ORGANIC CHEMISTRY

D0022815

ORGANIC CHEMISTRY

K. Peter C. Vollhardt

University of California, Berkeley

W. H. Freeman and Company

New York

Copyright © 1987 by
W. H. Freeman and Company

Printed in the United States
of America

Cover images
Computer-generated picture of the transition state of the S_N2 reaction of the enolate of menthone (a terpene found in nature) with bromomethane. Courtesy of Columbia University.

Periodic tables
Data taken from the "Periodic Table of the Elements," compiled by Fluck and Heumann, © VCH Verlagsgesellschaft 1986. The full table is available from VCH Publishers, New York.

Book and cover design
Lisa Douglis

Library of Congress Cataloging-in-Publication Data

Vollhardt, K. Peter C.
Organic chemistry.

Includes index.
1. Chemistry, Organic. I. Title.

QD251.2.V65 1987 547 86-31980
ISBN 0-7167-1786-7
ISBN 0-7167-1915-0 (international student ed.)

No part of this book may be reproduced by any mechanical, photographic, or electronic process, or in the form of a phonographic recording, nor may it be stored in a retrieval system, transmitted, or otherwise copied for public or private use, without the written permission of the publisher.

3 4 5 6 7 8 9 RRD 9 9 8 7 6 5 4 3 2 1

A Note to the Instructor

Organic chemistry is a rapidly changing field. In the past ten years, exciting technical and intellectual advances have been made in the development of synthetic methodology, mechanistic tools, and spectroscopic instrumentation. My conviction in writing this book is that such advances should be incorporated in the teaching of introductory organic chemistry even as basic rules and concepts are emphasized. I am fortunate that I have had the opportunity to teach a great variety of students at both undergraduate and graduate levels. Invariably, the beginning students arrive with diverse amounts of preparation, as well as different reasons for studying chemistry. Out of this teaching and my ongoing desire to communicate the excitement of research in organic chemistry to students, this book was born. My goal is to present the richness of modern organic chemistry in a simple and readily understandable manner. The effort has taken almost eight years and three drafts; I hope that you, my fellow teachers, will judge that effort to have been worthwhile.

Following the Functional-Group Approach

The unifying theme of this book is one that has proved to be instructionally most successful in the past: organic chemistry is best comprehended when divided into the chemistry of its functional groups. Thus, after a fairly comprehensive review of first-year chemistry (which in my experience most students have forgotten), the presentation follows a logical sequence starting with the alkanes (which have no functional group) and progressing to more and more highly functionalized classes of molecules. The physical properties of simple organic compounds are introduced early, as are such basic physical concepts as kinetics and thermodynamics. Within the functional-group approach, chemical reactions are routinely juxtaposed to the mechanisms by which they proceed. The point is made that, in this respect, organic chemistry is much like a language—the reactions are the vocabulary, the mechanistic descriptions the grammar. Finally, at strategic points throughout the book, the important spectroscopic techniques in organic chemistry are introduced.

Pedagogical Features

Several pedagogical features of this book distinguish it from others.

Learning Simplified by the Use of Color

One of the most important innovative features of this text is the way in which color is used. Color is used not only in marking centers of reactivity, as it is in many textbooks, but also in a functional sense. For example, the application of several colors is very effective in demonstrating how to name a functional molecule: the stem, the ancillary substituents, and the functional group of the structure can be differentiated by color, as can the corresponding components of the name itself. Whenever possible, *s* orbitals are shown in red, 2*p* orbitals in blue, and 3*p* orbitals in green. In the chapter on stereoisomerism (Chapter 5), the order of diminishing priority of substituents is indicated by the sequence red-blue-green-black. Most importantly, color is used to specify the type of reactivity of transforming centers in reactions and mechanisms. For example, all radicals, which are introduced in Chapter 3, are shown in green. In the mechanistic description of polar reactions, all nucleophiles are red, electrophiles are blue, and leaving groups are green. The various applications of color are explained to the student wherever necessary.

Illustrations

The thoughtful rendering of illustrations is essential in an organic chemistry textbook, because the massive scope of the subject dictates that all possible visual assistance be available to students. Color-coded airbrushed drawings throughout this book illustrate the three-dimensionality of orbitals and related spheres. The use of color in these drawings and in diagrams serves to reinforce key physical and chemical principles. A particularly nice example is found in sp^3 hybrids whose composition of *s* (red) and *p* (blue) orbitals is indicated by violet.

Nomenclature

A never-ending problem in teaching is that the naming of organic compounds, particularly the highly functionalized molecules, can be extremely confusing to students: this is because of the abundance of common names. This book introduces rigorous IUPAC nomenclature for the alkanes and then builds on this system for the naming of compounds having functional groups: the alkanols, alkanals, alkanones, alkanoic acids, alkanenitriles, and so forth. If a common name is firmly entrenched in the literature, it is given in parentheses following the systematic name. For example, the smallest ketone, propanone, is followed by (acetone). Common names are given in this way with particular frequency in the chapters on aromatic compounds and heterocycles.

Spectral Data

Almost all ^{1}H NMR and IR spectra presented in this book have been recorded on state-of-the-art equipment. Most of the ^{1}H NMR spectra were measured at

90 MHz (on a Varian Associates EM390 instrument), a frequency that is becoming the standard for routine use. IR spectra were recorded on a Perkin-Elmer 681 Spectrometer equipped with a 580B data station. The application of color in these spectra affords a unique opportunity to highlight spectral assignments.

Authenticity of Reactions

With the help of a generous grant from my publisher, all reactions in this book have been checked in the literature by undergraduate students to verify yields and conditions. Some of the reactions for which such data were unavailable were rerun in the laboratory.

Up-to-dateness of the Material

To the extent that they enhance the teaching value of the material presented and enable the student to perceive that organic chemistry is an advancing science, new developments at the forefront of organic research have been incorporated. These developments encompass natural products, medicinal compounds, coal chemistry, synthetic methodology, biochemical mechanisms, the mode of action of carcinogenic compounds, and energy-related matters.

Synthetic Strategy

The concepts of retrosynthetic analysis are introduced early so that the student can work backward in a synthetic scheme to unravel a target structure. These concepts are continually reinforced and reviewed.

Biological and Industrial Processes

Reactions and mechanisms pertinent to biochemistry and industrial chemistry are introduced to demonstrate the application of concepts developed throughout the text. Their incorporation serves to illustrate the relevance of organic chemistry to the life sciences, to our everyday lives, and to the economy.

End-of-Chapter Summaries

The text of each chapter ends with a summary of the important new concepts presented therein. Most chapters also include a summary of new reactions.

Cross-references

The book is extensively cross-referenced, to call the student's attention to the relation of a new concept or reaction to material covered in earlier chapters or to material that is to come. Basic principles are presented repeatedly within the framework of new reaction chemistry to insure that the student has an intellectual grasp of them and to emphasize their generality.

Exercises and Problems

The in-text exercises and the end-of-chapter problems are included not only for practice in the application of synthetic methodology, mechanisms, and new concepts, but also for the introduction of new material, whenever it is readily derived from the chemistry described in the text. Many of the exercises and problems have as their bases biochemical reactions, with a strong emphasis on the chemical aspects of these processes. Answers to all of the exercises can be found at the end of the book; those to the end-of-chapter problems are in a separate solutions manual.

Boxes

Many textbooks introduce ancillary or more-advanced material in the form of special topics, addenda, or appendices. In my experience, a two-semester course does not allow enough time to cover these topics. It is my belief that a discussion of this type of material, particularly if it concerns advanced concepts, should be left to advanced textbooks. In this book, such topics are presented briefly in boxes to inform the student not only of their existence, but also of the richness of organic chemistry beyond the introductory level. They are meant to stimulate the interested student, by adding depth or breadth to the subject at hand.

Ancillary Materials

The study guide and solutions manual was written by my colleague at the University of California at Davis, Professor Neil Schore, who presents the subject matter in a lively and easily understandable fashion. In this guide, Professor Schore summarizes the contents of each chapter again, but from a different perspective. Sample problems are worked out and the solutions to the end-of-chapter problems given. Finally, many hints to the student point out some of the pitfalls of applying faulty logic and help those who find it hard to visualize the steps and sequences that comprise solutions to various exercises.

A set of four-color overhead transparencies is available from the publisher. This selection of illustrations from the text includes reproductions of spectra, orbital pictures, diagrams, and mechanisms. It is designed to assist the instructor in presenting information in lectures.

K. Peter C. Vollhardt

January 1987

A Note to the Student

Organic chemistry is sometimes perceived by students as being a formidable subject, with an overwhelming number of facts to be memorized and with many seemingly difficult concepts. It does have a fairly rigid structure, with each new topic building on the preceding one. However, there is nothing inherently difficult about the subject, although it may be quite different from others that you have studied. Having spent much of my life in the study and teaching of organic chemistry, I have several bits of advice to offer that may be of help to you. Avoid falling behind. Make sure that your schedule allows you to set aside a short period every day for reading the book, for working the problems, for reviewing material presented in class, and for practicing the "vocabulary and grammar" of organic chemistry—that is, for learning the reactions and understanding their mechanisms. Make regular use of the office hours of your instructors and teaching assistants. They have specifically arranged their schedules so that they can help you deal with difficult material, instruct you on how to solve exercises, and inspire you to *think* organic chemistry. If you are not under pressure because of falling behind, you will enjoy organic chemistry as a learning experience that gives you a stimulating and different view of the chemical world that surrounds you.

To help you organize your thoughts and to provide you with an easy review, each chapter has a short introductory paragraph and an extensive summary section. The order of presentation of topics is the same in most chapters: you will learn, first, how to name the compounds to be covered; then, what their physical properties, in particular spectral characteristics, are; subsequently, the methods used to make these compounds; and, finally, how they react. The many transformations that you will encounter are reported in a consistent fashion: first, by an outline of the reagents, substrates, and reaction conditions; and, second, by the mechanistic details of the reaction. All topics are extensively reviewed in the study guide and solutions manual that accompanies this book.

The guide also summarizes new reactions and concepts, reminds you of material covered earlier, presents you with alternative explanations, and helps you to solve problems.

I urge you to acquire a molecular model-making kit (such as the Maruzen Molecular Structure Model set). It is an *invaluable* tool for dealing with stereochemistry, the shape of molecules, intra- and intermolecular interactions, and molecular mobility. Its utility is not limited to this course; it can also be used in other courses dealing with organic molecules.

Perhaps one of the strongest features of this book and one that most distinctively sets it apart from others is the use of color as a teaching tool. Color is employed not only as a ''marker'' to indicate the fate of individual atoms or molecular units in the course of a particular transformation, but also as a ''highlighter'' to show the relation of the names of organic molecules to their structures and the association of spectral features with certain molecular units. Most importantly, it is used in a functional sense: all electron-rich, nucleophilic moieties are in red, all electron-deficient, electrophilic parts blue, and all radicals and leaving groups green. The movement of electrons in a particular molecular transformation is indicated by red arrows. In descriptions of the mechanisms of organic reactions, the color reveals the reactivity of the functional groups. Because such reactivity may change from step to step, the color of such units may change as the overall reaction progresses. Do not be confused by these changes; they should allow you to visualize the metamorphosis of reactive centers as they transform starting materials into products. Although an attempt has been made to use color consistently within a section, you will note that its application may change from section to section (and chapter to chapter) because of emphasis on the material under discussion or the particular concept being highlighted. Frequently, color is applied only sparingly, and dropped when it serves no purpose. You will quickly grasp the basic idea behind the functional use of color, and you should be able to exploit it to your advantage. After you have done so, you should be able to apply what you have learned without the use of color. For this reason, the summary sections, as well as almost all exercises and all end-of-chapter problems, are presented without color, just as they would be in a test.

Enjoy organic chemistry and good luck!

K. Peter C. Vollhardt

January 1987

Acknowledgments

The book in its present form is the product of many contributors. I am indebted to the following reviewers for their invaluable criticisms and suggestions.

Harold Bell, *Virginia Polytechnic Institute*
Peter Bridson, *Memphis State University*
William Closson, *State University of New York at Albany*
Fred Clough, *formerly of University of Wisconsin, Parkside*
Otis Dermer, *Oklahoma State University*
Thomas Fisher, *Mississippi State University*
Marye Anne Fox, *University of Texas at Austin*
Raymond Funk, *University of Nebraska, Lincoln*
Roy Garvey, *North Dakota State University*
Edward Grubbs, *San Diego State University*
Gene Hiegel, *California State University at Fullerton*
Earl Huyser, *University of Kansas*
Taylor Jones, *The Master's College, Newhall, California*
George Kenyon, *University of California, San Francisco*
Robert Kerber, *State University of New York at Stony Brook*
Karl Kopecky, *University of Alberta*
James Moore, *University of Delaware*
Harry Pearson, *Bedales School, England*
William Pryor, *Louisiana State University*
William Rosen, *University of Rhode Island*
Neil Schore, *University of California, Davis*
Jay Siegel, *University of California, San Diego*
Richard Sundberg, *University of Virginia*
Michael Tempesta, *University of Missouri, Columbia*
Jack Timberlake, *University of New Orleans*
William Tucker, *North Carolina State University*
Desmond Wheeler, *University of Nebraska, Lincoln*
Joseph Wolinsky, *Purdue University*
Steven Zimmerman, *University of Illinois, Urbana*

I am grateful to Raymond Funk for his contributions in working out the color schemes and to Neil Schore for writing the end-of-chapter problems. I wish to thank Professor Schore and my graduate students Jim Drage, Ron Halterman, Joe King, and Patty McGovern, for recording spectral data, and Julie Bertuccelli, Harold Helson, and Eric Rouse, for spending many hours in the library researching the literature. I also wish to thank Juris Germanas, Douglas Grotjahn, and George Sheppard, also graduate students in my group, for their careful proofreading of galleys and page proof. Finally, my thanks go to Kim Denison, Lynn Goodman, Gene Sharp, Sandy Young, Janet Wortendyke, and, especially, Matt Pease, who were instrumental in translating my scribblings into a readable manuscript.

Contents

List of Topics

CHAPTER 1

Structure and Bonding in Organic Molecules

CHAPTER 2

Alkanes: Molecules Lacking Functional Groups

CHAPTER 3

The Reactions of Alkanes: Pyrolysis and Dissociation Energies, Combustion and Heat Content, Free-Radical Halogenation, and Relative Reactivity

CHAPTER 7

Further Reactions of Haloalkanes: Unimolecular Substitution and Pathways of Elimination

CHAPTER 8

The Formation of the Hydroxy Functional Group: Properties of the Alcohols and Synthetic Strategy in Their Preparation

CHAPTER 9

The Reactions of Alcohols and the Chemistry of Ethers

CHAPTER 10

The Use of Nuclear Magnetic Resonance Spectroscopy to Deduce Structure

CHAPTER 11

Alkenes: Hydrocarbons Containing Double Bonds

CHAPTER 12

The Reactions of Alkenes

CHAPTER 13

Alkynes: The Carbon–Carbon Triple Bond

CHAPTER 14

Delocalized Pi Systems and Their Investigation by Ultraviolet and Visible Spectroscopy

CHAPTER 18

Carboxylic Acid Derivatives and Mass Spectroscopy

CHAPTER 19

The Special Stability of the Cyclic Electron Sextet: Benzene and Electrophilic Aromatic Substitution

CHAPTER 20

Electrophilic and Nucleophilic Attack on Derivatives of Benzene: Substituents Control Regioselectivity

CHAPTER 21

Amines and Their Derivatives: New Functional Groups Containing Nitrogen

ORGANIC CHEMISTRY

CHAPTER 1

Structure and Bonding in Organic Molecules

1-1 Introduction and Overview

The subject of chemistry is a description of the structure of molecules and the rules that govern their interactions. Its subfields are known as analytical chemistry, biochemistry, inorganic, nuclear, organic, physical, polymer, and theoretical chemistry. What is organic chemistry? And what distinguishes it from the other disciplines of chemistry? A partial answer to these questions can be found in the most common definition of the subject: *organic chemistry is the chemistry of carbon and its compounds.*

How Organic Chemistry Affects Us

Organic molecules constitute the very essence of life. Proteins, the nucleic acids, sugars, and fats are compounds in which the principal component is carbon. We take for granted our everyday use of organic chemicals. Many of the clothes that we wear are made of organic molecules, some of natural origin such as cotton and silk, some artificial such as polyester. Toothbrushes, toothpaste, soaps, shampoos, deodorants, and perfumes—all contain organic compounds, as do furniture, carpets, the plastic in light fixtures and cooking utensils, paintings, food, liquor, and countless other items. Such organic substances as gasoline, coal, most medicines, vaccines, pesticides, and insect repellents very strongly influence if not control our lives. You may wonder whether this development is good or bad. On the one hand, organic chemicals have helped us to improve the quality of life. On the other hand, we have become dependent on them. On occasion, their uncontrolled disposal has seriously polluted the environment, causing the deterioration of animal and plant life, as well as injury, disease, and death to human beings. Whatever your feelings have been about organic chemistry, now is the time to delve into the subject, its underlying principles, and their applications.

What You Will Learn from This Book

This book begins by reviewing some basic principles of structure and bonding as they apply to organic molecules. A class of organic compounds composed of only hydrogen and carbon, the alkanes, is then used to introduce the systematic naming of organic molecules and to contrast this system with their traditional names. An example of an alkane is ethane, and its structural mobility is the starting point for a simple review of the thermodynamics and kinetics of reactions. This review is followed by a discussion of the strength of alkane bonds. We shall see how these bonds can be broken either by heat or by reagents, setting the stage for a description of the chlorination of methane to chloromethane.

$CH_3—CH_3$
Ethane

$CH_3—Cl$
Chloromethane

$$\begin{array}{ccc} & CH_2 & \\ H_2C & & CH_2 \\ | & & | \\ H_2C & & CH_2 \\ & CH_2 & \end{array}$$
Cyclohexane

After dealing with the special features of the cycloalkanes, particularly cyclohexane, we shall turn to the topic of stereoisomerism to review the variety of ways in which compounds of carbon can be arranged in space. This discussion is then followed by a study of two basic organic reactions: substitution and elimination.

A Substitution Reaction

$$CH_3—Cl + K^+I^- \longrightarrow CH_3—I + K^+Cl^-$$

An Elimination Reaction

$$\underset{\displaystyle H}{\underset{|}{CH_2}}—\underset{\displaystyle I}{\underset{|}{CH_2}} + K^{+\,-}OH \longrightarrow H_2C{=}CH_2 + HOH + K^+I^-$$

$H_2C{=}CH_2$
An alkene

$H_2C{=}O$
An aldehyde

$$CH_3—\overset{\displaystyle O}{\overset{\|}{C}}—CH_3$$
A ketone

$CH_3—NH_2$
An amine

At this stage, the foundation should have been laid for a presentation of the major classes of organic compounds, characterized by the molecular units that control their reactivity: the **functional groups.** For example, the carbon–carbon double bond is a functional group for the alkenes, the carbon–oxygen double bond fulfills this role for aldehydes and ketones, and amines contain nitrogen in their functional group. Whenever appropriate, biologically and industrially relevant examples will be introduced. Subsequent chapters will expand on this theme and will describe several important classes of molecules in nature that contain multiple functional groups, such as the carbohydrates, alkaloids, amino acids, peptides, and nucleic acids.

In the course of your studies, you will learn about the basic physical properties of organic molecules, how they are made, and how they can be interconverted. Several sections deal with methods for identifying molecules: elemental analysis and various forms of spectroscopy.

Synthesis: The Making of Molecules

A very important part of organic chemistry is the making of molecules: **synthesis.** Compounds that contain carbon were originally called "organic" because they were thought to exist only in living matter. However, it was soon found that organic compounds could be synthesized in the laboratory from carbon-containing substances obtainable from inanimate material. The first clear-cut report of such a feat dates to 1828. In that year, Friedrich Wöhler* succeeded in converting lead cyanate into urea by treatment with aqueous ammonia. Thus, an inorganic salt was converted into a natural product of human and animal protein metabolism. Urea is excreted in urine (hence the name urea). In the 150 years since Wöhler's discovery, more than six million organic substances have been synthesized in the laboratory from simpler organic and inorganic materials.

$$\underset{\textbf{Lead cyanate}}{Pb(OCN)_2} + \underset{\textbf{Water}}{2\,H_2O} + \underset{\textbf{Ammonia}}{2\,NH_3} \longrightarrow \underset{\textbf{Urea}}{2\,H_2N\overset{\overset{\displaystyle O}{\|}}{C}NH_2} + \underset{\textbf{Lead hydroxide}}{Pb(OH)_2}$$

These substances have included naturally occurring molecules, such as the penicillins, and entirely new ''unnatural'' ones, such as cubane and saccharin. The synthesis of cubane afforded an opportunity to study new fundamental modes of bonding and reactivity in organic compounds. The making of the artificial sweetener saccharin had more immediate practical application.

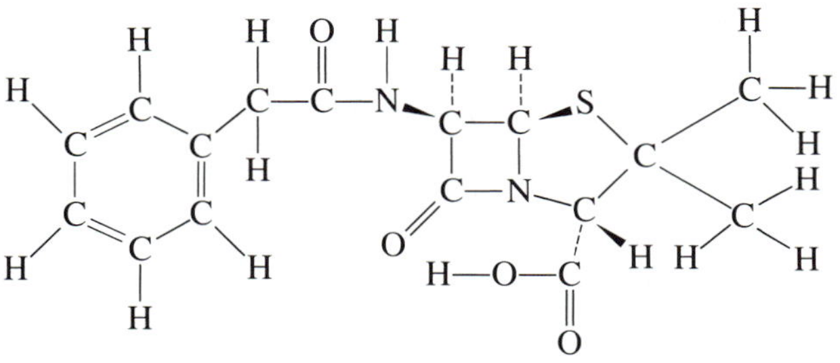

Benzylpenicillin 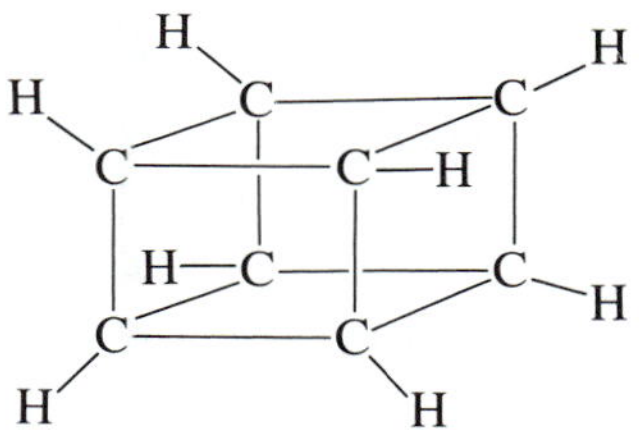 **Cubane** 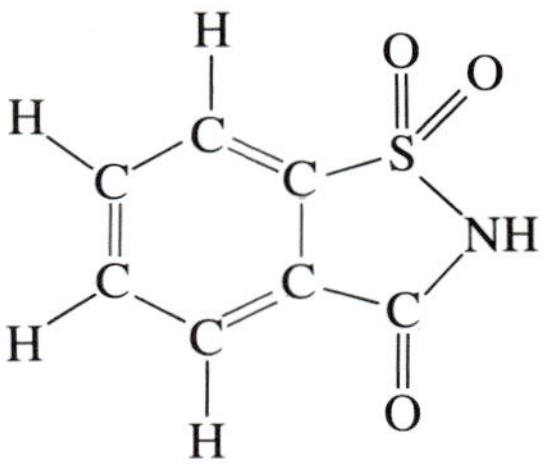 **Saccharin**

To be able to convert one type of molecule into another, a chemist must know organic reactions. He or she must also understand the physical characteristics that govern such processes—for example, the effects of solvents, temperature, and molecular structure. This knowledge is equally important in biological applications of organic chemistry, particularly the biosynthesis of organic molecules, their metabolism and degradation, and their interaction with the environment.

Reactions and Mechanism: The Vocabulary and Grammar of Organic Chemistry

A study of organic chemistry requires an awareness of the interplay between a **reaction**—namely, the overall conversion of one or more substances into others—and its **mechanism**—that is, its intricate details at the molecular level. For example, reagent A may undergo a reaction with reagent B to give C. The overall transformation would then be described as $A + B \rightarrow C$. However, in reality, the sequence might have been quite different: A + B could have first formed an unobserved intermediate D that very rapidly transformed into the observed product C. That reaction would therefore be described by $A + B \rightarrow D \rightarrow C$. Even more detail can be obtained by establishing how, when, and how fast certain bonds break and form, in which way they do so in three dimensions, and how subtle changes in the structure of the substrates affect the outcome of

*Professor Friedrich Wöhler, 1800–1882, University of Göttingen, Germany.

the reaction. In this respect, the ''learning'' and ''using'' of organic chemistry is much like learning and using a language. You need the vocabulary (i.e., the reactions) to be able to use the right words, but you also need the grammar (i.e., the mechanisms) to be able to converse intelligently. Neither one on its own gives complete knowledge and understanding, but together they form a powerful means of communication, rationalization, and predictive analysis.

To begin to understand the principles of organic chemistry, you should realize that organic molecules are no more than collections of bonded atoms. Let us review some of the elementary principles of bonding.

1-2 A Simplistic View of Bonding: Coulomb Forces

The bonds between atoms hold a molecule together. Let us examine the reasons why bonding exists. Later, we shall find these concepts useful in predicting reactivity.

Why Are There Bonds?

Two atoms form a bond only if their interaction is ''favorable.'' This qualitative term is used by chemists to indicate that energy—heat, for example—is released in the process. (Conversely, breaking that bond requires a corresponding amount of energy.)

What are the causes of this release of energy? There are two, based on some fundamental laws of physics:

1 Remote opposite charges are brought into closer proximity.
2 Electrons are spread out in space.

Bonds Are Made by Coulombic Attraction and Electron Exchange

Each atom consists of a nucleus, containing electrically neutral particles, or neutrons, and positively charged protons. Surrounding the nucleus are negatively charged electrons, numerically matched to the protons so that the net charge is zero. As two atoms approach each other, the positively charged nucleus of the first attracts the electrons of the second; similarly, the nucleus of the second attracts the electrons of the first. This sort of bonding is described by simple **Coulombic attraction.** Coulomb's law states that opposite charges attract each other with a force inversely proportional to the square of the distance between the centers of the charges.

Coulomb's Law

$$\text{Force} = \text{constant} \times \frac{(+)\text{charge} \times (-)\text{charge}}{\text{distance}^2}$$

Bonding begins at a certain distance between two atoms and increases until no more energy is released. The distance between the two nuclei at this point is called the **bond length** (Figure 1-1). Interestingly, bringing the two atoms even closer together results in a sharp increase in energy. Why? Because, if the atoms are too near one another, there are both electron-electron and nuclear-nuclear repulsions with forces that are stronger than those of the initial attraction. The picture that emerges, then, is one of a delicate balance of mutually attractive

FIGURE 1-1
The changes in energy, E, that result when two atoms are brought into close proximity. At the separation defined as bond length, maximum bonding is achieved.

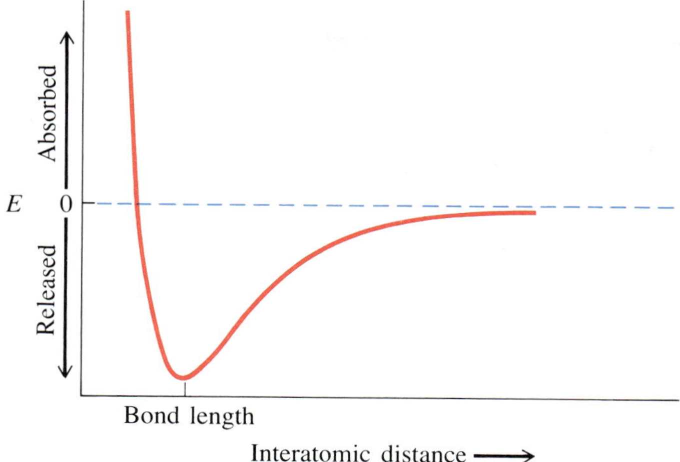

FIGURE 1-2
Attractive (solid line) and repulsive (dashed line) forces in the bonding between two atoms. The large circles represent areas in space in which the electrons are found around the nucleus. The small circle around the plus sign stands for the nucleus.

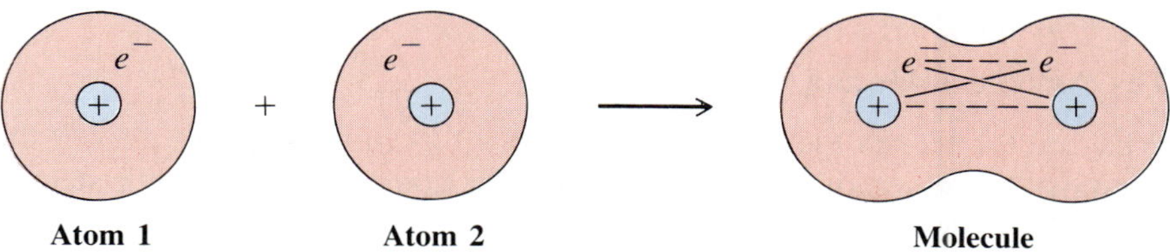

FIGURE 1-3
An alternative mode of bonding results from the complete transfer of an electron from atom 1 to atom 2.

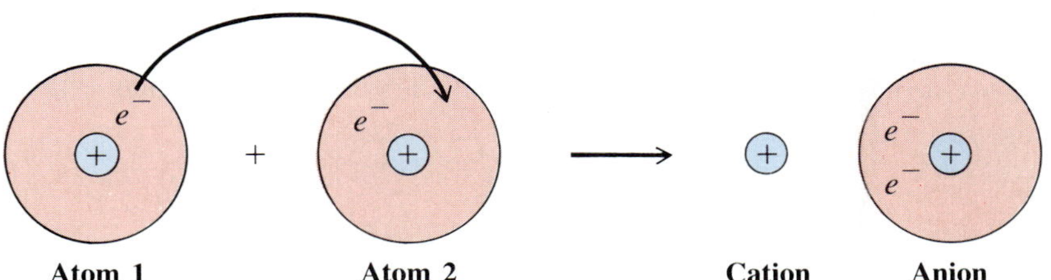

and repulsive forces keeping the two atoms at an equilibrium distance for maximum bonding (Figure 1-2).

An alternative to this type of bonding results from the complete transfer of an electron from one atom to the other. The result is two charged *ions:* one positively charged, a *cation,* and one negatively charged, an *anion* (Figure 1-3). Again, the bond formed is purely Coulombic, but now *full* charges attract each other.

The Coulombic bonding models of attracting and repelling charges shown in Figures 1-2 and 1-3 are highly simplified views of the interactions that take place in the bonding of atoms. The situation is not quite so simple, because, as mentioned, a bond can also be formed when the electrons from one atom exchange places with those of another: electrons are not localized but delocalized. Nevertheless, electrostatic phenomena are an integral part of all bonding. In fact, examination of the interaction of fully or partially charged particles does much to explain many of the structural and dynamic features of organic chemistry.

We have seen that attraction between negatively and positively charged particles is a basis for bonding. How does this concept work in real molecules?

1-3 Ionic and Covalent Bonds

There are two extreme types of bonding:

1 The **ionic bond** is formed by the transfer of one or more electrons from one atom to another (as shown in Figure 1-3), generating ions as in the inorganic salts.

2 The **covalent bond** is formed by the mutual sharing of electrons (not unlike that shown in Figure 1-2).

However, we will see that many atoms bind to carbon in a way that is not described by either extreme: some ionic bonds have covalent character, and some covalent bonds have ionic character.

The Periodic Table and the Octet Rule

What are the factors that account for the two types of bonds? To answer this question, let us return to the atoms and their composition. We will start by looking at the periodic table and at how the electronic make-up of the elements changes as the atomic number increases.

The partial periodic table depicted in Table 1-1 includes those elements of greatest importance to organic chemistry—mainly, the elements of the first three periods and the larger halogens. The elements most widely found in organic molecules are carbon (C), hydrogen (H), oxygen (O), nitrogen (N), sulfur (S), chlorine (Cl), bromine (Br), and iodine (I). Certain reagents, indispensable for synthesis and commonly used, contain such elements as lithium (Li), magnesium (Mg), boron (B), and phosphorus (P). If you are not familiar with these elements, you should memorize Table 1-1.

EXERCISE 1-1

(a) Redraw Figure 1-1 for a weaker bond than the one depicted; (b) write down Table 1-1 from memory.

The elements in the periodic table are listed according to their atomic number, or nuclear charge (number of protons). This number increases by one with each element listed. Recall that the number of electrons that an atom has matches the number of its protons. A defined number of electrons occupies each energy level, or ''shell.'' For example, the first shell has room for two electrons; the second, eight; and the third, eighteen. Some of the configurations

TABLE 1-1
Partial periodic table

Period							Halogens	Noble gases
First	H^{1}							He^{2}
Second	$Li^{2,1}$	$Be^{2,2}$	$B^{2,3}$	$C^{2,4}$	$N^{2,5}$	$O^{2,6}$	$F^{2,7}$	$Ne^{2,8}$
Third	$Na^{2,8,1}$	$Mg^{2,8,2}$	$Al^{2,8,3}$	$Si^{2,8,4}$	$P^{2,8,5}$	$S^{2,8,6}$	$Cl^{2,8,7}$	$Ar^{2,8,8}$
Fourth	$K^{2,8,8,1}$						$Br^{2,8,18,7}$	$Kr^{2,8,18,8}$
Fifth							$I^{2,8,18,18,7}$	$Xe^{2,8,18,18,8}$

Note: The superscripts indicate the number of electrons in each energy level of the atom.

determined by these numbers are especially stable: they are the configurations of the noble gases—that is, helium (He), neon (Ne), argon (Ar), krypton (Kr), and xenon (Xe). This class of elements lacks chemical reactivity. Helium has two electrons in its shell, neon has an additional eight electrons, argon eight more, and so on. Such patterns dominate the chemical behavior of most of the elements in organic chemistry. Indeed, as a rule of thumb, individual atoms *react in such a way as to fill the outer electron shell and attain a noble-gas electronic configuration.*

Electron Octets by the Transfer of Electrons: Ionic Bonds

Sodium (Na), a reactive metal, interacts with chlorine, a reactive gas, in a violent manner to produce a stable substance: sodium chloride. Similarly, sodium reacts with fluorine (F), bromine, or iodine to give the respective salts, reactions also undergone by other alkali metals, such as lithium and potassium (K). These transformations succeed because both reaction partners attain noble-gas character by the *transfer of outer-shell electrons* from the alkali metals on the left side of the periodic table to the halogens on the right.

$$Li^{2,1} \xrightarrow{-1\,e} [Li^{2}]^{+}$$ Helium configuration

Lithium cation

$$Na^{2,8,1} \xrightarrow{-1\,e} [Na^{2,8}]^{+}$$ Neon configuration

Sodium cation

$$Cl^{2,8,7} \xrightarrow{+1\,e} [Cl^{2,8,8}]^{-}$$ Argon configuration

Chloride anion

The outer-shell electrons are the **valence electrons;** they are the electrons that participate in bonding. In this way, both reaction partners achieve electronic shells with noble-gas character. The bonds thus formed are **ionic bonds,** in which the cation and anion are held together by electrostatic attraction.

The hydrogen atom may either lose an electron to become a bare nucleus, the *proton,* or accept an electron to form the *hydride* ion, which adopts the helium configuration. Indeed, the hydrides of lithium, sodium, and potassium are commonly used reagents.

$$\mathrm{H}^1 \xrightarrow{-1\,e} [\mathrm{H}^0]^+$$ **Proton** — Bare nucleus

$$\mathrm{H}^1 \xrightarrow{+1\,e} [\mathrm{H}^2]^-$$ **Hydride anion** — Helium configuration

More than one electron may be donated (or accepted) to achieve favorable electronic shells. Magnesium, for example, has two valence electrons. Donation to an appropriate acceptor produces the corresponding doubly charged cation (dication) with the electronic structure of neon. In this way, the ionic bonds of typical salts are formed.

Formation of Ionic Bonds by Electron Transfer

$$\mathrm{Na}^{2,8,1} + \mathrm{F}^{2,7} \longrightarrow [\mathrm{Na}^{2,8}]^+\,[\mathrm{F}^{2,8}]^-, \text{ or NaF}$$

Sodium fluoride

$$\mathrm{Mg}^{2,8,2} + 2\,\mathrm{Cl}^{2,8,7} \longrightarrow [\mathrm{Mg}^{2,8}]^{2+}\,[\mathrm{Cl}^{2,8,8}]^-{}_2, \text{ or } \mathrm{MgCl_2}$$

Magnesium chloride

A more convenient way of depicting valence electrons is by means of dots around the symbol for the element. In this case, the letters represent the nucleus and all the electrons in the inner shells, also called the **nuclear core.** The location of the dots shown here does not have any physical significance.

Valence Electrons as Electron Dots

Li·	Be:	·Ḃ·	·Ċ·	·N̈·	:Ö·	:F̤̈·
Na·	Mg:	·Ȧl·	·Ṡi·	·P̈·	:S̈·	:C̤̈l·

Electron-Dot Picture of Salts

$$\mathrm{Li}\cdot + \cdot\ddot{\underset{\cdot\cdot}{\mathrm{Cl}}}: \xrightarrow{e\text{ transfer}} \mathrm{Li}^+\,:\ddot{\underset{\cdot\cdot}{\mathrm{Cl}}}:^-$$

$$\mathrm{Mg}: + 2\,\cdot\ddot{\underset{\cdot\cdot}{\mathrm{Cl}}}: \longrightarrow \mathrm{Mg}^{2+}[:\ddot{\underset{\cdot\cdot}{\mathrm{Cl}}}:]^-{}_2$$

EXERCISE 1-2

Write electron-dot pictures for ionic LiBr, Na_2O, BeF_2, $AlCl_3$, and MgS.

Electron Octets by Sharing Electrons: Covalent Bonds

For elements at the center of the periodic table, the formation of ionic bonds is less favorable because it becomes more and more difficult to donate or accept enough electrons to attain the noble-gas configuration. Such is the case for carbon. This element would have to shed four electrons to reach the helium electronic structure or add four electrons for a neonlike arrangement. The large amount of charge that would develop means that these processes would be energetically unfavorable.

$$\mathrm{C}^{4+} \xleftarrow{-4\,e} \cdot\dot{\underset{\cdot}{\mathrm{C}}}\cdot \xrightarrow{+4\,e} :\ddot{\underset{\cdot\cdot}{\mathrm{C}}}:^{4-}$$

Helium configuration — **Neon configuration**

In this case, an alternative mode of bonding takes place, in which elements *share* electrons to attain octets around each nucleus. Hydrogen atoms may share electrons to maintain an electron duet around their core. Typical products of such sharing are H_2 and HF. In HF, the fluorine atom assumes an octet structure by sharing one of its valence electrons with that of hydrogen. Similarly, the fluorine molecule, F_2, is diatomic because both component atoms gain octets by sharing two electrons. Such bonds are called **covalent single bonds.**

Electron-Dot Picture of Covalent Bonds

$$\text{H}\cdot + \cdot\text{H} \longrightarrow \text{H:H}$$

$$\text{H}\cdot + \cdot\ddot{\underset{\cdot\cdot}{\text{F}}}: \longrightarrow \text{H:}\ddot{\underset{\cdot\cdot}{\text{F}}}:$$

$$:\ddot{\underset{\cdot\cdot}{\text{F}}}\cdot + \cdot\ddot{\underset{\cdot\cdot}{\text{F}}}: \longrightarrow :\ddot{\underset{\cdot\cdot}{\text{F}}}:\ddot{\underset{\cdot\cdot}{\text{F}}}:$$

Because carbon has four valence electrons, it must share four electrons through four single bonds to gain the neon configuration, as in methane. Nitrogen has five valence electrons and needs three to share, as in ammonia, and oxygen, with six valence electrons, requires only two to share, as in water.

It is possible for an atom to donate two electrons to the covalent bond to be formed. Examples are the ammonium ion, NH_4^+, and the hydronium ion, H_3O^+. Bonds formed in this manner are sometimes referred to as *dative bonds* (*dativus,* Latin, giving), or **coordinate covalent bonds.**

Besides two-electron (single) bonds, atoms may form four-electron (double) and six-electron (triple) bonds to gain noble-gas configurations. This is done by sharing more than one electron pair, as in ethylene or the nitrogen molecule.

H H
C::C
H H
Ethylene

:N:::N:
Nitrogen

H
H:C:H
H
Methane

H:N̈:H
H
Ammonia

H:Ö:H
Water

$$\text{H:}\underset{\text{H}}{\overset{\text{H}}{\text{N}}}\text{:} + \text{H}^+ \longrightarrow \left[\text{H:}\underset{\text{H}}{\overset{\text{H}}{\text{N}}}\text{:H}\right]^+$$

Ammonium ion

$$\text{H:}\underset{\text{H}}{\ddot{\text{O}}}\text{:} + \text{H}^+ \longrightarrow \left[\text{H:}\underset{\text{H}}{\ddot{\text{O}}}\text{:H}\right]^+$$

Hydronium ion

EXERCISE 1-3

Write electron-dot structures for Cl_2, SiF_4, CCl_4, PH_3, BrI, OH^-, NH_2^-, and NCl_3. (Where applicable, the first element is at the center of the molecule.) Make sure that all atoms have noble-gas electron structure.

Donating and Accepting Electrons: A Quantitative Measure

A more-quantitative measure of the ability of an element to donate electrons is its **ionization potential** (*IP*). Similarly, a measure of an element's ability to accept electrons is its **electron affinity** (*EA*). The ionization potential is the energy required to remove an electron from an atom in the gas phase. The energy needed to remove the first electron is called the *first ionization potential,* that required to remove the second electron is the *second ionization potential,* and so forth. Table 1-2 lists the ionization potentials of selected elements. The magnitude of the first and subsequent *IP*s depends strongly but not solely on the amount of positive nuclear charge present. A more-positive nuclear core frequently requires a higher *IP*. Compare, for example, *IP* (carbon) = 260 kcal $mole^{-1}$ with *IP* (fluorine) = 402 kcal $mole^{-1}$. However, the loss of an electron to create a noble-gas configuration is relatively easy (see *IP* for Li, Na).

Electron affinity is defined as the amount of energy *released* when an electron is absorbed by an atom. The electron affinity of an atom is equal to the

TABLE 1-2
Ionization potentials of common elements

Element	Ionization potential (kcal mole^{-1})[a,b] 1st	2nd	Element	1st
H	314		Na	119
He	567	1255	Mg	175
Li	124	1744	Al	138
Be	214	420	Si	188
B	191	580	P	242
C	260	562	S	239
N	335		Cl	300
O	314		Br	273
F	402		I	241
Ne	498			

[a] A kilocalorie (kcal) is the energy required to raise the temperature of one kilogram of water by 1°C. The traditional kcal mole^{-1} values for energy will be used in this book. The International Committee on Weights and Measures recommends the use of SI units (after Système International), in which energy is expressed in joules (kg m^2 sec^{-2}). One joule (J) is the energy required to accelerate 1 kilogram every second by 1 meter per second. It is also referred to as the mechanical equivalent of heat. Conversion factor: 1 kcal = 4187 J.
[b] Because energy is required to remove one or more electrons from the elements listed here, all *IP* values are given as positive quantities. However, *IP* values may be negative.

TABLE 1-3
Electron affinities of selected elements

Element	Electron affinity (kcal mole^{-1})[a]
H	−18
C	−25
O	−34
F	−80
Cl	−83
Br	−77
I	−71

[a] The sign of all the electron-affinity values is negative because energy is released on adding one electron to any one of the elements listed here. However, *EA* values may be positive.

ionization potential of the anion that results when the electron is absorbed. Table 1-3 gives some typical values of *EA*s.

The driving force to attain electron octets is reflected in the relatively high *EA*s of the halogens. On inspection of these tables, it becomes clear why ionic bonds are formed between elements lying at the extreme ends of the periodic table and covalent bonds between atoms located close to each other. However, what about those cases in between?

Not All Electrons Are Equally Shared: Polar Covalent Bonds

The elements on the left of the periodic table are often called *electropositive,* electron-donating or "electron pushing," because of their relatively low *IP*s. Conversely, the elements on the right are called *electronegative,* electron-accepting or "electron pulling," because of their relatively high *EA*s. Table 1-4 gives, on an arbitrary scale, the relative electronegativity of some elements. On this scale, fluorine, the most electronegative of them, is assigned the value 4. Covalent bonds between atoms of differing electronegativity are said to be **polarized.**

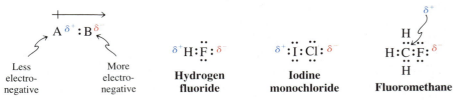

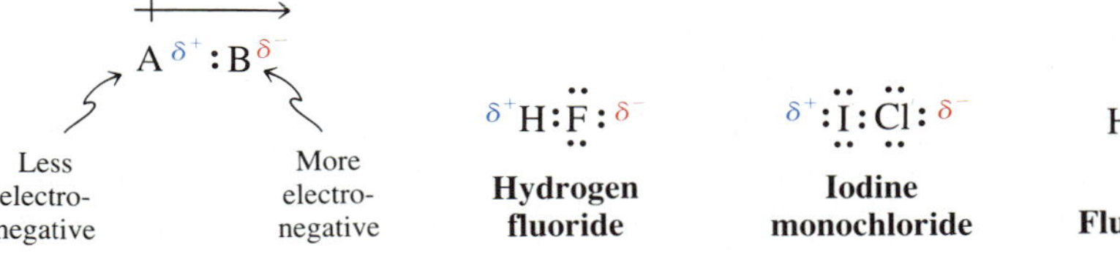

TABLE 1-4
Electronegativities of selected elements

H 2.2						
Li 1.0	Be 1.6	B 2.0	C 2.6	N 3.0	O 3.4	F 4.0
Na 0.9	Mg 1.3	Al 1.6	Si 1.9	P 2.2	S 2.6	Cl 3.2
K 0.8						Br 3.0
						I 2.7

In these bonds, the center of electron density provided by the shared electrons is shifted from the midpoint of the bond toward the more electronegative atom. Although the molecule as a whole stays electrically neutral, one end has a partial positive charge, designated δ^+, the other a partial negative charge, δ^-. A molecule A:B of this type is a *dipole,* characterized by a positive charge q at one end compensated by a negative charge q at the other, separated by the distance d. The product of these quantities, charge $q \times$ distance d, is the **dipole moment,** μ. It is measured by the tendency of the dipole to turn in an external electric field and is expressed in units of Debye (D), when the charge has electrostatic units (esu) and the distance is measured in Angstroms (Å, 1 Å$=10^{-8}$ cm). The dipole moment of HF is 1.82 D, that of HCl is smaller, 1.08 D. (Why?)

A Dipole

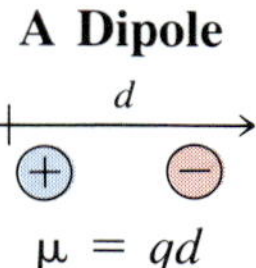

The polarization of a bond can be felt on adjacent groups. Thus, the electron-withdrawing character of the fluorine atom in fluoromethane also causes the hydrogens to be slightly positively charged (positively polarized). This phenomenon of the transmission of charge through a chain of atoms is called *inductive effect*. Its consequences will be discussed in Section 8-3.

Not all molecules with polar bonds are dipoles. A molecule is polarized only when the charge distribution is uneven, as in HF, HCl, and CH_3F, and the center of positive charge does not coincide with the center of negative charge. For example, in some symmetrical structures the polarizations of the individual bonds counteract each other to result in a net zero dipole moment. Examples are CO_2 and CCl_4. To know whether a structure is symmetrical, we have to know the shape of the molecule in question.

Molecules Can Have Polar Bonds but No Net Dipole

$2\delta^+$
δ^- :Ö::C::Ö: δ^-

δ^-
:C̈l:
δ^- :C̈l:C:C̈l: δ^- ($4\delta^+$ on C)
:C̈l: δ^-

EXERCISE 1-4

Show the bond polarization in CH_4, H_2O, SCO, SO, IBr, PCl_3, and BeH_2 by the use of arrows to indicate the displacement of electron density. (In the last two examples, place P and Be, respectively, in the center of the molecule.)

Electron Repulsion Controls the Shape of Molecules

The phenomenon of electron repulsion causes molecules to adopt a shape in which such repulsion is minimized. In diatomic species such as H_2 or LiH, there is only one bond and one possible arrangement of the two atoms. However, beryllium fluoride, BeF_2, is a triatomic species. Will it be bent or linear?

BeF_2 Is Linear

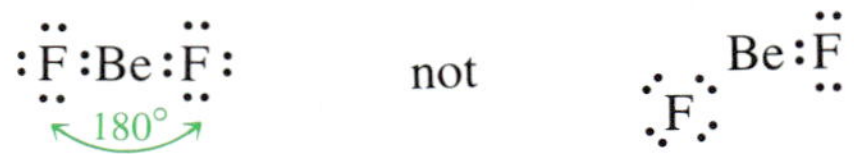

Electrons are farthest apart

Electrons are closer

Electron repulsion is at a minimum in a *linear* structure.* Linearity is also expected for other derivatives of beryllium, as well as other elements in the same column of the periodic table.

BCl_3 Is Trigonal

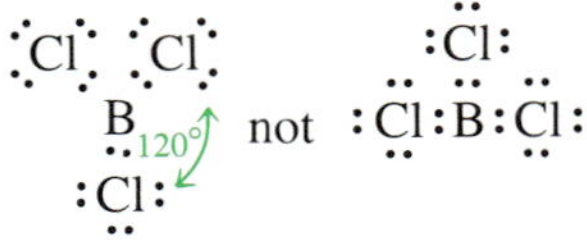

In boron trichloride, the three valence electrons of boron help to form covalent bonds with three chlorine atoms. Electron repulsion enforces a regular *trigonal* arrangement—that is, the three chlorine nuclei are at the corners of an equilateral triangle, the center of which is occupied by boron. Other derivatives of boron, and the analogous compounds with other elements in the same column of the periodic table, are again expected to adopt trigonal structures.

Applying this principle to carbon, we can see that methane, CH_4, has to be tetrahedral. Pointing its four valences toward the vertices of a tetrahedron is the best arrangement for minimizing electron repulsion.

CH_4 Is Tetrahedral

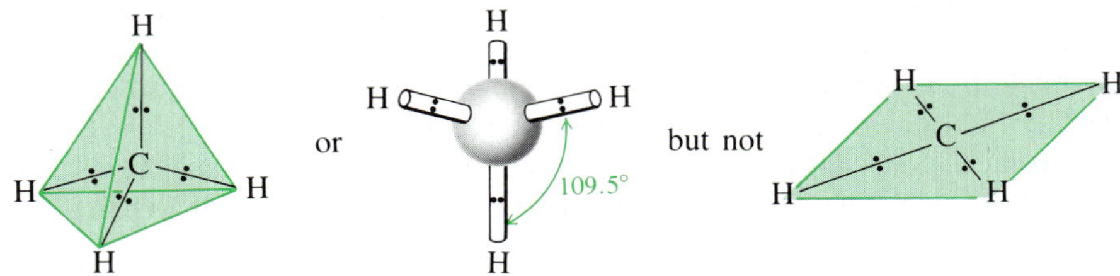

Ammonia, NH_3, and water, H_2O, nearly maintain such tetrahedral structures, in which the extra electron pairs can be thought of as occupying one (in NH_3) or two (in H_2O) of the vertices of the tetrahedron.

The Structures of NH_3 and H_2O Are Nearly Tetrahedral

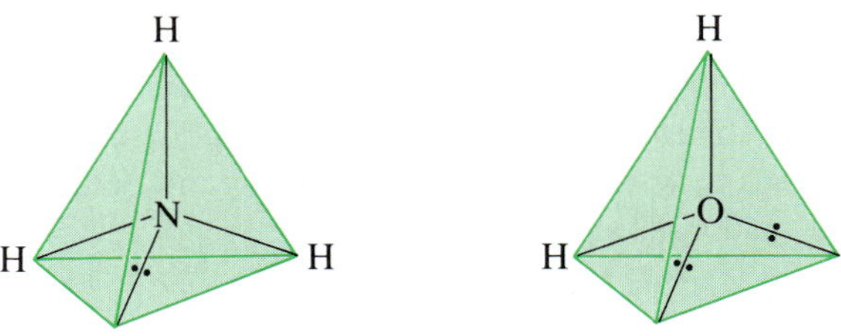

(Electron pairs of covalent bonds have been omitted for clarity.)

*This is true only for the gas. In the solid state, BeF_2 is aggregated and does not have a distinct linear triatomic structure.

To summarize, there are two extreme types of bonding, ionic and covalent. Both derive favorable energetics from Coulomb forces, electron delocalization, and the attainment of noble-gas electronic structures. Most bonds are better described as a combination of the two types: the polar covalent (or covalent ionic) bonds. Polarity in bonds may give rise to polar molecules. This depends on the shape of the molecule, which is derivable in a simple manner by arranging its bonds and nonbonding electrons so as to minimize electron repulsion.

1-4 The Electron-Dot Model of Bonding: Lewis Structures

The structures depicted in the preceding section using pairs of electron dots to represent bonds are also called **Lewis* structures.** In this section, rules are given for writing such structures correctly and for doing accurate electron bookkeeping.

Lewis Structures Follow Simple Rules

The procedure for writing correct electron-dot structures is straightforward, as long as the following rules are observed.

RULE 1 *Draw the molecular skeleton*. As an example, consider methane. The molecule has four hydrogen atoms bonded to one central carbon atom.

H
H C H
H
Correct

H H C H H
Incorrect

RULE 2 *Count the number of available valence electrons*. Simply add up all the valence electrons of the component atoms. Special care has to be taken with charged structures (anions or cations), in which case the appropriate number of electrons has to be added or subtracted to account for extra charges.

CH_4	4 H	4 × 1 electron	= 4 electrons
	1 C	1 × 4 electrons	= 4 electrons
		Total	8 electrons
HF	1 H	1 × 1 electron	= 1 electron
	1 F	1 × 7 electrons	= 7 electrons
		Total	8 electrons
H_2O	2 H	2 × 1 electron	= 2 electrons
	1 O	1 × 6 electrons	= 6 electrons
		Total	8 electrons
H_3O^+	3 H	3 × 1 electron	= 3 electrons
	1 O	1 × 6 electrons	= 6 electrons
	Charge		−1 electron
		Total	8 electrons
NH_2^-	2 H	2 × 1 electron	= 2 electrons
	1 N	1 × 5 electrons	= 5 electrons
	Charge		+1 electron
		Total	8 electrons

*Professor Gilbert N. Lewis, 1875–1946, University of California, Berkeley.

RULE 3 (The octet rule.) *Draw the covalent bonds between all the atoms, giving as many atoms as possible a surrounding electron octet, except for H, which requires a duet.* Make sure that the number of electrons used is *exactly* the number counted according to Rule 2. Elements at the right in the periodic table may have nonbonding valence electrons, also called "lone electron pairs."

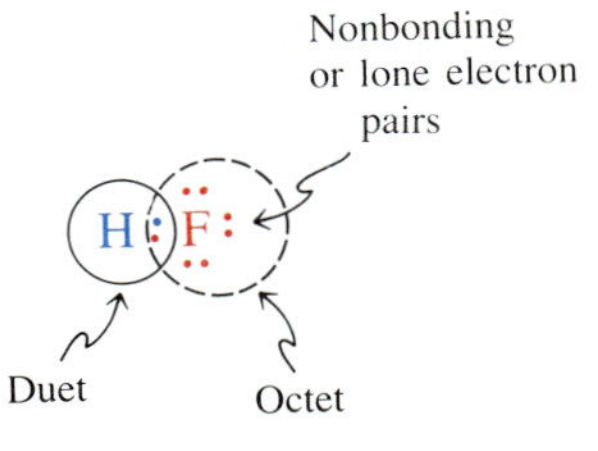

Consider, for example, hydrogen fluoride. The shared electron pair supplies the hydrogen atom with a duet, the fluorine with an octet, because the fluorine carries three lone electron pairs. On the other hand, in methane, the four C–H bonds satisfy the requirement of the hydrogens, and at the same time furnish the octet for carbon. Examples of correct and incorrect Lewis structures are shown below.

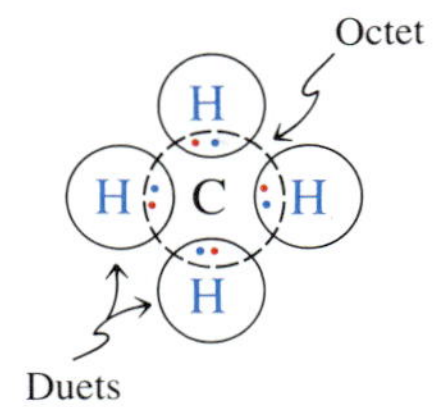

Correct Lewis Structures **Incorrect Lewis Structures**

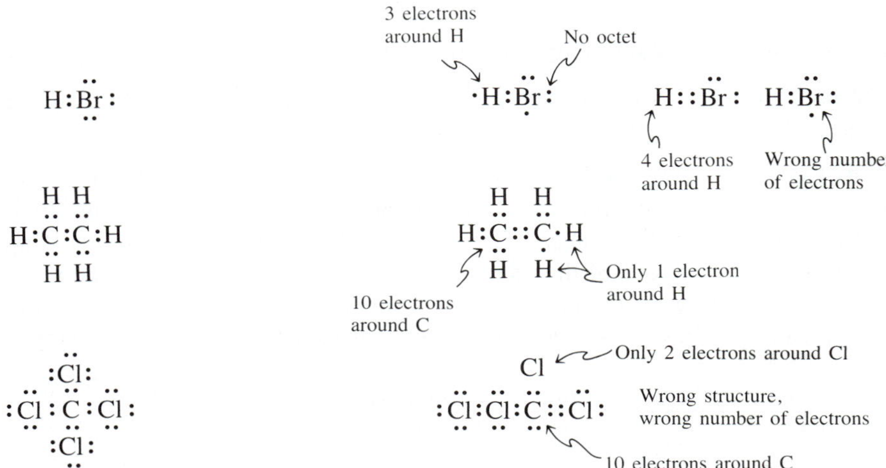

Frequently, as in unsaturated molecules, the number of valence electrons is not sufficient to allow the drawing of only single-bonded structures. In this event, double (two shared electron pairs) and even triple bonds (three shared pairs) are necessary to obtain octets. An example is the nitrogen molecule, N_2, which has ten valence electrons.

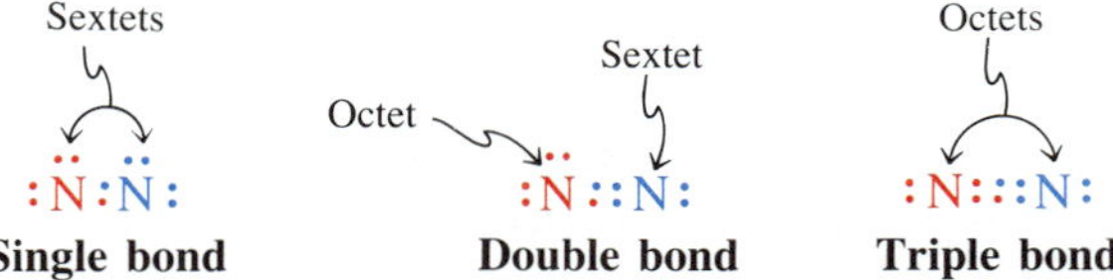

An N–N single bond would leave both atoms with electron sextets, and a double bond provides only one nitrogen atom with an octet. It is the molecule with a triple bond that satisfies both atoms. Further examples of such molecules are shown below and at the top of the facing page.

Correct Lewis Structures **Incorrect Lewis Structures**

H H
C::C
H H

6 electrons around C

H H
C:C:H
H

Correct Lewis Structures **Incorrect Lewis Structures**

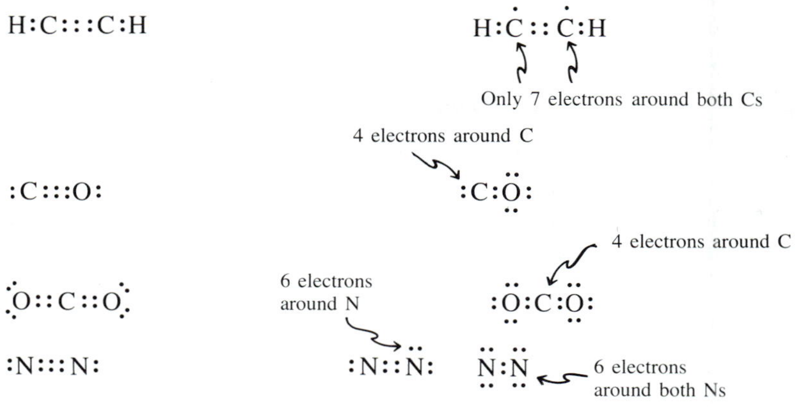
Only 7 electrons around both Cs
4 electrons around C
4 electrons around C
6 electrons
around N
6 electrons
around both Ns

EXERCISE 1-5

Write Lewis structures for the following molecules: HI, $CH_3CH_2CH_3$, CH_3OH, HSSH, SiO_2, O_2, CS_2.

RULE 4 *Assign charges in the molecule*. To place any charges requires identifying those atoms for which the surrounding number of electrons does not match the nuclear charge. To do so, add all nonbonding and core electrons around each atom to one-half the number of bonding ones. If that figure is equal to the proton count, the net charge is zero; if these numbers are not equal, the atom is charged.

The Formal Charge Formula

$$\begin{bmatrix}\text{Nuclear}\\ \text{charge}\end{bmatrix} - \begin{bmatrix}\text{Number of nonbonding}\\ \text{and core electrons}\end{bmatrix} - \frac{1}{2}\begin{bmatrix}\text{Number of bonding}\\ \text{electrons}\end{bmatrix} = \begin{matrix}\text{Formal}\\ \text{charge}\end{matrix}$$

For example, where does the positive charge reside in the hydronium ion? There are single bonds between the oxygen and the three hydrogen atoms; this leads to a net contribution of one electron to each hydrogen. Consequently, the charge at each hydrogen atom is zero. However, the electron count around the oxygen [three single bonds (3 electrons) plus one lone electron pair (2 electrons) plus core (2 electrons) gives a total of 7 electrons; oxygen's nuclear charge is +8] indicates the deficiency of one electron on this nucleus. Hence the positive charge is placed on oxygen.

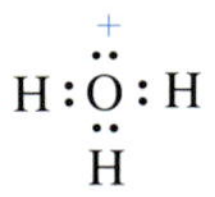

Hydronium ion

Nitrosyl cation

Another example is the nitrosyl cation, NO^+. The molecule bears a lone electron pair on nitrogen, which has two core electrons, in addition to a triple bond connecting the nitrogen to the oxygen atom. This gives nitrogen an electron count of seven, which matches its proton count; therefore, the nitrogen atom has no charge. The same electron count is found on oxygen, and so the positive charge is placed on this atom because of its higher nuclear charge. Other examples of charged molecules are shown below.

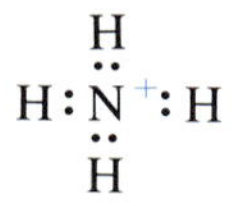

Ammonium ion

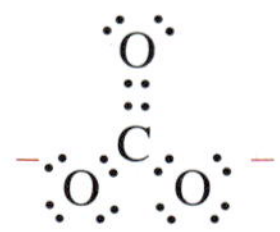

Carbonate ion

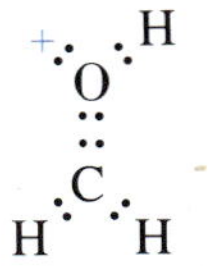

Protonated formaldehyde

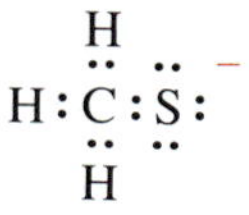

Methane-thiolate ion

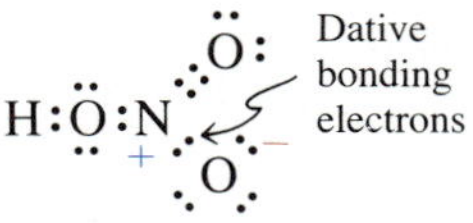

Dative bonding electrons

Nitric acid

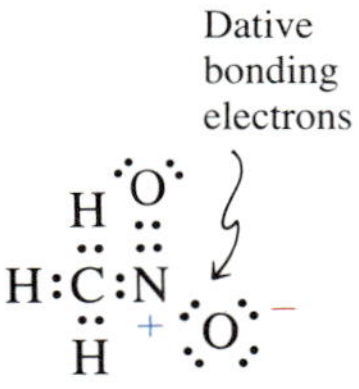

Dative bonding electrons

Nitromethane

Charge-Separated Lewis Structures

There are cases in which the writing of octet structures and the resulting electron count generate charge-separated Lewis structures, although the net charge on the molecule is zero. Examples may be found in some compounds containing nitrogen–oxygen bonds, as in nitric acid, HNO_3, and nitromethane, CH_3NO_2. Some of the N–O bonds in these molecules are thought of as dative (coordinate covalent, see Section 1-3); that is, formed by the addition of a lone pair on nitrogen to an electron-deficient oxygen. The simplest example of this type of arrangement is trimethylamine oxide.

Formation of a Dative Bond in Trimethylamine Oxide

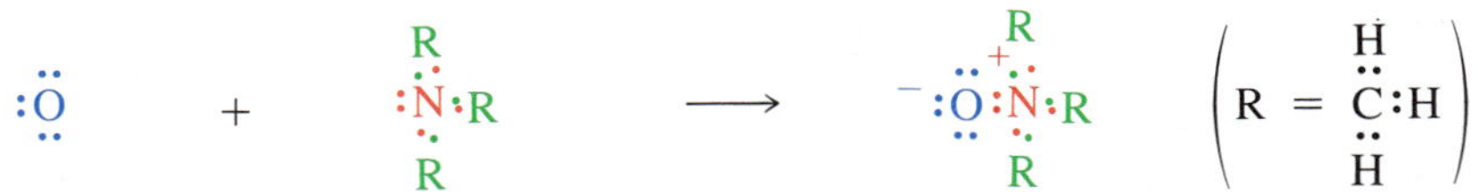

Electron-deficient oxygen + Electron-rich trimethylamine ⟶ Trimethylamine oxide (R = CH_3)

Covalent Bonds Can Be Depicted as Straight Lines

The writing of electron-dot structures can be cumbersome, particularly for larger molecules. A simpler way of showing a covalent bond is to connect the two bound atoms with a straight line, double bonds are represented by two such lines, and triple bonds by three. Lone electron pairs can either be shown as dots or simply omitted. The use of such notation was first suggested by the German chemist August Kekulé,* long before electrons were discovered, and structures of this type are therefore often called **Kekulé structures.**

Line Notation for the Covalent Bond

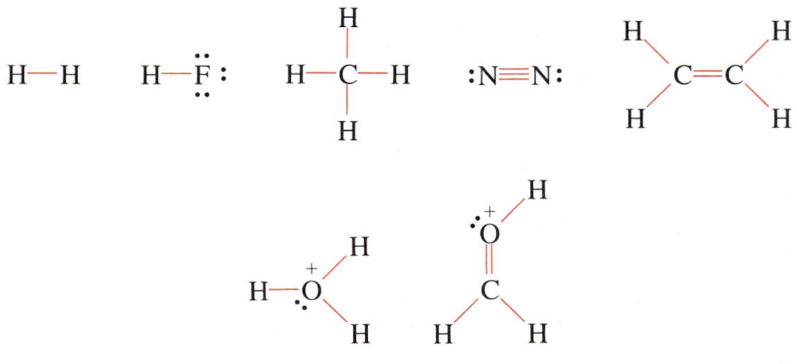

H—H H—F: H—C—H :N≡N: C=C

EXERCISE 1-6

Draw Lewis structures of the following molecules, assigning formal charges to atoms:

$$SO,\ F_2O,\ HNO_2,\ BF_3{-}NH_3,\ CH_3\overset{\overset{\displaystyle O}{\|}}{C}Cl,\ CN^-,\ C_2^{2-}.$$

*Professor August Kekulé, 1829–1896, University of Bonn, Germany.

You have now learned how to systematically derive electronic structures of molecules based on the total number of valence electrons and the octet rule. Because you also know how to predict the approximate shape of simple molecules by consideration of minimum electron repulsion, you should now be able to formulate a reasonably accurate picture of simple organic molecules.

1-5 Resonance Structures

We will now encounter a set of molecules for which there are no unique Lewis structures but *several* correct ones. How can that be?

The Carbonate Ion Has Several Correct Lewis Structures

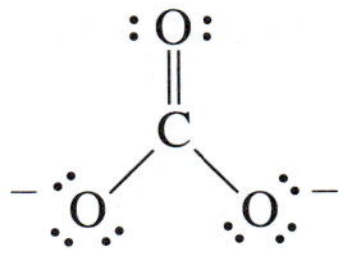

Let us return to the carbonate ion, CO_3^{2-}. Following the set of rules developed in the preceding section, we can derive a Lewis structure A in which every atom is surrounded by an electron octet. In it, the two negative charges are on the bottom two oxygen atoms, the third oxygen on top being neutral, connected to the central carbon atom by a double bond, and bearing two lone electron pairs. Why pick the bottom two oxygen atoms as the charge carriers? There is no particular reason at all for this choice—it is completely arbitrary. We could equally well have written structures B or C to describe the carbonate ion. All three Lewis pictures are *equivalent* and are called **resonance structures.** Resonance structures have the characteristic property of being interconvertible by *electron-pair movement only,* the nuclear positions in the molecule remaining *unchanged.* Note that, to convert A into B and then into C, we have to shift two electron pairs in each case. Such movement of electrons can be depicted by curved arrows, a procedure informally called "electron pushing."

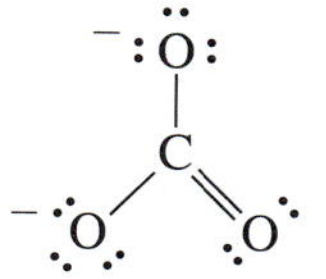

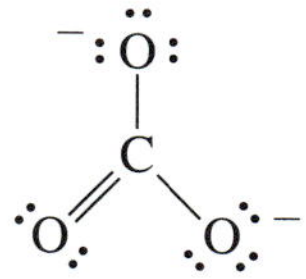

A

B

C

Arrow Signs for "Electron Pushing" of Electron Pairs

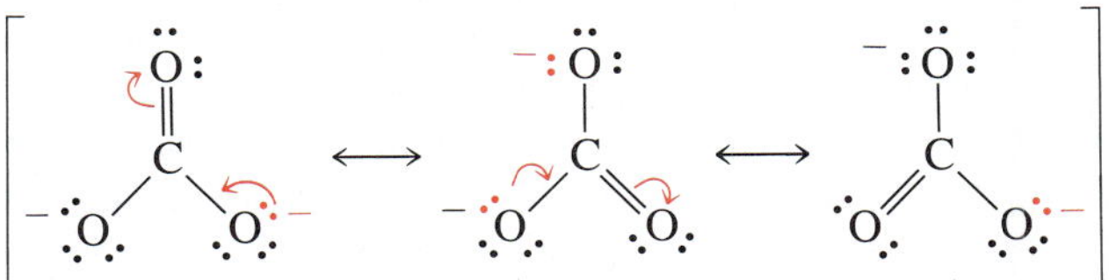

A ⟷ B ⟷ C

What Is the True Structure of the Carbonate Ion?

Does the carbonate ion have one uncharged oxygen atom bound to carbon through a double bond and two other oxygen atoms bound through a single bond each, both bearing a negative charge as suggested by the Lewis structures? The answer is no. Carbonate is perfectly symmetrical and contains a trigonal central carbon, all C–O bonds being of equal length, between that of a double and a single bond. The negative charge is evenly distributed over all three oxygens: it is said to be delocalized. In other words, none of the Lewis structures formulated for this molecule is structurally correct.

We can arrive at a correct description if *we take A, B, and C and conceptually create an average structure out of all three*. This picture is a **resonance hybrid.** Carbonate is a resonance hybrid of the three resonance structures A, B, and C. Because all three structures are equivalent, they contribute equally to the true structure of the molecule, but none of them accurately represents it; each is a distortion. Because resonance structures are hypothetical, they cannot be isolated or observed. To differentiate their interconversion from ordinary chemical equations, they are drawn connected by double-headed arrows and placed between square brackets.

The word resonance implies that the molecule vibrates or equilibrates from one form to the other. This is incorrect. The molecule has only one structure. The resonance approach is just one way of describing it. The term hybrid is better. A hybrid is defined as the offspring of genetically unequal parents. Carbonate has three "parents" (namely, A, B, and C) giving rise to one hybrid structure.

Dotted-Line Notation of Carbonate as a Resonance Hybrid

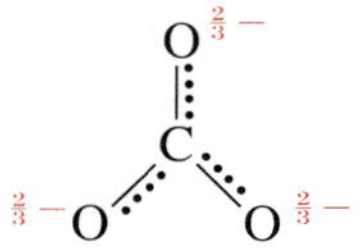

There are other conventions used to depict the structure of carbonate. One is to represent the carbon–oxygen bonds as a combination of a solid and a dotted line. Because only partial charges (on average $\frac{2}{3}$ of a negative charge) reside on the oxygen atoms, this fact is indicated by a $\frac{2}{3}-$ sign.

Other examples of resonance hybrids are the acetate and allyl ions.*

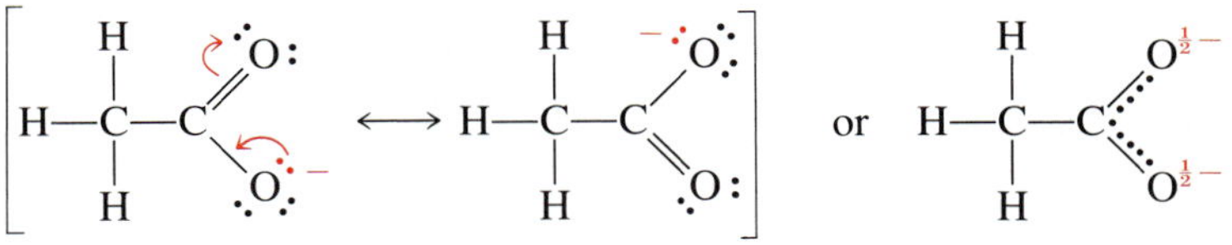

Acetate ion

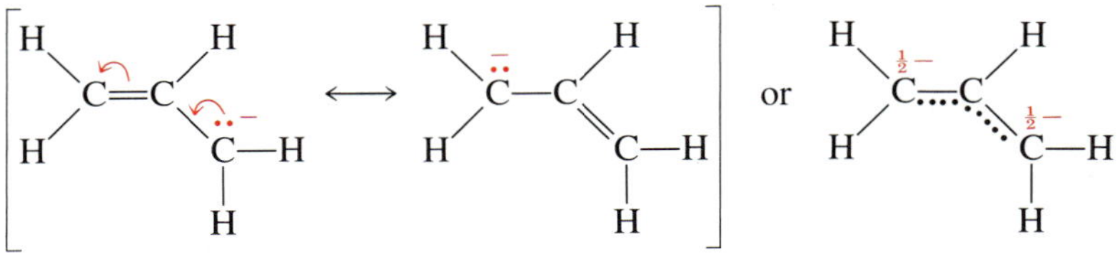

Allyl ion

When writing resonance structures, keep in mind that (1) pushing one electron pair toward one atom and away from another results in a movement of charge; (2) the relative positions of all the atoms stay unchanged; (3) the energies of all equivalent resonance structures are the same; and (4) the arrows connecting resonance structures are double-headed (↔).

EXERCISE 1-7

Write two resonance structures for nitrite ion, NO_2^-. What can you say about the geometry of this molecule?

How to Deal with Nonequivalent Resonance Structures

The resonance structures for the carbonate, acetate, and allyl anions are all equivalent. However, many molecules can be described by a set of Lewis structures that are not equivalent. An example is the enolate ion, an oxygen analog of

*These are common names. The systematic name for acetate is *ethanoate* (Section 17-1) and for allyl is 2-*propenyl* (Section 14-1). Because the systematic naming of organic molecules has not yet been described, common names are used in this discussion and throughout this chapter.

the allyl anion. The two resonance structures differ in the locations of both the double bonds and the charges.

The Two Nonequivalent Resonance Structures of the Enolate Ion

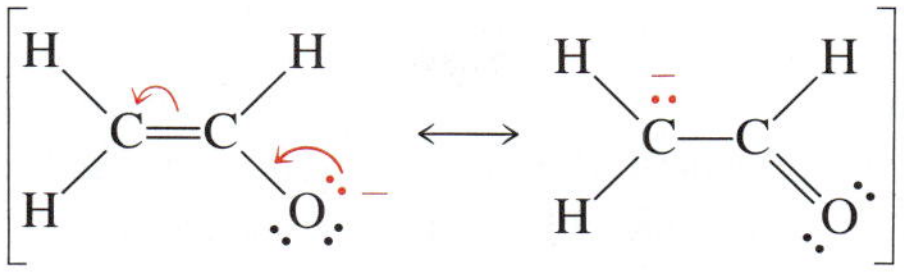

Because the two structures differ, one is closer to the truth than the other, although neither structure is quite correct. How can such structures be recognized, and what are the features contributing to their relative stability? Several guidelines help to resolve this problem.

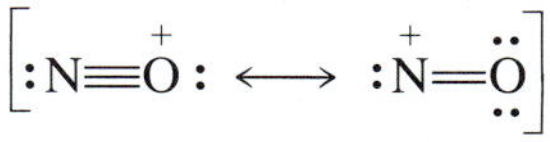

GUIDELINE 1 *Structures with a maximum of octets are preferred.* In the enolate ion, all component atoms in either structure are surrounded by octets. Consider, however, the nitrosyl cation, NO^+: one of the two resonance structures has a positive charge on oxygen with electron octets around both atoms; the other has the positive charge on nitrogen but has only one octet, which is on oxygen. Because of the octet rule, the second structure contributes less to the hybrid. Thus, the molecule is more likely to have a triple than a double bond, and more of the positive charge is on oxygen than on nitrogen.

Major resonance contributor **Minor resonance contributor**

GUIDELINE 2 *Charges should be preferentially located on atoms with compatible electronegativity.* Consider again the enolate ion. Which is the preferred resonance structure? Guideline 2 requires that it is the first, in which the negative charge resides on the more-electronegative oxygen atom.

Looking again at NO^+, you might find Guideline 2 confusing. The major resonance contributor to NO^+ has the positive charge on the more-electronegative oxygen. In cases such as this, the explanation for this seeming contradiction is that the octet rule overrides the electronegativity criterion; that is, Guideline 1 takes precedence over Guideline 2.

GUIDELINE 3 *Structures with a minimum of charge separation are preferred.* This rule is a simple consequence of Coulomb's law: separating charges requires energy; hence neutral structures are better than dipolar ones.

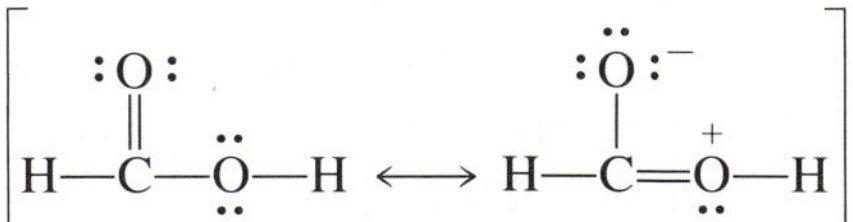

Major **Minor**

Formic acid

GUIDELINE 4 *Charge separation may be enforced by the octet rule.* In some cases, to ensure octet Lewis structures, charge separation is acceptable, that is, Guideline 1 takes precedence over Guideline 3. An example is carbon monoxide:

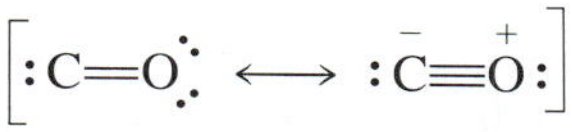

Minor **Major**

Carbon monoxide

$$\left[\begin{array}{c} \mathrm{H_2C{=}\overset{+}{N}{=}\overset{-}{\ddot{N}}{:}} \\ \textbf{Major} \\ \updownarrow \\ \mathrm{H_2\overset{-}{\ddot{C}}{-}\overset{+}{N}{\equiv}N{:}} \\ \textbf{Minor} \end{array}\right]$$

Diazomethane

If in such a case several charge-separated resonance structures are possible, the most favorable is the one in which the charge distribution best accommodates the relative electronegativities of the component atoms (Guideline 2). In diazomethane, for example, nitrogen is more electronegative than carbon.

EXERCISE 1-8

Write resonance structures for the following two molecules. Indicate the more favorable resonance contributor in each case. (a) CNO^-; (b) NO^-.

In summary, there are molecules that cannot be described accurately by one Lewis structure but exist as hybrids of several extreme resonance structures. To find the most important resonance contributor to a given molecule, consider the octet rule, make sure that there is a minimum of charge separation, and place on the relatively more electronegative atoms as much negative and as little positive charge as possible.

1-6 A Quantum-Mechanical Description of Electrons Around the Nucleus: Atomic Orbitals

We are now ready to delve a little deeper into the nature of the atom, particularly into the way in which the electrons are distributed around the nucleus, both spatially and energetically. The simplified treatment presented here has as its basis the theory of quantum mechanics developed independently in the 1920s by Heisenberg, Schrödinger, and Dirac.* In this theory, the mathematics describing the movement of an electron around the hydrogen nucleus is expressed in the form of equations that are very similar to those characteristic of waves. The solutions to these equations allow us to describe the probability of finding the electron in a certain region in space. The shape of these domains depends on the energy of the electron.

The Electron Is Described by Wave Equations

The classical picture of the atom (Bohr theory) assumed that electrons move on more or less defined trajectories around the nucleus. Their energy was thought to relate to the distance of their trajectory from the nucleus. We thus have a picture that resembles that of a planetary system in which orbiting planets (electrons) surround a central sun (nucleus). This view is intuitively appealing because it coincides with our physical understanding of classical mechanics. Yet this view must be incorrect for several reasons.

First, the classical picture of an electron moving in its orbit would require the emission of electromagnetic radiation. This emission should cause enough loss of energy from the system that the electron would spiral toward the nucleus, presumably resulting in eventual collapse. However, as we know, this is not observed—atoms (unless radioactive) are very stable entities.

*Professor Werner Heisenberg, 1901–1976, University of Munich, Nobel Prize 1932 (physics); Professor Erwin Schrödinger, 1887–1961, University of Dublin, Nobel Prize 1933 (physics); Professor Paul Dirac, 1902–1984, Florida State University, Nobel Prize 1933 (physics).

Second, in accord with classical mechanics, an atom should have an infinite number of electronic states with differing energies, corresponding to an infinite number of orbits of differing radii. This, again, is not what is observed. Rather, only certain defined energy states are possible for an electron occupying space around a nucleus. Thus, classical mechanics does not satisfactorily explain atomic structure and, ultimately, bonding.

A better model is afforded by considering the wave nature of moving particles. Matter of mass m that moves with velocity v has a wavelength λ:

$$\lambda = \frac{h}{mv}$$

in which h = Planck's constant. As a consequence, the movement and energetics of an orbiting electron can be described by equations that are the same as those used for the descriptions of waves (Figure 1-4). The latter have amplitudes with alternating positive and negative signs. Points at which the sign of the wave changes are called *nodes*. Waves that interact in phase reinforce each other, as in Figure 1-4B. Those out of phase interfere with each other to make smaller waves (and possibly even cancel each other), as in Figure 1-4C.

FIGURE 1-4

A. A wave. The signs of the amplitude are assigned arbitrarily. Points of zero amplitude are called nodes.

B. Waves with amplitude of equal sign (in phase) reinforce each other to make larger waves.

C. Waves out of phase subtract from each other to make smaller waves.

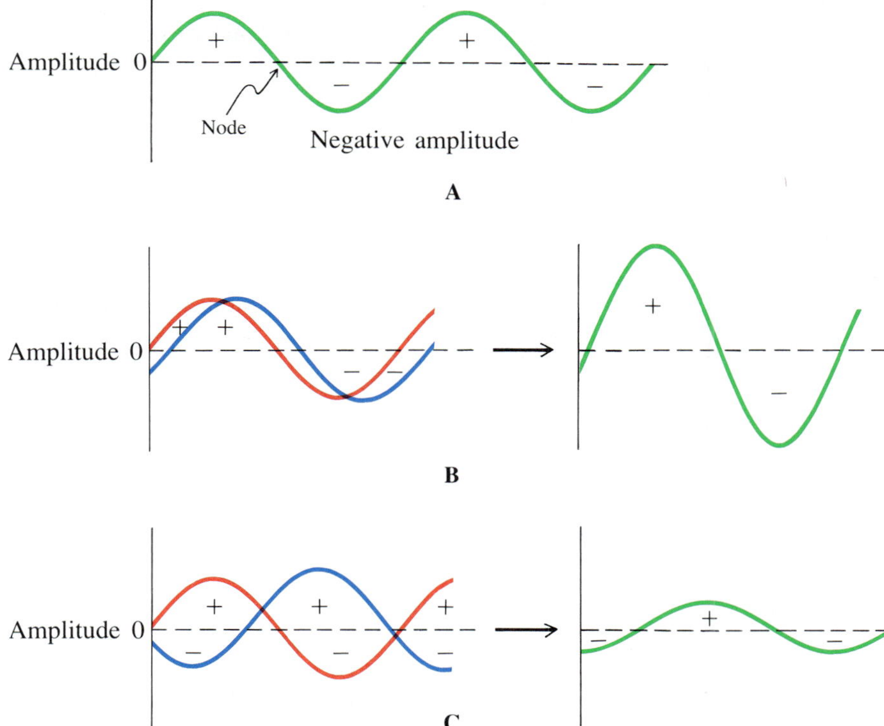

The mathematical theory describing the motion of the electron around the nucleus is called *quantum,* or *wave, mechanics*. The differential equations developed in this theory are the *wave equations*. These equations have a series of solutions, the *wave functions*. Wave functions are usually described by the Greek letter psi, ψ. Their values for each point in space around the nucleus are not directly identifiable with any observable property of the atom. However, *their squares* (ψ^2) *describe the probability of finding an electron at any given point*. Solutions of the wave equations are also referred to as **atomic orbitals.** The boundary conditions imposed by the physical realities of the atom make solutions attainable only for certain *specific energies* of the system. There is no continuum of electronic states with differing energies. Instead, there are only discrete states. The system is said to be *quantized*.

EXERCISE 1-9

Draw a picture similar to Figure 1-4 of two waves overlapping such that their amplitudes cancel each other.

Atomic Orbitals Have Characteristic Shapes

Wave functions when plotted have the appearance of spheres or lobes with positive and negative amplitudes and inflection points (nodes) where positive values change to negative and vice versa. High-energy solutions have more nodes than those of low energy.

Let us consider the shapes of the atomic orbitals for the simplest case, that of the hydrogen atom, consisting of a proton surrounded by an electron. The lowest-energy wave function is called the 1*s* orbital and constitutes a single first solution of the wave equation. We will see shortly that there may be several equivalent solutions of this equation at higher energy. The number in front of the letter refers to the first (lowest) energy level.

The label 1*s* denotes the shape and number of nodes of the orbital. The 1*s* orbital is *spherically symmetric* (Figure 1-5) and has no nodes. This orbital can be represented pictorially as a diffuse cloud, as in Figure 1-5A, or simply as a circle, as in Figure 1-5B.

FIGURE 1-5

Representations of a 1*s* orbital: (A) the three-dimensional nature of the orbital; (B) a simplified two-dimensional view. The plus sign denotes the sign of the wave function and *not a charge*.

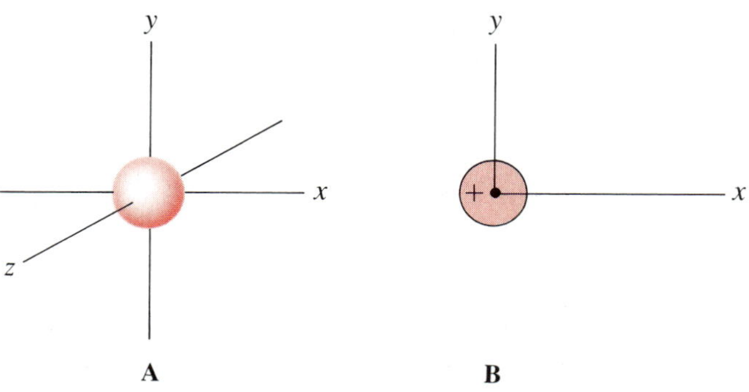

Because the square of the numerical value of the wave function at a certain point in space indicates the probability of finding the electron at that point, the atomic-orbital drawings define regions in space within which there is a high (typically about 90%) probability of locating the electron. The electron density outside this sphere (and toward its center) diminishes rapidly.

The next higher-energy solution of the wave equation is again unique, and it, too, is spherical: the 2*s* orbital. There are basically two differences between a 1*s* and a 2*s* orbital. First, the 2*s* orbital is larger, owing to the higher energy of the electron, which is on the average farther from the positive nucleus. Second, the 2*s* orbital has one node. Because the sign of the wave function changes on crossing this spherical surface (Figure 1-6), the probability of finding an electron on the node is zero. Like that of classical waves, the sign of the wave function on either side of the node is arbitrary, as long as the sign changes at the node. The sign of the wave function will become important in the discussion of bonding between atoms (Section 1-7).

FIGURE 1-6

Representations of a 2*s* orbital. Notice that it is larger than the 1*s* orbital and that a node is present. The plus and minus denote the sign of the wave function. Part A shows the orbital in three dimensions, with a section removed to allow the visualization of the node. Part B shows the more-conventional two-dimensional representation.

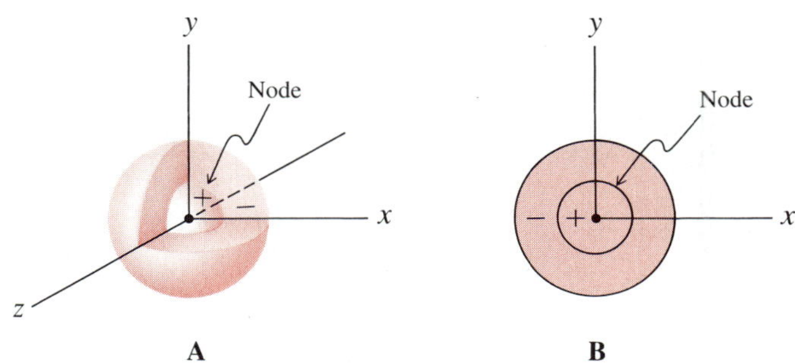

After the 2*s* orbital, the wave equations for the electron around a hydrogen atom furnish three energetically equivalent solutions, the $2p_x$, $2p_y$, and $2p_z$ orbitals. Such *p* orbitals consist of two touching spheres, or lobes, that look like a "spherical figure eight," as shown in Figure 1-7. Solutions of equal energy of this type are also called *degenerate* (*degenus,* Latin, without genus or kind). A *p* orbital is characterized by its directionality in space. The orbital axis can be aligned with any one of the *x, y,* and *z* axes of the rectangular Cartesian coordinate system, with the nucleus at the origin (hence the labels p_x, p_y, p_z). The two lobes of opposite sign of each orbital are separated by a nodal plane perpendicular to the orbital axis.

The next higher orbital levels are the 3*s* and 3*p* atomic orbitals. They are similar in shape but more diffuse than their lower-energy counterparts and have two nodes. Solutions of the wave equation for higher energy than 3*s* and 3*p* are designated *d* (and eventually *f*). These orbitals are characterized by an increasing number of nodes and a variety of shapes. They are of much less importance in organic chemistry than are the lower orbitals. The relative energies of the important atomic orbitals up to the 5*s* level are shown in Figure 1-8.

FIGURE 1-7

Representations of $2p$ orbitals: (A) in three dimensions; (B) in the conventional two dimensions.

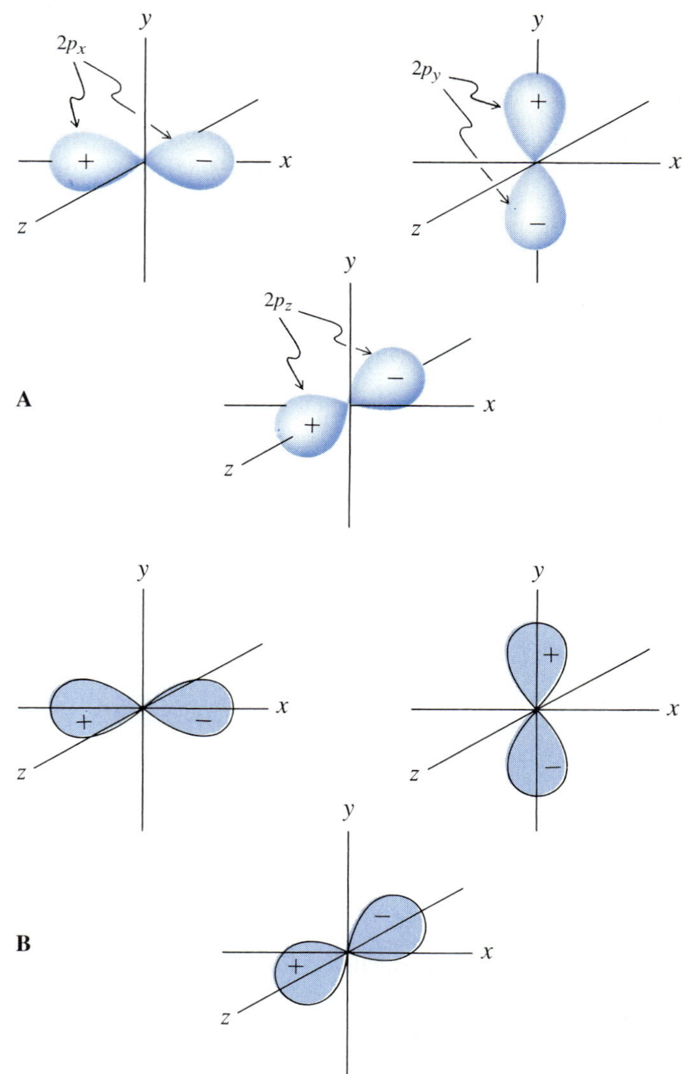

Atomic Orbitals for Elements Other Than Hydrogen

Would the solutions of the wave equations for the hydrogen atom be the same for other atoms? Strictly speaking, the answer is "no," because the hydrogen atom has only one proton and one electron, which interact mutually. The mathematics required for solving the quantum-mechanical equations for elements having more electrons would have to take into account the effect of the electrons on each other, and such calculations become very complicated. Fortunately, to a first approximation, the shapes and nodal properties of the atomic orbitals of other elements are very similar to those of hydrogen. Therefore, we may use *s* and *p* orbitals in a description of the electronic configuration of helium, lithium, and so forth. However, orbitals decrease in size from left to right along a period

FIGURE 1-8
Schematic energy diagram for the atomic orbitals in hydrogen. The energy scale is non-linear, the lower end being compressed relative to the higher end.

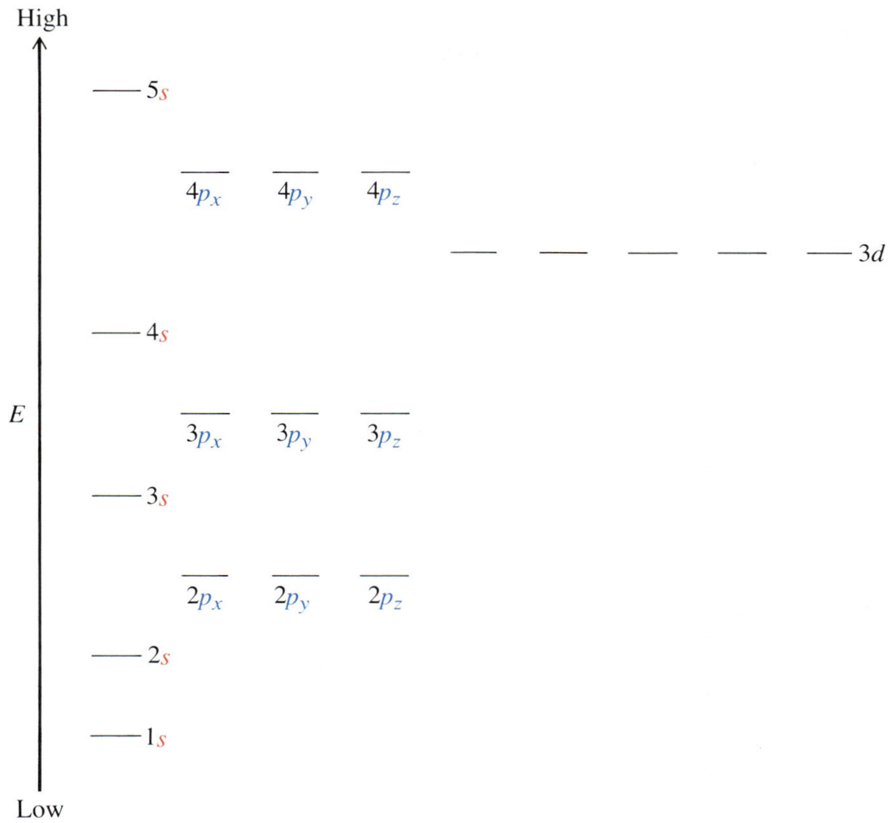

of the periodic table. This phenomenon is due to the increasing nuclear charge. We will see that it may have consequences in bonding.

How to Assign Electrons to Atomic Orbitals: The Aufbau Principle

With the help of Figure 1-8, we can give an electronic configuration to every atom in the periodic table. To do so, we need to follow three rules for assigning electrons to atomic orbitals:

1 Lower-energy orbitals are filled before those with higher energy.

2 The **Pauli exclusion principle** in effect states that any given orbital may not be occupied by more than two electrons and that, furthermore, these two electrons must differ in the orientation of their intrinsic angular momentum, their **spin.** There are two possible directions of the electronic spin, usually depicted by vertical arrows pointing in opposite directions. An orbital is filled when it is occupied by two electrons of opposing spin, frequently referred to as *paired electrons*.

3 **Hund's rule** gives instructions on the sequence in which degenerate orbitals, such as the *p* orbitals, are filled: each is first occupied by one electron, all of the same spin. Subsequently, three more, each of opposite spin, are added to the first set.

FIGURE 1-9

The most stable electronic configuration of elemental carbon, $(1s)^2(2s)^2(2p)^2$. Notice that the unpaired spins of both electrons in the p orbitals are in accord with Hund's rule, and the paired spins of the electrons in the filled $1s$ and $2s$ orbitals in accord with Pauli's principle and Hund's rule. The two unpaired electrons have been arbitrarily placed into the p_x, p_y orbitals. Any other combination of two $2p$ orbitals would have been equally correct.

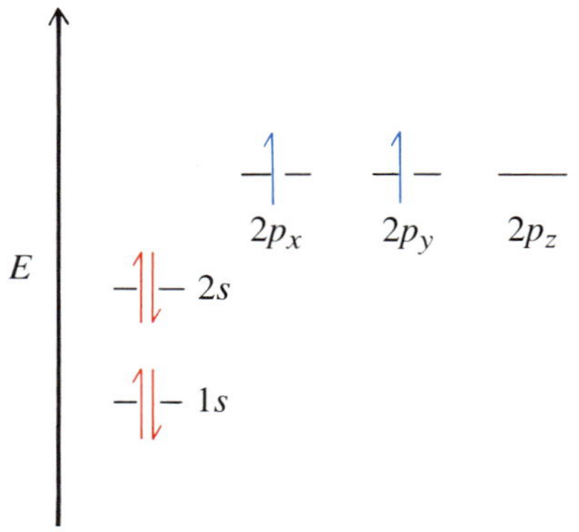

With these rules in hand, the determination of electronic configuration becomes simple. Helium has two electrons in the $1s$ orbital and its electronic structure is abbreviated $(1s)^2$. Lithium $[(1s)^2(2s)^1]$ has one and beryllium $[(1s)^2(2s)^2]$ two additional electrons in the $2s$ orbital. In boron $[(1s)^2(2s)^2(2p)^1]$, we begin to fill up the three degenerate $2p$ orbitals with one electron, followed by additional unpaired electrons for carbon $[(1s)^2(2s)^2(2p)^2]$ and nitrogen $[(1s)^2(2s)^2(2p)^3]$, and then pairing electrons for oxygen, fluorine, and neon to fill all p levels. The electronic configuration of carbon is depicted in Figure 1-9. Atoms with completely filled atomic orbitals are said to have a *closed shell*. For example, helium, beryllium, and neon have this attribute. Carbon, in contrast, has an *open-shell* configuration.

The process of adding electrons one by one to the orbital sequence calculated for the hydrogen atom (Figure 1-8) is called the *Aufbau principle* (*Aufbau,* German, build up). It is easy to see that the Aufbau principle affords a rationale for the stability of the electron octet and duet. These numbers are required to create closed-shell configurations. For helium, it signifies filling the $1s$ orbital with two electrons of opposite spin. In neon, the $2s$ and $2p$ orbitals are occupied by an additional eight electrons; in argon, the $3s$ and $3p$ levels again require eight more electrons (Figure 1-10).

EXERCISE 1-10

Using Figure 1-8, write the electronic configurations of sulfur and phosphorus.

To summarize, the motion of electrons around the nucleus is described by wave equations. Their solutions, atomic orbitals, can be symbolically represented as spheres in space, with each point given a positive, negative, or zero (at the node) numerical value, the square of which represents the probability of finding the electron there. The Aufbau principle allows us to assign electronic configurations to all atoms.

FIGURE 1-10
Closed-shell configuration of the noble gases helium, neon, and argon.

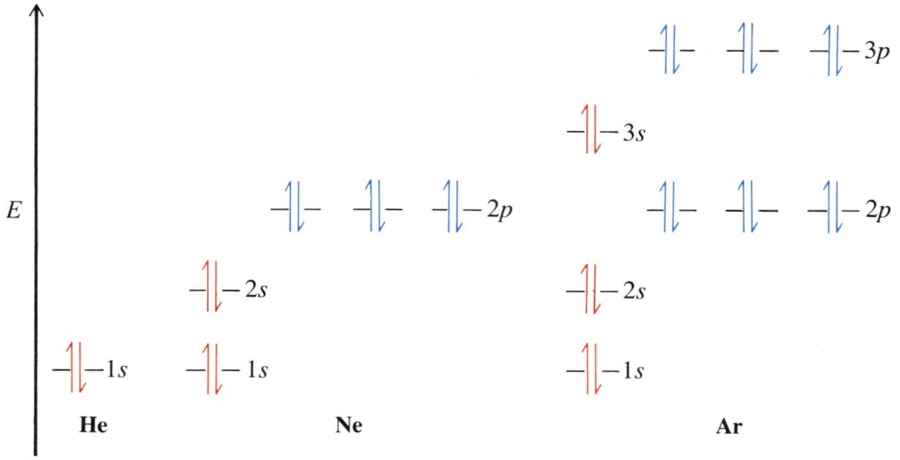

1-7
Bonding by Overlap of Atomic Orbitals: Molecular Orbitals

We will now see how covalent bonds are constructed by the in-phase overlap of atomic orbitals.

The Bond in the Hydrogen Molecule

Let us begin by looking at the simplest case: the bond between the two hydrogen atoms in H_2. In a Lewis structure of the hydrogen molecule, we would write the bond as an electron pair shared by both atoms to give each a helium configuration. How do we construct H_2 using atomic orbitals? An answer to this question was developed by Pauling*: *bonds are made by the in-phase overlap of atomic orbitals*. What is meant by that? Recall that atomic orbitals are solutions of wave equations. Like waves, they may interact in a reinforcing way (Figure 1-4) if the overlap is between areas of the wave function of the same sign. They may also interact in a destructive way if the overlap is between areas of opposite sign.

The in-phase overlap of the two 1*s* orbitals results in a new orbital of lower energy called the **bonding molecular orbital** (Figure 1-11). On the other hand, out-of-phase overlap between the same two orbitals results in a destabilizing interaction and formation of an **antibonding molecular orbital** (Figure 1-11). In the bonding combination, the wave function in the space between the nuclei is strongly reinforced. This also means that the probability of finding the electrons occupying this molecular orbital in that region is very high: a condition for bonding between the two atoms. This picture is strongly reminiscent of that shown in Figure 1-2. The use of two wave functions with *positive* signs for

*Professor Linus Pauling, b. 1901, Professor Emeritus, Stanford University, Nobel prizes 1954 and 1963 (peace).

FIGURE 1-11

In-phase (bonding) and out-of-phase (antibonding) combination of the orbitals of atomic hydrogen to give molecular hydrogen. The plus and minus signs denote the *sign* of the wave function and not charges. The dots represent the electrons. The antibonding molecular orbital has a node.

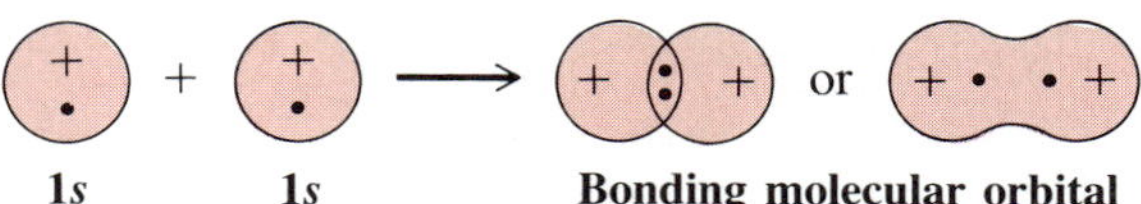

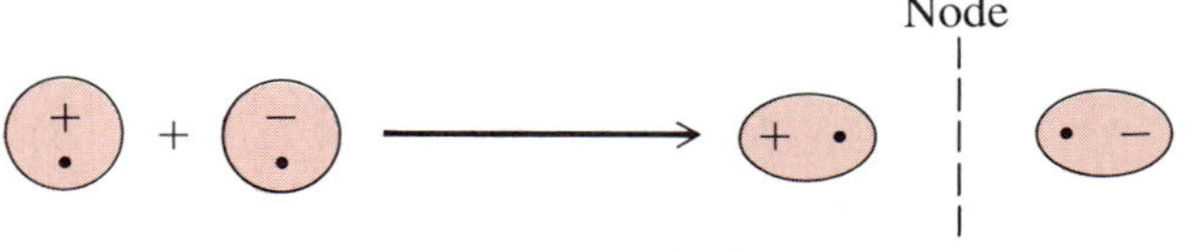

representing the in-phase combination of the two 1*s* orbitals in Figure 1-11 is arbitrary. Overlap between two *negative* orbitals would give identical results. In other words, it is overlap between *like* lobes that makes a bond, regardless of the sign of the wave function. In the antibonding molecular orbital, the amplitude of the wave function is cancelled in the space between the two atoms, giving rise to a node.

Thus, the net result of the interaction of the two 1*s* atomic orbitals of hydrogen is the generation of two molecular orbitals. One is bonding and lower in energy; the other is antibonding and higher in energy. Because the total number of electrons available to the system is only two, they are placed in the lower-energy molecular orbital. The result is a gain in energy compared with two noninteracting hydrogen atoms. This situation can be depicted schematically in an energy diagram, as in Figure 1-12A.

Is there any practical significance to the presence of an antibonding molecular orbital? Indeed there is. Under certain circumstances, bonding electrons may be activated by some form of energy (heat or light) in such a way that one of the bonding electrons is excited to the higher-energy antibonding level. In terms of bond strengths, this is approximately equivalent to having no bond at all, because the energy gained by placing one electron in the bonding molecular orbital is cancelled by the energy lost in placing the second electron into its antibonding counterpart. As a result such excitation may lead to the breaking of a bond.

It is readily understandable why hydrogen exists as H_2, whereas helium exists in the atomic form. The overlap of two completely filled atomic orbitals, as in helium, is energetically unprofitable, leading to a filled bonding as well as antibonding molecular orbital (Figure 1-12B). Therefore, no energy is gained by making a He–He bond.

Bonds Are Made by the Overlap of Atomic Orbitals: Sigma and Pi Bonds

The splitting of energy levels by interacting atomic orbitals is a general phenomenon that applies not only to the 1*s* orbitals of hydrogen, but also to other atomic orbitals. The extent of the splitting—that is, the amount of energy by

FIGURE 1-12
Schematic representation of the interaction of two (A) singly (as in H_2) and (B) doubly (as in He_2) occupied atomic orbitals to give two molecular orbitals (MO). (Not drawn to scale.)

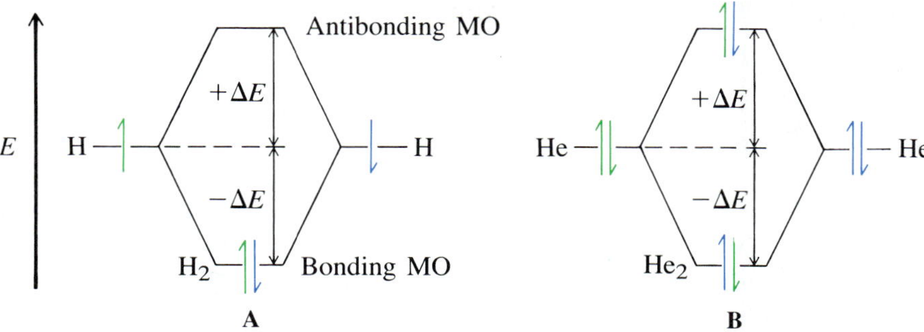

which the bonding level drops and the antibonding one is raised, which reflects the strength of the bond being made—depends on a variety of factors. One is how close in energy the starting atomic orbitals are before interaction. For example, two 1*s* orbitals will interact with each other more effectively than a 1*s* orbital and a 3*s* orbital. This result can be readily understood on inspection of the actual shape of the orbitals (Figure 1-13). Overlap is best between orbitals of similar size (and therefore similar energy) because it is related to the extent by which the same region in space is occupied by both bonding electrons. Therefore, overlap of a small orbital with a much larger one is poorer, because the large orbital is associated with a more-diffuse electron distribution.

Geometric factors also play a significant role in determining the degree of overlap. This is very important for orbitals with directionality in space, such as the *p* orbitals, which give rise to two types of bonds: one in which the atomic

FIGURE 1-13
Bonding between atomic orbitals: (A) 1*s* and 1*s* (e.g., H_2); (B) 1*s* and 2*p* (e.g., HF); (C) 2*p* and 2*p* aligned along internuclear axis (e.g., F_2); (D) 2*p* and 2*p* perpendicular to internuclear axis, a pi bond; (E) 2*p* and 3*p* (e.g., FCl). Note the arbitrary use of plus and minus signs to indicate in-phase interactions of the wave functions.

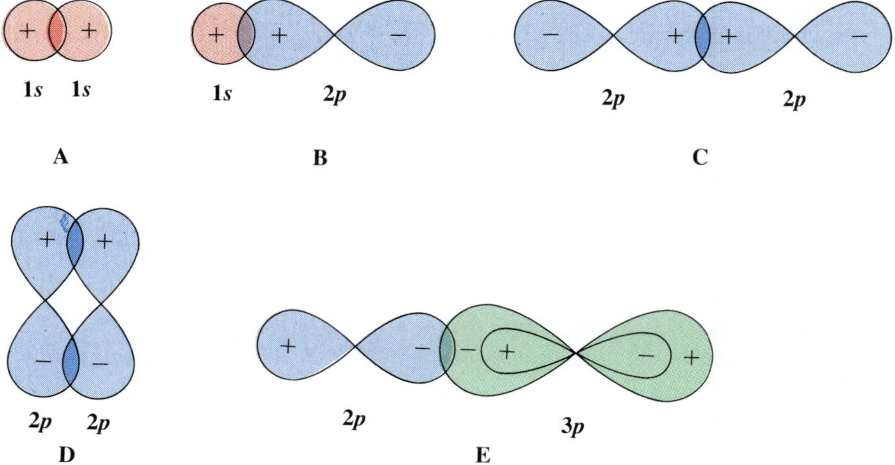

orbitals are aligned along the internuclear axis (parts A, B, C, and E in Figure 1-13) and one in which they are perpendicular (part D). The first is said to give rise to **sigma (σ) bonds,** the second to **pi (π) bonds.** All carbon–carbon single bonds are σ bonds; however, double and triple bonds also have π components (Chapters 11 and 13).

EXERCISE 1-11

Construct a molecular orbital and energy-level-splitting diagram of the bonding in He_2^+. Is it favorable?

We have come a long way in our description of bonding. First, we thought of bonds in terms of Coulombic forces, then in terms of covalency and shared electron pairs, and now we have a quantum-mechanical picture. Bonds are a result of the overlap of atomic orbitals. The two bonding electrons are placed in the bonding molecular orbital. Because this molecular orbital is lower in energy than were the two initial atomic orbitals, energy is given off on bond formation.

1-8 Bonding in Complex Molecules: Hybrid Orbitals

Let us now construct bonding schemes for more complex molecules by using quantum-mechanical techniques. How can we use atomic orbitals to build linear (as in BeH_2), trigonal (as in BH_3), and tetrahedral molecules (as in CH_4, NH_3, and H_2O)?

Mixing Orbitals in an Atom Gives Hybrid Orbitals: Beryllium Hydride

In general, the picture of the overlap of atomic orbitals satisfactorily illustrates bonding in simple diatomic molecules. However, this scheme does not work for larger systems.

Consider the triatomic molecule beryllium hydride, BeH_2. Application of the Aufbau principle (Figure 1-8) reveals that the electronic configuration of beryllium is closed shell, with two electrons in the 1*s* orbital and two electrons in the 2*s* orbital. This arrangement does not appear to allow for bonding.

However, it takes a relatively small amount of energy to promote one electron from the 2*s* orbital to one of the 2*p* levels (Figure 1-14). In the $1s^2 2s^1 2p^1$ configuration, beryllium is ready to enter into bonding, because there are now two singly filled atomic orbitals available for overlap. One could propose bond formation by overlap of the Be 2*s* orbital with the 1*s* orbital of one H, on the one hand, and the Be 2*p* orbital with the second H, on the other (Figure 1-15). This proposal has immediate consequences for the structure of the molecule. Namely, one would observe two different bonds of unequal length, probably at an angle. Unfortunately, BeH_2 does not exist as a single molecule (monomer) but as a collection of many (polymer), and hence this prediction cannot be verified experimentally. However, very similar compounds exist, such as gaseous dimethylberyllium, $Be(CH_3)_2$, that have been shown to be *linear* and to contain C–Be bonds of *equal* length. We know that this is the structure adopted as a consequence of electron repulsion (Section 1-3). How can we explain the bonding in these molecules using *s* and *p* orbitals? To answer this question, we use a quantum-mechanical approach called **orbital hybridization.**

FIGURE 1-14
Promotion of an electron in beryllium to allow the use of both valence electrons in bonding.

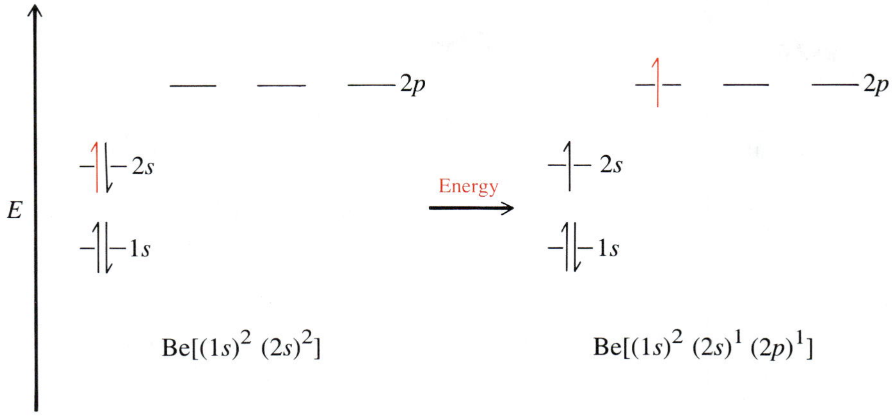

E

2p

2s

1s

Energy

2p

2s

1s

$Be[(1s)^2\ (2s)^2]$

$Be[(1s)^2\ (2s)^1\ (2p)^1]$

FIGURE 1-15
Possible bonding in BeH_2 by separate use of a 2*s* and a 2*p* orbital on beryllium. The nodal surface in the former is not shown. Moreover, the other two empty *p* orbitals and the lower-energy filled 1*s* orbital are omitted for clarity. The dots indicate the valence electrons.

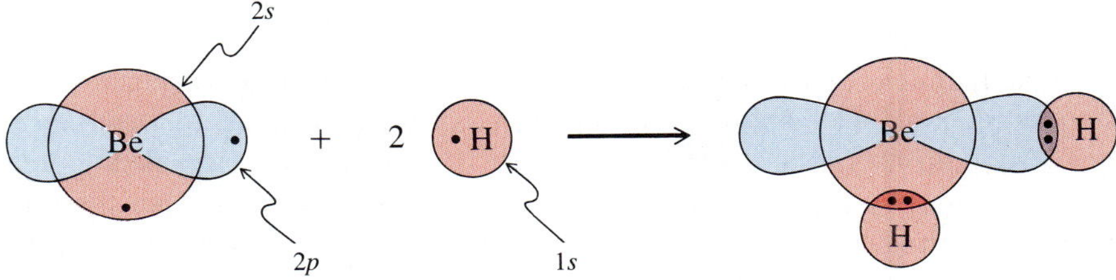

As in the mixing of atomic orbitals on different atoms to make covalent bonds, atomic orbitals on the *same* atom can be combined to form new **hybrid orbitals.** For beryllium, we mix the 2*s* and one of the 2*p* wave functions to give two new hybrids, called *sp* orbitals, made up of 50% *s* and 50% *p* character. This mixing rearranges the orbital lobes in space as shown in Figure 1-16: the major parts of the orbitals, also called front lobes, point away from each other at an angle of 180°. There are two additional minor back lobes (one for each *sp* hybrid) with opposite sign. The remaining two *p* orbitals are left unchanged.

This hybridization scheme minimizes electron repulsion and maximizes bonding, owing to the greater spatial availability of the *sp* lobes. Overlap with the two hydrogen 1*s* orbitals then occurs naturally to yield linear BeH_2. In the same manner, the two *sp* hybrids of beryllium can overlap the orbitals of other elements to form other beryllium compounds. Note that hybridization does not change the overall number of orbitals available for bonding. Hybridization of the four orbitals in beryllium gives a new total set of four: two *sp* hybrid orbitals and two relatively unchanged 2*p* orbitals. We will see in Section 13-2 that carbon utilizes *sp* hybrids to form triple bonds.

FIGURE 1-16
Hybridization in beryllium to create two *sp* hybrids and the resulting bonding for BeH_2. Again, both remaining *p* orbitals and the 1*s* orbital have been omitted for clarity. The sign of the wave function for the large *sp* lobes is opposite that for the small lobes.

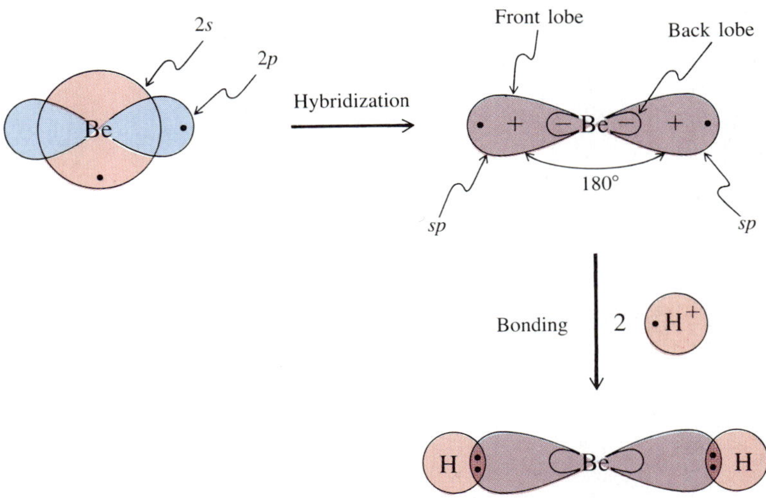

sp^2 Hybrids Create Trigonal Structures

Now let us consider the group of elements in the periodic table with three valence electrons. What bonding scheme can be derived for borane, BH_3, trimethylaluminum, $Al(CH_3)_3$, and similar molecules?

Using atomic orbitals to describe borane, we again resort to hybridization. In the present case, we combine one 2*s* and *two* of the 2*p* orbitals to create *three* new hybrid orbitals. These are designated sp^2 to indicate the component atomic orbitals (Figure 1-17). The third *p* orbital is left unchanged, and so the total number of orbitals stays the same: four. As with beryllium, it need not concern us that the Aufbau principle calls for a boron atom of configuration $(1s)^2(2s)^2(2p)^1$, and hence only one unpaired electron. Promotion of a 2*s* electron to one of the 2*p* levels gives the three singly filled atomic orbitals (one 2*s*, two 2*p*) needed for hybrid-orbital formation.

The three sp^2 orbitals of boron in BH_3 extend in space in a way that achieves the desired equilateral trigonal arrangement with minimum electron repulsion. Hybridization also enhances the possibility of overlap and therefore stronger bonds than those formed by *p* orbitals (Figure 1-17). The three front lobes have the same sign and overlap the respective 1*s* orbitals of the hydrogen atoms to form BH_3. The remaining unchanged *p* orbital is perpendicular to the plane incorporating the sp^2 hybrids. It is empty and does not enter significantly into bonding. As with *sp* hybrids, sp^2 hybrids can form bonds through overlap with atomic orbitals other than 1*s*. For example, boron trifluoride, BF_3, is constructed by interaction of the three sp^2 hybrids of boron with three singly filled 2*p* orbitals on fluorine to form three new doubly filled sp^2-*p* molecular orbitals.

The molecule BH_3 is *isoelectronic* with the methyl cation, CH_3^+; that is, they have the same number of electrons. Bonding in CH_3^+ also occurs by the use of three sp^2 hybrid orbitals, and we will see in Section 11-2 that carbon utilizes sp^2 hybrids in double-bond formation.

FIGURE 1-17
Hybridization in boron to yield three sp^2 hybrids, and the resulting bonding in BH_3. There are three front lobes of one sign and three back lobes of opposite sign. The remaining p orbital (p_z) extends perpendicular to the molecular plane (the plane of the page; one lobe is above, the other below that plane) and has been omitted.

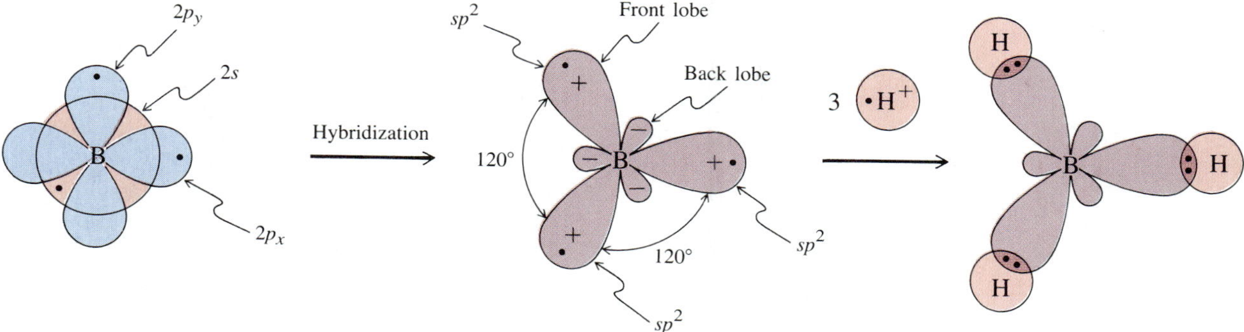

Hybridization in Saturated Carbon Compounds: Methane

Consider the element whose bonding is of most interest to us: carbon. Its electronic configuration is $(1s)^2(2s)^2(2p)^2$, with two unpaired electrons residing in two $2p$ orbitals. Promotion of one electron from $2s$ to $2p$ results in four singly filled orbitals for bonding, in accord with the tetravalent character of carbon. We have learned that the arrangement of the four C–H bonds of methane in space that would lead to least electron repulsion is tetrahedral, in which electron density is distributed along the four bonds pointed in the direction of the vertices of a tetrahedron (Section 1-3). To be able to achieve this geometry, the $2s$ orbital on carbon is hybridized with *all three* $2p$ orbitals to make *four* equivalent sp^3 orbitals with tetrahedral symmetry, each occupied by one electron. Overlap with four hydrogen $1s$ orbitals furnishes methane with four equal C–H bonds. The HCH bond angles are typical of a tetrahedron: 109.5° (Figure 1-18).

FIGURE 1-18
Hybridization in carbon to create four sp^3 hybrids, and the resulting bonding in CH_4. The sp^3 hybrids contain small back lobes of sign opposite that of the front lobes.

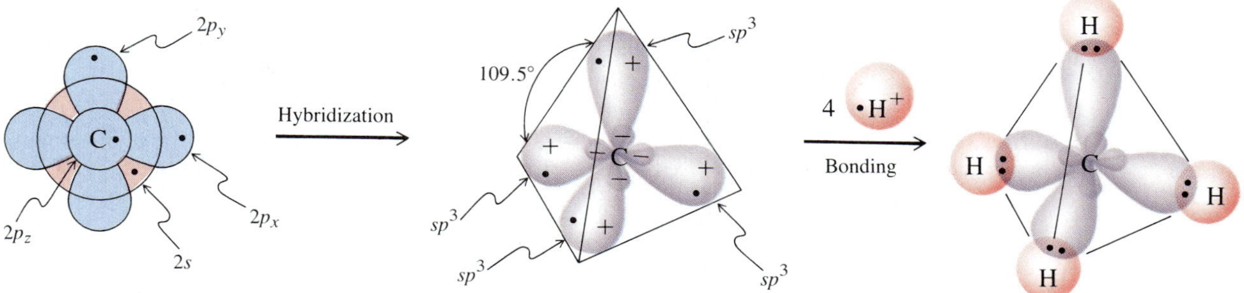

Like all other hybrid orbitals discussed so far, sp^3 hybrids may also overlap with other orbitals and hybrids. Overlap with, for example, four chlorine $3p$ orbitals would result in tetrachloromethane, CCl_4. Carbon–carbon bonds are formed by overlap between two hybrid orbitals. In ethane, CH_3–CH_3 (Figure

1-19), C–C overlap occurs between two sp^3 hybrids, one from each of two CH_3 units. Every hydrogen atom in methane and ethane may be replaced by CH_3 or other groups to give several new combinations. In all of these molecules, carbon is approximately tetrahedral. It is this ability of carbon to form chains of atoms bearing a variety of additional substituents that gives rise to the extraordinary diversity of organic chemistry.

FIGURE 1-19
Overlap of two sp^3 orbitals to form the carbon–carbon bond in ethane.

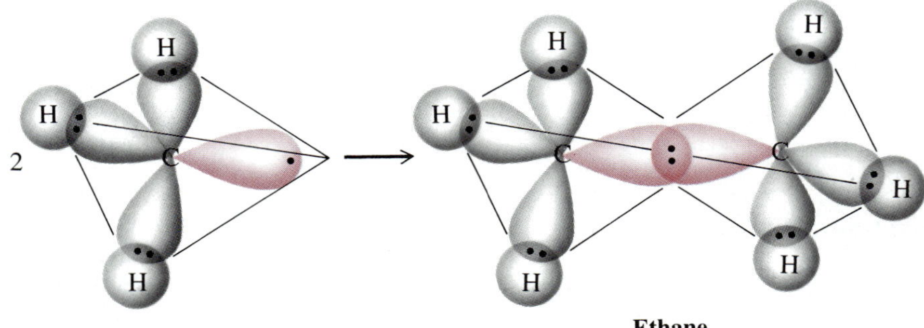

Lone Electron Pairs and Hybrids: Ammonia and Water

What sort of orbitals should be used to describe the bonding in ammonia and water? Let us begin with ammonia. The electronic configuration of nitrogen, $(1s)^2(2s)^2(2p)^3$, explains why nitrogen is trivalent, three covalent bonds being needed to facilitate electron octet formation. We could use pure *p* orbitals for overlap, leaving the nonbonding electron pair in the 2*s* level. However, this arrangement does not minimize electron repulsion. The best way to achieve this effect is by again using sp^3 hybrid orbitals. Three of the sp^3 orbitals are used to make bonds to the hydrogen atoms, whereas the fourth contains the lone electron pair. The HNH bond angles (107.3°) in ammonia are almost tetrahedral in accord with this picture (Figure 1-20).

Similarly, the bonding in water is best described by using sp^3 hybrid orbitals. The HOH bond angle is 104.5°, not quite but close to tetrahedral.

FIGURE 1-20
Bonding and electron repulsion in ammonia and water. The arcs indicate increased electron repulsion by the lone pairs located close to the central nucleus.

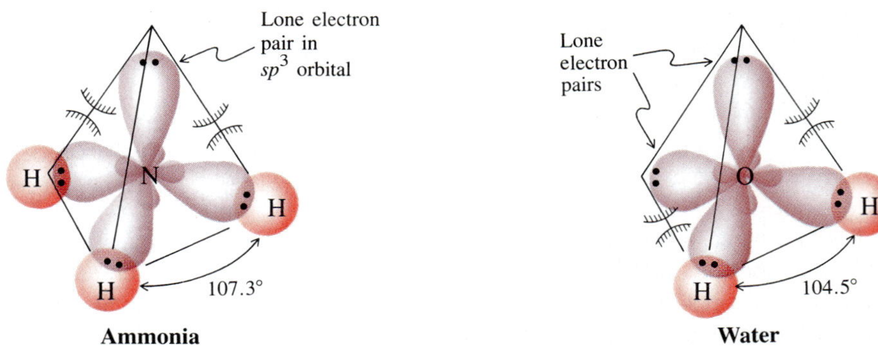

An explanation for these deviations in bond angles is found on consideration of the lone electron pairs in these systems (Figure 1-20). Because they are not used for bonding, lone electron pairs are not shared. As a result they are located closer to the single nucleus, nitrogen or oxygen, than are bonding electrons. This phenomenon causes increasing repulsion of the electrons in the respective bonds to hydrogen. Therefore, the bond angles in both ammonia and water are compressed from the tetrahedral value of 109.5°.

EXERCISE 1-12

Write a scheme for the hybridization and bonding in methyl cation, CH_3^+, and methyl anion, CH_3^-.

In summary, to minimize electron repulsion and maximize bonding in triatomic and larger molecules, the application of the concept of orbital hybridization to construct orbitals of appropriate shape is required. Combination of *s* and *p* atomic orbitals creates hybrids. Thus, a 2*s* and a 2*p* orbital mix to furnish two linear *sp* hybrids, the remaining two *p* orbitals being unchanged. Combination of the 2*s* with two *p* orbitals gives three sp^2 hybrids used in trigonal molecules. Finally, mixing the 2*s* with all three *p* levels results in the four sp^3 hybrids employed to describe the tetrahedral bonding in methane and similar saturated carbon compounds.

1-9 Composition, Structure, and Formulas of Organic Molecules

Now that we have a good understanding of bonding in organic molecules, let us find out how organic chemists go about determining their compositions and structures and, once the latter are established, how they apply various conventions for depicting these molecules on paper.

Elemental Analysis Reveals the Composition of Organic Compounds

How do we know the composition of a molecule? A classical way of obtaining initial information on the composition of an unknown structure is **elemental analysis.** In an elemental analysis, a small, accurately weighed amount of an unknown compound is converted into a measurable amount of a known substance containing the element to be determined. For example, the complete oxidation or combustion of a hydrocarbon to carbon dioxide and water can be accurately measured by using a machine called a C,H-analyzer. By knowing the amounts of the two products, one can very simply calculate the percentage of carbon and hydrogen in the original sample (this is in fact done automatically). Nitrogen-containing compounds can be made to produce measurable N_2 gas. Molecules incorporating chlorine are converted into solutions of chloride ion, which may be precipitated as silver chloride, AgCl, and then weighed, and so forth. Each element has its own way of yielding to quantitative analysis. Examples of the kinds of numbers obtained are shown on the next page.

Quantitative analysis reveals the number of atomic equivalents of each element contributing to the structure of a molecule. This assembled information is known as the **empirical formula,** which describes the ratio of the elements present. For example, the empirical formula of methane is CH_4. This is derived

Typical Elemental Analyses

CH_4: **Methane**	C 74.87%; H 25.13%
CCl_4: **Tetrachloromethane**	C 7.81%; Cl 92.19%.
$CH_3SH = CH_4S$: **Methanethiol**	C 24.97%; H 8.38%; S 66.65%.
$CH_3CH_2CH_3 = C_3H_8$: **Propane**	C 81.71%; H 18.29%.

from an elemental analysis, which indicates that there are four times as many hydrogen atoms present than there are carbons. How is this done in practice?

Let us assume that we obtained a sample of an unknown gas. To know the empirical formula, we would take a certain amount of the substance—say, 100 mg—and subject it to complete oxidation. This treatment would lead to measurable amounts of CO_2 (in this case, 275 mg) and H_2O (225 mg). These figures tell us that 100 mg of the gas furnish

$$\frac{\text{mg } CO_2}{\text{molecular weight } CO_2} = \frac{275}{44} = 6.25 \text{ mmoles of carbon (as } CO_2\text{),}$$

corresponding to 6.25 atomic equivalents (gram-atoms) of carbon

$$\frac{\text{mg } H_2O}{\text{molecular weight } H_2O} = \frac{225}{18} = 12.5 \text{ mmoles of } H_2 \text{ (as } H_2O\text{),}$$

corresponding to 25 atomic equivalents (gram-atoms) of (atomic) hydrogen

In this way, we can deduce the ratio of carbon to hydrogen in our sample as 1:4. Because 6.25 mmoles of carbon corresponds to $6.25 \times 12 = 75$ mg and 25 mmoles of hydrogen to 25 mg, no other elements can be present (total equals the total weight of our original sample), showing an empirical formula of CH_4. The true composition of the molecule, called the **molecular formula,** could, in principle, also be a multiple of CH_4, such as C_2H_8, C_3H_{12}, and so forth. The molecular formula must be derived with the help of other techniques. One such method is mass spectroscopy, to be discussed in Chapter 18. It relies on a sophisticated instrument with which one can measure molecular weights of molecules. In this case, it would be 16 for CH_4.

Molecular formulas are also established by the outcome of certain chemical reactions. For example, the introduction of one chlorine atom into C_3H_6 to give C_3H_5Cl allows one to distinguish this molecular formula from C_6H_{12}, which would give $C_6H_{11}Cl$, a substance of completely different elemental composition. For methane, however, the issue is simple: only CH_4 makes sense in terms of chemical bonding. Why?

Elemental analysis of oxygen is rather complicated and not usually done. Instead, oxygen is reported as the difference between the sum of the percentage values for the measured elements and 100%. For example, an elemental analysis of a liquid furnished carbon, 52.2%, hydrogen, 13.0%. No other element could be found; hence oxygen must be 34.8%. Therefore, the liquid contains $52.2/12 = 4.35$ atom equivalents of C; $13.0/1 = 13$ atom equivalents of H; and

34.8/16 = 2.175 atom equivalents of O. An empirical formula might then be $C_{4.35}H_{13}O_{2.175}$. This formula has to be normalized to account for the fact that molecules contain only whole atoms. Normalization by division through the smallest atom subscript (2.175) produces a molar ratio of C:H:O = 2:5.98:1, which may be rounded off to furnish an empirical formula of C_2H_6O. Two structures fit this formula: that of dimethyl ether, CH_3OCH_3, and that of ethanol, CH_3CH_2OH.

EXERCISE 1-13

Calculate the percentages of carbon and hydrogen in glucose, $C_6H_{12}O_6$ (see Chapter 23).

How to Establish the Identity and Structure of a Molecule

The establishment of an empirical formula does not reveal either the identity or the structure of the molecule. There are several ways in which we can distinguish between structural alternatives. For example, as already mentioned, the empirical formula C_2H_6O corresponds to two molecules: ethanol and dimethyl ether. One way of distinguishing between the two is to measure the difference in their respective physical properties—for example, their melting and boiling points, refractive indices, specific gravities, and so forth. Thus, ethanol is a liquid (b.p. 78.5°C) and dimethyl ether a gas (−23°C) at room temperature. There are also differences in chemical reactivity. For example, ethanol reacts with sodium to give hydrogen and sodium ethoxide, whereas dimethyl ether is inert. Molecules such as this, which have the same empirical formula but differ in the sequence *(connectivity)* in which the atoms are held together, are called **structural isomers.**

Ethanol and Dimethyl Ether: Two Isomers

```
   H  H
   |  |
H—C—C—O—H
   |  |
   H  H
```

Ethanol
(b.p. 78.5°C)

```
   H      H
   |      |
H—C—O—C—H
   |      |
   H      H
```

Dimethyl ether
(b.p. −23°C)

EXERCISE 1-14

Construct as many isomers with the molecular formula C_4H_{10} as you can.

Suppose now that the compound under investigation has never been studied before and therefore there are no data in the published research literature with which it can be compared. In this case, differentiation requires the use of other methods, most of which are various forms of **spectroscopy.** Spectroscopy is the branch of chemistry and physics concerned with the response of substances subjected to radiation of various wavelengths. These responses depend on the connectivity of the atoms (structure) in the molecule. The various spectroscopic methods and their application in organic structural elucidation will be dealt with in later chapters.

Apart from the connectivity of the atoms, we want to know their arrangement in space. Some information in this regard can again be obtained by spectroscopic techniques. The most complete method is that of **X-ray diffraction,** which allows the elucidation of the structure of a molecule in its crystalline state. X-ray crystallography supplies a detailed picture of the molecule and the exact position of every atom, as if viewed through an extremely powerful magnifying glass. Once the positions of all the atoms are known, one can deduce bond lengths, bond angles, and other geometrical features. The structural pictures that emerge in this way for the two isomers ethanol and dimethyl ether are exemplified in the form of ball-and-stick models in Figure 1-21. Note the tetrahedral bonding of the carbon atoms and the bent arrangement of the two bonds

FIGURE 1-21

Three-dimensional ball-and-stick representations of ethanol and dimethyl ether. Bond lengths are given in Angstrom units, bond angles in degrees.

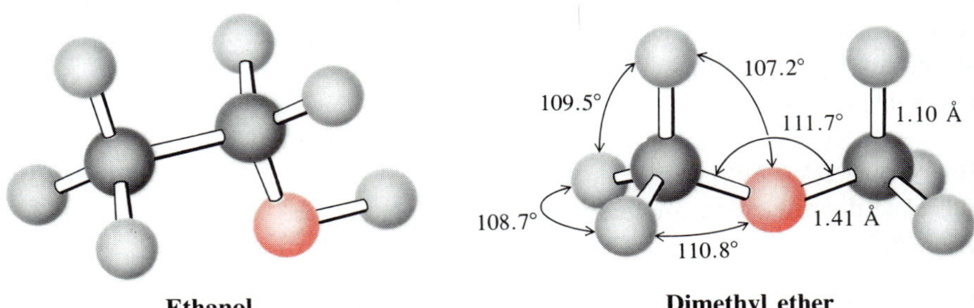

to oxygen, which is hybridized as in water. The figure also illustrates the kind of structural detail available through modern methods of structure determination.

Figure 1-21 calls attention to a problem to be faced throughout this book: how to draw the structure of organic molecules. Because it is exceedingly time consuming to represent structures as ball-and-stick models, organic chemists have designed various conventions.

Drawing Structures and Formulas

A particularly simple way of drawing organic molecules so as to indicate their three-dimensional arrangement (one that will be used extensively in this book) is the *dashed-wedged line notation*. In this picture, the carbon atoms in a chain are written in a zigzag manner and connected by straight lines to depict the tetrahedral bond angle around each carbon. Each carbon atom along the chain is then shown with dashed or wedged lines to represent the remaining two bonds: the dashed line means that the bond is *below the plane* of the sheet of paper; the wedge that it is *above that plane* (Figure 1-22). This method of illustrating three-dimensional structure requires that all carbon atoms in the chain be viewed as lying *in the plane* of the paper. The appropriate atoms or groups, such as H, OH, CH_3, and so forth, are written at the ends of the dashed or wedged lines. In a drawing of methane's structure, two hydrogen atoms take the place of two of the carbon atoms in a chain: for ethane, it is only one.

FIGURE 1-22

Dashed (red) and wedged (blue) line notation for: (A) a carbon chain; (B) methane; (C) ethane; (D) ethanol; and (E) dimethyl ether.

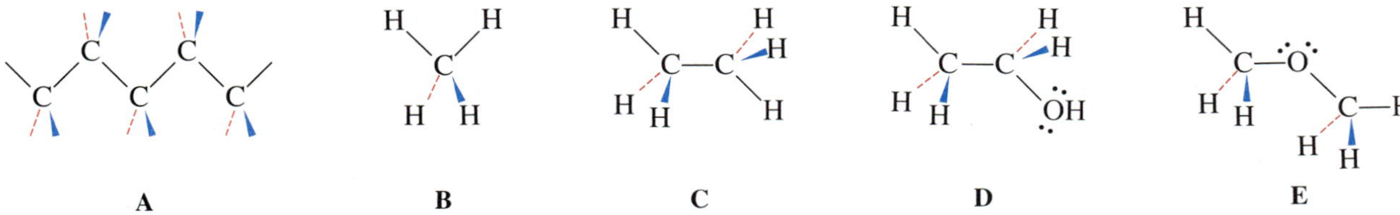

The perception of organic molecules in three-dimensions is essential for understanding their structure and frequently reactivity. You may find it difficult to visualize the spatial arrangements of the atoms in even very simple molecules. A good aid is a *molecular model kit,* which allows the construction of fairly accurate models of molecules. Such model sets are commercially available at a reasonable cost. You should acquire one of them and practice assembling organic structures.

It is often unnecessary to show molecules in three dimensions. In those cases, ordinary Lewis or line structures suffice. To simplify these even further, chemists use *condensed formulas* in which common groups are written as CH_3, CH_2, OH, and so forth. As shown, the atoms attached to each carbon are usually placed to the right. Lone electron pairs are usually omitted.

Condensed Formulas

H—C(H)(H)—C(H)(H)—$\ddot{O}$H equals CH_3CH_2OH

Ethanol

$H_2C=CH_2$ equals $CH_2{=}CH_2$

Ethylene

H—C≡C—H equals HC≡CH

Acetylene

H—C(H)(H)—C(=O)—C(H)(H)—H equals $CH_3\overset{\displaystyle O}{\overset{\|}{C}}CH_3$

Acetone

$^{-}$O—C(=O)—O^{-} equals CO_3^{2}

Carbonate ion

H—C(=O)—$\ddot{O}$—H equals $H\overset{\displaystyle O}{\overset{\|}{C}}OH$

Formic acid

In summary, you should now be able to determine molecular composition and deduce empirical formulas. The structures of molecules in three dimensions are best depicted by the dashed and wedged line notation. Condensed formulas are useful as a ‘‘shorthand’’ approach to drawing molecules.

Summary of Important Concepts

1 Organic chemistry is the chemistry of carbon.

2 Ionic bonds result because of Coulombic attraction of oppositely charged ions. These ions are formed by the complete transfer of electrons from one atom to another. Such electron transfer occurs frequently to achieve the noble-gas configuration.

3 Covalent bonds result because of electron sharing between two atoms. Again, electrons are shared to allow the atoms to attain noble-gas configurations. Dative (coordinate covalent) bonds can be thought of as covalent bonds in which one of the atoms has supplied both electrons.

4 Bond length is the average distance between two covalently bonded atoms. Bond formation releases energy. Breaking this bond requires energy to be added to the molecule. Pushing the atoms closer than the bonding distance also requires energy because of electron-electron and nuclear-nuclear repulsions.

5 Polar bonds are formed between atoms of differing electronegativity, which is a measure of an atom's ability to attract electrons.

6 The nuclear core contains the nucleus and all but the valence electrons.

7 The ionization potential is the energy required (positive sign) or released (negative sign) to remove an electron from an atom or molecule in the gas phase.

8 The electron affinity is the energy required or released on adding an electron to an atom or molecule in the gas phase.

9 A molecule that is a dipole contains an uneven charge distribution, with one positive and one negative end. The dipole moment $\mu = qd$.

10 The shape of molecules is strongly influenced by electron repulsion.

11 Lewis structures describe bonding by the use of electron dots. They are drawn so as to give hydrogen an electron duet, the other atoms electron octets (octet rule). Charge separation should be minimized but not at the expense of the octet rule.

12 If a molecule has two or more Lewis structures, they are called resonance forms. They are formally interconverted by movement of electrons only. None correctly describes the molecule, its true representation being the weighted average (hybrid) of all its Lewis structures. If the resonance forms of a molecule are unequal, those which best satisfy the rules for writing Lewis structures and the electronegativity requirements of the atoms are more important.

13 The motion of the electrons around the nucleus can be described by wave equations. The solutions to these equations are atomic orbitals, which roughly delineate regions in space in which there is a high probability of finding the electron.

14 An *s* orbital is spherical; a *p* orbital looks like two touching spheres, or a "spherical figure eight." The sign of the orbital can be positive, negative, or

zero (node). With increasing energy, there is an increasing number of nodes. Each orbital can be occupied by a maximum of two electrons of opposite spin (Pauli exclusion principle, Hund's rule).

15 The process of adding electrons one by one to the atomic orbitals, starting with those of lowest energy, is called the Aufbau principle.

16 A molecular orbital is formed when two atomic orbitals overlap to generate a bond. Atomic orbitals of the same sign overlap to give a bonding molecular orbital of lower energy. Atomic orbitals of opposite sign give rise to an antibonding molecular orbital of higher energy and containing a node. The number of molecular orbitals equals the number of atomic orbitals from which they derive.

17 Bonds made by overlap along the internuclear axis are called σ bonds; those made by overlap of p orbitals perpendicular to the internuclear axis are called π bonds.

18 The overlap, or mixing, of orbitals on the same atom results in new hybrid orbitals of different shape. One s and one p orbital mix to give two linear sp hybrids, used, for example, in the bonding of BeH_2. One s and two p orbitals result in three trigonal sp^2 hybrids, used, for example, in BH_3. One s and three p orbitals furnish four tetrahedral sp^3 hybrids, used, for example, in CH_4. The orbitals that are not hybridized stay unchanged. Hybrid orbitals may overlap with each other. Overlapping sp^3 hybrid orbitals on different carbon atoms forms the carbon–carbon bonds in ethane and other organic molecules. Hybrid orbitals may also be occupied by lone electron pairs, as in NH_3 and H_2O.

19 The composition (i.e., ratios of types of atoms) of organic molecules is revealed by elemental analysis. The empirical formula indicates the molecular composition; the molecular formula the number of atoms of each kind.

20 Molecules that have the same molecular formula but different spatial arrangements of their atoms, and hence different properties, are called structural isomers.

21 Dashed-wedged line drawings illustrate the structure of molecules in three dimensions. Condensed formulas are abbreviated forms of describing structures.

Problems

1 Using the data in Tables 1-2 and 1-3, calculate the energy change for each process shown below and decide whether each reaction is energetically favorable (i.e., energy is released) or unfavorable (i.e., energy is required). Assume for the purposes of this problem that the ions are formed far apart from one another, so that electrostatic forces between them can be ignored.

(a) $Na + Cl \longrightarrow Na^+ + Cl^-$
(b) $H + Cl \longrightarrow H^+ + Cl^-$
(c) $Li + F \longrightarrow Li^+ + F^-$
(d) $Be + 2\ Br \longrightarrow Be^{2+} + 2\ Br^-$
(e) $Be + 2\ H \longrightarrow Be^{2+} + 2\ H^-$

2 Repeat your calculation for each reaction in Problem 1, this time including the energy released as a result of electrostatic attraction of oppositely charged ions. For each plus-minus attraction, this energy can be calculated using the formula

$$E = \frac{(332\ \text{kcal/mole})(\text{charge on positive ion})(\text{charge on negative ion})}{\text{distance between positive ion and negative ion}}$$

The distance between ions is the sum of their ionic radii in Å. Use these values:

H^+ 0.0 Å	Na^+ 1.0 Å	H^- 2.1Å	Cl^- 1.8 Å
Li^+ 0.6 Å	Be^{2+} 0.3 Å	F^- 1.4 Å	Br^- 2.0 Å

Again, note whether each reaction is favorable or unfavorable.

3 Draw a Lewis structure for each of the following molecules and assign charges where appropriate. The order in which the atoms are connected is given in parentheses.

(a) ClF
(b) BrCN
(c) HOCl
(d) $SOCl_2$ ($OSCl_2$)
(e) CH_3NH_2
(f) $(CH_3)_2O$
(g) N_2H_2 (HNNH)
(h) CH_2CO
(i) HN_3 (HNNN)
(j) N_2O (NNO)

4 Draw a Lewis structure for each of the following species. Again, assign charges where appropriate.

(a) H^-
(b) CH_3^-
(c) CH_3^+
(d) CH_3
(e) $CH_3NH_3^+$
(f) CH_3O^-
(g) CH_2
(h) HC_2^- (HCC)
(i) H_2O_2 (HOOH)
(j) HO_2^- (HOO)

5 Several of the compounds in Problem 3 cannot be represented adequately by a single Lewis structure and are best described by resonance hybrids. Identify these molecules and write all additional resonance Lewis structures that you can for each of them. In each case, indicate the major contributor to the resonance hybrid.

6 Draw all resonance structures for each of the following species. Indicate the major contributor or contributors to the hybrid in each case.

(a) OCN^-
(b) CH_2CHNH^-
(c) $HCONH_2$ ($\overset{\text{O}}{\text{HC}}NH_2$)
(d) O_3 (OOO)
(e) $CH_2CHCH_2^+$
(f) SO_2 (OSO)
(g) $CH_2NH_2^+$
(h) $HOCHNH_2^+$
(i) $HONO_2$ ($\text{HON}\overset{\text{O}}{\text{O}}$)
(j) CH_3CNO

7 Two resonance forms can be written for a bond between trivalent boron and any atom with a lone pair of electrons.

(a) Write the two resonance forms for each of these three molecules: (i) R_2BNR_2, (ii) R_2BOR, and (iii) R_2BF (in each case R = CH_3).

(b) Using the guidelines from Section 1-5, predict the relative importance of the resonance forms to the resonance hybrids for each of the compounds in part **a**.

(c) How do the electronegativity differences between N, O, and F affect the relative importance of the resonance forms in each case?

(d) Predict the hybridization of N in (i) and O in (ii).

8 The unusual molecule [2.2.2]propellane is pictured below. Some of its geometrical features are also shown. On the basis of the geometry of the molecule, what hybridization scheme best describes the carbons marked with asterisks? (Make a model of the molecule to help you visualize its shape.) What orbitals are used in the bond between the carbons identified by asterisks? Would you expect the bond between these carbons to be stronger or weaker than an ordinary carbon–carbon single bond? (Ordinarily, these bonds are about 1.54 Å long.)

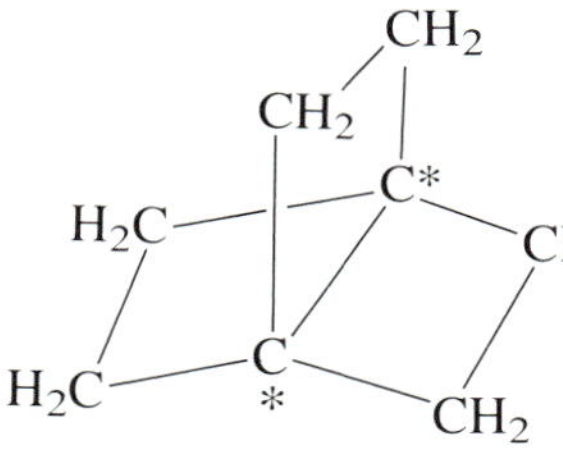

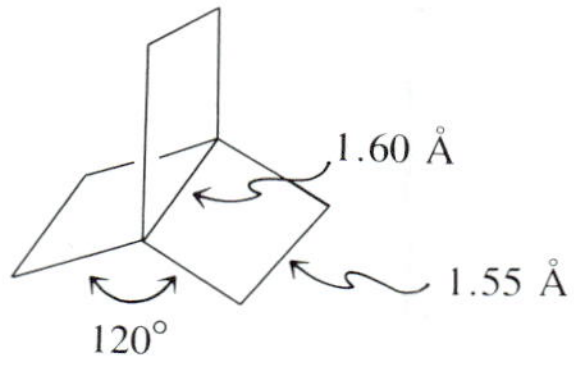

9 Use a molecular orbital analysis to predict which species in each of the following pairs has the stronger bonding between atoms.

(a) H_2 or H_2^+
(b) O_2 or O_2^+
(c) N_2 or N_2^+

10 Describe the hybridization of each carbon atom in each of the following structures. Base your answer on the geometry about the carbon atom.

(a) CH_3Cl
(b) CH_3OH
(c) $CH_3CH_2CH_3$
(d) $CH_2{=}CH_2$ (trigonal carbons)
(e) $HC{\equiv}CH$ (linear structure)
(f)

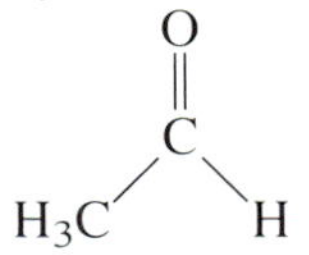

(g)

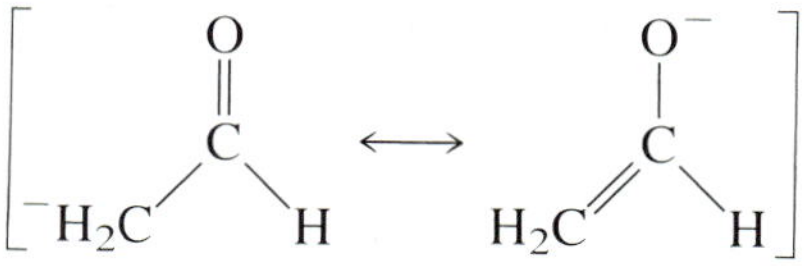

11 **(a)** Based on the information in Problem 10, give the likely hybridization of the orbital that contains the unshared pair of electrons (responsible for the negative charge) in each of the following anionic species: (i) $CH_3CH_2^-$; (ii) CH_2CH^-; (iii) HC_2^-.

(b) The closer electrons are to an atomic nucleus, the more stable they are. Rank the three species in part **a** in order of stability of the negative charge. (Hint: Which electrons are on average closer to the nucleus, those in an *s* orbital or those in a *p* orbital?)

(c) Given any acid, HA, its acid strength is related to the stability of its conjugate base, A^-. In other words, the ionization $HA \rightleftharpoons H^+ + A^-$ is favored for a more-stable A^-. Although CH_3CH_3, $CH_2{=}CH_2$, and $HC{\equiv}CH$ are weak acids, their acid strengths are not identical. Rank them in order of acid strength.

12 Calculate empirical formulas based on the following percentage compositions. If the total percentages do not add up to 100%, assume the remainder is due to oxygen.

- **(a)** C 92.31%; H 7.69%
- **(b)** C 62.07%; H 10.34%
- **(c)** C 71.11%; H 6.67%; N 10.37%
- **(d)** C 48.70%; H 2.90%; Cl 20.58%
- **(e)** C 83.72%; H 16.28%
- **(f)** C 57.14%; H 4.76%

13 Convert the following condensed formulas into line formulas.

- **(a)** CH_3CN
- **(b)** 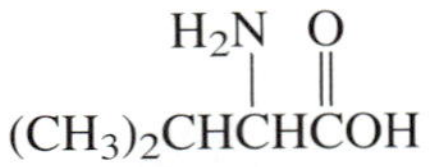 $(CH_3)_2CHCH(NH_2)COOH$ (drawn as $(CH_3)_2CHCHCOH$ with H_2N on the second CH and $=O$ on the last C) 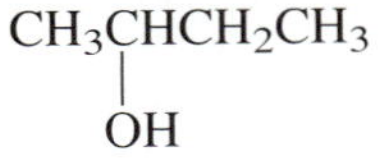
- **(c)** $CH_3CH(OH)CH_2CH_3$ (drawn as $CH_3CHCH_2CH_3$ with OH below the CH)
- **(d)** $CH_2BrCHBr_2$
- **(e)**  $CH_3CCH_2COCH_3$ (with O double-bonded above each of the first two C's)
- **(f)** $HOCH_2CH_2OCH_2CH_2OH$

14 Convert the following dashed-wedged line formulas into condensed formulas.

(a) 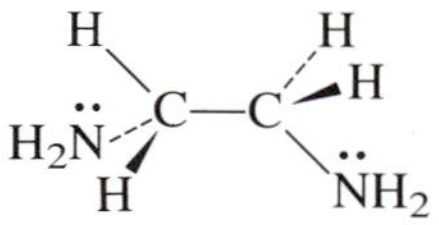
```
 H             H
  \        ,' ◄H
 ..  C —— C
H2N' ◄     \ ..
     H       NH2
```

(b)
```
               H
            H \        H
 H           ◄ C —— C /
   \        /       ◄ \ H
    C —— O:            H
 H' ◄   ..
     C
     ///
     N
     ..
```

(c)
```
         Br
        /
H —— C -- Br
       ◄
       Br
```

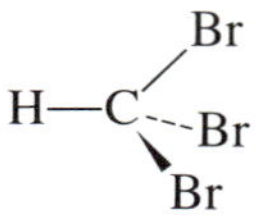

15 Convert the following line formulas into condensed formulas.

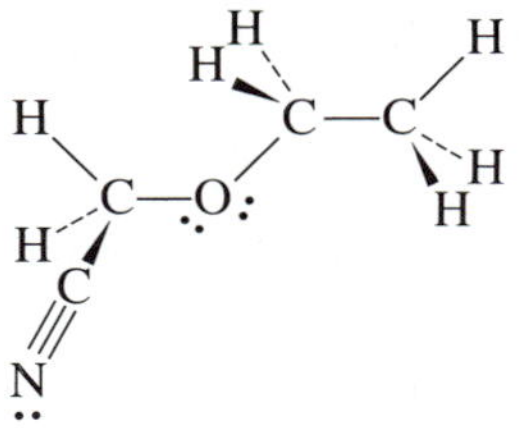

(a) 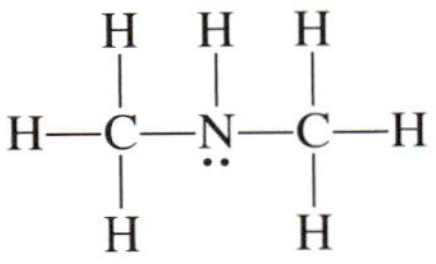
```
    H   H   H
    |   |   |
H — C — N — C — H
    |   ..  |
    H       H
```

(b)
```
    H   O   H   H   H
    |   ‖   |   |   |
H — C — C — N — C — C — H
    |       ..  |   |
    H           H   H
```

(c)
```
 ..     H   H  :Ö—H    H
 ..     |   |   |      |
H — S — C — C — C ———— C — H
 ..     |   |   |      |
        H   H   H      H
```

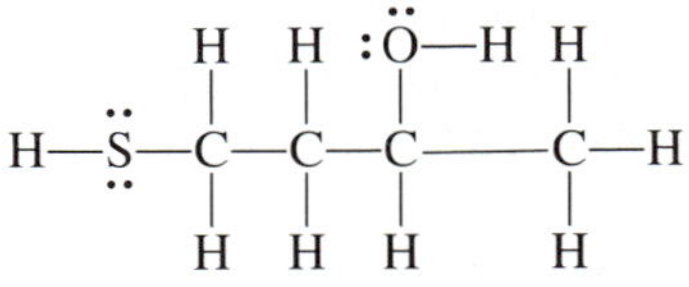

(d) 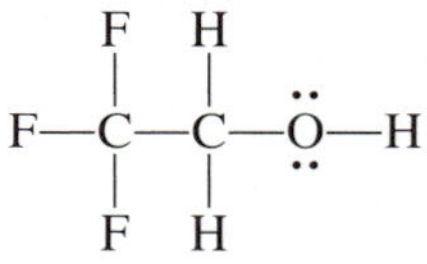
```
    F   H
    |   |   ..
F — C — C — O — H
    |   |   ..
    F   H
```

(e)
```
H       H  H       H
  \    /    \     /
   C         C
  /  \      /  \
H     C == C     H
     /      \    H
    H        \  /
              C
             / \
            H   H
```

(f)
```
H            H   H
  \         /    |
   C == C        C — H
  /         \   /|
H            C   H
             ‖
             O
```

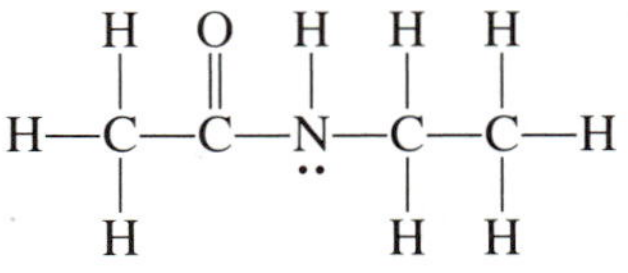

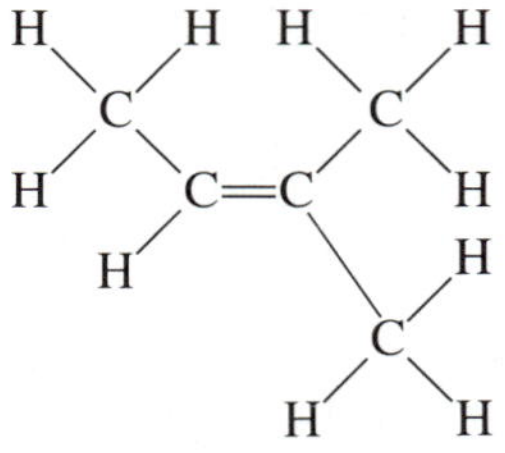

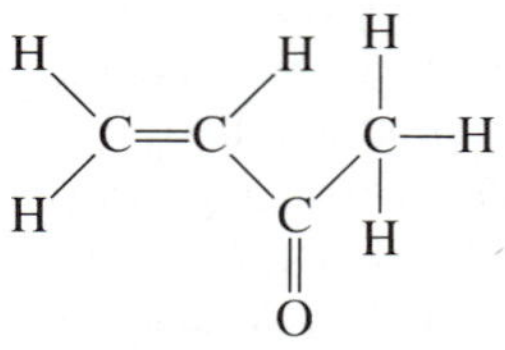

16 Convert the following condensed formulas into dashed-wedged line structures.

(a) CH_3CHOCH_3 (with CN on the central C)
(b) $CHCl_3$
(c) $(CH_3)_2NH$
(d) $CH_3CHCH_2CH_3$ (with SH on the second C)

17 Calculate percentage compositions for each of the following compounds.

Pain killers:
(a) Aspirin, $C_9H_8O_4$
(b) Acetaminophen, $C_8H_9NO_2$
(c) Ibuprofen, $C_{13}H_{18}O_2$

Artificial sweeteners:
(d) Saccharin, $C_7H_5NO_3S$
(e) Cyclamate, $C_6H_{13}NO_3S$
(f) Aspartame, $C_{14}H_{18}N_2O_5$

18 A number of substances containing carbon atoms at the positive ends of polarized bonds have been implicated as ''cancer suspect agents'' (i.e., suspected carcinogens, or cancer-inducing substances). Assuming that the degree of positive character at carbon is one factor related to carcinogenicity, rank the compounds below in order of increasing likelihood of being a cancer-causing substance.

(a) CH_3Cl
(b) $(CH_3)_4Si$
(c) $ClCH_2OCH_2Cl$
(d) CH_3OCH_2Cl
(e) $(CH_3)_3C^+$

Note: This is only one of many factors known to be related to carcinogenicity. Moreover, none of the factors can actually be applied in as straightforward a way as this question implies.

19 The structure of the substance lynestrenol, a component of certain oral contraceptives, is presented below.

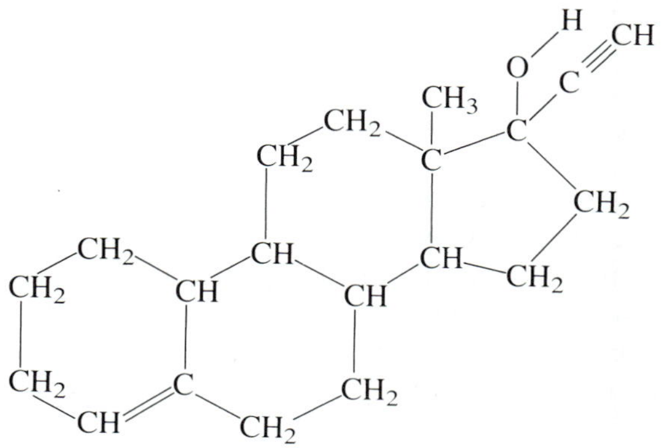

Lynestrenol

Locate in this structure an example of each of the following bonds or atoms:

(a) A highly polarized covalent bond.
(b) A nearly unpolarized covalent bond.
(c) An *sp*-hybridized carbon atom.
(d) An sp^2-hybridized carbon atom.
(e) An sp^3-hybridized carbon atom.
(f) A bond between atoms of different hybridization.

CHAPTER 2

Alkanes

MOLECULES LACKING FUNCTIONAL GROUPS

Having reviewed the basics of structure and bonding in organic molecules, we now turn to their chemical and physical properties. The classification that will be used relies on the recognition of functional groups. This chapter commences with a brief description of these groups. It then moves to the simplest class of molecules, the alkanes: their names; physical properties; and their mobility, a phenomenon based on the ability of tetrahedral carbon to rotate in space.

2-1 The Functional Group: Center of Reactivity

Organic molecules are composed of two parts: a backbone (skeleton) and functional groups. A simple system contains a backbone that is alkanelike—namely, it consists predominantly of sp^3-hybridized carbon atoms with hydrogen atoms attached. It is comparatively unreactive but may bear **functional groups,** which become the locations of any reactivity. Such groups have characteristic functions and control the reactivity of the molecule as a whole.

Attachment of Functional Groups (G) to an Alkane Chain

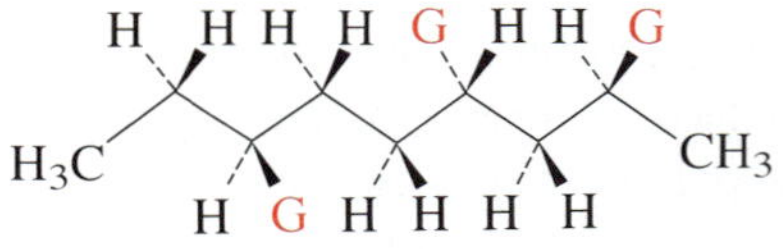

H H H H G H H G

H_3C CH_3

H G H H H H

One such functional characteristic is polarity, a property of the haloalkanes. Recall that polarity is due to a difference in the electronegativity of two atoms bound to each other. For example, the C–Cl bond in chloromethane is readily broken in certain chemical reactions because of this property.

$CH_3{-}\ddot{\underset{..}{Cl}}: + H\ddot{\underset{..}{O}}:^- \longrightarrow H\ddot{\underset{..}{O}}{-}CH_3 + :\ddot{\underset{..}{Cl}}:^-$

Chloromethane **Hydroxide ion** **Methanol** **Chloride ion**

In some molecules, the functional group is a double or triple bond, which allows the molecule to attack a reagent to form a new compound with only single bonds.

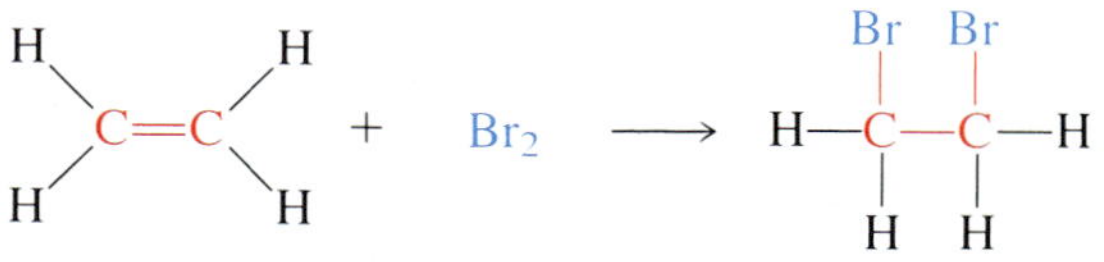

$H_2C{=}CH_2 + Br_2 \longrightarrow H{-}CH(Br){-}CH(Br){-}H$

Ethene (Ethylene)* **Bromine** **1,2-Dibromoethane**

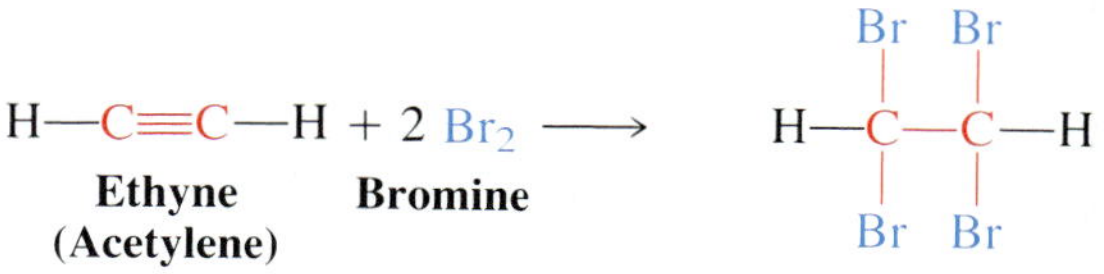

$H{-}C{\equiv}C{-}H + 2\,Br_2 \longrightarrow H{-}CBr_2{-}CBr_2{-}H$

Ethyne (Acetylene) **Bromine** **1,1,2,2-Tetrabromoethane**

Certain functional groups incorporate hydrogen atoms that are removable as protons, as in methanol, CH_3OH (illustrated at the right).

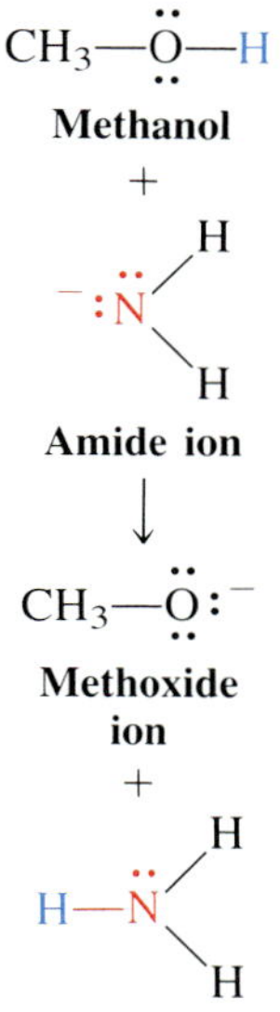

$CH_3{-}\ddot{\underset{..}{O}}{-}H$ **Methanol** + $^-:\ddot{N}H_2$ **Amide ion** $\downarrow$ $CH_3{-}\ddot{\underset{..}{O}}:^-$ **Methoxide ion** + $H{-}\ddot{N}H_2$ **Ammonia**

Functional groups carrying atoms with lone electron pairs can enter into dative (coordinate covalent) bonding.

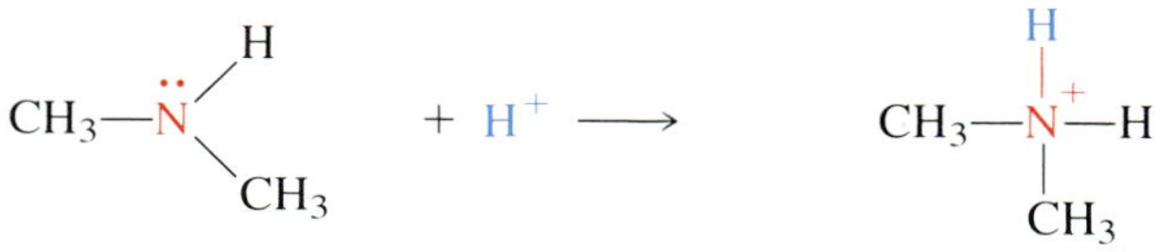

$CH_3{-}\ddot{N}(H){-}CH_3 + H^+ \longrightarrow CH_3{-}\overset{+}{N}H_2{-}CH_3$

***N*-Methylmethanamine (Dimethylamine)** ***N*-Methylmethanammonium ion (Dimethylammonium ion)**

Other functional groups are readily oxidized, reduced, or otherwise transformed.

Hydrocarbons Contain Only Hydrogen and Carbon

Molecules described by the general empirical formula C_xH_y are **hydrocarbons.** Representatives of this class of molecules containing only single bonds and no functional groups, such as methane, ethane, and propane, are **alkanes.** Cyclic analogs, the **cycloalkanes,** are exemplified by cyclohexane. **Alkenes** and **alkynes** are hydrocarbons that contain double and triple C–C bonds, respectively. Examples are ethene, propene, ethyne, and propyne.

Saturated Hydrocarbons: Alkanes

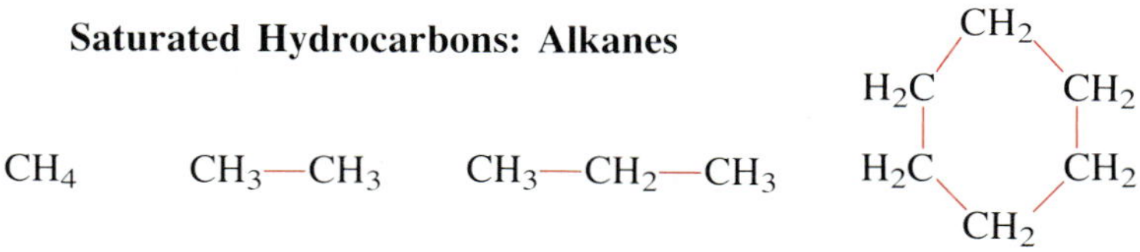

CH_4 **Methane**

$CH_3{-}CH_3$ **Ethane**

$CH_3{-}CH_2{-}CH_3$ **Propane**

$(CH_2)_6$ (ring of six CH_2 groups) **Cyclohexane**

*In the subsequent discussion, many compounds have more than one name. The systematic naming of molecules (Section 2-3) will be adhered to as much as possible, but common names will be given in parentheses if they are still frequently used.

Unsaturated Hydrocarbons: Alkenes and Alkynes

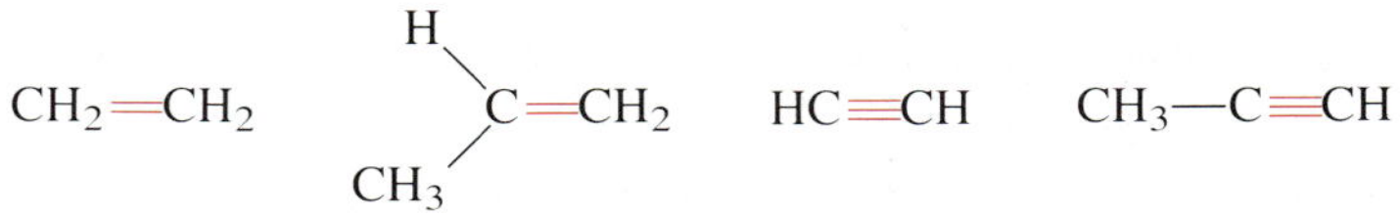

$CH_2{=}CH_2$ — **Ethene (Ethylene)**

$(H)(CH_3)C{=}CH_2$ — **Propene**

$HC{\equiv}CH$ — **Ethyne (Acetylene)**

$CH_3{-}C{\equiv}CH$ — **Propyne**

Because alkenes and alkynes can undergo addition reactions, they are said to be **unsaturated.** In contrast, the alkanes, which cannot enter into such reactions, are referred to as **saturated.** The chemistry and properties of the alkanes are described in the next section and in Chapters 3 and 4. The alkenes are the topic of Chapters 11 and 12, and the alkynes of Chapter 13.

A special unsaturated hydrocarbon is **benzene,** C_6H_6, in which three double bonds are incorporated into a six-membered ring. Benzene and its derivatives are **aromatic,** a term coined because some substituted benzenes have a strong fragrance. Nonaromatic compounds are **aliphatic.** Aromatic compounds are discussed in Chapters 14, 19, 20, 24, and 25.

Aromatic Compounds

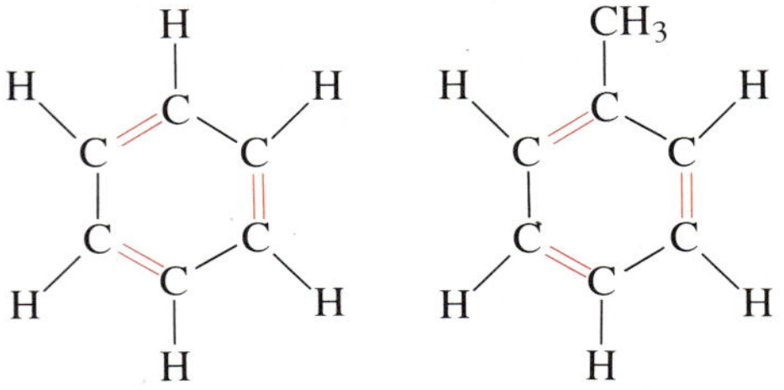

Benzene

Methylbenzene (Toluene)

Frequently, a hydrocarbon-derived unit replaces a hydrogen atom in an alkane chain. Such an appendage is generally called *alkyl* and is usually abbreviated by the symbol R (for ''radical'' or residue) if derived from an alkane and *phenyl* (C_6H_5) if derived from benzene. The general formula for an alkane is therefore often written as RH.

Hydrocarbon Substituents

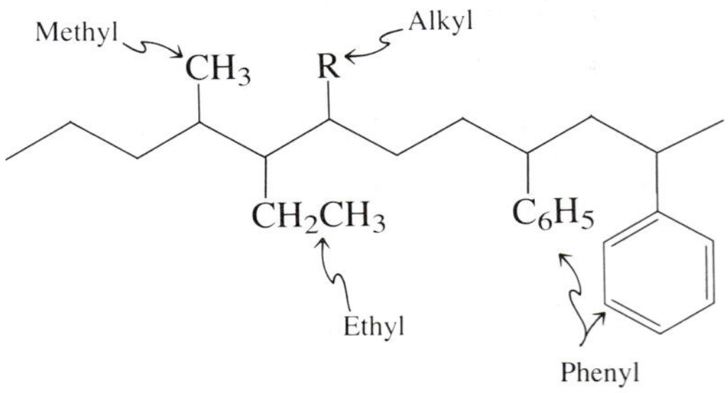

Methyl CH_3
Alkyl R
CH_2CH_3 Ethyl
C_6H_5 Phenyl

Hydrocarbons are abundant on earth. For example, methane is the chief component of natural gas. Petroleum (crude oil) consists mainly of saturated, unsaturated, and aromatic hydrocarbons. After processing and distillation, it supplies the precious gasoline with which we run our cars. Other fractions furnish kerosene (airplane fuel), fuel oil, diesel fuel, lubricating oil, paraffin wax, and other useful products.

Many Functional Groups Contain Oxygen and Nitrogen

A common functional group is the *hydroxy* group, —OH, obtained by removing a hydrogen atom from water and characteristic of the **alcohols.** An *alkoxy* group, —OR, in which R again signifies alkyl, is characteristic of **ethers.**

Alcohols

CH_3OH **Methanol**

CH_3CH_2OH **Ethanol**

Ethers

CH_3OCH_3 **Methoxymethane (Dimethyl ether)**

$CH_3CH_2OCH_2CH_3$ **Ethoxyethane (Diethyl ether)**

The simple representatives of these classes of compounds, such as methanol and ethanol, are frequently used as solvents in organic reactions. Ethanol is the "active" ingredient of alcoholic drinks; it is also used as a fuel additive ("gasohol"). The functional group in alcohols and those in some ethers can be converted into a large variety of other functional groups and are therefore important in synthetic transformations. This chemistry is the subject of Chapters 8 and 9.

The *carbonyl* function, C=O, is found in **aldehydes** and **ketones,** and, in conjunction with an attached —OH, in the **carboxylic acids.** Aldehydes and ketones are discussed in Chapters 15 and 16, the carboxylic acids and their derivatives in Chapters 17 and 18.

Aldehydes

Methanal (Formaldehyde)

$CH_3\overset{O}{\overset{\|}{C}}H$ or CH_3CHO **Ethanal (Acetaldehyde)**

Ketones

$CH_3\overset{O}{\overset{\|}{C}}CH_3$ **Propanone (Acetone)**

$CH_3CH_2\overset{O}{\overset{\|}{C}}CH_3$ **Butanone (Methyl ethyl ketone)**

Carboxylic Acids

$H\overset{O}{\overset{\|}{C}}OH$ or $HCOOH$ **Methanoic acid (Formic acid)**

$CH_3\overset{O}{\overset{\|}{C}}OH$ or CH_3COOH **Ethanoic acid (Acetic acid)**

Other elements give rise to further characteristic functional groups. For example, saturated alkyl nitrogen compounds are **amines.** The replacement of oxygen in alcohols by sulfur furnishes **thiols.**

Amines

CH_3NH_2 **Methanamine (Methylamine)**

$CH_3\overset{H}{\overset{|}{N}}CH_3$ or $(CH_3)_2NH$ ***N*-Methylmethanamine (Dimethylamine)**

A Thiol

CH_3SH **Methanethiol**

Table 2-1 (on the next two pages) depicts a selection of common functional groups (in the order in which we will encounter them), the class of compounds to which they give rise, a general structure, and an example.

TABLE 2-1
Common functional groups

Compound class	General structure	Functional group	Example
Alkanes	R—H	None	$CH_3CH_2CH_2CH_3$ **Butane**
Haloalkanes	R—X (X = F, Cl, Br, I)	—X	CH_3CH_2—I **Iodoethane**
Alcohols	R—OH	—OH	$(CH_3)_2CH$—OH **2-Propanol (Isopropyl alcohol)**
Ethers	R—O—R′	—O—	CH_3CH_2—O—CH_3 **Methoxyethane (Ethyl methyl ether)**
Thiols	R—SH	—SH	CH_3CH_2—SH **Ethanethiol**
Alkenes	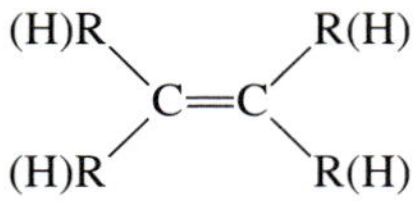 (H)R(H)RC=CR(H)R(H)	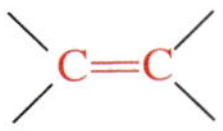 >C=C<	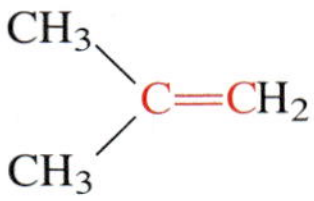 $(CH_3)_2C{=}CH_2$ **2-Methylpropene**
Alkynes	(H)R—C≡C—R(H)	—C≡C—	$CH_3C{\equiv}CCH_3$ **2-Butyne**
Aromatic compounds	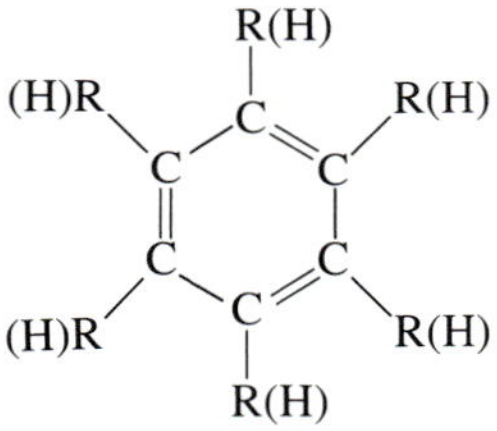 C_6R(H)$_6$ (benzene ring with six R(H) substituents)	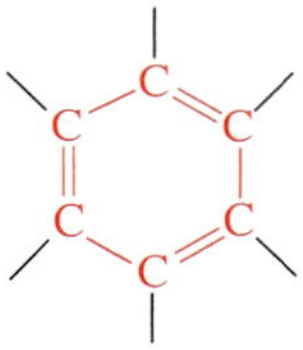 C_6 ring	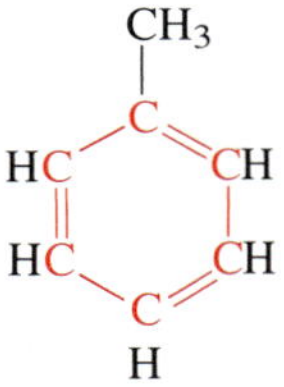 $C_6H_5CH_3$ **Methylbenzene (Toluene)**

Note: The letter R denotes an alkyl group. Different alkyl groups can be distinguished by adding primes to the letter R: R′, R″, and so forth.

2-2 The Straight-Chain and Branched Alkanes

As mentioned earlier, hydrocarbons that contain only single bonds are alkanes. They are classified into several types according to structure: the continuous *straight-chain* alkanes; the *branched* alkanes, in which the carbon chain contains one or several branching points; the cyclic alkanes, or *cycloalkanes;* and the more complicated bicyclic, tricyclic, and higher polycyclic alkanes. One example of each type is shown on the facing page.

Compound class	General structure	Functional group	Example
Aldehydes	$\mathrm{R{-}\overset{\overset{O}{\|}}{C}{-}H}$	$\mathrm{{-}\overset{\overset{O}{\|}}{C}{-}H}$	$\mathrm{CH_3CH_2\overset{\overset{O}{\|}}{C}H}$ **Propanal**
Ketones	$\mathrm{R{-}\overset{\overset{O}{\|}}{C}{-}R'}$	$\mathrm{{-}\overset{\overset{O}{\|}}{C}{-}}$	$\mathrm{CH_3CH_2\overset{\overset{O}{\|}}{C}CH_2CH_3}$ **3-Pentanone**
Carboxylic acids	$\mathrm{R{-}\overset{\overset{O}{\|}}{C}{-}O{-}H}$	$\mathrm{{-}\overset{\overset{O}{\|}}{C}{-}OH}$	$\mathrm{CH_3CH_2\overset{\overset{O}{\|}}{C}OH}$ **Propanoic acid**
Anhydrides	$\mathrm{R{-}\overset{\overset{O}{\|}}{C}{-}O{-}\overset{\overset{O}{\|}}{C}{-}R'\ (H)}$	$\mathrm{{-}\overset{\overset{O}{\|}}{C}{-}O{-}\overset{\overset{O}{\|}}{C}{-}}$	$\mathrm{CH_3CH_2\overset{\overset{O}{\|}}{C}O\overset{\overset{O}{\|}}{C}CH_2CH_3}$ **Propanoic anhydride**
Esters	$\mathrm{(H)R{-}\overset{\overset{O}{\|}}{C}{-}O{-}R'}$	$\mathrm{{-}\overset{\overset{O}{\|}}{C}{-}O{-}}$	$\mathrm{CH_3\overset{\overset{O}{\|}}{C}OCH_3}$ **Methyl ethanoate (Methyl acetate)**
Amides	$\mathrm{R{-}\overset{\overset{O}{\|}}{C}{-}\underset{\underset{R''\ (H)}{\|}}{N}{-}R'\ (H)}$	$\mathrm{{-}\overset{\overset{O}{\|}}{C}{-}\underset{\|}{N}{-}}$	$\mathrm{CH_3CH_2CH_2\overset{\overset{O}{\|}}{C}NH_2}$ **Butanamide**
Nitriles	$\mathrm{R{-}C{\equiv}N}$	$\mathrm{{-}C{\equiv}N}$	$\mathrm{CH_3C{\equiv}N}$ **Ethanenitrile (Acetonitrile)**
Amines	$\mathrm{R{-}\underset{\underset{R''}{\|}}{N}{-}R'}$	$\mathrm{{-}N{<}}$	$\mathrm{(CH_3)_3N}$ ***N,N*-Dimethylmethanamine (Trimethylamine)**

A Straight-Chain Alkane

$\mathrm{CH_3{-}CH_2{-}CH_2{-}CH_3}$

Butane, C_4H_{10}

A Branched Alkane

$\mathrm{CH_3{-}\underset{\underset{CH_3}{\|}}{\overset{\overset{CH_3}{\|}}{C}}{-}H}$

2-Methylpropane, C_4H_{10} (Isobutane)

A Cycloalkane

$$\begin{array}{ccc} \mathrm{CH_2} & - & \mathrm{CH_2} \\ | & & | \\ \mathrm{CH_2} & - & \mathrm{CH_2} \end{array}$$

Cyclobutane, C_4H_8

A Bicyclic Alkane

CH₂ CH₂ CH CH₂ CH CH₂ CH₂ CH₂

Bicyclo[2.2.2]octane, C_8H_{14}

Homologous and Isomeric Alkanes

The straight-chain alkanes consist of carbon chains in which each carbon is bound to its two neighbors and to two hydrogen atoms. Exceptions are the two terminal carbon nuclei, which are bound to only one carbon atom and three hydrogen atoms. Several general formulas may be written for the series:

$H—(CH_2)_n—H$ $\quad\quad$ $CH_3—(CH_2)_{n-1}—H$ $\quad\quad$ C_nH_{2n+2}

(Not limited to the straight-chain alkanes)

Each member of this series differs from the next one by the presence or absence of a methylene group, $—CH_2—$. Molecules that are related in this way are **homologs** of each other (*homos,* Greek, same as), and the series is a **homologous series.** Methane ($n = 1$) is the first member of the homologous series of the alkanes, ethane ($n = 2$) the second, and so forth.

Branched alkanes are derived from the straight-chain systems by removal of a hydrogen atom from a methylene group and replacement with an alkyl group. They have the same empirical formula as the straight-chain alkanes, C_nH_{2n+2}. The smallest example is 2-methylpropane, C_4H_{10}, with the same molecular formula as (and therefore isomeric to) butane. The two are related by simply changing the connectivity of the C–C and C–H bonds. In theory, butane could be turned into 2-methylpropane by exchanging an internal hydrogen atom with a terminal methyl group. In practice, this is not easy to do, because C–C and C–H bonds would have to be broken.

For the higher alkane homologs ($n > 4$), more than two isomers are possible. There are three pentanes, C_5H_{12}, as shown below. There are five hexanes, C_6H_{14}, nine heptanes, C_7H_{16}, and eighteen octanes, C_8H_{18}.

TABLE 2-2
Number of possible isomeric alkanes C_nH_{2n+2}

n	Isomers
1	1
2	1
3	1
4	2
5	3
6	5
7	9
8	18
9	35
10	75
15	4,347
20	366,319

Isomeric Pentanes

$CH_3—CH_2—CH_2—CH_2—CH_3$

Pentane

$$CH_3—CH_2—\overset{\overset{\large CH_3}{|}}{\underset{\underset{\large CH_3}{|}}{C}}—H$$

2-Methylbutane

$$CH_3—\overset{\overset{\large CH_3}{|}}{\underset{\underset{\large CH_3}{|}}{C}}—CH_3$$

2,2-Dimethylpropane (Neopentane)

The number of possibilities in connecting n carbon atoms to each other and to $2n + 2$ surrounding hydrogen atoms increases dramatically with the size of n (Table 2-2).

EXERCISE 2-1

(a) Write the structures of the five isomeric hexanes; (b) write the structures for all the possible next-higher and -lower homologs of 2-methylbutane.

The multiplicity of choices of assembling carbon atoms and attaching various substituents is the main reason why so many organic molecules exist. This diversity poses a problem: how can we systematically differentiate by name all of these compounds? Is it possible, for example, to name all of the C_6H_{14}

isomers, so that information on any of them (such as boiling points, melting points, chemical reactions) might easily be found in a compound index of an appropriate handbook? And is there a way to name a compound that you have never seen in such a way as to be able to draw the structure on a sheet of paper? Indeed, there is a precise system for naming the alkanes.

2-3 The System for Naming Alkanes

The problem of naming organic molecules has been with organic chemistry from its very beginning. It has been worsened by the multitude of new compounds and classes of compounds that are discovered every year. There are also national language problems [e.g., a German chemist would call tetrachloromethane (also known as carbon tetrachloride), CCl_4, *Tetrachlorkohlenstoff;* a Frenchman would call it *tetraclorure de carbone;* a Spaniard *tetracloruro de carbono;* a Dane *tetrachlorkulstof*). Some compounds have been named after their discoverers: for example, ''Nenitzescu's hydrocarbon,'' ''Prelog-Djerassi lactone,'' and ''Wieland-Miescher ketone.'' Localities can also come into play: ''münchnones,'' a certain class of compounds discovered in Munich; ''sydnones,'' prepared in Sydney. Quite a few names refer to the shape of a molecule: ''cubane,'' a hydrocarbon looking like a cube, (see Section 1-1); ''basketene,'' a basket-shaped hydrocarbon; ''barrelene''; ''tetrahedrane''; ''snoutane''; or to its origin: ''formic acid,'' isolated from ants (*formica,* Latin, ant); ''anisole,'' from anise oil; ''vanillin'' from vanilla. Such names have mainly historical significance although many are in common use **(common names),** especially when the structure-descriptive name is complex.

Systematic nomenclature was first introduced by a chemical congress in Geneva, Switzerland, in 1892 and has continually been revised since then, mostly by the International Union of Pure and Applied Chemistry (IUPAC). Rules for naming many organic molecules are called **IUPAC rules.** The names of the first twenty straight-chain alkanes are given in Table 2-3 (on the next page). Their stems are mainly of Latin or Greek origin revealing the number of carbon atoms in the chain. For example, the name heptadecane is composed of the Greek word *hepta,* seven, and the Latin word *decem,* ten. The first four alkanes have special names that have been accepted as part of the IUPAC system but also all end in **-ane.** You should study Table 2-3 closely, because it serves as the basis for naming a large fraction of all organic molecules.

$CH_3—$
Methyl

$CH_3—CH_2—$
Ethyl

$CH_3—CH_2—CH_2—$
Propyl

Alkane appendages are named by replacing the ending **-ane** with **-yl.** A few lower homologs of branched alkanes have common names that still have widespread use. They make use of the additional prefixes **iso-** and **neo-** as in isobutane, isopentane, and neohexane.

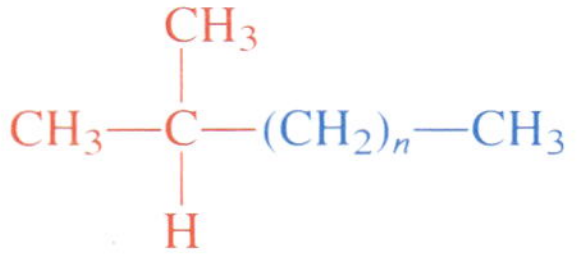

$$CH_3—\underset{\displaystyle H}{\overset{\displaystyle CH_3}{\underset{|}{\overset{|}{C}}}}—(CH_2)_n—CH_3$$

An isoalkane
(e.g., n = 1, Isopentane)

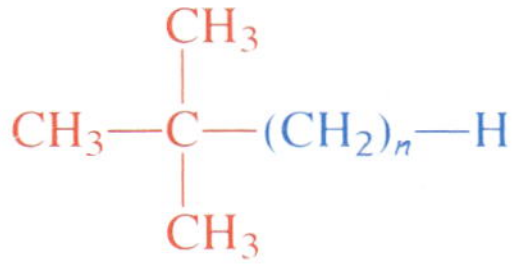

$$CH_3—\underset{\displaystyle CH_3}{\overset{\displaystyle CH_3}{\underset{|}{\overset{|}{C}}}}—(CH_2)_n—H$$

A neoalkane
(e.g., n = 2, Neohexane)

TABLE 2-3
Names of straight-chain alkanes, C_nH_{2n+2}

n	Name	Formula
1	Methane	CH_4
2	Ethane	CH_3CH_3
3	Propane	$CH_3CH_2CH_3$
4	Butane	$CH_3CH_2CH_2CH_3$
5	Pentane	$CH_3(CH_2)_3CH_3$
6	Hexane	$CH_3(CH_2)_4CH_3$
7	Heptane	$CH_3(CH_2)_5CH_3$
8	Octane	$CH_3(CH_2)_6CH_3$
9	Nonane	$CH_3(CH_2)_7CH_3$
10	Decane	$CH_3(CH_2)_8CH_3$
11	Undecane	$CH_3(CH_2)_9CH_3$
12	Dodecane	$CH_3(CH_2)_{10}CH_3$
13	Tridecane	$CH_3(CH_2)_{11}CH_3$
14	Tetradecane	$CH_3(CH_2)_{12}CH_3$
15	Pentadecane	$CH_3(CH_2)_{13}CH_3$
16	Hexadecane	$CH_3(CH_2)_{14}CH_3$
17	Heptadecane	$CH_3(CH_2)_{15}CH_3$
18	Octadecane	$CH_3(CH_2)_{16}CH_3$
19	Nonadecane	$CH_3(CH_2)_{17}CH_3$
20	Eicosane	$CH_3(CH_2)_{18}CH_3$

EXERCISE 2-2
Write the structures of isohexane and neopentane.

Many of the branched alkyl groups have common names, as shown in Table 2-4. Here, in addition to iso- and neo-, we use the prefixes *sec-* (or *s-*) for secondary and *tert-* (or *t-*) for tertiary. The classification terms **secondary** and **tertiary** are used in conjunction with the term **primary,** although this last term has no application as a prefix. A primary carbon is one attached to only one other carbon atom. For example, all carbon atoms at the end of alkane chains are primary. The hydrogen atoms attached to such a carbon are designated primary hydrogens. A secondary carbon is attached to two, and a tertiary to three other carbon nuclei.*

Alkyl groups are also labeled in this way. An alkyl group created by removing a primary hydrogen is called primary. Removal of a secondary hydrogen results in a secondary alkyl group, removal of a tertiary hydrogen in a tertiary alkyl group (Table 2-4).

---

*Although not used in the naming of organic molecules, the term **quaternary** is applied to carbon atoms bound to four other carbons.

TABLE 2-4
Branched alkyl groups

Structure	Common name	Systematic name	Derived from	Designation
$CH_3-CH(CH_3)-$	Isopropyl	1-Methylethyl	Propane	Secondary
$CH_3-CH(CH_3)-CH_2-$	Isobutyl	2-Methylpropyl	2-Methylpropane (Isobutane)	Primary
$CH_3-CH(CH_2CH_3)-$	*sec*-Butyl	1-Methylpropyl	Butane	Secondary
$CH_3-C(CH_3)_2-$	*tert*-Butyl	1,1-Dimethylethyl	2-Methylpropane (Isobutane)	Tertiary
$CH_3-C(CH_3)_2-CH_2-$	Neopentyl	2,2-Dimethylpropyl	2,2-Dimethylpropane (Neopentane)	Primary

Primary, Secondary, and Tertiary Carbons and Hydrogens

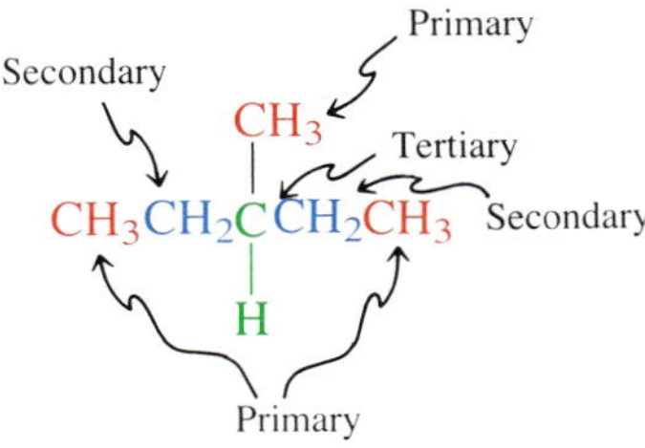

EXERCISE 2-3
Label the primary, secondary, and tertiary hydrogens in 2-methylpentane (isohexane).

Rules for Naming Branched Alkanes

Using the information in Table 2-3 allows you to name the first twenty straight-chain alkanes. How do we go about naming branched systems? IUPAC has supplied us with a set of rules that make this a relatively simple task, as long as they are carefully followed and in sequence.

RULE 1 *Find the longest chain in the molecule and name it.* This task is not as easy as it seems. The problem is that, in the condensed formula, complex alkanes may be written in ways that mask the identity of the longest chain. In the following examples, the longest chain, or stem, is clearly marked; the alkane stem gives the molecule its name.

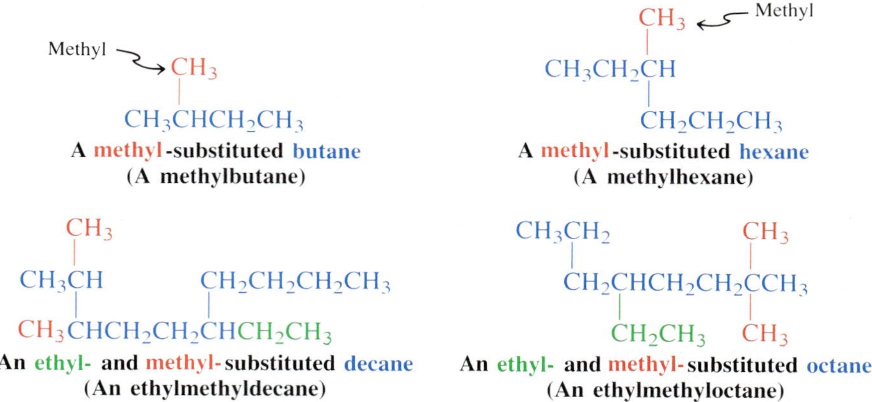

A methyl-substituted butane
(A methylbutane)

A methyl-substituted hexane
(A methylhexane)

An ethyl- and methyl-substituted decane
(An ethylmethyldecane)

An ethyl- and methyl-substituted octane
(An ethylmethyloctane)

If a molecule has two or more chains of equal length, the chain with the largest number of substituents is the base stem.

$$\begin{array}{c} CH_3\ \ CH_3 \\ CH_3CHCHCHCHCH_2CH_3 \\ CH_3\ \ CH_2 \\ CH_2 \\ CH_3 \end{array} \quad \text{not} \quad \begin{array}{c} CH_3\ \ CH_3 \\ CH_3CHCHCHCHCH_2CH_3 \\ CH_3\ \ CH_2 \\ CH_2 \\ CH_3 \end{array}$$

4 substituents
A heptane

3 substituents
A heptane

Before we move on to Rule 2, let us consider another, even more simplified way of picturing branched alkanes. This method is convenient because even condensed formulas can be cumbersome for complex alkanes. It consists of "zigzag" straight lines, the end of each line segment representing a carbon atom. All hydrogen atoms are omitted. The longest chain usually extends horizontally. Some of the alkanes pictured earlier are redrawn here in this manner:

A methylbutane

A methylhexane

Ethyl

An ethylmethyldecane

An ethylmethyloctane

Both straight-line and condensed formulas will be used in the following discussion.

RULE 2 *Name all groups attached to the longest chain as alkyl substituents*. For straight-chain substituents, Table 2-3 can be used to derive the alkyl name. However, what if the substituent chain is branched? In this case, the same IUPAC rules apply as did to the main chain: first, find the longest chain in the substituent; next, name all the appendages; and then proceed to Rule 3.

RULE 3 *Number the carbons of the longest chain beginning with the end that is closest to a substituent.*

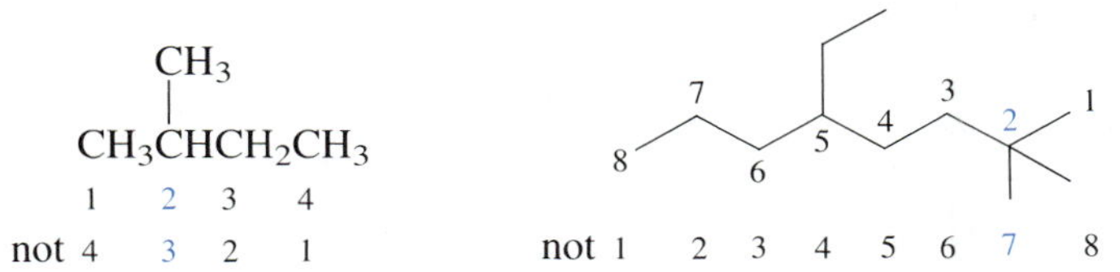

CH_3
$CH_3CHCH_2CH_3$
1 2 3 4
not 4 3 2 1

7 3 1
8 4 2
6 5
not 1 2 3 4 5 6 7 8

If there are two substituents at equal distance from the two ends of the chain, use the alphabet to decide how to number the stem. The substituent to come first in alphabetical order is attached to the carbon with the lower number.

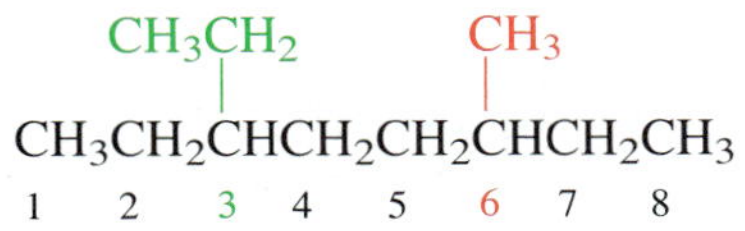

CH_3CH_2 CH_3
$CH_3CH_2CHCH_2CH_2CHCH_2CH_3$
1 2 3 4 5 6 7 8
Ethyl before methyl

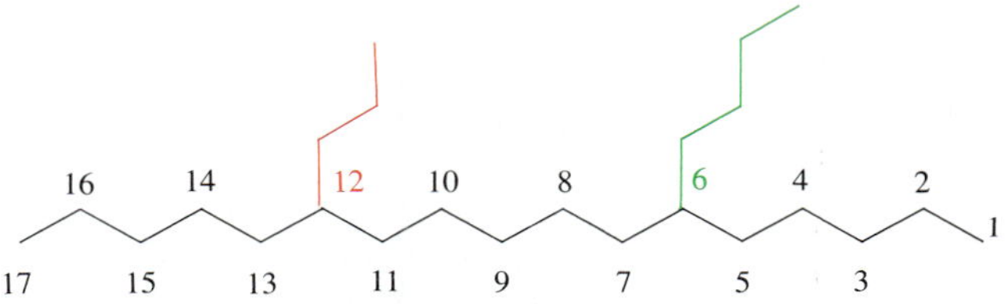

16 14 12 10 8 6 4 2
17 15 13 11 9 7 5 3 1
Butyl before propyl

In a substituent chain, the carbon numbered one is *always* the carbon atom bound to the principal chain.

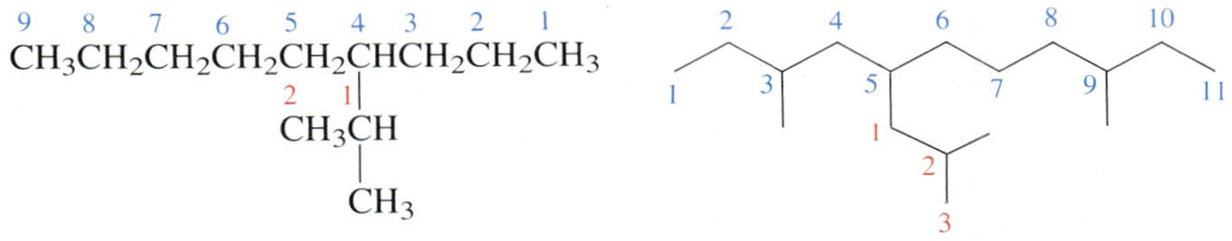

9 8 7 6 5 4 3 2 1
$CH_3CH_2CH_2CH_2CH_2CHCH_2CH_2CH_3$
2 1
CH_3CH
CH_3

2 4 6 8 10
1 3 5 7 9 11
1
2
3

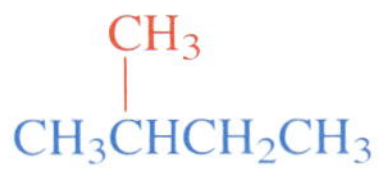

RULE 4 *Write the name of the alkane by first arranging all the substituents in alphabetical order (each preceded by the carbon number to which it is attached and a hyphen) and then add the name of the stem, as shown at the right.*

Should a molecule contain more than one of the same substituent, its alkyl name is preceded by the prefix di, tri, tetra, penta, and so forth (see Table 2-3). The positions of attachment to the stem are given collectively before the substituent name and are separated by commas. These prefixes, as well as *sec-* and *tert-*, are not considered in the alphabetical ordering except when they are part of a complex substituent name.

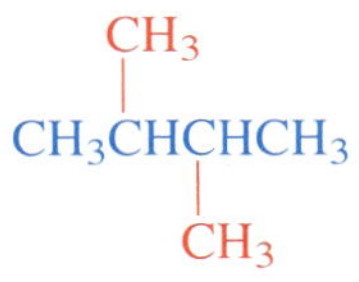

CH_3
$CH_3CHCH_2CH_3$
2-Methylbutane

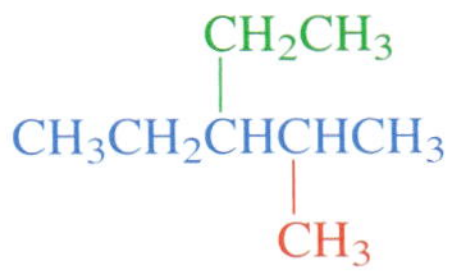

CH_2CH_3
$CH_3CH_2CHCHCH_3$
CH_3
3-Ethyl-2-methylpentane

CH_3
$CH_3CHCHCH_3$
CH_3
2,3-Dimethylbutane

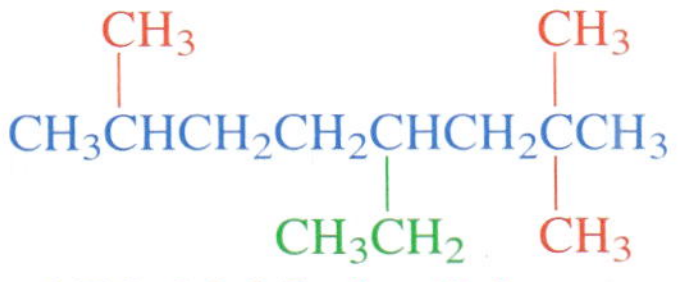

CH_3 CH_3
$CH_3CHCH_2CH_2CHCH_2CCH_3$
CH_3CH_2 CH_3
4-Ethyl-2,2,7-trimethyloctane

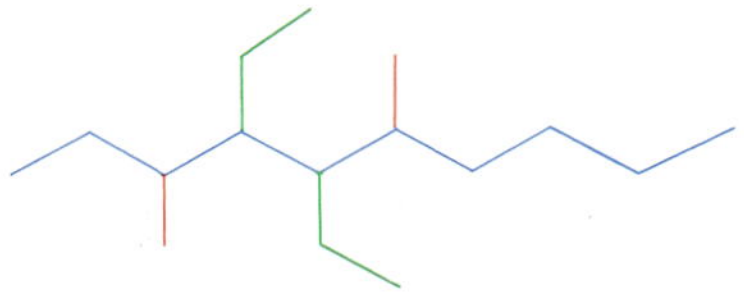

4,5-Diethyl-3,6-dimethyldecane

Although the common substituent names in Table 2-4 are permitted by IUPAC, it is preferable to use systematic naming. Such complex substituents are usually cited in parentheses, to avoid possible ambiguities. If they occur repeatedly, their names are preceded by the prefix bis, tris, tetrakis, pentakis, and so on.

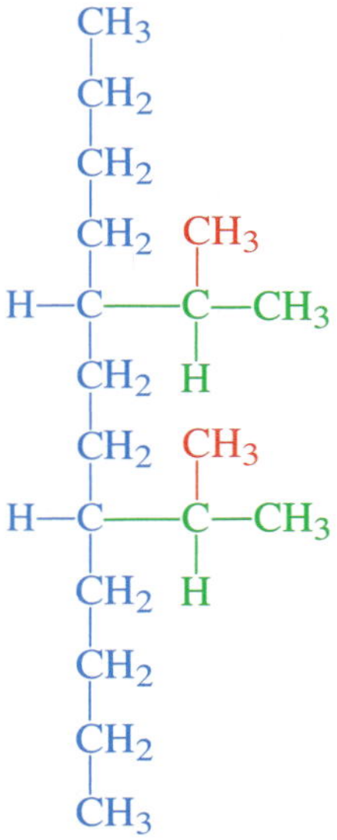

5,8-Bis(1-methylethyl)-dodecone

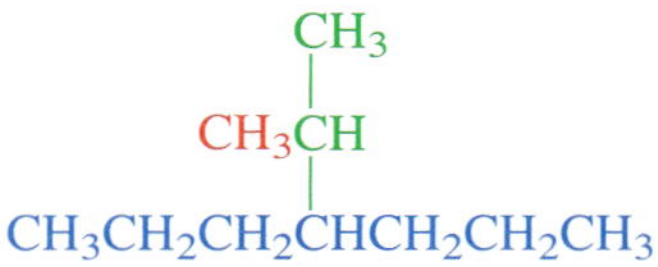

4-(1-Methylethyl)heptane (4-Isopropylheptane)

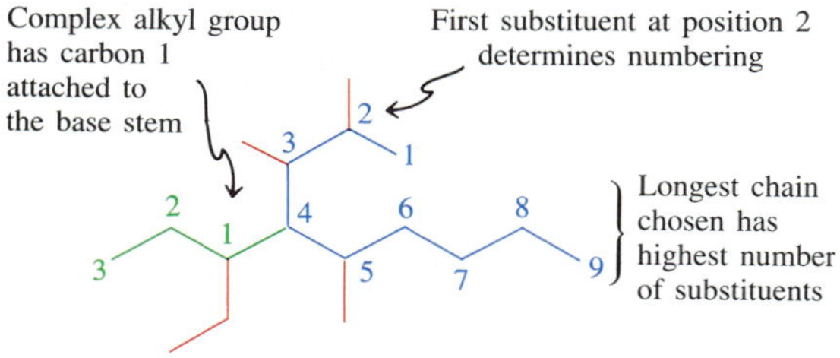

4-(1-Ethylpropyl)-2,3,5-trimethylnonane

Complex alkyl group has carbon 1 attached to the base stem

First substituent at position 2 determines numbering

Longest chain chosen has highest number of substituents

Further instructions on nomenclature will be presented when new classes of compounds, such as the cycloalkanes and haloalkanes, are introduced.

EXERCISE 2-4

Write down the names of the preceding eight branched alkanes, close the book, and reconstruct their structures from those names.

To summarize, four rules should be applied in sequence when naming a branched alkane: (1) find the longest chain; (2) find the names of all the alkyl groups attached to the stem; (3) number the chain; (4) name the alkane, with substituent names in alphabetical order and preceded by numbers to indicate their locations.

2-4 Physical Properties of the Alkanes

What do the structures of alkanes look like in three dimensions? What are their physical appearances, and what are their physical properties? These questions will be addressed next.

At room temperature, the lower homologs of the alkanes are gases or colorless liquids, the higher homologs solids. Their structural features are remarkably regular and can adopt the zigzag patterns (among others) used in the line notation for long hydrocarbon chains (Figure 2-1). The carbon atoms are tetrahedral, with bond angles approximating 109° and with regular C–H (~1.10 Å) and C–C (1.54 Å) bond lengths. To depict the three-dimensional structure of the hydrocarbon chain, we make use of the dashed-wedged line notation (Figure 1-22). In this case, the principal chain and one of the hydrogen atoms at each terminus are placed in the plane of the page (Figure 2-2).

EXERCISE 2-5

Draw zigzag dashed-wedged line structures for 2-methylbutane and 2,3-dimethylbutane.

The regularity in alkane structures suggests that their physical constants would reveal predictable trends. Indeed, inspection of the data presented in Table 2-5 reveals regular incremental increases along the homologous series. For example, from pentane to pentadecane, every additional CH_2 group causes an increase in boiling point ranging from 20°C to 30°C (Figure 2-3).

FIGURE 2-1
Two molecular models of hexane. (Model sets courtesy of Maruzen Co., Ltd., Tokyo.)

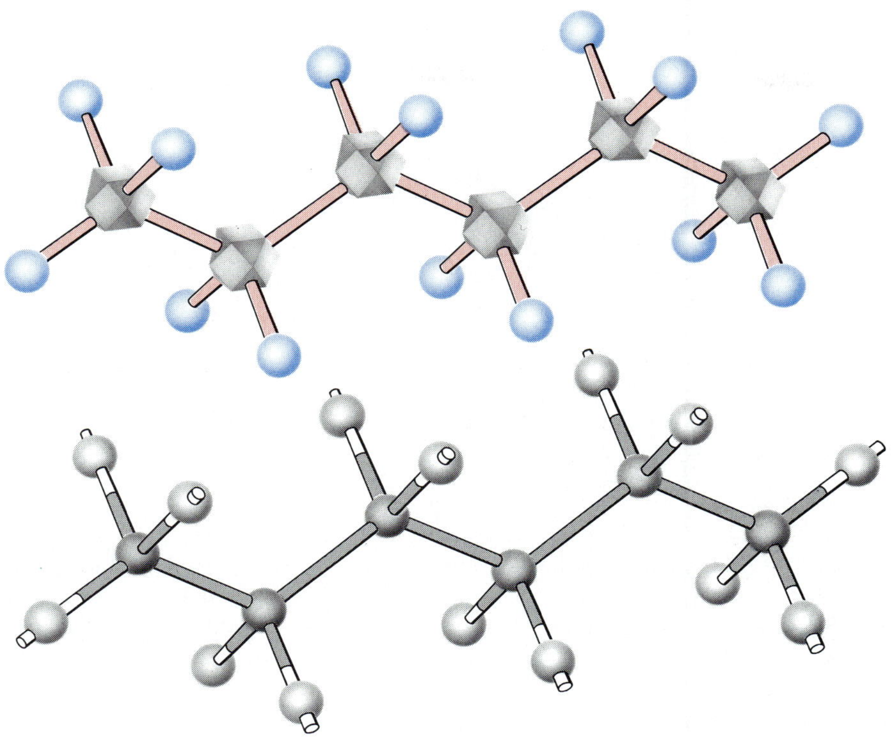

FIGURE 2-2
Dashed-wedged line structures of methane through pentane. Note the zigzag arrangement of the principal chain and the two unique terminal hydrogens, which are part of it.

109.5°
1.10 Å
109.5°
1.54 Å
1.095 Å

TABLE 2-5
Physical properties of straight-chain alkanes

Hydrocarbon	Boiling point (°C)	Melting point (°C)	Density at 20°C (g ml^{-1})
Methane	−161.7	−182.5	0.5547 (at 0°C)
Ethane	−88.6	−183.3	0.509 (at −60°C)
Propane	−42.1	−187.7	0.5005
Butane	−0.5	−138.3	0.5787
Pentane	36.1	−129.8	0.5572
Hexane	68.7	−95.3	0.6603
Heptane	98.4	−90.6	0.6837
Octane	125.7	−56.8	0.7026
Nonane	150.8	−53.5	0.7177
Decane	174.0	−29.7	0.7299
Undecane	195.8	−25.6	0.7402
Dodecane	216.3	−9.6	0.7487
Tridecane	235.4	−5.5	0.7564
Tetradecane	253.7	5.9	0.7628
Pentadecane	270.6	10	0.7685
Eicosane	343	36.8	0.7886

FIGURE 2-3
Increases in values of physical constants of straight-chain alkanes with increasing size.

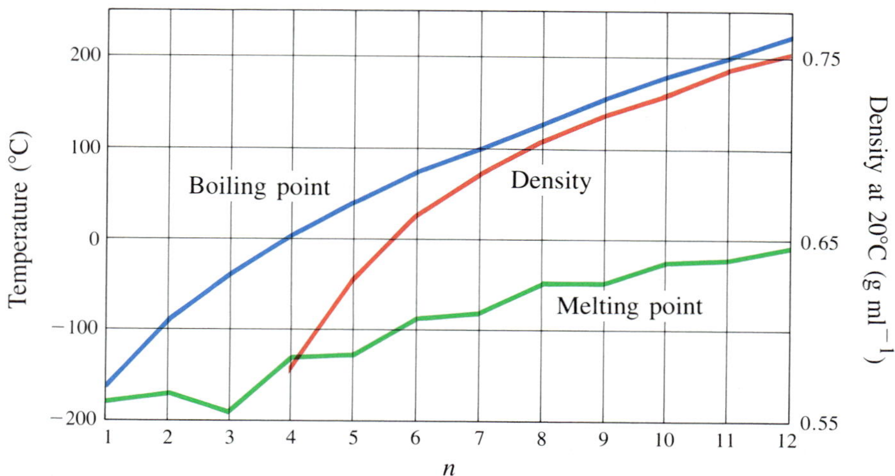

A plot of the melting points of alkanes versus increasing size, however, produces an irregular curve. This irregularity is due to the fact that odd-membered alkanes have slightly lower melting points than do even-membered systems. On the other hand, the values for the densities increase smoothly.

Molecules Are Attracted to Each Other by Forces

Why are there such trends? They exist because of **intermolecular,** or **van der Waals, forces.** Molecules exert attractive forces on each other causing aggregation into more highly organized arrangements as solids and liquids. The breaking up of such aggregates requires energy, usually supplied in the form of heat. Most organic molecules exist as crystals in the solid state, in which they are highly ordered. Ionic compounds, such as salts, are rigidly held in a crystal lattice, mainly by strong Coulombic forces. Nonionic but polar molecules, such as chloromethane (CH_3Cl), are held together by weaker dipolar forces—namely, dipole–dipole interactions, again of Coulombic nature (Sections 1-2 and 6-2). In contrast, the very nonpolar alkanes attract each other by **London forces.** These forces are due to *electron correlation*. A simplified explanation of this effect is to consider alkanes essentially electron clouds. As one alkane approaches a second, the electrons in the various bonds of the first are affected by the electrons of the second, and the electrons in the two individual molecules begin to correlate their movements. This correlation is one of attraction.

London forces are very weak but nevertheless have an appreciable effect on the physical constants of alkanes. Figure 2-4 is a simple picture comparing ionic, dipolar, and van der Waals attractions. In contrast with Coulombic forces, which change with the square of the distance between charges, London forces change with the distance between molecules to the power of six. However, there is a limit as to how close these forces can bring molecules together. At a certain proximity, nuclear-nuclear and electron-electron repulsions outweigh these attractive forces.

There are relations between these forces and the physical constants of elements and compounds. For example, to cause the melting of a compound, the

FIGURE 2-4
A. Coulombic attraction in sodium ethanoate (acetate) crystals.
B. Dipole–dipole interactions in solid chloromethane; the molecules arrange to allow for favorable Coulombic attraction.
C. London forces in crystalline pentane. In this simplified picture, the electron clouds as a whole mutually interact to produce partial charges of opposite sign. Note that this situation is not static but changes continually as the electrons continue to correlate their movement in the two molecules.

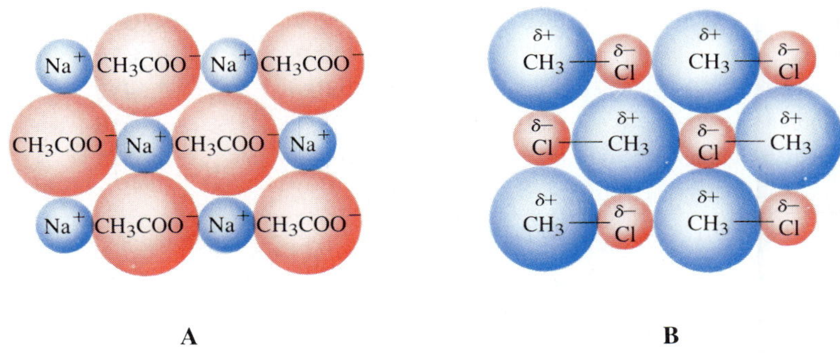

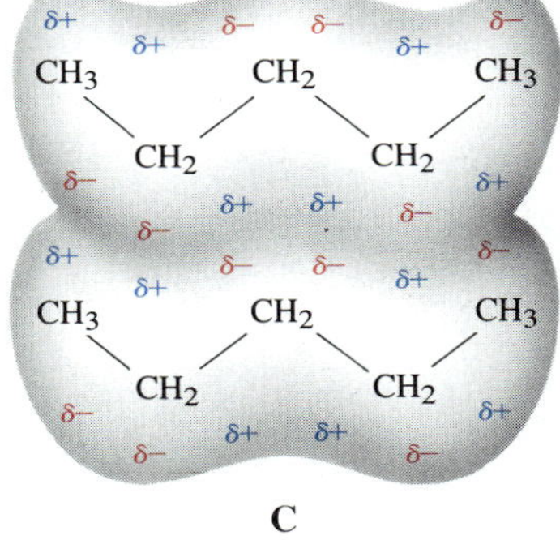

attractive forces that are responsible for the highly ordered crystalline state must be overcome. In an ionic compound, such as the sodium salt of ethanoic (acetic) acid, sodium ethanoate (acetate), which is shown in Figure 2-4A, the strong interionic forces require a relatively high temperature (322°–324°C) for the compound to melt.

In the liquid state, the same attractive forces are in effect as those in the solid state. For a molecule to escape these forces and enter the gas phase, again heat has to be applied. At the point at which the vapor pressure of the liquid equals the atmospheric pressure, boiling occurs. Like melting points, the boiling points of compounds are relatively high if the intermolecular forces are relatively large. The molecular weight also contributes to the value of the boiling point. A heavy compound requires more kinetic energy to leave the liquid phase than does, for example, its lower homolog.

Why do boiling and melting points increase as the size of the alkane increases? We now know the answer to this question: to a large extent, this effect may be ascribed to increasing London forces and molecular weights. With regard to the differences in melting points between even- and odd-membered straight-chain alkanes (Figure 2-3), the even-membered systems are more tightly packed in the crystalline state, allowing for better interactions. This is also the reason for the increasing density values recorded in Figure 2-3 and Table 2-5.

Knowing the effect of attractive forces allows predictions to be made with respect to the physical properties of branched alkanes. They have smaller surface areas and are unable to pack as well as their straight-chain isomers. This effect leads to diminished attractive forces and therefore to lower boiling and melting points. On the other hand, it should give them relatively high densities. Table 2-6 bears out that prediction. It also shows another interesting phenomenon: symmetrical branched alkanes have unusually high melting points (e.g., 2,2-dimethylpropane and 2,2,3,3-tetramethylbutane). Owing to their symmetry, these compounds readily form the regular lattices required for good crystal

TABLE 2-6
Physical properties of some branched alkanes

Alkane	Boiling point (°C)	Melting point (°C)	Density at 20°C (g ml^{-1})
2-Methylpropane	−11.7	−159.4	0.5572
2-Methylbutane	29.9	−159.9	0.6196
2,2-Dimethylpropane	9.4	−16.8	0.5904
2-Methylpentane	60.3	−153.6	0.6532
3-Methylpentane	63.3	−118.0	0.6644
2,2-Dimethylbutane	49.7	−100.0	0.6492
2,3-Dimethylbutane	58.0	−128.4	0.6616
2,2,3,3-Tetramethylbutane	106.3	100.6	0.6568

formation. Yet, their attractive forces are relatively small, resulting in low boiling points. Consequently, for some of these hydrocarbons, the difference between melting point and boiling point is small.

In summary, straight-chain alkanes have regular structures. Their melting points, boiling points, and densities increase with molecular weight because of increasing attraction between molecules. These relations do not hold for the branched isomers of straight-chain alkanes, because branched alkanes have smaller surface areas and, in some cases, form good crystals because of symmetry.

2-5 Molecules Are Not Rigid: Rotation Interconverts Conformational Isomers

We have considered how intermolecular forces can affect the physical properties of molecules. These forces act *between* molecules. In this section, we will examine the forces present *within* molecules (i.e., intramolecular forces), which make some arrangements in space energetically more favorable than others.

If you build a molecular model of ethane, you can see that the two methyl groups are readily rotated with respect to each other. For the molecule itself, the energy required to move the hydrogen atoms past each other, the *barrier to rotation,* is 3 kcal $mole^{-1}$. This turns out to be a very low figure, in fact so low that chemists speak of ''free rotation'' of the methyl groups. In general, there is free rotation around all single bonds.

Figure 2-5 depicts the rotational movement in ethane by the use of dashed-wedged line structures. There are two extreme ways of drawing ethane: the staggered conformation and the eclipsed one. If the **staggered conformation** is viewed along the C–C axis, each hydrogen atom on the first carbon is seen to be positioned perfectly between two hydrogen atoms on the second. The second extreme is derived from the first by a 60° turn of one of the methyl groups around the C–C bond. Now, if this conformation is viewed along the C–C axis, all hydrogen atoms on the first carbon are directly opposite those on the second—that is, those on the first **eclipse** those on the second. A further 60° turn converts the eclipsed form into a new but equivalent staggered arrangement. Between

FIGURE 2-5
Rotation in ethane: (A and C) staggered conformations; (B) eclipsed.

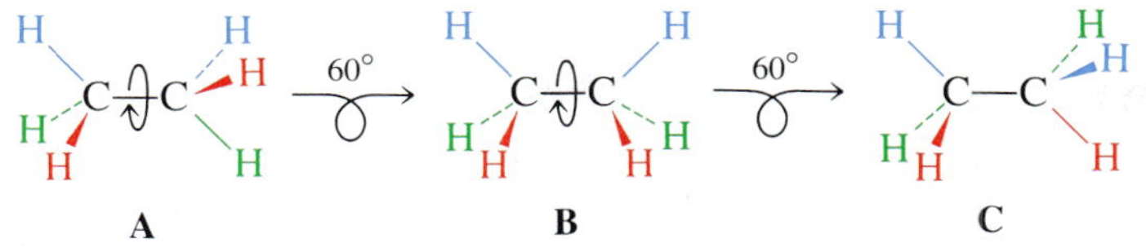

these two extremes, rotation of the methyl group results in numerous additional positions, referred to collectively as **skew conformations.**

The many forms of ethane (and, as we shall see, substituted analogs) created by such rotations are **conformational isomers.** Such isomers are also called **conformers** or **rotamers.** All of them rapidly interconvert at room temperature. The study of their thermodynamic and kinetic behavior is **conformational analysis.**

A Different Perspective: Newman Projections

A simple alternative to the dashed-wedged line structures for illustrating the conformers of ethane is the **Newman projection.** You can arrive at a Newman projection from the dashed-wedged line picture by turning the molecule out of the plane of the page toward you and viewing it along the C–C axis (Figure 2-6A and B). In this notation, the front carbon obscures the back carbon, but the bonds emerging from both are clearly seen. The front carbon is depicted as the point of juncture of the three bonds attached to it, one of them usually drawn vertically and pointing up. The rear carbon is a circle (Figure 2-6C). The bonds to this carbon project from the outer edge of the circle. The extreme conformational isomers of ethane are readily drawn in this way (Figure 2-7). To make the three rear hydrogen atoms more visible in eclipsed conformations, they are drawn somewhat rotated out of the perfectly eclipsing position.

FIGURE 2-6
Converting a dashed-wedged line structure (A) into a Newman projection (C).

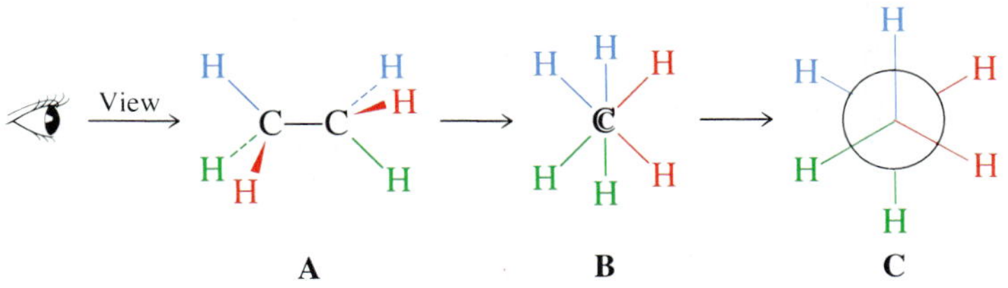

FIGURE 2-7
Newman projections of staggered and eclipsed rotamers of ethane. In these projections, the back carbon is rotated clockwise in increments of 60°.

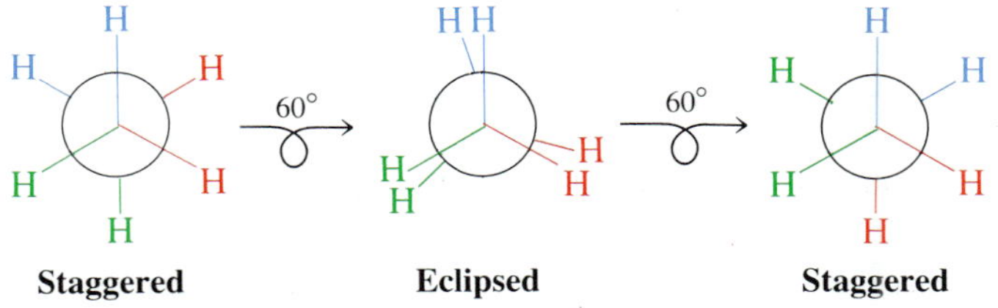

The Rotamers of Ethane Have Different Potential Energies

Not all rotamers of ethane have the same potential-energy content; the combined energy stored in the bonds of the various species differs. As mentioned earlier, about 3 kcal $mole^{-1}$ of heat is required to rotate the methyl groups in ethane. What is the reason for this requirement? A simple explanation is based on electron repulsion. As one methyl group turns around the C–C axis, starting from a staggered conformation, the distance between the hydrogen atoms of the respective methyl groups begins to diminish. This results in increasing interaction between the bonding pairs of electrons in the C–H bonds, an interaction that is repulsive. Thus, the potential energy of the system rises steadily as the methyl group rotates from staggered through skew to eclipsed conformations. At the point of eclipsing, the molecule has its highest energy content, because at this stage all six hydrogen atoms and the two sets of six bonding electrons are closest. This point is 3 kcal $mole^{-1}$ above the lowest energy state of the molecule, the staggered rotamer.

A Convenient Way of Depicting the Energy Changes During Bond Rotation: Potential-Energy Diagrams

The differences in potential energy between rotamers can be pictured by plotting the energy changes against the degree of rotation (Figure 2-8). Such a plot is called a **potential-energy diagram,** in this case drawn for the rotation of a methyl group in ethane. Potential-energy diagrams are also useful in the description of other chemical processes.

In potential-energy diagrams, the changes in potential energy during a process or reaction are plotted against a reaction progress coordinate, in this case the degree of rotation. Other alternative coordinates, such as the distance between opposing pairs of hydrogen atoms, would produce a plot very similar to Figure 2-8. In the staggered conformation, each hydrogen atom on one carbon atom has two closest-lying hydrogen neighbors on the other carbon, 2.55 Å away. The eclipsing hydrogens, on the other hand, have moved closer together, to a distance of 2.29 Å.

Ethane is best described in its staggered conformation. In fact, the eclipsed rotamer has only a fleeting lifetime (of the order of a molecular vibration, 10^{-12} sec) as the hydrogens rapidly move past each other equilibrating one staggered arrangement with another. Because eclipsed conformations have the highest energy in this process, they are positioned as maxima in potential-energy diagrams. Such points are called **transition states** (TS), marking the transition from one staggered rotamer to another. The energy of the transition state can be viewed as the barrier to be overcome when the molecule goes from one staggered arrangement to the next. This energy is called the **activation energy,** E_a, for the rotational process. The lower its value the faster the rotation.

How to Get Past the Activation-Energy Barrier

To pick up energy in excess of the activation energy, molecules have to collide with each other, with other solvent molecules, or with the walls of the container. Such collisions lead to energy transfer and are strongly affected by the *kinetic energy*. At room temperature, the average kinetic energy of organic compounds is only about 0.6 kcal $mole^{-1}$. A graph that depicts the distribution

FIGURE 2-8

Potential-energy diagram of the rotational isomerism in ethane; TS = transition state.

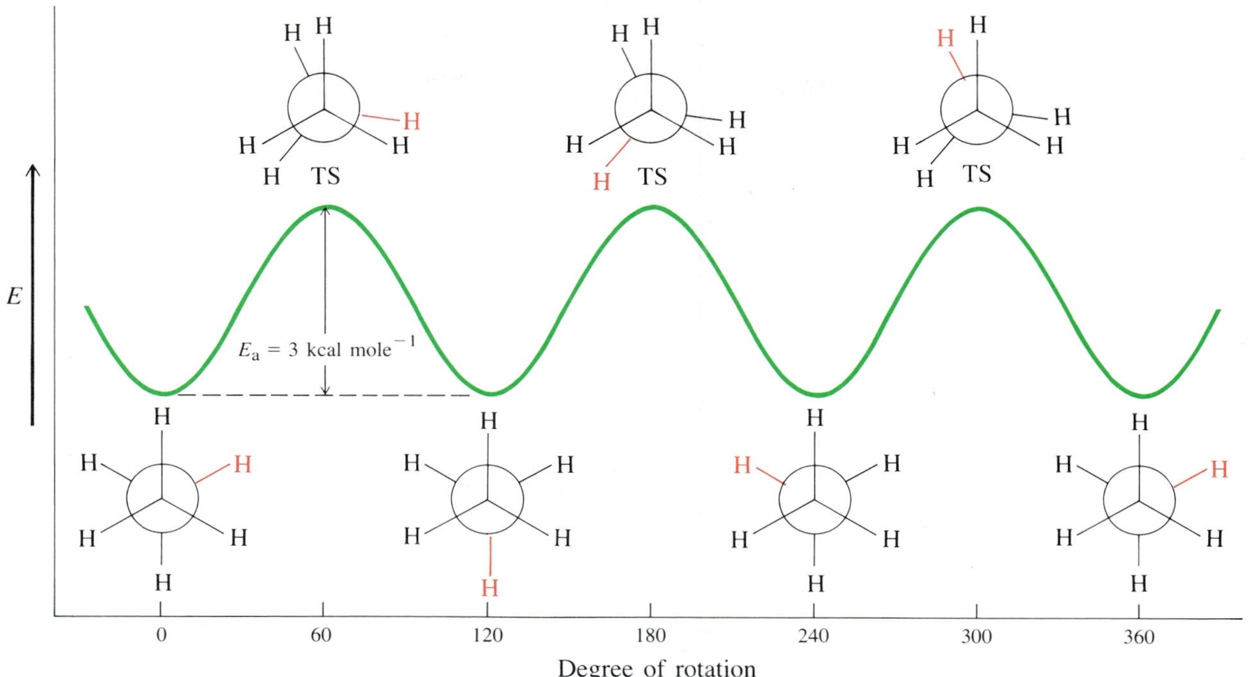

of kinetic energy at a given temperature is called a *Boltzmann* distribution curve* (Figure 2-9). It shows that, although most molecules have only average speed, some have substantially lower energy and others substantially higher. At room temperature, this may be as high as 25 kcal mole^{-1}. A good part of this

FIGURE 2-9

Boltzmann curves at two temperatures. At the higher temperature (green curve), there are more molecules of kinetic energy E than at the lower temperature (blue curve).

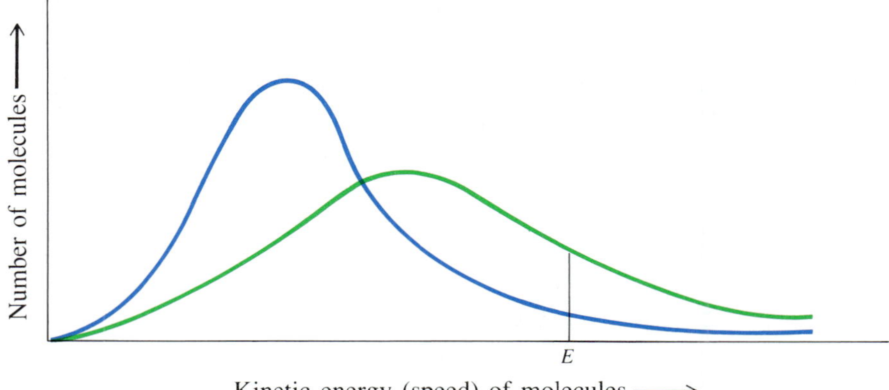

*Professor Ludwig Boltzmann, 1844–1906, University of Leipzig, Germany.

kinetic energy is translated in the molecule into the energy required to overcome the activation barrier for rotation in ethane. Because of the rapid redistribution of kinetic energy by continual collisions, all molecules will eventually overcome this barrier. This is why the qualitative term ''free rotation'' is used for ethane. At higher temperatures, the average kinetic energy increases and the Boltzmann curve is flattened and distorted toward the higher-energy side (Figure 2-9). Now, more molecules with energy higher than that required by the transition state are available; so the net speed of the rotational process increases. On the other hand, at lower temperatures, the rate of rotation decreases.

Rotation in Substituted Ethanes

How does the potential-energy diagram change when a substituent is added to ethane? Consider, for example, propane, whose structure is similar to that of ethane, except that a methyl group replaces one of ethane's hydrogen atoms.

A potential-energy diagram for the rotation around a C–C bond in propane is shown in Figure 2-10. The Newman projections of propane differ from those of ethane only by the substituting methyl group. Again, the extreme conformations are staggered and eclipsed. However, the activation barrier separating the two is 3.4 kcal mole^{-1}, slightly higher than that for ethane. This difference is due to an unfavorable steric interaction between the methyl substituent and the eclipsing hydrogen in the transition state, a phenomenon called **steric hindrance.** To a first approximation, this effect can be attributed simply to bulk: two molecular

FIGURE 2-10

Potential-energy diagram of the rotational isomerism in propane.

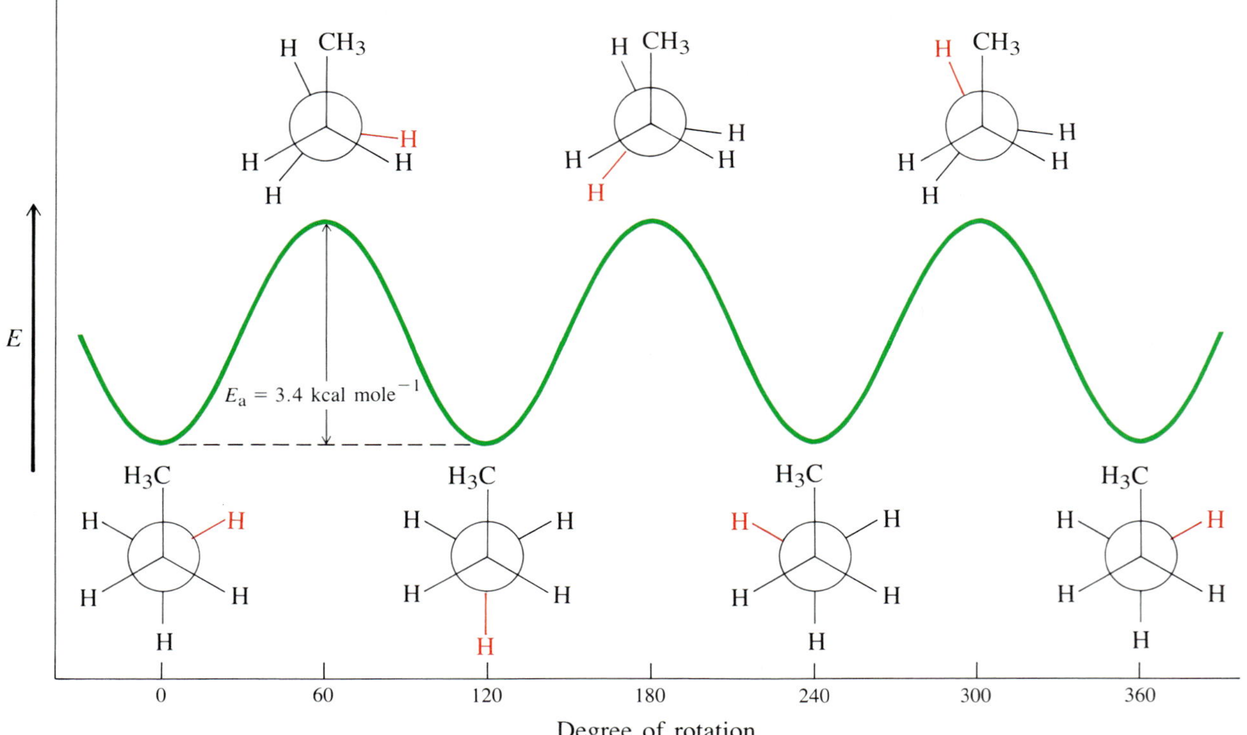

fragments cannot occupy the same region in space. Steric hindrance in propane is actually worse than what is indicated by the E_a value for rotation. Methyl substitution raises the energy not only of the eclipsed conformation, but also of the staggered (ground state) one, the latter to a lesser extent because of less steric interaction. The activation energy reveals only the *difference* between ground and transition states; the full effect of eclipsing a methyl group in propane compared with a hydrogen atom in ethane is not reproduced in a correspondingly higher E_a. In propane, *both* conformations are of higher energy, the eclipsed rotamer being affected somewhat more. The net result is only a minor increase in E_a.

Conformational Analysis of Butane: There Can Be More Than One Staggered and One Eclipsed Conformation

Build a model and look at the rotation around the central C–C bond of butane. You will find that there are more conformations than one staggered and one eclipsed (Figure 2-11). Start with the conformer in which the two methyl groups are as far away from each other as possible. This arrangement, called ***anti*** (i.e., opposed), is the most stable because steric hindrance is minimized. Rotation of the rear carbon in the Newman projections in either direction (in Figure 2-11, the direction is clockwise) produces an eclipsed conformation with two CH_3—H interactions. This rotamer is 3.8 kcal $mole^{-1}$ higher in energy than the *anti* precursor. Further rotation furnishes a *new* staggered structure in which the two methyl groups are in closer proximity than they are in the *anti* conformation. To distinguish this conformer from the others, it is named ***gauche*** (*gauche,* French, in the sense of awkward, clumsy). As a consequence of steric hindrance, the *gauche* conformer is higher in energy than the *anti* conformer by about 0.9 kcal $mole^{-1}$. Further rotation (Figure 2-11) results in a *new* eclipsed arrangement in which the two methyl groups are superposed. Because the two bulkiest substituents eclipse in this rotamer, it is energetically highest, 4.5 kcal $mole^{-1}$ higher than the most stable *anti* structure. Further rotation produces another *gauche* conformer. The activation energy for *gauche* $\rightleftharpoons$ *gauche* interconversion is 3.6 kcal $mole^{-1}$. A potential-energy diagram summarizes the energetics of the rotation (Figure 2-12). The most stable *anti* conformer is the most abundant in solution (about 80% at 25°C). Its less-stable *gauche* counterpart is present in lesser concentration (20%).

You can see from Figure 2-12 that knowing the difference in thermodynamic stability of two conformers (e.g., 0.9 kcal $mole^{-1}$ between the *anti* and *gauche*

FIGURE 2-11

Clockwise rotation of the rear carbon along the C-2—C-3 bond in a Newman projection of butane.

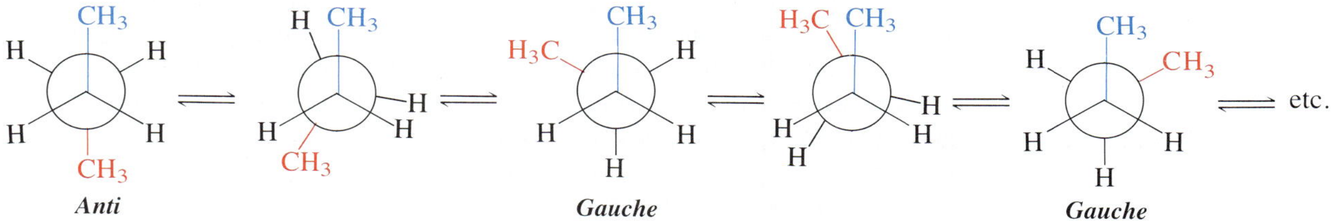

FIGURE 2-12

Potential-energy diagram of the rotation around the C-2—C-3 bond in butane. There are three processes: *anti* → *gauche* conversion with E_{a1} = 3.8 kcal mole^{-1}, *gauche* → *gauche* rotation with E_{a2} = 3.6 kcal mole^{-1}, and *gauche* → *anti* transformation with E_{a3} = 2.9 kcal mole^{-1}.

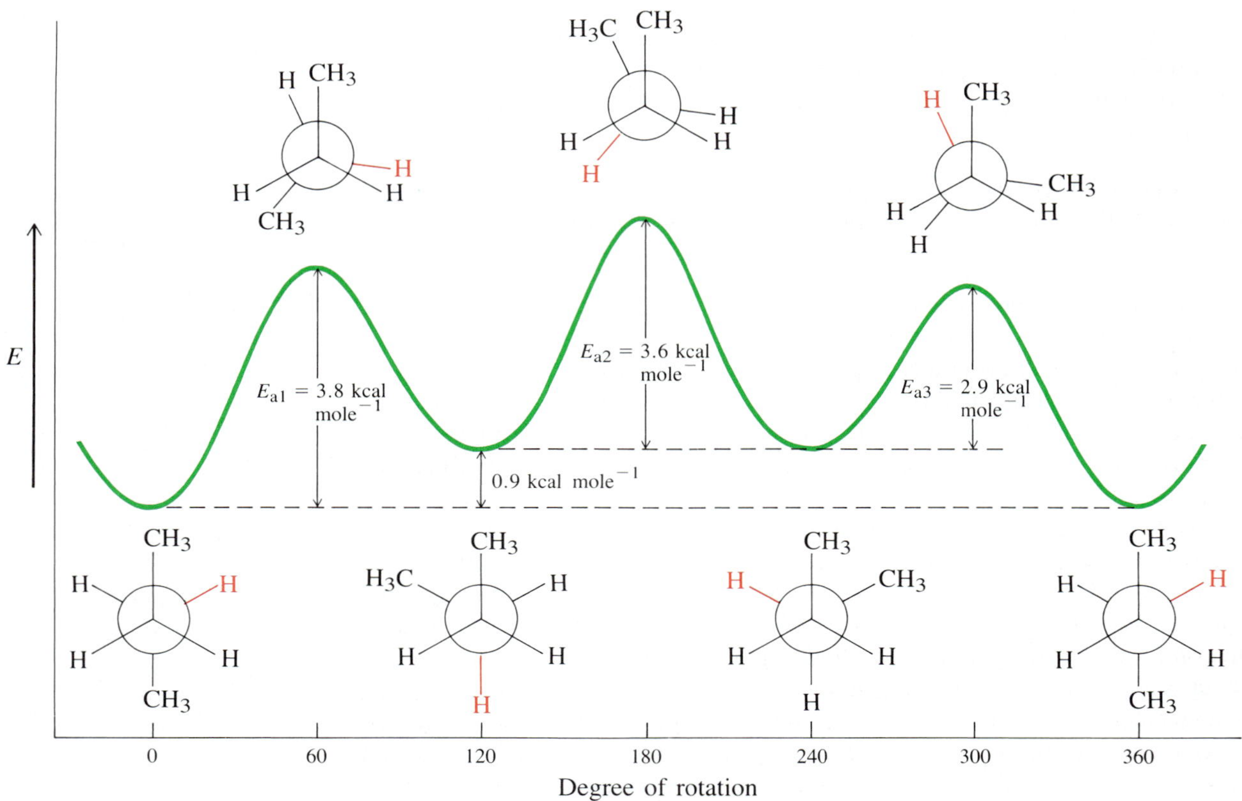

isomers) and the activation energy for proceeding from the first to the second (e.g., 3.8 kcal mole^{-1}) allows you to estimate the activation barrier of the reverse reaction. In this case, E_a for the *gauche*-to-*anti* conversion is 3.8 − 0.9 = 2.9 kcal mole^{-1}.

Qualitative Predictions Using Conformational Analysis

Conformational analysis helps in the understanding of reactivity. Because the most stable conformer is most abundant, it is frequently the species through which a reaction proceeds. However, in many cases, a reaction will occur through a less-stable conformation simply because a particular reagent selects for that conformation.

Conformational analysis is particularly useful in dealing with alkanes more complicated than butane. Consider, for example, the two staggered conformers of 2,3-dimethylbutane viewed along the C-2—C-3 bond. Counting the number of unfavorable *gauche* interactions in each of the two conformers allows a qualitative prediction about which structure should be more stable in solution or in the gas phase. However, such predictions may not hold for the solid state, because van der Waals forces can override conformational considerations.

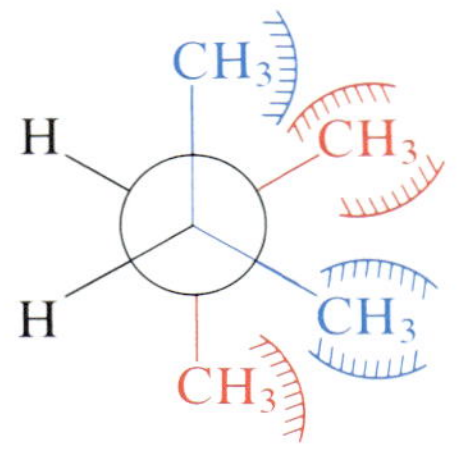

Three mutual *gauche* interactions: less stable

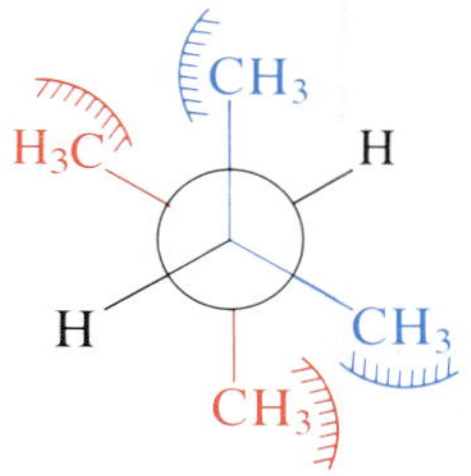

Two mutual *gauche* interactions: more stable

An even clearer case is presented by 2,2,4-trimethylpentane. Staggering the bulky 1,1-dimethylethyl (*tert*-butyl) between the two methyl groups is clearly less favorable than staggering it between only one methyl group and a hydrogen atom, as shown at the right.

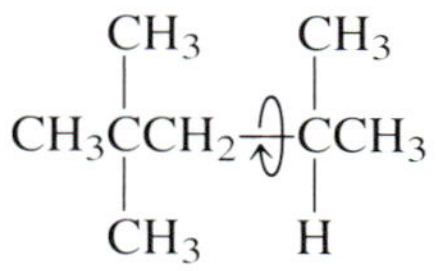

2,2,4-Trimethylpentane

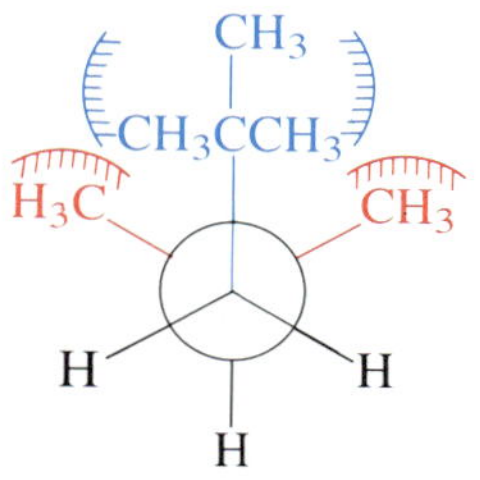

Less stable

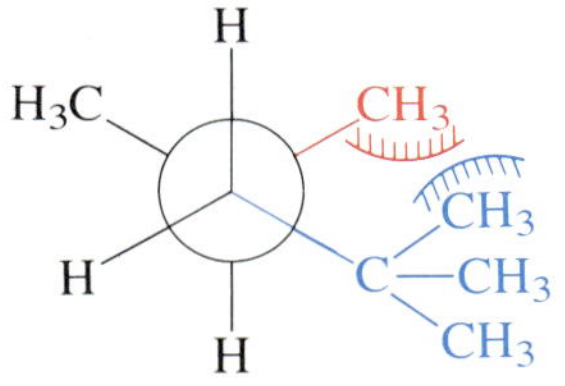

More stable

EXERCISE 2-6

Draw the expected potential-energy diagram for the rotation around the C-2—C-3 bond in 2,3-dimethylbutane.

In summary, intramolecular forces control the arrangement of substituents on neighboring and bonded saturated carbon atoms. In ethane and its substituted derivatives, the relatively stable staggered conformations are interconverted through higher-energy transition states in which substituents are eclipsed. To reach these transition states, molecules have to absorb the kinetic energy of others. The energy distribution of a collection of molecules at any given temperature is depicted by a Boltzmann curve. The energetics of rotation around the C–C bond is conveniently pictured in a potential-energy diagram. If two substituents (one on each carbon atom) are 180° apart in a staggered Newman projection, they belong to an *anti* conformer. If they are 60° apart, the conformer is *gauche*. *Gauche* conformations are usually less stable than their *anti* counterparts. Conformational analysis is the study of the changes in potential energy that take place on rotation around single bonds.

2-6 Kinetics and Thermodynamics of Conformational Isomerism and of Simple Reactions

The *anti* $\rightleftharpoons$ *gauche* conformational isomerism is a typical example of an equilibrium between two chemically distinct species. Although no bonds are broken or made, as in the usual chemical reaction, the process is controlled by the same physical criteria as are ordinary reactions.

This section reviews some of the basic principles governing chemical reactions:

1 **Chemical thermodynamics** deals with the changes in energy that take place when molecules react, a feature that controls the *extent* to which a reaction will go to completion.

2 **Chemical kinetics** concerns the velocity or rate at which the concentrations of reactants and products change, in other words the *speed* at which a reaction will go to completion.

The two phenomena are frequently related. Reactions that are thermodynamically very favorable often proceed faster than do less-favorable ones. On the other hand, some reactions are faster than others even though they result in a comparatively less stable product. A transformation that yields the most-stable products is said to be under **thermodynamic control.** Its outcome is determined by the net favorable change in energy in going from starting materials to products. A reaction that furnishes products that are less stable than others that could have formed is defined as being under **kinetic control.** Let us put these statements on a more quantitative footing.

Equilibria and the Thermodynamics of Chemical Transformations

All chemical reactions are reversible, and reactants and products interconvert to varying degrees. When the concentrations of reactants and products no longer change, the reaction is in a **state of equilibrium.** In many cases, equilibrium lies extensively (say, more than 99.9%) on the side of the products. When this occurs, the reaction is said to have *gone to completion*. (In such cases, the arrow indicating the reverse reaction is usually omitted). Equilibria are described by equilibrium constants, K. To find an equilibrium constant, divide the product of the concentrations of the components on the right side of the reaction by the product of the concentrations of the components on the left, all given in units of mole liter^{-1}. A large value for K indicates that a reaction will go to completion; it is said to have a large **driving force.**

Typical Chemical Equilibria

$$\mathrm{A} \underset{}{\overset{K}{\rightleftharpoons}} \mathrm{B} \qquad K = \frac{[\mathrm{B}]}{[\mathrm{A}]}$$

$$\mathrm{A} + \mathrm{B} \underset{}{\overset{K}{\rightleftharpoons}} \mathrm{C} \qquad K = \frac{[\mathrm{C}]}{[\mathrm{A}][\mathrm{B}]}$$

$$\mathrm{A} + \mathrm{B} \underset{}{\overset{K}{\rightleftharpoons}} \mathrm{C} + \mathrm{D} \qquad K = \frac{[\mathrm{C}][\mathrm{D}]}{[\mathrm{A}][\mathrm{B}]}$$

If a reaction has ''gone to completion'' or has ''a large driving force,'' a certain amount of energy has been released. The equilibrium constant can be related directly to the thermodynamic function of the **Gibbs standard free energy change,** ΔG°,* at equilibrium:

$$\Delta G^\circ = -RT \ln K = -2.303\ RT \log K \text{ (in kcal mole}^{-1}\text{)}$$

in which R is the gas constant (1.986 cal deg^{-1} mole^{-1}) and T is the absolute

*The descriptor ΔG° refers to the free energy of a reaction with the molecules in their standard states (e.g., ideal molar solutions) after it has reached equilibrium. The free energy change ΔG of such a system *before* equilibration is given by:

$$\Delta G = \Delta G^\circ + RT \ln K$$

In this book, only ΔG° values will be used.

temperature measured in kelvins (K). A negative value for ΔG° signifies a release of energy. The equation shows that a large value for K indicates a large favorable free energy change. At room temperature (25°C, 298 K), the preceding equation becomes

$$\Delta G^\circ = -1.36 \log K \text{ (in kcal mole}^{-1}\text{)}$$

This expression tells us that an equilibrium constant of 10 would have a ΔG° of -1.36 kcal mole^{-1}, and, conversely, a K of 0.1 would have a $\Delta G^\circ = +1.36$ kcal mole^{-1}. Because the relation is logarithmic, doubling the ΔG° value increases the K value exponentially. When $K = 1$, starting materials and products are present in equal concentrations and ΔG° is zero (Table 2-7).

TABLE 2-7
Equilibria and free energy

K	Percentage B	Percentage A	ΔG° (kcal mole^{-1} at 25°C)
0.01	0.99	99.0	+2.73
0.1	9.1	90.9	+1.36
0.33	25	75	+0.65
1	50	50	0
2	67	33	−0.41
3	75	25	−0.65
4	80	20	−0.82
5	83	17	−0.95
10	90.9	9.1	−1.36
100	99.0	0.99	−2.73
1000	99.9	0.1	−4.09
10000	99.99	0.01	−5.46

EXERCISE 2-7
Calculate the equilibrium concentration of *gauche*-butane at 25°C and at 100°C.

The Free Energy Change Is Related to Changes in Bond Strengths and the Degree of Order in the System

The Gibbs standard free energy change is related to two other thermodynamic quantities, the change in **enthalpy,** ΔH°, and in **entropy,** ΔS°:

$$\Delta G^\circ = \Delta H^\circ - T\Delta S^\circ$$

In this equation, T is again expressed in kelvins and ΔH° in kcal mole^{-1}, whereas ΔS° is in cal deg^{-1} mole^{-1}, also called entropy units (e.u.).

The enthalpy change, ΔH°, is defined as the heat of a reaction at constant pressure. Enthalpy changes in an organic chemical reaction relate mainly to changes in bond strengths in the course of the reaction. Thus, the value of ΔH°

can be estimated by subtracting the sum of the strengths of the bonds formed from those broken:

$$\begin{pmatrix}\text{Sum of strengths}\\ \text{of bonds broken}\end{pmatrix} - \begin{pmatrix}\text{Sum of strengths}\\ \text{of bonds formed}\end{pmatrix} = \Delta H^\circ$$

If the bonds formed are stronger than those broken, the value of ΔH° is negative and the reaction is defined as **exothermic** (releasing heat). In contrast, a positive ΔH° is characteristic of an **endothermic** process. An example of an exothermic reaction is the combustion of methane, the main component of natural gas, to carbon dioxide and liquid water. This process has a ΔH° value of -213 kcal mole^{-1}.

$$CH_4 + 2\ O_2 \longrightarrow CO_2 + 2\ H_2O_{liq} \qquad \Delta H^\circ = -213 \text{ kcal mole}^{-1}$$

The exothermic nature of this reaction is due to the very strong bonds formed in the products. Many hydrocarbons release a lot of energy on combustion and are therefore valuable fuels.

If the enthalpy of a reaction strongly depends on changes in bond strength, what is the significance of ΔS°? The entropy change, ΔS°, is a measure of the changes in the order of a system. The value of ΔS° increases with increasing disorder. Because the value of ΔS° in the expression for ΔG° is negative, a positive value for ΔS° makes a negative contribution to the free energy of the system. In other words, going from order to disorder is energetically favorable.

What is meant by order and disorder in a chemical reaction? Frequently, these terms describe the degree of free motion available to a molecule. Such motion may be translational—that is, related to how free a molecule is to move in space. For example, a reaction in which a compound in solution becomes immobilized as a solid has a relatively large negative ΔS°. On the other hand, a reaction in which more molecules are made from fewer, such as the cleavage of I_2 to atomic iodine, has a relatively large positive ΔS°. Molecular motion also includes vibrational and rotational freedom, terms that describe the ability of individual atoms and groups of atoms in a molecule to move relative to each other. For example, cyclohexane, because of its relatively increased rigidity, has fewer degrees of vibrational and rotational freedom than does straight-chain hexane.

When ΔH° and $-T\Delta S^\circ$ have opposing signs, whether ΔG° is positive or negative depends on the relative magnitude of each term. The driving force and hence the position of the equilibrium of many simple organic reactions are controlled primarily by ΔH° and to a lesser extent by ΔS°. Only if ΔH° is relatively small or if the temperature is high does the entropy of a reaction have much influence. For example, the chlorination of ethane with chlorine gas (see Section 3-5) to give chloroethane and hydrogen chloride has a ΔH° of -28 kcal mole^{-1} and only a small ΔS° of $+0.5$ e.u. This means that at room temperature (298 K) the contribution of $-T\Delta S^\circ$ to ΔG° is only on the order of -0.15 kcal mole^{-1}, essentially negligible. The considerable driving force for the chlorination lies mainly in the large negative ΔH° term.

$$CH_3CH_3 + Cl_2 \longrightarrow \underset{\textbf{Chloroethane}}{CH_3CH_2Cl} + HCl \qquad \begin{aligned}\Delta H^\circ &= -28 \text{ kcal mole}^{-1}\\ \Delta S^\circ &= +0.5 \text{ e.u.}\end{aligned}$$

In this reaction, the two starting materials convert into two products. The entropy of a reaction is considerably larger if the number of reacting molecules differs from the number formed. For example, the reaction of ethene (ethylene) with hydrogen chloride to give chloroethane is exothermic by -15.5 kcal mole^{-1}, but the entropy makes an unfavorable contribution to the $\Delta G°$, $\Delta S° = -31.3$ e.u.

$$CH_2{=}CH_2 + HCl \longrightarrow CH_3CH_2Cl \qquad \Delta H° = -15.5 \text{ kcal mole}^{-1}, \quad \Delta S° = -31.3 \text{ e.u.}$$

EXERCISE 2-8

Calculate the $\Delta G°$ at 25°C for the preceding reaction. Give an explanation for the relatively large negative entropy of the reaction.

How Fast Is Equilibrium Established? The Rate of a Chemical Reaction Depends on the Activation Energy

Does an understanding of the basic thermodynamic features of chemical reactions tell us anything about their rates? Let us return to the conformational analysis of butane (Figure 2-12). We note that it is thermodynamically favorable for the molecule to adopt the *anti* conformation. However, the driving force for *gauche-anti* conversion is minimal. The energy change is small, partly because bonds are neither broken nor formed. Nonetheless, *anti-gauche* equilibrium is established exceedingly rapidly even at very low temperatures. This contrasts with the sluggishness of reactions that are considerably more exothermic, such as the combustion of methane mentioned earlier. We know that methane does not spontaneously ignite in air at room temperature even though this process releases 213 kcal mole^{-1}. The reason for this seeming discrepancy is that the rate of a chemical reaction is controlled by the amount of its activation energy, E_a. This activation energy reflects the energy of the transition state. A high-lying transition state (e.g., one of high energy), as in methane oxidation, is associated with a low rate; a low-lying (low energy) transition state, as in the *gauche-anti* rotation of butane, indicates a fast reaction (Figure 2-13).

FIGURE 2-13

Comparison of potential-energy diagrams for (A) *gauche-anti* conversion in butane and (B) the combustion of methane.

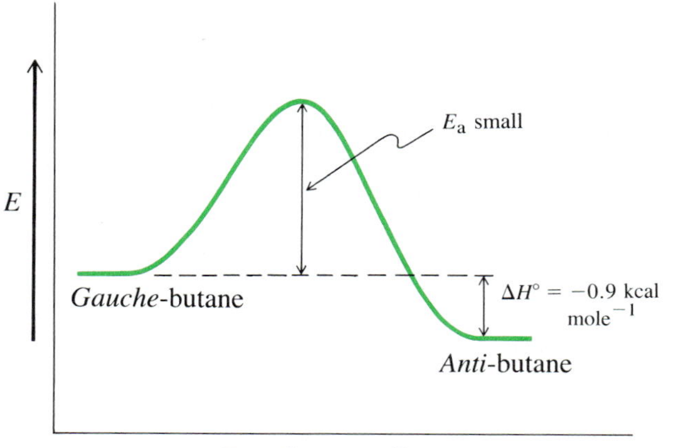

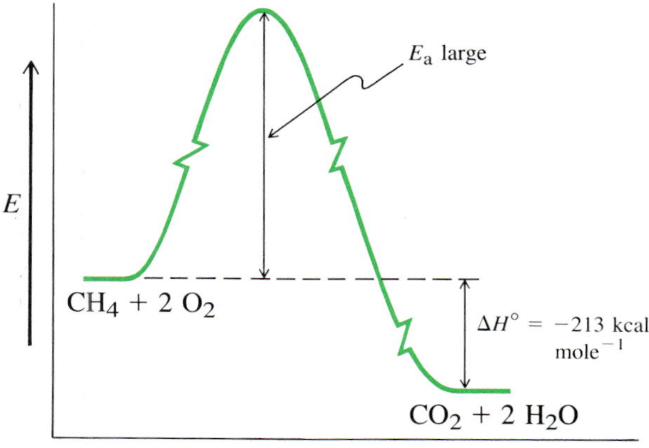

How can there be such high activation energies for exothermic reactions? A simple answer is that bond formation is usually preceded by bond breaking. Thus, before energy is gained through bonding, varying amounts of energy have to be expended to enable bond breaking. The point at which the initial energy loss is compensated by a corresponding gain is described by the transition state.

The Concentration of Reactants Can Affect Reaction Rates

The concentration of reactants can influence the rate of a reaction. Consider the addition of reagent A to reagent B to give C:

$$\mathrm{A + B \longrightarrow C}$$

In many transformations of this type, it is observed that increasing the concentration of either reactant increases the rate of the reaction. For some cases, the transition state is formed as the result of collision of molecules A and B. The rate can then be expressed by

$$\text{Rate} = k\,[\mathrm{A}][\mathrm{B}] \text{ in units of moles } \mathrm{l}^{-1}\ \mathrm{sec}^{-1}$$

in which the proportionality constant, k, is also called the **rate constant** of the reaction. The initial rate equals the rate constant when the two starting materials are at one molar concentration. A reaction for which the rate depends on the concentrations of two molecules in this way is said to be of **second order.**

There are processes whose rate depends on the concentration of only one reactant, such as in the hypothetical reaction:

$$\mathrm{A \longrightarrow B}$$
$$\text{Rate} = k[\mathrm{A}] \text{ in units of moles } \mathrm{l}^{-1}\ \mathrm{sec}^{-1}$$

A reaction of this type is said to be of **first order.** Rotation around a carbon–carbon bond follows such a rate law.

EXERCISE 2-9

Calculate the changes in rate of a first-order and a second-order reaction in which all starting materials have 1 molar concentration, after 50% conversion.

Temperature and Rates: The Arrhenius Equation

Temperature also greatly affects reaction rates. An increase in temperature leads to faster reactions. The reason for this finding lies in the increasing kinetic energy of molecules when they are heated, which means that a larger fraction of molecules are capable of overcoming the activation barrier (Figure 2-9). A useful rule of thumb that applies to many reactions is that raising the reaction temperature by 10°C causes the rate to increase by a factor of two to three.

The Swedish chemist Arrhenius* noticed the dependence of reaction rate on

---

*Professor Svante Arrhenius, 1859–1927, Technical Institute of Stockholm, Sweden, Nobel Prize 1903, director of the Nobel Institute from 1905 until shortly before his death.

temperature. He found that his measured data conformed to the equation

$$k = Ae^{-E_a/RT} \qquad \text{(Arrhenius equation)}$$

in which A is a constant characteristic of the reaction.

The Arrhenius equation also indicates in a quantitative manner how strongly reaction rates may be affected by activation energies. Thus, two reactions occurring at approximately 500 K (such that $RT \sim 1$ kcal mole^{-1}) with a difference in their activation energies of 10 kcal mole^{-1} and having the same A value will differ in rate by a factor of more than 20,000.

EXERCISE 2-10

Chloroethane, CH_3CH_2Cl, decomposes into ethene, $CH_2{=}CH_2$, and HCl at 500°C with an A value of 10^{14} and an activation energy of 58.4 kcal mole^{-1}. Estimate the rate of this transformation and its $\Delta G°$, using the $\Delta H°$ and $\Delta S°$ values given earlier for the reverse reaction.

How Do We Measure Activation Energies?

The Arrhenius equation can be reformulated by taking the natural logarithm of both sides and converting it into the logarithm of base 10:

$$\ln k = \ln (Ae^{-E_a/RT})$$

$$\ln k = \ln A - E_a/RT$$

$$\log k = \log A - \frac{E_a}{2.3RT}$$

This form shows how measuring the temperature dependence of the rate constant of a reaction allows us to derive its activation energy. If we plot log k against $1/T$, the slope of the ensuing straight line is equal to $-E/2.3R$, and the intercept at $1/T = 0$ is equal to log A.

This completes our brief review of the simple thermodynamic and kinetic relations governing many organic transformations. These concepts will be added to and elaborated on as necessary in subsequent chapters. Remember that all reactions are described by equilibrating starting materials and products. On which side the equilibrium lies depends on the size of the equilibrium constant, in turn related to the Gibbs free energy changes, $\Delta G°$. An increase in the equilibrium constant by a factor of 10 is associated with a change in $\Delta G°$ of about -1.36 kcal mole^{-1} at room temperature. The free energy change of a reaction is composed of changes in enthalpy, $\Delta H°$, and entropy, $\Delta S°$. Contributions to the former stem mainly from variations in bond strengths, to the latter from the relative degree of freedom in starting materials and products. Whereas these terms define the position of an equilibrium, the rate at which it is established depends on the concentration of starting materials, the activation barrier separating reactants and products, and the temperature. The relation between rate, E_a, and T is expressed by the Arrhenius equation.

Summary of Important Concepts

1 Organic molecules may be viewed as being composed of a carbon skeleton with attached functional groups.

2 Hydrocarbons are made up of carbon and hydrogen only. As a class of compounds, they are divided into saturated and unsaturated hydrocarbons. Saturated hydrocarbons are also called alkanes. They do not contain functional groups. They may exist as a single continuous chain or they may be branched or cyclic. The empirical formula for the straight-chain and branched alkanes is C_nH_{2n+2}.

3 Molecules that differ only by the number of methylene groups, CH_2, in the chain are called homologs and are said to belong to a homologous series.

4 A primary carbon is attached to only one other carbon. A secondary carbon is attached to two, a tertiary to three other carbon atoms. The hydrogen atoms bound to such carbon atoms are likewise designated primary, secondary, or tertiary.

5 The IUPAC rules for naming saturated hydrocarbons are:

(a) Find the longest chain in the molecule and name it.

(b) Name all groups attached to the longest chain as alkyl substituents.

(c) Number the carbon atoms of the longest chain.

(d) Write the name of the alkane, citing all substituents as prefixes arranged in alphabetical order and preceded by numbers designating their positions.

6 Alkanes attract each other through London forces, polar molecules through dipole–dipole interactions, and salts mainly through ionic interactions.

7 Rotation around carbon–carbon single bonds gives rise to conformational isomers (conformers, rotamers). Substituents on the carbon atoms may be staggered or eclipsed. The eclipsed conformation is a transition state between staggered conformers. The energy required to reach the eclipsed state is called the activation energy for rotation. When both carbons bear alkyl or other groups, there may be additional conformers: those in which the groups are in close proximity (60°) are *gauche;* those in which the groups are directly opposite (180°) one another are *anti*. Molecules tend to adopt conformations in which steric hindrance, as in *gauche* conformations, is minimized.

8 Chemical reactions can be described as equilibria controlled by thermodynamic and kinetic parameters. The change in the Gibbs free energy, ΔG°, is related to the equilibrium constant by: $\Delta G^\circ = -RT \ln K = -1.36 \log K$ (at 25°C). The free energy has contributions from changes in enthalpy, ΔH°, and entropy, ΔS°: $\Delta G^\circ = \Delta H^\circ - T\Delta S^\circ$. Changes in enthalpy are mainly due to differences between the strengths of the bonds made compared with those broken. A reaction is exothermic when the former is larger than the latter. It is endothermic when there is a net loss in combined bond strengths. Changes in entropy are controlled by the relative degree of order in starting materials compared with that in products. The greater the disorder, the larger a positive ΔS°. Order and disorder in organic molecules are mainly due to the relative degree of translational, vibrational, and rotational freedom.

9 The rate of a chemical reaction depends mainly on the concentrations of starting material(s), the activation energy, and temperature. These correlations are expressed in the Arrhenius equation: rate constant $k = Ae^{-E_a/RT}$.

10 If the rate depends on the concentration of only one starting material, the reaction is said to be of first order. If it depends on the concentrations of two reagents it is of second order.

Problems

1 For each example in Table 2-1, identify all polarized covalent bonds and label the appropriate atoms with partial positive or negative charges. (Do not consider carbon–hydrogen bonds.)

2 On the basis of electrostatics (Coulomb attraction), predict which atom in each of the following organic molecules is likely to react with the indicated reagent. Write "no reaction" if none seems likely. (See Table 2-1 for the structures of the organic molecules.)

- **(a)** Iodoethane, with the oxygen of HO^-.
- **(b)** Propanal, with the nitrogen of NH_3.
- **(c)** Methoxyethane, with H^+.
- **(d)** 3-Pentanone, with the carbon of CH_3^-.
- **(e)** Ethanenitrile, with the carbon of CH_3^+.
- **(f)** Butane, with HO^-.

3 Name the following molecules according to the IUPAC system of nomenclature.

(a) 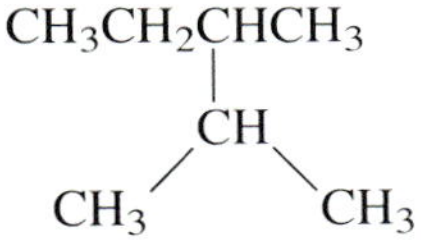
```
CH3CH2CHCH3
       |
       CH
      /  \
   CH3    CH3
```

(b) 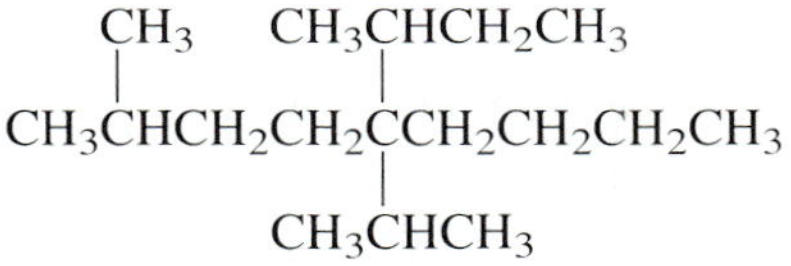
```
   CH3    CH3CHCH2CH3
   |         |
CH3CHCH2CH2CCH2CH2CH2CH3
             |
          CH3CHCH3
```

(c) 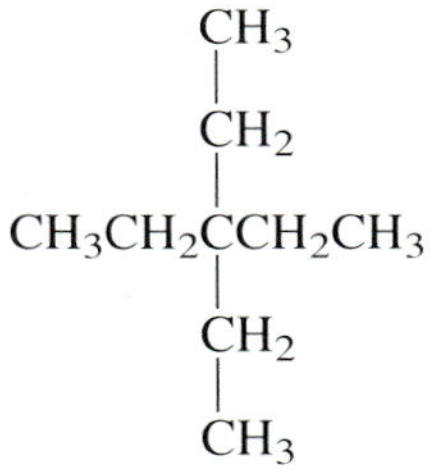
```
       CH3
       |
       CH2
       |
CH3CH2CCH2CH3
       |
       CH2
       |
       CH3
```

(d) $CH_3CH(CH_3)CH(CH_3)CH(CH_3)CH(CH_3)_2$

(e) 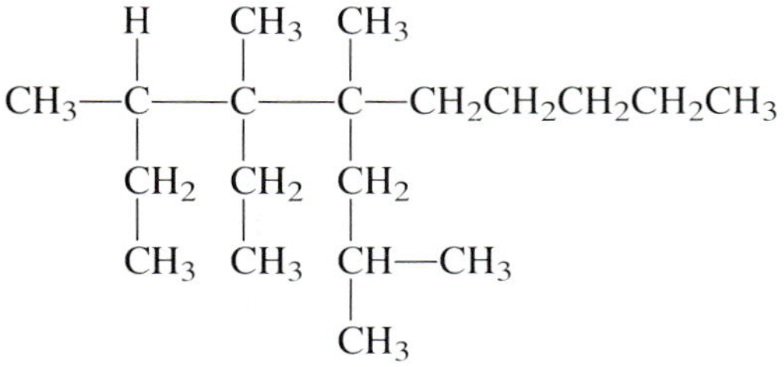
```
       H    CH3  CH3
       |    |    |
CH3—C——C——C—CH2CH2CH2CH2CH3
       |    |    |
       CH2  CH2  CH2
       |    |    |
       CH3  CH3  CH—CH3
                 |
                 CH3
```

(f) 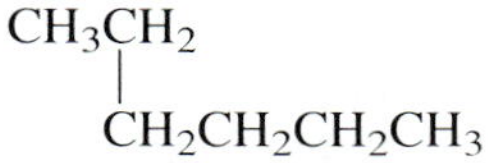
```
CH3CH2
   |
   CH2CH2CH2CH3
```

(g)  **(h)** 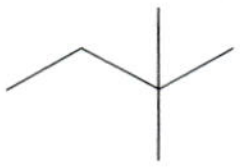 **(i)** 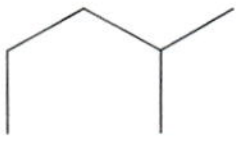 **(j)**

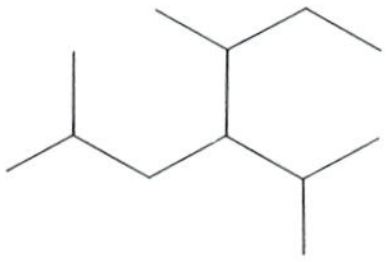

4 Convert the following names into the corresponding molecular formulas. After doing so, check to see if the name of each molecule as given here is in accord with the IUPAC system of nomenclature. If not, name the molecule correctly.

- **(a)** 2-Methyl-3-propylpentane.
- **(b)** 5-(1,1-Dimethylpropyl)nonane.

(c) 2,3,4-Trimethyl-4-butylheptane.
(d) 4-*tert*-Butyl-5-isopropylhexane.
(e) 4-(2-Ethylbutyl)decane.
(f) 2,4,4-Trimethylpentane.
(g) 4-*sec*-Butylheptane.

5 Draw and name all possible isomers of the formula C_7H_{16} (isomeric heptanes).

6 Identify the primary, secondary, and tertiary carbon atoms and hydrogen atoms in each of the following molecules.
(a) Ethane.
(b) Pentane.
(c) 2-Methylbutane.
(d) 3-Ethyl-2,2,3,4-tetramethylpentane.

7 Identify each of the following alkyl groups as being primary, secondary, or tertiary, and give it a systematic IUPAC name.

(a)
$$\begin{array}{ccccccc} & & CH_3 & & & & \\ & & | & & & & \\ -CH_2 & - & CH & - & CH_2 & - & CH_3 \end{array}$$

(b)
$$\begin{array}{cccccccc} & & CH_3 & & & & & \\ & & | & & & & & \\ CH_3 & - & CH & - & CH_2 & - & CH_2 & - \end{array}$$

(c)
$$\begin{array}{cccccc} & & CH_3 & & CH_3 & \\ & & | & & | & \\ CH_3 & - & CH & - & CH & - \end{array}$$

(d)
$$\begin{array}{cccccccc} & & CH_3-CH_2 & & & & & \\ & & | & & & & & \\ CH_3-CH_2 & - & CH & - & CH_2 & - & & \end{array}$$

(e)
$$\begin{array}{ccccc} & & CH_3-CH- & & \\ & & | & & \\ CH_3-CH_2 & - & CH & - & CH_3 \end{array}$$

(f)
$$\begin{array}{ccccc} & & CH_3-CH_2 & & \\ & & | & & \\ CH_3-CH_2 & - & C & - & CH_3 \\ & & | & & \end{array}$$

8 Rank the following molecules in order of increasing boiling point (*without* looking up the real values).
(a) 2-Methylhexane.
(b) Heptane.
(c) 2,2-Dimethylpentane.
(d) 2,2,3-Trimethylbutane.

9 Draw dashed-wedged line structures for the following molecules in the conformations indicated.
(a) Staggered propane.
(b) Eclipsed propane.
(c) *Anti*-butane.
(d) *Gauche*-butane.

10 At room temperature, 2-methylbutane exists primarily as two alternating conformations of rotation around the C-2—C-3 bond. About 90% of the molecules exist in the more-favorable conformation and 10% in the less-favorable one.

(a) Calculate the free energy change (ΔG°) between these two conformations.

(b) Draw a potential-energy diagram for rotation around the C-2—C-3 bond in 2-methylbutane. To the best of your ability, assign relative energy values to all the conformations on your diagram.

(c) Draw Newman projections for all staggered and eclipsed rotamers in part **b** and indicate the two most-favorable ones.

11 The equation relating ΔG° with K contains a temperature term. Refer to your answer to Problem 10, part **a,** to calculate the answers to the questions that follow. You will need to know that ΔS° for the more-stable conformer of 2-methylbutane is +1.4 cal deg^{-1} $mole^{-1}$, relative to the next most-stable conformer.

(a) Calculate the enthalpy difference (ΔH°) between these two conformers from the equation $\Delta G^\circ = \Delta H^\circ - T\Delta S^\circ$. How well does this agree with the ΔH° calculated on the basis of the number of *gauche* interactions in each conformer?

(b) Assuming that ΔH° and ΔS° do not change as a function of temperature, calculate ΔG° between these two conformations at the following three temperatures: (i) −250°C, (ii) −100°C, and (iii) +500°C.

(c) Calculate K for these two conformations at the same three temperatures.

12 Explain the following experimental results in terms of thermodynamics and kinetics.

A chemical reaction forms two products, A and B, and the ratio of these products depends on the conditions under which the reaction takes place. At −60°C, the yield is 80% A and 20% B. However, at or above room temperature, 25% of A and 75% of B form. Finally, if the low-temperature product mixture (80% A and 20% B) is allowed to warm up to 25°C, the product ratio changes to 25% A and 75% B.

13 The hydrocarbon propene ($CH_3—CH{=}CH_2$) can react in two different ways with bromine (Chapters 12 and 14):

(i) $CH_3—CH{=}CH_2 + Br_2 \longrightarrow CH_3—CH(Br)—CH_2(Br)$

(ii) $CH_3—CH{=}CH_2 + Br_2 \longrightarrow CH_2(Br)—CH{=}CH_2 + HBr$

Bond	Average strength
C—C	83
C═C	146
C—H	99
Br—Br	46
H—Br	87
C—Br	68

(a) Using the bond strengths (kcal $mole^{-1}$) given in the margin, calculate ΔH° for each of these reactions.

(b) $\Delta S^\circ \sim -35$ cal deg^{-1} $mole^{-1}$ for reaction (i); $\Delta S^\circ \sim 0$ for reaction (ii). Briefly explain the reason for this difference. Calculate ΔG° for each of these reactions at room temperature (25°C) and at 600°C.

(c) Both reactions have E_a values low enough to allow them to proceed at reasonable rates under ordinary conditions; however, E_a for reaction (i) is lower than E_a for reaction (ii). Using all the data given and the values that you have worked out for parts **a** and **b** of this problem, predict the product(s) of the reaction of propene with bromine at 25°C and at 600°C.

14 Using the Arrhenius equation, calculate the effect on k of 10°, 30°, and 50°C increases in temperature for the following activation energies. Use 300 K (approximately room temperature) as your initial T value.

(a) E_a = 15 kcal $mole^{-1}$.
(b) E_a = 30 kcal $mole^{-1}$.
(c) E_a = 45 kcal $mole^{-1}$.

15 For each of the following naturally occurring compounds, identify the compound class(es) to which it belongs, and circle all functional groups.

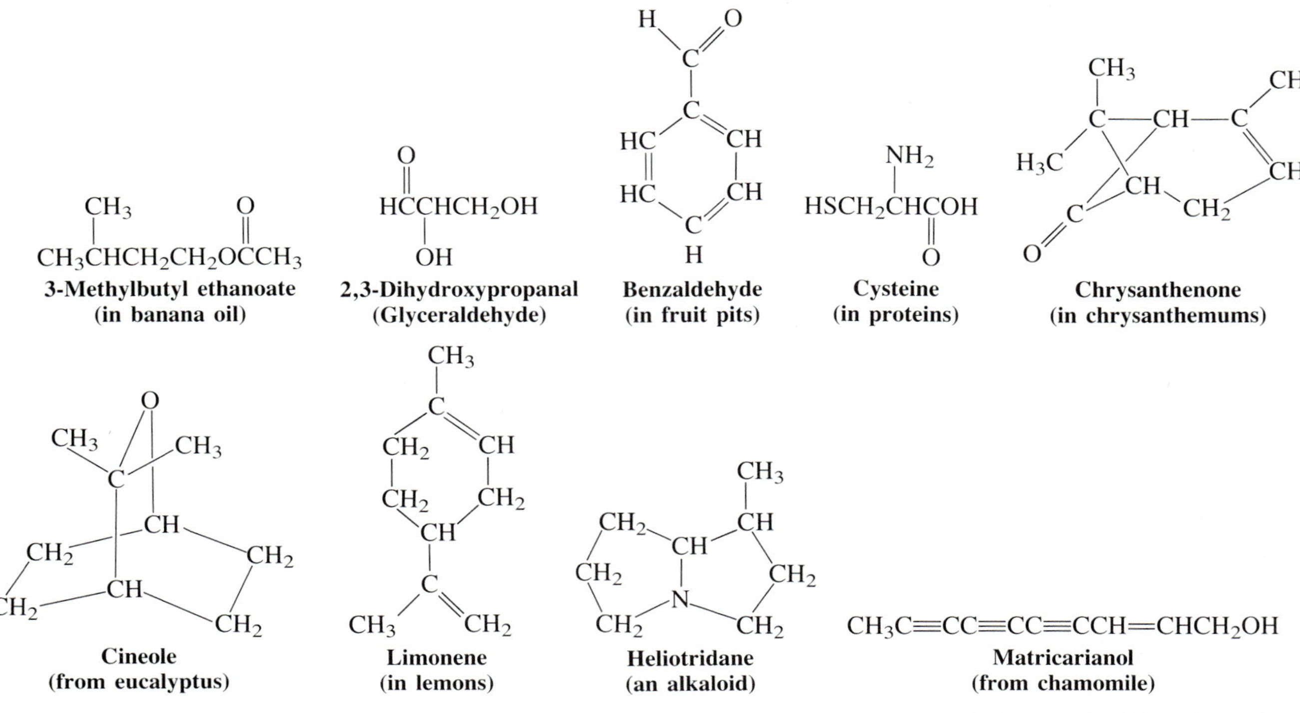

3-Methylbutyl ethanoate (in banana oil) · **2,3-Dihydroxypropanal (Glyceraldehyde)** · **Benzaldehyde (in fruit pits)** · **Cysteine (in proteins)** · **Chrysanthenone (in chrysanthemums)** · **Cineole (from eucalyptus)** · **Limonene (in lemons)** · **Heliotridane (an alkaloid)** · **Matricarianol (from chamomile)**

16 Give IUPAC names for all alkyl groups marked by dashed lines in each of the following biologically important compounds. Identify each group as a primary, secondary, or tertiary alkyl substituent.

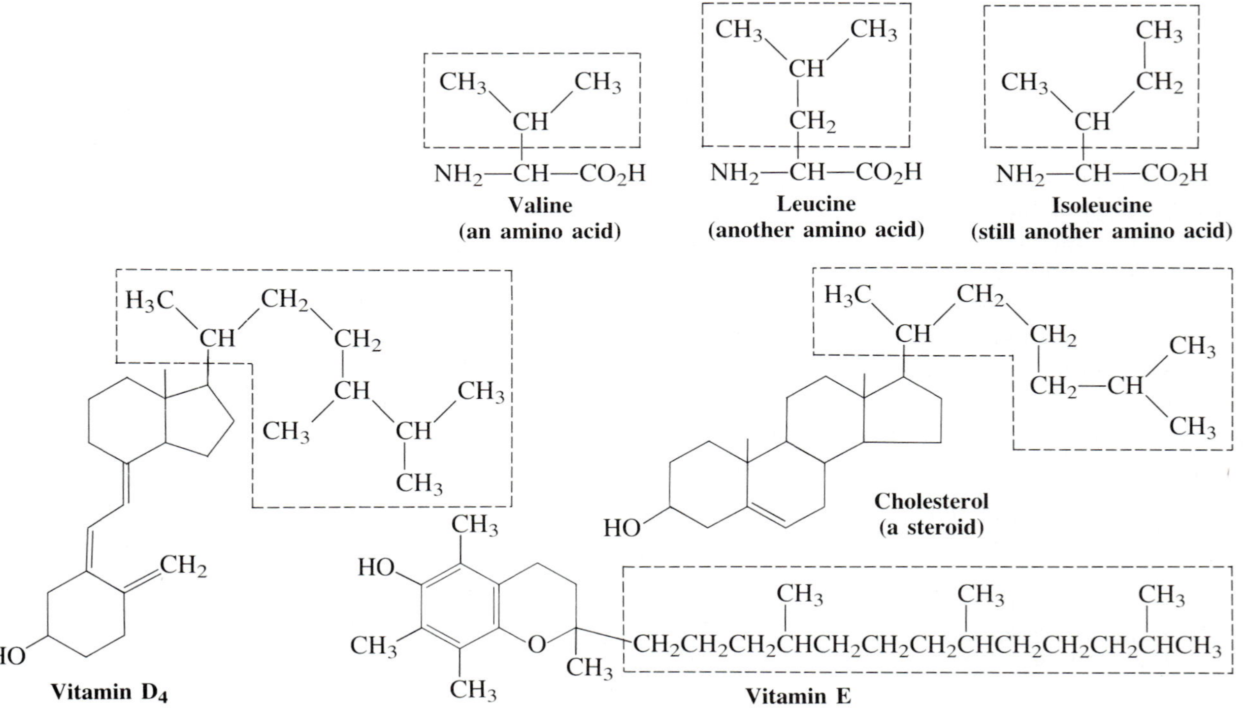

Valine (an amino acid) · **Leucine (another amino acid)** · **Isoleucine (still another amino acid)** · **Cholesterol (a steroid)** · **Vitamin D_4** · **Vitamin E**

CHAPTER 3

The Reactions of Alkanes

PYROLYSIS AND DISSOCIATION ENERGIES, COMBUSTION AND HEAT CONTENT, FREE-RADICAL HALOGENATION, AND RELATIVE REACTIVITY

As stated in Chapter 2, alkanes are organic chemicals that lack functional groups. To turn alkanes into synthetically useful compounds, it is necessary to introduce such groups, a process called **functionalization.** Many liquid and solid alkanes are obtained cheaply from petroleum by distillation and cracking (Section 3-3). Thus, nature has given us large quantities of hydrocarbons that can be used as chemical feedstocks, or starting materials, for the synthesis of other organic molecules. Natural alkanes are produced by the slow decomposition of animal and vegetable matter in the presence of water but in the absence of oxygen, a process lasting millions of years. The smaller alkanes—methane, ethane, propane, and butane—are gases and are present in natural gas, methane being by far its major component. In the United States, natural gas is a major source of energy, with annual production in hundreds of millions of tons.

This chapter begins by explaining what happens when alkanes are heated to high temperatures: carbon–carbon and carbon–hydrogen bonds are broken. This process is called **bond dissociation** and requires energy: bond strength or bond-dissociation energy, DH°. A discussion of hydrocarbon combustion leads to a description of the methods used to establish the heat content of molecules, the heats of formation, ΔH_f°. Finally, an important functionalization reaction of the alkanes is introduced: halogenation.

3-1
The Strength of Alkane Bonds: Pyrolysis

Chapter 1 explained why and how bonds are formed and that energy is released on bond formation. For example, bringing two hydrogen atoms into bonding distance produces 104 kcal mole^{-1} of heat (refer to Figures 1-1 and 1-12).

$$\text{H}^{\cdot} + \text{H}^{\cdot} \longrightarrow \text{H—H} \qquad \Delta H^{\circ} = -104 \text{ kcal mole}^{-1}$$

Consequently, breaking such a bond *requires* heat, in fact the same amount of heat that was released when the bond was made. This energy is called **bond-dissociation energy,** DH°.

$$\text{H—H} \longrightarrow \text{H}^{\cdot} + \text{H}^{\cdot} \qquad \Delta H^{\circ} = DH^{\circ} = 104 \text{ kcal mole}^{-1}$$

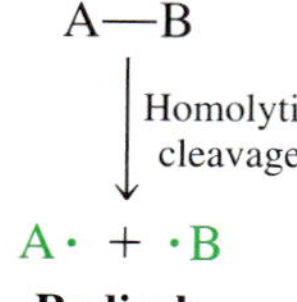

Radicals

$:\ddot{\underset{\cdot\cdot}{\text{Cl}}}\cdot$

Chlorine atom

$H_3C\cdot$ (H—C—H with H above, unpaired electron on C)

Methyl radical

$H_3C\text{—}\dot{C}H_2$

Ethyl radical

The breaking of a bond in such a way that the two bonding electrons divide equally between the two participating atoms (or fragments) is called **homolytic cleavage.** The fragments that form have unpaired electrons, such as $H^{\cdot}$, $Cl^{\cdot}$, $CH_3{}^{\cdot}$, and $CH_3CH_2{}^{\cdot}$. These species (unless they are atoms) are **radicals,** a term used in botany to refer to a branch of a plant attached to its stem, much like a substituent attached to an alkane chain. Because of the unpaired electron, a radical is very reactive and usually cannot be isolated. However, radicals can be present in low concentration as intermediates in reactions.

In an alternative way of breaking a bond, the entire bonding electron pair is donated to one of the atoms. This process is **heterolytic cleavage** and results in **ions.**

$$\text{A—B} \xrightarrow{\text{Heterolytic cleavage}} \text{A}^{+} + :\text{B}^{-}$$

Ions

Dissociation energies, DH°, *refer only to homolytic cleavages.* They have characteristic values for the various bonds that can be formed between the elements. Table 3-1 lists dissociation energies of some common bonds. Note the relatively strong bonds to hydrogen, as in H–F and H–OH. Even though these bonds have high DH° values, they readily undergo *heterolytic* cleavage in water to H^{+} and F^{-} or HO^{-}; do not mix up homolytic and heterolytic processes.

Bonds made by overlapping orbitals that are closely matched in energy and size are stronger than those not meeting these criteria. For example, the strength of the bonds between hydrogen and the halogens decreases in the order F > Cl > Br > I, because the *p* orbital of the halogen contributing to the bonding becomes larger and more diffuse with each element. Consequently, the degree to which overlap can be achieved with the relatively small 1*s* orbital on hydrogen diminishes along the series. A similar trend holds for bonding between the halogens and carbon.

EXERCISE 3-1

Compare the bond-dissociation energies of CH_3–F, CH_3–OH, and CH_3–NH_2. Why do the bonds get weaker along this series even though the orbitals participating in overlap become better matched in size and energy? (Hint: Consider Figure 1-2 and Table 1-4 for a simple explanation.)

TABLE 3-1

Bond-dissociation energies of some A–B bonds (DH° in kcal $mole^{-1}$)

	B						
A	**H**	**F**	**Cl**	**Br**	**I**	**OH**	**NH_2**
H	104	135	103	87	71	119	107
CH_3	105	110	85	71	57	93	80
CH_3CH_2	98	107	80	68	53	92	77
$CH_3CH_2CH_2$	98	107	81	68	53	91	78
$(CH_3)_2CH$	94.5	106	81	68	53	92	93
$(CH_3)_3C$	93	110	81	67	52	93	93

Note: These numbers are being revised continually because of improved methods for their measurement. Some of the values given here may be in (small) error.

The C–H Bond Strengths in Alkanes Vary with Structure

A C–H bond in an alkane has a strength of about 98 kcal $mole^{-1}$. The C—C bond strength is approximately 86 kcal $mole^{-1}$. The bond-dissociation energies of various alkane bonds are given in Table 3-2. Note that bond energies generally decrease with the progression from methane to primary, secondary, and tertiary carbon. For example, the C–H bond in methane has a high DH° value of 105 kcal $mole^{-1}$. In ethane, this bond energy is 7 kcal $mole^{-1}$ less: DH° = 98 kcal $mole^{-1}$. This number is typical for primary C–H bonds, as can be seen for those bonds in propane and 2-methylpropane. The secondary C–H bond is even weaker, with a DH° of 94.5 kcal $mole^{-1}$, and a tertiary carbon atom binds to hydrogen with only 93 kcal $mole^{-1}$.

$$CH_4 \longrightarrow CH_3\cdot + H\cdot \qquad DH^\circ = 105 \text{ kcal mole}^{-1}$$

$$R—H \longrightarrow R\cdot + H\cdot$$

$$\begin{array}{ll} \text{R primary} & DH^\circ = 98 \text{ kcal mole}^{-1} \\ \quad\text{secondary} & DH^\circ = 94.5 \text{ kcal mole}^{-1} \\ \quad\text{tertiary} & DH^\circ = 93 \text{ kcal mole}^{-1} \end{array}$$

TABLE 3-2

Bond-dissociation energies for some alkanes

Compound	**DH° (kcal $mole^{-1}$)**	**Compound**	**DH° (kcal $mole^{-1}$)**
CH_3–H	105	CH_3–CH_3	90
C_2H_5–H	98	C_2H_5–CH_3	86
C_3H_7–H	98	C_3H_7–CH_3	87
$(CH_3)_2CHCH_2$–H	98	C_2H_5–C_2H_5	82
$(CH_3)_2CH$–H	94.5	$(CH_3)_2CH$–CH_3	86
$(CH_3)_3C$–H	93	$(CH_3)_3C$–CH_3	84
		$(CH_3)_3C$–$C(CH_3)_3$	72

Note: See footnote for Table 3-1.

A slightly less pronounced but similar trend is seen for the relatively weaker C–C bonds, the extremes being the central bond in ethane (DH° = 90 kcal mole^{-1}) and 2,2,3,3-tetramethylbutane (DH° = 72 kcal mole^{-1}).

Why do all of these dissociations exhibit different DH° values? An explanation for this finding is *that the radicals formed have differing energies*. Their stability increases along the series from primary to secondary to tertiary; consequently, the energy required to create them decreases. From this observation, we may conclude that the primary radical has the highest energy, followed by the secondary, and then the tertiary radical (Figure 3-1).

Stability of Alkyl Radicals

$CH_3\cdot$ < primary < secondary < tertiary

What is the reason for this ordering? To answer this question, we need to inspect the structure of alkyl radicals more closely.

FIGURE 3-1
The different energy contents of primary, secondary, and tertiary radicals in an alkane: $CH_3CH_2CHR_2$.

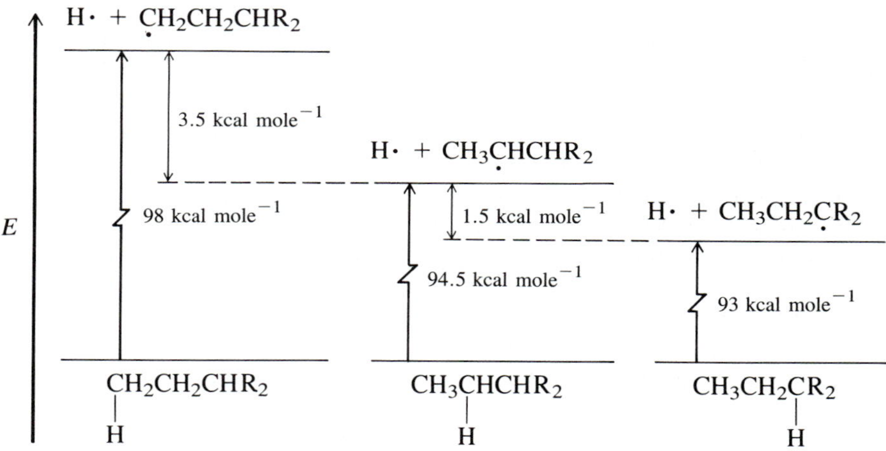

EXERCISE 3-2
In which order would you expect the C–C bonds to break in ethane and 2,2-dimethylpropane?

In summary, bond homolysis in alkanes yields radicals. The heat required to do so is called the bond-dissociation energy, DH°. Its value is characteristic for the bonds between the elements. Bond breaking that results in tertiary radicals necessitates relatively less energy than that furnishing secondary radicals; the latter are in turn formed more readily than primary radicals. The methyl radical is the most difficult to obtain in this way.

3-2 The Structure of Alkyl Radicals and Hyperconjugation

The answer to the question regarding the relative stability of radicals has as its

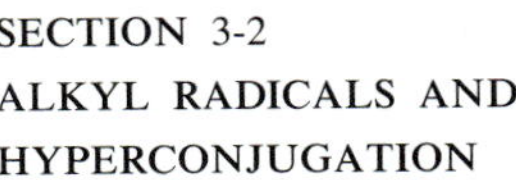

basis electron delocalization in which a neighboring C–H σ bond overlaps with the p orbital characteristic of a radical center.

Consider the structure of the methyl radical, formed by removal of a hydrogen atom from methane. Its bonding in principle could be described as involving an sp^3-hybridized carbon with three sp^3 C–H bonds, the odd electron occupying the fourth sp^3 molecular orbital. Spectral measurements, however, have shown that the methyl radical, and probably other alkyl radicals, rehybridize to adopt a nearly planar configuration, more accurately described by sp^2 hybridization (Figure 3-2). Here, the unpaired electron occupies the remaining p orbital perpendicular to the molecular plane.

FIGURE 3-2
Rehybridization of methane on methyl radical formation. The nearly planar arrangement is reminiscent of the hybridization in BH_3 (Figure 1-17).

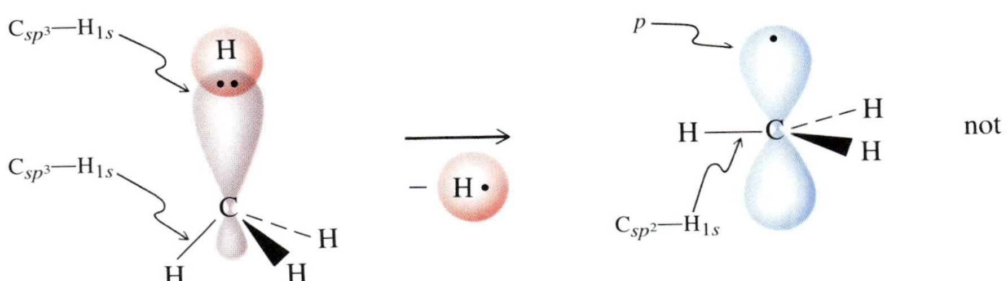

not

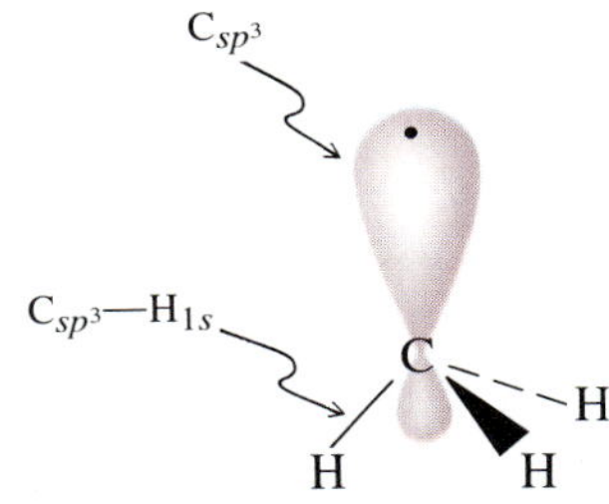

The planar structure of the alkyl radicals helps to explain their relative stabilities. The increasing stability is due to the increasing number of methyl groups replacing hydrogen atoms. For example, in the ethyl radical, a methyl group substitutes for one of the hydrogen atoms in the methyl radical. By making a model, you can see that there is a conformer in which a C–H bond in the methyl substituent eclipses one of the lobes of the singly filled p orbital (Figure 3-3). This configuration allows for a certain amount of delocalization of the bonding pair of electrons by overlap with the partly filled p lobe, a phenomenon called **hyperconjugation.** Recall (Exercise 1-11) that the interaction between a filled orbital and a singly occupied orbital has a net stabilizing effect (Figure 3-4).

FIGURE 3-3
Hyperconjugation (green dashed lines) between filled sp^3 hybrids and the partly filled p orbital in ethyl, 1-methylethyl, and 1,1-dimethylethyl radicals.

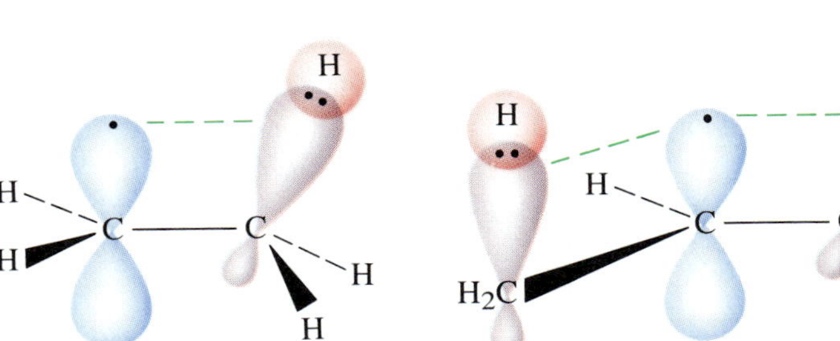

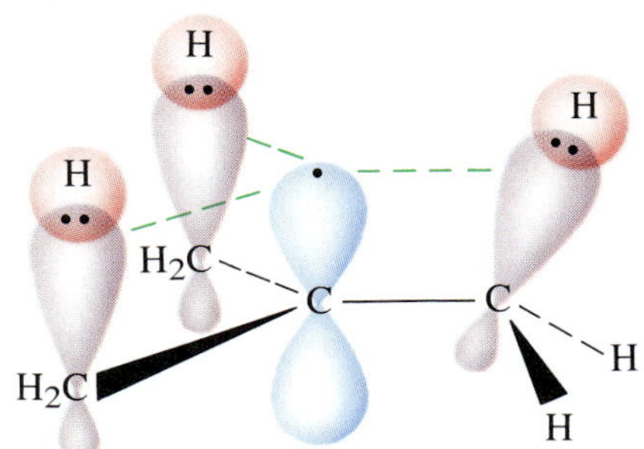

FIGURE 3-4

The interaction between a filled and a singly occupied orbital is stabilizing: the net energy change, to a first approximation, $\Delta E - 2\Delta E = -\Delta E$, is favorable. For the ethyl radical, the filled level represents the two electrons in the C–H bond, the singly occupied level a carbon $2p$ orbital. The filled orbital may represent any C–R bond.

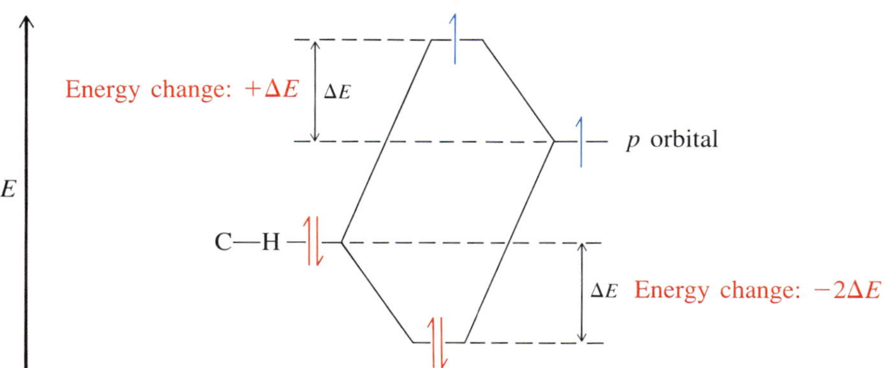

As further hydrogen atoms on the radical carbon are replaced successively by other alkyl substituents, the possibility for hyperconjugation doubles, as in the 1-methylethyl radical, and triples, as in the 1,1-dimethylethyl radical (Figure 3-3). In this way, the relative stability of radicals is readily understood to be a consequence of increasing hyperconjugative stabilization. An additional contribution to the stability of secondary and tertiary radicals is the decrease in steric interactions between the substituent groups as the system becomes planar.

3-3 Petroleum and Hydrocarbon Cracking: An Example of Pyrolysis

A knowledge of bond-dissociation energies and the stability of radicals is important for rationalizing the thermal chemistry of hydrocarbons. We will now see how this knowledge can be applied to understanding the conversion of crude petroleum oil into more volatile hydrocarbons.

When alkanes are heated to a high temperature, both C–H bonds and C–C bonds rupture, a process called **pyrolysis.** In the absence of oxygen, the resulting radicals can combine to form new higher or lower alkanes. They can also abstract hydrogen atoms from the carbon atom adjacent to another radical center to give alkenes. Clearly, very complicated mixtures of alkanes and alkenes form in pyrolyses.

Under certain conditions, the product distribution in these transformations can be controlled to obtain a large proportion of hydrocarbons of a defined chain

The Pyrolysis of Hexane

Cleavage to radicals:

$$\underset{\textbf{Hexane}}{\overset{1\ \ 2\ \ 3\ \ 4}{CH_3CH_2CH_2CH_2CH_2CH_3}} \xrightarrow{\text{C-1, C-2 cleavage}} CH_3\cdot + \cdot CH_2CH_2CH_2CH_2CH_3$$

$$\xrightarrow{\text{C-2, C-3 cleavage}} CH_3CH_2\cdot + \cdot CH_2CH_2CH_2CH_3$$

$$\xrightarrow{\text{C-3, C-4 cleavage}} CH_3CH_2CH_2\cdot + \cdot CH_2CH_2CH_3$$

Combination of radicals:

$$CH_3\cdot + \cdot CH_2CH_3 \longrightarrow \underset{\textbf{Propane}}{CH_3CH_2CH_3}$$

$$CH_3CH_2CH_2CH_2CH_2\cdot + \cdot CH_2CH_2CH_3 \longrightarrow \underset{\textbf{Octane}}{CH_3CH_2CH_2CH_2CH_2CH_2CH_2CH_3}$$

Hydrogen abstraction:

$$CH_3CH_2\cdot + CH_3\overset{H}{\overset{|}{C}}H{-}CH_2\cdot \longrightarrow \underset{\textbf{Ethane}}{CH_3\overset{H}{\overset{|}{C}}H_2} + \underset{\textbf{Propene}}{CH_3CH{=}CH_2}$$

$$\overset{H}{\overset{|}{C}}H_2CH_2\cdot + CH_3CH_2CH_2\cdot \longrightarrow \underset{\textbf{Ethene}}{CH_2{=}CH_2} + \underset{\textbf{Propane}}{CH_3CH_2\overset{H}{\overset{|}{C}}H_2}$$

length. Such conditions frequently include the use of special catalysts, examples being crystalline sodium aluminosilicates, also called zeolites. These catalysts form a three-dimensional network of tetrahedral SiO_4 and AlO_4 units. The composition of "zeolite A" is $Na_{12}(AlO_2)_{12}(SiO_2)_{12}(H_2O)_{27}$. Pyrolysis of dodecane over a catalyst of this type gives mainly products containing from three to six carbon atoms.

$$\text{Dodecane} \xrightarrow[\text{2 min}]{\text{Zeolite, 482°C}} \underset{17\%}{C_3} + \underset{31\%}{C_4} + \underset{23\%}{C_5} + \underset{18\%}{C_6}$$

BOX 3-1

The Function of a Catalyst

What is the function of the zeolite catalyst? A catalyst is a substance that speeds up a reaction; that is, it increases the rate at which equilibrium is established. It does so by allowing reactants and products to be interconverted by a new pathway in which the rate-determining step has an activation energy lower than the reaction would have in the absence of the catalyst (Figure 3-5). Apart from zeolites and other mineral-derived surfaces, many metals act as catalysts. In nature, *enzymes* usually fulfill this function (Chapter 27). The presence of catalysts allows many transformations to take place at lower temperatures and under generally milder conditions.

FIGURE 3-5

The difference in activation energies between a catalyzed (E_{cat}) and uncatalyzed (E_a) reaction.

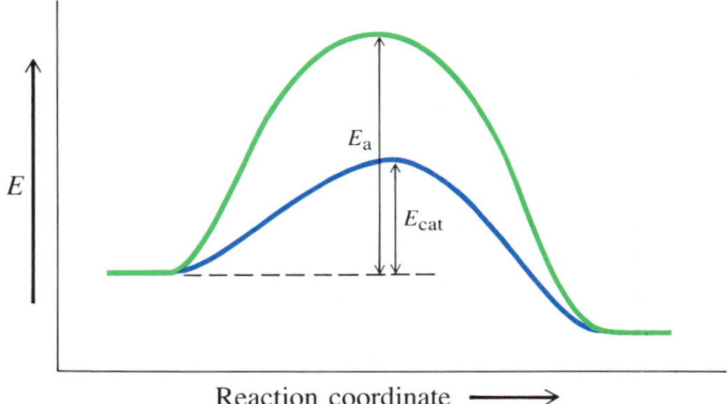

The Conversion of Petroleum

Breaking an alkane down into smaller fragments is also known as **cracking.** Such processes are important in the oil-refining industry for the production of gasoline and other liquid fuels from petroleum.

As mentioned in the introduction to this chapter, petroleum is believed to be the microbial-degradation product of living organic matter that existed several hundred million years ago. Crude oil, a dark viscous liquid, is a mixture of several hundred different hydrocarbons, particularly straight-chain alkanes, some branched alkanes, and varying quantities of aromatic hydrocarbons. In addition, there are oxygen-, sulfur-, and nitrogen-containing compounds. Distillation yields several fractions with a typical product distribution, as shown in Table 3-3. However, the composition of petroleum varies widely, depending on the origin of the oil. For example, Middle-Eastern oil tends to be rich in lower-boiling hydrocarbons. Mexican oil, on the other hand, is composed of higher-molecular-weight fractions.

TABLE 3-3

Product distribution in a typical distillation of crude petroleum

Amount (% of volume)	Boiling point (°C)	Carbon atoms	Products
1–2	<30	C_1–C_4	Natural gas, methane, propane, butane, liquefied petroleum gas (LPG)
15–30	30–200	C_4–C_{12}	Petroleum ether ($C_{5,6}$), ligroin (C_7), naphtha, straight-run gasoline[a]
5–20	200–300	C_{12}–C_{15}	Kerosene, heater oil
10–40	300–400	C_{15}–C_{25}	Gas oil, diesel fuel, lubricating oil, waxes, asphalt
8–69	>400 (Nonvolatiles)	>C_{25}	Residual oil, paraffin waxes, asphalt (tar)

[a]This refers to gasoline straight from petroleum, without having been treated in any way.

To increase the quantity of the much-needed gasoline fraction, the higher-boiling oils are cracked by pyrolysis. Originally (in the 1920s), this process required high temperatures (800°–1000°C). Modern cracking processes use catalysts, such as zeolites, at relatively low temperatures (500°C). Other catalysts employed are nickel or tungsten supported on a silica-alumina (SiO_2-Al_2O_3) surface. The reaction takes place in a hydrogen atmosphere (hydrocracking), which promotes the "capping off" of the radical centers by hydrogen to give lower-boiling products. It also serves to hydrogenate some alkenes to alkanes and to convert the polluting and catalyst-poisoning nitrogen- and sulfur-containing contaminants into readily removable ammonia, NH_3, and hydrogen sulfide,

H_2S. Cracking the residual oil from crude-petroleum distillation gives approximately 30% gas, 50% gasoline, and 20% higher-molecular-weight oils and a residue called coke.

Another process converts alkanes into aromatic hydrocarbons having approximately the same number of carbon atoms. The aromatics are highly efficient fuels and are used as feedstocks for the chemical industry. Because the process reforms a new hydrocarbon from an old one, it is referred to as **reforming,** and the product as **reformate.** Other terms refer to more-precise reaction conditions, such as platforming (platinum catalyst) or hydroforming (in the presence of hydrogen).

An example of reforming is the conversion of heptane into methylbenzene (toluene). Millions of barrels of reformate gasoline are produced in the United States alone.

$CH_3CH_2CH_2CH_2CH_2CH_2CH_3$
Heptane

$\downarrow$ $Pt\text{-}SiO_2\text{-}Al_2O_3$, 500°C, 20 atm H_2

CH_3-substituted benzene ring (H on the remaining five ring carbons) $+ 4\ H_2$

Methylbenzene (Toluene)

BOX 3-2

Petroleum and Gasoline Are Our Main Energy Sources

Oil and natural gas supply about 67% of the U.S. energy requirement. Other industrialized nations have a similar or even higher dependence on petroleum as a source of energy. U.S. yearly production of natural gas approximates 18 trillion cubic feet. In 1984, U.S. energy sources apart from oil (42%) and gas (25%) were coal (23%), hydroelectric power (5%) and nuclear power (5%). Domestic annual oil production peaked in 1971 at 4.2 billion barrels. In 1984, it was slightly more than 3 billion barrels, with almost 2 billion barrels imported.

The dependence of many countries on imported oil has had important economic and political consequences, as demonstrated by the severalfold increase in the price of crude oil in the seventies, the Arab oil embargo in 1973, and the recent plunge in oil prices. Strong efforts are being undertaken to decrease economic dependence on imported oil and to develop new energy sources to satisfy demand when oil and gas reserves are depleted. Potential contributors to the world's energy supply include controversial nuclear-breeder power and nuclear fusion, hydroelectric power, solar and geothermal energy, and coal. Other more-specialized and less-developed approaches rely on wind power, ocean-wave energy, and even the conversion of manure and plant wastes into fuels.

3-4 Combustion of Alkanes

How do we measure the energy contained in hydrocarbons (and therefore in some of the fuels that we use)? The answer is by burning, or **combustion.** Combustion is defined as *a chemical reaction with oxygen (usually at elevated temperatures) in which an alkane (or other reactant) is converted into carbon dioxide and water (or other oxidized products).* Both products in the combustion of alkanes have an exceedingly low energy content, and hence their formation is identified with a large negative ΔH° (ΔH°_{comb}), released as heat:

$$2\ C_nH_{2n+2} + (3n + 1)O_2 \longrightarrow 2nCO_2 + (2n + 2)H_2O + \text{heat}$$

The combustion of alkanes was considered earlier: in elemental analysis (Section 1-9) and in the discussion of ΔH° (Section 2-5). The heat released in the burning of an alkane is called its **heat of combustion.** The heats of combustion of many organic compounds have been measured with high accuracy. Table 3-4 lists them for several alkanes and other organic molecules.

TABLE 3-4

Heats of combustion (kcal mole^{-1} normalized to 25°C) of various organic compounds

Structure (state)	Name	ΔH°_{comb}
CH_4 (gas)	Methane	−212.8
C_2H_6 (gas)	Ethane	−372.8
$CH_3CH_2CH_3$ (gas)	Propane	−530.6
$CH_3(CH_2)_2CH_3$ (gas)	Butane	−687.4
$(CH_3)_3CH$ (gas)	2-Methylpropane	−685.4
$CH_3(CH_2)_3CH_3$ (gas)	Pentane	−845.2
$CH_3(CH_2)_3CH_3$ (liquid)	Pentane	−838.8
$CH_3(CH_2)_4CH_3$ (liquid)	Hexane	−995.0
(liquid)	Cyclohexane	−936.9
CH_3CH_2OH (gas)	Ethanol	−336.4
CH_3CH_2OH (liquid)	Ethanol	−326.7
$C_{12}H_{22}O_{11}$ (solid)	Cane sugar (sucrose)	−1348.2

Note: Combustion to CO_2 gas and liquid H_2O.

When giving the heat of combustion of a compound, it is customary to mention the physical state of the compound undergoing combustion (gas, liquid, solid). For example, the heat of combustion of liquid ethanol, $CH_3CH_2OH_{(l)}$, is −326.7 kcal mole^{-1}, of gaseous ethanol, $CH_3CH_2OH_{(g)}$, −336.4 kcal mole^{-1}. The difference is due to the heat of vaporization of ethanol, $\Delta H^\circ_{vap} = 9.7$ kcal mole^{-1}.

It is not surprising that heats of combustion increase as the number of carbon atoms increases, simply because there is more carbon and hydrogen to burn along the homologous series. What can be expected, however, for isomeric alkanes containing the same number of carbons and hydrogens?

The Relative Stability of Alkanes: Heats of Formation

A comparison of the heats of combustion of isomeric alkanes reveals that their values are usually *not* the same. Consider the two simple isomeric alkanes, butane and 2-methylpropane. The combustion of butane has a ΔH°_{comb} of −687.4 kcal mole^{-1}, whereas its isomer releases $\Delta H^\circ_{comb} = -685.4$ kcal mole^{-1}, 2 kcal mole^{-1} less (Table 3-4). This finding shows that 2-methylpropane has a *smaller* energy content than butane, because combustion yielding the identical kind and number of products (4 molecules of CO_2 and 5 molecules of H_2O) produces less energy (Figure 3-6). Butane is said to be *thermodynamically less stable* than its isomer.

FIGURE 3-6
Butane has a higher energy content than does 2-methylpropane, as measured by the release of energy on combustion.

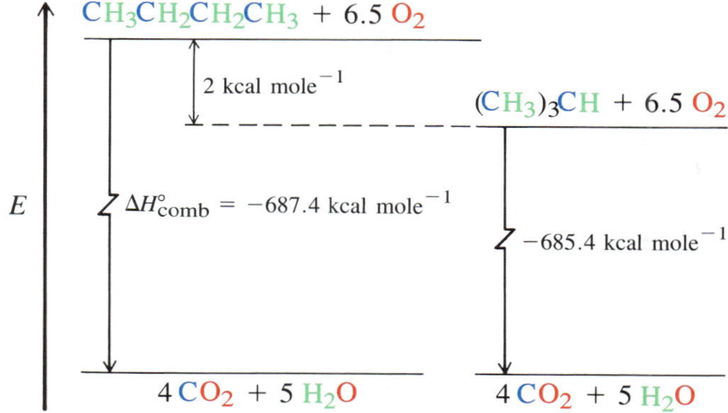

A useful way of tabulating the energy content of molecules is by their **heats of formation,** ΔH°_f. *The heat of formation of a molecule is the enthalpy for the production of one mole of a substance from its elements.* It relates the heat content of a molecule to a hypothetical reaction in which the molecule is formed from its component elements. This is in contrast with the heat of combustion, which relates the heat content of a molecule to its combustion products.

In an estimation of ΔH°_f, the elements are taken in their standard states. A standard state is defined as the stable form of the element at room temperature and normal pressure. The enthalpy of formation of an element in its standard state is defined as zero. Hydrogen, oxygen, and nitrogen have gaseous standard states, whereas the standard state of carbon is taken as graphite. The heats of formation of several organic molecules are given in Table 3-5.

TABLE 3-5
Heats of formation of selected elements and compounds (kcal mole^{-1} normalized to 25°C)

Structure (state)	ΔH°_f	Structure (state)	ΔH°_f
C (graphite)	0	$(CH_3)_3CH$ (gas)	−32.4
C (diamond)	0.45	$CH_3(CH_2)_3CH_3$ (gas)	−35.1
CO_2 (gas)	−94.1	$CH_3(CH_2)_3CH_3$ (liquid)	−41.4
H_2O (gas)	−57.8	H_2, O_2, N_2 (gases)	0
H_2O (liquid)	−68.3	H (atom)	52.1
CH_4 (gas)	−17.9	O (atom)	59.6
CH_3CH_3 (gas)	−20.2	C (atom)	171.3
$CH_3CH_2CH_3$ (gas)	−24.8	$CH_2{=}CH_2$ (gas)	12.5
$CH_3(CH_2)_2CH_3$ (gas)	−30.4	$HC{\equiv}CH$ (gas)	54.2

As you can see, the values for ΔH°_f can be positive or negative. For example, the diamond form of carbon has a ΔH°_f = 0.45 kcal mole^{-1}. Thus, the conversion of diamond into graphite is exothermic. (Fortunately, this process has a high activation energy, see Section 2-6). The heats of formation of atomic species as H and O are obtained by dividing the corresponding molecular bond strength in half (H—H: 104 kcal mole^{-1}, O—O: 119.2 kcal mole^{-1}). These values are positive because energy (DH°) must be expended to obtain such species from their standard molecular states.

Heats of Formation of H· and O·

$$\text{H—H} \longrightarrow \text{H}^\cdot + \text{H}^\cdot \qquad \Delta H^\circ = DH^\circ = +104 \text{ kcal mole}^{-1}$$
$$\Delta H_f^\circ(\text{H}^\cdot) = 52 \text{ kcal mole}^{-1}$$
$$\text{O—O} \longrightarrow \text{O}^\cdot + \text{O}^\cdot \qquad \Delta H^\circ = DH^\circ = +119.2 \text{ kcal mole}^{-1}$$
$$\Delta H_f^\circ(\text{O}^\cdot) = 59.6 \text{ kcal mole}^{-1}$$

Most organic molecules have negative heats of formation; that is, energy is released when they are formed from their component elements. For example, the reaction of graphite with two molecules of H_2 to give methane releases 17.9 kcal mole^{-1}.

$$C_{\text{graphite}} + 2\,H_2 \longrightarrow CH_4 \qquad \Delta H^\circ = -17.9 \text{ kcal mole}^{-1}$$

The different heat contents of butane and 2-methylpropane are again revealed by their differing heats of formation: -30.4 and -32.4 kcal mole^{-1}, respectively. Less heat is released in the formation of butane compared with 2-methylpropane even though the same number of graphite carbon atoms (4 C) and hydrogen molecules (5 H_2) are consumed in both reactions. Again, we can conclude that 2-methylpropane is slightly more stable than butane. The difference in the amount of energy released in both the formation and the combustion of butane and 2-methylpropane, then, is exactly the same: 2 kcal mole^{-1}. These relations are summarized in Figure 3-7.

You can see from Figure 3-7 that the heat of formation of a hydrocarbon may be calculated by measuring its heat of combustion and relating it to the heats of formation of CO_2 and H_2O, because the enthalpy changes for chemical reactions are additive (Hess's* Law). Let us attempt to estimate the heat of formation of methane in this way. Its heat of combustion (at 25°C) to give gaseous CO_2 and liquid H_2O (2 moles) is -212.8 kcal mole^{-1}.

$$\underset{\textbf{Gas}}{CH_4} + \underset{\textbf{Gas}}{2\,O_2} \longrightarrow \underset{\textbf{Gas}}{CO_2} + \underset{\textbf{Liquid}}{2\,H_2O} \qquad \Delta H^\circ = -212.8 \text{ kcal mole}^{-1}$$

Using Table 3-5, we can now estimate the combined heats of formation of the products formed on combustion:

$$\begin{array}{lr} 1 \times \Delta H_f^\circ (CO_2 \text{ gas}) & -94.1 \\ 2 \times \Delta H_f^\circ (H_2O \text{ liquid}) & \underline{-136.6} \\ & -230.7 \text{ kcal mole}^{-1} \end{array}$$

The observed heat of combustion of methane, however, amounts to only -212.8 kcal mole^{-1}. The difference between the total heat calculated and the heat measured is equal to the sum of the ΔH_f° for methane plus oxygen. Because

---

*Professor Germain Hess, 1802–1850, St. Petersburg, Russia.

FIGURE 3-7
Formation of butane and 2-methylpropane from the elements shows that the second is more stable than the first. Consequently, combustion of the latter releases less heat.

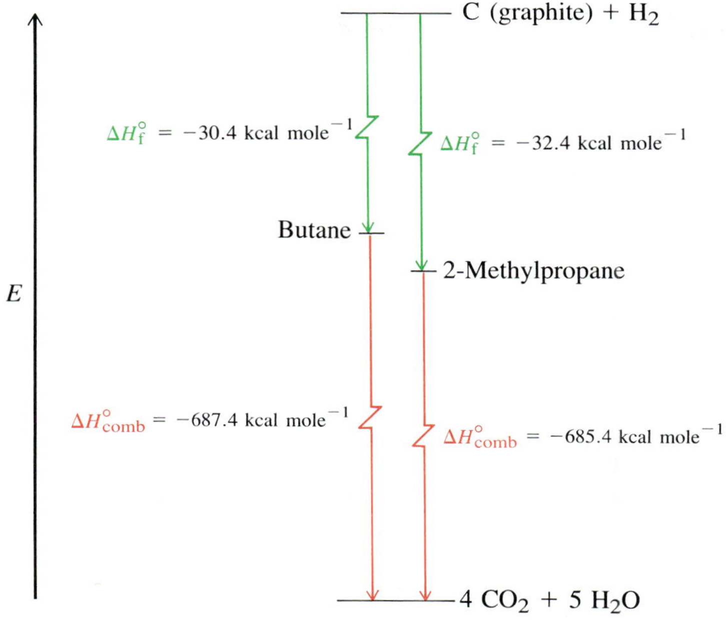

the latter is equal to zero, the difference is equal to the ΔH_f° of methane: $-230.7 - (-212.8) = -17.9$ kcal mole^{-1}.

In a similar way, knowledge of the heat of formation of a molecule allows the calculation of its heat of combustion.

EXERCISE 3-3

The heat of formation of cyclopropane

$$\begin{array}{c} CH_2 \\ / \quad \backslash \\ H_2C\text{———}CH_2 \end{array}$$

is $+12.7$ kcal mole^{-1}. What should its heat of combustion be?

BOX 3-3

Alkanes Are Functionalized by Catalytic Oxidation

Combustion is one way to activate the ordinarily quite unreactive alkane nucleus, but, unfortunately, because of the high temperatures, the reaction is relatively unselective and leads to the destruction of a major part of the starting material. An alternative and milder way to functionalize alkanes is by *enzymatic activation*. Enzymes are biocatalysts that effect a variety of transformations in living systems (see Chapter 27). The *mono-oxygenases* are a class of enzymes found in mammalian tissue, where they function in the oxidation of drugs, steroids (Section 4-7), and fatty acids (Section 17-12). In microbial systems, these enzymes also catalyze the oxidation of alkanes. For example, an enzyme from *Methylococcus capsulatus* inserts oxygen into a number of hydrocarbons to give alcohols. Bacteria have been suggested as a possible means of cleaning up oil spills by oxidative degradation.

Enzymatic Alkane Activation

R—H
Alkanes (C_1–C_8)

$\downarrow$ Enzyme, O_2

R—OH
Alcohols

Another approach to the controlled oxidation of the alkanes employs transition metal catalysts; for example, cobalt:

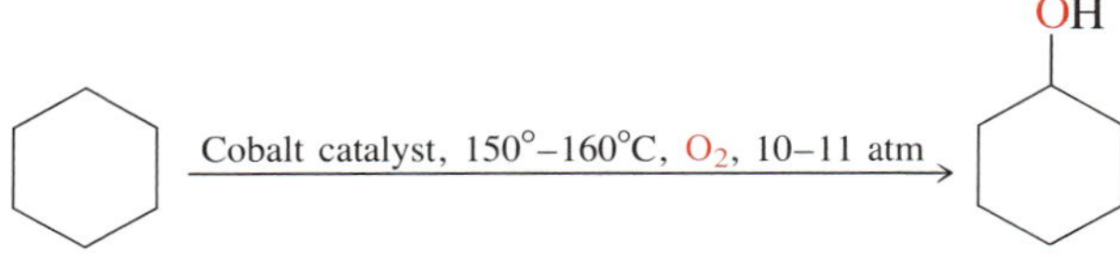

BOX 3-4

Fuel Additives Prevent Knocking

A problem with gasoline combustion is that occasionally the alkane mixture that constitutes the fuel explosively oxidizes (before the spark plug fires). This phenomenon is called preignition, or *knocking,* and is particularly pronounced with linear alkanes but less so with branched ones. A purpose of the reforming process is to isomerize linear alkanes to branched ones, thus improving the fuel quality. The amount of knocking is measured by the octane number. This is an arbitrary scale in which heptane is rated at zero and 2,2,4-trimethylpentane (isooctane) at 100.

The desire to improve the octane ratings of gasoline has led to the development of various additives. In the 1950s, most gasolines contained about 2.4 g per gallon of tetraethyllead, $Pb(CH_2CH_3)_4$, an organometallic compound that decomposes to give radicals that promote alkane combustion. The octane rating of 2,2,4-trimethylpentane-tetraethyllead is 120.3. To allow the lead to be removed from the combustion chamber as part of the exhaust gases, 1,2-dichloro- and 1,2-dibromoethane also were added; these compounds are sources of halogen, which scavenges the metal as lead halide. In 1967, the United States produced 685 million pounds of tetraethyllead, most of which was dissipated into the environment as oxides and halides through automotive exhaust. The environmental problems created by this practice have led to the phasing out of lead additives in gasoline. To worsen the problem, 1,2-dichloro- and 1,2-dibromoethane are suspected cancer-causing agents. Moreover, the catalytic converters used in newer cars are poisoned by lead. These converters have been added to improve the combustion process and to decrease the amount of toxic carbon monoxide and unburned hydrocarbon exhaust.

To summarize, the heat of combustion of alkanes and other organic molecules affords a quantitative estimate of their energy content. The same information is given by the heats of formation.

3-5 The Halogenation of Methane

We have seen that alkanes, despite their lack of chemical reactivity, undergo chemical transformations if subjected to pyrolysis and combustion. This section deals with the effect of exposing an alkane, methane, to halogens. A reaction takes place to produce a halomethane and a hydrogen halide. To establish its mechanism, each step in this transformation will be analyzed.

Chlorine Converts Methane into Chloromethane

When methane and chlorine gas are mixed in the dark at room temperature, there is no reaction. The mixture must be heated to temperatures above 300°C *or* irradiated with ultraviolet light before entering into reaction, sometimes violently. One of the two initial products is chloromethane, derived from methane in which a hydrogen atom is removed and replaced by chlorine. This process is referred to as a **substitution reaction.** The other product of this transformation is hydrogen chloride. Further substitution leads to dichloromethane (methylene chloride), CH_2Cl_2, trichloromethane (chloroform), $CHCl_3$, and tetrachloromethane (carbon tetrachloride), CCl_4.

Why should this reaction proceed? A clue may be obtained from a consideration of its ΔH°. Note that a C–H bond in methane ($DH^\circ = 105$ kcal mole^{-1}) and a Cl–Cl bond ($DH^\circ = 58$ kcal mole^{-1}) are lost, whereas the C–Cl bond of chloromethane ($DH^\circ = 85$ kcal mole^{-1}) and the substantial energy of the H–Cl linkage ($DH^\circ = 103$ kcal mole^{-1}) are gained. The net result is a gain in bond energies of 25 kcal mole^{-1}: the reaction is appreciably *exothermic*.

Chlorination of Methane

$$CH_3\text{—}H + :\ddot{\underset{..}{Cl}}\text{—}\ddot{\underset{..}{Cl}}: \xrightarrow{\Delta \text{ or } h\nu} CH_3\text{—}\ddot{\underset{..}{Cl}}: + H\text{—}\ddot{\underset{..}{Cl}}: \qquad \Delta H^\circ = -25 \text{ kcal mole}^{-1}$$

Chloromethane

DH° (kcal mole^{-1}): 105 (C–H), 58 (Cl–Cl), 85 (C–Cl), 103 (H–Cl)

Why then does the thermal chlorination of methane not occur at room temperature? We already know the answer to this question. The fact that a reaction is exothermic does not necessarily mean that it should proceed rapidly and spontaneously. Remember that the rate of a reaction is quite independent of its ΔH°; it is dependent on the activation parameters for the process, in this case evidently unfavorable. Why is this so, and what is the function of irradiation when the reaction *does* proceed at room temperature? Answering these questions requires an investigation of the mechanism of the reaction. This mechanism consists of three stages: initiation, propagation, and termination. (Note that, in this scheme and in those that follow, all radicals and atoms are in green.)

Stages in Mechanism of Chlorination of Methane:

Initiation:

$$Cl_2 \longrightarrow 2\,:\ddot{\underset{..}{Cl}}\cdot$$

Propagation:

$$CH_4 + :\ddot{\underset{..}{Cl}}\cdot \longrightarrow CH_3\cdot + H\ddot{\underset{..}{Cl}}:$$

$$CH_3\cdot + Cl_2 \longrightarrow CH_3\ddot{\underset{..}{Cl}}: + :\ddot{\underset{..}{Cl}}\cdot$$

Termination:

$$:\ddot{\underset{..}{Cl}}\cdot + :\ddot{\underset{..}{Cl}}\cdot \longrightarrow Cl_2$$

$$:\ddot{\underset{..}{Cl}}\cdot + CH_3\cdot \longrightarrow CH_3\ddot{\underset{..}{Cl}}:$$

$$CH_3\cdot + CH_3\cdot \longrightarrow CH_3\text{—}CH_3$$

Let us look at these stages in more detail.

The Chlorination of Methane: A Step by Step Analysis

The mechanism of the chlorination of methane includes the intermediate formation of radicals. In the first step of the reaction sequence, the weakest bond in the mixture, the Cl–Cl link, is broken. As mentioned earlier, this bond breaks thermally at about 300°C or by the absorption of a photon of light. When the latter happens, bonding electrons are excited to the antibonding level (Section

1-7), leading to bond rupture. Whatever the source of energy, the initial chlorine dissociation, also called the initiation step, requires a minimum of 58 kcal mole^{-1}.

STEP 1: Initiation

$$:\ddot{\underset{..}{\text{Cl}}}\text{—}\ddot{\underset{..}{\text{Cl}}}: \xrightarrow{\Delta \text{ or } h\nu} 2\ :\ddot{\underset{..}{\text{Cl}}}\cdot \qquad \Delta H^\circ = +58 \text{ kcal mole}^{-1}$$

Chlorine atom

In the next step, the chlorine atom attacks methane by abstracting a hydrogen atom. The resulting products are hydrogen chloride and a methyl radical.

STEP 2: Hydrogen abstraction

$$:\ddot{\underset{..}{\text{Cl}}}\cdot + \text{H—CH}_3 \longrightarrow :\ddot{\underset{..}{\text{Cl}}}\text{—H} + \cdot\text{CH}_3 \qquad \Delta H^\circ = +2 \text{ kcal mole}^{-1}$$

DH° (kcal mole^{-1}): 105 (H—CH_3), 103 (Cl—H)

Methyl radical

The ΔH° for this step is slightly positive but not sufficiently so to prevent appreciable equilibrium concentrations of products from being formed. What is the activation energy, E_a, for hydrogen abstraction? Is there enough external heat to overcome this barrier? The answer is yes, because radical abstraction generally requires very little activation. For the chlorine atom, a molecular-orbital description of the transition state of hydrogen abstraction from methane (Figure 3-8) reveals the simplicity of the process. The hydrogen being abstracted is positioned between the carbon and the chlorine, partially bound to both. The transition state is located only about 4 kcal mole^{-1} above the starting materials. Because this state cannot be isolated (it is only a point along the reaction coordinate), it is labeled by a special sign,‡. Note that, in the transition state, H–Cl bond formation has occurred to about the same extent as C–H bond breaking. Were the C–H bond to be broken to a substantially greater extent before hydrogen chloride formation, the E_a would be considerably higher. A potential-energy diagram describing this step is shown in Figure 3-9.

Step 2 gives one of the products of the chlorination reaction, HCl. What about the desired product, CH_3Cl? Chloromethane is formed in Step 3. Here the methyl radical abstracts a chlorine atom from one of the starting chlorine molecules, furnishing chloromethane and a new chlorine atom. The latter reenters Step 2, thus closing a cycle. Note the appreciable exothermicity of this transformation, −27 kcal mole^{-1}. This step supplies the overall driving force for the reaction of methane with chlorine.

STEP 3: Chloromethane formation

$$\text{H}_3\text{C}\cdot + :\ddot{\underset{..}{\text{Cl}}}\text{—}\ddot{\underset{..}{\text{Cl}}}: \longrightarrow \text{H}_3\text{C—Cl} + \cdot\ddot{\underset{..}{\text{Cl}}}: \qquad \Delta H^\circ = -27 \text{ kcal mole}^{-1}$$

DH° (kcal mole^{-1}): 58 (Cl—Cl), 85 (CH_3—Cl)

The exothermic nature of Step 3 allows for the unfavorable equilibrium in Step 2 to be pushed toward the product side by the rapid depletion of the methyl radical product in the subsequent reaction:

$$CH_4 + Cl\cdot \rightleftharpoons CH_3\cdot + HCl \xrightleftharpoons{Cl_2} CH_3Cl + Cl\cdot + HCl$$

Slightly unfavorable — Very favorable "Drives" first equilibrium

Steps 2 and 3 constitute the propagation cycle; they result in the formation of products. The potential-energy diagram in Figure 3-10 further illustrates this point by continuing the progress of the reaction begun in Figure 3-9. The second step clearly determines the rate of propagation. The diagram also shows that the overall ΔH° of the reaction is made up of two contributing values [ΔH° (Step 2)

FIGURE 3-8

Approximate molecular-orbital description of the abstraction of a hydrogen atom by a chlorine atom to give a methyl radical and hydrogen chloride. Notice the rehybridization at carbon on forming the planar methyl radical. The additional three nonbonded electron pairs on chlorine have been omitted. The orbitals are not drawn to scale.

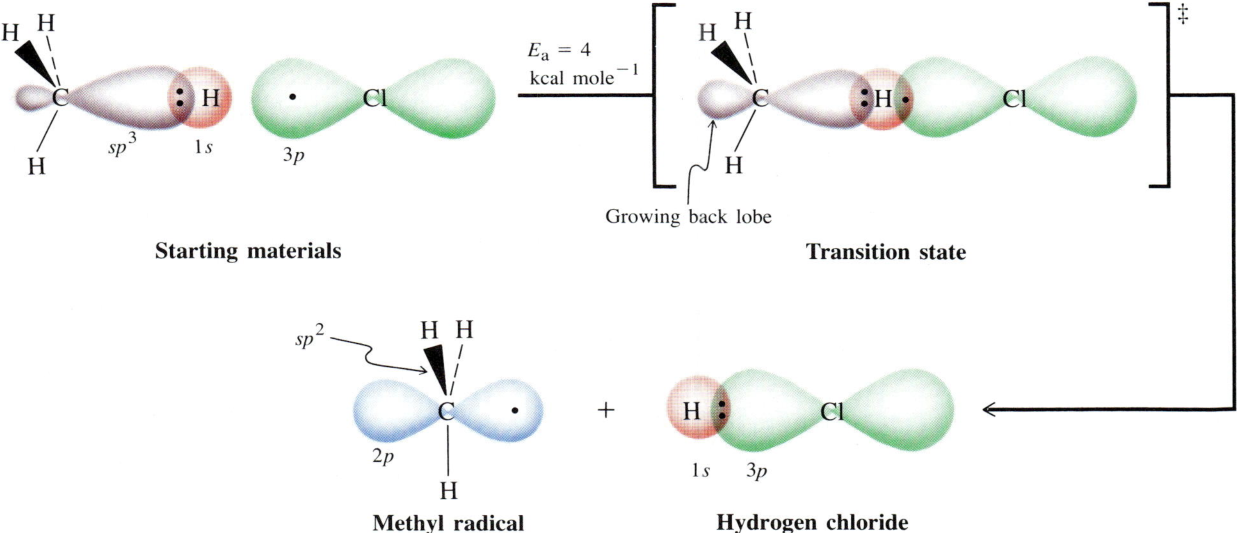

FIGURE 3-9

Potential-energy diagram of the reaction of methane with a chlorine atom. Partial bonds in the transition state are depicted by dotted lines.

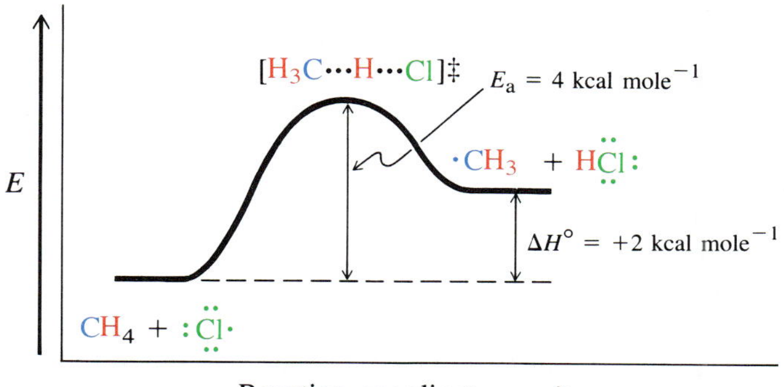

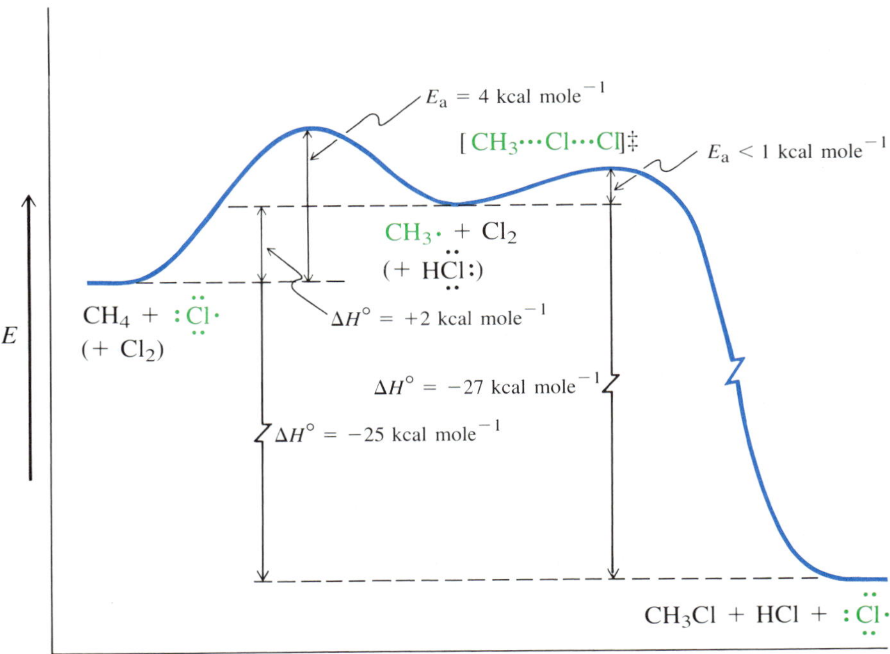

FIGURE 3-10
Potential-energy diagram for the formation of CH_3Cl from methane and chlorine. Note the ΔH° of the overall reaction $CH_4 + Cl_2 \rightarrow CH_3Cl + HCl$, which amounts to -25 kcal mole^{-1}.

and ΔH° (Step 3)]: $+2 - 27 = -25$ kcal mole^{-1}. You can see that this should be so by adding the equations in Steps 2 and 3:

	ΔH° (kcal mole^{-1})
$:\ddot{Cl}\cdot + CH_4 \longrightarrow CH_3\cdot + H\ddot{Cl}:$	$+2$
$CH_3\cdot + Cl_2 \longrightarrow CH_3\ddot{Cl}: + :\ddot{Cl}\cdot$	-27
$CH_4 + Cl_2 \longrightarrow CH_3\ddot{Cl}: + H\ddot{Cl}:$	-25

A General Radical Chain Mechanism:

Initiation:

$$X_2 \longrightarrow 2\ :\ddot{X}\cdot$$

Propagation:

$$:\ddot{X}\cdot + RH \longrightarrow R\cdot + H\ddot{X}:$$

$$X_2 + R\cdot \longrightarrow R\ddot{X}: + :\ddot{X}\cdot$$

Chain termination:

$$:\ddot{X}\cdot + :\ddot{X}\cdot \longrightarrow X_2$$

$$R\cdot + :\ddot{X}\cdot \longrightarrow RX$$

$$R\cdot + R\cdot \longrightarrow R_2$$

The mechanism for the chlorination of methane is an example of a **radical chain mechanism.**

Only a few halogen atoms are necessary for initiating the reaction, because the propagation steps are self-sufficient in $:\ddot{X}\cdot$. The first propagation step consumes a halogen atom, the second produces one. The newly generated halogen atom then reenters the propagation cycle in the first propagation step. In this way, a *radical chain* is set in motion that can drive the reaction for many thousands of cycles.

Can the chain be terminated? The answer is yes, mainly by the covalent bonding of radicals. However, the concentration of radicals in the reaction mixture is very small, and hence the chances of one radical finding another are also small. Therefore, chain termination is relatively infrequent.

EXERCISE 3-4

Chlorination of ethane furnishes chloroethane. Write a mechanism for this transformation and calculate ΔH° for each step (see Tables 3-1 and 3-2).

One of the practical problems of chlorinating methane is the control of product selectivity. As mentioned earlier, the reaction does not stop at the formation of chloromethane but continues to form di-, tri-, and tetrachloromethane by successive substitution. This problem is compounded by the increasingly weakened C–H bonds in the transformation of CH_3Cl (DH° = 101 kcal mole^{-1}) into CH_2Cl_2 (DH° = 99 kcal mole^{-1}) and then $CHCl_3$ (DH° = 96 kcal mole^{-1}), which favors competitive hydrogen abstraction from these compounds compared with methane. A simple solution to this problem is the use of a large excess of methane in the reaction. Under such conditions, the reactive intermediate chlorine atom is at any given moment surrounded by many more methane molecules than product CH_3Cl. Thus, the chances of $Cl^\cdot$ finding CH_3Cl to eventually make CH_2Cl_2 are greatly diminished, and product selectivity is achieved.

Other Halogenations: Fluorine Is Most, Iodine Least Reactive

Fluorine and bromine, but not iodine, also react with methane by free radical mechanisms to furnish the corresponding halomethanes. The dissociation energies of X_2 (X = F, Br, I) are lower than that of Cl_2, ensuring ready initiation of the radical chain.

	F_2	Cl_2	Br_2	I_2
DH° (X_2) in kcal mole^{-1}	37	58	46	36

It is interesting, however, to compare the enthalpies of the two propagation steps (Table 3-6). It is apparent that there are quite striking differences in the driving force for hydrogen abstraction. For fluorine, this step is exothermic by −30 kcal mole^{-1}. We have already seen that, for chlorine, the same step is slightly endothermic; for bromine, it is substantially so. Finally, atomic iodine shows about as much endothermicity (ΔH° = +34 kcal mole^{-1}) as fluorine shows exothermicity. This strong attenuation along the series is due to the decreased bond strengths of the hydrogen halides in going from fluorine to iodine (see Table 3-1). The fact that the fluorine atom forms a strong bond with hydrogen is demonstrated in its reactivity in hydrogen abstractions. Fluorine is more reactive than chlorine, chlorine is more reactive than bromine, and the least-reactive halogen atom is iodine.

Relative Reactivities of $X^\cdot$ in Hydrogen Abstractions

$$F^\cdot > Cl^\cdot > Br^\cdot > I^\cdot$$

However, we must be careful with the use of the word reactive. We need to distinguish *thermodynamic reactivity,* governed by the enthalpy, ΔH° (or, if we

TABLE 3-6
Enthalpies of the propagation steps in the halogenation of methane (kcal mole^{-1})

	F	Cl	Br	I
$:\ddot{X}\cdot + CH_4 \longrightarrow \cdot CH_3 + H\ddot{X}:$	−30	+2	+18	+34
$\cdot CH_3 + X_2 \longrightarrow CH_3\ddot{X}: + :\ddot{X}\cdot$	−73	−27	−25	−21
$CH_4 + X_2 \longrightarrow CH_3\ddot{X}: + H\ddot{X}:$	−103	−25	−7	+13

include the entropy contribution, the free energy change, ΔG°), from *kinetic reactivity,* controlled by the activation energy, which is in turn related to the structure of the transition state.

BOX 3-5

Early and Late Transition States

In some cases, a qualitative judgment can be made about the relative ease with which a reaction proceeds by inspection of the structure of the transition state. Consider the reaction of $F^\cdot$ with methane. The transition state is of the type shown in Figure 3-8: there is a partial H–F bond and a partial radical character on carbon. However, because the H–F bond strength is so much greater than that of H–Cl (see Table 3-1), less C–H bond breaking has to have occurred at this stage. Such a transition state is said to be *early*—that is, more to the left in the reaction-coordinate diagram in Figure 3-11. *Early transition states* are frequently characteristic of fast exothermic processes (Hammond* postulate).

FIGURE 3-11
Potential-energy diagrams of an early transition state (the reaction is exothermic) and a late transition state (the reaction is endothermic).

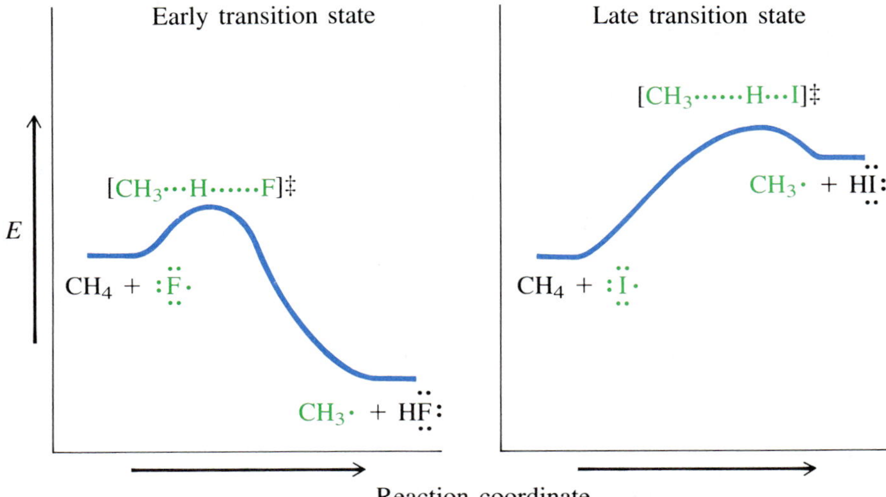

---

*Professor George S. Hammond, b. 1921, Allied Corporation.

On the other hand, in the hypothetical reaction of I· with methane, there has to be a considerable degree of H–I bond formation balancing the loss of a C–H bond in the transition state (Figure 3-11). Therefore, this transition state is substantially further along the reaction coordinate; it is said to be *late*. Late transition states are frequently typical of relatively slow, not very exothermic transformations or of endothermic ones.

The Thermodynamics and Kinetics of the Second Propagation Step

Let us now consider the second propagation step in Table 3-6. This process is exothermic for all the halogens. Again, the reaction is fastest and most exothermic for fluorine. The combined enthalpies of the two steps for the fluorination of methane result in a ΔH° of -105 kcal mole^{-1}. Indeed, this value is so large that, at sufficiently high concentrations of methane and fluorine gas, violent reaction occurs. The formation of chloromethane is less exothermic, and that of bromomethane even less so. In the latter case, the appreciably endothermic nature of the first step ($\Delta H^\circ = +18$ kcal mole^{-1}) is barely overcome by the enthalpy of the second ($\Delta H^\circ = -25$ kcal mole^{-1}), resulting in a gain in energy of only -7 kcal mole^{-1} for the overall substitution. Finally, inspection of the thermodynamics of iodination reveals why iodine does not react with methane to furnish methyl iodide and hydrogen iodide. The first step is so costly in energy that even the reasonably exothermic second step cannot force the reaction equilibrium to the product side.

EXERCISE 3-5

Predict the product distribution of the reaction of methane with an equimolar mixture of chlorine and bromine at low conversion.

In summary, halogens, with the exception of iodine, react with methane (and, as we shall see, other alkanes) to give halomethanes. The reaction proceeds through a mechanism in which a small amount of the halogen molecule is cleaved homolytically (initiation) by either heat or light to give halogen atoms. These are capable of maintaining a free-radical chain reaction (propagation) consisting of two steps: (1) hydrogen abstraction to give a methyl radical and HX, and (2) reaction of $CH_3\cdot$ to give CH_3X and regenerated $X\cdot$. The reaction is terminated by radical combination. The heats of reaction of the individual steps can be calculated by taking into account which bonds are being formed or broken. In these processes, the more exothermic steps also have lower activation energies. This explains the relative reactivities of the halogens, fluorine being the most reactive, iodine the least. Finally, an early transition state is often characteristic of a strongly exothermic reaction in which bonds that are formed are relatively strong. Conversely, a late transition state in many cases typifies an endothermic or relatively less exothermic process. Here, the bonds that are formed are less strong.

3-6
The Chlorination of Higher Alkanes: Relative Reactivity and Selectivity

What is the scope of the free-radical halogenation of methane when extended to other alkanes? Will the different types of R–H bonds—namely, primary, secondary, and tertiary—react in the same way? Let us address these questions by considering the chlorination of ethane, then propane, and finally 2-methylpropane.

The monochlorination of ethane gives chloroethane as the product.

The Chlorination of Ethane

$$CH_3CH_3 + Cl_2 \xrightarrow{\Delta \text{ or } h\nu} \underset{\textbf{Chloroethane}}{CH_3CH_2Cl} + HCl \qquad \Delta H^\circ = -27 \text{ kcal mole}^{-1}$$

This reaction proceeds by a free-radical chain mechanism analogous to the one observed for methane. The propagation steps include formation of the ethyl radical from ethane and the chlorine atom, followed by generation of the product and the release of another chlorine atom (Exercise 3-4). The greatest difference between the two mechanisms lies in the change in values for ΔH°. Thus, the abstraction of a hydrogen from the alkane ($DH^\circ = 98$ kcal mole^{-1}) is no longer endothermic, as in methane, but favorable by -5 kcal mole^{-1}. The reason for this finding is the weaker C–H bond of ethane.

Propagation Steps in the Chlorination of Ethane

$$CH_3CH_3 + :\ddot{\underset{..}{Cl}}\cdot \longrightarrow CH_3CH_2\cdot + H\ddot{\underset{..}{Cl}}: \qquad \Delta H^\circ = -5 \text{ kcal mole}^{-1}$$

$$CH_3CH_2\cdot + Cl_2 \longrightarrow CH_3CH_2\ddot{\underset{..}{Cl}}: + :\ddot{\underset{..}{Cl}}\cdot \qquad \Delta H^\circ = -22 \text{ kcal mole}^{-1}$$

What can be expected for the next homolog, propane?

Secondary C–H Bonds Are More Reactive Than Primary Ones

In propane, two kinds of hydrogen atoms are available for reaction with chlorine, six primary and two secondary. If the two types of hydrogen were to react at equal rates, we could expect three times as much 1-chloropropane as 2-chloropropane. This is a purely statistical ratio; there are three times as many primary hydrogen atoms as there are secondary.

The Chlorination of Propane

$$Cl_2 + CH_3CH_2CH_3 \xrightarrow{h\nu} \underset{\textbf{1-Chloropropane}}{CH_3CH_2CH_2Cl} + \underset{\textbf{2-Chloropropane}}{CH_3\overset{\overset{\displaystyle Cl}{|}}{C}HCH_3} + HCl$$

Expected statistical ratio	3	:	1
Expected C–H bond reactivity ratio	Less	:	More
Experimental ratio (25°C)	43	:	57
Experimental ratio (600°C)	3	:	1

On the other hand, because the secondary C–H bond is weaker than the primary one ($DH^\circ = 94.5$ versus 98 kcal mole^{-1}), more 2-chloro- than 1-chloropropane might be expected.

The Difference in ΔH° of the First Propagation Steps in the Chlorination of Propane

$$CH_3CH_2CH_3 + :\ddot{\underset{..}{Cl}}\cdot \longrightarrow \underset{\textbf{Propyl radical}}{CH_3CH_2CH_2\cdot} + H\ddot{\underset{..}{Cl}}: \qquad \Delta H^\circ = -5 \text{ kcal mole}^{-1}$$

$$CH_3CH_2CH_3 + :\ddot{\underset{..}{Cl}}\cdot \longrightarrow \underset{\textbf{1-Methylethyl (isopropyl) radical}}{CH_3\dot{C}HCH_3} + H\ddot{\underset{..}{Cl}}: \qquad \Delta H^\circ = -8.5 \text{ kcal mole}^{-1}$$

It is difficult to accurately predict a product ratio on the basis of these differences in DH°, because that ratio is determined not by the difference in the heat of formation of the resulting propyl and 1-methylethyl (isopropyl) radicals, but by the relative energies of the two transition states that lead to them. Because the species in these two transition states have only partial radical character, the difference in stability of the radical products will be only partly reflected in the relative energies of the transition states. Thus, the ΔE_a between the two modes of hydrogen abstraction is only about 1 kcal mole^{-1} (Figure 3-12).

FIGURE 3-12

Hydrogen abstraction by a chlorine atom from the secondary carbon in propane is more exothermic and faster than that from the primary carbon.

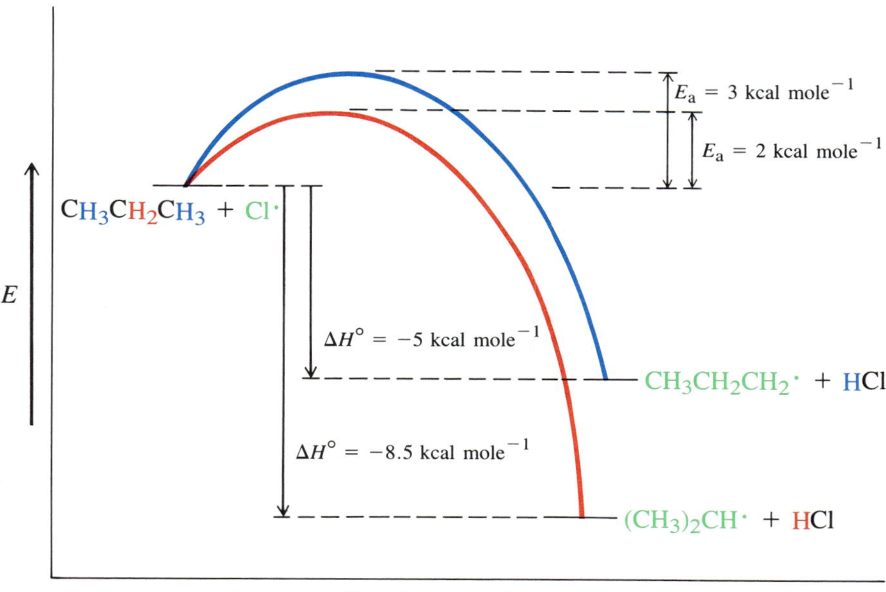

The experimental product ratio of 1-chloropropane:2-chloropropane in this reaction is found to be 43:57 at 25°C. This result indicates that both statistical and bond-energy factors play a role in product formation.

We can calculate the *relative reactivity of secondary and primary hydrogens* by factoring out the statistical contribution to the product ratio. For example, reaction of each of the six primary hydrogens contributes 43/6 = 7.2% to the total amount of the 1-chloropropane product. Similarly, each secondary hydrogen contributes 57/2 = 28.5% to the total amount of 2-chloropropane product. The relative reactivity of secondary:primary is consequently 28.5:7.2 = 4:1. In other words, the secondary hydrogen in the chlorination of propane at 25°C is four times as reactive as the primary one.

It would be tempting to conclude that, in general, *all* secondary positions are four times as reactive as all primary ones in *all* free-radical chain reactions. This generalization, however, is not quite correct. Thus, although we will see that secondary hydrogens generally react at a faster rate than do their primary counterparts, their relative reactivity very much depends on the nature of the attacking radical, X·, the strength of the resulting H–X bond, and even the temperature. For example, the chlorination of propane at 600°C results in the statistical distribution of products. At such a temperature, both reacting species have so much thermal energy that almost every collision leads to successful reaction. The chlorine radical is said to be less selective at higher temperatures, and statistical factors begin to control the outcome of the process. Conversely, at lower temperatures, there is more selectivity.

EXERCISE 3-6

What do you expect the products of monochlorination of butane to be? In what ratio will they be formed at 25°C?

Tertiary C–H Bonds Are More Reactive Than Secondary Ones

Let us now find the relative reactivity of a *tertiary* hydrogen in the chlorination of alkanes. For this purpose, we expose 2-methylpropane, a molecule containing nine primary hydrogens and a tertiary one, to the chlorination conditions at 25°C. The resulting two products, 2-chloro-2-methylpropane (*tert*-butyl chloride) and 1-chloro-2-methylpropane (isobutyl chloride) are formed in the relative yields of 36% and 64%, respectively.

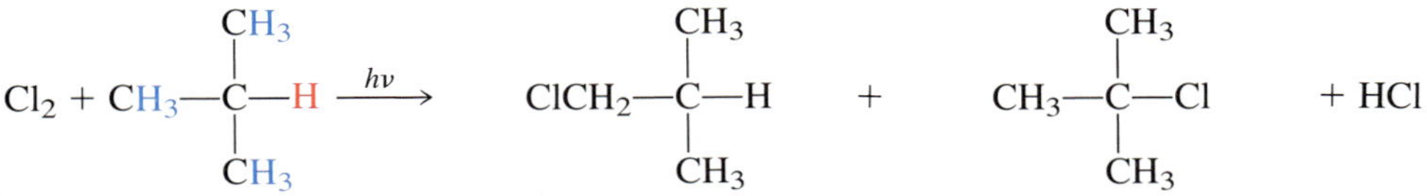

$$Cl_2 + CH_3-\underset{CH_3}{\overset{CH_3}{C}}-H \xrightarrow{h\nu} ClCH_2-\underset{CH_3}{\overset{CH_3}{C}}-H + CH_3-\underset{CH_3}{\overset{CH_3}{C}}-Cl + HCl$$

1-Chloro-2-methylpropane (Isobutyl chloride)   **2-Chloro-2-methylpropane (*tert*-Butyl chloride)**

Expected statistical ratio	9	:	1
Expected C–H bond reactivity ratio	Less	:	More
Experimental ratio (25°C)	64	:	36
Experimental ratio (600°C)	80	:	20

Factoring out the statistical contribution of nine gives a relative reactivity of tertiary : primary = 36/1 : 64/9 = 5.1 : 1. This selectivity, again, decreases at higher temperatures. However, we can say that, at 25°C, the relative reactivities of the various C–H bonds in chlorinations are:

$$\text{tertiary} : \text{secondary} : \text{primary} = 5 : 4 : 1$$

The result agrees well with the relative reactivity expected from consideration of bond strength: the tertiary C–H bond is weaker than the secondary, and the latter in turn weaker than the primary.

We can verify this ordering by looking at the competition between all three types of hydrogens within a single substrate. 2-Methylbutane, which contains nine primary hydrogens, two secondary hydrogens, and one tertiary hydrogen is an example. Because this molecule has two types of primary hydrogens, one set of six and one set of three, reaction with chlorine yields a total of four different monochlorination products.

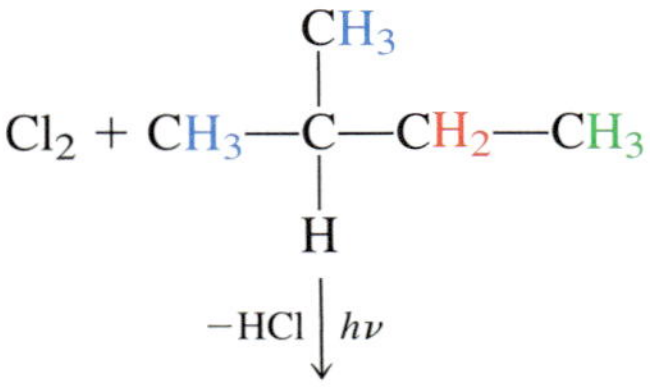

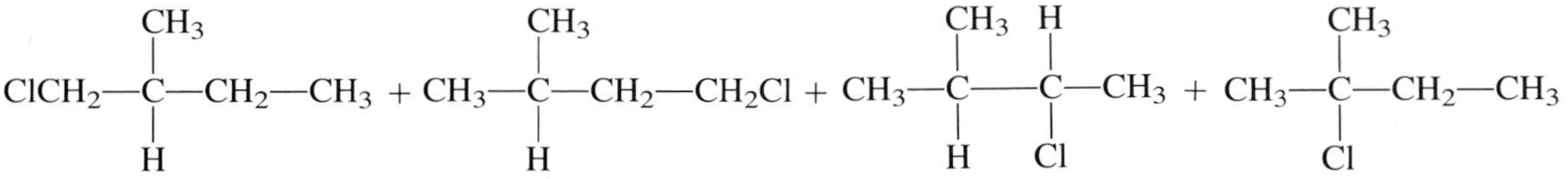

$$Cl_2 + CH_3-\underset{H}{\overset{CH_3}{C}}-CH_2-CH_3 \xrightarrow[-HCl]{h\nu}$$

$$ClCH_2-\underset{H}{\overset{CH_3}{C}}-CH_2-CH_3 + CH_3-\underset{H}{\overset{CH_3}{C}}-CH_2-CH_2Cl + CH_3-\underset{H}{\overset{CH_3}{C}}-\underset{Cl}{\overset{H}{C}}-CH_3 + CH_3-\underset{Cl}{\overset{CH_3}{C}}-CH_2-CH_3$$

27%	14%	36%	23%
1-Chloro-2-methylbutane	**1-Chloro-3-methylbutane**	**2-Chloro-3-methylbutane**	**2-Chloro-2-methylbutane**

The combined yield of the two primary halide products is 41% (1-chloro-2-methylbutane plus 1-chloro-3-methylbutane), the secondary halide 2-chloro-3-methylbutane is formed in 36%, and the tertiary halide in 23%. Therefore:

$$\begin{aligned} &\text{primary:secondary:tertiary halide} && = 41:36:23 \\ &\text{relative reactivity, primary:secondary:tertiary} && = 41/9:36/2:23/1 \\ &&& = 1:4:5 \end{aligned}$$

as expected.

EXERCISE 3-7

Give products and the ratio in which they are expected to form for the monochlorination of methylcyclohexane at 25°C.

To summarize, the relative reactivity of primary, secondary, and tertiary hydrogens follows the trend expected on the basis of their relative C–H bond strengths. Relative reactivity ratios can be calculated by factoring out statistical considerations. These ratios are temperature dependent, with greater selectivity at lower temperatures.

3-7 Selectivity in Halogenations of Alkanes with Fluorine and Bromine

What type of selectivity do halogens other than chlorine have in the halogenation of alkanes? Can we expect similar types of reactivity differences? We shall see in this section that the earlier observation that increased reactivity goes hand in hand with reduced selectivity can be generalized.

Table 3-6 showed that within the halogen series the fluorine atom is expected to be the most reactive, hydrogen abstraction from methane to form the very strong H–F bond being exothermic by −30 kcal mole^{-1}. This large and negative ΔH° gives rise to an early transition state and a low activation energy. On the other hand, the bromine radical is much less reactive, the same step having a large and positive ΔH° (18 kcal mole^{-1}), a relatively high activation barrier (E_a = 19 kcal mole^{-1}), and a late transition state. Does this difference result in changes in selectivity?

The answer to this question is given by the results of the reactions of 2-methylpropane with fluorine and bromine, respectively. Single fluorination at 25°C furnishes the two possible products: 2-fluoro-2-methylpropane (*tert*-butyl fluoride) and 1-fluoro-2-methylpropane (isobutyl fluoride) in the ratio 14:86, indicative of a relative reactivity of tertiary:primary hydrogens of 1.4:1. On the other hand, bromination of the same compound gives the tertiary bromide almost exclusively. Even at 98°C, the product ratio reveals a relative reactivity of 6300:1.

$$F_2 + (CH_3)_3CH \xrightarrow{h\nu} (CH_3)_3CF + FCH_2{-}C(CH_3)_2{-}H + HF$$

(CH₃)₃CF: 14%, **2-Fluoro-2-methylpropane (*tert*-Butyl fluoride)**

FCH₂–CH(CH₃)₂: 86%, **1-Fluoro-2-methylpropane (Isobutyl fluoride)**

Relative reactivity, tertiary:primary = 1.4:1 (at 25°C)

$$Br_2 + (CH_3)_3CH \xrightarrow{h\nu} (CH_3)_3CBr + BrCH_2-\underset{CH_3}{\overset{CH_3}{\underset{|}{\overset{|}{C}}}}-H + HBr$$

>99% 2-Bromo-2-methylpropane (*tert*-Butyl bromide)

<1% 1-Bromo-2-methylpropane (Isobutyl bromide)

Relative Reactivity, tertiary : primary = 6300 : 1 (at 98°C)

We can conclude that reactivity and selectivity in free-radical halogenations are properties that are related to each other *inversely*. The more-reactive halogens fluorine and chlorine are much less discriminating in regard to the various types of C–H bonds than is the less-reactive bromine. Table 3-7 summarizes these features.

TABLE 3-7

Relative reactivities of halogen atoms with alkane C–H bonds

	CH_3—H	RCH_2—H	R_2CH—H	R_3C—H
F· (25°C, gas)	0.5	1	1.2	1.4
Cl· (25°C, gas)	—	1	4	5
Cl· (100°C, liquid)	—	1	2.0	3.0
Br· (98°C, gas)	—	1	250	6300
Br· (150°C, gas)	0.002	1	80	1700

3-8 Synthetic Features of Free-Radical Halogenations

What considerations have to be made in devising a successful alkane halogenation? Selectivity, convenience, efficiency, and price.

Fluorinations are unattractive for several reasons. Fluorine is relatively expensive, corrosive, and, what is probably worse, quite unselective and dangerously reactive. Special conditions are necessary for the control of fluorination. At the opposite end of the scale, free-radical iodinations do not work because of unfavorable thermodynamics.

On the other hand, chlorinations are important, particularly in industry, simply because chlorine is cheap. Chlorine is prepared almost exclusively by the electrolysis of sodium chloride. The drawback in using chlorine for halogenations is the relatively low selectivity of the process, which gives mixtures of isomers that are very difficult to separate. Sometimes this problem is circumvented by the use of an alkane that contains only one type of hydrogen as a substrate, resulting (at least initially) in only one product. An example is cyclopentane.

$$\text{Cyclopentane} + Cl_2 \xrightarrow{h\nu} \text{Chlorocyclopentane} + HCl$$

Cyclopentane **Chlorocyclopentane**

Even in these cases, multiple substitution can complicate the outcome of the reaction. However, because the more-chlorinated products usually have higher boiling points (for every addition of chlorine, the boiling point of the haloalkane increases by about 60°C), they can be separated by distillation.

On an industrial scale, alkanes are chlorinated in large vessels fitted with elaborate controls to allow for smooth and safe operation. In the research laboratory, the use of chlorine gas is frequently avoided, because it is highly toxic and corrosive and because it is relatively difficult to weigh accurately. Several alternative chlorinating agents have been developed that accomplish the same task as chlorine but can be handled more safely and accurately. These compounds are usually either liquids or solids, such as sulfuryl chloride, SO_2Cl_2, and *N*-chlorobutanimide (*N*-chlorosuccinimide, NCS).

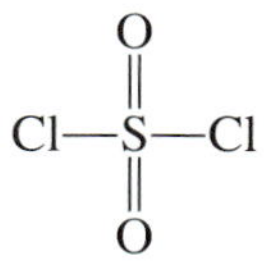

Sulfuryl chloride
(b.p. 69°C)

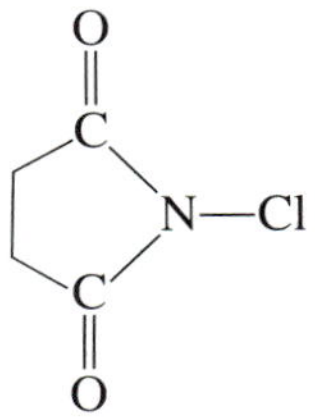

***N*-Chlorobutanimide**
(*N*-Chlorosuccinimide)
(m.p. 148°C)

BOX 3-6

The Lewis Structure of Sulfuryl Chloride

Sulfuryl chloride is a compound containing a third-row element for which Lewis structures with ten or more valence electrons are allowed. Such valence shell expansions are also found in other third- and higher-row elements, such as phosphorus, arsenic, and iodine. An octet structure for SO_2Cl_2 requires double charge separation (see also Section 6-7).

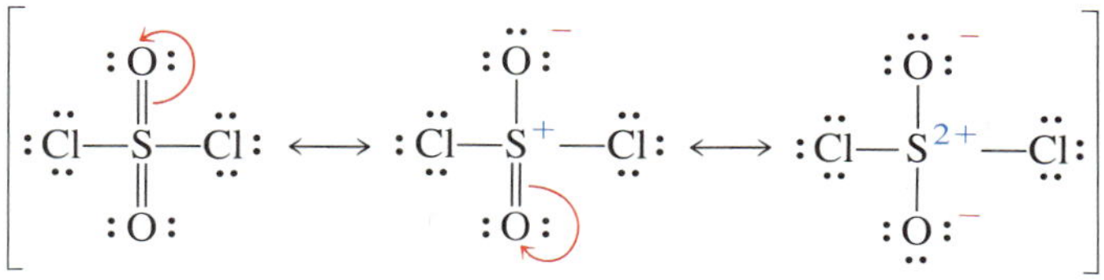

Examples:

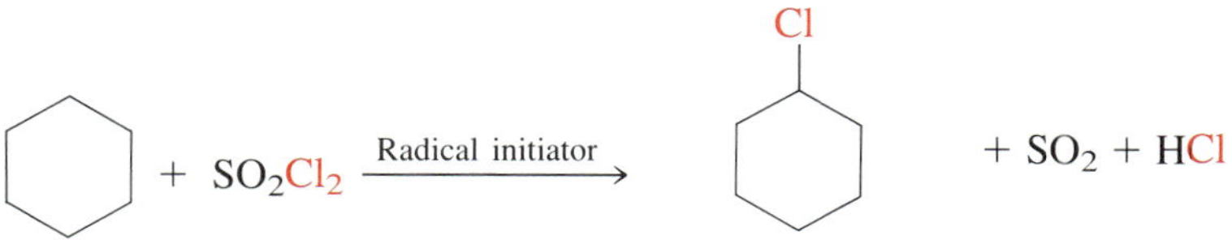

$$\text{Cyclohexane} + SO_2Cl_2 \xrightarrow{\text{Radical initiator}} \text{Chlorocyclohexane} + SO_2 + HCl$$

57%
Chlorocyclohexane

General NCS Reaction

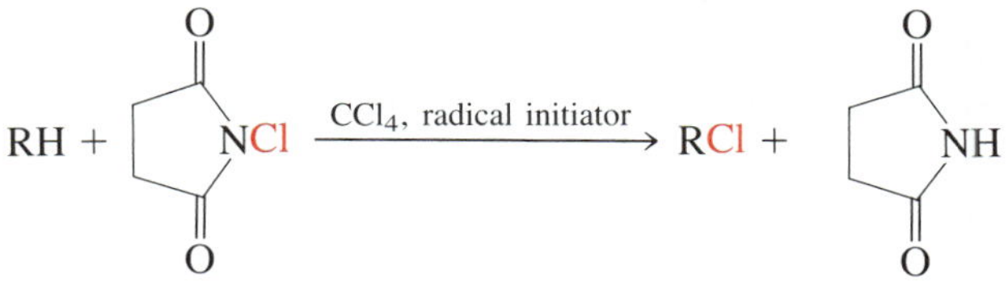

$$RH + \text{NCS} \xrightarrow{CCl_4,\ \text{radical initiator}} RCl + \text{Butanimide}$$

Butanimide
(Succinimide)

The use of sulfuryl chloride in alkane chlorinations requires catalytic amounts of a radical initiator. This is necessary because of the relatively high strength of the $Cl{-}SO_2Cl$ bond (estimated at 63 kcal mole^{-1}) compared with the Cl_2 bond, which makes it more difficult to break thermally or photolytically. The initiator fulfills the role of light in these chlorinations.

BOX 3-7

The Initiators in Free-Radical Reactions

Two common initiators in free-radical reactions are 2,2′-azodi(2-methylpropanenitrile), also called azodiisobutyronitrile, or AIBN, and dibenzoyl peroxide. (The latter compound is the active ingredient in some acne creams.) Both initiators readily generate radicals that react with SO_2Cl_2. The wavy lines in the structures below indicate the bonds in the initiators that are prone to breakage.

$$CH_3-\underset{CH_3}{\overset{CN}{C}}-\ddot{N}=\ddot{N}-\underset{CH_3}{\overset{CN}{C}}-CH_3 \xrightarrow{\Delta} 2\ CH_3-\dot{C}(CN)(CH_3) + :N\equiv N:$$

2,2′-Azodi(2-methylpropanenitrile) **2-Cyano-2-propyl radical**

$$C_6H_5-\overset{O}{\overset{\|}{C}}-\ddot{O}-\ddot{O}-\overset{O}{\overset{\|}{C}}-C_6H_5 \xrightarrow{\Delta} 2\ C_6H_5-\overset{O}{\overset{\|}{C}}-\ddot{O}\cdot \longrightarrow 2\ C_6H_5\cdot + 2\ CO_2$$

Dibenzoyl peroxide **Phenyl radical**

EXERCISE 3-8

Suggest mechanisms for the free radical chlorination of an alkane RH with (a) SO_2Cl_2 and (b) *N*-chlorobutanimide. Clearly delineate initiation, propagation, and termination steps.

EXERCISE 3-9

Which of the following compounds would give a monochlorination product with reasonable selectivity: propane, 2,2-dimethylpropane, cyclohexane, methylcyclohexane?

Because of its greater selectivity and because bromine is a liquid, brominations are frequently the method of choice for halogenating an alkane in the research laboratory. Halogenation occurs at the more-substituted carbon even in statistically unfavorable situations. Typical solvents in brominations are chlorinated methanes (CCl_4, $CHCl_3$, CH_2Cl_2), which are comparatively unreactive to bromine.

Bromine is obtained from aqueous sodium bromide solutions (found in natural brines) by treatment with chlorine. This process yields sodium chloride and bromine. The bromine vapors are swept out in a current of air and then condensed. Bromine is used less in industry because of its relatively greater cost and greater equivalent weight. Another popular and solid brominating agent is *N*-bromobutanimide (*N*-bromosuccinimide, NBS, see Section 14-2), an analog of NCS.

In summary, even though more expensive, bromine is the reagent of choice for selective free-radical halogenations. A solid brominating agent is *N*-bromobutanimide which, like *N*-chlorobutanimide, needs a radical initiator to enter into halogenation. Similarly, sulfuryl chloride may be used as a liquid chlorinating agent.

Summary of Important Concepts

1 The ΔH° for bond homolysis is defined as the bond-dissociation energy, DH°. Bond homolysis gives radicals.

2 The C–H bond strengths in the alkanes decrease in the order:

$$CH_3{-}H > RCH_2{-}H > R{-}\underset{\displaystyle H}{\overset{\displaystyle R}{C}}H > R{-}\underset{\displaystyle R}{\overset{\displaystyle R}{C}}{-}H$$

because the order of stability of the corresponding alkyl radical is:

$$CH_3\cdot < RCH_2\cdot < R{-}\overset{\displaystyle R}{C}H\cdot < R{-}\underset{\displaystyle R}{\overset{\displaystyle R}{C}}\cdot$$

The last sequence is due to increasing hyperconjugative stabilization.

3 Catalysts, such as zeolites or metal surfaces, speed up the establishment of an equilibrium between starting materials and products, such as in hydrocarbon rearrangements (reforming).

4 The ΔH° of the combustion of an alkane is defined as the heat of combustion, ΔH°_{comb}. Comparison of the heats of combustion of two isomeric hydrocarbons (and, more generally, compounds) allows an assessment of their relative stability.

5 The ΔH° of the formation of a molecule from its component elements is defined as its heat of formation ΔH°_f. Comparison of the heats of formation of two isomeric hydrocarbons (or compounds) also allows an assessment of their relative stability.

6 Given the heats of formation of water and carbon dioxide, knowledge of either the heat of combustion or the heat of formation of a molecule allows the calculation of the other quantity.

7 Alkanes react with halogens (except iodine) by a free-radical halogenation mechanism to give haloalkanes. The mechanism consists of initiation to create a halogen atom, two propagation steps, and various termination steps.

8 The first propagation step is rate and product determining. In it, a hydrogen atom is abstracted from the alkane chain resulting in an alkyl radical and HX. Hence, the order of reactivity increases from I_2 to F_2. The order of selectivity decreases along the same series, as well as with temperature.

9 The relative reactivities of the various types of alkane C–H bonds in halogenations can be estimated by factoring out statistical contributions. They are roughly constant under identical conditions and follow the order

$$CH_4 < \text{primary CH} < \text{secondary CH} < \text{tertiary CH}$$

In free-radical chlorinations of alkanes at 25°C, the approximate relative reactivity of the tertiary:secondary:primary positions is 5:4:1. In fluorinations, this ratio is about 1.4:1.2:1, whereas in brominations (98°C) it is 6300:250:1.

10 Chemists often use halogenating agents other than the halogens. Examples are sulfuryl chloride, SO_2Cl_2, and *N*-chloro- and *N*-bromobutanimide.

Problems

1 Label the primary, secondary, and tertiary hydrogens in each of the following compounds.

(a) $CH_3CH_2CH_3$

(b) $CH_3CH_2CH_2CH_3$

(c) 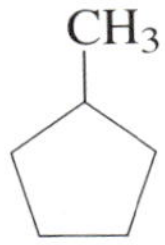 (cyclopentane ring bearing CH_3)

(d) 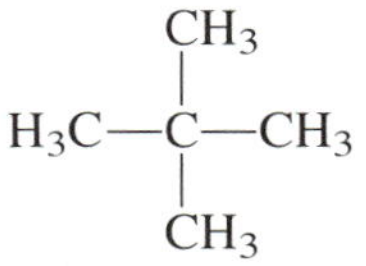 $H_3C—C(CH_3)_2—CH_3$ (central C with CH_3 above and below)

(e)  $(H_3C)_2CHCH_2CH_3$ (drawn as H_3C and H_3C joined to $CHCH_2CH_3$)

2 Write as many products as you can think of that might result from the pyrolytic cracking of propane. Assume that the only initial process is C–C bond cleavage.

3 Answer the question posed in Problem 2 for (a) butane and (b) 2-methylpropane. Use the data in Table 3-2 to determine the bond most likely to cleave homolytically, and use that bond cleavage as your first step.

4 Would you expect hyperconjugation to

(a) stabilize $CH_3CH_2^+$ relative to CH_3^+?

(b) stabilize $CH_3CH_2^-$ relative to CH_3^-?

5 Using the data in Section 3-4, calculate ΔH_f° values for cyclohexane (liquid), ethanol (gas), ethanol (liquid), and sucrose (solid).

6 The heats of combustion for several organic molecules are given below. Calculate ΔH_f° for each of them.

(a) C_6H_6, benzene (for structure, see Section 2-1), -781.0 kcal mole^{-1}

(b) C_3H_6O, propanone (acetone; for structure, see Section 15-1), -427.9 kcal mole^{-1}

(c) C_3H_6O, propanal (for structure, see Section 15-1), -434.1 kcal mole^{-1}

7 Referring to your results in Problem 6, which isomeric C_3H_6O compound is more stable, propanone (acetone) or propanal?

8 Referring to your results in Exercise 3-3 and Problem 5, calculate ΔH° for the hypothetical reaction

$$2\ \text{Cyclopropane} \longrightarrow \text{cyclohexane}$$

9 Calculate ΔH° for the reaction of ethyne (acetylene, HC≡CH) with hydrogen to produce ethane.

10 Ordinary glassblowing torches are fueled by natural gas. Welders, on the other hand, require much hotter temperatures for their work, and often use torches fueled by ethyne (acetylene, HC≡CH).

(a) Write a balanced equation for the combustion of ethyne and calculate its heat of combustion from data in Section 3-4.

(b) Compare your result with the heat of combustion of propane, a typical component of natural gas, both per mole and per gram. Does this explain the hotter flame of ethyne?

11 Section 3-4 tells how ΔH_f° for methane can be derived from the heat of combustion of methane, given the ΔH_f° values for liquid H_2O and gaseous CO_2. Suppose that, instead of ΔH_f° values for H_2O and CO_2, you were given heats of combustion for carbon and hydrogen. Show how you would then derive ΔH_f° for methane.

12 A hypothetical alternative mechanism for the halogenation of methane has the following propagation steps:

$$(1)\ X\cdot + CH_4 \longrightarrow CH_3X + H\cdot$$
$$(2)\ H\cdot + X_2 \longrightarrow HX + X\cdot$$

(a) Using either DH° or ΔH_f° values from appropriate tables, calculate ΔH° for these steps for any one of the halogens.

(b) Compare your ΔH° values with those for the accepted mechanism (Table 3-6). Do you expect this alternative mechanism to compete successfully with the accepted one?

13 Write a mechanism for the free-radical bromination of the hydrocarbon benzene, C_6H_6 (for structure, see Section 2-1). Use propagation steps similar to those in the halogenation of alkanes, as presented in Sections 3-5 through 3-7. Calculate ΔH° values for each step and for the reaction as a whole. How does this reaction compare thermodynamically with the bromination of other hydrocarbons? Data: $DH^\circ_{C_6H_5\text{–}H} = 111$ kcal mole^{-1}; $DH^\circ_{C_6H_5\text{–}Br} = 81$ kcal mole^{-1}.

14 The addition of certain materials called free-radical inhibitors to halogenation reactions causes the reactions to come virtually to a complete stop. An example is the inhibition by I_2 of the chlorination of methane. Explain how this inhibition might come about. Hint: Calculate ΔH° values for possible reactions of the various species present in the system with I_2, and evaluate the possible further reactivity of the products of these I_2 reactions.

15 Calculate ΔH° values for the following possible reactions.

(a) $H_2 + F_2 \longrightarrow 2\ HF$
(b) $H_2 + Cl_2 \longrightarrow 2\ HCl$
(c) $H_2 + Br_2 \longrightarrow 2\ HBr$
(d) $H_2 + I_2 \longrightarrow 2\ HI$

(e) $(CH_3)_3CH + F_2 \longrightarrow (CH_3)_3CF + HF$
(f) $(CH_3)_3CH + Cl_2 \longrightarrow (CH_3)_3CCl + HCl$
(g) $(CH_3)_3CH + Br_2 \longrightarrow (CH_3)_3CBr + HBr$
(h) $(CH_3)_3CH + I_2 \longrightarrow (CH_3)_3CI + HI$

16 Write the major organic product(s), if any, of each of the following reactions.

(a) $CH_3CH_3 + I_2 \xrightarrow{\Delta}$

(b) $CH_3CH_2CH_3 + F_2 \longrightarrow$

(c) 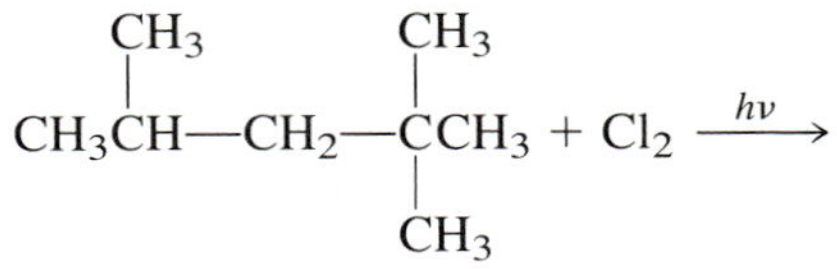 $\begin{array}{c} \quad CH_3 \qquad\qquad CH_3 \\ CH_3\overset{|}{C}H{-}CH_2{-}\overset{|}{\underset{|}{C}}CH_3 + Cl_2 \xrightarrow{h\nu} \\ \qquad\qquad\qquad CH_3 \end{array}$

(d) 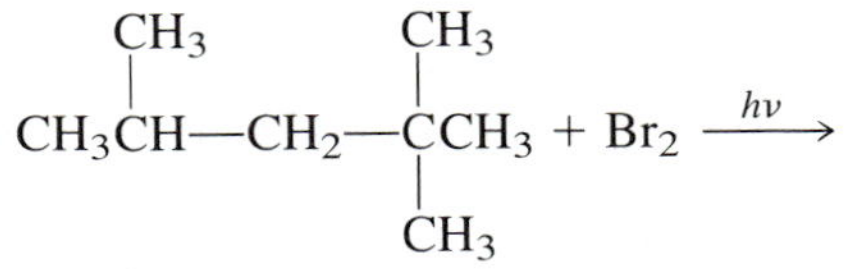 $\begin{array}{c} \quad CH_3 \qquad\qquad CH_3 \\ CH_3\overset{|}{C}H{-}CH_2{-}\overset{|}{\underset{|}{C}}CH_3 + Br_2 \xrightarrow{h\nu} \\ \qquad\qquad\qquad CH_3 \end{array}$

(e) 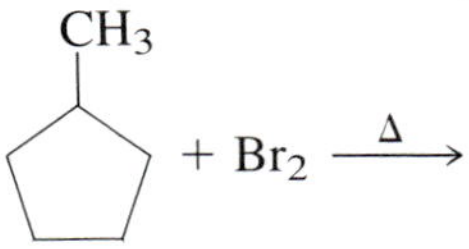 (methylcyclopentane, CH_3 on ring) $+ Br_2 \xrightarrow{\Delta}$

17 Calculate product ratios in each of the reactions in Problem 16. Use relative reactivity data for F_2 and Cl_2 at 25°C and for Br_2 at 150°C (Table 3-7).

18 Which, if any, of the reactions in Problem 16 give the major product with reasonable selectivity (i.e., are useful "synthetic methods")?

19 If a gaseous mixture of CH_3I and HI were to be heated, what would you expect to be the likely result? Suggest a detailed mechanism and calculate ΔH° for each required step. Hint: Begin by breaking the weakest bond of all those present in the starting materials, and then examine possible free-radical chain processes that might follow.

20 Predict the major product(s) of free-radical monobromination of each of the following compounds (identified by their common names). Point out any reaction that gives the major product with reasonable selectivity. All of the hydrocarbons shown are derived from molecules representative of the class of naturally occurring compounds called terpenes (see Section 4-7).

(a) 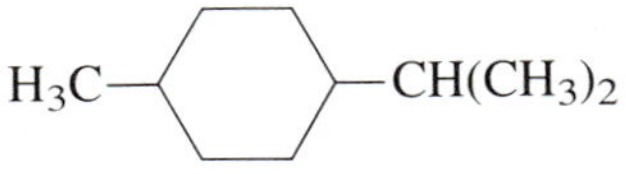 H_3C–(cyclohexane)–$CH(CH_3)_2$

(Menthane)

(b) 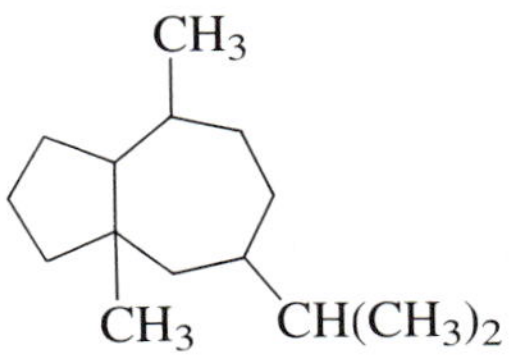 (structure with CH_3, CH_3, $CH(CH_3)_2$ substituents)

(Pseudoguaiane)

(c) (structure with CH_3, CH_3, CH_3 substituents)

(Bornane)

(d) 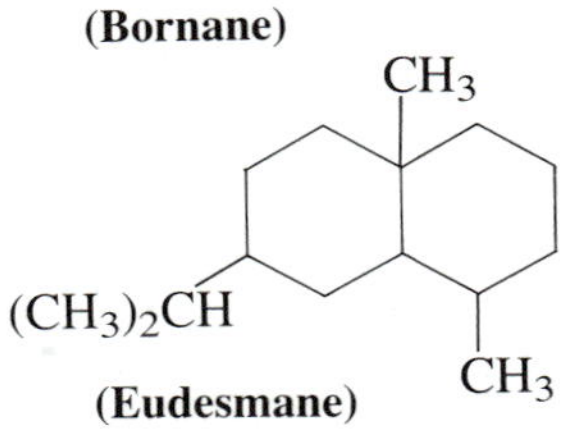 (structure with CH_3, $(CH_3)_2CH$, CH_3 substituents)

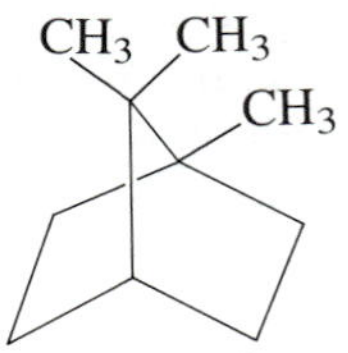

(Eudesmane)

21 The molecule commonly named *para*-cymene is a terpene-related molecule that is readily halogenated by a number of reagents.

(a) Write the major product of the reaction of *para*-cymene with (i) SO_2Cl_2, and (ii) *N*-bromobutanimide, each in the presence of dibenzoyl peroxide as a radical initiator. Refer to Problem 13 for relevant information.

(b) Write a detailed mechanism for the reaction of *para*-cymene with *N*-bromobutanimide (compare Exercise 3-8).

CH_3 / CH / H_3C CH_3

1-Methyl-4-(1-methylethyl)-benzene (*para*-Cymene)

22 Two simple organic molecules that have been recently considered for use as fuel additives are methanol (CH_3OH) and 2-methoxy-2-methylpropane [*tert*-butyl methyl ether, $(CH_3)_3COCH_3$]. The ΔH_f° values for these compounds in the gas phase are -48.1 kcal mole^{-1} for methanol and -70.6 kcal mole^{-1} for 2-methoxy-2-methylpropane.

(a) Write balanced equations for the complete combustion of each of these molecules to CO_2 and H_2O.

(b) Calculate ΔH°_{comb} for each of these molecules.

(c) Using Table 3-4, compare the ΔH°_{comb} values for these compounds with those of alkanes with similar molecular weights.

23 Typical hydrocarbon fuels (e.g., 2,2,4-trimethylpentane, a common component of gasoline) have very similar heats of combustion when calculated in kilocalories *per gram* (compare Problem 10).

(a) Calculate heats of combustion per gram for several representative hydrocarbons in Table 3-4.

(b) Make the same calculation for ethanol (Table 3-4) and for the two compounds discussed in Problem 22.

(c) In evaluating the feasibility of ''gasohol'' (90% gasoline and 10% ethanol) as a motor fuel, it has been estimated that an automobile running on pure ethanol would get approximately 40% fewer miles per gallon than would an identical automobile running on standard gasoline. Is this estimate consistent with the results in parts **a** and **b?** What can you say in general about the fuel capabilities of oxygen-containing molecules relative to hydrocarbons?

24 Figure 3-12 compares the reactions of $Cl^\cdot$ with the primary and secondary hydrogens of propane. Draw a similar diagram comparing the reactions of $Br^\cdot$ with the primary and secondary hydrogens of propane. Use the data given in the margin and answer the following questions.

(a) Calculate ΔH° for both the primary and the secondary hydrogen abstraction reactions.

(b) Which among the transition states of these reactions would you call ''early,'' and which ''late''?

(c) Judging from the locations of the transition states of these reactions along the reaction coordinate, should they show greater or lesser radical character than do the corresponding transition states for chlorination (Figure 3-12)?

(d) Is your answer to part **c** consistent with the selectivity differences between $Cl^\cdot$ reacting with propane and $Br^\cdot$ reacting with propane? Explain.

	ΔH_f° (kcal mole^{-1})
$CH_3CH_2CH_3$	-24.8
$CH_3CH_2CH_2^\cdot$	$+23$
$CH_3\dot{C}HCH_3$	$+18$
HBr	-8.7
$Br^\cdot$	$+26.7$

Approximate E_a for $Br^\cdot$ + primary C–H: 13 kcal mole^{-1}.

Approximate E_a for $Br^\cdot$ + secondary C–H: 10 kcal mole^{-1}.

CHAPTER 4

Cyclic Alkanes

This chapter deals with the names, physical properties, structural features, and conformational isomerism of the cycloalkanes. Members of this class of compounds will be used to review and amplify some of the principles presented in Chapter 2 regarding the linear alkanes.

4-1 Names and Physical Properties of Cycloalkanes

To begin, let us find out how the cycloalkanes differ in their names and simple physical properties from their noncyclic (also called *acyclic*) analogs containing the same number of carbons.

Naming Cycloalkanes

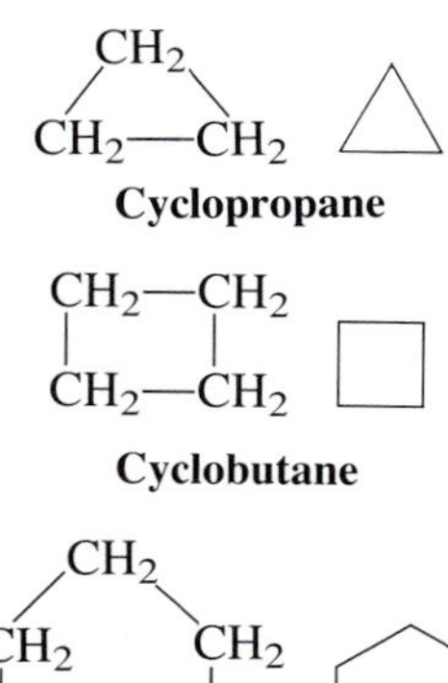

You can construct a molecular model of a cycloalkane by removing two terminal hydrogen atoms from a model of a straight-chain alkane and allowing the resulting two radical centers to form a bond. The empirical formula of a cycloalkane is C_nH_{2n}. The system for naming members of this class of compounds is straightforward: alkane names are preceded by the prefix **cyclo.** Three members in the homologous series, starting with the smallest, cyclopropane, are shown at the left, written both in condensed form and in line notation.

EXERCISE 4-1

Make molecular models of cyclopropane through cyclododecane. Compare the conformational flexibility of the rings within the series and with that of the corresponding straight-chain alkanes.

Naming a substituted cyclic alkane requires numbering of the individual ring carbons only if more than one substituent is attached to the ring. In monosubstituted systems, the carbon of attachment is defined as carbon 1 of the ring. For polysubstituted compounds, take care to provide the lowest-possible numbering sequence. If two such sequences are possible, the alphabetical order of the substituent names takes precedence. Radicals derived from cycloalkanes by abstraction of a hydrogen atom are **cycloalkyl radicals.** Substituted cycloalkanes are therefore sometimes named as cycloalkyl derivatives (large rings take precedence over small rings).

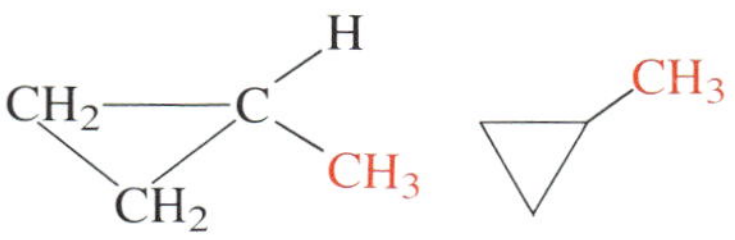

Methylcyclopropane

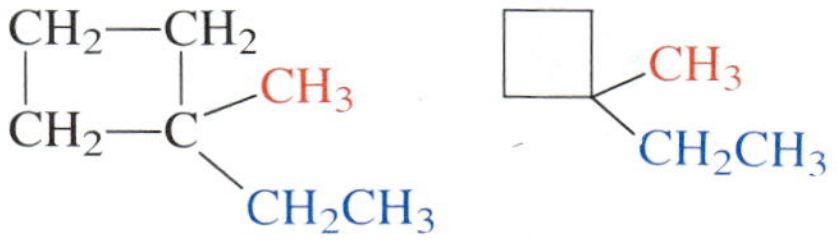

1-Ethyl-1-methylcyclobutane

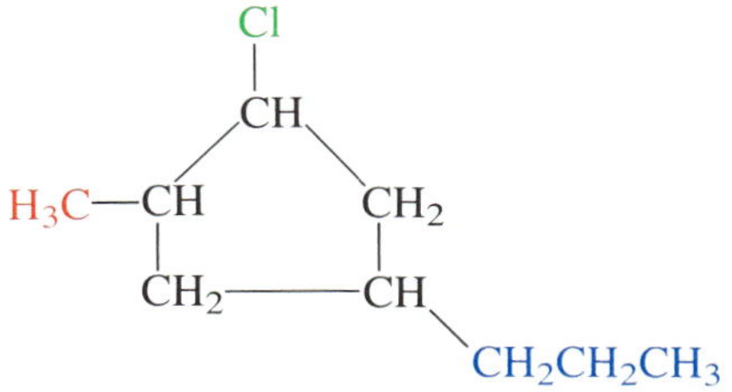

1-Chloro-2-methyl-4-propylcyclopentane
(Not 1-methyl-2-chloro-4-propylcyclopentane)

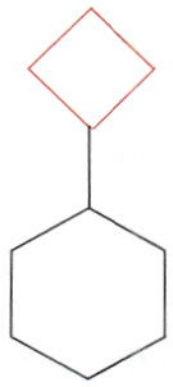

Cyclobutylcyclohexane

Inspection of molecular models of disubstituted cycloalkanes in which the two substituents are located on different carbons shows that *two isomers are possible* in each case. In one, the two substituents are positioned on the *same* face, or side, of the ring; in the other, on *opposite* faces. Substituents on the same face are called **cis** with respect to each other (*cis,* Latin, on the same side): those on opposite sides **trans** (*trans,* Latin, across).

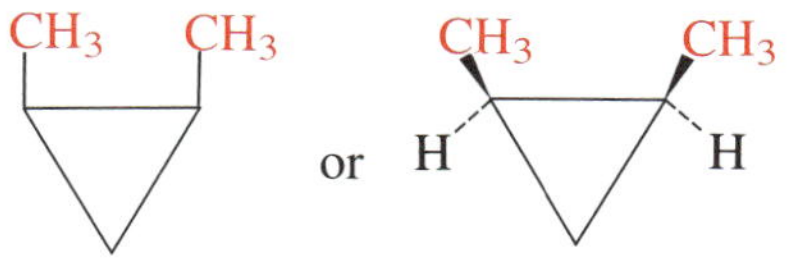

***cis*-1,2-Dimethylcyclopropane**

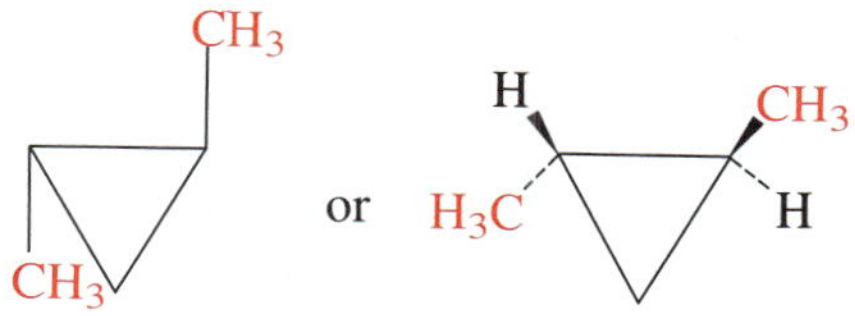

***trans*-1,2-Dimethylcyclopropane**

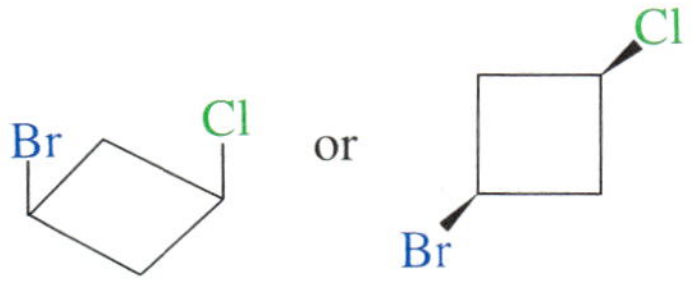

***cis*-1-Bromo-3-chlorocyclobutane**

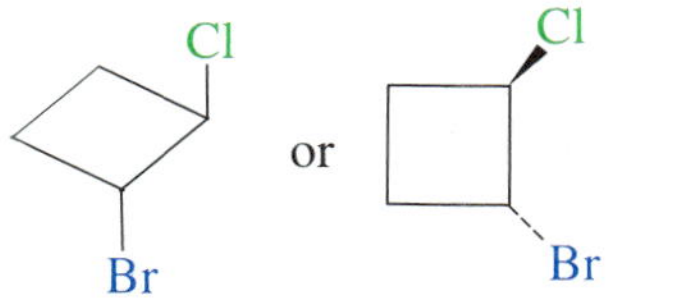

***trans*-1-Bromo-2-chlorocyclobutane**

Cis and trans isomers are **stereoisomers**—compounds that have the same sequence of bonds but differ in the arrangement of their atoms in space. They should be distinguished from structural isomers (Sections 1-9 and 2-2), which are compounds with differing bond sequences. Conformational isomers (Section 2-5), on the other hand, also are stereoisomers, by the above definition. However, unlike cis and trans isomers, which can be interconverted only by *breaking* bonds (try it on your models), conformers are readily equilibrated by *rotation* around bonds. The subject of stereoisomerism will be discussed in more detail in Chapter 5.

Dashed and wedged line structures can be used to depict the three-dimensional arrangement of the substituted cycloalkanes. The positions of any remaining hydrogens are not always shown. Structural and cis-trans isomerism gives rise to a variety of structural possibilities in substituted cycloalkanes. For example, there are eight isomeric bromomethylcyclohexanes (three of which are shown below), all with different and distinct physical and chemical properties.

$BrCH_2$ | H_3C Br | H_3C H Br H

(Bromomethyl)-cyclohexane | **1-Bromo-1-methyl-cyclohexane** | ***cis*-1-Bromo-2-methylcyclohexane**

EXERCISE 4-2

Give the structures and names of the other five isomeric bromomethylcyclohexanes.

Physical Properties of the Cycloalkanes

The physical properties of some of the cycloalkanes are recorded in Table 4-1. Note that, compared with the corresponding straight-chain alkanes (Table 2-5), the cycloalkanes have higher boiling and melting points, as well as higher densities. This is due in large part to increased London interactions of the relatively more rigid and more symmetric cyclic systems. In comparing lower cycloalkanes possessing an odd number of carbons with those having an even number, we find a pronounced alternation in their melting points. This phenomenon has been ascribed to differences in crystal packing forces between the two series.

TABLE 4-1

Physical properties of some cycloalkanes

	Boiling point (°C)	**Melting point (°C)**	**Density at 20°C ($g\ ml^{-1}$)**
Cyclopropane	−32.7	−127.6	
Cyclobutane	−12.5	−50.0	0.720
Cyclopentane	49.3	−93.9	0.7457
Cyclohexane	80.7	6.6	0.7786
Cycloheptane	118.5	−12.0	0.8098
Cyclooctane	148.5	14.3	0.8349
Cyclododecane	160 (100 torr)	64	0.861
Cyclopentadecane	110 (0.1 torr[a])	66	0.860

[a] Sublimation point.

In summary, names of the cycloalkanes are derived in a straightforward manner from those of the straight-chain alkanes. In addition, the position of a single substituent is defined as being at C-1. Disubstituted cycloalkanes can give rise to cis and trans isomers, depending on the location of the substituents. The simple physical properties parallel those of the straight-chain alkanes, except that the individual values for boiling and melting points, and for densities, are higher for the cyclic compounds of equal carbon number.

4-2 Ring Strain and the Structure of Cycloalkanes

Let us now compare the heat contents of the cycloalkanes with those of their open-chain analogs.

Before studying this section, recall Exercise 4-1. In making the molecular models, what obvious differences did you perceive between cyclopropane, cyclobutane, cyclopentane, and so forth, and the corresponding straight-chain alkanes? One notable feature of the first two members in the series is how difficult it is to close the ring without breaking the plastic tubes used to represent bonds. The reason for this problem lies in the tetrahedral carbon model. The C–C–C bond angles in, for example, cyclopropane and cyclobutane differ considerably from the tetrahedral value. As the ring size increases, this difference diminishes. Thus, cyclohexane can be assembled without distortion or strain. A second feature has to do with the eclipsing of the hydrogens: a large degree of eclipsing occurs in the lower cycloalkanes. As the ring size increases, however, the hydrogens can adopt staggered configurations.

Do these observations tell us anything about the relative stability of the cycloalkanes—for example, as measured by their heats of formation, ΔH_f°? How do the lower cycloalkanes accommodate the distortions at tetrahedral carbon that are necessary for the structures to exist? Are there conformational effects? This section and the following one address these questions.

The Heats of Formation of the Cycloalkanes Reveal the Presence of Ring Strain

In Section 3-4, methods were described by which the relative heat content of alkanes could be measured. One such method is to estimate the heats of formation, ΔH_f°, from the elements in their standard states, as shown in Table 3-5. Inspection of this table reveals that the (negative) ΔH_f° values for the homologous series of the straight-chain alkanes increase by approximately the same amount with each successive member—

The (Negative) ΔH_f° Values for the Homologous Alkanes Increase by Regular Increments

	ΔH_f°	Increment
$CH_3CH_2CH_3$	−24.8	
		−5.6
$CH_3CH_2CH_2CH_3$	−30.4	
		−4.7
$CH_3(CH_2)_3CH_3$	−35.1	
		−4.8
$CH_3(CH_2)_4CH_3$	−39.9	

kcal mole^{-1}

that is, by an increment of about 5 kcal mole^{-1} for every additional CH_2 group. When averaged over a larger number of alkanes, this value refines to 5.15 kcal mole^{-1}. Because cycloalkanes have the empirical formula $(CH_2)_n$, it should be possible to predict their approximate ΔH_f°: it should simply be

TABLE 4-2

Calculated and experimental heats of formation (kcal mole^{-1}) of some cycloalkanes

Ring size (C_n)	ΔH_f° (calculated)	ΔH_f° (experimental)	Total strain	Strain per CH_2 group
3	−15 .5	+12.7	28	9.4
4	−20.6	+6.8	27	6.9
5	−25.8	−18.4	7	1.5
6	−30.9	−29.5	1	0.2
7	−36.1	−28.2	8	1.1
8	−41.2	−29.7	12	1.4
9	−46.4	−31.7	15	1.6
10	−51.5	−36.9	15	1.5
11	−56.7	−42.9	14	1.3
12	−61.8	−55.0	7	0.6
17	−87.6	−87.1	1	0.3

Note: The calculated numbers are based on the value of −5.15 kcal mole^{-1} for a CH_2 group.

$n \times 5.15$ kcal mole^{-1} (Table 4-2, column 2). However, when the actual heats of formation are measured, a different set of values is obtained (Table 4-2, column 2). For example, cyclopropane should have a ΔH_f° of about −15.5 kcal mole^{-1}, but the experimental value is +12.73 kcal mole^{-1}. The difference between expected and observed values is 28 kcal mole^{-1} and is attributed to a property of cyclopropane of which you are already aware because of having built a model: **ring strain.** The strain per CH_2 group in this molecule is 9.4 kcal mole^{-1}. A similar calculation for cyclobutane (Table 4-2) reveals a ring strain of 27 kcal mole^{-1}, 6.9 kcal mole^{-1} per CH_2 group. In cyclopentane, this effect is significantly weakened, total strain amounting to only 7 kcal mole^{-1}, and cyclohexane is almost strain free. However, succeeding homologs, again, have considerable strain. Only the very large rings have strain-free structures. Because of this trend, organic chemists have loosely defined four groups of cycloalkanes:

1. *Small rings* (cyclopropane, cyclobutane).
2. *Common rings* (cyclopentane, cyclohexane, cycloheptane).
3. *Medium rings* (from eight- to twelve-membered).
4. *Large rings* (thirteen-membered and larger).

What factors contribute to strain in a ring? We shall see that they include: (1) **bond-angle strain,** the energy needed to distort tetrahedral carbon so as to close the cycle; (2) **eclipsing strain,** introduced by eclipsed hydrogen atoms; (3) ***gauche* interactions,** as in butane (Section 2-5); and (4) **transannular strain,** caused by the steric crowding of hydrogen atoms across the ring (*trans,* Latin, across; *annulus,* Latin, ring).

The Structures and Conformations of Cyclopropane, Cyclobutane, and Cyclopentane

The structure of the smallest cycloalkane, *cyclopropane,* is represented in Figure 4-1. Being triangular, this structure is by necessity flat and has C–C–C bond angles of 60°, a significant deviation from the "natural" tetrahedral bond angle of 109.5°. In addition, all methylene hydrogens in this molecule are eclipsed. As indicated in Table 4-2, the result is a heat of formation for this molecule that is considerably more positive than that of other cycloalkanes. In other words, cyclopropane is much less stable than expected for three methylene groups.

How is it possible for three supposedly tetrahedral carbon atoms to maintain a bonding relation at such highly distorted angles? The problem is perhaps best illustrated in Figure 4-2, in which the bonding in the strain-free "open cyclopropane," the trimethylene diradical $\cdot CH_2CH_2CH_2\cdot$, is compared with that in the closed form. You can see that the two ends of the trimethylene diradical cannot "reach" far enough to close the ring without "bending" the two C–C bonds already present. However, if all three C–C bonds in cyclopropane adopt a bent configuration (orbital angle 104°, see Figure 4-2B), overlap is sufficient for bond formation.

FIGURE 4-1
Cyclopropane:
(A) molecular model;
(B) bond lengths and angles.

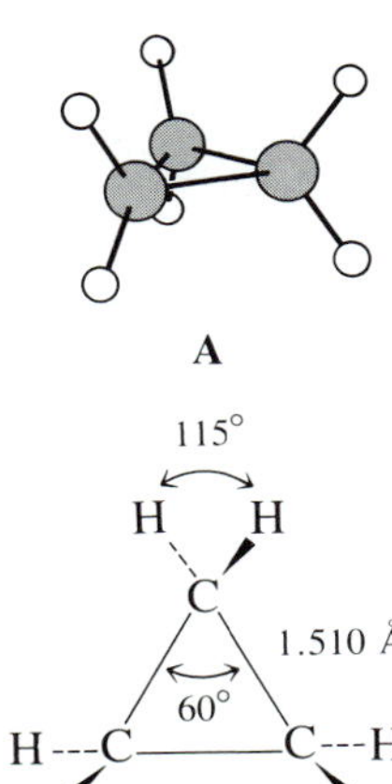

FIGURE 4-2
Molecular-orbital picture of (A) the trimethylene diradical and (B) the bent bonds in cyclopropane. Only the hybrid orbitals forming C–C bonds are shown. Note the interorbital angle of 104° in cyclopropane.

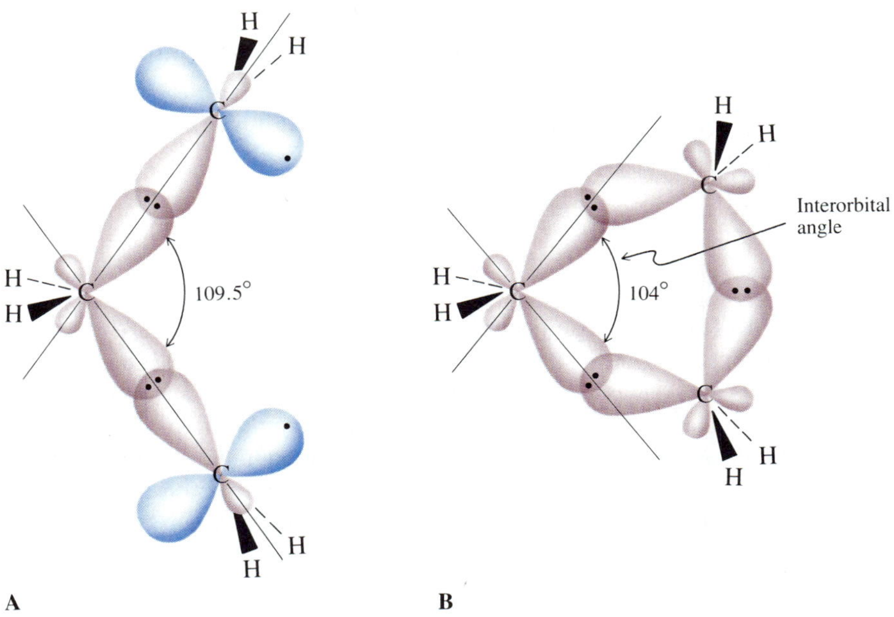

A consequence of this bonding scheme is that cyclopropane has relatively weak C–C bonds. The energy for opening the ring is 65 kcal mole^{-1}. This value is low (recall that the C–C bond strength in ethane is 90 kcal mole^{-1}) because of the release of ring strain. As a consequence, cyclopropane can undergo several unusual reactions. For example, on heating, the molecule rearranges to propene, and reaction with hydrogen in the presence of a palladium catalyst gives propane:

$$\text{cyclopropane} \xrightarrow{500°\text{C}} CH_2{=}CHCH_3 \quad \Delta H° = -7.9 \text{ kcal mole}^{-1}$$

Propene

$$\text{cyclopropane} + H_2 \xrightarrow{\text{Pd catalyst}} CH_3CH_2CH_3 \quad \Delta H° = -37.6 \text{ kcal mole}^{-1}$$

Propane

The structure of *cyclobutane* (Figure 4-3) reveals that this molecule is not planar but puckered, with an approximate angle of bend of 26°. The nonplanar structure of the ring, however, is not very rigid. The molecule ''flips'' rapidly from one puckered conformation to the other.

FIGURE 4-3

Cyclobutane: (A) molecular model; (B) bond lengths and angles.

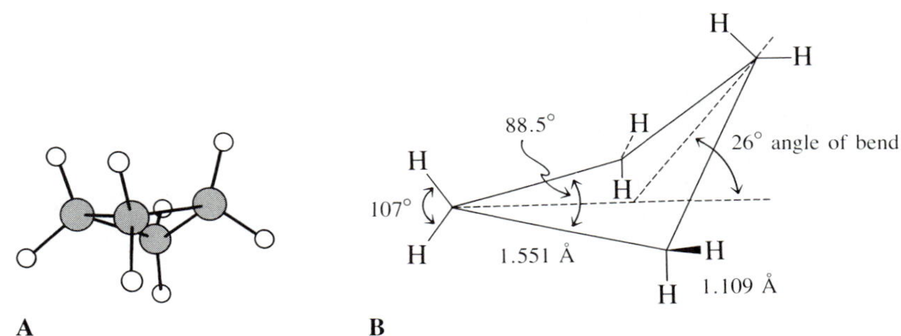

Construction of a molecular model shows why distorting the four-membered ring from planarity is favorable: it partly relieves the strain introduced by the eight eclipsing hydrogens. Moreover, bond-angle strain is considerably reduced compared with that in cyclopropane, although maximum overlap is, again, only possible by the use of bent bonds. The C–C bond strength in cyclobutane also is low (about 63 kcal mole^{-1}) because of the release of ring strain on ring opening and the consequences of relatively poor overlap in bent bonds. Cyclobutane is less reactive than cyclopropane, but undergoes similar ring-opening processes.

$$\text{cyclobutane} \xrightarrow{500°\text{C}} 2\ CH_2{=}CH_2$$

Ethene

$$\text{cyclobutane} + H_2 \xrightarrow{\text{Pd catalyst}} CH_3CH_2CH_2CH_3$$

Butane

Cyclopentane might be expected to be planar because the angles in a regular pentagon are 108°, close to tetrahedral. However, such a planar arrangement would have *ten* unfavorable eclipsing H–H interactions. The puckering of the ring prevents this from happening, as indicated in the structure of the molecule (Figure 4-4). Although puckering reduces eclipsing, it also increases bond-angle strain. The conformation of lowest energy is a compromise in which the energy of the system is minimized. There are two puckered conformations possible for cyclopentane: the "envelope" and the "half chair."

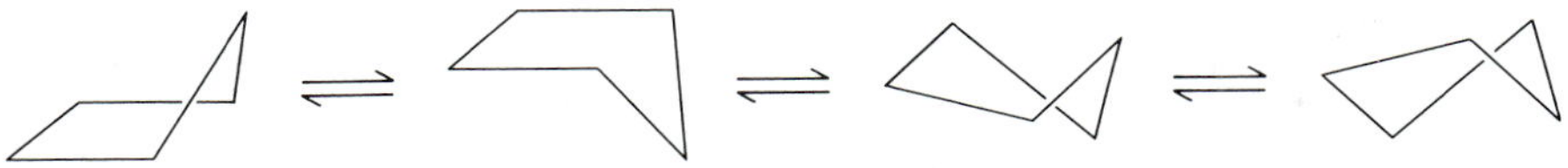

Envelope

Half chair

There is little difference in energy between them, and the activation barriers for rapid interconversion are low. Overall, cyclopentane has relatively little ring strain and hence is neither readily decomposed on heating nor hydrogenated to pentane.

FIGURE 4-4

Cyclopentane: (A) molecular model; (B) bond lengths and angles.

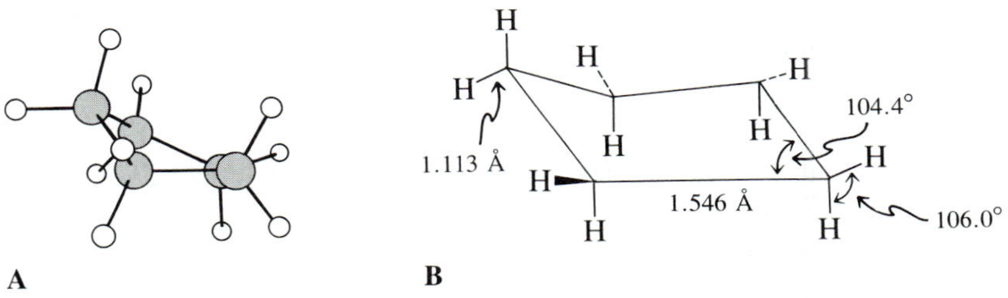

A

B

1.113 Å

1.546 Å

104.4°

106.0°

Table 4-2 reveals that cyclohexane is unusual in that it is almost strain free. Why?

4-3 Cyclohexane, a Strain-Free Cycloalkane?

The cyclohexane ring is one of the most abundant and important structural units in organic chemistry. Its substituted derivatives exist in many natural products (see Section 4-7) and an understanding of its conformational mobility is an important aspect of organic chemistry.

A hypothetical planar cyclohexane would suffer from twelve eclipsing H–H interactions and sixfold bond-angle strain. The strain would arise because a regular hexagon requires 120° bond angles. However, one conformation of cyclohexane, obtained by moving carbons 1 and 4 out of planarity in opposite directions, is in fact almost strain free (Figure 4-5). This structure is called the **chair conformation** of cyclohexane (because it resembles a chair), in which eclipsing is completely prevented and the bond angles are very nearly tetrahedral. As seen in Table 4-2, the calculated ΔH_f° of cyclohexane (−30.9 kcal mole^{-1}) based on a strain-free hexamethylene model is very close to the experimentally determined value (−29.5 kcal mole^{-1}).

FIGURE 4-5

Conversion of the (A) hypothetical planar cyclohexane into the chair conformation: (B) bond lengths and angles; (C) molecular model.

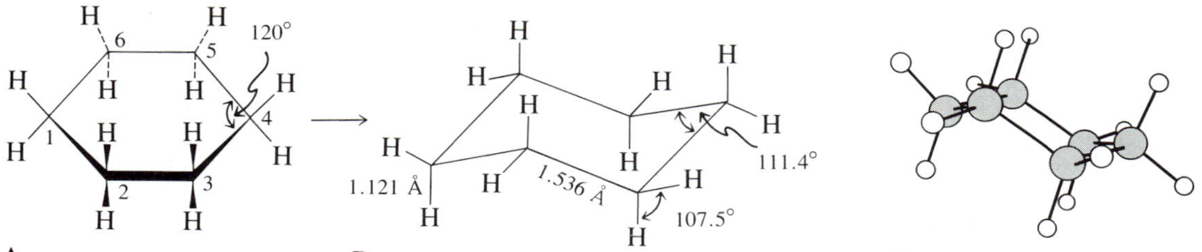

A **Planar cyclohexane** (120° bond angles; 12 eclipsing hydrogens)

B **Chair cyclohexane** (Nearly tetrahedral bond angles; no eclipsing hydrogens)

C

FIGURE 4-6

View along one of the C–C bonds in the chair conformation of cyclohexane. Note the staggered arrangement of all substituents and the *gauche* methylene groups.

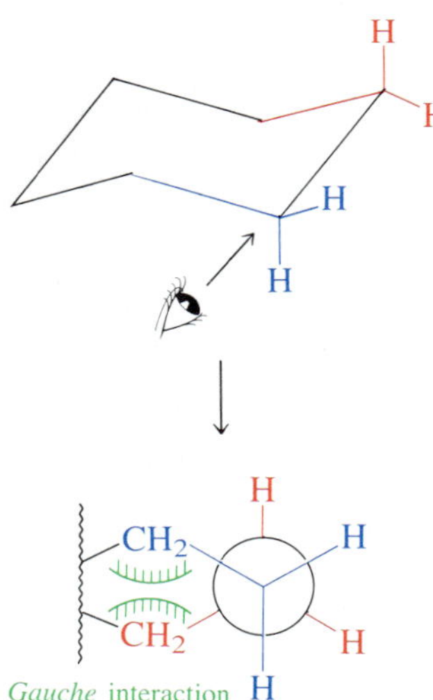

To follow the discussion, make a molecular model of cyclohexane, and use it to observe the conformational behavior of the molecule. If you view it along (any) one C–C bond, you can see the staggered arrangement of all substituent groups along it. You can visualize this arrangement by drawing a Newman projection of that view (Figure 4-6). Because of its lack of strain, cyclohexane is as inert as a straight-chain alkane.

EXERCISE 4-3

Figure 2-12 gives the energy difference between *gauche* and eclipsed butane. Assuming that the eclipsing of the C–C bonds in cyclohexane requires an equal amount of energy (and therefore neglecting bond-angle strain), calculate the energy difference between the planar and chair conformations of cyclohexane.

Table 4-2 shows that cyclohexane is not completely strain free; there is about 1 kcal $mole^{-1}$ residual strain. Where does it come from? The answer is most easily found by again looking at the Newman projection of the chair conformation of cyclohexane in Figure 4-6: all methylene groups function as *gauche* substituents to the adjacent C–C bond. The *anti* conformer is not attainable in a six-membered ring.

Cyclohexane Has More Than One Conformation

There are other, less-stable but nevertheless readily accessible, conformations of cyclohexane. One is the **boat form,** in which carbons 1 and 4 are out of the plane in the *same* direction (Figure 4-7). The boat is less stable than the chair form by 6.5 kcal $mole^{-1}$. The reasons for this difference are the eclipsing of eight hydrogen atoms at the base of the boat and the unfavorable steric hindrance due to the close proximity of the two inside hydrogens in the boat framework. The distance between these two hydrogens is only 1.83 Å, small enough to create an energy of repulsion of about 3 kcal $mole^{-1}$. This effect is an example of transannular strain.

Boat cyclohexane is fairly flexible. If one of the bonds is twisted relative to the second, the boat form can be somewhat stabilized by the partial removal of the inside hydrogen interactions. The new conformation obtained is called the

FIGURE 4-7

Conversion of the hypothetical planar cyclohexane into the boat form. Note the non-bonded interactions of the inside hydrogens at carbons 1 and 4 and the eclipsing of hydrogens at carbons 2, 3, 5, and 6.

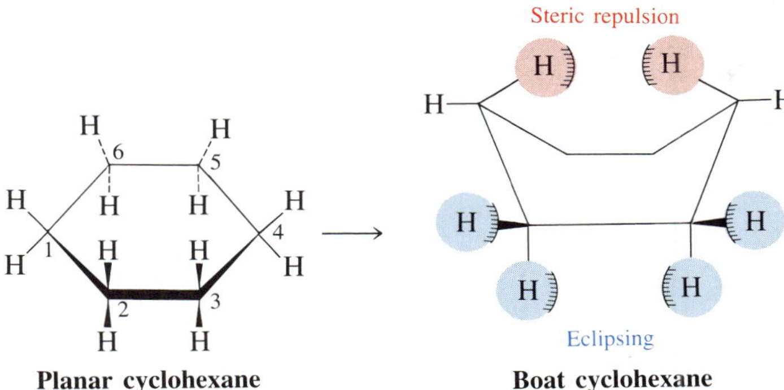

FIGURE 4-8

The conversion of a boat into a twist-boat cyclohexane.

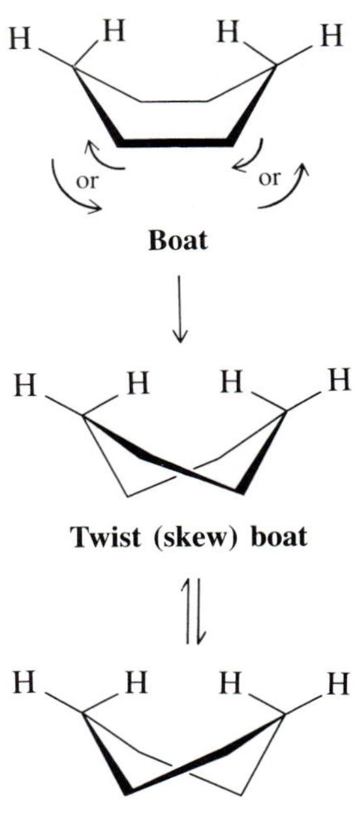

twist- (or **skew-**) **boat conformation** of cyclohexane (Figure 4-8). The stabilization relative to the boat form amounts to about 1.5 kcal mole^{-1}. As shown, there are two possible twist-boat forms. They interconvert rapidly through the boat conformer as a *transition state* (verify this with your model). Thus, the boat cyclohexane is not a normally isolable species, the twist boat is present in very small amounts, and the chair form is the major conformer (Figure 4-9). As we shall see, there are also two rapidly interconverting chair conformers of cyclohexane.

FIGURE 4-9

Potential-energy diagram for the interconversion of the various conformers of cyclohexane.

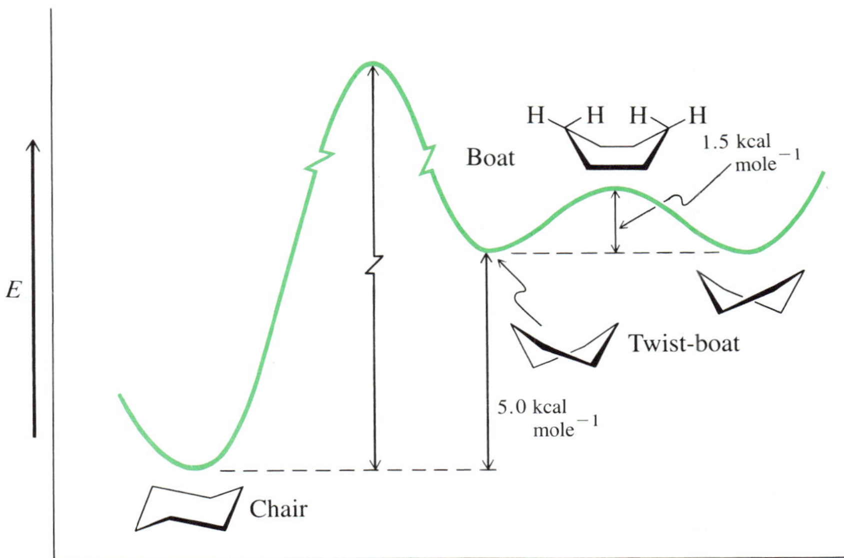

Cyclohexane Has Axial and Equatorial Hydrogen Atoms

By looking at the chair conformation of cyclohexane, you can see that the molecule has two types of hydrogens. Six carbon–hydrogen bonds are parallel to the molecular axis (Figure 4-10) and hence are **axial;** the other six are perpendicular to that axis and are therefore **equatorial.***

FIGURE 4-10

The axial and equatorial positions in the chair form of cyclohexane.

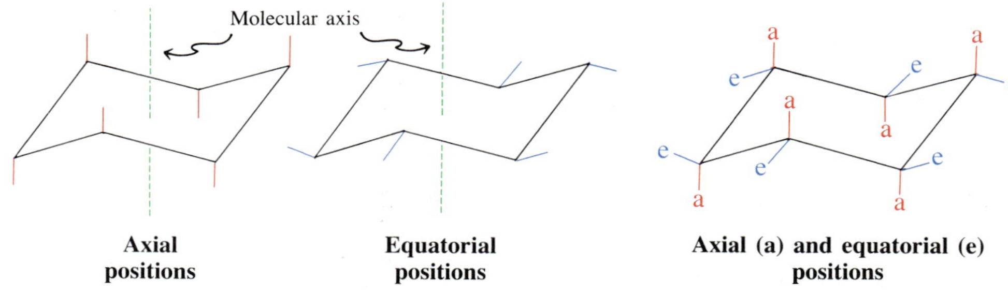

BOX 4-1

Rules for Drawing Cyclohexane Chair Conformations

To be able to draw cyclohexane chair conformations correctly is helpful for learning the chemistry of six-membered rings. Several rules might be useful:

1 Draw the chair so as to place the C-2 and C-3 atoms at the right of C-5 and C-6, with apex 1 pointing downward on the left and apex 4 pointing upward on the right.

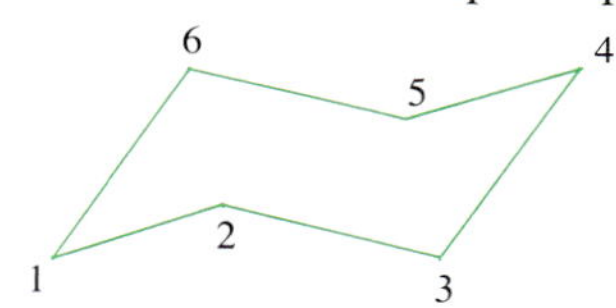

2 Add all the axial bonds as vertical lines, pointing downward at C-1, C-3, and C-5 and upward at C-2, C-4, and C-6.

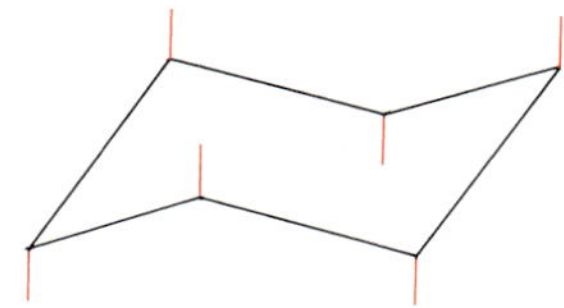

3 Draw the two equatorial bonds at C-1 and C-4 at a slight angle from horizontal, pointing upward at C-1 and downward at C-4, parallel to the bond between C-2 and C-3 (or C-5 and C-6).

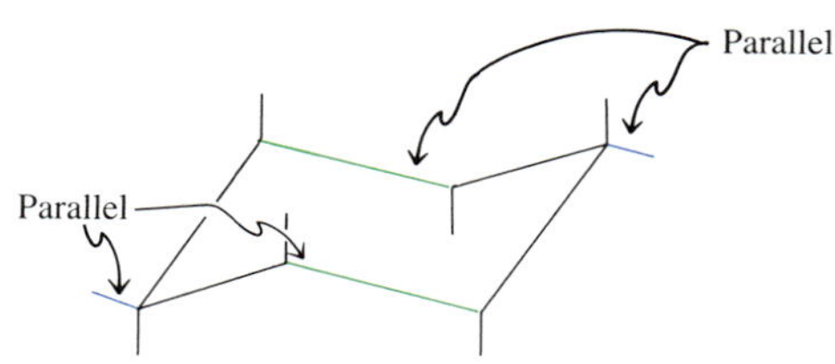

*An equatorial plane is defined as being perpendicular to the axis of rotation of a rotating body and equidistant from its poles, such as the equator of the planet Earth. The equatorial hydrogens in cyclohexane are in its equatorial plane.

4 This rule is the most difficult to follow: add the remaining equatorial bonds at C-2, C-3, C-5, and C-6 by aligning them *parallel* to the C–C bond "once removed," as shown.

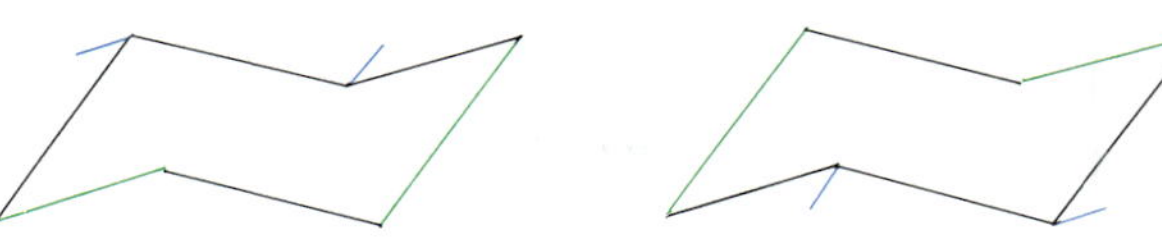

Conformational Flipping Interconverts Axial and Equatorial Hydrogens

Cyclohexane is not conformationally rigid. It is capable of interconverting one chair conformation with another, thus equilibrating axial with equatorial protons. In this process ("flipping"), all axial protons in one chair become equatorial in the other and vice versa (Figure 4-11). The activation energy for this process is 10.8 kcal $mole^{-1}$. As suggested in Section 2-5, this value is so low that at room temperature there is very rapid interconversion of the two equivalent chair forms (approximately 100,000 times per second). Only when solutions of the molecule are cooled to very low temperatures (about $-100°C$) is this equilibration stopped.

FIGURE 4-11

Chair–chair interconversion in cyclohexane.

$E_a = 10.8$ kcal $mole^{-1}$

To summarize Sections 4-2 and 4-3, the discrepancy between calculated and measured heats of formation in the cycloalkanes can be attributed to bond-angle, eclipsing, *gauche,* and transannular strain. Because of strain, the small cycloalkanes are chemically reactive, undergoing ring-opening reactions. Cyclohexane is almost strain free, except for *gauche* interactions. It has a lowest-energy chair, as well as additional higher-energy conformations, particularly the boat and twist-boat structures. Chair–chair interconversion is rapid at room temperature; it is a process in which equatorial and axial hydrogen atoms interchange their positions.

4-4 Substituted Cyclohexanes

We can now apply our knowledge of conformational analysis to substituted cyclohexanes. Let us look at the simplest alkylcyclohexane, methylcyclohexane.

The Difference Between Axial and Equatorial Methylcyclohexane

In methylcyclohexane, the methyl group occupies either an equatorial or an axial position:

$\Delta G^\circ = +1.7$ kcal mole^{-1}

Ratio 95:5

1,3-Diaxial interactions

Are the two forms equivalent? Clearly not. In the equatorial conformer, the methyl group extends into space away from the remainder of the molecule, whereas, in the axial conformer, the methyl substituent is *gauche* to two ring C–C bonds and close to the other two axial hydrogens on the same side of the molecule. The distance to these hydrogens is small enough (about 2.7 Å) to result in steric repulsion, another example of transannular strain. Because this effect is due to axial substituents on carbon atoms that have a 1,3-relation (in the drawing, 1,3 and 1′,3′), it is called a **1,3-diaxial interaction.**

The two forms of chair methylcyclohexane are in equilibrium, with the equatorial conformer favored by a ratio of 95:5. Using the expression $\Delta G^\circ = -1.36 \log K$ (at 25°C, Section 2-6), the difference in energy between the two species is 1.7 kcal mole^{-1}. The activation energy for chair–chair interconversion is similar to that in cyclohexane itself (about 11 kcal mole^{-1}), and equilibrium between the two conformers is established rapidly at room temperature.

The unfavorable *gauche* and 1,3-diaxial interactions to which an axial substituent is exposed are readily seen in Newman projections of the ring C–C bond bearing that substituent. In contrast with that in the axial form, the substituent in the equatorial conformer is *anti* to C-3 and C-5 (Figure 4-12) and away from the axial hydrogens.

FIGURE 4-12

Newman projections of a substituted cyclohexane. The equatorial substituent Y adopts an *anti* conformation, whereas the axial Y is *gauche*.

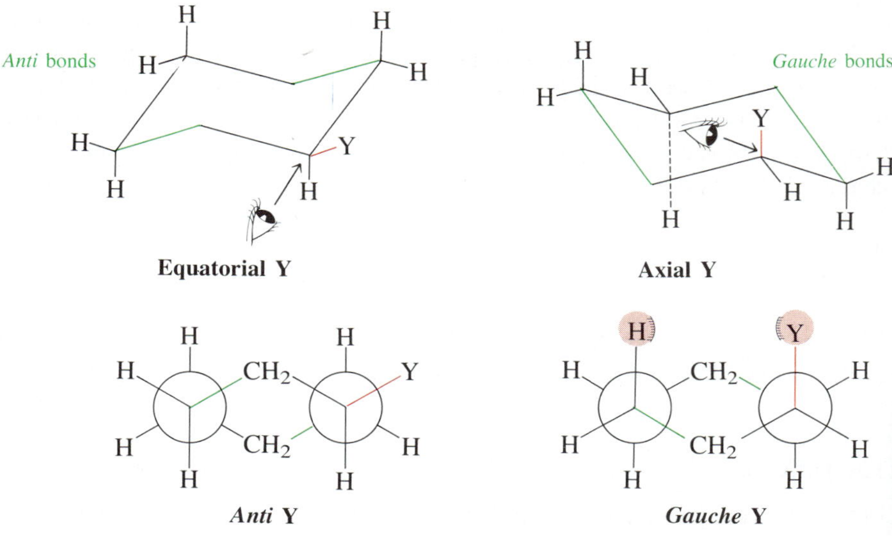

The energy difference, ΔG°, between the axial and the equatorial isomers of many monosubstituted cyclohexanes has been measured; some values are shown in Table 4-3. In many cases (but not all), particularly for alkyl substituents, the energy difference between the two forms increases with the size of the substituent, a direct consequence of increasing unfavorable *gauche* and 1,3-diaxial interactions. This is particularly pronounced in (1,1-dimethylethyl)-cyclohexane (*tert*-butylcyclohexane). The energy difference here is so large (about 5 kcal mole^{-1}) that very little (about 0.01%) of the axial isomer is present in solution. The substituent is said to ''lock'' the conformation.

TABLE 4-3

Free-energy differences between axial and equatorial cyclohexane conformers (in all examples, the equatorial form is more stable)

Substituent	ΔG° (kcal mole^{-1})	Substituent	ΔG° (kcal mole^{-1})
H	0	F	0.25
CH_3	1.70	Cl	0.52
CH_3CH_2	1.75	Br	0.55
$(CH_3)_2CH$	2.20	I	0.46
$(CH_3)_3C$	≈5	HO	0.94
$HO{-}C({=}O)$	1.41	CH_3O	0.75
$CH_3O{-}C({=}O)$	1.29	H_2N	1.4

Substituent Effects Can Be Additive

In disubstituted cyclohexanes, the conformation with the maximum number of equatorial substituents is generally the most abundant. Let us look at a few simple examples.

There are several isomers of dimethylcyclohexane. In 1,1-dimethylcyclohexane, one methyl group is always equatorial and the other axial. The two chair forms are identical, and hence the ΔG° for their interconversion is zero.

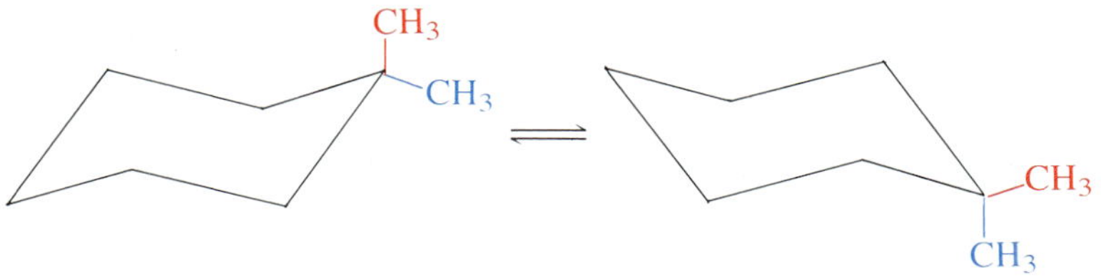

CH_3 CH_3 ⇌ CH_3 CH_3

$\Delta G^\circ = 0$ kcal mole^{-1}

1,1-Dimethylcyclohexane

For 1,2-, 1,3-, and 1,4-dimethylcyclohexane, there are cis and trans isomers of different conformations. For example, in *cis*-1,4-dimethylcyclohexane, both chairs have one axial and one equatorial substituent and are of equal energy.

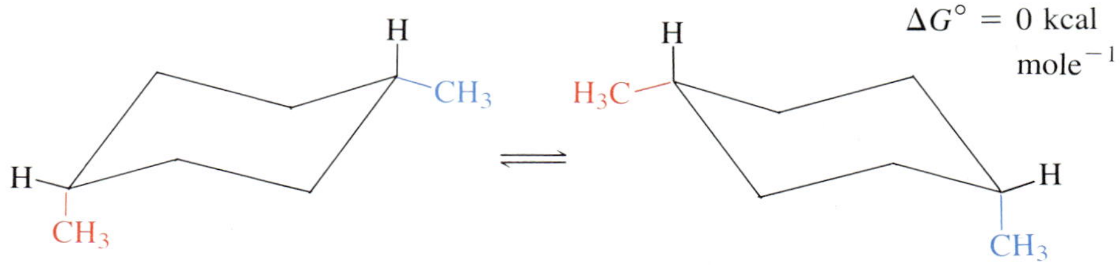

$\Delta G^\circ = 0$ kcal mole^{-1}

On the other hand, the trans isomer can exist in two different chair conformations: one with two axial methyl groups (diaxial) and the other with two equatorial ones (diequatorial).

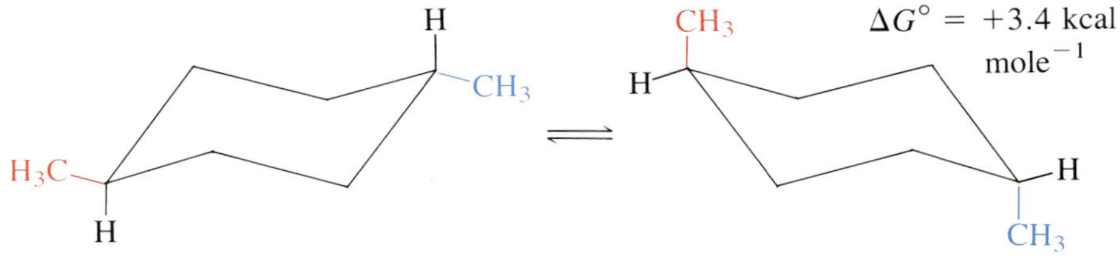

$\Delta G^\circ = +3.4$ kcal mole^{-1}

Diequatorial **Diaxial**

***trans*-1,4-Dimethylcyclohexane**

In the diaxial arrangement, both methyl groups are subject to an approximately equal degree of *gauche* and 1,3-diaxial strain. This form is therefore destabilized about twice as much ($\Delta G^\circ = +3.4$ kcal mole^{-1}) as the axial monomethylcyclohexane, relative to the stable conformer. This finding indicates that the substituent values in Table 4-3 are approximately additive and that they can be used to calculate ΔG° values for a number of chair–chair equilibria of substituted cyclohexanes.

EXERCISE 4-4

Calculate ΔG° for the equilibrium between the two chair conformers of: (a) 1-ethyl-1-methylcyclohexane; (b) *cis*-1-ethyl-4-methyl-cyclohexane; (c) *trans*-1-ethyl-4-methylcyclohexane.

EXERCISE 4-5

What is the most stable of the four *boat* conformations of methylcyclohexane, and why?

In *trans*-1,2-dimethylcyclohexane, the diequatorial conformer is more stable than the diaxial one by about 2 kcal mole^{-1} (Figure 4-13). We might have expected this difference to be a little larger, considering that in monomethylcyclohexane the equatorial isomer is 1.7 kcal mole^{-1} more stable than the axial isomer. With two such methyl groups, the difference should be approximately double. The primary reason for the failure of this argument is that there is an additional methyl-methyl *gauche* interaction in the diequatorial conformer (Figure 4-13), which raises its ground state in relation to its diaxial counterpart. Thus, the diaxial–diequatorial energy difference is smaller than expected.

In cis-1,2-disubstituted cyclohexanes, one substituent is always axial, the other equatorial. The predominance of one conformer over the other, then, depends on how likely each substituent is to occupy an equatorial or an axial position. For example, the size of the 1,1-dimethylethyl (*tert*-butyl) group in

FIGURE 4-13

Conformations of *trans*-1,2-dimethylcyclohexane. The Newman projections reveal a *gauche* interaction between the two methyl groups in the diequatorial isomer, whereas the diaxial isomer has the expected *gauche* and 1,3-diaxial interactions (a total of four each; not all are indicated).

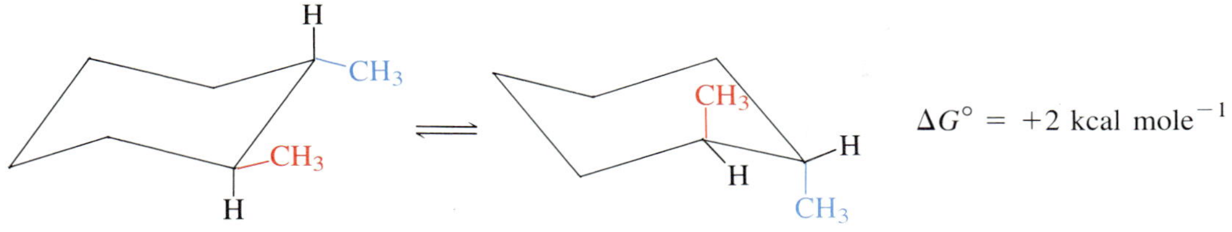

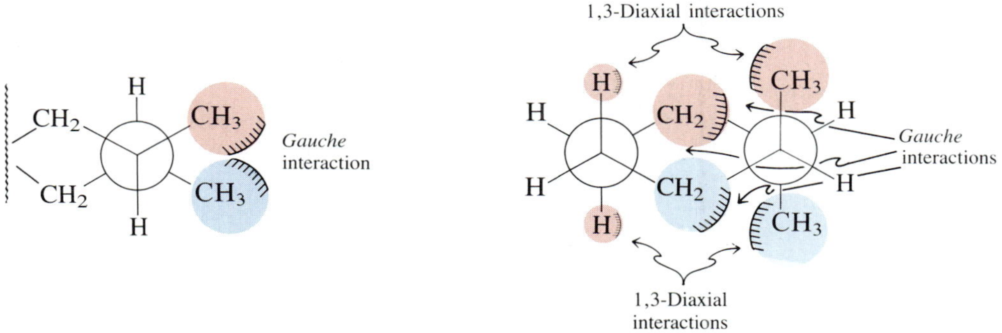

H
CH_3
CH_3
H
CH_3
H
H
CH_3

$\Delta G^\circ = +2$ kcal mole^{-1}

Diequatorial **Diaxial**

***trans*-1,2-Dimethylcyclohexane**

CH_2 H CH_3 CH_2 CH_3 H

Gauche interaction

1,3-Diaxial interactions

H H H H CH_2 CH_2 CH_3 CH_3 H H

Gauche interactions

1,3-Diaxial interactions

cis-1-(1,1-dimethylethyl)-2-methylcyclohexane forces the methyl group into the axial position to allow the larger group to be equatorial:

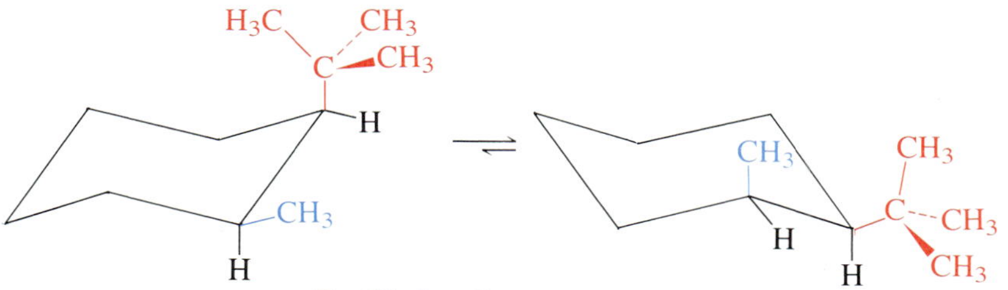

H_3C CH_3 C CH_3 H CH_3 H

CH_3 H CH_3 C CH_3 CH_3 H

Equilibrium lies to the right

Working through the various permutations of 1,2- 1,3-, and 1,4- substitution in either cis or trans configurations, you will identify two general features: (1) in cis-1,2-, trans-1,3-, and cis-1,4-disubstituted cyclohexanes, one substituent is always in the axial position, and the other is in the equatorial position; and (2) trans-1,2-, cis-1,3-, and trans-1,4-disubstituted cyclohexanes exist as equilibrating pairs of diaxial and diequatorial conformers.

EXERCISE 4-6

Draw all the possible all-chair conformers of cyclohexylcyclohexane.

In summary, the conformational analysis of cyclohexane allows us to predict the relative stability of its various conformers and even to approximate the energy differences between two chair configurations. Bulky substituents, particularly a 1,1-dimethylethyl group, tend to shift the chair–chair equilibrium toward the side in which the large substituent is equatorial.

Do similar relations hold for the larger cycloalkanes?

4-5 Larger Cycloalkanes

Table 4-2 shows that rings larger than cyclohexane also have more strain. This strain is due to a combination of bond-angle distortion, partial eclipsing of the hydrogens, *gauche* interactions, and nonbonded repulsions between hydrogen atoms on opposite sides of the ring. The individual molecules adopt a compromise solution for their most-stable conformation, which frequently results in several energetically very close lying geometries. Make molecular models to help in comprehending the following discussion.

Cycloheptane can be thought of as a cyclohexane in which one apex has been extended by one carbon atom (Figure 4-14). However, this is not the most-stable conformation because of excessive eclipsing of the hydrogens at positions C-4 and C-5. A slight twist along this C–C bond furnishes a more-favorable arrangement: "twist cycloheptane."

FIGURE 4-14
Two conformations of cycloheptane. The upper one is derived from chair cyclohexane by inserting carbon 5. Twisting removes some of the eclipsing strain along C-4 and C-5.

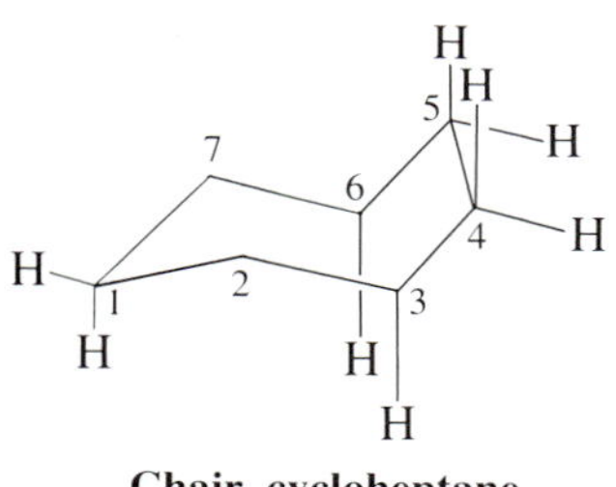

Chair cycloheptane

Twist

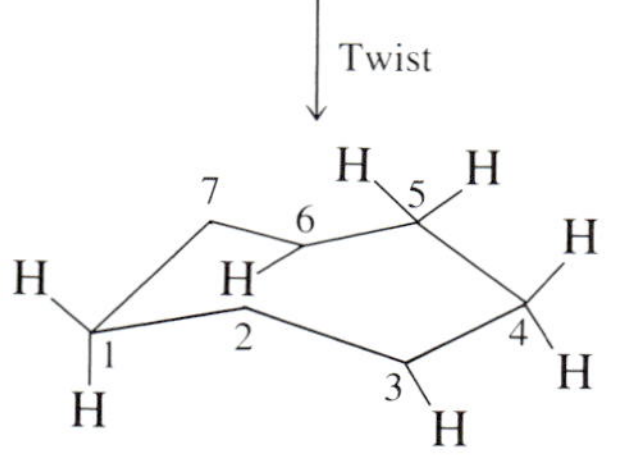

Twist cycloheptane

Cyclooctane exists mostly in a boat-chair conformation rather than the crown arrangement (named after its crown-shaped structure):

Boat part
Chair part

Boat-chair **Crown**

Cyclooctane

In *cyclodecane,* the ring is large enough to accommodate several *anti* conformations. However, the molecule is not strain free, partly because of nonbonded H–H repulsions across the carbon ring.

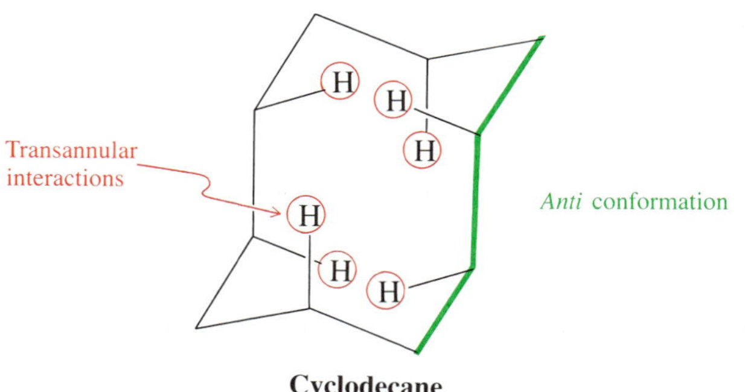

Cyclodecane

Essentially strain-free conformations are attainable only for large-ring cycloalkanes, such as cycloheptadecane (Table 4-2). In such rings, the carbon chain adopts a structure very similar to that of the straight-chain alkanes (Section 2-4), having staggered hydrogens and an all-*anti* configuration. However, even in these systems, the attachment of substituents usually introduces varying amounts of strain. Most cyclic molecules described in this book are not strain free.

The cycloalkanes discussed so far contain only one ring and therefore may be referred to as monocyclic alkanes. More complex structures in which two or more rings share carbon atoms—the bi-, tri-, tetra-, and higher polycyclic hydrocarbons—are the topic of the next section. We shall see the structural variety possible in these compounds, many of which, when bearing alkyl and functional groups, occur in nature.

4-6 Polycyclic Alkanes

Molecular models of polycyclic alkanes can be readily constructed by linking the carbon atoms of two alkyl substituents in a monocyclic alkane. For example, if two hydrogen atoms are removed from the methyl groups in 1,2-diethylcyclohexane, allowing a new C–C bond to be formed, the result is a new molecule with the common name decalin. In decalin, two cyclohexanes share two adjacent carbon atoms, and the two rings are said to be **fused,** or **annelated.**

CH_3 CH_3 $\xrightarrow{-2\,H}$

Decalin

If we treat a molecular model of *cis*-1,3-dimethylcyclopentane in the same way, we obtain another carbon skeleton, that of norbornane.

CH_3 CH_3 $\xrightarrow{-2\,H}$ same as

Norbornane

Compounds constructed in this way are *bicyclic* ring systems. They are characterized by two carbon atoms, the *bridgehead* carbons, being shared by two rings.

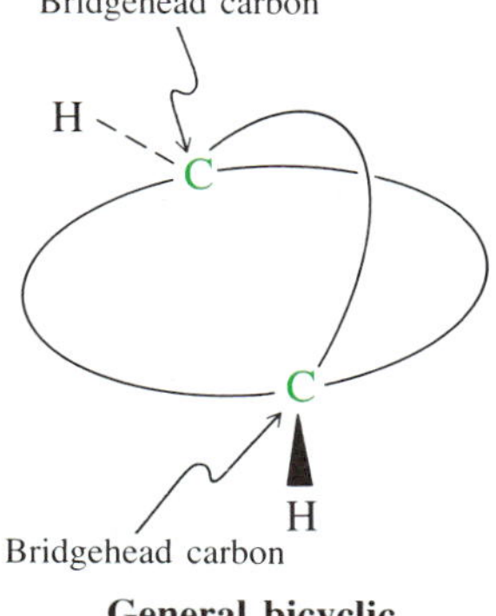

General bicyclic ring structure

Such molecules are named systematically as derivatives of the cycloalkane containing the same number of ring carbons. The alkane name is preceded by the prefix *bicyclo*. Decalin is then a *bicyclodecane*. The size of the bridge (number of carbons) between the two bridgehead carbons is indicated in the order of decreasing size by numerical inserts in square brackets between the prefix bicyclo and the alkane name. For example, norbornane is a bicyclic heptane with two two-carbon bridges, the third bridge containing only one carbon. It is therefore correctly named bicyclo[2.2.1]heptane. Decalin is a little different. Here, one of the bridges is simply a bond containing no atoms: a zero bridge. Consequently, its proper name is bicyclo[4.4.0]decane. Other examples of bicyclic hydrocarbons are *cis*-bicyclo[1.1.0]butane and *cis*-bicyclo[4.2.0]octane. In numbering the bicyclic framework, begin at one of the bridgehead carbons, proceed through the longest chain to the second bridgehead, continue through the second-longest chain back to the first bridgehead carbon, and finally complete numbering through the last bridge.

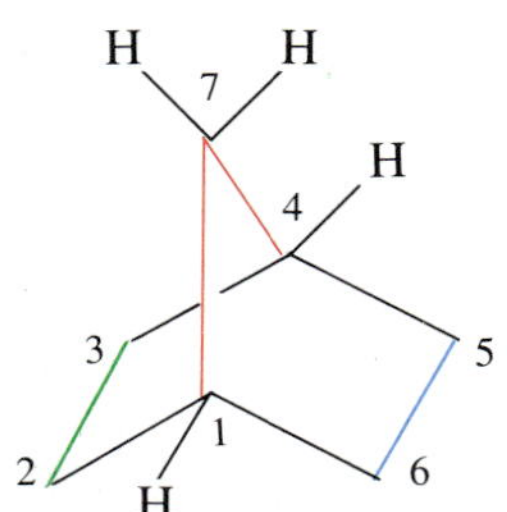

Bicyclo[2.2.1]heptane (Norbornane)

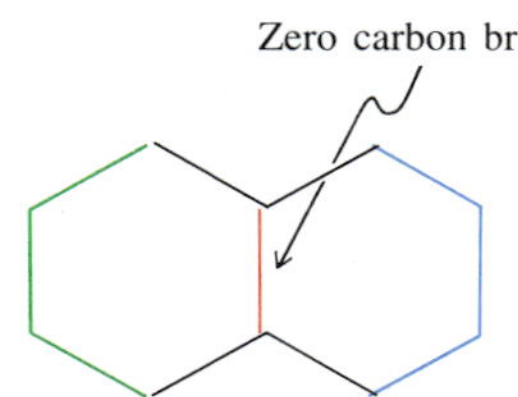

Bicyclo[4.4.0]decane (Decalin)

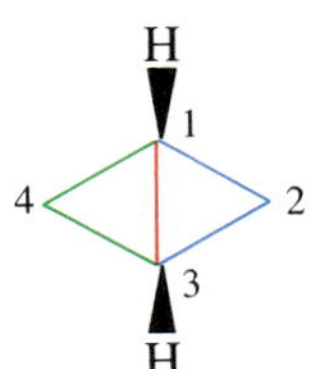

***cis*-Bicyclo[1.1.0]butane**

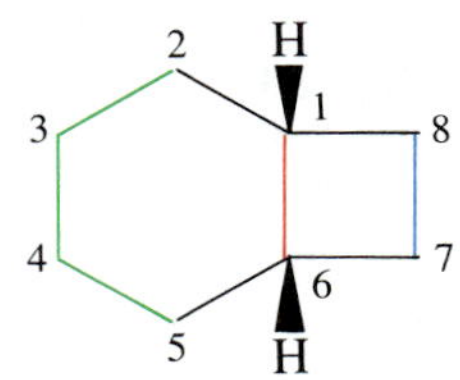

***cis*-Bicyclo[4.2.0]octane**

Bicycloalkanes can be cis or trans fused. Sometimes the latter is sterically unfeasible, as in norbornane or *cis*-bicyclo[1.1.0]butane. On the other hand, in bicyclo[4.4.0]decane both isomers are possible (Figure 4-15).

FIGURE 4-15

Conventional drawings and chair conformations of *trans*- and *cis*-bicyclo[4.4.0]decane.

H

H

H

H

Equatorial C–C bonds

***trans*-Bicyclo[4.4.0]decane**

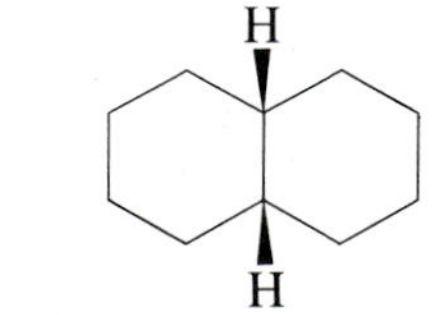

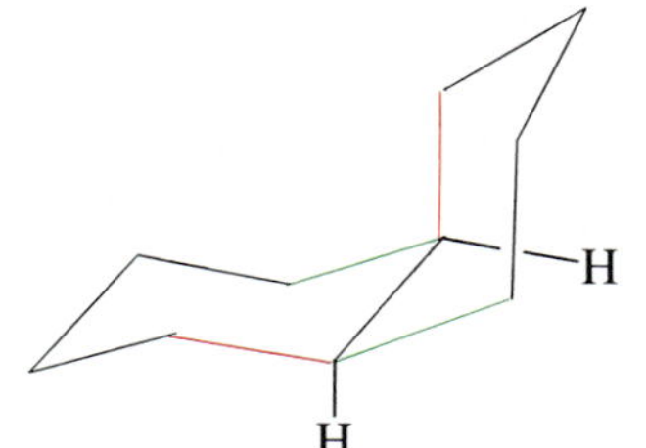

Axial C–C bonds Equatorial C–C bonds

***cis*-Bicyclo[4.4.0]decane**

EXERCISE 4-7

Construct molecular models of both *cis*- and *trans*- bicyclo[4.4.0]decane. What can you say about their conformational mobility?

Strained Hydrocarbons: What Is the Limit?

The search for the limits of strain in hydrocarbon bonds is an area of fascinating research that has resulted in the synthesis of many exotic molecules. What is surprising is how much bond-angle distortion a saturated carbon atom is able to tolerate. You have already made models of several molecules that are constructed only with difficulty. A case in point in the bicyclic series is bicyclo[1.1.0]butane, the strain energy of which is estimated to be 66.5 kcal mole^{-1}. Considering that its heat of formation is +51.9 kcal mole^{-1}, it is remarkable that the molecule exists at all.

A series of compounds attracting the attention of synthetic chemists possess a carbon framework geometrically equivalent to the Platonic solids: the *tetrahedron* (tetrahedrane), the *hexahedron* (cubane), and the pentagonal *dodecahedron* (dodecahedrane). Of them, the hexahedron was synthesized first in 1964, a C_8H_8 hydrocarbon shaped like a cube, and accordingly named cubane. The experimental strain energy (157 kcal mole^{-1}) is approximately equal to the total strain of six cyclobutanes. Although tetrahedrane itself is unknown, a tetra(1,1-dimethylethyl) derivative was synthesized in 1978. Despite the estimated strain ranging from 129 to 137 kcal mole^{-1}, the compound is stable and has a melting point of 135°C. The synthesis of dodecahedrane was achieved in 1982. It required twenty-three synthetic operations, starting from a simple cyclopentane derivative. The last step gave 1.5 mg of pure compound. Although small, this amount was sufficient to permit complete characterization of the molecule. Its melting point at 430°C is extraordinarily high for a C_{20} hydrocarbon; it is indicative of the symmetry of the compound. For comparison, eicosane melts at 36.8°C (Table 2-5).

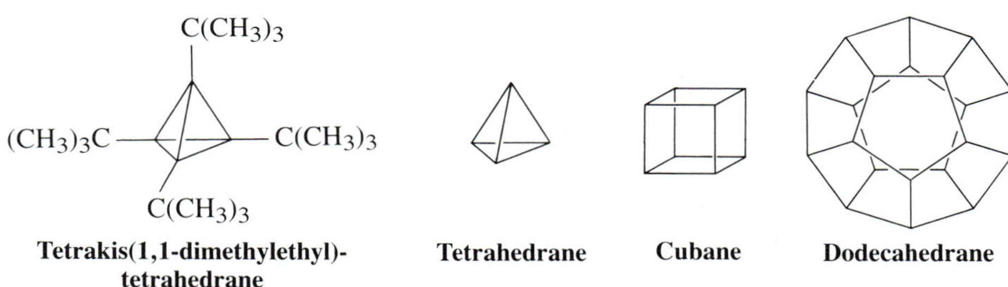

Tetrakis(1,1-dimethylethyl)-tetrahedrane **Tetrahedrane** **Cubane** **Dodecahedrane**

In summary, carbon in cyclic compounds tolerates a great amount of strain in its bonds to other nuclei, particularly to other carbon atoms. This capability has allowed the preparation of a series of cyclic molecules in which carbon is severely deformed from its tetrahedral shape.

4-7 Carbocyclic Products in Nature

Let us now take a brief look at the variety of cyclic molecules created in nature, cyclic **natural products.**

Natural products are organic compounds produced by living organisms. Some of these compounds are extremely simple, such as methane; others have great structural complexity. Scientists have attempted to classify the multitude of natural products in various ways. Generally, four schemes are followed, in which these products are classified according to: (1) chemical structure, (2) physiological activity, (3) organism or plant specificity (taxonomy), and (4) biochemical origin.

There are many reasons why organic chemists are interested in natural products. Many of these compounds are powerful drugs, others function as coloring or flavoring agents, and yet others are important raw materials. A study of animal secretions furnishes information concerning the ways in which animals use chemicals to mark trails, harm their predators, and attract the opposite sex. Investigations of the biochemical pathways by which an organism metabolizes and otherwise transforms a compound are sources of insight into the detailed workings of the organism's bodily functions. This section deals with two classes of natural products—terpenes and steroids—that, for the above reasons, have received a large amount of attention from organic chemists.

Terpenes

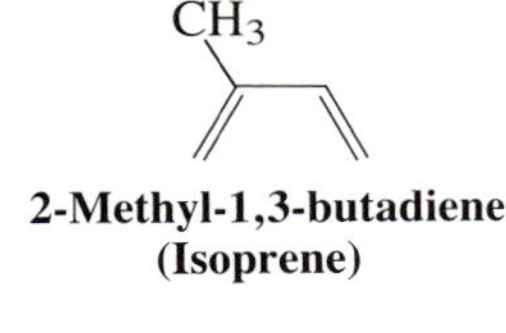

2-Methyl-1,3-butadiene (Isoprene)

Isoprene unit in terpenes
(Some contain double bonds)

Most of you have smelled the frequently strong odor emanating from freshly crushed plant leaves. The odor is due to the liberation of a mixture of fairly volatile compounds usually containing ten, fifteen, or twenty carbon atoms, the terpenes. These compounds are used as food flavorings (the extracts from cloves and peppermint), as perfumes (roses, lavender, sandalwood), and as solvents (turpentine).

Terpenes are biosynthesized in the plant by the linkage of at least two molecular units containing five carbon atoms. The structure of these units is like that of 2-methyl-1,3-butadiene (isoprene), and so they are referred to as "isoprene units." Depending on how many isoprene units are incorporated into the structure, terpenes are classified as mono- (C_{10}), sesqui- (C_{15}), and diterpenes (C_{20}). (The isoprene building units are shown in color in the examples of terpenes given here.)

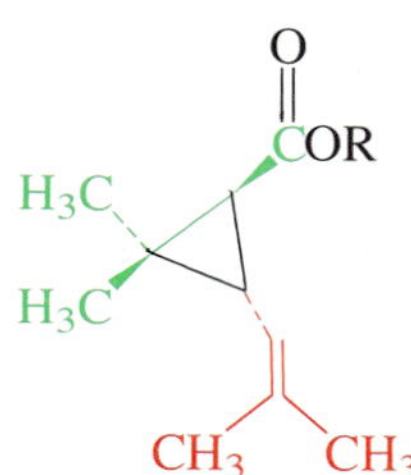

***trans*-Chrysanthemic acid (R = H)**
***trans*-Chrysanthemic esters (R ≠ H)**

Chrysanthemic acid is a monocyclic terpene containing a three-membered ring. Its esters are found in the flower heads of pyrethrum *(Chrysanthemum cinerariae-folium)*. These esters are naturally occurring insecticides, the cis isomers generally being more active than their trans counterparts. A cyclobutane is present in grandisol, the sex-attracting chemical used by the male boll weevil *(Anthonomus grandis)*. Menthol (peppermint oil) is an example of a substituted cyclohexane natural product, whereas camphor (from the camphor tree) and β-cadinene (from juniper and cedar) are simple bicyclic terpenes, the first a bicyclo[2.2.1] system, the second a bicyclo[4.4.0] (decalin) derivative.

Grandisol

Menthol

Camphor

β-Cadinene

EXERCISE 4-8

Draw the preferred chair conformation of menthol.

In contrast with the terpenes, which are classified according to their size and biological origins, the steroids are classified on the basis of their structures.

EXERCISE 4-9

After reviewing Section 2-1, specify the functional groups present in the terpenes shown in this section.

Steroids: Tetracyclic Natural Products with Powerful Physiological Activity

Steroids are abundant in nature and many derivatives have physiological activity. Steroids frequently function as *hormones,* which are regulators of biochemical activity. In the human body, for example, they control sexual development and fertility, in addition to other functions. Because of this feature, many steroids, often the products of laboratory synthesis, are used in medicines in, for example, the treatment of cancer, arthritis, or allergies, and in birth control.

In the steroids, three cyclohexane rings are fused to each other in an angular manner rather than linearly (i.e., the central ring is fused to its neighbors in such a way as to form an angle) and usually by trans-ring junctures, as in *trans*-bicyclo[4.4.0]decane. The fourth ring is a cyclopentane; its addition gives the typical tetracyclic structure. The four rings are labeled A, B, C, D, and the carbons are numbered according to a scheme specific to steroids. Many steroids have methyl groups attached at C-10 and C-13 and oxygen at C-3 and C-17. In addition, longer side chains may be found at C-17. The transfusion of the rings allows for a least-strained all-chair configuration in which the methyl groups and hydrogen atoms at the ring junctions occupy axial positions.

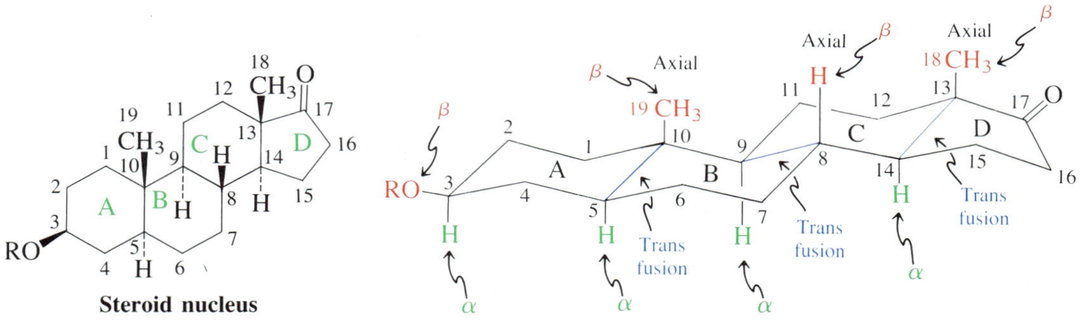

Steroid nucleus

Groups attached above the plane of the steroid molecule as written are β substituents, whereas those below are referred to as α. Thus, the structure shown above has a 3β-OR, 5α-H, 10β-CH_3 and so forth. The axial methyl groups are also called *angular* methyls, because they protrude sharply from the general framework (*angulus,* Latin, at an angle; being at a sharp corner).

Among the most abundant steroids is cholesterol. It is present in almost all human and animal tissue, particularly in the brain and the spinal cord. In fact, it is so concentrated in the spinal cord of cattle that it is commercially isolated from it by simple extraction. The adult human body contains from 200 to 300 g of cholesterol; gallstones may consist entirely of it. This steroid has been implicated as being responsible for several circulatory diseases because it precipitates in the arteries, causing arteriosclerosis and heart diseases. Although its biological function in the body is not completely understood, it is a precursor of steroid hormones and bile acids. Bile acids are produced in the liver as part of a fluid delivered to the duodenum to aid in the emulsification, digestion, and absorption of fats. An example is cholic acid.

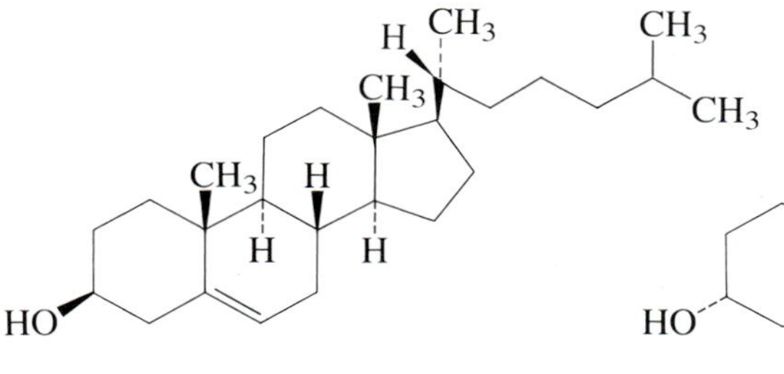

Cholesterol 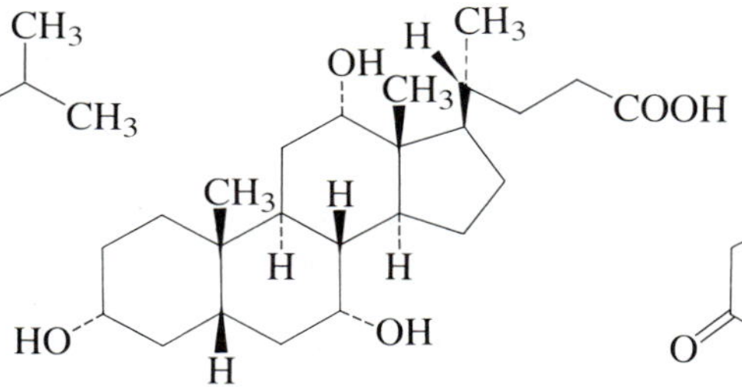 **Cholic acid** 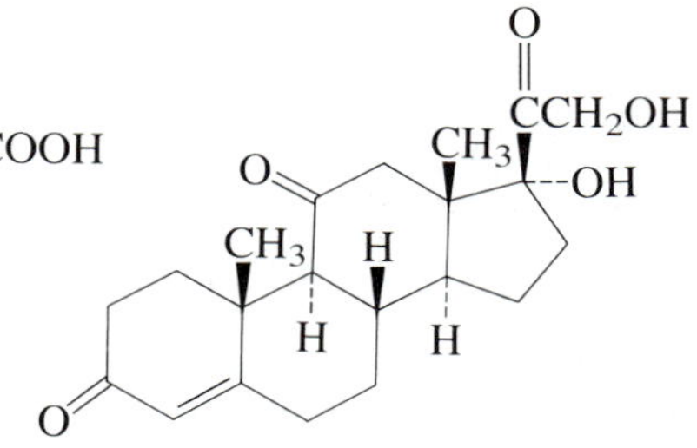 **Cortisone**

Cortisone, used extensively in the treatment of rheumatoid inflammations, is one of the adrenocortical hormones produced by the outer part (cortex) of the adrenal glands. These hormones participate in regulating the electrolyte and the water balance in the body, as well as protein and carbohydrate metabolism.

The sex hormones are divided into three groups: (1) the male sex hormones, or *androgens;* (2) the female sex hormones, or *estrogens;* and (3) the pregnancy hormones, or *progestins*.

Testosterone is the principal male sex hormone. Produced by the testes, it is responsible for male (masculine) characteristics (deep voice, facial hair, general physical constitution). Estradiol is the female sex hormone. It was first isolated by extraction of four tons of sow ovaries, yielding only several milligrams of pure steroid. Estradiol is responsible for the development of the secondary female characteristics and participates in the control of the menstrual cycle. An example of a progestin is progesterone, responsible for preparing the uterus for implantation of the fertilized egg.

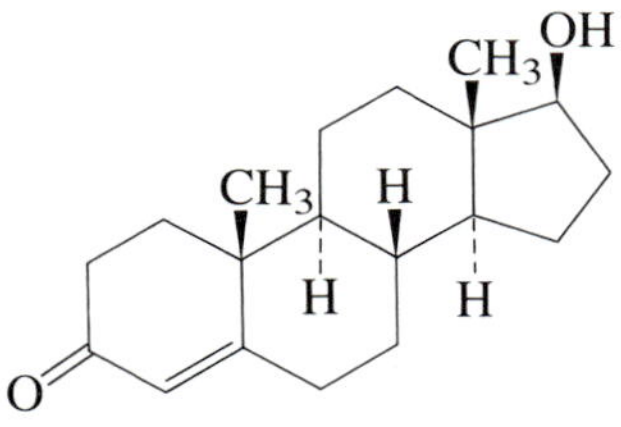

Testosterone 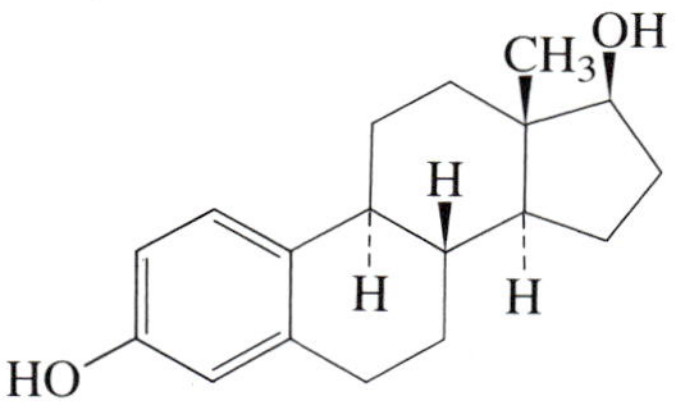 **Estradiol** 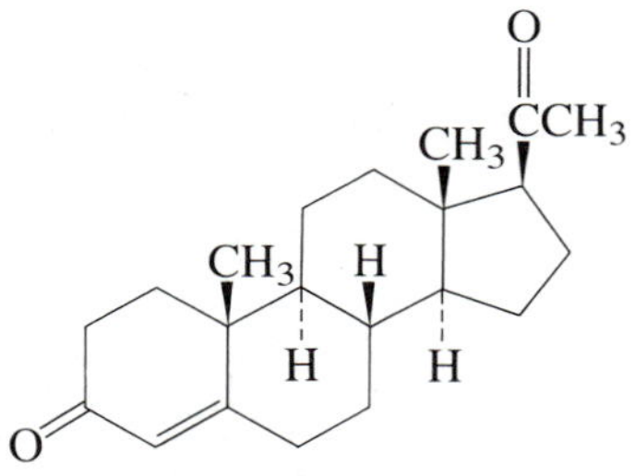 **Progesterone**

The structural similarity of the steroid hormones is remarkable considering their widely divergent activity. Steroids are the active ingredients of "the pill," functioning as an antifertility agent for the control of the female menstrual cycle and ovulation. At the peak of its use, more than 100 million women throughout the world took "the pill" as the primary form of contraception.

In summary, there is great variety in the structure and function of naturally occurring organic products, as manifested in the terpenes and the steroids. Natural products are frequently introduced in subsequent chapters to illustrate the presence and chemistry of a functional group, to demonstrate synthetic strategy or the use of reagents, to picture three-dimensional relations, and to exemplify

BOX 4-2

The Control of Fertility

The menstrual cycle is controlled by three protein hormones from the pituitary gland. The follicle-stimulating hormone (FSH) induces the growth of the egg; the luteinizing hormone (LH), its release from the ovaries. The third pituitary hormone (luteotropic hormone) induces formation of an ovarian tissue called the *corpus luteum*. As the cycle begins and egg growth is initiated, the tissue around the egg secretes increasing quantities of estrogens. Once a certain concentration of estrogen in the blood stream is reached, the production of FSH is turned off. The egg is released at this stage in response to LH. At the time of ovulation, LH also triggers the formation of the *corpus luteum*, which in turn begins to secrete increasing amounts of progesterone. The latter hormone suppresses any further ovulation by turning off the production of LH. If the egg is not fertilized, the *corpus luteum* and ovum are expelled (menstruation). Pregnancy, on the other hand, leads to increased production of estrogens and progesterone to prevent pituitary hormone secretion and thus renewed ovulation. The pill consists of a mixture of synthetic potent estrogen and progesterone derivatives (more potent than the natural hormones), which, when taken throughout most of the menstrual cycle, prevent both development of the ovum and ovulation by turning off production of both FSH and LH. The female body is essentially being tricked into believing that it is pregnant. One of the commercial pills contains a combination of norethynodrel (2.5 mg) and mestranol (0.1 mg). Other preparations consist of similar analogs with minor structural variations.

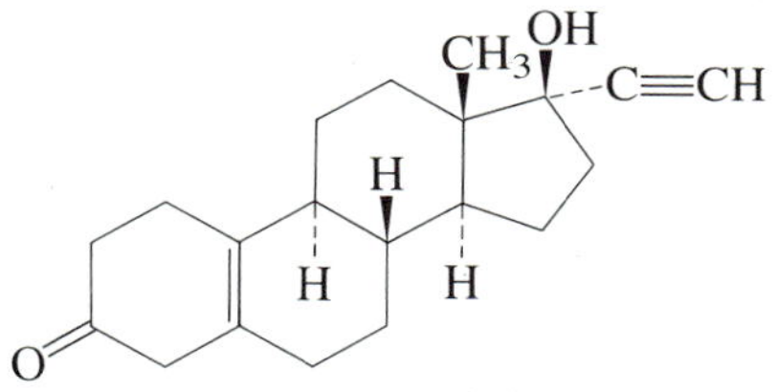

Norethynodrel

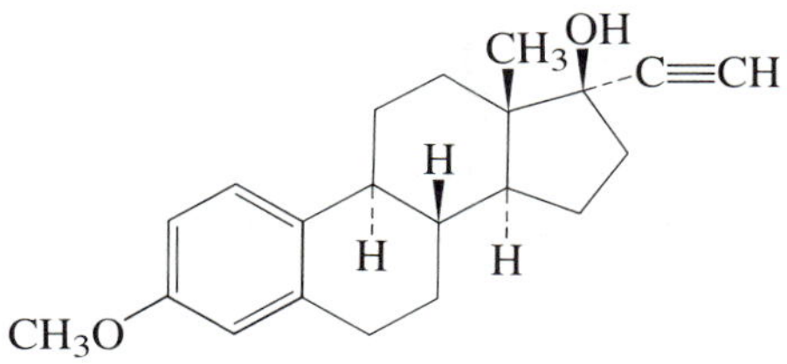

Mestranol

medicinal applications. Several classes of natural products will be discussed more extensively: fats (Sections 17-12 and 18-4), carbohydrates (Chapter 23), alkaloids (Section 26-7), and amino and nucleic acids (Chapter 27).

Summary of Important Concepts

1 Cycloalkane nomenclature is derived from that of the straight-chain alkanes.

2 All but the 1,1-disubstituted cycloalkanes exist as two isomers: if both substituents are on the same face of the molecule, they are cis; if they are on opposite faces, they are trans.

3 Some cycloalkanes are strained. Strain is due to bond-angle distortion, eclipsing, *gauche* interactions, and transannular interactions.

4 Cyclohexane is almost strain free, subjected only to *gauche* interactions.

5 Bond-angle strain in the small cycloalkanes is largely accommodated by the formation of bent bonds.

6 Bond-angle, eclipsing, and other strain in the cycloalkanes larger than cyclopropane (which is by necessity flat) can be relieved by deviations from planarity.

7 Ring strain in the small cycloalkanes gives rise to reactions that result in opening of the ring.

8 Deviations from planarity lead to conformationally mobile structures, such as chair, boat, and twist-boat cyclohexane.

9 Chair cyclohexane contains two types of hydrogens: axial and equatorial. These interconvert rapidly at room temperature by a conformational chair–chair (''flip'') interconversion, with an activation energy of 10.8 kcal $mole^{-1}$.

10 In monosubstituted cyclohexanes, the ΔG° of equilibration between the two chair conformations is substituent dependent. Axial substituents are exposed to *gauche* and 1,3-diaxial interactions.

11 In more highly substituted cyclohexanes, substituent effects are often additive, the bulkiest substituents being the most likely to emerge equatorially.

12 Completely strain-free cycloalkanes are those that can readily adopt an all-*anti* conformation and lack transannular interactions.

13 Bicycloalkanes contain two bridgehead carbons common to two rings. Fusion can be cis or trans.

14 Natural products are generally classified according to structure, physiological activity, taxonomy, and biochemical origin. Examples of the last class are the terpenes, of the first the steroids.

15 Terpenes are made up of isoprene units of five carbons.

16 Steroids contain three angularly fused cyclohexanes (A, B, C rings) attached to the cyclopentane D ring. Beta substituents are above the molecular plane, alpha substituents below.

17 An important class of steroids are the sex hormones, which have a number of physiological functions, including the control of fertility.

Problems

1 Write as many structures as you can that have the formula C_5H_{10} and contain one ring. Name them.

2 Name the following molecules according to the IUPAC nomenclature system.

(a)  I

(b) 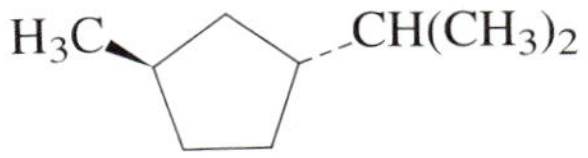 H_3C $CH(CH_3)_2$

(c) 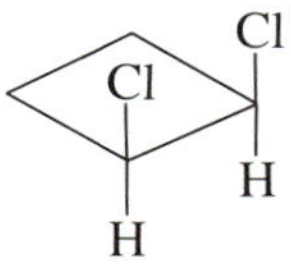 Cl Cl H H

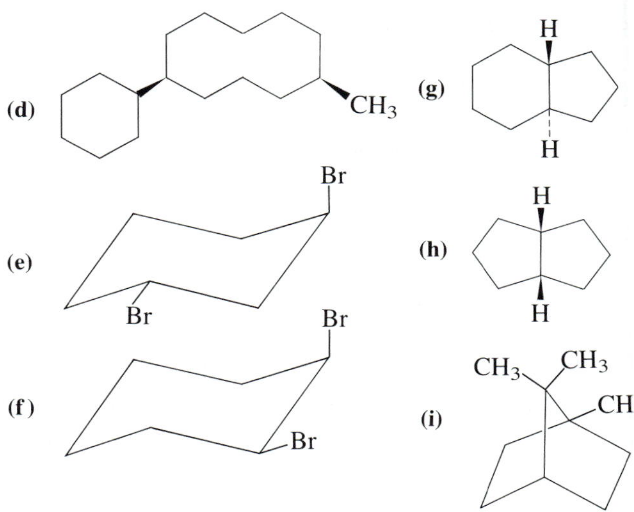

3 The kinetic data for the free-radical chlorination of several cycloalkanes shown at the right illustrate that the C–H bonds of cyclopropane and, to a lesser extent, cyclobutane are somewhat abnormal.

(a) What do these data tell you about (i) the strength of the cyclopropane C-H bond and (ii) the stability of the cyclopropyl radical?

(b) Suggest a reason for the stability characteristics of the cyclopropyl radical. Hint: Consider bond-angle strain in the radical relative to cyclopropane itself.

(c) Under the conditions of the experiment described in the table at the right, 1,1-dimethylcyclopropane has a relative reactivity per hydrogen toward Cl$^{\cdot}$ of 0.6. Given this information, predict the major monohalogenation product for each of the following reactions:

Reactivity per hydrogen toward Cl$^{\cdot}$, relative to cyclohexane = 1.0[a]

Cycloalkane	Reactivity
Cyclopentane	0.9
Cyclobutane	0.7
Cyclopropane	0.1

[a]Conditions: 68°C, $h\nu$, CCl_4 solvent.

H_3C CH_3 (cyclopropane) $\xrightarrow{Cl_2,\ h\nu}$ H CH_3 (cyclopropane) $\xrightarrow{Cl_2,\ h\nu}$

4 Use the data in Tables 3-2 and 4-2 to estimate the DH° value for a C–C bond in:

(a) Cyclopropane.
(b) Cyclobutane.
(c) Cyclopentane.
(d) Cyclohexane.

5 Propose a possible mechanism for the conversion of cyclopropane into propene at high temperature.

6 Calculate ΔH° values for the following possible transformations.

(a) (cyclohexane) + H_2 ⟶ hexane

(b) (cyclohexane) + HCl ⟶ 1-chloro-hexane

(c) (cyclopropane) + H_2 ⟶ propane

(d) (cyclopropane) + HCl ⟶ 1-chloro-propane

7 If cyclobutane were flat, it would have exactly 90° C–C–C bond angles and could conceivably use pure p orbitals in its C–C bonds. What would be a possible resultant hybridization for the carbon atoms of the molecule that would allow all the C-H bonds to be equivalent? Exactly where would the hydrogens on each carbon be located?

What are the real structural features of the cyclobutane molecule that contradict this hypothesis?

8 Estimate ΔH° for the interconversion

$$\text{Planar cyclopentane} \rightleftharpoons \text{envelope cyclopentane}$$

Consider only differences in eclipsing H–H interactions.

9 For each of the following cyclohexane derivatives, indicate (1) whether the molecule is a cis or trans isomer and (2) whether it is in its most stable conformation. If your answer to part 2 is "no," "flip" the ring and draw its most stable conformation.

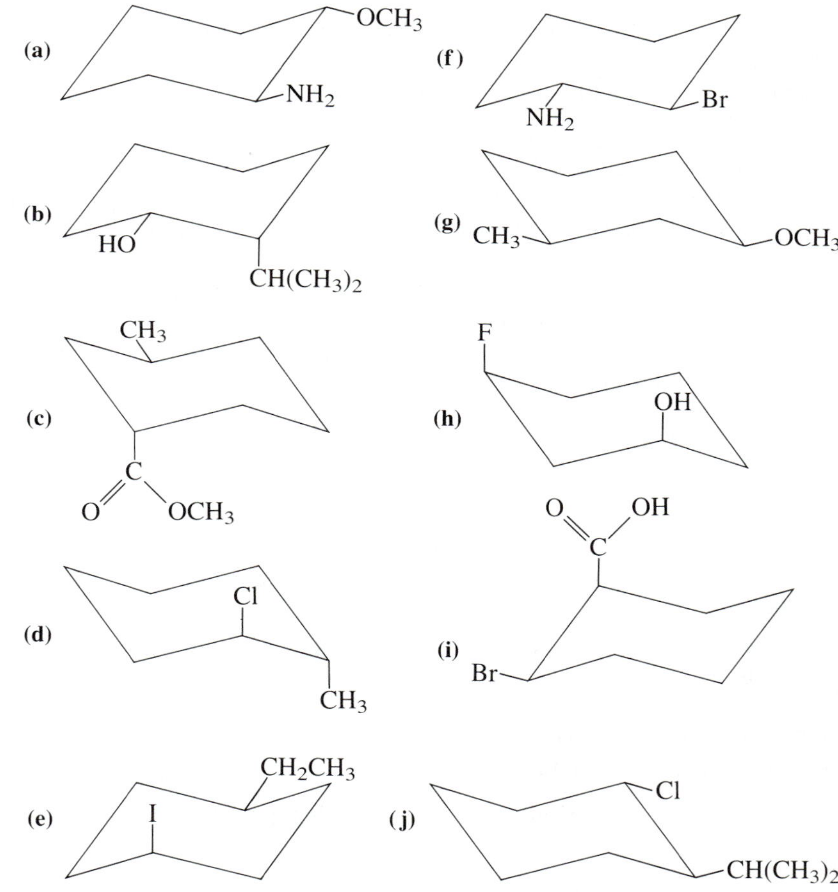

10 Draw the most-stable conformation for each of the following substituted cyclohexanes. Then, in each case, "flip" the ring and redraw the molecule in its next-best conformation.

(a) Cyclohexanol.
(b) *trans*-3-Methylcyclohexanol.
(c) *cis*-1-(1-Methylethyl)-2-methylcyclohexane.

(d) *cis*-1-Ethyl-2-methoxycyclohexane.
(e) *trans*-1-(1,1-Dimethylethyl)-4-chlorocyclohexane.

11 For each molecule in Problem 10, estimate the energy difference between the most-stable and next-best conformation. Calculate the approximate ratio of the two at 300 K.

12 The most-stable conformation of *trans*-1,3-bis(1,1-dimethylethyl)-cyclohexane is not a chair. What conformation would you predict for this molecule? Explain.

13 Compare the structure of cyclodecane in an all-chair conformation (illustrated below) with that of *trans*-bicyclo[4.4.0]decane (*trans*-decalin). Explain why the all-chair conformation is highly strained, and yet *trans*-bicyclo[4.4.0]-decane is a nearly strain-free molecule. Make models.

14 Suggest an explanation for the fact that *cis*-bicyclo[4.4.0]decane melts at $-43°C$, whereas the trans isomer melts at $-30°C$. Refer to your answer to Exercise 4-7 for relevant information.

15 Qualitatively, predict the relative stabilities of **(a)** *cis*- versus *trans*-bicyclo[4.4.0]decane; **(b)** *cis*- versus *trans*-bicyclo[4.3.0]nonane (indane); **(c)** *cis*- versus *trans*-bicyclo[4.2.0]octane. Again, make models.

16 Identify each of the molecules below as being either a monoterpene, a sesquiterpene, or a diterpene (all names are common).

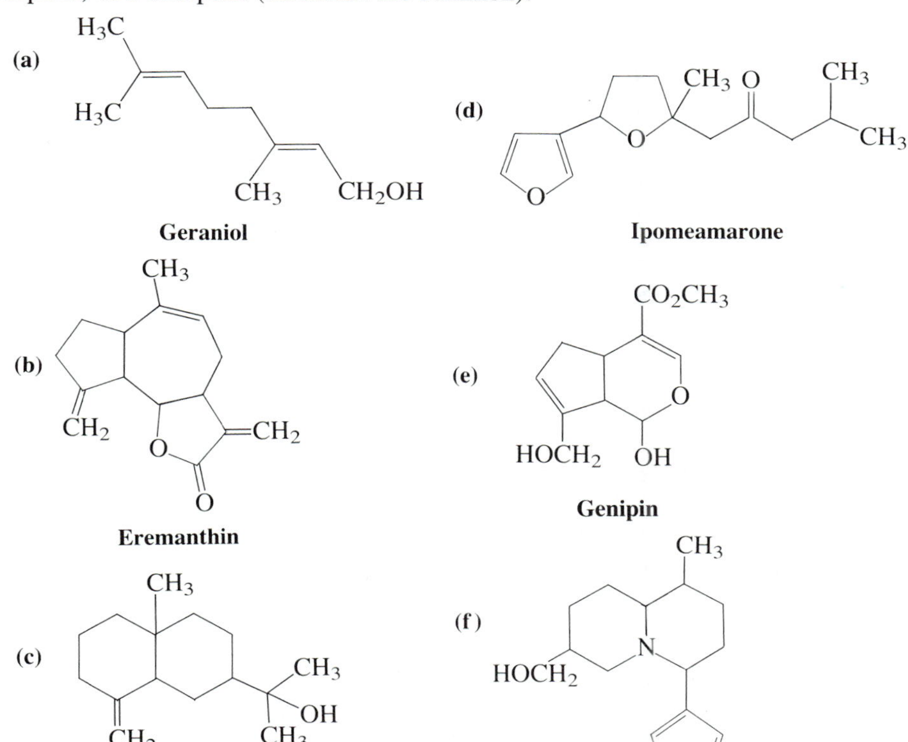

(g) Cantharidin

(h) Vitamin A

17 Find the 2-methyl-1,3-butadiene (isoprene) units in each of the naturally occurring organic molecules pictured in Problem 16.

18 Circle and identify by name all the functional groups in any three of the steroids illustrated in Section 4-7. Label any polarized bonds with partial positive and negative charges (δ^+ and δ^-).

19 Several additional examples of naturally occurring molecules with strained ring structures are shown below.

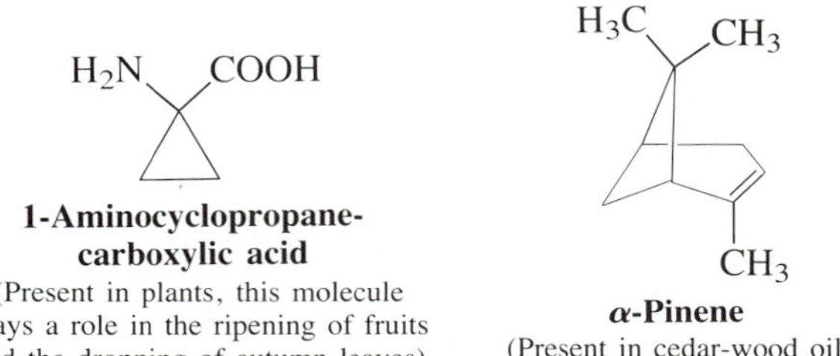

1-Aminocyclopropane-carboxylic acid
(Present in plants, this molecule plays a role in the ripening of fruits and the dropping of autumn leaves)

α-Pinene
(Present in cedar-wood oil)

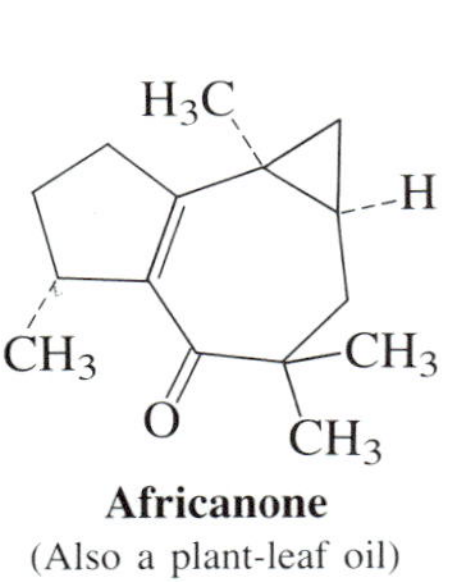

Africanone
(Also a plant-leaf oil)

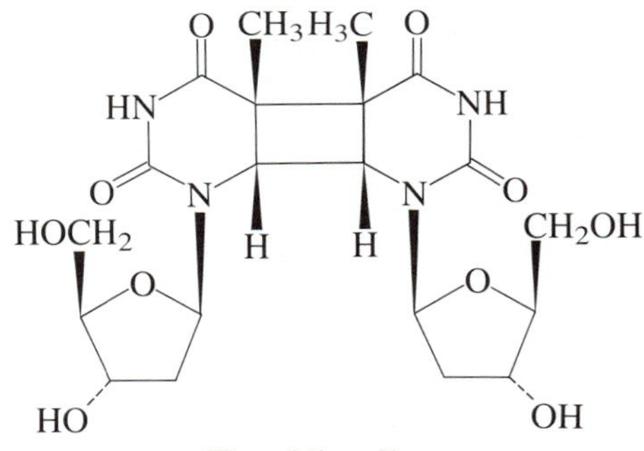

Thymidine dimer
(A component of DNA that has been exposed to ultraviolet light)

(a) Identify the terpenes (if any) in the preceding group of structures. Find the 2-methyl-1,3-butadiene units in each structure and classify as mono-, sesqui-, or diterpene.

(b) Supply IUPAC names for the bicyclic alkanes related to pinene and africanone shown below.

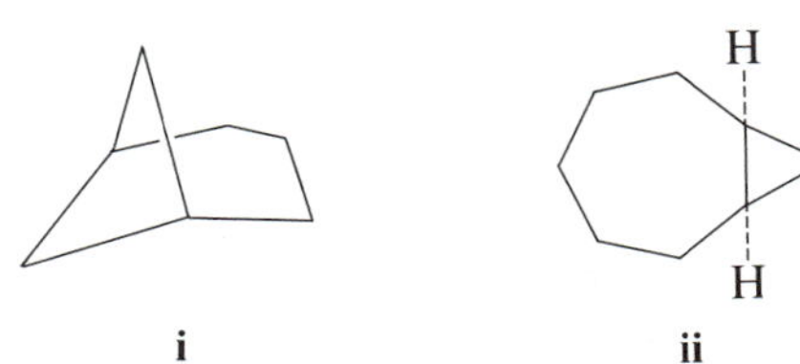

20 Fusidic acid is a steroidlike microbial product that is an extremely potent antibiotic with a broad spectrum of biological activity. Its molecular shape is most unusual (see below) and has supplied important clues to researchers investigating the methods by which steroids are synthesized in nature.

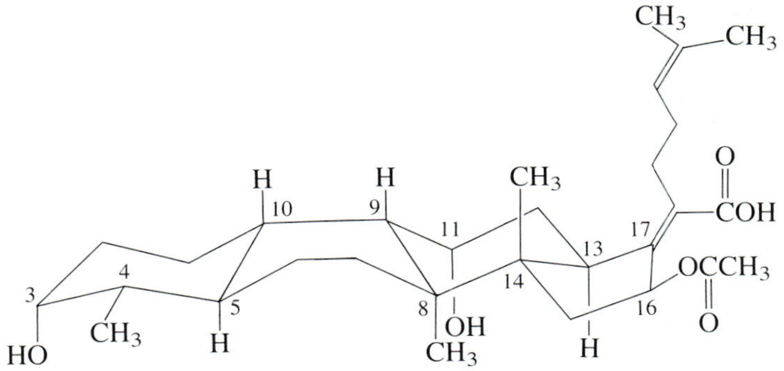

Fusidic acid

(a) Locate all the rings in fusidic acid and describe their conformations.

(b) Identify all ring fusions in the molecule as having either cis or trans geometry.

(c) Identify all groups attached to the rings as being either α- or β-substituents.

(d) Describe in detail how this molecule differs from the typical steroid in structure and stereochemistry.

As an aid to answering these questions, the carbon atoms of the framework of the molecule have been numbered.

21 The enzymatic oxidation of cyclohexane to produce cyclohexanol (Section 3-4) is a simplified version of the reactions that produce the adrenocortical steroid hormones. In the biosynthesis of corticosterone from progesterone (Section 4-7), two such oxidations take place successively:

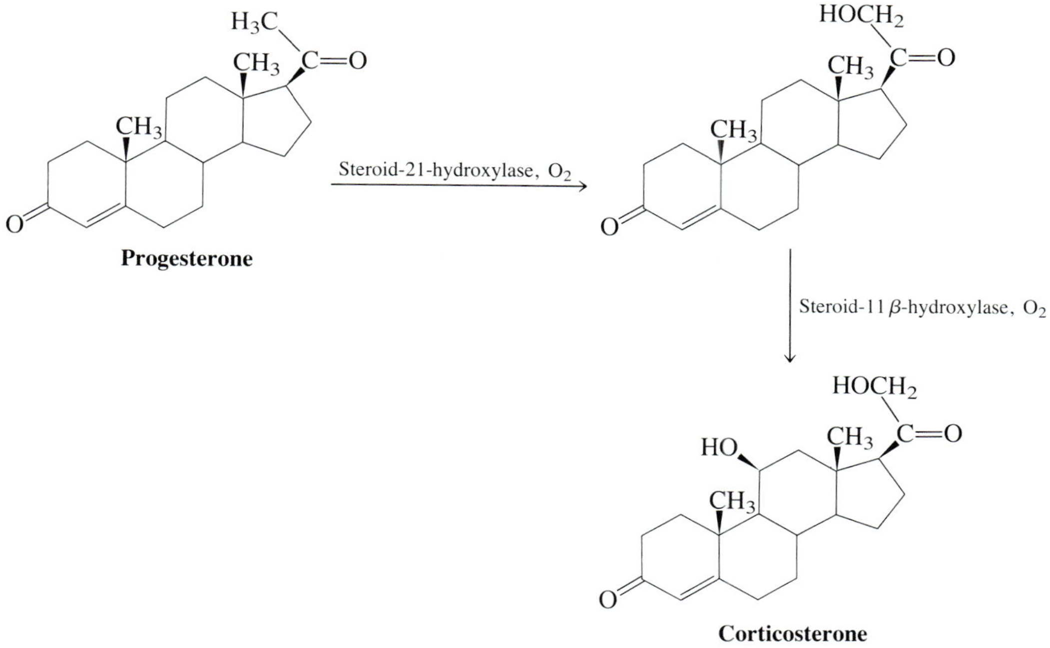

Progesterone

Steroid-21-hydroxylase, O_2

Steroid-11β-hydroxylase, O_2

Corticosterone

It is thought that the hydroxylase (or mono-oxygenase) enzymes act as complex oxygen-atom donors in these reactions. A suggested mechanism, as applied to cyclohexane, consists of the two steps shown below.

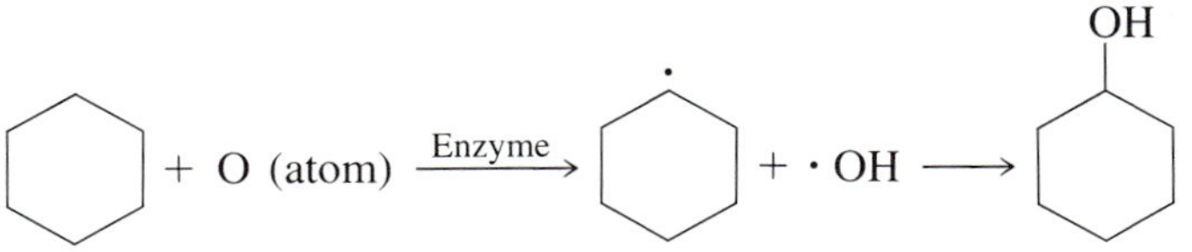

$$\text{cyclohexane} + \text{O (atom)} \xrightarrow{\text{Enzyme}} \text{cyclohexyl}\cdot + \cdot\text{OH} \longrightarrow \text{cyclohexanol (OH)}$$

Calculate ΔH° for each step and for the overall oxidation reaction. Use the data in Table 3-5, together with the following ΔH_f° values: cyclohexane, -29.5 kcal mole^{-1}; cyclohexanol, -68.4 kcal mole^{-1}; cyclohexyl radical, $+15$ kcal mole^{-1}; HO$\cdot$, $+9.4$ kcal mole^{-1}.

22 Iodobenzene dichloride, formed by the reaction of iodobenzene and chlorine, is, like sulfuryl chloride and NCS (Section 3-8), a reagent for the chlorination of alkane C-H bonds. Chlorinations using iodobenzene dichloride may be light-initiated and do not require the addition of chemical radical initiators.

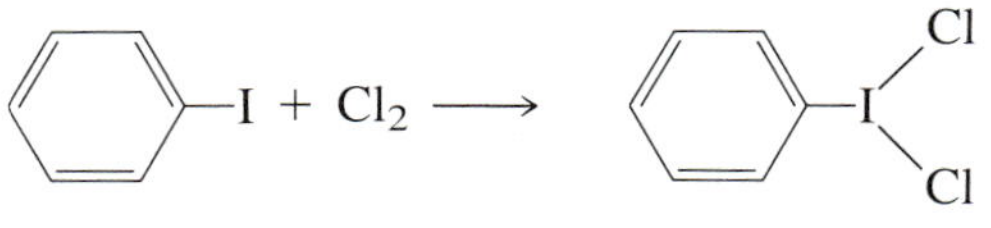

$$C_6H_5\text{—I} + Cl_2 \longrightarrow C_6H_5\text{—}ICl_2$$

Iodobenzene dichloride

(a) Propose a free-radical chain mechanism for the chlorination of a typical alkane RH by iodobenzene dichloride. The overall equation for the reaction is given below, as is the initiation step to get you started.

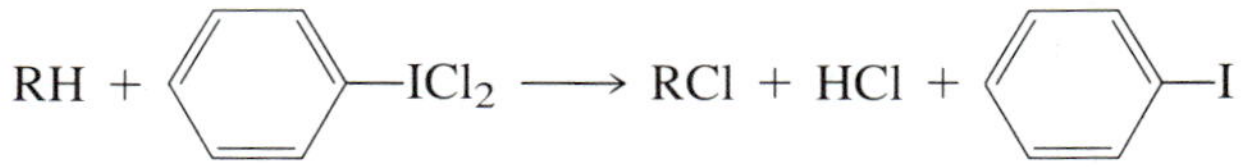

$$RH + C_6H_5\text{—}ICl_2 \longrightarrow RCl + HCl + C_6H_5\text{—I}$$

Initiation:

$$C_6H_5\text{—}ICl_2 \xrightarrow{h\nu} C_6H_5\text{—}\dot{I}\text{—Cl} + Cl\cdot$$

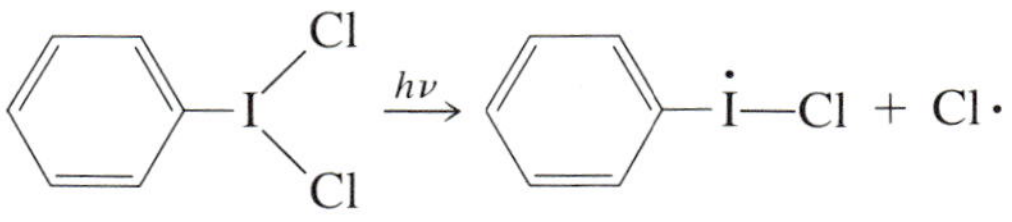

(b) Free-radical chlorination of typical steroids by iodobenzene dichloride gives, predominantly, three isomeric monochlorination products. On the basis of both reactivity (tertiary, secondary, primary) considerations and steric effects (which might hinder the approach of a reagent toward a C–H bond that might otherwise be reactive), predict the three major sites of chlorination in the steroid molecule. Either make a model or carefully analyze the drawings of the steroid nucleus in Section 4-7.

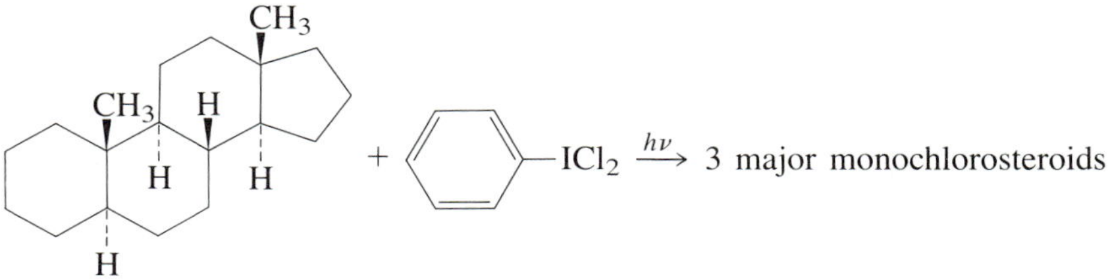

$$\text{steroid (CH}_3\text{, CH}_3\text{, H, H, H, H)} + C_6H_5\text{—}ICl_2 \xrightarrow{h\nu} \text{3 major monochlorosteroids}$$

23 As Problem 21 indicates, the enzymatic reactions that introduce functional groups to the steroid nucleus in nature are highly selective, unlike the laboratory chlorination described in Problem 22. However, by means of a clever adaptation of this reaction, it is possible to partly mimic nature's selectivity in the laboratory. Two such examples are illustrated below.

(a)

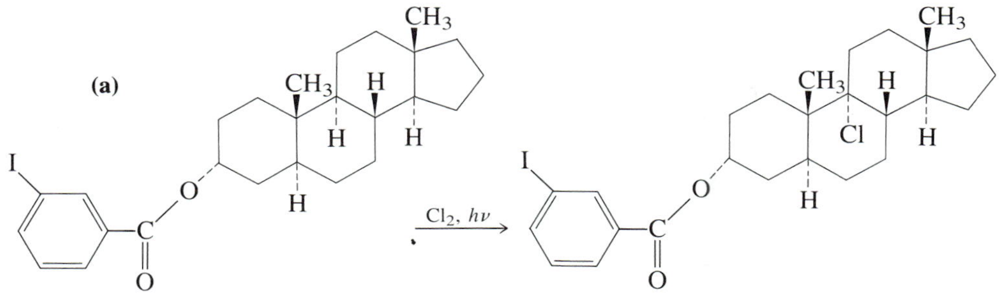

$\xrightarrow{Cl_2,\ h\nu}$

(b)

$\xrightarrow{Cl_2,\ h\nu}$

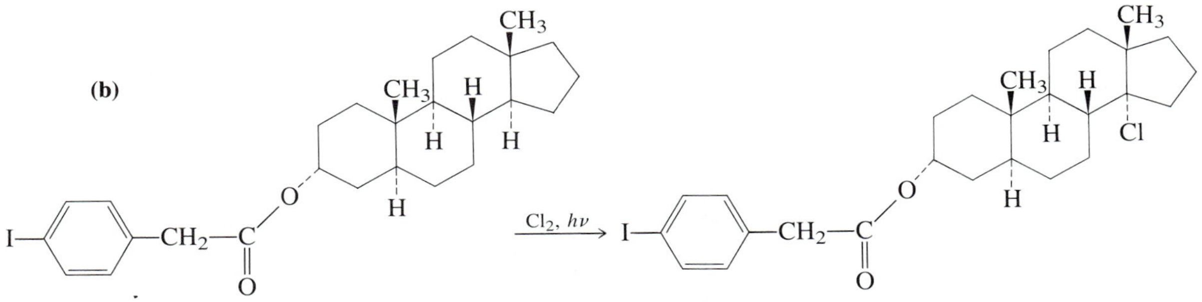

Propose reasonable explanations for the results of these two reactions. Make a model of the product of the addition of Cl_2 to each iodocompound (compare Problem 22) to help you analyze each system.

CHAPTER 5

Stereoisomerism

The preceding chapters have dealt with two kinds of isomerism: structural and stereo. *Structural isomerism* describes compounds of identical molecular formula but the order in which the individual atoms are connected differs.

Structural Isomers

C_4H_{10}	$CH_3CH_2CH_2CH_3$	$H_3C{-}CH(CH_3)_2$
	Butane	**2-Methylpropane**
C_2H_6O	CH_3CH_2OH	CH_3OCH_3
	Ethanol	**Methoxymethane (Dimethyl ether)**

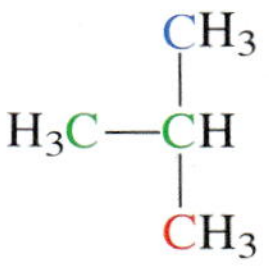

EXERCISE 5-1

Are bicyclo[4.2.0]octane and bicyclo[3.3.0]octane isomers?

Stereoisomerism describes isomers whose atoms are connected in the same order, but their spatial arrangement differs. Examples of stereoisomers include the relatively stable cis-trans isomers and the rapidly equilibrating conformational ones (Section 4-1).

Stereoisomers

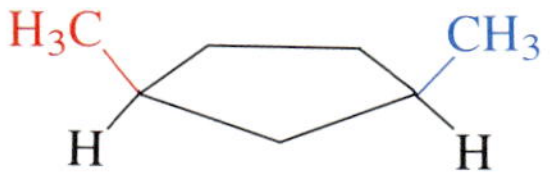

H_3C CH_3 H H

***cis*-1,3-Dimethylcyclopentane**

H CH_3 H_3C H

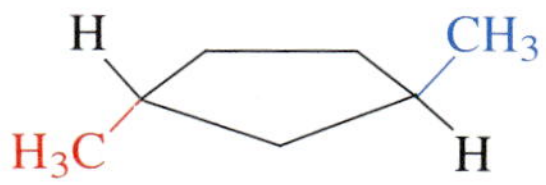

***trans*-1,3-Dimethylcyclopentane**

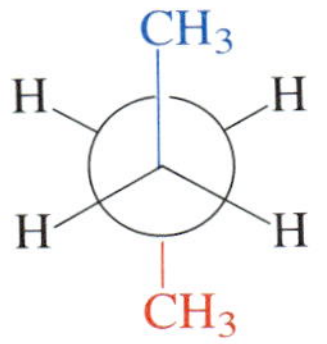

CH$_3$ H H H H CH$_3$ CH$_3$ H CH$_3$ H H H H CH$_3$ H CH$_3$

Anti rotamer of butane

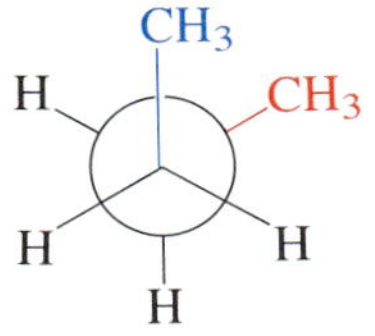

Gauche rotamer of butane

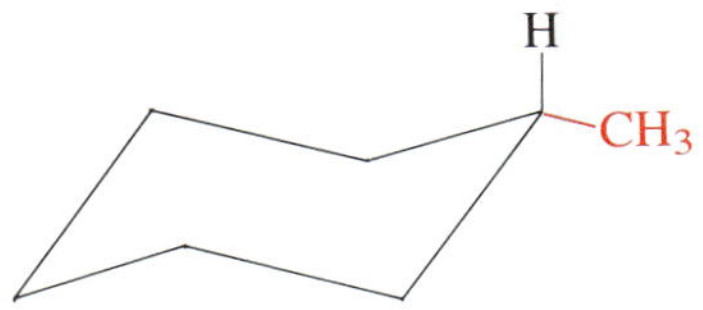

Equatorial methylcyclohexane

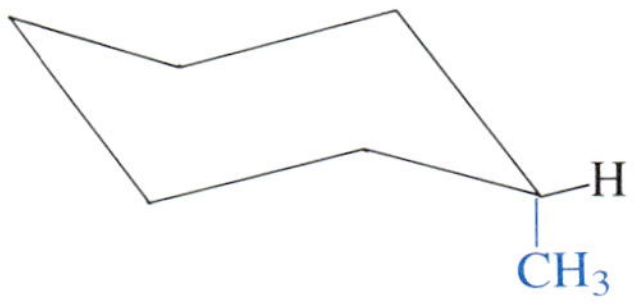

Axial methylcyclohexane

EXERCISE 5-2

Draw additional stereoisomers of methylcyclohexane.

This chapter introduces a third type of stereoisomerism, one that has as its basis the ''handedness'' of certain molecules. We shall see that there are structures whose mirror images are not superimposable on their images, just as your left hand is not superimposable on your right hand. The two are therefore different objects; they may have different properties and may react differently.

5-1 Chiral Molecules

How can a molecule exist as two nonsuperimposable isomers? Consider the free-radical bromination of butane. This reaction proceeds mainly at one of the secondary carbons to furnish 2-bromobutane. A molecular model of the starting material shows that two seemingly equivalent hydrogens on that carbon are replaceable to give only one ''possible'' 2-bromobutane (Figure 5-1). However, is this really true?

FIGURE 5-1

Replacement of one of the secondary hydrogens in butane results in two stereoisomeric forms of 2-bromobutane.

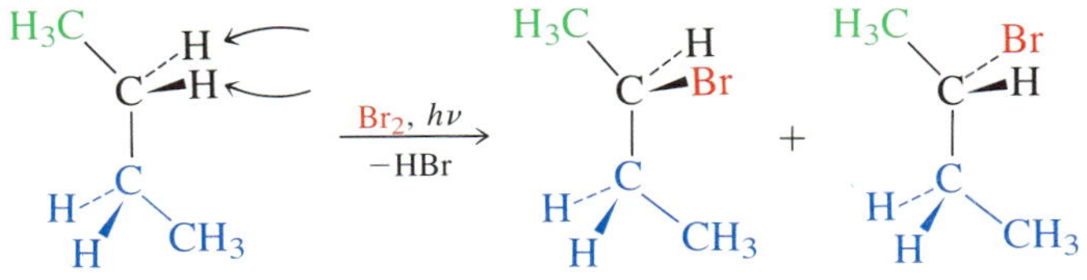

H_3C H H C C H H CH$_3$ $\xrightarrow[-HBr]{Br_2,\ h\nu}$ H_3C H Br C C H H CH$_3$ + H_3C Br H C C H H CH$_3$

If we look more closely at the two 2-bromobutanes obtained by replacing either of the methylene hydrogens with bromine, we see that the two structures are nonsuperimposable and therefore *not identical*. The two molecules are related as object and mirror image, but to convert one into the other would require the breaking of bonds. Pairs of molecules that exist as nonsuperimposable mirror images of each other are **enantiomers** (*enantios,* Greek, opposite). Such compounds, which lack *reflection symmetry,* are **chiral.** In our example of the bromination of butane, a 1 : 1 mixture of enantiomers is formed. Such a mixture is called a **racemate,** or **racemic mixture.**

In contrast with chiral molecules, structures that *are* superimposable with their mirror images are **achiral.** Examples of chiral and achiral molecules are shown at the top of the next page (the first two chiral structures are depicted as pairs of enantiomers).

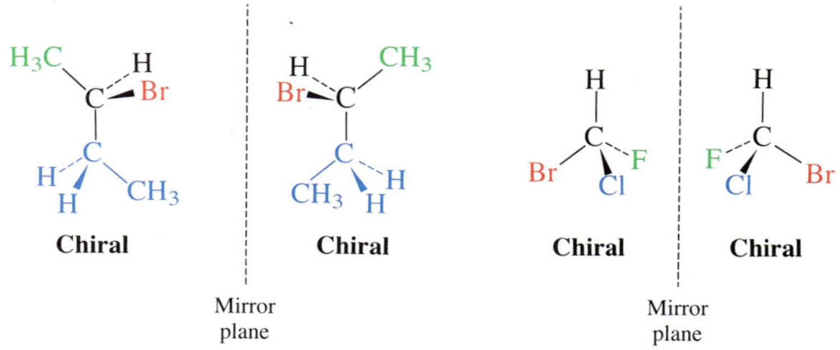

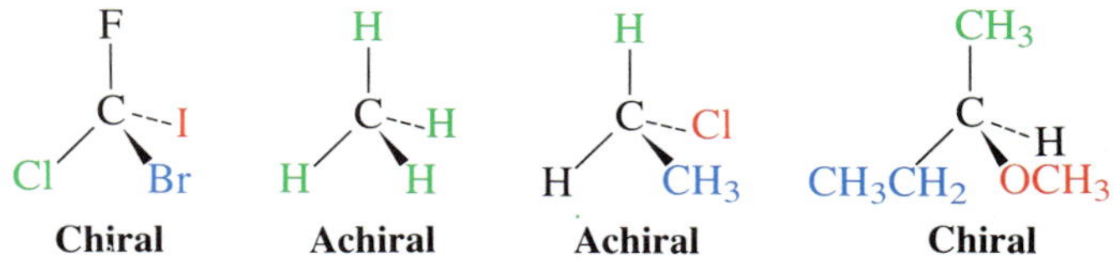

All the chiral examples contain an atom that is connected to four *different* substituent groups. Such a nucleus is called an **asymmetric atom** (e.g., asymmetric carbon), or **stereocenter.** Centers of this type are sometimes denoted by an asterisk. Molecules with one stereocenter are always chiral. We shall see in Section 5-5 that this statement is not necessarily true for structures incorporating more than one such center.

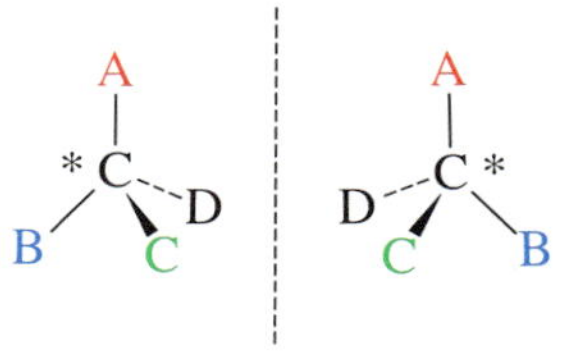

(C* = a stereocenter based on asymmetric carbon)

BOX 5-1

Chiral Natural Products

Many chiral natural products exist in nature as only one enantiomer, some as both. For example, natural *alanine* (systematic name: 2-aminopropanoic acid) is an abundant amino acid that is found in only one form. On the other hand, *lactic acid* (2-hydroxypropanoic acid) is present in blood and muscle fluid as one enantiomer but in sour milk and some fruits and plants as the racemate. A curious case is *carvone* [2-methyl-5-(1-methylethenyl)-2-cyclohexenone], which contains a stereocenter in a six-membered

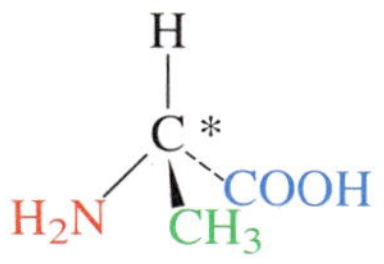

2-Aminopropanoic acid
(Alanine)

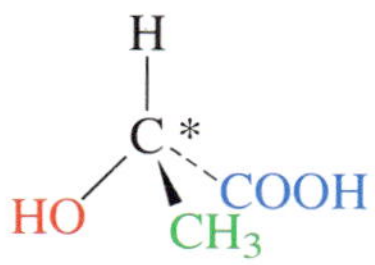

2-Hydroxypropanoic acid
(Lactic acid)

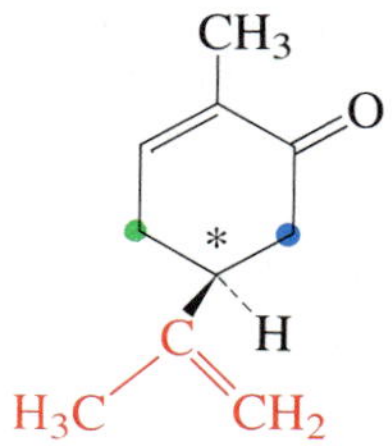

2-Methyl-5-(1-methylethenyl)-2-cyclohexenone
(Carvone)

ring. This carbon atom may be thought of as bearing four different groups if we consider the ring itself to be two separate and different substituents. They are different because, starting from the stereocenter, the clockwise sequence of atoms differs from the counterclockwise sequence. Carvone is found in nature in both enantiomeric forms. The characteristic odor of caraway and dill seed is due to the enantiomer shown, whereas the flavor of spearmint is due to the other enantiomer.

EXERCISE 5-3

Among the natural products shown in Section 4-7, which are chiral and which are achiral? Give the number of stereocenters in each case.

Molecules Can Be Chiral Without Containing an Asymmetric Carbon: Handedness

The word *chiral* is derived from the Greek *cheir,* meaning ''hand'' or ''handedness.'' Human hands have the mirror-image relation typical of enantiomers (Figure 5-2). Among the many other objects that are chiral are shoes, ears, screws, and spiral staircases. At one extreme, they may exist in one enantiomeric form, such as a collection of left shoes; at the other extreme, as a racemate, such as a collection of matching pairs of shoes. On the other hand, there are many achiral objects, such as balls, ordinary water glasses, hammers, and nails.

FIGURE 5-2
Left and right hands as models for enantiomeric relations.

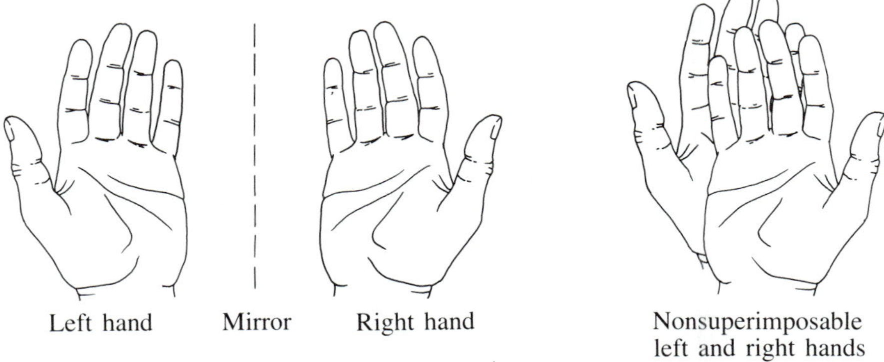

Many chiral objects, such as spiral staircases, do not have stereocenters. This is also true for many chiral molecules. *Remember that the only criterion for chirality is the nonsuperimposable nature of object and mirror image*. For example, the *gauche* conformer of 1,2-dichloroethane is chiral, whereas the *anti* stereoisomer is achiral (Figure 5-3). In contrast with 2-bromobutane, the enantiomers in this case undergo rapid interconversion at room temperature, and no bonds need to be broken.

We shall see that, among other examples of chiral molecules lacking a stereocenter, the more fascinating are those with spiral or helical shapes (*helix,* Greek, spiral), such as the helicenes (Section 25-4, Figure 25-5).

FIGURE 5-3

Gauche rotamers of 1,2-dichloroethane: rotation around the carbon–carbon bond interconverts the two enantiomeric forms.

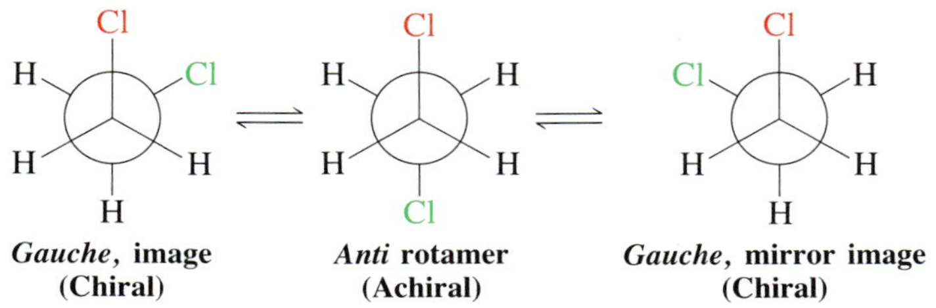

Gauche, **image (Chiral)** ⇌ ***Anti*** **rotamer (Achiral)** ⇌ ***Gauche,*** **mirror image (Chiral)**

How to Distinguish Chiral Structures from Achiral Ones: Symmetry in Molecules

As you have undoubtedly already noticed, it is not always easy to tell whether a molecule is chiral or not. A foolproof way is to construct molecular models of the molecule and its mirror image and look for superimposability. However, this procedure is very time consuming. Fortunately, two aids can be used to rapidly establish the presence or absence of chirality in a molecule. They are based on a molecule's properties of symmetry.

Certain manipulations, after being applied to the molecule under investigation, leave its structure and the position of its component atoms in space unchanged. We need to consider only two: the establishment of a plane of symmetry or a center of symmetry. A **plane of symmetry** is one that bisects the molecule so that the part of the structure lying on one side of the plane mirrors that lying on the other. For example, methane has six planes of symmetry, dichloromethane has two, bromochloromethane has one, and bromochlorofluoromethane has none (Figure 5-4, in which only one of those planes is shown for methane).

FIGURE 5-4

Examples of planes of symmetry in (A) methane (only one is shown), (B) dichloromethane, and (C) bromochloromethane. Bromochlorofluoromethane (D) has none.

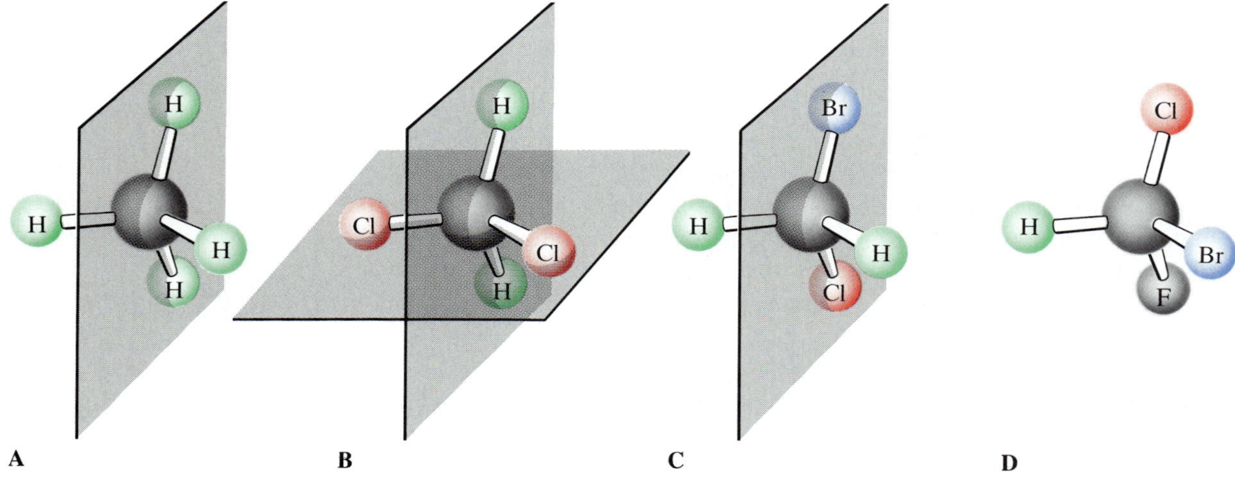

A **center of symmetry** is a point in a molecule that divides any straight line drawn through it into two sets of points on either side with the same environment. There can be only *one* such point. The molecule shown in Figure 5-5 has such a center of symmetry. There are far fewer structures having a center of symmetry than there are those having a plane of symmetry.

To distinguish a chiral molecule from an achiral one, we have to remember only that *chiral molecules cannot have either a plane of symmetry or a center of symmetry*. For example, the first three methanes in Figure 5-4 are clearly achiral because of the presence of mirror planes. The example in Figure 5-5 actually lacks such a plane, but the molecule is achiral by virtue of a center of symmetry. You will be able to classify most molecules in this book as chiral or achiral by simple recognition of the presence or absence of a plane of symmetry.

FIGURE 5-5
A molecule with a center of symmetry. The groups in color are in equal environments on various straight lines incorporating this center.

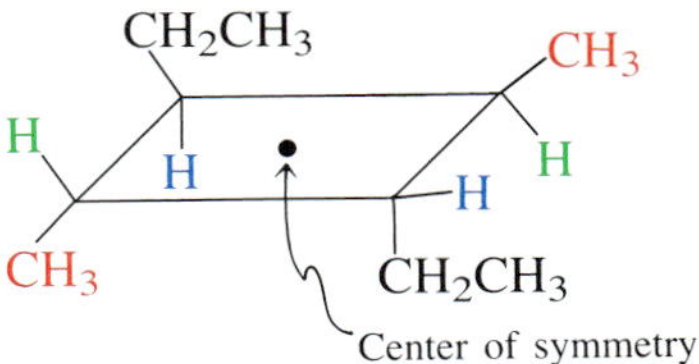

EXERCISE 5-4

Write the structures of all dimethylcyclobutanes. Specify those that are chiral.

To summarize, chiral molecules are stereoisomers in which image is not superimposable on mirror image. The two isomers are called enantiomers. Whereas many chiral organic molecules contain stereocenters, other chiral structures lack such centers. A chiral molecule has neither a plane nor a center of symmetry.

5-2 Optical Activity

Considering the close similarity of enantiomers (all bonds are identical; their heats of formation, ΔH_f, therefore are identical), you may wonder how it is possible to distinguish one enantiomer from another. This is indeed a very difficult task because most *physical* properties of enantiomers are identical. A notable exception is the interaction with a special type of light.

Recall our first example of a chiral molecule, the two enantiomers of 2-bromobutane, and assume that you have isolated each enantiomer in pure form. Comparison of their physical properties, such as boiling points, melting points, and densities, proves them to be indistinguishable. However, if **plane-polarized light** (which will be defined shortly) is passed through a sample of one of the enantiomers, the plane of polarization of the incident light is *rotated* by a certain amount in one direction (either clockwise or counterclockwise). If the same experiment is repeated with the mirror image of the first enantiomer, the plane of the polarized light is rotated by exactly the same amount, but in the opposite direction. An enantiomer that rotates the plane of light in a clockwise sense as the viewer faces the light source is **dextrorotatory** (*dexter,* Latin, right) and the compound is (arbitrarily) labeled the (+) enantiomer. Consequently, the other enantiomer, which will effect counterclockwise rotation is **levorotatory** (*laevus,* Latin, left) and labeled the (−) enantiomer. Because of this special interaction with light, enantiomers are frequently called **optical isomers** and the phenomenon **optical activity.**

The Measurement of Optical Rotation

What is plane-polarized light, and how is its rotation measured? Ordinary light may be thought of as electromagnetic radiation with oscillating electric and magnetic-field vectors at right angles and perpendicular to the light path (Figure

FIGURE 5-6

Representation of electromagnetic radiation.

A. One electric-field vector and the perpendicular magnetic field. In this representation, the path of the light is perpendicular to the plane of the page, and the field vectors extend into the plane.

B. Ordinary light has field vectors in all directions.

C. A different picture of part A in which the light travels from the left to the right side of the page.

Parts A and C are also representative of plane-polarized light.

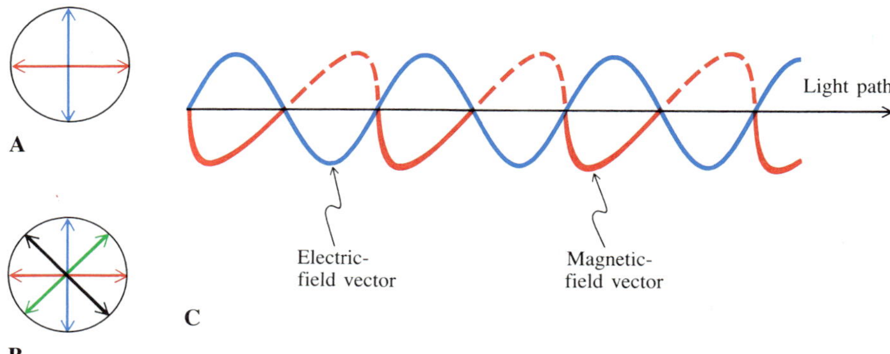

5-6) extending in all directions. The situation is different with plane-polarized light, in which the electric-field vectors lie in only *one* plane. Ordinary light becomes plane polarized if it is passed through a Nicol prism, which allows only one of the infinite number of planes of light to pass through. The resulting beam is said to be plane polarized (Figure 5-6). When light travels through a molecule, the electrons in the various bonds and around the nuclei interact with the electric field of the light beam. If the light is plane polarized, the electric-field vector may or may not change, depending on the sample. If the sample contains achiral molecules, the vector remains the same, and the sample is **optically inactive.** It is inactive because there is a mirror plane or a center of symmetry in such structures. In simplified terms, such symmetry cancels whatever effect one part of the molecule (or, rather, the component electrons) has on the light by its "mirrored" part. On the other hand, if a plane-polarized light beam is passed through a chiral substance, the electric field interacts differently with, say, the "left" half of the molecule compared with the "right" half (Figure 5-7). In solution, the chiral molecules have many different orientations; hence, the polarized light passes through them at all possible angles, with corresponding variations in rotation of the polarization plane. The resulting observed rotation is a macroscopic property—the net sum of many small molecular contributions. It is called **optical rotation,** and the sample giving rise to it is referred to as being **optically active.**

The instrument with which optical rotations are measured is a **polarimeter** (Figure 5-8). The light source used most frequently is a *monochromatic* (only one wavelength) sodium D lamp (λ = 5890 Å). The light is first plane polarized by a Nicol prism. It subsequently traverses a cell containing the sample. Rotation of the plane is detectable by another Nicol prism, the analyzer. In practice, this is simply done by rotating the prism to allow for maximum transmittance of the light beam to the eye of the observer. The measured difference

FIGURE 5-7
Uneven interaction of the oscillating electric field of plane-polarized light with a chiral molecule. "Left is not equal to right"; the consequence is a rotation of the plane of polarization.

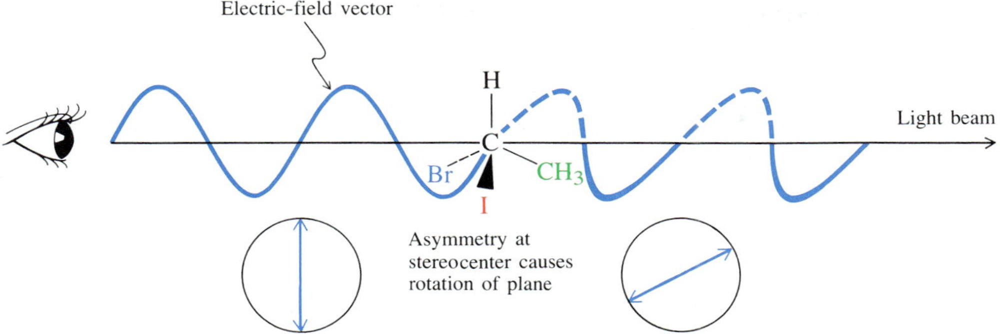

(in degrees) is the **observed optical rotation,** α, of the sample. Its value depends on the concentration and structure of the optically active molecule, the length of the sample cell, the wavelength of the light, the solvent, and the temperature. To avoid ambiguities, chemists have agreed upon the use of standard values of α, the **specific rotation,** $[\alpha]$, of a chiral compound. This quantity (which is solvent dependent) is defined as:

$$[\alpha]_{\lambda}^{t^{\circ}\mathrm{C}} = \frac{\alpha}{l \cdot c}$$

$[\alpha]$ = specific rotation
t = temperature in °C
λ = wavelength of incident light; for the sodium D lamp simply indicated by "D"
α = observed optical rotation
l = length of sample container in decimeters (tens of centimeters); its value is frequently 1 (i.e., 10 cm)
c = concentration (g ml^{-1} of solution)

FIGURE 5-8
A schematic diagram of a polarimeter.

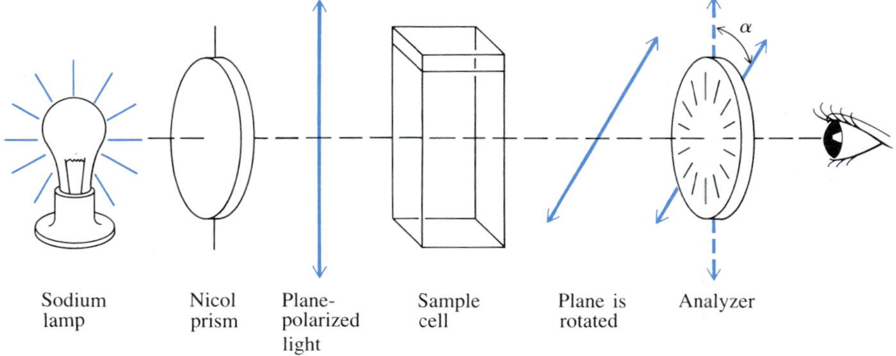

The specific rotation of an optically active molecule is a physical constant characteristic of that molecule, just like its melting point, boiling point, and density. Some specific rotations are recorded in Table 5-1.

TABLE 5-1
Specific rotations of some chiral molecules $[\alpha]_D^{25°C}$

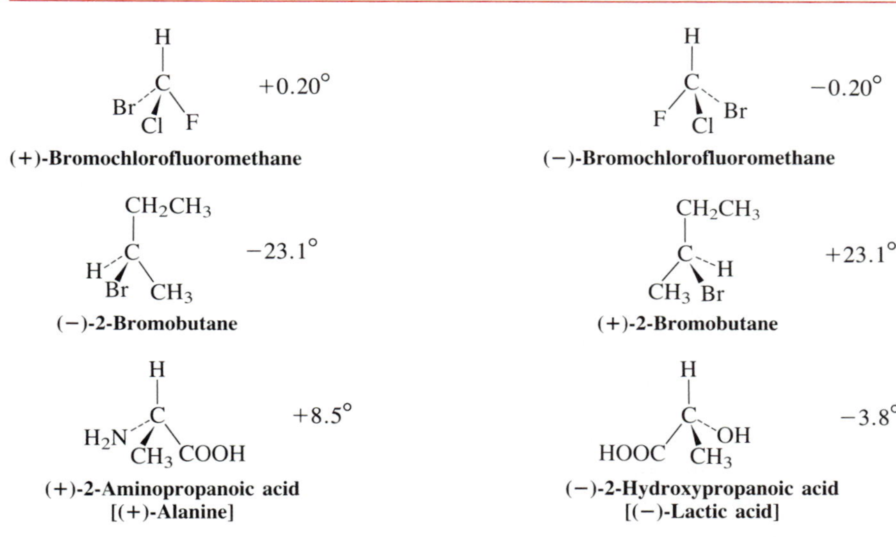

Note: Neat for the haloalkanes; in aqueous solution for the acids.

Specific Rotations in Mixtures of Enantiomers

As mentioned, enantiomers rotate plane-polarized light by equal amounts but in opposite directions. Thus, in 2-bromobutane the (−) enantiomer rotates this plane counterclockwise by 23.1°, its mirror image (+)-2-bromobutane clockwise by 23.1°. It follows that a racemate yields no rotation and is therefore optically inactive. [The term racemate is derived from *racemic acid,* an optically inactive tartaric acid (see Section 5-5) found in grapes; *racemus,* Latin, cluster of grapes.] To observe optical activity in a mixture of two enantiomers requires that more of one enantiomer be present than the other. Using the value of the specific rotation, we can calculate the composition of mixtures that are not enantiomerically pure. For example, if a solution of 2-bromobutane has an $[\alpha]$ value of only +11.55° (i.e., one-half the maximum possible for the pure dextrorotatory isomer), we can deduce that the mixture is composed of three parts of the (+) enantiomer and one part of the (−) one. The latter cancels the rotational effect of one part of its mirror image; thus, we can conclude that the mixture is 50% racemate (2 parts) and 50% pure (+) enantiomer (2 parts):

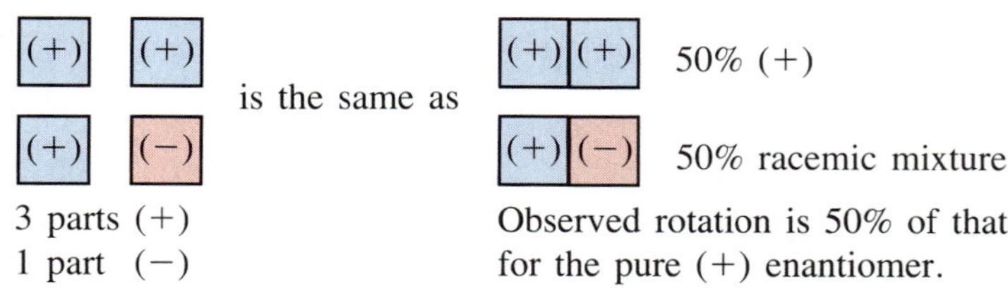

A mixture of this composition is said to be 50% *optically pure*. The specific rotation observed is only one-half of the maximum.

$$\% \text{ optical purity} = \left[\frac{[\alpha]_{\text{observed}}}{[\alpha]} \cdot 100\right]$$

If one enantiomer equilibrates with its mirror image by some process, it is said to undergo **racemization,** or to **racemize.** For example, racemization occurs by rotation in the *gauche* conformers of 1,2-dichloroethane (see Figure 5-3). We shall also look into examples of chemical reactions that lead to racemization.

In summary, two enantiomers can be distinguished by their optical activity—that is, their interaction with plane-polarized light as measured in a polarimeter. One enantiomer always rotates such light clockwise (dextrorotatory), the other counterclockwise (levorotatory) by the same amount. The specific rotation, $[\alpha]$, is a physical constant possible only for chiral molecules. The interconversion of enantiomers leads to racemization and the disappearance of optical activity.

5-3 Absolute Configuration: The *R-S* Sequence Rules

How do we determine which enantiomer is which? And, once we know the answer, is there a way in which we can unambiguously name an enantiomer and distinguish it from its mirror image? This section deals with these two questions.

The Absolute Configuration Is Not Related to the Sign of the Optical Rotation and Can Be Established by X-ray Analysis

How do we establish the structure of a pure enantiomer of one chiral compound? The relative arrangement of the atoms with respect to each other, the **relative configuration,** is the opposite for both enantiomers, but which is which? As already pointed out, virtually all the physical characteristics of one enantiomer are identical with those of its mirror image, except for the sign of optical rotation. Is there a correlation between the sign of optical rotation and the actual spatial arrangement of the substituent groups, the **absolute configuration?** Is it possible to determine the structure of an enantiomer by measurement of its $[\alpha]$ value? The answer to both questions is, unfortunately, no. There is no straightforward correlation between the sign of rotation and the structure of the particular enantiomer (other than the obvious one—namely, that it has to be the opposite of its mirror image). For example, although a specific stereostructure of 2-bromobutane has been assigned a positive $[\alpha]$ (Table 5-1), and its mirror image a negative $[\alpha]$, this assignment is not obvious and was in fact based on additional structural information. Such information can be obtained through a special type of *X-ray structural analysis* (anomalous dispersion) that directly furnishes the three-dimensional arrangement of the atoms of a molecule. Absolute configuration can also be established by *chemical correlation* with a structure whose own absolute configuration is known from an X-ray study.

BOX 5-2

Absolute Configuration: A Historical Note

Before the advent of X-ray crystallography, the absolute configurations of chiral molecules were unknown. It is amusing from a historical point of view that 2,3-dihydroxypropanal (glyceraldehyde), an important chemical relay point that may be converted into a variety of other chiral molecules, was arbitrarily assigned the correct absolute configuration, the dextrorotatory enantiomer being given a structure labeled ''D-glyceraldehyde.'' The label ''D'' did not refer to the sign of rotation of plane-polarized light but to the relative arrangement of the substituent groups, arbitrarily written as shown below.

H, O, HC, C, OH, CH_2OH $\quad [\alpha]_D^{25°C} = +8.7°$

D-(+)-2,3-Dihydroxypropanal
[D-(+)-Glyceraldehyde]

H, C, O, CH, HO, CH_2OH $\quad [\alpha]_D^{25°C} = -8.7°$

L-(−)-2,3-Dihydroxypropanal
[L-(−)-Glyceraldehyde]

The levorotatory isomer was called L-glyceraldehyde. All chiral compounds that could be chemically correlated with (+)-glyceraldehyde—that is, they could be converted into dextrorotatory glyceraldehyde by undergoing reactions that did not affect the configuration at the stereocenter—were assigned the D configuration, and their mirror images the L configuration. Examples of molecules with D and L configurations are shown below. Note that, although D-glyceraldehyde is dextrorotatory, other derivatives with the same stereostructure may be levorotatory. It was only in 1951 that the absolute configuration of D-glyceraldehyde became known and the original guess found to be correct.

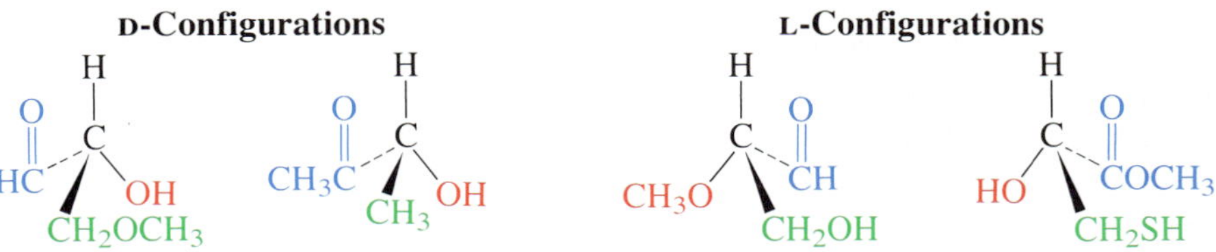

D,L nomenclature is still being used for sugars (Chapter 23) and amino acids (Chapter 27).

Naming Enantiomers: Handedness Is Labeled *R* and *S*

To unambiguously name enantiomers we need a system that allows us to indicate the handedness in the molecule, a sort of ''left hand'' versus ''right hand'' nomenclature. Such a system was developed by three chemists, R. S. Cahn, C. Ingold (both from London), and V. Prelog (Zürich).*

*Dr. Robert S. Cahn, 1899–1981, Fellow of the Royal Institute of Chemistry, London; Professor Christopher Ingold, 1893–1970, University College, London; Professor Vladimir Prelog, b. 1906, Swiss Federal Institute of Technology, Zürich, Nobel Prize 1975.

Although rules for naming chiral molecules without stereocenters also have been formulated, the rules presented here are restricted to stereoisomers containing unsymmetrically substituted tetrahedral carbon atoms, the situation commonly found in organic chemistry. The first step in establishing the handedness around such a carbon atom is to rank all four substituents as *a*, *b*, *c*, *d* in the order of decreasing priority. Priority is established by making use of sequence rules, as we shall see shortly. Substituent *a* has the highest priority, *b* the second highest, *c* the third, and *d* the lowest. Next, position the molecule (mentally, on paper, or by using your molecular model set) so that the lowest priority substituent is placed as far away from you as possible (Figure 5-9). This process results in two (and only two) possible arrangements of the remaining substituents. If the procession from *a* to *b* to *c* is counterclockwise, the configuration at the stereocenter is named *S* (*sinister,* Latin, left). On the other hand, if the procession is clockwise, the center is *R* (*rectus,* Latin, right). The symbol *R* or *S* is added as a prefix in parentheses to the name of the chiral compound, as in (*R*)-2-bromobutane and (*S*)-2,3-dihydroxypropanal. A racemic mixture is designated *R*,*S*, as in (*R*,*S*)-bromochlorofluoromethane. The sign of the rotation of plane-polarized light may be added (but is not necessary for the unambiguous labeling of a chiral compound), as in (*S*)-(+)-2-bromobutane and (*R*)-(+)-2,3-dihydroxypropanal. It is important to remember that the symbols *R* and *S* are *not* necessarily correlated with the sign of α.

FIGURE 5-9

Assignment of *R* or *S* configuration at a tetrahedral stereocenter. In many of the structural drawings in this chapter, the color scheme shown here is used to indicate the priority of substituents.

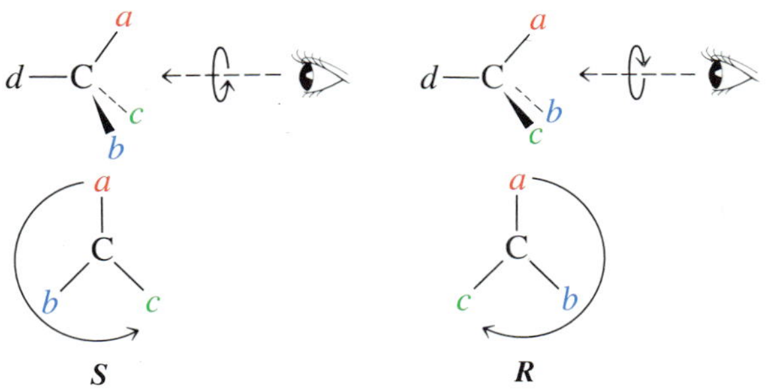

Assigning Priorities to Substituents: The Sequence Rules

We assign priorities to the substituents on a stereocenter by following the **sequence rules.**

RULE 1 Priority is established by the atomic numbers of the attached atoms. A substituent atom of higher atomic number takes precedence over one of lower atomic number. Consequently, the substituent of lowest priority is hydrogen.

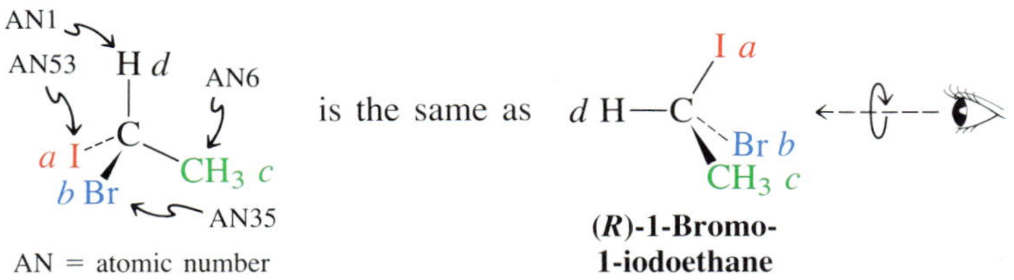

(*R*)-1-Bromo-1-iodoethane

RULE 2 If two substituents have the same rank when you consider the atoms directly attached to the stereocenter, rank the elements along the substituent chains until you reach a point of difference between the two chains at which a distinction in priority is possible.

For example, an ethyl substituent takes priority over methyl. Why? At the point of attachment to the stereocenter, each substituent has a carbon nucleus, equal in priority. Farther away from that center, however, methyl has only hydrogen atoms, but ethyl has a carbon atom (higher in priority).

$$\underset{\textbf{Methyl}}{-\overset{\overset{\displaystyle H}{|}}{\underset{\underset{\displaystyle H}{|}}{C}}-H} \quad \text{ranks lower in priority than} \quad \underset{\textbf{Ethyl}}{-\overset{\overset{\displaystyle H}{|}}{\underset{\underset{\displaystyle H}{|}}{C}}-\overset{\overset{\displaystyle H}{|}}{\underset{\underset{\displaystyle H}{|}}{C}}-H}$$

On the other hand, 1-methylethyl takes precedence over ethyl because, at the first carbon, ethyl bears only one other carbon substituent, but 1-methylethyl bears two. Similarly, 2-methylpropyl takes priority over butyl but ranks lower than 1,1-dimethylethyl.

$$\underset{\textbf{Ethyl}}{-\overset{\overset{\displaystyle H}{|}}{\underset{\underset{\displaystyle H}{|}}{C}}-CH_3} \quad \text{ranks lower in priority than} \quad \underset{\textbf{1-Methylethyl (Isopropyl)}}{-\overset{\overset{\displaystyle CH_3}{|}}{\underset{\underset{\displaystyle H}{|}}{C}}-CH_3}$$

$$\underset{\textbf{Butyl}}{-\overset{\overset{\displaystyle H}{|}}{\underset{\underset{\displaystyle H}{|}}{C}}-\overset{\overset{\displaystyle H}{|}}{\underset{\underset{\displaystyle H}{|}}{C}}-CH_2CH_3} \quad \text{ranks lower in priority than} \quad \underset{\textbf{2-Methylpropyl}}{-\overset{\overset{\displaystyle H}{|}}{\underset{\underset{\displaystyle H}{|}}{C}}-\overset{\overset{\displaystyle CH_3}{|}}{\underset{\underset{\displaystyle H}{|}}{C}}-CH_3}$$

$$\underset{\textbf{2-Methylpropyl}}{-\overset{\overset{\displaystyle H}{|}}{\underset{\underset{\displaystyle H}{|}}{C}}-\overset{\overset{\displaystyle CH_3}{|}}{\underset{\underset{\displaystyle H}{|}}{C}}-CH_3} \quad \text{ranks lower in priority than} \quad \underset{\textbf{1,1-Dimethylethyl (\textit{tert}-Butyl)}}{-\overset{\overset{\displaystyle CH_3}{|}}{\underset{\underset{\displaystyle CH_3}{|}}{C}}-CH_3}$$

Remember that the decision on priority is made at the *first* point of difference along otherwise similar substituent chains. Once that point is reached, the constitution of the remainder of the chain is irrelevant:

$$-\overset{\overset{\displaystyle H}{|}}{\underset{\underset{\displaystyle H}{|}}{C}}-CH_2OH \quad \text{ranks lower in priority than} \quad -\overset{\overset{\displaystyle CH_3}{|}}{\underset{\underset{\displaystyle H}{|}}{C}}-CH_3$$

$$-\overset{\overset{\displaystyle H}{|}}{\underset{\underset{\displaystyle H}{|}}{C}}-CH_2CH_2CCl_3 \quad \text{ranks lower in priority than} \quad -\overset{\overset{\displaystyle CH_3}{|}}{\underset{\underset{\displaystyle H}{|}}{C}}-CH_3$$

If you reach a point along a substituent chain at which it branches, choose the branch that is higher in priority. If two substituents have similar branches, rank the elements in those branches until you reach a point of difference:

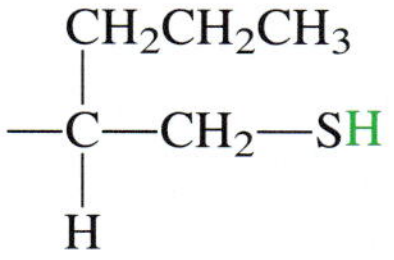

$$-\underset{\text{H}}{\overset{\text{CH}_2\text{CH}_2\text{CH}_3}{\text{C}}}-\text{CH}_2-\text{SH} \quad \text{ranks lower in priority than} \quad -\underset{\text{H}}{\overset{\text{CH}_2\text{CH}_2\text{CH}_3}{\text{C}}}-\text{CH}_2-\text{S}-\text{CH}_3$$

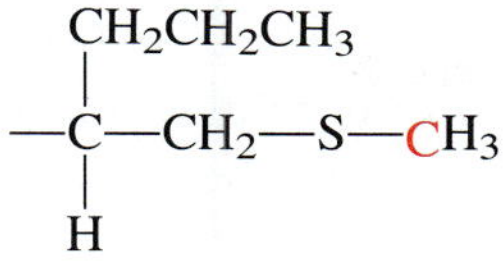

Examples:

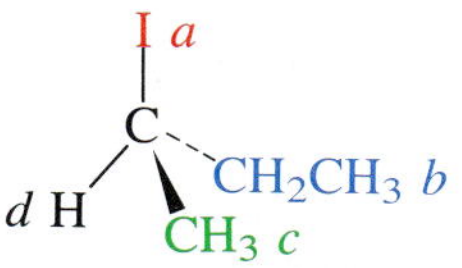

I *a*; CH_2CH_3 *b*; CH_3 *c*; *d* H
(*R*)-2-Iodobutane

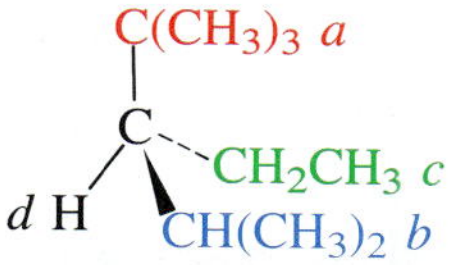

$C(CH_3)_3$ *a*; CH_2CH_3 *c*; $CH(CH_3)_2$ *b*; *d* H
(*S*)-3-Ethyl-2,2,4-trimethylpentane

RULE 3 Double and triple bonds are treated as if they were saturated, as shown:

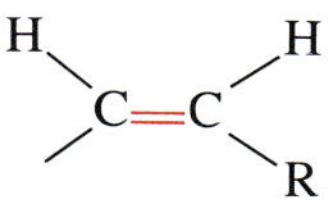

$$\text{H(H)C=C(H)R} \quad \text{is treated as} \quad -\underset{\text{C}}{\overset{\text{H}}{\text{C}}}-\underset{\text{C}}{\overset{\text{H}}{\text{C}}}-\text{R}$$

$$-\text{C}\equiv\text{C}-\text{R} \quad \text{is treated as} \quad -\underset{\text{C}}{\overset{\text{C}}{\text{C}}}-\underset{\text{C}}{\overset{\text{C}}{\text{C}}}-\text{R}$$

$$-\overset{\text{O}}{\overset{\|}{\text{C}}}-\text{H} \quad \text{is treated as} \quad -\underset{\text{H}}{\overset{\text{O}}{\text{C}}}-\text{O}-\text{C}$$

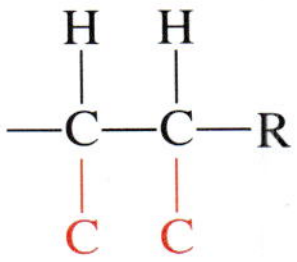

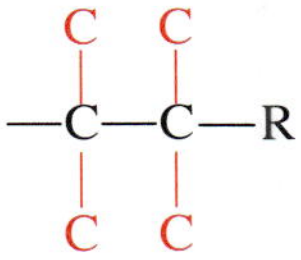

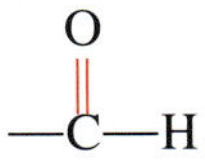

$$-\overset{\text{O}}{\overset{\|}{\text{C}}}-\text{OH} \quad \text{is treated as} \quad -\underset{\text{OH}}{\overset{\text{O}}{\text{C}}}-\text{O}-\text{C}$$

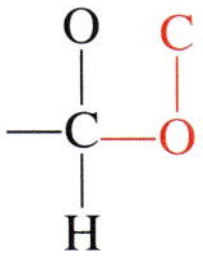

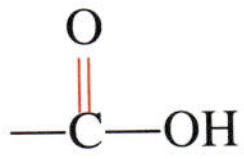

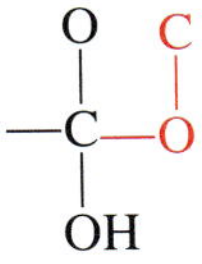

Examples:

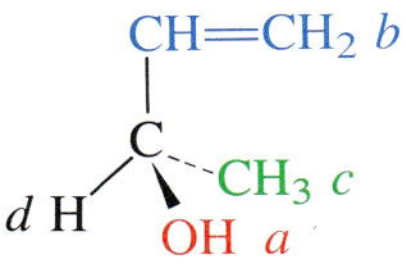

$CH{=}CH_2$ *b*; CH_3 *c*; OH *a*; *d* H
R

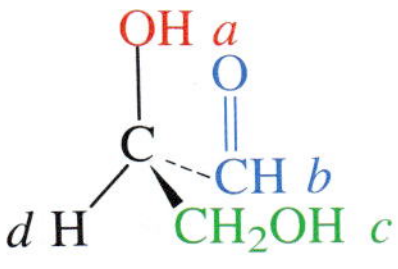

OH *a*; $\overset{\text{O}}{\|}$CH *b*; CH_2OH *c*; *d* H
R

EXERCISE 5-5

Assign the absolute configuration of the molecules depicted in Table 5-1.

EXERCISE 5-6

Draw one enantiomer (specify which) of your choice of: 2-chlorobutane, 2-chloro-2-fluorobutane, and $(HC{\equiv}C)(CH_2{=}CH)C(Br)(CH_3)$.

To correctly assign the stereostructure of stereoisomers, you must develop a fair amount of three-dimensional ''vision'' or ''stereoperception.'' In the structures that have been shown to illustrate the priority rules, the lowest-priority substituent has been located at the left of the carbon center and in the plane of the page and the remainder of the substituents at the right, the upper-right group also being positioned in this plane. However, this is not the only way of drawing dashed-wedged line structures; others are equally correct. Consider some of the structural drawings of (*S*)-2-bromobutane. These are simply different views of the same molecule.

Six Ways of Drawing (*S*)-2-Bromobutane

H—C(Br)(CH$_3$)(CH$_2$CH$_3$)   Br, H, C, CH$_2$CH$_3$, CH$_3$ Br—C(H)(CH$_2$CH$_3$)—CH$_3$ H$_3$C---C(H)(CH$_2$CH$_3$)---Br   Br, C—H, CH$_3$CH$_2$, CH$_3$ CH$_3$CH$_2$—C(Br)(H)—CH$_3$

It can be useful to employ chiral auxiliary aids to help you visualize stereochemistry. One such device is your left hand, which can function as a model for assigning absolute configuration (Figure 5-10). First, draw the molecule with the lowest-priority substituent pointing to the left, as was done in the presentation of the sequence rules. Then use your left hand as a model for the tetrahedral stereocenter to be named: the left wrist constitutes the lowest-priority group; the left index finger, middle finger, and thumb represent the various substituents. With the palm of your hand facing you, and your wrist (lowest-priority group) pointing away from you, mark the fingers with a pencil according to their priorities. The absolute configuration becomes clearly visible in this way.

FIGURE 5-10

The left hand as a model of a chiral molecule and the configurational assignment of (*R*)-2-bromobutane.

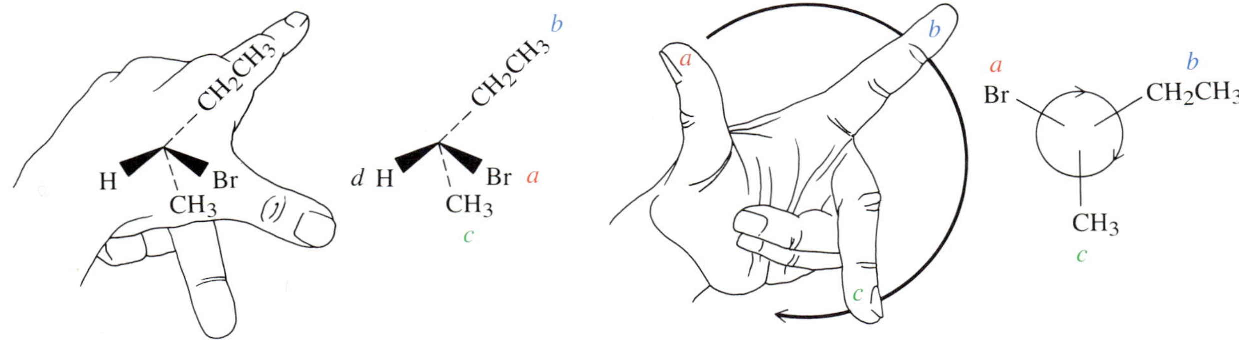

To summarize, the sign of optical rotation cannot be used to establish the absolute configuration of a stereoisomer. Instead, special methods of X-ray diffraction (or chemical correlations) must be used. We can express the absolute configuration of the chiral molecule as *R* or *S* by applying the sequence rules, which allows us to rank all substituents in order of decreasing priority. Turning the structures so as to place the lowest-priority group at the back causes the remaining substituents to be arranged in clockwise (*R*) or counterclockwise (*S*) fashion.

5-4 Fischer Projections

A **Fischer* projection** is a standard way of depicting tetrahedral carbon atoms and their substituents in two dimensions. With this method, the molecule is drawn in the form of a cross, the central carbon being at the point of intersection. The horizontal lines signify bonds directed *toward* the viewer; the vertical lines are pointing *away*. Dashed-wedged line structures have to be arranged in this way to facilitate their conversion into Fischer projections, as shown for 2-bromobutane.

Conversion of the Dashed-Wedged Line Structures of 2-Bromobutane into Fischer Projections

H
Br—C—CH_2CH_3
CH_3
Dashed-wedged line structure

H
Br—+—CH_2CH_3
CH_3
Fischer projection

(*R*)-2-Bromobutane

H
CH_3CH_2—C—Br
CH_3
Dashed-wedged line structure

H
CH_3CH_2—+—Br
CH_3
Fischer projection

(*S*)-2-Bromobutane

We must be very careful in the manipulation of Fischer projections. For example, rotation in the plane by 90° gives the structure of the enantiomer. It follows then that similar rotation by 180° gives the original enantiomer. You can verify this statement by drawing conventional dashed-wedged line structures or by looking at molecular models. It is best not to rotate Fischer projections in this manner because you may quickly lose the correct absolute configuration.

H / Br—+—CH_2CH_3 / CH_3 (*R*) $\xrightarrow{90° \text{ turn}}$ Br / H_3C—+—H / CH_2CH_3 (*S*) is the same as Br / H_3C—C—H / CH_2CH_3 (*S*)

H / CH_3CH_2—+—Br / CH_3 (*S*) $\xrightarrow{180° \text{ turn}}$ CH_3 / Br—+—CH_2CH_3 / H (*S*) is the same as CH_3 / Br—C—CH_2CH_3 / H (*S*)

*Professor Emil Fischer, 1852–1919, University of Berlin, Nobel Prize 1902.

EXERCISE 5-7

Draw Fischer projections for all the molecules in Exercises 5-5 and 5-6.

How to Interconvert Fischer Projections and Maintain Absolute Configuration

As is the case for dashed-wedged line structures, there are several Fischer projections of the same enantiomer, a situation that may lead to confusion. How can we quickly ascertain that two Fischer projections are depicting the same enantiomer rather than image and mirror image? We have to find a sure way to convert one Fischer projection into another in a manner that either leaves the absolute configuration unchanged or converts it into its opposite. It turns out that this task can be achieved by simply having substituent groups trade places. As you can readily verify by using molecular models, any single such exchange turns one enantiomer into its mirror image. Two such exchanges (you may select different substituents every time) produce the original absolute configuration.

Changes of Absolute Configuration on Switching Substituents in Fischer Projections

CH_3 (top), Cl (left), Br (right), CH_2CH_3 (bottom) — *S* $\xrightarrow{CH_3 \leftrightarrow Br}$ Br (top), Cl (left), CH_3 (right), CH_2CH_3 (bottom) — *R* $\xrightarrow{CH_3CH_2 \leftrightarrow Br}$ CH_2CH_3 (top), Cl (left), CH_3 (right), Br (bottom) — *S*

(The double arrow denotes two groups trading places.)

It is now easy to establish whether two different Fischer projections depict the same configuration or opposite ones. If the conversion of one structure into another takes an even number of exchanges, the structures are identical. If it requires an odd number of such moves, the structures are mirror images of each other.

Consider, for example, the two Fischer projections A and B. Do they represent molecules having the same configuration? The answer is found quickly. We convert A into B by two exchanges; so A equals B.

Cl (top), H (left), CH_3 (right), CH_2CH_3 (bottom) — **A** $\xrightarrow{H \leftrightarrow Cl}$ H (top), Cl (left), CH_3 (right), CH_2CH_3 (bottom) $\xrightarrow{Cl \leftrightarrow CH_3}$ H (top), H_3C (left), Cl (right), CH_2CH_3 (bottom) — **B**

BOX 5-3

A Simple Way to Establish *R* and *S* Configuration by the Use of Fischer Projections

Although there is no substitute for an accurate perception of space when dealing with stereochemical problems, Fischer projections allow you to assign absolute configurations without having to visualize the three-dimensional arrangement of the atoms. For this purpose, first draw the molecule as a (any) Fischer projection. Next, rank all the substituents in accord with the sequence rules. Finally, exchange two groups so that the lowest-priority substituent is at the top, and then exchange any other pair (to make sure that the absolute configuration stays unchanged from the original). On completion of these manipulations, you will find that the three groups of priority, *a*, *b*, and *c*, turn out to be arranged in either clockwise or counterclockwise fashion, in turn corresponding to either the *R* or the *S* configuration.

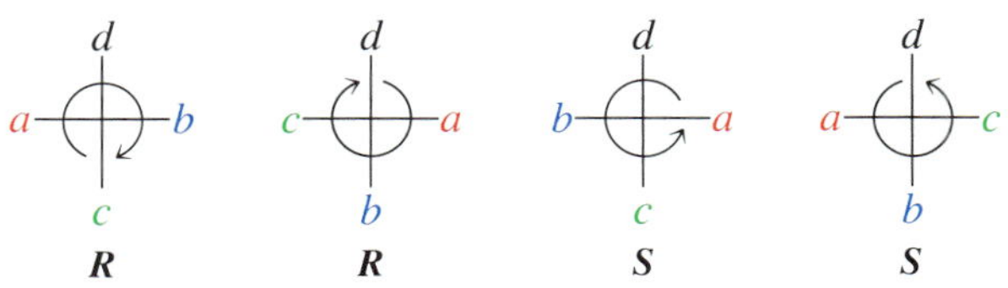

EXERCISE 5-8

What is the absolute configuration of the following molecules?

Br (top), H (left), D (right), CH_3 (bottom)

Cl (top), F (left), Br (right), I (bottom)

CH_3 (top), H_2N (left), COH with C=O (right), H (bottom)

In summary, a Fischer projection is a convenient way of drawing chiral molecules. Do not rotate such projections in the plane. Switching substituents reverses absolute configuration, if done an odd number of times, but leaves it intact when the number of such exchanges is even. By placing the lowest-priority substituent on top, the absolute configuration is readily assigned.

5-5
Molecules Incorporating Several Stereocenters: Diastereomers

Molecules containing several carbon atoms that bear all dissimilar substituents are the subject of this section. Because the configuration around each center can be *R* or *S*, a variety of possible structures emerge, all of which are isomeric.

The Chlorination of 2-Bromobutane at C-3 Produces Several Stereoisomers

Section 5-1 described how a carbon-based stereocenter can be created by the free-radical monohalogenation of butane. Let us now consider the chlorination of 2-bromobutane to give (among other products) 2-bromo-3-chlorobutane. Our starting material is chiral and, as a racemate, exists as an equimolar mixture of *R* and *S* isomers. The introduction of a chlorine atom at C-3 produces a new stereocenter in the molecule. This center may have either the *R* or the *S* configuration, assignable by using the same sequence rules as apply to molecules with only one such center.

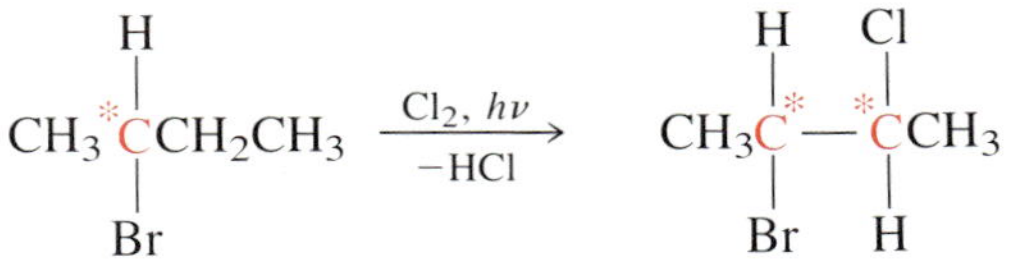

$$CH_3\overset{*}{C}(H)(Br)CH_2CH_3 \xrightarrow[-HCl]{Cl_2,\ h\nu} CH_3\overset{*}{C}(H)(Br)-\overset{*}{C}(Cl)(H)CH_3$$

One stereocenter **2-Bromo-3-chlorobutane Two stereocenters**

How many stereoisomers are possible for 2-bromo-3-chlorobutane? This question can be answered by completing a simple exercise in permutation: each stereocenter can be either *R* or *S*, and, hence, the possible combinations are: *RR*, *RS*, *SR*, and *SS*. There are four stereoisomers. These may be drawn as either dashed-wedged line structures or Fischer projections (Figure 5-11). When using Fischer projections in making stereochemical assignments, it is best to view each stereocenter separately and regard the substituent group containing the other as a simple appendage. This is particularly important when interconverting Fischer projections in an effort to assign the stereochemistry at any one center.

FIGURE 5-11

The four stereoisomers of 2-bromo-3-chlorobutane: (left) dashed-wedged line pictures in which all structures are drawn in their staggered conformations; (right) Fischer projections and the stereochemical relations of the four stereoisomers.

Mirror plane

(2*R*,3*R*)-2-Bromo-3-chlorobutane **(2*S*,3*S*)-2-Bromo-3-chlorobutane**

(2*R*,3*S*)-2-Bromo-3-chlorobutane **(2*S*,3*R*)-2-Bromo-3-chlorobutane**

Mirror plane

Enantiomers

Diastereomers

Enantiomers

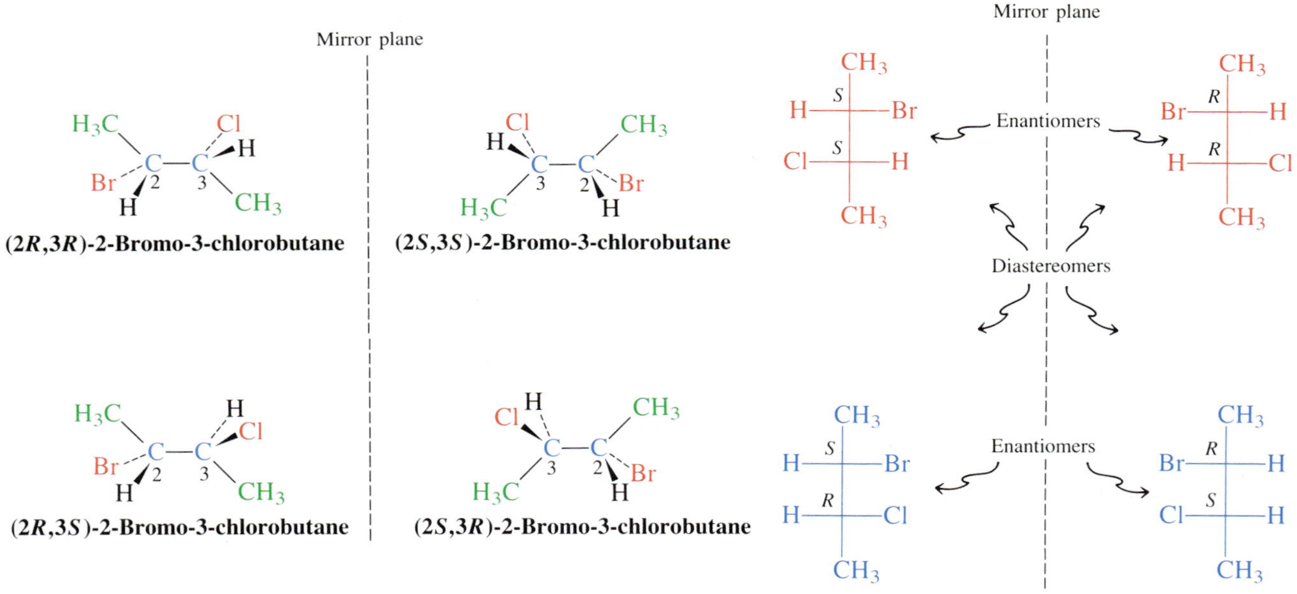

By looking closely at the structures of the four stereoisomers of 2-bromo-3-chlorobutane, we see that there are two related pairs of compounds: an *R,R*/*S,S* pair and an *R,S*/*S,R* pair. The members of each individual pair are mirror images of each other and therefore enantiomers. On the other hand, each member of one pair is not a mirror image of any member of the other pair; hence, the two pairs are not enantiomeric with respect to each other. Stereoisomers that are not related as object and mirror image are called **diastereomers** (*dia*, Greek, across). Figure 5-11 shows that 2-bromo-3-chlorobutane exists as two diastere-

BOX 5-4

Assigning the Absolute Configuration of a Stereocenter in a 2-Bromo-3-chlorobutane

CH_3; H—C—Br; Cl—C—H; CH_3 — Write as a tetrasubstituted methane → $CHBrCH_3$; Cl—C—H; CH_3

PROBLEM: *R* or *S*?

Double exchange ↓

H; H_3C—C—$CHBrCH_3$; Cl

S

SOLUTION: The center under scrutiny is *S*.

omers, each of which forms a pair of enantiomers. In contrast with enantiomers, diastereomers are distinct molecules with *different physical and chemical properties*. They can be separated from each other by fractional distillation or crystallization or by chromatography. They have different physical properties, such as melting and boiling points and densities, just as structural isomers do; for example, 1,2- and 1,3-bromochlorobutane. They also have different specific rotations.

EXERCISE 5-9

What are the stereochemical relations (identical, enantiomers, diastereomers) of the following four molecules? Assign absolute configurations at each stereocenter.

1: CH_3; H—C—F; H—C—CH_2CH_3; CH_3

2: H; H_3C—C—F; H—C—CH_2CH_3; CH_3

3: CH_3; F—C—H; CH_3CH_2—C—H; CH_3

4: H; H_3C—C—CH_2CH_3; F—C—H; CH_3

Diastereomers: The Cyclic Analogy

It is instructive to compare the stereoisomers of 2-bromo-3-chlorobutane with those of a cyclic analog, 1-bromo-2-chlorocyclobutane (Figure 5-12). In analogy to the open-chain derivative, there are four stereoisomers: *R,R; S,S; R,S;* and *S,R*. In the cyclic compound, however, the stereoisomeric relation of the first pair to the second is easily recognized: one pair has cis stereochemistry, the other trans. Cis and trans isomers (Section 4-1) in cycloalkanes are in fact diastereomers.

FIGURE 5-12
The diastereomeric relation of *cis*- and *trans*-1-bromo-2-chlorocyclobutane.

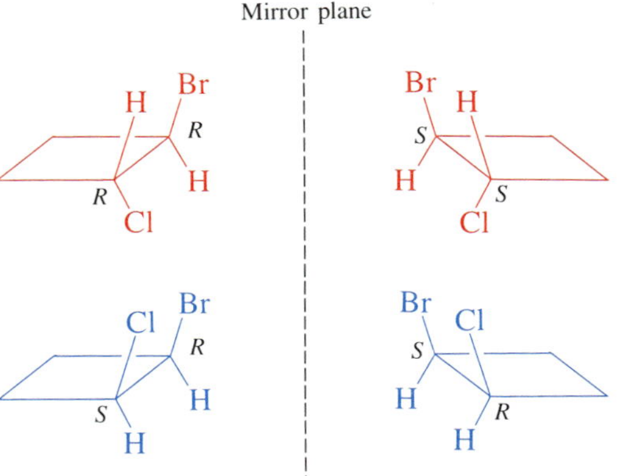

Two Equally Substituted Stereocenters Give Rise to Only Three Stereoisomers

The molecule 2-bromo-3-chlorobutane contains two stereocenters that differ in their substitution pattern, one being distinguished from the other by the difference in the nature of the halogen. How many stereoisomers are to be expected if both centers are equally substituted? Such a situation holds for 2,3-dibromobutane, which can be obtained by the free-radical bromination of 2-bromobutane. As we did for 2-bromo-3-chlorobutane, we have to consider four structures, resulting from the various permutations in *R* and *S* configurations (Figure 5-13).

$$CH_3\overset{*}{C}(H)(Br)CH_2CH_3 \xrightarrow[-HBr]{Br_2,\ h\nu} H_3C-\overset{*}{C}(H)(Br)-\overset{*}{C}(Br)(H)-CH_3$$

One stereocenter | **2,3-Dibromobutane Two stereocenters**

The first pair of stereoisomers, with *R,R* and *S,S* configurations, is clearly recognizable as a pair of enantiomers. However, a close look at the second pair reveals that image (*S,R*) and mirror image (*R,S*) are superimposable and therefore identical. Thus, the *S,R* diastereomer of 2,3-dibromobutane is achiral and not optically active, even though it contains two stereocenters. The identity of the two structures can be confirmed readily by using molecular models. A compound that contains two (or, as we shall see, even more than two) stereocenters but that is superimposable with its mirror image is a **meso compound** (*mesos,* Greek, middle). A characteristic feature of a meso compound is that it *contains a mirror plane,* which mirrors one stereocenter (or several of them) into the other. For example, in 2,3-dibromobutane, the 2*R* center is mirrored into the 3*S* center. This is best seen in an eclipsed dashed-wedged line structure (Figure 5-14). Recall that the presence of a mirror plane in a molecule makes it achiral (Section 5-1).

FIGURE 5-13

The stereochemical relations of the stereoisomers of 2,3-dibromobutane.

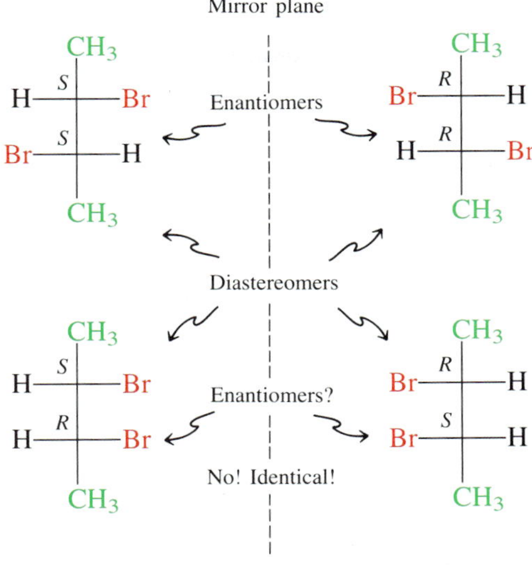

FIGURE 5-14

meso-2,3-Dibromobutane contains a mirror plane.

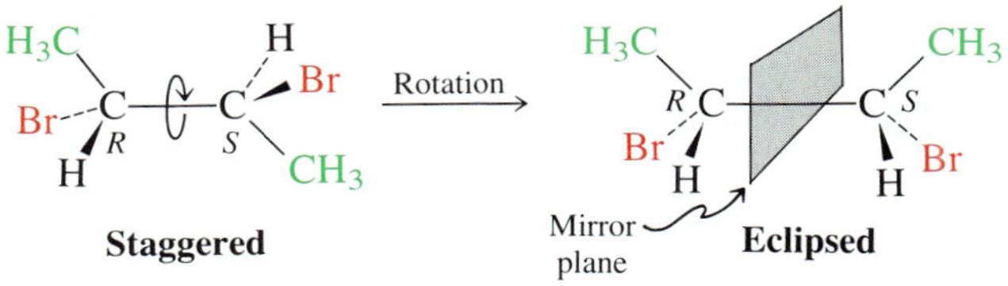

1,2-Dibromocyclobutane Is the Cyclic Analog of 2,3-Dibromobutane

It is again instructive to compare the stereochemical situation in 2,3-dibromobutane with that in an analogous cyclic molecule: 1,2-dibromocyclobutane. You can see that *trans*-1,2-dibromocyclobutane exists as two enantiomers (R,R and S,S) and may therefore be optically active. The cis isomer, on the other hand, has a mirror plane and is meso, achiral, and optically inactive (Figure 5-15).

FIGURE 5-15

Trans isomer of 1,2-dibromocyclobutane is chiral; cis isomer is a meso compound.

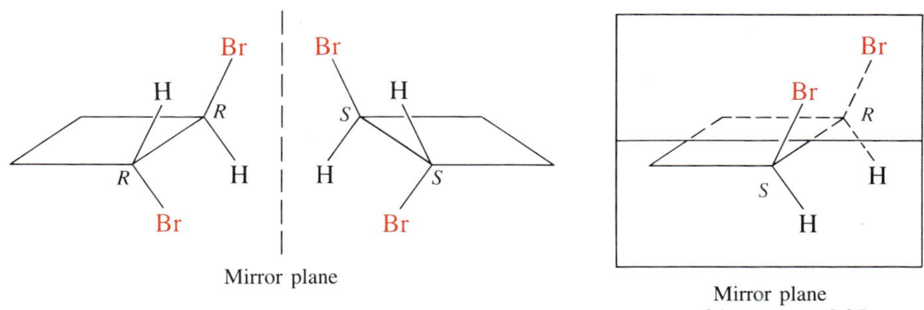

BOX 5-5

The Stereoisomers of Tartaric Acid

Tartaric acid (systematic name, 2,3-dihydroxybutanedioic acid) is a naturally occurring dicarboxylic acid containing two stereocenters with equal substitution patterns. Therefore it exists as a pair of enantiomers (which have identical physical properties but which rotate plane-polarized light in opposite directions) and as an achiral meso compound (with different physical and chemical properties from those of the chiral diastereomers).

COOH	COOH	COOH
H—$\overset{R}{\text{C}}$—OH	HO—$\overset{S}{\text{C}}$—H	H—$\overset{R}{\text{C}}$—OH
HO—$\overset{R}{\text{C}}$—H	H—$\overset{S}{\text{C}}$—OH	H—$\overset{S}{\text{C}}$—OH
COOH	COOH	COOH
(+)-Tartaric acid	**(−)-Tartaric acid**	***meso*-Tartaric acid**
$[\alpha]_D^{20°C} = +12.0°$	$[\alpha]_D^{20°C} = -12.0°$	$[\alpha]_D^{20°C} = 0°$
m.p. 168°–170°C	m.p. 168°–170°C	m.p. 146°–148°C
Density (g ml^{-1}) $d = 1.7598$	$d = 1.7598$	$d = 1.666$

The dextrorotatory enantiomer of tartaric acid is widely distributed in nature, being present in many fruits (fruit acid). The monopotassium salt is found as a deposit during the fermentation of grape juice. Levorotatory tartaric acid is rare. The racemate is also called racemic acid, as mentioned earlier. Like the levorotatory enantiomer, the meso isomer does not seem to be prevalent in nature.

Tartaric acid is of historical significance, because it was the first chiral molecule whose racemate was separated into the two enantiomers. This happened in 1848, long before it was recognized that carbon could be tetrahedral in organic molecules. By 1848, natural tartaric acid had been shown to be dextrorotatory, and the racemate was known. The French chemist Louis Pasteur* obtained a sample of the mixed sodium ammonium salt of racemic acid and noticed that there were two types of crystals: one set was the mirror image of the second. In other words, the crystals were chiral. By manually separating the two sets, dissolving them in water and measuring their optical rotation, he found one of the crystalline forms to be the pure salt of (+)-tartaric acid and the other to be the levorotatory form. Remarkably, the chirality of the individual molecules in this rare case had given rise to the macroscopic property of chirality in the crystal. Pasteur concluded from his observation that the molecules themselves must be chiral. These findings and others led in 1874 to the first proposal by van't Hoff and Le Bel,† independently, that saturated carbon has a tetrahedral—and not, for example, a square planar—bonding arrangement. (Why is the idea of a planar carbon incompatible with that of a stereocenter?)

---

*Professor Louis Pasteur, 1822–1895, Sorbonne, Paris.

†Professor Jacobus H. van't Hoff, 1852–1911, University of Amsterdam, Nobel Prize 1901; Dr. Joseph A. Le Bel, 1847–1930, Ph.D. from Sorbonne, Paris.

EXERCISE 5-10

Draw all the other possible isomeric dibromocyclobutanes. Which ones are chiral, which are achiral, and which are meso? Do the same for dibromocyclopentane.

Stereocenters Need Not Be Adjacent to Give Rise to Diastereomers

So far, the discussion in this section has centered on compounds having two adjacent carbon-based stereocenters. However, the same stereochemical relations hold for systems in which such centers are separated by one or more atoms: if the two stereocenters are different, four stereoisomers are again possible. However, if the substitution pattern of both stereocenters is the same, there will be three stereoisomers: two enantiomers and one meso compound. Examples are 2-bromo-4-chloro- and 2,4-dichloropentane. The first example has four stereoisomers, the second only three (only one enantiomer of each diastereomer is shown).

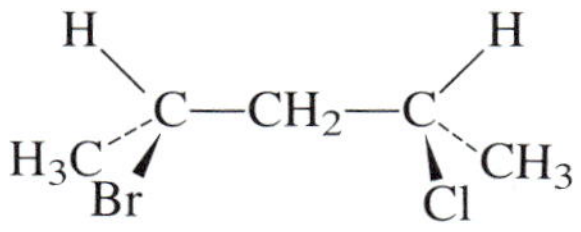

H, H_3C, Br — C—CH_2—C — H, CH_3, Cl

(2*R*,4*S*)-2-Bromo-4-chloropentane

H, H_3C, Br — C—CH_2—C — H, Cl, CH_3

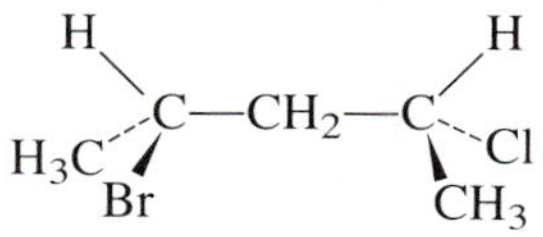

(2*R*,4*R*)-2-Bromo-4-chloropentane

Mirror plane

H, H_3C, Cl — C—CH_2—C — H, CH_3, Cl

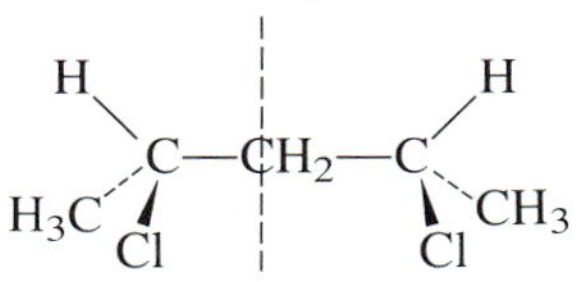

***meso*-2,4-Dichloropentane**

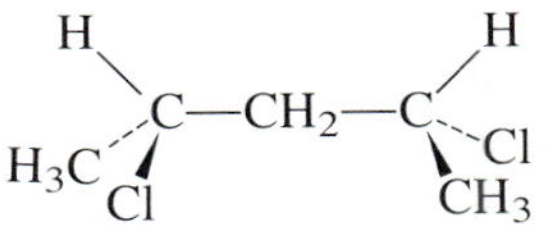

H, H_3C, Cl — C—CH_2—C — H, Cl, CH_3

(2*R*,4*R*)-2,4-Dichloropentane

More Than Two Stereocenters: More Stereoisomers

What structural variety do we expect for a compound having three stereocenters? We may again approach this problem by permuting the various possibilities. If we label the three centers consecutively as either *R* or *S*, the following sequence emerges:

RRR *RRS* *RSR* *SRR* *RSS* *SRS* *SSR* *SSS*

a total of eight stereoisomers. They can be arranged to reveal a division into four enantiomer pairs of diastereomers:

Image:	*RRR*	*RRS*	*RSS*	*SRS*
Mirror image:	*SSS*	*SSR*	*SRR*	*RSR*

The total number of stereoisomers decreases if there are meso forms. Generally, *a compound with* n *stereocenters can have a maximum of* 2^n *stereoisomers*. Therefore, a compound having three such centers gives rise to a maximum of eight stereoisomers; one having four produces sixteen; one having five, thirty-two; and so forth. The structural possibilities are quite staggering for larger systems.

EXERCISE 5-11

Draw all the stereoisomers of 2-bromo-3-chloro-4-fluoropentane.

In summary, the introduction of more than one stereocenter into a molecule gives rise to diastereomers. These are stereoisomers that are not related to each other as object and mirror image. In cyclic compounds, diastereomers are cis and trans isomers. The incorporation of n asymmetric centers gives rise to a maximum of 2^n stereoisomers. This number can decrease if the molecule has properties of symmetry, such as the presence of mirror planes.

5-6 Stereochemistry in Chemical Reactions

This section contains a detailed description of how a chemical reaction introduces chirality into a molecule. In particular, we shall understand more clearly why the conversion of achiral butane into chiral 2-bromobutane gives racemic material. We shall also see that the chiral environment of a stereocenter already present in the molecule exerts some control on the stereochemistry of a reaction that introduces a second one. Let us begin with another examination of the free-radical bromination of butane.

Why the Bromination of Butane Results in Racemic 2-Bromobutane

The free-radical bromination of butane at C-2 creates a chiral molecule (Section 5-1). This happens because one of the methylene hydrogens is replaced by a new substituent, furnishing a new stereocenter—a carbon atom with four different substituents. There are two such hydrogens; exchanging one for bromine gave one enantiomer, whereas exchanging the other gave the mirror image. Such a pair of hydrogens is called **enantiotopic,** because the environment around one is the mirror image of the environment around the other.

Because these hydrogens are chemically equivalent in this reaction, they will be abstracted by the attacking bromine atom at the same rate. However, consideration of the mechanism of free radical halogenation (Sections 3-5 and 3-6) shows that this step does not generate a new stereocenter, because it furnishes a planar and therefore achiral radical. The radical center also has two enantiotopic reaction sites—the two lobes of the *p* orbital (Figure 5-16); it is this species that gives rise to a chiral product. Because the two sites are equally prone to attack by bromine, the reaction leads to racemic 2-bromobutane. You can see that the two transition states resulting in the respective enantiomers are mirror images of each other. They are enantiomeric and therefore energetically equivalent. The rates of formation of *R* and *S* product are hence equal, and a racemate is formed. In general, *the formation of chiral compounds* (e.g., 2-bromobutane) *from achiral reactants* (e.g., butane and bromine) *yields racemates*. Or, *optically inactive reagents furnish optically inactive products*.*

---

*We shall see later that it *is* possible to generate optically active products from optically inactive starting materials, if we use an optically active reagent.

FIGURE 5-16

The creation of racemic 2-bromobutane from butane by free-radical bromination at C-2.

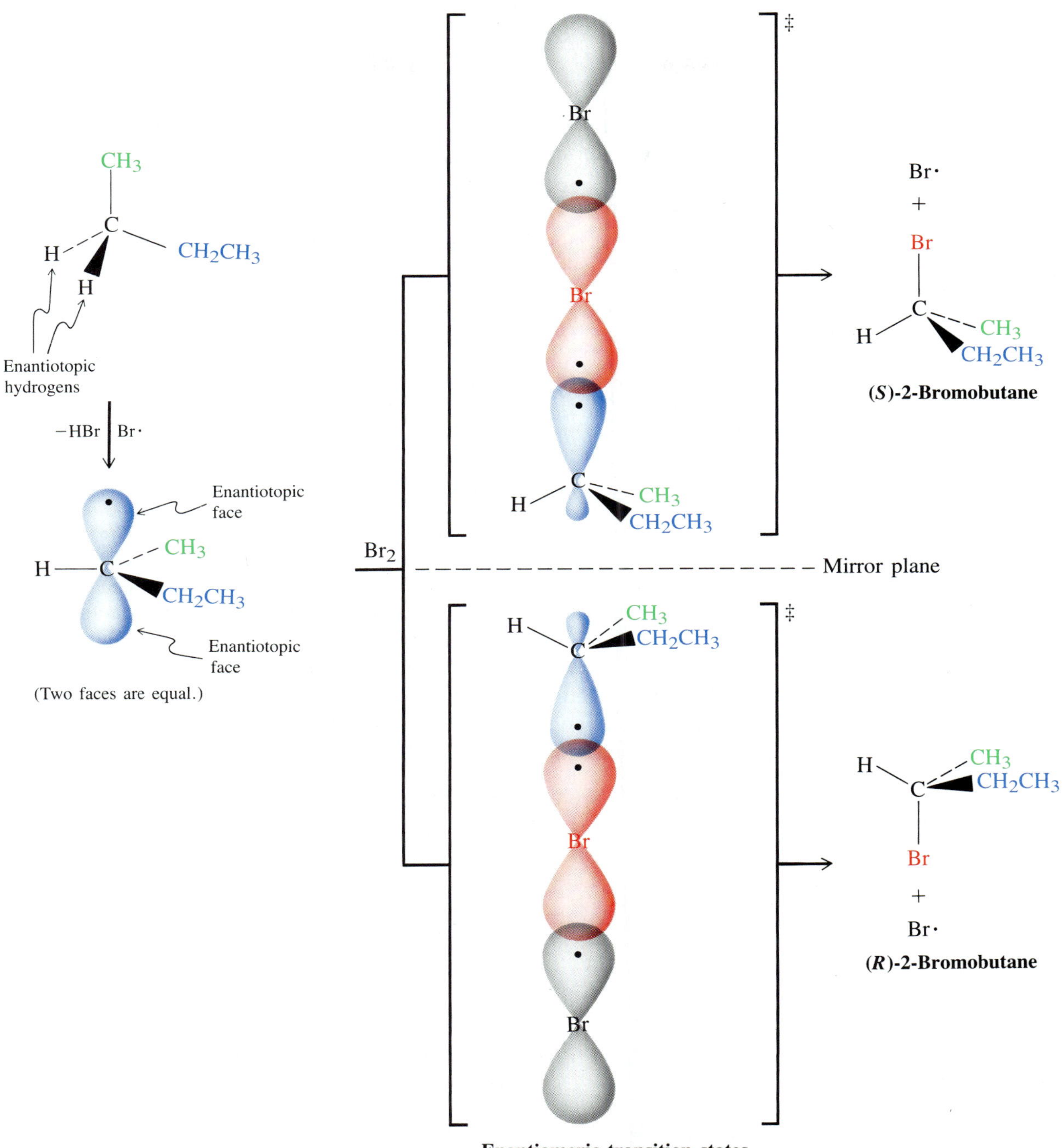

The Effect of the Presence of a Stereocenter: The Chlorination of (*S*)-2-Bromobutane

Now that we understand that the halogenation of an achiral molecule gives a racemic halide we may ask: What products can we expect from the halogenation of a chiral and enantiomerically pure molecule?

For example, consider the free-radical chlorination of the *S* enantiomer of 2-bromobutane. In contrast with the simplified picture presented in Section 5-5, in this case the chlorine atom has several options for attack: the two terminal methyl groups, the single hydrogen at C-2, and the two hydrogens on C-3. Let us examine each of these reaction paths.

Chlorination of (*S*)-2-Bromobutane at C-1

$^1CH_3$; H—^{2}C—Br; H—^{3}C—H; $^4CH_3$ $\xrightarrow[-HCl]{Cl_2,\ h\nu}$ $^1CH_2Cl$; H—^{2}C—Br; H—^{3}C—H; $^4CH_3$

2*S*
Optically active

2*R*
Optically active

Chlorination of the methyl group that includes C-1 is straightforward and proceeds through the primary radical to give 2-bromo-1-chlorobutane. The resulting compound is optically active because the original stereocenter is still present. Note, however, that the conversion of the methyl group into a chloromethyl one changes the sequence of priorities around C-2. Thus, although the stereocenter itself did not participate in the reaction, its designated configuration changed from *S* to *R*. This change is also indicated by the use of color: the priority sequence progresses from red (highest priority) to blue to green.

What about halogenation at C-2, the stereocenter?

Chlorination of (*S*)-2-Bromobutane at C-2

$^1CH_3$; H—^{2}C—Br; H—^{3}C—H; $^4CH_3$ $\xrightarrow[-HCl]{Cl_2,\ h\nu}$ $^1CH_3$; Cl—^{2}C—Br; H—^{3}C—H; $^4CH_3$ + $^1CH_3$; Br—^{2}C—Cl; H—^{3}C—H; $^4CH_3$

50:50

2*S*
Optically active

2*S* **2*R***
Racemic
(Optically inactive)

The product from chlorination at C-2 of (*S*)-2-bromobutane is 2-bromo-2-chlorobutane. The reaction takes place at the stereocenter, but the molecule remains chiral even though the substitution pattern has changed. However, an attempt to measure the α value for the product will reveal the absence of optical activity—the compound is racemic. How can this be explained?

A racemate forms because hydrogen abstraction from C-2 furnishes a planar and achiral radical. Chlorination from either side through enantiomeric transition states of equal energy, as in the bromination of butane (Figure 5-16), produces (*S*)- and (*R*)-2-bromo-2-chlorobutane at equal rates, giving a racemic mixture (Figure 5-17). The reaction is an example of a transformation in which an optically active compound leads to an optically inactive product (a racemate).

FIGURE 5-17
Chlorination of (*S*)-2-bromobutane at C-2 gives racemic product. Again, the color scheme indicates the priority of the groups in starting materials and products.

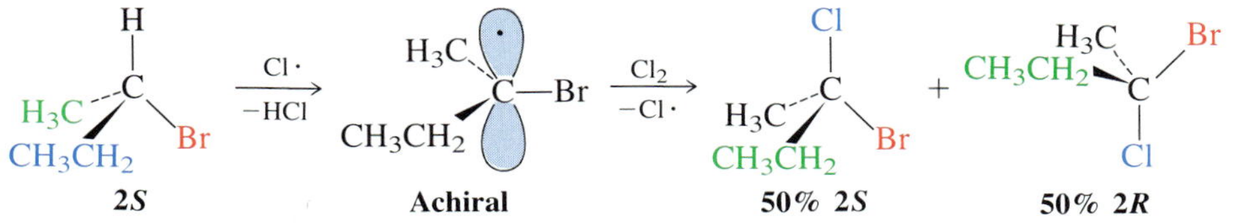

H, H_3C, CH_3CH_2, Br — **2*S*** $\xrightarrow[-HCl]{Cl\cdot}$ H_3C, CH_3CH_2, C—Br — **Achiral** $\xrightarrow[-Cl\cdot]{Cl_2}$ Cl, H_3C, CH_3CH_2, Br — **50% 2*S*** + H_3C, CH_3CH_2, Br, Cl — **50% 2*R***

EXERCISE 5-12

What other halogenations of (*S*)-2-bromobutane would furnish optically inactive products?

The chlorination of (*S*)-2-bromobutane at C-3 results in a second stereocenter and therefore gives rise to diastereomers. Specifically, replacement of the left-hand hydrogen at C-3 in the drawing below gives (2*S*,3*S*)-2-bromo-3-chlorobutane, whereas replacement of the right-hand hydrogen gives its diastereomer (2*S*,3*R*)-2-bromo-3-chlorobutane. Because the two hydrogens at C-3 are not equivalent, they are called **diastereotopic** and give rise to diastereomers upon substitution, in contrast with the enantiotopic methylene hydrogens in butane.

Chlorination of (*S*)-2-Bromobutane at C-3

1 CH_3; H—2—Br; H—3—H; 4 CH_3 — **2*S*** **Optically active** $\xrightarrow[-HCl]{Cl_2,\ h\nu}$ 1 CH_3; H—2—Br; H—3—Cl; 4 CH_3 — **2*S*,3*R*** **Optically active** + 1 CH_3; H—2—Br; Cl—3—H; 4 CH_3 — **2*S*,3*S*** **Optically active**

(Unequal amounts)

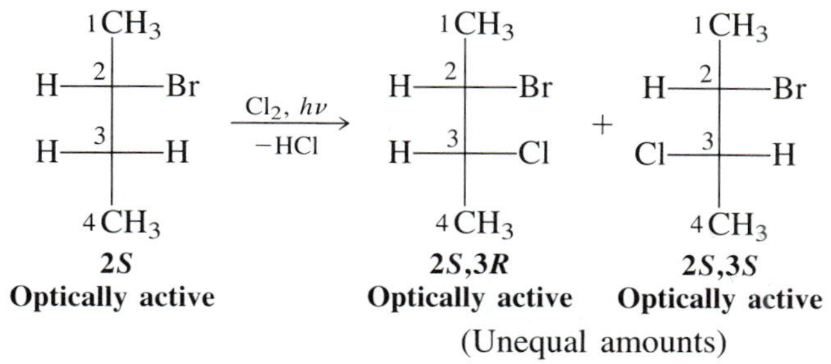

The chlorination at C-2 gives a 1:1 mixture of enantiomers. Does the reaction at C-3 also give an equimolar mixture of diastereomers? The answer is no. This finding is readily explained on inspection of the two transition states leading to the product (Figure 5-18). Abstraction of either one of the diastereotopic protons results in an approximately planar radical center at C-3. In contrast with the radical formed in the chlorination at C-2, however, the two faces of this radical are not mirror images of each other; they are not enantiotopic. The cause of this phenomenon is the retention of molecular chirality in the reactive spe-

FIGURE 5-18

The chlorination of (*S*)-2-bromobutane at C-3 produces the two diastereomers of 2-bromo-3-chlorobutane in unequal amounts.

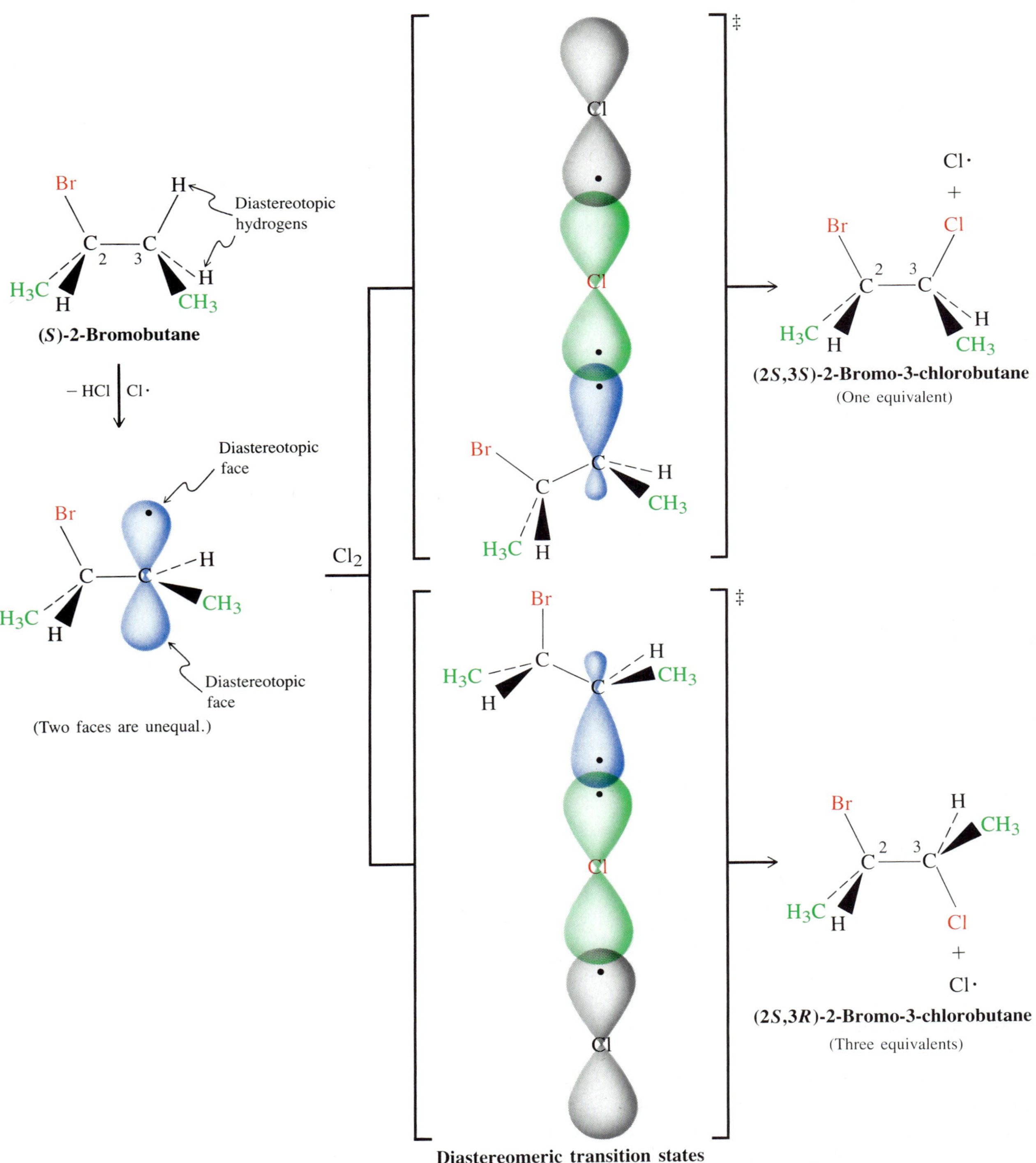

cies, which renders the two sides of the *p* orbital nonequivalent (regardless of the conformation of the molecule). The two faces of the radical are said to be diastereotopic.

What are the consequences of this nonequivalency? If the rate at which the two faces of the radical are attacked differs, as one would predict, then the rates of formation of the two diastereomers should be different, as is indeed found. The two transition states leading to products are not mirror images of each other and are not superimposable: they are diastereomeric. They therefore have different energies and represent different pathways.

One final product is possible in the chlorination of (*S*)-2-bromobutane, derived from substitution at C-4. In this reaction, neither is a new stereocenter created nor is the old one destroyed. The result is an optically active product. Chlorination at C-4 also does not alter the order of the substituents around (the original) C-2; hence its designated configuration remains *S*. However, to maintain the lowest possible substituent numbering, the carbon originally assigned C-4 has now become C-1. The compound is named (*S*)-3-bromo-1-chlorobutane.

Chlorination of (*S*)-2-Bromobutane at C-4

CH_3 / H—C—Br / H—C—H / CH_3 $\xrightarrow[-HCl]{Cl_2,\ h\nu}$ CH_3 / H—C—Br / H—C—H / CH_2Cl

2*S* **(Optically active)** — **(*S*)-3-Bromo-1-chlorobutane (Optically active)**

EXERCISE 5-13

Write the structures of the products of monobromination of (*S*)-2-bromopentane at each carbon atom. Name the products and specify whether they are chiral or achiral, whether they will be formed in equal or unequal amounts, and which will be in optically active form.

Stereoselectivity: The Preference for One Stereoisomer

The observation of an unequal distribution of diastereomers formed in the chlorination of 2-bromobutane at C-3 indicated that the presence or lack of symmetry can to a certain extent influence, perhaps even control, the stereochemical outcome of the reaction. A reaction that leads to the predominant (or exclusive) formation of one of several possible stereoisomeric products is **stereoselective.** For example, the chlorination of (*S*)-2-bromobutane at C-3 is stereoselective or, more specifically, *diastereoselective,* because the products are diastereomers. The corresponding chlorination at C-2, on the other hand, is not stereoselective (or, more specifically, not *enantioselective,* given that the products are enantiomers), a racemate being formed.

How much stereoselectivity is possible? The answer depends very much on substrate, reagents, the particular reaction in question, and conditions. Enzymes in nature manage to convert achiral compounds into chiral molecules with very high enantioselectivity. They are capable of this task because enzymes themselves have handedness and therefore convert achiral materials into those that are compatible with their own chirality. The situation is very similar to shaping flexible achiral objects with your hands. For example, clasping a piece of modeling clay with your left hand furnishes a shape that is the mirror image of that made with your right hand.

In chiral molecules, such as (*S*)-2-bromobutane, the molecular asymmetry may cause varying degrees of diastereoselectivity. In our example, (2*S*,3*R*)-2-bromo-3-chlorobutane is preferred over the 2*S*,3*S* isomer by a factor of 3 (Figure 5-18). Similarly, bromination of racemic 2-bromobutane at C-3 gives 2.5 times as much of the meso dibromide as of the *R*,*R*/*S*,*S* isomer.

EXERCISE 5-14

Draw all other possible products of the chlorination of (*S*)-1-bromo-2,2-dimethylcyclobutane. Specify whether they are chiral or achiral, whether they are formed in equal or unequal amounts, and which are optically active when formed.

In conclusion, chemical reactions, as exemplified by free-radical halogenation, can be stereoselective or not. Starting from achiral materials, such as butane, a racemic (nonstereoselective) product is formed (in butane by bromination at C-2). The two enantiotopic hydrogens at the methylene carbons of butane are equally susceptible to substitution, the halogenation step in the mechanism of free-radical bromination proceeding through two enantiomeric transition states of equal energy. Similarly, starting from chiral and enantiomerically pure 2-bromobutane, halogenation of the stereocenter also gives a racemic product through the intermediacy of an achiral radical. On the other hand, diastereoselectivity is possible in the formation of a new stereocenter, because the chiral environment retained by the molecule results in two unequal modes of attack on the intermediate radical. The two transition states have a diastereomeric relation, which leads to the formation of products at unequal rates.

5-7 Resolution: The Separation of Enantiomers

Although the physical and chemical characteristics of enantiomers have been discussed herein repeatedly, so far no mention has been made of their chemical preparation as isolated entities. As we know, the generation of a chiral structure from an achiral starting material furnishes a racemic mixture, and this raises the question of how to prepare pure enantiomers of a chiral compound.

A possible approach is to start with the racemate and separate one enantiomer from the other. This process is called the **resolution** of enantiomers. As mentioned earlier in regard to tartaric acid (Box 5-5), enantiomers occasionally crystallize into mirror-image shapes, which can be manually separated by "visual resolution." However, this process is time consuming, uneconomic for anything but minute-scale separations, and applicable only in rare cases.

A better strategy is based on the difference in the physical properties of diastereomers. If we can find a reaction for converting a racemate into a mixture of diastereomers by attachment of an enantiomerically pure chiral moiety, all the *R* forms of the original enantiomer mixture should be separable from the corresponding *S* forms by fractional crystallization, distillation, or chromatography of such diastereomers. For example, reaction of a racemate, $X_{R,S}$ (in which X_R and X_S are the two enantiomers), with an optically active compound, Y_S (arbitrarily given the S configuration—the mirror image could in principle work just as well), would form two optically active diastereomers, X_RY_S and X_SY_S, separable by standard techniques (Figure 5-19). If the bond between the two fragments is readily broken, X_R and X_S could be regenerated from the two diastereomers in the enantiomerically pure state. In addition, the optically active auxiliary agent Y_S might be recoverable in this manner and reused in other resolutions of this type.

FIGURE 5-19
Schematic flow sheet for the separation (resolution) of two enantiomers.

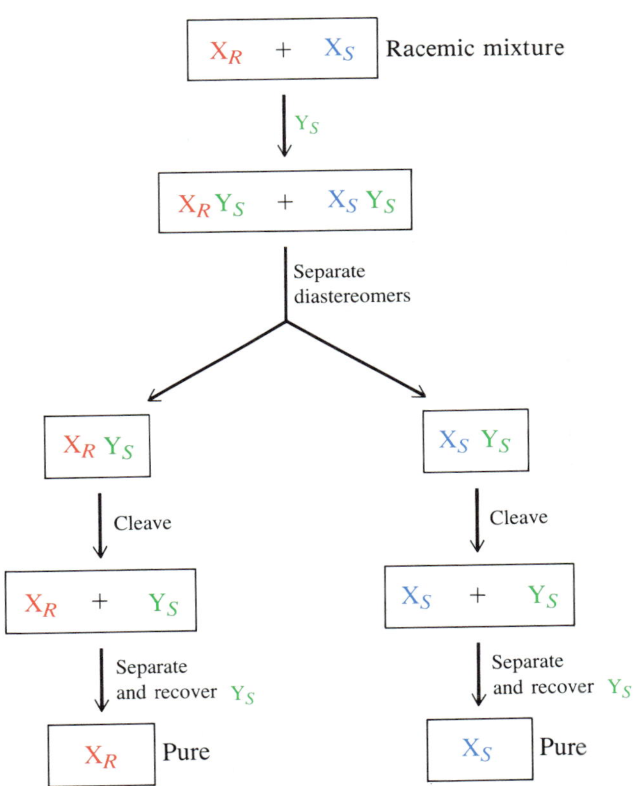

What we need, then, is a readily available enantiomerically pure compound, Y, that can be attached to the molecule to be resolved in a reversible chemical reaction. Frequently, natural products are used for this purpose, nature having provided us with a large number of pure optically active molecules, an example of which is (+)-2,3-dihydroxybutanedioic acid [(+)-(R,R)-tartaric acid]. A popular (and readily reversible) reaction employed in resolution is salt formation between acids and bases. For example, (+)-tartaric acid functions as an effective resolving agent of racemic amines. Figure 5-20 (on the next page) shows how 3-butyn-2-amine is resolved in this manner. The racemate is first treated with (+)-tartaric acid to form two diastereomeric tartrate salts. The dextrorotatory isomer crystallizes on standing and can be filtered away from the mother liquors containing the levorotatory tartrate. Treatment of the (+) salt with aqueous carbonate liberates the free amine, (+)-(*R*)-3-butyn-2-amine, which can be extracted with ether and purified by distillation. The aqueous layer contains potassium tartrate. Similar treatment of the foregoing mother liquors gives the (−)-*S* enantiomer (evidently slightly less pure because its optical rotation is slightly lower).

FIGURE 5-20

Resolution of 3-butyn-2-amine with (+)-2,3-dihydroxybutanedioic [(+)-tartaric] acid.

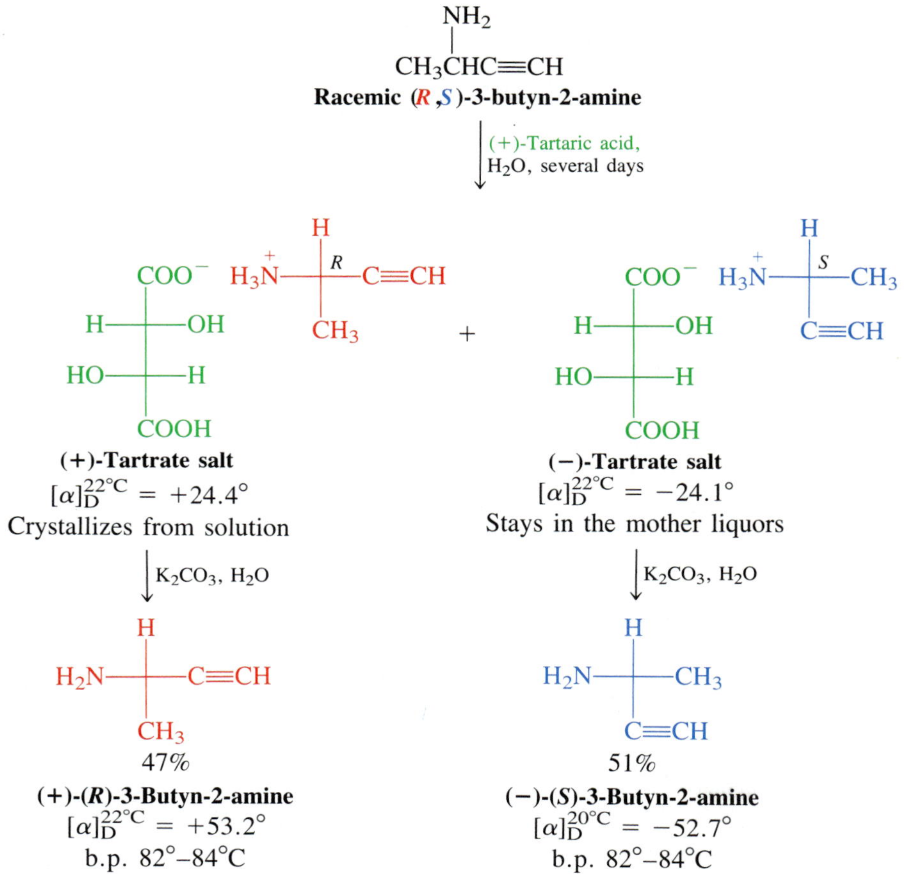

NH_2
$CH_3CHC{\equiv}CH$
Racemic (*R*,*S*)-3-butyn-2-amine

↓ (+)-Tartaric acid, H_2O, several days

COO^- / H—OH / HO—H / COOH ; $H_3\overset{+}{N}$—C(*R*)(H)(CH_3)—$C{\equiv}CH$
(+)-Tartrate salt
$[\alpha]_D^{22°C} = +24.4°$
Crystallizes from solution

+

COO^- / H—OH / HO—H / COOH ; $H_3\overset{+}{N}$—C(*S*)(H)($C{\equiv}CH$)—CH_3
(−)-Tartrate salt
$[\alpha]_D^{22°C} = -24.1°$
Stays in the mother liquors

↓ K_2CO_3, H_2O

H_2N—C(H)(CH_3)—$C{\equiv}CH$
47%
(+)-(*R*)-3-Butyn-2-amine
$[\alpha]_D^{22°C} = +53.2°$
b.p. 82°–84°C

↓ K_2CO_3, H_2O

H_2N—C(H)($C{\equiv}CH$)—CH_3
51%
(−)-(*S*)-3-Butyn-2-amine
$[\alpha]_D^{20°C} = -52.7°$
b.p. 82°–84°C

In this example, the pure enantiomer of an acid was used to resolve a racemic amine. The converse is also possible: the resolution of a racemic acid with an enantiomerically pure amine. There are many other ways in which the formation of diastereomers can be used in the resolution of racemates.

Summary of Important Concepts

1 Isomers have the same molecular formula but are different compounds. Structural isomers differ in the order in which the individual atoms are connected. Stereoisomers have the same connectivity but differ in the three-dimensional arrangement of the atoms. Stereoisomers include cis and trans isomers and enantio- and diastereomers. Cis and trans isomers constitute special cases of diastereoisomerism.

2 An object that is not superimposable with its mirror image is chiral. A carbon atom bearing four different substituents (asymmetric carbon) is an example of a stereocenter.

3 One stereocenter in a molecule gives rise to a pair of enantiomers. These are stereoisomers in which one member of the pair is related to the other as an object is to its mirror image.

4 Two stereocenters in a molecule give rise to four stereoisomers—namely, two enantiomer pairs of diastereomers. Diastereomers are stereoisomers that are not related to each other as object to mirror image. The maximum number of stereoisomers that a compound with n stereocenters can have is 2^n. The presence of some elements of symmetry in the molecule (e.g., as in meso compounds) reduces that number.

5 Chirality need not be due to the presence of stereocenters, because it is a property of handedness.

6 Chiral rotamers may undergo racemization by rotation.

7 For a molecule to be chiral, it must have neither a plane of symmetry nor a center of symmetry.

8 Most of the physical properties of enantiomers are the same. A major exception is the way in which they interact with plane-polarized light: one enantiomer will be dextrorotatory, the other will be levorotatory. This phenomenon is called optical activity. The extent of the rotation of the plane of plane-polarized light is measured in degrees and is expressed by the specific rotation, $[\alpha]$. A racemate shows zero rotation. The optical purity of a partial racemate is given by

$$\% \text{ optical purity} = \left[\frac{[\alpha]_{\text{observed}}}{[\alpha]} \cdot 100\right]$$

9 The absolute configuration at a stereocenter can be assigned R or S, using the sequence rules of Cahn, Ingold, and Prelog.

10 Fischer projections are used to simplify the depiction of stereoformulas and to help in describing their absolute configuration.

11 The chemical introduction of chirality into an achiral compound by free-radical halogenation leads to a racemate through enantiomeric (related as image and mirror image) transition states, in which both enantiotopic faces of the planar radical undergo reaction.

12 Free-radical halogenation of a chiral molecule containing one stereocenter will give a racemate if the reaction takes place at that center. Two diastereomers are formed in unequal amounts through diastereomeric transition states if two diastereotopic hydrogens participate in the reaction.

13 The preference for the formation of one stereoisomer, when several are possible, is called stereoselectivity.

14 The separation of enantiomers is called resolution. It is best achieved by the reaction of the racemate with the pure enantiomer of a chiral auxiliary compound to yield separable diastereomers. Cleavage of the chiral auxiliary frees both enantiomers of the original racemate.

Problems

1 Classify each of the common objects listed below as being either chiral or achiral. Assume in each case that the object is in its simplest form, without decoration or printed labels.

(a) A ladder.
(b) A door.
(c) An electric fan.
(d) A refrigerator.
(e) The earth.
(f) A baseball.
(g) A baseball bat.
(h) A baseball glove.
(i) A flat sheet of paper.
(j) A fork.
(k) A spoon.
(l) A knife.

2 Each part of this problem lists two objects or sets of objects. As precisely as you can, describe the relation between the two sets, using the terminology of this chapter; that is, specify whether they are identical, enantiomeric, diastereomeric, and so forth.

(a) An American toy car compared with a British toy car (same color and design but steering wheels on opposite sides).

(b) Two left shoes compared with two right shoes (same color, size, and style).

(c) A pair of skates compared with two left skates (same color, size, and style).

(d) A right glove on top of a left glove (palm to palm) compared with a left glove on top of a right glove (palm to palm; same color, size, and style).

3 For each pair of molecules below, indicate whether its members are (1) stereoisomers, (2) structural isomers, or (3) identical. If they are stereoisomers, indicate whether the two molecules are readily interconverted by bond rotations.

(a) $CH_3CH_2CH_2CH(CH_3)_2$ and $CH_3CH_2CH(CH_3)CH_2CH_3$

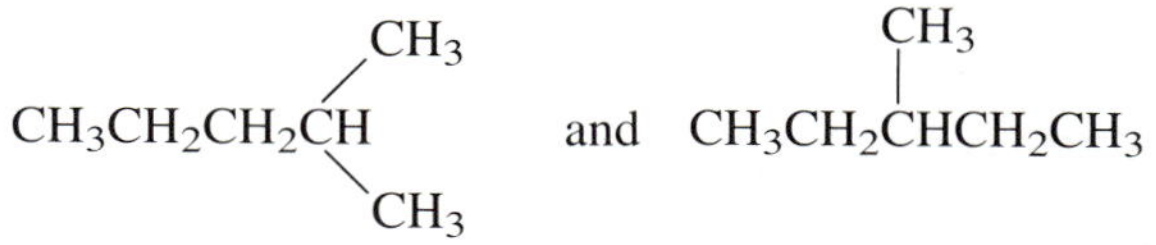

(b) H_3C, H on a ring carbon and H, CH_3 on a ring carbon

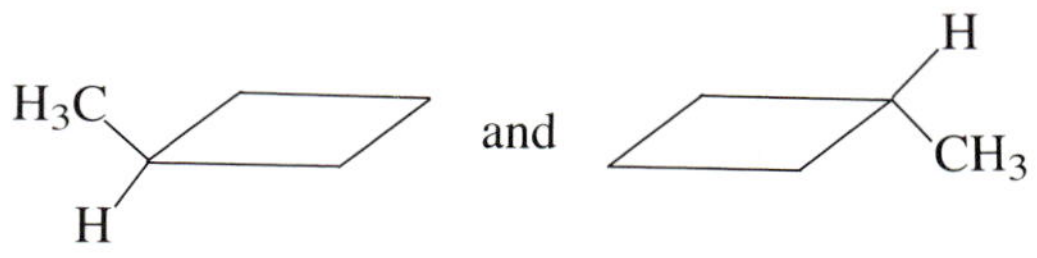

(c) Newman projections: front Br, H_3C, H; rear H, H, H and front CH_3, H, Br; rear H, H, H

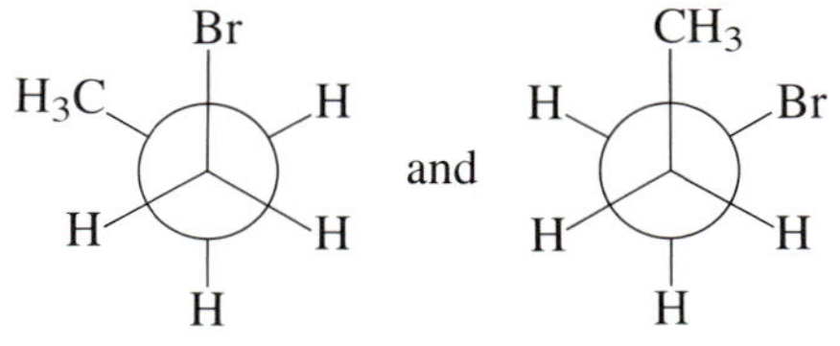

(d) 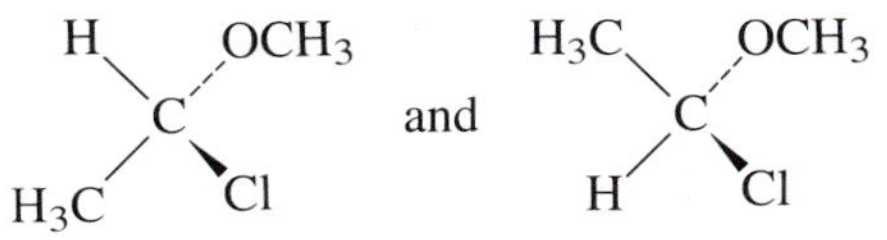 [structures: H, OCH_3, H_3C, Cl on C] and [structures: H_3C, OCH_3, H, Cl on C]

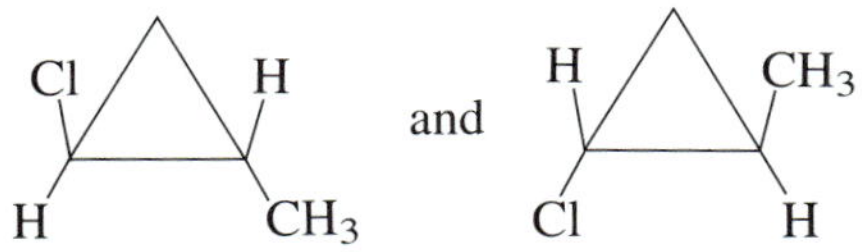

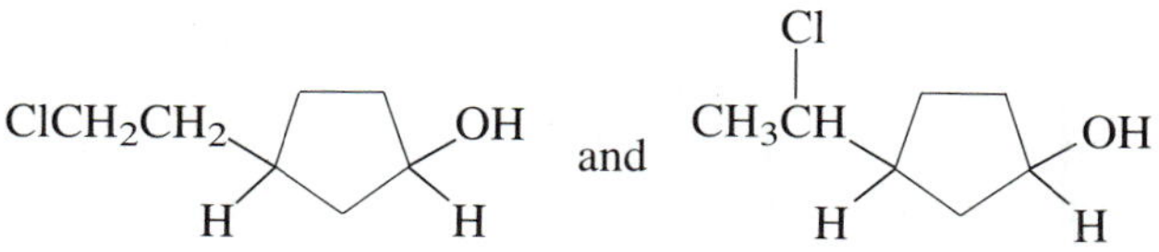

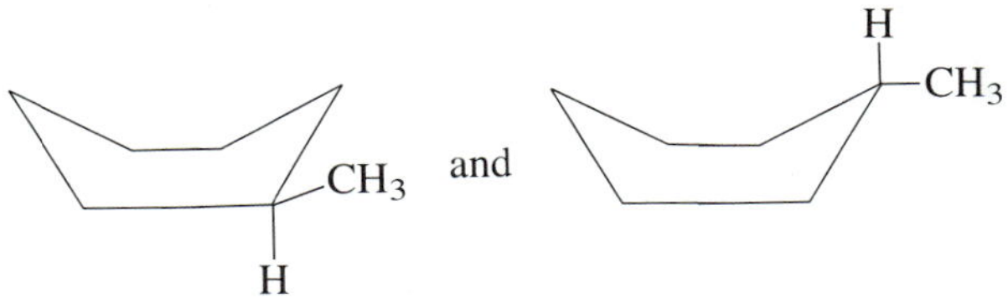

(e) [cyclopropane with Cl, H, H, CH_3] and [cyclopropane with H, CH_3, Cl, H]

(f) $ClCH_2CH_2$–[cyclopentane]–OH (H, H) and $CH_3CH(Cl)$–[cyclopentane]–OH (H, H)

(g) [cyclohexane chair with CH_3, H] and [cyclohexane chair with H, CH_3]

(h) $CH_3C(Cl)(Br)CH_2CH_2CH_3$ and $CH_3CH(Br)CH_2CH(Cl)CH_3$

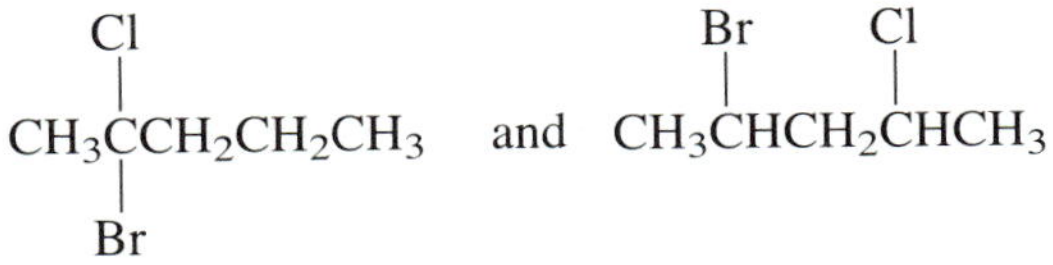

4 Which of the following compounds are chiral?

(a) 2-Methylheptane.
(b) 3-Methylheptane.
(c) 4-Methylheptane.
(d) 1,1-Dibromopropane.
(e) 1,2-Dibromopropane.
(f) 1,3-Dibromopropane.
(g) Ethene, $H_2C{=}CH_2$
(h) Ethyne, $HC{\equiv}CH$
(i) Benzene,

(Note: Like ethene, benzene contains all sp^2-hybridized carbons and therefore is planar.)

(j) Epinephrine (adrenalin),

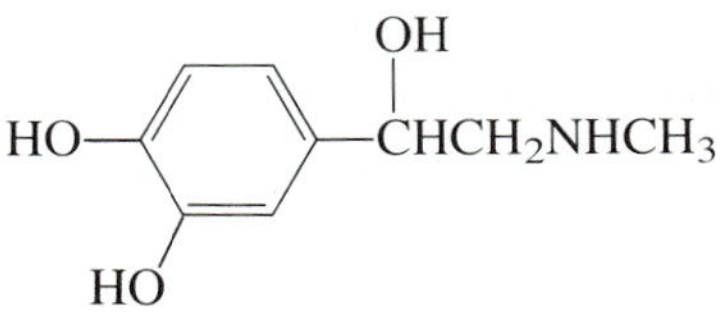

HO–[benzene ring]–$\overset{OH}{|}$CHCH$_2$NHCH$_3$ (HO)

(k) Vanillin,

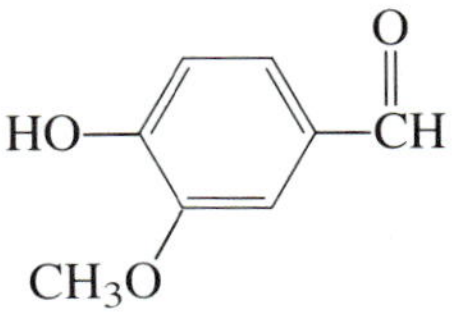

HO–[benzene ring]–CH(=O) (CH_3O)

(l) Citric acid,

$HOC(=O)CH_2C(OH)(COOH)CH_2C(=O)OH$

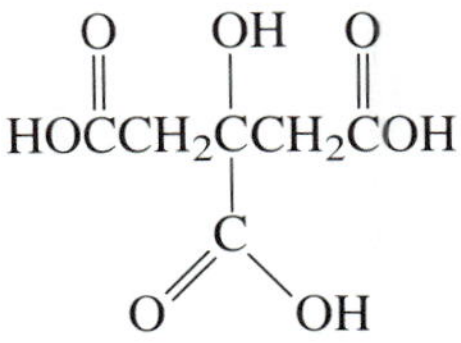

(m) Ascorbic acid,

$HOCH_2$–HOCH–[ring: O, C=O, H, HO, OH]

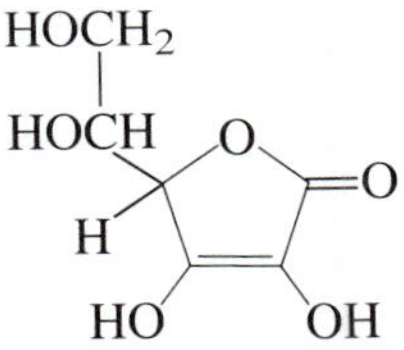

(n) *p*-Menthane-1,8-diol (Terpin hydrate),

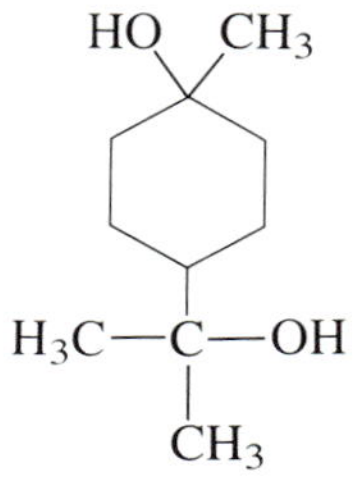

HO CH$_3$ / H$_3$C—C—OH / CH$_3$

(o) Meperidine (Demerol),

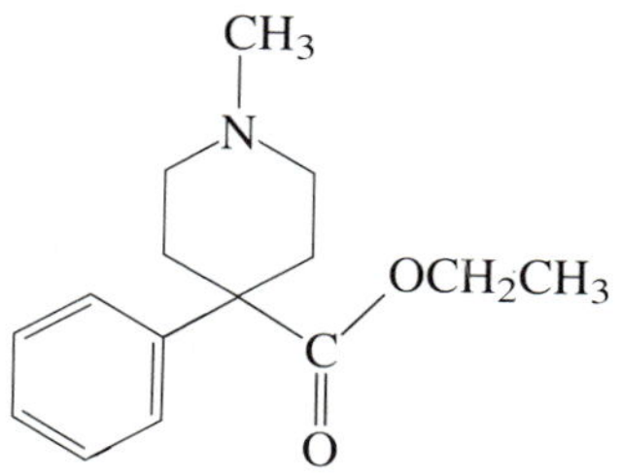

CH$_3$ / N / OCH$_2$CH$_3$ / C / O

5 Which of the following cyclohexane derivatives are chiral? For the purpose of determining the chirality of a cyclic compound, the ring may generally be treated as if it were planar.

(a) O, H$_3$C, CH$_3$, O

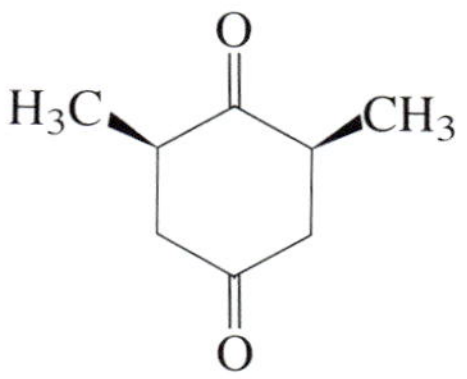

(c) O, CH$_3$, H$_3$C, O

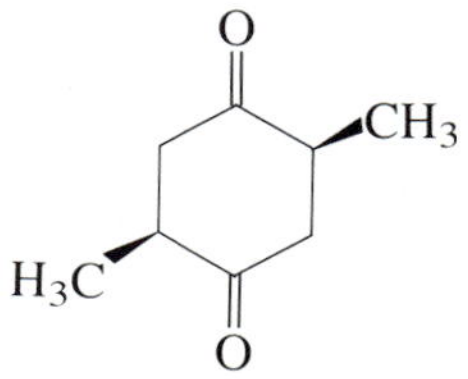

(b) O, H$_3$C, CH$_3$, O

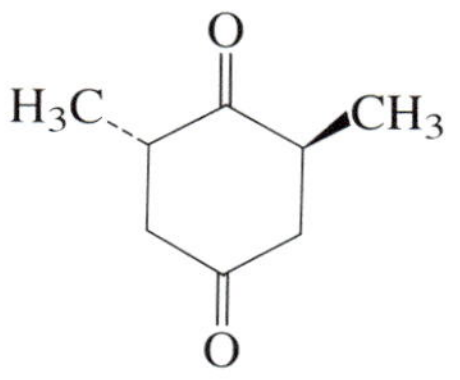

(d) O, CH$_3$, H$_3$C, O

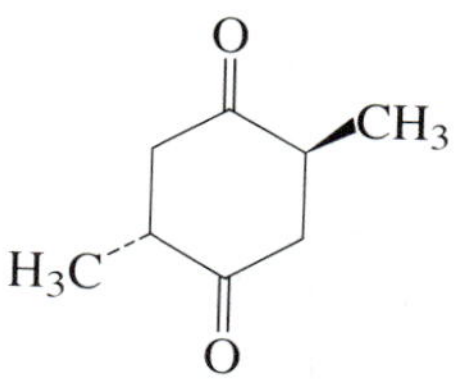

6 Which of the following cyclohexane derivatives are chiral? As in Problem 5, the ring may be treated as if it were planar.

(a) CH$_3$, CH$_3$

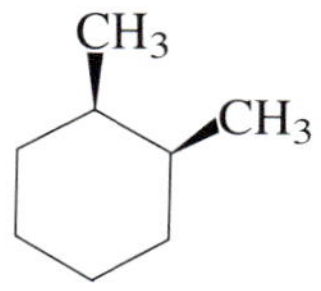

(c) CH$_3$, CH$_3$

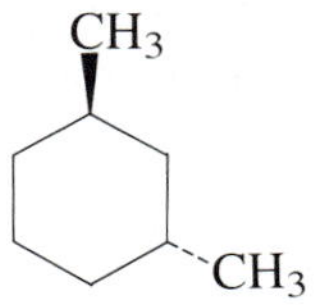

(b) CH$_3$, CH$_3$

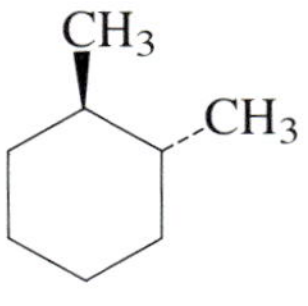

(d) CH$_3$, CH$_3$

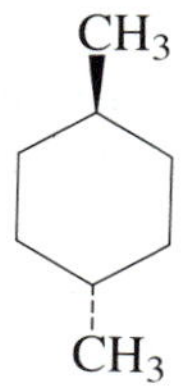

7 For each formula below, (1) identify every structural isomer containing one or more stereocenters, (2) give the number of stereoisomers for each, and (3) draw and fully name at least one of the stereoisomers in each case.

(a) C_7H_{16}
(b) C_8H_{18}
(c) C_5H_{10}, with one ring

8 Mark the stereocenters in each of the chiral molecules in Problem 4. Draw any single stereoisomer of each of these molecules, and assign the appropriate designation (*R* or *S*) to each stereocenter.

9 The two isomers of carvone [systematic name, 2-methyl-5-(1-methylethenyl)-2-cyclohexenone] are drawn below. Which is *R* and which is *S*?

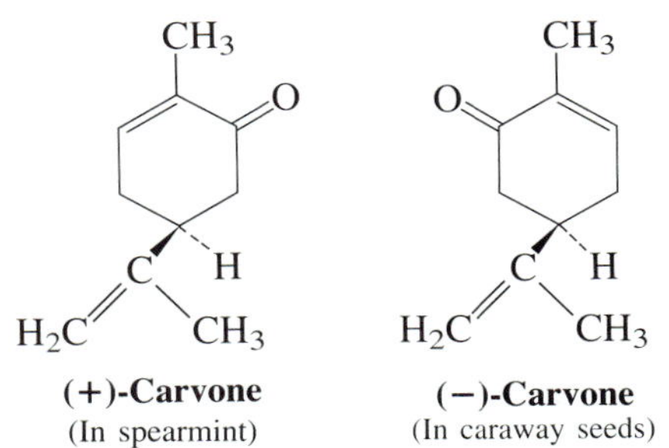

(+)-Carvone (In spearmint) **(−)-Carvone** (In caraway seeds)

10 Draw structural representations of each of the following molecules. Be sure that your structure clearly shows the configuration at each stereocenter.

(a) (*R*)-3-Bromo-3-methylhexane.
(b) (1*S*,2*S*)-1-Chloro-1-trifluoromethyl-2-methylcyclobutane.
(c) (3*R*,5*S*)-3,5-Dimethylheptane.
(d) (2*R*,3*S*)-2-Bromo-3-methylpentane.
(e) (*S*)-1,1,2-Trimethylcyclopropane.
(f) (1*R*,2*R*,3*S*)-1,2-Dichloro-3-ethylcyclohexane.

11 For each of the following questions, assume that all measurements are made in 10-cm polarimeter sample containers.

(a) A solution of 0.4 g of optically active 2-butanol in 10 ml of water displays an optical rotation of $-0.56°$. What is its specific rotation?

(b) The specific rotation of sucrose (common sugar) is $+66.4°$. What would be the observed optical rotation of a solution containing 3 g of sucrose in 10 ml of water?

(c) A solution of pure (*S*)-2-bromobutane in ethanol is found to have an observed $\alpha = 57.3°$. If $[\alpha]$ for (*S*)-2-bromobutane is 23.1°, what is the concentration of the solution?

12 Natural epinephrine, $[\alpha]_D^{25°C} = -50°$, is used medicinally. Its enantiomer is medically worthless and is, in fact, toxic. You, a pharmacist, are given a solution said to contain 1 g of epinephrine in 20 ml of liquid, but the optical purity is not specified. You place it in a polarimeter (10-cm tube) and get a reading of $-2.5°$. What is the optical purity of the sample? Is it safe to use medicinally?

13 Sodium hydrogen (*S*)-glutamate [(*S*)-monosodiumglutamate], $[\alpha]_D^{25°C} = +24°$, is the active flavor enhancer known as MSG. The condensed formula of MSG is shown below.

$$\mathrm{HO\overset{\overset{\displaystyle O}{\|}}{C}{-}\overset{\overset{\displaystyle NH_2}{|}}{C}HCH_2CH_2\overset{\overset{\displaystyle O}{\|}}{C}O^-Na^+}$$

(a) Draw the structure of the *S* enantiomer of MSG.

(b) If a commercial sample of MSG was found to have a $[\alpha]_D^{25°C} = +8°$, what would be its optical purity? What would the percentages of the *S* and *R* enantiomers be in the mixture?

(c) Answer the same questions for a sample with $[\alpha]_D^{25°C} = +16°$.

14 For each compound below, (1) mark each stereocenter, (2) assign an *R* or *S* designation, and (3) draw a clear picture of the molecule's enantiomer.

(a) H, HS, CH_3, C, OH, C, O

(b) CH_3, C, $C(CH_3)_3$, H, $CH_2CH_2OCH_2CH_3$

(c) CH_3, H_3C, C—C, Cl, H, H, $CH{=}CH_2$

(d) Br, H, Cl, OH

(e) $HC{\equiv}C$—C—C, H, CH_3, O, H

(f) O, H, Cl

(g) N, H, C, $CH_2CH_2N(CH_3)_2$, Cl

Chlorpheniramine
(As in Coricidin decongestant)

Note: The carbons in benzene or benzenelike rings are treated in the same way as those in alkenes (use Sequence Rule 3 from "Assigning Priorities," Section 5-3).

(h) CH_3, H, $C{=}CH_2$, CH_3

Limonene
(From trees, fruits, etc.)

15 For each pair of structures below, indicate whether the two compounds are identical or are enantiomers of each other.

(a) CH_2CH_3, Br, C, Cl, CH_2CH_3 and CH_2CH_3, C, Br, CH_3CH_2, Cl

(b) CH_3, H—C—Cl, Br and CH_3, C, Cl, H, Br

(c) 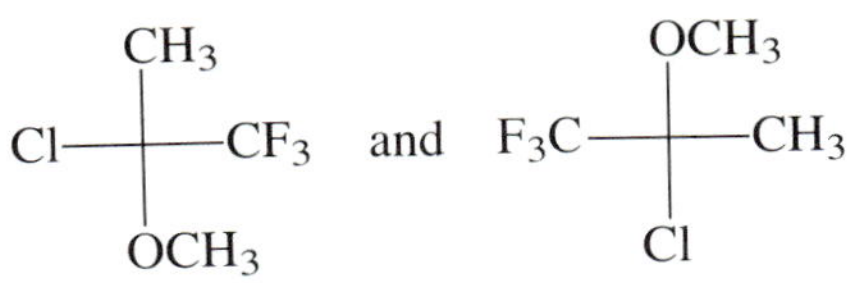 [Fischer projections: Cl, CF_3, CH_3, OCH_3 and F_3C, CH_3, OCH_3, Cl]

(d) 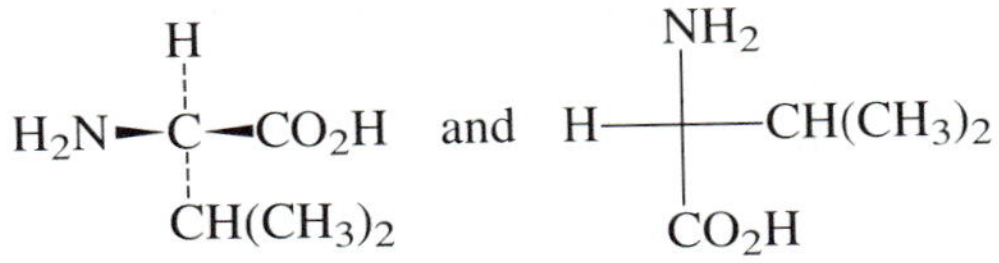 [structures: H_2N–C(H)(CH(CH_3)$_2$)–CO_2H and Fischer projection H, NH_2, CH(CH_3)$_2$, CO_2H]

16 Determine the *R* or *S* designation for each stereocenter in the structures in Problem 15.

17 Redraw each of the following molecules as a Fischer projection; then assign *R* or *S* designations to each stereocenter.

(a) 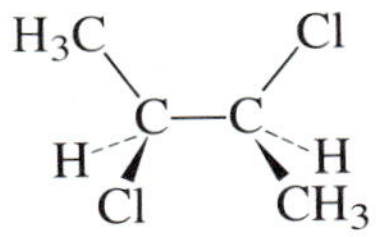 [structure: H_3C, H, Cl / Cl, H, CH_3]

(c) 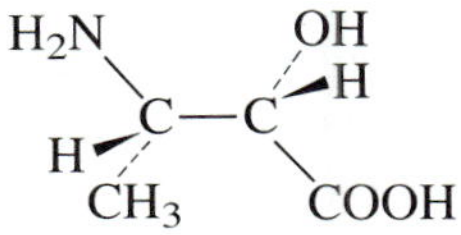 [structure: H_2N, H, CH_3 / OH, H, COOH]

(b) 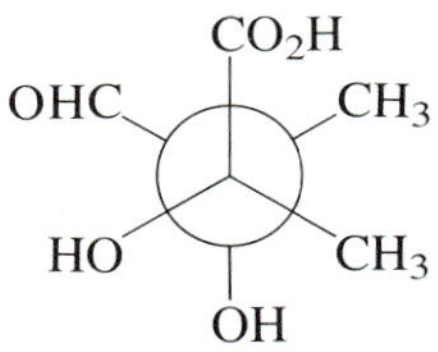 [Newman projection: CO_2H, OHC, CH_3, HO, CH_3, OH]

(d) 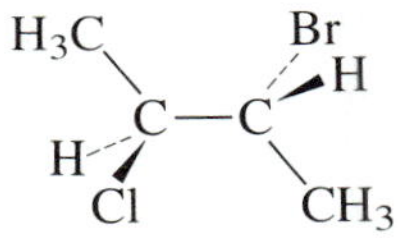 [structure: H_3C, H, Cl / Br, H, CH_3]

18 The compound pictured at the right is a sugar called (−)-arabinose. Its specific rotation is −105°.

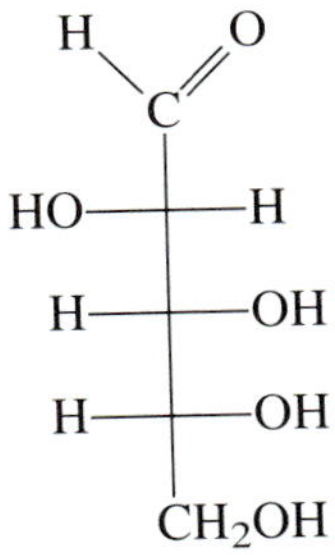

(a) Draw an enantiomer of (−)-arabinose.

(b) Does (−)-arabinose have any other enantiomers?

(c) Draw a diastereomer of (−)-arabinose.

(d) Does (−)-arabinose have any other diastereomers?

(e) If possible, predict the specific rotation of the structure that you drew for part **a.**

(f) If possible, predict the specific rotation of the structure that you drew for part **c.**

(g) Does (−)-arabinose have any optically inactive diastereomers? If it does, draw one.

[Fischer projection: CHO (H, O), HO–H, H–OH, H–OH, CH_2OH]

(−)-Arabinose

19 Write the complete IUPAC name of this compound (do not forget stereochemical designations).

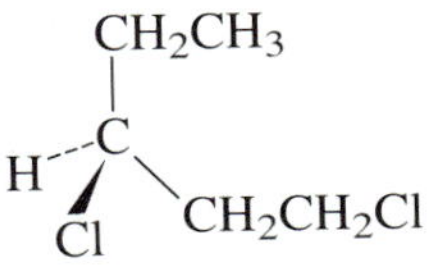

[structure: CH_2CH_3, H, Cl, CH_2CH_2Cl] $C_5H_{10}Cl_2$

Reaction of this compound with one mole of Cl_2 in the presence of light produces several isomers of the formula $C_5H_9Cl_3$. For each part of this problem, give the following information:

How many stereoisomers are formed?

If more than one is formed, are they formed in equal or unequal amounts?

Designate whether every stereocenter in each stereoisomer is *R* or *S*.

(a) Chlorination at C-3.
(b) Chlorination at C-4.
(c) Chlorination at C-5.

20 Monochlorination of methylcyclopentane can result in several products. Give the same information as that requested in Problem 19 for the monochlorination of methylcyclopentane at C-1, C-2, and C-3.

21 Illustrate how to resolve racemic 1-phenylethanamine, using the method of reversible conversion into diastereomers.

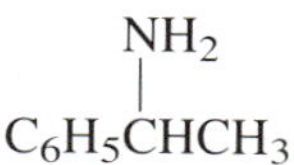

22 Draw a flow chart that diagrams a method for the resolution of racemic 2-hydroxypropanoic acid (lactic acid), using (*S*)-1-phenylethanamine (for structure, see Problem 21).

23 How many different stereoisomeric products are formed in the monobromination of

(a) racemic *trans*-1,2-dimethylcyclohexane?
(b) pure (*R*,*R*)-1,2-dimethylcyclohexane?
(c) For your answers to parts **a** and **b,** indicate whether you expect equal or unequal amounts of the various products to be formed. Indicate to what extent products can be separated on the basis of having different physical properties (e.g., solubility, boiling point, etc.).

24 Identify and label all enantiotopic pairs of atoms and all diastereotopic pairs of atoms in the molecules below.

(a) 2,2-Dimethylbutane.
(b) 3-Methylhexane.
(c) 3-Methylpentane.
(d) Cyclohexanone.

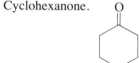

25 Make a model of *cis*-1,2-dimethylcyclohexane in its most stable conformation. If the molecule were rigidly locked into this conformation, would it be chiral? (Test your answer by making a model of the mirror image and checking for superimposability.)

Flip the ring of the model. What is the stereoisomeric relation between the original conformation and the conformation after flipping the ring? How do the results that you have obtained in this problem relate to your answer to Problem 6, part **a?**

26 Morphinane is the parent substance of the broad class of chiral molecules known as the morphine alkaloids. Interestingly, the (+) and (−) enantiomers of the compounds in this family have rather different physiological properties. The (−) compounds, such as morphine, are "narcotic analgesics" (pain killers), whereas the (+) compounds are "antitussives" (ingredients in cough syrup). Dextromethorphan is one of the simplest and most common of the latter.

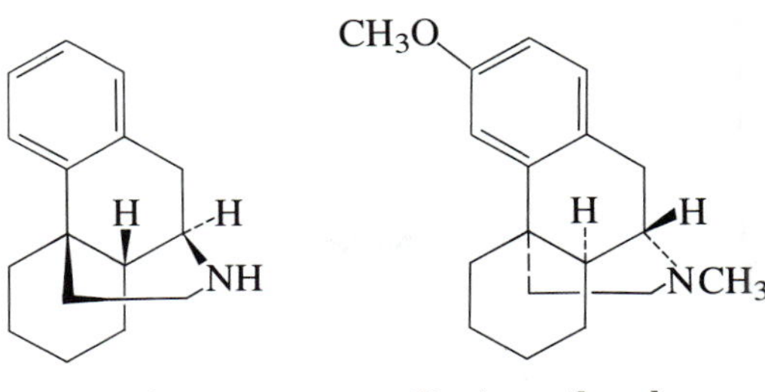

Morphinane **Dextromethorphan**

(a) Locate and identify all the stereocenters in dextromethorphan.

(b) Draw the enantiomer of dextromethorphan.

(c) As best you can (and it is not easy), assign *R* and *S* configurations to all the stereocenters in dextromethorphan.

27 The enzymatic introduction of a functional group into a biologically important molecule is not only specific with regard to the location of the reaction in the molecule (see Chapter 4, Problem 21), but also usually specific in the stereochemistry obtained. The biosynthesis of adrenalin first requires that a hydroxy group be introduced specifically to produce (−)-norepinephrine from the achiral substrate dopamine. (The completion of the synthesis of adrenalin will be presented in Problem 25 of Chapter 6.) Only the (−) enantiomer is functional in the appropriate physiological manner; so the synthesis must be highly stereoselective.

HO, HO, $CH_2CH_2NH_2$ — Dopamine hydroxylase, O_2 → HO, HO, HO, H, C, CH_2NH_2

Dopamine **(−)-Norepinephrine**

(a) Is the configuration of (−)-norepinephrine *R* or *S*?

(b) What term describes the relation between the two hydrogens on the methylene carbon of dopamine where this reaction takes place?

(c) In the absence of an enzyme, would the transition states of a free radical oxidation leading to (−)- and (+)-norepinephrine be of equal or unequal energy? What term describes the relation between these transition states?

(d) In your own words, describe how the enzyme must affect the energy of these transition states to favor production of the (−) enantiomer. Does the enzyme have to be chiral or can it be achiral?

CHAPTER 6

The Properties and Reactions of Haloalkanes

BIMOLECULAR NUCLEOPHILIC SUBSTITUTION

Having had occasion to examine the preparation of haloalkanes by the free-radical halogenation of alkanes in detail in Chapter 3 and the stereochemical consequences of this reaction in Chapter 5, we will now investigate the chemistry of the haloalkanes.

Let us start with the procedures for naming haloalkanes and briefly consider some of their physical characteristics.

6-1 The Naming of Haloalkanes

Similar to the alkanes, whose abbreviated molecular formula is R–H, haloalkanes are often depicted as R–X, in which X represents the halogen atom.

In the systematic (IUPAC) nomenclature, the halogen is treated as a substituent to the alkane framework; the halogenated hydrocarbon is named as a **haloalkane.**

CH_3I — **Iodomethane**

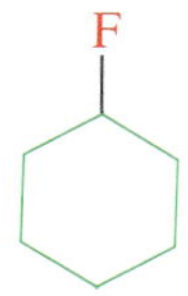

(cyclohexane ring bearing F) — **Fluorocyclohexane**

$CH_3C(CH_3)(Cl)CH_2CH_3$ — **2-Chloro-2-methylbutane**

The halogen is treated as if it were an alkyl substituent, without any priority. The longest alkane chain is numbered so as to give the first substituent from either end the lowest number. As usual, substituents are ordered according to the alphabet.

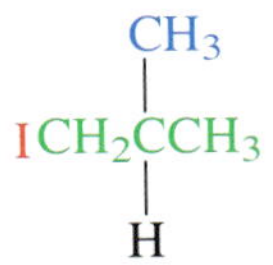

$ICH_2CH(CH_3)CH_3$ — **1-Iodo-2-methylpropane**

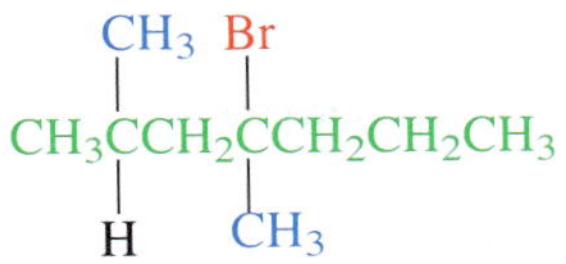

$CH_3CH(CH_3)CH_2CBr(CH_3)CH_2CH_2CH_3$ — **4-Bromo-2,4-dimethylheptane**

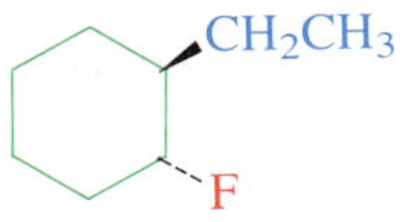

trans-1-Ethyl-2-fluorocyclohexane

Complex appendages are named according to the same rules.

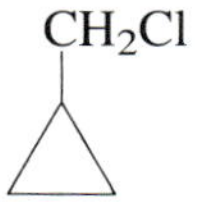

(Chloromethyl)-cyclopropane

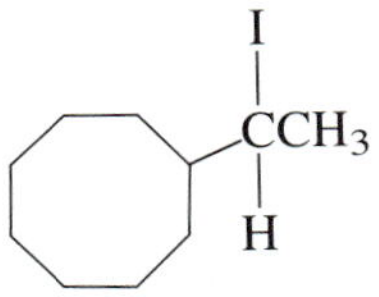

(1-Iodoethyl)cyclooctane

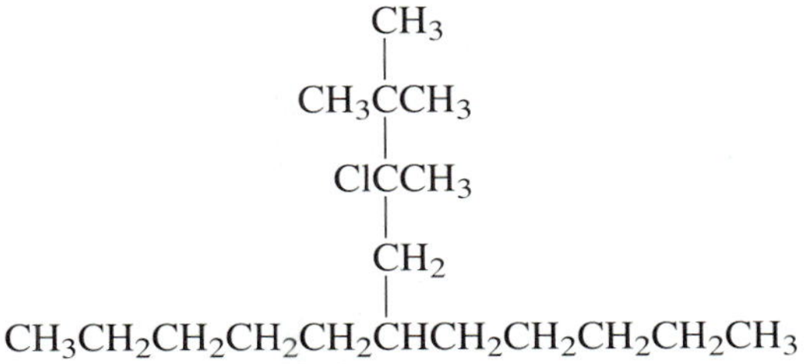

$CH_3CH_2CH_2CH_2CH_2CH[CH_2CCl(CH_3)C(CH_3)_3]CH_2CH_2CH_2CH_2CH_3$ — **6-(2-Chloro-2,3,3-trimethylbutyl)-undecane**

The common names of the haloalkanes are based on the alkyl halide stem.

CH_3I	CH_3CH_2F	$(CH_3)_3C—Br$
Methyl iodide	**Ethyl fluoride**	**_tert_-Butyl bromide**

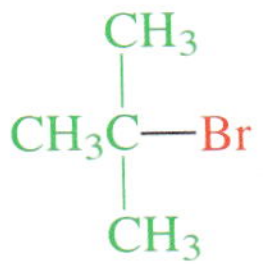

Some chlorinated solvents have common names: for example, carbon tetrachloride, CCl_4; chloroform, $CHCl_3$; and methylene chloride, CH_2Cl_2.

EXERCISE 6-1

Draw the structures of (2-iodoethyl)cyclooctane and 5-butyl-3-chloro-2,2,3-trimethyldecane.

In summary, haloalkanes are named in accord with the rules that apply to naming the alkanes (Section 2–3), the halo substituent being ranked equally with alkyl groups.

6-2 The Physical Properties of Haloalkanes

The physical properties of the haloalkanes are quite distinct from those of the corresponding alkanes. Their bond strengths, bond lengths, dipole moments, and boiling points are strongly affected by the differences in size of the halogen substituents and by the polarity of the carbon–halogen bond.

TABLE 6-1

Bond lengths and bond strengths in CH_3X

Halomethane	Bond length (Å)	Bond strength (kcal $mole^{-1}$)
CH_3F	1.385	110
CH_3Cl	1.784	85
CH_3Br	1.929	71
CH_3I	2.139	57

The Bond Strength of C–X Decreases As the Size of X Increases

The bond between carbon and a halogen is made up mainly by the overlapping of an sp^3 hybrid orbital on carbon with a nearly pure p orbital on the halogen (Figure 6-1). In the procession from fluorine to iodine in the periodic table, the size of the halogen p orbital increases, the electron cloud around the halogen atom becomes more diffuse, and the ability to overlap with the carbon orbitals diminishes. Consequently, there is a decrease in the C–X bond strength. For example, the C–X bond-dissociation energy in the halomethanes, CH_3X, decreases along the series at the same time as the C–X bond lengths increase (Table 6-1), as can be expected.

FIGURE 6-1

Bond between an alkyl carbon and a halogen. The size of the p orbital is substantially larger than shown for X = Cl, Br, or I.

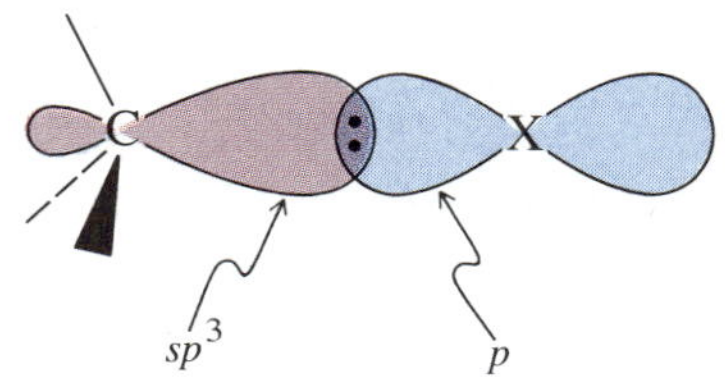

The C–X Bond Is Polarized

The Polar Character of the C–X Bond

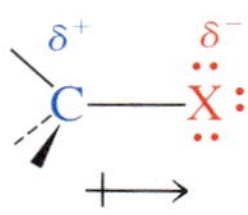

A characteristic physical property of the haloalkanes is the dipolar character of the C–X bond. Recall (Section 1-3, Table 1-4) that the halogens are more electronegative than is carbon. Thus, the electron density along the C–X bond is displaced in the direction of X, giving the halogen a partial negative charge (δ^-). The resulting dipole moment (Section 1-3) is appreciable and governs a substantial part of the chemical behavior of the haloalkanes. For example, we will see that anions and other electron-rich species are capable of attacking the positively polarized carbon atom. Cations and other electron-deficient species, on the other hand, can attack the halogen atom.

Does the dipolar character of the C–X bond affect other physical properties of the haloalkanes?

Boiling Points and Polarizability

The boiling points of haloalkanes are generally higher than those of the corresponding alkanes (Table 6-2), mainly because the dipolar structure of the haloalkanes causes dipole–dipole interactions in the liquid state. Boiling points also increase with increasing size of X. This result may be ascribed to increased molecular weight and to greater London interactions (Section 2-4) in the larger haloalkanes. Recall that London forces are due essentially to the mutual correlation of electrons among molecules. This effect is stronger for structures containing atoms in which the electrons are not held very tightly around the nuclear core, such as the atoms in the lower regions of the periodic table. This phenomenon is measurable as the **polarizability** of the atom. The polarizability (as

TABLE 6-2
Boiling points of haloalkanes (R–X)

		Boiling point (°C)				
R	**X =**	**H**	**F**	**Cl**	**Br**	**I**
CH_3		−161.7	−78.4	−24.2	3.6	42.4
CH_3CH_2		−88.8	−37.7	12.3	38.4	72.3
$CH_3(CH_2)_2$		−42.1	−2.5	46.6	71.0	102.5
$CH_3(CH_2)_3$		−0.5	32.5	78.4	101.6	130.5
$CH_3(CH_2)_4$		36.1	62.8	107.8	129.6	157.0
$CH_3(CH_2)_7$		125.7	142.0	182.0	200.3	225.5

opposed to polarity) is roughly a measure of the capability of the electrons around the nucleus to respond to a changing electric field. The more polarizable an atom, the more effectively it will enter into London interactions.

To summarize, the increasingly diffuse nature of the halogen orbitals along the series F, Cl, Br, I has several interrelated effects: (1) the C–X bond strength decreases; (2) the C–X bond becomes longer; (3) for the same R, the boiling points increase; (4) London interactions improve; and (5) the polarizability of R–X becomes greater. We shall see that these effects also play an important role in the chemistry of these compounds, as exemplified by nucleophilic substitution.

6-3
Nucleophilic Substitution: Introduction and Scope

The nucleophilic substitution of a haloalkane is described by either of two general equations:

$$\underset{\text{Nucleophile}}{\text{Nu}:^-} + \overset{\delta^+}{\underset{\text{Electrophile}}{\text{R}}}-\overset{\delta^-}{\ddot{\underset{..}{\text{X}}}}: \longrightarrow \text{R}-\text{Nu} + \underset{\text{Leaving group}}{:\ddot{\underset{..}{\text{X}}}:^-}$$

or

$$\underset{\text{Nucleophile}}{\text{Nu}:} + \underset{\text{Electrophile}}{\text{R}}-\ddot{\underset{..}{\text{X}}}: \longrightarrow [\text{R}-\text{Nu}]^+ + \underset{\text{Leaving group}}{:\ddot{\underset{..}{\text{X}}}:^-}$$

In this transformation, a reagent with an unshared electron pair (typically an anion, such as iodide, $:\ddot{\underset{..}{\text{I}}}:^-$, or a neutral species, such as ammonia, $:NH_3$) attacks the haloalkane to replace the halide substituent. This section explains why the reaction occurs and describes the types of molecules that will enter into it. In many of the equations and mechanisms that follow, nucleophiles, electrophiles, and leaving groups will be shown in red, blue, and green, respectively, as indicated in the preceding scheme.

Nucleophiles Attack Electrophilic Centers

As already noted, the polarization of a carbon–halogen bond causes the development of a positive charge on the carbon atom. Because of this charge, the carbon center is **electrophilic.** Such reactivity manifests itself toward species with available electron density (electrophilic = "electron loving"; *philos,* Greek, loving). In turn, electron-rich compounds that react with electrophilic nuclei are called nucleophilic species, or simply **nucleophiles,** Nu ("nucleus loving"). Nucleophiles typically have a negative charge or lone pairs of electrons or both. As shown at the beginning of this section, a negatively charged nucleophile reacts with a haloalkane by displacing the halogen as the anion to yield a neutral alkyl substitution product. A neutral nucleophile produces a positively charged product. In both cases, the group that has been replaced is $:\ddot{\underset{\cdot\cdot}{X}}:^-$. Because it may be thought of as having departed from the starting material, it is called the **leaving group.**

Note that the two equations formulated for nucleophilic substitution of a haloalkane are identical except for the difference in the charge distribution. In the first equation, charge neutrality is preserved by having one negative charge on each side. In the second, there are two neutral compounds on the left side, and on the right side the positive charge in the alkyl product is cancelled by the negative charge on the leaving group.

From the expression *nucleophilic substitution,* one might think that the nucleophile is the attacking species. The choice of words in this expression is somewhat arbitrary, because nucleophiles and electrophiles are mutually reactive. Thus, nucleophilic substitution can also be considered an electrophilic attack by the alkyl group; the haloalkane is said to *alkylate* the nucleophile.

Let us look at some typical nucleophiles and their reactions with several haloalkanes (Table 6-3). The reaction is fairly general with respect to both nucleophile and substrate.* However, in the examples, only primary or secondary halides serve as substrates. This is for good reason, because tertiary halides, even though they may (under certain conditions) undergo similar reactions, behave differently in the presence of the nucleophiles listed. Because of this behavior, they will be discussed separately in Chapter 7. Similarly, secondary halides entering into reaction with nucleophiles may give not only substitution, but also other products, again discussed in Chapter 7. The cleanest reactions of this type are obtained with primary haloalkanes.

Let us inspect these transformations in greater detail. In *reaction 1,* a hydroxide ion displaces chloride from chloromethane to give methanol. This is a general synthetic method for converting a haloalkane into an alcohol. Ordinarily, organic chemists use alkali metal hydroxides, such as sodium or potassium hydroxide, in this process.

A variation of this transformation is *reaction 2*. Methoxide ion reacts with iodoethane to give methoxyethane, an example of the Williamson ether synthesis (Section 9-2).

In both reaction 1 and reaction 2, the species attacking the haloalkane contains an anionic oxygen nucleophile. *Reaction 3* demonstrates the applicability of other nucleophiles and shows that a halide ion may function not only as a

---

*A *substrate* is a compound acted on in a reaction (*substratus,* Latin, to have been subjected).

TABLE 6-3
The diversity of nucleophilic substitution

	Substrate	Nucleophile		Product	Leaving group
1	CH_3Cl **Chloromethane**	$+\ HO^-$	$\longrightarrow$	CH_3OH **Methanol**	$+\ Cl^-$
2	CH_3CH_2I **Iodoethane**	$+\ CH_3O^-$	$\longrightarrow$	$CH_3CH_2OCH_3$ **Methoxyethane**	$+\ I^-$
3	$CH_3CH(Br)CH_2CH_3$ **2-Bromobutane**	$+\ I^-$	$\longrightarrow$	$CH_3CH(I)CH_2CH_3$ **2-Iodobutane**	$+\ Br^-$
4	$CH_3CH(CH_3)CH_2I$ **1-Iodo-2-methyl-propane**	$+\ N{\equiv}C^-$	$\longrightarrow$	$CH_3CH(CH_3)CH_2C{\equiv}N$ **3-Methylbutane-nitrile**	$+\ I^-$
5	Cyclohexyl–Br **Bromocyclohexane**	$+\ CH_3S^-$	$\longrightarrow$	Cyclohexyl–SCH_3 **Methylthiocyclohexane**	$+\ Br^-$
6	CH_3CH_2I **Iodoethane**	$+\ :NH_3$	$\longrightarrow$	$CH_3CH_2\overset{+}{N}H_3$ **Ethylammonium iodide**	$+\ I^-$
7	CH_3Br **Bromomethane**	$+\ :P(CH_3)_3$	$\longrightarrow$	$CH_3\overset{+}{P}(CH_3)_3$ **Tetramethylphosphonium bromide**	$+\ Br^-$

leaving group, but also as a nucleophile. The reverse reaction (the replacement of iodide in 2-iodobutane by bromide) is also possible. Thus, in reaction 3, the haloalkanes are in equilibrium. To get good yields of iodo product, propanone (acetone) is frequently used as a solvent. This choice is good because sodium iodide is soluble in propanone, whereas the product sodium bromide (or sodium chloride if chloroalkanes are used) is not. In this way, equilibrium can be shifted to the right by the precipitation of one of the products. Another option is to add excess sodium iodide.

Reaction 4 depicts a carbon nucleophile in the form of cyanide (e.g., sodium cyanide, $Na^{+-}CN$). It leads to the formation of a new carbon–carbon bond, an important means of lengthening the carbon chain.

Reaction 5 shows the sulfur analog of reaction 2, in which the nucleophile methanethiolate ion reacts with bromocyclohexane to produce a sulfide (Section 9-7). The success of this process demonstrates that nucleophiles in the same column of the periodic table react similarly to give analogous products. This conclusion is also borne out by *reactions 6* and *7*. However, these two reactions are qualitatively different from the other five, in which negatively charged nucleophiles displace negatively charged leaving groups. Amines, NR_3, and phosphines, PR_3, are neutral nucleophiles. In reactions with these neutral nucleophiles, the expulsion of the negatively charged leaving group results in a cationic species, an ammonium or phosphonium salt.

EXERCISE 6-2

What are the substitution products of the reaction of 1-bromobutane with:

(a) $:\ddot{\underset{\cdot\cdot}{I}}:^-$; (b) $CH_3CH_2\ddot{\underset{\cdot\cdot}{O}}:^-$; (c) $N_3:^-$; (d) $:As(CH_3)_3$; (e) $(CH_3)_2\ddot{\underset{\cdot\cdot}{Se}}$?

EXERCISE 6-3

Suggest starting materials for the preparation of: (a) $(CH_3)_4N^+I^-$; (b) $CH_3SCH_2CH_3$.

In summary, nucleophilic substitution is a fairly general reaction for primary and secondary haloalkanes. The halide functions as the leaving group, and there are several types of nucleophilic atoms that enter into the process.

6-4 A First Look at the Mechanism of Nucleophilic Substitution: Kinetics

Many questions can be raised at this stage. What are the kinetics of the reaction? Are they first order, second order (Section 2-6), or more complicated? What happens with optically active haloalkanes? Do other leaving groups exist? Do the relative rates of substitution predictably depend on the nature of the leaving group or that of the nucleophile, on the steric environment of the substrate or that of the solvent? These questions will be addressed one by one in the remainder of this chapter.

$$CH_3Cl + NaOH \xrightarrow{H_2O,\ \Delta} CH_3OH + NaCl$$

In this section, a study of the kinetics of the reaction will greatly assist us in ruling out possible mechanisms and in pointing out the most likely pathway. Our example will be the reaction of chloromethane with sodium hydroxide in water. Heating (indicated by the upper case Greek letter *delta,* Δ) will eventually result in a high yield of the products methanol and sodium chloride. However, this does not tell us anything about *how* starting materials are converted into products. Measurement of the changes in the speed of the reaction when the concentration of starting materials is changed will enable us to make certain predictions about the outcome of this experiment, based on our mechanistic hypotheses. By what means can we imagine methanol to be formed? We shall consider two possibilities.

A First Hypothetical Mechanism for the Conversion of Chloromethane into Methanol: Heterolytic Dissociation

Because the C–Cl bond is polarized, it may seem very reasonable to imagine an initial *heterolytic* dissociation of the molecule into an electron-deficient methyl cation and a chloride ion (Step 1). The highly electrophilic cation could subsequently join the electron-rich hydroxide ion to furnish the product (Step 2). This mechanism is controlled, therefore, by two steps, each having a rate constant, k_1 and k_2. If true, what should be the rate of the overall reaction? In transformations such as this, consisting of more than one step, the overall rate *is controlled by the slowest step in the sequence:* the **rate-determining step.** It seems intuitively sensible to assume that breaking the CH_3–Cl bond in our example would be rate determining ($k_1 < k_2$) because it requires the separation of opposite charges. The second step, in which a cation reacts with an anion to produce methanol, should be considerably faster.

STEP 1

$$CH_3Cl \xrightarrow{k_1} CH_3^+ + Cl^-$$

STEP 2

$$CH_3^+ + {}^-OH \xrightarrow{k_2} CH_3OH$$

Think of the rate-determining step as being analogous to a ‘‘bottleneck.’’ Imagine a water hose along which you have attached several clamps restricting the flow of the water (Figure 6-2). You can see that the rate at which the water will spew out of the end is controlled by the point of greatest constriction. If you were to reverse the water flow (to model the reversibility of a reaction), again the rate of flow would be controlled by this point. Such is the case in multistep reactions.

FIGURE 6-2

The rate at which water flows through a hose is controlled by the greatest constriction. The flow rate at various points is labeled k_1, k_2, and k_3; the overall flow stays constant (k_3) after this constriction.

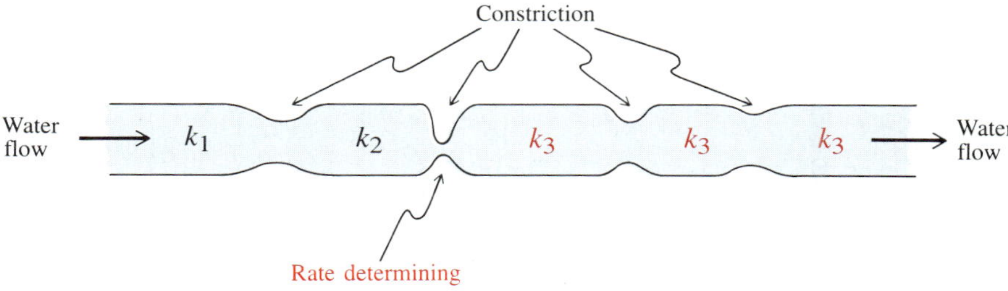

EXERCISE 6-4

Instead of the heterolytic dissociation in the preceding mechanism for methanol formation from chloromethane, consider an analogous homolytic pathway in which the CH_3–Cl bond is homolytically cleaved in the first step. Formulate this pathway and assign the rate-determining step.

A Second Possible Mechanism: Direct Displacement

The alternative to a dissociative mechanism is the direct attack of the haloalkane by the nucleophile with the simultaneous expulsion of the leaving group. In this process, also called **concerted,** bond-making would occur *at the same time* as bond-breaking (‘‘in concert’’).

The reaction can be thought of as proceeding in one of two extreme ways. The nucleophile approaches the substrate from the side of the leaving group and one group is exchanged for the other. This mechanism is **a frontside displace-**

ment (Figure 6-3). In its transition state, both oxygen and chlorine bear partially negative charges.

FIGURE 6-3

Frontside nucleophilic substitution. The concerted nature of bond-making (to OH) and bond-breaking (from Cl) is indicated by the dotted lines. Note how arrows are used in the first structure to indicate *electron flow*. This ''electron pushing'' technique (Section 1-5) will help you to keep track of the electrons in reactions. In the present case, an electron pair on the hydroxide ion is used to form a bond to carbon, and the electrons between carbon and chlorine depart with the leaving group.

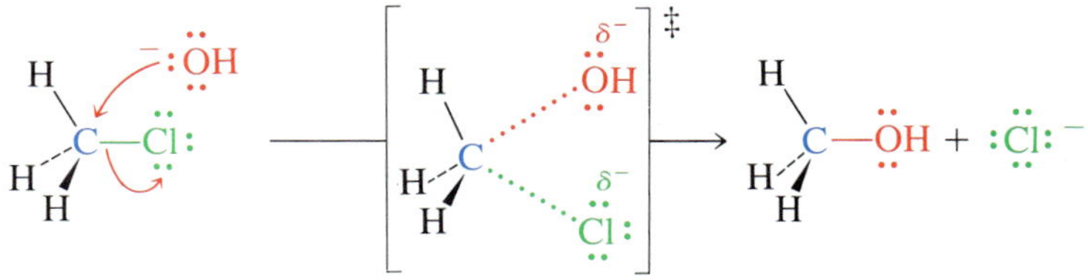

Another possibility is a concerted **backside displacement,** in which the nucleophile approaches carbon from the side opposite chlorine (Figure 6-4).

FIGURE 6-4

Backside nucleophilic substitution.

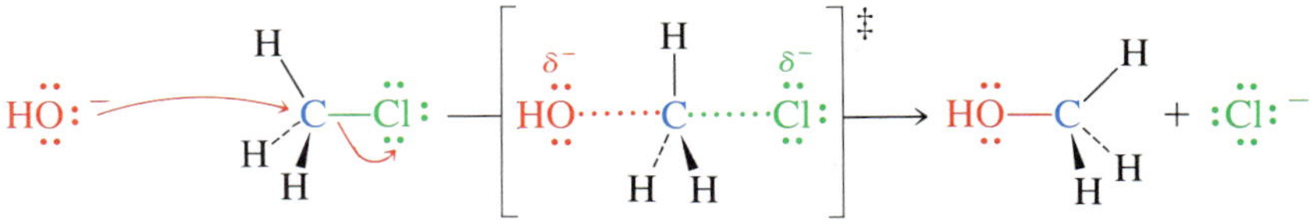

Is it possible to distinguish between the two sets of mechanism, two step or concerted? The answer is yes, by studying kinetics.

Kinetic Measurements Precisely Determine the Molecularity of the Substitution Reaction

A potential-energy diagram that qualitatively reflects the progress of our reaction according to the first proposed mechanism is shown in Figure 6-5. It indicates that the first step is rate determining by showing that its transition state is at a higher energy than that of the second step. Because the first transition state affects only the haloalkane, it follows that the rate of the reaction will be directly proportional only to the concentration of R–X and *independent of the concentration of hydroxide*. It should therefore be governed by the following rate equation:

$$\text{Rate} = k[\text{CH}_3\text{Cl}] \text{ moles } 1^{-1} \text{ sec}^{-1}$$

Recall that a reaction proceeding according to such a rate law is first order

FIGURE 6-5

Reaction profile for the hypothetical heterolytic dissociation mechanism of the substitution of hydroxide for chloride in chloromethane. The first step is slow and rate determining; the second is fast and does not affect the rate of the reaction.

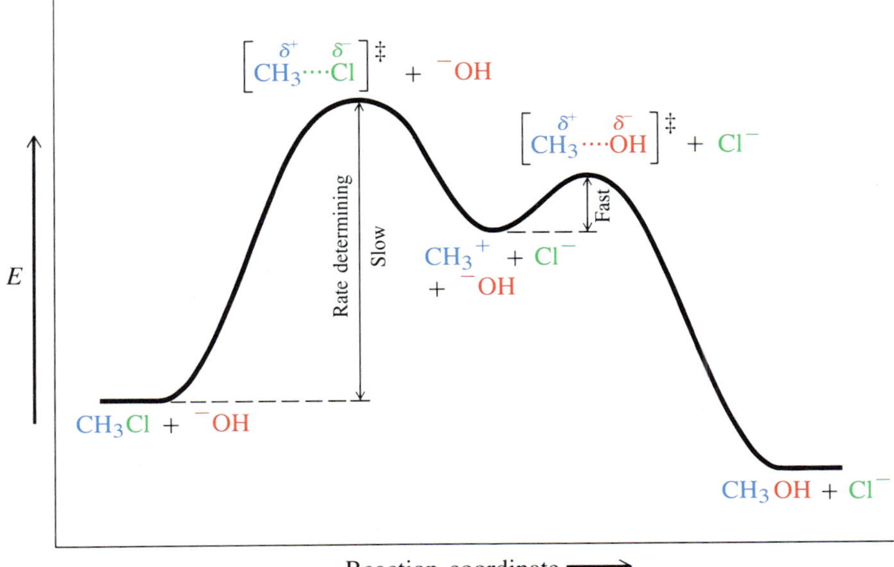

with respect to the reactant (Section 2-6). Because the mechanism is controlled by a rate-determining step affecting only this molecule, the reaction is **unimolecular.**

What kinetic results can we expect for the two concerted mechanisms? Because these mechanisms allow for the direct displacement of the leaving group, there is only one transition state to consider in each case, resulting from a direct collision between both reactants (Figure 6-6, on the next page). Here, the rate of the reaction will depend on the concentrations of *both* reaction partners, simply because the chances of one reactant molecule finding the other improve if the concentration of either one or both is increased. For example, doubling the concentration of the hydroxide should double the rate at which the haloalkane disappears. Likewise, at a given hydroxide concentration, doubling the concentration of chloromethane should have the same effect. Doubling the concentrations of both reactants should "doubly double" the rate; that is, increase it fourfold. The reaction, which is said to be of second order (Section 2-6), is governed by the following rate equation:

$$\text{Rate} = k[CH_3Cl][HO^-] \text{ mole } l^{-1} \text{ sec}^{-1}$$

A reaction following such a rate law is **bimolecular.**

What are the actual results of the kinetic experiments? In all cases, we find that, *in the reaction of chloromethane with hydroxide ion, the kinetics are second order*. We can monitor the progress of the reaction by following the disappearance of hydroxide ion with time by, for example, acid titration. Alternatively, the disappearance of chloromethane and the appearance of methanol

FIGURE 6-6

Reaction profile for the concerted replacement of chloride by hydroxide in chloromethane. There is only one transition state and it involves both reactant molecules.

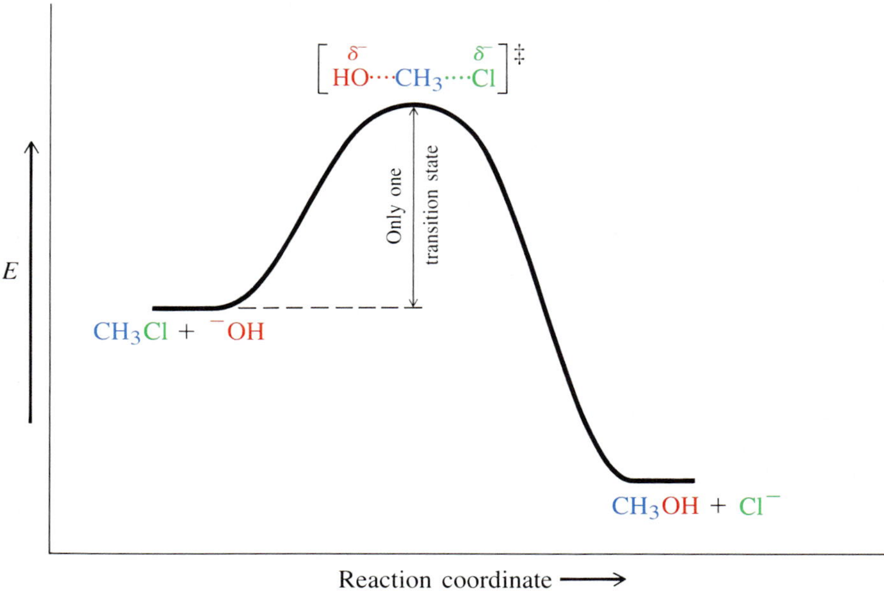

BOX 6-1

An Alternative but Chemically Unreasonable Mechanism for the S_N2 Reaction

The existence of second-order kinetics does not eliminate from consideration a dissociative mechanism in which the first step is *fast,* making the second bimolecular step rate determining. In this case, the reaction profile in Figure 6-5 would change so as to make the second transition state higher in energy than the first. The rate of the reaction would follow the rate law

$$\text{Rate} = k_2[CH_3{}^+][HO^-]$$

The concentration of $CH_3{}^+$ is related to the concentration of methyl chloride, as shown in the following equation:

$$\text{Rate (of } CH_3{}^+ \text{ formation)} = k_1[CH_3Cl]$$

Incorporation of this relation into the preceding one changes it as follows:

$$\text{Overall rate} = k_1 k_2[CH_3Cl][HO^-]$$

This rate law is consistent with a bimolecular reaction mechanism. Chemical reasoning suggests that it is not likely that the first step in the hypothetical mechanism would be faster than the second, and therefore we can drop this pathway from consideration. The stereochemical studies described in Section 6-5 offer an alternative way of eliminating it.

can be measured by gas-chromatographic analysis. Finally, we can also follow the rate of appearance of chloride ion by its precipitation as silver chloride. The results indicate that the substitution reaction proceeds according to a bimolecular mechanism, thus ruling out the dissociative pathway considered earlier. All the examples given in Section 6-3 follow such a mechanism. General bimolecular nucleophilic substitution is abbreviated as **S_N2,** in which **S** stands for substitution, **N** for nucleophilic, and **2** for bimolecular.

In summary, the S_N2 reaction follows second-order kinetics. It has a rate-determining transition state affecting both starting materials, which rules out a unimolecular pathway.

6-5 Frontside or Backside Attack? The Stereochemistry of the S_N2 Reaction

This section deals with an aspect of many organic reactions that provides us with additional mechanistic information: stereochemistry. Do the various mechanisms for the S_N2 reaction considered earlier have different consequences with respect to the three-dimensional arrangement of the groups around the reacting carbon center? To answer this question, we must examine the stereochemical fate of a chiral haloalkane.

The Carbocation Mechanism Gives Racemic Products

Consider the reaction of (*S*)-2-bromobutane with iodide ion. The various mechanisms of the displacement reaction considered earlier would lead to quite different stereochemical results. The dissociative pathway would have as an intermediate a planar (Section 1-8) and hence achiral cation; therefore, the stereochemical information present in the starting material would be lost in the products. Attack by iodide ion in the second step of the mechanism could occur from either side of the intermediate cation, affording both enantiomers of the product. Thus, the product 2-iodobutane would be racemic. In fact, it is found that the product from this S_N2 reaction is optically active.

Cationic Intermediates Lead to Racemates

H, H_3C, CH_2CH_3 on C—Br $\xrightarrow{-\,Br^-}$ planar cation (H, H_3C, CH_2CH_3, +) $\xrightarrow{I^-}$ $CH_3CHCH_2CH_3$ with I on C2

S
Chiral and optically active

Planar, achiral, and not optically active: attack from either side equally likely

R,S
Racemic and not optically active

The S_N2 Reaction Is Stereospecific

In the concerted mechanisms, both the frontside and the backside displacements on an optically active starting material should result in the formation of an optically active product. Frontside displacement leads to 2-iodobutane with the *same* configuration as the starting material, and the reaction is said to proceed with **retention of stereochemistry.** Backside displacement furnishes product of *opposite* configuration, and the reaction takes place with **inversion of stereochemistry.**

Frontside Displacement Gives Retention

S — Chiral and optically active → Frontside displacement → *S* — Chiral and optically active: retention; $+ Br^-$

Backside Displacement Gives Inversion

S — Chiral and optically active → Backside displacement → *R* — Chiral and optically active: inversion; $+ Br^-$

It is found that (*S*)-2-bromobutane gives (*R*)-2-iodobutane on treatment with iodide; in other words, the reaction proceeds with *inversion of configuration*. This finding, which applies to other S_N2 processes, rigorously rules out dissociative and frontside displacement mechanisms. It leaves as the most likely mechanistic candidate backside displacement.

BOX 6-2

Stereospecific and Stereoselective Reactions

A reaction such as the bimolecular substitution of a haloalkane, in which one stereoisomer (in the conversion of 2-bromobutane into 2-iodobutane, either the *R* or the *S* enantiomer of the starting material) is converted cleanly into another, is **stereospecific.** Here, this expression is preferable to the more-general term *stereoselective* (Section 5-6), which denotes a reaction that, regardless of the stereochemistry of the precursor, furnishes more of one stereoisomer than the other. A stereospecific reaction is also stereoselective, but not vice versa. These terms are often used interchangeably (and incorrectly) in the chemical literature. Sometimes the term stereoselective is incorrectly applied to reactions that are only partly stereospecific, whereas stereospecific is reserved for reactions in which the selectivity is quantitative. In this book, use of these terms will be restricted to the above definitions.

A Molecular-Orbital Picture of the Transition State of the S_N2 Reaction

The transition state for the S_N2 reaction can be described in molecular-orbital terms, as shown in Figure 6-7. As the nucleophile approaches the back lobe of the sp^3 hybrid orbital used by carbon to bind the halogen atom, the molecule becomes planar at the transition state, by rehybridization at carbon to sp^2. The negative charge is no longer located entirely on the nucleophile, but also partially on the leaving group. As the reaction proceeds to products, the inversion motion is completed, the carbon rehybridizes back to the tetrahedral sp^3 configuration, and the leaving group turns into a fully charged anion.

FIGURE 6-7
Molecular-orbital description of the S_N2 reaction. The inversion process is reminiscent of the "inversion" of an umbrella exposed to gusty winds.

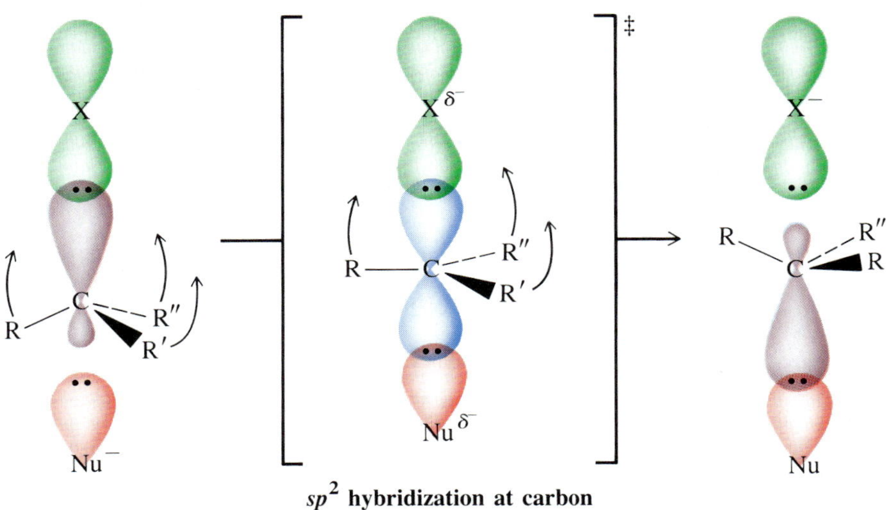

EXERCISE 6-5

Write the structures of the products of the S_N2 reactions of cyanide ion with: (a) *meso*-2,4-dibromopentane (double S_N2 reaction); (b) *trans*-1-iodo-4-methylcyclohexane.

To summarize, kinetic data indicate that the mechanism for nucleophilic substitution is bimolecular. Stereochemically, the reaction proceeds by inversion of configuration, through a concerted backside displacement.

6-6 Chemical Consequences of Inversion in S_N2 Reactions

What are the synthetic ramifications of the inversion of stereochemistry in the S_N2 reaction?

Enantiomers in S_N2 Reactions

Consider the conversion of 2-bromooctane into 2-octanethiol by reaction with hydrogen sulfide ion. If we were to start with optically pure *R* bromide, the only product obtainable would be *S* thiol, none of its *R* enantiomer. (The color scheme in this reaction is that used in Section 5-3 to indicate substituent priority.)

$$CH_3(CH_2)_4CH_2(CH_3)(H)C{-}Br \xrightarrow{HS^-} HS{-}C(H)(CH_3)CH_2(CH_2)_4CH_3 + Br^-$$

(*R*)-2-Bromooctane $[\alpha] = -34.6°$ → **(*S*)-2-Octanethiol** $[\alpha] = +36.4°$

What, however, if we wanted to convert (*R*)-2-bromooctane into the *R* thiol? A possible means of doing so would be a sequence of two S_N2 reactions, both of which would have to result in inversion of configuration at the stereocenter. For example, S_N2 reaction with iodide would first generate (*S*)-2-iodooctane, a halide with inverted configuration. Subsequent displacement of the iodide by HS^- ion would then furnish the *R* thiol. In this way, overall retention is attained by double inversion.

Retention by Double Inversion

$$CH_3(CH_2)_4CH_2(CH_3)(H)C{-}Br \xrightarrow[-\,Br^-]{I^-} I{-}C(H)(CH_3)CH_2(CH_2)_4CH_3 \xrightarrow[-\,I^-]{HS^-} CH_3(CH_2)_4CH_2(CH_3)(H)C{-}SH$$

(*R*)-2-Bromooctane $[\alpha] = -34.6°$ → **(*S*)-2-Iodooctane** $[\alpha] = +46.3°$ → **(*R*)-2-Octanethiol** $[\alpha] = -36.4°$

In these examples, the S_N2 reaction of an enantiomerically pure substrate changed its designated configuration from *R* to *S* or the reverse. Remember that this is not a necessary requirement of inversion, because the notation *R* and *S* is based strictly on priority rules. An example of an *S* enantiomer undergoing an S_N2 reaction to an *S* product is shown below.

$$CH_3CH_2O^- + {}^{b}CH_3S(H)(CH_3\,{}^{c})C{-}Br^{a} \longrightarrow CH_3CH_2O^{b}{-}C(SCH_3\,{}^{a})(H)(CH_3\,{}^{c}) + Br^-$$

S → ***S***

Diastereomers in S_N2 Reactions

So far, we have been concerned with S_N2 reactions of molecules containing only one stereocenter. What about molecules with several such centers? Reactions of these molecules are generally predictable: inversion will take place at all the primary and secondary carbons that undergo reaction with the incoming nucleophile. Note that the reaction of (2*S*,4*R*)-2-bromo-4-chloropentane with excess cyanide ion results in a meso product.

$$\begin{array}{c} CH_3 \\ H-\!\!\!-\!\!\!-Br \\ H-\!\!\!-\!\!\!-H \\ H-\!\!\!-\!\!\!-Cl \\ CH_3 \\ \mathbf{2S,4R} \end{array} \; + \; {}^-CN \; \xrightarrow[\text{Excess}]{\text{Ethanol}} \; \begin{array}{c} CH_3 \\ NC-\!\!\!-\!\!\!-H \\ H-\!\!\!-\!\!\!-H \\ NC-\!\!\!-\!\!\!-H \\ CH_3 \\ \mathbf{2R,4S: Meso} \end{array} \; + \; Br^- \; + \; Cl^-$$

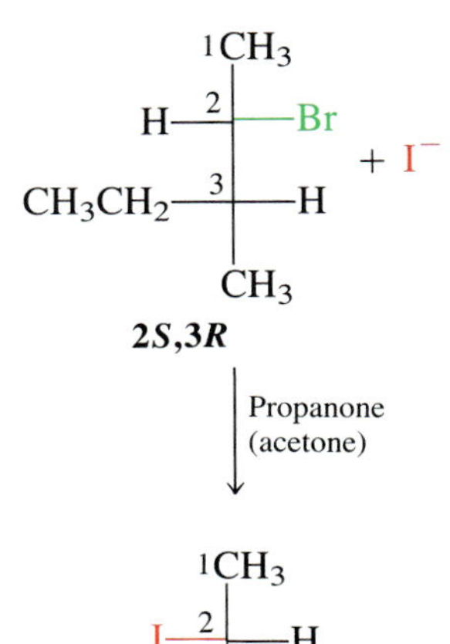

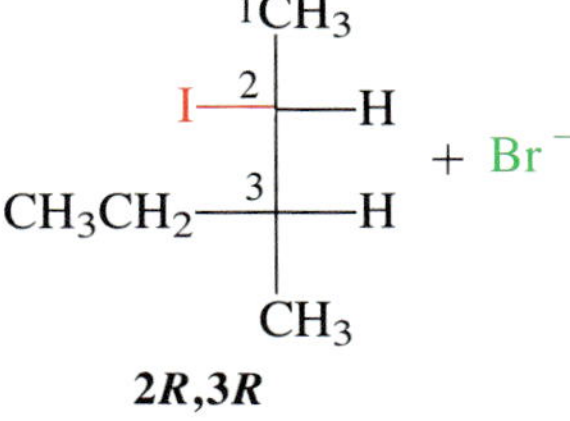

Ring Size Affects the Rate of the S_N2 Reaction: Halocycloalkanes

Halocycloalkanes also undergo S_N2 reactions, but with significant rate differences, depending on the size of the ring. For example, the strained halocyclopropanes are quite unreactive. An explanation for this finding lies in the structure of the transition state for the displacement. We have seen (Figure 6-7) that the reacting carbon adopts an sp^2 configuration as the nucleophile replaces the leaving group. The normal bond angle for such hybridization would be 120°, yet the cyclopropane ring cannot be distorted much from 60° (Figure 4-1). Therefore, much additional strain would have to be introduced into the transition state (Figure 6-8) to allow the S_N2 reaction to occur.

A similar explanation may be advanced for the relatively low reactivity of halocyclobutanes, although they do enter into S_N2 reactions, albeit only sluggishly.

Halocyclopentanes undergo substitution at rates comparable to their acyclic counterparts, such as 2-halopentanes. Halocyclohexanes, although seemingly more capable of attaining sp^2 hybridization at the reacting carbon, are slower to react. It appears that this inhibition is due to steric hindrance either to the approach of the attacking nucleophile if the leaving group is equatorial or to the departure of the leaving group if it is axial (Figure 6-9).

FIGURE 6-8

Hypothetical transition state for an S_N2 reaction of a halocyclopropane.

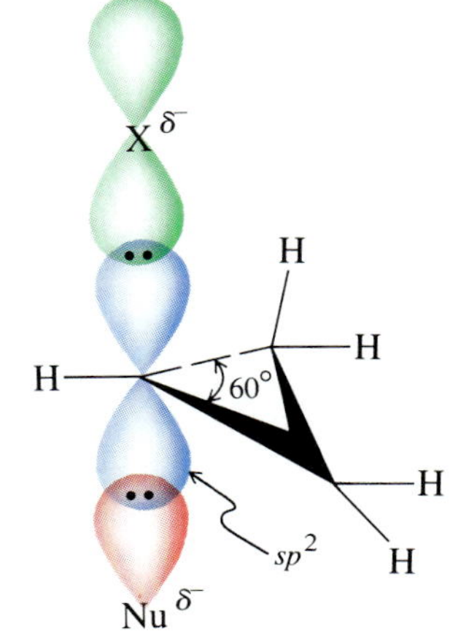

FIGURE 6-9

A. A nucleophile approaching an equatorial halide X. Steric hindrance by the two axial hydrogens at carbons 3 and 5 is apparent.
B. Replacement of an axial X. The leaving group encounters steric hindrance.

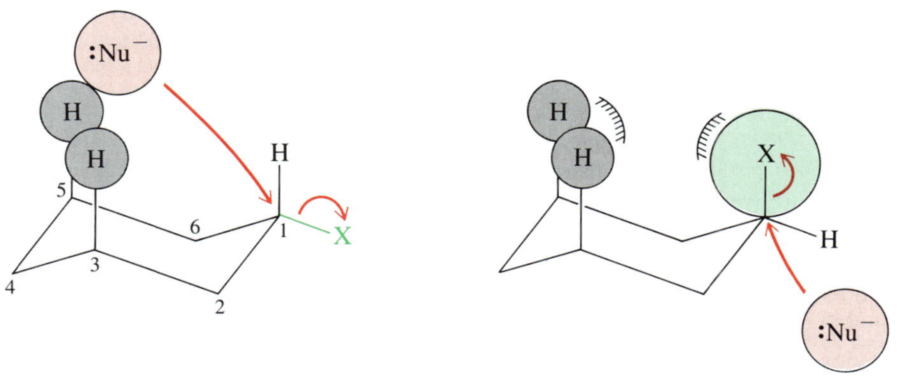

A

Inversion of stereochemistry at the reacting carbon in disubstituted cyclohexanes (and other cycloalkanes) is equivalent to changing the substituent pattern from cis to trans or trans to cis.

H, Br, H, CH_3 — Cis $\xrightarrow{\text{NaI, propanone (acetone)}}$ I, H, H, CH_3 — Trans + NaBr

In summary, inversion of configuration in the S_N2 reaction has distinct stereochemical consequences. Optical activity is retained, unless the nucleophile and the leaving group are the same or meso compounds are formed. In cyclic systems, cis and trans stereoisomers are interconverted. Halocyclopropanes do not undergo substitution because of prohibitive strain in the transition state.

6-7 Effect of the Structure of the Leaving Group on the Rate of Nucleophilic Displacement

As indicated earlier, the relative facility of S_N2 displacements depends on a variety of factors. Having examined the mechanistic details and some of the synthetic applications of this reaction, we should have a closer look at these factors. Let us start with the leaving group: are there any structural features that would allow us to predict, at least qualitatively, whether a leaving group is ''good'' or ''bad''?

The Leaving-Group Ability

The leaving group in nucleophilic substitution is frequently a negatively charged species. It is therefore not surprising that the relative ease with which it can be displaced, also called *leaving-group ability,* can be correlated with its capacity to accommodate a negative charge. Remember that a certain amount of negative charge is transferred to the leaving group in the transition state of the reaction (Figure 6-7).

Leaving Group Ability

$I^- > Br^- > Cl^- > F^-$

For the halogens, leaving-group ability increases in the procession from fluorine to iodine in the periodic table. Thus, iodide is regarded as a ''good'' leaving group; fluoride, on the other hand, is so ''bad'' that S_N2 reactions of fluoroalkanes are rarely observed.

EXERCISE 6-6

Predict the product of the reaction of 1-chloro-6-iodohexane with one equivalent of sodium methylselenide ($Na^{+\,-}SeCH_3$).

The quality of the leaving group should be related to its base strength; that is, the basicity of the leaving group should also depend on its capability to accommodate negative charge. What do we mean by these terms? Perhaps a brief review of acids and bases is in order.

Acids and Bases: A Brief Review

Brønsted and Lowry have given us a simple definition of acids and bases: an acid is a proton donor and a base is a proton acceptor. Acidity and basicity are commonly measured in water. An acid donates protons to water to give the hydronium ion, whereas a base removes them to give hydroxide ion. Examples are hydrogen chloride for the former and sodium methoxide for the latter.

$$\text{H—}\ddot{\underset{..}{\text{Cl}}}\text{:} + \text{H}\ddot{\underset{..}{\text{O}}}\text{H} \rightleftharpoons \text{H—}\ddot{\text{O}}^{+}\text{(H)}_2 + \text{:}\ddot{\underset{..}{\text{Cl}}}\text{:}^{-} \qquad \text{CH}_3\ddot{\underset{..}{\text{O}}}\text{:}^{-}\text{Na}^{+} + \text{H}\ddot{\underset{..}{\text{O}}}\text{H} \rightleftharpoons \text{CH}_3\text{OH} + \text{Na}^{+} \quad {}^{-}\text{:}\ddot{\underset{..}{\text{O}}}\text{H}$$

Hydronium ion **Hydroxide ion**

Water itself is neutral. It forms an equal number of hydronium and hydroxide ions by self-dissociation:

$$H_2O + H_2O \overset{K_w}{\rightleftharpoons} H_3O^+ + OH^-$$

Self-dissociation is governed by the equilibrium constant K_w, the self-ionization constant of water.

$$K_w = [H_3O^+][OH^-] = 10^{-14}\ \text{moles}^2\ l^{-2}$$

The concentration of water does not appear in this expression because it stays constant at its molarity: 1000 g l^{-1}/molecular weight (18 g mole^{-1}) = 55.5 moles l^{-1}. From the value for K_w, it follows that the concentration of H_3O^+ in pure water is 10^{-7} moles l^{-1}.

The pH is defined as the negative logarithm of the value for $[H_3O^+]$:

$$\text{pH} = -\log\ [H_3O^+]$$

For pure water, it is therefore +7. An aqueous solution with a pH lower than 7 is acidic; that with a pH higher than 7 is basic.

The acidity of a general acid, HA, is expressed by the chemical equation

$$HA + H_2O \overset{K}{\rightleftharpoons} H_3O^+ + A^-$$

which is described by the equilibrium constant K. According to the general definition of an equilibrium constant, K is expressed as

$$K = \frac{[H_3O^+][A^-]}{[HA][H_2O]}$$

Because, in aqueous solution, $[H_2O]$ will again be constant at 55 moles l^{-1}, that number may be incorporated into a new constant K_a, defined as the acidity constant:

$$K_a = \frac{[H_3O^+][A^-]}{[HA]}\ \text{moles}\ l^{-1}$$

As with pH, this measurement may be put on a logarithmic scale by the corresponding definition of pK_a:

$$pK_a = -\log K_a$$

An acid with a pK_a lower than 1 is defined as strong, one with a pK_a higher than 4 as weak. The acidities of some common acids are compiled in Table 6-4 and compared with those of compounds of high pK_a. It can be seen that the hydrogen halides (with the exception of HF) and sulfuric acid are very strong acids. Substances such as hydrogen cyanide, water, methanol, and methane are very weak acids.

TABLE 6-4
pK_a values for some common acids (25°C)

Name	Acid	pK_a
Hydrogen iodide	HI	−5.2
Hydrogen bromide	HBr	−4.7
Hydrogen chloride	HCl	−2.2
Sulfuric acid	H_2SO_4	−5[a]
Hydronium ion	H_3O^+	−1.7
Methanesulfonic acid	CH_3SO_3H	−1.2
Hydrogen fluoride	HF	3.2
Ethanoic (acetic) acid	CH_3COOH	4.7
Hydrogen cyanide	HCN	9.2
Methanethiol	CH_3SH	10.0
Methanol	CH_3OH	15.5
Water	H_2O	15.7
Ammonia	NH_3	35
Methane	CH_4	~50

[a] First dissociation equilibrium.

Like acid dissociation, the protonation of bases and their basicity can be described by a corresponding set of equations. The basicity of a base, A^-, can be expressed by the equation

$$A^- + H_2O \underset{}{\overset{K'}{\rightleftharpoons}} HO^- + HA$$

which is governed by the equilibrium constant K':

$$K' = \frac{[HO^-][HA]}{[A^-][H_2O]}$$

This transforms into a new equation by incorporation of the constant value for $[H_2O]$ into K':

$$K_b = \frac{[HO^-][HA]}{[A^-]} \text{ moles } l^{-1}$$

in which K_b is defined as the basicity constant, giving rise to a set of pK_b values.

The two constants, K_a and K_b, can be related to each other by simple multiplication:

$$K_a \times K_b = \frac{[H_3O^+][A^-]}{[HA]} \times \frac{[HO^-][HA]}{[A^-]} = [H_3O^+][HO^-] = K_w = 10^{-14}$$

We see that the product of the two is equal to the self-ionization constant of water. Hence,

$$pK_a + pK_b = 14$$

Therefore, if we know the pK_a of an acid HA, we automatically know the pK_b of A^-. Because of this relation, the species A^- derived from HA is frequently referred to as its **conjugate** base (*conjugatus,* Latin, joined together). Conversely, a species HA would be the conjugate acid of base A^-. For example, Cl^- is the conjugate base of HCl, and CH_3OH is the conjugate acid of CH_3O^-. Or, HCl may be viewed as the conjugate acid of Cl^-, CH_3O^- as the conjugate base of CH_3OH. It follows from this discussion that the conjugate base of a strong acid is weak, as is the conjugate acid of a strong base.

EXERCISE 6-7

Calculate the pK_b values for the conjugate bases of the acids in Table 6-4.

Are there structural features that allow us to predict, at least qualitatively, the relative strength of an acid HA (and hence the relative weakness of its conjugate base)? Yes, there are several. Prominent among them are two:

1 The strength of the H–A bond. The validity of this point is seen in the ordering of the acid strength of the hydrogen halides: HI > HBr > HCl > HF (for the DH° values of HX see Table 3-1).

2 The ability of A^- to accommodate the negative charge by either or both of two effects:

a The electronegativity of A. The more electronegative the atom to which the acidic proton is attached, the more acidic it will be. For example, in the first row of the periodic table, the decreasing order of acidity in the series HF > H_2O > H_3N > H_4C parallels the decreasing electronegativity of X (Table 1-4). In the hydrogen halides, this trend is outweighed by bond strength.

b The resonance in A allowing for delocalization of charge over several atoms. For example, ethanoic (acetic) acid is more acidic than methanol. In both cases, an O–H bond is broken heterolytically but, in contrast with methoxide, the ethanoate (acetate) ion has two resonance structures (Section 1-5).

Ethanoic (Acetic) Acid Is More Acidic Than Methanol Because of Resonance

$$CH_3\ddot{O}-H + H_2O \rightleftharpoons CH_3-\ddot{O}:^- + H_3O^+$$

$$CH_3C(=O)-\ddot{O}-H + H_2O \rightleftharpoons \left[CH_3C(=O)-\ddot{O}:^- \longleftrightarrow CH_3C(-\ddot{O}:^-)=\ddot{O}\right] + H_3O^+$$

The effect is even more pronounced in sulfuric and methanesulfonic acid. In these compounds, the availability of the *d* orbitals on sulfur enables us to write "valence shell expanded" Lewis structures containing as many as twelve electrons. Alternatively, charge-separated structures with one or two positive charges on sulfur can be used. Both representations indicate that the pK_a of these acids should be low:

$$\left[HO-S^{2+}(O^-)_3 \longleftrightarrow HO-S^{+}(=O)(O^-)_2 \longleftrightarrow HO-S(=O)_2-O^- \longleftrightarrow HO-S(=O)(O^-)=O \longleftrightarrow \text{etc.}\right]$$

Hydrogen sulfate ion

$$\left[CH_3-S^{2+}(O^-)_3 \longleftrightarrow CH_3-S^{+}(=O)(O^-)_2 \longleftrightarrow CH_3-S(=O)_2-O^- \longleftrightarrow CH_3-S(=O)(O^-)=O \longleftrightarrow \text{etc.}\right]$$

Methanesulfonate ion

As a rule of thumb, the acidity of an acid HA increases to the right and down in the periodic table.

Leaving-Group Ability Is Related to Base Strength

The leaving-group ability of the halides should be related to their base strength: the weaker that X^- is as a base, and therefore the stronger the acidity of its conjugate acid HX, the better it will function as a leaving group. A comparison of pK_a values (Table 6-4) and experimental evidence of the order of leaving-group abilities shows this postulate to be correct: HF is clearly the weakest acid, HCl is stronger, and HBr and HI are stronger still.

Table 6-4 also lists the pK_a values of other strong acids. Can their conjugate bases also function as good leaving groups in S_N2 reactions? The answer is yes. Halides are not the only groups that can be displaced by nucleophiles in S_N2 reactions. Other units that function as weak bases are also good leaving groups. Typical examples are sulfur derivatives of the type $ROSO_3^-$ and RSO_3^-, such as methyl sulfate ion, $CH_3OSO_3^-$, and various sulfonate ions. Their conjugate acids are very strong, comparable to methanesulfonic and sulfuric acid, mainly because of the resonance stabilization of the corresponding anions. Alkyl sulfate and sulfonate leaving groups are used so often that trivial names, such as mesylate, triflate, and tosylate, have found their way into the chemical literature.

Good Leaving Groups

Methyl sulfate ion	Methanesulfonate ion (Mesylate ion)	Trifluoromethanesulfonate ion (Triflate ion)	4-Methylbenzenesulfonate ion (*p*-Toluenesulfonate ion, tosylate ion)
$CH_3O{-}SO_2{-}O^-$	$CH_3{-}SO_2{-}O^-$	$CF_3{-}SO_2{-}O^-$	$CH_3{-}C_6H_4{-}SO_2{-}O^-$

For example, dimethyl sulfate, $(CH_3)_2SO_4$, a commercially available liquid, is often employed as a methylating agent because of its reactivity with oxygen and nitrogen nucleophiles. The nucleophile attacks the electrophilic methyl carbon atom, methyl sulfate ion functions as the leaving group, and the nucleophile emerges with an attached methyl group. The nucleophile is said to have been *methylated*.

$$CH_3CH_2\ddot{O}:^- + H_3C{-}O{-}SO_2{-}OCH_3 \xrightarrow{\text{Ethanol}} CH_3CH_2\ddot{O}CH_3 + {}^-:\ddot{O}{-}SO_2{-}OCH_3$$

Leaving group
Dimethyl sulfate

$$CH_3CH_2\ddot{N}H_2 + H_3C{-}O{-}SO_2{-}OCH_3 \xrightarrow{H_2O} CH_3CH_2\overset{+}{N}H_2(CH_3) + {}^-:\ddot{O}{-}SO_2{-}OCH_3$$

Alkyl sulfonates, such as mesylates and tosylates, are readily prepared from the corresponding sulfonyl chlorides and alcohols. This transformation will be discussed in more detail in Section 9-3.

$$CH_3CH_2CH_2OH + CH_3{-}SO_2{-}Cl \longrightarrow CH_3CH_2CH_2OSO_2CH_3 + HCl$$

Methanesulfonyl chloride (Mesyl chloride) → **Propyl methanesulfonate (Propyl mesylate)**

$$CH_3{-}CH(OH){-}CH_3 + Cl{-}SO_2{-}C_6H_4{-}CH_3 \longrightarrow CH_3{-}CH(CH_3){-}O{-}SO_2{-}C_6H_4{-}CH_3 + HCl$$

4-Methylbenzenesulfonyl chloride (*p*-Toluenesulfonyl chloride) → **1-Methylethyl 4-methylbenzenesulfonate (Isopropyl tosylate)**

The products react smoothly with a variety of nucleophiles.

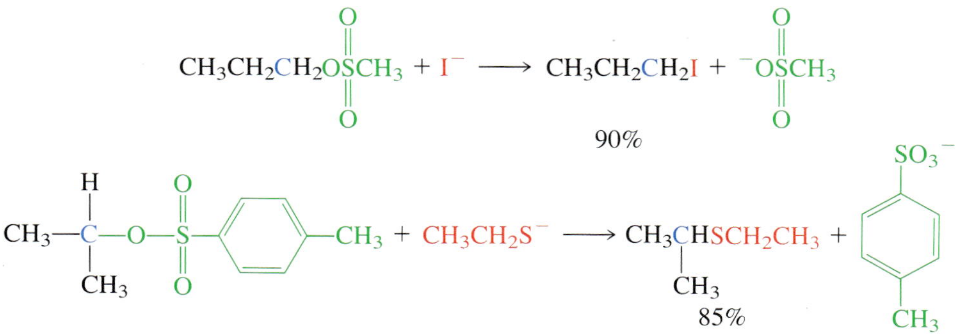

$$CH_3CH_2CH_2O\overset{O}{\underset{O}{\overset{\|}{\underset{\|}{S}}}}CH_3 + I^- \longrightarrow \underset{90\%}{CH_3CH_2CH_2I} + {}^-O\overset{O}{\underset{O}{\overset{\|}{\underset{\|}{S}}}}CH_3$$

$$CH_3-\overset{H}{\underset{CH_3}{\overset{|}{\underset{|}{C}}}}-O-\overset{O}{\underset{O}{\overset{\|}{\underset{\|}{S}}}}-C_6H_4-CH_3 + CH_3CH_2S^- \longrightarrow \underset{85\%}{CH_3\underset{CH_3}{\underset{|}{C}}HSCH_2CH_3} + {}^-O_3S-C_6H_4-CH_3$$

Sulfonate Intermediates in Nucleophilic Displacement of the Hydroxy Group in Alcohols

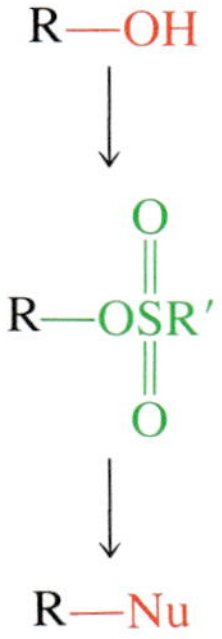

$$R-OH \longrightarrow R-O\overset{O}{\underset{O}{\overset{\|}{\underset{\|}{S}}}}R' \longrightarrow R-Nu$$

The foregoing reactions, when applied in sequence, afford a useful way of converting an alcohol into a new derivative in which the hydroxy group has been replaced by a nucleophile. This is important because, in contrast with the hydrogen halides and sulfonic acids, water has a very high pK_a; consequently, hydroxide is an exceedingly poor leaving group. However, its conversion into a sulfonate allows the ready formation of products of nucleophilic substitution.

Inspection of Table 6-4 indicates that $CH_3CO_2^-$, CN^-, CH_3S^-, HO^-, and H_2N^- should be increasingly poorer leaving groups. This is indeed found to be qualitatively the case. In fact, for practical purposes none of these substituents can function as leaving groups. However, remember that this is only a rough correlation. We cannot use the data in Table 6-4 to make any quantitative predictions about the exact relative rates of substitution of alkyl groups bearing various leaving groups.

EXERCISE 6-8

What is the product of the following reaction sequence?

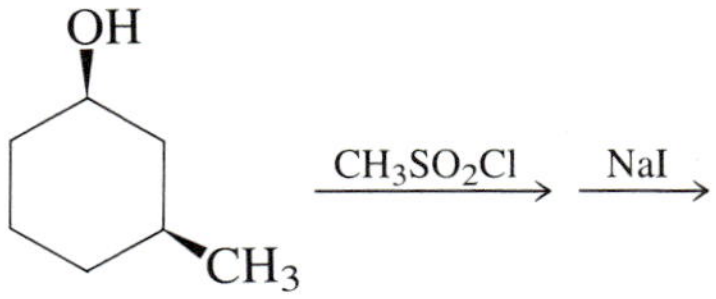

(cis-3-methylcyclohexanol structure with OH and CH_3) $\xrightarrow{CH_3SO_2Cl}$ $\xrightarrow{NaI}$

Water Can Be a Leaving Group

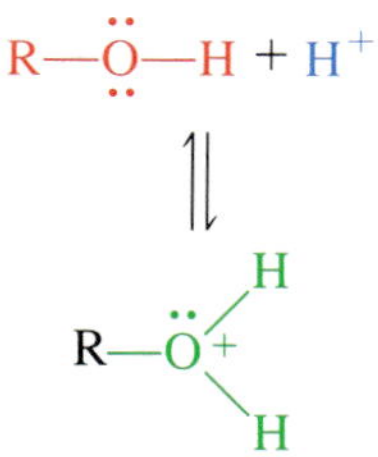

$$R-\ddot{\underset{..}{O}}-H + H^+ \rightleftharpoons R-\ddot{O}^+\!\!<^{H}_{H}$$

Oxonium ion

As mentioned, OH can be converted into a good leaving group by conversion into sulfonate. Another, and simpler, transformation of this kind is protonation. The attachment of a proton to the oxygen lone electron pair gives a species called an **oxonium ion.**

Because one of the lone pairs of electrons on the originally neutral OH group is now used for bonding to the proton, the positive charge resides on the oxygen atom. Protonation in this way turns OH from being a bad leaving group into a good leaving group: namely, neutral water. Water is a weak base, as exemplified by the very low pK_a (-1.7) of its conjugate acid, the hydronium ion H_3O^+.

$$pK_a\ (H_3O^+) = 14 - pK_a\ (H_2O) = 14 - 15.7 = -1.7$$

Nucleophilic displacement of H_2O can occur by the conjugate base of the acid employed to effect the initial protonation.

$$R\text{—}\ddot{\underset{\cdot\cdot}{O}}\text{—}H + HBr \longrightarrow R\text{—}\ddot{O}^{+}(H)(H) \quad Br^{-} \longrightarrow R\text{—}Br + H_2O$$

Thus, the treatment of some alcohols with concentrated hydrogen halides leads to the corresponding haloalkanes in good yield.

Examples:

$$CH_3(CH_2)_{10}CH_2OH + HBr \longrightarrow CH_3(CH_2)_{10}CH_2Br + HOH$$

1-Dodecanol → **1-Bromododecane** (91%)

$$HO(CH_2)_6OH + 2\,HI \longrightarrow I(CH_2)_6I + 2\,H_2O$$

1,6-Hexanediol → **1,6-Diiodohexane** (85%)

Similarly, the RO group in ethers, ROR, may be activated by very strong acid to become a leaving group by protonation (see Section 9-3). The mechanism of this reaction is analogous to that formulated for the alcohols.

$CH_3CH_2OCH_2CH_3$
Ethoxyethane
↓ HBr
CH_3CH_2Br
Bromoethane
+
CH_3CH_2OH
Ethanol

Mechanism:

[Note that color is used in a functional sense here: the nucleophilic group (red) turns into a leaving group (green).]

$$CH_3CH_2\ddot{\underset{\cdot\cdot}{O}}CH_2CH_3 \overset{H^+}{\rightleftharpoons} CH_3CH_2\text{—}\ddot{O}^{+}(H)(CH_2CH_3)$$

Oxonium ion

$$Br^{-} + CH_3CH_2\text{—}\ddot{O}^{+}(H)(CH_2CH_3) \longrightarrow CH_3CH_2Br + H\ddot{\underset{\cdot\cdot}{O}}CH_2CH_3$$

The alcohol formed as the second product may react to give more of the bromoalkane. The reactions of alcohols and ethers in the presence of acids will be discussed in more detail in Chapter 9.

EXERCISE 6-9

Reaction of oxacyclohexane (tetrahydropyran) with HI gives 1,5-diiodopentane. Give a mechanism for this reaction (write equations and use "electron pushing" arrows; do not use words).

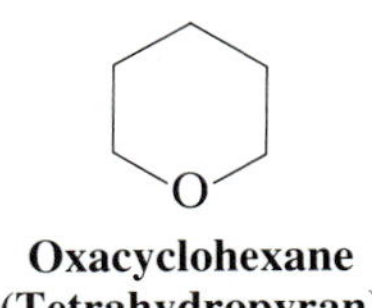

Oxacyclohexane (Tetrahydropyran)

In summary, the leaving-group ability of a substituent is roughly proportional to the strength of the conjugate acid. Both depend on the ability of the leaving group to accommodate negative charge. This section included a review of the equations by which the strengths of acids and bases are expressed and those that correlate the pK_a with the pK_b of the conjugate base ($pK_a + pK_b = 14$). Acid strength (and therefore leaving-group ability of the conjugate base) is related to the dissociation energy $DH°$(H–A), the electronegativity of A, and resonance effects. In addition to the halides (Cl^-, Br^-, and I^-), sulfonates (such as methane- or 4-methylbenzenesulfonates) are used as good leaving groups in synthetic applications. Particularly effective is the conversion of an alcohol into the corresponding sulfonate and its subsequent substitution, which allows the replacement of OH by a nucleophile in a two-step sequence. Similarly, the OH substituent can be changed into a good leaving group by protonation by use of a very strong acid HA. This procedure allows the substitution of OH by A in some cases (A = Br, I), chemistry that is also possible with ethers.

6-8 Effect of the Nature and Structure of the Nucleophile on the Rate of Nucleophilic Displacement

Let us now turn to a discussion of nucleophiles and their relative nucleophilic strength, their **nucleophilicity.** We shall see that nucleophilicity depends on a variety of factors: charge, basicity, solvent, polarizability, and the nature of substituents. To grasp the relative importance of these effects, let us consider a series of comparative experiments.

Increasing Negative Charge Increases Nucleophilicity

If the same nucleophilic atom is used, does charge play a role in the reactivity of a given nucleophile? The following experiments answer this question:

EXPERIMENT 1

$$CH_3Cl + HO^- \longrightarrow CH_3OH + Cl^- \quad \text{Fast}$$
$$CH_3Cl + H_2O \longrightarrow CH_3OH_2^+ + Cl^- \quad \text{Very slow}$$

EXPERIMENT 2

$$CH_3Cl + H_2N^- \longrightarrow CH_3NH_2 + Cl^- \quad \text{Fast}$$
$$CH_3Cl + H_3N \longrightarrow CH_3NH_3^+ + Cl^- \quad \text{Slower}$$

EXPERIMENT 3

$$CH_3Cl + HS^- \longrightarrow CH_3SH + Cl^- \quad \text{Fast}$$
$$CH_3Cl + H_2S \longrightarrow CH_3SH_2^+ + Cl^- \quad \text{Slower}$$

Conclusion: Of a pair of nucleophiles containing the same reactive atom, the species with a negative charge is the more powerful nucleophile. Or, of a base and its conjugate acid, the base is always more nucleophilic. This finding is intuitively very reasonable. Because nucleophilic attack is characterized by the formation of a bond with an electrophilic carbon center, the more negative the attacking species the faster the reaction should be.

EXERCISE 6-10

Predict which member in each of the following pairs of nucleophiles is better: (a) CH_3SCH_3 or CH_3S^-; (b) CH_3NH^- or CH_3NH_2; (c) HSe^- or H_2Se.

Nucleophilicity Decreases to the Right in the Periodic Table

Experiments 1 through 3 compared pairs of nucleophiles containing the same nucleophilic element (e.g., oxygen in H_2O versus HO^- and nitrogen in NH_3 versus H_2N^-). What about nucleophiles of similar structure but with different nucleophilic atoms? Let us examine the elements along one row of the periodic table.

EXPERIMENT 4

$$CH_3CH_2Br + H_3N \longrightarrow CH_3CH_2NH_3^+ + Br^- \quad \text{Fast}$$
$$CH_3CH_2Br + H_2O \longrightarrow CH_3CH_2OH_2^+ + Br^- \quad \text{Very slow}$$

EXPERIMENT 5

$$CH_3CH_2Br + H_2N^- \longrightarrow CH_3CH_2NH_2 + Br^- \quad \text{Fast}$$
$$CH_3CH_2Br + HO^- \longrightarrow CH_3CH_2OH + Br^- \quad \text{Slower}$$

EXPERIMENT 6

$$CH_3CH_2Br + CH_3S^- \longrightarrow CH_3CH_2SCH_3 + Br^- \quad \text{Fast}$$
$$CH_3CH_2Br + Cl^- \longrightarrow CH_3CH_2Cl + Br^- \quad \text{Very slow}$$

Conclusion: Nucleophilicity again appears to correlate with basicity: the more-basic species is seemingly the more-reactive nucleophile. Therefore, in procession from the left to the right of the periodic table, elements are observed to have decreasing nucleophilic character. The approximate order of reactivity for nucleophiles in the first row is

$$H_2N^- > HO^- > NH_3 > F^- > H_2O$$

We see that a negatively charged species is still more nucleophilic than a neutral analog, even though the latter might lie to its left in the periodic table (e.g., $HO^- > NH_3$ and $F^- > H_2O$), again reflecting the relatively higher basicity of the species.

EXERCISE 6-11

In each of the following pairs of molecules, predict which is the more nucleophilic: (a) $P(CH_3)_3$ or $S(CH_3)_2$; (b) $CH_3CH_2Se^-$ or Br^-; (c) H_2O or HF.

BOX 6-3

A Reflection on the Correlation Between Basicity and Nucleophilicity

There appears to be a fairly good correlation between basicity and nucleophilicity, in accord with intuitive expectation. However, consider the following arguments. Basicity (as well as acidity) is a measure of a *thermodynamic* phenomenon: namely, the equilibrium between a base and its conjugate acid in water:

$$A^- + H_2O \overset{K}{\rightleftharpoons} AH + HO^- \qquad K = \text{equilibrium constant}$$

On the other hand, nucleophilicity is a measure of a *kinetic* phenomenon: namely, the rate of reaction of a nucleophile with an electrophile.

$$Nu^- + R{-}X \xrightarrow{k} Nu{-}R + X^- \qquad k = \text{rate constant}$$

It is interesting to discover that, despite these inherently different features of basicity and nucleophilicity, such a good correlation exists in the cases examined so far: charged versus neutral nucleophiles, and nucleophiles along a row of the periodic table.

Nucleophilicity Is Impeded by Solvation

If it is a general rule that nucleophilicity correlates with basicity, then the elements from top to bottom of a column of the periodic table should have decreasing nucleophilic power. Recall (Section 6-7) that basicity decreases in an analogous fashion. To test this prediction, let us consider another series of experiments.

EXPERIMENT 7

$$CH_3CH_2CH_2O\overset{O}{\underset{O}{\overset{\|}{\underset{\|}{S}}}}CH_3 + Cl^- \xrightarrow{CH_3OH} CH_3CH_2CH_2Cl + {}^-O_3SCH_3 \qquad \text{Slow}$$

$$CH_3CH_2CH_2O\overset{O}{\underset{O}{\overset{\|}{\underset{\|}{S}}}}CH_3 + Br^- \xrightarrow{CH_3OH} CH_3CH_2CH_2Br + {}^-O_3SCH_3 \qquad \text{Faster}$$

$$CH_3CH_2CH_2O\overset{O}{\underset{O}{\overset{\|}{\underset{\|}{S}}}}CH_3 + I^- \xrightarrow{CH_3OH} CH_3CH_2CH_2I + {}^-O_3SCH_3 \qquad \text{Fastest}$$

EXPERIMENT 8

$$CH_3CH_2CH_2Br + CH_3O^- \xrightarrow{CH_3OH} CH_3CH_2CH_2OCH_3 + Br^- \qquad \text{Not very fast}$$

$$CH_3CH_2CH_2Br + CH_3S^- \xrightarrow{CH_3OH} CH_3CH_2CH_2SCH_3 + Br^- \qquad \text{Very fast}$$

EXPERIMENT 9

$$CH_3I + N(CH_3)_3 \xrightarrow{CH_3OH} {}^+N(CH_3)_4I^- \qquad \text{Fast}$$

$$CH_3I + P(CH_3)_3 \xrightarrow{CH_3OH} {}^+P(CH_3)_4I^- \qquad \text{Faster}$$

Conclusion: Nucleophilicity *increases* in procession down the periodic table, a trend directly opposing that expected on consideration of the basicity of the nucleophiles tested. Sulfur nucleophiles are more reactive than the analogous oxygen nucleophiles but less so than their selenium counterparts. The same is true for phosphorus versus nitrogen nucleophiles, the former being considerably more powerful. How can these trends be explained?

In the following discussion, a distinction is made between charged and neutral nucleophiles. A principal reason for the increasing nucleophilicity of *nega-*

tively charged nucleophiles from the top to the bottom of a column in the periodic table appears to be solvent effects.

When a solid dissolves, the intermolecular forces that held it together (Section 2-4) are replaced by molecule-solvent interactions, which may be of similar nature. This phenomenon is called **solvation.** A molecule in solution is surrounded by a solvent shell; it is said to be *solvated* by the solvent molecules. Salts require very polar solvents, such as water or alcohols, to enter the liquid phase. Solvents of this type, in which a hydrogen is attached to a strongly electronegative atom Y, contain a highly polarized $\overset{\delta^+}{H}-\overset{\delta^-}{Y}$ bond. Because of this effect the hydrogen has protonlike character and the solvent is called **protic.** Protic solvents solvate salts especially well because the small, positively charged hydrogen can interact strongly with the anion derived from the salt. This interaction is called **hydrogen bonding** and will be discussed in more detail in Section 8-2. Polar solvents not containing positively polarized hydrogens are called **aprotic** (see Section 6-10). The solvents used in many nucleophilic substitution reactions are protic.

How does the solvent affect the strength of the nucleophile? Generally, it weakens it. Moreover, smaller anions are more tightly solvated than larger ones because their charge is more concentrated. Their effective size becomes larger than that of their less well solvated counterparts. This phenomenon creates a solvent-induced barrier to attack at the nucleus carrying the leaving group. A pictorial representation of this effect, with methanol being the solvent, is shown in Figure 6-10. Evidence for this hypothesis is derived from the fact that many S_N2 reactions proceed considerably faster in aprotic (nonhydrogen-bonding) solvents (see Section 6-10). In these solvents, the order of reactivity may be reversed, following the trend originally expected on consideration of basicity.

FIGURE 6-10

Approximate schematic representation of the difference in solvation of (A) a small anion (F^-) and (B) a large anion (I^-).

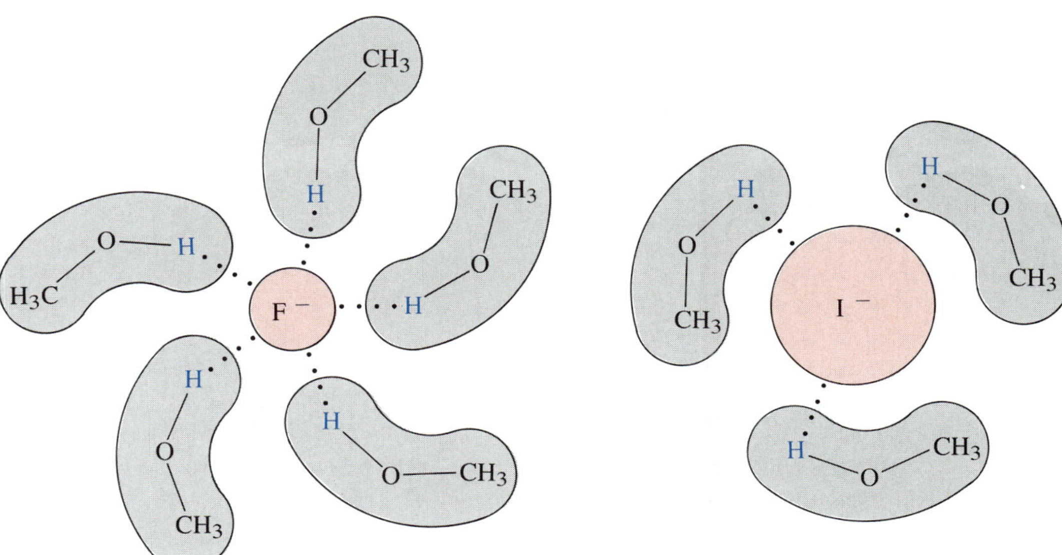

Increasing Polarizability Improves Nucleophilic Power

The solvation effects just described should be very pronounced only for charged nucleophiles. Nevertheless, the degree of nucleophilicity increases in procession down the periodic table even for *uncharged nucleophiles,* for which solvent effects should be much less strong: for example, $H_2Se > H_2S > H_2O$, and $PH_3 > NH_3$. Therefore, there must be an additional explanation for the observed nucleophilicity trends.

This explanation requires consideration of the polarizability of the nucleophile. Larger elements have larger, more-diffuse, and more-polarizable electron clouds. This added "softness" allows for more-effective overlap in the transition state, with the back lobe of the slowly rehybridizing sp^3 hybrid used to maintain bonding to the leaving group (Figure 6-11). For the same reason, the basicity of larger elements is relatively poor compared with that of the smaller elements, largely because overlap with the hydrogen $1s$ orbital is poor. A picture of this situation is shown in Figure 6-11, in which iodide and fluoride are compared in their reactivity in the S_N2 reaction.

EXERCISE 6-12

Which species is more nucleophilic: (a) CH_3SH or CH_3SeH; (b) $(CH_3)_2NH$ or $(CH_3)_2PH$?

FIGURE 6-11

Comparison of I^- and F^- in the S_N2 reaction: (A) the polarizable $5p$ orbital on iodide is distorted toward the electrophilic carbon atom; (B) the tight, less-polarizable $2p$ orbital on fluoride does not interact as effectively with the electrophilic carbon at a point along the reaction coordinate comparable to the one for part A.

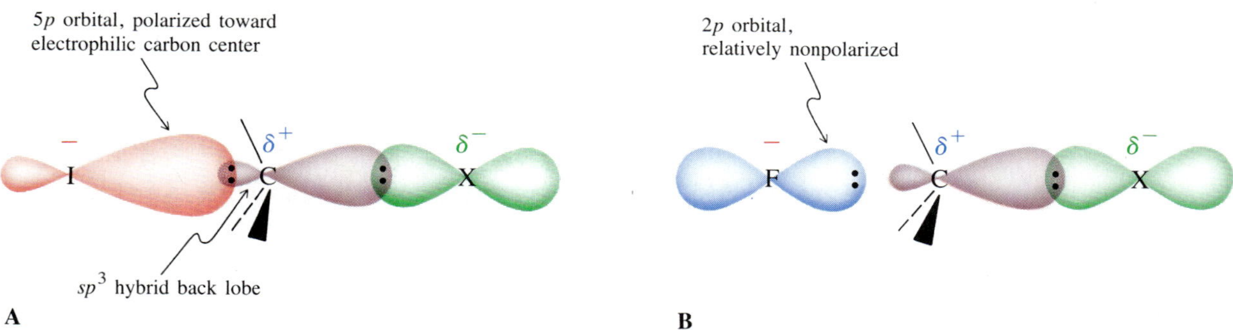

Sterically Hindered Nucleophiles Are Relatively Poor Reagents

We have seen that the steric bulk of the surrounding solvent shell may adversely affect the power of a nucleophile. Such steric hindrance may also be built into the molecule through its bulky substituents. The effect on the rate of reaction can be seen in Experiment 10.

EXPERIMENT 10

$$CH_3I + CH_3O^- \longrightarrow CH_3OCH_3 + I^- \quad \text{Fast}$$

$$CH_3I + (CH_3)_3CO^- \longrightarrow CH_3OC(CH_3)_3 + I^- \quad \text{Slower}$$

Conclusion: Sterically bulky nucleophiles react more slowly.

EXERCISE 6-13

Which of the two nucleophiles in parts a and b will react more rapidly with bromomethane?

(a) CH_3S^- or $CH_3CH(CH_3)S^-$; (b) $(CH_3)_2NH$ or $(CH_3CH(CH_3))_2NH$.

To summarize, nucleophilicity is controlled by a number of factors. Increased negative charge and procession from right to left and down the periodic table generally increase nucleophilic power. Table 6-5 compares the reactivity of a range of nucleophiles relative to that of methanol (arbitrarily set at 1). You can confirm the validity of the conclusion of this section by inspecting the various entries.

TABLE 6-5
Relative rates of reaction of various nucleophiles with iodomethane

Nucleophile	Relative rate	Nucleophile	Relative rate
CH_3OH	1	CH_3SCH_3	347,000
NO_3^-	~32	N_3^-	603,000
F^-	500	Br^-	617,000
SO_4^{2-}	3,160	CH_3O^-	1,950,000
		CH_3SeCH_3	2,090,000
$CH_3C(=O)O^-$	20,000	CN^-	5,010,000
Cl^-	23,500	$(CH_3CH_2)_3As$	7,940,000
$CH_3CH_2SCH_2CH_3$	219,000	I^-	26,300,000
NH_3	316,000	HS^-	100,000,000

6-9 Effect of the Structure of the Substrate on the Rate of Nucleophilic Displacement

Does the structure of the substrate, particularly in the vicinity of the center bearing the leaving group, influence the rate of nucleophilic attack?

As in the preceding sections, some indication of comparative reactivities can be gained from looking at the relative rates of reaction of representative haloalkanes with an appropriate nucleophile. A particularly easy way of obtaining information on *relative* reactivity is through a **competition experiment.** In this experiment, equal amounts of two substrates whose reactivity is of interest are exposed to a small amount (typically 0.05–0.1 molar equivalents) of reagent. Analysis of the ratio of the two possible products gives an immediate indication of the relative rates at which the two starting materials competed for the nucleophile.

The Effect of Chain Lengthening

Let us first investigate whether simply adding methylene groups to a halomethane affects the rate of its S_N2 reaction. For this purpose, we may start with a competition experiment in which a 1:1 mixture of chloromethane and chloroethane is exposed to a small amount of iodide ion.

EXPERIMENT 1

$$\underbrace{\boxed{CH_3Cl + CH_3CH_2Cl}}_{\text{Ratio 1:1}} + \underbrace{I^-}_{\text{Small amount}} \longrightarrow \underbrace{\boxed{CH_3I + CH_3CH_2I}}_{\text{Ratio 80:1}} + \text{starting chloroalkanes}$$

The outcome of this experiment tells us that chloromethane is approximately eighty times as reactive as chloroethane in its reaction with iodide. Will adding yet another methylene group increase the ratio? To find out, let us study the competition between chloromethane and 1-chloropropane in a similar experiment.

EXPERIMENT 2

$$\underbrace{\boxed{CH_3Cl + CH_3CH_2CH_2Cl}}_{\text{Ratio 1:1}} + \underbrace{I^-}_{\text{Small amount}} \longrightarrow \underbrace{\boxed{CH_3I + CH_3CH_2CH_2I}}_{\text{Ratio 150:1}} + \text{starting chloroalkanes}$$

We see that further chain extension reduces the reactivity of the chloroalkane by almost a factor of 2. This result can be verified by an independent control experiment in which chloroethane is found to be roughly twice as reactive as 1-chloropropane. Will this trend continue? The answer to this question is *no*. All of the higher haloalkanes have approximately the same reactivity toward nucleophiles. Is there an explanation for the observed results?

We can find a solution to this puzzle by inspecting the nucleophile's path of approach (Figure 6-12). A halomethane reacts essentially in the manner described for the reaction of chloromethane with hydroxide ion (Figures 6-7 and 6-12A). There appear to be no obstacles in the way of the nucleophile. However, replacement of a hydrogen atom by a methyl group, as in the haloethanes, creates a substrate in which the hydrogens of the additional methyl group sterically interfere with the incoming nucleophile (Figure 6-12B). This effect retards nucleophilic attack by a significant factor (80 in Experiment 1) when compared with analogous reactions of halomethanes. The 1-halopropanes have an additional methyl group in the vicinity of the reacting carbon center, but it seems to lead to only a small reduction in the rate of reaction. The effect is small because

FIGURE 6-12

S_N2 attack by Nu^- on (A) a halomethane, (B) a haloethane, and (C) three conformations of a 1-halopropane.

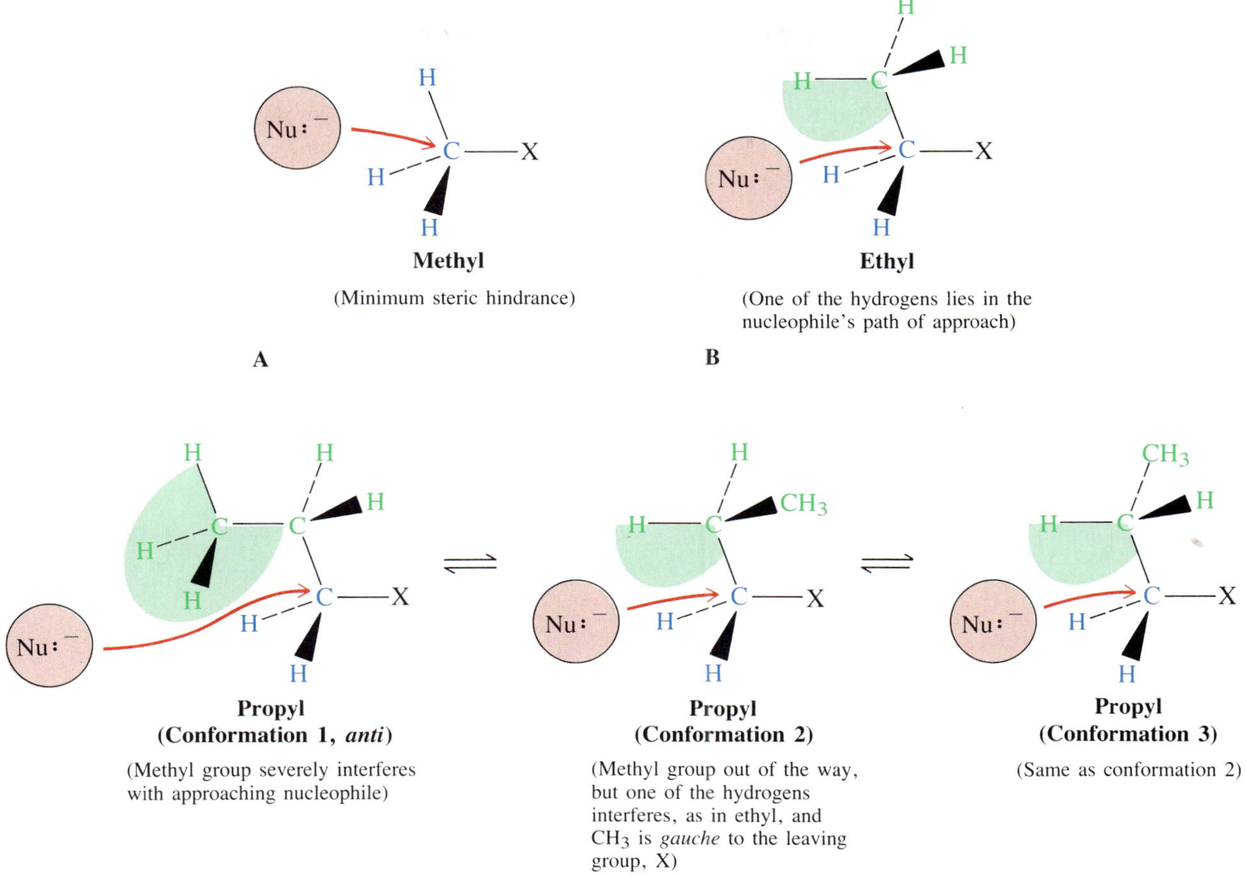

the molecule can adopt either of two staggered conformations (2 and 3 in Figure 6-12C) in which the extra methyl group is rotated out of the nucleophile's path of approach. However, these two conformations, which are best for the substitution reaction, are not the lowest-energy rotamers. The methyl group comes to lie *gauche* to the halogen in these structures. On the other hand, the most stable *anti* arrangement (conformation 1) places the methyl group right in the path of the approaching nucleophile, thus blocking reaction. As a result, the molecule that adopts conformation 1 has to proceed through a higher-energy *gauche* form (2 or 3) to undergo S_N2 displacement. This small energy expenditure results in the further decrease in rate in comparison with ethyl. Further chain elongation has no further effect, because the additional carbon atoms can be part of low-energy conformations that do not interfere with the incoming nucleophile.

Branching at the Reacting Carbon Decreases the Rate of Nucleophilic Substitution

As we have seen, the replacement of one hydrogen atom in a halomethane by a methyl group (Figure 6-12B) produces significant steric hindrance in S_N2 reac-

TABLE 6-6

Relative rates of S_N2 reaction of branched bromoalkanes with iodide

Bromoalkane	Rate
CH_3Br	145
CH_3CH_2Br	1
$CH_3CH(CH_3)Br$	0.0078
$CH_3C(CH_3)_2Br$	Negligible

tions. What happens if all *three* hydrogens are successively replaced by methyl groups? In other words, what are the relative bimolecular nucleophilic reactivities of methyl, primary, secondary, and tertiary halides? Competition experiments similar to the type described for chloroalkanes show that such reactivity rapidly decreases in the order shown in Table 6-6. Thus, the successive introduction of alkyl groups at the carbon bearing the leaving group has a cumulative effect on the steric hindrance of substitution (Figure 6-13). The results shown in Table 6-6 refer only to product formation by a bimolecular mechanism. As mentioned earlier, tertiary (and some secondary) halides may react by alternative pathways (Chapter 7).

EXERCISE 6-14

Predict the relative rates of the S_N2 reaction of cyanide with the following pairs of substrates:

(a) bromocyclohexane and 1-bromo-1-methylcyclohexane; (b) $CH_3CH_2C(CH_3)_2Br$ and $CH_3CH_2CH_2Br$.

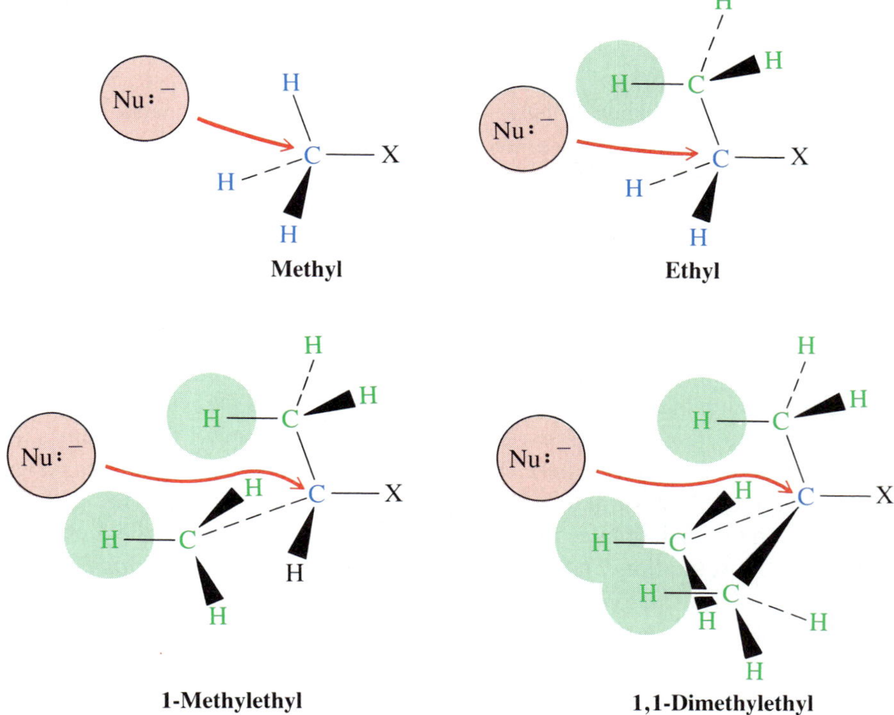

FIGURE 6-13

Steric effect of branching at the reacting carbon in S_N2 reactions.

Branching next to the Reacting Carbon

We can now consider a series of experiments probing the effect of multiple substitution at the position next to the electrophilic carbon and compare reactivity with that of bromoethane. The data obtained for the corresponding

TABLE 6-7
Relative reactivities of branched bromoalkanes with iodide

Bromoalkane	Relative rate	Bromoalkane	Relative rate
$H{-}C(H)(H)CH_2Br$	1		
$CH_3C(H)(H)CH_2Br$	0.8	$CH_3C(CH_3)(CH_3)CH_2CH_2Br$	0.03
$CH_3C(CH_3)(H)CH_2Br$	0.003	$CH_3C(CH_3)(CH_3)CH_2CH_2CH_2Br$	1
$CH_3C(CH_3)(CH_3)CH_2Br$	1.3×10^{-5}		

bromoalkanes are shown in Table 6-7, which includes two additional examples in which the branching is at a position removed from the bromine-bearing carbon. A dramatic decrease in reactivity is seen on further substitution of 1-bromopropane: 1-bromo-2-methylpropane is two orders of magnitude less reactive in its S_N2 reaction with iodide. 1-Bromo-2,2-dimethylpropane reacts so slowly that it may be regarded as being essentially inert. The general order of reactivity is then: methyl > ethyl > propyl > 2-methylpropyl > 2,2-dimethylpropyl halide.

Branching at carbon atoms removed from the bromine-bearing carbon has a much smaller effect. Can we rationalize these results?

Recall that, in Figure 6-12, the slight decrease in the reactivity of the halopropanes versus that of the haloethanes was explained by steric hindrance. We can use the same picture to understand the data in Table 6-7 (Figure 6-14).

A 1-halo-2-methylpropane molecule can adopt only one conformation (Figure 6-14C) in which the nucleophile can approach the backside of the reacting carbon. In this conformation, two methyl groups lie *gauche* to the halide substituent, a situation considerably more sterically constrained than that for the 1-halopropanes in which only one methyl group does so. The 1-halo-2,2-dimethylpropanes are almost completely blocked with respect to backside attack (Figure 6-14E and F).

EXERCISE 6-15

Predict the order of reactivity in the S_N2 reaction of:

Br versus CH_3 Br

FIGURE 6-14

Steric hindrance in the S_N2 reactions of (A through D) 1-halo-2-methylpropane systems and (E and F) 1-halo-2,2-dimethylpropane derivatives. In the 1-halo-2-methylpropane systems, conformations A and B contain methyl groups that block the path of the incoming nucleophile; only conformation C allows nucleophilic displacement. However, the two *gauche* methyl groups in conformation C result in considerable steric hindrance, as shown in part D, a Newman projection of this conformation. For the 1-halo-2,2-dimethylpropane, there is no conformation in which a methyl group does not shield the backside of the halogen-bearing carbon from an approaching nucleophile, as shown in parts E and F.

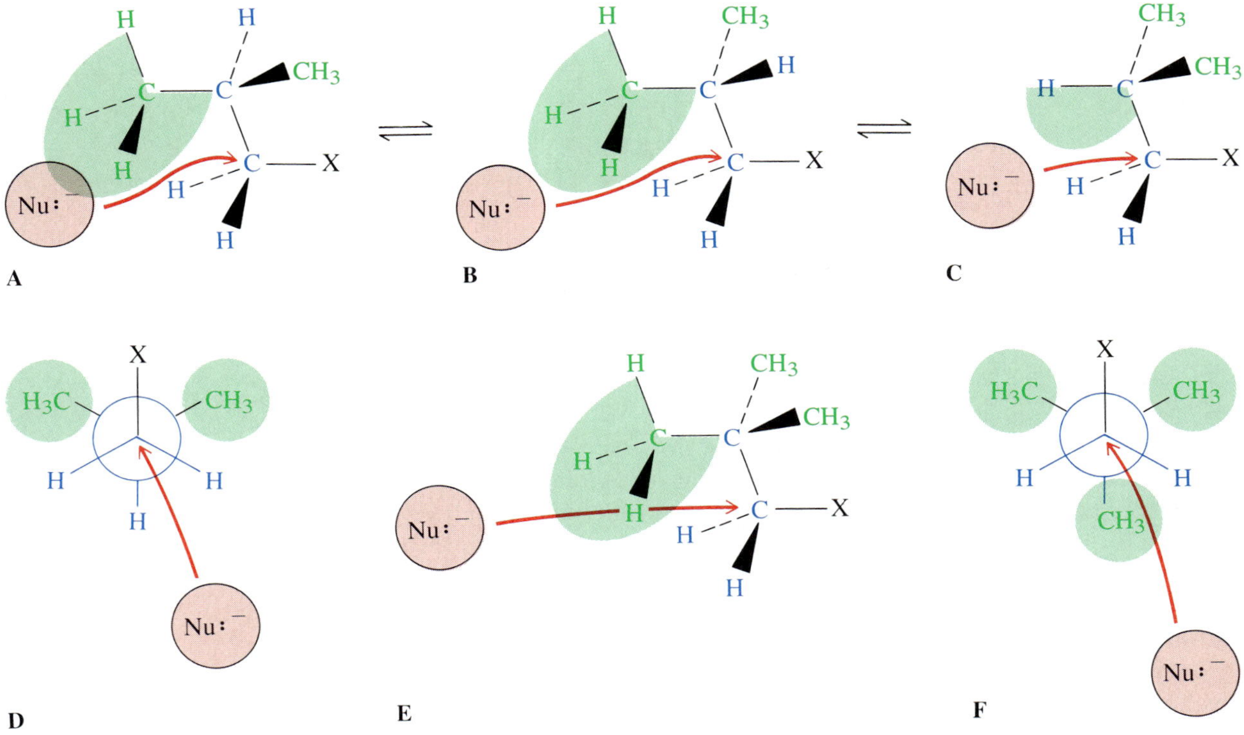

A B C D E F

In summary, the structure of the alkyl part of a haloalkane can have a pronounced effect on nucleophilic attack. Competition experiments have established that simple chain elongation has little influence on the rate of the S_N2 reaction. However, increased branching leads to strong steric hindrance and rate retardation.

6-10 Aprotic Solvent Effects

To close the chapter, this section will deal briefly with the effect of aprotic solvents on the rate of nucleophilic substitutions. As mentioned earlier (Section 6-8), the solvents frequently used in such reactions are protic, such as methanol, ethanol, or water. In these solvents, negatively charged nucleophiles are heavily solvated through hydrogen bonds. Aprotic media, such as propanone (acetone), also are useful solvents in these transformations. Because hydrogen bonds are not formed in such media, the nucleophiles are considerably less solvated, a condition sometimes termed ''naked.'' The reactivity of the naked nucleophile

TABLE 6-8
Polar aprotic solvents

Name	Structure	Boiling point (°C)	Dielectric constant	Dipole moment (D)
Propanone (Acetone)	$CH_3C(=O)CH_3$	56.5	20.70	2.88
Ethanenitrile (Acetonitrile)	$CH_3C\equiv N$	81.8	37.5	3.92
N,*N*-Dimethylmethanamide (*N*,*N*-Dimethylformamide, DMF)	$HC(=O)N(CH_3)_2$	39.9 (10 mm)	36.71	3.86
N-Methylmethanamide* (*N*-Methylformamide)	$HC(=O)N(H)CH_3$	103 (20 mm)	182.4	3.86
Methanamide* (Formamide)	$HC(=O)NH_2$	111 (20 mm)	111.0	3.73
Dimethyl sulfoxide (DMSO)	$CH_3S(=O)CH_3$	87 (20 mm)	46.6	3.9
Hexamethylphosphoric triamide (HMPA)	$O{=}P[N(CH_3)_2]_3$	100 (6 mm)	30	4.31
Sulfolane	five-membered ring with SO_2	283	43.3	4.81

*Still weakly protic.

is raised, sometimes drastically. For example, bromomethane is converted into iodomethane by potassium iodide in propanone (acetone) approximately 500 times as fast as in methanol.

Several polar aprotic solvents are listed in Table 6-8. They are characterized by the absence of protons capable of hydrogen bonding, large dipole moments, and large dielectric constants. Dipole moments in polar molecules were discussed in Section 1-3. The dielectric constant of a solvent is a measure of its ability to keep a positive ion, M^+, separated from its counterion, X^-. In nonpolar solvents, salts, even if they dissolve, tend to form ion pairs, M^+X^-, and aggregates, which are relatively unreactive in S_N2 reactions. This phenomenon explains why nucleophilic substitutions are relatively fast in solvents with a high dielectric constant. The importance of solvation in these reactions can be seen in a comparison of the rates of substitution of chloride for iodide in iodomethane in various solvents; Table 6-9 gives relative rates for methanol

TABLE 6-9

Relative rates of S_N2 reaction of iodomethane with chloride ion in various solvents

$$CH_3I + Cl^- \xrightarrow[k_{rel}]{\text{Solvent}} CH_3Cl + I^-$$

Solvent:	CH_3OH	$HCONH_2$	$HCONHCH_3$	$HCON(CH_3)_2$
k_{rel}:	1	12.5	45.3	1,200,000

(protic), methanamide (formamide, weakly protic), *N*-methylmethanamide (*N*-methylformamide, very weakly protic), and *N,N*-dimethylmethanamide (*N,N*-dimethylformamide, DMF, aprotic). This reaction is more than a million times as fast in DMF as in methanol.

Another manifestation of solvent effects is the finding that even 2,2-dimethylpropyl derivatives undergo efficient S_N2 reactions in a polar aprotic solvent:

$$(CH_3)_3CCH_2O\overset{O}{\underset{O}{\overset{\|}{\underset{\|}{S}}}}\text{—}C_6H_4\text{—}CH_3 + {}^-CN \xrightarrow{\text{HMPA, 100°C, 7 h}} \underset{90\%}{(CH_3)_3CCH_2CN}$$

2,2-Dimethylpropyl 4-methylbenzenesulfonate **3,3-Dimethylbutanenitrile**

Aprotic solvents also change the reactivity of nucleophiles that are normally heavily solvated through hydrogen bonds in protic solvents. As mentioned earlier, small ions are more encumbered by protic solvents than are large ones, which explains why the nucleophilicity of the halide ions increases in procession from top to bottom of the periodic table. In aprotic solvents, this ordering should change because solvation does not play such an overriding role. Indeed, in DMF, the order of nucleophilicity is $Cl^- > Br^- > I^-$, as originally expected from acid-base considerations. In propanone (acetone), the order of reactivity is the same as in alcohols, but the reactivity differences are much smaller.

Summary of Important Concepts

1 Compounds composed of an alkyl group and a halogen are commonly named as alkyl halides, systematically as haloalkanes. In IUPAC nomenclature, the halo substituent and an alkyl group have equal rank.

2 The physical properties of the haloalkanes are strongly affected by the polarization of the C–X bond and the polarizability of X.

3 Reagents bearing lone electron pairs are called nucleophilic when they attack positively polarized centers (other than a proton). Conversely, the latter are called electrophilic. When such a reaction leads to displacement of a substituent, it is a nucleophilic substitution. The group being displaced by the nucleophile is the leaving group.

4 The kinetics of the reaction of nucleophiles with primary (and most secondary) haloalkanes are second order, indicative of a bimolecular mechanism.

5 The bimolecular nucleophilic substitution reaction (S_N2) is stereospecific and proceeds with inversion of configuration.

6 A molecular-orbital description of the S_N2 transition state includes an sp^2-hybridized carbon center, partial bond-making between the nucleophile and the electrophilic carbon, and simultaneous partial bond-breaking between that carbon and the leaving group. Both the nucleophile and the leaving group bear partial charges.

7 Strained halocycloalkanes either do not undergo the S_N2 reaction (as is the case for three-membered rings) or do so only with difficulty because of bond-angle strain.

8 The leaving-group ability is roughly proportional to the strength of the conjugate acid. Especially good leaving groups are chloride, bromide, iodide, sulfonates, and water. Hydroxide is poor in that capacity but can be converted into a good leaving group by treatment with sulfonyl chlorides or by protonation.

9 Nucleophilicity increases with negative charge, for elements further to the left and down the periodic table, and in polar aprotic solvents.

10 Branching at the reacting carbon (or at the carbon next to it) of a haloalkane decreases the rate of bimolecular substitution. Tertiary halides do not enter into S_N2 processes but follow other reaction paths to be discussed in Chapter 7. 1-Halo-2,2-dimethylpropanes are normally quite unreactive in the presence of nucleophiles.

11 Polar aprotic solvents accelerate S_N2 reactions by: (a) allowing for the generation of "naked" nucleophilic reagents devoid of tight, bulky, hydrogen-bonded solvent shells; (b) being able to dissolve polar molecules (large dipole moment); and (c) being capable of keeping anionic nucleophiles away from their counterions (large dielectric constant).

Problems

1 Name the following molecules according to the IUPAC system of nomenclature.

(a) CH_3CH_2Cl

(b) $BrCH_2CH_2Br$

(c) $CH_3CH_2CHCH_2F$ (with CH_2CH_3 on the CH carbon)

(d) $(CH_3)_3CCH_2I$

(e) cyclohexane–CCl_3

(f) $CHBr_3$

2 Draw structures for each of the following molecules.

(a) 3-Ethyl-2-iodopentane.

(b) 3-Bromo-1,1-dichlorobutane.

(c) *cis*-1-(Bromomethyl)-2-(2-chloroethyl)cyclobutane.

(d) (Trichloromethyl)cyclopropane.

(e) 1,2,3-Trichloro-2-methylpropane.

3 Draw and name all possible structural isomers having the formula C_3H_6BrCl.

4 Draw and name all structurally isomeric compounds having the formula $C_5H_{11}Br$.

5 For each structural isomer in Problems 3 and 4, identify all stereocenters and give the total number of stereoisomers that can exist for the structure.

6 For each reaction in Table 6-3, identify (1) the nucleophile, (2) its nucleophilic atom (draw its Lewis structure first), (3) the electrophilic atom in the organic substrate, and (4) the leaving group.

7 A second Lewis structure can be drawn for one of the nucleophiles in Problem 6.

(a) Identify it and draw its alternate structure (which is simply a second resonance form).

(b) Does this second resonance form predict the presence of another nucleophilic atom in the nucleophile? If so, rewrite the reaction of Problem 6 using the new nucleophilic atom, and write a correct Lewis structure for the product.

8 In Section 6-4, four possible mechanisms for nucleophilic displacement are discussed: (1) heterolytic dissociation/combination, (2) homolytic dissociation (see Exercise 6-4), (3) frontside displacement, and (4) backside displacement.

For *each* of these mechanisms, predict the effect of the changes given below on the rate of the following reaction:

$$CH_3Cl + {}^-OCH_3 \xrightarrow{CH_3OH} CH_3OCH_3 + Cl^-$$

(a) Change substrate from CH_3Cl to CH_3I.
(b) Change nucleophile from CH_3O^- to CH_3S^-.
(c) Change substrate from CH_3Cl to $(CH_3)_2CHCl$.
(d) Change solvent from CH_3OH to $(CH_3)_2SO$.

9 The table below compares rate with concentration for the following reaction:

$$CH_3Cl + KSCN \xrightarrow{DMF} CH_3SCN + KCl$$

[CH_3Cl], *M*	**[KSCN], *M***	**Rate (moles l^{-1} sec^{-1})**
0.1	0.1	2×10^{-8}
0.2	0.1	
0.2	0.3	
0.4	0.4	

(a) Calculate the rate constant for this reaction.
(b) Fill in the rates that have been left out.

10 Determine the *R*/*S* designations for both starting materials and products in the following S_N2 reactions. Indicate whether the products are optically active.

(a) 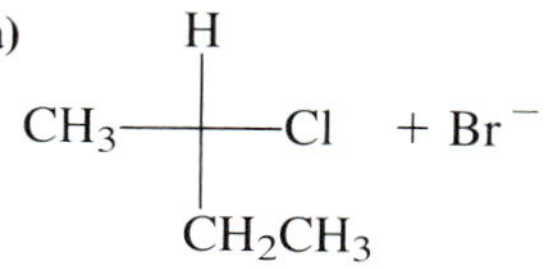 H / CH_3—C—Cl / CH_2CH_3 + Br^-

(c) cyclohexane with Cl and HO + $^-OCCH_3$ (C=O)

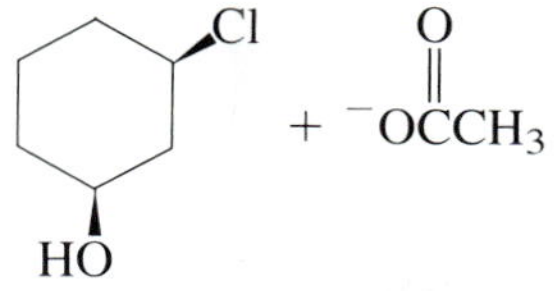

(b) 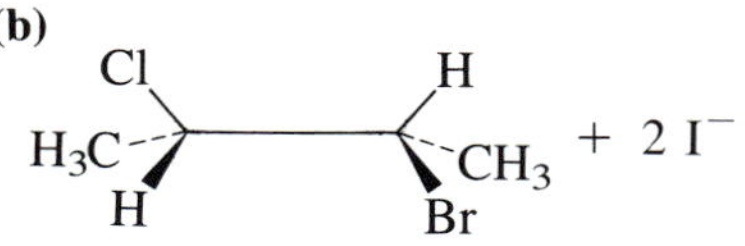 Cl, H_3C, H, H, CH_3, Br + 2 I^-

(d) 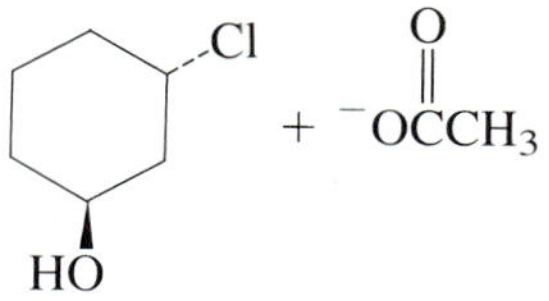 cyclohexane with Cl and HO + $^-OCCH_3$ (C=O)

11 Although all three bimolecular substitution reactions given below are comparably thermodynamically favorable, the reactions of the two cyclic substrates are *very slow* relative to the acyclic compound. Suggest an explanation.

REACTION 1

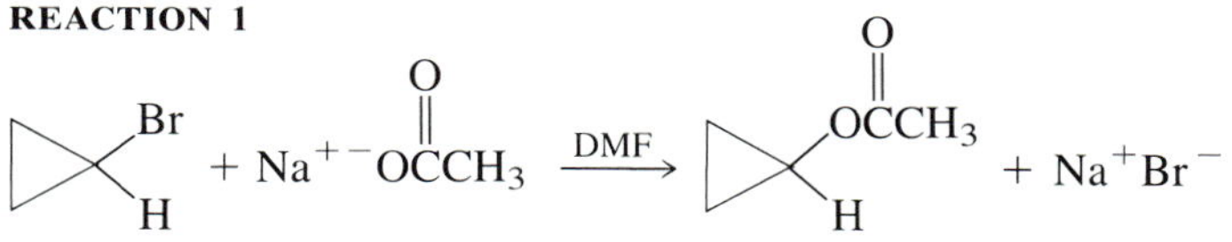

cyclopropyl (Br, H) + $Na^+ \ ^-OCCH_3$ (C=O) $\xrightarrow{DMF}$ cyclopropyl ($OCCH_3$, H) + Na^+Br^-

REACTION 2

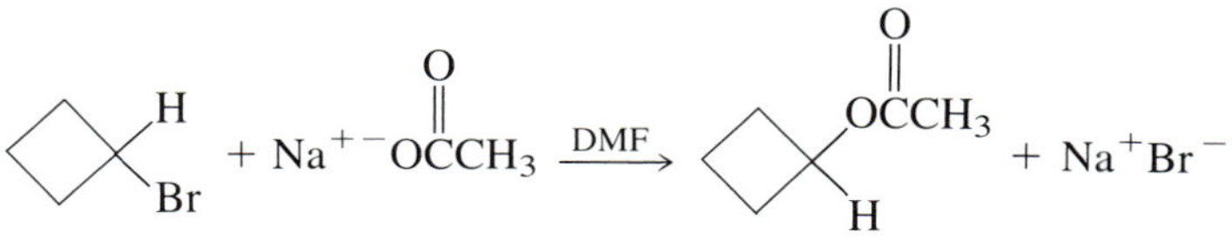

cyclobutyl (H, Br) + $Na^+ \ ^-OCCH_3$ (C=O) $\xrightarrow{DMF}$ cyclobutyl ($OCCH_3$, H) + Na^+Br^-

REACTION 3

$(H_3C)_2CHBr + Na^+ \ ^-OCCH_3$ (C=O) $\xrightarrow{DMF}$ $(H_3C)_2CHOCCH_3$ (C=O) + Na^+Br^-

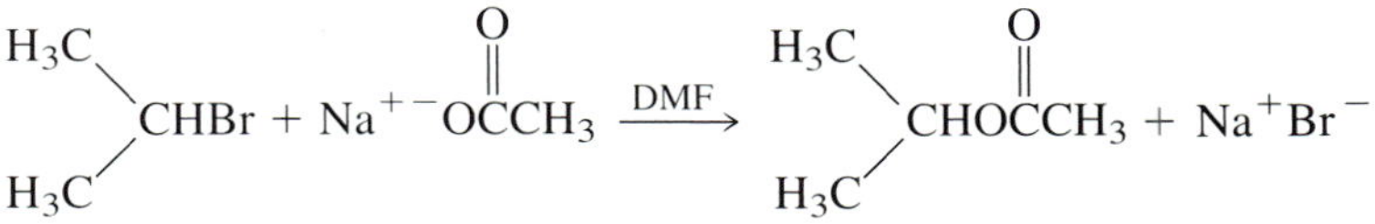

12 Write the product(s) of the reaction of 1-bromopropane with each of the following reagents. Write "no reaction" where appropriate.

(a)	H_2O	**(f)**	HCl
(b)	H_2SO_4	**(g)**	$(CH_3)_2S$
(c)	KOH	**(h)**	NH_3
(d)	CsI	**(i)**	Cl_2
(e)	NaCN	**(j)**	KF

13 Write the product of each of the following reactions using proper condensed formulas. If it does not seem that a reaction should occur, write "no reaction" as your answer.

(a) $CH_3CH_2CH_2CH_2Br + K^+ \ ^-OH \xrightarrow{CH_3CH_2OH}$

(b) $CH_3CH_2I + K^+Cl^- \xrightarrow{DMF}$

(c) 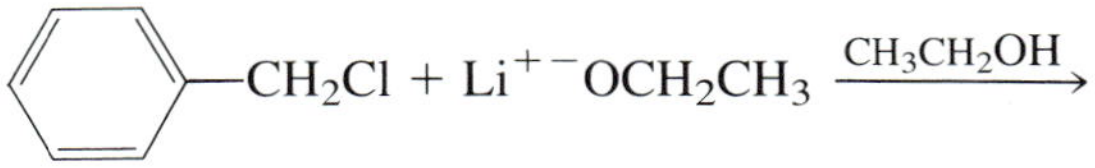 C_6H_5—$CH_2Cl + Li^+ \ ^-OCH_2CH_3 \xrightarrow{CH_3CH_2OH}$

(d) $(CH_3)_2CHCH_2Br + Cs^+I^- \xrightarrow{CH_3OH}$

(e) $CH_3CH_2CH_2Cl + K^{+-}SCN \xrightarrow{CH_3CH_2OH}$

(f) $CH_3CH_2F + Li^+Cl^- \xrightarrow{CH_3OH}$

(g) $CH_3CH_2CH_2OH + K^+I^- \xrightarrow{DMSO}$

(h) $CH_3I + Na^{+-}SCH_3 \xrightarrow{CH_3OH}$

(i) $CH_3CH_2OCH_2CH_3 + Na^{+-}OH \xrightarrow{H_2O}$

(j) $CH_3CH_2I + K^{+-}OC(=O)CH_3 \xrightarrow{DMSO}$

14 Write appropriate equations to show how each of the transformations given below might be accomplished. Use a different method from Section 6-7 for each one.

(a) N-methylpiperidine $\longrightarrow$ N,N-dimethylpiperidinium (N^+, H_3C, CH_3)

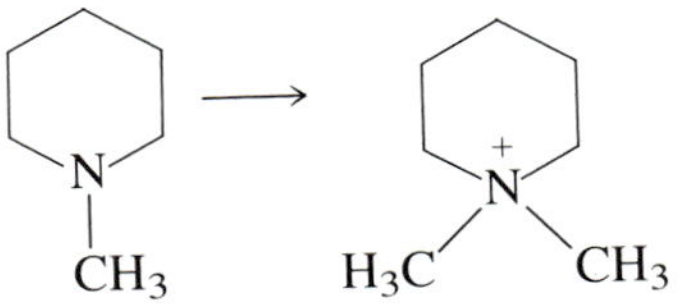

(b) (R)-$CH_3CH(OH)CH_2CH_3 \longrightarrow (S)$-$CH_3CH(SH)CH_2CH_3$

(c) Fischer projection: CH_3 (top), HO—H, CH_3O—H, CH_3 (bottom) $\longrightarrow$ CH_3 (top), H—CN, CH_3O—H, CH_3 (bottom)

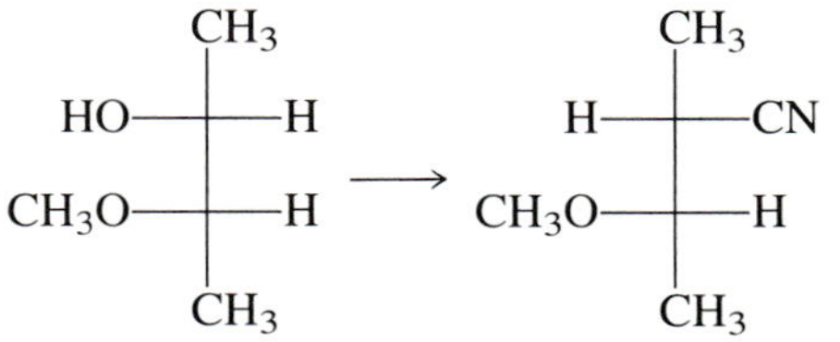

(d) bicyclic (H, H)---OH $\longrightarrow$ bicyclic (H, H)◀Br

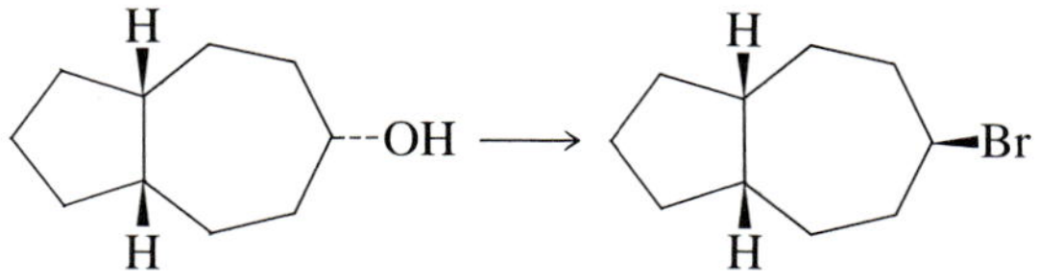

15 Write a detailed mechanism for the reaction that occurs when 1-propanol, Na^+Br^-, and concentrated H_2SO_4 are mixed. Write equations and use "electron pushing" arrows, not words.

16 Rings are readily prepared by means of intramolecular S_N2 reactions. Explain what is happening in each of the following reactions by a complete, step-by-step mechanism.

(a) $HOCH_2CH_2CH_2CH_2Cl + NaOH \xrightarrow{CH_3OH}$ (tetrahydrofuran, O)

(b) $HSCH_2CH_2Br + NaOH \xrightarrow{CH_3CH_2OH}$ (thiirane, S)

(c) $BrCH_2CH_2CH_2CH_2CH_2Br + NaOH \xrightarrow{CH_3OH}$ (tetrahydropyran, O)

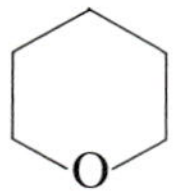

(d) $BrCH_2CH_2CH_2CH_2CH_2Br + NH_3 \xrightarrow{CH_3CH_2OH}$ (piperidine: six-membered ring with N–H)

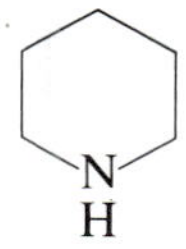

17 Rank the members of each group of species below in the order of (1) basicity, (2) nucleophilicity, and (3) leaving-group ability. Briefly explain your answers.

(a) H_2O, HO^-, $CH_3CO_2^-$
(b) Br^-, Cl^-, F^-, I^-
(c) $^-NH_2$, NH_3, $^-PH_2$
(d) ^-OCN, ^-SCN
(e) F^-, HO^-, $^-SCH_3$
(f) H_2O, H_2S, NH_3

18 Write the product(s) of each of the following reactions. Write "no reaction" as your answer if appropriate.

(a) $CH_3CH_2CH_2CH_3 + Na^+Cl^- \xrightarrow{CH_3OH}$

(b) $CH_3CH_2Cl + Na^{+\,-}OCH_3 \xrightarrow{CH_3OH}$

(c) 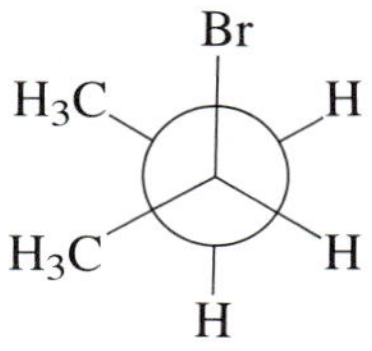 (Newman projection: front carbon Br, H, H; back carbon H_3C, H_3C, H) $+ Na^+I^- \xrightarrow{\text{Propanone (acetone)}}$

(d) 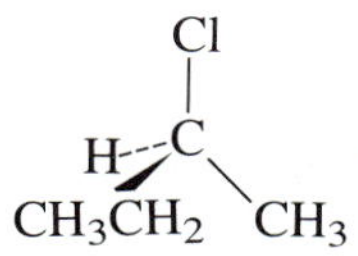 (Cl–C with H, CH_3CH_2, CH_3 in wedge/dash) $+ Na^{+\,-}SCH_3 \xrightarrow{\text{Propanone (acetone)}}$

(e) $CH_3CH(OH)CH_3 + Na^{+\,-}CN \longrightarrow$

(f) $CH_3CH(OH)CH_3 + HCN \longrightarrow$

(g) $CH_3CH(OH)CH_3 + HBr \longrightarrow$

(h) 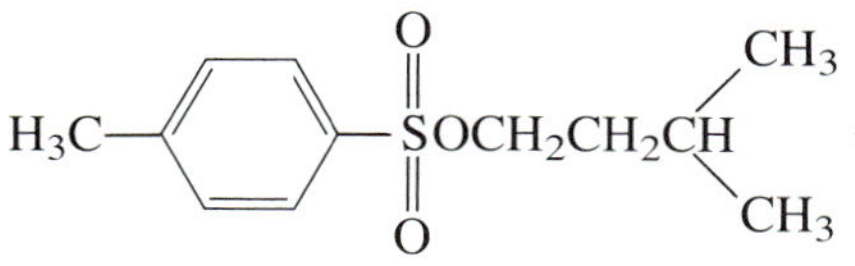 $H_3C-C_6H_4-SO_2-OCH_2CH_2CH(CH_3)_2 + K^{+\,-}SCN \xrightarrow{CH_3OH}$

(i) $CH_3CH_2NH_2 + Na^+Br^- \xrightarrow{DMSO}$

(j) $CH_3I + Na^{+\,-}NH_2 \xrightarrow{NH_3}$

(k) Product of part **j** + more $CH_3I \longrightarrow$

(l) 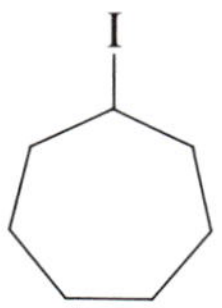 [iodocycloheptane] + $Na^{+}\,^{-}SH \xrightarrow{CH_3OH}$

(m) 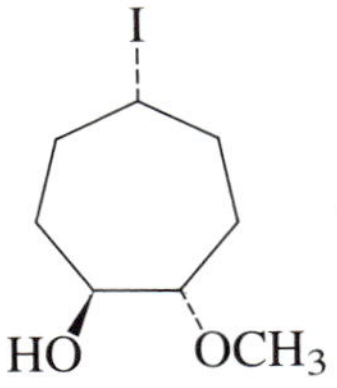 [iodocycloheptane bearing HO and OCH_3] + $Na^{+}\,^{-}SH \xrightarrow{CH_3OH}$

(n) $CH_3\underset{}{\overset{CH_3}{|}}CHCH_2Br$ + $(C_6H_5)_3P \xrightarrow{CH_3CH_2OH}$

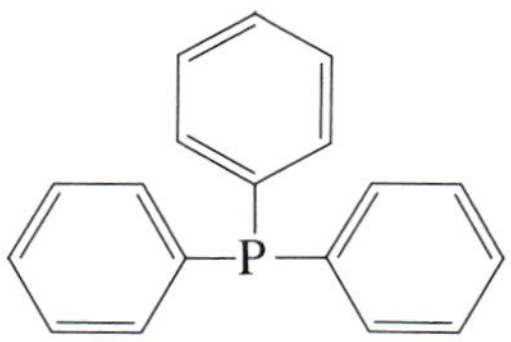

19 Using the information in Chapters 3 and 6, propose the best possible synthesis of each of the following compounds, using propane as your organic starting material and any other reagents you need. Note: Based on the information in Section 3-6, do not expect to find very good answers for parts a, c, and e. There is one general approach that is best, however.

(a) 1-Chloropropane. **(d)** 2-Bromopropane.
(b) 2-Chloropropane. **(e)** 1-Iodopropane.
(c) 1-Bromopropane. **(f)** 2-Iodopropane.

20 Propose four syntheses of *trans*-1-methyl-2-(methylthio)cyclohexane (shown below), beginning with each of the starting compounds indicated below.

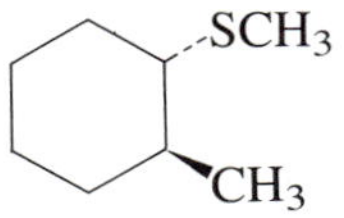

[cyclohexane with SCH_3 and CH_3 substituents]

(a) *cis*-1-Chloro-2-methylcyclohexane.
(b) *trans*-1-Chloro-2-methylcyclohexane.
(c) *cis*-2-Methylcyclohexanol.
(d) *trans*-2-Methylcyclohexanol.

21 Rank each group of molecules below in order of increasing S_N2 reactivity.

(a) CH_3CH_2Br, CH_3Br, $(CH_3)_2CHBr$.

(b) $(CH_3)_2CHCH_2CH_2Cl$, $(CH_3)_2CHCH_2Cl$, $(CH_3)_2CHCl$.

(c) CH_3CH_2Cl, CH_3CH_2I, [cyclohexyl]–Cl.

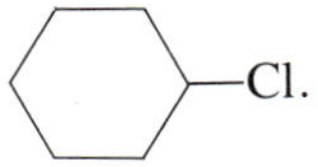

(d) $(CH_3CH_2)_2CHCH_2Br$, $CH_3CH_2CH_2\underset{CH_3}{\underset{|}{C}}HBr$, $(CH_3)_2CHCH_2Br$.

22 The table below presents rate data for the reactions of CH_3I with three different nucleophiles in two different solvents.

Nucleophile	k_{rel}, CH_3OH	k_{rel}, DMF
Cl^-	1	1.2×10^6
Br^-	20	6×10^5
^-SeCN	4×10^3	6×10^5

What is the significance of these results regarding relative reactivity of nucleophiles under different conditions?

23 The formidable looking molecule 5-methyltetrahydrofolic acid (abbreviated 5-methyl-FH_4) is the product of sequences of biological reactions that convert carbon atoms from a variety of simple molecules such as methanoic (formic) acid and the amino acid histidine into methyl groups:

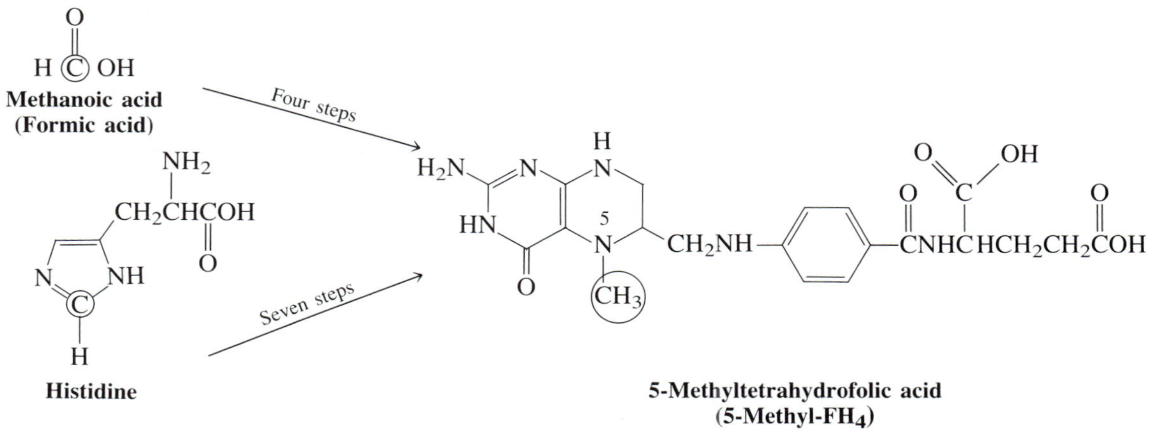

Methanoic acid (Formic acid)

Four steps

Seven steps

Histidine

5-Methyltetrahydrofolic acid (5-Methyl-FH_4)

The simplest synthesis of 5-methyltetrahydrofolic acid is from tetrahydrofolic acid (FH_4) and trimethylsulfonium ion, a reaction carried out by microorganisms in the soil:

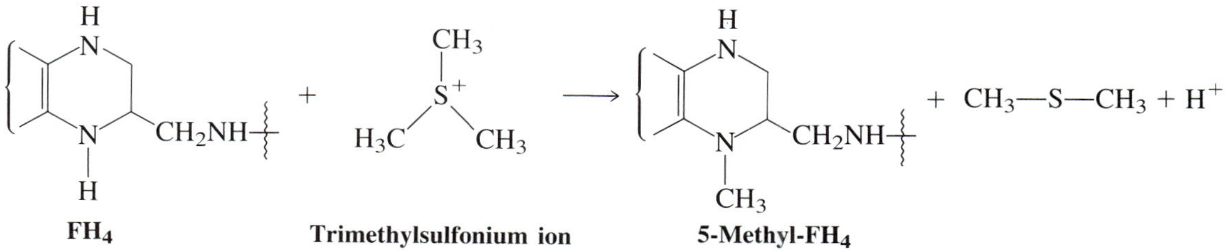

$$FH_4 + (CH_3)_3S^+ \longrightarrow \text{5-Methyl-}FH_4 + CH_3\text{—}S\text{—}CH_3 + H^+$$

FH_4 **Trimethylsulfonium ion** **5-Methyl-FH_4**

(a) Can this reaction be reasonably assumed to proceed through a nucleophilic substitution mechanism? Write the mechanism using "electron pushing" arrow notation.

(b) Identify the nucleophile, the nucleophilic and electrophilic atoms participating in the reaction, and the leaving group.

(c) Based on the concepts presented in Sections 6-7 through 6-9, are all the groups that you identified in part **b** behaving in a reasonable way in this reaction? Does it help to know that species such as H_3S^+ are very strong acids (e.g., pK_a of $CH_3SH_2^+$ is -7)?

24 The role of 5-methyl-FH_4 (Problem 23) in biology is to serve as a donor of methyl groups to small molecules. The synthesis of the amino acid methionine from homocysteine is perhaps the best-known example:

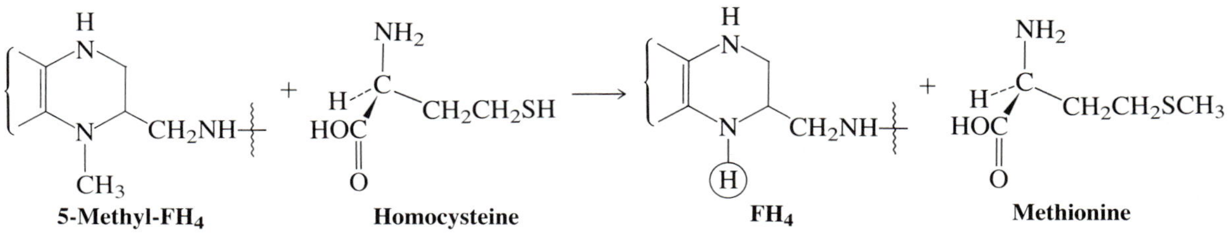

5-Methyl-FH_4 **Homocysteine** **FH_4** **Methionine**

For this problem, answer the same questions that were posed in Problem 23. The pK_a of the circled hydrogen in FH_4 is 5. Does this cause a problem with any feature of your mechanism? In fact, methyl transfer reactions of 5-methyl-FH_4 require a proton source. Review the material in Section 6-7, especially the subsection titled "Water Can Be a Leaving Group." Then suggest a useful role for a proton in the reaction above.

25 Adrenalin (epinephrine) is produced in your body in a two-step process that accomplishes the transfer of a methyl group from methionine (Problem 24) to norepinephrine (see Reaction 1 below and Reaction 2 on the next page).

(a) Explain in detail what is going on mechanistically in these two reactions, and analyze the role played by the molecule of ATP.

(b) Would you expect methionine to react directly with norepinephrine? Explain.

(c) Propose a laboratory synthesis of adrenalin from norepinephrine.

REACTION 1

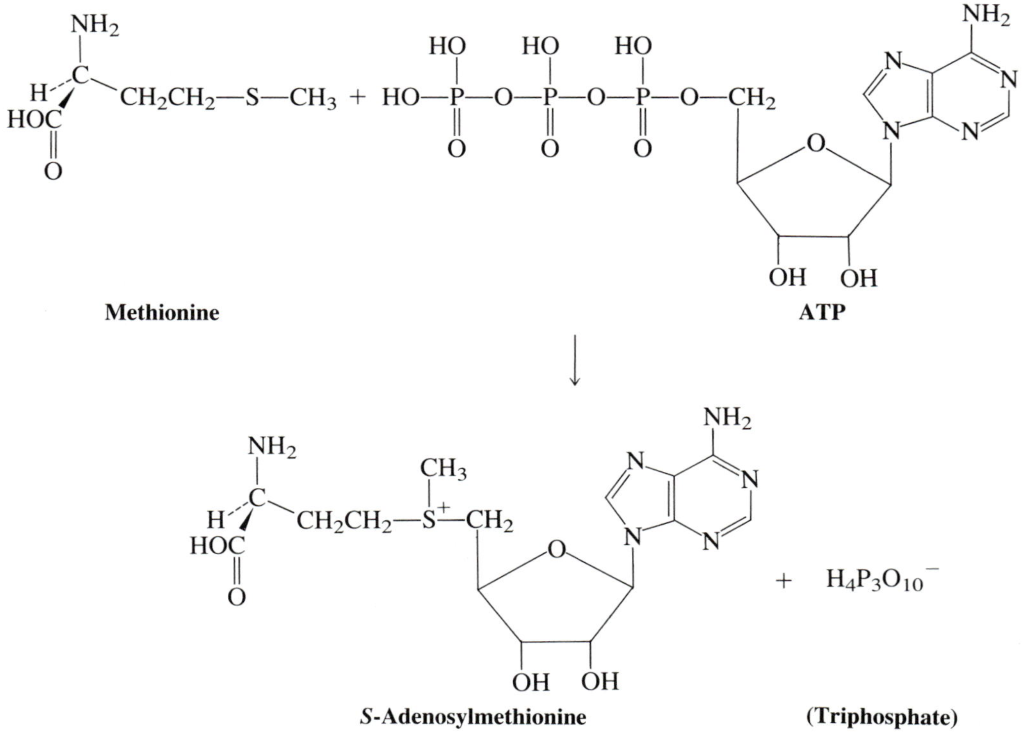

Methionine **ATP**

↓

***S*-Adenosylmethionine** **(Triphosphate)**

REACTION 2

S-Adenosylmethionine + Norepinephrine → S-Adenosylhomocysteine + Adrenalin + H^+

Norepinephrine

S-Adenosylhomocysteine

Adrenalin

CHAPTER 7

Further Reactions of Haloalkanes

UNIMOLECULAR SUBSTITUTION AND PATHWAYS OF ELIMINATION

As mentioned in Chapter 6, S_N2 reactions are not the only transformations that haloalkanes can undergo in the presence of nucleophiles, particularly when the haloalkanes are tertiary or secondary. We will see in this chapter that, indeed, bimolecular substitution is only one out of *four* possible modes of reaction.

7-1 Solvolysis of Tertiary Haloalkanes

Let us begin by looking at the aforementioned alternative mode of reactivity of tertiary haloalkanes in the presence of nucleophiles. We have seen that the rate of the S_N2 reaction diminishes drastically if the reacting center changes from primary to secondary and finally tertiary. Thus, whereas the S_N2 reactivity of bromomethane and bromoethane with iodide ion in propanone (acetone) is moderate, 2-bromopropane was relatively unreactive, and 2-bromo-2-methylpropane was extremely slow to transform by this pathway. However, this is true only for the bimolecular substitution reaction: secondary and tertiary halides do undergo substitution by another possible mechanism. This section will show that secondary and tertiary haloalkanes in fact react quite rapidly with nucleophiles to give the corresponding substitution products. For example, 2-bromo-2-methylpropane (*tert*-butyl bromide), when dissolved in aqueous propanone (acetone), is rapidly converted into 2-methyl-2-propanol (*tert*-butyl alcohol) and hydrogen bromide. The nucleophile here is water, which is normally quite unreactive in the S_N2 process.

An Example of Solvolysis: Hydrolysis

$$\underset{\substack{\textbf{2-Bromo-2-methylpropane}\\ (\textit{tert}\textbf{-Butyl bromide})}}{CH_3\overset{\displaystyle CH_3}{\underset{\displaystyle CH_3}{C}}Br} + H{-}OH \xrightleftharpoons{\text{Propanone (acetone)}} \underset{\substack{\textbf{2-Methyl-2-propanol}\\ (\textit{tert}\textbf{-Butyl alcohol})}}{CH_3\overset{\displaystyle CH_3}{\underset{\displaystyle CH_3}{C}}OH} + HBr$$

2-Bromopropane converts in a similar fashion, albeit more slowly, whereas 1-bromopropane, bromoethane, and bromomethane are relatively untouched by these conditions.

Such a transformation, in which a substrate undergoes substitution by molecules that are part of the solvent, is called **solvolysis.** When the solvent is water, the term **hydrolysis** is applied; for alcohols we use the word **alcoholysis** (more specifically, methanolysis, ethanolysis, and so on).

Methanolysis of 2-Chloro-2-methylpropane

$$\underset{\substack{\textbf{2-Chloro-}\\ \textbf{2-methylpropane}}}{CH_3\overset{\displaystyle CH_3}{\underset{\displaystyle CH_3}{C}}Cl} + \underset{\textbf{Solvent}}{CH_3OH} \rightleftharpoons \underset{\substack{\textbf{2-Methoxy-}\\ \textbf{2-methylpropane}}}{CH_3\overset{\displaystyle CH_3}{\underset{\displaystyle CH_3}{C}}OCH_3} + HCl$$

The relative rates of reaction of 2-bromopropane and 2-bromo-2-methylpropane with water to give the corresponding alcohols are shown in Table 7-1 and compared with the corresponding rates of hydrolysis of their unbranched counterparts. Although the process appears to give the products expected from an S_N2 reaction, the order of reactivity is *reversed* from that found under typical S_N2 conditions. Thus, primary halides are very slow in their reaction with water, secondary halides are somewhat more reactive, and tertiary systems are about *one million times* as fast as primary ones. These observations suggest that the mechanism of solvolysis of tertiary haloalkanes must be different from that of bimolecular substitution. In an effort to understand the details of this transformation, we can use techniques similar to those employed in the elucidation of the mechanistic features of S_N2 process: kinetics, stereochemistry, and the effect of substrate structure and solvent on reaction rates.

TABLE 7-1
Relative reactivities of various bromoalkanes with water

Bromoalkane	Relative rate
CH_3Br	1
CH_3CH_2Br	1
$(CH_3)_2CHBr$	12
$(CH_3)_3CBr$	1.2×10^6

EXERCISE 7-1

Whereas compound A (shown at the right) is completely stable in ethanol, B is rapidly converted into another compound. Explain.

H_3C CH_2Br (on cyclohexane) **A** H_3C Br (on cyclohexane) **B**

7-2 Mechanism of the Solvolysis of Tertiary Haloalkanes: Unimolecular Nucleophilic Substitution

In this section we will learn about a new mechanism of nucleophilic substitution. In constrast with the S_N2 reaction, which follows bimolecular kinetics, generates products stereospecifically, and speeds up in the order tertiary-secondary-primary substrate, this mechanism follows unimolecular kinetics, is not stereospecific, and, as already noted, is characterized by the opposite order of reactivity.

Solvolysis Follows Unimolecular Kinetics

In accord with a bimolecular rate-determining transition state, the rate of the S_N2 reaction is proportional to the concentration of both the haloalkane and the nucleophile. We can conduct similar rate studies for the reaction of 2-bromo-2-methylpropane with water. The results of these experiments show that the rate of hydrolysis of the bromide is proportional only to the concentration of starting halide.

$$\text{Rate} = k[(CH_3)_3CBr] \text{ mole } l^{-1} \text{ sec}^{-1}$$

The observed first-order kinetics therefore point to a unimolecular mechanism, that is, only the haloalkane participates in the rate-determining step. Such a mechanism was discussed as a hypothetical pathway for the conversion of chloromethane and hydroxide ion into methanol (Figure 6-5). Let us reconsider this mechanism for the hydrolysis of 2-bromo-2-methylpropane, using water as a nucleophile instead of hydroxide ion.

A Proposed Mechanism for the Hydrolysis of 2-Bromo-2-methylpropane

The initial and rate-determining step of the hydrolysis of 2-bromo-2-methylpropane may be envisaged to be dissociation to a cation and bromide ion (Step 1). This hypothesis alone would account for the kinetic results.

STEP 1: Dissociation

$$(CH_3)_3C{-}Br \xrightleftharpoons{\text{Rate determining}} (CH_3)_3C^+ + Br^-$$

1,1-Dimethylethyl cation (*tert*-Butyl cation)

The 1,1-dimethylethyl (*tert*-butyl) cation formed in this manner is a powerful electrophile, which is immediately trapped by the surrounding water (Step 2). This process can be viewed as a nucleophilic attack by the water molecule on the electron-deficient carbon.

STEP 2: Nucleophilic Attack by Water

$$(CH_3)_3C^+ + \ddot{O}H_2 \xrightleftharpoons{\text{Fast}} (CH_3)_3C{-}\overset{+}{\ddot{O}}H_2$$

The resulting oxonium ion (see Section 6-7) is rapidly deprotonated by the aqueous medium to furnish the product 2-methyl-2-propanol.

STEP 3: Deprotonation

$$CH_3\underset{CH_3}{\overset{CH_3}{|}}{C}\!-\!\overset{+}{O}H_2 + H_2\ddot{O}: \;\underset{}{\overset{\text{Fast}}{\rightleftharpoons}}\; CH_3\underset{CH_3}{\overset{CH_3}{|}}{C}\ddot{O}H + H\overset{+}{\ddot{O}}H_2$$

2-Methyl-2-propanol

A potential-energy diagram of the proposed mechanism is depicted in Figure 7-1.

The reaction is said to proceed by **unimolecular nucleophilic substitution,** abbreviated $\mathbf{S_N1}$. The number 1 indicates that only the haloalkane participates in the rate-determining transition state.

FIGURE 7-1
Reaction coordinate diagram for the hydrolysis of 2-bromo-2-methylpropane.

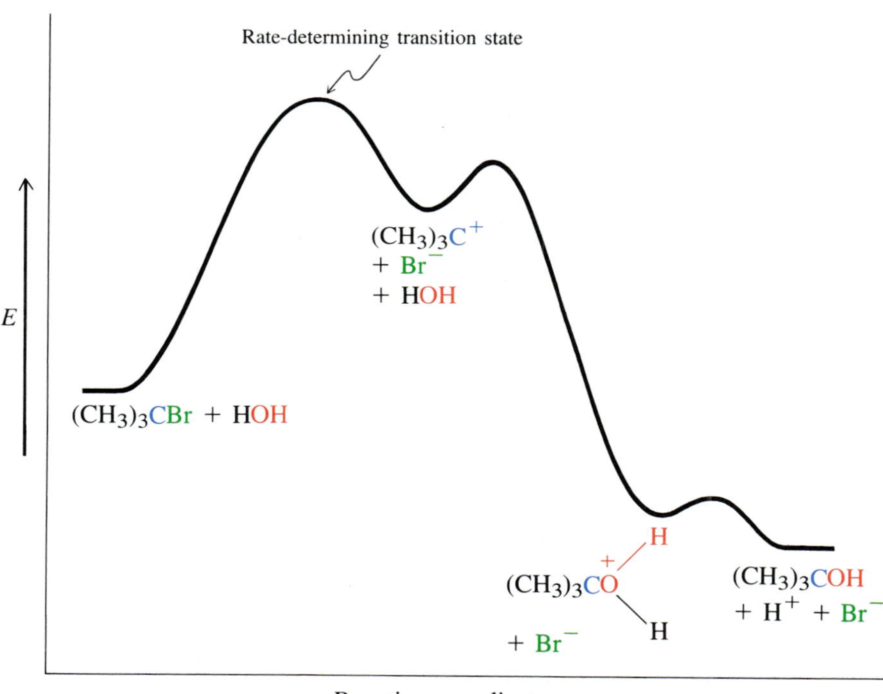

EXERCISE 7-2

Using the bond-strength data in Table 3-1, calculate the $\Delta H°$ for the hydrolysis of 2-bromo-2-methylpropane to 2-methyl-2-propanol and hydrogen bromide.

Solvolysis Is Reversible

All three steps of the mechanism of solvolysis are reversible, and the three equilibria are driven toward product by the favorable thermodynamics of the overall process. However, this situation can be reversed by employing a large excess of hydrogen bromide. Thus, a common method of preparing 2-bromo-2-methylpropane is to simply shake a solution of 2-methyl-2-propanol in concentrated aqueous hydrogen bromide in a separatory funnel. The product separates out as a water-immiscible layer.

$$(CH_3)_3COH + \underset{\text{Excess}}{HBr} \rightleftharpoons (CH_3)_3CBr + H_2O$$

Note that, although this process appears related to the formation of primary haloalkanes from primary alcohols on exposure to concentrated hydrogen halides (Section 6-7), the two mechanisms are fundamentally different. The intermediates in reactions of tertiary alcohols are **carbocations,** whereas those in the S_N2 reactions of primary systems are oxonium ions.

EXERCISE 7-3

Write a mechanism for the following transformation:

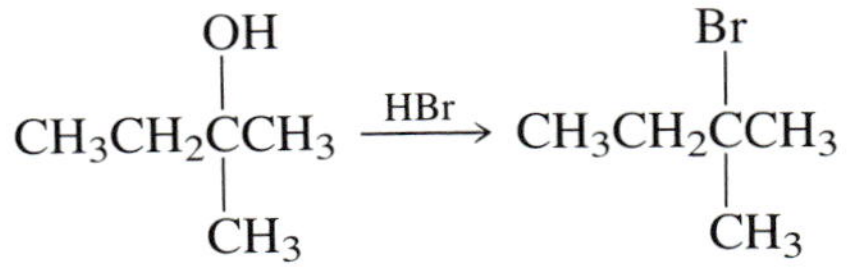

$$CH_3CH_2C(OH)(CH_3)CH_3 \xrightarrow{HBr} CH_3CH_2C(Br)(CH_3)CH_3$$

The S_N1 Reaction Leads to Racemic Products

The proposed mechanism of nucleophilic substitution in which the intermediate is a carbocation has predictable stereochemical consequences. Because the cation is planar, it is achiral. Hence, starting with an optically active tertiary haloalkane, we should obtain racemic S_N1 products (Figure 7-2). This result is observed in many solvolyses.

FIGURE 7-2

Hydrolysis of (R)-3-bromo-3-methylhexane. Initial ionization furnishes a planar and achiral carbocation. This ion, when trapped with water, results in racemic alcohol.

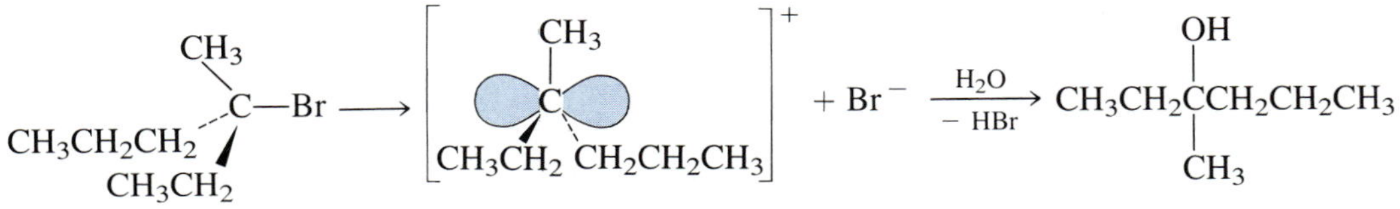

$$(CH_3)(CH_3CH_2CH_2)(CH_3CH_2)C{-}Br \longrightarrow \left[(CH_3)(CH_3CH_2)(CH_2CH_2CH_3)C \right]^+ + Br^- \xrightarrow[-\ HBr]{H_2O} CH_3CH_2C(OH)(CH_3)CH_2CH_2CH_3$$

(R)-3-Bromo-3-methylhexane — **Planar, achiral** — **Racemic**

EXERCISE 7-4

(R)-3-Bromo-2-methylhexane loses its optical activity when dissolved in propanone (acetone). Explain.

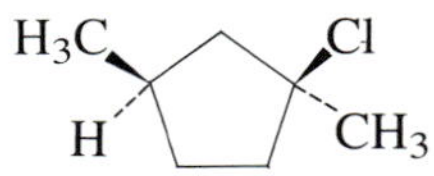

H_3C, H, Cl, CH_3

A

EXERCISE 7-5

Hydrolysis of molecule A (shown at the left) gives two alcohols. Explain.

Polar Solvents Accelerate the S_N1 Reaction

Heterolytic cleavage of a C–X bond in the rate-determining step of the S_N1 reaction implies a transition-state structure that is highly polarized (Figure 7-3A), because the reaction eventually leads to two fully charged ions. In contrast, in a typical S_N2 transition state (Figure 7-3B), the charge density on the incoming nucleophile decreases at the same time as that on the leaving group increases. Charges are not being created; in fact, quite the opposite: charge is dispersed.

FIGURE 7-3

Transition states for (A) the S_N1 reaction and (B) the S_N2 reaction, showing less net charge separation in transition state B than in A.

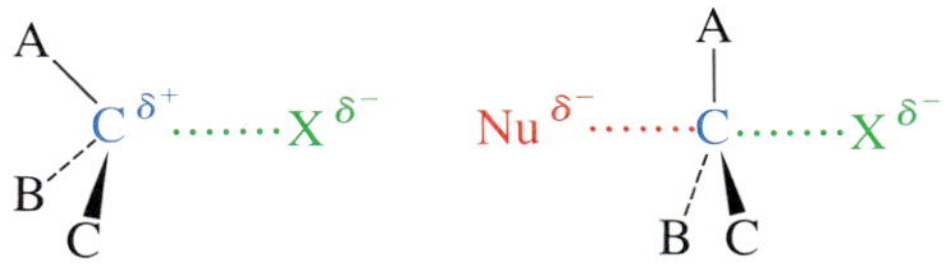

A: A, B, C attached to $C^{\delta+}$ ······ $X^{\delta-}$

B: $Nu^{\delta-}$ ······ C ······ $X^{\delta-}$ (C bearing A, B, C)

Because of the presence of such a polar transition state, the rate of an S_N1 reaction increases as solvent polarity is increased. A particularly striking effect is measured when the solvent is changed from aprotic to protic. For example, hydrolysis of 2-bromo-2-methylpropane is much faster in pure water than in a 9:1 mixture of propanone (acetone) and water. The protic solvent accelerates the S_N1 reaction because it stabilizes the transition state shown in Figure 7-3A by hydrogen bonding. Remember that, in contrast, the S_N2 reaction is accelerated in polar *aprotic* solvents, mainly because of a solvent effect on the reactivity of the nucleophile and *not* of the substrate.

	Relative rate
$(CH_3)_3CBr \xrightarrow{100\%\ H_2O} (CH_3)_3COH + HBr$	400,000
$(CH_3)_3CBr \xrightarrow{90\%\ \text{propanone (acetone)},\ 10\%\ H_2O} (CH_3)_3COH + HBr$	1

The S_N1 Reaction Speeds Up with Better Leaving Groups

Because the leaving group dissociates in the rate-determining step of the S_N1 reaction, it is not surprising that the rate of the reaction increases as the leaving-group ability of the dissociating group improves. Thus, tertiary iodoalkanes are more readily solvolyzed than the corresponding bromides, and these are in turn more reactive than the chlorides. Sulfonates are particularly prone to dissociation.

Relative Rate of Solvolysis of RX (R = Tertiary Alkyl)

$$X = \text{—}OSO_2R' > \text{—}I > \text{—}Br \sim \text{—}\overset{+}{O}H_2 > \text{—}Cl$$

The Strength of the Nucleophile Affects the Product Distribution but Not the Rate

What effect does changing the nature of the incoming nucleophile have on the rate of the solvolysis reaction? Recall that, in the S_N2 process, the rate of reaction increases significantly as the nucleophilicity of the attacking species improves. However, because the rate-determining step of the S_N1 process does not include the nucleophile, varying its nature should not affect the rate of disappearance of a tertiary haloalkane. On the other hand, when two or more nucleophiles compete for capture of the intermediate carbocation, the *product distribution* may be profoundly affected by their relative strengths and concentrations.

For example, solvolysis of a 0.1 molar solution of 2-chloro-2-methylpropane in methanol gives 2-methoxy-2-methylpropane with a rate constant k_1. However, in the presence of 0.1 mole of sodium azide, the product is 1,1-dimethylethyl (*tert*-butyl) azide, still formed at the *same* rate. In this case, the much more powerful nucleophile N_3^- (Table 6-5) wins out in competition with methanol. Thus, even though the rate of disappearance of 2-chloro-2-methylpropane (regardless of the product eventually formed) is determined by k_1, the relative rate of formation of product is governed by the relative nucleophilicities of the competing nucleophiles (k_{CH_3OH} is much smaller than $k_{N_3^-}$):

$$(CH_3)_3CCl + CH_3OH + NaN_3 \xrightarrow[\text{Rate determining}]{k_1} (CH_3)_3C^+ + Cl^- \xrightarrow{k_{CH_3OH}} (CH_3)_3COCH_3 + HCl$$

2-Methoxy-2-methylpropane

$$(CH_3)_3C^+ + Cl^- \xrightarrow{k_{N_3^-}} (CH_3)_3CN_3 + NaCl$$

1,1-Dimethylethyl azide (*tert*-Butyl azide)

Product determining

It is therefore useful to distinguish between the *rate-determining* and the *product-determining* transition states (Figure 7-4). Regardless of how many transition states (and hence intermediates) separate starting materials from products, the rate-determining transition state has the highest energy. However, if there are several available pathways after this step, the relative energies of the subsequent transition states determine the product distribution.

EXERCISE 7-6

A solution of 2-methyl-2-propanol in propanone (acetone) containing excess sodium chloride and sodium bromide was treated with sulfuric acid. The product was 2-bromo-2-methylpropane. Explain.

In this section, we have looked at the following pieces of evidence to establish the mechanism of the reaction of tertiary haloalkanes with certain nucleophiles: kinetics, stereochemistry, the nature of the solvent, leaving-group ability, and the strength of the nucleophile. This evidence strongly supports an

FIGURE 7-4

Difference between rate- and product-determining transition states in the S_N1 reaction of a tertiary haloalkane in the presence of two competing nucleophiles. The energy of activation for this reaction is given by E_a. The product distribution is given by the difference in energy between the two product-determining transition states.

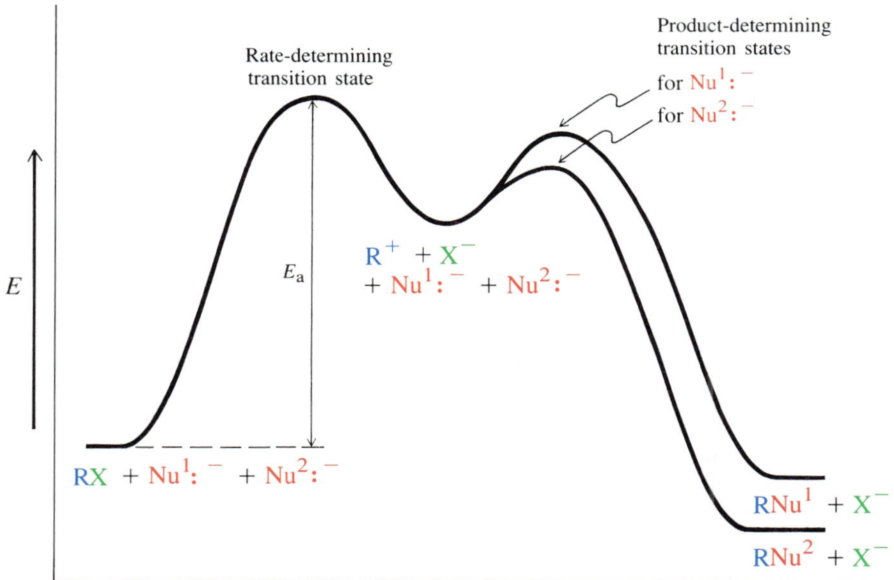

initial rate-determining dissociation as the crucial step in the mechanism. The next question to be answered is, Why? What is so special about tertiary haloalkanes that they undergo conversion by S_N1 reactions, whereas primary systems follow the S_N2 pathway? How do secondary haloalkanes fit into this scheme?

7-3 Effect of Substrate Structure on the Rate of the S_N1 Reaction: The Stability of Carbocations

We shall see in this section that the degree of branching at the reacting carbon controls the extent to which S_N2 or S_N1 pathways are followed in the reaction of haloalkanes (and related derivatives) with nucleophiles. The more-substituted carbon is more prone to forming carbocations; for this reason, tertiary systems transform by the S_N1 mechanism, primary haloalkanes by the S_N2, and secondary haloalkanes by either pathway, depending on reaction conditions.

The Stability of Carbocations Increases in Procession from Primary to Secondary to Tertiary

Of the two nucleophilic-substitution mechanisms, we have learned that primary haloalkanes undergo direct substitution, whereas, in secondary and tertiary systems, as steric hindrance builds up on the reacting carbon, a carbocation mechanism takes over. Why do primary haloalkanes not form carbocations? The answer lies in the relatively high energy of primary carbocations, too high to allow for their generation in solution under the conditions of the S_N1 reaction. From measurements in the gas phase, it has been established that carbocations *increase* in stability with increasing alkyl substitution (Table 7-2).

TABLE 7-2
Heats of formation of isomeric butyl cations in the gas phase

Carbocation		ΔH_f (kcal mole^{-1})
$CH_3CH_2CH_2\overset{+}{C}H_2$	(primary)	218
$CH_3CH_2\overset{+}{C}HCH_3$	(secondary)	192
$(CH_3)_3C^+$	(tertiary)	176

Note that this order of stability parallels that observed for the corresponding radicals (Section 3-2) and has its roots in the same phenomenon: *hyperconjugation* (see Figure 3-3). The only difference between a radical and a carbocation is that the latter has one less electron. Hyperconjugation in a carbocation occurs between an empty *p* orbital and an adjacent bond (usually C–H or C–C). This arrangement is illustrated in Figure 7-5, in which a relatively stable ion, the 1,1-dimethylethyl (*tert*-butyl) cation, is compared with methyl. As shown, hyperconjugation essentially amounts to electron donation from an alkyl substituent to the electron-deficient center. Because of this effect, organic chemists tend to think of an alkyl group as an "electron pusher," stabilizing adjacent positive charges.

FIGURE 7-5
Orbital picture of (A) the methyl cation, which lacks hyperconjugative stabilization, and (B) the 1,1-dimethylethyl (*tert*-butyl) cation, which has three hyperconjugating C–H bonds.

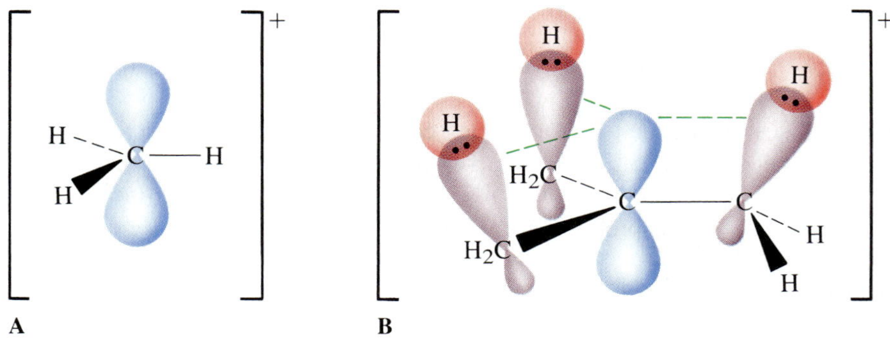

The finding that tertiary carbocations are more stable than their less-substituted relatives explains the S_N1 reactivity of tertiary haloalkanes: ionization is relatively facile. There is also a contributing steric effect, because on ionization the steric pressure caused by the close proximity of the three alkyl groups is removed. During dissociation, the reactive center rehybridizes from sp^3 to sp^2, allowing the substituents to separate from each other further in space and thus relieving the congestion initially present in starting material (Figure 7-6).

FIGURE 7-6
Relief of steric interactions on ionization of a tertiary haloalkane.

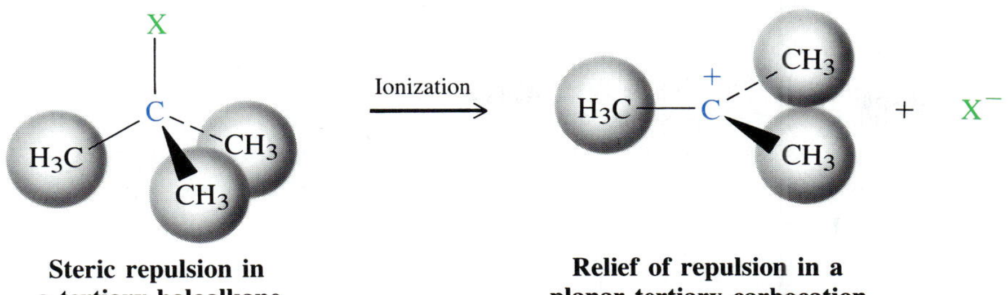

Steric repulsion in a tertiary haloalkane — **Relief of repulsion in a planar tertiary carbocation**

EXERCISE 7-7

Explain the approximate relative solvolysis rates of the two 4-methylbenzenesulfonates (tosylates) shown in the adjoining table.

$$(X = OS(=O)_2-C_6H_4-CH_3)$$

Structure	Relative rate
$(CH_3)_3C—X$	1
X	10^{-13}

A Comparison of the Reactivity of Methyl, Primary, and Tertiary Haloalkanes

Table 7-3 summarizes our observations about the nucleophilic reactivity of haloalkanes.

TABLE 7-3
General reactivity of R–X in nucleophilic substitutions:
$R—X + Nu^- \longrightarrow R—Nu + X^-$

R	S_N1	S_N2
CH_3	Not observed in solution (methyl cation too high in energy)	Frequent; fast with good nucleophiles and good leaving groups
Primary	Not observed in solution (primary carbocations too high in energy; exceptions are resonance-stabilized cations, see Chapter 14)	Frequent; fast with good nucleophiles and good leaving groups, slow when branching at C-2 is present in R
Tertiary	Frequent; particularly fast in polar protic solvents and with hindered R groups carrying good leaving groups	Extremely slow

Secondary Systems Undergo S_N1 and S_N2 Reactions

Are secondary systems prone to S_N2 reactivity or are they more likely to enter into carbocation formation?

From Chapter 6, we know that secondary haloalkanes and other secondary alkyl derivatives undergo S_N2 reactions. However, we also know that steric hindrance decreases the rate of these reactions compared with those of primary alkyl compounds by approximately two orders of magnitude. Does the fact that secondary carbocations are appreciably more stable than methyl and primary cations enable secondary systems to undergo S_N1 reactions? Yes, under some conditions. If we use a substrate carrying a very good leaving group, a nucleophile that is poor, and a polar protic solvent (S_N1 conditions), unimolecular substitution is favored. Only if we employ a good nucleophile and a polar aprotic solvent with a haloalkane bearing a reasonable leaving group (S_N2 conditions) is bimolecular substitution predominant. As in Table 7-3, these observations are summarized in Table 7-4.

S_N2 Conditions

$$(CH_3)_2CH\text{—}Br + CH_3S^- \xrightarrow{\text{Propanone (acetone)}} CH_3S\text{—}CH(CH_3)_2 + Br^-$$

S_N1 Conditions

$$(CH_3)_2CH\text{—}OSO_2CF_3 \xrightarrow{H_2O} (CH_3)_2CH\text{—}OH + CF_3SO_3H$$

TABLE 7-4
Reactivity of secondary halides in nucleophilic substitutions

R	S_N1	S_N2
Secondary	Relatively slow; best in polar protic solvents with good leaving groups and sterically hindering branching	Relatively slow; best with good nucleophiles in aprotic solvents

EXERCISE 7-8

Explain the following results:

(a) 2-chlorobutane (R) + CN^- $\xrightarrow{\text{Propanone (acetone)}}$ 2-methylbutanenitrile (S)

(b) 2-iodobutane (R) + CH_3OH $\longrightarrow$ 2-methoxybutane ($R + S$)

In contrast with S_N2 processes, S_N1 reactions are of limited use in synthesis because of the relative complexity of carbocation chemistry. As we shall see in Section 9-4, carbocations are prone to rearrangements, frequently resulting in complex mixtures of products. In addition, they undergo another reaction to be discussed in Section 7-4: *loss of a proton* to furnish a double bond.

To summarize, tertiary haloalkanes are reactive in the presence of nucleophiles even though they are too sterically hindered to undergo S_N2 reactions: the tertiary carbocation is readily formed because it is stabilized by hyperconjugation. Subsequent trapping by a nucleophile, such as a solvent (solvolysis), results in the product of nucleophilic substitution. Primary haloalkanes do not react in this manner—the primary cation is too highly energetic to be formed in solution. The reacting carbon is unhindered and transformation is directly through the S_N2 process. Whether secondary systems are converted into substitution products through one pathway or the other depends on the nature of the leaving group, the solvent, the nucleophile, and the substrate structure.

7-4 Unimolecular Elimination: E1

Although carbocations are reactive in the presence of nucleophiles and furnish the corresponding products of attack at the positively charged carbon, this is not their only mode of reaction. A second pathway open to this class of compounds results in the formation of alkenes by loss of a proton. For example, when 2-bromo-2-methylpropane is dissolved in methanol, the starting material rapidly disappears to give not only the expected S_N1 product, 2-methoxy-2-methylpropane, but also a significant amount of an alkene, 2-methylpropene. A kinetic analysis shows that the rate of alkene formation depends only on the concentration of starting halide. The rate-determining step is the same as in the S_N1 reaction: dissociation to the carbocation. This intermediate thus has two competing pathways at its disposal: loss of proton or nucleophilic trapping. The overall result of the first is loss of HX from RX. Such a process is termed an **elimination reaction,** abbreviated E. Because the kinetics of this particular elimination are unimolecular, **E1** is used to describe it.

$$(CH_3)_3CBr \underset{}{\overset{CH_3OH}{\rightleftharpoons}} H_3C\text{—}\overset{+}{C}(CH_3)_2 + Br^-$$

2-Bromo-2-methylpropane

E1: $H_2C{=}C(CH_3)_2 + H^+ + Br^-$ — 20% **2-Methylpropene**

S_N1, CH_3OH: $(CH_3)_3COCH_3 + H^+ + Br^-$ — 80% **2-Methoxy-2-methylpropane**

General Element Reaction

$$H\text{—}C\text{—}C\text{—}X \xrightarrow{\text{Base: }:B^-} C{=}C + H\text{—}B + X^-$$

Note:
While we depict the protons evolving in the scheme on the left as H^+, they are usually removed by some basic species—in aqueous solution by H_2O as H_3O^+ (Section 6-7)—here by CH_3OH as $CH_3\overset{+}{O}H_2$ (Figure 7-7). For simplicity, we will usually show the H^+ notation, but remember there is no such thing as a "free proton" in ordinary organic reactions.

FIGURE 7-7

A. A representation of the 1,1-dimethylethyl (*tert*-butyl) cation. The electron-withdrawing positive charge causes a weakening of the C–H bond, the hyperconjugating hydrogen assuming a partial positive charge.

B. A molecular orbital description of proton loss from 1,1-dimethylethyl (*tert*-butyl) cation to methanol (the solvent). The activated hydrogen atom is approached by the electron-rich oxygen of methanol. In the transition state, this hydrogen is partially bonded to the oxygen atom, and its bond with carbon is partially broken. The originally sp^3-hybridized orbital used to make the C–H bond rehybridizes to sp^2. The proton is removed, leaving an electron pair behind. These two electrons redistribute over the two p orbitals of the emerging double bond.

C. An electron-pushing scheme of the same process.

H^{δ^+} Weakened bond

H_3C $\overset{+}{C}$—C H

H_3C Electron pull H

A

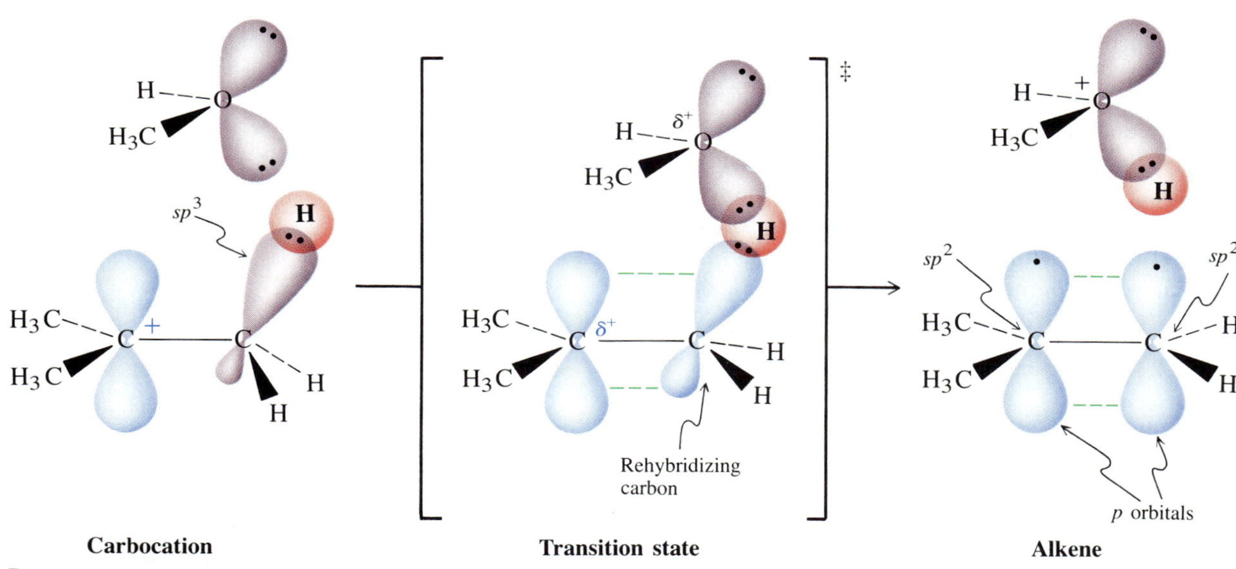

B

$HOCH_3$

H

H_3C $\overset{+}{C}$—C H

H_3C H

$\longrightarrow$ H_3C C=C H, H_3C H + $H\overset{+}{O}CH_3$, H

C

Nucleophiles May Act as Bases

Let us examine in more detail how the proton is lost (Figure 7-7). Recall that a substituted carbocation engages in hyperconjugation with neighboring C–H bonds. Through this mechanism, electron density is transferred from these bonds to the empty *p* orbital. Because of the removal of electron density, such bonds are weakened (Figure 7-7A). Moreover, the electron-withdrawing effect of the positively charged carbon causes the C–H bond to be polarized, placing a partial positive charge on the hydrogen atom. This hydrogen is therefore subject to removal as a proton. Deprotonation is caused by the solvent, acting in this case as a base.

Removal of the proton leaves behind a neutral fragment containing a double bond. This process is depicted in Figure 7-7, both as an orbital picture (Figure 7-7B) and, more simply, as an electron pushing scheme (Figure 7-7C).

In the mechanistic schemes presented in this book, proton abstraction from a carbocation will be written as simple proton dissociation. In such cases, it is understood, however, that proton removal is effected by a base.

Any hydrogen atom positioned on any carbon next to the center bearing the leaving group can participate in the E1 reaction. The 1,1-dimethylethyl (*tert*-butyl) cation has nine such hydrogens, each of which is equally reactive. In this case, the product is the same regardless of the identity of the proton lost. In other cases, more than one product may be obtained. An example is 3-chloro-3-methylhexane, which gives six different reaction products, five of which are the different isomeric alkenes. These pathways will be discussed in further detail in Chapter 11.

$$CH_3CH_2CH_2C(CH_3)(Cl)CH_2CH_3 \xrightarrow[-\ HCl]{CH_3OH,\ \Delta} CH_3CH_2CH_2C(CH_3)(OCH_3)CH_2CH_3 + (CH_3CH_2CH_2)(CH_3CH_2)C{=}CH_2$$

3-Chloro-3-methylhexane → **S_N1 product** + **E1**

$$+ (CH_3)(CH_3CH_2CH_2)C{=}C(H)(CH_3) + (CH_3)(CH_3CH_2CH_2)C{=}C(CH_3)(H)$$

E1 + **E1**

$$+ (CH_3CH_2)(H)C{=}C(CH_3)(CH_2CH_3) + (H)(CH_3CH_2)C{=}C(CH_3)(CH_2CH_3)$$

E1 + **E1**

Relative Rates of S_N1 and E1 Reactions

The partition between elimination and substitution in the solvolyses of tertiary haloalkanes should be the same irrespective of the nature of the leaving group, because in all cases the intermediate should be the same. This is indeed observed qualitatively (Table 7-5). The product ratio may be affected by the addition of a mild base, which will accelerate the deprotonation of the carbocation at the expense of nucleophilic attack. For example, adding an equivalent of NaOH to aqueous ethanol containing 2-chloro-2-methylpropane leads to the virtually

TABLE 7-5
Ratio of S_N1 to E1 product in the hydrolyses of 2-halo-2-methylpropanes at 25°C

X in $(CH_3)_3CX$	Ratio S_N1:E1
Cl	95:5
Br	95:5
I	96:4

exclusive formation of 2-methylpropene. In this case, hydroxide, a much stronger base than either water or ethanol, attacks the intermediate carbocation exclusively and rapidly at the hydrogen rather than the carbon atom. This change is, however, complicated by the fact that strong bases are capable of reacting with haloalkanes by a direct elimination pathway. This reaction will be the subject of the next section.

Preferential E1 Reaction from 2-Chloro-2-methylpropane in Ethanol in the Presence of Added Hydroxide

$$(CH_3)_3CCl \rightleftharpoons H_3C-\overset{+}{C}(CH_3)_2 + Cl^-$$

E1, Relatively fast, HO^-:

$$Cl^- + HOH + CH_2{=}C(CH_3)_2$$

2-Methylpropene

S_N1, Relatively slow, CH_3CH_2OH, H_2O:

$$(CH_3)_3COCH_2CH_3 + (CH_3)_3COH + H^+ + Cl^-$$

2-Ethoxy-2-methylpropane; 2-Methyl-2-propanol

EXERCISE 7-9

If 2-bromo-2-methylpropane is dissolved in aqueous ethanol at 25°C, a mixture of $(CH_3)_3COCH_2CH_3$ (30%), $(CH_3)_3COH$ (60%), and $(CH_3)_2C{=}CH_2$ (10%) is obtained. Explain.

To summarize, carbocations formed in solvolysis reactions are not only trapped by nucleophiles to give S_N1 products, but also deprotonated to result in elimination (E1). In this process, the nucleophile (usually the solvent) acts as a base.

7-5 Bimolecular Elimination: E2

Apart from S_N2, S_N1, and E1 reactions, there is a fourth mode of reactivity of haloalkanes with nucleophiles. In it the nucleophile again acts as a base and again effects elimination, but this time by a bimolecular mechanism. Such eliminations are abbreviated as **E2.**

Strong Bases Effect Bimolecular Elimination

From the preceding section, we know that tertiary haloalkanes undergo E1 reactions (in addition to substitution) in non-nucleophilic, weakly basic solvents, such as water and alcohols, and that the fraction of E1 products can be increased by adding some base. If we were to monitor the rate of the elimination reaction we would, however, note that beyond a certain concentration the rate of alkene formation would become proportional to the concentration of base: elimination is bimolecular.

Kinetics of the E2 Reaction of 2-Chloro-2-methylpropane

$$(CH_3)_3CCl + Na^{+-}OH \xrightarrow{k} CH_2{=}C(CH_3)_2 + NaCl + H_2O$$

$$\text{Rate} = k[(CH_3)_3CCl][^{-}OH] \text{ mole } l^{-1} \text{ sec}^{-1}$$

Evidently, strong bases are capable of attacking a hydrogen on the carbon next to that carrying the leaving group, before the formation of the carbocation. This reaction is general for haloalkanes, although in secondary and primary systems, there is competition by S_N2 processes.

$$CH_3CH_2CH_2Br \xrightarrow{CH_3O^-Na^+,\ CH_3OH} CH_3CH_2CH_2OCH_3 + (H_3C)HC{=}CH_2$$

92% **1-Methoxypropane** — 8% **Propene**

EXERCISE 7-10

What products do you expect from the reaction of bromocyclohexane with hydroxide ion?

EXERCISE 7-11

Give the products (if any) of the E2 reaction of the following substrates: CH_3CH_2I; CH_3I; $(CH_3)_3CCl$; $(CH_3)_3CCH_2I$.

The E2 Reaction Includes Concerted *Anti* Elimination of HX

The kinetics of the E2 reaction point to a bimolecular transition state. Pictures of its structure are shown in Figure 7-8, first by electron pushing and second through orbitals. Its characteristic features include removal of a proton that is preferentially located conformationally *anti* with respect to the leaving halide and three simultaneous events: bond formation between the hydroxide ion and the hydrogen; rehybridization of the sp^3 orbitals to sp^2 to furnish two p orbitals to form the new double bond; and bond breaking between the carbon and the leaving group.

A comparison with the transition state for the E1 reaction (Figure 7-7) shows that they are very similar and differ only in the sequence of events. Here proton abstraction and leaving-group dissociation occur at the same time. In the E1 reaction, the halide leaves first, to be followed by attack by a base. A good way of thinking about the difference between the two mechanisms is to consider the strong base necessary to effect the E2 reaction to be more "aggressive." It does not wait for the dissociation of a tertiary or secondary halide, but attacks the substrate directly.

FIGURE 7-8

The E2 reaction of 2-chloro-2-methylpropane with hydroxide ion: (A) electron-pushing description; (B) orbital description.

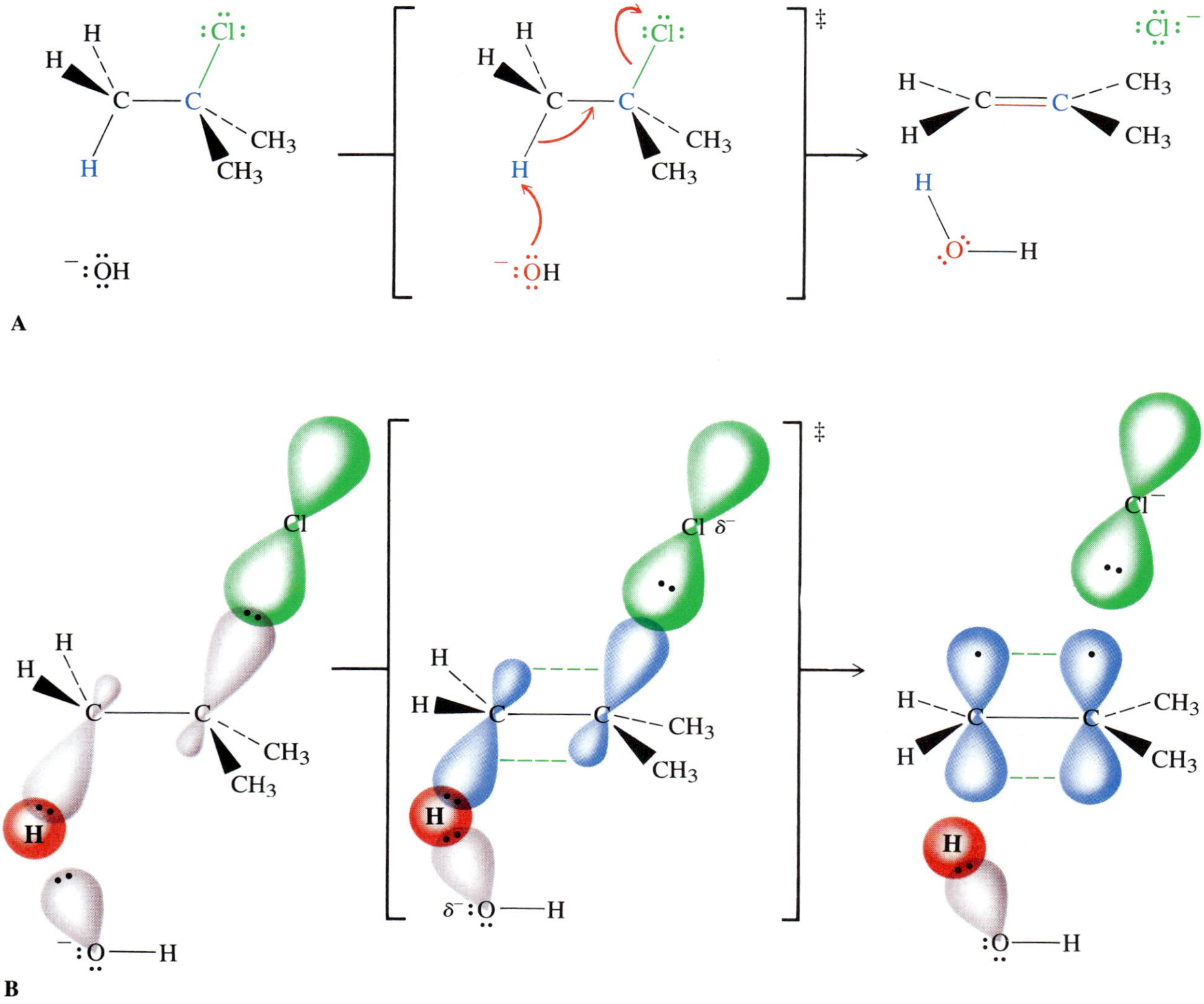

How to Obtain a Detailed Picture of the Transition State of the E2 Reaction

How do we know the transition state with such precision? Several different types of experiments have been carried out, each of which addresses a specific feature of its structure. As pointed out earlier, the kinetics clearly show that the haloalkane and the base have to be closely associated. It can also be shown that the bond to the leaving group is partially broken in the transition state by varying the nature of this group. Better leaving groups result in faster eliminations.

Order of Reactivity in the E2 Reaction

$$RCl < RBr < RI$$

EXERCISE 7-12

Explain the result in the reaction shown in the margin.

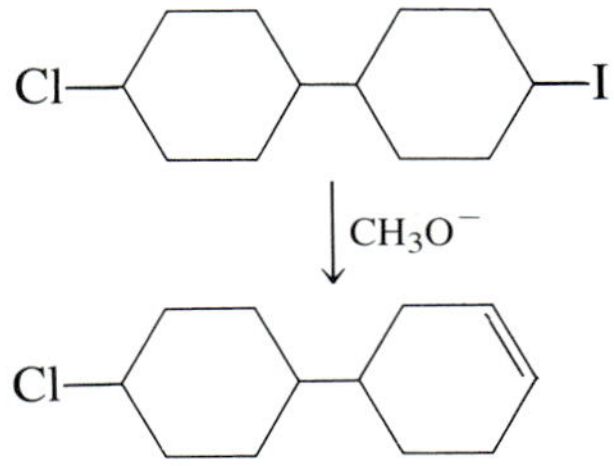

$$Cl\text{-}C_6H_{10}\text{-}C_6H_{10}\text{-}I \xrightarrow{CH_3O^-} Cl\text{-}C_6H_{10}\text{-}C_6H_9$$

Our transition-state structure (Figure 7-8) also shows that a C–H bond must partially break. To establish it with certainty, organic chemists have used isotopic labeling with deuterium substituted for hydrogen. It is known that the C–H bond is somewhat weaker than the C–D bond. If the C–H bond is cleaved in the rate-determining step of the E2 reaction, then replacing H by D should result in a decreased rate. This is indeed the case:

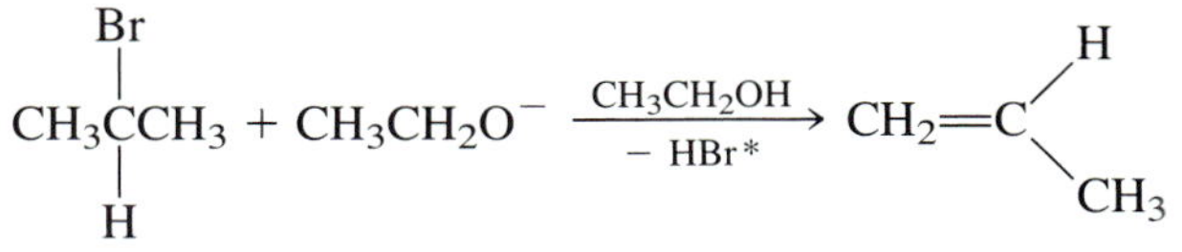

$$CH_3CHBrCH_3 + CH_3CH_2O^- \xrightarrow[-\ HBr^*]{CH_3CH_2OH} CH_2{=}C(H)CH_3 \qquad \text{Faster}$$

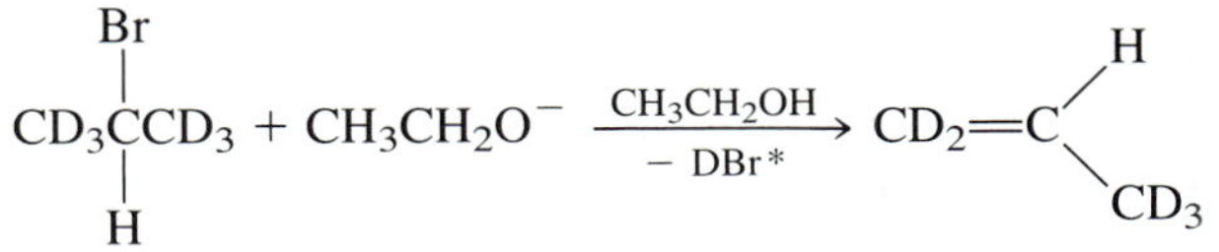

$$CD_3CHBrCD_3 + CH_3CH_2O^- \xrightarrow[-\ DBr^*]{CH_3CH_2OH} CD_2{=}C(H)CD_3 \qquad \text{Slower}$$

$$\frac{k_{CH_3CHBrCH_3}}{k_{CD_3CHBrCD_3}} = \text{deuterium isotope effect} = 6.7$$

Typical deuterium isotope effects in E2 reactions are on the order of k_H/k_D = 3–8.

How do we establish that the proton to be abstracted by a base should be preferentially *anti* to the leaving group? For this purpose, we can use the principles of stereochemistry. For example, in the all-trans stereoisomeric form of 1,2,3,4,5,6-hexachlorocyclohexane, there are no hydrogens located *anti* with respect to any of the chlorine atoms. Consequently, this compound undergoes the E2 reaction at a rate that is several thousand times lower than any of its other isomers, all of which have at least one *anti* hydrogen. Elimination still occurs, but the hydrogen is on the same side of the leaving group. Unlike *anti* elimination, this type of process is said to occur through a *syn* transition state (*syn*, Greek, together).

Neither Conformer of All-*trans*-1,2,3,4,5,6-hexachlorocyclohexane Bears *Anti* Hydrogens Conducive to Bimolecular Elimination

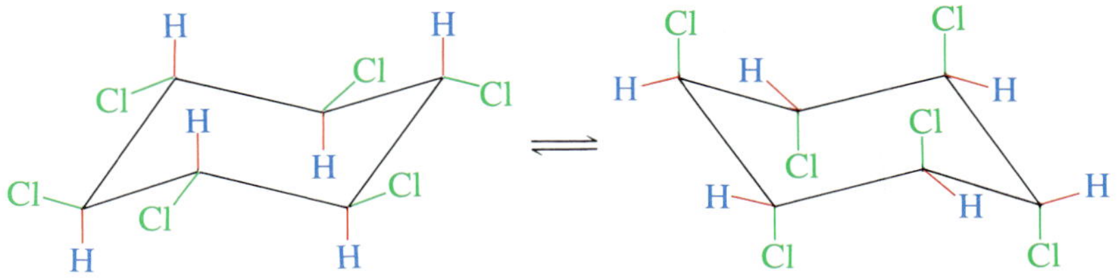

Equatorial ⇌ **Axial**

All-*trans*-1,2,3,4,5,6-hexachlorocyclohexane

*This notation indicates that the elements of the acid have been removed from the starting material. In reality, the proton ends up protonating the base. This system will be occasionally used in other elimination reactions in this book.

Another consequence of *anti* elimination is the defined stereochemistry of the alkenes produced. This subject and the reason for the preference for *anti* elimination will be discussed in Section 11-5.

EXERCISE 7-13

trans-1-Bromo-4-(1,1-dimethylethyl)cyclohexane undergoes the E2 reaction in the presence of sodium methoxide much more slowly than the corresponding cis isomer. Explain.

Steric Factors Control the Various Modes of Reactivity of Haloalkanes with Nucleophiles

At this stage, the variety of possible reactions that haloalkanes may undergo in the presence of nucleophiles may seem confusing: S_N2, S_N1, E2, E1. However, certain overriding factors allow you to predict the reactivity of a haloalkane. Among them, steric factors are dominant.

For example, the relatively unhindered primary haloalkanes usually undergo bimolecular nucleophilic substitution even with strong bases such as ^-OR or ^-OH, because of the accessibility of the reactive carbon to the incoming nucleophile. This situation changes when we employ a base that, by virtue of its structural bulk, is severely sterically hindered in attacking carbon. In this case, elimination may predominate, even with primary systems, because deprotonation by the base occurs at the less-hindered periphery of the molecule. Examples of sterically hindered bases are potassium *tert*-butoxide and lithium diisopropylamide (LDA) (these are common names).* Reactions in which these bases are used are usually carried out in a solvent that consists of the corresponding conjugate acid; for example, 2-methyl-2-propanol or *N*-(1-methylethyl)-1-methylethanamine (diisopropylamine).

Sterically Hindered Bases

$$(CH_3)_3C\ddot{\underset{\cdot\cdot}{O}}:^-K^+$$

Potassium *tert*-butoxide

$$[(CH_3)_2CH]_2\ddot{\underset{\cdot\cdot}{N}}^-\ Li^+$$

Lithium diisopropylamide (LDA)

$$CH_3CH_2CH_2CH_2Br \xrightarrow[-\ HBr]{(CH_3)_3CO^-K^+,\ (CH_3)_3COH} \underset{85\%}{CH_3CH_2CH{=}CH_2} + \underset{15\%}{CH_3CH_2CH_2CH_2OC(CH_3)_3}$$

$$(CH_3)_2CHCH_2CH_2CH_2Cl \xrightarrow[-\ HCl]{[(CH_3)_2CH]_2N^-Li^+,\ [(CH_3)_2CH]_2NH} \underset{87\%}{(CH_3)_2CHCH_2CH{=}CH_2} + \underset{13\%}{(CH_3)_2CHCH_2CH_2CH_2N[CH(CH_3)_2]_2}$$

---

*The name amide for the bases $^-:\ddot{N}R'R''$ should not be mistaken for the same name given to carboxylic amides, $RC(=O)NR'R''$ (Section 18-5).

Only when the leaving group is very good, such as a sulfonate (e.g., 4-methylbenzenesulfonate), does this principle not apply, and the S_N2 reaction is still favored:

$$CH_3CH_2CH_2CH_2OSO_2\text{—}C_6H_4\text{—}CH_3 \xrightarrow[-\ CH_3\text{—}C_6H_4\text{—}SO_3H]{(CH_3)_3CO^-K^+,\ (CH_3)_3COH} \underset{99\%}{CH_3CH_2CH_2CH_2OC(CH_3)_3} + \underset{1\%}{CH_3CH_2CH{=}CH_2}$$

Increasing the steric bulk of the haloalkane at the reacting carbon slows down substitution processes, leading to increasing elimination. Thus, secondary and tertiary haloalkanes undergo predominant bimolecular elimination with a strong base and virtually exclusive elimination with a sterically hindered strong base.

$$CH_3\overset{\overset{CH_3}{|}}{\underset{\underset{H}{|}}{C}}Br \xrightarrow[-\ HBr]{CH_3CH_2O^-Na^+,\ CH_3CH_2OH} \underset{87\%}{H_2C{=}CHCH_3} + \underset{13\%}{CH_3\overset{\overset{CH_3}{|}}{\underset{\underset{H}{|}}{C}}OCH_2CH_3}$$

Propene
Predominant elimination

2-Ethoxypropane

$$CH_3\overset{\overset{CH_3}{|}}{\underset{\underset{H}{|}}{C}}Br \xrightarrow[-\ HBr]{(CH_3)_3CO^-K^+,\ (CH_3)_3COH} \underset{98\%}{H_2C{=}CHCH_3}$$

Propene
Exclusive elimination

Other branching has a similar effect.

$$CH_3CH_2CH_2Br \xrightarrow[-\ HBr]{CH_3CH_2O^-Na^+,\ CH_3CH_2OH} CH_3CH_2CH_2OCH_2CH_3 + H_3CCH{=}CH_2$$

91% substitution 9% elimination

$$CH_3\overset{\overset{CH_3}{|}}{\underset{\underset{H}{|}}{C}}CH_2Br \xrightarrow[-\ HBr]{CH_3CH_2O^-Na^+,\ CH_3CH_2OH} CH_3\overset{\overset{CH_3}{|}}{\underset{\underset{H}{|}}{C}}CH_2OCH_2CH_3 + (H_3C)_2C{=}CH_2$$

40% substitution 60% elimination

However, good nucleophiles that are weak bases, such as I^-, Br^-, RS^-, N_3^-, and PR_3, give good yields of S_N2 products. For example, 2-bromopropane reacts with iodide ion cleanly through the S_N2 pathway:

$$CH_3\overset{\overset{CH_3}{|}}{\underset{\underset{H}{|}}{C}}CH_2Br + Na^+I^- \xrightarrow{\text{Propanone (acetone)}} CH_3\overset{\overset{CH_3}{|}}{\underset{\underset{H}{|}}{C}}CH_2I + Na^+\ Br^-$$

EXERCISE 7-14

Which nucleophile in each of the following pairs will give a higher E2/S_N2 product ratio in reaction with 1-bromo-2-methylpropane?

(a) $N(CH_3)_3$, $P(CH_3)_3$; (b) H_2N^-, $(CH_3\underset{}{\overset{CH_3}{|}}CH)_2N^-$; (c) I^-, Cl^-.

The Effect of Base Strength

As the basicity of the base decreases or its nucleophilicity increases, the amounts of substitution increase by both S_N2 and S_N1 mechanisms, depending on conditions, substrates, and the exact nature of the nucleophile. For example, ethanoate (acetate) is a weaker base than ethoxide. Thus, whereas ethoxide gives 87% elimination, ethanoate (acetate) furnishes 100% substitution in its reaction with 2-bromopropane in propanone (acetone):

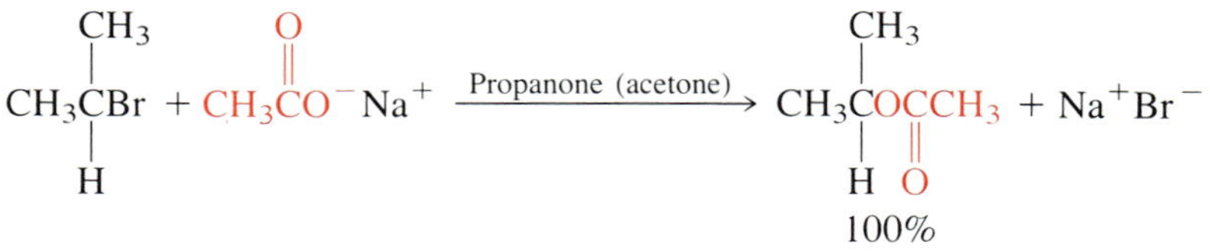

$$CH_3\overset{CH_3}{\underset{H}{C}}Br + CH_3\overset{O}{\overset{\|}{C}}O^-Na^+ \xrightarrow{\text{Propanone (acetone)}} \underset{100\%}{CH_3\overset{CH_3}{\underset{H}{C}}O\underset{O}{\underset{\|}{C}}CH_3} + Na^+Br^-$$

However, both give virtually exclusive elimination with 2-bromo-2-methylpropane.

The Versatile Secondary Haloalkanes

Secondary alkyl derivatives have the greatest versatility in their reactions with nucleophiles. Virtually exclusive bimolecular substitution may be attained with good nucleophiles that are weak bases. On the other hand, strong bases effect bimolecular elimination. When secondary haloalkanes are exposed to conditions that are neither strongly nucleophilic nor basic but include the use of a polar medium, such as a protic solvent, dissociation becomes competitive, simply because both S_N2 and E2 processes slow down sufficiently to allow the formation of the secondary cation. Depending on the nature of the medium, the relative amounts of substitution (S_N1) and elimination (E1) vary. Many secondary haloalkanes exhibit product formation following all four possible pathways: unimolecular and bimolecular substitution and elimination. It is therefore often impossible to foresee the outcome of the reaction of a secondary haloalkane with a nucleophile. Nevertheless, you should be able to predict, at least qualitatively, in which direction the product distribution will change upon changing substrate structures, leaving groups, solvents, and nucleophiles.

EXERCISE 7-15

Predict which reaction in each of the following pairs will have a higher E2/E1 product ratio and explain why.

(a)  $CH_3CH_2\overset{CH_3}{\overset{|}{C}}HBr \xrightarrow{CH_3OH}$ $\qquad$ $CH_3CH_2\overset{CH_3}{\overset{|}{C}}HBr \xrightarrow{CH_3O^-Na^+,\ CH_3OH}$

(b) Iodocyclohexane $\xrightarrow{(CH_3CH(CH_3))_2N^-Li^+,\ (CH_3CH(CH_3))_2NH}$ Iodocyclohexane $\xrightarrow{\text{Propanone (acetone)}}$

In summary, strong bases react with haloalkanes not only by substitution, but also by elimination. The kinetics of these reactions are second order, pointing to a bimolecular mechanism. An *anti* transition state is preferred, in which the base abstracts a proton at the same time as the leaving group departs. Steric effects that decrease the rate of substitution lead to increasing amounts of elimination products. Similarly, decreasing nucleophilic power and increasing basicity of the nucleophile favor E2 over S_N2. Secondary haloalkanes exhibit the greatest versatility in substitution and elimination reactions.

Summary of Important Concepts in Haloalkane Chemistry: Elimination or Substitution? Bimolecular or Unimolecular?

The preceding sections have demonstrated the complexity of the reactions of haloalkanes (and other alkyl derivatives containing suitable leaving groups) in the presence of nucleophiles. There are many possible variations of the parameters that enter into the reaction: substrate structure, nucleophiles, leaving groups, solvent effects, and base strength. The following guidelines are useful for comparing relative reactivity and for making at least qualitative predictions of reaction products.

Let us first summarize the major reaction types. Alkyl derivatives containing good leaving groups may undergo four different types of reaction with nucleophiles.

Reactions of Alkyl Derivatives with Nucleophiles

1 Ionization of secondary and tertiary derivatives to carbocations followed by nucleophilic trapping (S_N1):

$$CH_3CH_2C(CH_3)_2Br \xrightarrow[-\,Br^-]{} CH_3CH_2C^+(CH_3)_2 \xrightarrow{:Nu^-} CH_3CH_2C(CH_3)_2Nu$$

2 Dissociation of secondary and tertiary derivatives to carbocations followed by proton abstraction by the nucleophile acting as a base (E1):

$$CH_3C(CH_3)(H)\text{—}C(CH_3)(H)Cl \xrightarrow[-\,Cl^-]{} CH_3C(CH_3)(H)\text{—}C^+(CH_3)H \xrightarrow{:B^-} CH_3C(CH_3)(H)\text{—}C(=CH_2)H + (H_3C)_2C{=}C(CH_3)H + BH$$

3 Direct substitution of the leaving group by the incoming nucleophile with inversion of configuration in primary and secondary alkyl systems (S_N2):

$$\underset{\underset{CH_2CH_3}{H\cdots|}}{H_3C}\!\!>\!C{-}I \xrightarrow{:Nu^-} Nu{-}C\!\!<\!\!\underset{\underset{CH_2CH_3}{\cdots H}}{CH_3} + I^-$$

4 Simultaneous elimination of the leaving group and a neighboring proton to produce a double bond in primary, secondary, and tertiary derivatives (E2):

$$CH_3CH_2CH_2I \xrightarrow{:B^-} CH_3CH{=}CH_2 + BH + I^-$$

What generalizations can be made with respect to the relative reactivity of the various classes of haloalkanes in these reactions?

Primary Alkyl Derivatives

Unhindered primary alkyl derivatives always react in a bimolecular way and almost always give predominantly substitution products, except when sterically hindered strong bases, such as potassium *tert*-butoxide, are employed. In these cases, the S_N2 pathway is slowed down sufficiently for steric reasons to allow the E2 mechanism to take over.

Another way of reducing substitution is to introduce branching. However, even in these cases, good nucleophiles still furnish predominantly substitution products. Only strong bases, such as alkoxides, RO^-, or amides, R_2N^-, effect preferential elimination. Exceptions are the 2,2-dialkylpropyl and related systems. Here elimination is impossible because no hydrogen is available to permit it. Because the S_N2 reaction is also very slow, such derivatives remain uniquely unreactive in the presence of nucleophiles. The reactivity of primary derivatives is summarized below.

Reactivity of Primary Alkyl Derivatives R–X with Nucleophiles (Bases)

Unhindered R–X. S_N2 with good nucleophiles:

$$CH_3CH_2CH_2Br + {}^-CN \xrightarrow{\text{Propanone (acetone)}} CH_3CH_2CH_2CN + Br^-$$

Even with strong base:

$$CH_3CH_2CH_2Br + CH_3O^- \xrightarrow{CH_3OH} CH_3CH_2CH_2OCH_3 + Br^-$$

But E2 with strong hindered base:

$$CH_3CH_2CH_2Br + CH_3\overset{\overset{CH_3}{|}}{\underset{\underset{CH_3}{|}}{C}}O^- \xrightarrow[-\,HBr]{(CH_3)_3COH} CH_3CH{=}CH_2$$

No (or exceedingly slow) reaction with poor nucleophiles (CH_3OH).

Branched R–X. S_N2 with good nucleophiles (although slow compared with unhindered R–X):

$$\underset{\underset{H}{|}}{\overset{\overset{CH_3}{|}}{CH_3CCH_2Br}} + I^- \xrightarrow{\text{Propanone (acetone)}} \underset{\underset{H}{|}}{\overset{\overset{CH_3}{|}}{CH_3CCH_2I}} + Br^-$$

But E2 with strong base (not necessarily hindered):

$$\underset{\underset{H}{|}}{\overset{\overset{CH_3}{|}}{CH_3CCH_2Br}} + CH_3CH_2O^- \xrightarrow[-\,HBr]{CH_3CH_2OH} \overset{\overset{CH_3}{|}}{CH_3C{=}CH_2}$$

No (or exceedingly slow) reaction with poor nucleophiles.

Secondary Alkyl Derivatives

Secondary alkyl derivatives undergo, depending on conditions, both eliminations and substitutions by either possible pathway: uni- and bimolecular. Good nucleophiles favor S_N2, bases result in E2, and polar nonnucleophilic media give mainly S_N1 and E1 products. The reactivity of secondary derivatives is summarized below.

Reactivity of Secondary Alkyl Derivatives R–X with Nucleophiles (Bases)

When X is a good leaving group in a highly polar nonnucleophilic medium: S_N1 (+E1)

$$\underset{\underset{H}{|}}{\overset{\overset{CH_3}{|}}{CH_3CBr}} \xrightarrow[-\,HBr]{CH_3CH_2OH} \underset{\text{Major}}{\underset{\underset{H}{|}}{\overset{\overset{CH_3}{|}}{CH_3COCH_2CH_3}}} + \underset{\text{Minor}}{CH_3CH{=}CH_2}$$

With good nucleophiles: S_N2

$$\underset{\underset{H}{|}}{\overset{\overset{CH_3}{|}}{CH_3CBr}} + CH_3S^- \xrightarrow{CH_3CH_2OH} \underset{\underset{H}{|}}{\overset{\overset{CH_3}{|}}{CH_3CSCH_3}} + Br^-$$

With strong base: E2

$$\underset{\underset{H}{|}}{\overset{\overset{CH_3}{|}}{CH_3CBr}} + CH_3CH_2O^- \xrightarrow[-\,HBr]{CH_3CH_2OH} CH_3CH{=}CH_2$$

Tertiary Alkyl Derivatives

Tertiary systems eliminate (E2) with a strong base and are substituted in nonbasic media (S_N1). Bimolecular substitution is not observed, but elimination by E1 is a side reaction of S_N1. This reactivity is summarized on the next page.

Reactivity of Tertiary Alkyl Derivatives R–X with Nucleophiles (Bases)

When X is a good leaving group and no base is present, in polar solvents: S_N1 + E1

$$CH_3CH_2C(CH_3)_2Br \xrightarrow[-\ HBr]{HOH,\ \text{propanone (acetone)}} CH_3CH_2C(CH_3)_2OH + \text{alkenes}$$

With strong base: E2

$$CH_3CH_2C(CH_2CH_3)_2Cl \xrightarrow[-\ HCl]{CH_3O^-,\ CH_3OH} CH_3CH_2C(CH_2CH_3){=}CHCH_3$$

With weak base: E1

$$CH_3C(CH_3)_2Br \xrightarrow[-\ HBr]{CH_3CH_2OH,\ \text{weak base}} CH_2{=}C(CH_3)CH_3$$

Problems

1 Write the major substitution product of each of the following solvolysis reactions.

(a) $CH_3C(CH_3)_2Br \xrightarrow{CH_3CH_2OH}$

(b) $(CH_3)_2CBrCH_2CH_3 \xrightarrow{CF_3CH_2OH}$

(c) 1-chloro-1-ethylcyclopentane (CH_3CH_2, Cl on the same ring carbon) $\xrightarrow{CH_3OH}$

(d) cyclohexyl–$C(Br)(CH_3)$–CH_3 $\xrightarrow{CH_3\overset{O}{\overset{\|}{C}}OH}$

(e) $CH_3C(CH_3)_2Cl \xrightarrow{D_2O}$

(f) $CH_3C(CH_3)_2Cl \xrightarrow{\text{cyclohexane-C(H)OD}}$

2 Write the two major substitution products of the reaction shown at the right.

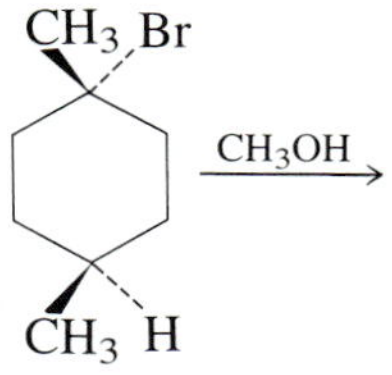

(a) Explain the formation of each of the products, using a mechanistic argument.

(b) If the reaction is interrupted before it goes to completion, an *isomer* of the starting material is found to be present in the reaction mixture. Draw its structure and explain how it is formed.

CH_3 Br, CH_3 H (cyclohexane) $\xrightarrow{CH_3OH}$

3 Write the two major substitution products of the reaction shown below.

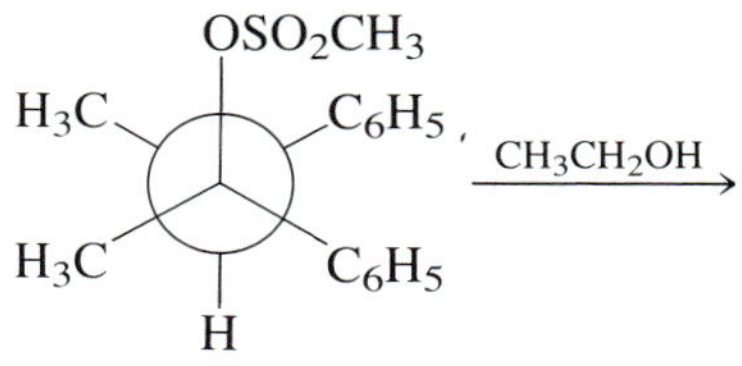

OSO_2CH_3, H_3C, C_6H_5, H_3C, C_6H_5, H $\xrightarrow{CH_3CH_2OH}$

4 How would each reaction in Problem 1 be affected by the addition of each of the following substances to the solvolysis mixture?

(a) H_2O
(b) H_2S
(c) KI
(d) NaN_3
(e) Propanone (acetone)
(f) $CH_3CH_2OCH_2CH_3$

5 Draw a qualitative potential-energy diagram for the following reaction.

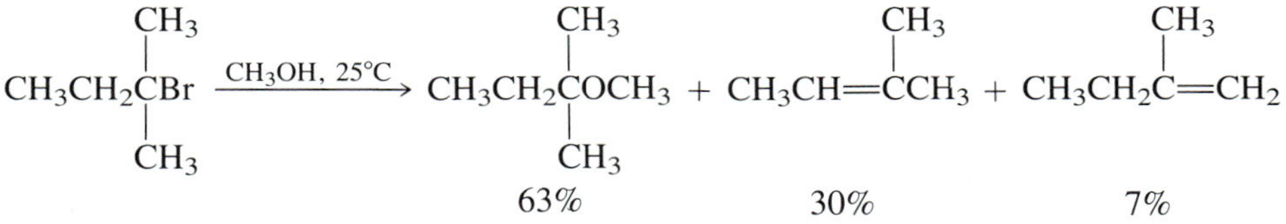

$$CH_3CH_2\overset{CH_3}{\underset{CH_3}{C}}Br \xrightarrow{CH_3OH,\ 25°C} CH_3CH_2\overset{CH_3}{\underset{CH_3}{C}}OCH_3 + CH_3CH{=}\overset{CH_3}{C}CH_3 + CH_3CH_2\overset{CH_3}{C}{=}CH_2$$

63%   30%   7%

Indicate the locations of the rate-determining and the product-determining transition states.

6 2-Bromo-2-phenylpropane (see margin) undergoes solvolysis in a unimolecular, strictly first order process. The reaction rate for [RBr] = 0.1 molar RBr in 9:1 propanone (acetone)-water is measured to be 2×10^{-4} mole l^{-1} sec^{-1}.

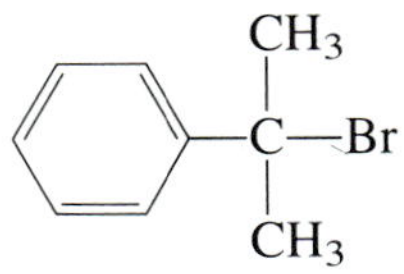

(a) Calculate the rate constant k from this data. What is the product of this reaction?

(b) In the presence of 0.1 molar LiCl, the rate is found to increase to 4×10^{-4} mole l^{-1} sec^{-1}, although the reaction still remains strictly first order. Calculate the new rate constant k_{LiCl} and suggest an explanation.

(c) If 0.1 molar LiBr is present instead of LiCl, the measured rate *drops* to 1.6×10^{-4} mole l^{-1} sec^{-1}. Explain this observation and write the appropriate chemical equations to describe the reactions.

RBr = C_6H_5–$\overset{CH_3}{\underset{CH_3}{C}}$–Br

7 Rank the following carbocations in decreasing order of stability.

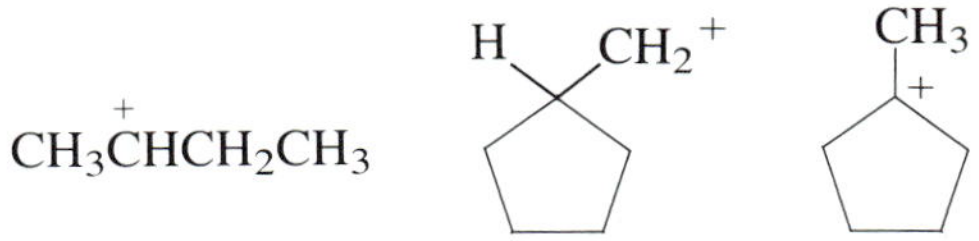

$CH_3\overset{+}{C}HCH_2CH_3$ H, CH_2^+ (cyclopentane) CH_3, + (cyclopentyl)

8 The three cyclic cations illustrated below have very different stabilities. Predict the order of stability of the three and suggest a rationalization for your prediction.

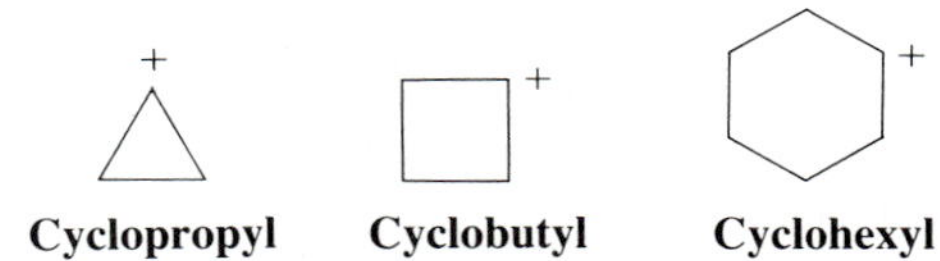

9 Rank the compounds in each group below in order of decreasing rate of solvolysis in aqueous propanone (acetone).

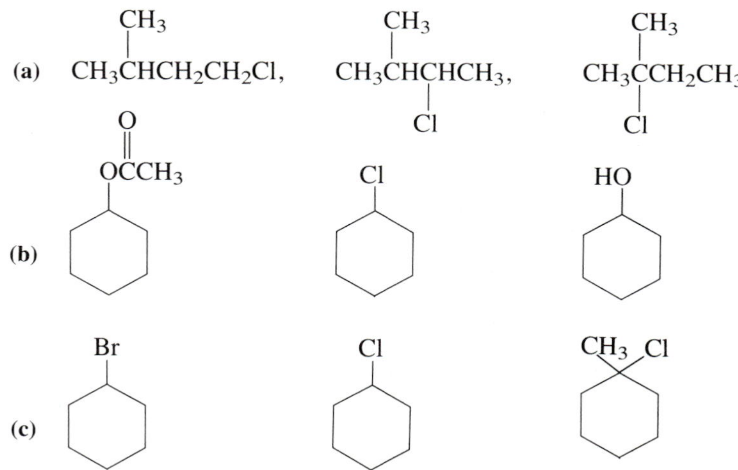

10 Write the products of the following substitution reactions. Indicate whether the reaction proceeds through the S_N1 or the S_N2 mechanism, and write the full mechanism in detail; that is, each step separately, using electron-pair arrows to denote all bonding changes in each step.

(a) $(CH_3)_2CHOSO_2CF_3 \xrightarrow{CH_3CH_2OH}$

(b) cyclopentane bearing CH_3 and OH on the same carbon $\xrightarrow{\text{Conc. HCl, } H_2O}$

(c) $CH_3CH_2CH_2CH_2Br \xrightarrow{(C_6H_5)_3P\text{, DMSO}}$

(d) $CH_3CH_2CHClCH_2CH_3 \xrightarrow{\text{NaI, propanone (acetone)}}$

11 Propose a synthesis of *(R)*-$CH_3\underset{\displaystyle N_3}{\underset{|}{C}}HCH_2CH_3$, starting from *(R)*-2-bromobutane.

12 Match each of the transformations below to the correct reaction profile shown on the next page, and draw the structures of the species present at each point on the energy curves marked by a capital letter.

(a) $(CH_3)_3CCl + (C_6H_5)_3P \longrightarrow$
(b) $(CH_3)_2CHI + KBr \longrightarrow$
(c) $(CH_3)_3COH + HBr \longrightarrow$
(d) $CH_3CH_2Br + NaOCH_2CH_3 \longrightarrow$

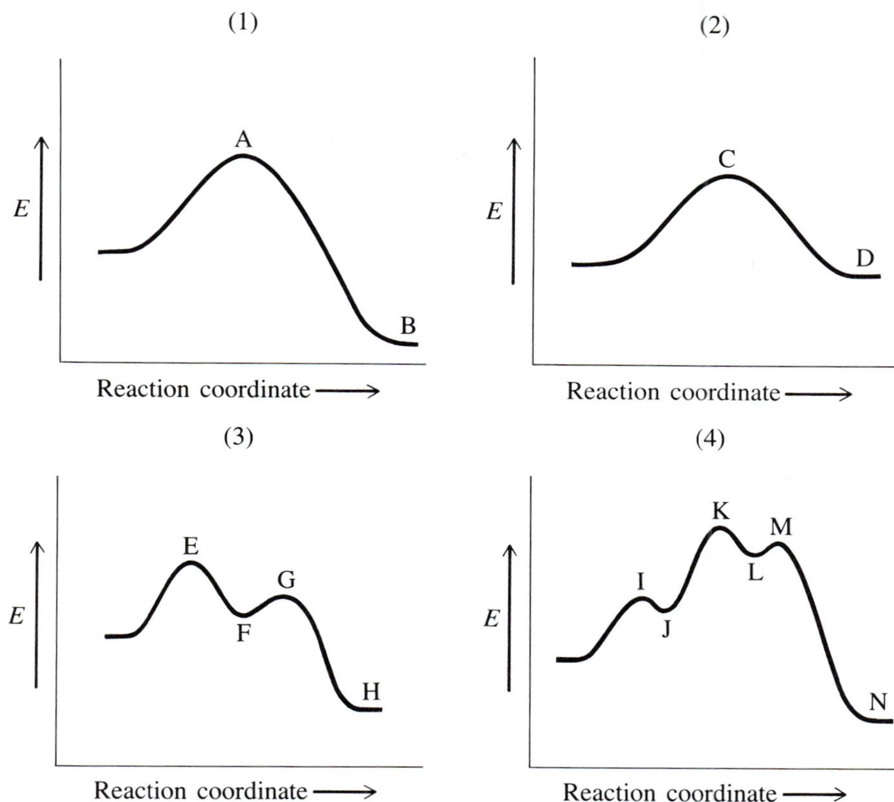

13 Two substitution reactions of *(S)*-2-bromobutane are shown below.

$$\text{(1)}\quad (S)\text{-}CH_3CH_2\underset{\substack{|\\ Br}}{C}HCH_3 \xrightarrow{\overset{\overset{O}{\|}}{HCOH}}$$

$$\text{(2)}\quad (S)\text{-}CH_3CH_2\underset{\substack{|\\ Br}}{C}HCH_3 \xrightarrow{\overset{\overset{O}{\|}}{HCO}{}^{-}Na^{+},\ DMSO}$$

Predict the comparative stereochemical results of these two reactions.

14 Write all possible products of E1 processes for each reaction in Problems 1, 2, and 3.

15 Write the structure of the most likely product of the following reaction of 4-chloro-4-methyl-1-pentanol in neutral, polar solution. The formula of the product is $C_6H_{12}O$.

$$(CH_3)_2\overset{\overset{Cl}{|}}{C}CH_2CH_2CH_2OH \longrightarrow HCl + ?$$

If the same molecule is dissolved in *basic* solution, the product still has the formula $C_6H_{12}O$, but it has a completely different structure. What is it?

Explain the difference between the two results.

16 Fill in the spaces in the table below with the major product(s) of reaction of each haloalkane with each of the reagents shown.

	Reagent			
Haloalkane	H_2O	$NaSCH_3$	$NaOCH_3$	$KOC(CH_3)_3$
CH_3Cl	______	______	______	______
CH_3CH_2Cl	______	______	______	______
$(CH_3)_2CHCl$	______	______	______	______
$(CH_3)_3CCl$	______	______	______	______

17 Indicate the major mechanism(s) (simply specify S_N2, S_N1, E2, or E1) required for the formation of each product that you wrote in Problem 16.

18 The reaction below can proceed through both E1 and E2 mechanisms.

$$C_6H_5CH_2C(CH_3)_2Cl \xrightarrow{NaOCH_3, CH_3OH} C_6H_5CH{=}C(CH_3)_2 + C_6H_5CH_2C(CH_3){=}CH_2$$

The E1 rate constant $k_{E1} = 1.4 \times 10^{-4}$ sec^{-1} and the E2 rate constant $k_{E2} = 1.9 \times 10^{-4}$ l mole^{-1} sec^{-1}; 0.02 molar haloalkane.

(a) What is the predominant elimination mechanism with 0.5 molar $NaOCH_3$?

(b) What is the predominant elimination mechanism with 2.0 molar $NaOCH_3$?

(c) At what concentration of base does exactly 50% of the starting material react by an E1 route and 50% by an E2 pathway?

19 Write the products of the following elimination reactions. Indicate whether the reaction proceeds predominantly through the E1 or E2 mechanism, and write the full mechanism in detail as described for Problem 10.

(a) $(CH_3)_3COH \xrightarrow{\text{Conc. } H_2SO_4,\ \Delta}$

(b) $CH_3CH_2CH_2CH_2Cl \xrightarrow{\text{LDA, } [(CH_3)_2CH]_2NH}$

(c) $C_6H_{11}\text{–}CH(Br)\text{–}C_6H_{11} \xrightarrow{\text{Excess KOH, } CH_3CH_2OH}$ (two cyclohexane rings joined through CHBr)

(d) 1-chloro-1-methylcyclohexane (chair; Cl, CH_3) $\xrightarrow{NaOCH_3,\ CH_3OH}$

20 Predict the mechanism and major product for the reaction of a secondary haloalkane in a polar aprotic solvent with each of the nucleophiles below. In each case, the pK_a of the conjugate acid of the nucleophile is given in parentheses.

(a)	N_3^- (4.6)	**(f)**	F^- (3.2)
(b)	H_2N^- (35)	**(g)**	$C_6H_5O^-$ (9.9)
(c)	NH_3 (9.5)	**(h)**	PH_3 (−12)
(d)	HSe^- (3.7)	**(i)**	NH_2OH (6.0)
(e)	H^- (38)	**(j)**	NCS^- (−0.7)

21 Three reactions of 2-chloro-2-methylpropane are shown below:

$$(1)\quad (CH_3)_3CCl \xrightarrow{H_2S,\ CH_3OH}$$

$$(2)\quad (CH_3)_3CCl \xrightarrow{CH_3\overset{\overset{O}{\|}}{C}O^-K^+,\ CH_3OH}$$

$$(3)\quad (CH_3)_3CCl \xrightarrow{CH_3O^-K^+,\ CH_3OH}$$

(a) Write the major product of each reaction.
(b) Compare the rates of the three reactions. Assume identical solution polarities and reactant concentrations. Explain mechanistically.
(c) Draw qualitative potential-energy diagrams for each reaction.

22 Write the major product(s) of the following reactions. Indicate whether the reaction proceeds through the appropriate mechanism(s): S_N1, S_N2, E1, or E2. (Do not write out the full mechanisms.) If no reaction occurs, write "no reaction."

(a) 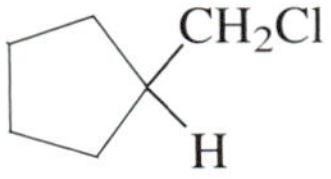 Cyclopentane bearing CH_2Cl and H $\xrightarrow{KOC(CH_3)_3,\ (CH_3)_3COH}$

(b) $CH_3\overset{\overset{F}{|}}{C}HCH_2CH_3 \xrightarrow{KBr,\ \text{propanone (acetone)}}$

(c) 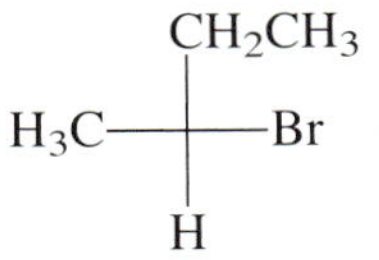 H_3C–C(–CH_2CH_3)(–Br)(–H) $\xrightarrow{H_2O}$

(d) 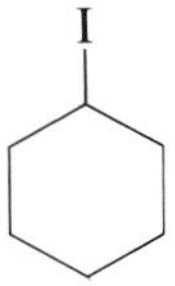 Iodocyclohexane $\xrightarrow{NaNH_2,\ \text{liquid } NH_3}$

(e) $(CH_3)_2CHCH_2CH_2CH_2Br \xrightarrow{NaOCH_2CH_3,\ CH_3CH_2OH}$

(f) 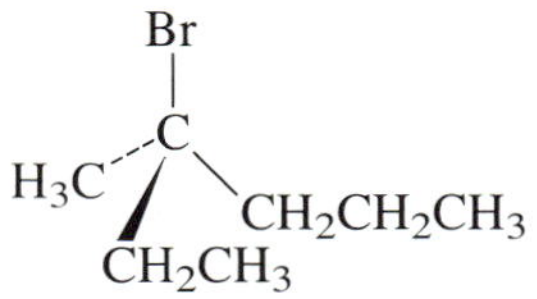 Br–C(H_3C)(CH_2CH_3)($CH_2CH_2CH_3$) $\xrightarrow{NaI,\ \text{propanone (acetone)}}$

(g)  Cyclopentane bearing OH and H $\xrightarrow{KOH,\ CH_3CH_2OH}$

(h) 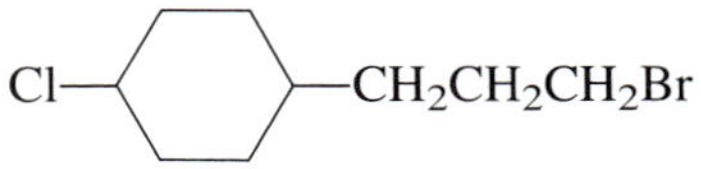 Cl–(cyclohexane)–$CH_2CH_2CH_2Br \xrightarrow{\text{Excess KCN},\ CH_3OH}$

(i) 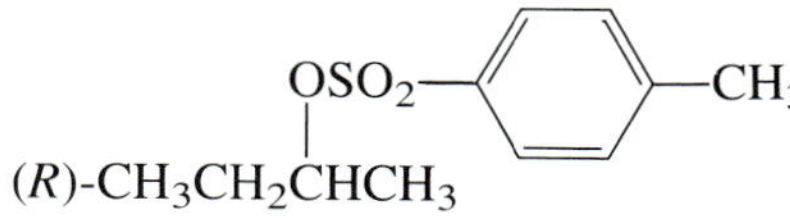 (*R*)-$CH_3CH_2\overset{\overset{OSO_2-C_6H_4-CH_3}{|}}{C}HCH_3 \xrightarrow{NaSH,\ CH_3CH_2OH}$

(j) $(CH_3CH_2)_3COH \xrightarrow{\text{Conc. HCl},\ H_2O}$

(k) $(CH_3)_3CCl \xrightarrow{NaNH_2,\ \text{liquid } NH_3}$

(l) 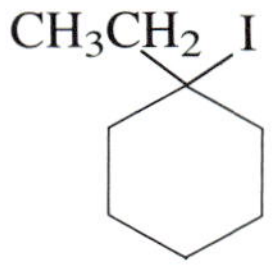 Cyclohexane bearing CH_3CH_2 and I on the same carbon $\xrightarrow{CH_3OH}$

(m) $(CH_3)_3C\overset{\overset{Br}{|}}{C}HCH_3 \xrightarrow{KOH,\ CH_3CH_2OH}$

(n) $CH_3CH_2Cl \xrightarrow{CH_3\overset{\overset{O}{\|}}{C}OH}$

23 For each equation below, indicate whether the reaction depicted would work well, poorly, or not at all. Write the actual major product for each reaction that would not work well as written.

(a) $CH_3CH_2CHCH_3$ (with Br on C-2) $\xrightarrow{\text{NaOH, propanone (acetone)}}$ $CH_3CH_2CHCH_3$ (with OH on C-2)

(b) CH_3CHCH_2Cl (with H_3C on C-2) $\xrightarrow{CH_3OH}$ $CH_3CHCH_2OCH_3$ (with CH_3 on C-2)

(c) Cyclohexane with H, Cl $\xrightarrow{\text{HCN, } CH_3OH}$ cyclohexane with H, CN

(d) $CH_3—C(CH_3)(OSO_2CH_3)—CH_2CH_2CH_2CH_2OH$ (drawn with CH_3 above and CH_3SO_2O below the C) $\xrightarrow{(CH_3CH_2)_2O}$ 2,2-dimethyltetrahydropyran (CH_3, CH_3, O in ring)

(e) Cyclopentane with CH_3, CH_2I $\xrightarrow{\text{NaSCH}_3\text{, } CH_3OH}$ cyclopentane with CH_3, CH_2SCH_3

(f) $CH_3CH_2CH_2Br \xrightarrow{NaN_3,\ CH_3OH} CH_3CH_2CH_2N_3$

(g) $(CH_3)_3CCl \xrightarrow{\text{NaI, propanone (acetone)}} (CH_3)_3CI$

(h) $(CH_3CH_2)_2O \xrightarrow{CH_3I} (CH_3CH_2)_2\overset{+}{O}CH_3 + I^-$

(i) $CH_3I \xrightarrow{CH_3OH} CH_3OCH_3$

(j) $CH_3CH_2CH_2CH_2Cl \xrightarrow{NaOCH_2CH_3,\ CH_3CH_2OH} CH_3CH_2CH{=}CH_2$

(k) $CH_3CHCH_2CH_2Cl$ (with CH_3 on C-2) $\xrightarrow{NaOCH_2CH_3,\ CH_3CH_2OH}$ $CH_3CHCH{=}CH_2$ (with CH_3 on C-2)

(l) $(CH_3CH_2)_3COCH_3 \xrightarrow{NaBr,\ CH_3OH} (CH_3CH_2)_3CBr$

24 Propose syntheses of the molecules below from the indicated starting materials. Make use of any other reagents or solvents that you need. In many cases, there may be no alternative but to employ a reaction that yields a mixture of products. In such cases, apply reagents and reaction conditions that will maximize the yield of the desired product (compare Problem 19 in Chapter 6).

(a) $CH_3CH_2CHCH_3$ (with I on C-2), from butane.
(b) $CH_3CH_2CH_2CH_2I$, from butane.
(c) $(CH_3)_3COCH_3$, from methane and 2-methylpropane.
(d) Cyclohexene, from cyclohexane.
(e) Cyclohexanol, from cyclohexane.
(f) Tetrahydrofuran (five-membered ring with O), from $HOCH_2CH_2CH_2CH_2OH$.
(g) Bicyclic ring with two S atoms, from 1,3-dibromopropane.

25 Cortisone is an important steroidal anti-inflammatory agent. Cortisone can be synthesized efficiently from the alkene shown below.

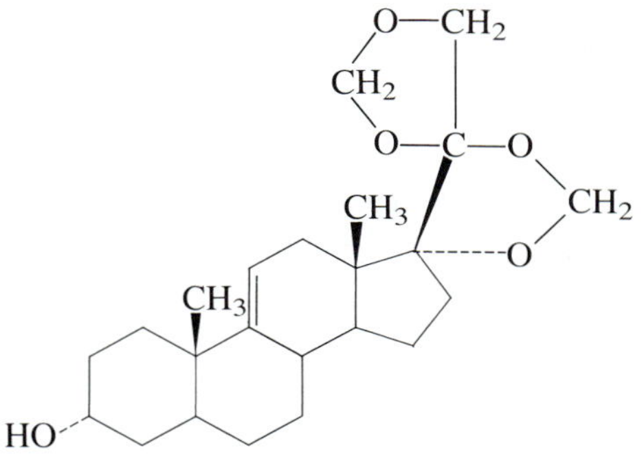

Alkene

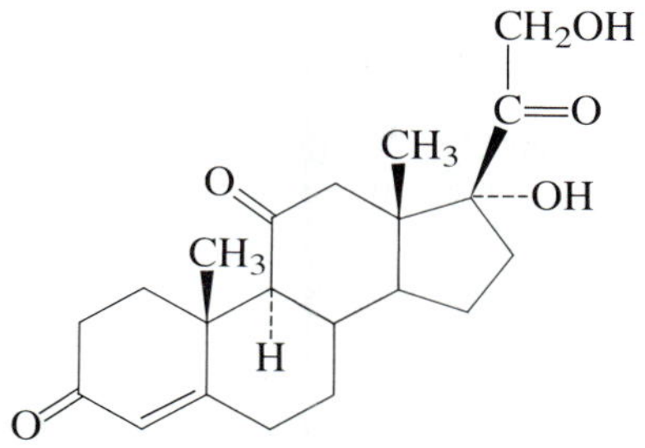

Cortisone

Of the three chlorinated compounds shown below, two give reasonable yields of the alkene shown above on E2 elimination with base, but one does not. Which one does not work well, and why? What does it give on attempted E2 elimination?

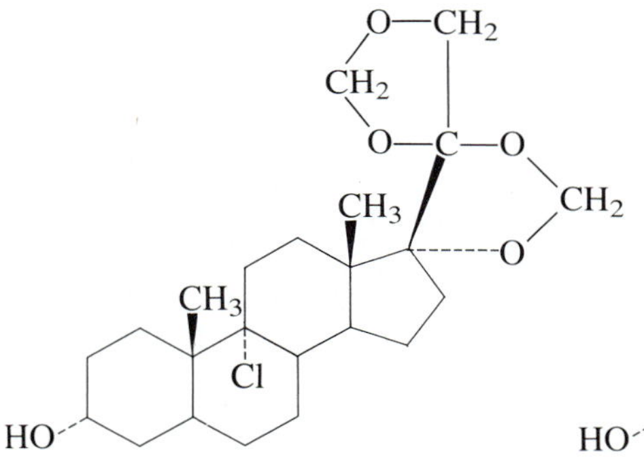

A

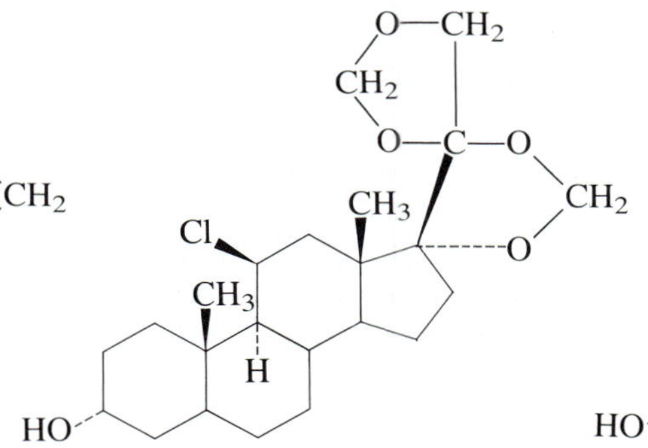

B

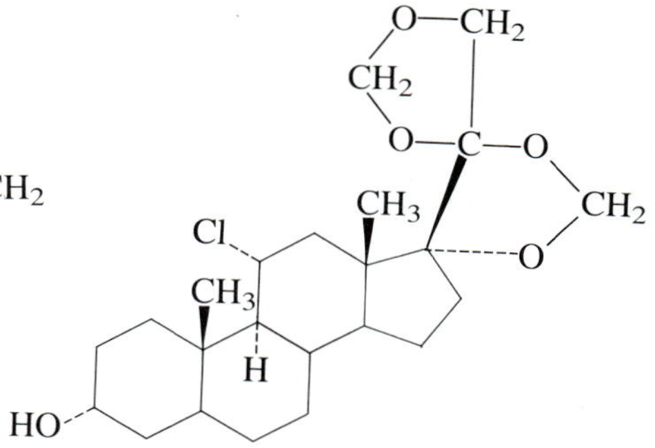

C

26 The chemistry of derivatives of *trans*-bicyclo[4.4.0]decane is of interest because this ring system is part of the structure of steroids. Make models of the brominated systems shown in the margin to help you answer the questions that follow.

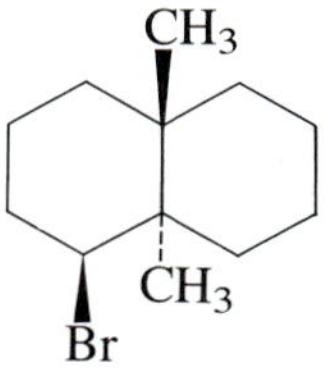

(a) One of the molecules shown at the right undergoes E2 reaction with $NaOCH_2CH_3$ in CH_3CH_2OH considerably faster than the other. Identify which is which and explain.

(b) If the carbon next to the one containing bromine has been specifically deuterated, the following results are obtained:

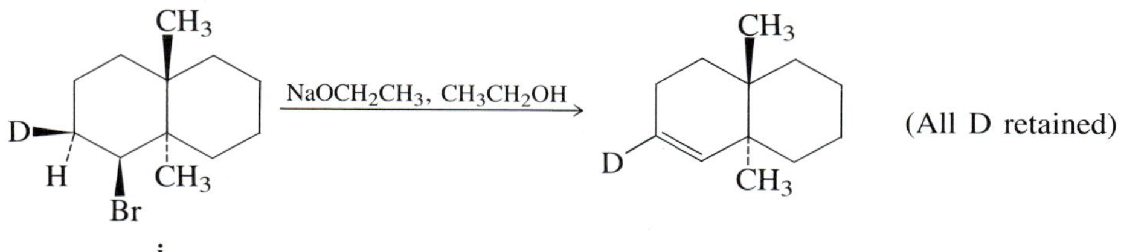

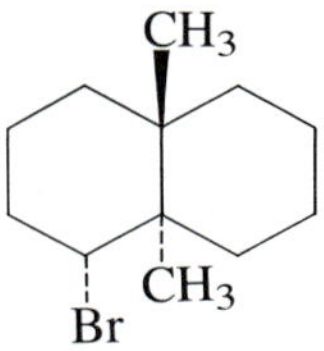

$NaOCH_2CH_3$, CH_3CH_2OH

(All D retained)

i

i

ii

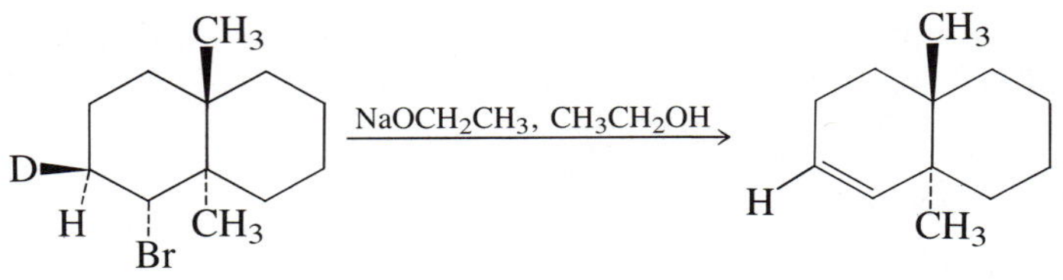

$$\text{ii} \xrightarrow{NaOCH_2CH_3,\ CH_3CH_2OH} \text{(All D lost)}$$

ii

Identify each of these as either an *anti* or a *syn* elimination. In each case, draw the conformation that the molecule must adopt for elimination to take place. Does your answer here help explain the situation described in part **a?** How?

(c) Does either reaction in part **b** exhibit a deuterium isotope effect? Explain.

27 The biosynthesis of unsaturated fatty acids is a multiple-step process that starts with small, saturated carboxylic acids and includes both chain-lengthening reactions and a double-bond–forming sequence. The synthesis of the most common example, 11-octadecenoic acid (vaccenic acid), utilizes 3-hydroxydecanoic acid as an intermediate.

$$CH_3(CH_2)_5CH{=}CH(CH_2)_9\overset{\overset{\displaystyle O}{\|}}{C}OH \qquad CH_3(CH_2)_5CH_2\overset{\overset{\displaystyle OH}{|}}{C}HCH_2\overset{\overset{\displaystyle O}{\|}}{C}OH$$

11-Octadecenoic acid **3-Hydroxydecanoic acid**

Even though the detailed chemistry of alcohols has not yet been covered, you already have seen methods that are applicable to the conversion of alcohols into alkenes through elimination.

Suggest reactions that would be capable of converting 3-hydroxydecanoic acid into one or more alkenes through **(a)** an E1 pathway and **(b)** an E2 pathway. Write the structure(s) of the likely product(s). Careful! Is the —OH group in the starting compound a good leaving group yet?

CHAPTER 8

The Formation of the Hydroxy Functional Group

PROPERTIES OF THE ALCOHOLS AND SYNTHETIC STRATEGY IN THEIR PREPARATION

Alcohols and ethers may be regarded as derivatives of water, in which one or two hydrogen atoms have been replaced by alkyl groups:

H—O—H	CH_3—O—H	CH_3—O—CH_3
Water	**Methanol**	**Methoxymethane**

Alcohols are abundant in nature, where they appear in a variety of complicated structures (see, e.g., Section 4-7). Simple alcohols have many practical uses as solvents and synthetic intermediates (see Section 9-8). This chapter is the first of two in which alcohols and ethers are discussed as classes of compounds whose chemistry is shaped by the presence of their functional groups. The naming of alcohols will be presented first, followed by a brief description of their structures and other physical properties, particularly in comparison with those of the alkanes and the haloalkanes.

8-1 The Naming of Alcohols

As with other compound classes, alcohols are named in accord with both systematic and common nomenclature.

Alcohols Are Alkanols in IUPAC Nomenclature

The systematic (IUPAC) way of naming alcohols is to treat them as derivatives of alkanes. The ending **-e** of the alkane is replaced by **-ol.** Thus, an alkane is converted into an **alkanol.** For example, the simplest alcohol is derived from methane: methanol. Ethanol stems from ethane, propanol from propane, and so on. In more complicated, branched systems, the name of the alcohol has as its derivation the longest chain *containing the OH substituent*. This chain may not always be the longest one in the molecule.

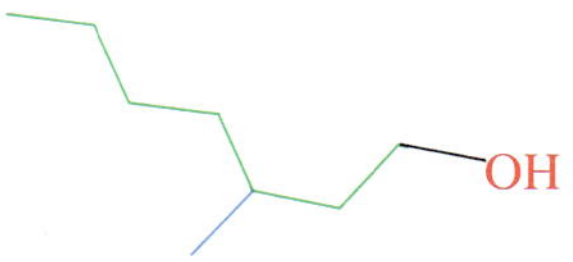

OH

A methylheptanol

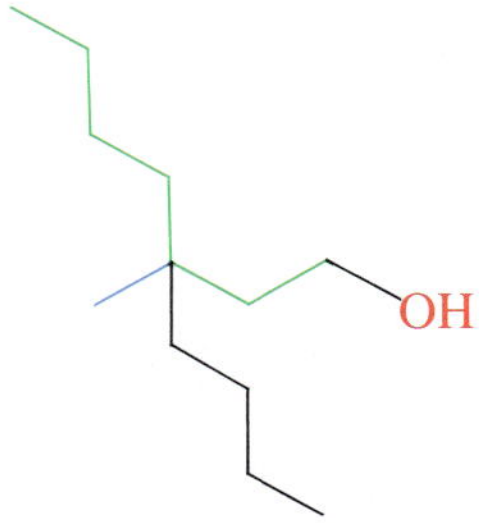

OH

A butylmethylheptanol

To find the functional group's position in the chain, number each carbon atom beginning from the closest terminus; this gives the position of the OH group the smallest possible number.

$\overset{3}{C}H_3\overset{2}{C}H_2\overset{1}{C}H_2OH$ — **1-Propanol**

$\overset{1}{C}H_3\overset{2}{C}H(OH)\overset{3}{C}H_3$ — **2-Propanol**

$\overset{1}{C}H_3\overset{2}{C}H(OH)\overset{3}{C}H_2\overset{4}{C}H_2\overset{5}{C}H_3$ — **2-Pentanol**

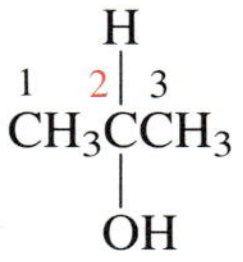

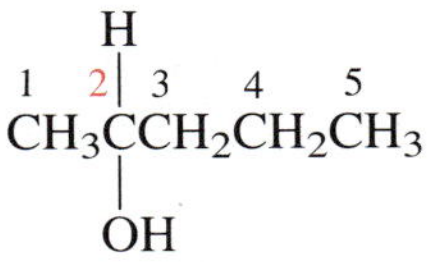

The numbers assigned to other substituents along the chain are then predetermined. The names of these substituents are added to the alkanol stem as prefixes. Complex alkyl appendages are named according to the IUPAC rules for hydrocarbons (Section 2-3).

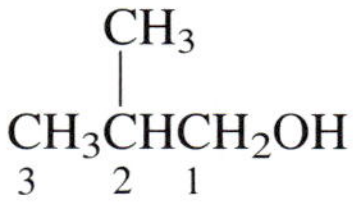

$\underset{3}{C}H_3\underset{2}{C}H(CH_3)\underset{1}{C}H_2OH$ — **2-Methyl-1-propanol**

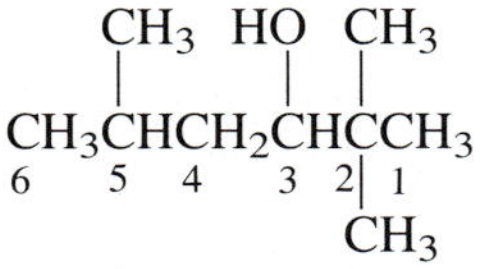

$\underset{6}{C}H_3\underset{5}{C}H(CH_3)\underset{4}{C}H_2\underset{3}{C}H(OH)\underset{2}{C}(CH_3)_2\underset{1}{C}H_3$ — **2,2,5-Trimethyl-3-hexanol**

If the OH group of an alcohol is attached to the alkane in such a way that either of two chains of equal length could be chosen for naming the molecule, the chain of greater complexity (i.e., containing the larger number of substituents) is used. Cyclic alcohols are called **cycloalkanols** and, when substituted, are numbered so that the carbon carrying the functional group automatically receives the number one:

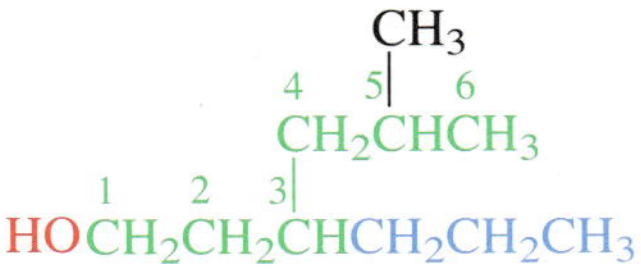

$HO\overset{1}{C}H_2\overset{2}{C}H_2\overset{3}{C}H(\overset{4}{C}H_2\overset{5}{C}H(CH_3)\overset{6}{C}H_3)CH_2CH_2CH_3$

5-Methyl-3-propyl-1-hexanol
[not 3-(2-Methylpropyl)-1-hexanol]

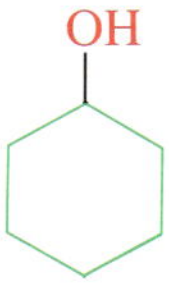

Cyclohexanol

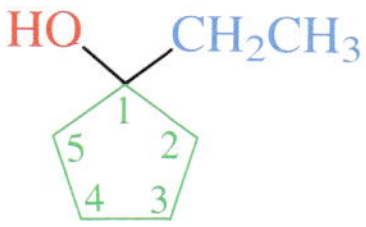

1-Ethylcyclopentanol

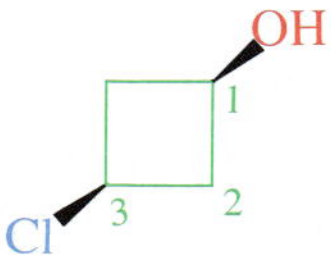

***cis*-3-Chlorocyclobutanol**

As a substituent, the OH group is called **hydroxy.** Similar to alkyl substituents and haloalkanes, the alcohols can be classified as primary, secondary, and tertiary:

RCH_2OH — **A primary alcohol**

OH on RCR′ with H (RCH(OH)R′) — **A secondary alcohol**

OH on RCR′ with R″ (RR′R″COH) — **A tertiary alcohol**

EXERCISE 8-1

Draw the structures of the following alcohols: (a) (*S*)-3-methyl-3-hexanol; (b) *trans*-2-bromocyclopentanol; (c) 2,2-dimethyl-1-propanol (neopentyl alcohol).

EXERCISE 8-2

Name the following compounds:

(a) $CH_3CH(CH_3)CH_2CH(OH)CH_3$ (b) CH_3CH_2–cyclohexane–OH (c) $CH_3CH(Br)CH(Cl)CH_2OH$

Alcohols Are Alkyl Derivatives in Common Nomenclature

In common nomenclature, the name of the alkyl group is followed by the word "alcohol," written separately. Such names are found in the older literature, and it is probably best not to use them.

CH_3OH — **Methyl alcohol**

$CH_3CH(CH_3)OH$ — **Isopropyl alcohol**

$CH_3C(CH_3)_2OH$ — ***tert*-Butyl alcohol**

Some alcohols have unique names:

$HOCH_2CH_2OH$ — **Ethylene glycol (Systematic name, 1,2-ethanediol)**

CH_2OH–HCOH–CH_2OH — **Glycerol (1,2,3-Propanetriol)**

C_6H_5OH — **Phenol (Benzenol)**

The position of a substituent is signaled by a Greek letter indicating its distance from the functional group. In alphabetical order, the carbon bearing the hydroxy group is labeled α, the neighboring carbon β, and so on.

$\overset{\delta}{Cl}CH_2\overset{\gamma}{C}H_2\overset{\beta}{C}H_2\overset{\alpha}{C}H_2OH$

δ-Chlorobutyl alcohol

In summary, alcohols can be named as alkanols (IUPAC) or alkyl alcohols. In IUPAC nomenclature, the name is derived from the chain bearing the functional group, whose position is given the lowest possible number. Common names incorporate the Greek alphabet to designate the position of additional substituents relative to the OH function.

8-2
Structure and Physical Properties of the Alcohols

The physical characteristics of the alcohols are strongly shaped by the presence of the hydroxy functional group. Its effect is revealed in their molecular structure, in their ability to enter into hydrogen bonding, and in the resulting changes in water solubility and boiling points.

The Structures of Alcohols Reveal a Nearly Tetrahedral Oxygen and a Short O–H Bond

The structural features of alcohols resemble those of water, from which they are formally derived. The structure of methanol is compared with that of water and methoxymethane (dimethyl ether) in Figure 8-1. Note their basic similarity. Minor differences in bond angles and bond lengths are due to the steric effect produced when the two hydrogen atoms in H_2O are sequentially replaced with alkyl groups. For example, the R–O–R angle increases along the series from $104.5°$ to $108.9°$ to $111.7°$. The O–H bond is considerably shorter than the C–H bond, owing to a large degree to the higher electronegativity of oxygen relative to carbon. Consistent with this bond shortening is the order of bond strengths: $DH°$ (OH) = 104 kcal $mole^{-1}$, $DH°$ (CH) = 98 kcal $mole^{-1}$.

FIGURE 8-1
The structures of (A) water, (B) methanol, and (C) methoxymethane. The carbon and oxygen are approximately sp^3-hybridized, so that all three adopt a nearly tetrahedral configuration. The tetrahedral nature of the oxygen may be visualized by remembering that it bears two lone electron pairs in two nonbonding sp^3 hybrid orbitals (see part B).

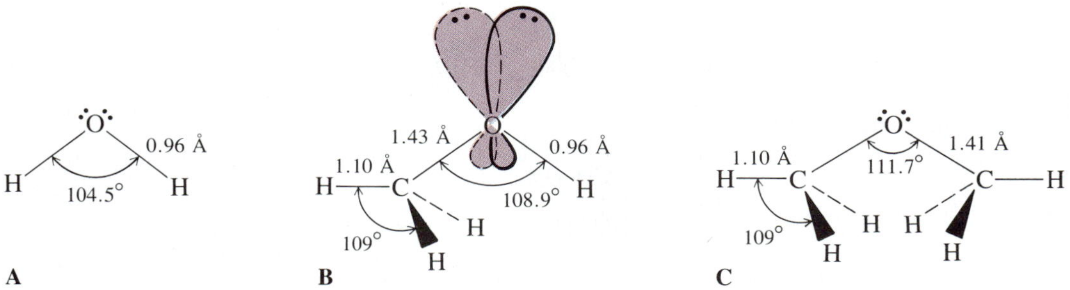

The electronegativity of oxygen causes an unsymmetrical distribution of charge in alcohols, creating a molecular dipole moment (Section 1-3) similar to that observed for water.

Dipole Moments of Water and Methanol

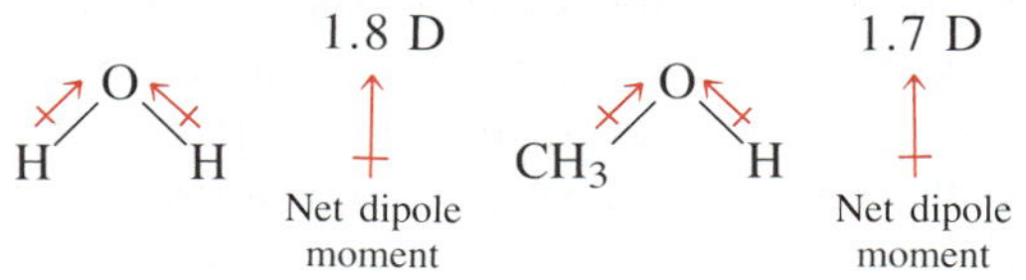

The Effect of Hydrogen Bonding on the Physical Properties of the Alcohols

In the discussion of the physical properties of the haloalkanes (Section 6-2), it was noted that polar molecules generally have higher boiling points than the corresponding nonpolar alkanes. Because alcohols have dipole moments similar to those of the haloalkanes (e.g., CH_3Cl, $\mu = 1.94$ D; CH_3OH, $\mu = 1.70$ D), it is reasonable to expect that the boiling points of similar members of these two classes of compounds would closely correspond. Inspection of Table 8-1 shows this statement to be incorrect. Alcohols are found to have unusually high boiling points, much higher than those of alkanes and haloalkanes of comparable size. How can we explain this observation?

TABLE 8-1

Physical properties of alcohols and selected analogous haloalkanes and alkanes

Compound	IUPAC name	Common name	Melting point (°C)	Boiling point (°C)	Solubility in H_2O at 23°C
CH_3OH	Methanol	Methyl alcohol	−97.8	65.0	Infinite
CH_3Cl	Chloromethane	Methyl chloride	−97.7	−24.2	0.74 g/100 ml
CH_4	Methane		−182.5	−161.7	3.5 ml (gas)/100 ml
CH_3CH_2OH	Ethanol	Ethyl alcohol	−114.7	78.5	Infinite
CH_3CH_2Cl	Chloroethane	Ethyl chloride	−136.4	12.3	0.447 g/100 ml
CH_3CH_3	Ethane		−183.3	−88.6	4.7 ml (gas)/100 ml
$CH_3CH_2CH_2OH$	1-Propanol	Propyl alcohol	−126.5	97.4	Infinite
$CH_3CHOHCH_3$	2-Propanol	Isopropyl alcohol	−89.5	82.4	Infinite
$CH_3CHClCH_3$	2-Chloropropane	Isopropyl chloride	−117.2	35.7	0.305 g/100 ml
$CH_3CH_2CH_3$	Propane		−187.7	−42.1	6.5 ml (gas)/100 ml
$CH_3CH_2CH_2CH_2OH$	1-Butanol	Butyl alcohol	−89.5	117.3	8.0 g/100 ml
$(CH_3)_3COH$	2-Methyl-2-propanol	*tert*-Butyl alcohol	25.5	82.2	Infinite
$CH_3(CH_2)_4OH$	1-Pentanol	Pentyl alcohol	−79	138	2.2 g/100 ml
$(CH_3)_3CCH_2OH$	2,2-Dimethyl-1-propanol	Neopentyl alcohol	53	114	Infinite

The answer lies in hydrogen bonding. In the alcohols, such bonds may be formed between the oxygen atoms of one alcohol molecule and the hydroxy hydrogen atoms of another. Alcohols build up an extensive network of these interactions (Figure 8-2A, on the next page). Although they are much weaker ($DH^\circ \sim 5$ kcal mole^{-1}) than the covalent O–H linkage ($DH^\circ = 104$ kcal mole^{-1}), a fact also indicated by the hydrogen bond's relatively extended length (Figure 8-2A), so many of them form that their combined strength makes volatilization difficult, resulting in correspondingly high boiling points.

The effect is even more pronounced in water, which has two hydrogens available for hydrogen bonding (Figure 8-2B). In pure liquid water, each molecule is bonded on the average to 3.4 neighbors through hydrogen bonds. Although these bonds are continually broken and reformed, the net result is an averaged highly ordered structure. This phenomenon explains why water, with a molecular weight of only 18 has an unusually high boiling point of 100°C. Without this property, water would be a gas at ambient temperatures. Considering the importance of water in all living organisms, you can imagine the effect that the absence of liquid water would have had on evolution on our planet.

FIGURE 8-2

Hydrogen bonding (A) in an alcohol and (B) between an alcohol and water.

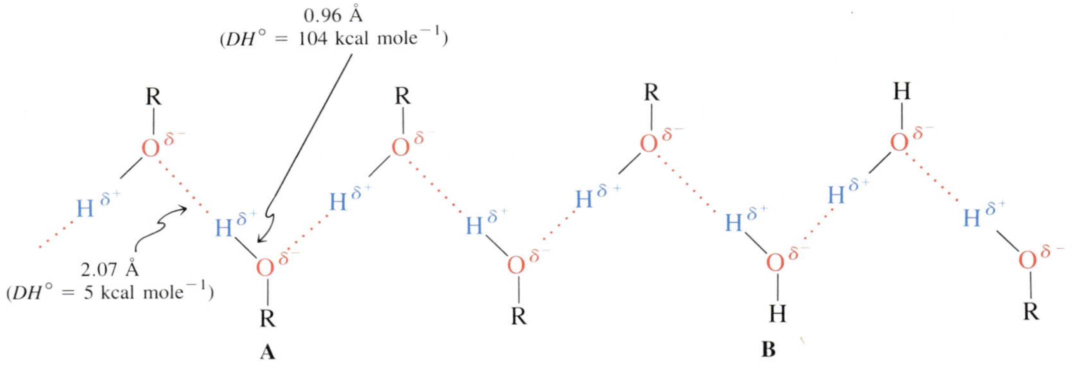

The hydrogen-bonding capacity of water and alcohols is responsible for another property: many alcohols are appreciably water soluble (Table 8-1). Solubility decreases as the size of the alkane chain increases. The nonpolar alkanes themselves are very poorly solvated by water and hence do not dissolve to any great extent in it. Because of this property alkanes are **hydrophobic** (*hydro,* Greek, water; *phobos,* Greek, fear). The same applies to alkyl chains in general. On the other hand, the readily solvated OH group and other polar substituents of this type (e.g., COOH and NH_2) are **hydrophilic** and enhance water solubility. As you can see in Table 8-1, an increase in the alkyl (hydrophobic) part of a molecule decreases its solubility in polar solvents. At the same time, it increases its solubility in nonpolar solvents. The introduction of polar (hydrophilic) groups into an alkane framework causes the opposite effect: solubility in polar solvents and relative insolubility in nonpolar solvents (Figure 8-3). The "waterlike" structure of the lower alcohols, particularly methanol and ethanol, makes them excellent solvents for polar compounds and even salts. It is not surprising, then, that alcohols are popular solvents in the S_N2 reaction (Chapter 6).

FIGURE 8-3

Schematic representation of the hydrophobic and hydrophilic parts of (A) ethanol and (B) 1-pentanol. The relatively increased size of the hydrophobic part in the higher alcohol leads to relatively decreased water solubility (Table 8-1).

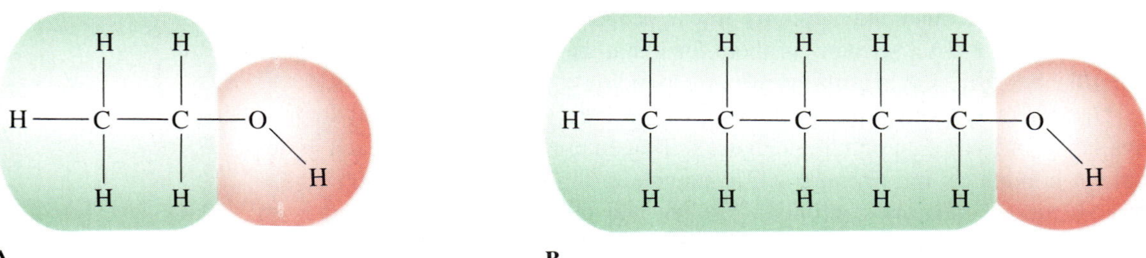

In summary, the oxygen in alcohols (and ethers) is tetrahedral and nearly sp^3-hybridized. The covalent O–H bond is stronger than the C–H bond. Because of the electronegativity of oxygen, alcohols, like water and ethers, have appreciable dipole moments. The hydroxy hydrogen enters into hydrogen bonding with other alcohol molecules. These properties lead to a substantial increase in the boiling points and in the solubilities of alcohols in polar solvents compared with those of the alkanes and haloalkanes.

8-3 Alcohols Are Acidic and Basic

Alcohols are acidic and give alkoxide salts on deprotonation (Section 6-7; Table 6-4). In this section, we shall see how inductive, polarizability, and steric effects influence the pK_a of alcohols. Alcohols also react as bases by utilizing the lone electron pairs on oxygen to form oxonium ions (Section 6-7).

The pK_a of Alcohols

The acidity of the alcohols in water is governed by the equilibrium constant K (Section 6-7):

$$\text{ROH} + \text{H}_2\text{O} \overset{K}{\rightleftharpoons} \text{H}_3\text{O}^+ + \text{RO}^-$$

Assuming that the concentration of water stays constant (55.5 moles l^{-1}) when it is the medium in which the equilibrium is measured, we can derive a new equilibrium constant K_a:

$$K_a = \frac{[\text{H}_3\text{O}^+][\text{RO}^-]}{[\text{ROH}]} \text{ moles } l^{-1} \qquad (\text{in which } K_a = K[\text{H}_2\text{O}])$$

and

$$\text{p}K_a = -\log K_a$$

Table 8-2 lists the pK_a values of several alcohols and related compounds. A comparison of these values with those given in Table 6-4 for mineral and other strong acids shows that alcohols, like water, are fairly weak acids.

TABLE 8-2
pK_a values of some alcohols and related compounds

Compound	pK_a	Compound	pK_a
H_2O	15.7	HOCl	7.53
CH_3OH	15.5	$ClCH_2CH_2OH$	14.3
CH_3CH_2OH	15.9	CF_3CH_2OH	12.4
$(CH_3)_2CHOH$	17.1	$CF_3CH_2CH_2OH$	14.6
$(CH_3)_3COH$	18	$CF_3CH_2CH_2CH_2OH$	15.4
H_2O_2	11.64		

Why are alcohols acidic, whereas alkanes and haloalkanes are not? The answer to this question is found in the relatively strong electronegativity of the oxygen to which the proton is attached. This effect serves to polarize the O–H bond to generate a partially positively charged hydrogen. In the alkoxide, it helps to stabilize the full negative charge.

Strong Bases Are Needed to Convert Alcohols into Their Conjugate Bases

To drive the equilibrium between alcohol and alkoxide to the side of the conjugate base, it is necessary to use a base *stronger* than the alkoxide formed or a base whose conjugate acid is *weaker* than the alcohol. An example is the reaction of sodium amide, $NaNH_2$, with methanol to furnish sodium methoxide and ammonia.

$$\underset{pK_a = 15.5}{CH_3OH} + Na^{+-}NH_2 \overset{K}{\rightleftharpoons} Na^{+-}OCH_3 + \underset{pK_a = 35}{NH_3}$$

This equilibrium lies well to the right ($K \sim 10^{35-15.5} = 10^{19.5}$) because methanol is a much stronger acid than ammonia or, alternatively, because amide is a much stronger base than methoxide. Section 9-1 describes how strong bases are utilized in synthetic applications of alkoxides.

In many synthetic applications of alkoxides, it is sufficient to generate them in equilibrium concentrations. For this purpose, we may add an alkali hydroxide to the alcohol.

$$\underset{pK_a = 15.9}{CH_3CH_2OH} + Na^{+-}OH \overset{K}{\rightleftharpoons} CH_3CH_2O^-Na^+ + \underset{pK_a = 15.7}{H_2O}$$

With this base present, approximately one-half of the alcohol will exist as the alkoxide (assuming equimolar concentrations of starting materials).

Table 8-2 shows quite a variation in the pK_a values of the alcohols listed there. What is the origin of this diversity?

Steric Hindrance, Polarizability, and Inductive Effects Control the pK_a Values of Alcohols

As indicated in the first column of Table 8-2, the pK_a of alcohols increases in procession from methanol to primary, secondary, and tertiary systems.

Order of pK_a Values of Alcohols (in Solution)

$$CH_3OH < \text{primary} < \text{secondary} < \text{tertiary}$$

This ordering has been ascribed to steric hindrance to solvation and hydrogen bonding in the alkoxide. Because solvation and hydrogen bonding stabilize the negative charge on oxygen, interference with these processes leads to an increase in pK_a.*

*In the gas phase, in which no solvation takes place, the acidity scale is reversed. The explanation for this observation appears to be polarizability. The larger branches are more polarizable and better stabilize the adjacent charge.

The second column in Table 8-2 reveals the consequences of placement of electron-withdrawing groups close to the OH group: the acidity increases. The reason is the inductive effect of these substituents (Section 1-3), which stabilizes the negative charge on oxygen in the alkoxide. The inductive effect increases with the number of electronegative groups but decreases with distance from the oxygen.

EXERCISE 8-3

Rank the following alcohols in order of increasing acidity:

OH OH OH Cl OH

Cl Cl

EXERCISE 8-4

Which side of the following equilibrium reaction is favored (assuming equimolar concentrations of starting materials)?

$$(CH_3)_3CO^- + CH_3OH \rightleftharpoons (CH_3)_3COH + CH_3O^-$$

The Lone Electron Pairs on Oxygen Make Alcohols Basic

Alcohols can be not only acidic, but also, because of the presence of the two lone electron pairs, basic. Alcohols as bases were discussed in Section 6-7, where it was shown that very strong acids effected protonation of the OH group. The alcohols are only very weakly basic, as indicated by the low pK_a values (strong acidity) of their conjugate acids, the oxonium ions (Table 8-3). Molecules that may be both acids and bases are called **amphoteric** (*ampho*, Greek, both).

TABLE 8-3
pK_a values of some protonated alcohols

Compound	pK_a
$CH_3\overset{+}{O}H_2$	−2.2
$CH_3CH_2\overset{+}{O}H_2$	−2.4
$(CH_3)_2CH\overset{+}{O}H_2$	−3.2
$(CH_3)_3C\overset{+}{O}H_2$	−3.8

The amphoteric nature of the hydroxy functional group characterizes the chemical reactivity of alcohols. In strong acids, they exist as oxonium ions, in neutral media as alcohols, and in strong bases as alkoxides. Amphoteric molecules have more than one pK_a value. This characteristic is reminiscent of some mineral acids, such as sulfuric and phosphoric acid, which may be deprotonated more than once. Normally, however, the pK_a of an alcohol is given in reference to its equilibrium with the alkoxide, and not with the oxonium ion.

$$R\overset{+}{O}H_2 \underset{\text{Mild base}}{\overset{\text{Strong acid}}{\rightleftharpoons}} ROH \underset{\text{Mild acid}}{\overset{\text{Strong base}}{\rightleftharpoons}} RO^-$$

Oxonium ion **Alcohol** **Alkoxide**

In summary, alcohols are amphoteric. They are weak acids by virtue of the electronegativity of the oxygen. In solution, the steric bulk of branching induces inhibition of solvation, leading to increased pK_a values, overriding the opposing effect of polarizability. Electron-withdrawing substituents close to the functional group lower the pK_a values. Alcohols are also weakly basic and can be protonated (or attacked by Lewis acids, such as carbocations) to furnish oxonium ions.

BOX 8-1

Oxonium Ions as Lewis Acid-Base Adducts

Oxonium ions are also formed when carbocations are trapped by water or alcohols (Section 7-1). In this process, the electron-deficient cation acts as an electrophile, the oxygen in the trapping agent as a nucleophile. The similarity between the action of the carbocation and that of a proton is taken into account in the general acid-base theory of G. N. Lewis. He defined acids as all species that can function as electron acceptors (i.e., all electrophiles) and bases as all molecules that act as electron donors (i.e., all nucleophiles). In accord with this definition, a carbocation is a Lewis acid and an alcohol is a Lewis base.

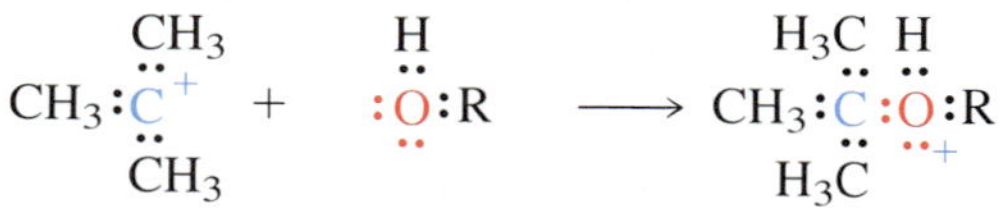

$$\underset{\textbf{Lewis acid}}{(CH_3)_3C^+} + \underset{\textbf{Lewis base}}{H\ddot{O}R} \longrightarrow (CH_3)_3C{-}\overset{+}{\ddot{O}}(H)R$$

Other examples of Lewis acid-base interactions were discussed earlier in connection with the concept of dative (coordinate covalent) bonds (Section 1-4).

8-4 The Synthesis of Alcohols

Let us now turn to the preparation of alcohols. A distinction will be made between special methods of industrial importance and approaches that allow the general introduction of a hydroxy function into organic molecules. In this section, we shall see that alcohols are made by the reduction of carbon monoxide, the hydrolysis of haloalkanes and related alkyl derivatives, the hydrogenation and hydride reduction of carbonyl compounds, and the hydride opening of strained cyclic ethers.

Several other procedures, which will be discussed later, convert a given functional group into that of an alcohol. They include:

1 Hydration of alkenes (Section 12-3).

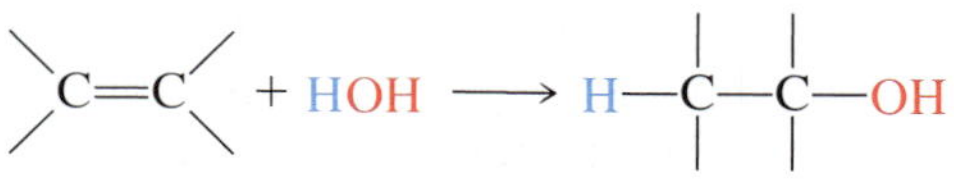

$$\gt C{=}C\lt + HOH \longrightarrow H{-}\overset{|}{\underset{|}{C}}{-}\overset{|}{\underset{|}{C}}{-}OH$$

2 Reaction of organometallic reagents with carbonyl compounds (Sections 8-6 and 13-4).

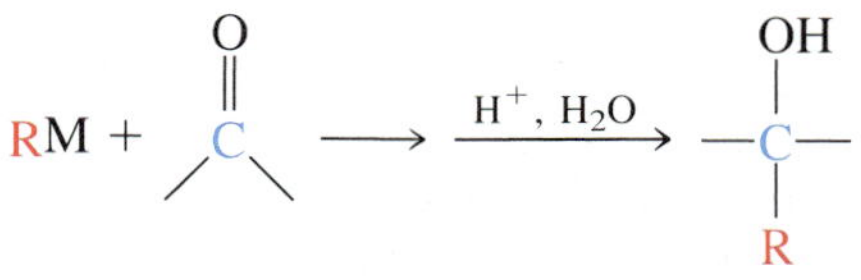

$$RM + \gt C{=}O \longrightarrow \xrightarrow{H^+,\ H_2O} {-}\overset{OH}{\underset{R}{C}}{-}$$

3 Aldol condensation (Section 15-7).

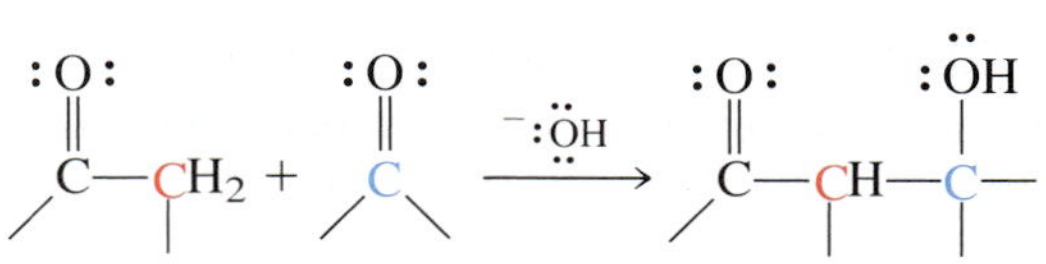

$$\gt C(=O){-}\overset{|}{C}H_2 + \gt C{=}O \xrightarrow{^-:\ddot{O}H} \gt C(=O){-}\overset{|}{C}H{-}\overset{OH}{\underset{|}{C}}{-}$$

Carbon Monoxide and Ethene: Convenient Industrial Sources of Alcohols

Methanol is made on a large scale (more than one billion gallons in the United States in 1984) from pressurized synthesis gas (a mixture of CO and H_2) by catalytic hydrogenation (reduction) of CO, the catalyst consisting of copper, zinc oxide, and chromium oxide.

$$\text{CO} + 2\,\text{H}_2 \xrightarrow{\text{Cu-ZnO-Cr}_2\text{O}_3,\ 250^\circ\text{C},\ 50\text{–}100\ \text{atm}} \text{CH}_3\text{OH}$$

When rhodium or ruthenium catalysts are heated in the presence of pressurized synthesis gas, 1,2-ethanediol (ethylene glycol) is formed selectively by reductive coupling of CO. 1,2-Ethanediol is an important industrial chemical, and its main method of preparation is the hydrolysis of oxacyclopropane (oxirane, ethylene oxide, Section 9-8).

$$2\,\text{CO} + 3\,\text{H}_2 \xrightarrow{\text{Rh or Ru, pressure, heat}} \underset{\text{OH}}{\text{CH}_2}\text{—}\underset{\text{OH}}{\text{CH}_2}$$

1,2-Ethanediol
(Ethylene glycol)

Reactions of synthesis gas that result in the selective formation of one product are the focus of much current research because of the direct availability of mixtures of CO and H_2 from the gasification of coal (for the approximate composition of coal, see Section 25-4) in the presence of water.

$$\text{Coal} \xrightarrow{\text{Air, H}_2\text{O, }\Delta} x\,\text{CO} + y\,\text{H}_2$$

There are abundant coal supplies in the United States and other parts of the world, which could be used as a source of liquid fuels, as well as industrial raw materials and feed stocks.

Another catalytic reaction of synthesis gas that furnishes alcohols, but only as by-products, is the cobalt- or iron-mediated formation of hydrocarbons usable as fuels and oils. This reaction was discovered at about the turn of the century and developed in Germany beginning in the 1920s. Its application enabled that country to supply its energy (particularly gasoline) needs from coal during the Second World War, when its supply of petroleum was virtually shut off. The process is known as the **Fischer-Tropsch reaction:**

$$n\,\text{CO} + (2n+1)\,\text{H}_2 \xrightarrow{\text{Co or Fe, pressure, }200^\circ\text{–}350^\circ\text{C}} \text{C}_n\text{H}_{2n+2} + n\,\text{H}_2\text{O}$$

At the height of production, in 1943, more than 500,000 tons of hydrocarbon and other products (gasoline, diesel fuel, oils, waxes, and detergents) were made in Germany by this process. Currently, South Africa is the only country that satisfies a substantial amount of its fuel needs by use of the Fischer-Tropsch reaction.

Ethanol is prepared in large quantities by fermentation of sugars (Chapter 23). Industrially it is made by the phosphoric acid-catalyzed hydration of ethene (Sections 9-8 and 12-9). The United States produced about 800 million pounds in this way in 1984.

$$CH_2{=}CH_2 + HOH \xrightarrow{H_3PO_4,\ 300^{\circ}C} \underset{\displaystyle H}{\underset{|}{C}H_2}{-}\underset{\displaystyle OH}{\underset{|}{C}H_2}$$

Other simple alcohols are also available by this route from the corresponding alkenes (see Sections 12-3 and 12-4).

Structurally more complex alcohols are often synthesized by converting other functional groups into the OH substituent. An example is nucleophilic substitution of a haloalkane.

Alcohols by Nucleophilic Substitution

A method that furnishes alcohols by conversion of other functional units—namely, the hydrolysis of alkyl derivatives containing suitable leaving groups by S_N2 and S_N1 mechanisms—was described in Chapters 6 and 7. These methods are not generally useful because, frequently, the required haloalkanes can be made only from the corresponding alcohols. They also suffer from the usual drawbacks of nucleophilic substitutions: bimolecular elimination can be a major side reaction with hindered primary systems and with secondary derivatives. With tertiary halides, carbocations are formed that may undergo E1 reactions or rearrange (see Section 9-2).

Alcohols by Nucleophilic Substitution

$$CH_3Br + HO^- \xrightarrow[S_N2]{H_2O,\ 100^{\circ}C} CH_3OH + Br^-$$

$$(CH_3)_3CCl \xrightarrow[S_N1]{HOH,\ \text{propanone (acetone)}} (CH_3)_3COH + Cl^- + H^+$$

A way around the problem of elimination is to use a less-basic oxygen nucleophile, such as ethanoate (acetate). As mentioned in Section 7-5, ethanoate is a weaker base than hydroxide and hence leads to substitution products in good yield with less contamination by elimination products. Once this step is accomplished (Step 1), the resulting alkyl ethanoate can be cleaved by aqueous hydroxide to give the desired alcohol (Step 2).

Alcohols from Haloalkanes Through Ethanoate (Acetate) Formation

STEP 1: Ethanoate (acetate) formation

$$CH_3CH_2\overset{\displaystyle CH_3}{\overset{|}{C}}HCH_2CH_2Br + CH_3\overset{\displaystyle O}{\overset{\|}{C}}O^-Na^+ \xrightarrow{DMF,\ 80^{\circ}C} \underset{95\%}{CH_3CH_2\overset{\displaystyle CH_3}{\overset{|}{C}}HCH_2CH_2O\overset{\displaystyle O}{\overset{\|}{C}}CH_3} + Na^+Br^-$$

1-Bromo-3-methylpentane **3-Methylpentyl ethanoate**

STEP 2: Cleavage

$$CH_3CH_2\overset{\displaystyle CH_3}{\overset{|}{C}}HCH_2CH_2O{\ddagger}\overset{\displaystyle O}{\overset{\|}{C}}CH_3 + Na^{+}\,{}^{-}OH \xrightarrow{H_2O} \underset{85\%}{CH_3CH_2\overset{\displaystyle CH_3}{\overset{|}{C}}HCH_2CH_2OH} + \left(HO\overset{\displaystyle O}{\overset{\|}{C}}CH_3 \xrightarrow[-\,H_2O]{HO^-} {}^{-}O\overset{\displaystyle O}{\overset{\|}{C}}CH_3\right)$$

3-Methyl-1-pentanol

The second reaction proceeds by a mechanism other than displacement, an *ester hydrolysis,* which occurs by cleavage of the carbonyl carbon–oxygen bond, and not by the S_N2 reaction. This process will be discussed in Section 18-4.

EXERCISE 8-5

Treatment of (*R*)-2-iodobutane sequentially with sodium ethanoate (acetate) followed by hydroxide gives an optically active alcohol. What is its absolute configuration?

Alcohols by Catalytic Hydrogenation of Aldehydes and Ketones

The carbon–oxygen double bonds of aldehydes and ketones (Section 2-1) react with hydrogen, H_2, by addition to furnish alcohols.

$$\underset{\text{Aldehyde (R'=H) or ketone (R' = alkyl)}}{RC(=O)R'} \xrightarrow{H_2,\ \text{catalyst}} \underset{\text{Alcohol}}{R-\overset{OH}{\underset{H}{C}}-R'}$$

The reaction requires a catalyst and, in many cases, the application of hydrogen pressure to proceed. Most of the catalysts used are **heterogeneous catalysts** (*heteros,* Greek, other; *genos,* Greek, kind). Most of these catalysts are composed of finely divided metal, such as palladium, platinum, or nickel, frequently deposited on some type of support, such as carbon, to maximize surface area. They are insoluble in the reaction solvent and used in suspension.

$$\underset{\text{3-Methylbutanal}}{(CH_3)_2CHCH_2CHO} \xrightarrow{H_2,\ \text{Pd-C}} \underset{\text{3-Methyl-1-butanol}}{(CH_3)_2CHCH_2CH_2OH}$$

$$\underset{\text{Cyclohexanone}}{C_6H_{10}O} \xrightarrow{H_2,\ \text{Pt}} \underset{\text{Cyclohexanol}}{C_6H_{11}OH}$$

The mechanism of catalytic hydrogenation (for alkenes) will be discussed in Section 12-2. It proceeds through hydrogen atoms and substrate, both of which are bound to the catalyst's surface. The reaction does not proceed without the catalyst because the energy of the H–H bond (DH° = 104 kcal mole^{-1}) is too high to allow the purely thermal formation of hydrogen atoms, which could then add sequentially to the double bond. Instead, the catalyst permits a different pathway requiring less activation energy (see also Figure 3-5).

Alcohols by Reduction of the Carbonyl Group with Hydrides

The electrons in the carbonyl group are not distributed evenly between the two component atoms because oxygen is more electronegative than carbon. Hence, the carbon of a carbonyl group is electrophilic, the oxygen nucleophilic. Moreover, the electrophilic nature of this carbon atom can be indicated by formulating a dipolar resonance structure in which the carbon assumes the role of a carbocation, the oxygen that of an alkoxide:

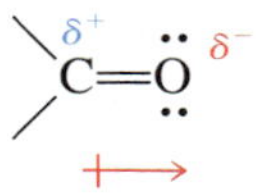

Polarization in a carbonyl group

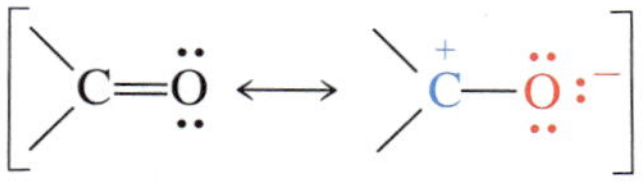

Resonance structures of a carbonyl group

The electrophilic nature of the carbonyl carbon can be used in reductions employing hydride reagents that deliver the equivalent of H^- to it. Two such reagents are sodium borohydride, $NaBH_4$, and lithium aluminum hydride, $LiAlH_4$.

General Hydride Reductions of Aldehydes and Ketones

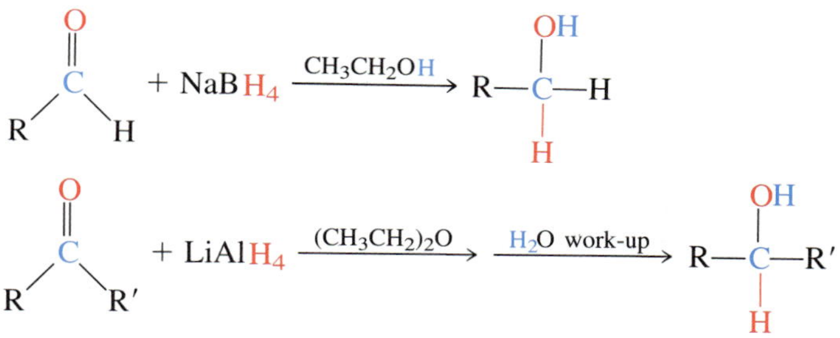

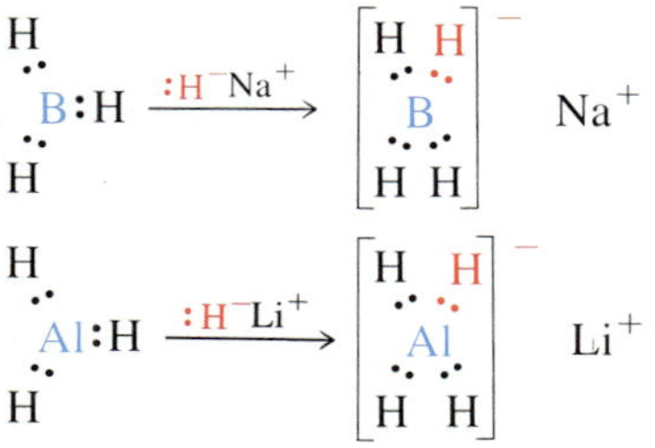

$$RCHO + NaBH_4 \xrightarrow{CH_3CH_2OH} R\text{—}CH(OH)\text{—}H$$

$$RCOR' + LiAlH_4 \xrightarrow{(CH_3CH_2)_2O} \xrightarrow{H_2O\ \text{work-up}} R\text{—}CH(OH)\text{—}R'$$

$$H_3B + {:}H^- Na^+ \longrightarrow [BH_4]^- \quad Na^+$$

$$H_3Al + {:}H^- Li^+ \longrightarrow [AlH_4]^- \quad Li^+$$

These compounds may be thought of as being formed by the reaction of an alkali metal hydride with borane, BH_3, or alane, AlH_3, respectively. The addition of a hydride, H^-, to the sp^2-hybridized electron-deficient BH_3 (see Figure 1-17) gives the boron a complete electron octet. The BH_4^- ion is tetrahedral, with four equivalent hydrogens and the same electronic structure as methane or the ammonium ion, NH_4^+. The same is true for AlH_4^-.

The chemistry of these hydride reagents is dominated by the hydridic (H^-) character of the hydrogen atoms. Sodium borohydride is less reactive than lithium aluminum hydride, but both are attacked by water or alcohols to give hydrogen.

$$NaBH_4 + 4\ HOH \xrightarrow{\text{Relatively slow}} NaOH + B(OH)_3 + 4\ H\text{—}H$$

$$LiAlH_4 + 4\ CH_3OH \xrightarrow{\text{Fast}} LiAl(OCH_3)_4 + 4\ H\text{—}H$$

In this transformation, the hydride ion acts as a base. Because $NaBH_4$ is less reactive, it can be used in either water or in alcohol as a solvent. Lithium aluminum hydride is too reactive in protic media, and ethoxyethane (diethyl ether) or another ether solvent must be used in this case. The work-up of the reaction mixture is done with aqueous acid to hydrolyze excess reagent and any boron and aluminum salts formed.

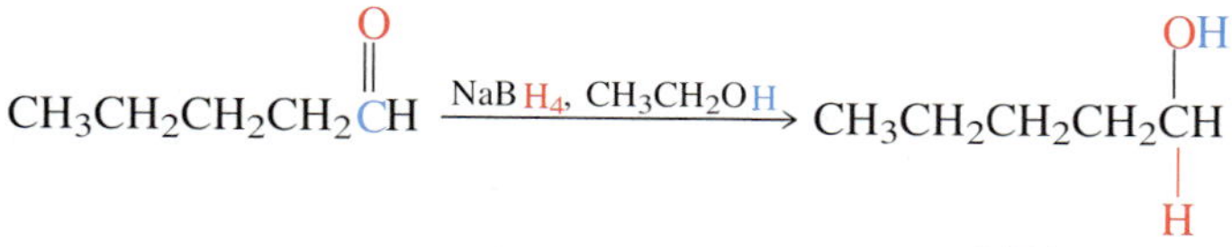

$$CH_3CH_2CH_2CH_2CHO \xrightarrow{NaBH_4,\ CH_3CH_2OH} CH_3CH_2CH_2CH_2CH_2OH$$

Pentanal → **1-Pentanol** 85%

Cyclobutanone $\xrightarrow{1.\ LiAlH_4,\ (CH_3CH_2)_2O \quad 2.\ H^+,\ H_2O}$ Cyclobutanol 90%

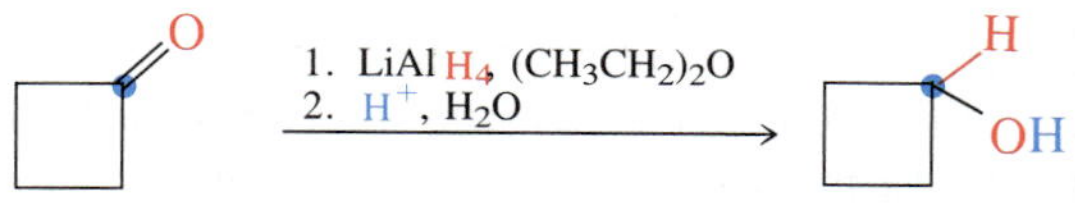

Cyclobutanone → **Cyclobutanol**

EXERCISE 8-6

Hydride reductions are often stereospecific, with the delivery of hydrogen from the less-hindered side of the substrate molecule. Predict the stereochemical outcome of the treatment of compound A, shown at the right, with $NaBH_4$.

A

As already mentioned, the mechanism of hydride reduction consists of nucleophilic addition to the electrophilic carbonyl carbon. If borohydride is used in a protic solvent (e.g., ethanol), the oxygen is thought to be protonated simultaneously with hydride delivery. The ethoxide combines with the boron fragment to form ethoxyborohydride.

Mechanism of $NaBH_4$ Reduction:

$$Na^+ \, {}^-H_3B{-}H \;\; {>}C{=}O \;\; H{-}\ddot{O}CH_2CH_3 \longrightarrow H{-}\overset{|}{\underset{|}{C}}{-}OH \; + \; Na^+ \, {}^-H_3BOCH_2CH_3$$

Ethanol solvent **Product alcohol** **Sodium ethoxyborohydride**

The resulting ethoxyborohydride is capable of attacking three more carbonyl substrates until all hydride atoms have been used up. The boron reagent is ultimately converted into tetraethoxyborate, $^-B(OCH_2CH_3)_4$. If lithium aluminum hydride is used, the initial product is an alkoxyaluminate, which is decomposed on aqueous work-up.

Mechanism of $LiAlH_4$ Reduction:

$${>}C{=}O + Li^+ \, {}^-AlH_4 \longrightarrow H{-}\overset{|}{\underset{|}{C}}{-}OAl^-H_3Li^+ \xrightarrow{\text{More}} \xrightarrow{\text{carbonyl}} \xrightarrow{\text{substrate}}$$

$$(H{-}\overset{|}{\underset{|}{C}}{-}O)_4Al^-Li^+ \xrightarrow{\text{HOH work-up}} 4\; H{-}\overset{|}{\underset{|}{C}}{-}OH + Al(OH)_3 + LiOH$$

Lithium tetraalkoxyaluminate

Lithium aluminum hydride is so powerful a reducing agent that it also reacts with carboxylic acids and esters to yield alcohols (see Sections 17-10 and 18-4):

$$C_6H_{11}CH_2\overset{O}{\overset{\|}{C}}OH \xrightarrow[2.\ H^+,\ H_2O]{1.\ LiAlH_4,\ (CH_3CH_2)_2O} C_6H_{11}CH_2CH_2OH$$

97%

2-Cyclohexylethanoic acid **2-Cyclohexylethanol**

$$CH_3CH_2\overset{O}{\overset{\|}{C}}OCH_3 \xrightarrow[2.\ H^+,\ H_2O]{1.\ LiAlH_4,\ (CH_3CH_2)_2O} CH_3CH_2CH_2OH + CH_3OH$$

92%

Methyl propanoate **1-Propanol**

It will also reduce haloalkanes to alkanes by standard S_N2 displacement.

$$\underset{\textbf{1-Bromononane}}{CH_3(CH_2)_7CH_2—Br} \xrightarrow[-\ LiBr]{LiAlH_4,\ (CH_3CH_2)_2O} \underset{\textbf{Nonane}}{\underset{96\%}{CH_3(CH_2)_7CH_2—H}}$$

Use of Ring Strain in Synthesis: Hydride Reduction of Strained Ethers to Alcohols

The nucleophilic power of lithium aluminum hydride also manifests itself in its ability to attack three-membered ring ethers, such as oxacyclopropane (oxirane, ethylene oxide):

$$\text{Oxacyclopropane (Oxirane, ethylene oxide)} \xrightarrow[2.\ H^+,\ H_2O]{1.\ LiAlH_4,\ (CH_3CH_2)_2O} \underset{\textbf{Ethanol}}{H—CH_2—CH_2—OH}$$

The reaction proceeds by nucleophilic attack of the hydride, with the ether oxygen functioning as an intramolecular leaving group.

Mechanism of Oxacyclopropane Opening by $LiAlH_4$:

$$H—Al^-H_3 + H_2C\overset{O}{—}CH_2 \longrightarrow H—CH_2—CH_2—\ddot{\underset{..}{O}}:^- \xrightarrow{HOH} HCH_2CH_2\ddot{\underset{..}{O}}H$$

This transformation is an unusual S_N2 reaction for two reasons: (1) alkoxides are usually very bad leaving groups; and (2) the leaving group does not actually "leave," it stays bound to the molecule. The driving force is the release of strain (about as much as in cyclopropane, 27 kcal mole^{-1}, see Section 4-2) on opening the ring. As in other S_N2 processes, the reduction of strained ethers occurs with inversion at the reacting carbon and is retarded by branching. For example, in unsymmetrical systems, the hydride attacks the less-substituted side. This aspect of ether chemistry is discussed further in Section 9-6.

$$\text{H}_2\text{C}\overset{O}{—}\text{CHR} \xrightarrow[2.\ H^+,\ H_2O]{1.\ LiAlH_4,\ (CH_3CH_2)_2O} \underset{H}{CH_2}\overset{OH}{CHR}$$

EXERCISE 8-7

Give four possible precursors of the alcohol 3-hexanol and the conditions by which you would convert them into product.

In summary, alcohols can be synthesized in several ways. Certain processes that include the use of transition metal catalysts allow the conversion of carbon monoxide into some of the smaller alcohols (methanol, 1,2-ethanediol) of industrial importance. More general approaches to the synthesis of alcohols are the transformation of haloalkanes (and other derivatives) into alcohols by S_N2 chemistry, hydrogenation and hydride reduction of carbonyl compounds, and the ring-opening of strained ethers.

8-5 Organometallic Reagents Containing Lithium and Magnesium: Sources of Nucleophilic Carbon

We have seen that the reduction of carbonyl compounds with hydride reagents is a useful way of synthesizing alcohols. This approach would be even more powerful if, instead of hydride, we could use a source of *nucleophilic carbon*. Attack on a carbonyl group would furnish an alcohol with the simultaneous formation of a carbon–carbon bond. To achieve such transformations, we need to find a way of making carbon-based nucleophiles, R^-. We shall see in this section that this goal can be reached through the action of metals, particularly lithium and magnesium, on haloalkanes to generate new compounds, called **organometallic reagents,** in which a carbon atom of an organic group is bound to a metal. These species are strong bases, as well as nucleophiles, and are extremely useful in organic synthesis.

Alkyllithium and Alkylmagnesium Reagents Are Prepared from Haloalkanes

Organometallic compounds of lithium and magnesium are most conveniently prepared by direct reaction of a haloalkane with the metal suspended in ethoxyethane (diethyl ether) or oxacyclopentane (tetrahydrofuran, THF). The reactivity of the haloalkane increases in the order Cl < Br < I; fluorides are not normally used as starting materials in these reactions.

Alkyllithium Synthesis

$$CH_3Br + 2\,Li \xrightarrow{(CH_3CH_2)_2O,\ 0°-10°C} CH_3Li + LiBr$$

Methyllithium

$$CH_3CH_2CH_2CH_2Br + 2\,Li \xrightarrow{THF,\ 0°-10°C} CH_3CH_2CH_2CH_2Li + LiBr$$

Butyllithium

THF = [five-membered ring with O] = oxacyclopentane (tetrahydrofuran)

Alkylmagnesium (Grignard) Synthesis

$$(CH_3)_2CHI + Mg \xrightarrow{THF,\ 20°C} (CH_3)_2CHMgI$$

1-Methylethylmagnesium iodide

Organomagnesium compounds, RMgX, are also called **Grignard reagents,** named after their discoverer, V. Grignard.* Their formation is possible starting with primary, secondary, and tertiary haloalkanes (and, as we shall see in later chapters, with haloalkenes and halobenzenes).

A typical experimental apparatus for the preparation of these compounds consists of a three-neck flask, a stirrer, an addition funnel, and a condenser, flushed with dry N_2, as shown in Figure 8-4. As the reaction proceeds, the suspended metal slowly dissolves. If lithium is being used, the halogen precipitates as lithium halide and can be removed by filtration. Alkyllithium compounds and Grignard reagents are rarely isolated; they are formed in solution and react directly with the substrate. Several alkyllithium reagents such as methyllithium, butyllithium, and 1,1-dimethylethyllithium (*tert*-butyllithium) are commercially available as standard solutions. The reagents are sensitive to air and moisture; thus, they must be prepared and handled under rigorously air-free and dry conditions. The structures of alkyllithium reagents differ from system to system and are not always known with certainty because these species are capable of aggregating into clusters of varying size.

FIGURE 8-4

Apparatus for preparing organometallic reagents by the addition of a haloalkane to suspended lithium or magnesium in an ether solvent.

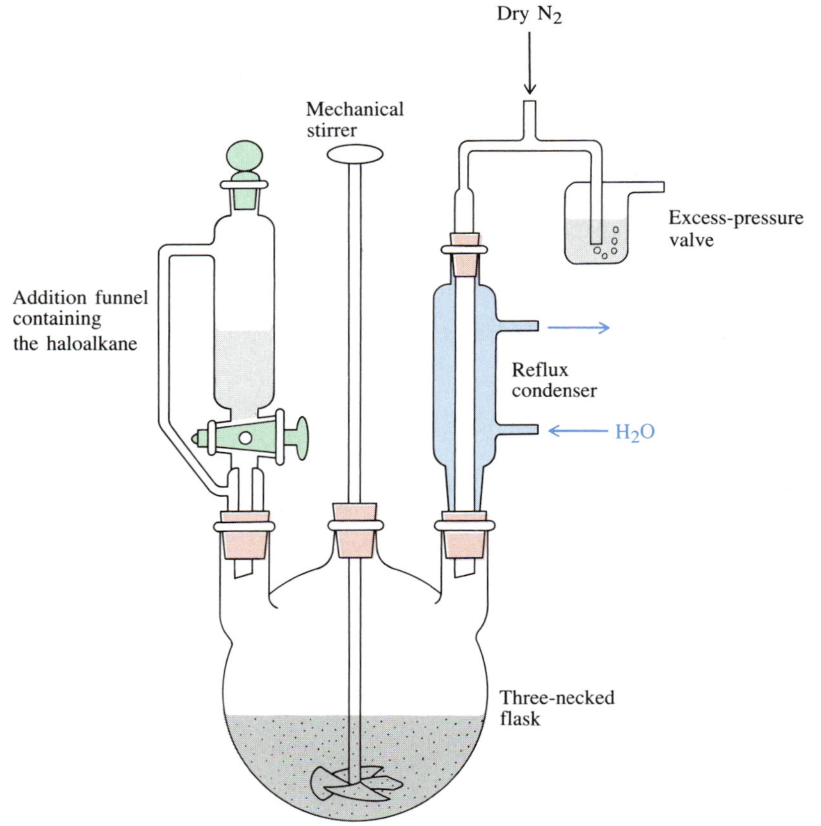

*Professor Victor Grignard, 1871–1935, University of Lyons, France, Nobel Prize 1912.

The composition of Grignard organometallics is somewhat different from that of alkyllithium reagents. The haloalkane reacts with the magnesium surface by essentially "extracting" a metal atom and inserting it into the R–X bond. Because magnesium is divalent, the resulting RMgX product satisfies the valence requirements of the metal. However, to gain an electron octet, Grignard reagents are coordinated to two solvent ether molecules. This interaction may be viewed as that typical of a dative (coordinate covalent) bond (Section 1-4) or of a Lewis acid-base reaction (Section 8-3). When the structures of Grignard reagents are written, coordination to the ether solvent is not shown. However, its importance is underlined by the necessity to have an ether present in their preparation; the reaction is very difficult in its absence.

Grignard Reagents Are Coordinated to Solvent

$$R{-}X + Mg \xrightarrow{(CH_3CH_2)_2O} \begin{array}{c} CH_3CH_2\ddot{\underset{\cdot\cdot}{O}}CH_2CH_3 \\ \downarrow \\ R{-}Mg{-}X \\ \uparrow \\ CH_3CH_2\ddot{\underset{\cdot\cdot}{O}}CH_2CH_3 \end{array}$$

Polarization in the Alkylmetal Bond

Alkyllithium and magnesium reagents have strongly polarized carbon–metal bonds, the strongly electropositive metal (Table 1-4) assuming the role of the positive end of the dipole. This polarization is sometimes described as "percent ionic bond character." The carbon–lithium bond, for example, has about 40% ionic character and the carbon–magnesium bond, 35%. Such systems react chemically as if they contained a negatively charged carbon, a **carbanion.** This characteristic can be symbolized in resonance terms by a structure that places the full negative charge on the carbon atom attached to the metal.

The Carbon–Metal Bond in Alkyllithium and Alkylmagnesium Compounds

$$\left[-\overset{|}{\underset{|}{C}}{}^{\delta^-}{-}M^{\delta^+} \longleftrightarrow -\overset{|}{\underset{|}{C}}{:}^- \ M^+ \right] \quad M = \text{metal}$$

Polarized   **Charge separated**

The preparation of alkylmetals from haloalkanes illustrates an important principle in synthetic organic chemistry: **reverse polarization.** In a haloalkane, the presence of the electronegative halogen turns the carbon into an electrophilic center. On treatment with a metal, the $C^{\delta^+}{-}X^{\delta^-}$ unit is converted into $C^{\delta^-}{-}M^{\delta^+}$, in which the direction of the original dipole moment is reversed. Therefore, metallation turns an electrophilic carbon into a nucleophilic (and basic) center with the corresponding characteristic reactivity.

The Alkyl Group in Alkylmetals Reacts as a Base

Organometallic reagents *must* be prepared in the absence of moisture because they react (in some cases, violently) with water; hydrolysis occurs, with the

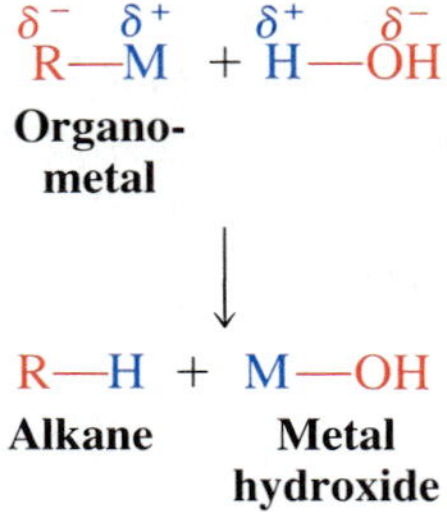

formation of the metal hydroxide and an alkane. The outcome of this transformation is predictable on purely electrostatic grounds: the negatively polarized alkyl substituent acts as a base and is protonated, whereas the positively polarized metal combines with the hydroxide group.

$$CH_3Li + HOH \longrightarrow \underset{100\%}{HCH_3} + LiOH$$

$$\underset{\textbf{3-Methylpentylmagnesium bromide}}{CH_3CH_2\overset{\overset{CH_3}{|}}{C}HCH_2CH_2MgBr} + HOH \longrightarrow \underset{\textbf{3-Methylpentane}}{\underset{100\%}{CH_3CH_2\overset{\overset{CH_3}{|}}{C}HCH_2CH_2H}} + BrMgOH$$

The method is not generally useful, but it affords a means by which a halogen substituent may be removed from an alkane chain. The reaction of a haloalkane with lithium aluminum hydride, discussed earlier, is a more direct way of achieving the same goal. Another possibility is the introduction of hydrogen isotopes, such as deuterium, into a molecule by the hydrolysis of the organometallic compound with labeled water.

$$(CH_3)_3CCl \xrightarrow[2.\ D_2O]{1.\ Mg} (CH_3)_3CD$$

EXERCISE 8-8

Show how you would prepare 1-deuteriocyclohexane from cyclohexane.

The Alkyl Group in Alkylmetals Reacts as a Nucleophile

Organometallic compounds containing metals other than lithium or magnesium can be prepared by nucleophilic reaction of alkyllithium or Grignard reagents with metal halides, a transformation called **transmetallation** because it converts one organometallic into another.

General Transmetallation

$$R—M + M'X \longrightarrow R—M' + MX$$

Examples:

$$3\ CH_3MgCl + SiCl_4 \longrightarrow \underset{\textbf{Chlorotrimethylsilane}}{(CH_3)_3SiCl} + 3\ MgCl_2$$

$$2\ CH_3\overset{\overset{CH_3}{|}}{C}HMgCl + CdCl_2 \longrightarrow \underset{\textbf{Di(1-methylethyl)cadmium}}{\left(CH_3\overset{\overset{CH_3}{|}}{C}H\right)_2Cd} + 2\ MgCl_2$$

$$3\ CH_3Li + AlCl_3 \longrightarrow \underset{\textbf{Trimethylaluminum}}{(CH_3)_3Al} + 3\ LiCl$$

$$CH_3CH_2CH_2CH_2Li + CuI \longrightarrow \underset{\textbf{Butylcopper}}{CH_3CH_2CH_2CH_2Cu} + LiI$$

$$CH_3CH_2CH_2CH_2Cu + CH_3CH_2CH_2CH_2Li \longrightarrow \underset{\substack{\textbf{Lithium dibutylcuprate} \\ \textbf{(A cuprate reagent)}}}{(CH_3CH_2CH_2CH_2)_2CuLi}$$

Several of these reagents are useful in organic synthesis. Examples of their applications are given in subsequent chapters in dealing with the appropriate functional-group transformations.

An example to be discussed here is carbon–carbon bond formation between cuprates, R_2CuLi, and (mainly) primary haloalkanes, $R'X$, to give coupled alkanes R–R′. Cuprates can be prepared directly from two equivalents of an alkyllithium reagent and one equivalent of cuprous iodide, CuI.

Direct General Cuprate Synthesis

$$2\ RLi + CuI \longrightarrow \underset{\textbf{A cuprate}}{R_2CuLi} + LiI$$

The detailed structure of these species is unknown and the reagents are usually generated in situ (*in situs,* Latin, on site), to be used immediately.

Alkanes from Cuprates and Haloalkanes

$$R_2CuLi + 2\ R'X \longrightarrow 2\ R{-}R' + LiX + \text{copper salts}$$

Example:

$$\underset{\substack{\textbf{Lithium} \\ \textbf{dimethylcuprate}}}{(CH_3)_2CuLi} + \underset{\textbf{1-Iododecane}}{CH_3(CH_2)_8CH_2I} \xrightarrow{(CH_3CH_2)_2O,\ 0°C} \underset{\substack{90\% \\ \textbf{Undecane}}}{CH_3(CH_2)_8CH_2CH_3}$$

Note that the inorganic products have been omitted from this equation. These products are removed by neutral or slightly acidic aqueous work-up and are normally of no interest to the organic chemist.

EXERCISE 8-9

Show how you would prepare 2,2,5,5-tetramethylhexane from 2,2-dimethylpropane.

As pointed out in the introduction to this section, among the most attractive applications of organometallic reagents of lithium and magnesium are those in which nucleophilic attack by the alkyl group results in the formation of alcohols. This topic is the subject of the next section.

In summary, haloalkanes can be converted into organometallic compounds of lithium or magnesium (Grignard reagents) by reaction with the respective metals in ether solvents. In these compounds, the alkyl group is negatively polarized, opposite to the polarization found in the haloalkane. Although the alkyl metal bond is to a large extent covalent, the carbon attached to the metal reacts as a carbanion. Thus, it can be protonated to give the alkane, and it can attack other metal halides and haloalkanes.

8-6 Organometallic Reagents Containing Magnesium and Lithium in the Synthesis of Alcohols

The nucleophilic property of the alkyl group in the organometallic reagents of lithium and magnesium can be used in the synthesis of alcohols by addition to carbonyl compounds and nucleophilic ring-opening of oxacyclopropanes.

Alcohols by the Addition of Organometallic Reagents to the Carbonyl Group of Aldehydes and Ketones

Organometallic reagents add to aldehydes and ketones in a manner very similar to that of hydrides to give alcohols. The difference is that a new carbon–carbon bond is formed in the process.

General Alcohol Syntheses from Aldehydes, Ketones, and Organometallics

$$\ddot{O}{:}^{\delta-}{=}C^{\delta+} \;+\; M^{\delta+}{-}R^{\delta-} \longrightarrow {-}\overset{\ddot{O}:^{-}M^{+}}{\underset{R}{C}}{-} \xrightarrow{H^+,\ H_2O} {-}\overset{OH}{\underset{R}{C}}{-}$$

The use of electron-pushing helps to understand the reaction: the alkyl group in the organometallic compound attacks the carbonyl carbon as a nucleophile with its electron pair, "pushing" two electrons from the double bond onto the oxygen and thus producing a metal alkoxide. The addition of a dilute aqueous acid furnishes the alcohol by hydrolyzing the metal–oxygen bond. As mentioned earlier, the work-up step is often omitted in the writing of synthetic sequences.

The reaction of organometallic compounds with *methanal* (formaldehyde) results in *primary alcohols:*

$$CH_3CH_2CH_2CH_2MgBr + H_2C{=}O \xrightarrow{(CH_3CH_2)_2O} \xrightarrow{H^+,\ H_2O} CH_3CH_2CH_2CH_2\overset{H}{\underset{H}{C}}OH$$

Butylmagnesium bromide — **Methanal (Formaldehyde)** — 93% **1-Pentanol**

On the other hand, *aldehydes* other than methanal are converted into *secondary alcohols:*

$$CH_3\overset{MgBr}{C}HCH_3 + CH_3\overset{O}{\overset{\|}{C}}H \xrightarrow{(CH_3CH_2)_2O} \xrightarrow{H^+,\ H_2O} (CH_3)_2CH\overset{OH}{\underset{H}{C}}CH_3$$

1-Methylethylmagnesium bromide — **Ethanal (Acetaldehyde)** — 54% **3-Methyl-2-butanol**

Ketones furnish *tertiary alcohols:*

$$\underset{\textbf{Butyllithium}}{CH_3CH_2CH_2CH_2Li} + \underset{\substack{\textbf{Propanone}\\\textbf{(Acetone)}}}{CH_3\overset{O}{\overset{\|}{C}}CH_3} \xrightarrow{THF} \xrightarrow{H^+,\ H_2O} \underset{\substack{95\%\\\textbf{2-Methyl-2-hexanol}}}{CH_3CH_2CH_2CH_2\overset{CH_3}{\overset{|}{\underset{CH_3}{\underset{|}{C}}}}OH}$$

EXERCISE 8-10

Suggest a synthetic scheme for the conversion of 2-propanol, $(CH_3)_2CHOH$, into 2-methyl-1-propanol, $(CH_3)_2CHCH_2OH$.

Alcohols by the Addition of Organometallic Reagents to Esters

Carboxylic esters react with two equivalents of alkyllithium or Grignard reagents to give tertiary (if R = alkyl) or secondary alcohols (if R = H). In these transformations, the alkoxy group is displaced from the ester by a mechanism in which an intermediate ketone or aldehyde is thought to react with the organometallic species faster than does the starting ester. A detailed discussion of this mechanism is presented in Section 18-4.

General Alcohol Synthesis from Esters

$$\underset{\textbf{Ester}}{R\overset{O}{\overset{\|}{C}}OR'} \xrightarrow[-\ LiOR']{R''Li} \underset{\substack{\text{Not isolable}\\\textbf{Aldehyde or ketone}}}{\left[R\overset{O}{\overset{\|}{C}}R''\right]} \xrightarrow[\text{Fast}]{R''Li} \xrightarrow{H^+,\ H_2O} \underset{\textbf{Alcohol}}{R\overset{OH}{\overset{|}{\underset{R''}{\underset{|}{C}}}}R''}$$

Examples:

$$\underset{\substack{\textbf{Methyl methanoate}\\\textbf{(Methyl formate)}}}{H\overset{O}{\overset{\|}{C}}OCH_3} + 2\ CH_3MgBr \xrightarrow{(CH_3CH_2)_2O} \xrightarrow{H^+,\ H_2O} \underset{\substack{90\%\\\textbf{2-Propanol}}}{H\overset{OH}{\overset{|}{\underset{CH_3}{\underset{|}{C}}}}CH_3}$$

$$\underset{\substack{\textbf{Ethyl ethanoate}\\\textbf{(Ethyl acetate)}}}{CH_3\overset{O}{\overset{\|}{C}}OCH_2CH_3} + 2\ CH_3CH_2MgBr \xrightarrow{(CH_3CH_2)_2O} \xrightarrow{H^+,\ H_2O} \underset{\substack{67\%\\\textbf{3-Methyl-3-pentanol}}}{CH_3\overset{OH}{\overset{|}{\underset{CH_2CH_3}{\underset{|}{C}}}}CH_2CH_3}$$

Alcohols by the Addition of Organometallic Reagents to Strained Cyclic Ethers

Although ordinary ethers are not readily attacked by Grignard reagents and alkyllithium compounds, strained cyclic ethers are sufficiently reactive to undergo additions. For example, as with a hydride reagent, the nucleophilic carbon in the organometallic species is able to open the ring of oxacyclopropane:

$$\underset{\text{O}}{\text{H}_2\text{C}\!-\!\text{CH}_2} + \text{CH}_3\text{CH}_2\text{CH}_2\text{CH}_2\text{MgBr} \xrightarrow{\text{THF}} \xrightarrow{\text{H}^+,\ \text{H}_2\text{O}} \underset{62\%}{\text{CH}_3\text{CH}_2\text{CH}_2\text{CH}_2\text{CH}_2\text{CH}_2\text{OH}}$$

1-Hexanol

$$\underset{\text{O}}{\text{H}_2\text{C}\!-\!\text{CH}_2} + (\text{CH}_3)_3\text{CLi} \xrightarrow{\text{THF}} \xrightarrow{\text{H}^+,\ \text{H}_2\text{O}} \underset{72\%}{(\text{CH}_3)_3\text{CCH}_2\text{CH}_2\text{OH}}$$

3,3-Dimethyl-1-butanol

EXERCISE 8-11

Propose efficient syntheses of the following products from starting materials containing no more than four carbons.

(a) $CH_3CH_2CH_2\overset{OH}{\overset{|}{C}}HCH_3$; (b) 1-$C(CH_3)_3$-cyclobutan-1-ol (OH and $C(CH_3)_3$ on the same ring carbon of a cyclobutane); (c) $CH_3\overset{OH}{\overset{|}{C}}(CH_3)_2$; (d) $CH_3(CH_2)_5OH$

In summary, alkyllithium and magnesium reagents add to aldehydes and ketones to give the corresponding alcohols in which the alkyl group has formed a bond to the original carbonyl carbon. With esters, such additions take place twice. Three-membered ring ethers open up to give alcohols in which the hydrocarbon part is extended by two methylene units.

8-7 Complex Alcohols: An Introduction to Synthetic Strategy

The new reactions introduced in this chapter may be viewed as part of the "vocabulary" of organic chemistry. Without knowing reactions that allow you to carry out specific molecular manipulations and functional-group interconversions, you cannot speak the language of organic chemistry. You should become familiar with these reactions; their types; the reagents used; the reaction conditions if they are specifically mentioned to be crucial to the success of the reaction; and the limitations of a method. Although this may seem to be a monumental task, an understanding of the mechanistic features of these reactions makes it easier. We have seen how reactivity can be predicted by simple considerations of electronegativity, Coulombic forces, and bond strengths. Examples are shown below and on the next page.

How to Predict the Outcome of a Reaction on Mechanistic Grounds

$$\underset{\text{Not formed}}{\text{ICH}_2\text{CH}_2\text{CH}_2\text{Br}} \xleftarrow[\times]{\text{I}^-,\ \text{propanone (acetone)}} \text{FCH}_2\text{CH}_2\text{CH}_2\text{Br} \xrightarrow{\text{I}^-,\ \text{propanone (acetone)}} \text{FCH}_2\text{CH}_2\text{CH}_2\text{I}$$

Explanation: Bromide is a better leaving group than fluoride.

$$\underset{\text{Not formed}}{CH_3\overset{OCH_3}{\underset{H}{C}}MgBr} \xleftarrow[\times]{(CH_3CH_2)_2O} \overset{O^{\delta-}}{\underset{H_3C \quad H}{\overset{\|}{C}^{\delta+}}} + \overset{\delta^-}{C}H_3\overset{\delta^+}{Mg}Br \xrightarrow{(CH_3CH_2)_2O} CH_3\overset{\overset{-}{O}\overset{+}{Mg}Br}{\underset{H}{C}}CH_3$$

Explanation: The positively polarized carbonyl carbon forms a bond to the negatively polarized alkyl group of the organometallic reagent.

Other bromides + [1-bromo-3-methylcyclohexane, Not formed] + [cyclohexylmethyl radical, CH_2Br] $\xleftarrow[\times]{Br_2,\ h\nu}$ [methylcyclohexane, CH_3] $\xrightarrow{Br_2,\ h\nu}$ [1-bromo-1-methylcyclohexane, H_3C, Br]

Explanation: The tertiary C–H bond is weaker than a primary or secondary C–H bond.

A good working knowledge of chemical reactions is a prerequisite to the successful execution of synthetic organic transformations. As mentioned at the beginning of this text (Section 1-1), synthesis is concerned with the making of molecules. It can be divided into two closely related subareas: (1) the development of new *synthetic methods;* and (2) *total synthesis,* in which the aim is to make a specific target molecule as efficiently as possible, usually by a series of reactions.

This section is a review of the reaction chemistry presented so far in the context of these two topics.

The Discovery and Application of Synthetic Methods

New reactions are found by design or by accident. For example, consider how two different students might discover the reactivity of a Grignard reagent with a ketone to give an alcohol. The first student, knowing about electronegativity and the electronic make-up of ketones, would predict that the nucleophilic alkyl group of the Grignard species should attach itself to the electrophilic carbonyl carbon. This student would be pleased by the successful outcome of the experiment, verifying chemical principles in practice. The second student, with less knowledge, might attempt to dilute a particularly concentrated solution of a Grignard reagent with what he might conceive to be a perfectly good polar solvent: propanone (acetone). A violent reaction would immediately reveal the powerful potential of the reagent in alcohol synthesis.

Once a reaction is discovered, it is important to show its scope and its limitations. For this purpose, many different substrates are tested, side products (if any) noted, new functional groups subjected to the reaction conditions, and mechanistic studies carried out. Should these investigations prove the new reaction to be generally applicable, it will be added as a new synthetic method to the organic chemist's arsenal.

Because a reaction leads to a very specific change in a molecule, it is frequently useful to emphasize this "molecular alteration." A simple example is the addition of a Grignard reagent to a three-membered-ring cyclic ether. What is the structural change in this transformation? A two-carbon unit is added to an

R—M
Alkyl group
+
H_2C—CH_2 (bridged by O)
Two-carbon unit
↓
R—CH_2—CH_2—OH

alkyl group. The method is valuable because it allows a straightforward two-carbon extension, also called a double homologation.

Because Grignard reagents are available from the reaction of haloalkanes and magnesium, and because haloalkanes can be prepared from the corresponding alcohols (e.g., by treatment with HX, Section 6-7) or alkanes (by free radical halogenation, Section 3-5), it is easy to conceive of further synthetic possibilities:

$$\text{ROH} \xrightarrow[\text{3. } H_2C\overset{O}{—}CH_2 \quad \text{4. } H^+,\ H_2O]{\text{1. HBr} \quad \text{2. Mg}} \text{RCH}_2\text{CH}_2\text{OH} \qquad \text{RH} \xrightarrow[\text{3. } H_2C\overset{O}{—}CH_2 \quad \text{4. } H^+,\ H_2O]{\text{1. } Br_2,\ h\nu \quad \text{2. Mg}} \text{RCH}_2\text{CH}_2\text{OH}$$

$$RCH_2CH_2OH \xrightarrow[-H_2O]{HBr} RCH_2CH_2Br \xrightarrow[-\text{LiBr}]{\text{LiAlH}_4} RCH_2CH_2H$$

Another aspect of this method is the generation of a new hydroxy group at the terminus of the elongated carbon chain. The use of this group allows for additional synthetic flexibility. Alternatively, if not wanted, it can be removed. You already know how to do that. For example, you can first convert the alcohol into a haloalkane and then reduce it with lithium aluminum hydride.

Why can OH not be replaced directly by H through the use of $LiAlH_4$? There are two reasons: first, hydroxide is a poor leaving group; second, hydride reagents react with the acidic hydroxy function to give alkoxides and hydrogen.

$$4\ RCH_2CH_2OH + LiAlH_4 \longrightarrow Li[Al(OCH_2CH_2R)_4] + 4\ H—H$$

The terminal hydroxy group is synthetically useful because it can be converted into a bromoalkane. Bromide is a good leaving group; therefore, nucleophilic displacement allows the introduction of many new functional groups at the terminus:

$$RCH_2CH_2OH \xrightarrow{\text{1. HBr} \quad \text{2. } :Nu^-} RCH_2CH_2Nu$$

Nu = OH, OR, I, CN, SR, NH_2, or R′ (from reaction with cuprates, etc.)

Bromoalkanes are also excellent substrates for the synthesis of organometallic reagents that, in turn, can be exploited to effect a variety of new molecular changes:

$$RCH_2CH_2Br \xrightarrow{Mg} RCH_2CH_2MgBr$$

- $RCH_2CH_2MgBr \xrightarrow{D_2O} RCH_2CH_2D$
- $RCH_2CH_2MgBr \xrightarrow{CH_3\overset{O}{\overset{\|}{C}}H} RCH_2CH_2\underset{H}{\overset{OH}{C}}CH_3$
- $RCH_2CH_2MgBr \xrightarrow{CH_3\overset{O}{\overset{\|}{C}}CH_3} RCH_2CH_2\underset{CH_3}{\overset{OH}{C}}CH_3$
- $RCH_2CH_2MgBr \xrightarrow{H\overset{O}{\overset{\|}{C}}OCH_3} H\overset{OH}{C}CH_2CH_2R$ (with CH_2CH_2R on the same carbon)
- $RCH_2CH_2MgBr \xrightarrow{H_2C\overset{O}{—}CH_2} RCH_2CH_2CH_2CH_2OH$

Each one of the products in the scheme can enter into further transformations of its own, leading to more complicated products. You can see how many synthetic permutations are already at your disposal, even with the few reagents that you know at this stage.

When we ask, ''What good is a reaction? What sort of structures can we make by applying it?'' we address a problem of synthetic methodology. Development of a synthetic method, finding the optimum conditions for its uses, defining its scope and limitations, and applying it to the synthesis of potentially valuable molecules are important aspects of synthetic organic research. Let us ask a different question. Suppose that we want to prepare a specific target molecule. How would we go about devising an efficient route to it? How do we find suitable starting materials? The problem with which we are dealing now is that of a total synthesis.

Total Synthesis of Organic Molecules

Organic chemists want to make complex molecules for specific purposes. For example, certain compounds might have valuable medicinal properties but are not readily available from natural sources. Biochemists need a particular isotopically labeled molecule to trace metabolic pathways. Physical organic chemists frequently design novel structures (recall the strained-ring hydrocarbons in Section 4-6) on which to measure physical constants, spectral characteristics, chemical behavior, and other properties. There are many reasons for the total synthesis of organic molecules.

Whatever the final target, a successful synthesis is characterized by brevity and high overall yield. The starting materials should be readily available, preferably commercial, and inexpensive. In industrial syntheses, the turnover of large quantities of chemicals requires special process technology. Moreover, plant safety and environmental concerns demand that the reagents used be relatively nontoxic and easily handled.

In some cases, it is worthwhile to accept a low-yield step if it allows a significant shortening of the synthetic sequence. For example (assuming all starting materials to be of comparable cost), a seven-step synthesis in which each step has an 85% yield is less productive than a four-step synthesis with three yields at 95% and one at 45%. The overall efficiency in the first sequence comes to $(0.85 \times 0.85 \times 0.85 \times 0.85 \times 0.85 \times 0.85 \times 0.85) \times 100 = 32\%$, whereas the second synthesis, in addition to being three steps shorter, gives $(0.95 \times 0.95 \times 0.95 \times 0.45) \times 100 = 39\%$.

Many compounds that are commercially available and inexpensive are also small, containing five or fewer carbon atoms. Therefore, the most frequent task facing the synthetic planner is that of building up a larger, complicated molecule from smaller, simple fragments. The best approach to the preparation of the target is to work its synthesis *backward,* an approach called **retrosynthetic analysis** (*retro,* Greek, reverse). In this analysis, strategic carbon–carbon bonds are broken at points at which their formation appears possible. For example, a retrosynthetic analysis of the synthesis of ethanol from two one-carbon units would suggest its formation from a methyl organometallic compound and methanal. A double-barreled arrow is used to indicate that the desired transformation (in this case, CH_3CH_2OH from $CH_3MgBr + CH_2O$) does not necessarily constitute a real reaction. It indicates a strategic disconnection; to achieve the connection might require several steps.

Retrosynthetic Analysis of Ethanol Synthesis from Two One-Carbon Fragments

$$CH_3CH_2OH \Longrightarrow CH_3MgBr + H_2C{=}O$$

Methylmagnesium bromide — Methanal (Formaldehyde)

Two alternative but inferior retrosyntheses of ethanol are:

$$CH_3CH_2OH \Longrightarrow CH_3\overset{O}{\overset{\|}{C}}H + NaBH_4$$

$$CH_3CH_2OH \Longrightarrow CH_3CH_2Br + OH^-$$

They are not as good as the first because they do not significantly simplify the target structure: no carbon–carbon bonds are broken.

Let us apply retrosynthetic analysis to the construction of a tertiary alcohol. There are several possible retrosynthetic steps. The strategic bond(s) to be cleaved in these steps are between the existing functional group and the reactive carbon center. For 4-methyl-4-heptanol, there are three disconnections leading to simpler precursors. Path *a* cleaves the methyl group from C-4, suggesting as the starting material for its construction methylmagnesium bromide and 4-heptanone. Cleavage *b* is an alternative possibility leading to a propyl Grignard reagent and 2-pentanone as precursors. Finally, a particularly efficient possibility is shown in path *c*. This alternative has as its basis the fact that the target alcohol contains two identical substituents at C-4: the propyl groups. We know that such alcohols can be prepared from esters and two equivalents of an organometallic compound. In this case, the molecule is readily made from methyl ethanoate (acetate) and propylmagnesium bromide.

Retrosynthetic Analysis of the Synthesis of 4-Methyl-4-heptanol

4-Methyl-4-heptanol: $CH_3C(OH)(CH_2CH_2CH_3)CH_2CH_2CH_3$

(a) $\Longrightarrow CH_3MgBr + CH_3CH_2CH_2\overset{O}{\overset{\|}{C}}CH_2CH_2CH_3$

Methylmagnesium bromide — 4-Heptanone

(b) $\Longrightarrow CH_3CH_2CH_2MgBr + CH_3\overset{O}{\overset{\|}{C}}CH_2CH_2CH_3$

Propylmagnesium bromide — 2-Pentanone

(c) $\Longrightarrow CH_3\overset{O}{\overset{\|}{C}}OCH_3 + 2\ CH_3CH_2CH_2MgBr$

Methyl ethanoate (Methyl acetate) — Propylmagnesium bromide

An evaluation of the relative merits of these three retrosyntheses suggests that path *c* is best: the starting materials are inexpensive and contain the least number of carbon atoms. The next-best method is path *b*. Its drawbacks relative to path *c* are that only one C–C bond is made in one step and that 2-pentanone is a large building block. Finally, path *a* is the worst alternative because the fragments that would form the target structure are very unequal: one of them contains more than five carbon atoms—namely, seven.

When considering the synthesis of a complex alcohol, remember that there are, in principle, several ways of putting it together by reaction of an appropriate organometallic reagent with an aldehyde or ketone.

A General Synthesis of Alcohols

$$\mathrm{R\overset{O}{\overset{\|}{C}}R' + R''M \Longleftarrow R\!-\!\underset{R''}{\underset{|}{\overset{OH}{\overset{|}{C}}}}\!-\!R' \Longrightarrow R\overset{O}{\overset{\|}{C}}R'' + R'M}$$

$$\Downarrow$$

$$\mathrm{R'\overset{O}{\overset{\|}{C}}R'' + RM}$$

EXERCISE 8-12

Give an economical retrosynthetic analysis of 2,3,3-trimethyl-2-heptanol.

EXERCISE 8-13

Devise a synthetic scheme that will convert 2-methyl-2-propanol into ether A.

$$(CH_3)_3COH \longrightarrow \underset{\textbf{A}}{(CH_3)_3CCH_2CH_2OCH_3}$$

Pitfalls in Synthetic Planning and Possible Solutions

There are several considerations to keep in mind when practicing synthetic chemistry that will help to avoid designing unsuccessful or low-yielding approaches to a target molecule.

Do not use reagents whose molecules have functional groups that would interfere with the desired reaction. For example, treating a hydroxy aldehyde with a Grignard reagent leads to alcoholysis and not carbon–carbon bond formation:

$$\mathrm{HOCH_2CH_2\underset{CH_3}{\underset{|}{\overset{OH}{\overset{|}{C}}}}H \;\not\longleftarrow\; HOCH_2CH_2\overset{O}{\overset{\|}{C}}H + CH_3MgBr \longrightarrow BrMgOCH_2CH_2\overset{O}{\overset{\|}{C}}H + \overset{H}{\overset{|}{C}}H_3}$$

A possible solution to this problem would be to add two equivalents of Grignard reagent, one to quench the acidic hydrogen and the second to achieve the desired reactivity at the carbonyl group.

Do not try to make a Grignard reagent from a bromo ketone. Such a reagent is not stable and will, as soon as it is formed, decompose by reacting with the carbonyl group of the starting material, as indicated on the next page.

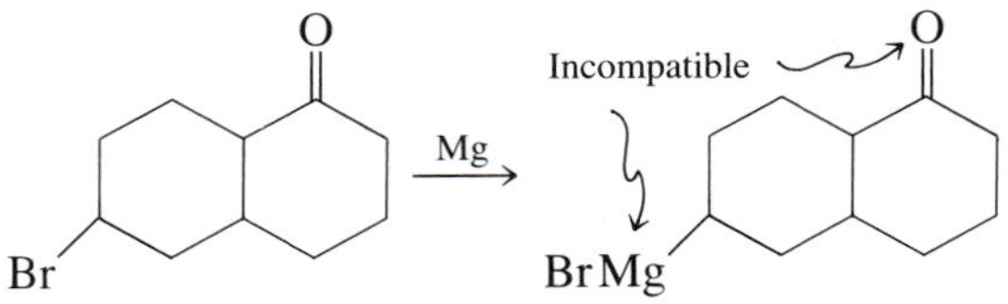

To remove the potentially complicating presence of reactive functionality, chemists use **protecting groups.** By this means, a functional group is converted into its unreactive form but in a way that allows the original unit to be regenerated. A way to protect an alcohol, for example, is in the form of a 1,1-dimethylethyl ether (Section 9-5). The ether is readily formed from the alcohol and a tertiary halide. The alcohol group can be regenerated by hydrolysis with acid (deprotection):

$$\mathrm{ROH} \xrightarrow[-\,\mathrm{HX,\ protection}]{(\mathrm{CH_3})_3\mathrm{CX}} \mathrm{ROC(CH_3)_3} \xrightarrow[\mathrm{Deprotection}]{\mathrm{H^+,\ H_2O}} \mathrm{ROH} + (\mathrm{CH_3})_3\mathrm{COH}$$

Other methods for protecting the various functional groups will be described later.

Examples of Neopentyl-like Hindered Haloalkanes

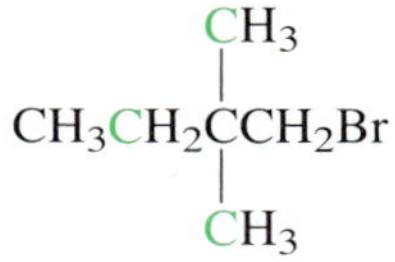

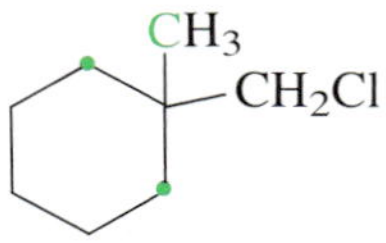

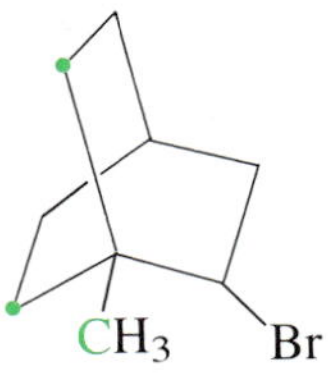

Take into consideration any mechanistic and structural constraints affecting the reactions that you want to use. For example, free-radical brominations are more selective than chlorinations. Keep in mind the structural limitations on nucleophilic reactions, and do not forget the lack of reactivity of the 2,2-dimethyl-1-halopropanes (neopentyl halides). Although sometimes difficult to recognize, many haloalkanes have "neopentyl-like" structures and are similarly unreactive. On the other hand, such systems do form organometallic reagents and may be further functionalized in this manner. For example, treatment of the Grignard reagent made from 1-bromo-2,2-dimethylpropane with methanal (formaldehyde) leads to the corresponding alcohol.

$$\underset{\textbf{1-Bromo-2,2-dimethylpropane}}{(\mathrm{CH_3})_3\mathrm{CCH_2Br}} \xrightarrow{\text{1. Mg} \quad \text{2. } \mathrm{CH_2{=}O}} \underset{\textbf{3,3-Dimethyl-1-butanol}}{(\mathrm{CH_3})_3\mathrm{CCH_2CH_2OH}}$$

Tertiary halides, if incorporated into a more complex framework, also are sometimes difficult to recognize. Remember that tertiary halides do not undergo S_N2 reactions but eliminate in the presence of bases.

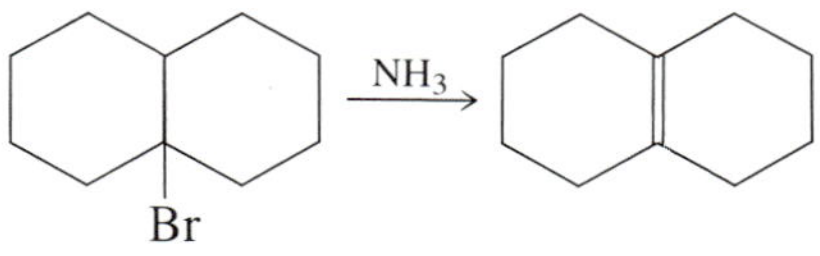

Expertise in synthesis, as in many other aspects of organic chemistry, develops largely from practice. Planning the synthesis of complex molecules requires that you review the reactions and mechanisms covered in earlier sections. The knowledge thus acquired can then be applied to the solution of synthetic problems.

Summary of New Reactions

1 ACID-BASE PROPERTIES OF ALCOHOLS

$$\underset{\text{Oxonium ion}}{R{-}\overset{+}{O}H_2} \underset{H^+}{\rightleftharpoons} \underset{\text{Alcohol}}{ROH} \xrightleftharpoons{\text{Base }:B^-} \underset{\text{Alkoxide}}{RO^-} + BH$$

2 SYNTHESIS GAS

$$\text{Coal} \xrightarrow{\text{Air, } H_2O,\ \Delta} \underset{\text{Synthesis gas}}{x\,CO + y\,H_2}$$

3 METHANOL SYNTHESIS FROM SYNTHESIS GAS

$$CO + 2\,H_2 \xrightarrow{\text{Cu-ZnO-}Cr_2O_3\text{, 250°C, 50–100 atm}} CH_3OH$$

4 FISCHER-TROPSCH SYNTHESIS

$$n\,CO + (2n + 1)\,H_2 \xrightarrow{\text{Co or Fe, pressure, 200°–350°C}} \underset{\text{Hydrocarbons}}{C_nH_{2n+2}} + n\,H_2O$$

5 ETHANOL BY HYDRATION OF ETHENE

$$CH_2{=}CH_2 + HOH \xrightarrow{H_3PO_4\text{, 300°C}} CH_3CH_2OH$$

Preparation of Alcohols

6 NUCLEOPHILIC DISPLACEMENT OF HALIDES AND OTHER LEAVING GROUPS BY HYDROXIDE ION

$$RCH_2X + HO^- \xrightarrow[S_N2]{} RCH_2OH + X^-$$

X = halide, sulfonate

$$R'CHBr + CH_3\overset{O}{\overset{\|}{C}}O^- \xrightarrow[S_N2]{} R'CHO\overset{O}{\overset{\|}{C}}CH_3 \xrightarrow[\text{Ester hydrolysis}]{HO^-} R'CHOH$$

$$RR'R''CX \xrightarrow[S_N1]{H_2O\text{, propanone (acetone)}} RR'R''COH$$

7 CATALYTIC HYDROGENATION OF ALDEHYDES AND KETONES

$$R\overset{O}{\overset{\|}{C}}H \xrightarrow{H_2\text{, Pt}} RCH_2OH \qquad R\overset{O}{\overset{\|}{C}}R' \xrightarrow{H_2\text{, catalyst}} RC(OH)(H)R'$$

8 REDUCTION OF ALDEHYDES AND KETONES BY HYDRIDES

$$R\overset{O}{\overset{\|}{C}}H \xrightarrow{NaBH_4\text{, }CH_3CH_2OH} RCH_2OH \qquad R\overset{O}{\overset{\|}{C}}R' \xrightarrow{\text{1. }LiAlH_4\text{, }(CH_3CH_2)_2O\quad \text{2. }H^+\text{, }H_2O} RC(OH)(H)R'$$

9 REDUCTION OF CARBOXYLIC ACIDS AND ESTERS BY LITHIUM ALUMINUM HYDRIDE

$$\underset{\text{Carboxylic acid}}{\mathrm{RCOH}\ (\mathrm{C{=}O})} \xrightarrow[\text{2. } H^+,\ H_2O]{\text{1. } LiAlH_4} RCH_2OH \qquad \underset{\text{Ester}}{\mathrm{RCOR'}\ (\mathrm{C{=}O})} \xrightarrow[\text{2. } H^+,\ H_2O]{\text{1. } LiAlH_4} RCH_2OH + R'OH$$

10 NUCLEOPHILIC OPENING OF OXACYCLOPROPANE BY LITHIUM ALUMINUM HYDRIDE

$$H_2C\overset{O}{—}CH_2 \xrightarrow[\text{2. } H^+,\ H_2O]{\text{1. } LiAlH_4} CH_3CH_2OH$$

Preparation of Organometallic Reagents

11 REACTION OF METALS WITH HALOALKANES

$$RX + Li \xrightarrow{(CH_3CH_2)_2O} \underset{\text{Alkyllithium reagent}}{RLi} + LiX$$

$$RX + Mg \xrightarrow{(CH_3CH_2)_2O} \underset{\text{Grignard reagent}}{RMgX}$$

12 TRANSMETALLATION

$$4\ RMgX + SiX_4 \longrightarrow \underset{\text{Tetraalkylsilane}}{SiR_4} + 4\ MgX_2$$

$$2\ RMgX + CdX_2 \longrightarrow \underset{\text{Dialkylcadmium}}{CdR_2} + 2\ MgX_2$$

$$3\ RLi + AlX_3 \longrightarrow \underset{\text{Trialkylaluminum}}{R_3Al} + 3\ LiX$$

$$RLi + CuI \longrightarrow \underset{\text{An alkylcopper reagent}}{RCu} + LiI$$

$$RCu + RLi \longrightarrow \underset{\text{Lithium dialkylcuprate (A cuprate reagent)}}{R_2CuLi}$$

Reactions of Organometallic Compounds

13 HYDROLYSIS

$$RLi + H_2O \longrightarrow RH + LiOH$$

$$RMgX + D_2O \longrightarrow RD + Mg(OD)X$$

14 ALKANES FROM CUPRATES AND (PRIMARY) HALOALKANES

$$R_2CuLi + 2\ R'X \longrightarrow 2\ R—R' + LiX + \text{copper salts}$$

15 ALKANES FROM HALOALKANES AND LITHIUM ALUMINUM HYDRIDE

$$RX + LiAlH_4 \longrightarrow RH$$

Alcohol Synthesis with Organometallic Reagents

16 ADDITION OF ORGANOMETALLIC COMPOUNDS TO ALDEHYDES AND KETONES

$$\text{RLi or RMgX} + \underset{\textbf{Methanal}}{CH_2{=}O} \longrightarrow \underset{\textbf{Primary alcohol}}{RCH_2OH}$$

$$\text{RLi or RMgX} + \underset{\textbf{Aldehyde}}{R'\overset{\overset{\large O}{\|}}{C}H} \longrightarrow \underset{\textbf{Secondary alcohol}}{R\overset{OH}{\underset{H}{C}}R'}$$

$$\text{RLi or RMgX} + \underset{\textbf{Ketone}}{R'\overset{\overset{\large O}{\|}}{C}R''} \longrightarrow \underset{\textbf{Tertiary alcohol}}{R\overset{OH}{\underset{R''}{C}}R'}$$

17 ADDITION OF ORGANOMETALLIC COMPOUNDS TO ESTERS

$$R\overset{\overset{\large O}{\|}}{C}OR' + 2\ R''MgX \longrightarrow \underset{\textbf{Tertiary alcohol}}{R\overset{OH}{\underset{R''}{C}}R''}$$

$$\underset{\textbf{Methanoate (formate) ester}}{H\overset{\overset{\large O}{\|}}{C}OR} + 2\ R'Li \longrightarrow \underset{\textbf{Secondary alcohol}}{H\overset{OH}{\underset{R'}{C}}R'}$$

18 NUCLEOPHILIC OPENING OF OXACYCLOPROPANE BY ORGANOMETALLIC COMPOUNDS

$$\text{RLi or RMgX} + H_2C\overset{O}{—}CH_2 \longrightarrow RCH_2CH_2OH$$

Summary of Important Concepts

1 Alcohols are alkanols in IUPAC nomenclature. The stem containing the functional group gives the alcohol its name. Alkyl and halo substituents are added as prefixes.

2 Alcohols have a polarized and short O—H bond. The proton is hydrophilic and enters into hydrogen bonding. The alkyl part is hydrophobic.

3 Alcohols are acidic and basic. Electron-withdrawing substituents increase the acidity (and lower the basicity). Complete deprotonation occurs with bases whose conjugate acid is considerably weaker than the alcohol.

4 An oxonium ion may be viewed as a Lewis acid-base complex between a carbocation and an alcohol.

5 The hydrogenation of aldehydes and ketones to alcohols requires a catalyst on which both hydrogen and the substrate are activated by surface adsorption.

6 The carbon atom in a carbonyl group is electrophilic and therefore subject to attack by nucleophiles, such as a hydride or the alkyl group in organometallic compounds. The products of such transformations are alcohols.

7 Strained cyclic ethers, such as oxacyclopropanes, are unusually reactive because nucleophilic ring-opening releases ring strain.

8 The conversion of the electrophilic alkyl group in a haloalkane into its nucleophilic analog in an organometallic compound is an example of reverse polarization.

Problems

1 Name the following molecules according to the IUPAC nomenclature system. Indicate in each case stereochemistry (if any) and whether the molecule is a primary, secondary, or tertiary alcohol.

(a) $CH_3CH_2CHOHCH_3$

(b) $CH_3CHBrCH_2CHOHCH_2CH_3$

(c) $HOCH_2CH(CH_2CH_2CH_3)_2$

(d) CH_2Cl, H, C, CH_3, OH

(e) CH_2CH_3, OH

(f) OH, Br

(g) $C(CH_2OH)_4$

(h) CH_2OH, H, OH, H, OH, CH_2OH

(i) OH, CH_2CH_2OH

(j) CH_3CH_2, Cl, CH_2OH, CH_3

2 Convert the following names into structural formulas.

(a) 2-(Trimethylsilyl)ethanol.
(b) 1-Methylcyclopropanol.
(c) 3-(1-Methylethyl)-2-hexanol.
(d) (*R*)-2-Pentanol.
(e) 3,3-Dibromocyclohexanol.

3 Predict the relative boiling points within each group of compounds below.

(a) Cyclohexane, cyclohexanol, chlorocyclohexane.
(b) 2,3-Dimethyl-2-pentanol, 2-methyl-2-hexanol, 2-heptanol.

4 Explain the relative water solubilities within each group of compounds below.

(a) Ethanol > chloroethane > ethane.
(b) Methanol > ethanol > 1-propanol.

5 Explain why 1,2-ethanediol exists in the *gauche* conformation to a much larger extent than does 1,2-dichloroethane.

Would you expect the *gauche*/*anti* conformational ratio of 2-chloroethanol to be similar to that of 1,2-dichloroethane or more like that of 1,2-ethanediol?

6 Predict the order of acidity in solution within each of the following groups of alcohols.

(a) $CH_3CHClCH_2OH$, $CH_3CHBrCH_2OH$, $ClCH_2CH_2CH_2OH$
(b) $CH_3CCl_2CH_2OH$, CCl_3CH_2OH, $(CH_3)_2CClCH_2OH$
(c) $(CH_3)_2CHOH$, $(CF_3)_2CHOH$, $(CCl_3)_2CHOH$

7 Write an appropriate equation to show how each of the alcohols below acts as (1) a base and (2) an acid in solution. In each case, compare the qualitative base and acid strengths relative to those of methanol.

(a) $(CH_3)_2CHOH$
(b) CH_3CHFCH_2OH
(c) CCl_3CH_2OH

8 Given the pK_a values of −2.2 for $CH_3\overset{+}{O}H_2$ and 15.5 for CH_3OH, calculate the pH at which

(a) methanol will contain exactly equal amounts of $CH_3\overset{+}{O}H_2$ and CH_3O^-.
(b) 50% CH_3OH and 50% $CH_3\overset{+}{O}H_2$ will be present.
(c) 50% CH_3OH and 50% CH_3O^- will be present.

9 Do you expect hyperconjugation to be important in the stabilization of oxonium ions (e.g., $R\overset{+}{O}H_2$, $R_2\overset{+}{O}H$, etc.)? Explain your answer.

10 Consider the reaction of bromocyclohexane with each of the four reagents below, and answer the questions that follow.

(i) H_2O; (ii) HO^-; (iii) $CH_3\overset{O}{\overset{\|}{C}}OH$; (iv) $CH_3\overset{O}{\overset{\|}{C}}O^-$

(a) What is the most important reaction mechanism in each case?
(b) Which reagent gives the most elimination product?
(c) Which reagent is most useful in synthesizing the alcohol?

11 Evaluate each of the possible alcohol syntheses below as being good (the desired alcohol is the major or only product), not so good (the desired alcohol is a minor product), or worthless.

(a) $CH_3CH_2Cl \xrightarrow{H_2O,\ CH_3\overset{O}{\overset{\|}{C}}CH_3} CH_3CH_2OH$

(b) $CH_3OSO_2\text{—}C_6H_4\text{—}CH_3 \xrightarrow{HO^-,\ H_2O,\ \Delta} CH_3OH$

(c) I $\xrightarrow{HO^-,\ H_2O,\ \Delta}$ OH

(d) $CH_3\overset{I}{\overset{|}{C}}HCH_2CH_2CH_3 \xrightarrow{H_2O,\ \Delta} CH_3\overset{OH}{\overset{|}{C}}HCH_2CH_2CH_3$

(e) $CH_3\overset{CN}{\overset{|}{C}}HCH_3 \xrightarrow{HO^-,\ H_2O,\ \Delta} CH_3\overset{OH}{\overset{|}{C}}HCH_3$

(f) $CH_3OCH_3 \xrightarrow{HO^-,\ H_2O,\ \Delta} CH_3OH$

(g) H_3C Br $\xrightarrow[2.\ HO^-,\ H_2O]{1.\ CH_3\overset{O}{\overset{\|}{C}}OH}$ H_3C OH

(h) $CH_3\overset{CH_3}{\overset{|}{C}}HCH_2Cl \xrightarrow{HO^-,\ H_2O,\ \Delta} CH_3\overset{CH_3}{\overset{|}{C}}HCH_2OH$

12 Write the major product(s) of each of the following reactions. It is implied that aqueous work-up has taken place in all those cases that require it to obtain the organic product.

(a) $CH_3CH{=}CHCH_3 \xrightarrow{H_3PO_4,\ H_2O,\ \Delta}$

(b) $(S)\text{-}CH_3(CH_2)_5\overset{OSO_2CH_3}{\overset{|}{C}}HCH_3 \xrightarrow{Na^{+}\ {}^{-}O\overset{O}{\overset{\|}{C}}CH_3,\ DMSO}$

(c) Product of part **b** $\xrightarrow{NaOH,\ CH_3OH,\ H_2O}$

(d) $CH_3\overset{O}{\overset{\|}{C}}CH_2CH_2\overset{O}{\overset{\|}{C}}CH_3 \xrightarrow{H_2,\ Pt}$

(e) Cyclohexane ring bearing H and a CHO group ($C(=O)H$) on the same carbon $\xrightarrow{NaBH_4,\ CH_3CH_2OH}$

(f) Bromo-substituted cyclopentane fused to an epoxide (Br at the ring carbon opposite the epoxide O) $\xrightarrow{LiAlH_4,\ (CH_3CH_2)_2O}$

(g) $CH_3O\overset{O}{\overset{\|}{C}}CH_2\text{—}$(tetrahydrofuran-2-yl) $\xrightarrow{LiAlH_4,\ (CH_3CH_2)_2O}$

(h) Cyclohexanone bearing a wedged CH_3 on the carbon adjacent to C=O and a wedged CH_3CH_2 on the ring $\xrightarrow{NaBH_4,\ CH_3CH_2OH}$

13 In which direction does the equilibrium below lie? Data: the pK_a for H_2 is about 38.

$$H^- + H_2O \rightleftharpoons H_2 + HO^-$$

14 Write the product of each of the following reactions.

(a) CH_3CH (C=O) $\xrightarrow[2.\ H^+,\ H_2O]{1.\ LiAlD_4}$

(b) 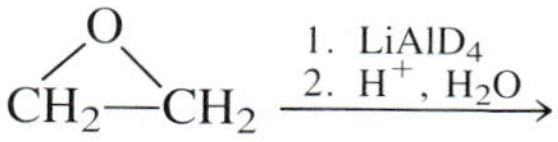 CH_2—CH_2 (epoxide, O bridging) $\xrightarrow[2.\ H^+,\ H_2O]{1.\ LiAlD_4}$

(c) 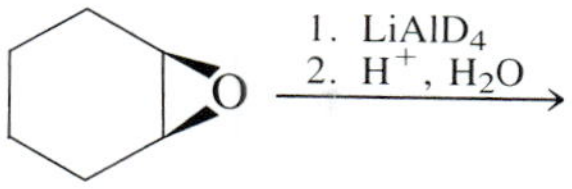 cyclohexene oxide $\xrightarrow[2.\ H^+,\ H_2O]{1.\ LiAlD_4}$

(d) 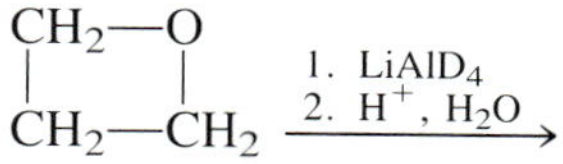 CH_2—O / CH_2—CH_2 (oxetane) $\xrightarrow[2.\ H^+,\ H_2O]{1.\ LiAlD_4}$

15 Write the major product(s) of each of the following reactions.

(a) $CH_3(CH_2)_5CHCH_3$ (Cl on CH) $\xrightarrow{Mg,\ (CH_3CH_2)_2O}$

(b) Product of part **a** $\xrightarrow{D_2O}$

(c) 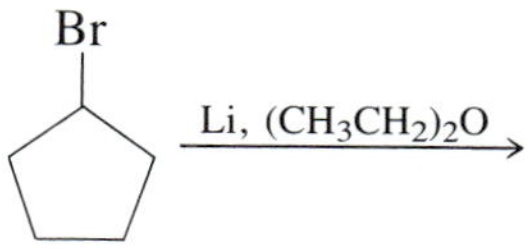 bromocyclopentane (Br) $\xrightarrow{Li,\ (CH_3CH_2)_2O}$

(d) Product of part **c** $\xrightarrow{ZnCl_2}$

16 Write the structure of the final product(s) of each reaction or reaction sequence shown below.

(a) $(CH_3)_3CCl \xrightarrow[2.\ (CH_3)_2SiCl_2]{1.\ Mg}$

(b) $(CH_3)_2CHBr + 2\ Li \xrightarrow{(CH_3CH_2)_2O}$

(c) Product of part **b** (2 equivalents) + CuI $\longrightarrow$

(d) Product of part **c** + $CH_3CHCH_2CH_2Br$ (CH_3 on CH) $\longrightarrow$

(e) $CH_3CH_2CH_2Cl + Mg \xrightarrow{(CH_3CH_2)_2O}$

(f) Product of part **e** + $C_6H_5CCH_3$ (C=O) $\longrightarrow$

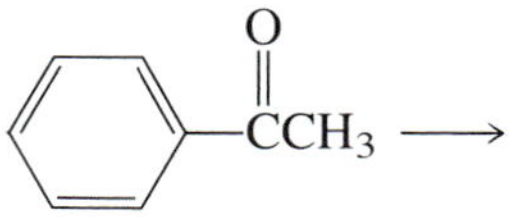

(g) 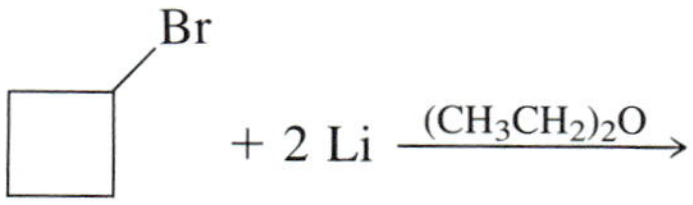 bromocyclobutane (Br) + 2 Li $\xrightarrow{(CH_3CH_2)_2O}$

(h) 2 moles of product of part **g** + 1 mole of $HCOCH_3$ (C=O) $\longrightarrow$

17 Organometallic reagents of the most-electropositive metals are difficult to prepare because they react rapidly with haloalkanes. For example, most attempts to prepare alkyl sodium reagents lead to alkanes by a reaction called *Wurtz coupling:*

$$2\ RX + 2\ Na \longrightarrow R{-}R + 2\ NaX$$

the result of

$$R{-}X + 2\ Na \longrightarrow R{-}Na + NaX$$

followed rapidly by

$$R{-}Na + R{-}X \longrightarrow R{-}R + NaX$$

When it was still in use, the Wurtz coupling reaction was employed mainly for the preparation of alkanes from the coupling of two identical alkyl groups (e.g., equation 1 below). Suggest a reason why Wurtz coupling might not be a useful method for coupling two different alkyl groups (equation 2 below).

$$2\ CH_3CH_2CH_2Cl + 2\ Na \longrightarrow CH_3CH_2CH_2CH_2CH_2CH_3 + 2\ NaCl \quad (1)$$

$$CH_3CH_2Cl + CH_3CH_2CH_2Cl + 2\ Na \longrightarrow CH_3CH_2CH_2CH_2CH_3 + 2\ NaCl \quad (2)$$

18 Write the major product(s) of each of the following reactions (after aqueous work-up). The solvent in each case is ethoxyethane (diethyl ether).

(a) 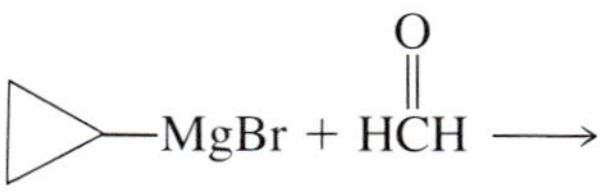 cyclopropyl–MgBr + $\overset{O}{\overset{\|}{HCH}}$ $\longrightarrow$

(b) $CH_3\overset{CH_3}{\overset{|}{C}}HCH_2MgCl + CH_3\overset{O}{\overset{\|}{C}}H \longrightarrow$

(c) $C_6H_5CH_2Li + C_6H_5\overset{O}{\overset{\|}{C}}H \longrightarrow$

(d) 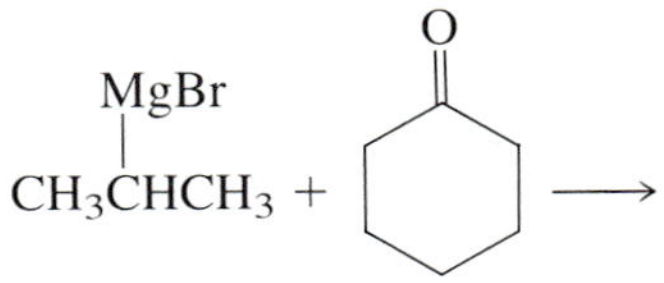 $CH_3\overset{MgBr}{\overset{|}{C}}HCH_3$ + cyclohexanone $\longrightarrow$

(e) $2\ CH_3MgI + C_6H_5\overset{O}{\overset{\|}{C}}OCH_3 \longrightarrow$

(f) $2\ CH_3CH_2Li + H\overset{O}{\overset{\|}{C}}OCH_3 \longrightarrow$

(g) 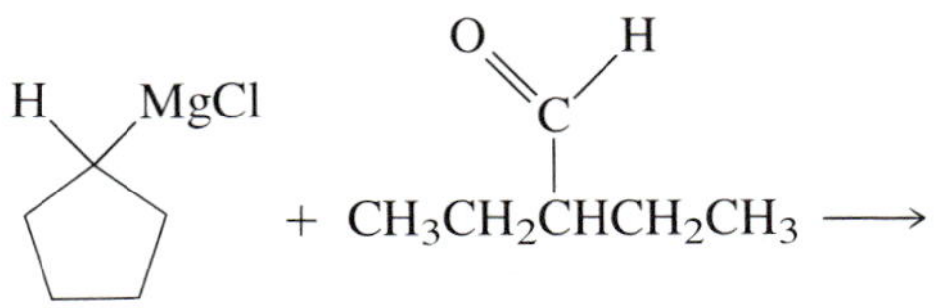 cyclopentane bearing H and MgCl on the same carbon + $CH_3CH_2\overset{CHO}{\overset{|}{C}}HCH_2CH_3 \longrightarrow$

(h) $C_6H_5Li + CH_2{-}CH_2$ (with O bridging, ethylene oxide) $\longrightarrow$

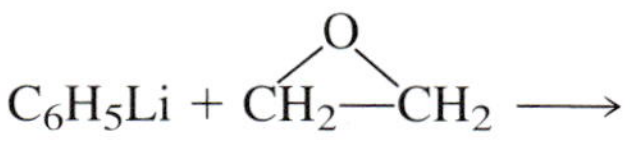

19 The reaction of two equivalents of Mg with 1,4-dibromobutane produces compound A. The reaction of A with two equivalents of CH_3CHO (ethanal), followed by work-up with dilute aqueous acid, produces compound B, having the formula $C_8H_{18}O_2$. What are the structures of A and B?

20 The reaction of compound B (Problem 19) with excess HI produces compound C ($C_8H_{16}I_2$). Treatment of C with two equivalents of Mg yields compound D. The reaction of D with two equivalents of HCHO (methanal) produces, after aqueous acid work-up, compound E ($C_{10}H_{22}O_2$). What are the structures of C, D, and E?

21 If compound D (Problem 20) undergoes reaction instead with the ester shown at the right, a completely different type of product is formed: compound F ($C_9H_{18}O$). Write a sensible structure of F and explain how it is formed, step by step.

$$\overset{\displaystyle O}{\overset{\|}{H\text{C}OCH_3}}$$

Methyl methanoate (Methyl formate)

22 Suggest the best synthetic route that you can devise for each of the simple alcohols below, using in each case a simple alkane as your initial starting molecule. What are the obvious problems in beginning syntheses with alkanes?

(a) Methanol.
(b) Ethanol.
(c) 1-Propanol.
(d) 2-Propanol.
(e) 1-Butanol.
(f) 2-Butanol.
(g) 2-Methyl-2-propanol.

23 For each alcohol in Problem 22, suggest (if possible) a synthetic route that starts with (1) an aldehyde, (2) a ketone, (3) an oxacyclopropane.

24 Which of the alcohols in Problem 22 can be synthesized from carboxylic acids or carboxylic acid esters? Write out the syntheses in full.

25 Suggest four different syntheses of 2-methyl-2-hexanol. Each synthesis should utilize one of the starting materials listed below. Then use any number of steps and any other reagents needed.

(a) $CH_3\overset{\overset{\displaystyle O}{\|}}{C}CH_3$

(b) $CH_3\overset{\overset{\displaystyle O}{\|}}{C}CH_2CH_2CH_2CH_3$

(c) $CH_3CH_2CH_2CH_2CO_2CH_3$

(d) $(CH_3)_2CHCH_2CH_2Br$

26 Propose at least two additional syntheses of 2-methyl-2-hexanol, using starting materials different from those in Problem 25. Evaluate these syntheses in light of the discussion of synthetic methodology.

27 Devise three different syntheses of 3-octanol starting with

(a) a ketone.
(b) an aldehyde.
(c) a different aldehyde.

28 Using the principles of retrosynthetic analysis, design two or three possible syntheses for each of the molecules below. Then, in each case, choose the best of the routes on the basis of the criteria presented in Section 8-7.

(a) $(CH_3)_3CCH_2CH_2OH$

(b) $CH_3CH_2\overset{\overset{\displaystyle CH_3}{|}}{\underset{\underset{\displaystyle CH_3CH_2}{|}}{C}}CH_2OH$

(c) 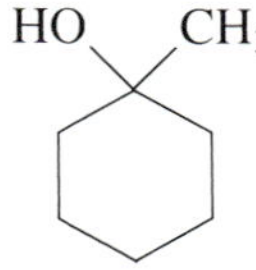 HO CH₃ on cyclohexane (1-methylcyclohexanol)

(d) $C_6H_5CH_2\overset{\overset{\displaystyle OH}{|}}{C}HCH_2C_6H_5$

(e) 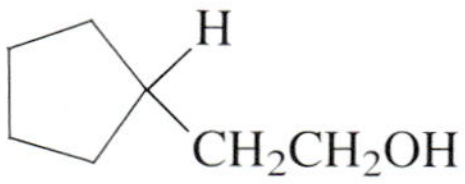 cyclopentane ring bearing H and CH_2CH_2OH

29 For each molecule in Problem 28, attempt to design a synthetic process that would replace the —OH group with each of the following atoms or groups.

(1) —Br
(2) —H
(3) $—CH_3$
(4) $—CH(CH_3)_2$
(5) $—C(CH_3)_3$

If in any case no synthetic sequence seems feasible, explain the nature of the problem.

$$CH_3(CH_2)_{14}\overset{O}{\overset{\|}{C}}O(CH_2)_{15}CH_3$$

1-Hexadecyl hexadecanoate

30 Waxes are naturally occurring esters (alkyl alkanoates) containing long, straight alkyl chains. Whale oil contains the wax 1-hexadecyl hexadecanoate:

(a) How would you synthesize this wax using an S_N2 reaction?

(b) What would the products of cleavage of this compound be with aqueous sodium hydroxide?

31 The B vitamin commonly known as niacin is used by the body to synthesize the coenzyme nicotinamide adenine dinucleotide (NAD, see Section 26-5). In the presence of a variety of enzyme catalysts, the reduced form of this substance (NADH) acts as a biological hydride donor, capable of reducing aldehydes and ketones to alcohols, according to the general formula

$$R\overset{O}{\overset{\|}{C}}R + NADH + H^+ \xrightarrow{\text{Enzyme}} R\overset{OH}{\overset{|}{C}}HR + NAD^+$$

Write the products of the NADH reduction of each of the molecules below.

(a) $CH_3\overset{O}{\overset{\|}{C}}H + NADH \xrightarrow{\text{Alcohol dehydrogenase}}$

(b) $CH_3\overset{O}{\overset{\|}{C}}\overset{O}{\overset{\|}{C}}OH + NADH \xrightarrow{\text{Lactate dehydrogenase}}$ **Lactic acid**

2-Oxopropanoic acid (Pyruvic acid)

(c) $HO\overset{O}{\overset{\|}{C}}CH_2\overset{O}{\overset{\|}{C}}\overset{O}{\overset{\|}{C}}OH + NADH \xrightarrow{\text{Malate dehydrogenase}}$ **Malic acid**

2-Oxobutanedioic acid (Oxaloacetic acid)

32 Reductions by NADH (Problem 31) are stereospecific, with the stereochemistry of the product controlled by an enzyme. The common forms of lactate and malate dehydrogenases produce exclusively the *S* stereoisomers of lactic and malic acids, respectively. Draw these stereoisomers.

33 Chemically modified steroids have become increasingly important in medicine. Write the product(s) of the reactions shown below. If more than one stereoisomer can be formed, predict the major product on the basis of delivery of the attacking reagent from the less-hindered side of the substrate molecule.

(a)

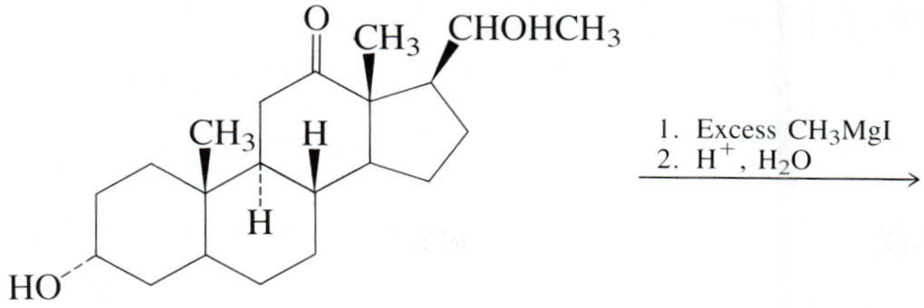

O, CH_3, $CHOHCH_3$, CH_3, H, H, HO

1. Excess CH_3MgI
2. H^+, H_2O →

(b)

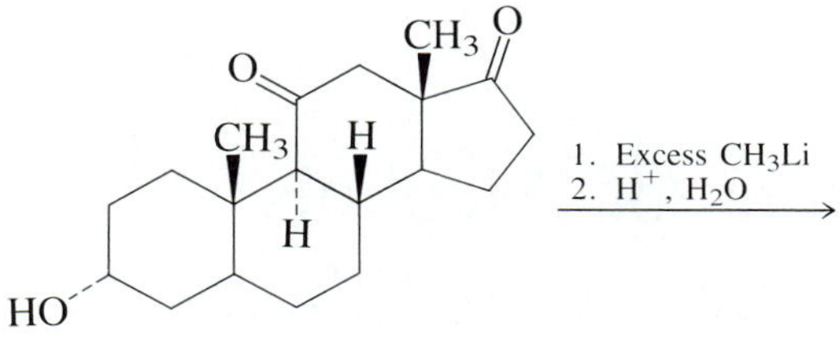

O, CH_3, O, CH_3, H, H, HO

1. Excess CH_3Li
2. H^+, H_2O →

(c)

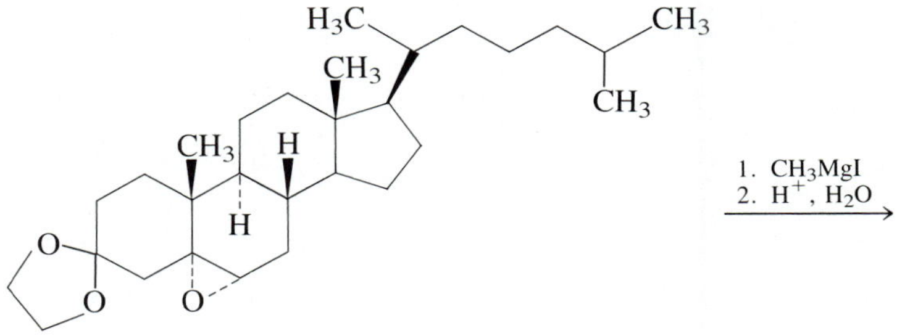

H_3C, CH_3, CH_3, CH_3, CH_3, H, H, O, O, O

1. CH_3MgI
2. H^+, H_2O →

CHAPTER 9

The Reactions of Alcohols and the Chemistry of Ethers

To explore the reactions of alcohols, we shall start with a review of some of the preparative aspects of their acidity and basicity and then examine the diverse chemistry of the carbocations formed on treatment of secondary and tertiary alcohols with acid. This is followed by an introduction to the preparation of esters from alcohols and a discussion of the oxidation of alcohols to aldehydes and ketones—the reverse of the reduction chemistry described in Section 8-4. The next section deals with alcohols as precursors to ethers through the intermediacy of their alkoxides or oxonium ions. The chemistry of ethers, particularly cyclic ethers, is the subject of subsequent sections.

The reactivity of alcohols expresses itself in a variety of ways, depending on which part of the OH group and its surroundings is attacked. Usually, at least one of the four bonds marked *a*, *b*, *c* and *d* in Figure 9-1 is cleaved.

9-1 The Preparation of Alkoxides and Carbocations

This section is a review of the preparative aspects of producing alkoxides by the deprotonation of alcohols and of synthesizing carbocations by the protonation of alcohols and the loss of water from the resulting oxonium ions.

Basic Reagents for the Production of Alkoxides

As discussed earlier (Section 8-3), alcohols are acidic and, to produce large quantities of an alkoxide, protons must be removed from the acid-base equilibrium by the use of a base that is stronger than the alkoxide. Such bases are potassium *tert*-butoxide, lithium diisopropylamide, or butyllithium (see the facing page).

FIGURE 9-1

Four general modes of reaction of alcohols: *a*, deprotonation by base; *b*, protonation by acid followed by uni- or bimolecular substitution; *b*, *c*, elimination; and *a*, *d*, oxidation.

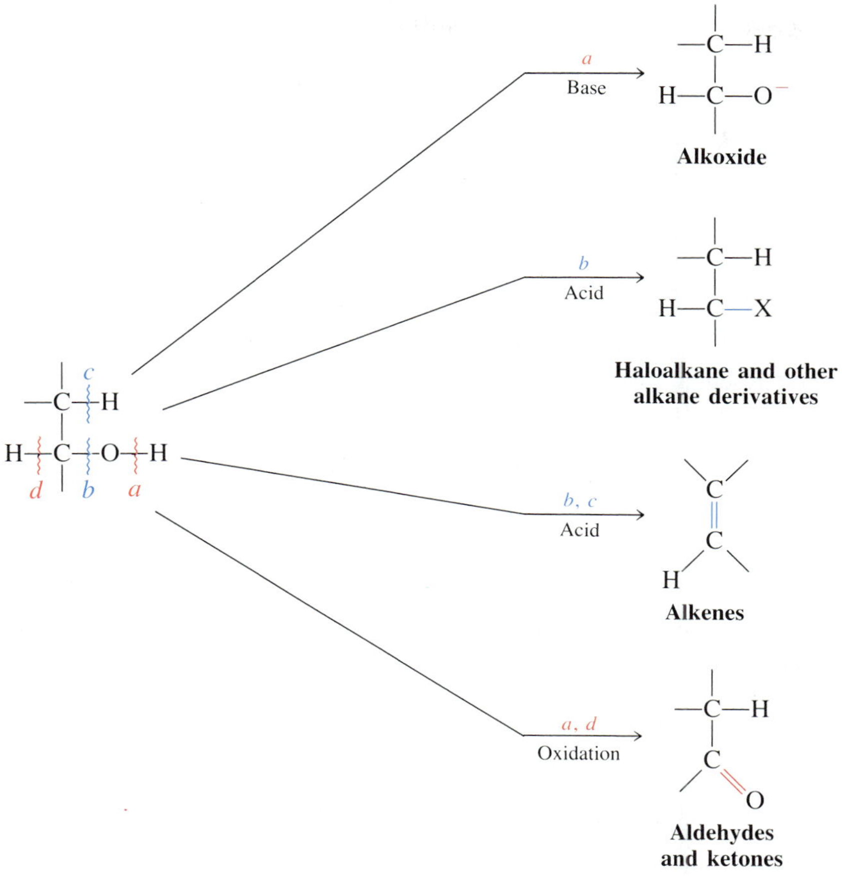

Three Ways of Making Methoxide from Methanol

$$\underset{pK_a = 15.5}{CH_3OH} + (CH_3)_3CO^-K^+ \underset{}{\overset{K = 300}{\rightleftharpoons}} CH_3O^-K^+ + \underset{pK_a = 18}{(CH_3)_3COH}$$

$$\underset{pK_a = 15.5}{CH_3OH} + Li^{+-}NCH(CH_3)_2\,[CH(CH_3)_2] \overset{K = 10^{24.5}}{\rightleftharpoons} CH_3O^-Li^+ + \underset{pK_a = 40}{HNCH(CH_3)_2\,[CH(CH_3)_2]}$$

$$\underset{pK_a = 15.5}{CH_3OH} + CH_3CH_2CH_2CH_2Li \overset{K = 10^{34.5}}{\rightleftharpoons} CH_3O^-Li^+ + \underset{pK_a = 50}{CH_3CH_2CH_2CH_2H}$$

Other bases employed to create alkoxides are sodium amide, $NaNH_2$, and hydrides, such as sodium hydride, NaH, or potassium hydride, KH. The hydrides are particularly useful because the only by-product of the reaction with alcohols is hydrogen gas.

$$\underset{\text{p}K_a = 18}{(CH_3)_3COH} + Na^{+}\,{}^{-}NH_2 \xrightleftharpoons{K = 10^{15}} (CH_3)_3CO^{-}Na^{+} + \underset{\text{p}K_a = 33}{HNH_2}$$

$$\underset{\text{p}K_a = 15.9}{CH_3CH_2OH} + K^{+}H^{-} \xrightleftharpoons{K = 10^{22.1}} CH_3CH_2O^{-}K^{+} + \underset{\text{p}K_a = 38}{H{-}H}$$

EXERCISE 9-1

Considering the pK_a data in Table 6-4, would you use sodium cyanide as a reagent to convert methanol into sodium methoxide? Explain your answer.

Alkali Metals React with Alcohols to Give Alkoxides

Another method frequently employed to obtain alkoxides is the exposure of alcohols to alkali metals. Such metals react with water as reducing agents, in some cases violently, to yield alkali hydroxides and hydrogen gas. With the more-reactive metals (sodium, potassium, and cesium), exposure to water in air may be explosive, because the hydrogen generated can ignite spontaneously and even detonate.

$$2\ H{-}OH + 2\ M\ (Li,\ Na,\ K) \longrightarrow 2\ M^{+}\ {}^{-}OH + H_2$$

The same chemical transformation takes place with alcohols to give alkoxides, but is less vigorous.

Alkoxides from Alcohols and Alkali Metals

$$2\ CH_3OH + 2\ Li \longrightarrow 2\ CH_3O^{-}Li^{+} + H_2$$

$$2\ CH_3CH_2OH + 2\ Na \longrightarrow 2\ CH_3CH_2O^{-}Na^{+} + H_2$$

$$2\ (CH_3)_3COH + 2\ K \longrightarrow 2\ (CH_3)_3CO^{-}K^{+} + H_2$$

The order of reactivity of the alcohols in this process decreases with increasing substitution, methanol being most reactive and tertiary alcohols least reactive.

Reactivity of ROH with Alkali Metals

$$R = CH_3 > \text{primary} > \text{secondary} > \text{tertiary}$$

2-Methyl-2-propanol reacts so slowly that it may be used safely to destroy potassium residues in the laboratory without any particular precautions.

What are alkoxides good for? We have already seen that they may be useful reagents in organic synthesis. For example, the reaction of hindered alkoxides with haloalkanes gives elimination.

$$CH_3CH_2CH_2CH_2Br \xrightarrow{(CH_3)_3CO^{-}K^{+},\ (CH_3)_3COH} CH_3CH_2CH{=}CH_2 + (CH_3)_3COH + K^{+}Br^{-}$$

Less-branched alkoxides attack primary and some secondary haloalkanes by the S_N2 reaction to give ethers. This method is described in Section 9-5.

Protonation of Alcohols and Carbocation Synthesis

The protonation of alcohols gives oxonium ions (Section 8-3). When derived from secondary and tertiary systems, these ions may lose water to form carbocations. Primary oxonium ions, however, are relatively stable in this respect because the resulting primary carbocations are too high in energy (Section 7-3). On the other hand, primary oxonium ions are subject to nucleophilic attack. For example, the oxonium ion resulting from the treatment of 1-butanol with HBr undergoes displacement by bromide to form 1-bromobutane (Section 6-7). Note that in the following scheme color is again used in a functional sense. The originally nucleophilic (red) oxygen is protonated by the electrophilic proton (blue) to give the oxonium ion containing an electrophilic (blue) carbon and H_2O as a leaving group (green). In the subsequent S_N2 reaction, bromide acts as a nucleophile.

$$CH_3CH_2CH_2CH_2OH + HBr \longrightarrow CH_3CH_2CH_2CH_2\overset{+}{O}H_2 + Br^- \longrightarrow CH_3CH_2CH_2CH_2Br + H_2O$$

Iodoalkanes also can be made in this fashion; however, primary chloroalkanes cannot, because chloride ion is too weak a nucleophile. The conversion of a primary alcohol into the corresponding chloroalkane requires other methods (Section 9-3).

In the absence of a good nucleophile, primary oxonium ions are stable enough to be observable. For example, Olah* discovered that protonated 1-butanol is stable after undergoing protonation by fluorosulfonic acid in liquid sulfur dioxide solvent.

Protonation of 1-Butanol by Fluorosulfonic Acid

$$CH_3CH_2CH_2CH_2OH + \underset{\textbf{Fluorosulfonic acid}}{FSO_3H} \xrightarrow{SO_2 \text{ solvent}} \underset{\textbf{Stable}}{CH_3CH_2CH_2CH_2\overset{+}{O}H_2} + \underset{\textbf{Extremely weak nucleophile}}{FSO_3^-}$$

Both the solvent and the conjugate base, the fluorosulfonate ion, FSO_3^-, are extremely weak nucleophiles, incapable of entering into S_N2 reactions with the protonated alcohol.

As stated earlier, in contrast with primary alcohols, their secondary and tertiary counterparts lose water after protonation to give the corresponding carbocations (Sections 7-2 and 7-4), which are reactive. For example, if nucleophiles are present, we observe S_N1 products and, in their absence and at elevated temperatures, E1 products. Is it possible to generate and observe such carbocations? Yes, but only if they are prepared at low temperature and in the strict absence of nucleophiles. For example, antimony pentafluoride as a Lewis acid strips 2-methyl-2-propanol of its hydroxy group. The resulting solution contains the 1,1-dimethylethyl (*tert*-butyl) cation and $HOSbF_5^-$ as a nonnucleophilic counterion in SO_2.

---

*Professor George A. Olah, b. 1927, University of Southern California, Los Angeles.

$$(CH_3)_3C\text{—}OH + SbF_5 \xrightarrow{SO_2\ \text{solvent}} CH_3\text{—}\overset{+}{C}(CH_3)_2 + HO\overline{S}bF_5$$

Since this initial discovery, many other carbocations, formerly invoked as reactive intermediates in mechanistic descriptions, have been produced in similar ways. Obtaining these intermediates as products enabled researchers to study, for the first time, their physical and chemical properties directly. Of particular interest were the kinetics of their decomposition and a reaction called rearrangement. This process is the subject of the next section.

In summary, the treatment of an alcohol with a strong base or an alkali metal gives an alkoxide. A strong acid leads to protonation to oxonium ions, which are relatively stable if they are primary but which convert into carbocations if secondary or tertiary.

9-2 Carbocation Rearrangements

Carbocations can rearrange to become other carbocations by both hydrogen and alkyl shifts. The new species may undergo S_N1 and E1 reactions, frequently giving complex mixtures unless there is a thermodynamic driving force toward one product.

Hydrogen Shifts Give New S_N1 Products

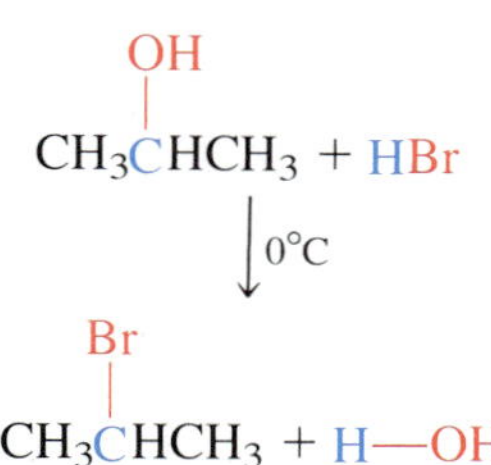

The treatment of 2-propanol with hydrogen bromide at low temperatures (to prevent elimination) gives 2-bromopropane, as expected. However, exposure of the more highly substituted secondary alcohol, 3-methyl-2-butanol, to the same reaction conditions produces a surprising result. The expected 2-bromo-3-methylbutane is only a minor product, the major component in the reaction mixture being 2-bromo-2-methylbutane. In this reaction scheme and in those that follow, color is used to indicate the origin of certain groups or atoms in a product. The migrating group or atom is in boldface type.

$$CH_3\overset{\mathbf{H}}{\underset{H_3C}{C}}\text{—}\overset{OH}{\underset{H}{C}}CH_3 \xrightarrow{HBr,\ 0^\circ C} CH_3\overset{\mathbf{H}}{\underset{H_3C}{C}}\text{—}\overset{Br}{\underset{H}{C}}CH_3 + CH_3\overset{Br}{\underset{H_3C}{C}}\text{—}\overset{\mathbf{H}}{\underset{H}{C}}CH_3 + H\text{—}OH$$

Minor product / Major product

3-Methyl-2-butanol | **2-Bromo-3-methylbutane** | **2-Bromo-2-methylbutane**

What is the mechanism of this transformation? The explanation for the observed result is found in the capability of carbocations to undergo rearrangements by **hydrogen shifts.** Initially, protonation of the alcohol followed by loss of water gives the expected secondary carbocation. A shift of the tertiary hydrogen to the electron-deficient neighbor then generates a tertiary cation, which is more stable. This species is finally trapped by bromide ion to give the rearranged S_N1 product. The mechanism for this rearrangement is illustrated below. (Recall that the arrows denote the flow of electron pairs.) In this mechanistic scheme and those that follow, color is used to indicate the electrophilic (blue), nucleophilic (red), and leaving-group (green) character of the reacting centers. Therefore, a color may again "switch" from one group or atom to another as the reaction proceeds.

Mechanism:

STEP 1: Protonation

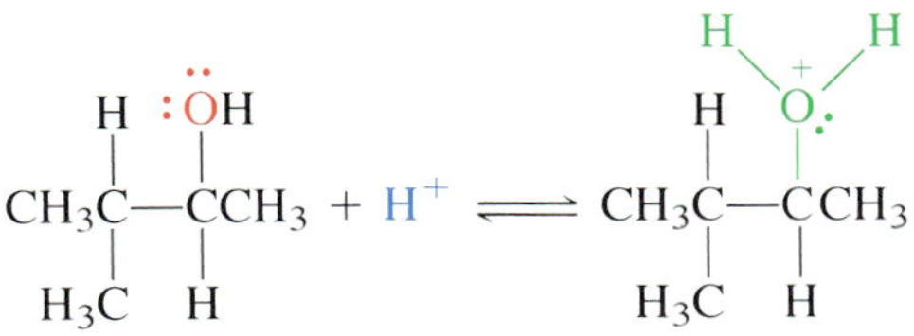

H :ÖH
$CH_3C—CCH_3 + H^+ \rightleftharpoons CH_3C—CCH_3$ (with $H\overset{+}{O}H$ on the second carbon)
H_3C H

STEP 2: Loss of water

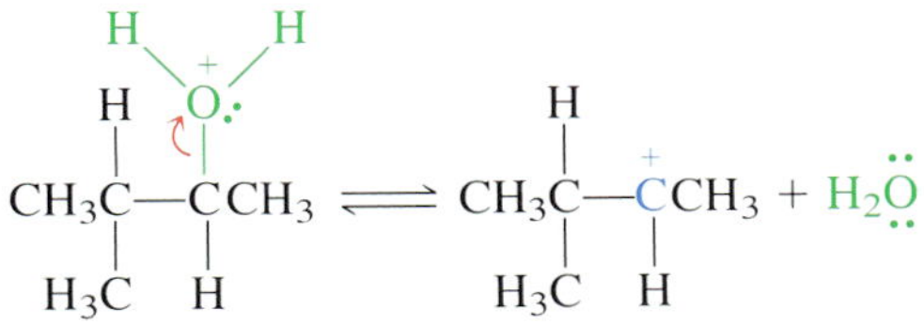

H $H\overset{+}{O}H$
$CH_3C—CCH_3 \rightleftharpoons CH_3C—\overset{+}{C}CH_3 + H_2\ddot{O}$
H_3C H

STEP 3: Hydrogen shift

H **H**
$CH_3C—\overset{+}{C}CH_3 \longrightarrow CH_3\overset{+}{C}—CCH_3$
H_3C H

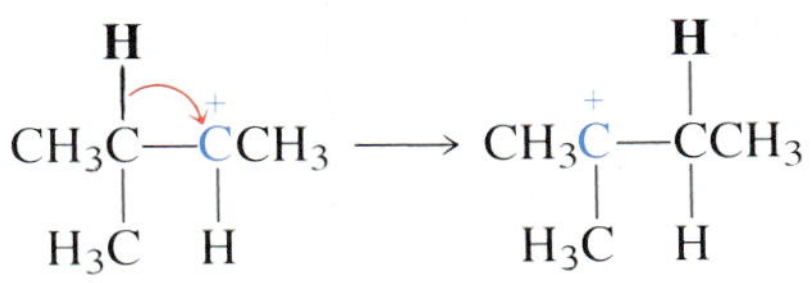

STEP 4: Trapping by bromide

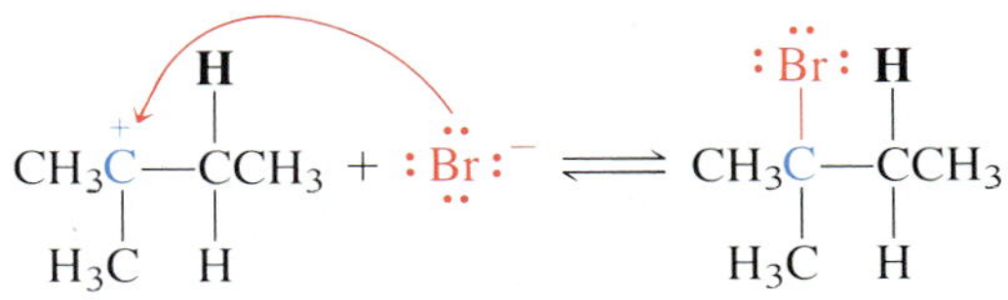

H :Br: **H**
$CH_3\overset{+}{C}—CCH_3 + :\ddot{Br}:^- \rightleftharpoons CH_3C—CCH_3$
H_3C H

The details of the transition state of the observed hydrogen shift are shown schematically in Figure 9-2. The process is another example of a concerted reaction in which a bond forms and a bond breaks simultaneously. As the hydrogen begins its migration, it increasingly interacts with the empty *p* orbital of the adjacent positively charged carbon. In the transition state, the orbitals of both carbons rehybridize; the center that was originally sp^3-hybridized eventually adopts an sp^2 configuration and assumes the positive charge when the hydrogen migration is complete. The hydrogen moves with both electrons to secure a bond to the neighboring carbon atom, a process sometimes called "hydride transfer" (although free H^- is not a participant). As the bond is formed between the migrating hydrogen and the originally cationic carbon, it rehybridizes from sp^2 to sp^3. A simple rule to remember when executing hydrogen shifts in carbocations is that the hydrogen and the positive charge exchange places between the two neighboring carbon atoms participating in the reaction.

Hydrogen shifts are generally very fast and particularly favorable when the new carbocation is more stable than the original one, as in the example depicted in Figure 9-2.

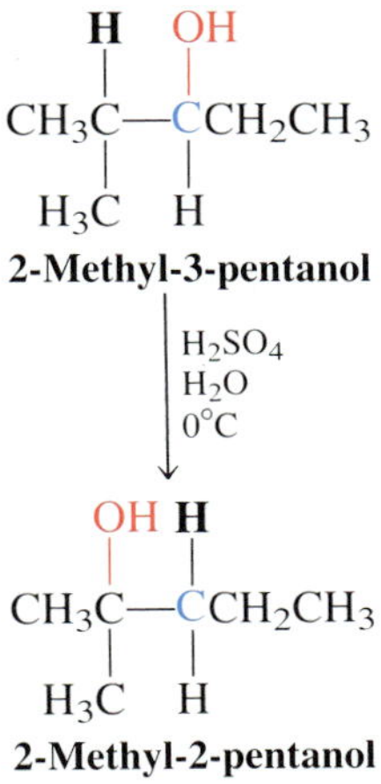

EXERCISE 9-2

2-Methylcyclohexanol on treatment with HBr gives 1-bromo-1-methylcyclohexane. Explain by a mechanism.

Rearranged carbocations can be trapped by any relatively good nucleophile in the reaction mixture. For example, treatment of 2-methyl-3-pentanol with aqueous sulfuric acid furnishes the rearranged alcohol 2-methyl-2-pentanol. The hydrogen sulfate ion in this case is too weakly nucleophilic to compete with water for the intermediate cation.

H OH
$CH_3C—CCH_2CH_3$
H_3C H
2-Methyl-3-pentanol

↓ H_2SO_4, H_2O, 0°C

OH **H**
$CH_3C—CCH_2CH_3$
H_3C H
2-Methyl-2-pentanol

FIGURE 9-2

The rearrangement of a carbocation by a hydrogen shift: (A) dotted-line notation; (B) orbital picture.

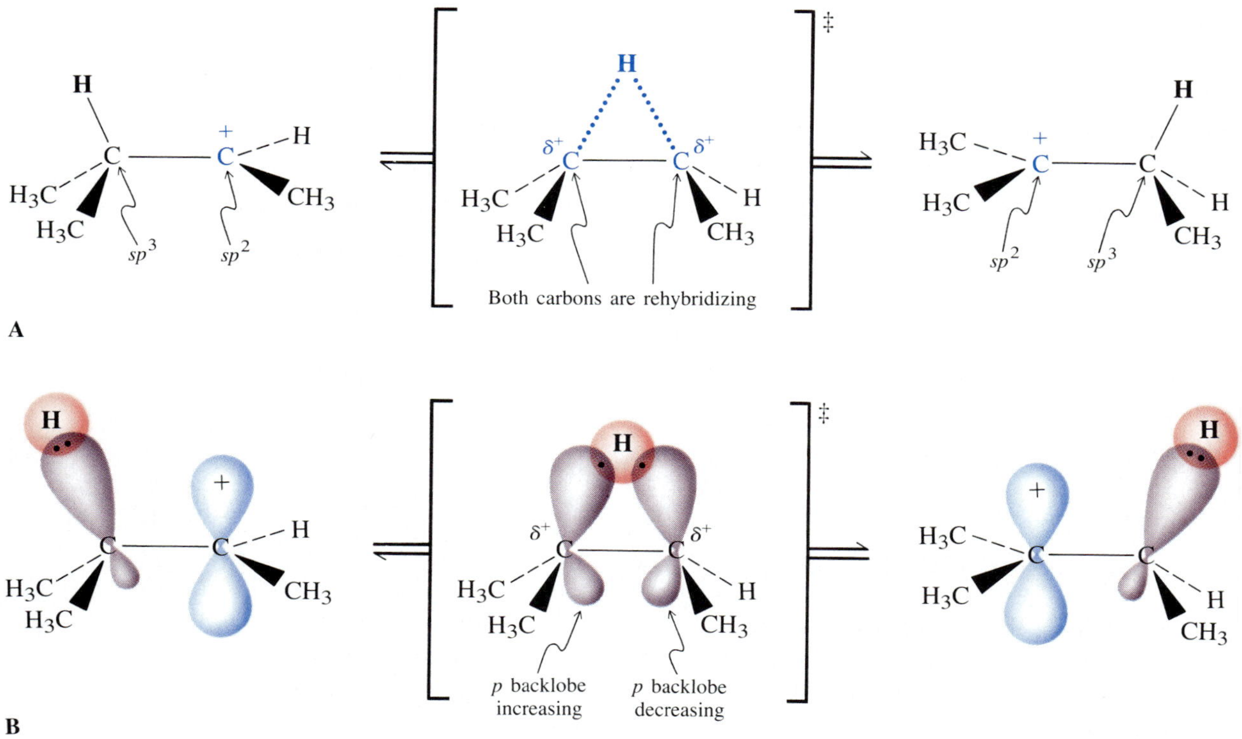

A

Both carbons are rehybridizing

B

p backlobe increasing

p backlobe decreasing

EXERCISE 9-3

Predict the major product from the following reactions: (a) 2-methyl-3-pentanol + H_2SO_4, CH_3OH solvent; (b) 1-cyclohexylethanol (cyclohexane ring bearing H and $CH_3CH(OH)$ on the same carbon) + HCl

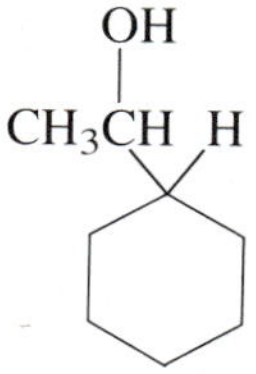

Primary carbocations are too unstable to be formed by rearrangement. However, carbocations of comparable stability—for example, secondary-secondary or tertiary-tertiary—equilibrate readily. In this case, any added nucleophile will trap all carbocations present, furnishing mixtures of products:

$$CH_3CH(OH)CH_2CH_2CH_3 \xrightarrow{HBr,\ 0^\circ C} CH_3CHBrCH_2CH_2CH_3 + CH_3CH_2CHBrCH_2CH_3$$

BOX 9-1

The Cyclopentyl Cation Undergoes a Degenerate Rearrangement

A particularly fascinating rearrangement is observed in the cyclopentyl cation. Here, successive hydrogen shifts are rapid, equilibrating all carbons. This phenomenon can be detected by the isotopic labeling of one carbon in the molecule. Thus, when ^{13}C-labeled cyclopentanol is treated with HBr, three bromides are formed in the statistical ratio of 1:2:2; they differ only with respect to the relative positions of label and halogen.

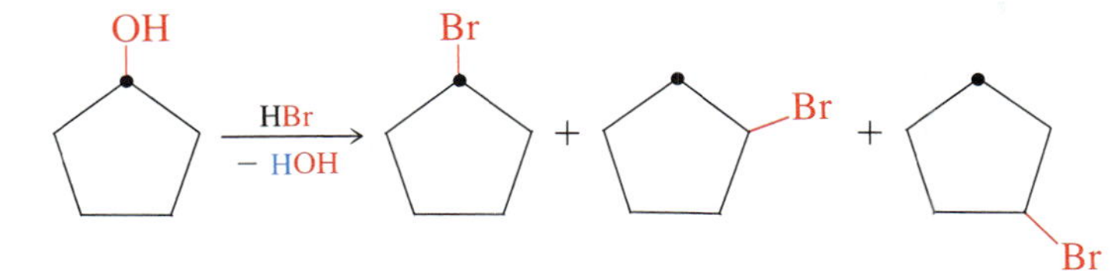

OH — HBr, − HOH → Br + Br + Br

• = ^{13}C label   1 : 2 : 2
Statistical ratio

Mechanism:

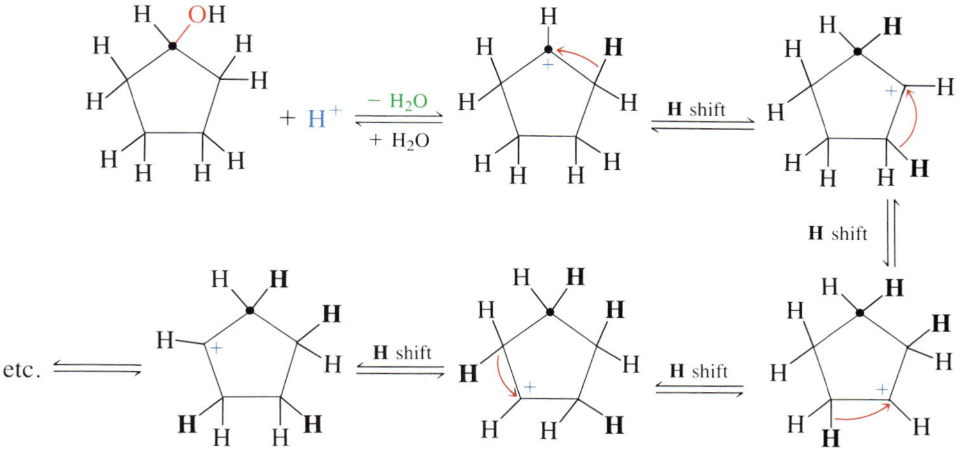

+ H^+ ⇌ (− H_2O / + H_2O) ⇌ **H** shift ⇌ **H** shift ⇌ **H** shift ⇌ **H** shift ⇌ etc.

Transformations of this type, in which a molecular reorganization leads to the same molecule, are ordinarily undetectable and are called **degenerate rearrangements.**

Under conditions favoring dissociation, haloalkanes also can undergo carbocation rearrangements. For example, ethanolysis of 2-bromo-3-ethyl-2-methylpentane gives the two possible tertiary ethers:

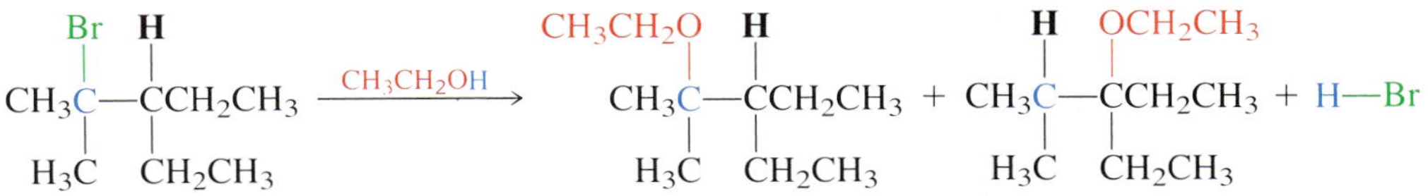

$$CH_3C(Br)(CH_3)—CH(CH_2CH_3)CH_2CH_3 \xrightarrow{CH_3CH_2OH} CH_3C(OCH_2CH_3)(CH_3)—CH(CH_2CH_3)CH_2CH_3 + CH_3CH(CH_3)—C(OCH_2CH_3)(CH_2CH_3)CH_2CH_3 + H—Br$$

2-Bromo-3-ethyl-2-methylpentane → **2-Ethoxy-3-ethyl-2-methylpentane** + **3-Ethoxy-3-ethyl-2-methylpentane**

EXERCISE 9-4

Predict the outcome of the reaction of 2-chloro-4-methylpentane with methanol.

Rearrangements to Hydroxy Carbocations

Resonance Structures of a Hydroxy Carbocation

$$\left[\ \overset{+}{C}-\ddot{O}H \longleftrightarrow C=\overset{+}{\ddot{O}}-H\ \right]$$

There is a strong driving force for a hydrogen shift that leads to a hydroxy-substituted carbocation. This cation is stabilized by resonance, in which a lone electron pair on oxygen is donated to the electron-deficient carbon atom. The deprotonation of such an ion produces a carbonyl compound. Hydroxy carbocations may be intermediates in transformations of 1,2-diols. For example, 2,3-butanediol on treatment with sulfuric acid yields 2-butanone.

$$CH_3CH(OH)CH(OH)CH_3 \xrightarrow[-\ H_2O]{H_2SO_4} CH_3COCH_2CH_3$$

2,3-Butanediol → 2-Butanone (95%)

Mechanism:

$$CH_3CH(\ddot{O}H)CH(\ddot{O}H)CH_3 \underset{-\ H^+}{\overset{+\ H^+}{\rightleftharpoons}} CH_3CH(\ddot{O}H)CH(\overset{+}{O}H_2)CH_3 \underset{+\ H_2\ddot{O}:}{\overset{-\ H_2\ddot{O}:}{\rightleftharpoons}} CH_3CH(\ddot{O}H)\overset{+}{C}HCH_3 \xrightarrow{\text{H shift}}$$

$$\left[\ CH_3\overset{+}{C}(\ddot{O}H)CH_2CH_3 \longleftrightarrow CH_3C(=\overset{+}{O}H)CH_2CH_3\ \right] \underset{+H^+}{\overset{-\ H^+}{\rightleftharpoons}} CH_3C(=\ddot{O}:)CH_2CH_3$$

When unsymmetrical 1,2-diols are heated with acid, the more-stable carbocation is formed first, controlling the outcome of the reaction as shown in the preparation of 2,3-dimethylbutanal from 2,3-dimethyl-1,2-butanediol.

$$(CH_3)_2CHC(OH)(CH_3)CH_2OH \xrightarrow{H^+,\ 100°C,\ 4\ h} (CH_3)_2CHCH(CH_3)CHO$$

2,3-Dimethyl-1,2-butanediol → 2,3-Dimethylbutanal (61%)

Mechanism:

$$RC(\ddot{O}H)(CH_3)CH_2\ddot{O}H + H^+ \underset{+\ H_2\ddot{O}:}{\overset{-\ H_2\ddot{O}:}{\rightleftharpoons}} R\overset{+}{C}(H_3C)-CH(H)-\ddot{O}H \longrightarrow$$

$$\left[\ RCH(H_3C)-\overset{+}{C}H(\ddot{O}H) \longleftrightarrow R-CH(H_3C)-CH=\overset{+}{O}-H\ \right] \underset{+\ H^+}{\overset{-\ H^+}{\rightleftharpoons}} RCH(H_3C)CH=\ddot{O}:$$

EXERCISE 9-5

Predict the product of the treatment of 1,2-cyclohexanediol with sulfuric acid.

Carbocation Rearrangements Also Give E1 Products

At elevated temperatures and in the presence of a relatively nonnucleophilic medium, rearranged carbocations furnish alkenes by the E1 mechanism. For example, treatment of 2-methyl-2-pentanol with sulfuric acid at 80°C gives the same major alkene product as that formed if the starting material is 4-methyl-2-pentanol. The conversion of the second alcohol includes a hydrogen shift of the initial carbocation, followed by deprotonation.

$$\underset{\textbf{2-Methyl-2-pentanol}}{CH_3\overset{OH}{\underset{CH_3}{C}}CH_2CH_2CH_3} \xrightarrow[-\,H_2O]{H_2SO_4,\ 80^\circ C} \underset{\textbf{2-Methyl-2-pentene}}{\underset{\text{Major product}}{(H_3C)_2C{=}C(H)CH_2CH_3}} \xleftarrow[-\,H_2O]{H_2SO_4,\ 80^\circ C} \underset{\textbf{4-Methyl-2-pentanol}}{CH_3\overset{CH_3}{\underset{H}{C}}CH_2\overset{OH}{\underset{H}{C}}CH_3}$$

Mechanism:

$$CH_3\overset{CH_3}{\underset{H}{C}}CH_2\overset{OH}{\underset{H}{C}}CH_3 + H^+ \underset{+\,H_2O}{\overset{-\,H_2O}{\rightleftharpoons}} CH_3\overset{H_3C}{\underset{H}{C}}{-}\overset{\mathbf{H}}{C}H\overset{+}{\underset{H}{C}}CH_3 \overset{\mathbf{H}\ \text{shift}}{\rightleftharpoons}$$

$$CH_3\overset{H_3C}{\underset{H}{C}}\overset{+}{C}H\overset{\mathbf{H}}{\underset{H}{C}}CH_3 \rightleftharpoons (H_3C)_2C{=}C(H)CH_2CH_3 + H^+$$

EXERCISE 9-6

Treatment of 4-methylcyclohexanol with hot acid gives 1-methylcyclohexene, Explain by a mechanism.

CH_3

Carbocation Rearrangements by Alkyl Shifts

Carbocations without suitable (secondary and tertiary) hydrogens next to the positively charged carbon can undergo another mode of rearrangement known as **alkyl group migration.** For example, 3,3-dimethyl-2-butanol on treatment with hydrogen bromide gives a high yield of 2-bromo-2,3-dimethylbutane, derived by methyl migration:

$$\underset{\textbf{3,3-Dimethyl-2-butanol}}{CH_3\overset{\mathbf{H_3C}}{\underset{H_3C}{C}}{-}\overset{CH_3}{\underset{H}{C}}OH} \xrightarrow[-\,HOH]{HBr} \underset{\textbf{2-Bromo-2,3-dimethylbutane}}{\underset{94\%}{CH_3\overset{Br}{\underset{H_3C}{C}}{-}\overset{\mathbf{CH_3}}{\underset{H}{C}}CH_3}}$$

Alkyl Group Migration

$$R{-}C{-}\overset{+}{C} \longrightarrow \left[C\overset{R}{\underset{+}{\cdots}}C \right]^{\ddagger} \longrightarrow \overset{+}{C}{-}C{-}R$$

General Carboxylic Ester Synthesis

$$RCOH(=O) + R'OH \overset{H^+}{\rightleftharpoons} RCOR'(=O) + HOH$$

The Synthesis of Organic Esters

Alcohols react with carboxylic acids in the presence of catalytic amounts of mineral acid to give esters and water. Starting materials and products in this transformation form an equilibrium that can be shifted in either direction by the use of an excess of one of the components. For example, use of the alcohol as solvent drives the reaction to the product side, whereas excess water favors the hydrolysis of the ester. Examples of both processes are shown below. The mechanism of ester formation is discussed in detail in Section 17-8.

Esterification

$$\underset{\textbf{Ethanoic (acetic) acid}}{CH_3\overset{O}{\overset{\|}{C}}OH} + \underset{\textbf{Ethanol solvent}}{CH_3CH_2OH} \xrightarrow{H^+} \underset{\textbf{Ethyl ethanoate (acetate)}}{CH_3\overset{O}{\overset{\|}{C}}OCH_2CH_3} + HOH$$

Ester Hydrolysis

$$C_6H_{11}\overset{O}{\overset{\|}{C}}OCH_3 + \underset{\text{Excess}}{H_2O} \xrightarrow{H^+} C_6H_{11}\overset{O}{\overset{\|}{C}}OH + CH_3OH$$

Inorganic Esters as Intermediates in Haloalkane Synthesis

Primary and secondary alcohols react with phosphorus tribromide, a readily available commercial compound, to give bromoalkanes and phosphorous acid. This method constitutes a general way of making bromoalkanes from alcohols. All three bromide atoms may be transferred from the phosphorus to the alkyl group.

$$\underset{\textbf{3-Pentanol}}{3\ CH_3CH_2\overset{CH_3CH_2}{\overset{|}{\underset{H}{\underset{|}{C}}}}OH} + \underset{\textbf{Phosphorus tribromide}}{PBr_3} \xrightarrow{(CH_3CH_2)_2O} \underset{\textbf{3-Bromopentane}}{\underset{47\%}{3\ CH_3CH_2\overset{CH_3CH_2}{\overset{|}{\underset{H}{\underset{|}{C}}}}Br}} + \underset{\textbf{Phosphorous acid}}{H_3PO_3}$$

What is the mechanism of action of PBr_3? In the first step, the alcohol and the phosphorus reagent form a species that can be viewed as a protonated inorganic ester derivative of phosphorous acid:

$$RCH_2\ddot{O}H + Br_2P{-}Br \longrightarrow RCH_2\overset{+}{\ddot{O}}(H){-}PBr_2 + Br^-$$

$HOPBr_2$ is a good leaving group and is displaced by the bromide generated in the ester-forming step:

$$:\ddot{\underset{..}{Br}}:^- + RCH_2-\overset{+}{\ddot{O}}(H)-PBr_2 \longrightarrow RCH_2\ddot{\underset{..}{Br}}: + H\ddot{\underset{..}{O}}PBr_2$$

$HOPBr_2$ will then react successively with two more molecules of alcohol by mechanisms analogous to that of the first ester-forming step:

$$RCH_2\ddot{\underset{..}{O}}H + H\ddot{\underset{..}{O}}PBr_2 \longrightarrow RCH_2\overset{+}{\ddot{O}}(H)-\underset{..}{\overset{Br}{P}}OH + :\ddot{\underset{..}{Br}}:^-$$

$$:\ddot{\underset{..}{Br}}:^- + RCH_2-\overset{+}{\ddot{O}}(H)-\underset{..}{\overset{Br}{P}}OH \longrightarrow RCH_2\ddot{\underset{..}{Br}}: + H\ddot{\underset{..}{O}}\overset{Br}{P}\ddot{\underset{..}{O}}H$$

$$RCH_2OH + (HO)_2PBr \longrightarrow \longrightarrow RCH_2\ddot{\underset{..}{Br}}: + H_3PO_3$$

Note that the success of this transformation depends on the initial conversion of the hydroxy substituent in the alcohol into a good leaving group, as in the transformation of alcohols into bromoalkanes by hydrogen bromide. The latter method, however, is encumbered by the strongly acidic reaction conditions and carbocation rearrangements.

If instead of a bromoalkane we want the corresponding iodoalkane, the required phosphorus triiodide, PI_3, is best generated in situ, because it is a reactive species. We do this by adding red elemental phosphorus and elemental iodine to the alcohol. This procedure allows any PI_3 formed to be consumed as soon as it is generated.

$$CH_3(CH_2)_{14}CH_2OH \xrightarrow{P,\ I_2,\ \Delta} \underset{85\%}{CH_3(CH_2)_{14}CH_2I} + H_3PO_3$$

To convert alcohols into chloroalkanes, we may use phosphorus pentachloride, PCl_5:

$$\text{(cyclohexyl)}CH_2OH + PCl_5 \longrightarrow \text{(cyclohexyl)}CH_2Cl + HCl + POCl_3$$

The mechanism of this reaction is similar to that of bromination by PBr_3, except that now the participating species contain pentavalent phosphorus. Initially, after consumption of one equivalent of alcohol, phosphorus oxychloride, $POCl_3$, is generated. Further reaction with additional alcohol molecules furnishes more chloroalkane and eventually phosphoric acid, H_3PO_4.

Mechanism:

$$RCH_2OH + PCl_5 \longrightarrow RCH_2OPCl_4 + H^+ + Cl^-$$

$$H^+ + Cl^- + RCH_2OPCl_4 \longrightarrow RCH_2Cl + HOPCl_4$$

$$HOPCl_3Cl \longrightarrow O{=}PCl_3 + HCl$$

$$3\ RCH_2OH + O{=}PCl_3 \longrightarrow 3\ RCH_2Cl + \underset{\textbf{Phosphoric acid}}{O{=}P(OH)_3}$$

Another useful chlorinating agent is thionyl chloride, $SOCl_2$. Simply warming an alcohol in its presence results in the evolution of SO_2 and HCl and the formation of the chloroalkane.

$$CH_3CH_2CH_2OH + SOCl_2 \longrightarrow \underset{91\%}{CH_3CH_2CH_2Cl} + O{=}S{=}O + HCl$$

Mechanistically, the alcohol RCH_2OH again first forms an inorganic ester, RCH_2O_2SCl. The chloride ion created in this process attacks the ester as a nucleophile, yielding one molecule each of SO_2 and HCl.

Mechanism:

$$RCH_2OH + Cl\overset{O}{\overset{\|}{S}}Cl \longrightarrow RCH_2O\overset{O}{\overset{\|}{S}}Cl + H^+ + Cl^-$$

$$H^+ + :\ddot{Cl}:^- + \underset{R}{\underset{|}{C}}H_2{-}\ddot{O}{-}\overset{:\ddot{O}:}{\overset{\|}{S}}{-}\ddot{Cl}: \longrightarrow :\ddot{Cl}CH_2R + \ddot{O}{=}S{=}\ddot{O} + H\ddot{Cl}:$$

$(CH_3CH_2)_3N:$
***N*,*N*-Diethylethanamine (Triethylamine)**
+
HCl
↓
$(CH_3CH_2)_3\overset{+}{N}HCl^-$

The reaction works better in the presence of an amine, which neutralizes the hydrogen chloride generated. An example is *N*,*N*-diethylethanamine (triethylamine), which forms the corresponding ammonium hydrochloride under these conditions.

The inorganic esters in these reactions are special examples of leaving groups derived from sulfur-based acids. They are related to the sulfonates discussed in Section 6-7, in which it was mentioned that sulfonates can be readily prepared from the corresponding sulfonyl chlorides and an alcohol.

$$\underset{\textbf{2-Methyl-1-propanol}}{CH_3\overset{CH_3}{\overset{|}{C}}HCH_2OH} + \underset{\textbf{Methanesulfonyl chloride}}{CH_3\overset{O}{\overset{\|}{\underset{O}{\underset{\|}{S}}}}Cl} + \underset{\textbf{Pyridine}}{C_5H_5N} \longrightarrow \underset{\textbf{2-Methylpropyl methanesulfonate}}{CH_3\overset{CH_3}{\overset{|}{C}}HCH_2O\overset{O}{\overset{\|}{\underset{O}{\underset{\|}{S}}}}CH_3} + \underset{\textbf{Pyridinium hydrochloride}}{C_5H_5\overset{+}{N}H\ Cl^-}$$

Like thionyl chloride, sulfonyl chlorides undergo reaction best in the presence of an amine, but, unlike RCH_2O_2SCl, sulfonates can be isolated.

$$\underset{\textbf{Cyclohexanol}}{C_6H_{11}OH} + \underset{\textbf{4-Methylbenzenesulfonyl chloride (p-Toluenesulfonyl chloride)}}{CH_3{-}C_6H_4{-}SO_2Cl} + (CH_3CH_2)_3N \longrightarrow \underset{\textbf{Cyclohexyl 4-methylbenzenesulfonate (Cyclohexyl tosylate)}}{C_6H_{11}OSO_2{-}C_6H_4{-}CH_3} + (CH_3CH_2)_3\overset{+}{N}H\,Cl^-$$

As we know, the displacement of sulfonate groups by halide ions (Cl^-, Br^-, I^-) readily yields the corresponding haloalkanes, particularly with primary and secondary systems, in which S_N2 reactivity is good. The preparation of tertiary halides by this method is inefficient because carbocation formation leads to elimination and rearrangement products. Tertiary chlorides and bromides are best made from the corresponding alcohols by treatment with the cold concentrated hydrogen halide.

EXERCISE 9-8

Supply reagents with which you would prepare the following haloalkanes from the corresponding alcohols: (a) $I(CH_2)_6I$; (b) $(CH_3CH_2)_3CCl$;

(c) Br

In summary, alcohols react with carboxylic acids by loss of water to furnish organic esters and with inorganic halides, such as PBr_3, PCl_5, $SOCl_2$, and RSO_2Cl, by loss of HX to produce inorganic esters. These inorganic esters contain good leaving groups in nucleophilic substitutions that are, for example, displaced by halide ions to furnish the corresponding haloalkanes.

9-4 Oxidation of Alcohols: Preparation of Aldehydes and Ketones

Section 8-4 described the synthesis of alcohols from aldehydes and ketones by reduction with hydrogen or hydride reagents. The reverse is also possible: oxidation of alcohols to aldehydes and ketones.

General Scheme for the Oxidation of Alcohols

$$RCH_2-OH \xrightarrow{\text{Oxidation}} RCH{=}O \qquad\qquad R\underset{H}{\overset{OH}{C}}R' \xrightarrow{\text{Oxidation}} R\overset{O}{\overset{\|}{C}}R'$$

Primary alcohol → **Aldehyde**; **Secondary alcohol** → **Ketone**

Chromium Reagents in Alcohol Oxidations

One of the preferred reagents for the oxidation of alcohols is a transition metal in a high oxidation state: chromium (VI). In this form, chromium usually has a yellow orange color. On exposure to an alcohol, the Cr(VI) species is reduced to the deep-green Cr(III). The reagent is usually sold as the dichromate salt ($K_2Cr_2O_7$ or $Na_2Cr_2O_7$) or as CrO_3, and the oxidation is performed in acidic aqueous solution, as shown at the right for 1-propanol.*

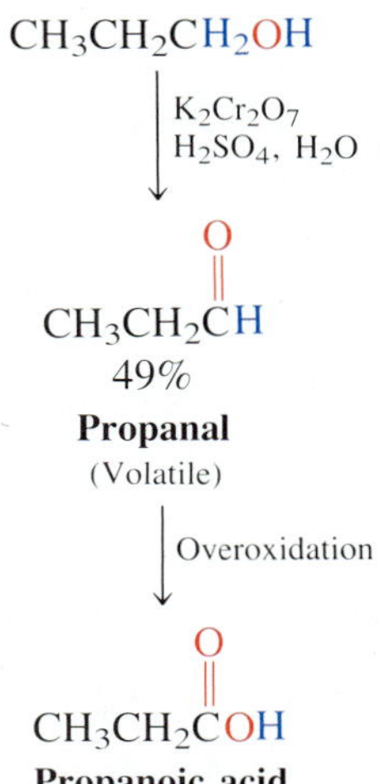

*The balanced equation for this reaction is:

$$7\,H_2SO_4 + 3\,CH_3CH_2CH_2OH + 2\,K_2Cr_2O_7 \longrightarrow 3\,CH_3CH_2\overset{O}{\overset{\|}{C}}H + 2\,Cr_2(SO_4)_3 + K_2SO_4 + 10\,H_2O$$

If you are unfamiliar with the system for balancing redox reactions, review this subject in a general chemistry text.

Under the reaction conditions, the oxidizing agent can further oxidize the aldehyde to the acid. Therefore, to isolate the aldehyde, it should be continuously removed as it is formed, usually by distillation. This technique can be used successfully only with relatively volatile aldehydes, and yields are moderate. Overoxidation can be prevented by the use of anhydrous conditions under which aldehydes are stable—for example, by employing the chromium trioxide(pyridine)$_2$ complex (see also Section 15-3):

$$\text{C}_5\text{H}_9\text{-CH}_2\text{CH}_2\text{CH}_2\text{CH}_2\text{OH} \xrightarrow{\text{CrO}_3(\text{pyridine})_2,\ \text{CH}_2\text{Cl}_2} \text{C}_5\text{H}_9\text{-CH}_2\text{CH}_2\text{CH}_2\text{CH=O} \quad 93\%$$

Ketones are more stable with Cr(VI) as a reagent because overoxidation requires C–C bond cleavage, and good yields are obtained in the oxidations of secondary alcohols:

$$\text{cyclohexanol} \xrightarrow{\text{Na}_2\text{Cr}_2\text{O}_7,\ \text{H}_2\text{SO}_4,\ \text{H}_2\text{O}} \text{cyclohexanone} \quad 96\%$$

A solution of chromium trioxide in aqueous sulfuric acid is known as the **Jones reagent.**

Oxidation with Jones Reagent

$$\text{norborneol} \xrightarrow{\text{CrO}_3,\ \text{H}_2\text{SO}_4,\ \text{H}_2\text{O}} \text{norbornanone} \quad 88\%$$

Chromium (VI) oxidations of alcohols proceed through the intermediacy of chromic esters. Their formation is formally analogous to that of organic esters from carboxylic acids. The oxidation state of chromium stays unchanged on ester formation:

$$\text{RCH}_2\ddot{\text{O}}\text{H} + \text{HO}-\overset{\text{O}}{\underset{\text{O}}{\text{Cr}^{\text{VI}}}}-\text{OH} \rightleftharpoons \text{RCH}_2\text{O}-\overset{\text{O}}{\underset{\text{O}}{\text{Cr}^{\text{VI}}}}-\text{OH} + \text{H}_2\text{O}$$

Chromic acid **Chromic ester**

The subsequent step is equivalent to an E2 reaction, with water acting as a mild base, removing the proton next to the alcohol oxygen, and $HCrO_3^-$ functioning as a leaving group. The donation of an electron pair to the chromium atom changes its oxidation state by two units, yielding Cr(IV):

$$\mathrm{H_2\ddot{O}:} \rightarrow \mathrm{H{-}C(R)(H){-}\ddot{O}{-}Cr^{VI}(=O)_2{-}\ddot{O}H} \longrightarrow \mathrm{H_2\ddot{O}^+{-}H} + \mathrm{RC(H){=}\ddot{O}} + \mathrm{{:}Cr^{IV}(=O)(\ddot{O}{:}^-)(\ddot{O}H)}$$

In contrast with the kinds of E2 reactions discussed so far, this elimination furnishes a carbon–oxygen instead of a carbon–carbon double bond. The Cr(IV) species formed may disproportionate to Cr(III) and Cr(V) or it may function as an oxidizing agent independently. Eventually all Cr(VI) is reduced to Cr(III).

Tertiary alcohols cannot be readily oxidized by chromate because they do not carry hydrogens next to the functional group.

Oxidation with Iodine: Iodoform Reaction

Secondary 2-alkanols (and ethanol) are attacked by iodine in aqueous sodium hydroxide to give a yellow precipitate of triiodomethane. This observation is characteristic for alcohols containing the structural unit $RCH(OH)CH_3$. The transformation **(iodoform reaction)** can be employed as an analytical test for the presence of this group. A carbon–carbon bond is cleaved and the remainder of the molecule converts into the corresponding carboxylate salt. The mechanism of this process will be discussed in Section 16–2.

General Iodoform Reaction

$$\mathrm{R{-}C(H)(OH){-}CH_3} \xrightarrow{I_2,\ NaOH} \underset{\text{Yellow precipitate}}{CHI_3} + \mathrm{RC(=O)O^-Na^+}$$

Synthetic Applications of Alcohol Oxidations

Having a convenient method for oxidizing alcohols to aldehydes and ketones at our disposal, we can ask how this new knowledge might be turned to synthetic advantage. A brief review of Section 8-7 will reveal immediate applications: every time a primary or secondary alcohol is generated (see Sections 8-4 and 8-6), it can be oxidized to a carbonyl compound, which in turn can be further transformed by the use of an organometallic reagent in a multitude of ways.

The Utility of Alcohol Oxidations in Synthesis

$$\mathrm{RC(OH)(H)R'} \xrightarrow{H^+,\ Na_2Cr_2O_7} \mathrm{RC(=O)R'} \xrightarrow[M\ =\ \text{metal}]{R''M} \xrightarrow{H^+,\ H_2O} \mathrm{RC(OH)(R'')R'}$$

EXERCISE 9-9

Show how you would prepare 2-bromo-2-methylpropane from methanol as the only organic starting material.

To summarize, chromium(VI) reagents oxidize primary alcohols to aldehydes and secondary alcohols to ketones. To prevent overoxidation of aldehydes, it is best to remove them as they are formed (when volatile) or to use chromium trioxide(pyridine)$_2$. Iodine in the presence of a base gives rise to the iodoform reaction with secondary 2-alkanols and ethanol. In the course of this reaction, the methyl carbon is cleaved from the chain, leaving behind a carboxylate salt.

9-5 Nucleophilic Displacements with the Alcohol Oxygen: Williamson and Other Ether Syntheses

Ethers have been mentioned on several occasions in preceding chapters. This section introduces this class of organic molecules in a systematic way, starting with the rules for naming them, continuing with a description of some of their physical properties, and ending with an outline of the methods used for their preparation.

The Naming of Ethers

$CH_3OCH_2CH_3$
Methoxyethane

$CH_3CH_2\ddot{O}C(CH_3)_3$
2-Ethoxy-2-methylpropane

cis-1-Ethoxy-2-methoxycyclopentane

The IUPAC system for naming ethers treats them as alkanes that bear an alkoxy substituent; that is, as alkoxyalkanes. The smaller substituent is considered part of the alkoxy group, and the larger defines the stem.

The alkoxyalkanes may be thought of as derivatives of alcohols in which the hydroxy proton has been replaced by an alkyl. Their common names are based on this picture: the names of the two alkyl groups are followed by the word ether. Hence, CH_3OCH_3 is dimethyl ether, $CH_3OCH_2CH_3$ is ethyl methyl ether, and so forth.

Physical Properties of Ethers

The molecular formula of simple alkoxyalkanes is $C_nH_{2n+2}O$, identical with that of the alkanols. However, because of the absence of hydrogen bonding, the boiling points of ethers are much lower than those of the corresponding isomeric alcohols (Table 9-1). The two smallest members of the series are water-miscible, but ethers become less water soluble as the hydrocarbon residues increase in size. For example, methoxymethane is completely water soluble, whereas ethoxyethane forms only an approximately 10% aqueous solution.

Ethers are generally fairly unreactive (except for strained cyclic derivatives, see Chapter 8 and Section 9-6) and are therefore frequently used as solvents in organic reactions. Some of these systems are cyclic (see Section 26-1) or contain several ether functions within their frameworks or both. All have common names.

Ether Solvents and Their Names

$CH_3CH_2OCH_2CH_3$	(six-membered ring, O at 1 and 4)	$CH_3OCH_2CH_2OCH_3$	(five-membered ring with O)
Ethoxyethane (Diethyl ether)	**1,4-Dioxacyclohexane (1,4-Dioxane)**	**1,2-Dimethoxyethane (Glycol dimethyl ether, glyme)**	**Oxacyclopentane (Tetrahydrofuran, THF)**

TABLE 9-1
The boiling points of ethers and the isomeric 1-alkanols

Ether	Name	Boiling point (°C)	1-Alkanol	Boiling point (°C)
CH_3OCH_3	Methoxymethane (Dimethyl ether)	−23	CH_3CH_2OH	78.5
$CH_3OCH_2CH_3$	Methoxyethane (Ethyl methyl ether)	10.8	$CH_3CH_2CH_2OH$	82.4
$CH_3CH_2OCH_2CH_3$	Ethoxyethane (Diethyl ether)	34.5	$CH_3(CH_2)_3OH$	117.3
$(CH_3CH_2CH_2CH_2)_2O$	1-Butoxybutane (Dibutyl ether)	142	$CH_3(CH_2)_7OH$	194.5

Ethers by S_N2 Reactions: The Williamson Ether Synthesis

The simplest way to synthesize an ether is to have an alkoxide undergo reaction with a primary haloalkane or a sulfonate ester under typical S_N2 conditions (Chapter 6), the **Williamson* ether synthesis.** The alcohol from which the alkoxide is derived can be used as the solvent (if inexpensive), but other polar molecules, such as dimethyl sulfoxide (DMSO) or hexamethylphosphoric triamide (HMPA), are equally or even more effective (Table 6-8).

Examples of the Williamson Ether Synthesis

$$CH_3CH_2CH_2CH_2O^-Na^+ + ClCH_2CH_2CH_2CH_3 \xrightarrow[\text{or DMSO, 9.5 h}]{CH_3CH_2CH_2CH_2OH,\ 14\ h} CH_3CH_2CH_2CH_2OCH_2CH_2CH_2CH_3 + Na^+Cl^-$$

60% (butanol solvent)
95% (DMSO solvent)
1-Butoxybutane

$$(CH_3)_2CHO^-Na^+ + CH_3CH_2Br \xrightarrow{HMPA} (CH_3)_2CHOCH_2CH_3 + Na^+Br^-$$

85%
2-Ethoxypropane

$$\text{cyclopentyl}(H)(O^-Na^+) + CH_3(CH_2)_{15}CH_2OSO_2CH_3 \xrightarrow{DMSO} \text{cyclopentyl}(H)(OCH_2(CH_2)_{15}CH_3) + Na^{+\,-}O_3SCH_3$$

91%
Cyclopentoxy-heptadecane

Because alkoxides are strong bases, their use in ether synthesis is restricted to primary unhindered alkylating agents; otherwise a significant amount of E2 product is formed (Section 7-5).

*Professor Alexander W. Williamson, 1824–1904, University College, London.

EXERCISE 9-10

Treatment of 4-bromo-1-butanol with aqueous sodium hydroxide gives oxacyclopentane (tetrahydrofuran). Explain.

Cyclic Ethers by Intramolecular Williamson Synthesis

The Williamson ether synthesis is also applicable to the preparation of cyclic ethers starting from halo alcohols and including an intramolecular S_N2 reaction.

General Cyclic Ether Synthesis

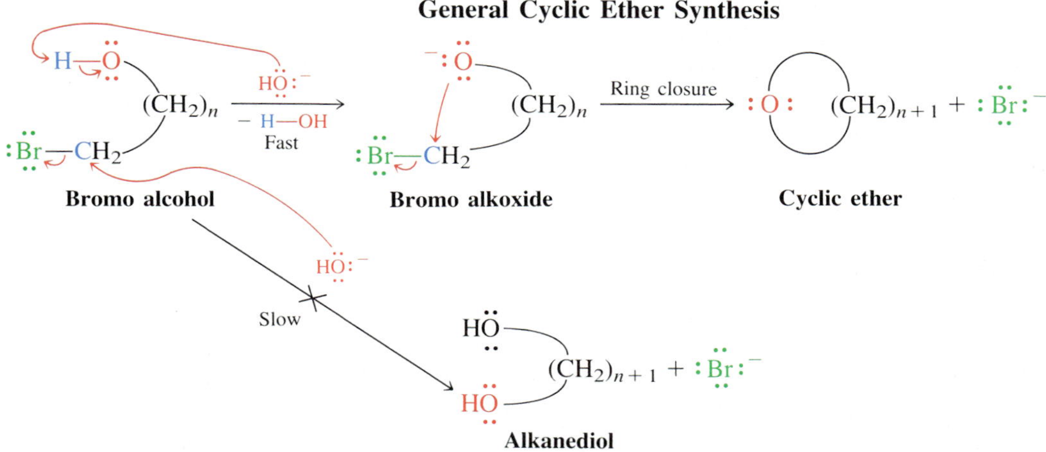

(The black curved lines denote a chain of carbon atoms.)

The mechanism of this reaction consists of the formation of an intermediate bromo alkoxide by fast proton transfer to the base, followed by ring closure to furnish the cyclic ether. The initial deprotonation of the alcohol and the subsequent step are much faster than the hydroxide's attack at the bromide terminus to give the alkanediol.

Intramolecular ring closures are usually carried out in dilute solutions to minimize the chances for intermolecular reaction. For example, in concentrated solutions of bromo alcohol, the intermediate bromo alkoxide might attack a molecule of starting material to give a new bromo alkoxy alkanol, which in turn could undergo further alkylation to give long-chain ethers. This reaction path is less likely in dilute solutions.

Intermolecular Bromo Alcohol Coupling

$$2\ \text{Br—CH}_2(\text{CH}_2)_n\text{—OH} \xrightarrow[-\ \text{Br}^-]{\text{HO}^-} \text{Br—CH}_2(\text{CH}_2)_n\text{—O—CH}_2(\text{CH}_2)_n\text{—OH} \xrightarrow{\text{HO}^-} \text{etc.}$$

Bromo alkoxy alkanol

The intramolecular Williamson synthesis allows the preparation of cyclic ethers of varying ring size, including small rings.

$$\text{HOCH}_2\text{CH}_2\text{Br} + \text{HO}^- \longrightarrow \text{(oxacyclopropane ring: O 1, C 2, C 3)} + \text{Br}^- + \text{HOH}$$

Oxacyclopropane
(Oxirane, ethylene oxide)

$$HO(CH_2)_2CH_2Br + HO^- \longrightarrow \text{Oxacyclobutane} + Br^- + HOH$$

Oxacyclobutane

$$HO(CH_2)_3CH_2Br + HO^- \longrightarrow \text{Oxacyclopentane} + Br^- + HOH$$

Oxacyclopentane
(Tetrahydrofuran)

$$HO(CH_2)_4CH_2Br + HO^- \longrightarrow \text{Oxacyclohexane} + Br^- + HOH$$

Oxacyclohexane
(Tetrahydropyran)

Cyclic ethers are members of a class of cycloalkanes in which one or more carbons have been replaced by a *heteroatom*—in this case, oxygen. A heteroatom is defined as any atom except carbon and hydrogen. Cyclic compounds of this type are called heterocycles (see Chapter 26).

The naming of cyclic ethers is diverse (see Section 26-1). The simplest system has as its basis the **oxacycloalkane stem,** in which the prefix *oxa* indicates the replacement of carbon by oxygen in the ring. Thus, three-membered cyclic ethers are oxacyclopropanes (other names used are oxiranes and ethylene oxides), four-membered systems are oxacyclobutanes, and the next two higher homologs are oxacyclopentanes (tetrahydrofurans) and oxacyclohexanes (tetrahydropyrans). The compounds are numbered by starting at oxygen and proceeding around the ring.

BOX 9-2

1,2-Dioxacyclobutanes Can Be Chemiluminescent

An intramolecular Williamson-type reaction of a special kind is that in which a 2-bromo hydroperoxide is the reactant. The peroxide product is a 1,2-dioxacyclobutane (1,2-dioxetane). This species is unusual because it decomposes to the corresponding carbonyl compounds with emission of light **(chemiluminescence)** by a process in which the O–O bond is cleaved to form a biradical, which then proceeds to product. Dioxacyclobutanes seem to be responsible for the luminescent character of certain species in nature **(bioluminescence),** although other mechanisms also contribute to this phenomenon. Terrestrial organisms, such as the firefly, the glowworm, and certain click beetles, are well known to be bioluminescent. However, most bioluminescent species live in the ocean; they range from microscopic bacteria and plankton to fish. The emitted light serves many purposes and seems to be important in courtship and communication, sex differentiation, finding prey, and hiding from or scaring off predators.

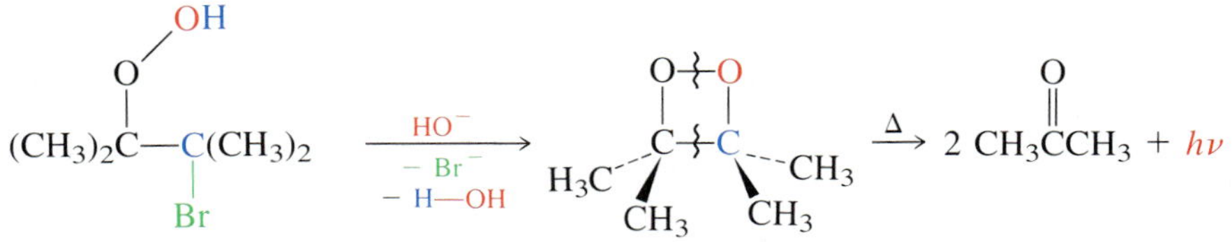

$(CH_3)_2C(OOH)—C(CH_3)_2Br \xrightarrow[-\,Br^-,\ -\,H—OH]{HO^-}$ 3,3,4,4-Tetramethyl-1,2-dioxacyclobutane $\xrightarrow{\Delta}$ $2\ CH_3CCH_3 + h\nu$

A 2-bromo hydroperoxide

3,3,4,4-Tetramethyl-1,2-dioxacyclobutane (A 1,2-dioxetane)

Propanone (Acetone)

An example of a chemiluminescent molecule in nature that causes bioluminescence is firefly luciferin. The basic oxidation of this molecule furnishes a dioxacyclobutanone intermediate that decomposes in a manner analogous to that of 3,3,4,4-tetramethyl-1,2-dioxacyclobutane to give a complex heterocycle, carbon dioxide, and emitted light.

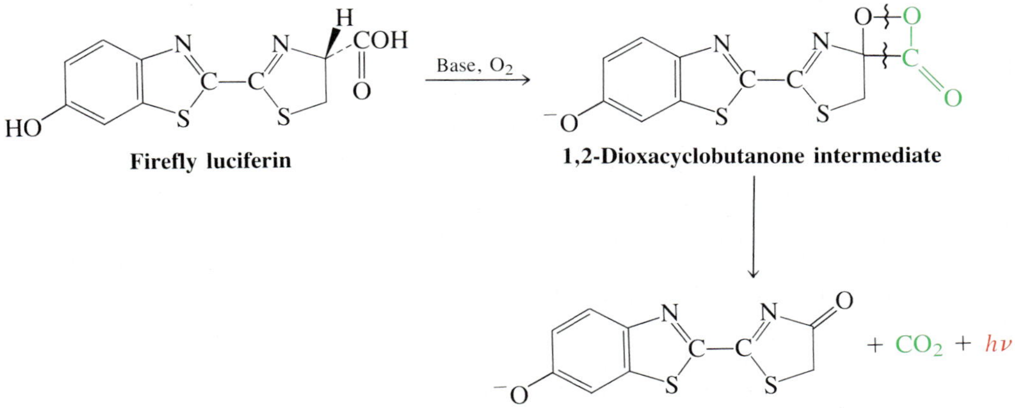

Firefly luciferin $\xrightarrow{\text{Base, } O_2}$ 1,2-Dioxacyclobutanone intermediate $\longrightarrow$ (heterocycle) $+ CO_2 + h\nu$

Firefly luciferin

1,2-Dioxacyclobutanone intermediate

Some Cyclic Polyethers Are Good Cation Binders

Certain cyclic ethers containing multiple ether functions have extraordinary solvating power. Those based on the 1,2-ethanediol unit are called **crown ethers,** so named because the molecules adopt a crownlike conformation in the crystalline state and, presumably, in solution. The polyether 18-crown-6 is shown in Figure 9-3. The number 18 refers to the total number of atoms in the ring, and 6 to the number of oxygens in the ring.

FIGURE 9-3

The crownlike structural arrangement of 18-crown-6.

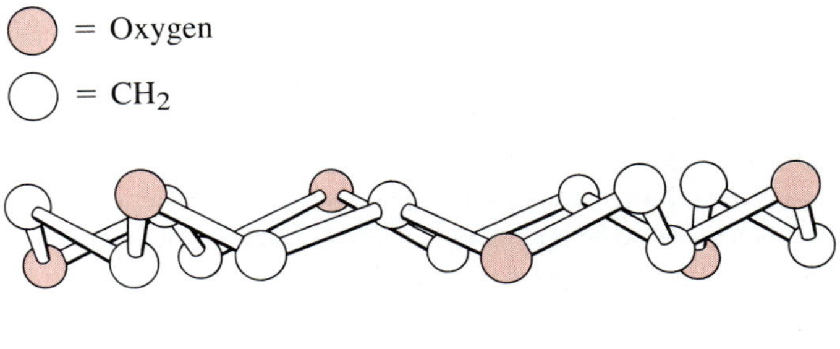

= Oxygen

= CH_2

Crown ethers have the remarkable capacity of strongly binding to cations, such as those found in ordinary salts, and rendering them soluble in organic solvents. For example, potassium permanganate, a deep-violet, completely insoluble solid in benzene, is readily dissolved in that solvent if 18-crown-6 is added to it. This solution is useful because it allows oxidations with potassium permanganate to be carried out in organic solvents. Dissolution is possible by effective solvation of the metal ion by the six crown oxygens:

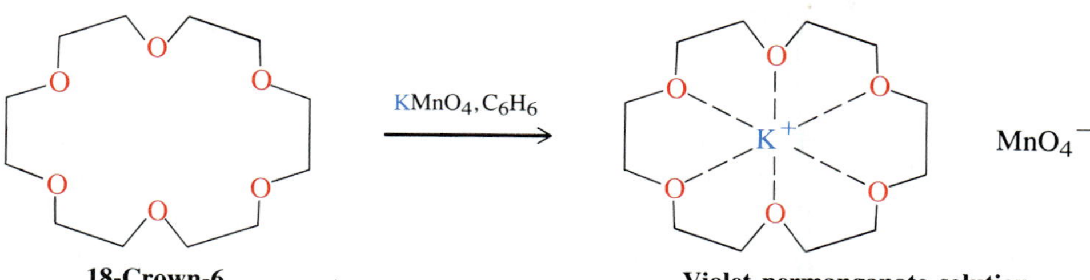

The size of the ''cavity'' in the crown ether can be tailored to allow for the selective binding of only certain cations—namely, those whose ionic radius is best accommodated by the polyether. This concept has been extended successfully into three dimensions by the synthesis of polycyclic ethers, also called **cryptands** (*kryptos,* Greek, hidden), which are highly selective in alkali and other metal binding (Figure 9-4).

FIGURE 9-4

The binding of a cation by a polycyclic ether (cryptand) to form a complex (cryptate). The system shown selectively binds the potassium ion, with a binding constant of $K = 10^{10}$. The order of selectivity is $K^+ > Rb^+ > Na^+ > Cs^+ > Li^+$. The binding constant for lithium is about 100. Thus, the total range within the series of alkali metals spans eight orders of magnitude.

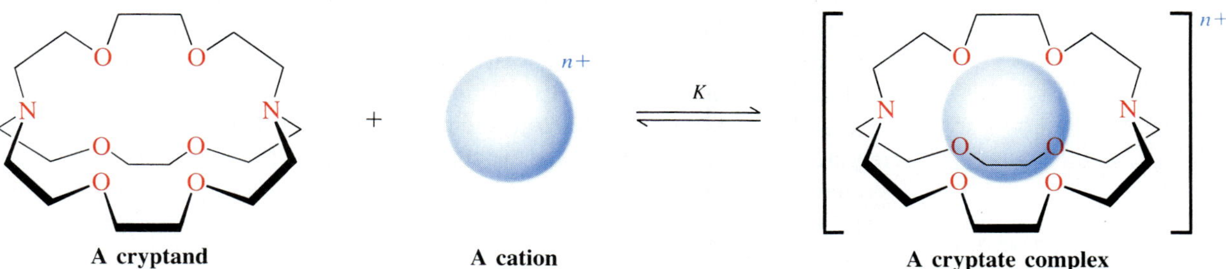

Ring Size Controls the Speed of Cyclization

As mentioned earlier, even strained-ring ethers are accessible by the intramolecular Williamson synthesis. This is so because the exothermicity of the S_N2 displacement by alkoxide is large enough to overcome the build-up of unfavorable ring strain present in the product. However, we would expect the formation of strained cyclic ethers to be relatively slow in comparison with that of larger systems, because the developing ring strain should affect the relative energy of the transition states. In contrast with this expectation, a comparison of the relative rates of cyclic ether formation reveals a surprising fact: three-membered rings form quickly, with rates comparable to five-membered ring closures, and greater than those observed for six-membered ring systems. Relatively slow to form are four-membered rings and the larger oxacycloalkanes. What effects are at work here?

Relative Rates of Cyclic Ether Formation

Ring size: $3 \geq 5 > 6 > 4 \geq 7 > 8$

To understand the relative rates of cyclic ether formation, recall that the activation energy of a reaction is composed of an enthalpy contribution, to which ring strain adds a positive (in other words, unfavorable) increment, and an entropy contribution, mainly influenced by the proximity of the reacting centers and relative rigidity of the transition state. Both factors vary systematically, but in opposite directions, from smaller to larger rings.

The preparation of an oxacyclopropane from a 2-bromo alcohol is entropically ideal, because nucleophile and leaving group are as close to each other as possible. Therefore, even though ring strain is worst in this case, the transition-state energy is relatively small, because a favorable entropy contribution allows relatively rapid ring construction. Oxacyclobutanes are generated much more slowly than their higher and lower homologs. The entropy factor here is considerably worse than that for the oxacyclopropanes, because the two reacting centers are separated by an extra methylene group. (Make a model of the preferred all *anti*-configuration.) Lengthening the chain also adds considerably more flexibility (degrees of freedom) to the starting material, flexibility that has to be given up in the transition state of the ring closure. The ring strain (see Table 4-2 for ring strains in comparable cycloalkanes), however, is about the same. The net result is a very low relative rate of formation. On the other hand, the synthesis of five-membered ring ethers is easy. Although the reacting centers are even farther apart, the unfavorable strain contribution to the enthalpy of activation is greatly diminished. Because of this effect, five-membered rings generally form very rapidly, even faster than six-membered rings, despite the fact that cyclohexanes and their derivatives are virtually strain free. The relatively greater distance between alkoxide and electrophilic carbon and the corresponding greater degree of freedom of the chain makes the difference in this case. The synthesis of larger rings is usually impeded by adverse entropy effects, eclipsing, and other strain factors (see Section 4-5).

Stereochemical Consequences of the Intramolecular Williamson Synthesis

The Williamson ether synthesis proceeds with inversion of configuration at the carbon bearing the leaving group. This finding is in accord with expectations based on an S_N2 mechanism. The attacking nucleophile approaches the electrophilic carbon from the opposite side of the leaving group. Only one conformation of the halo alkoxide can undergo efficient substitution. For example, oxacyclopropane formation requires an *anti* arrangement of the nucleophile and the leaving group. The alternative two *gauche* conformations cannot give the product (Figure 9-5).

FIGURE 9-5
Only the *anti* conformation (A) of a 2-bromo alkoxide allows for oxacyclopropane formation; the *gauche* conformers (B and C) cannot undergo intramolecular backside attack at the bromine-bearing carbon.

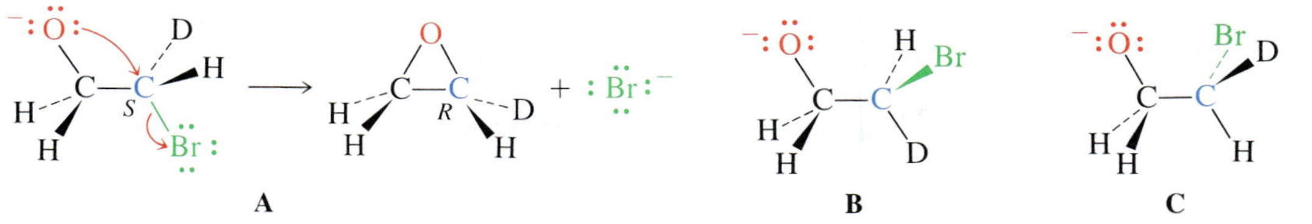

EXERCISE 9-11
(1*R*,2*R*)-2-Bromocyclopentanol undergoes rapid reaction with sodium hydroxide to yield an optically inactive product. In contrast, the (1*S*,2*R*)-enantiomer is much less reactive. Explain.

Ethers from Alcohols and Mineral Acid: S_N2 and S_N1 Mechanisms

Whereas treatment of primary alcohols with HBr or HI furnishes the corresponding haloalkanes through intermediate oxonium ions (Section 9-2), exposure to catalytic amounts of strong nonnucleophilic acids—for example, sulfuric acid—at elevated temperatures leads to ether formation. In this case, the strongest nucleophile present in solution is the unprotonated starting alcohol. As soon as one alcohol molecule has been protonated, nucleophilic attack begins, the ultimate products being an ether and water:

$$2\ CH_3CH_2OH \xrightarrow{H_2SO_4,\ 130^\circ C} CH_3CH_2OCH_2CH_3 + HOH$$

Mechanism:

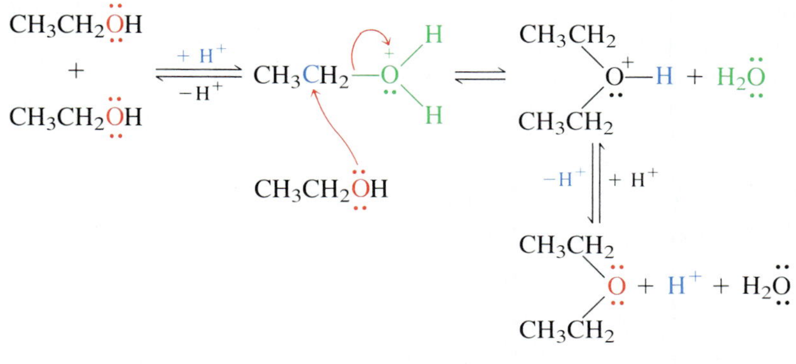

$$CH_3CH_2\ddot{O}H + CH_3CH_2\ddot{O}H \underset{-H^+}{\overset{+H^+}{\rightleftharpoons}} CH_3CH_2-\overset{+}{O}H_2 \ (+\ CH_3CH_2\ddot{O}H) \rightleftharpoons (CH_3CH_2)_2\overset{+}{O}-H + H_2\ddot{O} \underset{+H^+}{\overset{-H^+}{\rightleftharpoons}} (CH_3CH_2)_2\ddot{O} + H^+ + H_2\ddot{O}$$

Elimination of water can also be observed, but usually requires higher temperatures:

$$CH_3CH(H)CH_2OH \xrightarrow{H_2SO_4,\ 180^\circ C} CH_3CH{=}CH_2 + HOH$$

Similarly, secondary and tertiary ethers can be made from secondary and tertiary alcohols treated with acid. With these substrates, initial carbocation formation is prevalent, followed by trapping with alcohol (S_N1), as described in Section 9-2. The major side reaction follows the E1 pathway.

$$2\ CH_3\underset{\substack{|\\ H}}{\overset{\substack{OH\\ |}}{C}}CH_3 \xrightarrow{H^+} \underset{75\%}{(CH_3)_2CHOCH(CH_3)_2} + HOH + H^+$$

2-Propanol **2-(1-Methylethoxy)propane (Diisopropyl ether)**

Mixed ethers containing two different alkyl groups are accessible only with difficulty, because the mixing of two alcohols in the presence of an acid usually results in mixtures of all three possible products. However, mixed ethers containing one tertiary and one primary or secondary alkyl substituent can be prepared in good yield in the presence of dilute acid. Under these conditions, the much more rapidly formed tertiary carbocation is trapped by the other alcohol.

$$CH_3\underset{\substack{|\\ CH_3}}{\overset{\substack{CH_3\\ |}}{C}}OH + CH_3CH_2OH \xrightarrow[-\ HOH]{15\%\ \text{aqueous}\ NaHSO_4} \underset{80\%}{CH_3\underset{\substack{|\\ CH_3}}{\overset{\substack{CH_3\\ |}}{C}}OCH_2CH_3}$$

2-Ethoxy-2-methylpropane

1,1-Dimethylethyl (*tert*-butyl) ethers are readily hydrolyzed with dilute acid and therefore function as useful *protecting groups* for alcohols (Section 8-7). An example is 4-bromo-1-butanol, which can be protected to allow for Grignard formation, followed by reaction with methanal (formaldehyde), and finally deprotection. Deprotection is achieved with dilute aqueous acid, which leaves alkyl ethers other than tertiary systems untouched. The mechanism of deprotection is the reverse of that in the protection step. The 1,1-dimethylethyl (*tert*-butyl) cation rapidly decomposes by an E1 process to furnish 2-methylpropene.

$$\underset{\textbf{4-Bromo-1-butanol}}{BrCH_2CH_2CH_2CH_2OH} \xrightarrow[-\ H_2O,\ \text{protect}]{(CH_3)_3COH,\ H^+} \underset{\textbf{Protected 4-bromo-1-butanol}}{BrCH_2CH_2CH_2CH_2OC(CH_3)_3} \xrightarrow{Mg}$$

$$\underset{\substack{\textbf{Grignard reagent}\\ \textbf{(Protected from self-protonation)}}}{BrMgCH_2CH_2CH_2CH_2OC(CH_3)_3} \xrightarrow{\substack{1.\ CH_2{=}O\\ 2.\ H^+,\ H_2O}} \underset{\textbf{Half-protected 1,5-pentanediol}}{HOCH_2CH_2CH_2CH_2CH_2OC(CH_3)_3} \xrightarrow[\text{Deprotect}]{H^+,\ H_2O}$$

$$\underset{\textbf{1,5-Pentanediol}}{HO(CH_2)_5OH} + \underset{\textbf{2-Methylpropene}}{H_2C{=}C(CH_3)_2}$$

EXERCISE 9-12

Show how you would achieve the following interconversion (the dashed arrow indicates that several steps are required):

$$BrCH_2CH_2CH_2OH \dashrightarrow DCH_2CH_2CH_2OH$$

EXERCISE 9-13

Write mechanisms for the following two reactions: (a) 1,4-butanediol + $H^+ \rightarrow$ oxacyclopentane (tetrahydrofuran); (b) 5-methyl-1,5-hexanediol + $H^+ \rightarrow$ 2,2-dimethyloxacyclohexane (2,2-dimethyltetrahydropyran).

Ethers by Alcoholysis

As we know, tertiary and secondary ethers may also form by the alcoholysis of the corresponding haloalkanes or alkyl sulfonates (Section 7-2). The starting material is simply dissolved in an alcohol until the S_N1 process is complete (an example is shown at the right).

H_3C Cl

1-Chloro-1-methyl cyclohexane

↓ CH_3CH_2OH

H_3C OCH_2CH_3 $+ H^+ + Cl^-$

86%

1-Ethoxy-1-methylcyclohexane

EXERCISE 9-14

There are several ways of constructing an ether from an alcohol and a haloalkane. Which approach would you choose for the preparation of: (a) 2-methyl-2-(1-methylethoxy)butane; (b) 1-methoxy-2,2-dimethylpropane?

In summary, ethers can be named as alkoxyalkanes or as dialkyl ethers. They have lower boiling points than their comparable alcohols, because they cannot enter into hydrogen bonding with themselves. Ethers are prepared by the Williamson synthesis, which includes an S_N2 reaction of an alkoxide with a haloalkane. This reaction works best with primary halides or sulfonates that do not undergo ready elimination. Cyclic ethers are formed by the intramolecular version of this method but require dilute conditions. The relative rates of ring closure in this case are highest for three- and five- membered rings. Ethers can also be prepared by treatment of alcohols with acid through S_N2 and S_N1 pathways, with oxonium ions or carbocations as intermediates, and by alcoholysis of secondary or tertiary haloalkanes or alkane sulfonates.

9-6 The Reactions of Oxacyclopropanes

As mentioned earlier, ethers are rather unreactive. Some ethers, however, react slowly with oxygen to form hydroperoxides and peroxides by free-radical mechanisms. Peroxides are highly dangerous compounds because they may decompose explosively. Extreme care should therefore be taken with samples of ethers that have been exposed to air for several days.

Ethers are also attacked by strong acid to give alcohols or haloalkanes through the intermediacy of oxonium ions, but, because of the strongly acidic reaction conditions, this method is not generally useful (Section 6-7).

Peroxides from Ethers

$$2\ RO\overset{|}{\underset{|}{C}}H + O_2$$

$$\downarrow$$

$$2\ RO\overset{|}{\underset{|}{C}}—O—OH$$

An ether hydroperoxide

$$\downarrow$$

$$RO\overset{|}{\underset{|}{C}}—O—O—\overset{|}{\underset{|}{C}}OR$$

An ether peroxide

EXERCISE 9-15

Treatment of methoxymethane with hot HI gives iodomethane. Formulate a mechanism.

Although ordinary ethers are relatively inert, we know that the strained ring in oxacyclopropane undergoes a variety of ring-opening reactions with nucleophiles. This section presents further details of these processes.

Regioselectivity in Nucleophilic Oxacyclopropane Openings

Oxacyclopropane is subject to nucleophilic opening not only by hydride and organometallic reagents (Sections 8-4 and 8-6), but also by other anionic nucleophiles. These transformations are due to attack at either carbon of the ether, both equally reactive because of the symmetry of the substrate.

Anionic Nucleophilic Opening of Oxacyclopropane

$$\text{CH}_2\text{—CH}_2\text{(O)} \xrightarrow{:\text{Nu}^-} \xrightarrow[-\ \text{HO}^-]{\text{HOH}} \text{HO—CH}_2\text{—CH}_2\text{—Nu}$$

Example:

$$\text{CH}_2\text{—CH}_2\text{(O)} + \text{CH}_3\ddot{\text{S}}:^- \xrightarrow[-\ \text{HO}^-]{\text{HOH}} \text{H}\ddot{\text{O}}\text{CH}_2\text{CH}_2\ddot{\text{S}}\text{CH}_3$$

What is the situation with unsymmetrical systems? Consider, for example, the reaction of 2-methyloxacyclopropane with methoxide. There are two possible reaction sites: the primary carbon (*a*) to give 1-methoxy-2-propanol, and the methyl-substituted secondary center (*b*) resulting in 2-methoxy-1-propanol. The product analysis after reaction reveals the exclusive formation of 1-methoxy-2-propanol through path *a*. Is this surprising?

(*S*)-1-Methoxy-2-propanol $\xleftarrow[a]{\text{CH}_3\text{O}^-,\ \text{CH}_3\text{OH}}$ **(*S*)-2-Methyloxacyclopropane** $\xrightarrow[b]{\text{CH}_3\text{O}^-,\ \text{CH}_3\text{OH}}$ ✗ **(*R*)-2-Methoxy-1-propanol** (Not formed)

No, because, as we know, if there is more than one possibility, nucleophilic attack will be at the *less*-substituted carbon center (Section 6-9). If the starting material is optically pure (*S*)-2-methyloxacyclopropane, configuration at the stereocenter is retained, because this carbon does not participate in the reaction. Pathway *b*, on the other hand, would give inversion. Thus, the rules of nucleophilic substitution developed for simple alkyl derivatives also apply to strained cyclic ethers.

The selectivity in the nucleophilic opening of substituted oxacyclopropanes is referred to as **regioselectivity** because, of two possible and similar "regions," the nucleophile attacks only one.

As expected, ring-opening occurs with inversion at the target carbon:

Inversion on Oxacyclopropane Opening

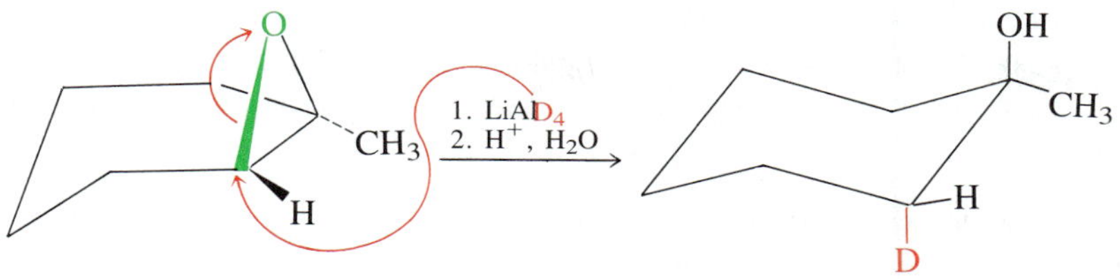
O
CH3
H
1. LiAlD4
2. H+, H2O
OH
CH3
H
D

99.4%

D and OH are trans, not cis.

In this example, $LiAlD_4$ reacts as if it were a source of D^-.

EXERCISE 9-16

Draw the initial conformation of the product of ring-opening of compound A by methyllithium. Will this conformation be stable? (Hint: Make a model.)

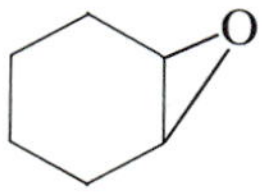
O

A

Ring-opening of oxacyclopropanes is catalyzed by acids. The reaction in this case proceeds through initial cyclic oxonium ion formation followed by ring-opening on nucleophilic attack.

Acid-Catalyzed Ring-Opening of Oxacyclopropane

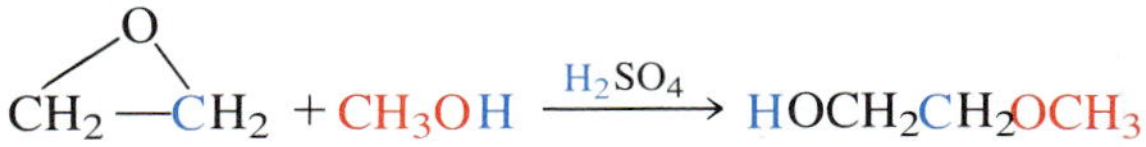
O
CH2—CH2 + CH3OH
H2SO4
HOCH2CH2OCH3

2-Methoxyethanol

Mechanism:

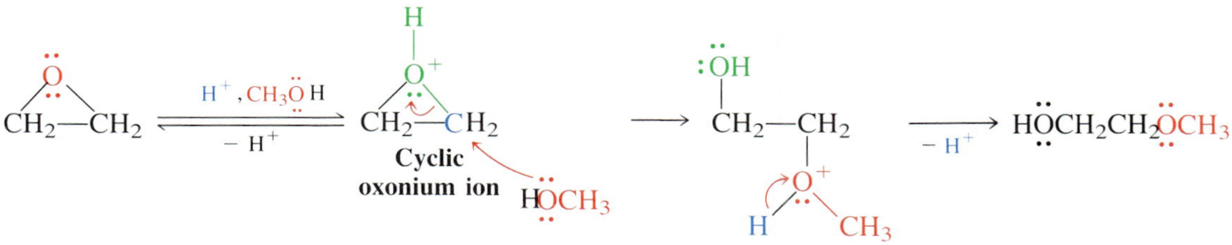
O
CH2—CH2
H+, CH3OH
− H+
H
O+
CH2—CH2
Cyclic
oxonium ion
HOCH3
:OH
CH2—CH2
O+
H
CH3
− H+
HOCH2CH2OCH3

Is this reaction also regioselective and stereospecific as in the anionic nucleophilic opening of oxacyclopropanes discussed first? The answer is yes, but not completely. Reconsider the opening of (*S*)-2-methyloxacyclopropane by methanol. Instead of being charged, the nucleophile is now neutral. The result is surprising: both possible regioisomeric products are formed. The reaction is only partly regioselective, with attack mainly at the primary carbon but with some product derived from attack at the secondary carbon. Inversion occurs at the reacting centers in both cases:

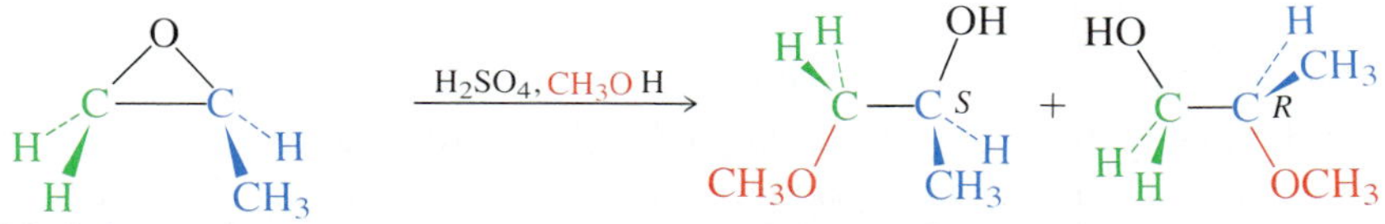
O
C
C
H
H
H
CH3
H2SO4, CH3OH
H
H
OH
C
C
S
CH3O
H
CH3
+
HO
H
CH3
C
C
R
H
H
OCH3

(*S*)-2-Methyloxacyclopropane **Major product** **Minor product**

Why is the more-hindered position attacked? A possible explanation is the special nature of the acid-catalyzed displacement reaction. Protonation at the oxygen of the ether generates a reactive intermediate oxonium ion with substantially polarized oxygen–carbon bonds. This polarization places partial positive charges on the ring carbons. Because alkyl groups act as electron donors (Section 7-3), more positive charge is located on the secondary than on the primary carbon.

Mechanism of Acid-Catalyzed Ring-Opening of (*S*)-2-Methyloxacyclopropane by Methanol:

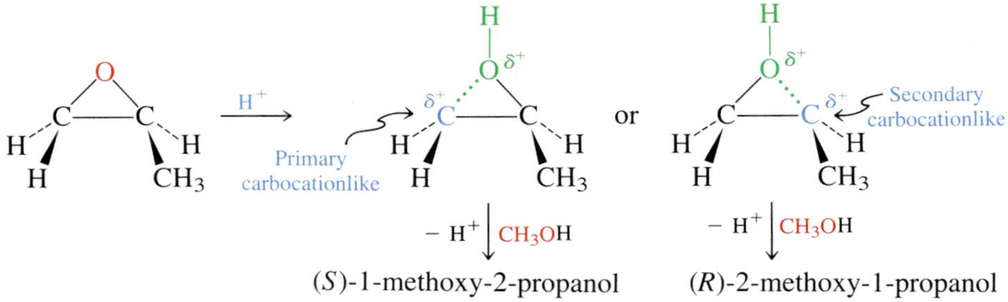

Primary carbocationlike

or

Secondary carbocationlike

$-H^+$ | CH_3OH

(*S*)-1-methoxy-2-propanol

(*R*)-2-methoxy-1-propanol

This uneven charge distribution counteracts (but does not completely override) the steric-hindrance effect, methanol being attracted by Coulombic forces more to the secondary than to the primary center. The influence of the charge seems to predominate in the acid-catalyzed ring-opening of 2,2-dimethyloxacyclopropane:

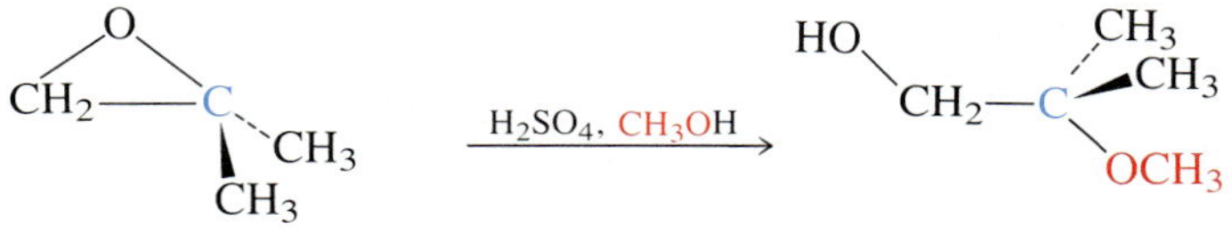

H_2SO_4, CH_3OH

2,2-Dimethyloxacyclopropane

Only one product

EXERCISE 9-17

Treatment of (2*R*,3*R*)-*trans*-2,3-dimethyloxacyclopropane with acidic water results in an optically inactive compound. Explain.

In summary, although ordinary ethers are relatively inert, the rings in oxacyclopropanes can be opened both regioselectively and stereospecifically. For anionic nucleophiles, the usual rules of bimolecular nucleophilic substitution hold: attack is at the less-hindered carbon center, which undergoes inversion. Acid catalysis, on the other hand, changes the regioselectivity—attack is at the more-hindered center.

9-7 Sulfur Analogs of Alcohols and Ethers: Thiols and Sulfides

The reactivity of the sulfur analogs of alcohols and ethers is in many respects similar to that of the oxygen compounds. Those differences that do exist are due to the fact that sulfur can undergo valence-shell expansion. This section compares the two classes of compounds.

Naming Sulfur Analogs of Alcohols and Ethers

The sulfur analogs of alcohols, R–SH, are called **thiols** in the IUPAC system (*theion,* Greek, brimstone—a common name for sulfur). The ending thiol is added to the alkane stem to yield the alkanethiol name. The SH group is referred to as **mercapto,** and its location is indicated by the appropriate numbering of the longest chain, as in alkanol nomenclature.

$CH_3\ddot{S}H$ — **Methanethiol**

$CH_3CH_2CH(CH_3)CH_2\ddot{S}H$ (carbons numbered 4, 3, 2, 1) — **2-Methyl-1-butanethiol**

$CH_3CH_2CH(\ddot{S}H)CH_2CH_3$ — **3-Pentanethiol**

$C_6H_{11}\ddot{S}H$ — **Cyclohexanethiol**

The sulfur analogs of ethers (common name, thioethers) are called **sulfides,** as in alkyl ether nomenclature.* The RS group is named **alkylthio,** the RS^- group **alkanethiolate.**

$CH_3\ddot{S}CH_2CH_3$ — **Ethyl methyl sulfide**

$CH_3C(CH_3)_2\ddot{S}(CH_2)_6CH_3$ — **Heptyl (1,1-dimethyl)-ethyl sulfide**

$CH_3\ddot{S}\colon^-$ — **Methanethiolate ion**

The Properties of Thiols: Less Hydrogen Bonded, More Acidic Than Alcohols

Sulfur, because of its large size, its diffuse orbitals, and the relatively nonpolarized S–H bond (Table 1-4), does not enter into hydrogen bonding very efficiently. Thus, thiols do not have the abnormally high boiling points common to alcohols (Table 9-2), but their volatility lies between those of the analogous haloalkanes and alcohols.

Partly because of the relatively weak hydrogen–sulfur bond, thiols are also more acidic than water, with pK_a values ranging from 9 to 12. They can therefore be more readily deprotonated by hydroxide and alkoxide anions.

Acidity of Thiols

$$R\ddot{S}H + H\ddot{O}\colon^- \rightleftharpoons R\ddot{S}\colon^- + HOH$$
$$pK_a = 9\text{–}12$$

TABLE 9-2
Comparison of the boiling points of thiols, alcohols, and haloalkanes

Compound	Boiling point (°C)
CH_3SH	6.2
CH_3Cl	−24.2
CH_3OH	65.0
CH_3CH_2SH	37
CH_3CH_2Cl	12.3
CH_3CH_2OH	78.5

The Chemical Reactivity of Thiols and Sulfides Can Be Similar to That of Alcohols and Ethers

Thiols and sulfides can react in ways that are very similar to those of the corresponding oxygen analogs. The sulfur in these compounds is even more nucleophilic than the oxygen in alcohols and ethers. Therefore, thiols and sulfides are readily prepared by nucleophilic attack of RS^- or HS^- on haloalkanes.

*A more-systematic nomenclature would be based on the alkylthioalkane system, as in the naming of alkoxyalkanes.

A large excess of the HS^- is used in the preparation of thiols to ensure that the product does not react with the starting halide to give the dialkyl sulfide.

$$CH_3\underset{|}{\overset{CH_3}{C}}HBr + Na^{+-}SH \xrightarrow{CH_3CH_2OH} CH_3\underset{|}{\overset{CH_3}{C}}HSH + Na^+Br^-$$

Excess — 2-Propanethiol

$$RSH + R'Br \xrightarrow{NaOH} RSR' + NaBr + H_2O$$

Sulfides are prepared in an analogous way by alkylation of thiols in the presence of base, such as hydroxide. The base generates the alkanethiolate, which reacts with the haloalkane by an S_N2 process. Because of the strong nucleophilicity of thiolates, there is no competition from hydroxide in this displacement.

The nucleophilicity of sulfur is also manifest in the capacity of sulfides to attack haloalkanes to furnish sulfonium ions:

$$(H_3C)_2\ddot{S}: + CH_3{-}\ddot{I}: \longrightarrow (H_3C)_2\overset{+}{\ddot{S}}{-}CH_3 + :\ddot{I}:^-$$

95%

Trimethylsulfonium iodide

Like their oxonium analogs, sulfonium salts are subject to nucleophilic attack at carbon, the sulfide functioning as the leaving group:

$$H\ddot{O}:^- + CH_3{-}\overset{+}{\ddot{S}}(CH_3)_2 \longrightarrow H\ddot{O}CH_3 + \ddot{S}(CH_3)_2$$

EXERCISE 9-18

(a) Sulfide A is a powerful poison used as a chemical-warfare agent (''mustard gas'') in the First World War. Propose a synthesis starting with oxacyclopropane. (b) Its mechanism of action is believed to involve sulfonium salt B, which is thought to react with nucleophiles in the body. How is B formed and how would it react with nucleophiles?

$$\underset{\textbf{A}}{ClCH_2CH_2SCH_2CH_2Cl} \qquad \underset{\textbf{B}}{ClCH_2CH_2\overset{+}{S}\langle{}^{CH_2}_{CH_2}|\ \ Cl^-}$$

Valence-Shell Expansion of Sulfur in Thiols and Thioethers Allows for Unique Reactivity

$$\underset{\textbf{Methanethiol}}{CH_3SH} \xrightarrow{KMnO_4} \underset{\textbf{Methanesulfonic acid}}{CH_3\overset{O}{\overset{\|}{\underset{\underset{O}{\|}}{S}}}OH}$$

Sulfur, being a third-row element with *d* orbitals, is capable of valence-shell expansion to accommodate more electrons than are allowed by the octet rule (see Section 6-7). We have already seen that, in some of its compounds, sulfur is surrounded by ten or even twelve valence electrons, and this capacity allows for reactions inaccessible to the corresponding oxygen analogs. For example, oxidation of thiols with strong oxidizing agents, such as hydrogen peroxide or potassium permanganate, gives the corresponding sulfonic acids. In this way methanethiol is converted into methanesulfonic acid.

Sulfonic acids give sulfonyl chlorides by treatment with PCl_5. The use of sulfonyl chlorides was demonstrated in sulfonate synthesis (Sections 6-7 and 9-4).

Milder oxidation of thiols, by the use of iodine, results in the formation of **disulfides,** which are readily reduced back to thiols by alkali metals.

The Thiol-Disulfide Redox Reaction

Oxidation:

$$2\ CH_3CH_2CH_2SH + I_2 \longrightarrow CH_3CH_2CH_2SSCH_2CH_2CH_3 + 2\ HI$$

1-Propanethiol **Dipropyl disulfide**

Reduction:

$$CH_3CH_2CH_2SSCH_2CH_2CH_3 \xrightarrow[2.\ H^+,\ H_2O]{1.\ Li,\ liquid\ NH_3} CH_3CH_2CH_2SH$$

Reversible disulfide formation from thiols is an important biological process. Many proteins and peptides contain free SH groups that form bridging disulfide linkages. Nature exploits this mechanism in order to link amino acid chains both intra- and intermolecularly (see Chapter 27).

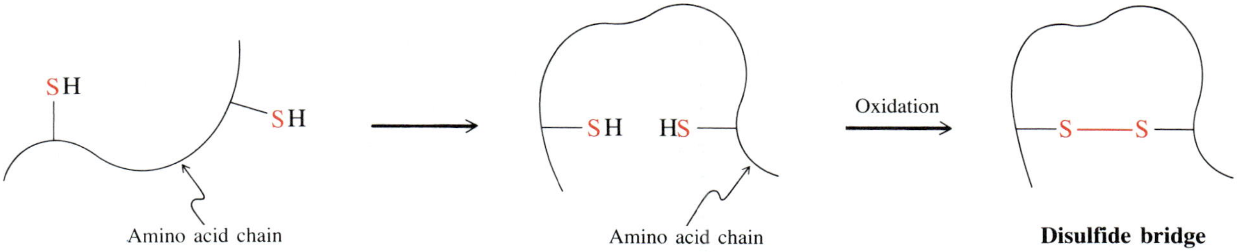

SH SH → SH HS —Oxidation→ S—S

Amino acid chain Amino acid chain **Disulfide bridge**

Sulfides are readily oxidized to **sulfones,** a transformation proceeding through the intermediacy of a **sulfoxide.** For example, oxidation of dimethyl sulfide first gives dimethyl sulfoxide, which subsequently furnishes dimethyl sulfone. Dimethyl sulfoxide has already been mentioned as a highly polar nonprotic solvent of great use in organic chemistry, particularly in nucleophilic substitutions (see Section 6-10, Table 6-8).

$CH_3\ddot{\underset{..}{S}}CH_3$
Dimethyl sulfide

↓ H_2O_2

:O:
‖
CH_3SCH_3
Dimethyl sulfoxide (DMSO)

↓ H_2O_2

:O:
‖
CH_3SCH_3
‖
:O:
Dimethyl sulfone

Unsymmetrical sulfoxides are chiral, the oxygen and the lone electron pair (which has the lowest priority) playing the role of substituents.

Chirality in Sulfoxides

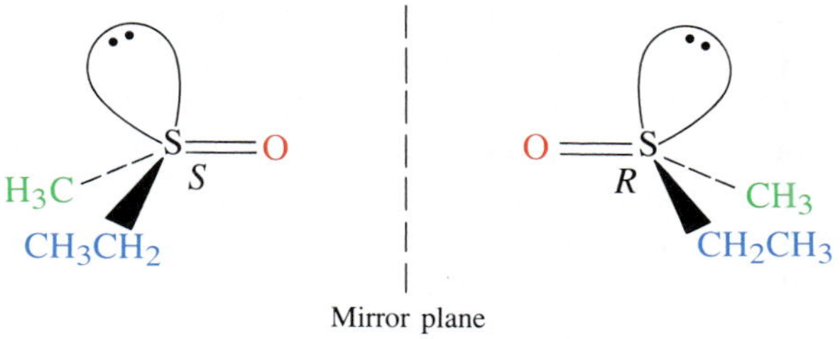

H_3C, CH_3CH_2, S=O, *S* | Mirror plane | O=S, *R*, CH_3, CH_2CH_3

Note: Colors indicate substituent priority (Section 5-3).

In summary, the naming of thiols and sulfides is related to the system used for alcohols and ethers. Thiols are more volatile, more acidic, and more nucleophilic than alcohols. Thiols and sulfides can be oxidized, the former to disulfides or sulfonic acids, the latter to sulfoxides and sulfones.

9-8 Physiological and Other Properties and Uses of Some Alcohols and Ethers

Modern industrial *methanol* synthesis uses the catalytic reduction of carbon monoxide with hydrogen at high pressures and temperatures (Section 8-4). Methanol is sold for a variety of purposes, examples being as a solvent for paint and other materials, as a fuel for camp stoves and soldering torches, and as a synthetic intermediate. It is highly poisonous, and ingestion or chronic exposure may lead to blindness. Death from ingestion of as little as 30 ml has been reported. It is sometimes added to commercial ethanol to render it unfit for consumption (denatured alcohol). The toxicity of methanol is thought to be due to metabolic oxidation to methanal (formaldehyde), $CH_2{=}O$, which seemingly interferes with the physiochemical processes of vision. Further oxidation to formic acid, HCOOH, causes acidosis, an unusual lowering of the blood pH. This condition leads to disruption of oxygen transport in the blood and eventually to coma.

Methanol has recently been studied as a possible precursor of gasoline. Mobil Corporation announced in 1976 that certain zeolite catalysts (Section 3-3) allow the conversion of methanol into a mixture of hydrocarbons, ranging in length from four-carbon chains to ten-carbon ones, with a composition that on distillation yields largely gasoline (see Table 3-3).

$$n\ CH_3OH \xrightarrow{\text{Zeolite, } 340°\text{–}375°C} \underset{67\%}{C_nH_{2n+2}} + \underset{6\%}{C_nH_{2n}} + \underset{27\%}{\text{aromatics}}$$

Ethanol diluted by various amounts of flavored water is an alcoholic beverage. It is classified pharmacologically as a general depressant, because it induces a nonselective, reversible depression of the central nervous system. Approximately 95% of the alcohol ingested is metabolized in the body (usually in the liver) to products that are eventually transformed into carbon dioxide and water. Although high in calories, ethanol has little nutritional value.

The rate of metabolism of most drugs in the liver increases with their concentration, but this is not true for alcohol, which is degraded linearly with time. An adult metabolizes about 10 ml of pure ethanol per hour, roughly the ethanol content of a cocktail, a shot of spirits, or a can of beer. Depending on a person's weight, the ethanol content of the drink, and the speed with which it is consumed, as few as two or three drinks can produce a level of alcohol in the blood that is more than 0.1%, the current legal limit beyond which the operation of a motorized vehicle is prohibited in the United States.

Ethanol is poisonous. Its lethal concentration in the blood stream has been estimated at 0.4%. Its effects are characterized by progressive euphoria, disinhibition, disorientation, and decreased judgment (drunkenness), followed by general anesthesia, coma, and death. It dilates the blood vessels, producing a "warm flush," but it actually decreases body temperature. Although long-term ingestion of moderate amounts (the equivalent of about two beers a day) does not appear to be harmful, larger amounts can be the cause of a variety of physical and psychological disorders, usually described by the general term *alcoholism*. These disorders include hallucinations, psychomotor agitation, liver diseases, dementia, gastritis, and psychological dependence.

Interestingly, a near-toxic dose of ethanol is applied in cases of acute methanol or 1,2-ethanediol (ethylene glycol) poisoning. This treatment prevents the metabolism of the more-toxic alcohols and allows their excretion before damaging concentrations of secondary products can accumulate.

Ethanol destined for human consumption is prepared by fermentation of sugars and starch (rice, potatoes, corn, wheat, flowers, fruit, etc., Chapter 23). Fermentation is catalyzed by enzymes in a multistep sequence that converts carbohydrates into ethanol and carbon dioxide:

$$\underset{\textbf{Starch}}{(C_6H_{10}O_5)_n} \xrightarrow{\text{Enzymes}} \underset{\textbf{Glucose}}{C_6H_{12}O_6} \xrightarrow{\text{Enzymes}} 2\ CH_3CH_2OH + 2\ CO_2$$

Commercial alcohol not intended as a beverage is made industrially by hydration of ethene (Sections 8-4 and 12-8). It is used, for example, as a solvent in perfumes, varnishes, and shellacs and as a synthetic intermediate, as demonstrated in earlier equations. Interest in ethanol production has surged recently because of its potential as a gasoline additive ("gasohol").

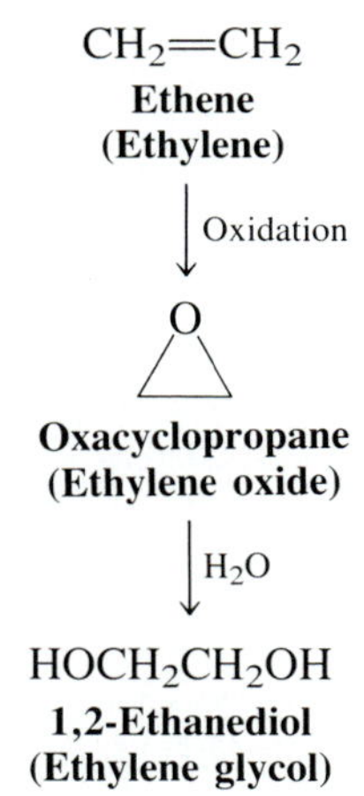

2-Propanol is toxic but (unlike methanol) not absorbed through the skin. Therefore it is popular as a rubbing alcohol and used as a solvent.

1,2-Ethanediol (ethylene glycol) is prepared by oxidation of ethene to oxacyclopropane, followed by hydrolysis in quantities exceeding two million tons in the United States per year. Its low melting point (−11.5°C), its high boiling point (198°C), and its complete miscibility with water make it a useful antifreeze agent.

1,2,3-Propanetriol (glycerol, glycerine), $HOCH_2CHOHCH_2OH$, is a viscous greasy substance, soluble in water, and nontoxic. It is obtained by alkaline hydrolysis of triglycerides, the major component of fatty tissue. The sodium and potassium salts of the long-alkyl-chain acids produced from fats ("fatty acids," Section 17-12) are sold as soaps:

$$\underset{\substack{\textbf{Triglyceride}\\ \textbf{("Fat")}\\ R\,=\,\text{long alkyl chain}}}{\begin{array}{l} CH_2OCR\ (C{=}O) \\ | \\ CHOCR\ (C{=}O) \\ | \\ CH_2OCR\ (C{=}O) \end{array}} \xrightarrow{H_2O,\ NaOH} \underset{\substack{\textbf{1,2,3-Propanetriol}\\ \textbf{(Glycerol, glycerine)}}}{\begin{array}{c} CH_2OH \\ | \\ HCOH \\ | \\ CH_2OH \end{array}} + \underset{\textbf{Soap}}{\overset{O}{\overset{\|}{RCO^-}}Na^+}$$

Phosphorus esters of 1,2,3-propanetriols (phosphoglycerides, Section 18-4) are primary components of cell membranes.

1,2,3-Propanetriol is used in lotions and other cosmetics, as well as in medicinal preparations. Treatment with nitric acid gives a trinitrate ester known as nitroglycerine, an extremely powerful explosive. The explosive potential of this substance results from its shock-induced highly exothermic decomposition to gaseous products (N_2, CO_2, H_2O gas, O_2), raising temperatures to more than 3,000°C and creating pressures higher than 2,000 atmospheres in a fraction of a second.

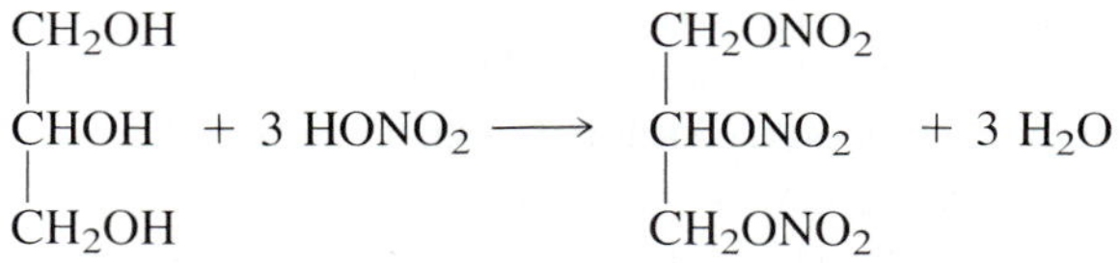

$$\begin{matrix} CH_2OH \\ | \\ CHOH \\ | \\ CH_2OH \end{matrix} + 3\ HONO_2 \longrightarrow \begin{matrix} CH_2ONO_2 \\ | \\ CHONO_2 \\ | \\ CH_2ONO_2 \end{matrix} + 3\ H_2O$$

Nitroglycerine

Cholesterol is an important steroid alcohol (Section 4-7).

Ethoxyethane (diethyl ether) was at one time used as a general anaesthetic. It produces unconsciousness by depressing central nervous activity. Because of adverse effects such as irritation of the respiratory tract and extreme nausea, its use has been discontinued, and 1-methoxypropane (methyl propyl ether, ''neothyl'') and other compounds have replaced it in such applications. Ethoxyethane and other ethers are explosive when mixed with air.

Oxacyclopropane (oxirane, ethylene oxide) is a fumigating agent for seeds and grains.

Many *natural products,* some of which are quite physiologically active, contain alcohol and ether groups.

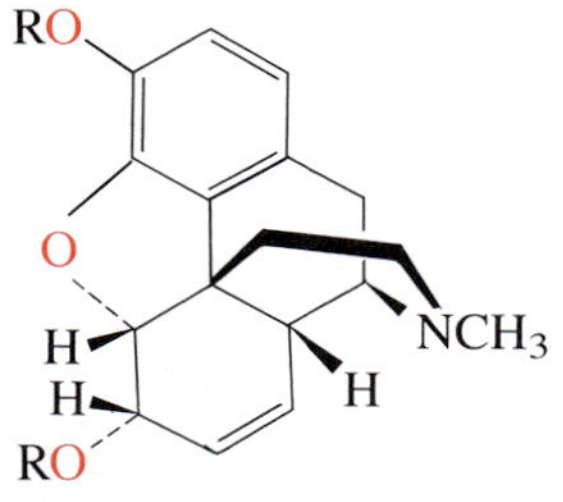

Morphine
(R = H)
Heroin
$\left(R = \underset{\underset{O}{\|}}{C}CH_3\right)$

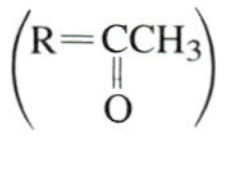

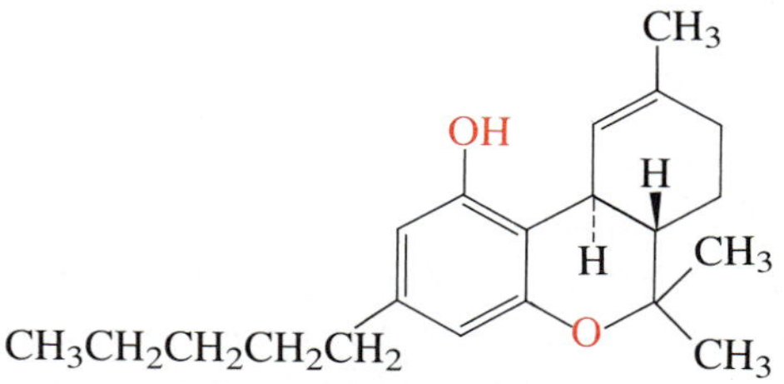

Tetrahydrocannabinol

The lower thiols and sulfides are most notorious for their foul smell. *Ethanethiol* is detectable by its odor even when diluted in fifty millions parts of air. The major volatile components of the skunk's defensive spray are *3-methyl-1-butanethiol,* trans-*2-butene-1-thiol,* and trans-*2-butenyl methyl disulfide*.

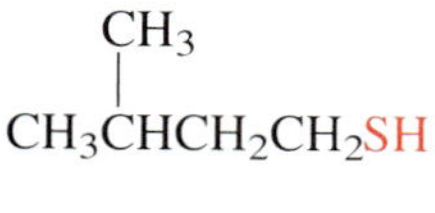

$$CH_3\underset{\overset{|}{CH_3}}{C}HCH_2CH_2SH$$

3-Methyl-1-butanethiol

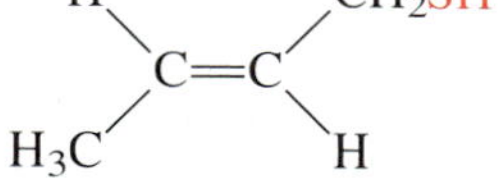

$$\begin{matrix} H & & CH_2SH \\ & C{=}C & \\ H_3C & & H \end{matrix}$$

***trans*-2-Butene-1-thiol**

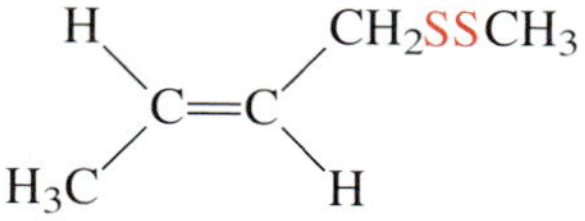

$$\begin{matrix} H & & CH_2SSCH_3 \\ & C{=}C & \\ H_3C & & H \end{matrix}$$

***trans*-2-Butenyl methyl disulfide**

Interestingly, when highly diluted, sulfur compounds can assume a rather pleasant odor. For example, the smell of freshly chopped onions or garlic is due to the presence of small thiols and sulfides. Dimethyl sulfide is a component of the aroma of black tea.

Many beneficial drugs contain sulfur in their molecular framework. Particularly well known are the *sulfonamide,* or *sulfa drugs,* powerful antibacterial agents (Section 19-6):

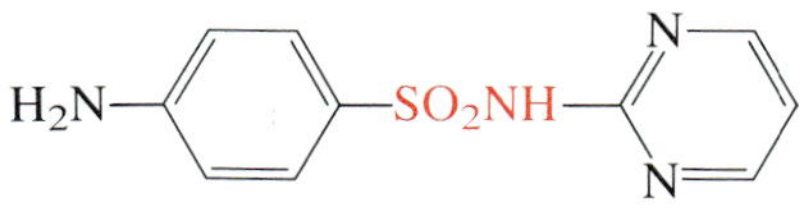

$H_2N-C_6H_4-SO_2NH-C_4H_3N_2$

Sulfadiazine
(An antibacterial drug)

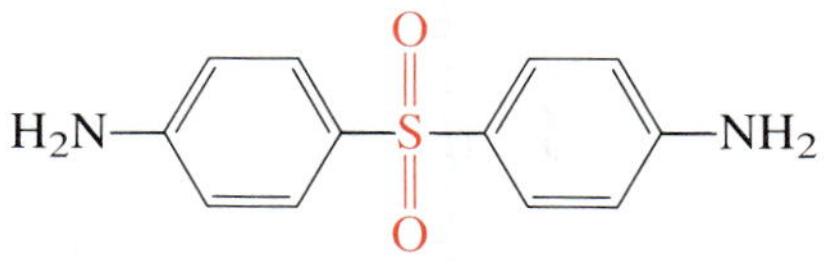

$H_2N-C_6H_4-S(=O)_2-C_6H_4-NH_2$

Diaminodiphenylsulfone
(An antileprotic drug)

To summarize, alcohols and ethers have varied uses, both as chemical raw materials and as medicinal agents. Many of their derivatives can be found in nature; others are readily synthesized.

Summary of New Reactions

The Reactions of Alcohols

1 ACID-BASE REACTIONS

$$RO^- \underset{}{\overset{-H^+}{\rightleftharpoons}} ROH \overset{+H^+}{\rightleftharpoons} R\overset{+}{O}H_2$$

2 CARBOCATION SYNTHESIS

$$RCH(OH)R \xrightarrow{SbF_5,\ SO_2} R\overset{+}{C}HR + HOS\overset{-}{b}F_5$$

3 CARBOCATION REARRANGEMENTS BY ALKYL AND HYDROGEN SHIFTS

$$-\overset{(R)H}{C}-\overset{OH}{C}- \xrightarrow[-H_2O]{H^+} -\overset{(R)H}{C}-\overset{+}{C}(H)< \longrightarrow >\overset{+}{C}-\overset{H(R)}{C}(H)-$$

$$\xrightarrow[-H^+]{E1} >C=C<\ (H(R)) \qquad \xrightarrow[S_N1]{Nu^-} -\overset{Nu}{C}-\overset{H(R)}{C}(H)-$$

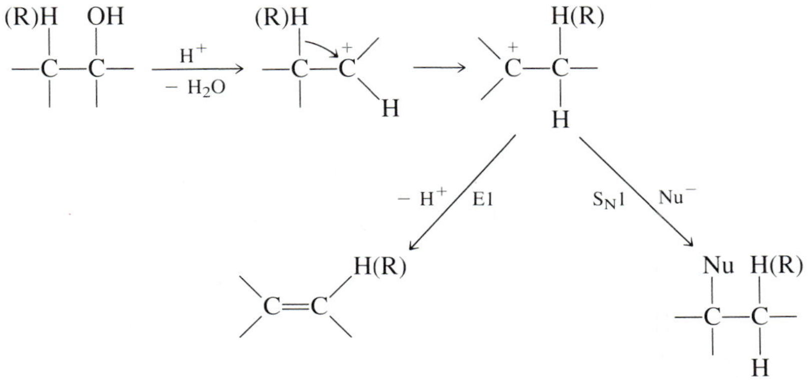

4 VICINAL DIOL REARRANGEMENTS BY ALKYL AND HYDROGEN SHIFTS

$$-\overset{OH}{\underset{(R)H}{C}}-\overset{OH}{C}- \xrightarrow[-H_2O]{H^+} -\overset{OH}{\underset{(R)H}{C}}-\overset{+}{C}< \longrightarrow \left[HO-\overset{+}{C}<-\underset{H(R)}{C}- \longleftrightarrow H\overset{+}{O}=C<-\underset{H(R)}{C}- \right] \xrightarrow{-H^+} O=C<-\underset{H(R)}{C}-$$

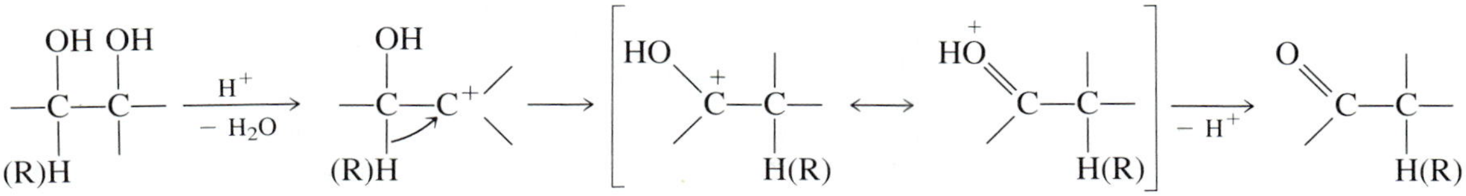

5 CONCERTED ALKYL SHIFTS FROM PRIMARY ALCOHOLS

$$R\!-\!\underset{R''}{\overset{R'}{C}}\!-\!CH_2OH \xrightarrow{H^+} R\!-\!\underset{R''}{\overset{R'}{C}}\!-\!CH_2\!-\!\overset{+}{O}H_2 \xrightarrow[-\,H_2O]{} \underset{R''}{\overset{R}{\ }}\overset{+}{C}\!-\!CH_2R' \longrightarrow \text{etc.}$$

6 ORGANIC ESTER FORMATION

$$R\overset{O}{\overset{\|}{C}}OH + R'OH \underset{}{\overset{H^+}{\rightleftharpoons}} R\overset{O}{\overset{\|}{C}}OR' + H_2O$$

Haloalkanes from Alcohols

7 PHOSPHORUS REAGENTS

$$3\ ROH + PBr_3 \longrightarrow 3\ RBr + H_3PO_3$$
$$6\ ROH + 6\ P + 3\ I_2 \longrightarrow 6\ RI + 6\ H_3PO_3$$
$$5\ ROH + PCl_5 \longrightarrow 5\ RCl + H_3PO_4 + H_2O$$

8 SULFUR REAGENTS

$$ROH + SOCl_2 \xrightarrow{N(CH_2CH_3)_3} RCl + SO_2 + (CH_3CH_2)_3\overset{+}{N}H\ Cl^-$$

$$ROH + R'SO_2Cl \longrightarrow \underset{\textbf{Alkane sulfonate}}{ROSO_2R'} \xrightarrow{X^-} RX + RSO_3^-$$

Oxidation of Alcohols

9 CHROMIUM REAGENTS

$$\underset{\textbf{Primary alcohol}}{RCH_2OH} \xrightarrow{CrO_3(\text{pyridine})_2} \underset{\textbf{Aldehyde}}{R\overset{O}{\overset{\|}{C}}H}$$

$$\underset{\textbf{Secondary alcohol}}{R\overset{OH}{\overset{|}{C}}HR} \xrightarrow{Na_2Cr_2O_7,\ H^+ \text{ or } CrO_3,\ H_2SO_4 \text{ (Jones reagent)}} \underset{\textbf{Ketone}}{R\overset{O}{\overset{\|}{C}}R}$$

10 IODOFORM REACTION

$$R\!-\!\underset{H}{\overset{OH}{C}}\!-\!CH_3 \xrightarrow{I_2,\ NaOH} R\overset{O}{\overset{\|}{C}}O^-Na^+ + \underset{\textbf{Yellow precipitate}}{CHI_3}$$

11 ETHERS FROM ALCOHOLS

Primary alcohols:

$$RCH_2OH \xrightarrow{H^+,\ \text{low temperature}} RCH_2\overset{+}{O}H_2 \xrightarrow[-\,H_2O]{RCH_2OH,\ \Delta} RCH_2OCH_2R$$

$$RCH_2OH \xrightarrow[2.\ R'X,\ S_N2]{1.\ \text{Base}} RCH_2OR'$$

R′ must be unhindered to prevent E2.

Secondary alcohols:

$$R_2CHOH \xrightarrow[-\,H_2O]{H^+} R_2CH{-}O{-}CHR_2 + \text{E1 products}$$

$$R_2CHOH \xrightarrow[2.\ R'X,\ S_N2]{1.\ \text{Base}} R_2CH{-}OR' + \text{E2 products}$$

R′ = primary

Tertiary alcohols:

$$R_3COH + R'OH \xrightarrow[S_N1,\ -\,H_2O]{NaHSO_4,\ H_2O} R_3C{-}OR' + \text{E1 products}$$

R′ = (mainly) primary

12 NUCLEOPHILIC OPENING OF OXACYCLOPROPANES

Anionic nucleophiles:

$$\text{Nu}^- + \text{oxacyclopropane (R-substituted)} \longrightarrow \xrightarrow{H^+,\ H_2O} NuCH_2CH(OH)R$$

Acid-catalyzed opening:

$$\text{protonated oxacyclopropane (R-substituted)} + \text{Nu} \longrightarrow HOCH_2CH(Nu)R + NuCH_2CH(OH)R$$

Major product (HOCH₂CH(Nu)R) — Minor product (NuCH₂CH(OH)R)

13 PREPARATION OF THIOLS AND SULFIDES

$$RX + \underset{\text{Excess}}{HS^-} \longrightarrow \underset{\textbf{Thiol}}{RSH}$$

$$RSH + R'X \xrightarrow{\text{Base}} \underset{\textbf{Alkyl sulfide}}{RSR'}$$

14 ACIDITY OF THIOLS

$$RSH + HO^- \rightleftharpoons RS^- + H_2O \qquad pK_a(RSH) = 9\text{–}12$$

15 NUCLEOPHILICITY OF SULFIDES

$$R_2\ddot{\underset{\cdot\cdot}{S}} + R'X \longrightarrow R_2\overset{+}{S}R'\ X^-$$

Sulfonium salt

16 OXIDATION OF THIOLS

$$RSH \xrightarrow{KMnO_4 \text{ or } H_2O_2} RSO_3H$$

Alkanesulfonic acid

$$RSH \underset{\text{Li, liquid } NH_3}{\overset{I_2}{\rightleftharpoons}} RS\text{—}SR$$

Dialkyl disulfide

17 OXIDATION OF SULFIDES

$$RSR' \xrightarrow{H_2O_2} \underset{\cdot\cdot}{\overset{\overset{\displaystyle O}{\|}}{RSR'}} \xrightarrow{H_2O_2} \underset{\underset{\displaystyle O}{\|}}{\overset{\overset{\displaystyle O}{\|}}{RSR'}}$$

Chiral

Dialkyl sulfoxide **Dialkyl sulfone**

Summary of Important Concepts

1 The reactivity of ROH with alkali metals to give alkoxides and hydrogen follows the order $R = CH_3 >$ primary $>$ secondary $>$ tertiary.

2 Treatment of primary alcohols with strong nonnucleophilic acid at a low temperature furnishes an observable oxonium ion. At a higher temperature, concerted loss of water with simultaneous hydrogen or alkyl shift creates a secondary or tertiary carbocation. In the presence of acid and a nucleophilic counterion, primary alcohols undergo S_N2 reactions. Secondary and tertiary alcohols tend to form carbocations in the presence of acid, capable of E1 and S_N1 product formation, before and after rearrangement.

3 The oxidation of alcohols to aldehydes and ketones opens up important synthetic possibilities, because carbonyl compounds can be modified further by organometallic reagents.

4 Intramolecular Williamson ether synthesis is favored by high-dilution reaction conditions, which slow down intermolecular reactions.

5 Crown ethers and ether cryptands are compounds capable of selective cation binding, allowing for the dissolution of inorganic salts in organic solvents.

6 The relative rates of cyclic ether ring closure follow the order of ring size: $3 \geq 5 > 6 > 4 \geq 7 > 8$.

7 Whereas nucleophilic ring opening of oxacyclopropanes by anions is at the less-substituted ring carbon according to the rules of the S_N2 reaction, acid-catalyzed opening may be at both centers, but often favors the more-substituted carbon, because of charge control of nucleophilic attack.

8 Sulfur has more-diffuse orbitals than does oxygen. In thiols, the S–H bond is less polarized than the O–H bond in alcohols, leading to diminished

hydrogen bonding. The S–H bond is also weaker than the O–H bond, resulting in the increased acidity of thiols.

Problems

1 On which side of the equation do you expect each of the following equilibria to lie (left or right)?

(a) $(CH_3)_3COH + K^{+\,-}OH \rightleftharpoons (CH_3)_3CO^-K^+ + H_2O$

(b) $CH_3OH + NH_3 \rightleftharpoons CH_3O^- + NH_4^+$ ($pK_a = 9.2$)

(c) CH_3CH_2OH + (piperidine ring with N^-Li^+) $\rightleftharpoons$ $CH_3CH_2O^-Li^+$ + (piperidine ring with N–H) ($pK_a = 40$)

(d) NH_3 ($pK_a = 35$) + $Na^+H^- \rightleftharpoons Na^{+\,-}NH_2 + H_2$ ($pK_a \sim 35$)

2 For each of the following alcohols: (1) write the structure of the oxonium ion produced after protonation by strong acid; (2) if the oxonium ion is capable of losing water readily, write the structure of the resulting carbocation; and (3) if the carbocation obtained is likely to be susceptible to rearrangement, write the structures of all new carbocations that might be reasonably expected to form.

(a) $CH_3CH_2CH_2OH$

(b) $CH_3CHOHCH_3$

(c) $CH_3CH_2CH_2CH_2OH$

(d) $(CH_3)_2CHCH_2OH$

(e) $(CH_3)_3CCH_2CH_2OH$

(f) (cyclopentane ring bearing OH and CH_3)

(g) (cyclobutane ring bearing H_3C and a CH(OH)CH_3 group)

(h) (cyclohexane ring bearing $(CH_3)_3C$ and OH on the same carbon)

(i) (cyclohexane ring bearing OH, with two H_3C groups on one adjacent carbon and two CH_3 groups on the other adjacent carbon)

(j) (cycloheptane ring bearing OH, with two CH_3 groups on the adjacent carbon)

3 Write all products of the reaction of each of the alcohols in Problem 2 with concentrated H_2SO_4 under elimination conditions.

4 Write all sensible products of the reaction of each of the alcohols in Problem 2 with concentrated aqueous HBr.

$CH_3CH_2CH_2CH_2OH$

↓ $NaBr, H_2SO_4$

$CH_3CH_2CH_2CH_2Br$

5 Primary alcohols are often converted into bromides by reaction with NaBr in aqueous H_2SO_4. Explain how this works and why it might be considered a superior method to the use of concentrated aqueous HBr.

6 In an attempt to make 1-chloro-1-cyclobutylpentane, the following reaction sequence was employed. The actual product isolated, however, was not the desired molecule, but an isomer of it. Suggest a structure for the product and give a mechanistic explanation for its formation.

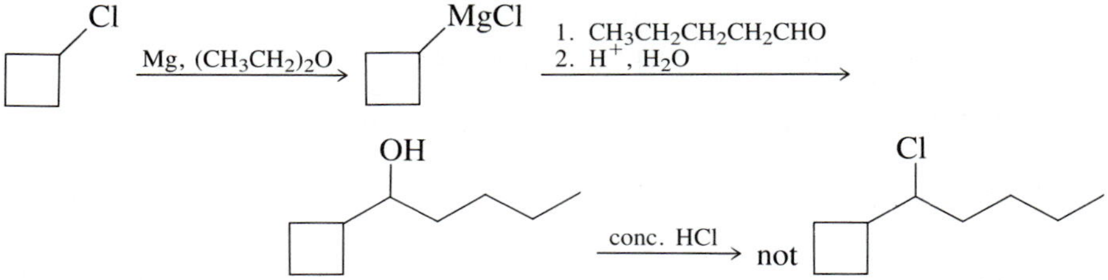

Cl — Mg, $(CH_3CH_2)_2O$ → MgCl — 1. $CH_3CH_2CH_2CH_2CHO$ 2. H^+, H_2O →

OH — conc. HCl → not Cl

7 Write the expected product(s) of the reactions of the following substrates with aqueous sulfuric acid.

(a) 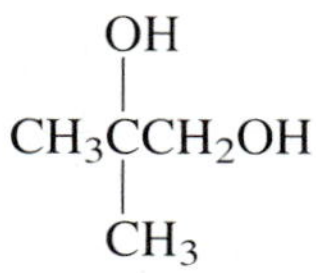 $CH_3C(OH)(CH_3)CH_2OH$

(b) 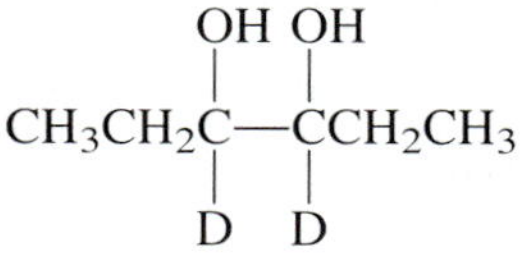 $CH_3CH_2C(OH)(D)$—$C(OH)(D)CH_2CH_3$

(c) 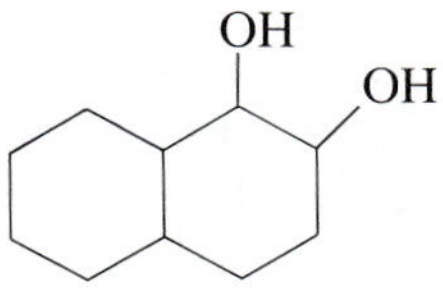 OH, OH

(d) OH, CH_2OH

CH₃, OH, OH, CH₃

Conc. H_2SO_4

O, H_3C, CH_3

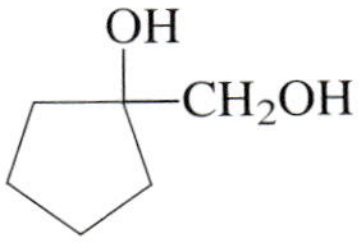

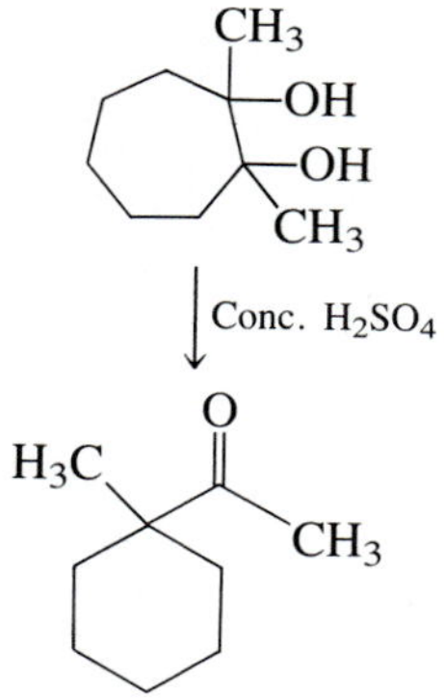

8 Suggest a mechanism for the reaction shown in the margin. What other product could be formed? Which of the two products would you expect to be the major one?

9 Write the most likely product(s) that can be expected from each of the following reactions.

(a) 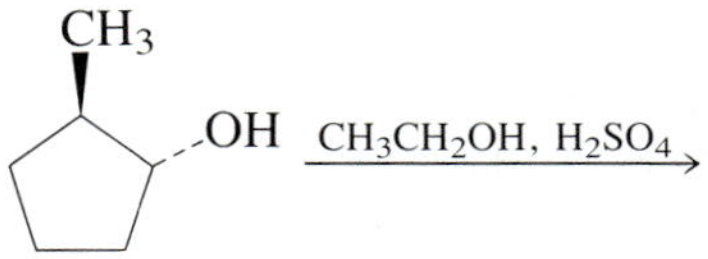 CH_3, OH — CH_3CH_2OH, H_2SO_4 →

(b) 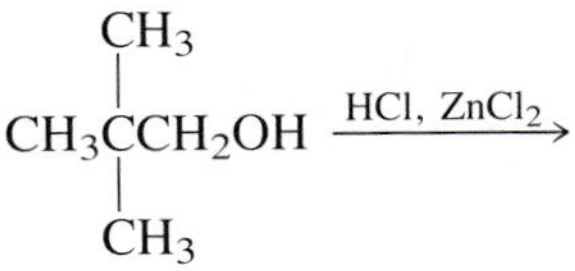 $CH_3C(CH_3)_2CH_2OH$ — HCl, $ZnCl_2$ →

(c) CH_2OH — Conc. H_2SO_4, 180°C 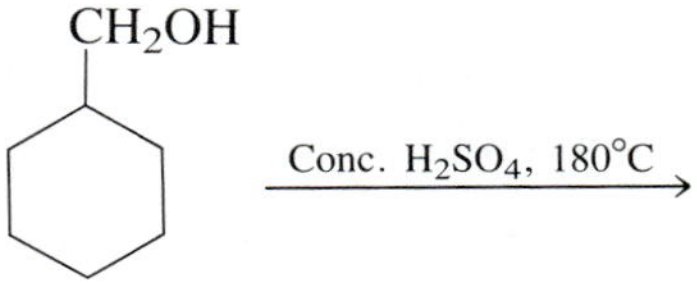 →

(d) 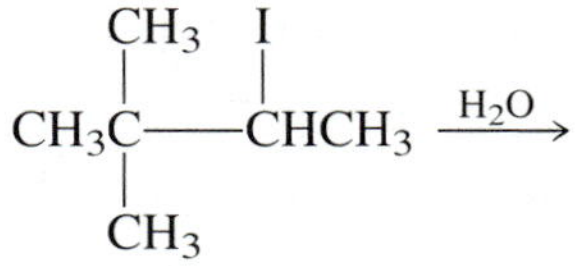 $CH_3C(CH_3)_2$—$CHICH_3$ — H_2O →

(e) 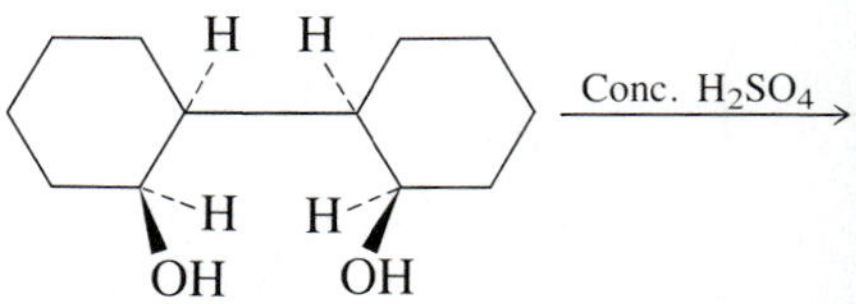 H H, H OH, H OH — Conc. H_2SO_4 →

(f) 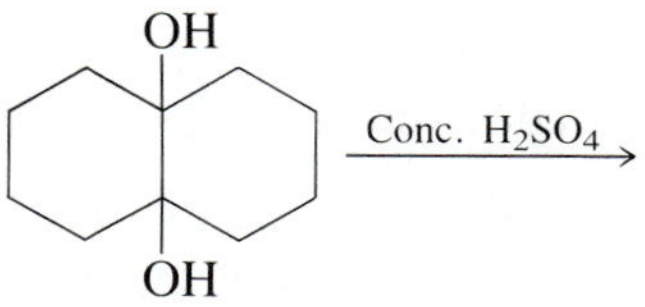 OH, OH — Conc. H_2SO_4 →

10 Write the expected main product of the reaction of each of the alcohols in Problem 2 with PBr_3. Compare and contrast the results with those of Problem 4.

11 Write the expected product(s) of the reaction of 1-pentanol with each of the following reagents.

(a) $K^+ \ {}^-OC(CH_3)_3$
(b) Sodium metal
(c) CH_3Li
(d) Concentrated HI
(e) $HCl + ZnCl_2$
(f) FSO_3H
(g) Concentrated H_2SO_4 at 130°C
(h) Concentrated H_2SO_4 at 180°C
(i) $(CH_3)_2CHCOOH +$ HCl (As catalyst)
(j) PBr_3
(k) $SOCl_2$
(l) $K_2Cr_2O_7 + H_2SO_4 + H_2O$
(m) $CrO_3(\text{pyridine})_2$
(n) $(CH_3)_3COH +$ H_2SO_4 (As catalyst)

12 Write the expected product(s) of the reaction of *trans*-3-methylcyclopentanol with each of the reagents in Problem 11.

13 Which of the following alcohols undergo the iodoform reaction when mixed with iodine in aqueous hydroxide? For each case in which the reaction takes place, write the products.

(a) $CH_3CH_2CH_2CH_2CH_2OH$
(b) $CH_3CH_2CH_2CH(OH)CH_3$
(c) $CH_3CH_2CH(OH)CH_2CH_3$
(d) 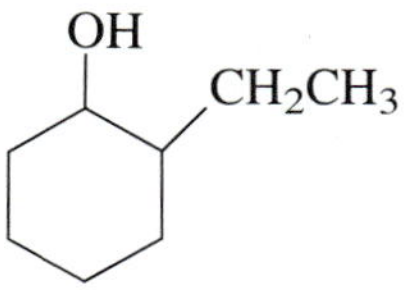 2-ethylcyclohexanol (cyclohexane ring bearing OH and, on the adjacent carbon, CH_2CH_3)
(e) 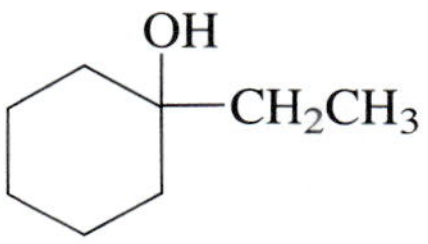 cyclohexane ring bearing OH and CH_2CH_3 on the same carbon
(f) 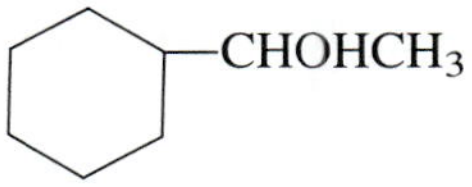 cyclohexyl–$CHOHCH_3$
(g) 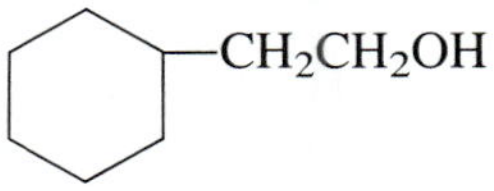 cyclohexyl–CH_2CH_2OH

14 Write IUPAC names for each of the molecules below.

(a) $(CH_3)_2CHOCH_2CH_3$
(b) $CH_3OCH_2CH_2OH$
(c) 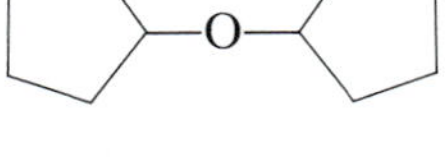 cyclopentyl–O–cyclopentyl
(d) $(ClCH_2CH_2)_2O$
(e) 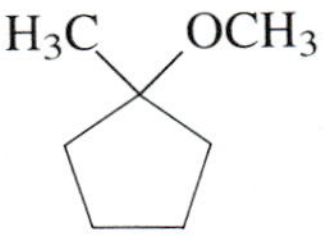 cyclopentane ring bearing H_3C and OCH_3 on the same carbon
(f) 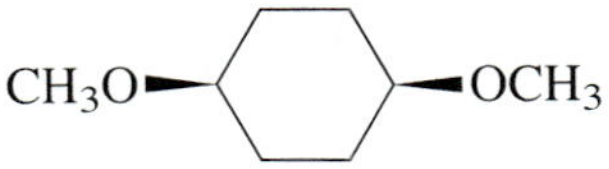 CH_3O–(cyclohexane)–OCH_3 (1,4-disubstituted, both substituents on wedge bonds)
(g) CH_3OCH_2Cl

15 Explain why the boiling points of ethers are so much lower than the boiling points of alcohols that have the same molecular weight.

How would you expect the water solubilities of ethers to compare with the water solubilities of alcohols that have the same molecular weight?

16 Write the expected major product(s) of each of the following attempted ether syntheses.

(a) $CH_3CH_2CH_2Cl + CH_3CH_2\underset{}{\overset{O^-}{\overset{|}{C}}}HCH_2CH_3 \xrightarrow{DMSO}$

(b) $CH_3CH_2CH_2O^- + CH_3CH_2\overset{Cl}{\overset{|}{C}}HCH_2CH_3 \xrightarrow{HMPA}$

(c) 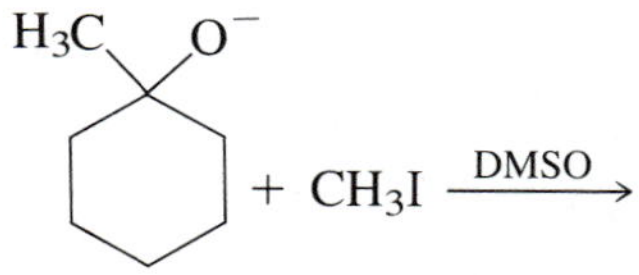 1-methylcyclohexoxide (H_3C, O^- on the same ring carbon) $+ CH_3I \xrightarrow{DMSO}$

(d) $(CH_3)_2CHO^- + (CH_3)_2CHCH_2CH_2Br \xrightarrow{(CH_3)_2CHOH}$

(e) 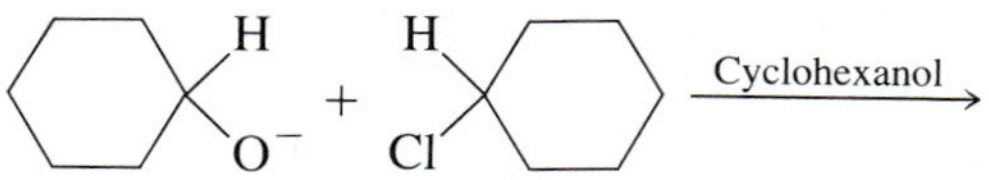 cyclohexyl (H, O^-) + cyclohexyl (H, Cl) $\xrightarrow{\text{Cyclohexanol}}$

(f) 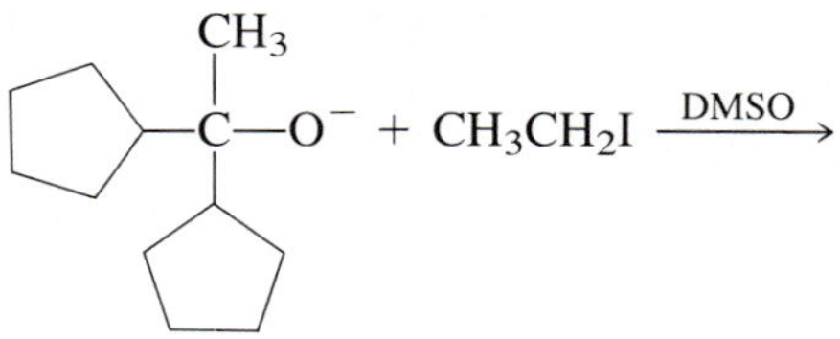 (cyclopentyl)$_2$–$\overset{CH_3}{\overset{|}{C}}$—$O^- + CH_3CH_2I \xrightarrow{DMSO}$

17 For each synthesis proposed in Problem 16 that is not likely to give a good yield of ether product, suggest an alternative synthesis beginning with suitable alcohols or haloalkanes that will give a superior result. Hint: See Problem 1 of Chapter 7.

18 Write the product(s) of reaction of each of the following molecules or pairs of molecules with NaOH in dilute solution in DMSO.

(a) $CH_3\overset{OH}{\overset{|}{C}}HCH_2CH_2\overset{Cl}{\overset{|}{C}}H_2$

(b) cyclohexane ring with H, OH on one carbon and H, CH_2CH_2Br on the adjacent carbon

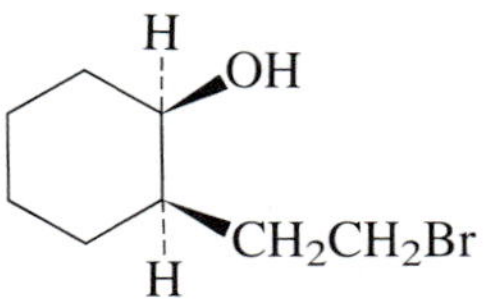

(c) 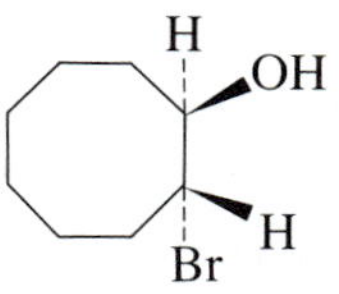 cyclooctane ring with H, OH on one carbon and Br, H on the adjacent carbon

(d) $HOCH_2CH_2OCH_2CH_2OH + BrCH_2CH_2OCH_2CH_2Br$

(e) 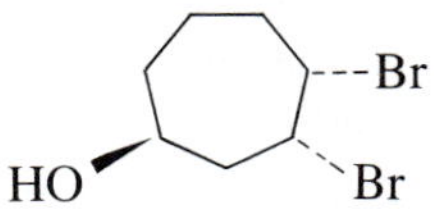 cycloheptane ring with HO, Br, Br substituents

19 What is the product of the reaction shown in the margin? (Pay attention to stereochemistry at the reacting centers.) What is the kinetic order of this reaction?

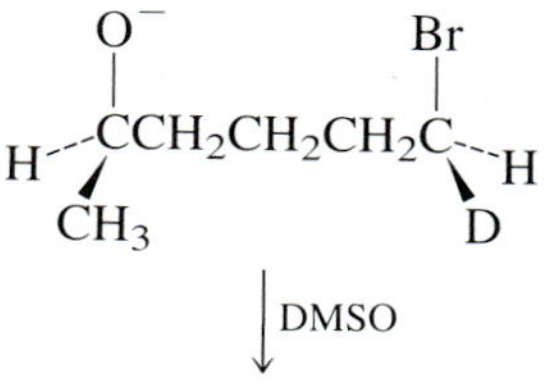

O^- ... Br; H--$CCH_2CH_2CH_2C$--H; CH_3 ... D; ↓ DMSO

20 Propose efficient syntheses for each of the following ethers, using haloalkanes or alcohols as starting materials.

(a) 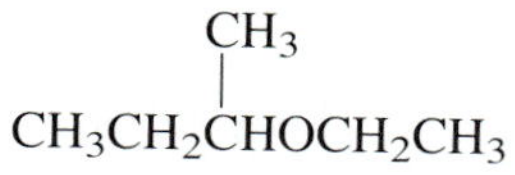 $CH_3CH_2CH(CH_3)OCH_2CH_3$

(b) 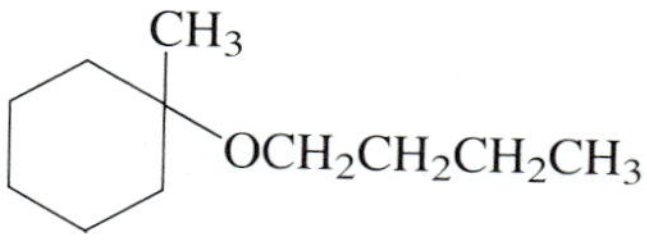 1-methylcyclohexyl–$OCH_2CH_2CH_2CH_3$ (CH_3)

(c) 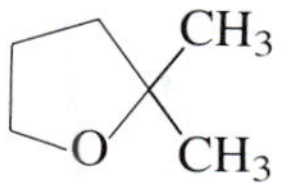 2,2-dimethyltetrahydrofuran (CH_3, CH_3, O)

(d) 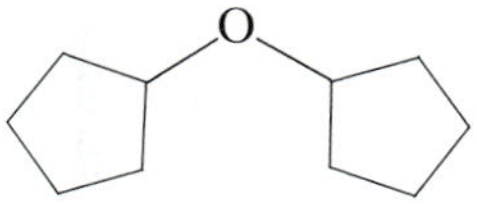 dicyclopentyl ether (O)

21 Write the major product(s) of each of the following reactions. Refer to Section 6-7 if necessary.

(a) $CH_3CH_2OCH_2CH_2CH_3 \xrightarrow{\text{Excess conc. HI}}$

(b) $CH_3OCH(CH_3)_2 \xrightarrow{\text{Excess conc. HBr}}$

(c) $CH_3OCH_2CH_2OCH_3 \xrightarrow{\text{Excess conc. HI}}$

(d) 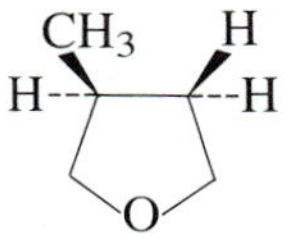 (CH_3, H--, H, --H; O ring) $\xrightarrow{\text{Excess conc. HBr}}$

(e) 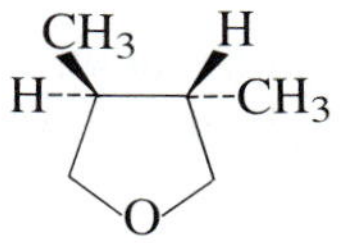 (CH_3, H--, H, --CH_3; O ring) $\xrightarrow{\text{Excess conc. HBr}}$

(f) 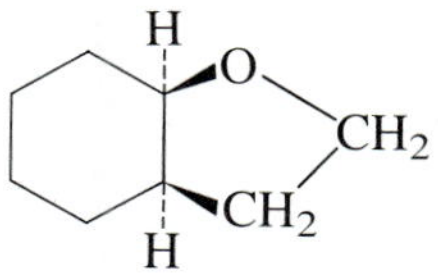 (H, O, CH_2, CH_2, H; cyclohexane-fused ring) $\xrightarrow{\text{Excess conc. HBr}}$

22 Write the major product(s) of each of the following reactions.

(a)  (oxirane, O) $\xrightarrow{Na^{+}\,{}^{-}NH_2,\ NH_3}$

(b) 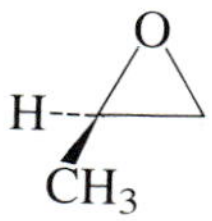 (O; H--, CH_3) $\xrightarrow{Na^{+}\,{}^{-}SCH_2CH_3,\ CH_3CH_2OH}$

(c) 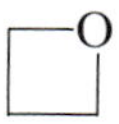 (oxetane, O) $\xrightarrow{\text{Excess conc. HBr}}$

(d) 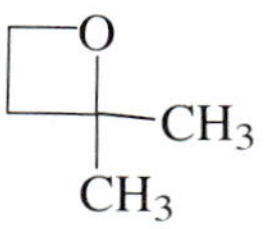 (O; CH_3, CH_3) $\xrightarrow{\text{Dilute HCl in } CH_3OH}$

(e) 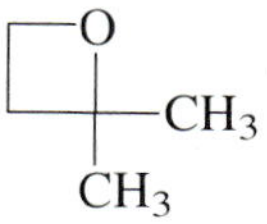 (O; CH_3, CH_3) $\xrightarrow{Na^{+}\,{}^{-}OCH_3 \text{ in } CH_3OH}$

(f) 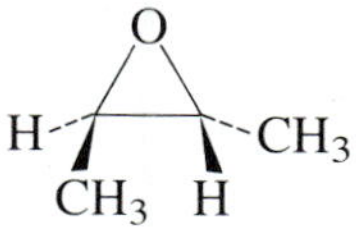 (O; H--, --CH_3; CH_3, H) 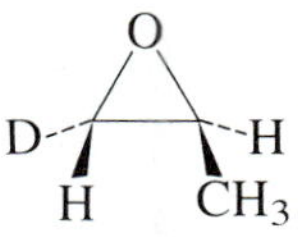 $\xrightarrow{\text{Dilute } H_2SO_4 \text{ in } CH_3CH_2OH}$

(g) (O; H--, --CH_3; CH_3, H) $\xrightarrow{LiAlD_4}$

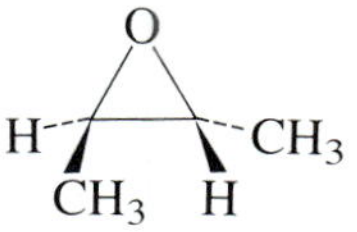

(h) (O; D--, --H; H, CH_3) $\xrightarrow{\text{Dilute } H_2SO_4 \text{ in } CH_3CH_2OH}$

23 Suggest a good synthetic method for preparing each of the following haloalkanes from the corresponding alcohols.

(a) $CH_3CH_2CH_2Cl$

(b) $CH_3CH_2CH(CH_3)CH_2Br$

(c) 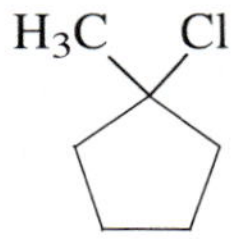 H_3C Cl

(d) $CH_3CHICH(CH_3)_2$

24 Suggest the best method for preparing each of the following compounds from an appropriate alcohol.

(a) 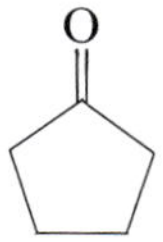 O

(b) $CH_3CH_2CH_2CH_2COOH$

(c) 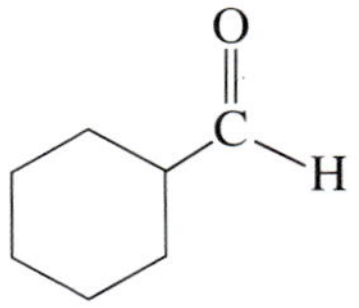 cyclohexanecarbaldehyde (cyclohexyl–CHO)

(d) $CH_3CH(CH_3)CCH_3$ with C=O (CH_3CHCCH_3, CH_3 on CH, O double-bonded to C)

(e) CH_3CHO ($CH_3\overset{O}{\overset{\|}{C}}H$)

25 Propose sensible synthetic schemes for the preparation of each of the following compounds, using only the organic starting material(s) indicated. Use any organic solvents or inorganic reagents necessary.

(a) $CH_3CH(CH_3)—C(CH_3)(OH)—CH_2CH_3$, from ethane and propane

(b) $CH_3CH_2CH(CH_3)C(=O)CH_2CH_3$, from butane, ethane, and methanal $\left(H\overset{O}{\overset{\|}{C}}H\right)$

$H\overset{O}{\overset{\|}{C}}OCH_3$
Methyl methanoate

(c) $(CH_3CH_2)_2CHCH_2COOH$, from ethane, methyl methanoate, and oxacyclopropane

(d) 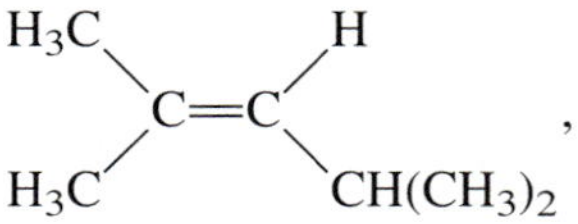 $(H_3C)_2C{=}CH{-}CH(CH_3)_2$, from propane and methyl methanoate

26 Propose syntheses of the following molecules, choosing reasonable starting materials on the basis of the principles of synthetic strategy introduced in preceding chapters, particularly in Section 8-7. Suggested positions for carbon–carbon bond formation are indicated by wavy lines.

(a) 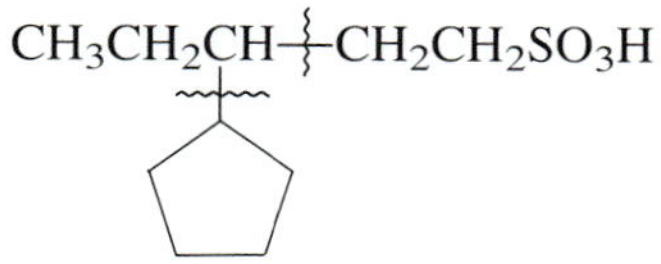 CH_3CH_2CH–$CH_2CH_2SO_3H$ (with a cyclopentyl group on CH)

(b) 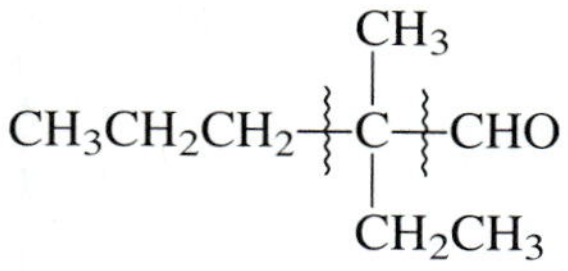 $CH_3CH_2CH_2$–$C(CH_3)(CH_2CH_3)$–CHO

27 Name each of the following compounds using the IUPAC system of nomenclature.

(a) 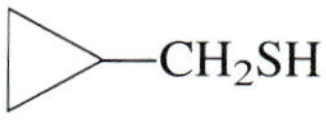 cyclopropyl–CH_2SH

(b) 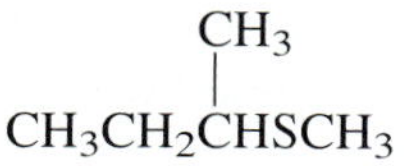 $CH_3CH_2CH(CH_3)SCH_3$

(c) $CH_3CH_2CH_2SO_3H$

(d) CF_3SO_2Cl

28 In each of the following pairs of compounds indicate (1) which is the stronger acid and (2) which is the stronger base.

(a) CH_3SH, CH_3OH
(b) HS^-, HO^-
(c) H_3S^+, H_2S

29 Write reasonable products for each of the following reactions.

(a) $ClCH_2CH_2CH_2CH_2Cl \xrightarrow{\text{Excess NaSH}}$

(b) $ClCH_2CH_2CH_2CH_2Cl \xrightarrow{\text{One equivalent } Na_2S}$

(c) 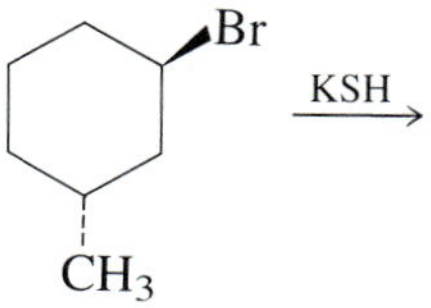 Br-substituted cyclohexane with CH_3 $\xrightarrow{\text{KSH}}$

(d) 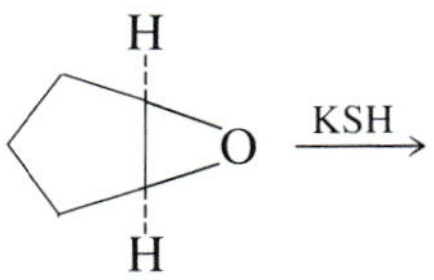 cyclopentene oxide (H, H) $\xrightarrow{\text{KSH}}$

(e) 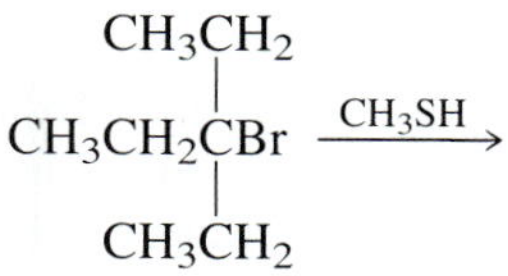 $(CH_3CH_2)_3CBr \xrightarrow{CH_3SH}$

(f) 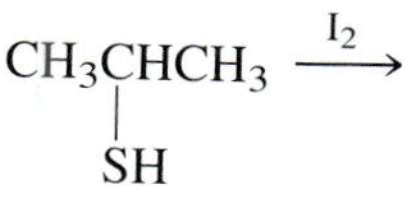 $CH_3CH(SH)CH_3 \xrightarrow{I_2}$

(g) 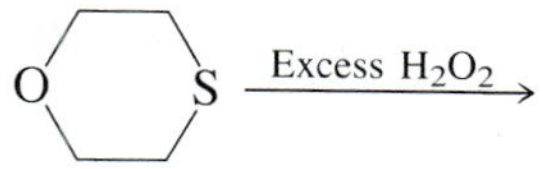 1,4-oxathiane (O, S ring) $\xrightarrow{\text{Excess } H_2O_2}$

30 Propose a synthesis for 1,4-oxathiane (O, S ring), beginning with 2-bromoethanol.

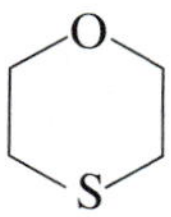

31 Compare the following methods of alkene synthesis from a general primary alcohol. State the advantages and disadvantages of each one.

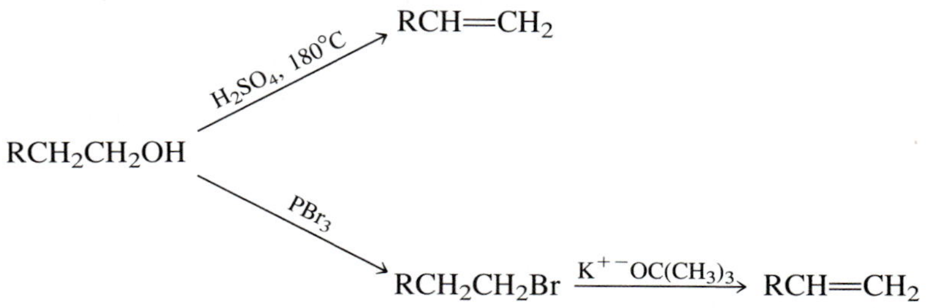

$$RCH_2CH_2OH \xrightarrow{H_2SO_4,\ 180°C} RCH{=}CH_2$$

$$RCH_2CH_2OH \xrightarrow{PBr_3} RCH_2CH_2Br \xrightarrow{K^{+}\,{}^{-}OC(CH_3)_3} RCH{=}CH_2$$

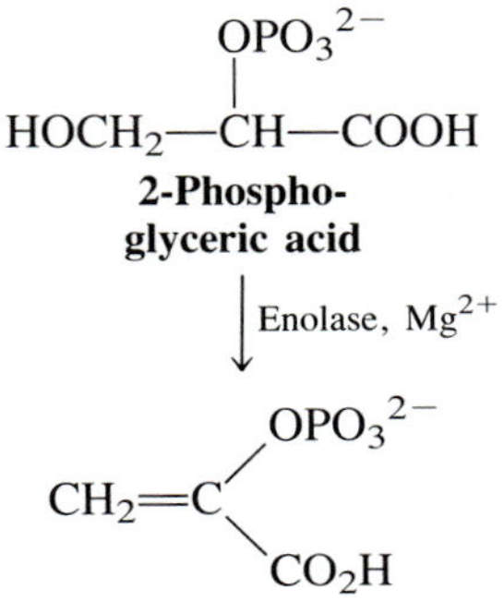

$$HOCH_2{-}CH(OPO_3^{2-}){-}COOH \xrightarrow{\text{Enolase, } Mg^{2+}} CH_2{=}C(OPO_3^{2-})CO_2H$$

2-Phosphoglyceric acid → **2-Phosphoenolpyruvic acid**

32 Sugars, being polyhydroxylic compounds (Chapter 23), undergo reactions characteristic of alcohols. In one of the later steps in glycolysis (the metabolism of glucose), one of the glucose metabolites with a remaining hydroxy group, 2-phosphoglyceric acid, is converted into 2-phosphoenolpyruvic acid; this reaction is catalyzed by the enzyme enolase in the presence of a Lewis acid such as Mg^{2+}.

(a) What kind of reaction is this?
(b) What possible role could the Lewis acidic metal ion be playing?

33 Several enzymes, together with certain derivatives of vitamin B_{12}, catalyze a class of reactions that includes the one shown below.

$$HOCH_2CH(OH)CH_2OH \longrightarrow HOCH_2CH_2CH{=}O$$

1,2,3-Propanetriol → **3-Hydroxypropanal**

(a) To what nonenzymatic reaction does this process bear a resemblance?
(b) Under what nonenzymatic conditions can this reaction take place (i.e., what would you do to obtain successful results in the laboratory)?
(c) Starting with $HOCD_2CH(OH)CD_2OH$, what product(s) would you expect to obtain? With $HOCD_2CH(OH)CH_2OH$?

$$RCH(OH)R + NAD^{+} \xrightarrow{\text{Enzyme}} RC({=}O)R + NADH + H^{+}$$

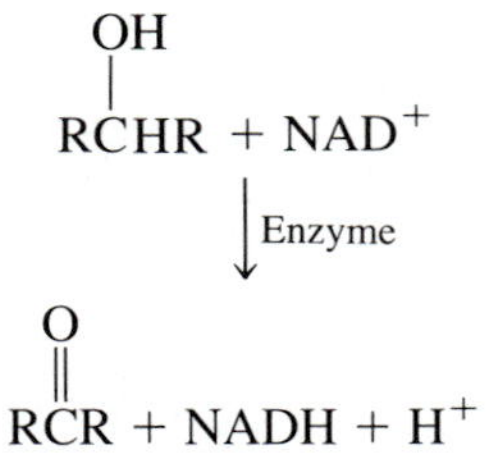

34 The reverse of NADH reduction of aldehydes and ketones (Chapter 8, Problems 31 and 32) also occurs biologically: oxidation of alcohols by NAD^{+}. When monodeuterated ethanal (acetaldehyde), CH_3CDO, is reduced by NADH in the presence of alcohol dehydrogenase (compare Chapter 8, Problem 31, part **a**), ethanol containing exactly one deuterium atom per molecule (CH_3CHDOH) is formed. If this ethanol is then oxidized by NAD^{+}, CH_3CDO is regenerated, containing all of the original deuterium.

(a) What stereochemical information does this tell you concerning NADH reduction of aldehydes and NAD^+ oxidation of alcohols? Hint: Note that CH_3CHDOH contains a stereocenter because of the isotopic difference between H and D.

(b) If deuterated NADH (i.e., NADD) is used to reduce CH_3CHO, monodeuterated ethanol, CH_3CHDOH, is also obtained. Is this ethanol identical with the ethanol formed by the reaction between CH_3CDO and NADH? If not, how does it differ?

35 **(a)** Only the trans isomer of 2-bromocyclohexanol can react with sodium hydroxide to form an oxacyclopropane-containing product. Explain the lack of reactivity of the cis isomer. Hint: Draw the available conformations of both the cis and trans isomers around the C-1–C-2 bonds (compare Figure 4-13). Use models if necessary.

(b) The synthesis of some oxacyclopropane-containing steroids has been achieved by use of a two-step procedure starting with steroidal bromo ketones. Suggest suitable reagents for accomplishing a conversion such as the following one.

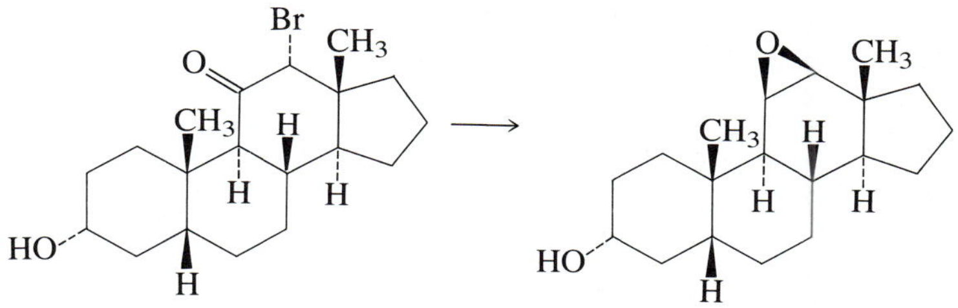

(c) Do any of the steps in your proposed sequence have specific stereochemical requirements in order for the oxacyclopropane-forming step to be successful?

36 Freshly cut garlic contains allicin, a compound responsible for the true garlic odor. Allicin also possesses substantial antibacterial properties and is apparently effective in limiting the increase in cholesterol levels for animals on high-cholesterol diets. Propose a short synthesis of allicin, starting with 3-chloropropene.

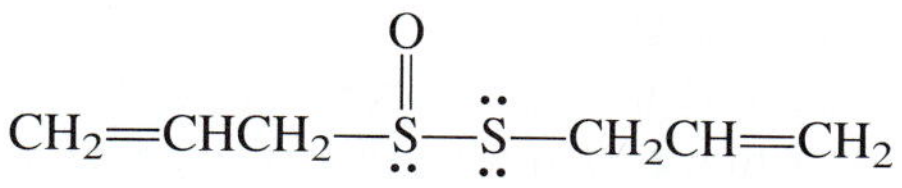

$$CH_2{=}CHCH_2{-}\overset{O}{\overset{\|}{S}}{-}S{-}CH_2CH{=}CH_2$$

Allicin

CHAPTER 10

The Use of Nuclear Magnetic Resonance Spectroscopy to Deduce Structure

Knowledge of the variety of organic reactions and functional groups presented in the preceding chapters should enable you to synthesize a reasonably complicated organic molecule in the laboratory. At this point, we need to return to a question raised in Section 1-9: how do we ascertain the structure of a molecule that we have synthesized? How do we know, for example, that a specific Grignard reagent has converted a ketone into the desired alcohol? To answer these questions, we have at our disposal a variety of measurements and tests. We begin by purifying our sample by means of chromatography, distillation, or recrystallization. Once it is pure, we can measure some of its physical properties—that is, melting point, boiling point, dipole moment, refractive index, and so forth. The values obtained can then be compared with data for known compounds, published in the literature or in appropriate handbooks. A match between several such measurements means that we can be reasonably certain of the structure of our molecule. However, this is true only if identification of the compound under investigation relies on an earlier structural proof. Many substances made in the laboratory are completely new; therefore, no data are available. What other options are there for solving such structural problems?

A useful device is elemental analysis (mentioned in Section 1-9) because it reveals the gross chemical composition of our molecule. A complementary approach consists of a series of tests aimed at elucidating the chemistry of the compound and hence its functional-group content. For example, we saw earlier how we can distinguish chemically between methoxymethane and ethanol by exposing them to sodium metal (Section 1-9). However, the problem becomes considerably more difficult for larger molecules with more structural variations. What if a reaction gave us an alcohol of molecular formula $C_7H_{16}O$? Even though the test with sodium metal would reveal the presence of a hydroxy functional group, this finding would not provide an unambiguous structure for the product. In fact, there are many possibilities, three of which are shown below.

Three Structural Possibilities for an Alcohol $C_7H_{16}O$

$$CH_3(CH_2)_5CH_2OH \qquad CH_3\overset{\displaystyle CH_3}{\underset{\displaystyle CH_3}{C}}CH_2CH_2CH_2OH \qquad CH_3\overset{\displaystyle CH_2CH_3}{\underset{\displaystyle CH_3}{C}}CH_2CH_2OH$$

EXERCISE 10-1

Write alternative structures of secondary and tertiary alcohols having the empirical formula $C_7H_{16}O$.

How do we differentiate between all of these alternatives? The modern organic chemist makes use of another tool to solve this problem: **spectroscopy** (*spectrum,* Latin, appearance, apparition, referring to the appearance of characteristic recorded lines).

There are many types of spectroscopy. Those most often used in organic chemistry fall into four categories: (1) *nuclear magnetic resonance* (NMR); (2) *infrared* (IR); (3) *ultraviolet* (UV); and (4) *mass spectroscopy* (MS). The most important of these methods is **NMR spectroscopy.** It provides a probe for the structural environment of individual nuclei (particularly hydrogens and carbons, but also certain other elements). Infrared spectroscopy (Section 17-3) affords a means for identifying certain functional groups (such as hydroxy and carbonyl) in a molecule. It also supplies an unambiguous ''fingerprint'' of its framework, which can be compared with ''fingerprints'' of known substances. Ultraviolet or, more generally, UV-visible spectra (Section 14-7) furnish information about the electronic structure of certain compounds. Finally, mass spectroscopy (Section 18-7) measures the molecular mass of a molecule. Its instrumentation and physical techniques are quite different from those of the other three types. This chapter deals primarily with the principles and applications of nuclear magnetic resonance spectroscopy.

10-1 What Is Spectroscopy?

We shall begin with a brief overview, in simple terms, of the type of spectroscopy relevant to NMR, IR and UV, followed by a description of how a spectrometer works in principle.

The Electromagnetic Spectrum

Organic molecules absorb radiation, in discrete ''packets'' of $\Delta E = h\nu$, which are also called **quanta** of energy (Figure 10-1). The absorbed energy causes some kind of electronic or mechanical ''motion'' in the molecule, a process also called *excitation*. This motion is quantized, and absorption occurs only when radiation supplying exactly the right ''packet'' of energy impinges on the compound under investigation.

FIGURE 10-1
The energy difference, ΔE, between the ground state and the excited state of a molecule is overcome by incident radiation of frequency ν matched exactly to equal ΔE.

ν frequency of absorbed radiation

h (Planck's constant) = 6.626×10^{-27} erg sec

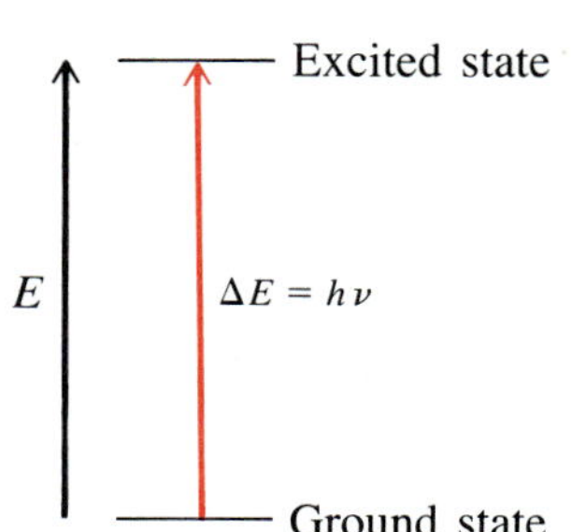

FIGURE 10-2

The spectrum of electromagnetic radiation. Energies are expressed in kcal $mole^{-1}$; wave numbers, $\tilde{\nu}$, in units of cm^{-1}; and wavelength, λ, in units of nanometers (1 nm = 10^{-9} m), micrometers (1 μm = 10^{-6} m), millimeters (mm), and meters (m). Wave number is defined as $\tilde{\nu} = 1/\lambda$, expressing the number of waves per centimeter. This quantity is related to (but should not be mistaken for) the frequency of radiation $\nu = c/\lambda$ (in cycles per second, or Hertz), in which c = velocity of light = 3×10^{10} cm sec^{-1}. A simple conversion between ΔE (kcal $mole^{-1}$) and λ (nm) is given by the equation $\Delta E = 28{,}600/\lambda$.

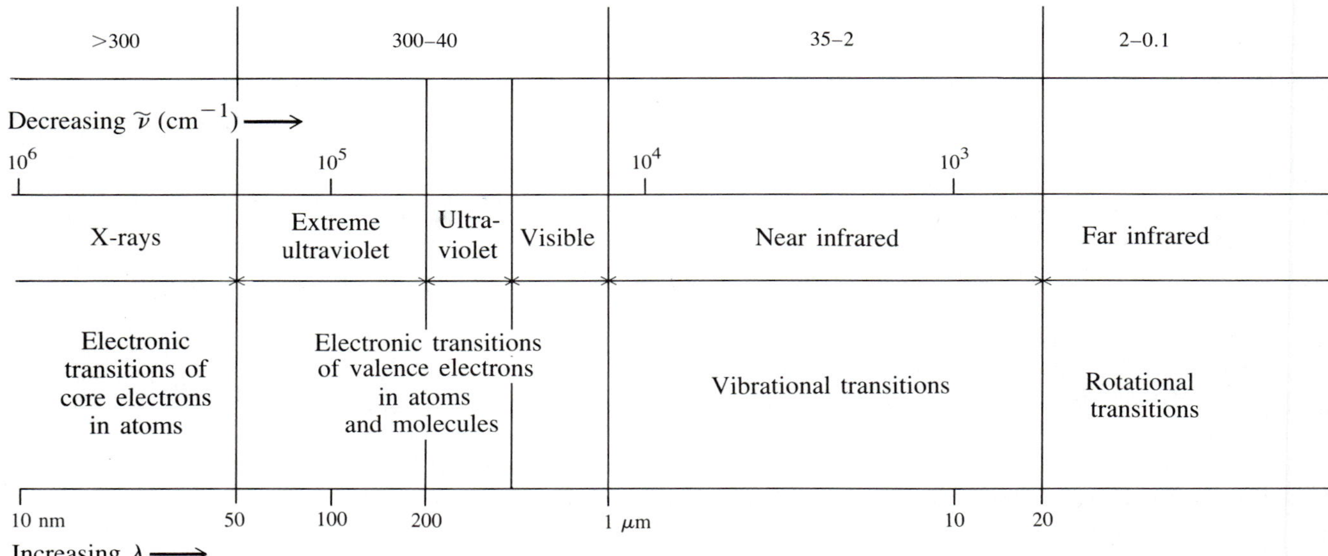

A molecule can undergo many different kinds of excitations, each of which requires its own distinctive energy, ΔE. X-rays, for example, are high-energy radiation that can move electrons in atoms from inner shells to outer ones, a change called an *electronic transition* and (in this case) requiring energy higher than 300 kcal $mole^{-1}$. Ultraviolet and visible light can move valence-shell electrons, typically from a filled bonding molecular orbital to an unfilled anti-bonding one (see Figure 1-12). The energy needed for this transfer ranges from 40 to 300 kcal $mole^{-1}$. Infrared radiation causes vibrational excitation of the molecular framework of a compound (ΔE is 1–10 kcal $mole^{-1}$); quanta of microwave radiation effect rotations around bonds (ΔE is about 10^{-4} kcal $mole^{-1}$); and radio waves reorient nuclear spins (ΔE is about 10^{-6} kcal $mole^{-1}$), a phenomenon that is the basis of nuclear magnetic resonance spectroscopy. The various forms of radiation, the energies, ΔE, related to them, and the corresponding wavelength ranges are depicted in Figure 10-2. Remember that radiation increases in energy with increasing frequency, ν, or wave number, $\tilde{\nu}$, but decreasing wavelength, λ.

EXERCISE 10-2

What type of radiation (in wavelength, λ) would be minimally required to initiate the free-radical chlorination of methane? (Refer to Section 3-5.)

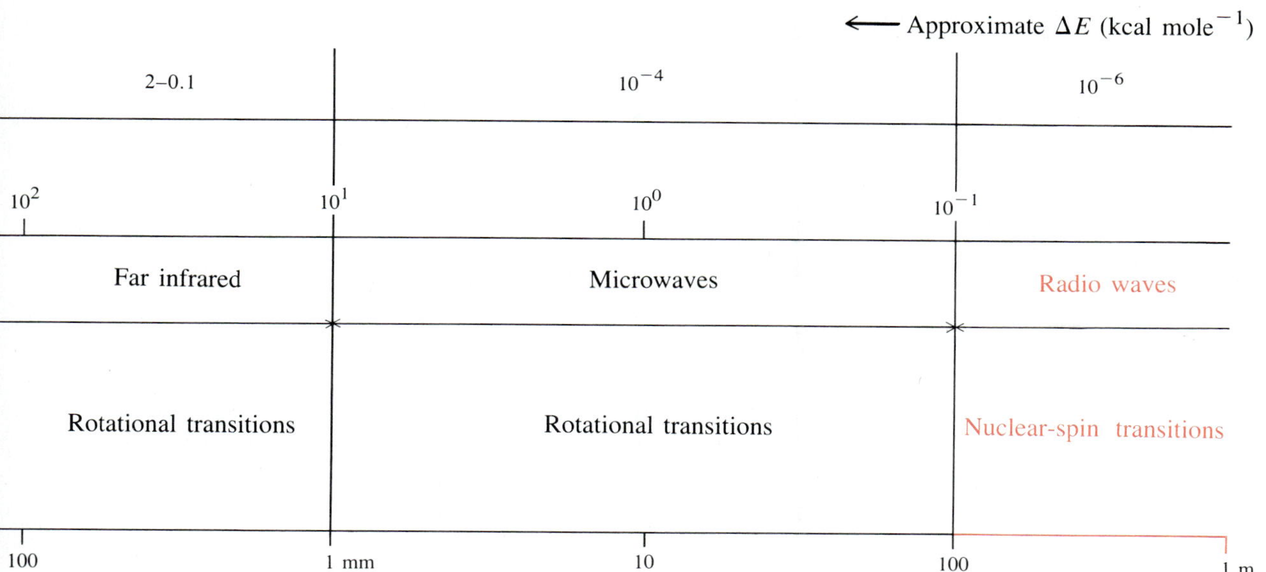

The Spectrometer

As illustrated in Figure 10-1, the absorption of a quantum of radiation by a molecule brings about a transition from its normal (ground) state to an excited state. Spectroscopy is essentially a technical procedure by which these absorptions can be mapped and recorded. The instruments used for these purposes are called spectrometers. A spectrometer has an appropriate source of electromagnetic radiation with a frequency in the region of interest (infrared, radio wave, etc.). This radiation is frequently split into an incident beam and a reference beam of equal intensity. The former travels through a sample tube, whereas the latter is unperturbed (Figure 10-3). A detector at the end of the spectrometer measures the intensity of both beams and records even slight deviations from equality. The apparatus is designed in such a way as to allow only radiation of a specific frequency (within the range of the source) to pass through the sample. A continuous change of the frequency of the incident radiation allows over time a sweep through the entire range of interest. This sweep is recorded on a piece of calibrated chart paper as a line. Whenever the sample absorbs incident light, the resulting intensity difference relative to the reference beam is measured by the detector and electronically relayed to the recorder to give a ‘‘peak,’’ a deviation from the straight line observed when both intensities are equal. The resulting line pattern, or plot, is called a *spectrum* of the sample.

FIGURE 10-3
General schematic diagram of a spectrometer. The source beam is split into reference and incident beams of equal intensity ($I_r = I_i$). The incident beam traverses the sample and emerges as the transmitted beam (I_t). When no absorption occurs, its intensity, I_t, will equal I_i and hence I_r. When $I_t = I_r$, the detector notes no difference and a straight line (zero or base line) is drawn by the recorder. On absorption, $I_t \neq I_r$ and a peak results. The resulting diagram is called a spectrum.

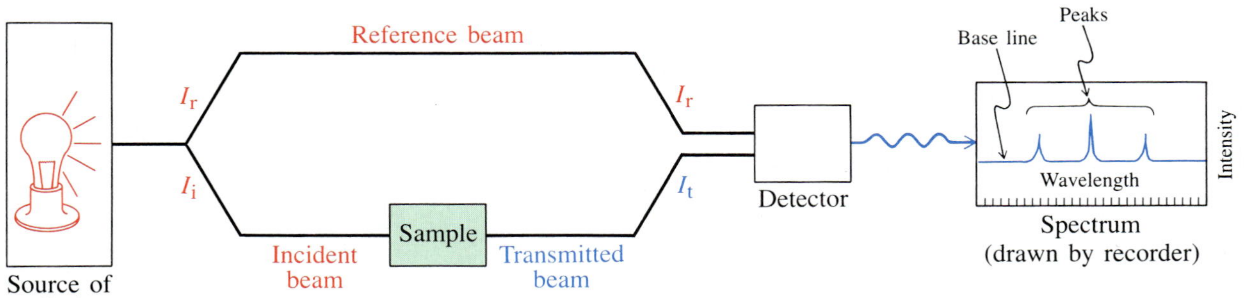

In summary, electromagnetic radiation is absorbed in discrete quanta of incident radiation measurable by spectroscopy. Nuclear magnetic resonance spectroscopy requires low-energy radiation in the radio-frequency range.

10-2 Proton Nuclear Magnetic Resonance

This section presents the elementary principles of nuclear magnetic resonance, particularly in regard to the hydrogen nucleus.

Nuclear Spins and the Absorption of Radio Waves

Many atomic nuclei behave as if they were spinning and are therefore said to have a **nuclear spin.** One of those nuclei is hydrogen, written as ^1H (the hydrogen isotope of mass one) to differentiate it from other isotopes (deuterium, tritium; Table 10-1). Let us consider the simplest form of hydrogen, the proton. Because the proton is positively charged, its spinning motion creates (as does any moving charged particle) a **magnetic moment.** The net result of this effect is that a proton may be viewed as a tiny bar magnet floating freely in solution or in space. When the proton is exposed to an external magnetic field, H_0, it may have one of two orientations: it may be aligned with H_0, an energetically favorable choice, or against H_0, a move that will cost energy. The two possibilities are designated spin α and spin β, respectively (Figure 10-4).

These two energetically different possibilities afford the necessary condition for spectroscopy: irradiation of the sample with a source of just the right frequency to bridge the difference in energy between the α and β states produces **resonance,** manifesting itself in absorption. Resonance is a general spectroscopic phenomenon that occurs when this condition is satisfied. It also occurs in a physical system, such as a weight on a spring, when an external force acting periodically on it does so with a frequency matching the natural oscillation frequency of the system. This phenomenon is illustrated schematically for a pair

FIGURE 10-4

Proton nuclei as tiny bar magnets with the two poles indicated by plus and minus signs (A) in the absence of a magnetic field (the positive charges on the nuclei have been omitted) and (B) in a magnetic field, H_0: the nuclear spins align with the field (α) or against the field (β).

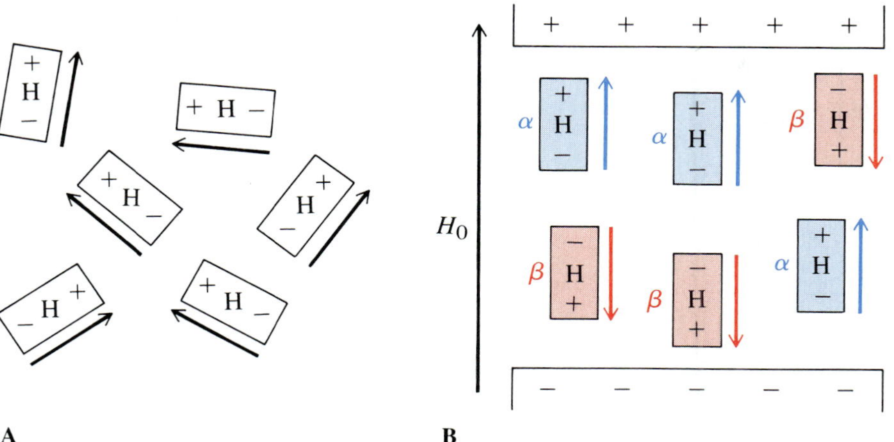

of protons in Figure 10-5. Nuclear magnetic resonance is unusual in that the different energetic states required for spectroscopy must be created by the applied external field. In other forms of spectroscopy—for example, UV and IR—these states are already present in the molecule. After excitation, the nuclei relax and return to their original states by a variety of pathways (which will not be discussed here), all of which release the absorbed energy as heat. At resonance, therefore, there is continuous excitation and relaxation.

As might be expected, the difference in energy, ΔE, between spin states α and β depends directly on the external field strength, H_0. The stronger the external field, the larger the difference in energy. This finding is simply a reflection of the relative difficulty of ''flipping'' an α to a β spin. The absorption frequency, ν, is directly proportional to H_0 with a proportionality constant k:

$$\nu = kH_0$$

The value of k is characteristic of the type of nucleus under investigation and its environment.

How much energy must be expended for the spin of a nucleus to flip from α to β? To answer this question, we need to know typical values for H_0. The commercial magnets employed today range in field strength from about 14,000 to 150,000 Gauss. The corresponding ν values needed to observe resonance lie in the radio-frequency range between 60 and 600 MHz (megaHertz, millions of Hertz, or cycles sec^{-1}). For example, at 21,150 Gauss, hydrogen nuclei require irradiation with radio waves of 90 MHz to cause resonance. Because $\Delta E_{\beta-\alpha} = h\nu$, we can calculate how much energy is being absorbed in this process. The amount is very small, on the order of 9×10^{-6} kcal $mole^{-1}$. Because so little energy is being absorbed, equilibration between the two states is fast, and on average only slightly more than one-half of all proton nuclei in a magnetic field will adopt the α state, the remainder having a β spin.

FIGURE 10-5

Generation of energetically different spin states by the exposure of protons to an external field. Irradiation with energy of frequency ν in which $h\nu = \Delta E$ causes absorption: (A) an α-to-β spin flip in the sample on absorption of $h\nu$; (B) energetics of excitation in the α-to-β spin flip shown in part A.

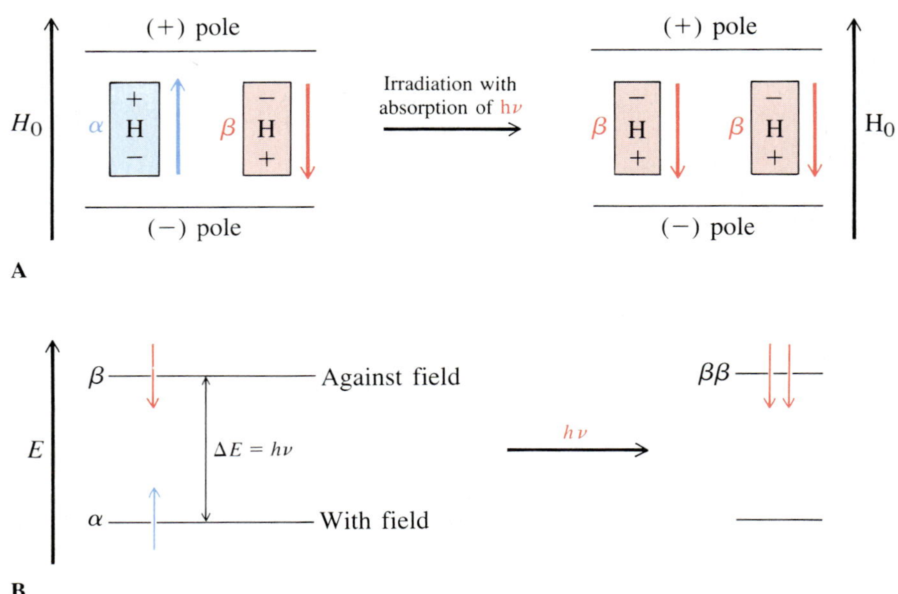

Nuclei Other Than Hydrogen Undergo Nuclear Magnetic Resonance

Hydrogen is not the only nucleus capable of nuclear magnetic resonance. Other nuclei responsive to NMR and of importance in organic chemistry are deuterium (D or ^{2}H = hydrogen isotope of mass 2), the ^{13}C isotope (but not the more-abundant ^{12}C nucleus), ^{14}N and ^{15}N, ^{17}O (but not ^{16}O), ^{19}F, ^{31}P, ^{35}Cl, and ^{37}Cl (Table 10-1).

TABLE 10-1

NMR activity and natural abundance of selected nuclei

Nucleus	NMR activity	Natural abundance (%)	Nucleus	NMR activity	Natural abundance (%)
^{1}H	Active	99.985	^{16}O	Inactive	99.759
^{2}H (D)	Active	0.015	^{17}O	Active	0.037
^{3}H (T)	Active	0	^{18}O	Inactive	0.204
^{12}C	Inactive	98.89	^{19}F	Active	100
^{13}C	Active	1.11	^{31}P	Active	100
^{14}N	Active	99.63	^{35}Cl	Active	75.53
^{15}N	Active	0.37	^{37}Cl	Active	24.47

FIGURE 10-6

The predicted NMR spectrum of CHDClF at 21,150 Gauss. Six lines should be observed: at 7.34 MHz (^{37}Cl), 8.82 MHz (^{35}Cl), 13.7 MHz (D), 22.6 MHz (^{13}C), 84.6 MHz (^{19}F), and 90 MHz (^{1}H). This is a simplified picture, because several of these nuclei (the chlorine isotopes, ^{2}H, and ^{13}C) require special techniques to be detectable by NMR spectroscopy owing to their low abundance or sensitivity or both.

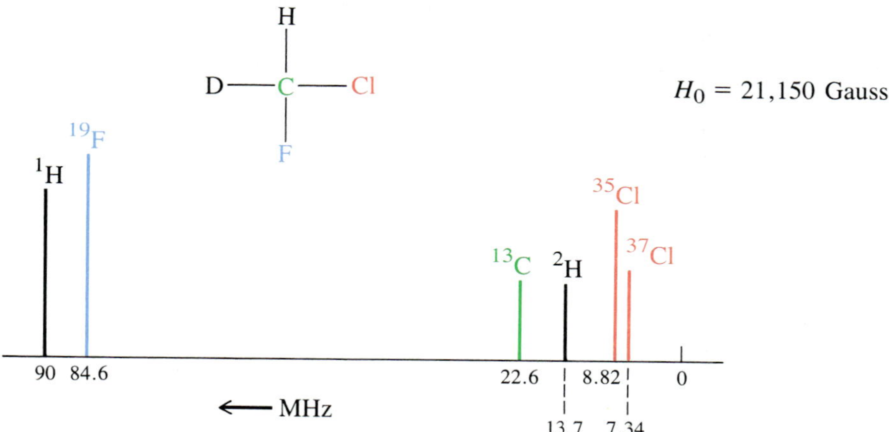

Each one of these nuclei has a different characteristic k value; hence, at *equal field strength, different nuclei will resonate at different values of ν.* This statement also applies to isotopes of the same element: hydrogen versus deuterium, chlorine-35 versus chlorine-37, and so forth. For example, if we were to scan a hypothetical spectrum of a sample of deuteriochlorofluoromethane, CHDClF, in a 21,150 Gauss magnet, we would observe six absorptions corresponding to the six NMR "active" nuclei in the molecule: the highly abundant ^{1}H, ^{19}F, ^{35}Cl, and ^{37}Cl, and the much-less plentiful ^{13}C (1.11%) and ^{2}H (0.015%), as shown in Figure 10-6.

The spectrum could be spread out further by scanning at higher field strengths. For example, at 42,300 Gauss all nuclei would resonate at double the frequency shown in Figure 10-6, ranging from a low of 14.68 MHz for ^{37}Cl, to a high of 180 MHz for the ^{1}H nucleus. Such a wider spectral range allows better separation (resolution) of the individual NMR absorptions, a factor which is very important in **high-resolution NMR spectroscopy.** As we shall see, this technique can also be used to differentiate nuclei of the same species.

High-Resolution Nuclear Magnetic Spectroscopy

Consider the NMR spectrum of chloro(methoxy)methane (chloromethyl methyl ether), $ClCH_2OCH_3$. A spectral sweep at 21,150 Gauss from 0 to 90 MHz would give large peaks for the most-abundant nuclei present (Figure 10-7A). High-resolution NMR allows us to study these peaks more closely by expanding the spectrum in the immediate vicinity of the main resonances. Modern instrumentation is advanced enough to enable us to look at an incredibly small part of this spectrum. At the hydrogen resonance, this part ranges from 90,000,000 to 90,000,900 Hz. In examining this expanded spectrum, we find that it does not

FIGURE 10-7

Hypothetical NMR spectrum of $ClCH_2OCH_3$ at 21,150 Gauss: (A) at low resolution, the spectrum shows four peaks for the four different most-abundant nuclei present in the molecule (weak lines due to the presence of small amounts of ^2H and ^{17}O have been omitted); (B) at high resolution, the hydrogen spectrum shows two peaks for the two sets of hydrogens (one shown in blue in the structure, the other in red). Note that the high-resolution sweep covers only 0.001% of that at low resolution. The high-resolution ^{13}C spectrum (C, measured by the use of a special technique described in Section 10-7), shows two peaks for the two different carbon atoms in the molecule.

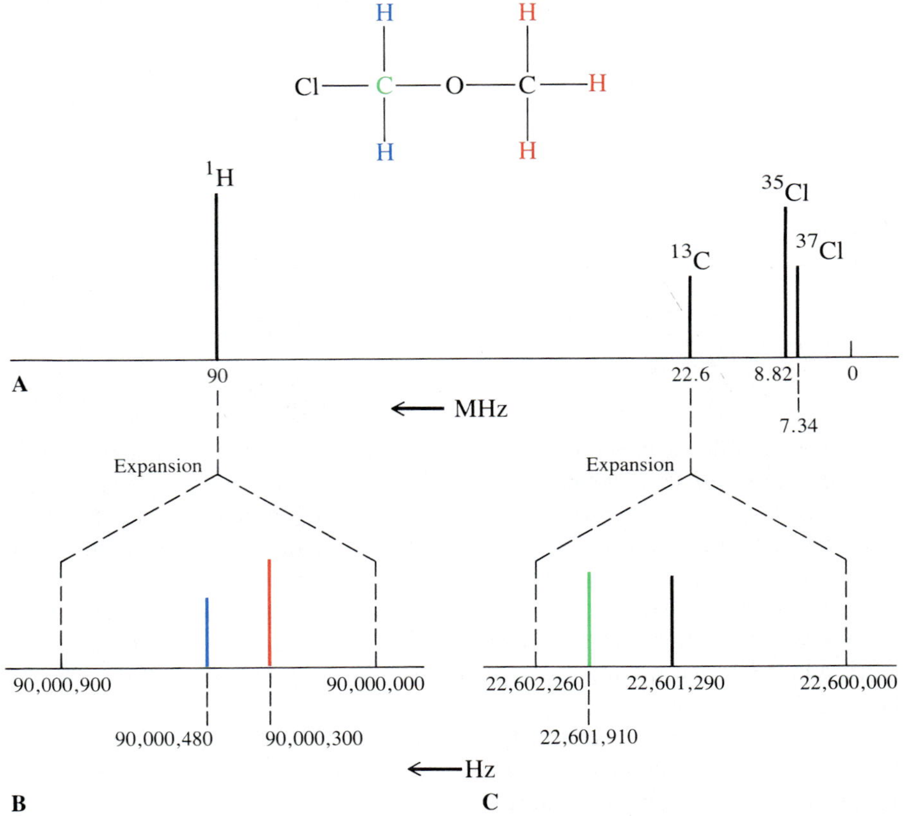

consist of only one peak in that region but two peaks that were not resolved at first (Figure 10-7B). Similarly, the high-resolution ^{13}C spectrum measured in the immediate vicinity of 22.6 MHz shows two peaks (Figure 10-7C). These absorptions reveal the presence of the two types of hydrogens and carbons. The finding that high-resolution NMR spectroscopy distinguishes both hydrogen and carbon atoms in different structural environments has made this method a most powerful tool for elucidating structures.

The organic chemist uses NMR spectroscopy more often than any other spectroscopic technique. The next section briefly describes its technical aspects before proceeding with its application.

How to Obtain an NMR Spectrum: The NMR Spectrometer

NMR spectroscopy has been routinely available since about 1960. The first mass-produced proton NMR spectrometers used magnets at 14,000 Gauss,

FIGURE 10-8
A 90-MHz ^{1}H NMR spectrometer (Varian Associates Model EM-390) is shown at the left below. The photograph at the right illustrates how an NMR sample tube is inserted into the magnet. (Reproduced with permission of Varian Associates.)

which set the resonance frequency of hydrogen nuclei at about 60 MHz. Most of the spectra published in the older literature are recorded at that frequency. Recent advances in computer technology, the advent of stable superconducting magnets, and other improvements in design have led to the widespread use of NMR spectrometers at higher field strength with hydrogen resonance frequencies as high as 500 MHz. This book will depict most ^{1}H NMR spectra at 90 MHz, with the occasional use of higher fields for comparative purposes.

What does a basic NMR spectrometer look like? A photograph of a machine in operation is reproduced in Figure 10-8, in which the control panel and the recorder on which the spectrum is being drawn can be clearly seen. The housing for the (covered) magnet with the inlet for the test sample is at the right of the spectrometer and shown separately in the margin.

The sample to be studied (a few milligrams) is usually dissolved in a solvent (0.3–0.5 ml) preferably not containing any atoms that themselves absorb in the NMR range under investigation. For proton NMR spectroscopy, these solvents are tetrachloromethane, CCl_4, or deuterated ones such as trichlorodeuteriomethane (deuteriochloroform), $CDCl_3$, hexadeuteriopropanone (hexadeuterioacetone), CD_3COCD_3, hexadeuteriobenzene, C_6D_6, and octadeuteriooxacyclopentane (octadeuteriotetrahydrofuran), C_4D_8O. The solution is transferred into an NMR sample container, a cylindrical precision-bore glass tube 18 cm in length and 5 mm in diameter, that is inserted into the magnet (Figure 10-9). To make sure that all molecules in the sample are rapidly averaged with respect to their position in the magnetic field, the NMR tube is

FIGURE 10-9
Schematic representation of the essentials of an NMR spectrometer. Energy from the radio-frequency (RF) power source is absorbed at resonance. The α-to-β spin flip generates a small electric current in a coil of wire around the sample, detectable through an RF detector and electronically translated into a signal on the recorder.

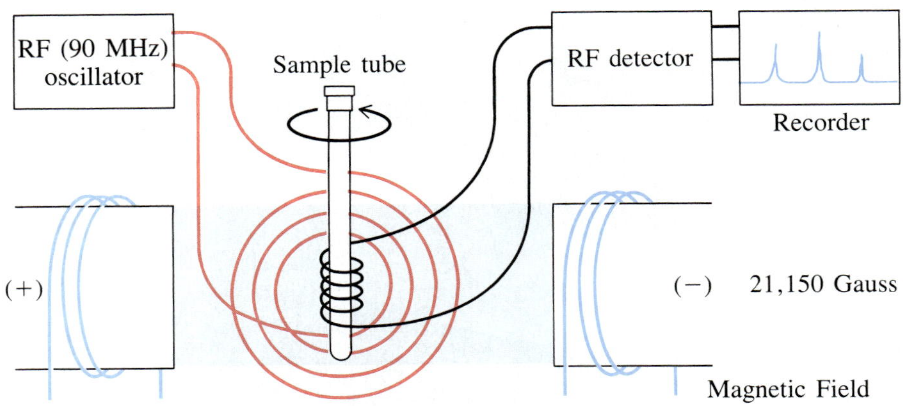

rapidly spun by an air jet. As indicated in Figure 10-7, to obtain a hydrogen NMR spectrum at 21,150 Gauss, the radio frequency (RF) could in principle be varied from 90,000,000 to 90,000,900 Hz. Most hydrogen nuclei in organic compounds resonate within this range. In practice, however, the radio frequency is left constant (e.g., at 90 MHz), and the field strength is varied to scan the spectrum. This practice may be a cause of confusion at first, because NMR spectral paper is printed with calibration in Hertz rather than Gauss. In other words, the spectra appear to be recorded as if the radio frequency had been varied at constant H_0. In actuality, they are not, but this difference is inconsequential in their interpretation. The ^{1}H NMR spectrum of $ClCH_2OCH_3$ is shown in Figure 10-10.

In summary, certain nuclei, such as ^{1}H and ^{13}C, can be viewed as tiny atomic magnets that, when exposed to a magnetic field, can align with it (α) or against it (β). These two states are of unequal energy, a condition giving rise to nuclear magnetic resonance spectroscopy. At resonance, radio-frequency radiation is absorbed by the nucleus to effect α-to-β transitions (excitation). The β state relaxes to the α state by giving off a small amount of heat. The resonance frequency is characteristic of the nucleus and its environment, and proportional to the strength of the external magnetic field.

10-3
Different Hydrogens Resonate at Different Field Strengths: The Proton Chemical Shift

Why do we observe different lines for the hydrogens in chloro(methoxy)-methane? This section will answer this question. We shall see that the position of an NMR absorption, also called the chemical shift, depends on the electron density around the hydrogen, which in turn is controlled by the structural environment of the nucleus observed. Therefore, *the NMR chemical shift is diagnostic of the structure of a molecule.*

FIGURE 10-10
90-MHz ^{1}H NMR spectrum of chloro(methoxy)methane. The zero Hertz line is set at exactly 90 MHz at the right-hand side of the spectral paper.

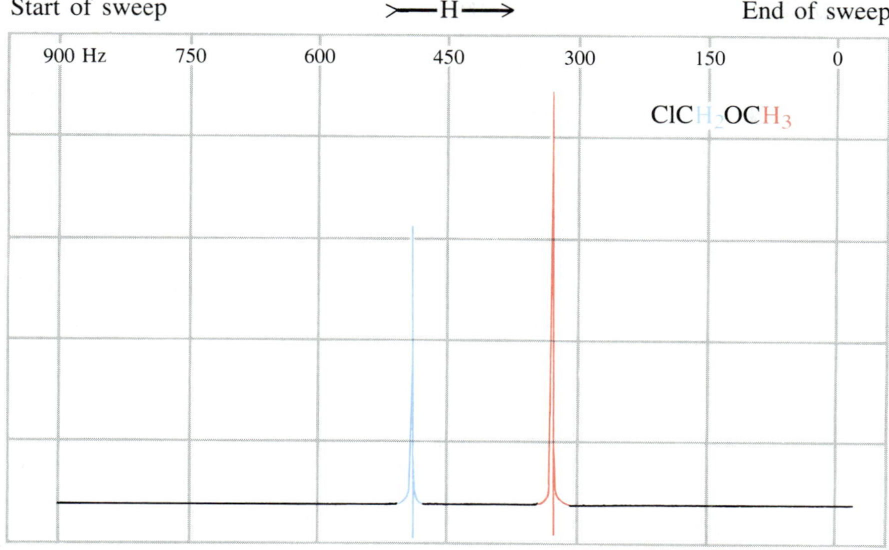

The Position of the NMR Peak Is Controlled by Electronic Shielding and Deshielding of the Nucleus

The high-resolution ^{1}H NMR spectrum of chloro(methoxy)methane depicted in Figure 10-10 reveals that the two kinds of hydrogens give rise to two separate resonance absorptions. What is the origin of this effect? It lies in the electronic environment of the hydrogen nucleus under examination. The free proton is a nucleus essentially unperturbed by outside electronic factors. When exposed to an external magnetic field, the nucleus becomes aligned with or against that field in the manner shown in Figure 10-4. Because organic molecules contain covalently bound hydrogen nuclei and not free protons,* we should expect such bonds to affect nuclear magnetic resonance absorptions. Bound hydrogens are surrounded by electronic shells whose electron density varies, depending on the polarity of the bond, the hybridization of the attached atom, and the presence of electron-donating or -accepting groups. When a nucleus surrounded by electrons is exposed to a magnetic field, *the electrons flow* around the nucleus, generating a small *local magnetic field,* h_{local}, *opposing* the external field, H_0. A consequence of this phenomenon is a reduction of the total field strength in the vicinity of the hydrogen nucleus, which is thus said to be **shielded** from H_0 by its electron cloud (Figure 10-11). The degree of shielding depends on the amount of electron density surrounding the nucleus. The addition of electrons increases shielding; their removal causes deshielding.

*In the discussion of NMR, the terms proton and hydrogen are interchanged (albeit incorrectly) frequently. ''Proton NMR'' and ''protons in molecules'' are used even in reference to covalently bound hydrogen.

FIGURE 10-11

The external field, H_0, causes an electronic current of the bonding electrons around a hydrogen nucleus, which in turn generates a local magnetic field opposing H_0 (Lenz's law).

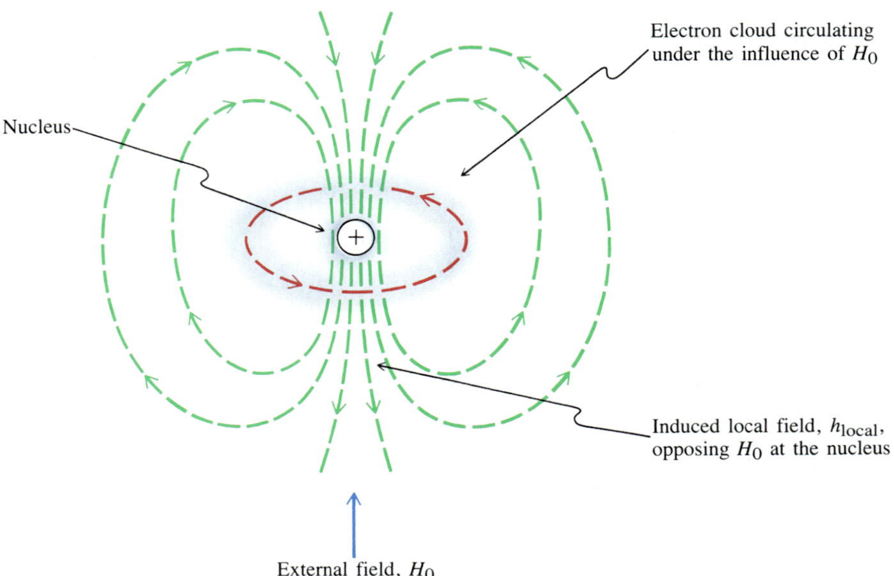

What is the effect of shielding on the relative position of an NMR absorption? If the spectrum is recorded at constant H_0, the fact that the local magnetic field at the nucleus is diminished to $H_0 - h_{local}$ means a decrease in the frequency necessary to cause resonance (Figure 10-12). Conversely, using the

FIGURE 10-12

Effect of shielding on the absorption of a "proton" when covalently bound. At constant RF of energy $h\nu$, the free proton resonates at H_0. Shielding causes the local field to decrease to $H_0 - h_{local}$. Hence, to "match" $h\nu$, the external field must be increased by an amount equal to h_{local}. At constant external H_0, the local field around the nucleus is $H_0 - h_{local}$; hence, a lower RF suffices to cause resonance. In both cases, the net result is that the peak shifts to the right-hand side of the spectrum.

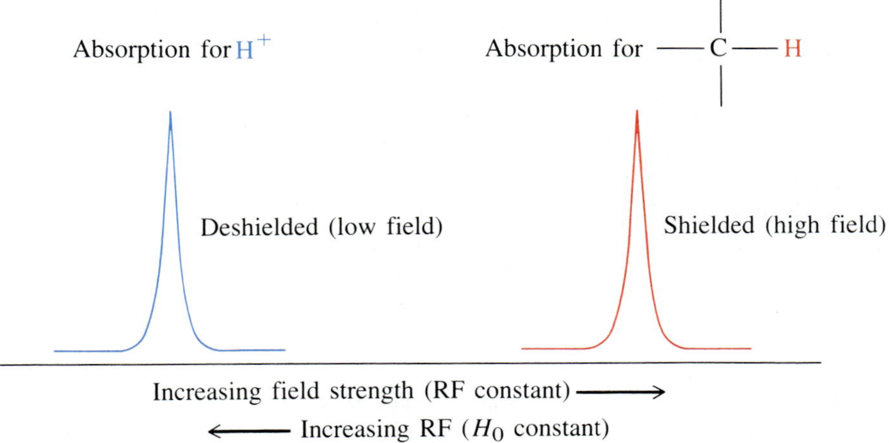

currently customary procedure of keeping the radio frequency constant and varying the magnetic field, we find that a higher external-field strength is required to overcome the shielding effect. The hydrogen signal is said to "occur at higher field" in the spectrum. Spectra are recorded with increasing field strength from left to right (or increasing frequency from right to left), and so a shifting to *higher field* means that the signal (peak) appears farther to the right. Because each chemically unique hydrogen has a unique electronic environment, it gives rise to a unique resonance. Chemically equivalent hydrogens show peaks at the same position. As examples, the NMR spectra of two familiar molecules, 1,2-dimethoxyethane and 2,2-dimethyl-1-propanol, are shown on the next page in Figures 10-13 and 10-14.

The Chemical Shift Describes the Position of an NMR Peak

In which manner are spectral data reported? As noted earlier, most hydrogen absorptions in 90-MHz ^{1}H NMR fall within a range of 900 Hz. Rather than measuring and recording the exact frequency of each resonance (which is difficult to do technically with accuracy), we add an internal standard relative to which the peak positions in the spectrum are reported. This compound is tetramethylsilane, $(CH_3)_4Si$ (b.p. 26.5°C). The twelve equivalent hydrogens in this compound are shielded relative to those in most organic compounds; therefore, they resonate at a position conveniently removed from the usual spectral range. Commercial solutions of common NMR solvents are available with tetramethylsilane (about 1%) already added. The NMR absorptions of a compound under investigation may then be measured in terms of their distance (in Hertz) from the signal of the internal standard. In this way, the signals of, for example, 2,2-dimethyl-1-propanol (Figure 10-14) can be reported as being located 78, 258, and 287 Hz downfield from $(CH_3)_4Si$.

A problem with these numbers, however, is that they vary with the strength of the applied magnetic field. Because field strength and resonance frequency are directly proportional, doubling or tripling the field strength will double or triple the distance (in Hertz) of the observed peaks relative to $(CH_3)_4Si$. To avoid this complication and to be able to compare reported literature spectra at different field strengths, we standardize the measured frequency by dividing the distance to $(CH_3)_4Si$ (in Hertz) by the frequency of the spectrometer. This procedure yields a *field-independent* number, the **chemical shift δ:**

$$\delta = \frac{\text{distance of peak from } (CH_3)_4Si \text{ in Hertz}}{\text{spectrometer frequency in megaHertz}} \text{ ppm}$$

The chemical shift is reported in units of parts per million (ppm), for hydrogen usually to three significant figures. For $(CH_3)_4Si$, δ is defined as 0.00. The NMR spectrum of 2,2-dimethyl-1-propanol in Figure 10-14 would then be reported in the following format: ^{1}H NMR (90 MHz, CCl_4) $\delta = 0.83, 2.87, 3.19$ ppm.

EXERCISE 10-3

The two NMR peaks of 1,2-dimethoxyethane (Figure 10-13) are 288 and 297 Hz away from $(CH_3)_4Si$ at 90 MHz. What is their chemical shift? How many Hertz would separate them from $(CH_3)_4Si$ when measured in a 100-MHz spectrometer?

FIGURE 10-13
90-MHz ^1H NMR spectrum of 1,2-dimethoxyethane, $CH_3OCH_2CH_2OCH_3$, in CCl_4. Two peaks are observed for the two sets of different hydrogens. The sharp peak at the right is due to a small amount of a reference compound, tetramethylsilane, $(CH_3)_4Si$. The scale at the top indicates in Hertz the distance of the signals from this reference.

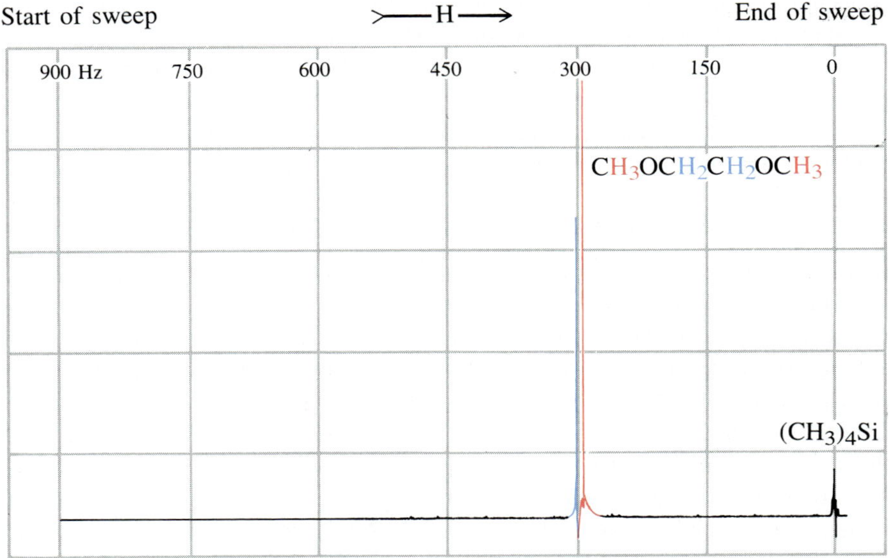

FIGURE 10-14
90-MHz ^1H NMR spectrum of 2,2-dimethyl-1-propanol in CCl_4. Three peaks are observed for the three sets of different hydrogens. (The scale at the bottom indicates the chemical shift in δ, a subject discussed in the next subsection.)

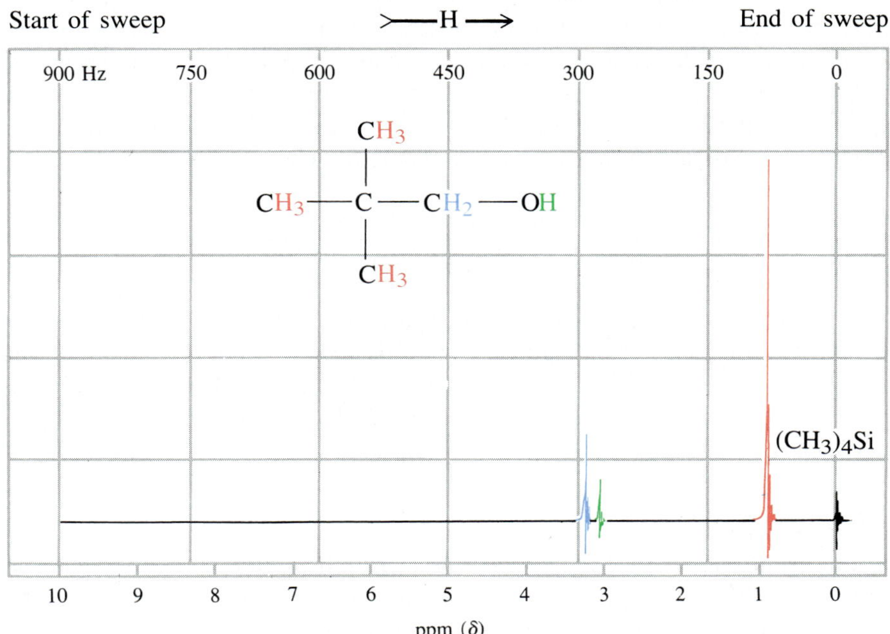

TABLE 10-2
Typical hydrogen chemical shifts in organic molecules

Type of hydrogen	Chemical shift δ in ppm
Primary alkyl, RCH_3	0.8–1.0
Secondary alkyl, RCH_2R'	1.2–1.4
Tertiary alkyl, R_3CH	1.4–1.7
Allylic (next to a double bond), $R_2C{=}C(CH_3)R'$	1.6–1.9
Benzylic (next to a benzene ring), $ArCH_2R$	2.2–2.5
Chloroalkane, RCH_2Cl	3.6–3.8
Bromoalkane, RCH_2Br	3.4–3.6
Iodoalkane, RCH_2I	3.1–3.3
Ether, RCH_2OR'	3.3–3.9
Alcohol, RCH_2OH	3.3–4.0
Ketone, $RC({=}O)CH_3$	2.1–2.6
Aldehyde, $RC({=}O)H$	9.5–9.6
Terminal alkene, $R_2C{=}CH_2$	4.6–5.0
Internal alkene, $R_2C{=}CHR'$	5.2–5.7
Aromatic, ArH	6.0–9.5
Alkyne, $RC{\equiv}CH$	1.7–3.1
Alcoholic hydroxy, ROH	0.5–5.0 (variable)
Amine, RNH_2	0.5–5.0 (variable)

Functional Groups Have Characteristic Chemical Shifts

A powerful use of NMR spectroscopy derives from the finding that the chemical shift is characteristic of the chemical (structural) environment around the nucleus measured. The hydrogen chemical shifts typical of standard organic structural units are listed in Table 10-2. They will be discussed in more detail in relation to specific functional groups in subsequent chapters. Acquaint yourself with the values shown in Table 10-2 for hydrogens adjacent to the structural types discussed so far: alkanes, haloalkanes, ethers, alcohols, aldehydes, and ketones.

Note that the chemical shifts of the alkane hydrogens are at relatively high field ($\delta = 0.8–1.7$ ppm) and exhibit a small but distinct downfield shift on increasing substitution; in other words, the trend δ_{R-H} is: tertiary > secondary > primary. A hydrogen close to an electron-withdrawing group or atom (such as halogen or oxygen) is shifted to relatively lower field, because such substituents deshield that neighboring hydrogen by removing electrons from its vicinity. Table 10-3 shows how adjacent heteroatoms affect the chemical shift of a methyl group. The more electronegative the atom, the more deshielded are the methyl hydrogens relative to methane. Several such substitu-

Cumulative Deshielding in Chloromethanes

CH_3Cl
$\delta = 3.05$ ppm

CH_2Cl_2
$\delta = 5.30$ ppm

$CHCl_3$
$\delta = 7.27$ ppm

ents exert a cumulative effect, as seen in the series of the three chlorinated methanes at the left. The deshielding effect of electron-withdrawing substituents diminishes rapidly with increasing distance from the carbon bearing the electronegative moiety:

$$\underset{\delta\,=\,1.06}{CH_3}—\underset{1.81}{CH_2}—\underset{3.47\text{ ppm}}{CH_2}—Br$$

$$CH_3—\overset{\overset{CH_3}{|}}{\underset{\underset{H}{|}}{C}}—I \qquad CH_3\text{: }1.89\text{ ppm};\quad H\text{: }4.24\text{ ppm}$$

TABLE 10-3

The deshielding effect of electronegative atoms

CH_3X	Electronegativity of X (from Table 1-4)	Chemical shift δ in ppm of CH_3 group
CH_3F	4.0	4.26
CH_3OH	3.4	3.40
CH_3Cl	3.2	3.05
CH_3Br	3.0	2.68
CH_3I	2.7	2.16
CH_3H	2.2	0.23

Another example is 1,2,2-trichloropropane, $CH_3CCl_2CH_2Cl$, which exhibits two absorptions, one at $\delta = 4.00$ ppm and the other at 2.23 ppm. The lower-field peak is assigned to the methylene group, which bears secondary hydrogens and is close to three electron-withdrawing chlorine atoms, one of which is located on the same carbon atom. The methyl group, on the other hand, is seen to absorb at higher field, because the hydrogens are primary and next to only two chlorines.

EXERCISE 10-4

Explain the assignment of the ^1H NMR signals of chloro(methoxy)methane (Figure 10-10) and 1,2-dimethoxyethane (Figure 10-13).

As noted in Table 10-2, hydroxy (and amine) hydrogens absorb over a range of frequencies. This variety is due to hydrogen bonding. The oxygen in an alcohol molecule has two lone electron pairs and consequently exerts a strong shielding effect on the hydrogen directly bound to it. (Compare, however, the inductively deshielding power of oxygen on an indirectly bound hydrogen in Table 10-3). Consequently, in dilute dry solutions, this nucleus resonates at fairly high field. In more-concentrated or wet samples, however, hydrogen bonding (between alcohol molecules or to water) diminishes the shielding power of the oxygen lone electron pairs and the chemical shift increases, its value depending on the composition of the solution (and temperature). In spectra of such samples, the absorption peak of the OH group is frequently relatively broad. This observation can be explained (in a simplified manner) by the fact that the different degrees of hydrogen bonding expose the hydrogen nuclei to a variety of magnetic environments at any given time. Thus, the peak "smears out" over a certain area, as observed. Similar considerations apply to amines, RNH_2, which also undergo hydrogen bonding and show the broad peaks characteristic of varying chemical shifts (Section 21-2). Because of this broadening, amine and alcohol hydrogen absorptions are usually readily recognized in NMR spectra, despite the fact that they may have ill-defined δ values.

In summary, the various hydrogen atoms present in an organic molecule can be recognized by their characteristic NMR peaks at certain chemical shifts δ. An electron-poor environment is deshielded and leads to low-field (high-δ) absorptions, whereas an electron-rich environment results in the opposite trend (shielded or high-field peaks). The chemical shift δ is measured in parts per million by dividing the difference between the measured resonance and that of the internal standard, tetramethylsilane, in Hertz by the spectrometer frequency in megaHertz. The NMR spectra for the OH groups of alcohols and the NH_2 groups of amines have characteristic broad peaks with concentration- and moisture-dependent δ values in a broad range

10-4
Chemically Equivalent Hydrogens Have the Same Chemical Shift

In the NMR spectra presented so far, two or more hydrogens occupying positions that are chemically equivalent give rise to only *one* NMR absorption. It can be said, in general, that chemically equivalent protons *have the same chemical shift*. The converse is not necessarily true: hydrogens having the same chemical shift are not necessarily chemically equivalent. They may just coincidentally occupy the same position in the NMR spectrum. We shall see that it is not always easy to establish the chemical, and therefore chemical-shift, equivalence of individual nuclei. We shall take recourse to the type of symmetry operations presented in Chapter 5 to help us decide on the expected NMR spectrum of a specific compound. The present section also introduces another NMR recording technique, called integration, which reveals the relative number of hydrogens in a molecule that give rise to an NMR peak.

Symmetry and Chemical-Shift Equivalence

To establish chemical equivalence, we have to recognize the symmetry of molecules and their substituent groups, a task similar to the one undertaken to recognize the presence or absence of chirality (Section 5-2). The molecule can be subjected to symmetry operations in such a way as to exchange the positions of two or more hydrogen atoms whose equivalence is under scrutiny. If the structure of the molecule stays the same, the nuclei in question are chemically equivalent and should show the same NMR absorption. As we know, two tests of symmetry are the presence of a mirror plane and a center of symmetry. A third test of this type is rotation. For example, Figure 10-15 demonstrates how two successive 120° rotations of a methyl group allows each hydrogen to occupy the positions of the other two without effecting any structural change. Thus, in a rapidly rotating methyl group all hydrogens should have the same chemical shift. We shall see shortly that this is indeed the case.

Application of the principles of rotational or mirror symmetry or both allows the assignment of equivalent nuclei in other compounds (Figure 10-16).

FIGURE 10-15
Counterclockwise rotation of a methyl group as a test of symmetry.

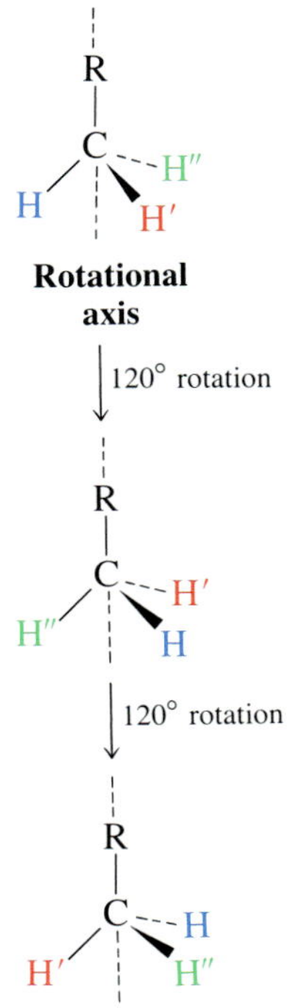

R, C, H, H′, H″
Rotational axis
120° rotation
R, C, H″, H′, H
120° rotation
R, C, H′, H, H″

FIGURE 10-16

Chemical-shift-equivalent hydrogens in a variety of organic molecules. The colors red and blue differentiate between nuclei giving rise to separate absorptions. All structures have rotational or mirror symmetry or both.

CH_3CH_3 (a a)

$(CH_3)_3CH$ (CH₃ a, CH b)

$CH_3OCH_2CH_2OCH_3$ (a b b a)

EXERCISE 10-5

How many NMR absorptions would you expect for: (a) 2,2,3,3-tetramethylbutane; (b) $CH_3OCH_2CH_2OCH_2CH_2OCH_3$; and (c) oxacyclopropane?

Another useful device for ascertaining the equivalence of hydrogens in a molecule is the execution of a (hypothetical or actual) chemical reaction that replaces one of these hydrogens by another group. If replacement of any of the other hydrogens by that group leads to the same product molecule or its enantiomer, the hydrogens have equivalent chemical shifts. This principle had application in the discussion of the stereochemistry of the free-radical halogenation of alkanes (Section 5-6).

Symmetry and the NMR Time Scale

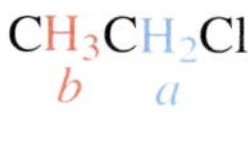

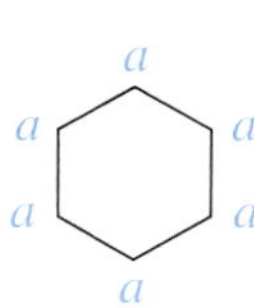

Let us look a little more closely at two additional examples, chloroethane and cyclohexane. The first should have two NMR peaks owing to the two sets of equivalent hydrogens; the second has twelve chemically equivalent hydrogen nuclei and is expected to show only one absorption. However, are these expectations really justified? Consider the possible conformations of these two molecules (Figure 10-17). Beginning with chloroethane, the most stable conformation is the staggered arrangement, in which one of the methyl hydrogens (H_{b_3}) is located *anti* with respect to the chlorine atom. We would expect this particular nucleus to have a chemical shift different from the two *gauche* hydrogens (H_{b_1} and H_{b_2}). In fact, however, the NMR spectrometer cannot resolve that difference, because the fast rotation of the methyl group averages out the signals for H_b. This rotation is said to be "fast on the NMR time scale." The spectrometer (like a camera or an eye) has only a limited ability to record individual events occurring in fast sequence. Beyond a certain speed, its "vision" becomes "blurred," and all events merge into one. For chloroethane, fast rotation "averages" the two signals of H_b into one. The resulting absorption appears at an average δ of the two signals expected for H_b.

FIGURE 10-17

A. Newman projections of chloroethane. H_{b_3} is located *anti* to the chlorine substituent and is therefore not in the same environment as H_{b_1} and H_{b_2}. However, fast rotation equilibrates all the methyl hydrogens on the NMR time scale. B. In any given conformation of cyclohexane, the axial hydrogens are different from the equatorial ones. However, conformational flip is rapid on the NMR time scale and equilibrates the two environments so that only one average signal is observed. Colors are used here to distinguish between magnetic environments and thus to indicate specific chemical shifts.

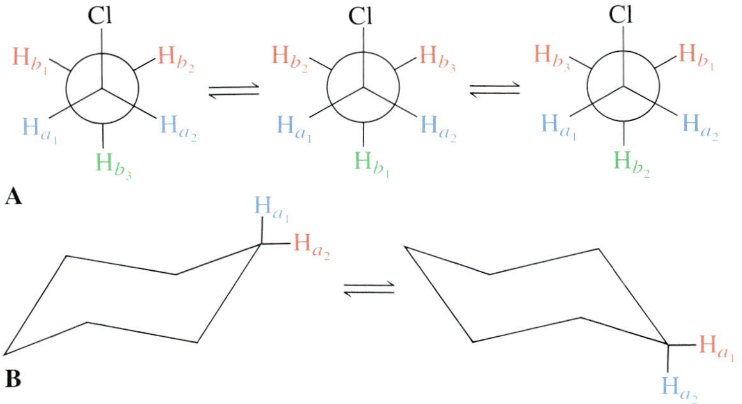

In theory, it should be possible to slow the rotational process in chloroethane by cooling the sample. Below a certain temperature, the originally averaged signal should split into two new peaks, corresponding to $H_{b_{1,2}}$ and H_{b_3}. In this particular example, this is technically very difficult to do, because the activation barriers for rotation are on the order of only a few kcal mole^{-1}. To "freeze out" the rotational process, we would have to cool the sample to a point ($\sim -180°C$) at which most solvents would crystallize, in which case ordinary NMR spectroscopy would not be possible.

A similar effect holds for the cyclohexane molecule. Here, fast conformational isomerism equilibrates the axial set of hydrogens with the equatorial one (Figure 10-17B); therefore, at room temperature, the NMR spectrum shows only one sharp line at $\delta = 1.36$ ppm. However, in contrast with that for chloroethane, the process is slow enough at $-90°C$ that, instead of a single absorption, two are observed: one for the six axial hydrogens at $\delta = 1.12$ ppm; the other for the six equatorial hydrogens at $\delta = 1.60$ ppm. The conformational isomerization in cyclohexane is frozen on the NMR time scale at this temperature because the activation barrier to ring flip is much higher ($E_a = 10.8$ kcal mole^{-1}, Section 4-3) than the barrier to rotation in chloroethane. In general, the lifetime of a species in such an equilibrium must be on the order of about a second to allow its well-resolved observation by NMR. If this period decreases substantially, an average spectrum is obtained. Such time- and temperature-dependent NMR phenomena are in fact used by chemists to estimate the activation parameters of chemical processes.

Enantiotopic and Diastereotopic Hydrogens in NMR Spectroscopy

Let us return to our example of chloroethane and its NMR spectrum. We have established that all H_b equilibrate by fast rotation. A Newman projection of the molecule also shows us that the two hydrogens labeled H_{a_1} and H_{a_2} adjacent to the chlorine substituent are equivalent. You can verify this statement by drawing a mirror plane bisecting the molecule and incorporating both carbon atoms, the chlorine nucleus, and H_{b_3}. Substitution of one H_a with another group will result in a pair of enantiomers (Section 5-6). The methylene hydrogens H_a are therefore called enantiotopic. They are chemically equivalent and consequently give rise to the same NMR signal.*

The Enantiotopic Hydrogens H_a in Chloroethane

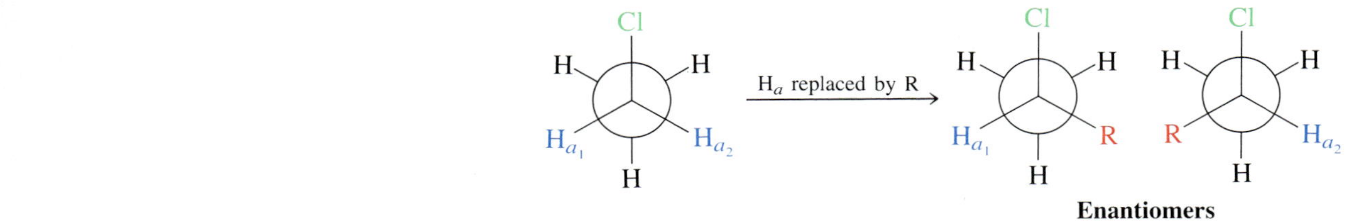

The situation is entirely different when such a methylene group is located close to a stereocenter. An example is 1-chloro-2-fluoropropane. This molecule has no mirror plane and, in fact, no symmetry operation allows for an interchange of the methylene hydrogens. Consequently, a chemical reaction that leads to substitution of one of the two hydrogen atoms at C-1 creates a new stereocenter next to the old one. The result is the formation of two diastereomers, and nuclei H_a and H_b are therefore diastereotopic (Section 5-6). Because they are chemically nonequivalent, these two nuclei will have different chemical shifts. This finding applies to all methylene groups bearing two (but not more) identical substituents *in any* chiral compound: the two substituents give rise to different sets of NMR absorptions.

The Diastereotopic Hydrogens in 1-Chloro-2-fluoropropane

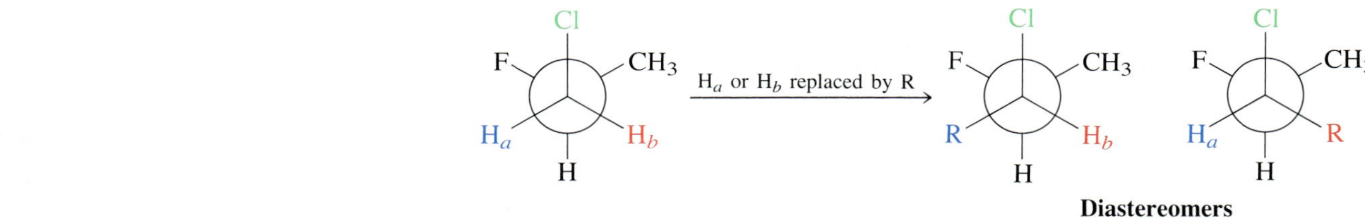

A molecule need not be chiral to have diastereotopic nuclei. For example, in a substituted cycloalkane, such as methylcyclohexane, the hydrogens located cis with respect to the substituent are different from those positioned trans. Similarly, 3-methylpentane contains two equivalent diastereotopic CH_2 groups. Even though the molecule has a mirror plane and is therefore achiral, no such plane bisects a CH_2 fragment, nor does any other test of symmetry prove the equivalency of the two hydrogens.

The chemical-shift differences between diastereotopic groups are often small, in which case they frequently give rise to only one absorption, as if they

*This is strictly true only with achiral reagents and in achiral solvents. In a chiral environment, the two hydrogens are no longer equivalent. You can establish this fact by enclosing a molecular model of chloroethane in one hand. The hand, being chiral, makes the two hydrogens in question diastereotopic.

were equivalent (at least at 90 MHz). We should keep in mind in these cases, however, that such simplicity is coincidental.

Diastereotopic Substituents (Indicated by Arrows)

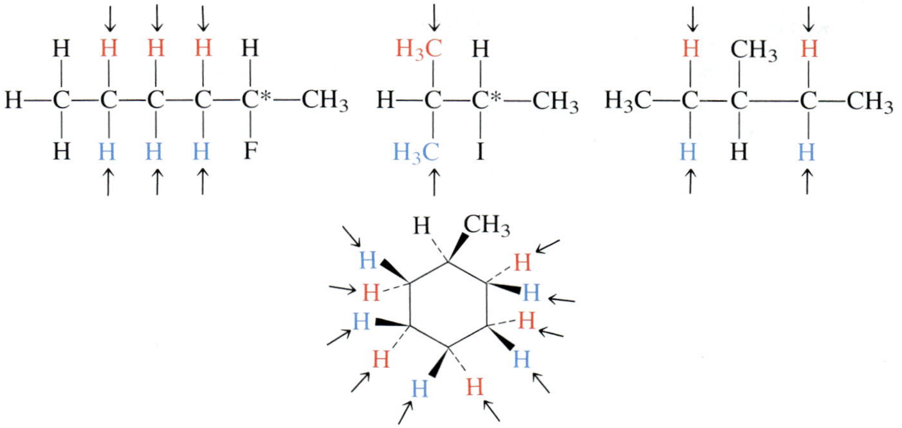

EXERCISE 10-6

Which of the CH_2 groups in the following compounds contains chemical-shift-non-equivalent hydrogens: (a) cyclopropane; (b) methylcyclopropane; (c)

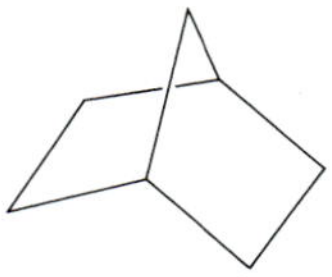

The Integration Mode Reveals the Relative Number of Hydrogens Recorded in an NMR Peak

Inspection of the heights of the peaks in the NMR spectra presented so far (see, e.g., Figures 10-13 and 10-14) reveals another useful feature of NMR spectroscopy: the relative intensity of a signal correlates with the relative abundance of the nucleus with which it corresponds. The more hydrogens of one kind there are in a molecule, the more intense the corresponding NMR absorption relative to the other signals. By measuring the area under a peak and comparing it with the corresponding peak areas of other signals, we can quantitatively estimate the nuclear ratios. For example, in the spectrum of 2,2-dimethyl-1-propanol (Figure 10-14), three signals are observed, with relative areas of 9:2:1.

To simplify this measurement, the spectrometer has an electronic feature called **integration.** At the push of a button, the instrument switches into the *integrator mode,* which allows a rescanning of the spectrum to specifically record the intensity ratio of the observed absorptions. This spectral integration is plotted (for clarity, displaced from the original spectral base line, Figure 10-18) as a straight, initially horizontal line (base line). Whenever a point in the scan is reached at which the normal recording mode would draw a peak, the pen in the integrator mode moves vertically upward until no further peak intensity is measurable. It then again moves horizontally until the next peak is reached and so forth. A ruler can then be used to measure the distance by which the horizontal line is displaced at every peak. The relative sizes of these displacements furnish the ratio of hydrogens giving rise to the various signals. Figure 10-18 depicts the ^1H NMR spectra of 2,2-dimethyl-1-propanol, chloro(methoxy)methane, and 1,2-dimethoxyethane, including plots of the integration.

FIGURE 10-18

Integrated 90-MHz spectra of (A) 2,2-dimethyl-1-propanol, (B, on the facing page) chloro(methoxy)methane, and (C) 1,2-dimethoxyethane, in CCl_4 with added $(CH_3)_4Si$. For example A, the integrated areas measured by a ruler are 11:6:52 (in mm). (Note that the integrator converts units of area into units of distance.) Normalization through division by the smallest number gives a peak ratio of 1.8:1:8.7. The slight deviations from integer numbers are due to small experimental errors; rounding off gives the expected ratio of 2:1:9. Note that the integration gives only *ratios,* not absolute values for the number of hydrogens present in the sample. Thus, in examples B and C, the integrated peak ratio is 3:2, yet the hydrogen ratio corresponds to this number only in example B. The compound in example C contains hydrogens in a ratio of 6:4.

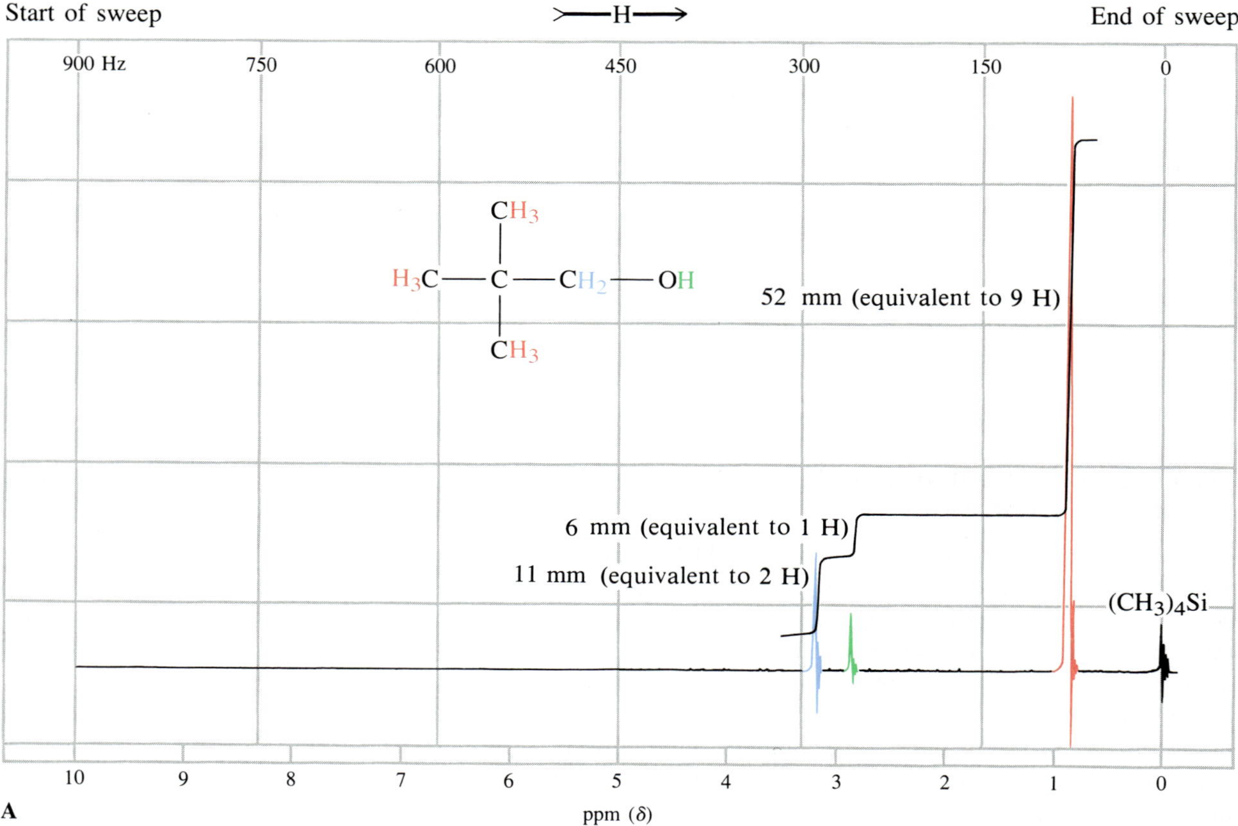

Integration Aids in Assigning the Structure of a Molecule

The assignment of chemical shifts in conjunction with peak integration is extremely useful in structure elucidation. Consider, for example, the three products obtained in the monochlorination of 1-chloropropane, $CH_3CH_2CH_2Cl$. All have the same empirical formula $C_3H_6Cl_2$ and very similar physical properties (such as boiling points):

$$CH_3CH_2CH_2Cl \xrightarrow[-\ HCl]{Cl_2,\ h\nu,\ 100°C} CH_3CH_2CHCl_2 + CH_3CHClCH_2Cl + ClCH_2CH_2CH_2Cl$$

10%	27%	14%
1,1-Dichloropropane	**1,2-Dichloropropane**	**1,3-Dichloropropane**
b.p. 87°–90°C	b.p. 96°C	b.p. 120°C

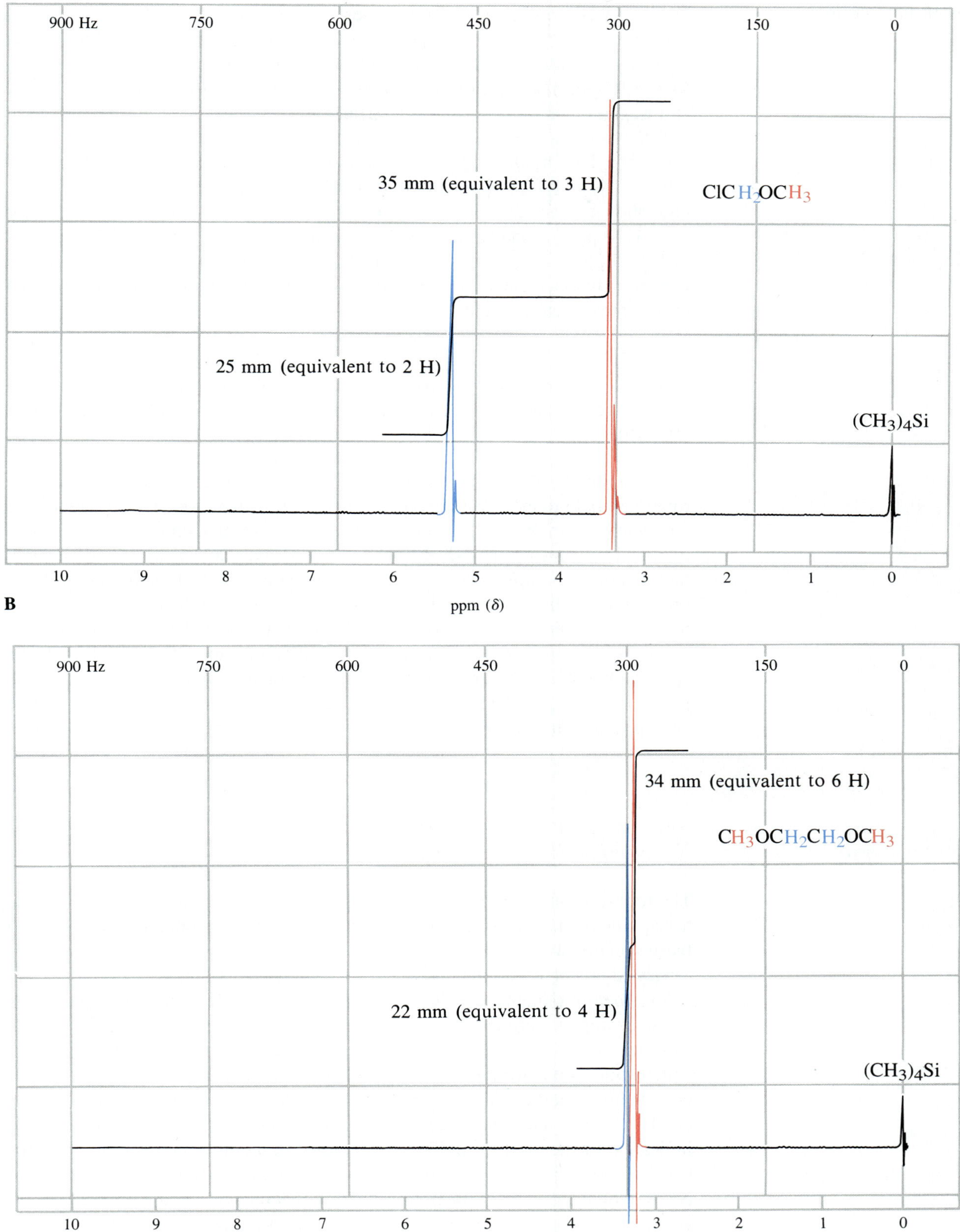

900 Hz
750
600
450
300
150
0
35 mm (equivalent to 3 H)
ClCH2OCH3
25 mm (equivalent to 2 H)
(CH3)4Si
10
9
8
7
6
5
4
3
2
1
0
B
ppm (δ)
900 Hz
750
600
450
300
150
0
34 mm (equivalent to 6 H)
CH3OCH2CH2OCH3
22 mm (equivalent to 4 H)
(CH3)4Si
10
9
8
7
6
5
4
3
2
1
0
C
ppm (δ)

NMR spectroscopy clearly distinguishes all three isomers. 1,1-Dichloropropane contains three types of nonequivalent hydrogens, giving rise to three NMR signals in the ratio of 3:2:1. The single hydrogen absorbs at relatively low field ($\delta = 5.93$ ppm) because of the cumulative deshielding effect of the two halogen atoms, the others at relatively high field ($\delta = 1.01$ and 2.34 ppm). The chiral 1,2-dichloropropane exhibits four different NMR peaks in the ratio of 3:1:1:1 (the methylene hydrogens are diastereotopic). In this case, three signals ($\delta =$ 2.62, 2.74, and 4.17 ppm), in the ratio of 1:1:1, appear at low field, and only one, shown by integration to represent three hydrogens, at relatively high field ($\delta = 1.70$ ppm). Finally, 1,3-dichloropropane shows only two peaks ($\delta = 3.71$ and 2.25 ppm) in a relative ratio of 2:1, clearly distinct from the other two isomers. By this means, the structures of the three products are readily assigned by a simple measurement.

EXERCISE 10-7

Chlorination of chlorocyclopropane gives three compounds of molecular formula $C_3H_4Cl_2$. Draw their structures and describe how you would differentiate them by ^1H NMR.

In summary, the properties of symmetry, particularly mirror images and rotations, help to establish the chemical-shift equivalence or nonequivalence of the hydrogens in organic molecules. Those structures that undergo rapid conformational changes on the NMR time scale show only averaged spectra at room temperature. Examples are rotamers and ring conformers. In some cases, these processes may be "frozen" at low temperatures to allow distinct absorptions to be observed. Diastereotopic hydrogens (and other groups) have different chemical shifts, whereas their enantiotopic counterparts do not. Recording NMR spectra in the integration mode reveals the relative areas under the various peaks, in turn equivalent to the relative number of hydrogens giving rise to these absorptions. This technique can be a powerful aid in structural elucidation.

10-5 Spin-Spin Splitting: The Influence of Nonequivalent Neighboring Hydrogens on Each Other

The high-resolution NMR spectra presented so far have rather simple line patterns: single sharp peaks, also called singlets. The compounds giving rise to these spectra have one feature in common: nonequivalent hydrogens are always separated by at least one hydrogen-free nucleus (carbon or oxygen), as in chloro(methoxy)methane, 1,2-dimethoxyethane, and 2,2-dimethyl-1-propanol. These examples were chosen for good reason, because the presence of magnetically nonequivalent neighboring nuclei causes a complication in the spectrum called **spin-spin coupling,** or **spin-spin splitting.** An example of this phenomenon is shown in the NMR spectrum of 1,1-dichloro-2,2-diethoxyethane in Figure 10-19. The multiplet patterns due to spin-spin splitting in such spectra are indicative of the number and kinds of hydrogen atoms adjacent to the nucleus (nuclei) giving rise to the absorption. In conjunction with the other features of NMR (chemical shifts, integration), this phenomenon frequently helps us to arrive at a complete structural assignment of an unknown compound.

FIGURE 10-19

90-MHz spectrum of 1,1-dichloro-2,2-diethoxyethane in CCl_4. The splitting patterns include two doublets, one triplet, and one quartet for the four types of protons.

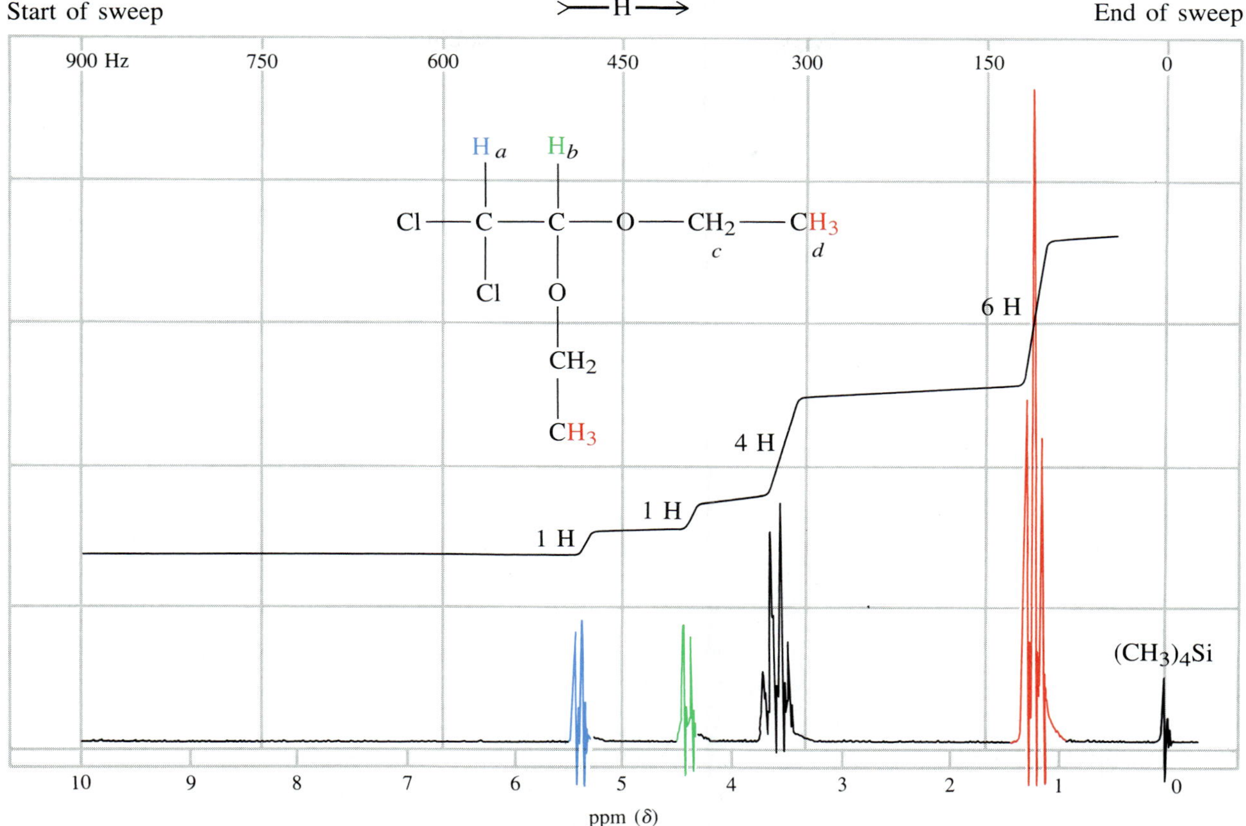

The Spin of the Neighboring Hydrogens Affects the Chemical Shift of the Resonating Nucleus

Figure 10-19 reveals four absorptions assignable to four sets of hydrogens H_a–H_d. Although the methylene protons designated H_c are diastereotopic, they have a nearly identical chemical shift, and we may treat them to a first approximation as being equal. Instead of appearing as single peaks, the NMR absorptions adopt more complex patterns: two two-peak patterns (doublets), one four-peak absorption (quartet), and one three-peak resonance (triplet). How can this be understood?

Let us first consider the two doublets of respective relative integration 1, assigned to the two single hydrogens H_a and H_b. The splitting of these peaks is explained by the behavior of nuclei in an external magnetic field: they are like tiny magnets aligned with (α) or against (β) the field. The energy difference between the two states is miniscule (see Section 10-2), and at room temperature their populations are nearly equal, with slightly more in the α state. In the case under consideration, this means that in essence there are two magnetic types of H_a: approximately half are next to H_b in the α configuration, the other half has a neighboring H_b in its β state. Conversely, H_b has two types of neighboring H_as,

FIGURE 10-20
The effect of a hydrogen nucleus on the chemical shift of its neighbor: two peaks are generated. External field contribution: H_0; local field contribution: h_{local}.

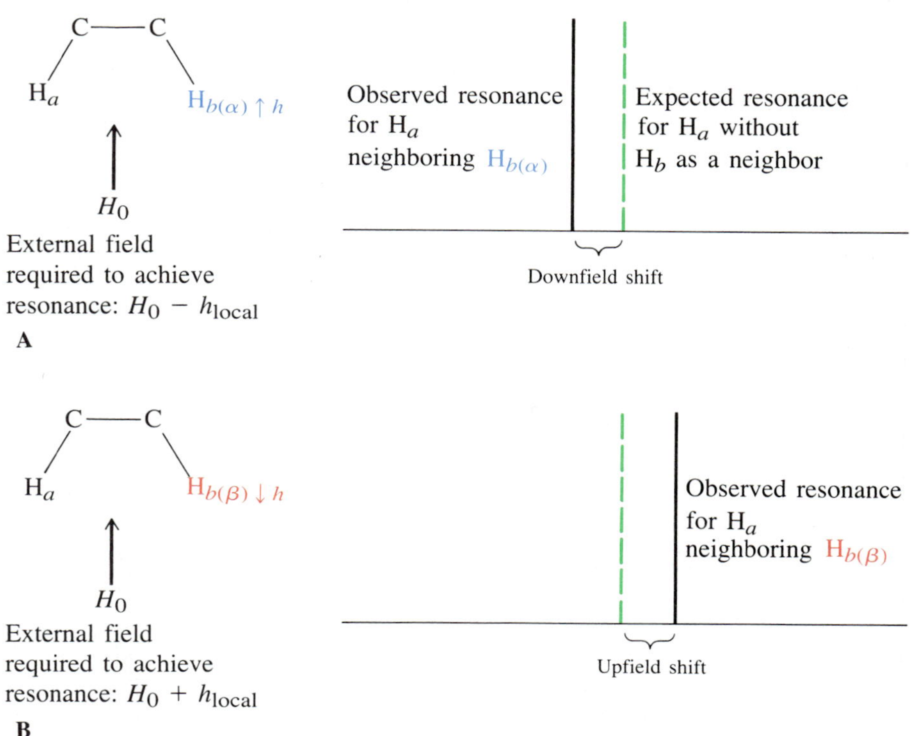

half of them in the α, half in the β state. What are the consequences of this phenomenon in the NMR spectrum?

Proton H_a, which has an H_b aligned *with* the field as its neighbor, is exposed not only to the external field H_0, but also to an additional *strengthening* local contribution (though very small) from the α spin of H_b. To achieve resonance for this type of H_a, a lesser field strength is required than that expected for H_a in the absence of a perturbing neighbor. A peak at lower field than that expected is indeed observed (Figure 10-20A). However, this absorption is due to only half the H_a protons. The other half has H_b in its β state as a neighbor. Because H_b in its β state is aligned *against* the external field, the local field around H_a in this case is *diminished*. To achieve resonance the external field has to be increased; an upfield shift is observed (Figure 10-20B).

Because the local contribution of H_b to the external field, whether positive or negative, is of the same magnitude, the downfield shift of the hypothetical signal equals the upfield shift. The single absorption expected for a neighbor-free H_a is said to be ''split'' into a doublet. Integration of each peak of this doublet shows a contribution of 0.5 hydrogens each, but the total integrated value of the doublet relative to the remainder of the integrated spectrum amounts to a full hydrogen. The chemical shift of this signal is reported as the center of the doublet (Figure 10-21).

The signal for H_b is subject to similar considerations. This hydrogen also has two types of hydrogen as neighbors: $H_{a(\alpha)}$ and $H_{a(\beta)}$. Consequently its absorp-

FIGURE 10-21
Spin-spin splitting between H_a and H_b in 1,1-dichloro-2,2-diethoxyethane. The coupling constant J_{ab} is the same for both doublets. The chemical shift is reported as the center of the doublet in the following format: δ_{H_a} = 5.36 ppm (d, J = 7 Hz, 1 H), δ_{H_b} = 4.39 ppm (d, J = 7 Hz, 1 H), in which "d" stands for the splitting pattern (doublet), and the last entry refers to the integrated value of the absorption.

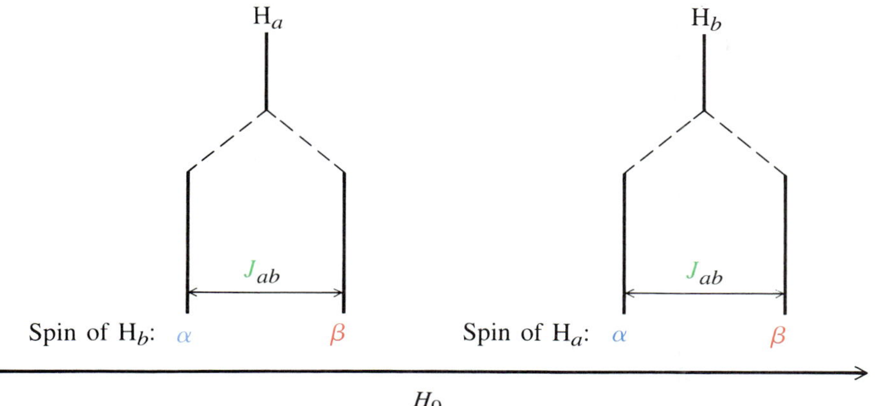

tion lines appear in the form of a doublet. In NMR jargon, H_a is said to be split by H_b and vice versa. The amount of this mutual splitting is equal; that is, the distance (in Hertz) between the individual peaks making up each doublet is identical. This distance is termed the **coupling constant, *J*.** In our example, J_{ab} = 7 Hz (Figure 10-21). The coupling constant is *independent* of the external-field strength. For example, J_{ab} is 7 Hz at 90 MHz, as well as at 180 MHz, even though the distance (in Hertz) between the two doublets doubles, as do the distances between the other absorptions.

Spin-spin splitting is generally observed only between hydrogens that are immediate neighbors, bound either to the same carbon atom [**geminal coupling** (*geminus*, Latin, twin)] or to two adjacent carbons [**vicinal coupling** (*vicinus*, Latin, neighbor)]. Hydrogen nuclei separated by a bridge larger than two atoms are usually too far apart to exhibit appreciable coupling.

Coupling Between Close-Lying Hydrogens

J_{ab}, geminal coupling, variable 0–18 Hz

J_{ab}, vicinal coupling, typically 6–8 Hz

J_{ab}, 1,3-coupling, usually negligible

Note that *chemical-shift-equivalent nuclei do not exhibit spin-spin splitting*. This statement makes intuitive sense. In a simplified way, it could be argued that, in this case, the α and β spins of all the hydrogen participants on the NMR time scale rapidly equilibrate because they resonate simultaneously; the spectrometer now "sees" only one average field around all of these nuclei.

Local-Field Contributions due to More Than One Hydrogen Are Additive

Let us return to the spectrum of 1,1-dichloro-2,2-diethoxyethane shown in Figure 10-19. In addition to showing the two doublets assignable to H_a and H_b, this spectrum records a triplet due to the methyl protons H_d and a quartet assignable to the methylene hydrogens H_c. Because these two nonequivalent sets of nuclei are next to each other, vicinal coupling is observed as expected. However, compared with the peak patterns for H_a and H_b, those for H_c and H_d are considerably more complicated. They can be understood by application of what we know in connection with the mutual coupling of H_a and H_b.

FIGURE 10-22

Nucleus H_d is represented by a three-peak pattern because of the presence of three magnetically nonequivalent neighbor combinations: $H_{c(\alpha\alpha)}$, $H_{c(\alpha\beta \text{ and } \beta\alpha)}$, and $H_{c(\beta\beta)}$. The chemical shift of the absorption is reported as that of the center line of the triplet: δ_{H_d} = 1.23 ppm (t, J = 8 Hz, 6 H), in which "t" stands for triplet.

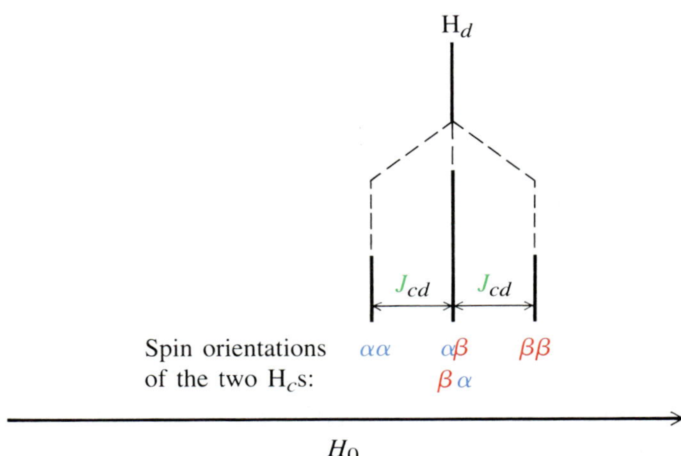

Let us begin with the triplet of relative integration 6. Its chemical shift and integrated value allow it to be assigned to the hydrogens of the two methyl groups. Instead of observing one peak, we observe three, in the approximate ratio of 1:2:1. The splittings must be due to coupling to the adjacent methylene groups in the following way. The three equivalent methyl hydrogens in each ethoxy group have two methylene counterparts as their neighbors, each of which may adopt an α or β configuration. Thus, H_d may "see" its neighbors as an $\alpha\alpha$, $\alpha\beta$, $\beta\alpha$, or $\beta\beta$ combination of the two H_cs. Those methyl hydrogens that are adjacent to the first possibility, $H_{c(\alpha\alpha)}$, are exposed to a strengthened local field and give rise to a lower-field absorption. In the $\alpha\beta$ or $\beta\alpha$ combination, one of the H_c nuclei is aligned with the external field and the other is opposed to it. The net result is a cancellation of the local-field contribution at H_d. In these cases, a spectral peak should appear at a chemical shift identical with the one expected if there were no coupling between H_c and H_d. Moreover, because two equivalent combinations of neighboring H_cs [$H_{c(\alpha\beta)}$ and $H_{c(\beta\alpha)}$] contribute to this peak (instead of only one, as did $H_{c(\alpha\alpha)}$ to the first peak), its relative integration (and approximate height) should be double that of the first peak, as observed. Finally, H_d may have the $H_{c(\beta\beta)}$ combination as its neighbor. In this case, the local field subtracts from the external field, and an upfield peak of relative intensity 1 ensues. The resulting pattern for H_d is the observed 1:2:1 triplet with a total integration corresponding to six hydrogens (because there are two methyl groups). The distance between each adjacent pair of peaks reveals that the coupling constant J_{cd} = 8 Hz. The splitting of the methyl group by the adjacent methylene group is represented in Figure 10-22.

The quartet observed for H_c can be analyzed in the same manner. At any given instant, this nucleus is exposed to four different types of H_d proton combinations as neighbors: one in which all protons are aligned with the field

FIGURE 10-23

Splitting of H_c into a quartet by the various spin combinations of H_d. The chemical shift of the quartet is reported as its midpoint: $\delta_{H_c} = 3.63$ ppm (q, $J = 8$ Hz, 4 H), in which "q" stands for quartet.

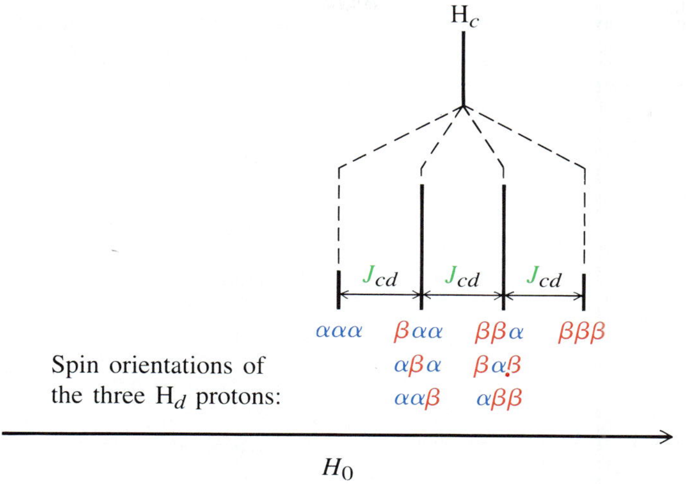

$[H_{d(\alpha\alpha\alpha)}]$, three equivalent arrangements in which one H_d is opposed to the external field and the other two are aligned with it $[H_{d(\beta\alpha\alpha,\ \alpha\beta\alpha,\ \alpha\alpha\beta)}]$, another set of three equivalent arrangements in which only one proton remains aligned with the field $[H_{d(\beta\beta\alpha,\ \beta\alpha\beta,\ \alpha\beta\beta)}]$, and a final possibility in which all H_ds oppose the external magnetic field $[H_{d(\beta\beta\beta)}]$. The resulting spectrum is predicted—and observed—to consist of four peaks in the ratio of 1:3:3:1 (integrated value 4) (Figure 10-19 and 10-23). The coupling constant J_{cd} is given as the distance between any two lines in the quartet and is identical with the coupling constant measured in the triplet for H_d (8 Hz).

Spin-Spin Splitting Simplified: The $N + 1$ Rule

The foregoing analyses afford a means of formulating a set of simple rules that hold for most organic molecules: nuclei in a given equivalent set located adjacent to a single neighboring hydrogen appear as a doublet, those adjacent to two neighbors of a second equivalent set form a triplet, and those next to three hydrogens resonate as a quartet. Table 10-4 shows the expected splitting patterns for nuclei adjacent to N equivalent neighbors. You see that they *split into* $N + 1$ *peaks*. Their relative ratio is given by a mathematical mnemonic device called Pascal's triangle. Each number in this triangle is the sum of the two numbers closest to it in the line above. Examples are shown in Figures 10-24, 10-25, and 10-26 (integration has been omitted). Figure 10-26, on page 391, depicts an example in which equivalent nuclei on two groups (the two equivalent methyls in 2-iodopropane) give rise to a septet ($N + 1 = 7$) for the central hydrogen. Conversely, the signal for the two methyl groups appears as a doublet owing to coupling to the tertiary hydrogen.

TABLE 10-4
NMR splittings of a set of hydrogens with N equivalent neighbors and their integrated ratios (Pascal's triangle)

Equivalent neighboring (N) hydrogens	Number of peaks ($N + 1$)	Name for peak pattern (abbreviation)	Integrated ratios of individual peaks
0	1	Singlet (s)	1
1	2	Doublet (d)	1:1
2	3	Triplet (t)	1:2:1
3	4	Quartet (q)	1:3:3:1
4	5	Quintet (quin)	1:4:6:4:1
5	6	Sextet (sex)	1:5:10:10:5:1
6	7	Septet (sep)	1:6:15:20:15:6:1

FIGURE 10-24
90-MHz NMR spectrum of bromoethane in CCl_4. The methylene group appears as a quartet at $\delta = 3.24$ ppm, $J = 7$ Hz. The methyl hydrogens, on the other hand, absorb as a triplet at $\delta = 1.58$ ppm, $J = 7$ Hz.

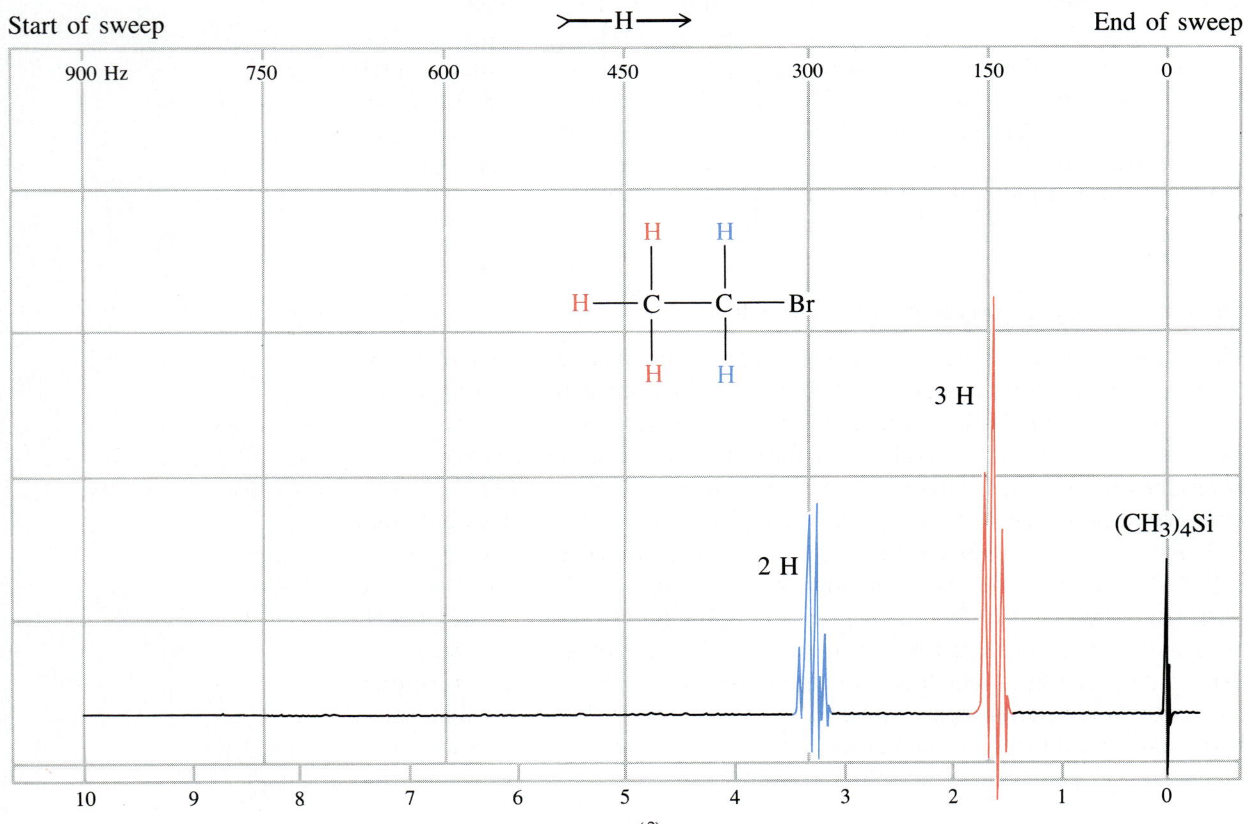

FIGURE 10-25

90-MHz NMR spectrum of 1,1,2-trichloroethane in CCl_4: $\delta = 5.58$ (t, $J = 7$ Hz, 1 H), 2.13 (d, $J = 7$ Hz, 2 H) ppm.

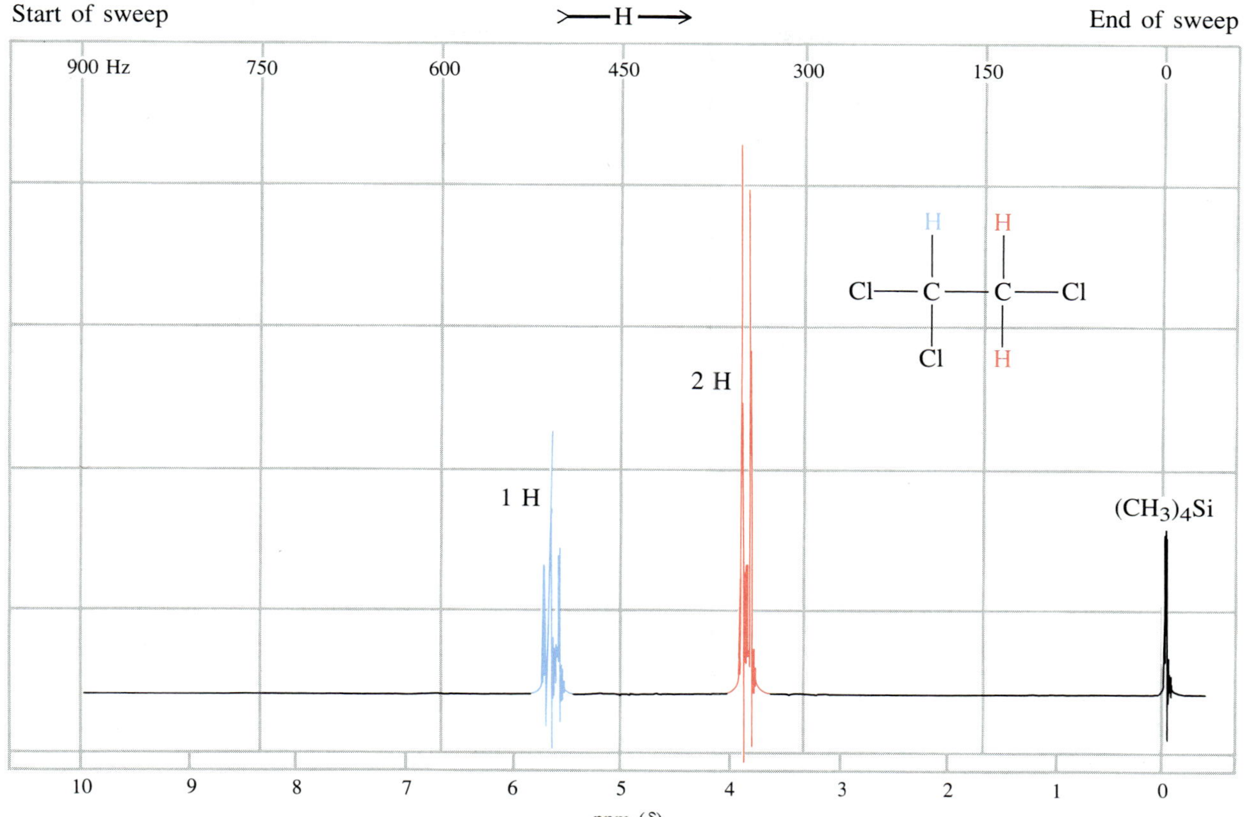

It is important to remember that nonequivalent nuclei mutually split each other. In other words, the observation of one split absorption necessitates the presence of another split signal in the spectrum. Moreover, the coupling constants for these patterns must be the same. Some frequently encountered multiplets and the corresponding structural units are shown in Table 10-5, on the next page.

EXERCISE 10-8

Predict the NMR spectra of: (a) ethoxyethane (diethyl ether); (b) 1,3-dibromopropane; (c) 2-methyl-2-butanol. Specify approximate chemical shifts and multiplicities.

To summarize, spin-spin splitting occurs between vicinal and geminal nonequivalent hydrogens following the $N + 1$ rule. Usually N equivalent neighbors will split the absorption of the observed hydrogen into $N + 1$ peaks, their relative intensities being in accord with Pascal's triangle. In this way, the common alkyl groups give rise to characteristic NMR patterns.

TABLE 10-5

Frequently observed spin-spin splittings in common alkyl groups

Splitting pattern for H_a	Structure	Splitting pattern for H_b

Note: H_a and H_b are assumed to have no other coupled nuclei in their vicinity.

FIGURE 10-26

90-MHz NMR spectrum of 2-iodopropane in CCl_4: $\delta = 4.12$ (sep, $J = 7.5$ Hz, 1 H), 1.82 (d, $J = 7.5$ Hz, 6 H), in which ''sep'' stands for septet. Note that the outer peaks of the septet are of such small intensity that they are difficult to see in the spectrum recorded on scale. It is therefore frequently advisable to ''blow up'' split peaks in intensity to clarify some of their features. Such an enlargement is shown in the inset in which the septet for the tertiary hydrogen has been rerecorded at higher sensitivity.

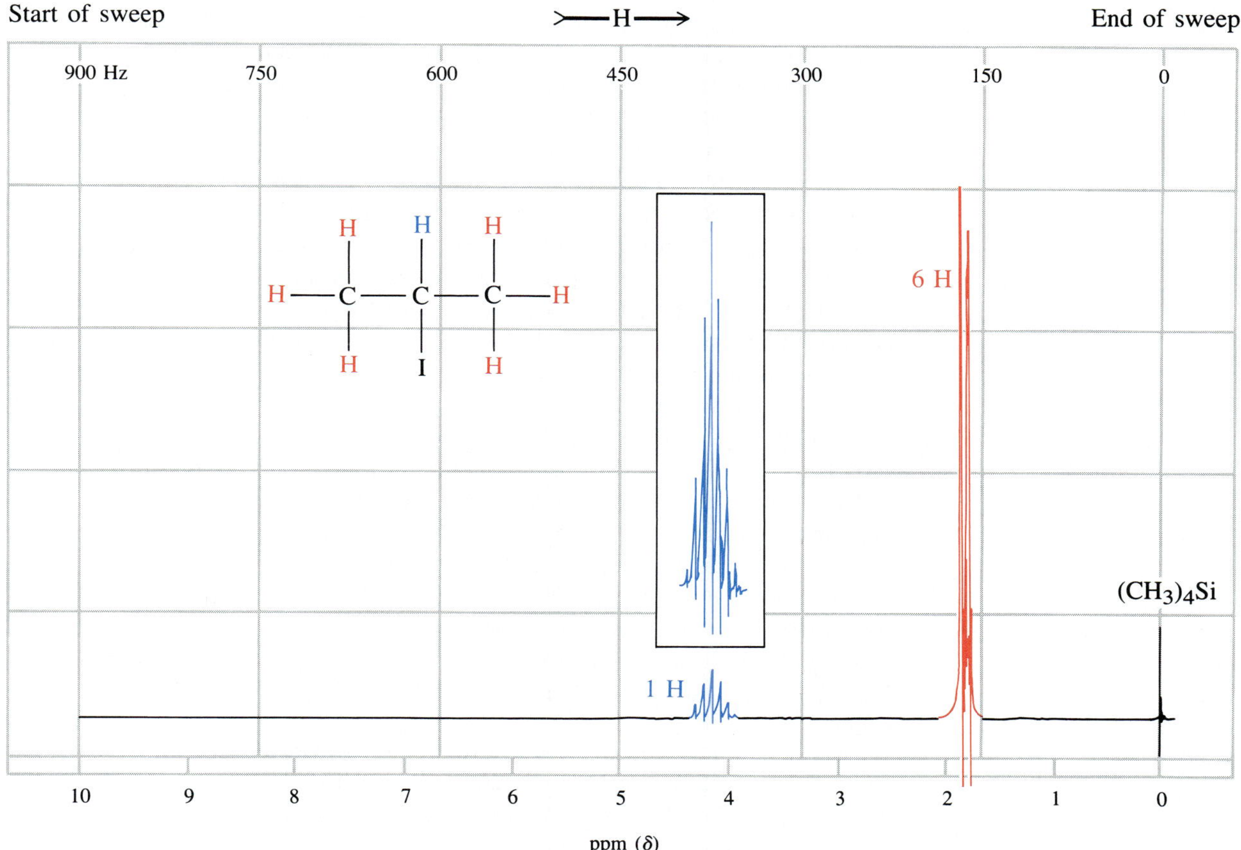

10-6 Spin-Spin Splitting: Complications

The rules governing the appearance of split peaks outlined in Section 10-5 are somewhat idealized. There are cases in which, because of a relatively small difference in δ between two absorptions, more-complex patterns (multiplets) are observed that are not interpretable without the use of computers. Moreover, the $N + 1$ rule may not be applicable in a direct way if two or more types of neighboring hydrogens are coupled to the resonating nucleus with fairly different coupling constants. Finally, the hydroxy proton may appear as a singlet (see Figure 10-14) even if vicinal hydrogens are present.

Close-Lying Peak Patterns May Give Rise to Non-First-Order Spectra

A careful look at the spectra shown in Figures 10-19, 10-25, and 10-26 shows that the relative intensities of the splitting patterns frequently do not conform to the idealized peak ratios expected from consideration of Pascal's triangle: the patterns are skewed. This is a general phenomenon that sometimes helps in interpreting a spectrum, because a multiplet is skewed in the direction of the resonance frequency of the nucleus responsible for its splitting. Near-perfect peak ratios are seen only when the chemical shifts of the coupled nuclei are very different. To be more specific, to observe intensity ratios dictated by Pascal's triangle and the $N + 1$ rule, the difference in chemical shift between coupled protons must be much larger than their coupling constant: $\Delta\delta \gg J$. Under these circumstances, the spectrum is said to be of *first order*.* On the other hand, when the difference in chemical shift between such hydrogens becomes smaller, the expected peak pattern is subject to increasing distortion. Initially, this distortion manifests itself in the skewing phenomenon. However, as the chemical shifts get closer, further complications arise. The simple rules devised in Section 10-5 do not apply any more, the resonance absorptions assume more complex shapes, and the spectra are said to be of *non-first order*. Although such spectra can be simulated with the help of computers, this treatment is beyond the scope of the present discussion.

Particularly striking examples of non-first-order spectra are of compounds containing alkyl chains, including even the simple straight-chain alkanes. Figure 10-27 shows an NMR spectrum of octane, which is obviously not of first order because all nonequivalent hydrogens (there are four types) have very similar chemical shifts. All methylenes absorb as one broad multiplet. In addition, there is a highly distorted triplet for the terminal methyl groups.

Figure 10-28A depicts the NMR spectrum of methylcyclohexane at 90 MHz, which is slightly more complex than that of octane because of the presence of eight different types of hydrogens. At higher field (hydrogen resonating at 250 MHz), the pattern is much better resolved (Figure 10-28B) but still not first order.

A case in which improved field strength has a more dramatic effect is found for 2-chloro-1-(2-chloroethoxy)ethane (Figure 10-29, on page 394). In this compound, the deshielding effect of the oxygen is about equal to that of the chlorine substituent. As a consequence, the two sets of methylene hydrogens give rise to very close lying peak patterns. The resulting absorption has a symmetrical shape, but is very complicated, exhibiting more than thirty-two peaks of various intensity. Because non-first-order spectra arise when $\Delta\delta \approx J$, it should be possible to "improve" the appearance of a multiplet by measuring a spectrum at higher field. Recall (Section 10-2) that the resonance frequency is proportional to the external-field strength (and vice versa) but that the coupling constant J is independent of field. Indeed, recording the NMR spectrum in a 500-MHz spectrometer (Figure 10-29B) produces a first-order pattern.

Coupling to Nonequivalent Neighbors: A Modification of the $N + 1$ Rule

A complex absorption does not always signify that the spectrum is non-first-order. Some hydrogens may be coupled to two nonequivalent neighbors, giving

*This expression derives from the definition of a "first order theory": in general, a theory that takes into account only the important terms and variables of a system.

FIGURE 10-27
90-MHz NMR spectrum of octane in CCl_4.

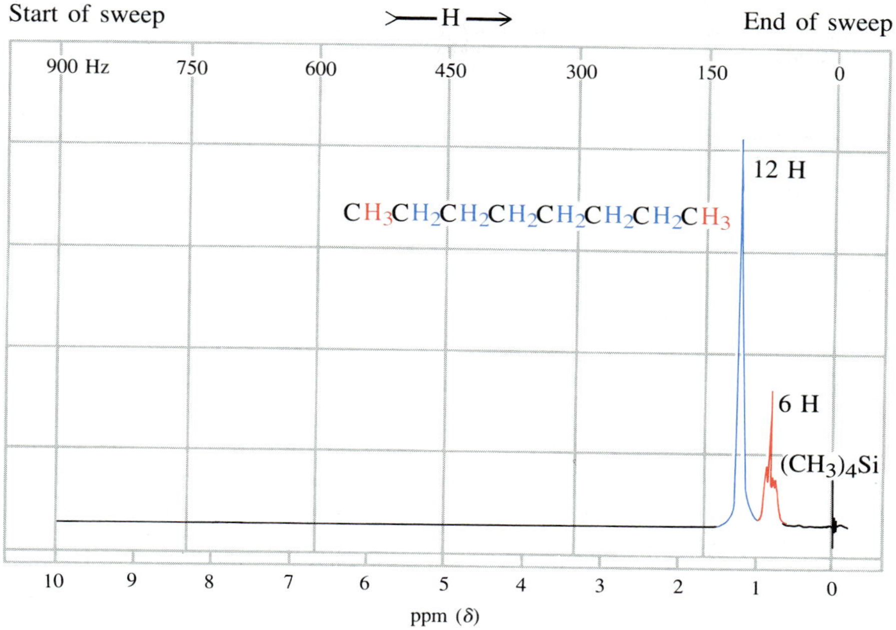

Start of sweep
$\succ$—H⟶
End of sweep

900 Hz 750 600 450 300 150 0

$CH_3CH_2CH_2CH_2CH_2CH_2CH_2CH_3$

12 H

6 H

$(CH_3)_4Si$

10 9 8 7 6 5 4 3 2 1 0

ppm (δ)

FIGURE 10-28
NMR spectrum of methylcyclohexane at (A) 90 MHz; (B) 250 MHz in CCl_4.

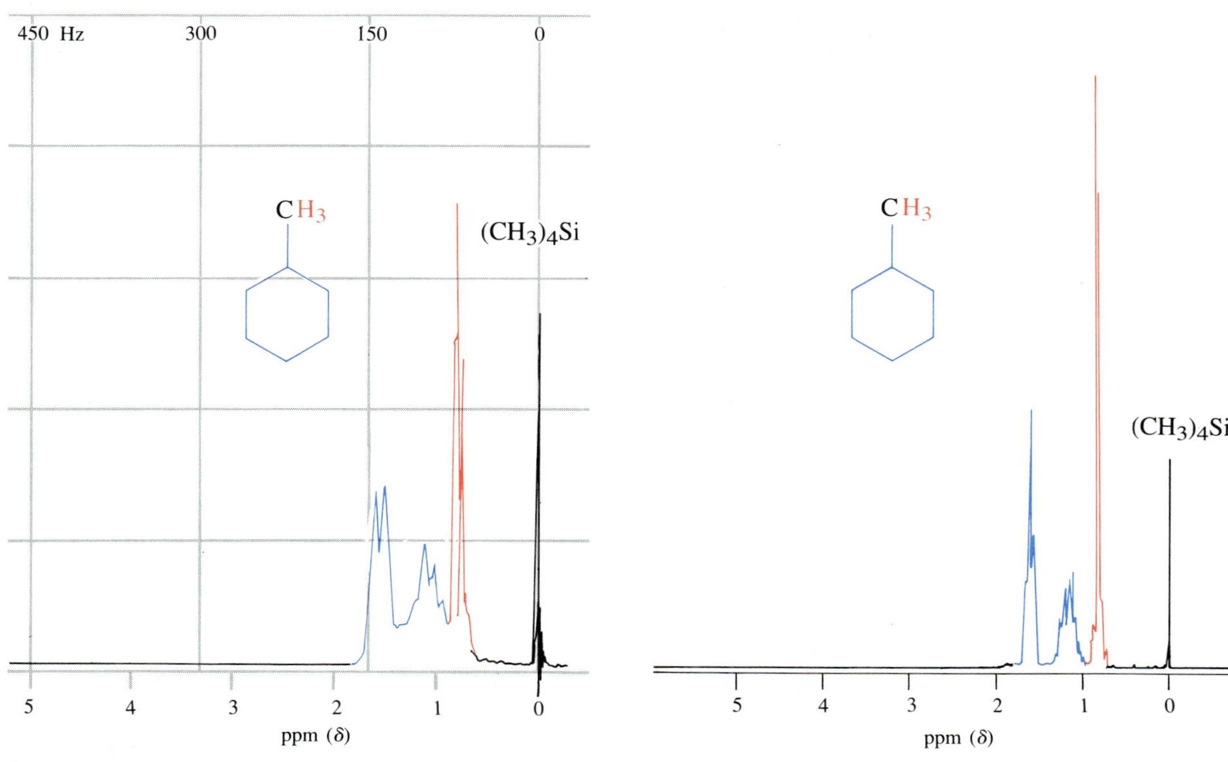

FIGURE 10-29

NMR spectrum of 2-chloro-1-(2-chloroethoxy)ethane at (A) 90 MHz; (B) 500 MHz in CCl_4. At high field, the complex multiplet observed at 90 MHz is simplified into two slightly distorted triplets, as might be expected for two mutually coupled CH_2 groups.

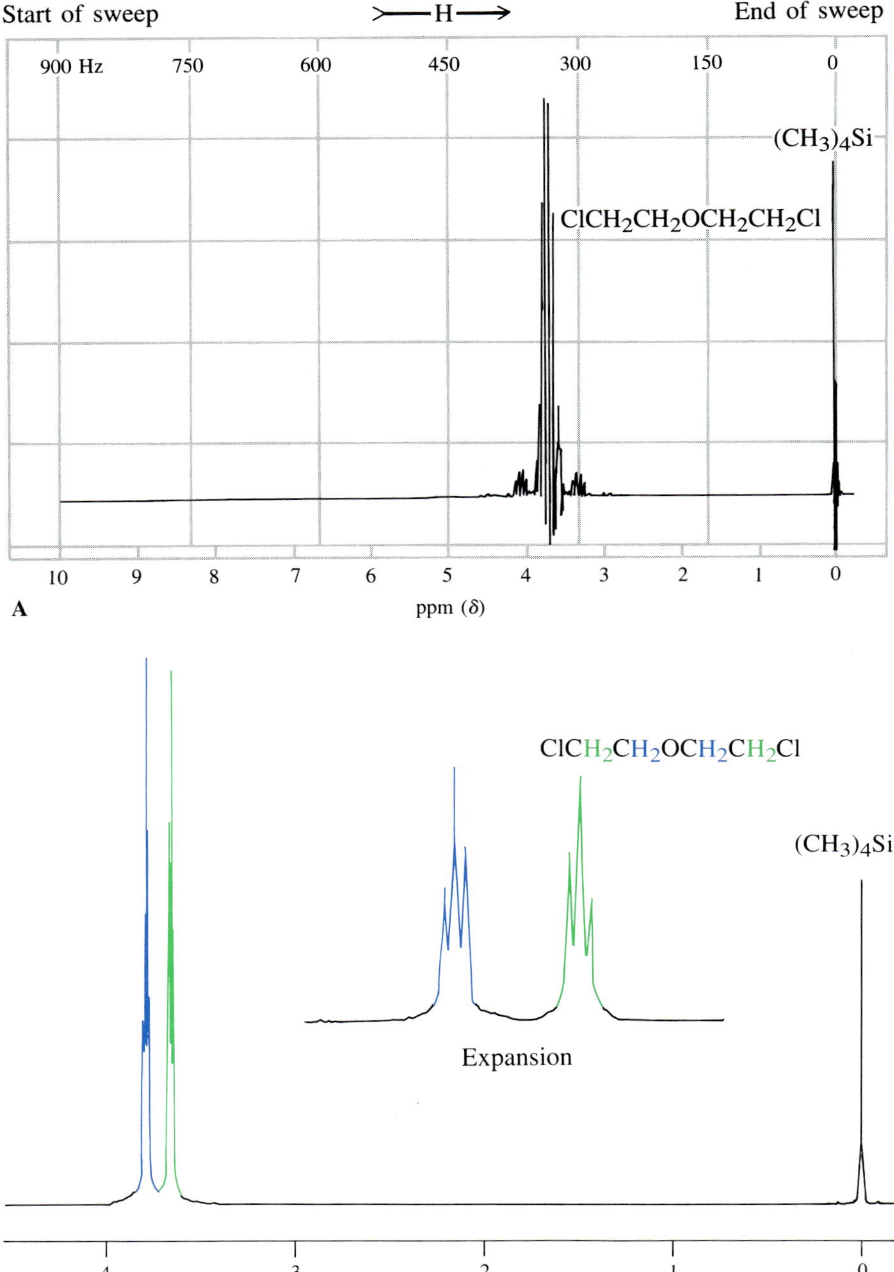

rise to relatively complicated but nevertheless first-order splitting patterns. An example is 1-bromopropane, whose spectrum is shown in Figure 10-30. In this compound, the hydrogens at C-2 are located between a methyl group and a methylene group, and are coupled to both independently.

FIGURE 10-30
90-MHz NMR spectrum of 1-bromopropane in CCl_4.

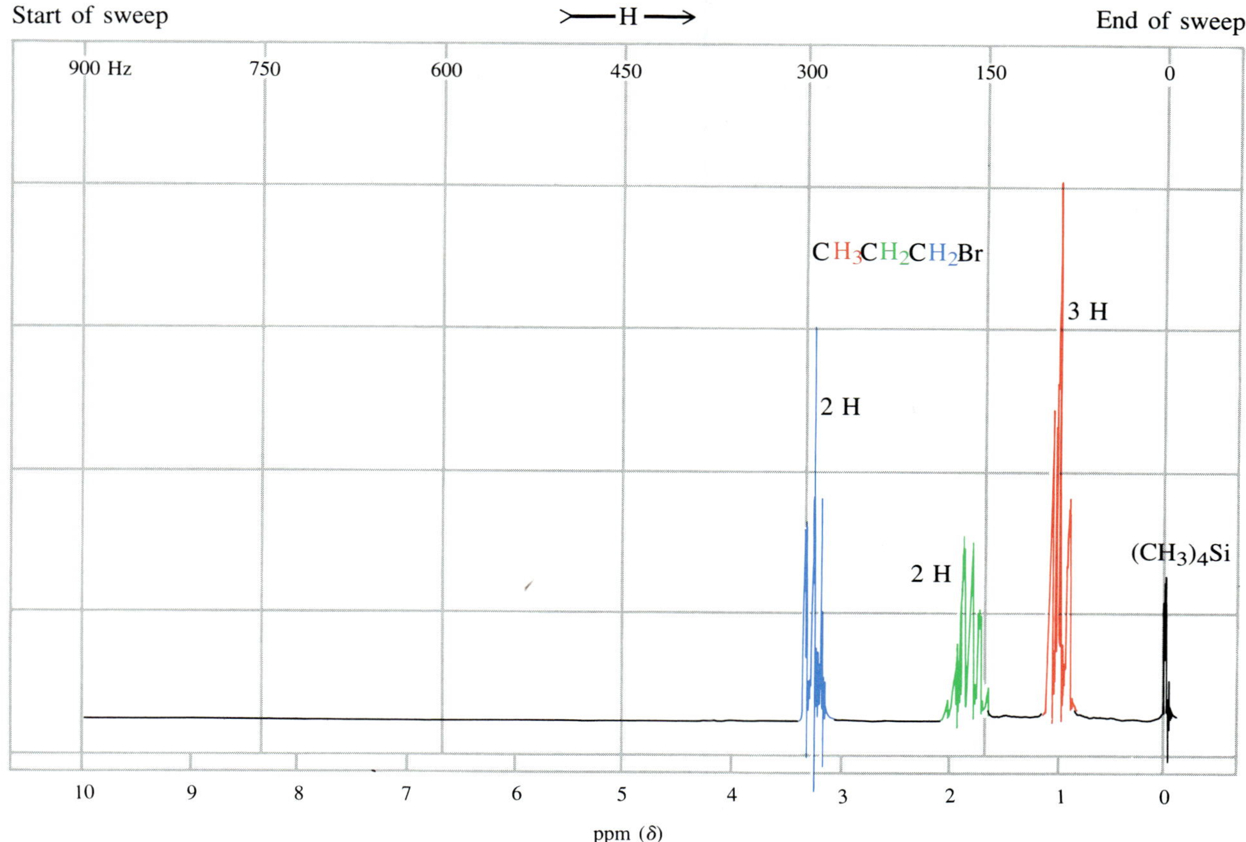

Let us analyze the spectrum in detail. We first notice two triplets, one at low-field ($\delta = 3.28$ ppm, $J = 6.0$ Hz, 2 H) and one at high-field ($\delta = 1.03$ ppm, $J = 6.5$ Hz, 3 H), assignable to the 1-methylene group and the methyl group, respectively. The low-field absorption is due to the hydrogens adjacent to the deshielding halogen, and the methyl group resonates as expected at highest field. The CH_2 group at C-2 gives rise to a pattern that appears to be a slightly distorted sextet ($\delta = 1.86$ ppm, $J \approx 6$–7 Hz, 2 H). Is this the splitting we would have expected?

From what we have observed, our answer to this question might be in the affirmative. The nuclei giving rise to this absorption have a total of five hydrogens as their neighbors: three from the methyl group, two from the 1-methylene group. Application of the $N + 1$ rule suggests that a sextet should be observed. This conclusion is, however, not entirely correct. The $N + 1$ rule applies strictly to splitting by magnetically *equivalent* neighbors. Here, we have two sets of different adjacent hydrogens, which should couple to the CH_2 group at C-2 with characteristic coupling constants. The effect of these couplings can be estimated separately, by sequential application of the $N + 1$ rule. First, the methyl group should cause a splitting of the resonance of its neighbors into a quartet. Subsequent consideration of the pair of hydrogens at C-1 suggests that each peak in this quartet should be further split into a triplet. The maximum number of peaks should therefore be twelve. The sequence of this analysis is not

FIGURE 10-31

Splitting pattern expected for H_b in a propyl derivative when (A) $J_{ab} > J_{bc}$, (B) $J_{bc} > J_{ab}$, and (C) $J_{ab} \sim J_{bc}$. In the last case, several of the peaks coincide, giving rise to a deceptively simple spectrum: a sextet.

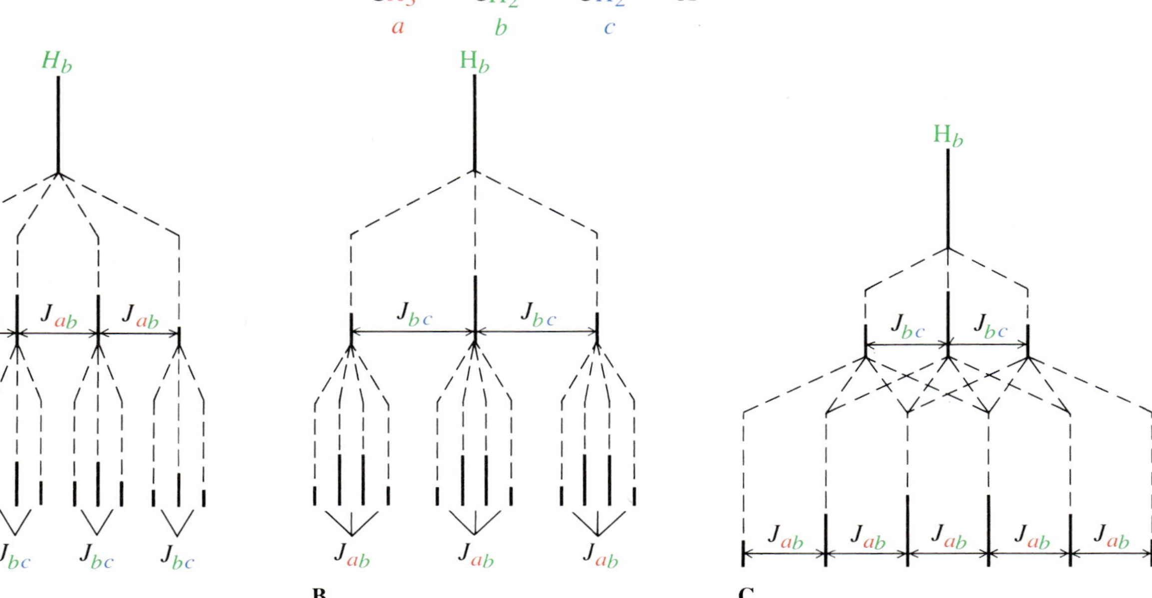

important: we could equally well start with the coupling to the hydrogens at C-1 to furnish a triplet, each peak of which should be split further into a quartet by coupling to the methyl group. The hydrogens at C-2 are said to be split into a triplet of quartets, or into a quartet of triplets. Why, then, do we observe the simpler pattern of a sextet?

The answer is that the final appearance of the multiplet depends very much on the coupling constants giving rise to it. If the J values differ appreciably, the expected twelve-peak pattern is readily constructed (Figure 10-31A and B). As the coupling constants become closer to being equal in magnitude, the spectrum simplifies to six lines because of the degeneracy of some resonances, as shown in Figure 10-31C. In alkyl derivatives such as 1-bromopropane, the magnitude of the coupling to both sides of H_b is very similar. Their spectra seem to obey the $N + 1$ rule. Remember, however, that this observation is by coincidence and not necessarily expected. The near equality of the two coupling constants J_{ab} and J_{bc} becomes apparent on inspection of the splitting in the triplets due to H_a and H_c. For 1-bromopropane, they are almost identical, 6 and 6.5 Hz, respectively.

EXERCISE 10-9

Explain the peak patterns observed in the NMR spectrum of 1-chloro-2-methylpropane (Figure 10-32).

FIGURE 10-32
90-MHz NMR spectrum of 1-chloro-2-methylpropane in CCl_4. The inset is an expanded (factor of two) rerecording of the multiplet at $\delta = 1.88$ ppm.

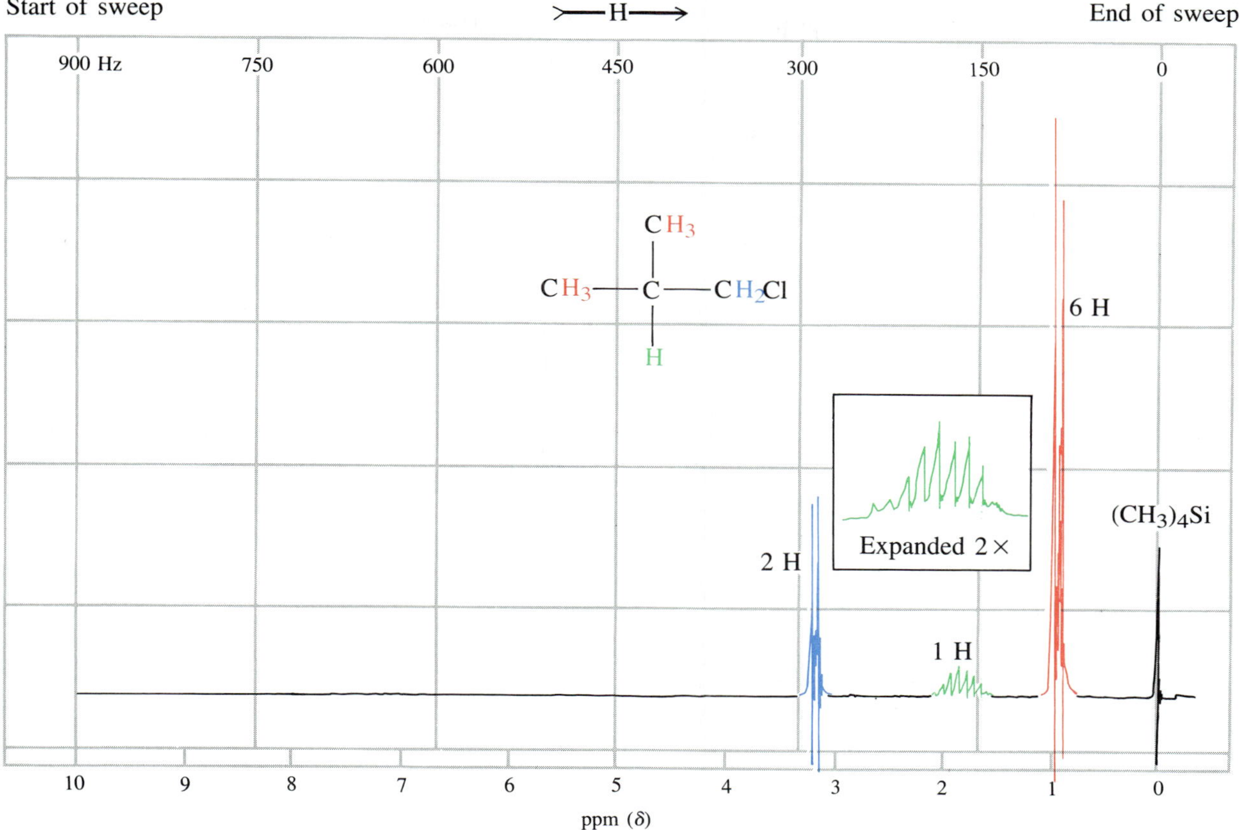

A case in which the coupling constants to two adjacent sets of hydrogens are sufficiently different to allow the observation of the theoretically expected spectrum is 1,1,2-trichloropropane (Figure 10-33).

Nucleus H_a ($\delta = 5.69$ ppm) and the higher-field methyl hydrogens H_c ($\delta = 1.64$ ppm) give rise to two doublets because of coupling to H_b. The two coupling constants are $J_{ab} = 3.6$ Hz and $J_{bc} = 6.8$ Hz. Consequently, H_b should appear as a doublet of quartets incorporating these two J values, as shown in the expanded inset. The chemical shift of H_b is the midpoint of the split absorption ($\delta = 4.18$ ppm).

Fast Proton Exchange Decouples Hydroxy Hydrogens

With our knowledge of vicinal coupling, let us now return to the NMR of alcohols. We note in the NMR spectrum of 2,2-dimethyl-1-propanol (Figure 10-14) that the OH absorption appears as a single peak, devoid of any splitting. This is curious, because the hydrogen is adjacent to two others, which should cause its appearance as a triplet. The CH_2 hydrogens that show up as a singlet should in turn appear as a doublet incorporating the same coupling constant. Why, then, do we not observe spin-spin splitting? Because there is fast proton exchange both between alcohol molecules and with traces of water under the

FIGURE 10-33

90-MHz NMR spectrum of 1,1,2-trichloropropane in CCl_4. Nucleus H_b gives rise to a doublet of quartets at $\delta = 4.18$ ppm: eight peaks.

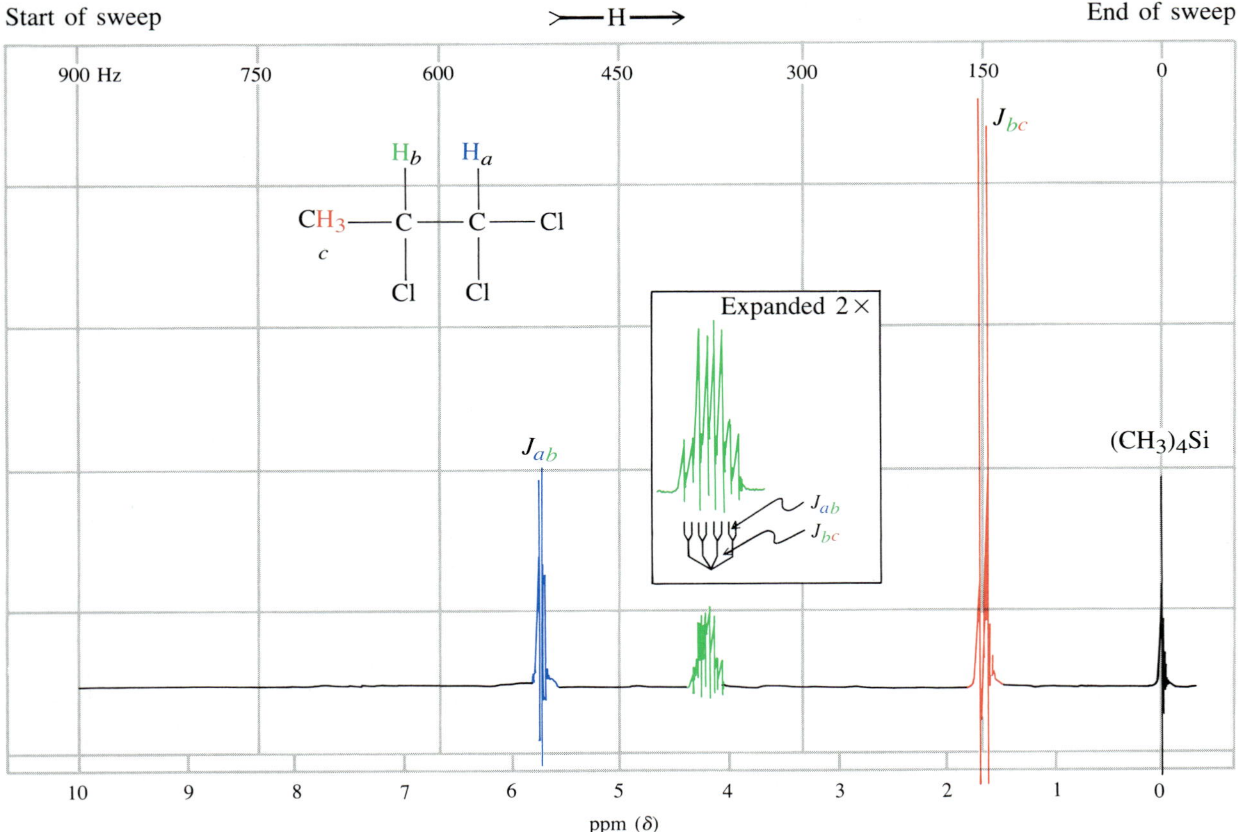

conditions of the NMR measurement. The rate of this process is fast on the NMR time scale, about 10^5 protons per second, a number that translates into an average residence time of 10^{-5} sec for an individual proton in the vicinity of the oxygen. As a consequence, the NMR spectrometer sees only an average signal for the OH hydrogen. No coupling is visible, because the binding time of the proton to the oxygen is too short. The three situations in which the OH proton is located next to the respective $\alpha\alpha$, $\alpha\beta$-$\beta\alpha$, $\beta\beta$ combinations of the neighboring CH_2 group interconvert so fast that all three chemical shifts merge into one. A similar argument holds for the lack of splitting in the CH_2 unit.

Absorptions of this type are said to be **decoupled** by proton exchange. The exchange may be slowed down by removal of traces of water or acid or by cooling, when the OH bond retains its integrity long enough to be observed on the NMR time scale. An example is shown in Figure 10-34 for methanol. At 37°C, two singlets are observed, corresponding to the two types of hydrogens, both devoid of spin-spin splitting. On the other hand, on cooling, the peaks begin to broaden, until at −65°C the expected coupling pattern is detectable: a quartet and a doublet.

FIGURE 10-34

Temperature dependence of spin-spin splitting in methanol (after H. Günther, *NMR-Spektroskopie,* Georg Thieme Verlag, Stuttgart, 1973).

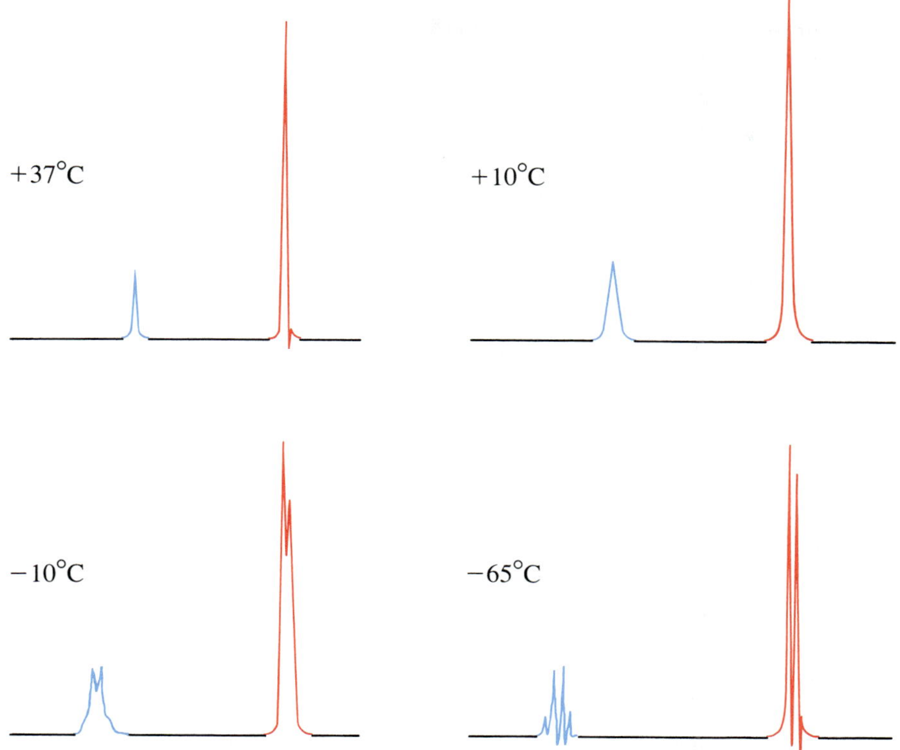

In summary, in many NMR spectra the peak patterns are not first order because the difference in chemical shift between the nonequivalent hydrogens is close to the value of the corresponding coupling constant. High-field NMR spectroscopy may improve the appearance of such spectra. Coupling of a hydrogen to different nonequivalent hydrogens occurs separately, with distinctly different coupling constants. In many simple alkyl derivatives, they are similar (J = 6–7 Hz), and so the spectrum simplifies to the one predicted in accord with the $N + 1$ rule. In other examples, this is not the case, but the resulting multiplets are nevertheless analyzable. Vicinal coupling through the oxygen in alcohols is frequently not observed because of decoupling by fast proton exchange.

10-7 Carbon-13 Nuclear Magnetic Resonance

Proton nuclear magnetic resonance is a powerful tool in the elucidation of organic structures because most organic compounds contain hydrogen nuclei. However, potentially even more powerful is NMR spectroscopy of carbon given that, by definition, *all* organic compounds contain this element as part of their structural framework. On its own or in combination with ^{1}H NMR, it can have many applications. This section is a brief introduction to the NMR of carbon and some of its applications.

Carbon NMR Utilizes an Isotope in Low Natural Abundance: ^{13}C

Carbon NMR is possible. However, there is a complication: the most abundant isotope of carbon, carbon-12, is not detectable by NMR. Fortunately, another isotope, carbon-13, is found in nature at a level of about 1.11%. Its behavior in the presence of a magnetic field is the same as that of hydrogen. One might therefore expect it to give spectra very similar to those observed in ^{1}H NMR spectroscopy. This expectation turns out to be only partly correct, because important (and in fact very useful) differences exist between the two types of NMR techniques.

Carbon-13 NMR (^{13}C NMR) spectra are much more difficult to record than are hydrogen spectra not only because of the low natural abundance of the nucleus under observation, but also because of the much weaker magnetic resonance of ^{13}C. Thus, under comparable conditions, ^{13}C signals are about 1/6,000 times as strong as those for hydrogen. For these weak signals to be observed, ^{13}C spectra are scanned repetitively and stored in a computer. The computer then adds up all of the signals to produce one composite average spectrum. To collect data at a reasonable rate (sometimes several thousand scans are necessary) requires the use of a special technique called Fourier Transform NMR (FT NMR), the details of which are beyond the scope of this introduction. Suffice it to say that development of this method is making ^{13}C NMR as routine a tool in the organic laboratory as ^{1}H NMR.

One advantage of the low abundance of ^{13}C is the absence of carbon–carbon coupling. Just as with hydrogen spectra, two adjacent carbons, if magnetically nonequivalent (as, e.g., in bromoethane), should split each other. In practice, such splitting is not observed. Why? Because coupling can occur only if two ^{13}C isotopes come to lie next to each other. With the abundance of ^{13}C in the molecule being 1.11%, this event has a very low probability. Most ^{13}C nuclei are surrounded by an excess of ^{12}C, which, having no spin, does not give rise to spin-spin splitting. Although this effect simplifies ^{13}C NMR spectra appreciably, coupling to any neighboring hydrogens must be taken into consideration.

Figure 10-35 depicts a ^{13}C NMR spectrum of bromoethane. The chemical shift δ is defined as in ^{1}H NMR and is determined relative to an internal standard, normally the carbon absorption in $(CH_3)_4Si$. Note that the chemical-shift range of carbon is much larger than that of hydrogen. For most organic compounds, it covers a distance of about 200 ppm, in contrast with the relatively narrow spectral ‘‘window’’ (10 ppm) of hydrogen. Figure 10-35 also reveals the relative complexity of the absorptions caused by extensive ^{13}C–H couplings. Directly bound hydrogens are coupled with relatively large coupling constants (~125–200 Hz). Coupling tapers off rapidly, however, with increasing distance from the ^{13}C nucleus under observation, such that the geminal coupling constant $J_{^{13}C\text{–}C\text{–}H}$ is in the range of only 0.7–6 Hz.

EXERCISE 10-10

Predict the ^{13}C NMR spectral pattern of 1-bromopropane.

FIGURE 10-35
62.8-MHz ^{13}C NMR spectrum of bromoethane. There is an upfield quartet (δ = 18.3 ppm) and a downfield triplet (δ = 26.6 ppm) resonance for the two carbon atoms. Note the large chemical-shift range. Tetramethylsilane, defined to be located at δ = 0 ppm as in ^1H NMR, absorbs as a quartet (J = 118 Hz; the two outside peaks are barely visible) because of coupling of each carbon to three equivalent hydrogens. The inset shows that the methyl carbon of bromoethane actually shows up as a quartet of triplets (J = 126 and 3 Hz) because of (the expectedly large) coupling to the immediately bound three methyl hydrogens, and further splitting by the adjacent methylenes. The methylene carbon absorbs as a triplet of quartets (J = 151 and 5 Hz).

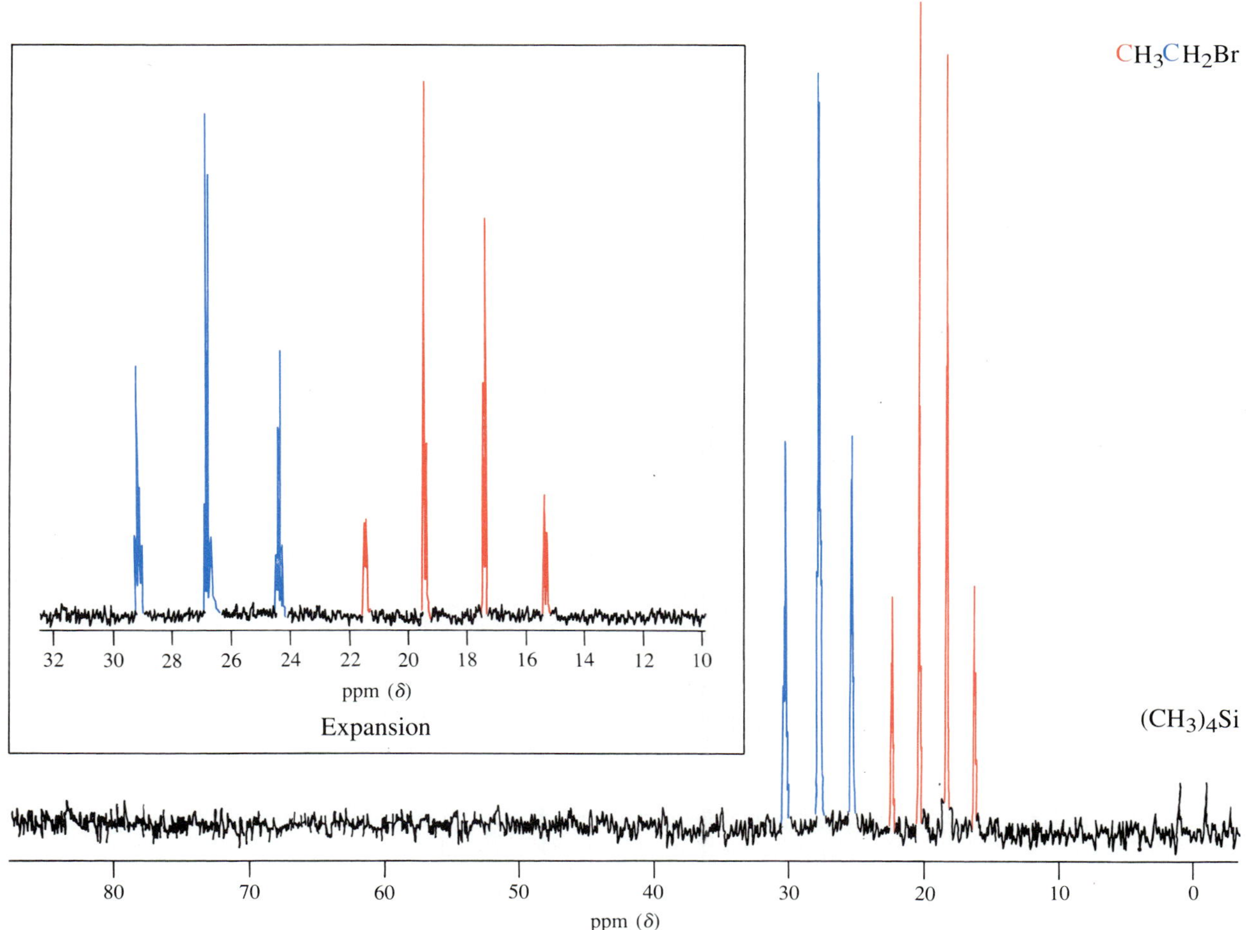

Hydrogen Decoupling Gives Single Lines

A technique that completely removes ^{13}C–H coupling is called **noise decoupling,** or **broad-band hydrogen** (or **proton) decoupling.** This method employs a strong, broad radio-frequency signal that covers the resonance frequencies of all the hydrogens and is applied at the same time as the ^{13}C spectrum is recorded. For example, in a magnetic field of 58,750 Gauss, carbon-13 resonates at 62.8 MHz (see spectrum in Figure 10-35), hydrogen at 250 MHz. In a proton-decoupled carbon spectrum, the sample is irradiated at both frequencies.

FIGURE 10-36
62.8-MHz ^{13}C NMR spectrum of bromoethane, with broad-band proton decoupling at 250 MHz. The field strength is 58,750 Gauss. All lines simplify to singlets, including the absorption for $(CH_3)_4Si$.

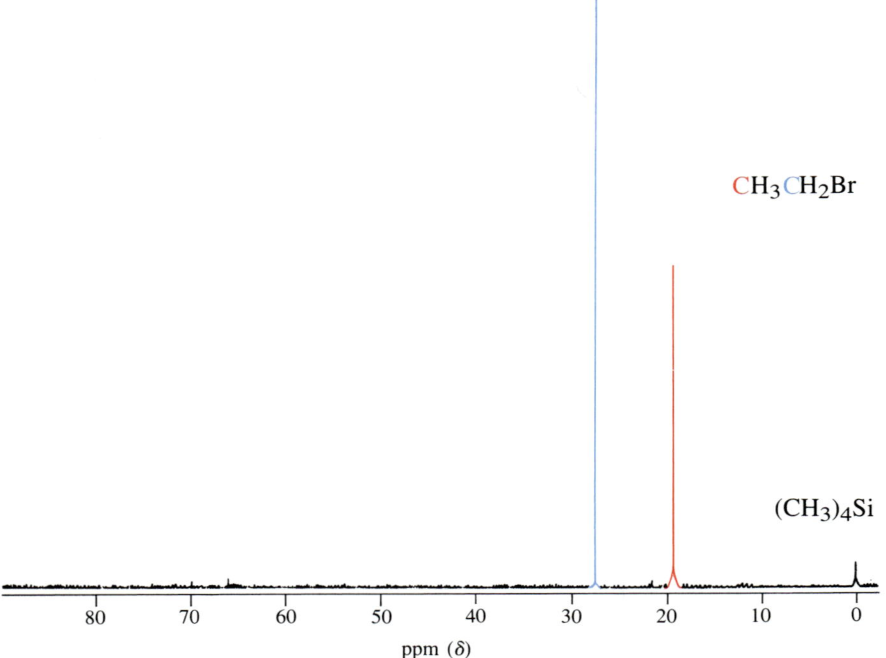

The first radio-frequency signal is used to effect carbon magnetic resonance. Simultaneous exposure to the second signal causes all the hydrogens to undergo rapid α-to-β spin flips, fast enough that the attached carbon atom is not exposed to distinct α and β spin states of the neighboring hydrogens but to an average single local magnetic field contribution: the net result is the absence of coupling. By use of this technique, the ^{13}C NMR spectrum of bromoethane simplifies to two single lines, shown in Figure 10-36.

The power of this technique becomes evident when spectra of relatively complex molecules are recorded. *Every magnetically distinct carbon gives only one single peak in the ^{13}C NMR spectrum.* Consider, for example, a hydrocarbon such as methylcyclohexane. Analysis by ^1H NMR is made very difficult by the close-lying chemical shifts of the eight different types of hydrogens (see Figure 10-28). The spectrum is exceedingly complex because of extensive H–H couplings. On the other hand, a proton-decoupled ^{13}C spectrum shows only five peaks, clearly depicting the presence of the five different types of carbons (Figure 10-37).

Table 10-6 shows that, similar to hydrogen (Table 10-2), carbon has characteristic chemical shifts depending on its structural environment. As in ^1H NMR, electron-withdrawing groups cause deshielding and the chemical shifts go up in the order primary < secondary < tertiary carbon. Apart from the diagnostic usefulness of such δ values, a knowledge of the number of different carbon atoms in the molecule can be an aid to structural identification. Consider, for example, the analytical differentiation of methylcyclohexane from other iso-

FIGURE 10-37
62.8-MHz ^{13}C NMR spectrum of methylcyclohexane with hydrogen decoupling (in C_6D_6). There are five magnetically different types of carbon in this compound, giving rise to five peaks: $\delta = 23.1$, 26.7, 26.8, 33.1, and 35.8 ppm.

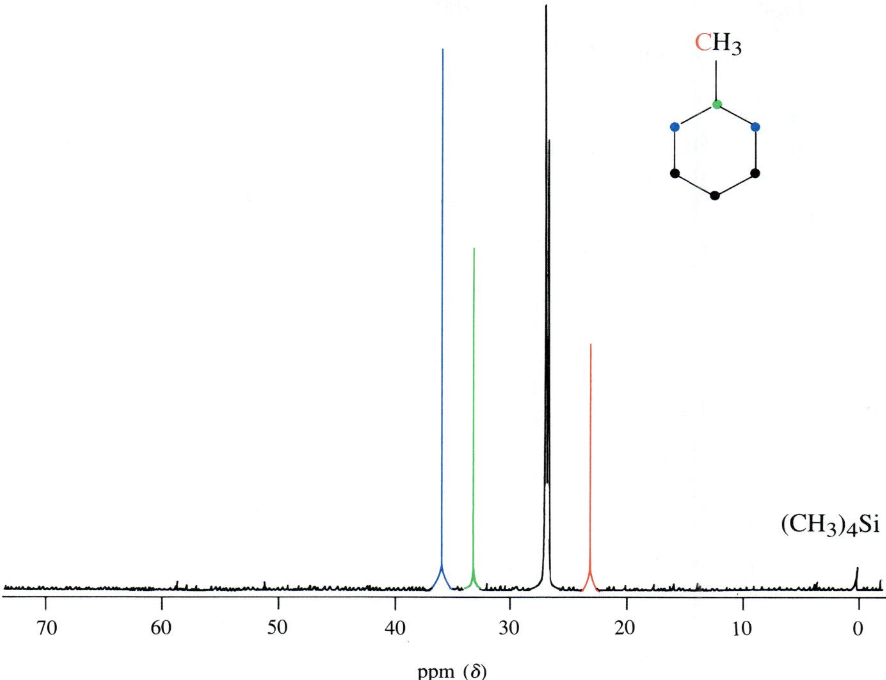

TABLE 10-6
Typical ^{13}C NMR chemical shifts

Type of carbon	Chemical shift δ in ppm
Primary alkyl, RCH_3	5–20
Secondary alkyl, RCH_2R'	20–30
Tertiary alkyl, R_3CH	30–50
Quaternary alkyl, R_4C	30–45
Allylic, $R_2C{=}C(R'')CH_2R'$	20–40
Chloroalkane, RCH_2Cl	25–50
Bromoalkane, RCH_2Br	20–40
Ether or alcohol, RCH_2OR' or RCH_2OH	50–90
Aldehyde or ketone, $RC(=O)H$ or $RC(=O)R'$	170–200
Alkene, aromatic, $R_2C{=}CR_2$	100–150
Alkyne, $RC{\equiv}CR$	50–95

mers with the same empirical formula C_7H_{14}. Many of the possibilities have a different number of nonequivalent carbons incorporated in their structure and therefore give distinctly different carbon spectra (find some with the same number of ^{13}C NMR peaks). Notice how much the (lack of) symmetry in a molecule influences the complexity of the carbon spectrum.

Number of ^{13}C Peaks in Some C_7H_{14} Isomers

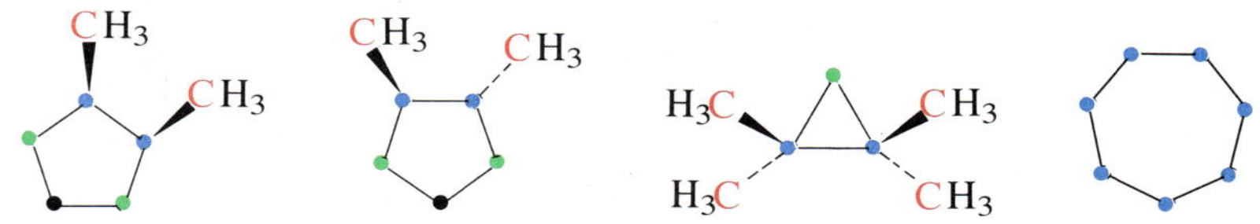

Four peaks — Four peaks — Three peaks — One peak

EXERCISE 10-11

How many peaks would you expect in the proton-decoupled ^{13}C NMR spectra of:

(a) 2,2-dimethyl-1-propanol;

(b) 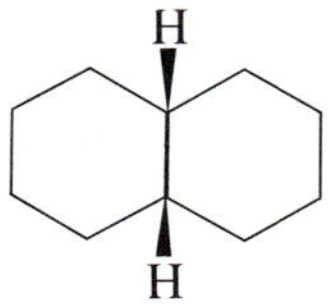 ; (c) 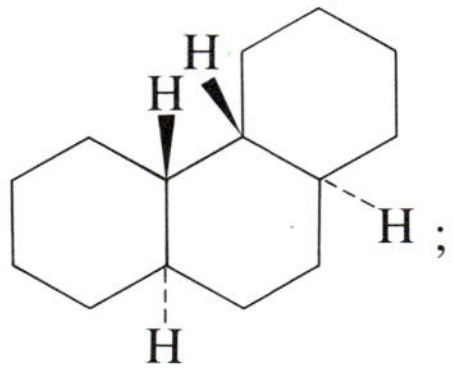 ; (d) 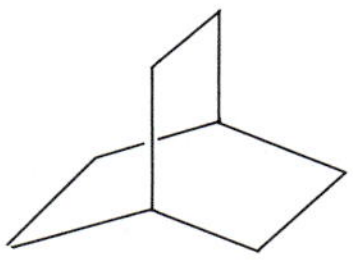 ?

Hint: Look for symmetry.

In summary, ^{13}C NMR is not quite as easy to measure as ^{1}H NMR because of the low natural abundance of the carbon-13 isotope and its intrinsically lower sensitivity in this experiment. Two additional potential complications are ^{13}C–^{1}H and ^{13}C–^{13}C coupling. The first is taken care of by broad-band proton decoupling, and the second never really presents a problem because of the scarcity of ^{13}C. The ^{13}C NMR chemical-shift range is large, about 200 ppm in organic molecules.

Summary of Important Concepts

1 NMR is the most important spectroscopic tool in the elucidation of the structures of organic molecules.

2 Spectroscopy is possible because molecules exist in various energetic forms, those at lower energy being convertible into states of higher energy by absorption of discrete quanta of electromagnetic radiation.

3 NMR is possible because certain nuclei, especially ^{1}H and ^{13}C, when exposed to a strong magnetic field, align with it (α) or against it (β). The α-to-β transition can be effected by radio-frequency radiation, leading to proton resonance and a spectrum with characteristic absorptions. The higher the external-field strength, the higher the resonance frequency. For example, a magnetic field of 21,150 Gauss causes hydrogen to absorb at 90 MHz.

4 High-resolution NMR allows for the differentiation of hydrogen and carbon nuclei in different chemical environments. Their characteristic position in the spectrum is measured as the chemical shift, δ, in ppm from an internal standard, tetramethylsilane.

5 The chemical shift is highly dependent on the presence (shielding) or absence (deshielding) of electron density. The former results in relatively high field [to the right, toward $(CH_3)_4Si$] peaks, the latter in low-field ones. Therefore, electron-donor substituents shield, and electron-withdrawing components deshield. The hydroxy proton of alcohols (and the NH proton in amines) shows a variable chemical shift and a broad absorption because of hydrogen bonding and proton exchange.

6 Chemically equivalent hydrogens and carbons have the same chemical shift. Equivalence is best established by the application of symmetry operations, such as those involving the use of mirror planes and rotations.

7 The number of hydrogens giving rise to a peak is measured by integration.

8 The number and kind of hydrogen neighbors is given by the spin-spin splitting patterns following the $N + 1$ rule. Equivalent hydrogens show no spin-spin splitting.

9 If the NMR absorptions are relatively close, non-first-order spectra with complicated coupling patterns are observed.

10 If the constants for coupling to nonequivalent types of neighboring hydrogens are different, the $N + 1$ rule is modified to apply to each type of hydrogen separately.

11 Carbon NMR utilizes the low-abundance ^{13}C isotope. Carbon–carbon coupling is not observed in ordinary ^{13}C spectra. Carbon–hydrogen coupling can be removed by proton decoupling, simplifying most ^{13}C spectra to a collection of single peaks.

Problems

1 Where on the chart presented in Figure 10-2 would the following be located: sound waves ($\nu \sim 1$ kHz $= 10^3$ Hz $= 10^3$ sec^{-1}, or cycles sec^{-1}); AM radio waves ($\nu \sim 1$ MHz $= 1000$ kHz $= 10^6$ Hz); FM and TV broadcast frequencies ($\nu \sim 100$ MHz $= 10^8$ sec^{-1})?

2 Convert each of the following quantities into the specified units.
(a) 1050 cm^{-1} into λ, in μm (micrometers).
(b) 510 nm (green light) into ν, in sec^{-1} (cycles sec^{-1}, or Hertz).
(c) 6.15 μm into $\tilde{\nu}$, in cm^{-1}.
(d) 2250 cm^{-1} into ν, in sec^{-1} (Hz).

3 Convert each of the following quantities into energies, in kcal mole^{-1}.
(a) A bond rotation of 750 wave numbers (cm^{-1}).
(b) A bond vibration of 2900 wave numbers (cm^{-1}).
(c) An electronic transition of 350 nm (ultraviolet light, capable of sunburn).
(d) A 20-Hz pipe-organ note.

(e) A 40,000-Hz dog-whistle note.

(f) The broadcast frequency of the audio signal of TV channel 6 (87.25 MHz).

(g) A "hard" X-ray with a 0.07-nm wavelength.

4 Calculate to three significant figures the amount of energy absorbed by a hydrogen when it undergoes an α-to-β spin flip in the field of a

(a) 21,150-Gauss magnet ($\nu = 90$ MHz).

(b) 137,500-Gauss magnet ($\nu = 500$ MHz).

5 The best modern NMR instruments can resolve resonance signals separated by as little as 0.15 Hz. How much energy does this correspond to in kcal $mole^{-1}$?

6 Sketch a hypothetical low-resolution NMR spectrum, showing the positions of the resonance peaks for all magnetic nuclei for each of the following molecules. Assume an external magnetic field of 21,150 Gauss.

(a) $CHCl_3$ (chloroform).

(b) $CFCl_3$ (Freon 11).

(c) $CF_3CHClBr$ (Halothane).

How would the spectra change if the magnetic field were 84,600 Gauss?

7 If the NMR spectra of the molecules in Problem 6 were recorded using high resolution for each nucleus, what differences would be observed?

8 The ^1H NMR spectrum of $CH_3COCH_2C(CH_3)_3$, 4,4-dimethyl-2-pentanone, taken at 90 MHz shows signals at the following positions: 92, 185, and 205 Hz, downfield from tetramethylsilane.

(a) What are the chemical shifts (δ) of these signals?

(b) What would their positions be in Hz, relative to tetramethylsilane, if the spectrum were recorded at 60 MHz? At 360 MHz?

(c) Assign each signal to a set of hydrogens in the molecule.

9 How many signals would be present in the ^1H NMR spectra of each molecule illustrated below? What would the *approximate* chemical shift be for each of these signals?

(a) $CH_3CH_2CH_2CH_3$

(b) $CH_3CHBrCH_3$

(c) 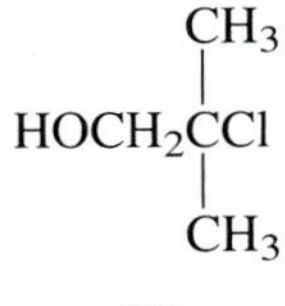 $HOCH_2CCl$ with two CH_3 groups on C: $HOCH_2C(CH_3)_2Cl$

(d) 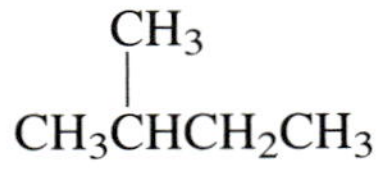 $CH_3CHCH_2CH_3$ with CH_3 on CH: $CH_3CH(CH_3)CH_2CH_3$

(e) 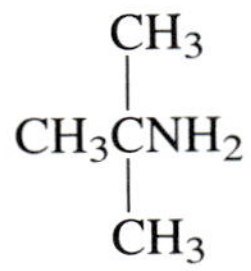 CH_3CNH_2 with two CH_3 groups on C: $CH_3C(CH_3)_2NH_2$

(f) $CH_3CH_2CH(CH_2CH_3)_2$

(g) $CH_3OCH_2CH_2CH_3$

(h) 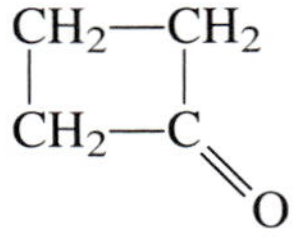 $CH_2—CH_2$ / $CH_2—C=O$ (four-membered ring)

(i) 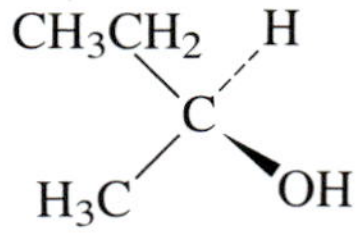 CH_3CH_2, H, H_3C, OH on C

(j) 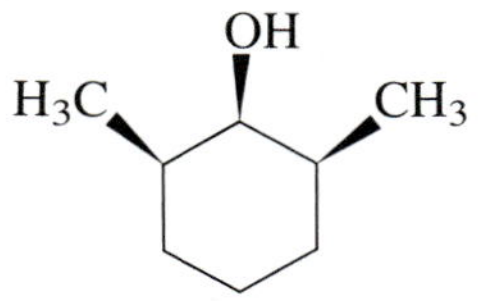 OH, H_3C, CH_3 on a cyclohexane ring

10 For each compound in each group of isomers shown below, indicate (1) the number of signals in the ^1H NMR spectrum, (2) the *approximate* chemical shift of each signal, and (3) the integration ratios for the signals. Finally, indicate whether all the isomers in each group can be distinguished from each other by means of these three pieces of information alone.

(a) $CH_3\overset{\overset{CH_3}{|}}{C}BrCH_2CH_3$, $BrCH_2\overset{\overset{CH_3}{|}}{C}HCH_2CH_3$, $CH_3\overset{\overset{CH_3}{|}}{C}HCH_2CH_2Br$

(b) $ClCH_2CH_2CH_2CH_2OH$, $CH_3\overset{\overset{CH_2Cl}{|}}{C}HCH_2OH$, $CH_3\overset{\overset{CH_3}{|}}{C}ClCH_2OH$

(c) $ClCH_2\overset{\overset{CH_3}{|}}{C}Br—\overset{\overset{CH_3}{|}}{C}HCH_3$, $ClCH_2\overset{\overset{CH_3}{|}}{C}H—\overset{\overset{CH_3}{|}}{C}BrCH_3$, $ClCH_2\underset{\underset{CH_3}{|}}{\overset{\overset{CH_3}{|}}{C}}—CHBrCH_3$, $ClCH_2CHBr\underset{\underset{CH_3}{|}}{\overset{\overset{CH_3}{|}}{C}}CH_3$

11 The ^1H NMR spectra for two haloalkanes are shown below. Propose structures for these compounds that are consistent with the spectra.

(a) $C_5H_{11}Cl$, spectrum A.

(b) $C_4H_8Br_2$, spectrum B.

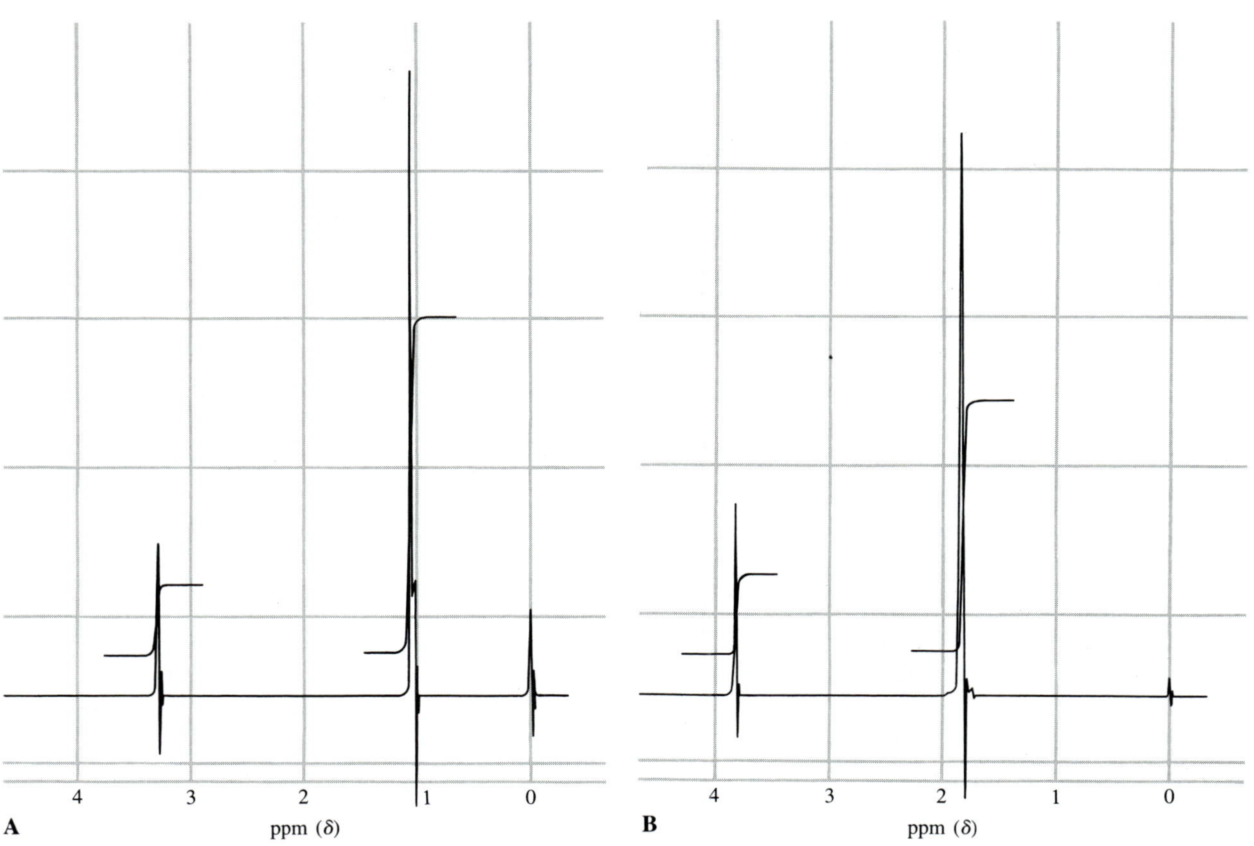

12 The ^1H NMR signals for three molecules with ether functional groups are given below. All the signals are single, sharp peaks. Propose structures for these compounds.

(a) $C_3H_8O_2$, $\delta = 3.3$ and 4.4 ppm (ratio 3:1).
(b) $C_4H_{10}O_3$, $\delta = 3.3$ and 4.9 ppm (ratio 9:1).
(c) $C_5H_{12}O_2$, $\delta = 1.2$ and 3.1 ppm (ratio 1:1).

Compare and contrast these spectra with that of 1,2-dimethoxyethane (Figure 10-13).

13 **(a)** The ^1H NMR spectrum of a ketone with the formula $C_6H_{12}O$ has $\delta = 1.2$ and 2.1 ppm (ratio 3:1). Propose a structure for this molecule.

(b) Each of two isomeric molecules related to the ketone in part **a** has the formula $C_6H_{12}O_2$. Their ^1H NMR spectra are as follows:

Isomer 1, $\delta = 1.5$ and 2.0 ppm (ratio 3:1).
Isomer 2, $\delta = 1.2$ and 3.6 ppm (ratio 3:1).

All signals in these spectra are sharp, single peaks. Propose structures for these two isomeric compounds. To what compound class do they belong?

14 ^1H NMR spectra C through G correspond to five isomeric alcohols with the formula $C_5H_{12}O$.

(a) As best you can, assign structures to the four alcohols represented by spectra C, D, E and F.

(b) Spectrum G is of 2-methyl-1-butanol. Identify any unusual characteristics of this spectrum; explain them on the basis of the molecule's structure.

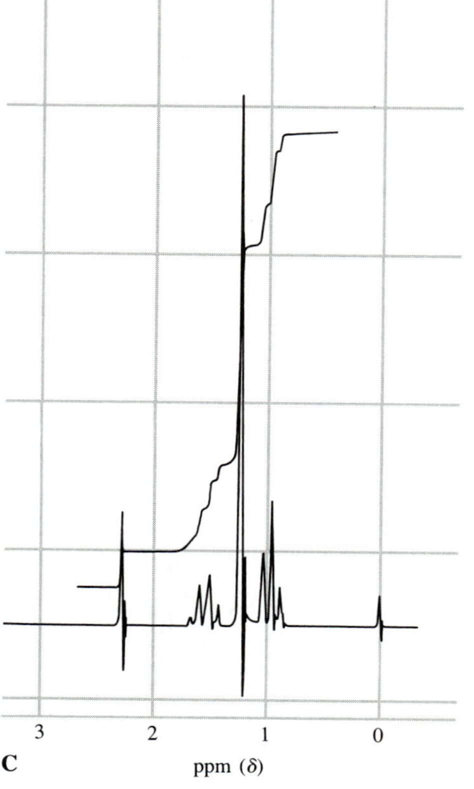

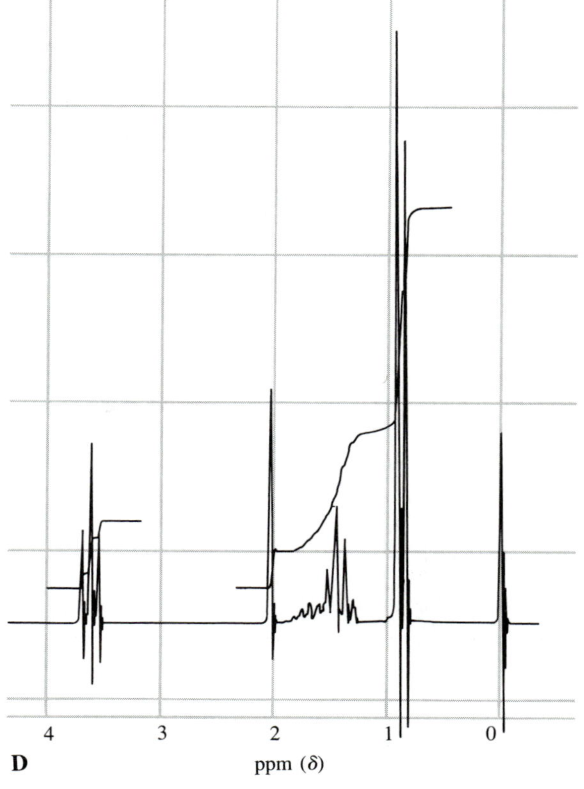

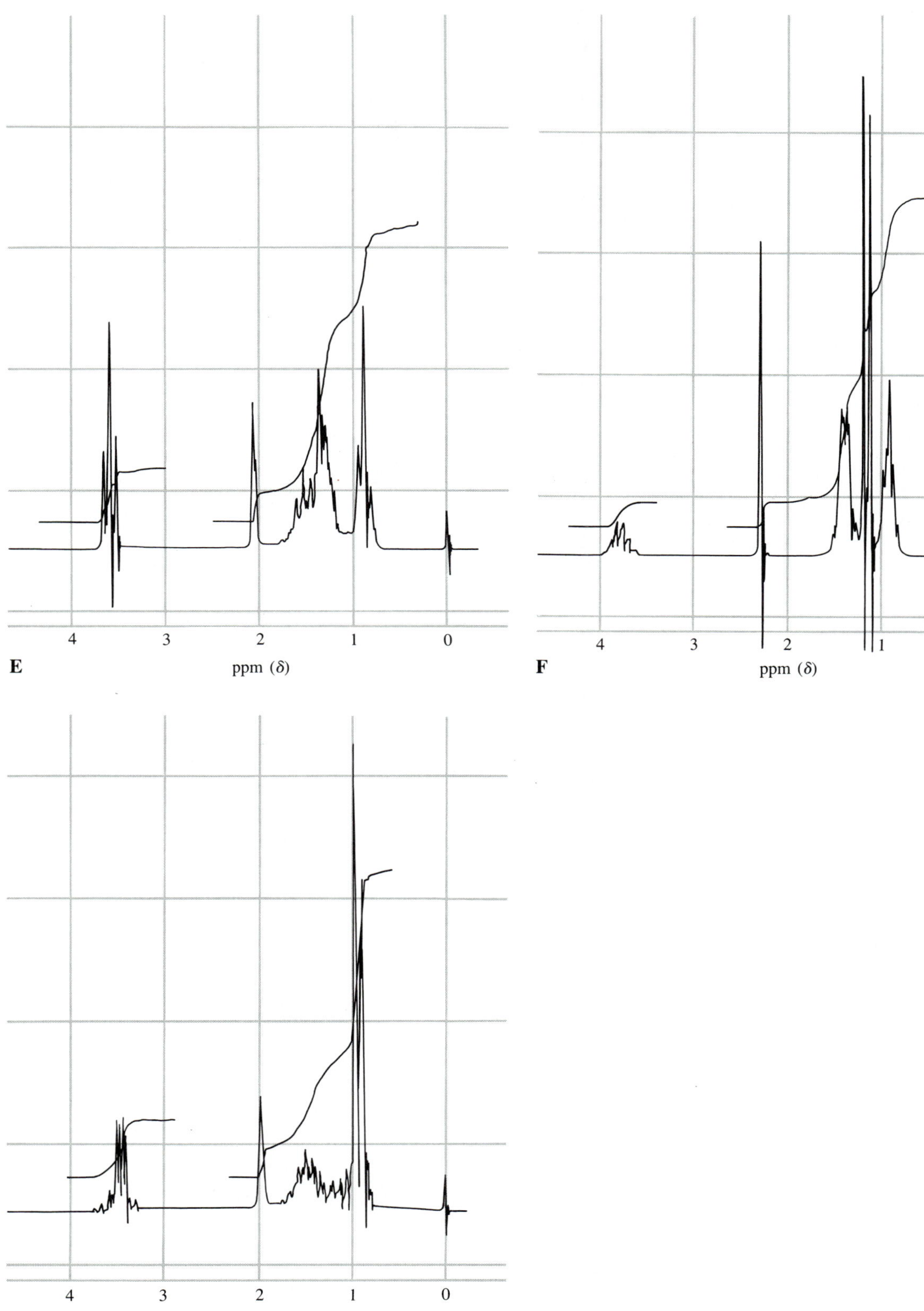
4
3
2
1
0
E
ppm (δ)
4
3
2
1
0
F
ppm (δ)
4
3
2
1
0
G
ppm (δ)

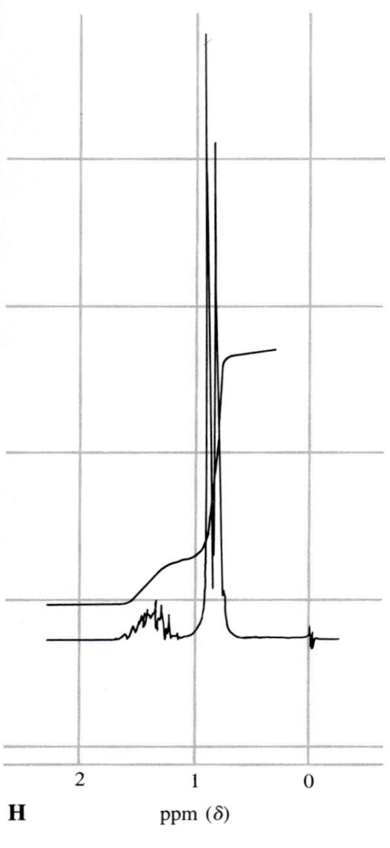

15 A hydrocarbon with the formula C_6H_{14} has ^1H NMR spectrum H. What is its structure? This molecule has a structural feature similar to another molecule whose spectrum is illustrated in this chapter. What molecule is that? Explain similarities and differences in the spectra of the two molecules.

16 The ^1H NMR spectrum of octane is shown at 500 MHz (spectrum I). Compare this with the 90-MHz spectrum, Figure 10-27, and explain the differences in appearance of the two spectra.

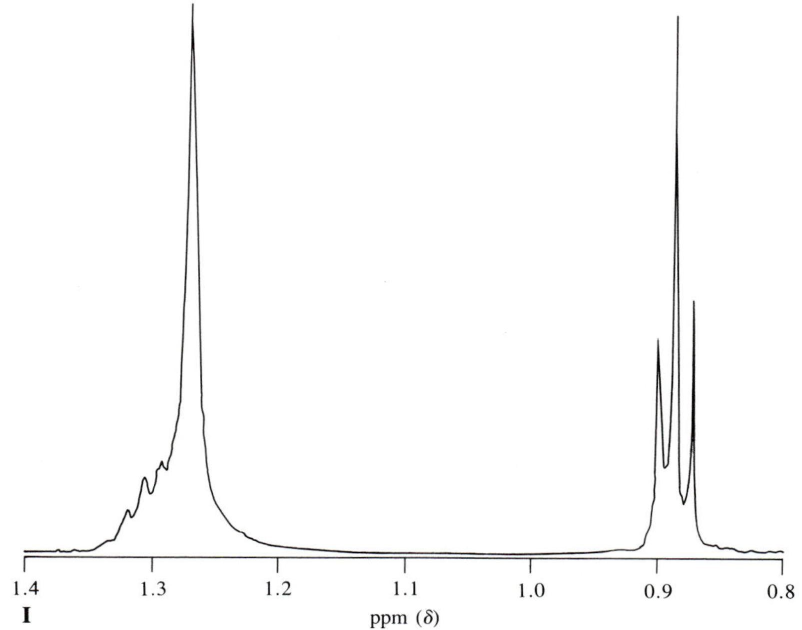

17 The ^1H NMR spectrum of 1-chloropentane is shown at 60 MHz (spectrum J) and 500 MHz (spectrum K). Explain the differences in appearance of the two spectra, and assign the signals to specific hydrogens in the molecule.

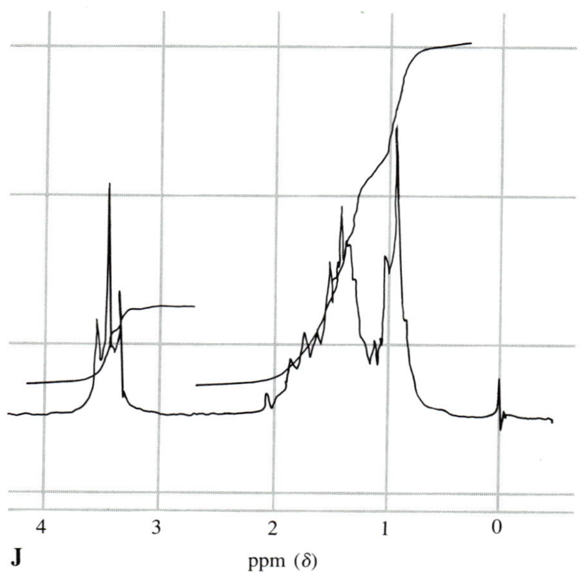

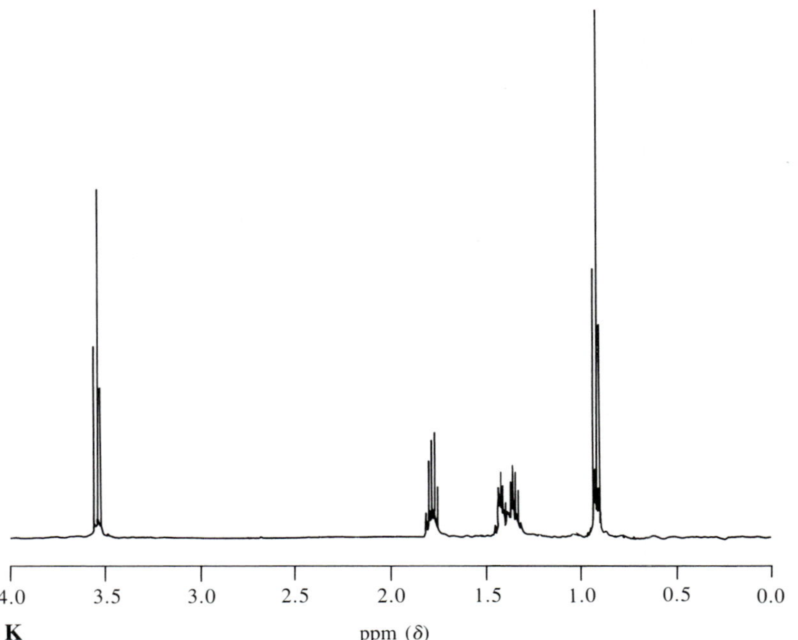

18 Can the three isomeric pentanes be distinguished unambiguously from their broad-band proton-decoupled ^{13}C NMR spectra *alone?* How about the five isomeric hexanes?

19 Predict the ^{13}C NMR spectra of the compounds in Problem 9, with and without proton decoupling.

20 For each group of isomeric compounds presented in Problem 10, answer the questions posed there on the basis of the ^{13}C NMR spectra that they should exhibit.

21 From each group of three molecules below, pick the one most consistent with the proton-decoupled ^{13}C NMR spectrum indicated. Explain your choices.

(a) $CH_3(CH_2)_4CH_3$, $(CH_3)_3CCH_2CH_3$, $(CH_3)_2CHCH(CH_3)_2$; $\delta = 19.5$ and 33.9 ppm.

(b) 1-Chlorobutane, 1-chloropentane, 3-chloropentane; $\delta = 13.2$, 20.0, 34.6, and 44.6 ppm.

(c) Cyclopentanone, cycloheptanone, cyclononanone; $\delta = 24.0$, 30.0, 43.5, and 214.9 ppm.

(d) $ClCH_2CHClCH_2Cl$, $CH_3CCl_2CH_2Cl$, $CH_2{=}CHCH_2Cl$; $\delta = 45.1$, 118.3, and 133.8 ppm.

22 Propose a reasonable structure for each molecule whose formula is given below, based on the indicated ^1H NMR and proton-decoupled ^{13}C NMR spectra.

(a) $C_7H_{16}O$, spectra L and M.
(b) $C_8H_{18}O_2$, spectra N and O.

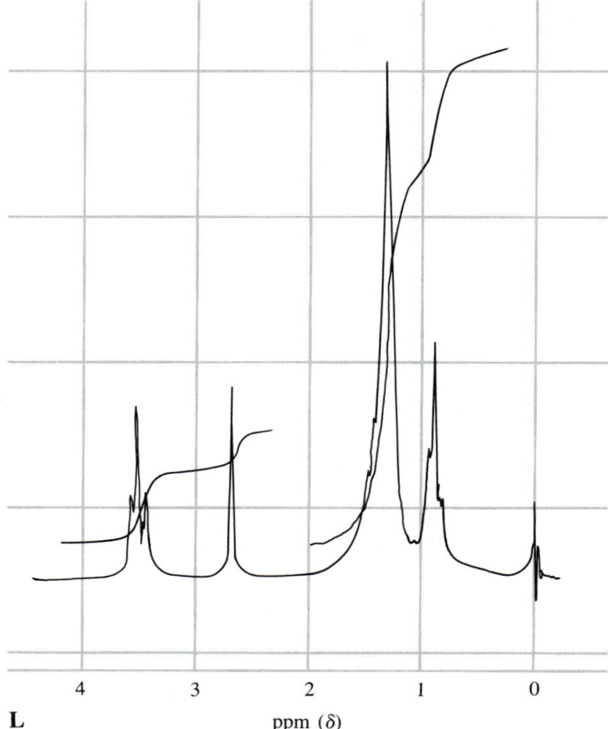

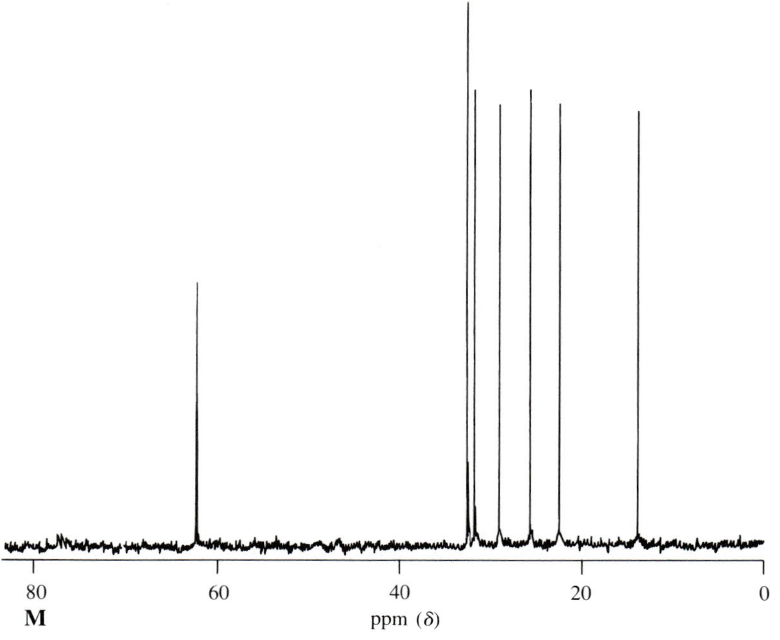

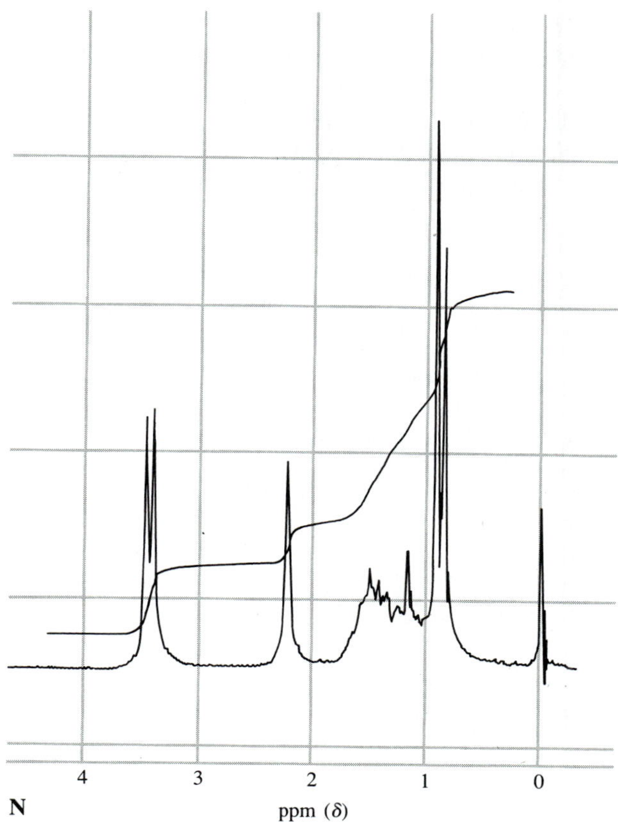

N

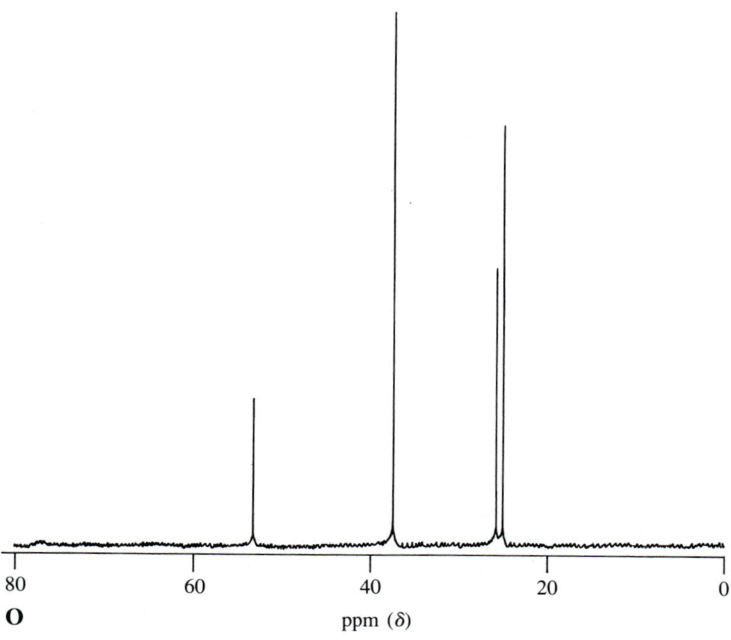

O

23 The ^{1}H NMR spectrum of cholesteryl benzoate, a derivative of a typical complex biological molecule, is shown in spectrum P. In spite of its complexity, a number of distinguishing features may be identified. Explain as many of the peaks or peak patterns in this spectrum as you can. The inset is a 2× expansion of the signal at $\delta = 4.8$ ppm. Why is this signal so complex? By comparison, can you explain the simplicity of the signal at $\delta = 5.4$ ppm?

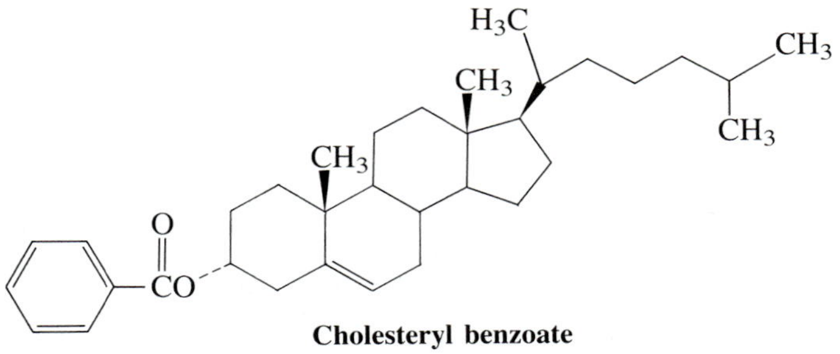

Cholesteryl benzoate

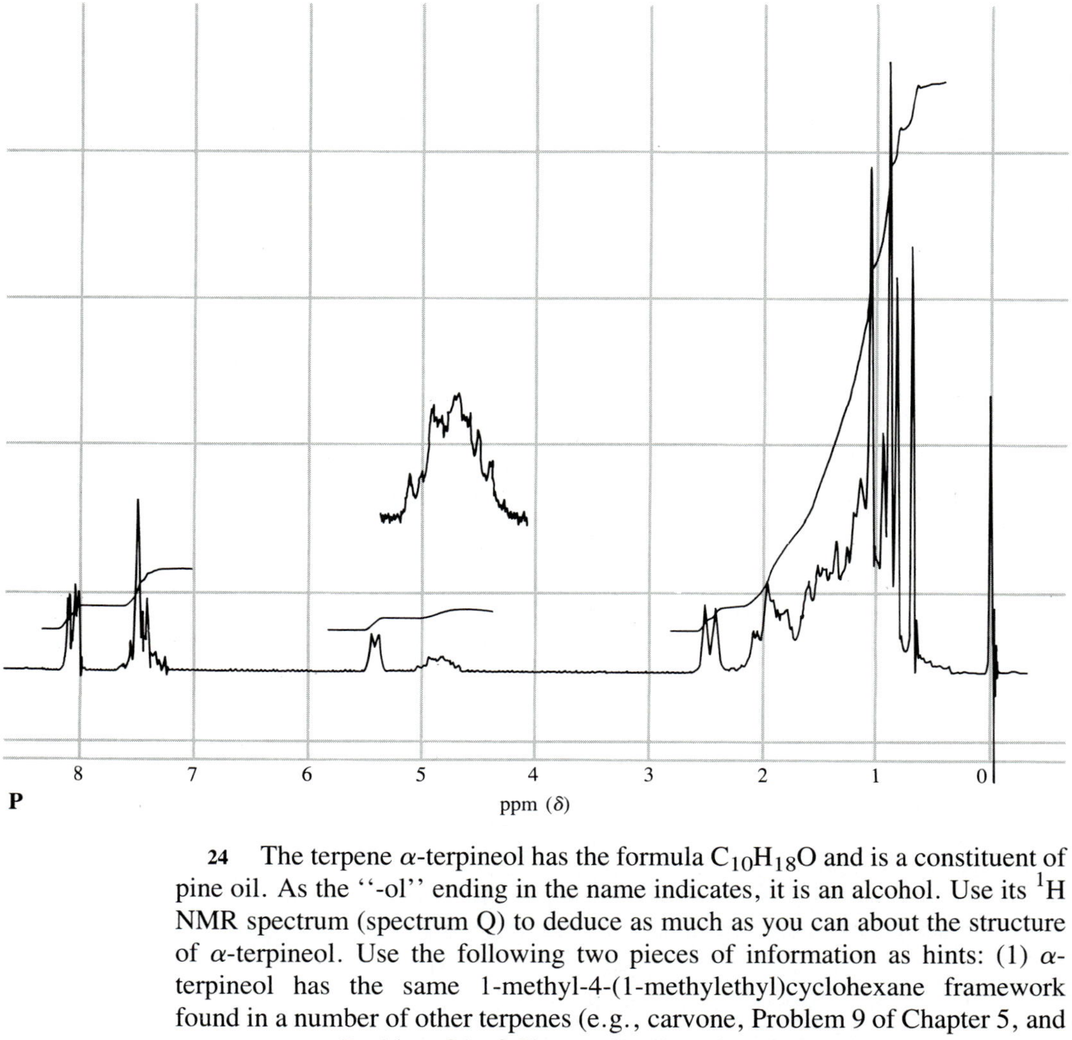

P

24 The terpene α-terpineol has the formula $C_{10}H_{18}O$ and is a constituent of pine oil. As the ''-ol'' ending in the name indicates, it is an alcohol. Use its ^{1}H NMR spectrum (spectrum Q) to deduce as much as you can about the structure of α-terpineol. Use the following two pieces of information as hints: (1) α-terpineol has the same 1-methyl-4-(1-methylethyl)cyclohexane framework found in a number of other terpenes (e.g., carvone, Problem 9 of Chapter 5, and *para*-cymene, Problem 21 of Chapter 3); (2) rather than trying to interpret this

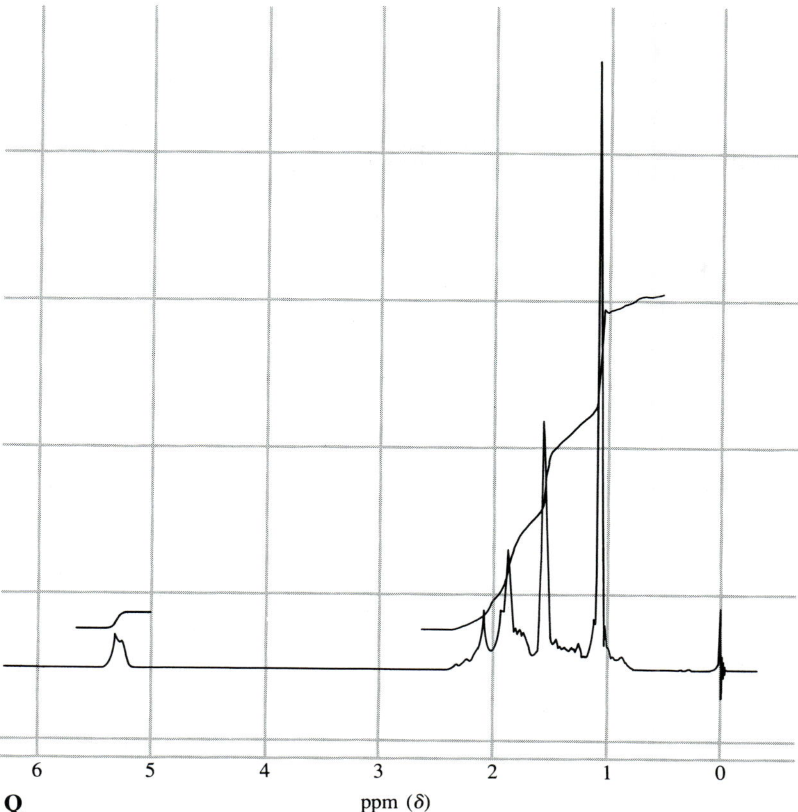

complicated NMR spectrum fully, concentrate on the most obvious features (peaks at δ = 1.1, 1.6, and 5.3 ppm) and use chemical shifts, integrations, and splitting patterns (if any) to help you.

25 Study of the solvolysis reactions of derivatives of menthol [5-methyl-2-(1-methylethyl)-1-cyclohexanol] has greatly enhanced our understanding of these types of reactions. Heating of the 4-methylbenzenesulfonate ester of the particular menthol isomer shown below in 2,2,2-trifluoroethanol (a highly ionizing solvent of low nucleophilicity) leads to two products with the formula $C_{10}H_{18}$.

CH_3

OSO_2—⟨benzene ring⟩—CH_3 $\xrightarrow{CF_3CH_2OH,\ \Delta}$ two $C_{10}H_{18}$ products

(D)H $CH(CH_3)_2$

(a) The major product displays ten different signals in its ^{13}C NMR spectrum. Two of the carbon signals are downfield of δ = 100 ppm, at about δ = 120 and δ = 145 ppm, respectively. The ^{1}H NMR spectrum has a one-hydrogen multiplet near δ = 5 ppm; otherwise all the signals are upfield of δ = 3 ppm. Identify the product.

(b) The minor product has only seven ^{13}C signals. Again, two are at low field ($\sim \delta$ = 125 ppm and δ = 140 ppm). However, no ^{1}H NMR signals are found at lower field than $\delta = 3$ ppm. Identify this product and explain its formation mechanistically.

(c) If the starting sulfonate ester has deuterium in place of hydrogen at C-2 of the cyclohexane ring, two products are again formed. The major product has virtually the same ^{13}C spectrum as described in part **a,** but the ^{1}H NMR signal at $\delta = 5$ ppm is much smaller than it should be. How might this result be explained? Hint: The mechanism of formation of the minor product (part **b**) plays a role.

CHAPTER 11

Alkenes

HYDROCARBONS CONTAINING DOUBLE BONDS

A carbon–carbon double bond is the functional group characteristic of the **alkenes** (see also Section 2-1). An older name for this class of compounds is **olefins** (*oleum,* Latin, oil; *facere,* Latin, to make). This name arose because many gaseous alkenes give oils by addition reactions to the double bond. The general formula of the alkenes is C_nH_{2n}, the same as that for the cycloalkanes.

This chapter deals with the naming of alkenes, their physical properties, how the stability of the double bond depends on its substitution pattern, and the preparation of alkenes. The last topic affords an opportunity to review and augment what we know about elimination reactions.

11-1 The Naming of Alkenes

As is true of other classes of organic compounds, the common names of some of the alkenes are still in use. In common names, the **-ane** ending of the alkane is replaced by **-ylene.** Substituent names are added as prefixes.

Common Names of Alkenes

$CH_2{=}CH_2$	$CH_2{=}C(CH_3)H$	$Cl_2C{=}C(Cl)H$
Ethylene	**Propylene**	**Trichloroethylene**

In IUPAC nomenclature, the simpler ending **-ene** is used instead of -ylene, as in ethene and propene. More-complicated systems require adaptations and extensions of the rules for naming alkanes (Section 2-3):

RULE 1 Find the longest chain that *includes* the functional group—in this case, *both* carbons making up the double bond. The molecule may have longer carbon chains, but they are ignored in the naming of the stem.

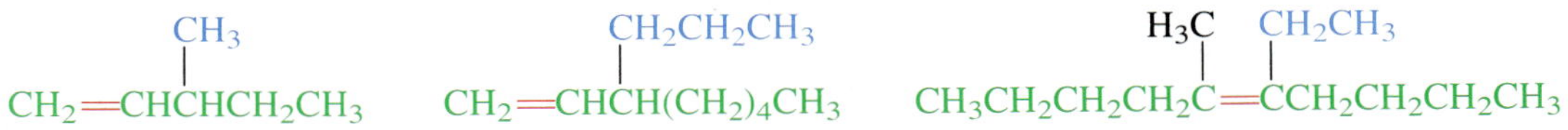

$CH_2{=}CHCH(CH_3)CH_2CH_3$
A methylpentene

$CH_2{=}CHCH(CH_2CH_2CH_3)(CH_2)_4CH_3$
A propyloctene
(Not a hexene derivative)

$CH_3CH_2CH_2CH_2C(CH_3){=}C(CH_2CH_3)CH_2CH_2CH_2CH_3$
An ethylmethyldecene
(Not a pentene or a heptene or an octene derivative)

RULE 2 The location of the double bond in the main chain is indicated by number. Numbering of the chain starts at the end *closer* to the double bond. In cycloalkenes, carbons 1 and 2 are defined to be part of the double bond. Alkenes of the same molecular formula differing only in the location of the double bond (as in 1-butene and 2-butene) are called *double-bond isomers*. A 1-alkene is also referred to as a *terminal* alkene; the others are labeled *internal*. Note that alkenes are easily depicted in line notation.

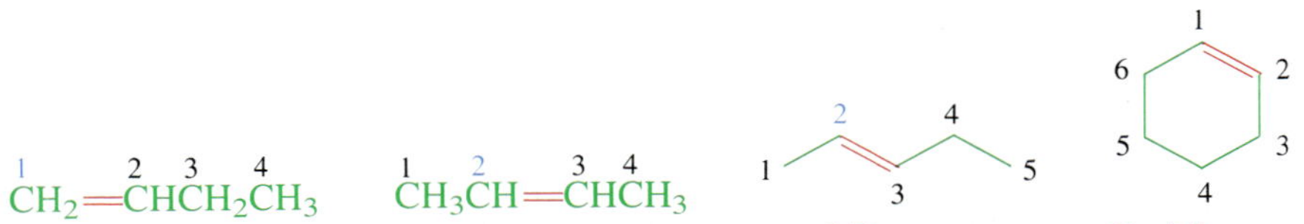

$\overset{1}{C}H_2{=}\overset{2}{C}H\overset{3}{C}H_2\overset{4}{C}H_3$
1-Butene
(Not 3-butene)
A terminal alkene

$\overset{1}{C}H_3\overset{2}{C}H{=}\overset{3}{C}H\overset{4}{C}H_3$
2-Butene
(An internal alkene and a double-bond isomer of 1-butene)

2-Pentene
(Not 3-pentene)

Cyclohexene

RULE 3 The substituents and their positions are added to the alkene name as prefixes. If the alkene stem is symmetrical, number in a way that gives the first substituent along the chain the lowest possible assignment.

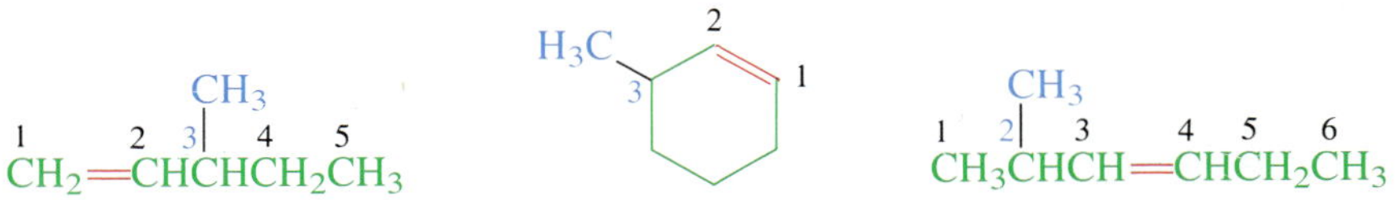

$\overset{1}{C}H_2{=}\overset{2}{C}H\overset{3}{C}H(CH_3)\overset{4}{C}H_2\overset{5}{C}H_3$
3-Methyl-1-pentene

3-Methylcyclohexene
(Not 6-methylcyclohexene)

$\overset{1}{C}H_3\overset{2}{C}H(CH_3)\overset{3}{C}H{=}\overset{4}{C}H\overset{5}{C}H_2\overset{6}{C}H_3$
2-Methyl-3-hexene
(Not 5-methyl-3-hexene)

EXERCISE 11-1

Name the following two alkenes:

(a) 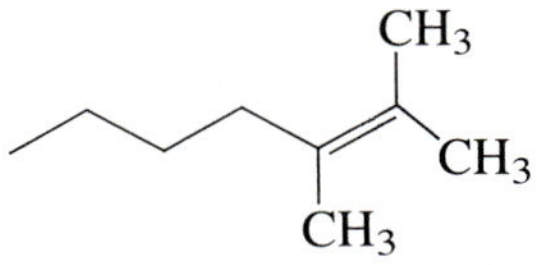 [line structure with CH_3, CH_3, CH_3 substituents]; (b) 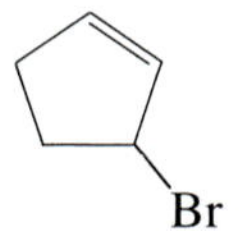 [cyclopentene with Br]

RULE 4 In a 1,2-disubstituted ethene, the two substituents may be on the same side of the double bond or on opposite sides. The first stereochemical arrangement is called cis, and the second trans, in analogy to the cis-trans names of the disubstituted cycloalkanes. Two alkenes of the same molecular formula differing only in their stereochemistry are called cis-trans isomers and are examples of structural isomers.

***cis*-2-Butene** | ***trans*-2-Butene** | **4-Chloro-*cis*-2-pentene**

EXERCISE 11-2

Name the following two alkenes:

(a) ; (b)

In the smaller substituted cycloalkenes, the double bond can exist only in the cis configuration. The trans arrangement is prohibitively strained (as building a model reveals). However, in larger cycloalkenes, trans isomers are stable (see Section 11-4).

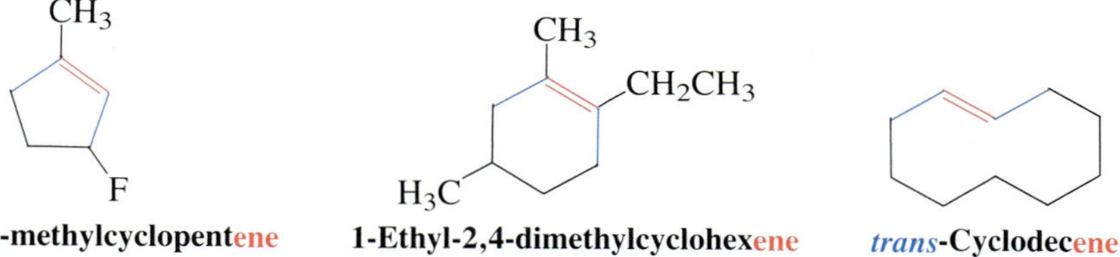

3-Fluoro-1-methylcyclopentene | **1-Ethyl-2,4-dimethylcyclohexene** | ***trans*-Cyclodecene**

(In both cases, only the cis isomer is stable.)

RULE 5 The labels cis and trans are more difficult to assign unambiguously to a compound in which each carbon participating in the double bond bears only one or no hydrogen. An alternative and more general system for naming substituted (particularly tri- and tetrasubstituted) alkenes has been adopted by IUPAC: the *E,Z system*. In this convention, the sequence rules devised for establishing priority in *R,S* names (Section 5-3) are applied separately to the two groups on each double-bonded carbon (the geminal groups). If the two groups of highest priority are on opposite sides of the double bond, the molecule is of the *E* configuration (E from *entgegen,* German, opposite). If the two substituents of highest priority appear on the same side of the double bond, the molecule is of the *Z* form (Z from *zusammen,* German, together).

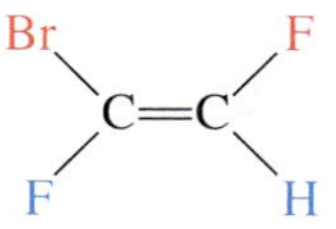

$$\underset{\text{(}Z\text{)-1-Bromo-1,2-difluoroethene}}{\begin{array}{ccc} Br & & F \\ & C{=}C & \\ F & & H \end{array}} \qquad \underset{\text{(}E\text{)-1-Chloro-3-ethyl-4-methyl-3-heptene}}{\begin{array}{ccc} CH_3CH_2 & & CH_2CH_2CH_3 \\ & C{=}C & \\ ClCH_2CH_2 & & CH_3 \end{array}}$$

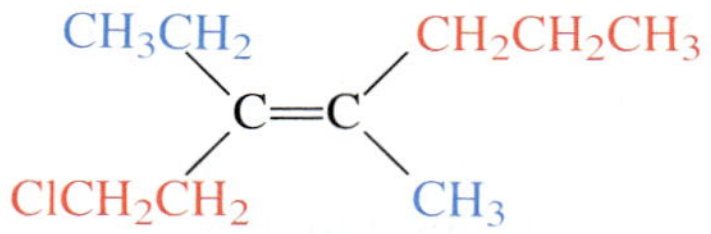

EXERCISE 11-3

Name the following two alkenes:

$$\text{(a)} \quad \begin{array}{ccc} D & & D \\ & C{=}C & \\ H_3C & & H \end{array} ; \qquad \text{(b)} \quad \begin{array}{ccc} F & & OCH_3 \\ & C{=}C & \\ H_3C & & CH_2CH_3 \end{array} .$$

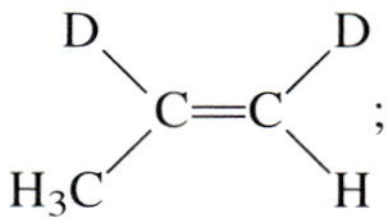

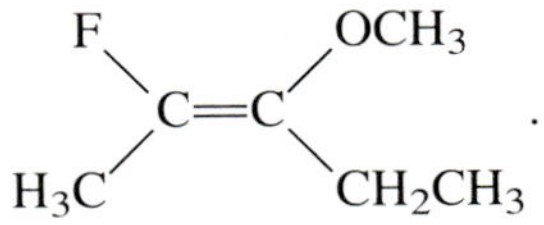

RULE 6 The hydroxy functional group has priority over the double bond. Alcohols containing double bonds in their molecular framework are named as alkenols, and the stem incorporating both functions is numbered to give the carbon bearing the OH group the lowest possible assignment. Note that the last *e* in alkene is dropped in the naming of alkenols.

$$\underset{\substack{\textbf{2-Propen-1-ol} \\ \text{(Not 1-propen-3-ol)}}}{CH_2{=}CHCH_2OH} \qquad \underset{\substack{\textbf{(Z)-5-Chloro-3-ethyl-4-hexen-2-ol} \\ \text{(The two stereocenters are} \\ \text{unspecified.)}}}{\begin{array}{ccc} & & OH \\ & & | \; 2 \\ & & \overset{1}{C}H_3CH \\ & & | \; 3 \\ Cl & & CHCH_2CH_3 \\ & \overset{5}{C}{=}\overset{4}{C} & \\ \overset{6}{H_3C} & & H \end{array}}$$

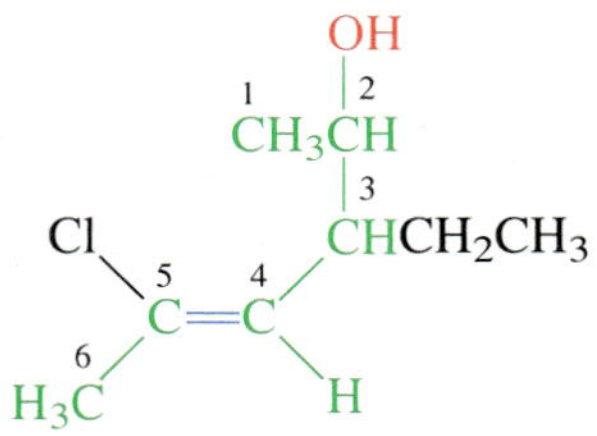

EXERCISE 11-4

Draw the structures of the following molecules: (a) *trans*-3-penten-1-ol; (b) 3-cyclohexenol.

RULE 7 Substituents containing a double bond are named alkenyl, such as ethenyl (common name, vinyl), 2-propenyl (allyl), and *cis*-1-propenyl.

$$\underset{\substack{\textbf{Ethenyl} \\ \textbf{(Vinyl)}}}{CH_2{=}CH{-}} \qquad \underset{\substack{\textbf{2-Propenyl} \\ \textbf{(Allyl)}}}{CH_2{=}CH{-}CH_2{-}} \qquad \underset{\textbf{\textit{cis}-1-Propenyl}}{\begin{array}{ccc} H & & H \\ & C{=}C & \\ H_3C & & \diagdown \end{array}}$$

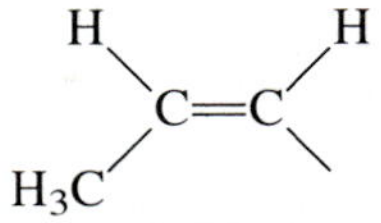

As usual, the numbering of a substituent chain begins at the point of attachment to the basic stem:

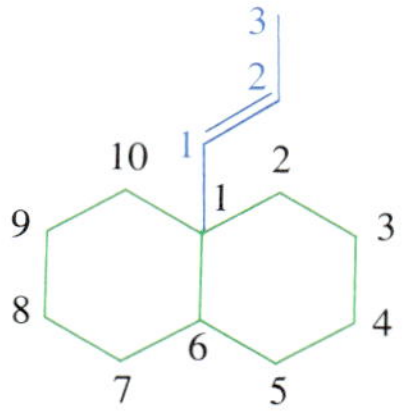

1-(*trans*-1-Propenyl)-bicyclo[4.4.0]decane

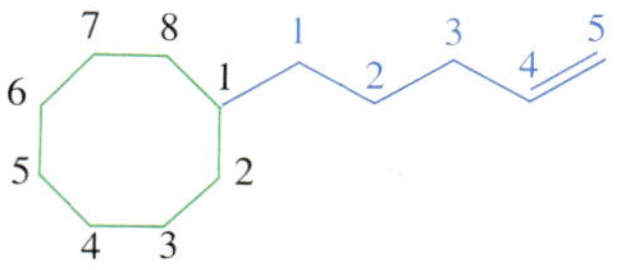

4-Pentenylcyclooctane

$\overset{1}{C}H_2\!=\!\overset{2}{C}H\overset{3}{C}H_2O\overset{1}{C}H_2\overset{2}{C}H\!=\!\overset{3}{C}H_2$

3-(2-Propenyloxy)-1-propene
(Diallyl ether)

EXERCISE 11-5

(a) Draw the structure of *trans*-1,2-diethenylcyclopropane; (b) name the structure shown in the margin.

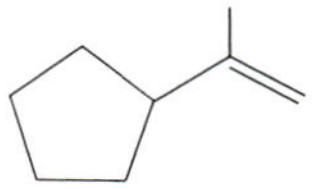

11-2 Structure and Bonding in the Alkenes

By virtue of their carbon–carbon double bonds, alkenes have special electronic and structural features. This section describes these aspects—in particular, hybridization; the nature of the two bonds defined as σ and π; the strength of the π bond; the molecular structure of ethene; and some physical properties of alkenes.

The Nature of the Double Bond in Ethene

In ethene, which contains the simplest double bond, both carbon atoms are sp^2 hybridized, as in the methyl radical (Section 3-2, Figure 3-2). In $CH_3\cdot$, three sp^2 hybrid orbitals overlap with the hydrogen 1*s* orbitals (Figure 11-1A) to form three σ bonds (Section 1-7). In addition, one singly occupied perpendicular *p*

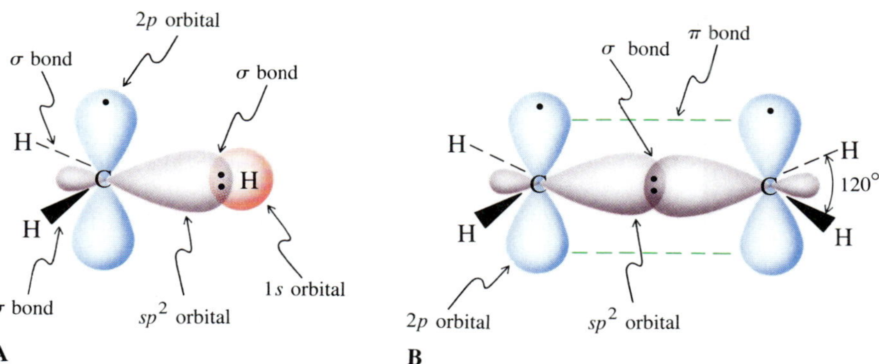

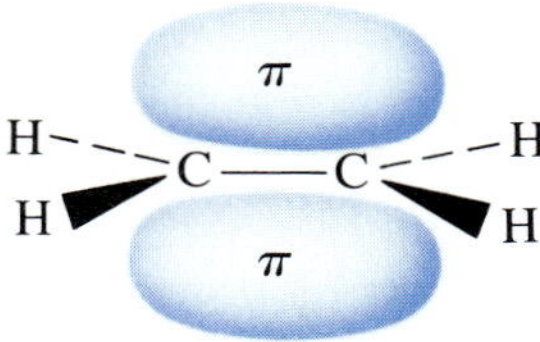

FIGURE 11-1
Molecular-orbital pictures of (A) the methyl radical, in which only one C–H bond is shown in orbital terms, and (B) ethene. The sp^2–sp^2 carbon–carbon bond is called σ; the pair of *p* orbitals perpendicular to the molecular plane overlap to form the additional π bond. For clarity, this overlap is indicated by the dashed green line, the orbital lobes shown as being some distance apart. Another way of presenting the π bond is shown in part C, in which the "π electron cloud" is above and below the molecular plane.

orbital extends in both directions. We can combine two sp^2-hybridized carbons by using two sp^2 hybrid orbitals for the carbon–carbon σ bond. This leaves two p orbitals that are aligned parallel but close enough to enter into overlap (Figure 11-1B). This type of interaction is called a π bond and is typical of the double bonds in alkenes (see also Figure 1-13D). The π electrons are delocalized over both carbons above and below the molecular plane, as indicated in Figure 11-1C.

FIGURE 11-2

Overlap between two sp^2 hybrid orbitals to form the σ bond of ethene. In-phase interaction, which is between regions of the wave function having the same sign, arbitrarily chosen as (+) [it could have been two (−) lobes; recall that this is not a charge], reinforces bonding (positive overlap). Both electrons end up occupying this level in a paired manner. The picture of the resulting molecular orbital σ is indicative of the relatively high probability of finding the σ electrons along the internuclear axis. Out-of-phase interaction—that is, between regions of the wave function of opposite sign—results in an unfilled antibonding orbital, usually designated by an asterisk, in this case σ^*. The antibonding molecular orbital is characterized by a node, across which the sign of the wave function changes, as shown in the drawing. The stabilization energy ΔE_σ on formation of the σ bond is larger than ΔE_π obtained in the generation of a π bond (see Figure 11-3).

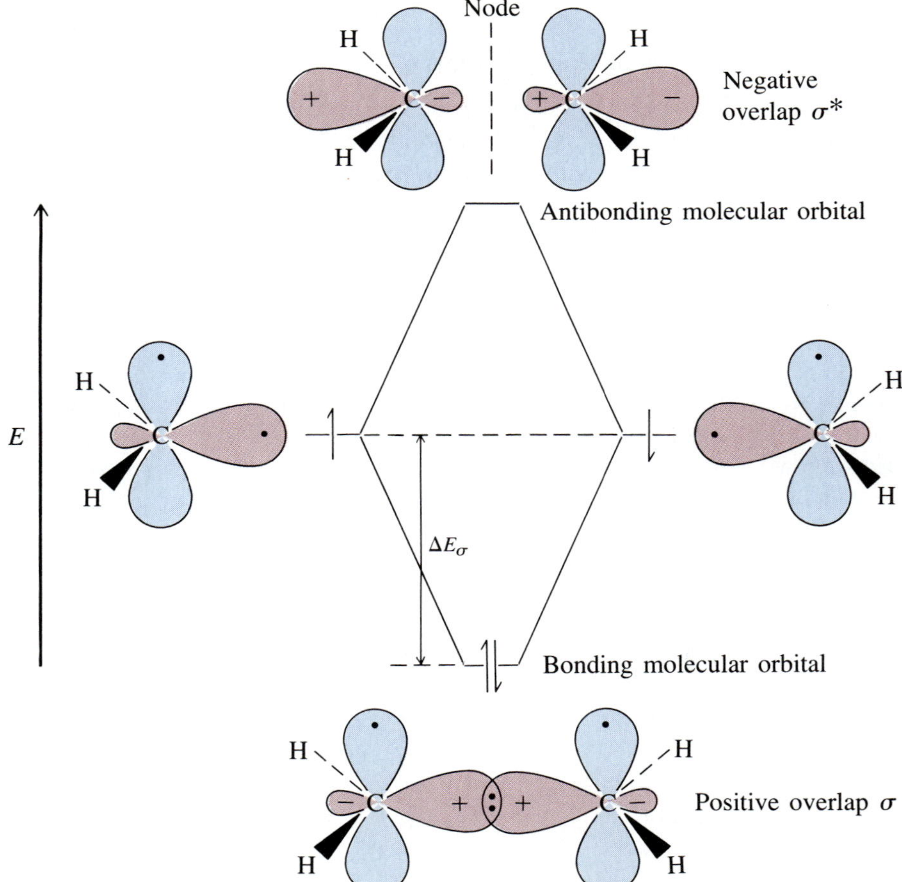

The Relative Strength of Sigma and Pi Bonds

As our orbital picture shows, the double bond is made up of two different types of bonds: a σ bond and a π bond. How much does each contribute to the total double-bond strength? We know from Section 1-7 that bonds are made by overlap of orbitals and that their relative strengths depend on the effectiveness of this overlap. Therefore, we can expect overlap in a σ bond to be considerably better than that in a π bond because of the directionality of the orbitals (Figure 11-1). Thus, a π bond should be weaker than a σ bond. This situation is illustrated in energy-level-interaction diagrams (Figures 11-2 and 11-3) analogous to those used to describe the bonding in the hydrogen molecule (Figures 1-11 and 1-12). There are two types of interactions for each of the two types of orbitals: bonding and antibonding. In the first, the two orbitals overlap in a reinforcing way (in phase) to form the σ and π bonds; in the second, they overlap in a counteracting way (out of phase) to give the antibonding combinations of σ^* and π^*. The shapes of the molecular orbitals in both bonding and antibonding interactions are shown in Figures 11-2 and 11-3. Because π overlap is less efficient than σ overlap, the difference in energy ΔE_{π} is smaller than ΔE_{σ}. The bonding molec-

FIGURE 11-3

Overlap of two p orbitals to form a π bond in ethene (the sp^2 orbitals are ignored). In-phase interaction results in positive overlap and the π molecular orbital being filled, whereas out-of-phase mixing in the empty π^* level contains a node. The picture of the bonding π orbital is indicative of the probability of finding the π electrons between the two sp^2 hybridized carbons above and below the molecular plane. The stabilization energy ΔE_{π} is smaller than ΔE_{σ} (see Figure 11-2).

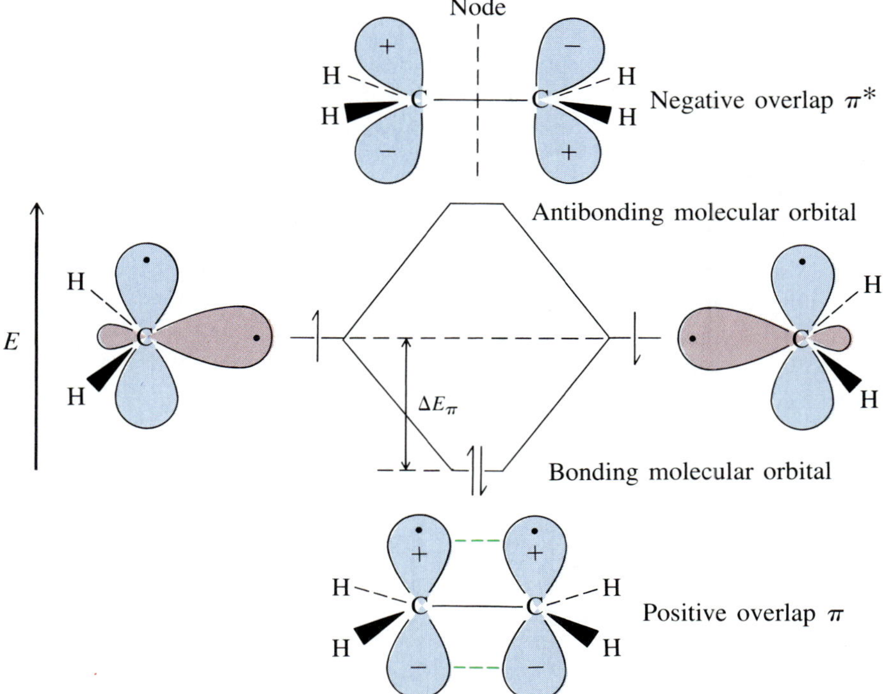

FIGURE 11-4
Energy ordering of the molecular orbitals making up the double bond. The four electrons occupy only bonding orbitals.

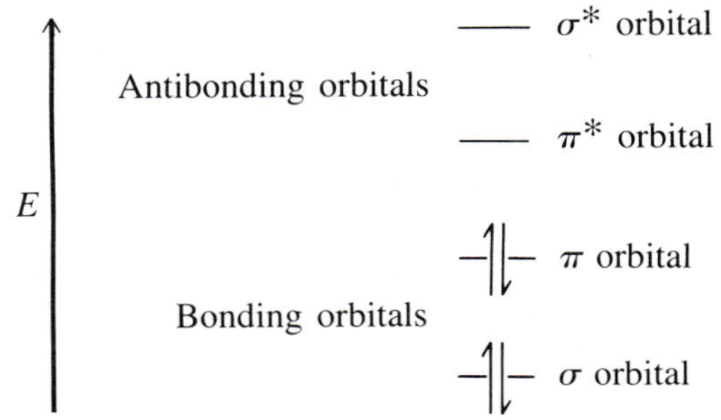

ular orbital of the σ bond is lower in energy than that of the corresponding π bond. This bonding difference has chemical and physical consequences. For example, ejection of an electron from an alkene, as measured by the ionization potential (Section 1-3), is from the higher-lying π level, because the π electrons are less strongly bound than the σ electrons. The molecular-orbital ordering in a double bond is given in Figure 11-4.

The relative strengths of the π and σ bonds can be estimated by measuring the energy required to effect thermal cis-trans isomerization of a specifically substituted alkene—for example, 1,2-dideuterated ethene. In this process, the two p orbitals responsible for the π bond are rotated 180°. At the midpoint of this rotation, 90°, the two p orbitals are perpendicular to one another and consequently are not overlapping; this is the transition state of the isomerization (Figure 11-5). At this stage, the π bond (but not the σ bond) has been broken; thus, the activation energy for the reaction can be thought of as being roughly equal to the π energy of the double bond.

Experimentally, such a purely thermal isomerization requires fairly high temperatures (400°–500°C) to take place at measurable rates. The activation energy is 65 kcal $mole^{-1}$, a value usually assigned to the strength of the π

FIGURE 11-5
Thermal isomerization of *cis*-dideuterioethene to the trans isomer. The reaction proceeds from starting material (A) through rotation around the C–C bond until it reaches the point of highest energy, the transition state (B). At this stage, the two p orbitals used to construct the π bond are perpendicular to each other. Further rotation in the same direction results in a product in which the two deuterium atoms are trans (C).

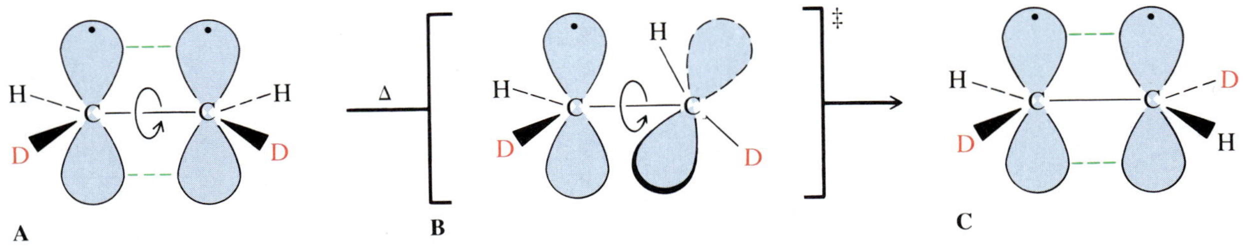

bond. At temperatures below 300°C, most double bonds are configurationally stable; that is, cis stays cis and trans remains trans. The strength of the double bond as a whole in ethene—in other words, the energy required for dissociation into two methylene fragments—is estimated to be 173 kcal $mole^{-1}$; consequently, the σ bond in this molecule has an energy content of about 108 kcal $mole^{-1}$ (Figure 11-6). The bond between an alkyl substituent or a hydrogen atom and the alkenyl carbon is unusually strong in comparison with the analogous bonds in alkanes (Table 3-2), the differences amounting to more than 10 kcal $mole^{-1}$. To a large extent, this effect is due to the improved overlap between the relatively tight sp^2 hybrids and the substituent orbitals. As a consequence, free-radical reactions of alkenes do not occur by abstraction of the strongly bound vinyl hydrogen. In fact, most of the chemistry of the double bond is characterized by the reactivity of the weaker bond: the π bond (Chapter 12).

FIGURE 11-6

Approximate bond strengths in an alkene (in kcal $mole^{-1}$).

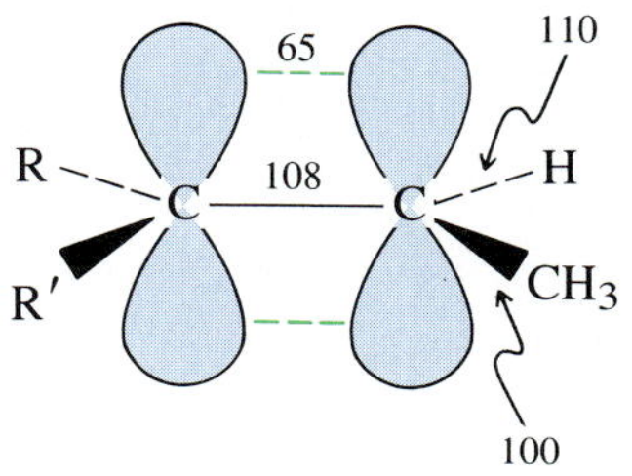

The Molecular Structure of Ethene Is due to Its Hybridization

The utilization of sp^2 orbital hybrids in the formation of a double bond has significant structural consequences. Ethene is planar, with two trigonal carbon atoms and bond angles close to 120° (Figure 11-7). Therefore, the double bond exhibits no three-dimensional stereochemistry but only that related to cis-trans isomerism.

FIGURE 11-7

The molecular structure of ethene.

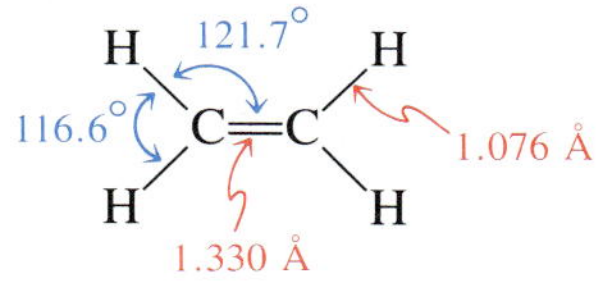

Some Physical Constants of Alkenes

There is nothing unusual about the boiling points of the alkenes, which lie very close to those of the corresponding alkanes. Ethene (b.p. −103.7°C), propene (b.p. −47.4°C), and the butenes are gases at room temperature. The pentenes boil just above room temperature (b.p. 30°–40°C).

Depending on their structure, alkenes may have dipole moments, because the alkenyl–carbon bond is polarized in the direction of the sp^2-hybridized carbon. Polarization is due to the fact that the percentage of s character in an sp^2 hybrid orbital is larger than that in an sp^3. On the average, electrons in orbitals with increased s character are held closer to the nucleus than those in orbitals containing more spatially directed p character. This effect makes the sp^2 carbon relatively electron withdrawing and creates a dipole along an alkenyl–carbon bond. Often, particularly in cis-disubstituted alkenes, a net molecular dipole is the result. In trans-disubstituted alkenes, such dipole moments are small, because the polarization of individual local bonds is in opposite directions and thus the dipoles may cancel each other.

Polarization in Alkenes

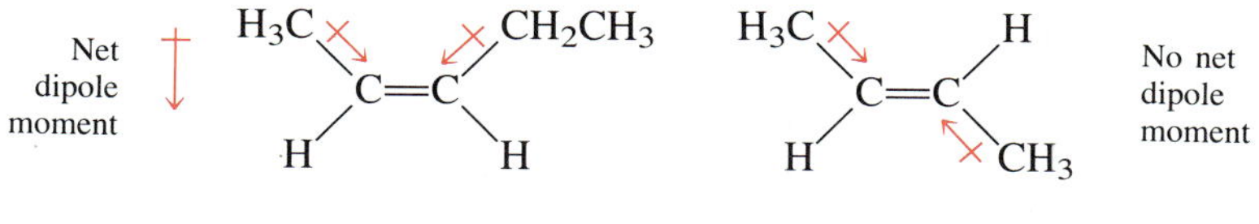

Another consequence of the electron-attracting property of the sp^2 carbon is the increased acidity of the vinyl hydrogen. Whereas ethane has an approximate pK_a of 50, ethene is somewhat more acidic with a pK_a of 44. Even so, ethene is a very poor source of protons compared with other compounds, such as the carboxylic acids or even alcohols.

Acidity of the Vinylic Hydrogen

$$CH_3\text{—}CH_2\text{—}H \underset{K\,\sim\,10^{-50}}{\rightleftharpoons} CH_3\text{—}\ddot{C}H_2^{\,-} + H^+$$

Ethyl anion

$$CH_2\text{=}CH_2 \underset{K\,\sim\,10^{-44}}{\rightleftharpoons} CH_2\text{=}\ddot{C}H^- + H^+$$

Ethenyl (vinyl) anion

EXERCISE 11-6

Ethenyllithium (vinyllithium) is not generally prepared by direct deprotonation of ethene but rather from chloroethene (vinyl chloride) or tetraethenyltin (tetravinyltin) by (trans)-metallation (Section 8-5):

$$CH_2\text{=}CHCl + 2\,Li \xrightarrow{(CH_3CH_2)_2O} CH_2\text{=}CHLi + LiCl \quad 60\%$$

$$4\,CH_3CH_2CH_2CH_2Li + (CH_2\text{=}CH)_4Sn \xrightarrow{THF} 4\,CH_2\text{=}CHLi + (CH_3CH_2CH_2CH_2)_4Sn$$

On treatment of ethenyllithium with propanone (acetone) followed by aqueous work-up, a colorless liquid is obtained in 74% yield. Propose a structure.

In summary, the characteristic hybridization scheme for the double bond of an alkene determines its physical and electronic features. This hybridization is responsible for the formation of a strong σ, as well as a weaker π, bond; for stable cis and trans isomers; and for the strength of the alkenyl–substituent bond. It gives rise to the planarity of the double bond, incorporating trigonal carbon atoms; the presence of polar bonds; and the somewhat increased acidity of the alkenyl hydrogen compared with an alkyl one.

11-3 The Double Bond Exerts Characteristic Deshielding: Nuclear Magnetic Resonance of Alkenes

The ^1H and ^{13}C NMR spectra of alkenes are highly diagnostic of the presence of the functional group (see Tables 10-2 and 10-6). We shall see how this feature can be exploited in the structural assignment of alkene isomers.

The Effects of Pi Electrons on Chemical Shifts

Hydrogen atoms bound to sp^2-hybridized carbons resonate at much lower field strengths than those in alkanes. Terminal alkene hydrogens (RR′C=CH$_2$) appear at $\delta = 4.6$–5.0 ppm, their internal counterparts (RCH=CHR′) at $\delta = 5.2$–5.7 ppm. A spectrum of *trans*-2,2,5,5-tetramethyl-3-hexene is shown in Figure 11-8. Only two signals are observed, one for the eighteen equivalent methyl hydrogens and one for the two alkene hydrogens. The absorptions ap-

FIGURE 11-8
90-MHz ^{1}H NMR spectrum of *trans*-2,2,5,5-tetramethyl-3-hexene in CCl_4. It reveals two sharp singlets for two sets of hydrogens, the six methyl groups at $\delta = 0.97$ ppm and two highly deshielded alkenyl hydrogens at $\delta = 5.30$ ppm.

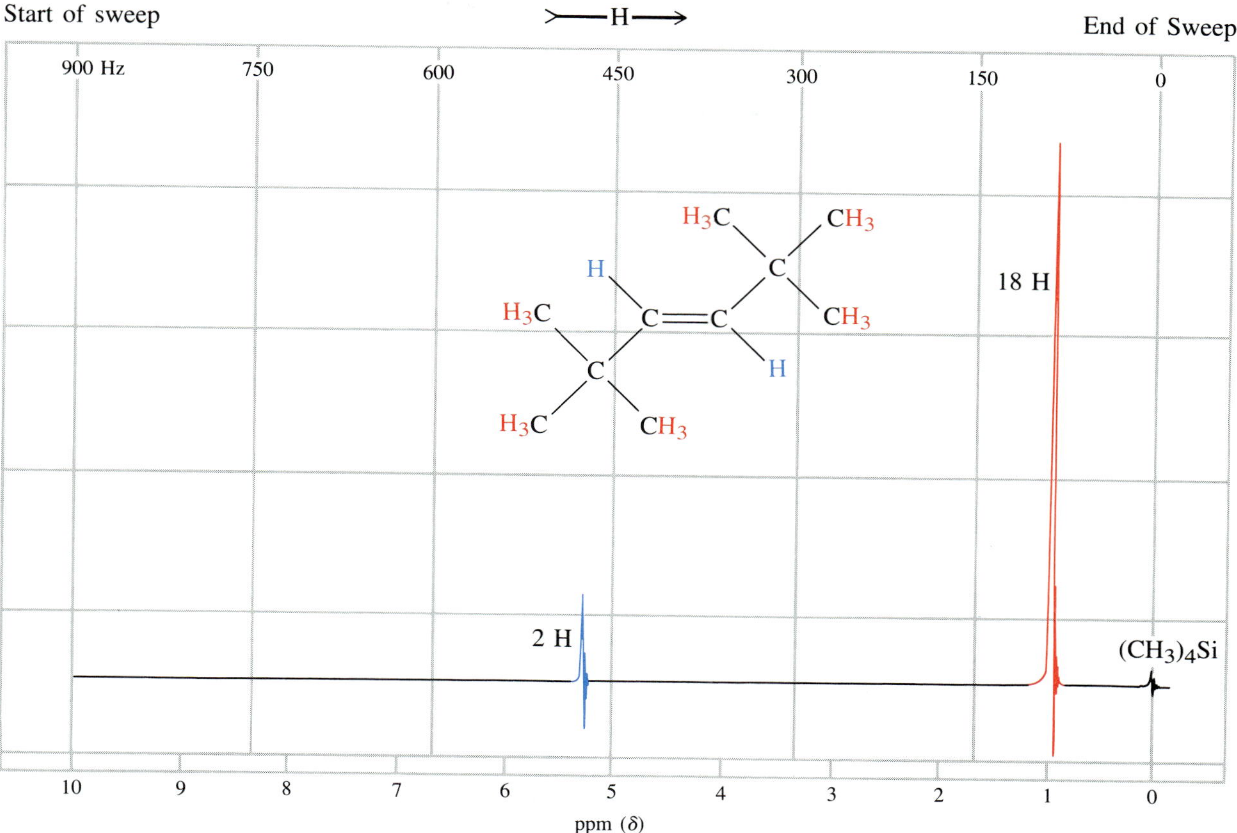

pear as singlets because no spin-spin splitting is observed between equivalent nuclei, and the methyl groups are far enough away from the alkenyl hydrogens to prevent detectable coupling.

Why is deshielding so pronounced for hydrogens bound to double bonds? One reason may be the electron-withdrawing character of the sp^2-hybridized carbon. Recall that, in NMR, withdrawing electron density from a nucleus leads to deshielding. Although this effect contributes to the observed low-field chemical shifts, another phenomenon is more important: *the movement of the electrons in the π bond*. When subjected to an external magnetic field perpendicular to the double-bond axis, these electrons enter into a circular motion that creates a local magnetic field opposed to the external field at the center of the π bond (Figure 11-9). This movement is similar to the electron current in a loop of copper wire when exposed to a magnet (see also Figure 10-12). At the edge of the double bond in the region of the alkenyl hydrogens, the local field *reinforces* the external field. As a consequence, less magnetic-field strength is required to bring these hydrogens to resonance: they are strongly deshielded.

In solution, not all the molecules will align in the way shown in Figure 11-9; that is, exactly perpendicular to the external field and therefore exposed to

FIGURE 11-9

An external field, H_0, induces a circular motion of the π electrons (shown in red) above and below the plane of a double bond. This motion in turn induces a local field (shown in green) that opposes H_0 at the center of the double bond but reinforces it in the regions occupied by the alkenyl hydrogens.

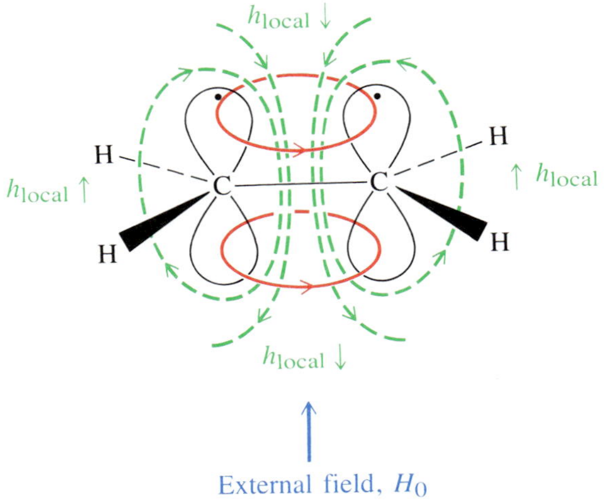

maximum deshielding. Most will be at some angle to it, rapidly changing positions as they move about in solution, and hence less deshielded. These changes are fast on the NMR time scale and therefore the observed chemical shift is an average of all the molecular contributions, not quite as large as the possible maximum, but still characteristically high, as observed.

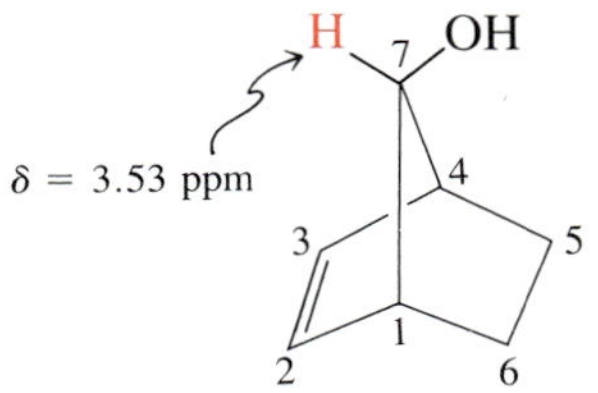

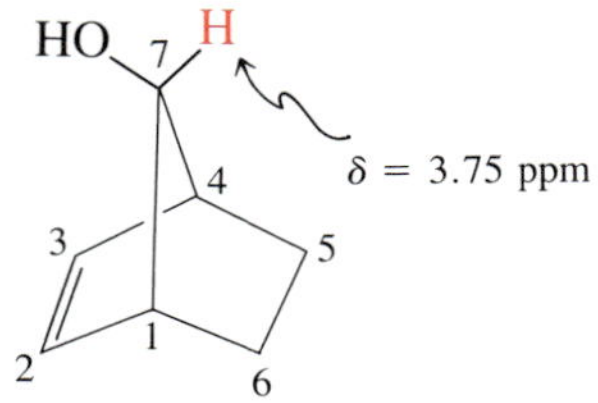

EXERCISE 11-7

Explain the differences in chemical shifts of the bridge hydrogens in the two bicyclo[2.2.1]hept-2-en-7-ols shown at the left. Hint: Try to apply the principles developed in Figure 11-9 to these compounds.

Coupling Through the Double Bond: Cis Is Different from Trans

Unsymmetrically substituted double bonds contain nonequivalent neighboring alkenyl hydrogens that are coupled to each other. Examples are shown in the spectra of *cis*- and *trans*-3-chloropropenoic acid (Figure 11-10). Note that the coupling constant for the hydrogens situated cis (J = 9 Hz) is different from those arranged trans (J = 14 Hz).

Table 11-1 is a compilation of the magnitude of the various possible couplings around a double bond. Although the range of J_{cis} overlaps that of J_{trans}, within a set of isomers the first is always smaller than the second. In this way cis and trans isomers can be distinguished readily.

FIGURE 11-10 (ON FACING PAGE)

90-MHz ^{1}H NMR spectra of (A) *cis*-3-chloropropenoic acid and (B) the corresponding trans isomer in CCl_4. The two alkenyl hydrogens are nonequivalent and coupled. The carboxylic acid proton (—CO_2H) resonates at δ = 12.35 ppm and is shown in the inset, offset by 3 ppm to higher field.

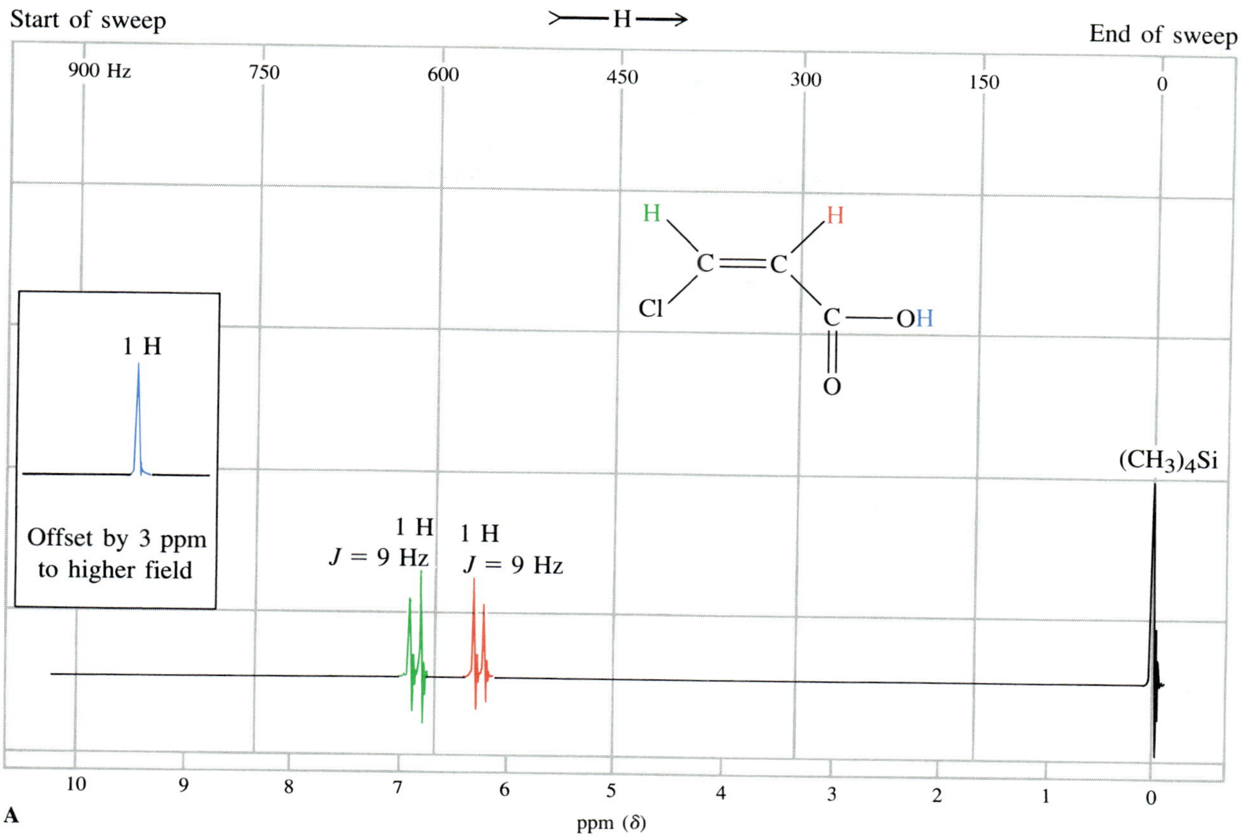

Start of sweep

>—H—>

End of sweep

900 Hz 750 600 450 300 150 0

H C=C H

Cl C—OH

O

1 H

Offset by 3 ppm to higher field

1 H J = 9 Hz

1 H J = 9 Hz

$(CH_3)_4Si$

10 9 8 7 6 5 4 3 2 1 0

ppm (δ)

A

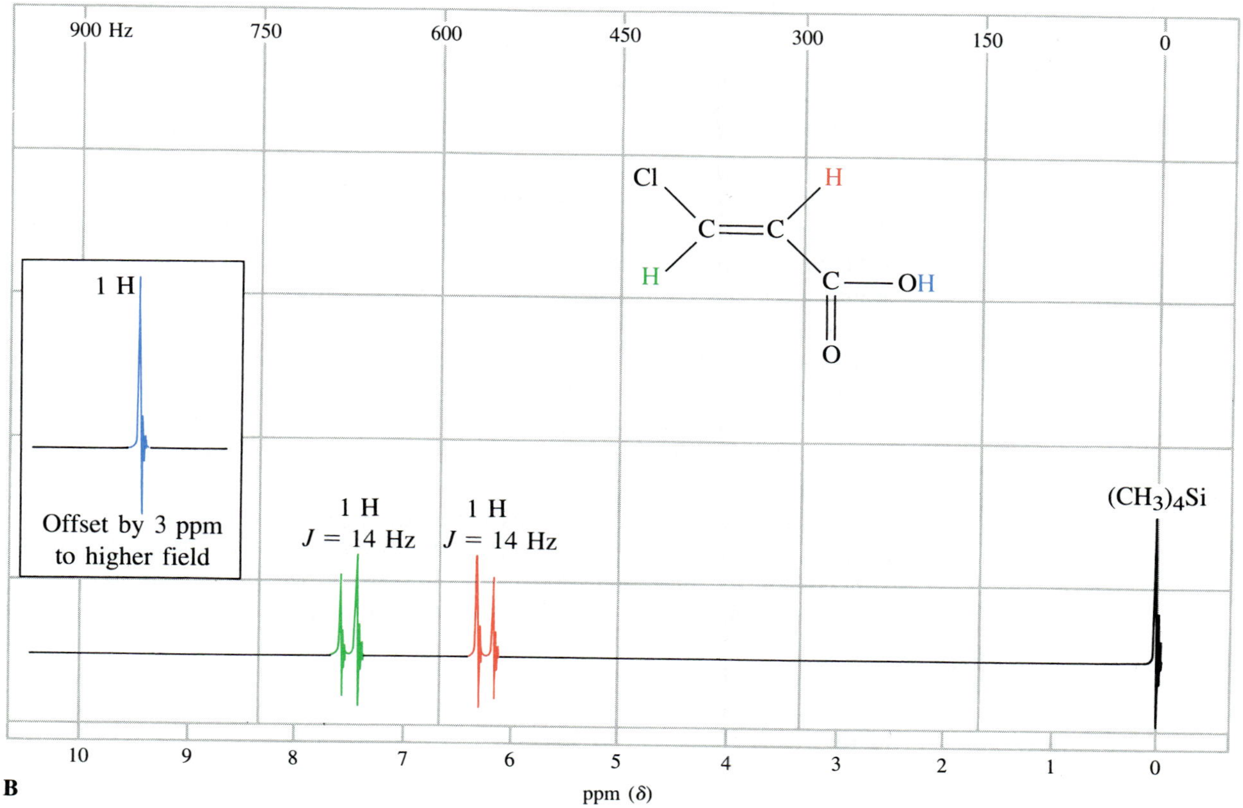

900 Hz 750 600 450 300 150 0

Cl C=C H

H C—OH

O

1 H

Offset by 3 ppm to higher field

1 H J = 14 Hz

1 H J = 14 Hz

$(CH_3)_4Si$

10 9 8 7 6 5 4 3 2 1 0

ppm (δ)

B

TABLE 11-1
Coupling constants around a double bond

Type of coupling	Name	J (Hz)
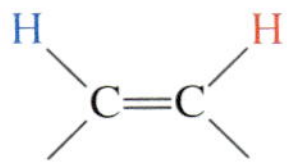 H(C=C)H, cis	Vicinal, cis	6–14
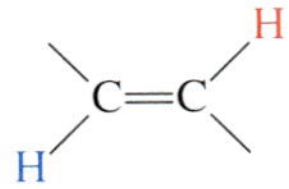 H(C=C)H, trans	Vicinal, trans	11–18
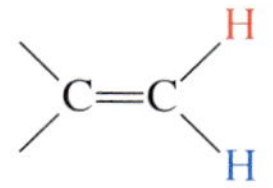 C=C(H)H	Geminal	0–3
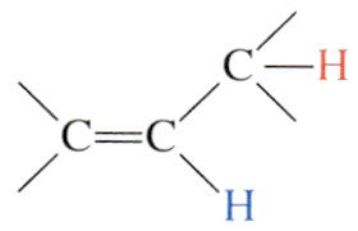 C=C(H)–C–H	None	4–10
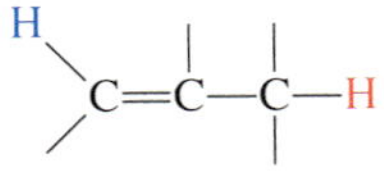 H–C=C–C–H	Allylic, (1,3)-cis or -trans	0.5–3
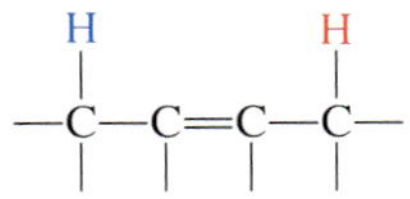 H–C–C=C–C–H	(1,4)- or long range	0–1.6

J_{cis} and J_{trans} are called **vicinal coupling constants.** Coupling between non-equivalent terminal hydrogens is small and labeled **geminal coupling** (Table 11-1). Coupling to neighboring alkyl hydrogens (**allylic,** see Section 11-1) and across the double bond (**1,4** or **long range**) is also possible, sometimes giving rise to complicated spectral patterns. Thus, the simple rule devised for saturated systems, discounting coupling between hydrogens farther than two intervening atoms apart, does not hold for alkenes.

The spectra of 3,3-dimethyl-1-butene and 1-pentene illustrate the potential complexity of the coupling patterns (Figure 11-11). In both spectra, the alkene hydrogens appear as complex multiplets. In 3,3-dimethyl-1-butene, H_a located on the more highly substituted carbon atom resonates at lower field (δ = 5.81 ppm) and in the form of a double doublet with two relatively large coupling constants (J_{ab} = 18 Hz, J_{ac} = 10.5 Hz). Each of hydrogens H_b and H_c also absorb as double doublets owing to their respective coupling to H_a and their mutual coupling (J_{bc} = 1.5 Hz). Because of the small chemical-shift difference between them, their signals overlap, but, as shown in the inset (twofold expansion), the coupling pattern can be readily analyzed and both hydrogens

FIGURE 11-11 (ON FACING PAGE)
90-MHz ^{1}H NMR spectra of (A) 3,3-dimethyl-1-butene and (B) 1-pentene in CCl_4.

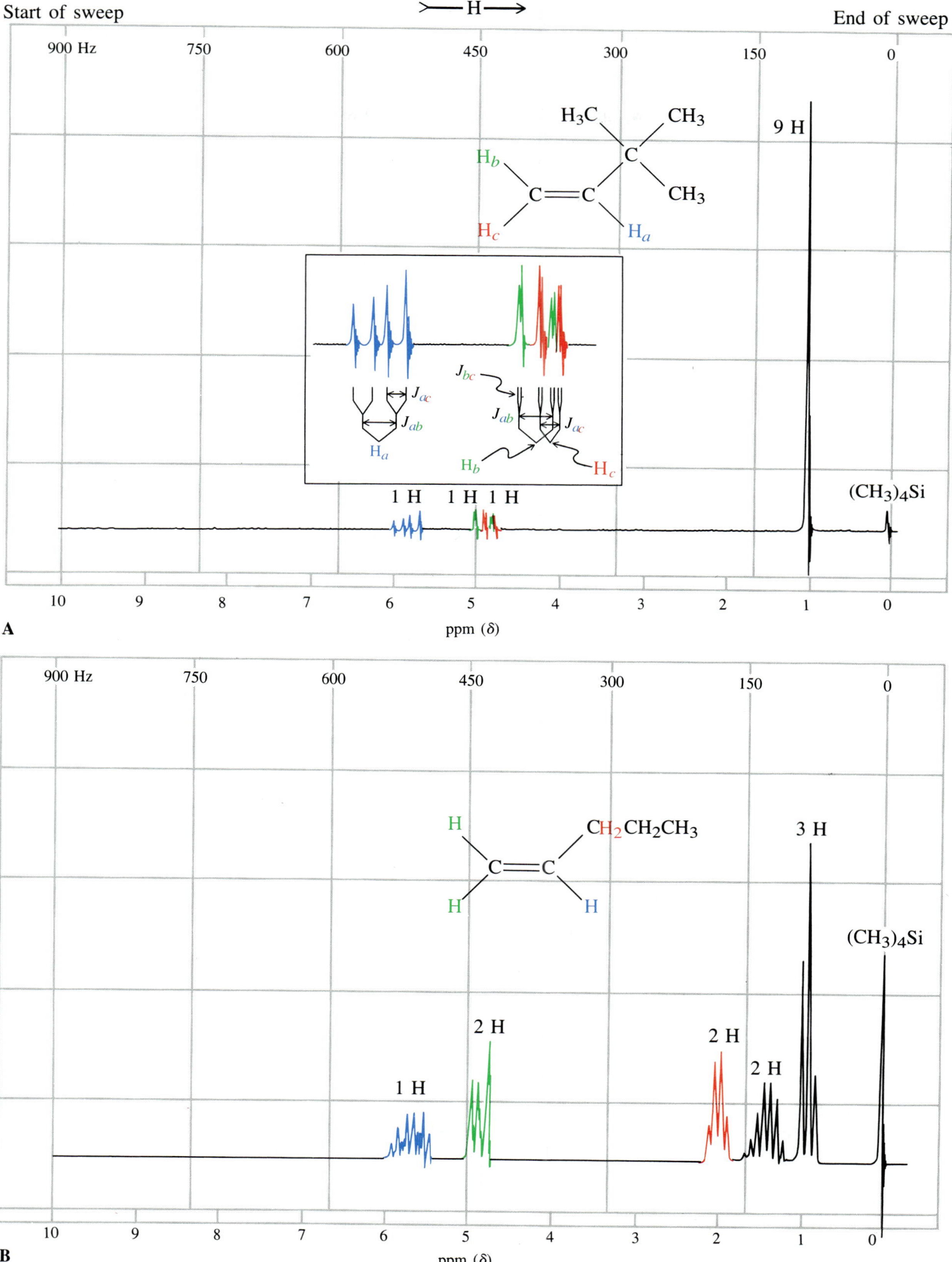

Start of sweep
H
End of sweep
900 Hz
750
600
450
300
150
0
H_3C
CH_3
H_b
C
C=C
CH_3
H_c
H_a
9 H
J_{ac}
J_{ab}
H_a
J_{bc}
J_{ab}
J_{ac}
H_b
H_c
1 H
1 H
1 H
$(CH_3)_4Si$
10
9
8
7
6
5
4
3
2
1
0
A
ppm (δ)
900 Hz
750
600
450
300
150
0
H
$CH_2CH_2CH_3$
C=C
H
H
3 H
$(CH_3)_4Si$
2 H
2 H
2 H
1 H
10
9
8
7
6
5
4
3
2
1
0
B
ppm (δ)

assigned. In the spectrum of 1-pentene, additional coupling due to the attached alkyl group (see Table 11-1) creates a situation that is too complex for a first-order analysis. However, the two sets of alkene hydrogens (terminal and internal) are clearly differentiated. In addition, the electron-withdrawing effect of the sp^2 carbon causes the directly attached (allylic) CH_2 group to be slightly deshielded, its multiplet appearing at lower field ($\delta = 1.94$ ppm) than that of the other alkyl absorptions.

EXERCISE 11-8

Ethyl 2-butenoate (ethyl crotonate), $CH_3CH{=}CHCO_2CH_2CH_3$, in CCl_4 has the following ^1H NMR spectrum: $\delta = 6.95$ (dq, $J = 16$, 6.8 Hz, 1 H), 5.81 (dq, $J = 16$, 1.7 Hz, 1 H), 4.13 (q, $J = 7$ Hz, 2 H), 1.88 (dd, $J = 6.8$, 1.7 Hz, 3 H), and 1.24 (t, $J = 7$ Hz, 3 H) ppm. Assign the various hydrogens and indicate whether the double bond is substituted cis or trans (consult Table 11-1).

^{13}C NMR of Alkenes

The carbon NMR absorptions of the alkenes are also highly diagnostic. Relative to alkanes, the corresponding alkene carbons (with similar substituents) absorb at about 100-ppm lower field (see Table 10-6). Two examples are shown in Table 11-2, in which the carbon chemical shifts of an alkene are compared with those of its saturated counterpart. Recall that, in broad-band decoupled ^{13}C NMR spectroscopy, all magnetically unique carbons absorb as sharp single lines. It is therefore very easy to determine the presence of sp^2 carbons by this method.

TABLE 11-2
Comparison of ^{13}C NMR absorptions of alkenes with the corresponding alkane carbon chemical shifts (in ppm)

Structure of alkene	Structure of alkane
$(CH_3)_2C{=}C(CH_3)_2$: C=C 122.8; CH_3 18.9	$(CH_3)_2CH{-}CH(CH_3)_2$: CH 34.0; CH_3 19.2
cis-$CH_3CH{=}CHCH_2CH_3$ (H, H on C=C): C bearing CH_3 123.7; C bearing CH_2CH_3 132.7; CH_3 12.3; CH_2 20.5; CH_3 (of ethyl) 14.0	$CH_3CH_2CH_2CH_2CH_3$: CH_3 13.5; C-2 22.2; C-3 34.1

In summary, NMR is highly effective in establishing the presence of double bonds in organic molecules. Alkenyl hydrogens and carbons are strongly deshielded. The order of coupling is $J_{trans} > J_{cis} > J_{gem}$. In addition, various coupling constants are typical for allylic substituents.

11-4 The Relative Stability of Double Bonds: Heats of Hydrogenation

We have seen that there are several possible substitution patterns around a double bond, depending on the number of substituents and their positions. Is there a difference in the relative stabilities of these molecular arrangements? For example, are there differences in the heats of formation of the three isomeric butenes: 1-butene, *cis*-2-butene, and *trans*-2-butene? This question is answered in the next subsection by consideration of the heats of hydrogenation of isomeric alkenes.

The Heats of Hydrogenation Reveal the Relative Stability of Alkene Isomers

How can we establish the relative energy of any compound? A method that might enable us to answer this question was presented in Section 3-4: measurement of the heat of combustion to estimate the heat of formation, $\Delta H_f°$. Depending on the relative energy content of the respective alkene, we expect more or less energy to be released in this experiment. Another possibility, which is particularly applicable to the isomeric alkenes, is the measurement of the heat of another reaction: *hydrogenation of the double bond.*

When an alkene and hydrogen gas are mixed in the presence of a catalyst (usually palladium or platinum), the two hydrogen atoms in elemental hydrogen add to the double bond to give the saturated alkane (see Section 12-2), much like the hydrogenation of a carbonyl compound to give an alcohol (Section 8-4). The heat of this reaction can be measured accurately and is approximately (depending on the alkene) -30 kcal mole^{-1} per double bond: the reaction is appreciably exothermic.

General Hydrogenation of an Alkene

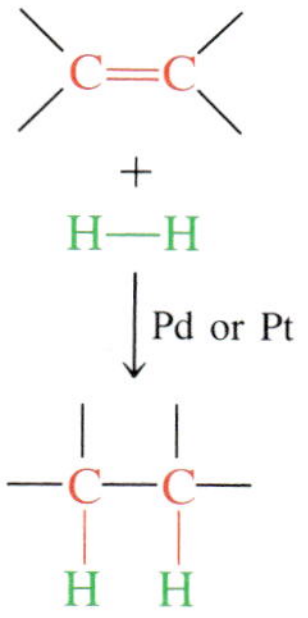

$$\text{>C=C<} + \text{H—H} \xrightarrow{\text{Pd or Pt}} \text{—}\overset{|}{\underset{\text{H}}{\text{C}}}\text{—}\overset{|}{\underset{\text{H}}{\text{C}}}\text{—}$$

$\Delta H° \sim -30$ kcal mole^{-1}

For the isomeric butenes, hydrogenation leads to the *same* product in each case: butane. If the respective heats of formation of the butenes were equal, their heats of hydrogenation should also be equal; however, as shown below, they are not:

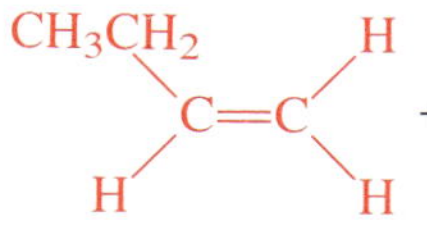

$$\text{(CH}_3\text{CH}_2\text{)(H)C=CH}_2 + H_2 \xrightarrow{Pt} CH_3CH_2CH_2CH_3 \qquad \Delta H° = -30.3 \text{ kcal mole}^{-1}$$

1-Butene → **Butane**

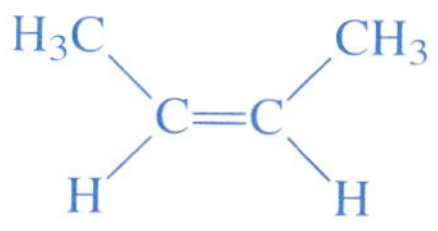

$$cis\text{-}CH_3CH{=}CHCH_3 + H_2 \xrightarrow{Pt} CH_3CH_2CH_2CH_3 \qquad \Delta H° = -28.6 \text{ kcal mole}^{-1}$$

***cis*-2-Butene** → **Butane**

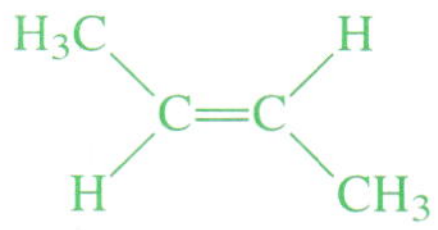

$$trans\text{-}CH_3CH{=}CHCH_3 + H_2 \xrightarrow{Pt} CH_3CH_2CH_2CH_3 \qquad \Delta H° = -27.6 \text{ kcal mole}^{-1}$$

***trans*-2-Butene** → **Butane**

FIGURE 11-12

The relative heats of formation of the isomeric butenes according to their heats of hydrogenation. The drawing is not to scale.

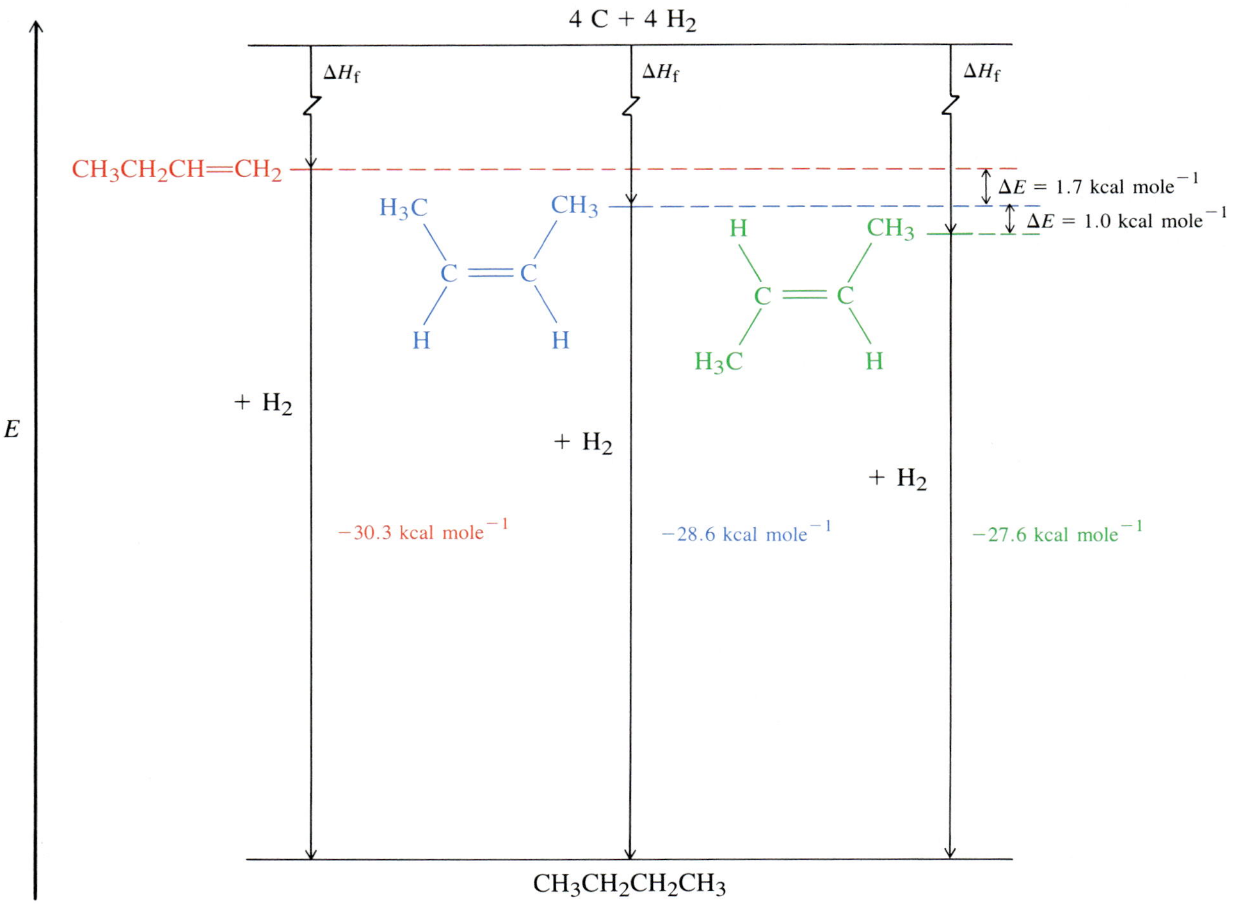

The largest amount of heat is evolved on hydrogenation of the terminal double bond; the next most exothermic reaction is that with *cis*-2-butene, and finally the *trans* isomer gives off the least amount of heat. This result leads to the conclusion that the thermodynamic stabilities of the butenes increase in the order 1-butene < *cis*-2-butene < *trans*-2-butene (Figure 11-12).

The preceding results may be generalized and expanded: the relative stability of the alkenes increases with increasing substitution, and trans isomers are usually more stable than their cis counterparts. The first trend cannot be explained in a straightforward way, but it is due in part to hyperconjugation. Just as the stability of a radical increases with increasing alkyl substitution (because of hyperconjugation of the singly filled *p* orbital with adjacent C–H bonds, Section 3-2), the *p* orbitals of a π bond may be stabilized by alkyl substituents. The second finding is more easily rationalized, particularly when we look at molecular models. In cis-disubstituted alkenes, the substituent groups frequently crowd each other.

Relative Stabilities of the Alkenes

$$CH_2{=}CH_2 < RCH{=}CH_2 < \underset{H}{\overset{R}{>}}C{=}C\underset{H}{\overset{R}{<}} < \underset{R}{\overset{H}{>}}C{=}C\underset{H}{\overset{R}{<}} < \underset{R}{\overset{R}{>}}C{=}C\underset{H}{\overset{R}{<}} < \underset{R}{\overset{R}{>}}C{=}C\underset{R}{\overset{R}{<}}$$

This steric interference is energetically disadvantageous and absent in the corresponding trans isomers (Figure 11-13).

FIGURE 11-13

Representation of (A) steric congestion in cis-disubstituted alkenes and (B) its absence in trans alkenes.

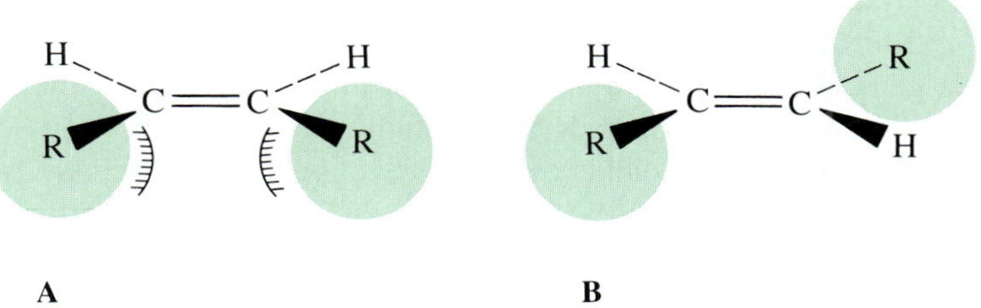

EXERCISE 11-9

Rank in order of stability of the double bond to hydrogenation (order of ΔH° of hydrogenation): 2,3-dimethyl-2-butene, *cis*-3-hexene, *trans*-4-octene, and 1-hexene.

Small Trans Cycloalkenes Are Destabilized by Strain

An exception to the general finding that trans alkenes are more stable than cis alkenes is found in cycloalkenes. In the small- and medium-ring (Section 4-2) members of this class of compounds, the trans isomers are much more strained (see Section 11-1). The smallest isolated unsubstituted trans cycloalkene is *trans*-cyclooctene, which is 9.2 kcal $mole^{-1}$ less stable than the cis isomer. Its structure is not planar but highly twisted. If you construct a model of this molecule, you will note that it has handedness and is therefore chiral. Resolution of *trans*-cyclooctene reveals that the activation barrier for racemization is relatively high: 36 kcal $mole^{-1}$. For this molecule to undergo conversion from one enantiomer into the other requires ring flip, the difficulty of which is readily seen by the use of molecular models.

Racemization of *trans*-Cyclooctene

E_a = 36 kcal $mole^{-1}$

EXERCISE 11-10

Alkene A hydrogenates to B with an estimated release of 65 kcal $mole^{-1}$, more than double the value of the hydrogenation shown in Figure 11-12. Explain.

A $\xrightarrow{H_2,\ \text{catalyst}}$ B $\qquad \Delta H^\circ = -65$ kcal $mole^{-1}$

In summary, the relative energies of various isomeric alkenes can be estimated by measuring their heats of hydrogenation. The more-energetic alkene has a higher ΔH° of hydrogenation. Stability increases with increasing substitution because of hyperconjugation. Trans alkenes are more stable than their cis isomers because of steric hindrance. The exceptions are the small- and medium-ring cycloalkenes, in which cis substitution is more stable than trans because of ring strain.

11-5 The Preparation of Alkenes from Haloalkanes and Alkyl Sulfonates: Bimolecular Elimination Revisited

General Scheme for Elimination

$$\underset{\mathrm{A}\ \ \ \mathrm{B}}{-\overset{|}{\mathrm{C}}-\overset{|}{\mathrm{C}}-} \longrightarrow \mathrm{C{=}C} + \mathrm{AB}$$

With the physical aspects of alkene structure and stability as a background, let us now concern ourselves with the various ways in which alkenes can be made. The most general approach is by **elimination,** in which two adjacent groups on a carbon framework are removed. Elimination reactions were discussed in the dehydrohalogenation of haloalkanes by E1 (Section 7-4) and E2 (Section 7-5). A special type of elimination was described in connection with hydrocarbon cracking (Section 3-3), the major industrial source of ethene (Section 12-8) and propene.

This section reviews the E2 reaction as the most common laboratory source of alkenes. Another method of alkene synthesis, the dehydration of alcohols, is described in Section 11-6. Other approaches to alkene synthesis are discussed in later chapters. They include:

1 Reduction of alkynes (Section 13-6)

$$H_2 + -C{\equiv}C- \longrightarrow \underset{H\qquad H}{C{=}C}$$

2 Condensation of carbonyl groups (Sections 15-4 and 15-7)

$$C{=}O + H_2C \longrightarrow C{=}C + H_2O$$

$$C{=}O + R_3P{=}C \longrightarrow C{=}C + O{=}PR_3$$

3 Ester pyrolysis (Section 18-4)

$$\underset{\text{H on one C, O–C(=O)R on the other}}{-\overset{|}{\underset{|}{C}}-\overset{|}{\underset{|}{C}}-} \xrightarrow{\Delta} C{=}C + R\overset{O}{\overset{\|}{C}}OH$$

4 Hofmann elimination (Section 21-5)

$$\mathrm{H{-}\overset{|}{\underset{|}{C}}{-}\overset{|}{\underset{|}{C}}{-}\overset{+}{N}R_3 + {}^{-}OH \longrightarrow \;\rangle C{=}C\langle\; + HOH + NR_3}$$

5 Cope elimination (Section 21-5)

$$\mathrm{-\overset{\overset{H}{|}}{\underset{|}{C}}-\overset{\overset{O^{-}-\overset{+}{N}R_2}{/}}{\underset{|}{C}}- \xrightarrow{\Delta} \;\rangle C{=}C\langle\; + R_2NOH}$$

Regioselectivity in E2 Reactions: Internal Compared with Terminal Alkene Formation

As discussed earlier, haloalkanes (or alkyl sulfonates) in the presence of base can undergo elimination of the elements of HX with simultaneous formation of a carbon–carbon double bond. Sterically hindered and strong bases, such as potassium *tert*-butoxide (Section 7-5), are best, to prevent the reactants from entering E1 pathways and the complications that arise from the formation of carbocations.

With many substrates, elimination can be in more than one direction of the chain, giving rise to double-bond isomeric alkenes. Can we control the *regioselectivity* of the reaction in such cases? The answer is yes, but only to a limited extent. A simple example is the dehydrobromination of 2-bromo-2-methylbutane. Reaction with sodium ethoxide in hot ethanol furnishes mainly 2-methyl-2-butene, but also some 2-methyl-1-butene:

$$\mathrm{CH_3CH_2\overset{\overset{CH_3}{|}}{\underset{\underset{Br}{|}}{C}}CH_3 \xrightarrow[-\,HBr]{CH_3CH_2O^-Na^+,\ CH_3CH_2OH,\ 70^\circ C} \underset{70\%}{(H_3C)(H)C{=}C(CH_3)_2} + \underset{30\%}{(CH_3CH_2)(H_3C)C{=}CH_2}}$$

2-Bromo-2-methylbutane → **2-Methyl-2-butene** (70%) + **2-Methyl-1-butene** (30%)

The first alkene contains a trisubstituted double bond and is thermodynamically more stable than the second. Indeed, many eliminations are regioselective in this way, with the thermodynamic product predominating. This result can be explained by analysis of the transition state of the reaction (Figure 11-14). Elimination of the elements of hydrogen bromide proceeds through attack by the base on one of the vicinal hydrogens situated *anti* to the leaving group. In the transition state, there is partial C–H bond rupture, partial C–C double-bond formation, and partial cleavage at C–Br. The transition state leading to 2-methyl-2-butene is slightly more stabilized than that generating 2-methyl-1-butene: the more-stable product is formed faster because the transition state of the reaction resembles the products to some extent. Elimination reactions of this

FIGURE 11-14

The two transition states leading to products in the dehydrobromination of 2-bromo-2-methylbutane. Transition state A is preferred over transition state B because there are more substituents around the partial double bond.

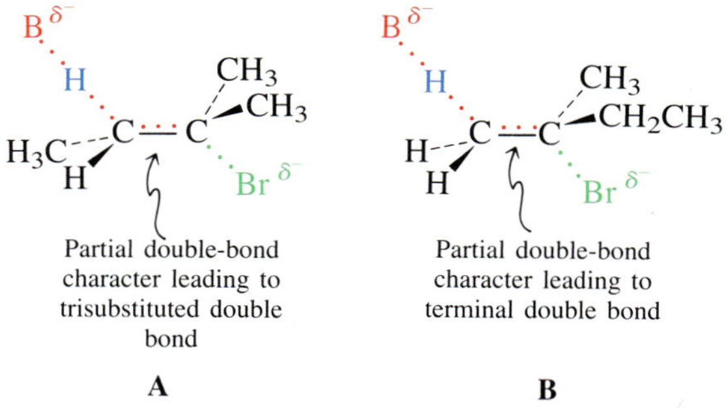

type that lead to the more highly substituted alkene are said to follow the **Saytzev rule.***

A different product distribution is obtained when a more-hindered base is used in the same reaction; more of the thermodynamically *less* favored terminal alkene is generated:

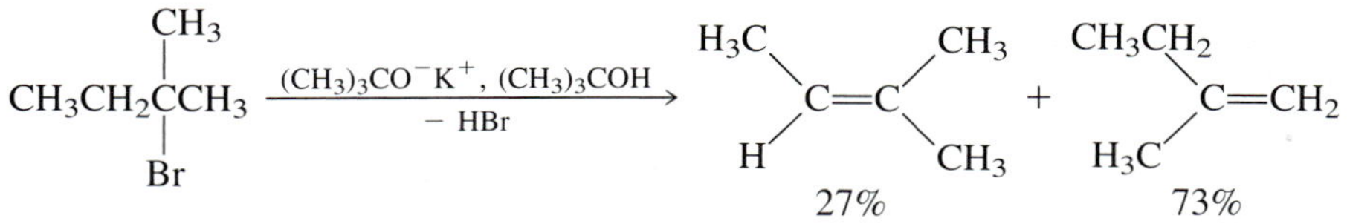

$$CH_3CH_2C(CH_3)(Br)CH_3 \xrightarrow[-\ HBr]{(CH_3)_3CO^-K^+,\ (CH_3)_3COH} \underset{27\%}{(H_3C)(H)C{=}C(CH_3)_2} + \underset{73\%}{(CH_3CH_2)(H_3C)C{=}CH_2}$$

This finding is again explicable by examination of the transition state. Removal of a secondary hydrogen (from C-3) in the starting bromide is sterically more difficult than abstracting one of the more-exposed primary methyl hydrogens. This effect is accentuated if the base used is bulky; here, because of steric constraints, the thermodynamically less favored isomer is generated faster, because the transition state leading to it has lower energy. Such a kinetically controlled regiochemical outcome of the E2 reaction is sometimes referred to as following the **Hofmann rule.**†

EXERCISE 11-11

The following reaction has been carried out with *tert*-butoxide in 2-methyl-2-propanol (*tert*-butyl alcohol) and with ethoxide in ethanol to give two products, A and B, in the ratio 23 : 77 under the first conditions and 82 : 18 under the second. What are A and B?

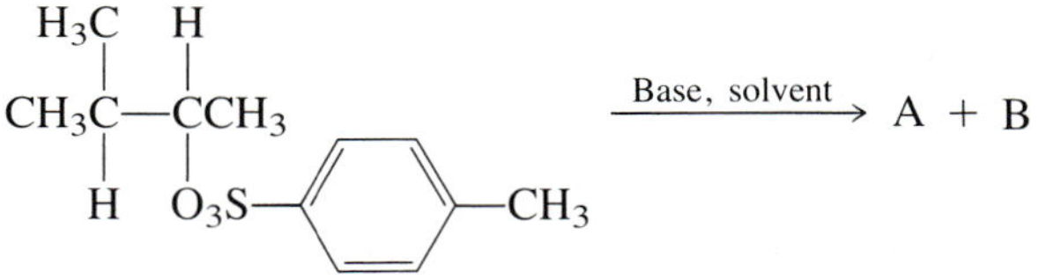

$$CH_3C(H_3C)(H)\text{—}C(H)(O_3S\text{—}C_6H_4\text{—}CH_3)CH_3 \xrightarrow{\text{Base, solvent}} A + B$$

*Alexander M. Saytzev (also spelled Zaitsev or Saytzeff), 1841–1910, Russian chemist.
†Professor August W. von Hofmann, 1818–1892, University of Berlin.

Stereoselectivity in the E2 Reaction: Cis or Trans?

Depending on the structure of the alkyl substrate, the E2 reaction can also lead to cis,trans alkene mixtures, in some cases with selectivity. For example, treatment of 2-bromopentane with sodium ethoxide furnishes 51% *trans*- and only 18% *cis*-2-pentene, the remainder of the product being the terminal regioisomer. According to our definitions of selectivity in stereochemical reactions leading to stereoisomers, this process would have to be classified as stereoselective because, regardless of the stereochemistry of the starting bromide (*R* or *S* in our example), one stereoisomer is formed in higher yields than the other (Section 5-6). The outcome of this and related reactions appears to be controlled again to some extent by the relative thermodynamic stabilities of the products, the most-stable trans double bond being formed preferentially.

Stereoselective Dehydrobromination of 2-Bromopentane

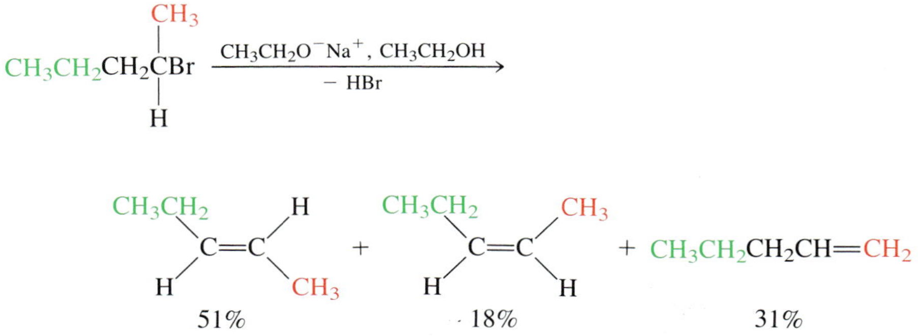

$$CH_3CH_2CH_2\underset{H}{\overset{CH_3}{C}}Br \xrightarrow[-\ HBr]{CH_3CH_2O^-Na^+,\ CH_3CH_2OH}$$

$$\underset{51\%}{(CH_3CH_2)(H)C{=}C(H)(CH_3)} + \underset{18\%}{(CH_3CH_2)(H)C{=}C(CH_3)(H)} + \underset{31\%}{CH_3CH_2CH_2CH{=}CH_2}$$

However, it also might be the preferred ground-state conformation of starting material that is responsible for the stereochemistry of the reaction. This conformation (Figure 11-15A) prevents *gauche* interactions between the ethyl and methyl groups, unlike the less-favorable alternative shown in Figure 11-15B. As elimination proceeds from the ground state shown in Figure 11-15A, the two alkyl substituents move into the unhindered trans positions. On the other hand, attempted elimination from the conformation shown in Figure 11-15B pushes the two *gauche* alkyl groups into even closer proximity. Elimination will thus proceed mainly through conformation A to give the trans alkene for two reasons: first, it predominates as the ground-state conformation; second, its transition state is of lower energy.

Unfortunately from a synthetic viewpoint, complete trans selectivity is rare in E2 reactions. Chapter 13 deals with alternative methods for the preparation of stereochemically pure cis and trans alkenes.

FIGURE 11-15 The two conformations in 2-bromopentane leading to either (A) *trans*-2-pentene or (B) *cis*-2-pentene.

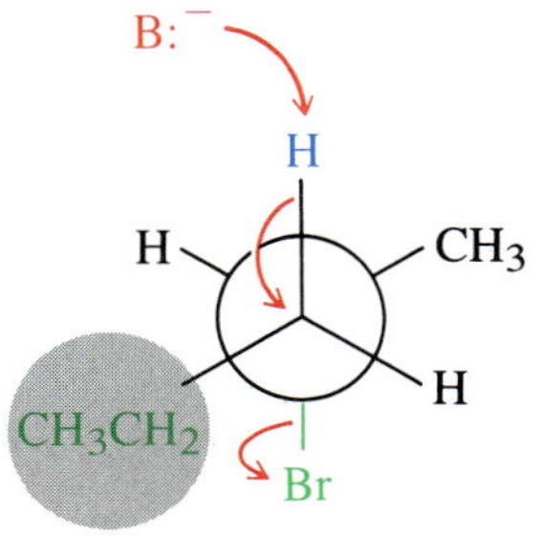

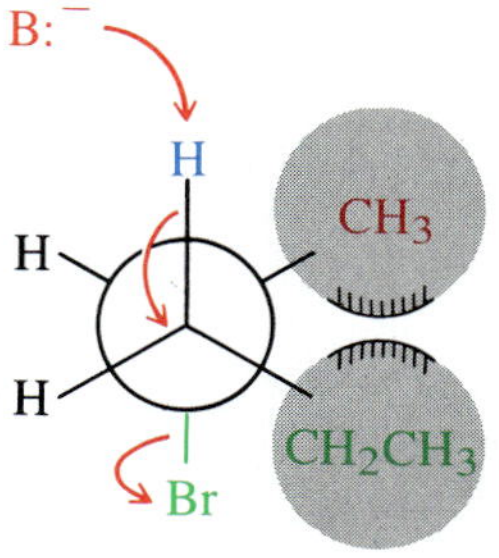

Stereospecificity in the E2 Reaction: *Z* or *E*?

The fact that the preferred transition state of elimination places the proton to be removed and the leaving group *anti* with respect to each other has additional stereochemical consequences. For example, the E2 reaction of the two diastereomers of 2-bromo-3-methylpentane to give 3-methyl-2-pentene is completely stereospecific. Employment of the (*R*,*R*) or the (*S*,*S*) isomer or both leads *exclusively* to the formation of the (*E*) isomer of the alkene. On the other hand, reaction of the (*R*,*S*) or (*S*,*R*) diastereomer or both gives only the (*Z*) alkene.

Stereospecificity in the E2 Reaction of 2-Bromo-3-methylpentane

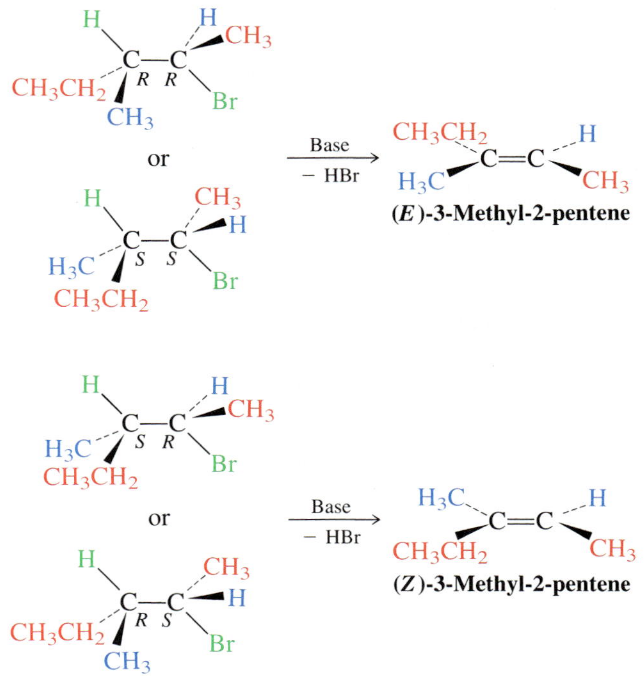

Base
$-$ HBr

(*E*)-3-Methyl-2-pentene

Base
$-$ HBr

(*Z*)-3-Methyl-2-pentene

As shown in these dashed-wedged line structures, *anti* elimination of HBr dictates the eventual configuration around the double bond. The reaction is stereospecific, one diastereomer (and its mirror image) resulting in only one stereoisomeric alkene, the other diastereomer furnishing the opposite configuration.

EXERCISE 11-12

Which diastereomer of 2-bromo-3-deuteriobutane gives (*E*)-2-deuterio-2-butene, and which diastereomer gives the *Z* isomer? What can you say about the isotopic purity of the respective products?

Why *Anti* Elimination?

Why is elimination from a conformation in which proton and leaving group are *anti* so much easier than elimination from a *gauche* conformation? The answer is that this is the only low-energy conformation that generates the two *p* orbitals of the π bond in the parallel alignment required for good overlap (Figure 11-16, see also Figure 7-8). Elimination from a *gauche* arrangement causes the two *p* orbitals to develop at a 60° angle, a less favorable situation. The third alternative, a *syn* elimination, with both H and X departing from the same side of the molecule, proceeds through a similarly unfavorable eclipsed rotamer, with less-good overlap between the developing *p* lobes. Thus, among these conformations, *anti* elimination is preferred. Only in special cases or if the *anti* pathway is not possible are other modes of elimination observed, and they frequently require more stringent reaction conditions (Section 7-5).

FIGURE 11-16

The various conceivable transition states for the E2 reaction: (A) H and X are located *anti,* the groups in the molecule are staggered, and the two new *p* orbitals being developed are perfectly parallel to form the π bond; (B) H and X are *gauche,* the conformation is staggered, but the two new *p* orbitals are at 60° angles; (C) H and X are *syn,* and the conformation is *eclipsed.* Although the two developing *p* orbitals are parallel in part C, overlap is worse than in part A, and the requirement for an eclipsed conformation makes this pathway unfavorable.

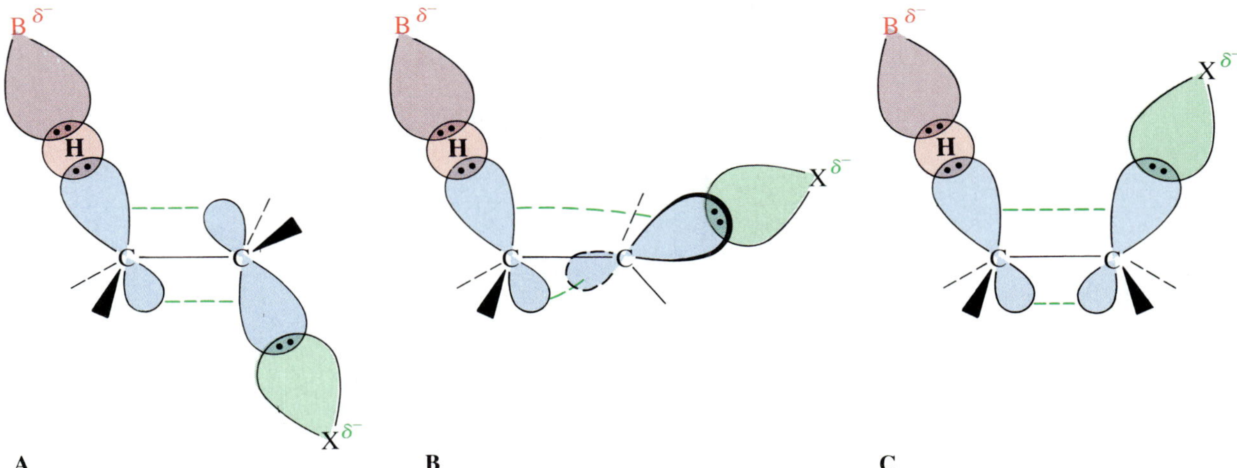

In summary, alkenes are most generally made by E2 reactions. To avoid S_N2 side products, particularly with primary substrates, the use of a bulky base is advisable. Generally, the thermodynamically more stable internal alkenes are formed faster than the terminal isomers (Saytzev rule). The reaction may be stereoselective, producing greater quantities of trans isomers than their cis counterparts from racemic starting materials. It is also stereospecific, certain haloalkane diastereomers furnishing only one out of the two possible stereoisomeric alkenes. Bulky bases may lead to more of the products with thermodynamically less stable (e.g., terminal) double bonds (Hofmann rule). *Anti* elimination is favored because overlap between the two developing *p* orbitals in π-bond formation is maximized.

11-6 The Preparation of Alkenes by Dehydration of Alcohols

Acid-Mediated Dehydration of Alcohols

$$\text{H}-\overset{|}{\underset{|}{C}}-\overset{|}{\underset{|}{C}}-\text{OH} \xrightarrow{\text{Acid, }\Delta} \text{C}=\text{C} + \text{HOH}$$

Treatment of alcohols with mineral acid at elevated temperatures results in loss of water, a process called **dehydration,** which proceeds by E1 or E2 pathways (Chapters 7 and 9). This section reviews some of the preparative and mechanistic aspects of this reaction.

Preparative Methods and Scope of Alcohol Dehydration

The usual method employed to dehydrate an alcohol is to heat it in the presence of sulfuric or phosphoric acid at relatively high temperatures (120°–170°C).

Another, sometimes more efficient, dehydration procedure requires the use of aluminum oxide (alumina), Al_2O_3, as a Lewis acid catalyst. In this case, vapors of the alcohol are passed over alumina powder at a temperature ranging from 350°C to 400°C, and the alkene is collected in a trap at the exit of the reaction vessel.

The ease of elimination of water from alcohols increases with increasing substitution of the hydroxy-bearing carbon:

Ease of Dehydration of Alcohols (ROH)

R = tertiary > secondary > primary

$$CH_3CH_2OH \xrightarrow[-\ HOH]{\text{Conc. } H_2SO_4,\ 170^\circ C} CH_2{=}CH_2$$

$$CH_3CH(OH)CH_2CH_3 \xrightarrow[-\ HOH]{50\%\ H_2SO_4,\ 100^\circ C} \underset{80\%}{CH_3CH{=}CHCH_3} + \underset{\text{Trace}}{CH_2{=}CHCH_2CH_3}$$

$$(CH_3)_3COH \xrightarrow[-\ HOH]{\text{Conc. } H_2SO_4,\ 50^\circ C} \underset{100\%}{H_2C{=}C(CH_3)_2}$$

There are two mechanisms to consider for these dehydration reactions: unimolecular and bimolecular elimination (see also Sections 7-4, 7-5, 9-2, and 9-5).

Unimolecular Dehydration

The unimolecular elimination pathway is followed by secondary and tertiary alcohols. Protonation of the weakly basic hydroxy oxygen, followed by loss of water, supplies the respective secondary or tertiary carbocations, which then deprotonate to give the alkene.

Mechanism of Unimolecular Dehydration:

$$R{-}\underset{|}{\overset{\ddot{O}H}{C}}{-}\underset{|}{\overset{H}{C}}{-} \underset{-\ H^+}{\overset{+\ H^+}{\rightleftharpoons}} R{-}\underset{|}{\overset{H_2\overset{+}{O}}{C}}{-}\underset{|}{\overset{H}{C}}{-} \underset{+\ H_2\ddot{O}}{\overset{-\ H_2\ddot{O}}{\rightleftharpoons}} \overset{+}{C}{-}\underset{|}{\overset{H}{C}}{-} \underset{+H^+}{\overset{-\ H^+}{\rightleftharpoons}} C{=}C$$

Alkene formation is particularly favored at higher temperatures, at which the large positive entropy of the reaction (one molecule turns into two) is dominant (recall that $\Delta G^\circ = \Delta H^\circ - T\Delta S^\circ$), and in the presence of nonnucleophilic media; otherwise S_N1 products are dominant. Because carbocations are intermediates, the reaction is nonstereoselective and subject to all the side reactions of which carbocations are capable, particularly hydrogen and alkyl shifts (Section 9-2).

Dehydration with Rearrangement

$$\mathrm{CH_3\overset{CH_3}{\underset{H}{C}}{-}CH_2{-}\overset{OH}{\underset{H}{C}}CH_3 \xrightarrow[-\,H_2O]{H_2SO_4,\ \Delta} (H_3C)_2C{=}C(H)CH_2CH_3\ (54\%) + CH_3\overset{CH_3}{\underset{H}{C}}CH{=}CHCH_3\ (8\%) + \text{other minor isomers}}$$

$$\mathrm{CH_3\overset{CH_3}{\underset{H_3C}{C}}{-}\underset{OH}{C}HCH_3 \xrightarrow[-\,H_2O]{85\%\ H_3PO_4,\ \Delta} (H_3C)_2C{=}C(CH_3)_2\ (64\%) + H_2C{=}C(\overset{CH_3}{C}HCH_3)(CH_3)\ (\text{Minor})}$$

EXERCISE 11-13

Referring to Sections 7-4 and 9-2, formulate mehanisms for the two preceding reactions.

Deprotonation Is Reversible: Kinetic Compared with Thermodynamic Control

Often acid-catalyzed dehydration gives product mixtures even without rearrangements. This is the case with unsymmetrical alcohols:

$$\text{2-methylcyclohexanol} \xrightarrow{H_3PO_4,\ 100^\circ C} \text{1-methylcyclohexene (Major)} + \text{3-methylcyclohexene (Minor)}$$

However, usually the thermodynamically most stable alkene or alkene mixture is formed. Thus, whenever possible, the most-highly-substituted system is generated; if there is a choice, trans-substituted alkenes predominate over the cis isomers. The reason for this finding is that the reaction is subject to thermodynamic control because deprotonation is *reversible*. Hence, even if in the early stages of the dehydration reaction, a thermodynamically less stable alkene builds up (kinetically), reprotonation will regenerate the original (or a new) carbocation, which ultimately will give the thermodynamic product.

Thermodynamic Control in Acid-Catalyzed Dehydrations

Initial proton loss might give terminal alkene:

$$\mathrm{CH_3CH_2\overset{OH}{\underset{H}{C}}CH_3 \underset{-\,H_2O}{\overset{H^+}{\rightleftharpoons}} CH_3CH_2\overset{+}{\underset{H}{C}}CH_3 \overset{-\,H^+}{\rightleftharpoons} CH_3CH_2CH{=}CH_2}$$

Reprotonation regenerates carbocation, which eventually leads to internal alkene:

$$\mathrm{CH_3CH_2CH{=}CH_2 \overset{H^+}{\rightleftharpoons} CH_3CH_2\overset{+}{\underset{H}{C}}CH_3 \overset{-\,H^+}{\rightleftharpoons} CH_3CH{=}CHCH_3}$$

The final product ratio therefore depends entirely on the relative heats of formation of the alkenes formed. For the isomeric butenes, as noted earlier (Figure 11-12), *cis*-2-butene is more stable than 1-butene by 1.7 kcal $mole^{-1}$, and *trans*-2-butene is more stable than the cis isomer by 1 kcal $mole^{-1}$. Applying our thermodynamic equation (Section 2-6), $\Delta G° = -RT \ln K$, we calculate that the equilibrated butene mixture at room temperature must consist of 74% *trans*-2-butene, 23% of the cis isomer, and only 3% of 1-butene:

$$CH_3CH(OH)CH_2CH_3 \xrightarrow[\text{85\% (Isolated yield)}]{60\%\ H_2SO_4,\ 25°C} \underset{74\%}{trans\text{-}CH_3CH{=}CHCH_3} + \underset{23\%}{cis\text{-}CH_3CH{=}CHCH_3} + \underset{3\%}{CH_3CH_2CH{=}CH_2}$$

(Relative yields)

Isolation of any one of the pure isomeric butenes and reexposure to the acidic reaction conditions will regenerate the thermodynamic equilibrium mixture.

Note that this procedure affords a catalytic method by which less-stable alkenes can be converted into their more-stable isomers:

$$\underset{\textbf{Cis}}{cis\text{-}(CH_3)_3CCH{=}CHC(CH_3)_3} \xrightarrow{\text{Catalytic } H^+} \underset{\textbf{Trans}}{trans\text{-}(CH_3)_3CCH{=}CHC(CH_3)_3}$$

$$\underset{\textbf{Terminal}}{H_2C{=}C_6H_{10}} \xrightarrow{\text{Catalytic } H^+} \underset{\textbf{Internal}}{1\text{-methylcyclohexene } (H_3C\text{–}C_6H_9)}$$

EXERCISE 11-14

Write a mechanism for the following rearrangement. What is the driving force for the reaction?

$$\text{1-ethenyl-2-methylcyclohexane (}H_3C\text{)} \xrightarrow{H^+} \text{1-ethyl-2-methylcyclohexene (}CH_2CH_3,\ H_3C\text{)}$$

BOX 11-1

The Acid-Catalyzed Dehydration of α-Terpineol

Acid-catalyzed dehydration is not generally useful for preparative purposes because product mixtures may be formed. An example is the dehydration of α-terpineol, a pleasant-smelling, naturally occurring, unsaturated terpene alcohol (see Section 4-7) isolated from pine oil:

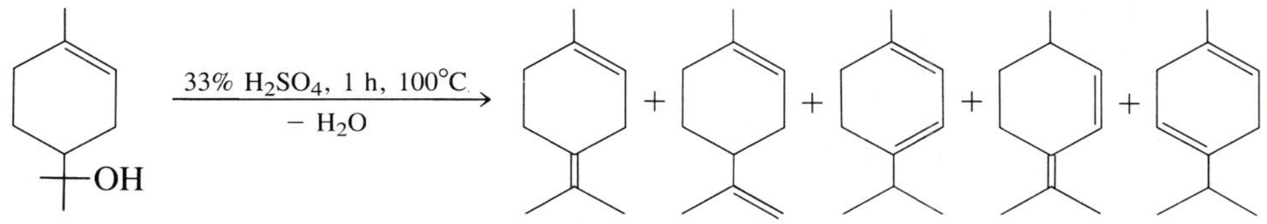

$$\text{α-Terpineol} \xrightarrow[-\,H_2O]{33\%\ H_2SO_4,\ 1\ h,\ 100^\circ C} 15\% + 9\% + 28.5\% + 18.5\% + 15\%$$

α-Terpineol (—OH)

Bimolecular Dehydration of Primary Alcohols

Treatment of primary alcohols with mineral acids at elevated temperatures also leads to alkenes; for example, ethanol gives ethene and 1-propanol yields propene (Section 9-5).

$$CH_3CH_2CH_2OH \xrightarrow{\text{Conc. } H_2SO_4,\ 180^\circ C} CH_3CH{=}CH_2$$

The mechanism of this reaction consists of the initial protonation of oxygen, which is then attacked by hydrogensulfate ion or another alcohol molecule to effect bimolecular elimination of the elements of H_3O^+.

Mechanism:

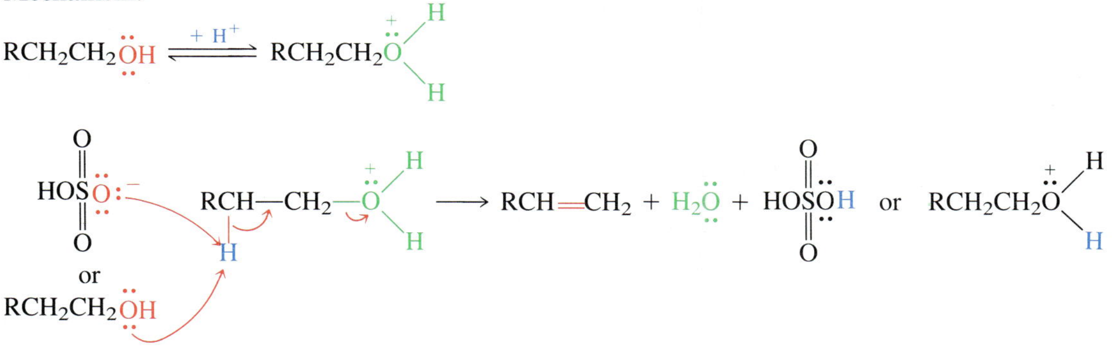

$$RCH_2CH_2\ddot{O}H \underset{}{\overset{+\,H^+}{\rightleftharpoons}} RCH_2CH_2\overset{+}{\ddot{O}}H_2$$

$$HOSO_2\ddot{O}:^- \text{ or } RCH_2CH_2\ddot{O}H + RCH(H){-}CH_2{-}\overset{+}{\ddot{O}}H_2 \longrightarrow RCH{=}CH_2 + H_2\ddot{O} + HOSO_2\ddot{O}H \text{ or } RCH_2CH_2\overset{+}{\ddot{O}}H_2$$

An S_N2 process can compete with the E2 reaction to produce ethers (Section 9-5). As we know, this pathway is predominant at lower temperatures. However, ethers also undergo elimination reactions at elevated temperatures by a mechanism similar to the mechanism of dehydration, and thus furnish the same alkenes.

Alkenes from Ethers

$$CH_3CH_2OCH_2CH_3 \xrightarrow{\text{Conc. } H_2SO_4,\ \Delta} 2\ CH_2{=}CH_2 + H_2O$$

EXERCISE 11-15

Formulate a mechanism for the formation of ethene from ethoxyethane (diethyl ether) on treatment with hot acid.

EXERCISE 11-16

One of the products obtained in the oxidation of cholesterol with acidic sodium dichromate is the ketone A. Can you suggest a mechanism for its formation? Hint: Refer to the discussion of hydroxy cations in Section 9-2.

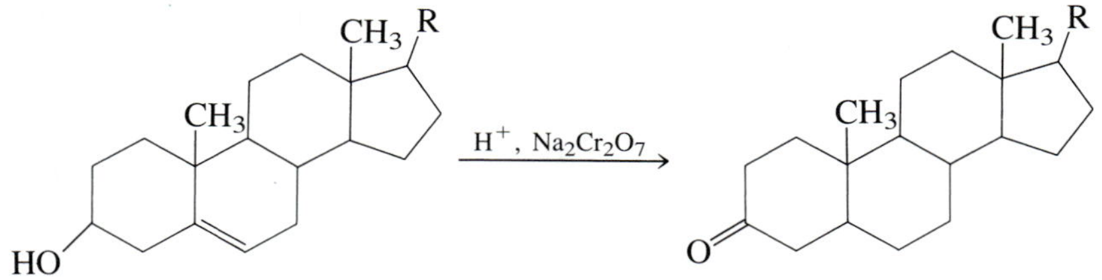

$$\text{Cholesterol} \xrightarrow{H^+,\ Na_2Cr_2O_7} \text{A}$$

Cholesterol **A**

In summary, alkenes can be made by dehydration of alcohols. Secondary and tertiary systems proceed through the intermediacy of carbocations, whereas primary alcohols can undergo E2 reactions from the intermediate oxonium ions. All systems are subject to rearrangement and thus frequently give mixtures.

Summary of New Reactions

1 HYDROGENATION OF ALKENES

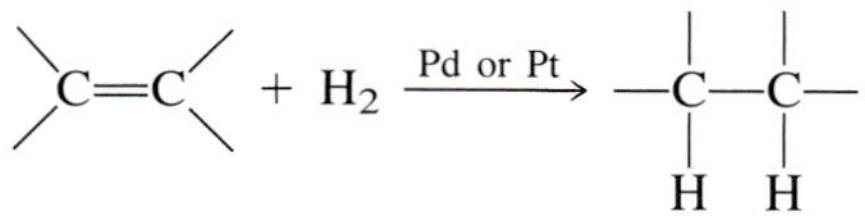

$$>C{=}C< + H_2 \xrightarrow{\text{Pd or Pt}} -\underset{H}{\overset{|}{C}}-\underset{H}{\overset{|}{C}}- \qquad \Delta H^\circ \sim -30 \text{ kcal mole}^{-1}$$

Order of stability of the double bond:

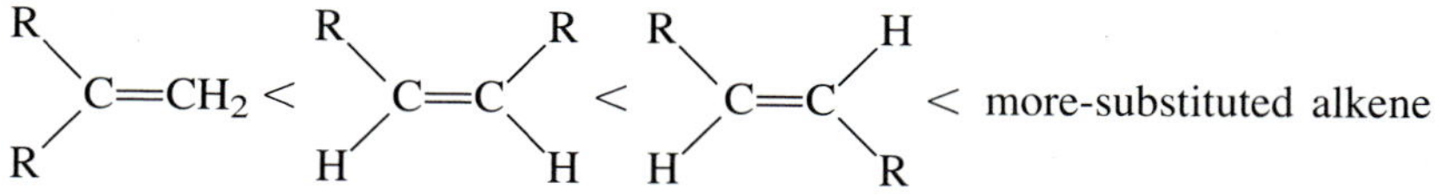

$$R_2C{=}CH_2 < RHC{=}CHR \ (cis) < RHC{=}CHR \ (trans) < \text{more-substituted alkene}$$

Preparation of Alkenes

2 ELIMINATION FROM HALOALKANES

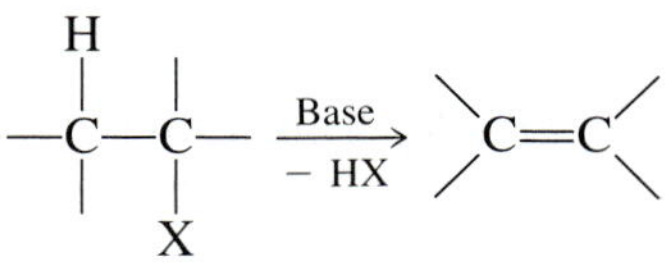

$$-\overset{H}{\underset{|}{\overset{|}{C}}}-\underset{X}{\underset{|}{\overset{|}{C}}}- \xrightarrow[-\ HX]{\text{Base}} >C{=}C<$$

3 **SAYTZEV ELIMINATION**

$$\text{H}-\overset{|}{\underset{|}{\text{C}}}-\overset{|}{\underset{\text{X}}{\text{C}}}-\text{CH}_3 \xrightarrow[-\text{HX}]{\text{Base}} \text{C}=\text{C}(\text{CH}_3)$$

More-substituted (more-stable) alkene

4 **HOFMANN ELIMINATION**

$$\text{H}-\overset{|}{\underset{|}{\text{C}}}-\overset{|}{\underset{\text{X}}{\text{C}}}-\text{CH}_3 \xrightarrow[-\text{HX}]{\text{Bulky base}} \text{C(H)}-\text{C}=\text{CH}_2$$

Less-substituted (less-stable) alkene

5 **STEREOCHEMISTRY OF E2 REACTION**

$$\text{H(R}'\text{)(R)C}-\text{C(R}''\text{)(R}'''\text{)X} \xrightarrow[-\text{HX}]{\text{Base}} \text{R}'\text{(R)C}=\text{C(R}''\text{)(R}'''\text{)}$$

Anti elimination

6 **DEHYDRATION OF ALCOHOLS**

$$-\overset{|}{\underset{\text{H}}{\text{C}}}-\overset{|}{\underset{\text{OH}}{\text{C}}}- \xrightarrow[-\text{H}_2\text{O}]{\text{H}^+} \text{C}=\text{C}$$

Order of reactivity: tertiary > secondary > primary

7 **ALKENE ISOMERIZATION**

$$\text{H(RCH}_2\text{)C}=\text{CH}_2 \xrightarrow{\text{H}^+} \text{RCH}=\text{CHCH}_3$$

Terminal **Internal**

$$\text{R(H)C}=\text{C(R)(H)} \xrightarrow{\text{H}^+} \text{R(H)C}=\text{C(H)(R)}$$

Cis **Trans**

Summary of Important Concepts

1 The IUPAC names of alkenes are derived from alkanes, the longest chain incorporating the double bond serving as the stem. Double-bond isomers include terminal, internal, cis, and trans arrangements. Tri- and tetrasubstituted

alkenes are named according to the E,Z system, in which the R,S priority rules apply.

2 The double bond is composed of a σ and a π part. The former is obtained by overlap of the two sp^2 hybrid lobes on carbon, the latter by interaction of the two remaining p orbitals. The π bond is weaker (~65 kcal mole^{-1}) than its σ counterpart (~108 kcal mole^{-1}) but strong enough to allow for the existence of stable cis and trans isomers.

3 The functional group in the alkenes is flat, sp^2 hybridization being responsible for the possibility of creating dipoles and for the relatively high acidity of the alkenyl hydrogen.

4 Alkenyl hydrogens appear at low field in ^{1}H NMR ($\delta = 4.6$–5.7 ppm) and ^{13}C NMR ($\delta = 100$–140 ppm) experiments. J_{trans} is larger than J_{cis}, $J_{geminal}$ is very small, and $J_{allylic}$ variable but small.

5 The relative stability of isomeric alkenes can be established by comparing heats of hydrogenation. It decreases with decreasing substitution; trans isomers are more stable than their cis analogs.

6 Elimination of haloalkanes (and other alkyl derivatives) may follow the Saytzev rule (nonbulky base, internal alkene formation) or the Hofmann rule (bulky base, terminal alkene formation). Trans alkenes as products predominate over cis alkenes. Elimination is stereospecific, as dictated by the *anti* transition state, which ensures the maximum overlap between the developing π orbital lobes.

Problems

1 Name each of the following molecules in accord with the IUPAC system of nomenclature.

(a) CH_3CH_2 / H on one carbon, CH_3 / H on the other: $(CH_3CH_2)HC{=}CH(CH_3)$, with CH_3CH_2 and CH_3 on the upper side, H and H on the lower side

(b) $CH_3CH_2CHCH{=}CH_2$ with CH_3CH_2 on the CH

(c) Cl, H on one carbon; H, $CH_2CH_2CHCH_3$ (with OH on the CH) on the other: Cl upper left, H lower left, H upper right, $CH_2CH_2CH(OH)CH_3$ lower right

(d) F upper left, Cl lower left, Br upper right, I lower right: C=C

(e) $HOCH_2$ upper left, CH_3CH_2 lower left, $CHCH_3$ (with CF_3) upper right, H lower right: C=C

(f) CH_3CH_2 upper left, H lower left, Cl upper right, Cl lower right: C=C

(g) CH_3O upper left, H_3C lower left, OCH_3 upper right, H lower right: C=C

(h)

$$\begin{array}{c} CH_3 \\ | \\ CH_3CH \qquad\qquad CH_2CH_2CH_3 \\ \diagdown \quad \diagup \\ C{=}C \\ \diagup \quad \diagdown \\ H_3C \qquad\qquad H \end{array}$$

(i) (cyclohexene ring bearing CH_3 and CH_2CH_3)

CH_3 CH_2CH_3

(j) (cyclohexane ring bearing $OCH_2CH{=}CH_2$)

$OCH_2CH{=}CH_2$

2 Determine structures for each of the following molecules from the indicated ^{1}H NMR spectra. Consider stereochemistry, where applicable.

(a) C_4H_7Cl, NMR spectrum A.
(b) $C_5H_8O_2$, NMR spectrum B (on page 450).
(c) $C_6H_{11}I$, NMR spectrum C (on page 451).
(d) Another $C_6H_{11}I$, NMR spectrum D (on page 452).
(e) $C_3H_4Cl_2$, NMR spectrum E (on page 453).

3 Explain the splitting patterns in ^{1}H NMR spectrum D in detail. Inset is $\delta = 5.7$–6.7 region expanded fivefold.

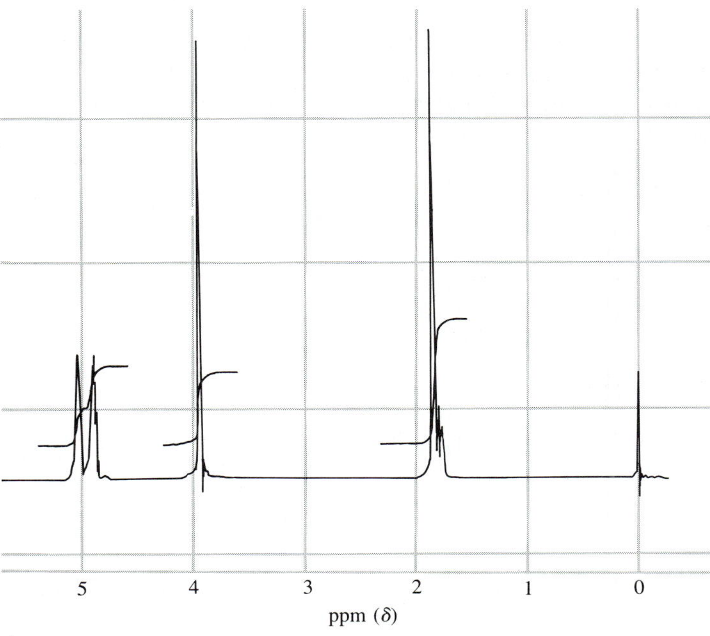

A

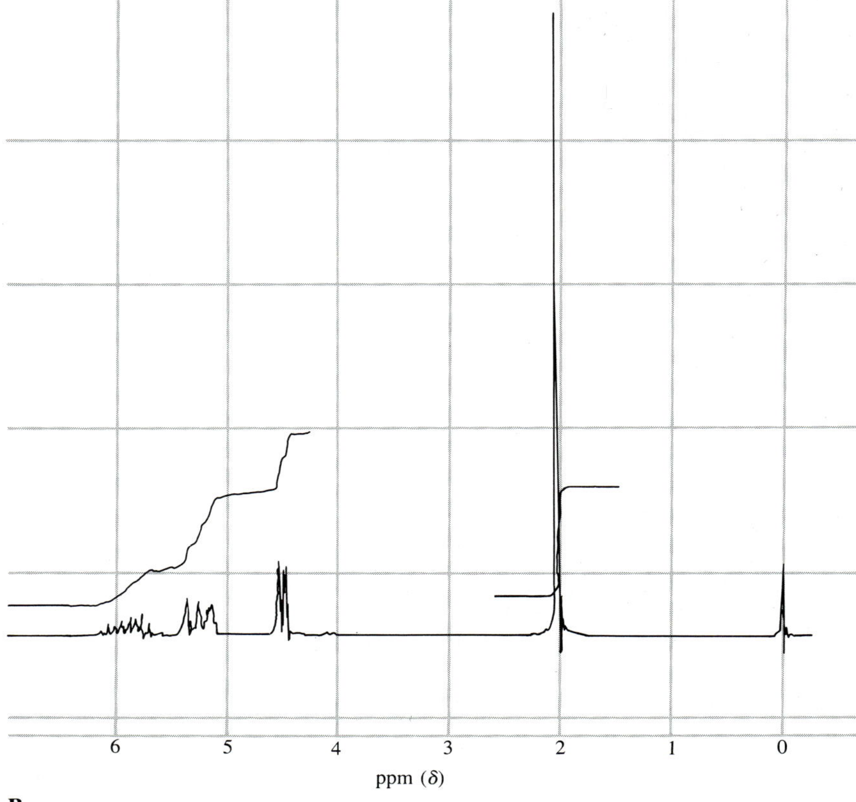

B

4 For each of the following pairs of alkenes, indicate whether measurements of dipole moments alone would be sufficient to distinguish the compounds from one another. Where possible, predict which compound would have the larger dipole moment.

(a) $\mathrm{H(H_3C)C{=}C(CH_3)H}$ — structure drawn with H and CH_3 above, H_3C and H below:

$$\begin{array}{ccc} \mathrm{H} & & \mathrm{CH_3} \\ & \mathrm{C{=}C} & \\ \mathrm{H_3C} & & \mathrm{H} \end{array} \quad \text{and } \mathrm{CH_3CH_2CH{=}CH_2}$$

(b)

$$\begin{array}{ccc} \mathrm{H_3C} & & \mathrm{CH_2CH_2CH_3} \\ & \mathrm{C{=}C} & \\ \mathrm{H} & & \mathrm{H} \end{array} \quad \text{and} \quad \begin{array}{ccc} \mathrm{CH_3CH_2} & & \mathrm{CH_2CH_3} \\ & \mathrm{C{=}C} & \\ \mathrm{H} & & \mathrm{H} \end{array}$$

(c)

$$\begin{array}{ccc} \mathrm{H_3C} & & \mathrm{CH_2CH_2CH_3} \\ & \mathrm{C{=}C} & \\ \mathrm{H} & & \mathrm{H} \end{array} \quad \text{and} \quad \begin{array}{ccc} \mathrm{H} & & \mathrm{CH_2CH_3} \\ & \mathrm{C{=}C} & \\ \mathrm{CH_3CH_2} & & \mathrm{H} \end{array}$$

5 The molecular formulas and ^{13}C NMR data (in ppm) for several compounds are given on the facing page. The splitting pattern of each signal, taken from the undecoupled spectrum, is given in parentheses. Deduce structures for each.

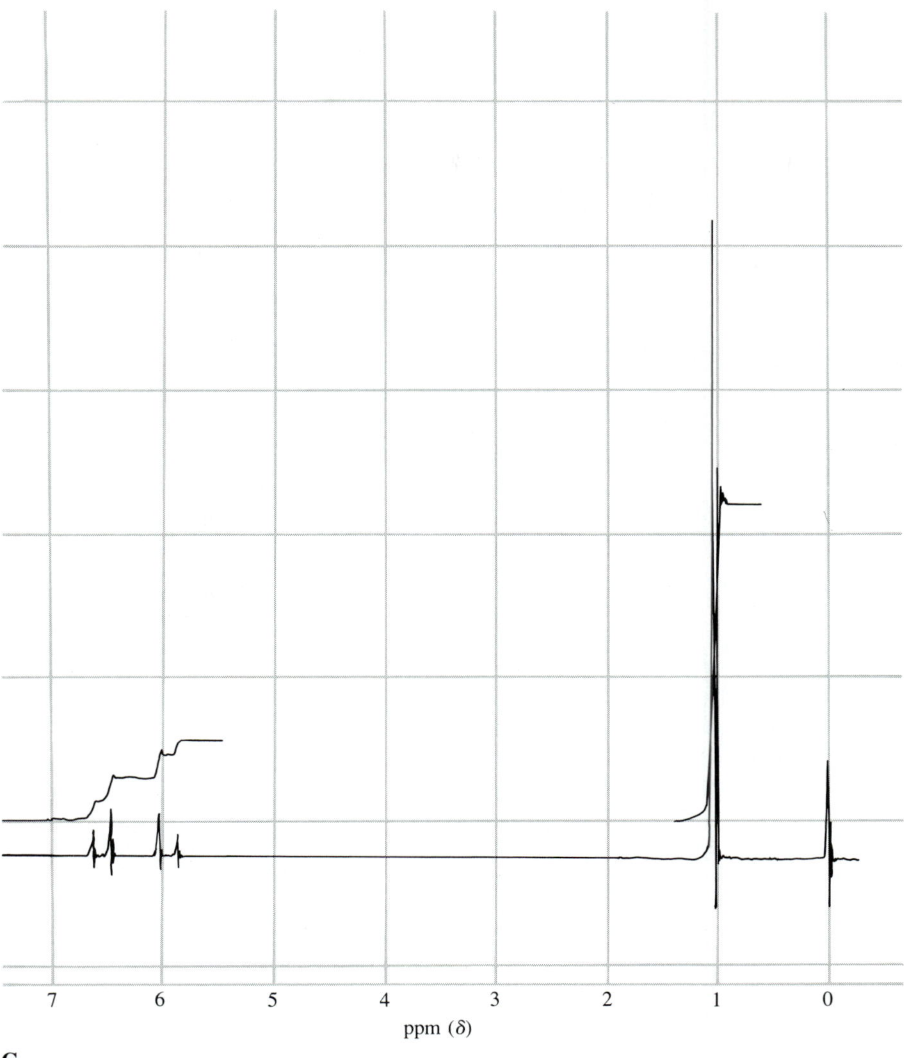

C

(a) C_4H_6: 30.2(t), 136.0(d).
(b) C_4H_6O: 18.2(q), 134.9(d), 153.7(d), 193.4(d).
(c) C_4H_8: 13.6(q), 25.8(t), 112.1(t), 139.0(d).
(d) $C_5H_{10}O$: 17.6(q), 25.4(q), 58.8(t), 125.7(d), 133.7(s).
(e) C_5H_8: 15.8(t), 31.1(t), 103.9(t), 149.2(s).
(f) C_7H_{10}: 25.2(t), 41.9(d), 48.5(t), 135.2(d). (This one is difficult. Hint: The molecule has one double bond. How many rings must it have?)

6 Data from the ^{13}C NMR spectra of several compounds, all with the formula C_5H_{10}, are given below. The splitting of each signal, taken from the undecoupled spectrum, is given in parentheses after the chemical-shift value. Propose structures for each.

(a) 25.3(t).
(b) 13.3(q), 17.1(q), 25.5(q), 118.7(d), 131.7(s).
(c) 12.0(q), 13.8(q), 20.3(t), 122.8(d), 132.4(d).

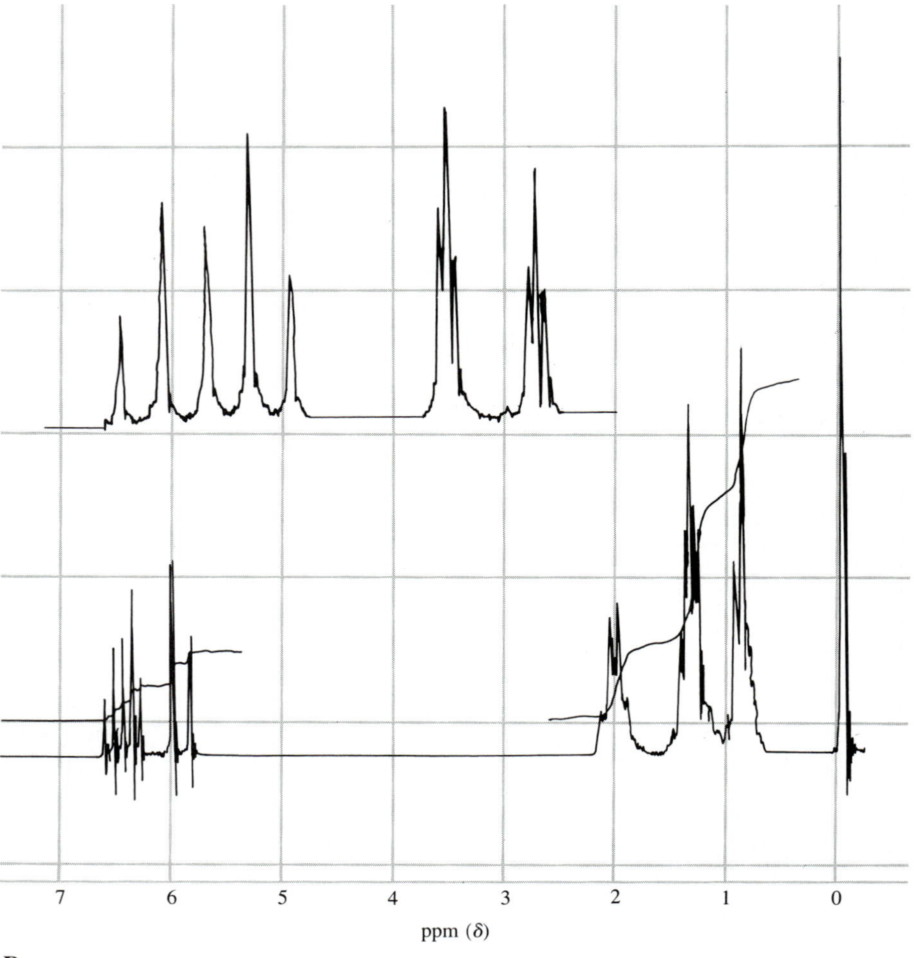

D

7 Rewrite each of the following groups of alkenes in order of (1) increasing stability of the double bond and (2) increasing heat of hydrogenation.

(a) $CH_2{=}CH_2$ $\quad (H_3C)_2C{=}C(CH_3)_2 \quad (H_3C)_2C{=}CH_2$

(b) 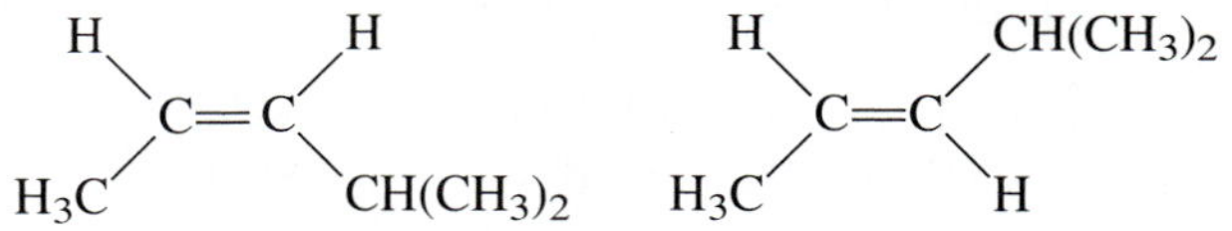 $H(H_3C)C{=}C(H)CH(CH_3)_2$ (cis: H, H on same side) $\quad H(H_3C)C{=}C(CH(CH_3)_2)H$ (trans)

$H((CH_3)_2CH)C{=}C(H)CH(CH_3)_2$ (cis)

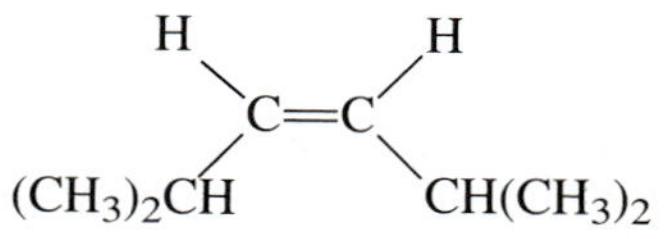

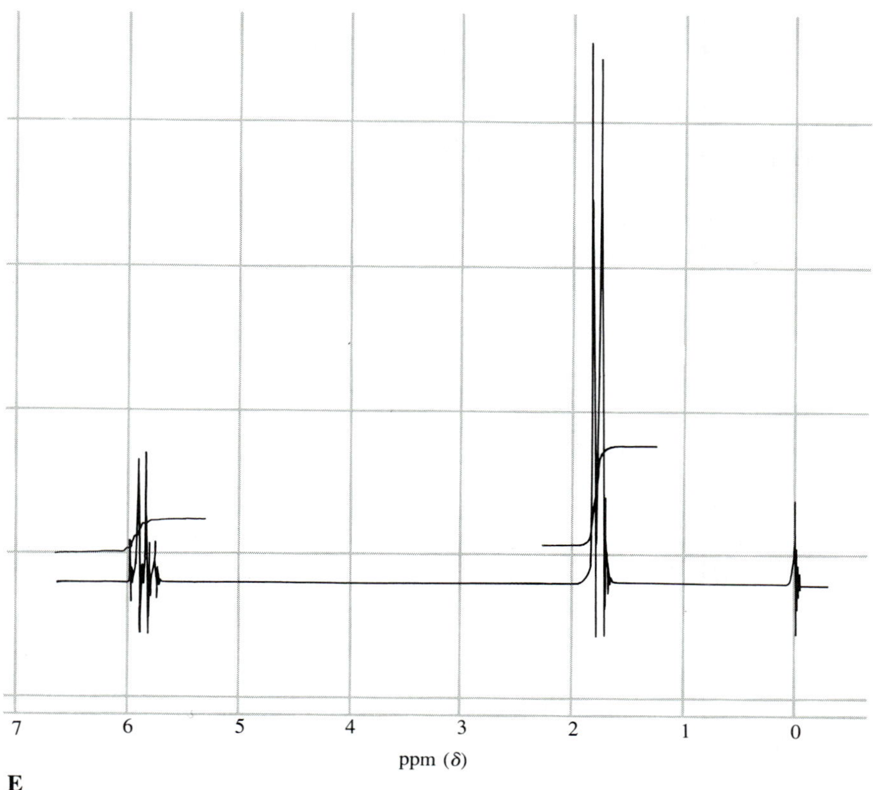

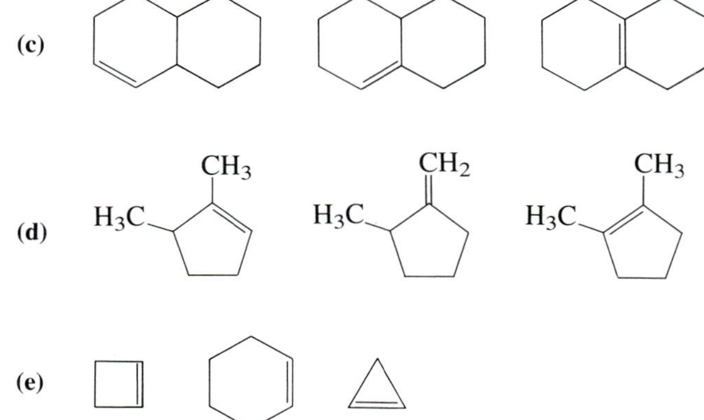

8 Using the heats of formation given (in kcal mole^{-1} at 25°C), calculate the equilibrium mixture compositions for each group of alkenes below. In each case, explain the order of stability.

(a) 1-Pentene (−5.3), *cis*-2-pentene (−7.0), *trans*-2-pentene (−7.9).

(b) 3-Methyl-1-butene (−6.9), 2-methyl-1-butene (−8.6), 2-methyl-2-butene (−10.1).

9 1-Methylcyclohexene is more stable than methylenecyclohexane (A, at the right), but methylenecyclopropane (B) is more stable than 1-methylcyclopropene. Explain.

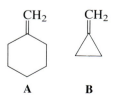

CH_2 CH_2

A **B**

10 The E2 elimination reaction between 2-bromobutane and sodium ethoxide in ethanol gives rise to three products. What are they? Draw Newman projections of the reactive conformations that are responsible for the formation of each of the elimination products. What do these conformations tell you about the relative ease of each product's formation?

11 What key structural feature distinguishes haloalkanes that give more than one stereoisomer on E2 elimination (e.g., 2-bromobutane, Problem 10) from those that give only a single isomer exclusively (e.g., 2-bromo-3-methylpentane, Section 11-5)?

12 Write the most likely major product(s) of each of the following haloalkanes with (1) sodium ethoxide in ethanol or (2) potassium *tert*-butoxide in 2-methyl-2-propanol (*tert*-butanol).

(a) Chloromethane.
(b) 1-Bromopentane.
(c) 2-Bromopentane.
(d) 1-Chloro-1-methylcyclohexane.
(e) (1-Bromoethyl)cyclopentane.
(f) (2*R*,3*R*)-2-Chloro-3-ethylhexane.
(g) (2*R*,3*S*)-2-Chloro-3-ethylhexane.
(h) (2*S*,3*R*)-2-Chloro-3-ethylhexane.

13 Draw Newman projections of the four stereoisomers of 2-bromo-3-methylpentane in the conformations required for E2 elimination. (See the structures labeled "Stereospecificity in the E2 Reaction of 2-Bromo-3-Methylpentane" on page 440). Are the reactive conformations also the most-stable conformations? Explain.

14 Write the products of E2 elimination from each of the following isomeric halogenated compounds.

(a) C_6H_5 / H—+—Br / H—+—CH_3 / C_6H_5

(b) C_6H_5 / H—+—Br / H_3C—+—H / C_6H_5

One of these compounds undergoes elimination at a rate fifty times as fast as the other. Which compound is it? Why? How does a factor of 50 in rate correspond to activation energy (use T = 298 K)? Does this energy difference make sense chemically? Hint: See Problem 13.

15 The following reaction gives two products whose ratio depends on the reaction conditions. Which will be the major product with dilute KOH in ethanol? With concentrated KOH in ethanol? Explain.

CH_3 CH_3 Br CH_3 H H ⟶ CH_3 CH_3 CH_3 H (a) + CH_3 CH_2 CH_3 H H (b)

16 Explain in detail the differences between the mechanisms giving rise to the following two experimental results.

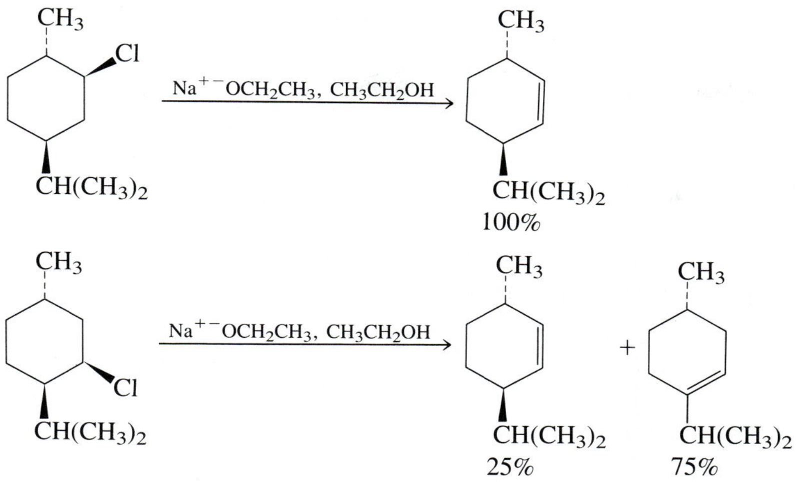

CH_3, Cl, $CH(CH_3)_2$ $\xrightarrow{Na^{+}\,{}^{-}OCH_2CH_3,\ CH_3CH_2OH}$ CH_3, $CH(CH_3)_2$ 100%

CH_3, Cl, $CH(CH_3)_2$ $\xrightarrow{Na^{+}\,{}^{-}OCH_2CH_3,\ CH_3CH_2OH}$ CH_3, $CH(CH_3)_2$ 25% + CH_3, $CH(CH_3)_2$ 75%

17 Write the structure of the haloalkane that will give each alkene below upon base-promoted elimination, with the highest possible stereoselectivity.

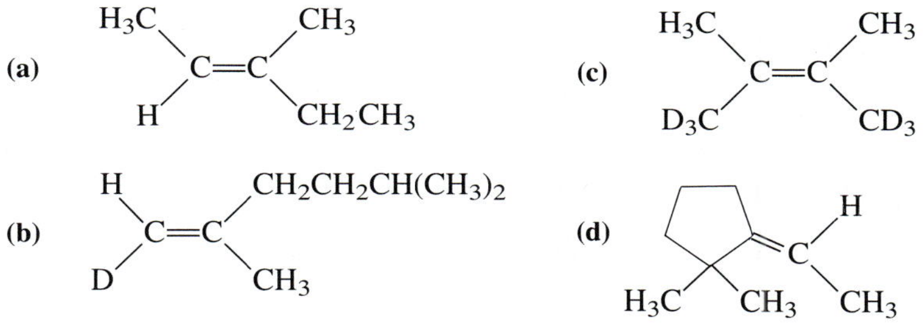

(a) H_3C, CH_3, C=C, H, CH_2CH_3

(b) H, $CH_2CH_2CH(CH_3)_2$, C=C, D, CH_3

(c) H_3C, CH_3, C=C, D_3C, CD_3

(d) H, C, H_3C, CH_3, CH_3

18 Referring to the answer to Problem 14 in Chapter 7, predict (qualitatively) the relative amounts of isomeric alkenes that are formed in the elimination reactions shown.

19 Referring to the answers to Problem 3 of Chapter 9, predict (qualitatively) the relative yields of all the alkenes formed in each reaction.

20 The elimination reactions of the following deuterium-labeled molecules proceed as shown. Explain, mechanistically.

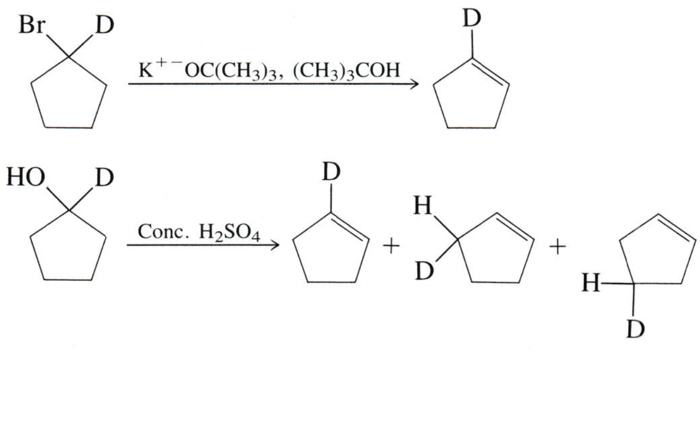

Br, D $\xrightarrow{K^{+}\,{}^{-}OC(CH_3)_3,\ (CH_3)_3COH}$ D

HO, D $\xrightarrow{\text{Conc. } H_2SO_4}$ D + H, D + H, D

21 Explain the following transformations from both energetic and mechanistic viewpoints.

(a) $(CH_3)_2CHCH_2CH_2OH \xrightarrow{H_2SO_4,\ \Delta} (CH_3)_2C{=}CHCH_3$

(b) $(CH_3)_3CCH{=}CH_2 \xrightarrow{H_2SO_4,\ \Delta} (CH_3)_2C{=}C(CH_3)_2$

(c) H_3C CH_2OH $\xrightarrow{H_2SO_4,\ \Delta}$ CH_3

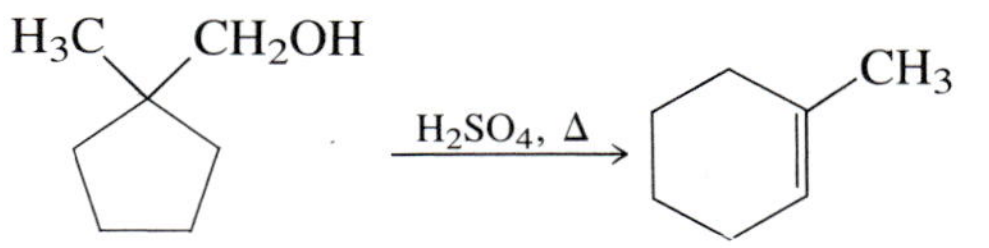

22 Write, in detail, the mechanism of the acid-catalyzed dehydration of 1-propanol, using unprotonated alcohol as the necessary base in the elimination step.

23 Referring to Problem 25 of Chapter 7, write the structure of the alkene that you would expect to be formed as the major product from E2 elimination of each of the chlorinated steroids shown.

24 An enzyme has been discovered in the bacteria *Escherichia coli* that catalyzes the dehydration of a thioester derivative of (−)-3-hydroxydecanoic acid to give a mixture of the corresponding thioester derivatives of *trans*-2-decenoic acid and *cis*-3-decenoic acid:

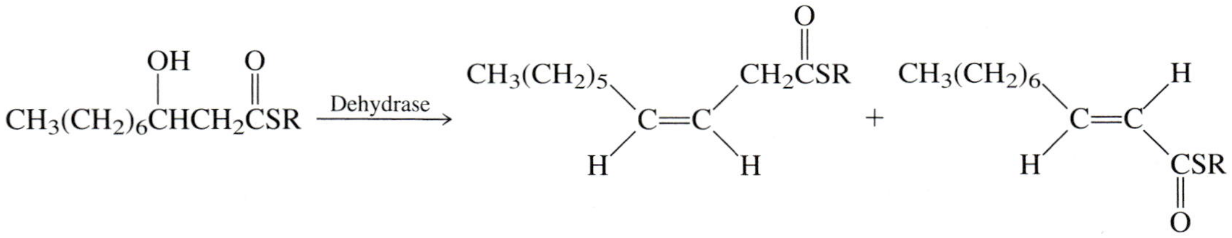

$$CH_3(CH_2)_6\overset{OH}{\overset{|}{C}}HCH_2\overset{O}{\overset{\|}{C}}SR \xrightarrow{\text{Dehydrase}} \underset{H}{\overset{CH_3(CH_2)_5}{}}C{=}C\underset{H}{\overset{CH_2\overset{O}{\overset{\|}{C}}SR}{}} + \underset{H}{\overset{CH_3(CH_2)_6}{}}C{=}C\underset{\underset{O}{\overset{\|}{C}SR}}{\overset{H}{}}$$

How does this result compare with those that can be expected from simple acid-catalyzed dehydration (e.g., H_2SO_4 and heat)?

25 The prostaglandins are a family of extremely potent hormonelike substances with many biological functions, including muscle stimulation, inhibition of platelet aggregation, lowering of blood pressure, enhancement of inflammatory reactions, and induction of labor in childbirth. The most-active mammalian prostaglandin is "prostaglandin E2." Illustrated below and at the top of the facing page are three members of the prostaglandin family, followed by the proton-decoupled ^{13}C NMR data for prostaglandin E2. On the basis of the NMR data, which of the three structures is that of prostaglandin E2?

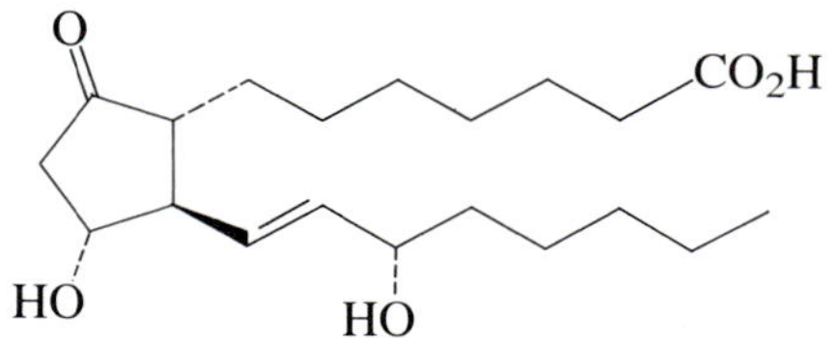

O, CO_2H, HO, HO

A

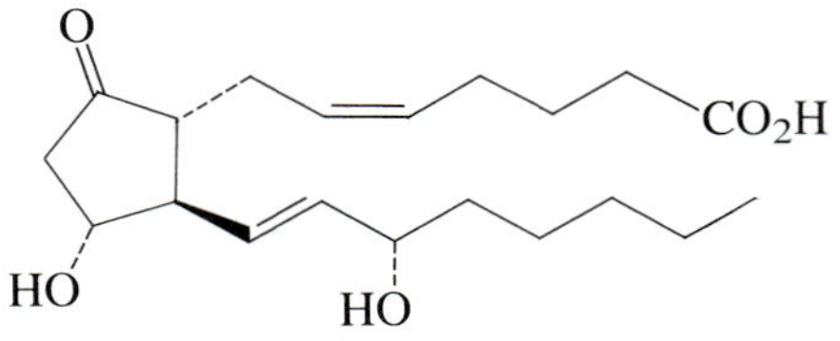

O, CO_2H, HO, HO

B

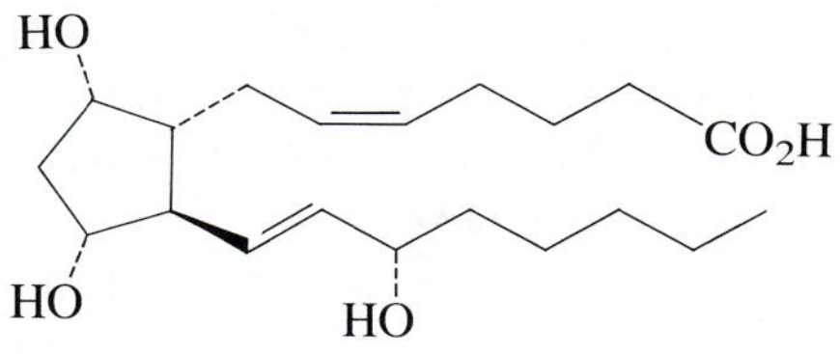

HO

CO_2H

HO HO

C

^{13}C NMR: $\delta = 14.1, 22.7, 24.6, 25.2, 26.5, 31.8, 33.4, 37.0, 46.2, 53.6, 54.6, 72.0, 73.2, 126.7, 131.0, 131.7, 136.7, 177.2$, and 215.4 ppm.

26 A very simple molecule that has been used effectively as a starting material for prostaglandin synthesis has the formula $C_5H_8O_2$. Deduce its structure from the following NMR data:

^1H NMR: $\delta = 1.54$ ppm (doublet of triplets; J for doublet splitting = 15 Hz; J for triplet splitting = 4 Hz; 1 H).

$\delta = 2.70$ ppm (doublet of triplets; J for doublet splitting = 15 Hz; J for triplet splitting = 7 Hz; 1 H).

$\delta = 3.20$ ppm (very broad signal; 2 H).

$\delta = 4.65$ ppm (doublet of doublets of doublets, J values = 1, 4, and 7 Hz; 2 H).

$\delta = 6.03$ ppm (doublet, $J = 1$ Hz, 2 H).

Proton-decoupled ^{13}C NMR: 3 signals (at approximately $\delta = 45$, 80, and 135 ppm).

Hints: See "Coupling Between Close-Lying Hydrogens" on page 385 for an explanation of the 15-Hz coupling between the signals at $\delta = 1.54$ and 2.70 ppm. The ^1H NMR spectrum exhibits the complications of coupling to nonequivalent neighbors (Section 10-6). The 1-Hz and 4-Hz J values are a bit lower than might be expected for the types of hydrogens resonating. Use the ^{13}C data to identify symmetry in the molecule.

27 The *citric acid cycle* is a series of biological reactions that play a central role in cell metabolism. The cycle includes dehydration reactions of both malic and citric acids, yielding fumaric and aconitic acids, respectively (all common names). Both proceed strictly by enzyme-catalyzed *anti*-elimination mechanisms.

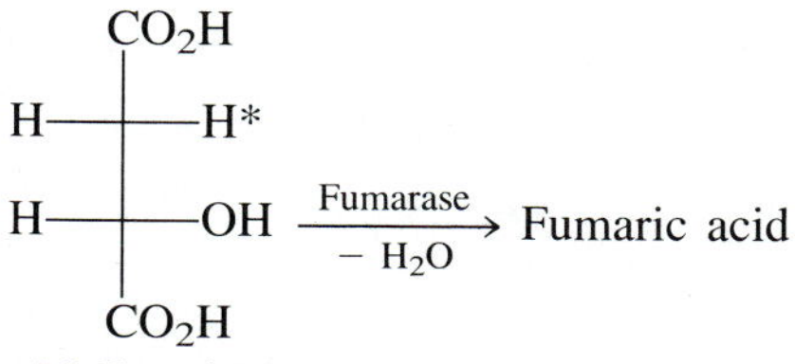

CO_2H
H—|—H*
H—|—OH $\xrightarrow[-\,H_2O]{\text{Fumarase}}$ Fumaric acid
CO_2H
Malic acid

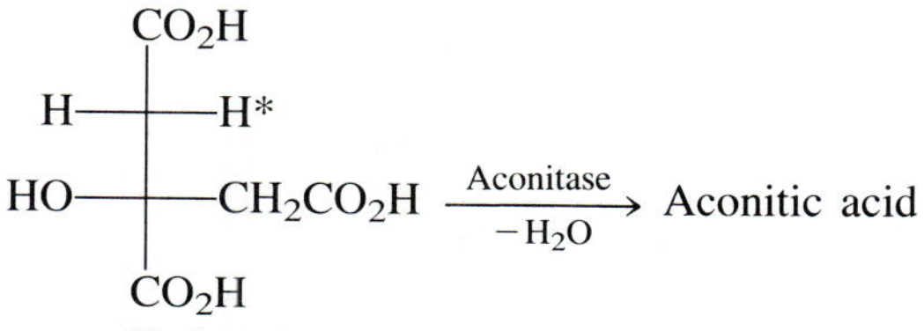

CO_2H
H—|—H*
HO—|—CH_2CO_2H $\xrightarrow[-\,H_2O]{\text{Aconitase}}$ Aconitic acid
CO_2H
Citric acid

(a) In each dehydration, only the hydrogen identified by an asterisk is removed, together with the —OH group on the carbon below. Write the structures for fumaric and aconitic acids as they are formed in these reactions. Make sure that the stereochemistry of each product is clearly indicated.

(b) Specify the stereochemistry of each of these products, using either cis-trans or *E*,*Z* notation, as appropriate.

(c) Are the hydrogens on the CH_2 group of malic acid enantiotopic or diastereotopic? Answer the same question for citric acid.

(d) Isocitric acid

$$\begin{array}{c} \text{OH} \\ | \\ HO_2CCHCHCH_2CO_2H \\ | \\ CO_2H \end{array}$$

is also dehydrated by aconitase. How many stereoisomers can exist for isocitric acid? Remembering that this reaction proceeds through *anti* elimination, write the structure of a stereoisomer of isocitric acid that will give on dehydration the same isomer of aconitic acid that is formed from citric acid. Label the chiral carbons in this isomer of isocitric acid using *R*,*S* notation.

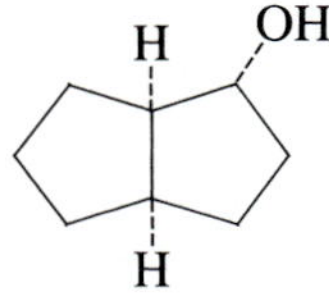

28 Identify the compounds labeled A, B, and C from the information given below, and explain the chemistry that is taking place.

Reaction of *exo*-bicyclo[3.3.0]octan-2-ol (margin) with 4-methylbenzenesulfonyl chloride in pyridine produced A ($C_{15}H_{20}SO_3$). Reaction of A with lithium diisopropylamide (LDA, Section 7-5) produces a single product, B (C_8H_{12}), which displays in its ^1H NMR a two-proton multiplet at about δ = 5.6 ppm. If, however, A is treated with NaI before the reaction with LDA, two products are formed: B and an isomer, C, whose NMR shows a multiplet at δ = 5.2 ppm that integrates as only one proton.

CHAPTER 12

The Reactions of Alkenes

The double bond can undergo a variety of reactions, many of which lead to saturated (Section 2-1) products by **addition.** Such additions include the participation of catalytically activated hydrogen, electrophilic reagents (among which are the proton, halogens, and mercuric ion), borane to give alkylboranes (which can be further functionalized), oxidizing agents (which may lead to diols or to the complete rupture of the double bond), and free-radical reagents.

This chapter begins by demonstrating that additions to the π bond are generally exothermic; they are certain to occur if suitable pathways can be found.

12-1 Thermodynamic Feasibility of Addition Reactions

The π bond is relatively weak, and the chemical characteristics of the alkenes are largely governed by its reactions. The most common transformation is addition of a reagent A–B to give a saturated compound. The thermodynamic feasibility of this process depends on the strength of the π bond, the dissociation energy, $DH^\circ{}_{A-B}$, and the strengths of the newly formed bonds of A and B to carbon. Remember that we can estimate the ΔH° of such reactions by subtracting the combined strength of the bonds made from that of the bonds broken (Section 2-6):

$$\Delta H^\circ = [DH^\circ\ (\pi\ \text{bond}) + DH^\circ\ (\text{A–B})] - [DH^\circ\ (\text{C–A}) + DH^\circ\ (\text{C–B})]$$

in which C stands for carbon.

Table 12-1 gives the DH° values (obtained by using the data from Table 3-1 and Section 3-5 and by equating the strength of the π bond to 65 kcal mole^{-1}) and the estimated ΔH° values for various additions to ethene. In all the examples, the combined strength of the bonds formed exceeds, sometimes significantly, that of the bonds broken. Therefore, if kinetically feasible, additions to alkenes should proceed to products with release of energy.

General Addition to the Alkene Double Bond

$$\text{>C=C<} + \text{A–B} \xrightarrow{\Delta H^\circ = ?} \text{–}\overset{\text{A}}{\overset{|}{\underset{|}{\text{C}}}}\text{–}\overset{\text{B}}{\overset{|}{\underset{|}{\text{C}}}}\text{–}$$

TABLE 12-1
Estimated $\Delta H°$ (all values in kcal mole^{-1}) for additions to ethene

$CH_2{=}CH_2$	+	A—B	⟶	H—C(A)—C(B)—H (with H, H below)		
$DH°$ (π bond)		$DH°$ (A–B)		$DH°$ (A–C)	$DH°$ (B–C)	~$\Delta H°$
Hydrogenation						
$CH_2{=}CH_2$	+	H—H	⟶	H–CH_2—CH_2–H		−27
65		104		98	98	
Bromination						
$CH_2{=}CH_2$	+	Br—Br	⟶	H—C(Br)(H)—C(Br)(H)—H		−25
65		46		~68	~68	
Hydrochlorination						
$CH_2{=}CH_2$	+	H—Cl	⟶	H—C(H)(H)—C(Cl)(H)—H		−10
65		103		~98	80	
Hydration						
$CH_2{=}CH_2$	+	H—OH	⟶	H—C(H)(H)—C(OH)(H)—H		−6
65		119		~98	92	

EXERCISE 12-1

Calculate the ΔH° for the addition of H_2O_2 to ethene to give 1,2-ethanediol (ethylene glycol) (DH°_{HO-OH} = 51 kcal mole^{-1}).

12-2 The Catalytic Hydrogenation of Alkenes

The simplest of the reactions of the double bond is its saturation with hydrogen. This transformation was discussed in Section 11-4 in connection with the use of heats of hydrogenation to estimate the relative stability of various substituted alkenes. Hydrogenation requires a catalyst, which may be either heterogeneous (Section 8-4) or homogeneous.

Heterogeneous Catalysis in Hydrogenation

The hydrogenation of an alkene to an alkane, although exothermic, will not take place even at elevated temperatures. Ethene and hydrogen can be heated in the gas phase to 200°C for prolonged periods without any measurable change. However, as soon as a catalyst is added, hydrogenation proceeds even at room temperature at a steady rate. The catalysts frequently are the same as those used for the catalytic hydrogenation of carbonyl compounds to alcohols (Section 8-4): heterogeneous materials such as palladium (e.g., dispersed on carbon, Pd-C), platinum (Adams's* catalyst, PtO_2, which is converted into colloidal platinum metal in the presence of hydrogen), and nickel (finely dispersed, as in a preparation called Raney† nickel, Ra-Ni). The major function of the catalyst is the activation of hydrogen to generate metal-bound hydrogen on the catalyst surface (Figure 12-1). Without the metal, thermal cleavage of the strong H–H bond is energetically prohibitive. The solvents commonly used in such hydrogenations are methanol, ethanol, ethanoic (acetic) acid, and ethyl ethanoate (ethyl acetate), among others.

FIGURE 12-1

Catalytic hydrogenation of ethene to ethane.

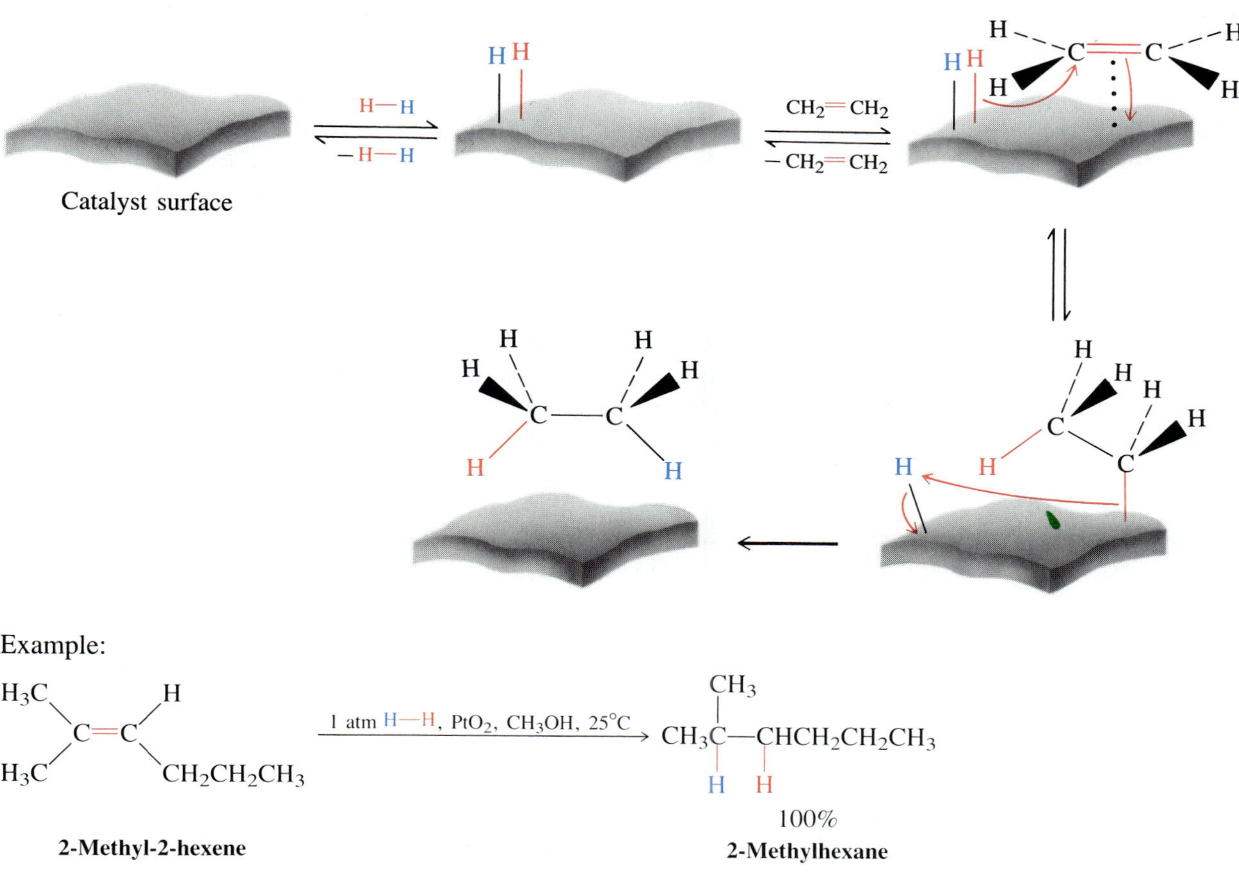

*Professor Roger Adams, 1889–1971, University of Illinois.

†Dr. Murray Raney, 1885–1966, Raney Catalyst Company.

Hydrogenation Is Stereospecific

An important feature of hydrogenation is *stereospecificity*. The two hydrogen atoms are added to the same side of the double bond (*syn* addition). For example, 1-ethyl-2-methylcyclohexene is hydrogenated over platinum to give specifically *cis*-1-ethyl-2-methylcyclohexane. Addition of hydrogen can be from above or from below the molecular plane with equal probability. Therefore, both stereocenters are generated as image and mirror image, and the product is racemic.

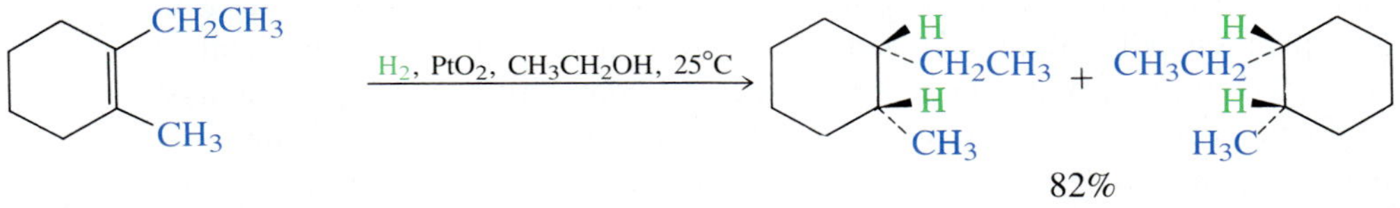

H_2, PtO_2, CH_3CH_2OH, 25°C

82%

1-Ethyl-2-methylcyclohexene

***cis*-1-Ethyl-2-methylcyclohexane (Racemic)**

Deuterium gas can be employed to effect *syn deuteration*. For example, deuteration of *cis*-2-butene gives *meso*-2,3-dideuteriobutane (note that simple addition of the two hydrogen or deuterium atoms *syn* to the double bond on paper results in the eclipsed conformation of the product):

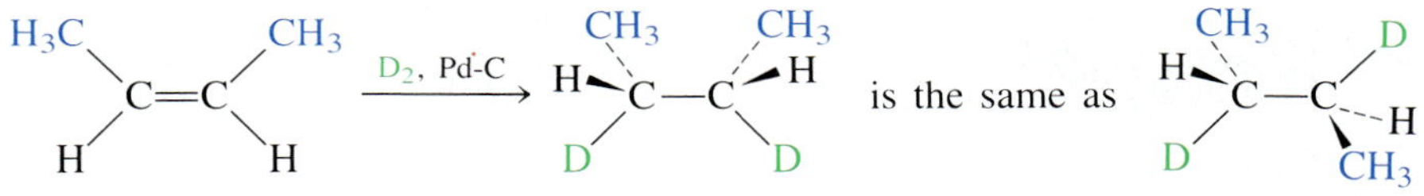

D_2, Pd-C

is the same as

Eclipsed **Staggered**

***meso*-2,3-Dideuteriobutane**

In this case, addition from either above or below the molecular plane makes the same stereoisomer. In contrast, deuteration of *trans*-2-butene leads to the racemic 2*R*,3*R* and 2*S*,3*S* mixture of 2,3-dideuteriobutane, one enantiomer being formed by *syn* addition of deuterium from one side of the π system, the other by addition from the opposite side.

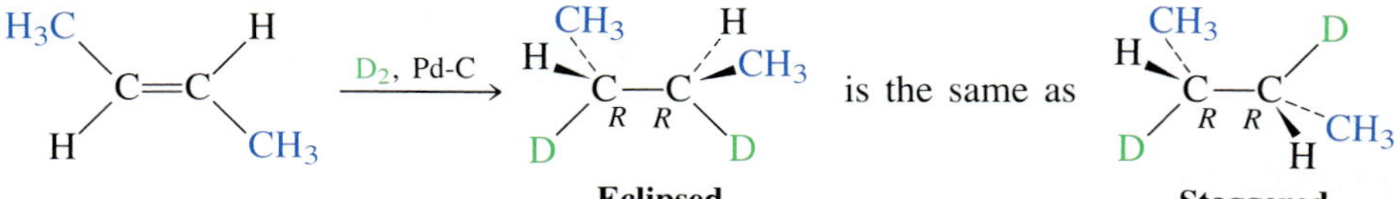

D_2, Pd-C

is the same as

Eclipsed **Staggered**

(2*R*,3*R*)-2,3-Dideuteriobutane

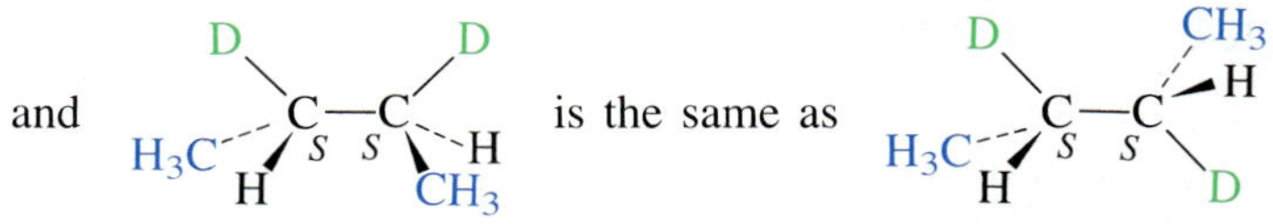

and

is the same as

Eclipsed **Staggered**

(2*S*,3*S*)-2,3-Dideuteriobutane

EXERCISE 12-2

Draw the products of hydrogenation of 2,3-dideuterio-*cis*-2-pentene. Clearly show the stereochemistry.

When the alkene is chiral, addition of hydrogen may still give only one diastereomer if it occurs with preference for one π face of the molecule. This effect is particularly pronounced in rigid bicyclic systems. For example, hydrogenation of the bicyclic alkene car-3-ene, a constituent of turpentine, over platinum gives only one saturated product, with the common name *cis*-carane. The prefix cis indicates that the methyl group and the cyclopropane ring are on the same side of the cyclohexane ring. This result shows that hydrogen has been added only from the less-hindered side of the double bond, opposite the three-membered ring, thus pushing the methyl group cis to that ring. (Make a model and use a table top to represent the catalyst surface, as in Figure 12-1).

CH_3 H_3C CH_3 — 100 atm H_2, PtO_2, CH_3CH_2OH, 25°C → H_3C H H H_3C CH_3 not H_3C H H H_3C CH_3

Car-3-ene 98% ***cis*-Carane**

EXERCISE 12-3

Explain the following result:

$C(CH_3)_3$ — D_2, Pd-C, CH_3OH → $C(CH_3)_3$ D H D H + $C(CH_3)_3$ H D H D

20% 80%

EXERCISE 12-4

Catalytic hydrogenation of (*S*)-2,3-dimethyl-1-pentene gives only one optically active product. Show the product and explain the result.

In summary, hydrogenation of the double bond in alkenes requires a catalyst. This transformation occurs stereospecifically by *syn* addition, from the least-hindered side of the molecule. Achiral alkenes may furnish chiral products but in racemic form. Chiral alkenes may give diastereomers; whether such reactions are stereoselective may depend on steric factors.

12-3 The Basic and Nucleophilic Character of the Pi Bond: Electrophilic Additions

As noted earlier, the π electrons of a double bond are not as strongly bound as those of a σ bond. The electron cloud above and below the molecular plane of the alkene is polarizable and subject to attack by electron-deficient species, as are the lone electron pairs in typical Lewis bases. The proton is not the only electrophile that attacks double bonds (Section 11-6); halogens and mercuric ion, among others, also have this capacity. As in hydrogenation, the double bond is altered by *addition,* but by different mechanisms. We shall see that these transformations can be regioselective and stereospecific. Let us begin with the simplest of all electrophiles, the proton.

Electrophilic Attack by Protons Gives Carbocations

The proton of a strong acid may add to a double bond to yield a carbocation (Section 11-6). The transition state for the process is the same as that formulated for the deprotonation step in the E1 reaction (Section 7-4, Figure 7-7). In the absence of a good nucleophile capable of trapping the carbocation, rearrangement is observed. However, in the presence of such a nucleophile, particularly at low temperatures, the carbocation is intercepted to give the product of an **electrophilic addition** to the double bond. Low temperatures are necessary because the reaction can be reversed on heating. For example, treatment of alkenes with hydrogen halides leads to such additions.

General Electrophilic Addition of HX to Alkenes

$$\text{C=C} + H^+ \xrightarrow[\text{Electrophilic attack}]{} \text{H–C–C}^+ \xrightarrow[\text{Nucleophilic trapping}]{X^-} \text{H–C–C–X}$$

Examples:

$$CH_3CH_2CH_2CH{=}CH_2 \xrightarrow{HBr,\ 0^\circ C} CH_3CH_2CH_2CH(Br)CH_2H$$

1-Pentene → **2-Bromopentane** ($> 84\%$)

$$\text{Cyclohexene} \xrightarrow{HI,\ 0^\circ C} \text{Iodocyclohexane}$$

Cyclohexene → **Iodocyclohexane** (90%)

In a typical experiment, the gaseous hydrogen halide is bubbled through neat or dissolved alkene. Or, it can be added to the alkene in a solvent, such as ethanoic (acetic) acid. Aqueous work-up furnishes the haloalkane in high yield. All the hydrogen halides can be used successfully in such addition reactions.

Regioselectivity in Electrophilic Additions: The Markovnikov Rule

Are additions of HX to unsymmetrical alkenes regioselective? To answer this question, let us consider the reaction of propene with hydrogen chloride. Two products are possible: 2-chloropropane and 1-chloropropane. However, the only product observed is 2-chloropropane.

Regioselective Electrophilic Addition to Propene

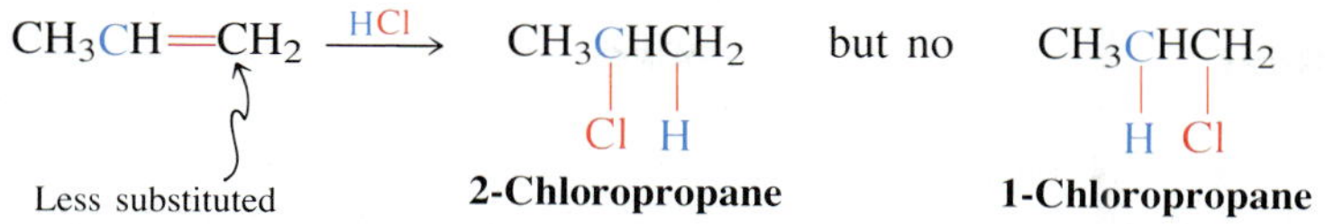

$$CH_3CH{=}CH_2 \xrightarrow{HCl} \underset{\textbf{2-Chloropropane}}{CH_3\underset{Cl}{C}H\underset{H}{C}H_2} \quad \text{but no} \quad \underset{\textbf{1-Chloropropane}}{CH_3\underset{H}{C}H\underset{Cl}{C}H_2}$$

Less substituted

Similarly, reaction of 2-methylpropene with hydrogen bromide gives only 2-bromo-2-methylpropane, and 1-methylcyclohexene combines with HI to furnish only 1-iodo-1-methylcyclohexane:

$$(H_3C)_2C{=}CH_2 \xrightarrow{HBr} CH_3\overset{CH_3}{\underset{Br}{C}}CH_2H$$

Less substituted

$$\text{1-methylcyclohexene} \xrightarrow{HI} \text{1-iodo-1-methylcyclohexane (CH}_3\text{, I, H)}$$

Less substituted

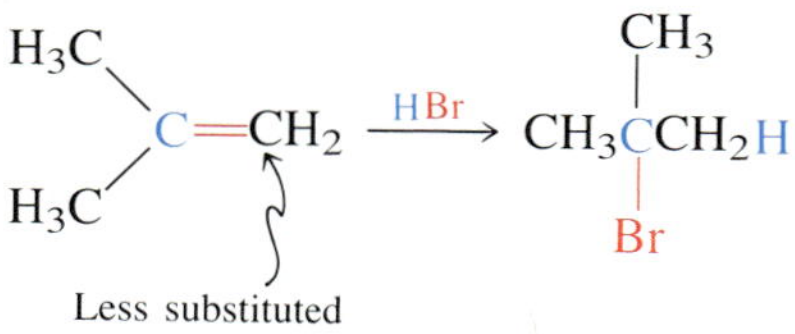

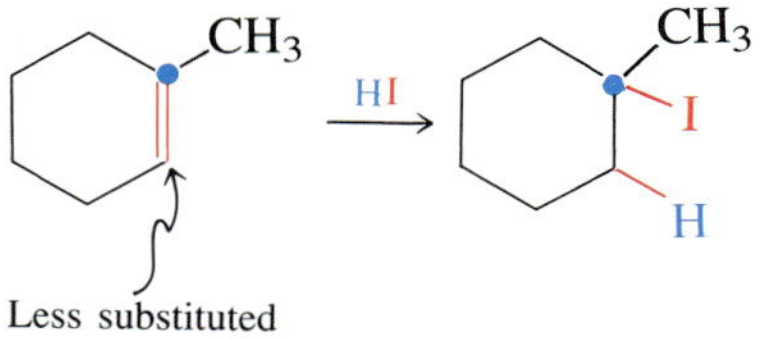

On the other hand, addition of HBr to *trans*-2-pentene gives a mixture of the two possible bromopentanes:

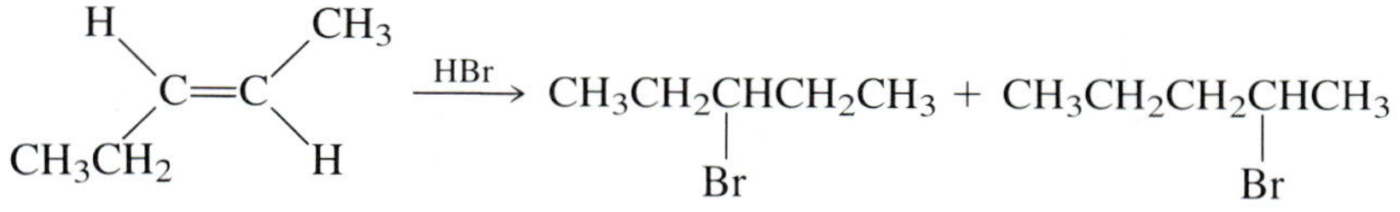

$$\text{H(CH}_3\text{CH}_2\text{)C{=}C(CH}_3\text{)H} \xrightarrow{HBr} CH_3CH_2\underset{Br}{C}HCH_2CH_3 + CH_3CH_2CH_2\underset{Br}{C}HCH_3$$

You can see from these first examples that, if the carbon atoms participating in the double bond are not equally substituted, *the proton from the hydrogen halide attaches itself to the less-substituted one*. As a consequence, the halogen tends to end up at the more-substituted carbon. This phenomenon is referred to as **Markovnikov's* rule,** named after its discoverer. This rule can be explained in terms of what we know about the mechanism of electrophilic additions of protons to alkenes and the relative stability of the resulting carbocations.

Consider the hydrochlorination of propene. In the first step, the proton attacks the π system to give an intermediate carbocation. The regiochemistry of the reaction is determined in this step because, once the carbocation is generated, trapping by chloride proceeds quickly. The proton attacks either of the two carbon atoms. Addition to the internal carbon leads to the primary propyl cation.

Protonation of Propene at C-2

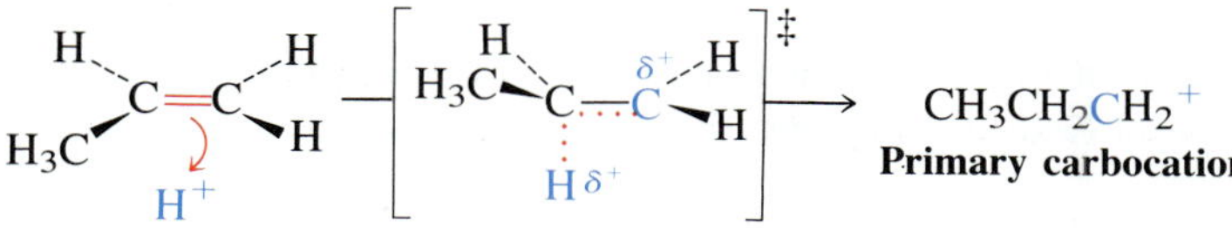

$$\text{H(H}_3\text{C)C{=}CH}_2 + H^+ \longrightarrow \left[H_3C(H)C\cdots C^{\delta+}H_2 \;\; \vdots\; H^{\delta+} \right]^{\ddagger} \longrightarrow CH_3CH_2CH_2^+$$

Primary carbocation

*Professor Vladimir V. Markovnikov, 1838–1904, formulated his rule in 1869, University of Moscow.

In contrast, protonation at the terminal carbon results in formation of the secondary 1-methylethyl (isopropyl) cation.

Protonation of Propene at C-1

$$\mathrm{CH_3(H)C{=}CH_2} + \mathrm{H^+} \longrightarrow \left[\mathrm{CH_3(H)\overset{\delta^+}{C}{\cdots}CH_2{\cdots}H^{\delta^+}}\right]^{\ddagger} \longrightarrow \mathrm{CH_3\overset{+}{C}HCH_3}$$

Secondary carbocation

The second species is more stable and, because the structure of the transition state for protonation resembles that of the resulting cation, is formed considerably faster. Figure 12-2 is a potential-energy diagram of this situation.

FIGURE 12-2
Potential-energy diagram for the two possible modes of HCl addition to propene. Transition state 1 (TS-1), which leads to the higher-energy primary propyl cation, is less favorable than transition state 2 (TS-2), which gives the 1-methylethyl (isopropyl) cation.

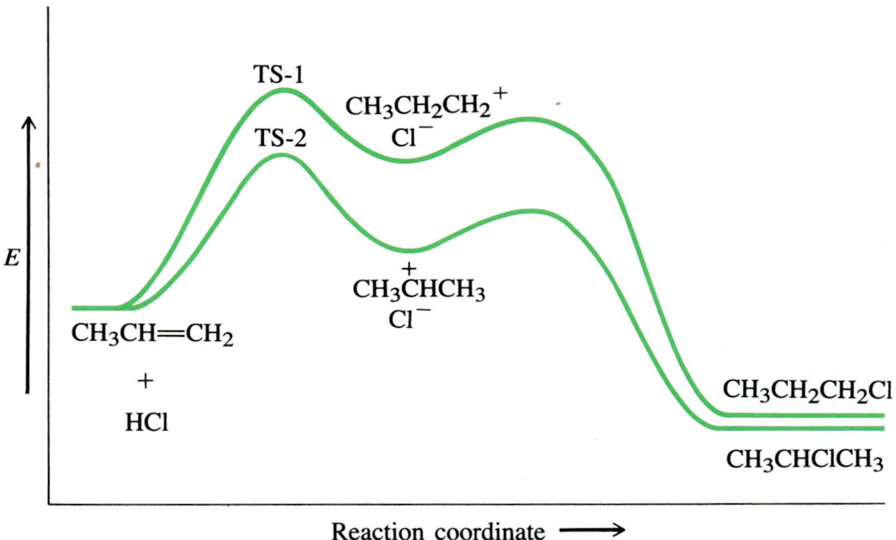

Given this analysis, the empirical Markovnikov rule can be rephrased as follows: HX adds to unsymmetrical alkenes in such a way that *the initial protonation gives the more-stable carbocation*. For alkenes that are similarly substituted at both sp^2 carbons, as in *trans*-2-pentene, product mixtures are to be expected, because carbocations of comparable stability are formed.

EXERCISE 12-5

Predict the outcome of the addition of HBr to: (a) 1-hexene; (b) 2-methyl-2-butene; (c) 4-methylcyclohexene. How many isomers can be formed in each case?

Alcohol Synthesis by Electrophilic Hydration

So far, the nucleophilic trapping agents have been the counterions to the protons of the attacking acids. What about other nucleophiles? When an alkene is exposed to an *aqueous* solution of an acid having a poor nucleophilic counterion, such as sulfuric acid, water plays the role of the trapping nucleophile and intercepts the carbocation formed by the initial protonation. In the overall reaction, the elements of water are added to the double bond, a **hydration.** The process is the reverse of the acid-induced dehydration of alcohols (Section 11-6), and its mechanism is the same in reverse, as illustrated in the hydration of 2-methylpropene, a reaction of industrial importance leading to 2-methyl-2-propanol (*tert*-butyl alcohol).

Electrophilic Hydration

$$\text{1-Methylcyclohexene} \xrightarrow{50\%\ \text{HOH},\ H_2SO_4} \text{1-Methylcyclohexanol}\ (50\%)$$

1-Methylcyclohexene → 1-Methylcyclohexanol (50%)

$$CH_3\overset{O}{\overset{\|}{C}}CH_2CH_2CH{=}C(CH_3)_2 \xrightarrow{HOH,\ H_2SO_4} CH_3\overset{O}{\overset{\|}{C}}CH_2CH_2\underset{H}{CH}{-}\overset{CH_3}{\underset{CH_3}{C}}OH \quad 85\%$$

6-Methylhept-5-en-2-one → 6-Hydroxy-6-methyl-2-heptanone

Mechanism of the Hydration of 2-Methylpropene:

$$(H_3C)_2C{=}CH_2 \underset{-H^+}{\overset{H^+}{\rightleftharpoons}} (H_3C)_2\overset{+}{C}CH_3 \underset{-\ H\ddot{O}H}{\overset{+\ H\ddot{O}H}{\rightleftharpoons}} CH_3\overset{CH_3}{\underset{CH_3}{C}}{-}\overset{+}{\ddot{O}}H_2 \underset{+H^+}{\overset{-H^+}{\rightleftharpoons}} CH_3\overset{CH_3}{\underset{CH_3}{C}}{-}\ddot{O}H$$

Hydration-Dehydration Equilibrium

$$RCH{=}CH_2 + H_2O \underset{}{\overset{\text{Catalytic } H^+}{\rightleftharpoons}} R\underset{OH}{CH}CH_3$$

As in the acid-mediated dehydration of alcohols, the proton acts only as a catalyst and is not consumed in the reaction. Indeed, without the acid, hydration would not occur; alkenes are stable in neutral water. The presence of acid, however, establishes an equilibrium between alkene and alcohol. This equilibrium may be driven in either direction, toward alkene (dehydration) or alcohol (hydration). For entropic reasons, higher temperatures generally favor the alkene (Section 11-6). If the alkene is volatile, the equilibrium may be shifted in its direction by direct distillation from the reaction mixture. At lower temperatures and particularly with a large excess of water present, the alcohol is the predominant product.

EXERCISE 12-6

Treatment of 2-methylpropene with catalytic deuterated sulfuric acid in D_2O gives $(CD_3)_3COD$. Explain by a mechanism.

Halogens Can Also Function as Electrophiles: Halogen Addition

Reagents that do not appear to contain electrophilic atoms also can attack double bonds electrophilically. An example is the halogenation of alkenes, which proceeds with addition of two halogen atoms to the double bond to give a vicinal dihalide. The reaction works best for chlorine and bromine. Fluorine

General Halogenation of Alkenes

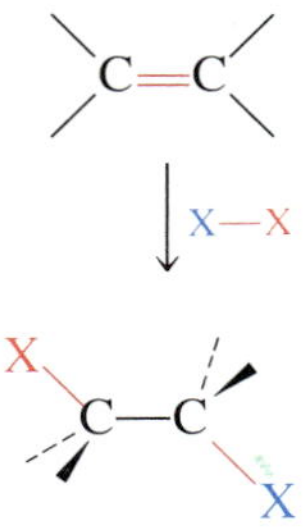

X—X

Vicinal dihalide

X=Cl, Br

adds too violently to alkenes, whereas diiodide formation is virtually thermoneutral.

EXERCISE 12-7

Calculate (as in Table 12-1) the ΔH° values for the addition of F_2 and I_2 to ethene. (For $DH^\circ_{X_2}$, see Section 3-5.)

Bromine addition is particularly easy to recognize because bromine solutions immediately change from red to colorless when exposed to an alkene. This phenomenon is sometimes used to test for unsaturation. Saturated systems do not react with bromine unless a free-radical initiator is present (Section 3-5).

Halogenations are best carried out at room temperature or, with cooling, in inert halogenated solvents such as the halomethanes.

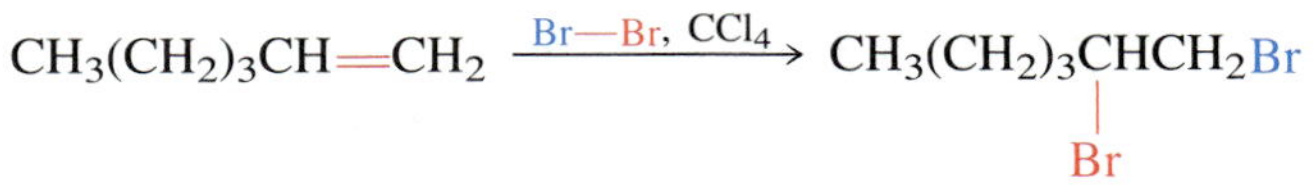

$$CH_3(CH_2)_3CH{=}CH_2 \xrightarrow{Br-Br,\ CCl_4} CH_3(CH_2)_3CHBrCH_2Br$$

1-Hexene → 90% **1,2-Dibromohexane**

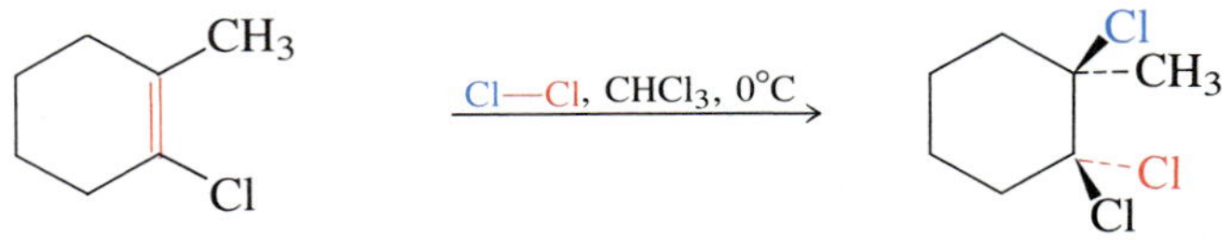

Cl—Cl, $CHCl_3$, 0°C

1-Chloro-2-methylcyclohexene → 30%–50% **1,1,2-Trichloro-2-methylcyclohexane**

On a superficial level, halogen additions to double bonds seem to be similar to hydrogenations. However, they are not. The mechanism of these reactions is quite different, as revealed by the stereochemistry of the process.

Stereochemistry and Mechanism of Halogenation: *Anti* Addition

The halogenation under discussion in this section is bromination, although similar arguments hold for other halogens.

What is the stereochemistry of bromination? Are the two bromine atoms added from the same side of the double bond (as in catalytic hydrogenation) or from opposite sides? Let us examine the bromination of cyclohexene. Double addition on the same side should give *cis*-1,2-dibromocyclohexane; the alternative would result in *trans*-1,2-dibromocyclohexane. The second alternative is borne out by experiment—only *anti* addition is observed. Because *anti* addition to the two reacting carbon atoms may occur with equal probability in two possible ways—in either case, from both above and below the π bond—the product is racemic.

Two Ways of Adding Bromine to a Double Bond in *Anti* Fashion

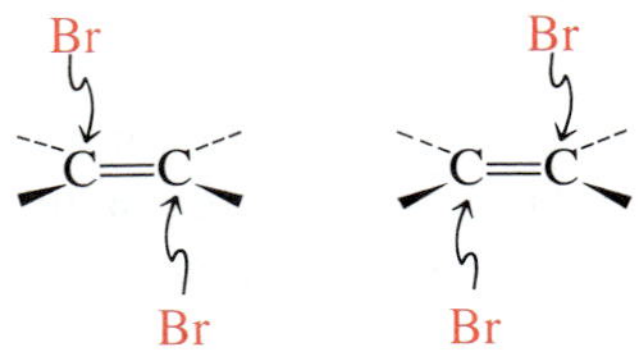

Br Br

Br Br

***Anti* Bromination of Cyclohexene**

Br_2, CCl_4

Br Br + Br Br

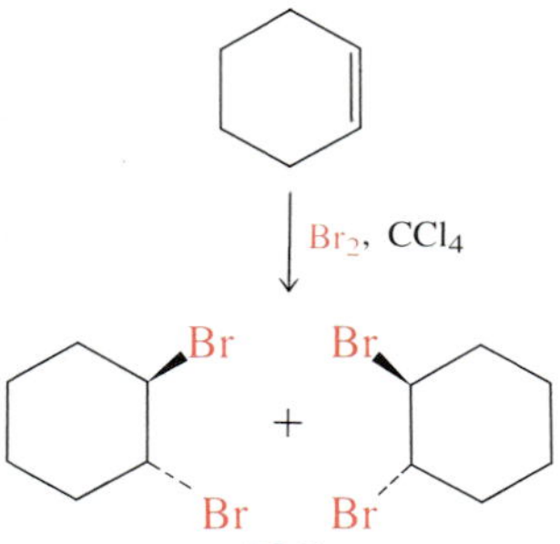

83%

Racemic *trans*-1,2-dibromocyclohexane

With acyclic alkenes, the reaction is also cleanly stereospecific. For example, *cis*-2-butene brominates to furnish a racemic mixture of (2*R*,3*R*)- and (2*S*,3*S*)-2,3-dibromobutane, the corresponding *trans*-2-butene resulting in the meso diastereomer.

Stereospecific 2-Butene Bromination

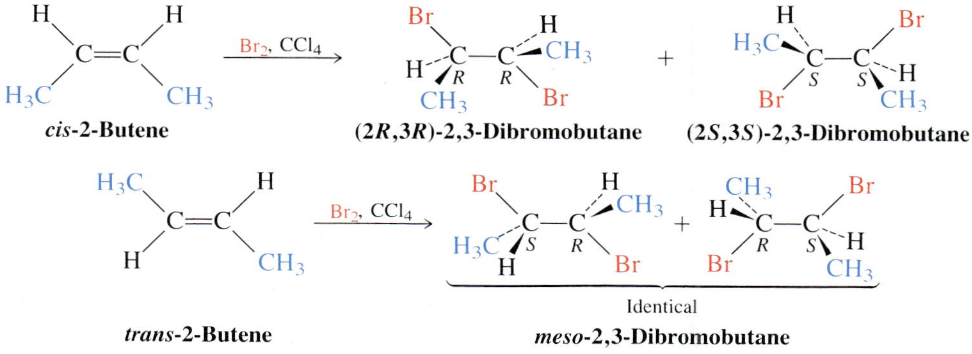

cis-2-Butene $\xrightarrow{Br_2,\ CCl_4}$ (2*R*,3*R*)-2,3-Dibromobutane + (2*S*,3*S*)-2,3-Dibromobutane

trans-2-Butene $\xrightarrow{Br_2,\ CCl_4}$ *meso*-2,3-Dibromobutane (Identical)

What mechanism explains the observed stereochemistry? How does bromine attack the electron-rich double bond even though it does not appear to contain an electrophilic center? The answers lie in the polarizability of the Br–Br bond, which is prone to heterolytic cleavage on reaction with a nucleophile. The π electron cloud of the alkene is nucleophilic and attacks one end of the bromine molecule with simultaneous displacement of the second bromine atom as bromide ion. The resulting intermediate is a **cyclic bromonium ion.** The characteristic feature of this species is that the bromine bridges both carbon atoms of the original double bond to form a three-membered ring. This species can be envisaged to arise through overlap between one of bromine's lone electron pairs and the two *p* orbitals of the original π bond (Figure 12-3, on the next page) in a kind of S_N2 reaction in which the π electrons act as nucleophiles and attack the bromine molecule, displacing Br^- as the leaving group.

An important feature of the intermediacy of the cyclic bromonium ion is that it explains the stereochemistry of bromination. The structure of this ion is rigid, and it may be attacked only on the side opposite the bromine. This is done by the bromide ion generated in the first step; the three-membered ring is thus opened stereospecifically. The second reaction is completely analogous to the nucleophilic ring-opening of oxacyclopropanes (Section 9-6). The leaving group is the bridging bromine. In symmetrical bromonium ions, attack is equally probable at either carbon atom, giving the racemic products observed when chiral bromoalkanes are made in this manner.

Nucleophilic Opening of a Cyclic Bromonium Ion

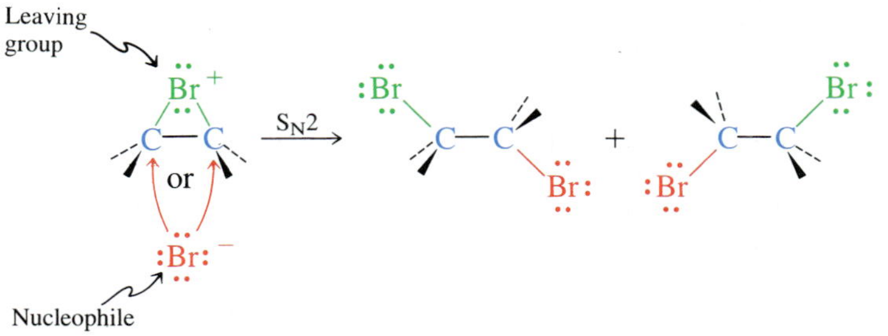

Leaving group
$\xrightarrow{S_N2}$
or
Nucleophile

FIGURE 12-3

A. Electron-pushing picture of cyclic bromonium ion formation. The alkene acts as a nucleophile to displace bromide ion from bromine. The molecular bromine behaves as if it were strongly polarized, one atom as a bromine cation, the other as an anion.
B. Molecular-orbital picture of bromonium ion formation.

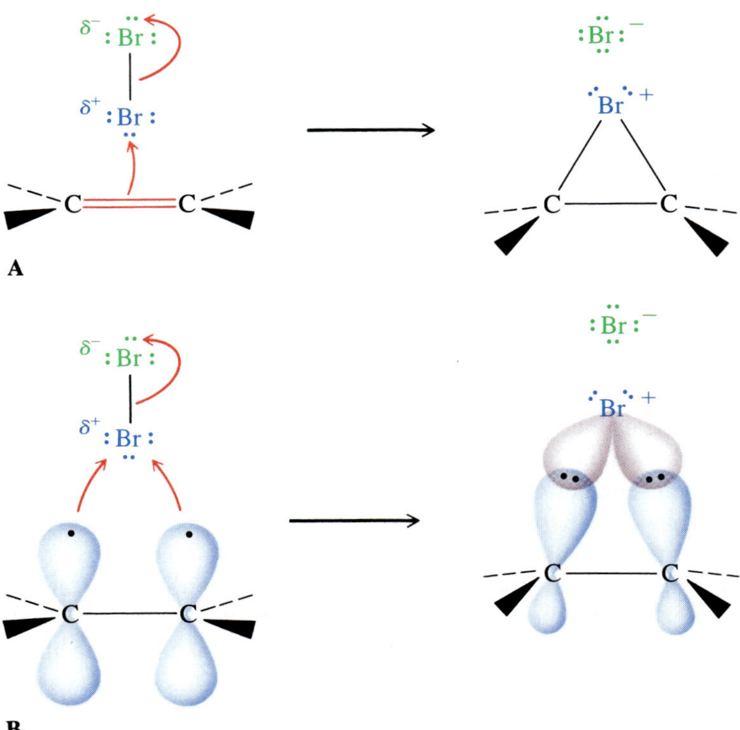

EXERCISE 12-8

Draw the intermediate in the bromination of cyclohexene, using the conformational picture below. Show why the product is racemic. What can you say about the initial conformation of the product?

Conformational flip in cyclohexene

The Bromonium Ion Can Be Trapped by Other Nucleophiles

The creation of a bromonium ion in alkene brominations suggests that, in the presence of other nucleophiles, competition with bromide might be observed in the trapping of the intermediate. Indeed, this is possible. For example, bromination of cyclopentene in the presence of excess chloride ion (added as a salt) gives a mixture of *trans*-1,2-dibromocyclopentane and *trans*-1-bromo-2-chlorocyclopentane.

Competitive Trapping of a Cyclic Bromonium Ion

(Although all products are racemic, only one enantiomer is shown in each case.)

Br_2, Cl^-, CCl_4, 0°C

$+ :Br:^- + :Cl:^-$

$:Br:^-$

$:Cl:^-$

A large excess of the competing nucleophile may prevent the formation of mixtures of products. For example, bromination of cyclopentene in water as solvent gives the vicinal bromo alcohol exclusively. In this case, the bromonium ion is attacked by water. The net transformation is the *anti* addition of the elements of Br–OH to the double bond. The other product formed is HBr. The corresponding chloro alcohols can be made from chlorine in water through the intermediacy of a chloronium ion.

Bromo Alcohol Synthesis

Br_2, H_2O, 0°C
$- Br^-$

$+ \ddot{O}H_2$

$- H^+$

***trans*-2-Bromocyclopentanol**

EXERCISE 12-9

Show the expected product from the reaction of (a) *trans*-2-butene and (b) *cis*-2-pentene with aqueous chlorine. Show the stereochemistry clearly.

2-Halo alcohols undergo intramolecular ring closure in the presence of base to give oxacyclopropanes (Section 9-5) and are therefore useful intermediates in organic synthesis. If alcohol is used instead of water as a solvent in these halogenations, the corresponding vicinal halo ethers are produced.

Vicinal Halo Ether Synthesis

Br—Br, H—OCH_3 / − HBr

H, OCH_3, Br H

76%

***trans*-1-Bromo-2-methoxycyclohexane**

Halonium Ion Opening Can Be Regioselective

In contrast with dihalogenations, mixed additions to double bonds can pose regiochemical problems. Is the addition of Br–OR to an unsymmetrical double bond selective? The answer is yes. For example, 2-methylpropene is converted by aqueous bromine only into 1-bromo-2-methyl-2-propanol, none of the alternative regioisomer, 2-bromo-2-methyl-1-propanol, being formed.

$$(H_3C)_2C{=}CH_2 \xrightarrow[-\ HBr]{Br—Br,\ H—OH} CH_3C(OH)(CH_3)CH_2Br \quad \text{but no} \quad CH_3C(Br)(CH_3)CH_2OH$$

82%

1-Bromo-2-methyl-2-propanol | **2-Bromo-2-methyl-1-propanol**

Similar selectivity is seen in other cases:

CH_2 (methylenecyclohexane) —Cl—Cl, H—OCH_3 / − HCl→ CH_3O, CH_2Cl (cyclohexane) but no Cl, CH_2OCH_3 (cyclohexane)

78%

The electrophilic halogen in the product always becomes linked to the less-substituted carbon of the original double bond, and the subsequently added nucleophile attaches to the more-highly-substituted center. This situation is reminiscent of the Markovnikov addition of the hydrogen halides to alkenes and has a similar origin. However, in contrast with the Markovnikov addition of acids, which yields intermediate carbocations, halonium ions are the intermediates. The question of regiochemistry is one of selectivity in the nucleophilic attack on this initial intermediate: the more-highly-substituted carbon is attacked preferentially.

How can this be explained? The situation is very similar to the acid-catalyzed nucleophilic ring-opening of oxacyclopropanes (Section 9-6), in which the intermediate contains a protonated oxygen in the three-membered ring. In both reactions, the nucleophile attacks the more-highly-substituted carbon of the ring, because this carbon is more positively polarized than the other.

Regioselective Opening of the Bromonium Ion Formed from 2-Methylpropene

$$(CH_3)_2\overset{Br^+}{C\!-\!-\!-}CH_2 + H\!-\!\ddot{O}H \longrightarrow (CH_3)_2\underset{H-\overset{+}{O}-H}{C}\!-\!CH_2\ddot{Br}: \xrightarrow{-H^+} CH_3\overset{CH_3}{\underset{:\ddot{O}H}{C}}\!-\!CH_2\ddot{Br}:$$

Attack at more-substituted carbon of bromonium ion

A simple rule of thumb is that electrophilic additions of unsymmetrical reagents of this type add in a regiochemical sense that is Markovnikov-like, the electrophilic unit emerging at the less-substituted carbon of the double bond. Mixtures are formed only when the two carbons are not sufficiently differentiated (see part b of Exercise 12-9).

EXERCISE 12-10

What are the product(s) of the following reactions?

(a) $CH_3CH{=}CH_2 \xrightarrow{Cl_2,\ CH_3OH}$ (b) 3-methylcyclohexene (ring with H_3C substituent) $\xrightarrow{Br_2,\ H_2O}$

EXERCISE 12-11

What would be a good alkene precursor for a racemic mixture of (2*R*,3*R*)- and (2*S*,3*S*)-2-bromo-3-methoxypentane? What other isomeric products might you expect to find from the reactions that you propose?

The Generality of Electrophilic Additions by Polarized or Polarizable Reagents A–B

Alkenes can undergo stereo- and regiospecific addition reactions with reagents of the type A–B, in which A acts as an electrophile, B as a nucleophile. Such reagents and how they will add to 2-methylpropene are shown in Table 12-2.

How can we tell which part of the reagent is electrophilic and which part nucleophilic? In other words, what determines the polarization of the A–B bond? Not unexpectedly, it is the relative electronegativity of A and B. In additions to an alkene the more-electronegative moiety acts as the nucleophile, the less-electronegative unit as the electrophile. For a qualitative idea of the relative electronegativities of the various structural units under consideration, see Table 1-4. The electron-attracting power of the elements increases from left to right and from the bottom to the top of the periodic table. For example, BrCl is polarized in the sense $\overset{\delta^+}{Br}\overset{\delta^-}{Cl}$, containing an electrophilic bromine. Similarly, polarization of ICl is $\overset{\delta^+}{I}\overset{\delta^-}{Cl}$, and RSCl is $\overset{\delta^+}{RS}\overset{\delta^-}{Cl}$.

Hybridization and the relative electronegativity of other attached atoms also contribute strongly to the sense of polarization. Cyanogen bromide, for example, is fairly strongly polarized as $\overset{\delta^+}{Br}\overset{\delta^-}{CN}$, even though the difference in electronegativity between bromine (3.0) and carbon (2.6) is minimal. The reasons for the observed sense of polarization are that the carbon is *sp*-hybridized and the nitrogen (electronegativity 3.0) attached to carbon is also relatively electron attracting.

TABLE 12-2
Reagents A–B that add to alkenes by electrophilic attack

$$(CH_3)_2C{=}CH_2 + {}^{\delta^+}A{-}B^{\delta^-} \longrightarrow H{-}\underset{A}{\overset{H}{C}}{-}\underset{B}{\overset{CH_3}{C}}{-}CH_3$$

Name	Structure	Addition product to 2-methylpropene
Bromine chloride	Br—Cl	$BrCH_2C(CH_3)_2$ with Cl on C
Cyanogen bromide	Br—CN	$BrCH_2C(CH_3)_2$ with CN on C
Iodine chloride	I—Cl	$ICH_2C(CH_3)_2$ with Cl on C
Sulfenyl chlorides	RS—Cl	$RSCH_2C(CH_3)_2$ with Cl on C
Mercuric salts	XHg—X	$XHgCH_2C(CH_3)_2$ with X on C

A Special Electrophilic Addition: Alcohols by Oxymercuration-Demercuration

The last example in Table 12-2 constitutes an electrophilic addition of a mercuric salt to an alkene. The reaction is called *mercuration,* and the resulting compound an alkylmercury derivative, in which the mercury can be removed in a subsequent step. Particularly useful is a reaction sequence titled **oxymercuration-demercuration,** which employs mercuric ethanoate (acetate) as the reagent. Treatment of an alkene with this species in the presence of water leads to the corresponding addition product (oxymercuration).

Oxymercuration

$$\rangle C{=}C\langle + \underset{\textbf{Mercuric ethanoate (acetate)}}{CH_3\overset{O}{\overset{\|}{C}}OHgO\overset{O}{\overset{\|}{C}}CH_3} + H{-}OH \xrightarrow{THF} -\underset{CH_3\underset{O}{\underset{\|}{C}}OHg}{\overset{|}{C}}-\overset{OH}{\underset{|}{C}}- + CH_3\overset{O}{\overset{\|}{C}}OH$$

Subsequently the mercury-containing substituent can be replaced by hydrogen on treatment with sodium borohydride in base (demercuration). The net result is the hydration of the double bond to give an alcohol.

Demercuration

$$-\overset{|}{\underset{\underset{\underset{O}{\|}}{CH_3COHg}}{C}}-\overset{OH}{\overset{|}{\underset{|}{C}}}- \xrightarrow{NaBH_4,\ NaOH} -\overset{|}{\underset{\underset{H}{|}}{C}}-\overset{OH}{\overset{|}{\underset{|}{C}}}- + Hg + CH_3\underset{\underset{O}{\|}}{C}O^-$$

Oxymercuration is predominantly *anti* stereospecific; it is also regioselective. This implies a mechanism similar to that for the electrophilic addition reactions discussed so far. The specificity of the reaction is illustrated in the oxymercuration-demercuration of 1-methylcyclopentene to furnish 1-methylcyclopentanol. The mercury reagent can be thought of as initially dissociating into a cationic mercury species and ethanoate (acetate) ion. The cation then attacks the alkene double bond, furnishing a mercurinium ion, probably of similar structure to a cyclic bromonium ion. The water present attacks the more-substituted carbon to give the alkylmercuric ethanoate (acetate) intermediate, which is reduced in a subsequent step by sodium borohydride. The mechanism of the reduction reaction (in which metallic mercury is generated) is complex and not completely understood. After reduction, the resulting product is the tertiary cyclopentanol, formally the product of Markovnikov hydration of the starting material. The entire oxymercuration-demercuration sequence conveniently takes place in the same flask, without the isolation of intermediates.

1-Methylcyclopentanol from 1-Methylcyclopentene by Oxymercuration-Demercuration

1 Dissociation:

$$CH_3\overset{\overset{O}{\|}}{C}OHgO\overset{\overset{O}{\|}}{C}CH_3 \rightleftharpoons CH_3\overset{\overset{O}{\|}}{C}O^- + {}^+HgO\overset{\overset{O}{\|}}{C}CH_3$$

2 Electrophilic attack:

1-Methylcyclopentene (CH$_3$) + $^+HgO\overset{\overset{O}{\|}}{C}CH_3$ ⟶ Mercurinium ion ($\overset{+}{H}gO\overset{\overset{O}{\|}}{C}CH_3$, CH$_3$)

Mercurinium ion

3 Nucleophilic opening:

Mercurinium ion ($\overset{+}{H}gO\overset{\overset{O}{\|}}{C}CH_3$, CH$_3$) + H—$\ddot{O}$H $\xrightarrow{-\,H^+}$ (CH$_3$, OH, H, $HgO\overset{\overset{O}{\|}}{C}CH_3$)

Alkylmercuric ethanoate (acetate)

4 Reduction:

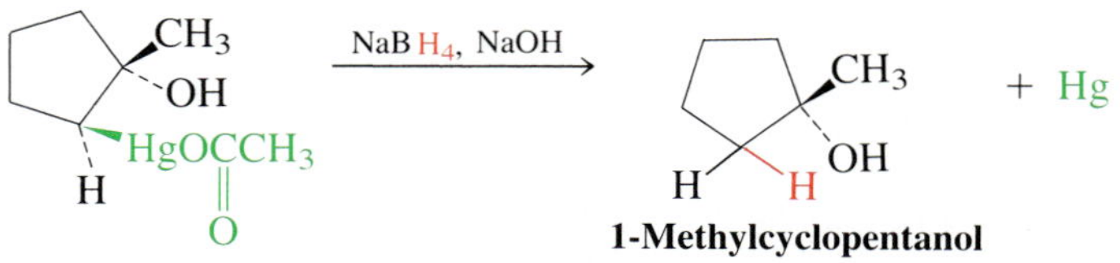
CH3
OH
HgOCCH3
O
H
NaB H4, NaOH
CH3
OH
H
H
+ Hg

1-Methylcyclopentanol

Ether Synthesis by Oxymercuration-Demercuration

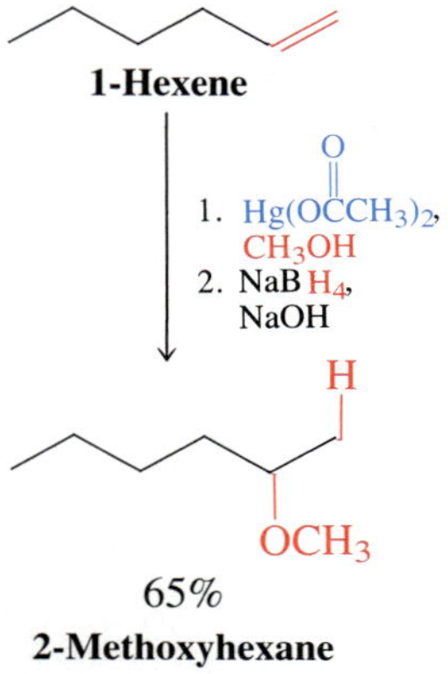
1-Hexene
1. Hg(OCCH3)2, CH3OH
2. NaB H4, NaOH
H
OCH3
65%
2-Methoxyhexane

When the oxymercuration of an alkene is executed in an alcohol solvent, demercuration gives an ether, as shown in the margin.

BOX 12-1

Oxymercuration-Demercuration in Juvenile-Hormone-Analog Synthesis

An application of the oxymercuration-demercuration reaction to the synthesis of an analog of juvenile hormone is shown below. Juvenile hormone is a substance that controls the larval metamorphosis in insects. It is produced by the male wild silk moth *Hyalophora cecropia L.*, and its presence prevents the maturation of insect larvae. The compound itself and modified analogs have been proposed as potential agents in insect control.

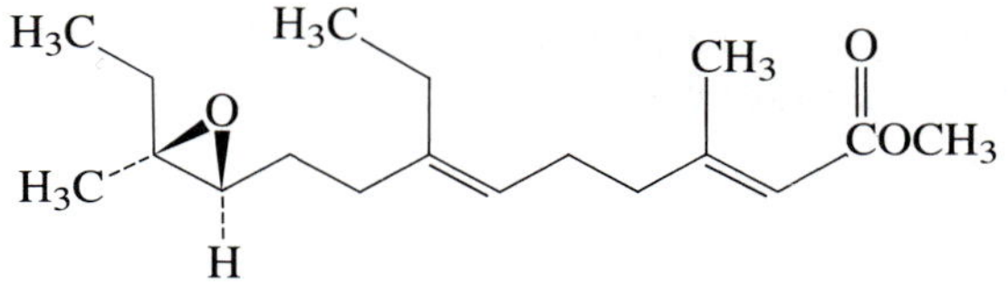
H3C
H3C
CH3
O
COCH3
O
H3C
H

Juvenile hormone

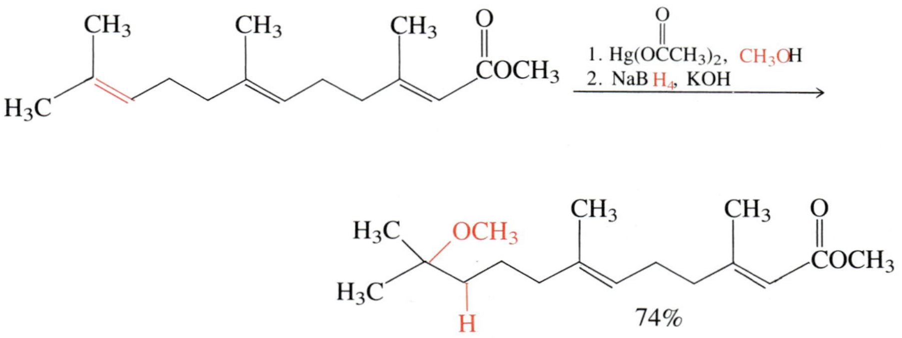
CH3
CH3
CH3
O
COCH3
H3C
1. Hg(OCCH3)2, CH3OH
2. NaB H4, KOH
H3C
OCH3
CH3
CH3
O
COCH3
H3C
H
74%

Analog of juvenile hormone

This example is noteworthy because the reaction can be controlled to take place only at the least-hindered electron-rich double bond. Unfortunately, the activity of the product is only 1/500 that of the natural compound.

EXERCISE 12-12

Explain the result shown at the right.

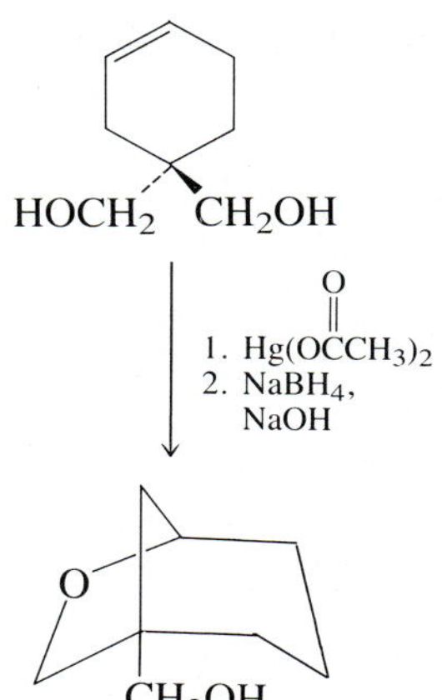

In summary, the new reactions in this section are mechanistically similar in that they constitute electrophilic additions to double bonds. The protonations of alkenes give carbocations that can subsequently be trapped by nucleophiles. These additions, obeying the Markovnikov rule, proceed through the intermediacy of the most-stable (most-highly-substituted) cations. In this way, alkenes can be regioselectively hydrohalogenated to haloalkanes and hydrated to alcohols. The additions of halogens furnish intermediate bridged halonium ions. These ions are subject to stereospecific and regioselective ring-opening in a manner mechanistically very similar to the nucleophilic opening of protonated oxacyclopropanes. Halonium ions can be trapped by halide ions, water, and alcohols to give vicinal dihaloalkanes, halo alcohols, and halo ethers, respectively. The principle of electrophilic addition can be applied to any reagent A–B containing a polarized or polarizable bond. Finally, a synthetically useful procedure for converting alkenes into Markovnikov alcohols or ethers is oxymercuration-demercuration.

12-4 Regioselective and Stereospecific Functionalization of Alkenes by Hydroboration

This section deals with a reaction that seems to lie mechanistically somewhere between hydrogenation and electrophilic addition: the hydroboration of double bonds. The resulting alkylboranes are very useful synthetically, because the boron can be replaced by functional groups, such as the hydroxy function and halogens.

The Boron–Hydrogen Bond Adds Across Double Bonds

Borane, BH_3, adds to double bonds without catalytic activation, a reaction called **hydroboration** by its discoverer, H. C. Brown.*

General Hydroboration

$$\text{>C=C<} + \underset{\textbf{Borane}}{BH_3} \longrightarrow \underset{\textbf{An alkylborane}}{\text{H–C–C–}BH_2} \xrightarrow{2\ \text{>C=C<}} \underset{\textbf{A trialkylborane}}{(\text{H–C–C})_3\text{–B}}$$

Borane (which by itself exists as a dimer, B_2H_6) is commercially available in ether solutions. In these solutions, borane exists as a Lewis acid-base complex with the ether oxygen, an aggregate that allows the boron to have an electron octet (for the molecular-orbital picture of BH_3, see Figure 1-17):

$$BH_3 + CH_3CH_2\ddot{\underset{..}{O}}CH_2CH_3 \longrightarrow \underset{\textbf{Borane-ether complex}}{H_3\overset{-}{B}\text{—}\overset{+}{\underset{..}{O}}(CH_2CH_3)_2}$$

*Professor Herbert C. Brown, b. 1912, Purdue University, Nobel Prize 1979.

How does the B–H unit add to the π bond? Because the π bond is electron rich and borane electron poor, it is reasonable to formulate an initial Lewis acid-base complex similar to that of a bromonium ion (Figure 12-3, without the positive charge), requiring the participation of the empty *p* orbital on BH_3, as in the borane-ether complex. Subsequently, one of the hydrogens is transferred by means of a four-center transition state to one of the alkene carbons, while the boron shifts to the other without any additional intermediates. The stereochemistry of the addition is *syn*. All three B–H bonds are reactive in this way.

Mechanism of Hydroboration:

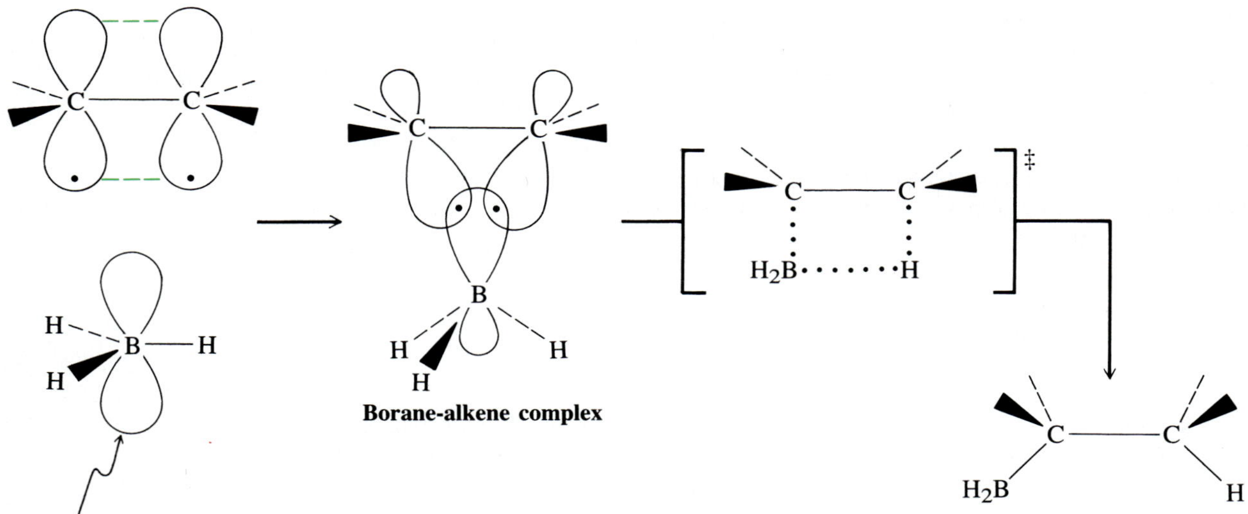

General Regioselective Hydroboration

$$3\ RCH{=}CH_2 + BH_3 \longrightarrow (RCH_2CH_2)_3B$$

Hydroboration is not only stereospecific (*syn* addition) but also regioselective. Unlike the electrophilic additions described in Section 12-3, steric and not electronic factors control the regioselectivity: the boron binds to the less-hindered (less-substituted) carbon. The reactivity of the trialkylboranes resulting from these hydroborations are of special interest to us.

Oxidation of Alkylboranes Gives Alcohols

Trialkylboranes can be oxidized with basic aqueous hydrogen peroxide to furnish alcohols in which the hydroxy function has replaced the boron atom. The net result of the two-step sequence, **hydroboration-oxidation,** is the addition of the elements of water to a double bond. In contrast with the hydrations described in Section 12-3, that involving borane is **anti-Markovnikov.**

General Hydroboration-Oxidation

$$3\ RCH{=}CHR \xrightarrow{BH_3} (RCH_2CHR)_3B \xrightarrow{H_2O_2,\ NaOH,\ H_2O} 3\ RCH_2CH(R)OH$$

Example:

$$(CH_3)_2CHCH_2CH{=}CH_2 \xrightarrow[2.\ H_2O_2,\ NaOH,\ H_2O]{1.\ BH_3} (CH_3)_2CHCH_2CH_2CH_2OH$$

80%

4-Methyl-1-pentene → **4-Methyl-1-pentanol**

In the mechanism of the oxidation of alkylboranes, the highly nucleophilic and electron-rich hydroperoxide ion attacks the electron-poor boron atom. The resulting boron species undergoes a rearrangement in which an alkyl group migrates with its electron pair (and with retention of its configuration) to the neighboring oxygen, expelling a hydroxide group in the process. Although hydroxide ion is normally a poor leaving group, the weak O–O bond and the intramolecular nature of this reaction allows it to depart as such. The initial product R_2BOR undergoes further oxidation to a trialkyl borate $(RO)_3B$ that is hydrolyzed by the basic medium to the alcohol and sodium borate.

Mechanism of Alkylborane Oxidation:

$$R_3B + {}^-:\ddot{O}-\ddot{O}H \longrightarrow R_3\bar{B}-\ddot{O}-\ddot{O}H \xrightarrow{-HO^-} R_2B-\ddot{O}R \xrightarrow[-HO^-]{HOO^-} \xrightarrow[-HO^-]{HOO^-}$$

Hydroperoxide ion

$$(RO)_3B \xrightarrow{H_2O,\ NaOH} HO-\bar{B}(OR)_3 \longrightarrow HOB(OR)_2 + {}^-OR \longrightarrow$$

A trialkyl borate

$$ROH + {}^-O-B(OR)_2 \longrightarrow \longrightarrow 3\ ROH + Na_3BO_3$$

Sodium borate

Rearrangement in the conversion of alkylboranes into alcohols is similar to alkyl-group migration on protonation of 2,2-dimethyl-1-propanol. In both cases, the migrating nucleus moves toward a relatively electron poor center.

$$(CH_3)_3C-CH_2-\overset{+}{O}H_2 \longrightarrow CH_3\overset{+}{C}(CH_3)-CH_2CH_3 + H_2\ddot{O}$$

Because borane addition to double bonds is so selective, subsequent oxidation allows the stereospecific and regioselective synthesis of alcohols.

Stereospecific and Regioselective Alcohol Synthesis by Hydroboration-Oxidation

3 1,2-Dideuteriocyclohexene $\xrightarrow{BH_3}$ (2-deuterio-2-boryl-1-deuteriocyclohexane, B and H cis) $\xrightarrow{H_2O_2,\ NaOH,\ H_2O}$ 3 *cis*-1,2-Dideuteriocyclohexanol (87%)

1-Methylcyclopentene $\xrightarrow{BH_3}$ (trans-2-methylcyclopentylborane) $\xrightarrow{H_2O_2,\ NaOH,\ H_2O}$ *trans*-2-Methylcyclopentanol (86%)

EXERCISE 12-13

Give the products of hydroboration-oxidation of: (a) propene and (b) (*E*)-2,3-dideuterio-2-butene. Show the stereochemistry clearly.

Hydroboration-Halogenation: Anti-Markovnikov Hydrohalogenation

Alkylboranes can also be precursors to haloalkanes. For example, exposure to bromine or iodine monochloride produces the corresponding haloalkanes. Again, the specificity of the initial hydroboration ensures that the halogen atom is added to only one side of the double bond. The net result of this **hydroboration-halogenation** sequence is the regioselective and stereospecific hydrohalogenation of an alkene, placing the halogen on the less-substituted carbon atom of the original double bond, in contrast with Markovnikov additions (Section 12-3).

Hydroboration-Halogenation

1. BH_3, THF, 0°C
2. Br_2

70%

2-Methyl-1-pentene → **1-Bromo-2-methylpentane**

1. BH_3
2. ICl

100%

Cyclohexene → **Iodocyclohexane**

EXERCISE 12-14

Draw the products of hydroboration-bromination of: (a) 1,2-dideuteriocyclopentene and (b) (*Z*)-2,3-dideuterio-2-butene. Show stereochemistry.

In summary, hydroboration constitutes another method by which to functionalize alkenes. The initial addition is *syn* and regioselective, the boron shifting to the less-hindered carbon. Oxidation of alkyl boranes with basic hydrogen peroxide gives anti-Markovnikov alcohols; halogenation results in the corresponding haloalkanes.

12-5 Oxidation of Alkenes by Electrophilic Oxidants

This section describes how electrophilic oxidizing agents are capable of delivering oxygen atoms to the π bond, furnishing oxacyclopropanes, vicinal *syn* and *anti* diols, and carbonyl compounds by complete cleavage of the double bond. Let us start with oxacyclopropane formation by peroxycarboxylic acids, a transformation that after hydrolytic ring-opening ultimately yields vicinal *anti* diols.

Peroxycarboxylic Acids Deliver Oxygen Atoms to Double Bonds: Oxacyclopropane and *Anti* Diol Formation

The OH group in peroxycarboxylic acids, $\overset{\overset{\displaystyle O}{\|}}{RC}OOH$, contains an electrophilic oxygen. These compounds react with alkenes by adding this oxygen to the double bond to form oxacyclopropanes. The other product of the reaction is a carboxylic acid, readily removed by extraction with aqueous base. The transformation is of value because, as we know, oxacyclopropanes are versatile synthetic intermediates (Section 9-6). It proceeds at room temperature in an inert solvent, such as chloroform, dichloromethane, or benzene. A frequently used commercial and crystalline peroxycarboxylic acid is *meta*-chloroperoxybenzoic acid (MCPBA).

General Oxacyclopropane Formation

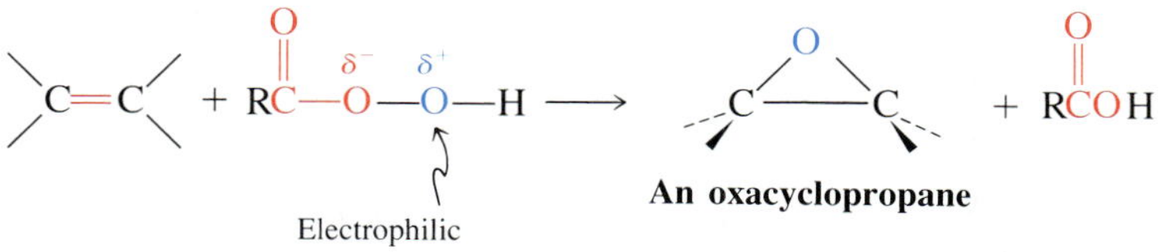

$$\text{>C=C<} + \text{RC(=O)}-\overset{\delta^-}{\text{O}}-\overset{\delta^+}{\text{O}}-\text{H} \longrightarrow \text{oxacyclopropane} + \text{RC(=O)OH}$$

Electrophilic

An oxacyclopropane

Examples:

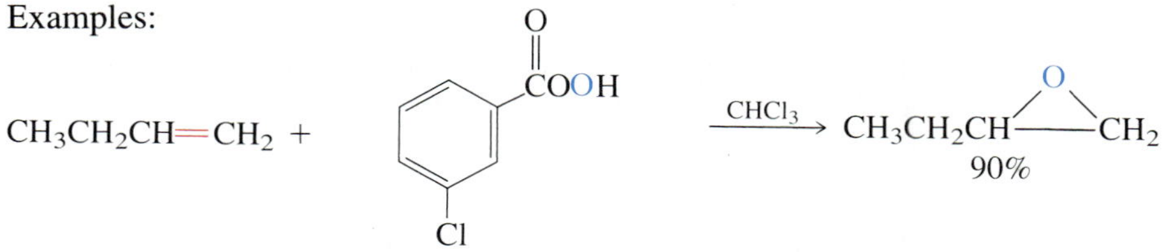

$$CH_3CH_2CH{=}CH_2 + \text{MCPBA} \xrightarrow{CHCl_3} CH_3CH_2CH\overset{O}{—}CH_2 \quad 90\%$$

1-Butene | ***meta*-Chloroperoxybenzoic acid (MCPBA)** | **Ethyloxacyclopropane**

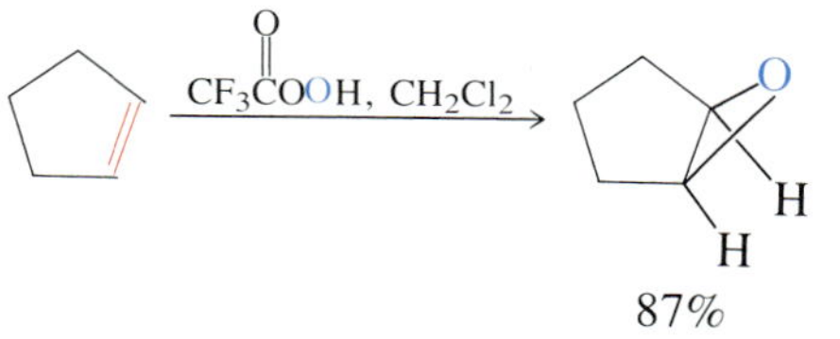

$$\text{Cyclopentene} \xrightarrow{CF_3COOH,\ CH_2Cl_2} \text{cyclopentene oxide} \quad 87\%$$

The transfer of oxygen is stereospecifically *syn,* the stereochemistry of the starting alkene being retained in the product. For example, *trans*-dideuterioethene gives *trans*-2,3-dideuteriooxacyclopropane; conversely, the *cis*-ethene yields *cis*-2,3-dideuteriooxacyclopropane.

$$\text{HDC=CHD} \xrightarrow{\text{MCPBA}} \text{oxacyclopropane} \quad 85\%$$

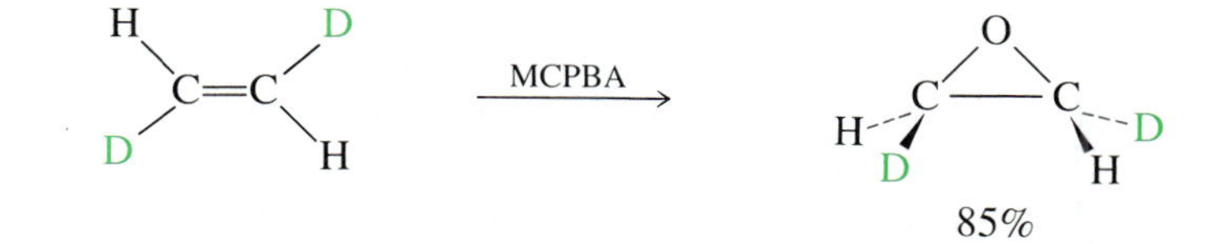

***trans*-2,3-Dideuterioethene** → ***trans*-2,3-Dideuteriooxacyclopropane**

$$\text{DHC=CHD} \xrightarrow{\text{MCPBA}} \text{oxacyclopropane} \quad 85\%$$

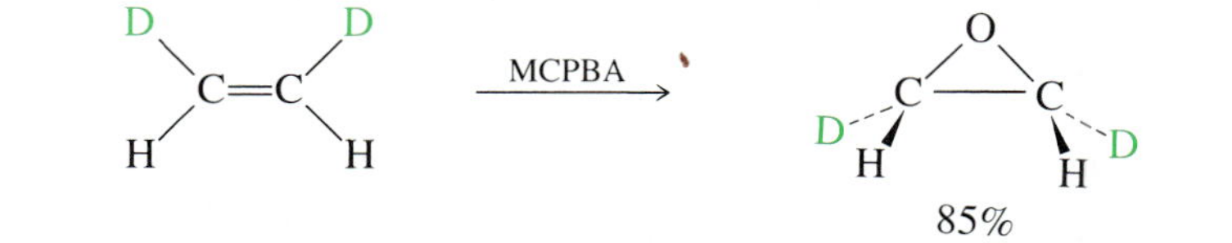

***cis*-2,3-Dideuterioethene** → ***cis*-2,3-Dideuteriooxacyclopropane**

What is the mechanism of this oxidation? We can write a cyclic transition state in which the peroxycarboxylic acid proton is transferred to its own carbonyl group at the same time as the electrophilic oxygen is added to the π bond.

Peroxycarboxylic Acids

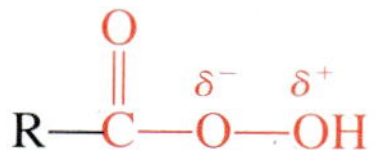

$$\text{R}-\text{C(=O)}-\overset{\delta^-}{\text{O}}-\overset{\delta^+}{\text{OH}}$$

A peroxycarboxylic acid

$$CH_3\overset{O}{\overset{\|}{C}}OOH$$

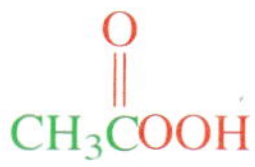

Peroxyethanoic (peracetic) acid

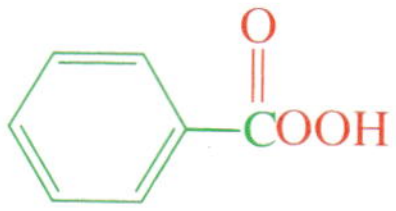

$$C_6H_5\overset{O}{\overset{\|}{C}}OOH$$

Peroxybenzoic (perbenzoic) acid

Formally, the electron pair of the alkene π system participates in the formation of one bond to the oxygen being added, whereas the electron pair responsible for the O–H linkage forms the other bond.

Mechanism of Oxacyclopropane Formation:

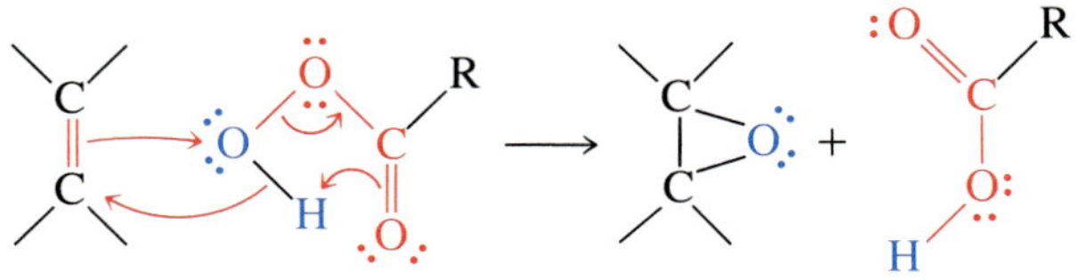

EXERCISE 12-15

Outline a short synthesis of *trans*-2-methylcyclohexanol from cyclohexene. Hint: Review the reactions of oxacyclopropanes (Section 9-6).

In accord with the electrophilic mechanism, the reactivity of alkenes to peroxycarboxylic acids increases with alkyl substitution, allowing for selective oxidations.

Relative Rates (in Parentheses) of Oxacyclopropane Formation

$CH_2{=}CH_2$ (1) < $RCH{=}CH_2$ (24) < $RCH{=}CHR$ ~ $R_2C{=}CH_2$ (500) < $R_2C{=}CHR$ (6500) < $R_2C{=}CR_2$ (very fast)

Example:

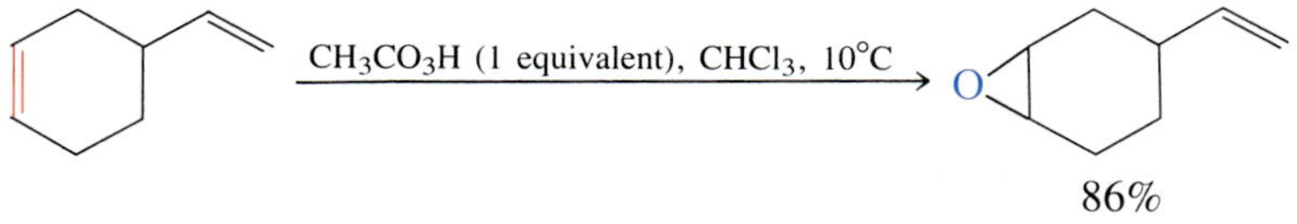

CH_3CO_3H (1 equivalent), $CHCl_3$, 10°C

86%

General Vicinal *Anti* Diol Formation from Alkenes

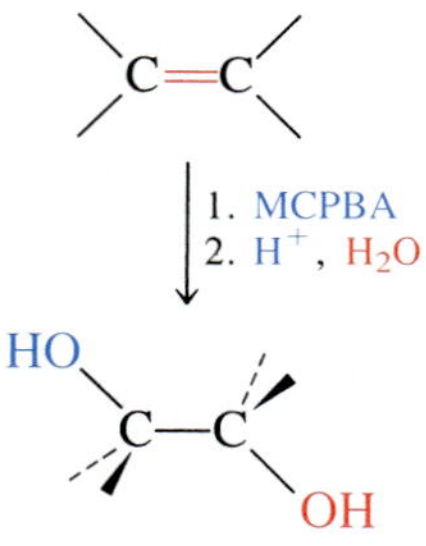

1. MCPBA
2. H^+, H_2O

The Hydrolysis of Oxacyclopropanes Furnishes the Products of *Anti* Dihydroxylation of an Alkene

Treatment of oxacyclopropanes with water in the presence of catalytic acid or base leads to ring-opening (Section 9-6) to the corresponding vicinal diols. In this reaction, water nucleophilically attacks the side opposite the oxygen in the three-membered ring, and so the net result of the oxidation-hydrolysis sequence constitutes an ***anti*** **dihydroxylation** of an alkene. In this way, *trans*-2-butene gives *meso*-2,3-butanediol, whereas *cis*-2-butene furnishes the racemic mixture of the 2*R*,3*R* and 2*S*,3*S* diastereomers.

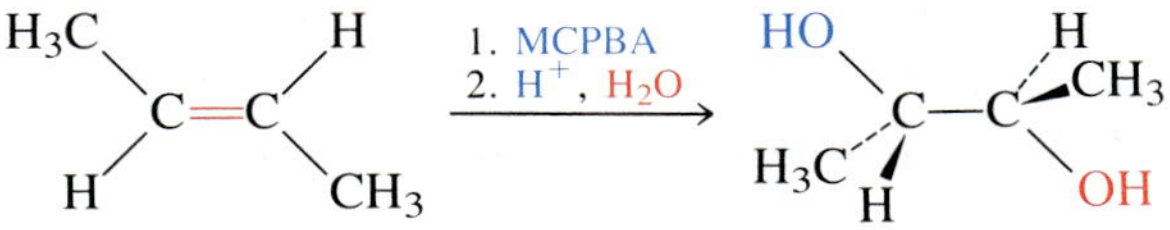

1. MCPBA
2. H^+, H_2O

***trans*-2-Butene** → ***meso*-2,3-Butanediol**

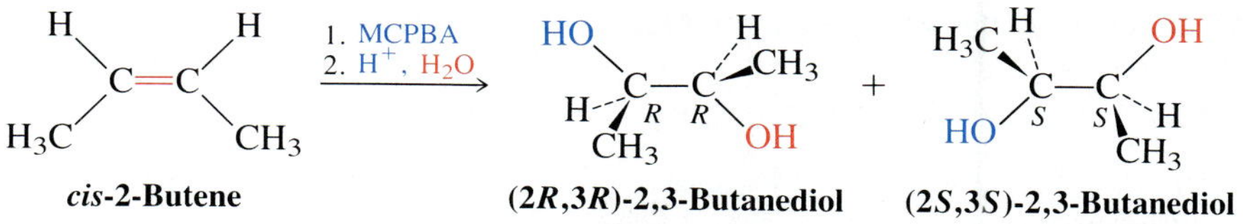

cis-2-Butene → (1. MCPBA; 2. H^+, H_2O) → (2*R*,3*R*)-2,3-Butanediol + (2*S*,3*S*)-2,3-Butanediol

EXERCISE 12-16

Give the products obtained by treating the following alkenes with MCPBA and then aqueous acid: (a) 1-hexene; (b) cyclohexene; (c) *cis*-2-pentene.

Oxidation of Alkenes to Vicinal *Syn* Diols

Potassium permanganate, in cold aqueous solution, reacts with alkenes under neutral conditions to give the corresponding vicinal *syn* diols. The other product in this reaction is insoluble brown manganese dioxide, MnO_2, itself a mild oxidizing agent. However, under neutral conditions, MnO_2 does not react further with either the alkene or the diol.

General Vicinal *Syn* Dihydroxylation with Permanganate

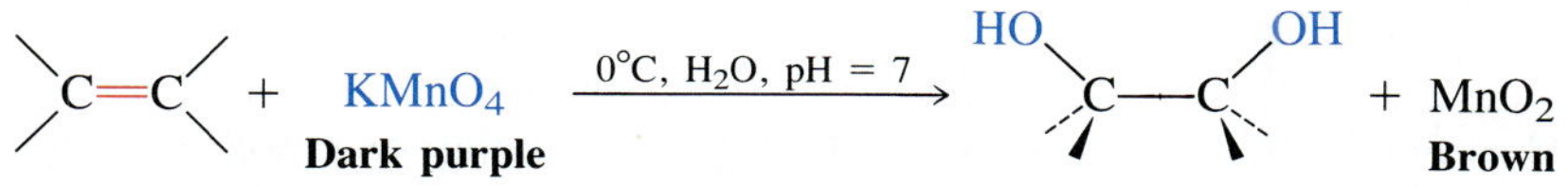

$$\text{C=C} + \underset{\textbf{Dark purple}}{KMnO_4} \xrightarrow{0°C,\ H_2O,\ pH\ =\ 7} \text{HO–C–C–OH} + \underset{\textbf{Brown}}{MnO_2}$$

Example:

Cyclohexene → ($KMnO_4$, H_2O, 0°C) → 37% *cis*-1,2-Cyclohexanediol

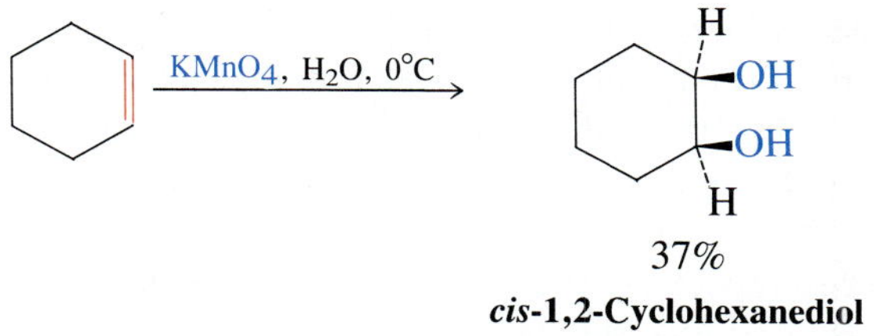

What is the mechanism of this transformation? The initial reaction of the π bond with permanganate constitutes a concerted addition in which three electron pairs move simultaneously to give a cyclic ester containing Mn(V). This process may be viewed as an electrophilic attack on the alkene—there is a net electron flow of two electrons from the alkene onto the metal that is reduced (Mn(VII) → Mn(V)). For steric reasons, the product can form only in a way that introduces the two oxygen atoms on the *same* side of the double bond: *syn*. This intermediate is reactive, hydrolyzing in the presence of water to give the free diol and an unstable Mn(V) species. Manganese in this oxidation state disproportionates into insoluble MnO_2 and (probably) a Mn(VI) species that further oxidizes the starting material. The details of these transformations are complex. The decolorization of a purple permanganate solution is a reaction diagnostic of alkenes. Other types of compounds studied so far, such as alkanes, halides, ethers, and ketones, are usually stable under these conditions.

Mechanism of the Permanganate Oxidation of Alkenes:

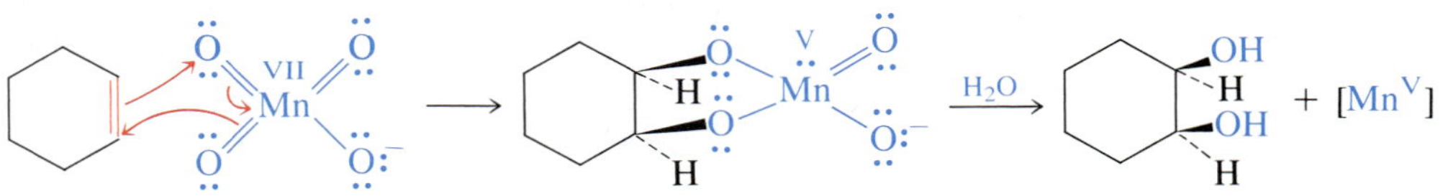

Ṡyn dihydroxylation can be achieved in better yields by using osmium tetroxide, OsO_4, which is quite similar to permanganate in its mechanism of action. The metal in OsO_4 has the same number of valence electrons as the manganese in MnO_4^-, osmium being located one column to the right in the periodic table. The reagent was originally employed in stoichiometric quantities and led to intermediate isolable cyclic esters. Typically, these intermediates were not isolated, but reductively hydrolyzed with H_2S or bisulfite, $NaHSO_3$.

Vicinal *Syn* Dihydroxylation with Osmium Tetroxide

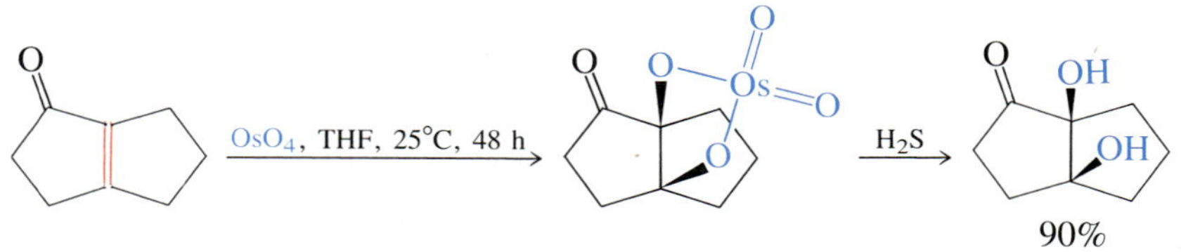

OsO_4, THF, 25°C, 48 h → H_2S → 90%

Oxidative Hydrolysis of the Intermediate Ester in Osmium Tetroxide Oxidations

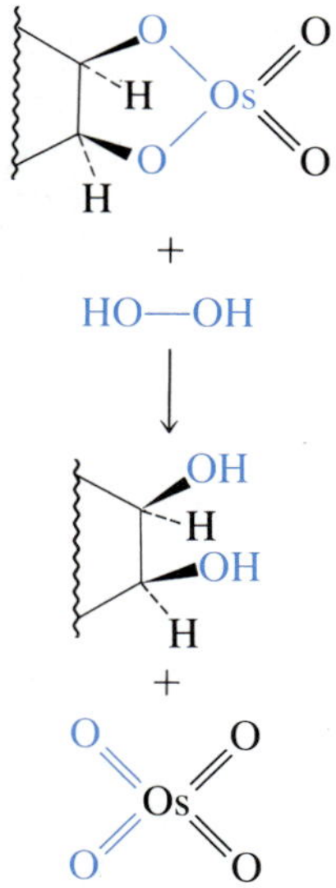

However, OsO_4 is expensive and highly toxic; therefore a modification calls for the use of only catalytic quantities of the osmium reagent and stoichiometric amounts of hydrogen peroxide (which does not react with alkenes under these conditions) as the oxidizing agent. In the catalytic cycle, OsO_4 first forms the intermediate ester, which is then presumably oxidatively hydrolyzed, as shown at the left, to free the vicinal *syn* diol and to regenerate OsO_4. This method furnishes the diol directly without requiring a second step.

An example of the application of the reaction in a steroid system reveals that dihydroxylation may be stereospecific with respect not only to the double bond but also to other stereocenters in the molecule. In this case, the α face (Section 4-7) of the steroid is less sterically hindered, resulting only in vicinal diol formation on the α side:

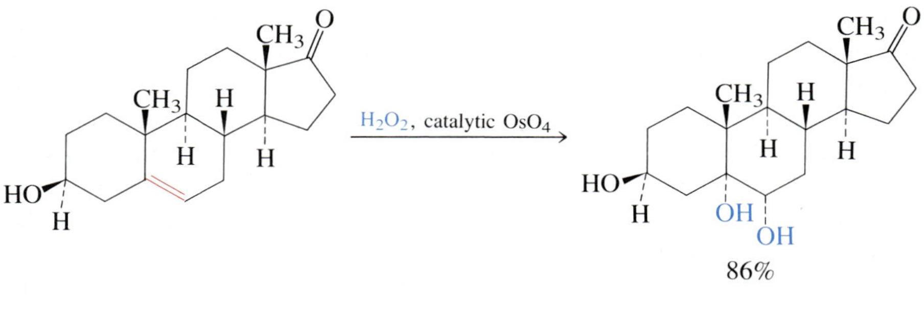

H_2O_2, catalytic OsO_4 → 86%

EXERCISE 12-17

The stereochemical consequences of the vicinal *syn* dihydroxylation of alkenes are complementary to those of vicinal *anti* dihydroxylation. Show the products (indicate stereochemistry) of the vicinal *syn* dihydroxylation of *cis*- and *trans*-2-butene.

The Complete Oxidative Cleavage of Alkenes: Ozonolysis

Although oxidation of alkenes with cold potassium permanganate or osmium tetroxide breaks only the π bond, other reagents may rupture the σ bond as well.

The most-general and mildest method of oxidatively cleaving alkenes to carbonyl compounds is **ozonolysis.** In this reaction, the alkene is treated with ozone, O_3, at low temperatures in, for example, methanol. Other common solvents are ethyl ethanoate (acetate) and dichloromethane. Ozone is produced in the laboratory in an instrument called an ozonator, in which an arc discharge generates from about 3% to about 4% of ozone in a dry oxygen stream. The gas mixture is passed through the solution containing the alkene. The first isolable intermediate is a species called an *ozonide,* which is reduced directly in a subsequent step to the two carbonyl products by various treatments, such as catalytic hydrogenation, exposure to zinc in ethanoic (acetic) acid, or by reaction with dimethyl sulfide. The net result of the ozonolysis-reduction sequence is the cleavage of the molecule at the C–C double bond and attachments of doubly bonded oxygen to each of the formerly mutually doubly bonded carbons.

General Ozonolysis Reaction

$$\text{>C=C<} \xrightarrow{O_3} \text{Ozonide} \xrightarrow[-[O]]{\text{Reduction}} \text{>C=O} + \text{O=C<}$$

Ozonide **Carbonyl products**

Examples:

$$CH_3CH_2(CH_3)C{=}C(H)CH_3 \xrightarrow[\text{2. Zn, } CH_3COOH]{\text{1. } O_3,\ CH_2Cl_2} CH_3CH_2\overset{O}{\overset{\|}{C}}CH_3 + CH_3\overset{O}{\overset{\|}{C}}H$$

(Z)-3-Methyl-2-pentene → **2-Butanone** (90%) + **Ethanal (Acetaldehyde)**

$$\text{1-Methylcyclohexene} \xrightarrow[\text{2. } H_2,\ Pt]{\text{1. } O_3,\ CH_2Cl_2} CH_3\overset{O}{\overset{\|}{C}}(CH_2)_4\overset{O}{\overset{\|}{C}}H$$

1-Methylcyclohexene → **6-Oxoheptanal** (85%)

$$\xrightarrow[\text{2. } (CH_3)_2S]{\text{1. } O_3,\ CH_3OH} \quad + \quad CH_2{=}O$$

(bicyclic H_3C, $=CH_2$ compound → bicyclic H_3C ketone, 76%) + $CH_2{=}O$ (Removed on work-up)

The mechanism of ozonolysis proceeds through initial electrophilic addition of ozone to the double bond, a transformation that yields the so-called *molozonide*. In this reaction, as in several others already presented, six electrons move in concerted fashion in a cyclic transition state. The molozonide is unstable and breaks apart into a carbonyl and a carbonyl oxide fragment through another cyclic six-electron rearrangement. Recombination of the two fragments in a head-to-tail manner yields the ozonide.

Mechanism of Ozonolysis:

STEP 1: Molozonide formation and cleavage

$C{=}C + O_3 \longrightarrow$ A molozonide $\longrightarrow C{=}O$ + A carbonyl oxide

A molozonide **A carbonyl oxide**

STEP 2: Ozonide formation and reduction

$C{=}O$ + carbonyl oxide $\longrightarrow$ **Ozonide**

$\xrightarrow{H_2,\ Pt}$ $C{=}O + H_2O$

$\xrightarrow{Zn,\ CH_3COOH}$ $C{=}O + ZnO$

$\xrightarrow{(CH_3)_2S}$ $C{=}O + (CH_3)_2S{=}O$

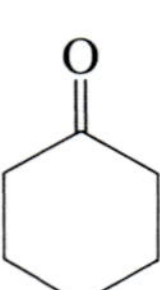

EXERCISE 12-18

An unknown hydrocarbon of the empirical (not molecular) formula C_3H_5 exhibited an ^1H NMR spectrum with a complex multiplet of signals between 1 and 2.2 ppm. Ozonolysis of this compound gave two equivalents of cyclohexanone, whose structure is shown at the left. What is the structure of the unknown?

Ozonolysis to Alcohols

Treatment of the ozonide with sodium borohydride leads to alcohols. In this way, a double bond can be oxidatively cleaved to produce two alcohols.

Alcohols from Alkenes

$$CH_3CH_2CH_2CH{=}CHCH_2CH_2CH_3 \xrightarrow[\text{2. NaBH}_4,\ CH_3OH]{\text{1. }O_3,\ CH_2Cl_2} 2\ CH_3CH_2CH_2CH_2OH$$

4-Octene — 95% **1-Butanol**

$$CH_3CH{=}CH(CH_2)_7COOCH_3 \xrightarrow[\text{2. NaBH}_4,\ CH_3OH]{\text{1. }O_3,\ CH_2Cl_2} CH_3CH_2OH + HOCH_2(CH_2)_7COOCH_3$$

Methyl 9-undecenoate — **Ethanol** — 91% **Methyl 9-hydroxynonanoate**

EXERCISE 12-19

Give the products of the following reactions:

(a) 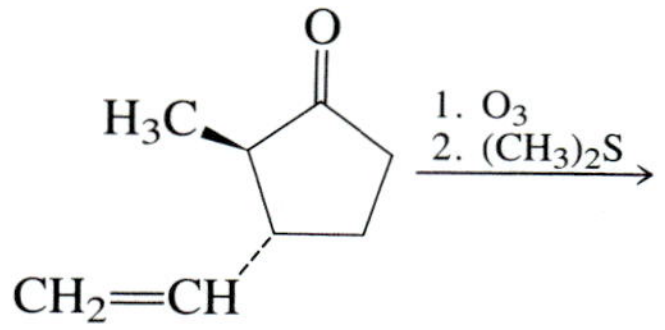 [2-methyl-3-vinylcyclopentanone; H_3C, $CH_2{=}CH$] $\xrightarrow[\text{2. }(CH_3)_2S]{\text{1. }O_3}$

(b) 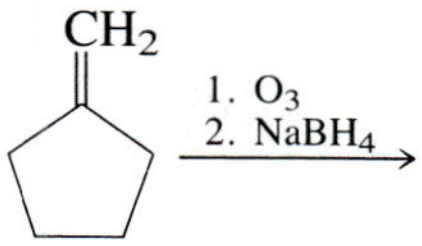 [methylenecyclopentane; CH_2] $\xrightarrow[\text{2. }NaBH_4]{\text{1. }O_3}$

EXERCISE 12-20

What is the structure of the following starting material?

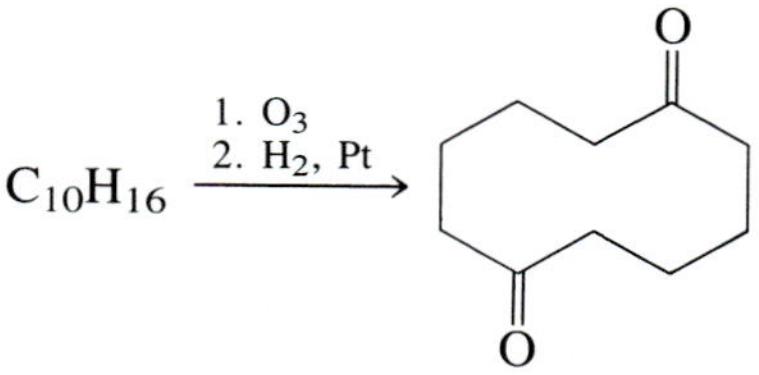

$$C_{10}H_{16} \xrightarrow[\text{2. } H_2,\ Pt]{\text{1. } O_3}$$ [cyclodecane-1,6-dione]

In summary, various electrophilic oxidizing agents convert alkenes into oxygenated products with or without cleavage of the molecule at the double bond. Peroxycarboxylic acids supply oxygen atoms to form oxacyclopropanes. Oxidation-hydrolysis reactions with peroxycarboxylic acids furnish vicinal *anti*-1,2-diols, whereas cold potassium permanganate or, better yet, catalytic osmium tetroxide in the presence of stoichiometric hydrogen peroxide leads to vicinal *syn*-1,2-diols. Finally, ozonolysis followed by reduction of the ozonide yields aldehydes and ketones by double-bond cleavage. Treatment of the ozonide with sodium borohydride gives the corresponding alcohols. Mechanistically, all these reactions can be related in that initial attack by an electrophilic agent on the π bond leads to its rupture and the two π electrons enter into bonding with the oxidizing agent.

12-6 Free-Radical Additions to Alkenes: Anti-Markovnikov Product Formation

This section deals with another mode of reactivity of the double bond: radical addition. In contrast with electrophilic reagents, which consume both electrons of the π bond on addition, a free radical requires only one electron for bond formation, so that an alkyl radical is formed. The consequences of this difference are anti-Markovnikov products.

Hydrogen Bromide Can Add to Alkenes in Anti-Markovnikov Fashion: A Change in Mechanism

When freshly distilled 1-butene is exposed to hydrogen bromide, clean Markovnikov addition to 2-bromobutane is observed. This result is in accord with the ionic mechanism discussed in Section 12-3. Curiously, the same reaction, when carried out with a sample of 1-butene that has been stored with some

Markovnikov Addition of HBr

$$CH_3CH_2CH{=}CH_2 \xrightarrow[24\text{ h}]{HBr} CH_3CH_2CH(Br)CH_2H$$

90%

Markovnikov product

Anti-Markovnikov Addition of HBr

$CH_3CH_2CH{=}CH_2$

Exposed to oxygen

$$\xrightarrow[4\text{ h}]{HBr} CH_3CH_2CH(H)CH_2Br$$

65%

Anti-Markovnikov product

exposure to air, proceeds much faster and gives an entirely different result. In this case, we isolate 1-bromobutane, formed by anti-Markovnikov addition.

This change caused considerable confusion in the early days of alkene chemistry, because one researcher would obtain only one hydrobromination product, whereas another would obtain a different product or mixtures from a seemingly identical reaction. The mystery was solved by Kharasch* in the 1930s, when it was discovered that the culprits responsible for anti-Markovnikov additions were free radicals formed from peroxides, ROOR, in alkene samples that had been stored in the presence of air.

The mechanism of the addition reaction under these conditions is not an ionic sequence; rather, it is a much faster **free-radical chain sequence.** (In this section, all radicals and single atoms are shown in green, as in Chapter 3.) The initiation steps are, first, the cleavage of the weak RO–OR bond ($DH^\circ \sim$ 35 kcal mole^{-1}) and, then, reaction of the resulting alkoxy radical (or subsequently formed radicals) with hydrogen bromide. The driving force for the second (exothermic) step is the formation of the strong O–H bond. The bromine atom so generated initiates chain propagation by attacking the double bond. One of the π electrons combines with the unpaired electron on the bromine atom to form the carbon–bromine bond. The other π electron remains on carbon, giving rise to a radical. The halogen atom's attack is *regioselective,* creating the relatively more stable secondary radical rather than the primary one. This result is reminiscent of the ionic additions of hydrogen bromide, except that the roles of the proton and bromine are reversed. In the ionic mechanism, a proton attacks first to generate the more stable carbocation, which is then trapped by bromide ion. In the free-radical mechanism, a bromine atom is the attacking species, creating the more-stable radical center. The latter subsequently reacts with HBr by abstracting a hydrogen and regenerating the chain-carrying bromine atom. Both propagation steps are exothermic, and the reaction proceeds rapidly. As usual, termination is by radical combination or by some other removal of the chain carriers (Section 3-5).

Mechanism of Free-Radical Hydrobromination:

Initiation:

$$R\ddot{\underset{..}{O}}{-}\ddot{\underset{..}{O}}R \xrightarrow{\Delta} 2\,R\ddot{\underset{..}{O}}\cdot \qquad \Delta H^\circ \sim +35 \text{ kcal mole}^{-1}$$

$$R\ddot{\underset{..}{O}}\cdot + H\ddot{\underset{..}{Br}}: \xrightarrow{\Delta} R\ddot{\underset{..}{O}}H + :\ddot{\underset{..}{Br}}\cdot \qquad \Delta H^\circ \sim -15.5 \text{ kcal mole}^{-1}$$

Chain propagation:

$$(CH_3CH_2)(H)C{=}CH_2 + :\ddot{\underset{..}{Br}}\cdot \longrightarrow CH_3CH_2\dot{C}H{-}CH_2\ddot{\underset{..}{Br}}: \qquad \Delta H^\circ \sim -3 \text{ kcal mole}^{-1}$$

Secondary radical

$$CH_3CH_2\dot{C}HCH_2Br + H:\ddot{\underset{..}{Br}}: \longrightarrow CH_3CH_2CH(H)CH_2\ddot{\underset{..}{Br}}: + :\ddot{\underset{..}{Br}}\cdot \qquad \Delta H^\circ \sim -7.5 \text{ kcal mole}^{-1}$$

*Professor M. S. Kharasch, 1895–1957, University of Chicago.

Termination:

$$:\ddot{\underset{..}{Br}}\cdot + \cdot\ddot{\underset{..}{Br}}: \longrightarrow Br_2$$

$$2\ CH_3CH_2\dot{C}HCH_2Br \longrightarrow \begin{array}{c} CH_3CH_2CHCH_2Br \\ | \\ CH_3CH_2CHCH_2Br \end{array}$$

Are Free-Radical Additions General?

Hydrogen chloride and hydrogen iodide do not give anti-Markovnikov addition products to alkenes because of unfavorable kinetics; in both cases, one of the propagating steps is endothermic and consequently so slow that the chain reaction terminates. For hydrogen iodide, the first step is uphill, because the strength of the newly formed carbon–iodine bond does not quite make up for the loss of the π bond.

$$CH_3CH_2CH{=}CH_2 + I\cdot \xrightarrow[\text{Endothermic}]{} CH_3CH_2\dot{C}H{-}CH_2I$$

For hydrogen chloride, the second step is endothermic; in this step, the strong hydrogen–chlorine bond must be broken, which renders the reaction unfavorably slow, although addition may be observed in certain cases.

$$CH_3CH_2\dot{C}HCH_2Cl + HCl \xrightarrow[\text{Endothermic}]{} CH_3CH_2\overset{\overset{H}{|}}{C}HCH_2Cl + Cl\cdot$$

EXERCISE 12-21

Calculate the DH° of the two preceding endothermic steps, using the data in Table 3-1 and a π-bond strength of 65 kcal mole^{-1}.

There are other reagents, however, that undergo successful free-radical additions to alkenes. Examples are thiols and some of the halomethanes.

Examples of Other Free-Radical Additions to Alkenes

$$CH_3CH{=}CH_2 + CH_3CH_2SH \xrightarrow{ROOR} CH_3\underset{\underset{H}{|}}{C}HCH_2SCH_2CH_3$$

Ethanethiol **Ethyl propyl sulfide**

1-Methylcyclohexene + HSH $\xrightarrow{ROOR}$ 2-methylcyclohexanethiol (H, CH_3, SH)

29%

2-Methylcyclohexanethiol
(Mixture of cis and trans isomers)

$$(H_3C)_2C{=}CH_2 + ClCCl_3 \xrightarrow{ROOR} CH_3\underset{CH_3}{\overset{Cl}{C}}CH_2CCl_3$$

78%

1,1,1,3-Tetrachloro-3-methylbutane

$$CH_3(CH_2)_5CH{=}CH_2 + HCCl_3 \xrightarrow{ROOR} CH_3(CH_2)_5\overset{H}{C}HCH_2CCl_3$$

22%

1,1,1-Trichlorononane

In most of these examples, the initiating alkoxy radical abstracts a hydrogen from the substrate to yield a chain carrier, because of the strength of the resulting OH bond. A typical example is trichloromethane (chloroform):

$$RO\cdot + CHCl_3 \longrightarrow ROH + \underset{\textbf{Chain carrier}}{\cdot CCl_3} \quad \text{not} \quad RO\cdot + Cl\underset{H}{\overset{Cl}{C}}{-}Cl \longrightarrow ROCl + \cdot CHCl_2$$

EXERCISE 12-22

Write a plausible mechanism for the free-radical addition of tetrabromomethane to propene.

Anti-Markovnikov additions are synthetically very useful because the derived products complement those obtained from ionic additions. This type of regiochemical control is an important feature in the development of new synthetic methods.

In summary, free-radical initiators alter the mechanism of the addition of HBr to alkenes from ionic to radical chain. The consequence of this change is anti-Markovnikov regioselectivity. Other species, most notably thiols and some halomethanes, are capable of undergoing similar reactions.

12-7 Dimerization, Oligomerization, and Polymerization of Alkenes

Is it possible for alkenes to react with each other? Indeed it is, but only in the presence of an appropriate catalyst—for example, an acid, a free radical, a base, or a transition metal. In this reaction, the unsaturated centers of the alkene monomer (*monos,* Greek, single; *meros,* Greek, part) are linked to form dimers, trimers, **oligomers** (*oligos,* Greek, few, small), and ultimately **polymers** (*polymeres,* Greek, of many parts), substances of great industrial importance.

Polymerization

$$\text{}{=}C\ \ C{=}C\ \ C{=}C\ \ C{=}\text{} \longrightarrow {-}C{-}C{-}C{-}C{-}C{-}C{-}$$

Monomers **Polymer**

Carbocations Attack Pi Bonds

Treatment of 2-methylpropene with hot aqueous sulfuric acid gives two dimers: 2,4,4-trimethyl-1-pentene and 2,4,4-trimethyl-2-pentene. This transformation is possible because 2-methylpropene can be protonated under the reaction conditions to furnish the 1,1-dimethylethyl (*tert*-butyl) cation. This species can attack the electron-rich double bond of 2-methylpropene with formation of a new carbon–carbon bond. Electrophilic addition proceeds according to the Markovnikov rule to generate the more-stable carbocation. Subsequent deprotonation in each of two directions furnishes a mixture of the two observed products.

Curiously, the terminal alkene predominates in the product mixture. This apparent violation of the rules of alkene stability seems to originate in the relatively large steric repulsion between the 1,1-dimethylethyl (*tert*-butyl) and the *Z*-methyl group in the minor isomer. Such steric hindrance renders the internal alkene less stable than the terminal one.

Dimerization of 2-Methylpropene

$CH_2{=}C(CH_3)_2$ + $CH_2{=}C(CH_3)_2$ $\xrightarrow{H^+}$ $CH_3C(CH_3)_2CH_2C(CH_3){=}CH_2$ Major **2,4,4-Trimethyl-1-pentene** + $CH_3C(CH_3)_2CH{=}C(CH_3)_2$ Minor **2,4,4-Trimethyl-2-pentene**

Mechanism of Dimerization of 2-Methylpropene:

$$CH_2{=}C(CH_3)_2 \xrightarrow{H^+} CH_3\overset{+}{C}(CH_3)_2 \quad CH_2{=}C(CH_3)_2 \longrightarrow$$

$$CH_3C(CH_3)_2{-}CH_2{-}\overset{+}{C}(CH_3)(CH_2H) \xrightarrow{-H^+} CH_3C(CH_3)_2CH_2C(CH_3){=}CH_2 + CH_3C(CH_3)_2CH{=}C(CH_3)_2$$

The two dimers of 2-methylpropene tend to react further with the starting alkene. For example, when 2-methylpropene is treated with mineral acid under more stringent conditions, trimers, tetramers, pentamers, and so forth, are formed by repeated electrophilic attack of intermediate carbocations on the double bond. This process, which leads to alkane chains of intermediate length, is called oligomerization.

Oligomerization of the 2-Methylpropene Dimers

$$CH_3C(CH_3)_2CH{=}C(CH_3)_2 + CH_3C(CH_3)_2CH_2C(CH_3){=}CH_2 \xrightarrow{H^+} CH_3C(CH_3)_2CH_2\overset{+}{C}(CH_3)_2 + CH_2{=}C(CH_3)_2 \longrightarrow$$

$$CH_3C(CH_3)_2CH_2C(CH_3)_2CH_2\overset{+}{C}(CH_3)_2 + CH_2{=}C(CH_3)_2 \longrightarrow CH_3C(CH_3)_2CH_2C(CH_3)_2CH_2C(CH_3)_2CH_2\overset{+}{C}(CH_3)_2 \quad \text{etc.}$$

BOX 12-2

Steroid Synthesis in Nature

A remarkable series of intramolecular alkene couplings is observed in nature as part of the biosynthetic pathway to the steroid nucleus. In this process, a molecule called squalene is enzymatically oxidized to the oxacyclopropane squalene oxide. Enzymatic acid-catalyzed oxacyclopropane ring-opening is followed by four sequential carbon–carbon bond-forming steps mechanistically related to alkene oligomerization. Further biochemical conversion yields lanosterol, a biological precursor to cholesterol. Such cyclizations (biomimetic alkene cyclizations) have also been carried out in the laboratory. These reactions are highly regioselective and stereospecific and afford a facile means of synthesizing a variety of steroids.

CH_3 H_3C CH_3 CH_3 H_3C CH_3 CH_3

Squalene

$\xrightarrow{O_2,\ \text{enzyme}}$

H_3C CH_3 CH_3 CH_3 CH_3 H_3C O CH_3 H^+ CH_3

Squalene oxide

$\xrightarrow{H^+,\ \text{enzyme}}$

H_3C CH_3 CH_3 CH_3 CH_3 H_3C HO + CH_3 CH_3

$\longrightarrow$

H_3C + H H CH_3 CH_3 CH_3 H_3C H CH_3 HO H H_3C CH_3

$\xrightarrow[-\ H^+]{}$

H H_3C H H_3C CH_3 CH_3 CH_3 H_3C HO H H_3C CH_3

Lanosterol

EXERCISE 12-23

Humulene and α-caryophyllene alcohol are terpene constituents of carnation extracts. The former is converted into the latter by acid-catalyzed hydration in one step. Write a mechanism. Hint: Follow the labeled carbon atoms retrosynthetically. The mechanism includes carbocation-induced cyclizations and hydrogen and alkyl-group migrations. This is difficult.

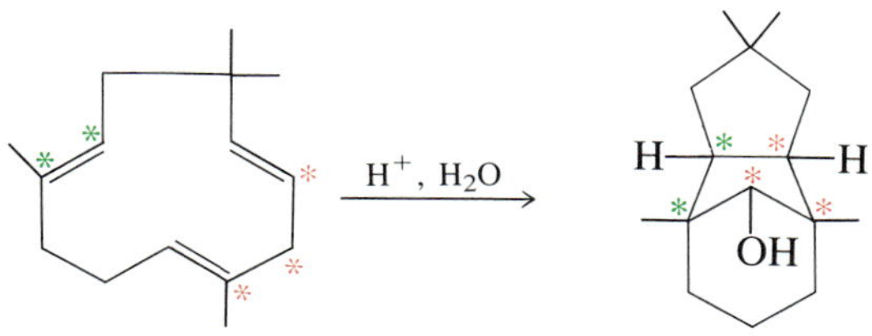

$$\text{Humulene} \xrightarrow{H^+,\ H_2O} \alpha\text{-Caryophyllene alcohol}$$

Humulene **α-Caryophyllene alcohol**

At higher temperatures, the oligomerization of alkenes continues to give polymers containing many subunits.

Polymerization of 2-Methylpropene

$$n\ CH_2{=}C(CH_3)_2 \xrightarrow{H^+,\ 200^\circ C} H{-}(CH_2{-}\underset{CH_3}{\overset{CH_3}{C}})_{n-1}{-}CH_2\overset{CH_3}{C}{=}CH_2$$

Poly-2-methylpropene (Polyisobutylene)

The Synthesis of Polymers

Many alkenes are suitable monomers for polymerization. Although polymerization can be an unwanted side reaction in their chemistry, it is exceedingly important in the chemical industry, because many polymers have desirable properties, such as durability, inertness to many chemicals, elasticity, transparency, and electrical and thermal resistance.

Names such as polyethylene, poly(vinyl chloride) (PVC), Teflon, polystyrene, Orlon, and Plexiglas (Table 12-3) have become household words. These substances have varied uses as synthetic fibers, films, pipes, coatings, molded articles, and so forth. Recently, energy conservation has led to the design and building of lighter automobiles. Plastics, because of their strength and light weight, have made a major contribution to this effort. In 1980, the incorporation of about 400 pounds of plastic resulted in, on the average, an overall *net* reduction of 1,000 pounds per car. However, despite the utility of polymers, their production has contributed to pollution—many of them are not biodegradable.

Acid-catalyzed polymerizations, such as that described for poly-2-methylpropene, are carried out with H_2SO_4, HF, or BF_3 as the initiators. Because they proceed through the intermediacy of carbocations, they are also called *cationic polymerizations*. Other mechanisms of polymerization are *free radical, anionic,* and *metal catalyzed*.

An example of the free-radical polymerization is that of ethene in the presence of an organic peroxide at 1000 atmospheres and temperatures in excess of 100°C. The reaction proceeds by a mechanism that, in its initial

Free-Radical Polymerization of Ethene

$$n\ CH_2{=}CH_2 \xrightarrow[>100^\circ C]{\text{ROOR, 1000 atm}} {-}(CH_2{-}CH_2)_n{-}$$

Polyethene (Polyethylene)

TABLE 12-3
Common polymers and their monomers

Monomer	Structure	Polymer (common name)
Ethene	$H_2C{=}CH_2$	Polyethylene
Chloroethene (vinyl chloride)	$H_2C{=}CHCl$	Poly(vinyl chloride) (PVC)
Tetrafluoroethene	$F_2C{=}CF_2$	Teflon
Phenylethene (styrene)	$C_6H_5CH{=}CH_2$	Polystyrene
Propenenitrile (acrylonitrile)	$H_2C{=}C(H)C{\equiv}N$	Orlon
Methyl 2-methylpropenoate (methyl methacrylate)	$H_2C{=}C(CH_3)COCH_3$ (with C=O)	Plexiglas

stages, resembles that of the free-radical addition to alkenes (Section 12-6). The peroxide initiators cleave into alkoxy radicals, which begin polymerization by addition to the double bond of ethene. The alkyl radical thus created attacks the double bond of another ethene molecule, furnishing another radical center, and so on. Termination of the polymerization can be by dimerization, disproportionation of the radical, or other radical-trapping reactions.

Mechanism of Radical Polymerization of Ethene:

Initiation:

$$RO{-}OR \longrightarrow 2\,RO\cdot$$

$$RO\cdot + CH_2{=}CH_2 \longrightarrow ROCH_2{-}\dot{C}H_2$$

Propagation:

$$ROCH_2CH_2\cdot + CH_2{=}CH_2 \longrightarrow ROCH_2CH_2CH_2CH_2\cdot$$

$$ROCH_2CH_2CH_2CH_2\cdot \xrightarrow{(n-1)\,CH_2=CH_2} RO{-}(CH_2CH_2)_n{-}CH_2CH_2\cdot$$

Termination:

$$2\,RO{-}(CH_2CH_2)_n{-}CH_2CH_2\cdot \longrightarrow RO{-}(CH_2CH_2)_{2n+2}{-}OR$$

$$2\,RO{-}(CH_2CH_2)_n{-}CH_2CH_2\cdot \longrightarrow RO{-}(CH_2CH_2)_n{-}CH_2CH_3 + RO{-}(CH_2CH_2)_n{-}CH{=}CH_2$$

Polyethene (polyethylene) produced in this way might be expected to have a linear structure. However, *branching* occurs by abstraction of a hydrogen along the growing chain by another radical center followed by chain growth originating from the new radical. The average molecular weight of polyethene is almost 1 million.

Branching in Ethene Polymerization

$$\sim\sim CH_2\underset{\displaystyle H}{\underset{|}{C}}HCH_2CH_2\sim\sim \xrightarrow[-\,RH]{R\cdot} \sim\sim CH_2\overset{\displaystyle H}{\overset{|}{\dot{C}}}CH_2CH_2\sim\sim \xrightarrow{CH_2=CH_2}$$

$$\sim\sim CH_2\overset{\displaystyle H}{\overset{|}{C}}(CH_2\dot{C}H_2)CH_2CH_2\sim\sim \xrightarrow{\text{Excess } CH_2=CH_2} \sim\sim CH_2\overset{\displaystyle H}{\overset{|}{C}}(CH_2CH_2\sim\sim)CH_2CH_2\sim\sim$$

Polychloroethene [poly(vinyl chloride), PVC] is made by similar radical polymerization. Interestingly, the reaction is regioselective. The peroxide initiator and the intermediate chain radicals add only to the unsubstituted end of the monomer, because the radical center formed next to chlorine is relatively stable. Thus, PVC has a very regular *head-to-tail structure* of molecular weight in excess of 1.5 million. Although PVC itself is fairly hard and brittle, it can be softened by addition of carboxylic acid esters (Section 18-4), called **plasticizers** (*plastikos,* Greek, to form). The resulting elastic material is used in "vinyl leather," plastic covers, and garden hoses.

$$CH_2=CHCl \xrightarrow{ROOR} -(CH_2\underset{\displaystyle Cl}{\underset{|}{C}}H)_n-$$

Polychloroethene (Polyvinyl chloride)

Exposure to chloroethene (vinyl chloride) has been linked to the incidence of a rare form of liver cancer (angiocarcinoma). The Occupational Safety and Health Administration (OSHA) has set limits to human exposure of less than an average of 1 ppm per eight-hour working day per worker.

An iron compound, $FeSO_4$, is used in the presence of hydrogen peroxide to effect the radical polymerization of propenenitrile (acrylonitrile). Polypropenenitrile (polyacrylonitrile), $-(CH_2CHCN)_n-$, also known as Orlon, is used to make fibers. Similar polymerizations of other monomers furnish Teflon and Plexiglas.

EXERCISE 12-24

Saran Wrap is made by radical copolymerization of 1,1-dichloroethene and chloroethene. Propose a structure.

Anionic polymerizations are initiated by strong bases such as alkyllithiums, Grignard reagents, amides, and alkoxides. For example, propenenitrile (acrylonitrile) undergoes rapid polymerization when treated with sodium amide in liquid ammonia. The amide ion attacks the methylene group to generate a carbanion, with its charge located next to the electron-withdrawing nitrile group. The latter is electron withdrawing because its carbon is *sp* hybridized, because the nitrogen polarizes the triple bond in the sense $-\overset{\delta^+}{C}\equiv\overset{\delta^-}{N}$, and because the negative charge can be delocalized.

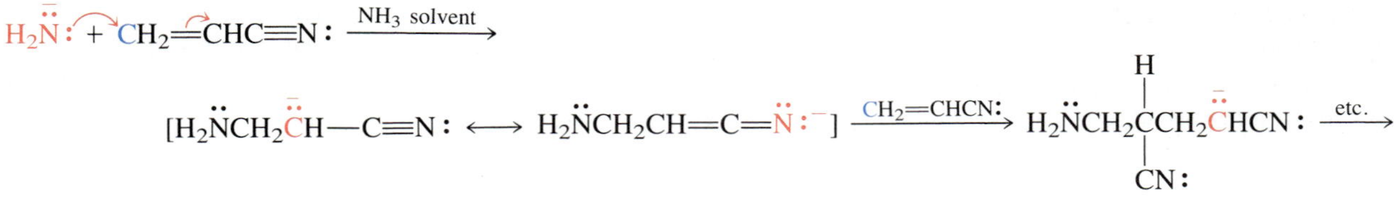

$$\mathrm{H_2\ddot{N}:^- + CH_2{=}CHC{\equiv}N: \xrightarrow{NH_3\ solvent}}$$

$$\mathrm{[H_2\ddot{N}CH_2\ddot{C}H{-}C{\equiv}N: \longleftrightarrow H_2\ddot{N}CH_2CH{=}C{=}\ddot{N}:^-] \xrightarrow{CH_2=CHCN:} H_2\ddot{N}CH_2\underset{CN:}{\overset{H}{C}}CH_2\ddot{C}HCN: \xrightarrow{etc.}}$$

An important *metal-catalyzed polymerization* is that initiated by Ziegler-Natta* catalysts. They are typically made from titanium tetrachloride and a trialkylaluminum, such as triethylaluminum, $Al(CH_2CH_3)_3$. The system polymerizes alkenes, particularly ethene, at relatively low pressures with remarkable ease and efficiency.

Ziegler-Natta Polymerization

$$\mathrm{R{-}\overset{R}{\underset{}{Ti}}{-}R \xrightarrow{CH_2=CH_2} R{-}\underset{CH_2=CH_2}{\overset{R}{Ti}}{-}R \longrightarrow R{-}\overset{R}{Ti}CH_2CH_2R \longrightarrow R{-}\underset{CH_2=CH_2}{\overset{R}{Ti}}CH_2CH_2R \xrightarrow{(n-2)CH_2=CH_2} R{-}(CH_2CH_2)_n{-}R}$$

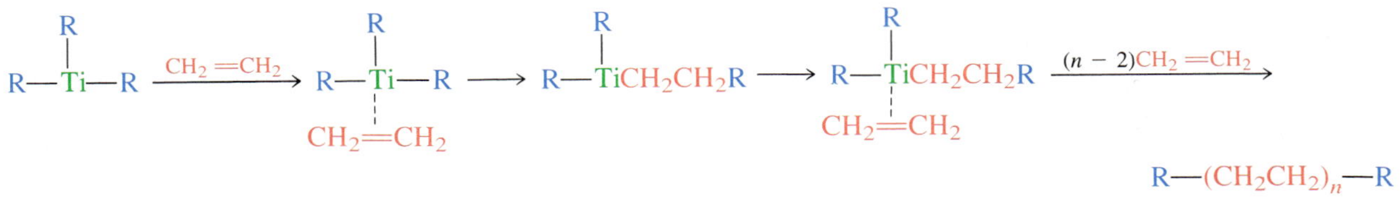

Although the mechanism of the reaction is a matter of controversy, it is not believed to proceed through any of the classical sequences discussed so far. Rather, polymerization is thought to occur by successive complexation of monomer to the transition metal bearing the growing chain, followed by repeated insertion. A feature of Ziegler-Natta polymerization is the regularity with which substituted alkane chains are constructed from substituted ethenes, such as propene. Thus, depending on reaction conditions, a terminal alkene is polymerized so as to place the substituent either on alternate sides of the chain *(syndiotactic polymer)* or on the same side *(isotactic polymer)*. This regularity gives the polymer advantageous properties, such as higher crystallinity, when compared with random polymers *(atactic polymers)*.

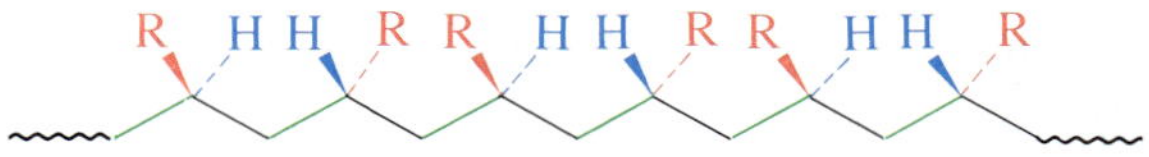

Syndiotactic polymer

Isotactic polymer

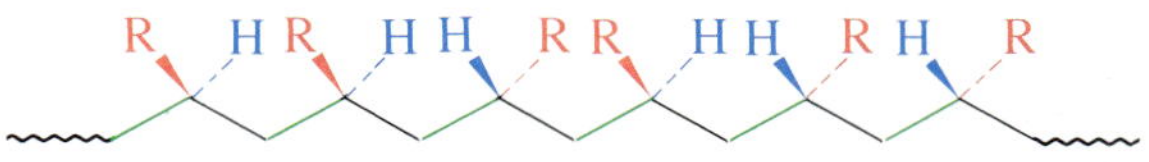

Atactic polymer

---

*Professor Karl Ziegler, 1898–1976, Max Planck Institute for Coal Research, Mülheim, W. Germany, Nobel Prize 1963; Professor Giulio Natta, 1903–1979, Polytechnic Institute of Milan, Nobel Prize 1963.

In summary, alkenes are subject to attack by carbocations, radicals, anions, and transition metals to give polymers. In principle, any alkene can function as a monomer. The intermediates are usually formed according to the rules that govern the stability of charges and radical centers.

12-8 Ethene: An Important Industrial Feedstock

In this section, ethene will serve as a case study of the importance of alkenes in industrial chemistry.

Ethene is an important monomer for polyethene production. More than 6 billion pounds of this polymer is produced in the United States annually. Currently, the major source of ethene is the pyrolysis of petroleum or hydrocarbons, such as ethane, propane, other alkanes, and cycloalkanes, derived from natural gas. Temperatures range from 750°C to 900°C and the yields of ethene from 20% to 30%. Cracking of larger alkanes typically proceeds through C–C bond breaking to alkyl radicals, the further fragmentation of which eliminates ethene (Section 3-3). In 1984, 31 billion pounds of ethene was made in this way. This amount is equivalent to about 10% of the total production of organic chemicals.

Apart from its direct use as a monomer, ethene is the starting material for many other industrial chemicals, some of which are themselves valuable monomers. For example, ethenyl ethanoate (vinyl acetate) is obtained in the reaction of ethene with ethanoic (acetic) acid in the presence of a palladium(II) catalyst, air, and cupric chloride.

$$CH_2{=}CH_2 \xrightarrow{CH_3\overset{O}{\overset{\|}{C}}OH,\ O_2,\ \text{catalytic } PdCl_2 \text{ and } CuCl_2} CH_2{=}CHO\overset{O}{\overset{\|}{C}}CH_3 + H_2O$$

Ethenyl ethanoate
(Vinyl acetate)

A similar reaction, in which water is used instead of ethanoic (acetic) acid, leads to the intermediate ethenol (vinyl alcohol). This species is unstable and spontaneously rearranges to ethanal (acetaldehyde, see Sections 13-6 and 16-2). The catalytic conversion of ethene into ethanal is also known as the *Wacker Process*.

The Wacker Process

$$CH_2{=}CH_2 \xrightarrow{H_2O,\ O_2,\ \text{catalytic } PdCl_2 \text{ and } CuCl_2} CH_2{=}CHOH \longrightarrow CH_3\overset{O}{\overset{\|}{C}}H$$

Ethenol **(Vinyl alcohol)** — **Ethanal** **(Acetaldehyde)**

Chloroethene (vinyl chloride) is made from ethene by a chlorination-dehydrochlorination sequence. Because chlorine is relatively expensive, an indirect process has been developed that uses HCl in the presence of oxygen and $CuCl_2$. These conditions lead to the same intermediate, 1,2-dichloroethane, which is converted into the desired product by elimination of HCl.

Chloroethene (Vinyl Chloride) Synthesis

$$CH_2{=}CH_2 \xrightarrow{Cl_2} \underset{\textbf{1,2-Dichloroethane}}{\underset{Cl\quad\;\; Cl}{CH_2{-}CH_2}} \xrightarrow[-\ HCl]{\Delta} \underset{\textbf{(Vinyl chloride)}}{\underset{\textbf{Chloroethene}}{CH_2{=}CHCl}}$$

Oxidation of ethene with oxygen in the presence of silver furnishes oxacyclopropane (ethylene oxide), the hydrolysis of which gives 1,2-ethanediol (ethylene glycol, Section 9-8). Hydration of ethene gives ethanol (Section 8-4).

$$CH_2{=}CH_2 \xrightarrow{O_2,\ \text{catalytic Ag}} \underset{\textbf{(Ethylene oxide)}}{\underset{\textbf{Oxacyclopropane}}{H_2C\overset{O}{\triangle}CH_2}} \xrightarrow{H^+,\ H_2O} \underset{\textbf{(Ethylene glycol)}}{\underset{\textbf{1,2-Ethanediol}}{\overset{OH\quad\; OH}{CH_2{-}CH_2}}}$$

Table 12-4 gives an idea of the sizable amount of ethene-derived raw materials produced in the United States in 1984.

TABLE 12-4
Chemicals from ethene

Chemical	**Millions of pounds**
1,2-Dichloroethane (ethylene dichloride)	7,330
Ethenylbenzene (styrene)	7,709
Ethylbenzene	7,562
Chloroethene (vinyl chloride)	6,085
Oxacyclopropane (ethylene oxide)	5,699
1,2-Ethanediol (ethylene glycol)	3,224
Ethanoic acid* (acetic acid)	2,618
Ethanol	1,060
Ethanoic anhydride* (acetic anhydride)	1,500
Ethanal (acetaldehyde)	892
Ethenyl ethanoate (vinyl acetate)	2,024

*Also made by other processes.

In summary, ethene is a valuable source of various industrial raw materials, particularly monomers, ethanol, and 1,2-ethanediol (ethylene glycol).

12-9 Alkenes in Nature: Insect Pheromones

Many natural products contain π bonds; several were mentioned in Sections 4-7 and 9-8. This section describes in more detail a specific group of naturally occurring alkenes, the **insect pheromones** (*pherein,* Greek, to bear; *hormon,* Greek, to stimulate).

Insect Pheromones

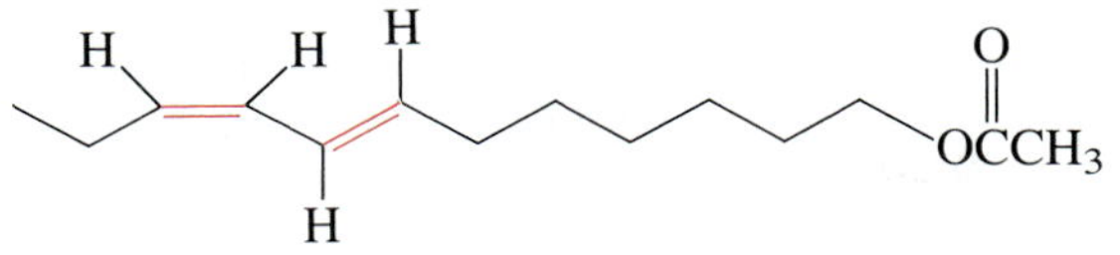

European vine moth

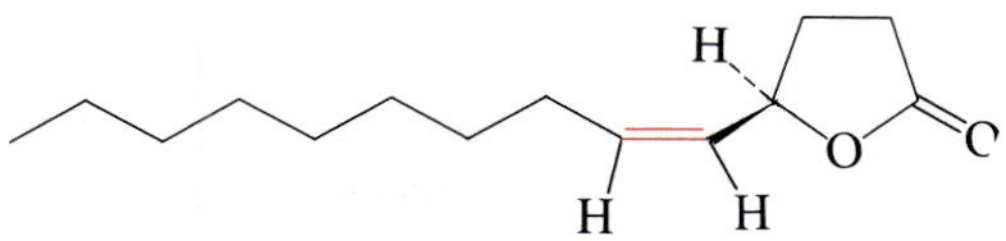

Japanese beetle

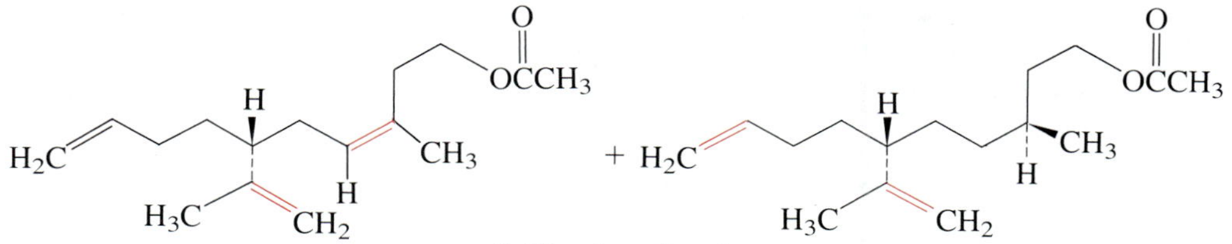

California red scale

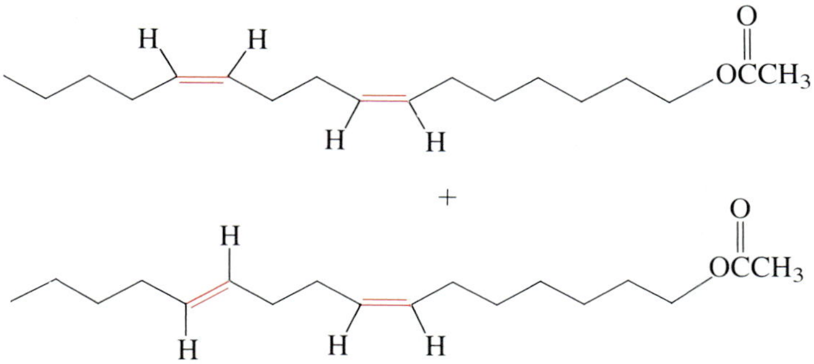

Pink bollworm moth

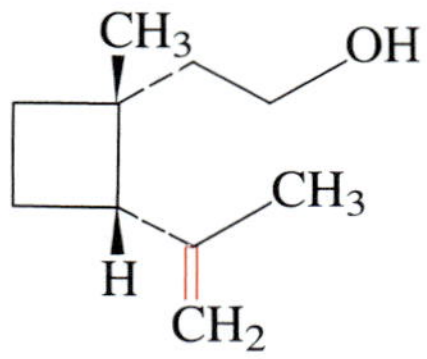

Male boll weevil (Grandisol)

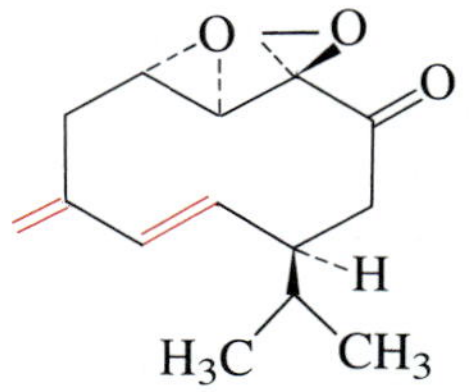

American cockroach (Periplanone-B)

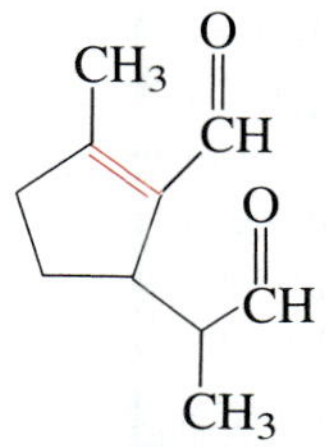

Defense pheromone of larvae of chrysomelid beetle

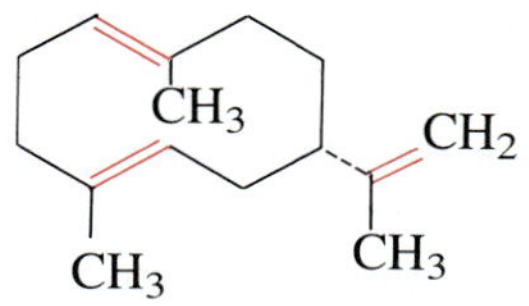

Alarm pheromone in some aphids

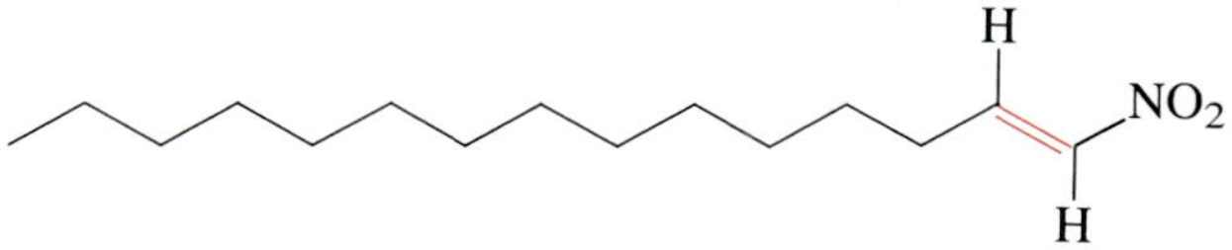

Termite defense pheromone

Pheromones are chemical substances used for communication within a living species. There are sex, trail, alarm, and defense pheromones, to mention a few. Many insect pheromones are simple alkenes; they are isolated by extraction of certain parts of the insect and separation of the resulting product mixture by chromatographic techniques. Often only minute quantities of the bioactive compound can be isolated, in which case the synthetic organic chemist can play a very important role in the design and execution of total syntheses. Interestingly, the specific activity of a pheromone frequently depends on the configuration around the double bond (e.g., *E* or *Z*), as well as on the absolute configuration of any chiral centers present (*R*,*S*) *and* the composition of isomer mixtures. For example, the male European corn borer is only weakly attracted to its pure purported sex pheromone, *cis*-3-tetradecenyl ethanoate (acetate). Another species, the red-banded leafroller, shows no response at all to this compound. Similarly, the trans isomer is completely inactive. However, when small amounts of this isomer are added to the cis compound, strong attraction in both species is observed.

Sex Pheromones of the European Corn Borer

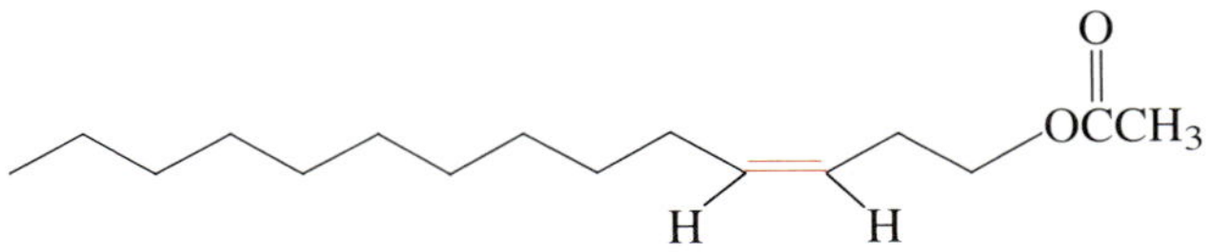

O
‖
OCCH$_3$
H H

***cis*-3-Tetradecenyl ethanoate (acetate)**

H
O
‖
OCCH$_3$
H

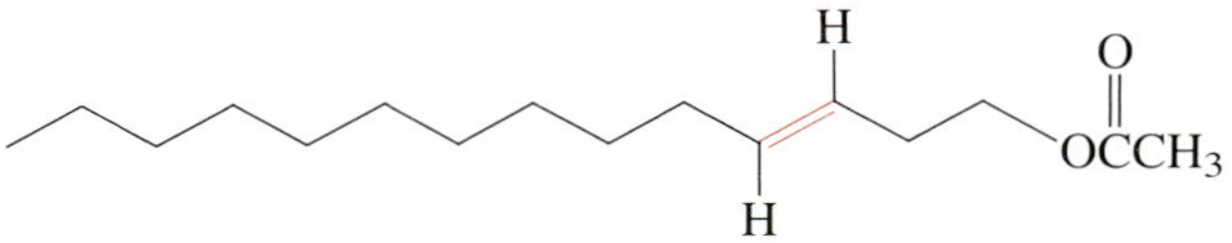

***trans*-3-Tetradecenyl ethanoate (acetate)**

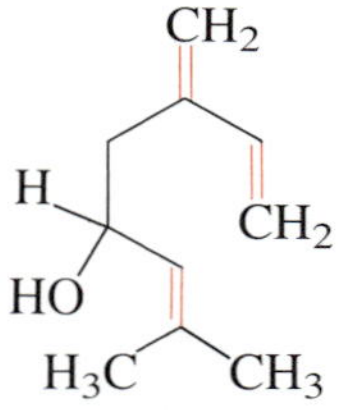

CH$_2$
H
CH$_2$
HO
H$_3$C CH$_3$
Ipsdienol

This phenomenon, in which a mixture of two or more compounds elicits more physiological activity than would be expected based on their individual effectiveness, is called **synergism** (*synergos,* Greek, to work together). Generally, synergism is defined as the simultaneous action of separate forces that together have greater total effect than the sum of their individual strengths (''the whole is greater than the sum of the parts''). Synergism is also observed with enantiomeric mixtures. For example, optimal response by the bark-boring beetle to the sex attractant ipsdienol is elicited by a mixture of the chemical that is 65% dextrorotatory and 35% levorotatory.

Pheromone research affords an important opportunity for achieving pest control. Minute quantities of sex pheromones can be used per acre of land to confuse male insects about the location of their female partners. These pheromones can thus serve as lures in traps to effectively remove insects without spraying crops with large amounts of other chemicals. It is clear that organic chemists in collaboration with insect biologists will make important contributions in this area in the years to come.

Summary of New Reactions

Addition Reactions

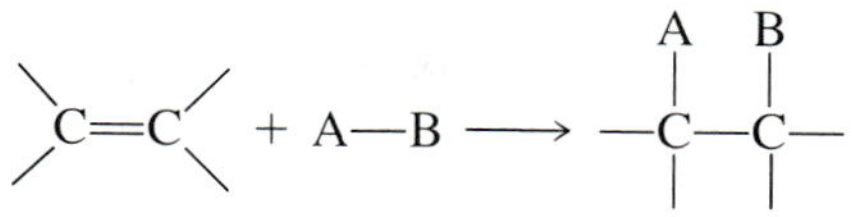

$$\mathrm{>C{=}C<} + \mathrm{A{-}B} \longrightarrow \mathrm{-\overset{A}{\underset{|}{\overset{|}{C}}}-\overset{B}{\underset{|}{\overset{|}{C}}}-}$$

1 HYDROGENATION

$$\mathrm{>C{=}C<} \xrightarrow{\mathrm{H_2,\ catalyst}} \mathrm{H{-}C{-}C{-}H}$$

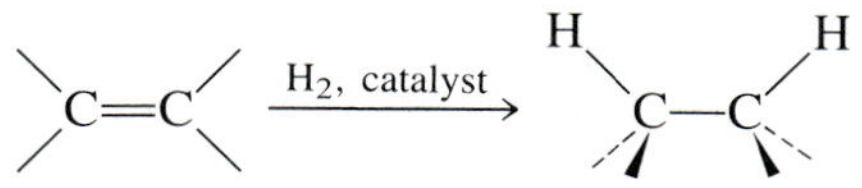

Cis addition

Electrophilic Additions

2 HYDROHALOGENATION

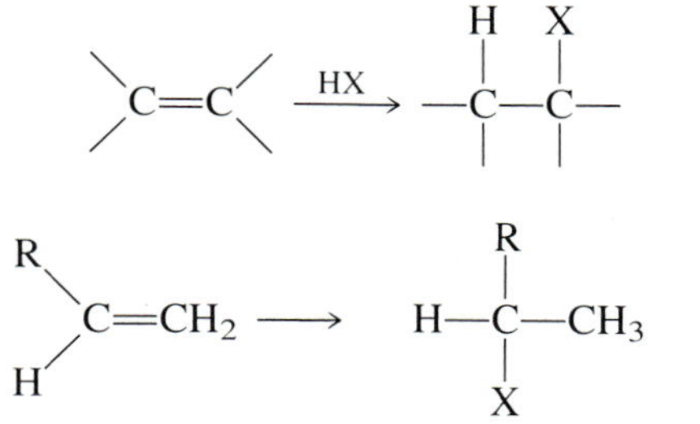

$$\mathrm{>C{=}C<} \xrightarrow{\mathrm{HX}} \mathrm{-\overset{H}{\underset{|}{\overset{|}{C}}}-\overset{X}{\underset{|}{\overset{|}{C}}}-}$$

$$\mathrm{R(H)C{=}CH_2} \longrightarrow \mathrm{H{-}\overset{R}{\underset{X}{C}}{-}CH_3}$$

Regiospecific (Markovnikov rule)

3 HYDRATION

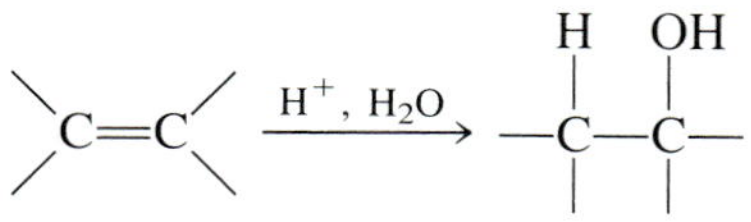

$$\mathrm{>C{=}C<} \xrightarrow{\mathrm{H^+,\ H_2O}} \mathrm{-\overset{H}{\underset{|}{\overset{|}{C}}}-\overset{OH}{\underset{|}{\overset{|}{C}}}-}$$

4 HALOGENATION

$$\mathrm{>C{=}C<} \xrightarrow{\mathrm{X_2}} \mathrm{X{-}C{-}C{-}X}$$

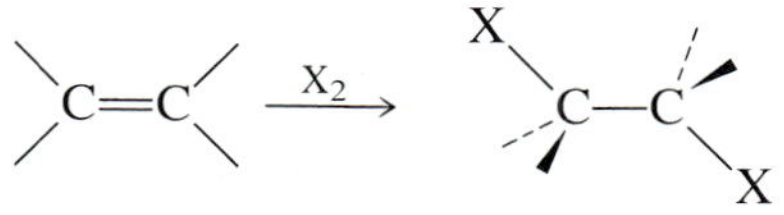

Stereospecific (*anti*)

5 VICINAL HALO ALCOHOL SYNTHESIS

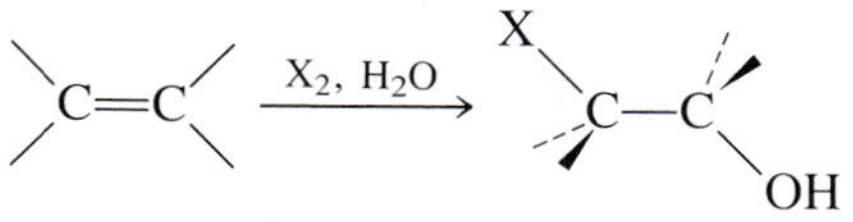

$$\mathrm{>C{=}C<} \xrightarrow{\mathrm{X_2,\ H_2O}} \mathrm{X{-}C{-}C{-}OH}$$

6 VICINAL HALO ETHER SYNTHESIS

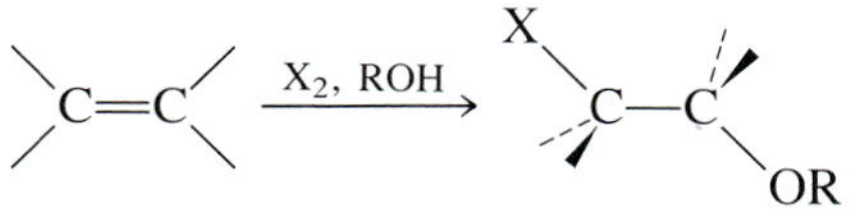

$$\mathrm{>C{=}C<} \xrightarrow{\mathrm{X_2,\ ROH}} \mathrm{X{-}C{-}C{-}OR}$$

7 GENERAL ELECTROPHILIC ADDITIONS

A = electropositive, B = electronegative

$$\mathrm{C{=}C} \xrightarrow{AB} \text{[C–C bridged by } A^{+}] \xrightarrow{B^-} A\text{–C–C–}B$$

8 OXYMERCURATION-DEMERCURATION

$$\mathrm{C{=}C} \xrightarrow[2.\ NaBH_4,\ NaOH]{1.\ Hg(O\overset{O}{\overset{\|}{C}}CH_3)_2,\ H_2O} \text{—}\overset{H}{\underset{|}{\overset{|}{C}}}\text{—}\overset{OH}{\underset{|}{\overset{|}{C}}}\text{—}$$

$$\mathrm{C{=}C} \xrightarrow[2.\ NaBH_4,\ NaOH]{1.\ Hg(O\overset{O}{\overset{\|}{C}}CH_3)_2,\ ROH} \text{—}\overset{H}{\underset{|}{\overset{|}{C}}}\text{—}\overset{OR}{\underset{|}{\overset{|}{C}}}\text{—}$$

9 HYDROBORATION

$$\mathrm{C{=}C} + BH_3 \longrightarrow (\text{—}\underset{H}{\underset{|}{\overset{|}{C}}}\text{—}\underset{|}{\overset{|}{C}})_3\text{—B}$$

$$\mathrm{RHC{=}CH_2} + BH_3 \longrightarrow (RCH_2CH_2)_3\text{—B}$$

Regiospecific

$$\text{1,2-dideuteriocyclohexene} + BH_3 \longrightarrow \text{cyclohexane with D, H on one carbon and D, B— on the adjacent carbon}$$

Stereospecific (*syn*) and anti-Markovnikov

10 HYDROBORATION-OXIDATION

$$\mathrm{C{=}C} \xrightarrow[2.\ H_2O_2,\ HO^-]{1.\ BH_3} \text{—}\overset{H}{\underset{|}{\overset{|}{C}}}\text{—}\overset{OH}{\underset{|}{\overset{|}{C}}}\text{—}$$

11 HYDROBORATION-HALOGENATION

$$\mathrm{C{=}C} \xrightarrow[2.\ X_2]{1.\ BH_3} \text{—}\overset{H}{\underset{|}{\overset{|}{C}}}\text{—}\overset{X}{\underset{|}{\overset{|}{C}}}\text{—}$$

Oxidation

12 OXACYCLOPROPANE FORMATION

$$>C{=}C< \xrightarrow{RCOOH} \text{oxacyclopropane} + RCOH$$

13 VICINAL ANTI DIHYDROXYLATION

$$>C{=}C< \xrightarrow[2.\ H^+,\ H_2O]{1.\ RCOOH} \text{HO–C–C–OH (anti)} + RCOH$$

14 VICINAL SYN DIHYDROXYLATION

$$>C{=}C< \xrightarrow{KMnO_4,\ H_2O,\ 0^\circ C,\ pH\ 7} \text{HO–C–C–OH (syn)}$$

$$>C{=}C< \xrightarrow{OsO_4,\ H_2S;\ \text{or}\ OsO_4,\ NaHSO_3;\ \text{or catalytic}\ OsO_4,\ H_2O_2} \text{HO–C–C–OH (syn)}$$

15 OZONOLYSIS

$$>C{=}C< \xrightarrow[2.\ (CH_3)_2S;\ \text{or Zn},\ CH_3COH;\ \text{or}\ H_2,\ Pt]{1.\ O_3} >C{=}O + O{=}C<$$

$$>C{=}C< \xrightarrow[2.\ NaBH_4]{1.\ O_3} -\underset{|}{\overset{H}{C}}-OH + HO-\underset{|}{\overset{H}{C}}-$$

Free-Radical Additions

16 FREE-RADICAL HYDROBROMINATION

$$>C{=}CH_2 \xrightarrow{HBr,\ ROOR} -\underset{|}{\overset{H}{C}}-\underset{H}{\overset{Br}{C}}-H$$

Anti-Markovnikov

17 OTHER FREE-RADICAL ADDITIONS

$$>C{=}C< \xrightarrow{RSH,\ ROOR} -\underset{|}{\overset{H}{C}}-\underset{|}{\overset{SR}{C}}-$$

$$>C{=}C< \xrightarrow{HCX_3,\ ROOR} -\underset{|}{\overset{H}{C}}-\underset{|}{\overset{CX_3}{C}}-$$

Monomers and Polymers

18 DIMERIZATION, OLIGOMERIZATION, AND POLYMERIZATION

$$n\ \rangle C{=}C\langle \xrightarrow{H^+ \text{ or } RO^\cdot \text{ or } B^-} -(\overset{|}{\underset{|}{C}}-\overset{|}{\underset{|}{C}})_n-$$

Ziegler-Natta polymerization:

$$n\ CH_2{=}CH_2 \xrightarrow{TiCl_4,\ AlR_3} -(CH_2CH_2)_n-$$

Syndiotactic or isotactic

19 ETHENE IN INDUSTRIAL PROCESSES

Synthesis by cracking:

$$R-CH_2-CH_2-R \xrightarrow{\Delta} CH_2{=}CH_2 + R-R + \text{other hydrocarbons}$$

Ethenyl ethanoate (vinyl acetate) synthesis:

$$CH_2{=}CH_2 \xrightarrow{CH_3\overset{O}{\overset{\|}{C}}OH,\ O_2,\ \text{catalytic } PdCl_2 \text{ and } CuCl_2} CH_2{=}C(H)(O\overset{\underset{\|}{O}}{C}CH_3)$$

Wacker Process:

$$CH_2{=}CH_2 \xrightarrow{H_2O,\ O_2,\ \text{catalytic } PdCl_2 \text{ and } CuCl_2} CH_3\overset{O}{\overset{\|}{C}}H$$

Chloroethene (vinyl chloride) synthesis:

$$CH_2{=}CH_2 \xrightarrow{Cl_2} \underset{Cl}{\underset{|}{CH_2}}-\underset{Cl}{\underset{|}{CH_2}} \xrightarrow[-\ HCl]{} CH_2{=}CHCl$$

Oxacyclopropane (ethylene oxide) and
1,2-ethanediol (ethylene glycol) synthesis:

$$CH_2{=}CH_2 \xrightarrow{O_2,\ \text{catalytic Ag}} H_2C\overset{O}{\diagup\diagdown}CH_2 \xrightarrow{H^+,\ H_2O} \overset{OH}{\overset{|}{CH_2}}-\overset{OH}{\overset{|}{CH_2}}$$

Summary of Important Concepts

1 The reactivity of the double bond manifests itself in exothermic addition reactions leading to saturated products.

2 The hydrogenation of alkenes is immeasurably slow unless a catalyst capable of splitting the strong H–H bond is used. Possible catalysts are palladium on carbon, platinum (as PtO_2), and Raney nickel. Addition of hydrogen is subject to steric control, the least-hindered side of the least-substituted double bond frequently being attacked preferentially.

3 Like a Lewis base, the π bond is subject to attack by acid and electrophiles, such as H^+, X^+, and Hg^{2+}. If the initial intermediate is a free carbocation, the more highly substituted carbocation is formed. Alternatively, a cyclic onium ion is generated subject to nucleophilic ring-opening at the more-substituted carbon. The former case leads to control of regiochemistry (Markovnikov rule), the latter to control of both regio- and stereochemistry.

4 Mechanistically, hydroboration seems to lie between hydrogenation and electrophilic addition. The first step is π complexation to the electron-deficient boron, whereas the second is a concerted transfer of the hydrogen to carbon. Hydroboration-oxidation or halogenation results in the anti-Markovnikov hydration and hydrohalogenation of alkenes.

5 Peroxycarboxylic acids may be thought of as containing an electrophilic oxygen atom, transferable as the electronic equivalent of $:\ddot{O}:$ to alkenes to give oxacyclopropanes.

6 Permanganate and osmium tetroxide act as electrophilic oxidants of alkenes; in the course of the reaction, the oxidation state of the metal is reduced by two units. Addition takes place in a concerted manner through cyclic six-electron transition states to give vicinal *syn*-diols.

7 Depending on the reducing agent, ozonolysis followed by reduction yields carbonyl compounds or alcohols derived by cleavage of the double bond.

8 In free-radical chain additions to alkenes, the chain carrier adds to the π bond to create the more-highly substituted radical. This method allows for the anti-Markovnikov hydrobromination of alkenes, as well as the addition of thiols and some halomethanes.

9 Alkenes react with themselves through initiation by charged species, free radicals, or some transition metals. The initial attack at the double bond yields a reactive intermediate that perpetuates carbon–carbon bond formation.

Problems

1 Using the appropriate data (e.g., DH° values in Table 3-1 and Section 3-5), calculate ΔH° values for addition of each of the following molecules to ethene, using 65 kcal mole^{-1} for the π-bond strength.

(a) Cl_2
(b) IF (DH° = 67 kcal mole^{-1})
(c) IBr (DH° = 43 kcal mole^{-1})
(d) HF
(e) HI
(f) Br—CN (DH° = 83 kcal mole^{-1}; DH° for $-\overset{|}{\underset{|}{C}}-CN$ = 124 kcal mole^{-1})
(g) HO—Cl (DH° = 60 kcal mole^{-1})
(h) CH_3S—H (DH° = 88 kcal mole^{-1}; DH° for $-\overset{|}{\underset{|}{C}}-S$ = 60 kcal mole^{-1})

2 Write the expected major product(s) of each of the following reactions.

(a) 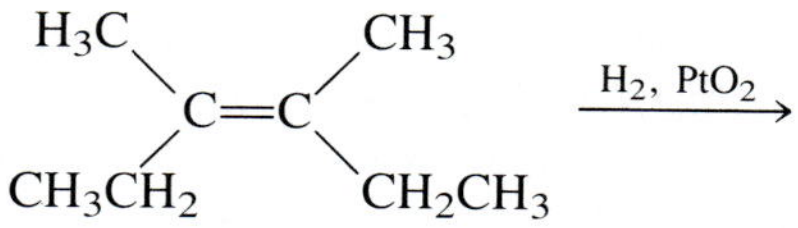 H_3C, CH_3, C=C, CH_3CH_2, CH_2CH_3 $\xrightarrow{H_2,\ PtO_2}$

(b) 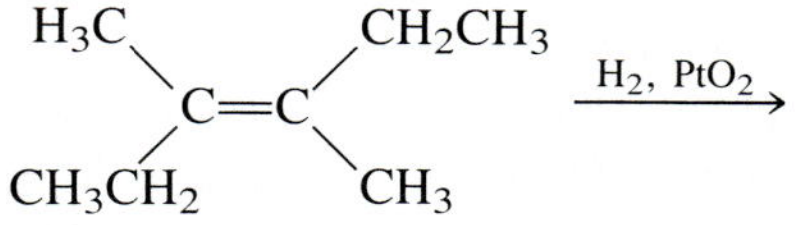 H_3C, CH_2CH_3, C=C, CH_3CH_2, CH_3 $\xrightarrow{H_2,\ PtO_2}$

(c) H, H, C=C, H_2C, CH_3, CH_3 $\xrightarrow{H_2,\ \text{Pd-C}}$

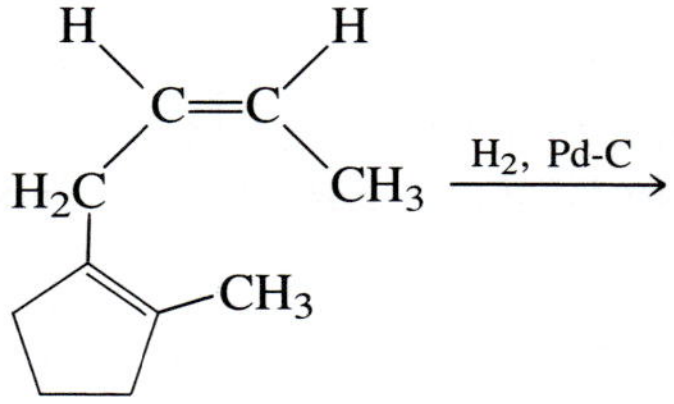

(d) 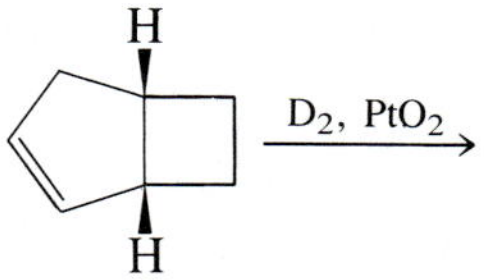 H, H $\xrightarrow{D_2,\ PtO_2}$

3 Write the expected major product of catalytic hydrogenation of each of the following alkenes. Clearly show and explain the stereochemistry of the product obtained in each case.

(a) 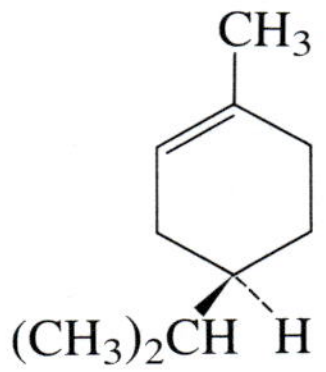 CH_3, $(CH_3)_2CH$ H (b) 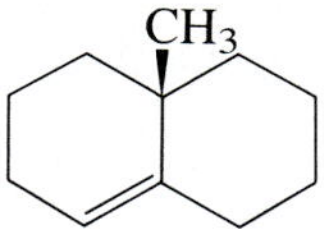 CH_3 (c) 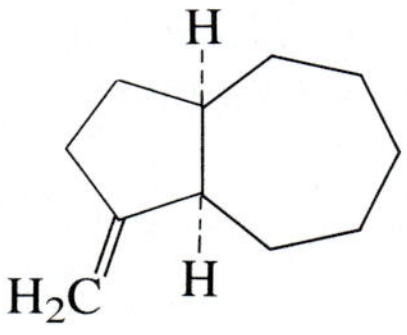 H, H, H_2C

4 Would you expect the catalytic hydrogenation of a small-ring cyclic alkene such as cyclobutene to be more or less exothermic than that of cyclohexene?

5 Based on elementary considerations of Lewis structure, electronegativity, and polarity, which of the following species might logically be considered a candidate for electrophilic addition to alkenes?

(a)	AlH_3	**(f)**	Li^+
(b)	NH_4^+	**(g)**	BH_4^-
(c)	O_2	**(h)**	NO^+
(d)	CH_3SeCl	**(i)**	$(CH_3)_3Si^{\cdot}$
(e)	$CH_2{=}\overset{+}{O}H$	**(j)**	HCN

6 Each of the following reactions is reversible. Give the typical reagents and experimental conditions necessary for each transformation to take place in the forward direction and in the backward one. Indicate which direction is

thermodynamically favored and how experimental conditions may allow the reaction to proceed to completion in either direction.

(a) $H_2O + CH_2{=}CH_2 \rightleftharpoons CH_3CH_2OH$
(b) $HBr + CH_2{=}CH_2 \rightleftharpoons CH_3CH_2Br$

7 Write the product(s) that you would expect to obtain from each of the following reactions. Show stereochemistry clearly.

(a) (cyclohexylidene)$CHCH_2CH_3$ $\xrightarrow{HCl}$

(b) *trans*-3-Heptene $\xrightarrow{Cl_2}$

(c) 1-Ethylcyclohexene $\xrightarrow{I_2,\ H_2O}$

(d) 3-methylcyclopentene (CH_3) $\xrightarrow[\text{2. } NaBH_4,\ CH_3OH]{\text{1. } Hg(OCCH_3)_2,\ CH_3OH}$ (O above C of OCCH₃)

(e) (bicyclic alkene with ring-fusion H) $\xrightarrow{I_2,\ \text{excess } Na^+Cl^-}$

(f) What would you expect to find as the product of the following hypothetical addition? Would you expect it to be faster or slower than similar addition reactions to "ordinary" alkenes (e.g., ethene)?

$$CH_2{=}CH{-}\overset{+}{N}(CH_3)_3 + HCl \longrightarrow$$

(g) (bicyclic alkene with ring-fusion CH_3) $\xrightarrow[\text{2. } H_2O_2,\ NaOH,\ H_2O]{\text{1. } BH_3}$

(h) (cyclohexene with CH_3 and CH_3 substituents) $\xrightarrow[\text{2. } ICl]{\text{1. } BH_3}$

8 Show how you would synthesize each of the following molecules from an alkene of appropriate structure (your choice).

(a) $(CH_3)_2CH\underset{}{\overset{OH}{C}}HCH_3$

(b) $ClCH_2\overset{CH_3}{C}HOCH(CH_3)_2$

(c) *meso*-$CH_3CH_2CH_2\underset{Br}{C}H\underset{Br}{C}HCH_2CH_2CH_3$ (i.e., the 4*R*,5*S* stereoisomer)

(d) $CH_3CH_2CH_2\underset{Br}{C}H\underset{Br}{C}HCH_2CH_2CH_3$ (racemic mixture of 4*R*,5*R* and 4*S*,5*S*)

(e) 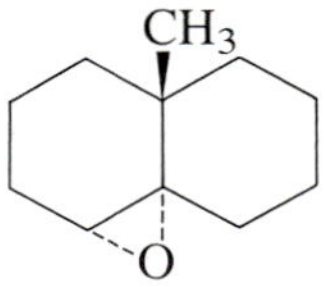 CH_3 O

(f) 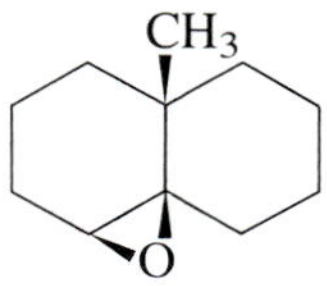 CH_3 O

9 Propose efficient methods for accomplishing each of the following transformations. Most will require more than one step.

(a) $CH_3CH_2CH(Br)CH_3 \longrightarrow CH_3CH_2CH_2CH_2I$

(b) $CH_3CH(OH)CH_2CH_3 \longrightarrow$ *meso*-$CH_3CH(OH)CH(OH)CH_3$ (i.e., 2*R*,3*S*)

(c) $CH_3CH(OH)CH_2CH_3 \longrightarrow CH_3CH(OH)CH(OH)CH_3$ (1:1 mixture of 2*R*,3*R* and 2*S*,3*S*)

(d) $(CH_3)_2C{=}CHCH_2CH_2CH{=}CH_2 \longrightarrow (CH_3)_2C\overset{O}{—}CHCH_2CH_2CH_2CH{=}O$

(e) *cis*-$(CH_3)_2CHCH_2CH{=}CHCH_3 \longrightarrow HC(=O)CH(CH_3)CH_2CH_2CH_2CH_3$

10 ^{1}H NMR spectrum A corresponds to a molecule with the formula C_3H_5Cl.

(a) Deduce the structure of the molecule.

(b) Assign each NMR signal to a hydrogen or group of hydrogens.

(c) The "doublet" at $\delta = 4$ ppm has $J = 6$ Hz. Is this in accord with your assignments in part **b**?

(d) This "doublet," on fivefold expansion, becomes a doublet of triplets (inset, spectrum A), with $J \sim 1$ Hz for the triplet splittings. What is the origin of this triplet splitting? Is it reasonable in light of your assignments in part **b**?

11 Reaction of C_3H_5Cl (Problem 10, spectrum A) with Cl_2 in H_2O gives rise to two products, both $C_3H_6OCl_2$, whose spectra are shown in B and C. Reaction of either of these products with KOH yields the same product, C_3H_5OCl, shown in spectrum D. The insets show expansions of some of the multiplets.

(a) Deduce the structures of the compounds giving rise to spectra B, C, and D.

(b) Why does reaction of the starting chloride compound with Cl_2 in H_2O give two isomeric products?

(c) Write mechanisms for the formation of the product C_3H_5OCl from both isomers $C_3H_6OCl_2$.

12 ^{1}H NMR spectrum E corresponds to a molecule with the formula C_4H_8O.

(a) Determine its structure.

(b) Assign each signal to a proton or group of protons in the molecule.

(c) Fully explain the splitting patterns for the signals at $\delta = 1.2$, 4.2, and 5.8 ppm (see inset for fivefold expansion).

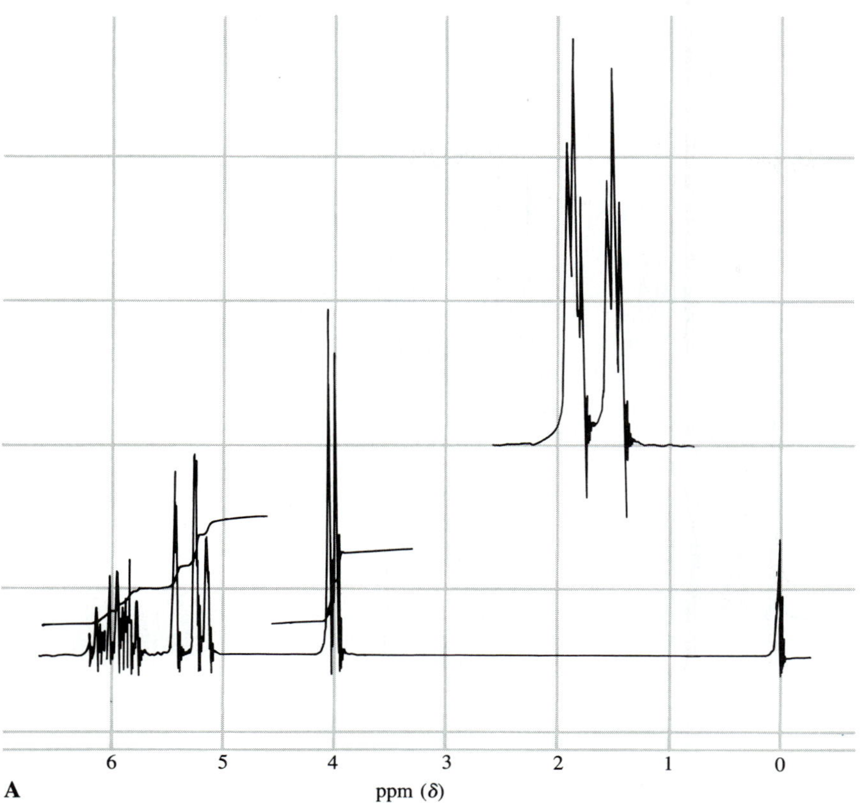

A

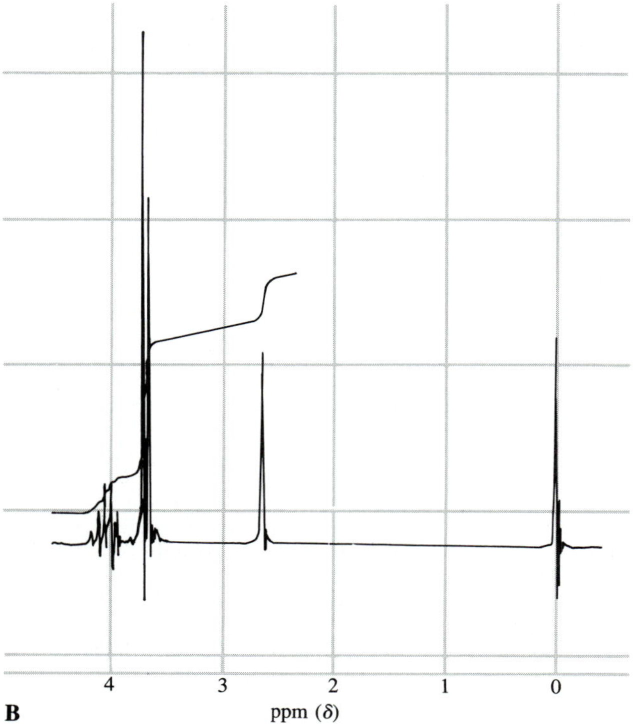

4 3 2 1 0
ppm (δ)

B

4 3 2 1 0
ppm (δ)

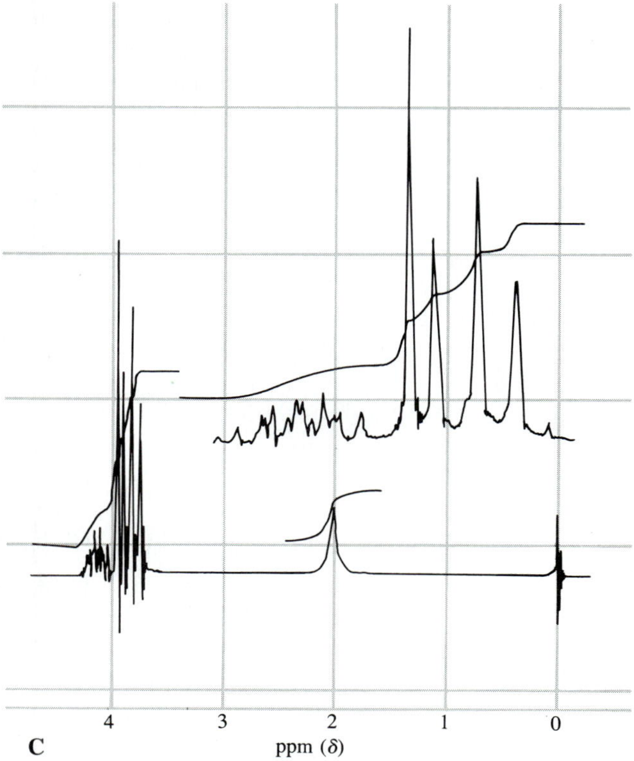

C

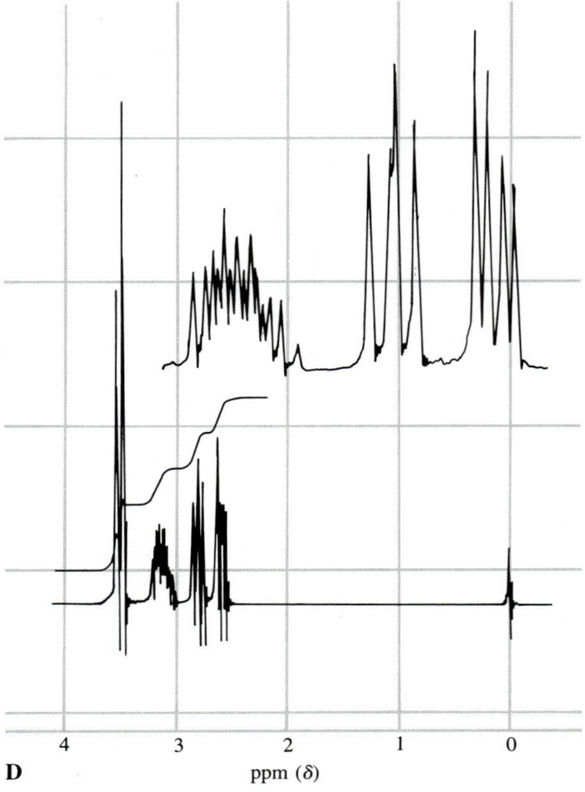

13 Reaction of the compound associated with spectrum E with $SOCl_2$ produces a compound, C_4H_7Cl, whose NMR spectrum is almost identical with E, except that the signal at $\delta = 3.5$ ppm is not present. Treatment of this compound with H_2 over PtO_2 results in C_4H_9Cl (spectrum F). Identify these two molecules.

14 Write the expected product of reaction of 2-methyl-1-pentene with each of the following reagents.

(a)	H_2, PtO_2	**(k)**	$Br_2 + CH_3CH_2OH$
(b)	D_2, Pd-C		
(c)	BH_3, then $NaOH + H_2O_2$	**(l)**	$Hg(O\overset{\overset{\displaystyle O}{\|}}{C}CH_3)_2 + H_2O$, then $NaBH_4$
(d)	HCl	**(m)**	MCPBA
(e)	HBr	**(n)**	Catalytic $OsO_4 + H_2O_2$
(f)	HBr + peroxides		
(g)	HI + peroxides	**(o)**	O_3, then $Zn + CH_3\overset{\overset{\displaystyle O}{\|}}{C}OH$
(h)	$H_2SO_4 + H_2O$	**(p)**	CH_3SH + peroxides
(i)	Cl_2	**(q)**	$CHBr_3$ + peroxides
(j)	ICl	**(r)**	Catalytic H_2SO_4 + heat

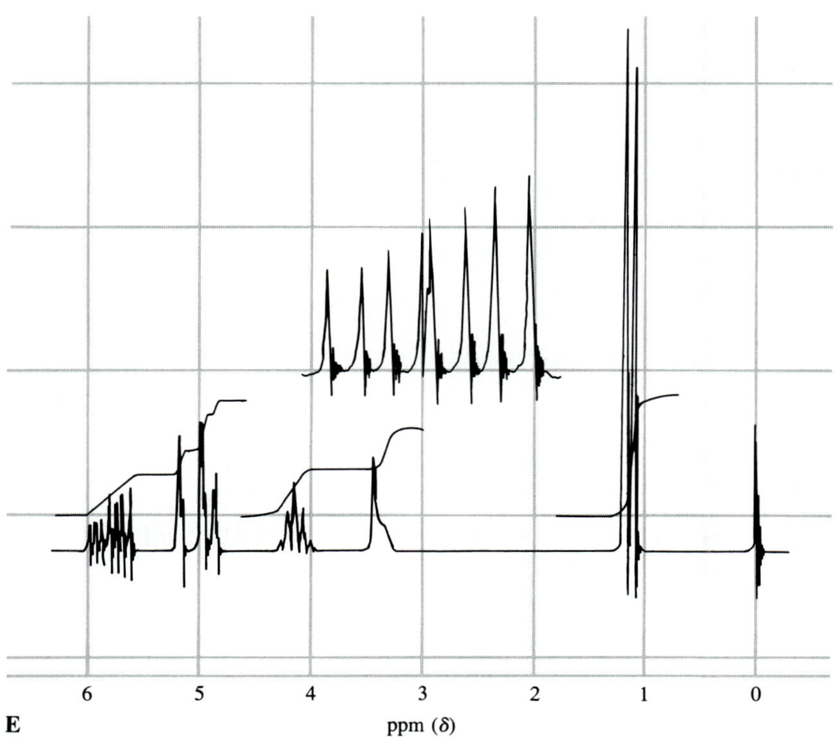

E

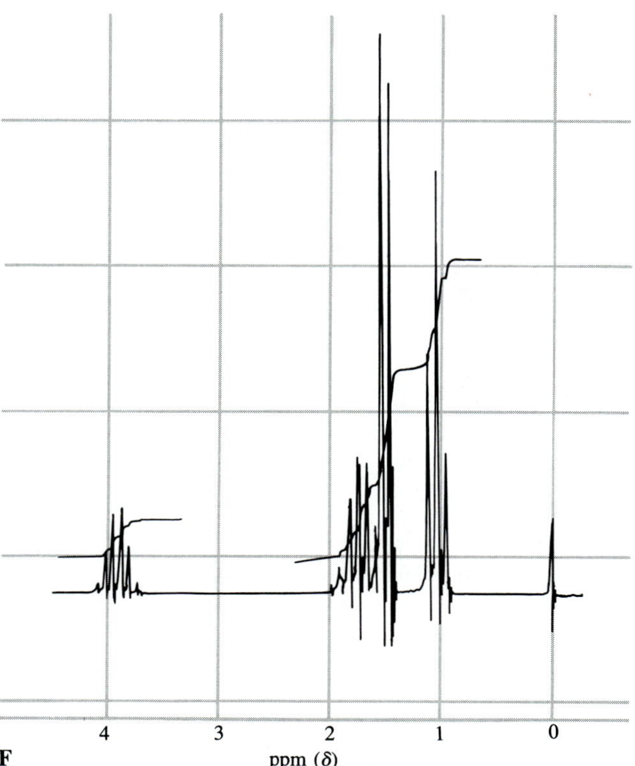

F

15 Write the expected product of reaction of (*E*)-3-methyl-3-hexene with each of the reagents in Problem 14.

16 Write the expected product of reaction of 1-ethylcyclopentene with each of the reagents in Problem 14.

17 (*E*)-5-Hepten-1-ol reacts with the following reagents to give products with the indicated formulas. In each case, write the structure of the product and describe its formation with a detailed mechanism.

(a) HCl, $C_7H_{14}O$ (no Cl!)
(b) Br_2, $C_7H_{13}OBr$ (no misprint!)

18 The following chemical reactions were compared in Problem 13 of Chapter 2.

$$\text{(i)} \quad CH_3{-}CH{=}CH_2 + Br_2 \longrightarrow CH_3{-}\underset{\displaystyle |}{\overset{\displaystyle Br}{\overset{|}{C}H}}{-}\overset{\displaystyle Br}{\overset{|}{C}H_2}$$

$$\text{(ii)} \quad CH_3{-}CH{=}CH_2 + Br_2 \longrightarrow \underset{\underset{\displaystyle Br}{|}}{CH_2}{-}CH{=}CH_2 + HBr$$

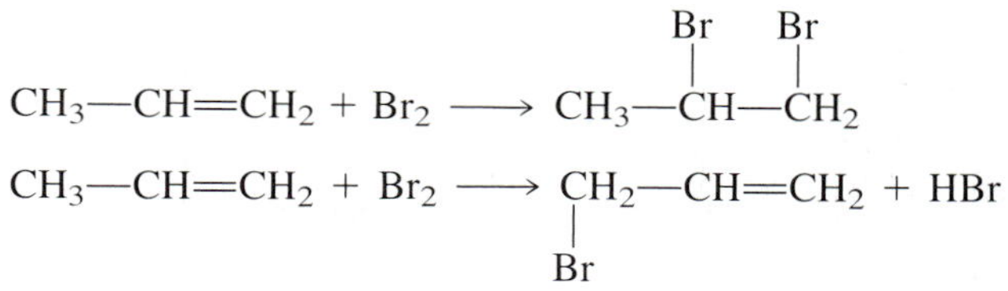

Given the material in Sections 3-5 through 3-7 and 12-6, suggest a mechanistic explanation for the results of reaction of propene with Br_2 in the gas phase at 600°C.

19 When a cis alkene is mixed with a small amount of I_2 in the presence of heat or light, it isomerizes to some trans alkene. Propose a detailed mechanism to account for this observation.

20 Plan syntheses of each of the following compounds, utilizing retrosynthetic-analysis techniques. Possible starting compounds are given in parentheses. However, other simple alkanes or alkenes may be used as starting materials, as long as you include at least one carbon–carbon bond-forming step in each synthesis.

(a) 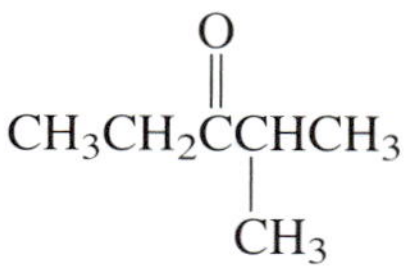 $CH_3CH_2\overset{\displaystyle O}{\overset{\|}{C}}\underset{\underset{\displaystyle CH_3}{|}}{C}HCH_3$ (propene)

(b) 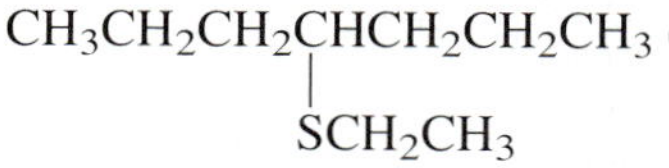 $CH_3CH_2CH_2\underset{\underset{\displaystyle SCH_2CH_3}{|}}{C}HCH_2CH_2CH_3$ (propene, again)

(c) 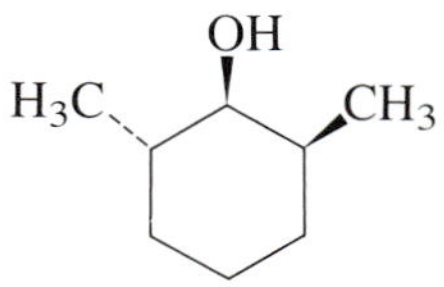 2,6-dimethylcyclohexanol structure: cyclohexane ring bearing OH (wedge), H_3C (dashed), and CH_3 (wedge) (cyclohexene)

(d) 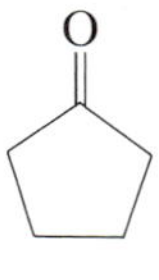 cyclopentanone structure (C=O on a five-membered ring) (cyclobutene, if you are ready for a challenge!)

21 Using the synthetic techniques presented so far, show how you would convert cyclopentane into each of the following molecules.

(a) *cis*-1,2-Dideuteriocyclopentane

(b) *trans*-1,2-Dideuteriocyclopentane

(c) 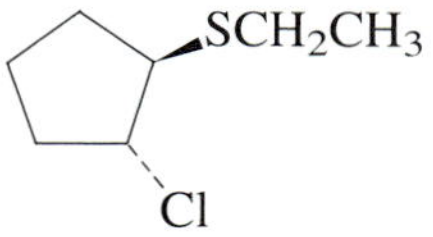 SCH_2CH_3 Cl

(e) 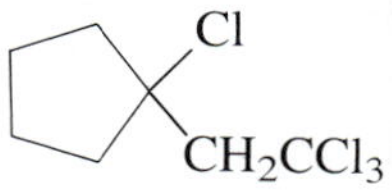 Cl CH_2CCl_3

(d) 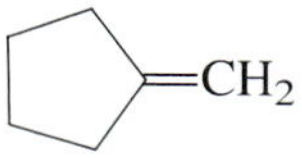 $=CH_2$

(f) 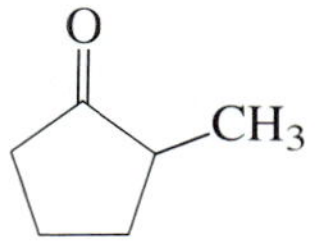 O CH_3

(g) 1,2-Dimethylcyclopentene

(h) *trans*-1,2-Dimethyl-1,2-cyclopentanediol

(i) 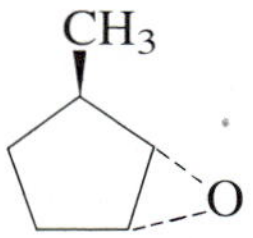 CH_3 O

22 Write the expected major product(s) of each of the following reactions.

(a) $CH_3OCH_2CH_2CH{=}CH_2 \xrightarrow[\text{2. NaBH}_4]{\text{1. Hg(OCCH}_3)_2\text{, CH}_3\text{OH}}$ (O above C)

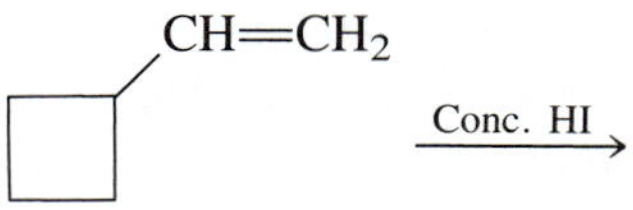

(b) $H_2C{=}C(CH_3)(CH_2OH) \xrightarrow[\text{2. H}^+\text{, H}_2\text{O}]{\text{1. CH}_3\text{COOH (with =O on C)}}$

(c) $CH{=}CH_2$ (on cyclobutane) $\xrightarrow{\text{Conc. HI}}$

(d) 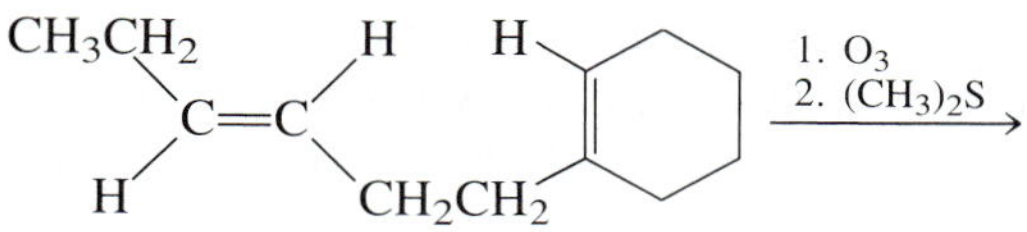 CH_3CH_2, H, H, H, C=C, CH_2CH_2 $\xrightarrow[\text{2. (CH}_3)_2\text{S}]{\text{1. O}_3}$

(e) 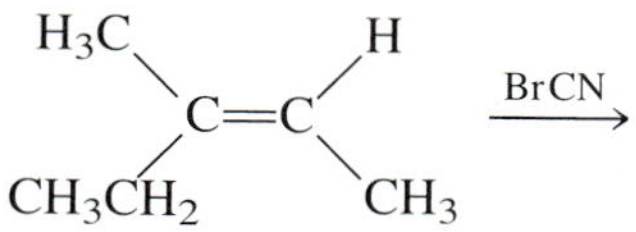 H_3C, H, C=C, CH_3CH_2, CH_3 $\xrightarrow{\text{BrCN}}$

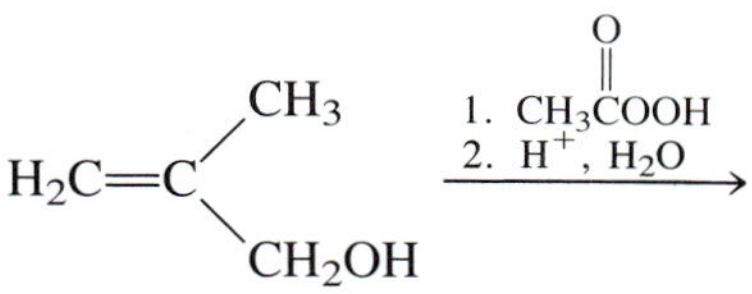

(f) [3-chlorocyclopentene] $\xrightarrow{\text{Cold KMnO}_4}$

(g) $(CH_3)_2C(OH)CH{=}CH_2 \xrightarrow{\text{HBr, ROOR}}$

(h) [cyclononene] $\xrightarrow{\text{Br}_2\text{, CCl}_4}$

(i) $CH_3CH{=}CH_2 \xrightarrow{\text{Catalytic HF}}$

(j) $CH_2{=}CHNO_2 \xrightarrow{\text{Catalytic KOH}}$

(Hint: Draw Lewis structures for the NO_2 group.)

23 Treatment of α-terpineol (Chapter 10, Problem 24) with aqueous mercuric ethanoate (acetate) followed by sodium borohydride reduction leads predominantly to an isomer of the starting compound ($C_{10}H_{18}O$) instead of a hydration product. This isomer is the chief component in oil of eucalyptus and, appropriately enough, is called eucalyptol. It is popularly used as a flavoring for otherwise foul-tasting medicines because of its pleasant spicy taste and aroma. Deduce a structure for eucalyptol on the basis of sensible mechanistic chemistry and the following proton-decoupled ^{13}C NMR data.

α-Terpineol [ring: CH_3 on alkene carbon; $(CH_3)_2COH$] $\xrightarrow[\text{2. NaBH}_4\text{, H}_2\text{O}]{\text{1. Hg(OCCH}_3)_2\text{, H}_2\text{O}}$ eucalyptol, ($C_{10}H_{18}O$) ^{13}C NMR: δ = 22.8, 27.5, 28.8, 31.5, 32.9, 69.6, and 73.5 ppm

24 Both borane and MCPBA react highly selectively with molecules such as limonene that contain double bonds in very different environments. Predict the products of reaction of limonene with **(a)** one equivalent of BH_3, followed by basic aqueous H_2O_2, and **(b)** one equivalent of MCPBA. Explain your answers.

Limonene [ring: CH_3 on alkene carbon; $C(CH_3){=}CH_2$; labels CH_3, H_3C, CH_2]

25 Oil of marjoram contains a pleasant, lemon-scented substance, $C_{10}H_{16}$ (compound G). Upon ozonolysis, G forms two products. One of them, H, has the formula $C_8H_{14}O_2$, and can be independently synthesized in the following way:

$(H_3C)_2C{=}CHCH_2Br \xrightarrow[\text{2. H}_2\text{C—CHCH}_3\text{ (epoxide, O)}]{\text{1. Mg}} \underset{\textbf{I}}{C_8H_{16}O} \xrightarrow{\text{CrO}_3\text{(pyridine)}_2} \underset{\textbf{J}}{C_8H_{14}O} \xrightarrow{\text{1. BH}_3\text{; 2. H}_2\text{O}_2\text{, NaOH; 3. CrO}_3\text{(pyridine)}_2} H$

From this information, propose reasonable structures for compound G through compound J.

26 Caryophyllene is an unusual sesquiterpene familiar to you as a major cause of the odor of cloves. Determine its structure from the information below. Caution: The structure is totally different from that of α-caryophyllene alcohol (Exercise 12-23).

Formula: $C_{15}H_{24}$

REACTION 1

$$\text{Caryophyllene} \xrightarrow{H_2,\ Pd\text{–}C} C_{15}H_{28}$$

REACTION 2

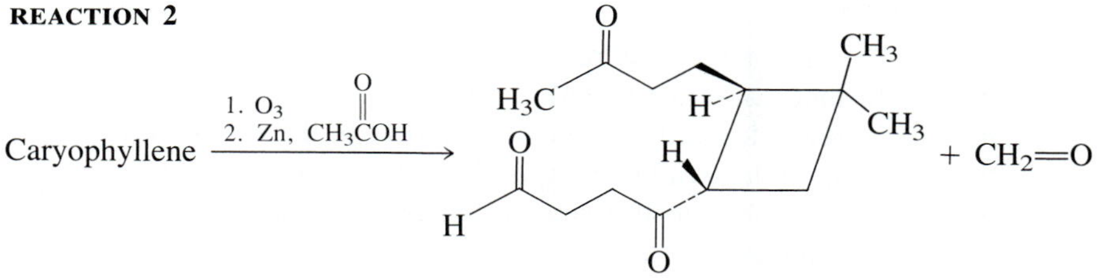

Caryophyllene $\xrightarrow{\text{1. } O_3 \quad \text{2. Zn, } CH_3COOH}$ [structure: H₃C–C(=O)–CH₂–CH₂–cyclobutane(H, gem-(CH₃)₂)–CH(H)–C(=O)–CH₂–CH₂–CHO] $+ CH_2{=}O$

REACTION 3

Caryophyllene $\xrightarrow{\text{1. One equivalent of } BH_3 \quad \text{2. } H_2O_2\text{, NaOH, } H_2O}$ $C_{15}H_{26}O$ $\xrightarrow{\text{1. } O_3 \quad \text{2. Zn, } CH_3COOH}$ [structure: H₃C–C(=O)–CH₂–CH₂–cyclobutane(H, gem-(CH₃)₂)–CH(H)–CH(HOCH₂)–CH₂–CH₂–CHO]

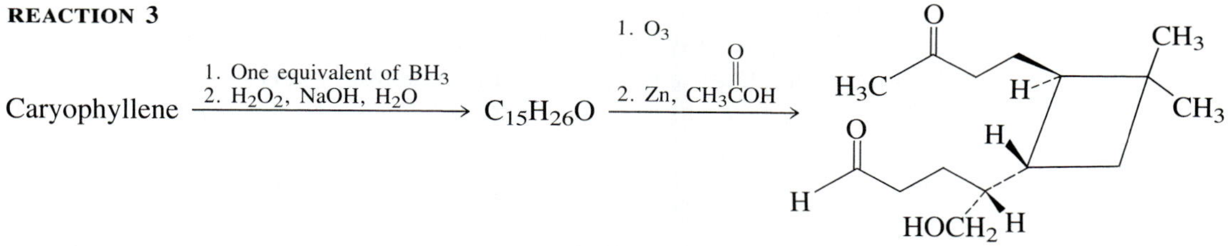

An isomer, isocaryophyllene, gives the same products as caryophyllene upon hydrogenation and ozonolysis. Hydroboration-oxidation of isocaryophyllene gives a $C_{15}H_{26}O$ product isomeric to the one shown in reaction 3; however, ozonolysis converts this $C_{15}H_{26}O$ compound into the same final product shown. In what way do caryophyllene and its isomer differ?

27 Juvenile hormone (JH, Section 12-3) itself has been synthesized in several ways. Three molecules from which it has been synthesized are shown below. Propose completions for syntheses of JH that start with each of them. Your syntheses for parts **a** and **b** should be stereospecific. Also for parts **a** and **b,** note that the double bond between C-10 and C-11 is the most reactive toward electrophilic reagents (compare the synthesis of the JH analog in Section 12-3).

(a) [structure with labels: H₃C, H₃C, H₃C, CH₃, O, COCH₃]

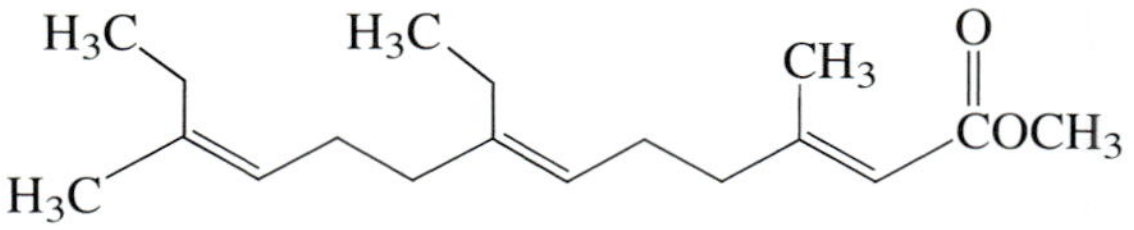

(b) [structure with labels: H₃C, CH₃, H₃C, CH₃, O, COCH₃]

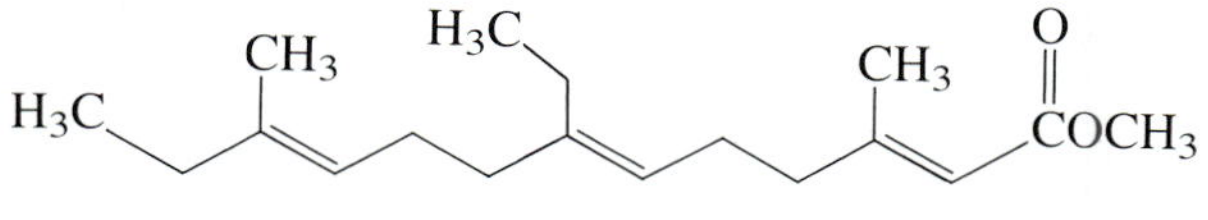

(c) [structure with labels: H₃C, H₃C, CH₃, O, COCH₃, O, Cl]

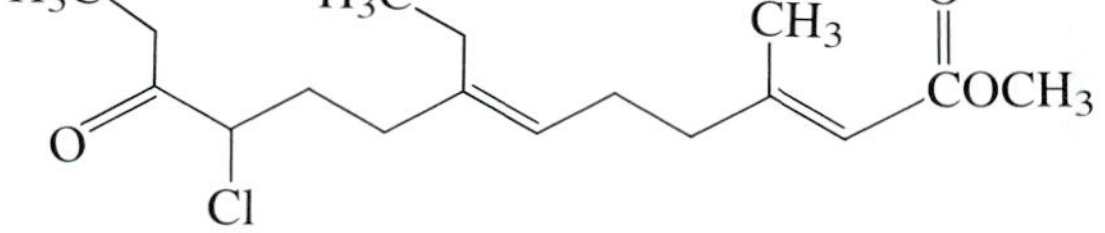

28 Many syntheses of complex natural products begin with simpler naturally occurring molecules. A plan to synthesize a complicated sesquiterpene aldehyde called sinensal requires that the monoterpene myrcene be converted into the seven-carbon iodide shown.

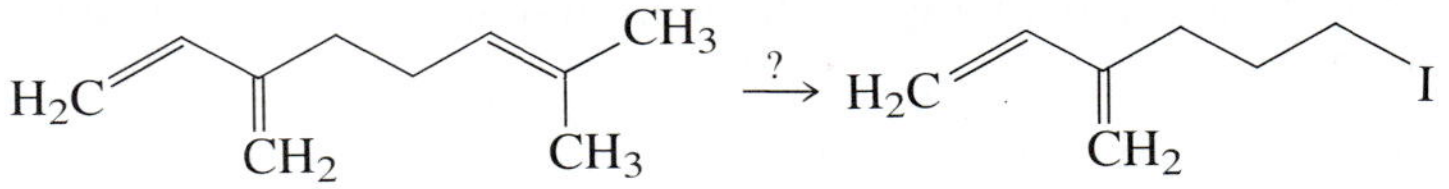

CH_3 H_2C CH_2 CH_3 $\xrightarrow{?}$ H_2C CH_2 I

Myrcene

Suggest a simple sequence of reactions that accomplishes this transformation. Hint: Make use of the differences in reactivity between the three double bonds of myrcene.

29 The monoterpene ketone camphor is used to make everything from plasticizers to mothballs to fireworks. Although it may be obtained readily from natural sources, demand is so great that most of the commercial supply is synthesized from more-available terpenes such as pinene and camphene. On treatment with ethanoic (acetic) acid in the presence of H_2SO_4, camphene gives isobornyl ethanoate (acetate), which, in turn, is converted into camphor in two subsequent steps.

Mechanistically explain the reaction that converts camphene into isobornyl ethanoate. Does any aspect of this reaction surprise you?

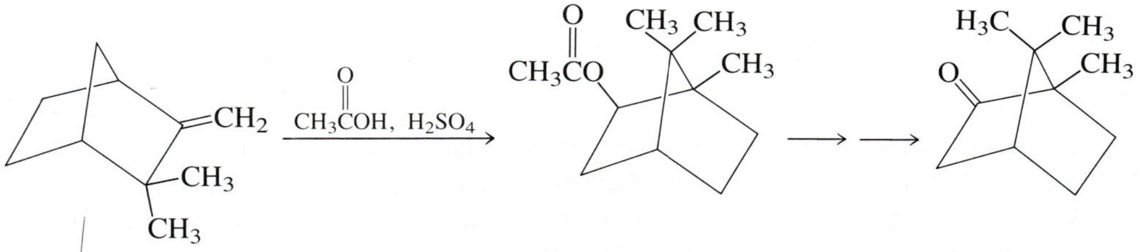

$=CH_2$ $-CH_3$ CH_3 $\xrightarrow{CH_3\overset{O}{\overset{\|}{C}}OH,\ H_2SO_4}$ $CH_3\overset{O}{\overset{\|}{C}}O$ CH_3 CH_3 CH_3 $\longrightarrow\ \longrightarrow$ O H_3C CH_3 CH_3

Camphene **Isobornyl ethanoate (acetate)** **Camphor**

CHAPTER 13

Alkynes

THE CARBON–CARBON TRIPLE BOND

A carbon–carbon triple bond is the functional group characteristic of the **alkynes.** The general formula for the alkynes is C_nH_{2n-2}, the same as that for the cycloalkenes.

The order of presentation in this chapter approximates that followed in the discussion of the structure, preparation, and reactions of alkenes. We shall see that the reactivity of the triple bond, with its two superimposed π bonds, is much like that of the double bond in many respects but quite different in others. For example, like alkenes, alkynes are electron rich and subject to attack by electrophiles and boranes, they can be made by elimination reactions, and they are more stable when the multiple bond is internal rather than terminal. In contrast with alkenyl hydrogens, alkynyl hydrogens are shielded in NMR, they are more acidic, and they can be removed in oxidative coupling reactions.

13-1 The Naming of Alkynes

The common names for alkynes, still in use for some molecules, include the ending *acetylene,* the common name of the smallest alkyne, C_2H_2. Other alkynes are treated as its derivatives—for example, the alkylacetylenes.

Common Names for Alkynes

$HC\equiv CH$	$CH_3C\equiv CCH_3$	$CH_3CH_2CH_2C\equiv CH$
Acetylene	**Dimethylacetylene**	**Propylacetylene**

The IUPAC rules for naming alkenes also apply to alkynes, the ending **-yne** replacing **-ene** and the position of the triple bond in the main chain indicated by a number.

$HC\equiv CH$	$CH_3C\equiv CCH_3$	$\overset{1}{C}H_3\overset{2}{C}\equiv\overset{3}{C}\overset{4}{C}H(Br)\overset{5}{C}H_2\overset{6}{C}H_3$	$\overset{4}{C}H_3\overset{3}{C}(CH_3)_2\overset{2}{C}\equiv\overset{1}{C}H$
Ethyne	**2-Butyne**	**4-Bromo-2-hexyne**	**3,3-Dimethyl-1-butyne**

Alkynes having the general structure $RC\equiv CH$ are *terminal,* whereas those with the structure $RC\equiv CR'$ are *internal*. The substituent with the structure $-C\equiv CH$ is named **ethynyl;** its homolog $-CH_2C\equiv CH$ is **2-propynyl** (propargyl). Like alkanes and alkenes, alkynes can be depicted in straight-line notation.

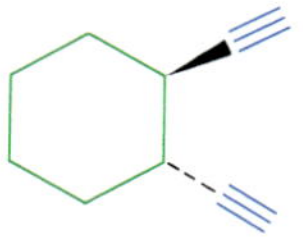

	$\triangleright$—$CH_2C\equiv CH$	$HC\equiv CCH_2OH$
***trans*-1,2-Diethynylcyclohexane**	**2-Propynylcyclopropane (Propargylcyclopropane)**	**2-Propyn-1-ol (Propargyl alcohol)**

In IUPAC nomenclature, the alkyne functional group has priority over that of the alkenes; thus, a hydrocarbon containing both double and triple bonds is called an *alkenyne*. The positions of the functional groups are assigned the lowest possible numbers. If a double and a triple bond are at equivalent locations in a molecule, the alkenyne is named so as to give the double bond the lower number. Alcohols have priority over alkynes; hence, alkynes incorporating the hydroxy function are named *alkynols*. Note the omission of the last letter of -ene in -enyne and of -yne in -ynol.

$\overset{6}{C}H_3\overset{5}{C}H_2\overset{4}{C}H=\overset{3}{C}H\overset{2}{C}\equiv\overset{1}{C}H$	$\overset{1}{C}H_2=\overset{2}{C}H\overset{3}{C}H_2\overset{4}{C}\equiv\overset{5}{C}H$	$\overset{6}{H}C\equiv\overset{5}{C}\overset{4}{C}H_2\overset{3}{C}H_2\overset{2}{C}H(OH)\overset{1}{C}H_3$
3-Hexen-1-yne	**1-Penten-4-yne**	**5-Hexyn-2-ol**
(Not 3-hexen-5-yne)	(Not 4-penten-1-yne)	(Not 1-hexyn-5-ol)

EXERCISE 13-1

Give the IUPAC names for (a) all the alkynes of composition C_6H_{10};

(b) 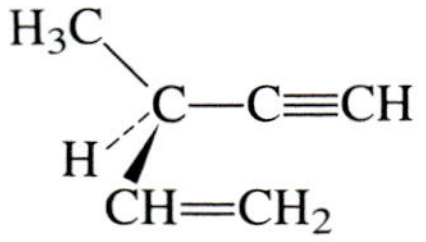 H_3C, H, $-C\equiv CH$, $CH=CH_2$ on a stereocenter C; (c) all butynols.

Remember to include and designate stereoisomers.

13-2 Structure and Bonding in the Alkynes

This section describes the triple bond in molecular-orbital terms: the carbons are *sp* hybridized and the four singly filled *p* orbitals form two perpendicular π bonds. This arrangement has a strong effect on the physical and chemical properties of the alkynes.

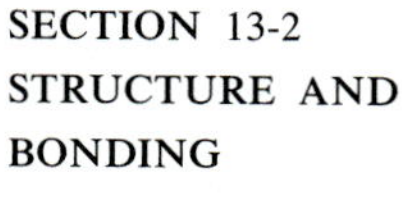

FIGURE 13-1

A. Molecular-orbital picture of *sp*-hybridized carbon.
B. Formation of the triple bond in ethyne by overlap between two *sp*-hybridized CH fragments to create a σ bond and two π bonds.
C. The two perpendicular π bonds in ethyne result in a cylindrical electron cloud around the molecular axis.

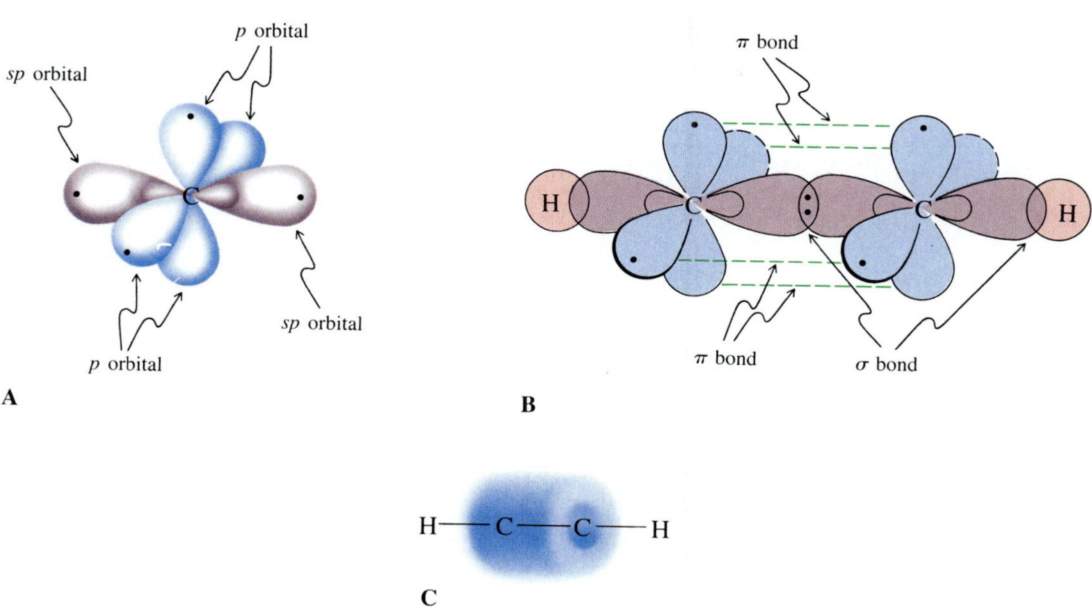

The Electronic Structure of Ethyne

In ethyne, the two carbons are *sp* hybridized (Figure 13-1A), as with beryllium in beryllium hydride (Figure 1-16). One of the hybrid orbitals on each carbon overlaps with hydrogen, the other forms a mutual σ bond (Figure 13-1B). The two perpendicular *p* orbitals on each carbon (in contrast with those of beryllium) contain one electron each. These two sets of *p* orbitals overlap to form two perpendicular π bonds between the two carbon atoms. Because of the diffuse character of the π bonds, the electronic distribution in the triple bond assumes the shape of a cylindrical cloud (Figure 13-1C).

As a consequence of hybridization and the two π interactions, the strength of the triple bond is quite high: about 200 kcal mole^{-1}. The C–H bond-dissociation energy of terminal alkynes is also high: 128 kcal mole^{-1}.

The Molecular Structure of Ethyne: A Linear Molecule with Short Bonds

Because both carbon atoms in ethyne are *sp* hybridized, its structure is linear (Figure 13-2). The carbon–carbon bond length is 1.203 Å, shorter than that of a double bond (1.33 Å, Figure 11-7). The carbon–hydrogen bond also is short because of the relatively large degree of *s* character in the *sp* hybrids used for bonding to hydrogen. The electrons in these orbitals (and in the bonds that they form by overlapping with other orbitals) reside relatively close to the nucleus and produce shorter (and stronger) bonds.

FIGURE 13-2

The molecular structure of ethyne.

Some Physical Constants of Alkynes

Alkynes have boiling points very similar to those of the corresponding alkenes and alkanes. Ethyne is unusual in that it has no boiling point at atmospheric pressure; rather it sublimes at −84°C. Propyne (b.p. −23.2°C) and 1-butyne (b.p. 8.1°C) are gases, whereas 2-butyne is barely a liquid (b.p. 27°C) at room temperature. The medium-sized alkynes are distillable liquids. Care must be taken in the handling of alkynes: they polymerize very easily—frequently with violence. Ethyne explodes under pressure but can be shipped in pressurized gas cylinders that contain propanone (acetone) and a porous filler (pumice).

In analogy to the alkenes, the relatively high *s* character in the carbon hybrid orbitals of substituted alkynes causes them to exhibit dipole moments, except when they are symmetrical.

Polarization in Alkynes

$$CH_3C{\equiv}CH \quad \mu = 0.74\ D \qquad CH_3CH_2C{\equiv}CH \quad \mu = 0.80\ D \qquad CH_3CH_2C{\equiv}CCH_2CH_3 \quad \mu = 0\ D$$

General Deprotonation of 1-Alkynes

$$RC{\equiv}C{-}H + :B^- \longrightarrow RC{\equiv}C:^- + HB$$

For the same reason, terminal alkynes are more acidic than their alkane and alkene analogs. The pK_a of ethyne, for example, is about 25, remarkably low compared with that of ethene ($pK_a \sim 44$) and ethane ($pK_a \sim 50$).

Order of Acidity of Alkynes, Alkenes, and Alkanes

	$HC{\equiv}CH$ >	$H_2C{=}CH_2$ >	H_3CCH_3
Hybridization:	sp	sp^2	sp^3
pK_a:	25	44	50

This effect turns out to be preparatively useful, because several strong bases can be used stoichiometrically to deprotonate terminal alkynes to the corresponding alkynyl anions. Examples are sodium amide in liquid ammonia, alkyllithiums, and Grignard reagents. The anions generated in this way react as bases and nucleophiles, as do other carbanions (Section 13-5).

Deprotonation of Alkynes

$$CH_3CH_2C{\equiv}CH + CH_3CH_2CH_2CH_2Li \xrightarrow{(CH_3CH_2)_2O} CH_3CH_2C{\equiv}CLi + CH_3CH_2CH_2CH_2{-}H$$

$$C_5H_9{-}C{\equiv}CH + CH_3CH_2MgBr \xrightarrow{THF} C_5H_9{-}C{\equiv}CMgBr + CH_3CH_2{-}H$$

$$HC{\equiv}CH + 2\ NaNH_2 \xrightarrow{\text{Liquid } NH_3} NaC{\equiv}CNa + 2\ H{-}NH_2$$

EXERCISE 13-2

Strong bases other than those mentioned here for the deprotonation of alkynes were introduced earlier. Two examples are potassium *tert*-butoxide and lithium diisopropyl amide (LDA). Would either (or both) of these compounds be suitable for making ethynyl anion from ethyne? Explain.

In summary, the characteristic hybridization scheme for the triple bond of an alkyne controls its physical and electronic features. It is responsible for strong bonds, the linear structure of alkynes, dipole moments, and the relatively acidic alkynyl hydrogen.

13-3 The Triple Bond Shields Alkyne Hydrogens: Nuclear Magnetic Resonance of the Alkynes

As discussed in Section 11-3, alkenyl hydrogens are unusually deshielded compared with the hydrogens in saturated alkanes because of the circular motion of the π electrons when exposed to a magnetic field. A similar (perhaps even stronger) effect might be expected for alkynyl hydrogens owing to the presence of two π bonds and to *sp* hybridization, which should cause deshielding. We shall see in this section that this expectation is not borne out by experiment and that these nuclei are in fact characteristically shielded.

The Chemical Shifts of Alkyne Hydrogens

The ^{1}H NMR absorptions of hydrogens bound to an *sp*-hybridized carbon atom are found at $\delta = 1.7$–3.1 ppm (see Table 10-2). An example is 3,3-dimethyl-1-butyne, shown in Figure 13-3. The alkynyl hydrogen is seen to absorb at $\delta = 1.74$ ppm, at considerably higher field than the alkenyl hydrogen in *trans*-2,2,5,5-tetramethyl-3-hexene ($\delta = 5.30$ ppm; see Figure 11-8). The 1,1-dimethylethyl (*tert*-butyl) group appears at $\delta = 1.13$ ppm, more or less in the expected position.

Why is the terminal alkyne hydrogen so shielded? Evidently, the electronic effect responsible for the low-field absorption of the alkene hydrogens is somehow canceled in the alkynes. The explanation for this observation lies in the cylindrical electron cloud around the molecular axis of the triple bond (see Figure 13-1C), which allows for a different type of electron movement. When the axis of the alkyne molecule is parallel to the external magnetic field, H_0, the electrons may flow relative to that axis in a way that *opposes* H_0 in the vicinity of the alkynyl hydrogen (Figure 13-4). The result is a strong *shielding* effect.

The normal influence of the π bond, depicted in Figure 11-9 for an alkene, is still in effect whenever the axis of an alkyne molecule is perpendicular (or at some angle) to the external field. In solution and in the spinning NMR tube, the molecules adopt all possible intermediate positions. Equilibration is rapid on the NMR time scale, giving only one averaged signal.

Coupling Through the Triple Bond

The alkyne functional group transmits coupling so well that the terminal hydrogen is split by the hydrogens across the triple bond, even though they are separated from them by three carbons, another example of long-range coupling.

FIGURE 13-3
90-MHz ^{1}H NMR spectrum of 3,3-dimethyl-1-butyne in CCl_4.

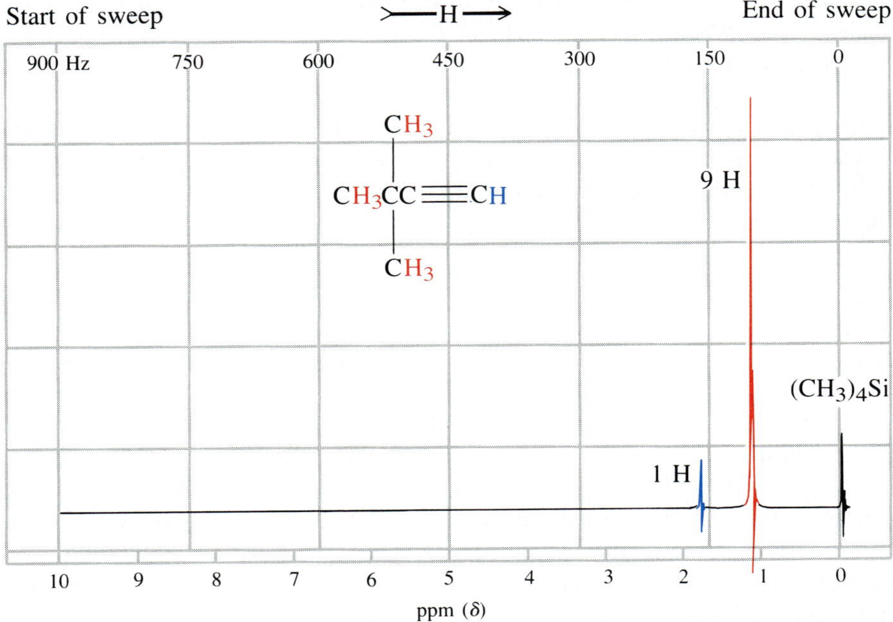

FIGURE 13-4
Electron current caused by the external field, H_0, being aligned with the molecular axis. This flow generates a local magnetic field, h_{local} (dashed lines), opposing the external field in the region of space occupied by the alkynyl hydrogen. A higher applied field is therefore necessary to cause this nucleus to resonate.

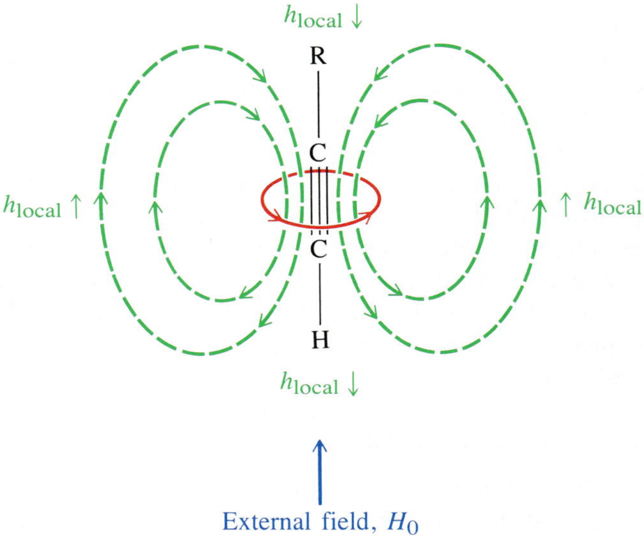

FIGURE 13-5
90-MHz ^1H NMR spectrum of 1-pentyne in CCl_4.

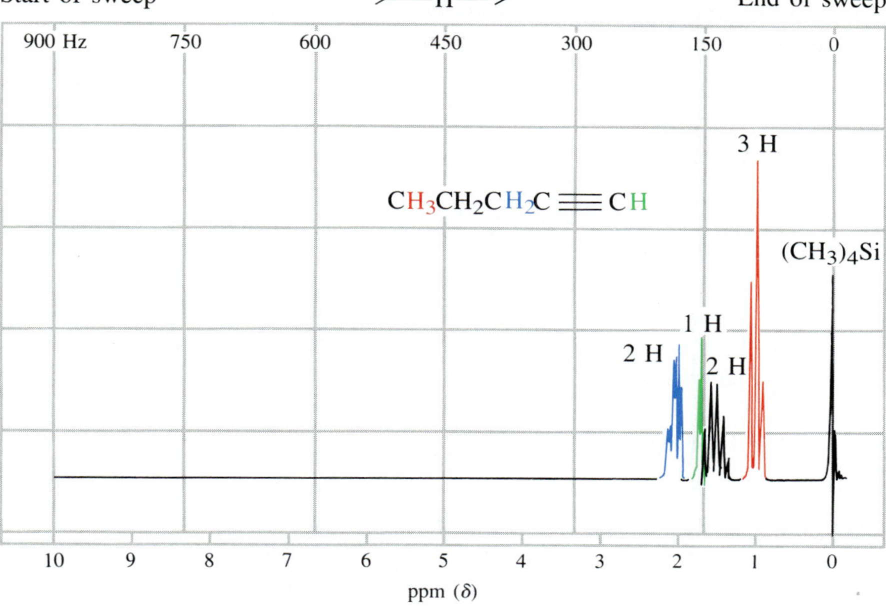

The coupling constants range from about 2 to 4 Hz. Figure 13-5 shows the NMR spectrum of 1-pentyne, clearly exemplifying this effect. The alkynyl hydrogen absorbs at $\delta = 1.71$ ppm and has the shape of a triplet ($J = 2.5$ Hz) because of coupling to the two equivalent hydrogens at C-3, which occur at $\delta = 2.07$ ppm as a doublet of triplets ($J = 2.5$, 6 Hz). Such long-range coupling ($J \sim 2$–3 Hz) is also observed across four carbons between two sets of nonequivalent hydrogens separated by a triple bond—for example, those present in 3-heptyne, $CH_3\mathbf{CH_2}C{\equiv}C\mathbf{CH_2}CH_2CH_3$.

Long-Range Coupling in Alkynes

H
|
—C—C≡C—H
|

$J = 2$–4 Hz

H H
| |
—C—C≡C—C—
| |

$J = 2$–3 Hz

EXERCISE 13-3

Predict the first-order splitting pattern in the ^1H NMR spectrum of *trans*-1-bromo-1-penten-4-yne.

Carbon-13 Nuclear Magnetic Resonance

Carbon-13 NMR spectroscopy also is useful in deducing the structure of alkynes. For example, the triple-bonded carbons in alkyl-substituted alkynes absorb in the narrow range of $\delta = 65$–85 ppm, quite separate from the chemical shifts of analogous alkane and alkene carbon atoms.

Typical Alkyne ^{13}C NMR Chemical Shifts

HC≡CH: $\delta = 71.9$

HC≡CCH₂CH₂CH₂CH₃: 68.6, 84.0, 18.6, 31.1, 22.4, 14.1

CH₃CH₂C≡CCH₂CH₃: 81.1, 15.6, 13.2 ppm

In summary, the cylindrical π cloud around the carbon–carbon triple bond induces local magnetic fields that lead to chemical shifts for alkynyl hydrogens at higher field than those of alkenyl hydrogens. Coupling through the triple bond is about as strong as that transmitted by the double bond.

13-4 The Stability of the Triple Bond

Does the heat content of the triple bond depend on the substituents, as it does for alkenes? Which type of alkyne is more stable—a terminal or an internal isomer? What about cycloalkynes? How large must a ring be to tolerate the presence of the linear triple bond, and how much distortion from linearity is possible? These questions can be addressed by studying the comparative heats of combustion and hydrogenation and the base-catalyzed equilibration of isomers.

Alkynes Are High-Energy Compounds

Alkynes have a high energy content. For example, the heat of formation of ethyne is $\Delta H_f^\circ = +54.3$ kcal mole^{-1}. In other words, ethyne is highly unstable relative to its component elements in their standard states. Because of this property, alkynes react in many cases with considerable release of energy. For example, when shocked by high pressure or exposed to catalytic amounts of copper, ethyne explosively decomposes to carbon and hydrogen. On combustion, it releases 317 kcal mole^{-1}. This energy is put to practical use in acetylene torches, employed in welding, which requires very hot flames (more than 2500°C).

Combustion of Ethyne

$$\text{HC}{\equiv}\text{CH} + 2.5\ \text{O}_2 \longrightarrow 2\ \text{CO}_2 + \text{H}_2\text{O} \qquad \Delta H^\circ = -317 \text{ kcal mole}^{-1}$$

The catalytic hydrogenation of ethyne proceeds in two stages: first to give ethene, then ethane. The respective heats of hydrogenation of these two steps show that the first π bond has a higher energy content than the second.

Hydrogenation of Ethyne

$$\text{HC}{\equiv}\text{CH} + \text{H}_2 \xrightarrow{\text{Catalyst}} \text{H}_2\text{C}{=}\text{CH}_2 \qquad \Delta H^\circ = -41.9 \text{ kcal mole}^{-1}$$

$$\text{H}_2\text{C}{=}\text{CH}_2 + \text{H}_2 \xrightarrow{\text{Catalyst}} \text{H}_3\text{CCH}_3 \qquad \Delta H^\circ = -32.7 \text{ kcal mole}^{-1}$$

To measure the relative stabilities of internal and terminal alkynes, we can compare their heats of hydrogenation. For example, both isomeric butynes hydrogenate to butane but on doing so release different amounts of heat:

$$\text{CH}_3\text{CH}_2\text{C}{\equiv}\text{CH} + 2\ \text{H}_2 \xrightarrow{\text{Catalyst}} \text{CH}_3\text{CH}_2\text{CH}_2\text{CH}_3 \qquad \Delta H^\circ = -69.9 \text{ kcal mole}^{-1}$$

$$\text{CH}_3\text{C}{\equiv}\text{CCH}_3 + 2\ \text{H}_2 \xrightarrow{\text{Catalyst}} \text{CH}_3\text{CH}_2\text{CH}_2\text{CH}_3 \qquad \Delta H^\circ = -65.1 \text{ kcal mole}^{-1}$$

Again, the internal π system is more stable than the terminal one, as might be expected on the basis of hyperconjugation (Sections 3-2 and 11-4).

Relative Stabilities of the Alkynes

$RC{\equiv}CH < RC{\equiv}CR'$

Cyclic Alkynes Are Strained

The energy content in alkynes may be increased further by bending the normally linear triple bond, as in the smaller cycloalkynes.

Calculated Strain Energies of Some Cycloalkynes

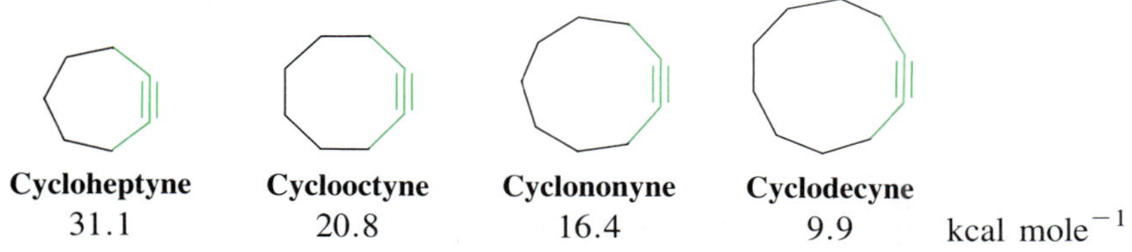

Cycloheptyne	Cyclooctyne	Cyclononyne	Cyclodecyne	
31.1	20.8	16.4	9.9	kcal mole^{-1}

The calculated strain energy in the cycloalkynes increases as the size of the ring decreases. The smallest unsubstituted cycloalkyne that has been isolated is cyclooctyne. Lower homologs are too reactive to be isolable. That strained cycloalkynes should be reactive becomes apparent on inspection of their molecular models. The distortion of the alkyne moiety reduces overlap between the p orbitals of the π bond in the molecular plane, leading to diradical character, as depicted in Figure 13-6. This radical character manifests itself in increased reactivity, particularly with respect to polymerization and reaction with oxygen.

FIGURE 13-6

A. Unstrained triple bond with good overlap between the two sets of p orbitals that form the two π bonds.

B. Strained triple bond forces the in-plane p orbitals to bend away from each other, causing decreased overlap.

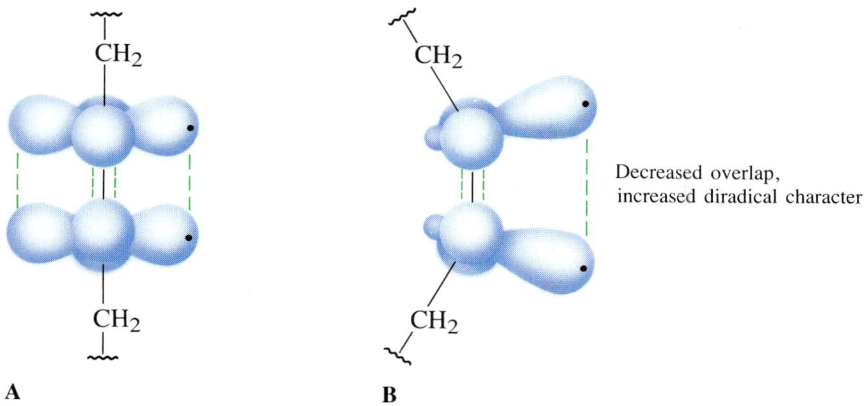

Bases Mediate the Isomerization of Alkynes

The relative stabilities of various isomeric alkynes can be estimated directly by chemical equilibration, as in the acid-catalyzed isomerization of alkenes (see

Section 12-3). In alkynes, isomerization is catalyzed by a base. For example, 1-alkynes are converted into 2-alkynes in the presence of hot alcoholic potassium hydroxide. At 25°C, this reaction is exothermic by about 4 kcal mole^{-1}, as can be expected based on the respective heats of formation of the two components.

Base-Catalyzed Isomerization of 1-Hexyne to 2-Hexyne

$$\underset{\textbf{1-Hexyne}}{CH_3CH_2CH_2CH_2C{\equiv}CH} \xrightleftharpoons{KOH,\ CH_3CH_2OH,\ 175°C} \underset{\textbf{2-Hexyne}}{CH_3CH_2CH_2C{\equiv}CCH_3} \qquad \Delta H° \sim -4 \text{ kcal mole}^{-1}$$

EXERCISE 13-4

The equilibrium ratio of 2- and 1-hexyne at 175°C was found to be 70:1. Calculate $\Delta G°$ at this temperature.

What is the mechanism of this transformation? When base attacks the terminal alkyne, it first detaches the most acidic terminal hydrogen to give an alkynyl anion.

Reversible Deprotonation of Terminal Alkynes

$$\underset{pK_a \sim 25}{RCH_2C{\equiv}CH} + CH_3CH_2\ddot{\underset{..}{O}}{:}^- \underset{K \sim 10^{-9}}{\rightleftharpoons} RCH_2C{\equiv}C{:}^- + \underset{pK_a \sim 16}{CH_3CH_2\ddot{\underset{..}{O}}H}$$

This process is reversible and produces only a small equilibrium concentration of the anion, because the pK_a of ethanol is lower than that of a terminal alkyne. The reaction is not only uphill thermodynamically, it is also a dead end mechanistically, given that it does not lead to the product. Although the C-3 position is considerably less acidic (pK_a ~ 35), the base will occasionally abstract a proton there to give the corresponding anion. This equilibrium does not favor the anion, but it serves to generate an intermediate that may proceed to product. A resonance structure can be written for this anion in which the negative charge resides at the terminal position. Reprotonation at this position gives a new isomer, an **allene.**

Mechanism of Terminal Alkyne Isomerization:

STEP 1: Anion formation at C-3

$$\underset{|\atop H}{RCHC{\equiv}CH} + CH_3CH_2\ddot{\underset{..}{O}}{:}^- \rightleftharpoons \left[\overset{H}{\underset{R}{>}}\bar{\ddot{C}}{-}C{\equiv}CH \longleftrightarrow \overset{H}{\underset{R}{>}}C{=}C{=}\bar{\ddot{C}}H \right] + CH_3CH_2\ddot{\underset{..}{O}}H$$

STEP 2: Reprotonation

$$\left[\overset{H}{\underset{R}{>}}\bar{\ddot{C}}{-}C{\equiv}CH \longleftrightarrow \overset{H}{\underset{R}{>}}C{=}C{=}\bar{\ddot{C}}H \right] + CH_3CH_2\ddot{\underset{..}{O}}H \longrightarrow \underset{\textbf{An allene}}{\overset{H}{\underset{R}{>}}C{=}C{=}C\overset{H}{\underset{H}{<}}} + CH_3CH_2\ddot{\underset{..}{O}}{:}^-$$

STEP 3: Second deprotonation-reprotonation

$$\underset{R}{\overset{H}{}}\!\!>C{=}C{=}C\!<\!\underset{H}{\overset{H}{}} + CH_3CH_2\ddot{\underset{\cdot\cdot}{O}}{:}^- \longrightarrow \left[R{-}\ddot{C}^-{=}C{=}C\!<\!\underset{H}{\overset{H}{}} \longleftrightarrow RC{\equiv}C{-}\ddot{C}^-H_2 \right] + CH_3CH_2\ddot{\underset{\cdot\cdot}{O}}H$$

$$\downarrow$$

$$RC{\equiv}CCH_2H + CH_3CH_2\ddot{\underset{\cdot\cdot}{O}}{:}^-$$

The isomerization of 1-alkynes to the corresponding allenes is usually slightly exothermic by about 1 kcal $mole^{-1}$. Continuing the isomerization, the base removes the internal allene proton to give another anion. Reprotonation provides the final rearranged alkyne.

EXERCISE 13-5

Predict which one of the following two compounds is deprotonated faster at the position printed in boldface: $CH_3\mathbf{CH_2}C{\equiv}CH$; $CH_3CH{=}C{=}\mathbf{CH_2}$.

Allenes Can Be Chiral

The allene intermediates in the alkyne rearrangement can sometimes be isolated, particularly if further rearrangement is blocked. For example, 3-methyl-1-butyne isomerizes to 1,1-dimethylallene on treatment with base.

1,1-Dimethylallene Synthesis

$$CH_3{-}\underset{CH_3}{\overset{H}{C}}{-}C{\equiv}CH \xrightarrow{\text{KOH, } CH_3CH_2OH\text{, } 170°C} \underset{CH_3}{\overset{CH_3}{}}\!\!>C{=}C{=}CH_2$$

3-Methyl-1-butyne → **3-Methyl-1,2-butadiene (1,1-Dimethylallene)**

Allenes have an unusual structure (Figure 13-7). The two π bonds are at right angles, because the central carbon atom is *sp* hybridized. Note that this arrangement places the four hydrogen atoms in the parent allene in the positions of an elongated tetrahedron. As a consequence, substituted allenes can be chiral even

FIGURE 13-7

The molecular-orbital structure of allene. The two π bonds are perpendicular to each other. Each double bond is 1.31 Å long (compare ethene, 1.34 Å; ethyne, 1.20 Å).

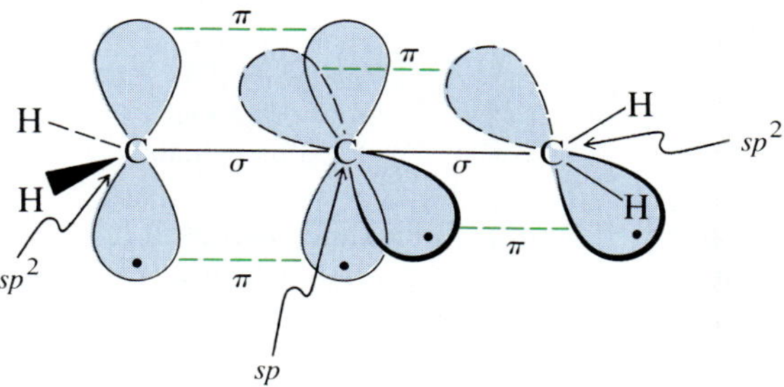

though they do not contain an asymmetric carbon. An example is 1,3-dichloro-1,2-propadiene. A molecular model will reveal that image and mirror image are not superimposable in this molecule.

Chirality in 1,3-Dichloro-1,2-propadiene

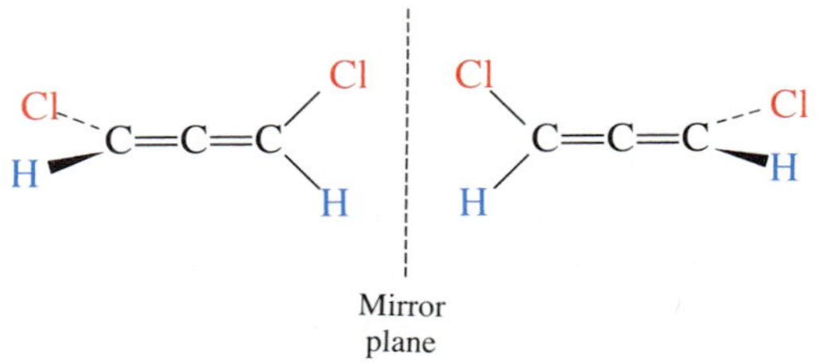

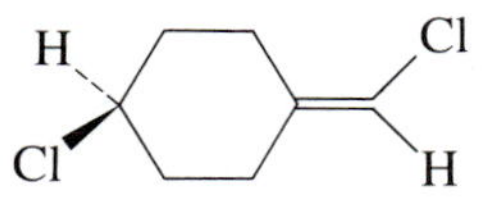

A

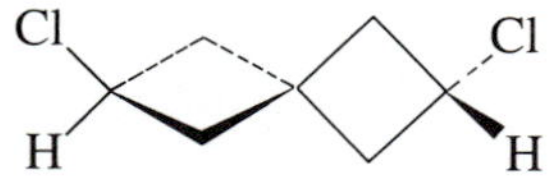

B

EXERCISE 13-6

Structures A and B, at the left, appear topologically similar to 1,3-dichloro-1,2-propadiene. Are they chiral?

In summary, alkynes are highly energetic, the first π bond releasing more energy on hydrogenation than the second. Cyclic alkynes can suffer from severe strain. Internal isomers are more stable than terminal ones, as demonstrated by their relative heats of hydrogenation and their base-mediated isomerization. The mechanism of isomerization proceeds through the intermediacy of allenes. Allenes can be chiral.

13-5 The Preparation of Alkynes

The two basic methods used to prepare alkynes are double elimination from 1,1- or 1,2-dihaloalkanes and the alkylation of terminal alkynyl anions.

Alkynes by Elimination

As discussed in Section 11-5, alkenes can be prepared by E2 reactions of haloalkanes. Application of this principle to alkyne synthesis suggests that treatment of vicinal or geminal dihaloalkanes with strong base should result in double elimination to furnish a triple bond.

General Double Elimination from Dihaloalkanes to Give Alkynes

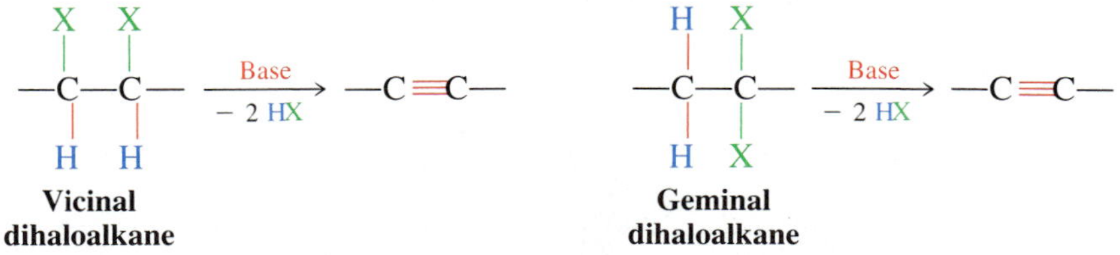

Indeed, addition of 1,2-dibromohexane (accessible by bromination of 1-hexene) to sodium amide in liquid ammonia followed by evaporation of solvent and aqueous work-up gives 1-hexyne:

$$\mathrm{CH_3CH_2CH_2CH_2\underset{\underset{\displaystyle Br}{|}}{C}H{-}CH_2Br \xrightarrow[-\ 2\ HBr]{1.\ NaNH_2,\ liquid\ NH_3\ \ 2.\ H_2O} CH_3CH_2CH_2CH_2C{\equiv}CH}$$

The other products in this reaction are the protonated base (i.e., NH_3), which is evaporated, and sodium bromide, which is removed in the work-up. Eliminations in liquid ammonia are usually carried out at the boiling point of ammonia, which is low (b.p. $-33°C$), and so the reaction conditions are mild and generally allow for the preparation of alkynes without isomerization (Section 13-4).

Because vicinal dihaloalkanes are readily available from alkenes by halogenation, this **halogenation–double-dehydrohalogenation** sequence is a ready means of converting alkenes into the corresponding alkynes.

Halogenation–Double-Dehydrohalogenation for Alkyne Synthesis

$$\underset{\textbf{2-Butene}}{\mathrm{CH_3CH{=}CHCH_3}} \xrightarrow{1.\ Br_2,\ CCl_4\ \ 2.\ NaNH_2,\ liquid\ NH_3} \underset{\textbf{2-Butyne}}{\mathrm{CH_3C{\equiv}CCH_3}}$$

$$\underset{\textbf{1,5-Hexadiene}}{\mathrm{CH_2{=}CHCH_2CH_2CH{=}CH_2}} \xrightarrow{1.\ Br_2,\ CCl_4\ \ 2.\ NaNH_2,\ liquid\ NH_3} \underset{\textbf{1,5-Hexadiyne}}{\underset{53\%}{\mathrm{HC{\equiv}CCH_2CH_2C{\equiv}CH}}}$$

Bases other than sodium amide, such as alkoxides, can be used in dehydrohalogenations. Their use is particularly effective if isomerization is structurally impossible. An example is the bromination-dehydrobromination of 3,3-dimethyl-1-butene with potassium *tert*-butoxide.

$$\underset{\textbf{3,3-Dimethyl-1-butene}}{\mathrm{CH_3\overset{\overset{\displaystyle CH_3}{|}}{\underset{\underset{\displaystyle CH_3}{|}}{C}}CH{=}CH_2}} \xrightarrow{1.\ Br_2,\ CCl_4\ \ 2.\ (CH_3)_3CO^-K^+,\ (CH_3)_3COH} \underset{\textbf{3,3-Dimethyl-1-butyne}}{\underset{95\%}{\mathrm{CH_3\overset{\overset{\displaystyle CH_3}{|}}{\underset{\underset{\displaystyle CH_3}{|}}{C}}C{\equiv}CH}}}$$

Geminal dihaloalkanes react like their vicinal counterparts:

$$\underset{\textbf{1,1-Dichloro-3-methylbutane}}{\mathrm{(CH_3)_2CHCH_2CHCl_2}} \xrightarrow{NaNH_2,\ 2,2,4\text{-}trimethylpentane,\ 170°C} \underset{\textbf{3-Methyl-1-butyne}}{\underset{75\%}{\mathrm{(CH_3)_2CHC{\equiv}CH}}}$$

Haloalkenes Are Intermediates in Alkyne Synthesis by Elimination

The mechanism of dehydrohalogenation of dihaloalkanes proceeds step by step through the intermediacy of haloalkenes, also called **alkenyl** or **vinyl halides.** Although mixtures of *E*- and *Z*-haloalkenes are in principle possible, with diastereomerically pure vicinal dihaloalkanes only one product is formed because elimination proceeds stereospecifically *anti* (Section 11-5).

Anti elimination → An alkenyl halide + BH + X^-

Anti elimination → An alkenyl halide + BH + X^-

We already know that stereoisomerically pure vicinal dihaloalkanes can be made by the addition of bromine to alkenes (see Section 12-3). For example, with *cis*-2-butene, bromine is added only to (2*R*,3*R*)- and (2*S*,3*S*)-2,3-dibromobutane (only the *S*,*S* isomer is shown below), which in turn dehydrobrominates to give only (*Z*)-2-bromo-2-butene. Conversely, *trans*-2-butene selectively furnishes (*E*)-2-bromo-2-butene when undergoing the same sequence.

Stereospecific Bromination-Dehydrobromination

***cis*-2-Butene** —Br_2→ **(2*R*,3*R*)- and (2*S*,3*S*)-2,3-Dibromobutane** —Rotation→ —$-$HBr→ **(*Z*)-2-Bromo-2-butene**

***trans*-2-Butene** —1. Br_2 2. Base→ **(*E*)-2-Bromo-2-butene**

The stereochemistry of the intermediate haloalkene is of no consequence to the outcome of the alkyne-forming reaction. Both isomers lead to the same product. However, starting from this intermediate, *anti* elimination is often faster than *syn* elimination, apparently for a reason similar to that making *anti* elimination the preferred pathway in the E2 reaction giving double bonds—overlap between the developing *p* orbitals of the new π bond is better in the *anti* transition state.

—$NaNH_2$, liquid NH_3, Fast→ $CH_3C{\equiv}CCH_3$ —$NaNH_2$, liquid NH_3, Slow→ $CH_3C{\equiv}CCH_3$

You will have noticed that elimination of HX from vicinal dihaloalkanes gives only alkenyl and not allylic (see Chapter 14) halides. Why does the base attack only the proton alpha to a halogen? An explanation in simple Coulombic terms is that the halogen is more electronegative than carbon; hence the proton next to it is relatively acidic. Thus, the E2 reaction leading to the alkenyl halide is faster than that furnishing the allylic isomer.

Regioselective Elimination from Vicinal Dihaloalkanes

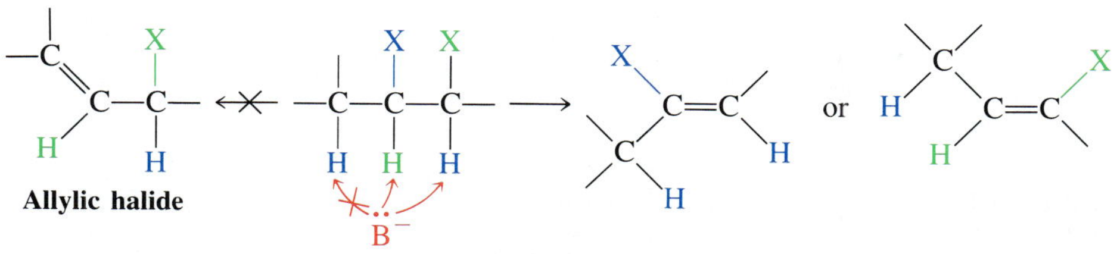

Allylic halide

EXERCISE 13-7

Predict the product of the reaction of sodium methoxide with 1-fluoro-2-bromoethane.

Alkenyl Halides Are Relatively Unreactive

Compared with haloalkanes, alkenyl halides are relatively unreactive toward nucleophiles. Although we have seen that, with strong bases, alkenyl halides undergo elimination reactions to give alkynes, they do not react with weak bases and relatively nonbasic nucleophiles, such as iodide. Similarly, S_N1 reactions do not normally take place because the intermediate alkenyl cations are species of high energy, partly owing to the minimal delocalization of electrons in this intermediate. The molecular-orbital structure of the ethenyl (vinyl) cation is shown in Figure 13-8. Note that the carbon adjacent to the positively charged one lacks a p orbital that would be parallel to the empty orbital on the carbon bearing the positive charge.

Alkenyl Halides Do Not Readily Undergo S_N2 or S_N1 Reactions

$$CH_2{=}CHBr \xrightarrow{I^-} \!\!\!\!\!\!\not\;\;\; CH_2{=}CHI + Br^- \qquad CH_2{=}CHBr \;\not\longrightarrow\; CH_2{=}\overset{+}{C}{-}H + Br^-$$

Ethenyl (vinyl) cation

FIGURE 13-8

Molecular-orbital picture of the ethenyl (vinyl) cation. One carbon is sp^2, the other sp hybridized.

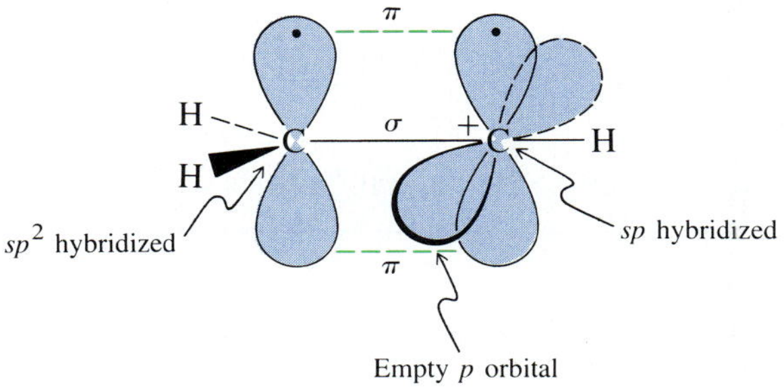

sp^2 hybridized

sp hybridized

Empty p orbital

Alkenyl halides, however, can be functional through the intermediate formation of alkenyl organometallics (see Exercise 11-6). These species allow access to a variety of specifically substituted alkenes.

Alkenyl Organometallics in Synthesis

$$CH_2{=}C(Br)H + Mg \xrightarrow{THF} CH_2{=}C(MgBr)H \xrightarrow[2.\ H^+,\ H_2O]{1.\ CH_3\overset{O}{\overset{\|}{C}}CH_3} CH_2{=}C(H)C(OH)(CH_3)_2$$

1-Bromoethene (Vinyl bromide) — 90% **Ethenylmagnesium bromide (A vinyl Grignard reagent)** — 65% **2-Methyl-3-buten-2-ol**

$$\underset{(Z)\text{-2-Bromo-2-pentene}}{CH_3CH_2(H)C{=}C(Br)CH_3} \xrightarrow[(CH_3CH_2)_2O,\ -78^\circ C]{CH_3CH_2CH_2CH_2Li,} \underset{(Z)\text{-2-Penten-2-yllithium}}{CH_3CH_2(H)C{=}C(Li)CH_3} \xrightarrow[-\ LiI]{CH_3I} \underset{\substack{72\% \\ \text{2-Methyl-2-pentene}}}{CH_3CH_2(H)C{=}C(CH_3)CH_3}$$

Alkynes by Alkylation of Alkynyl Anions

Alkynes can be prepared from other alkynes by the reaction of terminal alkynyl anions with alkylating agents, such as primary haloalkanes, oxacyclopropanes, aldehydes, or ketones. As we know, such anions are readily prepared from terminal alkynes by deprotonation with strong bases (mostly alkyllithium reagents, sodium amide in liquid ammonia, or Grignard reagents). Alkylation with primary haloalkanes is typically done in liquid ammonia or in ether solvents. These solvents sometimes contain 1,2-ethanediamine, $H_2NCH_2CH_2NH_2$, *N,N,N′,N′*-tetramethyl-1,2-ethanediamine, $(CH_3)_2NCH_2CH_2N(CH_3)_2$, or HMPA as cosolvents to increase the nucleophilic power of the anion (Section 6-10). The parent ethyne is alkylated in a series of steps through the selective formation of the monoanion to give mono- and dialkyl-derivatives.

Alkylation of Alkynyl Anions

$$CH_3C{\equiv}CH \xrightarrow[-\ NH_2H]{NaNH_2,\ \text{liquid}\ NH_3} CH_3C{\equiv}C{:}^{-}\,{}^{+}Na \xrightarrow[-\ NaI]{CH_3I} \underset{\substack{75\% \\ \text{2-Butyne}}}{CH_3C{\equiv}CCH_3}$$

$$\text{Cyclohexyl}{-}C{\equiv}CH \xrightarrow[-\ LiBr]{\substack{1.\ CH_3CH_2CH_2CH_2Li,\ THF,\ H_2NCH_2CH_2NH_2 \\ 2.\ CH_3CH_2CH_2Br}} \underset{\substack{85\% \\ \text{1-Pentynylcyclohexane}}}{\text{Cyclohexyl}{-}C{\equiv}CCH_2CH_2CH_3}$$

The ethynyllithium-1,2-ethanediamine complex is also commercially available. To obtain only monosubstituted ethyne, alkylation must take place at low temperatures. At high temperatures, singly metallated ethyne disproportionates to the insoluble dianion (or double Grignard reagent) and ethyne.

Disproportionation of Singly Metallated Ethyne

$$2\ HC{\equiv}CMgBr \xrightarrow{30^\circ C,\ (CH_3CH_2)_2O} HC{\equiv}CH + BrMgC{\equiv}CMgBr$$

Attempted alkylation of alkynyl anions with secondary and tertiary halides leads to E2 products because of the strongly basic character of the nucleophile.

Reactions with oxacyclopropanes or carbonyl compounds proceed in the same manner as similar transformations of other organometallic reagents.

$$HC{\equiv}CH \xrightarrow[-\ NH_2H]{LiNH_2\ (1\ equivalent),\ liquid\ NH_3} HC{\equiv}CLi \xrightarrow[-\ LiOH]{1.\ H_2C\overset{O}{—}CH_2\quad 2.\ HOH} HC{\equiv}CCH_2CH_2OH$$

92%

3-Butyne-1-ol

$$CH_3C{\equiv}CH \xrightarrow[-\ CH_3CH_2H]{CH_3CH_2MgBr,\ (CH_3CH_2)_2O,\ 20^\circ C} CH_3C{\equiv}CMgBr \xrightarrow{1.\ \text{cyclopentanone}\quad 2.\ H_2O} \text{1-(1-propynyl)cyclopentanol}$$

66%

EXERCISE 13-8

Suggest efficient and short syntheses of:

(a) $CH_3CH_2CH_2C{\equiv}CCH(OH)CH_2CH_3$ from ethyne; (b) $CH_3(CH_2)_3C{\equiv}CCH(OH)C{\equiv}C(CH_2)_3CH_3$ from $CH_3(CH_2)_3C{\equiv}CH$ (hint: review Section 8-6).

In summary, alkynes are made from vicinal and geminal dihaloalkanes by double stereospecific elimination through the intermediacy of alkenyl halides. These halides are unreactive in nucleophilic substitutions but can be metallated to organometallic reagents. Alkynes can also be prepared from other alkynes by alkylation with primary haloalkanes, oxacyclopropanes, or carbonyl compounds. Ethyne itself can be alkylated in a series of steps.

13-6 Reactions of the Alkynes: The Relative Reactivity of the Two Pi Bonds

In many respects, alkynes are like alkenes in their reactions, except for the availability of two π bonds, one generally being more reactive than the other. Thus, alkynes can undergo additions, such as hydrogenation, hydroboration, and other electrophilic attacks. Such reactions proceed in *syn* or *anti* manner and with or without regioselectivity.

General Addition of Reagents A–B to Alkynes

$$R{-}C{\equiv}C{-}R \xrightarrow{A-B} \underset{\text{A}\quad\quad\text{B}}{\overset{\text{R}\quad\quad\text{R}}{C{=}C}} \text{ or } \underset{\text{A}\quad\quad\text{R}}{\overset{\text{R}\quad\quad\text{B}}{C{=}C}} \xrightarrow{A-B} A{-}\underset{A}{\overset{R}{C}}{-}\underset{B}{\overset{R}{C}}{-}B \text{ or } A{-}\underset{B}{\overset{R}{C}}{-}\underset{A}{\overset{R}{C}}{-}B$$

The present section deals with this reactivity of the triple bond and introduces two new reactions: step-by-step hydrogenation and one-electron reduction by sodium to give cis and trans alkenes, respectively, and oxidative coupling to furnish diynes.

Hydrogenation: A Synthesis of Cis Alkenes

Alkynes can be hydrogenated under the same conditions used to hydrogenate alkenes. Typically, platinum or palladium on charcoal is suspended in a solution containing the alkyne, and the mixture is exposed to a hydrogen atmosphere. Under these conditions, the triple bond is saturated completely.

Complete Hydrogenation of Alkynes

$$\underset{\textbf{3-Heptyne}}{CH_3CH_2CH_2C{\equiv}CCH_2CH_3} \xrightarrow{H_2,\ Pt} \underset{\textbf{Heptane}}{\underset{100\%}{CH_3CH_2CH_2CH_2CH_2CH_2CH_3}}$$

$$\underset{\textbf{4-Methoxy-1-butyne}}{CH_3OCH_2CH_2C{\equiv}CH} \xrightarrow{H_2,\ Pd\text{-}C} \underset{\textbf{1-Methoxybutane}}{\underset{100\%}{CH_3OCH_2CH_2CH_2CH_2H}}$$

Hydrogenation is a step-by-step process, which means that, in some cases, the reaction can be stopped at the intermediate alkene stage. Alternatively, it is possible to prevent hydrogenation of the alkene by the use of modified catalysts. One such system is palladium that has been precipitated on calcium carbonate and treated with lead acetate and quinoline. This material is known as **Lindlar's* catalyst.** The surface of the metal in Lindlar's catalyst rearranges to a less-active configuration than that of palladium on carbon so that only the more-reactive first π bond of the alkyne is hydrogenated. Because the addition of hydrogen is *syn,* this method affords a stereoselective synthesis of cis alkenes.

Hydrogenations with Lindlar's Catalyst

Lindlar's catalyst: 5% Pd-CaCO$_3$, $Pb(O\overset{O}{\overset{\|}{C}}CH_3)_2$, quinoline

Quinoline

*Dr. Herbert H. M. Lindlar, b. 1909, Hoffmann La Roche and Co. A. G., Basel.

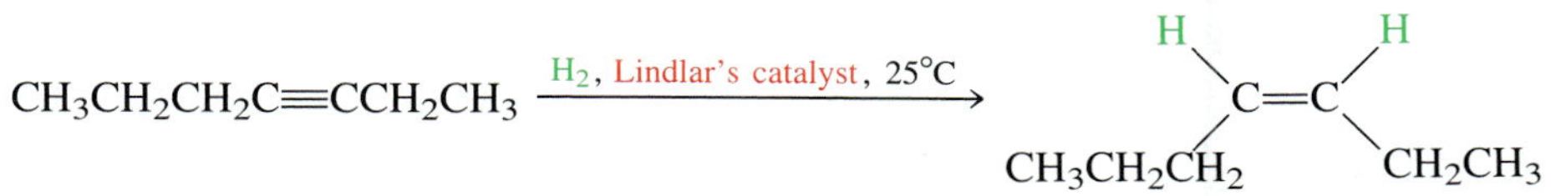

$$CH_3CH_2CH_2C{\equiv}CCH_2CH_3 \xrightarrow{H_2,\ \text{Lindlar's catalyst},\ 25°C}$$ (H)(CH_3CH_2CH_2)C=C(H)(CH_2CH_3)

100%

***cis*-3-Heptene**

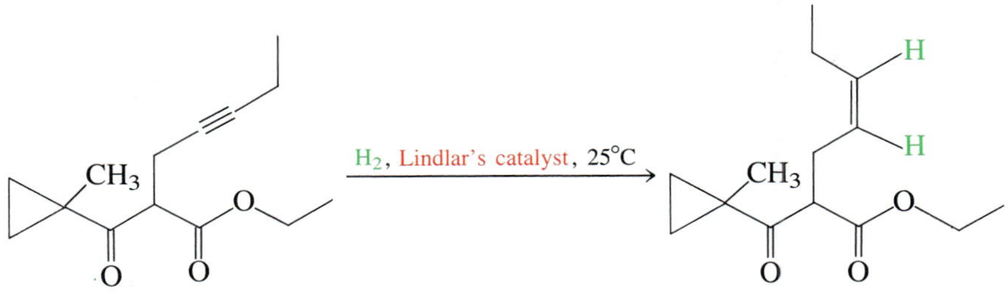

H₂, Lindlar's catalyst, 25°C

83% (after distillation)

With a method for the stereoselective construction of cis alkenes at our disposal, we may ask: Can we modify the reduction of alkynes so as to give only trans alkenes? The answer is yes, with a different reducing agent and through a different mechanism.

One-Electron Reduction of Alkynes to Trans Alkenes

If instead of using catalytically activated hydrogen we employ *sodium metal* dissolved in liquid ammonia as the reagent for the reduction of alkynes, we obtain trans alkenes as the products. For example, 3-heptyne is reduced to *trans*-3-heptene in this way. Unlike sodium amide in liquid ammonia, which functions as a strong base, sodium in liquid ammonia acts as a powerful electron donor (i.e., a reducing agent).

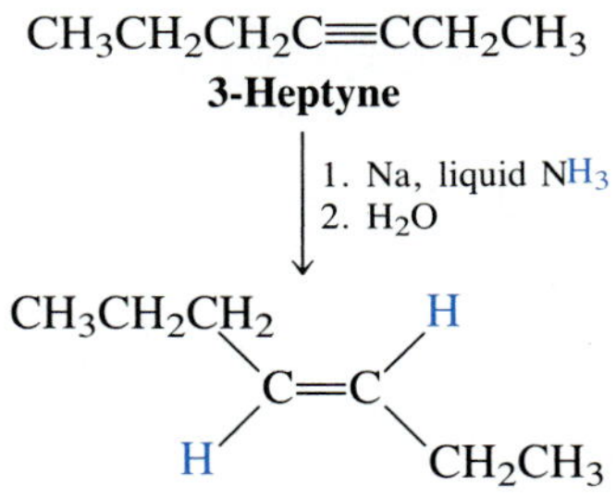

In the first step of the mechanism of this reduction, the π framework of the triple bond accepts one electron to give a radical anion. This anion is protonated by the ammonia solvent to give an alkenyl radical that is further reduced by accepting another electron to give an alkenyl anion. This species is again protonated to give the product alkene, stable to further reduction. The trans stereochemistry of the final alkene is due to the rapid equilibration of the intermediate alkenyl radical between the two possible cis and trans forms. The equilibrium lies on the side of the more-stable trans species.

$CH_3CH_2CH_2C{\equiv}CCH_2CH_3$

3-Heptyne

1. Na, liquid NH_3
2. H_2O

$CH_3CH_2CH_2$ / H — C=C — H / CH_2CH_3

86%

***trans*-3-Heptene**

Mechanism of the Reduction of Alkynes by Sodium in Liquid Ammonia:

STEP 1: One-electron transfer

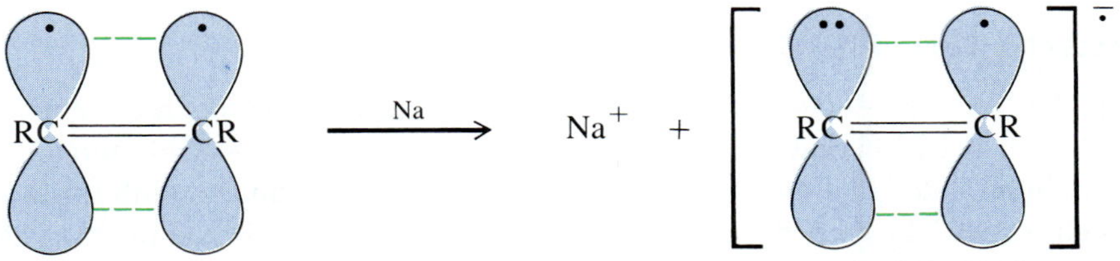

$$RC{\equiv}CR \xrightarrow{Na} Na^+ + [RC{\equiv}CR]^{\overset{.}{-}}$$

Alkyne radical anion

STEP 2: First protonation

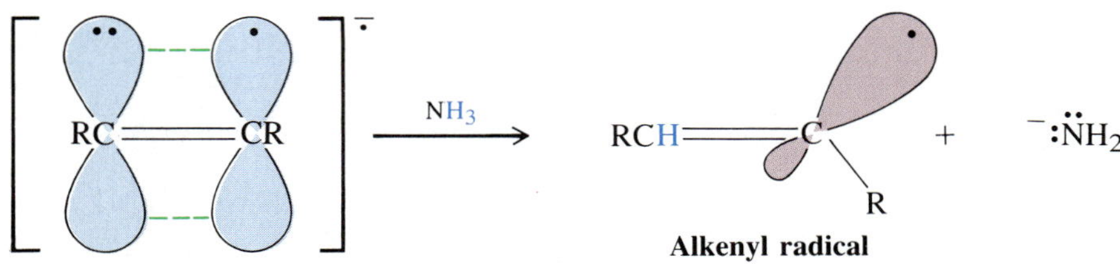

RC
CR
NH3
RCH
C
R
+
−:NH2
Alkenyl radical

STEP 3: Alkenyl radical equilibration

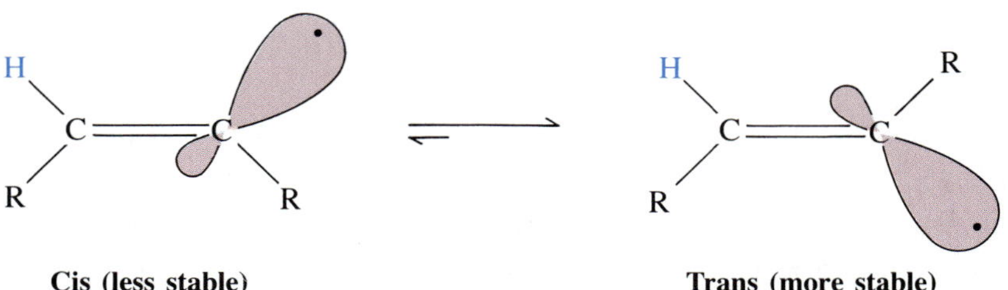

H
C
C
R
R
H
R
C
C
R
Cis (less stable)
Trans (more stable)

STEP 4: Second one-electron transfer

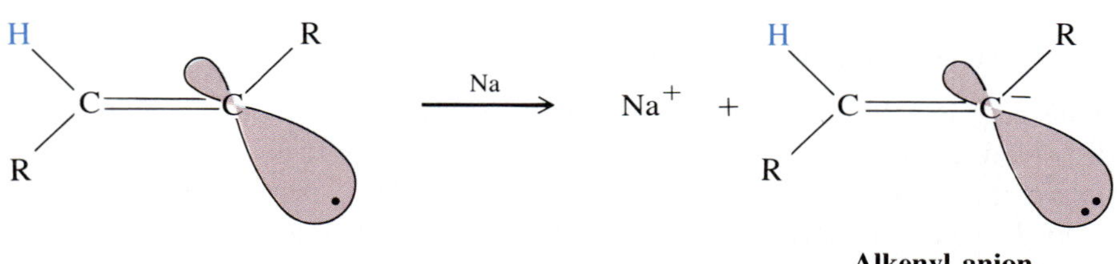

H
R
C
C
R
Na
Na+
+
H
R
C
C
R
Alkenyl anion

STEP 5: Second protonation

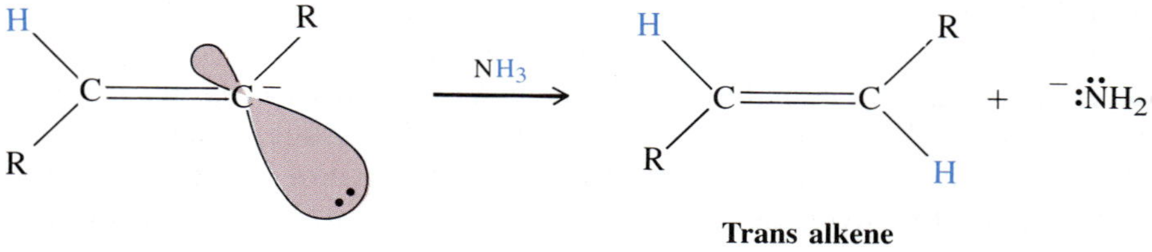

H
R
C
C
R
NH3
H
R
C
C
R
H
+
−:NH2
Trans alkene

EXERCISE 13-9

When 1,7-undecadiyne (11 carbons) was treated with a mixture of sodium *and* sodium amide in liquid ammonia, only the internal bond was reduced to give *trans*-7-undecen-1-yne. Explain.

Hydroboration of Alkynes

Another way of reducing a triple bond to a *cis* alkene that does not require use of hydrogen is hydroboration. The reaction of an internal alkyne with borane results in an alkenylborane that can be converted into the alkene with ethanoic (acetic) acid. The *cis* stereochemistry of the resulting alkene is controlled by the *syn* stereospecificity of the hydroboration reaction.

BOX 13-1

Synthesis of a Sex Pheromone

An application of the sequential one-electron reduction of alkynes to the synthesis of the sex pheromone (Section 12-9) of the spruce budworm is shown in the scheme below. The starting material, 10-bromo-1-decanol, is initially protected as a *tert*-butyl ether. This step prevents reaction of the hydroxy group with the organolithium reagent in the subsequent step. 1-Butynyllithium then displaces the bromide, the alcohol is deprotected, and the resulting alkynol reduced to the trans alkenol. Oxidation of the alcohol function gives the sex pheromone.

$$\underset{\textbf{10-Bromo-1-decanol}}{HO(CH_2)_{10}Br} \xrightarrow[\text{Protection}]{(CH_3)_2C{=}CH_2,\ H^+} (CH_3)_3CO(CH_2)_{10}Br \xrightarrow[-\,LiBr]{LiC{\equiv}CCH_2CH_3,\ THF,\ HMPA}$$

$$(CH_3)_3CO(CH_2)_{10}C{\equiv}CCH_2CH_3 \xrightarrow[\substack{-\,(CH_3)_2C{=}CH_2\\ \text{Deprotection}}]{H^+} \underset{\textbf{11-Tetradecyn-1-ol}}{HO(CH_2)_{10}C{\equiv}CCH_2CH_3} \xrightarrow[\text{Reduction}]{\text{Na, liquid } NH_3}$$

$$\underset{\textit{trans}\textbf{-11-Tetradecen-1-ol}}{\overset{HO(CH_2)_{10}\qquad H}{\underset{H\qquad CH_2CH_3}{C{=}C}}} \xrightarrow[\text{Oxidation}]{} \underset{\textbf{Sex pheromone of the spruce budworm}}{\overset{\overset{O}{\|}}{HC}(CH_2)_9\ \overset{\qquad H}{\underset{H\qquad CH_2CH_3}{C{=}C}}}$$

Hydroboration-Hydrolysis of Alkynes to Give Cis Alkenes

$$\underset{\textbf{3-Hexyne}}{3\ CH_3CH_2C{\equiv}CCH_2CH_3} \xrightarrow{BH_3,\ 0°C} \underset{\textbf{An alkenylborane}}{\left(\overset{CH_3CH_2\qquad CH_2CH_3}{\underset{H\qquad\quad}{C{=}C}}\right)_3 B} \xrightarrow{3\ CH_3\overset{O}{\overset{\|}{C}}OH}$$

$$\underset{\substack{80\%\\ \textit{cis}\textbf{-3-Hexene}}}{\overset{CH_3CH_2\qquad CH_2CH_3}{\underset{H\qquad\quad H}{C{=}C}}} + B(O\overset{O}{\overset{\|}{C}}CH_3)_3$$

Terminal alkynes are hydroborated in a regiospecific fashion, the boron attacking the less-hindered carbon. However, with borane itself, this reaction leads to hydroboration of both π bonds. To stop at the alkenylborane stage, less-reactive bulky borane reagents, such as di(1,2-dimethylpropyl) borane (diisoamylborane) or dicyclohexylborane, are used. These modified boranes are also remarkably regioselective in the hydroboration of unsymmetrical internal alkynes.

$$CH_3(CH_2)_3C{\equiv}CH + (CH_3)_2CHCH(CH_3)\text{–}BH\text{–}CH(CH_3)CH(CH_3)_2 \longrightarrow$$

Di(1,2-dimethylpropyl)borane (Diisoamylborane)

$$CH_3(CH_2)_3(H)C{=}C(H)B(CH(CH_3)CH(CH_3)_2)_2 \xrightarrow{CH_3COOH} CH_3(CH_2)_3CH{=}CH_2 \quad 90\%$$

$$(CH_3)_2CHC{\equiv}CCH_3 + (C_6H_{11})_2BH \longrightarrow (CH_3)_2CH(H)C{=}C(CH_3)B(C_6H_{11})_2$$

Dicyclohexylborane 92% (+ 8% of other regioisomer)

EXERCISE 13-10

Outline syntheses of *cis*- and *trans*-1-deuterio-1-hexene starting from 1-hexyne.

Similar to alkylboranes (Section 12-4), alkenylboranes can be oxidized to the corresponding alcohols—in this case, enols that spontaneously rearrange to the isomeric carbonyl compounds (Sections 12-8 and 16-2).

Hydroboration-Oxidation of Alkynes

$$CH_3CH_2C{\equiv}CCH_2CH_3 \xrightarrow[2.\ H_2O_2,\ HO^-]{1.\ BH_3} \left[CH_3CH_2(H)C{=}C(OH)CH_2CH_3\right] \xrightarrow{\text{Tautomerism}} CH_3CH_2CH_2\overset{O}{\overset{\|}{C}}CH_2CH_3$$

3-Hexyne — **Enol** — **3-Hexanone** 68%

1. Diisoamylborane
2. H_2O_2, HO^-

Tautomerism

OH

Enol

O

H H H

61%

Such a rearrangement is called **tautomerism;** the enol is said to *tautomerize* to the carbonyl compound, and the two species are referred to as *tautomers* (*tauto,* Greek, the same; *meros,* Greek, part). The term tautomerism is generally used when two isomers interconvert by the simultaneous shift of a proton and a double bond. In this way, internal alkynes are converted into ketones, whereas terminal alkynes give aldehydes.

EXERCISE 13-11

Outline a synthesis of the molecule in the margin from 3,3-dimethyl-1-butyne.

$$(CH_3)_3CCH_2CH{=}O$$

Borane is not the only electrophilic reagent that attacks triple bonds. Alkynes are also substrates for protonation, halogenation, and mercuration.

Protonation, Halogenation, and Mercuration of Alkynes

As a center of high electron density, the triple bond can be protonated readily. For example, hydrogen bromide added to 2-butyne yields the corresponding bromoalkene.

$$CH_3C{\equiv}CCH_3 \xrightarrow{HBr,\ Br^-} \underset{\substack{60\% \\ \textbf{(Z)-2-Bromobutene}}}{(H)(H_3C)C{=}C(CH_3)(Br)}$$

The stereochemistry of this type of addition is frequently (but not always) *anti*, particularly in the presence of excess bromide ion. Another molecule of hydrogen bromide added to the bromoalkene gives the geminal dihaloalkane with regioselectivity, in accord with Markovnikov's rule.

$$(H)(H_3C)C{=}C(CH_3)(Br) \xrightarrow{HBr} \underset{\substack{90\% \\ \textbf{2,2-Dibromobutane}}}{CH_3CHCBr_2CH_3}$$

The addition of hydrogen halides to terminal alkynes also proceeds in accord with Markovnikov's rule.

$$CH_3C{\equiv}CH \xrightarrow{HI,\ -70^\circ C} \underset{35\%}{(I)(H_3C)C{=}CH_2} + \underset{65\%}{CH_3CI_2{-}CH_2{-}H}$$

The mechanism of addition consists of initial protonation to form an alkenyl cation, followed by trapping by the counterion.

Mechanism of Hydrohalogenation of Alkynes:

$$RC{\equiv}CR \xrightarrow{H^+} \underset{\textbf{Alkenyl cation}}{R{-}\overset{+}{C}{=}C(H)(R)} \xrightarrow{X^-} (R)(X)C{=}C(H)(R) + (X)(R)C{=}C(H)(R)$$

EXERCISE 13-12

Reaction of 2-bromo-3-methyl-2-butene with hydrogen bromide gave only 2,2-dibromo-3-methylbutane. Explain. (Hint: Invoke resonance.)

As with alkenes, hydrogen bromide can add to triple bonds by a free-radical mechanism in an anti-Markovnikov fashion if light or other radical initiators are present. Both *syn* and *anti* additions are observed.

$$CH_3(CH_2)_3C{\equiv}CH \xrightarrow{HBr,\ ROOR} CH_3(CH_2)_3CH{=}CHBr \quad 74\%$$

1-Hexyne → ***cis-* and *trans*-1-Bromo-1-hexene**

$$(Br)(H_3C)C{=}CH_2 \xrightarrow{HBr,\ ROOR} CH_3CHBrCH_2Br \quad 66\%$$

2-Bromopropene → **1,2-Dibromopropane**

Electrophilic addition of halogen to alkynes proceeds through the intermediacy of isolable vicinal dihaloalkenes to eventually give tetrahaloalkanes.

$$CH_3CH_2C{\equiv}CCH_2CH_3 \xrightarrow{Br_2,\ CH_3COOH,\ LiBr} (CH_3CH_2)(Br)C{=}C(Br)(CH_2CH_3) \;(99\%) \xrightarrow{Br_2,\ CCl_4} CH_3CH_2CBr_2{-}CBr_2CH_2CH_3 \;(95\%)$$

3-Hexyne → **(*E*)-2,3-Dibromo-3-hexene** → **3,3,4,4-Tetrabromohexane**

$$CH_3CH_2C{\equiv}CH \xrightarrow{Cl_2,\ CCl_4} (CH_3CH_2)(Cl)C{=}C(Cl)(H) \;(\text{Mainly}) \xrightarrow{Cl_2,\ CCl_4} CH_3CH_2CCl_2{-}CHCl_2$$

(*E*)-1,2-Dichloro-1-butene → **1,1,2,2-Tetrachlorobutane**

In analogy to the hydration of alkenes, water can be added to alkynes in a Markovnikov sense to give ketones by tautomerism of the intermediate enols. The reaction is catalyzed by mercuric ions.

General Hydration of Alkynes

$$RC{\equiv}CR \xrightarrow{HOH,\ H^+,\ HgSO_4} RCH{=}C(OH)R \longrightarrow RCH_2{-}C(=O)R$$

Enol → **Ketone**

Symmetrical internal alkynes give only one carbonyl compound; unsymmetrical systems lead to mixtures of products.

$$CH_3CH_2C{\equiv}CCH_2CH_3 \xrightarrow{H_2SO_4,\ H_2O,\ HgSO_4} CH_3CH_2C(=O)CH_2CH_2CH_3 \quad 80\%$$

Only possible product

$$CH_3CH_2CH_2C{\equiv}CCH_3 \xrightarrow{H_2SO_4,\ H_2O,\ HgSO_4} \underset{50\%}{CH_3CH_2CH_2\overset{\overset{O}{\|}}{C}CH_2CH_3} + \underset{50\%}{CH_3CH_2CH_2CH_2\overset{\overset{O}{\|}}{C}CH_3}$$

Terminal alkynes furnish methyl ketones, and only ethyne gives an aldehyde—namely, ethanal (acetaldehyde).

$$\text{1-ethynylcyclohexanol} \xrightarrow{H_2SO_4,\ H_2O,\ HgSO_4} \underset{91\%}{\text{1-acetylcyclohexanol}}$$

$$HC{\equiv}CH \xrightarrow{H_2SO_4,\ H_2O,\ HgSO_4} \underset{H}{\overset{H}{}}\!\!>C{=}C<\!\!\underset{H}{\overset{OH}{}} \longrightarrow CH_3\overset{\overset{O}{\|}}{C}H$$

Ethanal
(Acetaldehyde)

Note that the regioselectivity of this hydration is opposite and hence complementary to that of the hydroboration-oxidation sequence.

EXERCISE 13-13

Propose a synthetic scheme that will convert compound A into B.

$$\underset{\textbf{A}}{CH_3\overset{\overset{O}{\|}}{C}CH_3} \dashrightarrow \underset{\textbf{B}}{CH_3\overset{\overset{OH}{|}}{\underset{\underset{CH_3}{|}}{C}}{-}\overset{\overset{O}{\|}}{C}CH_3}$$

Carbocations Attack Triple Bonds: Cationic Cyclizations

Triple bonds undergo electrophilic attack by carbocations. This reaction is sometimes preparatively useful, particularly if it is intramolecular. For example, heating the 4-methylbenzenesulfonate (tosylate) of 6-heptyn-2-ol in aqueous acid leads to 3-methylcyclohexanone.

$$CH_3\overset{\overset{OSO_2C_6H_4CH_3\text{-}p}{|}}{C}HCH_2CH_2CH_2C{\equiv}CH \xrightarrow{H^+,\ H_2O} \underset{91\%}{\text{3-methylcyclohexanone}}$$

3-Methylcyclohexanone

How does this remarkable transformation occur? Initially, a secondary carbocation is generated by ionization of the sulfonate as in an S_N1 reaction. The electrophilic carbon then attacks the electron-rich triple bond in Markovnikov fashion to furnish an alkenyl cation. This alkenyl cation is in turn trapped by water to give an intermediate enol, which rearranges to the ketone.

Mechanism:

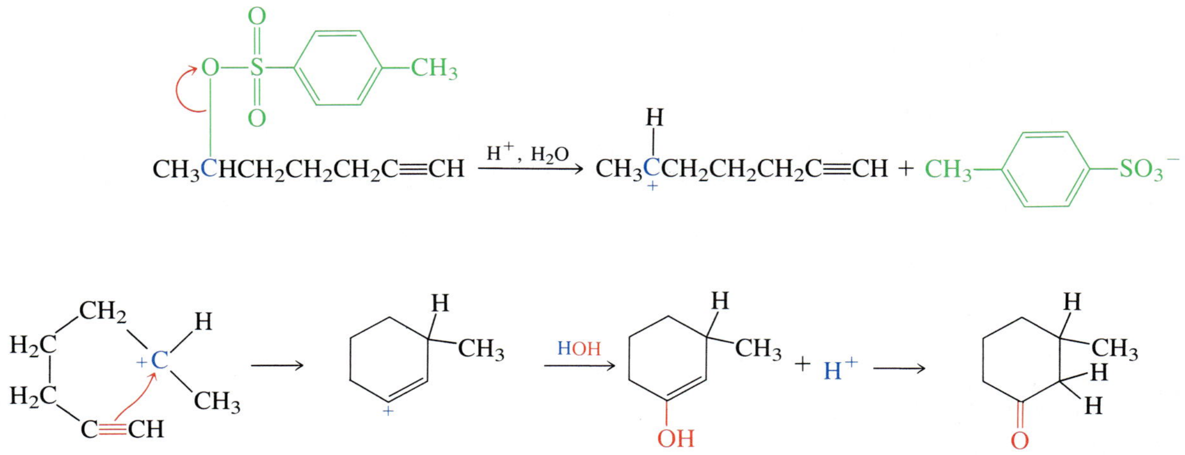

$$CH_3CH(OSO_2C_6H_4CH_3)CH_2CH_2CH_2C{\equiv}CH \xrightarrow{H^+,\ H_2O} CH_3\overset{+}{C}HCH_2CH_2CH_2C{\equiv}CH + CH_3C_6H_4SO_3^-$$

Cyclic alkenyl cation → (HOH) enol + H^+ → **3-Methylcyclohexanone**

Carbocations can also initiate polymerization of alkynes leading to polyalkynes. An unusual property of polyethyne (polyacetylene) is that it becomes a conductor, like a metal, when exposed to one-electron oxidants. Such treatment is called *doping*. Doped polyethyne is a sufficiently good conductor to be usable in rechargeable batteries.

Polymerization of Ethyne

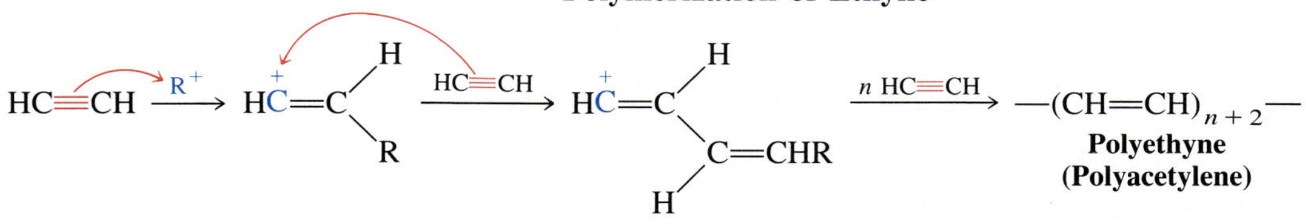

$$HC{\equiv}CH \xrightarrow{R^+} H\overset{+}{C}{=}C(H)R \xrightarrow{HC\equiv CH} H\overset{+}{C}{=}C(H)\text{–}C(H){=}CHR \xrightarrow{n\ HC\equiv CH} \text{—}(CH{=}CH)_{n+2}\text{—}$$

Polyethyne (Polyacetylene)

EXERCISE 13-14

Propose a mechanism for the following reaction:

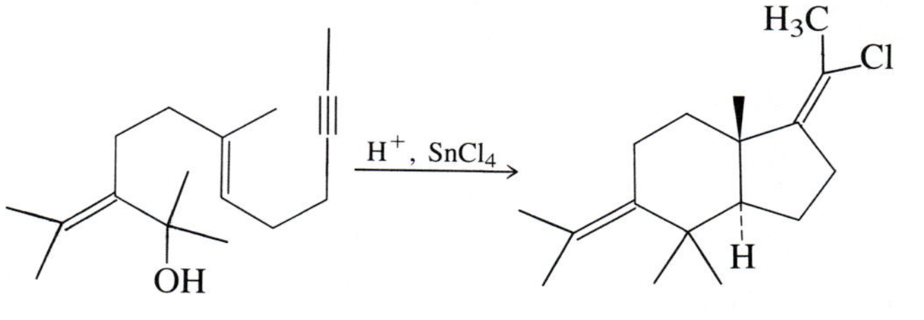

$$\xrightarrow{H^+,\ SnCl_4}$$

Oxidation of Alkynes: Activation of the Terminal C–H Bond

The oxidation of alkynes—for example, with a peroxycarboxylic acid—is not as straightforward as that of alkenes; alkyne oxidations generally give complex mixtures. Under controlled reaction conditions, carbon–carbon double bonds can be selectively oxidized in the presence of triple bonds:

$\xrightarrow{\text{MCPBA (1 equivalent)}}$

> 70%

A unique type of reactivity is observed when terminal alkynes are in the presence of a cuprous salt in an amine solvent with added oxygen. The two alkynyl hydrogens are removed with simultaneous coupling of the two triple bonds to give diynes. This reaction is called **oxidative coupling.**

Oxidative Coupling of Terminal Alkynes

$$RC{\equiv}CH + HC{\equiv}CR \xrightarrow{Cu^+,\ \text{amine},\ O_2} RC{\equiv}C{-}C{\equiv}CR + H_2O$$

A diyne

For example, 1-ethynylcyclohexanol can be coupled in the presence of cuprous chloride, *N,N,N′,N′*-tetramethyl-1,2-ethanediamine (TMEDA), and oxygen to furnish the coupled product in 93% yield.

OH, C≡CH $\xrightarrow{CuCl,\ (CH_3)_2NCH_2CH_2N(CH_3)_2,\ O_2,\ 28°C}$ OH, HO, C≡C—C≡C

1-Ethynylcyclohexanol

93%

Copper(II) is in fact the oxidant in these reactions, the oxygen serving to regenerate it from Cu(I).

Redox Reactivity of Copper in the Oxidative Coupling of Alkynes

$$2\ RC{\equiv}CH + 2\ Cu(II) \longrightarrow RC{\equiv}CC{\equiv}CR + 2\ H^+ + 2\ Cu(I)$$
$$2\ Cu(I) + \tfrac{1}{2}\ O_2 + 2\ H^+ \longrightarrow 2\ Cu(II) + H_2O$$

The method can be applied in an intramolecular fashion to give large rings, also called macrocycles (*makros,* Greek, long).

OH $\xrightarrow{CuCl,\ NH_4Cl,\ O_2,\ CH_3CH_2OH,\ 60°C}$ OH

40%

Similarly, diynes can be cyclodimerized, -trimerized, and -oligomerized to give analogous large-ring products. These macrocycles have been rearranged with the use of a strong base by Sondheimer* and his coworkers into cyclic polyenes called *annulenes* (Section 25-5).

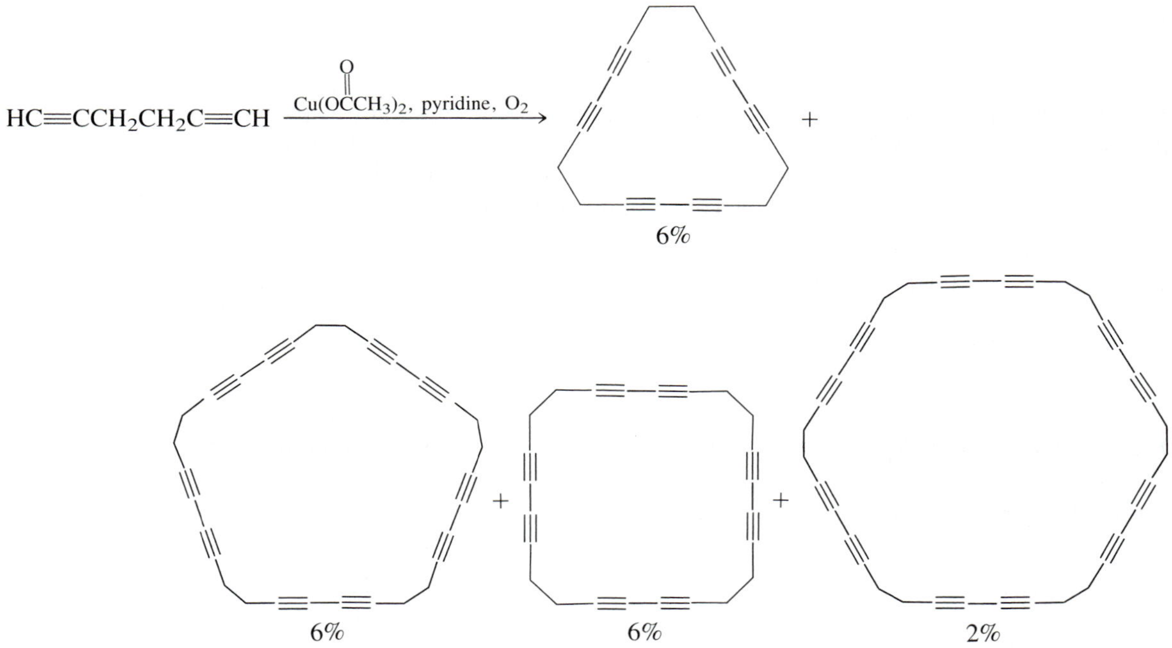

$$HC{\equiv}CCH_2CH_2C{\equiv}CH \xrightarrow{Cu(O\overset{O}{\overset{\|}{C}}CH_3)_2,\ pyridine,\ O_2}$$ +

6%

6% + 6% + 2%

EXERCISE 13-15

Propose a four-step synthesis of 1,8-octanediol from ethyne and other necessary reagents.

In summary, alkynes are very similar in reactivity to alkenes, except that they have two π bonds, both of which may be saturated by addition reactions. Hydrogenation of the first π bond, which gives cis alkenes, is more exothermic than that of the second. Alkynes are converted into trans alkenes by one-electron reduction with sodium in liquid ammonia. Electrophilic additions by boranes afford another pathway to cis alkenes by hydrolysis of the resulting alkenylboranes. Hydroboration-oxidation leads to anti-Markovnikov hydration and eventually to ketones or aldehydes by tautomerism of the corresponding enols. Reaction with electrophiles, such as protons, halogens, and mercuric ion, is in accord with Markovnikov's rule and gives products of *syn* and *anti* addition. Hydration catalyzed by mercuric ion results in ketones by Markovnikov addition. Free-radical addition of hydrogen bromide to terminal alkynes gives 1-bromoalkenes. The triple bond is subject to attack by carbocations. In intramolecular cases, this reaction gives rise to cyclic products; intermolecular attack may lead to polymers. Alkenes can be selectively oxidized in the presence of alkynes to give oxacyclopropanes. Oxidative coupling of terminal alkynes with cupric ion and oxygen is a unique carbon–carbon bond-forming reaction of terminal alkynes.

---

*Professor Franz Sondheimer, 1926–1981, University College, London.

13-7
Ethyne as an Industrial Starting Material

Ethyne was once one of the four or five major starting materials in the chemical industry for two reasons: addition reactions to one of the π bonds produces useful alkene monomers (Section 12-7), and it has a high heat content. Its industrial use has declined because of the availability of cheap ethene, propene, butadiene, and other hydrocarbons through oil-based technology. However, after the year 2000, oil reserves are expected to have dwindled to the point that other sources of energy will have to be developed. One such source is coal. There are currently no known processes for converting coal directly into the alkenes mentioned above; ethyne, however, can be produced from coal and hydrogen or from coke (a coal residue obtained after removal of volatiles) and limestone through the formation of calcium carbide. Consequently, it may once again become an important industrial raw material.

The Production of Ethyne

The high energy content of ethyne requires the use of methods that are costly in energy for its production. One process for making ethyne from coal uses hydrogen in an arc reactor at temperatures as high as several thousand degrees.

$$\text{Coal} + H_2 \xrightarrow{\Delta} \underset{33\%\ \text{conversion}}{HC{\equiv}CH} + \text{nonvolatile salts}$$

The oldest large-scale preparation of ethyne proceeds through calcium carbide. Limestone (calcium oxide) and coke are heated to about 2000°C, which results in the desired product and carbon monoxide.

$$\underset{\textbf{Coke}}{3\ C} + \underset{\textbf{Lime}}{CaO} \xrightarrow{2000^{\circ}C} \underset{\textbf{Calcium carbide}}{CaC_2} + CO$$

The calcium carbide is then treated with water at ambient temperatures to give ethyne and calcium hydroxide.

$$CaC_2 + 2\ H_2O \longrightarrow HC{\equiv}CH + Ca(OH)_2$$

Industrial Chemistry with Ethyne

Ethyne chemistry underwent important commercial development in the 1930s and 1940s in the laboratories of the Badische Anilin and Sodafabriken (BASF) in Ludwigshafen, Germany. Ethyne under pressure was brought into reaction with carbon monoxide, carbonyl compounds, alcohols, and acids in the presence of catalysts to give a multitude of valuable raw materials to be used in further transformations. For example, nickel carbonyl catalyzes the addition of carbon monoxide and water to ethyne to give propenoic (acrylic) acid. Similar exposure to alcohols or amines instead of water results in the corresponding acid derivatives. All of the products are valuable monomers (see Section 12-7). The addition of methanal (formaldehyde) to ethyne is achieved with high efficiency using copper acetylide as a catalyst.

Industrial Chemistry of Ethyne

Carbonylation:

$$HC{\equiv}CH + CO + H_2O \xrightarrow{Ni(CO)_4,\ 100\text{ atm},\ >250°C} H_2C{=}CHCOOH$$

Propenoic acid (Acrylic acid)

$$HC{\equiv}CH + CO + CH_3OH \xrightarrow{Ni(CO)_4,\ \Delta} H_2C{=}CH\overset{O}{\overset{\|}{C}}OCH_3$$

80%
Methyl propenoate (Methyl acrylate)

$$HC{\equiv}CH + CO + RNH_2 \xrightarrow{Ni(CO)_4,\ \Delta} H_2C{=}CH\overset{O}{\overset{\|}{C}}NHR$$

A propenamide (An acrylamide)

Methanal (formaldehyde) addition:

$$HC{\equiv}CH + CH_2{=}O \xrightarrow{Cu_2C_2\text{–}SiO_2,\ 125°C,\ 5\text{ atm}} HC{\equiv}CCH_2OH \text{ or } HOCH_2C{\equiv}CCH_2OH$$

2-Propyn-1-ol (Propargyl alcohol) — **2-Butyne-1,4-diol**

The resulting alcohols are useful synthetic intermediates. For example, 2-butyn-1,4-diol is a precursor for the production of oxacyclopentane (tetrahydrofuran) by hydrogenation, followed by acid-catalyzed dehydration.

Oxacyclopentane (Tetrahydrofuran) Synthesis

$$HOCH_2C{\equiv}CCH_2OH \xrightarrow{\text{Catalyst},\ H_2} HO(CH_2)_4OH \xrightarrow[260°\text{–}280°C,\ 90\text{–}100\text{ atm}]{H_3PO_4,\ pH\ 2,} \text{(cyclic } C_4H_8O) + H_2O$$

99%
Oxacyclopentane (Tetrahydrofuran, THF)

Ethyne can also be cyclooligomerized to its trimer, benzene, or to its tetramer, 1,3,5,7-cyclooctatetraene (see Section 25-5).

$$3\ HC{\equiv}CH \xrightarrow{Ni(CN)_2,\ P(C_6H_5)_3,\ THF,\ 70°C} C_6H_6$$

80%
Benzene

$$4\ HC{\equiv}CH \xrightarrow{Ni(CN)_2,\ THF,\ 70°C,\ 12–25\ atm}$$

70%
1,3,5,7-Cyclooctatetraene

Catalyzed Nucleophilic Addition Reactions

A number of technical processes have been developed in which reagents $\overset{\delta^+}{A}{-}\overset{\delta^-}{B}$ in the presence of a catalyst add to the triple bond. For example, ethyne can be hydrated (see Section 13-6) to ethanal [acetaldehyde; this production of ethanal became obsolete with the advent of the Wacker Process (see Section 12-8)]. Similarly, the catalyzed addition of ethanoic (acetic) acid furnishes vinyl ethanoate (acetate) and that of hydrogen chloride gives chloroethene (vinyl chloride). Chlorination results in tetrachloroethane, and addition of hydrogen cyanide produces propenenitrile (acrylonitrile).

Addition Reactions to Ethyne

$$HC{\equiv}CH + H_2O \xrightarrow{H_2SO_4,\ HgSO_4,\ FeSO_4,\ 95°C,\ 2\ atm} CH_3\overset{O}{\overset{\|}{C}}H$$

95%
Ethanal
(Acetaldehyde)

$$HC{\equiv}CH + CH_3\overset{O}{\overset{\|}{C}}OH \xrightarrow{Hg^{2+}} H_2C{=}CH{-}O\overset{O}{\overset{\|}{C}}CH_3$$

70%
Ethenyl ethanoate
(Vinyl acetate)

$$HC{\equiv}CH + HCl \xrightarrow{Hg^{2+},\ 100°–200°C} H_2C{=}CHCl$$

Chloroethene
(Vinyl chloride)

$$HC{\equiv}CH + HCN \xrightarrow{Cu^+,\ NH_4Cl,\ 70°–90°C,\ 1.3\ atm} H_2C{=}CHCN$$

80%–90%
Propenenitrile
(Acrylonitrile)

EXERCISE 13-16

Formulate a plausible mechanism for the hydration of ethyne in the presence of mercuric chloride. (Hint: Review the hydration of alkenes catalyzed by mercuric ion, Section 12-3.)

In summary, ethyne was once, and may again be in the future, a valuable industrial feedstock because of its ability to react with a large number of substrates to yield useful monomers and other compounds having functional groups. It can be made from coal and hydrogen at high temperatures or it can be liberated from calcium carbide by hydrolysis. Some of the industrial reactions that it undergoes are carbonylation, addition of methanal (formaldehyde), cyclooligomerization, and addition reactions with H_2O, HX, and X_2.

13-8 Naturally Occurring and Physiologically Active Alkynes

This section deals with the triple bond in natural products and compounds that have medicinal application.

Although alkynes are not very abundant in nature, they do exist in some plants and other organisms. The first such substance to be isolated, in 1826, was dehydromatricaria ester.

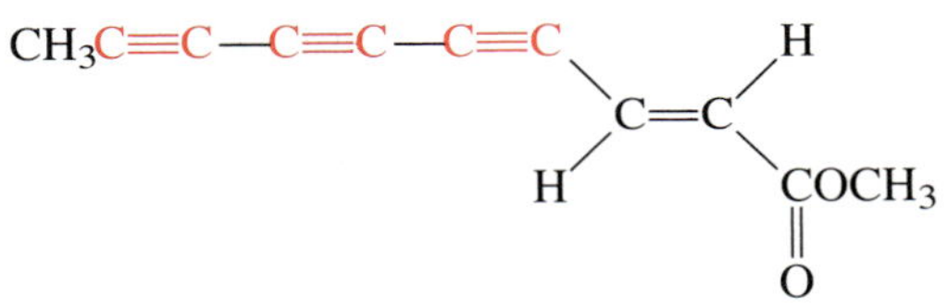

$CH_3C{\equiv}C{-}C{\equiv}C{-}C{\equiv}C{-}CH{=}CH{-}COCH_3$ (with C=O)

Dehydromatricaria ester

$CH_3C{\equiv}C{-}C{\equiv}C{-}C(=O){-}C_6H_5$

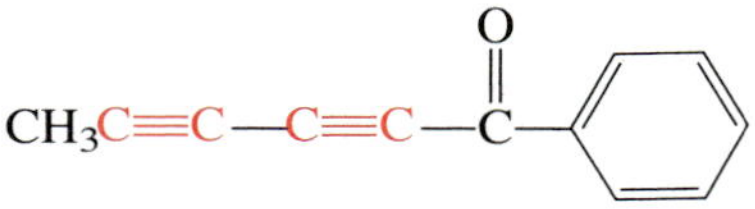

Capillin
(Active against skin fungi)

More than a thousand such compounds are now known, and some of them are physiologically active. For example, some naturally occurring ethynylketones, such as capillin, have fungicidal activity. The alkyne ichthyothereol is the active ingredient of a poisonous substance used by the Indians of the Lower Amazon Basin in arrowheads. It causes convulsions in mammals. Two enyne functional groups are incorporated in the compound hystrionicotoxin. It is one of the substances isolated from the skin of "arrow poison frog," a highly colorful species of the genus *Dendrobates*. The frog secretes this compound and similar ones as defensive venoms and mucosal-tissue irritants against both mammals and reptiles. How the alkyne units are constructed biosynthetically and their exact function is not clear.

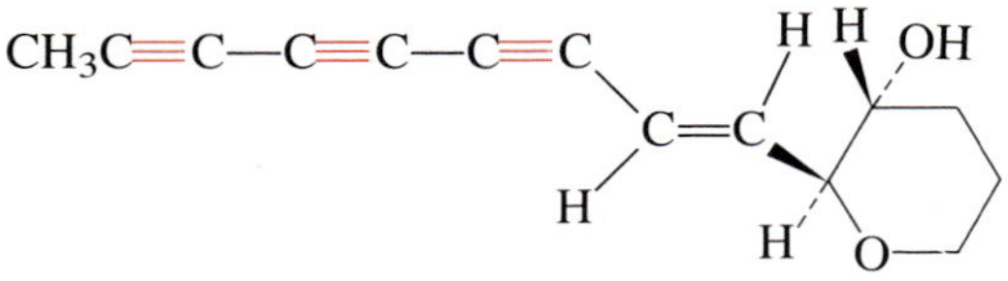

$CH_3C{\equiv}C{-}C{\equiv}C{-}C{\equiv}C{-}CH{=}CH{-}$(3-hydroxytetrahydropyran-2-yl)

Ichthyothereol
(A convulsant)

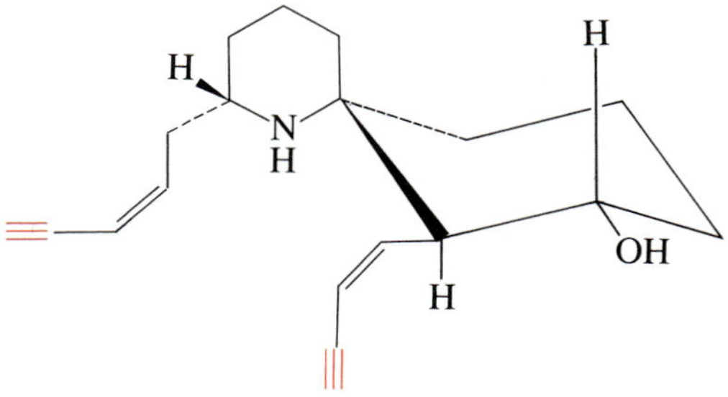

Hystrionicotoxin

Many drugs have been modified by synthesis to contain alkyne units, because such compounds are frequently more readily absorbed by the body, less toxic, and more active than the corresponding alkenes or alkanes. For example, 3-methyl-1-pentyn-3-ol is available as a nonprescription hypnotic, and several other alkynols are similarly effective. Certain chlorinated alkynylamines have been found to be anticancer agents. This activity is, however, thought to be due to their alkylating ability, not to the presence of the triple bond.

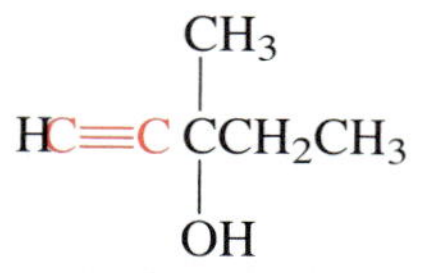

$$HC{\equiv}C\overset{\overset{CH_3}{|}}{\underset{\underset{OH}{|}}{C}}CH_2CH_3$$

3-Methyl-1-pentyn-3-ol
(Hypnotic)

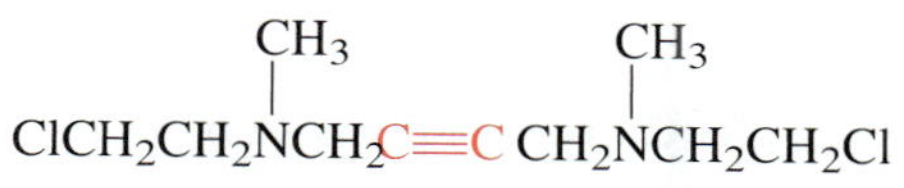

$$ClCH_2CH_2\overset{\overset{CH_3}{|}}{N}CH_2C{\equiv}CCH_2\overset{\overset{CH_3}{|}}{N}CH_2CH_2Cl$$

Anticancer agent

EXERCISE 13-17

Formulate a mechanism for the alkylating reactivity of the anticancer agent shown above. (Hint: Consult Exercise 9-18).

Ethynyl estrogens, such as 17-ethynylestradiol, are considerably more potent birth control agents than are the naturally occurring hormones (see Section 4-7). The diaminoalkyne tremorine induces symptoms characteristic of Parkinson's disease: spasms of uncontrolled movement. Interestingly, a simple cyclic homolog of tremorine acts as a muscle relaxant and counteracts the effect of tremorine. Compounds that cancel the physiological effects of other compounds are called *antagonists* (*antagonizesthai,* Greek, to struggle against). Finally, ethynyl analogs of amphetamine have been prepared in a search for alternative, more-active, more-specific, and less-addictive central-nervous-system stimulants.

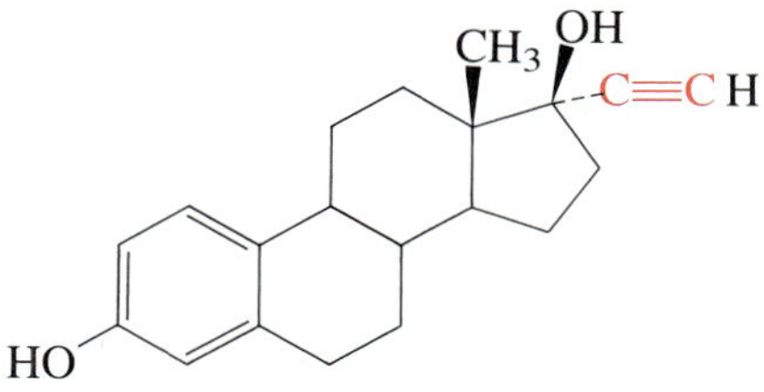

CH₃ OH, C≡CH, HO

17-Ethynylestradiol

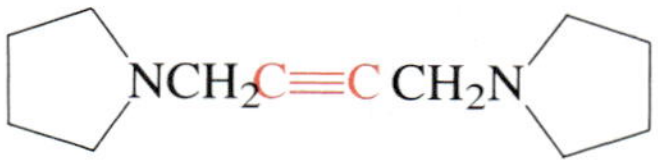

$$NCH_2C{\equiv}CCH_2N$$

Tremorine

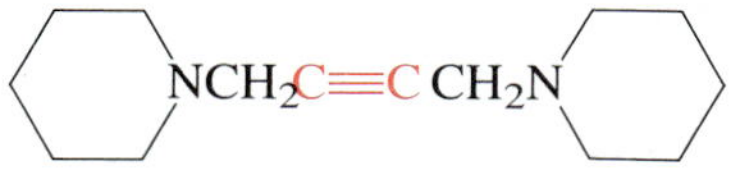

$$NCH_2C{\equiv}CCH_2N$$

Tremorine antagonist

$$-CH_2\overset{\overset{NH_2}{|}}{C}HC{\equiv}CH$$

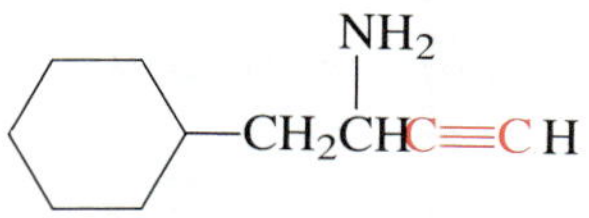

An amphetamine analog
(Active in the central nervous system)

In summary, there are a number of natural and synthetic compounds in which the alkyne unit is a necessary or auxiliary component for some type of physiological activity.

Summary of New Reactions

1 ACIDITY OF 1-ALKYNES

$$\underset{pK_a \sim 25}{RC{\equiv}CH} + :B^- \rightleftharpoons RC{\equiv}C:^- + BH$$

Base: $NaNH_2$-liquid NH_3, RLi-$(CH_3CH_2)_2O$, RMgX-THF

Conversion of Alkynes into Other Alkynes and Allenes

2 BASE-MEDIATED ISOMERIZATION

$$\underset{\textbf{1-Alkyne}}{RCH_2C{\equiv}CH} \xrightarrow{KOH,\ CH_3CH_2OH,\ \Delta} \underset{\textbf{2-Alkyne}}{RC{\equiv}CCH_3} \qquad \Delta H^\circ \sim -4 \text{ kcal mole}^{-1}$$

$$\underset{\textbf{1-Alkyne}}{R_2CHC{\equiv}CH} \xrightarrow{KOH,\ CH_3CH_2OH} \underset{\textbf{Allene}}{R_2C{=}C{=}CH_2}$$

3 ALKYLATION OF ALKYNYL ANIONS

$$RC{\equiv}CH \xrightarrow[2.\ R'X]{1.\ NaNH_2,\ \text{liquid } NH_3} RC{\equiv}CR'$$

$R' =$ primary

4 ALKYLATION WITH OXACYCLOPROPANE

$$RC{\equiv}CH \xrightarrow[2.\ H_2C\overset{O}{—}CH_2]{1.\ CH_3CH_2CH_2CH_2Li,\ THF} RC{\equiv}CCH_2CH_2OH$$

5 ALKYLATION WITH CARBONYL COMPOUNDS

$$RC{\equiv}CH \xrightarrow[2.\ R'\overset{O}{\overset{\|}{C}}R'']{1.\ \text{Base, THF}} RC{\equiv}C\underset{R''}{\overset{OH}{C}}R'$$

Preparation of Alkynes

6 ELIMINATION FROM DIHALOALKANES

$$\underset{\textbf{Vicinal dihaloalkane}}{R\overset{X}{\underset{H}{C}}—\overset{X}{\underset{H}{C}}R} \xrightarrow[-\ 2\ HX]{NaNH_2,\ \text{liquid } NH_3} RC{\equiv}CR$$

$$\underset{\text{Geminal dihaloalkane}}{RC(X)_2CH_2R} \xrightarrow[-\ 2\ HX]{NaNH_2,\ \text{liquid}\ NH_3} RC{\equiv}CR$$

Geminal dihaloalkane

7 FROM ALKENES BY HALOGENATION-DEHYDROHALOGENATION

$$RCH{=}CHR \xrightarrow{1.\ X_2,\ CCl_4;\ 2.\ NaNH_2,\ \text{liquid}\ NH_3} RCH{=}C(R)X \xrightarrow{NaNH_2,\ \text{liquid}\ NH_3} RC{\equiv}CR$$

Alkenyl halide intermediate

8 STEREOSPECIFIC ALKENYL HALIDE FORMATION

$$\underset{cis\text{-Alkene}}{(H)(R)C{=}C(H)(R)} \xrightarrow{1.\ Br_2,\ CCl_4;\ 2.\ \text{Base}} \underset{(Z)\text{-Alkenyl bromide}}{(R)(Br)C{=}C(H)(R)}$$

cis-Alkene → (*Z*)-Alkenyl bromide

$$\underset{trans\text{-Alkene}}{(H)(R)C{=}C(R)(H)} \xrightarrow{1.\ Br_2,\ CCl_4;\ 2.\ \text{Base}} \underset{(E)\text{-Alkenyl bromide}}{(R)(Br)C{=}C(R)(H)}$$

trans-Alkene → (*E*)-Alkenyl bromide

9 ALKENYL ORGANOMETALLICS

$$(R)(R')C{=}C(X)(R'') \xrightarrow{Mg,\ THF} (R)(R')C{=}C(MgX)(R'')$$

$$(R)(R')C{=}C(X)(R'') \xrightarrow{CH_3CH_2CH_2CH_2Li,\ THF} (R)(R')C{=}C(Li)(R'')$$

Reactions of Alkynes

10 HYDROGENATION

$$RC{\equiv}CR \xrightarrow{\text{Catalyst},\ H_2} RCH_2CH_2R \qquad \Delta H^\circ \sim -70\ \text{kcal mole}^{-1}$$

$$RC{\equiv}CR \xrightarrow{H_2,\ \text{Lindlar's catalyst}} (H)(R)C{=}C(H)(R) \qquad \Delta H^\circ \sim -40\ \text{kcal mole}^{-1}$$

Cis alkene

11 REDUCTION WITH SODIUM IN LIQUID AMMONIA

$$RC{\equiv}CR \xrightarrow[\text{2. }H^+,\ H_2O]{\text{1. Na, liquid }NH_3} \text{(H)(R)C=C(R)(H)}$$

Trans alkene

12 HYDROBORATION

$$RC{\equiv}CR \xrightarrow{BH_3} \text{(H)(R)C=C(B—)(R)}$$

$$RC{\equiv}CH \xrightarrow[\text{Diisoamyl- }[R' = (CH_3)_2CHCHCH_3]\text{ or dicyclohexylborane }(R' = C_6H_{11}\text{—})]{R'_2BH} \text{(R)(H)C=C(H)(BR}'_2)$$

13 REACTIONS OF ALKENYLBORANES

Conversion to alkenes:

$$\text{(H)(R)C=C(B—)(R)} \xrightarrow{CH_3\overset{O}{\overset{\|}{C}}OH} \text{(H)(R)C=C(H)(R)}$$

Cis alkene

Oxidation:

$$\text{(H)(R)C=C(B—)(R)} \xrightarrow{H_2O_2,\ HO^-} \left[\text{(H)(R)C=C(OH)(R)}\right] \xrightarrow[\text{Tautomerism}]{} RCH_2\overset{O}{\overset{\|}{C}}R$$

Enol

14 ELECTROPHILIC ADDITIONS

$$RC{\equiv}CR \xrightarrow{HX} RCH{=}CXR \xrightarrow{HX} RCH_2CX_2R$$

Geminal dihaloalkane

$$RC{\equiv}CH \xrightarrow{2\ HX} RCX_2CH_3$$

Markovnikov addition:

$$RC{\equiv}CR \xrightarrow{Br_2,\ Br^-} \text{(R)(Br)C=C(Br)(R)} \xrightarrow{Br_2} RCBr_2CBr_2R$$

Mainly trans

$$RC{\equiv}CR \xrightarrow{Hg^{2+},\ H_2O} RCH_2\overset{O}{\overset{\|}{C}}R$$

15 FREE-RADICAL ADDITION OF HYDROGEN BROMIDE

$$RC{\equiv}CH \xrightarrow{HBr,\ ROOR} RCH{=}CHBr$$

16 CATIONIC CYCLIZATION AND POLYMERIZATION

Cyclization:

$$\text{(cyclic) } RCH(O{-}SO_2{-}C_6H_4{-}CH_3)\text{ tethered to }C{\equiv}CH \xrightarrow{H^+,\ H_2O} \left[\text{ring }\overset{+}{C}{=}CH{-}CHR\right] \xrightarrow{H_2O} \text{ring } C({=}O){-}CH_2{-}CHR$$

Polymerization:

$$n\ RC{\equiv}CR \xrightarrow{Catalyst} {-}(RC{=}CR)_n{-}$$

Polyalkyne

17 OXIDATION

$$2\ RC{\equiv}CH \xrightarrow[\text{Oxidative coupling}]{Cu^+,\ amine,\ O_2} RC{\equiv}C{-}C{\equiv}CR$$

18 INDUSTRIAL PREPARATION AND USES OF ETHYNE

Preparation:

Directly from coal + H_2, Δ; or from coke + CaO $\longrightarrow$ $CaC_2 \xrightarrow{H_2O} HC{\equiv}CH$

Combustion (acetylene torch):

$$HC{\equiv}CH + 2.5\ O_2 \longrightarrow 2\ CO_2 + H_2O \qquad \Delta H^\circ = -317\ \text{kcal mole}^{-1}$$

Industrial chemistry:

$$C_2H_2 + CO + H_2O \xrightarrow{Ni(CO)_4} CH_2{=}CHCO_2H$$

$$C_2H_2 + CH_2O \xrightarrow{Cu_2C_2} HOCH_2C{\equiv}CH + HOCH_2C{\equiv}CCH_2OH$$

$$3\ C_2H_2 \xrightarrow{Ni(CN)_2,\ P(C_6H_5)_3} \text{benzene}$$

$$4\ C_2H_2 \xrightarrow{Ni(CN)_2} \text{cyclooctatetraene}$$

Additions:

Catalyzed by transition metal cations; addition of H_2O, $CH_3\overset{O}{\overset{\|}{C}}OH$, HX, X_2, HCN

Summary of Important Concepts

1 The rules for naming alkynes are essentially the same as those formulated for alkenes. The triple bond has priority over the double bond. Hydroxy groups have priority over both types of functions.

2 The electronic structure of the triple bond reveals two π bonds, perpendicular to each other, and a σ bond, formed by two overlapping *sp* hybrid orbitals. The strength of the triple bond is about 200 kcal $mole^{-1}$; that of the alkynyl C–H bond is 128 kcal $mole^{-1}$. Triple bonds form linear structures with respect to other attached atoms, with short C–C (1.20 Å) and C–H (1.06 Å) bonds.

3 The high *s* character of the terminal alkyne carbon makes the bound hydrogen relatively acidic ($pK_a \sim 25$).

4 The chemical shift of the alkynyl hydrogen is low ($\delta = 1.7$–3.1 ppm) compared with that of alkenyl hydrogens because of an induced electron current around the molecular axis caused by the external magnetic field. The shielding effect of this current cancels the deshielding effect of the "normal" π-bond current. The triple bond allows for long-range coupling.

5 Hydrogenation of the first π bond in an alkyne releases more heat than that of the second.

6 The smaller cyclic alkynes are strained because the normally linear arrangement around the functional group is distorted by bending.

7 Although the hydrogens bound to the carbon atoms adjacent to a triple bond are less acidic ($pK_a \sim 35$) than alkynyl hydrogens, they can be removed by bases to generate small equilibrium concentrations of the corresponding anions.

8 Allenes are slightly lower in energy (1–2 kcal $mole^{-1}$) than the isomeric terminal alkynes, but higher in energy (3 kcal $mole^{-1}$) than internal alkynes.

9 Allenes may show handedness and therefore may be chiral.

10 The elimination reaction with vicinal dihaloalkanes proceeds regioselectively to give only alkenyl halides because the proton alpha to the halogen is relatively acidic.

11 Selective cis dihydrogenation of alkynes is possible with Lindlar's catalyst, the surface of which is less active than palladium on carbon and therefore does not hydrogenate alkenes. Selective trans hydrogenation is possible with sodium metal dissolved in liquid ammonia because simple alkenes cannot be reduced by one-electron transfer. The stereochemistry is set by the greater stability of a trans disubstituted alkenyl radical intermediate.

12 To stop the hydroboration of terminal alkynes at the alkenylboron intermediate stage, modified dialkylboranes—particularly diisoamyl- or dicyclohexylborane—are used. Oxidation of the resulting alkenylboranes produces enols that are unstable with respect to rearrangement to carbonyl compounds (tautomerism).

Problems

1 Name each of the following compounds using the IUPAC system of nomenclature.

(a) $H_3C—C(CH_3)(Cl)—C{\equiv}CH$

(b) $H_3C—C(CH_3)(HO)—C{\equiv}CH$

(c) $CH_3CH_2CH_2CH(C{\equiv}CH)CH_2CH_2CH_2OH$

(d) $H_3C(H)C{=}C(H)C{\equiv}CH$ (H₃C and H on first carbon; H and C≡CH on second carbon; H₃C and C≡CH trans)

(e) $H_3C(CH_3CH_2)C{=}C(C{\equiv}CCH_3)(CHCH_2CH_2CH_3)$ with CH_3 on the CH (H₃C cis to C≡CCH₃; CH₃CH₂ cis to CH(CH₃)CH₂CH₂CH₃)

(f) Fischer projection: CH_3 (top), H (left), OH (right), $C{\equiv}CH$ (bottom)

(g) Cyclopentane with wedged $C{\equiv}CH$ and wedged $CH{=}CH_2$ on adjacent carbons

2 Compare C–H bond strengths in ethane, ethene, and ethyne. Reconcile these data with (1) hybridization, (2) polarity, and (3) acidity of the hydrogen in each of these bonds.

3 Compare the C-2–C-3 bonds in propane, propene, and propyne. Should they be any different with respect to either bond length or bond strength? If so, how should they vary?

4 Deduce structures for each of the following molecules from the NMR spectra on page 556. (Forewarning: The similarities between spectra B and C are deceptive.)

(a) C_6H_{10}, spectrum A.
(b) C_7H_{12}, spectrum B.
(c) C_7H_{12}, spectrum C.

5 Carbon-13 NMR signals for the four isomeric straight-chain octynes are listed below. Assign structures to each isomer on the basis of these data alone. Note: Signal intensities are not given; it is always possible that the signals for more than one carbon might coincidentally have the same chemical shift.

(a) δ = 12.7, 19.2, 21.4, and 79.0 ppm.
(b) δ = 2.9, 14.5, 18.4, 22.7, 29.5, 31.4, 74.8, and 78.5 ppm.
(c) δ = 14.9, 18.4, 23.4, 29.2, 32.0, 69.0, and 84.0 ppm.
(d) δ = 13.0, 14.6, 15.1, 18.7, 22.8, 31.9, 79.7, and 81.0 ppm.

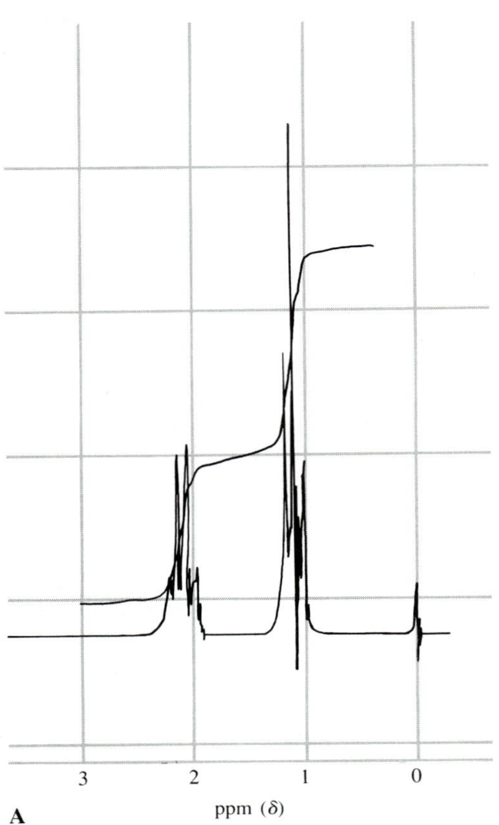

A

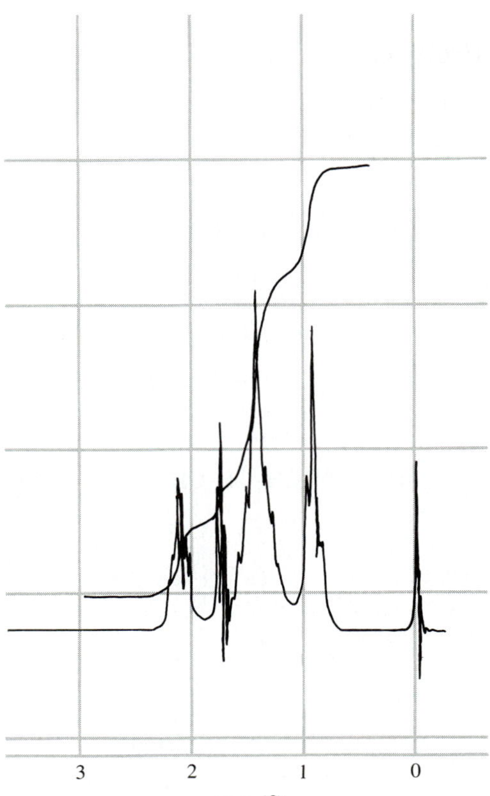

B

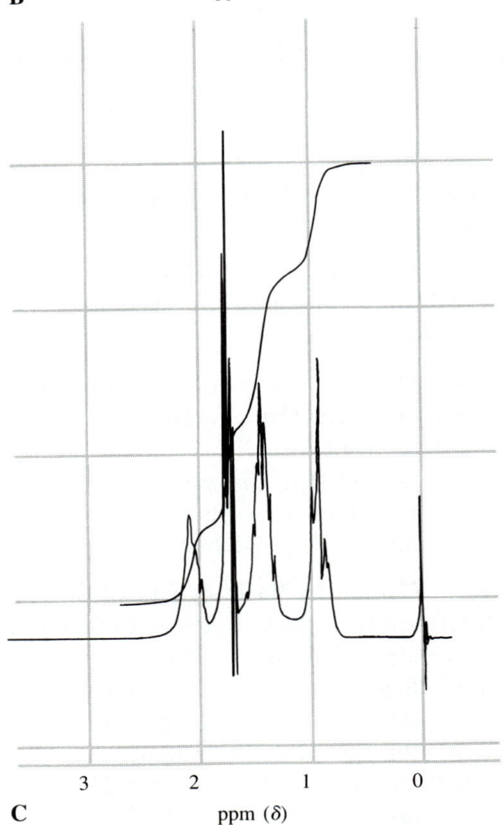

C

6 You are given the following approximate ΔH_f° values (in kcal mole^{-1}): butane, −30.4; *cis*-2-butene, −1.9; 2-butyne, 34.7; cycloheptane, −28.2; and cycloheptene, −2.2.

(a) Estimate ΔH_f° for cycloheptyne, taking the calculated strain energy into account (Section 13-4).

(b) Calculate ΔH° values for each of the following hydrogenations.

(i) Butyne + H_2 ⟶ *cis*-2-butene.

(ii) *cis*-2-Butene + H_2 ⟶ butane.

(iii) Cycloheptyne + H_2 ⟶ cycloheptene.

(iv) Cycloheptene + H_2 ⟶ cycloheptane.

7 Benzene is one of the most kinetically stable organic compounds known, capable of surviving heating to 1000°C. In contrast, benzyne can be observed as a stable species only in solid argon at 8 K and undergoes instant dimerization when ''warmed'' to 20 K. Explain the instability of benzyne.

Benzene, C_6H_6

Benzyne, C_6H_4

8 Predict the qualitative appearance of the ^{1}H NMR spectrum of 1,1-dimethylallene (chemical shifts, couplings, etc.).

9 Write the expected product(s) of each of the following reactions.

(a) $CH_3CH_2CH(CH_3)CH(Cl)CH_2Cl \xrightarrow{NaNH_2,\ liquid\ NH_3}$

(b) $CH_3OCH_2CH_2CH_2CH(Br)CH(Br)CH_3 \xrightarrow{NaNH_2,\ liquid\ NH_3}$

(c) *meso*-$CH_3CH(CH_3)CH_2CH(Cl)CH(Cl)CH_2CH(CH_3)CH_3 \xrightarrow{NaOCH_3,\ CH_3OH}$

(d) (4*R*,5*R*)-$CH_3CH(CH_3)CH_2CH(Cl)CH(Cl)CH_2CH(CH_3)CH_3 \xrightarrow{NaOCH_3,\ CH_3OH}$

(e) Which would react faster with $NaNH_2$: the product of reaction **c** or the product of reaction **d**?

10 Is it possible to form an alkyne from the treatment of 2,3-dibromo-2-methylbutane with $NaNH_2$ at high temperature? Write a detailed step-by-step mechanism describing the reaction.

11 Propose a mechanism for the formation of an alkenyl lithium reagent from an alkenyl bromide by transmetallation with $CH_3CH_2CH_2CH_2Li$ (Section 13-5). Predict the products of each of the following reactions.

(a) 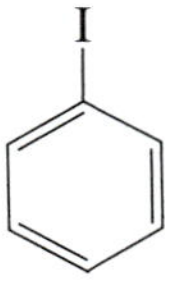 Iodobenzene + $CH_3CH_2CH_2CH_2Li \longrightarrow$

(b) $(CH_3)_3CC{\equiv}CBr + CH_3CH_2CH_2CH_2Li \longrightarrow$

12 Write the expected major product of reaction of 1-propynyllithium, $CH_3C{\equiv}C^-Li^+$, with each of the following molecules.

(a) CH_3CH_2Br

(b) $CH_3CH(CH_3)CH(Cl)CH(CH_3)_2$ (H₃C and Cl drawn above the second and third carbons: $CH_3\overset{H_3C}{\overset{|}{C}}H\overset{Cl}{\overset{|}{C}}HCH(CH_3)_2$)

(c) $CH_3\overset{Cl}{\overset{|}{C}}HCH_2OH$

(d) Cyclohexanone

(e) 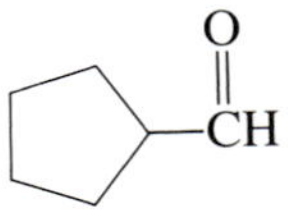 Cyclopentyl–$\overset{O}{\overset{\|}{C}}H$

(f) $CH_3CH{-}CH_2$ (epoxide, O bridging the two carbons)

(g) $CH_3CH_2CH_2\overset{O}{\overset{\|}{C}}OCH_3$

(h) 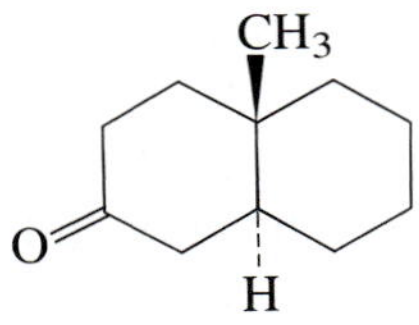 Bicyclic ketone (decalone) with angular CH_3 (wedge) and angular H (hashed), C=O on the ring

13 In an attempt to synthesize 6-heptyn-3-ol, the following reaction was carried out:

$$HC{\equiv}CCH_2CH_2\overset{O}{\overset{\|}{C}}H + CH_3CH_2MgBr \longrightarrow$$

The desired product was not obtained. Instead, a gas was observed to evolve from the reaction mixture, and work-up of the reaction gave only polymeric organic material. Explain these observations.

14 Propose reasonable syntheses of each of the following alkynes, using the principles of retrosynthetic analysis effectively. Each alkyne functional group in your synthetic target should come from a *separate* molecule, which may be any two-carbon compound (e.g., ethyne, ethene, ethanal, etc.).

(a) $CH_3CH_2C{\equiv}CCH_2CH_2CH_3$

(b) $(CH_3)_3CC{\equiv}CH$ (Be careful! What is wrong with $(CH_3)_3CCl + {}^-{:}C{\equiv}CH$?)

(c) $HC{\equiv}CCH_2CH_2C{\equiv}CH$

(d) $CH_3CH_2\underset{OH}{\underset{|}{\overset{CH_3}{\overset{|}{C}}}}C{\equiv}CH$

(e) HC≡CCH(OH)C≡CH

$$\begin{array}{c}\text{OH}\\ |\\ \text{HC}\equiv\text{CCHC}\equiv\text{CH}\end{array}$$

(f) 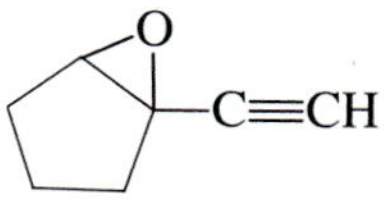 O —C≡CH

(g) $CH_3CH_2C{\equiv}C{-}C{\equiv}CCH_2CH_3$

15 Draw the structure of (*R*)-4-deuterio-2-hexyne. Propose an efficient strategy for the synthesis of this compound.

16 Compare the orbital pictures of the ethenyl (vinyl) radical (Section 13-6) and the ethenyl (vinyl) cation (Section 13-5). Why do you suppose that the cation carbon is *sp* hybridized, whereas the radical carbon is sp^2?

17 Write the expected major product of the reaction of propyne with each of the following reagents.

(a) D_2, Pd-$BaSO_4$, quinoline
(b) Na, ND_3
(c) $[(CH_3)_2CHCH_2CH_2]_2BD$, then CH_3CO_2H
(d) Diisoamylborane, then CH_3CO_2D
(e) 1 equivalent HI
(f) 2 equivalents HI
(g) 1 equivalent Br_2
(h) 1 equivalent ICl
(i) 2 equivalents ICl
(j) H_2O, $HgSO_4$, H_2SO_4
(k) Diisoamylborane, then NaOH, H_2O_2
(l) Cu_2Cl_2, O_2, pyridine

18 Write the product of the reaction of dicyclohexylethyne with each of the reagents in Problem 17.

19 Write the products of the reactions of your answers to parts **a** and **b** of Problem 18 with each of the following reagents.

(a) H_2, Pd-C
(b) Br_2
(c) BH_3, then NaOH, H_2O_2
(d) MCPBA
(e) Cold $KMnO_4$

20 Propose reasonable syntheses of each of the following molecules, using an alkyne at least once in each synthesis.

(a)
$$\begin{array}{c}\text{Br}\\ |\\ CH_3CH_2CCH_3\\ |\\ \text{Cl}\end{array}$$

(b) $CH_3CH_2CH_2CH_2CI_2CH_3$

(c) *meso*-2,3-Dibromobutane

(d) Racemic mixture of (2*R*,3*R*)- and (2*S*,3*S*)-2,3-dibromobutane

(e) Fischer projection: CH_3 at top; Br—C—Cl; H—C—Cl; CH_3 at bottom

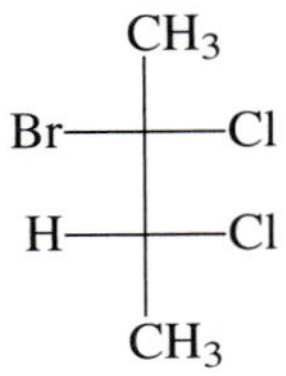

(f) $CH_3(CH_2)_3\overset{O}{\overset{\|}{C}}(CH_2)_4CH_3$

(g) $HOCH_2CH_2\overset{OH}{\overset{|}{C}}HCH_3$

(h) $CH_2{=}CHCH_2CH_2\overset{O}{\overset{\|}{C}}CH_3$

(i)

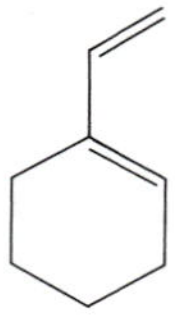

(j) H_3C CH_3 H O

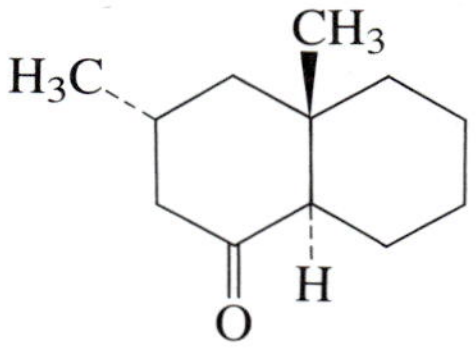

(k) O $CH_2CH_2CH_2CH_3$ CH_3

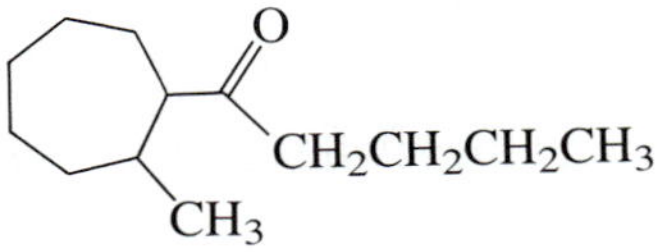

21 Cationic cyclization of 1-methyl-5-heptynyl 4-methylbenzenesulfonate (tosylate) gives a mixture of two major products:

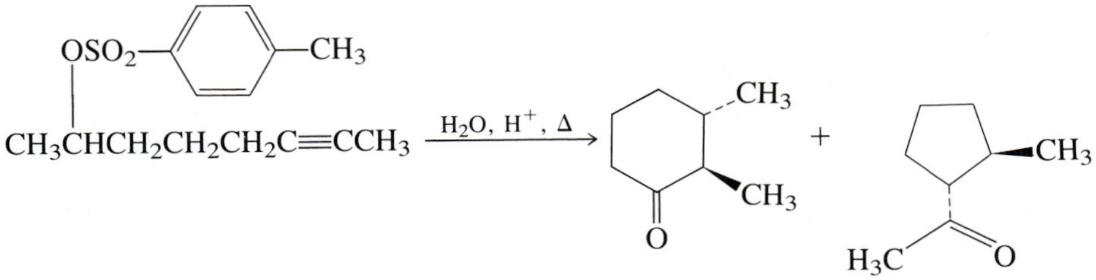

$$\underset{\displaystyle CH_3\overset{|}{C}HCH_2CH_2CH_2C{\equiv}CCH_3}{OSO_2\text{—}C_6H_4\text{—}CH_3} \xrightarrow{H_2O,\ H^+,\ \Delta}$$ CH_3 CH_3 O + CH_3 H_3C O

Write a detailed mechanism for this result, and compare this reaction with the cyclization of 1-methyl-5-hexynyl 4-methylbenzenesulfonate (tosylate) described in Section 13-6.

22 Predict the product of cationic cyclization of the tosylate of 6-heptyne-3-ol (conditions: H^+, H_2O, heating).

23 Propose a reasonable structure for calcium carbide, CaC_2, on the basis of its chemical reactivity. What might be a more-systematic name for it?

24 How would you carry out the final oxidation step in the synthesis of the spruce budworm pheromone (Section 13-6)?

25 Propose *two different* syntheses of linalool, a terpene found in cinnamon, sassafras, and orange flower oils. Start with the eight-carbon ketone shown below and use ethyne as your source of the necessary additional two carbons in both syntheses.

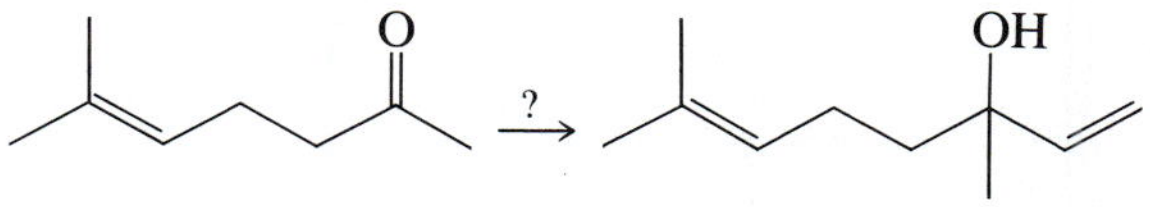

O ? OH

Linalool

26 A synthesis of the sesquiterpene farnesol requires the conversion of a dichloro compound into an alkynol. Suggest a way of achieving this transformation.

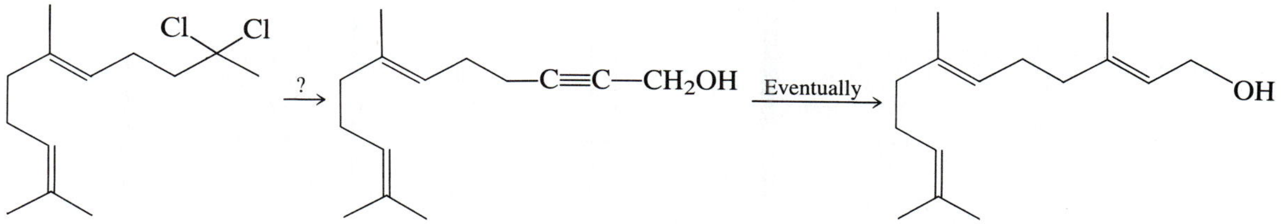

Cl Cl ? —C≡C—CH$_2$OH Eventually OH

Farnesol

27 The synthesis of chamaecynone, the essential oil of the Benihi tree, requires the conversion of a chloro alcohol into an alkynyl ketone. Propose a synthetic strategy to accomplish this task.

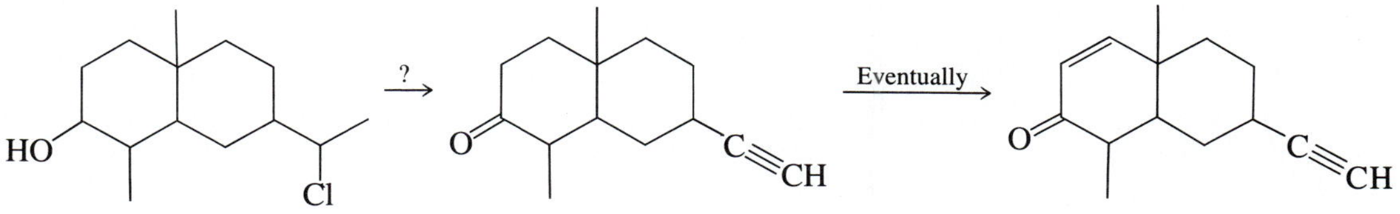

HO Cl ? O C≡CH Eventually O C≡CH

Chamaecynone

28 Synthesis of the sesquiterpene bergamotene proceeds from the alcohol shown below. Suggest a sequence to complete the synthesis.

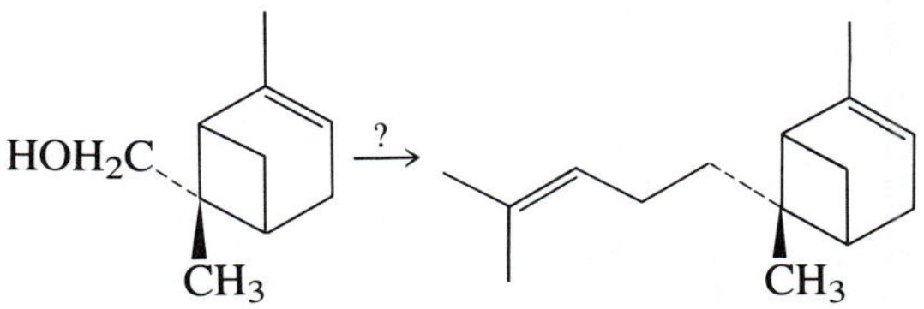

HOH$_2$C CH$_3$? CH$_3$

Bergamotene

29 An unknown molecule displays ^{1}H NMR spectrum D (on the next page). Reaction with H_2 in the presence of Lindlar's catalyst gives a compound that, after ozonolysis and treatment with Zn in aqueous acid, gives rise to one equivalent of $CH_3\overset{O}{\overset{\|}{C}}\overset{O}{\overset{\|}{C}}H$ and two of $H\overset{O}{\overset{\|}{C}}H$. What was the structure of the original molecule?

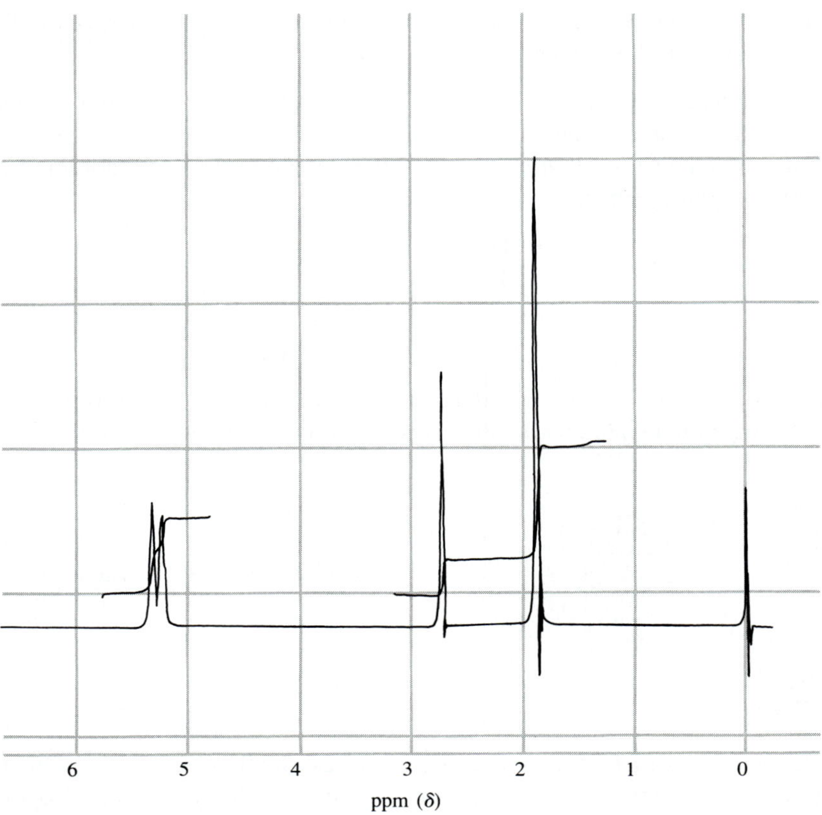

D

CHAPTER 14

Delocalized Pi Systems and Their Investigation by Ultraviolet and Visible Spectroscopy

Chapters 12 and 13 showed the importance of the overlap between two adjacent *p* orbitals that are parallel. This interaction produces energy and generates a new functional group, the π bond. If two overlapping *p* orbitals are energetically advantageous, are three or more such interactions better?

This chapter deals with compounds containing three or more contiguous overlapping *p* orbitals. Such systems are thermodynamically stabilized by extended overlap. Their chemical reactivity indicates the ready access to intermediates also stabilized by overlap. This chapter begins with the 2-propenyl system, also called allyl, containing three interacting *p* orbitals, and then proceeds to compounds that contain several adjacent double bonds. These molecules undergo not only the usual transformations of the alkenes, modified by the special situation of multiple overlap, but also unique thermal and photochemical cycloaddition reactions and ring closures. The extent of overlap can be probed and measured by ultraviolet and visible spectroscopy.

14-1 Overlap of Three Adjacent *p* Orbitals: Resonance in the 2-Propenyl (Allyl) System

Several apparently unrelated experimental facts about the 2-propenyl (allyl) system indicate some unusual features.

FACT 1 The primary carbon–hydrogen bond strength in propene is relatively low, only 87 kcal mole^{-1}.

$$H_2C{=}C(H)CH_2H \longrightarrow H_2C{=}C(H)CH_2\cdot + H\cdot \qquad DH^\circ = 87 \text{ kcal mole}^{-1}$$

Propene → **2-Propenyl radical**

Dissociation Energies of Various C–H Bonds

$CH_2{=}CHCH_2{-}H$
$DH^\circ = 87$ kcal mole^{-1}

$(CH_3)_3C{-}H$
$DH^\circ = 93$ kcal mole^{-1}

$(CH_3)_2CH{-}H$
$DH^\circ = 94.5$ kcal mole^{-1}

$CH_3CH_2{-}H$
$DH^\circ = 98$ kcal mole^{-1}

A comparison of this dissociation energy with those of other hydrocarbons (see Table 3-2), shows that the primary C–H bond in propene is even weaker than a tertiary C–H bond. Evidently, the 2-propenyl radical enjoys some type of special stability.

FACT 2 3-Chloro-1-propene ionizes relatively fast under S_N1 (solvolysis) conditions and undergoes rapid unimolecular substitution through a carbocation intermediate.

$$H_2C{=}C(H)CH_2Cl \xrightarrow[-\,Cl^-]{CH_3OH,\ \Delta} H_2C{=}C(H)CH_2^+ \xrightarrow[-\,H^+]{CH_3OH} H_2C{=}C(H)CH_2OCH_3$$

3-Chloro-1-propene → **2-Propenyl cation** → **3-Methoxy-1-propene (S_N1 product)**

This finding clearly contradicts what we would expect from our knowledge of the stability of primary cations. It appears that the cation derived from 3-chloro-1-propene is somehow more stable than other primary carbocations. By how much? The ease of formation of the 2-propenyl cation in solvolysis reactions has been measured to be roughly equal to that of a secondary carbocation.

FACT 3 The pK_a of propene is about 40.

$$H_2C{=}C(H)CH_2H \underset{K\,\sim\,10^{-40}}{\rightleftharpoons} H_2C{=}C(H)CH_2^- + H^+$$

2-Propenyl anion

Thus, propene is considerably more acidic than propane (p$K_a \sim 50$), and the formation of the propenyl anion by deprotonation appears unusually favored. How can we explain these three observations?

The 2-Propenyl (Allyl) System Is Stabilized by Conjugation

Each of the preceding three processes generates a moiety—a radical, a carbocation, and a carbanion, respectively—that is adjacent to the π framework of a double bond. This arrangement seems to impart special stability. Why? The reason is electron delocalization. Molecules for which several Lewis structures can be drawn have unusual stability (see Section 1-5). Here, each product may be described by a pair of equivalent resonance structures. It is because of their special electronic arrangement, and the resulting unusual chemistry, that these three-carbon intermediates have been given the common name **allyl** followed by the appropriate derivative term—radical, cation, or anion. Generally, carbon atoms and their bonds adjacent to a double bond are called *allylic*.

Resonance in the 2-Propenyl (Allyl) System

$[CH_2{=}CH{-}\dot{C}H_2 \longleftrightarrow \dot{C}H_2{-}CH{=}CH_2]$ or H, C, H_2C, CH_2

Radical

$[CH_2{=}CH{-}\overset{+}{C}H_2 \longleftrightarrow \overset{+}{C}H_2{-}CH{=}CH_2]$ or H, C, H_2C, +, CH_2

Cation

$[CH_2{=}CH{-}\overset{-}{\ddot{C}}H_2 \longleftrightarrow \overset{-}{\ddot{C}}H_2{-}CH{=}CH_2]$ or H, C, H_2C, −, CH_2

Anion

When inspecting the resonance forms of the 2-propenyl (allyl) systems, remember that resonance structures are *not* isomers, but extreme representations of the molecule. The true structure lies somewhere between these extreme views. The dotted-line drawings shown at the right of the classical pictures better reveal the symmetry of the allyl unit.

The Three Molecular Orbitals of the 2-Propenyl (Allyl) Framework

The stabilization of the 2-propenyl (allyl) system by resonance can also be described in terms of molecular orbitals, a more sophisticated (and complementary) quantum-mechanical approach. To what extent does the interaction with a p orbital modify the molecular-orbital picture of a simple π bond (Section 11-2, Figures 11-1, 11-2, and 11-3)?

The molecular framework consists of three carbons, each one of which is sp^2 hybridized and bears a p orbital perpendicular to the molecular plane. It may be regarded as a double bond to which an additional sp^2 carbon is attached (Figure 14-1). The molecule, however, is symmetrical, with equal C–C bond lengths.

FIGURE 14-1

The three p orbitals in the 2-propenyl (allyl) group. The σ framework is shown as solid lines.

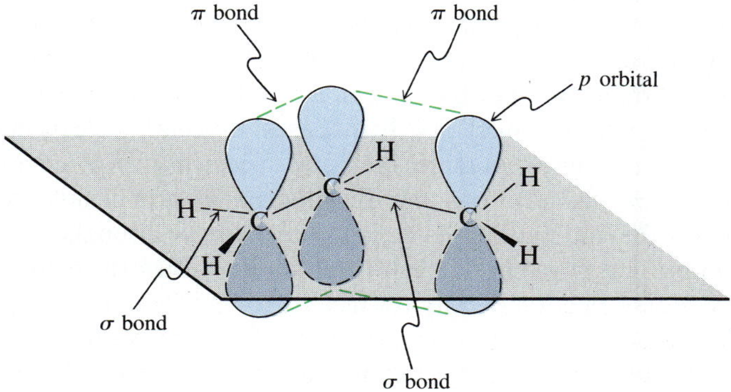

FIGURE 14-2
The three molecular π orbitals of 2-propenyl (allyl). Note that the size of the various lobes is not the same.

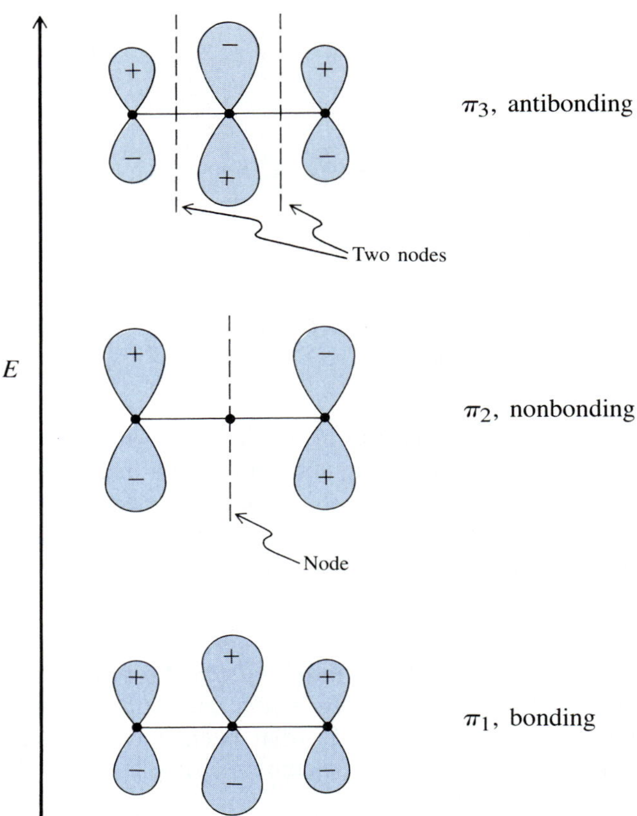

Ignoring the σ framework, we can combine the three *p* orbitals mathematically to give three π molecular orbitals. This process is analogous to mixing two atomic orbitals to give two molecular orbitals describing a π bond (Figures 11-1, 11-2, and 11-3), except that there is an extra orbital. Among the three new molecular orbitals, one is bonding and has no nodes (π_1), one is *nonbonding* (in other words, it has the same energy as the noninteracting *p* orbital) and has one node (π_2), and one is antibonding, with two nodes (π_3), as shown in Figure 14-2. Once the three molecular orbitals are derived, we can use the Aufbau principle to fill in the appropriate number of electrons (Figure 14-3). We are dealing with π electrons only; hence the cation, with a total of two, contains only one filled orbital, π_1. For the radical and, subsequently, the anion, we fill up the second molecular orbital, π_2. In all cases, the total π-electron energy of the system is lower (more favorable) than that expected from three noninteracting *p* orbitals—π_1 has dropped significantly and is completely filled in all cases, whereas the antibonding level, π_3, stays empty throughout.

The node passing through the central carbon in π_2 has an important consequence: any excess electron density (or lack thereof) will show up primarily on the two terminal carbons, as we would expect from the resonance structures. Therefore, on average, there is roughly half of a positive charge located at these carbons in the 2-propenyl (allyl) cation, half of a negative charge in the anion.

FIGURE 14-3
Filling up the 2-propenyl (allyl) π molecular orbitals for the cation, radical, and anion according to the Aufbau principle.

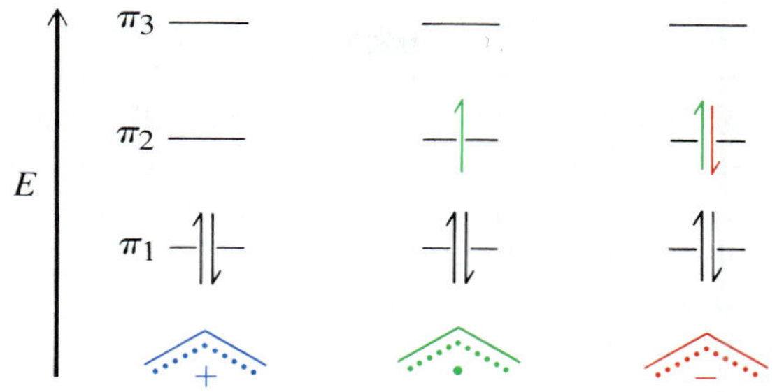

The central position stays roughly neutral, maintaining an octet. In the neutral radical, each carbon bears one p electron, and the terminal positions exhibit 50% radical character.

Partial Electron Density Distribution in the 2-Propenyl (Allyl) System

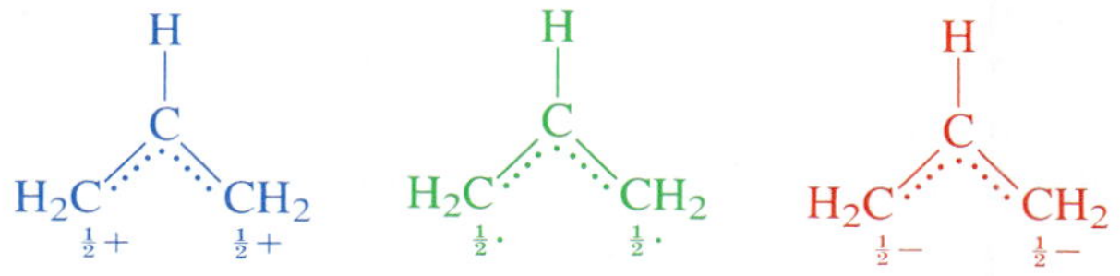

How Strong Is the Pi Bond in the 2-Propenyl (Allyl) Radical?

To estimate the strength of the π bond in the 2-propenyl (allyl) radical, an experiment can be carried out very similar to that used to measure the strength of the π bond in ethene (Section 11-2). That experiment determined the barrier to cis-trans isomerization in 1,2-dideuterioethene; for our problem, we can look at the corresponding reaction of the 1-deuterio-2-propenyl radical. The activation energy for this isomerization is about 15.7 kcal mole^{-1}, giving an estimate of the strength of the interaction of an electron in a p orbital with a double bond. Note that the interaction is appreciable but much weaker than the π bond in ethene.

Cis-Trans Isomerization of the 1-Deuteriopropenyl (1-Deuterioallyl) Radical

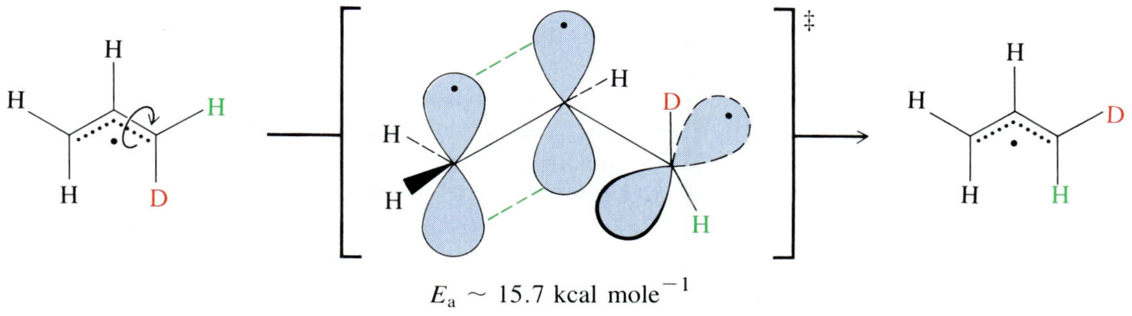

$E_a \sim 15.7$ kcal mole^{-1}

In summary, some experimental facts clearly implicate participation of the double bond in the stabilization of allylic radicals, cations, and anions. In Lewis terms, this participation is readily explained by resonance structures. In a molecular-orbital description, the three interacting *p* orbitals form three new molecular orbitals: one is considerably lower in energy than the *p* level, another one stays at the same level, and a third moves up. Because in the 2-propenyl (allyl) system only the lower two are populated with electrons, the total π energy of the system is lower than before such interaction. The distribution of charge or radical density is controlled by the nonbonding molecular orbital, with a node passing through the central carbon. The barrier to isomerization of the 1-deuterio-2-propenyl radical is from about 15 to 16 kcal $mole^{-1}$, a measure of the strength of the allylic π bond.

14-2 The Consequences of Delocalization: The Chemistry of the 2-Propenyl (Allyl) System

Allylic cations can be trapped by nucleophiles at either terminal carbon atom in S_N1 reactions, giving rise to kinetic product mixtures whose composition may change on heating to result in the product ratios expected by thermodynamic considerations. Allylic halides also undergo S_N2 and a new type of direct displacement reaction—S_N2'. In the latter, the nucleophile attacks the double bond, pushing the π-electron pair toward the leaving group. Free-radical halogenation of alkenes may occur at the allylic position, leaving the double bond untouched. Double bonds substituted by atoms carrying lone electron pairs can be viewed as the neutral analogs of 2-propenyl (allyl) anions.

Nucleophilic Substitution of 3-Halo-1-propenes (Allyl Halides)

The ready ionization of allylic halides has important chemical consequences. For example, the hydrolysis of either 1-chloro-2-butene or 3-chloro-1-butene results in the same alcohol mixture. The reason is the intermediacy of the same allylic cation.

Hydrolysis of Isomeric Allylic Chlorides

$$CH_3CH{=}CHCH_2Cl \xrightarrow[-\,Cl^-]{} \left[\overset{4}{C}H_3\overset{3}{C}H{=}\overset{2}{C}H\overset{1}{C}H_2^{+} \longleftrightarrow \underset{4}{C}H_3\underset{3}{\overset{+}{C}}H\underset{2}{C}H{=}\underset{1}{C}H_2\right] \xleftarrow[-\,Cl^-]{} CH_3\underset{\displaystyle |}{\overset{Cl}{C}}HCH{=}CH_2$$

1-Chloro-2-butene **Allylic cation** **3-Chloro-1-butene**

$$\downarrow HOH$$

$$\underset{\text{Minor}}{CH_3CH{=}CHCH_2OH} + \underset{\text{Major}}{CH_3\overset{OH}{\overset{|}{C}}HCH{=}CH_2} + H^+$$

2-Buten-1-ol **3-Buten-2-ol**

EXERCISE 14-1

Hydrolysis of (*R*)-3-chloro-1-butene gives exclusively racemic alcohols. Explain.

Interestingly, the major product from this reaction is 3-buten-2-ol, despite the fact that the thermodynamically less favored (terminal) double bond is created. This must be a kinetic effect; that is, the less-stable isomer must be formed faster. Why? The difference is due to the electronic makeup of the allylic cation intermediate in the reaction. This system is unsymmetrical; therefore, we would expect an unequal charge distribution between carbons 1 and 3. Indeed, more charge resides at C-3, the more-substituted carbon. In other words, in our scheme the second resonance structure contributes more to the ground state of the system than the first. Because nucleophilic trapping of the cationic intermediate is largely charge controlled, water will attack at C-3 faster, giving the initially observed main product.

Kinetic Compared with Thermodynamic Control

$$CH_3CH(OH)CH{=}CH_2 + H^+ \underset{\text{but reversible)}}{\overset{\text{Kinetic control (fast}}{\rightleftharpoons}} \left[CH_3CH{=}CH\overset{+}{C}H_2 \longleftrightarrow CH_3\overset{+}{C}HCH{=}CH_2\right] + HOH \overset{\text{Thermodynamic control (slow)}}{\rightleftharpoons} CH_3CH{=}CHCH_2OH + H^+$$

Less-stable product, predominates when reaction time is short or at low temperatures

More-stable product, predominates when reaction time is long or at higher temperatures

That the outcome of the hydrolysis is indeed determined kinetically can be demonstrated by equilibration of the products. Heating a mixture of the butenol products for a prolonged time gives the thermodynamic mixture, with the predominance of 2-buten-1-ol. The former conditions are said to exert kinetic, the latter thermodynamic control.

Why do prolonged reaction times give thermodynamic product ratios? The nucleophilic trapping of the intermediate allylic cation is reversible. Initially, the cation is trapped to form the kinetic product, stable at lower temperatures. However, on heating, the alcohols regenerate the allylic cation and give the slower nucleophilic reaction leading to the thermodynamically more stable isomer a chance to succeed.

This situation may be illustrated by a potential-energy diagram (Figure 14-4, on the next page). The kinetic alcohol is formed first, but its formation is reversible, allowing for the eventual (and slower) generation of the thermodynamic product.

EXERCISE 14-2

Treatment of 3-buten-2-ol with cold hydrogen bromide gives 1-bromo-2-butene and 3-bromo-1-butene in a 15:85 ratio. On heating, this ratio changes to give mainly 1-bromo-2-butene. Explain.

Allylic halides can also undergo S_N2 reactions. They are in fact faster than the equivalent S_N2 reactions of ordinary primary haloalkanes, apparently because the *p* orbital in the transition state of the displacement overlaps with the double bond (Figure 14-5, on page 571).

The S_N2 Reactions of 3-Chloro-1-propene and 1-Chloropropane

Reaction	Relative rate
$CH_2{=}CHCH_2Cl + I^- \xrightarrow{\text{Propanone (acetone), 50°C}} CH_2{=}CHCH_2I + Cl^-$	73
$CH_3CH_2CH_2Cl + I^- \xrightarrow{\text{Propanone (acetone), 50°C}} CH_3CH_2CH_2I + Cl^-$	1

FIGURE 14-4

Kinetic compared with thermodynamic control in the reaction of 1-methyl-2-propenyl (1-methylallyl) cation with water. The oxonium ion intermediates have been omitted from the diagram for clarity. At low temperatures, the distribution of the products is governed by their relative rates of formation (k_1, k_2), that is, the relative heights of the activation energy barriers (favoring the formation of 3-buten-2-ol), because under these conditions the rates of the reverse reactions, k_{-1} and k_{-2}, are negligible. At higher temperature, k_{-1} becomes competitive, regenerating the allylic cation, which slowly forms more and more of the thermodynamic product, 2-buten-1-ol. The rate of its back reaction, k_{-2}, is the lowest in the entire system. The green arrows indicate the direction along the reaction coordinate with which each rate k is associated.

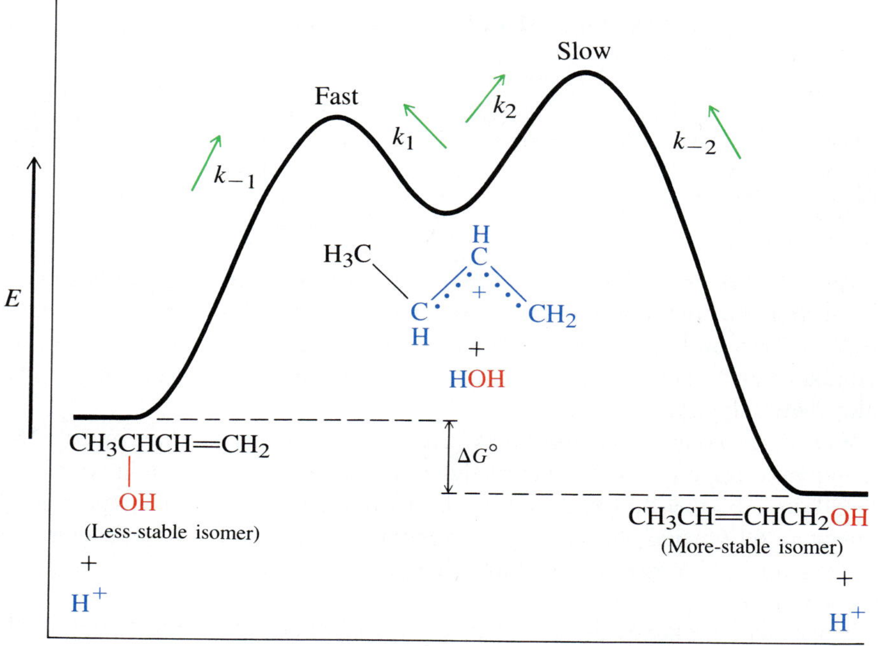

These bimolecular displacement processes can be complicated by another kind of attack of the nucleophile, in which the double bond is the target. As the new bond forms, the double bond moves toward the leaving group; in the ultimate product, the positions of the allylic substituent and the double bond have been exchanged. Because of its similarity to the S_N2 reaction, this displacement is called an **S_N2'** (S_N2 ''prime'') **reaction.**

The S_N2' Reaction

$$H_2C{=}CH{-}C(CH_3)(H){-}Cl \;+\; (CH_3CH_2)_2\ddot{N}H \longrightarrow (CH_3CH_2)_2\overset{+}{N}H{-}CH_2{-}CH{=}C(H)CH_3 + Cl^- \longrightarrow$$

$$(CH_3CH_2)_2NCH_2CH{=}C(H)CH_3 + H^+ + Cl^-$$

FIGURE 14-5

Delocalized transition state of the S_N2 reaction of 3-chloro-1-propene with iodide ion.

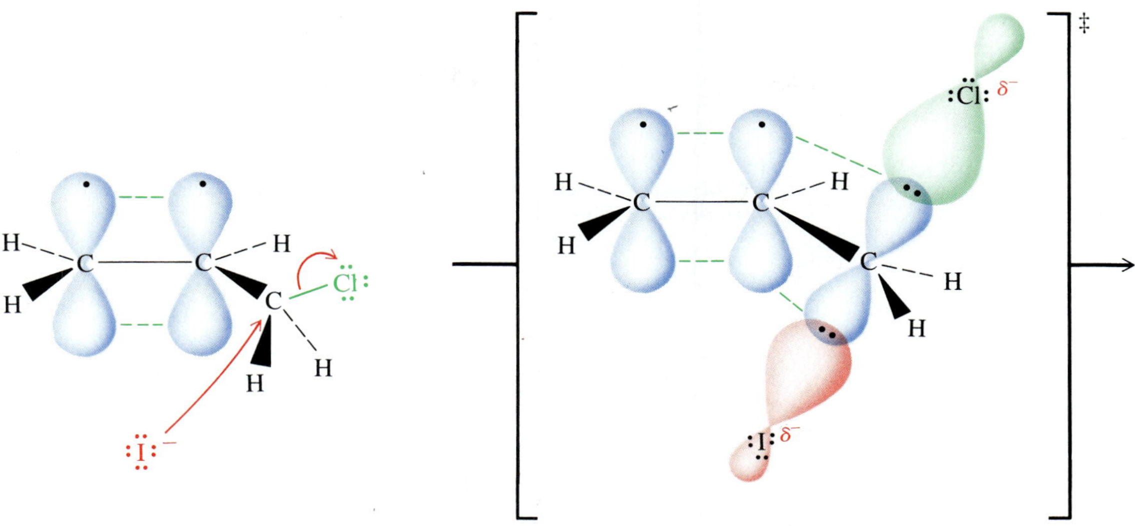

EXERCISE 14-3

Explain the following transformation.

HS⁀⁀⁀⁀(CH₃)OCR(=O) $\xrightarrow{HO^-}$ 2-(1-propenyl)thiacyclopentane

CH_3 O   OCR   HS HO^- S CH_3

Free-Radical Halogenation of Allylic Positions

Although halogens can add to alkenes to give the corresponding vicinal dihalides (Section 12-3), the course of this reaction is changed when the halogen is present in only low concentrations. These conditions favor free-radical chain mechanisms and lead to *free-radical allylic substitution*.*

Free-Radical Allylic Halogenation

$$CH_2{=}CHCH_3 \xrightarrow{X_2\text{ (low conc.)}} CH_2{=}CHCH_2X + HX$$

A reagent frequently used in allylic brominations is *N*-bromobutanimide (*N*-bromosuccinimide, NBS, Section 3-8) suspended in tetrachloromethane. This species is insoluble in CCl_4 and is the steady source of very small amounts of bromine formed by reaction with traces of hydrogen bromide.

*The reasons for this change require a detailed kinetic analysis of the three possible processes—allylic substitution, ionic addition, and free-radical addition—a discussion that is beyond the scope of this book. Suffice it to say that at low bromine concentrations the competing addition processes are slowed down to the point that substitution wins out.

NBS as a Source of Bromine

$$\text{N-Bromobutanimide (N-Bromosuccinimide, NBS)} + HBr \rightleftharpoons \text{Butanimide} + Br_2$$

N-Bromobutanimide
(N-Bromosuccinimide, NBS)

Butanimide

Example:

$$\text{Cyclohexene} + \text{NBS} \xrightarrow{h\nu,\ CCl_4} \text{3-Bromocyclohexene (85\%)} + \text{butanimide}$$

85%
3-Bromocyclohexene

The bromine then reacts with the alkene by a free-radical chain mechanism (Section 3-5) that begins when bromine is dissociated by radiation or traces of free radicals.

Mechanism of Allylic Brominations:

Chain initiation:

$$Br_2 \xrightarrow{h\nu \text{ or free-radical initiators}} 2\ Br\cdot$$

Chain propagation:

STEP 1

$$RCH{=}CHCH(R){-}H + Br\cdot \longrightarrow \left[RCH{=}CH\dot{C}H(R) \longleftrightarrow \dot{C}H(R)CH{=}CHR\right] + H{-}Br$$

$DH^\circ \sim 87$ kcal mole^{-1} (C–H); $DH^\circ = 87$ kcal mole^{-1} (H–Br)

STEP 2

$$RCH{=}CH\dot{C}H(R) + Br{-}Br \longrightarrow RCH{=}CHCH(R){-}Br + Br\cdot$$

$DH^\circ = 46$ kcal mole^{-1} (Br–Br); $DH^\circ = 55$ kcal mole^{-1} (C–Br)

The bromine atom abstracts the more weakly bound allylic hydrogen in the first chain-propagation step to generate the allylic radical. This reaction is virtually thermoneutral because the H–Br bond formed is as strong as the allylic C–H bond broken. In the next propagation step, the allylic radical reacts with more bromine to give the allylic bromide and another bromine atom, which then reenters the first propagation step. Step 2 is exothermic. The hydrogen bromide formed in the first propagation step reacts with more *N*-bromobutanimide, generating more Br_2, and so on. The reaction is readily monitored because NBS is heavier than CCl_4 and stays at the bottom of the flask. In contrast, the butanimide product, also insoluble in CCl_4, is lighter than the solvent and floats to the surface.

Allylic brominations with NBS can give mixtures of products when the intermediate allylic radical is not symmetrical.

$$CH_2{=}CH(CH_2)_5CH_3 \xrightarrow[-\,HBr]{NBS,\ h\nu} BrCH_2CH{=}CH(CH_2)_4CH_3 + CH_2{=}CH\underset{}{\overset{Br}{C}}H(CH_2)_4CH_3$$

1-Octene — **1-Bromo-2-octene** (44%) — **3-Bromo-1-octene** (17%)

However, secondary positions frequently react more rapidly than primary.

$$CH_3CH_2CH_2CH_2CH{=}CHCH_3 \xrightarrow{NBS,\ ROOR,\ CCl_4,\ 2\ h} CH_3CH_2CH_2\overset{Br}{C}HCH{=}CHCH_3 + HBr$$

2-Heptene — **4-Bromo-2-heptene** (76%)

Allylic chlorinations are important in industry because chlorine is relatively cheap. For example, 3-chloropropene (allyl chloride) is made commercially by the gas-phase chlorination of propene at 400°C.

$$CH_3CH{=}CH_2 + Cl_2 \xrightarrow{400^\circ C} ClCH_2CH{=}CH_2 + HCl$$

3-Chloropropene (Allyl chloride)

EXERCISE 14-4

Predict the outcome of the allylic bromination of the following substrates with NBS (1 equivalent): (a) cyclohexene; (b) [bicyclic structure: two fused six-membered rings sharing a central C=C double bond]; (c) 1-methylcyclohexene.

Allylic Organometallic Reagents: Useful Three-Carbon Nucleophiles

Because of the relative stability of the resulting conjugated carbanion, propene is appreciably more acidic than propane. Therefore, allylic lithium reagents can be made by the abstraction of protons from propene derivatives with an alkyllithium in the presence of *N,N,N′,N′*-tetramethylethane-1,2-diamine (tetramethylethylenediamine, TMEDA), a good solvating agent.

$$CH_3CH_2CH_2CH_2Li + H_2C{=}C(CH_3)_2 \xrightarrow{(CH_3)_2NCH_2CH_2N(CH_3)_2\ (TMEDA)} H_2C{=}C(CH_3)(CH_2Li) + CH_3CH_2CH_2CH_2{-}H$$

Allylic organometallics can also be prepared by transmetallation (Section 8-5). For example, treatment of tetra(2-propenyl)tin (tetraallyltin) with butyllithium results in 2-propenyllithium (allyllithium) and tetrabutyltin.

$$4\ CH_3CH_2CH_2CH_2Li + (CH_2{=}CHCH_2)_4Sn \xrightarrow{THF} 4\ CH_2{=}CHCH_2Li + (CH_3CH_2CH_2CH_2)_4Sn$$

2-Propenyllithium (Allyllithium) — **Tetrabutyltin**

An even more-straightforward way of producing an allylic organometallic is Grignard formation.

$$CH_2{=}CHCH_2Br \xrightarrow{Mg,\ THF,\ 0^\circ C} CH_2{=}CHCH_2MgBr$$

1-Bromo-2-propene → **2-Propenylmagnesium bromide**

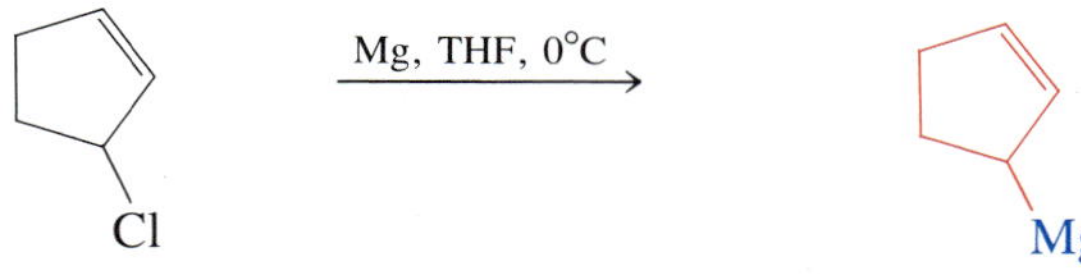

3-Chlorocyclopentene $\xrightarrow{Mg,\ THF,\ 0^\circ C}$ 2-Cyclopentenylmagnesium chloride

Cl, MgCl

3-Chlorocyclopentene **2-Cyclopentenylmagnesium chloride**

Like their alkyl counterparts, allylic lithium and Grignard reagents can function as nucleophiles.

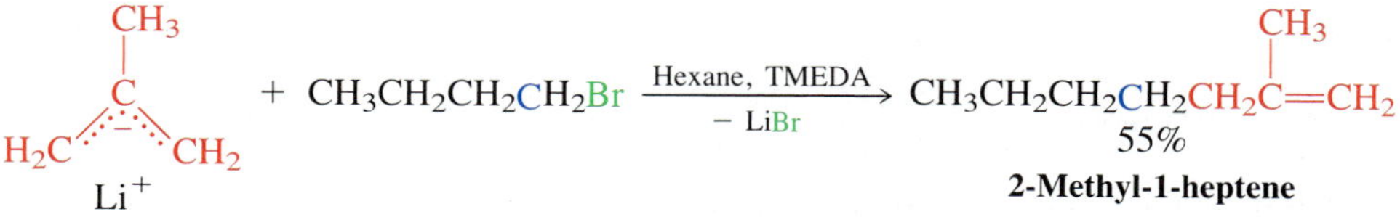

$$[H_2C{\cdots}C(CH_3){\cdots}CH_2]^- Li^+ + CH_3CH_2CH_2CH_2Br \xrightarrow[-\ LiBr]{Hexane,\ TMEDA} CH_3CH_2CH_2CH_2CH_2C(CH_3){=}CH_2$$

55%

2-Methyl-1-heptene

$$CH_2{=}CHCH_2MgBr + CH_3\overset{O}{\overset{\|}{C}}CH_3 \xrightarrow{(CH_3CH_2)_2O} \xrightarrow{H^+,\ H_2O} CH_2{=}CHCH_2\underset{CH_3}{\overset{OH}{C}}CH_3$$

85%

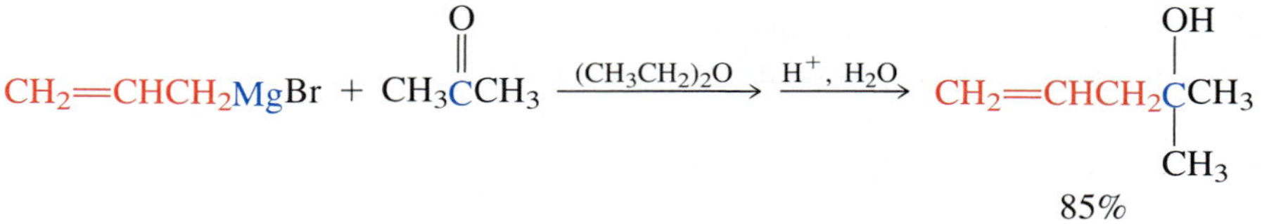

2-Methyl-4-penten-2-ol

EXERCISE 14-5

Show how to accomplish the following conversion in as few steps as possible:

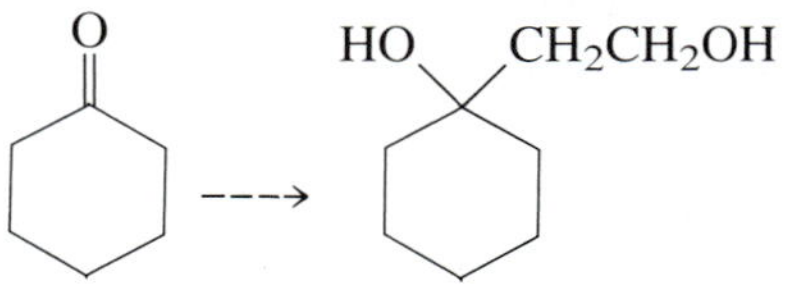

Cyclohexanone (O) ---→ 1-(2-hydroxyethyl)cyclohexanol (HO, CH_2CH_2OH)

Some Electron-Rich Double Bonds

$R_2C{=}C(H)\ddot{O}R'$

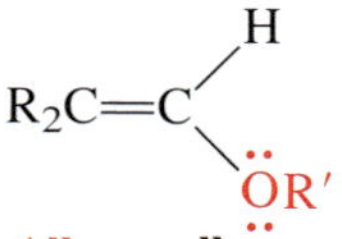

Alkoxy alkene (Enol ether)

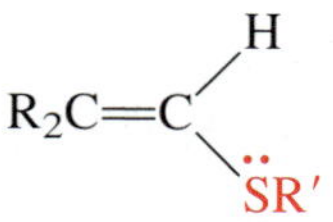

$R_2C{=}C(H)\ddot{S}R'$

Alkyl alkenyl sulfide

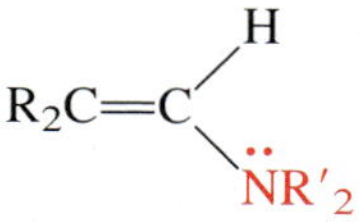

$R_2C{=}C(H)\ddot{N}R'_2$

***N,N*-Dialkylalkenamines (Enamine)**

Neutral Allylic Anion Equivalents: Delocalized Lone Electron Pairs

Allylic anions have special stability because they can delocalize the electron pair next to the double bond. A similar effect is manifest in some neutral species, such as alkenyl halides. The halogen atom can align the p orbital containing one of its lone electron pairs parallel to the p orbitals of the π bond, so that the nonbonding electrons can delocalize into the π bond (Figure 14-6). The result is a transfer of electron density to the π system. In other words, delocalization into a double bond makes the halogen substituent *electron donating* by resonance, even though inductively it is electron attracting. This effect can be emphasized by drawing a dipolar resonance structure (Figure 14-6B). A double bond bearing such electron-donating substituents is said to be *electron rich*. Other electron-rich double bonds are found in alkoxyalkenes, alkyl vinyl sulfides, and alkenamines (enamines, Section 15-6).

In summary, delocalization in the 2-propenyl (allyl) system gives rise to the unusual reactivity of allylic bonds. Allylic halides undergo relatively rapid S_N1 and S_N2 reactions, as well as a new type of bimolecular substitution, S_N2'. The

FIGURE 14-6

A. One of the nonbonding electron pairs on the halogen aligns with the π bond; this delocalizes the pair toward the two sp^2-hybridized carbons.
B. A resonance-structural description of delocalization in an alkenyl halide. Because it contains separated charges, the second structure is less important than the first.

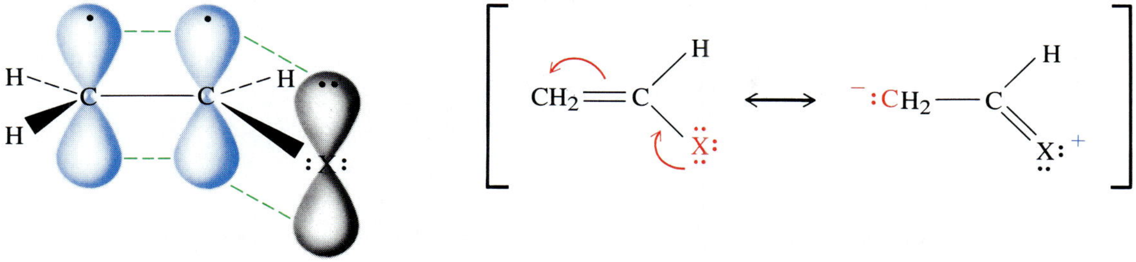

A B

S_N1 reaction leads to kinetic products at low temperatures but gives thermodynamically equilibrated mixtures at higher temperatures. Under free-radical conditions, alkenes containing allylic hydrogens enter into allylic halogenations. A particularly good reagent for allylic bromination is *N*-bromobutanimide, a source of dilute bromine. Alkenes tend to be deprotonated at the allylic position, resulting in the corresponding delocalized anions. Allylic Grignard reagents are made from the corresponding halides. Like their alkyl analogs, allyl organometallics function as nucleophiles. Like allylic anions, some neutral species are stabilized by delocalization of an electron pair into a double bond. These are alkenes bearing heteroatoms in which lone electron pairs are in resonance with the π electrons. This interaction makes the π bond electron rich.

14-3 Two Neighboring Double Bonds: Conjugated Dienes

This section goes one step beyond the allylic framework to consider the addition of yet another *p* orbital to the terminus of the π system, as is seen in structural units in which two double bonds are linked by a single bond. Systems containing alternating double and single bonds are called **conjugated** (*conjugatio,* Latin, union), in this case, conjugated dienes. In conjugated dienes, too, delocalization results in stability, as measured by their heats of hydrogenation. The interaction between the two double bonds is reflected in the molecular and electronic structures of these dienes and in their addition reactions with electrophiles and free radicals.

The Naming of Dienes: Hydrocarbons with Two Double Bonds

Conjugated dienes have to be contrasted with their nonconjugated isomers, in which the two double bonds are separated by saturated carbons, and the allenes (Section 13-4), in which π overlap between the two π bonds is impossible because of their perpendicular arrangement.

The names of conjugated and nonconjugated dienes are derived from those of the alkenes in a straightforward manner. The longest chain incorporating both double bonds is found, then numbered to indicate the positions of the functional and substituent groups. If necessary, cis-trans or *E,Z* prefixes indicate the geometry around the double bonds. Cyclic dienes are named accordingly.

The Simplest Conjugated and Nonconjugated Dienes

$CH_2{=}CH{-}CH{=}CH_2$
1,3-Butadiene

$CH_2{=}CHCH_2CH{=}CH_2$
1,4-Pentadiene

$CH_2{=}C{=}CH_2$
1,2-Propadiene
(Allene)

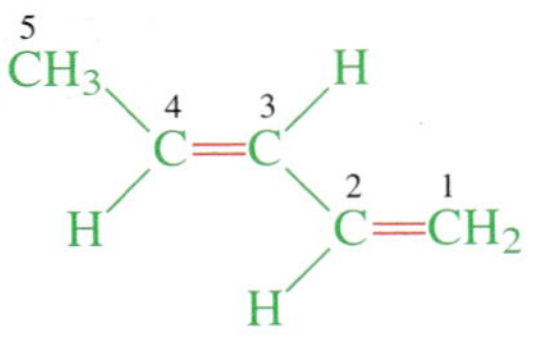

trans-1,3-Pentadiene

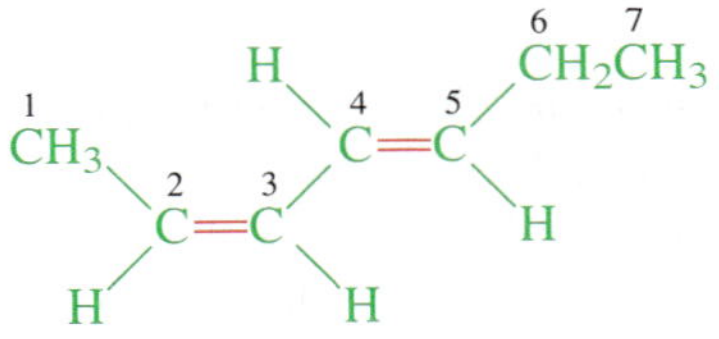

cis-2-*trans*-4-Heptadiene

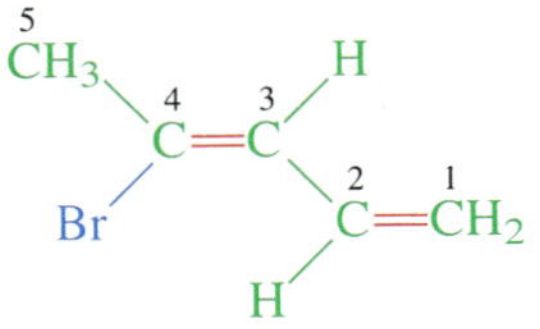

(*Z*)-4-Bromo-1,3-pentadiene

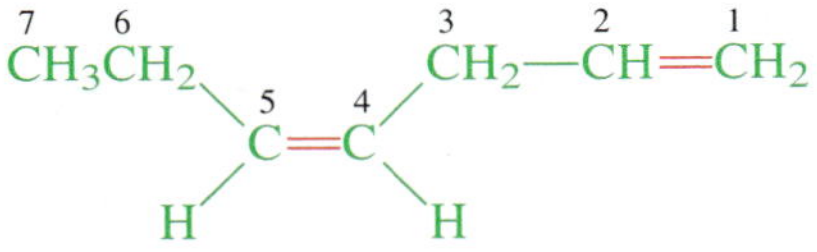

cis-1,4-Heptadiene (A nonconjugated diene)

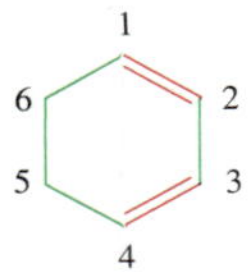

1,3-Cyclohexadiene

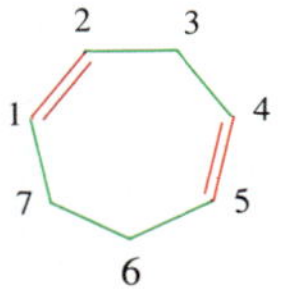

1,4-Cycloheptadiene (A nonconjugated cyclic diene)

EXERCISE 14-6

Suggest names or draw structures, as appropriate, for the following compounds: (a) 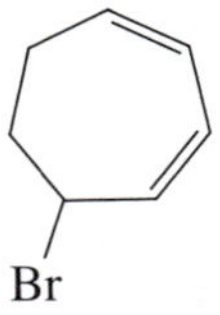 ; (b) 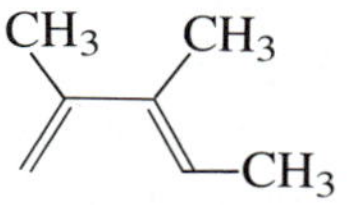 ; (c) *cis*-3,6-dimethyl-1,4-cyclohexadiene; (d) *cis,cis*-1,4-dibromo-1,3-butadiene.

Are Conjugated Dienes More Stable Than Nonconjugated Dienes?

The preceding sections noted that delocalization of electrons makes the allylic system especially stable. If a conjugated diene has the same property, it should be manifest in its heat of hydrogenation. Because the heat of hydrogenation of a terminal alkene such as 1-hexene is about -30 kcal mole^{-1} (see Section 11-4), a compound containing two noninteracting terminal double bonds should exhibit a heat of hydrogenation roughly twice this value, about -60 kcal mole^{-1}. Indeed, catalytic hydrogenation of either 1,5-hexadiene or 1,4-pentadiene releases just about that amount of energy.

The Heats of Hydrogenation of Nonconjugated Dienes and That of an Alkene

$$CH_3CH_2CH{=}CH_2 + H_2 \xrightarrow{Pt} CH_3CH_2CH_2CH_3 \qquad \Delta H^\circ = -30.3 \text{ kcal mole}^{-1}$$

$$CH_2{=}CHCH_2CH_2CH{=}CH_2 + 2\,H_2 \xrightarrow{Pt} CH_3(CH_2)_4CH_3 \qquad \Delta H^\circ = -60.5 \text{ kcal mole}^{-1}$$

$$CH_2{=}CHCH_2CH{=}CH_2 + 2\,H_2 \xrightarrow{Pt} CH_3(CH_2)_3CH_3 \qquad \Delta H^\circ = -60.8 \text{ kcal mole}^{-1}$$

However, when the same experiment is carried out with the conjugated diene 1,3-butadiene, less energy is produced.

Heat of Hydrogenation of 1,3-Butadiene

$$CH_2{=}CH{-}CH{=}CH_2 + 2\,H_2 \xrightarrow{Pt} CH_3CH_2CH_2CH_3 \qquad \Delta H^\circ = -57.1 \text{ kcal mole}^{-1}$$

FIGURE 14-7

The heats of hydrogenation of two equivalents of a monoalkene (1-butene) and a conjugated diene (1,3-butadiene). To make the two processes energetically comparable, the heat of formation of one equivalent of butane must be added to that of 1,3-butadiene.

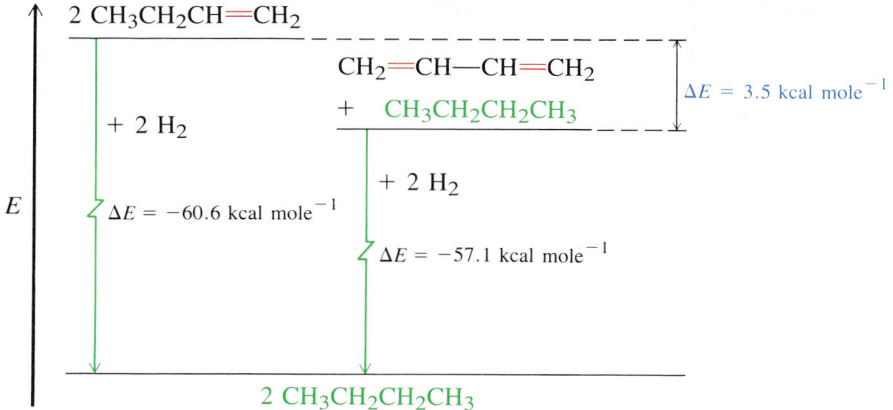

The difference in the heats of hydrogenation amounts to about 3.5 kcal mole^{-1}. In other words, the conjugated diene is more stable by about 3.5 kcal mole^{-1} than a system in which there is no stabilizing interaction between the two double bonds. This result is illustrated in Figure 14-7 (see also Figure 11-12). This stabilizing interaction between two adjacent double bonds is referred to as the **resonance energy.**

EXERCISE 14-7

The heat of hydrogenation of *trans*-1,3-pentadiene is 54.2 kcal mole^{-1}, 6.6 kcal mole^{-1} less than that of 1,4-pentadiene, even less than that expected from the resonance energy of 1,3-butadiene. Explain. Hint: *trans*-1,3-Pentadiene contains an internal double bond.

The Structure of 1,3-Butadiene Reveals the Effect of Conjugation

The structure of 1,3-butadiene is shown in Figure 14-8, including the relevant *p* orbitals, each containing one electron. The two π bonds are in parallel alignment, and so the two *p* orbitals on C-2 and C-3 may overlap. The resulting π interaction is weak but nevertheless amounts to a few kilocalories per mole. As expected, the barrier to rotation is lower than that for a normal π bond (Figure 14-8B).

An inspection of models shows that the molecule can adopt two possible extreme coplanar conformations. In one, designated *s*-cis, the two π bonds lie on the same side of the C-2–C-3 axis; in the other, called *s*-trans, the π bonds are on opposite sides (Figure 14-8B). The prefix *s* refers to the fact that the bridge between C-2 and C-3 constitutes essentially a *single* bond. The *s*-cis form is almost 3 kcal mole^{-1} less stable than the *s*-trans configuration because of the steric interference between the two hydrogens on the inside of the diene unit. Therefore, the *s*-cis conformer adopts a slightly nonplanar conformation, the two double bonds arranged *gauche*.

What is the π-bond strength of the two double bonds in 1,3-butadiene? The same type of experiment used for ethene (Section 11-2) can be applied to *cis*-1-

FIGURE 14-8

A. The structure of 1,3-butadiene. The central bond is shorter than that in an alkane (1.54 Å for the central C–C bond in butane). The p orbitals aligned perpendicular to the molecular plane form a contiguous interacting array.
B. 1,3-Butadiene can exist in two conformations.

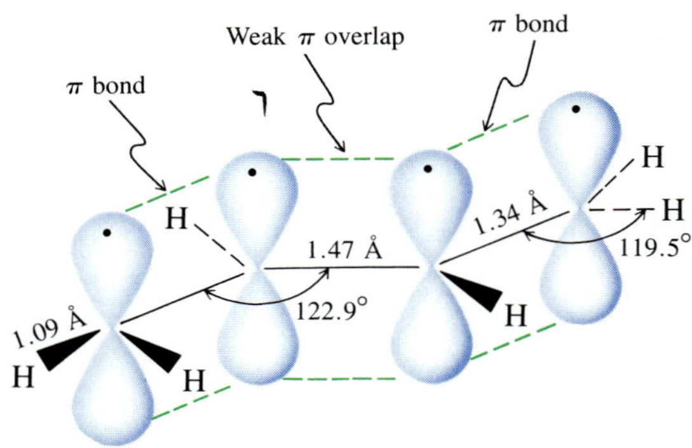

A

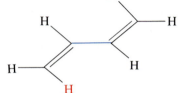

s-cis $\underset{}{\overset{E_a\,=\,3.9\text{ kcal mole}^{-1}}{\rightleftharpoons}}$ s-trans $\qquad \Delta H^\circ = -2.8\text{ kcal mole}^{-1}$

B

deuterio-1,3-butadiene. The result is a much lower energy of isomerization: only about 52 kcal mole^{-1}, 13 kcal mole^{-1} less than in ethene. Why should this be so?

An answer is seen in the transition state of such an isomerization. On one side of the original double bond appears an isolated radical center, but on the other a delocalized allylic radical. The allylic resonance energy is responsible for the lower barrier to isomerization.

EXERCISE 14-8

The dissociation energy of the central C-H bond in 1,4-pentadiene is only 71 kcal mole^{-1}. Explain.

The Molecular-Orbital Picture of 1,3-Butadiene

The π-electronic structure of 1,3-butadiene may be described by constructing four molecular orbitals out of the four p atomic orbitals (Figure 14-9, on page 580). The molecular orbitals increase in energy with increasing number of nodes. The molecular orbital lowest in energy, π_1, has no nodes and exhibits only bonding interactions between the various lobes. Orbital π_2 has one node and therefore one antibonding interaction. This molecular orbital is still bonding

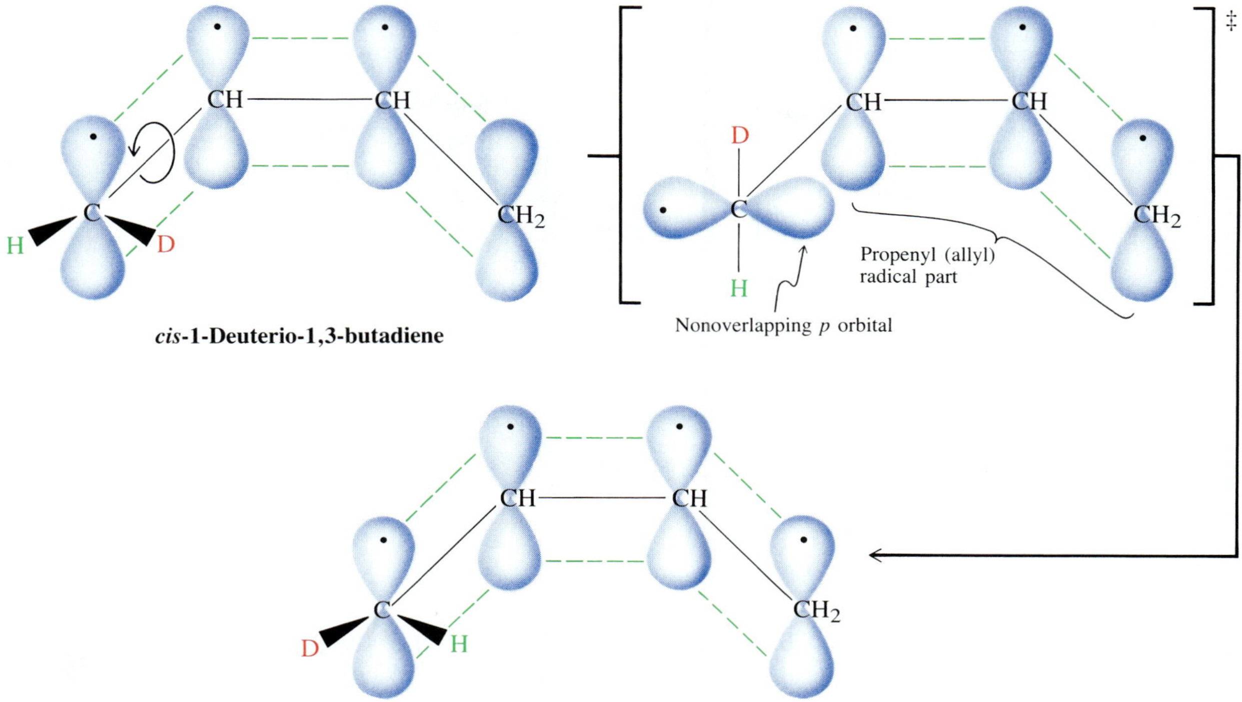

cis-1-Deuterio-1,3-butadiene

trans-1-Deuterio-1,3-butadiene

overall, because the antibonding relation of the two central lobes is made up for by two bonding interactions at the ends. Both π_1 and π_2 are lower in energy than isolated *p* orbitals. Orbital π_3 has two nodes and is antibonding overall (one bonding, two antibonding interactions), whereas π_4 is completely antibonding. The four π electrons are placed into the two bonding molecular orbitals (Figure 14-10, on the next page), revealing the net stabilization of the system when compared with four noninteracting *p* orbitals.

Conjugated Dienes Are Attacked by Electrophiles and Free Radicals

Conjugated dienes are centers of high electron density because of the presence of the π electrons. In fact, although more stable thermodynamically than dienes with isolated double bonds, conjugated dienes are actually *more reactive* in the presence of electrophiles and other reagents. 1,3-Butadiene, for example, readily absorbs one mole of gaseous hydrogen chloride. Two isomeric addition products are formed: 3-chloro-1-butene and 1-chloro-2-butene.

$$CH_2{=}CH{-}CH{=}CH_2 + HCl \xrightarrow{25^\circ C} \underset{\substack{\textbf{3-Chloro-1-butene} \\ 80\%}}{CH_2{=}CH{-}\overset{\overset{\displaystyle Cl}{|}}{C}H{-}CH_2H} + \underset{\substack{\textbf{1-Chloro-2-butene} \\ 20\%}}{HCH_2{-}CH{=}CH{-}CH_2Cl}$$

The generation of the first product is readily understood in terms of ordinary alkene chemistry. It is the result of a Markovnikov addition to one of the double bonds. What about the second product?

FIGURE 14-9

A π-molecular-orbital description of 1,3-butadiene. Note that the size of the various orbital lobes varies somewhat from orbital to orbital.

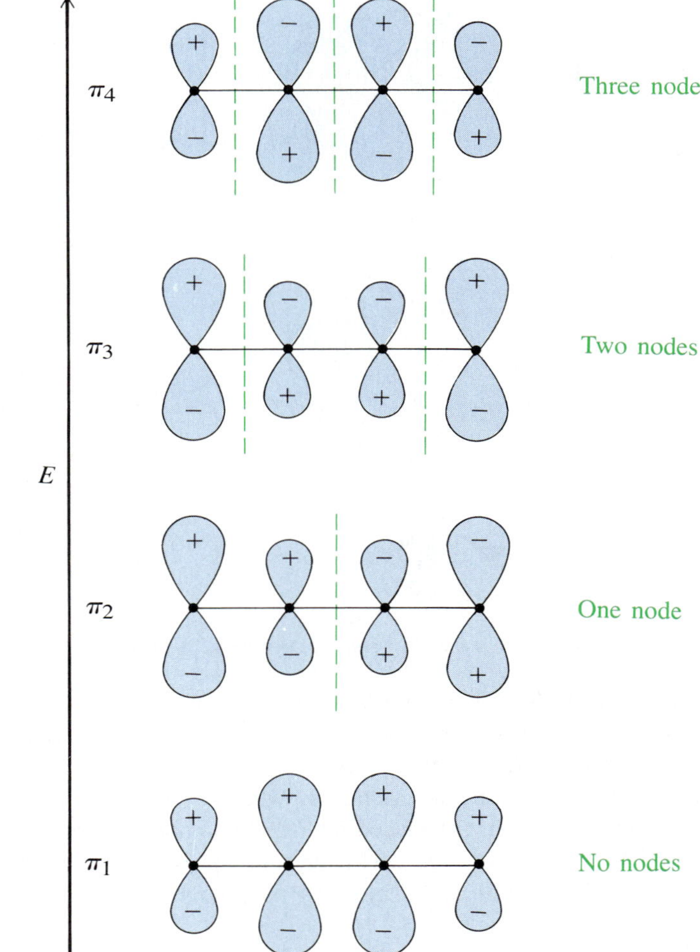

FIGURE 14-10

1,3-Butadiene has four π electrons, which are placed into the two lowest π molecular orbitals, π_1 and π_2. Both of them are bonding.

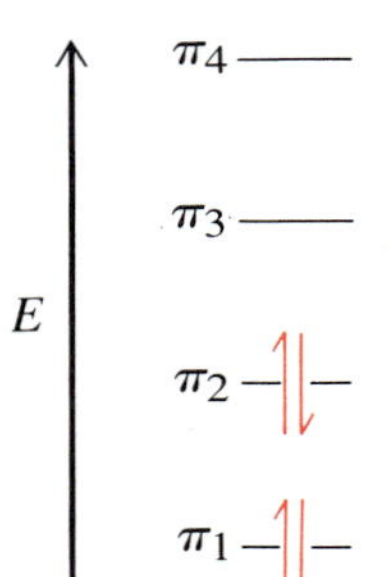

The presence of 1-chloro-2-butene is explained on consideration of the reaction mechanism. Initial protonation occurs at C-1 to give the thermodynamically most favored carbocation, namely an allylic cation.

Protonation of 1,3-Butadiene

$$\overset{+}{C}H_2\text{—}\underset{}{\overset{H}{\overset{|}{C}}}H\text{—}CH\text{=}CH_2 \;\xleftarrow[\text{Attack at C-2}]{H^+}\!\!\!\!\!\!\times\;\; \overset{1}{C}H_2\text{=}\overset{2}{C}H\text{—}\overset{3}{C}H_2\text{=}\overset{4}{C}H_2 \xrightarrow[\text{Attack at C-1}]{H^+}$$

Primary nondelocalized cation not formed

$$\left[CH_3\text{—}\overset{+}{C}H\text{—}CH\text{=}CH_2 \longleftrightarrow CH_3\text{—}CH\text{=}CH\text{—}\overset{+}{C}H_2\right]$$

Delocalized allylic cation formed exclusively

It can be trapped by chloride in two possible ways, to yield the two observed products: at the terminal carbon, it yields 1-chloro-2-butene; and at the internal carbon, it yields 3-chloro-1-butene. The 1-chloro-2-butene is said to form by **1,4-addition** of hydrogen chloride to butadiene because reaction has occurred at C-1 and C-4 of the original diene. The other product arises by the normal 1,2-addition. Many electrophilic additions to dienes give rise to product mixtures by both modes of addition.

Nucleophilic Trapping of the Allylic Cation Formed on Protonation of 1,3-Butadiene

$$CH_3\underset{\underset{Cl}{|}}{C}HCH{=}CH_2 \xleftarrow[\text{Attack at internal carbon}]{Cl^-} CH_3CH\overset{\overset{H}{|}}{\overset{+}{\cdots C \cdots}}CH_2 \xrightarrow[\text{Attack at terminal carbon}]{Cl^-} CH_3CH{=}CH{-}CH_2Cl$$

1,2-Addition **1,4-Addition**

Examples:

$$CH_2{=}CH{-}CH{=}CH_2 + Br{-}Br \xrightarrow{CCl_4} \underset{\underset{Br}{|}}{C}H_2{-}\underset{\underset{Br}{|}}{C}H{-}CH{=}CH_2 + \underset{\underset{Br}{|}}{C}H_2{-}CH{=}CH{-}\underset{\underset{Br}{|}}{C}H_2$$

54% **3,4-Dibromo-1-butene** 46% **1,4-Dibromo-2-butene**

1,3-Cyclohexadiene + DBr ⟶ (D, Br-substituted cyclohexene) + (D, Br-substituted cyclohexene)

90% **3-Bromo-4-deuteriocyclohexene** 10% **3-Bromo-6-deuteriocyclohexene**

These reactions are subject to the rules of thermodynamic and kinetic control. The 1,2-adducts frequently rearrange on heating to give the thermodynamically more stable 1,4-isomers.

Kinetic Compared with Thermodynamic Control in Electrophilic Additions to 1,3-Butadiene

$$BrCH_2\underset{\underset{Br}{|}}{C}HCH{=}CH_2 \overset{\Delta}{\rightleftharpoons} BrCH_2CH\overset{\overset{H}{|}}{\overset{+}{\cdots C \cdots}}CH_2 \;\; Br^- \rightleftharpoons BrCH_2CH{=}CHCH_2Br$$

Kinetic product **Thermodynamic product**

Conjugated dienes also function as monomers in polymerizations induced by electrophiles, free radicals, and other initiators (see Section 14-6).

EXERCISE 14-9

Conjugated dienes can be made by the methods applied to the preparation of ordinary alkenes. Propose syntheses of: (a) 2,3-dimethylbutadiene from 2,3-dimethyl-1,4-butanediol; (b) 1,3-cyclohexadiene from cyclohexane.

In summary, dienes are named according to the rules formulated for ordinary alkenes. Conjugated dienes are more stable than dienes containing two isolated double bonds, as measured by their heats of hydrogenation; the difference is their resonance energy. Conjugation is reflected in the molecular structure of 1,3-butadiene, revealing a relatively short central carbon–carbon bond with a small barrier to rotation of about 4 kcal $mole^{-1}$. The two rotamers *s*-trans and *s*-cis differ in energy by about 3 kcal $mole^{-1}$. The barrier to cis-trans isomerization is 52 kcal $mole^{-1}$, lower than that of ethene because of the allylic stabilization of one of the fragments in the transition state. The molecular-orbital picture of the π system in 1,3-butadiene shows two bonding and two antibonding orbitals. The four electrons are placed in the first two bonding levels, revealing the electronic stabilization of the system. Finally, conjugated dienes are electron rich and are attacked by electrophiles to give intermediate allylic cations on the way to 1,2- or 1,4-addition products.

14-4 Delocalization Among More Than Two Pi Bonds: Extended Conjugation and Benzene

What happens if a molecule contains more than two conjugated double bonds? Are cyclic conjugated systems different from their acyclic analogs? This section will begin to answer these questions.

Extended π Systems Are Thermodynamically Stable but Kinetically Reactive

When more than two double bonds are in conjugation, the molecule is called an **extended π system.** An example is 1,3,5-hexatriene, the next higher double-bond homolog (also called vinylog, because of the added vinyl group) of 1,3-butadiene. This compound is quite reactive and readily polymerizes, particularly in the presence of electrophiles. Despite its reactivity as a delocalized π system, it is also relatively stable thermodynamically. Its increased reactivity is due to the low activation barriers for electrophilic additions, which proceed through highly delocalized carbocations. For example, the bromination of 1,3,5-hexatriene produces a substituted pentadienyl cation intermediate that can be described by three resonance structures.

Bromination of 1,3,5-Hexatriene

$$\underset{\textbf{1,3,5-Hexatriene}}{CH_2{=}CH{-}CH{=}CH{-}CH{=}CH_2} \xrightarrow{Br_2} \left[\begin{array}{c} BrCH_2{-}\overset{+}{C}H{-}CH{=}CH{-}CH{=}CH_2 \\ \updownarrow \\ BrCH_2{-}CH{=}CH{-}\overset{+}{C}H{-}CH{=}CH_2 \\ \updownarrow \\ BrCH_2{-}CH{=}CH{-}CH{=}CH{-}\overset{+}{C}H_2 \end{array}\right] + Br^-$$

$$\downarrow$$

$$\underset{\substack{\textbf{5,6-Dibromo-1,3-hexadiene,}\\ \textbf{a 1,2-addition product}}}{BrCH_2\overset{\overset{Br}{|}}{C}HCH{=}CHCH{=}CH_2} + \underset{\substack{\textbf{3,6-Dibromo-1,4-hexadiene,}\\ \textbf{a 1,4-addition product}}}{BrCH_2CH{=}CH\overset{\overset{Br}{|}}{C}HCH{=}CH_2} + \underset{\substack{\textbf{1,6-Dibromo-2,4-hexadiene,}\\ \textbf{a 1,6-addition product}}}{BrCH_2CH{=}CHCH{=}CHCH_2Br}$$

The final mixture is the result of 1,2-, 1,4-, and 1,6-additions, the last product being the most favored thermodynamically, because it retains an internal conjugated diene system. 1,3,5-Hexatriene is also susceptible to attack by radicals and anionic reagents, giving similarly delocalized intermediate pentadienyl radicals and anions.

EXERCISE 14-10

On treatment with two equivalents of bromine, 1,3,5-hexatriene has been reported to give moderate amounts of 1,2,5,6-tetrabromo-3-hexene. Write a mechanism for the formation of this product.

Some highly extended π systems are found in nature. An example is β-carotene, the orange coloring agent in carrots, and its biological degradation product, vitamin A.

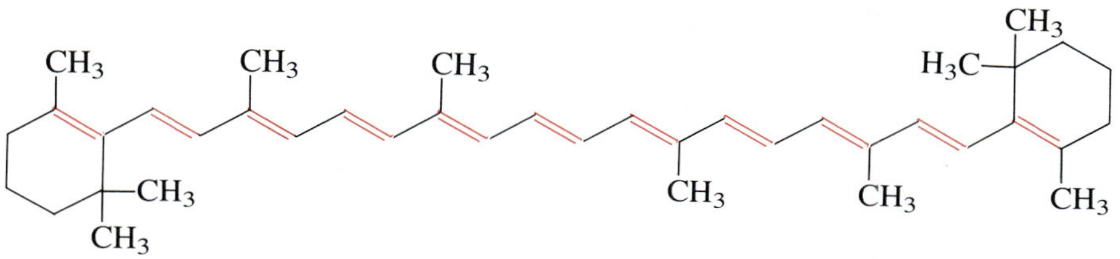

CH_3 CH_3 CH_3 CH_3 H_3C CH_3 CH_3 CH_3 CH_3 CH_3

β-Carotene

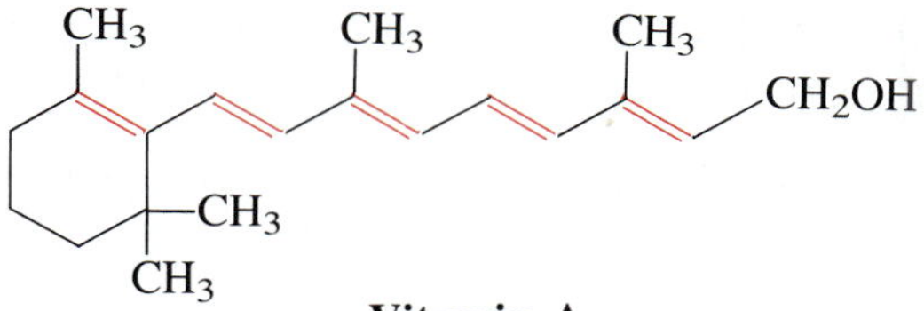

CH_3 CH_3 CH_3 CH_2OH CH_3 CH_3

Vitamin A

Compounds of this type can be very reactive because there are many potential sites for attack by reagents that add to double bonds. Such behavior contrasts with that of some cyclic conjugated systems, which may be considerably less reactive, depending on the number of π electrons (see Chapters 19 and 25). The most-striking example of this effect is benzene, the cyclic analog of 1,3,5-hexatriene.

Benzene and Its Resonance Structures

H H H H H C C C C C C

Benzene

The Special Stability of Benzene, a Cyclic Triene

Cyclic conjugated systems are special cases. The most common examples are the cyclic triene C_6H_6, better known as benzene, and its derivatives (Chapters 19, 20, 24, and 25). In contrast with hexatriene, benzene is unusually stable both thermodynamically and kinetically, because of its special electronic makeup (see Chapter 19). That benzene is unusual can be seen by drawing its resonance structures: there are two *equally* contributing forms. Benzene does not readily undergo addition reactions typical of unsaturated systems, such as catalytic hydrogenation, hydration, halogenation, and oxidation. In fact, because of its low reactivity, benzene can be used as a solvent in organic reactions.

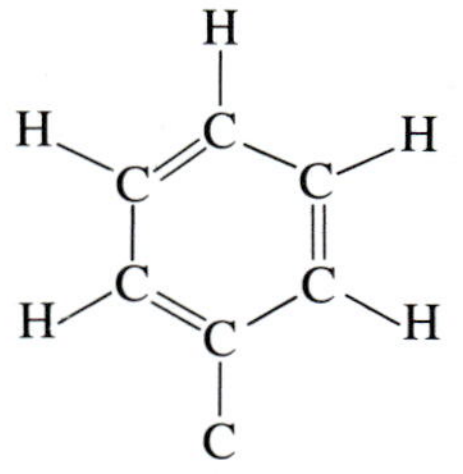

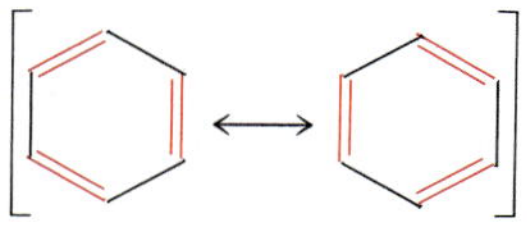

Benzene Is Unusually Unreactive

Exceedingly slow reaction ← H_2, Pd — benzene — $KMnO_4$, 25°C → No reaction

No reaction ← H^+, 25°C — benzene — Br_2, 25°C → No reaction

Although relatively inert, benzene can be attacked by electrophiles, but only under special conditions. Such transformations do not lead to addition, as with ordinary conjugated systems, but rather to products of *substitution*. For example, treatment of benzene with bromine in the presence of catalytic amounts of iron tribromide yields bromobenzene.

Bromination of Benzene Leads to Substitution, Not Addition

Benzene (H) —Br_2, catalytic $FeBr_3$→ Bromobenzene (Br) + HBr, no (Br, Br addition product)

80%

Benzene **Bromobenzene**

The reason for this behavior is that the system strives to retain its stable cyclic six-electron structure. Further examples of this tendency are found in the nitration and sulfonation of the benzene nucleus under conditions that would lead to the complete polymerization of an open conjugated polyene.

Nitration and Sulfonation of Benzene

Benzene —HNO_3, conc. H_2SO_4, Δ→ NO_2

85%

Nitrobenzene

Benzene —SO_3, conc. H_2SO_4, Δ→ SO_3H

95%

Benzenesulfonic acid

Many derivatives of benzene have strong aromas; so they are called aromatic compounds (Section 2-1). The chemistry of benzene and the mechanism of its substitution reactions will be discussed in more detail in Chapter 19.

To summarize, acyclic extended conjugated systems show increasing reactivity because of the many sites open to attack by reagents and the ease of formation of delocalized intermediates. The conjugated system of benzene, however, is unusually unreactive because of its cyclic form.

14-5 Conjugated Pi Systems Can Undergo Unusual Transformations: Cycloadditions and Electrocyclic Reactions

Conjugated double bonds are capable of more than just the types of reactions already encountered in the chemistry of the alkenes, such as electrophilic and free-radical additions. For example, conjugated dienes are capable of thermal cycloadditions to other double bonds to give substituted cyclohexenes. This reaction occurs by simultaneous and stereospecific formation of two carbon–carbon bonds with the nuclei at the ends of the diene. Conjugated π systems also undergo stereospecific thermal and photochemical ring closures called electrocyclic reactions. Cycloadditions and electrocyclic reactions belong to a new class of transformations called **pericyclic** (*peri,* Greek, around), because their transition states have a cyclic array of nuclei and electrons.

The Cycloaddition of Dienes to Alkenes Gives Cyclohexenes: The Diels-Alder Reaction

When a mixture of 1,3-butadiene and ethene is heated in the gas phase, a remarkable reaction takes place in which cyclohexene is formed by the simultaneous generation of two new carbon–carbon bonds.

$$CH_2{=}CH{-}CH{=}CH_2 + CH_2{=}CH_2 \xrightarrow{200°C} \text{cyclo-}(CH_2CH{=}CHCH_2CH_2CH_2)$$

20%

1,3-Butadiene, four π electrons — **Ethene, two π electrons** — **Cyclohexene, a cycloadduct**

This is the simplest example of the **Diels-Alder reaction,*** in which a conjugated diene adds to an alkene to yield cyclohexene derivatives. The Diels-Alder reaction is in turn a special case of the more-general class of **cycloaddition reactions** between π systems. In the Diels-Alder reaction, an assembly of four conjugated atoms containing four π electrons reacts with a double bond containing two π electrons. For that reason, the reaction is also called a [4+2]cycloaddition.

The prototype reaction of butadiene and ethene actually does not work very well and gives only low yields of cyclohexene. It is much better to use an electron-poor alkene with an electron-rich diene. Substitution of the alkene with electron-attracting groups and of the diene with electron-donating groups therefore creates excellent reaction partners.

How do we recognize such groups? Certain substituents exert an inductive effect that changes the electron density of an appended group. For example, trifluoromethyl is inductively electron withdrawing because of the relatively

*Professor Otto Diels, 1876–1954, University of Kiel, Germany, Nobel Prize 1950; Professor Kurt Alder, 1902–1958, University of Köln, Germany, Nobel Prize 1950.

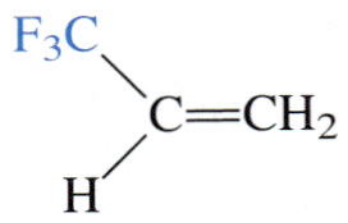

F_3C—CH=CH_2

An electron-poor alkene

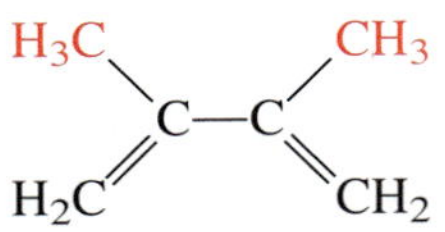

H_2C=C(CH_3)—C(CH_3)=CH_2

An electron-rich diene

high electronegativity of fluorine, whereas alkyl groups are electron pushers because of hyperconjugation (Section 7-3).

There are also many substituents that may donate or withdraw electron density by resonance. For example, carbonyl-containing groups, nitriles, and nitro substituents are good electron acceptors. This is evident from the formulation of charge-separated resonance structures that place positive charges on the alkene carbons and negative charges on the more-electronegative substituent atoms.

Groups That Are Electron Withdrawing by Resonance

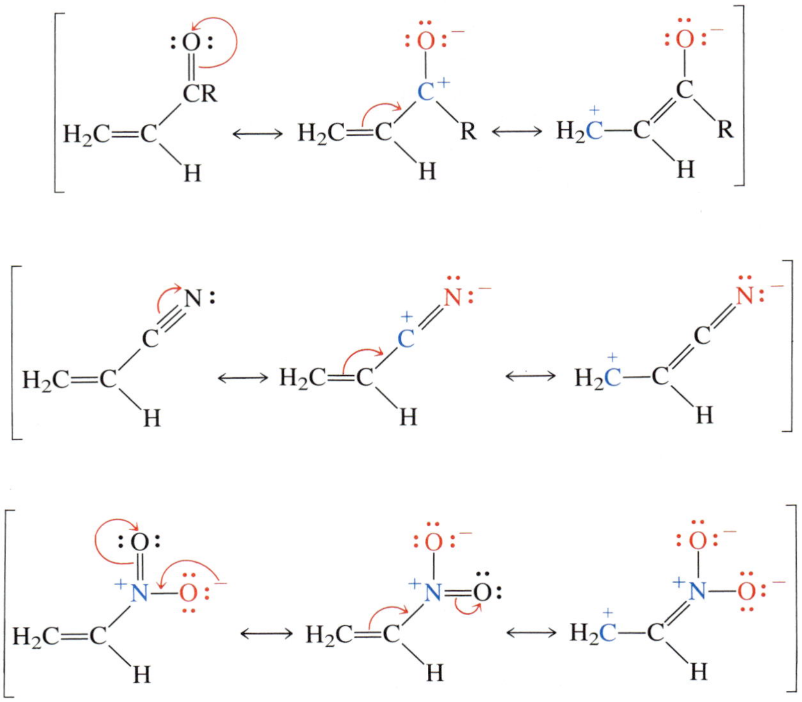

$$\left[H_2C{=}CH{-}C(R){=}\ddot{O}: \longleftrightarrow H_2C{=}CH{-}\overset{+}{C}(R){-}\ddot{O}:^- \longleftrightarrow H_2\overset{+}{C}{-}CH{=}C(R){-}\ddot{O}:^- \right]$$

$$\left[H_2C{=}CH{-}C{\equiv}N: \longleftrightarrow H_2C{=}CH{-}\overset{+}{C}{=}\ddot{N}:^- \longleftrightarrow H_2\overset{+}{C}{-}CH{=}C{=}\ddot{N}:^- \right]$$

$$\left[H_2C{=}CH{-}\overset{+}{N}(=\ddot{O}:){-}\ddot{O}:^- \longleftrightarrow H_2C{=}CH{-}\overset{+}{N}(-\ddot{O}:^-){=}\ddot{O}: \longleftrightarrow H_2\overset{+}{C}{-}CH{=}\overset{+}{N}(-\ddot{O}:^-){-}\ddot{O}:^- \right]$$

An example of the effect of such electronic "push pull" substitution on the efficiency of the Diels-Alder reaction is the cycloaddition of 2,3-dimethyl-1,3-butadiene to propenal (acrolein).

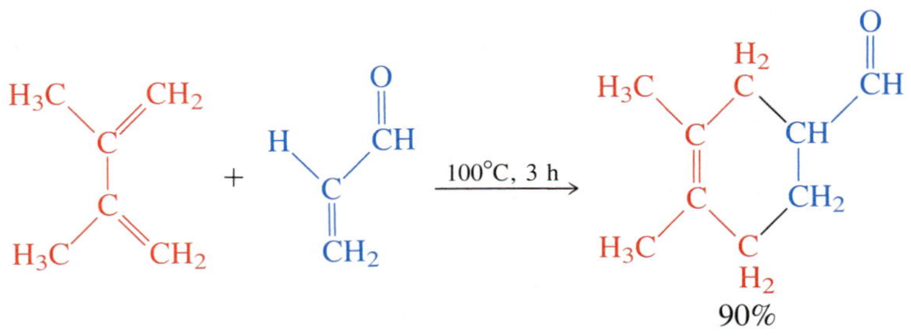

$$\underset{\textbf{2,3-Dimethyl-1,3-butadiene}}{CH_2{=}C(CH_3){-}C(CH_3){=}CH_2} + \underset{\textbf{Propenal}}{CH_2{=}CH{-}CHO} \xrightarrow{100^\circ C,\ 3\ h} \underset{\textbf{Diels-Alder adduct}}{\text{90\%}}$$

The double bond in the product is electron rich and therefore does not compete with the propenal for the diene:

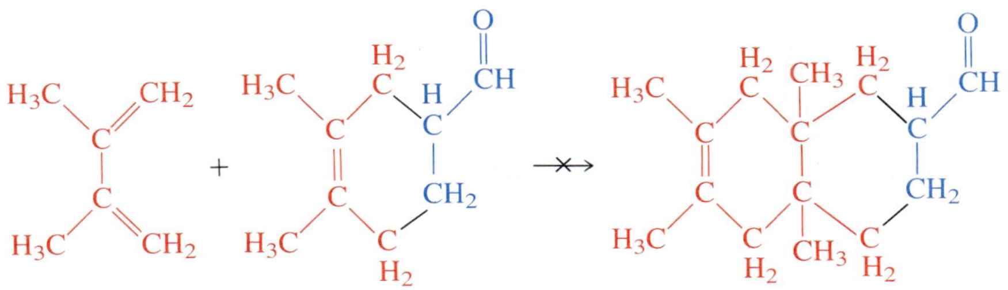

The parent 1,3-butadiene, without additional substituents, is electron rich enough to undergo [4+2]cycloadditions with electron-poor alkenes:

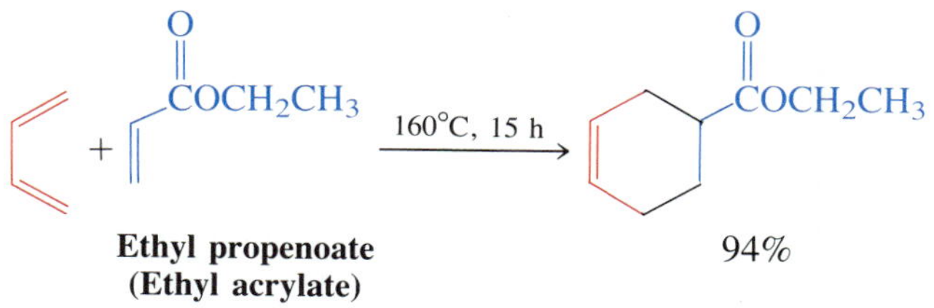

160°C, 15 h

Ethyl propenoate (Ethyl acrylate)

94%

In [4+2]cycloadditions, the substituted ethene is frequently called the **dienophile** (''diene loving''), to contrast it with the diene component. Typical dienes and dienophiles, many of which have common names, are shown in Table 14-1.

EXERCISE 14-11

Formulate the products of [4+2]cycloaddition of tetracyanoethene with: (a) 1,3-butadiene; (b) cyclopentadiene; (c) 1,2-dimethylenecyclohexane.

The Diels-Alder Reaction Is Concerted

The Diels-Alder reaction takes place in one step. Both new carbon–carbon single bonds and the new π bond are formed simultaneously, just as the three π bonds in the starting materials break. As already mentioned, such one-step mechanisms, in which bond breaking happens at the same time as bond making, are *concerted*. The concerted nature of the transformation can be shown as a delocalized transition state in which all six π electrons are indicated by a dotted circle or by the electron-pushing technique.

Two Pictures of the Transition State of the Diels-Alder Reaction

or

Dotted-line picture

Electron-pushing picture

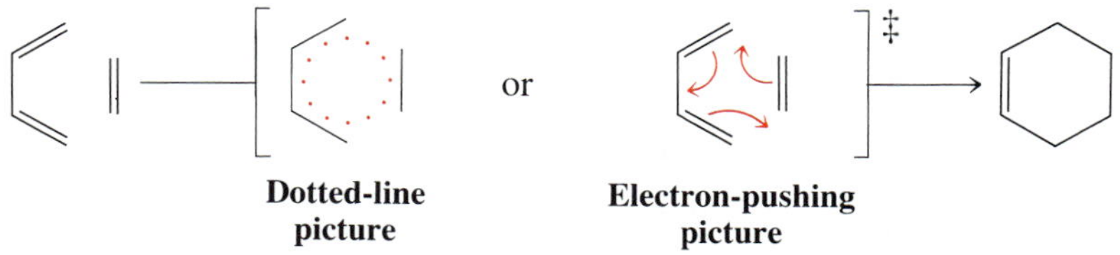

A molecular-orbital representation (Figure 14-11) shows bond formation clearly by overlap of the p orbitals of the dienophile with the terminal p orbitals of the diene. In this way, all of these four carbons in the reaction rehybridize to sp^3. The remaining two internal diene p orbitals are left to form the new π bond.

TABLE 14-1

Some dienes and dienophiles in the Diels-Alder reaction

Dienes		Dienophiles	
	1,3-Butadiene	$(NC)_2C{=}C(CN)_2$	**Tetracyanoethene**
H_3C, H_3C	**2,3-Dimethyl-1,3-butadiene**	H, CN, H, CN	***cis*-1,2-Dicyanoethene**
CH_3, CH_3	***trans,trans*-2,4-Hexadiene**	H, CO_2CH_3, H, CO_2CH_3	**Dimethyl *cis*-2-butenedioate (Dimethyl maleate)**
	1,3-Cyclopentadiene	H, CO_2CH_3, CH_3CO_2, H	**Dimethyl *trans*-2-butenedioate (Dimethyl fumarate)**
	1,3-Cyclohexadiene	O, O, O	**2-Butenedioic anhydride (Maleic anhydride)**
	5-Methylene-1,3-cyclopentadiene (Fulvene)	$CH_3O_2C{-}C{\equiv}C{-}CO_2CH_3$	**Dimethyl butynedioate (Dimethyl acetylene-dicarboxylate)**
	1,2-Dimethylenecyclohexane	$H_2C{=}CHCH{=}O$	**Propenal (Acrolein)**
		$H_2C{=}CHC(=O)OCH_3$	**Methyl propenoate (Methyl acrylate)**

FIGURE 14-11

Molecular-orbital representation of the Diels-Alder reaction between 1,3-butadiene and ethene. The two *p* orbitals at C-1 and C-4 of the former and the two *p* orbitals of the latter interact, as the reacting carbons rehybridize to sp^3 to maximize overlap in the two resulting new single bonds. At the same time, π overlap between the two *p* orbitals on C-2 and C-3 of the diene increases to create a full double bond.

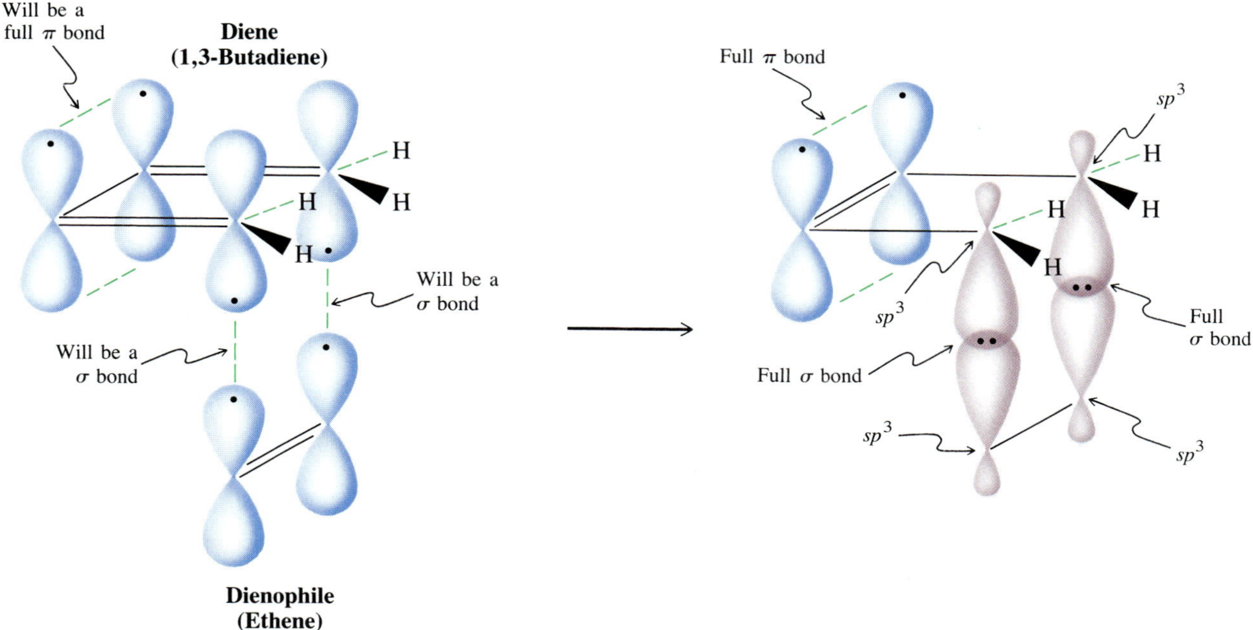

The Diels-Alder Reaction Is Stereospecific

As a consequence of the concerted mechanism, the Diels-Alder reaction is *stereospecific*. For example, reaction of 1,3-butadiene with dimethyl *cis*-2-butenedioate (dimethyl maleate, a cis alkene) gives dimethyl *cis*-4-cyclohexene-1,2-dicarboxylate. In other words, the stereochemistry at the original double bond of the dienophile is retained in the product. In the complementary reaction, dimethyl *trans*-2-butenedioate (dimethyl fumarate, a trans alkene) gives the trans adduct.

In the Diels-Alder Reaction, the Stereochemistry of the Dienophile Is Retained

$$\text{1,3-butadiene} + \text{HC(COOCH}_3\text{)=CH(COOCH}_3\text{)} \xrightarrow{150°\text{–}160°\text{C, 20 h}} \text{product (68\%)}$$

150°–160°C, 20 h

68%

Dimethyl *cis*-2-butenedioate (Dimethyl maleate) Cis starting material

Dimethyl *cis*-4-cyclohexene-1,2-dicarboxylate Cis product

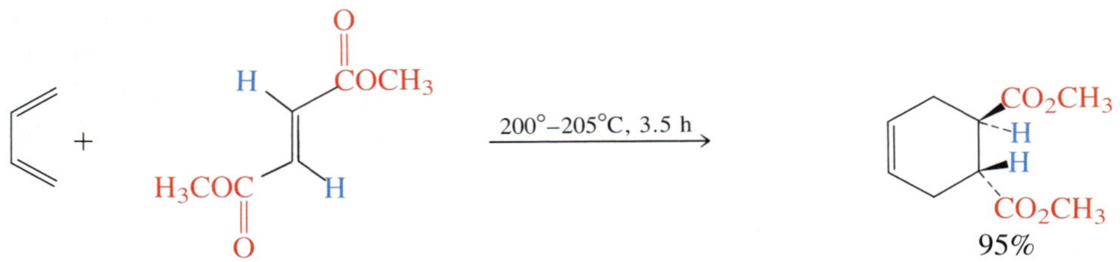

H, $COCH_3$ (=O); H_3COC (=O), H

+ → 200°–205°C, 3.5 h

CO_2CH_3, H, H, CO_2CH_3

95%

Dimethyl *trans*-2-butenedioate (Dimethyl fumarate) Trans starting material

Dimethyl *trans*-4-cyclohexene-1,2-dicarboxylate Trans product

Similarly, the stereochemistry of the diene is also retained.

In the Diels-Alder Reaction, the Stereochemistry of the Diene Is Retained

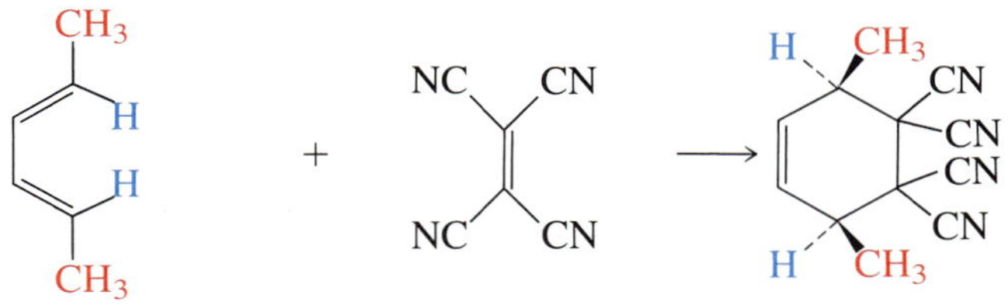

CH_3, H, H, CH_3 + NC, CN, NC, CN → H, CH_3, CN, CN, CN, CN, H, CH_3

***trans,trans*-2,4-Hexadiene** **Tetracyanoethene** **Cis product**

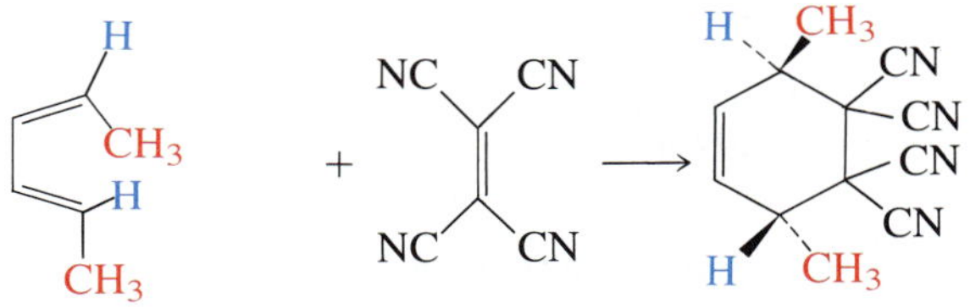

H, CH_3, H, CH_3 + NC, CN, NC, CN → H, CH_3, CN, CN, CN, CN, H, CH_3

***cis,trans*-2,4-Hexadiene** **Trans product**

EXERCISE 14-12

Add the missing products or starting materials to the following schemes.

(a) H_3C, H_3C + CN, NC →

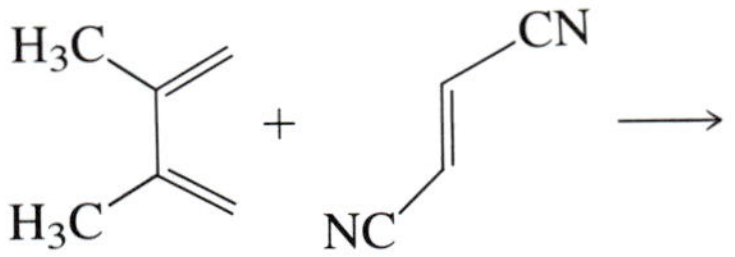

(b) + → H_3C, CH_3, F, F, F, F, CH_3

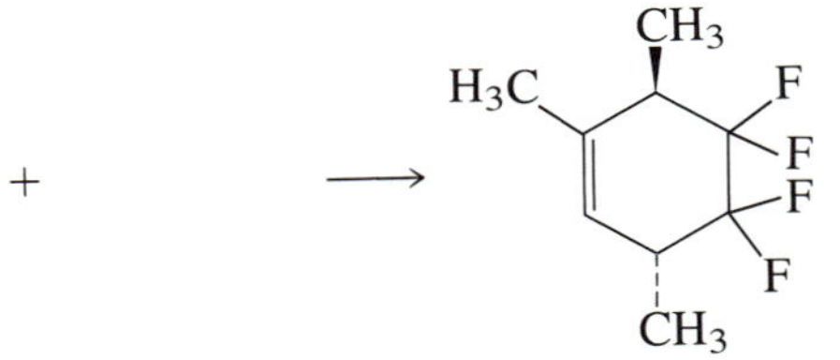

EXERCISE 14-13

cis-trans-2,4-Hexadiene reacts very sluggishly in [4+2]cycloadditions; the trans,trans isomer does so much more rapidly. Explain. Hint: The Diels-Alder reaction requires the *s*-cis arrangement of the diene (Figure 14-11, and see Figure 14-8).

Diels-Alder Cycloadditions Follow the Endo Rule

The Diels-Alder reaction is stereospecific not only with respect to the substitution pattern of the original double bonds, but also with respect to the orientation of the starting materials relative to each other. Consider the reaction of 1,3-cyclopentadiene with dimethyl *cis*-2-butenedioate. Two products are conceivable, one in which the two ester substituents on the bicyclic frame are on the same side (cis) as the bridge, the other in which they are on the side opposite (trans) from the bridge. The first is called the **exo adduct,** the second the **endo adduct** (*exo,* Greek, outside; *endo,* Greek, within). The terms refer to the position of groups in bridged systems. Exo substituents are placed cis with respect to the shorter bridge; endo substituents are positioned trans to this bridge.

Exo and Endo Cycloadditions to Cyclopentadiene

H H H H H CO_2CH_3 H CO_2CH_3 Exo addition CO_2CH_3 H CO_2CH_3 H

H H H H Endo addition CH_3O_2C CH_3O_2C H H H H CO_2CH_3 CO_2CH_3

The Diels-Alder reaction usually proceeds with endo selectivity; that is, only the *endo* product is formed, a result referred to as the **endo rule.**

The Endo Rule

+ $\underset{\underset{O}{\|}}{C}OCH_3$ 80°C H $\underset{\underset{O}{\|}}{C}OCH_3$

Methyl propenoate

91%
Endo **product**

+ O O O 25°C H H O O O

2-Butenedioic anhydride (Maleic anhydride)

100%
Endo **product**

EXERCISE 14-14

Predict the products of the following reactions (show the stereochemistry clearly): (a) *trans,trans*-2,4-hexadiene with methyl propenoate; (b) *trans*-1,3-pentadiene with 2-butenedioic anhydride (maleic anhydride); (c) 1,3-cyclopentadiene with dimethyl *trans*-2-butenedioate (dimethyl fumarate).

EXERCISE 14-15

The Diels-Alder reaction can also occur in an intramolecular fashion. Draw the two transition states leading to products in the following reaction:

O
‖
COCH₃ → 180°C, 5 h → [COCH₃ product, H, H] + [COCH₃ product, H, H]

65 : 35

75% (combined yield)

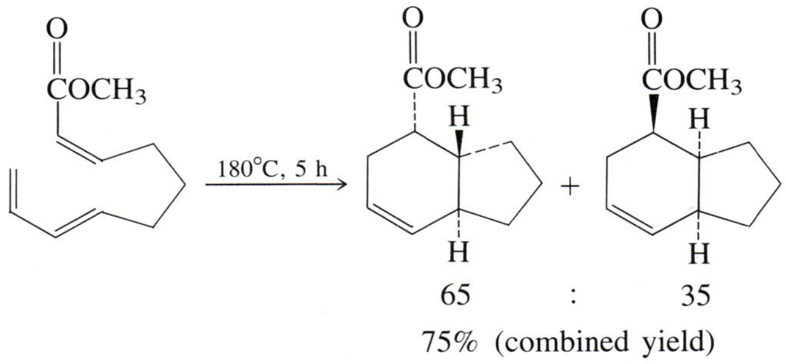

Alkynes as Dienophiles Produce 1,4-Cyclohexadienes

Alkynes also can function as dienophiles in [4+2]cycloadditions. Either one or both of the two π bonds in the alkyne may react. Single addition leads to a 1,4-cyclohexadiene derivative.

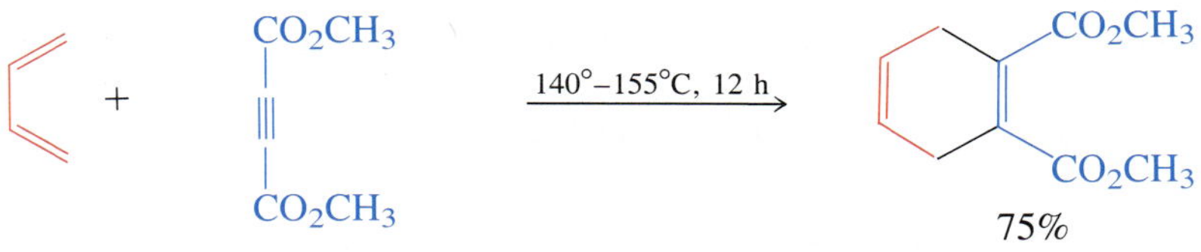

CO_2CH_3 140°–155°C, 12 h CO_2CH_3

CO_2CH_3 CO_2CH_3

75%

Dimethyl butynedioate (Dimethyl acetylene-dicarboxylate)

Dimethyl 1,4-cyclohexadiene-1,2-dicarboxylate

The electron-poor double bond of this product may react with another molecule of the diene.

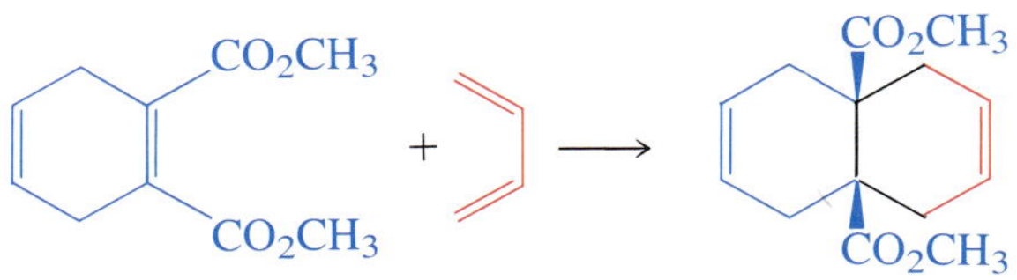

CO_2CH_3 + → CO_2CH_3

CO_2CH_3 CO_2CH_3

A *cis*-bicyclo[4.4.0]-decadiene derivative

Cycloaddition of an alkyne to a cyclic diene results in a bicyclic diene.

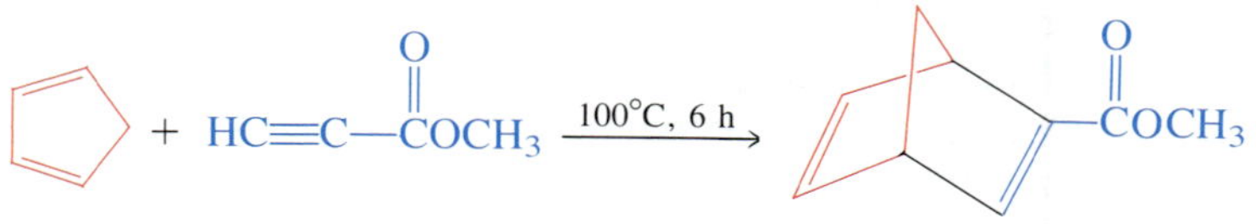

+ HC≡C—COCH₃ $\xrightarrow{100°C,\ 6\ h}$ —COCH₃

EXERCISE 14-16

The reaction of A with dimethyl butynedioate gives the cycloadducts B and C. Explain by a mechanism.

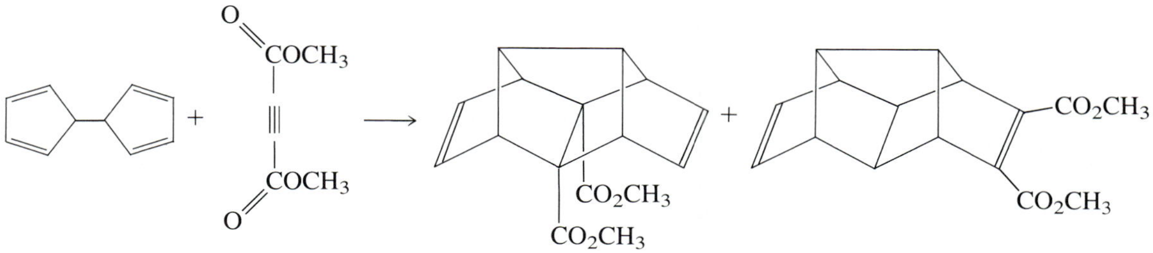

A + $CH_3OC(O)C{\equiv}CC(O)OCH_3$ ⟶ B + C

Other Cycloaddition Reactions

The Diels-Alder reaction is only one of several possible types of cycloadditions. For example, the reaction of ozone with alkenes to form a molozonide (Section 12-5), the first intermediate of ozonolysis, falls into the same category. Related to this reaction are the first steps of the vicinal dihydroxylations of alkenes with permanganate and osmium tetroxide (Section 12-5).

Although isolated double bonds do not normally undergo cycloadditions to each other when heated, they do so in the presence of light to give four-membered rings. Such reactions are called [2+2]cycloadditions.

Photochemical [2+2]Cycloadditions

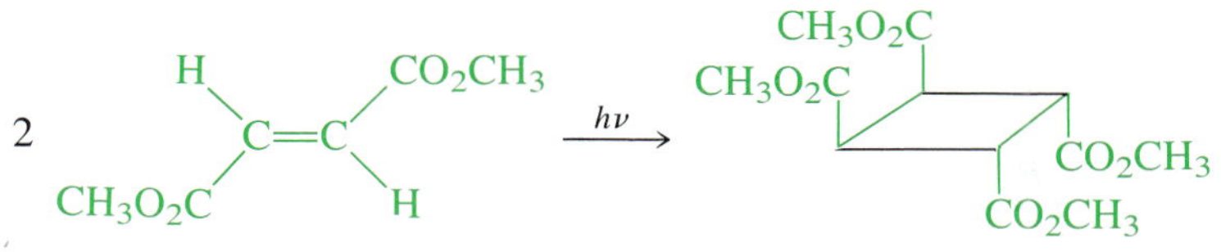

2 $CH_3O_2C(H)C{=}C(H)CO_2CH_3$ $\xrightarrow{h\nu}$ tetrakis(CO_2CH_3)cyclobutane 60%

2 cyclopentenone $\xrightarrow{h\nu}$ 49% + 45%

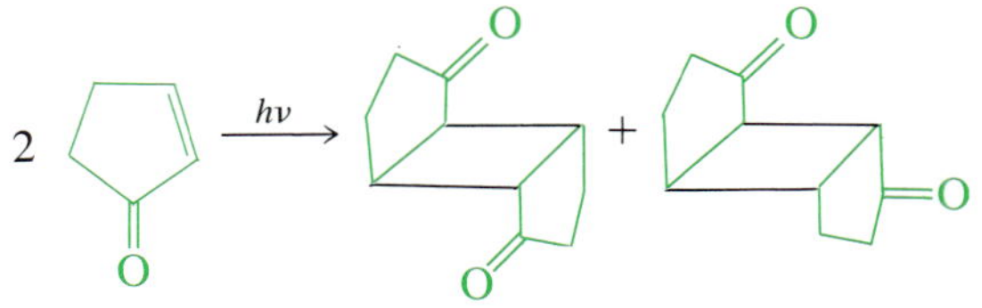

An unusual version of this reaction is the intramolecular photochemical conversion of bicyclo[2.2.1]hepta-2,5-diene (norbornadiene) into quadricyclane (common name). The reaction is thermally unfavorable, and the light energy is needed to drive it to the strained product. This product rapidly undergoes the reverse reaction in the presence of metal catalysts, releasing 26 kcal mole^{-1} of strain energy. A system of this type may become important in efforts to convert the energy of sunlight into a chemically storable and transportable form.

A Photochemical-Energy-Storage System

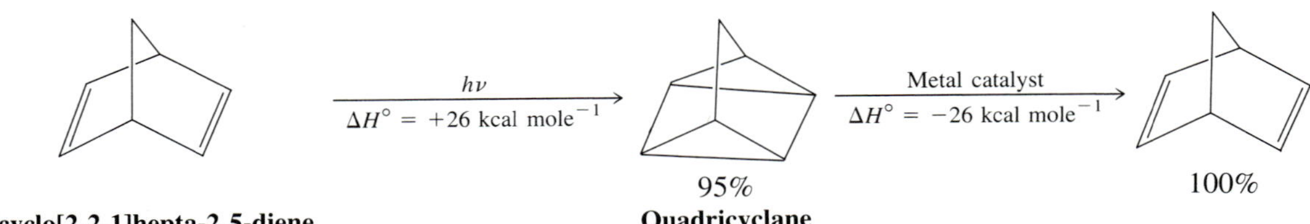

Bicyclo[2.2.1]hepta-2,5-diene (Norbornadiene)

Quadricyclane

Electrocyclic Reactions

In the absence of other reaction partners, conjugated di-, tri- and higher polyenes can isomerize by **electrocyclic reactions.** In these processes, the polyene is converted into a cyclic isomer by linking its two ends and rearranging the double bonds. Alternatively, a cyclic system may be ring-opened by the reverse process. The position of the equilibrium is governed by the relative heats of formation of the isomers. For example, heated *cis*-1,3,5-hexatriene undergoes electrocyclic closure to 1,3-cyclohexadiene. On the other hand, 1,3-butadiene ring closure to cyclobutene is endothermic because of ring strain; so the cyclobutene ring *opens* on heating. If light is used, these simple thermodynamic considerations no longer hold, and the energy of the light can be used to drive the 1,3-butadiene ring closure.

Examples of Electrocyclic Reactions

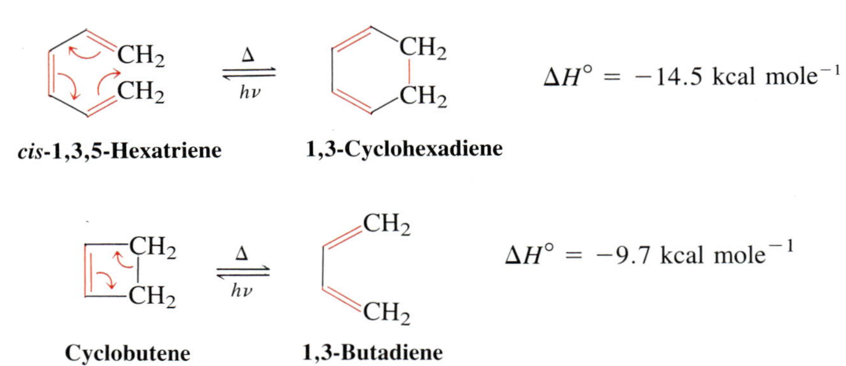

EXERCISE 14-17

Benzocyclobutene A can be heated in the presence of dimethyl-2-butenedioate B to give C. Explain. Hint: Combine an electrocyclic with a Diels-Alder reaction.

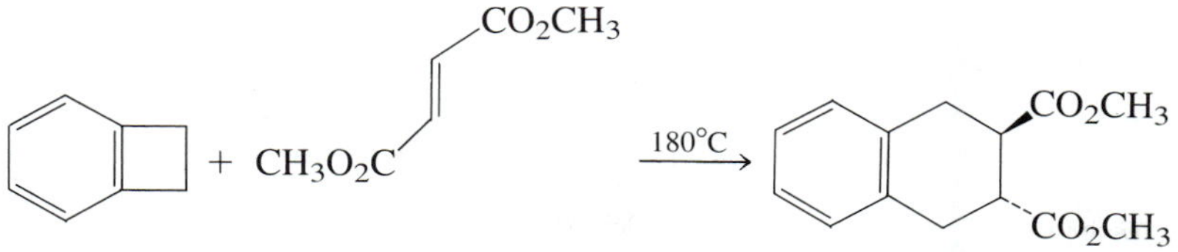

A + B (CH_3O_2C–CH=CH–CO_2CH_3) $\xrightarrow{180°C}$ C (CO_2CH_3, CO_2CH_3)

The Stereochemistry of Electrocyclic Reactions: The Dimethylcyclobutene-Hexadiene Equilibrium

Like the Diels-Alder cycloaddition, electrocyclic reactions are concerted and highly stereospecific. For example, heated *cis*-3,4-dimethylcyclobutene opens to give only *cis,trans*-2,4-hexadiene:

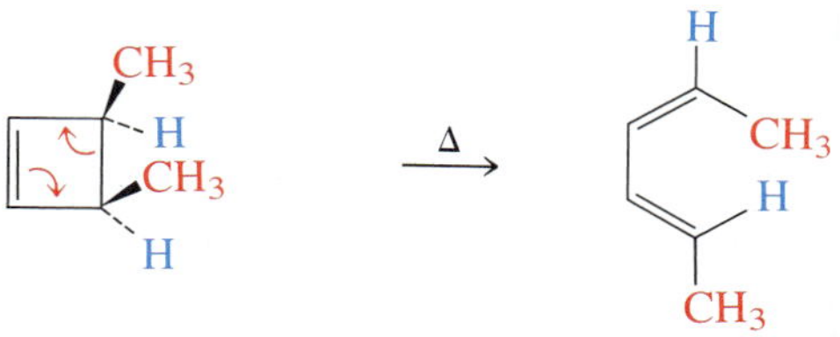

CH_3, H, CH_3, H $\xrightarrow{\Delta}$ H, CH_3, H, CH_3

***cis*-3,4-Dimethylcyclobutene** ***cis,trans*-2,4-Hexadiene**

On the other hand, heated *trans*-3,4-dimethylcyclobutene opens to *trans,trans*-2,4-hexadiene.

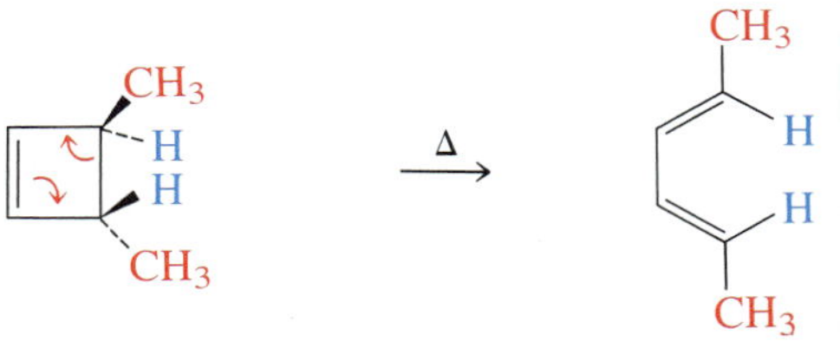

CH_3, H, H, CH_3 $\xrightarrow{\Delta}$ CH_3, H, H, CH_3

***trans*-3,4-Dimethylcyclobutene** ***trans,trans*-2,4-Hexadiene**

We can explain the stereochemistry of these two products by supposing that, as the bond between carbons C-3 and C-4 in cyclobutene is broken, the carbon atoms rotate in the *same* (clockwise or counterclockwise) direction. This mode is called a **conrotatory ring-opening.** During this rotation, the sp^3-hybridized carbons change to sp^2, and the resulting p orbitals overlap with the existing p orbitals of the original cyclobutene double bond to give the π framework of the diene (Figure 14-12A). Note that *trans*-3,4-dimethylcyclobutene has a choice of two conrotatory modes of ring-opening, one in which the two substituents move away from each other and from the center of the resulting diene, and one in which they move toward each other. The first mode, called **conrotatory outward,** results in the observed product (Figure 14-12B). The second, called **conrotatory inward,** would give *cis,cis*-2,4-hexadiene (Figure 14-12C), which is highly congested sterically and therefore not observed.

It is curious that light-induced electrocyclic reactions have different stereochemical results. For example, photochemical closure of butadiene to cyclobutene proceeds with stereochemistry *opposite* to that observed in the ther-

FIGURE 14-12

A. Conrotatory ring-opening of *cis*-3,4-dimethylcyclobutene. Both of the reacting carbons rotate clockwise. The sp^3 hybrid lobes in the ring change into pure *p* orbitals, the carbons becoming sp^2 hybridized. Overlap of these *p* orbitals with those already present in the cyclobutene starting material creates the two double bonds of the cis,trans diene. B. Similar conrotatory opening of *trans*-3,4-dimethylcyclobutene proceeds to the trans,trans diene. This mode of opening, in which the two substituents are placed away from the center of the molecule, is called conrotatory outward and is sterically favored. C. The alternative conrotatory inward process is prohibitive, because of steric congestion in the transition state.

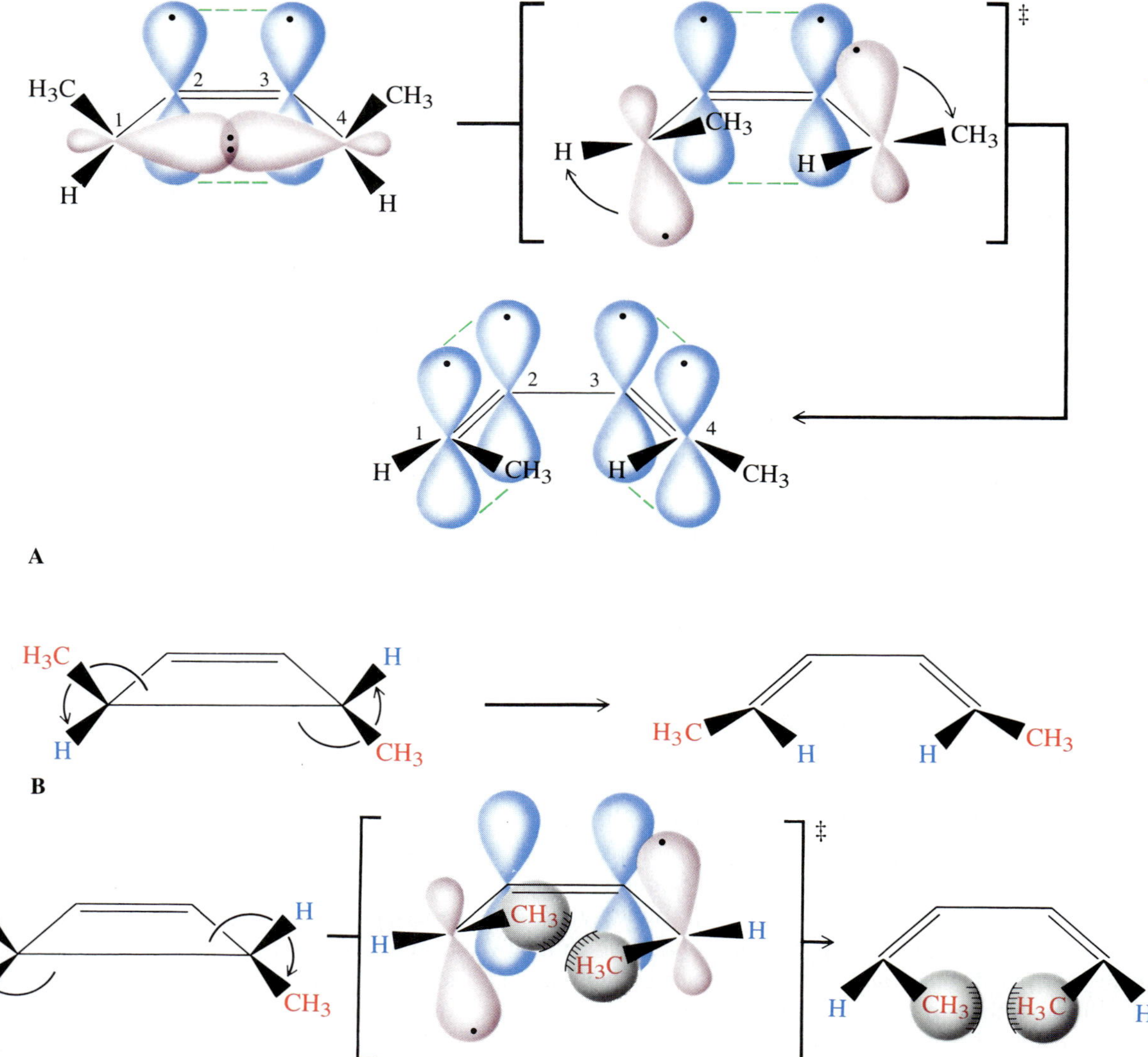

FIGURE 14-13

Disrotatory photochemical ring closure of *cis,trans-* and *trans,trans-*2,4-hexadiene. In the disrotatory mode, one carbon rotates clockwise, the other counterclockwise.

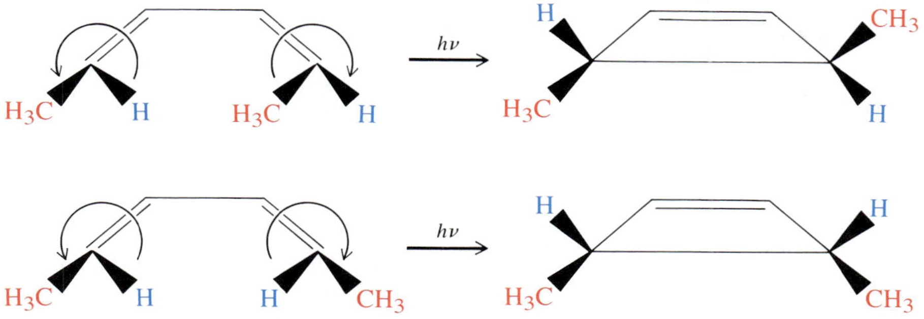

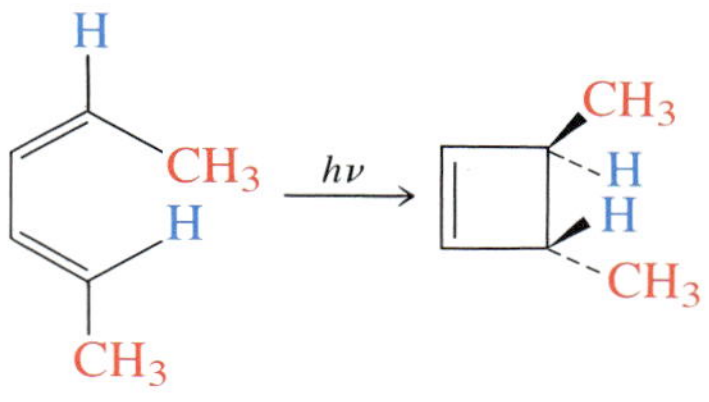

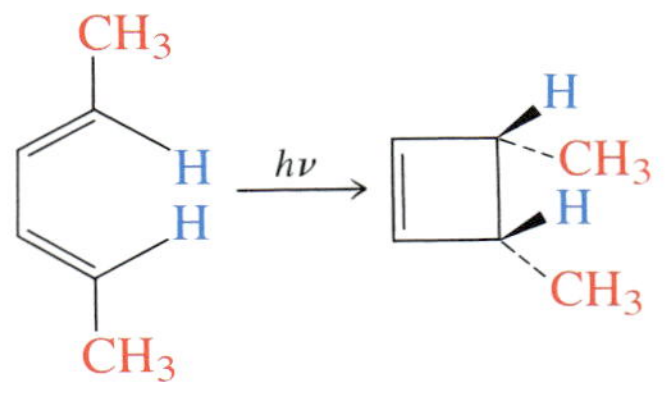

mal opening. In this case, the products arise by rotation of the two reacting carbons in *opposite* directions. In other words, if one rotates clockwise, the other does so counterclockwise. This mode of movement is called **disrotatory** (Figure 14-13).

Thus, whereas the thermal cyclobutene ring opening is conrotatory, the photochemical ring closure is disrotatory.

The Stereochemistry of the Cyclohexadiene–1,3,5-Hexatriene Interconversion

Heating *cis*-1,3,5-hexatriene converts it into cyclohexadiene. (*trans*-1,3,5-Hexatriene cannot undergo such a ring closure, because the two termini are sterically prevented from forming a bond.) Is the stereochemistry of this transformation the same as in the thermal cyclobutene–hexadiene interconversion? The answer, perhaps surprising, is no. The six-membered ring is formed thermally by the disrotatory mode, as may be shown by using substituted derivatives. For example, heated *trans,cis,trans*-2,4,6-octatriene gives *cis*-5,6-dimethyl-1,3-cyclohexadiene, and *cis,cis,trans*-2,4,6-octatriene closes to *trans*-5,6-dimethyl-1,3-cyclohexadiene, both disrotatory closures.

Stereochemistry of the Thermal 1,3,5-Hexatriene Ring Closure

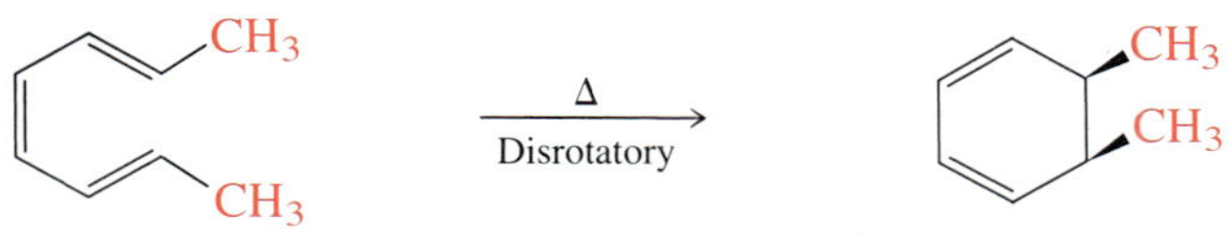

$\xrightarrow[\text{Disrotatory}]{\Delta}$

***trans,cis,trans*-2,4,6-Octatriene** → ***cis*-5,6-Dimethyl-1,3-cyclohexadiene**

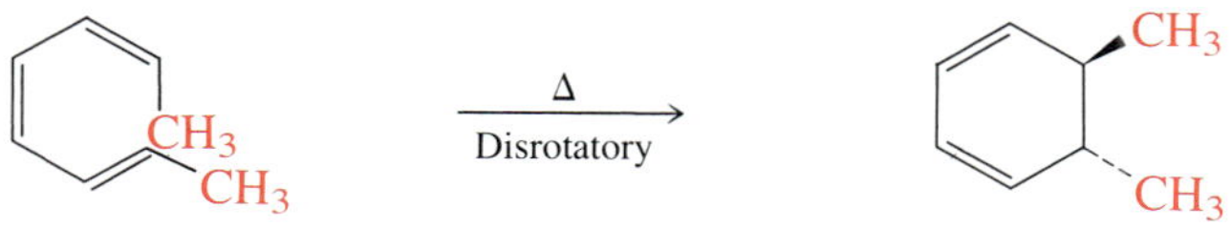

***cis,cis,trans*-2,4,6-Octatriene** → ***trans*-5,6-Dimethyl-1,3-cyclohexadiene**

In contrast, the corresponding photochemical reactions occur in conrotatory fashion.

Stereochemistry of the Photochemical 1,3,5-Hexatriene Ring Closure

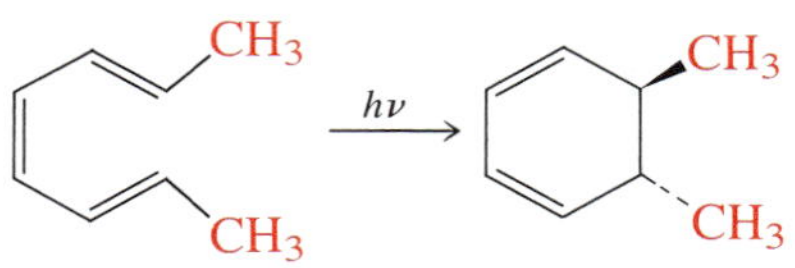

CH_3 $h\nu$ CH_3
CH_3 CH_3

Thus, whereas heated conjugated trienes undergo disrotatory ring closure to the isomeric 1,3-cyclohexadienes, irradiation causes conrotatory closure exactly opposite to the stereochemistry of the 1,3-diene–cyclobutene interconversions.

The guidelines for the stereochemistry of the preceding electrocyclic reactions are two examples of the **Woodward-Hoffmann* rules** governing such processes. These rules were derived from the symmetry of the molecular orbitals [i.e., the sequence of (+) and (−) lobes] primarily participating in these transformations. A more-complete description of the Woodward-Hoffmann rules and their theoretical analysis may be found in advanced organic chemistry textbooks.

EXERCISE 14-18

Photolysis of ergosterol gives provitamin D_2, a precursor of vitamin D_2 (an antirachitic agent). Is the ring-opening conrotatory or disrotatory?

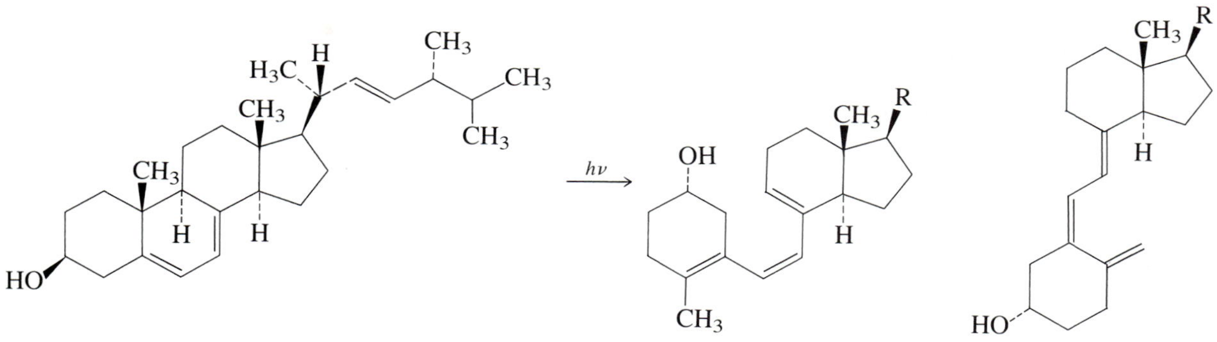

H H_3C CH_3 CH_3 CH_3 CH_3 CH_3 H H HO $h\nu$ OH CH_3 R H CH_3 CH_3 R H HO

Ergosterol **Provitamin D_2** **Vitamin D_2**

In summary, conjugated π systems undergo concerted cycloadditions and electrocyclic reactions. The Diels-Alder reaction is a [4+2]cycloaddition that proceeds best between electron-rich 1,3-dienes and electron-poor dienophiles and that leads to cyclohexenes. The reaction is stereospecific with respect to the stereochemistry of the double bonds and with respect to the arrangements of the

---

*Professor R. B. Woodward, 1917–1979, Harvard University, Nobel Prize 1965; Professor R. Hoffmann, b. 1937, Cornell University, Nobel Prize 1981.

substituents on diene and dienophile: it follows the *endo* rule. Electron-poor alkynes produce 1,4-cyclohexadienes. Photochemical [2+2]cycloadditions lead to substituted cyclobutanes. Electrocyclic reactions of dienes and hexatrienes are (reversible) ring closures to cyclobutenes and 1,3-cyclohexadienes, respectively. The hexadiene-cyclobutene system prefers thermal conrotatory and photochemical disrotatory modes. The hexatriene-cyclohexadiene manifold reacts in the opposite way, proceeding through thermal disrotatory and photochemical conrotatory rearrangements. The stereochemistry of such electrocyclic reactions is governed by the Woodward-Hoffmann rules.

14-6 Polymerization of Conjugated Dienes

Just as simple alkenes can be polymerized (Section 12-7), so can conjugated dienes. However, because they have four unsaturated centers instead of only two, there are more ways of linking the individual diene units. The elasticity of the resulting materials has led to their use as synthetic rubbers. The biochemical way by which natural rubber is made illuminates the structure of terpenes—in particular, the recurrence of the five-carbon unit of 2-methyl-1,3-butadiene (isoprene, see Section 4-7).

The Polymerization of 1,3-Butadiene

1,3-Butadiene may be polymerized by various initiators. Polymerization at carbons 1 and 2 yields a polyethenylethene (polyvinylethylene). Depending on the initiator (cation, radical, anion, organometal), syndio-, iso-, and atactic polymers may be formed, all with different properties (Section 12-8).

1,2-Polymerization of 1,3-Butadiene

$$2n\ CH_2{=}CH{-}CH{=}CH_2 \xrightarrow{\text{Initiator}} {-}(\underset{\substack{|\\ CH\\ \|\\ CH_2}}{CH}{-}CH_2{-}\underset{\substack{|\\ CH\\ \|\\ CH_2}}{CH}{-}CH_2)_n{-}$$

Alternatively, polymerization at carbons 1 and 4 gives either *trans*-polybutadiene, *cis*-polybutadiene, or a mixed polymer.

1,4-Polymerization of 1,3-Butadiene

$$n\ CH_2{=}CH{-}CH{=}CH_2 \xrightarrow{\text{Initiator}} {-}(CH_2{-}CH{=}CH{-}CH_2)_n{-}$$

***cis*- or *trans*-Polybutadiene**

Butadiene polymerization is unique in that the product itself may be unsaturated. The double bonds in this initial polymer may be linked by further treatment with added chemicals, such as radical initiators, or by radiation. In this way, **cross-linked polymers** arise, in which individual chains have been connected into a more rigid framework (Figure 14-14). Cross-linking generally increases the density and hardness of a polymer. It also greatly affects a property characteristic of butadiene polymers: **elasticity.** The individual chains in most polymers can be moved past one another, so that an external deforming force causes an irreversible change in the shape of a polymer bead. In cross-

FIGURE 14-14
The effect of cross-linking of polybutadiene chains.

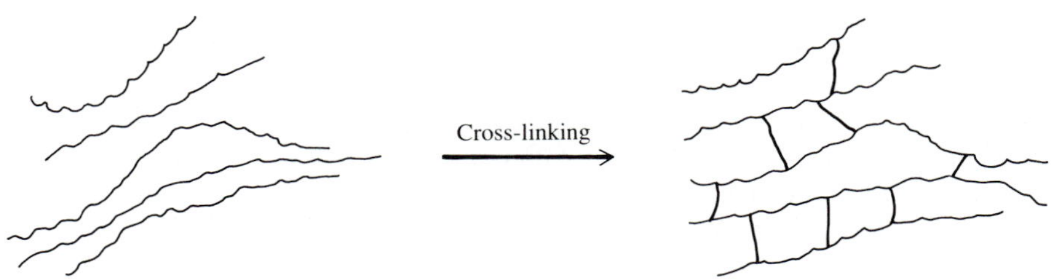

linked polymers, however, with the removal of the external force, the polymer chains snap back to their original positions (more or less). Such elasticity is typical of rubbers.

Synthetic Rubber

Polymerization of 2-methyl-1,3-butadiene (isoprene, Section 4-7) by a Ziegler-Natta catalyst (Section 12-8) results in a synthetic rubber *(polyisoprene)* of almost 100% (*Z*)-1,4-polybutadiene configuration. Similarly, 2-chloro-1,3-butadiene can be polymerized into an elastic, heat- and oxygen-resistant polymer called neoprene, again almost exclusively in the *Z* form.

$$n\ \mathrm{H_2C{=}\overset{\overset{\displaystyle CH_3}{|}}{C}{-}CH{=}CH_2} \xrightarrow{\mathrm{TiCl_4,\ AlR_3}} \mathrm{{-}(H_2C{-}C(CH_3){=}C(H){-}CH_2)_n{-}}$$

2-Methyl-1,3-butadiene → **(*Z*)-Polyisoprene**

$$n\ \mathrm{H_2C{=}\overset{\overset{\displaystyle Cl}{|}}{C}{-}CH{=}CH_2} \xrightarrow{\mathrm{TiCl_4,\ AlR_3}} \mathrm{{-}(H_2C{-}C(Cl){=}C(H){-}CH_2)_n{-}}$$

2-Chloro-1,3-butadiene → **Neoprene**

Many useful types of synthetic rubber are obtained by copolymerization of butadiene with other alkenes. For example, butadiene and phenylethene (styrene) form a rubber called Buna S or GRS (Government Rubber-Styrene). This substance is derived by linking all of the double bonds in the starting materials. The butadiene participates in both 1,2- and 1,4-fashion, the latter randomly giving *Z* and *E* double bonds, which are further cross-linked.

Natural *Hevea* rubber is a 1,4-polymerized (*Z*)-poly(2-methyl-1,3-butadiene) similar in structure to polyisoprene. To increase its elasticity, it is treated with hot elemental sulfur in a process called **vulcanization** (*Vulcanus,* Latin, the Roman god of fire), which creates sulfur cross-links. This reaction was discovered by Goodyear* in 1839.

---

*Charles Goodyear, 1800–1860, American inventor.

Biosynthesis of Natural Rubber

How is rubber made in nature? Plants construct the polyisoprene framework of natural rubber by using as a building block isopentenyl pyrophosphate or 3-methyl-3-butenyl pyrophosphate. This molecule is an ester of pyrophosphoric acid and 3-methyl-3-buten-1-ol. An enzyme equilibrates a small amount of this material with the 2-butenyl isomer, an allylic pyrophosphate.

Biosynthesis of the Two Isomers of 3-Methylbutenyl Pyrophosphate

$$H_2C{=}C(CH_3)CH_2CH_2OH \quad \text{3-Methyl-3-buten-1-ol}$$

$$HO{-}P(=O)(OH){-}OH + HO{-}P(=O)(OH){-}OH \quad \text{Phosphoric acid} \xrightarrow{-\,HOH} HO{-}P(=O)(OH){-}O{-}P(=O)(OH){-}OH \quad \text{Pyrophosphoric acid}$$

$$\xrightarrow{-\,HOH} H_2C{=}C(CH_3)CH_2CH_2O{-}P(=O)(OH){-}O{-}P(=O)(OH){-}OH \underset{}{\overset{\text{Enzyme}}{\rightleftharpoons}} (H_3C)_2C{=}CHCH_2O{-}P(=O)(OH){-}O{-}P(=O)(OH){-}OH$$

3-Methyl-3-butenyl pyrophosphate — 3-Methyl-2-butenyl pyrophosphate

Although the subsequent processes are enzymatically controlled, they can be formulated simply in terms of familiar mechanisms (OPP = pyrophosphate).

Mechanism of Natural Rubber Synthesis:

1 Ionization

$$\text{(CH}_3)_2C{=}CHCH_2OPP \xrightarrow{-\,OPP^-} \left[(CH_3)_2C{=}CHCH_2^+ \longleftrightarrow (CH_3)_2C^+{-}CH{=}CH_2\right]$$

2 Electrophilic attack

$$(CH_3)_2C{=}CHCH_2^+ + H_2C{=}C(CH_3)CH_2CH_2OPP \longrightarrow (CH_3)_2C{=}CHCH_2CH_2C^+(CH_3){-}CH(H)CH_2OPP$$

3 Proton loss

$$(CH_3)_2C{=}CHCH_2CH_2C^+(CH_3){-}CH_2CH_2OPP \xrightarrow{-\,H^+} (CH_3)_2C{=}CHCH_2CH_2C(CH_3){=}CHCH_2OPP$$

Geranyl pyrophosphate

4 Second oligomerization

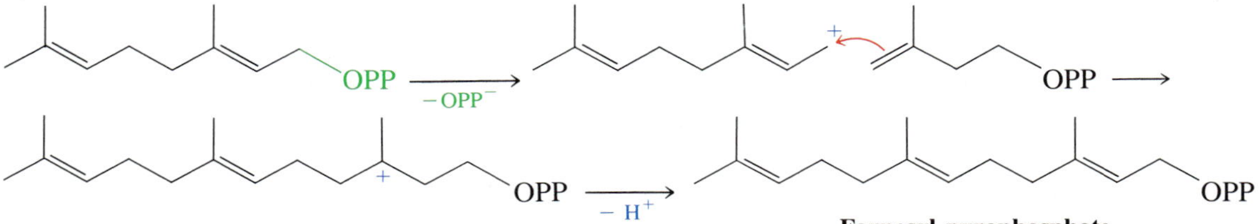

OPP —— $-OPP^-$ → OPP →

OPP —— $-H^+$ → OPP

Farnesyl pyrophosphate

Thus, ionization of the allylic pyrophosphate to the allylic cation, followed by attack on the initial 3-methyl-3-butenyl pyrophosphate and then proton loss, yields a dimer called geranyl pyrophosphate. Repetition of this process with another molecule of 3-methyl-3-butenyl pyrophosphate yields farnesyl pyrophosphate. Further oligomerization eventually leads to a polymer containing an (*E*)-poly-2-methyl-butadiene [(*E*)-polyisoprene] framework, a natural rubber known as guttapercha. Many plants form rubbers containing *Z* double bonds by a similar enzymatic process.

3-Methyl-3-butenyl Pyrophosphate Is an Important Five-Carbon Building Block in Biosynthesis

Many natural products are derived from 3-methyl-3-butenyl pyrophosphate, including the terpenes first discussed in Section 4-7. Terpenes are composed of five-carbon units ultimately traceable to 2-methyl-1,3-butadiene. Indeed, terpenes are made by joining several 3-methyl-3-butenyl pyrophosphate molecules in a variety of ways. The monoterpene geraniol and the sesquiterpene farnesol form by hydrolysis of their corresponding pyrophosphates.

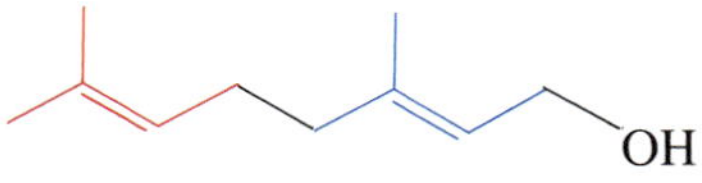

OH

Geraniol

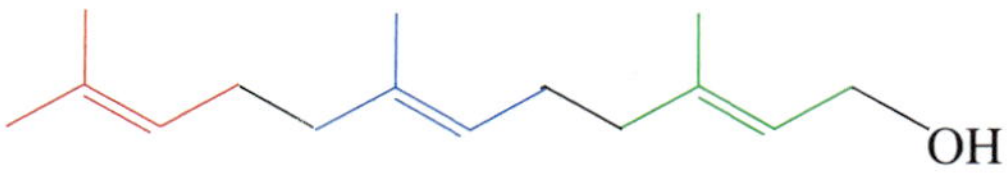

OH

Farnesol

Reductive coupling of two molecules of farnesyl pyrophosphate leads to squalene, a biosynthetic precursor of the steroid nucleus (Section 12-7).

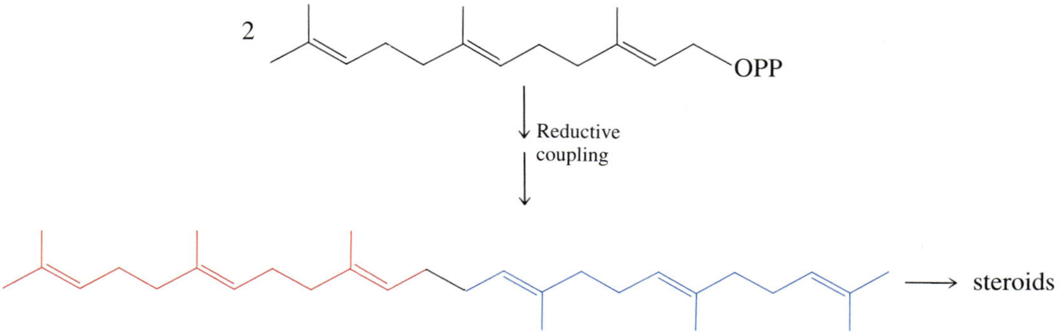

2 OPP

↓ Reductive coupling

→ steroids

Squalene

Bicyclic diterpenes of the camphor variety are built up from geranyl pyrophosphate by enzymatically controlled electrophilic carbon–carbon bond formations.

Camphor Biosynthesis from Geranyl Pyrophosphate

OPP — Cis, trans isomerization → OPP → $-$ OPP$^-$ → + →

Geranyl pyrophosphate

[+ is the same as +] → + → → **Camphor** (O)

Other higher terpenes are constructed by similar cyclization reactions.

To summarize, 1,3-butadiene polymerizes in a 1,2 or 1,4 manner to give polybutadienes with various amounts of cross-linking and therefore variable elasticity. Synthetic rubber can be made by polymerizing 2-methyl-1,3-butadiene into polymers containing various amounts of *E* and *Z* double bonds. Natural rubber is constructed by isomerization of 3-methyl-3-butenyl pyrophosphate to the 2-butenyl system, ionization, and electrophilic (step-by-step) polymerization. Similar mechanisms account for the incorporation of isoprene units into the polycyclic structure of terpenes.

14-7
Electronic Spectra: Ultraviolet and Visible Spectroscopy

Section 10-1 explained that organic molecules may absorb radiation at various wavelengths. Spectroscopy is possible because molecules must absorb quanta of specific energies, $h\nu$, to undergo certain excitations with energy change ΔE.

$$\Delta E = h\nu = \frac{hc}{\lambda} \qquad (c = \text{velocity of light})$$

In nuclear magnetic resonance, electromagnetic energy flips the spin of an atomic nucleus from alignment with a magnetic field to alignment against it. The energy required is very small, only a fraction of a calorie per mole. Consequently, radiowaves are sufficient to cause these transitions.

This section will discuss a form of spectroscopy that requires electromagnetic radiation of appreciably higher energy, within the wavelength ranges of 200 to 400 nm, called **ultraviolet spectroscopy,** and 400 to 800 nm, **visible**

FIGURE 14-15
Excitation of an electron from a bonding to an antibonding (excited) state.

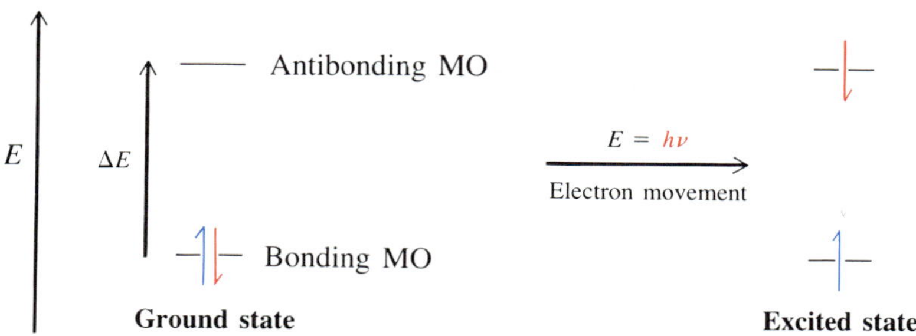

spectroscopy (see Figure 10-2). This type of spectroscopy is useful for investigating the electronic structures of unsaturated molecules and for measuring the extent of their conjugation. The spectra so produced are called **electronic spectra.** The event triggered by electromagnetic radiation at these wavelengths is the excitation of electrons from filled bonding (and nonbonding) to unfilled antibonding molecular orbitals.

Electronic Excitations are the Source of Electronic Spectra

Bonds are made by positive and negative (in-phase and out-of-phase) overlap of atomic orbitals (Section 1-7). Positive overlap gives bonding molecular orbitals occupied by the bonding electrons. Negative overlap gives empty, antibonding molecular orbitals of higher energy (Figure 14-15). When both electrons of a simple bond are located in the bonding orbital, the molecule is said to be in its ground state. Transfer of one of these electrons to an antibonding orbital by energy from incident radiation raises a molecule to an excited state. This electronic excitation is the basis of ultraviolet and visible spectroscopy. The wavelength of the absorbed light depends on the difference in energy between the two states; this difference in turn depends on that between the occupied and unoccupied molecular orbitals. Carbon–carbon and carbon–hydrogen σ bonds have a large gap between those orbitals, an indication of good overlap. Excitations of electrons in such bonds are called **$\sigma \rightarrow \sigma^*$ transitions,** the wavelengths absorbed are usually in the **extreme** or **vacuum ultraviolet spectral region,** below 200 nm. Special equipment is required to measure spectra in this range (e.g., vacuum pumps to remove air, which absorbs radiation below 200 nm). On the other hand, excitation of electrons in π bonds requires less energy and may be observed by scanning the region above 200 nm. The electrons being excited are said to undergo **$\pi \rightarrow \pi^*$ transitions.** It is also possible for nonbonding electrons to be excited to higher states. Such events are called **$n \rightarrow \pi^*$** or **$n \rightarrow \sigma^*$ transitions** (n stands for nonbonding).

A conjugated molecule may have a variety of bonding, nonbonding, and antibonding molecular orbitals. Therefore, a variety of transitions may occur, and several absorption peaks may be observed in their electronic spectra (Figure 14-16). These peaks will be found at the short-wavelength end of the spectrum for the energetically costly excitations, at the long-wavelength end for the easier ones. Conjugated systems that absorb light at wavelengths longer than about 400 nm are colored; hence this range is called the visible range. For example, molecules in which transitions require 450-nm light are orange-red, those that absorb at 550 nm are violet, at 650 nm blue-green (see Table 14-2).

FIGURE 14-16

Various possible electronic transitions in a conjugated molecule.

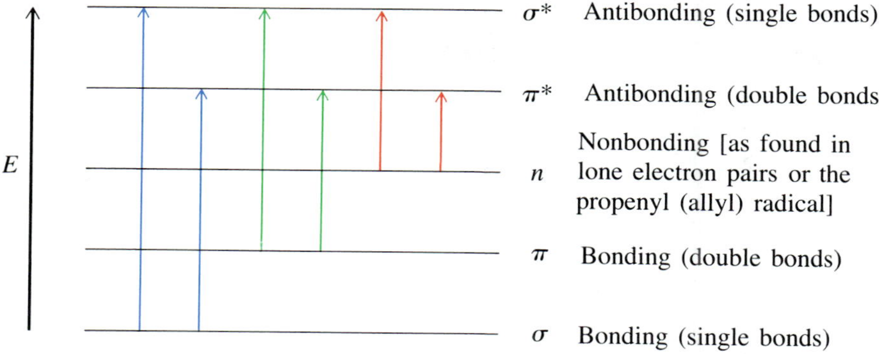

The UV-Visible Spectrometer and the Recorded Spectrum

A UV-visible spectrometer is constructed according to the general scheme in Figure 10-3. As in NMR, samples are usually dissolved in solvents that do not absorb in the spectral region under scrutiny. Examples are ethanol, methanol, and cyclohexane, none of which have absorption peaks above 200 nm. A typical UV spectrum is that of 2-methylbutadiene (isoprene), shown in Figure 14-17. A peak in a spectrum is quoted as a wavelength, a λ_{max} value (in nm). The height (absorbance) of such a peak is reported as the **extinction coefficient, ϵ,** which is characteristic of the molecule. The value of ϵ is calculated by dividing the measured peak height (absorbance, A) by the molar concentration, C, of the sample (assuming a standard sample cell of length 1 cm):

$$\epsilon = \frac{A}{C}$$

FIGURE 14-17

Ultraviolet spectrum of 2-methyl-1,3-butadiene in methanol, λ_{max} = 222.5 nm (ϵ = 10,800). The two indentations at the sides of the main peak are called shoulders.

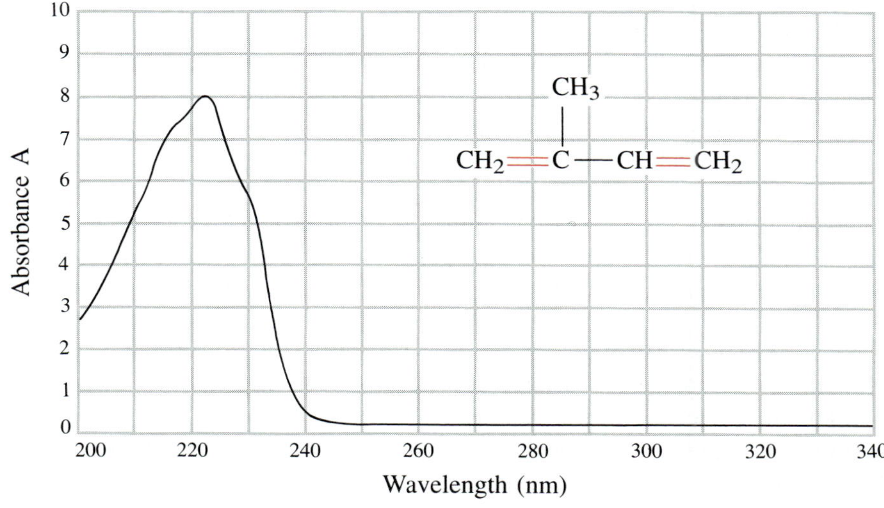

TABLE 14-2

Some λ_{max} values for lowest-energy transitions in ethene and conjugated pi systems

Alkene structure	Name	λ_{max} (nm)	ϵ
	Ethene	171	15,500
CH_2	1,4-Pentadiene	178	Not known
	1,3-Butadiene	217	21,000
	2-Methyl-1,3-butadiene	222.5	10,800
	trans-1,3,5-Hexatriene	268	36,300
	trans,trans-1,3,5,7-Octatetraene	330	Not known
	2,5-Dimethyl-2,4-hexadiene	241.5	13,100
	1,3-Cyclopentadiene	239	4,200
	1,3-Cyclohexadiene	259	10,000

Extinction coefficients typically range from several hundred to several hundred thousand. Electronic spectral peaks are frequently broad, as in Figure 14-17, not the sharp lines typical of many NMR spectra.

What Can the Electronic Spectrum Tell Us?

Electronic spectra can show how extensively conjugated an organic molecule is. The more double bonds there are in conjugation, the longer will be the wavelength for the lowest-energy excitation (and the more peaks will appear in the spectrum). For example, ethene absorbs at $\lambda_{max} = 171$ nm, and an unconjugated diene, such as 1,4-pentadiene, at $\lambda_{max} = 178$ nm. On the other hand, a conjugated diene, such as 1,3-butadiene, absorbs at much higher wavelength ($\lambda_{max} = 217$ nm), corresponding to a lower energy of excitation. Adding conjugated double bonds leads to corresponding incremental increases in the λ_{max} values, as shown in Table 14-2. As the table indicates, other factors influence the position of the peak of longest wavelength, particularly the way in which the π system is incorporated into a carbocyclic frame and the extent of substitution.

Some molecules that exhibit complicated electronic spectra are benzene (see Section 19-3, Figure 19-7) and the deep blue-violet azulene (Figure 14-18).

Alkene structure	Name	λ_{max} (nm)	ϵ
	A steroid diene	282	Not known
	A steroid triene	324	Not known
	A steroid tetraene	355	Not known
(For structure, see Section 14-4)	β-Carotene (Vitamin A precursor)	497 (orange)	133,000
	Azulene, a cyclic conjugated hydrocarbon	696 (blue-violet)	150

Why would the number of conjugated double bonds have such a strong effect on the electronic spectrum? As the degree of conjugation increases, the number

FIGURE 14-18

UV-visible spectrum of azulene in cyclohexane. The absorbance is plotted as log ϵ to compress the scale. The horizontal axis, representing wavelength, also is nonlinear.

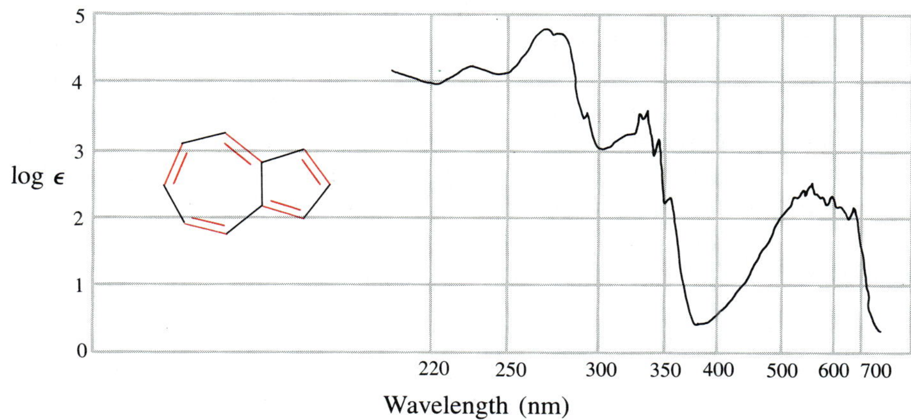

FIGURE 14-19
The π-molecular-orbital levels in ethene, the 2-propenyl (allyl) radical, and butadiene. The distance between the highest occupied molecular orbital (HOMO) and the lowest unoccupied molecular orbital (LUMO) decreases with increasing conjugation.

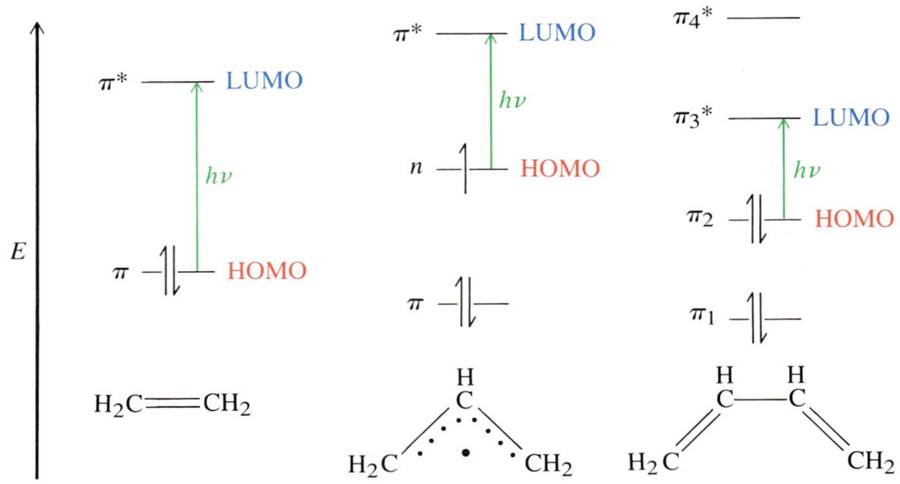

of energy levels corresponding to the respective *p* molecular orbitals also increases (Figure 14-19); likewise, the energy difference between the highest occupied and lowest unoccupied levels decreases. It is this difference that corresponds to the lowest-energy electronic excitation, which therefore requires radiation of less energy (longer wavelength) in the more extended π systems.

EXERCISE 14-19

Each substitution of a hydrogen by an alkyl group at an sp^2 carbon in a conjugated system causes the wavelength associated with the lowest-energy $\pi \rightarrow \pi^*$ transition to increase by 5 nm. Use this fact and the measured value of λ_{max} for 1,3-butadiene to calculate λ_{max} for 2-methylbutadiene and 2,5-dimethyl-2,4-hexadiene. Compare your results with the measured values in Table 14-2.

In summary, UV and visible spectroscopy can be used to detect electronic excitations in conjugated molecules. With increasing number of molecular orbitals there is an increasing variety of possible transitions and hence of absorption bands in the observed spectra. The absorption peak of longest wavelength in a UV-visible spectrum is typically associated with the movement of an electron from the highest occupied to the lowest unoccupied molecular orbital. This wavelength value increases, indicating a lesser difference between these orbitals with increasing conjugation and on substitution by alkyl groups. Its value is also characteristic of certain structural types.

Summary of New Reactions

1 DISSOCIATION ENERGY OF THE ALLYLIC HYDROGEN IN PROPENE

$$CH_3CH{=}CH_2 \longrightarrow H\cdot + \cdot CH_2CH{=}CH_2 \qquad DH^\circ = 87 \text{ kcal mole}^{-1}$$

2 S_N1 REACTIVITY OF ALLYLIC HALIDES

$$XCH_2CH{=}CH_2 \longrightarrow X^- + {}^+CH_2CH{=}CH_2 \xrightarrow{Nu^-} X^- + NuCH_2CH{=}CH_2$$

3 pK_a OF PROPENE

$$CH_3CH{=}CH_2 \rightleftharpoons H^+ + {}^-{:}CH_2CH{=}CH_2 \qquad pK_a \sim 40$$

4 THERMODYNAMIC COMPARED WITH KINETIC CONTROL IN S_N1 REACTIONS OF ALLYLIC DERIVATIVES

$$\underset{\textbf{More stable}}{CH_3CH{=}CHCH_2X} \xleftarrow{\text{Slow}} CH_3CH{=}CHCH_2{}^+ + X^- \underset{\text{Reversible}}{\overset{\text{Fast}}{\rightleftharpoons}} \underset{\textbf{Less stable}}{CH_3\overset{\overset{\displaystyle X}{|}}{C}HCH{=}CH_2}$$

5 S_N2 REACTIVITY OF ALLYLIC HALIDES

$$CH_2{=}CHCH_2X + Nu{:}^- \longrightarrow CH_2{=}CHCH_2Nu + X^-$$

S_N2' REACTIVITY OF ALLYLIC HALIDES

$$CH_2{=}CH\overset{\overset{\displaystyle R}{|}}{C}HX + Nu{:}^- \longrightarrow NuCH_2CH{=}CHR + X^-$$

6 FREE-RADICAL ALLYLIC HALOGENATION

$$RCH_2CH{=}CH_2 \xrightarrow{\text{NBS}, h\nu} R\overset{\overset{\displaystyle Br}{|}}{C}HCH{=}CH_2$$

7 ALLYLIC GRIGNARD REAGENTS

$$CH_2{=}CHCH_2Br \xrightarrow{Mg} CH_2{=}CHCH_2MgBr$$

8 ALLYLLITHIUM REAGENTS

$$RCH_2CH{=}CH_2 \xrightarrow{CH_3CH_2CH_2CH_2Li,\ TMEDA} R\overset{\overset{-}{\ddot{}}}{C}HCH{=}CH_2\ Li^+$$

$$(CH_2{=}CHCH_2)_4Sn \xrightarrow{CH_3CH_2CH_2CH_2Li} CH_2{=}CHCH_2{:}^-Li^+$$

9 HYDROGENATION OF CONJUGATED DIENES

$$CH_2{=}CH{-}CH{=}CH_2 \longrightarrow CH_3CH_2CH_2CH_3 \qquad \Delta H^\circ = -57.1 \text{ kcal mole}^{-1}$$

but compare:

$$CH_2{=}CH{-}CH_2{-}CH{=}CH_2 \longrightarrow CH_3(CH_2)_3CH_3 \qquad \Delta H^\circ = -60.8 \text{ kcal mole}^{-1}$$

10 ELECTROPHILIC REACTIONS OF 1,3-DIENES

$$CH_2{=}CH{-}CH{=}CH_2 \xrightarrow{HX} \underset{\textbf{Kinetic product}}{CH_2{=}CH\overset{\overset{\displaystyle X}{|}}{C}HCH_3} + \underset{\textbf{Thermodynamic product}}{XCH_2CH{=}CHCH_3}$$

$$CH_2{=}CH{-}CH{=}CH_2 \xrightarrow{X_2} CH_2{=}CH\overset{\overset{\displaystyle X}{|}}{C}HCH_2X + XCH_2CH{=}CHCH_2X$$

11 ELECTROPHILIC SUBSTITUTION OF BENZENE

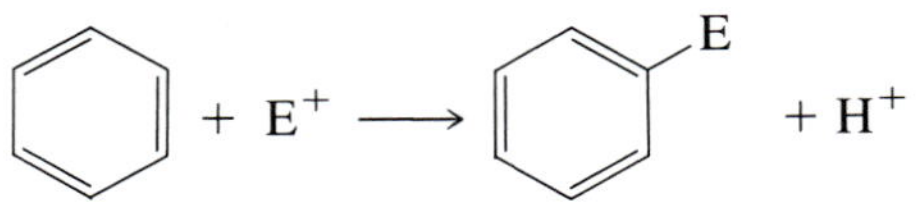

$+ E^+ \longrightarrow$ E $+ H^+$

E = Br (needs $FeBr_3$ catalyst), NO_2, SO_3H

12 DIELS-ALDER REACTION (CONCERTED AND STEREOSPECIFIC, ENDO RULE)

Alkenes:

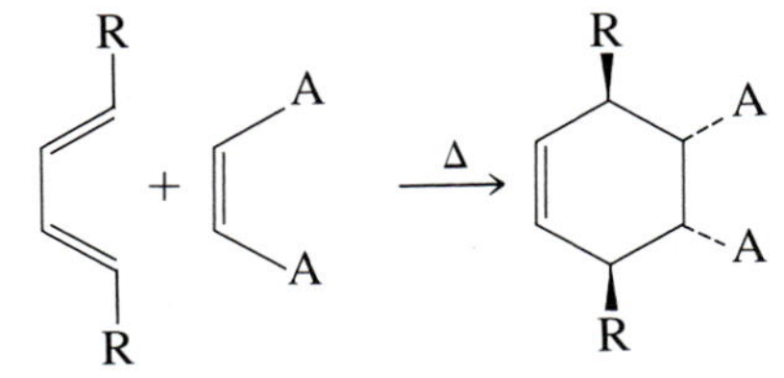

R R A A Δ R A A R

A = electron acceptor

Alkynes:

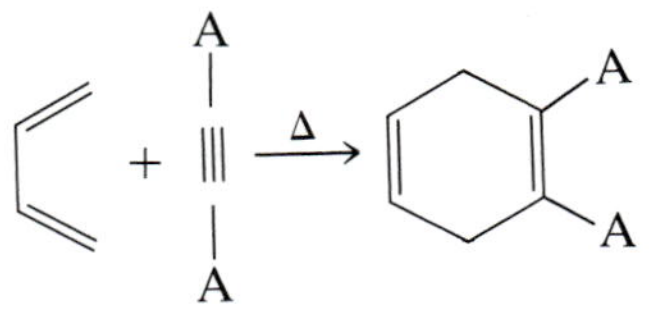

A A Δ A A

13 [2+2]PHOTOCYCLOADDITIONS

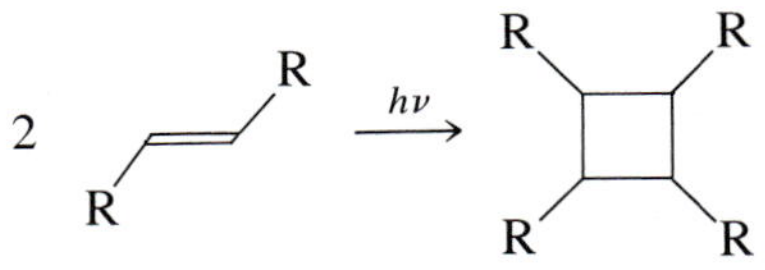

2 R R $h\nu$ R R R R

14 ELECTROCYCLIC REACTIONS

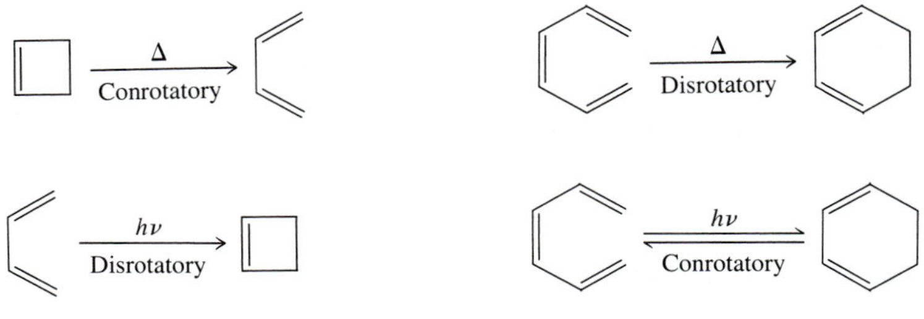

Δ Conrotatory

Δ Disrotatory

$h\nu$ Disrotatory

$h\nu$ Conrotatory

15 POLYMERIZATION OF 1,3-DIENES

1,2-Polymerization:

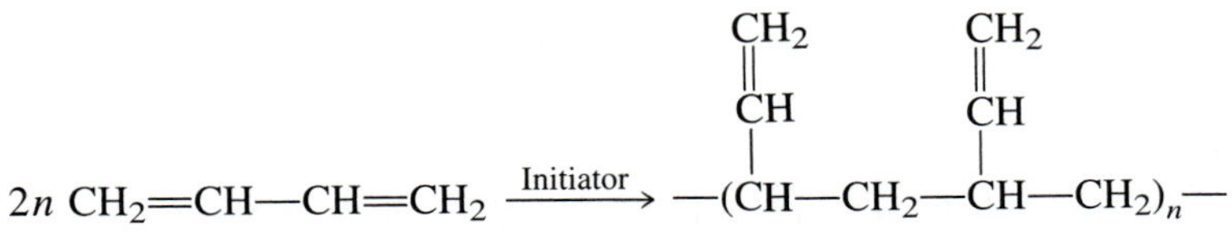

$$2n\ CH_2{=}CH{-}CH{=}CH_2 \xrightarrow{\text{Initiator}} {-}(\underset{\underset{\underset{CH_2}{\|}}{CH}}{CH}{-}CH_2{-}\underset{\underset{\underset{CH_2}{\|}}{CH}}{CH}{-}CH_2)_n{-}$$

1,4-Polymerization:

$$n\ CH_2{=}CH{-}CH{=}CH_2 \xrightarrow{\text{Initiator}} {-}(CH_2{-}CH{=}CH{-}CH_2)_n{-}$$

Cis or trans

16 3-METHYL-3-BUTENYL PYROPHOSPHATE AS A BIOCHEMICAL BUILDING BLOCK

$$CH_2{=}C(CH_3){-}CH_2CH_2OPP \underset{}{\overset{\text{Enzyme}}{\rightleftharpoons}} (CH_3)_2C{=}CHCH_2OPP \longrightarrow (CH_3)_2C{=}CHCH_2^{+} + {}^{-}OPP$$

3-Methyl-3-butenyl pyrophosphate — **Allylic cation** — **Pyrophosphate ion**

C–C bond formation:

$$\text{(allylic cation)} + \text{3-methyl-3-butenyl-OPP} \longrightarrow \text{(tertiary cation)-OPP} \xrightarrow[-\,H^+]{} \text{(geranyl)-OPP} \xrightarrow[-\,(^{-}OPP)]{} \text{(geranyl cation)}^{+} \longrightarrow \text{etc.}$$

Summary of Important Concepts

1 The 2-propenyl (allyl) system is stabilized by resonance. Its molecular-orbital description shows the presence of three π molecular levels: one bonding, one nonbonding, and one antibonding. Its structure is symmetrical, any charges or odd electrons being equally distributed between the two end carbons. The π-bond strength in the 2-propenyl (allyl) radical is about 15 kcal mole^{-1}.

2 The chemistry of the 2-propenyl (allyl) cation is subject to both thermodynamic and kinetic control. Nucleophilic trapping may occur more rapidly at an internal carbon that bears relatively more positive charge, giving the thermodynamically less stable product. The kinetic product may rearrange to its thermodynamic isomer by dissociation followed by eventual thermodynamic trapping.

3 The S_N2 reaction of allylic halides is accelerated by conjugation in the transition state.

4 The special stability of allylic radicals allows free-radical halogenations of alkenes at the allylic position.

5 The special stability of allylic anions allows allylic deprotonation by a strong base, such as butyllithium-TMEDA.

6 Conjugation of heteroatoms bearing lone electron pairs with a double bond makes this bond electron rich.

7 1,3-Dienes reveal the effects of conjugation by their resonance energy, a relatively short internal bond (1.47 Å), and a barrier to rotation (3.9 kcal mole^{-1}). Cis-trans isomerization around the double bonds is made relatively easy ($E_a \sim$ 52 kcal mole^{-1}) by the allylic radical nature of the transition state.

8 Electrophilic attack on 1,3-dienes leads to the preferential formation of allylic cations.

9 Extended conjugated systems are reactive because they have many sites for possible attack and the resulting intermediates are stabilized by resonance.

10 Benzene has special stability because of cyclic delocalization. Electrophilic attack results in substitution and not addition or polymerization.

11 The Diels-Alder reaction is a stereospecific concerted [4+2]cycloaddition reaction between an *s*-cis diene and a dienophile; it leads to cyclohexene or 1,4-cyclohexadiene derivatives.

12 Conjugated dienes, trienes, and their respective cyclic isomers undergo electrocyclic reactions, which result in concerted and stereospecific ring closures and openings. Conrotatory modes of opening and closing switch to disrotatory when thermal sources of energy are replaced by photochemical ones.

13 Polymerization of 1,3-dienes results in 1,2- or 1,4-additions to give polymers that are capable of further cross-linking. Synthetic rubbers can be synthesized in this way. Natural rubber is made by electrophilic carbon–carbon bond formation involving biosynthetic five-carbon cations derived from 3-methyl-3-butenyl pyrophosphate.

14 Ultraviolet and visible spectroscopy gives a way of estimating the extent of conjugation in a molecule. Its spectra usually have broad peaks, reported as λ_{max} (nm). Their intensity is given by the extinction coefficient ϵ.

Problems

1 Draw all resonance forms and a representation of the appropriate resonance hybrid for each of the following species.

(a) H_3C, H, C, CH_3, C, $\overset{+}{C}$, CH_3, CH_3

(b) H_3C, H, $\overset{+}{C}$, CH_3, C, CH, CH_2, CH_3

(c) CH_3

(d)

(e)

2 Illustrate by means of appropriate structures (including all relevant resonance forms) the initial species formed by

(a) breaking the weakest C-H bond in 1-butene.

(b) treating 4-methylcyclohexene with a powerful base (e.g., butyllithium-TMEDA).

(c) heating a solution of 3-chloro-1-methylcyclopentene in aqueous ethanol.

3 Rank primary, secondary, tertiary, and allylic radicals in order of decreasing stability. Do the same for the corresponding carbocations. Do the results indicate something about the relative ability of hyperconjugation and resonance to stabilize free-radical and cationic centers?

4 Write the major product(s) of each of the following reactions. If more than one product forms, indicate which is kinetic (i.e., formed fastest at low temperature and short reaction time) and which is thermodynamic (i.e., formed in highest yield at higher temperature after longer reaction times).

(a) $H_3C(CH_3)_2CH$–C=C–$(H)(CH_2OH)$ $\xrightarrow{\text{Conc. HBr}}$

(b) 1-chloro-1-methyl-2-cyclohexene (ring C bearing CH_3 and Cl) $\xrightarrow{H_2O}$

(c) 1-bromo-1-vinylcyclopentane (Br, $CH{=}CH_2$) $\xrightarrow{CH_3CH_2OH}$

(d) cyclohexylidene=C(CH_3)(CH_2I) $\xrightarrow{CH_3\overset{O}{\overset{\|}{C}}OH}$

(e) cyclohexylidene=C(CH_3)(CH_2I) $\xrightarrow{KSCH_3,\ DMSO}$

(f) $(H)(ClCH_2)$C=C$(H)(CH_2CH_2CH_2OH)$ $\xrightarrow{(CH_3CH_2)_2O,\ \Delta}$

5 Write detailed mechanisms for the following reactions in Problem 4: parts **a**, **c**, **e**, and **f**.

6 Rank primary, secondary, tertiary, and (primary) allylic chlorides in approximate order of (1) decreasing S_N1 reactivity and (2) decreasing S_N2 reactivity.

7 Rank the following six molecules in approximate order of (1) decreasing S_N1 reactivity and (2) decreasing S_N2 reactivity.

(a) $CH_3\overset{Cl}{\overset{|}{C}}HCH{=}CH_2$

(b) $(H_3C)(H)$C=C$(H)(CH_2Cl)$

(c) $(H_3C)(H)$C=C$(CH_3)(CH_2Cl)$

(d) $(H_3C)(H_3C)$C=C$(H)(CH_2Cl)$

(e) $(CH_3)_2C(Cl)CH{=}CH_2$

(f) $CH_2{=}CHCH_2Cl$

8 Each of the following reactions gives only the single substitution product shown. Indicate the mechanism for each case. Suggest a reason why rearrangement occurs in parts **a** and **b** but not in parts **c** and **d.**

(a) $CH_2{=}CHCHCl(C(CH_3)_3) \xrightarrow{NaOCH_2CH_3}$ $(CH_3CH_2OCH_2)(H)C{=}C(H)(C(CH_3)_3)$

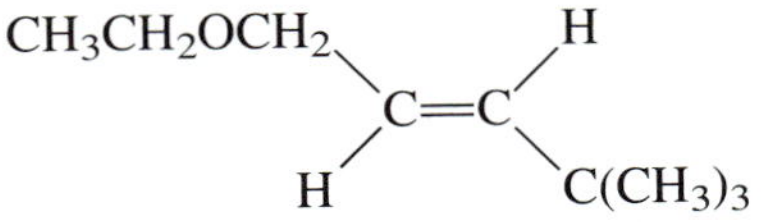

(b) $CH_2{=}CHC(CH_3)_2Cl \xrightarrow{NaSCH_2CH_3} CH_3CH_2SCH_2CH{=}C(CH_3)_2$

(c) $CH_3CH{=}CHCH_2I \xrightarrow{NaBr} CH_3CH{=}CHCH_2Br$

(d) $(CH_3)_2CH{=}CHCH_2Br \xrightarrow{NaCN} (CH_3)_2CH{=}CHCH_2CN$

9 Write the major product(s) of each of the following reactions.

(a) 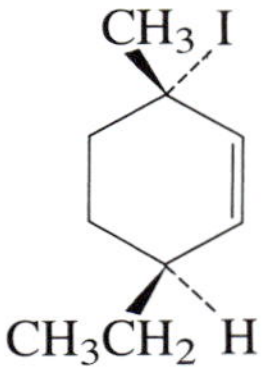 Cyclohexene ring bearing CH_3 and I on one carbon and CH_3CH_2 and H on the opposite carbon $\xrightarrow{H_2O}$

(b) 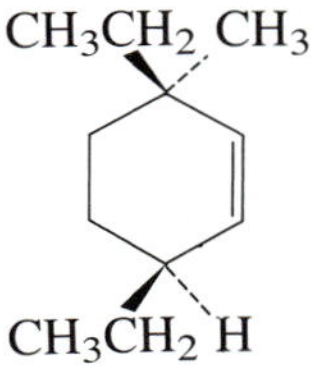 Cyclohexene ring bearing CH_3CH_2 and CH_3 on one carbon and CH_3CH_2 and H on the opposite carbon $\xrightarrow{NBS,\ CCl_4,\ ROOR}$

(c) $CH_3CH_2CH(CH_3){-}CH{=}CH_2 \xrightarrow{NBS,\ CCl_4,\ ROOR}$

(d) 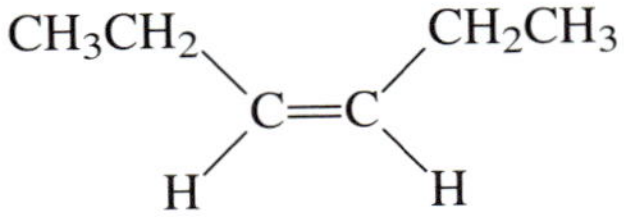 $(CH_3CH_2)(H)C{=}C(CH_2CH_3)(H) \xrightarrow{CH_3CH_2CH_2CH_2Li,\ TMEDA}$

(e) Product of part **d** $\xrightarrow[2.\ H^+,\ H_2O]{1.\ CH_3CHO}$

(f) 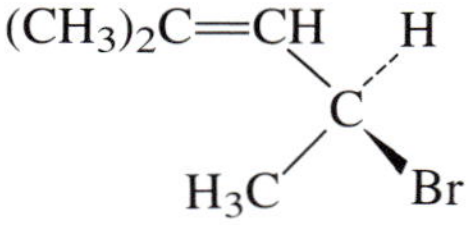 $(CH_3)_2C{=}CH{-}C(H)(H_3C)(Br) \xrightarrow{KSCH_3,\ DMSO}$

10 The following reaction sequence gives rise to two isomeric products. What are they? Explain the mechanism of their formation.

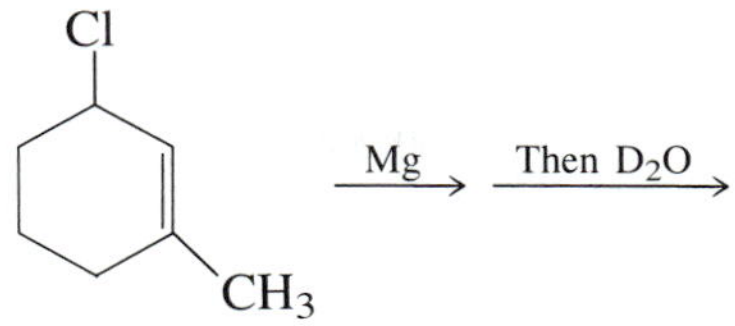

Cl, CH_3 $\xrightarrow{\text{Mg}}$ $\xrightarrow{\text{Then } D_2O}$

11 Starting with cyclohexene, propose a reasonable synthesis of the cyclohexene derivative shown in the margin.

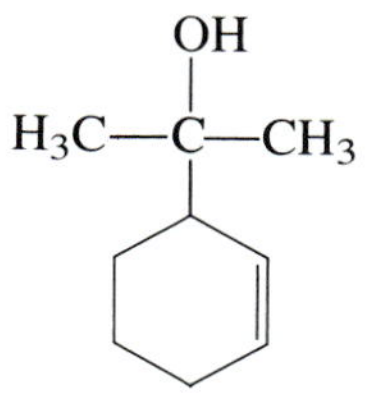

OH, $H_3C—C—CH_3$

12 Explain how the following reaction must take place.

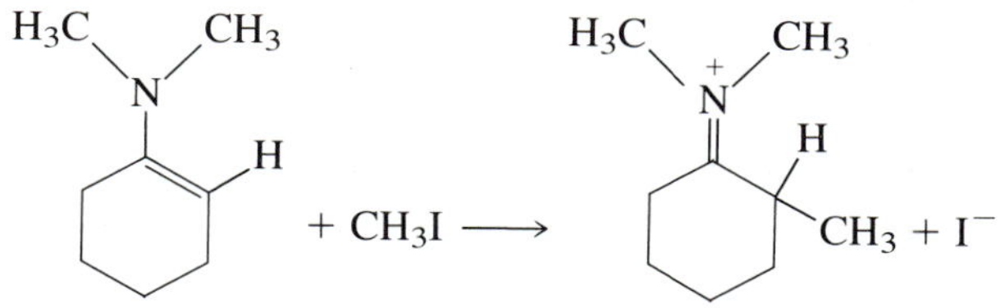

H_3C, CH_3, N, H $+ CH_3I \longrightarrow$ H_3C, CH_3, $\overset{+}{N}$, H, $CH_3 + I^-$

13 The ^1H NMR spectrum of ethenyl ethanoate (vinyl acetate, $CH_3CO_2CH{=}CH_2$) exhibits the following signals: $\delta = 2.05$(s, 3 H), 4.45(doublet of doublets, $J = 1$ and 6 Hz, 1 H), 4.77(doublet of doublets, $J =$ 1 and 14 Hz, 1 H), 7.23(doublet of doublets, $J = 6$ and 14 Hz, 1 H) ppm. Using the δ and J values as guides, assign each signal to a specific hydrogen or set of hydrogens in the molecule. Explain the large difference in chemical shift between the signals in the range $\delta = 4$ to $\delta = 5$ and the signal at $\delta = 7.23$ ppm. Hint: Refer to Figure 14-6 for useful information.

14 Give a systematic name to each of the following molecules.

(a) 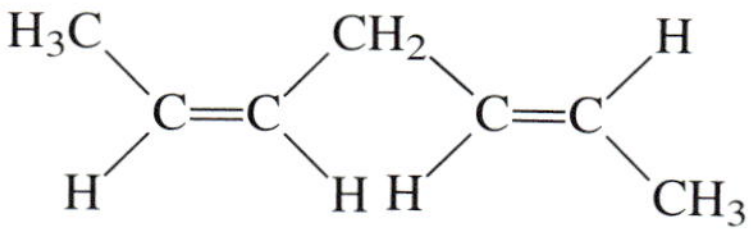 H_3C, CH_2, H, C=C, C=C, H, H H, CH_3

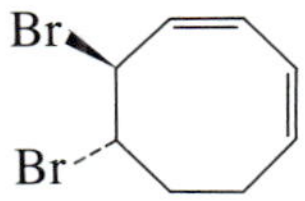

(b) $CH_2{=}CH—CH{=}CH—CH_2OH$

(c) Br, Br

(d)

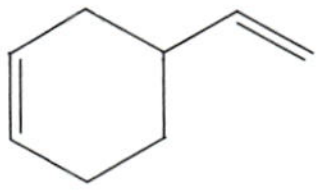

15 Compare the allylic bromination reactions of 1,3-pentadiene and 1,4-pentadiene. Which should be faster? Which is more energetically favorable? How do the product mixtures compare?

$$CH_2{=}CH—CH{=}CH—CH_3 \xrightarrow{\text{NBS, ROOR, } CCl_4}$$

$$CH_2{=}CH—CH_2—CH{=}CH_2 \xrightarrow{\text{NBS, ROOR, } CCl_4}$$

16 Compare the addition of H^+ to 1,3-pentadiene and 1,4-pentadiene (see Problem 15). Write the structures of the products. Draw a qualitative reaction profile showing both dienes and both proton addition products on the same graph. Which diene adds the proton faster? Which one gives the more-stable product?

17 What products would you expect from the electrophilic addition by each of the following reagents to 1,3-cycloheptadiene?

(a) HI
(b) Br_2 in H_2O
(c) IN_3
(d) H_2SO_4 in CH_3CH_2OH
(e) HBr + ROOR

18 Write the products of the reaction of *trans*-1,3-pentadiene with each of the reagents in Problem 17.

19 Write the products of reaction of 2-methyl-1,3-pentadiene with each of the reagents in Problem 17.

20 For each reaction in Problem 19, indicate which is the kinetic and which is the thermodynamic product.

21 Arrange the following carbocations in order of decreasing stability. Draw all other possible resonance forms for each of them.

(a) $CH_2{=}CH{-}\overset{+}{C}H_2$
(b) $CH_2{=}CH^+$
(c) $CH_3CH_2^{\,+}$
(d) $CH_3{-}CH{=}CH{-}\overset{+}{C}H{-}CH_3$
(e) $CH_2{=}CH{-}CH{=}CH{-}CH_2^{\,+}$

22 Sketch the molecular orbitals for the pentadienyl system in order of ascending energy (see Figures 14-2 and 14-9). Indicate how many electrons are present, and in which orbitals, for **(a)** the free radical, **(b)** the cation, and **(c)** the anion (see Figures 14-3 and 14-10). Draw all reasonable resonance forms for any one of these three species.

23 Dienes may be prepared by elimination reactions of substituted allylic compounds. For example,

$$H_3C{-}\underset{}{\overset{CH_3}{\overset{|}{C}}}{=}CH{-}CH_2OH \xrightarrow{\text{Catalytic } H_2SO_4,\ \Delta} H_2C{=}\overset{CH_3}{\overset{|}{C}}{-}CH{=}CH_2 \xleftarrow{\text{LDA}} H_3C{-}\overset{CH_3}{\overset{|}{C}}{=}CH{-}CH_2Cl$$

Propose detailed mechanisms for each of these 2-methyl-1,3-butadiene (isoprene) syntheses.

24 Write the structures of all possible products of the acid-catalyzed dehydration of vitamin A.

25 In a published synthetic procedure, propanone (acetone) is treated with ethenyl (vinyl) magnesium bromide, and the reaction mixture is then neutralized with strong aqueous acid. If the procedure is carried out properly, the product isolated exhibits ^1H NMR spectrum A. What is the product's structure?

If the reaction mixture is (improperly) allowed to remain in contact with aqueous acid for too long, a mixture of products is obtained. This mixture gives rise to the ^1H NMR spectrum B. New signals are present at $\delta = 1.77$ (several lines), 4.10 (a doublet with $J = 8$ Hz), and 5.45 (a broad triplet with $J = 8$ Hz) ppm. The relative intensities of these three signals are 6:2:1. What other compound is in the mixture besides the original product? How did it get there?

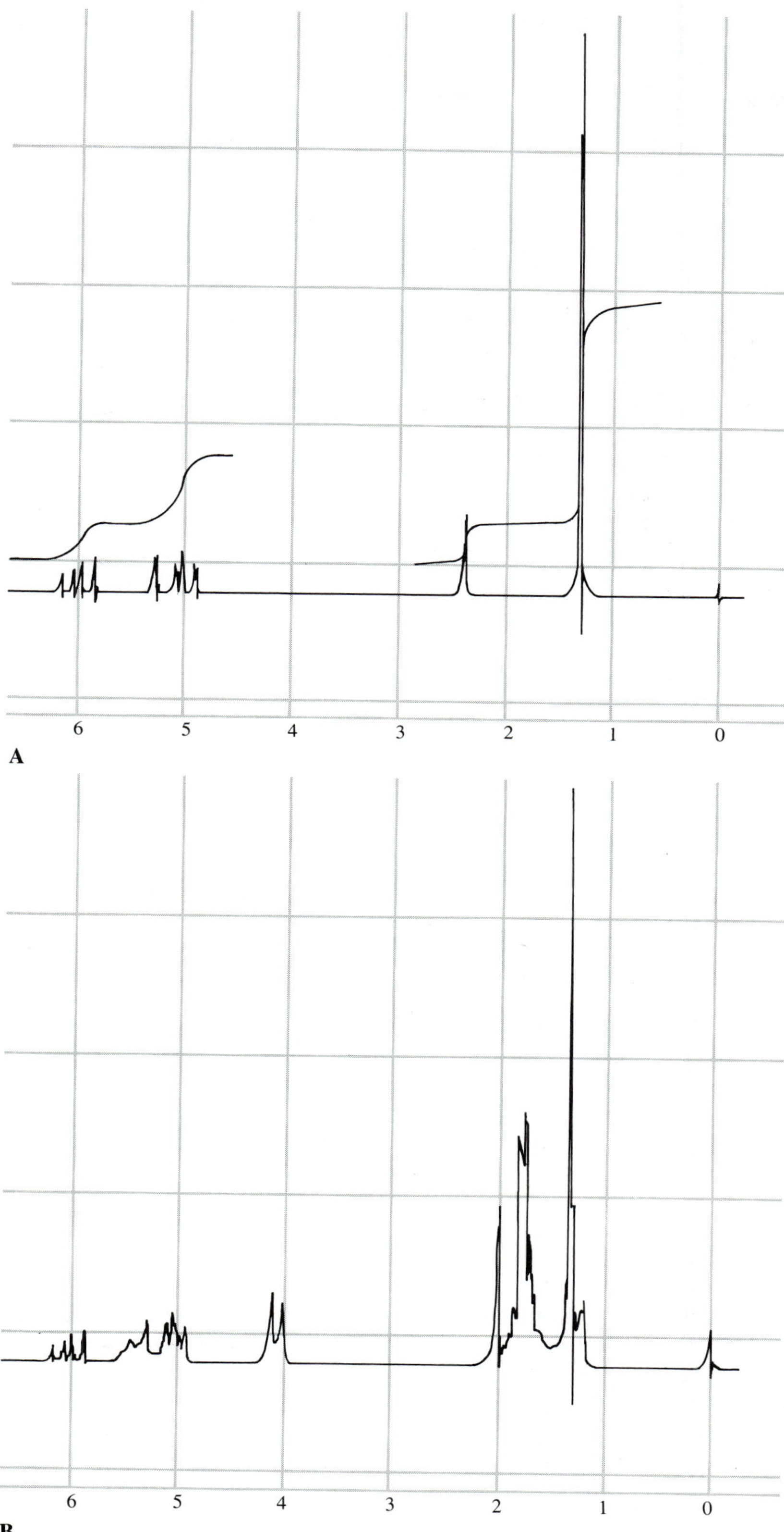
6
5
4
3
2
1
0
A
6
5
4
3
2
1
0
B

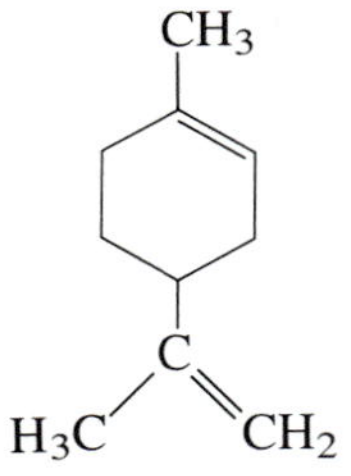

Limonene

26 The structure of the terpene limonene is shown in the margin. Show the two 2-methyl-1,3-butadiene (isoprene) units in limonene.

(a) Treatment of isoprene with catalytic amounts of acid leads to a variety of oligomeric products, one of which is limonene. Devise a detailed mechanism for the acid-catalyzed conversion of two molecules of isoprene into limonene. Take care to use sensible intermediates in each step.

(b) Two molecules of isoprene may also be converted into limonene by a completely different mechanism, which takes place in the strict absence of catalysts of any kind. Describe this mechanism. What is the name of this reaction?

27 Farnesol is a molecule that makes flowers smell nice (lilacs, for instance). Treatment with hot concentrated H_2SO_4 converts farnesol first into bisabolene and finally into cadinene, a compound of the essential oils of junipers and cedars. Propose detailed mechanisms for these conversions.

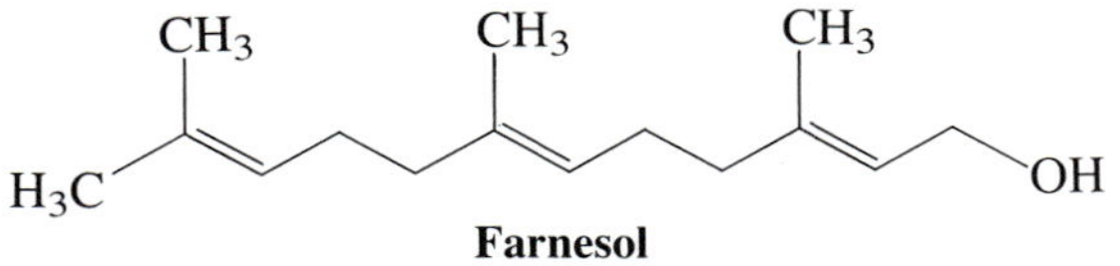

Farnesol

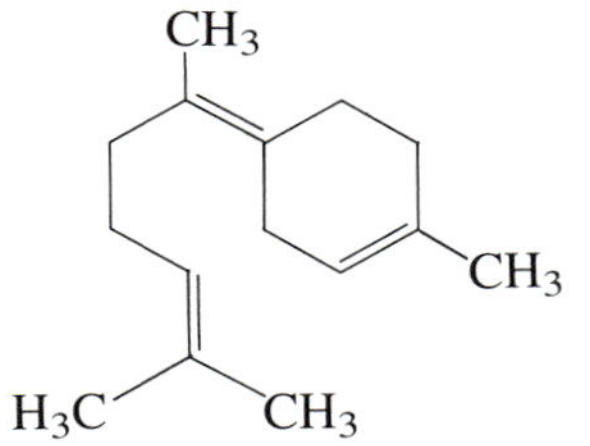

Bisabolene

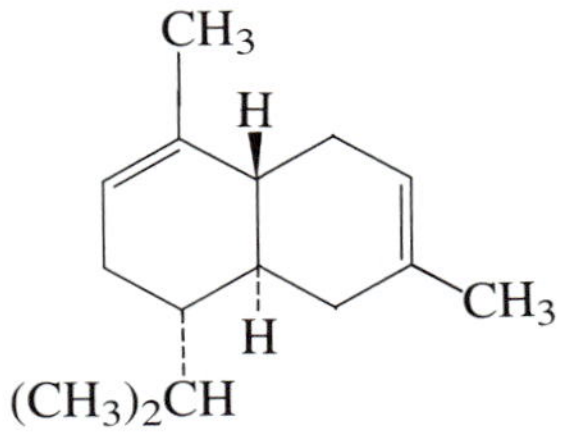

Cadinene

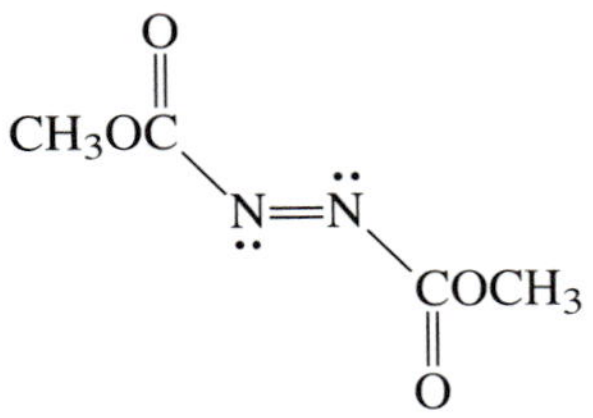

Dimethyl azodicarboxylate

28 Dimethyl azodicarboxylate takes part in the Diels-Alder reaction as a dienophile. Write the structure of the product of cycloaddition of this molecule with each of the following dienes.

(a) 1,3-Butadiene.
(b) *trans,trans*-2,4-Hexadiene.
(c) 5,5-Dimethoxycyclopentadiene.
(d) 1,2-Dimethylenecyclohexane.

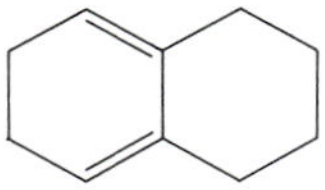

29 Bicyclic diene A reacts readily with appropriate alkenes by the Diels-Alder reaction, whereas diene B is totally unreactive. Explain.

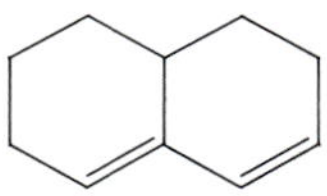

A

B

30 Propose a synthesis of each of the following molecules by Diels-Alder reactions.

(a) **(b)** 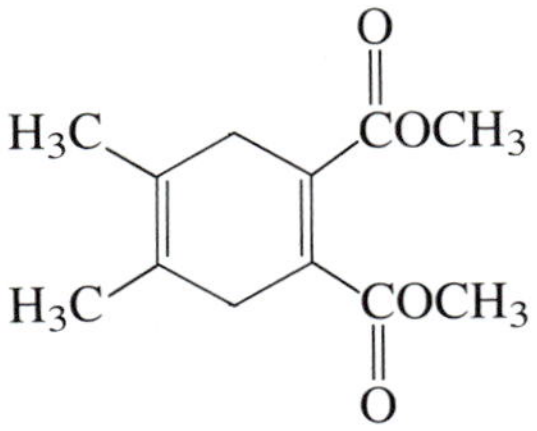 **(c)** 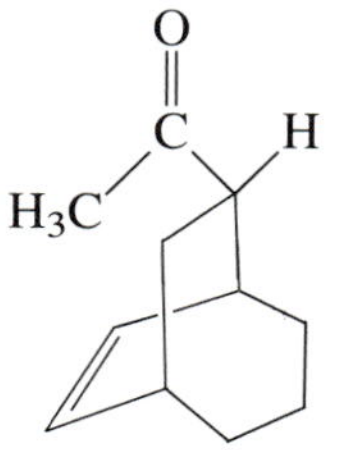 **(d)**

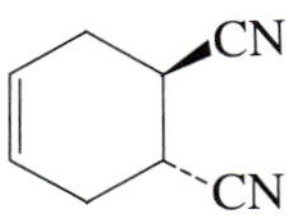

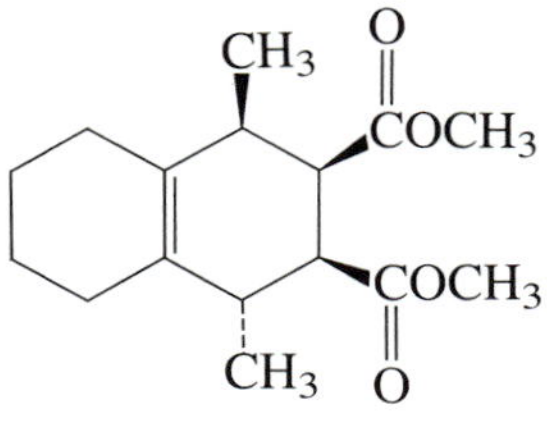

(e) 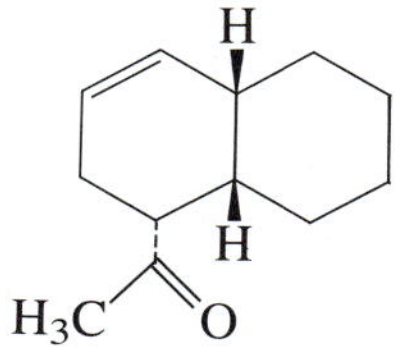 (Hint: See Exercise 14-15.)

31 Write reasonable products for each of the following reactions.

(a) 2 $\xrightarrow{h\nu}$

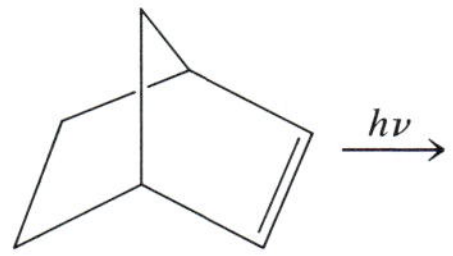

(b) 1,6-Heptadiene $\xrightarrow{h\nu}$

(c) Can you write a second, isomeric product of reaction b?

(d) 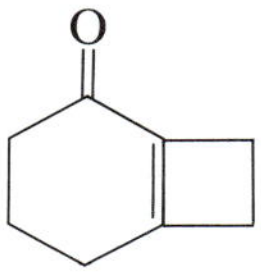 + $CH_2{=}CH_2$ $\xrightarrow{h\nu}$

32 Write the expected product of each of the following reactions.

(a) 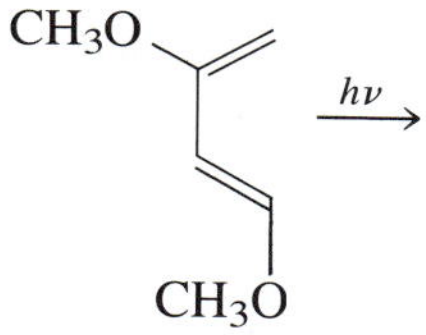 CH_3O ... CH_3O $\xrightarrow{h\nu}$

(d) 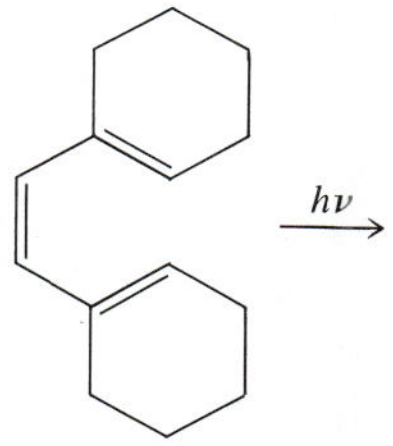 $\xrightarrow{h\nu}$

(b) 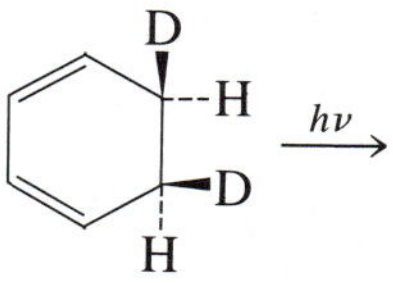 $\xrightarrow{h\nu}$

(e) 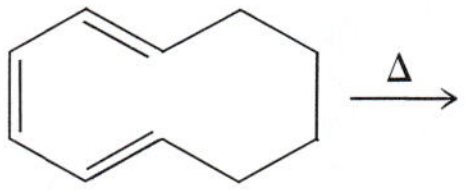 $\xrightarrow{\Delta}$

(c) 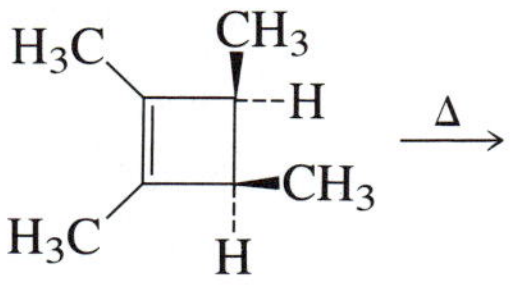 H_3C, CH_3, H, CH_3, H_3C, H $\xrightarrow{\Delta}$

33 When heated, bicyclo[2.2.0]hexa-2,5-diene (Dewar Benzene) isomerizes to benzene in a reaction that is exothermic by about 60 kcal mole^{-1}. This reaction, however, has an extraordinarily high activation energy of 37 kcal mole^{-1}. Suggest an explanation for the high barrier to reaction.

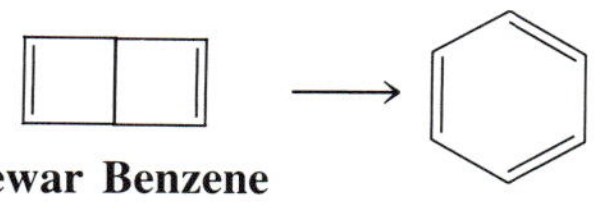

Dewar Benzene $\longrightarrow$

34 Explain each of the transformations labeled with a letter below.

Cl, Cl, + ; $h\nu$ (a); Cl, Cl, H H, H H; Na, naphthalene*; H H, H H; $h\nu$ (b); H, H; $h\nu$ (c); Δ (d); H, H; $h\nu$ †; Δ (e); $h\nu$ (f); H, H

*A special reagent combination used specifically for this reaction.
†Causes slow trans → cis isomerization of trans double bond.

35 Write abbreviated structures of each of the following:

(a) (E)-1,4-Poly-2-methyl-1,3-butadiene [(E)-1,4-polyisoprene].
(b) 1,2-Poly-2-methyl-1,3-butadiene (1,2-polyisoprene).
(c) 3,4-Poly-2-methyl-1,3-butadiene (3,4-polyisoprene).
(d) Copolymer of 1,3-butadiene and ethenylbenzene (styrene; $C_6H_5CH{=}CH_2$, SBR, used in automobile tires).
(e) Copolymer of 1,3-butadiene and propenenitrile (acrylonitrile; $CH_2{=}CHCN$, Latex).
(f) Copolymer of 2-methyl-1,3-butadiene (isoprene) and 2-methylpropene (butyl rubber, for inner tubes).

36 The carbocation derived from geranyl pyrophosphate (Section 14-6) not only is transformed biochemically into camphor, it is the biosynthetic precursor of limonene (Problem 26) and α-pinene (Chapter 4, Problem 19) as well. Write mechanisms for the formation of the second and third terpenes from this carbocation.

37 Using the labels introduced in Section 14-7 (particularly Figure 14-16), identify the longest-wavelength electronic transition in each of the following species.

(a) CH_4
(b) N_2
(c) H_2O
(d) Benzene (C_6H_6)
(e) Formaldehyde ($H_2C{=}O$)
(f) 2-Propenyl (allyl) cation

38 In each of the following groups of molecules, indicate which molecule would show the longest-wavelength electronic absorption.

(a) *cis,cis*-1,4,7-Nonatriene; *trans*-1,3,8-nonatriene; *trans,cis,trans*-2,4,6-nonatriene.

(b) 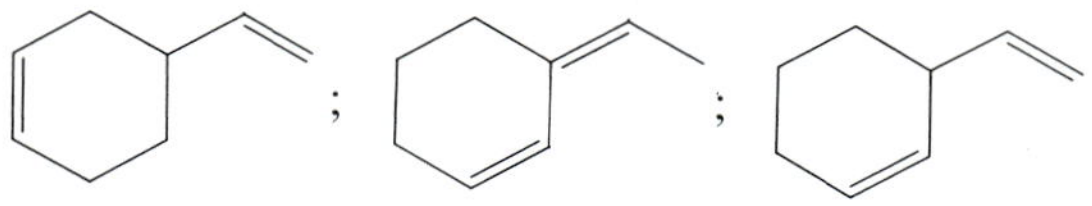 ; ;

39 Determine as best you can the structures of four unknown molecules from the following information. Each one reacts with excess H_2 over Pd to give hexane as the only product.

(a) NMR: $\delta = 1.6$–2.9 (5 H), 4.5–6.5 (5 H) ppm.
UV: $\lambda_{max} = 182$ nm.

(b) NMR: $\delta = 2.02$ (4 H), 4.5–6.5 (6 H) ppm.
UV: $\lambda_{max} = 177$ nm.

(c) NMR: $\delta = 1.68$ (6 H), 5.5–6.5 (4 H) ppm.
UV: λ_{max} 227 nm.

(d) NMR: $\delta = 0.8$–2.0 (5 H), 4.5–6.5 (5 H) ppm.
UV: $\lambda_{max} = 222$ nm.

CHAPTER 15

Aldehydes and Ketones

THE CARBONYL GROUP

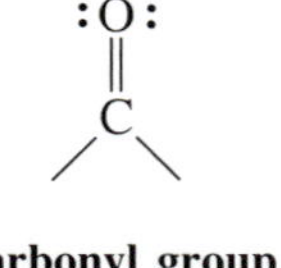

Carbonyl group

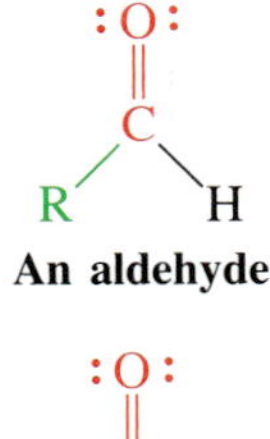

An aldehyde

A ketone

A carboxylic acid

Having dealt extensively with the chemistry of the carbon–carbon π bond, we now turn to the chemistry of the carbon–oxygen π bond, or the **carbonyl group.** This function is probably the most important in organic chemistry. This and the next chapter deal with the chemistry of **aldehydes,** in which the carbonyl carbon is bound to carbon and hydrogen [except for methanal (formaldehyde), in which it is bound to two hydrogens], and **ketones,** in which it is bound to two carbons. Chapters 17 and 18 will describe the chemistry of carboxylic acids and their derivatives, in which the carbonyl carbon is bound to carbon and heteroatoms, such as another oxygen, halogens, or nitrogen.

Aldehydes and ketones may be thought of as being derived from alcohols by removal of two hydrogen atoms, one from the hydroxy function and one from the neighboring carbon.

The Redox Relation Between Alcohols and Carbonyl Compounds

$$\text{R—CH(OH)—R}' \underset{+\,2\,\text{H}}{\overset{-\,2\,\text{H}}{\rightleftharpoons}} \text{RC(=O)R}'$$

As in the alcohols, the carbonyl oxygen bears two lone electron pairs, and therefore carbonyl groups are slightly basic. In addition, the C–O bond is polarized, making the carbonyl carbon electrophilic. Both of these properties shape the chemical behavior of this functional group. After explaining how to name aldehydes and ketones, this chapter considers their physical and structural properties and some of their spectral characteristics. It proceeds to the methods used in their preparation, primarily a brief review of reactions that we have encountered earlier, and finishes with the versatile addition chemistry of the carbon–oxygen double bond.

15-1 Naming Aldehydes and Ketones

Members of this class of compounds are called, like others, both by common and by systematic names.

The Common Names of Aldehydes and Ketones

For historical reasons, many of the simpler aldehydes and ketones have retained their common names. Many of the names of aldehydes have been derived from the stem of the common name of the corresponding carboxylic acid (see Section 17-1) by replacing the ending **-ic acid** with **aldehyde.**

$HCOOH$ **Formic acid** — $HCHO$ **Formaldehyde**

CH_3COOH **Acetic acid** — CH_3CHO **Acetaldehyde**

CH_3COCH_3 **Dimethyl ketone (Acetone)**

$CH_3COCH_2CH_3$ **Ethyl methyl ketone**

$CH_3CH_2COCH_2CH_3$ **Diethyl ketone**

Common names of ketones first give the two substituent groups and then add the word **ketone** at the end, similar to dialkyl ether nomenclature (Section 9-5). Dimethyl ketone is better known as acetone.

Systematic Names of Aldehydes and Ketones

Systematic names of aldehydes treat them as derivatives of the alkanes. The ending **-e** of the alkane name is replaced by **-al.** Thus, an alkane turns into an alkanal. For example, the simplest aldehyde is derived from methane—methanal. Ethanal stems from ethane, propanal from propane, and so on.

$HCHO$ **Methanal** — CH_3CHO **Ethanal** — CH_3CH_2CHO **Propanal**

$\overset{4}{Cl}CH_2\overset{3}{C}H_2\overset{2}{C}H_2\overset{1}{C}HO$ **4-Chlorobutanal**

4,6-Dimethylheptanal (skeletal structure, carbons numbered 1–7 from the CHO carbon)

The position of the carbonyl group does not have to be specified; its carbon is defined as carbon 1. This procedure automatically defines the numbering along the substituent chain as long as the aldehyde function is incorporated into the longest chain. Notice that the names of the aldehydes directly parallel those of the alkanols (Section 8-1). Systems that cannot be named readily by this use of the *-al* suffix are described as **carboxaldehydes.** The aldehyde functional group as a substituent should be called **methanoyl** (because of its relation to methane); nevertheless, the term **formyl** (derived from the common name of formic acid) has been retained by IUPAC and Chemical Abstracts.

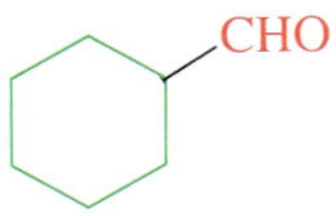

Cyclohexanecarboxaldehyde

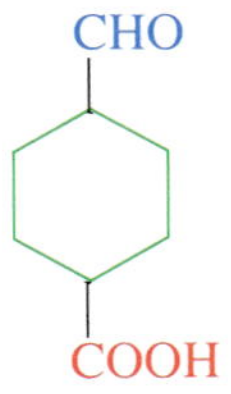

4-Methanoylcyclohexanecarboxylic acid (4-Formylcyclohexanecarboxylic acid)

According to IUPAC, ketones are called **alkanones,** the ending **-e** in the alkane stem being replaced by **-one.** You can see why the name of the common solvent acetone should be propanone. The position of the carbonyl function in the longest chain is indicated by numbering the chain to assign the lowest possible number to the carbonyl carbon. Whereas the methanoyl (formyl) group cannot be a part of a ring, ketones can be cyclic. Such compounds are called **cycloalkanones.** The carbonyl carbon is defined as carbon 1. In more complicated structures (e.g., in which another group has priority) and in conjunction with common names, the term **oxo** is used to indicate the presence of a carbonyl group. Various carbonyl-bearing appendages also have special substituent names.

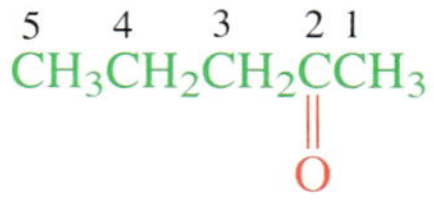

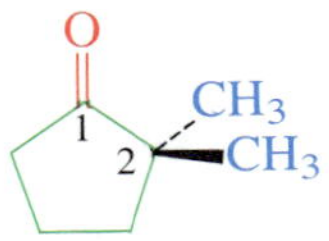

$CH_3CH_2CH_2C(=O)CH_3$

2-Pentanone

$CH_3CH(CH_3)CH_2CH(Cl)C(=O)CH_2CH_3$

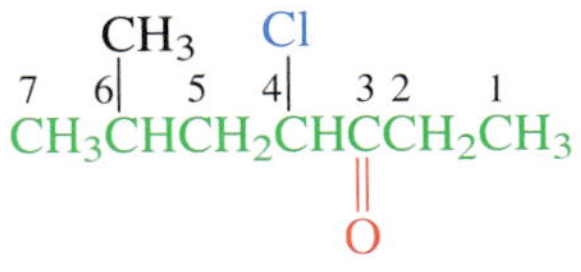

4-Chloro-6-methyl-3-heptanone

2,2-Dimethylcyclopentanone

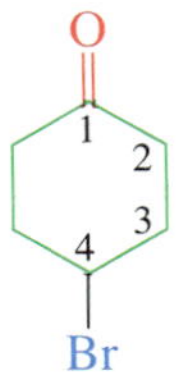

4-Bromocyclohexanone

The substituent $CH_3\overset{O}{\overset{\|}{C}}—$ should be called **ethanoyl,** but the IUPAC and Chemical Abstracts systems use the term **acetyl.** In this book, the general fragment $R\overset{O}{\overset{\|}{C}}—$ will be called **alkanoyl,** although **acyl** is preferred in other systems of naming compounds.

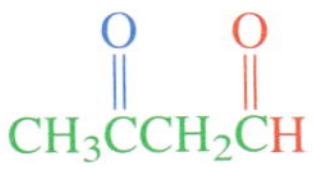

$CH_3\overset{O}{\overset{\|}{C}}CH_2\overset{O}{\overset{\|}{C}}H$

3-Oxobutanal

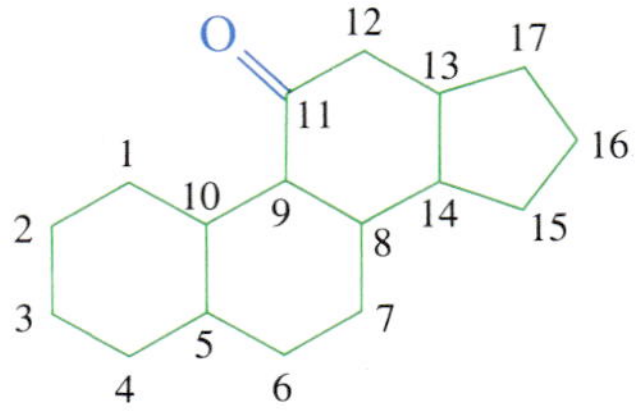

An 11-oxosteroid

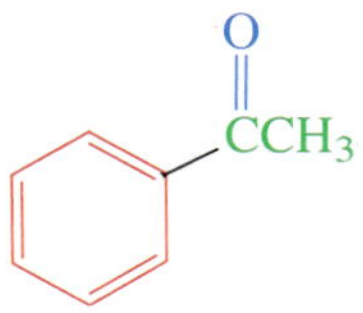

Ethanoylbenzene (Acetylbenzene, acetophenone)

As has been done throughout this book, the naming of organic molecules will be related to alkane nomenclature as much as possible, but other popular names will be given in parentheses. The carbonyl group has priority in names over all the other functional groups considered so far.

Aldehydes and Ketones with Other Functional Groups

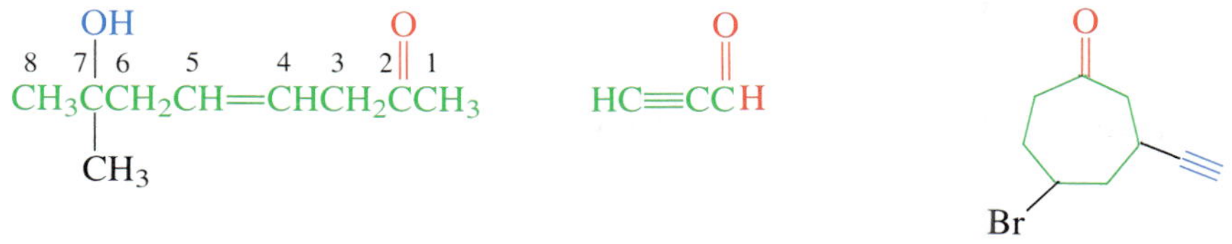

$$\overset{8}{CH_3}\overset{7}{\underset{\underset{CH_3}{|}}{\overset{\overset{OH}{|}}{C}}}\overset{6}{C}H_2\overset{5}{C}H{=}\overset{4}{C}H\overset{3}{C}H_2\overset{2}{\overset{\overset{O}{\|}}{C}}\overset{1}{C}H_3 \qquad H C{\equiv}C\overset{\overset{O}{\|}}{C}H$$

Br

7-Hydroxy-7-methyl-4-octen-2-one **Propynal** **5-Bromo-3-ethynylcycloheptanone**

(Note that the *e* in *-ene* and *-yne* is dropped in *enone* and *ynal*.)

EXERCISE 15-1

Name or draw the structures of the following compounds:

(a) 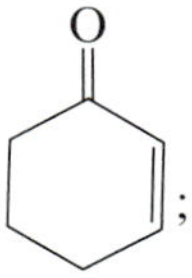 [structure: O]; (b) 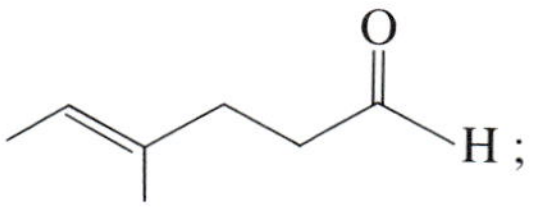 [structure: O, H]; (c) 4-octyn-3-one;

(d) 3-hydroxybutanal.

There are various ways of drawing aldehydes and ketones. As usual, condensed formulas or the zigzag notation may be used. Note that the condensed formulas for aldehydes are written as RCHO rather than RCOH to prevent confusion with the hydroxy group of alcohols.

Various Ways of Writing Aldehyde and Ketone Structures

Butanal: $CH_3CH_2CH_2\overset{\overset{O}{\|}}{C}H$ $\qquad CH_3CH_2CH_2CHO$ (Not a hydroxy group) $\qquad$ [zigzag structure: O, H]

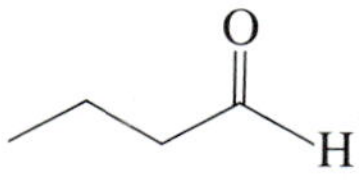

Butanone: $CH_3CH_2\overset{\overset{O}{\|}}{C}CH_3$ $\qquad CH_3CH_2COCH_3$ $\qquad$ [zigzag structure: O]

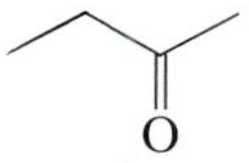

In summary, aldehydes and ketones are named systematically as alkanals and alkanones. The carbonyl functional group has priority over the hydroxy function and C–C double and triple bonds. With these rules, the usual guidelines for numbering the stem and labeling the substituents are followed.

15-2 The Physical Properties of Aldehydes and Ketones

To a certain extent, the carbonyl group is an oxygen analog of the alkene functional group. This similarity is reflected in its molecular-orbital description, in the structure of aldehydes and ketones, and in some of their physical and spectral properties. On the other hand, there are pronounced differences between the two characteristic double bonds because of the electronegativity of the oxygen,

FIGURE 15-1
Orbital picture of the carbonyl group. The orbital arrangement is similar to the one shown in Figure 11-1 for ethene.

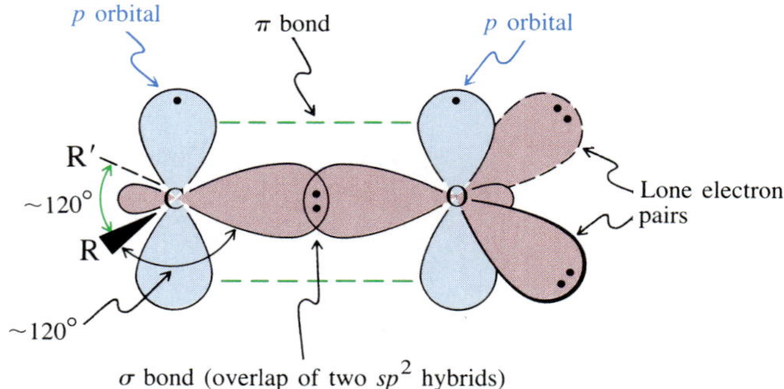

the absence of substituents on it, and the presence of its two lone pairs of electrons.

The Electronic Structure of the Carbonyl Group

Both the carbon and the oxygen of the carbonyl group are sp^2 hybridized and therefore lie in the same plane as the two additional groups on carbon, with bond angles approximating 120°. Perpendicular to the molecular frame are two p orbitals, one on carbon and one on oxygen, making up the π bond (Figure 15-1).

Comparison with the electronic structure of an alkene double bond reveals two important differences. First, the oxygen atom does not bear any additional groups, but only two lone electron pairs located in two nearly sp^2 hybrid orbitals. These point into that direction in space in which one would normally find the substituents of an sp^2-hybridized carbon. Second, oxygen is more electronegative than carbon. This property affects the π cloud to cause an appreciable polarization of the carbon–oxygen double bond, with a partial positive charge on carbon and an equal amount of negative charge on oxygen. In this way, the carbon is rendered electrophilic, the oxygen nucleophilic and slightly basic. This polarization can be described either by writing a polar resonance structure for the carbonyl moiety or by indicating partial charges. The carbonyl group has a substantial dipole moment (on the order of 2.7 D), comparable to those in the haloalkanes (Section 6-2).

Descriptions of a Carbonyl Group

$$\left[\ \rangle C{=}\ddot{\underset{..}{O}} \longleftrightarrow \rangle \overset{+}{C}{-}\ddot{\underset{..}{O}}{:}^{-}\ \right] \quad \text{or} \quad \rangle \overset{\delta^+}{C}{=}\overset{\delta^-}{\ddot{\underset{..}{O}}}$$

Electrophilic (C⁺) — Nucleophilic and basic (O⁻)

$\mu \sim 2.7$ D

Molecular Structure of Ethanal, a Typical Aldehyde

Figure 15-2 shows some of the structural features of the carbonyl compound ethanal (acetaldehyde). As expected, the molecule is planar, with a trigonal carbonyl carbon and a short carbon–oxygen bond, indicative of its double bond character. Not surprisingly, this bond is also rather strong, ranging from 175 to 180 kcal $mole^{-1}$.

FIGURE 15-2

The molecular structure of ethanal.

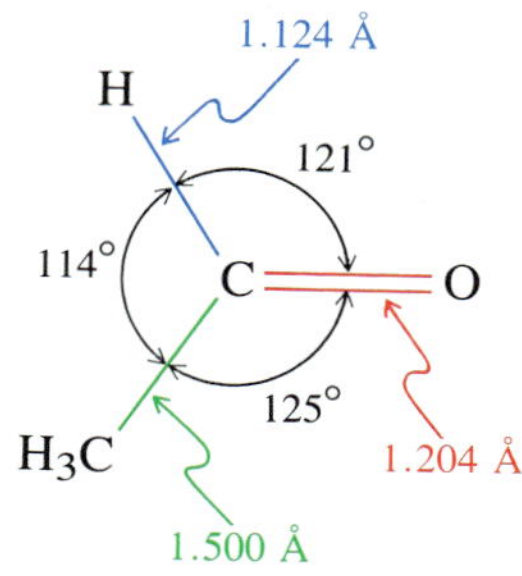

Some Physical Constants of Aldehydes and Ketones

The polarization of the carbonyl functional group makes the boiling points of aldehydes and ketones higher than those of the corresponding hydrocarbons (Table 15-1). However, the boiling points of the more-polar haloalkanes (Table 6-2) are remarkably close to those of aldehydes and ketones of similar size. The differences between isomeric aldehydes and ketones are negligible. Homologs with more than twelve carbons are solids at room temperature. Because of their dipole moments, the smaller carbonyl compounds are soluble in water. For example, ethanal (acetaldehyde) and propanone (acetone) are completely miscible with water. As the hydrophobic hydrocarbon part of the molecule increases in size, however, water solubility decreases. Carbonyl compounds with more than six carbons are fairly insoluble in water.

TABLE 15-1

Boiling points of aldehydes and ketones

Formula	Name	Boiling point (°C)
HCHO	Methanal (formaldehyde)	−21
CH_3CHO	Ethanal (acetaldehyde)	21
CH_3CH_2CHO	Propanal (propionaldehyde)	49
CH_3COCH_3	Propanone (acetone)	56
$CH_3CH_2CH_2CHO$	Butanal (butyraldehyde)	76
$CH_3CH_2COCH_3$	Butanone (ethyl methyl ketone)	80
$CH_3CH_2CH_2CH_2CHO$	Pentanal	102
$CH_3COCH_2CH_2CH_3$	2-Pentanone	102
$CH_3CH_2COCH_2CH_3$	3-Pentanone	102

Spectroscopic Properties of Aldehydes and Ketones

The carbonyl group gives rise to characteristic spectra. In ^1H NMR spectroscopy, the methanoyl (formyl) hydrogen of the aldehydes is very strongly deshielded, appearing between 9 and 10 ppm, which is unique for this class of compounds. The reason for this effect is twofold. First, the movement of the π electrons, as in alkenes (Section 11-3), causes a local magnetic field strengthening the external field. Second, the charge on the positively polarized carbon exerts an additional deshielding effect. Figure 15-3 (on the next page) shows the ^1H NMR spectrum of propanal with the methanoyl (formyl) hydrogen resonating at $\delta = 9.89$ ppm, split into a triplet ($J = 2$ Hz) because of a small coupling to the methylene hydrogens on the other side of the carbonyl function. The latter are also slightly deshielded relative to alkane hydrogens because of the electron-withdrawing character of the functional group. This effect is also seen in the ^1H NMR spectra of ketones.

^1H NMR Deshielding in Aldehydes and Ketones

$RCH_2C(=O)H$

$\delta \sim 2.5 \quad \sim 9.8$ ppm

$RCH(R')C(=O)CH_3$

$\delta \sim 2.6 \quad \sim 2.0$ ppm

Carbon-13 NMR spectra are diagnostic of both aldehydes *and* ketones, because of the characteristic chemical shift of the carbonyl carbon. Recall (Section

FIGURE 15-3
90-MHz ^{1}H NMR spectrum of propanal in CCl_4.

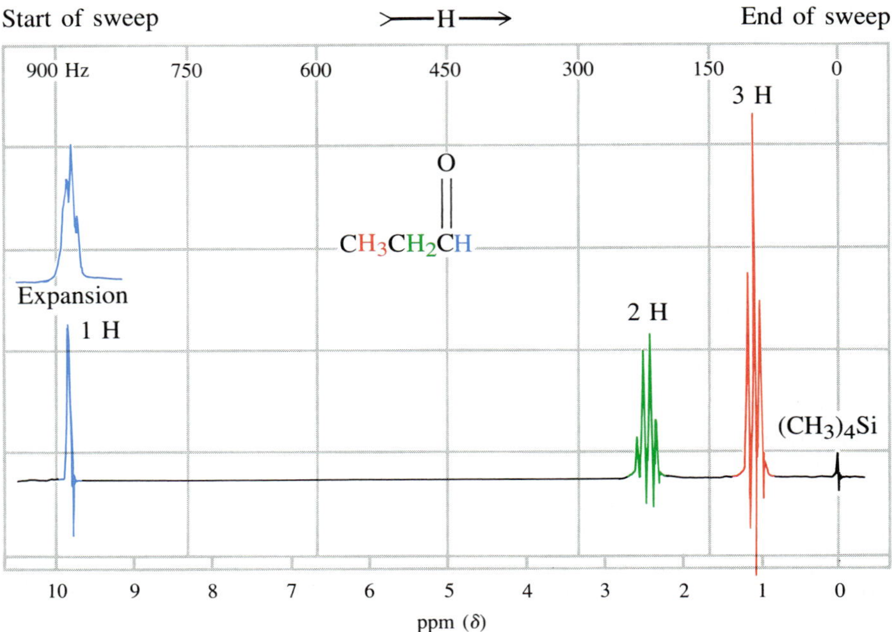

FIGURE 15-4
^{13}C NMR spectrum of cyclohexanone at 75.5 MHz in $CDCl_3$. The carbonyl carbon at 211.8 ppm is strongly deshielded relative to the other carbons. Because of symmetry, the molecule exhibits only four peaks; the three methylene carbon resonances absorb at increasingly lower field the closer they are to the carbonyl group. The triplet at 77 ppm is due to the carbon in the solvent, $CDCl_3$, split by deuterium. (The rules for spin-spin splitting by deuterium are different from those for hydrogen and will not be discussed here.)

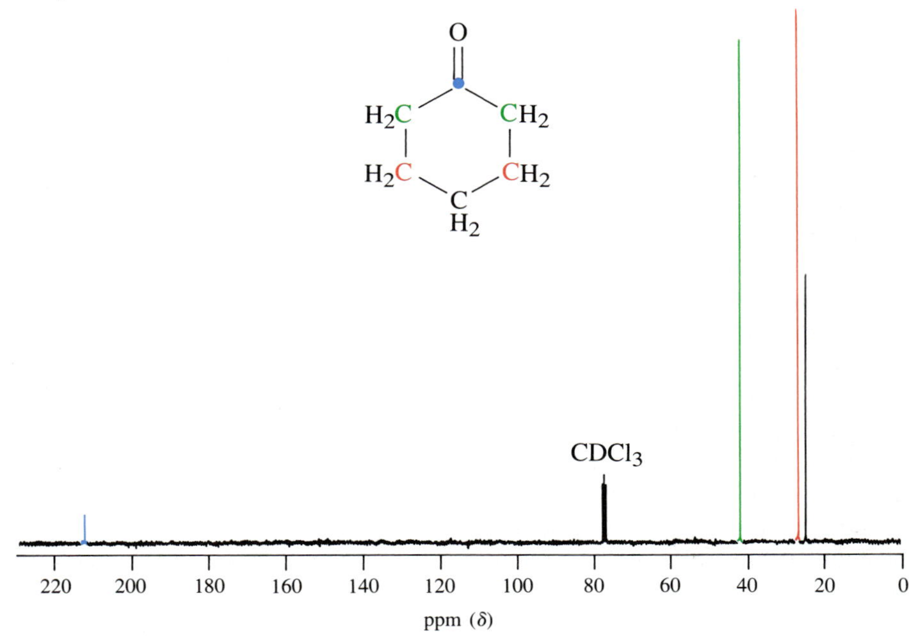

FIGURE 15-5
The $\pi \rightarrow \pi^*$ and $n \rightarrow \pi^*$ transitions in propanone (acetone).

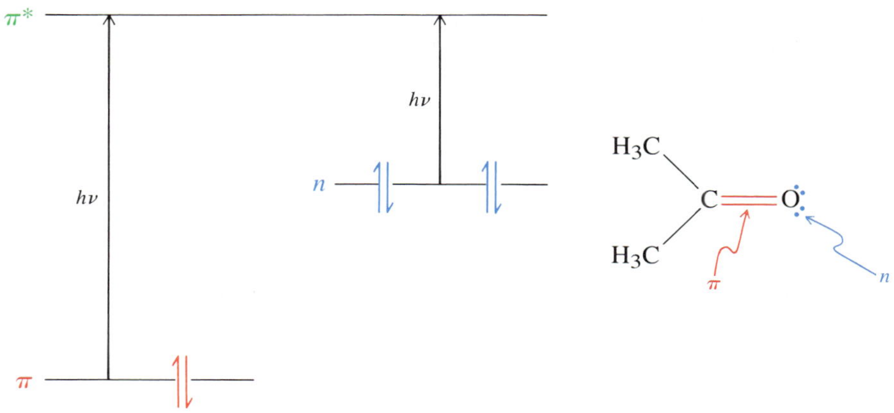

11-3) that the sp^2-hybridized carbons of an alkene resonate at low field (~120–130 ppm). Partly because of the electronegativity of the directly bound oxygen, the carbonyl carbons in aldehydes and ketones appear at even lower field (~200 ppm). The carbons next to the carbonyl group are also deshielded compared with those located farther away. The ^{13}C NMR spectrum of cyclohexanone is shown in Figure 15-4.

^{13}C NMR Chemical Shifts of Some Aldehydes and Ketones

CH_3—$\overset{O}{\overset{\|}{C}}H$
δ = 31.2 199.6 ppm

CH_3—CH_2—$\overset{O}{\overset{\|}{C}}H$
δ = 5.2 36.7 201.8 ppm

$CH_3\overset{O}{\overset{\|}{C}}CH_3$
δ = 30.2 205.1 ppm

$CH_3\overset{O}{\overset{\|}{C}}$—$CH_2$—$CH_2$—$CH_3$
δ = 29.3 206.6 45.2 17.5 13.5 ppm

EXERCISE 15-2

A student found an old bottle of propanone (acetone) that had been exposed to air. Gas-chromatography revealed a small amount of a new compound. Its ^1H NMR (CCl_4) spectrum had peaks at δ = 2.11 (s, 3 H) and 2.60 (s, 2 H) ppm. Its ^{13}C NMR spectrum exhibited only three lines, one at 206.8 ppm. Elemental analysis gave the composition C_3H_5O. Suggest a structure for this compound.

Carbonyl groups also exhibit characteristic electronic spectra, because the nonbonding lone electron pairs on the oxygen atom undergo low-energy $n \rightarrow \pi^*$, as well as $\pi \rightarrow \pi^*$, transitions (Figure 15-5). As a consequence, ordinary aldehydes and ketones exhibit absorption bands in the ultraviolet range between about 275 and 295 nm, much longer wavelengths (lower energy) than those observed in simple alkenes (Section 14-7). For example, propanone (acetone) shows an $n \rightarrow \pi^*$ band at 280 nm (ϵ = 15) in hexane. The corresponding $\pi \rightarrow \pi^*$ transition appears at about 190 nm (ϵ = 1100). Resonance with a carbon–carbon double bond has the same effect on the electronic spectra of carbonyl compounds as it has on the spectra of alkenes: absorptions are shifted to longer wavelengths. For example, the electronic spectrum of 3-buten-2-one, $CH_2{=}CHCOCH_3$, has peaks at 324 nm ($\epsilon = 24$, $n \rightarrow \pi^*$) and 219 nm ($\epsilon = 3600$, $\pi \rightarrow \pi^*$).

Electronic Transitions of Propanone (Acetone) and 3-Buten-2-one

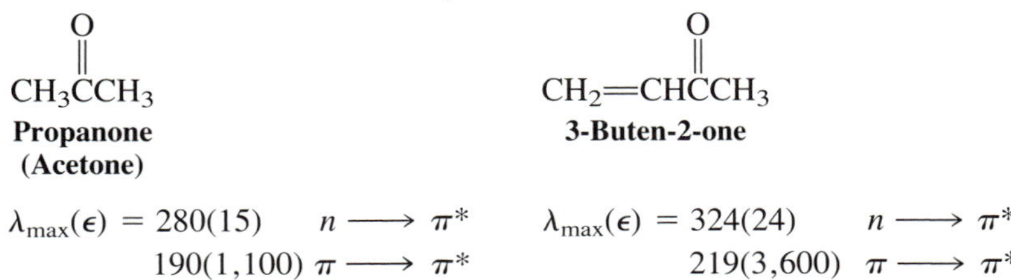

$\lambda_{max}(\epsilon) = 280(15) \quad n \longrightarrow \pi^*$

$\qquad\qquad 190(1{,}100) \quad \pi \longrightarrow \pi^*$

$\lambda_{max}(\epsilon) = 324(24) \quad n \longrightarrow \pi^*$

$\qquad\qquad 219(3{,}600) \quad \pi \longrightarrow \pi^*$

EXERCISE 15-3

An unknown C_4H_6O exhibited the following spectral data: ^1H NMR (CCl_4) $\delta = 2.03$ (dd, $J = 6.7$, 1.6 Hz, 3 H), 6.06 (ddq, $J = 16.1$, 7.7, 1.6 Hz, 1 H), 6.88 (dq, $J = 16.1$, 6.7 Hz, 1 H), 9.47 (d, $J = 7.7$ Hz, 1 H) ppm; ^{13}C NMR (CCl_4) $\delta = 18.4$, 132.8, 152.1, 191.4 ppm; UV $\lambda_{max}(\epsilon) = 220(15{,}000)$ and 314(32) nm. Suggest a structure.

To summarize, the carbonyl group in aldehydes and ketones is an oxygen analog of the carbon–carbon double bond. However, the electronegativity of oxygen polarizes the π bond, generating a strong dipole moment and rendering the alkanoyl substituent electron withdrawing. The arrangement of bonds around the carbon and oxygen is planar, a consequence of sp^2 hybridization. Methanoyl (formyl) hydrogens and the carbonyl carbons are strongly deshielded in NMR. Finally, the ability of nonbonding electrons to be excited into the π^* molecular orbitals causes the carbonyl group to exhibit characteristic relatively long wavelength UV absorptions.

15-3 The Preparation of Aldehydes and Ketones

There are many ways to prepare aldehydes and ketones, and the most important ones have already been described in connection with the chemistry of other functional groups: oxidation of alcohols (Section 9-4), oxidative cleavage of alkenes (Section 12-3), and hydroboration-oxidation or hydration of alkynes (Section 13-6). This section introduces some industrial preparations and reviews the above methods, pointing out some special features and additional examples. Other routes to aldehydes and ketones will be described in later chapters. These routes include

1 Reaction of carboxylic acids with alkyllithium reagents (Section 17-10)

$$\underset{\text{RCOH}}{\overset{\text{O}}{\|}} + \text{R}'\text{Li} \longrightarrow \xrightarrow{\text{H}^+,\ \text{H}_2\text{O}} \text{R}\overset{\text{O}}{\overset{\|}{\text{C}}}\text{R}'$$

2 Reaction of alkanoyl (acyl) chlorides with organometallic reagents (Section 18-2)

$$\mathrm{R\overset{O}{\overset{\|}{C}}Cl + R'M \longrightarrow \xrightarrow{H^+,\ H_2O} R\overset{O}{\overset{\|}{C}}R'}$$

3 Reduction of alkanoyl (acyl) chlorides with a modified hydride or catalytically activated hydrogen (Section 18-2)

$$\mathrm{R\overset{O}{\overset{\|}{C}}Cl + MH \longrightarrow \xrightarrow{H^+,\ H_2O} R\overset{O}{\overset{\|}{C}}H} \qquad \mathrm{R\overset{O}{\overset{\|}{C}}Cl + H_2 \xrightarrow{Pd\text{-}BaSO_4} R\overset{O}{\overset{\|}{C}}H}$$

4 Reduction of esters and amides with diisobutylaluminum hydride (Sections 18-4 and 18-5)

$$\mathrm{R\overset{O}{\overset{\|}{C}}OR' \text{ or } R\overset{O}{\overset{\|}{C}}NR'_2 + (CH_3\overset{CH_3}{\overset{|}{C}}HCH_2)_2AlH \longrightarrow \xrightarrow{H^+,\ H_2O} R\overset{O}{\overset{\|}{C}}H}$$

5 Reaction of nitriles with organometallic reagents (Section 18-6)

$$\mathrm{RC{\equiv}N + R'M \longrightarrow \xrightarrow{H^+,\ H_2O} R\overset{O}{\overset{\|}{C}}R'}$$

6 Reduction of nitriles with modified hydrides (Section 18-6)

$$\mathrm{RC{\equiv}N + MH \longrightarrow \xrightarrow{H^+,\ H_2O} R\overset{O}{\overset{\|}{C}}H}$$

7 Friedel-Crafts alkanoylation (acylation) (Section 19-7)

$$\mathrm{C_6H_6 + R\overset{O}{\overset{\|}{C}}Cl \xrightarrow{AlCl_3} \xrightarrow{H^+,\ H_2O} C_6H_5\overset{O}{\overset{\|}{C}}R}$$

8 Reactions of diazomethane with alkanoyl (acyl) chlorides (Section 21-5)

$$\mathrm{R\overset{O}{\overset{\|}{C}}Cl + CH_2N_2 \longrightarrow R\overset{O}{\overset{\|}{C}}CH_2Cl + N_2} \qquad \mathrm{R\overset{O}{\overset{\|}{C}}Cl + 2\ CH_2N_2 \longrightarrow R\overset{O}{\overset{\|}{C}}CHN_2 + HCl}$$

9 Alkylation of alkanoyl (acyl) anion equivalents (Section 22-2)

$$\mathrm{RX + \text{(2-R'-1,3-dithiane anion, } \ddot{C}^-\text{–R')} \longrightarrow \xrightarrow{H^+,\ HgCl_2} R\overset{O}{\overset{\|}{C}}R'}$$

10 Decarboxylation of 3-keto acids (Section 22-4).

$$\mathrm{R\overset{\overset{\displaystyle O}{\|}}{C}\underset{\underset{\displaystyle R'}{|}}{C}HCOOH \xrightarrow{\Delta} R\overset{\overset{\displaystyle O}{\|}}{C}CH_2R' + CO_2}$$

Some Industrial Preparations of Aldehydes and Ketones

The most important aldehyde in industry is methanal (formaldehyde); the most important ketone is propanone (acetone). About 6 billion pounds of methanal is made yearly in the United States by oxidation of methanol:

$$\mathrm{CH_3OH \xrightarrow{O_2,\ 600^\circ\text{–}650^\circ C,\ catalytic\ Ag} CH_2{=}O}$$

In aqueous solution (''formalin''), it has applications as a disinfectant, a germicide, and a fungicide. Its greatest use is in the preparation of phenolic resins (Section 24-4).

Propanone (acetone) is a valuable by-product of the cumene-hydroperoxide process (Section 24-3) and is sold as a solvent and starting material for the production of other industrial materials. Annual production in the United States alone is about 2 billion pounds.

Butanal is made by a process called hydroformylation, in which propene is exposed to synthesis gas ($CO + H_2$, Section 8-4) in the presence of a cobalt or rhodium catalyst. The reaction is general and can be used to prepare other aldehydes.

$$\mathrm{CH_3CH{=}CH_2 + CO + H_2 \xrightarrow{Co\ or\ Rh,\ \Delta,\ pressure} CH_3CH_2CH_2CHO}$$

Aldehydes and Ketones by the Oxidation of Alcohols

We have seen (Section 9-4) that chromium(VI) reagents can oxidize alcohols to carbonyl compounds. Secondary alcohols give ketones and primary alcohols give aldehydes, but, in the latter case, only if chromium trioxide is mixed with pyridine, which prevents overoxidation to carboxylic acids. The chromium oxidant is selective even in the presence of alkene and alkyne units.

Selective Alcohol Oxidations

$$\mathrm{CH_3\overset{\overset{\displaystyle OH}{|}}{C}HC{\equiv}C(CH_2)_3CH_3 \xrightarrow{CrO_3,\ H_2SO_4,\ propanone\ (acetone),\ 0^\circ C} \underset{80\%}{CH_3\overset{\overset{\displaystyle O}{\|}}{C}C{\equiv}C(CH_2)_3CH_3}}$$

3-Octyn-2-ol → **3-Octyn-2-one**

$$\mathrm{HC{\equiv}CCH{=}CHCH_2OH \xrightarrow{CrO_3,\ H_2SO_4,\ propanone\ (acetone),\ 0^\circ C} \underset{60\%}{HC{\equiv}CCH{=}CH\overset{\overset{\displaystyle O}{\|}}{C}OH}}$$

2-Penten-4-yn-1- ol → **2-Penten-4-ynoic acid**

$$\xrightarrow{CrO_3\,(pyridine)_2,\ HCl,\ CH_2Cl_2,\ Na^{+\,-}OOCCH_3}$$

85%

The successful selective oxidation of primary alcohols to aldehydes by chromium trioxide with pyridine requires the absence of water. In an aqueous medium, the aldehyde product can be hydrated to the geminal diol (Section 15-5), which is further oxidized.

Water Causes the Overoxidation of Primary Alcohols

$$RCH_2OH \xrightarrow{Cr(VI),\ H^+} RCH{=}O \xrightarrow{H_2O} RC(OH)_2H \xrightarrow{Cr(VI)} RC(=O)OH$$

Another mild reagent that specifically oxidizes allylic alcohols to aldehydes is manganese dioxide. Ordinary alcohols are not attacked at room temperature.

Selective Allylic Oxidations with Manganese Dioxide

$$HC{\equiv}CCH{=}C(CH_3)CH_2OH \xrightarrow{MnO_2,\ propanone\ (acetone),\ 25^\circ C} HC{\equiv}CCH{=}C(CH_3)CHO$$

35%

2-Methyl-2-penten-4-yn-1-ol **2-Methyl-2-penten-4-ynal**

$$\xrightarrow{MnO_2,\ CHCl_3,\ 25^\circ C}$$

62%

EXERCISE 15-4

Design a synthesis of cyclohexyl 1-propynyl ketone starting from cyclohexane. You may use any other reagents.

Aldehydes and Ketones by Oxidative C–C Bond Cleavage

Exposure to ozone followed by treatment with a mild reducing agent, such as catalytically activated hydrogen or dimethyl sulfide, cleaves alkenes to give aldehydes and ketones (Section 12-5).

Ozonolysis

$$>C{=}C< \xrightarrow{1.\ O_3\quad 2.\ Reducing\ agent} >C{=}O + O{=}C<$$

Aldehydes and Ketones from Alkynes

Addition of water to the carbon–carbon triple bond yields enols that tautomerize to carbonyl compounds (Section 13-6). Markovnikov addition of water takes place in aqueous acid in the presence of mercuric ion.

Markovnikov Hydration of Alkynes

$$\mathrm{RC{\equiv}CH} \xrightarrow{\mathrm{HOH,\ H^+,\ Hg^{2+}}} \left[\begin{array}{c}\mathrm{HO}\quad\quad\mathrm{H}\\ \mathrm{C{=}C}\\ \mathrm{R}\quad\quad\mathrm{H}\end{array}\right] \longrightarrow \mathrm{R\overset{O}{\overset{\|}{C}}CH_3}$$

Anti-Markovnikov addition is observed in hydroboration-oxidation.

Anti-Markovnikov Hydration of Alkynes

$$\mathrm{RC{\equiv}CH} \xrightarrow{(\text{cyclohexyl})_2\mathrm{BH}} \begin{array}{c}\mathrm{R}\quad\quad\mathrm{H}\\ \mathrm{C{=}C}\\ \mathrm{H}\quad\quad\mathrm{B(cyclohexyl)_2}\end{array} \xrightarrow{\mathrm{H_2O_2,\ HO^-}} \left[\begin{array}{c}\mathrm{R}\quad\quad\mathrm{H}\\ \mathrm{C{=}C}\\ \mathrm{H}\quad\quad\mathrm{OH}\end{array}\right] \longrightarrow \mathrm{RCH_2\overset{O}{\overset{\|}{C}}H}$$

In summary, there are three general methods of synthesizing aldehydes and ketones: oxidation of alcohols, oxidative cleavage of alkenes, and hydration of alkynes. Many other approaches exist and will be discussed in later chapters.

15-4 The Reactivity of the Carbonyl Group: Mechanisms of Addition

This section begins the discussion of the chemistry of the carbonyl group in aldehydes and ketones. Like the π bond in the alkenes, the carbon–oxygen double bond is prone to additions. For example, catalytic hydrogenation yields alcohols. Electrophiles attack at oxygen, nucleophiles at carbon.

The Three Regions of Reactivity in Aldehydes and Ketones

Aldehydes and ketones contain three regions at which most reactions occur: the oxygen, the carbonyl carbon, and the carbon adjacent to it:

:O: ← Attack by electrophiles
C ← Attack by nucleophiles
H ← Acidic

The remainder of this chapter is concerned with the first two areas of reactivity. The other reactions will be the subject of Chapter 16.

The Carbonyl Pi Bond Undergoes Hydrogenation

Like C–C π bonds, the carbonyl group is susceptible to catalytic hydrogenation (Section 8-4), which results in the formation of alcohols. Depending on the catalyst, aldehydes and ketones may be more sluggish than alkenes in this reaction, requiring pressure or elevated temperature to proceed at a useful rate.

$$CH_3CH_2CH_2\overset{O}{\overset{\|}{C}}H \xrightarrow{H_2,\ Ru,\ 160^\circ C,\ 30\ atm} CH_3CH_2CH_2\overset{H}{\overset{|}{C}}HOH$$

$$CH_3\overset{O}{\overset{\|}{C}}CH_2CH_3 \xrightarrow{H_2,\ Raney\ Ni,\ 80^\circ C,\ 5\ atm} CH_3\underset{H}{\underset{|}{\overset{OH}{\overset{|}{C}}}}CH_2CH_3$$

This difference in reactivity can be exploited in selective hydrogenations of unsaturated aldehydes or ketones. In difficult cases, hydrogen uptake can be monitored and the reaction stopped once the required amount of hydrogen gas has been absorbed.

The catalytic hydrogenations of aldehydes and ketones are addition reactions that proceed through surface intermediates. Additions to the carbon–oxygen double bond may also occur by ionic mechanisms that exploit the dipolar nature of the functional group.

Selective Enone Hydrogenation

H_2, Pt, 1 atm, 25°C

H H

100%

The Carbonyl Group Undergoes Ionic Additions

Polar reagents add to the dipolar carbonyl group according to Coulomb's law. Nucleophiles bond to the carbon, electrophiles to the oxygen. Sections 8-4 and 8-6 described a number of such additions by organometallic and hydride reagents to give alcohols. Because of the nucleophilic strength of the reagents, those additions were irreversible. This and the following two sections consider ionic additions of milder nucleophiles Nu–H, such as water, alcohols, thiols, and amines. These processes are not strongly exothermic but instead establish equilibria that can be pushed in either direction by the appropriate choice of reaction conditions.

General Additions to the Carbonyl Group

$$\overset{\delta^-}{O}{=}\overset{\delta^+}{C}< \; + \; \overset{\delta^+}{X}{-}\overset{\delta^-}{Y} \longrightarrow -\underset{Y}{\underset{|}{\overset{OX}{\overset{|}{C}}}}-$$

What is the mechanism of the ionic addition of these reagents to the C–O double bond? Two pathways can be formulated. The first begins with nucleophilic attack and proceeds in the absence of acid. As the nucleophile approaches the electrophilic carbon, the latter rehybridizes, and the electron pair of the π bond moves over to the oxygen, producing an alkoxide ion. Subsequent protonation, usually from a protic solvent such as water or alcohol, yields the final addition product.

Nucleophilic Addition-Protonation

$$\overset{\delta^+}{C}{=}\overset{\delta^-}{\ddot{O}}: + Nu:^- \rightleftharpoons Nu{-}C{-}\ddot{O}:^- \underset{-\,H_2\ddot{O}}{\overset{H\ddot{O}H}{\rightleftharpoons}} Nu{-}C{-}\ddot{O}H + H\ddot{O}:^-$$

Alkoxide ion

Note that the new Nu–C bond is made up entirely of the electron pair of the nucleophile. The entire process is reminiscent of an S_N2 reaction. However, instead of a leaving group, an electron pair is displaced from a position in which it is shared between two nuclei to one in which it is localized on one of them (the oxygen atom).

The second mechanism predominates in an acidic medium and begins with electrophilic attack. Here the polarization of the carbonyl function and the presence of lone electron pairs on the oxygen atom facilitate protonation. Protonated carbonyl groups were discussed in Section 9-2. The basicity of the oxygen is very weak, as shown by the acidity of its conjugate acid, which has a pK_a ranging from about -7 to -8. Thus, in a dilute acidic medium, such as the one in which many carbonyl addition reactions are carried out, most of the carbonyl compound will stay unprotonated. However, the small amount of protonated material behaves like a very reactive carbon electrophile. Nucleophilic attack by the nucleophile completes the addition process and shifts the first, unfavorable equilibrium.

Electrophilic Protonation-Addition

$$\overset{\delta^+}{C}=\overset{\delta^-}{\ddot{O}} + H^+ \rightleftharpoons \left[\overset{+}{C}-\ddot{O}H \longleftrightarrow C=\overset{+}{\ddot{O}}H \right] \underset{-:Nu^-}{\overset{:Nu^-}{\rightleftharpoons}} (Nu)C-\ddot{O}H$$

Protonated carbonyl group
p$K_a \sim -8$

In summary, there are three regions of reactivity in aldehydes and ketones. The first two are the two atoms of the carbonyl group and are the subject of the remainder of this chapter. A discussion of the third is deferred to Chapter 16. The reactivity of the carbonyl group is governed by addition processes. Catalytic hydrogenation gives alcohols. Ionic additions of NuH (Nu = OH, OR, SR, NR_2) are reversible and may begin with nucleophilic attack at the carbonyl carbon, followed by electrophilic trapping of the alkoxide anion so generated. Alternatively, in acidic media, protonation precedes the addition of the nucleophile.

15-5 Aldehydes and Ketones Add Water and Alcohols to Form Hydrates and Acetals

This section introduces the reactions of aldehydes and ketones with water and alcohols. These compounds attack the carbonyl group according to the general mechanisms outlined in Section 15-4, through both acid and base catalysis. In acid catalysis, the simple addition product is transformed further by replacement of the hydroxy group with alkoxy to give acetals. Similar reactivity is observed in thiols.

Hydration of the Carbonyl Group

$$C=O + HOH \underset{}{\overset{K}{\rightleftharpoons}} (HO)C-OH \quad (H^+ \text{ or } HO^-)$$

Geminal diol

Water Hydrates the Carbonyl Group

One of the reagents capable of attacking the carbonyl function of aldehydes and ketones is water. This transformation, which is catalyzed by either acid or base, leads to equilibration with the corresponding **geminal diols,** also called **carbonyl hydrates.** In the base-catalyzed mechanism, the hydroxide functions as the nucleophile; water then traps the intermediate adduct, a hydroxy alkoxide, to give the product diol and to regenerate the catalyst.

Mechanism of Base-Catalyzed Hydration:

$$C=\ddot{O} + {}^-:\ddot{O}H \rightleftharpoons (H\ddot{O})C-\ddot{O}:^- \overset{H\ddot{O}H}{\rightleftharpoons} (H\ddot{O})C-\ddot{O}H + H\ddot{O}:^-$$

Hydroxy alkoxide

In the acid-catalyzed mechanism, the sequence of events is reversed. Here, initial protonation is followed by nucleophilic attack of water and subsequent loss of the catalytic proton, which reenters the catalytic cycle.

Mechanism of Acid-Catalyzed Hydration:

$$>C{=}\ddot{\underset{..}{O}} + H^+ \rightleftharpoons >C{=}\overset{+}{\ddot{O}}{-}H \xrightleftharpoons{H_2\ddot{O}} H_2\overset{+}{\ddot{O}}{-}C{-}\ddot{\underset{..}{O}}H \rightleftharpoons H\ddot{\underset{..}{O}}{-}C{-}\ddot{\underset{..}{O}}H + H^+$$

As these equations indicate, the hydration of aldehydes and ketones is reversible. The equilibrium lies to the left for ketones, and to the right for formaldehyde and aldehydes bearing electron-withdrawing substituents. Ordinary aldehydes adopt an intermediate position, with equilibrium constants approaching unity.

How can these trends be explained? Generally, additions to carbonyl groups become more favored the more electrophilic the carbonyl carbon. The electrophilic power of this center can be roughly correlated with the stability of the carbocation formulated in the dipolar resonance structure. The more alkylated carbonyl carbon is the more stable, and its reactivity decreases along the series methanal, ethanal, propanone. For example, methanal (formaldehyde) has a dipolar resonance form incorporating a primary carbocation; in ethanal (acetaldehyde), this carbon is secondary; and propanone (acetone) has a resonance structure reminiscent of a tertiary cation. Electron-withdrawing substituents destabilize the positively polarized carbon, rendering it more reactive in nucleophilic attack.

Equilibrium Constants K for the Hydration of Some Carbonyl Compounds

$Cl_3CCH(=O)$ $K > 10^4$

$H_2C{=}O$ $K > 10^3$

$CH_3CH(=O)$ $K \sim 1$

$CH_3C(=O)CH_3$ $K < 10^{-2}$

Order of Reactivity of Carbonyl Groups

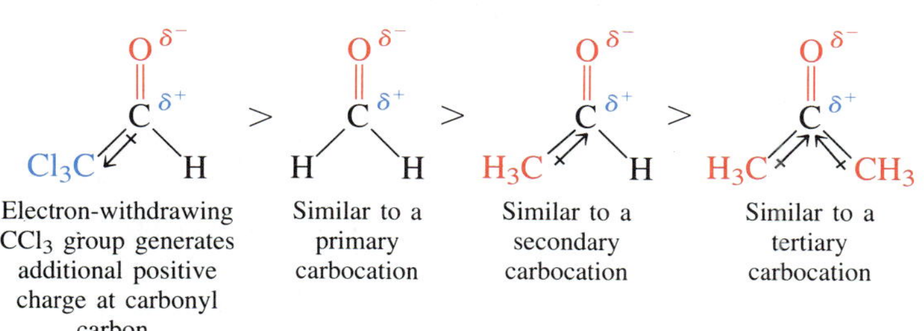

EXERCISE 15-5

(a) Rank in order of increasing reactivity in hydration: $Cl_3CCH(=O)$, $Cl_3CC(=O)CH_3$, $Cl_3CC(=O)CCl_3$

(b) Treatment of propanone (acetone) with $H_2{}^{18}O$ results in the formation of labeled propanone, $CH_3C(={}^{18}O)CH_3$. Explain.

Alcohols Add to Aldehydes and Ketones to Form Hemiacetals

Not surprisingly, alcohols also undergo addition to aldehydes and ketones, by a mechanism virtually identical with that outlined for water. The adducts formed are called **hemiacetals** (*hemi,* Greek, half), because they are intermediates on the way to acetals.

General Hemiacetal Formation

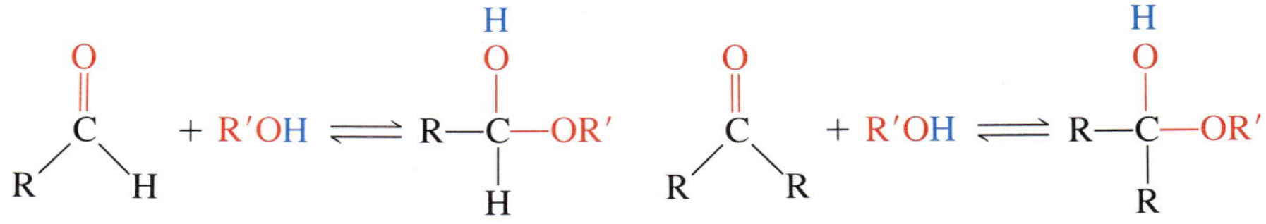

$$\underset{\text{R}\quad\text{H}}{\overset{\text{O}}{\overset{\|}{\text{C}}}} + \text{R'OH} \rightleftharpoons \text{R—}\underset{\text{H}}{\overset{\text{HO}}{\text{C}}}\text{—OR'} \qquad \underset{\text{R}\quad\text{R}}{\overset{\text{O}}{\overset{\|}{\text{C}}}} + \text{R'OH} \rightleftharpoons \text{R—}\underset{\text{R}}{\overset{\text{HO}}{\text{C}}}\text{—OR'}$$

A hemiacetal **A hemiacetal**

These addition reactions, too, are governed by equilibria that usually favor the starting carbonyl compound. Hemiacetals are therefore usually not isolable. Exceptions are those formed from reactive carbonyl compounds such as methanal (formaldehyde) or 2,2,2-trichloroethanal (2,2,2-trichloroacetaldehyde). Hemiacetals are also isolable from hydroxy aldehydes and ketones when cyclization leads to the formation of relatively strain free five- and six-membered rings.

Intramolecular Hemiacetal Formation

$$HOCH_2CH_2CH_2CH_2CH{=}O \xrightarrow[]{H^+} \text{cyclic hemiacetal}$$

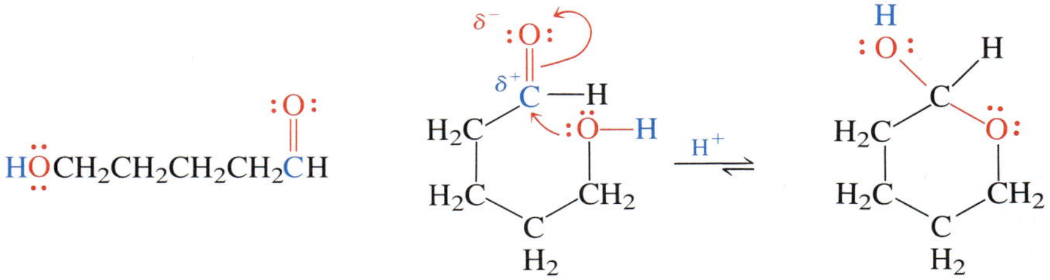

5-Hydroxypentanal **A cyclic hemiacetal, stable**

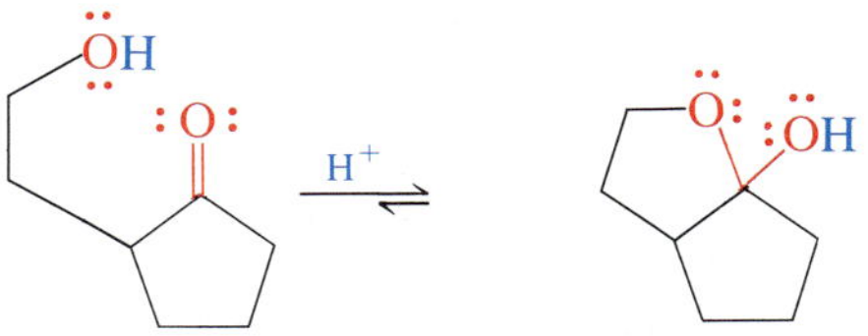

A cyclic hemiacetal, stable

Entropy explains why. In the intermolecular reaction, *two* molecules combine to form *one* new structure. This is costly in entropic terms, ΔS being very negative. On the other hand, the reverse reaction is entropically favored. In contrast, in the intramolecular version, *one* molecule transforms into *one* other. Although some freedom is lost (mainly rotational) in ''fixing'' the alkyl chain bearing the OH group, this entropy change is much less, and so ΔS is negative but small. The reaction proceeds to product because the enthalpy is favorable. In hemiacetal formation, the entropy term may control the outcome of the reaction, tipping the balance between a positive (unfavorable) or negative (favorable) ΔG°. Intramolecular hemiacetal formation is common in sugar chemistry (see Chapter 23).

Although cyclic hemiacetals are more stable than the hydroxy carbonyl compounds from which they are derived, remember that they are isomers in equilibrium. Therefore, they will show reactivity typical of an alcohol in the hemiacetal form and typical of the two functional groups in the hydroxy aldehyde or hydroxy ketone form.

Reactivity of a Cyclic Hemiacetal

Reduction occurs at the carbonyl group of the hydroxy aldehyde. $NaBH_4$

CrO_3 Oxidation occurs at the hemiacetal hydroxy group.

Acid Catalyzes Acetal Formation

In the presence of excess alcohol, the acid-catalyzed reaction of aldehydes and ketones proceeds beyond the hemiacetal stage. Under these conditions, the hydroxy function of the initial adduct is replaced by another alkoxy unit derived from the alcohol. The resulting compounds are called acetals. (An older term for acetals derived from ketones is ketals.)

General Acetal Synthesis

$$\text{RCR}(\text{C=O}) + 2\,\text{R'OH} \xrightleftharpoons{\text{H}^+} \text{R—C(OR')(R)—OR'} + \text{H}_2\text{O}$$

An acetal

The net change is the replacement of the carbonyl oxygen by two alkoxy groups and the formation of one equivalent of water.

Let us examine the mechanism of this transformation for an aldehyde. The initial reaction is the ordinary acid-catalyzed addition of the first molecule of alcohol. The resulting hemiacetal can be protonated at the hydroxy group, changing this substituent into water, a good leaving group. On loss of water, the resulting carbocation is stabilized by resonance with a lone electron pair on oxygen. A second molecule of alcohol now adds to the electrophilic carbon, giving initially the protonated acetal, which is then deprotonated to the final product.

Mechanism of Acetal Formation:

STEP 1: Hemiacetal generation

$$\text{RCH(=O)} \underset{-\text{H}^+}{\overset{+\text{H}^+}{\rightleftharpoons}} \left[\text{RC(=\overset{+}{O}H)H} \longleftrightarrow \text{R}\overset{+}{\text{C}}\text{(OH)H}\right] \underset{-\text{R'OH}}{\overset{\text{R'OH}}{\rightleftharpoons}} \text{R—C(OH)(H)—}\overset{+}{\text{O}}\text{(H)R'} \underset{+\text{H}^+}{\overset{-\text{H}^+}{\rightleftharpoons}} \text{R—C(OH)(H)—OR'}$$

Hemiacetal

STEP 2: Acetal generation

$$R{-}\underset{\substack{|\\ :\ddot{O}R'}}{\overset{\substack{:\ddot{O}H\\ |}}{C}}{-}H \underset{-\,H^+}{\overset{+\,H^+}{\rightleftharpoons}} R{-}\underset{\substack{|\\ :\ddot{O}R'}}{\overset{\substack{H\\ |\\ +:\ddot{O}H\\ |}}{C}}{-}H \underset{+\,H_2\ddot{O}}{\overset{-\,H_2\ddot{O}}{\rightleftharpoons}} \left[\begin{array}{c} R \quad H \\ \overset{+}{C} \\ | \\ :\ddot{O}R' \end{array} \longleftrightarrow \begin{array}{c} R \quad H \\ C \\ \| \\ {}^{+}\ddot{O}R' \end{array} \right] \underset{-\,R'\ddot{O}H}{\overset{R'\ddot{O}H}{\rightleftharpoons}}$$

$$R{-}\underset{\substack{|\\ :\ddot{O}R'}}{\overset{\substack{H\\ |}}{C}}{-}\overset{+}{\ddot{O}}\!:\!{\substack{R'\\ H}} \underset{+\,H^+}{\overset{-\,H^+}{\rightleftharpoons}} R{-}\underset{\substack{|\\ :\ddot{O}R'}}{\overset{\substack{H\\ |}}{C}}{-}\ddot{O}R'$$

Each step is reversible; the entire sequence, starting from carbonyl compound and ending with acetal, is an equilibrium process. As in hemiacetal formation, the overall equilibrium may lie to the right (mainly with aldehydes) or to the left (with ketones). Its position may be shifted in either direction by manipulating the reaction conditions: toward acetal, by using excess alcohol or by continually removing water from the reaction medium; toward aldehyde or ketone by using excess water. The latter process is called **acetal hydrolysis.**

Cyclic Acetals Are Useful Protecting Groups

1,2-Ethanediol and related diols react with aldehydes and ketones in the presence of catalytic acid to form cyclic acetals.

$$CH_3\overset{\substack{O\\ \|}}{C}H + HOCH_2CH_2OH \xrightarrow[-\,H_2O]{H^+} \begin{array}{c} H_2C{-}CH_2 \\ O \quad\; O \\ C \\ H_3C \quad\; H \end{array}$$

73%

A cyclic acetal

Use of a diol helps, because only *two* molecules of starting material (carbonyl compound and diol) are converted into two molecules of product (acetal and water). This reaction contrasts with that with ordinary alcohols, which is not as favorable entropically because *three* molecules (the carbonyl compound and two equivalents of alcohol) transform to the two products. Cyclic acetals readily revert to aldehydes and ketones in the presence of excess acidic water.

A useful aspect of acetalization is the relative inertness of the resulting groups. Thus, even though they are readily hydrolyzed by aqueous acid, they are not attacked by many basic, organometallic, and hydride reducing agents. This finding is not too surprising, considering that acetals are essentially cyclic ethers. We may regard an acetal as a ''masked'' aldehyde or ketone; the cyclic ethers act as protecting groups for the carbonyl function.

An application of cyclic acetals as protecting groups is the alkylation of an alkynyl anion with 3-iodopropanal 1,2-ethanediol acetal. The isolated product is readily hydrolyzed to the aldehyde. If the alkylation is attempted with unprotected 3-iodopropanal, the alkynyl anion attacks the carbonyl group.

The Use of a Protected Aldehyde in Synthesis

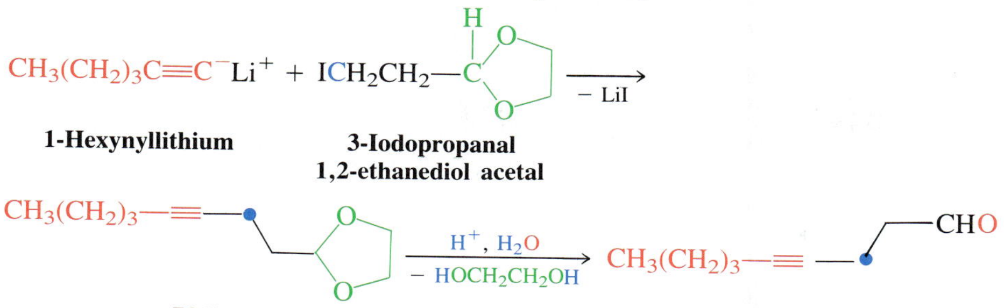

$CH_3(CH_2)_3C{\equiv}C^-Li^+ + ICH_2CH_2$—C(H)(OCH₂CH₂O) $\xrightarrow{-\,LiI}$

1-Hexynyllithium **3-Iodopropanal 1,2-ethanediol acetal**

$CH_3(CH_2)_3$—≡—(CH₂)₂-dioxolane $\xrightarrow[-\,HOCH_2CH_2OH]{H^+,\,H_2O}$ $CH_3(CH_2)_3$—≡—CH_2CH_2CHO

70% **4-Nonynal 1,2-ethanediol acetal** 90% **4-Nonynal**

EXERCISE 15-6

Suggest a convenient way of converting compound A (in the margin) into B.

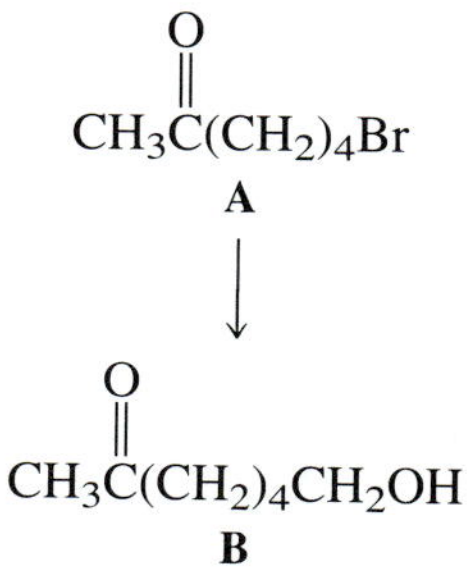

$CH_3\overset{O}{\overset{\|}{C}}(CH_2)_4Br$ **A** → $CH_3\overset{O}{\overset{\|}{C}}(CH_2)_4CH_2OH$ **B**

EXERCISE 15-7

Compound C isomerized to D on standing in slightly acidic solution. Write a mechanism.

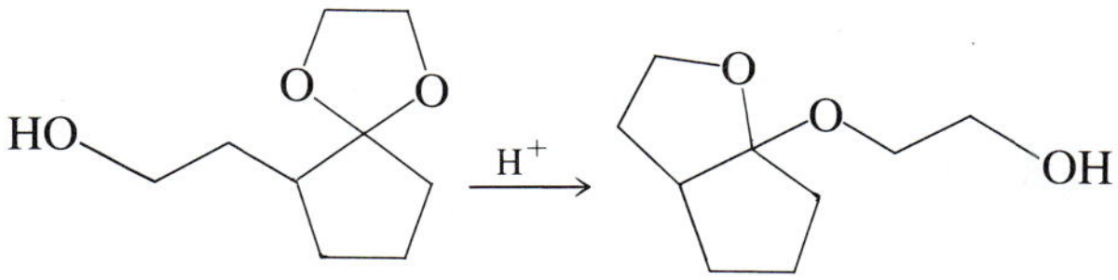

HO ... **C** $\xrightarrow{H^+}$ ... OH **D**

If a diol may be used to protect a carbonyl function, then a carbonyl compound should be able to protect a diol. This is indeed the case. For example, propanone (acetone) blocks the acidic sites in vicinal diols by forming an acetal with them. The protection of diols as propanone (acetone) acetals is an important reaction in sugar chemistry (Chapter 23).

Protection of a Vicinal Diol as the Propanone (Acetone) Acetal

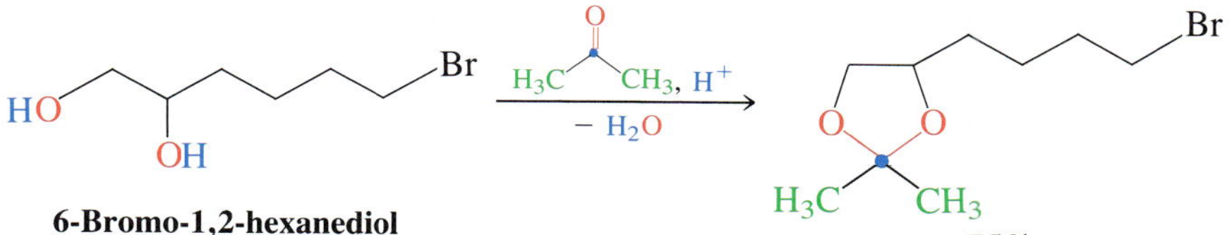

HO, OH, Br **6-Bromo-1,2-hexanediol** $\xrightarrow[-\,H_2O]{H_3CCOCH_3,\ H^+}$ O, O, Br, H_3C, CH_3 75%

EXERCISE 15-8

The reaction of the sugar sorbose A with propanone (acetone) in the presence of acid gives the polyether B. Explain how. Hint: Make models and work backward.

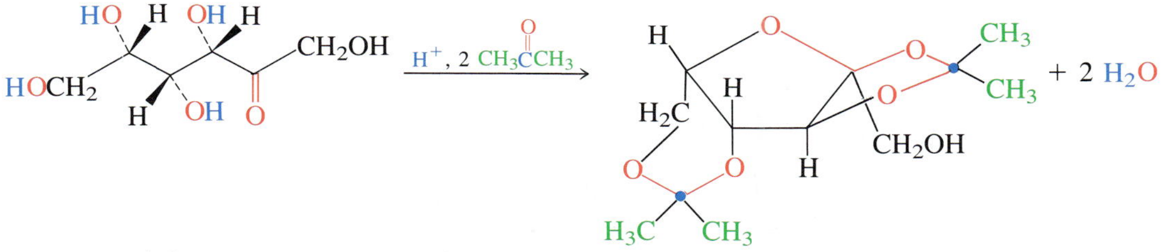

$HOCH_2$, HO, H, OH, H, CH_2OH, H, OH, O **A (sorbose)** $\xrightarrow{H^+,\ 2\ CH_3\overset{O}{\overset{\|}{C}}CH_3}$ H, O, O, CH_3, CH_3, H, H_2C, O, CH_2OH, O, O, H, H_3C, CH_3 **B** $+\ 2\ H_2O$

Thiols React with the Carbonyl Group as Alcohols Do

Thiols, the sulfur analogs of alcohols (see Section 9-7), also react with aldehydes and ketones by a mechanism identical with the one described for alcohols to form **thioacetals** in high yield. Instead of a proton catalyst, a Lewis acid, such as BF_3 or $ZnCl_2$, is often used.

Formation of Thioacetals

$$CH_3CH_2CH_2\overset{\overset{\displaystyle O}{\|}}{C}H \xrightarrow[-\ H_2O]{CH_3SH,\ BF_3,\ 20°C,\ 30\ min} CH_3CH_2CH_2\underset{\underset{\displaystyle SCH_3}{|}}{\overset{\overset{\displaystyle SCH_3}{|}}{C}}H$$

70%

A thioacetal

$$\xrightarrow[-\ H_2O]{HSCH_2CH_2SH,\ ZnCl_2,\ (CH_3CH_2)_2O,\ 25°C}$$

95%

A cyclic thioacetal

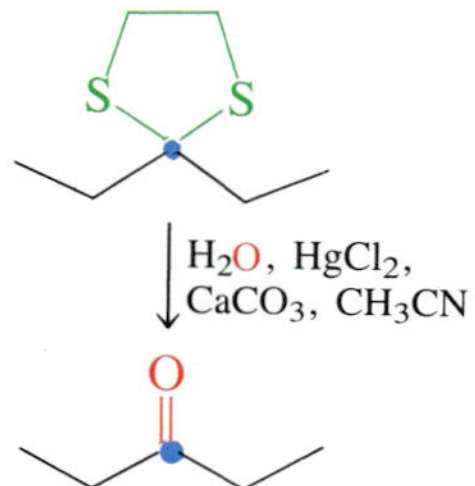

These sulfur derivatives are stable in aqueous acid, a medium that hydrolyzes ordinary acetals. The difference in reactivity may be useful in synthesis if it is necessary to differentiate two different carbonyl groups in the same molecule. Hydrolysis of thioacetals is carried out by using mercuric chloride in aqueous acetonitrile. The driving force is the formation of insoluble mercuric sulfides.

Thioacetals are desulfurized to the corresponding hydrocarbon by treatment with Raney nickel. Thioacetal generation followed by desulfurization is used to convert a carbonyl into a methylene group.

$$\xrightarrow{\text{Raney Ni}}$$

65%

$$\xrightarrow{1.\ HSCH_2CH_2SH,\ BF_3,\ 0°C \quad 2.\ \text{Raney Ni},\ CH_3CH_2OH,\ \Delta}$$

74%

EXERCISE 15-9

Suggest a possible synthesis of cyclodecane from 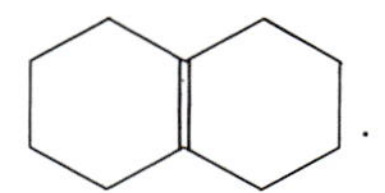.

In summary, the carbonyl group of the aldehydes and ketones is hydrated by water and converted into hemiacetals by alcohols. Aldehydes are more reactive than ketones. Electron-withdrawing substituents render the carbonyl carbon more electrophilic. All reactions are equilibria, which may be shifted either way by starting with excess water or alcohol. Both acid and base catalysis occurs. Intramolecular (cyclic) hemiacetal formation is favored over its intermolecular counterpart by entropy. Hemiacetals are converted into acetals by acid and excess alcohol. Cyclic acetals are good protecting groups for the carbonyl function and for diols. Thiols show similar reactivity. The formation of thioacetals is usually catalyzed by Lewis acids. Their hydrolysis requires the presence of mercuric salts. Thioacetals are reduced by Raney nickel to the corresponding hydrocarbons.

15-6 The Nucleophilic Addition of Amines to Aldehydes and Ketones: Condensation to Imines

Amines may be regarded as the nitrogen analogs of alcohols. The nitrogen is more nucleophilic than oxygen; so amines add very effectively to the carbonyl group in aldehydes and ketones. This section delineates the reactions of several amine derivatives with the carbonyl function.

On exposure to an amine, aldehydes and ketones form **hemiaminals,** the nitrogen analogs of hemiacetals. Hemiaminals of primary amines readily lose water to form a carbon–nitrogen double bond. This functional group is called an **imine** (an older name is *Schiff base*) and is the nitrogenous analog of an aldehyde or ketone.

Imine Formation from Amines and Aldehydes or Ketones

R—$\ddot{N}H_2$ + C=O —Addition⇌ $R\overset{+}{N}H_2$—C—O^- —Deprotonation at N, reprotonation at O⇌ RNH—C—OH —($- H_2O$ / $+ H_2O$, Elimination of water)⇌ C=N—R

A hemiaminal   An imine

The mechanism of the elimination of water from the hemiaminal is the same as that for the decomposition of a hemiacetal to the carbonyl compound and alcohol. It begins with protonation of the hydroxy group. (Protonation of the more-basic nitrogen just leads back to the carbonyl compound.) Dehydration follows, and then deprotonation of the intermediate iminium ion.

Mechanism of Hemiaminal Dehydration:

RNH—C—OH ($+ H^+$ / $- H^+$)⇌ RNH—C—$\overset{+}{O}H_2$ ($-$ HOH / $+$ HOH)⇌ [RNH—C^+ ⟷ RHN^+=C] ($- H^+$ / $+ H^+$)⇌ RN=C

Processes such as imine formation from a primary amine and an aldehyde or ketone, in which two molecules are joined with the elimination of water, are called **condensations.** Imine formation is reversible, and the usual measures have to be employed to shift the equilibrium in the desired direction. Imines can exist as *Z* and *E* isomers.

Condensation of a Ketone with a Primary Amine

$$RNH_2 + O{=}C(R')(R'') \rightleftharpoons RN{=}C(R')(R'') + H_2O$$

(Two isomers can be formed.)

Examples:

$$CH_3\overset{O}{\overset{\|}{C}}H + H_2NCH_2CH_2CH_2CH_3 \xrightarrow{KOH} CH_3CH{=}NCH_2CH_2CH_2CH_3 + H_2O$$

83%

$$CH_3\overset{O}{\overset{\|}{C}}CH_3 + C_6H_{11}NH_2 \xrightarrow{H^+} (H_3C)_2C{=}N{-}C_6H_{11} + H_2O$$

95%

EXERCISE 15-10

Reagent A has been used with aldehydes to prepare crystalline imidazolidine derivatives such as B, for the purpose of their isolation and structural identification. Write a mechanism for the formation of B.

$$C_6H_5NHCH_2CH_2NHC_6H_5 + H\overset{O}{\overset{\|}{C}}CH_3 \xrightarrow{CH_3OH,\ H^+} \text{B} + H_2O$$

A
***N,N′*-Diphenyl-1,2-ethanediamine**

B
2-Methyl-1,3-diphenyl-1,3-diazacyclopentane (2-Methyl-1,3-diphenylimidazolidine)
m.p. 102°C

Several amine derivatives condense with aldehydes and ketones to form products that are highly crystalline. For example, *hydroxylamine,* H_2NOH, in the form of its hydrochloride, condenses to **oximes.**

Oxime Synthesis

$$CH_3(CH_2)_5\overset{O}{\overset{\|}{C}}H + H_3\overset{+}{N}OHCl^- \xrightarrow[-\,H_2O]{H^+} CH_3(CH_2)_5(H)C{=}NOH$$

Hydroxylamine hydrochloride

93%
Heptanal oxime

Hydrazine, H_2NNH_2, and some of its derivatives condense to **hydrazones.** Reaction at both ends of the hydrazine produces **azines.**

Hydrazone and Azine Synthesis

$$CH_3\overset{O}{\overset{\|}{C}}CH_3 + H_2N{-}NH_2 \xrightarrow[-\ H_2O]{} (CH_3)_2C{=}N{-}NH_2$$

Hydrazine **Propanone (acetone) hydrazone**

$$2\ C_6H_5\overset{O}{\overset{\|}{C}}H + H_2N{-}NH_2 \xrightarrow[-\ 2\ H_2O]{} C_6H_5(H)C{=}N{-}N{=}C(H)C_6H_5$$

94%

Benzenecarboxaldehyde (Benzaldehyde) **Benzalazine**

Phenylhydrazines, $C_6H_5NHNH_2$ and, particularly, 2,4-dinitrophenylhydrazine, $(NO_2)_2C_6H_3NHNH_2$, have been employed traditionally in the synthesis of crystalline derivatives of liquid aldehydes and ketones for the purposes of structural identification or isolation.

$$2,4\text{-}(NO_2)_2C_6H_3NHNH_2 + \text{cyclopentanone} \xrightarrow[-\ H_2O]{} \text{cyclopentylidene}{=}N{-}NH{-}C_6H_3\text{-}2,4\text{-}(NO_2)_2$$

85%

2,4-Dinitrophenylhydrazine **Cyclopentanone 2,4-dinitrophenylhydrazone** m.p. 142°C

Finally, the hydrazine derivative *semicarbazide,* reacts with aldehydes and ketones to give **semicarbazones.**

Semicarbazone Synthesis

$$C_6H_{11}\overset{O}{\overset{\|}{C}}H + H_2N{-}NH\overset{O}{\overset{\|}{C}}NH_2 \xrightarrow[-\ H_2O]{H^+} C_6H_{11}(H)C{=}N{-}NH\overset{O}{\overset{\|}{C}}NH_2$$

90%

Semicarbazide **Cyclohexanecarboxaldehyde semicarbazone**

Deoxygenation of the Carbonyl Group Through Imines

In the presence of base at elevated temperatures, simple hydrazones decompose with evolution of nitrogen to give the corresponding hydrocarbon. This reaction, called the **Wolff-Kishner* reduction,** complements thioketal desulfurization (Section 15-5) as a method of deoxygenating an aldehyde or ketone.

Wolff-Kishner Reduction

$$\underset{\text{RCR}'}{\overset{\text{N—NH}_2}{\|}} + \text{NaOH} \xrightarrow{(\text{HOCH}_2\text{CH}_2)_2\text{O},\ 180°\text{–}200°\text{C}} \text{RCH}_2\text{R}' + \text{N}_2$$

The mechanism of nitrogen elimination includes a sequence of base-mediated hydrogen shifts. The base first removes a proton from the hydrazone to give the corresponding delocalized anion. This species may be viewed as an allylic anion in which two carbon atoms have been replaced by nitrogen. Reprotonation may occur on nitrogen, regenerating the starting material, or on carbon, leading on to the product. The base removes another proton from nitrogen on the new intermediate to generate a new anion, which rapidly decomposes irreversibly, with the extrusion of nitrogen gas. The alkyl anion so formed is rapidly protonated to the hydrocarbon.

Mechanism of Nitrogen Elimination in the Wolff-Kishner Reduction:

$$\text{R}_2\text{C=N–NH}_2 + \text{HO}^- \underset{+\text{H}_2\text{O}}{\overset{-\text{HOH}}{\rightleftharpoons}} \left[\text{RR}'\text{C=N–}\ddot{\text{N}}\text{H}^- \longleftrightarrow \text{RR}'\ddot{\text{C}}^-\text{–N=NH}\right] \underset{\text{Reprotonation}}{\overset{\text{HOH}}{\rightleftharpoons}}$$

(Deprotonation)

$$\text{RR}'\text{CH–N=NH} + \text{HO}^- \underset{+\text{H}_2\text{O}}{\overset{-\text{HOH}}{\rightleftharpoons}} \text{RR}'\text{CH–N=N}^- \xrightarrow[\text{loss of N}_2]{\text{Irreversible}} \text{RR}'\text{CH}^- \xrightarrow{\text{H}_2\text{O, protonation}} \text{RCH}_2\text{R}' + \text{HO}^-$$

(Deprotonation)

In practice, the Wolff-Kishner reduction is carried out without isolating the intermediate hydrazone. Hydrazine, commercially available as an 85% aqueous solution (hydrazine hydrate), is added to the carbonyl compound in the high-boiling alcohol $HOCH_2CH_2OCH_2CH_2OH$ (b.p. 245°C), containing sodium hydroxide, and the mixture is heated. Aqueous work-up yields the hydrocarbon.

In a variation of this method, dimethyl sulfoxide (DMSO) is used as a solvent and potassium *tert*-butoxide as base; these conditions allow the reaction to be carried out at lower temperatures (20°–100°C).

---

*Professor Ludwig Wolff, 1857–1919, University of Jena, Germany; Professor N. M. Kishner, 1867–1935, University of Moscow.

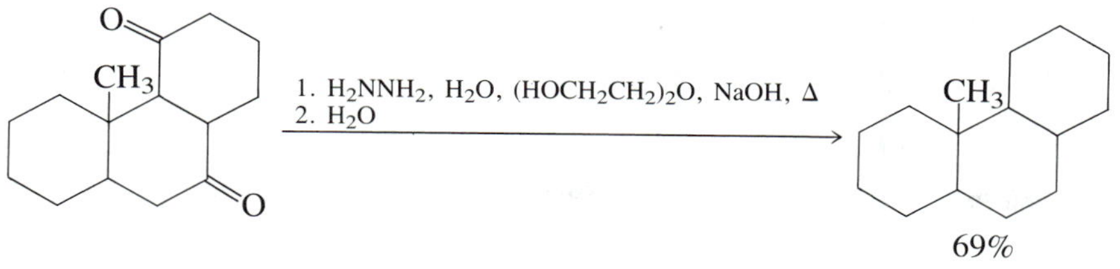

1. H_2NNH_2, H_2O, $(HOCH_2CH_2)_2O$, NaOH, Δ
2. H_2O

CH_3 CH_3

69%

Condensations with Secondary Amines Give Enamines

The condensations of amines described so far are possible only for primary derivatives, because the nitrogen of the amine has to supply both of the protons necessary to form water. Reaction with a secondary amine therefore takes a different course. After the initial addition, water is eliminated by deprotonation at *carbon* to produce an **enamine.** This functional group incorporates both the *ene* function of an alkene and the *amino* group of an amine. Enamine formation is reversible, and hydrolysis occurs readily in the presence of acidic water. Enamines are useful substrates in alkylations (Section 16-1).

Enamine Formation

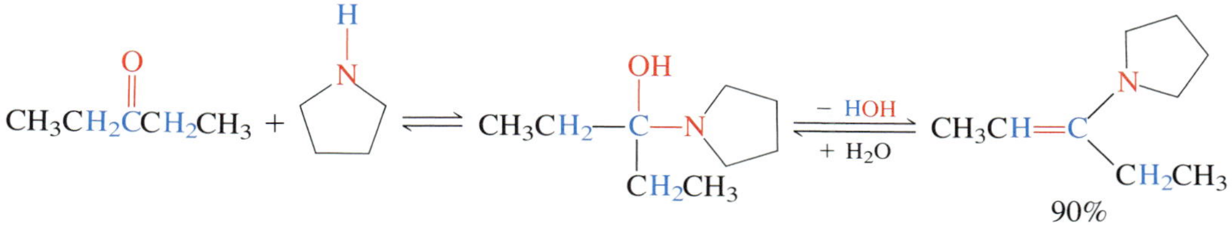

$$CH_3CH_2\overset{O}{\overset{\|}{C}}CH_2CH_3 + \text{pyrrolidine (HN)} \rightleftharpoons CH_3CH_2-\underset{CH_2CH_3}{\overset{OH}{C}}-N(\text{pyrrolidinyl}) \underset{+\,H_2O}{\overset{-\,HOH}{\rightleftharpoons}} CH_3CH{=}C(N\text{-pyrrolidinyl})CH_2CH_3$$

90%

An enamine

EXERCISE 15-11

Write the products of the following reactions:

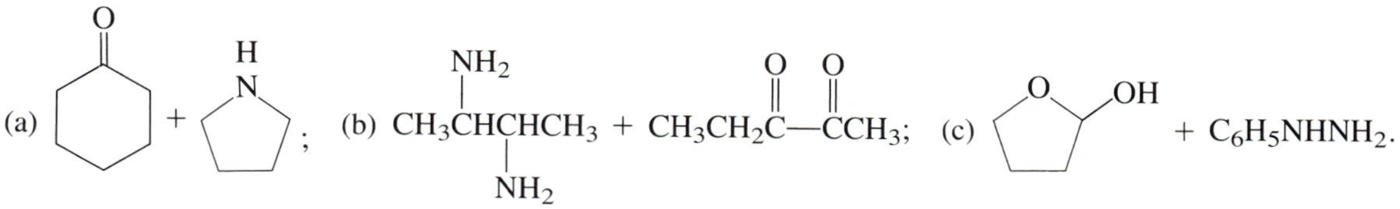

(a) cyclohexanone + pyrrolidine (HN); (b) $CH_3\underset{NH_2}{\overset{NH_2}{CHCH}}CH_3 + CH_3CH_2\overset{O}{\overset{\|}{C}}-\overset{O}{\overset{\|}{C}}CH_3$; (c) 2-hydroxytetrahydrofuran (O, OH) $+ C_6H_5NHNH_2$.

In summary, amines attack aldehydes and ketones to form imines by condensation. Hydroxylamine gives oximes, hydrazines lead to hydrazones and azines, and semicarbazide results in semicarbazones. The Wolff-Kishner reduction is the decomposition of a hydrazone by a base. It is the second part of a method of deoxygenating carbonyl compounds. Secondary amines react with aldehydes and ketones to give enamines.

15-7
Addition of Carbon Nucleophiles to Aldehydes and Ketones

Besides alcohols and amines, several other nucleophilic reagents may attack the carbonyl group. Particularly important are carbon nucleophiles, because new carbon–carbon bonds can be made in this way. Section 8-6 explained that organometallic compounds, such as Grignard and alkyllithium reagents, add to aldehydes and ketones to produce alcohols. In contrast with additions by alcohols and amines, these are normally *not* reversible. This section deals with the behavior of less-reactive carbon nucleophiles that are not organometallic reagents—the additions of cyanide ion and of a new class of compounds called ylides. Cyanide additions offer another way of protecting the carbon–oxygen double bond of aldehydes and ketones. The addition of ylides, on the other hand, leads to alkenes or oxacyclopropanes.

Addition of Hydrogen Cyanide to Carbonyl Compounds Gives Cyanohydrins

Hydrogen cyanide adds to the carbonyl group to form hydroxy alkanenitrile adducts commonly called **cyanohydrins.** This process is reversible because the negative charge on the cyanide ion is fairly well stabilized (Table 6-4). The equilibrium may be shifted toward the adduct by the use of liquid HCN as solvent. However, it is dangerous to use such large amounts of HCN, which is volatile and highly toxic. Alternatively, HCN may be generated *in situ* from a cyanide salt by the slow addition of an acid.

Cyanohydrin Formation

$$CH_3\overset{O}{\overset{\|}{C}}H + HCN \rightleftharpoons CH_3\overset{OH}{\underset{H}{C}}CN$$

70%

2-Hydroxypropanenitrile (Ethanal cyanohydrin)

$$\text{cyclohexanone} + Na^{+}\,{}^{-}CN \xrightarrow[-\,NaCl]{HCl} \text{1-hydroxycyclohexanecarbonitrile}$$

60%

1-Hydroxycyclohexanecarbonitrile (Cyclohexanone cyanohydrin)

The mechanism of cyanohydrin formation begins with nucleophilic attack by cyanide ion and ends with protonation at oxygen.

Mechanism of Cyanohydrin Formation:

$$>C{=}\ddot{O}: + {}^{-}:C{\equiv}N: \rightleftharpoons (NC)C{-}\ddot{O}:^{-} \underset{-\,H^{+}}{\overset{+\,H^{+}}{\rightleftharpoons}} (NC)C{-}\ddot{O}H$$

The reaction is readily reversed by the addition of base, which shifts the equilibrium to the free cyanide side by removing protons from the equation.

$$CH_3\overset{OH}{\underset{H}{C}}CN \rightleftharpoons CH_3\overset{O}{\overset{\|}{C}}H + HCN \xrightarrow[\text{Shifts equilibrium}]{HO^{-}} CH_3\overset{O}{\overset{\|}{C}}H + HOH + {}^{-}CN$$

Cyanohydrins are useful intermediates because the nitrile group can be modified by further reaction. For example, treatment of a cyanohydrin with aqueous acid gives the corresponding 2-hydroxy carboxylic acid (see Section 17-5); alternatively, reduction (see Section 18-6) of the cyanohydrin gives a hydroxy amine.

Hydrolysis and Reduction of 2-Hydroxybutanenitrile (Propanal Cyanohydrin)

$$\underset{\substack{70\% \\ \textbf{2-Hydroxybutanoic acid}}}{CH_3CH_2\overset{OH}{\underset{H}{C}}COOH} \xleftarrow{H^+,\ H_2O} \underset{\substack{\textbf{2-Hydroxybutanenitrile} \\ \textbf{(Propanal cyanohydrin)}}}{CH_3CH_2\overset{OH}{\underset{H}{C}}C{\equiv}N} \xrightarrow[2.\ H^+,\ H_2O]{1.\ LiAlH_4} \underset{\substack{80\% \\ \textbf{1-Amino-2-butanol}}}{CH_3CH_2\overset{OH}{\underset{H}{C}}CH_2NH_2}$$

Addition of Phosphorus Ylides to Aldehydes and Ketones: The Wittig Reaction

In the **Wittig* reaction,** a special reagent is used in which a carbanion is stabilized by an adjacent, positively charged phosphorus-containing substituent. Such a species is called a **phosphorus ylide** (another name is phosphorane). Delocalization of the negative charge onto the phosphorus allows the formulation of another resonance structure with expanded (pentavalent) phosphorus valence shells and a carbon–phosphorus double bond.

$$\left[(R)(H)\ddot{C}^{-}\text{—}\overset{+}{P}(C_6H_5)_3 \longleftrightarrow (R)(H)C{=}P(C_6H_5)_3 \right]$$

Ylide

In the most commonly used ylides, the other substituents on phosphorus are phenyl groups. Ylides are most conveniently prepared from haloalkanes by a two-step sequence; the first step is the nucleophilic displacement of halide by triphenylphosphine to furnish an alkyltriphenylphosphonium salt.

Phosphonium Salt Synthesis

$$\underset{\textbf{Triphenylphosphine}}{(C_6H_5)_3P\!:} + \overset{R}{C}H_2\text{—}\ddot{\underset{..}{X}}\!: \longrightarrow \underset{\substack{\textbf{An alkyltriphenylphosphonium} \\ \textbf{halide}}}{RCH_2\overset{+}{P}(C_6H_5)_3} + :\ddot{\underset{..}{X}}:^-$$

The positively charged phosphorus atom renders any neighboring proton acidic. In the second step, deprotonation by bases, such as alkoxides, sodium hydride, or butyllithium, gives the ylide. Ylides can be isolated, but they are usually produced and used as reagents *in situ*.

Ylide Formation

$$RCH_2\overset{+}{P}(C_6H_5)_3X^- + CH_3CH_2CH_2CH_2Li \xrightarrow{THF} \underset{\textbf{Ylide}}{RCH{=}P(C_6H_5)_3} + CH_3CH_2CH_2CH_2H + LiX$$

---

*Professor Georg Wittig, 1897–1987, University of Heidelberg, Germany, Nobel Prize 1979.

When an ylide is exposed to an aldehyde or ketone, their reaction ultimately produces an alkene by coupling the ylide carbon with that of the carbonyl. The other product of this reaction is triphenylphosphine oxide.

The Wittig Reaction

$$\rangle C{=}P(C_6H_5)_3 + O{=}C\langle \longrightarrow \rangle C{=}C\langle + (C_6H_5)_3P{=}O$$

Ylide **Aldehyde or ketone** **Alkene** **Triphenylphosphine oxide**

Examples:

$$\text{Cyclohexanone} + CH_2{=}P(C_6H_5)_3 \xrightarrow{(CH_3CH_2)_2O,\ \Delta} \text{Methylenecyclohexane} + (C_6H_5)_3PO$$

40%
Methylene-cyclohexane

$$CH_3CH_2CH_2\overset{O}{\overset{\|}{C}}H + CH_3CH_2\overset{CH_3}{\overset{|}{C}}{=}P(C_6H_5)_3 \xrightarrow{(CH_3CH_2)_2O,\ 10^\circ C} CH_3CH_2CH_2CH{=}\overset{CH_3}{\overset{|}{C}}CH_2CH_3 + (C_6H_5)_3PO$$

66%
3-Methyl-3-heptene

The Wittig reaction is a valuable addition to our synthetic arsenal because it forms carbon–carbon double bonds. In contrast with eliminations (Sections 11-5 and 11-6), it gives rise to alkenes in which the position of the newly formed double bond is unambiguous. Compare, for example, two syntheses of 2-ethyl-1-butene, one by the Wittig reaction, the other by elimination.

Syntheses of 2-Ethyl-1-butene

By Wittig reaction:

$$CH_3CH_2\overset{O}{\overset{\|}{C}}CH_2CH_3 + CH_2{=}P(C_6H_5)_3 \longrightarrow CH_3CH_2\overset{CH_2}{\overset{\|}{C}}CH_2CH_3 + (C_6H_5)_3P{=}O$$

Only one isomer

By elimination:

$$CH_3CH_2\underset{Br}{\underset{|}{\overset{CH_3}{\overset{|}{C}}}}CH_2CH_3 \xrightarrow{\text{Base}} CH_3CH_2\overset{CH_2}{\overset{\|}{C}}CH_2CH_3 + CH_3CH_2\overset{CH_3}{\overset{|}{C}}{=}CHCH_3$$

Mixture of isomers

What is the mechanism of the Wittig reaction? The negatively polarized carbon in the ylide is nucleophilic and can attack the carbonyl group. The result is a **phosphorus betaine,*** a dipolar species of the kind called a *zwitterion* (*Zwitter,* German, hybrid). The betaine is short-lived and rapidly forms a neutral heterocycle, an **oxaphosphacyclobutane (oxaphosphetane),** characterized by a four-membered ring containing phosphorus and oxygen. This substance then decomposes to the product alkene and triphenylphosphine oxide. The driving force for the last step is the formation of the very strong phosphorus–oxygen double bond.

Mechanism of the Wittig Reaction:

$$\left[RCH{=}P(C_6H_5)_3 \longleftrightarrow R\ddot{C}H{-}\overset{+}{P}(C_6H_5)_3\right] + \begin{matrix}R'\\ \diagdown\\ C{=}\ddot{\underset{..}{O}}\\ \diagup\\ R''\end{matrix} \longrightarrow \begin{matrix} & & R' & \\ & & | & \\ RCH & - & C & -R''\\ | & & | & \\ (C_6H_5)_3P^{+} & & :\ddot{\underset{..}{O}}:^{-} & \end{matrix} \longrightarrow$$

A phosphorus betaine

$$\begin{matrix} RCH & - & C(R')(R'') \\ | & & | \\ (C_6H_5)_3P & - & \ddot{\underset{..}{O}}: \end{matrix} \longrightarrow RCH{=}C(R')(R'') + (C_6H_5)_3P{=}O$$

An oxaphosphacyclobutane
(Oxaphosphetane)

Wittig reactions can be carried out in the presence of ether, ester, halogen, alkene, and alkyne functions. However, they are only sometimes stereoselective, and mixtures of Z and E alkenes may form.

$$CH_3\overset{O\quad O}{C}CH_2CH_2CH{=}P(C_6H_5)_3 + CH_2{=}CH\overset{O}{\overset{\|}{C}}H \xrightarrow[-\ (C_6H_5)_3PO]{\text{Benzene, }25°C} \begin{matrix} CH_3\overset{O\quad O}{C}CH_2CH_2 & & CH{=}CH_2\\ & C{=}C & \\ H & & H\end{matrix}$$

50%
Major isomer

$$CH_3CH_2CH_2CH{=}P(C_6H_5)_3 + CH_3(CH_2)_4CHO \xrightarrow[-\ (C_6H_5)_3PO]{\text{THF}}$$

$$\underset{60\%}{\begin{matrix} H & & H\\ & C{=}C & \\ CH_3(CH_2)_2 & & (CH_2)_4CH_3\end{matrix}} + \underset{10\%}{\begin{matrix} CH_3(CH_2)_2 & & H\\ & C{=}C & \\ H & & (CH_2)_4CH_3\end{matrix}}$$

The Wittig reaction has been employed extensively in total synthesis. For example, the following total synthesis of the pheromone (see Section 12-9) bombykol, the sex attractant of the silkworm moth, employs the Wittig reaction, as well as various other reactions that were described in preceding chapters.

* Betaine is the name of an amino acid, $(CH_3)_3\overset{+}{N}CH_2COO^-$, which is found in beet sugar (*beta,* Latin, beet) and exists as a zwitterion.

The Total Synthesis of Bombykol

$$CH_3CH_2CH_2Br \xrightarrow{NaC{\equiv}CH,\ \text{liquid } NH_3} CH_3CH_2CH_2C{\equiv}CH \xrightarrow[2.\ CH_2{=}O]{1.\ CH_3CH_2MgBr} CH_3CH_2CH_2C{\equiv}CCH_2OH \xrightarrow{PBr_3}$$

$$CH_3CH_2CH_2C{\equiv}CCH_2Br \xrightarrow{P(C_6H_5)_3} CH_3CH_2CH_2C{\equiv}CCH_2\overset{+}{P}(C_6H_5)_3Br^- \xrightarrow{CH_3CH_2O^-Na^+}$$

$$CH_3CH_2CH_2C{\equiv}CCH{=}P(C_6H_5)_3 \xrightarrow{CH_3CH_2O_2C(CH_2)_8CHO}$$

$$CH_3CH_2CH_2C{\equiv}CCH{=}CH(CH_2)_8CO_2CH_2CH_3 \xrightarrow{H_2,\ \text{Lindlar's catalyst (on the trans isomer)}}$$

Cis and trans isomers

$$\text{(cis)-}CH_3CH_2CH_2CH{=}CH\text{-(trans)-}CH{=}CH(CH_2)_8\overset{O}{\overset{\|}{C}}OCH_2CH_3 \xrightarrow{LiAlH_4} \text{(cis)-}CH_3CH_2CH_2CH{=}CH\text{-(trans)-}CH{=}CH(CH_2)_8CH_2OH$$

Bombykol

The first step is an alkylation of an alkynyl anion (Section 13-5) to give 1-pentyne. Then 1-pentyne is converted into its corresponding Grignard reagent by treatment with ethylmagnesium bromide (Section 13-5). This compound then reacts with methanal (formaldehyde) to give the corresponding alcohol (Section 8-6), which is converted into the bromide with PBr_3 (Section 9-3). Nucleophilic displacement with triphenylphosphine gives the phosphonium salt, which then undergoes the Wittig reaction with ethyl 10-oxodecanoate. Only the aldehyde function reacts, giving a mixture of cis and trans enynes. After separation, the trans isomer is hydrogenated stereospecifically by Lindlar's catalyst to give a new cis double bond (Section 13-6). The ester function may then be reduced to the corresponding alcohol by lithium aluminum hydride (Section 8-4).

The Wittig reaction has also found extensive application in industry. The chemical company Badische Anilin und Soda Fabriken (BASF) in West Germany synthesizes vitamin A_1 (Section 14-4) by a Wittig reaction in the crucial step. In this case, the reaction is stereoselective, giving only the trans alkene.

BASF Vitamin A_1 Synthesis

$$\text{R-}CH_2\overset{+}{P}(C_6H_5)_3X^- \xrightarrow[CH_3OH]{CH_3O^-} \text{R-}CH{=}P(C_6H_5)_3 \xrightarrow{O{=}CH\text{-}C(CH_3){=}CH\text{-}CH_2O\overset{O}{\overset{\|}{C}}CH_3}$$

$$\text{R-}CH{=}CH\text{-}C(CH_3){=}CH\text{-}CH_2O\overset{O}{\overset{\|}{C}}CH_3 \xrightarrow[\text{Ester hydrolysis}]{HO^-,\ H_2O} \text{R-}CH{=}CH\text{-}C(CH_3){=}CH\text{-}CH_2OH$$

Vitamin A_1

EXERCISE 15-12

Develop concise synthetic connections between starting material and product. You may use any material in addition to the given compound (more than one step will be required):

(a) cyclohexene ------→ $CH_2{=}CH(CH_2)_4CH{=}CH_2$

(b) 1,3-butadiene ------→ (cyclohexylidenemethyl)cyclohexene

Sulfur Ylides Lead to Oxacyclopropanes

Whereas phosphorus ylides react with aldehydes and ketones by nucleophilic attack followed by elimination of the phosphine oxide to give a double bond, **sulfur ylides,** derived by deprotonation of the corresponding sulfonium salts, react differently to give oxacyclopropanes.

Sulfur Ylide Formation

$$CH_3I + CH_3\ddot{S}CH_3 \longrightarrow CH_3\overset{+}{S}(CH_3)_2\ I^- \xrightarrow[-\ CH_3CH_2CH_2CH_2H,\ LiI]{CH_3CH_2CH_2CH_2Li} \left[\ :\overset{-}{C}H_2{-}\overset{+}{S}(CH_3)_2 \longleftrightarrow CH_2{=}\ddot{S}(CH_3)_2\ \right]$$

Trimethylsulfonium iodide — **Sulfur ylide**

Sulfur ylides are nucleophilic; as in the first step of the Wittig reaction, they attack the carbonyl carbon to give sulfur betaines. However, unlike the phosphorus betaine, the dimethyl sulfide in this adduct functions as a leaving group. It is displaced by intramolecular nucleophilic attack of the alkoxide ion, leading to the formation of an oxacyclopropane. Thus, the sulfur ylide acts as a *methylene transfer reagent,* adding a methylene group to the carbonyl double bond. Recall that oxacyclopropanes may be made by an alternative route (Section 12-5), employing a peroxycarboxylic acid to transfer oxygen to an alkene.

Oxacyclopropane Formation with Sulfur Ylides

$$\text{cyclohexanone} + :\overset{-}{C}H_2{-}\overset{+}{S}(CH_3)_2 \longrightarrow \text{sulfur betaine} \longrightarrow \text{oxacyclopropane} + CH_3\ddot{S}CH_3$$

82%

Sulfur betaine — **An oxacyclopropane**

EXERCISE 15-13

Use a sulfur ylide to prepare:

(a) 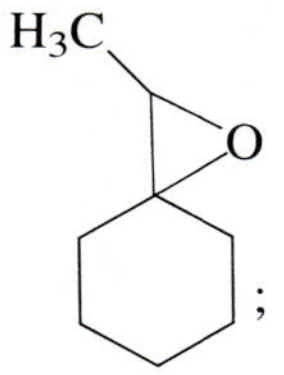 [1-methyl-spiro oxacyclopropane–cyclohexane structure]; (b) 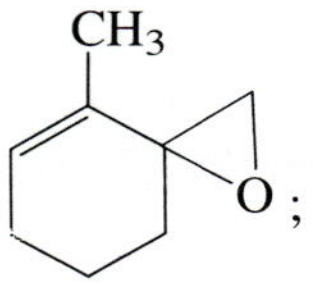 [spiro oxacyclopropane with 2-methylcyclohexene]; (c) 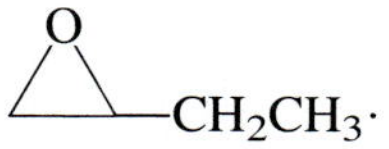 [oxacyclopropane]$-CH_2CH_3$.

EXERCISE 15-14

Explain the following transformation by a mechanism. Hint: The mechanism includes a carbocation rearrangement.

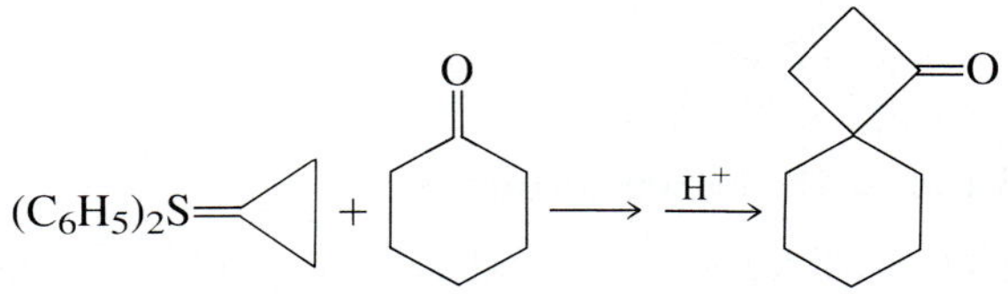

$$(C_6H_5)_2S{=}\text{cyclopropylidene} + \text{cyclohexanone} \longrightarrow \xrightarrow{H^+} \text{spiro[3.5]nonan-1-one}$$

EXERCISE 15-15

Diazomethane reacts as sulfur ylides do with aldehydes and ketones. Formulate a mechanism for the following result:

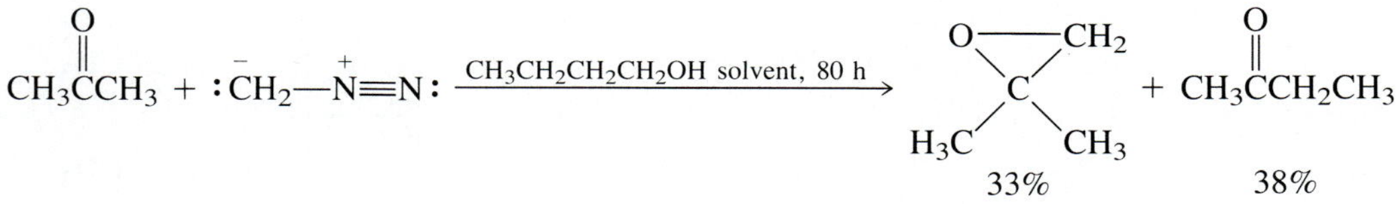

$$CH_3\overset{\overset{\displaystyle O}{\|}}{C}CH_3 + :\overset{-}{C}H_2{-}\overset{+}{N}{\equiv}N: \xrightarrow{CH_3CH_2CH_2CH_2OH\text{ solvent, 80 h}} \underset{33\%}{(CH_3)_2C\overset{O}{—}CH_2} + \underset{38\%}{CH_3\overset{\overset{\displaystyle O}{\|}}{C}CH_2CH_3}$$

In summary, the carbonyl group in aldehydes and ketones can be attacked by carbon-based nucleophiles. Organometallic reagents give alcohols, cyanide gives cyanohydrins, phosphorus ylides add to give betaines that decompose by forming carbon–carbon double bonds, and sulfur ylides give oxacyclopropanes. The Wittig reaction affords a means of synthesizing alkenes from carbonyl compounds and haloalkanes by way of the corresponding phosphonium salts.

15-8 Special Oxidations and Reductions of Aldehydes and Ketones

This section shows how the hydroperoxy group of peroxycarboxylic acids adds to the carbonyl group of aldehydes or ketones to eventually give acids or esters by rearrangement. It also describes a couple of oxidative tests for the aldehyde function. In contrast with these oxidations are some special methods for reducing aldehydes and ketones by the transfer of single electrons.

Peroxycarboxylic Acids Oxidize Aldehydes and Ketones

The carbonyl carbon in aldehydes and ketones may be attacked by the hydroxy group of a peroxycarboxylic acid (Section 12-5) as if the latter came from an alcohol. The result is a peroxide analog of a hemiacetal.

Peroxycarboxylic Acid Addition to the Carbonyl Group of Aldehydes and Ketones

$$\mathrm{R'\overset{O}{\overset{\|}{C}}H + R\overset{O}{\overset{\|}{C}}OOH \longrightarrow R'\overset{OH}{\overset{|}{C}}O\overset{O}{\overset{\|}{C}}R}$$
(with H on the central carbon of the product)

Peroxycarboxylic acid

$$\mathrm{R''\overset{O}{\overset{\|}{C}}R' + R\overset{O}{\overset{\|}{C}}OOH \longrightarrow R''\overset{OH}{\overset{|}{C}}O\overset{O}{\overset{\|}{C}}R}$$
(with R′ on the central carbon of the product)

These products are not stable and decompose through a cyclic eight-electron transition state by substituent shifts. In the adduct from an aldehyde, it is a hydrogen that migrates to give two carboxylic acids. One is derived by oxidation of the aldehyde, the other from the original peroxycarboxylic acid.

Decomposition of Aldehyde Peroxycarboxylic Acid Adduct

$$\mathrm{R'-\underset{H}{\overset{O(H\cdots O{=}C(R))}{C}}-O-O \longrightarrow R'\overset{O}{\overset{\|}{C}}OH + HO\overset{O}{\overset{\|}{C}}R}$$

In the adduct from a ketone, an alkyl group shifts in analogous manner to give an ester. This transformation is called the **Baeyer-Villiger* oxidation.**

Baeyer-Villiger Oxidation

$$\mathrm{R''\overset{O}{\overset{\|}{C}}R' + R\overset{O}{\overset{\|}{C}}OOH \longrightarrow R''-\underset{R'}{\overset{O(H\cdots O{=}C(R))}{C}}-O-O \longrightarrow R''\overset{O}{\overset{\|}{C}}OR' + HO\overset{O}{\overset{\|}{C}}R}$$

Example:

$$\mathrm{CH_3\overset{O}{\overset{\|}{C}}CH_2CH_3 \xrightarrow{CF_3\overset{O}{\overset{\|}{C}}OOH,\ CH_2Cl_2} CH_3\overset{O}{\overset{\|}{C}}OCH_2CH_3}$$

2-Butanone → **Ethyl ethanoate (acetate)** 72%

*Professor Adolf von Baeyer, 1835–1917, University of Munich, Nobel Prize 1905; V. Villiger, 1868–1934, BASF, Ludwigshafen, Germany.

Cyclic ketones are converted into cyclic esters:

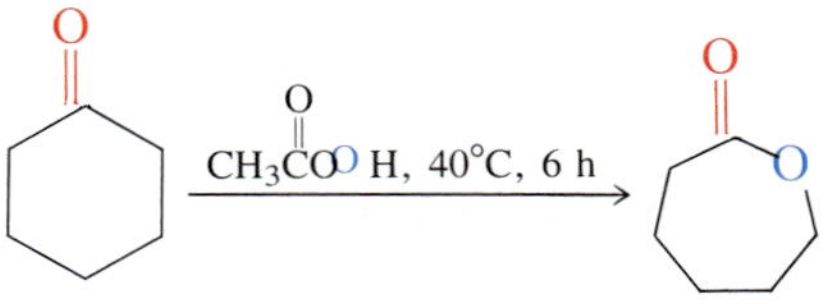

$$\text{cyclohexanone} \xrightarrow{CH_3\overset{O}{\overset{\|}{C}}OOH,\ 40^\circ C,\ 6\ h} \text{caprolactone (90\%)}$$

$$\text{bicyclic ketone} \xrightarrow{CH_3\overset{O}{\overset{\|}{C}}OOH,\ CHCl_3} \text{bicyclic lactone (56\%)}$$

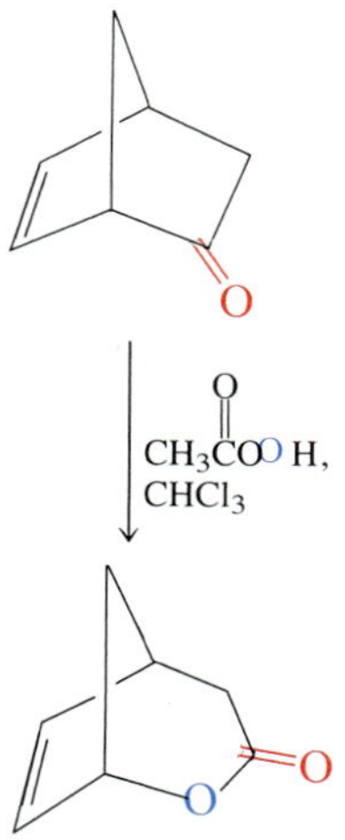

Attack is at the carbonyl rather than at the carbon–carbon double bond.

With unsymmetrical ketones, it is in principle possible to obtain two different esters. As the examples show, however, only one is formed in each case. By a series of experiments, a list of ''migratory aptitudes'' has been established, indicating the relative ease of migration of various substituents. This ordering indicates that there may be carbocation character of the migrating carbon in the transition state for rearrangement.

Migratory Aptitudes in the Baeyer-Villiger Reaction

hydrogen > tertiary > cyclohexyl > secondary ~ phenyl > primary > methyl

EXERCISE 15-16

Predict the outcome of the following oxidations with a peroxycarboxylic acid:

(a) $CH_2{=}CHCH_2CH_2\overset{O}{\overset{\|}{C}}CH_3$; (b) 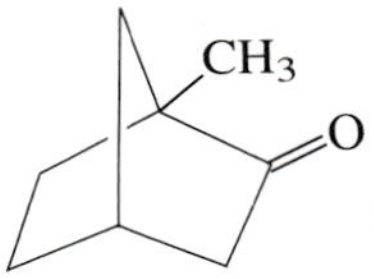 (bicyclic ketone with CH_3); (c) $(CH_3)_3C\overset{O}{\overset{\|}{C}}CH_2CH_3$.

Oxidative Chemical Tests for Aldehydes

Although the advent of NMR and other spectroscopy has made chemical tests for functional groups a rarity, they are still used in special cases in which other analytical tests may fail. Two characteristic simple tests for aldehydes will again turn up in the discussion of sugar chemistry in Chapter 23; they make use of the ready oxidation of aldehydes to carboxylic acids. The first is Fehling's* test, in which cupric ion is the oxidant. In a basic medium, the precipitation of cuprous oxide indicates the presence of an aldehyde function.

Fehling's Test

$$R\overset{O}{\overset{\|}{C}}H + Cu^{2+} \xrightarrow{NaOH,\ tartrate,\ H_2O} \underset{\textbf{Brick-red}}{Cu_2O} + R\overset{O}{\overset{\|}{C}}OH$$

The second is Tollens's† test, in which a solution of silver ion precipitates a silver mirror on exposure to an aldehyde.

*Professor Hermann C. von Fehling, 1812–1885, Polytechnic School of Stuttgart.

†Professor Bernhard Tollens, 1841–1918, University of Göttingen, Germany.

Tollens's Test

$$\mathrm{R\overset{\overset{\Large O}{\|}}{C}H} + \mathrm{Ag^+} \xrightarrow{\mathrm{NH_3,\ H_2O}} \underset{\text{Mirror}}{\mathrm{Ag}} + \mathrm{R\overset{\overset{\Large O}{\|}}{C}OH}$$

The Reduction of Aldehydes and Ketones by Zinc Amalgam: Clemmensen Reduction

The carbonyl group of aldehydes and ketones may be deoxygenated indirectly by thioacetal desulfurization (Section 15-5) and Wolff-Kishner reduction (Section 15-6). It may be deoxygenated directly by zinc amalgam, Zn(Hg), with concentrated hydrogen chloride. The reaction is known as the **Clemmensen* reduction.**

Clemmensen Reduction

$$\underset{\textbf{2-Octanone}}{\mathrm{CH_3(CH_2)_5\overset{\overset{\Large O}{\|}}{C}CH_3}} \xrightarrow{\mathrm{Zn(Hg),\ HCl,\ \Delta}} \underset{\textbf{Octane}}{\underset{62\%}{\mathrm{CH_3(CH_2)_5CH_2CH_3}}}$$

Zinc amalgam is an alloy of zinc and mercury and essentially serves to provide a highly active zinc surface. The mechanism of this reduction is not completely understood, but it probably involves electron transfers from the metal into the carbonyl π bond. The Clemmensen reduction has limited application because of the drastic conditions (strong acid) under which it is performed.

Some Metals Reduce Aldehydes and Ketones by Coupling

The carbonyl bond may be reduced by hydrides and by catalytic hydrogenation (Sections 8-4 and 15-4). Aldehydes and ketones are also subject to *one-electron reduction* by metals, which leads to the formation of a carbon–carbon bond between two carbonyl carbons. For example, when propanone (acetone) is treated with magnesium in benzene, a strongly exothermic reaction takes place, which, after aqueous work-up, gives 2,3-dimethylbutane-2,3-diol. This compound appeared in Section 9-2 as pinacol in the pinacol rearrangement. The reduction leading to its formation is called the **pinacol reaction.**

Pinacol Reaction

$$2\ \mathrm{CH_3\overset{\overset{\Large O}{\|}}{C}CH_3} + \mathrm{Mg} \xrightarrow{\text{Benzene}} \mathrm{CH_3\underset{\underset{\Large H_3C}{|}}{\overset{\overset{\Large ^-:\ddot{O}:}{|}}{C}}-\underset{\underset{\Large CH_3}{|}}{\overset{\overset{\Large :\ddot{O}:^-}{|}}{C}}CH_3} + \mathrm{Mg^{2+}} \xrightarrow{\mathrm{H^+,\ H_2O}} \mathrm{(CH_3)_2\overset{\overset{\Large HO}{|}}{C}-\overset{\overset{\Large OH}{|}}{C}(CH_3)_2}$$

40%
2,3-Dimethyl-2,3-butanediol (Pinacol)

The mechanism of the pinacol reaction proceeds through one-electron transfer steps. Initially, the carbonyl group accepts an electron to give a radical anion.

---

*E. C. Clemmensen, 1876–1941, President of Clemmensen Chemical Corporation, Newark, N.J.

Subsequently, two of the radicals couple to give a dianion that is protonated on aqueous work-up.

Mechanism of the Pinacol Reaction:

$$2\ CH_3\overset{O}{\overset{\|}{C}}CH_3 \xrightarrow[\text{One-electron transfer}]{Mg} 2\ CH_3\overset{:\ddot{O}:^-}{\overset{|}{\dot{C}}}CH_3 + Mg^{2+} \xrightarrow{\text{Coupling}} CH_3\overset{^-:\ddot{O}:}{\overset{|}{\underset{H_3C}{\underset{|}{C}}}}-\overset{:\ddot{O}:^-}{\overset{|}{\underset{CH_3}{\underset{|}{C}}}}CH_3 + Mg^{2+}$$

Radical anion

Reductive carbonyl coupling is also caused by a low-valence titanium species formed in the reduction of titanium trichloride with agents such as an alkali metal, lithium aluminum hydride, or a zinc-copper couple. However, instead of vicinal diol formation, complete deoxygenation to an alkene occurs.

2 (cyclohexanone) $\xrightarrow[-\ TiO_2]{TiCl_3,\ Li,\ THF}$ (cyclohexylidenecyclohexane) 79%

$$2\ CH_3CH_2CH_2CH_2\overset{O}{\overset{\|}{C}}H \xrightarrow[-\ TiO_2]{TiCl_3,\ K,\ THF} CH_3CH_2CH_2CH_2CH{=}CHCH_2CH_2CH_2CH_3$$

Pentanal — **5-Decene**, 77% (cis : trans = 3 : 7)

The mechanism of this transformation appears to begin with pinacol coupling followed by oxygen removal in a step like the reverse of the cis dihydroxylation of alkenes (Section 12-5). The titanium functions as a reducing agent and is oxidized to titanium dioxide.

Possible Mechanism of Titanium-mediated Deoxygenation:

(Ti–O–C–C–O five-membered ring) $\longrightarrow$ C=C $+\ TiO_2$

Mixed couplings are possible if one of the carbonyl components is present in excess.

(3,5,5-trimethylcyclohex-2-enone) $+\ CH_3\overset{O}{\overset{\|}{C}}CH_3 \xrightarrow[-\ TiO_2]{TiCl_3,\ Li,\ THF}$ (isopropylidene product) 63%

1 : 4

Intramolecular coupling leads to cycloalkenes.

$$CH_3(CH_2)_3\overset{\overset{\displaystyle O}{\|}}{C}(CH_2)_8\overset{\overset{\displaystyle O}{\|}}{C}(CH_2)_3CH_3 \xrightarrow[-\ TiO_2]{TiCl_3,\ Zn\text{-}Cu,\ THF}$$

75%

EXERCISE 15-17

Devise a rapid synthesis of 2-methyl-2-pentene starting from 1-butene.

In summary, aldehydes can be oxidized with peroxycarboxylic acids (and other oxidizing agents) to give carboxylic acids. Similarly, ketones give esters; with unsymmetrical ketones, the esters can be formed selectively, by migration of only one of the substituents. Aldehydes and ketones can be reduced with metals to the corresponding alkanes, to coupled vicinal diols, or to coupled alkenes.

Summary of New Reactions

Synthesis of Aldehydes and Ketones

1 OXIDATION OF ALCOHOLS

$$R\overset{\overset{\displaystyle R'}{|}}{C}HOH \xrightarrow{CrO_3} R\overset{\overset{\displaystyle O}{\|}}{C}R'$$

$$RCH_2OH \xrightarrow{CrO_3(pyridine)_2} R\overset{\overset{\displaystyle O}{\|}}{C}H$$

Stable to oxidizing agent

Allylic oxidation:

$$RCH{=}CHCH_2OH \xrightarrow{MnO_2} RCH{=}CH\overset{\overset{\displaystyle O}{\|}}{C}H$$

2 OZONOLYSIS OF ALKENES

$$\gt C{=}C\lt \xrightarrow{\text{1. } O_3 \quad \text{2. Reduction}} \gt C{=}O + O{=}C\lt$$

3 HYDRATION OF ALKYNES

$$RC{\equiv}CH \xrightarrow{H_2O,\ Hg^{2+},\ H_2SO_4} R\overset{\overset{\displaystyle O}{\|}}{C}CH_3$$

$$RC{\equiv}CR' \xrightarrow{H_2O,\ Hg^{2+},\ H_2SO_4} R\overset{\overset{\displaystyle O}{\|}}{C}CH_2R' + RCH_2\overset{\overset{\displaystyle O}{\|}}{C}R'$$

4 HYDROBORATION-OXIDATION OF ALKYNES

$$\text{RC}\equiv\text{CH} \xrightarrow[\text{2. } H_2O_2,\ HO^-]{\text{1. } HBR_2'} \text{RCH}_2\overset{\overset{\text{O}}{\|}}{\text{C}}\text{H}$$

$$\text{RC}\equiv\text{CR}' \xrightarrow[\text{1. } H_2O_2,\ HO^-]{\text{1. } HBR_2''} \text{RCH}_2\overset{\overset{\text{O}}{\|}}{\text{C}}\text{R}' + \text{R}\overset{\overset{\text{O}}{\|}}{\text{C}}\text{CH}_2\text{R}'$$

Reactions of Aldehydes and Ketones

5 HYDROGENATION

$$\text{R}\overset{\overset{\text{O}}{\|}}{\text{C}}\text{R}' \xrightarrow{H_2,\ \text{Ru or Raney Ni, pressure}} \text{R}\overset{\overset{\text{OH}}{|}}{\underset{\underset{\text{H}}{|}}{\text{C}}}\text{R}'$$

$$\text{RCH}{=}\text{CHCH}_2\text{CH}_2\overset{\overset{\text{O}}{\|}}{\text{C}}\text{R}' \xrightarrow{H_2,\ \text{Pt}} \text{R(CH}_2)_4\overset{\overset{\text{O}}{\|}}{\text{C}}\text{R}'$$

6 ADDITION OF WATER AND ALCOHOLS—HEMIACETALS

$$\text{R}\overset{\overset{\text{O}}{\|}}{\text{C}}\text{R}' \xrightleftharpoons{H_2O,\ H^+ \text{ or } HO^-} \text{R}\overset{\overset{\text{OH}}{|}}{\underset{\underset{\text{OH}}{|}}{\text{C}}}\text{R}'$$

Carbonyl hydrate
(A geminal diol)

$$\text{R}\overset{\overset{\text{O}}{\|}}{\text{C}}\text{R}' \xrightleftharpoons{R''OH,\ H^+ \text{ or } HO^-} \text{R}\overset{\overset{\text{OH}}{|}}{\underset{\underset{\text{OR}''}{|}}{\text{C}}}\text{R}'$$

Hemiacetal

Intramolecular addition:

$$\text{HOCH}_2\text{CH}_2\text{CH}_2\text{CH}_2\overset{\overset{\text{O}}{\|}}{\text{C}}\text{R} \xrightarrow{H^+ \text{ or } HO^-} \text{(2-hydroxy-2-R-tetrahydropyran ring: HO, R on ring C adjacent to O)}$$

Cyclic hemiacetal

7 ACID-CATALYZED ADDITION OF ALCOHOLS—ACETALS

$$\mathrm{R\overset{O}{\overset{\|}{C}}R' + 2\,R''OH \overset{H^+}{\rightleftharpoons} R\underset{OR''}{\underset{|}{\overset{OR''}{\overset{|}{C}}}}R' + H_2O}$$

Acetal

Cyclic acetals:

$$\mathrm{R\overset{O}{\overset{\|}{C}}R' + HOCH_2CH_2OH \overset{H^+}{\rightleftharpoons}}$$ (1,3-dioxolane ring: O–CH₂CH₂–O bonded to C bearing R and R′) $$+\ H_2O$$

Ketone, protected as cyclic acetal

$$\mathrm{R\text{–}CH(OH)\text{–}CH(OH)\text{–}R + CH_3\overset{O}{\overset{\|}{C}}CH_3 \overset{H^+}{\rightleftharpoons}}$$ (dioxolane ring: $(H_3C)(CH_3)C$ bonded to two O, which are bonded to the CH(R)–CH(R) carbons) $$+\ H_2O$$

Diol, protected as propanone (acetone) acetal

8 THIOACETALS

$$\mathrm{R\overset{O}{\overset{\|}{C}}R' + R''SH \xrightarrow{BF_3\ \text{or}\ ZnCl_2} (R''S)(R''S)C(R)(R')}$$

Hydrolysis of thioacetals:

$$\mathrm{(R''S)(R''S)C(R)(R') \xrightarrow{H_2O,\ HgCl_2,\ CaCO_3,\ CH_3CN} R\overset{O}{\overset{\|}{C}}R'}$$

9 RANEY NICKEL DESULFURIZATION

$$\mathrm{(1,3\text{-dithiolane: S–CH_2CH_2–S})C(R)(R') \xrightarrow{Raney\ Ni} RCH_2R'}$$

10 ADDITION OF AMINE DERIVATIVES

$$\text{RCR}' \ (\text{C=O}) \xrightarrow{R''NH_2,\ H^+} \text{R(R')C=N–R}'' + H_2O$$

Imine

$$\text{RCR}' \ (\text{C=O}) \xrightarrow[\text{Hydroxylamine}]{H_2NOH,\ H^+} \text{R(R')C=N–OH} + H_2O$$

Oxime

$$\text{RCR}' \ (\text{C=O}) \xrightarrow[\text{Hydrazine}]{H_2NNH_2,\ H^+,\ H_2O} \text{R(R')C=N–NH}_2 + H_2O$$

Hydrazone

$$2\ \text{RCR}' \ (\text{C=O}) \xrightarrow{H_2NNH_2,\ H^+,\ H_2O} \text{R(R')C=N–N=C(R)R}' + 2\ H_2O$$

Azine

$$\text{RCR}' \ (\text{C=O}) \xrightarrow[\text{2,4-Dinitrophenylhydrazine}]{(O_2N)_2C_6H_3NHNH_2,\ H^+} \text{R(R')C=N–NH–C}_6\text{H}_3(\text{NO}_2)_2 + H_2O$$

2,4-Dinitrophenylhydrazone

$$\text{RCR}' \ (\text{C=O}) \xrightarrow[\text{Semicarbazide}]{H_2NNHCNH_2\ (\text{C=O}),\ H^+} \text{R(R')C=N–NHCNH}_2\ (\text{C=O}) + H_2O$$

Semicarbazone

11 WOLFF-KISHNER REDUCTION

$$RC(=O)R' \xrightarrow{H_2NNH_2,\ H_2O,\ HO^-,\ \Delta} RCH_2R'$$

12 ENAMINES

$$RCH_2\overset{O}{\overset{\|}{C}}R' + R''R'''NH \rightleftharpoons RCH{=}C(R')N(R'')R''' + H_2O$$

Secondary amine — Enamine

13 CYANOHYDRINS

$$R\overset{O}{\overset{\|}{C}}R' + HCN \rightleftharpoons (HO)(CN)C(R)(R') \xrightarrow{LiAlH_4} RC(OH)(R')CH_2NH_2$$

Cyanohydrin

$$(HO)(CN)C(R)(R') \xrightarrow{H_2O,\ H^+} RC(OH)(R')COOH$$

14 WITTIG REACTION

$$R''CH_2X + P(C_6H_5)_3 \longrightarrow R''CH_2\overset{+}{P}(C_6H_5)_3X^-$$

Triphenylphosphine — Phosphonium halide

$$R''CH_2\overset{+}{P}(C_6H_5)_3X^- \xrightarrow{Base} R''CH{=}P(C_6H_5)_3$$

Ylide

$$R\overset{O}{\overset{\|}{C}}R' + R''CH{=}P(C_6H_5)_3 \longrightarrow (R)(R')C{=}CHR'' + (C_6H_5)_3P{=}O$$

Not always stereoselective

15 SULFUR YLIDES AS METHYLENE TRANSFER AGENTS

$$CH_3I + CH_3SCH_3 \longrightarrow (CH_3)_3S^+I^-$$

Trimethylsulfonium iodide

$$(CH_3)_3S^+I^- \xrightarrow{CH_3CH_2CH_2CH_2Li} H_2C{=}S(CH_3)_2$$

Sulfur ylide

$$RC(=O)R' + H_2C{=}S(CH_3)_2 \longrightarrow \underset{\text{epoxide: } RR'C\text{—}CH_2\text{—}O}{} + CH_3SCH_3$$

16 BAEYER-VILLIGER OXIDATION

$$\underset{\textbf{Aldehyde}}{RC(=O)H} + \underset{\textbf{Peroxycarboxylic acid}}{R'C(=O)OOH} \longrightarrow \underset{\textbf{Carboxylic acids}}{RC(=O)OH + R'C(=O)OH}$$

$$\underset{\textbf{Ketone}}{RC(=O)R'} + R''C(=O)OOH \longrightarrow \underset{\textbf{Ester}}{RC(=O)OR'} + R''C(=O)OH$$

Migratory aptitudes in Baeyer-Villiger oxidation:

H > tertiary > cyclohexyl > secondary ~ phenyl > primary > methyl

17 TEST FOR ALDEHYDES

$$RC(=O)H \xrightarrow{Cu^{2+} \text{ or } Ag^+} RC(=O)OH + \underset{\textbf{Red precipitate}}{Cu_2O} \text{ or } \underset{\textbf{Mirror}}{Ag}$$

18 CLEMMENSEN REDUCTION

$$RC(=O)R' \xrightarrow{Zn(Hg),\ HCl,\ \Delta} RCH_2R'$$

19 PINACOL REACTION

$$2\ CH_3C(=O)CH_3 \xrightarrow[2.\ H^+,\ H_2O]{1.\ Mg} \underset{\textbf{Pinacol}}{(CH_3)_2C(OH)\text{—}C(OH)(CH_3)_2}$$

20 TITANIUM COUPLING

$$2\ \overset{O}{\underset{/\ \backslash}{\overset{\|}{C}}} \xrightarrow{TiCl_3,\ \text{reducing agent}} \gt C{=}C \lt$$

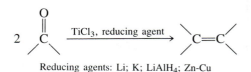

Reducing agents: Li; K; $LiAlH_4$; Zn-Cu

Summary of Important Concepts

1 The carbonyl group is the functional group of the aldehydes (alkanals) and ketones (alkanones). It has precedence over the hydroxy, alkenyl, and alkynyl groups in the naming of organic molecules.

2 The carbon–oxygen double bond and its attached two nuclei in aldehydes and ketones form a plane. The C=O unit is polarized, with a partial negative charge on oxygen.

3 The ^{1}H NMR spectra of aldehydes exhibit a peak at $\delta \sim 9.8$ ppm. The carbonyl carbon absorbs at ~200 ppm. Because of the availability of low-energy $n \rightarrow \pi^*$ transitions, the electronic spectra of aldehydes and ketones have relatively long wavelength bands.

4 The carbon–oxygen double bond undergoes catalytic hydrogenation and ionic additions. The catalysts for the former are heterogeneous transition metal surfaces; for the latter, acid or base.

5 The reactivity of the carbonyl group increases with increasing electrophilic character of the carbonyl carbon. Therefore, aldehydes are more reactive than ketones.

6 Intramolecular additions to the carbonyl group are entropically favored over their intermolecular variants.

7 Primary amines undergo condensation reactions with aldehydes and ketones to imines; secondary amines condense to enamines.

8 The Wittig reaction is an important carbon–carbon bond-forming reaction that produces alkenes directly from aldehydes and ketones.

9 Oxacyclopropanes are formed through the methylenation of aldehydes and ketones by sulfur ylides.

10 The addition of peroxycarboxylic acids to the carbonyl group of aldehydes and ketones produces carboxylic acids and esters, respectively.

Problems

1 Name or draw the structure of each of the following compounds.

(a) $(CH_3)_2CH\overset{O}{\overset{\|}{C}}CH(CH_3)_2$ **(b)** $(CH_3)_2CHCH_2CH_2CHO$

(c) 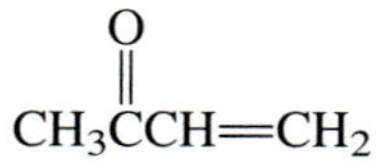 $CH_3\overset{\overset{\displaystyle O}{\|}}{C}CH{=}CH_2$

(d) 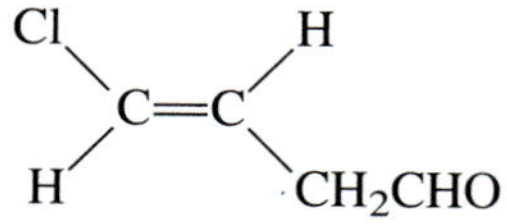 Cl(H)C=C(H)CH_2CHO

(e) 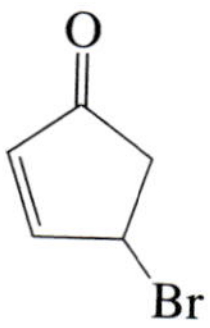 4-Bromo-2-cyclopentenone (ring structure: C=O at top, Br substituent)

(f) 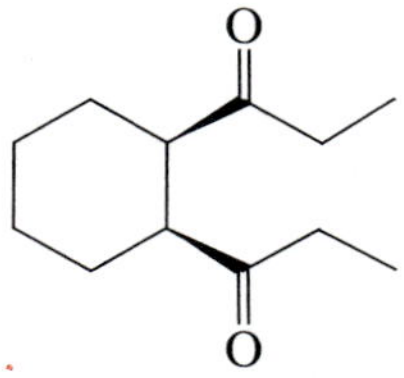 1,2-Dipropanoylcyclohexane (ring structure)

(g) (*Z*)-2-Ethanoyl-2-butenal

(h) *trans*-3-Chlorocyclobutanecarboxaldehyde

2 Spectroscopic data for two carbonyl compounds with the formula $C_8H_{12}O$ are given below. Suggest a structure for each. The letter "m" stands for the appearance of this particular part of the spectrum as an uninterpretable multiplet.

(a) ^{1}H NMR: δ = 1.6 (m, 4 H), 2.15 (s, 3 H), 2.19 (m, 4 H), and 6.78 (t, 1 H) ppm.

^{13}C NMR: δ = 21.8, 22.2, 23.2, 25.0, 26.2, 139.8, 140.7, and 198.6 ppm.

(b) ^{1}H NMR: δ = 0.94 (t, 3 H), 1.48 (sex, 2 H), 2.21 (q, 2 H), 5.8–7.1 (m, 4 H), and 9.56 (d, 1 H) ppm.

^{13}C NMR: δ = 13.6, 21.9, 35.2, 129.0, 135.2, 146.7, 152.5, and 193.2 ppm.

3 The compounds described in Problem 2 have very different ultraviolet spectra. One has $\lambda_{max}(\epsilon)$ = 232(13,000) and 308(1450) nm, whereas the other has $\lambda_{max}(\epsilon)$ = 272(35,000) nm and a weaker absorption near 320 nm (this value is hard to determine accurately owing to the intensity of the stronger absorption). Using the structures that you determined in Problem 2, match the compounds to these UV spectral data. Explain the spectra in terms of the structures.

4 Indicate which reagent or combination of reagents is best suited for each of the following reactions.

(a) 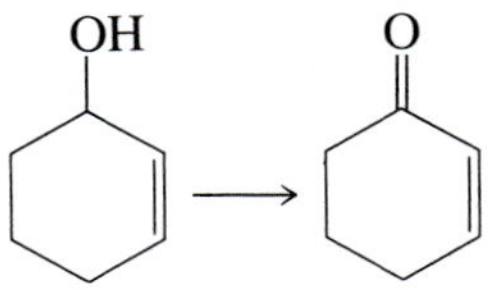 2-Cyclohexen-1-ol (OH) ⟶ 2-Cyclohexen-1-one (O)

(b) $CH_3CH_2CHCH_2CH_3$ (with CH_2OH on C-3) $\longrightarrow$ $CH_3CH_2CHCH_2CH_3$ (with CHO on C-3)

(c) 1-methylcyclopentene $\longrightarrow$ $CH_3\overset{O}{\overset{\|}{C}}CH_2CH_2CH_2CHO$

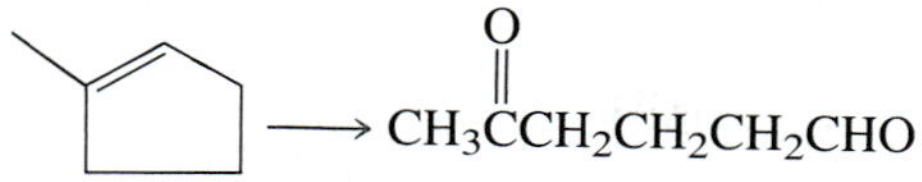

(d) 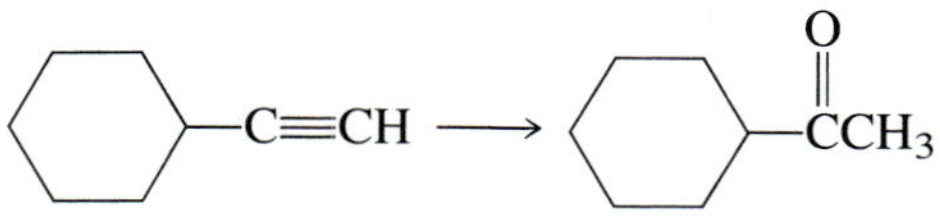 cyclohexyl–$C{\equiv}CH$ $\longrightarrow$ cyclohexyl–$\overset{O}{\overset{\|}{C}}CH_3$

(e) $(CH_3)_2CHC{\equiv}CCH(CH_3)_2 \longrightarrow (CH_3)_2CH\overset{O}{\overset{\|}{C}}CH_2CH(CH_3)_2$

5 Overoxidation of primary alcohols to carboxylic acids is caused by the water present in the usual aqueous acidic Cr(VI) reagents. The water adds to the initial aldehyde product to form a hydrate, which is further oxidized (Section 15-3). In view of these facts, explain the following two observations:

(a) Water adds to ketones to form hydrates, but no overoxidation follows the conversion of a secondary alcohol into a ketone.

(b) Successful oxidation of primary alcohols to aldehydes by the water-free CrO_3-pyridine reagent requires that the alcohol be added slowly to the CrO_3 reagent. If, instead, the CrO_3 reagent is added to the alcohol, a new side reaction forms an ester. This is illustrated below for 1-butanol:

$$CH_3CH_2CH_2CH_2OH \xrightarrow{CrO_3(pyridine)_2} CH_3CH_2CH_2\overset{O}{\overset{\|}{C}}OCH_2CH_2CH_2CH_3$$

6 Write the expected products of ozonolysis (followed by mild reduction—e.g., by Zn) of each of the following molecules.

(a) $CH_3CH_2CH_2CH{=}CH_2$

(b) 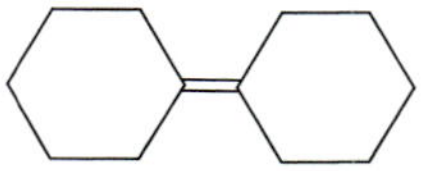 cyclohexylidenecyclohexane

(c) 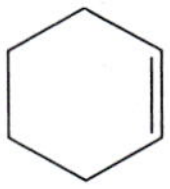 cyclohexene

(d)

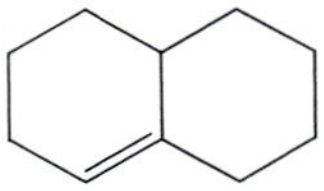

7 For each of the following groups, rank the molecules in decreasing order of reactivity toward addition of a nucleophile to the most electrophilic carbon.

(a) $(CH_3)_2C{=}O$, $(CH_3)_2C{=}NH$, $(CH_3)_2C{=}\overset{+}{O}H$

(b) 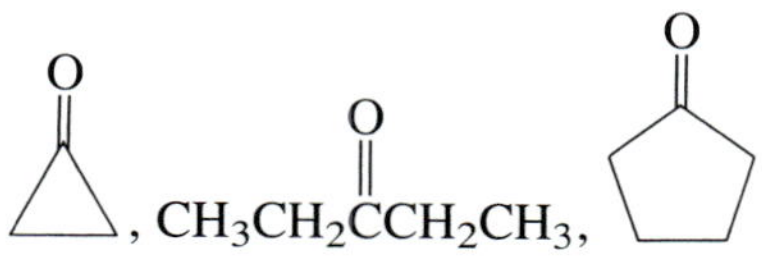 cyclopropanone, $CH_3CH_2\overset{O}{\overset{\|}{C}}CH_2CH_3$, cyclopentanone

(c) $CH_3\overset{O}{\overset{\|}{C}}CH_3$, $CH_3\overset{O}{\overset{\|}{C}}\overset{O}{\overset{\|}{C}}CH_3$, $CH_3\overset{O}{\overset{\|}{C}}\overset{O}{\overset{\|}{C}}\overset{O}{\overset{\|}{C}}CH_3$

8 Write in detail the mechanism of the BF_3-catalyzed reaction of CH_3SH with butanal (Section 15-5).

9 The commercial synthesis of vitamin C requires the following oxidation of sorbose to a carboxylic acid:

HO H HO H; $HOCH_2$; CH_2OH; H OH O → HO H HO H; $HOCH_2$; COOH; H OH O

Sorbose

Suggest a synthetic sequence capable of achieving this oxidation efficiently.

10 Write the expected product(s) of each of the following reactions.

(a) cyclohexanone $+$ excess $CH_3OH \xrightarrow{^-OH}$

(b) cyclohexanone $+$ excess $CH_3OH \xrightarrow{H^+}$

(c) 2-methylcyclopentanone (CH_3) $+ H_3C$–C_6H_4–$S(=O)_2$–$NHNH_2 \xrightarrow{H^+}$

(d) $CH_3\overset{O}{\overset{\|}{C}}CH_3 + HOCH_2\overset{OH}{\overset{|}{C}}HCH_2CH_2CH_3 \xrightarrow{H^+}$

(e) bicyclic enone with CH_3, O $+ 2\,CH_3CH_2SH \xrightarrow{BF_3}$

(f) cyclopentanone $+ (CH_3CH_2)_2NH \longrightarrow$

(g) 2-methylcyclohexanone (CH_3) $+$ morpholine (O, N, H) $\longrightarrow$

11 There are two isomers of ethanal oxime, $CH_3CH{=}NOH$, but only one of propanone oxime, $(CH_3)_2C{=}NOH$. Explain.

12 The rate of the reaction of NH_2OH with aldehydes and ketones is very sensitive to pH. It is very slow in solutions more acidic than pH 2 or more basic than pH 7. It is fastest in moderately acidic solution (pH ~ 4). Suggest explanations for these observations.

13 The formation of imines, oximes, hydrazones, and related derivatives from carbonyl compounds is reversible. Write a detailed mechanism for the acid-catalyzed hydrolysis of cyclohexanone semicarbazone back to cyclohexanone and semicarbazide.

$$C_6H_{10}{=}N{-}NHCONH_2 + H_2O \xrightarrow{H^+} C_6H_{10}{=}O + NH_2NHCONH_2$$

14 Propose reasonable syntheses of each of the following molecules, beginning with the indicated starting material.

(a) $HOCH_2CH(OH)CH_2CH_2CH_2C(OH)(CH_3)CH_3$ from $HOCH_2CH(OH)CH_2CH_2CH_2OH$

(b) $C_6H_5N{=}C(CH_2CH_3)_2$ from 3-pentanol

(c) 2,6-dihydroxytetrahydropyran (HO, O, OH) from 1,5-pentanediol

(d) 4-(3-hydroxybutyl)cyclohexanone (OH, O) from 4-vinylcyclohexanone (O)

(e) 1-hydroxycyclopentanecarboxylic acid (HO, COOH) from cyclopentane

15 The UV absorptions and colors of 2,4-dinitrophenylhydrazone derivatives of aldehydes and ketones depend sensitively on the structure of the carbonyl compound. Suppose that you are asked to identify the contents of three bottles whose labels have fallen off. The labels indicate that one bottle contained butanal, one contained *trans*-2-butenal, and one contained *trans*-3-phenyl-2-propenal. The 2,4-dinitrophenylhydrazones prepared from the contents of the bottles have the following characteristics:

Bottle 1: m.p. 187°–188°C; λ_{max} = 377 nm; orange color
Bottle 2: m.p. 121°–122°C; λ_{max} = 358 nm; yellow color
Bottle 3: m.p. 252°–253°C; λ_{max} = 394 nm; red color

Match up the hydrazones with the aldehydes (*without* first looking up the melting points of these derivatives), and explain your choices.

16 Indicate the reagent(s) best suited to effect these transformations.

(a) 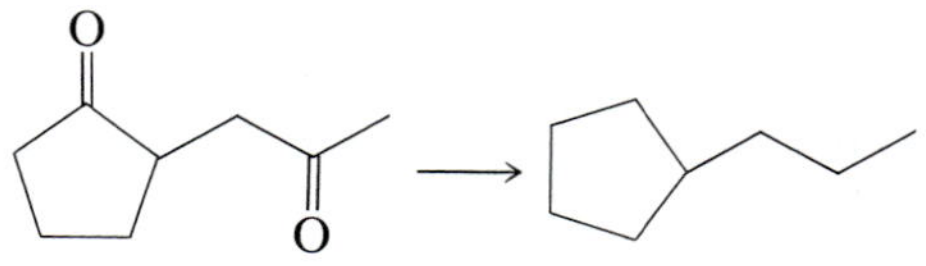 2-(2-oxopropyl)cyclopentanone $\longrightarrow$ propylcyclopentane

(b) $CH_3CH{=}CHCH_2CH_2\overset{O}{\overset{\|}{C}}H \longrightarrow CH_3CH_2CH_2CH_2CH_2\overset{O}{\overset{\|}{C}}H$

(c) $CH_3CH{=}CHCH_2CH_2\overset{O}{\overset{\|}{C}}H \longrightarrow CH_3CH{=}CHCH_2CH_2CH_2OH$

(d) 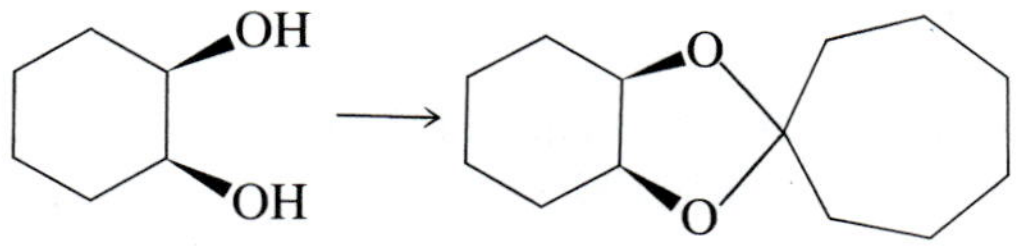 cis-1,2-cyclohexanediol (OH, OH) $\longrightarrow$ spiro ketal with cycloheptane (O, O)

17 The *Strecker amino acid synthesis* is written below. It is one of the oldest synthetic methods in organic chemistry, dating to 1850. It was the first method developed for the laboratory preparation of 2-amino acids, the basic constituents of all proteins.

$$RCHO \xrightarrow[2.\ H^+,\ H_2O]{1.\ NH_3,\ HCN;} R\overset{NH_2}{\overset{|}{C}}HCO_2H$$

Explain the steps of this synthesis in full mechanistic detail. Hint: The synthesis combines characteristics of both imine- and cyanohydrin-forming reactions.

18 For each of the following molecules, propose *two* methods of synthesis, from the different precursor molecules indicated.

(a) 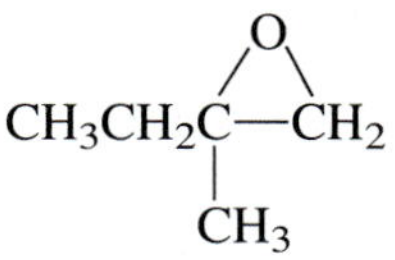 $CH_3CH_2\overset{O}{C}\text{—}CH_2$ (epoxide, with CH_3 on C) from (1) an alkene and (2) a ketone.

(b) $CH_3CH{=}CHCH_2CH(CH_3)_2$ from (1) an aldehyde and (2) a different aldehyde.

(c) 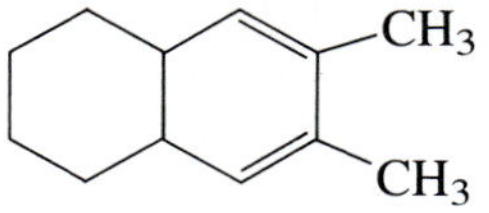 2,3-dimethyl bicyclic diene (CH_3, CH_3) from (1) a dialdehyde and (2) a diketone.

19 Three isomeric compounds with the formula $C_5H_{10}O$ are converted into pentane by Clemmensen reduction. Compounds A and C give a single product on Baeyer-Villiger oxidation; compound B gives two different products in a 1 : 1 ratio. The Fehling's test for compound C is positive; for compounds A and B, it is not. Identify compounds A, B, and C.

20 Compound D, formula $C_8H_{14}O$, is converted by $CH_2{=}P(C_6H_5)_3$ into E, C_9H_{16}. Treatment of D with $LiAlH_4$ yields *two* isomeric products F and G, both $C_8H_{16}O$, in unequal yield. Heating either F or G with concentrated H_2SO_4 produces H, with the formula C_8H_{14}. Ozonolysis of H produces a keto aldehyde after Zn—H^+, H_2O treatment. Oxidation of this keto aldehyde with aqueous Cr(VI) produces

$$CH_3\overset{\overset{\displaystyle O}{\|}}{C}CH_2CH_2CH_2\overset{\overset{\displaystyle CH_3}{|}}{C}HCO_2H$$

Identify compounds D through H. Pay particular attention to the stereochemistry of D.

21 Write the product(s) of reaction of hexanal with each of the following reagents.

(a) $HOCH_2CH_2OH$, H^+

(b) $LiAlH_4$

(c) NH_2OH, H^+

(d) NH_2NH_2, KOH, heat

(e) $(CH_3)_2CHCH_2CH{=}P(C_6H_5)_3$

(f) 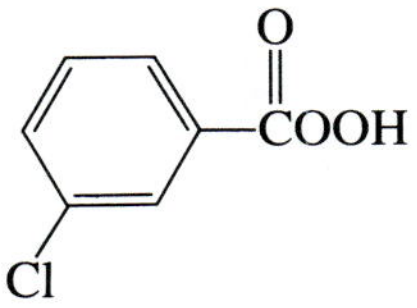 3-ClC_6H_4–$\overset{\overset{\displaystyle O}{\|}}{C}OOH$ (benzene ring bearing Cl)

(g) 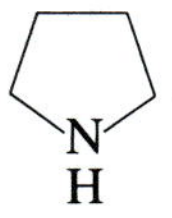 pyrrolidine (five-membered ring with N–H), H^+

(h) $CH_2{=}S(CH_3)_2$

(i) Ag^+, NH_3

(j) CrO_3, H_2SO_4, H_2O

(k) HCN, then H^+, H_2O, and heat

(l) Mg, then H^+, H_2O

22 Write the product(s) of reaction of cycloheptanone with each of the reagents in Problem 21.

23 Synthesize each of the following molecules, beginning with the indicated starting materials.

(a) 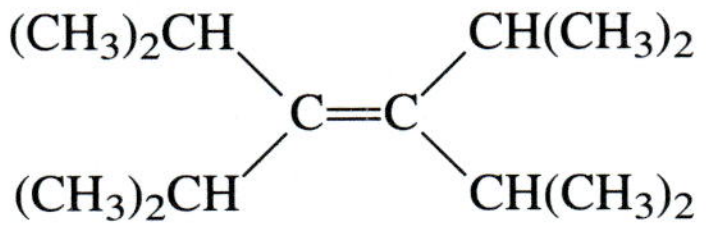 $[(CH_3)_2CH]_2C{=}C[CH(CH_3)_2]_2$ from 2,4-dimethyl-2-pentene

(b) Pentanal, from ethyne and any haloalkane of four carbons or less

(c) 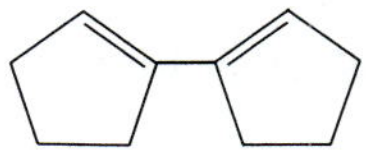 1,1′-bi(cyclopentenyl) (two cyclopentene rings joined) from cyclopentane

(d) 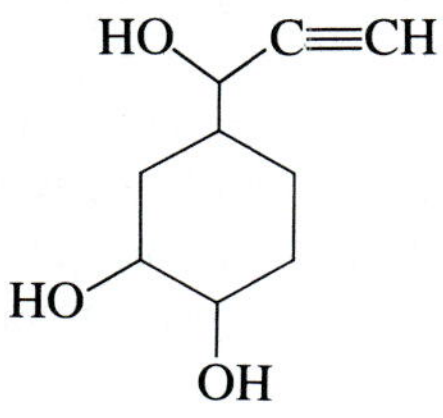 Cyclohexane bearing $CH(OH)C{\equiv}CH$ at C-1 and HO groups at C-3 and C-4 (HO, C≡CH; HO; OH) from 1,3-butadiene and $CH_2{=}CHCHO$

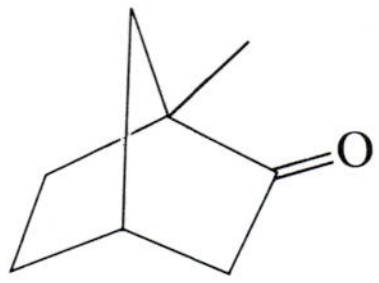

24 Write a detailed mechanism for the Baeyer-Villiger oxidation of the ketone shown in the margin (refer to Exercise 15-16).

25 Write the two theoretically possible Baeyer-Villiger products from each of the following compounds. Indicate which one is preferentially formed.

(a) 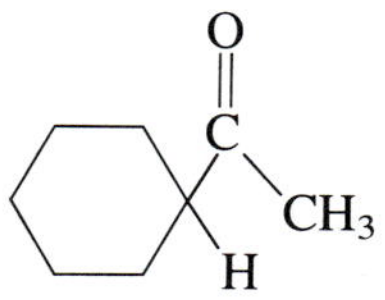 (cyclohexyl methyl ketone: cyclohexane ring bearing H and $C(=O)CH_3$)

(b) (2-methylcyclopentanone)

(c) $(CH_3)_2CHCCH_2CH(CH_3)_2$ (C=O at the central carbon)

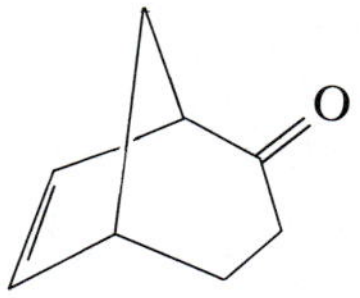

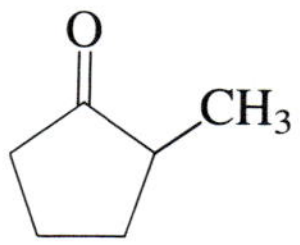

(d) (bicyclic unsaturated ketone)

(e) C_6H_5CH (with C=O; C_6H_5CHO)

(f) $C_6H_5CCH_3$ (with C=O)

26 Propose efficient syntheses of each of the following molecules, beginning with the indicated starting materials.

(a) 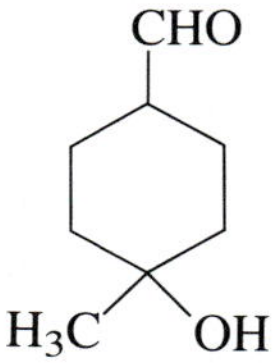 4-formyl-1-methylcyclohexanol (CHO on ring; H_3C and OH on C-1) from 4-hydroxycyclohexanecarbaldehyde (CHO; H and OH)

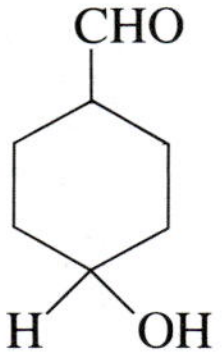

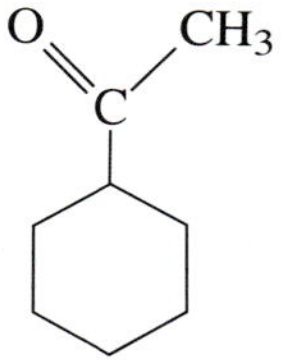

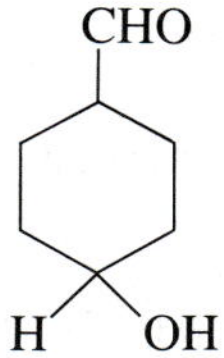

(b) cyclohexyl methyl ketone ($C(=O)CH_3$ on cyclohexane) from 4-hydroxycyclohexanecarbaldehyde (CHO; H and OH)

(c) 2-methyltetrahydrofuran-2-ol (OH, CH_3) from $ClCH_2CH_2CH_2OH$

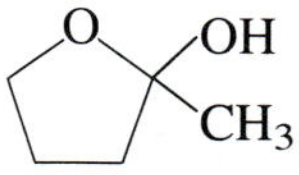

27 In 1862, it was discovered that cholesterol (for structure, see Section 4-7) is converted into a new substance named coprostanol by the action of bacteria in the human digestive tract. Make use of the following information to deduce the structure of coprostanol. Identify the structures of unknowns J through M as well.

(i) Coprostanol, on treatment with Cr(VI) reagents, gives compound J, UV $\lambda_{max}(\epsilon) = 281(22)$ nm.

(ii) Treatment of cholesterol with H_2 over Pt gives compound K, a stereoisomer of coprostanol. Treatment of K with the Cr(VI) reagent gives compound L, which has a UV peak very similar to that of J, $\lambda_{max}(\epsilon) = 285(23)$ nm, and turns out to be a stereoisomer of J.

(iii) Careful treatment of cholesterol with Cr(VI) reagent produces M: UV $\lambda_{max}(\epsilon) = 286(109)$ nm. Treatment of M with H_2 over Pt gives compound L described above.

28 Three reactions involving compound M (see Problem 27) are described below. Answer the questions that follow the descriptions.

(a) Treatment of M with catalytic amounts of acid in ethanol solvent causes isomerization to N: UV $\lambda_{max}(\epsilon) = 241(17{,}500)$ and $310(72)$ nm. Propose a structure for N.

(b) Hydrogenation of N (H_2-Pd, ether solvent) produces J (Problem 27). Is this the result that you would have predicted, or is there anything unusual about it?

(c) Wolff-Kishner reduction of N (H_2NNH_2, H_2O, HO^-, Δ) produces 3-cholestene. Propose a mechanism for this transformation.

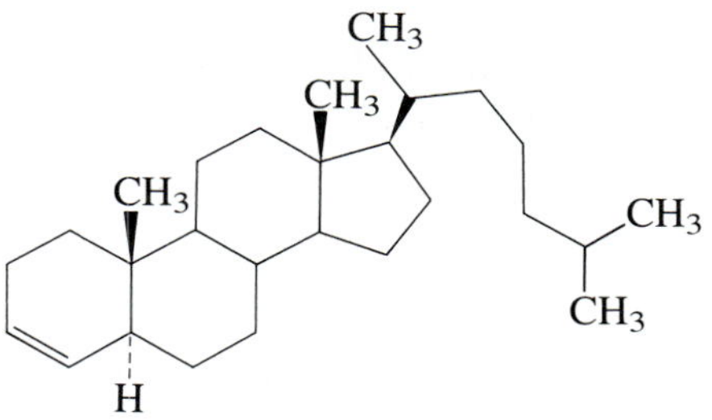

CH_3 CH_3 CH_3 CH_3 CH_3 H

3-Cholestene

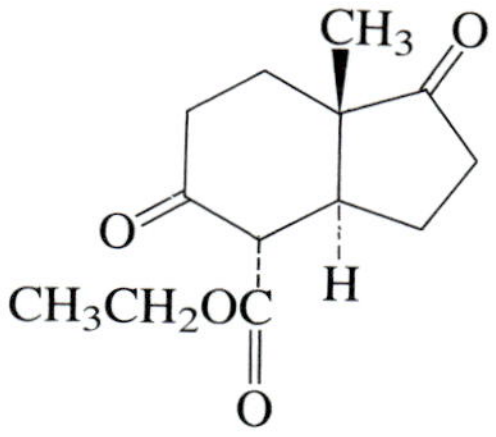

29 An efficient synthesis of the hormone estrone has been devised, starting from diketoester A (shown in the margin) and proceeding through compound B. Propose a synthetic scheme for the conversion of A into B. Hint: Use a Wittig reaction for your carbon–carbon bond-forming step.

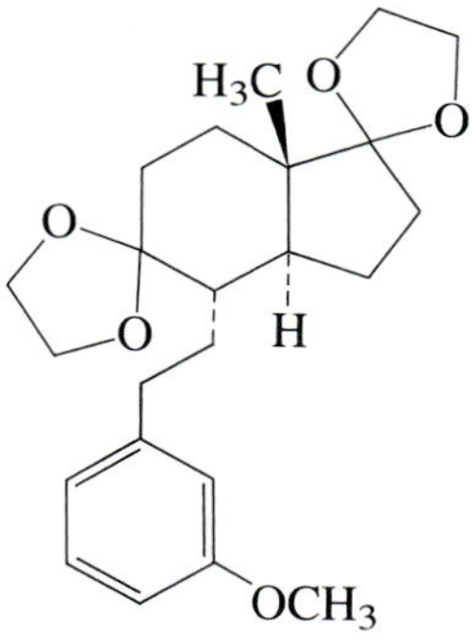

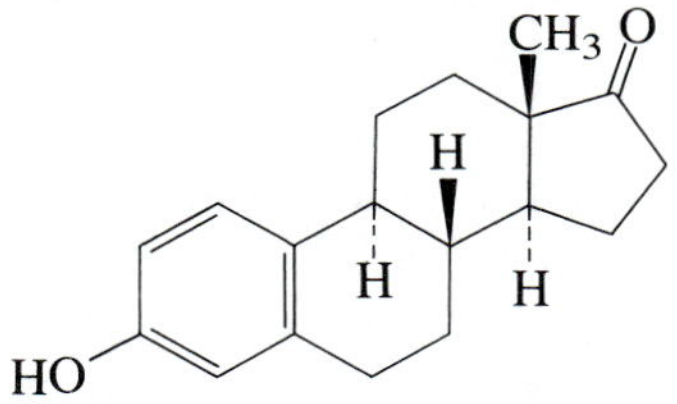

CH_3 O H H H HO

Estrone

CH_3 O O CH_3CH_2OC H O

A

Several steps

H_3C O O O O H OCH_3

B

30 Devise an efficient sequence for converting polyene acetal A into oxacyclopropane derivative B. The latter has been used in studies of steroid-producing cyclization reactions.

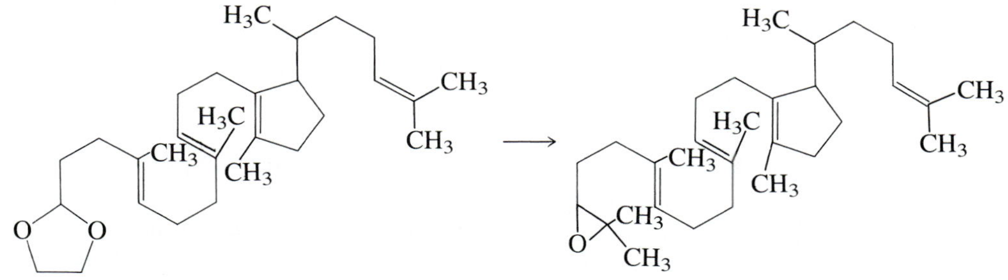

H_3C CH_3 H_3C CH_3 CH_3 CH_3 O O

A

H_3C CH_3 H_3C CH_3 CH_3 CH_3 CH_3 CH_3 O

B

CHAPTER 16

Enols and Enones

α,β-UNSATURATED ALCOHOLS, ALDEHYDES, AND KETONES

This chapter discusses another site of reactivity in aldehydes and ketones: the carbon next to the functional groups, in what is called the α position. The special reactivity of this carbon was pointed out in Sections 15-4 through 15-8. The polarized carbonyl group has an acidifying effect on α-hydrogens, allowing the formation of α,β-unsaturated alcohols (enols, for short) and their corresponding anions (enolate ions). These electron-rich species can be subjected to electrophilic attack by protons, alkylating agents, halogens, and the positively polarized carbon of other carbonyl compounds. Similarly, α,β-unsaturated aldehydes and ketones can enter into electrophilic additions to the carbon–carbon double bond. They also undergo nucleophilic attack, which, in analogy to the chemistry of conjugated dienes, may occur in 1,2 or 1,4 fashion, depending on the nature of the nucleophile.

16-1
The Acidity of α-Hydrogens in Aldehydes and Ketones: Enolate Ions

Sections 15-4 and 15-5 showed that the carbonyl group is subject to attack by electrophiles (most commonly, protons) at oxygen, and by nucleophiles at carbon. A third mode of reactivity is caused by strong bases. When these are hindered to minimize nucleophilic addition, they deprotonate the carbon at the α position. This section describes the synthetic utility of the resulting species.

General Deprotonation of a Carbonyl Compound

$$\text{H}-\overset{|}{\underset{|}{\text{C}}}-\text{C}(=\ddot{\text{O}}:) + \text{B}:^- \longrightarrow \left[{}^-\text{C}-\text{C}=\ddot{\text{O}}: \longleftrightarrow \text{C}=\text{C}-\ddot{\text{O}}:^- \right] + \text{BH}$$

$pK_a \sim 19–21$

The pK_as of aldehydes and ketones range from 19 to 21, considerably lower than the pK_as of ethene (44) or ethyne (25), but higher than those of alcohols (15–18). Why are aldehydes and ketones relatively acidic? There are two reasons: the inductive electron-withdrawing effect of the positively polarized carbonyl carbon and, more importantly, resonance. The anions formed by deprotonation, the **enolate ions,** or **enolates,** are stabilized by resonance.

To generate stoichiometric amounts of an enolate from an aldehyde is difficult because of side reactions (see Section 16-3). Ketones, however, can be deprotonated by lithium diisopropylamide or potassium hydride. The hydride atom in the latter acts as a base rather than a nucleophile and therefore does not give alcohols, as observed with sodium borohydride or lithium aluminum hydride (Sections 8-4 and 15-8).

Enolate Preparation

$$CH_3\overset{:O:}{\overset{\|}{C}}CH_3 + KH \xrightarrow{THF} K^+ \left[{}^-:CH_2-C(=\ddot{O}:)CH_3 \longleftrightarrow CH_2=C(-\ddot{O}:^-)CH_3 \right] + H-H$$

Propanone (acetone) enolate ion

$$\text{Cyclohexanone} + (CH_3\overset{CH_3}{\overset{|}{C}}H)_2NLi \xrightarrow{THF} \text{Cyclohexanone enolate ion } (:\ddot{O}:^- Li^+) + (CH_3\overset{CH_3}{\overset{|}{C}}H)_2NH$$

LDA **Cyclohexanone enolate ion**

Both resonance structures of the enolate contribute to the characteristics of the anion. Placing the negative charge on carbon keeps the relatively strong carbon–oxygen double bond intact. On the other hand, oxygen is more electronegative than carbon; we would therefore expect the negative charge to prefer to reside on oxygen. Both forms are important in the nucleophilic reactions of enolate ions, and electrophilic attack can be either at carbon or at oxygen. Although in most cases carbon is attacked, in others oxygen is the target. A species that may react at two different sites to give two different products is called **ambident** (''two fanged'': from *ambi,* Latin, both; *dens,* Latin, tooth). The enolate ion is an ambident anion. For example, alkylation of cyclohexanone enolate with 3-chloropropene occurs at carbon (C-alkylation), but protonation takes place at oxygen (O-protonation). The product of protonation is an enol, which is unstable and rapidly tautomerizes to the ketone (Section 13-6).

Ambident Behavior of Cyclohexanone Enolate Ion

:O:

$CH_2CH{=}CH_2$

$+ Cl^-$ $\xleftarrow[\text{C-alkylation}]{CH_2{=}CHCH_2Cl,\ THF}$ [Cyclohexanone enolate ion: :Ö:⁻ ⟷ :O:] $\xrightarrow[\text{O-protonation}]{H^+,\ H_2O}$

62%

2-(2-Propenyl)cyclohexanone

Cyclohexanone enolate ion

:ÖH → :O: H

Cyclohexanone enol **Cyclohexanone**

$C_6H_5\overset{O}{\overset{\|}{C}}CH(CH_3)_2$

2-Methyl-1-phenyl-1-propanone

$- H{-}H$, $- NaBr$ | 1. NaH, benzene, Δ 2. $(CH_3)_2C{=}CHCH_2Br$

$C_6H_5\overset{O}{\overset{\|}{C}}C(CH_3)_2$

$CH_2CH{=}C(CH_3)_2$

88%

2,2,5-Trimethyl-1-phenyl-4-hexen-1-one

Most alkylations occur at carbon. This is a general way to introduce an alkyl substituent next to the carbonyl group of a ketone, as shown at the left. A problem in these transformations is the control of dialkylation. Under the reaction conditions, a monoalkylated ketone may become deprotonated by the starting enolate and would therefore become subject to further alkylation.

Single and Double Alkylation of Enolates

$\xrightarrow[-\ H{-}H,\ -\ NaI]{1.\ NaH,\ CH_3OCH_2CH_2OCH_3\quad 2.\ CH_3I}$

O, H, H → O, CH_3, H (27%) + O, CH_3, CH_3 (38%)

Another complication arises in the alkylation of unsymmetrical ketones: both α positions are subject to electrophilic attack.

H H O H H $\xrightarrow[-\ (CH_3CH)_2NH\ (CH_3)]{(CH_3CH)_2NLi\ (CH_3),\ 20°C}$ $\ddot{O}{:}^-Li^+$ (53) : + $\ddot{O}{:}^-Li^+$ (47)

$- LiI$ | CH_3I

CH_3, O + O, CH_3

EXERCISE 16-1

The reaction of compound A with base gives three isomeric products $C_8H_{12}O$. What are they? (Hint: Try intramolecular alkylations.)

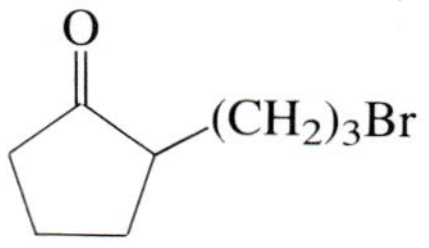

$(CH_2)_3Br$

A

EXERCISE 16-2

How would you convert cycloalkyne B into ketone C?

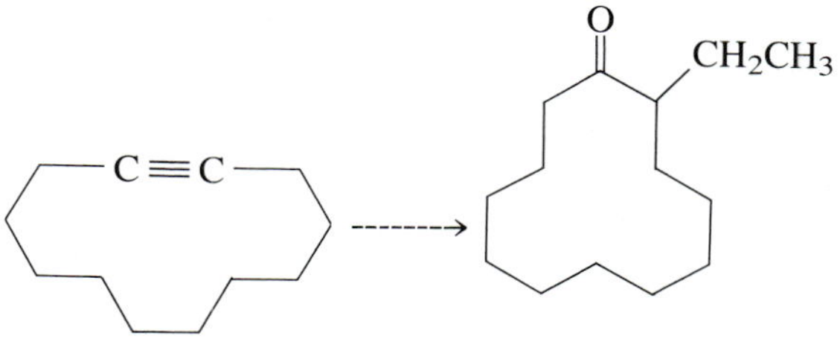

$C{\equiv}C$ - - - → CH_2CH_3

B **C**

Enamines Afford an Alternative Route for the Alkylation of Aldehydes and Ketones

Section 15-6 showed that the reaction of secondary amines with aldehydes or ketones produced enamines. Even though they are neutral, enamines are electron rich because of the presence of the nitrogen substituent (Section 15-6). They may be attacked by electrophiles at the carbon–carbon double bond because of the consequences of resonance of this bond with the lone electron pair of the nitrogen atom.

The dipolar resonance structure shows that an enamine has the same ambident potential as an enolate ion. Indeed, exposure to haloalkanes alkylates enamines at carbon to produce iminium salts. On aqueous work-up, iminium salts hydrolyze by a mechanism that is the reverse of the one formulated for imine formation in Section 15-6. The results are a new alkylated ketone and the original secondary amine.

Resonance in Enamines

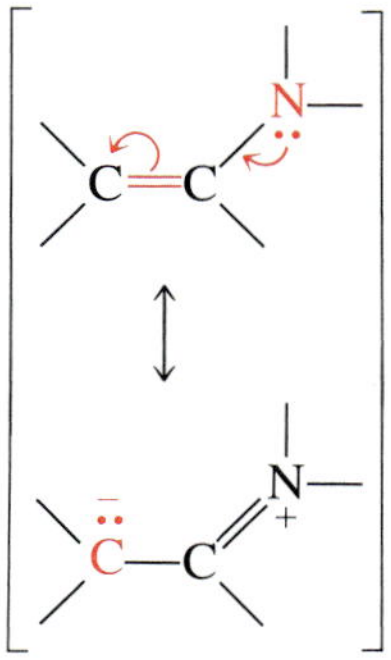

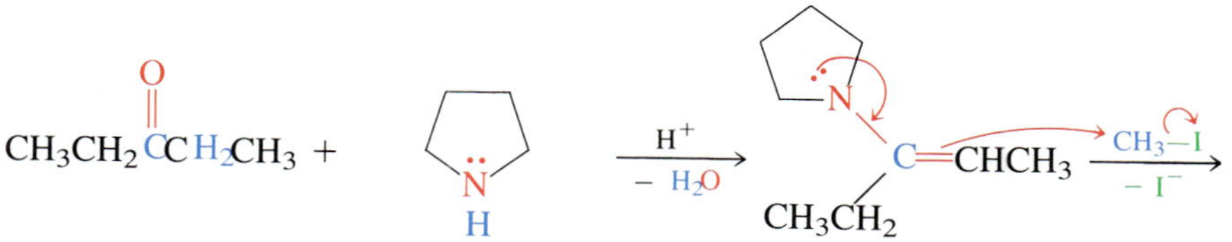

$$\left[>C{=}C{-}\ddot{N}< \longleftrightarrow >\overset{-}{\ddot{C}}{-}C{=}\overset{+}{N}< \right]$$

$$CH_3CH_2\overset{O}{\overset{\|}{C}}CH_2CH_3 + \text{pyrrolidine (N–H)} \xrightarrow[-\,H_2O]{H^+} \text{(pyrrolidin-1-yl)}C(CH_2CH_3){=}CHCH_3 \xrightarrow[-\,I^-]{CH_3{-}I}$$

3-Pentanone **Azacyclopentane (Pyrrolidine)**

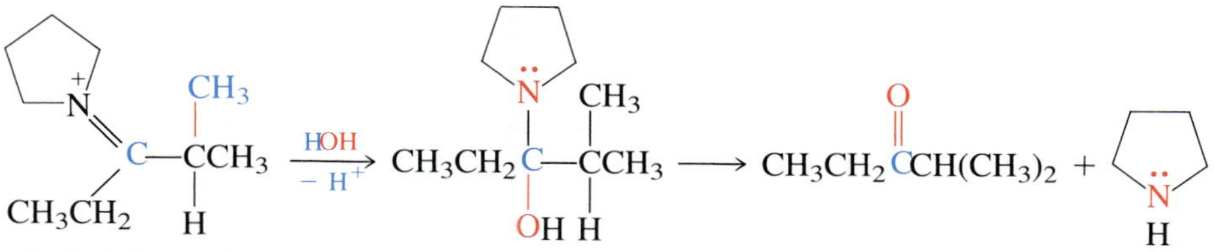

$$\text{(pyrrolidinium)}\overset{+}{N}{=}C(CH_2CH_3){-}CH(CH_3)_2 \xrightarrow[-\,H^+]{HOH} CH_3CH_2C(OH)(\text{N-pyrrolidinyl}){-}CH(CH_3)_2 \longrightarrow CH_3CH_2\overset{O}{\overset{\|}{C}}CH(CH_3)_2 + \text{pyrrolidine (N–H)}$$

An iminium salt **2-Methyl-3-pentanone**

This enamine-formation–alkylation sequence is complementary to the enolate-formation–alkylation procedure for the preparation of alkylated ketones.

1. $CH_3CH_2CH_2CH_2I$
2. H_2SO_4, H_2O

$-H_2O$

44%
2-Butylcyclo-hexanone

$(CH_3)_2CHCH$ (=O) → 1. pyrrolidine (N–H) 2. CH_3CH_2Br 3. H^+, H_2O → $(CH_3)_2C(CH_2CH_3)CH$ (=O)

2-Methylpropanal **2,2-Dimethylbutanal** 67%

EXERCISE 16-3

Explain the following result.

CH_3I, K_2CO_3, CH_3OH, H_2O, Δ

CH_3 … I^- + H_3C CH_3 … I^- + CH_3 … CH_3 I^-

In summary, the hydrogens on the carbon next to the carbonyl group in aldehydes and ketones are acidic, with pK_as ranging from 19 to 21. Deprotonation leads to the corresponding enolate ions, which may be attacked by electrophilic reagents at either oxygen or carbon. Alkylation is mainly at carbon, giving rise to alkylated derivatives. In these reactions, control of the extent and the position of alkylation, when there is a choice, may be a problem. Enamines derived from aldehydes and ketones undergo alkylation to the corresponding iminium salts, which may hydrolyze to the alkylated carbonyl compounds.

16-2 Keto-Enol Equilibria

Protonation of an enolate ion occurs on oxygen to furnish the enol, which tautomerizes to the carbonyl system. The reaction is an equilibrium process, and the aldehyde or ketone is said to have **keto** and **enol** forms. Normally, the keto form is by far the most predominant species. This section describes the factors that influence keto-enol equilibria, what the exact mechanism is of the tautomerization process, and what are its chemical consequences.

How Does an Enol Equilibrate with Its Keto Form?

Enol-keto tautomerization proceeds by either acid or base catalysis. In the base-catalyzed reaction, the proton is simply removed from the oxygen, reversing the initial protonation. Subsequent (and slower) attack at carbon furnishes the thermodynamically more stable keto form.

Base-Catalyzed Enol-Keto Equilibration

$$\text{Enol form} + :B^- \rightleftharpoons [\text{Enolate ion}] + BH \rightleftharpoons \text{Keto form} + :B^-$$

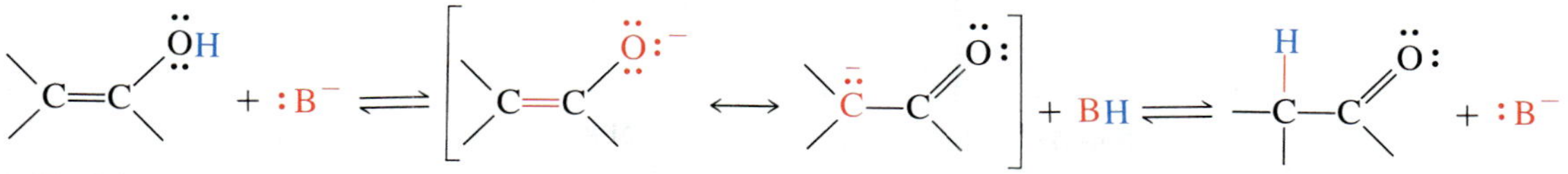

Enol form | Enolate ion | Keto form

In the acid-catalyzed process, the enol form is protonated at the double bond to give the resonance-stabilized carbocation next to oxygen. This species is simply a protonated carbonyl function. Its deprotonation then gives the keto form.

Acid-Catalyzed Enol-Keto Equilibration

$$\text{Enol} + H^+ \rightleftharpoons [\text{protonated carbonyl}] \rightleftharpoons \text{Keto} + H^+$$

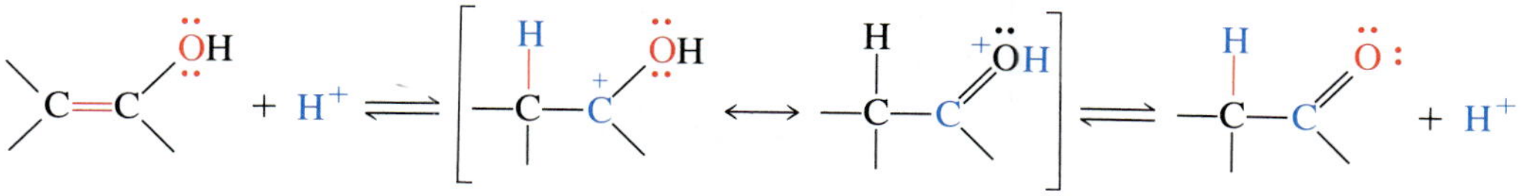

Both the acid- and base-catalyzed enol-keto interconversions occur relatively fast in solution, where there are traces of the required catalysts. Is it possible to produce isolable and observable quantities of simple enols? Yes, but only under special conditions, when acid and base can be rigorously excluded. For example, ethenol (vinyl alcohol) can be made by vacuum pyrolytic dehydration of 1,2-ethanediol.

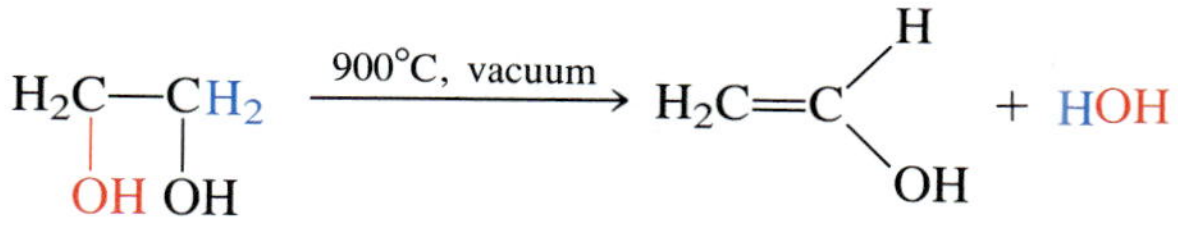

$$H_2C(OH){-}CH_2(OH) \xrightarrow{900^\circ C,\ \text{vacuum}} H_2C{=}CH(OH) + HOH$$

1,2-Ethanediol | **Ethenol (Vinyl alcohol)**

Ethenol has a half-life of about 30 min at room temperature, long enough to allow structural characterization (Figure 16-1). The barrier for the uncatalyzed ethenol (vinyl alcohol) → ethanal (acetaldehyde) interconversion in the gas phase has been calculated to be very high, about 74 kcal mole^{-1}. This result underscores the importance of catalysis in the rearrangement.

FIGURE 16-1

The molecular structure of ethenol (vinyl alcohol) is a composite of that of ethene (ethylene, Figure 11-7) and methanol (Figure 8-1B). However, the C–O bond is relatively short, a result of resonance. (Draw the dipolar resonance structures.)

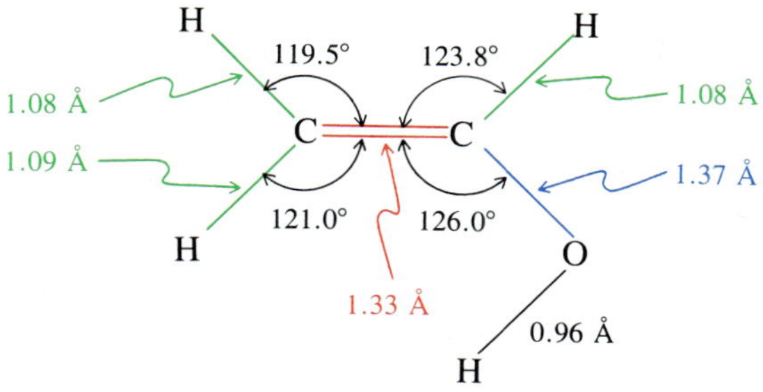

EXERCISE 16-4

Pyrolysis of cyclobutanol at 950°C gave a new compound that exhibited an ^{1}H NMR spectrum with the following parameters: $\delta = 3.91$ (dd, $J = 6.5$, 1.8 Hz, 1 H), 4.13 (dd, $J = 14.0$, 1.8 Hz, 1 H), 6.27 (dd, $J = 14.0$, 6.5 Hz, 1 H), 7.12 (broad s, 1 H) ppm. What is the structure of this compound? Assign the parts of the NMR spectrum to the individual hydrogens in the structure. Is there another compound that should be formed? If so, what is it?

Substituent Effects on Keto-Enol Equilibria

The equilibrium constants for the conversion of the keto into the enol forms are very small for ordinary aldehydes and ketones, only traces of enol being present. The enol of ethanal (acetaldehyde) is about a thousand times as stable relative to its keto form as the enol of propanone (acetone), perhaps because of the lesser stability of the less-substituted aldehyde group.

Keto-Enol Equilibria

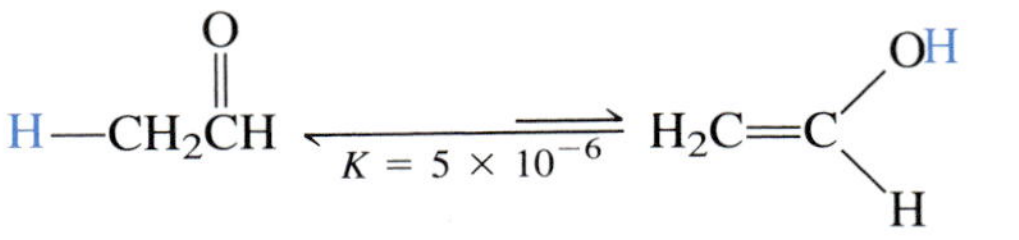

$$\text{H—CH}_2\overset{\overset{\text{O}}{\|}}{\text{C}}\text{H} \underset{K = 5 \times 10^{-6}}{\rightleftharpoons} \text{H}_2\text{C=C(OH)H} \qquad \Delta G^\circ \sim +9 \text{ kcal mole}^{-1}$$

Ethanal (Acetaldehyde) — **Ethenol (Vinyl alcohol)**

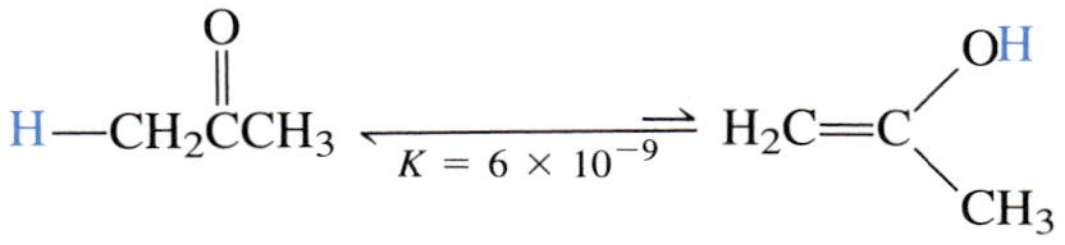

$$\text{H—CH}_2\overset{\overset{\text{O}}{\|}}{\text{C}}\text{CH}_3 \underset{K = 6 \times 10^{-9}}{\rightleftharpoons} \text{H}_2\text{C=C(OH)CH}_3 \qquad \Delta G^\circ \sim +11 \text{ kcal mole}^{-1}$$

Propanone (Acetone) — **2-Propenol**

Some special carbonyl compounds exist mostly in the enol form. Examples are β-diketones, such as 2,4-pentanedione (acetylacetone) (Section 22-4). Contributing to the stability of the enol form of 2,4-pentanedione is the formation of a conjugated enone and of an intramolecular hydrogen bond between the enol proton and the remaining carbonyl oxygen. Note that the resulting structure contains a relatively strain free six-membered ring.

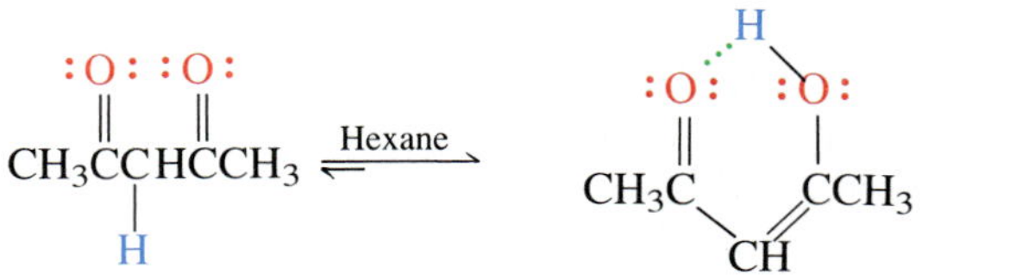

$$\text{CH}_3\overset{\overset{:\text{O}:}{\|}}{\text{C}}\underset{\underset{\text{H}}{|}}{\text{C}}\text{H}\overset{\overset{:\text{O}:}{\|}}{\text{C}}\text{CH}_3 \underset{}{\overset{\text{Hexane}}{\rightleftharpoons}} \text{CH}_3\text{C(=O:)–CH=C(–O–H}\cdots\text{)CH}_3$$

10% **2,4-Pentanedione (Acetylacetone)** — 90% **Z-4-Hydroxypent-3-en-2-one**

Consequences of Enol Formation: Deuterium Exchange and Stereoisomerization

Treatment of a ketone with traces of acid or base in D_2O as solvent leads to the complete exchange of *all* the hydrogens that can take part in keto-enol tautomerism. This reaction can be conveniently followed by ^{1}H NMR, because the signal for these hydrogens slowly disappears as each hydrogen is sequentially replaced by deuterium.

Hydrogen-Deuterium Exchange of Enolizable Hydrogens

$$\underset{\textbf{2-Butanone}}{CH_3\overset{\overset{\displaystyle O}{\|}}{C}CH_2CH_3} \xrightarrow{D_2O,\ DO^-} \underset{\textbf{1,1,1,3,3-Pentadeuterio-2-butanone}}{CD_3\overset{\overset{\displaystyle O}{\|}}{C}CD_2CH_3}$$

The mechanism of this exchange reaction is the same as that formulated for the acid- and base-catalyzed tautomerism, except that after enolization, or enolate formation, the original carbonyl compound is reformed through attack by D^+, instead of by H^+.

Mechanism of Single H–D Exchange:

Base catalysis:

$$\text{H}-\overset{|}{C}-C(=\ddot{O}:) + D\ddot{O}:^- \rightleftharpoons C{=}C-\ddot{O}:^- + D\ddot{O}H$$

$$C{=}C-\ddot{O}:^- + D-\ddot{O}D \rightleftharpoons D-\overset{|}{C}-C(=\ddot{O}:) + D\ddot{O}:^-$$

Acid catalysis:

$$H-\overset{|}{C}-C(=\ddot{O}:) + D^+ \rightleftharpoons H-\overset{|}{C}-C(=\overset{+}{\ddot{O}}-D) \rightleftharpoons C{=}C-\ddot{O}D + H^+$$

$$C{=}C-\ddot{O}D + D^+ \rightleftharpoons D-\overset{|}{C}-C(=\overset{+}{\ddot{O}}-D) \longrightarrow D-\overset{|}{C}-C(=\ddot{O}:) + D^+$$

Because D_2O is the solvent, the chances of reprotonation are small compared with the probability of deuteration. The proton essentially gets ''lost'' in the large excess of D_2O. Multiple execution of these exchange mechanisms leads to complete replacement of hydrogen by deuterium.

EXERCISE 16-5

Write the products (if any) of deuterium incorporation by the treatment of the following compounds with D_2O–NaOD: (a) cycloheptanone; (b) 2,2-dimethylpropanal; (c) 3,3-dimethyl-2-butanone; (d) [bicyclic ketone structure: O= on decalin ring, CHO substituent]

O
CHO

Another consequence of enolization is the stereochemical lability of stereocenters bearing enolizable hydrogens. For example, treatment of cis-2,3-disubstituted cyclopentanones with mild base furnishes the corresponding trans isomers. For steric reasons, these are thermodynamically more stable.

O
$CH_2CH{=}CH_2$
CH_3
10% KOH, CH_3CH_2OH
O^-K^+
$CH_2CH{=}CH_2$
CH_3
O
$CH_2CH{=}CH_2$
CH_3
>95%

The reaction proceeds through the enolate ion, in which the α-carbon is no longer a stereocenter. Reprotonation from the side cis to the 3-methyl group results in the observed product.

For the same reason, it is difficult to maintain optical activity in a compound whose stereocenter is next to a carbonyl group. For example, at room temperature, optically active 3-phenyl-2-butanone racemizes with a half-life of minutes in basic ethanol.

Racemization of Optically Active 3-Phenyl-2-butanone

C_6H_5, O, C—CCH_3, H, H_3C
(*S*)-3-Phenyl-2-butanone
$CH_3CH_2O^-$, CH_3CH_2OH
C_6H_5, O^-, C=C, H_3C, CH_3
Achiral
$CH_3CH_2O^-$, CH_3CH_2OH
C_6H_5, O, C—CCH_3, H_3C, H
(*R*)-3-Phenyl-2-butanone

Halogenation of Aldehydes and Ketones Through the Intermediacy of Enolates or Enols

Aldehydes and ketones react with halogens at the carbon next to the carbonyl group. In contrast with deuteration, the extent of halogenation depends on whether acid or base catalysis has been used. For example, in the presence of acid, halogenation usually stops after the first halogen has been introduced.

$$H{-}CH_2\overset{O}{\overset{\|}{C}}CH_3 \xrightarrow{Br{-}Br,\ CH_3CO_2H,\ H_2O,\ 70^\circ C} Br\,CH_2\overset{O}{\overset{\|}{C}}CH_3 + HBr$$

44%

Bromopropanone (Bromoacetone)

O, CH_3, H
$Cl{-}Cl,\ H^+,\ CCl_4$
O, CH_3, Cl
+ HCl
85%

2-Methylcyclohexanone → **2-Chloro-2-methylcyclohexanone**

The rate of the acid-catalyzed halogenation is independent of the halogen concentration, suggesting a rate-determining first step involving the carbonyl substrate. This step is enolization. The halogen then attacks the double bond to give an intermediate oxygen-stabilized halo carbocation (Section 9-2). Subsequent deprotonation of this species furnishes the product.

Mechanism of the Acid-Catalyzed Bromination of Propanone (Acetone):

STEP 1: Enolization

$$CH_3\overset{\overset{:O:}{\|}}{C}CH_3 \overset{H^+}{\rightleftharpoons} H_2C{=}C(\ddot{O}H)CH_3$$

STEP 2: Halogen attack

$$H_2C{=}C(\ddot{O}H)CH_3 + Br{-}Br \longrightarrow \left[BrH_2C{-}\overset{+}{C}(\ddot{O}H)CH_3 \longleftrightarrow BrH_2C{-}C(={\overset{+}{\ddot{O}}}H)CH_3 \right] + Br^-$$

STEP 3: Deprotonation

$$BrCH_2\overset{\overset{+:O{-}H}{\|}}{C}CH_3 \longrightarrow BrCH_2\overset{\overset{:O:}{\|}}{C}CH_3 + H^+$$

Why is halogenation slower after the first halogen has been introduced? The answer lies in the requirement for enolization. To repeat halogenation, the halo carbonyl compound must enolize again by the usual acid-catalyzed mechanism. However, the electron-withdrawing power of the halogen makes protonation, the initial step in enolization, *more difficult* than in the original carbonyl compound.

Halogenation Slows Down Enolization

Less basic than unsubstituted ketone

$$BrCH_2\overset{\overset{:O:}{\|}}{C}CH_3 \overset{H^+}{\rightleftharpoons} BrCH_2\overset{\overset{+:O{-}H}{\|}}{C}CH_3$$

Electron withdrawing

Therefore, the singly halogenated product is not attacked by additional halogen until the starting aldehyde or ketone has been used up. Only then does further halogenation occur.

$$C_6H_5\overset{\overset{O}{\|}}{C}CH_3 \xrightarrow[-\,2\,HCl]{Cl{-}Cl\text{ (excess)},\ CH_3CO_2H,\ 60^\circ C} C_6H_5\overset{\overset{O}{\|}}{C}CHCl_2$$

94%

1-Phenylethanone (Acetophenone) → **2,2-Dichloro-1-phenylethanone**

Base-mediated halogenation is entirely different. Here the reaction is difficult to stop and therefore is not generally useful. However, with methyl ketones the resulting trihalomethyl substituent functions as a leaving group under the basic conditions, and the ultimate product in many cases is a carboxylic acid and the trihalomethane (common name, haloform). This process appears in the iodoform test for secondary 2-alkanols and methyl ketones (Section 9-4).

$$(CH_3)_3C\overset{O}{\overset{\|}{C}}CH_3 \xrightarrow[-\,3\,NaBr,\ -\,3\,H_2O]{3\,Br{-}Br,\ NaOH,\ H_2O,\ 0^\circ C} (CH_3)_3C\overset{O}{\overset{\|}{C}}CBr_3 \xrightarrow{1.\ HO^-,\ H_2O\quad 2.\ H^+,\ H_2O} (CH_3)_3C\overset{O}{\overset{\|}{C}}OH + HCBr_3$$

3,3-Dimethyl-2-butanone → **1,1,1-Tribromo-3,3-dimethyl-2-butanone** → **2,2-Dimethylpropanoic acid** (74%) + **Tribromomethane (Bromoform)**

How does this reaction work? The base first forms the enolate ion. This species is nucleophilic enough to attack the halogen molecule to form the monohalo compound directly. The electron-withdrawing power of the halogen has an *acidifying* effect on the neighboring hydrogen, leading to accelerated enolate formation and hence further halogenation. The trihalomethyl group is cleaved off by nucleophilic attack of hydroxide on the carbonyl function, as in the base-mediated hydration of aldehydes and ketones (Section 15-5). Rather than leading to protonation, however, the negative charge departs along with the leaving group, thus regenerating the carbon–oxygen double bond. This process is very similar to cyanohydrin formation in reverse (Section 15-7).

Mechanism of the Base-Catalyzed Bromination of a Methyl Ketone:

STEP 1: Enolate formation

$$R\overset{:O:}{\overset{\|}{C}}{-}CH_2{-}H + {}^-{:}\ddot{O}H \rightleftharpoons \underset{R}{\overset{{}^-{:}\ddot{O}}{}}\!\!>C{=}CH_2 + H\ddot{O}H$$

STEP 2: Nucleophilic attack on bromine

$$\underset{R}{\overset{{}^-{:}\ddot{O}}{}}\!\!>C{=}CH_2 + Br{-}Br \longrightarrow R\overset{:O:}{\overset{\|}{C}}CH_2Br + Br^-$$

STEP 3: Complete bromination

$$R\overset{:O:}{\overset{\|}{C}}{-}\overset{H}{\overset{|}{C}}HBr \xrightarrow[-\,HOH]{{}^-{:}\ddot{O}H} \underset{R}{\overset{{}^-{:}\ddot{O}}{}}\!\!>C{=}CHBr \xrightarrow[-\,Br^-]{Br{-}Br} R\overset{:O:}{\overset{\|}{C}}CHBr_2 \xrightarrow[-\,HOH]{HO^-} \xrightarrow[-\,Br^-]{Br{-}Br} R\overset{:O:}{\overset{\|}{C}}CBr_3$$

STEP 4: Carboxylic acid formation

$$\mathrm{RC(=O)CBr_3} + {}^{-}\mathrm{:\ddot{O}H} \longrightarrow \mathrm{RC(O^-)(OH)}\!-\!\mathrm{CBr_3} \longrightarrow \mathrm{RC(=O)OH} + {}^{-}\mathrm{:CBr_3} \longrightarrow$$

$$\mathrm{RC(=O)\ddot{O}:^-} + \mathrm{HCBr_3} \xrightarrow{\mathrm{H^+,\ H_2O}} \mathrm{RC(=O)OH} + \mathrm{HCBr_3}$$

EXERCISE 16-6

Write the products of the acid- and base-catalyzed bromination of ethanoylcyclopentane (acetylcyclopentane).

EXERCISE 16-7

Explain mechanistically the outcome of the following transformation:

5,5-Dimethylcyclohexane-1,3-dione (ring with two C=O, H_3C and CH_3 on C5) $\xrightarrow[\text{2. HCl, } H_2O]{\text{1. } Cl_2\text{, NaOH, 35°–40°C}}$ $(CH_3)_2C(CH_2CO_2H)_2$
96%

(Hint: The ring opens at an intermediate stage through a mechanism similar to Step 4 of the base-catalyzed halogenation of a 2-alkanone.)

In summary, aldehydes and ketones are in equilibrium with their enol forms, which are roughly 10 kcal mole^{-1} less stable. Keto-enol equilibration is catalyzed by acid or base. Enolization allows for facile H–D exchange in D_2O and causes stereoisomerization at stereocenters next to the functional group. Halogenation in acid can proceed selectively to the monohalo carbonyl compounds. In base, halogenation is complete. With methyl ketones loss of the trihalomethyl group gives rise to carboxylic acids in the haloform reaction.

16-3 Attack by Enolates on the Carbonyl Function: The Aldol Condensation

Enolates may attack the carbonyl carbon to give hydroxy carbonyl compounds. Subsequent elimination of water leads to α,β-unsaturated aldehydes and ketones, the overall two-step sequence constituting a condensation reaction.

Aldehydes Undergo Base-Catalyzed Condensations

The smallest aldehyde with enolizable hydrogens, ethanal (acetaldehyde), is unstable in the presence of aqueous sodium hydroxide. Apart from being in rapid equilibrium with its hydrate, it is slowly (and, on heating, rapidly) converted into the condensation product, the α,β-unsaturated aldehyde *trans*-2-butenal (crotonaldehyde).

Aldol Condensation of Ethanal (Acetaldehyde)

$$(H)(H_3C)C{=}O + H_2C(H)CH{=}O \longrightarrow (H)(H_3C)C{=}C(H)CH{=}O + H_2O$$

***trans*-2-Butenal (Crotonaldehyde)**

This reaction is an example of the **aldol condensation.** The aldol condensation is general for aldehydes and, as we shall see, may also succeed with ketones.

The mechanism of this reaction is a straightforward example of enolate chemistry. Under the basic conditions employed, an equilibrium between the aldehyde and the corresponding enolate ion is set up. The latter, being surrounded by excess aldehyde, uses its nucleophilic carbon to attack the carbonyl group of another molecule of aldehyde. Protonation of the alkoxide ion furnishes the initial aldol adduct, 3-hydroxybutanal, which has been given the common name aldol.

Mechanism of Aldol Formation:

STEP 1: Enolate generation

$$HC(=O)—CH_2—H + {}^-{:}\ddot{O}H \rightleftharpoons H_2C{=}C(H){-}\ddot{O}{:}^- + H\ddot{O}H$$

Small equilibrium concentration of enolate

STEP 2: Nucleophilic attack

$$CH_3CH{=}O + CH_2{=}C(H){-}\ddot{O}{:}^- \rightleftharpoons CH_3CH(O^-)CH_2CH{=}O$$

STEP 3: Protonation

$$CH_3CH(O^-)CH_2CH{=}O + HOH \rightleftharpoons CH_3CH(OH)CH_2CH{=}O + HO^-$$

50%–60%

3-Hydroxybutanal ("Aldol")

Note that hydroxide ion functions as a catalyst in this reaction. The last two steps of the sequence drive the initially unfavorable equilibrium toward product, but the overall reaction is not very exothermic and therefore is quite reversible.

The aldol is formed in 50%–60% yield and does not react further if its preparation is carried out at low temperature (5°C).

EXERCISE 16-8

Commercial ''aldol'' has been reported to exhibit an NMR spectrum consistent with structure A, at the right. How can A be formed from ethanal (acetaldehyde)?

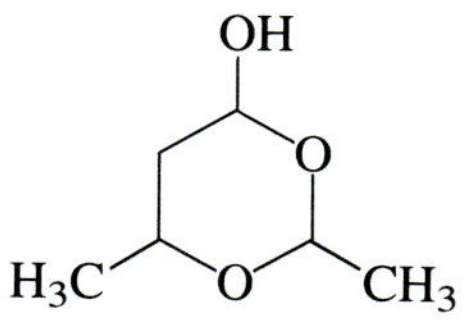

OH
O
H_3C O CH_3
A

At elevated temperature the aldol is converted into its enolate ion, which eliminates hydroxide ion to yield the final product. The net result of this second sequence is a hydroxide-catalyzed dehydration of the aldol.

Mechanism of Dehydration:

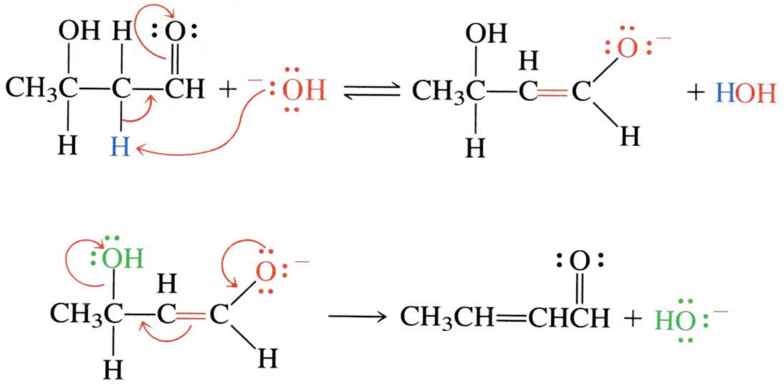

$$CH_3\overset{OH}{\underset{H}{C}}-\overset{H}{\underset{H}{C}}-\overset{O}{\overset{\|}{C}}H + {}^-:\ddot{O}H \rightleftharpoons CH_3\overset{OH}{\underset{H}{C}}-\overset{H}{C}=C(O^-)H + HOH$$

$$CH_3\overset{OH}{\underset{H}{C}}-\overset{H}{C}=C(O^-)H \longrightarrow CH_3CH=CH\overset{O}{\overset{\|}{C}}H + H\ddot{O}:^-$$

The synthetic usefulness of the aldol reaction stems from the fact that a new carbon–carbon bond is formed, producing either a hydroxy carbonyl or an α,β-unsaturated carbonyl moiety. For example,

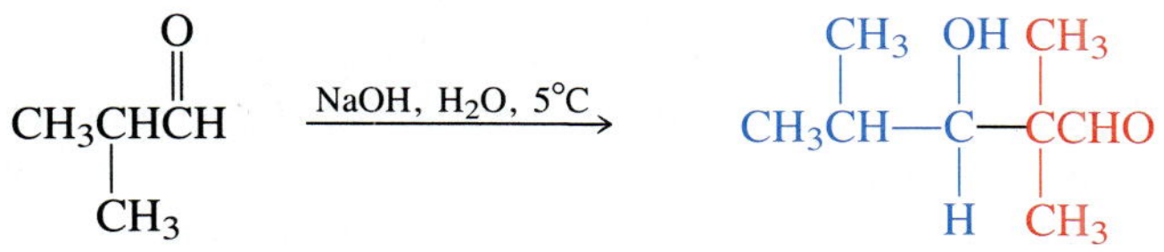

$$CH_3\underset{CH_3}{C}H\overset{O}{\overset{\|}{C}}H \xrightarrow{NaOH,\ H_2O,\ 5°C} CH_3\overset{CH_3}{C}H-\overset{OH}{\underset{H}{C}}-\overset{CH_3}{\underset{CH_3}{C}}CHO$$

2-Methylpropanal → 85% **3-Hydroxy-2,2,4-trimethylpentanal**

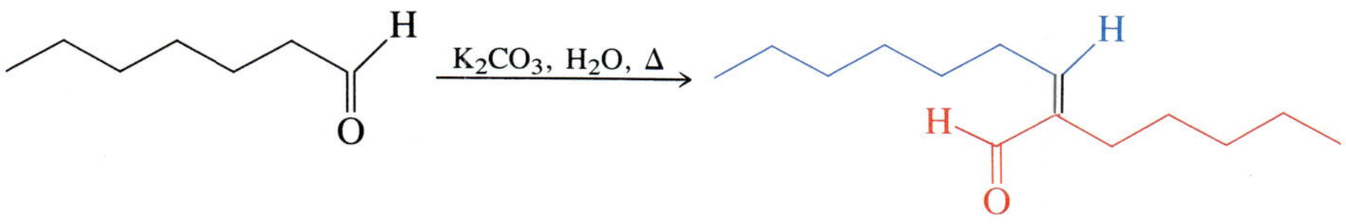

Heptanal $\xrightarrow{K_2CO_3,\ H_2O,\ \Delta}$ Z-2-Pentyl-2-nonenal (80%)

Heptanal **Z-2-Pentyl-2-nonenal**

A drawback of the aldol reaction is the lack of selectivity when two different aldehydes are employed **(crossed aldol condensation).** For example, a 1:1 mixture of ethanal (acetaldehyde) and propanal gives the four possible aldol addition products (or α,β-unsaturated aldehydes at higher temperatures) in essentially statistical ratios.

Nonselective Crossed Aldol Reaction of Ethanal (Acetaldehyde) and Propanal

(All four reactions occur simultaneously.)

$$CH_3\overset{\overset{O}{\|}}{C}H + CH_3\overset{\overset{O}{\|}}{C}H + \underset{\text{Does not participate}}{CH_3CH_2\overset{\overset{O}{\|}}{C}H} \longrightarrow CH_3\underset{H}{\overset{OH}{C}}-CH_2\overset{\overset{O}{\|}}{C}H$$

3-Hydroxybutanal

$$CH_3\overset{\overset{O}{\|}}{C}H + CH_3CH_2\overset{\overset{O}{\|}}{C}H \longrightarrow CH_3\underset{H}{\overset{OH}{C}}-\underset{CH_3}{C}H\overset{\overset{O}{\|}}{C}H$$

3-Hydroxy-2-methylbutanal

$$CH_3CH_2\overset{\overset{O}{\|}}{C}H + CH_3\overset{\overset{O}{\|}}{C}H \longrightarrow CH_3CH_2\underset{H}{\overset{OH}{C}}-CH_2\overset{\overset{O}{\|}}{C}H$$

3-Hydroxypentanal

$$CH_3CH_2\overset{\overset{O}{\|}}{C}H + CH_3CH_2\overset{\overset{O}{\|}}{C}H + \underset{\text{Does not participate}}{CH_3\overset{\overset{O}{\|}}{C}H} \longrightarrow CH_3CH_2\underset{H}{\overset{OH}{C}}-\underset{CH_3}{C}H\overset{\overset{O}{\|}}{C}H$$

3-Hydroxy-2-methylpentanal

This problem is minimized when one of the aldehydes does not have any enolizable hydrogens, because two of the normally expected products do not form. To ensure efficient one-to-one reactivity in this case, the enolizable aldehyde is added slowly to the nonenolizable reactant (usually used in excess). As soon as the enolate ion is formed, it will react preferentially with the other aldehyde. For example,

$$\underset{\textbf{2,2-Dimethylpropanal}}{CH_3\underset{CH_3}{\overset{CH_3}{C}}CHO} + \underset{\textbf{Propanal}}{\underset{\text{Added slowly}}{CH_3CH_2CHO}} \xrightarrow{NaOH,\ H_2O,\ \Delta} \underset{\textbf{2,4,4-Trimethyl-2-pentenal}}{\underset{65\%}{CH_3\underset{CH_3}{\overset{CH_3}{C}}CH{=}\overset{CH_3}{C}CHO}} + H_2O$$

$$\underset{\textbf{3-Phenyl-2-propenal}}{C_6H_5CH{=}CHCHO} + \underset{\textbf{Butanal}}{CH_3CH_2CH_2CHO} \xrightarrow[-H_2O]{NaOH,\ H_2O} \underset{\textbf{2-Ethyl-5-phenyl-2,4-pentadienal}}{\underset{55\%}{C_6H_5CH{=}CHCH{=}\overset{CH_2CH_3}{C}CHO}}$$

EXERCISE 16-9

Show the likely products of the following aldol condensations:

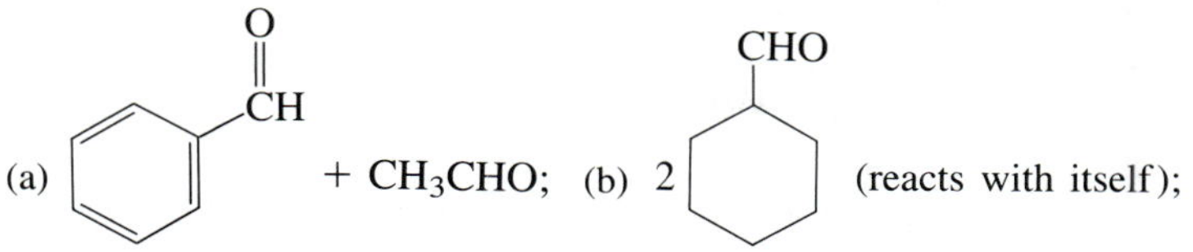

(a) C_6H_5CHO + CH_3CHO; (b) 2 cyclohexanecarbaldehyde (reacts with itself);

(c) $CH_2{=}CHCHO + CH_3CH_2CHO$.

Selectivity can also be achieved in some intramolecular aldol condensations, in which random intermolecular reactions are minimized.

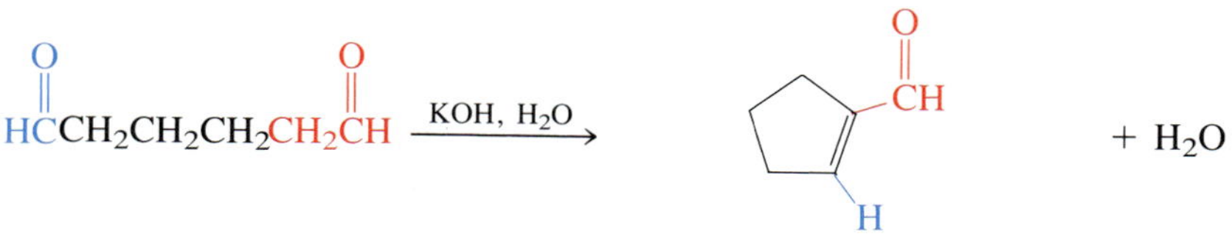

$$\overset{O}{\overset{\|}{H C}}CH_2CH_2CH_2CH_2\overset{O}{\overset{\|}{C}}H \xrightarrow{KOH,\ H_2O} \text{1-cyclopentene-CHO} + H_2O$$

62%
1-Cyclopentenecarboxaldehyde

Ketones in the Aldol Condensation

So far, only aldehydes have been discussed as substrates in the aldol condensation. What about ketones? Treatment of propanone (acetone) with base does indeed lead to some 4-hydroxy-4-methyl-2-pentanone, but only in a few percent yield and in equilibrium with starting material.

Aldol Formation from Propanone (Acetone)

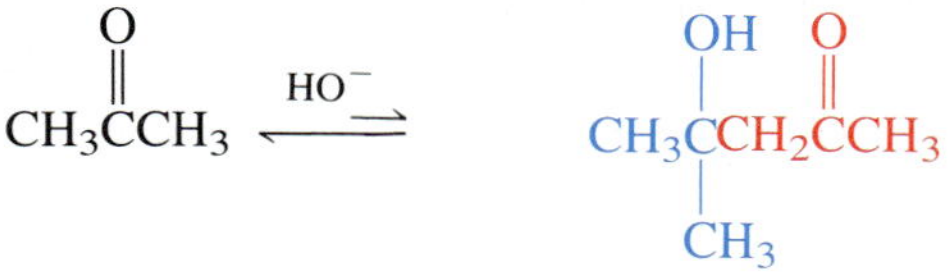

$$CH_3\overset{O}{\overset{\|}{C}}CH_3 \underset{}{\overset{HO^-}{\rightleftharpoons}} CH_3\overset{OH}{\underset{CH_3}{C}}CH_2\overset{O}{\overset{\|}{C}}CH_3$$

94% 6%
4-Hydroxy-4-methyl-2-pentanone

That this is a true equilibrium, which may be approached from either side of the equation, is demonstrated by treating 4-hydroxy-4-methyl-2-pentanone with base. This causes the reverse aldol **(retro-aldol)** reaction to take place rapidly, yielding propanone (acetone).

Retro-Aldol Reaction

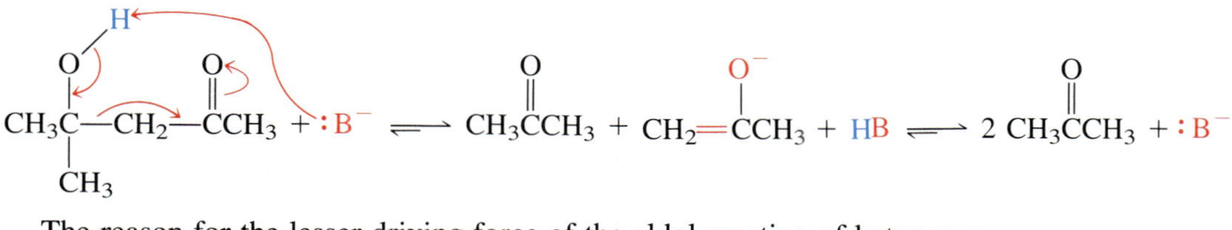

$$CH_3\overset{OH}{\underset{CH_3}{C}}{-}CH_2{-}\overset{O}{\overset{\|}{C}}CH_3 + :B^- \rightleftharpoons CH_3\overset{O}{\overset{\|}{C}}CH_3 + CH_2{=}\overset{O^-}{\overset{|}{C}}CH_3 + HB \rightleftharpoons 2\ CH_3\overset{O}{\overset{\|}{C}}CH_3 + :B^-$$

The reason for the lesser driving force of the aldol reaction of ketones compared with that of aldehydes is the presence of a slightly stronger (about 3 kcal mole^{-1}) carbonyl bond in the former. Thus, whereas the aldol addition of aldehydes is slightly exothermic, the same reaction of ketones is somewhat endothermic. The aldol reaction of ketones can, however, be driven forward by extracting the product alcohol continually from the reaction mixture as it is formed. Alternatively, under more vigorous reaction conditions, dehydration and removal of water move the equilibrium toward the α,β-unsaturated ketone, as shown at the left.

OH O
$CH_3CCH_2CCH_3$ (CH_3)
NaOH, H_2O, Δ
$(H_3C)_2C{=}CHCCH_3$ (O)
80%
4-Methyl-3-penten-2-one
+
H_2O (removed)

Intramolecular ketone condensations are a ready source of cyclic and bicyclic enones. Depending on the reaction conditions and the substrates, either the aldol or the condensation product is isolated.

$CH_3CCH_2CH_2CCH_3$ (O, O) —NaOH, H_2O→ 3-methyl-2-cyclopentenone (CH_3) + H_2O
42%
2,5-Hexanedione **3-Methyl-2-cyclopentenone**

CH_3 —KOH, H_2O, 20°C→ HO —Δ→ + H_2O
90%
2-(3-Oxobutyl)cyclohexanone

In principle, the intramolecular aldol condensation can in many cases give several products with different ring sizes. However, usually the least-strained ring is formed, typically five or six membered. For example, in one of the preceding examples, 2,5-hexanedione gives only 3-methyl-2-cyclopentenone, and none of the alternative three-membered ring product.

$CH_3CCH_2CH_2CCH_3$ (O, O) ⇌ (HO^-, H_2O) H_3C, OH, CCH_3 (O)
2,5-Hexanedione **2-Ethanoyl-1-methylcyclopropanol**

Similarly, 2,7-octanedione cyclizes to a five-membered ring enone, not the seven-membered alternative.

H_2O + (O) —CH_3 ⇌ (KOH, H_2O, *b*) $CH_3CCH_2CH_2CH_2CH_2CCH_3$ (O, O; *a*, *b*) ⇌ (KOH, H_2O, *a*) CH_3, CCH_3 (O) + H_2O
83%
3-Methyl-2-cycloheptenone **2,7-Octanedione** **1-Ethanoyl-2-methylcyclopentene**

The reason for these results is that the aldol addition is reversible and therefore leads to the thermodynamic products.

EXERCISE 16-10

The aldol reaction of 2-(3-oxobutyl)cyclohexanone could have lead to three other aldol adducts. Draw them (ignoring stereochemistry).

EXERCISE 16-11

Predict the outcome of the intramolecular aldol condensations of

(a) cyclodecane-1,5-dione; (b) $C_6H_5\overset{O}{\overset{\|}{C}}(CH_2)_2\overset{O}{\overset{\|}{C}}CH_3$;

(c) 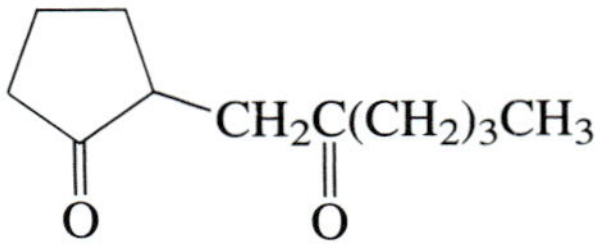 2-[$CH_2\overset{O}{\overset{\|}{C}}(CH_2)_3CH_3$]cyclopentanone

The Aldol Reaction Occurs in Nature

Aldol condensations occur in natural systems. For example, collagen fibers are strengthened by chemical cross-linking of aldehyde units through aldol condensations. Collagen is the most-abundant fibrous protein in mammals, being the major fibrous component of skin, bone, tendon, cartilage, and teeth. One of its functions is to hold cells together in discrete units. Its structure is basically a staggered array of *tropocollagen* molecules, which consist of triply stranded helical polypeptide chains (Section 27-3). Cross-linking (see Sections 12-7 and 14-6 for cross-linking of polymers) of these chains is by aldol condensations catalyzed by enzymes. First, lysine residues (Section 27-1) in the chain are enzymatically oxidized to aldehyde derivatives. Then, an aldol condensation cross-links two chains (Figure 16-2, on the next page).

The extent of cross-linking depends on the function of the tissue. For example, the collagen in the Achilles' tendon of rats is highly cross-linked, but, that of the more flexible tail tendon is less so.

EXERCISE 16-12

Prepare the following compounds from any starting material, using aldol reactions in the crucial step:

(a) 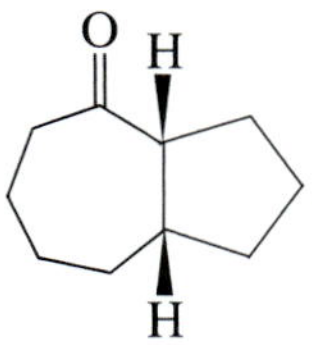 [bicyclic ketone with H, H] ; (b) 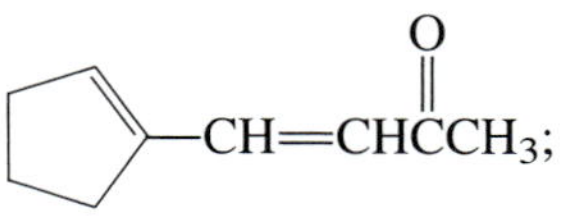 1-cyclopentenyl$-CH{=}CH\overset{O}{\overset{\|}{C}}CH_3$; (c) [tricyclic diol ketone with OH, OH, =O]

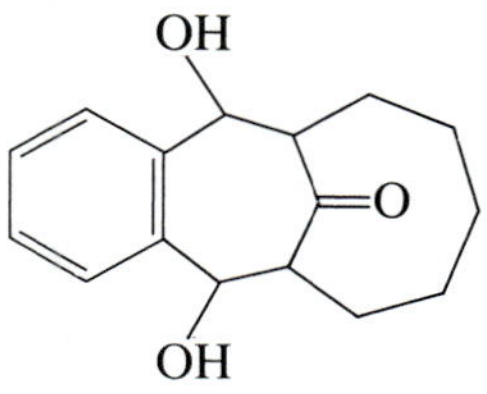

Hint: The last preparation requires a double aldol addition.

In summary, aldehydes and ketones undergo aldol condensation reactions in the presence of catalytic base to give α,β-unsaturated aldehydes and ketones. These reactions proceed by enolate attack on the carbonyl function, resulting first in a hydroxy carbonyl derivative, which is then dehydrated on heating.

FIGURE 16-2

Formation of an aldol cross-link from two lysine side chains.

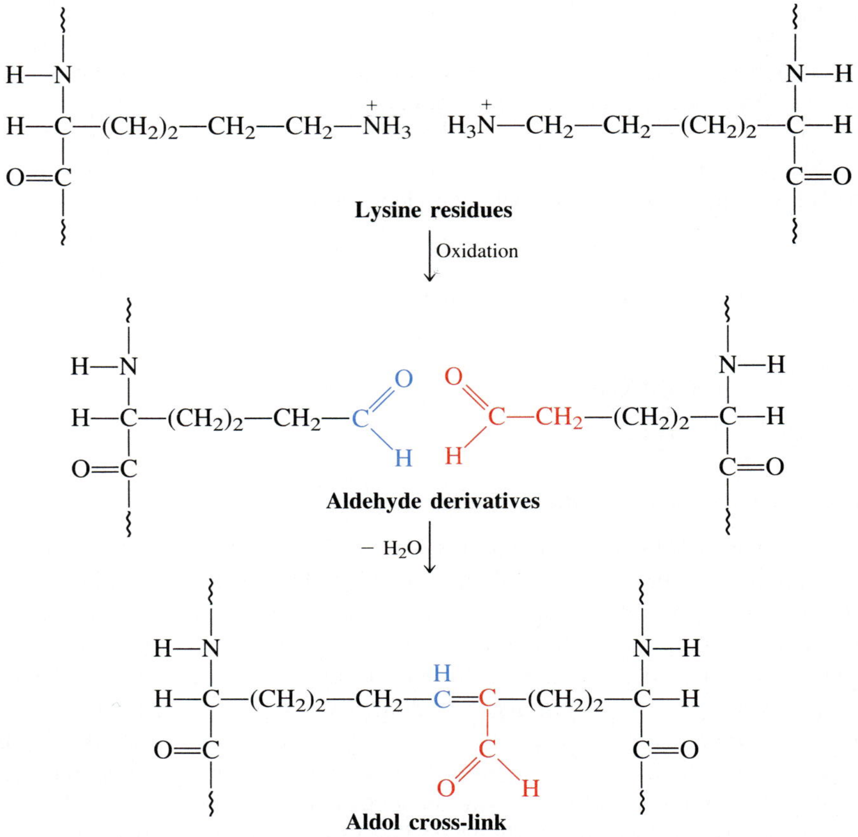

$$\text{H—C(—N—H)(—C=O)—(CH}_2)_2\text{—CH}_2\text{—CH}_2\text{—}\overset{+}{\text{N}}\text{H}_3 \quad \text{H}_3\overset{+}{\text{N}}\text{—CH}_2\text{—CH}_2\text{—(CH}_2)_2\text{—C(—N—H)(—C=O)—H}$$

Lysine residues

Oxidation

$$\text{H—C(—N—H)(—C=O)—(CH}_2)_2\text{—CH}_2\text{—C(=O)H} \quad \text{HC(=O)—CH}_2\text{—(CH}_2)_2\text{—C(—N—H)(—C=O)—H}$$

Aldehyde derivatives

$-\ H_2O$

$$\text{H—C(—N—H)(—C=O)—(CH}_2)_2\text{—CH}_2\text{—CH=C(—CHO)—(CH}_2)_2\text{—C(—N—H)(—C=O)—H}$$

Aldol cross-link

Crossed aldol condensations furnish product mixtures unless one of the reaction partners cannot enolize. Aldol addition to a ketone carbonyl group is energetically unfavorable. To drive the aldol condensation of ketones to product, special conditions have to be used, such as removal of the water or the aldol formed in the reaction. Intramolecular aldol condensation can be highly selective and gives the least-strained cycloalkenones. An example of the aldol reaction in nature is the cross-linking of collagen fibers.

16-4 The Preparation and Chemistry of α,β-Unsaturated Aldehydes and Ketones

α,β-Unsaturated aldehydes and ketones contain two functional groups. As in other difunctional compounds (Chapter 22), their chemistry may be a simple composite of the individual behavior of the two types of double bonds or, as described in Section 16-5, may involve the enone function as a whole. This section begins with a review of the preparation of these molecules.

You Can Prepare α,β-Unsaturated Aldehydes and Ketones by the Use of Familiar Reactions

Section 16-3 showed how α,β-unsaturated aldehydes and ketones are prepared by the aldol condensation. There are other synthetic routes to this class of molecules. For example, the carbon–carbon double bond can be introduced next to the carbonyl function by halogenation (Section 16-2) followed by base-mediated dehydrohalogenation, as shown at the right.

Another way of introducing the double bond is by a Wittig reaction (Section 15-7) of carbonyl ylides. For example, 2-chloroethanal (2-chloroacetaldehyde) can be converted into the phosphonium salt and subsequently deprotonated to the corresponding ylide.

Cyclopentanone $\xrightarrow{1.\ Cl_2;\ 2.\ Na_2CO_3}$ 2-Cyclopentenone (73%)

Synthesis of a Stabilized Ylide

$$ClCH_2\overset{O}{\overset{\|}{C}}H \xrightarrow{P(C_6H_5)_3} Cl^-\,(C_6H_5)_3\overset{+}{P}CH_2\overset{O}{\overset{\|}{C}}H \xrightarrow[-\,HOH,\ -\,NaCl]{NaOH}$$

$$\left[(C_6H_5)_3P{=}CH{-}\overset{:O:}{\overset{\|}{C}}H \longleftrightarrow (C_6H_5)_3\overset{+}{P}{-}\overset{-}{\ddot{C}}H{-}\overset{:O:}{\overset{\|}{C}}H \longleftrightarrow (C_6H_5)_3\overset{+}{P}{-}CH{=}\overset{:\ddot{O}:^-}{\overset{|}{C}}H\right]$$

A stabilized ylide

Deprotonation is facile because the product is stabilized by resonance. Such **stabilized ylides** are comparatively unreactive and can be readily isolated and stored. For example, they do not undergo the Wittig reaction with ketones. However, they do react with aldehydes to form the corresponding α,β-unsaturated aldehydes. Analogous reactions are possible with other alkanoyl ylides.

$$(C_6H_5)_3P{=}CH\overset{O}{\overset{\|}{C}}H + CH_3(CH_2)_5CHO \xrightarrow[-\,(C_6H_5)_3PO]{(CH_3CH_2)_2O,\ \Delta} CH_3(CH_2)_5CH{=}CHCHO$$

Heptanal → 81% ***trans*-2-Nonenal**

A fourth way to prepare α,β-unsaturated aldehydes and ketones is by oxidation of allylic alcohols. A specific reagent for this transformation is manganese dioxide, MnO_2 (Section 15-3). Vitamin A, for example, can be oxidized in this way to all-*trans*-retinal, a molecule of importance in the chemistry of vision (see page 696).

Vitamin A (…CH_2OH) $\xrightarrow{MnO_2,\ \text{propanone (acetone)}}$ (…CHO) 80% **All-*trans*-retinal**

Conjugated Unsaturated Aldehydes and Ketones Are More Stable than Their Unconjugated Isomers

Like conjugated dienes (Section 14-3), α,β-unsaturated aldehydes and ketones are stabilized by resonance.

Resonance in 2-Butenal

$$\left[CH_3CH{=}CH{-}\overset{\overset{\displaystyle :O:}{\|}}{C}H \longleftrightarrow CH_3CH{=}CH{-}\overset{\overset{\displaystyle :\ddot{O}:^-}{|}}{\overset{+}{C}}H \longleftrightarrow CH_3\overset{+}{C}H{-}CH{=}\overset{\overset{\displaystyle :\ddot{O}:^-}{|}}{C}H \right]$$

Thus, unconjugated β,γ-unsaturated carbonyl compounds rearrange readily to their conjugated isomers. The carbon–carbon double bond is said to "move into conjugation" with the carbonyl group.

Isomerization of β,γ-Unsaturated Carbonyl Compounds to Conjugated Systems

$$\underset{\textbf{3-Butenal}}{CH_2{=}CHCH_2\overset{\overset{\displaystyle O}{\|}}{C}H} \xrightarrow{H^+ \text{ or } HO^-,\ H_2O} \underset{\textbf{2-Butenal}}{CH_3CH{=}CH\overset{\overset{\displaystyle O}{\|}}{C}H}$$

$$\textbf{3-Cyclohexenone} \xrightarrow{H^+ \text{ or } HO^-,\ H_2O} \textbf{2-Cyclohexenone}$$

The isomerization can be acid or base catalyzed. The acid-catalyzed pathway proceeds through the conjugated dienol. Protonation at the terminus away from the hydroxy group generates a resonance-stabilized carbocation that is deprotonated at oxygen to give the product.

Mechanism of Acid-Mediated Isomerization of β,γ-Unsaturated Carbonyl Compounds:

$$CH_2{=}CHCH_2\overset{\overset{\displaystyle :O:}{\|}}{C}H \overset{H^+}{\rightleftharpoons} \underset{\textbf{Dienol}}{CH_2{=}CHCH{=}C(\ddot{O}H)H} \overset{H^+}{\rightleftharpoons}$$

$$\left[H{-}CH_2\overset{+}{C}H{-}CH{=}C(\ddot{O}H)H \longleftrightarrow CH_3CH{=}CH{-}C(\overset{+}{\ddot{O}}H)H \right] \overset{-H^+}{\rightleftharpoons} CH_3CH{=}CH\overset{\overset{\displaystyle O}{\|}}{C}H$$

In the base-catalyzed reaction, the intermediate is the conjugated dienolate ion, which is reprotonated at the carbon terminus.

Mechanism of Base-Mediated Isomerization of β,γ-Unsaturated Carbonyl Compounds:

$$CH_2{=}CHCH_2\overset{\overset{\displaystyle :O:}{\|}}{C}H + H\ddot{O}:^- \rightleftharpoons$$

$$\left[CH_2{=}CH{-}\ddot{C}^-H{-}\overset{\overset{\displaystyle :O:}{\|}}{C}H \longleftrightarrow CH_2{=}CH{-}CH{=}\overset{\overset{\displaystyle :\ddot{O}:^-}{|}}{C}H \longleftrightarrow :\bar{C}H_2{-}CH{=}CH{-}\overset{\overset{\displaystyle :O:}{\|}}{C}H\right] + H\ddot{O}H \rightleftharpoons$$

Dienolate ion

$$CH_3CH{=}CH\overset{\overset{\displaystyle :O:}{\|}}{C}H + H\ddot{O}:^-$$

α,β-Unsaturated Aldehydes and Ketones Undergo the Reactions Typical of Their Component Functional Groups

α,β-Unsaturated aldehydes and ketones undergo many reactions that are perfectly predictable from the known chemistry of the carbon–carbon and carbon–oxygen double bonds. For example, hydrogenation by palladium on carbon gives the saturated carbonyl compound.

H_2, Pd-C, $CH_3CO_2CH_2CH_3$ (ethyl ethanoate solvent) →

95%

Certain special catalysts cause the selective reduction of the carbonyl group without affecting the alkene double bond.

$$(H_3C)_2C{=}CH\overset{\overset{\displaystyle O}{\|}}{C}H \xrightarrow{H_2,\ PtO_2,\ FeSO_4,\ Zn(O\overset{O}{\|}{C}CH_3)_2,\ 30\ atm} (H_3C)_2C{=}CH\overset{\overset{\displaystyle H}{|}}{C}HOH$$

3-Methyl-2-butenal → **3-Methyl-2-buten-1-ol**

Electrophilic attack is at the carbon–carbon π system. For example, bromination furnishes a dibromocarbonyl compound.

$$CH_3CH{=}CH\overset{\overset{\displaystyle O}{\|}}{C}CH_3 \xrightarrow{Br{-}Br,\ CCl_4} CH_3\underset{\underset{\displaystyle Br}{|}}{C}H\overset{\overset{\displaystyle Br}{|}}{C}H\overset{\overset{\displaystyle O}{\|}}{C}CH_3$$

60%

3-Penten-2-one → **3,4-Dibromo-2-pentanone**

The carbonyl function undergoes the usual addition reactions. Thus, addition of amines results in the expected condensation products (see, however, Section 16-5).

4-Phenylbut-3-en-2-one $\xrightarrow[-\,H_2O]{NH_2OH,\ H^+}$ Oxime, m.p. 115°C

BOX 16-1

Reactions of Unsaturated Aldehydes in Nature: The Chemistry of Vision

Vitamin A (retinol) is an important nutritional factor in vision. Vitamin A deficiency causes night blindness. Living organisms use an enzyme called *retinol dehydrogenase* to oxidize the vitamin to *trans*-retinal. This molecule is present in the light receptor cells of the human eye, but before it can fulfill its biological function it has to be isomerized by another enzyme, *retinal isomerase,* to give *cis*-retinal.

trans-Retinal $\xrightarrow{\text{Retinal isomerase}}$ *cis*-Retinal

This molecule fits well into the active site of a protein called *opsin* (approximate molecular weight 38,000). *cis*-Retinal reacts with one of the amine substituents of opsin to form the imine *rhodopsin,* the light-sensitive chemical unit in the eye. The electronic spectrum of rhodopsin, with a λ_{max} at 506 nm ($\epsilon = 40{,}000$), has been interpreted as being indicative of the presence of a protonated imine group.

$$RCHO + H_2\ddot{N}{-}(CH_2)_4{-}\text{Protein} \xrightarrow[-\,H_2O]{H^+,\ H_2O} RCH{=}\overset{+}{N}(H){-}(CH_2)_4{-}\text{Protein}$$

cis-Retinal　　Opsin　　Rhodopsin

When a photon strikes rhodopsin, the *cis*-retinal part isomerizes extremely rapidly, in only picoseconds (10^{-12} sec), to the trans isomer. This isomerization induces a tremendous geometric change, which appears to severely disrupt the snug fit of the original molecule in the protein cavity. Within nanoseconds (10^{-9} sec), a series of new intermediates form from this photoproduct, accompanied by conformational changes in the protein structure, followed by eventual hydrolysis of the ill-fitting retinal unit. This sequence initiates a nerve impulse perceived by us as light. The *trans*-retinal is then reisomerized to the cis form by retinal isomerase and reforms rhodopsin, ready for another photon. What is extraordinary about this mechanism is its sensitivity, which allows the eye to register as little as one photon impinging on the retina. Curiously, all known visual systems in nature, even though they might have a completely different evolutionary history, use the retinal system for visual excitation. Evidently, this molecule offers an optimal solution to the problem of vision.

EXERCISE 16-13

Propose a synthesis of 1-pentanol starting from propanal.

In summary, this section reviewed synthetic methods of preparing α,β-unsaturated aldehydes and ketones. These are aldol condensations, halogenation-dehydrohalogenation of saturated aldehydes and ketones, Wittig reactions with stabilized ylides, isomerization of β,γ-unsaturated carbonyl systems, and MnO_2 oxidations of allylic alcohols. The resulting systems undergo the reactions typical of alkenes and carbonyl compounds. Finally, a particular unsaturated aldehyde, retinal, takes part in the chemistry of vision by imine formation and photochemical cis-trans isomerization.

16-5 1,4-Additions to α,β-Unsaturated Aldehydes and Ketones

This section shows how the conjugated carbonyl group of α,β-unsaturated aldehydes and ketones can enter into reactions that involve the entire functional system. These are 1,4-additions of the type encountered with 1,3-butadiene (Section 14-3). Depending on the reagents, the reactions proceed by acid-catalyzed, radical, or nucleophilic addition mechanisms.

1,2- Compared with 1,4-Additions to Conjugated Aldehydes and Ketones

The reactions of α,β-unsaturated aldehydes and ketones described in Section 16-4 can be classified as 1,2-addition to either of the π bonds in the system.

1,2-Addition of a Polar Reagent A–B to a Conjugated Enone

However, several reagents add to the conjugated π system in a 1,4 manner, a result also called **conjugate addition.** In these transformations, the nucleophilic part of a reagent attaches itself to the β-carbon, and the electrophilic part (commonly, a proton) binds to the carbonyl oxygen. The initial product is an enol, which subsequently rearranges to its keto form.

1,4-Addition of a Polar Reagent A–B to a Conjugated Enone

$$\mathrm{>C{=}C{<}C(=O)-} \xrightarrow{\text{A–B}} \mathrm{-C(B)-C{=}C(OA)-} \xrightarrow{\text{A = H}} \mathrm{-C(B)-C(H)-C(=O)-}$$

$$\mathrm{C_6H_5CCH{=}CH_2}\ (\text{C=O})$$
1-Phenylpropenone

$\downarrow$ KCN, H^+

$$\mathrm{C_6H_5C(=O)CH(H)CH_2CN}$$
67%
4-Oxo-4-phenylbutanenitrile

Hydrogen Cyanide Attacks Conjugated Carbonyl Compounds to Give β-Cyanocarbonyls

Treatment of a conjugated aldehyde or ketone with cyanide in the presence of acid may result in attack by cyanide at the β carbon, in contrast with cyanohydrin formation (Section 15-7). Although this transformation appears to give a 1,2-adduct to the C–C double bond, it proceeds through a 1,4-addition pathway, beginning with protonation of the oxygen, then nucleophilic β-attack, and finally enol-keto tautomerization.

Mechanism of Hydrogen Cyanide Addition to an α,β-Unsaturated Carbonyl Compound:

STEP 1: Protonation

$$\mathrm{C_6H_5C(=\ddot{O}:)CH{=}CH_2} \underset{}{\overset{H^+}{\rightleftharpoons}} \left[\mathrm{C_6H_5C(=\overset{+}{O}H)-CH{=}CH_2} \longleftrightarrow \mathrm{C_6H_5\overset{+}{C}(\ddot{O}H)-CH{=}CH_2} \longleftrightarrow \mathrm{C_6H_5C(\ddot{O}H){=}CH-\overset{+}{C}H_2}\right]$$

STEP 2: Cyanide attack

$$\mathrm{C_6H_5C(=\overset{+}{O}H)-CH{=}CH_2} + \mathrm{:\overset{-}{C}{\equiv}N:} \rightleftharpoons \mathrm{C_6H_5C(\ddot{O}H){=}CH-CH_2CN}$$

STEP 3: Enol-keto tautomerization

$$\mathrm{C_6H_5C(\ddot{O}H){=}CHCH_2CN} \rightleftharpoons \mathrm{C_6H_5C(=\ddot{O}:)CH(H)CH_2CN}$$

Conjugate Additions of Oxygen and Nitrogen Nucleophiles

Water, alcohols, and amines can be induced to undergo 1,4-additions. Although these reactions can be catalyzed by acid or base, the products are usually formed faster and in higher yields with base.

$$\underset{\textbf{3-Buten-2-one}}{\mathrm{CH_2{=}CHC(=O)CH_3}} \underset{}{\overset{\text{HOH, Ca(OH)}_2,\ -5^\circ\text{C}}{\rightleftharpoons}} \underset{\textbf{4-Hydroxy-2-butanone}}{\mathrm{HOCH_2CH(H)C(=O)CH_3}}$$

$$\mathrm{CH_3CH{=}CH\overset{O}{\overset{\|}{C}}H \xrightleftharpoons{CH_3OH,\ CH_3O^-K^+} CH_3\underset{H}{\underset{|}{C}}H\overset{CH_3O}{\overset{|}{C}}H\overset{O}{\overset{\|}{C}}H}$$

50%

2-Butenal **3-Methoxybutanal**

$$\mathrm{(H_3C)_2C{=}CH\overset{O}{\overset{\|}{C}}CH_3 \xrightleftharpoons{CH_3NH_2,\ H_2O} (CH_3)_2\overset{CH_3NH}{\overset{|}{C}}\underset{H}{\underset{|}{C}}H\overset{O}{\overset{\|}{C}}CH_3}$$

75%

4-Methyl-3-penten-2-one **4-Methyl-4-(methylamino)-2-pentanone**

Note that the hydration of an α,β-unsaturated carbonyl compound is the reverse of the second step of the aldol condensation. Indeed, at elevated temperatures, the 1,4-addition becomes reversible, and other products may be formed—for example, derived from aldol or amine condensation reactions (Section 16-4).

The mechanism of the base-catalyzed addition to conjugated aldehydes and ketones is direct nucleophilic attack at the β-carbon to give the enolate ion, which is subsequently protonated.

Mechanism of Base-Catalyzed Hydration of α,β-Unsaturated Aldehydes and Ketones:

$$\mathrm{HO^- + \;>C{=}C{<}C(=O) \rightleftharpoons HO{-}\overset{|}{C}{-}\overset{|}{C}{=}C{-}O^- \xrightleftharpoons{HOH} HO{-}\overset{|}{C}{-}\overset{|}{C}H{-}\overset{O}{\overset{\|}{C}}{-} + {}^-OH}$$

EXERCISE 16-14

Treatment of 3-chloro-2-cyclohexenone with sodium methoxide in methanol gave 3-methoxy-2-cyclohexenone. Write the mechanism of this reaction.

EXERCISE 16-15

Suggest a mechanism for the following reaction:

$$\mathrm{CH_2{=}CH\overset{O}{\overset{\|}{C}}CH_3 + NH_2NH_2 \xrightarrow{HCl,\ H_2O}}$$ 3-methyl-2-pyrazoline (ring with CH$_3$, N, N–H) $+ \; H_2O$

Organometallic Reagents Add in 1,2 or 1,4 Manner

Organometallic reagents may add to the α,β-unsaturated carbonyl function in either 1,2 or 1,4 fashion. Organolithium reagents, for example, react preferentially by direct nucleophilic attack at the carbonyl carbon.

$$\mathrm{(H_3C)_2C{=}CHCCH_3}\ (\text{C}{=}\text{O}) + \xrightarrow[\text{2. } H^+,\ H_2O]{\text{1. } CH_3Li,\ (CH_3CH_2)_2O} \mathrm{(H_3C)_2C{=}CHC(OH)(CH_3)CH_3}$$

4-Methyl-3-penten-2-one → 81% **2,4-Dimethyl-3-penten-2-ol**

On the other hand, cuprates give only products of conjugate addition.

$$\mathrm{CH_3(CH_2)_5CH{=}C(CH_3)CH{=}O} \xrightarrow[\text{2. } H^+,\ H_2O]{\text{1. } (CH_3)_2CuLi,\ THF,\ -78^\circ C,\ 4\ h} \mathrm{CH_3(CH_2)_5CH(CH_3)CH(CH_3)CH{=}O}$$

2-Methyl-2-nonenal → 40% **2,3-Dimethylnonanal**

$$\text{2-Cyclohexenone} \xrightarrow[\text{2. } H^+,\ H_2O]{\text{1. } (CH_2{=}CH)_2CuLi,\ THF,\ -78^\circ C} \text{3-}(CH{=}CH_2)\text{cyclohexanone}$$

2-Cyclohexenone → 65% **3-Ethenylcyclohexanone (3-Vinylcyclohexanone)**

The copper-mediated 1,4-addition reactions are thought to proceed through rapid and complex electron transfer mechanisms involving radicals and, possibly, additional organocopper species. The first isolable intermediate is an enolate ion, which can be trapped by alkylating species as shown in Section 16-1. Conjugate addition followed by alkylation constitutes a useful sequence for α,β-dialkylation of unsaturated aldehydes and ketones.

General α,β-Dialkylation of Unsaturated Carbonyl Compounds

$$\mathrm{{>}C{=}C{<}C(=O){-}} \xrightarrow[\text{2. } R'X]{\text{1. } R_2CuLi} \mathrm{{-}C(R){-}C(R'){-}C(=O){-}}$$

Example:

$$\text{2-Cyclohexenone} \xrightarrow[\text{2. } CH_3I]{\text{1. } (CH_3CH_2CH_2CH_2)_2CuLi,\ THF} \text{trans-2-}CH_3\text{-3-}CH_2CH_2CH_2CH_3\text{-cyclohexanone} + \text{cis isomer}$$

84%, 4:1

***trans*- and *cis*-3-Butyl-2-methylcyclohexanone**

BOX 16-2

α,β-Dialkylation in the Synthesis of Natural Products

The α,β-dialkylation procedure has been exploited in the total synthesis of **prostaglandins,** powerful physiologically active compounds (see Section 17-12). They appear to regulate a remarkable variety of bodily functions, including those of the endocrine, reproductive, nervous, digestive, hemostatic, respiratory, cardiovascular, and renal systems. Because of these properties, they are potential drugs in the treatment of hypertension, asthma, fever, inflammations, and ulcers. One of the commercially available prostaglandins induces labor in pregnant women. Others have applications in animal breeding by controlling the day in which the animal goes into heat.

A synthesis of prostaglandin $PGF_{2\alpha}$, developed by Stork,* includes two conjugate additions as well as an aldol reaction.

A Prostaglandin Synthesis

(R, R′, R″ are protecting groups; C_5H_{11} = pentyl)

O RO + LiCu (C_5H_{11} OR′)$_2$ $\xrightarrow[\text{1,4-Addition}]{(CH_3CH_2)_2O,\ -78°C}$ O^-Li^+ RO C_5H_{11} OR′ $\xrightarrow[\text{Aldol condensation}]{1.\ CH_2{=}O,\ -50°C \quad 2.\ H^+,\ H_2O}$

O CH_2 RO C_5H_{11} OR′ $\xrightarrow[\text{1,4-Addition}]{1.\ [R''O(CH_2)_4\ \ ]_2CuLi \quad 2.\ H^+,\ H_2O}$

O CH_2 $(CH_2)_4OR''$ RO C_5H_{11} OR′ ⟶ ⟶ OH CH_2 $(CH_2)_3CO_2H$ HO C_5H_{11} OH

$PGF_{2\alpha}$

EXERCISE 16-16

Show how you might synthesize the following compounds from 3-methyl-2-cyclohexenone:

(a) O, CH_3, CH_3, CH_3 (b) O, CH_3

Hints: Work backward, and the last step in part b is an intramolecular aldol condensation.

*Professor Gilbert Stork, b. 1921, Columbia University.

Enolate Ions Enter into Conjugate Additions: The Michael Reaction and Robinson Annelation

Like other nucleophiles, enolate ions undergo conjugate additions to α,β-unsaturated aldehydes and ketones, in a reaction known as the **Michael* reaction.** This transformation works best with enolates derived from β-dicarbonyl compounds (Section 22-3), but it also works with simpler systems.

$$CH_3\overset{O}{\overset{\|}{C}}CH_2\overset{O}{\overset{\|}{C}}CH_3 + CH_2{=}CH\overset{O}{\overset{\|}{C}}H \xrightarrow{\text{Pyridine}} (CH_3\overset{O}{\overset{\|}{C}})_2CHCH_2CH_2\overset{O}{\overset{\|}{C}}H \quad 27\%$$

$$\text{2-methylcyclohexanone} + CH_2{=}CH\overset{O}{\overset{\|}{C}}C_6H_5 \xrightarrow{CH_3CH_2O^-K^+,\ CH_3CH_2OH,\ (CH_3CH_2)_2O} \text{2-methyl-2-}(CH_2CH_2\overset{O}{\overset{\|}{C}}C_6H_5)\text{cyclohexanone} \quad 64\%$$

The mechanism of the Michael reaction includes nucleophilic attack by the enolate ion at the β-carbon of the unsaturated carbonyl compound (the Michael "acceptor"), followed by protonation.

Mechanism of the Michael Reaction:

$$^-{:}\ddot{O}{:}\text{–}C{=}C\langle \; + \; \rangle C{=}C\text{–}C{=}\ddot{O}{:} \longrightarrow {:}\ddot{O}{=}C\text{–}C\text{–}C\text{–}C{=}C\text{–}\ddot{O}{:}^- \xrightarrow[-\ HO^-]{HOH} {:}\ddot{O}{=}C\text{–}C\text{–}C\text{–}C(H)\text{–}C{=}\ddot{O}{:}$$

As the mechanism indicates, the reaction works because of the nucleophilic potential of the β-carbon of an enolate and the electrophilic reactivity of the β-carbon of an α,β-unsaturated carbonyl compound.

With some Michael acceptors, such as 3-buten-2-one, the products of the initial addition are capable of a subsequent intramolecular aldol condensation, which creates a new ring.

$$\text{2-methylcyclohexanone} + \underset{\textbf{3-Buten-2-one}}{CH_2{=}CH\overset{O}{\overset{\|}{C}}CH_3} \xrightarrow[\text{Michael addition}]{CH_3CH_2O^-Na^+,\ CH_3CH_2OH,\ (CH_3CH_2)_2O,\ -10^\circ C}$$

$$\text{2-methyl-2-(3-oxobutyl)cyclohexanone} \xrightarrow{\text{Intramolecular aldol condensation}} \text{bicyclic hydroxy ketone (CH}_3\text{, OH)}\ 54\% \xrightarrow[-\ HOH]{\Delta,\ HO^-} \text{bicyclic enone (CH}_3\text{)}\ 86\%$$

*Professor Arthur Michael, 1853–1942, Harvard University.

The synthetic sequence of a Michael addition followed by an intramolecular aldol condensation is also called a Robinson* annelation.

General Robinson Annelation

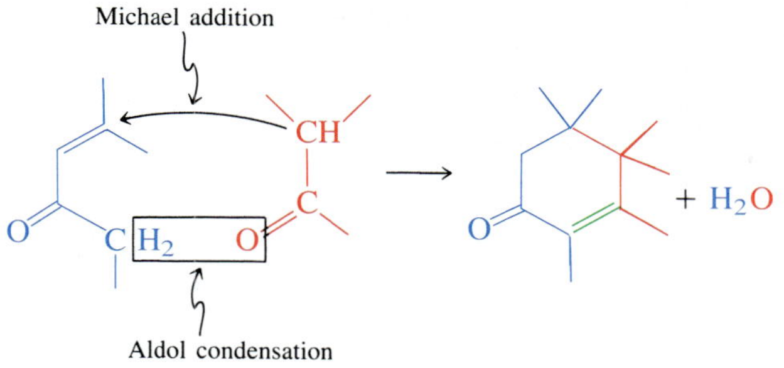

Michael addition

$+ H_2O$

Aldol condensation

The Robinson annelation has found extensive use in the synthesis of polycyclic ring systems, including steroids.

Steroid Synthesis by Robinson Annelation

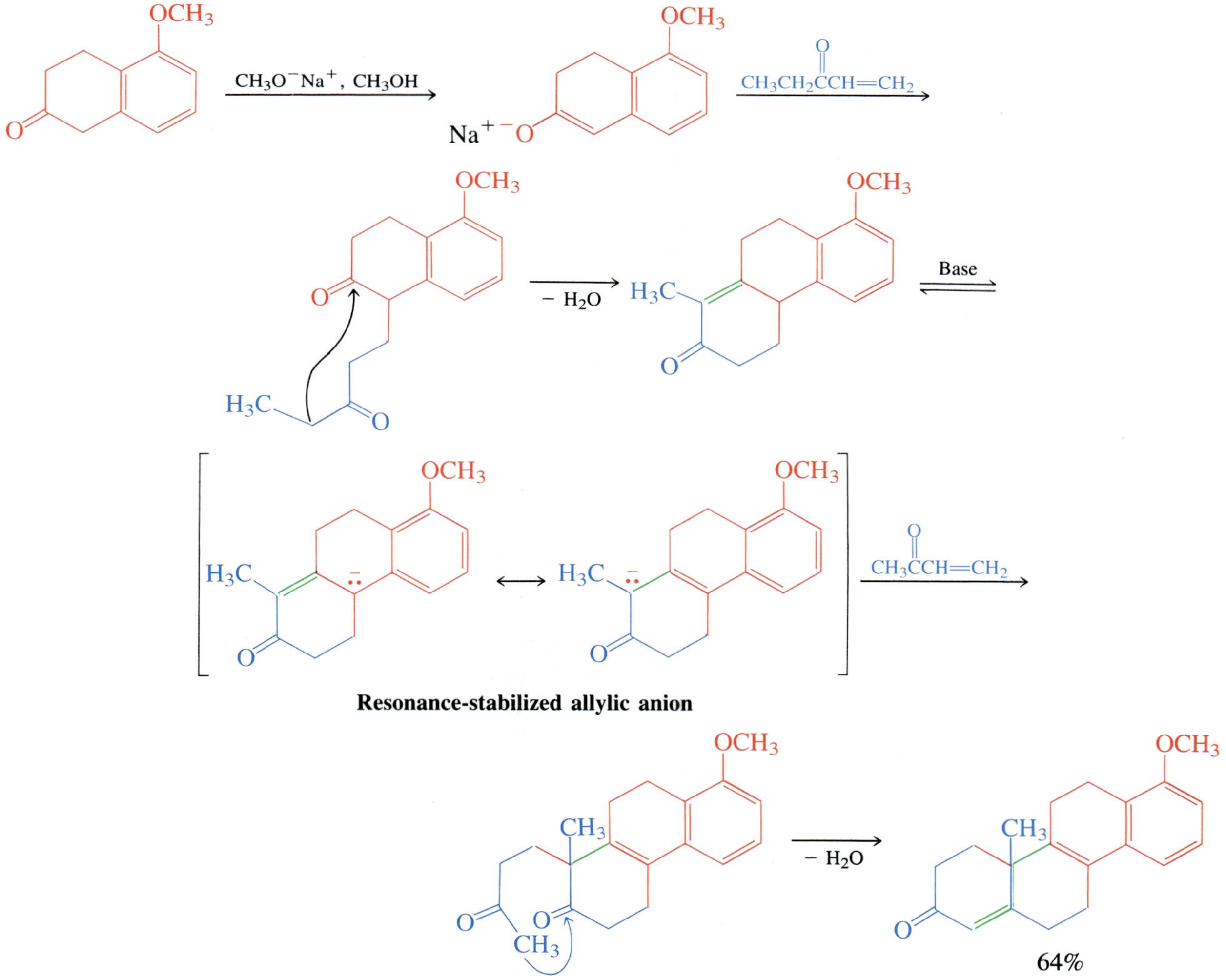

OCH_3

$CH_3O^-Na^+$, CH_3OH

$Na^+ {}^-O$

$CH_3CH_2CCH{=}CH_2$

$-H_2O$

Base

$CH_3CCH{=}CH_2$

Resonance-stabilized allylic anion

$-H_2O$

64%

*Sir Robert Robinson, 1886–1975, Oxford University, Nobel Prize 1947.

EXERCISE 16-17

Enamines also enter into Michael reactions. Explain the following transformation by a mechanism.

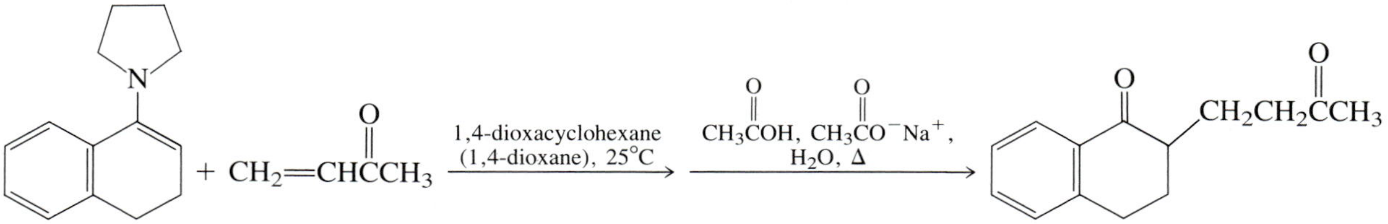

$$\text{(1-pyrrolidinyl-3,4-dihydronaphthalene)} + CH_2{=}CH\overset{O}{\overset{\|}{C}}CH_3 \xrightarrow[\text{(1,4-dioxane), 25°C}]{\text{1,4-dioxacyclohexane}} \xrightarrow[H_2O,\ \Delta]{CH_3\overset{O}{\overset{\|}{C}}OH,\ CH_3\overset{O}{\overset{\|}{C}}O^-Na^+,} \text{(2-tetralone derivative with } CH_2CH_2\overset{O}{\overset{\|}{C}}CH_3\text{)}$$

EXERCISE 16-18

Propose syntheses of the following compounds by Michael or Robinson reactions:

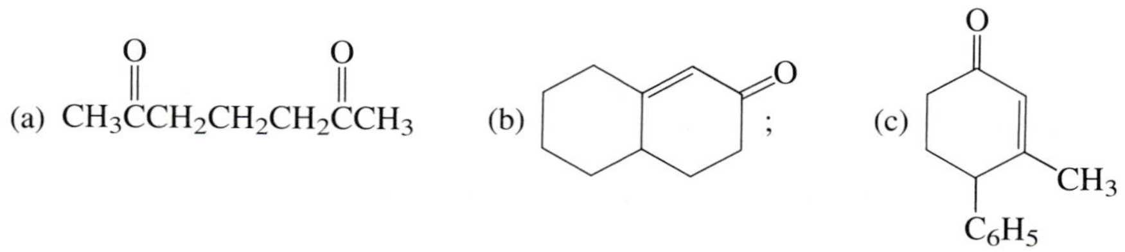

(a) $CH_3\overset{O}{\overset{\|}{C}}CH_2CH_2CH_2\overset{O}{\overset{\|}{C}}CH_3$ (b) [bicyclic enone] ; (c) [cyclohexenone with CH_3 and C_6H_5]

One-Electron Reduction of α,β-Unsaturated Aldehydes and Ketones

The double bond of some conjugated enones and enals can undergo ''conjugate reduction'' by a reducing system described earlier in the conversion of alkynes into trans alkenes: an alkali metal in liquid ammonia (Section 13-6).

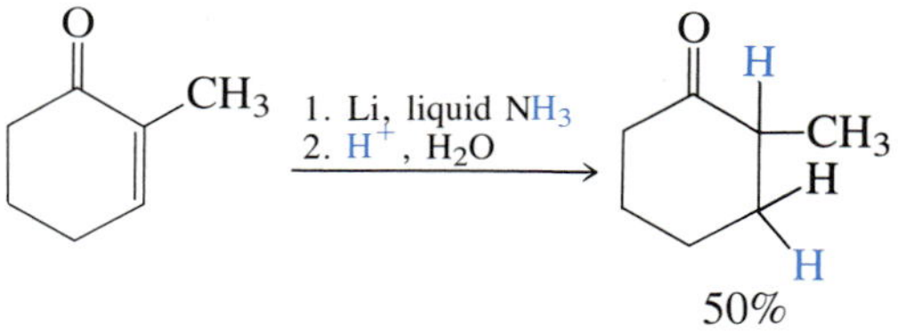

$$\text{2-methylcyclohexenone} \xrightarrow[2.\ H^+,\ H_2O]{1.\ Li,\ liquid\ NH_3} \text{2-methylcyclohexanone (H added at C2 and C3)}\quad 50\%$$

The mechanisms of these reductions are similar and include two one-electron transfers and two protonations. In the reduction of enones, the first electron transfer gives a resonance-stabilized enolate radical ion that is basic enough to be protonated by the ammonia solvent to furnish the corresponding radical. Further reduction gives the enolate ion, which is protonated on aqueous work-up. Note that, in contrast with the pinacol reaction (Section 15-8), there is no coupling in this reduction.

Mechanism of the Metal-Ammonia Reduction of Conjugated Enones and Enals:

(Note: The stereochemistry around the double bonds is not indicated.)

STEP 1: First one-electron transfer

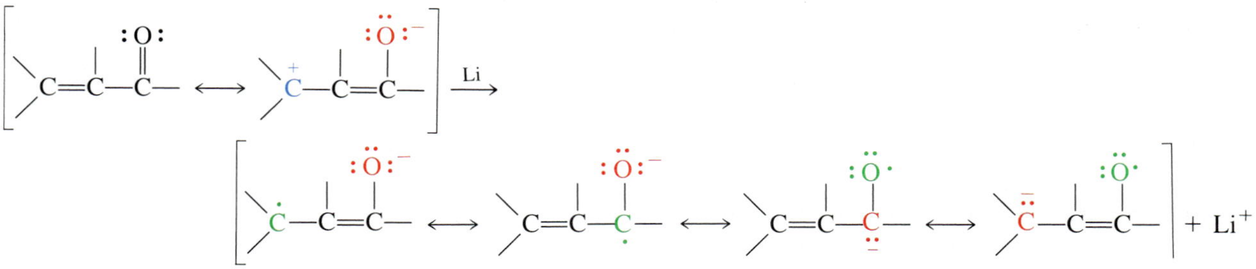

$$\left[>C{=}C{-}\overset{:O:}{\overset{\|}{C}}{-} \longleftrightarrow >\overset{+}{C}{-}C{=}\overset{:\ddot{O}:^-}{C}{-} \right] \xrightarrow{Li}$$

$$\left[>\dot{C}{-}C{=}\overset{:\ddot{O}:^-}{C}{-} \longleftrightarrow >C{=}C{-}\overset{:\ddot{O}:^-}{\dot{C}}{-} \longleftrightarrow >C{=}C{-}\overset{:\dot{O}\cdot}{\ddot{C}^-}{-} \longleftrightarrow >\ddot{C}^-{-}C{=}\overset{:\dot{O}\cdot}{C}{-} \right] + Li^+$$

STEP 2: Protonation

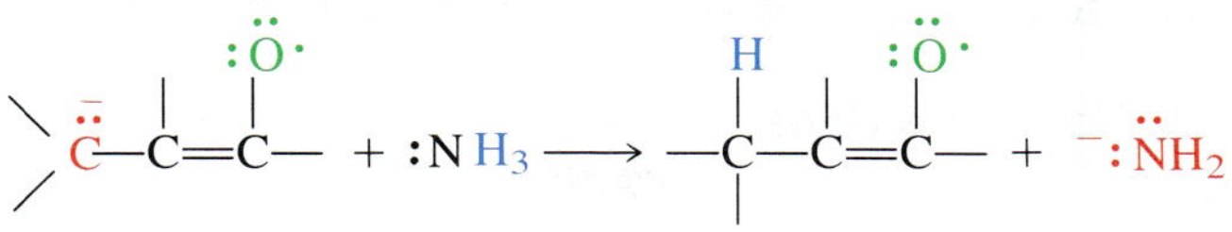

STEP 3: Second one-electron transfer

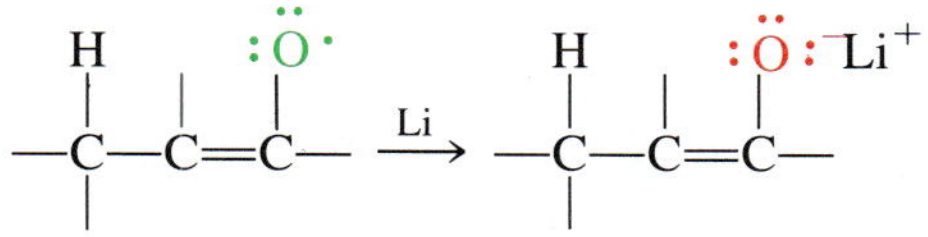

STEPS 4 AND 5: Protonation and enol-keto tautomerization by aqueous work-up

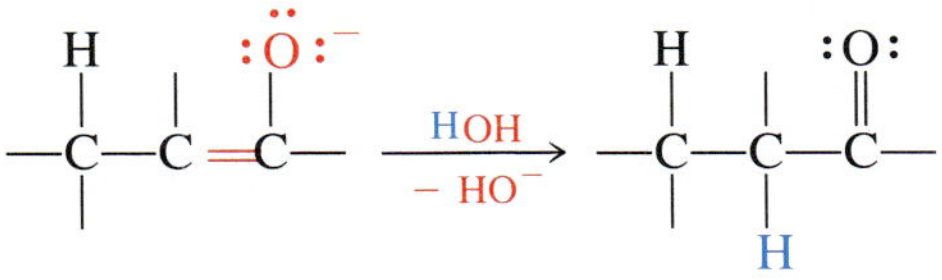

The method allows for the selective reduction of the conjugated double bond in the presence of unconjugated ones.

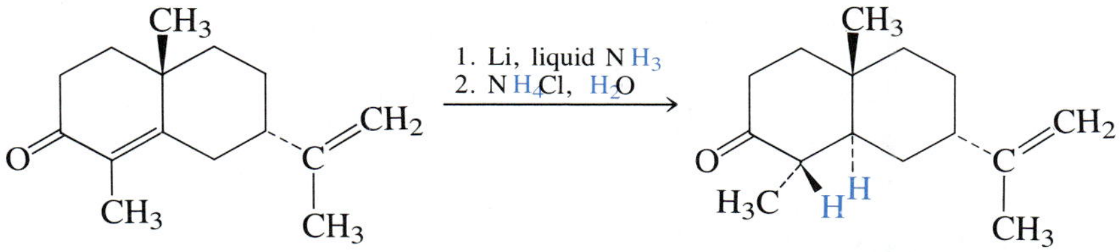

1. Li, liquid NH_3
2. NH_4Cl, H_2O

98%

Another useful feature of this reduction is the initial generation of an enolate ion. An alternative to working it up by protonation is to quench the reaction mixture with an alkylating agent.

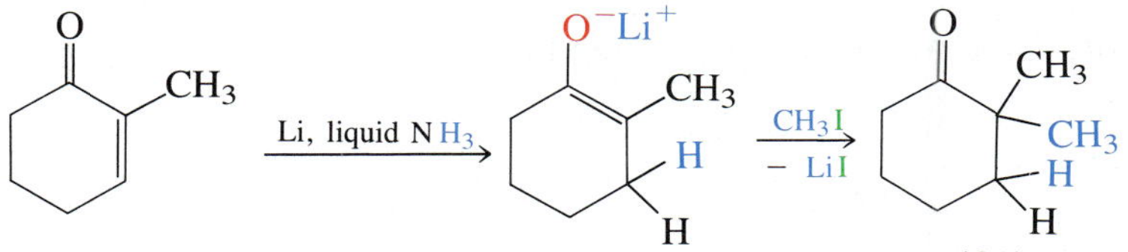

Li, liquid NH_3

CH_3I, $-$ LiI

40%

2-Methyl-2-cyclohexenone

2,2-Dimethylcyclohexanone

In summary, α,β-unsaturated aldehydes and ketones are synthetically useful building blocks in organic synthesis because of their ability to undergo 1,4-additions. Hydrogen cyanide addition leads to β-cyano carbonyl compounds; oxygen and nitrogen nucleophiles can add to the β-carbon; and organocuprates furnish β-alkylated derivatives after aqueous work-up, or α,β-dialkylated aldehydes and ketones after alkylation with a haloalkane. The Michael reaction results in the conjugate addition of an enolate ion to give dicarbonyl compounds. The Robinson annelation reaction combines a Michael addition with a subsequent intramolecular aldol condensation to give new cyclic enones. Finally, the reduction of conjugated aldehydes and ketones may proceed in 1,4-

manner with lithium in liquid ammonia through one-electron transfer steps and the intermediate formation of enolate ions before aqueous work-up.

Summary of New Reactions

Synthesis and Reactions of Enolates and Enols

1 ENOLATE IONS

$$\mathrm{RCH_2\overset{O}{\overset{\|}{C}}R'} \xrightarrow[\text{or other strong base, } -78^\circ\mathrm{C}]{\mathrm{LDA\ or\ KH\ or\ (CH_3)_3CO^-K^+}} \mathrm{RCH{=}C(O^-)R'}$$

Enolate ion

2 ENOLATE ALKYLATION

$$\mathrm{RCH{=}C(O^-)R'} \xrightarrow[-\,X^-]{R''X} \mathrm{RCH(R'')\overset{O}{\overset{\|}{C}}R'}$$

3 ENAMINE ALKYLATION

$$\mathrm{{>}C{=}C(NR_2){<}} \xrightarrow{R'X} \mathrm{{-}C(R'){-}C{=}\overset{+}{N}R_2}\ \ X^- \xrightarrow{H^+,\ H_2O} \mathrm{{-}C(R'){-}C{=}O} + \mathrm{R_2NH}$$

4 KETO-ENOL EQUILIBRIA

$$\mathrm{RCH_2\overset{O}{\overset{\|}{C}}R'} \xrightleftharpoons{\text{Catalytic } H^+ \text{ or } HO^-} \mathrm{RCH{=}C(OH)R'}$$

5 HYDROGEN-DEUTERIUM EXCHANGE

$$\mathrm{RCH_2\overset{O}{\overset{\|}{C}}R'} \xrightarrow{D_2O,\ DO^- \text{ or } D^+} \mathrm{RCD_2\overset{O}{\overset{\|}{C}}R'}$$

6 STEREOISOMERIZATION

$$\mathrm{(R)(H)(R')C{-}\overset{O}{\overset{\|}{C}}R''} \xrightleftharpoons{H^+ \text{ or } HO^-} \mathrm{(R)(R')(H)C{-}\overset{O}{\overset{\|}{C}}R''}$$

7 HALOGENATION

$$\mathrm{RCH_2\overset{O}{\overset{\|}{C}}R'} \xrightarrow[-\,HX]{X_2,\ H^+} \mathrm{RCH(X)\overset{O}{\overset{\|}{C}}R'}$$

Haloform reaction:

$$\mathrm{R\overset{O}{\overset{\|}{C}}CH_3 \xrightarrow{X_2,\ HO^-} R\overset{O}{\overset{\|}{C}}CX_3 \xrightarrow{HO^-} R\overset{O}{\overset{\|}{C}}O^- + CHX_3}$$

8 ALDOL CONDENSATIONS

$$\mathrm{2\ RCH_2\overset{O}{\overset{\|}{C}}H \underset{}{\overset{HO^-}{\rightleftharpoons}} \underbrace{RCH_2\overset{OH}{\overset{|}{\underset{H}{\underset{|}{C}}}}-\overset{}{\underset{R}{\underset{|}{C}}}H\overset{O}{\overset{\|}{C}}H}_{\textbf{Aldol adduct}} \overset{HO^-,\ \Delta}{\rightleftharpoons} \underbrace{RCH_2CH{=}C(CHO)(R)}_{\textbf{Condensation product}} + H_2O}$$

Mixed aldol condensation (one aldehyde not enolizable):

$$\mathrm{R\overset{O}{\overset{\|}{C}}H + R'CH_2\overset{O}{\overset{\|}{C}}H \xrightarrow[-\ H_2O]{HO^-,\ \Delta} RCH{=}C(R')(CHO)}$$

Ketones:

$$\mathrm{R\overset{O}{\overset{\|}{C}}CH_2R' \overset{HO^-}{\rightleftharpoons} R\overset{OH}{\overset{|}{\underset{CH_2R'}{\underset{|}{C}}}}-\overset{R'}{\overset{|}{C}}H-\overset{O}{\overset{\|}{C}}R \xrightarrow[\text{Drive equilibrium}]{-\ H_2O} R\underset{CH_2R'}{\underset{|}{C}}{=}\overset{R'}{\overset{|}{C}}-\overset{O}{\overset{\|}{C}}R}$$

Retro-aldol condensation:

$$\mathrm{{>}C{=}C{<}C(=O){-} \overset{HO^-,\ H_2O}{\rightleftharpoons} -\underset{HO}{\underset{|}{\overset{|}{C}}}-\underset{H}{\underset{|}{\overset{|}{C}}}-C(=O){-} \overset{HO^-,\ H_2O}{\rightleftharpoons} {>}C{=}O + H-\underset{H}{\underset{|}{\overset{|}{C}}}-C(=O){-}}$$

Intramolecular aldol condensation:

$$\mathrm{(CH_2)_n\text{-linked } \overset{O}{\overset{\|}{C}}CH_2R \text{ and } \overset{O}{\overset{\|}{C}}R' \xrightarrow[-\ H_2O]{HO^-} \text{cyclic enone } (CH_2)_n,\ \text{with } R \text{ and } R'}$$

Unstrained rings preferred

9 SYNTHESIS OF α,β-UNSATURATED ALDEHYDES AND KETONES

Aldol condensation: see preceding reactions

Bromination-dehydrobromination of aldehydes and ketones:

$$\mathrm{RCH_2CH_2\overset{O}{\overset{\|}{C}}R' \xrightarrow{1.\ X_2,\ H^+ \quad 2.\ Base} RCH{=}CH\overset{O}{\overset{\|}{C}}R'}$$

Wittig reaction with stabilized ylides:

$$\mathrm{RCH{=}O} + (C_6H_5)_3P{=}CHC(=O)R' \longrightarrow (R)(H)C{=}CHC(=O)R' + (C_6H_5)_3P{=}O$$

Oxidation of allylic alcohols:

$$>C{=}C<\!\!-\!C(HO)(H)\!- \xrightarrow{MnO_2} >C{=}C<\!\!-\!C(=O)\!-$$

Isomerization of β,γ-unsaturated aldehydes and ketones to conjugated carbonyl compounds:

$$RCH{=}CHCH_2CH(=O) \xrightarrow{H^+ \text{ or } HO^-} RCH_2CH{=}CHCH(=O)$$

Reactions of α,β-Unsaturated Aldehydes and Ketones

10 REDUCTIONS

Hydrogenation:

$$>C{=}C<\!\!-\!C(=O)\!- \xrightarrow{H_2,\ Pd} -\overset{|}{\underset{H}{C}}-\overset{|}{\underset{H}{C}}-C(=O)-$$

$$>C{=}C<\!\!-\!C(=O)\!- \xrightarrow{H_2,\ PtO_2,\ FeSO_4,\ Zn(OC(=O)CH_3)_2,\ 30\ atm} >C{=}C<\!\!-\!C(HO)(H)\!-$$

One-electron transfer reduction:

$$>C{=}C<\!\!-\!C(=O)\!- \xrightarrow{Li,\ liquid\ NH_3} -\overset{|}{\underset{H}{C}}-\overset{|}{\underset{H}{C}}-C(=O)-$$

11 ADDITION OF HALOGEN

$$>C{=}C<\!\!-\!C(=O)\!- \xrightarrow{X_2} -\overset{|}{\underset{X}{C}}-\overset{|}{\underset{X}{C}}-C(=O)-$$

12 CONDENSATIONS WITH AMINE DERIVATIVES

$$>C{=}C{<}C({=}O){-} \xrightarrow{ZNH_2} >C{=}C{<}C({=}N{-}Z){-}$$

Z = OH, NH_2, RNH, R, etc.

Conjugate Additions to α,β-Unsaturated Aldehydes and Ketones

13 HYDROGEN CYANIDE ADDITION

$$>C{=}C{<}C({=}O){-} \xrightarrow{KCN,\ H^+} -C(CN){-}C(H){-}C({=}O){-}$$

14 WATER, ALCOHOLS, AMINES

$$>C{=}C{<}C({=}O){-} \xrightarrow{H_2O} -C(OH){-}C(H){-}C({=}O){-}$$

$$>C{=}C{<}C({=}O){-} \xrightarrow{ROH} -C(OR){-}C(H){-}C({=}O){-}$$

$$>C{=}C{<}C({=}O){-} \xrightarrow{RNH_2} -C(RNH){-}C(H){-}C({=}O){-}$$

15 ORGANOMETALLIC REAGENTS

$$>C{=}C{<}C({=}O){-} \xrightarrow[\text{2. } H^+,\ H_2O]{\text{1. RLi}} >C{=}C{<}C(HO)(R){-}$$

1,2-Addition

$$>C{=}C{<}C({=}O){-} \xrightarrow[\text{2. } H^+,\ H_2O]{\text{1. } R_2CuLi} -C(R){-}C(H){-}C({=}O){-}$$

1,4-Addition

Cuprate additions followed by enolate alkylations:

$$\mathrm{>C{=}C(-)-C(=O)-} \xrightarrow[\text{2. R'X}]{\text{1. } R_2CuLi} \mathrm{-C(R)-C(R')-C(=O)-}$$

16 MICHAEL REACTION

$$\mathrm{^{-}O-C(R){=}C<} + \mathrm{>C{=}C-C(=O)R'} \longrightarrow \xrightarrow{H^+} \mathrm{RC(=O)-C-C-C-C(=O)R'}$$

17 ROBINSON ANNELATION

$$\xrightarrow{HO^-}$$

18 REDUCTION-ALKYLATION

$$\mathrm{>C{=}C-C(=O)-} \xrightarrow[\text{2. RX}]{\text{1. Li, liquid } NH_3} \mathrm{-C(H)-C(R)-C(=O)-}$$

Summary of Important Concepts

1 Hydrogens next to the carbonyl group are acidic because of the electron-withdrawing nature of the functional group and because the resulting enolate ion is resonance stabilized.

2 Electrophilic attack on enolates may occur at both the carbon and the oxygen. Haloalkanes usually prefer the former.

3 Enamines are neutral analogs of enolates. They can be β-alkylated to give iminium cations that hydrolyze to aldehydes and ketones on aqueous work-up.

4 Aldehydes and ketones are in equilibrium with their enol forms; the conversion is catalyzed by acid or base. This allows for facile α-deuteration and stereochemical equilibration.

5 α-Halogenation of carbonyl compounds may be acid or base catalyzed. With acid, the enol is halogenated by attack at the double bond; subsequent renewed enolization is slowed down by the halogen substituent. With base, the enolate is attacked at carbon and subsequent enolate formation is accelerated by the halogens introduced.

6 Enolates are nucleophilic and reversibly attack the carbonyl carbon of aldehydes and ketones in the aldol reaction, and the β-carbon of α,β-unsaturated carbonyl compounds in the Michael reaction.

7 Carbonyl ylides are stabilized because of resonance.

8 α,β-Unsaturated aldehydes and ketones show the normal chemistry of each individual double bond, but the entire conjugated system may react as a whole, as revealed by the ability of these compounds to undergo acid- and base-mediated 1,4-additions. Cuprates alkylate at the β-position, presumably through an electron-transfer process. β-Protonation in reductions with lithium in liquid ammonia may be followed by α-alkylation of the resulting enolate ions.

Problems

1 Write the structures of (i) every enol and (ii) enolate ion that can arise from each of the carbonyl compounds below.

(a) $CH_3CH_2\overset{O}{\overset{\|}{C}}CH_2CH_3$

(b) $CH_3\overset{O}{\overset{\|}{C}}CH(CH_3)_2$

(c) 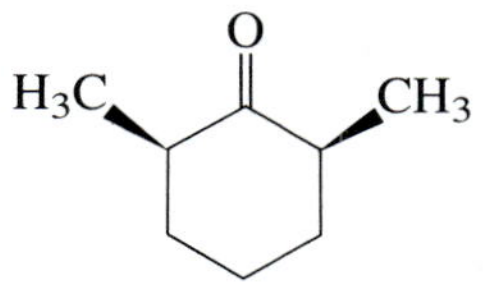 [cyclohexanone with H_3C and CH_3 on both α-carbons, both wedged]

(d) 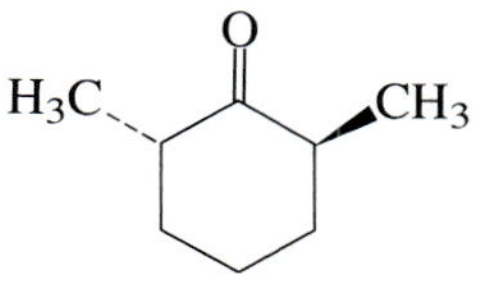 [cyclohexanone with H_3C (hashed) and CH_3 (wedged) on the α-carbons]

(e) 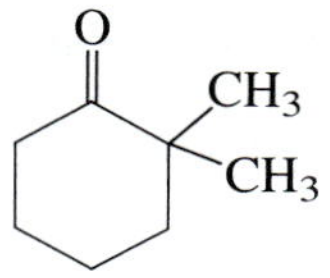 [cyclohexanone with two CH_3 groups on one α-carbon]

(f) 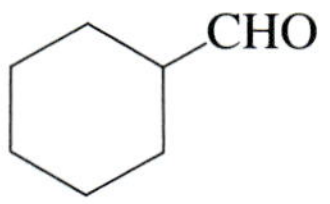 CHO

(g) $(CH_3)_3C\overset{O}{\overset{\|}{C}}H$

(h) $(CH_3)_3CCH_2\overset{O}{\overset{\|}{C}}H$

2 Write the product(s) that would be expected on reaction of cyclohexanone with one equivalent of LDA, followed by addition of 1 equivalent of

(a) CH_3CH_2Br

(b) $(CH_3)_2CHCl$

(c) 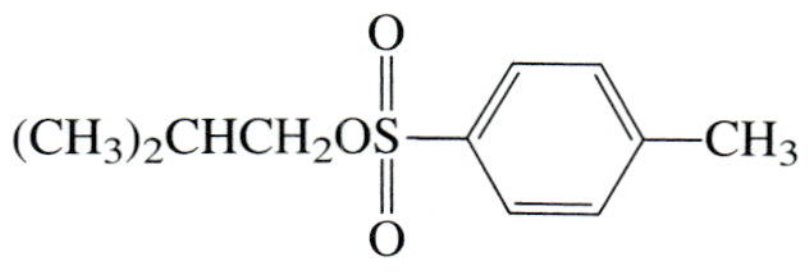 $(CH_3)_2CHCH_2OSO_2$–C_6H_4–CH_3 (p-toluenesulfonate)

(d) $(CH_3)_3CCl$

3 Write the product(s) of the following reaction sequences.

(a) 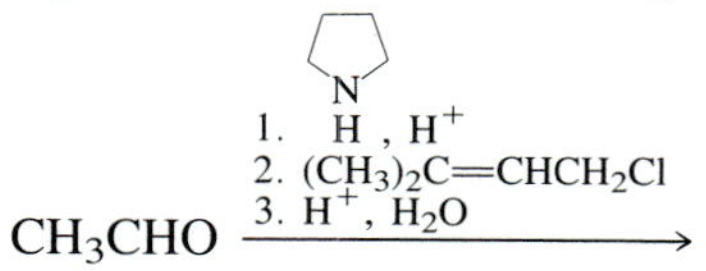 $CH_3CHO \xrightarrow[\text{3. } H^+,\ H_2O]{\text{1. pyrrolidine (N–H), } H^+;\ \text{2. } (CH_3)_2C{=}CHCH_2Cl}$

(b) $C_6H_5CH_2CHO \xrightarrow[\text{3. } H^+,\ H_2O]{\text{1. pyrrolidine, } H^+ \quad \text{2. } C_6H_5CH_2Br}$

4 The problem of double compared with single alkylation of ketones by iodomethane and base is mentioned in Section 16-1. Write a detailed mechanism showing how some double alkylation occurs even when only one equivalent each of the iodide and base is used.

Would you expect the use of the enamine alkylation procedure to solve this problem? Explain.

5 Would the use of an enamine instead of an enolate improve the likelihood of successful alkylation of a ketone by a secondary haloalkane?

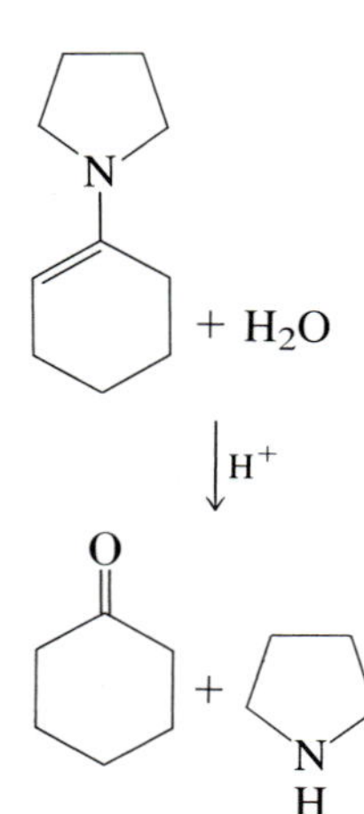

6 Propose a mechanism for the acid-catalyzed hydrolysis of the pyrrolidine enamine of cyclohexanone (shown at the left).

7 Which of the carbonyl compounds in Problem 1 would give a positive iodoform test?

8 What product(s) would form if each carbonyl compound in Problem 1 were treated with

(a) alkaline D_2O.
(b) 1 equivalent of Br_2 in ethanoic (acetic) acid.
(c) excess Cl_2 in aqueous base.

9 Propanedial, $OHCCH_2CHO$, in nonpolar solvent exists predominantly in an isomeric structure. Draw the more-stable isomer. Would you expect the same for butanedial, $OHCCH_2CH_2CHO$?

10 Describe the experimental conditions that would be best suited for the efficient synthesis of each of the following compounds from the corresponding nonhalogenated ketone.

(a) $C_6H_5CHBrCOCH_3$ (Br on CH, O on C)

(b) $CH_3CCl_2COCCl_2CH_3$ ($CH_3\overset{Cl}{\underset{Cl}{C}}\overset{O}{C}\overset{Cl}{\underset{Cl}{C}}CH_3$)

(c) CH_3COCH_2Cl

11 Write the structures of the likely products of each of the following aldol addition reactions.

(a) $2\ C_6H_5CH_2CHO \xrightarrow{NaOH}$

(b) $C_6H_5CHO + (CH_3)_2CHCHO \xrightarrow{NaOH}$

(c) 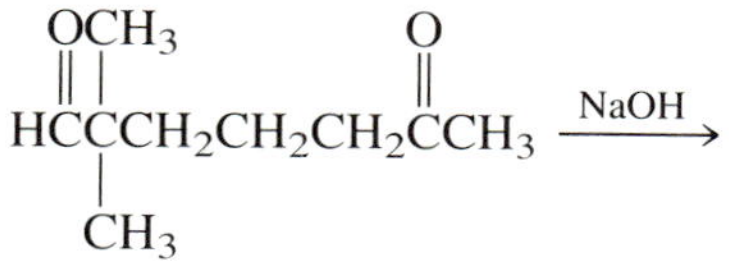 [structure: HC(=O)C(OCH3)(CH3)CH2CH2CH2C(=O)CH3] $\xrightarrow{\text{NaOH}}$

(d) 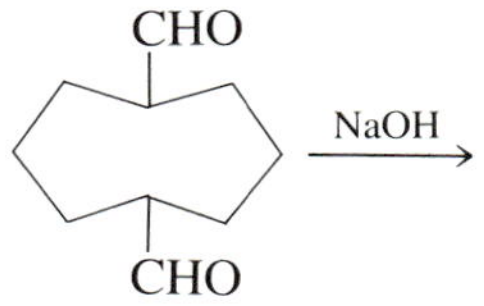 [structure: cyclooctane-1,5-dicarbaldehyde, CHO / CHO] $\xrightarrow{\text{NaOH}}$

12 Describe how you would prepare each of the following compounds, using an aldol condensation.

(a) 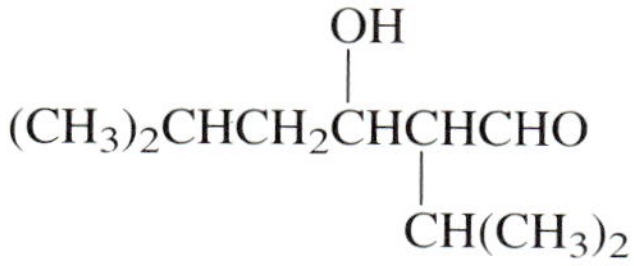 $(CH_3)_2CHCH_2CH(OH)CH(CH(CH_3)_2)CHO$

(b) 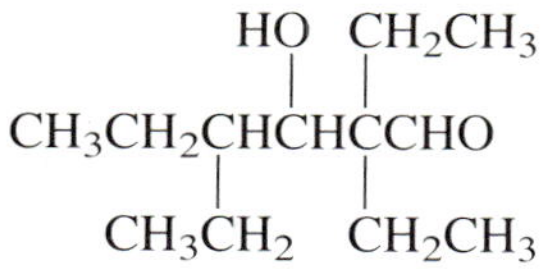 $CH_3CH_2CH(CH_2CH_3)CH(OH)C(CH_2CH_3)_2CHO$

(c) $(CH_3)_3CCH(OH)CH(CH_2CH_2CH_2CH_3)CHO$

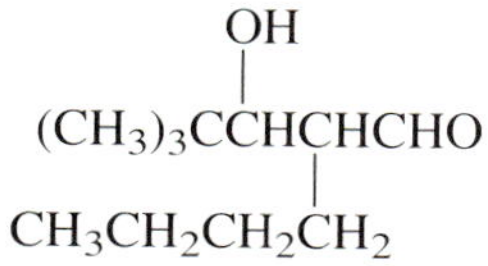

(d) 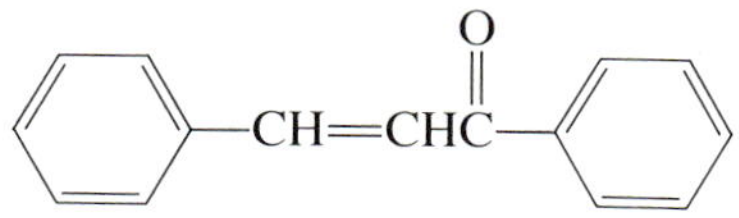 C_6H_5—CH=CHC(=O)—C_6H_5

(e) 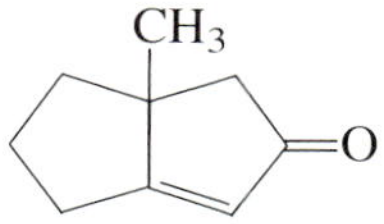 [structure: bicyclic enone with CH_3 and =O]

(f) 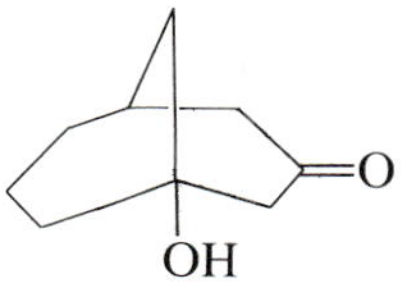 [structure: bridged bicyclic ketone with OH and =O]

13 Aldol condensations may be catalyzed by acids. Suggest a role for H^+ in the acid-catalyzed aldol condensation. (Hint: Consider what kind of nucleophile might exist in acid solution, where enolate ions are unlikely to be present.)

14 **(a)** The enzymatic oxidation of a lysine group to give an aldehyde is described in Section 16-3. What sort of intermediate(s) might appear in this oxidation? Hints: Refer to Problem 21 of Chapter 4 and to Sections 3-4 and 15-6.

(b) A similar enzyme-catalyzed oxidation is the first step in the metabolism of amphetamine, which takes place in the endoplasmic reticulum of the liver. Write the structures of both the final product of this oxidation and the intermediate that immediately precedes its formation.

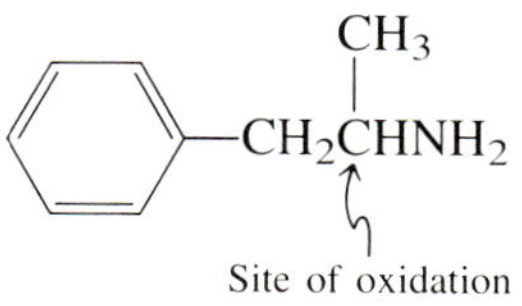

[structure: C_6H_5—$CH_2CH(CH_3)NH_2$; arrow labeled "Site of oxidation" pointing to CH]

Site of oxidation

Amphetamine

15 A recent (1981) and very clever synthesis of steroids in the cortisone family includes the sequence of compounds shown below. Describe how (reagents, reaction conditions) each of the three transformations **(a, b, c)** shown below and at the top of the next page might be carried out.

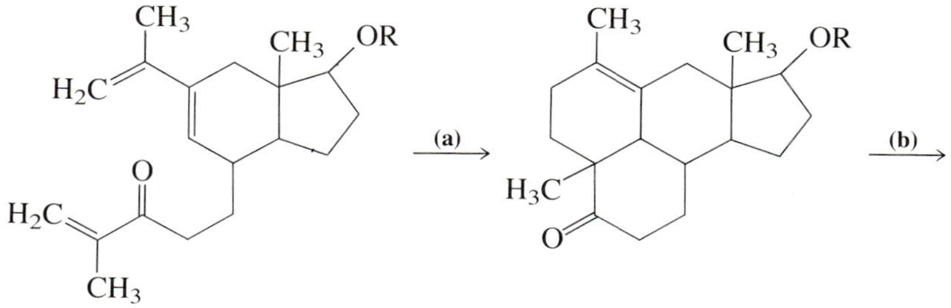

[structure: starting compound with H_2C, CH_3, CH_3, OR, O, H_2C, CH_3] $\xrightarrow{\text{(a)}}$ [structure: tetracyclic compound with CH_3, CH_3, OR, H_3C, O] $\xrightarrow{\text{(b)}}$

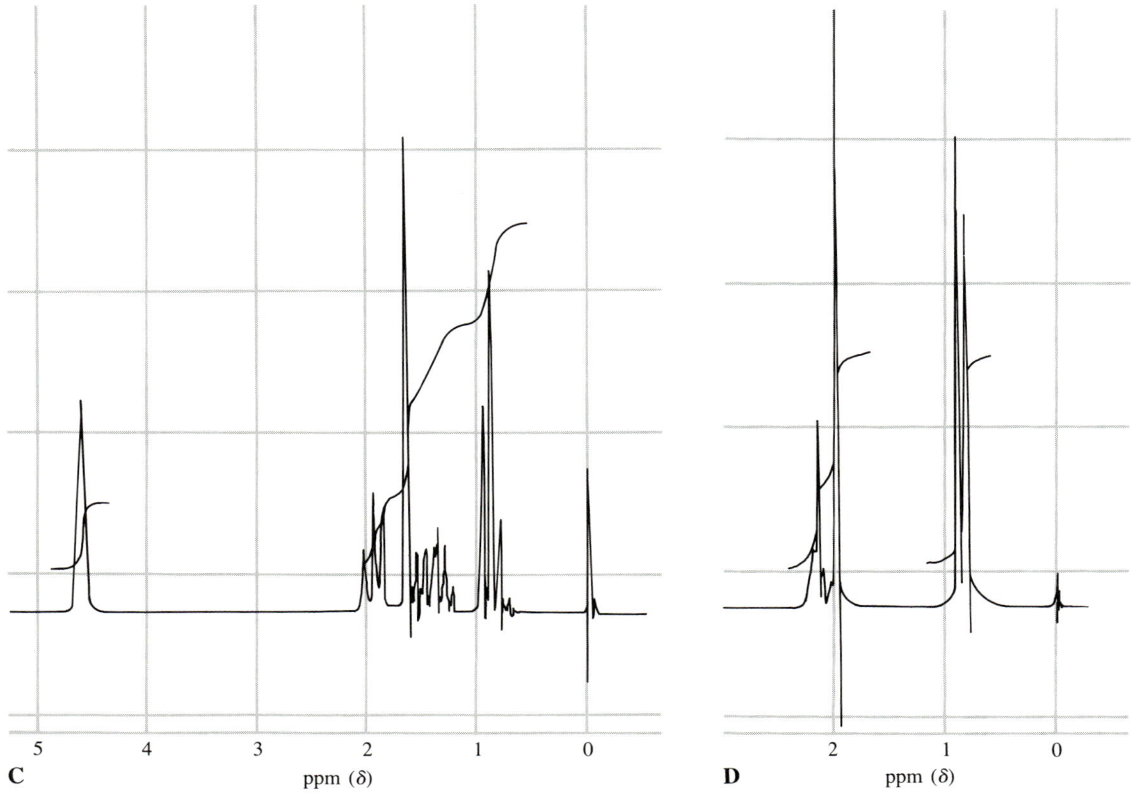

Next, for each of the following reactions, name an appropriate reagent for the indicated interconversion. (The letters refer to the compounds giving rise to NMR spectra A through D.)

(e) A $\longrightarrow$ C
(f) B $\longrightarrow$ D
(g) B $\longrightarrow$ A

21 Treatment of cyclopentane-1,3-dione with iodomethane in the presence of base leads mainly to a mixture of three products:

NaOH, CH_3I → **i** (CH_3, H) + **ii** (CH_3, CH_3) + **iii** (H, OCH_3)

(a) Give a mechanistic description of how these three products are formed.

(b) Reaction of product (iii) with a cuprate reagent results in loss of the methoxy group. For example,

iii (OCH_3) → 1. $(CH_3CH_2CH_2CH_2)_2CuLi$ 2. H^+, H_2O → **iv** ($CH_2CH_2CH_2CH_3$)

Suggest a mechanism for this reaction, which is another synthetic route to enones substituted at the β-carbon. Hint: See Exercise 16-14.

22 **(a)** Treatment of 2-cyclopentenone with NaOD in D_2O leads to complete disappearance of all the ^1H NMR signals, although the ^{13}C spectrum still shows resonances for all the original carbons at, essentially, the original chemical-shift positions. Explain why by a detailed mechanism.

(b) Keto aldehyde (i) was heated with aqueous base in an attempt to synthesize a specific cyclic α,β-unsaturated ketone. In fact, the expected product was not obtained. Instead, an isomer, enone (ii), was formed.

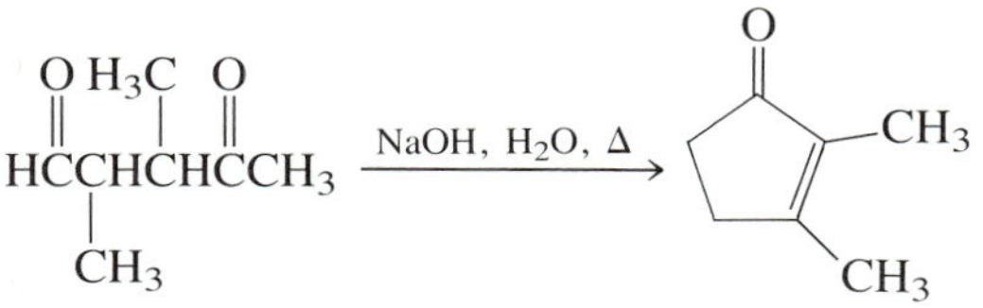

$$\underset{\textbf{i}}{\mathrm{HC(=O)CH(CH_3)CH(CH_3)C(=O)CH_3}} \xrightarrow{\text{NaOH, } H_2O,\ \Delta} \underset{\textbf{ii}}{\text{2,3-dimethyl-2-cyclopentenone}}$$

Write the structure of the initial enone product that you would expect to form on base treatment of (i). Then propose a mechanism to explain the formation of (ii). Hint: Consider whether the expected enone could be an intermediate in the formation of enone (ii).

23 A somewhat unusual synthesis of cortisone-related steroids includes the following two reactions:

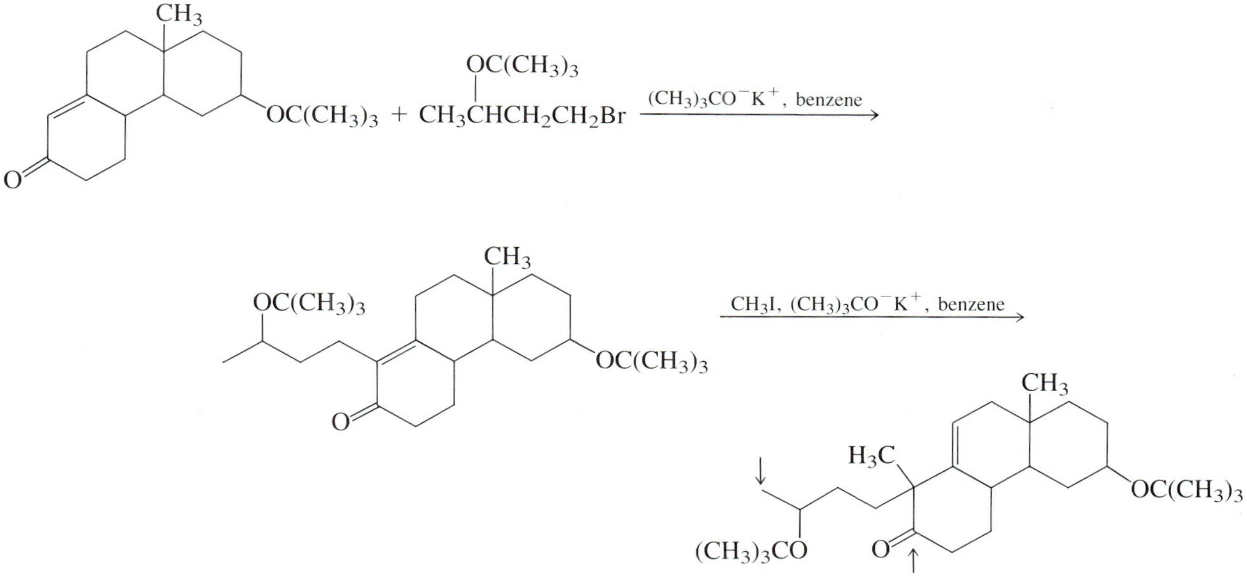

Tricyclic enone (CH_3, $OC(CH_3)_3$) + $CH_3CH(OC(CH_3)_3)CH_2CH_2Br$ $\xrightarrow{(CH_3)_3CO^-K^+,\ \text{benzene}}$

Alkylated enone (side chain with $OC(CH_3)_3$; ring CH_3; $OC(CH_3)_3$) $\xrightarrow{CH_3I,\ (CH_3)_3CO^-K^+,\ \text{benzene}}$ product (CH_3, H_3C, $(CH_3)_3CO$, O, $OC(CH_3)_3$; two carbons marked by arrows)

(a) Propose mechanisms for these two transformations. Be careful in choosing the initial site of deprotonation in the starting enone. The alkenyl hydrogen, in particular, is *not* the one initially removed by base in this reaction.

(b) Propose a sequence of reactions that will connect the carbons marked by arrows in the structure above to form another six-membered ring.

24 Write the expected product(s) of each of the following reactions.

(a) 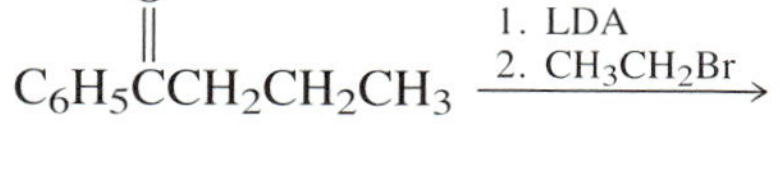 $C_6H_5\overset{\displaystyle O}{\overset{\|}{C}}CH_2CH_2CH_3 \xrightarrow[\text{2. } CH_3CH_2Br]{\text{1. LDA}}$

(b) 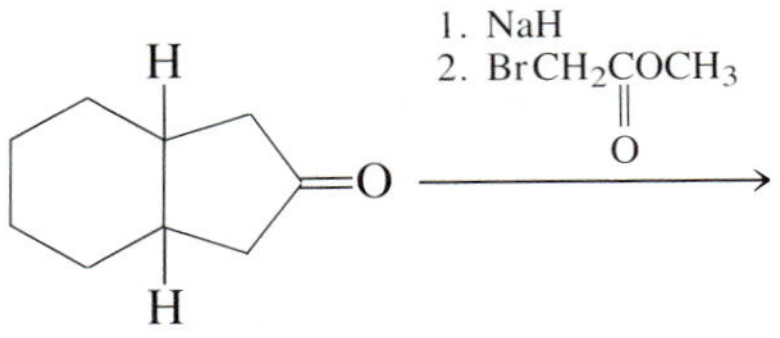 [cis-fused bicyclic ketone with H, H] $\xrightarrow[\text{2. } BrCH_2COCH_3]{\text{1. NaH}}$

(c) 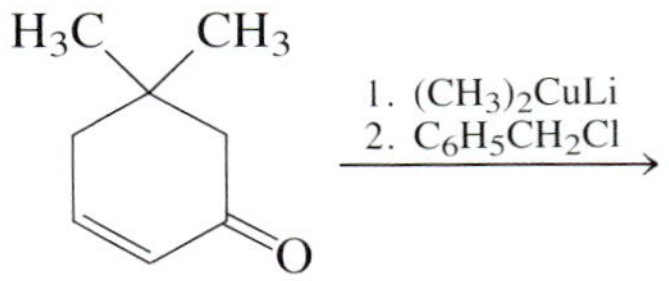 [3,3-dimethylcyclohex-2-enone; H_3C, CH_3] $\xrightarrow[\text{2. } C_6H_5CH_2Cl]{\text{1. } (CH_3)_2CuLi}$

(d) 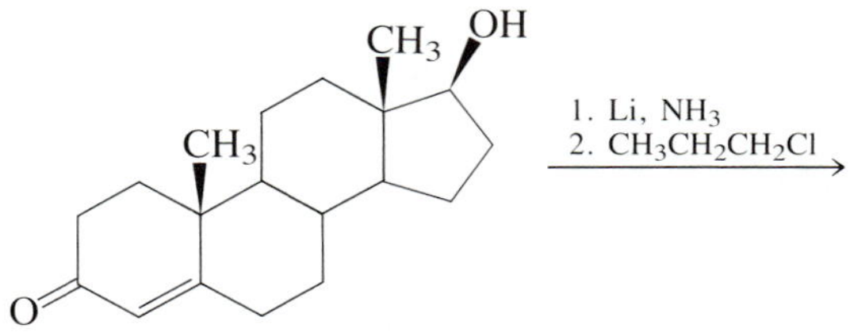 [steroid enone with CH_3, CH_3, OH] $\xrightarrow[\text{2. } CH_3CH_2CH_2Cl]{\text{1. Li, } NH_3}$

(e) 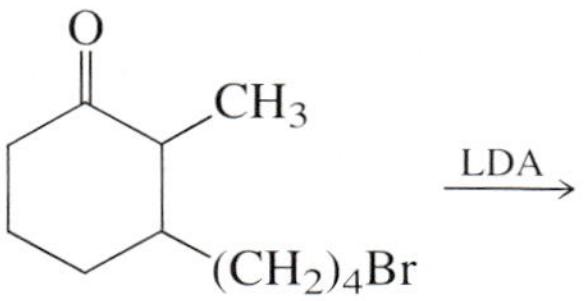 [2-methyl-3-(4-bromobutyl)cyclohexanone; CH_3, $(CH_2)_4Br$] $\xrightarrow{\text{LDA}}$

(f) Write a detailed mechanism for reaction **e.**

25 Write the products of each of the following reactions after aqueous work-up.

(a) $C_6H_5\overset{O}{\overset{\|}{C}}CH_3 + CH_2{=}CH\overset{O}{\overset{\|}{C}}C_6H_5 \xrightarrow{\text{LDA}}$

(b) 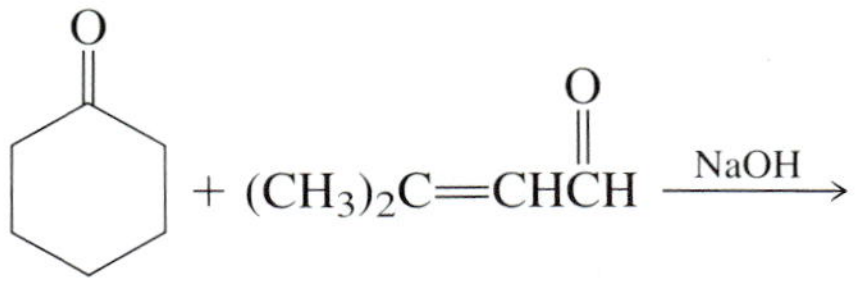 [cyclohexanone] $+ (CH_3)_2C{=}CH\overset{O}{\overset{\|}{C}}H \xrightarrow{\text{NaOH}}$

(c) [cyclopentenone] $\xrightarrow[\text{2. } CH_2{=}CH\overset{O}{\overset{\|}{C}}CH_3]{\text{1. } (CH_2{=}CH)_2CuLi}$

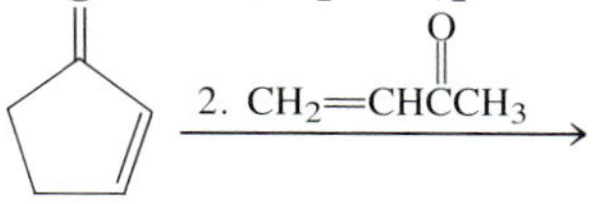

(d) 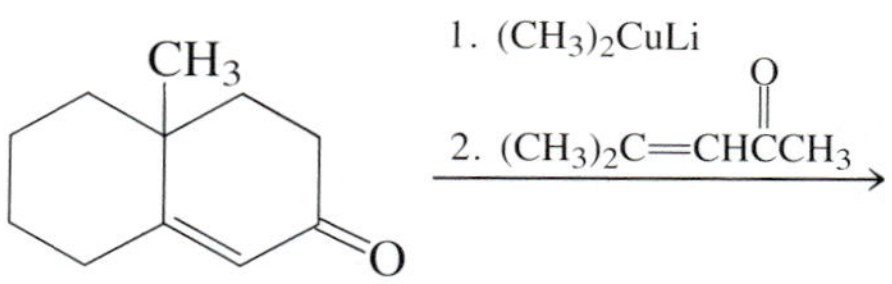 [bicyclic enone with angular CH_3] $\xrightarrow[\text{2. } (CH_3)_2C{=}CH\overset{O}{\overset{\|}{C}}CH_3]{\text{1. } (CH_3)_2CuLi}$

(e) Write the results that you expect from base treatment of the products of reactions **c** and **d.**

26 Write the final products of the following reaction sequences.

(a) 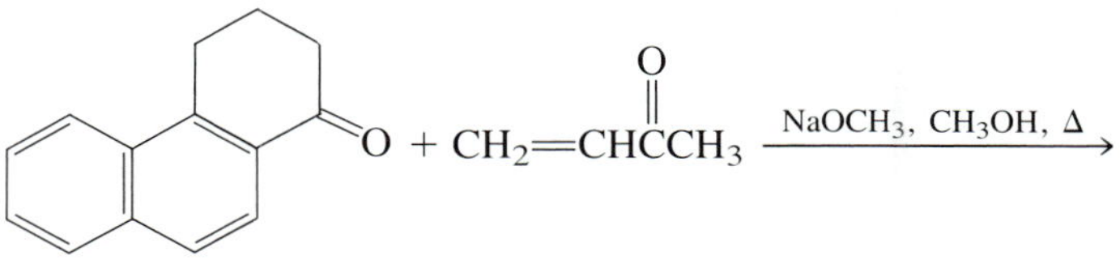 [structure] $\text{O} + CH_2{=}CH\overset{O}{\overset{\|}{C}}CH_3 \xrightarrow{NaOCH_3,\ CH_3OH,\ \Delta}$

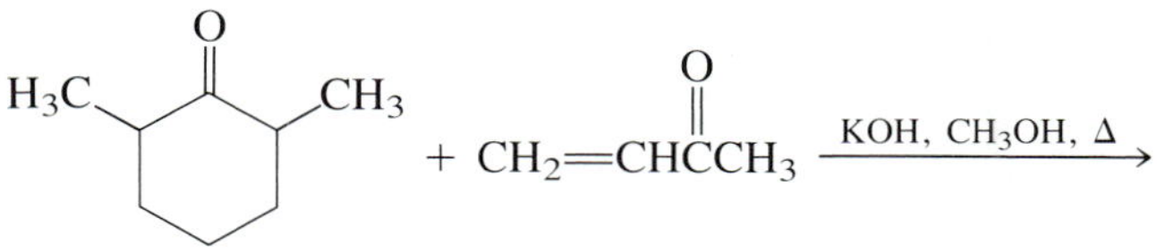

(b) H_3C ... CH_3 [structure] $+ CH_2{=}CH\overset{O}{\overset{\|}{C}}CH_3 \xrightarrow{KOH,\ CH_3OH,\ \Delta}$

(c) [structure] $\xrightarrow[]{1.\ NaH,\ (CH_3CH_2)_2O \quad 2.\ HC{\equiv}C\overset{O}{\underset{\|}{C}}CH_3}$

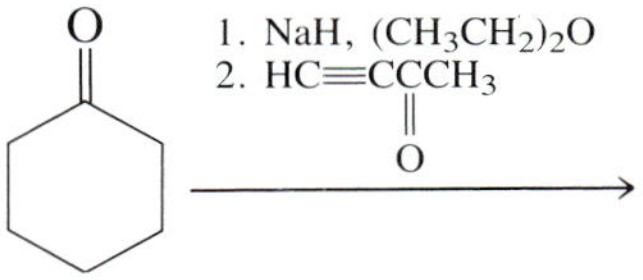

(d) Write a detailed mechanism for reaction sequence **c.**

27 Propose syntheses of the following compounds using Michael additions followed by aldol condensations (i.e., Robinson annelation). Each of the compounds shown has been instrumental in one or more total syntheses of steroidal hormones.

(a) 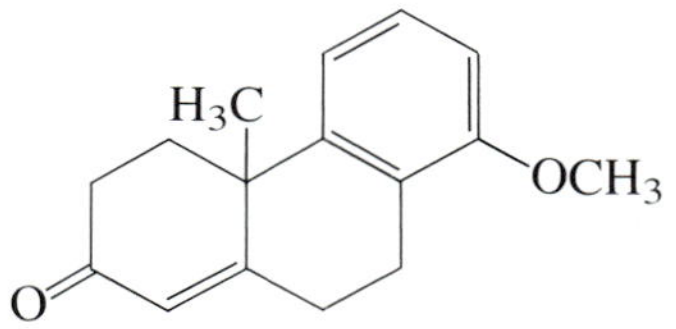 H_3C OCH_3 O

(b) 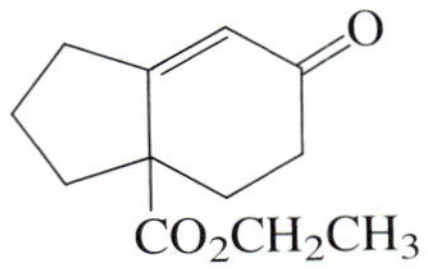 O $CO_2CH_2CH_3$

(c) 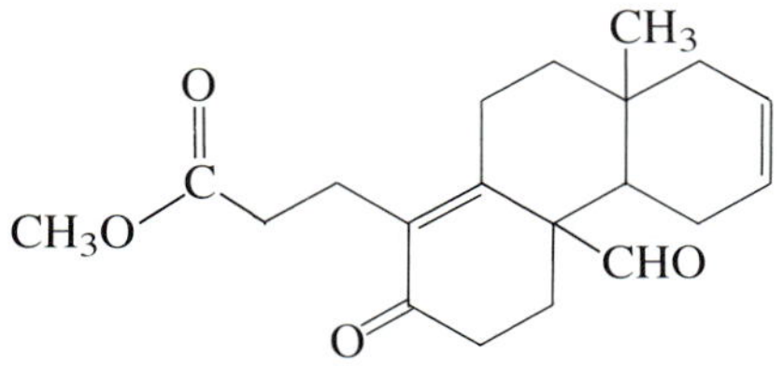 CH_3O C(=O) CH_3 CHO O

(d) 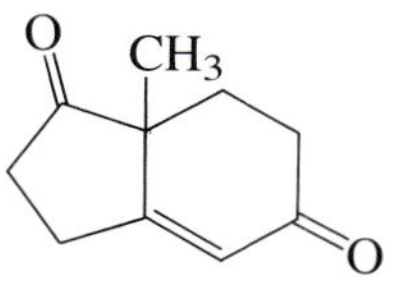 O CH_3 O

28 The following steroid synthesis contains modified versions of two key types of reactions presented in this chapter. Identify these reaction types, and give detailed mechanisms for each of the transformations shown.

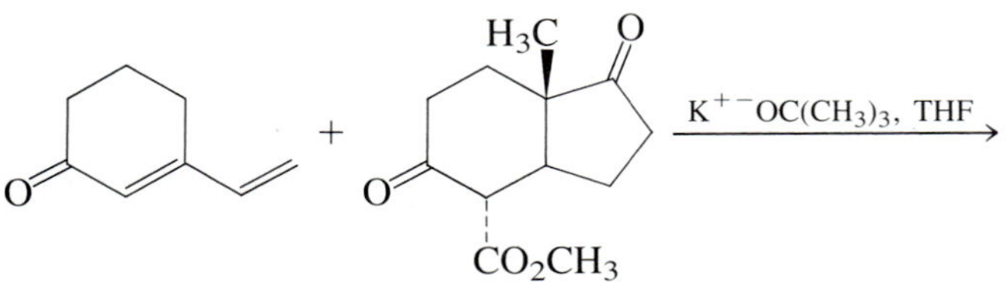

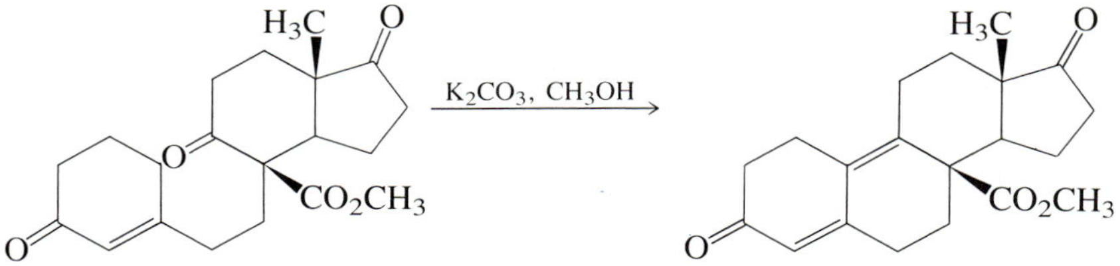

$K^{+}\,^{-}OC(CH_3)_3$, THF →

K_2CO_3, CH_3OH →

29 Devise reasonable plans for carrying out the following syntheses. Ignore stereochemistry in your strategies.

(a) 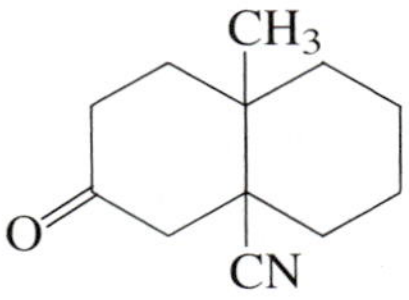 , starting from cyclohexanone

(b) 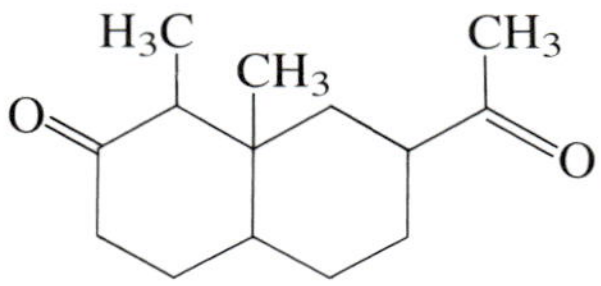 , starting from 2-cyclohexenone

(Hint: Prepare 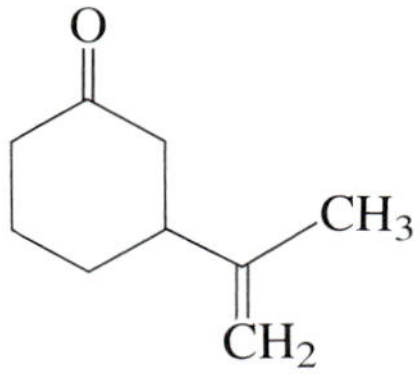 in your first step.)

30 Write reagents **a, b, c, d,** and **e** where they have been omitted from the following synthetic sequence. Each letter may correspond to one or more reaction steps. (This is the beginning of a synthesis of germanicol, a naturally occurring triterpene.)

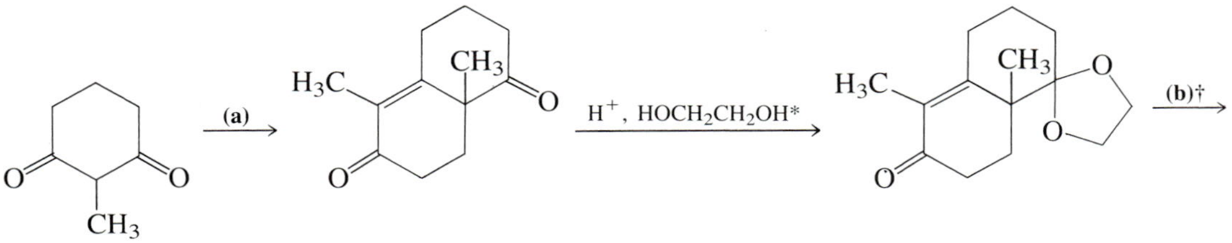

(a) →

H^{+}, $HOCH_2CH_2OH$* →

(b)† →

* Selective protection of the more-reactive carbonyl group.

† Hint: See Problem 23.

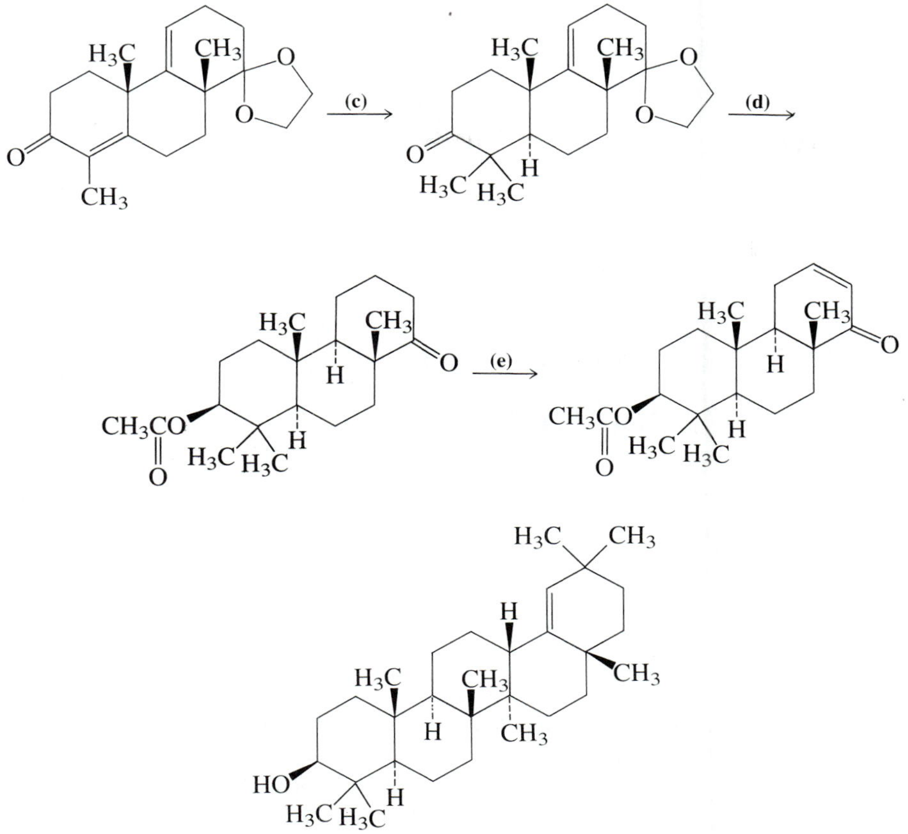

Germanicol

CHAPTER 17

Carboxylic Acids and Infrared Spectroscopy

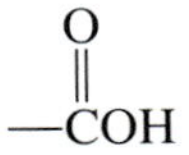
The carboxy group

When a hydroxy group is attached to the carbonyl function, a new functional group is formed, the **carboxy group** characteristic of the **carboxylic acids.** This substituent is usually written —COOH or —CO_2H, and both of these conventions will be used in the following sections. To a certain extent, carboxylic acids may be regarded as hydroxy carbonyl derivatives, because they show some of the reactivity of both the alcohols and the ketones. Thus, they are both acidic and basic: the OH proton as well as the entire OH group may be replaced by other substituents, and the carbonyl function is subject to nucleophilic attack at carbon. However, because of the close proximity of the two functional groups, the carboxy group also has its own distinct and unique chemistry.

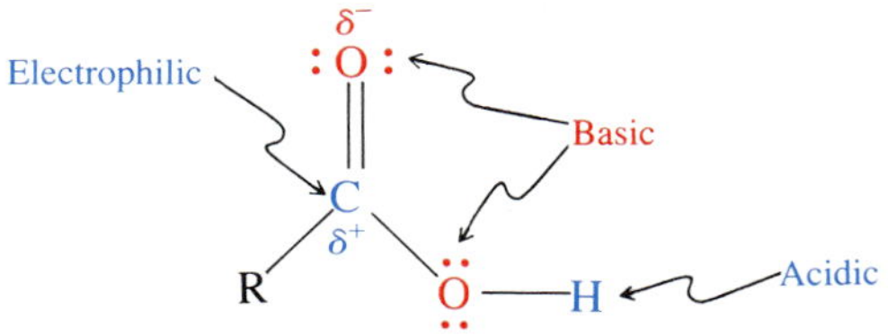

This chapter introduces the system of naming carboxylic acids. It then examines some of their physical properties, including their NMR spectra, and another analytical technique used by organic chemists: infrared spectroscopy. The remainder of the chapter describes the preparation and reactivity of carboxylic acids, and some of their naturally occurring representatives.

17-1 The System for Naming Carboxylic Acids

As usual, carboxylic acids have many common names. Many of them indicate the natural sources from which the acids were originally derived (Table 17-1)

TABLE 17-1
Names and natural sources of carboxylic acids

Structure	IUPAC name	Common name	Natural source
HCOOH	Methanoic acid	Formic acid	From the "destructive distillation" of ants (*formica,* Latin, ant)
CH_3COOH	Ethanoic acid	Acetic acid	Vinegar (*acetum,* Latin, vinegar)
CH_3CH_2COOH	Propanoic acid	Propionic acid	Dairy products (*pion,* Greek, fat)
$CH_3CH_2CH_2COOH$	Butanoic acid	Butyric acid	Butter (particularly if rancid) (*butyrum,* Latin, butter)
$CH_3(CH_2)_3COOH$	Pentanoic acid	Valeric acid	Valerian root
$CH_3(CH_2)_4COOH$	Hexanoic acid	Caproic acid	Odor of goats (*caper,* Latin, goat)

and are (particularly for the lower members of this class) used frequently in the literature.

The IUPAC system derives the names of the carboxylic acids by replacing the ending **-e** in the name of the alkane by **-oic acid.** The alkanoic acid stem is numbered by assigning the number 1 to the carboxy carbon, and labeling any substituents along the longest chain incorporating the functional group accordingly.

$CH_3CH(Br)COOH$
2-Bromopropanoic acid
(α-Bromopropionic acid)

$CH_2{=}CHCOOH$
Propenoic acid
(Acrylic acid)

$CH_3CH_2CH(CH_3)CH(CH_3)COOH$
2,3-Dimethylpentanoic acid
(α,β-Dimethylvaleric acid)

The carboxy function has priority over any functional group discussed so far. In multiply functionalized carboxylic acids, the longest chain is chosen to include other functional groups as much as possible.

$CH_2{=}CHCH(CH_2CH_2CH_2CH_3)CH_2CH_2CH_2COOH$
5-Butyl-6-heptenoic acid
(Better than 5-ethenylnonanoic acid)

$CH_3C({=}O)CH(CH_2CH_2CH_3)CH_2CH_2COOH$
5-Oxo-4-propylhexanoic acid

1-Bromo-2-chlorocyclopentanecarboxylic acid

Saturated cyclic acids are named as **cycloalkanecarboxylic acids.** In these compounds, the carbon attached to the functional group is C-1.

Dicarboxylic acids may be referred to as **alkanedioic acids,** as well as by their traditional common names. The following examples indicate the variety of ways in which the carboxy group can be depicted.

$HOOC{-}COOH$ (HOCCOH, with two C=O)
Ethanedioic acid
(Oxalic acid)

$HOOCCH_2COOH$ (HOCCH$_2$COH, with two C=O)
Propanedioic acid
(Malonic acid)

$HOOCCH_2CH_2COOH$
Butanedioic acid
(Succinic acid)

$HOOC(CH_2)_3COOH$
Pentanedioic acid
(Glutaric acid)

$HOOC(CH_2)_4COOH$
Hexanedioic acid
(Adipic acid)

$HO_2CCH{=}CHCO_2H$
***cis*-2-Butenedioic acid**
(Maleic acid)
or
***trans*-2-Butenedioic acid**
(Fumaric acid)

EXERCISE 17-1

Give systematic names or write the structure, as appropriate, of the following compounds:

(a) [skeletal structure: seven-carbon chain ending in C(=O)OH, with Br and Cl substituents] ; (b) [cyclohexanone ring bearing COOH para to C=O] ;

(c) 2,2-dibromohexanedioic acid; (d) 4-hydroxypentanoic acid.

In summary, the systematic naming of the carboxylic acids is based on the alkanoic acid stem. Cyclic derivatives are called cycloalkanecarboxylic acids, and dicarboxylic systems are labeled alkanedioic acids.

17-2 The Physical Properties of Carboxylic Acids

What is the structure of a typical carboxylic acid? Do carboxylic acids have characteristic physical constants? This section answers these questions. In particular, it describes the structure of methanoic (formic) acid. Carboxylic acids exist mainly as hydrogen-bonded dimers, and they exhibit characteristic NMR spectra.

The Structure of Methanoic (Formic) Acid

The molecular structure of methanoic (formic) acid is shown in Figure 17-1. It is roughly that expected for a ''hydroxymethanal'' with an approximately trigonal planar carbonyl carbon. (See the structure of methanol, Figure 8-1B, and the structure of ethanal, Figure 15-2.)

Physical Constants Reveal the Polarity and Hydrogen-Bonding Ability of the Carboxy Group

The carboxy function is strongly polar because of the presence of the polarizable carbonyl double bond and the hydroxy group, which forms hydrogen bonds to other polarized molecules, such as water, alcohols, and other carboxylic acids. Not surprisingly, therefore, the lower carboxylic acids are completely soluble in water (up to butanoic acid). As neat liquids and even in fairly dilute solutions (in nonhydroxylic solvents), carboxylic acids exist to a large extent as hydrogen-bonded dimers, each O–H···O interaction ranging in strength from about 6 to 8 kcal mole^{-1}.

FIGURE 17-1
The molecular structure of methanoic (formic) acid.

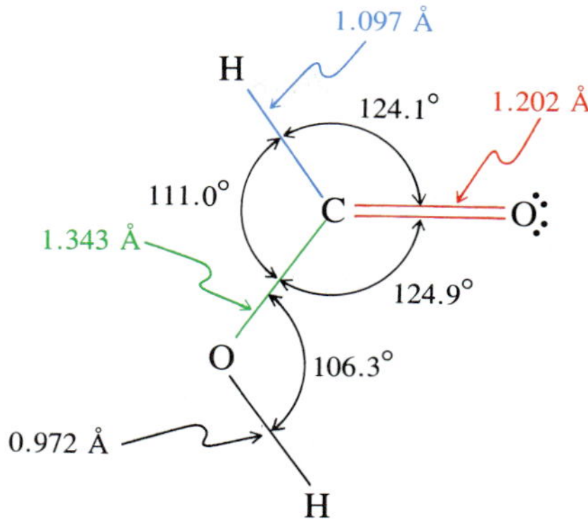

TABLE 17-2
Melting and boiling points of functional alkane derivatives with various chain lengths

Derivative	Melting point, °C	Boiling point, °C
CH_4	−182.5	−161.7
CH_3Cl	−97.7	−24.2
CH_3OH	−97.8	65.0
HCHO	−92.0	−21.0
HCOOH	8.4	100.6
CH_3CH_3	−183.3	−88.6
CH_3CH_2Cl	−136.4	12.3
CH_3CH_2OH	−114.7	78.5
CH_3CHO	−121.0	20.8
CH_3COOH	16.7	118.2
$CH_3CH_2CH_3$	−187.7	−42.1
$CH_3CH_2CH_2Cl$	−122.8	46.6
$CH_3CH_2CH_2OH$	−126.5	97.4
CH_3COCH_3	−95.0	56.5
CH_3CH_2CHO	−81.0	48.8
CH_3CH_2COOH	−20.8	141.8

Carboxylic Acids Form Dimers Readily

2 RCOOH ⟶ R—C(=O···H—O)(—O—H···O=)C—R

Two hydrogen bonds

They have relatively high melting and boiling points (Table 17-2), owing to their ability to hydrogen bond in the solid state, as well as in liquid form.

Nuclear Magnetic Resonance: The Carboxylic Acid Proton and the Carbonyl Carbon Absorb at Low Field

As in aldehydes and ketones, the hydrogens positioned on the carbon next to the carbonyl group are slightly deshielded in ^{1}H NMR spectra, partly because of the local magnetic field generated by π-electron movement and partly because of the inductive effect of the positively polarized carbonyl carbon. The effect diminishes rapidly with increasing distance from the functional group. A special case is methanoic (formic) acid. This compound has one "aldehydic" proton, giving rise to a signal at characteristically low field. The hydroxy proton resonates at very low field (δ = 10–13 ppm). As in the NMR spectra of alcohols, its chemical shift varies strongly with concentration, solvent, and temperature, because of the strong ability of the OH group to enter into hydrogen bonding. The free (non-hydrogen-bonded) acid proton has a chemical shift δ = 5.7 ppm. The ^{1}H NMR spectrum of pentanoic acid is shown in Figure 17-2.

^{1}H NMR Chemical Shifts of Alkanoic Acids

CH_3COOH	CH_3CH_2COOH	$(CH_3)_2CHCOOH$	$HCOOH$
δ = 2.08 (CH_3)	1.16 (CH_3), 2.36 (CH_2)	1.21 (CH_3), 2.56 (CH)	8.08 ppm (H)

The ^{13}C NMR chemical shifts of carboxylic acids are also similar to those of the aldehydes and ketones, with moderately deshielded carbons next to the carbonyl group and the typically low field carbonyl absorptions. However, the amount of deshielding is not quite as large, because the positive polarization of the carboxy carbon is somewhat attenuated by the presence of the extra OH group.

Typical ^{13}C NMR Chemical Shifts of Alkanoic Acids

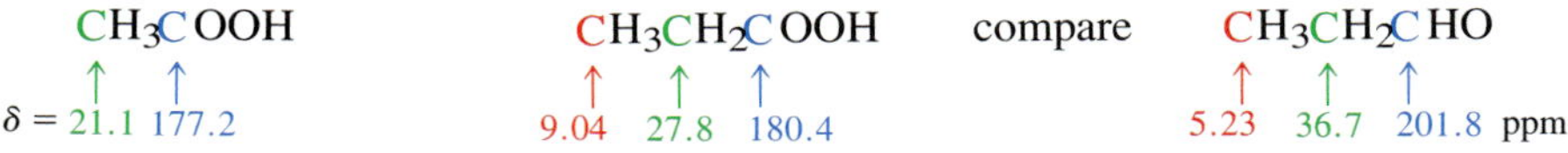

This attenuation is best visualized by writing dipolar resonance structures. For aldehydes and ketones, only one such structure is a reasonable contributor—namely, that with carbocation (hence, deshielding) character.

Resonance in Aldehydes and Ketones

$$\left[H_3C{-}C({=}\ddot{O}{:}){-}R \longleftrightarrow H_3C{-}\overset{+}{C}({-}{:}\ddot{O}{:}^{-}){-}R \right]$$

The contribution of the second resonance structure, though minor, explains the strong deshielding of the carbonyl and adjacent carbons.

However, in carboxylic acids there is an additional strongly contributing resonance structure, in which the hydroxy oxygen has donated an electron pair as in the oxonium ion resonance form of a hydroxy carbocation. In this way, the amount of positive charge on the carbonyl carbon is reduced.

FIGURE 17-2
90-MHz ^{1}H NMR spectrum of pentanoic acid in CCl_4. The scale has been expanded to 20 ppm, to allow the signal for the acid proton at $\delta = 11.83$ ppm to be shown. The methylene hydrogens at C-2 absorb at the next-lowest field as a triplet ($\delta = 2.25$ ppm, $J = 7$ Hz), followed by a four-hydrogen multiplet for the next two sets of methylenes. The methyl group appears as a distorted triplet at highest field ($\delta = 0.90$ ppm, $J =$ 6 Hz).

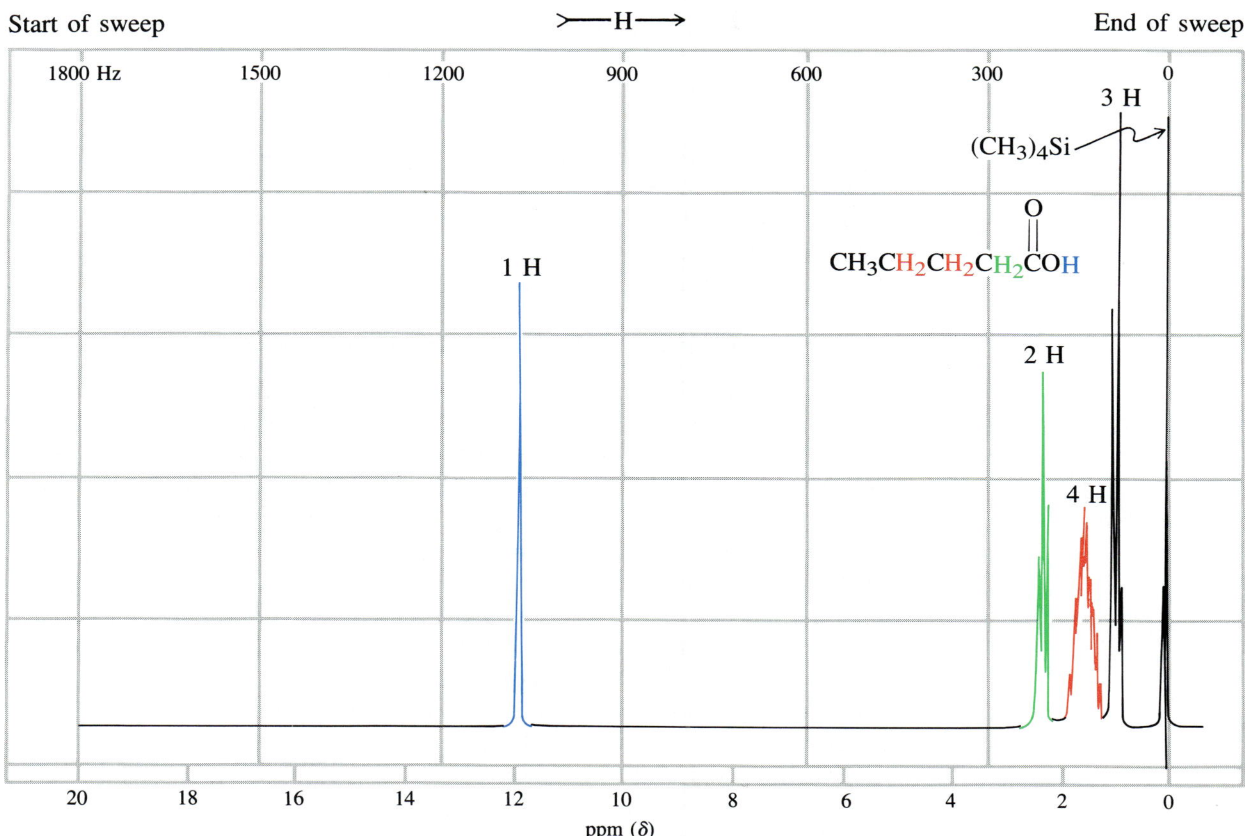

Resonance in Carboxylic Acids

The third resonance structure explains the attenuated deshielding effect of the carbonyl carbon compared with aldehydes and ketones.

EXERCISE 17-2

A foul-smelling carboxylic acid with b.p. 164°C gave the following NMR data: ^{1}H NMR (CCl_4) $\delta = 1.00$ (t, $J = 7.4$ Hz, 3 H), 1.65 (sex, $J = 7.5$ Hz, 2 H), 2.31 (t, $J = 7.4$ Hz, 2 H), and 11.68 (s, 1 H) ppm; ^{13}C NMR (CS_2) $\delta = 13.4$, 18.5, 36.3, and 179.6 ppm. Assign a structure to it.

In summary, the physical properties of the carboxylic acids show the presence of a polarizable carbonyl group, as well as the hydroxy function, which is capable of hydrogen bonding. Thus, carboxylic acids exhibit unusually high melting and boiling points, highly deshielded acid proton and carbonyl carbon signals, and moderately deshielded nuclei next to the functional group. The positive polarization at carbon is, however, somewhat attenuated by the contribution of an oxonium ion resonance structure.

17-3 Another Technique for Identifying Functional Groups: Infrared Spectroscopy

Another method of identifying carboxylic acids and, indeed, other functional groups is **infrared spectroscopy,** which measures the vibrational excitation of atoms around the bonds that connect them. The position of the absorption lines depends on the types of functional groups present, and the spectrum as a whole is a unique ''fingerprint'' of the entire molecule.

Absorption of Infrared Light Causes Molecular Vibrations

In nuclear magnetic resonance, radiowaves cause nuclear spins to change their alignment with the magnetic field ($\Delta E \sim 10^{-6}$ kcal mole^{-1}; Chapter 10). Ultraviolet–visible spectroscopy is performed with higher-energy light, which induces electronic transitions ($\Delta E \sim$ 40–300 kcal mole^{-1}; Section 14-7). At energies slightly lower than those of visible radiation, molecules absorb light by undergoing **vibrational excitation:** this is the **infrared,** or **IR, region** of the electromagnetic spectrum (see Figure 10-2). The intermediate range, or *middle infrared,* is most useful to the organic chemist. IR absorption bands are described by either the wavelength, λ, of the absorbed light in micrometers ($\lambda \sim$ 2.5–16.7 μm; see Figure 10-2) or its reciprocal value, called *wave number,* $\tilde{\nu}$ (in units of cm^{-1}; $\tilde{\nu} = 1/\lambda$). Thus, a typical infrared spectrum ranges from 600 to 4000 cm^{-1}, and the energy changes associated with absorption of this radiation range from 1 to 10 kcal mole^{-1}.

A simple IR spectrometer can be described by the general picture given in Figure 10-3. Modern systems use sophisticated rapid-scan techniques and are linked with computers. This allows for data storage, spectra manipulation, computer library searches, and matching of unknown compounds with stored spectra.

Vibrational excitation can be envisioned simply by thinking of two atoms A and B linked by a bond as two weights on a spring that stretches and compresses at a certain frequency, ν (Figure 17-3). In this picture, the frequency of the vibrations between two atoms depends both on the strength of the bond between them and on their atomic weights. In fact, it can be shown to be governed by Hooke's law, derived for the motions of a spring.

FIGURE 17-3
Two unequal weights on an oscillating (''vibrating'') spring: a model for vibrational excitation of a bond.

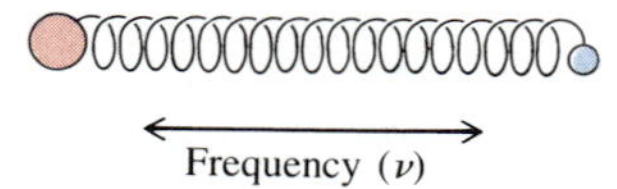

Hooke's Law and Vibrational Excitation

$$\tilde{\nu} = k\sqrt{f\frac{(m_1 + m_2)}{m_1 m_2}}$$

$\tilde{\nu}$ = vibrational frequency in wave numbers (cm^{-1})
k = constant

f = force constant, indicating the strength of the spring (bond)
m_1, m_2 = masses of attached weights (atoms)

This equation might lead us to expect every individual bond in a molecule to show a specific absorption band in the infrared spectrum. For example, ethanoic (acetic) acid should show five peaks. We might even be able to predict the relative positions of these bands. For example, the O–H bond strength is higher than that of C–H; therefore the O–H bond should vibrate at a frequency of higher wavenumber than the C–H bond does. A similar argument could be made for the C–O double bond compared with the single bond. The IR spectrum of ethanoic (acetic) acid is shown in Figure 17-4.

Possible Vibrational Excitations in Ethanoic (Acetic) Acid

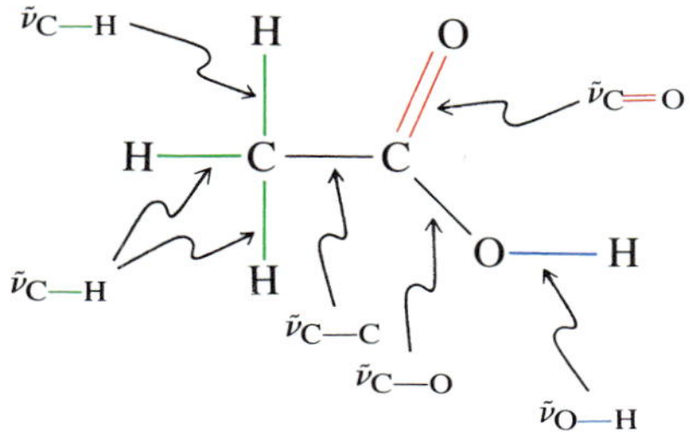

FIGURE 17-4

Infrared spectrum of ethanoic (acetic) acid. Note the format of the spectrum. The reciprocal wavelength is plotted, highest wave number to the left, against the percent transmittance. A 100% transmittance means *no* absorption.

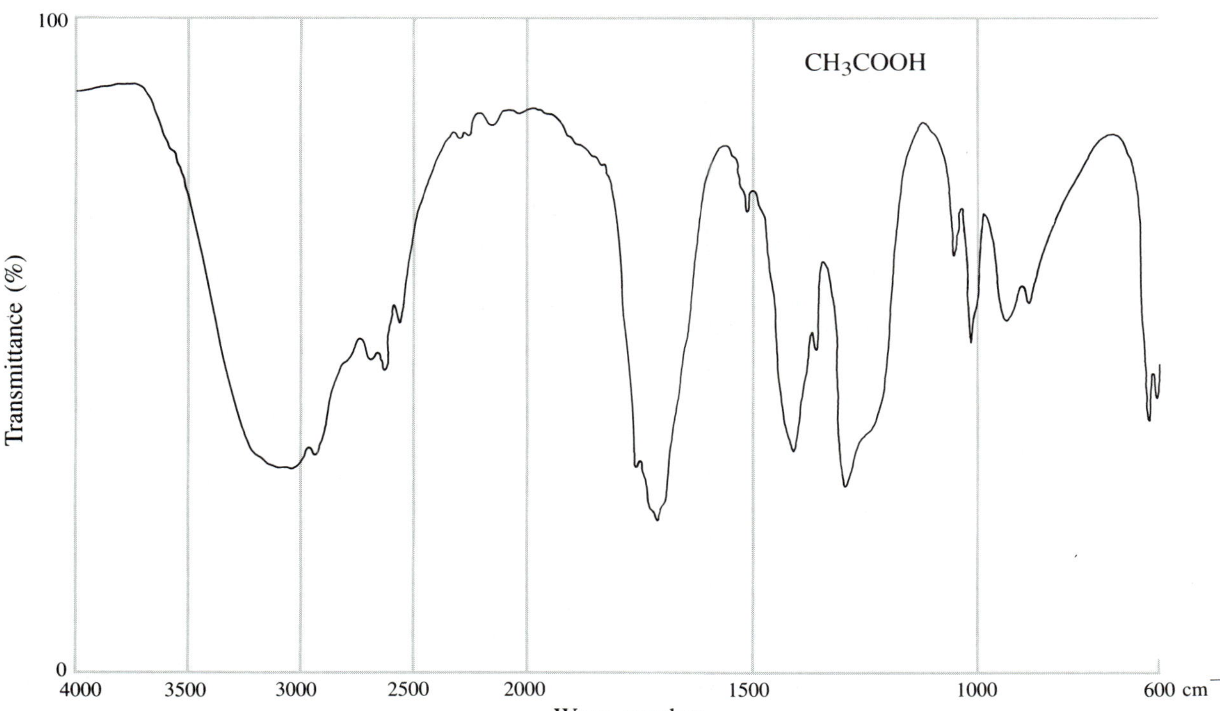

It is immediately apparent that things are not as simple as expected; clearly, more than five bands (some of them appearing as poorly resolved shoulders) can be discerned in this spectrum. How can this observation be explained?

Complex Vibrations and Couplings

The primary reason for the complexity of infrared spectra is the variety of vibrations possible in a molecule and the fact that many vibrations are mechanically coupled. This coupling is reminiscent of the coupling between nuclei in NMR that gives rise to complicated and non-first-order spectra, although it is of different origin. Thus, molecules that absorb infrared light undergo not only stretching but also various bending motions, and combinations of the two as well. The vibrational modes possible around tetrahedral carbon are shown in Figure 17-5. They include movements labeled symmetric and asymmetric stretching, rocking, twisting, and wagging.

FIGURE 17-5
The various vibrational modes possible around tetrahedral carbon.

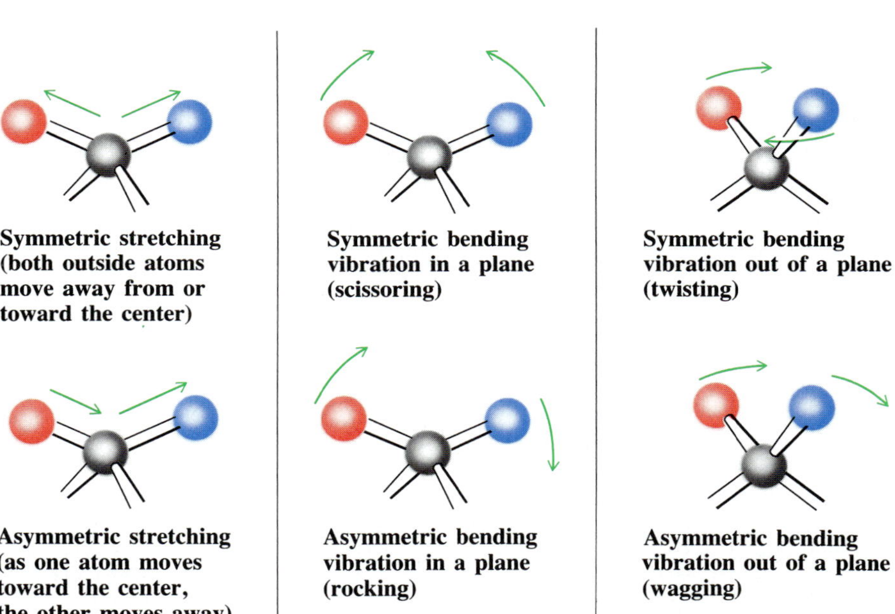

The complexity introduced by these possibilities makes a straightforward interpretation of the entire infrared spectrum very difficult. The practicing organic chemist can, however, find some good use for IR spectroscopy for two reasons: first, the vibrational bands of several functional groups appear at characteristic wave numbers and, second, the entire infrared spectrum may be used as a unique fingerprint of a compound.

Functional Groups Have Typical Infrared Absorptions

Table 17-3 lists some of the important characteristic stretching wave numbers for the functional groups and bonds encountered thus far: alkanes, alkenes,

TABLE 17-3
Characteristic infrared stretching wave number ranges of organic molecules

Bond or functional group	$\tilde{\nu}$ (cm^{-1})
RO—H (alcohols)	3200–3650
RC(=O)O—H (carboxylic acids)	2500–3300
R_2N—H (amines)	3300–3500
RC≡C—H (alkynes)	3260–3330
>C=C(—H)< (alkenes)	3050–3150
—C(—)(—)—H (alkanes)	2840–3000
RC≡CH (alkynes)	2100–2260
RC≡N (nitriles)	2220–2260
RC(=O)H, RC(=O)R′ (aldehydes, ketones)	1690–1750
RC(=O)OR′ (esters)	1735–1750
RC(=O)OH (carboxylic acids)	1710–1760
>C=C< (alkenes)	1620–1680
RC(—)(—)—OR′ (alcohols, ethers)	1000–1260

alkynes, alcohols, ethers, aldehydes and ketones, and (in this chapter) carboxylic acids. The following discussion will mention bending vibrations only briefly, because they are mostly of weaker intensity, they overlap with other absorptions, and they may show complicated patterns.

The Fingerprint Region

Figures 17-6 through 17-9 show the IR spectra of pentane and hexane at two different attenuations. Although these spectra have similar features, their fine structures are different. These differences become even clearer at higher recorder sensitivity, particularly in the range between 600 and 1500 cm^{-1}, called the *fingerprint region*. The typical C–H stretching absorptions for the alkanes are seen in the range from 2840 to 3000 cm^{-1}. Three other bands, due to bending motions, stand out at about 1460, 1380, and 730 cm^{-1}. All saturated hydrocarbons (including cycloalkanes) show similar absorptions.

FIGURE 17-6

IR spectrum of pentane: $\tilde{\nu}_{\text{C–H stretch}}$ = 2960, 2930, and 2870 cm^{-1}; $\tilde{\nu}_{\text{C–H bend}}$ = 1460, 1380, and 730 cm^{-1}.

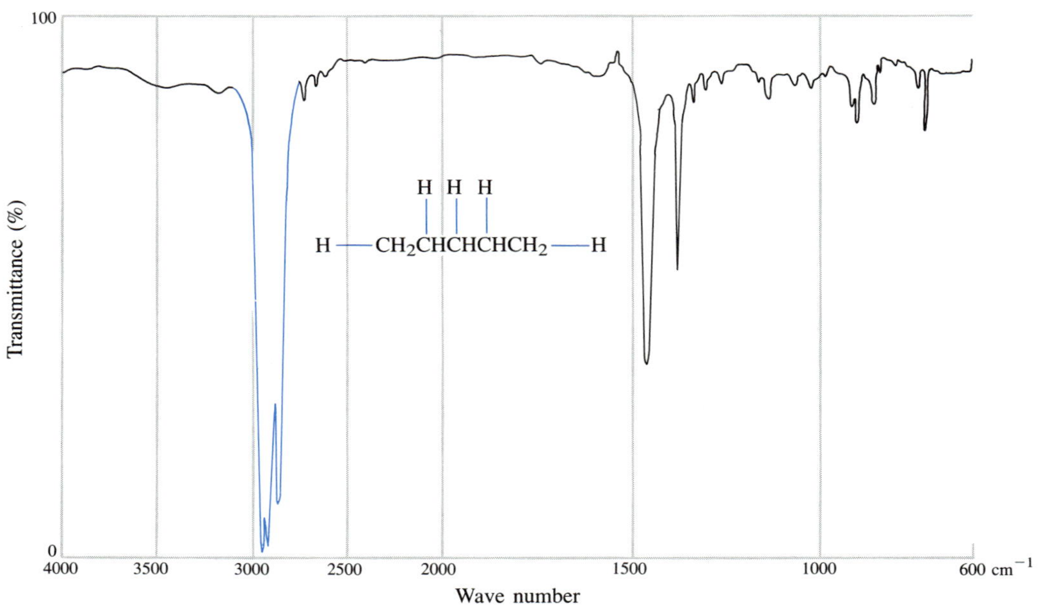

FIGURE 17-7

IR spectrum of hexane. Note the similarity of the location of the major bands to those in the IR spectrum of pentane.

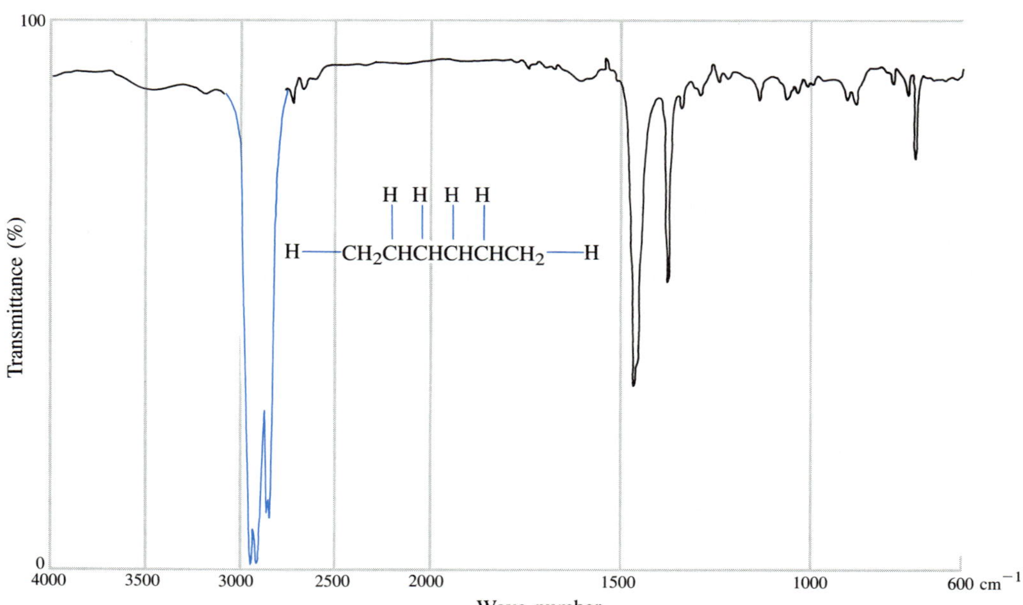

FIGURE 17-8

IR spectrum of a sample of pentane at higher recorder sensitivity (compared with Figure 17-6). Note the emergence of the fingerprint pattern between 600 and 1500 cm^{-1}, different from that pattern in the analogous spectrum of hexane (Figure 17-9).

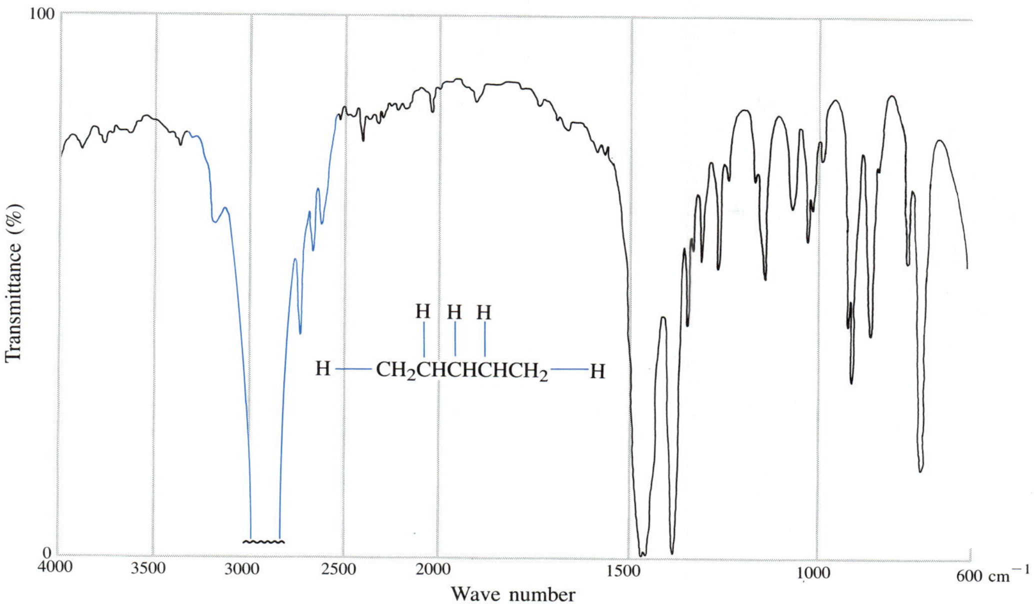

FIGURE 17-9

IR spectrum of a sample of hexane at higher recorder sensitivity (compared with Figure 17-7).

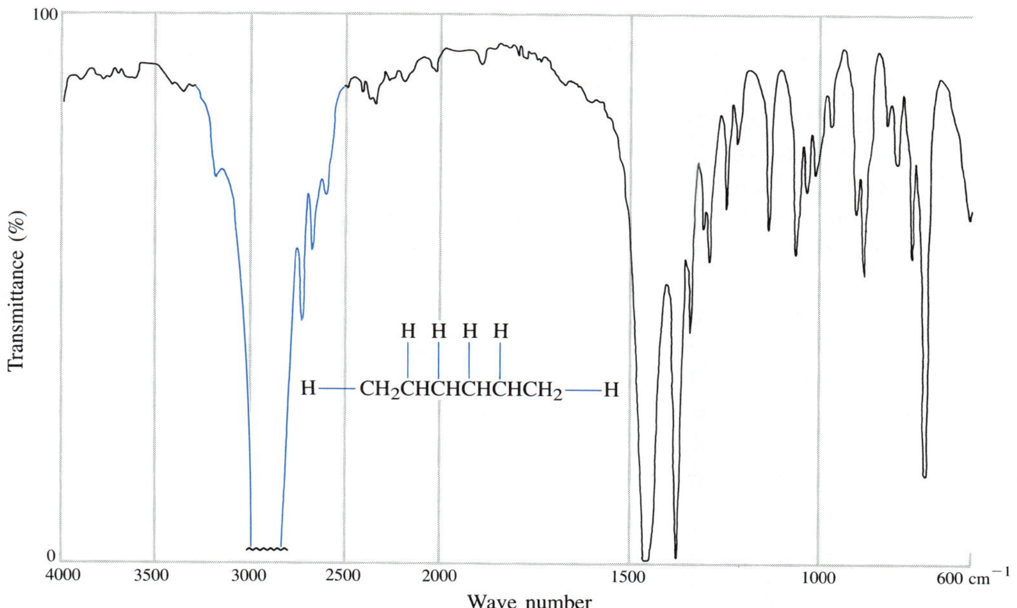

FIGURE 17-10

IR spectrum of 1-hexene: $\tilde{\nu}_{C_{sp^2}\text{–H stretch}} = 3080\ \text{cm}^{-1}$; $\tilde{\nu}_{C{=}C\text{ stretch}} = 1640\ \text{cm}^{-1}$; $\tilde{\nu}_{C_{sp^2}\text{–H bend}} = 995$ and $915\ \text{cm}^{-1}$.

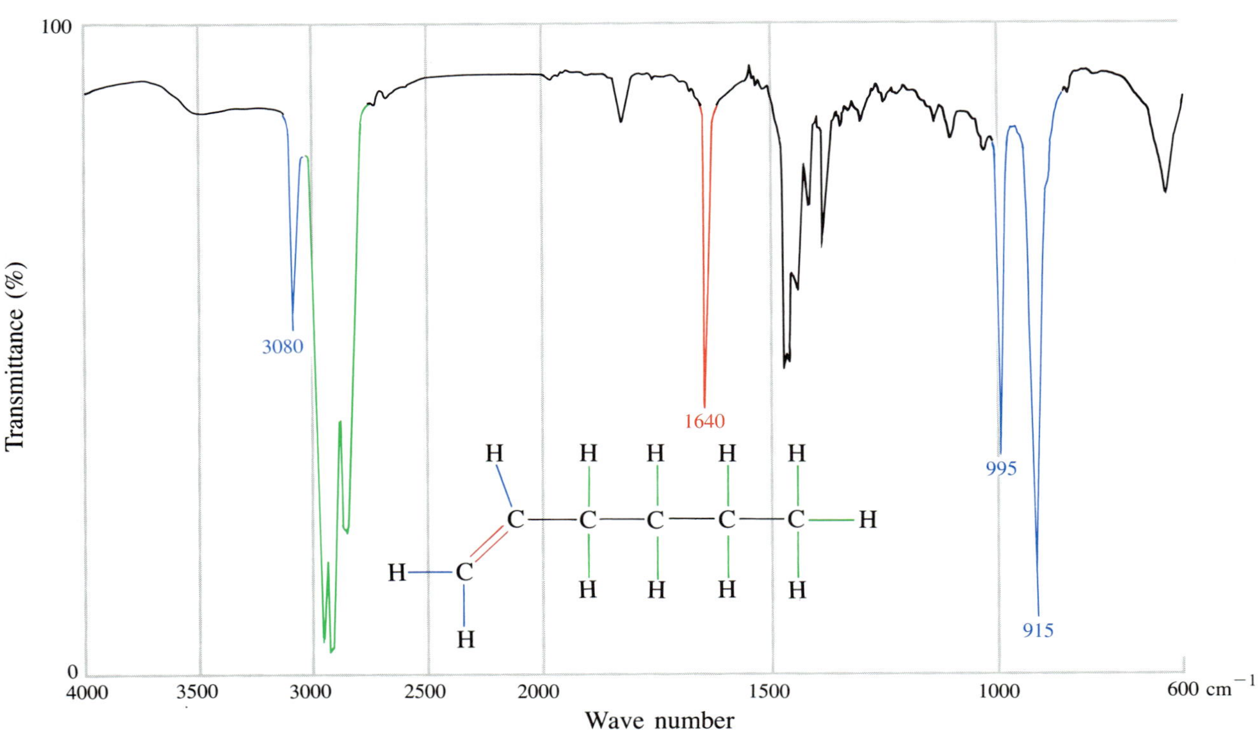

Strong Bending Modes for Alkenes

$RHC{=}CH_2$ (R, H / H, H)

$915, 995\ \text{cm}^{-1}$

$R_2C{=}CH_2$ (R, R / H, H)

$890\ \text{cm}^{-1}$

trans-$RHC{=}CHR$ (R, H / H, R)

$970\ \text{cm}^{-1}$

Alkenes

Figure 17-10 shows the IR spectrum of 1-hexene. A characteristic feature of alkenes when compared with alkanes is the stronger C_{sp^2}–H bond, which should therefore have a higher energy peak in the IR spectrum. Indeed, as the figure shows, there is a sharp spike at 3080 cm^{-1} due to this stretching mode, at slightly higher wave number than the remainder of the C–H stretching absorptions. According to Table 17-3, the C=C stretching band should appear between about 1620 and 1680 cm^{-1}. Figure 17-10 shows a relatively strong and sharp band at 1640 cm^{-1} assigned to this vibration. The other strong peaks are the result of bending motions. For example, the two signals at 915 and 995 cm^{-1} are typical of a terminal alkene.

Two other strong bending modes may be used as a diagnostic tool for the substitution pattern in alkenes. One results in a single band at 890 cm^{-1} and is characteristic of 1,1-dialkylethenes; the other gives a sharp peak at 970 cm^{-1} and is produced by the C_{sp^2}–H bending mode of a trans double bond. The presence or absence of such bands is often corroborating evidence for the presence of specifically substituted double bonds. This measurement, in conjunction with NMR (Section 11-3), allows for fairly certain structural assignments.

Alkynes

The most characteristic IR stretching bands of the alkynes (Table 17-3) are due to the alkynyl hydrogen (3260–3330 cm^{-1}) and the C≡C triple bond (2100–

FIGURE 17-11

IR spectrum of 1,7-octadiyne: $\tilde{\nu}_{C_{sp}\text{–H stretch}} = 3300\ cm^{-1}$; $\tilde{\nu}_{C\equiv C\text{ stretch}} = 2120\ cm^{-1}$; $\tilde{\nu}_{C_{sp}\text{–H bend}} = 640\ cm^{-1}$.

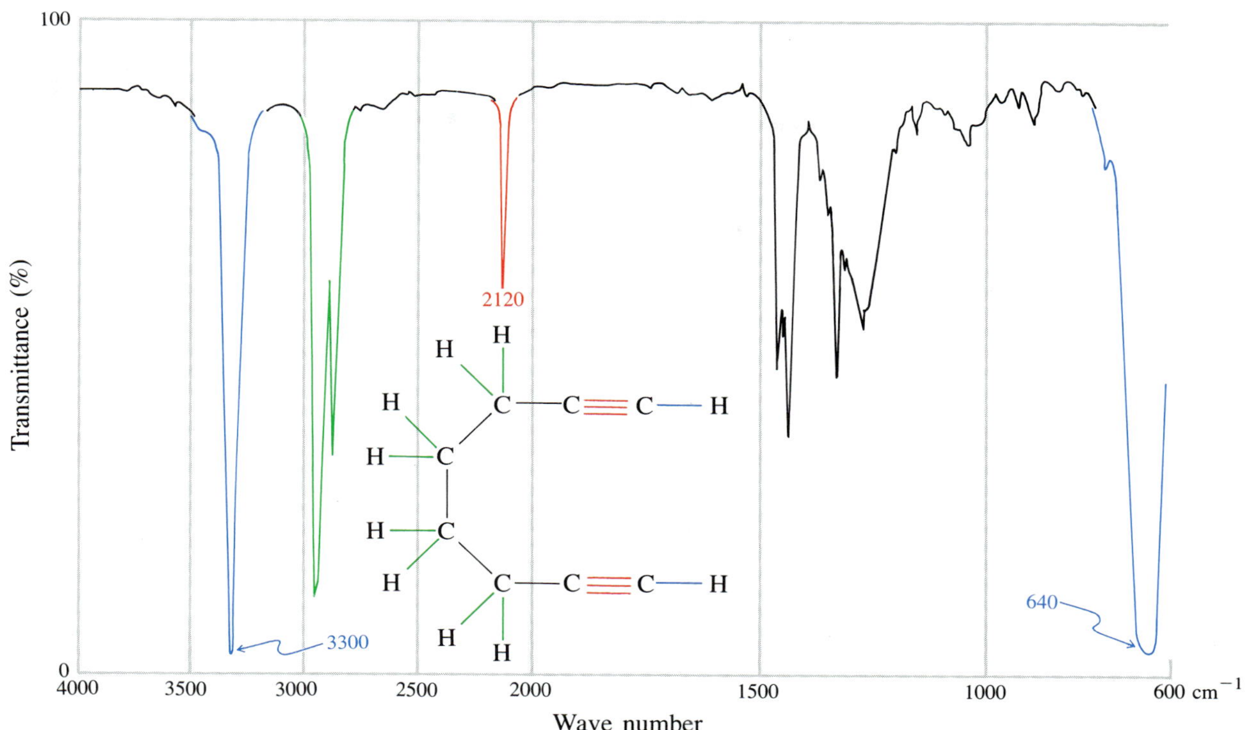

2260 cm^{-1}, Figure 17-11). Both appear at higher wave numbers than the corresponding vibrations in alkenes. A broad C_{sp}–H bending peak is found at about 640 cm^{-1}.

Alcohols

The O–H stretching absorption is the most characteristic band in the IR spectra of alcohols, appearing as a broad peak over a fairly wide range (3200–3650 cm^{-1}, Figure 17-12). The broadness of this peak is due to hydrogen bonding to other alcohol molecules or to water. Dry alcohols in dilute solution exhibit sharper bands in a narrower range (3620–3650 cm^{-1}).

Aldehydes and Ketones

Before the advent of routine ^{13}C NMR spectroscopy, which reveals the presence of the carbonyl group in aldehydes and ketones by the characteristic low-field chemical shifts of the carbonyl carbon, IR spectroscopy was the only reliable way of directly detecting this functionality. The C=O stretching frequency is unusually strong and typically appears in a relatively narrow range (1690–1750 cm^{-1}, Figure 17-13).

Carboxylic Acids

The carboxy group consists of a carbonyl group and an attached hydroxy substituent. Consequently, both characteristic stretching frequencies are seen in the

FIGURE 17-12

IR spectrum of cyclohexanol $\tilde{\nu}_{\text{O–H stretch}} = 3345\ \text{cm}^{-1}$; $\tilde{\nu}_{\text{C–O stretch}} = 1070\ \text{cm}^{-1}$. Note the broad O–H peak.

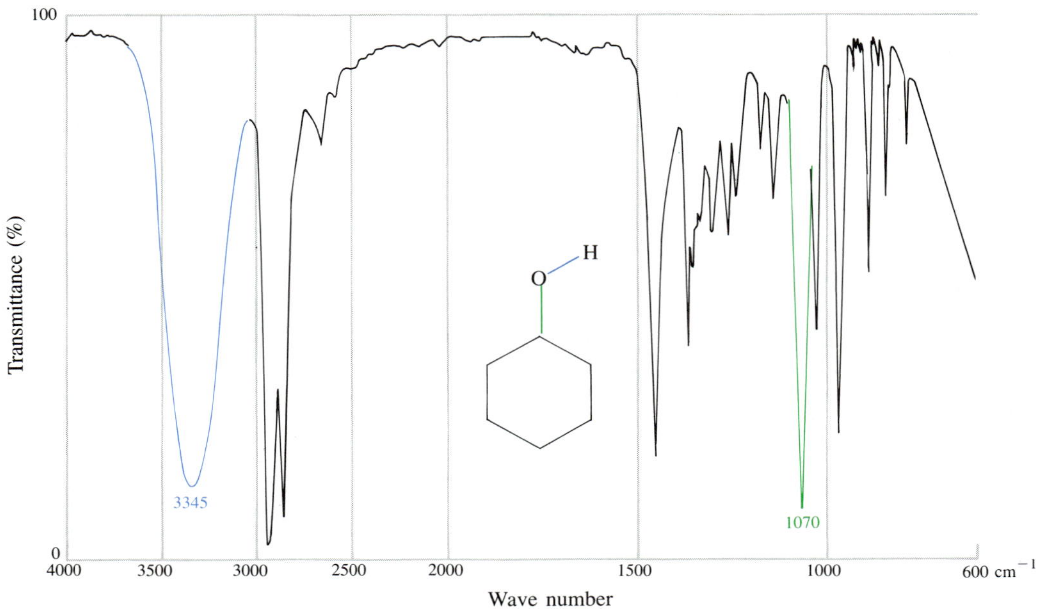

FIGURE 17-13

IR spectrum of 3-pentanone: $\tilde{\nu}_{\text{C=O stretch}} = 1715\ \text{cm}^{-1}$.

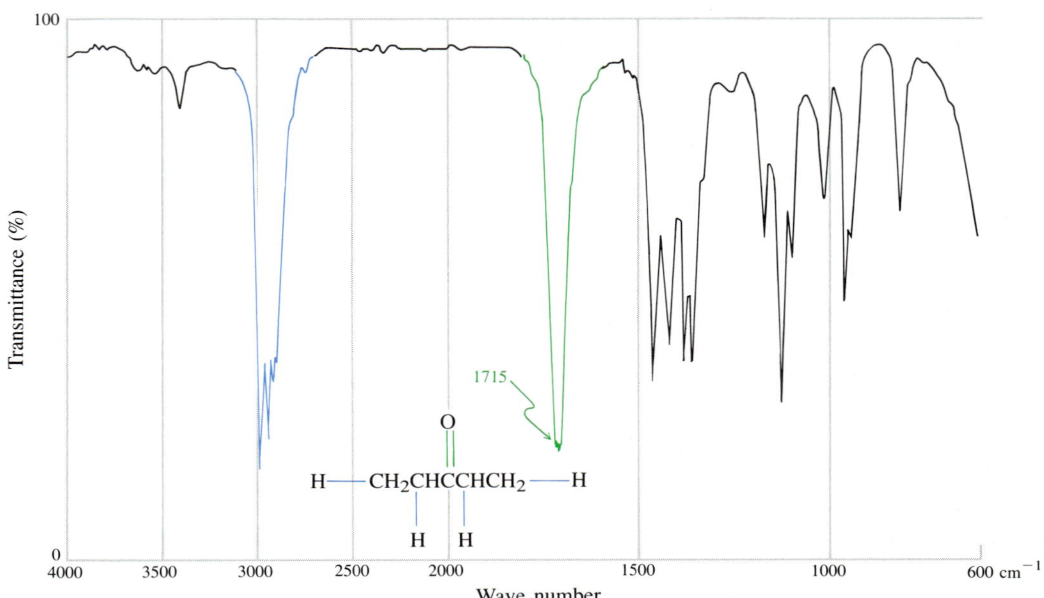

FIGURE 17-14

IR spectrum of propanoic acid: $\tilde{\nu}_{\text{O-H stretch}} = 3000\ \text{cm}^{-1}$; $\tilde{\nu}_{\text{C=O stretch}} = 1715\ \text{cm}^{-1}$.

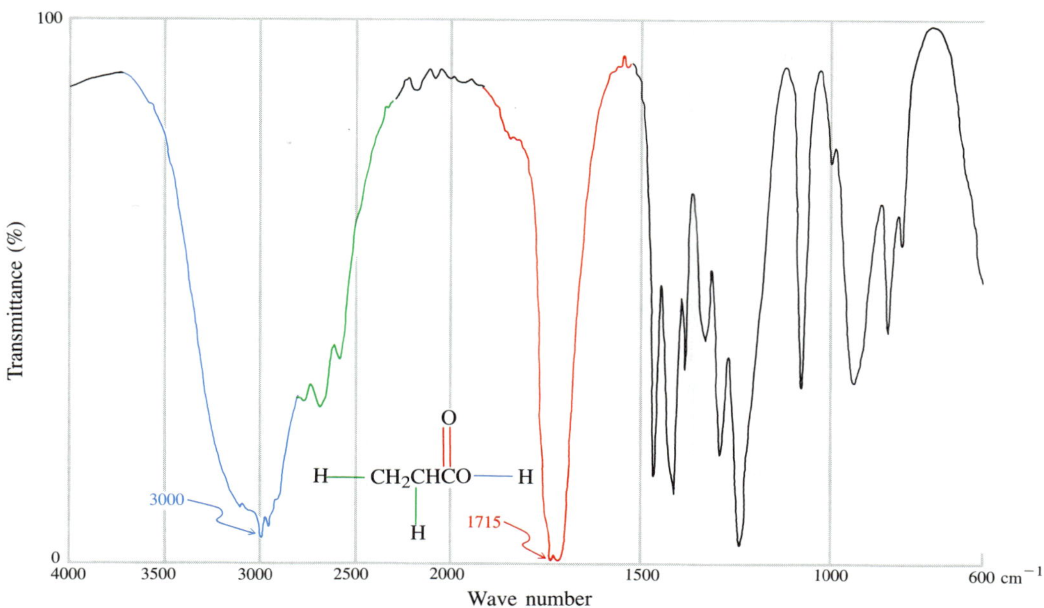

infrared spectrum (Figures 17-4 and 17-14). The O–H bond gives rise to a broad band at lower wave number (2500–3300 cm^{-1}) than is observed for alcohols, because of strong hydrogen bonding. The IR spectrum of propanoic acid is shown in Figure 17-14; it is instructive to compare it with the spectrum of ethanoic (acetic) acid (Figure 17-4). These acids have very similar absorption patterns for the O—H stretching motion (overlapping with the C—H stretching bands) and for the C=O stretching motion, but clearly different fingerprint regions.

EXERCISE 17-3

A colorless liquid gave an elemental analysis with the following values: C, 25.41%; H, 3.18%; Cl, 37.53%. The spectral data were as follows: ^{1}H NMR (CD_3COCD_3) δ = 10.35 (s, 1 H) and 5.36 (s, 2 H) ppm; ^{13}C NMR (CS_2) δ = 173.8 and 41.3 ppm; IR (neat) 3080 and 1728 cm^{-1}. What is the structure of this compound?

In summary, the presence of specific functional groups can be ascertained by infrared spectroscopy. Infrared light causes the vibrational excitation of bonds in molecules. Strong bonds and light atoms vibrate at relatively high stretching frequencies measured in wave numbers (reciprocal wavelengths). Conversely, weak bonds and heavy atoms absorb at lower wave numbers, as would be expected from Hooke's law. Because of the variety of stretching and bending modes and because of mechanical coupling, infrared spectra usually show complicated patterns. These are, however, diagnostic fingerprints for particular compounds. The presence of variously substituted alkenes may be detected by stretching signals at about 3080 (C—H) and 1640 (C=C) cm^{-1}, and bending modes between 890 and 990 cm^{-1}. Alkynes show characteristic bands at about 3300 (C—H) and 2120 (C≡C) cm^{-1}, alcohols at about 3345 (O—H), aldehydes and ketones at about 1715 (C=O), and carboxylic acids at about 3000 (O—H) and 1710 (C=O) cm^{-1}.

17-4 The Acidity and Basicity of Carboxylic Acids

Like alcohols (Section 8-3), carboxylic acids are both acidic and basic: deprotonation to carboxylate ions is relatively easy, protonation more difficult.

Deprotonation of Carboxylic Acids Gives Resonance-Stabilized Anions

As the name implies, carboxylic acids are acidic, much more so than alcohols even though the acidic proton is in both cases derived from a hydroxy group.

Carboxylic Acids Are Relatively Strong Acids

$$\mathrm{RC(=O)\ddot{O}H} + \mathrm{H_2\ddot{O}} \rightleftharpoons \mathrm{RC(=O)\ddot{O}:^-} + \mathrm{H\overset{+}{\ddot{O}}H_2}$$

$$K_a \sim 10^{-4}\text{–}10^{-5}$$
$$pK_a \sim 4\text{–}5$$

Why should this be so? The difference lies in the presence of the carbonyl group to which the hydroxy substituent is attached. The positively polarized carbon exerts an inductive electron-withdrawing effect on the neighboring groups and, more importantly, allows for resonance stabilization of the resulting carboxylate ion. These effects are the same as in aldehydes and ketones, in which the acidifying influence of the carbonyl group on its neighboring C–H units gives rise to resonance-stabilized enolate ions (Chapter 16).

Resonance in Carboxylate and Enolate Ions

Carboxylate ion:

$$\mathrm{B:^-} + \mathrm{RC(=\!\ddot{O}\!:)\ddot{O}H} \rightleftharpoons \mathrm{BH} + \left[\mathrm{RC(=\!\ddot{O}\!:)\!-\!\ddot{O}:^-} \longleftrightarrow \mathrm{RC(\!-\!\ddot{O}:^-)\!=\!\ddot{O}}\right]$$

Enolate ion:

$$\mathrm{B:^-} + \mathrm{R'C(=\!\ddot{O}\!:)CH_2R} \rightleftharpoons \mathrm{BH} + \left[\mathrm{R'C(=\!\ddot{O}\!:)\!-\!\ddot{C}^-HR} \longleftrightarrow \mathrm{R'C(\!-\!\ddot{O}:^-)\!=\!CHR}\right]$$

In contrast with enolates, the two resonance structures in carboxylate ions are equivalent (Section 1-5). As a result, carboxylates are symmetrical, with equal carbon–oxygen bond lengths (1.26 Å), in between the lengths typical of the carbon–oxygen double (1.20 Å) and single (1.34 Å) bonds in the corresponding acids (Figure 17-1). They may be viewed as oxygen analogs of the 2-propenyl (allyl) anion, which is also described by equivalent resonance structures (Section 14-1).

Electron-Withdrawing Substituents Increase the Acid Strength of Carboxylic Acids

As with alcohols, aldehydes, and ketones, inductively electron-withdrawing substituents situated close to the carboxy group cause an increase in acidity. The pK_as of selected carboxylic acids are shown in Table 17-4. The presence of several electron-withdrawing substituents can result in carboxylic acids that are as strong as typical mineral acids. The inductive effect is much less when the substituents are at some distance from the functional group.

TABLE 17-4

pK_a Values of some carboxylic and other acids

Compound	pK_a	Compound	pK_a
CH_3COOH	4.74	$CH_3CH_2CH_2COOH$	4.9
$ClCH_2COOH$	2.86	$CH_3CH_2CHClCOOH$	3.8
$Cl_2CHCOOH$	1.26	$CH_3CHClCH_2COOH$	4.1
Cl_3CCOOH	0.64	$ClCH_2CH_2CH_2COOH$	4.5
F_3CCOOH	0.23		
		H_3PO_4	2.15 (first pK_a)
HOOCCOOH	2.77, 5.81	HNO_3	−1.3
$HOOCCH_2COOH$	3.15, 6.30	HCl	−2.2
$HOOCCH_2CH_2COOH$	5.84, 6.34	H_2SO_4	−5.2 (first pK_a)
$HOOC(CH_2)_4COOH$	5.57, 6.38		
		H_2O	15.7
		CH_3OH	15.5

EXERCISE 17-4

In the following sets of acids, rank the components in order of *decreasing* acidity:

(a) CH_3CH_2COOH $\quad$ $CH_3CHBrCOOH$ (Br on C-2) $\quad$ CH_3CBr_2COOH;

(b) CH_3CHFCH_2COOH (F on C-3) $\quad$ $CH_3CHBrCH_2COOH$ (Br on C-3);

(c) COOH $\quad$ F COOH $\quad$ COOH F

The dioic acids have two pK_a values, one for each of the functional groups. In ethanedioic (oxalic) and propanedioic (malonic) acid, the first pK_a is much lower than the second, because of the mutual electron-withdrawing effect of the two carboxy groups. In the higher dioic acids, the two pK_a values are very close.

The relatively strong acidity of carboxylic acids means that their corresponding **carboxylate salts** are readily made by treatment of the acid with base, such as NaOH. These salts are named by specifying the metal and replacing the ending **-ic acid** in the acid name by **-ate**. Thus, $HCOO^-Na^+$ is called sodium methanoate (formate), $CH_3COO^-Li^+$ is named lithium ethanoate (acetate), and so on. Carboxylate salts are much more water soluble than the corresponding acids, because the polar anionic group is readily solvated.

Carboxylate Salt Formation

$$\underset{\substack{\textbf{4,4-Dimethylpentanoic acid}\\\textbf{(Slightly water-soluble)}}}{CH_3\overset{CH_3}{\underset{CH_3}{C}}CH_2CH_2COOH} \xrightarrow{NaOH,\ H_2O} \underset{\substack{\textbf{Sodium 4,4-dimethylpentanoate}\\\textbf{(Water soluble)}}}{CH_3\overset{CH_3}{\underset{CH_3}{C}}CH_2CH_2COO^-Na^+} + HOH$$

Protonation of Carboxylic Acids Gives Resonance-Stabilized Cations

The lone electron pairs on both of the oxygen atoms in the carboxy group can, in principle, be protonated, just as alcohols are protonated by strong acids to give oxonium ions (Section 8-3). Which oxygen is more basic? The available evidence indicates that it is the carbonyl oxygen. Why? We can answer this question by considering resonance. Protonation at the hydroxy group would generate an oxonium ion for which only one reasonable structure can be written. On the other hand, protonation at the carbonyl oxygen gives a molecule with three contributing structures.

Protonation of a Carboxylic Acid

$$RC(=O)OH + H^+ \rightleftharpoons RC(=O)\overset{+}{O}H_2 \quad \text{Not observed}$$

$$RC(=O)OH + H^+ \underset{K\,\sim\,10^{-6}}{\rightleftharpoons} \left[RC(=\overset{+}{O}H)OH \longleftrightarrow R\overset{+}{C}(OH)OH \longleftrightarrow RC(OH)=\overset{+}{O}H \right]$$

$pK_a \sim -6$

Note that protonation is nevertheless very difficult, as shown by the very high acidity ($pK_a \sim -6$) of the conjugate acid. It is in fact more difficult than protonation of an alcohol (pK_a of oxonium ion ~ -3), revealing that the carboxy oxygen has less available electron density. We shall see, however, that such protonations are important in many reactions of the carboxylic acids and their derivatives.

EXERCISE 17-5

The pK_a of protonated propanone (acetone) is -7.2, and that of protonated ethanoic (acetic) acid is -6.1. Explain.

In summary, carboxylic acids are acidic because deprotonation gives resonance-stabilized anions. Electron-withdrawing groups increase acidity, although this effect wears off rapidly with increasing distance from the carboxy group. Protonation is difficult, but possible, and occurs on the carbonyl oxygen to give a resonance-stabilized cation.

17-5 The Preparation of Carboxylic Acids

This section describes methods for the preparation of carboxylic acids. Some of these procedures have appeared in the description of the chemistry of other functional groups: oxidation of primary alcohols and aldehydes (Sections 9-4, 15-3 and 15-8) and the haloform reaction (Sections 9-4 and 16-2). Others are new: basic permanganate oxidations of alkenes, the carbonation of organometallic reagents, and the hydrolysis of nitriles. Still other preparations of carboxylic acids will be described in later chapters, including

1 Hydrolysis of other carboxylic acid derivatives (Chapter 18)

$$\mathrm{R\overset{O}{\overset{\|}{C}}X + HOH \longrightarrow RCOOH + HX}$$

2 Benzilic acid rearrangement (Section 22-1)

$$\mathrm{R\overset{O}{\overset{\|}{C}}-\overset{O}{\overset{\|}{C}}R \xrightarrow[2.\ H^+,\ H_2O]{1.\ HO^-} R\underset{OH}{\overset{R}{C}}COOH}$$

3 Malonic and related ester syntheses (Section 22-4)

$$\mathrm{(H)(R)C(COOR')_2 \xrightarrow[-\,2\,R'OH]{1.\ HO^-;\ 2.\ H^+,\ H_2O} (H)(R)C(COOH)_2 \xrightarrow{\Delta} (H)(R)C(H)COOH + O{=}C{=}O}$$

4 Friedel-Crafts alkanoylation with anhydrides (Sections 25-3 and 25-4)

$$\mathrm{R{-}C_6H_4{-}H + \text{succinic anhydride} \xrightarrow{AlCl_3} C_6H_5COCH_2CH_2COOH}$$

5 Aromatic side-chain oxidation (Section 24-2)

$$\mathrm{C_6H_5CH_2R \xrightarrow{KMnO_4,\ HO^-} C_6H_5COOH}$$

6 Kolbe reaction (Section 24-4)

$$\text{C}_6\text{H}_5\text{OH} + CO_2 \xrightarrow[\text{2. } H^+,\ H_2O]{\text{1. } HO^-} \text{o-HOC}_6\text{H}_4\text{COOH}$$

7 Amino acid synthesis (Section 27-2)

Oxidation of Alkenes with Permanganate Ion Gives Carboxylic Acids

The treatment of alkenes with basic potassium permanganate can result in complete oxidative cleavage at the double bond with the formation of carboxylic acids. (Under neutral conditions, oxidation stops at the vicinal diol stage; see Section 12-5.) Such oxidations work best with terminal alkenes. For example, 3,7-dimethyl-1-octene yields 2,6-dimethylheptanoic acid (and CO_2). The procedure is potentially useful for the degradation of terminal alkenes to the homologous acids with one less carbon.

$$CH_3\overset{CH_3}{\overset{|}{C}}HCH_2CH_2CH_2\overset{CH_3}{\overset{|}{C}}HCH{=}CH_2 \xrightarrow[\text{propanone (acetone), 7°C}]{KMnO_4,\ NaHCO_3,} CH_3\overset{CH_3}{\overset{|}{C}}HCH_2CH_2CH_2\overset{CH_3}{\overset{|}{C}}HCOOH + CO_2$$

45%

2,6-Dimethylheptanoic acid

Oxidation of Primary Alcohols and Aldehydes to Carboxylic Acids

Primary alcohols oxidize to aldehydes, which in turn may readily oxidize further to the corresponding carboxylic acids (Sections 9-4, 15-3, and 15-8).

$$RCH_2OH \xrightarrow{\text{Oxidation}} R\overset{O}{\overset{\|}{C}}H \xrightarrow{\text{Oxidation}} R\overset{O}{\overset{\|}{C}}OH$$

The oxidizing agents can be CrO_3, $KMnO_4$, nitric acid, or others.

$$CH_3(CH_2)_3\overset{CH_2CH_3}{\overset{|}{C}}HCH_2OH \xrightarrow{KMnO_4,\ H_2O,\ NaOH,\ 12\ h} CH_3(CH_2)_3\overset{CH_2CH_3}{\overset{|}{C}}HCOOH$$

74%

2-Ethyl-1-hexanol **2-Ethylhexanoic acid**

Nitric acid, HNO_3, is one of the cheapest strong oxidants. Its oxidizing power is due to its ready reduction to NO_2:

$$2\ HNO_3 + ClCH_2CH_2\overset{O}{\overset{\|}{C}}H \xrightarrow{25°C} ClCH_2CH_2\overset{O}{\overset{\|}{C}}OH + 2\ NO_2 + H_2O$$

79%

3-Chloropropanal **3-Chloropropanoic acid**

In the presence of vanadium oxide, nitric acid may even oxidize secondary alcohols and ketones with simultaneous cleavage of C–C bonds.

$$\text{Cyclohexanol} \xrightarrow{50\%\ HNO_3,\ V_2O_5,\ 60^\circ C} \underset{60\%}{\text{Hexanedioic acid (Adipic acid)}}$$

Cyclohexanol

60%
Hexanedioic acid
(Adipic acid)

The Haloform Reaction Yields Carboxylic Acids by Degradation

The haloform reaction (Sections 9-4 and 16-2) is used occasionally for synthetic purposes.

$$CH_3\overset{O}{\overset{\|}{C}}CH_2C(CH_3)_2C_6H_5 \xrightarrow[2.\ H^+,\ H_2O]{1.\ I_2,\ HO^-} HO\overset{O}{\overset{\|}{C}}CH_2C(CH_3)_2C_6H_5 + CHI_3$$

4-Methyl-4-phenyl-2-pentanone

84%
3-Methyl-3-phenylbutanoic acid

Triiodomethane (Iodoform)

Organometallic Compounds React with Carbon Dioxide to Give Carboxylic Acids

Carbon dioxide may be regarded as a ''diketone'' of carbon. As such, organometallic reagents attack it much as they would attack aldehydes or ketones, except that a carboxylate salt is formed. Acidic aqueous work-up yields the acid.

Carbonation of Organometallics

$$R^{\delta-}\text{—}Mg^{\delta+}X + {}^{\delta-}{:}\ddot{O}{=}\overset{\delta+}{C}{=}\ddot{O}{:}^{\delta-} \longrightarrow R\text{—}C(=\ddot{O}{:})\text{—}\ddot{\underset{..}{O}}{:}^-\ {}^+MgX \xrightarrow[-\ XMgOH]{H^+,\ HOH} RCOOH$$

$$RLi + CO_2 \longrightarrow RCOO^-Li^+ \xrightarrow[-\ LiOH]{H^+,\ HOH} RCOOH$$

Because organometallic reagents can be made from the corresponding haloalkanes, this procedure allows for the preparation of the homologous acids with one more carbon: RX $\rightarrow$ RCOOH.

$$CH_3CH_2\overset{Cl}{\overset{|}{C}}HCH_3 + Mg \longrightarrow CH_3CH_2\overset{MgCl}{\overset{|}{C}}HCH_3 \xrightarrow[2.\ H^+,\ H_2O]{1.\ CO_2} CH_3CH_2\overset{COOH}{\overset{|}{C}}HCH_3$$

2-Chlorobutane

86%
2-Methylbutanoic acid

$$CH_3C{\equiv}CH \xrightarrow[3.\ H^+,\ H_2O]{1.\ CH_3CH_2CH_2CH_2Li,\ THF,\ hexane,\ -30^\circ C;\ 2.\ CO_2,\ 0^\circ C} CH_3C{\equiv}CCOOH$$

Propyne

98%
2-Butynoic acid

Nitriles Hydrolyze to Carboxylic Acids

Another method for producing the next higher carboxylic acid from a haloalkane is through the hydrolysis of an intermediate nitrile. Recall (Section 6-3) that cyanide is a good nucleophile and may be used to synthesize nitriles. The nitrile reacts with hot acidic or basic water to furnish the corresponding carboxylic acid and ammonia.

Carboxylic Acids from Haloalkanes Through Nitriles

$$\text{RX} \xrightarrow[-\,\text{X}^-]{^-\text{CN}} \text{RC}{\equiv}\text{N} \xrightarrow{\text{1. HO}^-\ \ \text{2. H}^+,\ \text{H}_2\text{O}} \text{RCOOH} + \text{NH}_3$$

The mechanism of this reaction will be discussed in Section 18-6. Although this procedure at first glance appears to have no particular advantage over Grignard carbonation, it is in fact complementary. For example, hydroxy and carboxy groups present in the molecule do not have to be protected.

$$\text{CH}_3(\text{CH}_2)_{15}\text{CN} \xrightarrow{\text{1. KOH, CH}_3\text{CH}_2\text{OH, H}_2\text{O, 31 h, }\Delta\ \ \text{2. HCl, H}_2\text{O}} \text{CH}_3(\text{CH}_2)_{15}\text{COOH}\ \ 79\%$$

Heptadecanenitrile → **Heptadecanoic acid**

$$\text{ClCH}_2\text{COOH} \xrightarrow{\text{1. NaCN, Na}_2\text{CO}_3\ \ \text{2. NaOH, }60^\circ\text{–}70^\circ\text{C}\ \ \text{3. HCl, H}_2\text{O}} \text{HOOCCH}_2\text{COOH}\ \ 80\%$$

Chloroethanoic acid → **Propanedioic acid (Malonic acid)**

$$\text{HOCH}_2\text{CH}_2\text{Cl} \xrightarrow{\text{1. NaCN}\ \ \text{2. HO}^-,\ \text{H}_2\text{O, }\Delta\ \ \text{3. H}^+,\ \text{H}_2\text{O}} \text{HOCH}_2\text{CH}_2\text{COOH}\ \ 65\%$$

2-Chloroethanol → **3-Hydroxypropanoic acid**

EXERCISE 17-6

Suggest ways to effect the following conversions (more than one step will be required):

(a) Cyclohexane-CHO ----→ Cyclohexane-HOCHCOOH

(b) Methylenecyclohexane ----→ 1-methylcyclohexane-1-carboxylic acid (H_3C, COOH)

(c) Br / OCH_3 substituted cyclohexane ----→ COOH / OCH_3 substituted cyclohexane

In summary, there are several synthetic methods for making carboxylic acids by oxidation, carbonation, and hydrolysis of appropriate precursors. For example, alkenes, secondary alcohols, and ketones can be oxidatively cleaved, and primary alcohols and aldehydes may be oxidized to carboxylic acids. The haloform reaction leads to loss of a carbon to give the next lower homologous acid, whereas carbonation of organometallic compounds or hydrolysis of nitriles, both accessible through haloalkanes, gives the next higher homologous acid.

17-6 Reactivity of the Carboxy Group: The Mechanism of Addition-Elimination

Apart from their acid-base properties, carboxylic acids show reactivity at the carbonyl nucleus similar to that in aldehydes and ketones (Section 15-4): the carbonyl carbon is subject to nucleophilic attack, the oxygen is the target of electrophiles, and the neighboring hydrogens are acidic and enolizable. However, the course of nucleophilic addition is different from that to aldehydes and ketones. The hydroxy substituent (by itself or in a modified form) can function as a leaving group, giving rise to new carbonyl derivatives.

The Carbonyl Carbon in Carboxylic Acids Is Attacked by Nucleophiles

The carbonyl carbon in carboxylic acids is electrophilic and can be attacked by nucleophiles. In contrast to the addition products of aldehydes and ketones, *the intermediate alkoxide can decompose by eliminating hydroxide*. The process by which the nucleophile replaces the hydroxy group is called **addition-elimination.**

Addition-Elimination of a Carboxylic Acid

$$\mathrm{RC(=O)OH} + \mathrm{Nu}^- \underset{}{\overset{\text{Addition}}{\rightleftharpoons}} \underset{\textbf{Tetrahedral intermediate}}{\mathrm{R{-}C(O^-)(Nu){-}OH}} \overset{\text{Elimination}}{\rightleftharpoons} \mathrm{RC(=O)Nu} + {}^-\mathrm{OH}$$

The species formed first in this reaction contains (in contrast with both starting material and product) a tetrahedral carbon center. It is therefore called the **tetrahedral intermediate.** Remember, however, that the hydroxy proton is acidic and that most nucleophiles are bases. Therefore, an acid-base reaction can interfere with nucleophilic attack:

$$\mathrm{R{-}C(O^-)(Nu){-}OH} \underset{\substack{\text{Reversible}\\ \text{nucleophilic}\\ \text{addition}\\ (\text{path } a)}}{\overset{\mathrm{Nu}:^-\ (a)}{\rightleftharpoons}} \mathrm{RC(=O){-}O{-}H} \underset{\substack{\text{Reversible}\\ \text{acid-base}\\ \text{reaction}\\ (\text{path } b)}}{\overset{:\mathrm{Nu}^-\ (b)}{\rightleftharpoons}} \mathrm{RC(=O)O^-} + \mathrm{NuH}$$

If the nucleophile is very basic (an alkoxide, e.g.), the formation of the carboxylate ion will be essentially irreversible. Addition of an excess of nucleophilic reagent to force addition to the carboxylate ion is futile, because carboxylates are virtually inert to nucleophilic attack. It takes a very strong nucleophile to add to a negatively charged species. Such reactions take place only with organometallic nucleophiles (Section 17-10).

If the nucleophile is less basic, carboxylate formation is reversible, and nucleophilic attack may become competitive. Even so, because of complications due to deprotonation when the nucleophile acts as a base, carboxylic acids themselves undergo addition-elimination reactions only under special circumstances (Sections 17-9 and 17-10). However, such transformations are common for other carboxylic acid derivatives, RCOL, in which L does not bear an acidic proton (e.g., L = Cl, Br, OR, NR_2). The result is substitution of the leaving group L by the nucleophile Nu.

Nucleophilic Substitution of Carboxylic Acid Derivatives

$$\mathrm{R\overset{O}{\overset{\|}{C}}L} + {}^{-}\!:\mathrm{Nu} \longrightarrow \mathrm{R\overset{O}{\overset{\|}{C}}Nu} + {}^{-}\!:\mathrm{L}$$

A special example of addition-elimination has already been described in Section 16-2. In the last step of the haloform reaction, hydroxide functioned as the nucleophile, and the leaving group was a trihalomethyl anion, ${}^{-}\!:CX_3$.

Addition-Elimination Is Catalyzed by Base or Acid

Additions to the carbonyl group in carboxylic acids and their derivatives are catalyzed by either base or acid. Base ensures the maximum concentration of negatively charged (deprotonated) nucleophile (such as HO^-, RO^-, and RS^-) when it is the attacking species.

Base-Catalyzed Addition-Elimination

(L = leaving group, B = base)

STEP 1: Deprotonation of NuH

$$\mathrm{Nu{-}H} + {}^{-}\!:\mathrm{B} \rightleftharpoons {}^{-}\!:\mathrm{Nu} + \mathrm{BH}$$

STEP 2: Addition-elimination

$$\mathrm{R\overset{:O:}{\overset{\|}{C}}L} + {}^{-}\!:\mathrm{Nu} \rightleftharpoons \mathrm{R{-}\underset{Nu}{\overset{:\ddot{O}:^{-}}{C}}{-}L} \rightleftharpoons \mathrm{R\overset{:O:}{\overset{\|}{C}}Nu} + {}^{-}\!:\mathrm{L}$$

STEP 3: Regeneration of catalyst

$${}^{-}\!:\mathrm{L} + \mathrm{H{-}B} \rightleftharpoons \mathrm{LH} + {}^{-}\!:\mathrm{B}$$

(Alternatively, ${}^{-}$:L may act as a base in Step 1.)

Acid protonates the oxygen to activate the carbonyl carbon making attack by a neutral nucleophile feasible.

Acid-Catalyzed Addition-Elimination

STEP 1: Protonation

$$\mathrm{RC(=\ddot{O}\!:)L} + H^+ \rightleftharpoons \mathrm{RC(=\overset{+}{O}H)L}$$

STEP 2: Addition-elimination

$$\mathrm{RC(=\overset{+}{O}H)L} + :\mathrm{NuH} \rightleftharpoons \mathrm{R{-}C(\ddot{O}H)(L){-}\overset{+}{Nu}H} \rightleftharpoons \mathrm{R{-}C(\ddot{O}H)(Nu){-}\overset{+}{L}H} \rightleftharpoons \mathrm{RC(=\overset{+}{O}H)Nu} + \mathrm{LH}$$

STEP 3: Deprotonation

$$\mathrm{RC(=\overset{+}{O}H)Nu} \rightleftharpoons \mathrm{RC(=\ddot{O}\!:)Nu} + H^+$$

The overall result of addition-elimination is the displacement of the leaving group by the nucleophile. You will see in the following sections and in Chapter 18 how important this pathway is for the interconversion of carboxylic acid derivatives.

In summary, potential nucleophilic attack on the carbonyl carbon of carboxylic acids may be complicated by a side reaction in which the nucleophile acts as a base to deprotonate the acid. With less basic nucleophiles and, more generally, with carboxylic acid derivatives lacking acidic hydrogens but bearing potential leaving groups L, nucleophilic attack can occur, to be followed by elimination of L. This addition-elimination sequence is catalyzed by base and acid.

17-7 Transformation of Carboxylic Acids into Their Derivatives: Alkanoyl (Acyl) Halides and Anhydrides

The potential complications arising from the acidity of the carboxy function may be bypassed by protecting or otherwise transforming the hydroxy group. This treatment gives rise to a variety of carboxylic acid derivatives, which are described here and in subsequent sections, and includes alkanoyl (acyl) halides, in which the hydroxy group of the acid has been replaced by a halogen, and anhydrides, in which it has been replaced by RCOO.

Alkanoyl Chlorides Through Anhydrides

$$\mathrm{RCOH} + \underset{\text{Phosgene}}{\mathrm{ClC(=O)Cl}} \xrightarrow[-\,\mathrm{HCl}]{} \underset{\text{An anhydride}}{\mathrm{RCOCCl}} \xrightarrow[-\,\mathrm{CO_2}]{} \mathrm{RCCl}$$

$$\mathrm{RCOH} + \underset{\text{Oxalyl chloride}}{\mathrm{ClC{-}CCl}} \xrightarrow[-\,\mathrm{HCl}]{} \underset{\text{An anhydride}}{\mathrm{RCOC{-}CCl}} \longrightarrow \mathrm{RCCl} + \mathrm{CO_2} + \mathrm{CO}$$

EXERCISE 17-7

Propose a mechanism for the decomposition of the anhydrides derived from (a) phosgene and (b) oxalyl chloride. Hint: Review the mechanism of the reaction of a carboxylic acid and thionyl chloride.

With alkanoyl chlorides other than phosgene and oxalyl chloride, the carboxylic acids can form stable anhydrides.

$$\underset{\text{Butanoic acid}}{\mathrm{CH_3CH_2CH_2COH}} + \underset{\text{Butanoyl chloride}}{\mathrm{ClCCH_2CH_2CH_3}} \xrightarrow{\Delta,\ 8\ h} \underset{\substack{85\% \\ \text{Butanoic anhydride}}}{\mathrm{CH_3CH_2CH_2COCCH_2CH_2CH_3}} + \mathrm{HCl}$$

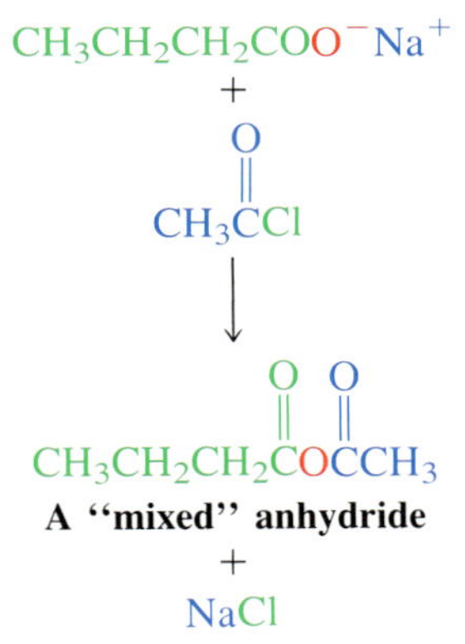

The reaction of an alkanoyl halide with a carboxylate salt produces the anhydride under neutral conditions, as shown at the left.

As the name indicates, carboxylic anhydrides are formally derived from the corresponding acids by loss of water. They can actually be made in this way by thermal dehydration, although this procedure is normally not a very effective way to prepare anhydrides from two acid molecules. On the other hand, the intramolecular version of this approach can be carried out by simply heating a dicarboxylic acid to give a cyclic anhydride. A condition for the success of this transformation is that the ring closure lead to a five- or six-membered ring product.

Cyclic Anhydride Formation

$$\underset{\substack{\text{Butanedioic acid} \\ \text{(Succinic acid)}}}{\mathrm{HOOC{-}CH_2{-}CH_2{-}COOH}} \xrightarrow{300^\circ\mathrm{C}} \underset{\substack{95\% \\ \text{Butanedioic anhydride} \\ \text{(Succinic anhydride)}}}{\text{cyclic } \mathrm{H_2C{-}C(=O){-}O{-}C(=O){-}CH_2}} + \mathrm{H_2O}$$

Because the halogen in the alkanoyl halide and the RCO_2 substituent in the anhydride are good leaving groups, and because they activate the adjacent carbonyl function, these carboxylic acid derivatives are useful synthetic intermediates for the preparation of other compounds, to be discussed in Sections 18-2 and 18-3.

EXERCISE 17-8

Suggest two preparations, starting from carboxylic acids or their derivatives, for each of the following compounds:

(a) $CH_3\overset{O}{\overset{\|}{C}}O\overset{O}{\overset{\|}{C}}CH_2CH_3$; (b) $CH_3\overset{H_3C}{\overset{|}{C}}H\overset{O}{\overset{\|}{C}}Cl$.

EXERCISE 17-9

Propose a mechanism for the thermal formation of butanedioic anhydride from the dioic acid.

In summary, the hydroxy group in COOH can be replaced by halogen using the same reagents employed in the conversion of alcohols into haloalkanes—$SOCl_2$, PCl_5, or PBr_3. In addition, phosgene and oxalyl chloride furnish alkanoyl chlorides from carboxylic acids by decomposition of intermediate anhydrides. Anhydrides are stable when formed from an ordinary alkanoyl chloride and an acid or its salt. Carboxylic anhydrides can also be made from their component acids by dehydration, a reaction that is most successful if carried out in an intramolecular manner, when it gives cyclic anhydrides.

17-8 Transformation of Carboxylic Acids into Their Derivatives: Ester Synthesis

Esters have the general formula $R\overset{O}{\overset{\|}{C}}OR'$. They are the most important of the carboxylic acid derivatives. This section describes how esters are made from carboxylic acids and elaborates the mechanistic details of one of these methods, the mineral-acid-catalyzed esterification of carboxylic acids with alcohols.

General Acid-Catalyzed Esterification

$$\underset{\textbf{Carboxylic acid}}{R\overset{O}{\overset{\|}{C}}OH} + \underset{\textbf{Alcohol}}{R'OH} \overset{H^+}{\rightleftharpoons} \underset{\textbf{Ester}}{R\overset{O}{\overset{\|}{C}}OR'} + H_2O$$

Carboxylic Acids React with Alcohols to Form Esters

When a carboxylic acid and an alcohol are mixed together, no reaction takes place. However, upon addition of catalytic amounts of a mineral acid, such as sulfuric acid or hydrogen chloride, the two components combine gradually to give an **ester** and water (see Section 9-3).

This transformation is not very exothermic, and the ester is formed in equilibrium concentrations. The equilibrium may be shifted toward the ester product side by using an excess of either one of the two starting materials or by selectively removing the ester or the water from the reaction mixture. For example, esterifications are often carried out by using the alcohol as a solvent:

$$\underset{\substack{\textbf{Ethanoic acid}\\\textbf{(Acetic acid)}}}{CH_3\overset{\displaystyle O}{\overset{\|}{C}}OH} + \underset{\textbf{Solvent}}{CH_3OH} \xrightleftharpoons{H_2SO_4,\ \Delta} \underset{\substack{85\%\\\textbf{Methyl ethanoate}\\\textbf{(Methyl acetate)}}}{CH_3\overset{\displaystyle O}{\overset{\|}{C}}OCH_3} + H_2O$$

The opposite of esterification is **ester hydrolysis.** This reaction is carried out under the same conditions as esterification, but, to shift the equilibrium, an excess of water is used in a water-miscible solvent.

$$\underset{\textbf{Ethyl 2,2-dimethylhexanoate}}{CH_3CH_2CH_2CH_2\overset{\displaystyle CH_3}{\underset{\displaystyle CH_3}{C}}COOCH_2CH_3} \xrightarrow{H_2SO_4,\ HOH,\ \text{propanone (acetone)},\ \Delta}$$

$$\underset{\substack{85\%\\\textbf{2,2-Dimethylhexanoic acid}}}{CH_3CH_2CH_2CH_2\overset{\displaystyle CH_3}{\underset{\displaystyle CH_3}{C}}COOH} + CH_3CH_2OH$$

The Mechanism of Esterification

The esterification of a carboxylic acid with methanol can be followed by labeling the alcohol oxygen with the ^{18}O isotope. The label allows differentiation between two mechanistic possibilities: it may appear either in the ester or in the water product.

Two Options in Esterification with Labeled Methanol

Label appears in the ester:

$$R\overset{\displaystyle O}{\overset{\|}{C}}OH + H^{18}OCH_3 \xrightleftharpoons{H^+} R\overset{\displaystyle O}{\overset{\|}{C}}{}^{18}OCH_3 + H_2O$$

Label appears in the water:

$$R\overset{\displaystyle O}{\overset{\|}{C}}OH + H^{18}OCH_3 \underset{\text{Not observed}}{\overset{H^+}{\not\rightleftharpoons}} R\overset{\displaystyle O}{\overset{\|}{C}}OCH_3 + H_2{}^{18}O$$

The result of this experiment shows that the first option is followed; that is, the alcohol oxygen is incorporated into the ester. This and other observations have led to the formulation of the following mechanism for acid-catalyzed esterification and its reverse, hydrolysis.

Mechanism of Acid-Catalyzed Esterification and Ester Hydrolysis:

STEP 1: Protonation of the carboxy group

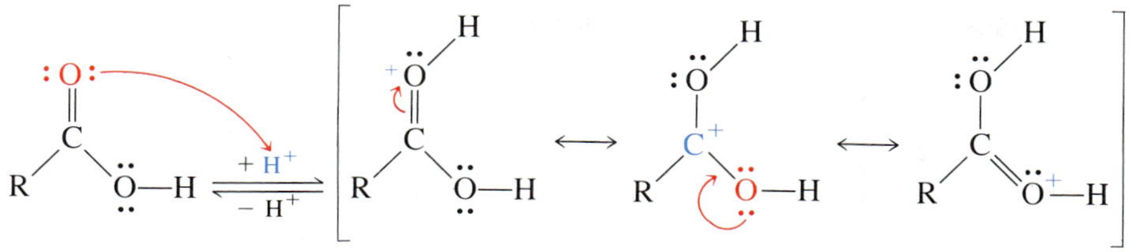

Dihydroxycarbocation

STEP 2: Attack by methanol

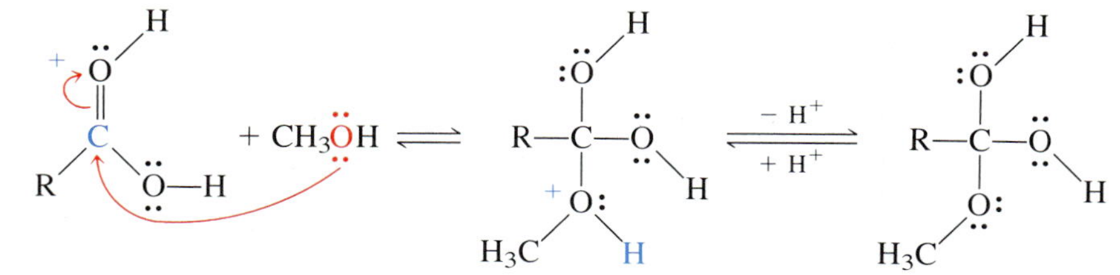

Tetrahedral intermediate

STEP 3: Elimination of water

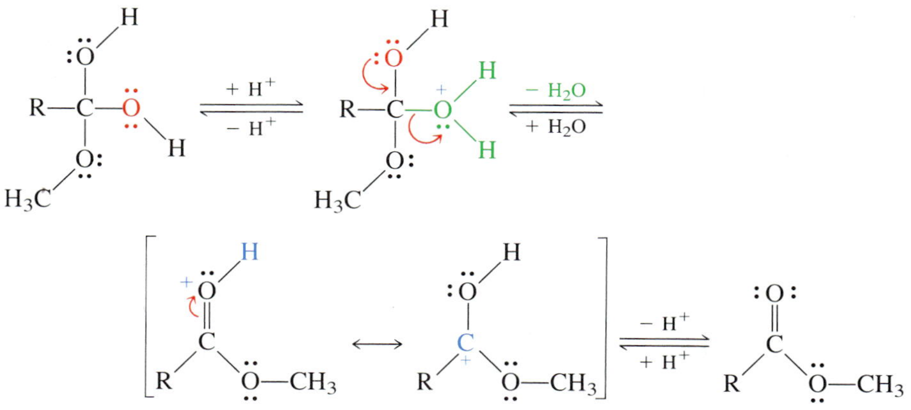

Initially, protonation of the acid gives a delocalized dihydroxycarbocation (Step 1). This formation renders the carbonyl carbon susceptible to nucleophilic attack by methanol. Proton loss from the initial adduct gives the tetrahedral intermediate of nucleophilic addition (Step 2). This species is a crucial relay point, because it undergoes acid-catalyzed decomposition in both the forward and the backward direction. In the latter, protonation at the methoxy oxygen induces elimination of methanol by the reverse of the sequence of Steps 1 and 2. However, protonation at either of the hydroxy oxygens leads to elimination of water and to the ester product (Step 3). This mechanism nicely explains why acid catalysis is necessary and why the ^{18}O label in the methanol ends up in the ester.

EXERCISE 17-10

Incomplete hydrolysis of methyl ethanoate (methyl acetate) with $H_2{}^{18}O$ leads to recovery of partly labeled starting material, $CH_3\overset{^{18}O}{\overset{\|}{C}}OCH_3$. Explain.

Hydroxy Acids May Undergo Intramolecular Esterification to Lactones

When hydroxy acids are treated with catalytic amounts of mineral acid, intramolecular esters can form. Such cyclic esters are called **lactones** (Section 18-4).

Lactone Formation

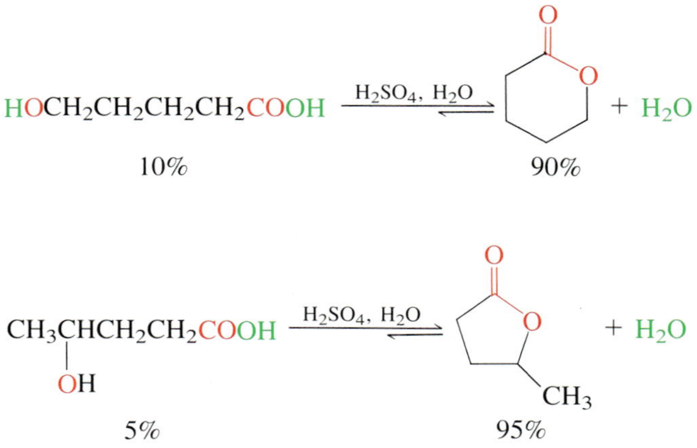

$$HOCH_2CH_2CH_2CH_2COOH \underset{}{\overset{H_2SO_4,\ H_2O}{\rightleftharpoons}} \text{(six-membered lactone)} + H_2O$$

10% 90%

$$CH_3\underset{OH}{\underset{|}{C}}HCH_2CH_2COOH \overset{H_2SO_4,\ H_2O}{\rightleftharpoons} \text{(five-membered lactone, } CH_3\text{)} + H_2O$$

5% 95%

The synthesis of lactones is favored only for five- and six-membered rings, and for larger rings in which ring strain and transannular interactions are negligible (see Sections 4-2 and 9-5).

EXERCISE 17-11

Explain the following result by a mechanism:

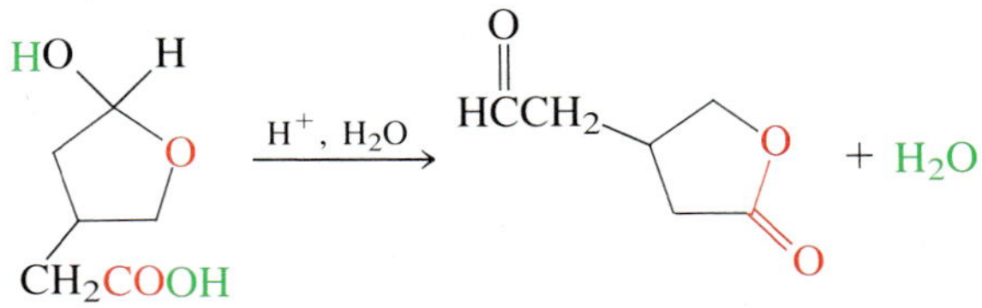

HO, H, O, CH_2COOH $\xrightarrow{H^+,\ H_2O}$ $H\overset{O}{\overset{\|}{C}}CH_2$—(lactone), O, O $+ H_2O$

The preparation of large-ring (macrocyclic) lactones requires high-dilution techniques to ensure that esterification is intramolecular. Nevertheless, polymer formation is frequently extensive, because lactone formation is reversible.

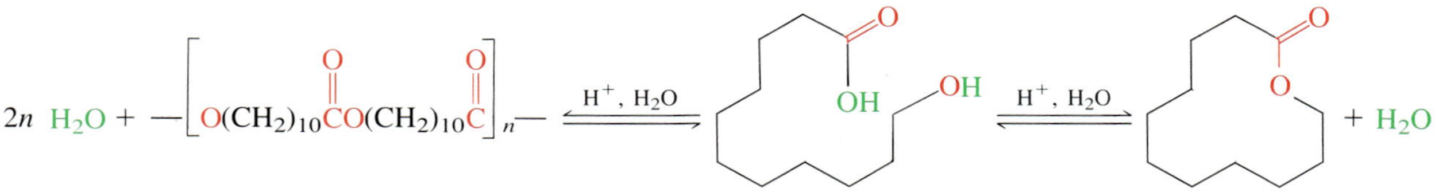

$$2n\ H_2O + -\left[O(CH_2)_{10}\overset{O}{\overset{\|}{C}}O(CH_2)_{10}\overset{O}{\overset{\|}{C}}\right]_n- \overset{H^+,\ H_2O}{\rightleftharpoons} \text{(hydroxy acid: OH, OH, O)} \overset{H^+,\ H_2O}{\rightleftharpoons} \text{(macrolactone: O, O)} + H_2O$$

Polymer **Low yield**

BOX 17-1

Enantioselective Lactone Synthesis in Nature

Like many other chiral natural products, most macrolide lactones are formed in nature as only one enantiomer. An example of how such selectivity may be achieved is the enantioselective lactone synthesis catalyzed by the enzyme horse liver alcohol dehydrogenase, starting from meso diols:

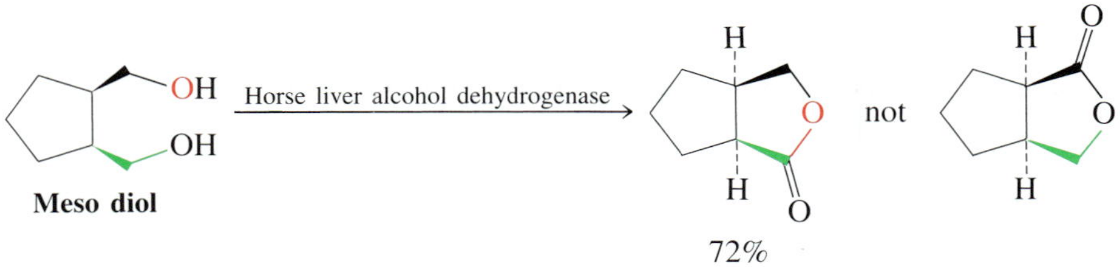

Meso diol —Horse liver alcohol dehydrogenase→ (lactone) 72% not (lactone)

Macrocyclic lactone synthesis is a topic of synthetic interest because the basic framework of many antibiotics (the *macrolide antibiotics*) consists of a substituted large-ring lactone.

Macrolide Antibiotics

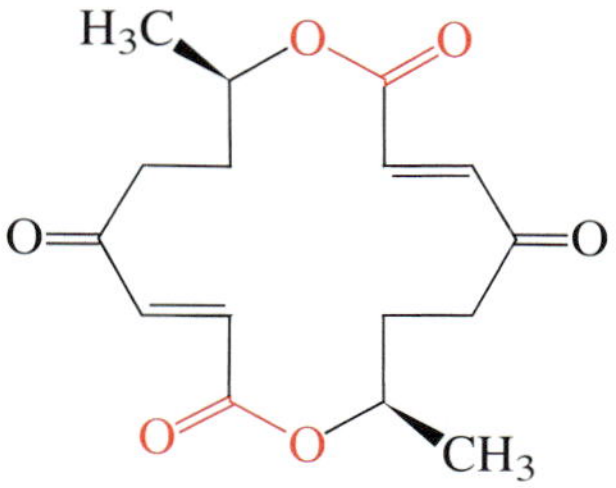

Pyrenophorin 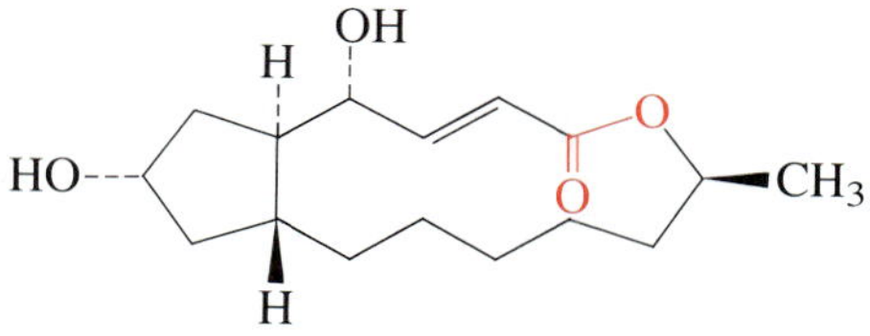 **Brefeldin A**

Carboxylic Acids Form Esters by Other Mechanisms

In addition to acid-catalyzed esterification, other reactions can transform carboxylic acids into esters. Two, in particular, are nucleophilic substitution of haloalkanes with carboxylate ions and methyl ester formation by the reaction of carboxylic acids with diazomethane, CH_2N_2.

The first of these methods has already been described, in connection with the synthesis of alcohols (Section 8-4). Carboxylate ions are nucleophiles that give esters by S_N2 processes, particularly when the substrates are primary haloalkanes. For example, iodobutane furnishes butyl ethanoate (acetate) on treatment with sodium ethanoate (acetate).

$$CH_3CH_2CH_2CH_2I + CH_3\overset{\overset{O}{\|}}{C}O^-Na^+ \xrightarrow{CH_3CO_2H,\ 100°C} CH_3CH_2CH_2CH_2O\overset{\overset{O}{\|}}{C}CH_3 + Na^+I^-$$

Iodobutane — **Sodium ethanoate (acetate)** — **Butyl ethanoate (acetate)** 95%

This method can be used to prepare macrocyclic lactones.

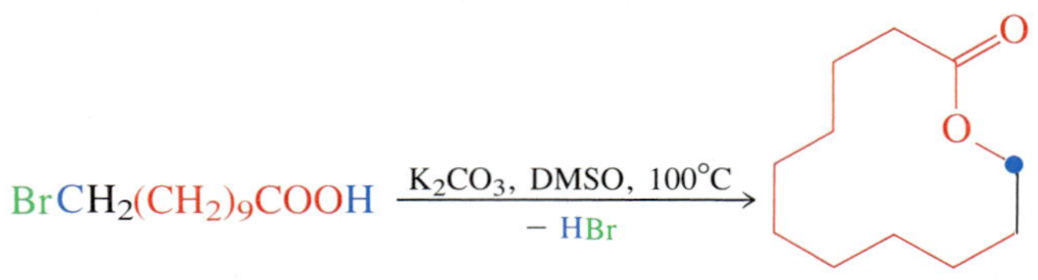

$$BrCH_2(CH_2)_9COOH \xrightarrow[-\ HBr]{K_2CO_3,\ DMSO,\ 100°C}$$ (macrocyclic lactone) 89%

EXERCISE 17-12

Propose a synthesis of lactone B from A. Hint: Convert the bromine substituent (more reactive) into a carboxy group.

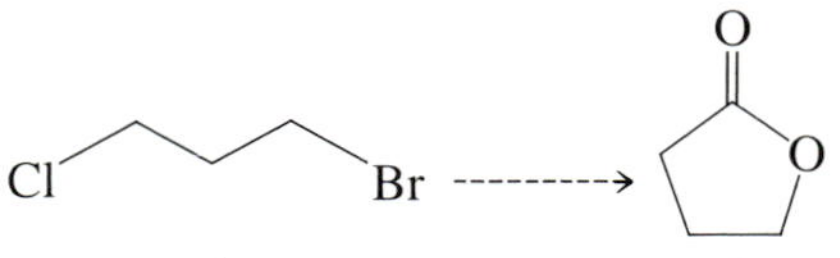

Cl–CH₂CH₂CH₂–Br (**A**) ----→ (five-membered lactone) (**B**)

EXERCISE 17-13

Explain the following stereochemical results:

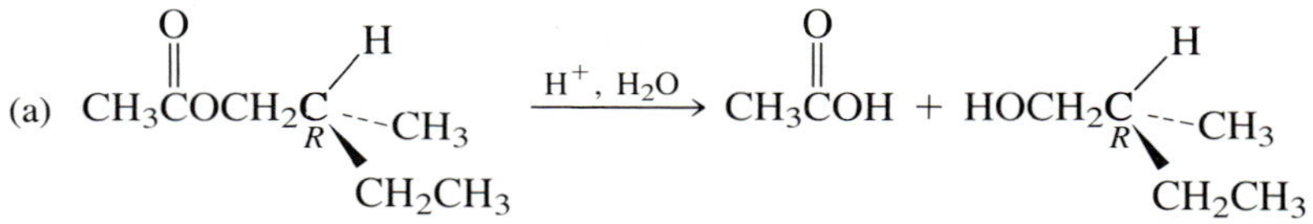

(a) $CH_3COCH_2C(H)(CH_3)(CH_2CH_3)$ (*R*) $\xrightarrow{H^+,\ H_2O}$ $CH_3COH + HOCH_2C(H)(CH_3)(CH_2CH_3)$ (*R*)

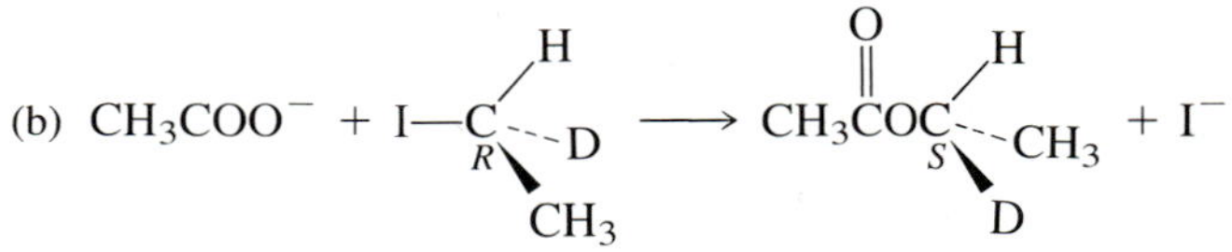

(b) $CH_3COO^- + I{-}C(H)(D)(CH_3)$ (*R*) $\longrightarrow$ $CH_3COC(H)(CH_3)(D)$ (*S*) $+ I^-$

The second method of esterification is used only on a small scale for the conversion of an acid into its methyl ester and requires the use of diazomethane, CH_2N_2 (Section 21-5), a highly reactive and toxic species. This transformation is driven forward by the production of nitrogen gas.

$$CH_3C{\equiv}CCOOH + CH_2N_2 \xrightarrow{(CH_3CH_2)_2O} CH_3C{\equiv}CCOOCH_2H + N_2$$

80%

2-Butynoic acid → **Methyl 2-butynoate**

In summary, carboxylic acids react with alcohols to form esters, as long as a catalytic amount of mineral acid is present. This reaction is only slightly exothermic and is characterized by an equilibrium that may be shifted in either direction by the appropriate choice of reaction conditions. The reverse of ester formation is ester hydrolysis. The mechanism of esterification is acid-catalyzed addition of alcohol to the carbonyl group followed by acid-catalyzed dehydration. Intramolecular ester formation results in lactones, favored only when five-

or six-membered rings are produced. Esters can be formed from carboxylic acids by other mechanisms, for example, the reaction of carboxylate ions with (primary) haloalkanes, and, for methyl alkanoates, of carboxylic acids with diazomethane.

17-9 Transformation of Carboxylic Acids into Their Derivatives: Amide Synthesis

This section shows that amines are also capable of attacking the carbonyl function in carboxylic acids to form another class of derivatives, **carboxylic amides.** The mechanism of this reaction is very similar to that of the acid-catalyzed esterification.

$$RC(=O)NR'_2$$

Carboxylic amide

Amines React with Carboxylic Acids as Bases and as Nucleophiles

Because nitrogen is not as electronegative as oxygen, amines are both more basic and more nucleophilic than alcohols (Chapter 21). They can react in either mode with carboxylic acids. Thus, exposure of the acid to the amine initially leads to the rapid formation of ammonium salts.

Ammonium Salts from Carboxylic Acids

$$RC(=O)O{-}H + :NH_3 \rightleftharpoons RC(=O)O:^- \; H\overset{+}{N}H_3$$

Ammonia — **An ammonium alkanoate**

On heating, salt formation is reversed and a slower but thermodynamically more favored process, in which the nitrogen attacks the carbonyl carbon, takes over. Addition-elimination leads to an **amide,*** in which NR_2 has replaced OH.

Mechanism of Amide Formation:

$$RC(=O)OH + :NH_3 \rightleftharpoons R{-}C(O^-)(\overset{+}{N}H_3){-}OH \xrightleftharpoons[\text{transfer}]{\text{Proton}} R{-}C(O^-)(NH_2){-}\overset{+}{O}H_2 \rightleftharpoons H_2O: + RC(=O)NH_2$$

An amide

Example:

$$CH_3CH_2CH_2COOH + (CH_3)_2NH \xrightarrow{155°C} CH_3CH_2CH_2C(=O)N(CH_3)_2 + HOH$$

84%

***N,N*-Dimethylbutanamide**

Amide formation is reversible. Thus, treatment of amides with hot acidic or basic water regenerates the component carboxylic acids and amines (see Section 18-5).

---

*Remember not to confuse the names of carboxylic amides with those of the alkali salts of amines, also called amides (e.g., lithium amide, $LiNH_2$, etc.).

Dicarboxylic Acids React with Amines to Give Imides

Dicarboxylic acids may react twice with the amine nitrogen of ammonia or of a primary amine. This sequence gives rise to **imides,** the nitrogen analogs of cyclic anhydrides.

$$\underset{\textbf{Butanedioic acid}}{\begin{matrix}CH_2COOH\\|\\CH_2COOH\end{matrix}} \xrightarrow{NH_3} \begin{matrix}CH_2COO^-NH_4^+\\|\\CH_2COO^-NH_4^+\end{matrix} \xrightarrow{290°C} \underset{\substack{83\%\\ \textbf{Butanimide}\\ \textbf{(Succinimide)}}}{\text{cyclic } H_2C(C{=}O)\ddot{N}H(C{=}O)CH_2} + 2\,H_2O + NH_3$$

Recall the use of *N*-halobutanimides in halogenations (Sections 3-8 and 14-2).

Amino Acids Cyclize to Lactams

In analogy to hydroxy carboxylic acids, some amino acids undergo cyclization to the corresponding cyclic amides, called lactams (Section 18-5).

$$H_2\overset{+}{N}(H)CH_2CH_2CH_2\overset{O}{\overset{\|}{C}}O^- \rightleftharpoons H_2\ddot{N}CH_2CH_2CH_2\overset{O}{\overset{\|}{C}}OH \xrightarrow{\Delta} \underset{\substack{86\%\\ \textbf{A lactam}}}{\text{(five-membered ring C(=O)–}\ddot{N}H)} + HOH$$

EXERCISE 17-14

Formulate a detailed mechanism for the formation of butanimide from butanedioic acid and ammonia.

In summary, amines react with carboxylic acids to form amides in much the same way that alcohols transform carboxylic acids into esters, except that the initial products are ammonium salts. Amines also attack dicarboxylic acids to give imides, the nitrogen analogs of cyclic carboxylic anhydrides. Finally, amino acids form intramolecular amides, called lactams.

17-10 Reactions of Carboxylic Acids with Organolithium Reagents and Lithium Aluminum Hydride: Nucleophilic Attack at the Carboxylate Group

This section points out that carboxylate groups, even though they are deactivated by the presence of the negative charge, can be attacked by strong nucleophiles, such as organolithium reagents and lithium aluminum hydride. The results of these transformations are, respectively, ketones and alcohols.

Organolithium Reagents Turn Carboxylate Ions into Ketones

Although nucleophilic attack on a negatively charged species is difficult, organolithium reagents are capable of nucleophilic addition to the carbonyl group of a carboxylate ion. For example, treatment of a carboxylate salt with methyllithium gives an intermediate that is essentially the dianion of a geminal diol. Aqueous work-up presumably gives the ketone hydrate (Section 15-5), which immediately dehydrates to the ketone.

$$\text{C}_6\text{H}_{11}\text{CH}_2\overset{\text{O}}{\overset{\|}{\text{C}}}\text{O}^-\text{Li}^+ \xrightarrow{\text{CH}_3\text{Li, THF}} \text{C}_6\text{H}_{11}\text{CH}_2\text{—}\underset{\text{CH}_3}{\underset{|}{\overset{\text{O}^-\text{Li}^+}{\overset{|}{\text{C}}}}}\text{—O}^-\text{Li}^+ \xrightarrow[-\,2\,\text{LiOH}]{\text{H}^+,\ \text{H}_2\text{O}} \text{C}_6\text{H}_{11}\text{CH}_2\text{—}\underset{\text{CH}_3}{\underset{|}{\overset{\text{OH}}{\overset{|}{\text{C}}}}}\text{—OH} \xrightarrow[-\,\text{H}_2\text{O}]{} \text{C}_6\text{H}_{11}\text{CH}_2\overset{\text{O}}{\overset{\|}{\text{C}}}\text{CH}_3$$

Lithium cyclohexylethanoate

76%
1-Cyclohexyl-2-propanone

This method constitutes a useful general synthesis of methyl ketones. For convenience, you can start with a carboxylic acid and simply add two equivalents of methyllithium. The first equivalent reacts with the acidic proton to give methane and the carboxylate salt. The second equivalent then attacks the salt to give the methyl ketone on work-up.

Step-by-Step Methyl Ketone Synthesis

$$\text{CH}_2\text{=CH}\underset{\text{CH}_3}{\underset{|}{\overset{\text{CH}_3}{\overset{|}{\text{C}}}}}\text{COOH} \xrightarrow[-\,\text{CH}_3\text{H}]{\text{CH}_3\text{Li, THF}} \text{CH}_2\text{=CH}\underset{\text{CH}_3}{\underset{|}{\overset{\text{CH}_3}{\overset{|}{\text{C}}}}}\text{COO}^-\text{Li}^+ \xrightarrow[-\,2\,\text{LiOH}]{\text{1. CH}_3\text{Li, THF} \quad \text{2. H}^+,\ \text{H}_2\text{O}} \text{CH}_2\text{=CH}\underset{\text{H}_3\text{C}}{\underset{|}{\overset{\text{H}_3\text{C}}{\overset{|}{\text{C}}}}}\text{—}\overset{\text{O}}{\overset{\|}{\text{C}}}\text{CH}_3$$

2,2-Dimethyl-3-butenoic acid

55%
3,3-Dimethyl-4-penten-2-one

Other alkyllithium reagents react like methyllithium to give other ketones.

EXERCISE 17-15

Treatment of butyllithium with carbon dioxide at elevated temperatures gives 5-nonanone on aqueous work-up. Explain.

Reduction of Carboxylic Acids with Lithium Aluminum Hydride Gives Alcohols

Another extremely strong nucleophile is lithium aluminum hydride. This reagent reduces carboxylic acids all the way to the corresponding alcohols, which are obtained by aqueous work-up:

$$\text{RCOOH} \xrightarrow{\text{1. LiAlH}_4,\ \text{THF} \quad \text{2. H}^+,\ \text{H}_2\text{O}} \text{RCH}_2\text{OH}$$

Example:

$$\text{(bicyclo[2.2.1]heptane-2-COOH)} \xrightarrow[\text{2. } H^+,\ H_2O]{\text{1. } LiAlH_4,\ THF} \text{(bicyclo[2.2.1]heptane-2-}CH_2OH\text{)}\quad 65\%$$

The first step in this transformation is the formation of the lithium salt of the acid and the generation of hydrogen gas. Subsequently, another hydride equivalent reduces the carbonyl function by addition, probably to give an aluminum-complexed dianion of a geminal diol. Another hydride equivalent displaces one of the oxide substituent groups; finally, the resulting alkoxide gives alcohol on acidic work-up.

Probable Mechanism of Reduction of Carboxylic Acids by Lithium Aluminum Hydride:

STEP 1: Salt formation

$$RC(=O)OH + LiAlH_4 \longrightarrow RC(=O)O^-Li^+ + H{-}H + AlH_3$$

STEP 2: Hydride addition

$$RC(=O)O^-Li^+ + LiAlH_4 \longrightarrow RC(H)(OAl\langle)O^-Li^+$$

STEP 3: Displacement by hydride

$$RC(H)(OAl\langle)O^-Li^+ + LiAlH_4 \longrightarrow RC(H)(H)O^-Li^+ + {}^-OAl\langle$$

STEP 4: Hydrolysis

$$RCH_2O^-Li^+ \xrightarrow{HOH} RCH_2OH + Li^{+-}OH$$

EXERCISE 17-16

Propose synthetic schemes that produce compound B from compound A: (a) **A:** $CH_3CH_2CH_2CN$, **B:** $CH_3CH_2CH_2CH_2OH$;

(b) **A:** cyclopropyl$-CH_2COOH$, **B:** cyclopropyl$-CH_2CD_2OH$.

In summary, the strongly nucleophilic character of organolithium reagents and lithium aluminum hydride allows these species to add to the carbonyl group of carboxylates. Alkyllithium species turn the carboxy function into the carbonyl group of a ketone, and lithium aluminum hydride reduces carboxylic acids to alcohols.

17-11 Reactions of Carboxylic Acids: Substitution next to the Carboxy Group and Decarboxylation

Like other carbonyl compounds, carboxylic acids (as the carboxylate salts) form enolate ions that can undergo nucleophilic substitution reactions. Enols are also intermediates in the bromination of the carbon next to the carboxy group. Finally, decarboxylation of carboxylic acids proceeds by one-electron oxidations of their carboxylate salts.

The Position next to the Carbonyl Group in Carboxylic Acids Is Acidic

The carbonyl group in aldehydes and ketones has an acidifying effect on adjacent hydrogens (Section 16-1). This is true also of carboxylic acids (and their derivatives, Chapter 18). On treatment with base, they first form carboxylate salts. In the presence of another equivalent of a strong base, such as lithium diisopropylamide (LDA), and a highly polar aprotic cosolvent, such as hexamethylphosphoric triamide, HMPA (Section 6-10, Table 6-8), they can deprotonate again to give a **carboxylic acid dianion,** which is a powerful nucleophile.

Dianion Formation from Nonanoic Acid

$$CH_3(CH_2)_6CH_2COOH \xrightarrow[-\,2\,(CH_3CH)_2NH,\ \text{with } CH_3 \text{ on CH}]{2\text{ LDA, THF, HMPA, 25°C, 30 min}} \left[CH_3(CH_2)_6\ddot{C}H^- \text{—}C(=\ddot{O})\ddot{O}:^- \longleftrightarrow CH_3(CH_2)_6\ddot{C}H^- \text{—}C(=\ddot{O})\ddot{O}:^- \longleftrightarrow CH_3(CH_2)_6CH{=}C(\ddot{O}:^-)_2 \right] + 2\,Li^+$$

Nonanoic acid — Nonanoic acid dianion

Its nucleophilic reactions are the same as those of other enolate ions (Sections 16-1 and 16-3)—namely, alkylation, oxacyclopropane opening, and aldol addition.

STEP 3: Bromination

$$RCH{=}C(OH)Br \xrightarrow{Br{-}Br} RCH(Br)\overset{O}{\overset{\|}{C}}Br + HBr$$

STEP 4: Exchange

$$RCH(Br)\overset{O}{\overset{\|}{C}}Br + RCH_2\overset{O}{\overset{\|}{C}}OH \rightleftharpoons RCH(Br)\overset{O}{\overset{\|}{C}}OH + RCH_2\overset{O}{\overset{\|}{C}}Br$$

The bromo carboxylic acids formed in the Hell-Volhard-Zelinsky reaction can be further functionalized. For example, treatment with aqueous base gives the 2-hydroxy acid, whereas an amine yields an amino acid (Section 27-2). Substitution with cyanide furnishes a 2-cyano carboxylic acid, which subsequently may be hydrolyzed to the dicarboxylic acid.

Further Functionalization of 2-Bromo Carboxylic Acids

$$CH_3\underset{}{C}H(CH_3)CH_2CH(Br)COOH \xrightarrow[-\ KBr]{1.\ HOH,\ K_2CO_3,\ \Delta;\ 2.\ H^+,\ H_2O} CH_3CH(CH_3)CH_2CH(OH)COOH \quad 72\%$$

2-Bromo-4-methylpentanoic acid → **2-Hydroxy-4-methylpentanoic acid**

$$CH_3CH(Br)COOH \xrightarrow[-\ Br^-]{NH_3,\ H_2O,\ 23^\circ C,\ 4\ days} CH_3CH(NH_2)COOH \quad 56\%$$

2-Bromopropanoic acid → **2-Aminopropanoic acid (Alanine)**

$$BrCH_2COOH \xrightarrow[-\ KBr,\ -\ H_2O]{KCN,\ NaOH} NCCH_2COO^-Na^+ \xrightarrow{1.\ H_2O,\ HO^-;\ 2.\ H^+,\ H_2O} HO\overset{O}{\overset{\|}{C}}CH_2\overset{O}{\overset{\|}{C}}OH$$

Bromoethanoic acid → **Propanedioic acid (Malonic acid)**

One-Electron Transfer Leads to the Decarboxylation of Carboxylic Acids

The one-electron oxidation of a carboxylate ion produces a very unstable RCOO· radical, which decomposes with evolution of carbon dioxide. The resulting alkyl radical then reacts further by dimerization or by abstracting a suitable atom from its environment. The net result is the **decarboxylation** of the carboxylic acid and the formation of an alkane or an alkene derivative.

Generation and Decomposition of an RCOO· Radical

$$\mathrm{RC(=O)\ddot{O}:^-} \xrightarrow[\text{Oxidation}]{-1e} \mathrm{RC(=O)\ddot{O}\cdot} \longrightarrow \mathrm{R\cdot} + \ddot{O}{=}C{=}\ddot{O}$$

In the **Kolbe* electrolysis,** the electron is removed at the anode of an electrolysis apparatus. The resulting radicals dimerize to give an alkane.

$$\mathrm{CH_3CH_2CH_2CH_2COO^-Na^+} \xrightarrow[\text{Anode oxidation}]{} \mathrm{CH_3CH_2CH_2CH_2COO\cdot} \longrightarrow \mathrm{CO_2} + \mathrm{CH_3CH_2CH_2\dot{C}H_2}$$

$$2\ \mathrm{CH_3CH_2CH_2\dot{C}H_2} \xrightarrow[\text{Dimerization}]{} \underset{90\%}{\mathrm{CH_3(CH_2)_6CH_3}}$$

BOX 17-2

Organic Electrochemical Reactions

Kolbe electrolysis is an organic redox reaction that is carried out in an electrochemical cell. The advantages of this method lie in the control over the electrode potential and in the simplicity of the reaction, because no chemical oxidants or reductants are needed. Electrochemical transformations have been used in the reduction of haloalkanes to alkanes, carbonyl compounds to alcohols, and nitro compounds to amines. Oxidations have included the conversion of alcohols into ketones, amines into imines, and sulfides into sulfoxides and sulfones, to name just a few. An important industrial application of this method is the electrochemical hydrodimerization of propenenitrile (acrylonitrile) to butanedinitrile (adiponitrile), which is discussed in Section 21-6.

In the **Hunsdiecker† reaction,** the oxidation is performed chemically: a silver salt of the carboxylic acid is treated with a halogen, usually bromine. Silver bromide precipitates, carbon dioxide is evolved, and a bromoalkane is formed in which the bromine has taken the position of the carboxy function.

General Hunsdiecker Reaction

$$\mathrm{RCOO^-Ag^+} + \mathrm{Br{-}Br} \longrightarrow \mathrm{RBr} + \mathrm{CO_2} + \mathrm{AgBr}$$

The reaction allows for the conversion of a carboxylate salt into a bromoalkane containing one less carbon atom. For example,

$$\underset{\textbf{Dodecanoic acid}}{\mathrm{CH_3(CH_2)_{10}COOH}} \xrightarrow{\text{1. } AgNO_3,\ KOH,\ H_2O \quad \text{2. } Br{-}Br,\ CCl_4} \underset{\textbf{1-Bromoundecane}}{\underset{67\%}{\mathrm{CH_3(CH_2)_{10}Br}}}$$

*Professor Hermann Kolbe, 1818–1884, University of Leipzig, Germany.

†Dr. Heinz Hunsdiecker, b. 1904, Vogt and Co., Köln-Braunsfeld, Germany.

The mechanism of this degradation reaction includes an alkanoyl hypobromite, RCOOBr, which decomposes to the RCOO· radical. After decarboxylation, the alkyl radical abstracts bromine from another molecule of hypobromite to yield the bromoalkane and another RCOO· radical, which reenters the reaction.

Mechanism of the Hunsdiecker Reaction:

STEP 1: Hypobromite formation

$$\mathrm{RC(=\ddot{O}:)\ddot{\underset{..}{O}}:^{-}\,Ag^{+}} \xrightarrow{\mathrm{Br{-}Br}} \mathrm{RC(=\ddot{O}:)\underset{..}{\ddot{O}}Br} + \mathrm{AgBr}$$

Hypobromite

STEP 2: RCOO· radical formation

$$\mathrm{RC(=\ddot{O}:)\underset{..}{\ddot{O}}{-}\underset{..}{\ddot{Br}}:} \longrightarrow \mathrm{RC(=\ddot{O}:)\underset{..}{\ddot{O}}\cdot} + :\underset{..}{\ddot{Br}}\cdot$$

STEP 3: RCOO· radical decomposition

$$\mathrm{R{-}C(=\ddot{O}:)\underset{..}{\ddot{O}}\cdot} \longrightarrow \mathrm{R\cdot} + \mathrm{\underset{..}{\ddot{O}}{=}C{=}\underset{..}{\ddot{O}}}$$

STEP 4: Haloalkane and RCOO· formation

$$\mathrm{R\cdot} + \mathrm{RC(=\ddot{O}:)\underset{..}{\ddot{O}}Br} \longrightarrow \mathrm{RBr} + \mathrm{RC(=\ddot{O}:)\underset{..}{\ddot{O}}\cdot}$$

Other examples of decarboxylation appear in the enzymatic degradation of 2-oxopropanoic (pyruvic) acid (Section 22-1) and in the reactions of 3-keto carboxylic acids (Section 22-4).

EXERCISE 17-18

Outline syntheses of the following compounds starting from butanoic acid:

(a) $CH_3CH_2CH(CH_3)Br$; (b) CH_3COOH.

In summary, the 2-position in carboxylate salts is weakly acidic and can be deprotonated with LDA in the presence of HMPA to give carboxylic acid dianions. The resulting enolate carbon is nucleophilic and can be alkylated in reactions like those of ordinary enolates. With catalytic amounts of phosphorus, carboxylic acids are brominated at C-2 in a transformation (the Hell-Volhard-Zelinsky reaction) that proceeds through the intermediacy of 2-bromoalkanoyl bromides. Electrochemical oxidation or halogenation of a carboxylate produces an unstable RCOO· radical, which decomposes to carbon dioxide and an alkyl radical. The radical can either couple to an alkane (Kolbe electrolysis) or be halogenated to the corresponding halo- (usually bromo-) alkane (the Hunsdiecker reaction).

17-12 The Distribution and Function of Some Carboxylic Acids

Considering the variety of reactions that carboxylic acids can undergo, it is no wonder that they are very important, not only as synthetic intermediates in the laboratory, but also in nature.

As Table 17-1 indicates, even the simplest carboxylic acids are abundant in nature. *Methanoic (formic) acid* is present not only in ants, where it functions as an alarm pheromone, but also in plants. For example, one reason why human skin hurts after it touches the stinging nettle is that methanoic (formic) acid is injected. This acid is prepared conveniently on a large scale by reaction of powdered sodium hydroxide with carbon monoxide under pressure. This transformation proceeds by nucleophilic addition followed by protonation.

Methanoic (Formic) Acid Synthesis

$$\text{NaOH} + \text{CO} \xrightarrow{150°\text{C, 100 psi}} \text{HCOO}^-\text{Na}^+ \xrightarrow{\text{H}^+,\ \text{H}_2\text{O}} \text{HCOOH}$$

Ethanoic (acetic) acid is formed in nature through the enzymatic oxidation of ethanol produced by fermentation. The acid and its anhydride are major industrial chemicals, used to manufacture monomers for polymerization, such as ethenyl ethanoate (vinyl acetate, Sections 12-8 and 13-7), as well as pharmaceuticals, dyes, and pesticides. There are three industrial preparations of importance that annually furnish more than 1 million tons of ethanoic (acetic) acid in the United States alone: ethene oxidation through ethanal (acetaldehyde, Section 12-8); air oxidation of butane; and carbonylation of methanol. The mechanisms of these reactions are complex.

Ethanoic (Acetic) Acid by Oxidation of Ethene

$$\text{CH}_2\text{=CH}_2 \xrightarrow[\text{Wacker process}]{\text{O}_2,\ \text{H}_2\text{O, catalytic PdCl}_2\text{ and CuCl}_2} \text{CH}_3\text{CHO} \xrightarrow{\text{O}_2,\text{ catalytic Co}^{3+}} \text{CH}_3\text{COOH}$$

Ethanoic (Acetic) Acid by Oxidation of Butane

$$\text{CH}_3\text{CH}_2\text{CH}_2\text{CH}_3 \xrightarrow{\text{O}_2,\text{ catalytic Co}^{3+},\ 15\text{–}20\text{ atm, }180°\text{C}} \text{CH}_3\text{COOH}$$

Ethanoic (Acetic) Acid by Carbonylation of Methanol (Monsanto Process)

$$\text{CH}_3\text{OH} \xrightarrow{\text{CO, catalytic Rh}^{3+},\ \text{I}_2,\ 30\text{–}40\text{ atm, }180°\text{C}} \text{CH}_3\text{COOH}$$

Ethanoic (acetic) acid is also important in nature. For example, it is the building block in fatty acid biosynthesis (see the next subsection), and it has been identified as the defense pheromone in some ants and scorpions.

A very important class of natural carboxylic acids is the *amino acids,* the monomeric units of important biological polymers such as peptides and proteins. These compounds will be discussed in Chapter 27.

Natural Fats Hydrolyze to Long-Chain Carboxylic Acids

The natural **fats** are esters of long-chain carboxylic acids (see Section 18-4). Hydrolysis or *saponification* (so called because the corresponding salts form soaps—see the next subsection; *sapo,* Latin, soap) yields the corresponding **fatty acids.** The most-important fatty acids—for example, hexadecanoic (palmitic) and *cis*-9-octadecenoic (oleic) acid—are between twelve and twenty-two carbons long and may be unsaturated.

$CH_3(CH_2)_{14}COOH$

Hexadecanoic acid (Palmitic acid)

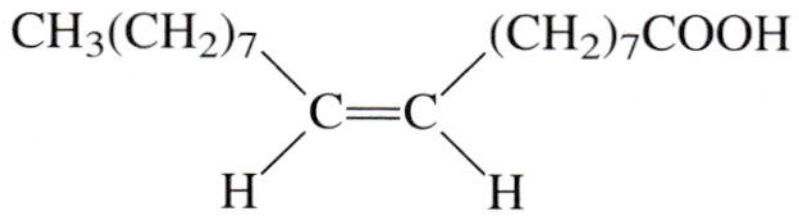

$CH_3(CH_2)_7(H)C{=}C(H)(CH_2)_7COOH$

***cis*-9-Octadecenoic acid (Oleic acid)**

Fatty acids consist mostly of even-numbered carbon chains, because they are derived biologically from ethanoic (acetic) acid. This relation has been demonstrated by a very elegant experiment in which singly labeled radioactive (^{14}C) ethanoic (acetic) acid was fed to several organisms. The resulting fatty acids were labeled only at every other carbon atom:

$$CH_3{}^{14}COOH \xrightarrow{\text{Organism}} CH_3{}^{14}CH_2CH_2{}^{14}CH_2CH_2{}^{14}CH_2CH_2{}^{14}CH_2CH_2{}^{14}CH_2CH_2{}^{14}CH_2CH_2{}^{14}CH_2CH_2{}^{14}COOH$$

Labeled hexadecanoic (palmitic) acid

The way in which this polymerization takes place may be schematized as follows. At first, ethanoic (acetic) acid is activated by formation of a reactive ester (acetyl coenzyme A) with the mercapto group (therefore, a thioester) of an important biological relay compound called coenzyme A (abbreviated HSCoA; Figure 17-15). This ester is then carboxylated with the help of an enzyme called acetyl CoA carboxylase to give malonyl coenzyme A. Now a new enzyme system comes into the act, fatty acid synthetase, which facilitates chain elongation. First, the two alkanoyl groups in both the acetyl and malonyl coenzyme A (these are all common names) are enzymatically transferred independently to

FIGURE 17-15

Structure of coenzyme A. For this discussion, the important part is the mercapto function. For convenience, the molecule may be abbreviated HSCoA.

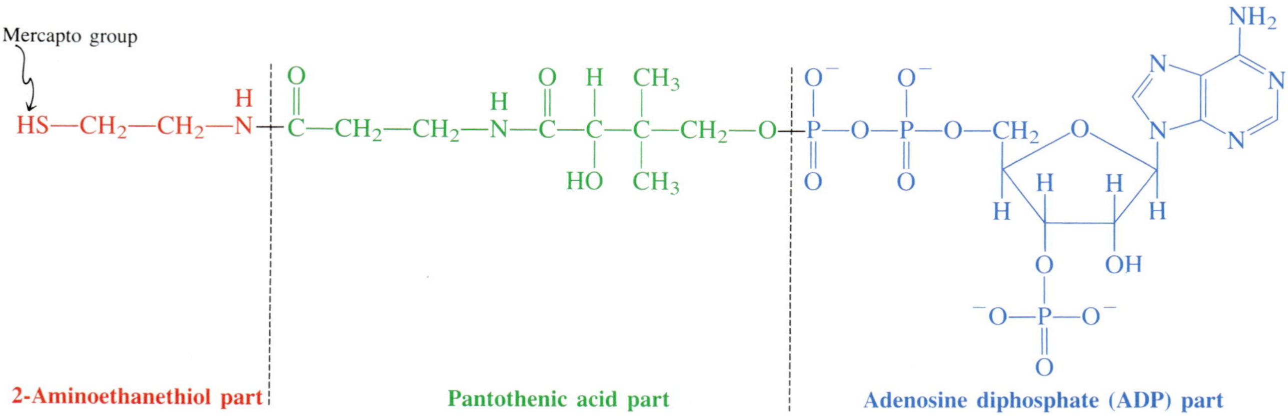

Mercapto group

$HS{-}CH_2{-}CH_2{-}NH{-}C({=}O){-}CH_2{-}CH_2{-}NH{-}C({=}O){-}CH(OH){-}C(CH_3)_2{-}CH_2{-}O{-}P(O^-)({=}O){-}O{-}P(O^-)({=}O){-}O{-}CH_2$–ribose(3′-$OPO_3^{2-}$, 2′-OH)–adenine($NH_2$)

2-Aminoethanethiol part | **Pantothenic acid part** | **Adenosine diphosphate (ADP) part**

two molecules of another mercapto-containing protein, the acyl carrier protein. The resulting two species then combine with loss of CO_2 to form a 3-oxobutanoic thioester. (The coupling step is similar to malonic ester synthesis, to be discussed in Section 22-4). The 3-oxo group of the thioester is subsequently reduced, first to the alcohol, which is dehydrated to the α,β-unsaturated ester, and then to the saturated ester. The resulting butanoic ester repeatedly undergoes a similar sequence of reactions elongating the chain, always by two carbons, to eventually furnish the fatty acid.

Mechanism of Fatty Acid Biosynthesis:

STEP 1: Thioester formation

$$\underset{\text{Ethanoic (acetic) acid}}{CH_3COOH} + \underset{\text{Coenzyme A}}{HSCoA} \longrightarrow \underset{\text{Acetyl coenzyme A}}{CH_3\overset{O}{\overset{\|}{C}}SCoA} + HOH$$

STEP 2: Carboxylation

$$CH_3\overset{O}{\overset{\|}{C}}SCoA + CO_2 \xrightarrow{\text{Acetyl CoA carboxylase}} \underset{\text{Malonyl CoA}}{HO\overset{O}{\overset{\|}{C}}CH_2\overset{O}{\overset{\|}{C}}SCoA}$$

STEP 3: Alkanoyl group transfer

$$CH_3\overset{O}{\overset{\|}{C}}SCoA + \underset{\text{Acyl carrier protein}}{HS\text{—}\boxed{\text{protein}}} \longrightarrow CH_3\overset{O}{\overset{\|}{C}}S\text{—}\boxed{\text{protein}} + HSCoA$$

$$HO\overset{O}{\overset{\|}{C}}CH_2\overset{O}{\overset{\|}{C}}SCoA + \underset{\text{Acyl carrier protein}}{HS\text{—}\boxed{\text{protein}}} \longrightarrow HO\overset{O}{\overset{\|}{C}}CH_2\overset{O}{\overset{\|}{C}}S\text{—}\boxed{\text{protein}} + HSCoA$$

STEP 4: Coupling

$$HO\overset{O}{\overset{\|}{C}}CH_2\overset{O}{\overset{\|}{C}}S\text{—}\boxed{\text{protein}} \xrightarrow[-CO_2]{CH_3\overset{O}{\overset{\|}{C}}S\text{—}\boxed{\text{protein}}} \underset{\text{A 3-oxobutanoic thioester}}{CH_3\overset{O}{\overset{\|}{C}}CH_2\overset{O}{\overset{\|}{C}}S\text{—}\boxed{\text{protein}}}$$

STEP 5: Reduction

$$CH_3\overset{O}{\overset{\|}{C}}CH_2\overset{O}{\overset{\|}{C}}S\text{—}\boxed{\text{protein}} \xrightarrow{\text{Reduction}} \underset{\text{A butanoic thioester}}{CH_3CH_2CH_2\overset{O}{\overset{\|}{C}}S\text{—}\boxed{\text{protein}}}$$

SUBSEQUENT STEPS: Repetition of steps 1 through 5 followed by hydrolysis

$$CH_3CH_2CH_2\overset{\overset{O}{\|}}{C}S\text{—}\boxed{\text{protein}} \xrightarrow{H_2O} \underset{\textbf{A long-chain fatty acid}}{CH_3(CH_2)_nCOOH} + HS\text{—}\boxed{\text{protein}}$$

The energetics of each step are carefully controlled, and the heats of the reactions usually drive other processes. The system has to be extremely selective. Thus, the enzyme fatty acid synthetase builds up the sixteen-carbon chain of hexadecanoic (palmitic) acid in one sequence of operations—never releasing any intermediates—until the final product is assembled. Although the structure of the "assembly line" is not exactly known, the enzyme is thought to consist of a cluster of several proteins with an estimated molecular weight of 2,300,000.

Long-Chain Carboxylates Make Soaps

The sodium and potassium salts of the fatty acids have the interesting property of aggregating as spherical clusters called **micelles** in aqueous solution (Figure 17-16). In such aggregates, all the hydrophobic alkyl chains of the acids attempt to occupy the same region in space because of their attraction by London forces and their tendency to be exposed to as little water as possible. The polar, water-solvated carboxylate "head groups" form a spherical wall around the hydrocarbonlike center. Carboxylate salts also create films on aqueous surfaces, in which the polar groups stick into the water while the alkyl chains assemble into a hydrophobic layer. This construction reduces surface tension and permits the foaming that is typical of soaps. Indeed, carboxylate salts were among the first soaps to be used. They basically act by dissolving ordinarily water-insoluble materials (oils, fats) in the hydrocarbon interior of the micelles and by reducing the surface tension of water, allowing it to permeate cloth and other fabrics.

FIGURE 17-16

Schematic representation of a micelle formed by adding a fatty acid to water.

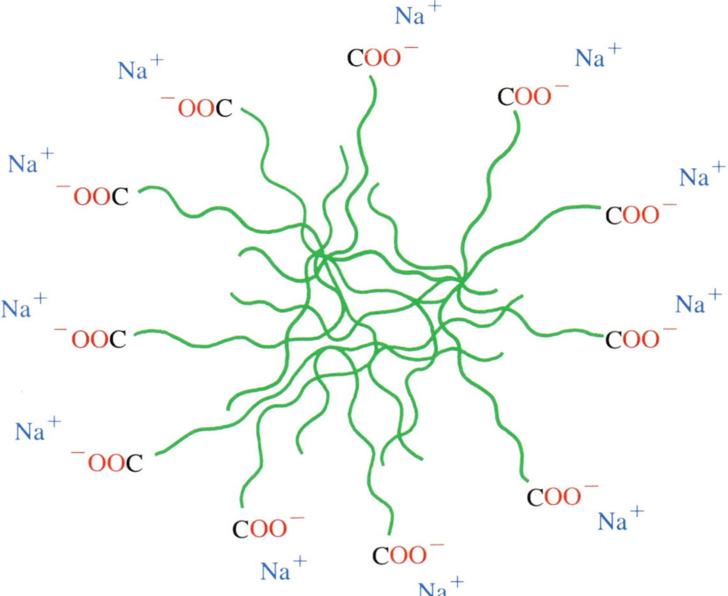

Modern detergents are based on alkanesulfonates, $CH_3(CH_2)_nCH_2SO_3^-Na^+$, which, like soaps, are biodegradable, but do not form precipitates with the calcium and magnesium ions in ''hard'' water, a problem with fatty acid anions.

An important natural long-chain carboxylic acid is *arachidonic acid*. This substance appears to be the biological precursor to a multitude of important chemicals in the human body, such as prostaglandins (Section 16-5), thromboxanes, prostacyclins, and leukotrienes (for an exercise, try to discern the structural origin of these compounds from arachidonic acid).

Arachidonic acid

Prostaglandin $F_{2\alpha}$

(Induces labor, abortion, menstruation)

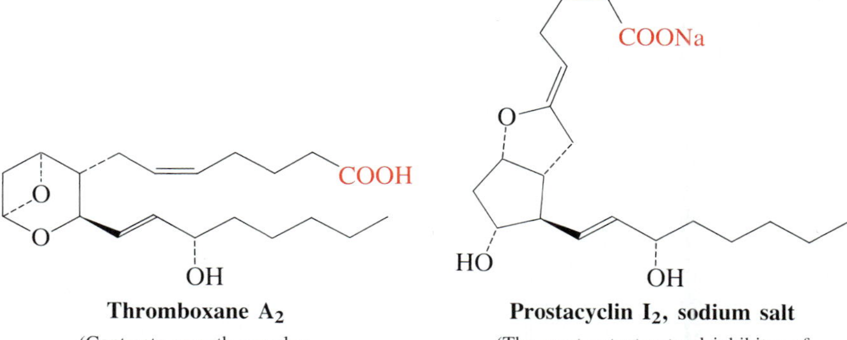

Thromboxane A_2

(Contracts smooth muscles; aggregates blood platelets)

Prostacyclin I_2, sodium salt

(The most-potent natural inhibitor of platelet aggregation; vasodilator, used in heart-bypass operations, in kidney patients, and in others)

Leukotriene B_4

(Potent chemotactic factor; e.g., causes cell migrations)

These molecules have powerful pharmacological properties and are the targets of a variety of synthetic approaches.

Apart from the long-chain carboxylic acids, many natural products have carboxy groups as substituents of a more complicated framework. In a biological environment, the function of the carboxy group may be to impart water solubility, to allow for salt formation or ion transport, and to enable micellar-type aggregation. *Gibberellic acid* is one of a group of plant growth-promoting substances manufactured by fermentation.

Lysergic acid is a major product of hydrolyzed extracts of ergot, a fungal parasite that lives on grasses, including rye. Many lysergic acid derivatives possess powerful psychotomimetic activity. In the Middle Ages, thousands who ate rye bread contaminated by ergot experienced the poisonous effects characteristic of these compounds: hallucinations, convulsions, delirium, epilepsy, and death. The synthetic lysergic acid diethylamide (LSD) is one of the most powerful hallucinogens known. The effective oral dose for humans is only about 0.05 mg.

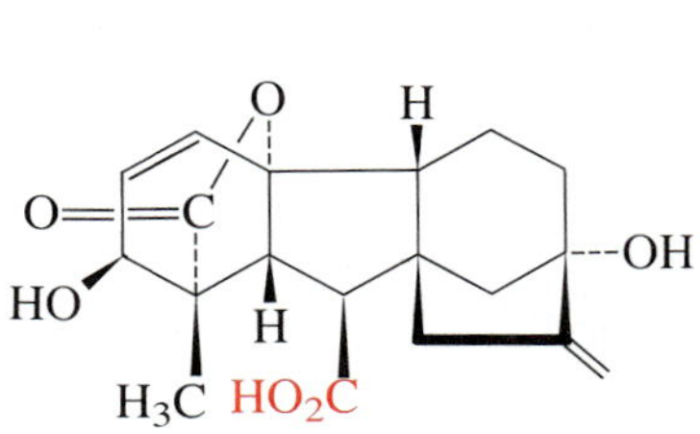

Gibberellic acid 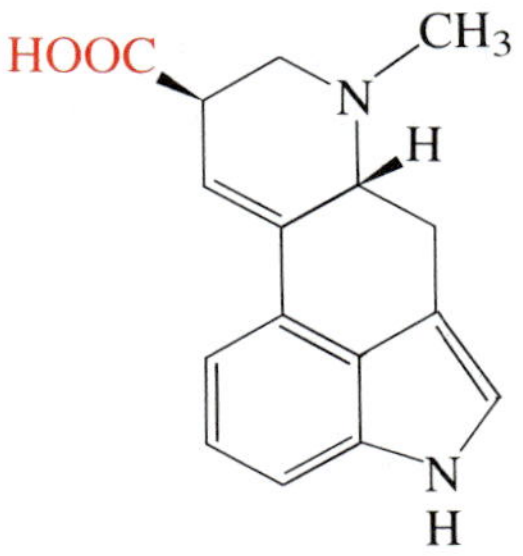 **Lysergic acid**

Certain steroid "bile acids," such as *cholic acid* (Section 4-7), are found in the bile duct and are useful in the emulsification and bodily absorption of fats through the formation of micelles. The *penicillins* are antibiotics derived from 2-amino acids (see Section 18-5).

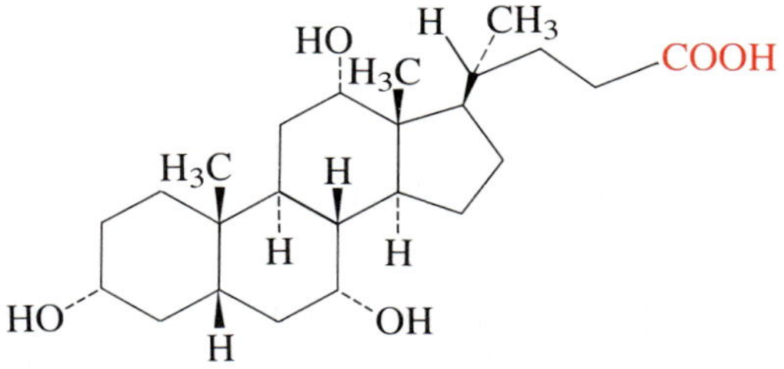

Cholic acid (A bile acid) **Penicillins (R variable)**

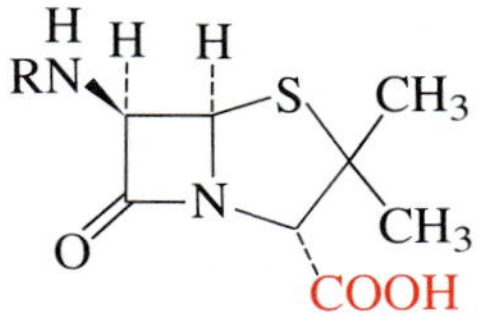

In summary, this section has given examples of the structural diversity of carboxylic acids, particularly as they occur in living organisms, as well as a glimpse of their biological functions.

Summary of New Reactions

1 ACIDITY OF CARBOXYLIC ACIDS

$$RCOOH + H_2O \longrightarrow [RCOO^- \longleftrightarrow RCOO^-] + H_3O^+$$

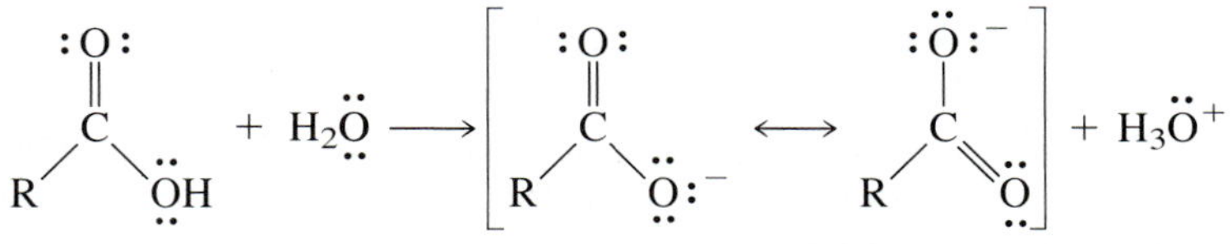

Resonance-stabilized carboxylate ion

$K_a = 10^{-4}$–10^{-5}, pK_a ~ 4–5

Salt formation:

$$\text{RCOOH} + \text{NaOH} \longrightarrow \text{RCOO}^-\text{Na}^+ + H_2O$$

2 BASICITY OF CARBOXYLIC ACIDS

$$\text{R—C}(\text{=}\ddot{\text{O}}\text{:})\text{—}\ddot{\text{O}}\text{H} + \text{H}^+ \rightleftharpoons \left[\text{R—C}(\text{=}\overset{+}{\ddot{\text{O}}}\text{—H})\text{—}\ddot{\text{O}}\text{—H} \longleftrightarrow \text{R—C}(\text{—}\ddot{\text{O}}\text{—H})\text{=}\overset{+}{\ddot{\text{O}}}\text{—H} \right]$$

Resonance-stabilized protonated carboxylic acid

Preparation of Carboxylic Acids

3 OXIDATIVE CLEAVAGE OF ALKENES

$$\text{RCH=CH}_2 \xrightarrow{\text{KMnO}_4,\ \text{HO}^-} \text{RCOOH} + CO_2$$

4 OXIDATION OF PRIMARY ALCOHOLS AND ALDEHYDES

$$\text{RCH}_2\text{OH} \xrightarrow{\text{Oxidizing agent}} \text{RCOOH}$$ Oxidizing agents: CrO_3, $KMnO_4$, HNO_3

$$\text{RCHO} \xrightarrow{\text{Oxidizing agent}} \text{RCOOH}$$ Oxidizing agents: CrO_3, $KMnO_4$, Ag^+, H_2O_2, HNO_3

5 OXIDATION OF KETONES

$$\text{R}\overset{\text{O}}{\overset{\|}{\text{C}}}\text{CH}_2\text{R}' \xrightarrow{50\%\ \text{HNO}_3,\ \Delta,\ \text{V}_2\text{O}_5} \text{RCOOH} + \text{R}'\text{COOH}$$

Haloform reaction:

$$\text{R}\overset{\text{O}}{\overset{\|}{\text{C}}}\text{CH}_3 \xrightarrow[\text{2. H}^+,\ \text{H}_2\text{O}]{\text{1. X}_2,\ \text{HO}^-} \text{RCOOH} + \text{HCX}_3$$

6 CARBOXYLATION OF ORGANOMETALLIC REAGENTS

$$\text{RMgX} + CO_2 \longrightarrow \text{RCOO}^{-\,+}\text{MgX} \xrightarrow{\text{H}^+,\ \text{H}_2\text{O}} \text{RCOOH}$$

$$\text{RLi} + CO_2 \longrightarrow \text{RCOO}^-\text{Li}^+ \xrightarrow{\text{H}^+,\ \text{H}_2\text{O}} \text{RCOOH}$$

7 HYDROLYSIS OF NITRILES

$$\text{RC}\equiv\text{N} \xrightarrow{\text{H}_2\text{O},\ \Delta,\ \text{H}^+\ \text{or}\ \text{HO}^-} \text{RCOOH} + NH_3$$

Imides:

$$(CH_2)_n(COOH)_2 + R'NH_2 \xrightarrow{\Delta} (CH_2)_n(C{=}O)_2NR' + 2\ H_2O$$

Cyclic amides (lactams):

$$(CH_2)_n(CH_2NH_2)(COOH) \xrightarrow{\Delta} (CH_2)_n(CH_2)(NH)(C{=}O) + H_2O$$

Lactam

13 REACTION WITH ORGANOMETALLIC REAGENTS

$$RCOOH \xrightarrow[2.\ H^+,\ H_2O]{1.\ 2R'Li} RC(=O)R'$$

14 REDUCTION WITH LITHIUM ALUMINUM HYDRIDE

$$RCOOH \xrightarrow[2.\ H^+,\ H_2O]{1.\ LiAlH_4} RCH_2OH$$

15 CARBOXYLIC ACID DIANION FORMATION AND REACTIONS

$$RCH_2COOH \xrightarrow{2\ LDA,\ THF,\ HMPA} R\ddot{C}^{-}HC(=O)\ddot{O}:^{-}\ Li^+\ (Li^+)$$

$$R\ddot{C}^{-}HC(=O)\ddot{O}:^{-} \xrightarrow[2.\ H^+,\ H_2O]{1.\ R'X} RCH(R')COOH$$

$$R\ddot{C}^{-}HC(=O)\ddot{O}:^{-} \xrightarrow[2.\ H^+,\ H_2O]{1.\ H_2C\text{—}CH_2\ (\text{epoxide, O})} RCH(CH_2CH_2OH)COOH$$

$$R\ddot{C}^{-}HC(=O)\ddot{O}:^{-} \xrightarrow[2.\ H^+,\ H_2O]{1.\ R'C(=O)R''} RCH(C(OH)R'R'')COOH$$

16 BROMINATION: HELL-VOLHARD-ZELINSKY REACTION

$$RCH_2COOH \xrightarrow{Br_2,\ \text{catalytic P}} RCH(Br)COOH$$

17 DISPLACEMENT REACTIONS OF 2-BROMO CARBOXYLIC ACIDS

$$RCH(Br)COOH \xrightarrow{HO^-,\ H_2O} RCH(OH)COOH$$

$$RCH(Br)COOH \xrightarrow{NH_3,\ \Delta} RCH(NH_2)COOH$$

$$RCH(Br)COOH \xrightarrow{NaCN,\ NaOH} RCH(C{\equiv}N)COO^-Na^+ \xrightarrow[2.\ H^+,\ H_2O]{1.\ H_2O,\ HO^-} RCH(COOH)COOH$$

One-Carbon Degradation of Carboxylic Acids

18 KOLBE ELECTROLYSIS

$$RCOO^-Na^+ \xrightarrow[-\,e^-\ \text{(anode)}]{} RCOO\cdot \xrightarrow[-\,CO_2]{} R\cdot \xrightarrow{2\times} R{-}R$$

19 HUNSDIECKER REACTION

$$RCOO^-Ag^+ \xrightarrow{Br_2} RBr + CO_2 + AgBr$$

Special Preparations

20 METHANOIC (FORMIC) ACID

$$CO + NaOH \longrightarrow HCOO^-Na^+ \xrightarrow{H^+,\ H_2O} HCOOH$$

21 ETHANOIC (ACETIC) ACID

Ethene oxidation:

$$CH_2{=}CH_2 \xrightarrow{O_2,\ H_2O,\ \text{catalytic } Pd^{2+} \text{ and } Cu^{2+}} CH_3CHO \xrightarrow{O_2,\ \text{catalytic } Co^{3+}} CH_3COOH$$

Butane oxidation:

$$CH_3CH_2CH_2CH_3 \xrightarrow{O_2,\ \text{catalytic } Co^{3+},\ \Delta,\ \text{pressure}} CH_3COOH$$

Carbonylation of methanol:

$$CH_3OH \xrightarrow{CO,\ \text{catalytic } Rh^{3+},\ I_2,\ \Delta,\ \text{pressure}} CH_3COOH$$

Summary of Important Concepts

1 Carboxylic acids are named as alkanoic acids. The carbonyl carbon is numbered 1 in the longest chain incorporating the carboxy group. Dicarboxylic acids are called alkanedioic acids.

2 The carboxy group is approximately trigonally planar. Except in very dilute solution, carboxylic acids form dimers by hydrogen bonding.

3 The carboxylic acid proton chemical shift is variable but relatively high ($\delta = 10–13$), because of hydrogen bonding. The carbonyl carbon is also relatively deshielded, but not as much as in aldehydes and ketones because of the resonance contribution of the hydroxy group.

4 Infrared spectroscopy measures vibrational excitation. The energy of the incident radiation ranges from about 1 to 10 kcal mole^{-1} ($\Delta\lambda \sim 2.5–16.7\ \mu$m; $\Delta\tilde{\nu} \sim 600–4000$ cm^{-1}). Characteristic peaks are observed for certain functional groups, a consequence of stretching, bending, and other modes of vibration, and their combination. Moreover, each molecule exhibits a characteristic infrared spectral pattern, called a fingerprint.

5 The carbonyl group in carboxylic acids is subject to addition by nucleophiles to give an unstable tetrahedral intermediate. This intermediate may decompose by elimination of the hydroxy group to give a carboxylic acid derivative.

6 Alkanoyl halides are formed by the action of inorganic halides ($SOCl_2$, PCl_5, PBr_3) on carboxylic acids. Alkanoyl chlorides can also be made by treating carboxylic acids with phosgene or oxalyl chloride. These processes generate intermediate, unstable, mixed carboxylic anhydrides, which decompose to the products with evolution of CO_2 and (with oxalyl chloride) CO.

7 The mechanism of acid-catalyzed esterification can be probed with labeled oxygen, which reveals that the alkoxy group is derived from the alcohol.

8 Organolithium reagents and lithium aluminum hydride are strong enough nucleophiles to add to the carbonyl group of carboxylate ions. These processes allow the syntheses of ketones and the reduction of carboxylic acids to primary alcohols.

9 The 2-position of carboxylic acid dianions enters into S_N2 reactions just as the corresponding position in an enolate ion does.

Problems

1 Name (IUPAC system) or draw the structure of each of the following compounds.

(a) 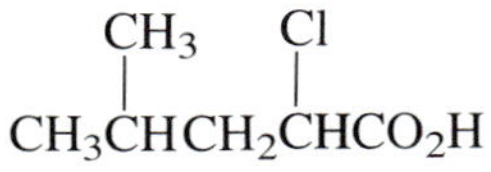 $CH_3CH(CH_3)CH_2CH(Cl)CO_2H$ (CH$_3$ on C-4, Cl on C-2)

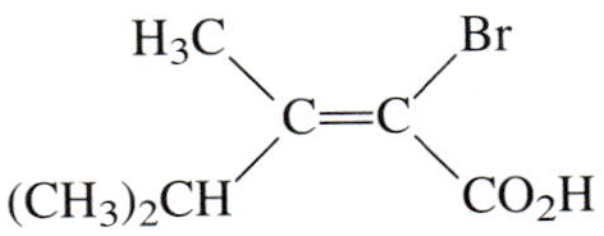

(b) 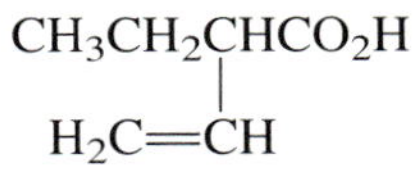 $CH_3CH_2CH(CH{=}CH_2)CO_2H$ (H$_2$C=CH substituent on the CH)

(c) $(H_3C)((CH_3)_2CH)C{=}C(Br)(CO_2H)$ (H$_3$C and Br on the same side; (CH$_3$)$_2$CH and CO$_2$H on the same side)

(d) 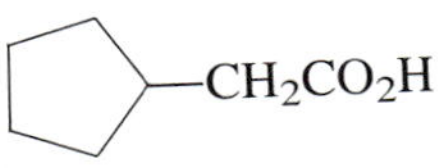 cyclopentyl–CH_2CO_2H

(e) OH CO_2H

(f) Cl(HO$_2$C)C=C(H)(CO$_2$H)

(g) 4-Aminobutanoic acid (also known as "GABA," a critical participant in brain biochemistry)

(h) *meso*-2,3-Dimethylbutanedioic acid

(i) 2-Oxopropanoic acid (pyruvic acid)

(j) *trans*-2-Methanoylcyclohexanecarboxylic acid

2 Rank the group of molecules shown in the margin in decreasing order of (1) boiling point and (2) solubility in water. Explain your answers.

COOH CHO CH_2OH CH_3

3 From the Hooke's Law equation, would you expect the C–X bonds of common haloalkanes (X = Cl, Br, I) to have IR bands at higher or lower wavenumbers than are typical for bonds between carbon and lighter elements (e.g., oxygen)?

4 Convert each of the following IR frequencies into micrometers.

(a) 1720 cm^{-1} (C=O) **(d)** 890 cm^{-1} (alkene bend)

(b) 1650 cm^{-1} (C=C) **(e)** 1100 cm^{-1} (C–O)

(c) 3300 cm^{-1} (alkyne C–H) **(f)** 2260 cm^{-1} (C≡N)

5 The IR spectrum of 1,4-nonadiyne displays a strong, sharp band at 3300 cm^{-1}. What is the origin of this band? Treatment of 1,4-nonadiyne with $NaNH_2$, then with D_2O, causes replacement of the band at 3300 cm^{-1} by a band at 2580 cm^{-1}.

(a) What is the product of this reaction?

(b) What new bond is responsible for the IR absorption at 2580 cm^{-1}?

(c) Using Hooke's law, calculate the approximate expected position of this new band from the structure of the original molecule and its IR spectrum. Assume that *k* and *f* have not changed.

6 You have just entered the chemistry stockroom to look for several isomeric bromopentanes. There are three bottles on the shelf marked $C_5H_{11}Br$, but their labels have fallen off. The NMR machine is broken, and so you devise the following experiment in an attempt to determine which isomer is in which bottle: a sample of the contents in each bottle is treated with NaOH in aqueous ethanol, and then the IR spectrum of each product or product mixture determined. Here are the results:

(i) $C_5H_{11}Br$ isomer in bottle A $\xrightarrow{\text{NaOH}}$ IR bands at 1660, 2850–3020, and 3350 cm^{-1}.

(ii) $C_5H_{11}Br$ isomer in bottle B $\xrightarrow{\text{NaOH}}$ IR bands at 1670 and 2850–3020 cm^{-1}.

(iii) $C_5H_{11}Br$ isomer in bottle C $\xrightarrow{\text{NaOH}}$ IR bands at 2850–2960 and 3350 cm^{-1}.

Answer the following questions.

(a) What do the data tell you about each product or product mixture?

(b) Suggest a structure for the contents of each bottle. Is more than one structure consistent with the data for any of the three bottles?

7 Match each structure at the left in the table below with one of the sets of IR bands given at the right. Only bands of specific diagnostic value have been listed (br stands for broad band).

Compounds	Infrared bands
(a) $CH_3(CH_2)_3O(CH_2)_3CH_3$	(1) 3060, 2925, 1646, 1438, 1380, 888 cm^{-1}
(b) 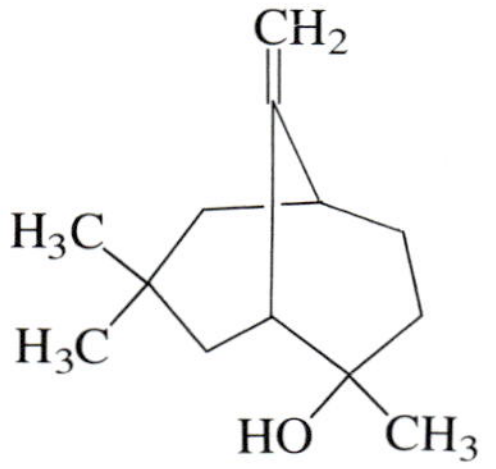 bicyclic structure bearing $=CH_2$, two H_3C groups, HO and CH_3	(2) 3000, 2915, 1460, 1370, 1120 cm^{-1}
(c) $CH_3CH_2CH_2\overset{\overset{\large O}{\|}}{C}OCH_2CH_3$	(3) 3080, 3020, 2960, 2900, 2260, 1647, 1418, 990, 935 cm^{-1}
(d) 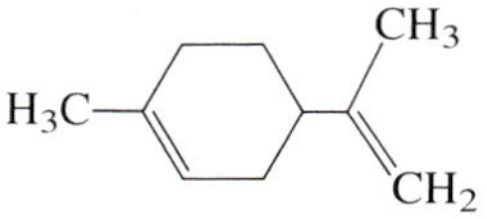 H_3C–cyclohexene ring–$C(CH_3)=CH_2$	(4) 2975, 2880, 2260, 1464, 1433 cm^{-1}
(e) $CH_3\overset{\overset{\large CH_3}{\vert}}{C}H\underset{\underset{\large O}{\|}}{C}CH_3$	(5) 3300–2500 (br), 2905, 2840, 1708, 1465, 1430, 1410 cm^{-1}
(f) $CH_2{=}CHCH_2CN$	(6) 3300–2500 (br), 1730 (br), 1710, 1400, 1366, 1163 cm^{-1}
(g) $CH_3(CH_2)_{14}CO_2H$	(7) 3330 (br), 3300, 2937, 2878, 2120, 1360, 1025 cm^{-1}
(h) $CH_3CH_2CH_2CN$	(8) 3356 (br), 2900, 1666, 1466, 1380, 1130, 896 cm^{-1}
(i) $HC{\equiv}CCH_2OH$	(9) 2970, 2880, 1718, 1470, 1370 cm^{-1}
(j) $CH_3\overset{\overset{\large O}{\|}}{C}CH_2CH_2CO_2H$	(10) 2955, 2930, 2870, 1737, 1461, 1372, 1183 cm^{-1}

8 **(a)** An unknown compound D has the formula $C_7H_{12}O_2$ and the infrared spectrum shown on the facing page. To which class does this compound belong? (Refer to the sample IR spectra in Section 17-3.)

(b) Use the NMR and IR spectra on pages 781 through 784 and spectroscopic and chemical information in the reaction sequence at the top of the facing page to determine the structures of compound D and the other unknown substances E through I. References are made to relevant sections of earlier chapters, but do not look them up before you have tried to solve the problem without the extra help.

(c) Another unknown compound, J, has the formula $C_8H_{14}O_4$ and the NMR and IR spectra on page 784. Propose a structure for this compound. Note: Table 17-3 may be helpful.

$$\underset{\textbf{E}}{C_6H_{10}} \xrightarrow[\text{Section 12-3}]{\text{1. } Hg(O\overset{O}{\overset{\|}{C}}CH_3)_2,\ H_2O \quad \text{2. } NaBH_4} \underset{\textbf{F}}{C_6H_{12}O} \xrightarrow[\text{Section 9-4}]{CrO_3,\ H_2SO_4,\ \text{propanone (acetone)},\ 0°C}$$

E: ^{13}C NMR: δ = 22.1, 24.5, 126.2 ppm

^1H NMR-E

F: ^{13}C NMR: δ = 24.4, 25.9, 35.5, 69.5 ppm

$$\underset{\textbf{G}}{C_6H_{10}O} \xrightarrow[\text{Section 15-7}]{CH_2{=}P(C_6H_5)_3} \underset{\textbf{H}}{C_7H_{12}} \xrightarrow[\text{Section 12-4}]{\text{1. } BH_3 \quad \text{2. } HO^-,\ H_2O_2} \underset{\textbf{I}}{C_7H_{14}O} \xrightarrow[\text{Section 9-4}]{Na_2Cr_2O_7,\ H_2O,\ H_2SO_4} \textbf{D}$$

G: ^{13}C NMR: δ = 23.8, 26.5, 40.4, 208.5 ppm

IR-G

H: IR-H

I: IR-I, ^1H NMR-I

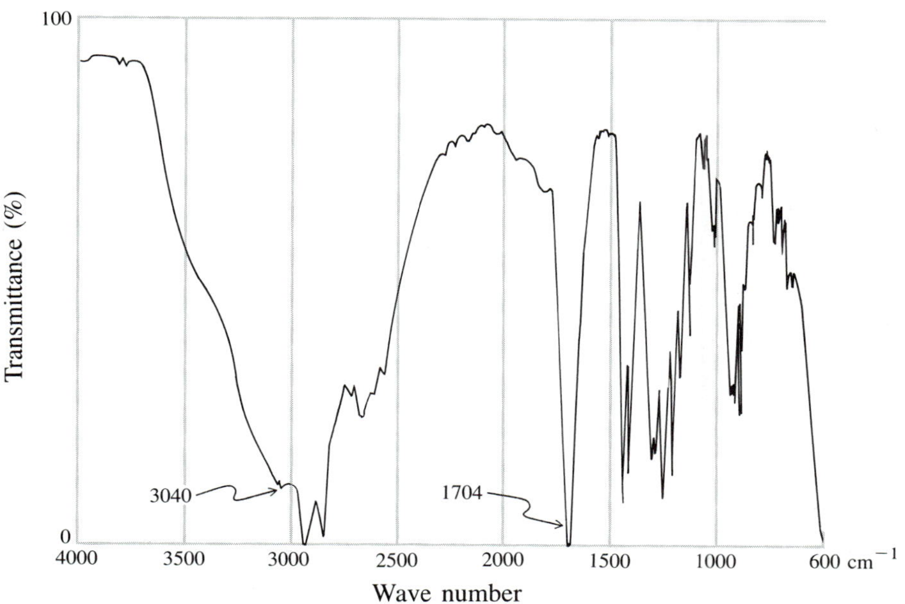

IR-D

(d) Compound J may be readily synthesized from E. Propose a sequence that accomplishes this efficiently.

(e) Propose a completely different sequence from that shown in part **b** for the conversion of F into D.

(f) Finally, construct a synthetic scheme that is the reverse of that shown in part **b;** namely, the conversion of D into E.

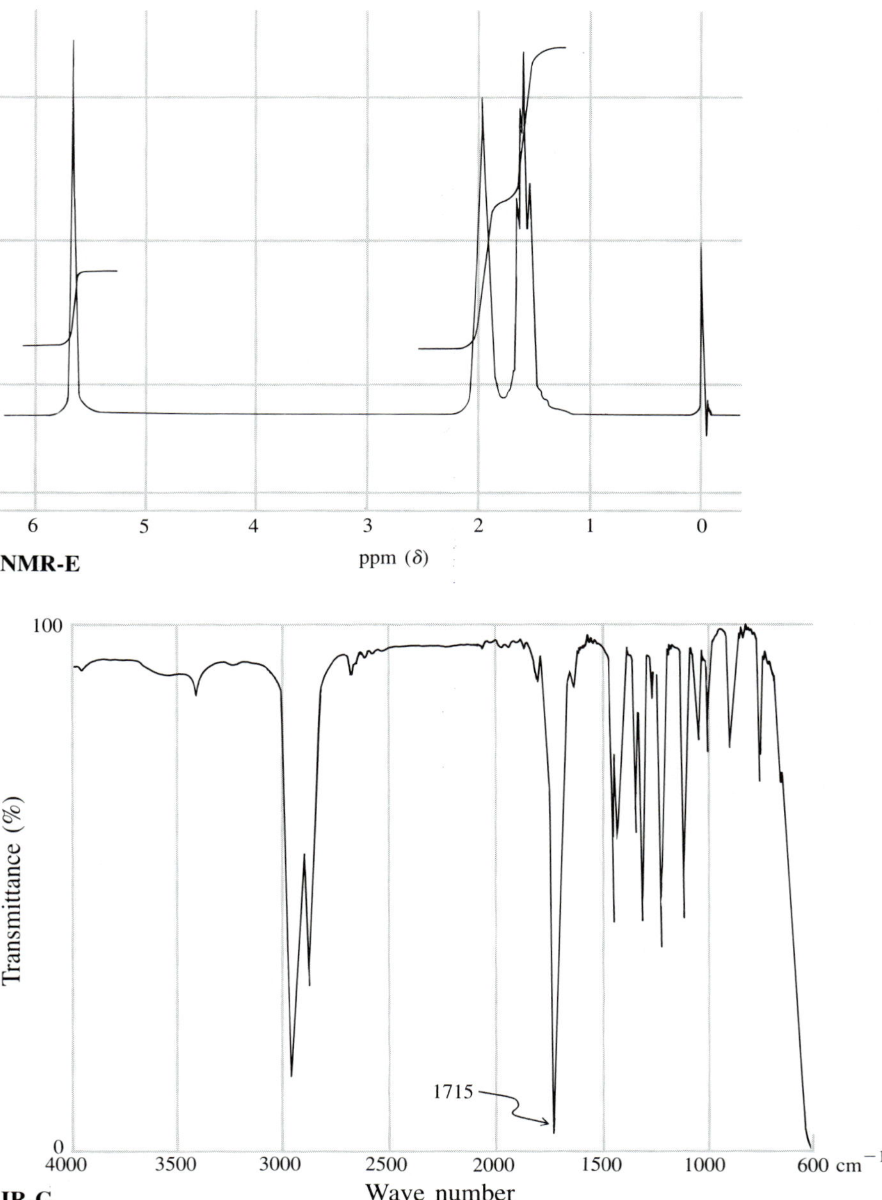

9 Rank each of the following groups of organic compounds in order of decreasing acidity.

(a) $CH_3CH_2CO_2H$, $CH_3\overset{O}{\overset{\|}{C}}CH_2OH$, $CH_3CH_2CH_2OH$

(b) $BrCH_2CO_2H$, $ClCH_2CO_2H$, FCH_2CO_2H

(c) $CH_3\underset{}{\overset{Cl}{\overset{|}{C}}}HCH_2CO_2H$, $ClCH_2CH_2CH_2CO_2H$, $CH_3CH_2\overset{Cl}{\overset{|}{C}}HCO_2H$

(d) CF_3CO_2H, CBr_3CO_2H, $(CH_3)_3CCO_2H$

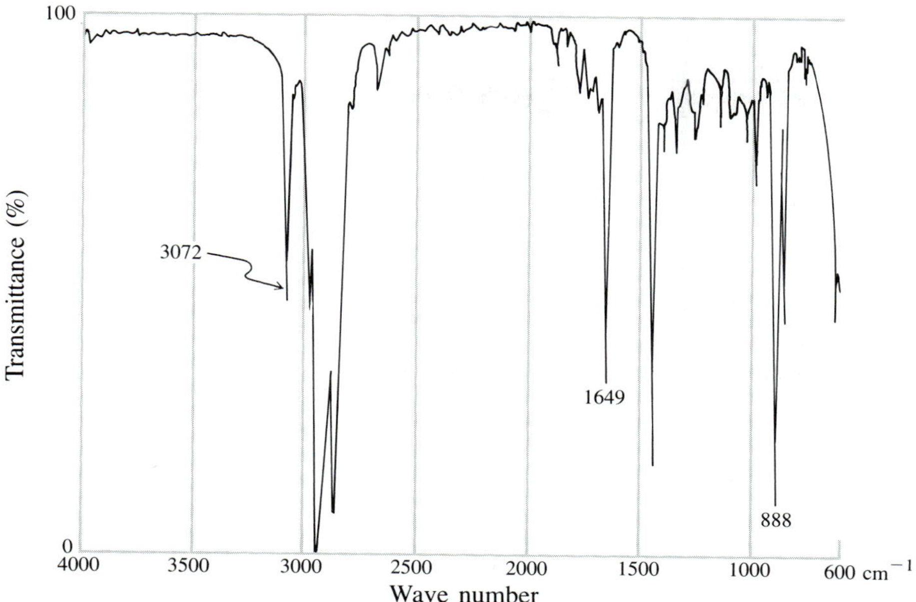

IR-H

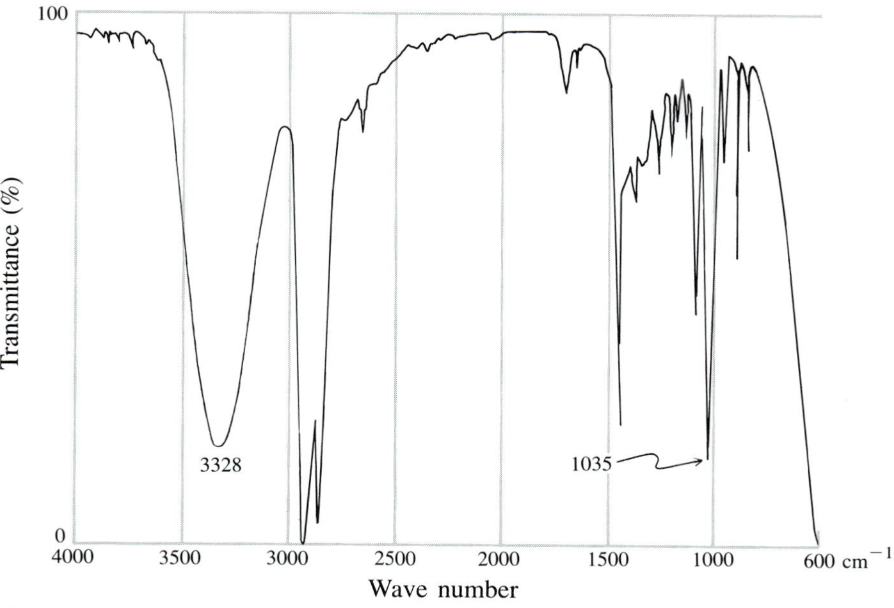

IR-I

10 How would you expect the acidity of ethanamide (acetamide) to compare with that of ethanoic (acetic) acid? With that of propanone (acetone)? Which protons in ethanamide (acetamide) are the most acidic? Where would you expect ethanamide (acetamide) to be protonated by very strong acid?

$$CH_3\overset{\overset{\displaystyle O}{\|}}{C}NH_2$$

Ethanamide (Acetamide)

11 Fill in suitable reagents to carry out the following transformations.

(a) $(CH_3)_2CHCH_2CHO \longrightarrow (CH_3)_2CHCH_2CO_2H$

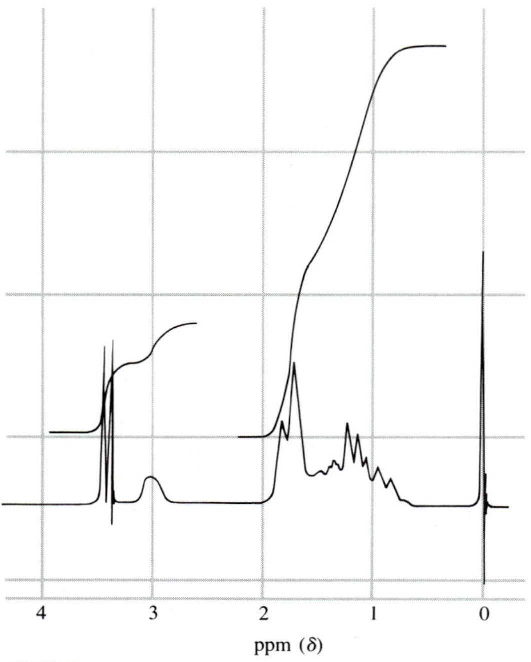

NMR-I

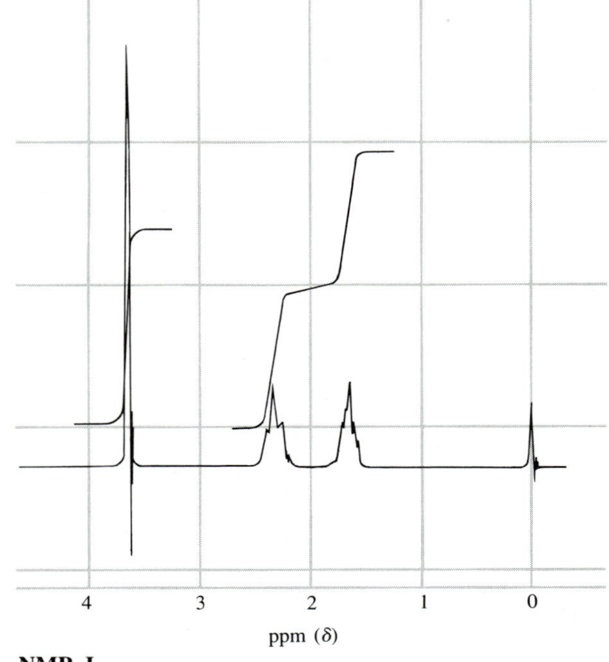

NMR-J

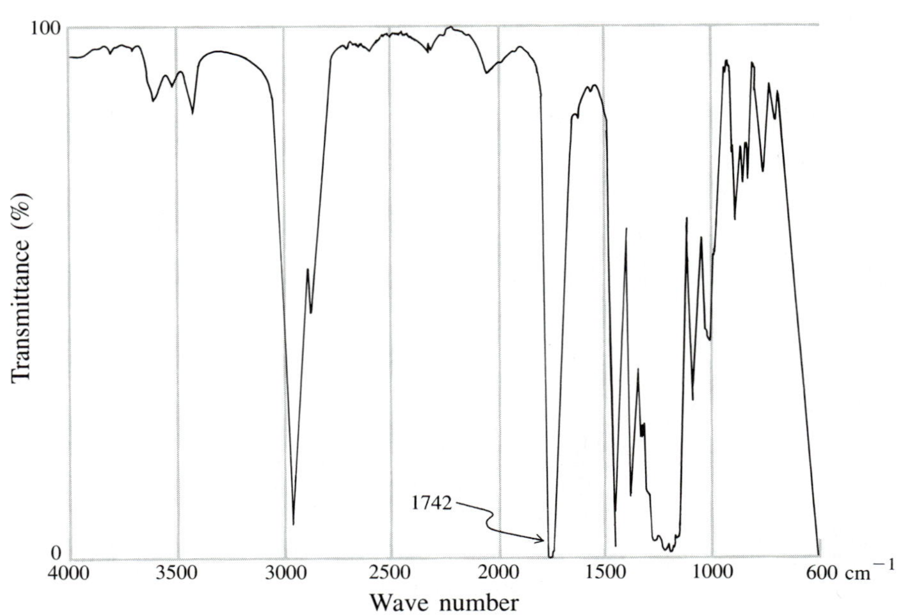

IR-J

(b) cyclopentyl—CH=CH$_2$ ⟶ cyclopentyl—CO$_2$H

(c) decalin-2-yl—Br ⟶ decalin-2-yl—CO$_2$H

(d) $CH_3CH(OH)CH_2CH_2Cl \longrightarrow CH_3CH(OH)CH_2CH_2CO_2H$

(e) $CH_3CH_2CH(CO_2H)CH_3 \longrightarrow CH_3CH_2CH(CH_3)—\overset{O}{\overset{\|}{C}}O\overset{O}{\overset{\|}{C}}—CH(CH_3)CH_2CH_3$

(f) $(CH_3)_3CCO_2H \longrightarrow (CH_3)_3CCO_2C(CH_3)_3$

(g) 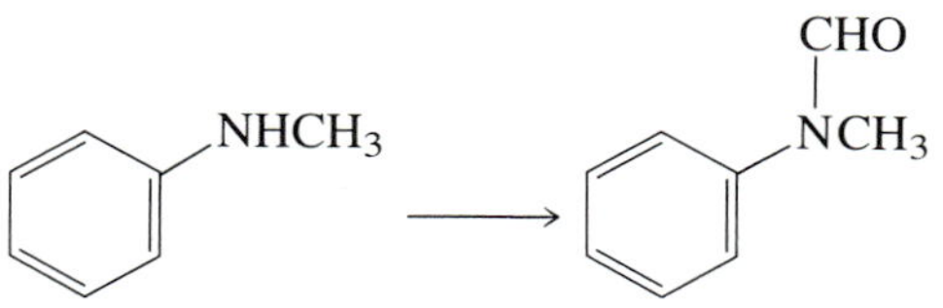 $C_6H_5NHCH_3 \longrightarrow C_6H_5N(CHO)CH_3$

(h)  $\text{cyclopropyl}—CH_2CO_2H \longrightarrow \text{cyclopropyl}—CH_2Br$

12 Attempted CrO_3 oxidation of 1,4-butanediol to butanedioic acid results in significant yields of "γ-butyrolactone." Explain mechanistically.

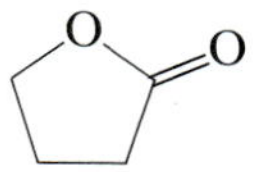

γ-Butyrolactone

13 Show how you might synthesize each of the following carboxylic acids, without using simple oxidations of primary alcohols or aldehydes. Otherwise, you may use any starting materials you wish.

(a) $CH_3CH_2CH_2CH_2CH_2CH_2CO_2H$

(b) $CH_3CH(OH)CH_2CO_2H$

(c) $H_3C—C(CH_3)_2—CO_2H$

(d) 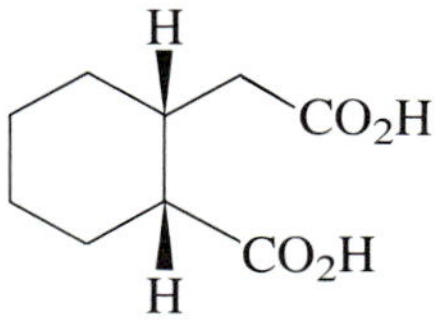 cyclohexane bearing CH_2CO_2H and CO_2H on adjacent carbons, with H atoms shown (wedged)

14 Following the general mechanistic schemes given in Section 17-6, write detailed mechanisms for each of the following substitution reactions. Note: These anticipate reactions that you will see in detail in Chapter 18, but do not refer to them.

(a) 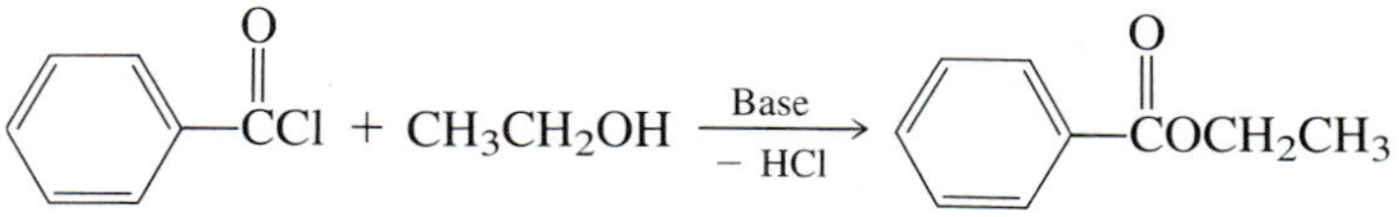 $C_6H_5\overset{O}{\overset{\|}{C}}Cl + CH_3CH_2OH \xrightarrow[-\ HCl]{\text{Base}} C_6H_5\overset{O}{\overset{\|}{C}}OCH_2CH_3$

(b) $CH_3\overset{O}{\overset{\|}{C}}NH_2 + H_2O \xrightarrow{\text{Acid}} CH_3\overset{O}{\overset{\|}{C}}OH + \overset{+}{N}H_4$

15 **(a)** Write a mechanism for the esterification of propanoic acid with ^{18}O-labeled ethanol. Show clearly the fate of the ^{18}O label.

(b) Acid-catalyzed hydrolysis of an unlabeled ester with ^{18}O-labeled water ($H_2{}^{18}O$) leads to incorporation of some ^{18}O into *both* oxygens of the carboxylic acid product. Explain by a mechanism. Hint: You must use the fact that all steps in the mechanism are reversible.

16 Suggest structures for the products of each reaction in the following synthetic sequence.

Starting ketone (IR: 1745 cm^{-1}) $\xrightarrow{\text{1. pyrrolidine, } H^+;\ \text{2. } (CH_3)_2C{=}CHCH_2Br;\ \text{3. } H^+, H_2O}$ $C_{14}H_{22}O$ **K** (IR: 1675 and 1745 cm^{-1}) $\xrightarrow{HOCH_2CH_2OH,\ H^+}$ $C_{16}H_{26}O_2$ **L** (IR: 1670 cm^{-1}) $\xrightarrow{KMnO_4,\ NaHCO_3,\ \text{propanone (acetone)}}$

$C_{13}H_{20}O_4$ **M** (IR: 1715 and 3000 (br) cm^{-1}) $\xrightarrow{\text{1. } H^+, H_2O;\ \text{2. } NaBH_4}$ $C_{11}H_{18}O_3$ **N** (IR: 1715, 3000 (br), and 3350 cm^{-1}) $\xrightarrow{\text{Catalytic } H^+,\ \Delta}$ $C_{11}H_{16}O_2$ **O** (IR: 1770 cm^{-1})

17 S_N2 reactions of simple carboxylate ions with haloalkanes in aqueous solution generally give very poor results.

(a) Explain why this is so.

(b) In Section 17-8, the reaction of 1-iodobutane with sodium ethanoate (acetate) in ethanoic (acetic) acid solvent is shown. Why is ethanoic (acetic) acid a better solvent for this reaction than water?

(c) The reaction of 1-iodobutane with sodium dodecanoate proceeds surprisingly well in aqueous solution, much better than the reaction with sodium ethanoate (acetate) (see the following equation). Explain this observation. (Hint: Sodium dodecanoate is a soap and forms micelles in water.)

$$CH_3CH_2CH_2CH_2I + CH_3(CH_2)_{10}CO_2{}^-Na^+ \xrightarrow{H_2O} CH_3(CH_2)_{10}\overset{O}{\overset{\|}{C}}OCH_2CH_2CH_2CH_3$$

18 **(a)** Diazomethane, CH_2N_2, is usually represented as a resonance hybrid of two contributing Lewis structures. Write them.

(b) Propose a mechanism for the formation of a methyl ester from diazomethane and a carboxylic acid.

19 Propose *two* possible mechanisms for the following reaction. Hint: Consider the possible sites of protonation in the molecule and the mechanistic consequences of each.

C_6H_5–CH(OH)–C_6H_4–COOH (ortho) $\xrightarrow{HCl,\ \Delta}$ lactone (C_6H_5–CH–O–C(=O)– fused to benzene ring)

Devise an isotope-labeling experiment that might distinguish your two mechanisms.

20 Write the products of reaction of propanoic acid with each of the following reagents.

(a) $SOCl_2$
(b) PBr_3
(c) CH_3CH_2COBr + pyridine
(d) $(CH_3)_2CHOH$ + HCl
(e) CH_2N_2
(f) KOH, CH_3CH_2I, DMSO
(g) C_6H_5—CH_2NH_2
(h) Product of part **g**, heated strongly
(i) $LiAlH_4$
(j) C_6H_5—Li
(k) 2 LDA in HMPA
(l) Product of part **k** + $CH_2{=}CHCH_2Cl$

21 The starting material for Problem 15 of Chapter 16 is prepared in two steps from the keto acid shown at the right. Suggest how this might be accomplished.

22 Beginning with cyclohexanol, design several alternative syntheses of 4-cyclohexyl-2-butanone, using in each case one of the following reaction types as a key carbon–carbon bond-forming step.

(a) Conjugate addition.
(b) Three-membered ring ether (oxacyclopropane) ring-opening.
(c) Wittig reaction.

Which of these routes is the most efficient?

23 Propose a synthesis of dihydrotagetone, a naturally occurring essence in Japanese Ho leaf oil, starting exclusively with compounds containing four carbons or less.

Dihydrotagetone

24 Suggest a preparation of hexanoic acid from pentanoic acid.

25 Give reagents and reaction conditions that would allow efficient conversion of 2-methylbutanoic acid into

(a) the corresponding alkanoyl chloride.
(b) the corresponding methyl ester.
(c) the corresponding ester with 2-butanol.
(d) the mixed anhydride with ethanoic (acetic) acid.
(e) the *N*-methylamide.
(f) $CH_3CH_2CH(CH_3)COCH_3$
(g) $CH_3CH_2CH(CH_3)CH_2OH$
(h) $CH_3CH_2CBr(CH_3)CO_2H$
(i) $CH_3CH_2C(CH_3)_2CO_2H$
(j) 2-bromobutane.

26 Show how the Hell-Volhard-Zelinsky reaction might be used in the synthesis of each of the following compounds, beginning in each case with a simple monocarboxylic acid. Write detailed mechanisms for all the reactions in *one* of your syntheses.

(a) 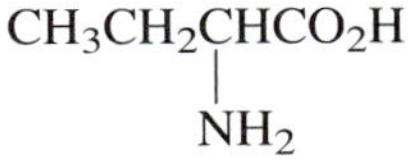 $CH_3CH_2CHCO_2H$
NH_2

(b) 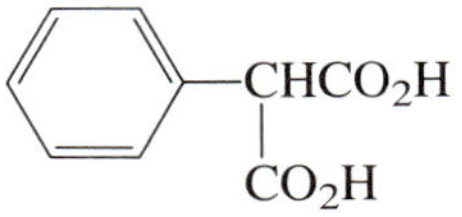 C_6H_5–$CHCO_2H$
CO_2H

(c) CH_3
$CH_3CH_2CHCH_2CHCO_2H$
OH

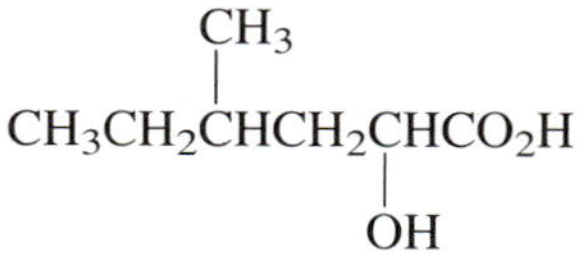

(d) $HO_2CCH_2SSCH_2CO_2H$

(e) $(CH_3CH_2)_2NCH_2CO_2H$

(f) 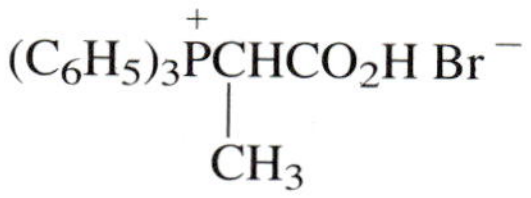 $(C_6H_5)_3\overset{+}{P}CHCO_2H\ Br^-$
CH_3

27 Draw and compare the resonance structures of the dianions of carbonic acid, H_2CO_3, and ethanoic (acetic) acid.

(a) Compare the basicities of these dianions. Explain the difference.

(b) Propose structures for dianions that might be derived from (1) propanone (acetone), and (2) 2-methylpropene. Draw all resonance forms for each.

(c) Suggest methods by which each of these dianions might be synthesized.

(d) The product of the following reaction shows only two signals in the ^1H NMR spectrum: a singlet near $\delta = 0.5$ and another singlet near $\delta = 4.0$ ppm, in an intensity ratio of 9:1. What is this product?

$$CH_3CO_2H \xrightarrow[\text{2. 2 }(CH_3)_3SiCl]{\text{1. 2 LDA, HMPA}}$$

28 Suggest ways of synthesizing each of the following molecules from the indicated starting materials. You may use any other reagents that you find helpful, as well.

(a) 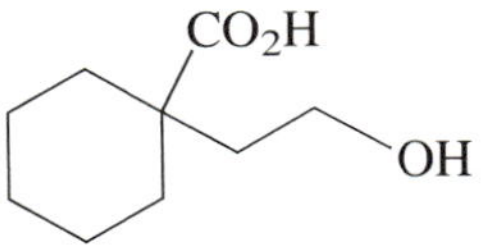 CO_2H / OH (cyclohexane structure) from cyclohexanone

(b) 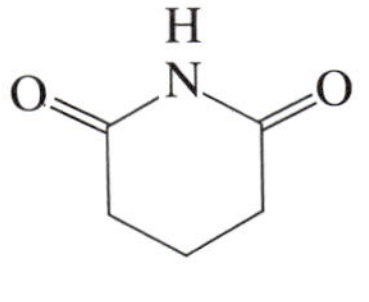 H N, O, O (ring structure) from cyclopentene

(c) 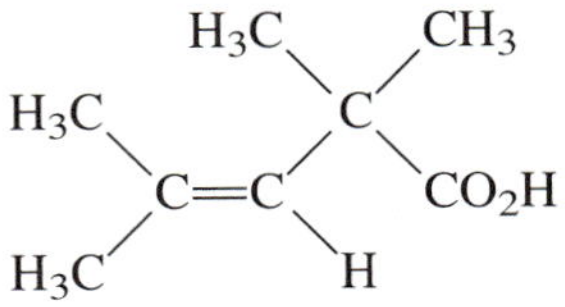 H_3C CH_3
H_3C C
C=C CO_2H from 2-propanol
H_3C H

(d) [structure: six-membered lactone, δ-valerolactone] from ethanoic (acetic) acid and 3-bromopropene

29 Hunsdiecker decarboxylation of cyclopropanecarboxylic acids is a method of choice for synthesis of bromocyclopropanes. Explain why this is so by describing the problems with other methods for the synthesis of haloalkanes as applied to cyclopropyl compounds.

30 The "iridoids" are a class of monoterpenes with powerful and varied biological activities. They include insecticides, agents of defense against predatory insects, and animal attractants. Following is a synthesis of neonepetalactone, one of the nepetalactones, which are primary constituents of catnip. Use the information given to deduce the structures that have been left out, including that of neonepetalactone itself.

$C_{10}H_{16}O_2$ **P** $\xrightarrow{\text{Base}}$ [cyclopentene with methyl, –CHO, and isopropenyl substituents] $\xrightarrow{CrO_3,\ H_2SO_4,\ 0°C}$

IR: 890, 1645, 1725 (very strong), and 2705 cm^{-1}

$C_{10}H_{14}O_2$ **Q** $\xrightarrow{CH_2N_2}$ $C_{11}H_{16}O_2$ **R** $\xrightarrow{\text{1. Diisoamylborane; 2. } HO^-,\ H_2O_2}$

Q — IR: 890, 1630, 1640, 1720, and 3000 (very broad) cm^{-1}

R — IR: 890, 1630, 1640, and 1720 cm^{-1}

$C_{11}H_{18}O_3$ **S** $\xrightarrow{H^+,\ H_2O,\ \Delta}$ $C_{10}H_{14}O_2$ **Neonepetalactone**

S — IR: 1630, 1720, and 3335 cm^{-1}

Neonepetalactone — IR: 1645 and 1710 cm^{-1}; UV λ_{max} = 241 nm

resonating lone electron pairs on the central oxygen of an anhydride have to be shared with two carbonyl groups.

TABLE 18-1

C–L bond lengths in $\mathrm{R\overset{O}{\overset{\|}{C}}L}$ compared with R–L single bond distances

L	Bond length (Å) in R—L	Bond length (Å) in $\mathrm{R\overset{O}{\overset{\|}{C}}—L}$
Cl	1.78	1.79
OCH_3	1.43	1.36
NH_2	1.47	1.36

The Extent of Resonance Affects the Structure of the Functional Group

The extent of resonance shows itself in the structural parameters of carboxylic acid derivatives. In particular, in procession from the most-reactive (alkanoyl halides) to the least-reactive systems (esters and amides), the C–L bond becomes progressively shorter than the normal C–L single bond (Table 18-1). The dipolar resonance contribution in amides is strong enough to impart considerable rigidity to the amide linkage. This rigidity is seen in the NMR spectra of amides. For example, *N,N*-dimethylmethanamide (*N,N*-dimethylformamide) at room temperature exhibits two singlets for the two methyl groups, because of the slowness of rotation around the C–N bond on the NMR time scale. The barrier to rotation (E_a = 21 kcal mole^{-1}) is due to the strength of the π bond between the carbonyl carbon and the nitrogen.

Slow Rotation in *N,N*-Dimethylmethanamide

$$(H_3C)_2N\text{—}CH{=}O \xrightleftharpoons{\text{Rotation}} (H_3C)_2N\text{—}CH{=}O$$

$$E_a = 21 \text{ kcal mole}^{-1}$$

EXERCISE 18-1

The methyl group in the ^{1}H NMR spectrum of 1-ethanoyl 2-phenylhydrazide, $\mathrm{CH_3\overset{O}{\overset{\|}{C}}NHNHC_6H_5}$, exhibits two singlets at δ = 2.02 and 2.10 ppm at room temperature. On heating to 100°C in the NMR probe, the same compound gives rise to only one signal in that region. Explain.

Another spectral technique for probing the extent of resonance in carboxylic acid derivatives is infrared spectroscopy. An increased contribution of the dipolar resonance structure weakens the C=O bond and causes a corresponding decrease in the carbonyl stretching frequency (Table 18-2).

A similar diagnostic trend is not seen in the ^{13}C NMR chemical shifts of the carbonyl carbon in carboxylic acid derivatives—all appear at about 170 ppm (see Section 17-2).

TABLE 18-2

Carbonyl stretching frequencies of carboxylic acid derivatives $\mathrm{R\overset{O}{\overset{\|}{C}}L}$

L	$\tilde{\nu}_{C=O}$ (cm^{-1})	
Cl	1790–1815	
$\mathrm{O\overset{O}{\overset{\|}{C}}R}$	1740–1790 1800–1850	Because of mechanical coupling, two bands are observed, corresponding to asymmetric and symmetric stretching motions.
OR	1735–1750	
NR'_2	1650–1690	

^{13}C NMR Chemical Shifts of the Carbonyl Carbon in Carboxylic Acid Derivatives

$\mathrm{CH_3\overset{O}{\overset{\|}{C}}Cl}$ $\delta = 170.3$; $\mathrm{CH_3\overset{O}{\overset{\|}{C}}O\overset{O}{\overset{\|}{C}}CH_3}$ 166.9; $\mathrm{CH_3\overset{O}{\overset{\|}{C}}OH}$ 177.2; $\mathrm{CH_3\overset{O}{\overset{\|}{C}}OCH_3}$ 170.7; $\mathrm{CH_3\overset{O}{\overset{\|}{C}}NH_2}$ 172.6 ppm

Carboxylic Acid Derivatives Are Basic and Acidic

The extent of resonance in carboxylic acid derivatives is also seen in their basicity (protonation at the carbonyl oxygen) and acidity (enolate formation). In all cases, protonation requires strong acid, but it gets easier as the resonating ability of the L group increases.

Protonation of a Carboxylic Acid Derivative

$$\mathrm{R\overset{:O:}{\overset{\|}{C}}\ddot{L}} \xrightarrow{H^+} \left[\mathrm{R\overset{^+\ddot{O}H}{\overset{\|}{C}}\ddot{L}} \longleftrightarrow \mathrm{R\overset{:\ddot{O}H}{\overset{|}{C^+}}\ddot{L}} \longleftrightarrow \mathrm{R\overset{:\ddot{O}H}{\overset{|}{C}}{=}L^+} \right]$$

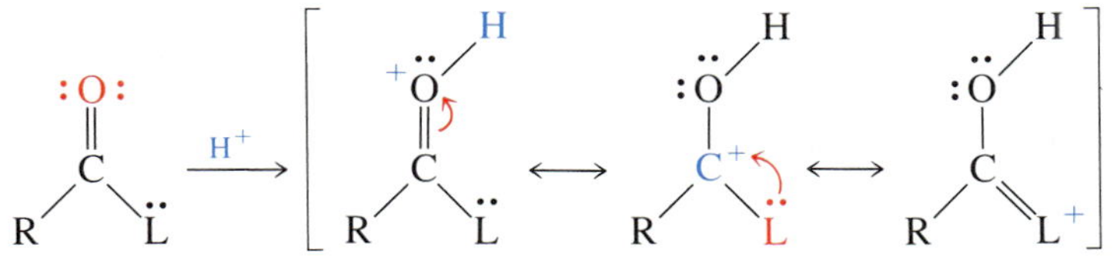

A relatively strong contribution of this resonance structure stabilizes the protonated species.

The pK_a values for the conjugate acids of carboxylic acid derivatives reveal that alkanoyl halides are the weakest bases (i.e., their conjugate acids are the strongest acids and have the lowest pK_as). Esters are about as basic as carboxylic acids, whereas amides are the most basic.

pK_a Values for Some Oxygen-Protonated Carboxylic Acid Derivatives

$$\mathrm{R\overset{^+\ddot{O}H}{\overset{\|}{C}}\ddot{\underset{..}{Cl}}:} < \mathrm{R\overset{^+\ddot{O}H}{\overset{\|}{C}}\ddot{\underset{..}{O}}R'} < \mathrm{R\overset{^+\ddot{O}H}{\overset{\|}{C}}\ddot{N}H_2}$$

pK_a ~ -9 $\quad$ ~ -6 $\quad$ ~ 0

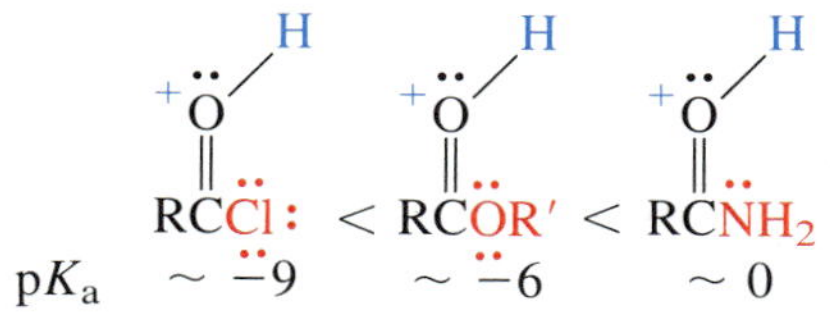

EXERCISE 18-2

Explain with resonance structures why ethanoyl (acetyl) chloride is a much weaker base than ethanamide (acetamide).

For related reasons, the acidity of the hydrogens next to the carbonyl group decreases along the series. The acidity of a ketone lies between those of an alkanoyl chloride and an ester.

Acidity of α-Hydrogens in Carboxylic Acid Derivatives in Comparison with Propanone (Acetone)

$$\underset{\sim 16}{CH_3\overset{O}{\overset{\|}{C}}Cl} > \underset{\sim 20}{CH_3\overset{O}{\overset{\|}{C}}CH_3} > \underset{\sim 25}{CH_3\overset{O}{\overset{\|}{C}}OCH_3} > \underset{\sim 30}{CH_3\overset{O}{\overset{\|}{C}}N(CH_3)_2}$$

pK_a: ~16, ~20, ~25, ~30

Amides, in which the nitrogen bears one or two protons, are deprotonated at *nitrogen* to form an **amidate ion.** The pK_a at this position is relatively low because, like a carboxylate ion, the amidate ion is stabilized by resonance. Because nitrogen is much less electronegative than oxygen, the pK_a of an amide is more than ten units larger than that of a carboxylic acid.

Resonance in Amidate Ions

$$R\overset{:O:}{\overset{\|}{C}}\ddot{N}H_2 \rightleftharpoons H^+ + \left[R\overset{:O:}{\overset{\|}{C}}\ddot{\underset{..}{N}}H^- \longleftrightarrow R\overset{:\ddot{O}:^-}{\overset{|}{C}}{=}\ddot{N}H \right]$$

$pK_a \sim 15$

An amidate ion

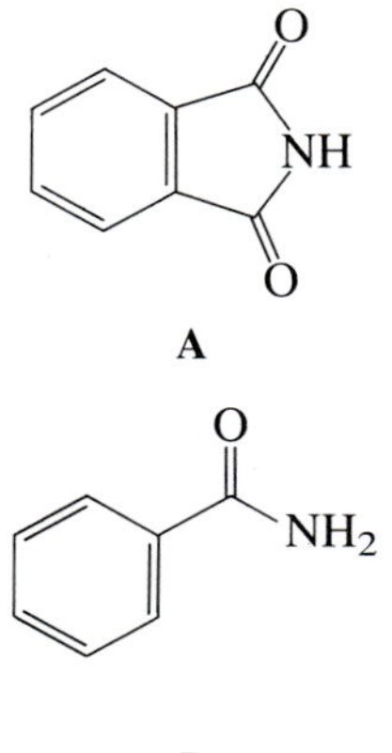

EXERCISE 18-3

The pK_a of 1,2-benzenedicarboxímide (phthalimide, A) is 8.3, considerably lower than the pK_a of benzenecarboxamide (benzamide, B). Why?

In summary, the relative electronegativity of L in $R\overset{O}{\overset{\|}{C}}L$ controls the extent of resonance of the lone electron pair(s) and therefore the relative reactivity of a carboxylic acid derivative in nucleophilic addition-elimination reactions. This effect manifests itself in structural and spectroscopic measurements, as well as in the relative acidity and basicity of the α-hydrogen and the carbonyl oxygen, respectively. Amides that bear hydrogens on the nitrogen atoms are deprotonated there to give the resonance-stabilized amidate ions.

18-2 The Chemistry of Alkanoyl Halides

After a brief description of nomenclature, this section presents the reactions of alkanoyl halides, particularly those in which nucleophiles displace the halide leaving group to give anhydrides, alcohols, esters, amides, ketones, and aldehydes.

The Names of Alkanoyl Halides

Compounds RCOX are named by changing the name of the alkano**ic acid** from which they are derived to alkano**yl halide.**

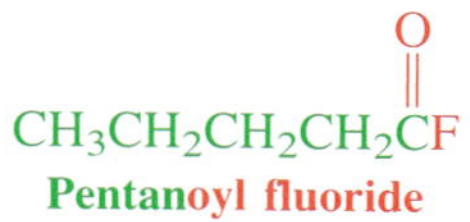

Alkanoyl Halides React with Nucleophiles

Alkanoyl halides undergo addition-elimination reactions with nucleophiles.

General Addition-Elimination Reactions of Alkanoyl Halides

$$\mathrm{R\overset{\overset{\Large :O:}{\|}}{C}\ddot{X}: + :Nu^- \longrightarrow R{-}\underset{\underset{\Large Nu}{|}}{\overset{\overset{\Large :\ddot{O}:^-}{|}}{C}}{-}\ddot{X}: \longrightarrow R\overset{\overset{\Large :O:}{\|}}{C}Nu + :\ddot{X}:^-}$$

The nucleophilic agents may be water (which gives the corresponding carboxylic acid), carboxylate ions (which give anhydrides), alcohols (which give esters), amines (which give amides), and organometallic reagents (which can give ketones). In addition, hydride reducing agents and hydrogen (which functions by the activation of catalysts) can yield aldehydes. It is because of this wide range of reactivity that alkanoyl halides are useful synthetic intermediates. Many of these reactions are also possible for the other carboxylic acid derivatives.

Let us consider these transformations one by one (except for anhydride formation, which was covered in Section 17-7). Examples will be restricted to alkanoyl chlorides, which are the most readily accessible, but their transformations can be generalized to a considerable extent to the other alkanoyl halides.

Nucleophilic Addition-Elimination Reactions of Alkanoyl Halides

$$\mathrm{R\overset{\overset{O}{\|}}{C}X \xrightarrow{HOH} R\overset{\overset{O}{\|}}{C}OH + HX}$$

$$\mathrm{R\overset{\overset{O}{\|}}{C}X \xrightarrow{R'\overset{\overset{O}{\|}}{C}O^-M^+} R\overset{\overset{O}{\|}}{C}O\overset{\overset{O}{\|}}{C}R' + MX}$$

$$\mathrm{R\overset{\overset{O}{\|}}{C}X \xrightarrow{R'OH} R\overset{\overset{O}{\|}}{C}OR' + HX}$$

$$\mathrm{R\overset{\overset{O}{\|}}{C}X \xrightarrow{R'\overset{\overset{H}{|}}{N}H} R\overset{\overset{O}{\|}}{C}NHR' + HX}$$

$$\mathrm{R\overset{\overset{O}{\|}}{C}X \xrightarrow{R'MgX} R\overset{\overset{O}{\|}}{C}R' + XMgX}$$

$$\mathrm{R\overset{\overset{O}{\|}}{C}X \xrightarrow{LiAl(OR)_3H} R\overset{\overset{O}{\|}}{C}H + LiX + Al(OR)_3}$$

$$\mathrm{R\overset{\overset{O}{\|}}{C}X \xrightarrow{H{-}H,\ catalyst} R\overset{\overset{O}{\|}}{C}H + HX}$$

$$\mathrm{CH_3CH_2\overset{O}{\overset{\|}{C}}Cl}$$
Propanoyl chloride
+
HOH
↓
$$\mathrm{CH_3CH_2\overset{O}{\overset{\|}{C}}OH}$$
100%
Propanoic acid
+
HCl

Water Hydrolyzes Alkanoyl Chlorides to Carboxylic Acids

Alkanoyl chlorides react with water, often violently, to give the corresponding carboxylic acids and hydrogen chloride. The mechanism of this transformation is a simple variation of the general addition-elimination scheme.

Mechanism of Alkanoyl Chloride Hydrolysis:

$$\mathrm{CH_3CH_2\overset{:O:}{\overset{\|}{C}}{-}\ddot{C}l: + H_2\ddot{O} \longrightarrow CH_3CH_2{-}\underset{:\overset{+}{O}H_2}{\overset{:\ddot{O}:^-}{C}}{-}\ddot{C}l: \xrightarrow{-\,:\ddot{C}l:^-} CH_3CH_2\overset{:O:}{\overset{\|}{C}}{-}\overset{+}{\ddot{O}}H_2 \longrightarrow CH_3CH_2\overset{:O:}{\overset{\|}{C}}\ddot{O}H + H^+}$$

Alcohols Convert Alkanoyl Chlorides into Esters

The reaction of alkanoyl chlorides with alcohols is mechanistically quite analogous to their reaction with water and is a highly effective way of producing esters. A base is usually added to neutralize the hydrogen chloride by-product. Because alkanoyl chlorides are readily made from the corresponding carboxylic acids (Section 17-7), the sequence RCOOH → RCOCl → RCOOR′ is a good method for esterification that avoids the equilibrium problem of acid-catalyzed ester formation (Section 17-8). Bases that may be used are alkali metal hydroxides, pyridine, or amines.

General Ester Synthesis from Carboxylic Acids Through Alkanoyl Chlorides

$$\mathrm{R\overset{O}{\overset{\|}{C}}OH \xrightarrow[-\,HCl]{SOCl_2} R\overset{O}{\overset{\|}{C}}Cl \xrightarrow[-\,HCl]{R'OH,\ base} R\overset{O}{\overset{\|}{C}}OR'}$$

Examples:

$$\mathrm{CH_3\overset{O}{\overset{\|}{C}}Cl + HOCH_2CH_2CH_3 \xrightarrow{N(CH_2CH_3)_3} CH_3\overset{O}{\overset{\|}{C}}OCH_2CH_2CH_3 + H\overset{+}{N}(CH_2CH_3)_3Cl^-}$$

Ethanoyl chloride (Acetyl chloride) — **1-Propanol** — **Propyl ethanoate** 75% — **Triethylammonium chloride**

Cyclobutanecarbonyl chloride + HO–cyclopropyl →(pyridine) cyclobutyl–C(=O)O–cyclopropyl (55%) + pyridinium $\mathrm{\overset{+}{N}{-}H\ Cl^-}$

Pyridinium hydrochloride

EXERCISE 18-4

You have learned that 2-methyl-2-propanol (*tert*-butyl alcohol) dehydrates in the presence of acid. Suggest a synthesis of 1,1-dimethylethyl ethanoate (*tert*-butyl acetate) from ethanoic (acetic) acid.

Amines React with Alkanoyl Chlorides to Give Amides

Secondary and primary amines, as well as ammonia, convert alkanoyl chlorides into amides. Again, the hydrogen chloride formed is neutralized by added base (which can be excess amine).

$$(CH_3)_2CHCCl(=O) + \underset{\text{Excess}}{NH_3} \xrightarrow{H_2O} \underset{70\%}{(CH_3)_2CHCNH_2(=O)} + \overset{+}{N}H_4Cl^-$$

2-Methylpropanoyl chloride → **2-Methylpropanamide**

$$CH_2{=}CHCCl(=O) + 2\,CH_3NH_2 \xrightarrow{\text{Benzene, 5°C}} \underset{68\%}{CH_2{=}CHCNHCH_3(=O)} + CH_3\overset{+}{N}H_3Cl^-$$

Propenoyl chloride → ***N*-Methylpropenamide**

The mechanism of this transformation is a straightforward modification of that formulated for the conversion of alkanoyl chlorides by water or alcohols.

Mechanism of Amide Formation from Alkanoyl Chlorides:

$$RC(=\ddot{O}:)\ddot{Cl}: + H_2\ddot{N}R' \longrightarrow R{-}C(-\ddot{O}:^-)(-\ddot{Cl}:)(-\overset{+}{N}H_2R') \xrightarrow{-:\ddot{Cl}:^-} RC(=\ddot{O}:)\overset{+}{N}H_2R' \longrightarrow RC(=\ddot{O}:)\ddot{N}HR' + H^+$$

Note that, in the last step, a proton is lost from nitrogen. Consequently, tertiary amines cannot form amides, although they do react to form **alkanoyl ammonium salts.** The reactivity of these salts with nucleophiles is like that of alkanoyl halides, because the tertiary amine is an excellent leaving group.

Alkanoyl Ammonium Salt Formation and Reactivity

$$RCCl(=O) + (CH_3)_3N: \longrightarrow RC(=O)\overset{+}{N}(CH_3)_3Cl^- \xrightarrow{:NuH} RCNu(=O) + H\overset{+}{N}(CH_3)_3Cl^-$$

An alkanoyl ammonium chloride

EXERCISE 18-5

Some amide preparations require the reaction of an alkanoyl halide and an *expensive* primary or secondary amine, so that the use of the latter as a base to neutralize the hydrogen halide is prohibitive. Suggest a potential solution to this problem.

*Professor Karl W. Rosenmund, 1884–1964, University of Kiel, Germany.

18-3 The Chemistry of Carboxylic Anhydrides: Slightly Less Reactive Analogs of Alkanoyl Halides

This section briefly examines the system of naming carboxylic anhydrides, a method for their synthesis through ketenes, and the scope of their reactions. Although less reactive, carboxylic anhydrides behave in a manner very similar to that of alkanoyl halides.

The Names of Anhydrides

Carboxylic anhydrides can be derived from carboxylic acids by dehydration (Section 17-7). Accordingly, they are named by simply adding the term **anhydride** to the acid name (or names in the cases of mixed anhydrides). This method also applies to cyclic derivatives.

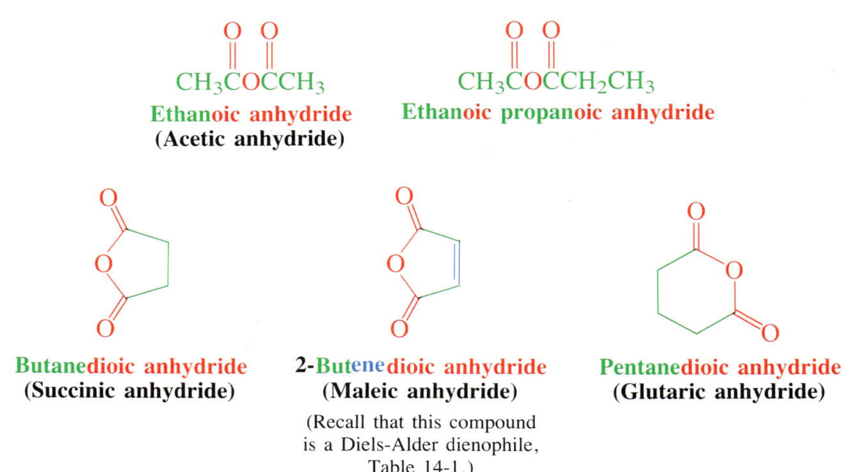

How to Prepare Anhydrides from Ketenes

Carboxylic anhydrides can be prepared not only from the reactions of alkanoyl halides with carboxylic acids or carboxylates (Section 17-7), but also from ketenes. Ketene itself, $CH_2{=}C{=}O$, may be regarded as the oxygen analog of allene, $CH_2{=}C{=}CH_2$ (see Section 13-4, Figure 13-7), or as the methylene analog of carbon dioxide, $O{=}C{=}O$. Treatment of ketene with ethanoic (acetic) acid furnishes ethanoic (acetic) anhydride.

$$\underset{\textbf{Ketene}}{CH_2{=}C{=}O} + \underset{\textbf{Ethanoic (acetic) acid}}{CH_3\overset{\overset{\displaystyle O}{\|}}{C}OH} \longrightarrow \underset{\textbf{Ethanoic (acetic) anhydride}}{CH_3\overset{\overset{\displaystyle O}{\|}}{C}O\overset{\overset{\displaystyle O}{\|}}{C}CH_3}$$

This procedure is used commercially to satisfy a large part of the U.S. demand for ethanoic (acetic) anhydride (0.75 million tons per year). The required ketene is generated on a large scale by the high-temperature pyrolysis of propanone (acetone), a reaction that also produces methane.

$$\mathrm{CH_3\overset{\overset{\displaystyle O}{\|}}{C}CH_3} \xrightarrow{700^\circ\mathrm{C}} \mathrm{CH_4} + \underset{\textbf{Ketene}}{\mathrm{CH_2{=}C{=}O}}$$

A general preparation of substituted ketenes is based on the dehydrohalogenation of alkanoyl halides or the dehalogenation of 2-haloalkanoyl halides. The latter starting materials are readily made by carrying out the Hell-Volhard-Zelinsky reaction (Section 17-11) with a stoichiometric amount of PBr_3.

General Preparation of Ketenes

$$\mathrm{RCH_2\overset{\overset{\displaystyle O}{\|}}{C}Cl} \xrightarrow{\mathrm{N(CH_2CH_3)_3}} \mathrm{RCH{=}C{=}O} + \mathrm{H\overset{+}{N}(CH_2CH_3)_3Cl^-}$$

$$\mathrm{RC\underset{\underset{\displaystyle Br}{|}}{H}\overset{\overset{\displaystyle O}{\|}}{C}Br} \xrightarrow{\mathrm{Zn}} \mathrm{RCH{=}C{=}O} + \mathrm{BrZnBr}$$

Somewhat like alkanoyl halides, ketenes are reactive by virtue of their electrophilic carbonyl carbon. They add water to give carboxylic acids, alcohols to give esters, amines to give amides, and, as we have seen, carboxylic acids to give anhydrides.

$$\mathrm{RR'CH\overset{\overset{\displaystyle O}{\|}}{C}OH} \xleftarrow{\mathrm{H_2O}} \mathrm{RR'C{=}C{=}O} \xrightarrow{\mathrm{R''OH}} \mathrm{RR'CH\overset{\overset{\displaystyle O}{\|}}{C}OR''}$$

$$\mathrm{RR'C{=}C{=}O} \xrightarrow{\mathrm{R''NH_2}} \mathrm{RR'CH\overset{\overset{\displaystyle O}{\|}}{C}NHR''}$$

$$\mathrm{RR'C{=}C{=}O} \xrightarrow{\mathrm{R''\overset{\overset{\displaystyle O}{\|}}{C}OH}} \mathrm{RR'CH\overset{\overset{\displaystyle O}{\|}}{C}O\overset{\overset{\displaystyle O}{\|}}{C}R''}$$

Carboxylic Anhydrides Undergo Nucleophilic Addition-Eliminations Analogous to Those of Alkanoyl Halides

The reactions of carboxylic anhydrides with nucleophiles, although less vigorous, are completely analogous to those of the alkanoyl halides. The leaving group is a carboxylate instead of a halide ion.

General Nucleophilic Addition-Elimination of Anhydrides

$$\mathrm{R\overset{\overset{\displaystyle :O:}{\|}}{C}{-}\ddot{\underset{..}{O}}{-}\overset{\overset{\displaystyle :O:}{\|}}{C}R} + \mathrm{:NuH} \longrightarrow \mathrm{R\overset{\overset{\displaystyle :\ddot{O}:^-}{|}}{\underset{\underset{\displaystyle ^+NuH}{|}}{C}}{-}\ddot{\underset{..}{O}}{-}\overset{\overset{\displaystyle :O:}{\|}}{C}R} \longrightarrow \mathrm{R\overset{\overset{\displaystyle :O:}{\|}}{C}{-}Nu\overset{+}{H}} + \mathrm{{}^-:\ddot{\underset{..}{O}}\overset{\overset{\displaystyle :O:}{\|}}{C}R} \longrightarrow \mathrm{R\overset{\overset{\displaystyle :O:}{\|}}{C}Nu} + \mathrm{H\ddot{\underset{..}{O}}\overset{\overset{\displaystyle :O:}{\|}}{C}R}$$

Examples:

$$\underset{\textbf{Ethanoic (acetic) anhydride}}{CH_3\overset{O}{\overset{\|}{C}}O\overset{O}{\overset{\|}{C}}CH_3} \xrightarrow{HOH} \underset{\textbf{Ethanoic (acetic) acid}}{\underset{100\%}{CH_3\overset{O}{\overset{\|}{C}}OH + HO\overset{O}{\overset{\|}{C}}CH_3}}$$

$$\underset{\textbf{Propanoic anhydride}}{CH_3CH_2\overset{O}{\overset{\|}{C}}O\overset{O}{\overset{\|}{C}}CH_2CH_3} \xrightarrow{CH_3OH} \underset{\textbf{Methyl propanoate}}{\underset{83\%}{CH_3CH_2\overset{O}{\overset{\|}{C}}OCH_3}} + \underset{\textbf{Propanoic acid}}{HO\overset{O}{\overset{\|}{C}}CH_2CH_3}$$

$$\underset{\textbf{Cyclohexanecarboxylic anhydride}}{C_6H_{11}\overset{O}{\overset{\|}{C}}-O-\overset{O}{\overset{\|}{C}}C_6H_{11}} \xrightarrow{(CH_3)_2CHNH_2} \underset{\textbf{\textit{N}-(1-Methylethyl)-cyclohexanecarboxamide}}{\underset{73\%}{C_6H_{11}\overset{O}{\overset{\|}{C}}NHCH(CH_3)_2}} + \underset{\textbf{Cyclohexanecarboxylic acid}}{HO\overset{O}{\overset{\|}{C}}C_6H_{11}}$$

Except for hydrolysis, the carboxylic acid side product is usually undesired, and is removed by work-up with basic water. Cyclic anhydrides undergo similar nucleophilic addition-elimination reactions that lead to ring-opening.

Nucleophilic Ring-Opening of Cyclic Anhydrides

$$\textbf{Butanedioic (succinic) anhydride} \xrightarrow{CH_3OH,\ 100^\circ C} \underset{96\%}{HO\overset{O}{\overset{\|}{C}}CH_2CH_2\overset{O}{\overset{\|}{C}}OCH_3}$$

$$\textbf{3,3-Dimethylpentanedioic anhydride} \xrightarrow{2\ NH_3} \underset{85\%}{{}^{+}NH_4{}^{-}O\overset{O}{\overset{\|}{C}}CH_2\underset{CH_3}{\overset{CH_3}{C}}CH_2\overset{O}{\overset{\|}{C}}NH_2}$$

EXERCISE 18-7

Treatment of butanedioic (succinic) anhydride with ammonia at elevated temperatures leads to a compound $C_4H_5NO_2$. What is its structure?

EXERCISE 18-8

Formulate the mechanisms for the reaction of ethanoic (acetic) anhydride with methanol in the presence of sulfuric acid or sodium methoxide.

In summary, anhydrides can be made from alkanoyl halides or ketenes by treatment with a carboxylic acid. They react with nucleophiles in the same way as alkanoyl halides do, except that the leaving group is a carboxylic acid or a carboxylate. Cyclic anhydrides furnish dicarboxylic acid derivatives.

18-4 Esters: Moderately Reactive but with an Extensive Chemistry

As mentioned in Section 17-8, esters constitute the most important class of carboxylic acid derivatives. After a brief introduction to the naming of esters, this section describes some of their properties and uses, their occurrence in nature, and their chemistry.

The Names of Esters and Some of Their Properties

Esters are named as alkyl alkanoates. The ester grouping, $-\overset{\overset{\displaystyle O}{\|}}{C}OR$, as a substituent is called **alkoxycarbonyl.** A cyclic ester is called a lactone (the common name, Section 17-8; a systematic name would be oxa-2-cycloalkanone, Section 26-1). Depending on ring size, its name may be preceded by the prefix α, β, γ, δ, and so on.

$CH_3\overset{\overset{\displaystyle O}{\|}}{C}OCH_3$
Methyl ethanoate
(Methyl acetate)

$CH_3CH_2\overset{\overset{\displaystyle O}{\|}}{C}OCH_2CH_3$
Ethyl propanoate
(Ethyl propionate)

$CH_3\overset{\overset{\displaystyle O}{\|}}{C}OCH_2CH_2\overset{\overset{\displaystyle CH_3}{|}}{C}HCH_3$
3-Methylbutyl ethanoate
(Isopentyl acetate, component of banana flavor)

$CH_3CH_2CH_2CH_2\overset{\overset{\displaystyle O}{\|}}{C}OCH_2CH_2\overset{\overset{\displaystyle CH_3}{|}}{C}HCH_3$
3-Methylbutyl pentanoate
(Isopentyl valerate, component of apple flavor)

$CH_3CH_2\overset{\overset{\displaystyle O}{\|}}{C}OCH_2\overset{\overset{\displaystyle CH_3}{|}}{C}HCH_3$
2-Methylpropyl propanoate
(Isobutyl propionate, component of rum flavor)

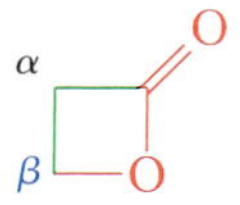

α, β, O, O

β-Propiolactone
(This compound is a carcinogen and would be systematically called oxa-2-cyclobutanone; see Section 26-1.)

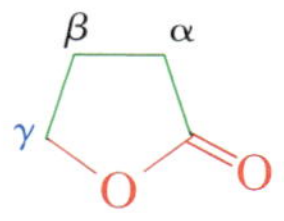

β, α, γ, O, O

γ-Butyrolactone
(Better: oxa-2-cyclopentanone)

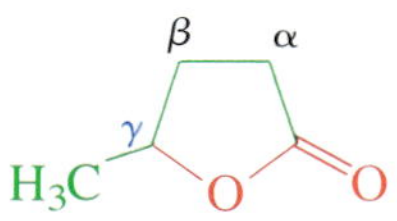

β, α, γ, H_3C, O, O

γ-Valerolactone
(Better: 5-methyloxa-2-cyclopentanone)

EXERCISE 18-9

Name the following esters:

(a) $CH_3CH_2\overset{\overset{\displaystyle O}{\|}}{C}OCH_2CH_2CH_3$; (b) $CH_3O\overset{\overset{\displaystyle O}{\|}}{C}CH_2CH_2\overset{\overset{\displaystyle O}{\|}}{C}OCH_3$; (c) $CH_2{=}CHCO_2CH_3$.

Many esters have characteristically pleasant odors. They are important contributors to natural and artificial fruit flavors. Lower esters, such as ethyl ethanoate (ethyl acetate, b.p. 77°C) and butyl ethanoate (butyl acetate, b.p. 127°C), are used as solvents. Higher nonvolatile esters are applied as softeners (called plasticizers, see Section 12-7) for brittle polymers—such as in flexible tubing (e.g., Tygon tubing), rubber pipes, and upholstery.

Some Waxes, Oils, and Fats Are Esters

Esters made up of long-chain carboxylic acids and long-chain alcohols are the main constituents of animal- and plant-derived **waxes.**

$$CH_3(CH_2)_{14}\overset{O}{\overset{\|}{C}}O(CH_2)_{15}CH_3 \qquad CH_3(CH)_n\overset{O}{\overset{\|}{C}}O(CH)_mCH_3$$

$n = 24, 26;\ m = 29, 31$

Hexadecyl hexadecanoate (Cetyl palmitate) (Wax from the sperm whale)

Beeswax

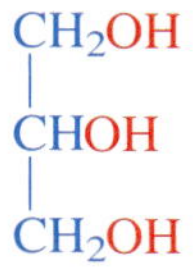

$$\begin{array}{c} CH_2OH \\ | \\ CHOH \\ | \\ CH_2OH \end{array}$$

1,2,3-Propanetriol (Glycerol)

$$\begin{array}{l} CH_2O\overset{O}{\overset{\|}{C}}R \\ | \\ HCO\overset{O}{\overset{\|}{C}}R' \\ | \\ CH_2O\overset{O}{\overset{\|}{C}}R'' \end{array}$$

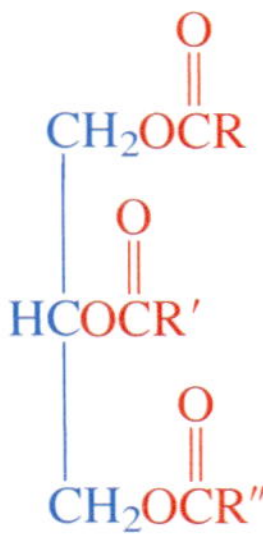

1,2,3-Propanetriol triester (Triglyceride)

Triesters of 1,2,3-propanetriol (glycerol) constitute the **oils** and **fats** found in plants and animals (see Section 17-12). They are also called triglycerides. The carboxylic acid parts of these esters are fatty acids.

There is no essential chemical difference between fats and oils; fats just happen to be solids at room temperature. However, oils are usually more unsaturated. They may be converted into solid fats by catalytic hydrogenation. A variety of cooking fats and margarines are produced in this way. Saturated fats have been implicated as a dietary factor in atherosclerosis (hardening of the arteries). Thus, for considerations of health, vegetable oils, which are highly unsaturated, are becoming increasingly popular. Biologically, fats are used as a source of energy, their metabolism leading ultimately to CO_2 and water.

Waxes and fats are constituents of **lipids,** which are defined as water-insoluble biomolecules highly soluble in organic solvents such as chloroform. They serve as molecular "fuel" and energy stores, as well as components of *membranes*. An important class of membrane lipids are the **phospholipids,** which are di- and triesters derived from carboxylic acids and phosphoric acid. In the **phosphoglycerides,** glycerol is esterified by two adjacent fatty acids and a phosphate unit, which bears another ester substituent, such as that derived from choline, $[HOCH_2CH_2N(CH_3)_3]^+HO^-$.

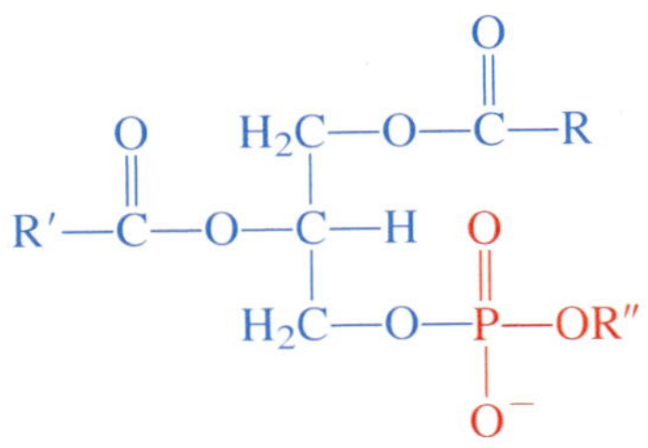

$$\begin{array}{l} \qquad\qquad\qquad H_2C-O-\overset{O}{\overset{\|}{C}}-R \\ \qquad\qquad\qquad\ \ | \\ R'-\overset{O}{\overset{\|}{C}}-O-C-H \\ \qquad\qquad\qquad\ \ | \\ \qquad\qquad\qquad H_2C-O-\overset{O}{\overset{\|}{P}}-OR'' \\ \qquad\qquad\qquad\qquad\qquad\ \ | \\ \qquad\qquad\qquad\qquad\qquad O^- \end{array}$$

A phosphoglyceride

Hexadecanoic (palmitic) acid unit

$$\begin{array}{l} \qquad\qquad\qquad\qquad\qquad\qquad\qquad CH_2O\overset{O}{\overset{\|}{C}}(CH_2)_{14}CH_3 \\ \qquad\qquad\qquad\qquad\qquad\qquad\qquad\ | \\ CH_3(CH_2)_7CH\overset{cis}{=}CH(CH_2)_7\overset{O}{\overset{\|}{C}}OCH \\ \qquad\qquad\qquad\qquad\qquad\qquad\qquad\ | \\ \qquad\qquad\qquad\qquad\qquad\qquad\qquad CH_2O\overset{O}{\overset{\|}{P}}O(CH_2)_2\overset{+}{N}(CH_3)_3 \\ \qquad\qquad\qquad\qquad\qquad\qquad\qquad\qquad\quad | \\ \qquad\qquad\qquad\qquad\qquad\qquad\qquad\qquad\quad O^- \end{array}$$

cis-9-Octadecenoic (oleic) acid unit

Choline unit

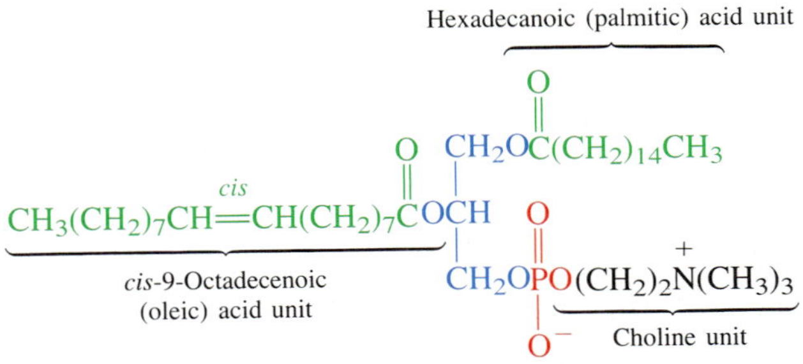

Palmitoyloleoylphosphatidyl choline

Because these molecules carry two long hydrophobic fatty acid chains and a polar head group (the phosphate and choline substituent), they are capable of forming micelles in aqueous solution (see Section 17-12, Figure 17-16). In the micelles, the phosphate unit is solubilized by water, and the ester chains are clustered inside the hydrophobic micellar sphere (Figure 18-1A). Phosphoglyc-

FIGURE 18-1

A. A micelle formed from phospholipid molecules.

B. A lipid bilayer formed from phospholipid molecules.

C. The polar heads and nonpolar tails in phospholipids. (After *Biochemistry*, 2d ed., by Lubert Stryer. W. H. Freeman and Company. Copyright © 1975, 1981.)

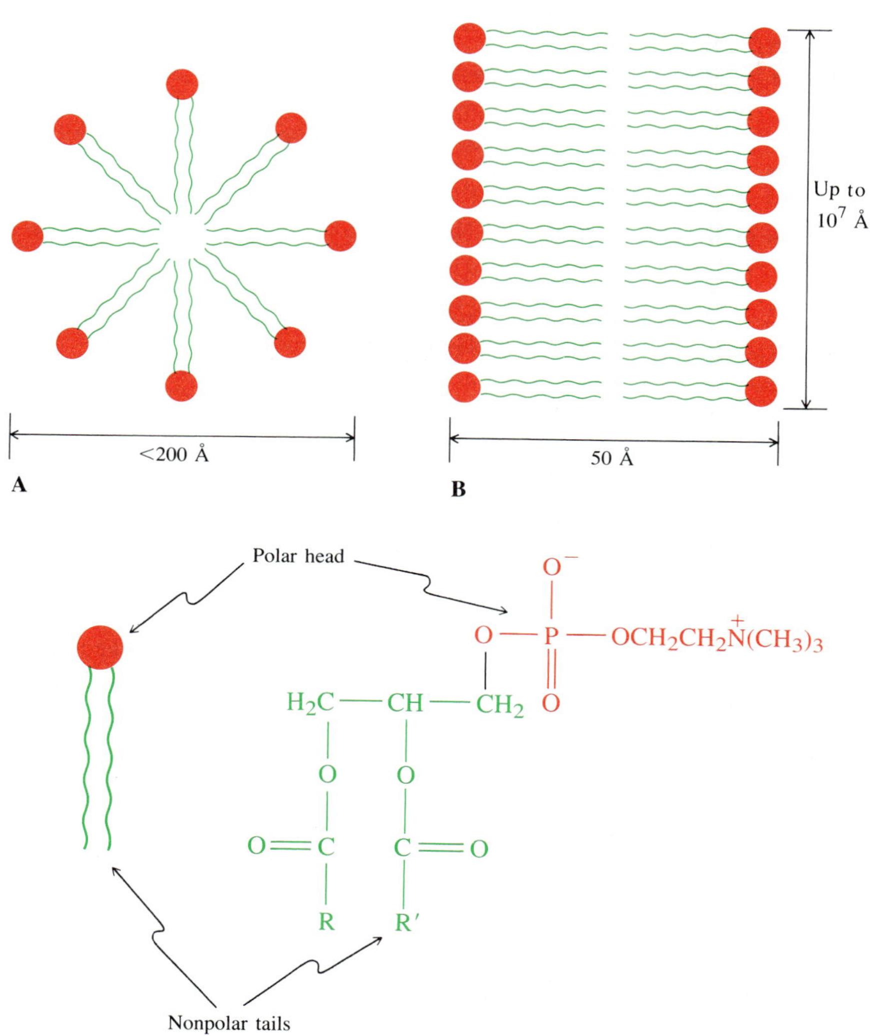

erides can also aggregate in a different way: they may form a unimolecular sheet called a **lipid bilayer** (Figure 18-1B). This capability is significant because, whereas micelles usually are limited in size (<200 Å in diameter), bilayers may be as much as 1 mm (10^7 Å) in length. This makes them ideal constituents of cell membranes, which act as permeability barriers regulating molecular transport into and out of the cell. Lipid bilayers are relatively stable molecular assemblies. The forces that drive their formation are similar to those at work in micelles: London interactions between the hydrophobic alkane chains, and Coulombic and solvation forces among the polar head groups between each other and water.

Esters Hydrolyze to Carboxylic Acids

In contrast with alkanoyl halides and carboxylic anhydrides, esters do not react with water and alcohols unless catalysts are present. Esters hydrolyze when they are heated with excess water in the presence of mineral acid. The mechanism of this reaction is the reverse of acid-catalyzed esterification (Section 17-8).

EXERCISE 18-10

Formulate a mechanism for the acid-catalyzed hydrolysis of γ-butyrolactone.

Ester hydrolysis is also catalyzed by base and proceeds by simple addition-elimination.

Mechanism of Base-Mediated Ester Hydrolysis:

$$RCOCH_3 \text{ (with C=O)} + {}^{-}\!:\!\ddot{O}H \rightleftharpoons R{-}\underset{\underset{:OH}{|}}{\overset{\overset{:\ddot{O}:^{-}}{|}}{C}}{-}\ddot{O}CH_3 \rightleftharpoons RC\ddot{O}{-}H \text{ (with C=O)} + {}^{-}\!:\!\ddot{O}CH_3 \longrightarrow$$

$$RC\ddot{O}:^{-} \text{ (with C=O)} + H\ddot{O}CH_3$$

Carboxylate ion

In contrast with acid-catalyzed hydrolysis, the reaction is not an equilibrium process. The last step, in which the acid is converted into the carboxylate salt, is essentially irreversible. Consequently, at least stoichiometric hydroxide is required (but, frequently, an excess is used).

$$\underset{\text{Methyl 3-methylbutanoate}}{CH_3\overset{CH_3}{\overset{|}{C}}HCH_2\overset{O}{\overset{\|}{C}}OCH_3} \xrightarrow[2.\ H^+,\ H_2O]{1.\ KOH,\ H_2O,\ CH_3OH,\ \Delta} \underset{\text{3-Methylbutanoic acid}}{\underset{100\%}{CH_3\overset{CH_3}{\overset{|}{C}}HCH_2\overset{O}{\overset{\|}{C}}OH}} + CH_3OH$$

EXERCISE 18-11

Formulate a mechanism for the base-mediated hydrolysis of γ-butyrolactone.

The Reaction of Esters with Alcohols: Transesterification

Esters react with alcohols in an acid- or base-catalyzed transformation called **transesterification.** It allows for the direct conversion of one ester into another without proceeding through the free acid. Transesterification is an equilibrium reaction. To shift the equilibrium, a large excess of the alcohol is usually employed, sometimes in the form of solvent.

General Transesterification

$$R\overset{O}{\overset{\|}{C}}OR' + R''OH \underset{H^+ \text{ or } {}^{-}OR''}{\rightleftharpoons} R\overset{O}{\overset{\|}{C}}OR'' + R'OH$$

$$\underset{\text{Ethyl octadecanoate}}{C_{17}H_{35}\overset{O}{\overset{\|}{C}}OCH_2CH_3} + \underset{\text{Solvent}}{CH_3OH} \xrightarrow{H^+ \text{ or } {}^{-}OCH_3} \underset{\text{Methyl octadecanoate}}{\underset{90\%}{C_{17}H_{35}\overset{O}{\overset{\|}{C}}OCH_3}} + CH_3CH_2OH$$

Lactones are opened to hydroxy esters by transesterification:

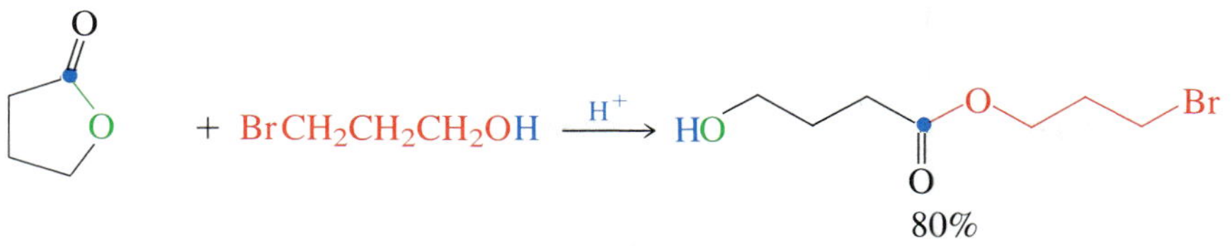

$$\gamma\text{-Butyrolactone} + BrCH_2CH_2CH_2OH \xrightarrow{H^+} HOCH_2CH_2CH_2COOCH_2CH_2CH_2Br \quad 80\%$$

γ-Butyrolactone **3-Bromopropanol** **3-Bromopropyl 4-hydroxybutanoate**

The mechanisms of transesterification by acid and base are straightforward permutations of the mechanisms of the corresponding hydrolyses to the carboxylic acids.

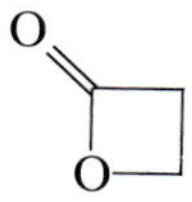

EXERCISE 18-12

β-Propiolactone A (margin) reacts with basic methanol to give the expected methyl 3-hydroxypropanoate B. However, treatment with acidic methanol gives 3-methoxypropanoic acid C. Can you formulate mechanisms that explain this divergence?

A

$$HOCH_2CH_2\overset{O}{\overset{\|}{C}}OCH_3$$
B

$$CH_3OCH_2CH_2\overset{O}{\overset{\|}{C}}OH$$
C

Amides from Esters

Esters react with amines, which are more nucleophilic than alcohols, to form amides; no catalyst is needed.

General Amide Formation from Methyl Esters

$$R\overset{O}{\overset{\|}{C}}OCH_3 + R'\overset{H}{N}H \longrightarrow R\overset{O}{\overset{\|}{C}}NHR' + CH_3OH$$

Example:

$$CH_3(CH_2)_7CH{=}CH(CH_2)_7\overset{O}{\overset{\|}{C}}OCH_3 + CH_3(CH_2)_{11}\overset{H}{N}H \xrightarrow{230°C}$$

Methyl 9-octadecenoate **1-Dodecanamine**

$$CH_3(CH_2)_7CH{=}CH(CH_2)_7\overset{O}{\overset{\|}{C}}NH(CH_2)_{11}CH_3 + CH_3OH$$

69%

***N*-Dodecyl-9-octadecenamide**

The mechanism of this reaction, too, includes addition-elimination.

EXERCISE 18-13

Formulate a mechanism for the formation of ethanamide (acetamide), $CH_3\overset{O}{\overset{\|}{C}}NH_2$, from methyl ethanoate (acetate) and ammonia.

Grignard Reagents Transform Esters into Alcohols

Esters can be converted into alcohols by using two equivalents of a Grignard reagent (Section 8-6). In this way, ordinary esters are transformed into tertiary alcohols, whereas methanoic (formic) esters furnish secondary alcohols.

Alcohols from Esters and Grignard Reagents

$$CH_3CH_2\overset{O}{\overset{\|}{C}}OCH_2CH_3 + 2\ CH_3CH_2CH_2MgBr \xrightarrow[-\ CH_3CH_2OH]{1.\ (CH_3CH_2)_2O \quad 2.\ H^+,\ H_2O} CH_3CH_2\underset{CH_2CH_2CH_3}{\overset{OH}{C}}CH_2CH_2CH_3 \quad 69\%$$

Ethyl propanoate **Propylmagnesium bromide** **4-Ethyl-4-heptanol**

$$H\overset{O}{\overset{\|}{C}}OCH_3 + 2\ CH_3CH_2CH_2CH_2MgBr \xrightarrow[-\ CH_3OH]{1.\ (CH_3CH_2)_2O \quad 2.\ H^+,\ H_2O} H\underset{CH_2CH_2CH_2CH_3}{\overset{OH}{C}}CH_2CH_2CH_2CH_3 \quad 85\%$$

Methyl methanoate (Methyl formate) **Butylmagnesium bromide** **5-Nonanol**

The reaction probably begins with addition of the organometallic to the carbonyl function in the usual manner to give the magnesium salt of a hemiacetal (Section 15-5). At room temperature, rapid elimination results in the formation of an intermediate ketone [or aldehyde, with methanoates (formates)]. The resulting carbonyl group then immediately adds a second equivalent of Grignard reagent. Subsequent acidic work-up leads to the observed alcohol.

Mechanism of the Alcohol Synthesis from Esters and Grignard Reagents:

$$R\overset{O}{\overset{\|}{C}}OCH_3 + R'MgBr \longrightarrow R-\underset{R'}{\overset{O-MgBr}{C}}-OCH_3 \xrightarrow[-\ CH_3OMgBr]{}$$

$$R\overset{O}{\overset{\|}{C}}R' \xrightarrow{R'MgBr} R-\underset{R'}{\overset{OMgBr}{C}}-R' \xrightarrow{H^+,\ H_2O} R-\underset{R'}{\overset{OH}{C}}-R'$$

Dialkyl carbonates, $RO\overset{O}{\overset{\|}{C}}OR$, the diesters of carbonic acid, $HO\overset{O}{\overset{\|}{C}}OH$, are special substrates in reactions with Grignard species, giving tertiary alcohols in which three alkyl groups have been introduced by the organometallic species. Here, both alkoxy substituents act as leaving groups.

$$CH_3O\overset{O}{\overset{\|}{C}}OCH_3 + 3\ CH_3CH_2MgBr \xrightarrow[-\ 2\ CH_3OH]{1.\ THF,\ 25^\circ C \quad 2.\ H^+,\ H_2O} CH_3CH_2-\underset{CH_2CH_3}{\overset{OH}{C}}-CH_2CH_3 \quad 85\%$$

Dimethyl carbonate **Ethylmagnesium bromide** **3-Ethyl-3-pentanol**

Dialkyl carbonates are readily made from phosgene by reaction with alcohols.

$$\underset{\text{Phosgene}}{\overset{\text{O}}{\overset{\|}{\text{ClCCl}}}} + 2\,\text{CH}_3\text{OH} \longrightarrow \underset{\text{Dimethyl carbonate}}{\text{CH}_3\text{O}\overset{\text{O}}{\overset{\|}{\text{C}}}\text{OCH}_3} + 2\,\text{HCl}$$

Esters Are Reduced by Hydride Reagents to Give Alcohols or Aldehydes

The reduction of esters to alcohols by lithium aluminum hydride was mentioned in Section 8-4. The process requires 0.5 equivalent of the hydride, because only two hydrogens are needed per ester function:

$$\text{(pyrrolidinyl)NCH(CH}_3\text{)}\overset{\text{O}}{\overset{\|}{\text{C}}}\text{OCH}_2\text{CH}_3 \xrightarrow[-\ \text{CH}_3\text{CH}_2\text{OH}]{\text{1. LiAlH}_4\ \text{(0.5 equivalent), (CH}_3\text{CH}_2\text{)}_2\text{O}\quad \text{2. H}^+,\ \text{H}_2\text{O}} \underset{90\%}{\text{(pyrrolidinyl)NCH(CH}_3\text{)CH}_2\text{OH}}$$

A milder reducing agent allows the reaction to be stopped at the aldehyde oxidation stage. Such a reagent is bis(2-methylpropyl)aluminum hydride (diisobutylaluminum hydride), when used at low temperatures in toluene.

$$\underset{\text{Ethyl 2-Methylpropanoate}}{\text{CH}_3\overset{\text{H}_3\text{C}}{\overset{|}{\text{C}}}\text{H}\overset{\text{O}}{\overset{\|}{\text{C}}}\text{OCH}_2\text{CH}_3} + \underset{\substack{\text{Bis(2-methylpropyl)aluminum hydride}\\ \text{(Diisobutylaluminum hydride)}}}{(\text{CH}_3\overset{\text{CH}_3}{\overset{|}{\text{C}}}\text{HCH}_2)_2\text{AlH}} \xrightarrow[-\ \text{CH}_3\text{CH}_2\text{OH}]{\text{1. Toluene, }-60^\circ\text{C}\quad \text{2. H}^+,\ \text{H}_2\text{O}} \underset{\text{2-Methylpropanal}}{\text{CH}_3\overset{\text{CH}_3}{\overset{|}{\text{C}}}\text{HCHO}}$$

With bis(2-methylpropyl)aluminum hydride instead of lithium aluminum hydride, the reaction proceeds only to the initial addition step. Acidic aqueous work-up of the resulting alkoxyaluminum compound furnishes the hemiacetal of the aldehyde, which rapidly decomposes to product. The method is successful even with complex molecules.

Mechanism of Bis(2-methylpropyl)aluminum (Diisobutylaluminum) Hydride Reduction:

$$\text{R}\overset{\text{O}}{\overset{\|}{\text{C}}}\text{OCH}_3 + \text{R}'_2\text{AlH} \longrightarrow \text{R}-\underset{\text{H}}{\underset{|}{\overset{\text{OAlR}'_2}{\overset{|}{\text{C}}}}}-\text{OCH}_3 \xrightarrow[-\ \text{HOAlR}'_2]{\text{H}^+,\ \text{H}_2\text{O}} \underset{\text{Hemiacetal}}{\text{R}-\underset{\text{H}}{\underset{|}{\overset{\text{OH}}{\overset{|}{\text{C}}}}}-\text{OCH}_3} \xrightarrow[-\ \text{CH}_3\text{OH}]{} \text{R}\overset{\text{O}}{\overset{\|}{\text{C}}}\text{H}$$

Example:

$$\text{(CH}_3\text{O)}_2\text{-benzoquinolizidine-CH}_2\text{CO}_2\text{CH}_2\text{CH}_3\ (\text{with } -\text{CH}_2\text{CH}_3) \xrightarrow{(\text{CH}_3\overset{\text{CH}_3}{\overset{|}{\text{C}}}\text{HCH}_2)_2\text{AlH, toluene, }-60^\circ\text{C}} \underset{83\%}{\text{(CH}_3\text{O)}_2\text{-benzoquinolizidine-CH}_2\text{CHO}\ (\text{with } -\text{CH}_2\text{CH}_3)}$$

Esters Form Enolates

The acidity of the α-hydrogens in esters is sufficiently high that **ester enolates** are formed by treatment of esters with strong base at low temperatures. Ester enolates react like ketone enolates, undergoing alkylations, oxacyclopropane openings, and aldol reactions.

$$CH_3CO_2CH_2CH_3 \xrightarrow{\text{LDA, THF, } -78^\circ C} CH_2{=}C(O^-Li^+)(OCH_2CH_3)$$
$pK_a \sim 25$

Ethyl ethanoate (acetate) enolate ion

$$\xrightarrow[\text{HMPA}]{CH_2{=}CHCH_2Br,\ -\ LiBr} CH_2{=}CHCH_2CH_2COCH_2CH_3 \ (\text{C=O})$$
97%
Ethyl 4-pentenoate

$$\xrightarrow{H_2C\text{—}CH_2 \ (\text{O bridge}),\ \text{HMPA}} Li^{+\,-}OCH_2CH_2CH_2COCH_2CH_3 \ (\text{C=O})$$

$$\xrightarrow[\text{See Exercise 18-14}]{-\ Li^{+\,-}OCH_2CH_3} \gamma\text{-butyrolactone}$$
55%
γ-Butyrolactone

$$\xrightarrow{1.\ (CH_3)_3CCHO \quad 2.\ H^+,\ H_2O} (CH_3)_3C\text{—}CH(OH)\text{—}CH_2COCH_2CH_3 \ (\text{C=O})$$
97%
Ethyl 3-hydroxy-4,4-dimethyl 1-1 pentanoate

The pK_a of esters is five units higher than that of aldehydes and ketones. Consequently, ester enolates are more basic than ketone enolates. Thus, they exhibit the typical side reactions of strong bases: E2 processes (particularly, with secondary, tertiary, and branched halides) and deprotonations.

EXERCISE 18-14

Explain how the reaction of the ethyl ethanoate (acetate) enolate ion with oxacyclopropane leads to γ-butyrolactone.

Ester enolates attack not only the carbonyl group of aldehydes and ketones, but also that of esters. In this transformation, known as the **Claisen* condensation,** the enolate ion undergoes an addition-elimination reaction with the ester function, furnishing a 3-keto ester. Here, the enolate species does not have to be present in stoichiometric amounts but may be produced in equilibrium concentrations as in the aldol reaction (Section 15-7). Both the alkoxide and the ester used should be derived from the same alcohol to prevent complications arising from transesterification.

*Professor Ludwig Claisen, 1851–1930, University of Berlin.

$$CH_3\overset{O}{\overset{\|}{C}}OCH_2CH_3 + CH_3\overset{O}{\overset{\|}{C}}OCH_2CH_3 \xrightarrow[-\ CH_3CH_2OH]{Na^{+}\ {}^{-}OCH_2CH_3,\ CH_3CH_2OH} CH_3\overset{O}{\overset{\|}{C}}CH_2\overset{O}{\overset{\|}{C}}OCH_2CH_3$$

75%

Ethyl 3-oxobutanoate

Mechanism of the Claisen Condensation:

STEP 1: Ester enolate formation

$$CH_3\overset{O}{\overset{\|}{C}}OCH_2CH_3 \underset{}{\overset{Na^{+}\ {}^{-}OCH_2CH_3}{\rightleftharpoons}}$$

$$Na^{+}\left[{}^{-}\!:\!CH_2\!-\!C(=\ddot{O}\!:)\!-\!\ddot{O}CH_2CH_3 \longleftrightarrow CH_2\!=\!C(-\ddot{O}\!:^{-})\!-\!\ddot{O}CH_2CH_3\right] + CH_3CH_2OH$$

STEP 2: Nucleophilic addition

$$CH_3\overset{:\ddot{O}:}{\overset{\|}{C}}OCH_2CH_3 + {}^{-}\!:\!CH_2\overset{O}{\overset{\|}{C}}OCH_2CH_3 \rightleftharpoons CH_3\!-\!\underset{CH_2\overset{\|}{\underset{O}{C}}OCH_2CH_3}{\overset{:\ddot{O}:^{-}}{C}}\!-\!OCH_2CH_3$$

STEP 3: Elimination

$$CH_3\underset{CH_2\underset{O}{\underset{\|}{C}}OCH_2CH_3}{\overset{:\ddot{O}:^{-}}{C}}\!-\!OCH_2CH_3 \rightleftharpoons CH_3\overset{O}{\overset{\|}{C}}CH_2\overset{O}{\overset{\|}{C}}OCH_2CH_3 + {}^{-}\!:\!\ddot{O}CH_2CH_3$$

3-Keto ester

STEP 4: Deprotonation

$$CH_3\overset{O}{\overset{\|}{C}}CH_2\overset{O}{\overset{\|}{C}}OCH_2CH_3 + {}^{-}\!:\!\ddot{O}CH_2CH_3 \longrightarrow$$

Acidic, pK_a ~ 11

$$\left[CH_3\overset{O}{\overset{\|}{C}}\!-\!\overset{-}{\ddot{C}}H\!-\!\overset{O}{\overset{\|}{C}}OCH_2CH_3 \longleftrightarrow CH_3\overset{:\ddot{O}:^{-}}{\overset{|}{C}}\!=\!CH\!-\!\overset{O}{\overset{\|}{C}}OCH_2CH_3 \longleftrightarrow CH_3\overset{O}{\overset{\|}{C}}\!-\!CH\!=\!\overset{:\ddot{O}:^{-}}{\overset{|}{C}}OCH_2CH_3\right] + CH_3CH_2OH$$

STEP 5: Protonation on work-up

$$CH_3\overset{O}{\overset{\|}{C}}\overset{-}{\ddot{C}}H\overset{O}{\overset{\|}{C}}OCH_2CH_3 \xrightarrow{H^{+},\ H_2O} CH_3\overset{O}{\overset{\|}{C}}CH_2\overset{O}{\overset{\|}{C}}OCH_2CH_3$$

The Claisen condensation is essentially the ester analog of the aldol condensation. It is an equilibrium reaction that is endothermic at the 3-keto ester stage.

The equilibrium is shifted to the product by irreversible conversion of the 3-keto ester into the corresponding enolate ion. This last step is highly favored, because the acidity of the protons flanked by the two carbonyl groups is much enhanced ($pK_a \sim 11$) by resonance stabilization in the anion. As a result of this effect, 3-keto esters—and, more generally, β-dicarbonyl compounds—are useful synthetic intermediates, to be discussed in detail in Section 22-4. In the Claisen condensation, it is the final acidic work-up that results in the 3-keto ester product.

Pyrolysis of Esters Furnishes Alkenes

General Ester Pyrolysis

$$\mathrm{RCOCH_2CH_2R'}\ (\text{C=O}) \xrightarrow{300^\circ\mathrm{C}} \mathrm{RCOH}\ (\text{C=O}) + \mathrm{CH_2{=}CHR'}$$

Esters are thermally quite stable. However, when heated to more than 300°C, they decompose into a carboxylic acid and an alkene. The reaction is concerted and includes a six-membered transition state. As the carbonyl oxygen begins to abstract a β-hydrogen, the alkoxy oxygen–carbon bond in the ester starts to break. In an electron-pushing picture of this transformation, three electron pairs can be seen shifting, somewhat as in the Diels-Alder and *retro*-Diels-Alder reactions (Section 14-5).

Mechanism of Ester Pyrolysis:

Six-electron, six-membered cyclic transition state

The hydrogen and the ester group have to depart from the same side of the developing π bond, in a process called ***syn* elimination.** It contrasts with the *anti* transition state of most E2 reactions (Section 7-5), which also involves six electrons but not a cyclic arrangement.

EXERCISE 18-15

Give the products of the reaction of methyl cyclohexanecarboxylate with the following compounds or under the following conditions (and followed by aqueous work-up, if necessary): (a) H^+, H_2O; (b) HO^-, H_2O; (c) $CH_3CH_2O^-$, CH_3CH_2OH; (d) NH_3, Δ; (e) 2 CH_3MgBr; (f) $LiAlH_4$; (g) 1. LDA, 2. CH_3I; (h) 300°C.

In summary, esters are named as alkyl alkanoates. Many of them have a pleasant odor and occur in nature as fragrant agents, waxes, oils, and fats. With acidic or basic water they hydrolyze to the corresponding carboxylic acids or carboxylates, with alcohols they undergo transesterification, and with amines at elevated temperatures they furnish amides. Grignard reagents add twice to give tertiary alcohols [or secondary alcohols, from methanoates (formates)]. Lithium aluminum hydride reduces esters all the way to the alcohols, whereas bis(2-methylpropyl)aluminum (diisobutylaluminum) hydride permits stopping at the aldehyde stage. With LDA it is possible to form ester enolates, which can be alkylated by electrophiles. If the latter is another ester, a 3-keto ester is formed in a reaction called the Claisen condensation. Finally, when heated to more than

300°C, esters pyrolyze to alkenes and carboxylic acids through a concerted elimination process.

18-5
Amides: The Least-Reactive Carboxylic Acid Derivatives

Carboxylic amides are carboxylic acid derivatives in which the carbonyl group is the least susceptible to nucleophilic attack. After a brief introduction to the naming of amides and an excursion into the activity of the β-lactam system of the penicillins, this section describes the reactions of amides.

The Names of Amides

Amides are called **alkanamides,** the ending **-e** of the alkane stem having been replaced by **-amide.** In common names, the ending **-ic** of the acid name is replaced by the **-amide** suffix. Substituents on the nitrogen are indicated by the prefix *N*- or *N,N*-, depending on the number of substituents. Accordingly, there are primary, secondary, and tertiary amides.

$HC(=O)NH_2$ — **Methanamide (Formamide)** — A primary amide

$CH_3C(=O)NHCH_3$ — ***N*-Methylethanamide (*N*-Methylacetamide)** — A secondary amide

$CH_3CH(Br)CH_2CH_2C(=O)N(CH_2CH_3)CH_3$ — **4-Bromo-*N*-ethyl-*N*-methylpentanamide** — A tertiary amide

There are several amide derivatives of carbonic acid, H_2CO_3 (Section 18-4): ureas, carbamic acids, and carbamic esters (urethanes).

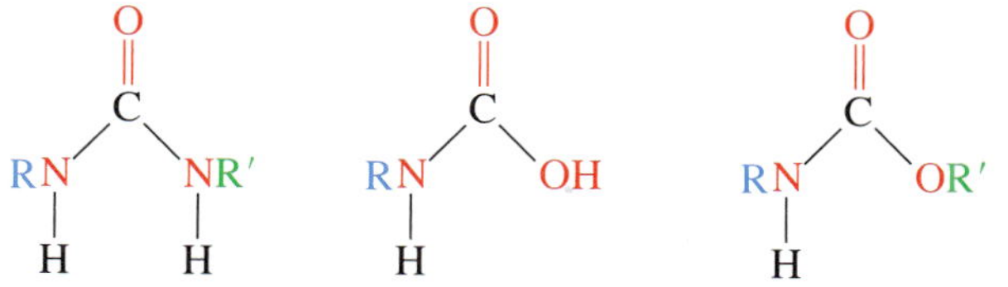

$RNH-C(=O)-NHR'$ — **A urea**

$RNH-C(=O)-OH$ — **A carbamic acid**

$RNH-C(=O)-OR'$ — **A carbamic ester (Urethane)**

Cyclic amides are called **lactams** (Section 17-9; a systematic name would be aza-2-cycloalkanones, Section 26-1), and the rules for naming them follow those used for lactones. The penicillins are annelated β-lactams.

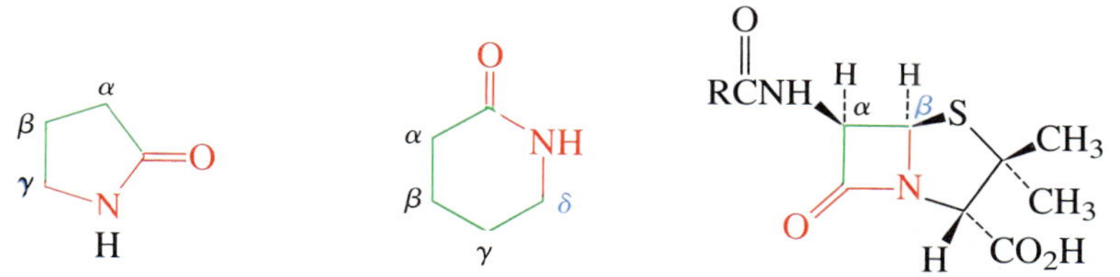

γ-Butyrolactam (Systematic name: aza-2-cyclopentanone)

δ-Valerolactam (Systematic name: aza-2-cyclohexanone)

Penicillin (A β-lactam derivative)

BOX 18-1

Penicillin, an Antibiotic Containing a β-Lactam Ring

The discovery of penicillin as a powerful broad-spectrum antibiotic was one of the milestones in medicinal chemistry. As with many such advances, serendipity played a major role. In 1928, the British bacteriologist Alexander Fleming noted that several *Staphylococcus* cultures set aside on a laboratory bench had been contaminated by microorganisms from the laboratory atmosphere. A green mold, *Penicillinum notatum,* had grown in some places, and the *Staphylococcus* in its vicinity was disintegrating. The substance causing this antibiotic activity was called penicillin, but it was not available in pure form until about 10 years later. Many different penicillins have subsequently been synthesized with different R groups. *Penicillin G* has a phenylmethyl (benzyl, $C_6H_5CH_2$) group attached to the amide function; *ampicillin* has a phenylaminomethyl ($C_6H_5CHNH_2$) substituent. Structurally and biologically related are the cephalosporins, important antibiotics that are frequently active when the penicillins are not.

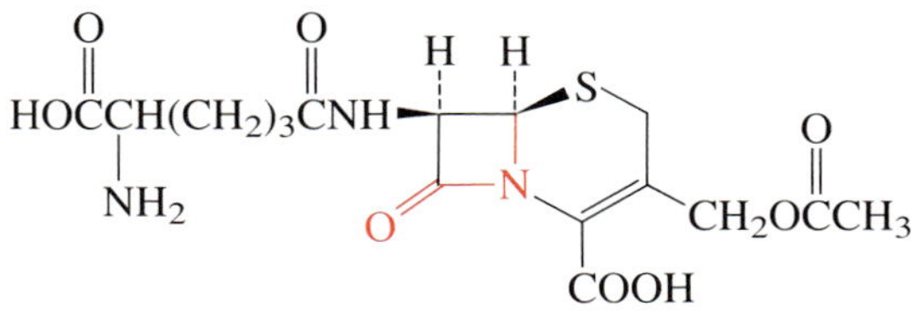

HOCCH(CH$_2$)$_3$CNH (with NH$_2$, C=O groups), H H, S, N, O, CH$_2$OCCH$_3$ (C=O), COOH

Cephalosporin C

The strained β-lactam ring is responsible for the antibiotic activity of these drugs. Because ring strain is relieved on opening, β-lactams are unusually reactive compared with ordinary amides. The enzyme *transpeptidase,* which catalyzes a crucial cross-linking reaction in the biosynthesis of bacterial cell walls, accepts penicillin as a substrate. The penicillin then alkanoylates a nucleophilic oxygen of the enzyme, rendering it inactive. Cell-wall construction stops, and the organism dies. The reaction is the reverse of amide formation from esters (Section 18-4).

Penicillin in Action

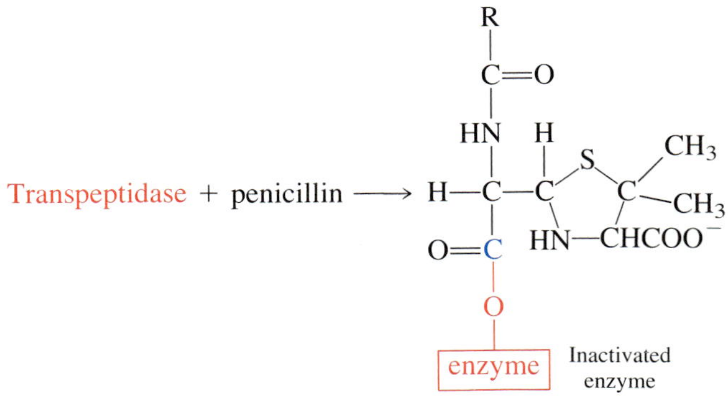

Transpeptidase + penicillin ⟶ R–C(=O)–HN–CH(–C(=O)–O–enzyme)–CH(S–C(CH$_3$)$_2$)–HN–CHCOO$^-$

Inactivated enzyme

Some bacteria are resistant to penicillin because they produce an enzyme, *penicillinase,* that hydrolyzes the β-lactam ring before it can attach itself to transpeptidase. The rate of this hydrolysis depends on the structure of the β-lactam. Cephalosporins are not affected by penicillinase. Nevertheless, the continual emergence of new, antibiotic-resistant bacterial strains, aided by the frequently indiscriminate prescription of penicillin and other antibiotics, has spurred intensive ongoing efforts to discover novel, more-active, and more-selective systems.

Nucleophilic Additions to Amides: Water, Alcohols, and Hydrides

The amides are the least reactive of the carboxylic acid derivatives, mainly because of the extra resonance capability of the nitrogen lone electron pair. As a consequence, their nucleophilic addition-eliminations require relatively harsh conditions. For example, hydrolysis occurs only on prolonged heating in strongly acidic or basic water. Acid hydrolysis liberates the amine in the form of the corresponding ammonium salt.

Acid Hydrolysis of an Amide

$$CH_3CH_2CH(CH_3)CH_2C(=O)NH_2 \xrightarrow{H_2SO_4,\ H_2O,\ \Delta,\ 3\ h} CH_3CH_2CH(CH_3)CH_2C(=O)OH + (NH_4)_2SO_4$$

3-Methylpentanamide → **3-Methylpentanoic acid** (88%)

Mechanism:

$$RC(=O)NH_2 \underset{-H^+}{\overset{+H^+}{\rightleftharpoons}} RC(=\overset{+}{O}H)NH_2 \underset{-H_2O}{\overset{+H_2O}{\rightleftharpoons}} R-C(OH)(\overset{+}{O}H_2)-NH_2 \underset{+H^+}{\overset{-H^+}{\rightleftharpoons}} R-C(OH)(OH)-NH_2 \underset{-H^+}{\overset{+H^+}{\rightleftharpoons}}$$

$$R-C(OH)(OH)-\overset{+}{N}H_3 \rightleftharpoons RC(=\overset{+}{O}H)OH + {:}NH_3 \longrightarrow RC(=O)OH + NH_4^+$$

Base hydrolysis initially furnishes the carboxylate salt and the amine. Acidic work-up then produces the acid.

Base Hydrolysis of an Amide

$$CH_3CH_2CH_2C(=O)NHCH_3 \xrightarrow{HO^-,\ H_2O,\ \Delta} CH_3CH_2CH_2C(=O)O^- + CH_3NH_2 \xrightarrow{H^+,\ H_2O} CH_3CH_2CH_2C(=O)OH + CH_3\overset{+}{N}H_3$$

***N*-Methylbutanamide** → **Butanoic acid** (87%)

Mechanism of Hydrolysis of Amides by Aqueous Base:

$$RC(=O)NH_2 \xrightarrow{^-:\ddot{O}H} R{-}C(O^-)(OH){-}NH_2 \rightleftharpoons RC(=O)OH + :NH_2^- \longrightarrow RC(=O)O^- + HNH_2$$

Similar reactions may be carried out with alcohols to give esters.

$$C_6H_{11}CNHCH_3\ (C{=}O) \xrightarrow{CH_3CH_2OH,\ HCl} C_6H_{11}COCH_2CH_3\ (C{=}O) + CH_3NH_3{}^+Cl^-$$

85%

N-Methylcyclohexanecarboxamide → **Ethyl cyclohexanecarboxylate**

On treatment with lithium aluminum hydride, amides are converted into the corresponding amines in high yield.

$$(CH_3)_2CHCH_2CH_2\overset{O}{\overset{\|}{C}}N(CH_2CH_3)_2 \xrightarrow[2.\ H^+,\ H_2O]{1.\ LiAlH_4,\ (CH_3CH_2)_2O} (CH_3)_2CHCH_2CH_2CH_2N(CH_2CH_3)_2$$

85%

N,N-Diethyl-4-methylpentanamide → **N,N-Diethyl-4-methylpentanamine**

$$\text{lactam} \xrightarrow[2.\ H^+,\ H_2O]{1.\ LiAlH_4,\ (CH_3CH_2)_2O} \text{pyrrolidine}$$

γ,γ-Dimethyl-γ-lactam (Systematic name: 5,5-dimethylaza-2-cyclopentanone)

80%
2,2-Dimethylazacyclopentane (2,2-Dimethylpyrrolidine)

In contrast with the reactions of carboxylic acids and esters with lithium aluminum hydride, this treatment does not produce alcohols and is special for amides. The mechanism of the reduction is thought to include hydride addition, followed by aluminate elimination. The resulting iminium ion intermediate (see Section 15-6) is then reduced by further hydride attack.

Possible Mechanism of Amide Reduction by Hydride:

$$R\overset{:O:}{\overset{\|}{C}}\ddot{N}R'_2 \xrightarrow{LiAlH_4} R{-}C(H)(\ddot{O}{-}Al){-}\ddot{N}R'_2 \xrightarrow[-\ {-}Al\ddot{O}:^-]{} RCH{=}\overset{+}{N}R'_2 \xrightarrow{LiAlH_4} RCH_2\ddot{N}R'_2$$

Modified hydride reducing agents allow the reduction of amides to be stopped at the aldehyde stage, probably by intermediate hemiaminal formation (Section 15-6). Such reductions generally work best with *N,N*-dialkylalkanamides. Frequently used hydrides are bis(2-methylpropyl)aluminum (diisobutylaluminum) hydride and lithium trialkoxyaluminum hydrides.

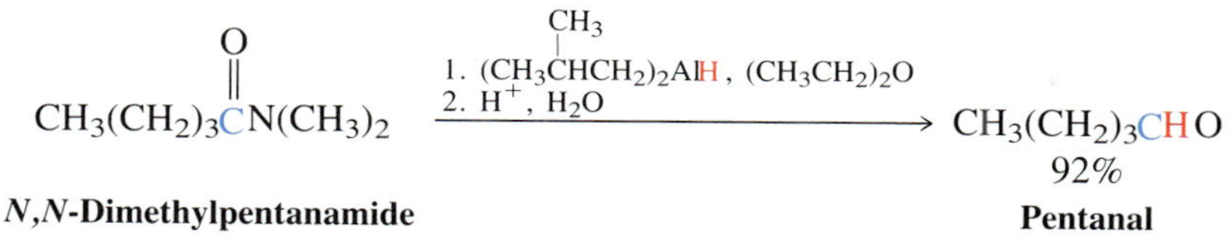

Mechanism of Reduction of Amides to Aldehydes:

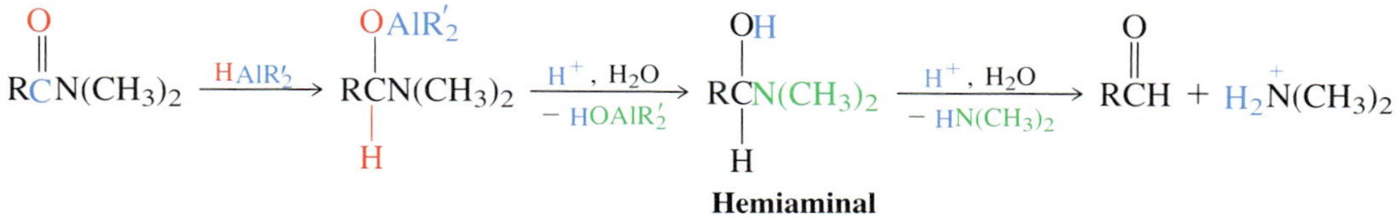

EXERCISE 18-16

Treatment of amide A with $LiAlH_4$, followed by acidic aqueous work-up, gave B. Explain. (Hint: Review Sections 15-5 and 15-6).

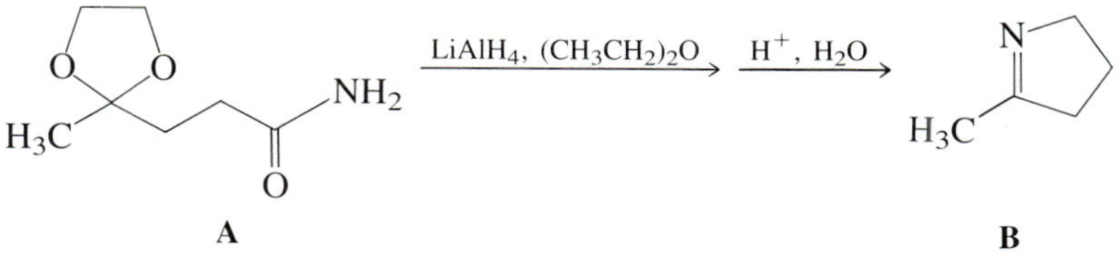

Amide Enolates and Amidates

When bearing hydrogens, both positions next to the carbonyl group in amides are acidic; the NH hydrogen has a pK_a of about 15, whereas the CH hydrogen is less acidic, with a pK_a of about 30 (Section 18-1).

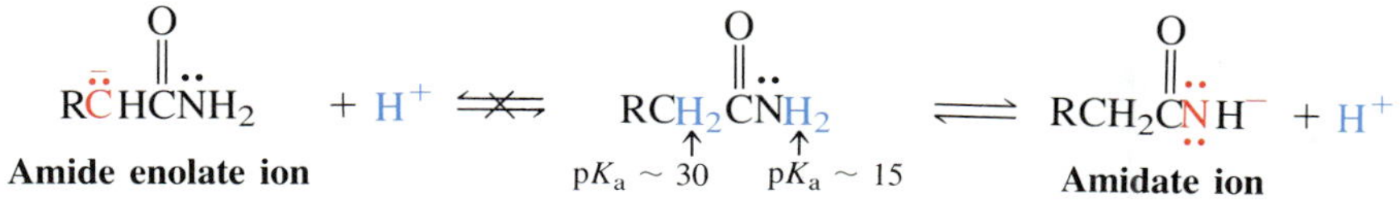

Practically speaking, therefore, a proton may be removed from carbon only with tertiary amides, in which the nitrogen is blocked. Both resulting anions act as nucleophiles and allow the synthesis of substituted amides.

Amidate and Amide Enolate Ions in Synthesis

$$\text{(steroid)}\text{-C(=O)NH}_2 \xrightarrow[-\,\text{H—H}]{\text{NaH, }C_6H_6} \text{(steroid)}\text{-C(=O)}\ddot{N}H^- Na^+ \xrightarrow[-\,\text{NaI}]{CH_3I\text{, DMF}} \text{(steroid)}\text{-C(=O)NHCH}_3$$

Amidate ion

>66%

$$CH_3CH_2\overset{O}{\overset{\|}{C}}N(CH_3)_2 \xrightarrow[-\,NH_2H]{NaNH_2,\ NH_3} CH_3\overset{..}{\underset{-}{C}}H\overset{Na^+O}{\overset{\|}{C}}N(CH_3)_2 \xrightarrow[-\,NaBr]{CH_3CH_2Br} CH_3\underset{CH_3CH_2}{\underset{|}{C}}H\overset{O}{\overset{\|}{C}}N(CH_3)_2$$

62%

***N,N*-Dimethylpropanamide** — **Amide enolate ion** — ***N,N*-Dimethyl-2-methylbutanamide**

Halogenation of Primary Amides Results in Amines

In the presence of base, primary amides undergo a special halogenation reaction, the **Hofmann* rearrangement,** in which the carbonyl group is expelled from the molecule to give a primary amine with one carbon less in the chain.

General Hofmann Rearrangement

$$R\overset{O}{\overset{\|}{C}}NH_2 \xrightarrow{X_2,\ NaOH,\ H_2O} RNH_2 + O{=}C{=}O$$

Example:

$$\underset{\textbf{Nonanamide}}{CH_3(CH_2)_6CH_2CONH_2} \xrightarrow{Cl_2,\ NaOH} \underset{\textbf{Octanamine}}{CH_3(CH_2)_6CH_2NH_2} + O{=}C{=}O$$

66%

Mechanism:

STEP 1: Amidate formation

$$R\overset{:O:}{\overset{\|}{C}}\ddot{N}H_2 + {}^-:\ddot{O}H \rightleftharpoons R\overset{:O:}{\overset{\|}{C}}\ddot{N}H^- + H\ddot{O}H$$

STEP 2: Halogenation

$$R\overset{:O:}{\overset{\|}{C}}\ddot{N}H^- + :\ddot{X}{-}\ddot{X}: \longrightarrow R\overset{:O:}{\overset{\|}{C}}\underset{:\ddot{X}:}{\underset{|}{\ddot{N}}}H + :\ddot{X}:^-$$

*This is the Hofmann of the Hofmann rule of E2 reactions (Section 11-5).

STEP 3: *N*-Halo amidate formation

$$\mathrm{RC(=O)NHX} + {}^{-}\!:\!\ddot{\mathrm{O}}\mathrm{H} \rightleftharpoons \mathrm{RC(=O)\ddot{N}^{-}\!-\!\ddot{X}\!:} + \mathrm{H\ddot{O}H}$$

An *N*-halo amidate

STEP 4: Halide elimination

$$\mathrm{RC(=O)\ddot{N}^{-}\!-\!\ddot{X}\!:} \longrightarrow \mathrm{RC(=O)\ddot{N}} + :\!\ddot{X}\!:^{-}$$

An acyl nitrene

STEP 5: Rearrangement

$$\mathrm{RC(=O)\ddot{N}\!:} \longrightarrow \ddot{\mathrm{O}}\!=\!\mathrm{C}\!=\!\ddot{\mathrm{N}}\!-\!\mathrm{R}$$

An isocyanate

Recall:

$$-\overset{|}{\underset{\mathrm{R}}{\mathrm{C}}}-\overset{+}{\mathrm{C}}\!\lessgtr \;\longrightarrow\; \gtrless\!\overset{+}{\mathrm{C}}-\overset{|}{\underset{\mathrm{R}}{\mathrm{C}}}-$$

STEP 6: Hydrolysis to the carbamic acid and decomposition

$$\mathrm{R}\!-\!\ddot{\mathrm{N}}\!=\!\mathrm{C}\!=\!\ddot{\mathrm{O}} \xrightarrow{H_2O} \mathrm{R(H)N\!-\!C(=O)\!-\!\ddot{O}H} \longrightarrow \mathrm{R\ddot{N}H_2} + \mathrm{CO_2}$$

A carbamic acid

The reaction begins with amidate ion formation, followed by α-halogenation, much like the base-catalyzed halogenation of aldehydes and ketones (Section 16-2). Subsequently, the second proton on nitrogen is abstracted by additional base to give an *N*-halo amidate, which spontaneously eliminates halide ion. The net result of these two steps is an unusual α-elimination of HX, both proton and leaving group coming from the same atom. The species formed contains an electron-deficient nitrogen atom surrounded by only an electron sextet. Such compounds are called nitrenes, in analogy to carbenes, $:CR_2$ (Section 21-5). In the Hofmann rearrangement, an acyl nitrene is a reactive intermediate. The acyl nitrene stabilizes itself by a 1,2-shift of the alkyl group to give an **isocyanate.** This transformation is related to the 1,2-shift of alkyl substituents in carbocations: in both cases, the migrating group moves with its electron pair toward the electron-deficient center. Isocyanates may be regarded as nitrogen analogs of ketenes, $R_2C{=}C{=}O$ (Section 18-3), or carbon dioxide, $O{=}C{=}O$. As in ketenes, the *sp*-hybridized carbonyl carbon is highly electrophilic. In the aqueous medium usually employed in the Hofmann reaction, hydration of the isocyanate produces a **carbamic acid.** Carbamic acids are unstable, and decompose to carbon dioxide and the product amine.

The Hofmann rearrangement is usually carried out by adding the amide to cold aqueous hypohalites (produced by the reaction $X_2 + HO^- \longrightarrow HOX + X^-$) and then warming the mixture.

$$\text{Cyclohexanecarboxamide} + \underset{\text{Sodium hypobromite}}{NaOBr} \xrightarrow[-\ NaBr]{75°C} \underset{\substack{65\% \\ \text{Cyclohexanamine}}}{C_6H_{11}NH_2} + O{=}C{=}O$$

In some cases, the procedure is modified by use of an alcohol solvent, such as methanol. Then the product of alcohol addition to the intermediate isocyanate is a carbamic ester (urethane), which is stable under the reaction conditions and may be isolated. Subsequent hydrolysis of the ester leads to decarboxylation and the desired amine.

Carbamic Ester Formation in the Hofmann Rearrangement

$$CH_3(CH_2)_{29}CH_2\overset{O}{\overset{\|}{C}}NH_2 \xrightarrow[-\ 2\ NaBr,\ -\ 2\ CH_3OH]{Br_2,\ 2\ Na^{+\ -}OCH_3,\ CH_3OH} CH_3(CH_2)_{29}CH_2N{=}C{=}O \xrightarrow{CH_3OH}$$

$$\underset{\substack{\text{A carbamic ester} \\ \text{(A urethane)}}}{CH_3(CH_2)_{29}CH_2NH\overset{O}{\overset{\|}{C}}OCH_3} \xrightarrow[-CH_3O^-]{^-OH,\ H_2O} \underset{100\%\ \text{overall}}{CH_3(CH_2)_{29}CH_2NH_2} + CO_2$$

The reaction of methyl isocyanate with various alcohols and amines is used in the industrial preparation of several powerful herbicides and insecticides.

$$\underset{\text{Methyl isocyanate}}{CH_3N{=}C{=}O} + \underset{\substack{\text{1-Naphthalenol} \\ \text{(1-Naphthol)}}}{C_{10}H_7OH} \longrightarrow \underset{\substack{\text{1-Naphthyl } N\text{-methylcarbamate} \\ \text{(Sevin, an insecticide)}}}{C_{10}H_7O\overset{O}{\overset{\|}{C}}NHCH_3}$$

U.S. consumption of methyl isocyanate has been estimated at 30 to 35 million pounds per year. In late 1984 in the city of Bhopal, India, a massive leak of this substance, used in the preparation of the insecticide Sevin, is estimated to have resulted in the deaths of more than 2,000 people; at least 300,000 more were exposed to it. This catastrophe, the worst chemical industrial accident in history, has led to a reappraisal of the safety measures for the handling of large quantities of toxic chemicals.

EXERCISE 18-17

Write a detailed mechanism for the addition of water to an isocyanate and for the decarboxylation of the resulting carbamic acid.

EXERCISE 18-18

Suggest a sequence by which you could convert ester A (margin) into amine B.

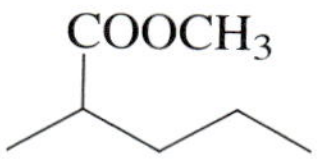

$COOCH_3$ **A** → NH_2 **B**

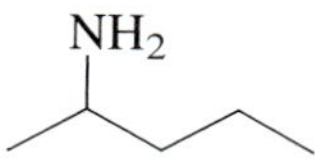

In summary, carboxylic amides are named as alkanamides, or lactams when cyclic. They can be hydrolyzed to carboxylic acids by acid or base catalysis, converted into esters by alcohols in the presence of acid or base, and reduced to amines by lithium aluminum hydride. Modified hydrides stop the reduction of amides at the aldehyde oxidation stage. Treatment of amides with base leads to deprotonation at nitrogen or, for tertiary amides, at carbon to the respective amidate or enolate ions. Finally, in the Hofmann rearrangement, amides react with basic halogen to yield amines containing one less carbon.

18-6 A Special Class of Carboxylic Acid Derivatives: Alkanenitriles

Nitriles, RC≡N, are considered derivatives of carboxylic acids because the nitrile carbon is in the same oxidation state as the carboxy carbon and because nitriles are readily interconverted with other carboxylic acid derivatives. This section describes the rules for naming nitriles, the structure and bonding in the nitrile group, and some of its spectral characteristics. Then it compares the chemistry of the nitrile group with that of other carboxylic acid derivatives.

The Names of Nitriles

A systematic way of naming this class of compounds is as **alkanenitriles.** In common names the ending **-ic acid** of the carboxylic acid is usually replaced with **-nitrile.** The chain is numbered as in the naming of carboxylic acids. Similar rules apply to dinitriles derived from dicarboxylic acids. The substituent —CN is called **cyano.** Cyanocycloalkanes are labeled cycloalkanecarbonitriles.

$CH_3C{\equiv}N$
Ethanenitrile
(Acetonitrile)

$CH_3CH_2C{\equiv}N$
Propanenitrile
(Propionitrile)

$CH_3CH(CH_3)CH_2C{\equiv}N$
3-Methylbutanenitrile

$CH_2C{\equiv}N$ / $CH_2C{\equiv}N$
Butanedinitrile
(Succinonitrile)

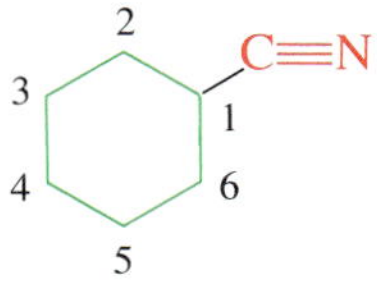

Cyclohexane ring (C1–C6) with C≡N at C1
Cyclohexanecarbonitrile

$N{\equiv}CCH_2COCH_2CH_3$ (C=O)
Ethyl cyanoethanoate
(Ethyl cyanoacetate)

Bonding, Molecular Structure, and Spectral Features of the Alkanenitriles

In the nitriles, both atoms in the functional group are *sp* hybridized, and there is a lone electron pair on nitrogen occupying an *sp* hybrid orbital pointing away from the molecule along the C–N axis. The hybridization and structure of the nitrile functional group very much resemble those of the alkynes (Figure 18-2; see also Figures 13-1 and 13-2).

In the infrared spectrum, the C≡N stretching vibration appears at about 2250 cm^{-1}, in the same range as the C≡C absorption. The ^1H NMR spectra of nitriles indicate that protons near the nitrile group are deshielded about as much as those in other carboxylic acid and alkyne derivatives (Table 18-3).

FIGURE 18-2

A. Molecular-orbital picture of the nitrile group.

B. Molecular structure of ethanenitrile (acetonitrile).

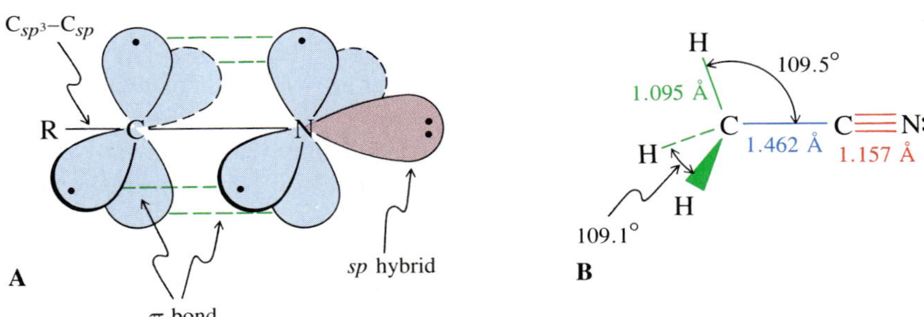

TABLE 18-3

^{1}H NMR chemical shifts of substituted methanes CH_3X

X	δ_{CH_3}
—H	0.23
—Cl	3.06
—OH	3.39
—CH(=O)	2.18
—COOH	2.08
—$CONH_2$	2.02
—C≡N	1.98
—C≡CH	1.80

The ^{13}C NMR absorption for the nitrile carbon appears at lower field ($\delta \sim$ 112–126 ppm) than that of the alkynes ($\delta \sim$ 65–85 ppm), owing to the greater electronegativity of nitrogen compared with carbon.

EXERCISE 18-19

1,3-Dibromopropane was treated with sodium cyanide in dimethyl sulfoxide-d_6 and the mixture monitored by ^{13}C NMR. After a few minutes, four new intermediate peaks appeared, one of which was located well downfield from the others at δ = 117.6 ppm. Subsequently, another three peaks began growing at δ = 119.1, 22.6, and 17.6 ppm, at the expense of the signals of starting material and the intermediate. Explain.

Nitriles Are Basic and Acidic

The electron-withdrawing power of the nitrogen in the nitrile group can be pictured in a dipolar resonance structure:

$$[\mathrm{R{-}C{\equiv}N:} \longleftrightarrow \mathrm{R{-}\overset{+}{C}{=}\ddot{N}:^-}]$$

The lone electron pair on the nitrogen makes the nitrile function slightly basic, like the carbonyl oxygen of carboxylic acids (Section 17-4). However, compared with an imine or amine, the *sp*-hybridized nitrile system is much less readily protonated.

Basicity of the Nitrogen Lone Electron Pair in Various Hybridization States

$$\underset{sp^3}{\mathrm{R_3N:}} > \underset{sp^2}{\mathrm{R_2C{=}\ddot{N}R}} > \underset{sp}{\mathrm{R{-}C{\equiv}N:}}$$

Recall that, in going from sp^3 to sp hybridization, the electron-withdrawing power of an atom is increased (Section 13-2). Therefore, the protonated nitrile, even though stabilized by resonance, is much more acidic than, for example, the ammonium ion, as seen in their pK_a values.

Ammonia Is Much More Basic Than an Alkanenitrile

$$\underset{\substack{\uparrow\\ sp}}{R{-}C{\equiv}N:} + H^+ \underset{K \sim 10^{-10}}{\rightleftharpoons} \left[R{-}C{\equiv}\overset{+}{N}H \longleftrightarrow R{-}\overset{+}{C}{=}\ddot{N}H\right] \quad pK_a \sim -10$$

$$\underset{\substack{\uparrow\\ sp^3}}{H_3N:} + H^+ \underset{K = 10^{9.5}}{\rightleftharpoons} NH_4^+ \quad pK_a = 9.24$$

Nitriles bearing hydrogens next to the cyano group are also acidic, their pK_as of deprotonation being in the range of those of esters. The negative charge of the resulting anions can be delocalized.

$$\underset{pK_a \sim 25}{RCH_2C{\equiv}N:} + :B^- \rightleftharpoons \left[R\overset{..}{\bar{C}}H{-}C{\equiv}N: \longleftrightarrow RCH{=}C{=}\ddot{N}:^-\right] + BH$$

As with other carboxylic acid derivatives, this deprotonation allows for the introduction of substituents by alkylation.

Alkylation of Nitriles

$$\underset{\textbf{Butanenitrile}}{CH_3CH_2CH_2CN} \xrightarrow[-\ LiBr]{\text{1. LDA, } (CH_3CH_2)_2O \quad \text{2. } CH_3CH_2Br} \underset{\textbf{2-Ethylbutanenitrile}}{\underset{75\%}{CH_3CH_2\overset{\overset{CH_3CH_2}{|}}{C}HCN}}$$

Nitriles Undergo Hydrolysis to Carboxylic Acids

As mentioned in Sections 15-7 and 17-5, nitriles can be hydrolyzed by aqueous acid or base to give the corresponding carboxylic acids. The mechanisms of these reactions proceed through the intermediate amide and include addition-elimination steps.

Mechanism of the Acid-catalyzed Hydrolysis of Nitriles:

$$R{-}C{\equiv}N: \underset{-\ H^+}{\overset{+\ H^+}{\rightleftharpoons}} \left[R{-}C{\equiv}\overset{+}{N}{-}H \longleftrightarrow R{-}\overset{+}{C}{=}\ddot{N}{-}H\right] \underset{-\ H_2\ddot{O}}{\overset{+\ H_2\ddot{O}}{\rightleftharpoons}} R{-}C(={\ddot{N}H}){-}\overset{+}{\ddot{O}H_2} \underset{+\ H^+}{\overset{-\ H^+}{\rightleftharpoons}}$$

$$\underset{\textbf{Tautomer of amide}}{R{-}C(={\ddot{N}H}){-}\ddot{O}H} \underset{-\ H^+}{\overset{+\ H^+}{\rightleftharpoons}} \left[R{-}C(={\overset{+}{N}H_2}){-}\ddot{O}H \longleftrightarrow R{-}C(\ddot{N}H_2){=}\overset{+}{\ddot{O}H}\right] \underset{+\ H^+}{\overset{-\ H^+}{\rightleftharpoons}} R{-}C(\ddot{N}H_2){=}\ddot{O}: \xrightarrow{H^+} \xrightarrow{H_2O} RCOOH + NH_4^+$$

In the acid-catalyzed process, initial protonation on nitrogen is followed by nucleophilic attack of water. On proton loss, a neutral intermediate is formed, a tautomer of the amide. Rearrangement of this species gives the amide, which may then be hydrolyzed as described in Section 18-5.

In base-catalyzed nitrile hydrolysis, direct attack of hydroxide gives the anion of the amide tautomer. Subsequent protonation followed by a proton shift results in the amide, which hydrolyzes with more base as described in Section 18-5.

Mechanism of the Base-catalyzed Hydrolysis of Nitriles:

$$R{-}C{\equiv}N: + {}^{-}:\ddot{O}H \rightleftharpoons R\overset{\displaystyle :\ddot{N}^{-}}{\overset{\|}{C}}\ddot{O}H \overset{HOH}{\rightleftharpoons} R\overset{\displaystyle :N{-}H}{\overset{\|}{C}}\ddot{O}{-}H + {}^{-}:\ddot{O}H \underset{+\,H\ddot{O}H}{\overset{-\,H\ddot{O}H}{\rightleftharpoons}}$$

$$\left[R\overset{\displaystyle :N{-}H}{\overset{\|}{C}}{-}\ddot{O}:^{-} \longleftrightarrow R\overset{\displaystyle {}^{-}:\ddot{N}{-}H}{\overset{|}{C}}{=}\ddot{O}:\right] \overset{HOH}{\rightleftharpoons} R\overset{\displaystyle \ddot{N}H_2}{\overset{|}{C}}{=}\ddot{O}: + {}^{-}:\ddot{O}H \longrightarrow \xrightarrow{H^+} R\overset{O}{\overset{\|}{C}}OH + NH_4^+$$

The conditions for the hydrolysis of nitriles are usually stringent, requiring concentrated acid or base at high temperatures.

$$N{\equiv}C(CH_2)_4C{\equiv}N \xrightarrow{H^+,\ H_2O,\ 300^\circ C} HOOC(CH_2)_4COOH \quad 97\%$$

Hexanedinitrile (Adiponitrile) → **Hexanedioic acid (Adipic acid)**

$$\text{1-(3-methylphenyl)cyclohexane-}C{\equiv}N \xrightarrow{KOH,\ H_2O,\ \Delta,\ HOCH_2CH_2OCH_2CH_2OH} \text{1-(3-methylphenyl)cyclohexane-}COOH \quad 93\%$$

Organometallic Reagents Attack Nitriles to Give Ketones

Strong nucleophiles, such as organometallic reagents, add to nitriles to give anionic imine salts. Work-up with acidic water gives the neutral imine, which is rapidly hydrolyzed to the ketone (Section 15-6).

General Ketone Synthesis from Nitriles

$$R{-}C{\equiv}N + R'M \longrightarrow R\overset{\displaystyle :\ddot{N}^{-}M^{+}}{\overset{\|}{C}}R' \underset{-\,MOH}{\xrightarrow{H^+,\ HOH}} R\overset{\displaystyle \ddot{N}H}{\overset{\|}{C}}R' \xrightarrow{H^+,\ H_2O} R\overset{O}{\overset{\|}{C}}R' + NH_4^+$$

Examples:

$$\text{(4-methylcyclohex-3-enyl)}\underset{}{\overset{CH_3}{\overset{|}{C}}}HCH_2CN \xrightarrow[\text{2. } H^+,\ H_2O]{\text{1. } CH_3\overset{CH_3}{\overset{|}{C}}HCH_2Li,\ (CH_3CH_2)_2O,\ \text{heptane}} \text{(4-methylcyclohex-3-enyl)}\overset{CH_3}{\overset{|}{C}}HCH_2\overset{O}{\overset{\|}{C}}CH_2CH(CH_3)_2 \quad 76\%$$

$$CH_3CN \xrightarrow[\text{2. } H^+,\ H_2O]{\text{1. } CH_3(CH_2)_3CH_2MgBr,\ THF} CH_3\overset{\overset{\displaystyle O}{\|}}{C}(CH_2)_4CH_3$$

44%

Ethanenitrile (Acetonitrile) → **2-Heptanone**

Reduction of Nitriles by Hydride Reagents: Synthesis of Aldehydes and Amines

Nucleophilic additions to the nitrile carbon may also be effected by hydride reagents. Like organometallic compounds, modified lithium aluminum hydride, $LiAlH(OCH_2CH_3)_3$, adds to nitriles only once to give the imine anion, probably complexed with aluminum. Aqueous work-up then produces aldehydes.

General Aldehyde Synthesis from Nitriles

$$R{-}C{\equiv}N + LiAlH(OCH_2CH_3)_3 \longrightarrow R{-}C(H){=}N{-}\overset{Li^+}{Al^-}(OCH_2CH_3)_3 \xrightarrow{H^+,\ H_2O} R{-}\overset{\overset{\displaystyle O}{\|}}{C}{-}H$$

Example:

$$CH_3CH_2CH_2C{\equiv}N \xrightarrow[\text{2. } H_2SO_4,\ H_2O]{\text{1. } LiAlH(OCH_2CH_3)_3,\ (CH_3CH_2)_2O} CH_3CH_2CH_2\overset{\overset{\displaystyle O}{\|}}{C}H$$

69%

Butanenitrile → **Butanal**

Another reagent that reduces nitriles to aldehydes is bis(2-methylpropyl)-aluminum (diisobutylaluminum) hydride.

1-(1,3-benzodioxol-5-yl)cyclopropane-CN $\xrightarrow[\text{2. } H^+,\ H_2O]{\text{1. } (CH_3\overset{CH_3}{\overset{|}{C}}HCH_2)_2AlH}$ 1-(1,3-benzodioxol-5-yl)cyclopropane-CHO

85%

Treatment of nitriles with strong hydride reducing agents results in double hydride addition, giving the amine on aqueous work-up. The best reagent for this purpose is lithium aluminum hydride:

$$CH_3CH_2CH_2CN \xrightarrow[\text{2. } H^+,\ H_2O]{\text{1. } LiAlH_4} CH_3CH_2CH_2CH_2NH_2$$

86%

Butanenitrile → **Butanamine**

EXERCISE 18-20

The reduction of a nitrile by $LiAlH_4$ to give an amine adds four hydrogen atoms to the C–N triple bond: two from the reducing agent and two from the water in the aqueous work-up. Formulate a mechanism for this transformation.

Like the triple bond of alkynes (Section 13-6), the nitrile group is hydrogenated by catalytically activated hydrogen. The result is the same as that with reduction by lithium aluminum hydride: amine formation. All four hydrogens are from the hydrogen gas.

$$CH_3CH_2CH_2C{\equiv}N \xrightarrow{H_2,\ PtO_2,\ CH_3CH_2OH,\ CHCl_3} CH_3CH_2CH_2CH_2NH_2 \quad 96\%$$

Butanenitrile → **Butanamine**

$$CH_3(CH_2)_5OCH_2CH_2C{\equiv}N \xrightarrow{H_2,\ Rh\text{–}Al_2O_3,\ NH_3,\ CH_3CH_2OH} CH_3(CH_2)_5OCH_2CH_2CH_2NH_2 \quad 100\%$$

3-Hexoxypropanenitrile → **3-Hexoxypropanamine**

EXERCISE 18-21

Show how you would prepare the following compounds from pentanenitrile:

(a) $CH_3CH_2CH_2CH(CH_3)CN$; (b) $CH_3(CH_2)_3COOH$; (c) $CH_3(CH_2)_3C(=O)CH_3$;

(d) $CH_3(CH_2)_3C(=O)(CH_2)_3CH_3$; (e) $CH_3(CH_2)_3CH(=O)$; (f) $CH_3(CH_2)_3CD_2ND_2$.

In summary, nitriles are named as alkanenitriles. Both atoms making up the C–N unit are *sp* hybridized, the nitrogen bearing a lone electron pair occupying an *sp* hybrid orbital. The nitrile stretching vibration appears at 2250 cm^{-1}, the ^{13}C NMR absorption at about 120 ppm. The electron pair on nitrogen is extremely weakly basic, whereas the hydrogens next to the functional group are about as acidic as the corresponding hydrogens in esters. Acid- or base-catalyzed hydrolysis of nitriles gives carboxylic acids, and organometallic reagents (RLi, RMgBr) add to give ketones after hydrolysis. With modified hydride reagents, addition and hydrolysis furnishes aldehydes, whereas $LiAlH_4$ or catalytically activated hydrogen transforms the nitrile function to the amine.

18-7 Measuring the Molecular Weight of Organic Compounds: Mass Spectroscopy

The preceding section completed the introduction to all major functional groups in organic chemistry (including amines, although a systematic discussion of nitrogen-containing compounds follows in Chapters 21, 26, and 27). This section introduces one last important physical technique used by organic chemists to characterize organic molecules: **mass spectroscopy,** which is employed to measure molecular weights. The section begins with a description of the apparatus used and the physical principles on which it is based. Organic molecules fragment under the conditions necessary for measuring molecular weights, giving rise to characteristic recorded patterns called mass spectra.

The Mass Spectrometer Separates Molecular Ions by Weight

Mass spectroscopy is not spectroscopy in the conventional sense, because no incident radiation is absorbed (Section 10-1). It is made possible by the fact that

FIGURE 18-3
Schematic diagram of a mass spectrometer.

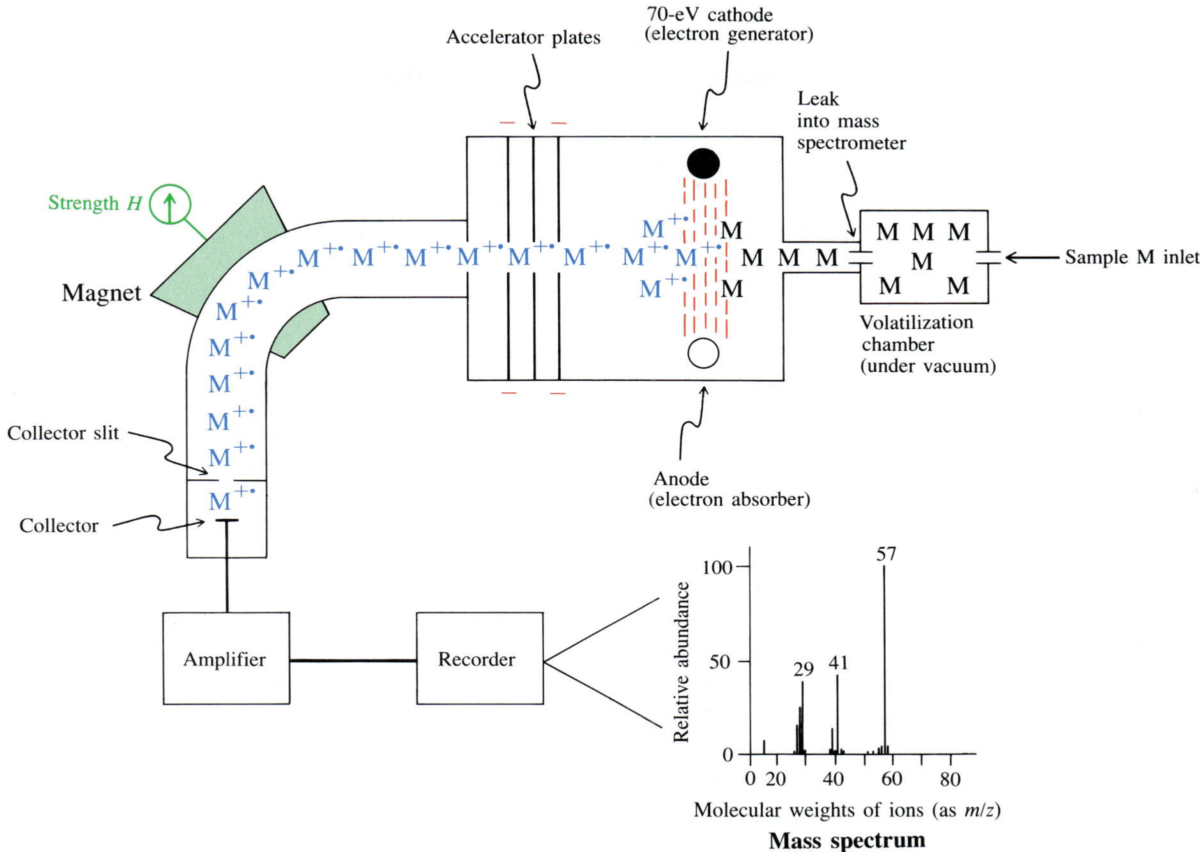

charged particles traveling through a magnetic field are deflected from a linear path, the lighter species being deflected more than the heavier ones. In this way, ions of different mass can be separated (Figure 18-3).

The sample is introduced into an inlet chamber and volatilized (unless it is already a gas). Subsequently, a small quantity is allowed to leak into the high-vacuum chamber of the spectrometer. At this stage, the neutral molecules (M) pass through a beam of electrons, usually accelerated to 70 eV (slightly less than 1600 kcal $mole^{-1}$). On electron impact, the molecules are energized sufficiently to eject an electron. A radical cation ($M^{+\cdot}$) is formed, also called the **molecular ion.**

Ionization of a Molecule on Electron Impact

M	+ *e* (70 eV) ⟶	$M^{+\cdot}$	+ 2*e*
Neutral molecule	**Ionizing beam**	**Radical cation (Molecular ion)**	

Most organic molecules undergo only single ionization. As a charged particle, the molecular ion can be accelerated by an electric field. The accelerated $M^{+\cdot}$ now enters a magnetic field, where it is deflected into a circular path whose radius depends largely on the mass of the ion and the strength of the field. As the strength of the electric field is changed, only ions of certain mass

Molecular Weights of Organic Molecules

CH_4
$m/z = 16$

CH_3OH
$m/z = 32$

$CH_3\overset{\overset{\displaystyle O}{\|}}{C}OCH_3$
$m/z = 74$

are allowed to pass through the collector slit, an event that is translated electronically into a signal that is recorded as a peak on a chart. As field strength sweeps through a range, a mixture of compounds gives rise to several such peaks, each one at a specific position on the chart corresponding to a particular molecular weight. Neutral molecules are not accelerated or deflected and are therefore lost in the instrument chamber, eventually to be pumped out.

The molecular weights of organic molecules are reported as mass-to-charge ratios, m/z. If only singly charged species are considered, the m/z value equals the mass of the ion in question. Mass spectra are plotted as m/z values (on the abscissa) versus peak height (on the ordinate), the latter being a measure of the relative number of ions with this molecular weight.

EXERCISE 18-22

Three unknown compounds containing only C, H, and O gave rise to the following molecular weights. Draw as many reasonable structures as you can: (a) $m/z = 46$; (b) $m/z = 30$; (c) $m/z = 56$.

Molecular Ions Undergo Fragmentations

Mass spectroscopy gives information not only about the molecular ion, but, owing to **fragmentation,** also about its component structural units. Because the energy of the ionizing beam far exceeds that required to break typical organic bonds, some of the ionized molecules break apart into virtually all possible pairs of fragments, giving rise to a number of additional mass-spectral peaks, *all of lower mass* than the molecular ion (also called the *parent ion*) from which they are derived. The spectrum that results is called the **mass-spectral fragmentation pattern.**

For example, the mass spectrum of methane contains, in addition to the parent ion peak, lines for $CH_3^{+\cdot}$, $CH_2^{+\cdot}$, $CH^{+\cdot}$, and $C^{+\cdot}$ (Figure 18-4). These can be formed as radical cations or carbocations, depending on the mode of fragmentation (which is left unspecified here). The relative abundance of these species, as indicated by the height of the peaks, gives a useful indication of the relative ease of their formation. It can be seen that the first C–H bond is cleaved

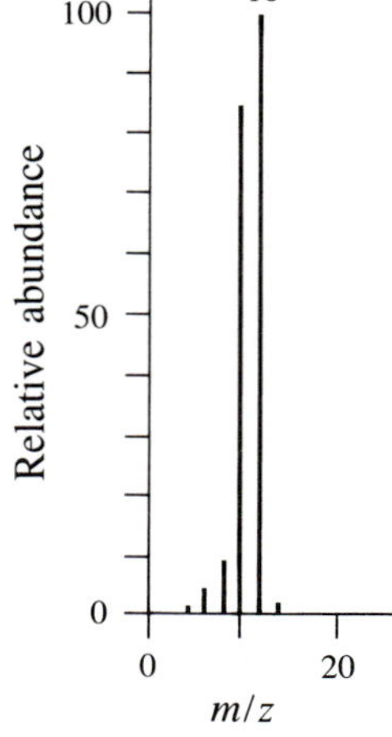

FIGURE 18-4

The mass spectrum of methane. At the left is the spectrum actually recorded, at the right is the tabulated form, the largest peak **(base peak)** being defined as 100%. For methane, the base peak at $m/z = 16$ is due to the parent ion.

Tabulated spectrum

m/z	Relative abundance (%)	Molecular or fragment ion
17	1.1	$(M + 1)^{+\cdot}$
16	100.0 (base peak)	$M^{+\cdot}$ (parent ion)
15	85.0	$(M - 1)^{+\cdot}$
14	9.2	$(M - 2)^{+\cdot}$
13	3.9	$(M - 3)^{+\cdot}$
12	1.0	$(M - 4)^{+\cdot}$

readily, the $m/z = 15$ peak reaching 85% of the abundance of the parent ion. Breaking two, three, or four C–H bonds is more difficult, and the corresponding ions have lower relative abundance.

Mass Spectra Reveal the Presence of the Isotopes of the Elements

An unusual feature in the mass spectrum of methane is a small (1.1%) peak at $m/z = 17$; it is designated $(M + 1)^{+\bullet}$. How is it possible to have a molecular ion present that has an extra mass unit? The answer lies in the fact that carbon is not isotopically pure. About 1.1% of natural carbon is the ^{13}C isotope (see Table 10-1), giving rise to the additional peak.

Hydrogen, too, has a naturally occurring higher isotope: deuterium, with about 0.015% abundance. This proportion is so small that deuterium is normally ignored when mass-spectral patterns are considered. However, other elements common in organic molecules, such as nitrogen (0.366% ^{15}N), oxygen (negligible ^{17}O; 0.204% ^{18}O), and sulfur (0.76% ^{33}S; 4.22% ^{34}S; negligible ^{36}S), have higher isotopes that may give rise to molecular ions of higher mass than the most common one. Spectra get more complicated and require careful statistical analysis when several such atoms are present. There are tables that list the predicted peak distribution for many isotopic combinations, but a detailed discussion is not needed here. There is a simple rule of thumb for many organic compounds (not containing Cl, S, Br, and other isotopically impure atoms): the $(M + 1)^{+\bullet}$ peak has a height (relative to $M^{+\bullet}$) of n times 1.1%, in which n = number of carbon atoms in the molecule. In the ethane spectrum, for example, the height of the $(M + 1)^{+\bullet}$ peak, at $m/z = 31$, is about 2.2% that of the parent ion. The reason for this finding is statistical. The chance of finding a ^{13}C atom in a compound containing two carbons is double that expected of a one-carbon molecule. For a three-carbon moiety, it would be threefold, and so on. A mass spectrum of the eighteen-carbon steroid estrone (see Section 4-7) is shown in Figure 18-5 (on the next page).

Halogen-containing compounds may also give rise to isotopic patterns. Whereas fluorine and iodine are isotopically pure, chlorine (75.53% ^{35}Cl; 24.47% ^{37}Cl) and bromine (50.54% ^{79}Br; 49.46% ^{81}Br) exist as a mixture of two isotopes in comparable proportions, giving rise to characteristic molecular ion patterns. For example, on the basis of the average atomic weights of the elements recorded in the periodic table, the weight of 1-bromopropane is 123. However, in its mass spectrum, there is no peak at this position (Figure 18-6). The answer to this puzzle lies in the true isotopic composition of this molecule: a nearly 1:1 mixture of $CH_3CH_2CH_2{}^{79}Br$ and $CH_3CH_2CH_2{}^{81}Br$. As a consequence, the mass spectrum shows two molecular ions, at $m/z = 122$ and $m/z = 124$. Similarly, the spectra of monochloroalkanes exhibit two molecular ions two mass units apart, for $R^{35}Cl$ and $R^{37}Cl$, but in this case in a 3:1 ratio, because of the different isotopic composition of this element. These peak patterns may be diagnostic for the presence of chlorine or bromine in a compound.

EXERCISE 18-23

What peak pattern do you expect for the molecular ion of dibromomethane?

EXERCISE 18-24

Nonradical compounds containing C, H, and O have even molecular weights, those containing C, H, O and an odd number of N atoms have odd molecular weights, but those with an even number of N atoms are even again. Explain.

FIGURE 18-5

Mass spectrum (below and on the facing page) of the female sex hormone estrone. The molecule contains eighteen carbon atoms and thus the $(M + 1)^{+\cdot}$ peak height is predicted to be approximately 18% of the intensity of the $M^{+\cdot}$ peak, a value close to that observed. Note the extensive and complex fragmentation pattern. The base peak is found at $m/z = 44$.

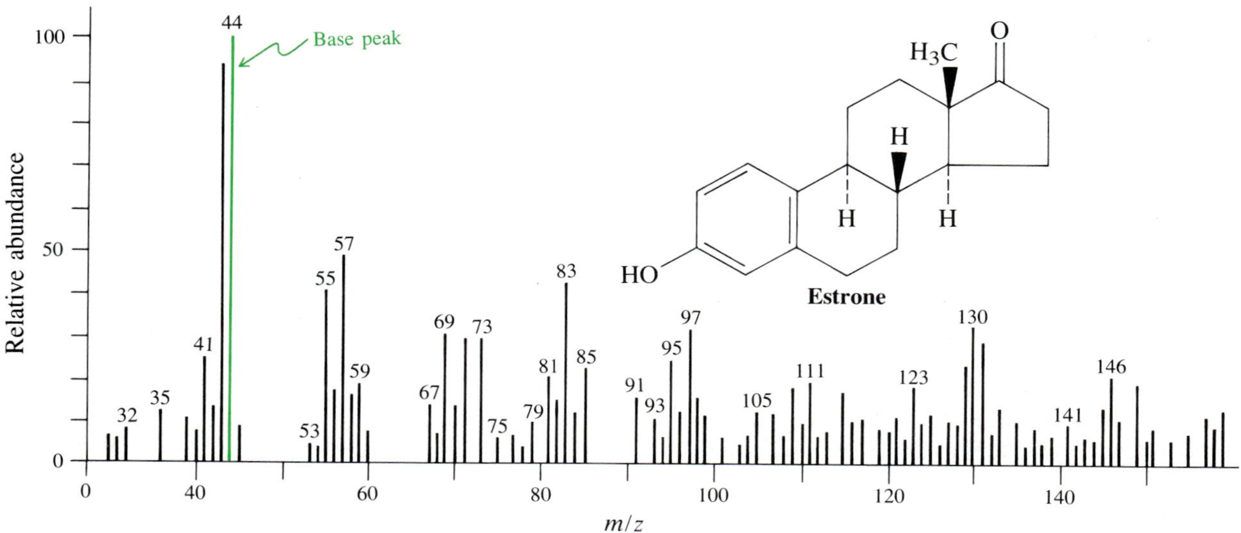

FIGURE 18-6

Mass spectrum of 1-bromopropane.

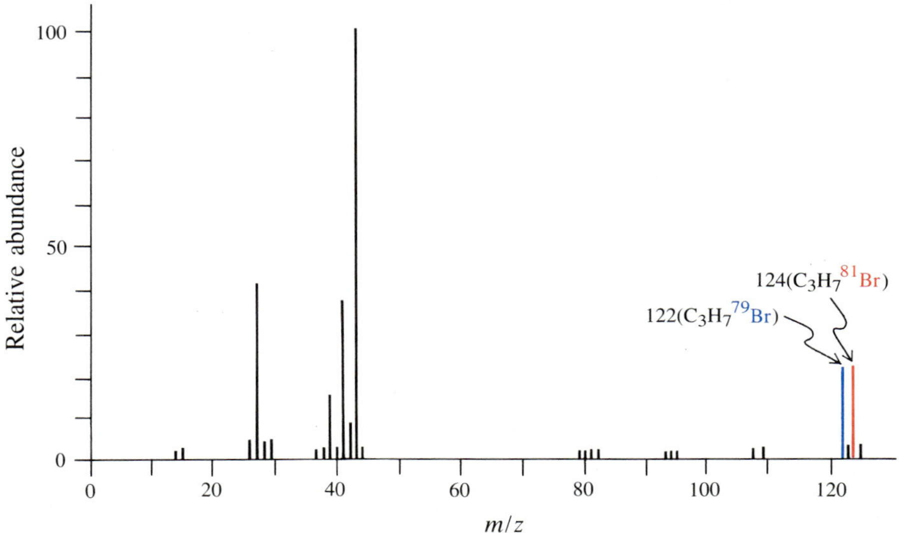

Molecules Fragment in Predictable Fashion

On electron impact, molecules fragment by the cleavage of weaker bonds before stronger ones. The resulting fragments may themselves fall apart into

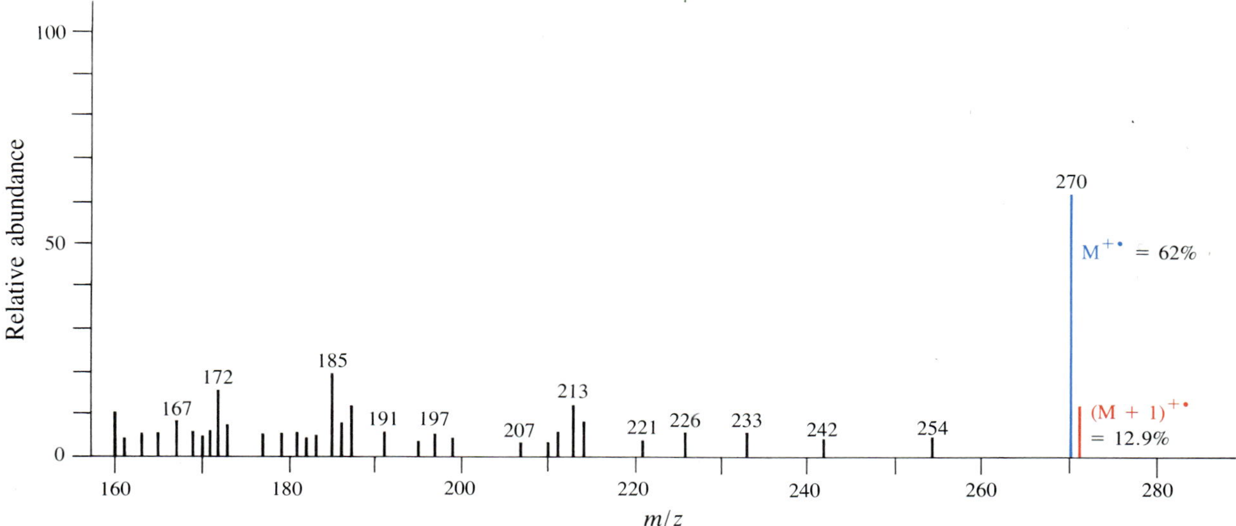

smaller pieces. The results of this phenomenon can be seen in the mass spectra of the isomeric hydrocarbons pentane, 2-methylbutane, and 2,2-dimethylpropane (Figures 18-7, 18-8, and 18-9, on pages 832 and 833). In each case, the molecular ion ($m/z = 72$) produces a relatively small peak, because fragmentation is rapid and extensive. However, the fragmentation patterns are very different for the three compounds. The mass spectrum of pentane (Figure 18-7) reveals a series of peaks consistent with more or less random C–C bond breaking, giving rise to charged C_4, C_3, C_2, and C_1 fragments (radical cations or carbocations):*

$$\underset{m/z\,=\,72}{[CH_3{-}CH_2{-}CH_2{-}CH_2{-}CH_3]^{+\cdot}} \longrightarrow \underset{m/z\,=\,15}{[CH_3]^+} + \underset{m/z\,=\,29}{[CH_3CH_2]^+} + \underset{m/z\,=\,43}{[CH_3CH_2CH_2]^+} + \underset{m/z\,=\,57}{[CH_3CH_2CH_2CH_2]^+}$$

Thus, in addition to the molecular ion signal, there is a line at $m/z = 57$ $(M - CH_3)^+$ followed by peaks indicating progressive loss of CH_2 units: $m/z = 43$ $(M - CH_2CH_3)^+$, and $m/z = 29$ $[CH_3CH_2]^+$. These fragment peaks are surrounded by clusters of smaller lines because of the presence of ^{13}C $(M + 1)^{+\cdot}$ and the loss of hydrogens [$(M - 1)^+$, $(M - 2)^+$, etc.].

The mass spectrum of 2-methylbutane (Figure 18-8) shows a fragmentation pattern similar to that of pentane; however, the relative intensities of the various peaks differ. Thus, there is a larger $(M - 1)^+$ peak at $m/z = 71$, and strong $(M - \text{alkyl})^+$ signals at $m/z = 57$ and 43, because of the relative stability of the

*This equation is neither mass nor charge balanced. All the fragment ions shown here and in subsequent mass-spectral equations can be cations or radical cations but, for convenience, are usually shown as cations only.

FIGURE 18-7
The mass spectrum of pentane.

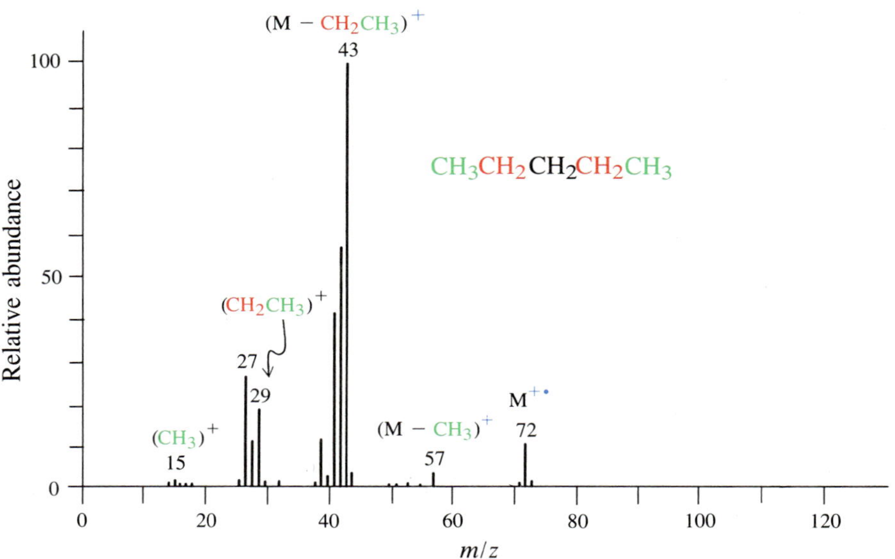

FIGURE 18-8
The mass spectrum of 2-methylbutane.

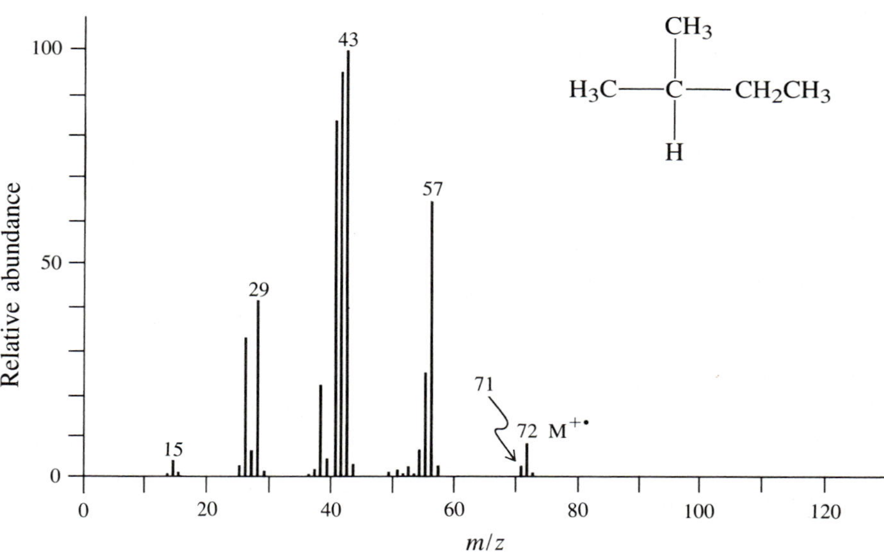

cations produced by preferred fragmentation at the more-highly-substituted carbon centers.

$$\left[H_3C{-}\underset{\displaystyle H}{\overset{\displaystyle CH_3}{C}}{-}CH_2CH_3 \right]^{+\bullet} \longrightarrow \underset{H_3C\quad CH_2CH_3}{\overset{CH_3}{\overset{+}{C}}} + \overset{H_3C}{\underset{H}{{}^{+}C}}{-}CH_2CH_3 + H_3C{-}\overset{CH_3}{\underset{H}{C^{+}}}$$

$m/z = 72$ $\qquad$ $m/z = 71$ $\qquad$ $m/z = 57$ $\qquad$ $m/z = 43$

FIGURE 18-9
The mass spectrum of 2,2-dimethylpropane.

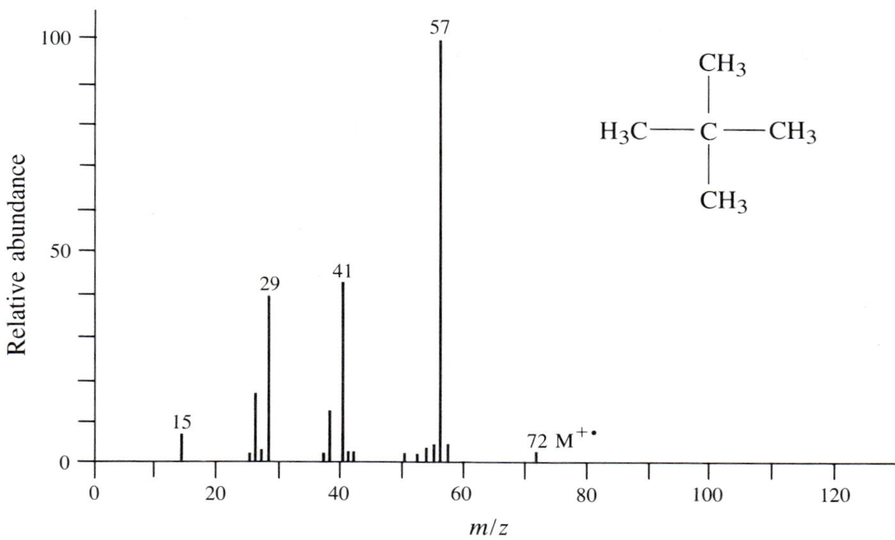

This effect is even more pronounced in the mass spectrum of 2,2-dimethylpropane (Figure 18-9). Here, loss of a methyl radical from the molecular ion produces the 1,1-dimethylethyl (*tert*-butyl) cation as the base peak at $m/z = 57$. This fragmentation is so facile that the molecular ion is barely visible. The spectrum also reveals peaks at $m/z = 41$ and 29, even though the molecule cannot directly fragment into pieces corresponding to these molecular weights. Fragment ions of this type are usually the result of complex structural reorganizations, some of which are like the carbocation rearrangements discussed in Section 9-2.

$$\left[(CH_3)_3C{-}CH_3\right]^{+\cdot} \quad m/z = 72 \quad \longrightarrow \quad (H_3C)_2C^{+}{-}CH_3 \quad m/z = 57$$

Fragmentations at Functional Groups

Particularly easy fragmentation of relatively weak bonds is also seen in the mass spectra of the haloalkanes. The fragment ion $(M - X)^+$ is frequently the base peak in these spectra, as illustrated in the mass spectrum of 1-bromopropane (Figure 18-6).

A similar phenomenon is observed in the mass spectra of alcohols, which eliminate water to give a large $(M - H_2O)^{+\cdot}$ peak (Figure 18-10). Loss of an 18-unit mass can be diagnostic for the presence of an alcohol function in a molecule.

Alcohol Fragmentation by Dehydration

$$\underset{M^{+\cdot}}{\left[R{-}\overset{HO}{\underset{H}{C}}{-}\overset{H}{C}HR'\right]^{+\cdot}} \longrightarrow \underset{(M-18)^{+\cdot}}{[RCH{=}CHR']^{+\cdot}} + H_2O$$

EXERCISE 18-25
Try to predict the appearance of the mass spectrum of 3-methyl-3-heptanol.

FIGURE 18-10

The mass spectrum of 1-butanol. The molecular ion, at $m/z = 74$, gives rise to a small peak because of ready loss of water to give the ion at $m/z = 56$. Other fragment ions are probably propyl ($m/z = 43$), 2-propenyl (allyl) ($m/z = 41$), and hydroxymethyl ($m/z = 31$).

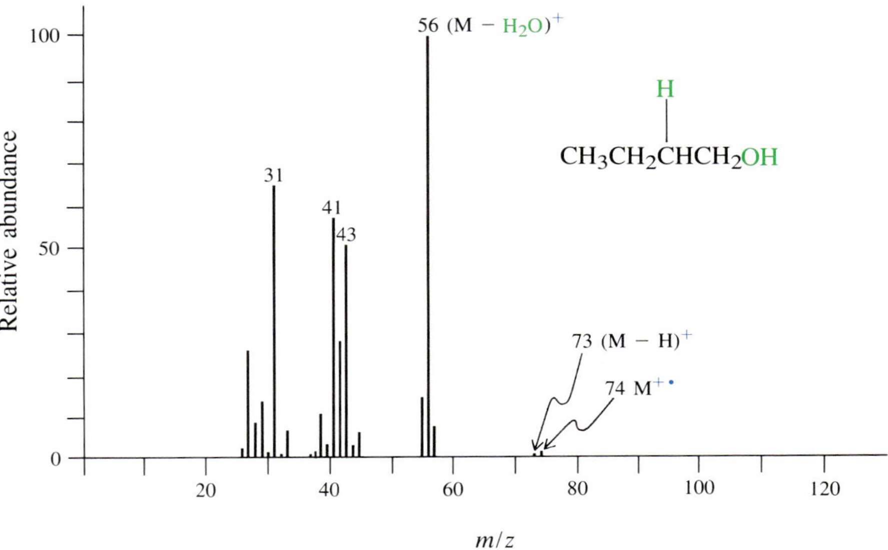

The fragmentation patterns of carbonyl compounds can also be very diagnostic. Consider, for example, the isomeric ketones 2-pentanone, 3-pentanone, and 3-methyl-2-butanone. Their mass spectra (Figure 18-11, on pages 836 and 837) reveal very clean and distinct fragmentation ions. The predominant pathway of decomposition is ***α* cleavage.** In this process, either of the alkyl bonds to the carbonyl function is severed to give the corresponding **acylium cation** and an alkyl fragment.

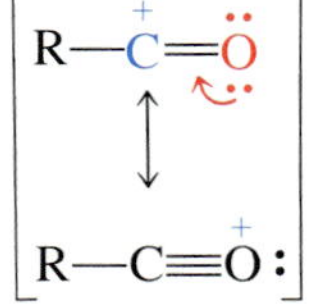

Acylium cation

α Cleavage of Carbonyl Compounds

$$\left[\mathrm{R}\!-\!\overset{\overset{\displaystyle :\mathrm{O}:}{\|}}{\mathrm{C}}\!-\!\mathrm{R}'\right]^{+\cdot} \xrightarrow{\alpha\text{ cleavage}} \mathrm{R}\overset{+}{\mathrm{C}}{\equiv}\mathrm{O}: + \mathrm{R}'^{+} + \mathrm{R}^{+} + :\overset{+}{\mathrm{O}}{\equiv}\mathrm{CR}'$$

The acylium cation (shown in the margin) forms easily because of resonance stabilization.

These fragment ions are valuable for structural assignments, because they allow the gross composition of the two alkyl groups in a ketone to be read from the spectrum. In this way, 2-pentanone is readily differentiated from 3-pentanone: α cleavage of 2-pentanone gives four fragment ions (two of which are coincidental), at $m/z = 15$, 43, and 71, but 3-pentanone gives only two, at $m/z = 29$ and $= 57$:

α Cleavage in 2-Pentanone

$$\underset{\substack{m/z\,=\,86\\ \textbf{2-Pentanone}}}{\mathrm{H_3C}\!-\!\overset{\overset{\displaystyle :\mathrm{O}:}{\|}}{\mathrm{C}}\!-\!\mathrm{CH_2CH_2CH_3}} \longrightarrow \underset{m/z\,=\,15}{\mathrm{CH_3}^{+}} + \underset{m/z\,=\,71}{:\overset{+}{\mathrm{O}}{\equiv}\mathrm{CCH_2CH_2CH_3}} + \underset{m/z\,=\,43}{\mathrm{CH_3C}{\equiv}\overset{+}{\mathrm{O}}:} + \underset{m/z\,=\,43}{{}^{+}\mathrm{CH_2CH_2CH_3}}$$

α Cleavage in 3-Pentanone

$$\underset{\substack{m/z\,=\,86\\ \textbf{3-Pentanone}}}{CH_3CH_2{-}\overset{\overset{:\ddot{O}:}{\|}}{C}{-}CH_2CH_3} \longrightarrow \underset{m/z\,=\,29}{CH_3CH_2^{+}} + \underset{m/z\,=\,57}{CH_3CH_2C{\equiv}\overset{+}{O}:}$$

Can 2-pentanone be distinguished from 3-methyl-2-butanone? Not by the observation of α cleavage—in both molecules, the substituent groups are CH_3 and C_3H_7. However, comparison of the mass spectra of the two compounds (Figure 18-11 A and C) reveals an additional prominent peak for 2-pentanone at $m/z = 58$, signifying the loss of a molecular fragment of weight $m/z = 28$. This fragment is absent from the spectra of both other isomers and is characteristic of the presence of hydrogens located gamma to the carbonyl group. Compounds with this structural feature and with sufficient flexibility in the chain to allow the γ-hydrogen to be close to the carbonyl oxygen undergo decomposition by the **McLafferty rearrangement.*** In this reaction, the molecular ion of the starting ketone splits into two pieces (a neutral fragment, which can, however, be ionized in the mass-spectral experiment to show up as a peak, and a radical cation) in a unimolecular process like the mechanism of ester pyrolysis (Section 18-4).

Similarity of Ester Pyrolysis and McLafferty Rearrangement

Ester pyrolysis:

$$\text{RHC(H)–H}_2\text{C–O–C(=O)R}' \xrightarrow{\Delta} \underset{\textbf{Alkene}}{RCH{=}CH_2} + \underset{\textbf{Carboxylic acid}}{O{=}C(OH){-}R'}$$

McLafferty rearrangement:

$$\left[\overset{\gamma}{\text{RHC(H)}}\text{–}\overset{\beta}{\text{H}_2\text{C}}\text{–}\overset{\alpha}{\text{CH}_2}\text{–C(=O)R}'\right]^{+\cdot} \longrightarrow RCH{=}CH_2 + \left[H_2C{=}C(OH){-}R'\right]^{+\cdot}$$

In contrast with ester pyrolysis, the McLafferty rearrangement does not produce a carboxylic acid; rather it yields the enol form of a new ketone. Note that the McLafferty rearrangement cannot take place in the absence of a γ-hydrogen, hence:

$$\underset{m/z\,=\,86}{\left[CH_3CH_2\overset{\overset{O}{\|}}{C}\underset{\alpha}{C}H_2\underset{\beta}{C}H_3\right]^{+\cdot}} \longrightarrow \text{no McLafferty rearrangement}$$

*Professor F. W. McLafferty, b. 1923, Cornell University, Ithaca, New York.

FIGURE 18-11

The mass spectra of: (A) 2-pentanone; (B) 3-pentanone; and (C) 3-methyl-2-butanone.

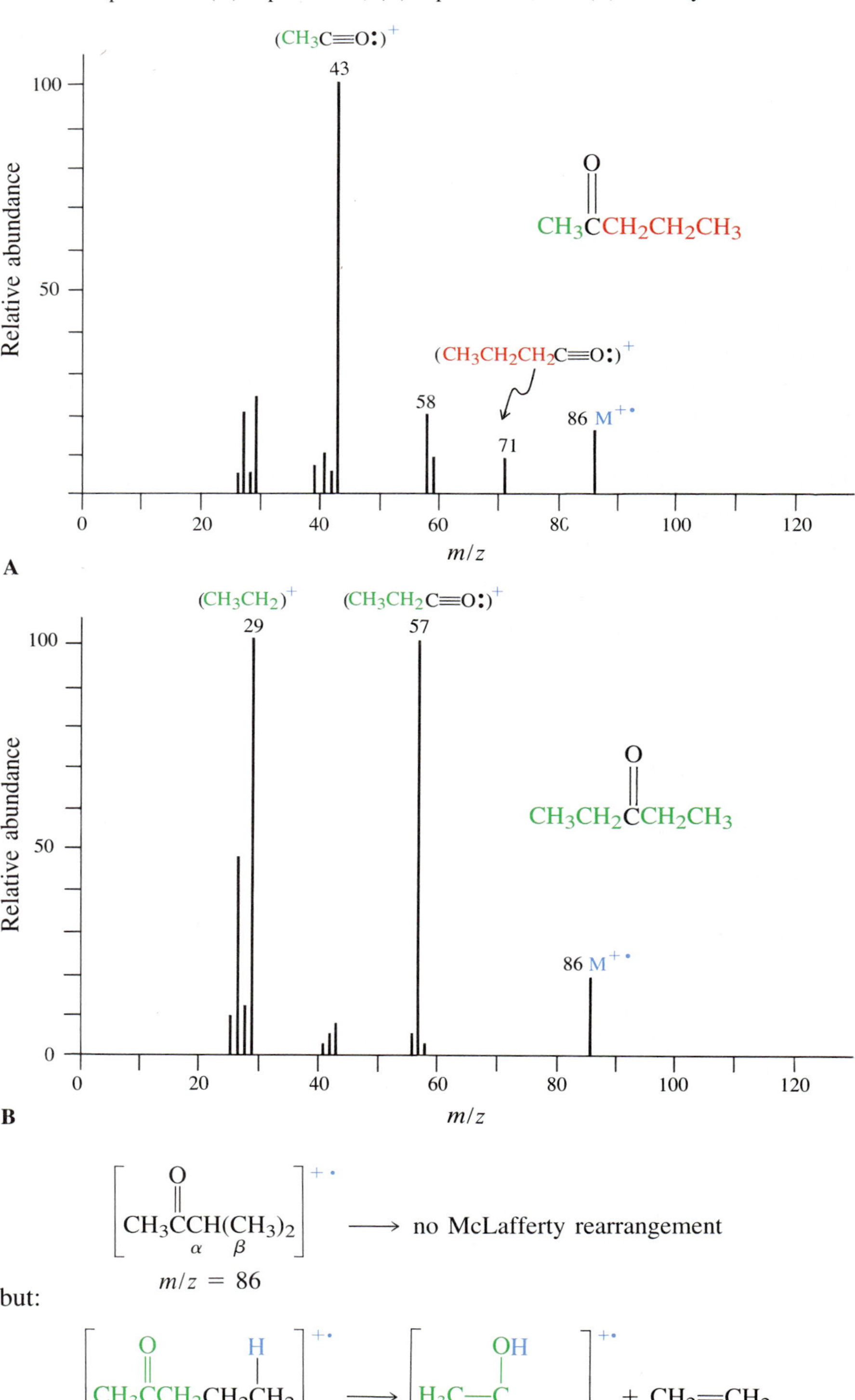

$$\left[\mathrm{CH_3\overset{O}{\overset{\|}{C}}\underset{\alpha}{C}H\underset{\beta}{(CH_3)_2}}\right]^{+\cdot} \longrightarrow \text{no McLafferty rearrangement}$$

$$m/z = 86$$

but:

$$\left[\mathrm{CH_3\overset{O}{\overset{\|}{C}}\underset{\alpha}{C}H_2\underset{\beta}{C}H_2\underset{\gamma}{\overset{H}{\overset{|}{C}}}H_2}\right]^{+\cdot} \longrightarrow \left[\mathrm{H_3C{-}\overset{OH}{\overset{|}{C}}{=}CH_2}\right]^{+\cdot} + \mathrm{CH_2{=}CH_2}$$

$$m/z = 86 \qquad\qquad m/z = 58$$

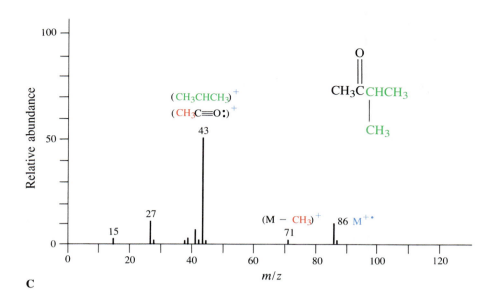

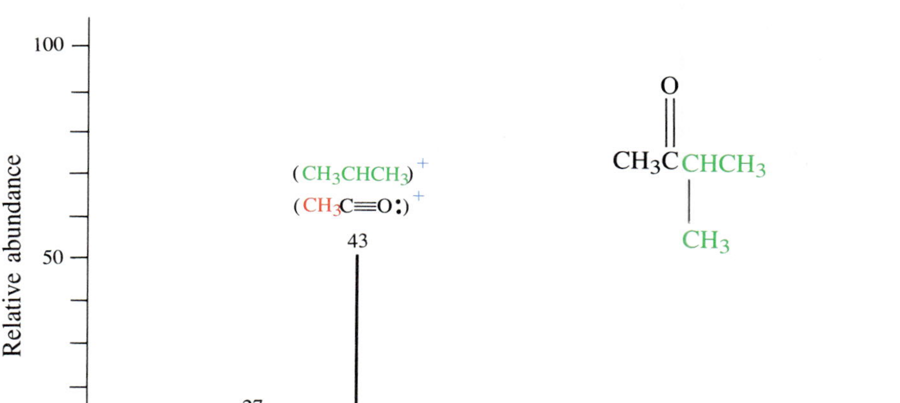

C

EXERCISE 18-26

How would you tell the difference between: (a) 3-methyl-2-pentanone and 4-methyl-2-pentanone, and (b) 2-ethylcyclohexanone and 3-ethylcyclohexanone, using only mass spectroscopy?

Similar rearrangements as well as α cleavages can be seen in the mass spectra of aldehydes and carboxylic acid derivatives.

EXERCISE 18-27

Interpret the labeled peaks in the mass spectra of pentanal, pentanoic acid, and methyl pentanoate shown in Figures 18-12, 18-13, and 18-14.

FIGURE 18-12

The mass spectrum of pentanal.

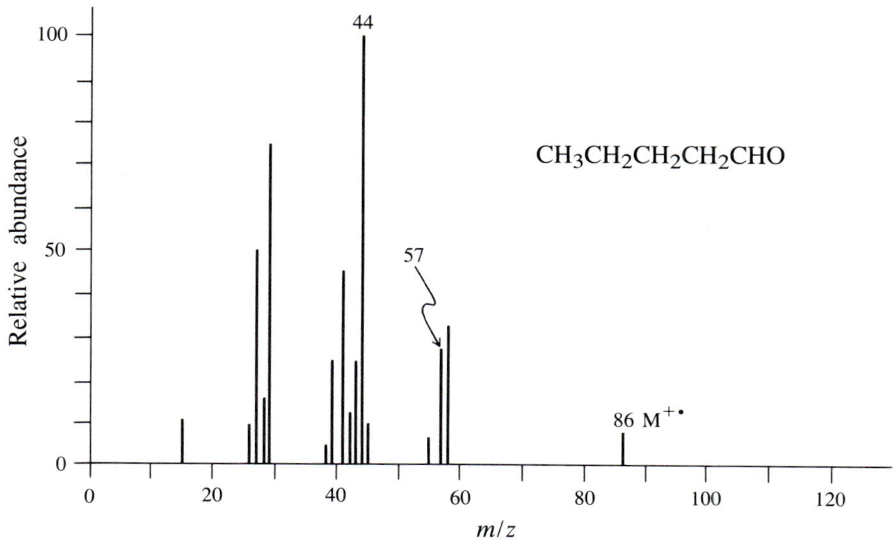

FIGURE 18-13
The mass spectrum of pentanoic acid.

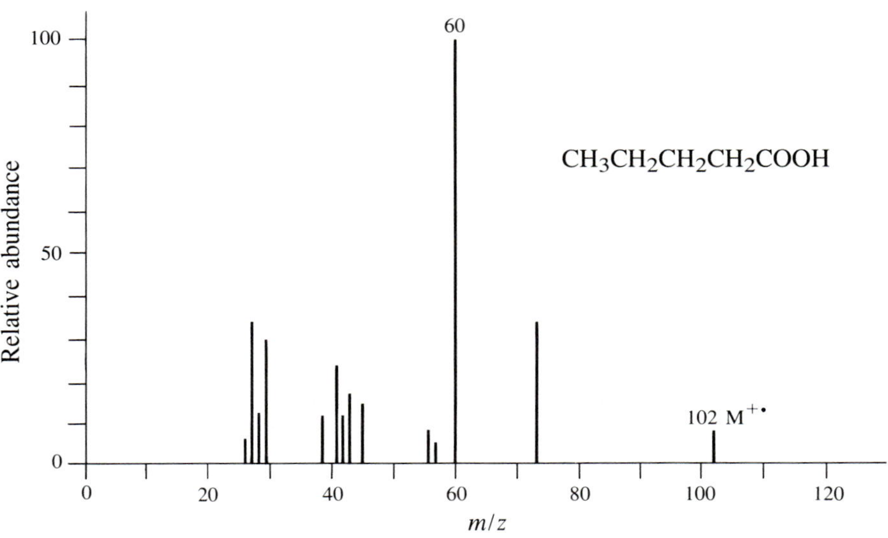

FIGURE 18-14
The mass spectrum of methyl pentanoate.

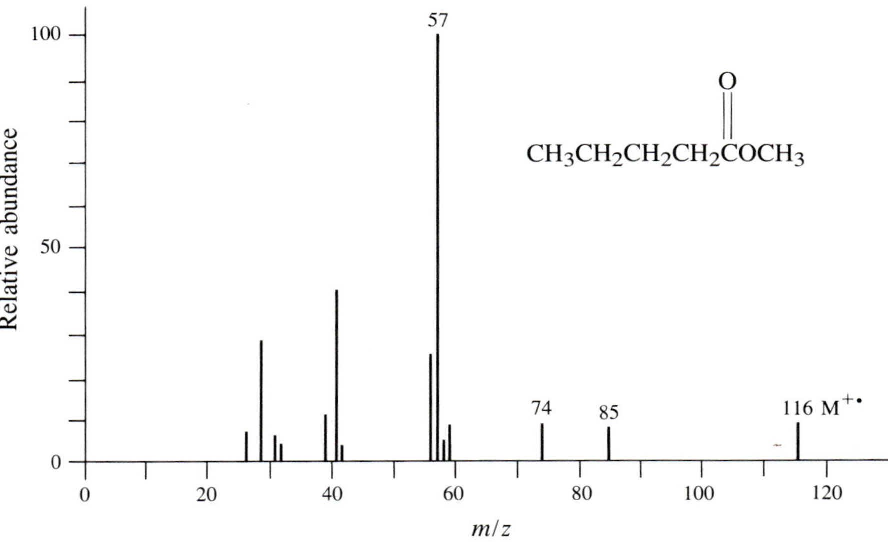

In summary, molecules can be ionized by an electron beam at 70 eV to give radical cations that are accelerated by an electric field and then separated by the different deflections that they undergo in a magnetic field. In a mass spectrometer, this effect is used to measure the molecular weights of molecules. The molecular ion is usually accompanied by less-massive fragments and isotopic ''satellites'' due to the presence of less-abundant isotopes. In some cases, such as with Cl and Br, more than one isotope may be present in substantial quantities. Fragmentation patterns can be interpreted for structural elucidation. For example, the radical cations of alkanes cleave to form the most-stable positively charged fragments, haloalkanes fragment by rupture of the carbon–halogen bond, alcohols readily dehydrate, and carbonyl compounds undergo α cleavage and the McLafferty rearrangement.

Summary of New Reactions

1 ORDER OF REACTIVITY OF CARBOXYLIC ACID DERIVATIVES

$$\underset{\textbf{Alkanoyl halide}}{\overset{\overset{\large O}{\|}}{RCX}} > \underset{\textbf{Anhydride}}{\overset{\overset{\large O\ \ O}{\|\ \ \|}}{RCOCR}} > \underset{\textbf{Ester}}{\overset{\overset{\large O}{\|}}{RCOR'}} > \underset{\textbf{Amide}}{\overset{\overset{\large O}{\|}}{RCNH_2}}$$

2 BASICITY OF THE CARBONYL OXYGEN

$$\text{:O: double bonded to C(R)(L:)} \xrightleftharpoons{H^+} \left[\underset{\textbf{A}}{{}^{+}\text{:O(H)=C(R)(L:)}} \longleftrightarrow \underset{\textbf{B}}{\text{:}\ddot{O}\text{(H)–}\overset{+}{C}\text{(R)(L:)}} \longleftrightarrow \underset{\textbf{C}}{\text{:}\ddot{O}\text{(H)–C(R)=}\overset{+}{L}} \right]$$

Basicity increases with increasing contribution of resonance structure C.

3 ENOLATE FORMATION

$$RCH_2\text{–C(=O:)–L:} \xrightleftharpoons{\text{Base}} \left[\underset{\textbf{A}}{RCH{=}C(\text{:}\ddot{O}\text{:}^-)(L\text{:})} \longleftrightarrow \underset{\textbf{B}}{R\overset{-}{\ddot{C}}H\text{–C(=}\ddot{O}\text{:)(L:)}} \longleftrightarrow \underset{\textbf{C}}{R\overset{-}{\ddot{C}}H\text{–C(:}\ddot{O}\text{:}^-\text{)=}L^+} \right]$$

Acidity of the neutral generally increases with decreasing contribution of resonance structure C in the anion.

4 AMIDATE FORMATION

$$\underset{pK_a \sim 15}{\overset{\overset{:O:}{\|}}{RC}\text{—}\ddot{N}HR'} \longrightarrow \left[\overset{\overset{:O:}{\|}}{RC}\text{—}\overset{-}{\ddot{\underset{\cdot\cdot}{N}}}R' \longleftrightarrow \overset{\overset{:\ddot{O}:^-}{|}}{RC}{=}\ddot{N}R' \right] + H^+$$

Reactions of Alkanoyl Halides

5 WATER

$$\overset{\overset{O}{\|}}{RCX} + H_2O \longrightarrow \underset{\textbf{Carboxylic acid}}{\overset{\overset{O}{\|}}{RCOH}} + HX$$

6 CARBOXYLIC SALTS

$$\mathrm{RCX} + \mathrm{R'CO_2^-Na^+} \longrightarrow \mathrm{RCOCR'} + \mathrm{Na^+X^-}$$

(RCX has C=O; RCOCR' has two C=O groups.)

Carboxylic anhydride

7 ALCOHOLS

$$\mathrm{RCX} + \mathrm{R'OH} \longrightarrow \mathrm{RCOR'} + \mathrm{HX}$$

Ester — HX (Removed with pyridine, triethylamine, or other base)

8 AMINES

$$\mathrm{RCX} + \mathrm{R'NH_2} \longrightarrow \mathrm{RCNHR'} + \mathrm{HX}$$

Amide — HX (Removed with pyridine, triethylamine, excess $R'NH_2$, or other base)

9 ORGANOMETALLIC REAGENTS

$$\mathrm{RCX} \xrightarrow[\text{2. } H^+,\ H_2O]{\text{1. } R'_2CuLi} \mathrm{RCR'} + \mathrm{CuX} + \mathrm{LiX}$$

Ketone

10 HYDROGEN (ROSENMUND REDUCTION)

$$\mathrm{RCX} + \mathrm{H_2} \xrightarrow{\text{Pd-BaSO}_4\text{, quinoline}} \mathrm{RCH} + \mathrm{HX}$$

Aldehyde

11 HYDRIDES

$$\mathrm{RCX} \xrightarrow[\text{2. } H^+,\ H_2O]{\text{1. LiAl[OC(CH}_3)_3]_3\text{H}} \mathrm{RCH} + \mathrm{LiX} + \mathrm{Al[OC(CH_3)_3]_3}$$

Aldehyde

12 DEHYDROHALOGENATION TO KETENES

$$\mathrm{RCH_2CX} \xrightarrow{\mathrm{N(CH_2CH_3)_3}} \mathrm{RCH{=}C{=}O} + \mathrm{(CH_3CH_2)_3\overset{+}{N}HX^-}$$

Ketene

13 DEHALOGENATION OF HALOALKANOYL HALIDES TO KETENES

$$\mathrm{RCHCX}\ (\text{with X on the CH}) \xrightarrow{\mathrm{Zn}} \mathrm{RCH{=}C{=}O} + \mathrm{ZnX_2}$$

14 REACTIONS OF KETENES

$$\text{>C=C=O} \xrightarrow{H_2O} -\overset{H}{\underset{|}{\overset{|}{C}}}-\overset{O}{\overset{\|}{C}}OH \quad \textbf{Carboxylic acid}$$

$$\text{>C=C=O} \xrightarrow{R\overset{O}{\overset{\|}{C}}OH} -\overset{H}{\underset{|}{\overset{|}{C}}}-\overset{O}{\overset{\|}{C}}O\overset{O}{\overset{\|}{C}}R \quad \textbf{Carboxylic anhydride}$$

$$\text{>C=C=O} \xrightarrow{ROH} -\overset{H}{\underset{|}{\overset{|}{C}}}-\overset{O}{\overset{\|}{C}}OR \quad \textbf{Ester}$$

$$\text{>C=C=O} \xrightarrow{RNH_2} -\overset{H}{\underset{|}{\overset{|}{C}}}-\overset{O}{\overset{\|}{C}}NHR \quad \textbf{Amide}$$

15 COMMERCIAL KETENE SYNTHESIS

$$CH_3\overset{O}{\overset{\|}{C}}CH_3 \xrightarrow{700^\circ C} CH_2{=}C{=}O + CH_4$$

16 COMMERCIAL ETHANOIC (ACETIC) ANHYDRIDE SYNTHESIS

$$CH_2{=}C{=}O + CH_3\overset{O}{\overset{\|}{C}}OH \longrightarrow CH_3\overset{O}{\overset{\|}{C}}O\overset{O}{\overset{\|}{C}}CH_3$$

Reactions of Carboxylic Acid Anhydrides

17 WATER

$$R\overset{O}{\overset{\|}{C}}O\overset{O}{\overset{\|}{C}}R + H_2O \longrightarrow \underset{\textbf{Carboxylic acid}}{2\ R\overset{O}{\overset{\|}{C}}OH}$$

18 ALCOHOLS

$$R\overset{O}{\overset{\|}{C}}O\overset{O}{\overset{\|}{C}}R + R'OH \longrightarrow \underset{\textbf{Ester}}{R\overset{O}{\overset{\|}{C}}OR'} + R\overset{O}{\overset{\|}{C}}OH$$

19 AMINES

$$\mathrm{R\overset{\overset{O}{\|}}{C}O\overset{\overset{O}{\|}}{C}R + R'NH_2 \longrightarrow \underset{\textbf{Amide}}{R\overset{\overset{O}{\|}}{C}NHR'} + R\overset{\overset{O}{\|}}{C}OH}$$

Reactions of Esters

20 WATER (ESTER HYDROLYSIS)

Acid catalysis:

$$\mathrm{R\overset{\overset{O}{\|}}{C}OR' + H_2O \xrightarrow{\text{Catalytic } H^+} \underset{\textbf{Carboxylic acid}}{R\overset{\overset{O}{\|}}{C}OH} + R'OH}$$

Base catalysis:

$$\mathrm{R\overset{\overset{O}{\|}}{C}OR' + \underset{\text{1 equivalent}}{{}^{-}OH} \xrightarrow{H_2O} \underset{\textbf{Carboxylate ion}}{R\overset{\overset{O}{\|}}{C}O^-} + R'OH}$$

21 ALCOHOLS (TRANSESTERIFICATION)

$$\mathrm{R\overset{\overset{O}{\|}}{C}OR' + R''OH \xrightarrow{H^+ \text{ or } {}^{-}OR''} \underset{\textbf{Ester}}{R\overset{\overset{O}{\|}}{C}OR''} + R'OH}$$

22 ORGANOMETALLIC REAGENTS

$$\mathrm{R\overset{\overset{O}{\|}}{C}OR'' \xrightarrow[\text{2. } H^+,\ H_2O]{\text{1. 2 } R'MgX} \underset{\textbf{Tertiary alcohol}}{R{-}\overset{\overset{OH}{|}}{\underset{\underset{R'}{|}}{C}}{-}R'} + R''OH}$$

Methyl methanoate (formate):

$$\mathrm{H\overset{\overset{O}{\|}}{C}OCH_3 \xrightarrow[\text{2. } H^+,\ H_2O]{\text{1. 2 } R'MgX} \underset{\textbf{Secondary alcohol}}{H{-}\overset{\overset{OH}{|}}{\underset{\underset{R'}{|}}{C}}{-}R'} + CH_3OH}$$

Dimethyl carbonate:

$$\mathrm{CH_3O\overset{\overset{O}{\|}}{C}OCH_3 \xrightarrow[\text{2. } H^+,\ H_2O]{\text{1. 3 } R'MgX} \underset{\textbf{Tertiary alcohol}}{R'{-}\overset{\overset{OH}{|}}{\underset{\underset{R'}{|}}{C}}{-}R'} + 2\ CH_3OH}$$

23 HYDRIDES

$$\mathrm{R\overset{O}{\overset{\|}{C}}OR'} \xrightarrow[\text{2. } H^+,\ H_2O]{\text{1. } LiAlH_4} \mathrm{RCH_2OH}$$

$$\mathrm{R\overset{O}{\overset{\|}{C}}OR'} \xrightarrow[\text{2. } H^+,\ H_2O]{\text{1. } (CH_3\overset{CH_3}{\overset{|}{C}}HCH_2)_2AlH} \mathrm{R\overset{O}{\overset{\|}{C}}H}$$

24 ENOLATES (PRODUCTS OBTAINED AFTER ACIDIC WORK-UP)

$$\mathrm{RCH_2\overset{O}{\overset{\|}{C}}OR'} \xrightarrow{\text{LDA}} \left[\mathrm{R\ddot{\bar{C}}H{-}\overset{:O:}{\overset{\|}{C}}OR'} \longleftrightarrow \mathrm{RCH{=}\overset{^-:\ddot{O}:}{\overset{|}{C}}OR'}\right]$$

Ester enolate ion

With $R''CHO$: $\mathrm{R''\underset{OH}{\underset{|}{C}}H\overset{R}{\overset{|}{C}}H\overset{O}{\overset{\|}{C}}OR'}$

With ethylene oxide: α-R-substituted γ-butyrolactone + R′OH

With $R''X$: $\mathrm{R\overset{R''}{\overset{|}{C}}H\overset{O}{\overset{\|}{C}}OR'}$

25 CLAISEN CONDENSATION

$$2\ \mathrm{CH_3\overset{O}{\overset{\|}{C}}OCH_2CH_3} \xrightarrow[\text{2. } H^+,\ H_2O]{\text{1. } CH_3CH_2O^-,\ CH_3CH_2OH} \mathrm{CH_3\overset{O}{\overset{\|}{C}}CH_2\overset{O}{\overset{\|}{C}}OCH_2CH_3} + \mathrm{CH_3CH_2OH}$$

↑ $pK_a \sim 11$ (at the CH_2)

26 ESTER PYROLYSIS

$$\mathrm{R\overset{O}{\overset{\|}{C}}OCH_2CH_2R'} \xrightarrow[\textit{Syn}\text{ elimination}]{300^\circ C} \mathrm{R\overset{O}{\overset{\|}{C}}OH} + \mathrm{CH_2{=}CHR'}$$

Reactions of Amides

27 WATER

$$\mathrm{R\overset{O}{\overset{\|}{C}}NHR'} + H_2O \xrightarrow{H^+,\ \Delta} \mathrm{R\overset{O}{\overset{\|}{C}}OH} + R'\overset{+}{N}H_3$$

Carboxylic acid

$$\mathrm{RCNHR' + H_2O \xrightarrow{HO^-,\ \Delta} RCO^- + R'NH_2}$$

(C=O in RCNHR′ and RCO⁻)

28 HYDRIDES

$$\mathrm{RC(=O)NHR' \xrightarrow[\text{2. } H^+,\ H_2O]{\text{1. } LiAlH_4} RCH_2NH_2}$$

Amine

$$\mathrm{RC(=O)NHR' \xrightarrow[\text{2. } H^+,\ H_2O]{\text{1. } (CH_3CH(CH_3)CH_2)_2AlH} RC(=O)H}$$

Aldehyde

$$\mathrm{RC(=O)NHR' \xrightarrow[\text{2. } H^+,\ H_2O]{\text{1. } LiAl(OCH_2CH_3)_3H} RC(=O)H}$$

Aldehyde

29 ENOLATES AND AMIDATES

$$\mathrm{RCH_2C(=\ddot{O}:)NR'_2 \xrightarrow{Base} RCH{=}C(\ddot{O}:^-)(NR'_2) \xrightarrow{R''X} RCH(R'')C(=O)NR'_2}$$

$pK_a \sim 30$ (at the CH_2 hydrogen)

Amide enolate ion

$$\mathrm{RCH_2C(=\ddot{O}:)\ddot{N}HR' \xrightarrow{Base} RCH_2C(\ddot{O}:^-){=}\ddot{N}R' \xrightarrow{R''X} RCH_2C(=\ddot{O}:)\ddot{N}(R'')R'}$$

$pK_a \sim 15$ (at the NH hydrogen)

Amidate ion

30 HOFMANN REARRANGEMENT

$$\mathrm{RC(=O)NH_2 \xrightarrow{Br_2,\ NaOH,\ H_2O,\ 75^\circ C} RNH_2 + CO_2}$$

Amine

$$\mathrm{RC(=O)NH_2 \xrightarrow{Br_2,\ NaOCH_3,\ CH_3OH} RNHC(=O)OCH_3 \xrightarrow{NaOH,\ H_2O} RNH_2 + CO_2}$$

Carbamic ester (Urethane) — **Amine**

Reactions of Nitriles

31 PROTONATION

$$\mathrm{R{-}C{\equiv}N: + H^+ \longrightarrow [R{-}C{\equiv}\overset{+}{N}{-}H \longleftrightarrow R{-}\overset{+}{C}{=}\ddot{N}{-}H]}$$

$pK_a \sim -10$

32 DEPROTONATION

$$\underset{pK_a \sim 25}{RCH_2C{\equiv}N\!:} + :B^- \longrightarrow [R{-}\ddot{\bar{C}}H{-}C{\equiv}N\!: \longleftrightarrow R{-}CH{=}C{=}\ddot{\bar{N}}\!:] \xrightarrow{R'X} R{-}\overset{R'}{\overset{|}{C}}HC{\equiv}N\!:$$

33 WATER

$$RC{\equiv}N + H_2O \xrightarrow{H^+ \text{ or } HO^-,\ \Delta} \underset{\textbf{Amide}}{R\overset{O}{\overset{\|}{C}}NH_2} \xrightarrow{H^+ \text{ or } HO^-,\ \Delta} \underset{\textbf{Carboxylic acid}}{R\overset{O}{\overset{\|}{C}}OH}$$

34 ORGANOMETALLIC REAGENTS

$$RC{\equiv}N \xrightarrow[2.\ H^+,\ H_2O]{1.\ R'MgX \text{ or } R'Li} \underset{\textbf{Ketone}}{R\overset{O}{\overset{\|}{C}}R'}$$

35 HYDRIDES

$$RC{\equiv}N \xrightarrow[2.\ H^+,\ H_2O]{1.\ LiAlH_4} \underset{\textbf{Amine}}{RCH_2NH_2}$$

$$RC{\equiv}N \xrightarrow[2.\ H^+,\ H_2O]{1.\ LiAlH(OCH_2CH_3)_3} \underset{\textbf{Aldehyde}}{R\overset{O}{\overset{\|}{C}}H}$$

$$RC{\equiv}N \xrightarrow[2.\ H^+,\ H_2O]{1.\ (CH_3\overset{CH_3}{\overset{|}{C}}HCH_2)_2AlH} \underset{\textbf{Aldehyde}}{R\overset{O}{\overset{\|}{C}}H}$$

36 CATALYTIC HYDROGENATION

$$RC{\equiv}N \xrightarrow{H_2,\ PtO_2} \underset{\textbf{Amine}}{RCH_2NH_2}$$

Mass-Spectroscopy Fragmentation

(All fragments can be cations, radical cations, or radicals.)

37 ALKANES

$$[RCH_2CH_2R]^{+\cdot} \longrightarrow RCH_2{}^+ + R\overset{+}{C}HCH_2R + RCH_2CH_2{}^+ + R^+ + H^+$$

$$\left[R{-}\overset{R}{\overset{|}{\underset{R}{\underset{|}{C}}}}{-}H\right]^{+\cdot} \longrightarrow R_3C^+ + R_2CH^+ + R^+ + H^+$$

38 HALOALKANES

$$[R{-}X]^{+\cdot} \longrightarrow R^+ + X^+$$

39 ALCOHOLS

$$\left[\begin{array}{c} \text{H}\quad\text{OH} \\ | \quad\; | \\ -\text{C}-\text{C}- \\ | \quad\; | \end{array}\right]^{+\cdot} \longrightarrow \left[\;\rangle\text{C}{=}\text{C}\langle\;\right]^{+\cdot} + H_2O$$

40 CARBONYL COMPOUNDS

α Cleavage:

$$\left[\begin{array}{c} \text{O} \\ \| \\ \text{RCR}' \end{array}\right]^{+\cdot} \longrightarrow \text{RC}{\equiv}\overset{+}{\text{O}}\text{:} + \text{R}'^{+} + \text{R}^{+} + \text{:}\overset{+}{\text{O}}{\equiv}\text{CR}'$$

McLafferty rearrangement:

$$\left[\underset{\gamma\quad\beta\quad\alpha}{\text{RCH}_2\text{CH}_2\text{CH}_2}\overset{\text{O}}{\overset{\|}{\text{C}}}\text{R}'\right]^{+\cdot} \longrightarrow \text{RCH}{=}\text{CH}_2 + \left[\text{CH}_2{=}\text{C}\begin{array}{l}\diagup\text{OH} \\ \diagdown\text{R}'\end{array}\right]^{+\cdot}$$

Summary of Important Concepts

1 The electrophilic reactivity of the carbonyl carbon in carboxylic acid derivatives is weakened by good electron-donating substituents. This effect, measurable by IR spectroscopy, is responsible not only for the decrease in the reactivity with nucleophiles and acid, but also for the increased basicity along the series: alkanoyl halides–anhydrides–esters–amides. Electron donation by the nitrogen in amides is so pronounced that there is hindered rotation around the amide bond on the NMR time scale.

2 Carboxylic acid derivatives are named as alkanoyl halides, carboxylic anhydrides, alkyl alkanoates, alkanamides, and alkanenitriles, depending on the functional group.

3 Carboxylic acid derivatives generally react with water (often with acid or base catalysis) to hydrolyze to the corresponding carboxylic acid; they combine with alcohols to give esters, and with amines to furnish amides. With Grignard and other organometallic reagents, they form ketones; esters may react further to form the corresponding alcohols. Reduction by catalytically activated hydrogen or hydrides gives products in various oxidation states: aldehydes, alcohols, or amines.

4 Long-chain esters are the constituents of animal and plant waxes. Triesters of glycerol are contained in natural oils and fats. Their hydrolysis gives soaps. Triglycerides containing phosphoric acid ester subunits belong to the class of phospholipids. Because they carry a highly polar head group and hydrophobic tails, they form micelles and lipid bilayers.

5 Transesterification can be used to convert one ester into another.

6 The functional group of nitriles has some similarity to that of the alkynes. The two component atoms are *sp* hybridized, causing the nitrogen lone electron pair to be relatively nonbasic. The IR stretching vibration appears at about 2250 cm^{-1}. The hydrogens next to the cyano group are acidic and are deshielded in ^{1}H NMR. The ^{13}C NMR absorptions for nitrile carbons are at relatively low field ($\delta \sim 112$–126 ppm), a consequence of the electronegativity of nitrogen.

7 Mass spectroscopy is a technique for ionizing molecules and separating the resulting ions magnetically by molecular weight. Because the ionizing beam has high energy, the ionized molecules also fragment to smaller particles, all of which are separated and recorded as the mass spectrum of a compound. The presence of certain elements (such as Cl, Br) can be detected by their isotopic patterns. The presence of fragment-ion signals in mass spectra can be used to deduce the structure of a molecule.

Problems

1 Use resonance structures to explain in detail the relative order of acidity of carboxylic acid derivatives, as presented in Section 18-1.

2 In each of the following pairs of compounds, decide which possesses the indicated property to the greater degree:

(a) Length of C–X bond: ethanoyl fluoride or ethanoyl chloride.

(b) Acidity of the boldface H: $\mathbf{CH_2}(COCH_3)_2$ or $\mathbf{CH_2}(COOCH_3)_2$.

(c) Reactivity toward addition of a nucleophile: an amide—for example, $CH_3\overset{O}{\overset{\|}{C}}N(CH_3)_2$—or an imide—for example, $CH_3\overset{O}{\overset{\|}{C}}N(CH_3)\overset{O}{\overset{\|}{C}}CH_3$

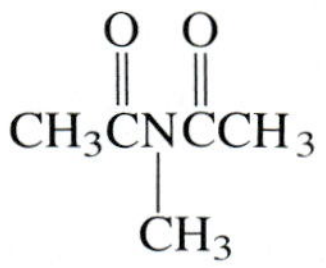

(d) High infrared carbonyl stretching frequency: ethyl ethanoate or ethenyl ethanoate.

3 On treatment with strong base followed by protonation, compounds (i) and (ii) undergo cis-trans isomerization, but compound (iii) does not. Explain.

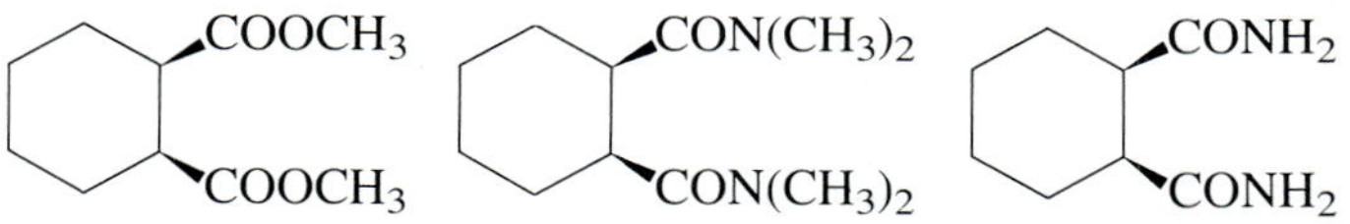

i: cyclohexane bearing two $COOCH_3$ groups on adjacent carbons

ii: cyclohexane bearing two $CON(CH_3)_2$ groups on adjacent carbons

iii: cyclohexane bearing two $CONH_2$ groups on adjacent carbons

4 Write the product(s) of each of the following reactions.

(a) 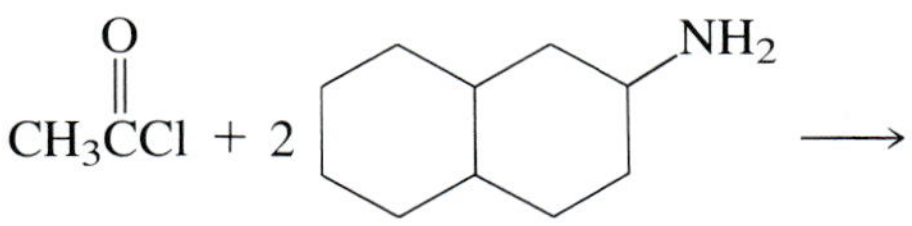 $CH_3\overset{O}{\overset{\|}{C}}Cl + 2$ (decahydronaphthalen-2-amine, NH_2) $\longrightarrow$

(b) 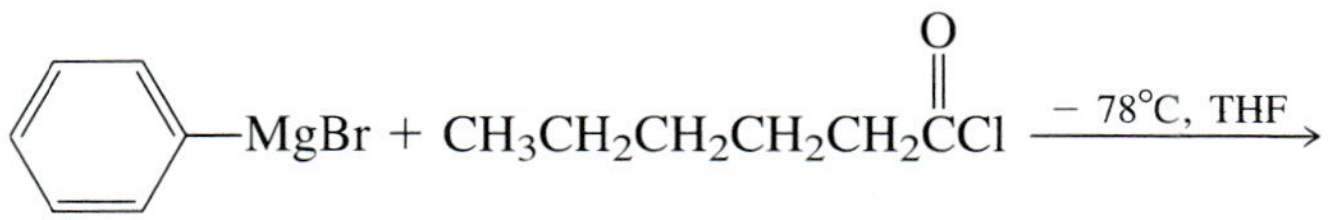 $C_6H_5MgBr + CH_3CH_2CH_2CH_2CH_2\overset{O}{\overset{\|}{C}}Cl \xrightarrow{-78°C,\ THF}$

(c) $(H_3C)_3C\overset{O}{\overset{\|}{C}}Cl \xrightarrow{LiAl[OC(CH_3)_3]_3H,\ THF,\ -78°C}$

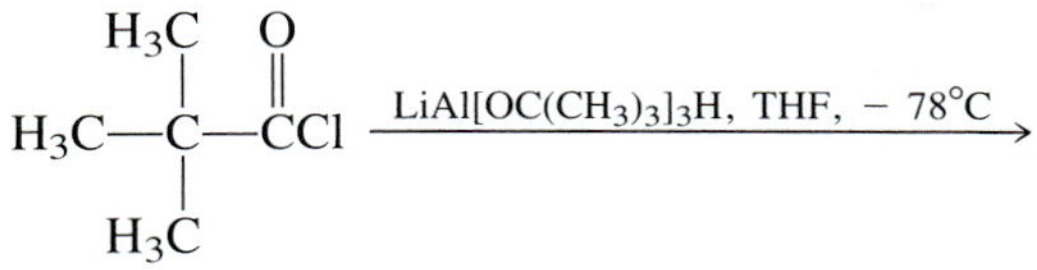

(d) CH_2—$\overset{O}{\overset{\|}{C}}$Cl, CH_2—$\underset{\underset{O}{\|}}{C}$Cl $\xrightarrow{2\ C_6H_5CH_2OH,\ \text{pyridine}}$

(e) (structure with OCH_3, H_3C, H_3C, $\underset{\underset{O}{\|}}{C}$Cl) $\xrightarrow{H_2,\ Pd\text{-}BaSO_4,\ \text{quinoline}}$

5 **(a)** Propose mechanisms for (1) the formation of methylketene, $CH_3CH{=}C{=}O$, from propanoyl chloride and $N(CH_2CH_3)_3$, and (2) the reaction of methylketene with water.

(b) Suggest factors that could influence whether an alkanoyl halide reacts with a tertiary amine to form an alkanoyl ammonium halide (Section 18-2) or a ketene (Section 18-3).

6 Write the product(s) that you would expect from the reactions of ethanoic (acetic) anhydride with each of the reagents below. Assume in all cases that the reagent is present in large excess.

(a) $(CH_3)_2CHOH$

(b) NH_3

(c) C_6H_5—MgBr

(d) $LiAlH_4$

7 Write the product(s) of the reaction of butanedioic (succinic) anhydride with each of the reagents in Problem 6.

8 On completing a synthetic procedure, every chemist is faced with the job of cleaning glassware. Because the compounds present may be dangerous in some way or have unpleasant properties, a little serious chemical thinking is often beneficial before "doing the dishes." Suppose that you have just completed a synthesis of hexanoyl chloride, perhaps to carry out the reaction in part **b** of Problem 4; first, however, you must clean the glassware contaminated with this alkanoyl halide. Both hexanoyl chloride and hexanoic acid have terrible odors.

(a) Would cleansing the glassware with soap and water be a good idea? Explain.

(b) Suggest a more-pleasant alternative, based on the chemistry of alkanoyl halides and the physical properties (particularly the odors) of the various carboxylic acid derivatives.

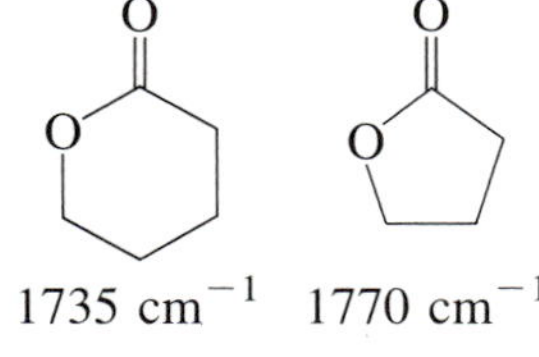

1840 cm^{-1}

9 Although esters typically have carbonyl stretching frequencies at about 1740 cm^{-1} in the infrared spectrum, the corresponding band for lactones can vary greatly with ring size. Three examples are shown in the margin. Propose an explanation for the IR bands of these smaller-ring lactones.

10 Write out a detailed mechanism for the acid-catalyzed transesterification of ethyl 2-methylpropanoate (ethyl isobutyrate) into the corresponding methyl ester. Your mechanism should clearly illustrate the catalytic role of the proton.

11 Show how you would carry out the following transformation, in which the ester at the lower left of the molecule is converted into an alcohol but the ester at the upper right is preserved. (Hint: Do not try ester hydrolysis. Look carefully at how the ester groups are linked to the steroid and consider an approach based upon transesterification.)

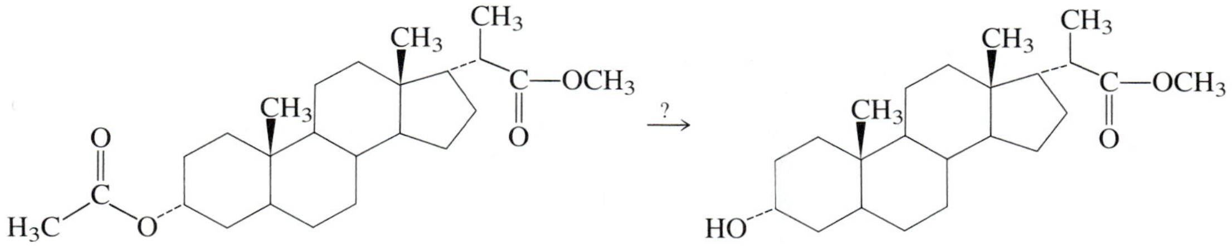

H_3C O C O CH$_3$ CH$_3$ CH$_3$ C—OCH$_3$ O $\xrightarrow{?}$ HO CH$_3$ CH$_3$ CH$_3$ C—OCH$_3$ O

12 Write the product of each of the following reactions.

(a) COOCH$_3$ COOCH$_3$ $\xrightarrow{\text{1. KOH, } H_2O\text{; 2. } H^+\text{, } H_2O}$

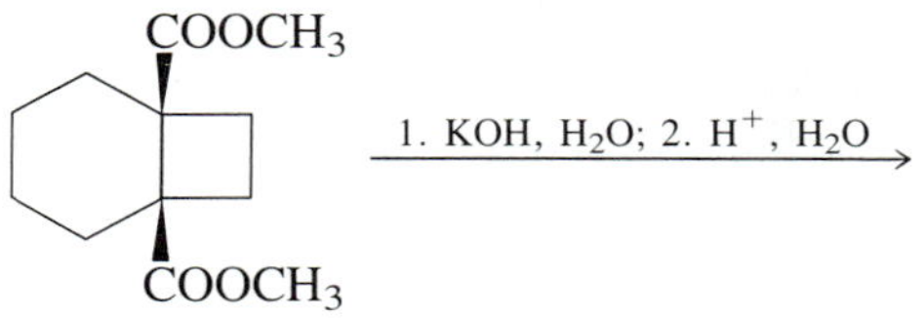

(b) H_3C O =O CH_3CH_2 $\xrightarrow{(CH_3)_2CHNH_2\text{, } CH_3OH\text{, } \Delta}$

(c) $CH_3\overset{O}{\overset{\|}{C}}OCH_3$ + excess (cyclopentyl)MgBr $\xrightarrow[\text{2. } H^+\text{, } H_2O]{\text{1. } (CH_3CH_2)_2O\text{, 20°C}}$

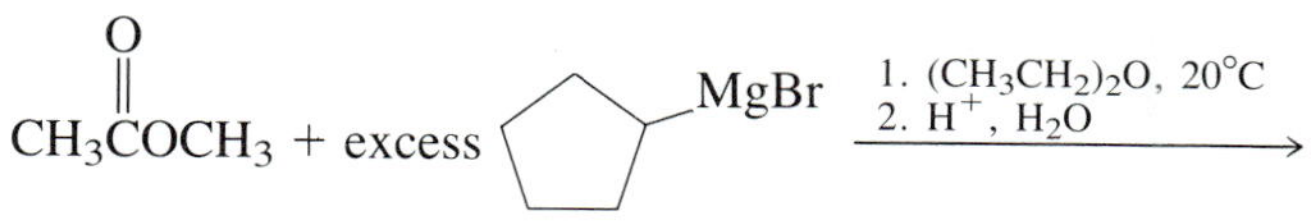

(d) $CH_3O\overset{O}{\overset{\|}{C}}OCH_3$ + excess (cyclopropyl)—MgBr $\xrightarrow[\text{2. } H^+\text{, } H_2O]{\text{1. } (CH_3CH_2)_2O\text{, 20°C}}$

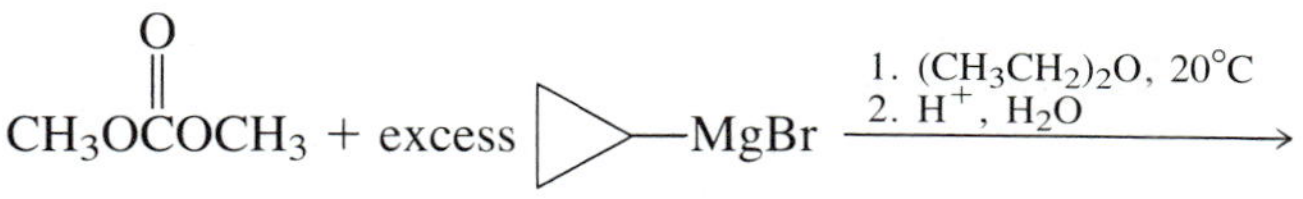

(e) (phenyl)—CH(CH$_3$)COOCH$_2$CH$_3$ $\xrightarrow[\text{3. } H^+\text{, } H_2O]{\text{1. LDA, THF, } -78°C\text{; 2. } \overset{O}{CH_2—CH_2}\text{, HMPA}}$

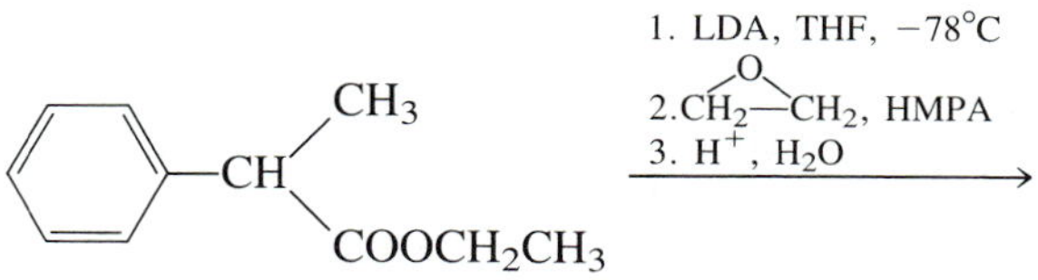

(f) (cyclohexyl)COOCH$_3$ $\xrightarrow[\text{2. } H^+\text{, } H_2O]{\text{1. } (CH_3\overset{CH_3}{\overset{|}{C}}HCH_2)_2AlH\text{, toluene, } -60°C}$

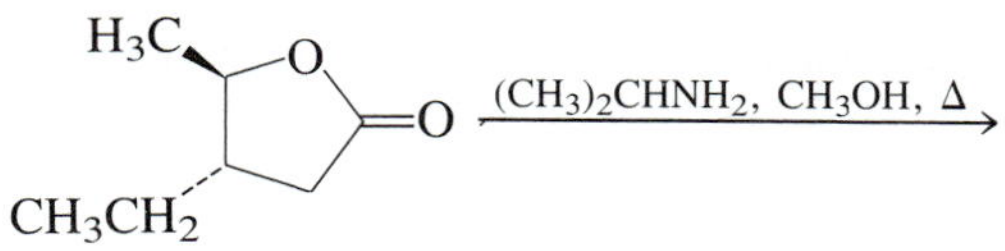

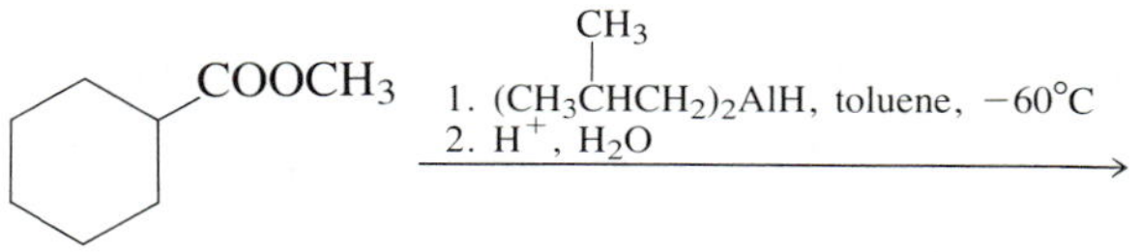

13 The removal of the C-17 side chain of certain steroids is a critical element in the synthesis of a number of hormones, such as testosterone, from steroids in the more-readily-available pregnane family:

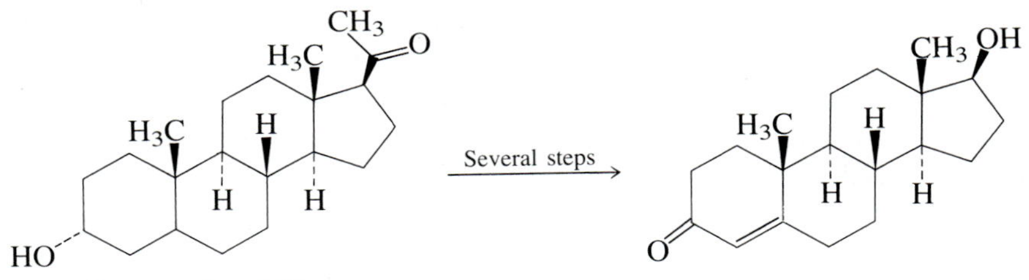

Pregnan-3α-ol-20-one $\xrightarrow{\text{Several steps}}$ **Testosterone**

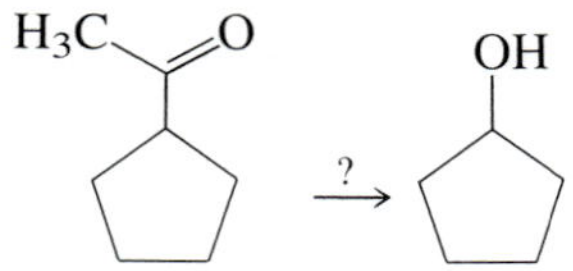

How would you carry out the comparable transformation, shown in the margin, of ethanoylcyclopentane into cyclopentanol? Note: In this and subsequent synthetic problems you may need to use reactions from several areas of carbonyl chemistry (Chapters 15–18).

14 A useful synthesis of certain types of diols includes the reaction of a "bis-Grignard" reagent with a lactone:

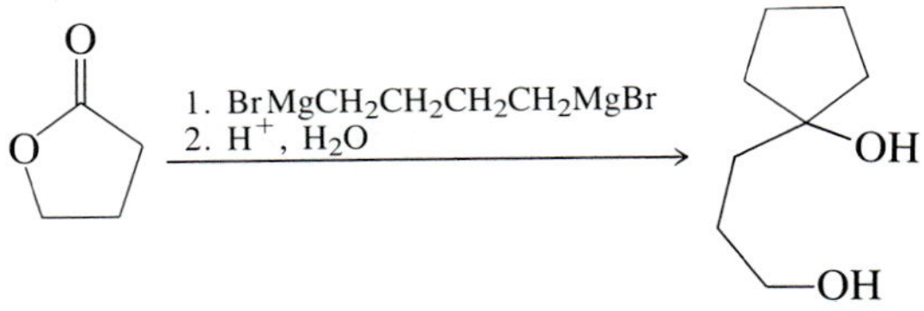

1. $BrMgCH_2CH_2CH_2CH_2MgBr$
2. H^+, H_2O

(a) Formulate a mechanism for this transformation.

(b) Show how you would apply this general method to the synthesis of diols (i) and (ii).

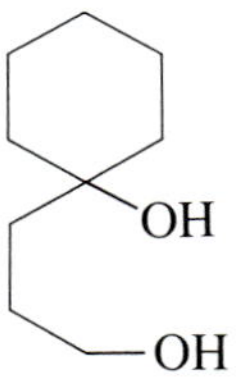

i 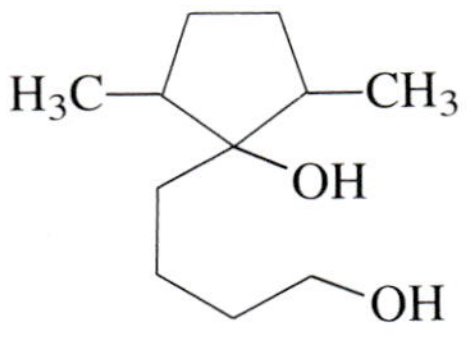 ii

15 The reaction between the ethyl ethanoate (ethyl acetate) enolate and 2,2-dimethylpropanal is given in Section 18-4 as an example of ester enolate addition to a ketone or aldehyde carbonyl group. Would the following corresponding reaction with propanal itself also work? Why or why not?

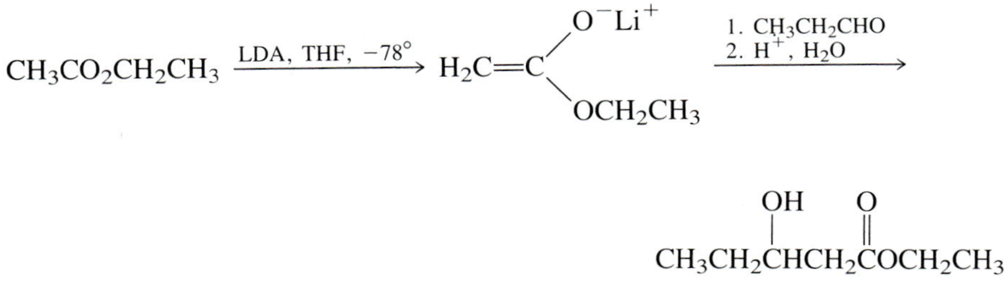

$$CH_3CO_2CH_2CH_3 \xrightarrow{\text{LDA, THF, } -78^\circ} H_2C{=}C(O^-Li^+)(OCH_2CH_3) \xrightarrow[\text{2. } H^+,\ H_2O]{\text{1. } CH_3CH_2CHO} CH_3CH_2CH(OH)CH_2C(=O)OCH_2CH_3$$

16 Propose a synthetic sequence to convert carboxylic acid (i), shown in the margin, into the naturally occurring sesquiterpene α-curcumene.

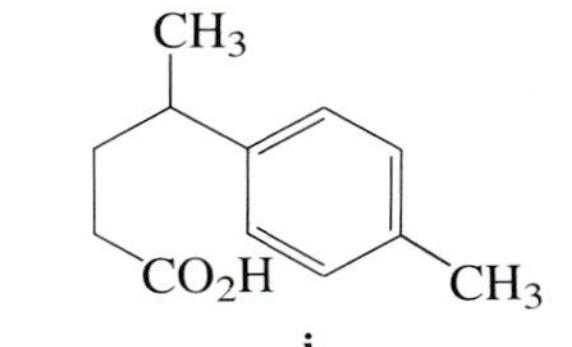

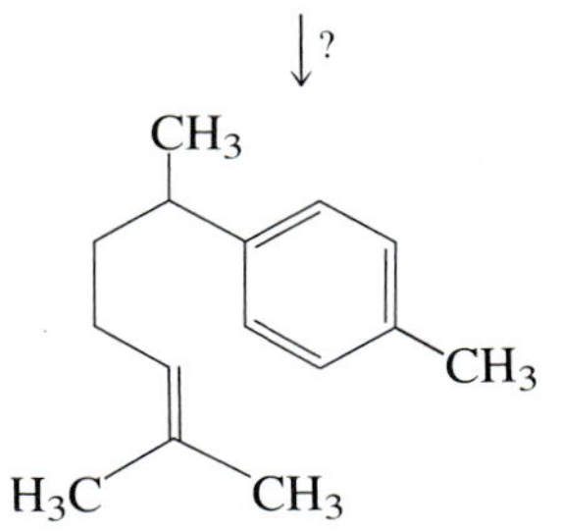

17 Propose a synthesis of oleuropic acid (which is found in olive trees), starting with keto acid (i), below.

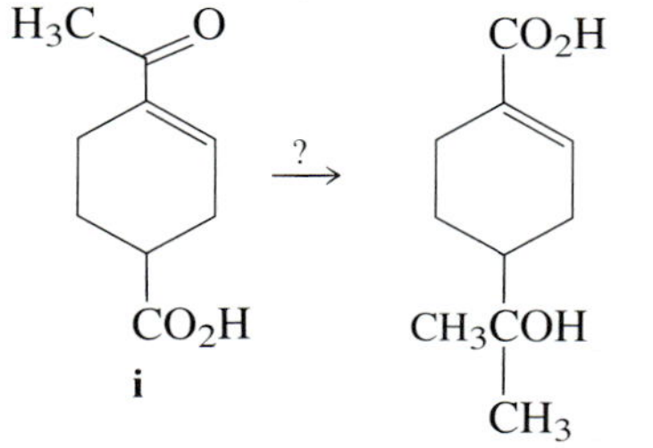

H_3C O CO_2H ? → CO_2H i CH_3COH CH_3

Oleuropic acid

CH_3 CO_2H CH_3 i ? CH_3 CH_3 H_3C CH_3

α-Curcumene

18 A synthesis of camphene from camphenilone starts with a Grignard reaction (the Wittig reaction does not work):

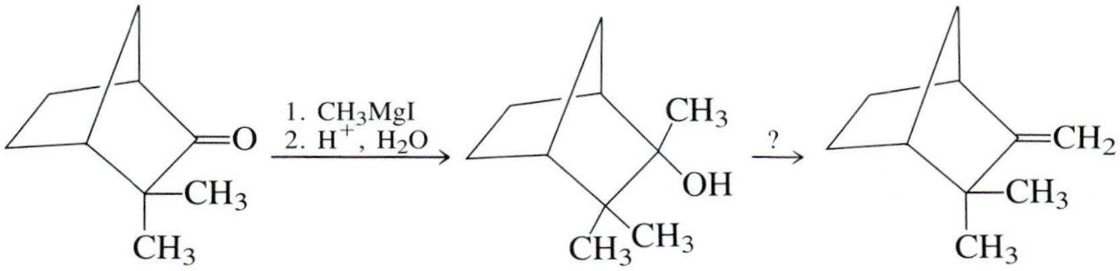

=O —CH_3 CH_3 1. CH_3MgI 2. H^+, H_2O → CH_3 OH CH_3 CH_3 ? → =CH_2 —CH_3 CH_3

Camphenilone **Camphene**

Unfortunately, normal methods of dehydration of the resulting tertiary alcohol are unsuccessful. Acid treatment and other attempts to convert the alcohol into a better leaving group lead to a tertiary carbocation that is very prone to structural rearrangement.

Propose a reaction sequence that converts this tertiary alcohol into camphene, avoiding conditions that would lead to carbocation rearrangement.

19 The following outline is part of a synthesis of chrysanthemic acid, a naturally occurring insecticide. Several compounds in the sequence are pictured, separated by arrows. Suggest reasonable reaction sequences (one or more steps may be required) for each transformation.

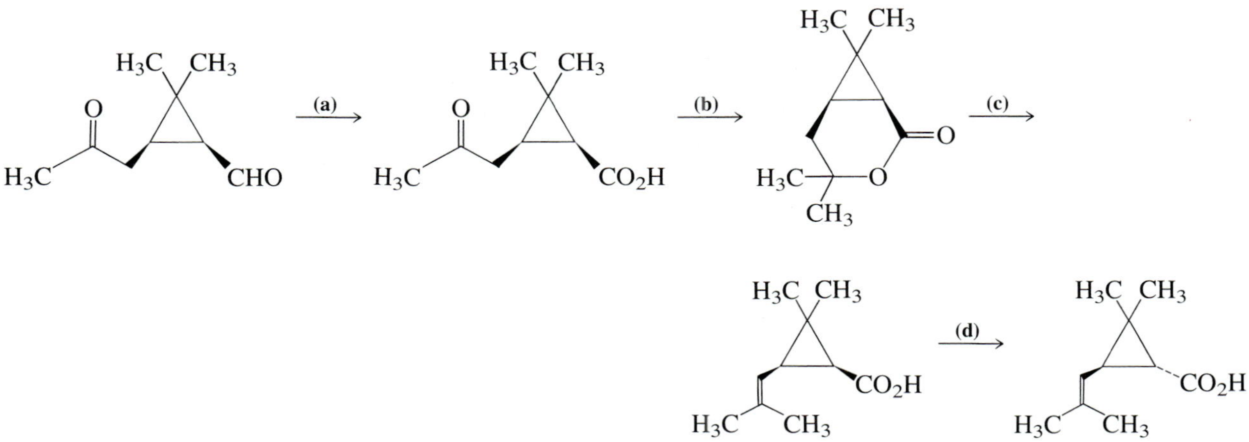

H_3C CH_3 O H_3C CHO (a) → H_3C CH_3 O H_3C CO_2H (b) → H_3C CH_3 =O H_3C O CH_3 (c) →

H_3C CH_3 CO_2H H_3C CH_3 (d) → H_3C CH_3 CO_2H H_3C CH_3

Chrysanthemic acid

20 Show how each of the following molecules can be prepared by a Claisen condensation.

(a) $CH_3CH_2C(=O)CH(CH_3)C(=O)OCH_2CH_3$

(b) $(CH_3)_2CHCH(C(=O)CH_2CH(CH_3)_2)CO_2CH_2CH_3$

(c) $C_6H_5CH_2C(=O)CH(C_6H_5)CO_2CH_2CH_3$

(d) (cyclopentyl)$CH_2C(=O)CH$(cyclopentyl)$C(=O)OCH_2CH_3$

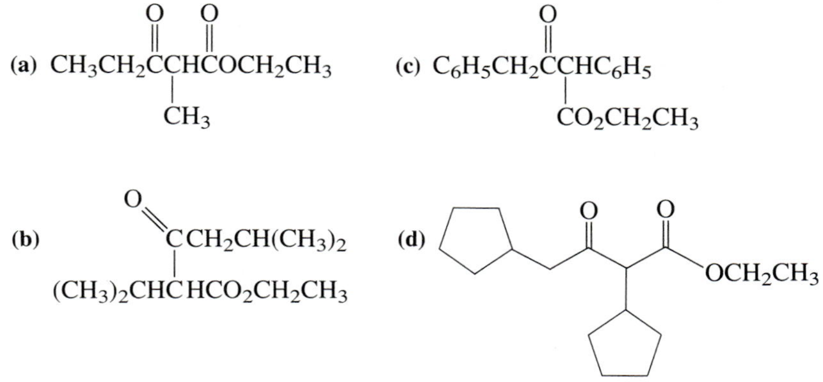

21 Would you expect a Claisen condensation between the enolate ion of one ester and the carbonyl group of another *(crossed Claisen condensation)* in general to be easy to accomplish in good yield? Explain.

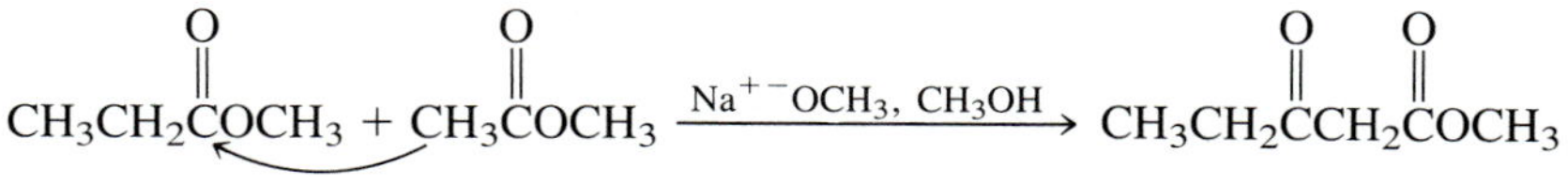

$$CH_3CH_2\overset{O}{\overset{\|}{C}}OCH_3 + CH_3\overset{O}{\overset{\|}{C}}OCH_3 \xrightarrow{Na^{+}\,^{-}OCH_3,\ CH_3OH} CH_3CH_2\overset{O}{\overset{\|}{C}}CH_2\overset{O}{\overset{\|}{C}}OCH_3$$

22 Crossed Claisen condensations (Problem 21) are useful for 3-keto ester synthesis if one of the esters lacks α-hydrogens. Why? Using crossed Claisen condensations, show how you would prepare each of the following 3-keto esters.

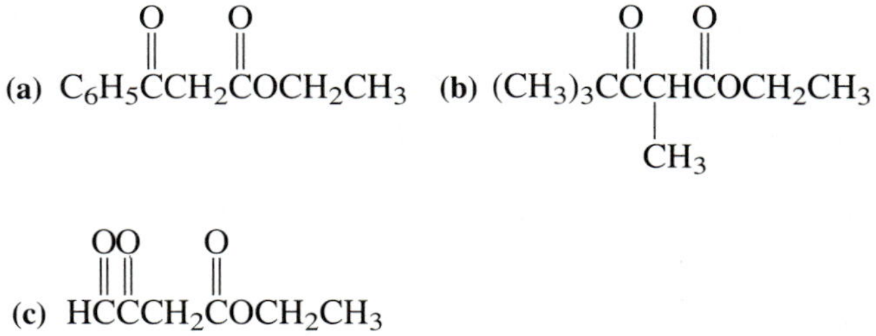

(a) $C_6H_5C(=O)CH_2C(=O)OCH_2CH_3$

(b) $(CH_3)_3CC(=O)CH(CH_3)C(=O)OCH_2CH_3$

(c) $HC(=O)C(=O)CH_2C(=O)OCH_2CH_3$

23 Propose a synthetic scheme for the conversion of lactone (i) into amine (ii), a precursor to the naturally occurring monoterpene (iii).

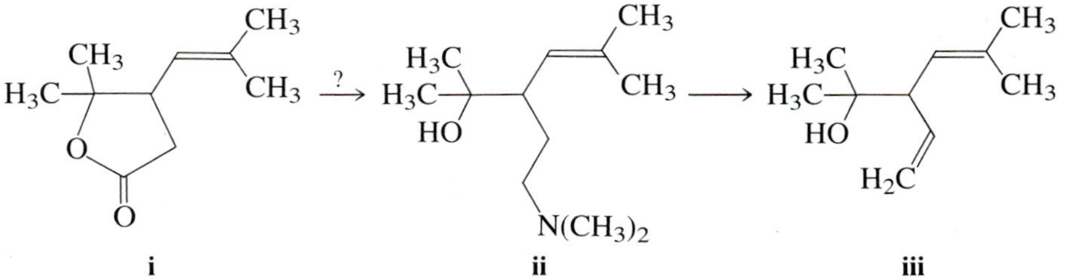

i $\xrightarrow{?}$ ii $\longrightarrow$ iii

i: lactone bearing H_3C, CH_3 and a $CH{=}C(CH_3)_2$ group; ii: $(H_3C)_2C(HO)$–, $CH{=}C(CH_3)_2$, $CH_2CH_2N(CH_3)_2$; iii: $(H_3C)_2C(HO)$–, $CH{=}C(CH_3)_2$, $CH{=}CH_2$ (H_2C)

24 Show how you might synthesize chlorpheniramine, a powerful antihistamine used in several decongestants, from each of carboxylic acids (i) and (ii) shown below. Use a different carboxylic amide in each synthesis.

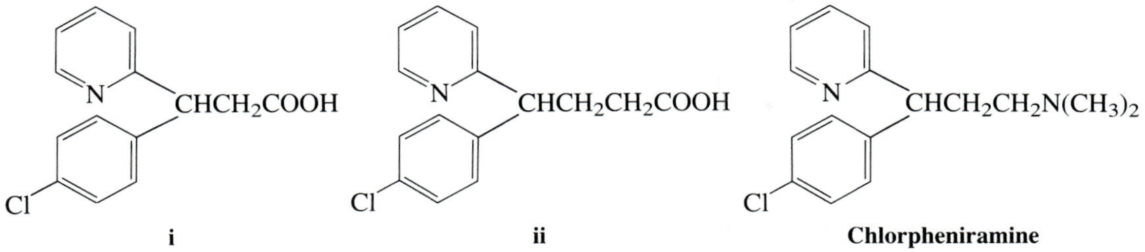

$CHCH_2COOH$ (i) $CHCH_2CH_2COOH$ (ii) $CHCH_2CH_2N(CH_3)_2$

i ii **Chlorpheniramine**

25 Write the intermediate products and explain each reaction step in the following conversion of cyclohexanone into 1-cyclohexenecarboxylic acid.

KCN, $(CH_3\overset{O}{\overset{\|}{C}})_2O$ → **(a)** 600°C → **(b)** 1. KOH, H_2O, Δ 2. H^+, H_2O →

1-Cyclohexene-carboxylic acid

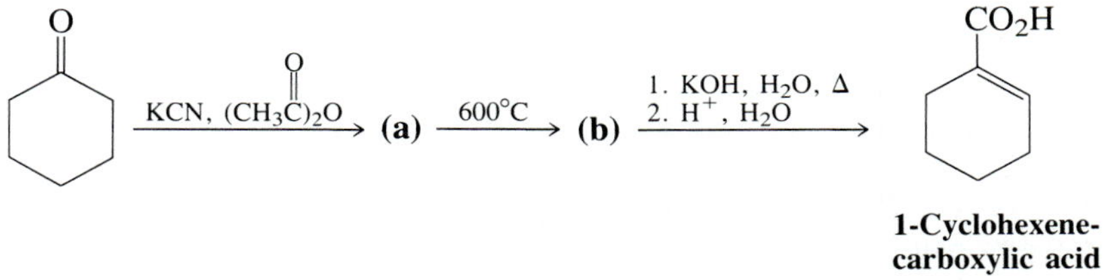

26 Propose a synthesis of β-selinene, a member of a very common family of sesquiterpenes, beginning with the alcohol shown here. Use a nitrile in your synthesis.

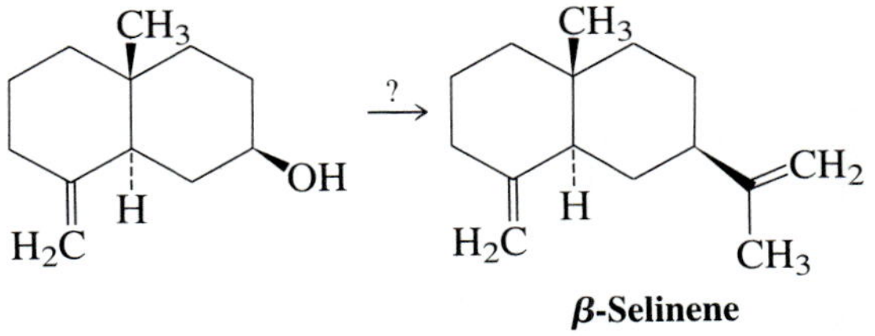

β-Selinene

27 Write the structure of the product of the first of the following reactions, and then propose a scheme that will ultimately convert it into the methyl-substituted ketone at the end of the scheme. This example illustrates a common method for introduction of "angular methyl groups" into synthetically prepared steroids. Note: It will be necessary to protect the ketone carbonyl.

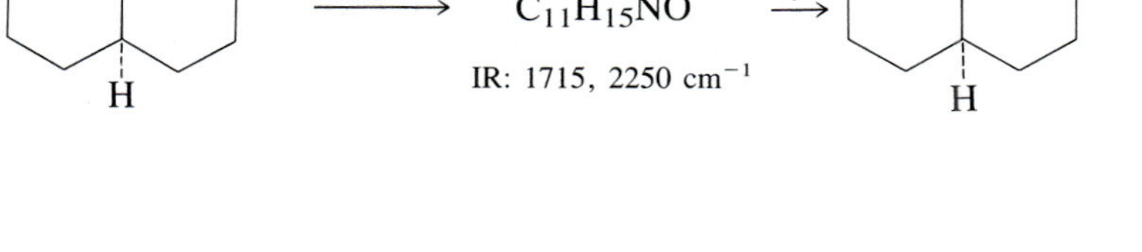

HCN → $C_{11}H_{15}NO$?→

IR: 1715, 2250 cm^{-1}

28 **(a)** The Dieckmann reaction is the intramolecular variant of the Claisen condensation. Show how keto esters (i) and (ii) may be formed by this reaction.

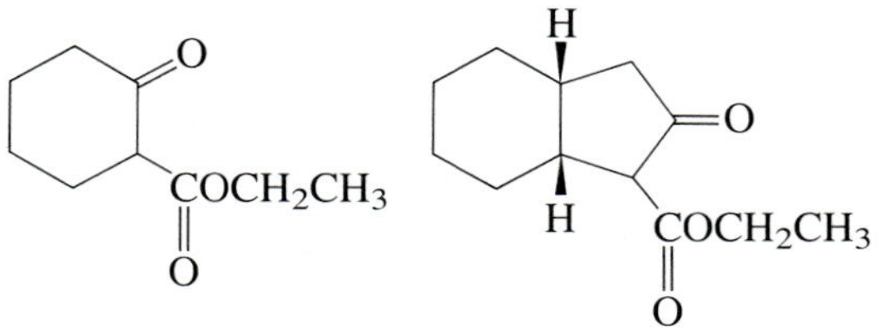

i **ii**

(b) A related condensation of dinitriles (the Thorpe-Ziegler reaction) is shown here. Formulate a detailed mechanism for this synthesis of cyclic α-cyano ketones.

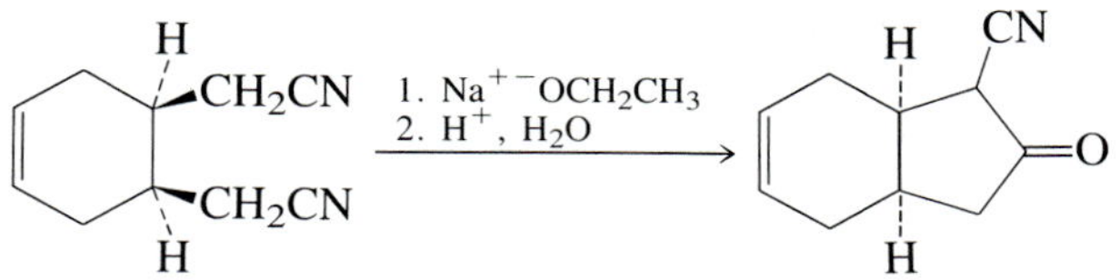

1. $Na^{+}\,{}^{-}OCH_2CH_3$
2. H^+, H_2O

29 Assign as many peaks as you can in the mass spectrum of 1-bromopropane (Figure 18-6).

30 The following table lists selected mass-spectral data for three isomeric alcohols with the formula $C_5H_{12}O$. On the basis of the peak positions and intensities, suggest structures for each of the three isomers. A dash means that the peak is very weak or absent entirely.

Relative peak intensities

m/z	Isomer A	Isomer B	Isomer C
88 M^+	–	–	–
87 $(M-1)^+$	2	2	–
73 $(M-15)^+$	–	7	55
70 $(M-18)^+$	38	3	3
59 $(M-29)^+$	–	–	100
55 $(M-15-18)^+$	60	17	33
45 $(M-43)^+$	5	100	10
42 $(M-18-28)^+$	100	4	6

31 Following are spectroscopic and analytical characteristics of two unknown compounds. Propose a structure for each compound.

(a) Analysis: 74.94% C, 12.58% H (remainder is O).
^{1}H NMR: δ = 0.90(t, 3 H), 1.0–1.6(m, 8 H), 2.05(s, 3 H), and 2.25(t, 2 H) ppm.
IR: 1715 cm^{-1}
UV: $\lambda_{max}(\epsilon)$ = 280(15) nm.

MS: m/z for molecular ion is 128; intensity of $(M + 1)^+$ peak is 9% of $M^{+\cdot}$ peak; important fragments are at $m/z = 113$ $(M - 15)^+$, $m/z = 85$ $(M - 43)^+$, $m/z = 71$ $(M - 57)^+$, $m/z = 58$ $(M - 70)^+$ (the second largest peak), and $m/z = 43$ $(M - 79)^+$ (the base peak).

(b) Analysis: 88.16% C, 11.84% H.
^{1}H NMR: spectrum B.
^{13}C NMR (undecoupled from H): $\delta = 20.5$(q), 23.8(q), 28.0(t), 30.6(t), 30.9(t), 41.2(d), 108.4(t), 120.8(d), 133.2(s), and 149.7(s) ppm.
IR: significant bands at 3060 (medium), 3010 (medium), 1680 (weak), 1646 (medium), and 880 (very strong) cm^{-1}.
UV: $\lambda_{max} < 200$ nm.
MS: m/z for molecular ion is 136; intensity of $(M + 1)^+$ peak is 11% of $M^{+\cdot}$ peak; important fragments are at $m/z = 121$ $(M - 15)^+$, $m/z = 95$ $(M - 41)^+$, $m/z = 68$ $(M - 68)^+$ (the base peak), and $m/z = 41$ $(M - 95)^+$.

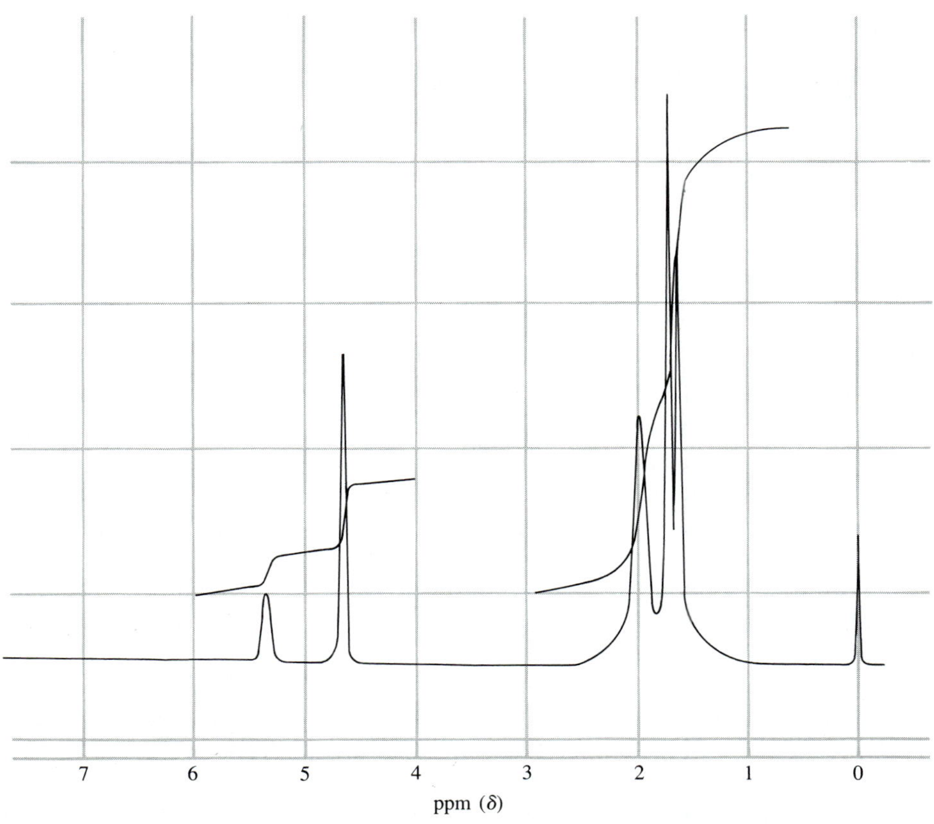

B

CHAPTER 19

The Special Stability of the Cyclic Electron Sextet

BENZENE AND ELECTROPHILIC AROMATIC SUBSTITUTION

Proposed Benzene Structures

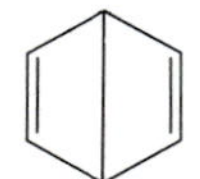

Dewar benzene

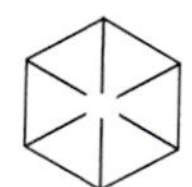

Claus benzene

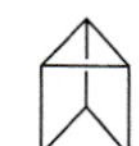

Ladenburg prismane

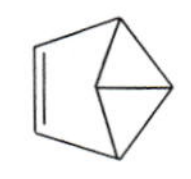

Benzvalene

In 1825, the English scientist Faraday* pyrolyzed whale oil to obtain a colorless liquid (b.p. 80.1°C, m.p. 5.5°C) that had the empirical formula CH. This compound posed a problem to the theory that all carbons had to have four valences to other atoms and was of particular interest because of its unusual stability and chemical inertness (Section 14-4). The molecule was named **benzene,** and it was eventually established that the molecular formula was C_6H_6. Various investigators proposed many incorrect structures, such as Dewar benzene, Claus benzene, Ladenburg prismane, and benzvalene. In fact, since then, Dewar benzene, prismane, and benzvalene (but not Claus benzene) have been synthesized. These compounds are unstable and isomerize to benzene in very exothermic transformations. It was Kekulé† who suggested, in 1865, that benzene should be viewed as a set of equilibrating cyclohexatriene *isomers*. We now know that this notion was not quite right. In terms of modern electronic theory, benzene is best described by two equivalent cyclohexatrienic *resonance* structures (Section 14-4).

This chapter introduces the physical and chemical properties of benzene. It begins with the system of naming substituted benzenes, then looks at the electronic and molecular structure of the parent molecule and reviews some of the evidence for a special stabilizing energy, the aromaticity of benzene. This aromaticity and the special structure of benzene affect its spectral properties. Fi-

---

*Professor Michael Faraday, 1791–1867, Royal Institution of Chemistry, London.

†Professor F. August Kekulé, see Section 1-4.

nally, the chapter elaborates on the mechanism of electrophilic aromatic substitution, a class of reactions introduced briefly in Section 14-4 that affords a ready synthetic path to substituted benzenes.

19-1 The System for Naming Benzenes

Because of their strong aroma, many derivatives of benzene are called **aromatic compounds** (Section 14-4). Therefore, benzene, even though its odor is not particularly pleasant, is viewed as the ''parent'' aromatic molecule. A number of fragrant aromatic compounds have common names, many of which refer to their natural sources. Several such common names have been accepted by IUPAC. As before, a consistent logical naming of these compounds will be adhered to as much as possible, with common names mentioned in parentheses.

Some Aromatic Flavoring Agents

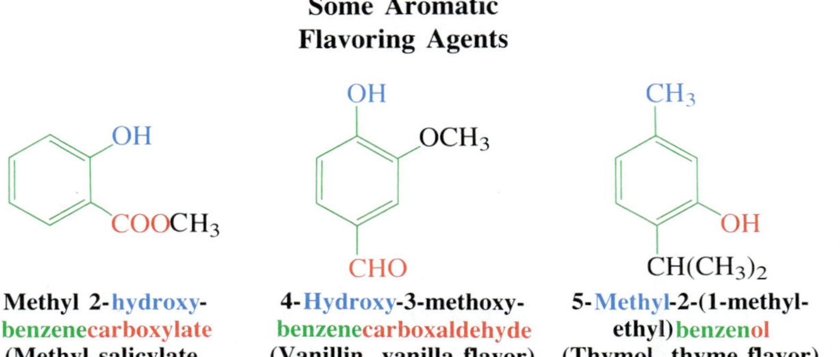

Methyl 2-hydroxybenzenecarboxylate (Methyl salicylate, oil of wintergreen flavor)

4-Hydroxy-3-methoxybenzenecarboxaldehyde (Vanillin, vanilla flavor)

5-Methyl-2-(1-methylethyl)benzenol (Thymol, thyme flavor)

The generic term for substituted benzenes is **arene.** An arene as a substituent is referred to as an **aryl group,** abbreviated **Ar.** The parent aryl substituent is **phenyl,** C_6H_5 (Section 14-4).

Whenever the symbol for the benzene ring with its three double bonds is written, it should be understood to represent only one of a pair of resonance structures. Alternatively, the ring is sometimes drawn as a regular hexagon with an inscribed circle.

is the same as

Many monosubstituted benzenes are named by simply adding a substituent prefix to the word benzene:

Fluorobenzene **Nitrobenzene** **(1,1-Dimethylethyl)benzene (*tert*-Butylbenzene)**

There are three possible arrangements of disubstituted benzenes. These are designated by the prefixes **1,2-** (**ortho-,** or ***o*-,** Greek, straight) for adjacent substituents, **1,3** (**meta-,** or ***m*-,** Greek, transposed) for 1,3-disubstitution, and **1,4** (**para-,** or ***p*-,** Greek, beyond) for 1,4-disubstitution. The substituents are listed in alphabetical order.

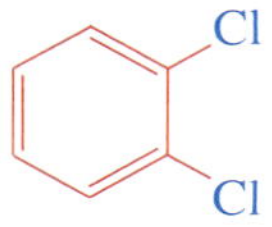

1,2-Dichlorobenzene (*o*-Dichlorobenzene)

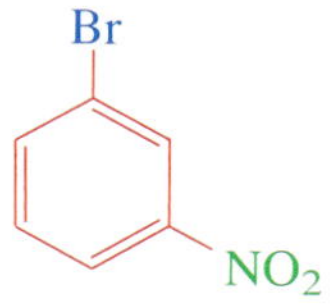

1-Bromo-3-nitrobenzene (*m*-Bromonitrobenzene)

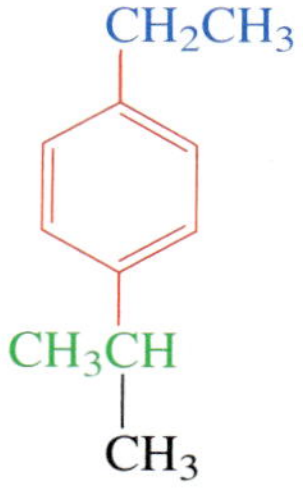

1-Ethyl-4-(1-methylethyl)benzene (*p*-Ethylisopropylbenzene)

To name tri- and more-highly-substituted derivatives, the six carbons of the ring are numbered using the lowest set of locants, and the substituents are labeled accordingly, as in cyclohexane nomenclature.

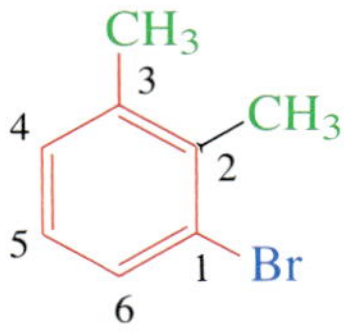

1-Bromo-2,3-dimethylbenzene

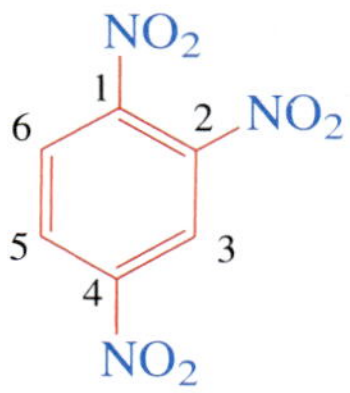

1,2,4-Trinitrobenzene

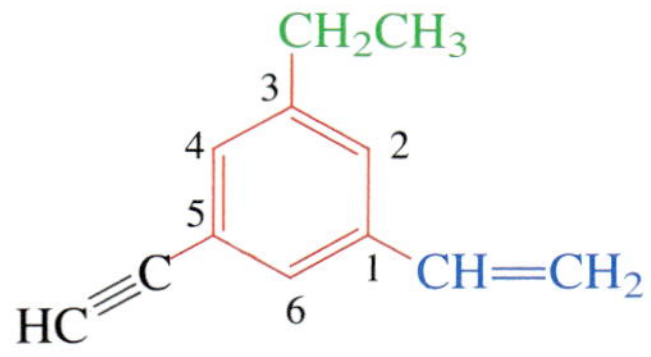

1-Ethenyl-3-ethyl-5-ethynylbenzene

The following series of benzene derivatives will be encountered in this book. Note that the presence of functional groups of higher priority than the aromatic ring is revealed in their names (e.g., benzenol rather than hydroxybenzene; benzenecarboxylic acid rather than hydroxycarbonylbenzene).

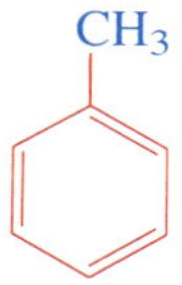

Methylbenzene (Toluene)

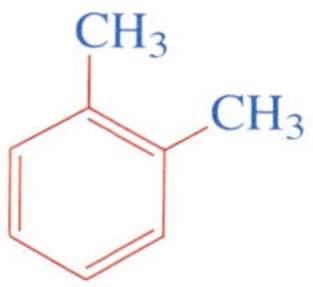

1,2-Dimethylbenzene (*o*-Xylene)

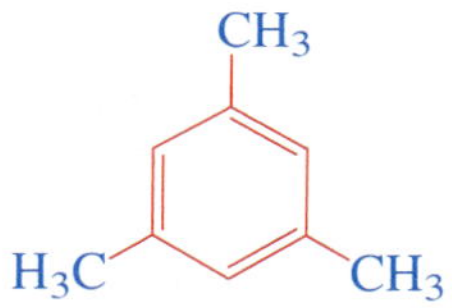

1,3,5-Trimethylbenzene (Mesitylene)

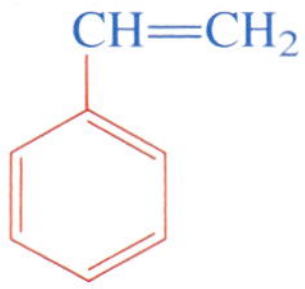

Ethenylbenzene (Styrene)

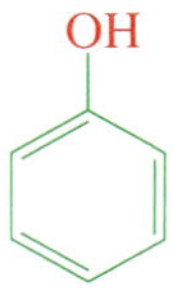

Benzenol (Phenol)

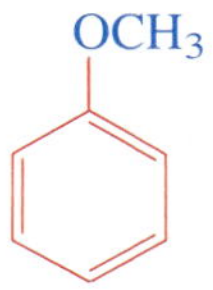

Methoxybenzene (Anisole)

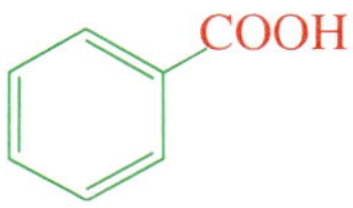

Benzenecarboxylic acid (Benzoic acid)

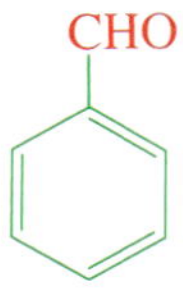

Benzenecarboxaldehyde (Benzaldehyde)

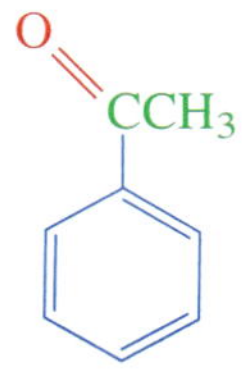

1-Phenylethanone (Acetophenone)

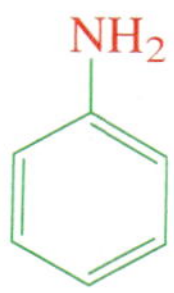

Benzenamine (Aniline)

Ring-substituted derivatives of these compounds are named by numbering the ring positions or by using the prefixes *o*-, *m*-, and *p*-. The substituent that gives the compound its base name is placed at carbon 1.

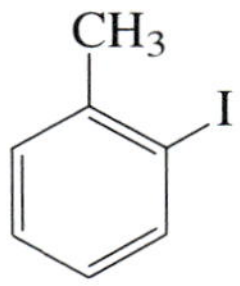

1-Iodo-2-methylbenzene (*o*-Iodotoluene) **2,4,6-Tribromobenzenol (2,4,6-Tribromophenol)** **1-Bromo-3-ethenylbenzene (*m*-Bromostyrene)**

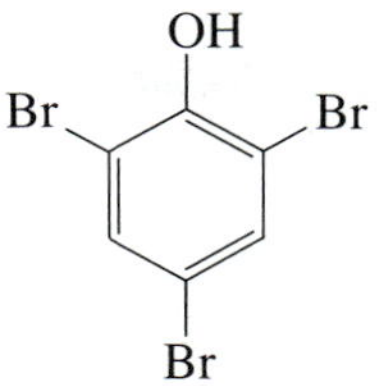

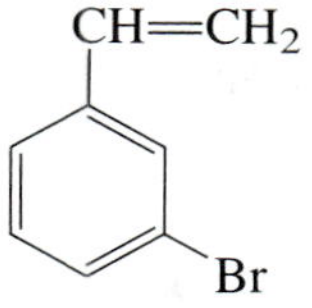

The C_6H_5 substituent, when attached to a chain carrying functional groups of higher priority or when bound to a complex molecule, is labeled as phenyl. The $C_6H_5CH_2$— group, which is related to the 2-propenyl (allyl) substituent (Section 24-1), is called **phenylmethyl (benzyl).**

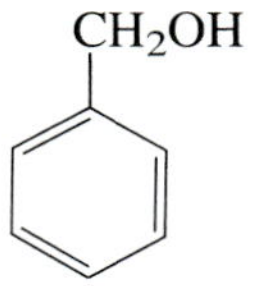

CH_2OH

Phenylmethanol (Benzyl alcohol)

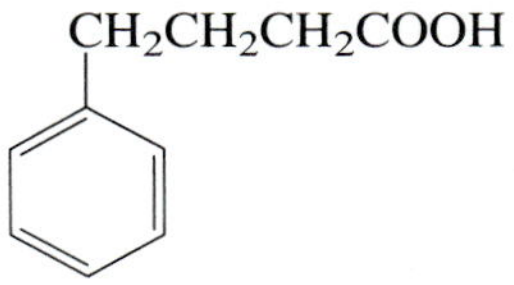

$CH_2CH_2CH_2COOH$

4-Phenylbutanoic acid

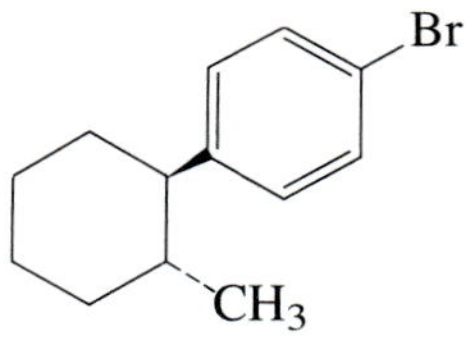

trans-1-(4-Bromophenyl)-2-methylcyclohexane

EXERCISE 19-1

Write systematic and common names of the following substituted benzenes:

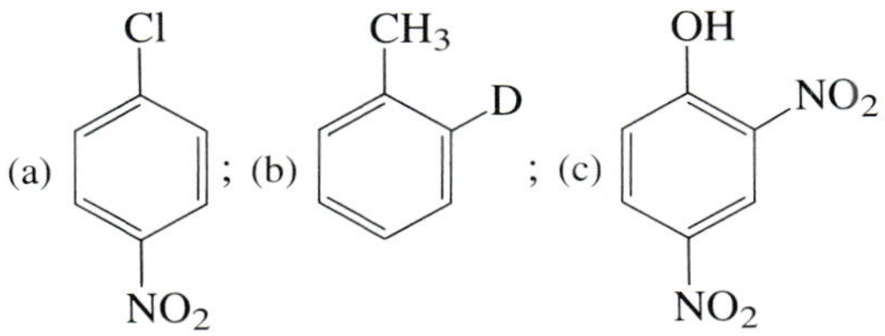

(a) ; (b) ; (c)

EXERCISE 19-2

Draw the structures of the following molecules: (a) (1-methylbutyl)benzene; (b) 1-ethenyl-4-nitrobenzene (*p*-nitrostyrene); (c) 2-methyl-1,3,5-trinitrobenzene (2,4,6-trinitrotoluene—the explosive TNT).

EXERCISE 19-3

The following names are incorrect. Write the correct form: (a) 3,5-dichlorobenzene; (b) *o*-aminophenyl fluoride; (c) *p*-fluorobromobenzene.

In summary, simple singly substituted benzenes are named by placing the substituent name before the word benzene. For more highly substituted systems, 1,2-, 1,3-, and 1,4- or *ortho, meta,* and *para* prefixes indicate the positions of disubstitution or the ring is numbered and the so-labeled substituents are named in alphabetical order. Many simple substituted benzenes have common names.

19-2 The Structure of Benzene: A First Look at Aromaticity

Benzene is unusually unreactive; it does not undergo the addition reactions common to alkenes (Section 14-4). You will see in this section why that is so: the cyclic six-electron arrangement confers a special stability. The extra energy

FIGURE 19-1

Orbital picture of the bonding in benzene. The σ framework is depicted as straight lines except for the bonding to one carbon, in which the p orbital and the sp^2 hybrids are shown explicitly.

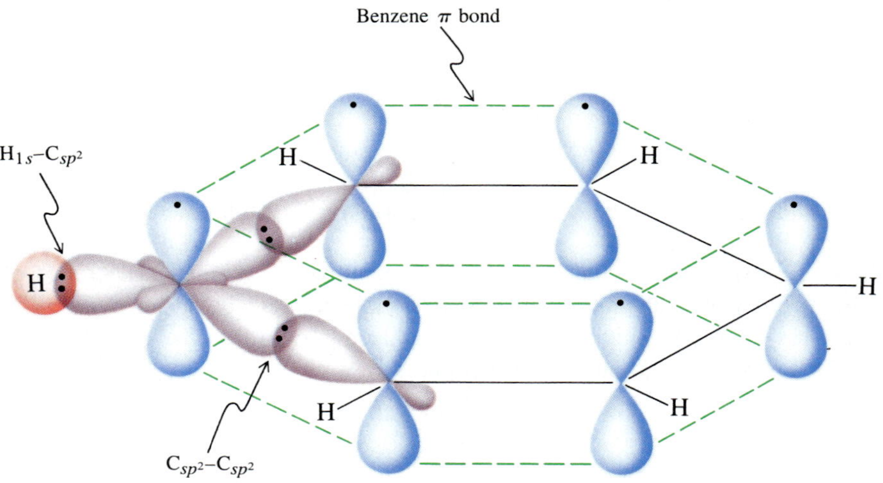

FIGURE 19-2

The six overlapping p orbitals in benzene form a π electron cloud located above and below the molecular plane.

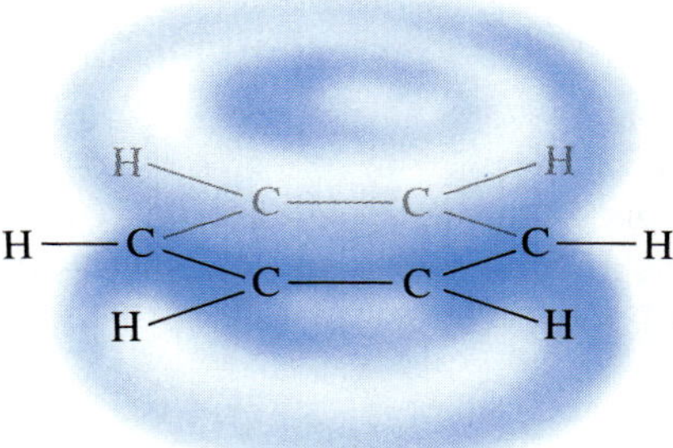

gained by delocalizing six π electrons in this way is called the resonance energy of benzene. An estimate of its size is obtained by comparing the heat of hydrogenation of benzene with those of model systems, such as 1,3-cyclohexadiene, devoid of cyclic conjugation.

The Benzene Ring Contains Six Equally Overlapping p Orbitals

The electronic structure of the benzene ring is shown in Figure 19-1. All carbons are sp^2 hybridized, and each p orbital overlaps to an equal extent with its two neighbors. The consequently delocalized electrons form a π *cloud* above and below the ring (Figure 19-2).

According to this picture, the benzene molecule should be a completely symmetrical hexagon with equal C–C bond lengths. Such is in fact the experimentally determined structure (Figure 19-3), revealing the absence of alternation between single and double bonds. This type of alternation would have been expected if benzene were a conjugated triene, a "cyclohexatriene." The C–C bond length in benzene is 1.39 Å, between the values found for the single (1.47 Å) and the double bond (1.34 Å) in 1,3-butadiene (Figure 14-8).

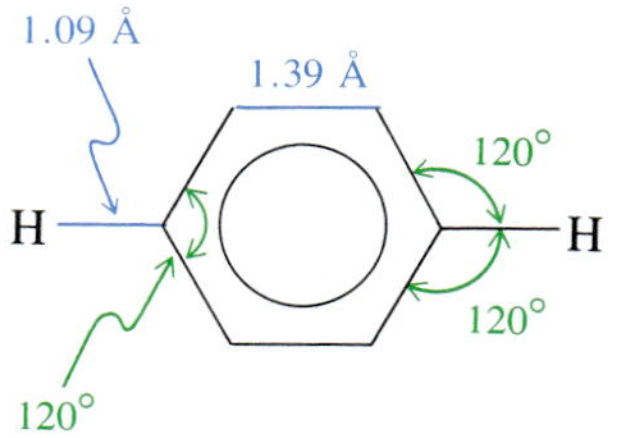

FIGURE 19-3

The molecular structure of benzene. All six C–C bonds are equal; all bond angles are 120°.

Benzene Is Especially Stable: Heats of Hydrogenation

A way to establish the relative stability of a series of alkenes is to measure their heats of hydrogenation (Sections 11-4 and 14-3). We may carry out a similar experiment with benzene, relating its heat of hydrogenation to those of 1,3-cyclohexadiene and cyclohexene. These molecules are conveniently compared because hydrogenation changes all three into cyclohexane.

The hydrogenation of cyclohexene is exothermic by -28.6 kcal mole^{-1}, a value expected for the hydrogenation of a cis double bond (Section 11-4). The heat of hydrogenation of 1,3-cyclohexadiene ($\Delta H^\circ = -54.9$ kcal mole^{-1}) is slightly less than double that of cyclohexene, because of the resonance stabilization in a conjugated diene (Section 14-3); in this case, $2 \times (28.6) - 54.9 = 2.3$ kcal mole^{-1}.

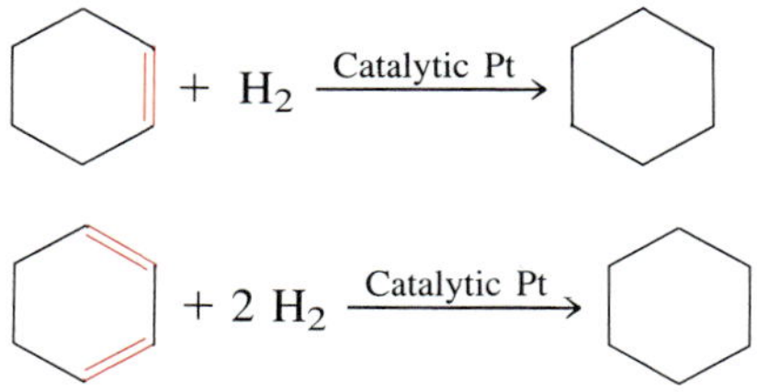

$$\text{cyclohexene} + H_2 \xrightarrow{\text{Catalytic Pt}} \text{cyclohexane} \qquad \Delta H^\circ = -28.6 \text{ kcal mole}^{-1}$$

$$\text{1,3-cyclohexadiene} + 2\,H_2 \xrightarrow{\text{Catalytic Pt}} \text{cyclohexane} \qquad \Delta H^\circ = -54.9 \text{ kcal mole}^{-1}$$

What do we expect for the heat of hydrogenation of benzene? In the absence of any special resonance effects, the molecule can be viewed as having three cyclohexene-type double bonds, all in conjugation but alternating with single bonds. (This is accentuated in the structure at the left below in which the regular hexagon usually written for benzene is distorted.) If the resonance effect in benzene were similar to that in 1,3-cyclohexadiene, we could estimate its ΔH° of hydrogenation as follows:

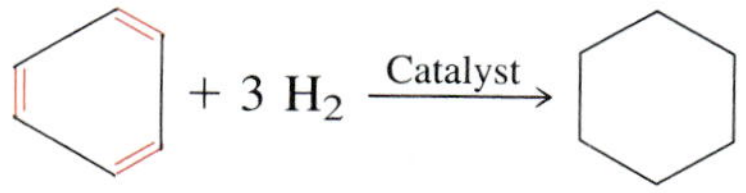

$$\text{cyclohexatriene} + 3\,H_2 \xrightarrow{\text{Catalyst}} \text{cyclohexane} \qquad \Delta H^\circ = ?$$

$$\begin{aligned}\Delta H^\circ &= 3\,(\Delta H^\circ \text{ of hydrogenation of cyclohexene}) + 3\,(\text{resonance correction in 1,3-cyclohexadiene})\\ &= (3 \times -28.6) + (3 \times 2.3) \text{ kcal mole}^{-1}\\ &= -85.8 + 6.9 \text{ kcal mole}^{-1}\\ &= -78.9 \text{ kcal mole}^{-1}\end{aligned}$$

Now let us look at the experimental data. Although benzene is hydrogenated only with difficulty (Section 14-4), special catalysts carry out this reaction, and so the heat of hydrogenation of benzene can be measured: $\Delta H^\circ = -49.3$ kcal mole^{-1}, much less than the -78.9 kcal mole^{-1} calculated.

Figure 19-4 summarizes the results of the heats-of-hydrogenation determinations in the three experiments and compares them with the calculated estimate for a hypothetical cyclohexatriene. It is immediately apparent that benzene is much more stable than a cyclic triene containing alternating single and double

FIGURE 19-4

The heats of hydrogenation of cyclohexene, 1,3-cyclohexadiene, the hypothetical 1,3,5-cyclohexatriene (calculated), and benzene. All four processes lead to the same product, cyclohexane. The resonance energy for benzene is estimated to be 29.6 kcal mole^{-1}.

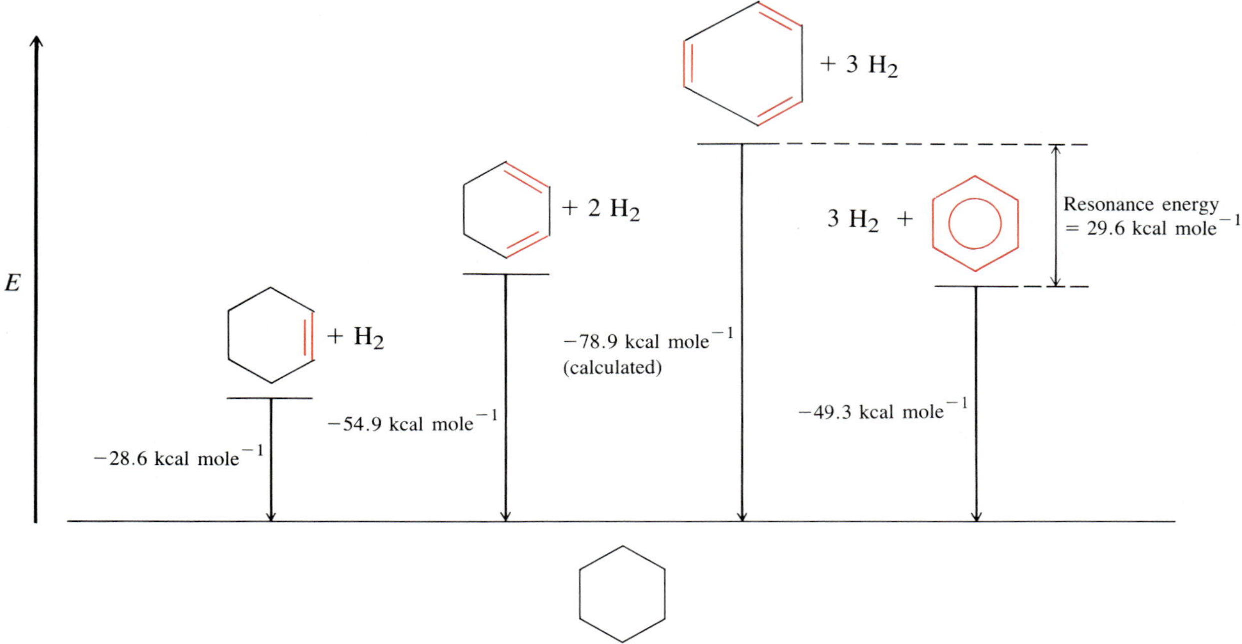

bonds. The difference in the heats of hydrogenation (and in the heats of formation) is about 30 kcal mole^{-1} and is called the **resonance energy** of benzene. Other terms used are *delocalization energy, aromatic stabilization,* or simply the **aromaticity** of benzene. The original meaning of this word has changed quite a bit with time, now referring to a thermodynamic property rather than to odor.

The Molecular Orbitals of Benzene

The six overlapping *p* orbitals in a cyclic arrangement give rise to a set of six molecular orbitals for benzene. Figure 19-5 compares these with the molecular orbitals derived in a similar manner for the open-chain analog, 1,3,5-hexatriene. As in 1,3-butadiene (Figure 14-9), molecular orbitals of increasing energy have an increasing number of nodes. Because of benzene's symmetry, a single node or two nodes can be added each in two possible ways, giving rise to pairs of molecular orbitals of equal energy (designated $\psi_{2,3}$ and $\psi_{4,5}$). These are said to be *degenerate* and constitute the highest occupied and lowest unoccupied orbitals of benzene. On the other hand, 1,3,5-hexatriene is described by a set of six distinct (nondegenerate) molecular orbitals. The energy levels of the molecular orbitals for both systems are depicted in Figure 19-6 (compare Figure 14-10). Each system contains six *p* electrons, giving rise to three filled bonding molecular orbitals. Benzene is more stable than 1,3,5-hexatriene in this picture, because two of the three filled molecular orbitals are of lower energy.

FIGURE 19-5

Molecular orbitals of benzene (A) compared with those of 1,3,5-hexatriene (B). The orbitals are shown at equal size for simplicity. Favorable overlap (bonding) takes place between orbital lobes of equal sign. A sign change is indicated by a node (dashed line). As the number of nodes increases, so does the energy of the orbitals. Note that benzene has two degenerate sets of orbitals, the lower-energy one occupied (ψ_2, ψ_3), the other not (ψ_4, ψ_5).

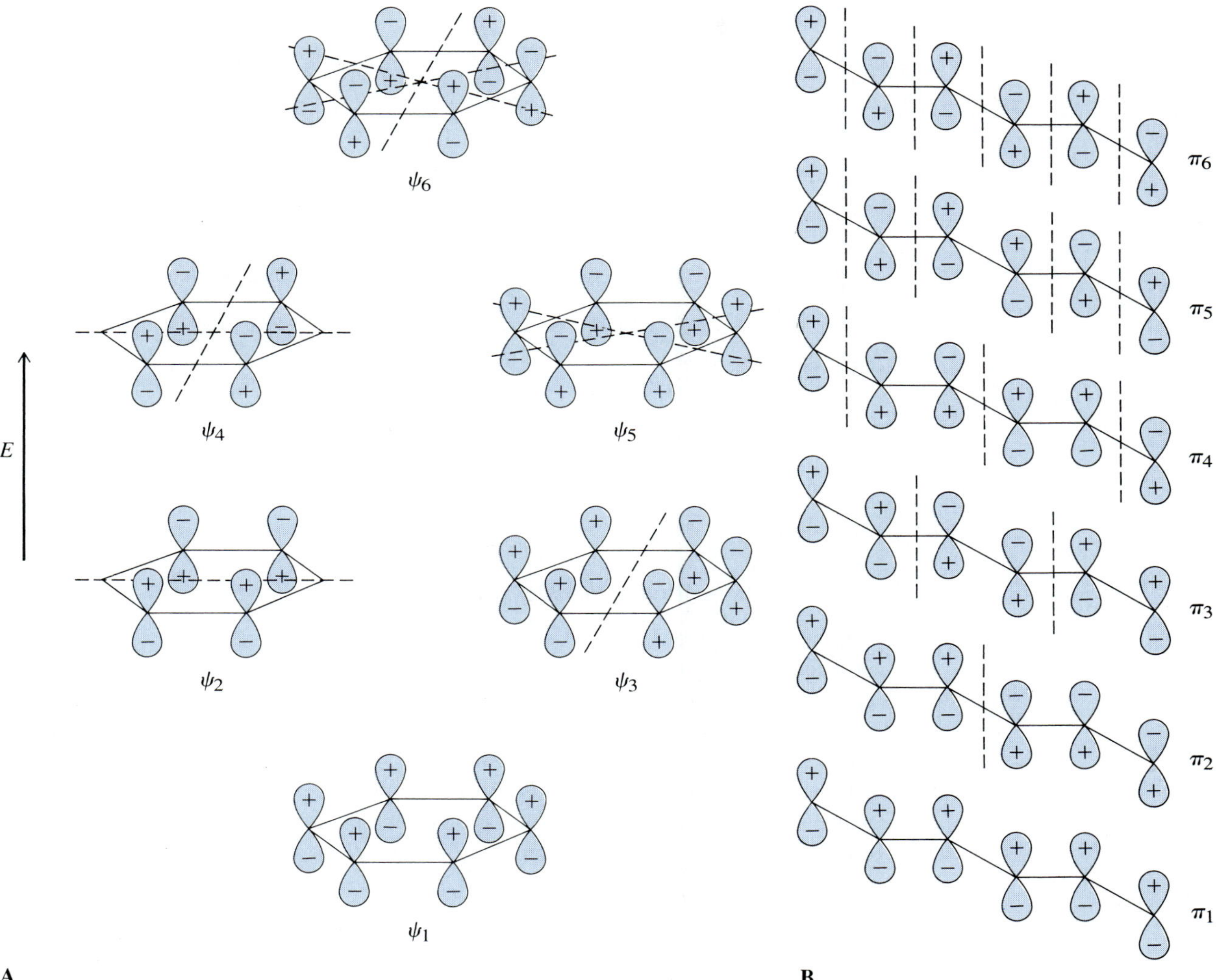

FIGURE 19-6

Energy levels of the molecular orbitals in benzene and 1,3,5-hexatriene. In both systems, the π electrons occupy only bonding molecular orbitals. In benzene, two filled molecular orbitals are lower in energy and one is higher than the corresponding orbitals in 1,3,5-hexatriene. The net result is a more-stable system of six π electrons.

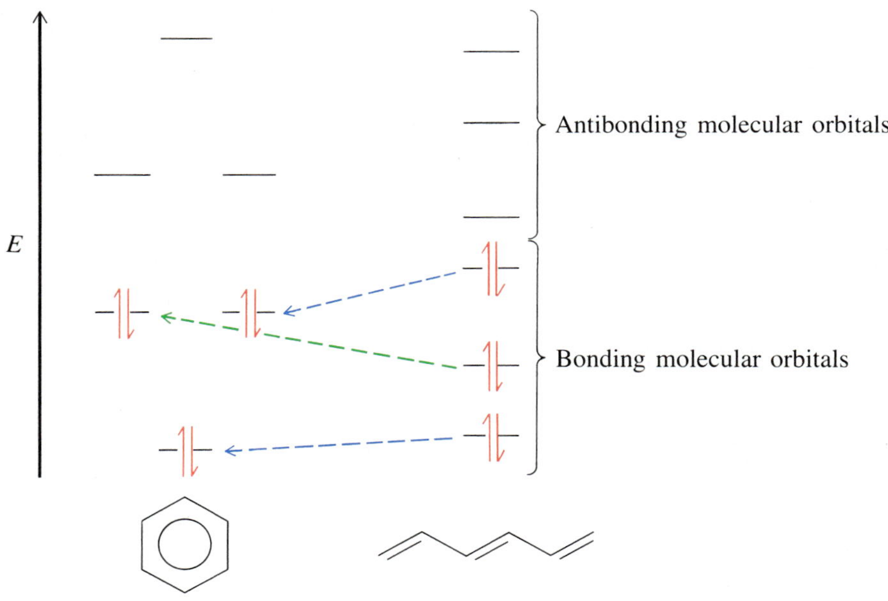

Some Reactions Have Aromatic Transition States

The structure of benzene accounts in a simple way for several reactions that proceed by what has seemed to be a complicated, concerted movement of three electron pairs: the Diels-Alder reaction (Section 14-5), ester pyrolysis (Section 18-4), and the McLafferty rearrangement (Section 18-7). In all three processes, there is a transition state with cyclic overlap of six electrons in π orbitals (or orbitals with π character). This electronic arrangement is similar to that in benzene and is energetically more favored than the alternative, sequential bond breaking and bond making. Transition states of this type are called aromatic.

Aromatic Transition States

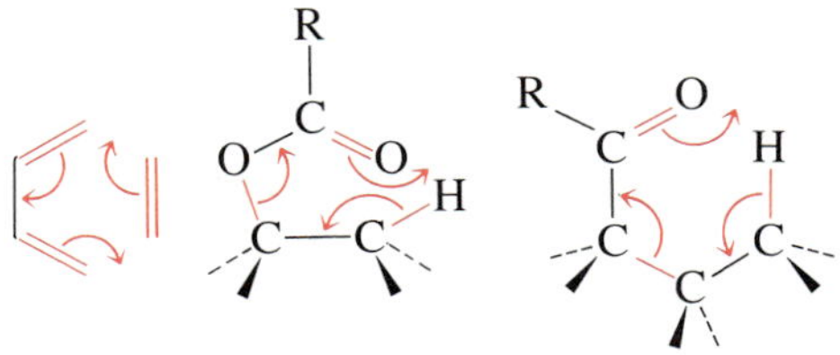

EXERCISE 19-4

If benzene were a cyclohexatriene, 1,2-dichloro- and 1,2,4-trichlorobenzene should exist as two isomers. Draw them.

EXERCISE 19-5

The thermal ring-opening of cyclobutene to 1,3-butadiene is exothermic by about 10 kcal $mole^{-1}$ (Section 14-5). On the other hand, the same reaction for benzocyclobutene, A, to B is *endothermic* by the same amount. Explain.

A

$\rightleftharpoons$

B

$\Delta H° \sim +10$ kcal $mole^{-1}$

To summarize, the structure of benzene is a regular hexagon made up of six sp^2-hybridized carbons. The aromatic C–C bond length is between those of a single and a double bond. The electrons occupying the p orbitals form a π cloud above and below the plane of the ring. The structure of benzene can be represented by two equally contributing cyclohexatriene resonance structures. Hydrogenation of benzene to cyclohexane releases about 30 kcal $mole^{-1}$ less energy than would be expected from the heats of hydrogenation of nonaromatic models. This difference is called the resonance energy of benzene. The molecular-orbital picture of benzene consists of three bonding and three antibonding orbitals. Two of the orbitals in each set are degenerate.

19-3 The Spectral Characteristics of Benzene

To what extent are cyclic conjugation and its related resonance energy manifest in the spectra of benzene and its derivatives? This section shows that the special electronic arrangement in these molecules gives rise to characteristic ultraviolet spectra, that the hexagonal structure is manifest in specific infrared bands, and, what is most striking, cyclic delocalization causes induced ring currents in NMR spectroscopy, resulting in unusually large deshielding of hydrogens attached to the aromatic ring. Moreover, the coupling constants between 1,2- *(ortho),* 1,3- *(meta),* and 1,4- *(para)* hydrogens in substituted benzenes are different and diagnostic of the substitution pattern.

The Electronic Spectrum of Benzene

Ultraviolet and visible spectroscopy is based on the ability of light of certain energy to effect electronic excitations—that is, to promote electrons from filled molecular orbitals to unfilled ones (Section 14-7). The rules that govern the relative ease with which such transitions occur are complicated. To a first approximation, however, we can assume that the band of longest wavelength corresponds to the lowest-energy electronic transition between the highest occupied molecular orbital and its lowest unoccupied neighbor. Figure 19-6 shows that this distance is smaller for hexatriene than for benzene; this is frequently true of aromatic compounds compared with their open-chain nonaromatic analogs. Hence, the λ_{max} values of 1,3,5-hexatriene are at a higher wavelength than those of benzene (Figure 19-7).

The ultraviolet and visible spectra of aromatic compounds vary characteristically with the introduction of substituents; this phenomenon has been exploited in the tailored synthesis of dyes (Section 24-6). Simple substituted benzenes absorb between 250 and 290 nm. For example, the water-soluble 4-aminobenzenecarboxylic acid (p-aminobenzoic acid, or PABA), 4-H_2N-C_6H_4-COOH, has a λ_{max} at 289 nm, with a rather high extinction coefficient of 18,600.

FIGURE 19-7

The ultraviolet spectra of benzene: $\lambda_{max}(\epsilon)$ = 234(30), 238(50), 243(100), 249(190), 255(220), 261(150) nm; and 1,3,5-hexatriene: $\lambda_{max}(\epsilon)$ = 247(33,900), 258(43,700), 268(36,300) nm. The extinction coefficients of the absorptions of 1,3,5-hexatriene are very much larger than those of benzene; therefore the spectrum at the right was taken at lower concentration.

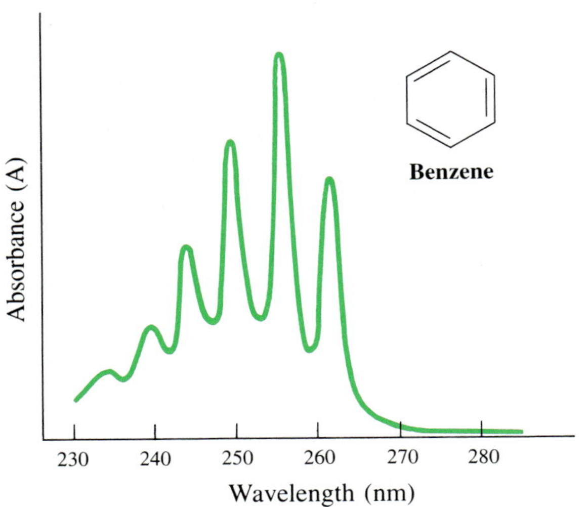

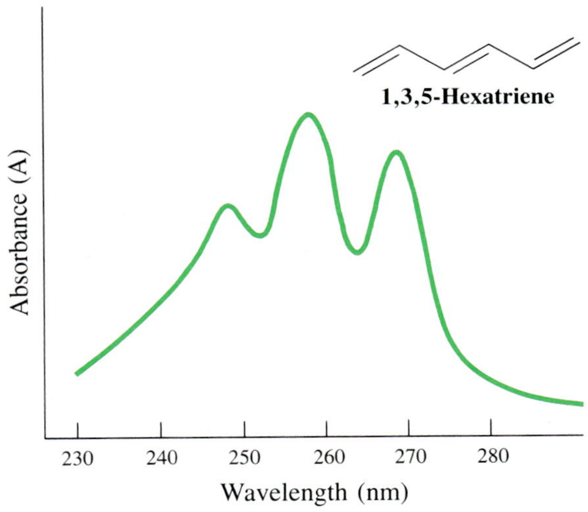

Because of this property (and because it is nontoxic), it is used in suntan lotions, in which it acts as a filter to block out the dangerous ultraviolet radiation emanating from the sun in this wavelength region.

Vibrational Absorptions of the Benzene Ring

The infrared spectra of benzene and its derivatives show characteristic bands in three regions. The first is at 3030 cm^{-1} for the phenyl-hydrogen stretching mode. This value is very close to the wave number of alkenyl-hydrogen stretching, as might be expected for a C_{sp^2}–H bond. The second region ranges from 1500 to 2000 cm^{-1} and records aromatic ring C–C stretching. Finally, a useful set of bands due to C–H out-of-plane bending motions is found between 650 and 1000 cm^{-1}.

Typical Infrared C–H Out-of-plane Bending Vibrations for Substituted Benzenes (cm^{-1})

Monosubstituted (R)	1,2-Disubstituted (R, R)	1,3-Disubstituted (R, R)	1,4-Disubstituted (R, R)
690–710 730–770	735–770	690–710 750–810	790–840

Their precise location is indicative of the specific substitution pattern. For example, 1,2-dimethylbenzene (*o*-xylene) has this band at 738 cm^{-1}, the 1,4

FIGURE 19-8

The infrared spectrum of 1,3-dimethylbenzene (*m*-xylene). There are two C–H stretching absorptions, one due to the aromatic bonds (3030 cm^{-1}), the other to saturated C–H bonds (2920 cm^{-1}). The two bands at 690 and 765 cm^{-1} are typical of 1,3-disubstituted benzenes.

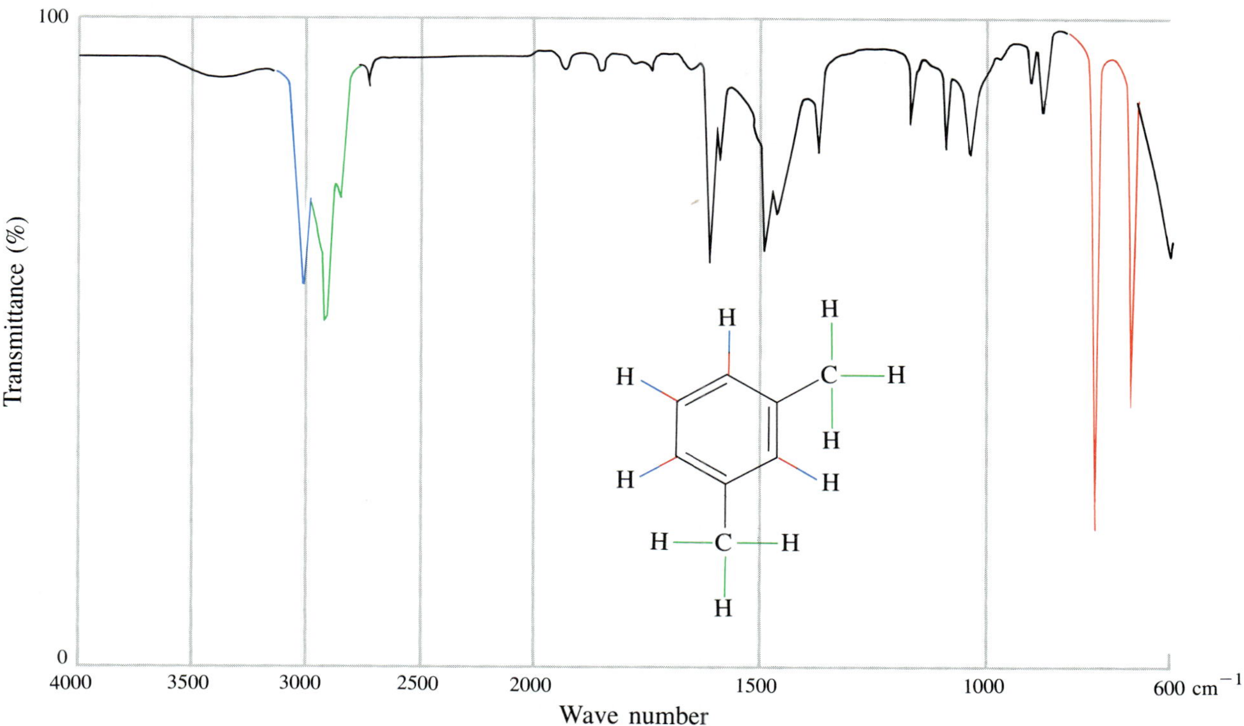

isomer at 793 cm^{-1}, and the 1,3 isomer (Figure 19-8) shows two absorptions in this range, at 690 and 765 cm^{-1}.

Nuclear Magnetic Resonance in Aromatic Compounds

A powerful spectroscopic technique for the identification of benzene and its derivatives is based on ^{1}H NMR. The cyclic delocalization of the aromatic ring gives rise to an unusual deshielding effect, which causes the ring hydrogens to resonate at very low field ($\delta \sim 6.5$–8.5 ppm), even lower than the already rather deshielded alkenyl hydrogens ($\delta \sim 4.6$–5.7 ppm, see Section 11-3).

The ^{1}H NMR spectrum of benzene, for example, exhibits a sharp singlet for the six equivalent hydrogens at $\delta = 7.27$ ppm. How can this strong deshielding be explained? In a simplified picture, the cyclic π system with its delocalized electrons may be compared to a loop of conducting metal. Exposure of such a loop to a perpendicular magnetic field (H_0) causes an electric current (called a ring current) to flow in the loop, in turn generating a new local magnetic field (h_{local}) that opposes H_0 on the inside of the loop (Figure 19-9). On the other hand, the induced local field h_{local} reinforces H_0 on the outside, just where the aromatic hydrogens are located. This results in a local field in their vicinity equal to $H_0 + h_{local}$. Therefore, to cause resonance at constant radio frequency ν, the applied field strength has to be reduced to ($H_0 - h_{local}$): the nuclei are deshielded. This effect is strongest close to the ring and diminishes with in-

FIGURE 19-9

The effect of an external magnetic field H_0 on the π electrons in benzene. A ring current in the six-membered ring generates a local magnetic field opposing H_0 inside the ring but reinforcing it on the outside.

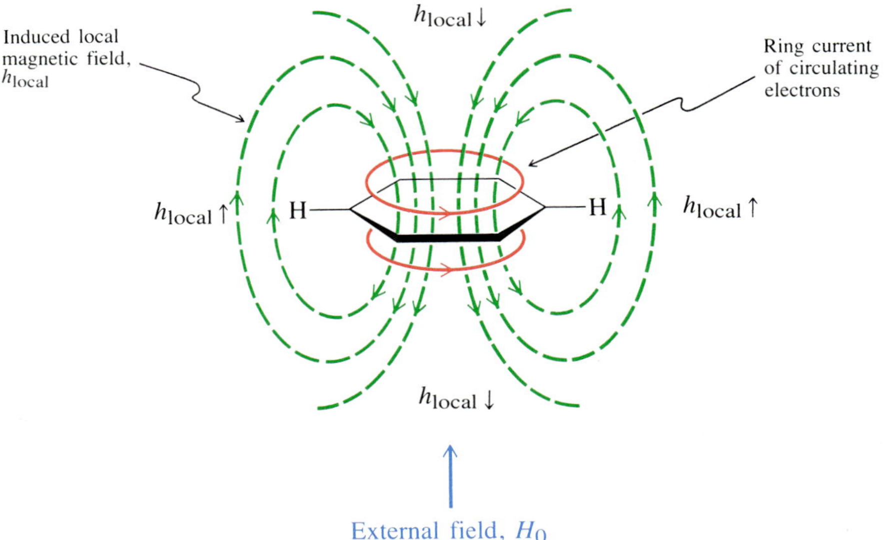

creasing distance from it. Thus, benzylic nuclei are deshielded only about 0.4 to 0.5 ppm more than their allylic counterparts, and hydrogens further away from the π system have chemical shifts that do not differ much from each other and are similar to those in the alkanes.

The ring-current model predicts not only deshielding of hydrogens located outside the π-electronic cycle, but also *shielding* of nuclei located above it. This phenomenon may be observed in special substituted benzenes. For example, the *para*-bridged benzene derivative [10]cyclophane (*phainein,* Greek, to resemble; i.e., resembling a ring) exhibits widely different chemical shifts for the methylene hydrogens, because some of them are located in the deshielding region, whereas others are in the shielding area of the induced field (Figure 19-10).

Even though benzene exhibits only a sharp singlet in its NMR spectrum, substituted derivatives may have more complicated patterns. For example, in a monosubstituted system, the hydrogens positioned *ortho, meta,* and *para* to the substituent are spectroscopically different and may have different chemical shifts. Moreover, coupling across the benzene ring takes place between all nonequivalent nuclei, frequently giving rise to non-first-order spectra. An example is the NMR spectrum of bromobenzene (Figure 19-11). The bromine causes a downfield shift ($\Delta\delta = 0.22$ ppm) of the neighboring hydrogens and a slight upfield shift of the *meta* ($\Delta\delta = -0.13$ ppm) and *para* hydrogens ($\Delta\delta = -0.03$ ppm) relative to the chemical shift of benzene. Moreover, all the protons are coupled to each other, giving rise to a complicated spectral pattern.

Some substituents do not perturb the chemical shifts of the benzene hydrogens by much, so that all absorb at (nearly) the same position, in the form of a broadened singlet. Such an absorption is found in the NMR spectrum of methylbenzene (toluene, Figure 19-12).

FIGURE 19-10

The alkane region of the 220-MHz ^{1}H NMR spectrum of [10]cyclophane in CCl_4. (Courtesy of Professor Michael J. McGlinchey, McMaster University, Hamilton, Ontario, Canada.)

α
δ
(CH3)4Si
ε
γ
δ
β
α
β
γ
ε
3
2
1
0
ppm (δ)

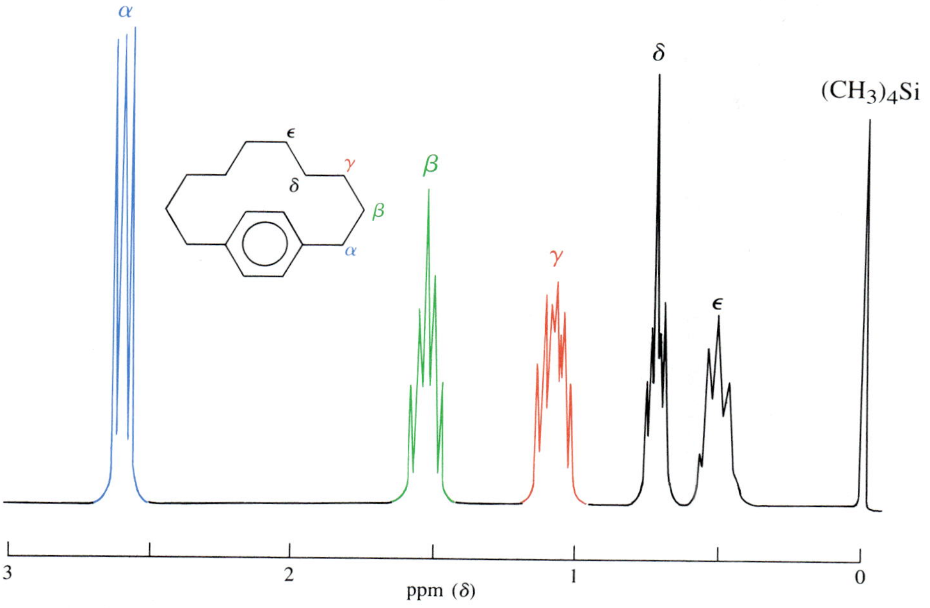

FIGURE 19-11

The 90-MHz ^{1}H NMR spectrum of bromobenzene in CCl_4.

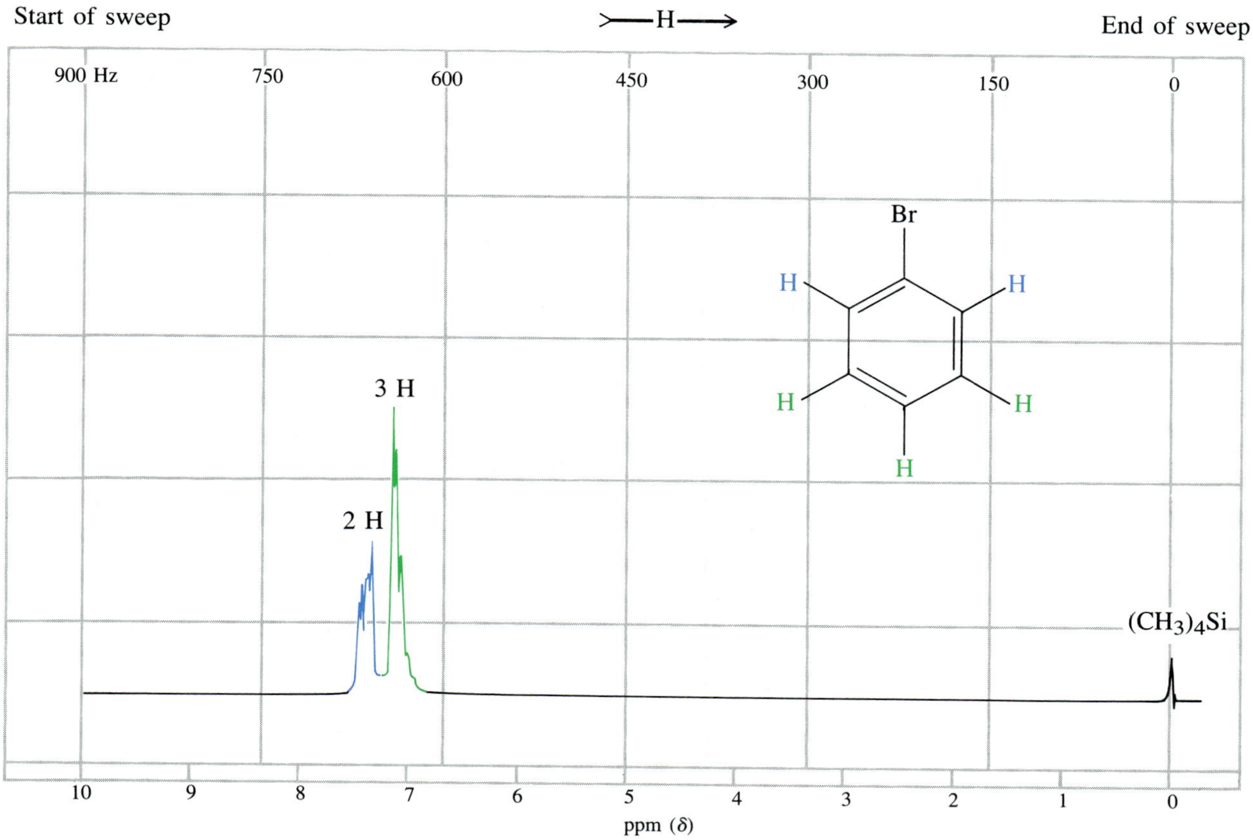

Start of sweep
H
End of sweep
900 Hz
750
600
450
300
150
0
Br
H
H
H
H
H
3 H
2 H
(CH3)4Si
10
9
8
7
6
5
4
3
2
1
0
ppm (δ)

FIGURE 19-12
The 90-MHz ^{1}H NMR spectrum of methylbenzene (toluene) in CCl_4.

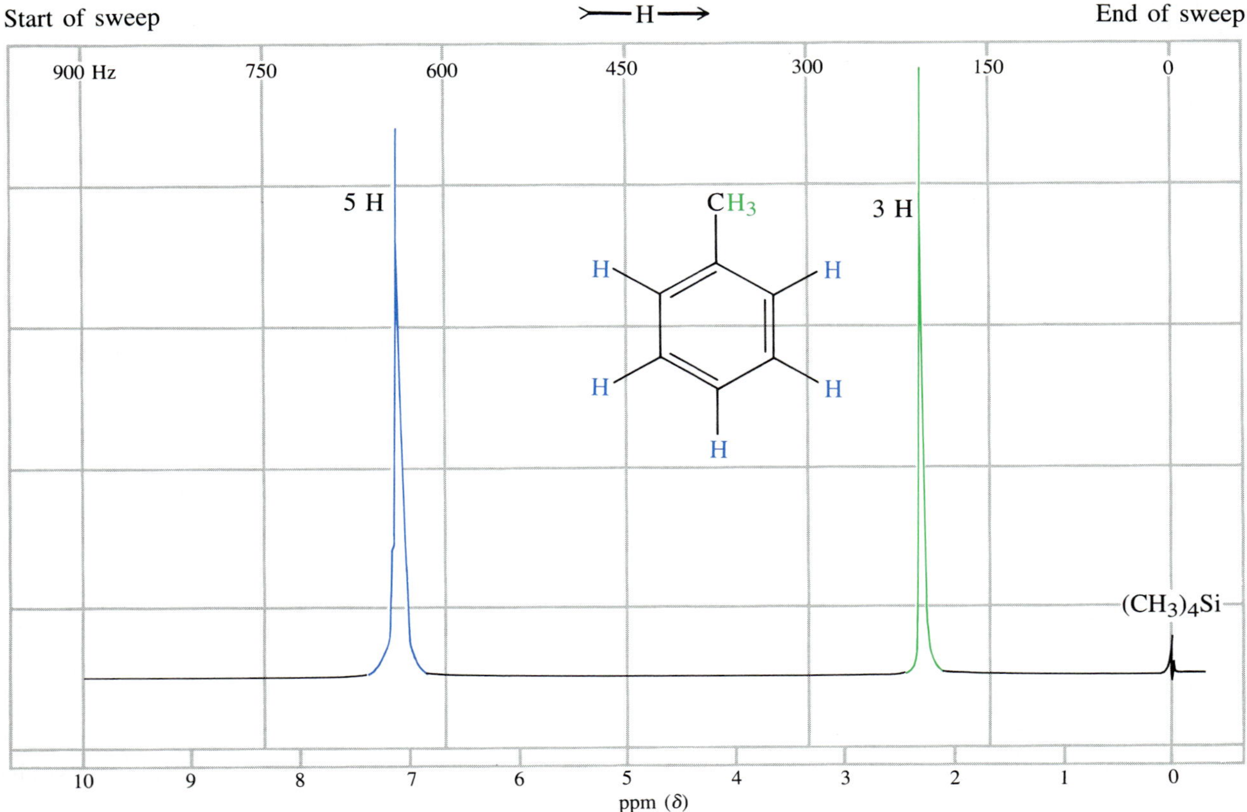

Sometimes the chemical-shift differences introduced by substituents are large enough to produce first-order spectra, especially for benzene derivatives that bear both electron-withdrawing (deshielding) and electron-donating (shielding) groups. For example, the ^{1}H NMR spectrum of 4-*N*,*N*-dimethylaminobenzene-carboxaldehyde (*p*-dimethylaminobenzaldehyde) reveals a set of two doublets for the aromatic hydrogens (Figure 19-13). The higher-field absorptions at δ = 6.69 ppm are due to the hydrogen nuclei *ortho* to the electron-rich amino substituent, whereas the doublet at δ = 7.76 is assigned to the hydrogens *ortho* to the electron-withdrawing methanoyl (formyl) group. The coupling constant between each of the two hydrogens next to the amino group and its respective neighbor next to the methanoyl (formyl) substituent is 9 Hz, which is typical for *ortho* coupling. Other couplings in benzene are smaller, from 2 to 3 Hz for *meta* and about 1 Hz for *para* coupling.

A first-order spectrum revealing all three types of couplings is shown in Figure 19-14. 1-Methoxy-2,4-dinitrobenzene (2,4-dinitroanisole) bears three ring hydrogens, with rather different chemical shifts and distinct splitting patterns. The *para* coupling is so small here that it manifests itself only as slight broadening. Hence, the hydrogen *ortho* to the methoxy group appears as a

FIGURE 19-13

The 90-MHz ^{1}H NMR spectrum of 4-*N*,*N*-dimethylaminobenzenecarboxaldehyde (*p*-dimethylaminobenzaldehyde) in $CDCl_3$. In addition to the two aromatic doublets (J = 9 Hz), there are singlets for the methyl (δ = 3.07 ppm) and the methanoyl (formyl) hydrogens (δ = 9.80 ppm).

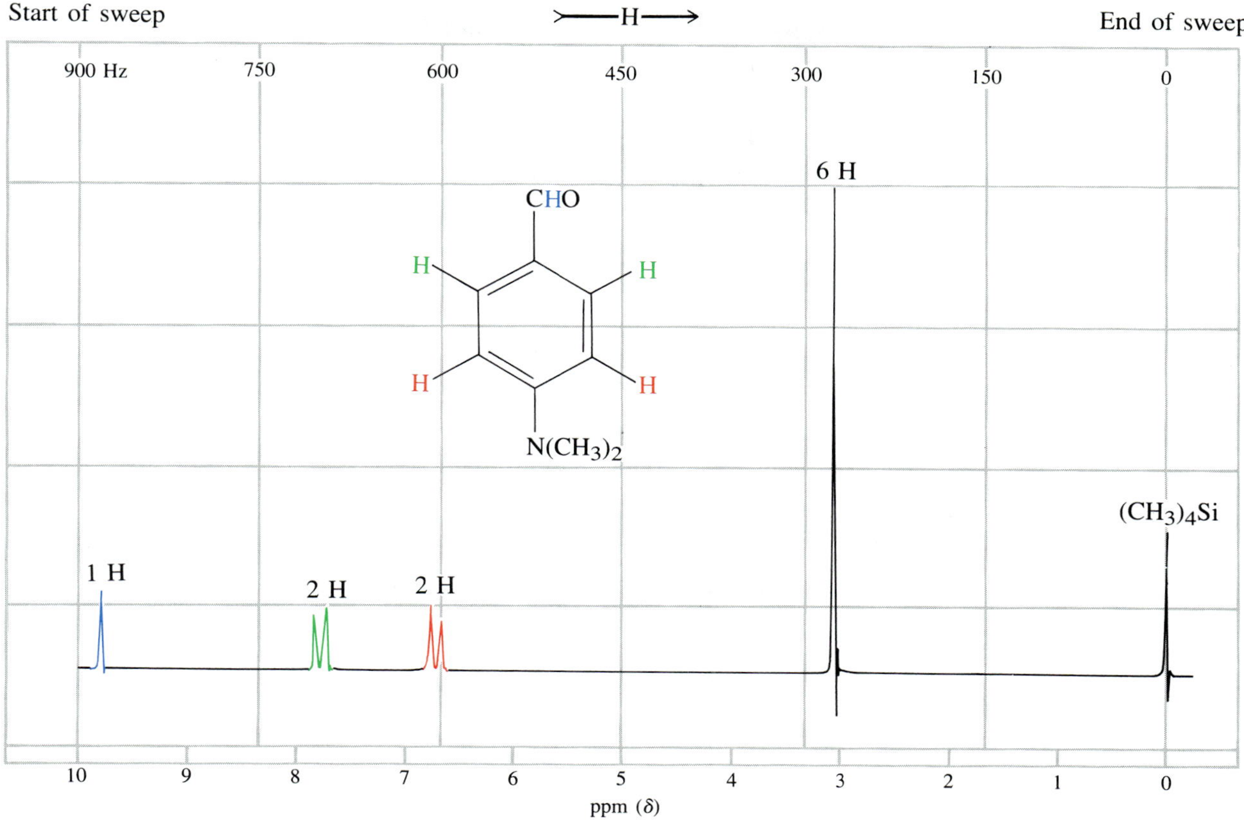

broadened doublet with *ortho* coupling (J = 9 Hz) at δ = 7.30 ppm. The hydrogen flanked by the two nitro substituents is at lowest field (δ = 8.73 ppm); it, too, appears as a doublet, with a small *meta* coupling (J = 3 Hz), because of spin-spin splitting by the nucleus *meta* to the methoxy group. This hydrogen gives rise to a double doublet at δ = 8.50 ppm because of simultaneous coupling to the other two ring hydrogens.

In contrast with ^{1}H NMR, ^{13}C NMR spectroscopy of benzene and its derivatives does not show the ring-current effect, because the chemical shifts of carbon are largely controlled by the influence of charge and hybridization. In addition, the carbons in the ring lie exactly midway between the shielding and deshielding areas around the benzene nucleus, diminishing any consequences of the ring current. Typical chemical shifts for unsubstituted benzene carbons range from 120 to 135 ppm, similar to those of the analogous alkene carbons. Alkyl-substituted nuclei resonate at lower field, and peaks for benzene carbons substituted by polar groups may be shifted in either direction from their original position (Table 19-1). Symmetry greatly simplifies the ^{13}C spectra of substituted benzenes. For example, 1,3,5-trimethylbenzene shows only three lines, whereas its 1,2,3- (6 lines) and 1,2,4-isomers (9 lines) have more.

FIGURE 19-14
^1H NMR spectrum of 1-methoxy-2,4-dinitrobenzene (2,4-dinitroanisole) in $CDCl_3$.

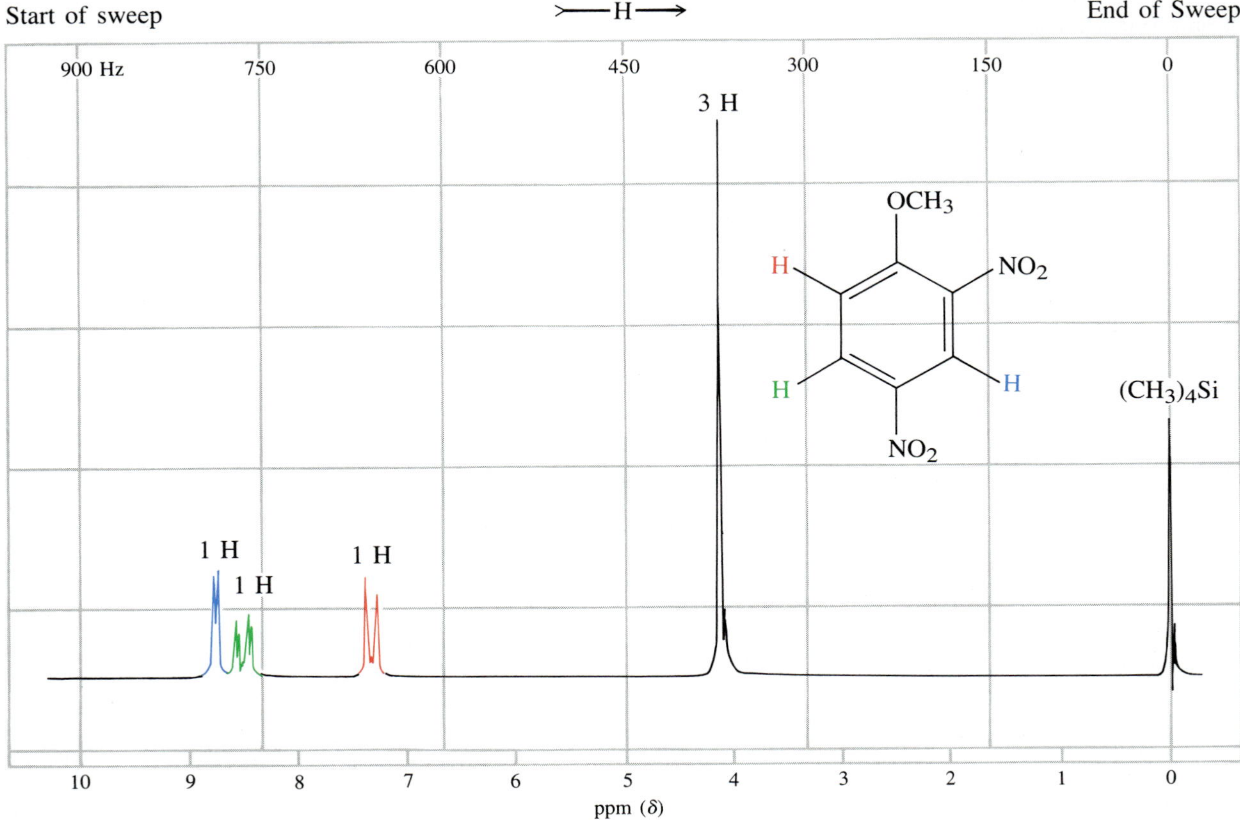

TABLE 19-1
^{13}C NMR data of selected benzene derivatives (ppm)

Benzene	Methylbenzene (H_3C)	Phenol (OH)	Iodobenzene (I)
128.7	CH_3: 21.3		
	C1: 137.8	C1: 155.6	C1: 96.7
	129.3	116.0	138.9
	128.5	130.5	131.6
	125.6	120.8	129.7

EXERCISE 19-6

A hydrocarbon was found to give the following elemental analysis: 89.55% C; 10.45% H. The spectral data for this compound were as follows: ^1H NMR (90 MHz) $\delta = 7.02$ (s, 4 H), 2.82 (heptet, $J = 7.0$ Hz, 1 H), 2.28 (s, 3 H), and 1.22 ppm (d, $J = 7.0$ Hz, 6 H); ^{13}C NMR $\delta = 21.3$, 24.2, 38.9, 126.6, 128.6, 134.8, and 145.7 ppm; mass spectrum $m/z = 134$ (M^+), 119, and 77; IR $\tilde{\nu} = 3030$, 2970, 2880, 1515, 1465, and 813 cm^{-1}; UV $\lambda_{max}(\epsilon) = 265(450)$. What is its structure?

In summary, benzene and its derivatives can be recognized and structurally characterized by their spectral data. Electronic absorptions take place between 250 and 290 nm. The infrared vibrational bands are found at 3030 ($C_{aromatic}$–H), from 1500 to 2000 (C–C), and from 650 to 1000 (C–H out-of-plane bending). Most informative is NMR, with low-field resonances for the aromatic hydrogens and carbons. Coupling is largest between the *ortho* hydrogens and is diminished progressively in their *meta* and *para* counterparts.

19-4 The Synthesis of Benzene Derivatives: General Electrophilic Aromatic Substitution

Benzene, although generally unreactive, can be attacked by electrophiles (Section 14-4). In contrast with the corresponding reactions of alkenes, this reaction results in *substitution* of hydrogens, *not addition* to the ring.

Electrophilic Aromatic Substitution

$$C_6H_6 + E^+X^- \longrightarrow C_6H_5E + H^+X^-$$

Electrophile

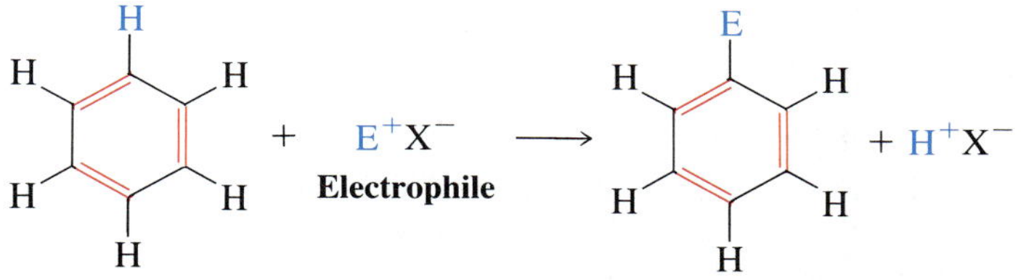

The conditions employed to effect these transformations would readily polymerize nonaromatic conjugated systems. However, the stability of the benzene moiety allows it to survive these reactions. This section deals with the general mechanism of electrophilic aromatic substitution.

Electrophilic Substitution in Benzene Proceeds by Addition of the Electrophile Followed by Proton Loss

The mechanism of electrophilic aromatic substitution requires the electrophile E^+ to attack the benzene nucleus, much as it would attack an ordinary double bond.

Mechanism of Electrophilic Aromatic Substitution:

STEP 1: Electrophilic attack

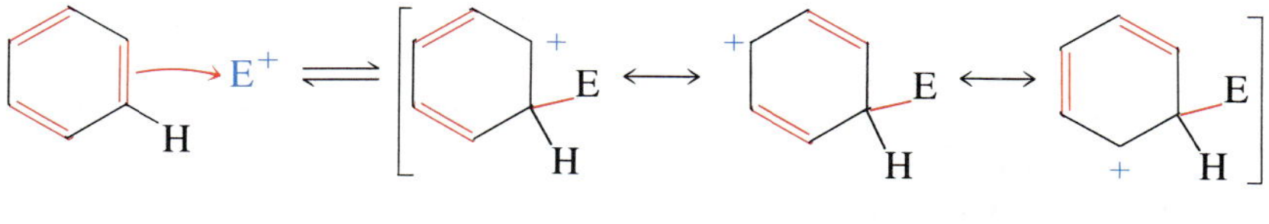

$$C_6H_6 + E^+ \rightleftharpoons [C_6H_6E^+ \longleftrightarrow C_6H_6E^+ \longleftrightarrow C_6H_6E^+]$$

STEP 2: Proton loss

$$[C_6H_6E^+ \longleftrightarrow C_6H_6E^+ \longleftrightarrow C_6H_6E^+] \rightleftharpoons C_6H_5E + H^+$$

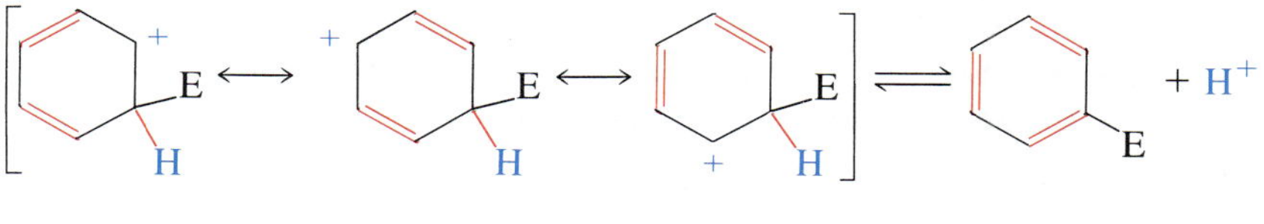

FIGURE 19-15

A. Orbital picture of the hexadienyl cation resulting from attack by an electrophile on the benzene ring. Cyclic conjugation is interrupted by the sp^3-hybridized carbon. The four electrons in the π system are not shown.
B. Dotted-line notation to indicate delocalized nature of the charge in the hexadienyl cation.

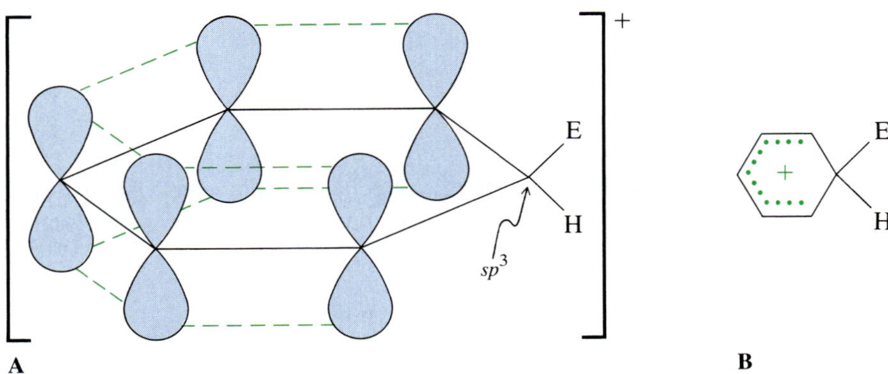

The first step is not favored thermodynamically, because cyclic delocalization is lost and, with it, the aromatic character of benzene. Some of this loss is made up by the delocalization of the charge in the resulting hexadienyl cation (Figure 19-15).

After this step, the aromatic ring is regenerated by extrusion of the proton at the sp^3-hybridized carbon. This is energetically more favored than nucleophilic trapping by the anion that accompanies E^+, a process that would give the addition product.

Both steps in the mechanism of electrophilic aromatic substitution are in principle reversible. However, the overall reaction is exothermic, the new bonds being stronger than the old ones. A potential-energy diagram illustrates the course of the electrophilic aromatic substitution of benzene (Figure 19-16). The first step in this reaction is rate determining; it leads to a transition state in which there is partial bond formation between the electrophile and a benzene carbon. Proton loss is much faster, because it leads to the aromatic product in an exothermic step, which furnishes the driving force for the overall sequence.

The mechanism of electrophilic aromatic substitution described here is general; its details depend on the electrophile. The following sections will look more closely at the most common reagents employed in this transformation.

EXERCISE 19-7

Review Section 12-1 and explain why addition reactions to benzene do not occur. (Calculate the ΔH° for some additions.)

In summary, the general mechanism of electrophilic aromatic substitution begins with electrophilic attack by E^+ to give an intermediate, charge-delocalized hexadienyl cation in a rate-determining step. Subsequent fast proton loss regenerates the (now substituted) aromatic ring.

FIGURE 19-16
Potential-energy diagram describing the course of the reaction of benzene with an electrophile. The first transition state is rate determining. Proton loss is relatively fast. The overall rate of the reaction is determined by E_a, the exothermicity by $\Delta H°$.

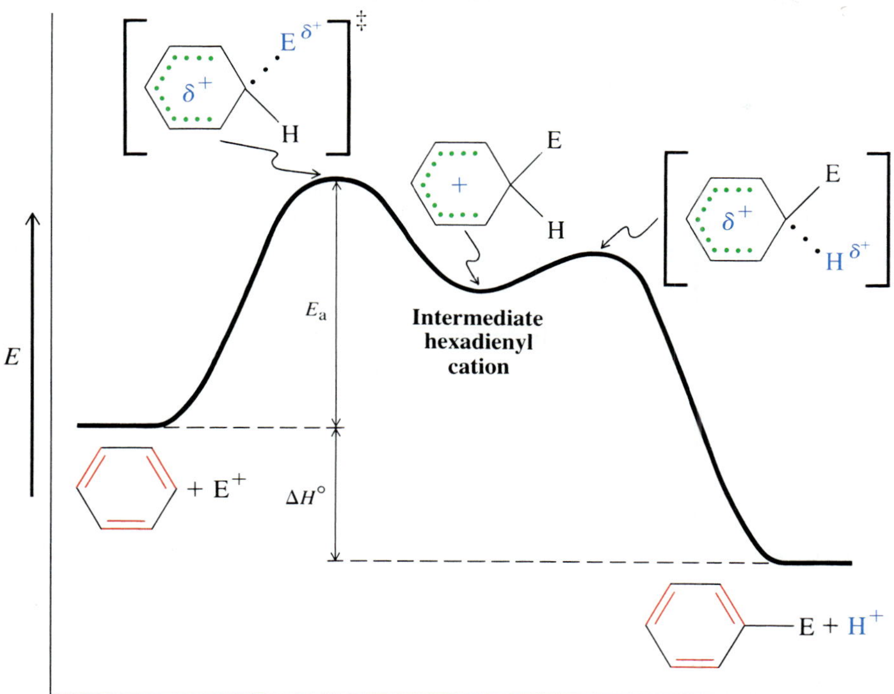

19-5 Halogenation of Benzene: The Need for a Catalyst

This section inspects an example of electrophilic aromatic substitution: halogenation.

Benzene is normally unreactive in the presence of halogens, because halogens are not electrophilic enough to disrupt its aromaticity. However, the halogen may be activated by Lewis acids, such as ferric halides, FeX_3, or aluminum halides, AlX_3, to become a much more powerful electrophile. How does this activation work? The characteristic of Lewis acids is their ability to accept electron pairs (Section 8-3). Halogens are relatively electron rich because of the presence of lone electron pairs. When a halogen such as bromine is exposed to $FeBr_3$, the two molecules combine in a Lewis acid-base sense much as a Grignard reagent (Lewis acid) complexes to the ether solvent in which it is made (Section 8-6).

Activation of Bromine by the Lewis Acid $FeBr_3$

$$:\ddot{\underset{..}{Br}}-\ddot{\underset{..}{Br}}: \curvearrowright FeBr_3 \longrightarrow \left[:\ddot{\underset{..}{Br}}-\overset{+}{\ddot{\underset{..}{Br}}}-\overset{-}{Fe}Br_3 \longleftrightarrow \overset{+}{\ddot{\underset{..}{Br}}}=\ddot{\underset{..}{Br}}-\overset{-}{Fe}Br_3\right]$$

This complex serves to polarize the Br–Br bond, imparting electrophilic character to one of the bromine atoms. Electrophilic attack on benzene occurs as described in the preceding section.

Electrophilic Attack on Benzene by Activated Bromine

$$\text{C}_6\text{H}_6 + :\ddot{\text{Br}}-\overset{+}{\ddot{\text{Br}}}-\overset{-}{\text{FeBr}_3} \longrightarrow \text{hexadienyl cation (H, Br)} + \text{FeBr}_4^-$$

The $FeBr_4^-$ formed in this step now functions as a base, abstracting a proton from the hexadienyl cation intermediate. This transformation not only furnishes the two products of the reaction, bromobenzene and hydrogen bromide, but also regenerates the $FeBr_3$ catalyst.

Bromobenzene Formation

$$\text{hexadienyl cation (H, Br)} + :\ddot{\text{Br}}-\overset{-}{\text{FeBr}_3} \longrightarrow \text{C}_6\text{H}_5\text{Br} + \text{HBr} + \text{FeBr}_3$$

A quick calculation confirms the exothermicity of the electrophilic bromination of benzene. The phenyl-hydrogen bond is about as strong as the alkenyl-hydrogen bond (Figure 11-6), approximately 111 kcal mole^{-1} (Table 19-2), and the dissociation energy of the bromine molecule into its component atoms is 46 kcal mole^{-1}. Therefore, a total of 157 kcal mole^{-1} in bond enthalpy is lost in the process. Counterbalancing this loss, a phenyl–bromine bond (DH° = 81 kcal mole^{-1}) and an H–Br bond (DH° = 87.5 kcal mole^{-1}) are formed, totaling 168.5 kcal mole^{-1}. Thus, the overall reaction is exothermic by 11.5 kcal mole^{-1}.

TABLE 19-2
Bond strengths DH° of bonds A–B (kcal mole^{-1})

	B				
A	**F**	**Cl**	**Br**	**I**	**H**
C_6H_5	126	96	81	65	111
F	37				135.8
Cl		58			103.2
Br			46		87.5
I				36	71.3

EXERCISE 19-8

Using the numbers in Table 19-2, calculate the ΔH° for the electrophilic fluorination, chlorination, and iodination of benzene.

As in the free-radical halogenation of alkanes (Section 3-7), the exothermicity of aromatic halogenation decreases down the periodic table. Fluorination is so exothermic that direct reaction of fluorine with benzene is explosive. Chlorination, on the other hand, is controllable, but requires the presence of an activating catalyst, such as aluminum chloride or ferric chloride. The mechanism of this reaction is identical with that of bromination. Finally, electrophilic iodination with iodine is endothermic and thus not normally possible.

The unfavorable thermodynamics of iodination can be changed by addition of a silver salt to the reaction mixture, which activates the iodine and also removes one of the products (iodide) from the reaction by precipitation.

$$C_6H_5\text{–}H + I\text{–}I + Ag^+NO_3^- \longrightarrow C_6H_5\text{–}I + AgI + HNO_3$$

Another method converts the iodine into the much more reactive iodonium ion by addition of an oxidizing agent such as dilute nitric acid (see Section 17-5).

$$I_2 \xrightarrow{HNO_3} 2\ I^+NO_3^-$$

Iodonium nitrate

$$C_6H_5\text{–}H + I^+NO_3^- \longrightarrow C_6H_5\text{–}I + HNO_3$$

EXERCISE 19-9

When benzene is dissolved in D_2SO_4, its ^1H NMR absorption at $\delta = 7.27$ ppm disappears and a new compound is formed having a molecular ion of $m/z = 84$. What is it? Propose a mechanism for its formation.

EXERCISE 19-10

Professor G. Olah (see Section 9-2) and his colleagues exposed benzene to the especially strong acid system HF-SbF_5 in the nonnucleophilic solvent SO_2ClF-SO_2F_2 in an NMR tube and observed a new ^1H NMR spectrum with absorptions at $\delta = 5.69$ (2 H), 8.22 (2 H), 9.42 (1 H), and 9.58 (2 H) ppm. Propose a structure for this species.

In summary, the halogenation of benzene increases in exothermicity from I_2 (endothermic) to F_2 (explosive). Chlorinations and brominations are achieved with the help of Lewis acid catalysts, which polarize the X–X bond and activate the halogen by increasing its electrophilic power.

19-6 The Nitration and Sulfonation of Benzene

In two typical electrophilic substitutions of benzene, the electrophiles are the nitronium ion, NO_2^+, leading to nitrobenzene, and sulfur trioxide, SO_3, giving benzenesulfonic acid. These reactions and their mechanisms are the subject of this section.

Nitration of Benzene Through Electrophilic Attack by the Nitronium Ion

Treatment of benzene with concentrated nitric acid in the presence of concentrated sulfuric acid at moderate temperatures leads to nitration of the ring (Section 14-4). Because the nitrogen in the nitrate group of HNO_3 has no electrophilic power, it must be somehow activated. This occurs by the action of the added sulfuric acid. Protonation of nitric acid followed by loss of water yields the nitronium ion, NO_2^+, a strongly electrophilic species.

Activation of Nitric Acid by Sulfuric Acid

$$HO{-}\overset{+}{N}(=O){-}O^- + H{-}OSO_3H \rightleftharpoons H\overset{+}{O}(H){-}\overset{+}{N}(=O){-}O^- + HSO_4^-$$

Nitric acid

$$H\overset{+}{O}(H){-}\overset{+}{N}(=O){-}O^- \longrightarrow H_2O + O{=}\overset{+}{N}{=}O$$

Nitronium ion

Some of its salts, such as $NO_2^{+}\,{}^{-}PF_6$, can even be isolated and used as such to nitrate benzene. Electrophilic attack on benzene proceeds by a simple variation of the general mechanism discussed earlier.

Mechanism of Aromatic Nitration:

$$C_6H_6 + \overset{+}{N}O_2 \longrightarrow [C_6H_6NO_2]^+ \xrightarrow{^-:OSO_3H} C_6H_5NO_2 + HOSO_3H$$

Nitrobenzene

Sulfonation: A Reversible Substitution

Except for effecting protonation (see Exercise 19-9), concentrated sulfuric acid does not react with benzene at room temperature. However, a more-reactive form, called "fuming sulfuric acid," undergoes electrophilic attack by SO_3 (Section 14-4). Commercial fuming sulfuric acid is made by adding about 8% of sulfur trioxide, SO_3, to the concentrated acid. This added ingredient is the reactive species. Because of the strong electron-withdrawing effect of the three oxygens, the sulfur in SO_3 is electrophilic enough to attack benzene directly. Subsequently proton transfer results in the sulfonated product, benzenesulfonic acid.

Mechanism of Aromatic Sulfonation:

Benzenesulfonic acid

The aromatic sulfonation reaction is nearly thermoneutral; it forms an equilibrium that may be shifted to the side of the starting materials by removing the sulfur trioxide with water to give sulfuric acid, a very exothermic process. In practice, heating benzenesulfonic acid in aqueous sulfuric acid completely reverses sulfonation.

Hydration of SO_3

$$SO_3 + HOH \longrightarrow O{=}S(=O)(OH)-OH + \text{heat}$$

Reverse Sulfonation

$$C_6H_5SO_3H \xrightarrow{H_2O,\ \text{catalytic } H_2SO_4,\ 100°C} C_6H_5H + HOSO_3H$$

Reversible sulfonation is of some use in synthetic applications of electrophilic aromatic substitution, because the sulfonic acid substituent may be employed as a directing blocking group (Section 20-5).

Benzenesulfonic Acids Have Important Uses

Sulfonation is also important in the synthesis of *detergents* (Section 17-12). Until recently, long-chain branched alkylbenzenes were sulfonated to the corresponding sulfonic acids, then converted into their sodium salts. Because such detergents are not readily biodegradable, they have been replaced by long-chain alkanesulfonates (Section 17-12).

Aromatic Detergent Synthesis

$$R{-}C_6H_5 \xrightarrow{SO_3,\ H_2SO_4} R{-}C_6H_4{-}SO_3H \xrightarrow[-\ H_2O]{NaOH} R{-}C_6H_4{-}SO_3^-\ Na^+$$

R = branched alkyl group

Another application of sulfonation is to the manufacture of dyes, because the sulfonic acid group imparts water solubility to colored organic molecules (Section 24-6).

Sulfonyl chlorides are the acid chlorides of sulfonic acids (see Sections 6-7 and 9-3). Like alkanoyl chlorides, they are usually prepared by reaction of their sodium salts with PCl_5 or $SOCl_2$.

$$2\ C_6H_5SO_3^-Na^+ + PCl_5 \longrightarrow 2\ C_6H_5SO_2Cl + POCl_3 + 2\ Na^+Cl^-$$

Although not quite as reactive as alkanoyl chlorides, they are synthetically useful. For example, recall that the hydroxy group of an alcohol may be turned into a good leaving group by conversion of the alcohol into the 4-methylbenzenesulfonate (*p*-toluenesulfonate, tosylate; Sections 6-7, 9-3).

Sulfonyl chlorides are important precursors of **sulfonamides,** many of which are chemotherapeutic agents, such as the *sulfa drugs* discovered in 1932 (Section 9-8). Sulfonamides are derived from the reaction of a sulfonyl chloride with an amine. Sulfa drugs specifically contain the 4-aminobenzenesulfonamide (sulfanilamide) function. Their mode of action is to interfere with the bacterial enzymes that help to synthesize folic acid (Section 26-6).

Some Sulfa Drugs

RHN–C_6H_4–SO_2NHR'

General structure

H_2N–C_6H_4–SO_2NH–(5-methylisoxazol-3-yl)

Sulfamethoxazole (Gantanol; antibacterial, used in urinary infections)

H_2N–C_6H_4–SO_2NH–(pyrimidin-2-yl)

Sulfadiazine (antimalarial)

H_2N–C_6H_4–SO_2NH–(3-CH_3O-pyrazin-2-yl)

Sulfalene (Kelfizina; antilepral)

About 15,000 sulfa derivatives have been synthesized and screened for antibacterial activity; some have become new drugs. With the advent of antibiotics, the medicinal use of sulfa drugs has greatly diminished, but their discovery was a milestone in the systematic development of medicinal chemistry.

EXERCISE 19-11

Formulate mechanisms for (a) the reverse of sulfonation and (b) the hydration of SO_3.

In summary, nitration of benzene requires the generation of the nitronium ion, NO_2^+, which functions as the active electrophile. The nitronium ion is formed by the loss of water from protonated nitric acid. Sulfonation is achieved with fuming sulfuric acid, in which sulfur trioxide, SO_3, is the electrophile. Sulfonation is reversible by the action of hot aqueous acid. Benzenesulfonic acids are used in the preparation of detergents, dyes, compounds containing leaving groups, and sulfa drugs.

19-7
Electrophilic Aromatic Substitution with Carbon–Carbon Bond Formation: The Friedel-Crafts Reactions

None of the electrophilic substitutions mentioned so far have led to carbon–carbon bond formation, one of the primary challenges in organic chemistry. In principle, such reactions could be carried out with benzene in the presence of a sufficiently electrophilic carbon-based electrophile. This section introduces two such transformations, named after their discoverers, the **Friedel-Crafts* reactions.** The secret to the success of these processes is the use of a Lewis acid, usually aluminum chloride. In the presence of this reagent, haloalkanes attack benzene to form alkylbenzenes, and alkanoyl halides give alkanoylbenzenes.

Electrophilic Alkylation of Benzene

In 1877, Friedel and Crafts discovered that a haloalkane reacts with benzene in the presence of an aluminum halide. The resulting products are the alkylbenzene and hydrogen halide. This reaction, which can be carried out in the presence of other Lewis acid catalysts, is called the **Friedel-Crafts alkylation** of benzene.

General Friedel-Crafts Alkylation

$$C_6H_5H + RX \xrightarrow{AlX_3} C_6H_5R + HX$$

The reactivity of the haloalkane decreases in the order RF > RCl > RBr > RI. Typical Lewis acids are (in the order of decreasing activity) $AlBr_3$, $AlCl_3$, $FeCl_3$, $SbCl_5$, and BF_3.

$$CH_3CH_2Cl + C_6H_5H \xrightarrow{AlCl_3,\ 25^\circ C} C_6H_5CH_2CH_3 + HCl$$

27.5%
Ethylbenzene

$$(CH_3)_3CCl + C_6H_5H \xrightarrow{AlCl_3,\ 0^\circ C} C_6H_5C(CH_3)_3 + HCl$$

60%
(1,1-Dimethylethyl)benzene
(*tert*-Butylbenzene)

*Professor Charles Friedel, 1832–1899, Sorbonne, Paris; Professor James M. Crafts, 1839–1917, Massachusetts Institute of Technology, Boston.

With primary halides, the reaction begins with coordination of the Lewis acid to the halogen of the haloalkane, much as in the activation of halogens in electrophilic halogenation. This places a partial positive charge on the halogen-bearing carbon, rendering it more electrophilic. Attack on the benzene ring is followed by proton loss in the usual manner, giving the observed product.

Mechanism of Friedel-Crafts Alkylation with Primary Haloalkanes:

STEP 1: Haloalkane activation

$$RCH_2\text{—}\ddot{\underset{..}{X}}\!: + AlX_3 \rightleftharpoons R\overset{\delta^+}{C}H_2\!:\!\overset{+}{\underset{..}{\ddot{X}}}\!:\!\overset{-}{Al}X_3$$

STEP 2: Electrophilic attack

$$C_6H_6 + \overset{\delta^+}{H_2C}(R)\text{—}\overset{+}{\ddot{X}}\text{—}\overset{-}{Al}X_3 \longrightarrow C_6H_6^+(H)(CH_2R) + AlX_4^-$$

STEP 3: Proton loss

$$C_6H_6^+(H)(CH_2R) + :\ddot{\underset{..}{X}}\text{—}\overset{-}{Al}X_3 \longrightarrow C_6H_5CH_2R + HX + AlX_3$$

With secondary and tertiary halides, free carbocations are usually formed as intermediates, which attack the benzene ring in the same way as the cation NO_2^+.

EXERCISE 19-12

Write a mechanism for the formation of (1,1-dimethylethyl)benzene (*tert*-butylbenzene) from 2-chloro-2-methylpropane (*tert*-butyl chloride) and benzene in the presence of $AlCl_3$.

EXERCISE 19-13

When benzene is treated with 1-bromo-3-fluoropropane in the presence of the Lewis acid boron tribromide, BBr_3, at $-10°C$ for 30 min, a new compound is obtained that has the molecular formula $C_9H_{11}Br$ and shows seven peaks in the ^{13}C NMR spectrum. Suggest a structure for this product.

Intramolecular Friedel-Crafts alkylations can be used to fuse a new ring onto the benzene nucleus.

$$C_6H_5(CH_2)_4Cl \xrightarrow[-\ HCl]{AlCl_3,\ CS_2\ \text{and}\ CH_3NO_2\ \text{(solvents)},\ 25°C,\ 72\ h} \text{Tetralin}$$

31%
Tetralin
(common name)

Friedel-Crafts alkylations can be carried out with any starting material, such as an alcohol or alkene, that functions as a precursor to a carbocation (Sections 9-2 and 12-3).

$$C_6H_6 + CH_3CH_2CH(OH)CH_3 \xrightarrow[-\ HOH]{BF_3,\ 60^\circ C,\ 9\ h} C_6H_5CH(CH_3)CH_2CH_3$$

36%
(1-Methylpropyl)benzene

$$C_6H_6 + \text{cyclohexene} \xrightarrow{HF,\ 0^\circ C} C_6H_5\text{-}C_6H_{11}$$

62%
Cyclohexylbenzene

EXERCISE 19-14

In 1984, more than 3.7 billion pounds of (1-methylethyl)benzene (isopropylbenzene or cumene), an important industrial intermediate in the manufacture of benzenol (phenol, Section 24-3), was synthesized in the United States from propene and benzene in the presence of phosphoric acid. Write a mechanism for its formation in this reaction.

Friedel Crafts Alkylations Have Limitations

The alkylation of benzenes under Friedel-Crafts conditions is accompanied by two important limiting reactions: one is *polyalkylation;* the second, *carbocation rearrangements*. Both cause the yield of the desired products to diminish and lead to mixtures that are difficult to separate.

For example, reaction of benzene with 2-bromopropane in the presence of iron tribromide as a catalyst furnishes a mixture of (1-methylethyl)benzene and 1,4-bis(1-methylethyl)benzene, both in relatively low yield because of the formation of many by-products.

$$C_6H_6 + (CH_3)_2CHBr \xrightarrow[-\ HBr]{FeBr_3} C_6H_5CH(CH_3)_2 + 1,4\text{-}[(CH_3)_2CH]_2C_6H_4$$

25%
(1-Methylethyl)benzene (Isopropylbenzene)

15%
1,4-Bis(1-methylethyl)benzene (*p*-Diisopropylbenzene)

In the preceding examples, electrophilic aromatic substitution could be stopped at the monosubstitution stage. Why do Friedel-Crafts alkylations have the problem of multiple electrophilic substitution? Because the substituents differ in electronic structure. Bromination, nitration, and sulfonation introduce an electron-withdrawing group into the benzene ring, which renders the product *less* susceptible to electrophilic attack than the starting material. In contrast, an alkylated benzene is more electron rich than unsubstituted benzene and thus *more* susceptible to electrophilic attack.

EXERCISE 19-15

Treatment of benzene with chloromethane in the presence of aluminum chloride results in a complex mixture of tri-, tetra-, and pentamethylbenzenes. One of the components in this mixture crystallizes out selectively: m.p. = 80°C; mass m/z = 134 $(M)^+$; ^1H NMR δ = 2.27 (s, 12 H) and 7.15 (s, 2 H) ppm; ^{13}C NMR δ = 19.2, 131.2, and 133.8 ppm. Draw a structure for this product.

The second complication in aromatic alkylation is skeletal rearrangement (Section 9-2). For example, the attempted propylation of benzene with 1-bromopropane and $AlCl_3$ produces (1-methylethyl)benzene.

$$C_6H_5\text{—}H + CH_3CH_2CH_2Br \xrightarrow[-\ HBr]{AlCl_3} C_6H_5\text{—}CH(CH_3)_2$$

The starting haloalkane rearranges to the thermodynamically favored 1-methylethyl (isopropyl) cation in the presence of the Lewis acid.

Rearrangement of 1-Bromopropane to 1-Methylethyl (Isopropyl) Cation

$$CH_3\overset{H}{\overset{|}{C}}H\text{—}CH_2\text{—}Br + AlCl_3 \longrightarrow CH_3\overset{+}{C}HCH_3 + Br\overset{-}{A}lCl_3$$

1-Methylethyl (isopropyl) cation

EXERCISE 19-16

Attempted alkylation of benzene with 1-chlorobutane in the presence of $AlCl_3$ gave not only the expected butylbenzene, but, as a major product, (1-methylpropyl) benzene. Write a mechanism for this reaction.

Because of these limitations, Friedel-Crafts alkylations are used rarely in synthetic chemistry. Is there a way to improve this process? It would require finding an electrophilic carbon species that cannot rearrange and that would, moreover, deactivate the ring to prevent further substitution. There is such a species: an **acylium cation equivalent,** used in the **Friedel-Crafts alkanoylation.**

Friedel-Crafts Alkanoylation to Alkanoylbenzenes (Acylbenzenes)

The prototype of the Friedel-Crafts alkanoylation is the ethanoylation (acetylation) of benzene with ethanoyl (acetyl) chloride in the presence of aluminum chloride to give 1-phenylethanone (acetophenone). The reaction proceeds through the intermediacy of an ethanoyl cation, $CH_3C{\equiv}O{:}^+$.

Friedel-Crafts Alkanoylation

$$C_6H_5H + CH_3\overset{O}{\overset{\|}{C}}Cl \xrightarrow{AlCl_3} C_6H_5\overset{O}{\overset{\|}{C}}-CH_3 + HCl$$

61%
1-Phenylethanone
(Acetophenone)

In general, acylium cations are formed by the reaction of alkanoyl halides with aluminum chloride. Carboxylic anhydrides react with Lewis acids in a similar way. The Lewis acid forms a complex by coordination with the carbonyl oxygen. This complex is in equilibrium with another species in which the aluminum chloride is bound to the halogen (in alkanoyl halides) or the bridging oxygen atom (in carboxylic anhydrides).

Lewis Acid Complexation with Alkanoyl Halides

$$RC(=\ddot{O}:)\ddot{X}: + AlCl_3 \rightleftharpoons \left[RC(={}^{+}\!:\!O\!-\!\overset{-}{A}lCl_3)\ddot{X}: \longleftrightarrow R\overset{+}{C}(-\ddot{O}\!-\!\overset{-}{A}lCl_3)\ddot{X}: \longleftrightarrow RC(-\ddot{O}\!-\!\overset{-}{A}lCl_3)=\ddot{X}:^{+} \right]$$

$$RC(={}^{+}\!:\!O\!-\!\overset{-}{A}lCl_3)\ddot{X}: \rightleftharpoons RC(=\ddot{O}:)\overset{+}{\ddot{X}}-\overset{-}{A}lCl_3$$

Lewis Acid Complexation with Carboxylic Anhydrides

$$RC(=\ddot{O}:)-\ddot{O}-C(=\ddot{O}:)R + AlCl_3 \rightleftharpoons \left[RC(={}^{+}\!:\!O\!-\!\overset{-}{A}lCl_3)-\ddot{O}-C(=\ddot{O}:)R \longleftrightarrow R\overset{+}{C}(-\ddot{O}\!-\!\overset{-}{A}lCl_3)-\ddot{O}-C(=\ddot{O}:)R \longleftrightarrow RC(-\ddot{O}\!-\!\overset{-}{A}lCl_3)=\overset{+}{\ddot{O}}-C(=\ddot{O}:)R \right]$$

$$RC(={}^{+}\!:\!O\!-\!\overset{-}{A}lCl_3)-\ddot{O}-C(=\ddot{O}:)R \rightleftharpoons RC(=\ddot{O}:)-\overset{+}{\ddot{O}}(-\overset{-}{A}lCl_3)-C(=\ddot{O}:)R$$

These adducts may dissociate to produce the acylium ion in small equilibrium concentrations. This process is relatively easy because the product is stabilized by resonance (Section 18-7).

Acylium Ion Generation

$$\mathrm{RC(=O)}-\ddot{\underset{..}{X}}^{+}-\overset{-}{A}lCl_3 \rightleftharpoons [\mathrm{RC}\equiv\overset{+}{O}: \longleftrightarrow \mathrm{R}\overset{+}{C}=\ddot{O}] + :\ddot{\underset{..}{X}}\overset{-}{A}lCl_3$$

Acylium ion

$$\mathrm{RC(=O)}-\overset{+}{\ddot{O}}(\overset{-}{A}lCl_3)-\mathrm{C(=O)R} \rightleftharpoons [\mathrm{RC}\equiv\overset{+}{O}: \longleftrightarrow \mathrm{R}\overset{+}{C}=\ddot{O}] + \mathrm{Cl_3}\overset{-}{A}l\ddot{O}\mathrm{C(=O)R}$$

The acylium ion is sufficiently electrophilic to attack benzene by the usual aromatic substitution mechanism.

Electrophilic Alkanoylation

$$\mathrm{C_6H_5H} + \mathrm{R}-\overset{+}{C}=\ddot{O}: \longrightarrow [\text{cyclohexadienyl cation bearing H and } \mathrm{C(=O)R}] \longrightarrow \mathrm{C_6H_5C(=O)R} + \mathrm{H^+}$$

Because the newly introduced alkanoyl substituent is electron withdrawing, it deactivates the ring and protects it from further substitution. The effect is accentuated by the formation of a strong complex between the aluminum chloride catalyst and the carbonyl function of the product ketone. (Recall that the carbonyl oxygen in ketones is basic; Section 15-2.)

Lewis Acid Complexation with 1-Phenylalkanones

$$\mathrm{C_6H_5C(=\ddot{O}:)R} + \mathrm{AlCl_3} \rightleftharpoons \mathrm{C_6H_5C(=\overset{+}{O}:-\overset{-}{A}lCl_3)R}$$

This complexation removes the $AlCl_3$ from the reaction mixture and necessitates the use of at least one full equivalent of the Lewis acid to allow the reaction to go to completion. Aqueous work-up is necessary to liberate the ketone by hydrating the complexed aluminum chloride.

Aqueous Work-up of Friedel-Crafts Alkanoylations

$$\mathrm{C_6H_5C(=\overset{+}{O}:-\overset{-}{A}lCl_3)R} + 3\,\mathrm{HOH} \longrightarrow \mathrm{C_6H_5C(=\ddot{O}:)R} + \mathrm{Al(OH)_3} + 3\,\mathrm{HCl}$$

Examples (all require aqueous work-up):

$$C_6H_5\text{—H} + CH_3CH_2\overset{\overset{\displaystyle O}{\|}}{C}Cl \xrightarrow[-\ HCl]{AlCl_3\ (1.7\ \text{equivalents})} C_6H_5\overset{\overset{\displaystyle O}{\|}}{C}CH_2CH_3$$

84%
1-Phenyl-1-propanone
(Propiophenone)

$$C_6H_5\text{—H} + CH_3\overset{\overset{\displaystyle O}{\|}}{C}O\overset{\overset{\displaystyle O}{\|}}{C}CH_3 \xrightarrow[-\ CH_3COOH]{AlCl_3\ (2.4\ \text{equivalents})} C_6H_5\overset{\overset{\displaystyle O}{\|}}{C}CH_3$$

85%
1-Phenylethanone
(Acetophenone)

EXERCISE 19-17

The Gattermann-Koch reaction enables the direct introduction of the methanoyl (formyl) group, —CHO, into the benzene ring by treatment with CO under pressure, in the presence of hydrogen chloride and Lewis acid catalysts. For example, methylbenzene (toluene) can be methanoylated (formylated) at the *para* position in this way in 51% yield. Write a mechanism for this reaction.

$$CH_3C_6H_4\text{—H} + CO + HCl \xrightarrow{AlCl_3,\ CuCl} p\text{-}CH_3C_6H_4CHO$$

Because the carbonyl group in the products of Friedel-Crafts alkanoylations may be completely reduced by a variety of methods (Clemmensen reduction, Wolff-Kishner reduction, Raney Ni desulfurization of a thioketal; see Sections 15-5, 15-6, and 15-8), the sequence alkanoylation–reduction gives convenient synthetic access to alkylbenzenes, circumventing the difficulties encountered in Friedel-Crafts alkylations.

Wolff-Kishner Reduction of a Friedel-Crafts Alkanoylation Product

$$C_6H_5\overset{\overset{\displaystyle O}{\|}}{C}CH_3 \xrightarrow{NH_2NH_2,\ H_2O,\ KOH,\ HOCH_2CH_2OCH_2CH_2OCH_2CH_2OH,\ \Delta} C_6H_5CH_2CH_3$$

95%

EXERCISE 19-18

Propose a synthesis of hexylbenzene from hexanoic acid.

EXERCISE 19-19

Intramolecular alkanoylations occur in a variety of carboxylic acid derivatives (including esters) in the presence of various Lewis and mineral acids. Explain the following transformation by a mechanism:

$$\xrightarrow{\text{HF}} \quad + \ CH_3OH$$

73%

1-Indanone

(common name)

In summary, the Friedel-Crafts alkylation produces carbocations (or their equivalents) capable of electrophilic aromatic substitution by formation of aryl–carbon bonds. Haloalkanes, alkenes, and alcohols can be used to achieve aromatic alkylation in the presence of a Lewis or mineral acid. The problems with this method are overalkylation and skeletal rearrangements by both hydrogen and alkyl shifts. These drawbacks are avoided in Friedel-Crafts alkanoylations, in which an alkanoyl halide or carboxylic acid anhydride is the reaction partner, in the presence of a Lewis acid. The intermediate acylium cations undergo electrophilic aromatic substitution to yield the corresponding aromatic ketones, which can be manipulated further.

Summary of New Reactions

1 HYDROGENATION OF BENZENE

$$C_6H_6 \xrightarrow{H_2,\ \text{catalyst}} C_6H_{12} \qquad \Delta H^\circ = -49.3 \text{ kcal mole}^{-1}$$

Resonance energy: ~ -30 kcal mole^{-1}

2 AROMATIC TRANSITION STATES

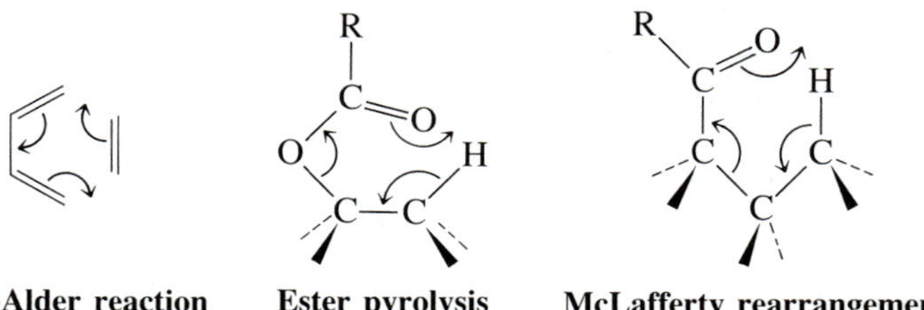

Diels-Alder reaction **Ester pyrolysis** **McLafferty rearrangement**

Electrophilic Aromatic Substitution

3 CHLORINATION, BROMINATION, NITRATION, AND SULFONATION

$$C_6H_6 \xrightarrow{X_2,\ FeX_3} C_6H_5X + HX \qquad X = Cl, Br$$

$$C_6H_6 \xrightarrow{HNO_3,\ H_2SO_4} C_6H_5NO_2 + H_2O$$

$$C_6H_6 \underset{H_2SO_4,\ H_2O,\ \Delta}{\overset{SO_3,\ H_2SO_4}{\rightleftharpoons}} C_6H_5SO_3H$$

4 IODINATION

$$C_6H_6 + I_2 + AgNO_3 \longrightarrow C_6H_5I + AgI + HNO_3$$

or

$$2\ C_6H_6 + I_2 + 2\ HNO_3 \longrightarrow 2\ C_6H_5I + 2\ NO_2 + 2\ H_2O$$

5 BENZENESULFONYL CHLORIDES

$$C_6H_5SO_3Na + PCl_5 \longrightarrow C_6H_5SO_2Cl + POCl_3 + NaCl$$

6 FRIEDEL-CRAFTS ALKYLATION

$$C_6H_6 + RX \xrightarrow{AlCl_3} C_6H_5R + HX + \text{overalkylated product}$$

R is subject to rearrangement

Intramolecular:

$$C_6H_5CH_2CH_2CH_2CH_2Cl \xrightarrow{AlCl_3} \text{tetrahydronaphthalene} + HCl$$

Alcohols and alkenes as substrates:

$$RCH(OH)R' + C_6H_6 \xrightarrow[-\ H_2O]{BF_3,\ 60^\circ C} C_6H_5CH(R')R$$

$$RCH{=}CH_2 + C_6H_6 \xrightarrow{HF,\ 0^\circ C} C_6H_5CH(R)CH_3$$

7 FRIEDEL-CRAFTS ALKANOYLATION

$$C_6H_6 + RCCl(=O) \xrightarrow{AlCl_3} C_6H_5C(=O)R + HCl$$

Anhydrides:

$$C_6H_6 + CH_3C(=O)OC(=O)CH_3 \xrightarrow{AlCl_3} C_6H_5C(=O)CH_3 + CH_3COOH$$

Summary of Important Concepts

1 Substituted benzenes are named by adding prefixes or suffixes to the word *benzene*. Disubstituted systems are labeled as 1,2- 1,3- and 1,4- or *ortho, meta,* and *para,* depending on the location of the substituents. Many benzene derivatives have common names, sometimes used as bases for naming their substituted analogs. As a substituent, an aromatic system is called aryl; the parent aryl substituent, C_6H_5, is called phenyl; its homolog $C_6H_5CH_2$ is named phenylmethyl (benzyl).

2 Benzene is not a cyclohexatriene but a delocalized cyclic system of six π electrons. It is a regular hexagon of six sp^2-hybridized carbons. All six *p* orbitals overlap equally with their neighbors. Its unusually low heat of hydrogenation indicates a resonance energy of about 30 kcal $mole^{-1}$. The stability imparted by aromatic delocalization is also evident in the transition state of some reactions, such as the Diels-Alder cycloaddition, ester pyrolysis, and the McLafferty rearrangement.

3 The special structure of benzene gives rise to unusual UV, IR, and NMR spectral data. ^1H NMR spectroscopy is particularly diagnostic because of the unusual deshielding of aromatic hydrogens by an induced ring current. Moreover, the substitution pattern is revealed by examination of the *o, m,* and *p* coupling constants.

4 The most-important reaction of benzene is electrophilic aromatic substitution. The rate-determining step is addition by the electrophile to give a delocalized hexadienyl cation in which the aromatic character of the original benzene ring has been lost. Fast deprotonation restores the aromaticity of the (now-substituted) benzene ring. Exothermic substitution is preferred over endothermic addition. The reaction can lead to halo- and nitrobenzenes, benzenesulfonic acids, and alkylated and alkanoylated derivatives. When necessary, Lewis acid (chlorination, bromination, Friedel-Crafts reaction), mineral acid (nitration, sulfonation), or other mediators (iodination) are applied. These catalysts enhance the electrophilic power of the reagents (chlorination, bromination, primary alkylation, sulfonation), generate strong, positively charged electrophiles (iodination, I^+; nitration, $NO_2{}^+$; alkylation, R^+; alkanoylation, RCO^+), or drive the process of substitution in some other ways (iodination).

5 Benzenesulfonic acids are precursors of benzenesulfonyl chlorides. The chlorides react with alcohols to form sulfonic esters containing useful leaving groups and with amines to give sulfonamides, some of which are medicinally important (sulfa drugs).

6 In contrast with other electrophilic substitutions, Friedel-Crafts alkylations activate the aromatic ring to further electrophilic substitution, leading to product mixtures.

Problems

1 Name each of the following compounds using the IUPAC system and, if possible, by a reasonable common alternative.

(a) 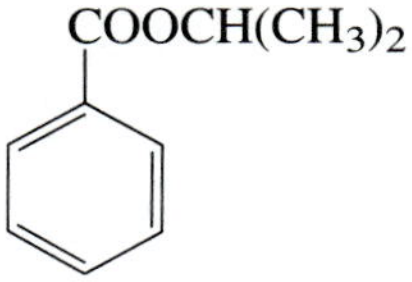 $COOCH(CH_3)_2$ on benzene ring

(f) 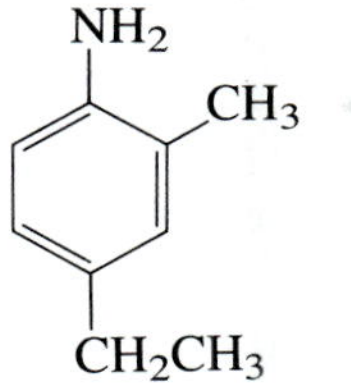 NH_2, CH_3, CH_2CH_3 on benzene ring

(b) 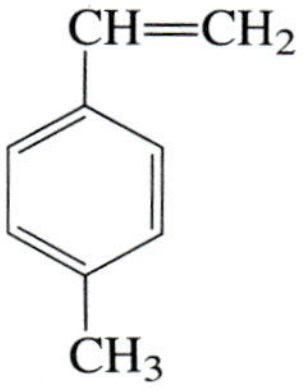 $CH{=}CH_2$, CH_3 on benzene ring

(g) 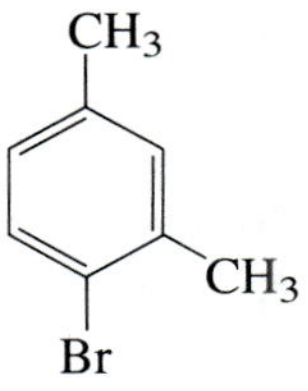 CH_3, CH_3, Br on benzene ring

(c) 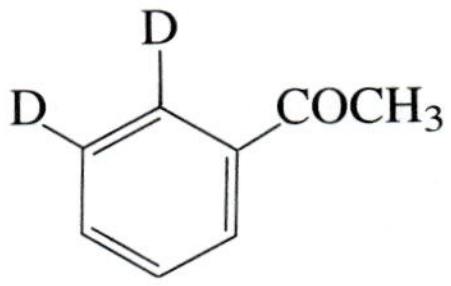 D, D, $COCH_3$ on benzene ring

(h) OH, CH_3O, OCH_3, Br on benzene ring

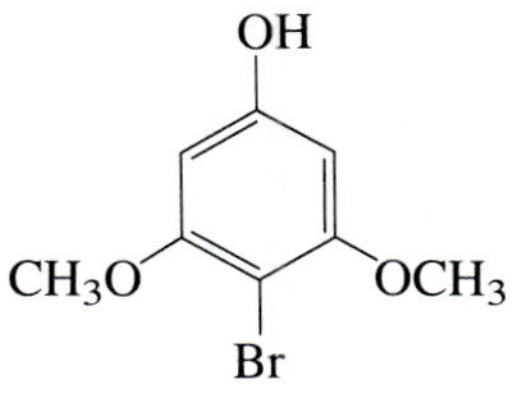

(d) OH, CHO on benzene ring

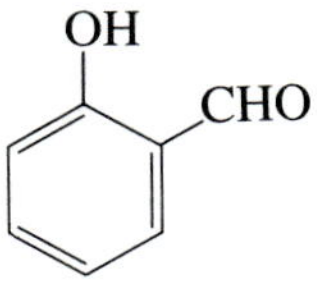

(i) 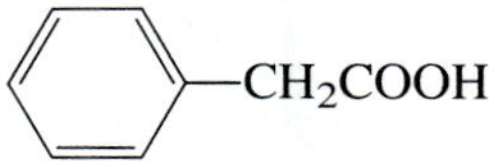 benzene ring —CH_2COOH

(e) 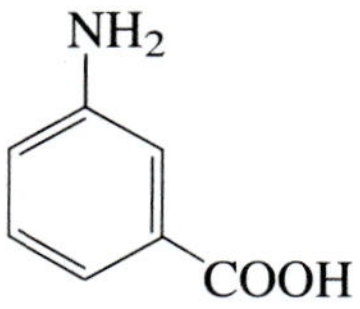 NH_2, COOH on benzene ring

(j) 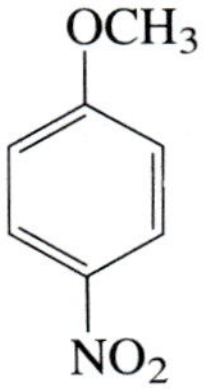 OCH_3, NO_2 on benzene ring

2 Give a proper IUPAC name for each of the commonly named substances below and at the top of the next page.

(a) 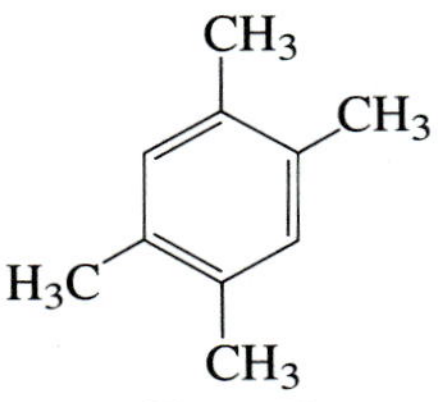 CH_3, CH_3, H_3C, CH_3 on benzene ring
(Durene)

(b) 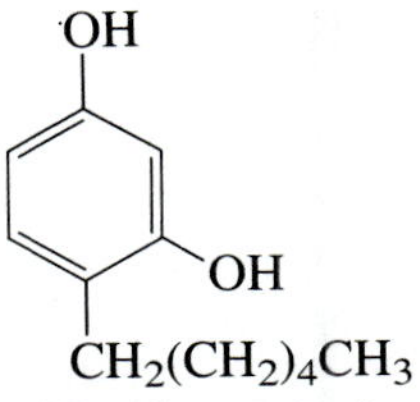 OH, OH, $CH_2(CH_2)_4CH_3$ on benzene ring
(Hexylresorcinol)

(c) $CH_3\overset{\overset{\large O}{\|}}{C}NH$–$C_6H_4$–OH

(Acetaminophen)

(d) $(CH_3)_2CHCH_2$–C_6H_4–$\overset{\overset{\large CH_3}{|}}{C}HCOOH$

(Ibuprofen)

3 Draw the structure of each of the following compounds. If the name itself is incorrect, give a correct systematic alternative.

(a) *o*-Chlorobenzyl alcohol.
(b) 2,4,6-Trihydroxybenzene.
(c) 4-Nitro-*o*-xylene.
(d) *m*-(Dimethylamino)benzoic acid.
(e) 4,5-Dibromoaniline.
(f) *p*-Methoxy-*m*-nitroacetophenone.

4 The complete combustion of benzene is exothermic by approximately -789 kcal mole^{-1}. What would this value be if benzene lacked aromatic stabilization?

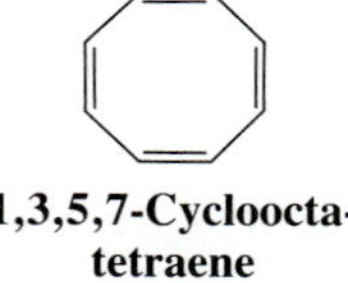

1,3,5,7-Cycloocta-tetraene

5 Complete hydrogenation of 1,3,5,7-cyclooctatetraene is exothermic by -101 kcal mole^{-1}. Hydrogenation of cyclooctene proceeds with $\Delta H^\circ = -23$ kcal mole^{-1}. On the basis of these data, would you call cyclooctatetraene aromatic?

6 From your answer to Problem 5, would you expect more than one distinct isomer of 1,2-dimethylcyclooctatetraene to exist? If so, draw their structures and give their complete IUPAC names.

7 The energy levels of the 2-propenyl (allyl) and cyclopropenyl π systems are compared qualitatively in the following diagram.

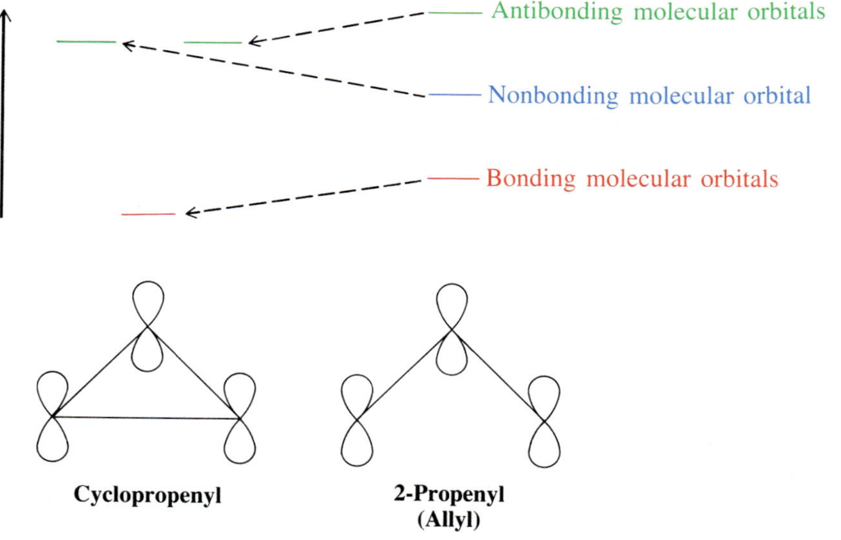

(a) Draw the three molecular orbitals of each system, using plus and minus signs and dotted lines to indicate bonding overlap and nodes, as in Figure 19-5. Does either of these systems possess degenerate molecular orbitals?

(b) How many π electrons would give rise to the maximum stabilization of the cyclopropenyl system, relative to 2-propenyl (allyl)? (Compare Figure 19-6, for benzene.) Draw Lewis structures for both systems with this number of π electrons and any appropriate formal charges.

(c) Could the cyclopropenyl system drawn in part **b** qualify as "aromatic"? Explain.

8 Referring to the description of aromatic transition states in Section 19-2, find as many additional reactions as you can among those presented in Section 14-5 that also have aromatic transition states.

9 Following are spectroscopic and other data for several compounds. Propose a structure for each of them.

(a) Analysis: 30.55% C, 1.71% H, 67.75% Br. ^{1}H NMR spectrum A (on page 894). ^{13}C NMR: 3 peaks. IR: $\tilde{\nu} = 745$ (s, broad) cm^{-1}. UV: $\lambda_{max}(\epsilon) =$ 263(150), 270(250), and 278(180) nm.

(b) Analysis: 79.98% C, 6.71% H. ^{1}H NMR spectrum B (on page 894). ^{13}C NMR: $\delta = 26.3$(q), 128.3(d), 128.6(d), 133.0(d), 137.3(s), and 197.4(s) ppm. IR: $\tilde{\nu} = 1680$(s), 755(s), and 690(s) cm^{-1}. UV: $\lambda_{max}(\epsilon) = 240(13000)$, 278(1100), and 319(50) nm.

(c) Analysis: 70.57% C, 5.92% H. ^{1}H NMR spectrum C (on page 895). IR: $\tilde{\nu} = 1690$(s) and 825(s) cm^{-1}.

(d) Analysis: 44.95% C, 3.78% H, 42.72% Br. ^{1}H NMR spectrum D (on page 895). ^{13}C NMR: 7 peaks. IR: $\tilde{\nu} = 765$(s) and 680(s) cm^{-1}.

(e) Analysis: 54.29% C, 5.57% H, 40.14% Br. ^{1}H NMR spectrum E (on page 896). ^{13}C NMR: $\delta = 20.6$(q), 23.6(q), 124.2(s), 129.0(d), 136.0(s), and 137.7(s) ppm.

10 The species resulting from the addition of benzene to HF-SbF_5 (Exercise 19-10) shows the following ^{13}C NMR absorptions: $\delta = 52.2$(t), 136.9(d), 178.1(d), and 186.6(d) ppm. The signals at $\delta = 136.9$ and $\delta = 186.6$ are twice the intensity of the other signals. Assign the signals in this spectrum.

11 Write the expected major product that should form with the addition to benzene of each of the following reagent mixtures.

(a) $Cl_2 + AlCl_3$

(b) $T_2O + BF_3$ (T = tritium, ^{3}H)

(c) $ICl + FeCl_3$ (Careful! $DH^\circ_{ICl} = 50$ kcal mole^{-1}. Is this reaction exothermic?)

(d) N_2O_5 (which tends to dissociate into NO_2^+ and NO_3^-)

(e) $(CH_3)_2C{=}CH_2 + H_3PO_4$

(f) $(CH_3)_3CCH_2CH_2Cl + AlCl_3$

(g) $(CH_3)_2C(Br)CH_2CH_2C(Br)(CH_3)_2 + AlBr_3$

(h) H_3C—⟨benzene ring⟩—$COCl + SbCl_5$

12 Write mechanisms for reactions **c** and **f** in Problem 11.

13 Propose a mechanism for the direct chlorosulfonylation of benzene (at the right), an alternative synthesis of benzenesulfonyl chloride.

Benzene + 2 $ClSO_3H$ → $C_6H_5SO_2Cl$ + HCl + H_2SO_4

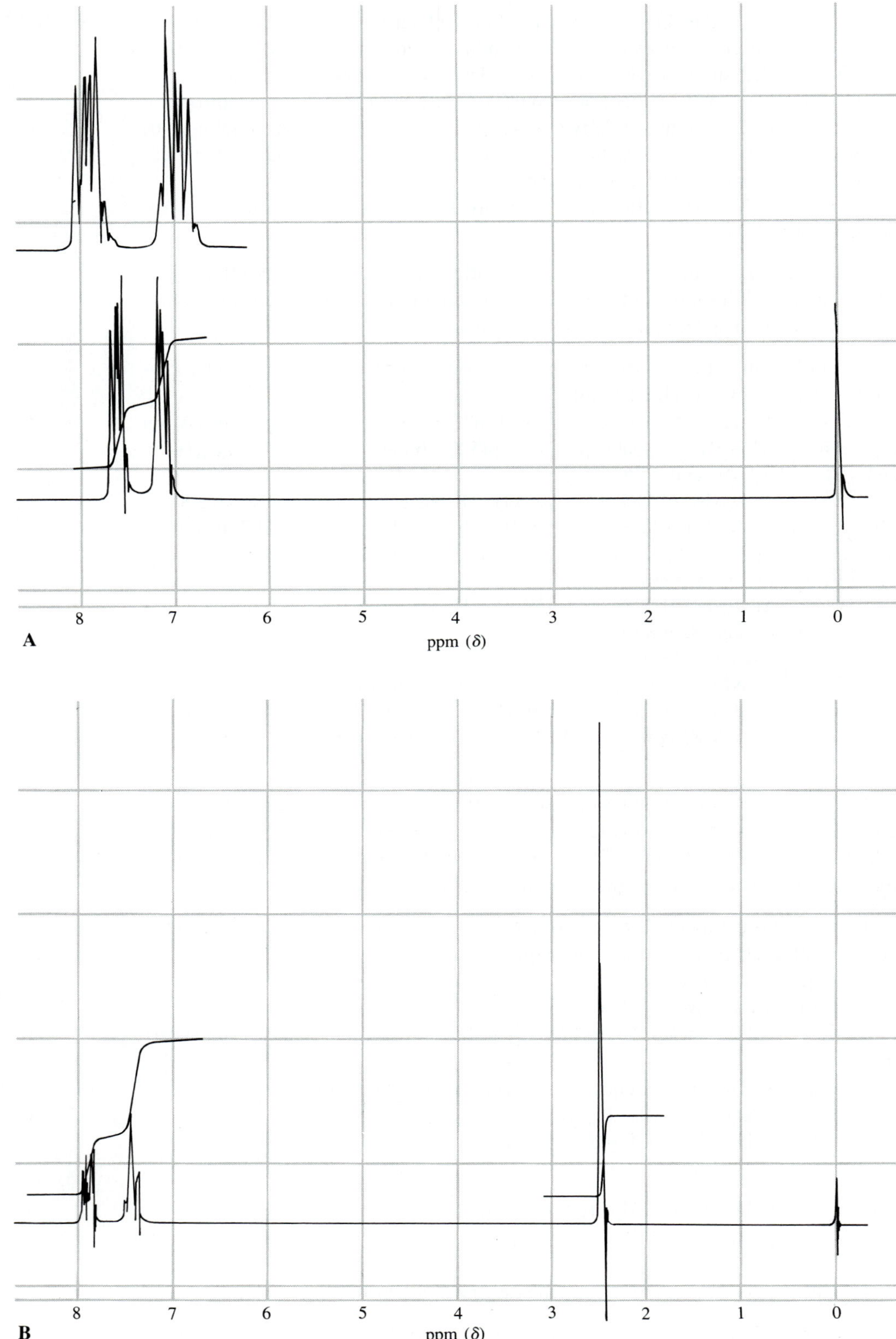
8
7
6
5
4
3
2
1
0
A
ppm (δ)
8
7
6
5
4
3
2
1
0
B
ppm (δ)

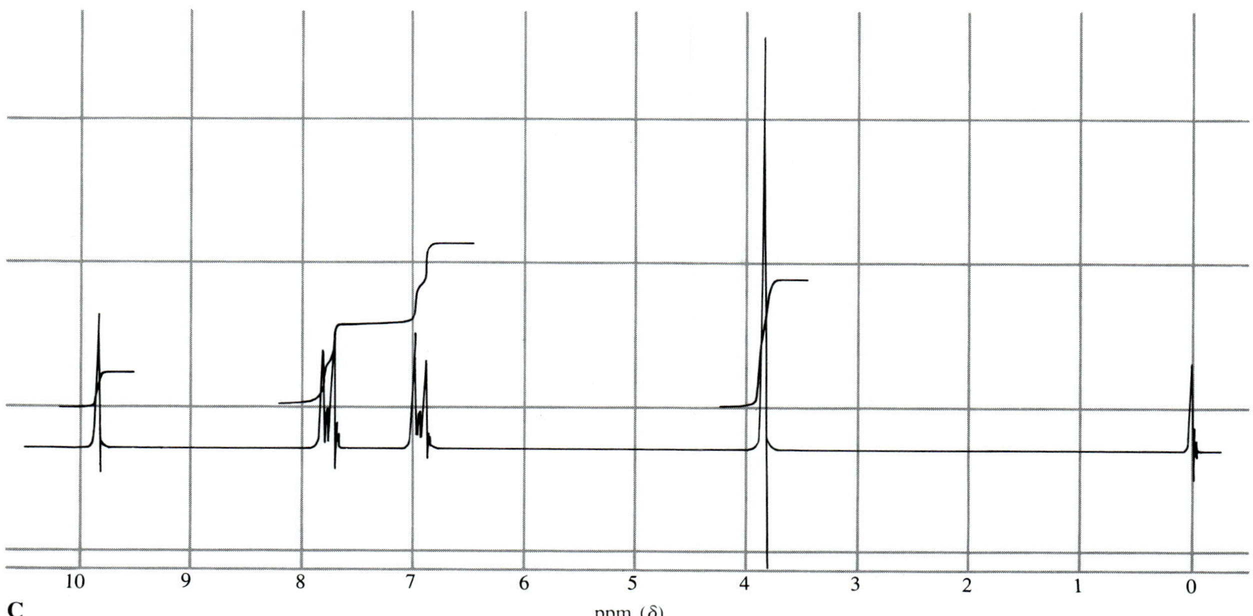
10
9
8
7
6
5
4
3
2
1
0
C
ppm (δ)

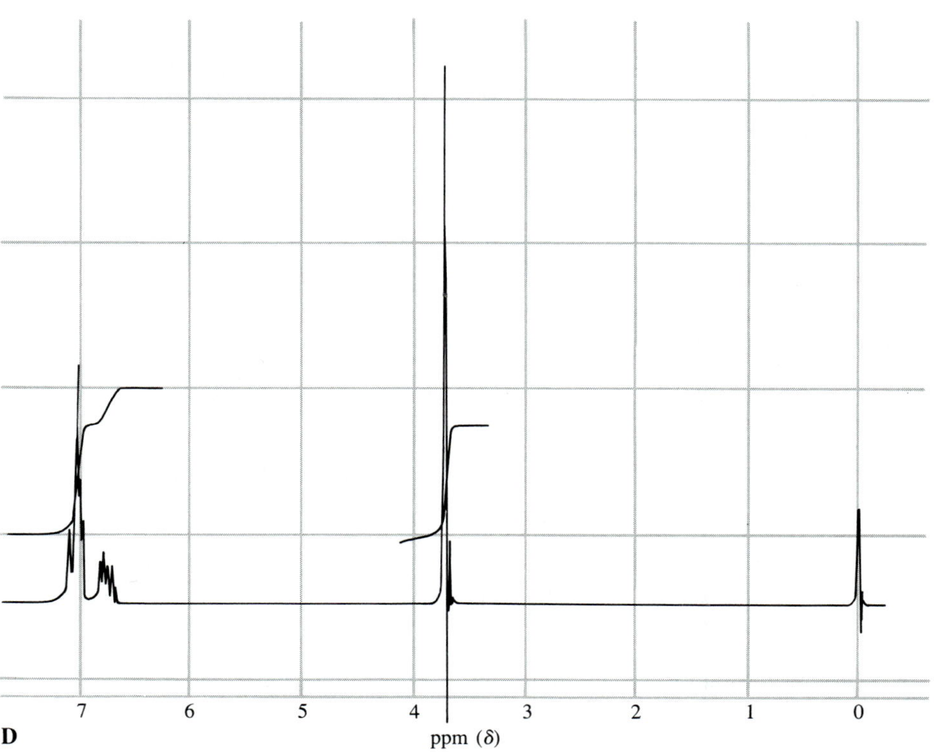
7
6
5
4
3
2
1
0
D
ppm (δ)

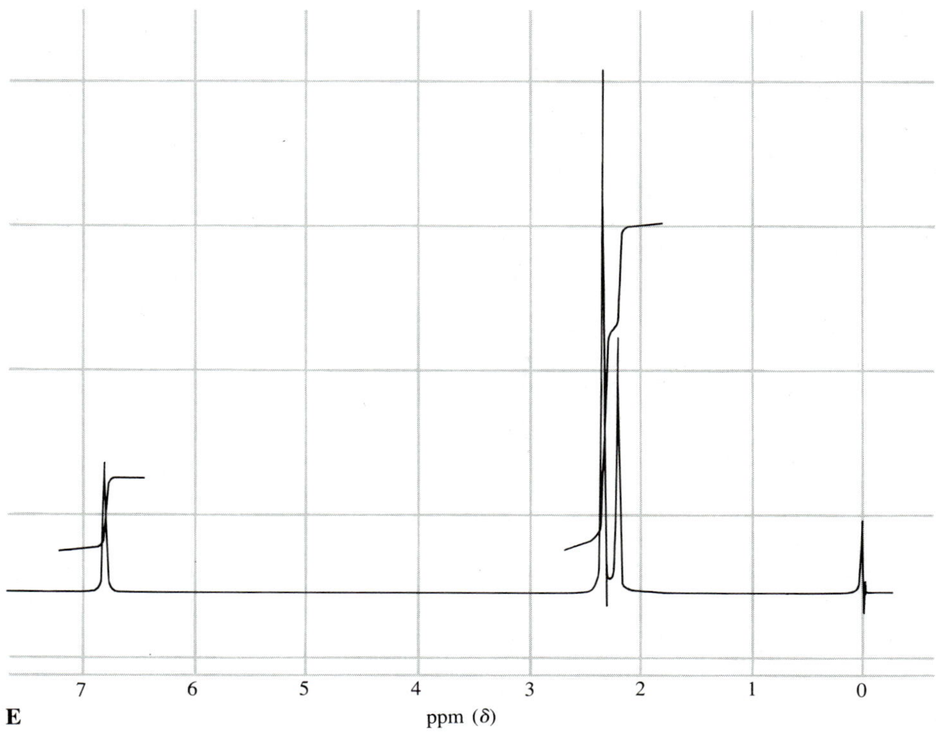

E

14 Besides the Gattermann-Koch reaction (Exercise 19-17), another method for direct introduction of the methanoyl (formyl) group into benzene is the Vilsmeier-Haack reaction:

$$C_6H_6 \xrightarrow[2.\ H_2O]{1.\ HC(=O)N(CH_3)_2,\ POCl_3} C_6H_5CHO$$

Propose a mechanism for this process.

15 Explain the following reaction and the indicated stereochemical result mechanistically.

$$C_6H_6 + H_2C\overset{O}{—}C(H)CH_3 \xrightarrow[2.\ H^+,\ H_2O]{1.\ AlCl_3} C_6H_5C(H)(CH_3)CH_2OH$$

16 A commonly used intramolecular variation of the Friedel-Crafts reaction with ketones gives rise to rings:

$$C_6H_5CH_2CH_2CH_2C(=O)CH_3 \xrightarrow[-\ H_2O]{H_3PO_4,\ \Delta} \text{4-methyl-1,2-dihydronaphthalene } (CH_3)$$

(a) Propose a mechanism for this reaction.

(b) Outline a possible synthesis of the starting ketone, beginning with either benzene or methylbenzene (toluene).

(c) Write the structure of the expected product of the first of the following reactions, and propose two different schemes for carrying out the subsequent transformation.

$$\text{C}_6\text{H}_6 + \text{succinic anhydride} \xrightarrow[2.\ H^+,\ H_2O]{1.\ AlCl_3} C_{10}H_{10}O_3 \xrightarrow{?} \text{(1-methyl-3,4-dihydronaphthalene, } CH_3\text{)}$$

IR: $\tilde{\nu}$ = 3000 (very broad), 1710, 1680, 755, and 690 cm^{-1}
UV: λ_{max} = 240, 278, and 319 nm

17 Propose a reasonable synthesis of each of the following compounds, using benzene as your starting material.

(a) C_6H_5—$CH_2CH_2CH_2CH_3$ (Careful! Will Friedel-Crafts *alkylation* work?)

(b) C_6H_5—$C(CH_2CH_3)_3$

(c) C_6H_5—$\overset{O}{\overset{\|}{C}}CH_2CH_2\overset{O}{\overset{\|}{C}}$—$C_6H_5$

(d) Two benzene rings bridged at ortho positions by CH_2CH_2 and CH_2CH_2 (In addition to benzene, you will need another aromatic starting material.)

18 The text states that alkylated benzenes are more susceptible to electrophilic attack than benzene itself. Draw a graph like Figure 19-16 to show how the energy profile of electrophilic substitution of methylbenzene (toluene) would differ qualitatively from that of benzene.

19 Starting with benzene and anything else you need, formulate syntheses of cinnamic acid and cinnamaldehyde, both constituents of natural cinnamon oils. (Hint: see Exercise 19-17.) Give a systematic name for each molecule.

C_6H_5—CH=CHCOOH
Cinnamic acid

C_6H_5—CH=CH$\overset{O}{\overset{\|}{C}}$H
Cinnamaldehyde

HgOCCH$_3$ (with C=O)

Phenylmercury ethanoate

20. Metal-substituted benzenes have a long history of use in medicine. Before antibiotics were discovered, phenylarsenic derivatives were the only treatment for a number of diseases. Phenylmercury compounds continue to be used as fungicides and antimicrobial agents to the present day. Based on the general principles explained in this chapter and your knowledge of the characteristics of compounds of Hg^{2+} (see Section 12-3), propose a sensible synthesis of phenylmercury ethanoate.

21 Like haloalkanes, haloarenes are readily converted into organometallic reagents, which are sources of nucleophilic carbon:

$$C_6H_5\text{-}Br \xrightarrow{Mg,\ (CH_3CH_2)_2O,\ 25°C} C_6H_5^{\delta-}\text{-}\overset{\delta+}{MgBr}$$

Phenylmagnesium bromide

$$C_6H_5\text{-}Cl \xrightarrow{Mg,\ THF,\ 50°C} C_6H_5^{\delta-}\text{-}\overset{\delta+}{MgCl}$$

Phenylmagnesium chloride

Grignard reagents

$$C_6H_5\text{-}Br \xrightarrow{Li,\ (CH_3CH_2)_2O,\ 25°C} C_6H_5^{\delta-}\text{-}\overset{\delta+}{Li} \xrightarrow{CuI,\ (CH_3CH_2)_2O,\ 0°C} (C_6H_5)_2CuLi$$

Phenyllithium **Lithium diphenylcuprate**

The chemical behavior of these reagents is very similar to that of their alkyl counterparts. Write the main product of each of the following reaction sequences:

(a) $C_6H_5Br \xrightarrow[\text{3. } H^+,\ H_2O]{\text{1. Li, } (CH_3CH_2)_2O \quad \text{2. } CH_3CHO}$

(b) $C_6H_5Cl \xrightarrow[\text{3. } H^+,\ H_2O]{\text{1. Mg, THF} \quad \text{2. } CH_2\text{—}CH_2 \text{ (epoxide, O bridging)}}$

(c) $2\ C_6H_5Br \xrightarrow[\text{3. } H^+,\ H_2O]{\text{1. 2 Mg, } (CH_3CH_2)_2O \quad \text{2. } CH_3CH_2COOCH_3}$

(d) $C_6H_5Br \xrightarrow[\text{4. } H^+,\ H_2O]{\text{1. Li, } (CH_3CH_2)_2O \quad \text{2. CuI, } (CH_3CH_2)_2O \quad \text{3. cyclohexenone}}$

22 Starting with benzene, design reasonable syntheses of each of the following compounds. Make use of the chemistry outlined in Problem 21.

(a) C_6H_5—CH_2OH

Phenylmethanol
(Benzyl alcohol)
Used as a protecting group; Sections 24-2 and 27-5

(b) C_6H_5—$CH_2CH(CH_3)NH_2$

Amphetamine
The prototypical central nervous system stimulant

(c) $(C_6H_5)_2C(OH)C(=O)OCH_2CH_2\overset{+}{N}H(CH_2CH_3)_2Br^-$

Benzactyzine
One of a recently developed family of drugs custom designed to pass through membranes separating brain cells from the blood; this one is a tranquilizer

CHAPTER 20

Electrophilic and Nucleophilic Attack on Derivatives of Benzene

SUBSTITUENTS CONTROL REGIOSELECTIVITY

Chapter 19 introduced a new reaction: the electrophilic replacement of a hydrogen on a benzene ring. With most electrophiles, this process stops at the stage of monosubstitution because the newly introduced group deactivates the benzene ring, protecting it from further electrophilic attack. This effect does not completely prevent further substitution; it simply slows it down. Thus, more-highly-substituted benzenes do arise from this reaction, if more of the first electrophile (or another one) reacts with the initial product. The same is true of substitutions in which the substituent group is electron donating, as in the Friedel-Crafts alkylation—here, in fact, higher substitution is facilitated. This chapter will more clearly define the nature of activating and deactivating groups on the benzene nucleus. The substituted benzene ring can also be attacked by nucleophiles and bases, through other mechanisms for further substituting the aromatic ring. With this chapter, all the major types of mechanisms that control the fate of reacting organic molecules will have been covered. Section 20-7 will review them.

20-1 Activation and Deactivation of the Benzene Ring

What are the factors that control the activation or deactivation of the benzene nucleus with respect to electrophilic aromatic substitution? A substituent on the benzene ring can have either an activating effect, by donating electron density, or a deactivating effect, by withdrawing it. In this way, a monosubstituted benzene may be more or less susceptible than benzene to further electrophilic attack. Donation and withdrawal of electron density can be due to either inductive effects or resonance.

Inductive Activation and Deactivation by Alkyl Groups

Consider, for example, methylbenzene (toluene). The alkyl group is by induction (and by hyperconjugation, Section 7-3) electron donating; therefore, the molecule is more reactive than benzene to further substitution. On the other hand, the strongly electronegative fluorine atoms in (trifluoromethyl)benzene render the trifluoromethyl group electron withdrawing. Consequently, this benzene derivative is less reactive than benzene in the presence of electrophiles.

Effect of Electron Donor and Acceptor Substituents on the Reactivity of the Benzene Ring

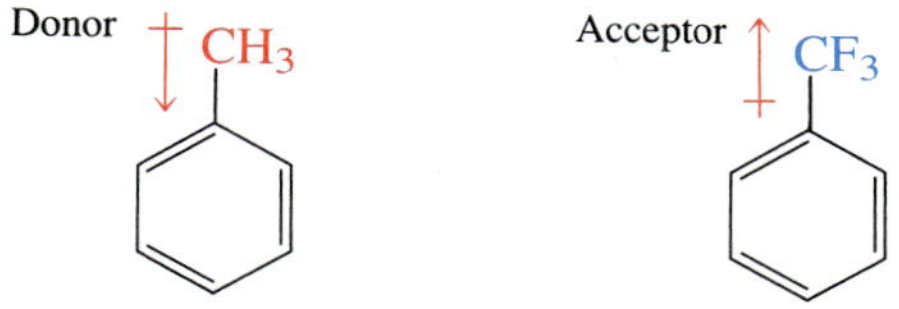

Donor CH_3 — Acceptor CF_3

Relatively electron-rich ring (more reactive) — Relatively electron-poor ring (less reactive)

EXERCISE 20-1

Rank the compounds at the right in order of decreasing activity in electrophilic substitution:

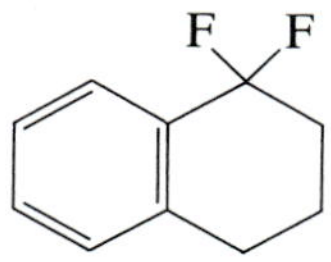

F F

A

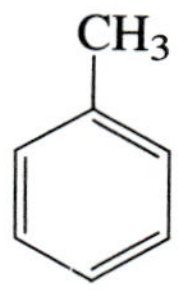

CH_3

B

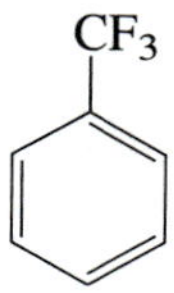

CF_3

C

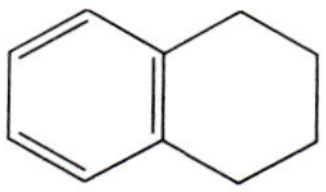

D

Resonance Contributions of the Amino and Hydroxy Groups Override Their Inductive Influence

Let us now consider substituents capable of participating in resonance, such as the amino group in benzenamine (aniline), $C_6H_5\ddot{N}H_2$. Nitrogen is more electronegative than carbon, and therefore the amino group is inductively electron withdrawing. However, the lone electron pair on the nitrogen atom may delocalize into the aromatic π system, adding negative charge to the ring. As with enamines (Section 15-6), this resonance contribution by far outweighs the inductive effect. Consequently, benzenamine (aniline) is activated to further substitution.

Inductive and Resonance Effects in Benzenamine (Aniline)

NH_2 — [$:NH_2$ ⟷ $^+NH_2$ ⟷ $^+NH_2$ ⟷ $^+NH_2$]

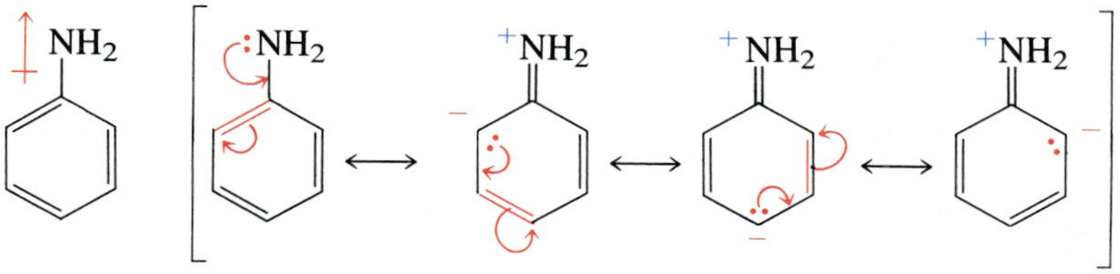

Inductive effect (minor) — **Resonance effect (major)**

In benzenol (phenol), C_6H_5OH, the electronegative oxygen atom also tends to withdraw electrons inductively. Nevertheless, resonance again has an overriding influence, leading to activation of the benzene nucleus.

Inductive and Resonance Effects in Benzenol (Phenol)

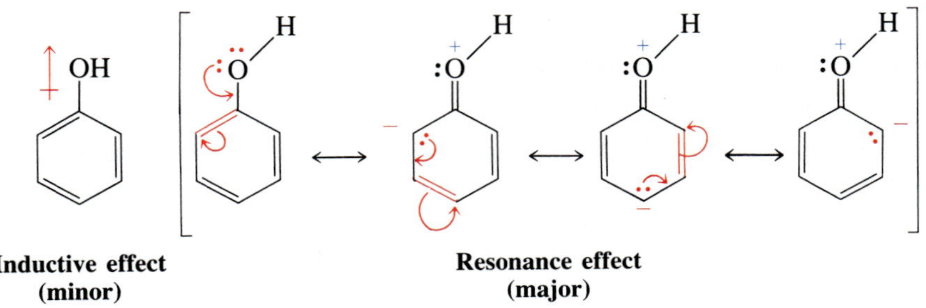

Inductive effect (minor)

Resonance effect (major)

EXERCISE 20-2

The pK_a of benzenol (phenol) is 10, much lower than that of an alkanol. Explain this phenomenon in terms of resonance structures of the molecule as well as of the corresponding anion. Similarly, benzenamine (aniline) is much less basic than methanamine (Section 21-3). Explain.

Halogen Substituents: Inductive Effects Override Resonance

In the haloarenes, both inductive and resonance contributions play an important role. The fluorine substituent is so strongly electronegative that, despite resonance, fluorobenzene is slightly deactivated compared with benzene. This deactivating tendency might be expected to become weaker in progression down the periodic table, because the electronegativity of the halogens decreases (see Table 1-4). However, all the halobenzenes are deactivated because of the diminishing ability of the bulky chlorine, bromine, and iodine atoms to enter into resonance with the π system of the aromatic ring. Remember that resonance is caused by π orbital overlap, which is most effective between orbitals of comparable size and energy (Section 1-7). This condition is fulfilled less and less in chloro-, bromo-, and iodobenzene (Figure 20-1); that is, in progression down the periodic table, both inductive and resonance contributions diminish simultaneously, the first always being slightly stronger than the second.

Deactivation by Resonance

Resonance may lead to the deactivation of the benzene ring. A good example is benzenecarboxylic (benzoic) acid, C_6H_5COOH, in which the carboxy group removes electron density by resonance. Because of the polarization of the C–O double bond (Section 15-2), dipolar resonance structures place a positive charge in the ring. Nitroarenes, aromatic carbonyl compounds, arenenitriles, and arenesulfonic acids are similarly affected.

Electron Withdrawal by Resonance in Benzenecarboxylic (Benzoic) Acid

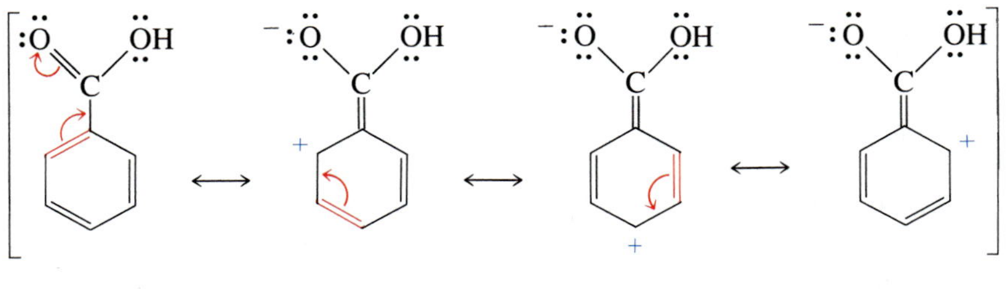

FIGURE 20-1

Overlap between the $2p$ orbital on fluorine and the π system of benzene (A). The orbitals are well matched; hence, the resulting resonance donation compensates for the inductive electron-attracting power of fluorine. With the larger halogens, overlap becomes progressively worse (B). Because electronegativity simultaneously decreases, the net effect—weak deactivation—stays approximately constant.

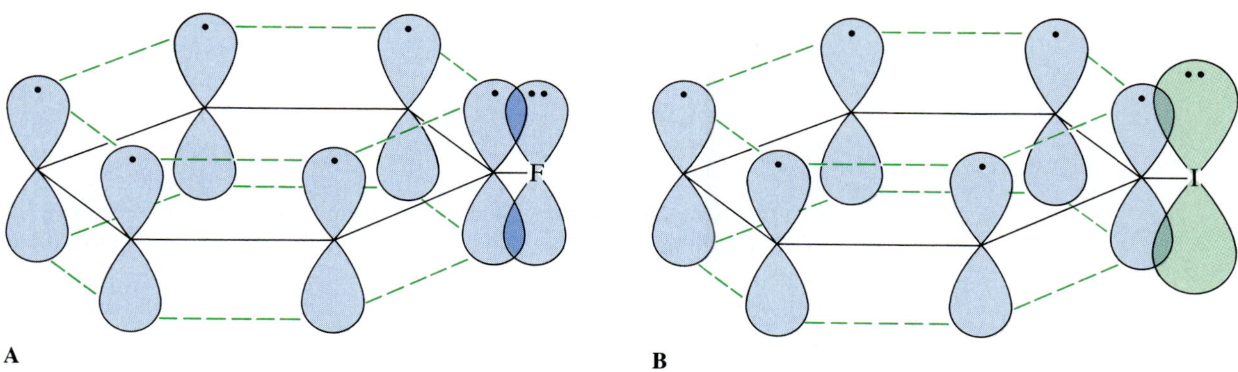

EXERCISE 20-3

Specify whether the benzene rings in the following compounds are activated or deactivated:

A: CH_2CH_3 / CH_2CH_3 (para-substituted benzene)

B: (bicyclic ketone, O)

C: COOH / CF_3

D: OCH_3 / $N(CH_3)_2$

In summary, induction and resonance determine whether a substituent activates or deactivates the benzene ring. Except for halogens, resonance effects typically are stronger than inductive ones. Table 20-1 lists several common substituents in aromatic compounds and their effects on the benzene ring.

TABLE 20-1

Activating and deactivating substituents on the benzene ring

Strong activators	**Strong deactivators**
$-\ddot{N}H_2$, $-\ddot{N}HR$, $-\ddot{N}R_2$,	$-NO_2$, $-CF_3$, $-\overset{+}{N}R_3$, $-\overset{O}{\overset{\Vert}{C}}OH$,
$-\ddot{\underset{\cdot\cdot}{O}}H$, $-\ddot{\underset{\cdot\cdot}{O}}R$	$-\overset{O}{\overset{\Vert}{C}}OR$, $-\overset{O}{\overset{\Vert}{C}}R$, $-SO_3H$, $-C\equiv N$
Weak activators	**Weak deactivators**
Alkyl, phenyl	$-\ddot{\underset{\cdot\cdot}{F}}:$, $-\ddot{\underset{\cdot\cdot}{Cl}}:$, $-\ddot{\underset{\cdot\cdot}{Br}}:$, $-\ddot{\underset{\cdot\cdot}{I}}:$

EXERCISE 20-4

Explain why (a) —NO_2, (b) —$\overset{+}{N}R_3$, and (c) —SO_3H are strong deactivators. (d) Why should phenyl be a weak activator? (Hint: Draw dipolar resonance structures for as many of the substituted benzenes in question as you can.)

20-2 Where Does a Second Group Go? Directing Inductive Effects by Alkyl Substituents

Does a substituent already present exert any kind of control over where an electrophile will attack next? Let us inspect the distributions of products derived from some of the substrates discussed in the preceding section. We shall begin with methylbenzene (toluene), in which the benzene ring bears an inductively activating methyl group.

Inductively Activating Substituents Direct *Ortho* and *Para*

Electrophilic bromination of methylbenzene (toluene) results mainly in *para* (60%) and *ortho* (40%) substitution, with virtually no *meta* product.

Electrophilic Bromination of Methylbenzene (Toluene) Gives *Ortho* and *Para* Substitution

CH_3 / Br—Br, $FeBr_3$, CCl_4 / − HBr

40%	<1%	60%
2-Bromo-1-methylbenzene (*o*-Bromotoluene)	**3-Bromo-1-methylbenzene (*m*-Bromotoluene)**	**4-Bromo-1-methylbenzene (*p*-Bromotoluene)**

Is bromination a special case? The answer is no; nitration and sulfonation give the same qualitative results: mainly *ortho* and *para* substitution. Evidently, the nature of the attacking electrophile has little influence on the isomer distribution.

CH_3 / HNO_3, H_2SO_4, 30°C / − H_2O

60%	5%	35%
1-Methyl-2-nitrobenzene (*o*-Nitrotoluene)	**1-Methyl-3-nitrobenzene (*m*-Nitrotoluene)**	**1-Methyl-4-nitrobenzene (*p*-Nitrotoluene)**

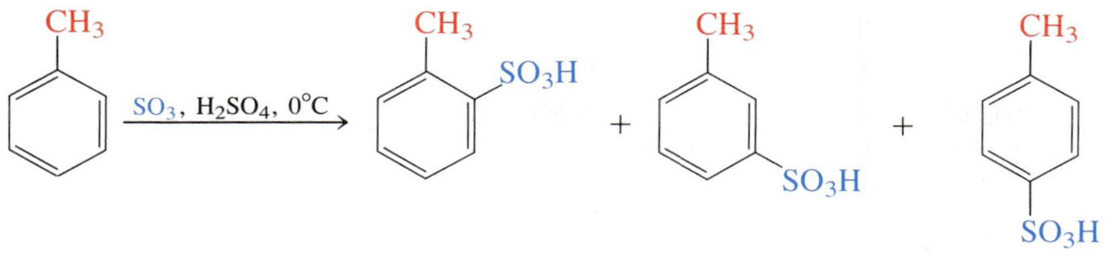

CH_3-benzene $\xrightarrow{SO_3,\ H_2SO_4,\ 0°C}$ 2-SO_3H + 3-SO_3H + 4-SO_3H

43%	4%	53%
2-Methylbenzene-sulfonic acid (*o*-Toluene-sulfonic acid)	**3-Methylbenzene-sulfonic acid (*m*-Toluene-sulfonic acid)**	**4-Methylbenzene-sulfonic acid (*p*-Toluene-sulfonic acid)**

Can we explain this selectivity by a mechanism? Let us inspect the possible resonance structures of the cations formed after the electrophile, E^+, has attacked the ring.

Consequences of *Ortho, Meta,* and *Para* Attack on Methylbenzene (Toluene)

Ortho attack: (E^+ = electrophile)

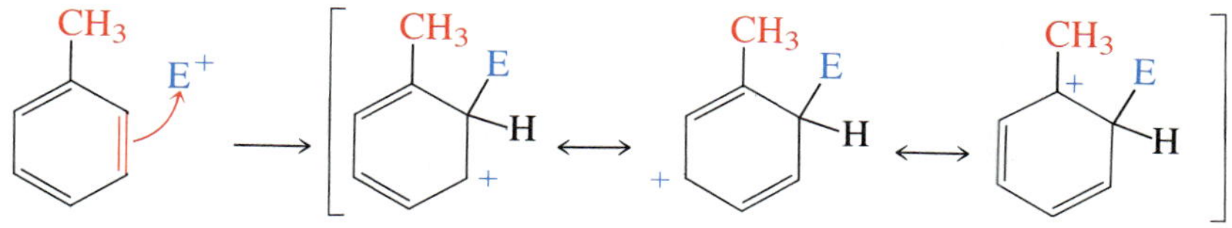

Most-significant resonance structure

Meta attack:

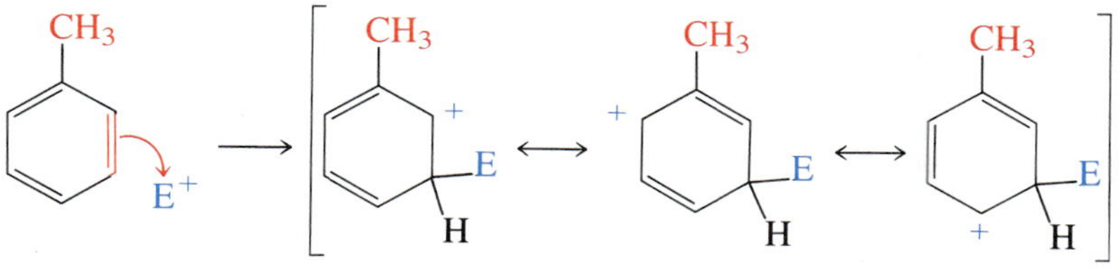

Para attack:

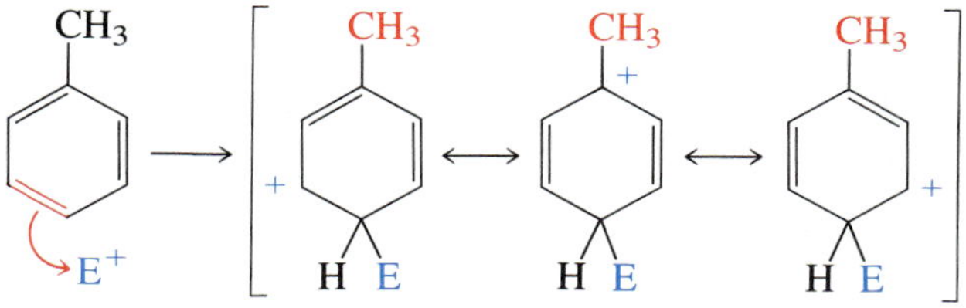

Most-significant resonance structure

Only attack at the *ortho* and *para* positions produces a hexadienyl cation in which a resonance structure places the positive charge *next* to the alkyl substituent. Because that structure has tertiary carbocation character, it is more important than the others, in which the positive charge is at a secondary carbon. On the other hand, *meta* attack produces an intermediate in which *none* of the resonance structures benefits from such tertiary carbocation stabilization. Thus, electrophilic attack on a carbon located *ortho* or *para* to the methyl (or other alkyl) group leads to a cationic intermediate that is more stable than the one derived by attack at the *meta* carbon. This intermediate forms relatively rapidly, through a transition state that is of relatively low energy.

EXERCISE 20-5

Propose a possible synthesis of 1,4-diethylbenzene from ethylbenzene. Hint: This synthesis may require separation of isomeric intermediates.

Inductively Deactivating Substituents Direct *Meta*

What about (trifluoromethyl)benzene? This compound is deactivated with respect to electrophilic attack, and therefore reacts only sluggishly. Nevertheless, under stringent conditions, substitution does take place but *only* at the *meta* positions.

Electrophilic Nitration of (Trifluoromethyl)benzene Gives *Meta* Substitution

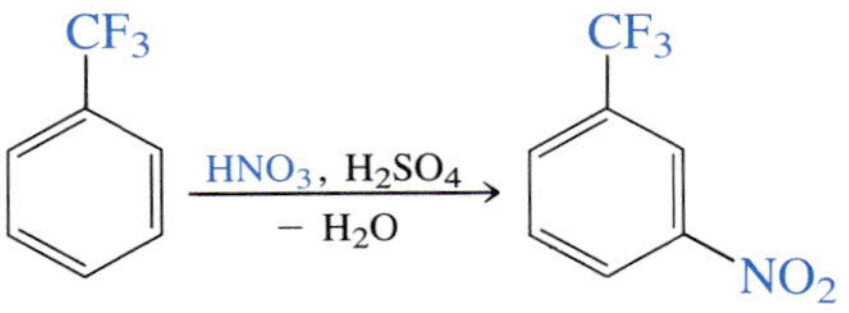

CF_3 — $\xrightarrow[-\ H_2O]{HNO_3,\ H_2SO_4}$ — CF_3, NO_2

Only product

Once again, the explanation lies in the various resonance structures produced by *ortho, meta,* and *para* attack.

Consequences of *Ortho, Meta,* and *Para* Attack on (Trifluoromethyl)benzene

Ortho attack:

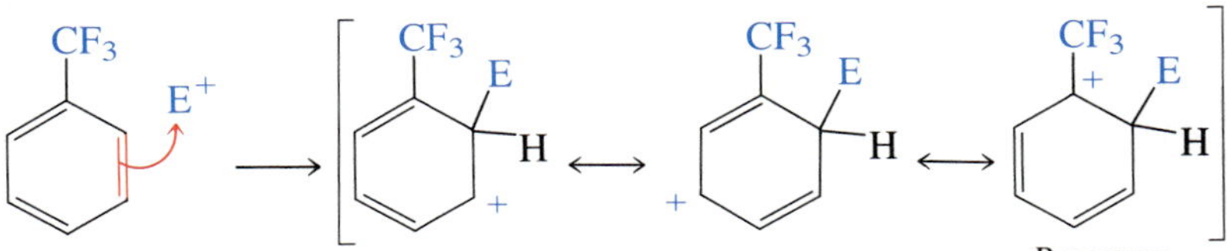

Resonance structure with little contribution

Meta attack:

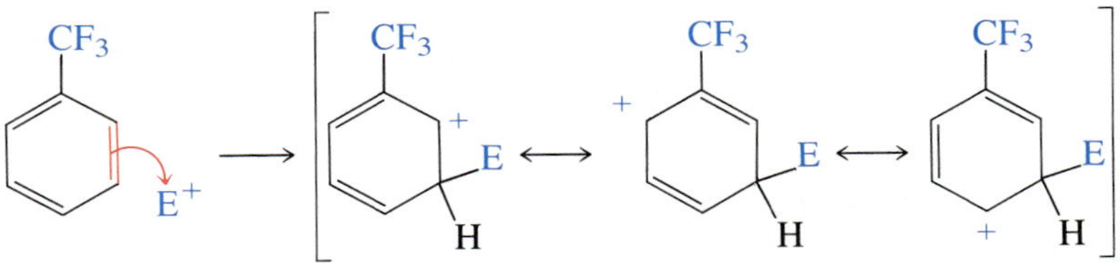

Para attack:

Ortho and *para* attack are relatively unfavored for the same reasons that they are favored with methylbenzene (toluene): in each case, one of the resonance structures in the intermediate cation places the positive charge next to the substituent. This structure is stabilized by an electron-donating group, but it is destabilized by an electron-withdrawing substituent—removing electron density from a positively charged center is energetically unfavored. *Meta* attack avoids this situation. Therefore, the trifluoromethyl group directs substitution *meta*, or, more accurately, *away* from the *ortho* and *para* carbons.

EXERCISE 20-6

Electrophilic bromination of an equimolar mixture of methylbenzene (toluene) and (trifluoromethyl)benzene with one equivalent of bromine gives only 2- and 4-bromo-1-methylbenzene. Explain.

In summary, electron-donating substituents inductively activate the benzene ring and direct electrophiles *ortho* and *para;* their electron-accepting counterparts deactivate the benzene ring and direct electrophiles to the *meta* positions.

20-3 Directing Effects by Substituents in Resonance with the Benzene Ring

What directing power do substituents have whose electrons are in resonance with those of the benzene ring? This section answers this question by again comparing the resonance structures of the intermediates formed by the various modes of electrophilic attack.

Groups That Activate by Resonance Direct *Ortho* and *Para*

Electrophilic attack on resonance-activated benzene rings does not even always require a catalyst. It proceeds rapidly and in a completely regioselective manner to give (often repeated) *ortho* and *para* substitution.

Electrophilic Bromination of Benzenamine (Aniline) and Benzenol (Phenol) Gives *Ortho* and *Para* Substitution

NH_2 — $\xrightarrow[-\ 3\ HBr]{3\ Br{-}Br,\ H_2O}$ — NH_2, Br, Br, Br — 100%

Benzenamine (Aniline) → **2,4,6-Tribromobenzenamine (2,4,6-Tribromoaniline)**

OH — $\xrightarrow[-\ 3\ HBr]{3\ Br{-}Br,\ H_2O}$ — OH, Br, Br, Br — 100%

Benzenol (Phenol) → **2,4,6-Tribromobenzenol (2,4,6-Tribromophenol)**

Derivatives of benzenamine (aniline) and benzenol (phenol), such as *N*-phenylethanamide (acetanilide) and methoxybenzene (anisole), react in a more controlled manner to give monosubstituted products.

N-Phenylethanamide (Acetanilide) $\xrightarrow[-\ H_2O]{HNO_3,\ H_2SO_4,\ 20°C}$

- 21% **N-(2-Nitrophenyl)-ethanamide (*o*-Nitroacetanilide)**
- + Trace **N-(3-Nitrophenyl)-ethanamide (*m*-Nitroacetanilide)**
- + 79% **N-(4-Nitrophenyl) ethanamide (*p*-Nitroacetanilide)**

Methoxybenzene (Anisole) $\xrightarrow[-\ H_2O]{HNO_3,\ 45°C}$

- 40% **1-Methoxy-2-nitrobenzene (*o*-Nitroanisole)**
- + 2% **1-Methoxy-3-nitrobenzene (*m*-Nitroanisole)**
- + 58% **1-Methoxy-4-nitrobenzene (*p*-Nitroanisole)**

Again, the observed regioselectivity can be explained by writing resonance structures for the various intermediate cations.

Consequences of *Ortho, Meta,* and *Para* Attack on Benzenamine (Aniline)

Ortho attack:

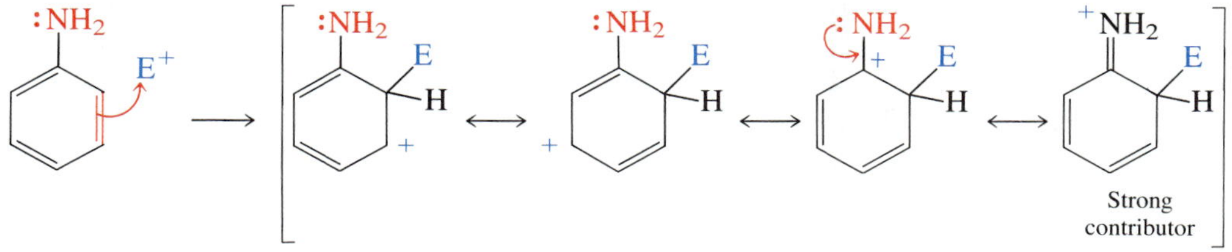

Strong contributor

Meta attack:

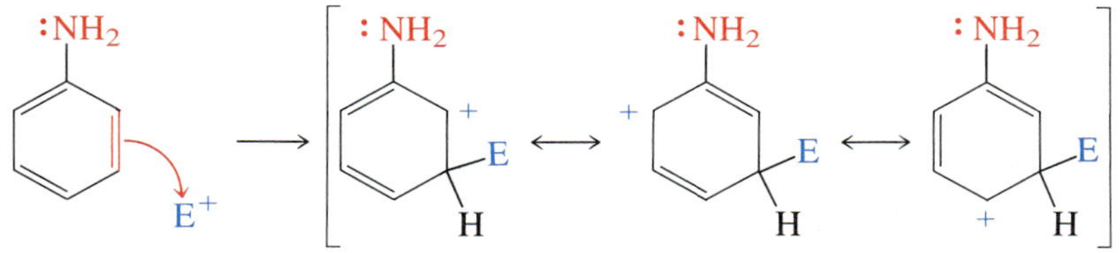

Para attack:

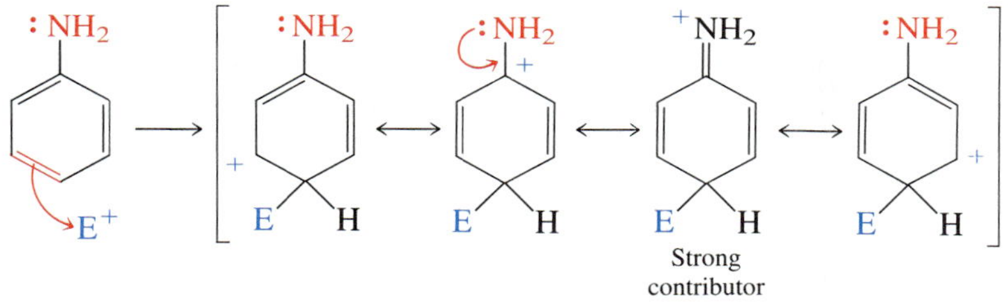

Strong contributor

Ortho and *para* substitutions are favored because they furnish cations for which four resonance structures can be formulated, rather than only three from *meta* attack. In benzenamine (aniline), the extra structure is due to electron donation by the heteroatom, which creates an iminium ion.

EXERCISE 20-7

Formulate resonance structures for the various modes of electrophilic attack on methoxybenzene (anisole).

EXERCISE 20-8

N-Phenylethanamide (acetanilide) is brominated more slowly than benzenamine (aniline). Explain. [Hint: Consider the effect of the ethanoyl (acetyl) substituent on the ability of the nitrogen lone electron pair to delocalize into the benzene ring.]

EXERCISE 20-9

In strongly acidic solution, benzenamine (aniline) becomes quite unreactive to electrophilic attack, and increased *meta* substitution is observed. Explain.

Groups That Deactivate by Resonance Direct *Meta*

Let us now turn to benzene derivatives that bear groups that deactivate by resonance. Benzenecarboxylic (benzoic) acid, for example, undergoes nitration at only about one-thousandth the rate of benzene. For comparison, methoxybenzene (anisole) enters into the same reaction about a thousand times as fast as benzene. However, whereas methoxybenzene gives almost exclusively *ortho* and *para* substitution, nitration of benzenecarboxylic acid gives predominantly *meta* substitution.

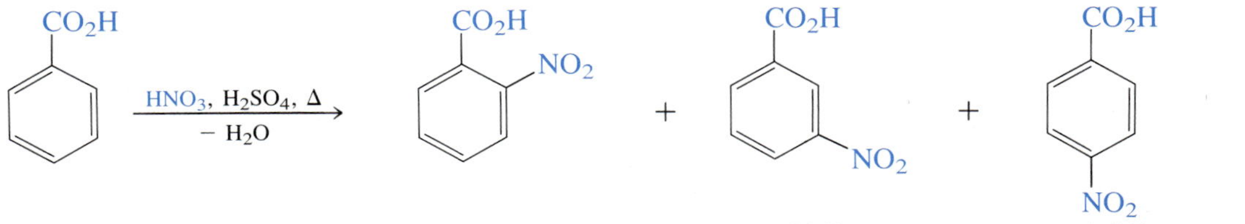

CO_2H — $\xrightarrow[-\ H_2O]{HNO_3,\ H_2SO_4,\ \Delta}$ — CO_2H, NO_2 + CO_2H, NO_2 + CO_2H, NO_2

18.5%
2-Nitrobenzene-carboxylic acid (*o*-Nitrobenzoic acid)

80%
3-Nitrobenzene-carboxylic acid (*m*-Nitrobenzoic acid)

1.5%
4-Nitrobenzene-carboxylic acid (*p*-Nitrobenzoic acid)

Resonance structures again explain this selectivity.

Consequences of *Ortho*, *Meta*, and *Para* Attack on Benzenecarboxylic (Benzoic) Acid

Ortho attack:

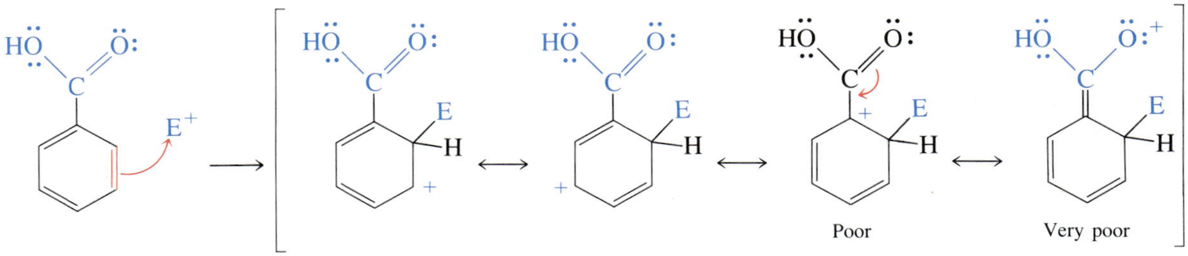

HO, O:, C, E^+ ⟶ [HO, O:, C, E, H, + ⟷ HO, O:, C, E, H, + ⟷ HO, O:, C, +, E, H ⟷ HO, $O:^+$, C, E, H]

Poor

Very poor

Meta attack:

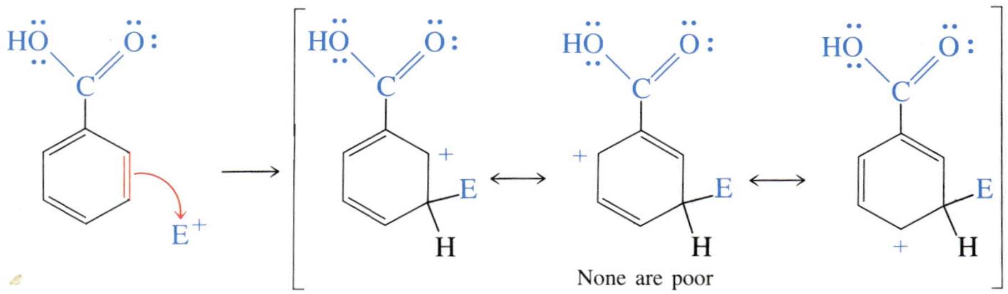

HO, O:, C, E^+ ⟶ [HO, O:, C, +, E, H ⟷ HO, O:, C, +, E, H ⟷ HO, O:, C, E, +, H]

None are poor

Para attack:

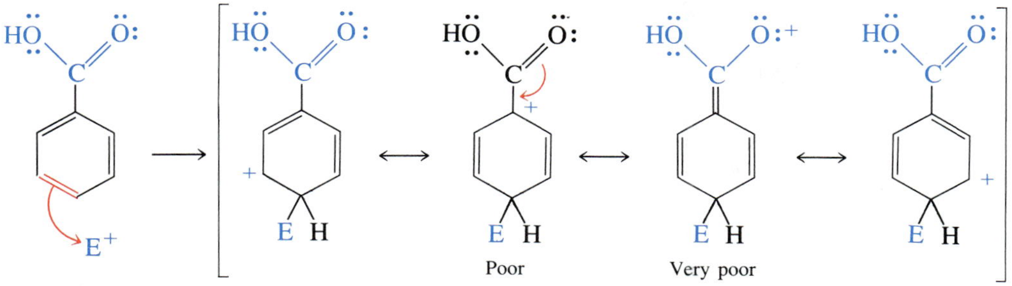

Poor

Very poor

Attack at the *meta* position avoids placing the positive charge next to the electron-withdrawing carboxy group, whereas *ortho* and *para* attacks necessitate the formulation of poor resonance structures. Similar situations arise with other deactivating substituents, such as —SO_3H, —C≡N, —CHO, —$COOCH_3$, and —NO_2.

Hence, it appears that deactivating groups, whether operating by induction or resonance, direct incoming electrophiles to the *meta* position, whereas activating groups direct to the *ortho* and *para* carbons. This statement is true for all classes of substituents except one: the halogens.

There Is Always an Exception: Halogen Substituents, Although Deactivating, Direct *Ortho* and *Para*

Halogen substituents inductively withdraw electron density, although they are donors by resonance. On average, the inductive effect wins out, rendering haloarenes deactivated (Section 20-1). Nevertheless, electrophilic substitution occurs mainly at the *ortho* and *para* positions.

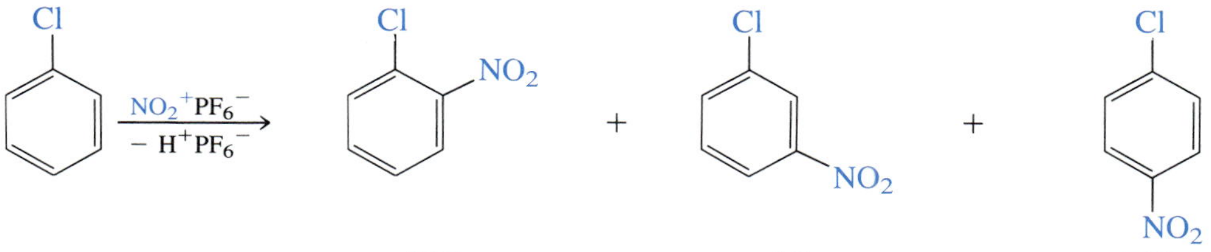

$NO_2^+PF_6^-$, $-\ H^+PF_6^-$

29% **1-Chloro-2-nitrobenzene (*o*-Chloronitrobenzene)**

1% **1-Chloro-3-nitrobenzene (*m*-Chloronitrobenzene)**

70% **1-Chloro-4-nitrobenzene (*p*-Chloronitrobenzene)**

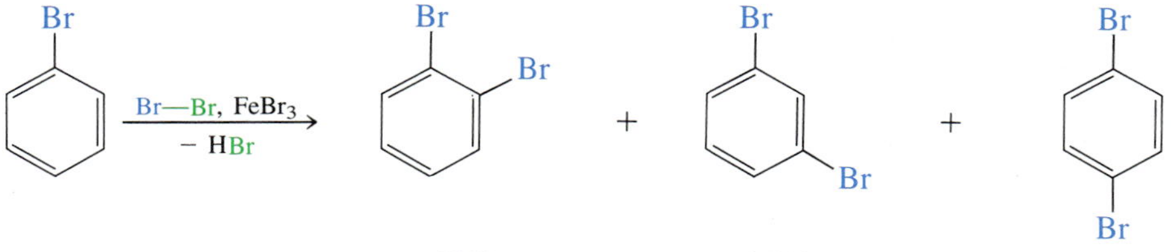

Br—Br, $FeBr_3$, − HBr

13% **1,2-Dibromobenzene (*o*-Dibromobenzene)**

2% **1,3-Dibromobenzene (*m*-Dibromobenzene)**

85% **1,4-Dibromobenzene (*p*-Dibromobenzene)**

The resonance structures for the various possible intermediates explain this seemingly contradictory reactivity.

Consequences of *Ortho*, *Meta*, and *Para* Attack on a Halobenzene

Ortho attack:

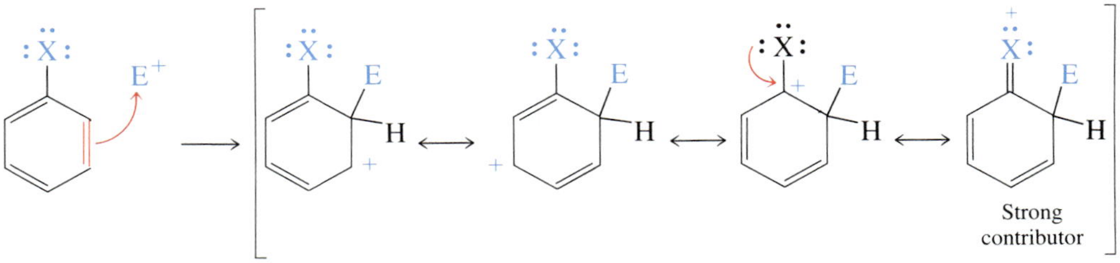

Strong contributor

Meta attack:

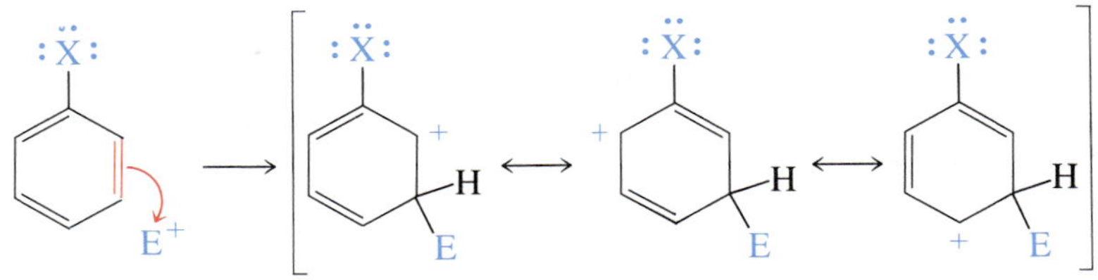

Para attack:

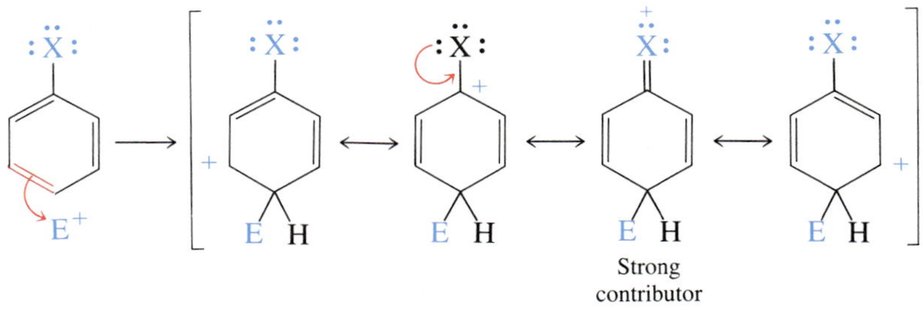

Strong contributor

Note that *ortho* and *para* attack leads to resonance structures in which the positive charge is placed next to the halogen substituent. Although this might be expected to be unfavorable, because the halogen is inductively electron withdrawing, resonance with the lone electron pairs prevails because it allows the charge to be delocalized. Therefore, *ortho* and *para* substitution becomes the preferred mode of reaction. In other words, in the hexadienyl cation (and in the transition state leading to it), resonance *outweighs* the inductive effect of the halogen, even though in the neutral starting material the opposite is true. The resonance effect is stronger in the charged system because of the delocalization of the charge.

This section completes the survey of the regioselectivity of electrophilic attack on monosubstituted benzenes, summarized in Table 20-2. Table 20-3 ranks various substituents by their activating power and lists the product distributions obtained on electrophilic nitration of the benzene ring.

TABLE 20-2

Direction by substituents in electrophilic aromatic substitution

Ortho and *para* directors	*Meta* directors
Strong activators:	**Strong deactivators:**
$—\ddot{N}H_2$, $—\ddot{N}HR$, $—\ddot{N}R_2$, $—\ddot{O}H$, $—\ddot{O}R$	$—NO_2$, $—CF_3$, $—\overset{+}{N}R_3$, $—\overset{O}{\overset{\|}{C}}OH$,
Weak activators:	$—\overset{O}{\overset{\|}{C}}OR$, $—\overset{O}{\overset{\|}{C}}R$, $—SO_3H$, $—C{\equiv}N$
Alkyl, phenyl	
Weak deactivators:	
$—\ddot{F}:$, $—\ddot{Cl}:$, $—\ddot{Br}:$, $—\ddot{I}:$	

TABLE 20-3

Relative rates and orientational preferences in the nitration of some monosubstituted benzenes, RC_6H_5

R	Relative rate	Percentage of isomer		
		Ortho	*Meta*	*Para*
OH	1000	40	<2	58
CH_3	25	58	4	38
H	1			
CH_2Cl	0.71	32	15.5	52.5
I	0.18	41	<0.2	59
Cl	0.033	31	<0.2	69
$CO_2CH_2CH_3$	0.0037	24	72	4
CF_3	2.6×10^{-5}	6	91	3
NO_2	6×10^{-8}	5	93	2
$\overset{+}{N}(CH_3)_3$	1.2×10^{-8}	0	89	11

20-4 Electrophilic Attack on Disubstituted Benzenes

Do the rules developed so far in this chapter predict the regioselectivity of higher substitution? Let us investigate the reactivity of disubstituted benzenes toward electrophiles.

Substituent Effects Are Additive

The effects of two substituents on the relative rate and orientation of electrophilic substitution of the benzene ring are additive. For example, the dimethylbenzenes (xylenes) are several times as reactive as toluene. All positions that are either *ortho* or *para* to a methyl group will be the subject of attack, regardless of whether they are also *meta* to a substituent. Remember that the effect of

FIGURE 20-2

Electrophilic attack on the dimethylbenzenes (xylenes). All positions *ortho* or *para* to a methyl group are activated. The *meta* carbon is neither activated nor deactivated.

an alkyl group on a *meta* carbon is negligible (i.e., neither activating nor deactivating). On the other hand, a doubly activated position (e.g., *ortho* to one, *para* to a second alkyl group) is more reactive than a singly activated one (Figure 20-2). For example, sulfonation of 1,3-dimethylbenzene (*m*-xylene) occurs only in the 2- and 4- (which is the same as 6-) positions, both being doubly (*ortho* and *para*) activated, whereas C-5 is *meta* to both substituents and hence is not as reactive. On the other hand, 1,2- and 1,4-dimethylbenzene (*o*- and *p*-xylene) are susceptible to attack at all positions, albeit more slowly, because each carbon is only singly (either *ortho* or *para*) activated.

Similar arguments apply to other activating substituents. Resonating activators usually override the effect of an inductively acting donor, such as an alkyl group. For example, the disinfectant 4-methylbenzenol (*p*-cresol) is brominated to give mainly 2-bromo-4-methylbenzenol (2-bromo-4-methylphenol).

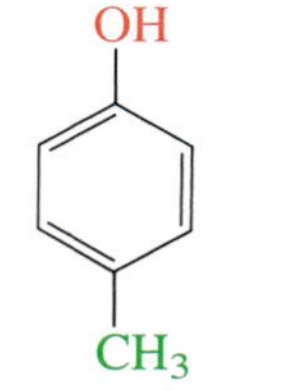

4-Methylbenzenol (*p*-Cresol)

$-$ HBr | Br—Br, $CHCl_3$, 0°C

80%

2-Bromo-4-methylbenzenol (2-Bromo-4-methylphenol)

EXERCISE 20-10

Predict the result of mononitration of:

(a) (b) (c) (d)

EXERCISE 20-11

The food preservative BHT (*tert*-butylated hydroxytoluene) has the structure shown here. Suggest a synthesis starting from 4-methylbenzenol (*p*-cresol).

4-Methyl-2,6-bis(1,1-dimethylethyl)benzenol (2,6-Di-*tert*-butyl-4-methylphenol)

When all substituents are deactivating, reaction is sluggish, and attack is directed to the position that is *meta* to most of them (the *ortho, para* positions are more deactivated). This effect can be seen in the nitration of 1,3-benzenedicarboxylic (isophthalic) and 1,2-benzenedicarboxylic (phthalic) acid.

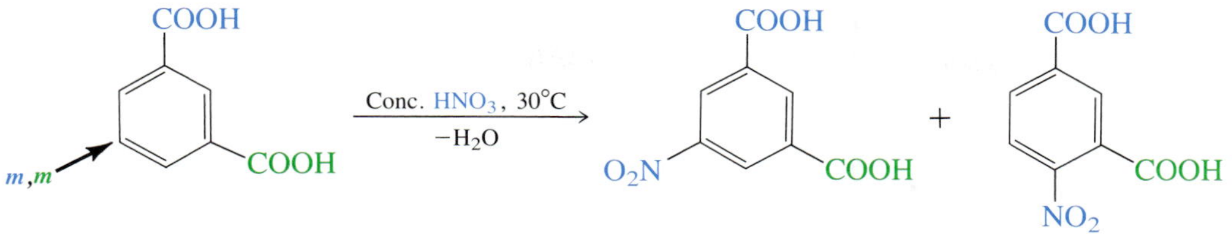

Conc. HNO_3, 30°C
$-H_2O$

m,m

96.9% | 3.1%

1,3-Benzenedicarboxylic acid (Isophthalic acid)

5-Nitro-1,3-benzenedicarboxylic acid (5-Nitroisophthalic acid)

4-Nitro-1,3-benzenedicarboxylic acid (4-Nitroisophthalic acid)

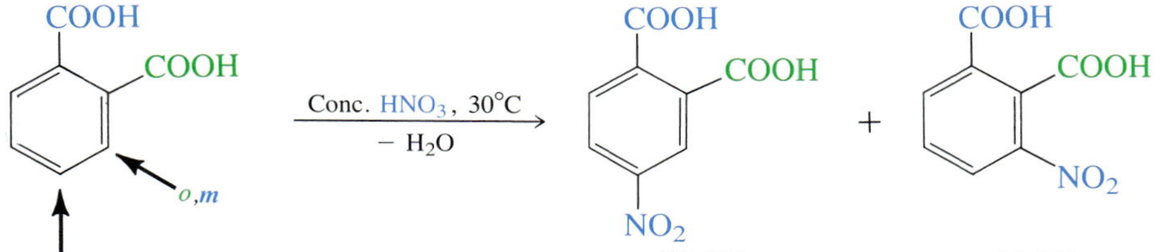

Conc. HNO_3, 30°C
$-H_2O$

o,m *m,p*

50.5% | 49.5%

1,2-Benzenedicarboxylic acid (Phthalic acid)

4-Nitro-1,2-benzenedicarboxylic acid (4-Nitrophthalic acid)

3-Nitro-1,2-benzenedicarboxylic acid (3-Nitrophthalic acid)

EXERCISE 20-12

Predict the result of the mononitration of:

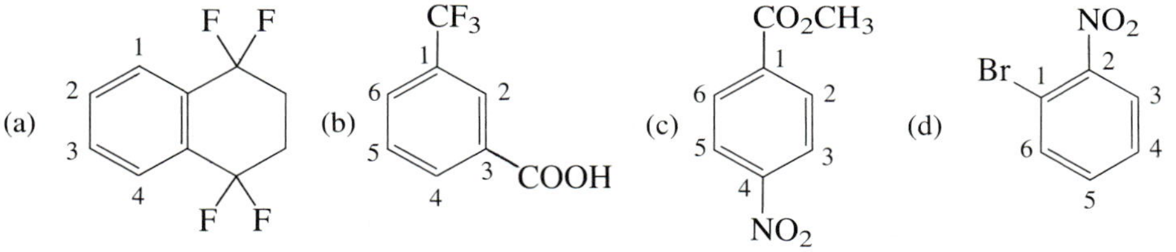

(a) (b) (c) (d)

The Activating Substituent Wins Out

The situation becomes more complicated when there are two groups with opposing directing (*ortho, para* versus *meta*) power. Usually each substituent acts independently. If the groups reinforce each other in controlling the position(s) of electrophilic attack, relatively clean results may be obtained.

Electrophilic Substitution with Reinforcing *o*, *m*, and *p* Directors

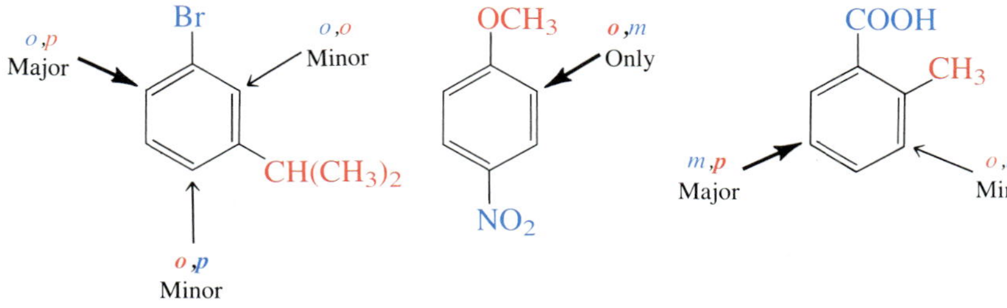

o,p Major | *o,o* Minor | *o,p* Minor

o,m Only

m,p Major | *o,m* Minor

Steric effects may render substitution at a position between two groups less favored, particularly if the groups are bulky. If the substituents compete with each other in controlling the site of reaction, the stronger activator wins out, by necessity an *ortho, para* director. If there is competition between two *ortho, para* directors, the better electron donor generally has the upper hand. In most other cases, product mixtures will ensue.

Regiocontrol by the Stronger Activator in Electrophilic Aromatic Substitution

CF_3 o,p Minor Br o,p Major

OCH_3 Cl o,m m,p

$NHCCH_3$ (O) o,m F

$NHCCH_3$ (O) CH_3 o,m m,p

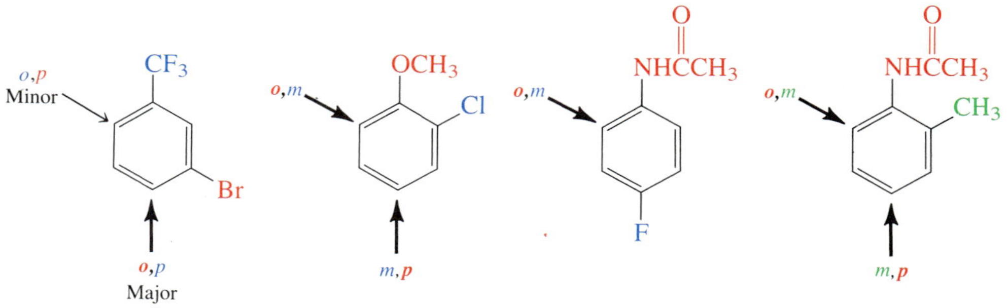

EXERCISE 20-13

Predict the site of electrophilic aromatic substitution in:

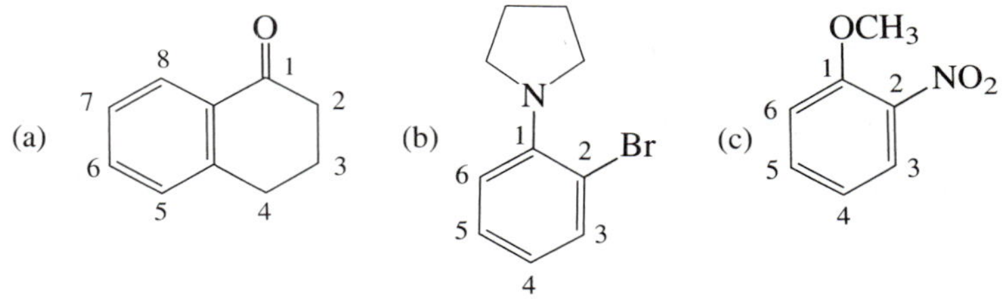

(a) 1-tetralone (positions 1–8) (b) 1-(2-bromophenyl)pyrrolidine, N at 1, Br at 2, positions 3–6 (c) OCH_3 at 1, NO_2 at 2, positions 3–6

In summary, electrophilic aromatic substitution of multiply substituted benzenes is controlled by the strongest activator and, to a certain extent, by steric effects. The greatest product selectivity appears if there is only one activator and all other groups deactivate or if all groups cooperate in their directing effects.

20-5 Synthetic Aspects of Benzene Chemistry

This section examines some of the industrial aspects of benzene production, how hydroaromatics may be ''aromatized'' in the laboratory, and, finally, the strategy of planning the synthesis of specifically substituted aromatic molecules.

Industrial Production of Benzene

Benzene is obtained commercially from aromatic distillation fractions in oil refineries, the steam cracking of alkenes, the so-called hydrodealkylation of methylbenzene (toluene), and the pyrolysis of coal. The United States produces about 1.6 billion gallons of benzene yearly, mainly by catalytic reforming of gasoline fractions (see Section 3-3), which is also a source of methylbenzene (toluene) and the dimethylbenzenes (xylenes). In this process, C_{6-8} hydrocarbons are dehydrogenated to simple aromatic compounds. Typically, the starting material in gaseous form reacts over a platinum-on-alumina catalyst at temperatures ranging from 450° to 550°C and a pressure between 10 and 50 atm of hydrogen. (This process is called platforming; Section 3-3).

Platforming to Aromatics

CH_3, CH_3 —(Pt-Al_2O_3, 450°–550°C, H_2 pressure)→ CH_3, CH_3 + 3 H_2

1,2-Dimethylcyclohexane **1,2-Dimethylbenzene (*o*-Xylene)**

If hydrogen is produced in these reactions, why is more added? The answer is, to suppress the formation of high-molecular-weight carbonaceous material (ultimately, coke), which tends to clog and deactivate the catalyst surface.

Benzene, methylbenzene (toluene), and the dimethylbenzenes (xylenes) are added to gasoline because their octane ratings (Section 3-4) are higher than 100.

Dehydrogenation of Hydroaromatic Compounds in the Laboratory

Dehydrogenation of cyclohexane, cyclohexene, and cyclohexadiene derivatives ("hydroaromatic" compounds) is a laboratory method for the preparation of substituted benzenes. The transformation is usually carried out at elevated temperatures with platinum or palladium metal as a catalyst, either as a fine powder or as a deposit on activated charcoal. The mechanism of the reaction is probably the reverse of the mechanism of hydrogenation of double bonds (Section 12-2). Dehydrogenations of this type have found some use in the synthesis of fused benzenes such as naphthalene (Section 25-3).

—(Pt or Pd-C, 350°C)→ + 3 H_2

—(Pd-C, 300°C)→ + 4 H_2

82%
Naphthalene

Alternatively, hydroaromatics are aromatized by chemical oxidation, using, for example, sulfur, selenium, or $KMnO_4$.

Aromatics by Oxidation of Hydroaromatics

—(S, Δ)→ + 3 H_2S

CH_3 —(Se, Δ)→ CH_3 + H_2Se

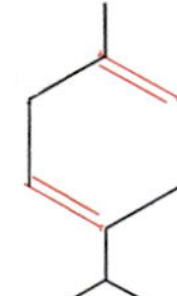

$\xrightarrow{\text{KMnO}_4\text{, benzene, 18-crown-6 (see Section 9-5)}}$

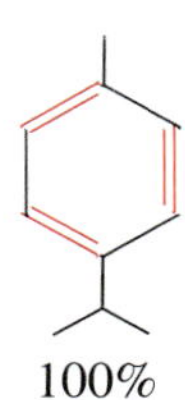

100%

1-Methyl-4-(1-methylethyl)-1,4-cyclohexadiene

1-Methyl-4-(1-methylethyl)benzene (*p*-Methylisopropylbenzene)

Specifically Substituted Benzenes Can Be Constructed by Carefully Planned Electrophilic Substitutions

The synthesis of a specific benzene derivative can be a major undertaking if the desired substitution pattern is ''wrong'' for the directing power of the first group introduced. For this reason, some useful reactions reverse the directing power of a substituent. An example of such a change is the conversion of the nitro group in nitrobenzenes (*meta* director) to an amino substituent (*ortho, para* director) and its reverse. Simple reagents accomplish this task. Reduction is effected by catalytic hydrogenation, oxidation by treatment with trifluoroperoxyethanoic (trifluoroperacetic) acid.

Reversible Conversion of a *Meta* into an *Ortho, Para* Director

$C_6H_5NO_2 \underset{CF_3CO_3H}{\overset{H_2,\ Ni \text{ or } Fe,\ HCl}{\rightleftharpoons}} C_6H_5NH_2$

For an example of the application of this strategy, consider the preparation of 3-ethylnitrobenzene. Direct nitration of ethylbenzene furnishes a mixture of 2- and 4-nitro-1-ethylbenzene and is therefore useless. An alternative approach, the Friedel-Crafts ethylation of nitrobenzene also fails, because nitrobenzene is too unreactive toward the relatively weak electrophilic Friedel-Crafts reagents. However, the target product is available in a roundabout way from 1-(3-nitrophenyl)ethanone (*m*-nitroacetophenone), prepared by nitration of 1-phenylethanone (acetophenone).

$m\text{-}CH_3CO\text{-}C_6H_4\text{-}NO_2 \xrightarrow{HCl,\ Zn(Hg)} m\text{-}CH_3CH_2\text{-}C_6H_4\text{-}NH_2 \xrightarrow{CF_3CO_3H,\ CH_2Cl_2} m\text{-}CH_3CH_2\text{-}C_6H_4\text{-}NO_2$

1-(3-Nitrophenyl)ethanone (*m*-Nitroacetophenone) — **3-Ethylbenzenamine (*m*-Ethylaniline)** — **3-Ethylnitrobenzene**

What is required is reduction of the carbonyl (*meta* director) to a methylene group (an *ortho, para* director). Clemmensen conditions can be used for this reduction (Section 15-8), but they lead to overreduction to 3-ethylbenzenamine (*m*-ethylaniline). The synthetic task can be completed by reoxidation of the amino group. This route also constitutes a ready synthesis of 3-ethylbenzenamine (*m*-ethylaniline), a compound in which two *ortho, para* directors emerge positioned *meta* to each other.

A similar problem arises in the attempt to prepare an *o*-disubstituted benzene in which both groups are *meta* directors. For example, what would be a rational synthesis of *o*-dinitrobenzene? A successful sequence starts with benzenamine (aniline) and protects the amino function by ethanoylation (acetylation; Section 18-5). Nitration gives a mixture of *o*- and *p*-nitro products, from which the former is separated. Deprotection by amide hydrolysis (Section 18-5) followed by oxidation of the 2-nitrobenzenamine (*o*-nitroaniline) then results in the desired product.

A Synthesis of 1,2-Dinitrobenzene (*o*-Dinitrobenzene)

NH_2 | $CH_3\overset{O}{\overset{\|}{C}}Cl$, pyridine → $HN\overset{O}{\overset{\|}{C}}CH_3$ | HNO_3 → $HN\overset{O}{\overset{\|}{C}}CH_3$, NO_2 + $HN\overset{O}{\overset{\|}{C}}CH_3$, NO_2 | Separate *ortho* isomer →

Benzenamine (Aniline) | ***N*-Phenyl-ethanamide (Acetanilide)** | ***N*-(2-Nitrophenyl)-ethanamide (*o*-Nitroacetanilide)** | ***N*-(4-Nitrophenyl)-ethanamide (*p*-Nitroacetanilide)**

H^+, H_2O → H_2N, NO_2 | $CF_3\overset{O}{\overset{\|}{C}}OOH$, CH_2Cl_2 → NO_2, NO_2

2-Nitrobenzenamine (*o*-Nitroaniline) | **1,2-Dinitrobenzene (*o*-Dinitrobenzene)**

Significant losses in this sequence arise in the second step, which furnishes substantial quantities of the wrong isomer and poses a difficult separation task. This problem can be prevented by yet another clever device: sulfonation as a blocking procedure. For steric reasons, *N*-phenylethanamide (acetanilide) is sulfonated mainly *para,* blocking this carbon from further electrophilic attack. Nitration now can occur only *ortho* to the nitrogen, and double deprotection by heating in aqueous acid completes the synthesis.

Reversible Sulfonation as a Blocking Procedure

$HN\overset{O}{\overset{\|}{C}}CH_3$ | SO_3, conc. H_2SO_4 → $HN\overset{O}{\overset{\|}{C}}CH_3$, SO_3H | HNO_3 → $HN\overset{O}{\overset{\|}{C}}CH_3$, NO_2, SO_3H | 1. H^+, H_2O, Δ 2. ^-OH, H_2O → NH_2, NO_2

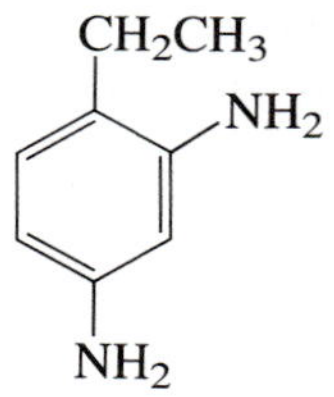

CH_2CH_3, NH_2, NH_2

A

EXERCISE 20-14

Suggest a synthetic route to compound A, starting from benzene.

In summary, benzene is obtained commercially mainly by reforming of petroleum distillates. In the laboratory, hydroaromatics can be aromatized by Pd-C, S, Se, or $KMnO_4$. Finally, by careful choice of the sequence in which new substituents are introduced, it is possible to devise specific syntheses of multiply substituted benzenes.

20-6 Attack at an Aromatic Carbon Already Bearing a Substituent: *Ipso* Substitution

In the preceding sections, we have tacitly assumed that an incoming reagent has only three choices in attacking a monosubstituted benzene: at the *o*, *m*, or *p* positions. We assumed that, for some reason (perhaps steric), attack at the substituted carbon is not competitive. Although this is generally true, it is not *always:* in some cases, aromatic substitution replaces a substituent already present. This process is called ***ipso* substitution** (*ipso,* Latin, on itself); it may follow various mechanisms and replace various leaving groups.

Electrophilic *Ipso* Substitution

Ipso substitution is observed in the **protodealkylation** of alkylbenzenes, a reaction that reverses the Friedel-Crafts alkylation. Tertiary alkyl groups are most easily cleaved off.

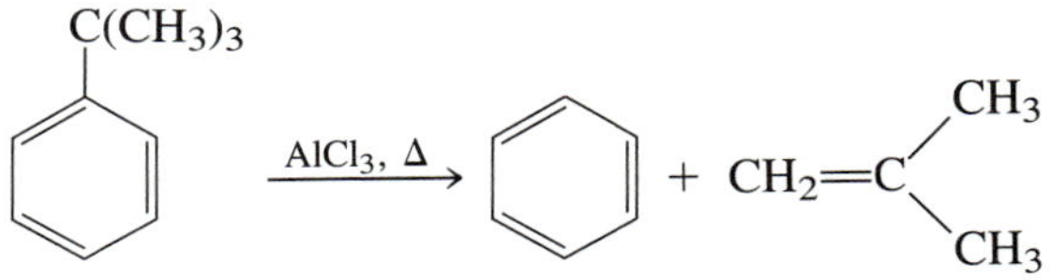

$$C_6H_5C(CH_3)_3 \xrightarrow{AlCl_3,\ \Delta} C_6H_6 + CH_2{=}C(CH_3)_2$$

The mechanism of this process probably begins with protonation by traces of HCl, followed by extrusion of the 1,1-dimethylethyl (*tert*-butyl) cation and its decomposition, regenerating the proton and furnishing 2-methylpropene.

Mechanism of Protodealkylation of (1,1-Dimethylethyl)benzene:

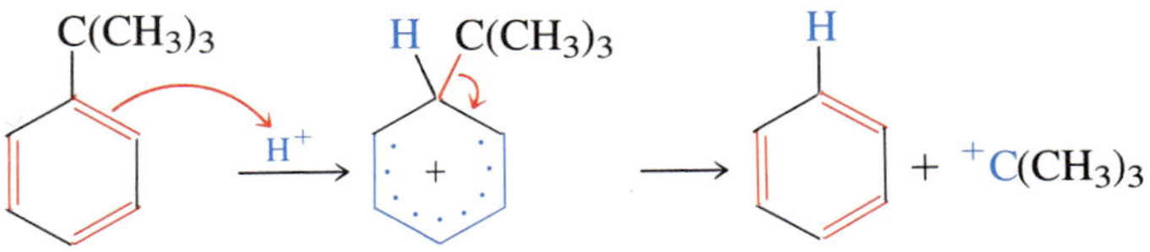

$$C_6H_5C(CH_3)_3 \xrightarrow{H^+} [C_6H_5(H)C(CH_3)_3]^+ \longrightarrow C_6H_6 + {}^+C(CH_3)_3$$

$$^+C(CH_3)_3 \longrightarrow CH_2{=}C(CH_3)_2 + H^+$$

Ipso attack is also responsible for the rearrangement of alkylbenzenes to their isomers; here the alkyl group undergoes migration from one carbon atom on the ring to another one, a typical carbocation rearrangement (Section 9-2).

Rearrangement of 1,2- to 1,3-Dimethylbenzene (*o*-Xylene to *m*-Xylene)

CH_3 CH_3 $\xrightleftharpoons{H^+}$ CH_3 CH_3

Mechanism:

CH_3 CH_3 H $+ H^+ \rightleftharpoons$ CH_3 H CH_3 + H $\rightleftharpoons$ CH_3 H + CH_3 H $\rightleftharpoons$ CH_3 H CH_3 $+ H^+$

EXERCISE 20-15

When 2,4,6-tribromobenzenol (2,4,6-tribromophenol) is treated with bromine in water, the red bromine color rapidly disappears and a neutral nonaromatic ketone of molecular formula $C_6H_2Br_4O$ is generated. Give a structure for this compound and a mechanism for its formation.

Nucleophilic Aromatic *Ipso* Substitution

Treatment of 1-chloro-2,4-dinitrobenzene with nucleophiles such as hydroxide ion, methoxide ion, or ammonia replaces the halogen with the reagent.

General Nucleophilic Aromatic *Ipso* Substitution

Cl NO_2 NO_2 $+ :Nu^- \longrightarrow$ Nu NO_2 NO_2 $+ Cl^-$

Examples:

Cl NO_2 NO_2 $\xrightarrow{Na_2CO_3,\ HOH,\ 100^\circ C}$ OH NO_2 NO_2 $+ NaCl$

90%

1-Chloro-2,4-dinitrobenzene → **2,4-Dinitrobenzenol (2,4-Dinitrophenol)**

Cl NO_2 NO_2 $\xrightarrow{:NH_3,\ \Delta}$ NH_2 NO_2 NO_2 $+ NH_4Cl$

85%

2,4-Dinitrobenzenamine (2,4-Dinitroaniline)

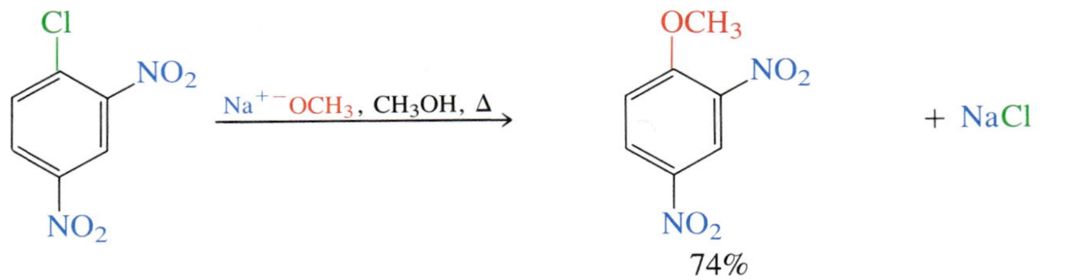

Cl, NO_2, NO_2 $\xrightarrow{Na^{+}\,{}^{-}OCH_3,\ CH_3OH,\ \Delta}$ OCH_3, NO_2, NO_2 + NaCl

74%

1-Methoxy-2,4-dinitrobenzene
(2,4-Dinitroanisole)

The reaction is called **nucleophilic aromatic substitution.** The key to its success is the presence of at least one strongly electron withdrawing substituent, preferably more, on the benzene ring located *ortho* or *para* to the leaving group. The effect of such substituents is, first, to decrease the electron density in the benzene ring, making it more favorable for nucleophilic attack, and, second, to stabilize the intermediate hexadienyl anion by resonance. In contrast with the S_N2 reaction of haloalkanes, substitution in these reactions takes place by a *two-step mechanism,* an *addition-elimination sequence* similar to the mechanism of nucleophilic substitution of carboxylic acid derivatives (Chapter 18).

Mechanism of Nucleophilic Aromatic Substitution:

STEP 1: Addition

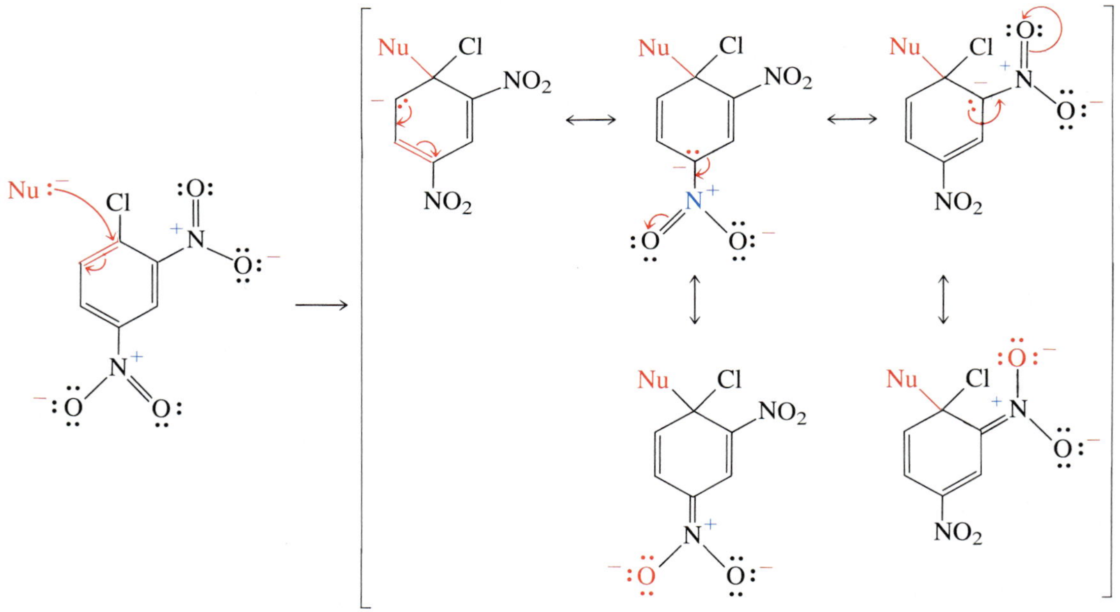

Nu :⁻ Cl :O: N⁺ :O:⁻ N⁺ ⁻:O: O: ⟶ [Nu Cl NO_2 NO_2 ⟷ Nu Cl NO_2 N⁺ :O :O:⁻ ⟷ Nu Cl :O: N⁺ :O:⁻ NO_2; Nu Cl NO_2 N⁺ ⁻:O :O:⁻; Nu Cl :O:⁻ N⁺ :O:⁻ NO_2]

STEP 2: Elimination (only one resonance structure is shown)

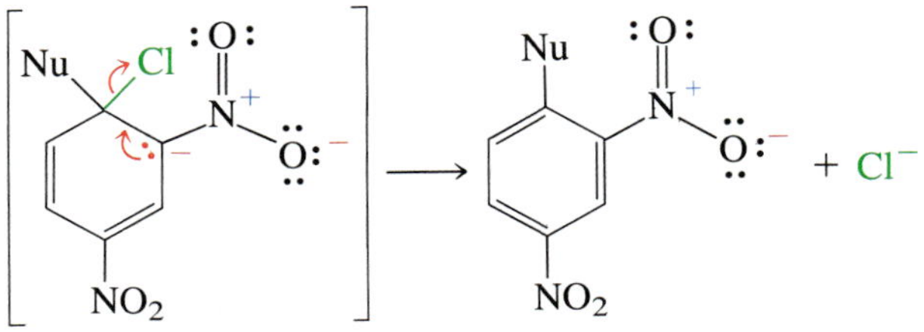

[Nu Cl :O: N⁺ :O:⁻ NO_2] ⟶ Nu :O: N⁺ :O:⁻ NO_2 + Cl^-

In the first step, *ipso* attack by the nucleophile produces an anion with a highly delocalized charge, for which several resonance structures may be written; five are shown. The important feature of this intermediate is the ability of the negative charge to be delocalized into the electron-withdrawing groups. For example, such delocalization is not possible in 1-chloro-3,5-dinitrobenzene, in which these groups are located *meta;* so this compound does not undergo *ipso* substitution under the conditions employed.

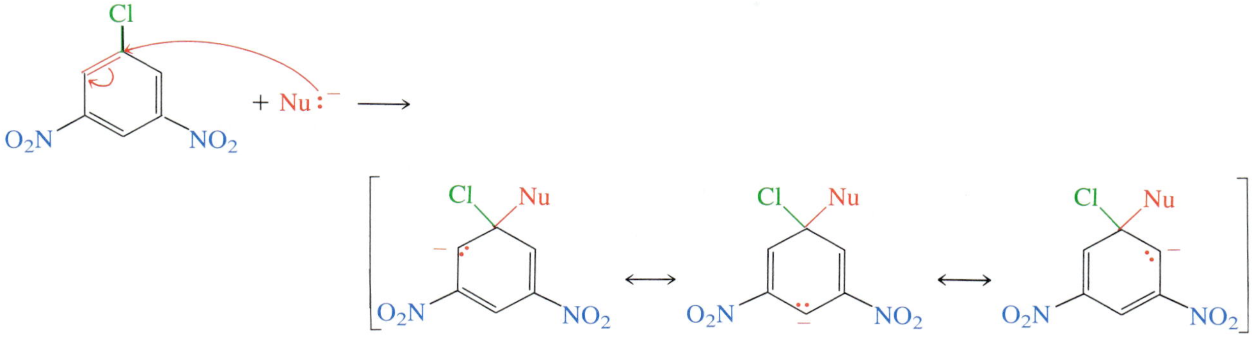

In the second step, the leaving group is expelled to regenerate the aromatic ring. Predictably, the reactivity of haloarenes in nucleophilic substitutions increases with the nucleophilic power of the reagent and the number of electron-withdrawing groups on the ring, particularly if they are in the *ortho* and *para* positions.

EXERCISE 20-16

Formulate a mechanism for the following conversion. Considering that the first step is rate determining, draw a potential-energy diagram depicting the progress of the reaction.

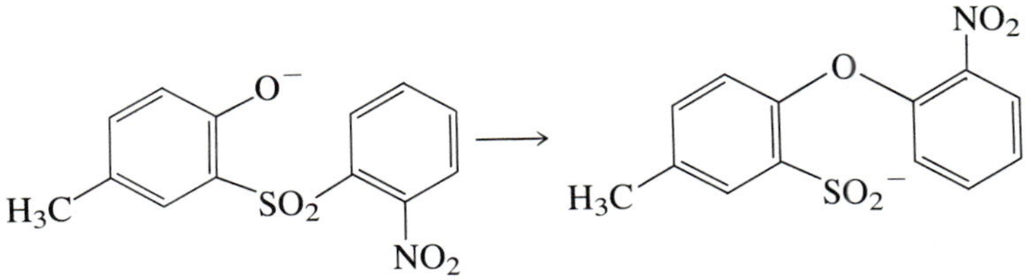

Nucleophilic Aromatic Substitution by Elimination-Addition: The Benzyne Mechanism

Haloarenes do not undergo simple S_N2 or S_N1 reactions. However, at highly elevated temperatures and pressures, it is possible to effect nucleophilic substitution. For example, if exposed to hot sodium hydroxide followed by neutralizing work-up, chlorobenzene furnishes benzenol (phenol).

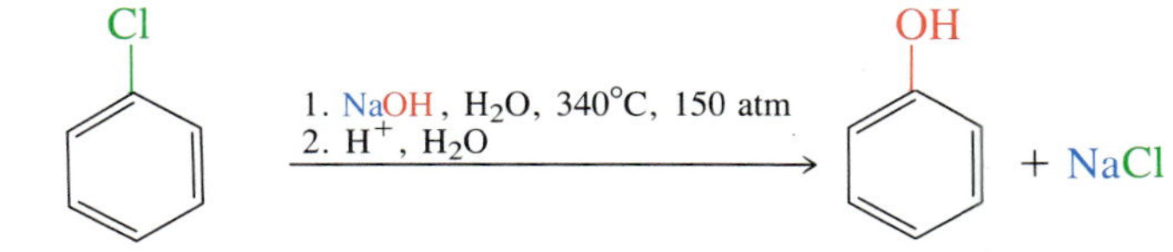

$$\text{Chlorobenzene} \xrightarrow[\text{2. } H^+,\ H_2O]{\text{1. NaOH, } H_2O,\ 340°C,\ 150\ \text{atm}} \text{Benzenol (Phenol)} + NaCl$$

Chlorobenzene **Benzenol (Phenol)**

Similar treatment with potassium amide results in benzenamine (aniline).

Cl → 1. KNH_2, liquid NH_3 2. H^+, H_2O → NH_2 + KCl

Benzenamine (Aniline)

It is tempting to assume that these substitutions are accomplished by mechanisms similar to those formulated for the haloalkanes. However, when the last reaction is performed with radioactively labeled chlorobenzene (^{14}C at C-1), a very curious result is obtained: only half of the product is substituted at the labeled carbon; in the other half, the nitrogen is located at the *neighboring* position.

Cl → KNH_2, liquid NH_3 / − KCl → NH_2 (50%) + NH_2

Chlorobenzene-1-^{14}C **Benzenamine-1-^{14}C** **Benzenamine-2-^{14}C**

Similar results are obtained from the reaction of substituted halobenzenes with strong bases. For example, 1-halo-2-methylbenzenes furnish *ortho*- and *meta*-substituted products. 1-Halo-4-methylbenzenes give *meta* and *para* substitution.

CH_3, Cl → $NaNH_2$, liquid NH_3 / − NaCl → CH_3, NH_2 + CH_3, NH_2 + CH_3, NH_2

Ratio: 45 : 55 : 0

66%

CH_3, Cl → $NaNH_2$, liquid NH_3 / − NaCl → CH_3, NH_2 + CH_3, NH_2 + CH_3, NH_2

Ratio: 0 : 62 : 38

58%

On the other hand, 1-halo-3-methylbenzenes give all possible products.

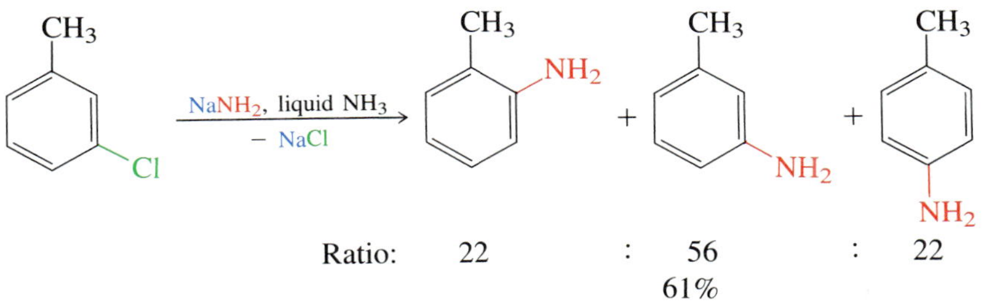

Ratio: 22 : 56 : 22
61%

Direct substitution mechanisms do not seem to operate on these compounds. What, then, is the answer to this puzzle? A clue is the emergence of the incoming nucleophile only at the *ipso* or at the *ortho* position relative to the leaving group. This and other observations can be accounted for by an initial base-induced elimination of HX from the benzene ring, a process reminiscent of the dehydrohalogenation of alkenyl halides to give alkynes (Section 13-5). In the present case, step-by-step elimination through a phenyl anion intermediate gives a highly strained and reactive species called **benzyne,** or **1,2-dehydrobenzene.**

Mechanism of Nucleophilic Substitution of Simple Haloarenes:

STEP 1: Elimination

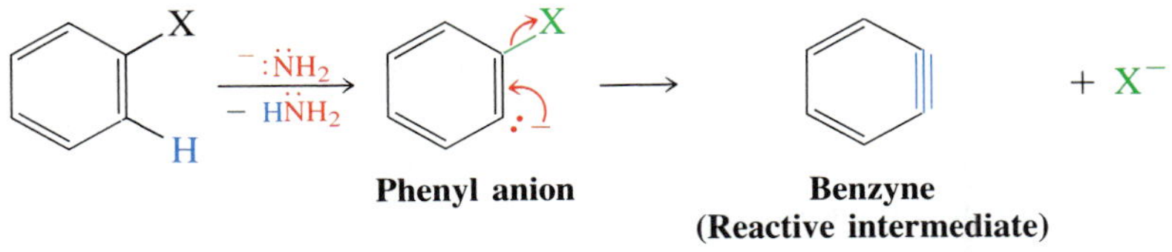

Phenyl anion **Benzyne (Reactive intermediate)**

STEP 2: Addition

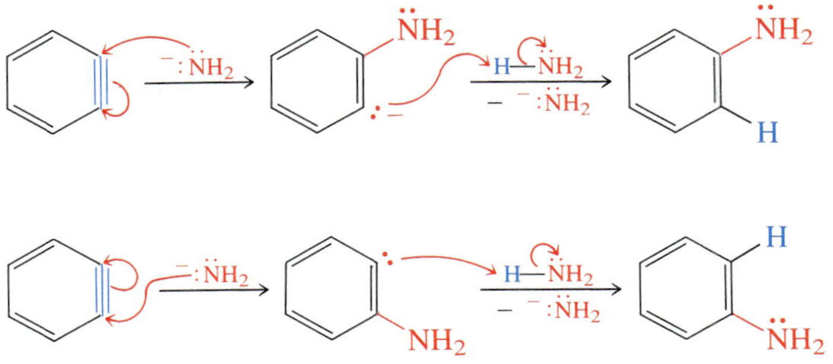

The reason for the reactivity of benzyne is the normal requirement for alkynes to adopt a linear rather than a bent structure, a consequence of the linear *sp* hybridization of the carbons making up the triple bond (see Section 13-2). Because of its cyclic structure, benzyne cannot meet that requirement. The molecule has only a fleeting life and is rapidly trapped by any nucleophile present. For example, ammonia solvent adds to furnish the product benzenamine (aniline). Because the two ends of the triple bond are equally reactive, nucleophilic addition may be at either carbon, explaining the product mixtures observed from labeled chlorobenzene and the (halo)methylbenzenes.

EXERCISE 20-17

Explain the regioselectivity observed in the following reaction. (Hint: Consider the effect of the methoxy group on the selectivity of attack on the benzyne by amide ion.)

Br, OCH_3 $\xrightarrow[-\text{ KBr}]{KNH_2,\ \text{liquid } NH_3}$ H_2N, OCH_3 (Major) + NH_2, OCH_3 (Minor)

Benzyne Is a Strained Cycloalkyne

Matrix Generation of Benzyne, a Reactive Intermediate

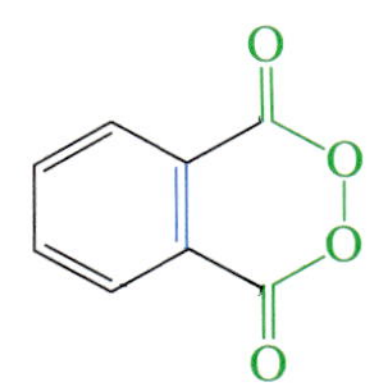

1,2-Benzenedicarbonyl peroxide (Phthaloyl peroxide)

$\downarrow h\nu$, 8 K

$+ 2\ CO_2$

Although benzyne is too reactive to be isolated and stored in a bottle, it can be observed spectroscopically under special conditions. One successful approach is to irradiate a specific precursor, 1,2-benzenedicarbonyl (phthaloyl) peroxide, at low temperature. Photolysis of this compound at 8 K (−265°C) in a solid argon matrix (m.p. = −189°C) produces a species whose IR and UV spectra are assignable to benzyne, formed by elimination of two equivalents of carbon dioxide.

From a detailed analysis of these spectra, it was concluded that benzyne has a cycloalkyne rather than a cycloallene structure (Figure 20-3A). The normally symmetrical benzene ring is distorted by introduction of the strained triple bond. Three short and three longer bonds can be detected, a phenomenon called *bond fixation* or *bond alternation*. The destabilization imposed on benzyne by these changes has been estimated to be 44 kcal $mole^{-1}$.

In summary, *ipso* substitution on the benzene ring may be due to either electrophilic or nucleophilic (basic) pathways. An example of the first is protodealkylation, which can also cause the redistribution of substituents on the benzene ring. If the ring bears enough strongly electron withdrawing groups

FIGURE 20-3

Representations of benzyne: (A) of the two possible resonance structures, benzyne prefers the cycloalkyne form; (B) the orbital picture of benzyne reveals the aromatic framework of six π electrons arranged perpendicular to the two reactive and poorly overlapping additional hybrid orbitals making up the distorted triple bond. The molecular structure of benzyne is quite deformed from regular hexagonal symmetry, because of the bond alternation imposed by the presence of the triple bond.

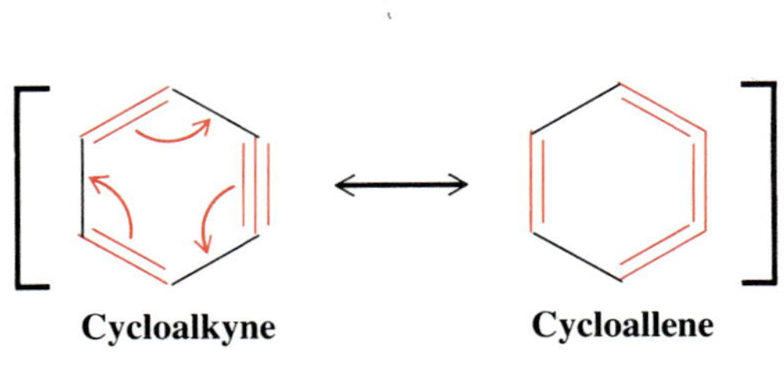

A

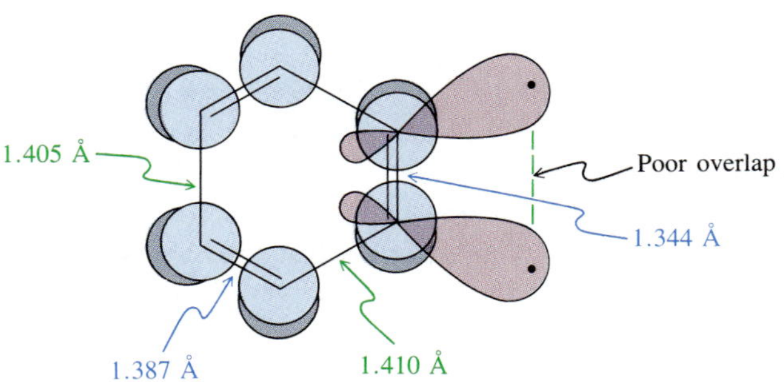

B

and a leaving group, addition to give an intermediate anion with delocalized charge becomes feasible, followed by elimination of the leaving group (nucleophilic aromatic substitution). Finally, very strong bases are capable of eliminating HX from haloarenes to form the reactive intermediate benzyne, which is subject to nucleophilic attack to give new substituted products.

20-7 A Summary of Organic Reaction Mechanisms: Substitution, Elimination, Addition, and Pericyclic Reactions

This chapter completes our survey of the major classes of organic reactions and their mechanisms. Let us review.

Substitution Reactions Allow the Introduction of New Groups

Perhaps the most frequently discussed reactions are those in which one atom or group is replaced by another, the **substitution reactions.** Chapter 3 showed that alkanes can be functionalized by *free-radical halogenation*. Chapter 6 described the two modes of *nucleophilic substitution* of the resulting haloalkanes: bimolecular (S_N2) and unimolecular (S_N1). In the S_N2 reaction, the leaving group is displaced directly by the incoming nucleophile, whereas the S_N1 process requires that dissociation to a carbocation precedes nucleophilic trapping (Chapter 7). A variant of these options is the nucleophilic addition-elimination sequence in the reactions of carboxylic acid derivatives (Chapters 17 and 18) and in nucleophilic aromatic substitution (Section 20-6). Finally, Chapters 19 and 20 pointed out that *electrophilic substitution* is also possible with benzene and its derivatives. The major modes of substitution are summarized in Table 20-4.

TABLE 20-4
The major modes of substitution

1 FREE-RADICAL HALOGENATION

Alkanes:

$$RH + X_2 \xrightarrow{h\nu \text{ or peroxide}} RX + HX \qquad \text{(Section 3-4)}$$

Allylic systems:

$$CH_3CH{=}CH_2 + X_2 \xrightarrow{h\nu \text{ or peroxide}} XCH_2CH{=}CH_2 + HX \qquad \text{(Section 14-2)}$$

2 NUCLEOPHILIC SUBSTITUTION

S_N2 reaction:

$$Nu{:}^- + CH_3{-}X \longrightarrow Nu{-}CH_3 + X^- \qquad \text{(Sections 6-3 through 6-11)}$$

S_N1 reaction:

$$(CH_3)_3CX \rightleftharpoons (CH_3)_3C^+ + X^- \xrightarrow{^-:Nu} (CH_3)_3CNu + X^- \qquad \text{(Sections 7-1 through 7-3)}$$

TABLE 20-4 (CONTINUED)

The major modes of substitution

2 NUCLEOPHILIC SUBSTITUTION (CONTINUED)

Addition-elimination of carboxylic acid derivatives:

$$\mathrm{R\overset{O}{\overset{\|}{C}}L} + {}^{-}\!:\!\mathrm{Nu} \longrightarrow \mathrm{R\underset{Nu}{\underset{|}{\overset{O^{-}}{\overset{|}{C}}}}\!-\!L} \longrightarrow \mathrm{R\overset{O}{\overset{\|}{C}}Nu} + \mathrm{L^{-}}$$

(Sections 17-6 through 17-10, and 18-1 through 18-6)

Nucleophilic aromatic substitution:

O_2N, NO_2, X, NO_2 (aryl halide) + $^{-}:Nu$ ⟶ (addition intermediate with X and Nu, O_2N, NO_2, NO_2) ⟶ (Nu, O_2N, NO_2, NO_2 substituted arene) + X^{-} (Section 20-6)

3 ELECTROPHILIC AROMATIC SUBSTITUTION

Benzene + E^{+} ⟶ (cation intermediate with E, H, +) ⟶ (benzene–E) + H^{+} (Chapters 19 and 20)

Elimination Reactions Result in Unsaturation

In another major pathway of organic reactions, two moieties are removed from adjacent atoms to produce an unsaturated molecule, such as an alkene, an alkyne, or a carbonyl compound. Two mechanisms for such eliminations, bimolecular (E2) and unimolecular (E1), were described in Chapter 7. A third was discussed in connection with carbonyl chemistry and the generation of benzyne: proton abstraction followed by extrusion of the leaving group (Chapter 15 and Section 20-6). The major modes of elimination are summarized in Table 20-5.

Addition Reactions Remove Unsaturation

A third type of reaction is addition to unsaturated moieties, such as alkenes (Chapters 12 and 14), alkynes (Chapter 13), carbonyl functions (Chapters 15–17), and benzyne (Section 20-6). Such additions may be electrophilic, nucleophilic, or radical. The major modes are summarized in Table 20-6.

Pericyclic Reactions Have No Intermediates

A final class of reactions are called pericyclic because they are characterized by cyclic transition states in which there is continuous overlap of a cyclic array of orbitals. Pericyclic transformations may produce new rings, as in the Diels-Alder reaction (Section 14-5); ring-openings and -closures, as in the 1,3-butadiene-cyclobutene interconversion (Section 14-5); and fragmentations, as in ester pyrolysis (Section 18-4) and the McLafferty rearrangement (Section 18-7). Examples are shown in Table 20-7.

TABLE 20-5

The major modes of elimination

1 BIMOLECULAR ELIMINATION

$$B:^- + RCH(H)—CH_2X \longrightarrow BH + RCH{=}CH_2 + X^-$$ (Section 7-5)

$$B:^- + RHC{=}CHX \longrightarrow BH + RC{\equiv}CH + X^-$$ (Section 13-5)

2 UNIMOLECULAR ELIMINATION

$$CH_3C(CH_3)_2X \rightleftharpoons (CH_3)_3C^+ + X^- \longrightarrow CH_2{=}C(CH_3)_2 + H^+ + X^-$$ (Section 7-4)

3 DEPROTONATION-EXTRUSION

Carbonyl compounds:

$$R—CH(OH)—X + {}^-:B \xrightleftharpoons{-\,BH} R—CH(O^-)—X \rightleftharpoons RCH{=}O + X^-$$ (Sections 15-4 through 15-7)

Benzyne formation:

$$C_6H_5X + {}^-:B \xrightarrow{-\,BH} C_6H_4X^- \longrightarrow C_6H_4\ (\text{benzyne}) + X^-$$ (Section 20-6)

TABLE 20-6

The major modes of addition

1 ELECTROPHILIC ADDITIONS

Alkenes:

$$RCH_2{=}CH_2 + HX \longrightarrow RCH(X)CH_2(H)$$ (Section 12-3)

Markovnikov product

$$RCH{=}CH_2 + XY \longrightarrow RCH(Y)CH_2X$$ (Section 12-3)

TABLE 20-6 (CONTINUED)
The major modes of addition

1 ELECTROPHILIC ADDITIONS (CONTINUED)

$$3\ RCH{=}CH_2 + BH_3 \longrightarrow (R\underset{}{\overset{H}{C}}HCH_2)_3B \qquad \text{(Section 12-4)}$$

$$RCH{=}CH_2 + H_2O_2 \xrightarrow{OsO_4} R\overset{OH}{C}HCH_2OH \qquad \text{(Section 12-5)}$$

Alkynes:

$$RC{\equiv}CH + HX \longrightarrow R\overset{X}{C}{=}\overset{H}{C}H \qquad \text{(Section 13-6)}$$

$$RC{\equiv}CH + XY \longrightarrow R\overset{Y}{C}{=}CHX \qquad \text{(Section 13-6)}$$

$$3\ RC{\equiv}CH + BH_3 \longrightarrow (RCH{=}CH)_3B \qquad \text{(Section 13-6)}$$

Carbonyl compounds:

$$R\overset{O}{\overset{\|}{C}}R + H^+ \longrightarrow R\overset{^+O-H}{\overset{\|}{C}}R \quad + \ ^-{:}Nu \longrightarrow R\underset{Nu}{\overset{OH}{C}}R \qquad \text{(Sections 15-4 through 15-7)}$$

2 NUCLEOPHILIC ADDITIONS

Carbonyl compounds:

$$R\overset{O}{\overset{\|}{C}}R + \ ^-{:}Nu \longrightarrow R\underset{Nu}{\overset{O^-}{C}}R \xrightarrow{HOH} R\underset{Nu}{\overset{OH}{C}}R + HO^- \qquad \text{(Sections 15-4 through 15-7, and 16-3)}$$

$$R\overset{O}{\overset{\|}{C}}L + \ ^-{:}Nu \longrightarrow R\underset{Nu}{\overset{O^-}{C}}L \xrightarrow{HOH} R\underset{Nu}{\overset{OH}{C}}L + HO^- \qquad \text{(Sections 17-6 through 17-10, and 18-1 through 18-6)}$$

α,β-Unsaturated carbonyl compounds:

$$RCH{=}CH{-}\overset{O}{\overset{\|}{C}}R + \ ^-{:}Nu \longrightarrow R\underset{Nu}{C}HCH{=}C\begin{smallmatrix}O^-\\R\end{smallmatrix} \xrightarrow{HOH} R\underset{Nu}{C}H\overset{H}{C}H\overset{O}{\overset{\|}{C}}R + HO^-$$

(Section 16-5)

Benzyne additions:

$$C_6H_4\ (\text{benzyne}) + \ ^-{:}Nu \longrightarrow C_6H_4(Nu)^{:-} \xrightarrow{NuH} C_6H_5Nu\ (\text{H ortho to Nu}) + \ ^-{:}Nu \qquad \text{(Section 20-6)}$$

TABLE 20-6 (CONTINUED)
The major modes of addition

3 FREE-RADICAL ADDITIONS

Alkenes:

$$\mathrm{RCH{=}CH_2 + HBr \xrightarrow{ROOR} R\overset{H}{\overset{|}{C}}HCH_2Br}$$
Anti-Markovnikov product (Section 12-6)

Alkynes:

$$\mathrm{RC{\equiv}CH + HBr \xrightarrow{ROOR} RCH{=}CHBr}$$ (Section 13-6)

Radical coupling:

$$\mathrm{R\cdot + R\cdot \longrightarrow R{-}R}$$ (Sections 3-3 and 3-5 through 3-7)

$$\mathrm{2\ R\overset{O}{\overset{\|}{C}}R \xrightarrow{2e} 2\ R\overset{O^-}{\overset{|}{\dot{C}}}R \longrightarrow R\overset{O^-}{\overset{|}{\underset{R}{\underset{|}{C}}}}{-}\overset{O^-}{\overset{|}{\underset{R}{\underset{|}{C}}}}R}$$ (Section 15-8)

TABLE 20-7
Examples of pericyclic reactions

1 DIELS-ALDER CYCLOADDITION

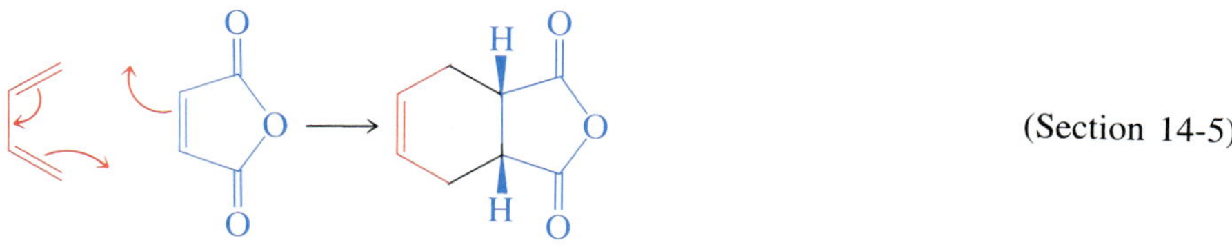

(Section 14-5)

2 OTHER CYCLOADDITIONS

Photodimerization of alkenes:

$$\mathrm{RCH{=}CH_2 + RCH{=}CH_2 \xrightarrow{h\nu}}$$ 1,2-disubstituted cyclobutane (R, R) (Section 14-5)

Ozone addition to alkenes:

$$\mathrm{O{=}\overset{+}{O}{-}O^- + RCH{=}CHR \longrightarrow}$$ five-membered ring O–O–O / RCH—CHR (Section 12-5)

Molozonide

TABLE 20-7 (CONTINUED)
Examples of pericyclic reactions

2 OTHER CYCLOADDITIONS (CONTINUED)

Cyclic osmate formation from alkenes:

(Section 12-5)

3 ELECTROCYCLIC REACTIONS

Cyclobutene—1,3-butadiene:

Δ or $h\nu$

(Section 14-5)

1,3-Cyclohexadiene—1,3,5-hexatriene:

Δ or $h\nu$

(Section 14-5)

4 FRAGMENTATIONS

Ester pyrolysis:

R O O H ⟶ R O OH +

(Section 18-4)

McLafferty rearrangement:

[H O R]$^{+\cdot}$ ⟶ [OH R]$^{+\cdot}$ +

(Section 18-7)

Summary of New Reactions

Synthesis of Benzene and Derivatives

1 INDUSTRIAL BENZENE SYNTHESIS BY PLATFORMING

$$\xrightarrow{\text{Pt-Al}_2\text{O}_3,\ 450\text{–}550^\circ\text{C},\ \text{H}_2\ \text{pressure}} + 3\ \text{H}_2$$

2 DEHYDROGENATION OF HYDROAROMATICS

Pt or Pd-C, Δ
$- H_2$

S or Se with Δ, or $KMnO_4$, C_6H_6, crown ether

R

R

Synthetic Planning: Switching and Blocking of Directing Power

3 INTERCONVERSION OF NITRO AND AMINO GROUPS

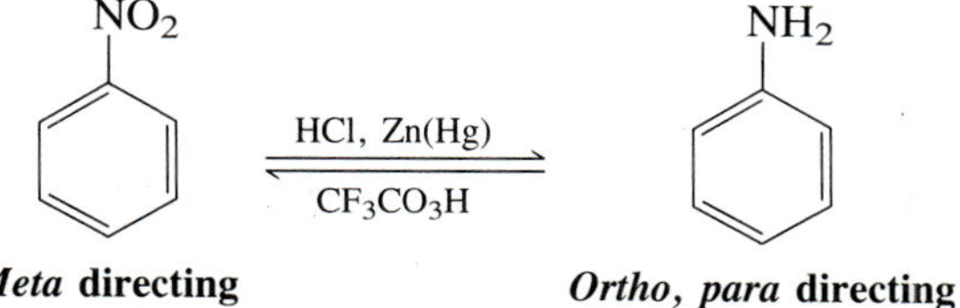

Meta directing — *Ortho, para* directing

4 CONVERSION OF ALKANOYL INTO ALKYL

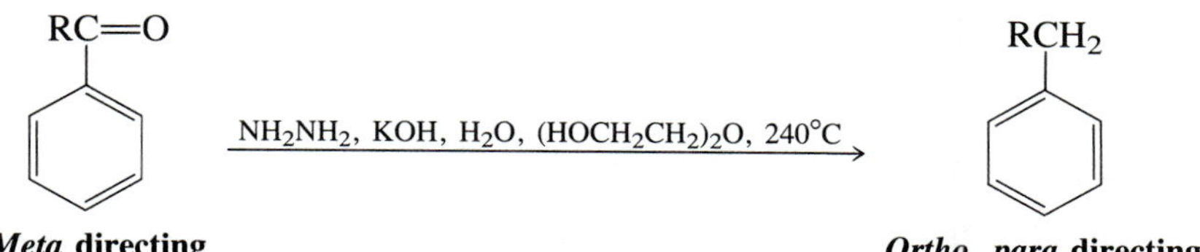

Meta directing — *Ortho, para* directing

5 BLOCKING BY SULFONATION

R → (SO_3, Block) → R, SO_3H → (E^+) → R, E, SO_3H → (H_2O, Δ; $- H_2SO_4$, Deblock) → R, E

Ipso Substitution

6 PROTODEALKYLATION

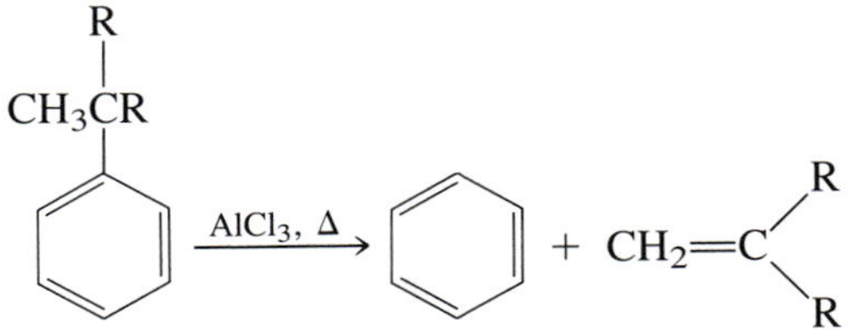

7 REARRANGEMENT

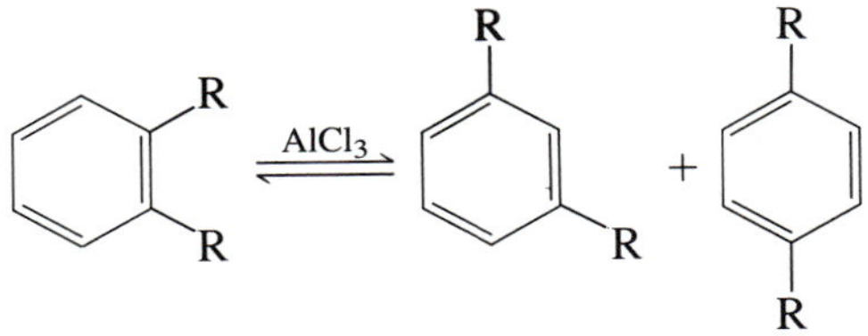

8 NUCLEOPHILIC AROMATIC SUBSTITUTION

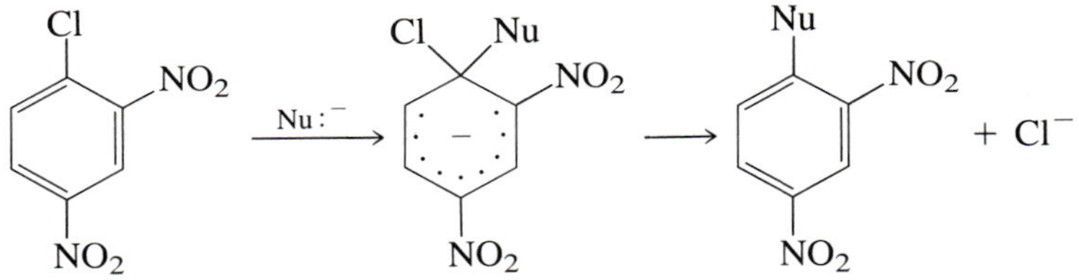

9 AROMATIC SUBSTITUTION THROUGH BENZYNE INTERMEDIATES

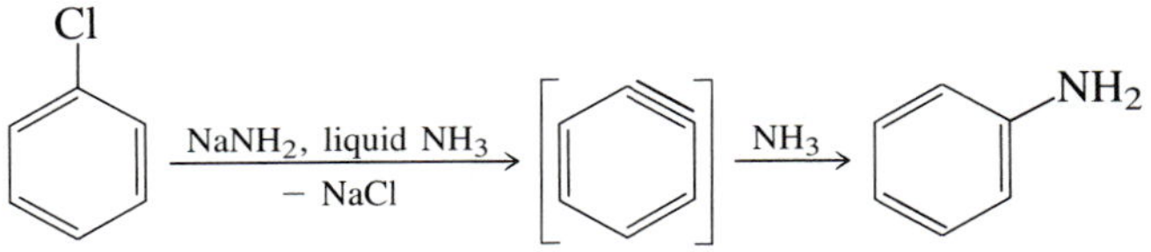

Summary of Important Concepts

1 Substituents on the benzene ring can be divided into two classes: those that activate the ring by electron donation and those that deactivate it by electron attraction. The mechanisms of donation and attraction are based on either induction or resonance. These effects may operate simultaneously to either reinforce or oppose each other. Amino and alkoxy substituents are strongly activating, alkyl and phenyl weakly so; nitro, trifluoromethyl, sulfonic and carboxylic acid, nitrile, and cationic groups are strongly deactivating, whereas halogens are weakly so.

2 Activators direct electrophiles *ortho* and *para;* deactivators direct *meta,* although at a much lower rate. The exceptions are the halogens, which direct *ortho* and *para*.

3 If there are several substituents, electrophilic aromatic substitution is governed by the activating power of each group. Generally, the stronger activator (or weaker deactivator) controls the regioselectivity of attack, in the following order: strongly activating *ortho,para* director > weakly activating *ortho,para* director > deactivating *ortho,para* director > weakly deactivating *meta* director > strongly deactivating *meta* director. For example,

$$OH > CH_3 > Br > CHO > NO_2$$

4 Strategies for the synthesis of highly substituted benzenes rely on the directing power of the substituents, the synthetic ability to change the sense of direction of these substituents by chemical manipulation, and the use of blocking groups.

5 Nucleophilic aromatic *ipso* substitution accelerates with the nucleophilic power of the attacking species and with the number of electron-withdrawing substituents on the benzene ring, particularly if they are located *ortho* or *para* to the point of attack.

6 Benzyne is destabilized by the strain imposed by the two *sp*-hybridized carbons forming the triple bond.

7 The major modes of organic reactivity are substitution, elimination, addition, and pericyclic reactions.

Problems

1 Rank the compounds in each of the following groups in order of decreasing reactivity toward electrophilic substitution. Explain your rankings.

(a) $C_6H_5CCl_3$ $C_6H_5CH_3$ $C_6H_5CHCl_2$ $C_6H_5CH_2Cl$

(b) $C_6H_5CH_2CH_3$ $C_6H_5CH_2CCl_3$ $C_6H_5CH_2CF_3$ $C_6H_5CF_2CH_3$

(c) $C_6H_5OCH_3$ $C_6H_5O^-Na^+$ $C_6H_5OC(=O)CH_3$

(d) $C_6H_5COCH_3$ $C_6H_5COO^-Na^+$ $C_6H_5CONH_2$

2 Specify whether you expect the benzene rings in the following compounds to be activated or deactivated.

(a) 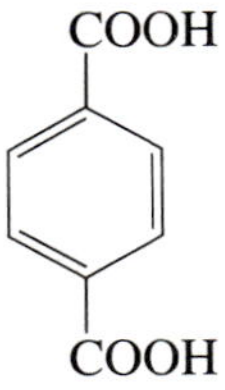 (b) 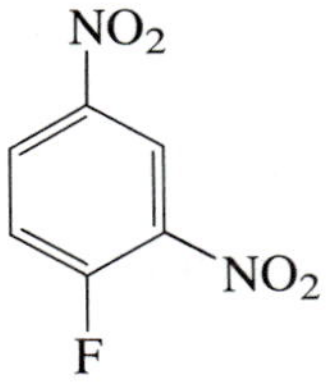 (c)

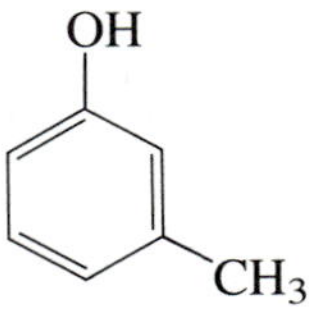

(d) 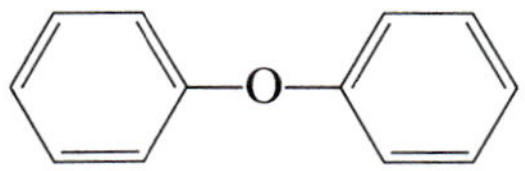 (e) 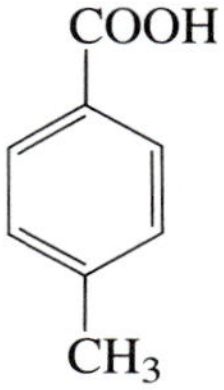 (f)

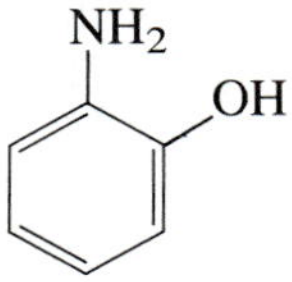

(g) 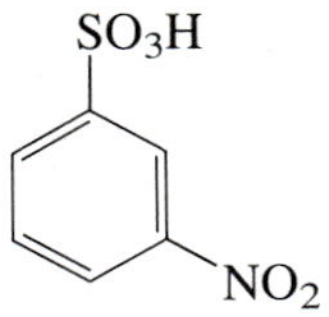 (h)

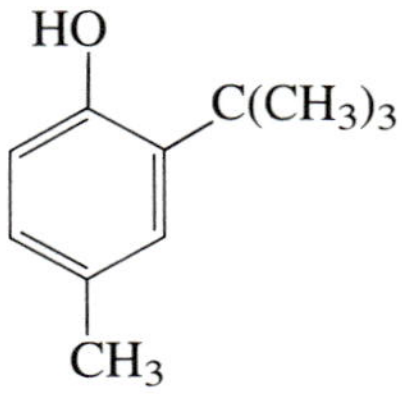

3 Draw appropriate resonance forms to explain the deactivating *meta*-directing character of the $—SO_3H$ group in benzenesulfonic acid.

4 Do you agree with the following statement? Explain your answer.

"Strongly electron withdrawing substituents on benzene rings are *meta* directing because they deactivate the *meta* positions less than they deactivate the *ortho* and *para* positions."

5 Write the expected major product(s) of each of the following electrophilic substitution reactions.

(a) 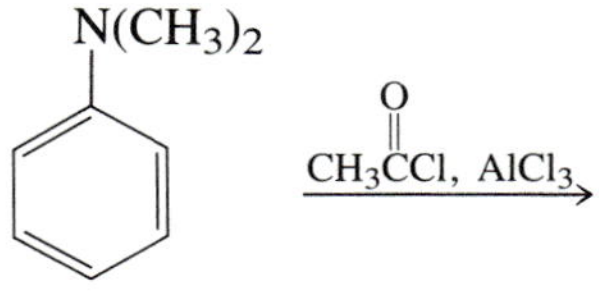 $\xrightarrow{CH_3\overset{O}{\overset{\|}{C}}Cl,\ AlCl_3}$

(b) $\xrightarrow{Br_2,\ Fe}$

(c) $\xrightarrow{HNO_3,\ H_2SO_4}$

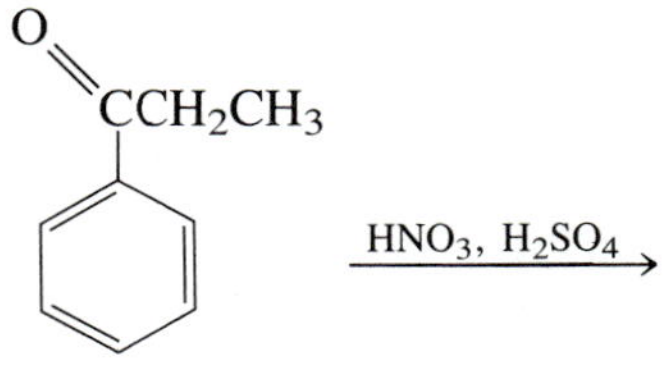

(d) 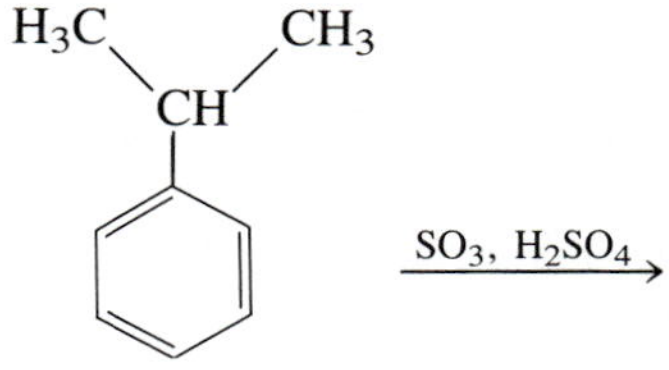 $\xrightarrow{SO_3,\ H_2SO_4}$

(e) 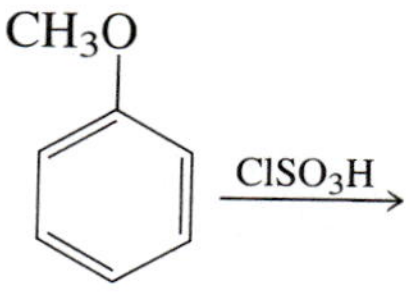 $\xrightarrow{ClSO_3H}$

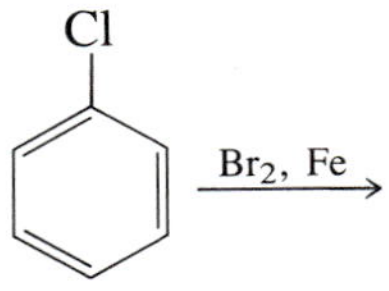

(f) [acetanilide, $C_6H_5NHCOCH_3$] $\xrightarrow{I_2,\ HNO_3}$

(g) [biphenyl] $\xrightarrow{CH_3COCCH_3 \text{ (O O)},\ AlCl_3}$

(h) [nitrobenzene, $C_6H_5NO_2$] $\xrightarrow{HNO_3,\ H_2SO_4,\ \Delta}$

6 Electrophilic substitution on the benzene ring of benzenethiol (thiophenol, C_6H_5SH) is not possible. Why? What do you think happens when benzenethiol is allowed to react with an electrophile? (Hint: Recall Chapters 6 and 9 regarding sulfur compounds.)

7 Although methoxy is a strongly activating (and *ortho,para*-directing) group, the *meta* positions in methoxybenzene (anisole) are actually slightly *deactivated* toward electrophilic substitution relative to benzene. Explain.

8 Write the expected major product(s) of each of the following reactions.

(a) [benzene ring: NO_2, CH_3 (meta)] $\xrightarrow{Cl_2,\ FeCl_3}$

(b) [benzene ring: CH_3O, OCH_3 (ortho)] $\xrightarrow{HNO_3,\ H_2SO_4}$

(c) [benzene ring: Cl, CH_3 (para)] $\xrightarrow{SO_3,\ H_2SO_4}$

(d) [benzene ring: COOH, CH_3CH_2 (ortho)] $\xrightarrow{HNO_3,\ H_2SO_4}$

(e) [benzene ring: NH_2, H_2N (meta)] $\xrightarrow{Br_2,\ H_2O}$

(f) [benzene ring: SO_3H, CCH_3 (=O) (meta)] $\xrightarrow{Br_2,\ FeBr_3}$

(g) [indane ring with O_2N] $\xrightarrow{HNO_3,\ H_2SO_4}$

(h) [benzene ring: $NHCCH_3$ (=O), NO_2 (para)] $\xrightarrow{Cl_2,\ Fe}$

(i) 4-chloronitrobenzene (NO_2, Cl) $\xrightarrow{CH_3Cl,\ AlCl_3}$

(j) (CH_3, NO_2, OCH_3) $\xrightarrow{SO_3,\ H_2SO_4}$

CH_3, CH_3, CH_3
1,2,3-Trimethylbenzene

CH_3, CH_3, CH_3
1,2,4-Trimethylbenzene

CH_3, H_3C, CH_3
1,3,5-Trimethylbenzene

9 Which of the three trimethylbenzenes, shown in the margin, would you expect to be most reactive toward electrophilic substitution?

10 The *nitroso* group, —NO, as a substituent on a benzene ring acts as an *ortho,para*-directing group but is deactivating. Explain this finding by the Lewis structure of the nitroso group and its inductive and resonance interactions with the benzene ring. (Hint: Consider possible similarities to another type of substituent that is *ortho,para*-directing but deactivating.)

11 Typical conditions for nitrosation are illustrated in the equation below. Propose a detailed mechanism for this reaction.

Phenol (OH) $\xrightarrow{NaNO_2,\ HCl,\ H_2O}$ o-nitrosophenol (OH, NO) + p-nitrosophenol (OH, NO)

12 Identify compounds A through D from the elemental analyses given below and the spectra presented below and on the next three pages. Then propose syntheses for compounds B, C, and D, each starting from compound A.
A: 45.90 C; 3.21 H; 50.89% Br. ^{1}H NMR and IR spectra A.
B: 41.89 C; 3.52 H; 46.45 Br; 8.14% N. ^{1}H NMR and IR spectra B.
C: Same analysis as B. ^{1}H NMR and IR spectra C.
D: 28.72 C; 2.01 H; 63.69 Br; 5.58% N. ^{1}H NMR and IR spectra D.

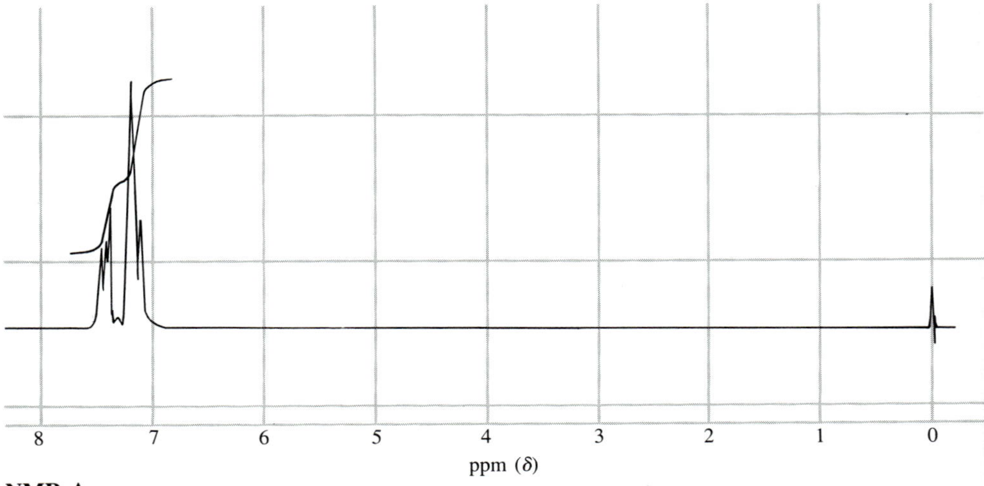

NMR-A

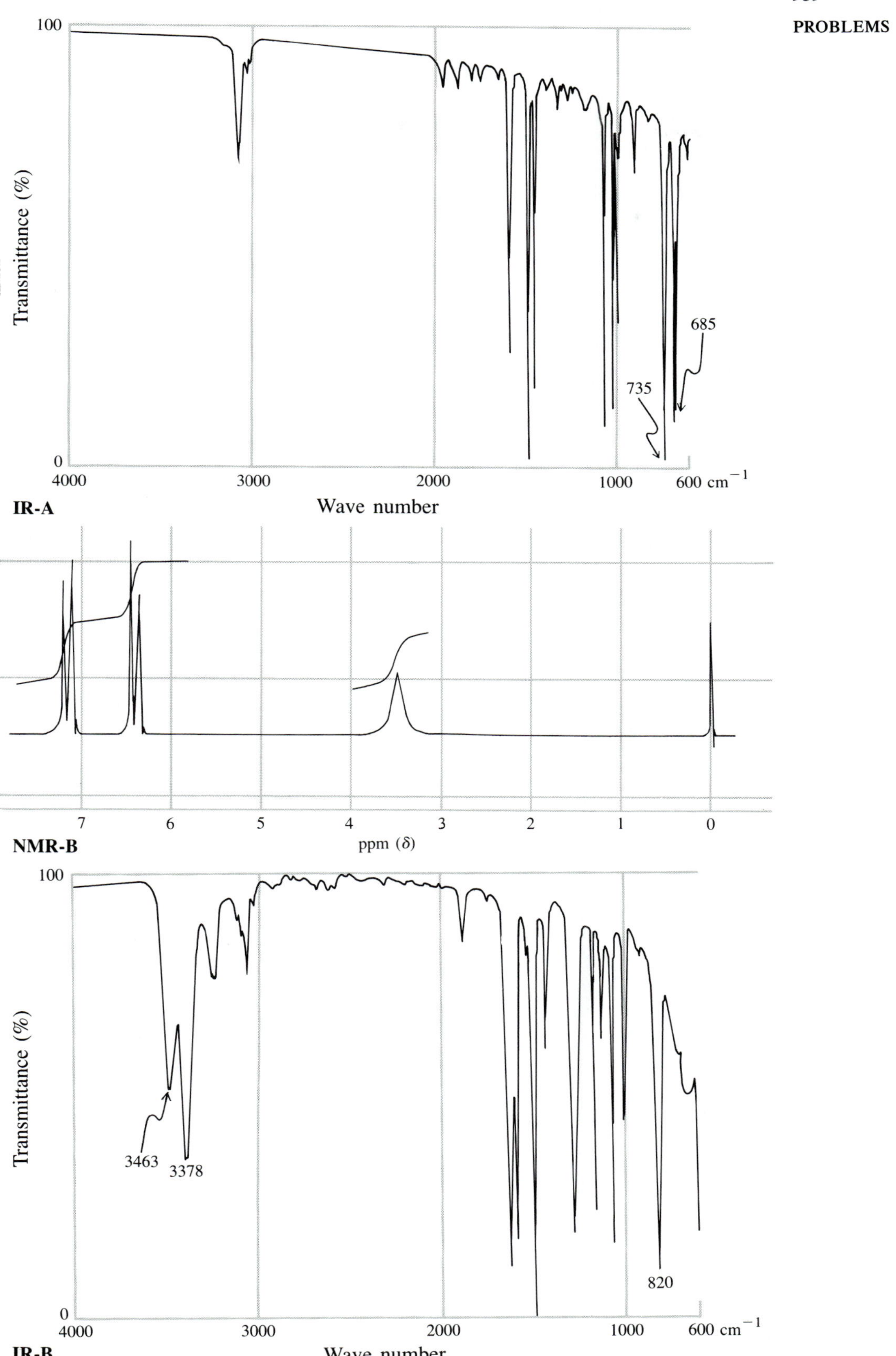
100
Transmittance (%)
685
735
0
4000
3000
2000
1000
600 cm⁻¹
IR-A
Wave number
7
6
5
4
3
2
1
0
NMR-B
ppm (δ)
100
Transmittance (%)
3463
3378
820
0
4000
3000
2000
1000
600 cm⁻¹
IR-B
Wave number

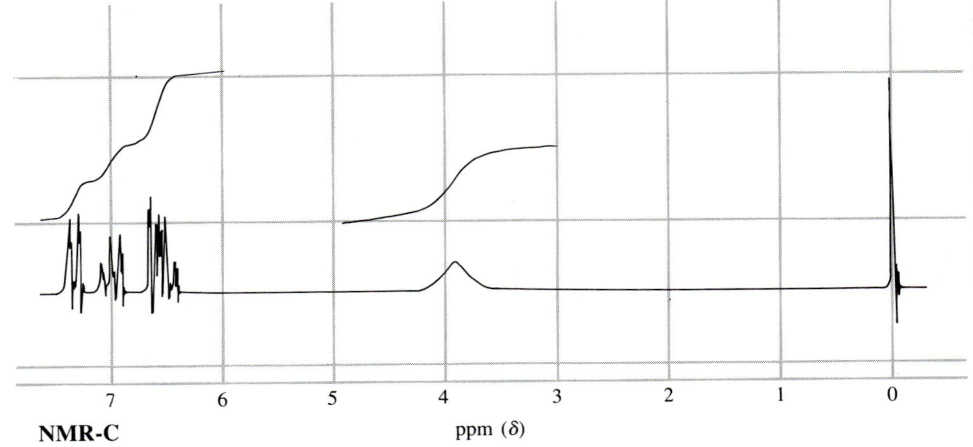
7
6
5
4
3
2
1
0
NMR-C
ppm (δ)

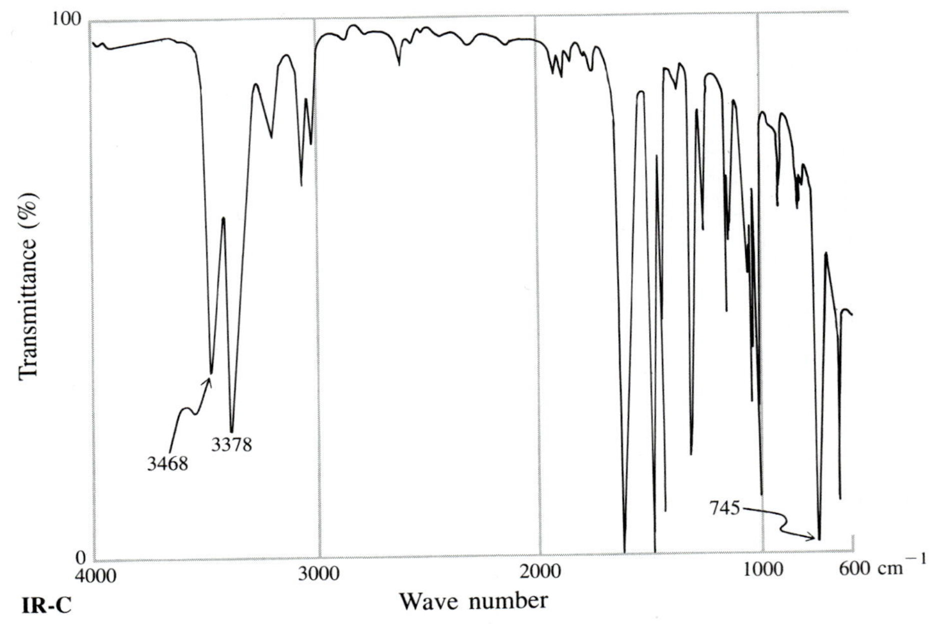
100
Transmittance (%)
3468
3378
745
0
4000
3000
2000
1000
600 cm⁻¹
Wave number
IR-C

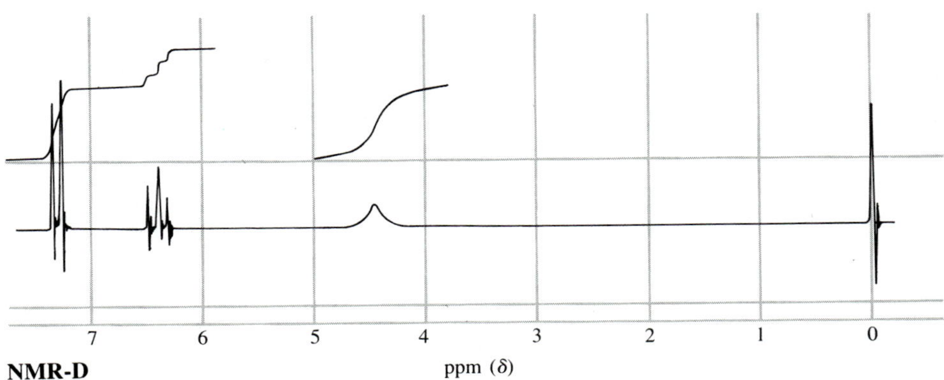
7
6
5
4
3
2
1
0
NMR-D
ppm (δ)

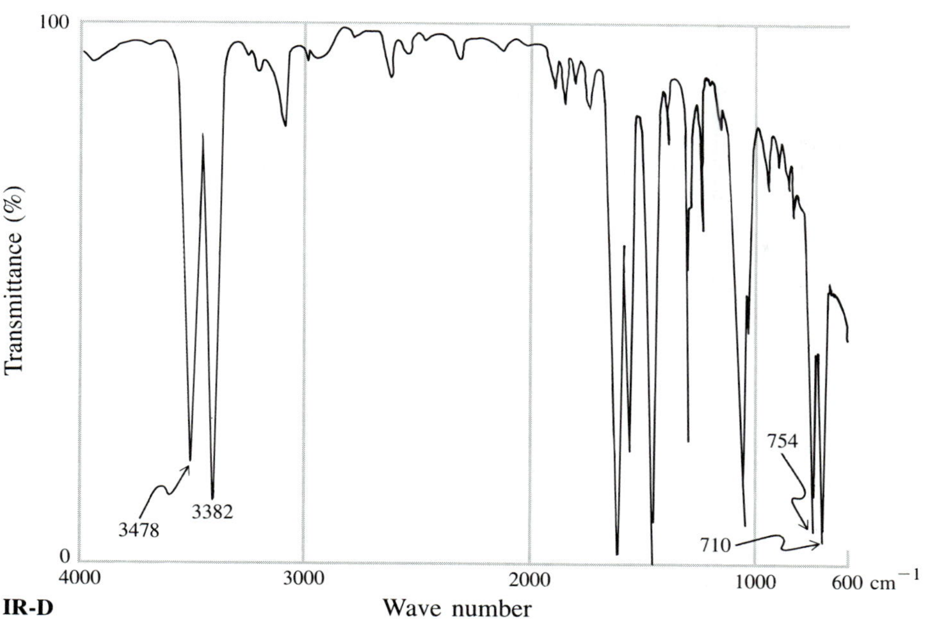

IR-D

13 Propose a reasonable synthesis of each of the following multiply substituted arenes from benzene.

(a) CH_2CH_3; $C(=O)CH_3$

(b) NO_2; Cl

(c) CH_2CH_3; SO_3H

(d) NO_2; NH_2

(e) Cl; NO_2; NO_2

(f) Cl; Br; O_2N

(g) Br; Cl

(h) CH_3; Br; Br

(i) NO_2; $CH_2CH_2CH_3$; Cl

(j) Br; Cl

14 4-Methoxyphenylmethanol (anisyl alcohol) contributes both to the flavor of licorice and to the fragrance of lavender. Propose a synthesis of this compound from methoxybenzene (anisole).

CH_2OH; OCH_3

4-Methoxyphenylmethanol (Anisyl alcohol)

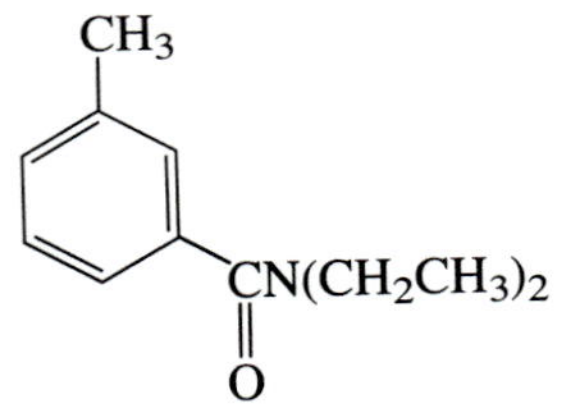

CH_3 / $CN(CH_2CH_3)_2$ / O

***N,N*-Diethyl-3-methyl-benzenecarboxamide**

15 The active ingredient in virtually all insect repellents is *N,N*-diethyl-3-methylbenzenecarboxamide (*N,N*-diethyl-*m*-toluamide). Show how you would prepare this compound from benzene. (Hint: First prepare 1-bromo-3-methylbenzene.)

16 Propose an efficient synthesis of 2,5-dimethylbenzenamine (2,5-dimethylaniline), an important precursor of a number of synthetic dyes. Start with methylbenzene (toluene).

17 Rank the following compounds in descending order of reactivity toward aqueous base.

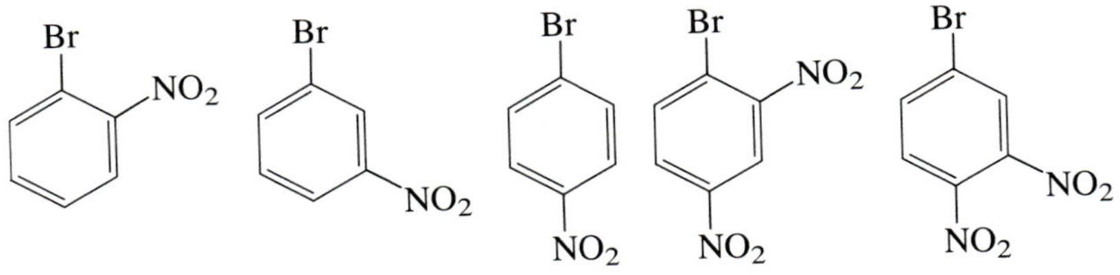

Br, NO_2 Br, NO_2 Br, NO_2 Br, NO_2, NO_2 Br, NO_2, NO_2

18 Predict the main product(s) of each of the following reactions. In each case, describe the mechanism(s) in operation.

(a) Cl, NO_2, NO_2 $\xrightarrow{NH_2NH_2}$

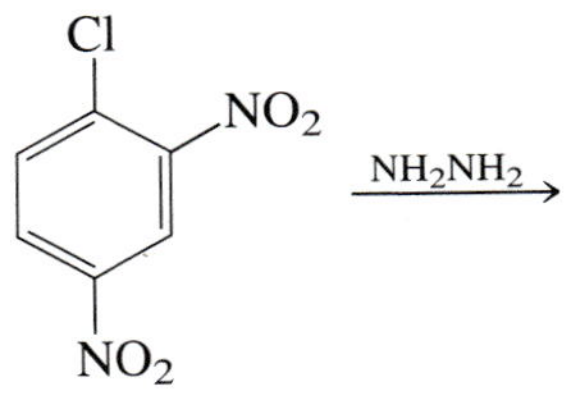

(b) Cl, Cl, O_2N, NO_2 $\xrightarrow{NaOCH_3,\ CH_3OH}$

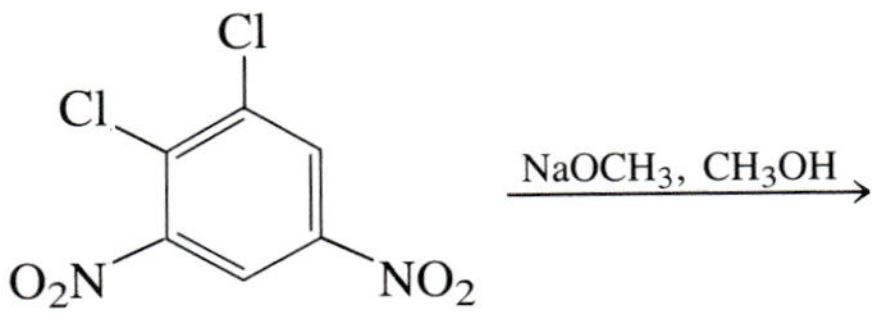

(c) Cl, CH_3 $\xrightarrow{KOH,\ H_2O,\ \Delta}$

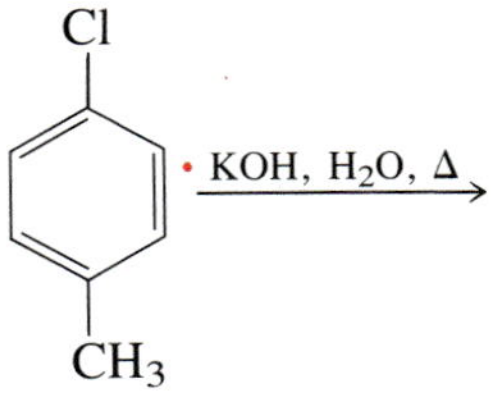

(d) CH_3, Br $\xrightarrow{LiN(CH_2CH_3)_2,\ (CH_3CH_2)_2NH}$

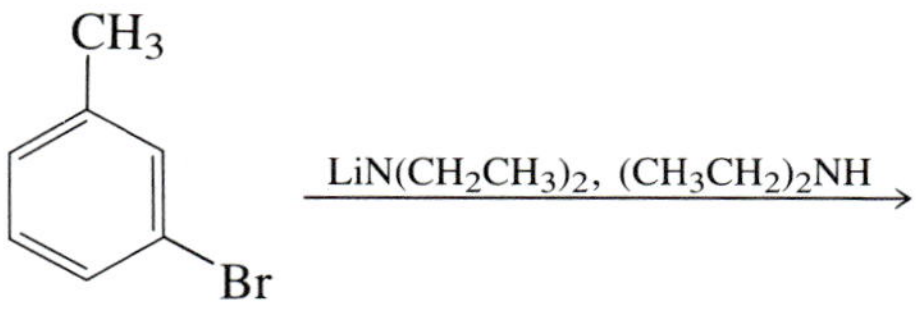

19 Following is an outline of a synthesis of the herbicide propalin. Fill in the missing reagents or synthetic intermediates. Determine the proper IUPAC name for propalin. (Hint: Consider the name of the parent compound carefully.)

Benzene —(a) 2 steps→ 1-bromo-4-chlorobenzene (Cl, Br) —(b) 2 steps→ 4-chlorobenzoic acid (Cl, COOH) —(c)→ (d) —$(CH_3CH_2CH_2)_2NH$→ [$N(CH_2CH_2CH_3)_2$, O_2N, NO_2, COOH] —(e) 3–4 steps→ [$N(CH_2CH_2CH_3)_2$, O_2N, NO_2, CH_3]

Propalin

20 Section 24-3 contains information about the herbicides 2,4-D and 2,4,5-T. The latter is notorious because it is commercially prepared from 2,4,5-trichlorobenzenol (2,4,5-trichlorophenol), which, on treatment with base, is partly converted into the extraordinarily toxic and infamous compound TCDD (2,3,7,8-tetrachlorodibenzo[*b,e*][1,4]dioxin, or just *dioxin,* for short). Explain this transformation by means of a reasonable mechanism.

[OCH_2COOH, Cl, Cl, Cl]

2,4,5-T

2 [OH, Cl, Cl, Cl] ⇌(Base) 2 [O^-, Cl, Cl, Cl] → [Cl, Cl, O, O, Cl, Cl]

TCDD (Dioxin)

21 Starting with benzenamine, propose a synthesis of aklomide, an agent used to treat certain exotic fungal and protozoal infections in veterinary medicine. As in Problem 19, several intermediates are already shown to give you the general route. Fill in the blanks that remain; each may require as many as three sequential reactions. Give the IUPAC name of aklomide when you are done.

[NH_2] —(a)→ [NH_2, Br] —(b)→ [NO_2, Cl, Br] —(c)→ [NO_2, Cl, CN] —(d)→ [NO_2, Cl, $CONH_2$]

Aklomide

22 Explain the mechanism of the following synthetic transformation. (Hint: Two equivalents of butyllithium are used.)

OCH_3, F (benzene) $\xrightarrow[3.\ H^+,\ H_2O]{1.\ CH_3CH_2CH_2CH_2Li\quad 2.\ H_2C{=}O}$ OCH_3, CH_2OH, $CH_2CH_2CH_2CH_3$ (benzene)

NC, CN, C=C, NC, CN

TCNE

NaOH, H_2O, 100°C

NC, CN, C=C, NC, OH

23 In a rather unusual reaction, tetracyanoethene (TCNE) is converted by boiling in aqueous base into tricyanoethenol (whose enol form is stabilized by the three nitrile groups). Suggest a mechanism for this transformation. (Hint: This reaction occurs for the same reason that one form of nucleophilic aromatic substitution takes place.)

24 Propose a continuation of the synthesis of estrone, beginning with structure B in Problem 29 of Chapter 15. If you need a hint, refer to Problem 16 of Chapter 19. You should be able to get as far as an estrone derivative containing methoxy instead of hydroxy on the aromatic ring.

25 Propose structures for the products A and B in the following scheme:

$\xrightarrow{h\nu}$ A ($C_{14}H_{12}$) $\xrightarrow{S,\ \Delta}$ B ($C_{14}H_{10}$) + H_2S

26 Place each of the following reactions into one or more of the categories described in Section 20-7.

(a) $RI \xrightarrow{LiAlD_4} RD$

(b) $RI \xrightarrow[2.\ H_2C{=}O]{1.\ Mg} RCH_2OH$

(c) $ROH \xrightarrow{SOCl_2} RCl$

(d) $RCH{=}CHR \xrightarrow{MCPBA} RCH{-}CHR$ (with O bridging, epoxide)

(e) $RC{\equiv}CR \xrightarrow{H^+,\ H_2O,\ Hg^{2+}} RCCH_2R$ (with C=O)

(f) $RC{\equiv}CH \xrightarrow[2.\ R'X]{1.\ CH_3CH_2CH_2CH_2Li} RC{\equiv}CR'$

(g) RCR (with C=O) $\xrightarrow{NH_2OH}$ RCR (with C=N–OH)

(h) $RCH_2\overset{\overset{O}{\|}}{C}H \xrightarrow{NaOH} RCH_2\overset{\overset{OH}{|}}{C}H\underset{\underset{R}{|}}{C}H\overset{\overset{O}{\|}}{C}H$

(i) $RCH_2COOCH_2CH_3 \xrightarrow{NaOCH_2CH_3} RCH_2\overset{\overset{O}{\|}}{C}\underset{\underset{R}{|}}{C}HCOOCH_2CH_3$

(j) $C_6H_5C(CH_3)_3 \xrightarrow{H^+} C_6H_6 + CH_2{=}C(CH_3)_2$

CHAPTER 21

Amines and Their Derivatives

NEW FUNCTIONAL GROUPS CONTAINING NITROGEN

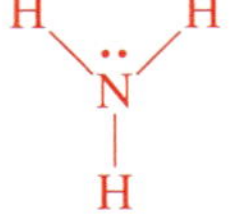

Ammonia

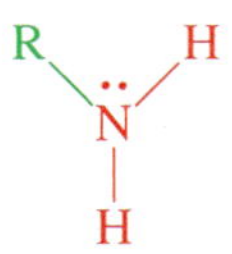

Primary amine

Secondary amine

Tertiary amine

Amines are derivatives of ammonia, in which one (primary), two (secondary), or three (tertiary) of the hydrogens have been replaced by alkyl or aryl groups. Therefore, in a way, amines are related to ammonia in the same sense as ethers and alcohols are related to water. The amines and other nitrogen-bearing compounds are among the most-abundant organic molecules; they are, for example, components of the amino acids, peptides, proteins (see Chapter 27), and alkaloids (Chapter 26). Many are medicinally active (Section 21-6).

In many respects, the chemistry of the amines is analogous to that of the alcohols and ethers (Chapters 8 and 9). For example, all amines are basic (although primary and secondary ones can also behave as acids), they form hydrogen bonds, and they act as nucleophiles in substitution reactions. However, there are some differences in reactivity, because nitrogen is less electronegative than oxygen. Thus, primary and secondary amines are less acidic and form weaker hydrogen bonds than alcohols and ethers, and they are more basic and more nucleophilic.

This chapter begins by naming amines. There follows a discussion of the physical, chemical, and physiological properties of this class of molecules.

21-1 The Naming of Amines

As for the other functional groups, the system for naming amines is confused by the variety of common names in the literature. Probably the best way to name aliphatic amines is that used by *Chemical Abstracts*—that is, as **alkanamines,** in which the name of the alkane stem is modified by replacing the ending **-e** by **-amine.** The position of the functional group is indicated by a prefix designating the carbon atom to which it is attached, as in the alcohols (Section 8-1).

CH_3NH_2
Methanamine

$CH_3CH(CH_3)CH_2NH_2$
2-Methyl-1-propanamine

$CH_3CH(NH_2)CH{=}CHCH_3$
3-Penten-2-amine

In this system, anilines are called benzenamines (Section 19-1). For secondary and tertiary amines, the largest alkyl substituent on nitrogen is chosen as the alkanamine stem, and the other groups are named by using the letter *N*-, followed by the name of the additional substituent(s).

$C_6H_5NH_2$
Benzenamine
(Aniline)

$CH_3N(H)CH_2CH_3$
***N*-Methylethanamine**

$CH_3N(CH_3)CH_2CH_2CH_3$
***N*,*N*-Dimethyl-1-propanamine**

An alternative way to name amines treats the functional group, called *amino-*, merely as a substituent of the alkane stem. This procedure would be equivalent to naming alcohols as hydroxy alkanes.

$CH_3CH_2NH_2$
Aminoethane

$(CH_3)_2NCH_2CH_2CH_3$
***N*,*N*-Dimethylaminopropane**

$FCH_2CH_2CH(CH_3)N(H)CH_2CH_3$
3-(Ethylamino)-1-fluorobutane

Many common names are based on the label *alkyl amine* (see margin), as in the naming of alkyl alcohols.

CH_3NH_2
Methylamine

$(CH_3)_3N$
Trimethylamine

$CH_3N(C_6H_{11})CH_2–C_6H_5$
Benzylcyclohexylmethylamine

EXERCISE 21-1

Name each of the following molecules twice, first as an alkanamine, then as an alkyl amine.

(a) $CH_3CH(NH_2)CH_2CH_3$; (b) $C_6H_5N(CH_3)_2$; (c) $BrCH_2CH_2CH_2CH_2CH(CH_3)NH_2$.

EXERCISE 21-2

Draw structures for the following compounds (common name in parentheses): (a) 2-propynamine (propargylamine); (b) (*N*-2-propenyl)phenylmethanamine (*N*-allylbenzylamine); (c) *N*-methyl-1,1-dimethylethanamine (*tert*-butylmethylamine); (d) *N*,2-dimethyl-2-propanamine (*tert*-butylmethylamine).

In summary, there are several systems for naming amines. *Chemical Abstracts* uses names of the type *alkanamine* and *benzenamine*. Alternatives are based on the labels *amino alkane, aniline,* and *alkyl amine*.

FIGURE 21-1

The nearly tetrahedral structure of methanamine (methylamine).

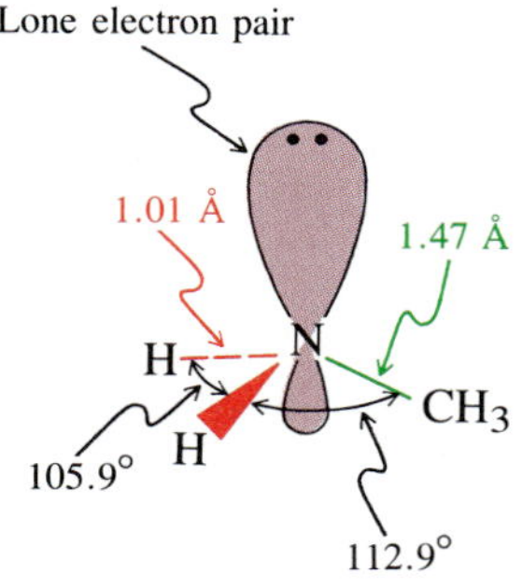

21-2 Physical and Acid-Base Properties of the Amines

Let us now look at some of the physical characteristics of simple amines. Amines adopt a tetrahedral structure around the heteroatom, but this arrangement is not rigid because of rapid inversion at nitrogen. Configurational stability is attained by protonation or alkylation to ammonium salts that can be resolved when chiral. After a brief discussion of some physical constants, this section gives examples of the behavior of amines in IR, NMR, and mass spectroscopy.

The Molecular Structure of Alkanamines Is Tetrahedral

The nitrogen orbitals in amines are very nearly sp^3 hybridized (see Section 1-8, Figure 1-20), forming an approximately tetrahedral structure. Three vertices of the tetrahedron are occupied by the three substituents; the fourth is the center of the lone electron pair. The structure of methanamine (methylamine) is depicted in Figure 21-1.

EXERCISE 21-3

On inspection of Figure 21-1, you will notice that the bonds to nitrogen in methanamine (methylamine) are slightly longer than those in the corresponding structure of methanol (Figure 8-1B). Explain.

The tetrahedral arrangement around an amine nitrogen suggests that it should be chiral if it bears three different substituents, the lone electron pair functioning as the fourth. The image and mirror image of such a compound would not be superimposable, by analogy with carbon-based stereocenters (Section 5-1). That this is so can be illustrated with the simple chiral alkanamine *N*-methylethanamine (ethylmethylamine).

Image and Mirror Image of *N*-Methylethanamime (Ethylmethylamine)

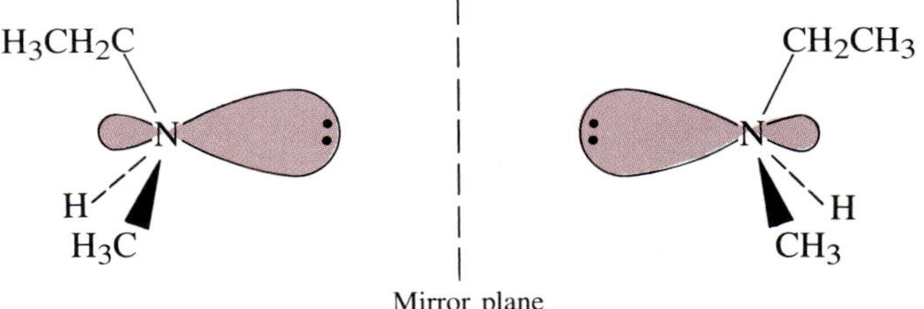

However, amines are not configurationally stable at nitrogen, because of rapid isomerization by a process called **inversion.** This transformation is somewhat like the inversion at carbon in the S_N2 reaction of haloalkanes (Section 6-5, Figure 6-7). However, inversion of amines does not require the presence of an additional reagent. The molecule can be visualized as passing through a transition state incorporating an sp^2-hybridized nitrogen atom, as illustrated in Figure 21-2. The barrier to this motion in ordinary small amines has been measured by spectroscopic techniques and found to be between about 5 and 7 kcal $mole^{-1}$. It is therefore impossible to keep an enantiomerically pure, simple di- or trialkylamine from racemizing at room temperature.

FIGURE 21-2
Inversion at nitrogen interconverts the two enantiomers of *N*-methylethanamine.

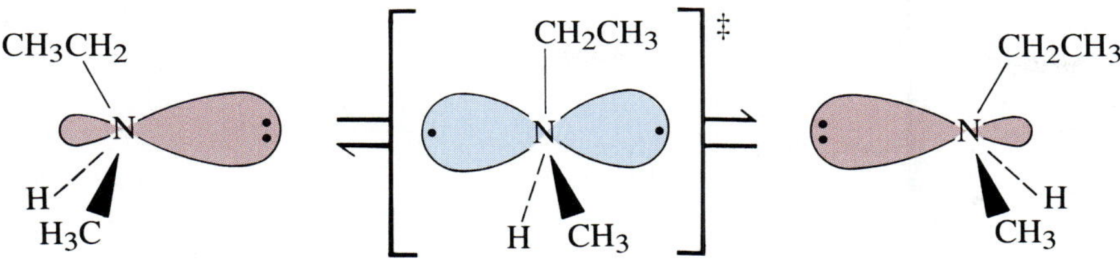

EXERCISE 21-4

The methylene hydrogens in *N*-methylethanamine (ethylmethylamine) are diastereotopic (review Section 10-4). Do you expect two different absorptions in the ^1H NMR spectrum at room temperature?

The Effect of the Lone Electron Pair: A First Look at Protonation and Alkylation

The basicity and nucleophilicity of the nitrogen in amines allow for its protonation and alkylation, giving rise to **substituted ammonium salts.** Depending on the number of substituents on nitrogen, they can be primary, secondary, tertiary, or quaternary.

$$R\ddot{N}H_2 + H^+Cl^- \longrightarrow RNH_3{}^+Cl^-$$
Primary ammonium chloride

$$R\ddot{N}H_2 + RCl \longrightarrow R_2NH_2{}^+Cl^-$$
Secondary ammonium chloride

$$R_3N\colon + H^+I^- \longrightarrow R_3NH^+I^-$$
Tertiary ammonium iodide

$$R_3N\colon + RBr \longrightarrow R_4N^+Br^-$$
Quaternary ammonium bromide

Ammonium salts can be named by attaching the substituent names to the ending **-ammonium** followed by the name of the counter ion.

$CH_3NH_3{}^+Cl^-$	$(CH_3CH_2)_4N^+I^-$	$[C_6H_5CH_2N(CH_3)_3]_2{}^{2+}SO_4{}^{2-}$
Methylammonium chloride	**Tetraethylammonium iodide**	**Benzyltrimethylammonium sulfate**

If all four substituents are different from each other, the ammonium ion is chiral. Because the lone electron pair originally present in the neutral amine is tied up by bonding to the fourth substituent, quaternary ammonium salts are configurationally stable and, if chiral, can be resolved. This can be achieved by adding enantiomerically pure chiral sulfonate counter ions (see Section 5-7) and fractionally crystallizing the diastereomeric salts.

Resolution of a Chiral Ammonium Salt

$$\left[\begin{array}{c} CH_2CH{=}CH_2 \\ | \\ CH_3NCH_2C_6H_5 \\ | \\ C_6H_5 \end{array}\right]^+ I^- \xrightarrow[\substack{-\ NaI \\ (R^* = \text{optically} \\ \text{pure chiral} \\ \text{substituent})}]{\substack{R^*SO_3^-Na^+ \\ \text{(optically pure)}}} \left[\begin{array}{c} CH_2CH{=}CH_2 \\ | \\ CH_3NCH_2C_6H_5 \\ | \\ C_6H_5 \end{array}\right]^+ {}^-O_3SR^*$$

A racemic ammonium iodide

Two diastereomers of a chiral ammonium sulfonate

1. Resolve (separate) by fractional crystallization
2. Add I^- (to replace $R^*SO_3^-$)

$\longrightarrow$ (H₃C, C₆H₅, CH₂C₆H₅, CH₂CH=CH₂ on N⁺ I⁻) + (C₆H₅, CH₃, CH₂C₆H₅, CH₂CH=CH₂ on N⁺ I⁻)

Two separated enantiomers of chiral ammonium iodide

EXERCISE 21-5

Assign the absolute configuration to the two enantiomers of the chiral ammonium salt whose resolution has been described.

Amines Form Weaker Hydrogen Bonds Than Alcohols

Because of the special ability of alcohols to enter into hydrogen bonding (Section 8-2, Figure 8-2, Table 8-1), they have unusually high boiling points. In principle, similar properties might be expected of amines, and indeed Table 21-1 bears out this expectation. However, because amines form weaker hydrogen bonds* than alcohols, their boiling points are lower and their solubility in water less. In general, the boiling points of the amines lie between those of the corresponding alkanes and alcohols. The smaller amines are soluble in water and in alcohols because they can form hydrogen bonds to the solvent. If the hydrophobic part of an amine exceeds six carbons, the solubility in water decreases strongly; the larger amines are essentially insoluble in water.

An Amino Group Can Be Detected by Spectroscopy

Primary and secondary amines can be recognized by infrared spectroscopy, because they exhibit a characteristic broad N–H stretching absorption in the range between 3300 and 3500 cm^{-1}. Primary amines show two strong peaks in this range, whereas secondary amines give rise to only a very weak single line. Tertiary amines do not give rise to signals at these wave numbers because they do not have a hydrogen that is directly bound to nitrogen. Figure 21-3 shows the infrared spectrum of cyclohexanamine.

Nuclear magnetic resonance spectroscopy is also useful for detecting the presence of amino groups. Amine hydrogens resonate to give (sometimes broadened) peaks almost anywhere in the normal hydrogen range, like the OH

---

*It should be pointed out that, whereas *all* amines can act as proton acceptors in hydrogen bonding, only primary and secondary amines can function as proton donors, because tertiary amines lack such protons.

TABLE 21-1
Comparative physical properties of amines, alcohols, and alkanes (°C)

Compound	Melting point	Boiling point	Compound	Melting point	Boiling point
CH_4	−182.5	−161.7	$(CH_3)_2NH$	−93	7.4
CH_3NH_2	−93.5	−6.3	$(CH_3)_3N$	−117.2	2.9
CH_3OH	−97.5	65.0	$(CH_3CH_2)_2NH$	−48	56.3
CH_3CH_3	−183.3	−88.6	$(CH_3CH_2)_3N$	−114.7	89.3
$CH_3CH_2NH_2$	−81	16.6	$(CH_3CH_2CH_2)_2NH$	−40	110
CH_3CH_2OH	−114.1	78.5	$(CH_3CH_2CH_2)_3N$	−94	155
$CH_3CH_2CH_3$	−187.7	−42.1	NH_3	−77.7	−33.4
$CH_3CH_2CH_2NH_2$	−83	47.8	H_2O	0	100
$CH_3CH_2CH_2OH$	−126.2	97.4			
$CH_3CH_2CH_2CH_2NH_2$	−49.1	77.8			

FIGURE 21-3
Infrared spectrum of cyclohexanamine.

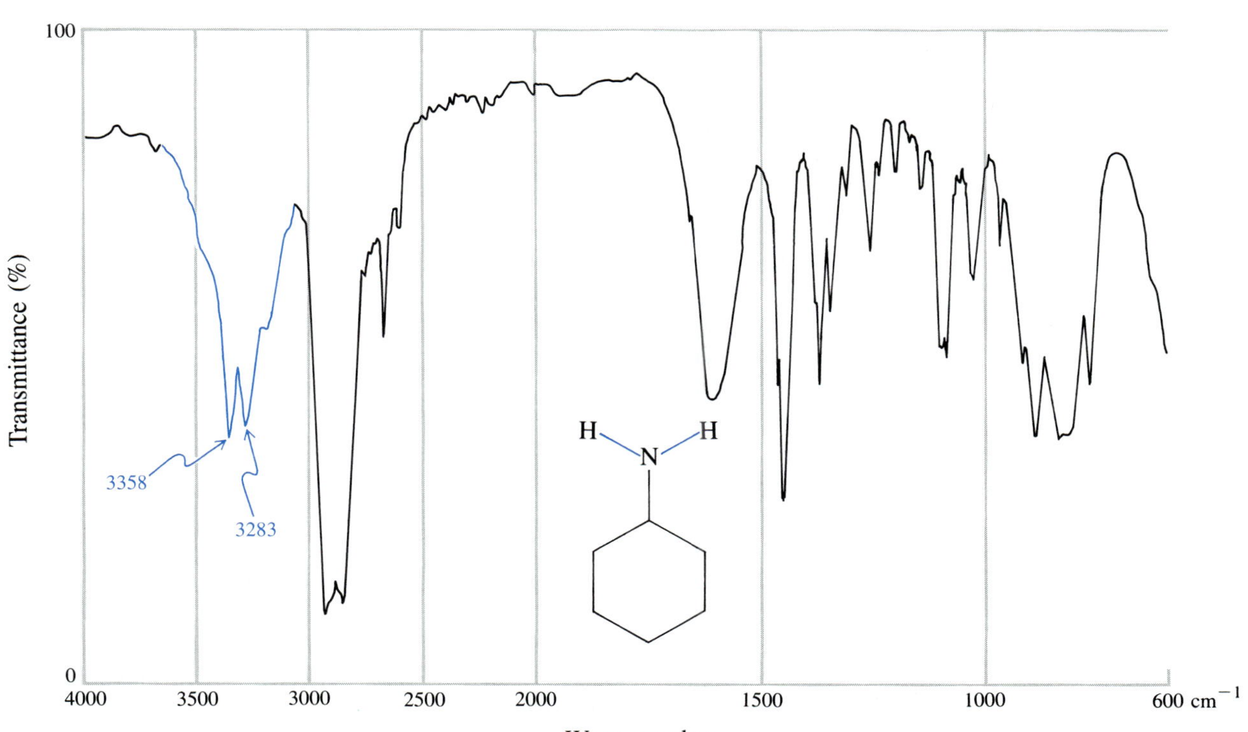

FIGURE 21-4

90-MHz ^{1}H NMR spectrum of azacyclohexane (piperidine) in dry CCl_4. Because of the use of dry solvent, the NH absorption is sharp.

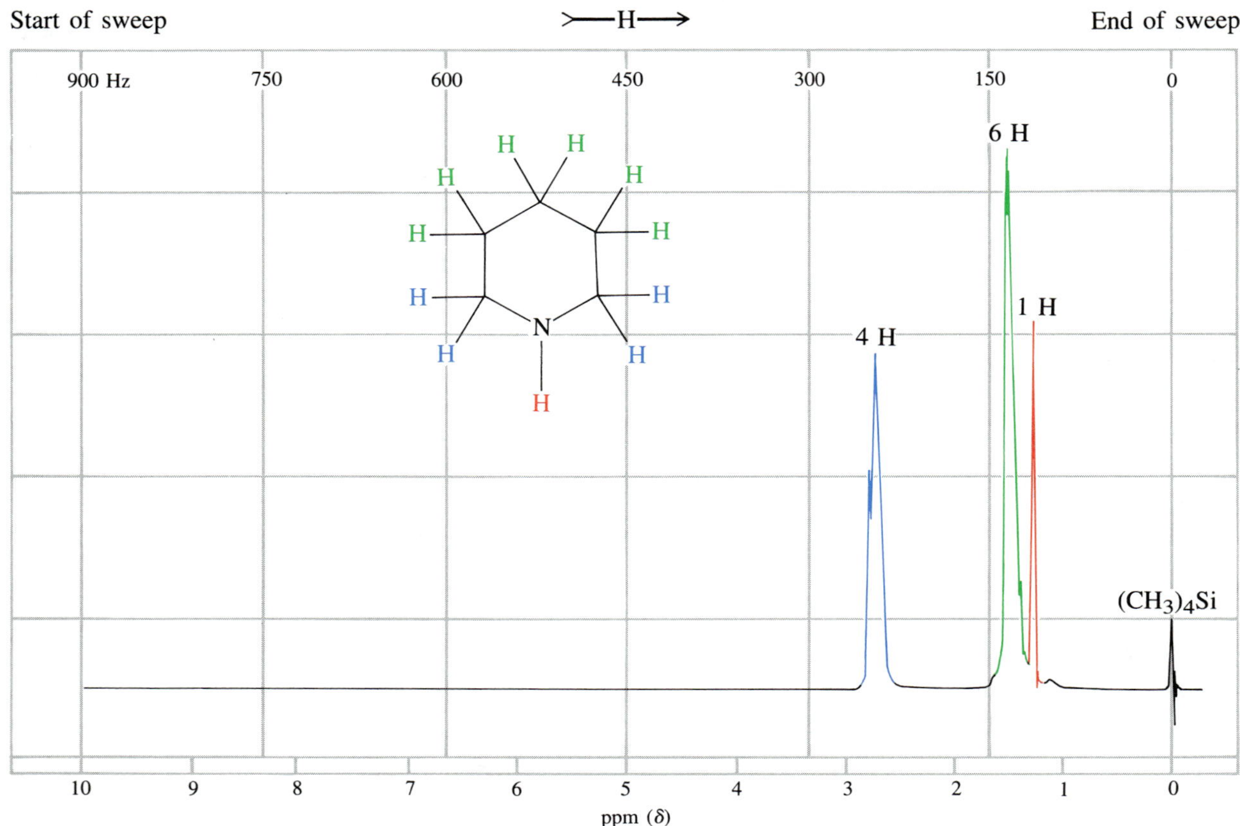

signal in the NMR spectra of alcohols. Their chemical shift depends mainly on the rate of exchange of protons with water in the solvent and the degree of hydrogen bonding. Figure 21-4 shows the ^{1}H NMR spectrum of azacyclohexane (piperidine), a cyclic secondary amine. The amine hydrogen appears at $\delta = 1.29$ ppm, and there are two other sets of signals, at $\delta = 1.52$ and 2.73 ppm. The absorption at lowest field can be assigned to the hydrogens neighboring the nitrogen, which must be deshielded because of the proximity of the electronegative nitrogen atom.

EXERCISE 21-6

Would you expect the hydrogens next to the heteroatom in an amine, $RC\mathbf{H_2}NH_2$, to be more or less deshielded than those in an alcohol, $RC\mathbf{H_2}OH$? Explain.

The ^{13}C NMR spectra of amines show a similar trend: carbons directly bound to nitrogen resonate at considerably lower field than the carbon atoms in alkanes. However, as in the hydrogen spectra (Exercise 21-6), nitrogen has a lesser deshielding effect than oxygen.

FIGURE 21-5

Mass spectrum of *N,N*-diethylethanamine (triethylamine), showing a molecular ion peak at $m/z = 101$. The base peak is due to loss of a methyl group: $m/z = 86$.

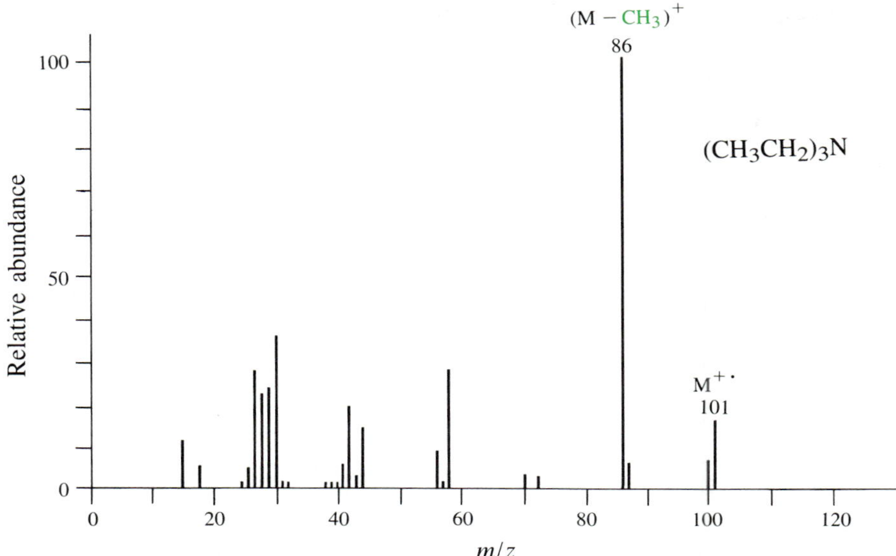

^{13}C Chemical Shifts in Some Amines (ppm)

Piperidine: 25.5, 27.2, 47.5 compared with tetrahydropyran: 23.6, 26.6, 68.0

Cyclohexanamine: 25.1, 25.7, 36.7, 50.4

$(CH_3CH_2)_2NH$: 15.4, 44.2

$CH_3N(H)CH(CH_3)_2$: 33.6, 50.9, 22.6

$H_2NCH_2CH_2CH_2CH_2CH_2NH_2$: 42.2, 33.4, 25.1

The isotopes ^{14}N and ^{15}N are directly observable by NMR (see Table 10-1). This technique is becoming increasingly useful for deducing the structure of nitrogen-containing compounds, but discussion of this prospect is better left to specialized texts.

Mass spectroscopy also can be used to establish the presence of nitrogen in an organic compound. Unlike carbon, which is tetravalent, nitrogen is trivalent. Because of these valence requirements and because nitrogen has an even atomic weight (14), molecules incorporating one nitrogen (or any odd number of nitrogens) have an *odd* molecular weight (recall Exercise 18-24). For example, the mass spectrum of *N,N*-diethylethanamine (triethylamine) shows the peak of a molecular ion at $m/z = 101$ (Figure 21-5). The base peak, at $m/z = 86$, is caused by the loss of a methyl group. Such a fragmentation is favored because it results in a resonance-stabilized **iminium ion.**

FIGURE 21-6
Mass spectrum of 1-hexanamine.

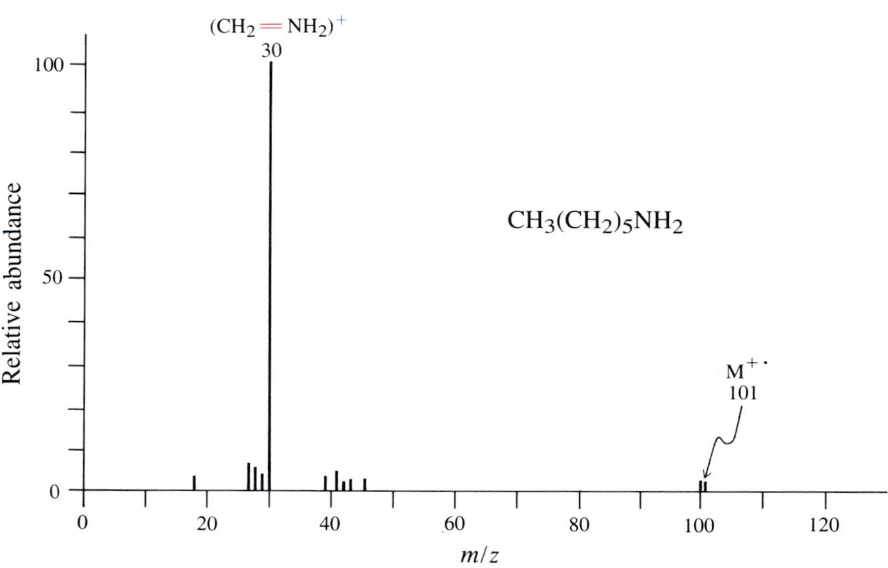

Mass-Spectral Fragmentation of *N,N*-Diethylethanamine

$$[(CH_3CH_2)_2\ddot{N}CH_2 \wr CH_3]^{+\cdot} \longrightarrow CH_3\cdot + [(CH_3CH_2)_2\ddot{N}{-}CH_2{}^+ \longleftrightarrow (CH_3CH_2)_2\overset{+}{N}{=}CH_2]$$

***N,N*-Diethylethanamine (Triethylamine)** — **An iminium ion**

The rupture of the C–C bond next to nitrogen is frequently so easy that the molecular ion cannot be observed. For example, in the mass spectrum of 1-hexanamine (an isomer of *N,N*-diethylethanamine), the molecular ion ($m/z = 101$) is barely visible; the dominating peak corresponds to the methyleneiminium fragment $[CH_2{=}NH_2]^+$ ($m/z = 30$; Figure 21-6).

EXERCISE 21-7

What approximate spectral data (IR, NMR, m/z) would you expect for *N*-ethyl-2,2-dimethylpropanamine?

$$CH_3\overset{\displaystyle CH_3}{\underset{\displaystyle CH_3}{\overset{|}{\underset{|}{C}}}}CH_2NHCH_2CH_3$$

***N*-Ethyl-2,2-dimethylpropanamine**

To summarize, amines adopt an approximately tetrahedral structure in which the lone electron pair occupies one vertex of the tetrahedron. They can, in principle, be chiral but are difficult to maintain in enantiomerically pure form because of fast nitrogen inversion. However, protonation and alkylation results in ammonium salts that are configurationally stable and, if chiral, can be resolved. Amines have boiling points higher than those of alkanes of similar size. Boiling points are lower than those of the analogous alcohols because of weaker hydrogen bonding. This attenuating effect also controls their water solubility, which is between that of comparable alkanes and alcohols. The IR stretching absorption of the N–H bond ranges between 3300 and 3500 cm^{-1}; the corresponding ^1H NMR peak can be found at variable δ. The electron-withdrawing power of the nitrogen deshields neighboring carbons and hydrogens, although to a lesser extent than is observed in alcohols and ethers. The mass spectra of simple alkanamines that contain only one nitrogen atom have odd-numbered molecular ion peaks, because of the trivalent character of nitrogen. Fragmentation occurs in such a way as to produce resonance-stabilized iminium ions.

21-3 The Acidity and Basicity of Amines

Like the alcohols (Section 8-3), amines are not only basic, but also acidic. Because of the lesser electronegativity of nitrogen compared with oxygen, the acidity of amines is about 20 orders of magnitude less than that of comparable alcohols. On the other hand, the lone electron pair is much more available for protonation, causing amines to be good bases.

Acidity and Basicity of Amines

$$R\ddot{N}H(H) + {}^-{:}B \underset{}{\overset{K_a}{\rightleftharpoons}} R\ddot{\underset{\cdot\cdot}{N}}H^- + HB$$

$$R\ddot{N}H_2 + HA \underset{}{\overset{K_b}{\rightleftharpoons}} R\overset{+}{N}H_3 + {}^-{:}A$$

Amines Are Poor Acids

We have seen evidence that amines are less acidic than alcohols: amide ions are used to deprotonate alcohols (Sections 8-3 and 9-1). The equilibrium of this proton transfer is shifted to the side of the alkoxide ion with an equilibrium constant of about 10^{20}. This high value is due to the strong basicity of amide ions, which is reflected, in turn, in the low acidity of amines. The pK_a of ammonia and alkanamines is on the order of 35.

The pK_a of Amines

$$R\ddot{N}H(H) + H_2\ddot{\underset{\cdot\cdot}{O}} \underset{}{\overset{K_a}{\rightleftharpoons}} R\ddot{\underset{\cdot\cdot}{N}}H^- + H_2\ddot{O}H^+ \qquad K_a = \frac{[R\ddot{N}H^-][H_2\ddot{O}H^+]}{[R\ddot{N}H(H)]} \sim 10^{-35}$$

$$pK_a \sim 35$$

The deprotonation of amines requires exceedingly strong bases, such as alkyllithium reagents. For example, lithium diisopropylamide, the special sterically hindered base used in some bimolecular elimination reactions (Section 7-5), is made in the laboratory by treatment of *N*-(1-methylethyl)-2-propanamine (diisopropylamine) with butyllithium.

Preparation of LDA

$$CH_3\underset{|}{\overset{CH_3}{C}}H\ddot{N}H\overset{CH_3}{C}HCH_3 \xrightarrow[-\,CH_3CH_2CH_2CH_2H]{CH_3CH_2CH_2CH_2Li} (CH_3\overset{CH_3}{C}H)_2\ddot{\underset{\cdot\cdot}{N}}^-\,Li^+$$

_N_-(1-Methylethyl)-2-propanamine (Diisopropylamine) — **Lithium diisopropylamide, LDA**

An alternative synthesis of amide ions is the treatment of amines with alkali metals. Alkali metals dissolve in amines (albeit relatively slowly) with the evolution of hydrogen and the formation of amine salts. For example, sodium amide can be made in liquid ammonia from sodium metal in the presence of catalytic amounts of Fe^{3+}, which facilitates electron transfer to the amine. In the absence of such a catalyst, sodium simply dissolves in ammonia to form a strongly reducing solution (Section 13-6).

Preparation of Sodium Amide

$$2\ Na + 2\ NH_3 \xrightarrow{\text{Catalytic } Fe^{3+}} 2\ NaNH_2 + H_2$$

Amines Are Bases

Amines deprotonate water to a small extent to form ammonium and hydroxide ions. Thus, as expected, amines are much more strongly basic than alcohols. On a quantitative level, the base strength of amines can be expressed by their pK_b values (review Section 6-7), which are about 4.

$$\underset{\textbf{Amine}}{R\ddot{N}H_2} + HOH \underset{}{\overset{K_b}{\rightleftharpoons}} R\overset{+}{N}H_3 + HO:^- \qquad K_b = \frac{[R\overset{+}{N}H_3][HO:^-]}{[R\ddot{N}H_2]} \sim 10^{-4}$$

$$pK_b \sim 4$$

Alkanamines are slightly more basic than ammonia but less basic than hydroxide ion ($pK_b = -1.7$).

pK_b Values of a Series of Simple Amines

	NH_3	CH_3NH_2	$(CH_3)_2NH$	$(CH_3)_3N$
$pK_b =$	4.76	3.38	3.27	4.21

Ammonium Salts Are Poor Acids

It is useful to view the basicity of a compound as a reflection of the acidity of its conjugate acid (Section 6-7). The conjugate acids of amines are the corresponding ammonium ions.

The Acidity of Ammonium Ions

$$R\overset{+}{N}H_3 + H_2\ddot{O} \overset{K_a}{\rightleftharpoons} R\ddot{N}H_2 + H_2\overset{+}{\ddot{O}}H \qquad K_a = \frac{[R\ddot{N}H_2][H_2\overset{+}{\ddot{O}}H]}{[R\overset{+}{N}H_3]} \sim 10^{-10}$$

$$pK_a \sim 10$$

The pK_b value is easily converted into the pK_a of the conjugate acid by the equation $pK_a + pK_b = 14$ (Section 6-7).

pK_a Values of a Series of Simple Ammonium Salts*

	$\overset{+}{N}H_4$	$CH_3\overset{+}{N}H_3$	$(CH_3)_2\overset{+}{N}H_2$	$(CH_3)_3\overset{+}{N}H$
$pK_a =$	9.24	10.62	10.73	9.79

*A confusing practice in the literature is to refer to the pK_a value of an ammonium salt as being that of the neutral amine. In the statement "the pK_a of methanamine is 10.62," what is given is the pK_a of the methylammonium ion. The pK_a of methanamine is actually 35.

In summary, amines are poor acids and require alkyllithium reagents or alkali metal treatment to form amide ions. In contrast, they are good bases, although they are weaker than hydroxide ion.

21-4 The Synthesis of Amines

This section introduces the variety of ways in which amines can be made. Some of them make use of transformations that have already been covered: alkylation of simple amines to give more-highly-substituted products (Section 21-2); alkylation of azide ion (Section 6-3), followed by reduction; alkylation of 1,2-benzenedicarboxylic imide (phthalimide), followed by liberation of the amine by hydrolysis; reductive amination of aldehydes and ketones; reduction of carboxylic amides by lithium aluminum hydride (Section 18-5); and Hofmann and related rearrangements (Section 18-5).

Amines Can Be Derived from Other Amines by Alkylation

Amines are nucleophilic and react with haloalkanes to give ammonium salts (Sections 6-3 and 21-2). Unfortunately, this reaction is not clean because the resulting amine product usually undergoes further alkylation. How does this complication arise?

Consider the alkylation of ammonia with bromomethane. When this transformation is carried out with equimolar quantities of starting materials, the product methylammonium bromide, as soon as it is formed, undergoes proton exchange with the starting ammonia. The small quantities of methanamine generated in this way then compete effectively with the ammonia for the alkylating agent. Further methylation leads to the generation of a dimethylammonium salt capable of donating a proton to either of the other two nitrogen bases present, in this way furnishing *N*-methylmethanamine (dimethylamine). This compound constitutes yet another nucleophile competing for bromomethane, leading to *N,N*-dimethylmethanamine (trimethylamine) and, eventually, tetramethylammonium bromide. The result is that a mixture of alkylammonium salts and alkanamines is formed.

Methylation of Ammonia

$$H_3N\colon + CH_3Br \longrightarrow CH_3\overset{+}{N}H_3\ Br^-$$

Methylammonium bromide

$$CH_3\overset{+}{N}H_2(H)\ Br^- + \colon NH_3 \rightleftharpoons CH_3\ddot{N}H_2 + H\overset{+}{N}H_3\ Br^-$$

Methanamine

$$CH_3\ddot{N}H_2 + CH_3Br \longrightarrow (CH_3)_2\overset{+}{N}H_2\ Br^-$$

Dimethylammonium bromide

$$(CH_3)_2\overset{+}{N}H(H)\ Br^- + \colon NH_3 \text{ or } CH_3\ddot{N}H_2 \rightleftharpoons (CH_3)_2\ddot{N}H + H\overset{+}{N}H_3\ Br^- \text{ or } CH_3\overset{+}{N}H_2(H)\ Br^-$$

***N*-Methylmethanamine (Dimethylamine)**

$$(CH_3)_2\ddot{N}H + CH_3Br \longrightarrow (CH_3)_3\overset{+}{N}H\ Br^-$$

Trimethylammonium bromide

$$(CH_3)_3\overset{+}{N}H\ Br^- + \text{amine} \rightleftharpoons (CH_3)_3N\!: + \text{amine hydrobromide}$$

N,N-Dimethylmethanamine (Trimethylamine)

$$(CH_3)_3N\!: + CH_3Br \longrightarrow (CH_3)_4N^+Br^-$$

Tetramethylammonium bromide

It is possible to minimize overalkylation by using a large excess of ammonia or other starting amine, which limits the ability of the product to compete for the alkylating agent. Nevertheless, even with this improvement, the process gives only moderate yields in many cases.

$$\underset{\text{Excess}}{(CH_3CH_2CH_2)_2NH} + CH_3CH_2I \longrightarrow \underset{40\%}{(CH_3CH_2CH_2)_2NCH_2CH_3} + (CH_3CH_2CH_2)_2\overset{+}{N}H_2I^-$$

N-Propylpropanamine (Dipropylamine) — **N-Ethyl-N-propylpropanamine (Ethyldipropylamine)**

Moreover, it is quite impractical if a relatively valuable amine is employed as a substrate. In such cases, indirect methods give better results.

EXERCISE 21-8

Like other amines, benzenamine (aniline) can be benzylated with chloromethylbenzene (benzyl chloride), $C_6H_5CH_2Cl$. In contrast with the reaction with alkanamines, which proceeds at room temperature, this transformation requires heating to between 90° and 95°C. Explain. (Hint: Review Section 20-1).

Indirect Alkylation Methods Afford an Effective Access to Amines

Controlled alkylation of amines requires a nitrogen-containing nucleophile that will undergo reaction only once. Cyanide ion, for example, turns haloalkanes into nitriles, which can be reduced subsequently to the corresponding amines (Section 18-6). This sequence allows the conversion $RX \rightarrow RCH_2NH_2$.

Conversion of a Haloalkane into the Homologous Amine by Cyanide Displacement-Reduction

$$RX + {}^-CN \longrightarrow RC{\equiv}N + X^-$$

$$RC{\equiv}N \xrightarrow{LiAlH_4 \text{ or } H_2,\ PtO_2} RCH_2NH_2$$

Example:

$$\underset{\textbf{1,8-Dibromooctane}}{Br(CH_2)_8Br} + NaCN \xrightarrow[-\,2NaBr]{DMSO} \underset{\substack{93\% \\ \textbf{Decanedinitrile}}}{NC(CH_2)_8CN} \xrightarrow{H_2,\ \text{Raney Ni},\ 100\ \text{atm}} \underset{\substack{80\% \\ \textbf{1,10-Decanediamine}}}{H_2NCH_2(CH_2)_8CH_2NH_2}$$

To introduce an amino group without additional carbons requires a modified nitrogen nucleophile, which should be unreactive after the first alkylation. Such a nucleophile is the azide ion, N_3^-, which reacts with haloalkanes to furnish **alkyl azides,** which in turn are reduced by catalytic hydrogenation (Pd-C) or by lithium aluminum hydride to the corresponding primary amines.

$$\text{(cyclopentyl)CH}_2\text{CH}_2\text{CH}_2\text{Br} + Na^+N_3^- \xrightarrow[-\ Na^+Br^-,\ S_N2\ \text{reaction}]{CH_3CH_2OH}$$

$$\text{(cyclopentyl)CH}_2\text{CH}_2\text{CH}_2\text{–N=}\overset{+}{N}\text{=N}^- \xrightarrow{1.\ LiAlH_4,\ (CH_3CH_2)_2O\quad 2.\ H^+,\ H_2O} \text{(cyclopentyl)CH}_2\text{CH}_2\text{CH}_2\text{NH}_2$$

91%
3-Cyclopentylpropyl azide

89%
3-Cyclopentylpropanamine

In a variant of this indirect method, nitrite reacts with a haloalkane to give the corresponding nitroalkane, which can be reduced to the amine by, for example, iron.

Amine Synthesis by Nitroalkane Formation and Reduction

$$CH_3(CH_2)_5\underset{}{\overset{Br}{C}}HCH_3 \xrightarrow[-\ NaBr]{NaNO_2,\ DMF,\ 45\ h} CH_3(CH_2)_5\overset{NO_2}{C}HCH_3 \xrightarrow{Fe,\ FeSO_4,\ H^+} CH_3(CH_2)_5\overset{NH_2}{C}HCH_3$$

2-Bromooctane — 58% **2-Nitrooctane** — 90% **2-Octanamine**

This sequence resembles aromatic nitration followed by reduction to the arenamine (aniline, Section 20-5), except that nitroarenes are made by electrophilic and not nucleophilic nitration.

Arenamines (Anilines) from Aromatics

$$C_6H_5CH_3 \xrightarrow[-\ 2\ H_2O]{HNO_3,\ H_2SO_4} CH_3C_6H_3(NO_2)_2 \xrightarrow{H_2,\ Ni,\ \Delta} CH_3C_6H_3(NH_2)_2$$

Methylbenzene (Toluene)

87%
1-Methyl-2,4-dinitrobenzene (2,4-Dinitrotoluene)

75%
4-Methyl-1,3-benzenediamine (2,4-Diaminotoluene)

EXERCISE 21-9

Nitromethane has the unusually low pK_a of about 10 (a property that has been exploited in the synthesis of alkylated derivatives and the corresponding amines). Explain why it is so low.

A nonreductive approach to synthesizing primary amines makes use of the anion of 1,2-benzenedicarboximide (phthalimide), the cyclic imide of 1,2-benzenedicarboxylic (phthalic) acid. This process is also known as the **Gabriel* synthesis.**

*Professor Siegmund Gabriel, 1851–1924, University of Berlin.

1,2-Benzenedicarboxylic acid (Phthalic acid) $\xrightarrow[-\ H_2O]{:NH_3,\ 300°}$ 1,2-Benzenedicarboxyimide (Phthalimide), 97% $\xrightarrow{K_2CO_3,\ H_2O}$ $:\ddot{N}:^- K^+$ $\xrightarrow[-\ KBr]{HC{\equiv}CCH_2Br,\ DMF,\ 100°C}$

N-2-Propynl-1,2-benzenedicarboximide (N-Propargylphthalimide), $:NCH_2C{\equiv}CH$, 93% $\xrightarrow{H_2SO_4,\ H_2O,\ 120°C}$ phthalic acid + $HC{\equiv}CCH_2\overset{+}{N}H_3HSO_4^-$ $\xrightarrow{NaOH,\ H_2O}$

$HC{\equiv}CCH_2\ddot{N}H_2$ + disodium 1,2-benzenedicarboxylate (O^-Na^+, O^-Na^+)

73%

2-Propynamine (Propargylamine)

Removed in aqueous work-up

Because the nitrogen in this system is adjacent to two carbonyl functions, the acidity of the NH group ($pK_a = 8.3$) is much greater than that of an ordinary amide ($pK_a = 15$; see Exercise 18-3). Deprotonation can therefore be achieved with as mild a base as carbonate ion, and the resulting anion monoalkylated in good yield. The amine subsequently can be liberated by acidic hydrolysis, initially as the ammonium salt. Base treatment of the salt then produces the free amine.

EXERCISE 21-10

The cleavage of an *N*-alkyl-1,2-benzenedicarboximide (*N*-alkyl phthalimide) is frequently carried out with base or with hydrazine, H_2NNH_2. The respective products of these two treatments are the 1,2-benzenedicarboxylate A or hydrazide B. Write mechanisms for these two transformations.

A (COO^-, COO^-) $\xleftarrow[-\ RNH_2]{^-OH}$ N-alkyl phthalimide (NR) $\xrightarrow[-RNH_2]{H_2NNH_2}$ **B** (NH–NH hydrazide)

Amine Syntheses Through Condensation-Reduction Sequences with Carbonyl Compounds

Like the carbon–oxygen double bond in aldehydes and ketones, the carbon–nitrogen double bond in imines can be reduced by catalytic hydrogenation or by hydride reagents. The products are the corresponding amine derivatives. Because imines are made by condensation of amines with carbonyl compounds, this method is a way of causing the **reductive amination** of an aldehyde or ketone.

General Reductive Amination of a Ketone

$$\mathrm{RR'C{=}O} + \mathrm{H_2NR''} \underset{\text{Condensation}}{\rightleftharpoons} \mathrm{RR'C{=}NR''} + \mathrm{H_2O}$$

$$\mathrm{RR'C{=}NR''} \xrightarrow[\text{Reduction}]{} \mathrm{R'{-}C(R)(H){-}NHR''}$$

This reaction succeeds because of the selectivity of the reducing agents: catalytically activated hydrogen or sodium cyanoborohydride, $Na^{+}\,{}^{-}BH_3CN$. Both react faster with the imine double bond than with that of the carbonyl group. In a typical procedure, the carbonyl component and the amine are allowed to equilibrate with the imine and water in the presence of the reductant.

$$\mathrm{C_6H_5CHO} + \mathrm{NH_3} \underset{+\,H_2O}{\overset{-\,H_2O}{\rightleftharpoons}} \mathrm{C_6H_5CH{=}NH} \xrightarrow{\mathrm{H_2,\ Ni,\ CH_3CH_2OH,\ 70^\circ C,\ 90\ atm}} \mathrm{C_6H_5CH(H)NH(H)}$$

Benzenecarboxaldehyde (Benzaldehyde) — Not isolated — 89% **Phenylmethanamine (Benzylamine)**

$$\text{Cyclohexanone} + \mathrm{NH_3} \underset{+\,H_2O}{\overset{-\,H_2O}{\rightleftharpoons}} \text{Cyclohexanimine} \xrightarrow{\mathrm{NaBH_3CN,\ CH_3CH_2OH,\ pH\ =\ 2\text{–}3}} \text{Cyclohexanamine}$$

Cyclohexanone — Not isolated — 61% **Cyclohexanamine**

Sodium cyanoborohydride is prepared by treating sodium borohydride, $NaBH_4$, with hydrogen cyanide. The electron-withdrawing cyanide group decreases the ability of the borohydride ion to deliver H^- to the functional group being reduced. Consequently, cyanoborohydride is more discriminating in its reducing power. In contrast with sodium borohydride, it is also fairly stable in protic acidic media (pH = 2–3), making it ideal for reductive aminations.

Reductive amination of methanal (formaldehyde) is a convenient method for the methylation of secondary amines. Primary amines are dimethylated by this procedure.

$$\text{C}_5\text{H}_9\text{–N(H)CH}_2\text{C}_6\text{H}_5 + \text{CH}_2\text{=O} \xrightarrow{\text{NaBH}_3\text{CN, CH}_3\text{OH}} \text{C}_5\text{H}_9\text{–N(CH}_3\text{)CH}_2\text{C}_6\text{H}_5 \quad 100\%$$

N-(Phenylmethyl)cyclopentanamine (Benzylcyclopentylamine) → **N-Methyl-N-(phenylmethyl)cyclopentanamine (Benzylcyclopentylmethylamine)**

$$(CH_3)_3CCH_2NH_2 + 2\ CH_2{=}O \xrightarrow{NaBH_3CN,\ CH_3OH} (CH_3)_3CCH_2N(CH_3)_2 \quad 84\%$$

2,2-Dimethylpropanamine → **N,N,2,2-Tetramethylpropanamine**

Reductive aminations with secondary amines give the corresponding *N,N*-dialkylamino derivatives.

$$\text{Ketone} \xrightarrow{(CH_3)_2NH,\ NaBH_3CN,\ CH_3OH} \text{H, N(CH}_3)_2 \text{ product} \quad 89\%$$

EXERCISE 21-11

Reductive amination with a secondary amine proceeds through the intermediacy of an iminium ion. Write a mechanism for the reaction.

EXERCISE 21-12

Explain the following transformation by a mechanism:

$$\text{HCCH}_2\text{CH}_2\text{CH(NH}_2\text{)CH}_2\text{CH}_2\text{CH (O=, =O)} \xrightarrow{NaBH_3CN,\ CH_3OH} \text{pyrrolizidine} \quad 35\%$$

In a reaction that is related to reductive amination, oximes (which are formed by condensation of carbonyl compounds with hydroxylamine, Section 15-6) are reduced by lithium aluminum hydride to give primary amines. The oxime is usually isolated before the reduction step.

Amines from Oximes

$$\text{Cyclohexanone} \xrightarrow[-\ H_2O]{NH_2OH} \text{Cyclohexanone oxime (N–OH)} \xrightarrow[2.\ H^+,\ H_2O]{1.\ LiAlH_4,\ (CH_3CH_2)_2O} \text{Cyclohexanamine (NH}_2) \quad 85\%$$

Treatment with the milder reducing agent sodium cyanoborohydride allows the isolation of the intermediate hydroxylamine.

Hydroxylamines from Oximes

$$\text{cyclohexanone oxime (C=N–OH)} \xrightarrow{NaBH_3CN,\ CH_3COOH} \text{N-cyclohexylhydroxylamine (HNOH)}\quad 81\%$$

Amines Can Be Made from Amides by Reduction and by Oxidation

Carboxylic amides can be versatile precursors of amines (Section 18-5). Recall that amides are readily available by reaction of alkanoyl halides with amines, and that the corresponding amidate ions are easily alkylated. Reduction with lithium aluminum hydride then converts them into the corresponding amines.

Utility of Amides in Amine Synthesis

$$\overset{O}{\overset{\|}{RCCl}} + H_2\ddot{N}R' \xrightarrow[-\ HCl]{Base} \overset{O}{\overset{\|}{RC\ddot{N}HR'}} \xrightarrow[-\ H—H]{NaH} \overset{O}{\overset{\|}{RC\ddot{\bar{N}}R'}}\ Na^+ \xrightarrow[-\ NaX]{R''X} \overset{O}{\overset{\|}{RC\ddot{N}R'R''}} \xrightarrow{LiAlH_4,\ (CH_3CH_2)_2O} RCH_2\ddot{N}R'R''$$

Amidate ion

Primary amides can be turned into amines also by oxidation with bromine or chlorine in the presence of sodium hydroxide, in other words, by the Hofmann rearrangement (Section 18-5). Recall that in this transformation the carbonyl group is extruded as carbon dioxide, so that the resulting amine bears one carbon less than the starting material.

Amines by Hofmann Rearrangement

$$\overset{O}{\overset{\|}{RCNH_2}} \xrightarrow{Br_2,\ NaOH,\ H_2O} RNH_2 + O{=}C{=}O$$

Similar transformations start with an alkanoyl halide, in the **Curtius* rearrangement,** and a carboxylic acid, in the **Schmidt† rearrangement.** In both cases, the additional reagent is sodium azide.

In the Curtius rearrangement, the azide ion first displaces the halide ion by addition-elimination. The resulting alkanoyl azide decomposes by nitrogen extrusion to give the same acyl nitrene intermediate postulated in the Hofmann rearrangement. Migration of the R group then results in an isocyanate that, in the presence of water, hydrolyzes to the amine.

---

*Professor Theodor Curtius, 1857–1928, University of Heidelberg, Germany.

†Dr. Karl F. Schmidt, 1887–1971, Knoll AG, Ludwigshafen, Germany.

Mechanism of the Curtius Rearrangement:

$$\mathrm{R\overset{O}{\overset{\|}{C}}Cl} + \mathrm{Na^+N_3^-} \xrightarrow[-\ \mathrm{NaCl}]{\mathrm{CHCl_3}} \mathrm{R\overset{O}{\overset{\|}{C}}{-}\ddot{\underset{\cdot\cdot}{N}}^{-}{-}\overset{+}{N}{\equiv}N{:}} \xrightarrow[-\ \mathrm{N_2}]{\Delta}$$

Alkanoyl azide

$$\mathrm{R\overset{O}{\overset{\|}{C}}N{:}} \xrightarrow[\text{Rearrangement}]{} \mathrm{RN{=}C{=}O} \xrightarrow{\mathrm{H_2O}} \mathrm{RNH_2} + \mathrm{O{=}C{=}O}$$

Acyl nitrene

The unique feature of the Curtius rearrangement is that it can be stopped at the isocyanate stage when inert solvents are used, such as methanenitrile (acetonitrile):

$$\text{(ethylene ketal)-CH}_2\text{CH}_2\mathrm{\overset{O}{\overset{\|}{C}}Cl} \xrightarrow[-\ \mathrm{N_2},\ -\ \mathrm{NaCl}]{\mathrm{Na^+N_3^-},\ \mathrm{CH_3CN},\ 82^\circ\mathrm{C}} \text{(ethylene ketal)-CH}_2\text{CH}_2\mathrm{N{=}C{=}O}$$

84%

The Schmidt rearrangement proceeds through the same sequence, but starts with a carboxylic acid and leads directly to the product amine. The initial addition-elimination that produces the alkanoyl azide is catalyzed by sulfuric acid, which is neutralized in a basic work-up.

$$\mathrm{CH_3(CH_2)_{16}COOH} \xrightarrow[\text{2. NaOH, } \mathrm{H_2O}]{\text{1. } \mathrm{Na^+N_3^-},\ \mathrm{H_2SO_4},\ \mathrm{C_6H_6}} \mathrm{CH_3(CH_2)_{15}CH_2NH_2}$$

96%

Octadecanoic acid — **Heptadecanamine**

EXERCISE 21-13

Suggest synthetic methods for the preparation of *N*-methylhexanamine from hexanamine (two syntheses) and from *N*-hexylmethanamide.

To summarize, amines can be made from ammonia or other amines by simple alkylation, but this method gives mixtures and poor yields. Better in the laboratory are indirect methods that use nitrile, azide, and nitro groups or protected systems, such as 1,2-benzenedicarboxylic imide (phthalimide) in the Gabriel synthesis. Reductive amination furnishes alkanamines by reductive condensation of amines with aldehydes and ketones. Finally, amides can be reduced to amines by hydrides or oxidized to them through intermediate isocyanates, as in the Hofmann rearrangement.

21-5 The Reactions of Amines Are Governed by the Presence of the Lone Electron Pair

The reactions of amines are controlled by the nucleophilic potential of the nitrogen atom. Earlier sections have touched on some of this reactivity. For example, amines react with haloalkanes to give ammonium salts (Sections 21-2 and 21-4). Amines also condense with aldehydes and ketones to produce imines and enamines (Section 15-6), and they enter into addition-elimination sequences with carboxylic acid derivatives to furnish amides (Section 18-5).

This section introduces a variety of other modes of nucleophilic reactivity of amines and the unique chemistry of the resulting nitrogen-containing compounds. This chemistry is characterized by several reactions that bear the names of their discoverers, such as the Hofmann elimination, the Mannich reaction, and the Cope elimination.

Quaternary Ammonium Salts Eliminate Tertiary Amines to Give Alkenes

Nucleophilic attack of amines on haloalkanes results in ammonium ions. If these are quaternary, further alkylation is impossible because there are no more replaceable protons on the nitrogen. Quaternary ammonium salts are nevertheless unstable, particularly in the presence of strong base, because of a bimolecular elimination reaction that furnishes alkenes. The hydrogen beta to the nitrogen is attacked by the base, and the leaving group is a trialkylamine departing as a neutral molecule with its electron pair. This reaction is known as the **Hofmann* elimination.** It resembles the acid-catalyzed dehydration of alcohols, in which water is the leaving group.

General Hofmann Elimination

$$\text{H–C–C–}\overset{+}{\text{N}}\text{R}_3 \quad (\text{with } ^{-}\text{:}\ddot{\text{O}}\text{H}) \longrightarrow \text{C=C} + \text{HOH} + \text{:NR}_3$$

Alkene

In the procedure of Hofmann elimination, the amine is first completely methylated with excess iodomethane *(exhaustive methylation)* and then treated with wet silver oxide (a source of HO^-) to produce the ammonium hydroxide. Heating degrades this salt to the alkene. The advantage of using iodomethane instead of other haloalkanes in this procedure is clear: from an alkyltrimethylammonium salt, only one product of decomposition can be formed.

Hofmann Elimination of Butanamine

$$\underset{\textbf{Butanamine}}{CH_3CH_2CH_2CH_2NH_2} \xrightarrow{\text{Excess } CH_3I,\ K_2CO_3,\ H_2O} \underset{\textbf{Butyltrimethylammonium iodide}}{CH_3CH_2CH_2CH_2\overset{+}{N}(CH_3)_3I^-} \xrightarrow[-\ AgI]{Ag_2O,\ H_2O}$$

$$\underset{\textbf{Butyltrimethylammonium hydroxide}}{CH_3CH_2\underset{}{\overset{H}{\overset{|}{C}}}HCH_2\overset{+}{N}(CH_3)_3\ HO^-} \xrightarrow{\Delta} \underset{\textbf{1-Butene}}{CH_3CH_2CH{=}CH_2} + N(CH_3)_3 + HOH$$

*This is the Hofmann of the Hofmann rule for E2 reactions (Section 11-5) and the Hofmann rearrangement (Section 18-5).

EXERCISE 21-14

Give the structures of the possible alkene products of the Hofmann elimination of (a) *N*-ethylpropanamine (ethylpropylamine) and (b) 2-butanamine.

The Hofmann elimination of amines has been applied to the elucidation of the structures of nitrogen-containing natural products (alkaloids, Section 26-7). Particularly if the nitrogen atom is part of a ring, repeated Hofmann eliminations allow the location of the heteroatom to be pinpointed.

N-Methylazacycloheptane —(1. CH_3I; 2. Ag_2O, H_2O; 3. Δ)→ **N,N-Dimethyl-5-hexenamine** ($N(CH_3)_2$) —(1. CH_3I; 2. Ag_2O, H_2O; 3. Δ)→ **1,5-Hexadiene** + $N(CH_3)_3$

EXERCISE 21-15

An unknown amine of the molecular formula $C_7H_{13}N$ has a ^{13}C NMR spectrum containing only three lines of $\delta = 21.0$, 26.8, and 47.8 ppm. Three cycles of Hofmann elimination are required to form 3-ethenyl-1,4-pentadiene (trivinylmethane) and its double-bond isomers (as side products arising from base-catalyzed isomerization). Propose a structure for the unknown.

Iminium Ions Alkylate Enols

The condensation of methanal (formaldehyde) with ammonia or primary or secondary amines gives iminium ions, which are electrophilic enough to react with enolizable aldehydes and ketones in a process related to the aldol reaction (Section 15-7). In this case, the enol attacks a carbon–nitrogen double bond and not a carbonyl function. The method is known as the **Mannich* reaction;** it leads to the corresponding *β-N*-methylamino carbonyl compounds. Typically, the enolizable carbonyl component is heated in the presence of methanal (formaldehyde), the amine, and HCl to give the hydrochloride salt of the product. The free amine (called a *Mannich base*) can be obtained on treatment with base.

Mannich Reaction

$$\underset{\textbf{2-Methylpropanal}}{CH_3\overset{CH_3}{\overset{|}{C}}HCH{=}O} + CH_2{=}O + CH_3NH_2 \xrightarrow[2.\ HO^-,\ H_2O]{1.\ HCl,\ CH_3CH_2OH,\ \Delta} (CH_3)_2\overset{CHO}{\overset{|}{C}}CH_2NHCH_3$$

70%

2-Methyl-(2-*N*-methylaminomethyl)-propanal
Mannich base

$$\text{Cyclohexanone} + CH_2{=}O + (CH_3)_2NH \xrightarrow{HCl,\ CH_3CH_2OH,\ \Delta} \text{2-}[CH_2\overset{+}{N}H(CH_3)_2Cl^-]\text{cyclohexanone}$$

85%

Salt of Mannich base

*Professor Carl Mannich, 1877–1947, University of Berlin.

Mechanism of the Mannich reaction:

STEP 1: Iminium ion formation

$$CH_2{=}O + (CH_3)_2\overset{+}{N}H_2Cl^- \longrightarrow CH_2{=}\overset{+}{N}(CH_3)_2Cl^- + H_2O$$

STEP 2: Enolization

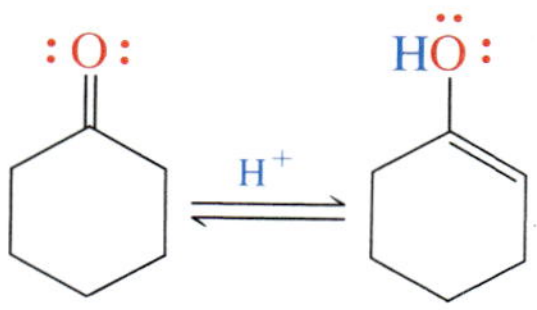

H⁺

STEP 3: Carbon–carbon bond formation

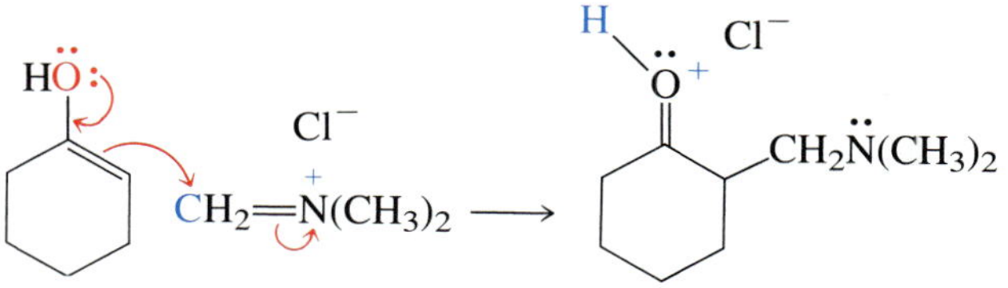

$$Cl^- \quad CH_2{=}\overset{+}{N}(CH_3)_2 \longrightarrow$$

$$Cl^- \quad CH_2\ddot{N}(CH_3)_2$$

STEP 4: Hydrochloride salt formation

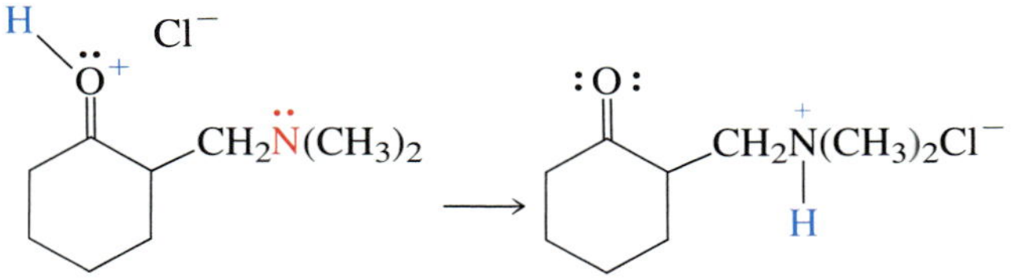

$$Cl^- \quad CH_2\ddot{N}(CH_3)_2 \longrightarrow CH_2\overset{+}{N}(CH_3)_2Cl^-$$

Salt of Mannich base

Nucleophilic Attack by Nitrogen on Oxygen: Oxidation of Amines to Amine Oxides

Amines are sensitive to oxidizing agents; in their presence, primary and secondary amines usually yield complex mixtures. Tertiary amines, however, are oxidized by aqueous hydrogen peroxide or peroxycarboxylic acids (Section 12-5) to the corresponding **amine oxides.** (For the bonding in amine oxides, see Section 1-4.)

$$(CH_3)_3N\colon + 30\%\ H_2O_2 \xrightarrow{CH_3OH,\ H_2O} (CH_3)_3\overset{+}{N}{-}\ddot{\underset{..}{O}}\colon^-$$

95%

Trimethylamine oxide

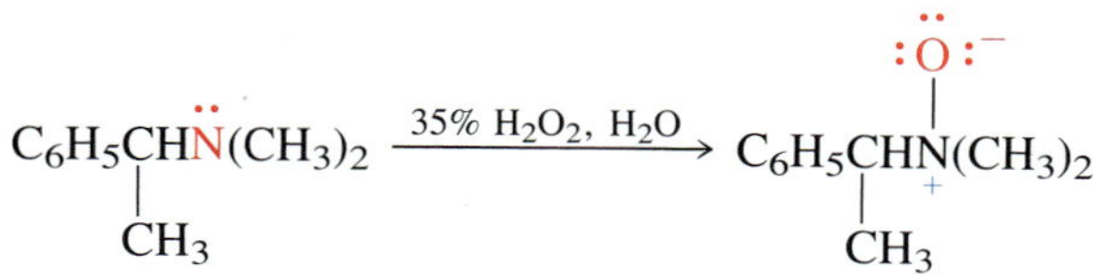

$$C_6H_5\underset{CH_3}{\underset{|}{C}}H\ddot{N}(CH_3)_2 \xrightarrow{35\%\ H_2O_2,\ H_2O} C_6H_5\underset{CH_3}{\underset{|}{C}}H\overset{\colon\ddot{O}\colon^-}{\overset{|}{\underset{+}{N}}}(CH_3)_2$$

98%

Heated above 100°C, an amine oxide bearing a β-hydrogen decomposes into an *N,N*-dialkylhydroxylamine and an alkene.

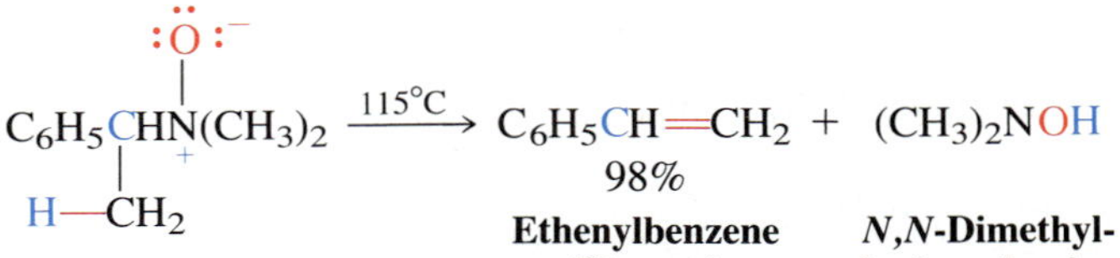

$$C_6H_5CH(CH_2\text{—}H)\overset{+}{N}(O^-)(CH_3)_2 \xrightarrow{115°C} C_6H_5CH{=}CH_2 + (CH_3)_2NOH$$

98%

Ethenylbenzene (Styrene) **N,N-Dimethylhydroxylamine**

This process, known as the **Cope* elimination,** is similar to ester pyrolysis (Section 18-4), but it proceeds at lower temperature and thus is more useful for synthesizing alkenes. Its mechanism is a *syn* elimination through a cyclic transition state.

Syn Elimination from Amine Oxides

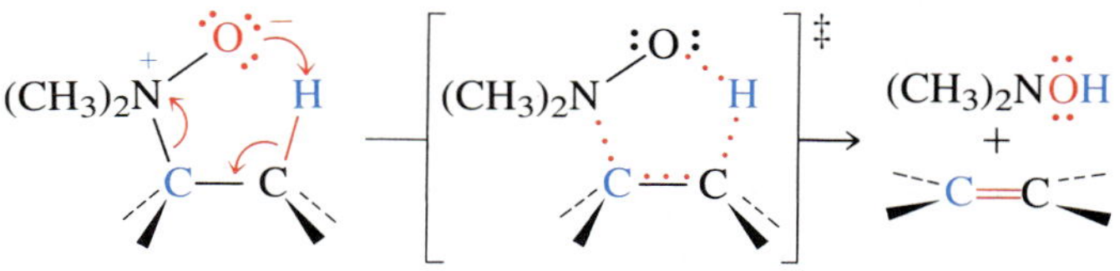

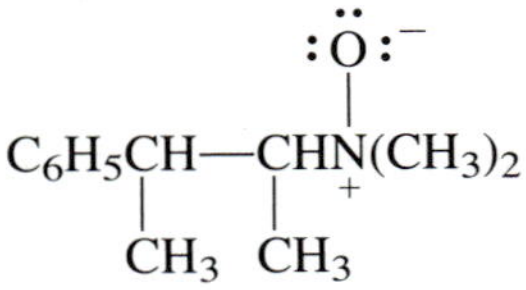

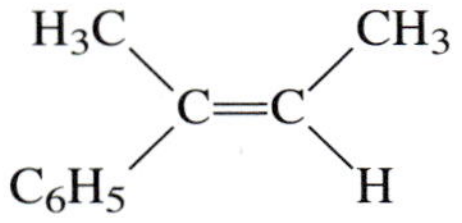

$$(CH_3)_2\overset{+}{N}(O^-)\text{–C–C–H} \longrightarrow [\,(CH_3)_2N\cdots O\cdots H\cdots C\cdots C\,]^{\ddagger} \longrightarrow (CH_3)_2NOH + C{=}C$$

EXERCISE 21-16

Draw the diastereomer of compound A (in the margin) that will give rise to (*E*-1-methyl-1-propenyl) benzene, B, by Cope elimination.

$$C_6H_5CH(CH_3)\text{—}CH(CH_3)\overset{+}{N}(O^-)(CH_3)_2$$

A

$$(H_3C)(C_6H_5)C{=}C(CH_3)(H)$$

B

EXERCISE 21-17

Formulate a concise synthesis (not more than five steps) of methylenecyclohexane, A (below), from cyclohexanecarboxylic acid that uses a Cope elimination in the last step.

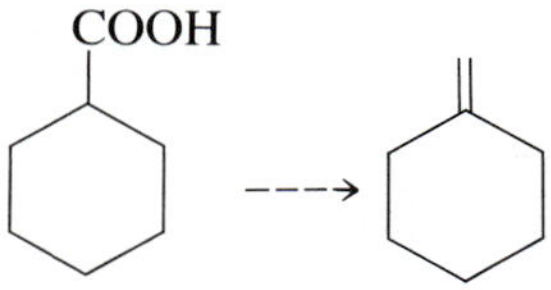

COOH (cyclohexane) - - -→ (methylenecyclohexane) **A**

Nucleophilic Attack by Nitrogen on Nitrogen: *N*-Nitrosamines and Diazonium Ions

The reactions of amines with nitrous acid proceed by nucleophilic attack on (or electrophilic attack by) a transient nitrosyl cation, NO^+. The product of such a transformation depends very much on whether the reactant is an alkanamine or a benzenamine (aniline) and whether it is tertiary, secondary, or primary. Because of their special structure, aromatic amines will be discussed in Section 24-4. This section deals only with simple alkanamines.

Nitrous acid is usually prepared *in situ* by the treatment of sodium nitrite with aqueous hydrogen chloride. Once made in such acid solution, it establishes an equilibrium with the nitrosyl cation. (Compare this sequence with the preparation of the nitronium cation from nitric acid, Section 19-6.)

---

*Professor Arthur C. Cope, 1909–1966, Massachusetts Institute of Technology.

Mechanism of Nitrosyl Cation Formation from Nitrous Acid:

STEP 1: Nitrous acid formation

$$Na^{+}\,{}^{-}:\ddot{O}-\ddot{N}=\ddot{O} \xrightarrow[-\,NaCl]{HCl} H\ddot{O}-\ddot{N}=\ddot{O}$$

Sodium nitrite **Nitrous acid**

STEP 2: Protonation

$$H\ddot{O}-\ddot{N}=\ddot{O} \overset{H^+}{\rightleftharpoons} H_2\overset{+}{\ddot{O}}-\ddot{N}=\ddot{O}$$

STEP 3: Loss of water

$$H_2\overset{+}{\ddot{O}}-\ddot{N}=\ddot{O} \rightleftharpoons [{}^{+}:N=\ddot{O} \longleftrightarrow :N\equiv\overset{+}{O}:] + H_2\ddot{O}$$

Nitrosyl cation

The nitrosyl cation is electrophilic and attacks amines to form an *N*-nitrosammonium salt.

$$-\ddot{N}< + {}^{+}\ddot{N}=\ddot{O} \longrightarrow -\overset{+}{N}-\ddot{N}=\ddot{O}$$

***N*-Nitrosammonium salt**

Depending on whether the amine nitrogen bears no, one, or two hydrogens, the reaction subsequently takes a specific course. Tertiary *N*-nitrosammonium salts are stable at low temperatures but decompose on heating to give a mixture of compounds. On the other hand, secondary *N*-nitrosammonium salts are simply deprotonated to furnish the relatively stable ***N*-nitrosamines** as the major products:

$$(CH_3)_2NH \xrightarrow{NaNO_2,\ HCl,\ H_2O,\ 0^\circ C} (CH_3)_2\overset{H}{\overset{|}{N}}{}^{+}-N=O\ Cl^- \xrightarrow[-\,HCl]{} (H_3C)_2N-N=O$$

88%–90%

***N*-Nitrosodimethylamine**

Similar treatment of primary amines initially gives the analogous monoalkyl-*N*-nitrosamines. However, these are unstable and rapidly decompose to complex mixtures. The lability of primary *N*-nitrosamines is due to the remaining proton on the nitrogen. By a series of hydrogen shifts, these compounds first rearrange to the corresponding diazohydroxides. Then, protonation followed by loss of water gives the highly reactive **diazonium ions.** The reason for the reactivity of these salts is their propensity to lose nitrogen and form the corresponding carbocations. These carbocations may rearrange, deprotonate, or undergo nucleophilic trapping (Section 9-2) to yield the mixtures of compounds that are usually observed.

Mechanism of Decomposition of Primary *N*-Nitrosamines:

STEP 1: Rearrangement to a diazohydroxide

$$\begin{matrix}R\\ \ \ \diagdown\\ \ \ \ddot{N}-\ddot{N}=\ddot{O}\\ \ \ \diagup\\ H\end{matrix} \underset{-\,H^+}{\overset{+\,H^+}{\rightleftharpoons}} \left[\begin{matrix}R\\ \diagdown\\ \ddot{N}-\ddot{N}=\overset{+}{\ddot{O}}H\\ \diagup\\ H\end{matrix} \longleftrightarrow \begin{matrix}R\\ \diagdown\\ \overset{+}{N}=\ddot{N}-\ddot{O}H\\ \diagup\\ H\end{matrix}\right] \underset{+\,H^+}{\overset{-\,H^+}{\rightleftharpoons}} R-\ddot{N}=\ddot{N}-\ddot{O}H$$

Diazohydroxide

STEP 2: Loss of water to give a diazonium ion

$$R-\ddot{N}=\ddot{N}-\ddot{O}H \underset{-\,H^+}{\overset{+\,H^+}{\rightleftharpoons}} R-\ddot{N}=\ddot{N}-\overset{+}{\ddot{O}}H_2 \underset{+\,H_2O}{\overset{-\,H_2O}{\rightleftharpoons}} R-\overset{+}{N}\equiv N:$$

Diazonium cation

STEP 3: Nitrogen loss to give a carbocation

$$R-\overset{+}{N}\equiv N: \xrightarrow[-\,N_2]{} R^+ \longrightarrow \text{product mixtures}$$

BOX 21-1

N-Nitrosodialkanamines Are Carcinogens

N-Nitrosodialkanamines are notorious as potent carcinogens in a variety of animals. Although there is no direct evidence, they are suspected of causing cancer in humans as well. Most nitrosamines appear to cause liver cancer, but certain of them are very organ-specific in their carcinogenic potential (bladder, lungs, esophagus, nasal cavity, etc.).

Their mode of carcinogenic action is not understood, but it is proposed that initial enzymatic oxidation of one of the α positions allows eventual formation of a monoalkyl-*N*-nitrosamine. This compound then decomposes to a carbocation that, as a powerful electrophile, is thought to attack one of the bases in DNA to inflict the kind of genetic damage that seems to lead to cancerous cell behavior.

Nitrosamines have been detected in a variety of cured meats, such as smoked fish, frankfurters (*N*-nitrosodimethylamine), and fried bacon [*N*-nitrosoazacyclopentane (*N*-nitrosopyrrolidine), see Exercise 26-11]. Moreover, they can be formed from natural amines and added nitrite ion at physiological pH in the stomach of test animals. The level of nitrite salts in the body depends on environmental factors, such as water supply and consumption of natural foods as well as food additives. For example, fresh spinach contains about 5 mg kg^{-1} of nitrite. On storing the vegetable in a refrigerator, its nitrite level may increase to 300 mg kg^{-1}. Nitrite has been widely used to preserve meats and enhance their color and flavor. In 1976, the FDA limited the amount of nitrite permissible for such purposes to between 50 and 125 ppm. More recently, this rule has been relaxed somewhat because of the lack of data pointing to nitrite as a carcinogenic additive.

N
|
NO

_N_-Nitrosoaza-cyclopentane (_N_-Nitroso-pyrrolidine)

Diazomethane from *N*-Methyl-*N*-nitrosamides

Nitrosation of *N*-methylamides gives rise to *N*-methyl-*N*-nitrosamides.

$$\mathrm{RC(=O)NHCH_3} \xrightarrow[-\,H^+]{NO^+} \mathrm{RC(=O)N(NO)CH_3}$$

An *N*-methyl-*N*-nitrosamide

These compounds are precursors of **diazomethane,** a useful synthetic intermediate (Section 17-8). For example, treatment of *N*-methyl-*N*-nitrosourea in ether with a 40% aqueous solution of KOH at 0°C produces a yellow solution of diazomethane.

Making Diazomethane

$$\mathrm{CH_3N(N{=}O)C(=O)NH_2} \xrightarrow{KOH,\ H_2O,\ (CH_3CH_2)_2O,\ 0^\circ C} \mathrm{CH_2{=}\overset{+}{N}{=}\ddot{N}{:}^-} + NH_3 + K_2CO_3 + H_2O$$

***N*-Methyl-*N*-nitrosourea** **Diazomethane**

The mechanism of this reaction probably begins with hydroxide ion addition. Then, diazoxide ion elimination gives an intermediate diazohydroxide, which, under the strongly basic conditions, eliminates water to furnish the product.

Mechanism of Diazomethane Formation from *N*-Methyl-*N*-nitrosourea:

STEP 1: Hydroxide addition

$$\mathrm{CH_3N(N{=}O)C(=O)NH_2} + {}^-{:}\ddot{O}H \longrightarrow \mathrm{CH_3N(N{=}O)\!-\!C(O^-)(OH)\!-\!NH_2}$$

STEP 2: Elimination

$$\mathrm{CH_3N(N{=}O)\!-\!C(O^-)(OH)\!-\!NH_2} \longrightarrow \mathrm{CH_3N{=}N{-}\ddot{O}{:}^-} + \mathrm{HOC(=O)NH_2}$$

$$\mathrm{CH_3N{=}N{-}O^-} \xrightarrow{HOH} \mathrm{CH_3N{=}NOH} + {}^-{:}\ddot{O}H \qquad \mathrm{HOC(=O)NH_2} \longrightarrow CO_2 + HNH_2$$

Diazohydroxide

STEP 3: Loss of water

$$\mathrm{H{-}CH_2N{=}NOH} + {}^-{:}\ddot{O}H \xrightarrow[-\,HOH]{} {}^-{:}\mathrm{CH_2{-}N{=}N{-}\ddot{O}H} \xrightarrow[-\ {}^-{:}\ddot{O}H]{} \left[{}^-{:}\mathrm{CH_2{-}\overset{+}{N}{\equiv}N{:}} \longleftrightarrow \mathrm{CH_2{=}\overset{+}{N}{=}\ddot{N}{:}^-}\right]$$

Diazomethane is exceedingly toxic and highly explosive in the gaseous state (b.p. −24°C) and in concentrated solutions. It is usually allowed to react as soon as it is generated. Residues may be destroyed by adding ethanoic (acetic) acid, which leads to methyl ester formation (Section 17-8).

$$CH_3\overset{O}{\overset{\|}{C}}OH + CH_2N_2 \longrightarrow CH_3\overset{O}{\overset{\|}{C}}OCH_2H + N_2$$

Diazomethane attacks alkanoyl chlorides to give two different types of products, depending on whether one or two equivalents of the reagent are employed. With one equivalent, a chloromethyl ketone is obtained:

$$CH_3CH_2\overset{O}{\overset{\|}{C}}Cl + CH_2N_2 \xrightarrow{(CH_3CH_2)_2O,\ 5°C} \underset{\substack{50\% \\ \textbf{1-Chloro-2-butanone}}}{CH_3CH_2\overset{O}{\overset{\|}{C}}CH_2Cl} + N_2$$

The reaction begins with nucleophilic attack on the carbonyl group by the carbon atom in diazomethane. Elimination of chloride ion then yields a diazonium ion. Finally, nucleophilic displacement of N_2 by the halide ion gives the product.

Mechanism of Chloromethyl Ketone Synthesis from Alkanoyl Chlorides and Diazomethane:

STEP 1: Nucleophilic addition

$$CH_3CH_2\overset{:O:}{\overset{\|}{C}}\ddot{C}l: + {}^{-}:CH_2—\overset{+}{N}{\equiv}N: \rightleftharpoons CH_3CH_2\underset{:\ddot{C}l:}{\overset{:\ddot{O}:^{-}}{\overset{|}{\underset{|}{C}}}}CH_2—\overset{+}{N}{\equiv}N:$$

STEP 2: Elimination of chloride

$$CH_3CH_2\underset{:\ddot{C}l:}{\overset{:\ddot{O}:^{-}}{\overset{|}{\underset{|}{C}}}}CH_2—\overset{+}{N}{\equiv}N: \rightleftharpoons \underset{\textbf{A diazonium ion}}{CH_3CH_2\overset{:O:}{\overset{\|}{C}}CH_2—\overset{+}{N}{\equiv}N:} + :\ddot{C}l:^{-}$$

STEP 3: Displacement of nitrogen by chloride

$$CH_3CH_2\overset{:O:}{\overset{\|}{C}}CH_2—\overset{+}{N}{\equiv}N: + :\ddot{C}l:^{-} \longrightarrow CH_3CH_2\overset{:O:}{\overset{\|}{C}}CH_2\ddot{C}l: + :N{\equiv}N:$$

On the other hand, with two equivalents of diazomethane, the intermediate diazonium ion is intercepted and deprotonated to give an **α-diazo ketone.**

Generation of an α-Diazo Ketone in the Reaction of Alkanoyl Chlorides with Diazomethane

$$CH_3CH_2\overset{\overset{:O:}{\|}}{C}\underset{|}{C}H\!-\!\overset{+}{N}\equiv N: + Cl^- + \ ^-:CH_2\!-\!\overset{+}{N}\equiv N: \longrightarrow H\!-\!CH_2\!-\!\overset{+}{N}\equiv N\ (+\ Cl^- \rightarrow CH_3Cl + N_2) + \left[CH_3CH_2\overset{\overset{:O:}{\|}}{C}\!-\!CH\!=\!\overset{+}{N}\!=\!\ddot{N}^- \longleftrightarrow CH_3CH_2\overset{\overset{:O:}{\|}}{C}\!-\!\overset{-}{\ddot{C}}H\!-\!\overset{+}{N}\equiv N: \longleftrightarrow CH_3CH_2\overset{:\ddot{O}:^-}{C}\!=\!CH\!-\!\overset{+}{N}\equiv N:\right]$$

(H on the α-carbon shown below CH)

An α-diazo ketone

Because of resonance including the carbonyl group, α-diazo ketones are more stable than diazoalkanes and may be isolated and stored. (Compare the stability of the corresponding keto ylides due to similar resonance, Section 16-4.)

Diazoalkanes decompose by extruding nitrogen on exposure to light, heat, or catalytic copper, forming highly reactive carbenes. Carbenes are electrophilic and can be trapped by added compounds containing double bonds to furnish cyclopropanes, usually in a stereospecific manner.

$$H_2\overset{-}{\ddot{C}}\!-\!\overset{+}{N}\equiv N: \xrightarrow{h\nu \text{ or } \Delta \text{ or Cu}} H_2C: \quad + :N\equiv N:$$

Methylene

Carbene Additions to Double Bonds

$$\text{Cyclohexene} + CH_2N_2 \xrightarrow{h\nu} \text{Bicyclo[4.1.0]heptane} + N_2$$

40%

Bicyclo[4.1.0]heptane

$$CH_3CH_2CH\!=\!CHCH_2CH_3\ (cis) + CH_2N_2 \xrightarrow{\text{Cu}, \Delta} cis\text{-}(CH_2CH_3)_2C_3H_4 + N_2$$

50%–70%

***cis*-Diethylcyclopropane**

EXERCISE 21-18

Irradiation of diazo compound A in heptane at $-78°C$ gave rise to a hydrocarbon C_4H_6, exhibiting three signals in ^1H NMR and two signals in ^{13}C NMR spectroscopy, all in the aliphatic region. Suggest a structure for this molecule.

$$CH_2{=}CHCH_2CH{=}\overset{+}{N}{=}\ddot{N}{:}^-$$
A

EXERCISE 21-19

What synthetic procedures would convert hexanoic acid into the following compounds?

(a) 1-Chloro-2-heptanone (b) $CH_3(CH_2)_4\overset{O}{\overset{\|}{C}}CHN_2$ (c) $CH_3(CH_2)_4\overset{O}{\overset{\|}{C}}-$(bicyclo[4.1.0]hept-7-yl)

Carbenes can also be prepared from halomethanes. For example, treatment of trichloromethane (chloroform) with strong base causes an unusual elimination reaction in which both the proton and the leaving group are removed from the same carbon to give dichlorocarbene. Similarly, exposure of dichloromethane (methylene chloride) to butyllithium furnishes chlorocarbene. Another source of the carbene methylene is diiodomethane reduced with zinc powder (usually activated with copper), a mixture called the Simmons-Smith reagent.* All of these species can be produced only as reactive intermediates, which are usually trapped by alkenes to give the corresponding cyclopropanes.

Dichlorocarbene from Chloroform and Its Trapping by Cyclohexene

$$HCCl_3 + (CH_3)_3CO^- \xrightarrow[-\ (CH_3)_3COH]{} {}^-{:}CCl_2(\text{—}Cl) \longrightarrow \underset{\textbf{Dichlorocarbene}}{:CCl_2} + Cl^-$$

$$\text{Cyclohexene} + :CCl_2 \longrightarrow \underset{59\%}{\text{7,7-dichlorobicyclo[4.1.0]heptane (cis H, H)}}$$

Chlorocarbene from Dichloromethane and Its Trapping by 2-Methylpropene

$$(H)_2C{=}C(CH_3)_2 + HCCl(H)(Cl) \xrightarrow[\substack{-\ CH_3CH_2CH_2CH_2H \\ -\ LiCl}]{CH_3CH_2CH_2CH_2Li} \text{1-chloro-2,2-dimethylcyclopropane (HCCl; H, H; CH}_3\text{, CH}_3)$$

---

*Dr. Howard E. Simmons, b. 1929, and Dr. Ronald D. Smith, b. 1930, both with E. I. du Pont de Nemours and Company, Wilmington, Delaware.

Simmons-Smith Reagent in Cyclopropane Synthesis

$$\underset{\text{H}_3\text{C}}{\overset{\text{H}}{}}\!\!\text{C}=\text{C}\!\!\underset{\text{H}}{\overset{\text{CH}_3}{}} + \text{CH}_2\text{I}_2 \xrightarrow[\text{− Metal iodide}]{\text{Zn-Cu, }(\text{CH}_3\text{CH}_2)_2\text{O}} \text{cis-1,2-dimethylcyclopropane product (CH}_2\text{ bridging; H, H}_3\text{C; CH}_3\text{, H)}$$

To summarize, the reactions of amines proceed by nucleophilic attack at carbon, oxygen, and nitrogen. Quaternary ammonium salts, synthesized by amine alkylation, undergo Hofmann elimination in the presence of base to give alkenes, probably through an E2 mechanism. On the other hand, amine oxides, which are made by oxidation from the corresponding tertiary amines, decompose to alkenes in the Cope elimination, a process more closely related to ester pyrolysis. Iminium ions produced by condensation of methanal (formaldehyde) with an amine react in aldol fashion with enolizable aldehydes and ketones to give aminomethylated carbonyl derivatives (Mannich reaction). Nitrous acid attacks amines by nitrosation. Secondary amines and amides give *N*-nitrosamines, notorious for their carcinogenicity, and nitrosamides. Primary *N*-nitrosamines decompose through carbocations to a variety of products. *N*-Methylnitrosamides release diazomethane on treatment with hydroxide. Diazomethane is a useful synthetic intermediate in the methyl esterification of carboxylic acids, in the conversion of alkanoyl chlorides into chloromethyl ketones, in the production of α-diazo ketones, and as a methylene source for forming cyclopropanes from alkenes. Carbenes as reactive intermediates can also be made from halomethanes by dehydrohalogenation or dehalogenation.

21-6 Some Uses of Amines

Amines have diverse uses. This section describes some medicinally active amines, points out the utility of optically active amines in resolution, and discusses the ability of quaternary ammonium salts to function as catalysts bridging the interface between organic and aqueous solvents (phase-transfer catalysts). It concludes with the chemistry of an industrially important amine, 1,6-hexanediamine (hexamethylenediamine), needed in the manufacture of nylon.

Many Amines Are Physiologically Active

A large number of physiologically active compounds contain nitrogen. Several simple amines are used as drugs. Examples are epinephrine, propylhexedrine, hexamethylenetetramine, amphetamine, and mescaline (shown at the top of the next page). A recurring pattern in many (but not all) of these compounds is the 2-phenylethanamine (β-phenethylamine) unit. Although the mechanism of action of these substances is not understood, it seems that such a structural feature is required for their binding to a receptor site.

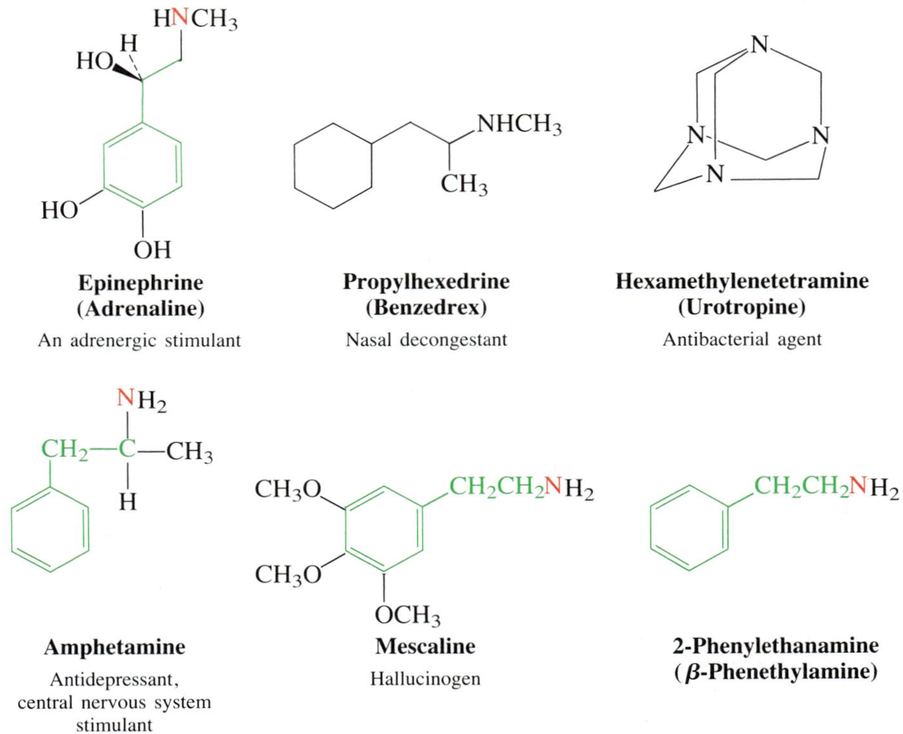

Epinephrine (Adrenaline)
An adrenergic stimulant

Propylhexedrine (Benzedrex)
Nasal decongestant

Hexamethylenetetramine (Urotropine)
Antibacterial agent

Amphetamine
Antidepressant, central nervous system stimulant

Mescaline
Hallucinogen

2-Phenylethanamine (β-Phenethylamine)

The molecules not only are potent central nervous system stimulants, but they also increase cardiovascular activity, raise body temperature, and cause loss of appetite. The last function is one reason for their extensive use in diets and the treatment of obesity. Unfortunately, they can cause psychic dependence and are potentially dangerous, particularly when taken without discretion.

Many amines in which the nitrogen is part of a ring (nitrogen heterocycles, Chapter 26) also have powerful physiological effects.

Optically Active Amines Can Be Used as Resolving Agents

Some naturally occurring optically active amines, particularly alkaloids (Chapter 26), are useful resolving agents (Sections 5-7 and 21-2). For example, they form readily crystallized diastereomeric ammonium salts with chiral racemic carboxylic acids. Fractional crystallization followed by acidification separates the pure enantiomers of the acid and allows the recovery of the resolving agent.

Quaternary Ammonium Salts Are Phase-Transfer Catalysts

Quaternary ammonium salts that bear hydrophobic alkyl substituents are mediators for reactions between species dissolved in immiscible liquids—usually, water and an organic solvent. Their action is called **phase-transfer catalysis,** because they transfer one of the species into the solvent containing the other.

When designing an organic reaction, chemists frequently face the problem of finding a solvent that will dissolve reagents having quite different solubility characteristics. A good example is the S_N2 reaction, in which an organic compound, such as a haloalkane, typically is brought to reaction with an inorganic material, a salt. Most organic solvents do not dissolve salts, whereas water, which does so, is unsuitable as a solvent for the haloalkane. In order to get

FIGURE 21-7
Phase-transfer catalysis of the S_N2 reaction of a chloroalkane with cyanide ion.

Organic phase $R_4N^+CN^- + R'Cl \rightleftharpoons R_4N^+Cl^- + R'CN$

Phase boundary

Aqueous phase $R_4N^+CN^- + Na^+Cl^- \rightleftharpoons R_4N^+Cl^- + Na^+CN^-$

around this problem, alcohols frequently are used as solvents in these reactions because they are both polar and protic and hence can dissolve both reactants. Phase-transfer catalysis is an alternative that circumvents the problem by allowing the transformation to proceed between two immiscible phases. Catalytic quantities of a quaternary ammonium salt (e.g., a tetrabutylammonium, hexadecyltrimethylammonium, or benzyltriethylammonium halide) enable one of the reactants to be brought into the solution containing the other.

For example, a heated mixture of aqueous sodium cyanide and 1-chlorooctane in decane solvent shows no sign of undergoing S_N2 reaction. However, addition of a small amount of benzyltriethylammonium chloride results in a quantitative yield of nonanenitrile in a few hours.

$$\underset{\textbf{1-Chlorooctane}}{CH_3(CH_2)_7Cl} + Na^+CN^- \longrightarrow \underset{\textbf{Nonanenitrile}}{\underset{100\%}{CH_3(CH_2)_7CN}} + Na^+Cl^-$$

This reaction succeeds because the quaternary ammonium ion, owing to the simultaneous presence of hydrophobic substituents and the polar end group, is soluble in both the aqueous *and* the organic phase. Whenever it passes into the organic phase, it carries chloride or cyanide as a counterion. The chloride causes no reaction, but the cyanide rapidly displaces chloride. The resulting quaternary ammonium chloride returns to the aqueous layer, where it exchanges its counterion for cyanide, and so on (Figure 21-7). The quaternary ammonium ion in essence acts as a shuttle carrying cyanide into the organic layer and chloride into the aqueous layer. Phase-transfer catalysis has been applied to a variety of organic reactions to achieve marked rate acceleration and increased selectivity and yield. The technique is used not only for S_N2 reactions, but also for carbene additions, oxidations, and reductions, among others.

Phase-Transfer Catalysis (Aqueous-Organic Phase)

Ether synthesis:

$$CH_3CH_2CH_2CH_2OH + C_6H_5CH_2Cl \xrightarrow[HCl]{(CH_3CH_2CH_2CH_2)_4\overset{+}{N}\ HSO_4^-,\ 35°C,\ 1.5\ h} \underset{92\%}{C_6H_5CH_2OCH_2CH_2CH_2CH_3}$$

Nitrile alkylation:

$$C_6H_5CH(H)CN + CH_3I + NaOH \xrightarrow[-\ NaI,\ -\ HOH]{R\overset{+}{N}(CH_2CH_3)_3Br^-\ (R\ =\ octyl)} \underset{66\%}{C_6H_5CH(CH_3)CN}$$

Carbene addition:

$$CHCl_3 + \text{cyclohexene} \xrightarrow[-\ NaCl,\ -\ HOH]{C_6H_5CH_2\overset{+}{N}(CH_2CH_3)_3Cl^-,\ NaOH} \text{7,7-dichlorobicyclo[4.1.0]heptane}$$

70%

Oxidation:

$$\text{cyclooctene} + KMnO_4 \xrightarrow{C_6H_5CH_2\overset{+}{N}(CH_2CH_3)_3Cl^-,\ NaOH} \text{cis-cyclooctane-1,2-diol (OH, OH)}$$

50%

Reduction:

$$\text{2-octanone} + NaBH_4 \xrightarrow{R_4N^+X^-} \text{2-octanol (OH, H)}$$

100%

An Industrial Amine: 1,6-Hexanediamine (Hexamethylenediamine)

One of the most-important industrially produced amines is 1,6-hexanediamine (hexamethylenediamine, HMDA). This compound is copolymerized with hexanedioic (adipic) acid, a process that leads to nylon 6,6, out of which hosiery, gears, and billions of pounds of textile fiber are made.

Copolymerization of Adipic Acid with HMDA

$$\underset{\substack{\textbf{Hexanedioic acid}\\ \textbf{(Adipic acid)}}}{HO\overset{O}{\overset{\|}{C}}(CH_2)_4\overset{O}{\overset{\|}{C}}OH} + \underset{\substack{\textbf{1,6-Hexanediamine}\\ \textbf{(Hexamethylenediamine)}}}{H_2N(CH_2)_6NH_2} \longrightarrow \underset{\textbf{Double salt}}{\begin{matrix} {}^-O\overset{O}{\overset{\|}{C}}(CH_2)_4\overset{O}{\overset{\|}{C}}O^- \\ H_2\overset{+}{N}H(CH_2)_6\overset{+}{N}H_3 \end{matrix}} \xrightarrow[\substack{-\ H_2O\\ \text{Polymerization}}]{270^\circ C,\ 250\ psi}$$

$$-\!\!\left[NH(CH_2)_6NH\overset{O}{\overset{\|}{C}}(CH_2)_4\overset{O}{\overset{\|}{C}}NH(CH_2)_6NH\overset{O}{\overset{\|}{C}}(CH_2)_4\overset{O}{\overset{\|}{C}}\right]_n\!\!-$$

Nylon 6,6

Nylon 6,6 is a polyamide formed by condensation of the acid with the diamine under pressure. The high demand for nylon has stimulated the development of several ingenious cheap syntheses of the monomeric precursors. Originally, the diamine was made by du Pont from hexanedioic (adipic) acid, which in turn was derived from benzene. In this process, benzene was first hydrogenated to cyclohexane, which was then oxidized with air to furnish a mixture of cyclohexanone and cyclohexanol. Oxidative ring cleavage then yielded the desired acid.

Hexanedioic (Adipic) Acid from Benzene

$$\text{Benzene} \xrightarrow{H_2,\ Pt\text{-}Al_2O_3,\ 200^\circ C,\ 500\ psi} \text{Cyclohexane} \xrightarrow{O_2,\ \text{catalytic } Co^{3+}}$$

$$\text{Cyclohexanone} + \text{Cyclohexanol} \xrightarrow{O_2,\ HNO_3,\ \text{catalytic Cu-V}} HO\overset{O}{\overset{\|}{C}}(CH_2)_4\overset{O}{\overset{\|}{C}}OH$$

Hexanedioic acid (Adipic acid)

The diacid was turned into hexanedinitrile (adiponitrile) by treatment with ammonia. Finally, catalytic hydrogenation furnished the diamine.

$$HO\overset{O}{\overset{\|}{C}}(CH_2)_4\overset{O}{\overset{\|}{C}}OH \xrightarrow[-\ 4\ H_2O]{NH_3,\ \Delta} N{\equiv}C(CH_2)_4C{\equiv}N \xrightarrow{H_2,\ Ni,\ 130^\circ C,\ 2000\ psi} H_2N(CH_2)_6NH_2$$

Hexanedinitrile (Adiponitrile) — **1,6-Hexanediamine (Hexamethylene-diamine)**

An improvement was achieved (again by du Pont) when it was realized that 1,3-butadiene could be a starting material for a shorter hexanedinitrile synthesis. Chlorination of butadiene furnished a mixture of 1,2- and 1,4-dichlorobutene (Section 14-3). This mixture could be directly converted into the dinitrile with sodium cyanide in the presence of cuprous cyanide. Selective hydrogenation then furnished the desired product.

Hexanedinitrile (Adiponitrile) from 1,3-Butadiene

$$CH_2{=}CH{-}CH{=}CH_2 \xrightarrow{Cl{-}Cl} ClCH_2CH{=}CHCH_2Cl + ClCH_2\overset{Cl}{\overset{|}{C}}HCH{=}CH_2 \xrightarrow[-\ NaCl]{CuCN,\ NaCN}$$

$$NCCH_2CH{=}CHCH_2CN \xrightarrow{H_2,\ \text{catalyst}} NC(CH_2)_4CN$$

In the mid-1960s, Monsanto developed a process that, even though it used a more-expensive starting material, was attractive because it was done in one step: the electrolytic hydrodimerization of propenenitrile (acrylonitrile).

Electrolytic Hydrodimerization of Propenenitrile (Acrylonitrile)

$$CH_2{=}CHC{\equiv}N \xrightarrow{1\ e} \dot{C}H_2{-}\ddot{C}H^{-}{-}C{\equiv}N \xrightarrow{H^+} \dot{C}H_2{-}CH_2{-}C{\equiv}N \xrightarrow{1\ e}$$

Propenenitrile (Acrylonitrile) — **Radical anion** — **Radical**

$$\ddot{C}H_2^{-}{-}CH_2{-}C{\equiv}N \xrightarrow{CH_2{=}CH{-}C{\equiv}N} N{\equiv}CCH_2CH_2CH_2\ddot{C}H^{-}C{\equiv}N \xrightarrow{H^+} NC(CH_2)_4CN$$

Anion

A suggested mechanism for this process, as shown, is cathodic reduction of propenenitrile to give its radical anion, followed by protonation to a radical, which is rapidly reduced further on the electrode surface to the anion. This species then attacks more starting material, as in a Michael reaction (Section 16-5), to give the dinitrile by protonation.

To counter Monsanto's challenge, du Pont devised yet another synthesis, again starting with 1,3-butadiene, but now avoiding the consumption of chlorine, removing the toxic-waste problems of the disposal of copper salts, and using cheaper hydrogen cyanide rather than sodium cyanide. The synthesis uses the conceptually simplest approach: direct regioselective addition of two molecules of hydrogen cyanide to butadiene. A transition metal catalyst is needed, such as iron, cobalt, or nickel. Typically required also are Lewis acids and phosphines, usually triphenylphosphine, $P(C_6H_5)_3$.

Hydrogen Cyanide Addition to 1,3-Butadiene

$$CH_2{=}CHCH{=}CH_2 + 2\ HCN \xrightarrow{\text{Catalyst}} NC(CH_2)_4CN$$

In summary, amines and their salts have uses as drugs, resolving agents, phase-transfer catalysts, and industrial chemicals, the last being exemplified by 1,6-hexanediamine.

Summary of New Reactions

1 ACIDITY OF AMINES AND AMIDE FORMATION

$$RNH_2 + H_2O \rightleftharpoons R\bar{N}H + H_3O^+ \qquad K_a \sim 10^{-35}$$

$$R_2NH + CH_3CH_2CH_2CH_2Li \rightleftharpoons \underset{\textbf{Lithium dialkylamide}}{R_2N^-Li^+} + CH_3CH_2CH_2CH_3$$

$$2\ NH_3 + 2\ Na \xrightarrow{\text{Catalytic } Fe^{3+}} 2\ NaNH_2 + H_2$$

2 BASICITY OF AMINES

$$RNH_2 + H_2O \rightleftharpoons R\overset{+}{N}H_3 + HO^- \qquad K_b \sim 10^{-3.5}$$

$$R\overset{+}{N}H_3 + H_2O \rightleftharpoons RNH_2 + H_3O^+ \qquad K_a \sim 10^{-10}$$

Salt formation:

$$RNH_2 + HCl \longrightarrow \underset{\textbf{An alkylammonium chloride}}{R\overset{+}{N}H_3Cl^-}$$

Preparation of Amines

3 AMINES BY ALKYLATION

$$R\ddot{N}H_2 + R'X \longrightarrow R\overset{+}{N}(R')H_2\,X^-$$

Drawback: multiple alkylation

$$R\overset{+}{N}(R')H_2X^- + R'X \longrightarrow \longrightarrow \longrightarrow R\overset{+}{N}R'_3X^-$$

4 PRIMARY AMINES FROM NITRILES

$$RX + {}^-CN \xrightarrow[- X^-]{S_N2} RCN \xrightarrow{LiAlH_4 \text{ or } H_2, \text{ catalyst}} RCH_2NH_2$$

5 PRIMARY AMINES FROM AZIDES

$$RX + N_3^- \xrightarrow[- X^-]{S_N2} RN_3 \xrightarrow{LiAlH_4} RNH_2$$

6 PRIMARY AMINES FROM NITROCOMPOUNDS

Nitroalkanes:

$$RX + NO_2^- \xrightarrow[- X^-]{S_N2} RNO_2 \xrightarrow{Fe, FeSO_4, H^+} RNH_2$$

Nitroarenes:

$$R\text{-benzene} \xrightarrow[- H_2O]{HNO_3, H_2SO_4} 4\text{-}R\text{-}C_6H_4\text{-}NO_2 \xrightarrow{H_2, Ni, \Delta} 4\text{-}R\text{-}C_6H_4\text{-}NH_2$$

7 PRIMARY AMINES BY GABRIEL SYNTHESIS

$$\text{phthalimide (NH)} \xrightarrow{1.\ K_2CO_3\quad 2.\ RX} \text{N-R phthalimide} \xrightarrow{1.\ H_2SO_4, H_2O, 120^\circ C\quad 2.\ NaOH} RNH_2$$

8 AMINES BY REDUCTIVE AMINATION

$$RCR' (C{=}O) \xrightarrow{NH_3, NaBH_3CN} R\text{—}\underset{H}{\overset{NH_2}{C}}\text{—}R'$$

Reductive methylation with methanal (formaldehyde):

$$R_2NH + CH_2{=}O \xrightarrow{NaBH_3CN} R_2NCH_3$$

9 PRIMARY AMINES AND HYDROXYLAMINES FROM OXIMES

$$\mathrm{RC(=O)R'} \xrightarrow{\mathrm{NH_2OH}} \underset{\textbf{Oxime}}{\mathrm{RC(=N{-}OH)R'}} \xrightarrow{\mathrm{LiAlH_4}} \underset{\textbf{Amine}}{\mathrm{RC(NH_2)(H)R'}}$$

$$\underset{\textbf{Oxime}}{\mathrm{RC(=N{-}OH)R'}} \xrightarrow{\mathrm{NaBH_3CN}} \underset{\textbf{Hydroxylamine}}{\mathrm{RCH(NHOH)R'}}$$

10 AMINES FROM AMIDES

$$\mathrm{RC(=O)NR'R''} \xrightarrow{\mathrm{LiAlH_4}} \mathrm{RCH_2NR'R''}$$

11 HOFMANN REARRANGEMENT

$$\mathrm{RC(=O)NH_2} \xrightarrow{\mathrm{Br_2,\ NaOH,\ H_2O}} \mathrm{RNH_2} + \mathrm{CO_2}$$

12 CURTIUS REARRANGEMENT

$$\mathrm{RC(=O)Cl} + \mathrm{Na^+N_3^-} \xrightarrow[-\ \mathrm{NaCl},\ -\ \mathrm{N_2}]{} \mathrm{RN{=}C{=}O} \xrightarrow{\mathrm{H_2O}} \mathrm{RNH_2} + \mathrm{CO_2}$$

13 SCHMIDT REARRANGEMENT

$$\mathrm{RC(=O)OH} \xrightarrow{\text{1. } \mathrm{Na^+N_3^-,\ H_2SO_4}\ \text{2. } \mathrm{NaOH}} \mathrm{RNH_2} + \mathrm{CO_2}$$

Reactions of Amines

14 HOFMANN ELIMINATION

$$\mathrm{RCH_2CH_2NH_2} \xrightarrow{\text{Excess } \mathrm{CH_3I,\ K_2CO_3}} \mathrm{RCH_2CH_2\overset{+}{N}(CH_3)_3I^-} \xrightarrow[-\ \mathrm{AgI}]{\mathrm{Ag_2O,\ H_2O}}$$

$$\mathrm{RCH_2CH_2\overset{+}{N}(CH_3)_3\ {}^-OH} \xrightarrow{\Delta} \mathrm{RCH{=}CH_2} + \mathrm{N(CH_3)_3} + \mathrm{H_2O}$$

15 MANNICH REACTION

$$\mathrm{RC(=O)CH_2R'} + \mathrm{CH_2{=}O} + \mathrm{(CH_3)_2NH} \xrightarrow{\text{1. HCl}\ \text{2. } \mathrm{HO^-}} \mathrm{RC(=O)CHR'(CH_2N(CH_3)_2)}$$

16 AMINE OXIDE SYNTHESIS

$$R_3N: + H_2O_2 \longrightarrow R_3\overset{+}{N}-\ddot{\underset{..}{O}}:^-$$

17 COPE ELIMINATION

$$\underset{RCH_2CHR'}{\overset{^-O-\overset{+}{N}(CH_3)_2}{|}} \xrightarrow{\Delta} RCH{=}CHR' + (CH_3)_2NOH$$

18 NITROSATION OF AMINES

Tertiary amines:

$$R_3N \xrightarrow{NaNO_2,\ H^+X^-} R_3\overset{+}{N}-NO\ X^-$$

Tertiary *N*-nitrosammonium salt

Secondary amines:

$$RR'NH \xrightarrow{NaNO_2,\ H^+} RR'N-N{=}O$$

***N*-Nitrosamine**

Primary amines:

$$RNH_2 \xrightarrow{NaNO_2,\ H^+} RN{=}NOH \xrightarrow[-\,H_2O]{H^+} RN_2^+ \xrightarrow[-N_2]{} R^+ \longrightarrow \text{mixture of products}$$

19 DIAZOMETHANE

$$CH_3N(N{=}O)\overset{O}{\overset{\|}{C}}R \xrightarrow{KOH} CH_2{=}\overset{+}{N}{=}\ddot{N}:^-$$

Reactions of diazomethane:

$$R\overset{O}{\overset{\|}{C}}OH + CH_2N_2 \longrightarrow R\overset{O}{\overset{\|}{C}}OCH_3 + N_2$$

$$R\overset{O}{\overset{\|}{C}}Cl + \underset{\text{1 Equivalent}}{CH_2N_2} \longrightarrow \underset{\textbf{Chloromethyl ketone}}{R\overset{O}{\overset{\|}{C}}CH_2Cl} + N_2$$

$$R\overset{O}{\overset{\|}{C}}Cl + \underset{\text{2 Equivalents}}{CH_2N_2} \longrightarrow \underset{\boldsymbol{\alpha}\textbf{-Diazo ketone}}{R\overset{O}{\overset{\|}{C}}CH{=}\overset{+}{N}{=}\ddot{N}:^-} + CH_3Cl + N_2$$

$$\text{cis-}RCH{=}CHR' + CH_2N_2 \xrightarrow{h\nu \text{ or } \Delta \text{ or Cu}} \text{cis-1-R-2-R'-cyclopropane}$$

Other sources of carbenes:

$$CHCl_3 \xrightarrow{\text{Base}} :CCl_2$$

$$CH_2Cl_2 \xrightarrow{\text{Base}} :CHCl$$

$$CH_2I_2 \xrightarrow{\text{Zn—Cu}} :CH_2$$

20 PHASE-TRANSFER CATALYSIS

$$R'X + {}^{-}:Nu \xrightarrow{R_4\overset{+}{N}X^-} R'Nu + X^-$$

21 NYLON 6,6

$$\underset{\substack{\textbf{Hexanedioic acid}\\\textbf{(Adipic acid)}}}{HO\overset{O}{\overset{\|}{C}}(CH_2)_4\overset{O}{\overset{\|}{C}}OH} + \underset{\substack{\textbf{1,6-Hexanediamine}\\\textbf{(Hexamethylene-}\\\textbf{diamine)}}}{H_2N(CH_2)_6NH_2} \xrightarrow[-\,H_2O]{} \underset{\textbf{Nylon 6,6}}{—\left[NH(CH_2)_6NH\overset{O}{\overset{\|}{C}}(CH_2)_4\overset{O}{\overset{\|}{C}}\right]_n—}$$

22 HEXANEDIOIC ACID SYNTHESIS

$$C_6H_6 \xrightarrow[\text{3. } O_2\text{, } HNO_3\text{, catalytic Cu-V}]{\substack{\text{1. } H_2\text{, Pt-}Al_2O_3\text{, 200°C, 500 psi}\\ \text{2. } O_2\text{, catalytic } Co^{3+}}} HOOC(CH_2)_4COOH$$

23 1,6-HEXANEDIAMINE SYNTHESIS

From hexanedioic acid:

$$HOOC(CH_2)_4COOH \xrightarrow{\substack{\text{1. } NH_3\text{, }\Delta\\ \text{2. } H_2\text{, Ni, 130°C, 2000 psi}}} H_2N(CH_2)_6NH_2$$

From 1,3-butadiene:

$$CH_2{=}CH{—}CH{=}CH_2 \xrightarrow{\substack{\text{1. } Cl_2\\ \text{2. CuCN, NaCN}\\ \text{3. } H_2\text{, catalyst}}} H_2N(CH_2)_6NH_2$$

$$CH_2{=}CH{—}CH{=}CH_2 \xrightarrow{\substack{\text{1. HCN, catalyst}\\ \text{2. } H_2\text{, catalyst}}} H_2N(CH_2)_6NH_2$$

From propenenitrile:

$$2\ CH_2{=}CH{—}C{\equiv}N \xrightarrow{2\,e,\ 2\,H^+} NC(CH_2)_4CN \xrightarrow{H_2\text{, catalyst}} H_2N(CH_2)_6NH_2$$

Summary of Important Concepts

1 Amines can be viewed as derivatives of ammonia just as ethers and alcohols can be regarded as derivatives of water.

2 *Chemical Abstracts* names amines as alkanamines (and benzenamines), alkyl substituents on the nitrogen being designated as *N*-alkyl. Another system is based on the label aminoalkane. Common names are based on the label alkylamine.

3 The nitrogen in amines is sp^3 hybridized, the nonbonding electron pair functioning as the equivalent of a substituent. This tetrahedral arrangement inverts rapidly through a planar transition state. However, when the lone electron pair is tied up in a bond to a fourth substituent, the tetrahedral structure is rigid and resolvable, if chiral.

4 The lone electron pair in amines is less tightly held than in alcohols and ethers, because nitrogen is less electronegative than oxygen. The consequences of this difference manifest themselves in a diminished capability for hydrogen bonding, higher basicity and nucleophilicity, and lower acidity.

5 Infrared spectroscopy helps to differentiate between primary and secondary amines. Nuclear magnetic resonance spectroscopy indicates the presence of nitrogen-bound hydrogens; both hydrogen and carbon atoms are deshielded in the vicinity of the nitrogen. Mass spectra are characterized by iminium ion fragments.

6 The NR_3 group in a quaternary amine, $R'{-}\overset{+}{N}R_3$, is a good leaving group in E2 reactions; this enables the Hofmann elimination to take place.

7 The Cope elimination of amine oxides proceeds through a cyclic transition state to give alkenes and alkylhydroxylamines.

8 The nucleophilic reactivity of amines manifests itself in reactions with electrophilic carbon, as in haloalkanes, aldehydes and ketones, and carboxylic acids and their derivatives. Amines also attack electrophilic oxygen, as in hydrogen peroxide or peroxycarboxylic acids, and nitrogen, as in the nitrosyl cation.

9 Phase-transfer catalysts act through their ability to move readily from an aqueous to an organic phase. This property enables associated anions to function as dissolved nucleophiles, bases, and redox reagents in both phases.

Problems

1 Give at least two names for each of the following amines.

(a) 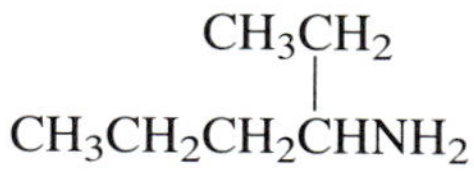 $CH_3CH_2CH_2CH(CH_2CH_3)NH_2$ (drawn as $CH_3CH_2CH_2\underset{}{C}HNH_2$ with CH_3CH_2 on the CH carbon)

(b) $(H_3C)_2CHNHCH_3$ (two H_3C groups on CHNHCH_3)

(c) Benzene ring bearing $-NH_2$, with Cl on the adjacent (ortho) carbon

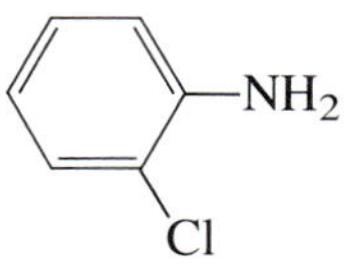

(d) 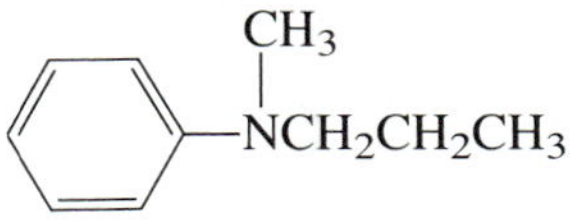 Benzene ring bearing $-N(CH_3)CH_2CH_2CH_3$

(e) $(CH_3)_3N$

(f) 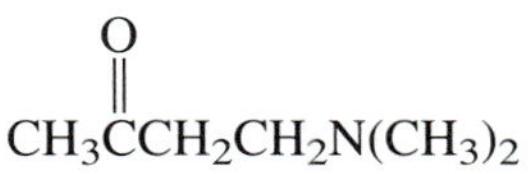 $CH_3\overset{O}{\overset{\|}{C}}CH_2CH_2N(CH_3)_2$

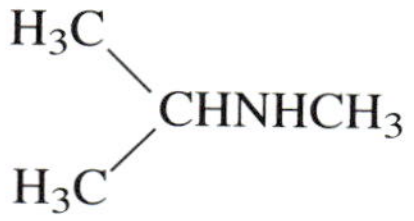

(g) C_5H_9—$N(CH_3)CH_2CH_2CH_2CH_2CH(CH_3)CH_2Cl$

(h) $(CH_3CH_2)_2NCH_2CH{=}CH_2$

2 Write structures that correspond to each of the following names.

(a) *N,N*-Dimethyl-3-cyclohexenamine.
(b) *N*-Ethyl-2-phenylethylamine.
(c) 2-Aminoethanol.
(d) *m*-Chloroaniline.

3 As mentioned in Section 21-2, the inversion of nitrogen requires a change of hybridization.

(a) What is the approximate energy difference between pyramidal nitrogen (sp^3 hybridized) and trigonal planar nitrogen (sp^2 hybridized) in ammonia and simple amines?

(b) Compare the nitrogen atom in ammonia with the carbon atom in each of the following: methyl cation, methyl radical, and methyl anion. Compare the most-stable geometries and the hybridizations of each of these species. Using fundamental notions of orbital energies and bond strengths, explain the similarities and differences among them.

4 Use the following NMR- and mass-spectral data to identify the structures of the two unknown compounds.

Compound A: ^1H NMR $\delta = 0.92$ (t, $J = 6$ Hz, 3 H), 1.32 (broad s, 12 H), 2.28 (broad s, 2 H), and 2.69 (t, $J = 7$ Hz, 2 H) ppm. Mass spectrum m/z (relative intensity) = 129(0.6) and 30(100).

Compound B: ^1H NMR $\delta = 1.00$ (s, 9 H), 1.17 (s, 6 H), 1.28 (s, 2 H), and 1.42 (s, 2 H) ppm. Mass spectrum m/z (relative intensity) = 129(0.05), 114(3), 72(4), and 58(100).

5 Spectroscopic data (^{13}C NMR and IR) are presented below for several isomeric amines of the formula $C_6H_{15}N$. Propose a structure for each compound.

(a) ^{13}C NMR: $\delta = 23.7$ (q) and 45.3 (d) ppm.
IR: 3300 cm^{-1}.

(b) ^{13}C NMR: $\delta = 12.6$ (q) and 46.9 (t) ppm.
IR: no bands in 3300–3500 cm^{-1} range.

(c) ^{13}C NMR: $\delta = 12.0$ (q), 23.9 (t), and 52.3 (t) ppm.
IR: 3280 cm^{-1}.

(d) ^{13}C NMR: $\delta = 14.2$ (q), 23.2 (t), 27.1 (t), 32.3 (t), 34.6 (t), and 42.7 (t) ppm.
IR: 3280 and 3365 cm^{-1}.

(e) ^{13}C NMR: $\delta = 25.6$ (q), 38.7 (q), and 53.2 (s) ppm.
IR: no bands in 3300–3500 cm^{-1} region.

6 Mass-spectral data for two of the compounds in Problem 5 are given below. Match the mass spectrum with a compound.

(a) m/z (relative intensity) = 101(8), 86(11), 72(79), 58(10), 44(40), and 30(100).

(b) m/z (relative intensity) = 101(3), 86(30), 58(14), and 44(100).

7 Is a molecule with a high pK_b value a stronger or weaker base than a molecule with a low pK_b value? Explain by using a general equilibrium equation.

8 How would you expect the following classes of compounds to compare with simple primary amines as (i) bases and (ii) acids?

(a) Carboxamides; for example, CH_3CONH_2

(b) Imides; for example, $CH_3CONHCOCH_3$

(c) Enamines; for example, $CH_2{=}CHN(CH_3)_2$

(d) Benzenamines; for example, $C_6H_5{-}NH_2$

9 Several functional groups containing nitrogen are considerably stronger bases than are ordinary amines. One is the amidine group found in DBN and DBU, both of which are widely used as bases in a variety of organic reactions.

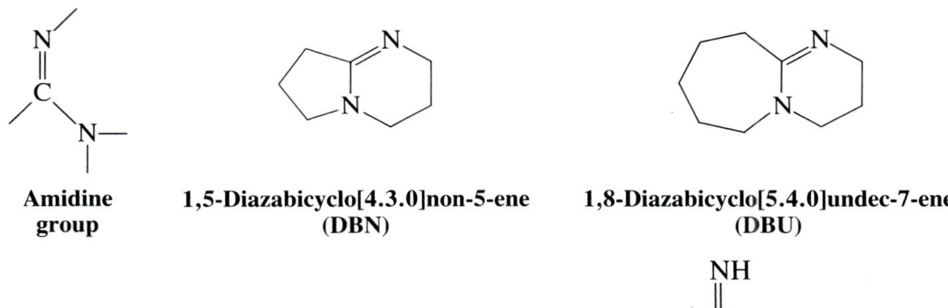

Amidine group | **1,5-Diazabicyclo[4.3.0]non-5-ene (DBN)** | **1,8-Diazabicyclo[5.4.0]undec-7-ene (DBU)**

Another unusually strong organic base is guanidine, $H_2N\overset{\overset{NH}{\|}}{C}NH_2$.

(a) Indicate which nitrogen in each of these bases is the one most likely to be protonated and explain the enhanced base strength of these systems relative to simple amines.

(b) Of which common organic compound class are amidines derivatives? Answer the same question for guanidine.

10 The reaction of (*R*)-*N*-benzyl-*N*-ethyl-1-phenylethanamine with iodomethane gives a high yield of two isomeric products, each with the formula $C_{18}H_{23}NI$. The two isomers can be separated by careful recrystallization. They have similar but not identical NMR spectra and different melting points.

(a) Draw the structure of the starting amine. Be sure to indicate stereochemistry clearly.

(b) What are the two products of this compound's reaction with iodomethane? Draw and name each of them.

(c) What is the relation between these two compounds?

11 The following are proposed syntheses of amines. In each case, indicate whether the synthesis will work (i) well, (ii) poorly, or (iii) not at all. If a synthesis will not work well, explain why.

(a) $CH_3CH_2CH_2CH_2Cl \xrightarrow[\text{2. LiAlH}_4]{\text{1. KCN}} CH_3CH_2CH_2CH_2NH_2$

(b) $(CH_3)_3CCl \xrightarrow[\text{2. LiAlH}_4]{\text{1. NaN}_3} (CH_3)_3CNH_2$

(c) $CONH_2$ $\xrightarrow{Br_2,\ NaOH,\ H_2O}$ NH_2

(d) Br $\xrightarrow{CH_3NH_2}$ $NHCH_3$

(e) H, H, CH_2Br, CH_3 $\xrightarrow{\text{1. } N^-K^+,\ DMF \quad \text{2. } H^+,\ H_2O,\ \Delta}$ H, H, CH_2NH_2, CH_3

(f) COCl $\xrightarrow{\text{1. } C_6H_5CH_2NH_2,\ HO^- \quad \text{2. } LiAlH_4}$ $(C_6H_5CH_2)_2NH$

(g) CHO $\xrightarrow{\text{1. } (CH_3)_3CNH_2 \quad \text{2. } NaBH_3CN}$ $CH_2NHC(CH_3)_3$

(h) $NH_2CH_2CH_2CHO$ $\xrightarrow{NaBH_3CN}$ NH

(i) Br $\xrightarrow{\text{1. } HNO_3,\ H_2SO_4 \quad \text{2. } Fe,\ H^+}$ Br, NH_2

(j) NH, O $\xrightarrow{\text{1. } NaH \quad \text{2. } CH_3I \quad \text{3. } LiAlH_4}$ N, CH_3

12 For each synthesis in Problem 11 that does not work well, propose an alternative synthesis of the final amine, starting either with the same material or with a material of similar structure and functionality.

13 On completion of most amine syntheses, the product must be separated from various organic and inorganic reagents, by-products, and impurities. A typical work-up procedure is outlined in the following flow chart:

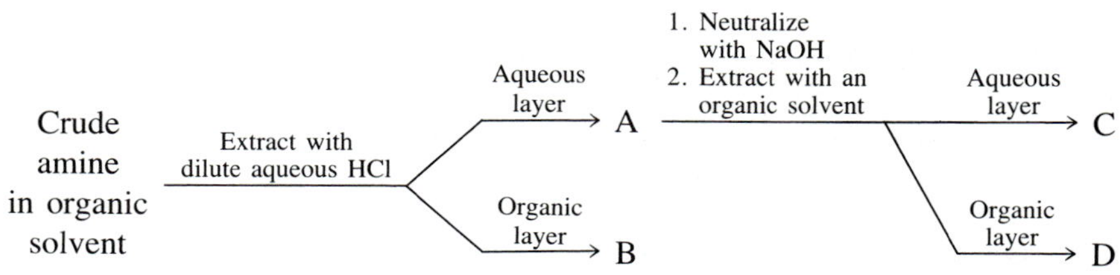

Assuming that the crude amine contains both inorganic (water-soluble) and organic (water-insoluble) impurities, explain the purpose of each step in the flow chart, and identify the materials present in the stages marked A, B, C and D. (Hint: First consider the effect that reaction with HCl should have on the water solubility of a typical amine.)

14 Write out detailed, step-by-step mechanisms for each of the following reactions (taken from Section 21-4). Use H^- as the active nucleophile derived from the hydride reagents.

(a) $(CH_3)_3CCH_2NH_2 \xrightarrow{CH_2{=}O,\ NaBH_3CN} (CH_3)_3CCH_2N(CH_3)_2$

(b) cyclopentyl–$CH_2CH_2CH_2CH_2N_3$ $\xrightarrow{1.\ LiAlH_4;\ 2.\ H^+,\ H_2O}$ cyclopentyl–$CH_2CH_2CH_2CH_2NH_2$

(c) cyclohexanone oxime ($C{=}N{-}OH$) $\xrightarrow{1.\ LiAlH_4;\ 2.\ H^+,\ H_2O}$ cyclohexylamine (NH_2)

(d) $CH_3(CH_2)_{16}COOH \xrightarrow{1.\ NaN_3,\ H_2SO_4;\ 2.\ NaOH} CH_3(CH_2)_{15}CH_2NH_2$

15 Nitroalkane anions (see Exercise 21-9) undergo reactions similar to those of enolates: alkylations and condensations. Write the products that would be expected from each of the following reaction sequences.

(a) $CH_3CH_2CH_2CH_2Br \xrightarrow{1.\ NaNO_2,\ DMF;\ 2.\ NaOCH_3;\ 3.\ CH_2{=}CHCH_2Cl}$

(b) $CH_3NO_2 \xrightarrow{1.\ C_6H_5CHO,\ NaOH;\ 2.\ H^+,\ H_2O}$

(c) chlorocyclohexane (Cl) $\xrightarrow{1.\ NaNO_2,\ DMF;\ 2.\ CH_2{=}CHCOCH_3,\ NaOH}$

16 In the past several years, pseudoephedrine has gradually replaced phenylpropanolamine as the favored decongestant in over-the-counter cold remedies.

$C_6H_5{-}CH(OH){-}CH(NH_2){-}CH_3$

Phenylpropanolamine

$C_6H_5{-}CH(OH){-}CH(NHCH_3){-}CH_3$

Pseudoephedrine

Suppose that you are the director of a major pharmaceutical laboratory with a huge stock of phenylpropanolamine on hand and the president of the company issues the order, "Pseudoephedrine from now on!" Analyze all your options, and propose the best solution that you can find for the problem.

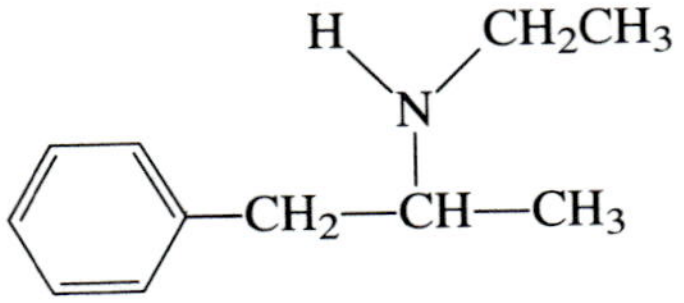

H CH$_2$CH$_3$ N — C$_6$H$_5$—CH$_2$—CH—CH$_3$

Apetenil

17 Apetenil, an appetite suppressant (i.e., diet pill), has the structure shown at the left. Is it a primary, a secondary, or a tertiary amine? Propose an efficient synthesis of apetenil from each of the following starting materials. Try to use a variety of methods.

(a) $C_6H_5CH_2COCH_3$

(b) $C_6H_5CH_2CH(Br)CH_3$

(c) $C_6H_5CH_2CH(CH_3)COOH$

18 Several of the natural amino acids are synthesized from 2-keto carboxylic acids by an enzyme-catalyzed reaction with a special coenzyme called pyridoxamine. Use electron-pushing arrows to describe each step in the following synthesis of phenylalanine from phenylpyruvic acid.

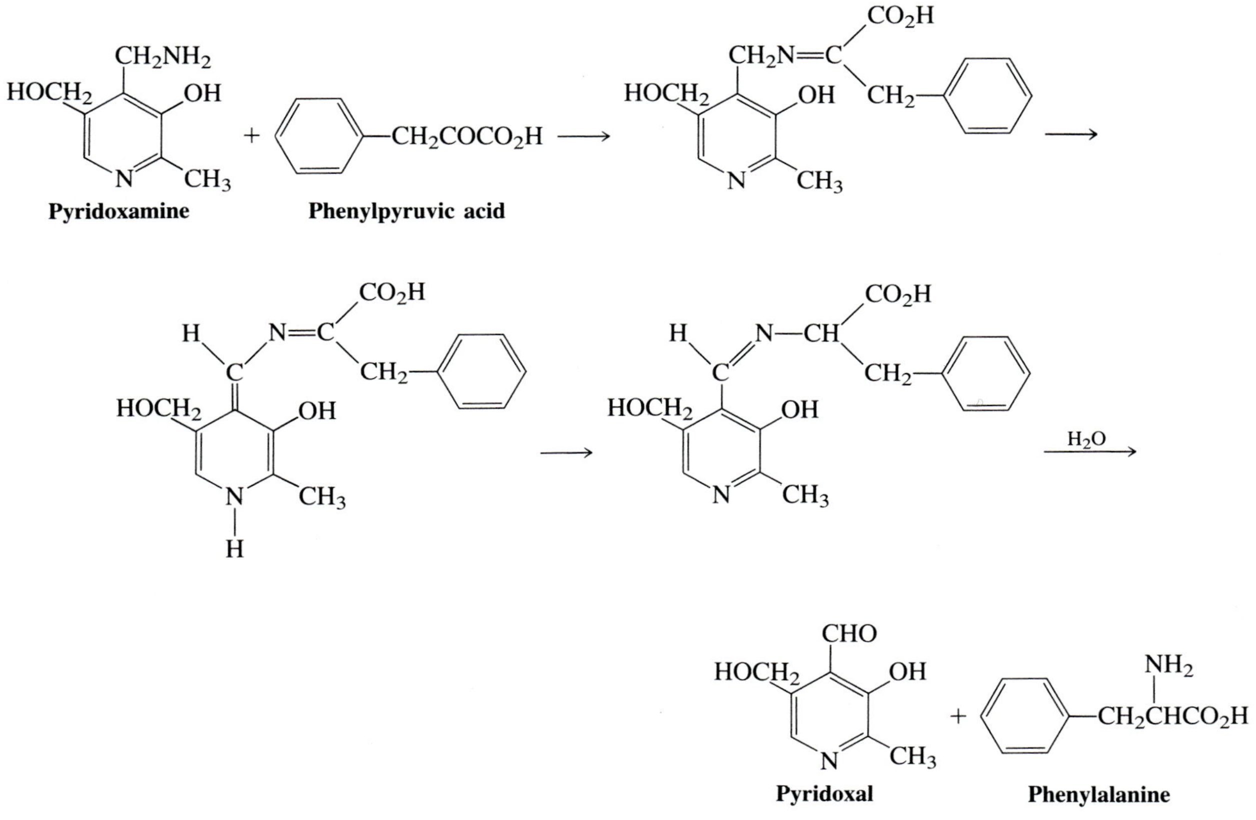

Pyridoxamine (CH$_2$NH$_2$, HOCH$_2$, OH, N, CH$_3$) + Phenylpyruvic acid (C_6H_5—CH$_2$COCO$_2$H) ⟶ (HOCH$_2$, OH, N, CH$_3$; CH$_2$N=C(CO$_2$H)CH$_2$—C_6H_5) ⟶

(H, C=, N=C(CO$_2$H)CH$_2$—C_6H_5; HOCH$_2$, OH, N—H, CH$_3$) ⟶ (H, C=N—CH(CO$_2$H)CH$_2$—C_6H_5; HOCH$_2$, OH, N, CH$_3$) $\xrightarrow{H_2O}$

Pyridoxal (CHO, HOCH$_2$, OH, N, CH$_3$) + **Phenylalanine** (C_6H_5—CH$_2$CH(NH$_2$)CO$_2$H)

19 Write the structures of the possible alkene products of Hofmann elimination of each of the following amines. If a compound can be cycled through multiple eliminations, give the products of each cycle.

(a) 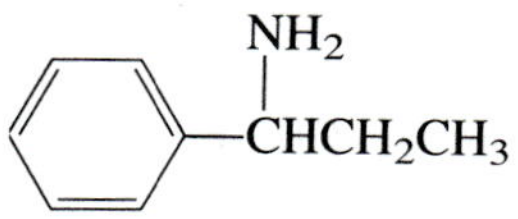 C_6H_5—CH(NH$_2$)CH$_2$CH$_3$ **(b)** 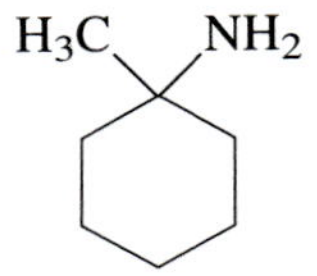 1-methylcyclohexylamine (H$_3$C, NH$_2$ on cyclohexane)

(c) 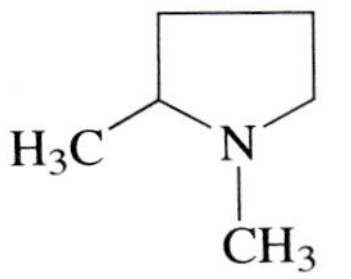 [structure: 1,2-dimethylpyrrolidine; labels H_3C, N, CH_3] (d) 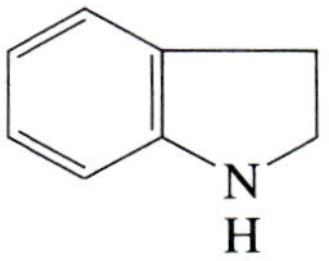 [structure: indoline; labels N, H] (e) 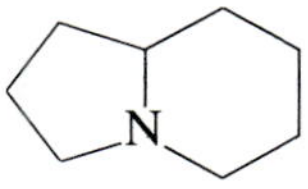 [structure: bicyclic amine; label N]

20 From the following information, deduce the structure of coniine, an amine found in poison hemlock, which, deservedly, has a very bad reputation.

IR: 3330 cm^{-1}.
^1H NMR: $\delta = 0.91$ (t, $J = 7$ Hz, 3 H), 1.33 (s, 1 H), 1.52 (m, 10 H), 2.70 (t, $J = 6$ Hz, 2 H), and 3.0 (m, 1 H) ppm.
Mass spectrum: molecular ion $m/z = 127$; also, $m/z = 84(100)$ and 56(20).

$$\text{Coniine} \xrightarrow[\text{3. }\Delta]{\text{1. CH}_3\text{I; 2. Ag}_2\text{O, H}_2\text{O}} \text{mixture of three compounds} \xrightarrow[\text{3. }\Delta]{\text{1. CH}_3\text{I; 2. Ag}_2\text{O, H}_2\text{O}} (CH_3)_3N + \text{mixture of 1,4-octadiene and 1,5-octadiene}$$

21 Pethidine, the active ingredient in the narcotic analgesic Demerol, was subjected to two successive exhaustive methylations with Hofmann eliminations, and then ozonolysis, with the following results:

(a) Propose a structure for pethidine based on this information.

$$\underset{\textbf{Pethidine}}{C_{15}H_{21}NO_2} \xrightarrow[\text{3. }\Delta]{\text{1. CH}_3\text{I; 2. Ag}_2\text{O, H}_2\text{O}} C_{16}H_{23}NO_2 \xrightarrow[\text{3. }\Delta]{\text{1. CH}_3\text{I; 2. Ag}_2\text{O, H}_2\text{O}}$$

$$(CH_3)_3N + C_{14}H_{16}O_2 \xrightarrow[\text{2. Zn, H}_2\text{O}]{\text{1. O}_3} 2\ CH_2O + C_6H_5C(CO_2CH_2CH_3)(CHO)(CHO)$$

[structure: C bearing phenyl, $CO_2CH_2CH_3$, OHC, and CHO]

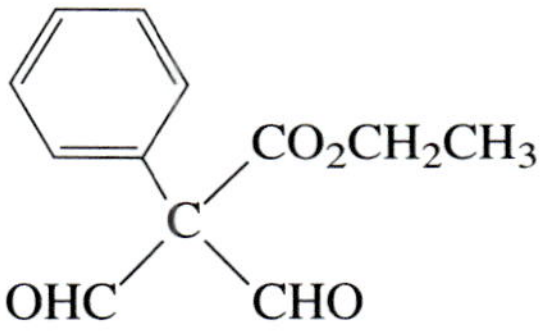

(b) Propose a synthesis of pethidine that begins with ethyl phenylethanoate and *cis*-1,4-dibromo-2-butene. (Hint: First prepare the dialdehyde ester shown in the margin, and then convert it into pethidine.)

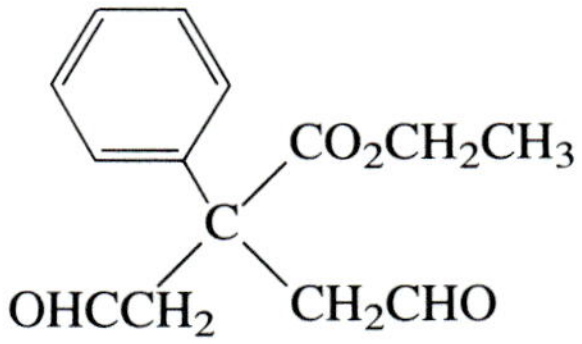

[margin structure: C bearing phenyl, $CO_2CH_2CH_3$, $OHCCH_2$, and CH_2CHO]

22 Skytanthine is a monoterpene alkaloid with the following properties.
Analysis: $C_{11}H_{21}N$.
^1H NMR: two CH_3 doublets ($J = 7$ Hz) at $\delta = 1.20$ and 1.33 ppm; one CH_3 singlet at $\delta = 2.32$ ppm; other hydrogens give rise to broad signals at $\delta = 1.3$–2.7 ppm.
IR: no bands ≥ 3100 cm^{-1}.

$$\text{Skytanthine} \xrightarrow[\text{3. }\Delta]{\text{1. CH}_3\text{I; 2. Ag}_2\text{O, H}_2\text{O}} \underset{\substack{\textbf{A}\\ \text{IR: } \tilde{\nu} = 1646\ cm^{-1}}}{C_{12}H_{23}N} \xrightarrow[\text{2. Zn, H}_2\text{O}]{\text{1. O}_3} CH_2{=}O + \underset{\substack{\textbf{B}\\ \text{IR: } \tilde{\nu} = 1715\ cm^{-1}}}{C_{11}H_{21}NO} \xrightarrow[\text{2. KOH, H}_2\text{O}]{\text{1. } m\text{-ClC}_6\text{H}_4\text{CO}_3\text{H}}$$

$$CH_3COOH + \underset{\substack{\textbf{C}\\ \text{IR: } \tilde{\nu} = 3620\ cm^{-1}}}{C_9H_{19}NO} \xrightarrow{\text{Careful oxidation}}$$

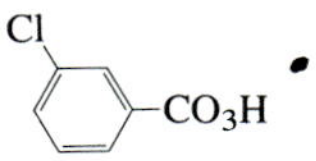

[structure: cyclopentanone (O=) bearing CH_3 and $CH_2N(CH_3)_2$ substituents; IR: $\tilde{\nu} = 1745\ cm^{-1}$]

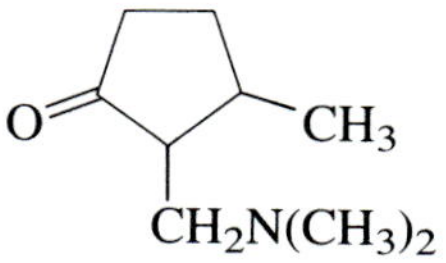

Deduce the structures of skytanthine and degradation products A, B, and C from this information.

23 Reaction of the tertiary amine tropinone with (bromomethyl)benzene (benzyl bromide) gives not one but two quaternary ammonium salts, A and B.

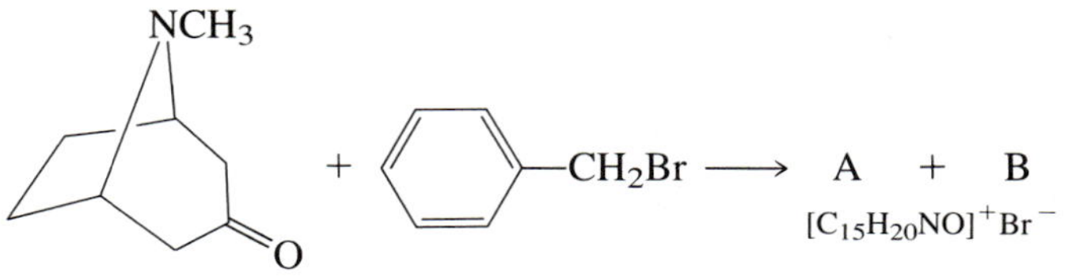

NCH_3 + (benzene ring)$-CH_2Br \longrightarrow$ A + B

$[C_{15}H_{20}NO]^+Br^-$

Tropinone

($C_8H_{13}NO$)

Compounds A and B are stereoisomers that are interconverted by base; that is, base treatment of either purified isomer leads to an equilibrium mixture of the two.

(a) Propose structures for A and B.
(b) What kind of stereoisomers are A and B?
(c) Suggest a mechanism for the equilibration of A and B by base.

24 Attempted Hofmann elimination of an amine containing a hydroxy group on the β-carbon gives an oxacyclopropane product instead of an alkene:

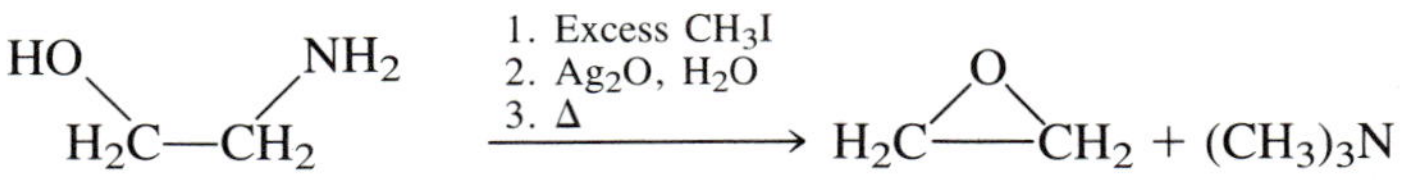

$$HO-H_2C-CH_2-NH_2 \xrightarrow[3.\ \Delta]{1.\ \text{Excess } CH_3I \quad 2.\ Ag_2O,\ H_2O} H_2C\overset{O}{—}CH_2 + (CH_3)_3N$$

(a) Propose a sensible mechanism for this transformation.

(b) Pseudoephedrine (see Problem 16) and ephedrine are closely related, naturally occurring compounds, as the similar names imply. In fact, they are stereoisomers. From the results of the following reactions, deduce the precise stereochemistries of ephedrine and pseudoephedrine.

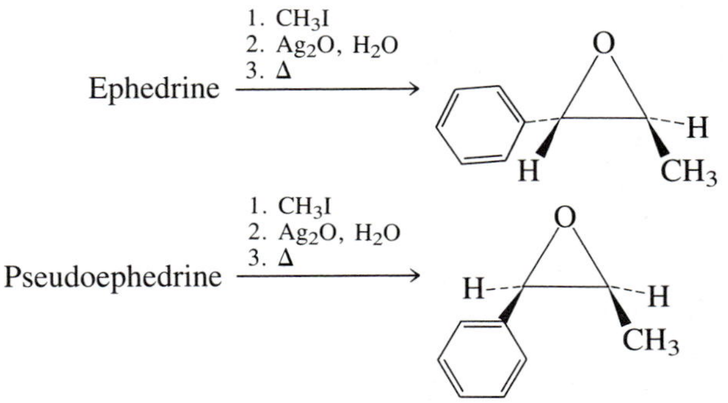

Ephedrine $\xrightarrow[3.\ \Delta]{1.\ CH_3I \quad 2.\ Ag_2O,\ H_2O}$ (oxacyclopropane: O; H; H; CH_3)

Pseudoephedrine $\xrightarrow[3.\ \Delta]{1.\ CH_3I \quad 2.\ Ag_2O,\ H_2O}$ (oxacyclopropane: O; H; H; CH_3)

25 Show how each of the following molecules might be synthesized by Mannich or Mannich-like reactions. (Hint: Work backward, identifying the bond made in the Mannich reaction.)

(a) $CH_3COCH_2CH_2N(CH_2CH_3)_2$

(b) 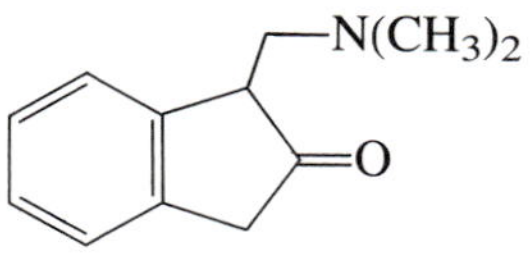 (indanone structure with $-N(CH_3)_2$ and $=O$)

(c) 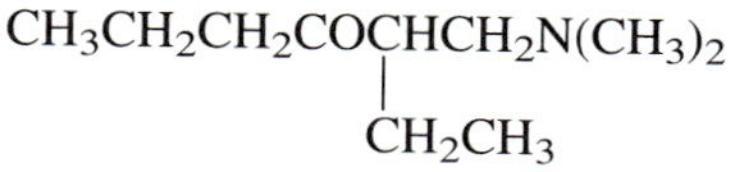 $CH_3CH_2CH_2COCH(CH_2CH_3)CH_2N(CH_3)_2$

(d) 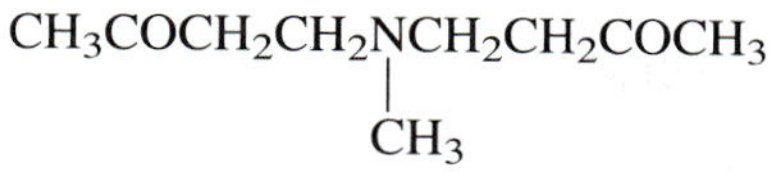 $CH_3COCH_2CH_2N(CH_3)CH_2CH_2COCH_3$

(e) $(CH_3CH_2)_2NCH_2CH_2NO_2$

(f) 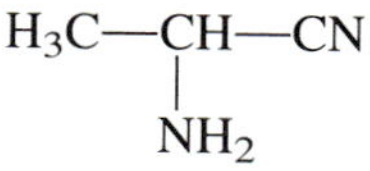 $H_3C—CH(NH_2)—CN$

26 Tropinone (Problem 23) was first synthesized by Sir Robert Robinson (famous for the Robinson annelation reaction, Section 16-5), in 1917, by the following reaction.

$$OHC–CH_2–CH_2–CHO + CH_3NH_2 + (H_3C)_2C{=}O \longrightarrow \text{tropinone (bicyclic ketone with } NCH_3 \text{ bridge)}$$

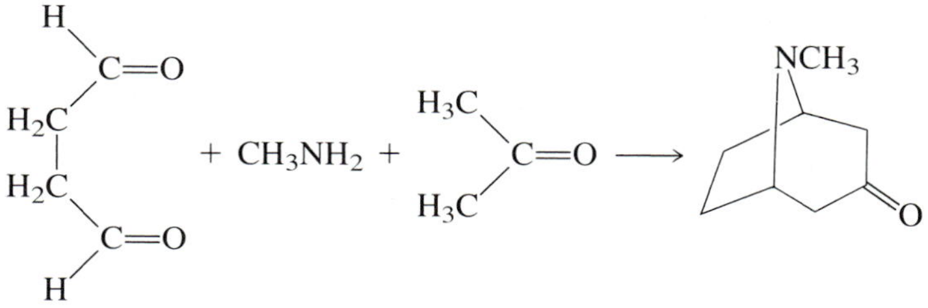

Show a mechanism for this transformation.

27 Many alkaloids are synthesized in nature from a precursor molecule called norlaudanosoline which, in turn, appears to be derived from the condensation of amine (i) with aldehyde (ii):

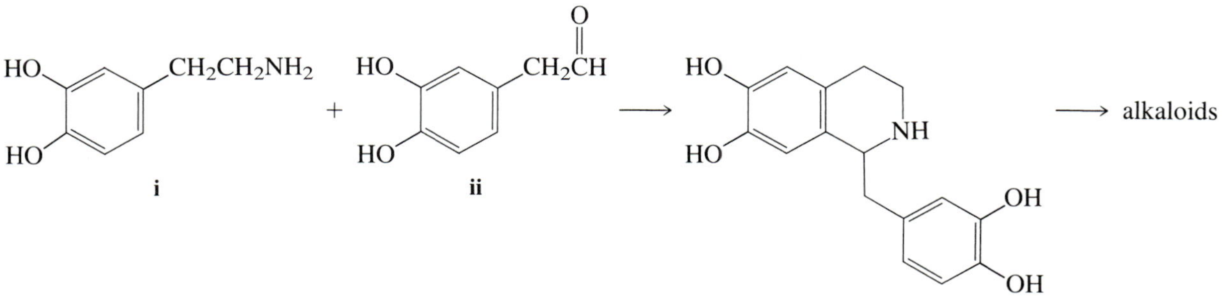

$$\underset{\textbf{i}}{(HO)_2C_6H_3CH_2CH_2NH_2} + \underset{\textbf{ii}}{(HO)_2C_6H_3CH_2CHO} \longrightarrow \textbf{Norlaudanosoline} \longrightarrow \text{alkaloids}$$

Formulate a mechanism for this transformation. Note that a carbon–carbon bond is formed in the process. Name a reaction presented in this chapter that is closely related to this carbon–carbon bond formation.

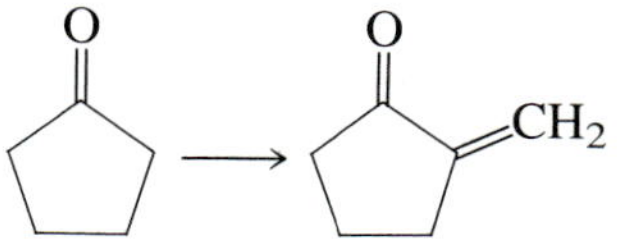

28 Illustrate two methods for achieving the synthetic transformation shown in the margin, using combinations of reactions presented in Section 21-5.

(Margin: cyclopentanone → 2-methylenecyclopentanone, $=CH_2$)

29 Write the expected product(s) of each of the following reactions.

(a) 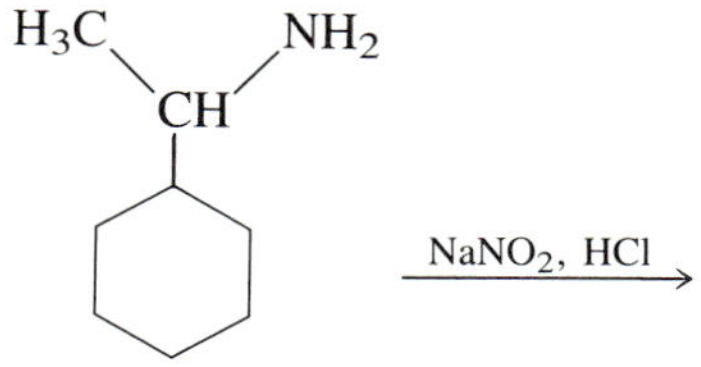 $H_3C–CH(NH_2)–C_6H_{11}$ (cyclohexyl) $\xrightarrow{NaNO_2,\ HCl}$

(b) Pyrrolidine (N–H) $\xrightarrow{NaNO_2,\ HCl,\ 0°C}$

(c) $H_2NCH_2COOCH_2CH_3 \xrightarrow{NaNO_2,\ HCl,\ 0°C}$

(d) Product of part **c** $\xrightarrow{\text{cyclohexene},\ CuSO_4,\ \Delta}$

(e) Cyclopentenyl–$CH_2COCl \xrightarrow{1\ \text{equivalent}\ CH_2N_2}$

(f) Cyclopentenyl–$CH_2COCl \xrightarrow[2.\ CuSO_4,\ \Delta]{1.\ 2\ \text{equivalents}\ CH_2N_2}$ $C_8H_{10}O$
A ketone

30 Describe how phase-transfer catalysis of dichlorocarbene addition to cyclohexene probably occurs, using a diagram similar to Figure 21-7. (Hint: Identify a reasonable anion for the quaternary ammonium cation to transfer from the aqueous phase into the organic phase.)

CHAPTER 22

Difunctional Compounds

We tend to view organic molecules in terms of the chemistry of their functional groups. What happens if there are different functional units in the same molecule? You have seen that such moieties may react independently of each other, according to their particular characteristics. For example, 5-hexenoic acid may be deprotonated with base (as a carboxylic acid) or hydrogenated in the presence of a catalyst (as an alkene).

$$\text{Hexanoic acid } (\ldots\overset{O}{\overset{\|}{C}}OH) \xleftarrow{H_2,\ Pt} \text{5-Hexenoic acid } (\ldots\overset{O}{\overset{\|}{C}}OH) \xrightarrow[-\ HOH]{NaOH} \text{Sodium 5-hexenoate } (\ldots\overset{O}{\overset{\|}{C}}O^-\ Na^+)$$

Hexanoic acid **5-Hexenoic acid** **Sodium 5-hexenoate**

In both of these processes, the reacting centers are converted independently into products, according to their established functional-group chemistry.

On the other hand, in some transformations, the chemical behavior typical of one functional group is drastically altered by the presence of another. For example, α,β-unsaturated carbonyl compounds (Sections 16-4 and 16-5) may undergo nucleophilic addition to the C–C double bond (1,4-addition) instead of attack at the C–O double bond (1,2-addition).

$$CH_3CH_2CH_2\overset{O}{\overset{\|}{C}}H \xrightarrow{KCN,\ H^+} CH_3CH_2CH_2\underset{CN}{\underset{|}{\overset{OH}{\overset{|}{C}}}}H$$

Butanal **2-Hydroxypentanenitrile (Butanal cyanohydrin)**

$$CH_3CH{=}CH\overset{O}{\overset{\|}{C}}H \xrightarrow{KCN,\ H^+} CH_3\overset{NC}{\overset{|}{C}}H\overset{H}{\overset{|}{C}}H\overset{O}{\overset{\|}{C}}H$$

2-Butenal **2-Methyl-4-oxobutanenitrile (3-Cyanobutanal)**

The presence of more than one function in a molecule can also preclude intermolecular reactions:

HO~~~~Br **5-Bromo-1-pentanol** $\xrightarrow[-\text{NaBr},\ -\text{HOH}]{\text{NaOH}}$ **Oxacyclohexane (Tetrahydropyran)**

HO~~~~Br $\xrightarrow[-\text{NaBr}]{\text{NaOH}}$ (does not occur) HO~~~~OH **1,5-Pentanediol**

The reactivity of benzene, with aromatic double bonds, is drastically different from that of cyclohexene.

$$C_6H_6 \xrightarrow[-\text{HBr}]{\text{Br—Br, FeBr}_3} C_6H_5Br$$

Benzene → **Bromobenzene**

Cyclohexene $\xrightarrow{\text{Br—Br}}$ ***trans*-1,2-Dibromocyclohexane**

EXERCISE 22-1

In each of the following pairs of compounds, the second can undergo reactions that the first cannot. For each pair, give at least one example of such reactions: (a) benzene and 1,3-cyclohexadiene; (b) propanal and propenal; (c) cyclohexanol and *cis*-1,2-cyclohexanediol.

This chapter presents the chemistry of difunctional compounds containing carbonyl and hydroxy functions. Earlier chapters dealt with the reactivity of the individual groups (Chapters 8, 9, and 15–18). Here, new reactions will illustrate some of the unique aspects of difunctional systems.

22-1 α-Dicarbonyl Derivatives and Their Precursors: α-Hydroxy Carbonyl Compounds

Some members of the general class of dicarbonyl compounds are shown below. Several of these have common names. Greek letters designate whether the two carbonyl carbons are adjacent or separated by one or more carbon atoms.

$HC(=O)—C(=O)H$	$C_6H_5C(=O)—C(=O)C_6H_5$	$CH_3C(=O)CH_2C(=O)CH_3$
Ethanedial (Glyoxal, an α-dicarbonyl compound)	**Diphenylethanedione (Benzil, an α-dicarbonyl compound)**	**2,4-Pentanedione (Acetylacetone, a β-dicarbonyl compound)**

$$\underset{\text{(Methyl acetoacetate, a }\beta\text{-dicarbonyl compound)}}{\underset{\textbf{Methyl 3-oxobutanoate}}{CH_3\overset{O}{\overset{\|}{C}}CH_2\overset{O}{\overset{\|}{C}}OCH_3}} \qquad \underset{\text{(a }\gamma\text{-dicarbonyl compound)}}{\underset{\textbf{4-Oxopentanal}}{CH_3\overset{O}{\overset{\|}{C}}CH_2CH_2\overset{O}{\overset{\|}{C}}H}} \qquad \underset{\text{(Malonic acid, a }\beta\text{-dicarbonyl compound)}}{\underset{\textbf{Propanedioic acid}}{HO\overset{O}{\overset{\|}{C}}CH_2\overset{O}{\overset{\|}{C}}OH}}$$

If the two carbonyl functions are separated by more than one carbon, such as in 4-oxopentanal, their chemistry can be independent. For example, reduction leads to the diol, condensation with 2,4-dinitrophenylhydrazine results in double hydrazone formation, and oxidation with silver oxide furnishes the keto acid.

The Chemistry of 4-Oxopentanal

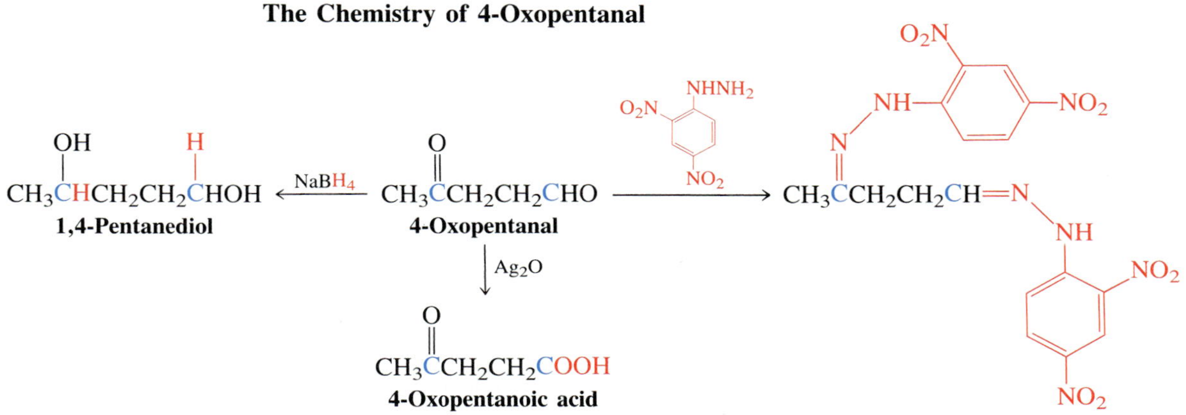

$$\underset{\textbf{1,4-Pentanediol}}{CH_3\overset{OH}{\overset{|}{C}}HCH_2CH_2\overset{H}{\overset{|}{C}}HOH} \xleftarrow{NaBH_4} \underset{\textbf{4-Oxopentanal}}{CH_3\overset{O}{\overset{\|}{C}}CH_2CH_2CHO} \xrightarrow{\text{2,4-}(O_2N)_2C_6H_3NHNH_2} CH_3\overset{N-NH-C_6H_3(NO_2)_2}{\overset{\|}{C}}CH_2CH_2CH{=}N{-}NH{-}C_6H_3(NO_2)_2$$

$$\underset{\textbf{4-Oxopentanal}}{CH_3\overset{O}{\overset{\|}{C}}CH_2CH_2CHO} \xrightarrow{Ag_2O} \underset{\textbf{4-Oxopentanoic acid}}{CH_3\overset{O}{\overset{\|}{C}}CH_2CH_2COOH}$$

On the other hand, mutual reaction of both carbonyl groups is also possible, as in intramolecular aldol condensations.

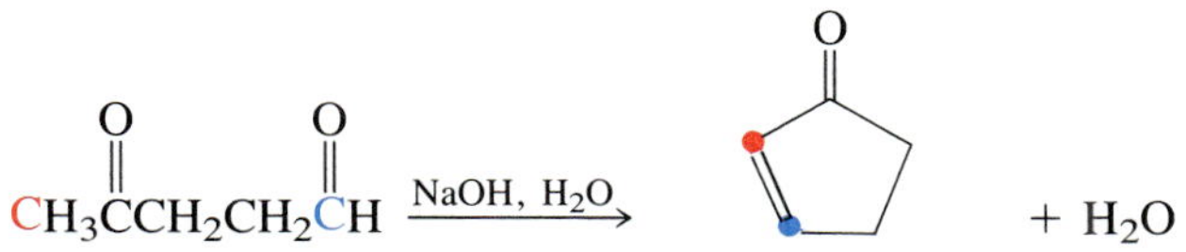

$$\underset{\textbf{4-Oxopentanal}}{CH_3\overset{O}{\overset{\|}{C}}CH_2CH_2\overset{O}{\overset{\|}{C}}H} \xrightarrow{NaOH,\ H_2O} \underset{\textbf{2-Cyclopentenone}}{\text{(cyclopentenone)}} + H_2O$$

In α-dicarbonyl compounds, the carbonyl groups are so close that they strongly influence each other's reactivity, which leads to novel behavior. The remainder of this section will deal with the preparation and reactions of these compounds.

α-Diketones and α-Keto Aldehydes Are Made by Oxidation of α-Hydroxy Carbonyl Compounds

α-Diketones and α-keto aldehydes are frequently obtained by oxidation of α-hydroxy carbonyl precursors. Because of the sensitivity of the products, special reagents and reaction conditions are required to prevent overoxidation to carboxylic acids through C–C bond breaking (Section 17-5). A relatively simple reagent is $KMnO_4$ in ethanoic (acetic) anhydride solvent, which converts α-hydroxy ketones into the desired products with relatively little overoxidation, as shown in the conversion of 2-hydroxy-1,2-diphenylethanone (benzoin) into diphenylethanedione (benzil).

$$\text{C}_6\text{H}_5\text{CH(OH)}\overset{\text{O}}{\overset{\|}{\text{C}}}\text{C}_6\text{H}_5 \xrightarrow{\text{KMnO}_4,\ \text{CH}_3\overset{\text{O}}{\overset{\|}{\text{C}}}\overset{\text{O}}{\overset{\|}{\text{C}}}\text{CH}_3\ \text{(solvent)},\ 5°\text{C}} \text{C}_6\text{H}_5\overset{\text{O}}{\overset{\|}{\text{C}}}\overset{\text{O}}{\overset{\|}{\text{C}}}\text{C}_6\text{H}_5$$

2-Hydroxy-1,2-diphenylethanone (Benzoin) → 73% **Diphenylethanedione (Benzil)**

A milder oxidizing agent is cupric ethanoate (acetate) in aqueous ethanoic (acetic) acid. The mechanism of its action is complex and is likely to include the transfer of electrons to the metal.

$$\xrightarrow{\text{Cu(O}\overset{\text{O}}{\overset{\|}{\text{C}}}\text{CH}_3)_2,\ 50\%\ \text{CH}_3\text{COOH, H}_2\text{O, CH}_3\text{OH},\ 75°\text{C}}$$

2-Hydroxycyclononanone → 75% **1,2-Cyclononanedione**

How to Obtain α-Hydroxy Ketones: Acyloin Condensation of Esters

The α-hydroxy ketones (also called *acyloins*) required for the synthesis of α-diketo compounds are readily obtained by the **acyloin condensation** of esters. In fact, the process is not a condensation, but rather a reductive dimerization similar to the pinacol reaction (Section 15-8). When esters are heated with sodium metal in ethoxyethane (diethyl ether) or benzene (followed by aqueous work-up), reductive coupling that forms the corresponding α-hydroxy ketone is observed.

Acyloin "Condensation"

$$2\ \text{R}\overset{\text{O}}{\overset{\|}{\text{C}}}\text{OR}' \xrightarrow[\text{2. H}^+,\ \text{H}_2\text{O}]{\text{1. Na, (CH}_3\text{CH}_2)_2\text{O}} \text{R}\overset{\text{HO}}{\overset{|}{\text{C}}}\text{H}\overset{\text{O}}{\overset{\|}{\text{C}}}\text{R} + 2\ \text{R}'\text{OH}$$

An acyloin

For example, this treatment converts ethyl butanoate into 5-hydroxy-4-octanone in 80% yield.

$$\text{CH}_3\text{CH}_2\text{CH}_2\overset{\text{O}}{\overset{\|}{\text{C}}}\text{OCH}_2\text{CH}_3 \xrightarrow[\text{2. H}^+,\ \text{H}_2\text{O}]{\text{1. Na, (CH}_3\text{CH}_2)_2\text{O}} \text{CH}_3\text{CH}_2\text{CH}_2\overset{\text{O}}{\overset{\|}{\text{C}}}\underset{\text{OH}}{\underset{|}{\text{C}}}\text{HCH}_2\text{CH}_2\text{CH}_3 + 2\ \text{CH}_3\text{CH}_2\text{OH}$$

Ethyl butanoate → 80% **5-Hydroxy-4-octanone**

A possible mechanism for this reaction is like that of the pinacol coupling. First, the ester is converted into the radical anion by electron transfer from the metal, perhaps directly onto the positively polarized carbonyl carbon. The resulting species dimerizes to the dianion of the corresponding diol.

Mechanism of the Acyloin Condensation:

$$\left[:\!O\!: \;\; \overset{\|}{\mathrm{RCOR'}} \longleftrightarrow :\!\ddot{O}\!:^{-} \; \mathrm{R\overset{+}{C}OR'}\right] \xrightarrow{\mathrm{Na}} \mathrm{Na^+}\; :\!\ddot{O}\!:^{-}\; \mathrm{R\dot{C}OR'} \xrightarrow{\text{Dimerization}} \mathrm{R{-}C(O^-Na^+)(OR'){-}C(O^-Na^+)(OR'){-}R} \xrightarrow{-\,2\,\mathrm{Na^{+}\,^{-}OR'}}$$

$$\left[\mathrm{RC(=O){-}C(=O)R} \longleftrightarrow \mathrm{R\overset{+}{C}(O^-){-}\overset{+}{C}(O^-)R}\right] \xrightarrow{2\,\mathrm{Na}} \underset{\textbf{Enediolate}}{\mathrm{Na^+Na^+}\;\mathrm{RC(O^-){=}C(O^-)R}} \xrightarrow[-\,2\,\mathrm{NaOH}]{2\,\mathrm{HOH}} \underset{\textbf{Enediol}}{\mathrm{RC(OH){=}C(OH)R}} \longrightarrow \underset{\textbf{Acyloin}}{\mathrm{RC(=O){-}CH(OH)R}}$$

In the pinacol coupling, the reaction ends at this stage, but, in the acyloin condensation, the dianion can extrude two alkoxide ions to furnish an α-diketone. Further reduction by transfer of two electrons results in an enediolate. Aqueous work-up then produces the α-hydroxy ketone through the intermediacy of the enediol. Although this sequence appears plausible, the actual mechanism of the reaction is thought to be more complicated, because other products do occasionally appear.

When dicarboxylic esters are subjected to the acyloin condensation, cyclic α-hydroxy ketones are formed. This process is one of the most-general ring-forming reactions, yielding both strained and large rings.

Intramolecular Acyloin Condensation

Dimethyl *cis*-1,2-dimethylcyclohexane-1,2-dicarboxylate $\xrightarrow[2.\ H^+,\ H_2O]{1.\ \mathrm{Na,\ NH_3,\ (CH_3CH_2)_2O,\ -78°C}}$ **8-Hydroxy-1,6-dimethylbicyclo[4.2.0]-7-octanone** (70%)

$\mathrm{CH_3OC(=O)(CH_2)_4C(=O)OCH_3}$ (**Dimethyl hexanedioate**) $\xrightarrow[2.\ H^+,\ H_2O]{1.\ \text{Na, methylbenzene (toluene)}}$ **2-Hydroxycyclohexanone** (57%)

Alkene, alkyne, and ether functional groups are unaffected by the reaction conditions:

$$CH_3O\overset{O}{\overset{\|}{C}}(CH_2)_4C{\equiv}C(CH_2)_4\overset{O}{\overset{\|}{C}}OCH_3 \xrightarrow[\text{2. } H^+,\ H_2O]{\text{1. Na, 1,2-dimethylbenzene (}o\text{-xylene)}}$$

73%

Dimethyl 6-dodecynedioate **2-Hydroxy-7-cyclododecynone**

EXERCISE 22-2

Propose the synthesis of each of the following molecules from an appropriate diester:

(a) $CH_3CH_2CH_2\overset{O}{\overset{\|}{C}}-\overset{O}{\overset{\|}{C}}CH_2CH_2CH_3$

(b) OH, O

(c) OH, O, CH_3O

BOX 22-1

The Synthesis of Catenanes by the Acyloin Condensation

A catenane

The intramolecular acyloin condensation has been used to prepare the first member of an unusual class of compounds: the **catenanes** (*catena,* Latin, chain). A catenane is a compound consisting of two interlocking rings, arranged like two links of a chain, its unusual feature being the absence of any covalent bonds holding the two rings together.

A catenane containing two rings of thirty-four carbons each was made in the following way. First, acyloin cyclization of diethyl tetratriacontanedioate gave the cyclic hydroxy ketone, which was then reduced to the corresponding cycloalkane by Clemmensen reaction (Section 15-8), but with deuterium chloride to incorporate deuterium label (about five D per molecule). The acyloin condensation of the original diester was then repeated, but in the presence of the deuterated macrocycloalkane.

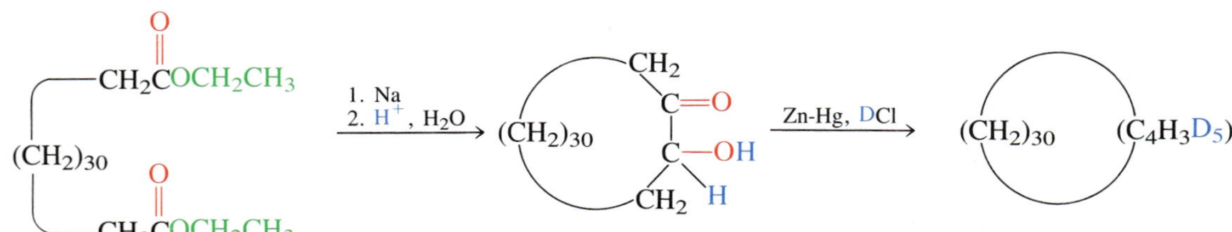

Diethyl tetratriacontanedioate **Pentadeuteriocyclotetratriacontane**

This reaction gave mainly the normal condensation product, but also a new material containing deuterium, as confirmed by infrared spectroscopy ($\tilde{\nu}_{C-D}$ = 2105, 2160, and 2200 cm^{-1}). The new structure was the catenane, formed by the threading of the diester chain through the large cycloalkane loop before the second cyclization.

Catenane Formation

$(CH_2)_{30}$, $(CH_2)_{32}$, $(C_4H_3D_5)$, $\overset{O}{\overset{\|}{C}}OCH_2CH_3$, $CH_3CH_2O\underset{O}{\underset{\|}{C}}$ $\xrightarrow[\text{2. } H^+,\ H_2O]{\text{1. Na}}$ $(CH_2)_{30}$, $(CH_2)_{32}$, $(C_4H_3D_5)$, $\underset{O}{\underset{\|}{C}}-\underset{OH}{\underset{|}{C}}-H$

1%–2%

Catenane

The structural integrity of the two rings was demonstrated by oxidative cleavage of the hydroxy ketone function, which generated tetratriacontanedioic acid bearing no deuterium and liberated the intact deuterated cycloalkane $C_{34}H_{63}D_5$. The term *topological isomer* has been coined to differentiate the catenane from the mixture of noninterlocked rings.

α-Hydroxy Ketones from Aldehydes: Benzoin and Related Condensations

A second approach to the preparation of α-hydroxy ketones is the dimerization of aldehydes in the presence of a suitable catalyst. For example, treatment of benzenecarboxaldehyde (benzaldehyde) with a catalytic amount of sodium cyanide in aqueous ethanol gives 2-hydroxy-1,2-diphenylethanone in high yield.

$$2\ C_6H_5CHO \xrightarrow{Na^{+\,-}CN,\ CH_3CH_2OH,\ H_2O} C_6H_5\overset{O}{\overset{\|}{C}}-\overset{OH}{\overset{|}{C}H}C_6H_5$$

95%

Benzenecarboxaldehyde (Benzaldehyde) — **2-Hydroxy-1,2-diphenylethanone (Benzoin)**

The common name of the product is *benzoin,* and the reaction is known as the **benzoin condensation** (although it is not strictly a condensation according to the definition in Section 15-6).

This cyanide ion catalysis works only for aromatic aldehydes. However, aliphatic aldehydes undergo the same coupling in the presence of thiazolium salts as catalysts. A thiazolium salt is derived from thiazole by alkylation at nitrogen. Thiazole is a heterocyclic compound containing sulfur and nitrogen (a systematic name would be 3-aza-1-thia-2,4-cyclopentadiene; see Section 26-4). An example of such a catalyst is *N*-dodecylthiazolium bromide, in which the long-chain alkyl substituent makes the salt soluble in organic solvents:

$$2\ CH_3CH_2CH_2\overset{O}{\overset{\|}{C}}H \xrightarrow[\textit{N}\text{-Dodecylthiazolium bromide}]{\text{thiazolium}-(CH_2)_{11}CH_3\ Br^-} CH_3CH_2CH_2\overset{O}{\overset{\|}{C}}-\overset{OH}{\overset{|}{C}}HCH_2CH_2CH_3$$

76%

Butanal — **5-Hydroxy-4-octanone**

Catalytic Dimerization of Aldehydes to α-Hydroxy Ketones

$$R\overset{O}{\overset{\|}{C}}H + H\overset{O}{\overset{\|}{C}}R \xrightarrow{\text{Catalyst}} R\overset{O}{\overset{\|}{C}}-\underset{H}{\underset{|}{\overset{OH}{\overset{|}{C}}}}R$$

Thiazole

R, X^-

Thiazolium salt

Mechanistically, these aldehyde dimerizations could in principle proceed by deprotonation of one aldehyde to form a benzoyl or alkanoyl anion intermediate, followed by nucleophilic addition to the carbonyl group of a second.

Plausible (but Wrong) Mechanism of α-Hydroxy Ketone Formation from Aldehydes:

$$\mathrm{RCH(=O)} \xrightleftharpoons{\text{Base}} \mathrm{R\ddot{C}(=O)^-} \xrightarrow{\mathrm{HC(=O)R}} \mathrm{RC(=O)-CHR(O^-)} \xrightarrow[-\,\mathrm{HO^-}]{\mathrm{HOH}} \mathrm{RC(=O)-CHR(OH)}$$

However, such anions cannot be generated in this way (base either adds to the carbonyl group or produces the corresponding enolates; see Chapters 15 and 16), and these steps do not account for the need for a special catalyst. Indeed, the reaction takes a different mechanistic course.

Mechanism of Benzoin Condensation:

STEP 1: Cyanohydrin formation

$$\mathrm{C_6H_5CH(=O)} + {}^-\mathrm{:CN:} \rightleftharpoons \mathrm{C_6H_5-C(CN)(O^-)-H} \xrightleftharpoons{\mathrm{H^+}} \mathrm{C_6H_5-C(CN)(OH)-H}$$

Cyanohydrin

STEP 2: Deprotonation at benzylic position

$$\mathrm{C_6H_5-C(CN)(OH)-H} \xrightleftharpoons{^-\mathrm{OH}} \mathrm{C_6H_5-\ddot{C}(CN)(OH)^-} + \mathrm{HOH}$$

Benzylic anion (A masked benzoyl anion)

STEP 3: Nucleophilic attack on initial aldehyde and protonation

$$\mathrm{C_6H_5-\ddot{C}(CN)(OH)^-} + \mathrm{HC(=O)C_6H_5} \rightleftharpoons \mathrm{C_6H_5C(CN)(OH)-C(O^-)(H)C_6H_5} \xrightleftharpoons{\mathrm{HOH}} \mathrm{C_6H_5C(CN)(OH)-C(OH)(H)C_6H_5} + \mathrm{HO^-}$$

STEP 4: Deprotonation and loss of cyanide ion

$$\mathrm{C_6H_5C(CN)(OH)-C(OH)(H)C_6H_5} \xrightleftharpoons{^-\mathrm{OH}} \mathrm{C_6H_5C(CN)(O^-)-C(OH)(H)C_6H_5} + \mathrm{HOH} \rightleftharpoons \mathrm{C_6H_5C(=O)-C(OH)(H)C_6H_5} + {}^-\mathrm{:CN:} + \mathrm{HOH}$$

With cyanide ion as the catalyst, a few molecules of the starting aromatic aldehyde are equilibrated with the corresponding cyanohydrin (Section 15-7). Although in the presence of base the cyanohydrin is more readily deprotonated at the hydroxy group (the first step of its conversion back into aldehyde), the benzylic position is sufficiently acidic (see Section 24-1) that every now and then proton abstraction will give the benzylic anion. This deprotonation is possible only because the resulting anion is stabilized by resonance both with the benzene ring and with the attached nitrile function (Section 18-6). Nucleophilic attack by this reactive species (which functions as a masked benzoyl anion) on the initial aldehyde gives, after protonation, an intermediate hydroxy cyanohydrin, which in turn loses hydrogen cyanide to regenerate the carbonyl function in the product.

EXERCISE 22-3

Formulate all possible resonance structures for the benzylic anion derived from benzenecarboxaldehyde (benzaldehyde) cyanohydrin.

Alkanals do not react in the same way with each other because, without the aromatic ring, generation of the masked alkanoyl anion is too sluggish. This problem is solved by the use of thiazolium salt catalysts. These salts have an unusual feature—namely, a relatively acidic proton located between the two heteroatoms (at C-2).

Thiazolium Cations Are Acidic

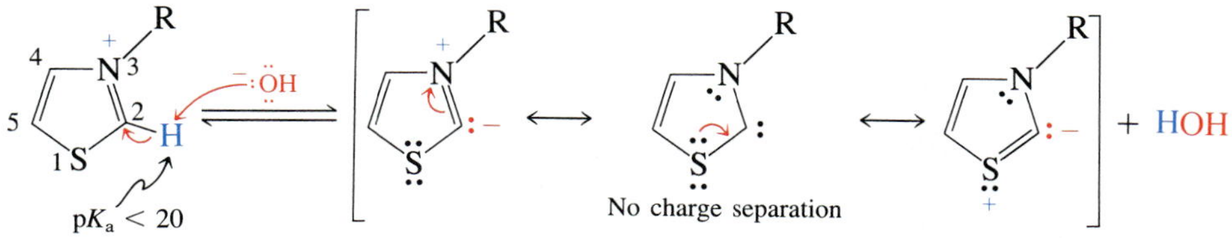

R, 4, N3, +, 5, 2, 1 S, H, $^{-}$:ÖH, $pK_a < 20$, + HOH, No charge separation

Its acidity can be explained by the activating effect of the adjacent positive charge on nitrogen and the possibility of writing several resonance structures (one of which exhibits no charge separation although it contains an electron-deficient carbon at position 2) for the deprotonated species. C-2 in the deprotonated molecule is nucleophilic and reversibly adds to the carbonyl function of an aldehyde, just as cyanide ion does.

Mechanism of Thiazolium Ion Catalysis in Aldehyde Coupling:

STEP 1: Deprotonation of thiazolium ion

STEP 2: Nucleophilic attack by catalyst

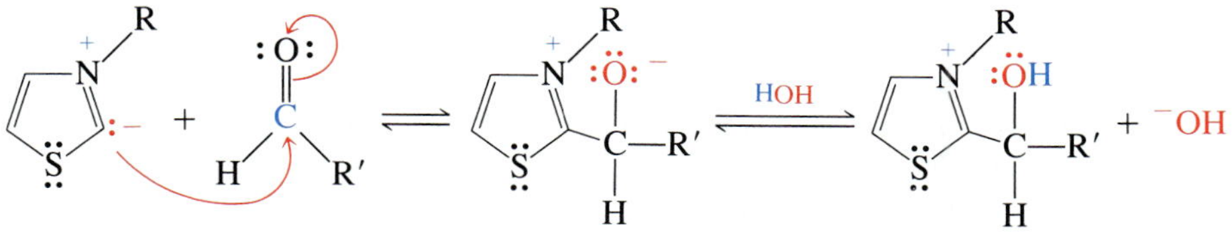

R, N, S, :O:, C, H, R′, :Ö:$^{-}$, C—R′, H, HOH, :ÖH, + $^{-}$OH

STEP 3: Masked alkanoyl anion formation

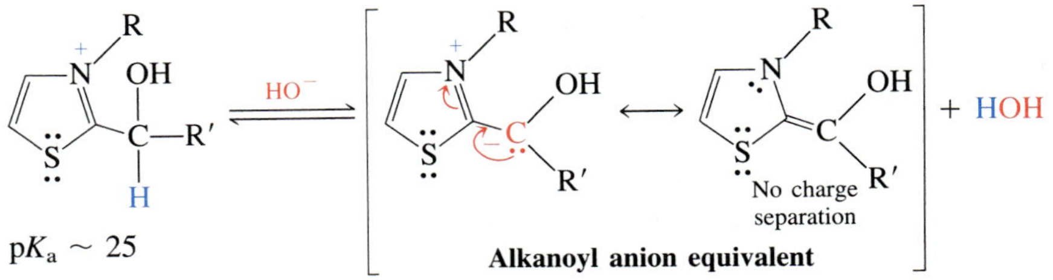

$pK_a \sim 25$

No charge separation

Alkanoyl anion equivalent

STEP 4: Nucleophilic attack on initial aldehyde

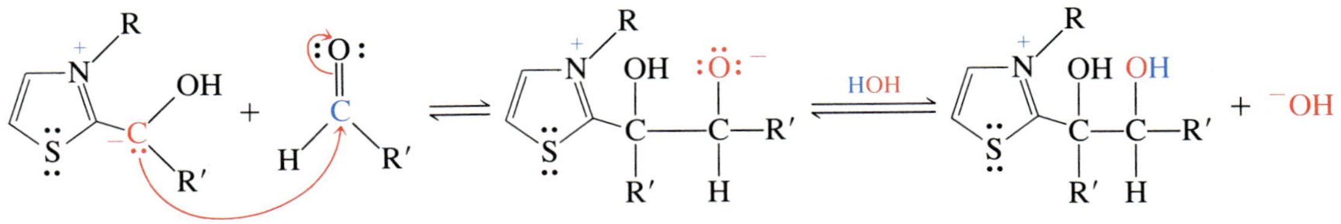

STEP 5: Liberation of α-hydroxy ketone

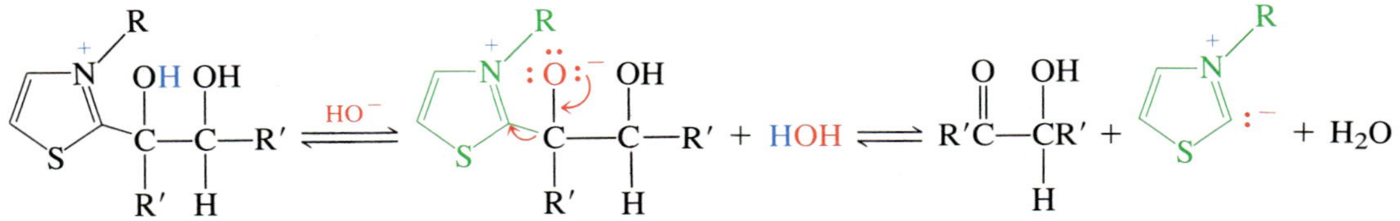

The product alcohol of Step 2 is unique in that the thiazolium unit is a substituent. This moiety is strongly electron withdrawing, both inductively and by resonance, and causes a strong increase in the acidity of the adjacent proton. Deprotonation leads to a masked alkanoyl anion of unusual stability because it can be described by several resonance structures, one of which again shows no charge separation. Nucleophilic attack by this anion on another molecule of aldehyde, followed by loss of the thiazolium substituent, liberates the α-hydroxy ketone. Overall, thiazolium salts react like cyanide. In both cases, the trick is to convert the aldehyde into an adduct that is more prone to lose a proton at carbon, furnishing the corresponding masked alkanoyl anion.

BOX 22-2

Thiamine: A Natural Catalytically Active Thiazolium Ion

The catalytic activity of thiazolium salts in aldehyde dimerization has a natural analogy, the action of *thiamine,* or vitamin B_1. Thiamine, in the form of its pyrophosphate, is a coenzyme for several biochemical transformations, including the transketolase reaction and the decarboxylation of 2-oxopropanoic (pyruvic) acid to ethanal (acetaldehyde). These processes include intermediates of the type that appear in the synthesis of α-hydroxy ketones catalyzed by thiazolium ion.

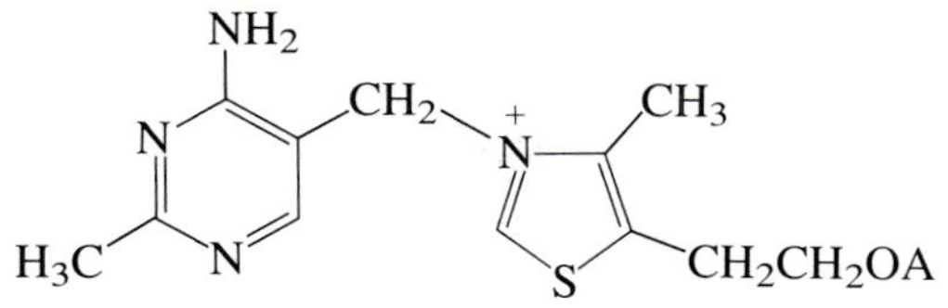

Thiamine

A = H

Thiamine pyrophosphate (TPP)

A = —P(=O)(OH)—O—P(=O)(OH)—OH

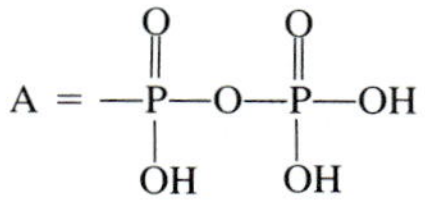

The active site of the enzyme *transketolase* includes a thiamine pyrophosphate (TPP), which enables it to transfer aldehyde units from a donor sugar to an acceptor to give a new sugar (see Chapter 23). The deprotonated thiazolium salt first attacks the donor molecule at a carbonyl function to form an addition compound, in a way completely analogous to addition to aldehydes.

Sugar Activation

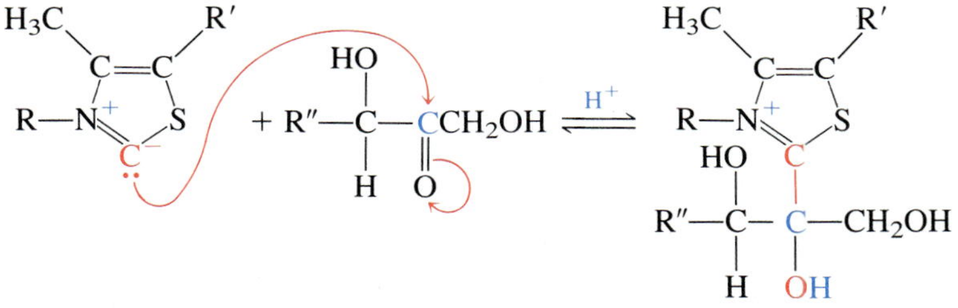

Deprotonated thiamine pyrophosphate + **Donor sugar** $\underset{}{\overset{H^+}{\rightleftharpoons}}$ **Addition compound**

Because the donor sugar contains a hydroxy group next to the reaction site, the initial addition compound can decompose by the reverse of the addition process to an aldehyde and a new thiamine intermediate, ready for the next step.

Removal of Old Aldehyde

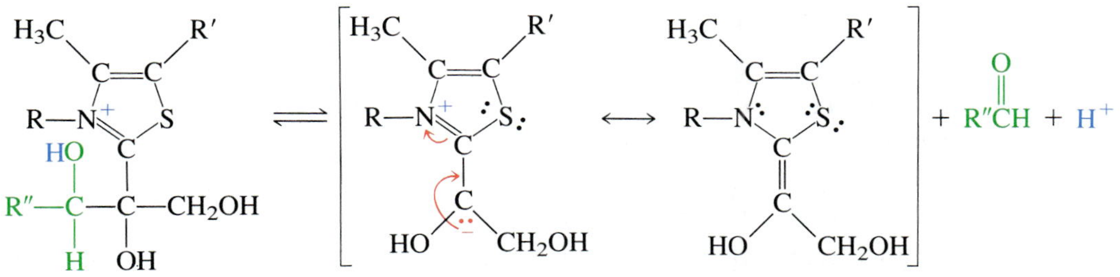

$\rightleftharpoons$ [... $\longleftrightarrow$ ...] + R″CHO + H^+

This next step is an attack on another aldehyde to produce a new addition product. The catalyst then dissociates as thiamine pyrophosphate, releasing a new sugar molecule.

Introduction of New Aldehyde

$$H^+ + R'''CH{=}O + \text{(thiazolium–}\ddot{C}(OH)CH_2OH) \rightleftharpoons HO{-}C(H)(R''')\text{–}C(OH)(CH_2OH)\text{–thiazolium} \rightleftharpoons HO{-}C(H)(R''')\text{–}C(=O)\text{–}CH_2OH + TPP$$

New sugar

Pyruvate dehydrogenase is responsible for catalyzing the decarboxylation of 2-oxopropanoic (pyruvic) acid. Its active site also is equipped with a thiamine unit capable of nucleophilic addition to the ketone function in the 2-oxopropanoate (pyruvate) ion. This addition triggers the loss of carbon dioxide.

2-Oxopropanoate (Pyruvate) + Thiamine part in pyruvate dehydrogenase ⟶ Addition compound ⟶ [resonance structures] + CO_2

The remaining product is subsequently oxidized to the ethanoyl (acetyl) derivative, the ethanoyl (acetyl) group eventually being transferred to coenzyme A to give acetyl CoA (see Section 17-12, Figure 17-15).

EXERCISE 22-4

Write a four-step synthesis of 4-hydroxy-2,5-dimethyl-3-hexanone from 2-bromopropane.

2-Oxopropanoic (Pyruvic) Acid, a Natural α-Keto Acid

2-Oxopropanoic (pyruvic) acid is an α-keto acid. It can be prepared by treatment of 2,3-dihydroxybutanedioic (tartaric) acid (an α-hydroxy acid; see Section 5-5) with potassium hydrogen sulfate as a dehydrating agent. The intermediate 2-oxobutanedioic acid loses carbon dioxide (this is a reaction typical of 3-keto acids; see Section 22-4) to give the final product.

2-Oxopropanoic (Pyruvic) Acid Synthesis

$$\underset{\substack{\textbf{2,3-Dihydroxybutanedioic acid}\\\textbf{(Tartaric acid)}}}{\text{HOOCCH(OH)CH(OH)COOH}} \xrightarrow[-\,H_2O]{KHSO_4,\ \Delta} \left[\underset{\text{Enol form}}{\text{HOOCH=C(OH)COOH}} \rightleftharpoons \underset{\text{Keto form}}{\text{HOOCCH}_2\text{C(=O)COOH}}\right] \xrightarrow{-\,CO_2} \underset{\substack{\textbf{2-Oxopropanoic acid}\\\textbf{(Pyruvic acid)}}}{CH_3\overset{O}{\overset{\|}{C}}COOH}$$

2-Oxobutanedioic acid

An alternative preparation is based on the hydrolysis of 2-oxopropanenitrile, which is available by the treatment of ethanoyl (acetyl) chloride with sodium cyanide:

$$CH_3\overset{O}{\overset{\|}{C}}Cl + Na^{+\,-}CN \xrightarrow[-\,NaCl]{} \underset{\substack{95\%\\\textbf{2-Oxopropanenitrile}}}{CH_3\overset{O}{\overset{\|}{C}}CN} \xrightarrow{\text{Conc. HCl, }0^{\circ}C} \underset{\substack{100\%\\\textbf{2-Oxopropanoic acid}\\\textbf{(Pyruvic acid)}}}{CH_3\overset{O}{\overset{\|}{C}}COOH}$$

Its ethyl ester is made by oxidation of ethyl 2-hydroxypropanoate (ethyl lactate):

$$\underset{\substack{\textbf{Ethyl 2-hydroxypropanoate}\\\textbf{(Ethyl lactate)}}}{CH_3\overset{OH}{\overset{|}{\underset{H}{\underset{|}{C}}}}CO_2CH_2CH_3} \xrightarrow{KMnO_4} \underset{\substack{54\%\\\textbf{Ethyl 2-oxopropanoate}\\\textbf{(Ethyl pyruvate)}}}{CH_3\overset{O}{\overset{\|}{C}}CO_2CH_2CH_3}$$

The two molecules 2-hydroxypropanoic (lactic) and 2-oxopropanoic (pyruvic) acid are interconverted in the body. This task is carried out by an enzyme in the muscle, *lactic acid dehydrogenase,* which reduces pyruvic to lactic acid during physical exercise. The enzyme reverses the process when muscles rest.

$$CH_3\overset{O}{\overset{\|}{C}}COOH \xrightleftharpoons{\text{Lactic acid dehydrogenase}} \text{(HO)(H)(H}_3\text{C)C—COOH}$$

(*S*)-(+)-2-Hydroxypropanoic (lactic) acid

The reduction is stereospecific, giving only the (*S*)-(+) acid. In nature, 2-oxopropanoic (pyruvic) acid is made by the enzymatic degradation of carbohydrates (Section 23-4).

α-Dicarbonyl Compounds Are Reactive

The reactivity of α-dicarbonyl compounds derives from the closeness of the two carbonyl double bonds, which cause mutual activation of the two functional groups to nucleophilic attack.

Activation of Dicarbonyl Compounds

$$\left[\underset{R}{\;}\overset{:O:\,:O:}{\overset{\|\quad\|}{C-C}}\underset{R}{\;} \longleftrightarrow \overset{^{-}:\ddot{O}:\,:O:}{\overset{|\quad\|}{^{+}C-C}} \longleftrightarrow \overset{:O:\,:\ddot{O}:^{-}}{\overset{\|\quad|}{C-C^{+}}} \longleftrightarrow \overset{^{-}:\ddot{O}:\,:\ddot{O}:^{-}}{\overset{|\quad|}{^{+}C-C^{+}}} \right]$$

Subject to nucleophilic attack

Ethanedial (glyoxal), for example, is readily hydrated, and it is quite difficult to completely dehydrate this compound, unlike normal aldehydes and ketones. Treatment with sodium hydroxide induces a rearrangement that produces the sodium salt of hydroxyethanoic acid (glycolic acid):

$$\underset{\textbf{Ethanedial (Glyoxal)}}{HC(=O)CH(=O)} \xrightarrow{NaOH,\ \Delta} \underset{\textbf{Sodium hydroxyethanoate (Glycolic acid sodium salt)}}{HOCH(H)CO^{-}Na^{+}}$$

This reaction is general for other α-dicarbonyl derivatives, such as α-keto aldehydes and α-diketones. It has been named **benzilic acid rearrangement,** because diphenylethanedione (benzil) is converted into diphenylhydroxyethanoic (benzilic) acid under these conditions.

Benzilic Acid Rearrangement

$$\underset{\textbf{Diphenylethanedione (Benzil)}}{C_6H_5C(=O)C(=O)C_6H_5} \xrightarrow{KOH,\ H_2O,\ CH_3CH_2OH,\ 100^{\circ}C} \underset{\textbf{Potassium diphenylhydroxyethanoate (Benzilic acid)}}{\underset{95\%}{(C_6H_5)_2C(OH)COO^{-}K^{+}}}$$

How does this rearrangement proceed? The mechanism of the benzilic acid synthesis begins with nucleophilic addition of hydroxide ion to one of the activated carbonyl carbons. Subsequently, the molecule reorganizes by migration of the substituent on the alkoxide carbon to the neighboring carbonyl function. The migrating group takes with it its electron pair, which is donated to the neighboring carbonyl double bond. Proton transfer completes the sequence.

Mechanism of the Benzilic Acid Rearrangement:

STEP 1: Addition of hydroxide ion

$$\overset{:O:\,:O:}{\overset{\|\quad\|}{RC-CR}} + {}^{-}:\ddot{\underset{..}{O}}H \rightleftharpoons \overset{^{-}:\ddot{O}:\,:O:}{\overset{|\quad\|}{H\ddot{\underset{..}{O}}\underset{\underset{R}{|}}{C}-CR}}$$

STEP 2: Rearrangement

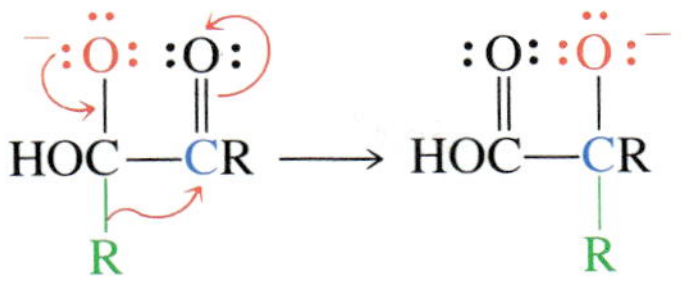

$$\text{HO}\overset{\overset{\ddot{O}:^-}{|}}{C}\text{—}\overset{\overset{:O:}{\|}}{C}\text{R} \ (\text{with R on first C}) \longrightarrow \text{HO}\overset{\overset{:O:}{\|}}{C}\text{—}\overset{\overset{:\ddot{O}:^-}{|}}{C}\text{R} \ (\text{with R on second C})$$

STEP 3: Proton transfer

$$\text{H}\ddot{O}\overset{\overset{:O:}{\|}}{C}\text{—}\overset{\overset{:\ddot{O}:^-}{|}}{C}\text{R (R)} \rightleftharpoons {}^-:\ddot{O}\overset{\overset{:O:}{\|}}{C}\text{—}\overset{\overset{:\ddot{O}H}{|}}{C}\text{R (R)}$$

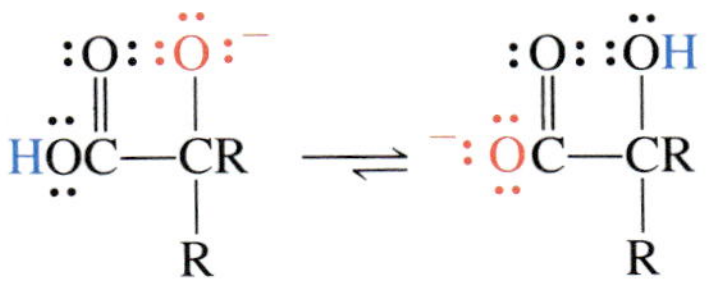

When cyclic α-diketones undergo this transformation, the ring contracts.

Benzilic Acid Rearrangement of 1,2-Cyclohexanedione

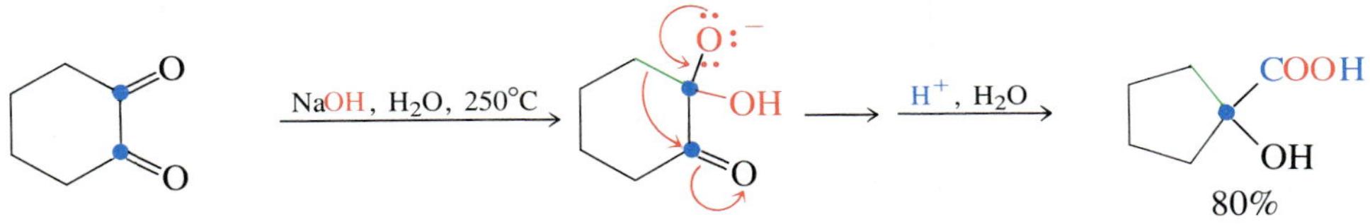

1,2-Cyclohexanedione $\xrightarrow{\text{NaOH, } H_2O, \ 250°C}$ [intermediate: O⁻, OH adduct] $\longrightarrow \xrightarrow{H^+, \ H_2O}$ 1-Hydroxycyclopentanecarboxylic acid (COOH, OH) 80%

1,2-Cyclohexanedione

1-Hydroxycyclopentane-carboxylic acid

EXERCISE 22-5

Explain the following rearrangement by a mechanism. Hint: Start the ring expansion with the nitrogen lone electron pair as shown.

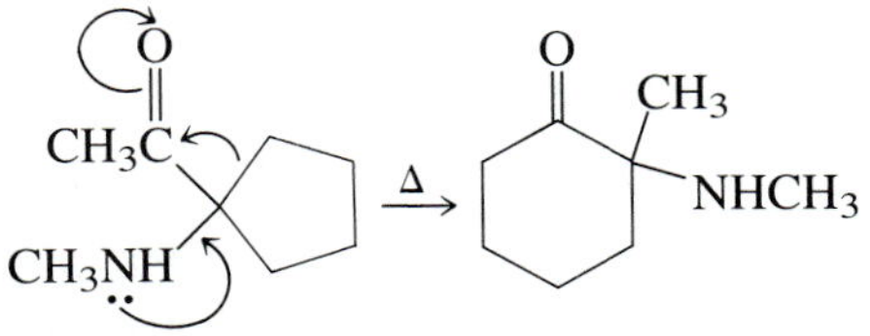

1-Acetyl-1-(methylamino)cyclopentane ($CH_3C(=O)$, $CH_3\ddot{N}H$) $\xrightarrow{\Delta}$ 2-methyl-2-(methylamino)cyclohexanone (CH_3, $NHCH_3$)

In summary, α-hydroxy ketones (also called acyloins) are available from esters by the acyloin condensation and from aldehydes by dimerization catalyzed by cyanide (aromatic aldehydes) or thiazolium ion (aliphatic aldehydes). The latter process has analogies in nature in thiamine-mediated aldehyde transfers and pyruvate decarboxylations. The acyloin condensation is a reductive coupling by a sequence of electron transfers. The aldehyde dimerization proceeds through masked benzoyl or alkanoyl anions. Oxidation of α-hydroxy ketones results in α-diketones. 2-Oxopropanoic (pyruvic) acid, an α-keto acid, is made by dehydration of 2,3-dihydroxybutanedioic (tartaric) acid or by hydrolysis of 2-oxopropanenitrile. It is formed in resting muscle by the (reversible) enzymatic oxidation of (*S*)-(+)-2-hydroxypropanoic (lactic) acid. The α-diketo function is more reactive than an isolated carbonyl group. Hydration occurs readily, and base causes the benzilic acid rearrangement.

22-2 1,3-Dithiacyclohexane (1,3-Dithiane) Anions: Stoichiometric Alkanoyl (Acyl) Anion Equivalents

The dimerization of aldehydes catalyzed by cyanide and thiazolium ion proceeds through benzoyl and alkanoyl anion equivalents as *reactive intermediates:* they are not isolable. There are, however, isolable compounds that function as such masked anions in a variety of stoichiometric reactions. The development of such reagents is important, because in them the normally electrophilic nature of a carbonyl carbon has been changed to nucleophilic. Such reverse polarizations appear also in the preparation of organometallic reagents from haloalkanes (Section 8-5). This section will describe the use of 1,3-dithiacyclohexane (1,3-dithiane) anions as masked alkanoyl anions.

1,3-Dithiacyclohexanes (1,3-Dithianes): Masked Alkanoyl Anion Precursors

When dimethoxymethane [the dimethyl acetal of methanal (formaldehyde)] is exposed to 1,3-propanedithiol in the presence of acid, a *transacetalization* takes place to give 1,3-dithiacyclohexane (1,3-dithiane), a cyclic dithioacetal of methanal (formaldehyde; see Section 15-5).

$$HSCH_2CH_2CH_2SH + H_2C(OCH_3)_2 \xrightarrow{BF_3,\ (CH_3CH_2)_2O,\ CHCl_3,\ \Delta} \text{1,3-dithiacyclohexane} + 2\ CH_3OH$$

1,3-Propane-dithiol — Dimethoxymethane — 77%–84% **1,3-Dithiacyclohexane (1,3-Dithiane)**

EXERCISE 22-6

Formulate a reasonable mechanism for the acid-catalyzed transacetalization of dimethoxymethane to 1,3-dithiacyclohexane.

The hydrogens on the methylene group positioned between the two sulfur atoms in 1,3-dithiacyclohexane are relatively acidic, exhibiting a pK_a of 31. This value is sufficiently low to enable strong bases such as butyllithium to deprotonate the compound to the corresponding anion. The reason for the acidity of these protons is the polarizability (see Figure 6-11) of sulfur, which allows for the stabilization of an adjacent negative charge.

Deprotonation of 1,3-Dithiacyclohexane

$$\text{1,3-dithiacyclohexane } (pK_a = 31.1) \xrightarrow{CH_3CH_2CH_2CH_2Li,\ THF} \text{2-lithio-1,3-dithiacyclohexane } (Li^+) + CH_3CH_2CH_2CH_2\text{–}H$$

Aldehydes other than methanal (formaldehyde) also can be converted into such thioacetals and subsequently deprotonated, furnishing the corresponding 2-alkyl-1,3-dithiacyclohexane anions.

$$CH_3CH_2CH(CH_3)CHO \xrightarrow[-\ H_2O]{HS(CH_2)_3SH,\ BF_3} \text{2-(1-methylpropyl)-1,3-dithiacyclohexane (76\%)} \xrightarrow[-\ CH_3CH_2CH_2CH_2H]{CH_3CH_2CH_2CH_2Li,\ THF} Li^+\ \text{2-lithio-2-(1-methylpropyl)-1,3-dithiacyclohexane}$$

2-Methylbutanal — **2-(1-Methylpropyl)-1,3-dithiacyclohexane** (76%)

These anions can be alkylated by a variety of reagents, including primary and secondary haloalkanes, aldehydes and ketones, and oxacyclopropanes. Thus, the electrophilic carbonyl carbon has undergone reverse polarization in the thioacetal to become nucleophilic.

General Alkylation of a 1,3-Dithiacyclohexane Anion

$$\text{2-R-1,3-dithiacyclohexane anion} \xrightarrow{R'X} \text{2-R-2-R'-1,3-dithiacyclohexane} + X^-$$

Examples:

$$\text{1,3-dithiacyclohexane anion } Li^+ + CH_3CH(I)CH_2CH_3 \xrightarrow{THF} \text{2-(1-methylpropyl)-1,3-dithiacyclohexane} + LiI$$

53%

2-(1-Methylpropyl)-1,3-dithiacyclohexane

$$\text{1,3-dithiacyclohexane anion} + C_6H_5CHO \longrightarrow \xrightarrow{H^+,\ H_2O} \text{2-[CH(OH)C}_6H_5\text{]-1,3-dithiacyclohexane}$$

91%

2-(1-Hydroxyphenylmethyl)-1,3-dithiacyclohexane

$$\text{1,3-dithiacyclohexane anion} + \text{oxacyclopropane} \longrightarrow \xrightarrow{H^+,\ H_2O} \text{2-(}CH_2CH_2OH\text{)-1,3-dithiacyclohexane}$$

75%

2-(2-Hydroxyethyl)-1,3-dithiacyclohexane

The thioacetal function is hydrolyzed by mercuric salts (Section 15-5) to yield the corresponding carbonyl compounds. The overall sequence of thioacetalization, anion formation, alkylation, and hydrolysis thus constitutes a general ketone synthesis from aldehydes.

General Hydrolysis of 1,3-Dithianes

$$\text{2-R-2-R'-1,3-dithiane} \xrightarrow{HgCl_2,\ HgO,\ \text{alcohol solvent},\ H_2O} RC(=O)R'$$

Example:

$$\underset{\textbf{2-Methyl-2-(1-methylpropyl)-1,3-dithiacyclohexane}}{\text{2-}H_3C\text{-2-}CH(CH_3)CH_2CH_3\text{-1,3-dithiane}} \xrightarrow{HgCl_2,\ HgO,\ HOCH_2CH_2OCH_2CH_2OCH_2CH_2OH,\ H_2O,\ 90°C} \underset{\substack{63\% \\ \textbf{3-Methyl-2-pentanone}}}{CH_3C(=O)CH(CH_3)CH_2CH_3}$$

EXERCISE 22-7

What is the structure of the product of the following sequence of reactions?

$$\text{1,3-dithiane} \xrightarrow[\substack{3.\ CH_3CH_2CH_2CH_2Li \\ 4.\ HgCl_2,\ HCl,\ HOCH_2CH_2OH,\ H_2O,\ 90°C}]{\substack{1.\ CH_3CH_2CH_2CH_2Li \\ 2.\ Cl(CH_2)_3Br}} \underset{42\%}{C_4H_6O}$$

In summary, the phenomenon of reverse polarization is seen in the conversion of aldehydes into the anions of the corresponding 1,3-dithiacyclohexanes (1,3-dithianes). The electrophilic carbonyl carbon changes into a nucleophilic center; this is the basis of a synthetic method for converting aldehydes into ketones.

22-3 Preparation of β-Dicarbonyl Compounds: The Unusual Acidity of Methylene Hydrogens Flanked by Two Carbonyl Groups

As Section 22-4 will show, β-dicarbonyl compounds, such as β-diketones and β-(or 3-)keto esters, are versatile synthetic intermediates. This section will describe how these systems are made.

β-Dicarbonyl Compounds Are Made by Claisen and Related Condensations

β-Keto esters can be made by the Claisen condensation (Section 18-4). For example, ethyl ethanoate (acetate) reacts with sodium ethoxide to give, after acidic work-up, ethyl 3-oxobutanoate (ethyl acetoacetate). The reaction before work-up generates the enolate ion of the keto ester, a step needed to drive the reaction to completion. This enolate enjoys unusual resonance stabilization, because both carbonyl groups participate in conjugation. The effect of reso-

BOX 22-3

1,3-Dithiacyclohexane Anion in the Synthesis of a Natural Product

A 1,3-dithiacyclohexane is used as an alkanoyl anion equivalent in the laboratory preparation of a sex attractant of the bark beetle *Ips confusus*. 2-(2-Methylpropyl)-1,3-dithiacyclohexane anion is first alkylated with 2-(bromomethyl)-1,3-butadiene. Hydrolysis of the thioacetal group in the product and reduction of the carbonyl function then gives the natural product.

Total Synthesis of the Sex Attractant of the Bark Beetle

2-(2-Methylpropyl)-1,3-dithiane anion ($CH_2CH(CH_3)_2$) + 2-(Bromomethyl)-1,3-butadiene ($BrCH_2$) $\xrightarrow{-\,Br^-}$ 2-alkylated dithiane ($(CH_3)_2CHCH_2$, CH_2) 51% $\xrightarrow{HgCl_2}$ ketone 59% $\xrightarrow{NaBH_4}$ alcohol (OH, H) 66%

Sex attractant of the bark beetle

Claisen Condensation of Ethyl Ethanoate (Acetate)

$$2\ CH_3\overset{O}{\overset{\|}{C}}OCH_2CH_3 \xrightarrow{Na^{+\,-}OCH_2CH_3,\ CH_3CH_2OH} \left[\begin{array}{c} CH_3\overset{:\ddot{O}:^-}{\overset{|}{C}}{=}CH{-}\overset{:O:}{\overset{\|}{C}}OCH_2CH_3 \\ \updownarrow \\ CH_3\overset{:O:}{\overset{\|}{C}}{-}\overset{..}{\bar{C}}H{-}\overset{:O:}{\overset{\|}{C}}OCH_2CH_3 \\ \updownarrow \\ CH_3\overset{:O:}{\overset{\|}{C}}{-}CH{=}\overset{:\ddot{O}:^-}{\overset{|}{C}}OCH_2CH_3 \end{array}\right] \xrightarrow{H^+,\ H_2O} CH_3\overset{O}{\overset{\|}{C}}\underset{H}{\underset{|}{C}}H\overset{O}{\overset{\|}{C}}OCH_2CH_3$$

Ethyl ethanoate (Ethyl acetate)

Ethyl 3-oxobutanoate (Ethyl acetoacetate)

TABLE 22-1
pK_a values for β-dicarbonyl and related compounds

Name	Structure	pK_a
2,4-Pentanedione (Acetylacetone)	$CH_3C(=O)CH_2C(=O)CH_3$	9
Methyl 2-cyanoethanoate (Methyl 2-cyanoacetate)	$NCCH_2C(=O)OCH_3$	9
Ethyl 3-oxobutanoate (Ethyl acetoacetate)	$CH_3C(=O)CH_2C(=O)OCH_2CH_3$	11
Propanedinitrile (Malonodinitrile)	$NCCH_2CN$	13
Diethyl propanedioate (Diethyl malonate)	$CH_3CH_2OC(=O)CH_2C(=O)OCH_2CH_3$	13

nance manifests itself in the unusually low pK_a values of ethyl 3-oxobutanoate and other β-dicarbonyl compounds, listed in Table 22-1. This list includes related systems, such as methyl 2-cyanoethanoate and propanedinitrile.

The importance of the final deprotonation step in the Claisen condensation is clearly apparent if the ester bears only one α-hydrogen. The product of this reaction would be a 2,2-disubstituted 3-keto ester lacking any of the acidic protons necessary to drive the equilibrium. Hence, no Claisen condensation product is observed.

An Example of the Failure of a Claisen Condensation

$$2\ (CH_3)_2CHC(=O)OCH_2CH_3 \underset{}{\overset{Na^{+}\,{}^{-}OCH_2CH_3,\ CH_3CH_2OH}{\rightleftharpoons}} (CH_3)_2CHC(=O)\text{—}C(CH_3)_2\text{—}C(=O)OCH_2CH_3 + CH_3CH_2OH$$

Ethyl 2-methylpropanoate → **Ethyl 2,2,4-trimethyl-3-oxopentanoate**

That this result is due to an unfavorable equilibrium may be demonstrated by treating a 2,2-disubstituted 3-keto ester with base: a **retro-Claisen condensation** takes place, proceeding through a mechanism that is the exact reverse of the forward reaction.

Retro-Claisen Condensation

$$(CH_3)_2CHC(=\ddot{O}\!:)\text{—}C(CH_3)_2\text{—}C(=\ddot{O}\!:)\ddot{O}CH_2CH_3 + CH_3CH_2\ddot{O}\!:^{-} \xrightarrow{CH_3CH_2\ddot{O}H} (CH_3)_2CHC(\text{—}\ddot{O}\!:^{-})(\text{—}\ddot{O}CH_2CH_3)\text{—}C(CH_3)_2\text{—}C(=\ddot{O}\!:)OCH_2CH_3 \longrightarrow$$

$$(CH_3)_2CHC(=\ddot{O}\!:)\ddot{O}CH_2CH_3 + (CH_3)_2C{=}C(\text{—}\ddot{O}\!:^{-})\ddot{O}CH_2CH_3 \overset{CH_3CH_2\ddot{O}H}{\rightleftharpoons} (CH_3)_2CHC(=\ddot{O}\!:)\ddot{O}CH_2CH_3 + CH_3CH_2\ddot{O}\!:^{-}$$

EXERCISE 22-8

Explain the following observation:

$$2\ CH_3\overset{O}{\overset{\|}{C}}-\underset{CH_3}{\underset{|}{\overset{CH_3}{\overset{|}{C}}}}-COOCH_3 \xrightarrow[2.\ H^+,\ H_2O]{1.\ CH_3O^-Na^+,\ CH_3OH} CH_3\overset{O}{\overset{\|}{C}}CH_2COOCH_3 + 2\ (CH_3)_2CHCOOCH_3$$

Claisen Condensations Between Two Different Esters

Mixed Claisen condensations frequently furnish, like mixed aldol reactions (Section 16-3), product mixtures.

Mixed Claisen Condensations Are Often Nonselective

$$CH_3CH_2\overset{O}{\overset{\|}{C}}OCH_2CH_3 + CH_3\overset{O}{\overset{\|}{C}}OCH_2CH_3 \xrightarrow[2.\ H^+,\ H_2O]{1.\ CH_3CH_2O^-Na^+,\ CH_3CH_2OH}$$

$$CH_3CH_2\overset{O}{\overset{\|}{C}}\underset{CH_3}{\underset{|}{C}}H\overset{O}{\overset{\|}{C}}OCH_2CH_3 + CH_3CH_2\overset{O}{\overset{\|}{C}}CH_2\overset{O}{\overset{\|}{C}}OCH_2CH_3 + CH_3\overset{O}{\overset{\|}{C}}\underset{CH_3}{\underset{|}{C}}H\overset{O}{\overset{\|}{C}}OCH_2CH_3 + CH_3\overset{O}{\overset{\|}{C}}CH_2\overset{O}{\overset{\|}{C}}OCH_2CH_3$$

However, a selective mixed condensation is possible when one of the reacting partners has no α-hydrogens, as in ethyl methanoate (formate) or ethyl benzenecarboxylate (benzoate).

Selective Mixed Claisen Condensations

$$H\overset{O}{\overset{\|}{C}}OCH_2CH_3 + CH_3\overset{O}{\overset{\|}{C}}OCH_2CH_3 \xrightarrow[2.\ H^+,\ H_2O]{1.\ CH_3CH_2O^-Na^+,\ CH_3CH_2OH} H\overset{O}{\overset{\|}{C}}CH_2\overset{O}{\overset{\|}{C}}OCH_2CH_3$$

Ethyl methanoate (Ethyl formate) → 80% **Ethyl 3-oxopropanoate**

$$C_6H_5CO_2CH_2CH_3 + CH_3CH_2\overset{O}{\overset{\|}{C}}OCH_2CH_3 \xrightarrow[2.\ H^+,\ H_2O]{1.\ CH_3CH_2O^-Na^+,\ CH_3CH_2OH} C_6H_5\overset{O}{\overset{\|}{C}}\underset{CH_3}{\underset{|}{C}}H\overset{O}{\overset{\|}{C}}OCH_2CH_3$$

Ethyl benzenecarboxylate (Ethyl benzoate) → 71% **Ethyl 2-methyl-3-oxo-3-phenylpropanoate**

Intramolecular Claisen Condensations Result in Cyclic Keto Esters

The intramolecular version of the Claisen reaction is called the **Dieckmann* condensation,** and produces cyclic 3-keto esters. As expected (Section 9-5), it works best for the formation of five- and six-membered rings.

---

*Professor Walter Dieckmann, 1869–1925, University of Munich.

$$CH_3CH_2O\overset{O}{\overset{\|}{C}}CH_2CH_2CH_2CH_2CH_2\overset{O}{\overset{\|}{C}}OCH_2CH_3 \xrightarrow[2.\ H^+,\ H_2O]{1.\ CH_3CH_2O^-Na^+,\ CH_3CH_2OH}$$ Ethyl 2-oxocyclohexanecarboxylate (with $CO_2CH_2CH_3$), 60%

Diethyl heptanedioate → **Ethyl 2-oxocyclohexanecarboxylate**

$$CH_3CH_2O\overset{O}{\overset{\|}{C}}\underset{CH_3}{\underset{|}{C}}HCH_2CH_2CH_2\overset{O}{\overset{\|}{C}}OCH_2CH_3 \xrightarrow[2.\ H^+,\ H_2O]{1.\ CH_3CH_2O^-Na^+,\ CH_3CH_2OH}$$ H_3C / $CO_2CH_2CH_3$ cyclopentanone, 70%

Diethyl 2-methylhexanedioate → **Ethyl 3-methyl-2-oxocyclopentanecarboxylate**

Cyclic compounds may also be obtained by (intermolecular followed by intramolecular) **double Claisen condensations** with diesters such as diethyl ethanedioate (diethyl oxalate).

$$CH_3CH_2O\overset{O}{\overset{\|}{C}}\overset{O}{\overset{\|}{C}}OCH_2CH_3 + CH_3CH_2O\overset{O}{\overset{\|}{C}}(CH_2)_3\overset{O}{\overset{\|}{C}}OCH_2CH_3 \xrightarrow[2.\ H^+,\ H_2O]{1.\ CH_3CH_2O^-Na^+,\ CH_3CH_2OH}$$ $CH_3CH_2O_2C$ / $CO_2CH_2CH_3$ dioxocyclopentane, 80%

Diethyl ethanedioate (Diethyl oxalate) + **Diethyl pentanedioate** → **Diethyl 4,5-dioxo-1,3-cyclopentanedicarboxylate**

EXERCISE 22-9

Formulate a mechanism for the reaction of diethyl ethanedioate with diethyl pentanedioate.

EXERCISE 22-10

Formulate a mechanism for the following reaction:

$$\text{Diethyl phthalate} + CH_3CO_2CH_2CH_3 \xrightarrow[2.\ H^+,\ H_2O]{1.\ CH_3CH_2O^-Na^+,\ CH_3CH_2OH}$$ 1,3-indandione-2-$CO_2CH_2CH_3$, 60%–80%

Diethyl 1,2-benzenedicarboxylate (Diethyl phthalate) (with two $CO_2CH_2CH_3$ groups)

Ketones Undergo Mixed Claisen Reactions

Ketones can participate in the Claisen condensation. Because they are more acidic than esters, they are deprotonated before the ester has a chance to undergo self-condensation. The products of this reaction (after acidic work-up) may be β-diketones, β-keto aldehydes, or other β-dicarbonyl compounds. The reaction can be carried out with a variety of ketones and esters both inter- and intramolecularly. Frequently, the use of a base stronger than alkoxide gives better yields.

$$CH_3\overset{O}{\overset{\|}{C}}OCH_2CH_3 + CH_3\overset{O}{\overset{\|}{C}}CH_3 \xrightarrow[2.\ H^+,\ H_2O]{1.\ NaH,\ (CH_3CH_2)_2O} CH_3\overset{O}{\overset{\|}{C}}CH_2\overset{O}{\overset{\|}{C}}CH_3$$

85%

Methyl 5-oxohexanoate $\xrightarrow[2.\ H^+,\ H_2O]{1.\ (C_6H_5)_3CO^-K^+,\ 1,2\text{-dimethylbenzene }(o\text{-xylene}),\ \Delta}$ 1,3-Cyclohexanedione (100%)

Methyl 5-oxohexanoate

100%

1,3-Cyclohexanedione

Carbonates furnish 3-keto esters, whereas methanoates (formates) lead to 3-keto aldehydes.

$$BrC_6H_4COCH_3 + CH_3CH_2O\overset{O}{\overset{\|}{C}}OCH_2CH_3 \xrightarrow[2.\ H^+,\ H_2O]{1.\ NaH,\ (CH_3CH_2)_2O,\ 25°C,\ 66\ h} BrC_6H_4\overset{O}{\overset{\|}{C}}CH_2\overset{O}{\overset{\|}{C}}OCH_2CH_3$$

1-(4-Bromophenyl)-ethanone (*p*-Bromoacetophenone)

Diethyl carbonate

25%

Ethyl 3-oxo-3-(4-bromophenyl)-propanoate

$$\text{Cyclohexanone} + H\overset{O}{\overset{\|}{C}}OCH_2CH_3 \xrightarrow[2.\ H^+,\ H_2O]{1.\ CH_3CH_2O^-Na^+,\ CH_3CH_2OH} \text{2-formylcyclohexanone}$$

Ethyl methanoate (Ethyl formate)

75%

2-Methanoylcyclohexanone (2-Formylcyclohexanone)

The mechanism of these reactions begins with the generation of the ketone enolates.

Mechanism of the Claisen Condensation with Ketones:

STEP 1: Ketone deprotonation

$$CH_3\overset{O}{\overset{\|}{C}}CH_3 \underset{\text{Reversible}}{\overset{CH_3CH_2O^-Na^+,\ CH_3CH_2OH}{\rightleftharpoons}} CH_2{=}C(O^-Na^+)CH_3 + CH_3CH_2OH$$

STEP 2: Nucleophilic addition-elimination

$$CH_3\overset{O}{\overset{\|}{C}}OCH_2CH_3 + CH_2{=}C(O^-)CH_3 \underset{\text{Reversible}}{\rightleftharpoons} CH_3\overset{O}{\overset{\|}{C}}CH_2\overset{O}{\overset{\|}{C}}CH_3 + {}^-OCH_2CH_3$$

2,4-Pentanedione (Acetylacetone)

STEP 3: Deprotonation of β-diketone

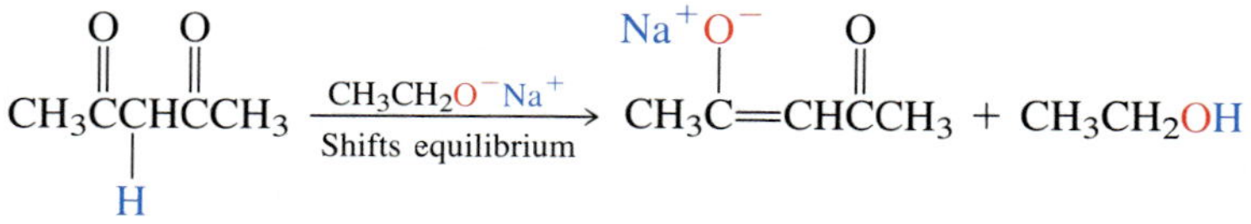

$$CH_3\overset{O}{\overset{\|}{C}}\underset{H}{\underset{|}{C}}H\overset{O}{\overset{\|}{C}}CH_3 \xrightarrow[\text{Shifts equilibrium}]{CH_3CH_2O^-Na^+} CH_3\overset{Na^+O^-}{\overset{|}{C}}{=}CH\overset{O}{\overset{\|}{C}}CH_3 + CH_3CH_2OH$$

EXERCISE 22-11

Suggest syntheses of the following molecules by Claisen or Dieckmann condensations:

(a) 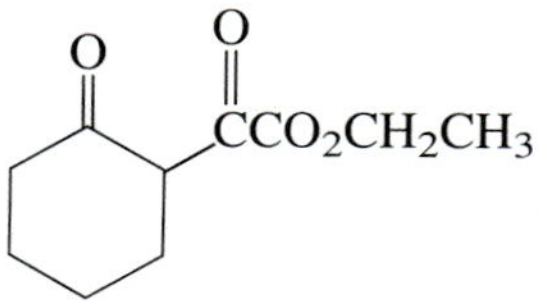 2-oxocyclohexyl–$\overset{O}{\overset{\|}{C}}CO_2CH_2CH_3$

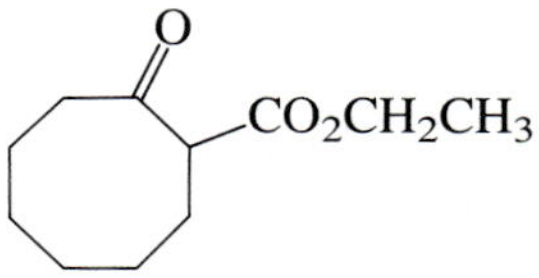

(b) 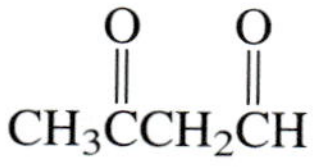 $CH_3\overset{O}{\overset{\|}{C}}CH_2\overset{O}{\overset{\|}{C}}H$

(c) 2-oxocyclooctyl–$CO_2CH_2CH_3$

(d) 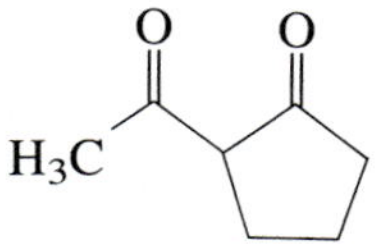 $H_3C\overset{O}{\overset{\|}{C}}$–(2-oxocyclopentyl)

In summary, Claisen condensations are endothermic and therefore would not occur without a stoichiometric amount of base strong enough to deprotonate the resulting 3-keto ester. Mixed Claisen condensations between two esters are nonselective, unless they are intramolecular (Dieckmann condensation) or one of the components is devoid of α-hydrogens. Ketones also participate in selective mixed Claisen reactions because they are more acidic than esters.

22-4
β-Dicarbonyl Compounds as Synthetic Intermediates

Having seen how to prepare β-dicarbonyl compounds, let us explore their synthetic utility. This section will show that the corresponding anions are readily alkylated and that 3-keto esters are hydrolyzed to the corresponding acids, which can be decarboxylated to give ketones or new carboxylic acids. These transformations open up versatile synthetic routes to other functionalized molecules.

β-Dicarbonyl Anions Are Nucleophilic

The unusual acidity of β-keto carbonyl compounds may be used to synthetic advantage, because the enolate ions obtained by deprotonation can be alkylated to give substituted derivatives. For example, in this way ethyl 3-oxobutanoate is readily converted into alkylated analogs.

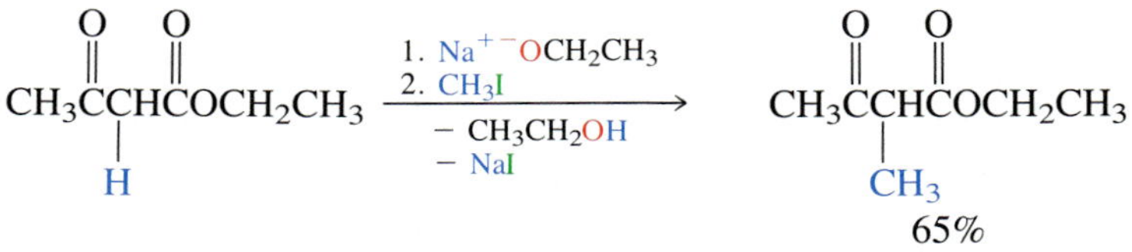

$$CH_3\overset{O}{\overset{\|}{C}}\underset{H}{\underset{|}{C}}H\overset{O}{\overset{\|}{C}}OCH_2CH_3 \xrightarrow[\substack{-\ CH_3CH_2OH \\ -\ NaI}]{\substack{1.\ Na^{+\,-}OCH_2CH_3 \\ 2.\ CH_3I}} CH_3\overset{O}{\overset{\|}{C}}\underset{CH_3}{\underset{|}{C}}H\overset{O}{\overset{\|}{C}}OCH_2CH_3$$

65%

Ethyl 3-oxobutanoate **Ethyl 2-methyl-3-oxobutanoate**

$$\mathrm{CH_3\overset{O}{\overset{\|}{C}}\underset{CH_3}{\underset{|}{C}H}\overset{O}{\overset{\|}{C}}OCH_2CH_3} \xrightarrow[-\,(CH_3)_3COH,\ -\,KBr]{1.\ K^{+}\,{}^{-}OC(CH_3)_3;\ 2.\ C_6H_5CH_2Br} \mathrm{CH_3\overset{O}{\overset{\|}{C}}-\overset{CH_3}{\overset{|}{\underset{CH_2C_6H_5}{\underset{|}{C}}}}-\overset{O}{\overset{\|}{C}}OCH_2CH_3}$$

77%

Ethyl 2-methyl-2-(phenylmethyl)-3-oxobutanoate

Other β-dicarbonyl compounds undergo similar reactions:

$$\mathrm{CH_3\overset{O}{\overset{\|}{C}}\underset{H}{\underset{|}{C}H}\overset{O}{\overset{\|}{C}}CH_3} \xrightarrow[-\,KHCO_3,\ -\,KI]{K_2CO_3,\ CH_3I,\ \text{propanone (acetone)},\ \Delta} \mathrm{CH_3\overset{O}{\overset{\|}{C}}\underset{CH_3}{\underset{|}{C}H}\overset{O}{\overset{\|}{C}}CH_3}$$

2,4-Pentanedione → 77% **3-Methyl-2,4-pentanedione**

Ethyl 2-oxocyclohexanecarboxylate (2-oxocyclohexane ring with H and $CO_2CH_2CH_3$ at C1)
$\xrightarrow[-\,CH_3CH_2OH,\ -\,NaBr]{1.\ Na^{+}\,{}^{-}OCH_2CH_3;\ 2.\ CH_3CH_2CH_2CH_2Br}$
Ethyl 1-butyl-2-oxocyclohexane-carboxylate (butyl and $CO_2CH_2CH_3$ at C1), 80%

$$\mathrm{CH_3CH_2O\overset{O}{\overset{\|}{C}}\underset{H}{\underset{|}{C}H}\overset{O}{\overset{\|}{C}}OCH_2CH_3} \xrightarrow[-\,CH_3CH_2OH,\ -\,NaBr]{1.\ Na^{+}\,{}^{-}OCH_2CH_3;\ 2.\ CH_3CH_2CH(CH_3)Br} \mathrm{CH_3CH_2O\overset{O}{\overset{\|}{C}}\underset{CH_3CH_2CHCH_3}{\underset{|}{C}H}\overset{O}{\overset{\|}{C}}OCH_2CH_3}$$

Diethyl propanedioate → 84% **Diethyl 2-(1-methylpropyl)-propanedioate**

EXERCISE 22-12

Give a synthesis of 2,2-dimethyl-1,3-cyclohexanedione from methyl 5-oxohexanoate.

EXERCISE 22-13

Explain the following observation:

2-Methyl-1,3-cyclohexanedione $\xrightarrow{Ba(OH)_2,\ H_2O}\ \xrightarrow{H^+,\ H_2O}$ $\mathrm{HOOC(CH_2)_3\overset{O}{\overset{\|}{C}}CH_2CH_3}$ 78%

Hint: Instead of deprotonation at position 2, consider nucleophilic attack by HO^- on one of the carbonyl groups.

3-Keto Acids Readily Undergo Decarboxylation: A New Ketone Synthesis

The synthetic usefulness of the alkylation reactions of 3-keto esters stems from their ready decarboxylation after hydrolysis, affording access to new ketones and carboxylic acids.

General Decarboxylation of 3-Keto Acids

$$\mathrm{RC(=O)CH(R')C(=O)OCH_2CH_3} \xrightarrow{\text{Hydrolysis}} \mathrm{RC(=O)CH(R')C(=O)OH} \xrightarrow{\Delta} \mathrm{RC(=O)CH(H)R'} + \mathrm{CO_2}$$

Ketone

$$\mathrm{CH_3CH_2OC(=O)CH(R)C(=O)OCH_2CH_3} \xrightarrow{\text{Hydrolysis}} \mathrm{HOC(=O)CH(R)C(=O)OH} \xrightarrow{\Delta} \mathrm{RCH(H)C(=O)OH} + \mathrm{CO_2}$$

Carboxylic acid

Examples:

$$\mathrm{CH_3C(=O)CH[(CH_2)_3CH_3]C(=O)OCH_2CH_3} \xrightarrow[\substack{-\,CH_3CH_2OH \\ -\,CO_2}]{\substack{1.\ NaOH,\ H_2O \\ 2.\ H_2SO_4,\ H_2O,\ 100^{\circ}C}} \mathrm{CH_3C(=O)CH(H)(CH_2)_3CH_3}$$

Ethyl 2-butyl-3-oxobutanoate → 60% **2-Heptanone**

$$\mathrm{CH_3CH_2OC(=O)CH[CH(CH_3)CH_2CH_3]C(=O)OCH_2CH_3} \xrightarrow[\substack{-\,CH_3CH_2OH \\ -\,CO_2}]{H_2SO_4,\ H_2O,\ \Delta} \mathrm{CH_3CH_2CH(CH_3)CH(H)COOH}$$

Diethyl 2-(1-methylpropyl)propanedioate → 65% **3-Methylpentanoic acid**

The decarboxylation step proceeds by a concerted mechanism and includes a cyclic transition state somewhat reminiscent of that formulated in the ester pyrolysis (Section 18-4) or the McLafferty rearrangement (Section 18-7).

Mechanism of Decarboxylation of 3-Keto Acids:

$$\mathrm{H_3C{-}C(=\ddot{O}{:})CH_2C(=\ddot{O}{:})\ddot{O}H} \longrightarrow \mathrm{H_3C{-}C(\ddot{O}H){=}CH_2} + \mathrm{{:}O{=}C{=}O{:}} \longrightarrow \mathrm{H_3C{-}C(=O)CH_2H} + \mathrm{CO_2}$$

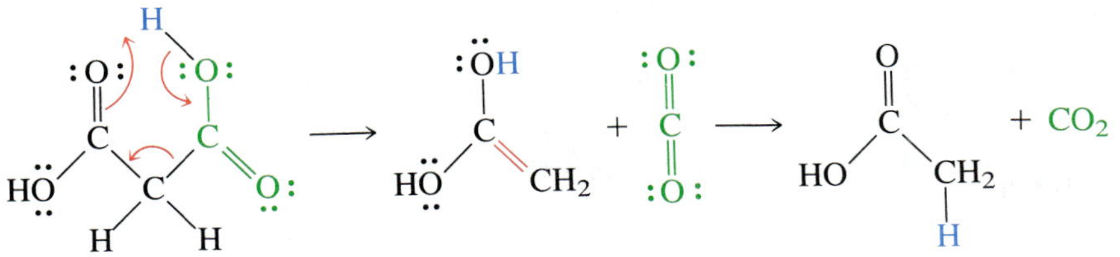

$$\text{HOOC–CH}_2\text{–COOH (cyclic H-transfer)} \longrightarrow \text{HO(C=CH}_2\text{)OH} + \text{O=C=O} \longrightarrow \text{HOC(=O)CH}_2\text{H} + CO_2$$

Loss of CO_2 can occur readily only from the free carboxylic acid. If the ester is hydrolyzed under basic conditions, the resulting carboxylate salt is usually neutralized with acid to enable subsequent decarboxylation. The decarboxylation process allows compounds such as ethyl 3-oxobutanoate to be converted ultimately into 3,3-disubstituted methyl ketones (a strategy called **acetoacetic ester synthesis**). Similarly, propanedioic (malonic) esters are good starting materials for 2,2-disubstituted carboxylic acids **(malonic ester synthesis).**

General Acetoacetic Ester Synthesis

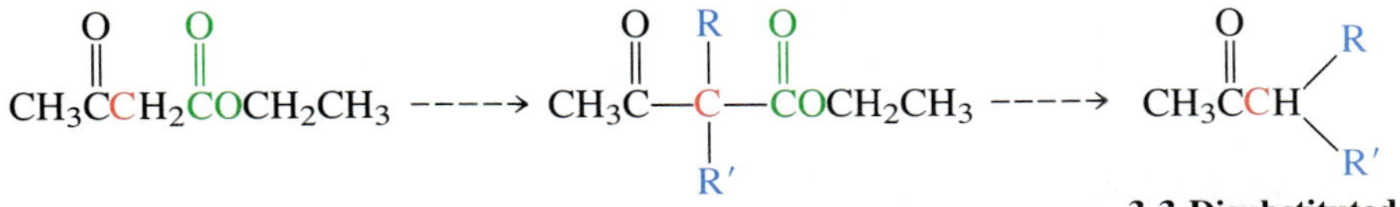

$$CH_3\overset{O}{\overset{\|}{C}}CH_2\overset{O}{\overset{\|}{C}}OCH_2CH_3 \dashrightarrow CH_3\overset{O}{\overset{\|}{C}}—\underset{R'}{\overset{R}{C}}—\overset{O}{\overset{\|}{C}}OCH_2CH_3 \dashrightarrow CH_3\overset{O}{\overset{\|}{C}}CH\begin{matrix}R\\R'\end{matrix}$$

3,3-Disubstituted methyl ketone

General Malonic Ester Synthesis

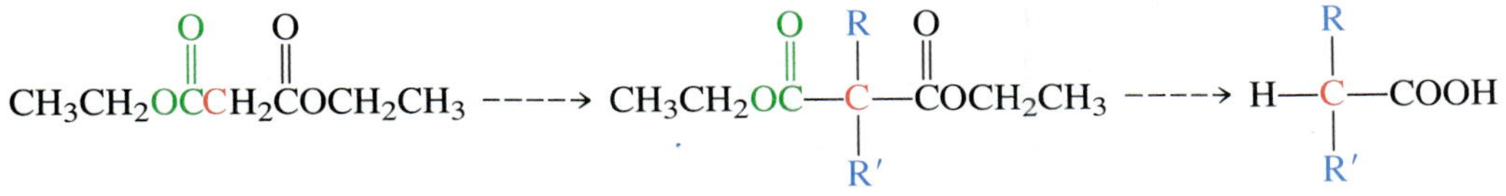

$$CH_3CH_2O\overset{O}{\overset{\|}{C}}CH_2\overset{O}{\overset{\|}{C}}OCH_2CH_3 \dashrightarrow CH_3CH_2O\overset{O}{\overset{\|}{C}}—\underset{R'}{\overset{R}{C}}—\overset{O}{\overset{\|}{C}}OCH_2CH_3 \dashrightarrow H—\underset{R'}{\overset{R}{C}}—COOH$$

2,2-Disubstituted carboxylic acid

The rules and limitations governing S_N2 reactions are still in effect in the alkylation steps. Thus, tertiary haloalkanes exposed to β-dicarbonyl anions give mainly elimination products. However, the anions can be successfully attacked by alkanoyl halides, α-bromo esters, α-bromo ketones, and oxacyclopropanes.

EXERCISE 22-14

The first-mentioned compound in each of the following parts is treated with the subsequent series of reagents; give the final products:

(a) $CH_3CH_2O_2C(CH_2)_5CO_2CH_2CH_3$: (1) $NaOCH_2CH_3$, (2) $CH_3(CH_2)_3I$, (3) NaOH, and (4) H^+, H_2O, Δ;

(b) $CH_3CH_2O_2CCH_2CO_2CH_2CH_3$: (1) $NaOCH_2CH_3$, (2) CH_3I, (3) KOH, and (4) H^+, H_2O, Δ;

(c) $CH_3\overset{O}{\overset{\|}{C}}\underset{CH_3}{\underset{|}{C}}HCO_2CH_3$: (1) NaH, C_6H_6, (2) $C_6H_5\overset{O}{\overset{\|}{C}}Cl$, and (3) H^+, H_2O, Δ;

(d) $CH_3\overset{O}{\overset{\|}{C}}CH_2CO_2CH_2CH_3$: (1) $NaOCH_2CH_3$, (2) $BrCH_2CO_2CH_2CH_3$, (3) NaOH, and (4) H^+, H_2O, Δ;
(e) $CH_3CH_2CH(CO_2CH_2CH_3)_2$: (1) $NaOCH_2CH_3$, (2) $BrCH_2CO_2CH_2CH_3$, and (3) H^+, H_2O, Δ;
(f) $CH_3\overset{O}{\overset{\|}{C}}CH_2CO_2CH_2CH_3$: (1) $NaOCH_2CH_3$, (2) $BrCH_2\overset{O}{\overset{\|}{C}}CH_3$, and (3) H^+, H_2O, Δ.

EXERCISE 22-15

Propose a synthesis of cyclohexanecarboxylic acid from diethyl propanedioate (malonate), $CH_2(CO_2CH_2CH_3)_2$, and 1-bromo-5-chloropentane, $Br(CH_2)_5Cl$. Hint: Consult Exercise 22-8.

EXERCISE 22-16

Explain the following result by a mechanism:

$$CH_3\overset{O}{\overset{\|}{C}}CH_2CO_2CH_2CH_3 \xrightarrow[\text{3. } H^+,\ H_2O]{\text{1. NaOH, } H_2O,\ 0^\circ C;\ \text{2. } H_2C\overset{O}{-}CH_2} H_3C\overset{O}{\overset{\|}{C}}\text{-(}\gamma\text{-butyrolactone-2-yl)}$$

60%

2-Ethanoyl-γ-butyrolactone

In summary, β-dicarbonyl compounds such as ethyl 3-oxobutanoate and diethyl propanedioate (malonate) are versatile synthetic building blocks for elaborating more-complex molecules. Their unusual acidity makes it easy to form the corresponding anions, which can be used in nucleophilic displacement reactions with a wide variety of substrates. Their hydrolysis produces 3-keto acids that are unstable and undergo decarboxylation on heating.

22-5 Extensions of β-Dicarbonyl Anion Chemistry: The Knoevenagel Condensation and Michael Additions

The stabilized anions derived from β-dicarbonyl compounds and related analogs (Table 22-1) undergo another type of reaction typical of enolate ions: nucleophilic attack on carbonyl compounds. This transformation leads to aldol products (Section 16-3) and is called the **Knoevenagel* condensation.** With α,β-unsaturated aldehydes and ketones, addition occurs in a 1,4 mode; these are further examples of the Michael addition (Section 16-5).

In the Knoevenagel condensation, the β-dicarbonyl compound is treated with a catalytic amount of a weak amine base, such as *N*-ethylethanamine (diethylamine), in the presence of an aldehyde or ketone to furnish the aldol condensation product.

---

*Professor Emil Knoevenagel, 1865–1921, University of Heidelberg.

Knoevenagel Condensation

$$CH_3CH_2CH_2CH_2\overset{\overset{\displaystyle O}{\|}}{C}H + H_2C(CO_2CH_2CH_3)_2 \xrightarrow[C_6H_6,\ \Delta]{(CH_3CH_2)_2NH,} CH_3(CH_2)_3CH{=}C(CO_2CH_2CH_3)_2 + H_2O$$

Pentanal — 80% — Ethyl 2-ethoxycarbonyl-2-heptenoate

The Knoevenagel condensation can be carried out with several of the species shown in Table 22-1.

EXERCISE 22-17

Write the products of the Knoevenagel condensation of the following reactants: (a) Cyclohexanone + $N{\equiv}CCH_2C{\equiv}N$; (b) $CH_3CH_2CH_2CHO + N{\equiv}CCH_2CO_2CH_2CH_3$; (c) $CH_2{=}O + C_6H_5\overset{\overset{\displaystyle O}{\|}}{C}CH_2\overset{\overset{\displaystyle O}{\|}}{C}C_6H_5$; (d) $C_6H_5CHO + CH_3\overset{\overset{\displaystyle O}{\|}}{C}CH_2CO_2CH_2CH_3$.

EXERCISE 22-18

The mechanism of the Knoevenagel condensation is analogous to that of the aldol reaction. Formulate it.

Exposure of β-dicarbonyl anions to α,β-unsaturated carbonyl compounds leads to 1,4-additions (Michael addition). The reaction works with α,β-unsaturated ketones, aldehydes, nitriles, and carboxylic acid derivatives (called Michael acceptors) and again requires only a catalytic amount of base.

Michael Addition

$$CH_2(CO_2CH_2CH_3)_2 + CH_2{=}CH\overset{\overset{\displaystyle O}{\|}}{C}CH_3 \xrightarrow[CH_3CH_2OH,\ -10^\circ \text{ to } 25^\circ C]{\text{Catalytic } CH_3CH_2O^-Na^+,} (CH_3CH_2O_2C)_2CHCH_2CH_2\overset{\overset{\displaystyle O}{\|}}{C}CH_3$$

3-Buten-2-one (Methyl vinyl ketone, Michael acceptor) — 71% — Diethyl 2-(3-oxobutyl)propanedioate

EXERCISE 22-19

Give the products of the following Michael additions [base in square brackets]:

(a) $CH_3CH_2CH(CO_2CH_2CH_3)_2 + CH_2{=}CH\overset{\overset{\displaystyle O}{\|}}{C}H\ [Na^{+-}OCH_2CH_3]$;

(b) 5,5-dimethylcyclohexane-1,3-dione (structure) $+ CH_2{=}CHC{\equiv}N\ [Na^{+-}OCH_3]$;

(c) 2-methyl-5-(ethoxycarbonyl)cyclopentanone with H_3C and $CO_2CH_2CH_3$ substituents (structure) $+ CH_3CH{=}CHCO_2CH_2CH_3\ [K^{+-}OCH_2CH_3]$.

EXERCISE 22-20

Explain the following observation:

$$\text{5,5-dimethyl-1,3-cyclohexanedione} + 2\ CH_2{=}CHC{\equiv}N \xrightarrow{Na^{+}\,{}^{-}OCH_3,\ CH_3OH} \text{2,2-bis(2-cyanoethyl)-5,5-dimethyl-1,3-cyclohexanedione}\ (81\%)$$

Hint: Consider proton transfer in the first Michael adduct.

EXERCISE 22-21

Explain the outcome of the following transformation:

$$\text{ethyl 2-oxocyclohexanecarboxylate} + CH_2{=}CH\overset{O}{\overset{\|}{C}}CH_3 \xrightarrow{Na^{+}\,{}^{-}OCH_2CH_3,\ CH_3CH_2OH} \text{bicyclic enone}-CO_2CH_3\ (70\%)$$

Hint: Review the Robinson annelation (Section 16-5), a combination of Michael addition to α,β-unsaturated ketones with the aldol condensation leading to cyclohexenones.

In summary, β-dicarbonyl anions, like ordinary enolate ions, undergo aldol-type condensations (Knoevenagel condensation) and Michael additions to α,β-unsaturated carbonyl compounds.

Summary of New Reactions

1 PREPARATION OF α-DIKETONES

$$R-\underset{H}{\underset{|}{\overset{OH}{\overset{|}{C}}}}-\overset{O}{\overset{\|}{C}}R' \xrightarrow[\text{or } Cu(O\overset{O}{\overset{\|}{C}}CH_3)_2,\ CH_3COOH]{KMnO_4,\ CH_3\overset{O}{\overset{\|}{C}}O\overset{O}{\overset{\|}{C}}CH_3} R\overset{O}{\overset{\|}{C}}-\overset{O}{\overset{\|}{C}}R'$$

2 ACYLOIN CONDENSATION

$$2\ R\overset{O}{\overset{\|}{C}}OR' \xrightarrow[2.\ H^+,\ H_2O]{1.\ Na,\ (CH_3CH_2)_2O} R\overset{OH}{\overset{|}{C}H}-\overset{O}{\overset{\|}{C}}R + 2\ R'OH$$

Intramolecular variant:

$$RO\overset{O}{\overset{\|}{C}}\frown\overset{O}{\overset{\|}{C}}OR \xrightarrow[2.\ H^+,\ H_2O]{1.\ Na,\ (CH_3CH_2)_2O} \text{cyclic } \overset{O}{\overset{\|}{C}}-\underset{H}{C}-OH + 2\ ROH$$

3 CATENANES

$$(CH_2)_n \text{ (ring)} + \text{ROC(=O)}\frown\text{C(=O)OR} \xrightarrow[2.\ H^+,\ H_2O]{1.\ Na} (CH_2)_n \text{ ring interlocked with ring containing } -C(=O)-CH(OH)-$$

Catenane

4 BENZOIN CONDENSATION

$$2\ C_6H_5CHO \xrightarrow{\text{Catalytic NaCN}} C_6H_5CH(OH)-C(=O)C_6H_5$$

5 THIAZOLIUM SALT CATALYSIS IN THE COUPLING OF ALDEHYDES

$$2\ RCHO \xrightarrow{\text{Catalytic } N\text{-}(CH_2)_{11}CH_3 \text{ thiazolium } Br^-} RCH(OH)-C(=O)R$$

6 2-OXOPROPANOIC (PYRUVIC) ACID, AN α-KETO ACID

Preparation:

$$HOOC-CH(OH)-CH(OH)-COOH \xrightarrow[-\ H_2O]{KHSO_4,\ \Delta} \left[HOOCCH{=}C(OH)COOH \rightleftharpoons HOOCCH_2C(=O)COOH\right] \xrightarrow{-\ CO_2} CH_3C(=O)COOH$$

2,3-Dihydroxybutanedioic acid (Tartaric acid) → **2-Oxopropanoic acid (Pyruvic acid)**

$$CH_3C(=O)Cl + NaCN \xrightarrow[-\ NaCl]{} CH_3C(=O)CN \xrightarrow{HCl,\ H_2O,\ 0°C} CH_3C(=O)COOH$$

Pyruvates by oxidation of lactates:

$$CH_3CH(OH)CO_2CH_2CH_3 \xrightarrow{KMnO_4,\ H_2O} CH_3C(=O)CO_2CH_2CH_3$$

Ethyl 2-hydroxypropanoate (Ethyl lactate) → **Ethyl 2-oxopropanoate (Ethyl pyruvate)**

Biological reduction:

$$CH_3C(=O)COOH \xrightarrow[\text{Stereospecific}]{\text{Lactic acid dehydrogenase}} (S)\text{-}CH_3CH(OH)COOH$$

(*S*)-(+)-2-Hydroxypropanoic (lactic) acid

7 ACTIVATION OF α-DICARBONYL COMPOUNDS

$$\underset{R}{}\overset{O^{\delta-}}{\overset{\|}{C}}{}^{\delta+}\!-\!\overset{O^{\delta-}}{\overset{\|}{C}}{}^{\delta+}\underset{R}{}$$

Nucleophilic attack

8 BENZILIC ACID REARRANGEMENT

$$R\overset{O}{\overset{\|}{C}}-\overset{O}{\overset{\|}{C}}R \xrightarrow{NaOH} R-\underset{OH}{\overset{R}{C}}-\overset{O}{\overset{\|}{C}}O^-Na^+$$

9 ALKANOYL (ACYL) ANION EQUIVALENTS

$$R-\underset{A}{\overset{A'}{C}}:^- \quad \text{synthetic equivalent of} \quad R-\overset{O}{\overset{\|}{C}}:^-$$

A = electron withdrawing, conjugating, or polarizable group

1,3-Dithiacyclohexanes (1,3-dithianes):

$$\text{1,3-dithiane} \xrightarrow[\text{2. RX}]{\text{1. } CH_3CH_2CH_2CH_2Li} \text{2-R-2-H-1,3-dithiane} \xrightarrow[\text{2. R'X}]{\text{1. } CH_3CH_2CH_2CH_2Li} \text{2-R-2-R'-1,3-dithiane} \xrightarrow{HgCl_2,\ HgO,\ CH_3OH} R\overset{O}{\overset{\|}{C}}R'$$

$$\text{1,3-dithiane} \xrightarrow[\text{2. } R\overset{O}{\overset{\|}{C}}R' \text{ 3. } HgCl_2,\ HgO]{\text{1. } CH_3CH_2CH_2CH_2Li} R\underset{R'}{\overset{OH}{C}}-\overset{O}{\overset{\|}{C}}H$$

$$\text{1,3-dithian-2-yl anion} \xrightarrow[\text{2. } HgCl_2,\ HgO,\ CH_3OH]{\text{1. } H_2C\overset{O}{-}CH_2} H\overset{O}{\overset{\|}{C}}CH_2CH_2OH$$

Synthesis of β-Dicarbonyl Compounds

10 CLAISEN AND RETRO-CLAISEN CONDENSATIONS

$$2\ R\overset{O}{\overset{\|}{C}}OR' \underset{}{\overset{Na^{+-}OR'}{\rightleftharpoons}} R\overset{O}{\overset{\|}{C}}CH_2\overset{O}{\overset{\|}{C}}OR' + R'OH$$

Acidic

11 DIECKMANN CONDENSATION

$$(CH_2)_n(CH_2CO_2R)(CO_2R) \xrightarrow{Na^{+}\,{}^{-}OR} (CH_2)_n\text{-ring: }C(=O)\text{—}CHCO_2R + ROH$$

12 DOUBLE CLAISEN CONDENSATION

$$RO_2CCH_2CH_2CH_2CO_2R + ROCCOR\ (\text{with two C=O}) \xrightarrow{Na^{+}\,{}^{-}OR} \text{cyclopentane-1,2-dione with two } RO_2C \text{ groups} + 2\ ROH$$

13 β-DIKETONE SYNTHESIS

$$\overset{O}{\overset{\|}{R C}}OR' + CH_3\overset{O}{\overset{\|}{C}}CH_3 \xrightarrow{Na^{+}\,{}^{-}OR} R\overset{O}{\overset{\|}{C}}CH_2\overset{O}{\overset{\|}{C}}CH_3 + R'OH$$

Intramolecular:

$$(CH_2)_n(\overset{O}{\overset{\|}{C}}CH_3)(CO_2R) \xrightarrow{Na^{+}\,{}^{-}OR} (CH_2)_n\text{-ring: }C(=O)\text{—}CH_2\text{—}C(=O) + ROH$$

14 3-KETO ESTER SYNTHESIS

$$R\overset{O}{\overset{\|}{C}}CH_3 + R'O\overset{O}{\overset{\|}{C}}OR' \xrightarrow{Na^{+}\,{}^{-}OR'} R\overset{O}{\overset{\|}{C}}CH_2\overset{O}{\overset{\|}{C}}OR' + R'OH$$

15 3-KETO ALDEHYDE SYNTHESIS

$$R\overset{O}{\overset{\|}{C}}CH_3 + H\overset{O}{\overset{\|}{C}}OR' \xrightarrow{Na^{+}\,{}^{-}OR'} R\overset{O}{\overset{\|}{C}}CH_2\overset{O}{\overset{\|}{C}}H + R'OH$$

3-Keto Esters as Synthetic Building Blocks

16 ENOLATE ALKYLATION

$$R\overset{O}{\overset{\|}{C}}CH_2CO_2R' \xrightarrow[2.\ R''X]{1.\ Na^{+}\,{}^{-}OR'} R\overset{O}{\overset{\|}{C}}\underset{R''}{\underset{|}{C}}HCO_2R'$$

17 3-KETO ACID DECARBOXYLATION

$$RC(=O)CH_2C(=O)OR' \xrightarrow[-\,R'O^-]{HO^-} RC(=O)CH_2C(=O)O^- \xrightarrow{H^+} RC(=O)CH_2C(=O)OH \xrightarrow[-\,CO_2]{\Delta} RC(=O)CH_3$$

18 ACETOACETIC ESTER SYNTHESIS

$$RC(=O)CH_2C(=O)OR' \xrightarrow{\text{1. NaOR}'\ \text{2. R}''\text{X}\ \text{3. HO}^-\ \text{4. H}^+,\,\Delta} RC(=O)CH_2R''$$

R″ = alkyl, alkanoyl (acyl), $CH_2C(=O)OR'''$, $CH_2C(=O)R'''$
R″X = oxacyclopropane

19 MALONIC ESTER SYNTHESIS

$$ROC(=O)CH_2C(=O)OR \xrightarrow{\text{1. NaOR}\ \text{2. R}'\text{X}\ \text{3. HO}^-\ \text{4. H}^+,\,\Delta} R'CH_2C(=O)OH$$

R′ = alkyl, alkanoyl (acyl), $CH_2C(=O)OR''$, $CH_2C(=O)R''$
R′X = oxacyclopropane

20 KNOEVENAGEL CONDENSATION

$$RC(=O)H + H_2C(CO_2R')_2 \xrightarrow{R_2NH} RCH{=}C(CO_2R')_2 + H_2O$$

21 MICHAEL ADDITION

$$CH_2{=}CHC(=O)R + H_2C(CO_2CH_3)_2 \xrightarrow{Na^{+}\,{}^{-}OCH_3} RC(=O)CH_2CH_2CH(CO_2CH_3)_2$$

Summary of Important Concepts

1 Molecules containing more than one functional group may react according to the individual properties of the reactive centers (particularly if they are well separated) or according to a combination of these properties.

2 Because alkanoyl (acyl) anions are not directly available by deprotonation of aldehydes, they have to be made as masked reactive intermediates or stoichiometric reagents by transformations of functional groups.

3 Thiamine (vitamin B_1) effects reactions in nature that are analogous to the coupling of alkanals to α-hydroxy carbonyl compounds catalyzed by thiazolium ion and the cyanide-catalyzed coupling of aromatic aldehydes.

4 β-Dicarbonyl compounds contain acidic hydrogens at the carbon between the two carbonyl groups because of the inductive electron-withdrawing effect of the two neighboring carbonyl functions and because the anions resulting from deprotonation are resonance stabilized.

5 The Claisen condensation is driven by the stoichiometric generation of a stable β-dicarbonyl anion.

6 Although mixed Claisen condensations between esters are usually not selective, they can be so with certain substrates (nonenolizable esters, intramolecular versions, ketones).

7 3-Keto acids are unstable; they decarboxylate in a concerted process through an aromatic transition state. This property, in conjunction with the nucleophilic reactivity of 3-keto ester anions, is a good basis for synthesizing substituted ketones and acids.

Problems

1 When an apparently ordinary reaction is attempted on one functional group in a difunctional compound, interference by the second functional group may cause it to fail. For each of the following equations, indicate whether the reaction is likely to proceed as written. If not, suggest what might occur instead.

(a) O ; Br_2 ; Br ; Br ; O

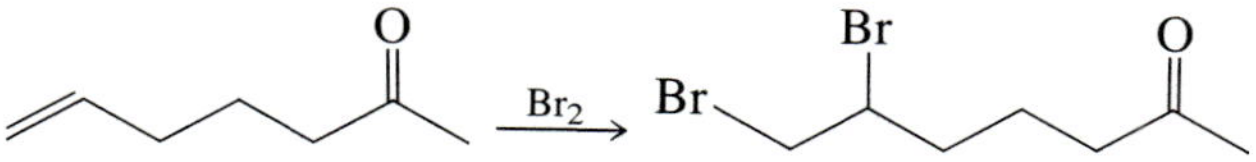

(b) O ; Br ; KCN ; O ; CN

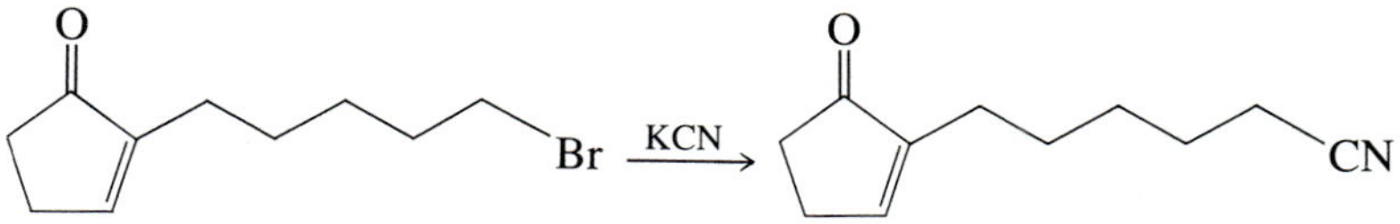

(c) COOH ; $LiAlH_4$; CH_2OH

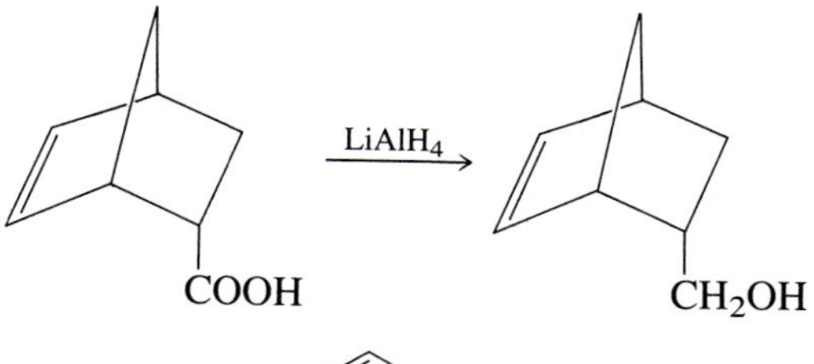

(d) OH ; HCl ; Cl

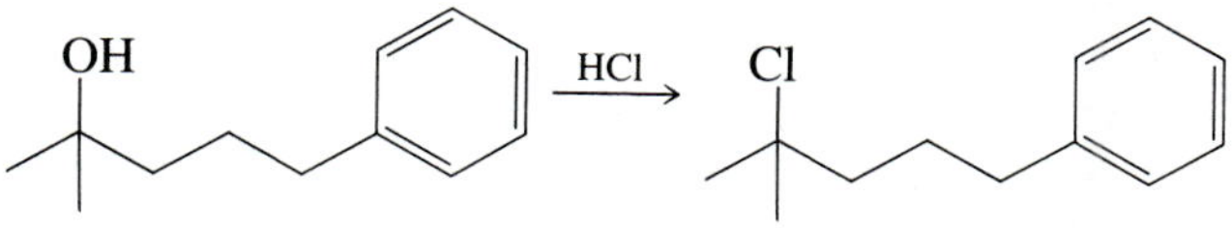

(e) O ; I ; LDA ; O

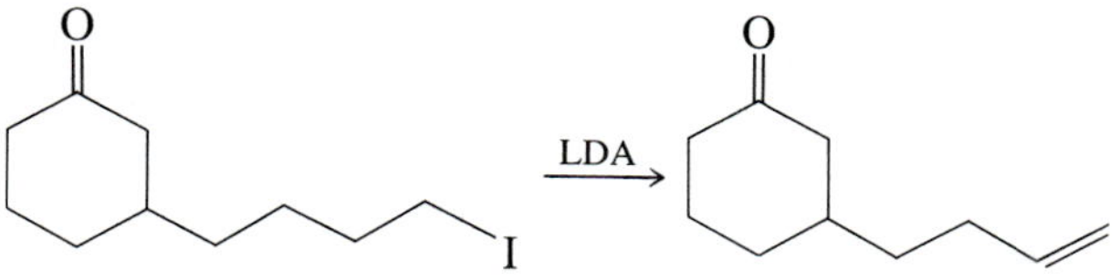

(f) H N ; COOH ; 1. $SOCl_2$ 2. $(CH_3)_2CHOH$; H N ; O ; O

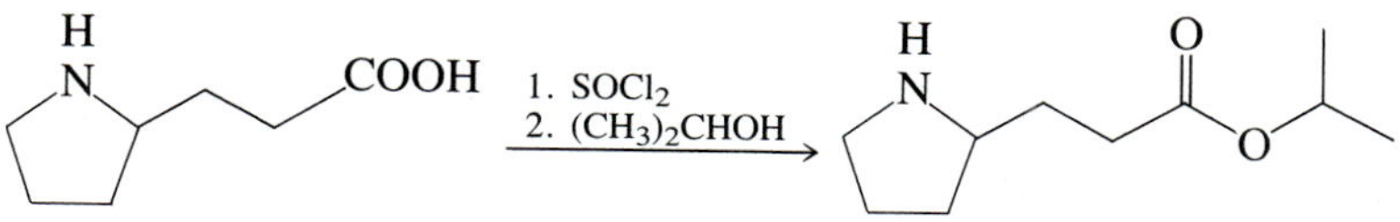

(g) 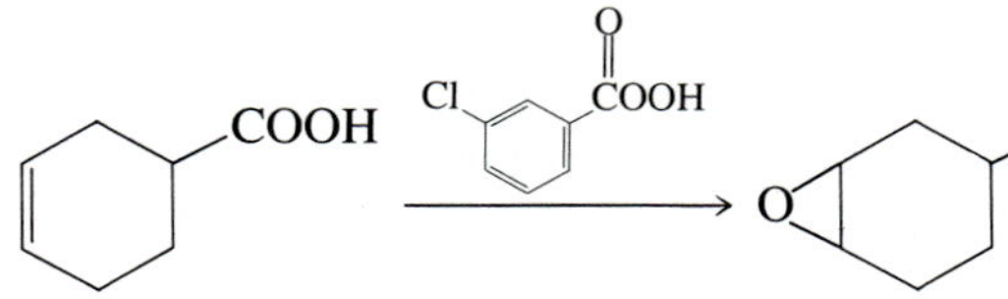

(h) 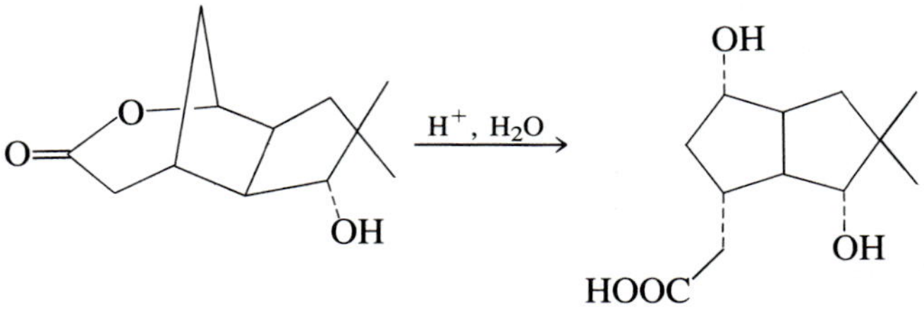

2 The sesquiterpene maaliol may be synthesized from alkene (i) through the intermediacy of hydroxy ketone (ii). Propose synthetic sequences that convert (i) into (ii) and then (ii) into maaliol.

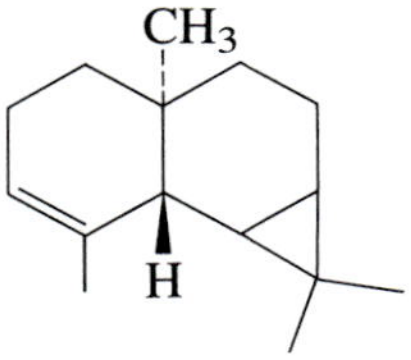

i

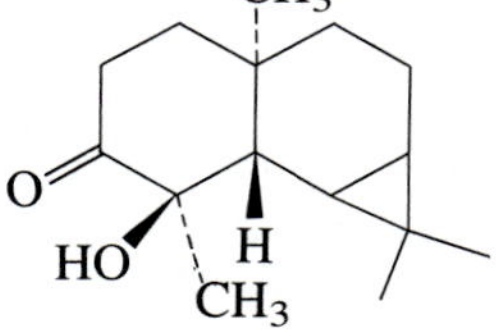

ii

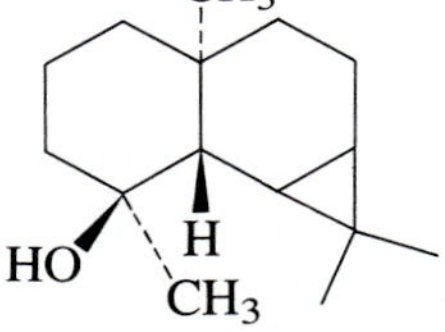

Maaliol

3 **(a)** Identify the unknowns A, found in fresh cream prior to churning, and B, possessor of the characteristic yellow color and buttery odor of butter. Here are some data:

A: MS m/z (relative abundance) = 88(molecular ion, weak), 45(100), and 43(80).
^{1}H NMR δ = 1.36(d, J = 7 Hz, 3 H), 2.18(s, 3 H), 3.73(broad s, 1 H), 4.22(q, J = 7 Hz, 1 H) ppm.
IR $\tilde{\nu}$ = 1718 and 3430 cm^{-1}.

B: MS m/z (relative abundance) = 86(17) and 43(100).
^{1}H NMR δ = 2.29(s) ppm.
IR $\tilde{\nu}$ = 1708 cm^{-1}.

(b) What kind of reaction is the conversion of A into B? Does it make sense that this should take place in the churning of cream to make butter? Explain.

(c) Outline laboratory syntheses of A and B, starting only with compounds containing two carbons.

(d) The UV spectrum of A has a λ_{max} of 271 nm, whereas that of B has a λ_{max} of 290 nm. (Extension of the latter absorption into the violet region of the visible spectrum is responsible for the yellow color of B.) Explain the difference in λ_{max}.

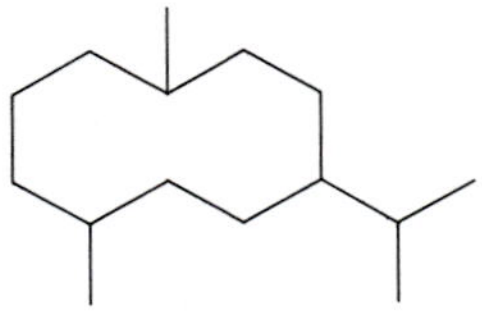

Germacrane

4 Illustrated at the left is germacrane, a naturally occurring cyclodecane derivative. Propose a short synthesis of germacrane starting from an appropriate acyclic diester.

5 Write the expected products of each of the following reactions.

(a) $CH_3OOCCH_2C(CH_3)_2CH_2CH_2C(CH_3)_2CH_2COOCH_3 \xrightarrow[2.\ H^+,\ H_2O]{1.\ Na,\ 1,2\text{-dimethylbenzene}\ (o\text{-xylene})}$

(b)  $(CH_3)_2CHCHO \xrightarrow{\text{thiazolium } N^+(CH_2)_{11}CH_3Br^-}$

(c) 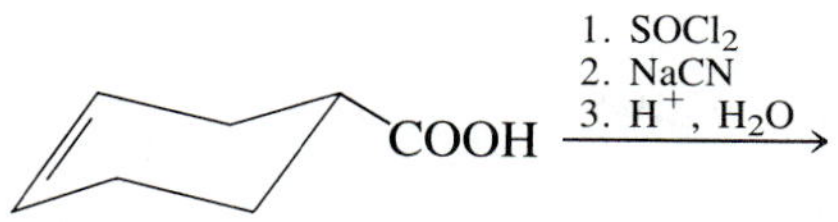 (cyclohexenyl)COOH $\xrightarrow{\text{1. } SOCl_2 \text{ 2. NaCN 3. } H^+, H_2O}$

(d) 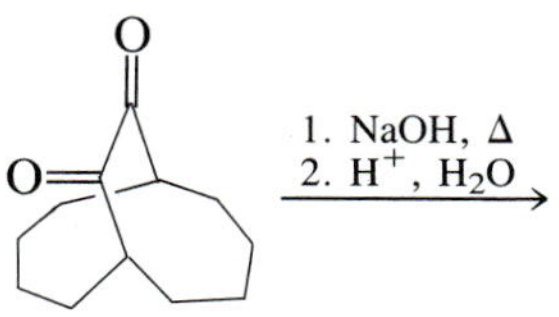 (bicyclic diketone) $\xrightarrow{\text{1. NaOH, } \Delta \text{ 2. } H^+, H_2O}$

6 Mixed ketone-ester coupling reactions are possible under the conditions of acyloin condensation. Write the product that you would expect from the reaction of sodium with the keto ester shown, followed by aqueous work-up.

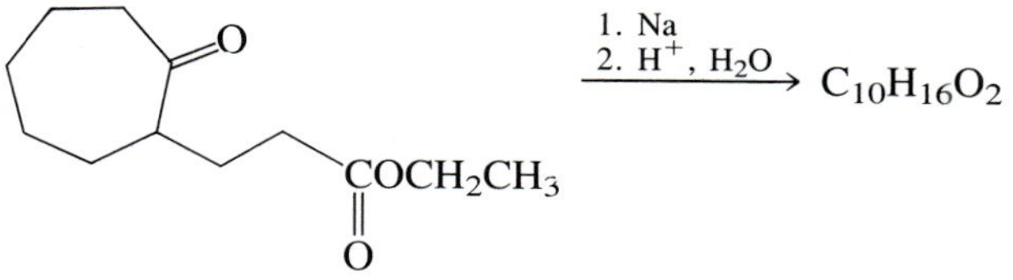

(2-oxocycloheptyl)$CH_2CH_2COCH_2CH_3$ (ester C=O) $\xrightarrow{\text{1. Na 2. } H^+, H_2O}$ $C_{10}H_{16}O_2$

7 The reaction of diphenylethanedione (benzil) with 4-methoxyphenylmagnesium chloride gives primarily the anomalous keto alcohol shown below. Explain this result mechanistically. Do you think that thermodynamics plays a role in the formation of this product?

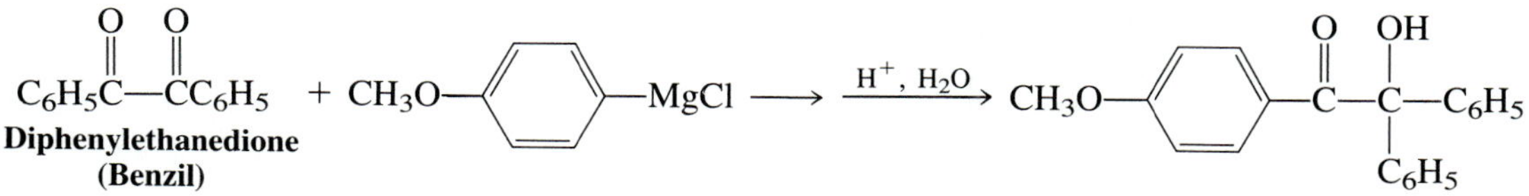

$C_6H_5\overset{O}{\overset{\|}{C}}-\overset{O}{\overset{\|}{C}}C_6H_5$ + $CH_3O-C_6H_4-MgCl \longrightarrow \xrightarrow{H^+, H_2O} CH_3O-C_6H_4-\overset{O}{\overset{\|}{C}}-C(OH)(C_6H_5)-C_6H_5$

Diphenylethanedione
(Benzil)

8 One commercial preparation of the analgesic pethidine (see Problem 21, Chapter 21) starts from carboxylic acid (i), through the intermediacy of chloro ketone (ii) and carboxylic acid (iii). Answer the questions that follow.

(a) Propose a synthetic sequence that converts (i) into (ii).

(b) The rearrangement of (ii) to (iii) in the presence of base is mechanistically related to the benzilic acid rearrangement (Section 22-2). Propose a mechanism for this rearrangement.

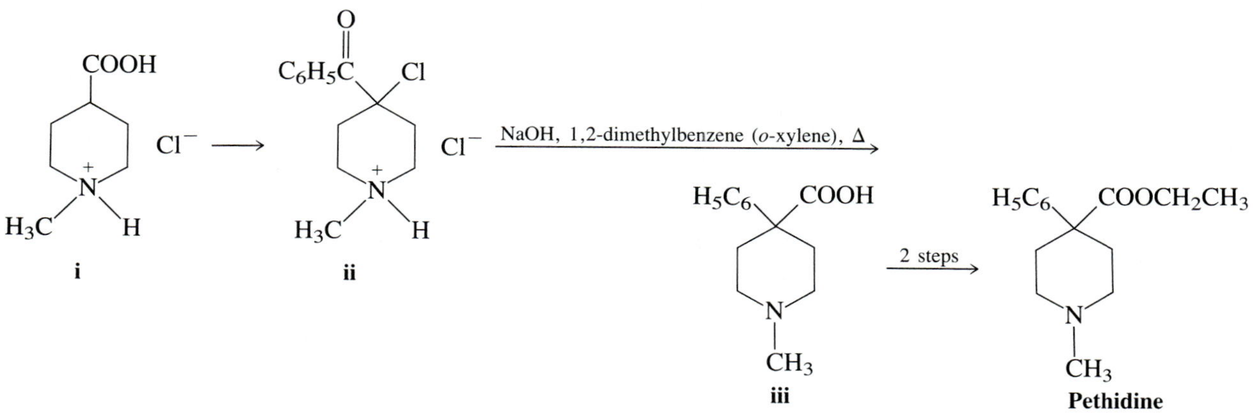

i (4-COOH, N-H, N-H_3C piperidinium Cl^-) $\longrightarrow$ **ii** (4-$C_6H_5\overset{O}{\overset{\|}{C}}$, 4-Cl, N-H, N-$H_3C$ piperidinium Cl^-) $\xrightarrow{\text{NaOH, 1,2-dimethylbenzene (}o\text{-xylene), } \Delta}$ **iii** (4-H_5C_6, 4-COOH, N-CH_3 piperidine) $\xrightarrow{\text{2 steps}}$ **Pethidine** (4-H_5C_6, 4-$COOCH_2CH_3$, N-CH_3 piperidine)

9 Write chemical equations to illustrate all primary reaction steps that can occur between a base such as ethoxide ion and a carbonyl compound such as ethanal (acetaldehyde). Explain why the carbonyl carbon is not deprotonated to any appreciable extent in this system.

10 Propose short syntheses of each of the following molecules, starting with the material indicated and making use of the benzilic acid rearrangement.

(a) 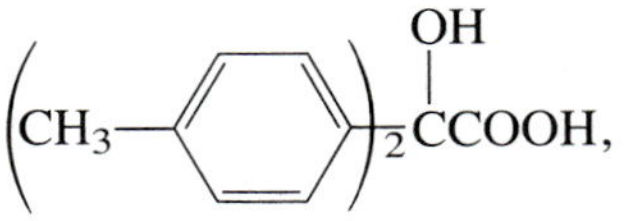 $(CH_3-C_6H_4)_2C(OH)COOH$, starting from $H_3C-C_6H_4-COOH$

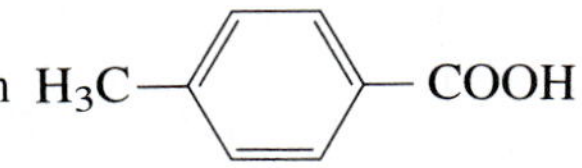

(b) $C_6H_5-CH(OH)-COOH$, starting from C_6H_5CHO

(c) $C_6H_5CH_2-C(OH)(C_6H_5)-COOH$, starting from C_6H_5CHO

11 Using methods described in section 22-3 (i.e., reverse polarization) propose a simple synthesis of each of the following molecules.

(a) $CH_2{=}CHCH(OH)C(=O)CH_2C_6H_5$

(b) 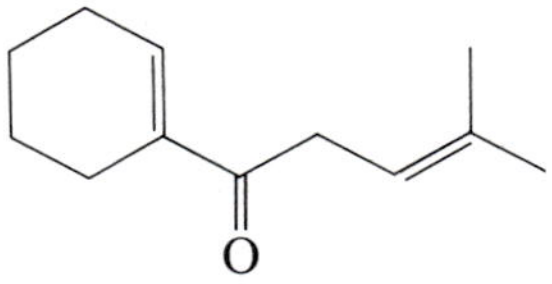 1-cyclohexenyl–C(=O)–CH$_2$–CH=C(CH$_3$)$_2$

(c) $CH_3CH(OH)CH(CH_3)CHO$

12 Propose a synthesis of ketone (iii), which was central in attempts to synthesize several antitumor agents. Start with aldehyde (i), lactone (ii), and anything else you need.

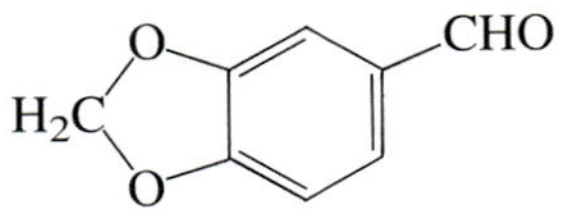

i: 3,4-methylenedioxybenzaldehyde (H$_2$C bridging two O on the ring, CHO)

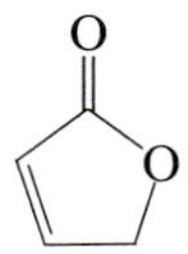

ii: unsaturated five-membered lactone (C=O, O in ring)

iii: H$_2$C(O)$_2$-aryl–C(=O)–(γ-butyrolactone ring)

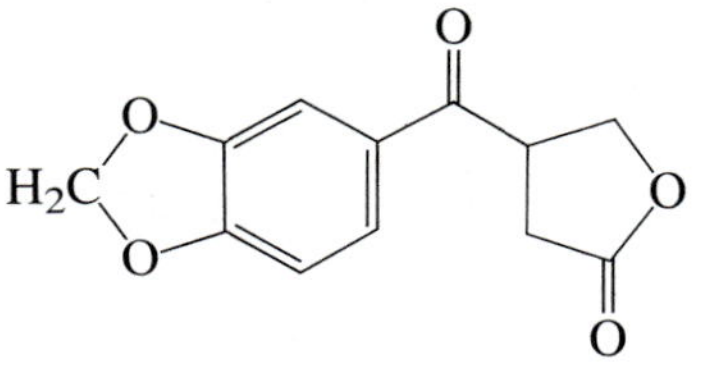

13 Write the expected results of reaction of each of the following molecules (or combinations of molecules) with $NaOCH_2CH_3$ in CH_3CH_2OH.

(a) $CH_3CH_2CH_2COOCH_2CH_3$

(b) $C_6H_5CH(CH_3)CH_2COOCH_2CH_3$

(c) $C_6H_5CH_2CH(CH_3)COOCH_2CH_3$

(d) $CH_3CH_2OC(=O)(CH_2)_4C(=O)OCH_2CH_3$

(e) $CH_3CH_2OC(=O)CH(CH_3)(CH_2)_4C(=O)OCH_2CH_3$

(f) $C_6H_5CH_2CO_2CH_2CH_3 + HCO_2CH_2CH_3$

(g) $C_6H_5CO_2CH_2CH_3 + CH_3CH_2CH_2CO_2CH_2CH_3$

(h) Cyclobutane with two $CO_2CH_2CH_3$ groups (on adjacent carbons, wedged) $+ CH_3CH_2OC(=O)CH_2CH_2C(=O)OCH_2CH_3$

(i) Benzene ring with two adjacent $CH_2CO_2CH_2CH_3$ groups $+ CH_3CH_2OC(=O)—C(=O)OCH_2CH_3$

14 The following mixed Claisen condensation works best when one of the starting materials is present in large excess.

$$CH_3CH_2C(=O)OCH_3 + (CH_3)_2CHC(=O)OCH_3 \xrightarrow{NaOCH_3,\ CH_3OH} (CH_3)_2CHC(=O)CH(CH_3)C(=O)OCH_3$$

Which of the two starting materials should be present in excess? Why? What side reaction will compete if the reagents are present in comparable amounts?

15 Suggest a synthesis of each of the following β-dicarbonyl compounds by Claisen or Dieckmann condensations.

(a) (cyclopentyl)$-CH_2C(=O)CH(C_5H_9)C(=O)OCH_2CH_3$ (CH bearing a cyclopentyl ring)

(b) $C_6H_5C(=O)CH(C_6H_5)C(=O)OCH_2CH_3$

(c) Cyclohexanone with H_3C on one α-carbon and $CO_2CH_2CH_3$ on the other α-carbon

(d) Cycloheptane-1,4-dione ring bearing two adjacent H_3C groups and two $CO_2CH_2CH_3$ groups

(e) $HC{-}CCH_2COCH_2CH_3$ (with C=O at each of the first, second and third carbonyl carbons)

(f) $C_6H_5CCH_2CC_6H_5$ (with C=O at both carbonyl carbons)

(g) $CH_3CH_2OCCH_2COCH_2CH_3$ (with C=O at both carbonyl carbons)

(h) cyclopropyl–CCH_2CCH_3 (with C=O at both carbonyl carbons)

(i) 2-formylcycloheptanone (CHO)

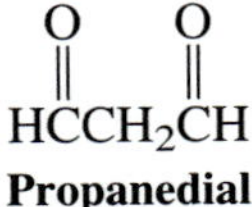

Propanedial

16 Do you think that propanedial can be easily prepared by a simple Claisen condensation? Why or why not?

17 Nootkatone is found in grapefruit. Fill in the necessary steps in the following scheme to make nootkatone from 4-(1-methylethenyl)cyclohexanone.

Nootkatone

18 Devise a preparation of each of the following ketones, using the acetoacetic ester synthesis.

(a) $CH_3CCH_2CH_2CHCH_3$ (C=O at C-2; CH_3 on C-5)

(b) $CH_3CCHCH_2CH{=}CH_2$ (C=O at C-2; H_2C–phenyl on C-3)

(c) cyclobutyl methyl ketone

(d) $CH_3CCHCH_2CH_3$ (C=O at C-2) with $CH_2COCH_2CH_3$ (C=O) on C-3

19 The following ketones cannot be synthesized by the acetoacetic ester method (Why?), but they can be prepared by a modified version of it. The modification includes the preparation (by Claisen condensation) and use of an appropriate 3-keto ester, $RCCH_2COCH_2CH_3$ (with C=O at both carbonyl carbons), containing an R group that appears in the final ketone product.

Synthesize each of the following ketones. For each, show the structure and synthesis of the necessary 3-keto ester as well.

(a) $CH_3CH_2\overset{O}{\overset{\|}{C}}CH_2CH_3$

(b) 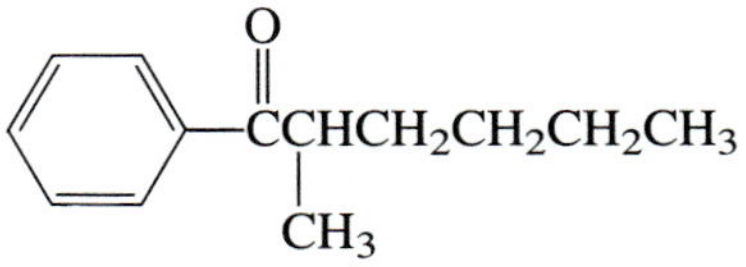 $C_6H_5\overset{O}{\overset{\|}{C}}\underset{CH_3}{\underset{|}{C}}HCH_2CH_2CH_2CH_3$

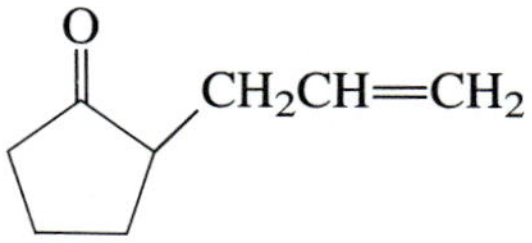

(c) 2-($CH_2CH{=}CH_2$)cyclopentanone (Use a Dieckmann condensation.)

(d) 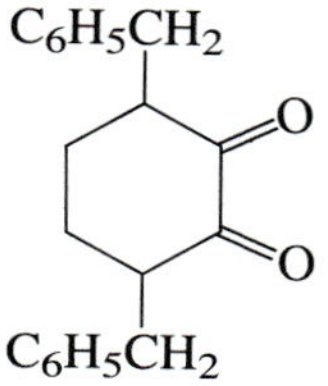 cyclohexane-1,2-dione with $C_6H_5CH_2$ groups at C3 and C6 (Use a double Claisen condensation.)

20 Devise a synthesis for each of the four following compounds, using the malonic ester synthesis.

(a) 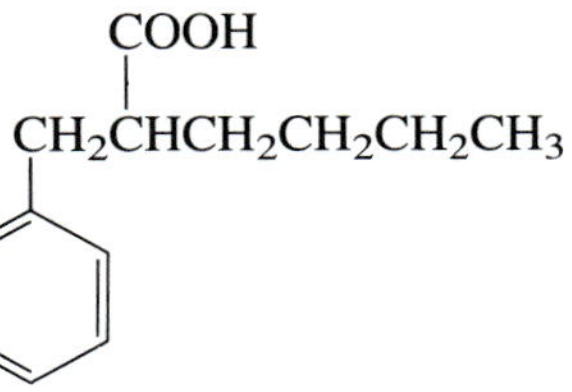 $C_6H_5CH_2\underset{}{\overset{COOH}{\overset{|}{C}}}HCH_2CH_2CH_2CH_3$

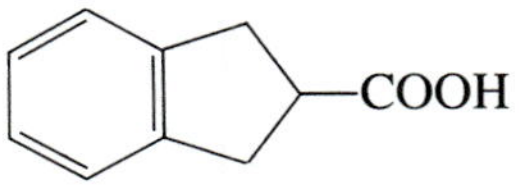

(b) $H_2C{-}COOH$ / $H_2C{-}COOH$ (two H_2C joined by a bond)

(c) indan-2-yl–COOH

(d) 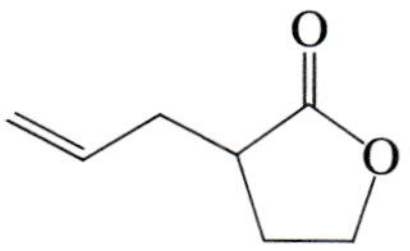 2-allyl-γ-butyrolactone (See Exercise 22-16.)

21 What would you expect if the product of Exercise 22-16 (2-ethanoyl-γ-butyrolactone) were heated in the presence of aqueous acid for an extended period of time?

22 What class of compounds results from treatment of the anion of a substituted malonic ester with an alkanoyl (acyl) halide, followed by acid hydrolysis and heating?

$$RCH(CO_2CH_2CH_3)_2 \xrightarrow[]{\text{1. NaH, } C_6H_6 \quad \text{2. } R'COCl \quad \text{3. } H^+, H_2O, \Delta}$$

Illustrate for $R = C_6H_5CH_2$ and $R' = (CH_3)_3C$. Indicate which parts of the final product molecule come from which starting compounds. (For a related reaction, see Exercise 22-14c.)

23 In Section 17-12, the biosynthesis of long-chain carboxylic acids (fatty acids) through the intermediate acetyl coenzyme A (acetyl CoA) was presented. From the information in the present chapter, formulate a plausible mechanism that explains one of the key steps in the process, the coupling of protein-bound alkanoyl and malonyl units (the step that elongates the chain by two carbons):

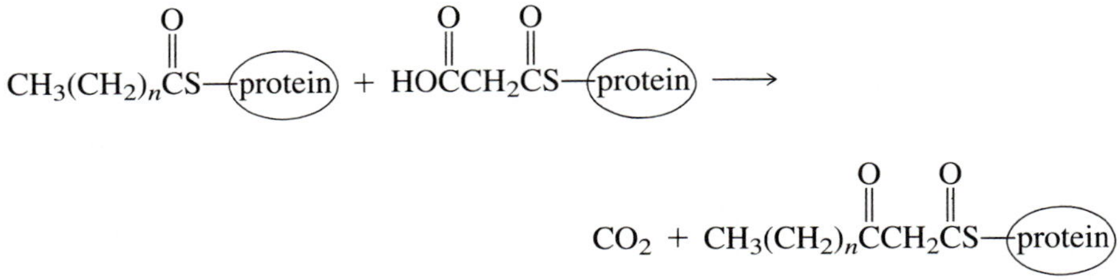

$$CH_3(CH_2)_n\overset{O}{\overset{\|}{C}}S\text{—(protein)} + HO\overset{O}{\overset{\|}{C}}CH_2\overset{O}{\overset{\|}{C}}S\text{—(protein)} \longrightarrow$$

$$CO_2 + CH_3(CH_2)_n\overset{O}{\overset{\|}{C}}CH_2\overset{O}{\overset{\|}{C}}S\text{—(protein)}$$

24 Use the methods described in Section 22-5, with other reactions if necessary, to synthesize each of the following compounds. In each case, your starting materials should include one aldehyde or ketone and one β-dicarbonyl compound.

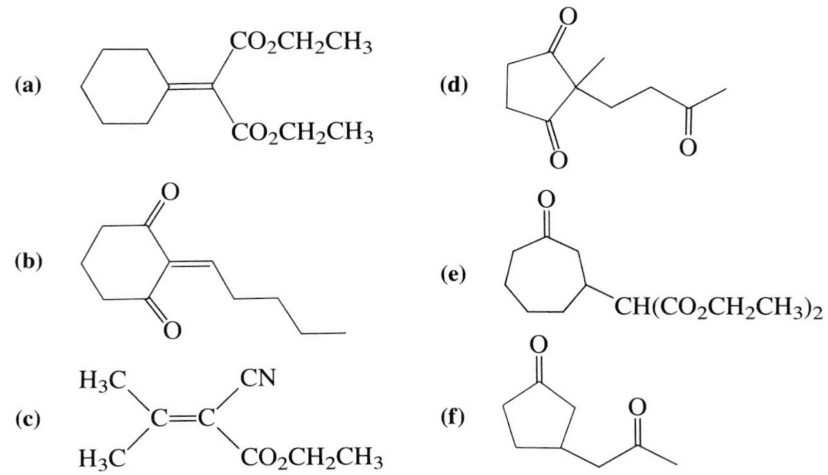

(a) Cyclohexylidene with $C(CO_2CH_2CH_3)_2$

(b) 2-Butylidene-1,3-cyclohexanedione

(c) $(H_3C)_2C{=}C(CN)CO_2CH_2CH_3$

(d) 2-Methyl-2-(3-oxobutyl)-1,3-cyclopentanedione

(e) 2-Oxocycloheptyl–$CH(CO_2CH_2CH_3)_2$

(f) 2-(2-Oxopropyl)cyclopentanone

(Hint: A decarboxylation is necessary.)

25 Write out, in full detail, the mechanism of the Michael addition of malonic ester to 3-buten-2-one in the presence of ethoxide ion. Be sure to indicate all steps that are reversible. Does the overall reaction appear to be exo- or endothermic? Explain why only a catalytic amount of base is necessary.

26 Some of the most important building blocks for synthesis are very simple molecules. Although cyclopentanone or cyclohexanone are readily commercially available, an understanding of how they might be made from simpler molecules is instructive. The following are possible retrosynthetic analyses for both of these ketones. Using them as a guide, write out a synthesis of each ketone from the indicated starting materials.

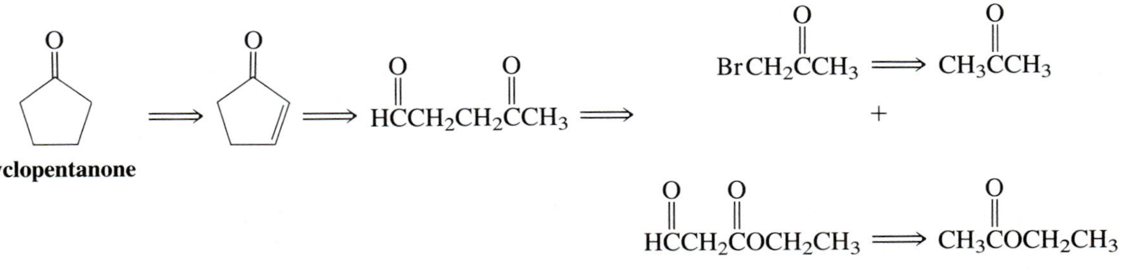

Cyclopentanone $\Longrightarrow$ 2-cyclopentenone $\Longrightarrow H\overset{O}{\overset{\|}{C}}CH_2CH_2\overset{O}{\overset{\|}{C}}CH_3 \Longrightarrow$

$$BrCH_2\overset{O}{\overset{\|}{C}}CH_3 \Longrightarrow CH_3\overset{O}{\overset{\|}{C}}CH_3$$

$$+$$

$$H\overset{O}{\overset{\|}{C}}CH_2\overset{O}{\overset{\|}{C}}OCH_2CH_3 \Longrightarrow CH_3\overset{O}{\overset{\|}{C}}OCH_2CH_3$$

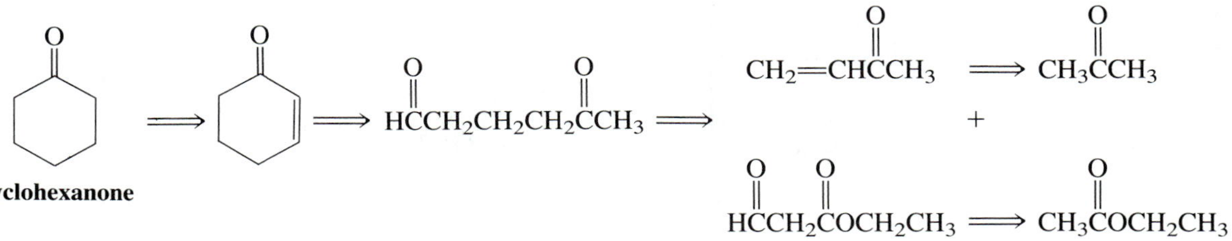

Cyclohexanone $\Longrightarrow$ 2-cyclohexenone $\Longrightarrow$ $HCCH_2CH_2CH_2CCH_3$ (O above each C) $\Longrightarrow$ $CH_2{=}CHCCH_3$ (C=O) $\Longrightarrow$ CH_3CCH_3 (C=O) + $HCCH_2COCH_2CH_3$ (O O) $\Longrightarrow$ $CH_3COCH_2CH_3$ (C=O)

27 Using the methods described in this chapter, design a multistep synthesis of each of the following molecules, making use of the indicated building blocks as the sources of all the carbon atoms in your final product.

(a) 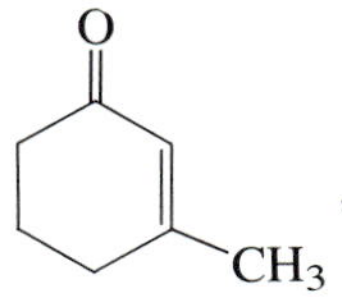 3-methyl-2-cyclohexenone (CH_3), from $CH_3CO_2CH_2CH_3$ and $CH_3COCH{=}CH_2$

(b) 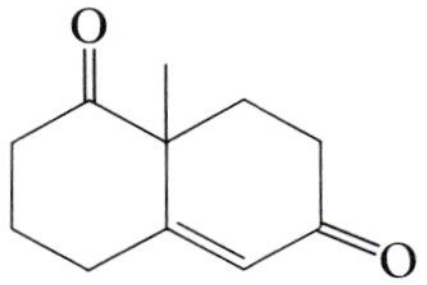 bicyclic enedione (O, O), from CH_3I, $CH_2(CO_2CH_2CH_3)_2$, and $CH_3CCH{=}CH_2$ (C=O)

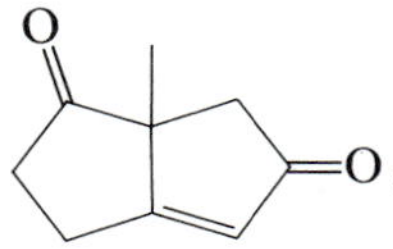

(c) bicyclic enedione (O, =O), from CH_3I, $CH_2(CO_2CH_2CH_3)_2$, and $BrCH_2CCH_3$ (C=O)

Hints for parts **b** and **c:** First make 1,3-cyclohexanedione and 1,3-cyclopentanedione, respectively.

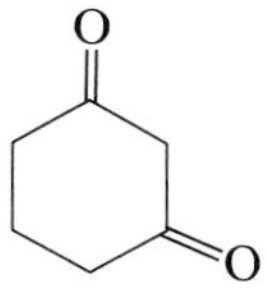

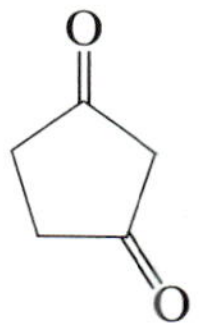

28 By means of either general equations or suitable examples, illustrate how the techniques presented in this chapter can be used to synthesize

(a) 1,4-dicarboxylic acids,
(b) 4-keto acids,
(c) 1,4-diketones,
(d) 1,5-dicarboxylic acids,
(e) 5-keto acids, and
(f) 1,5-diketones.

29 A short construction of the steroid skeleton (part of a total synthesis of the hormone estrone) is shown here. Formulate mechanisms for each of the steps.

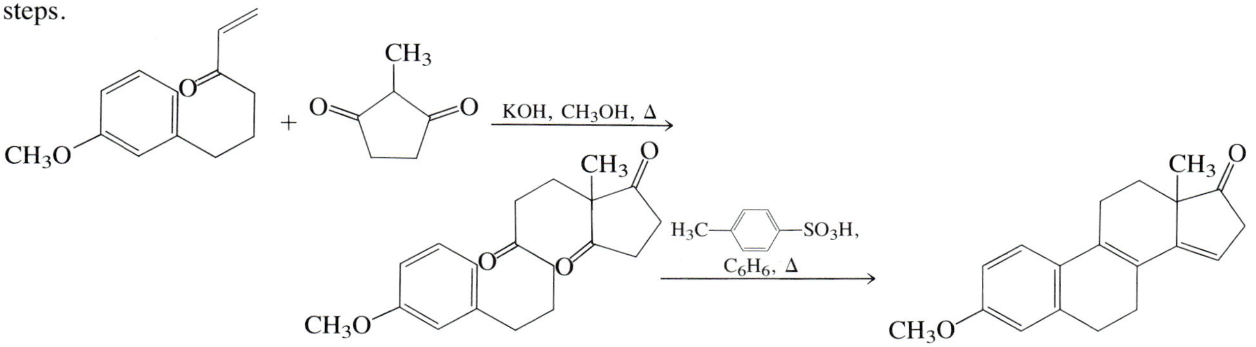

CH_3O–(vinyl ketone) + 2-methyl-1,3-cyclopentanedione (CH_3, O, O) $\xrightarrow{KOH,\ CH_3OH,\ \Delta}$ (CH_3, O, O, O; CH_3O) $\xrightarrow[C_6H_6,\ \Delta]{H_3C-C_6H_4-SO_3H,}$ (CH_3, O; CH_3O)

Hint: Processes similar to those taking place in the second step are presented in Problem 13 of Chapter 16 and in Problem 16 of Chapter 19.

30 The Knoevenagel condensation, combined with decarboxylation, is a simple way to make α,β-unsaturated acids, aldehydes, ketones, and nitriles. Show how you would use this method to convert benzenecarboxaldehyde (benzaldehyde) into each of the following molecules. Use a different Knoevenagel condensation for each example.

(a) C_6H_5—CH=CHCO$_2$H

Cinnamic acid

(b) C_6H_5—CH=CHCHO

Cinnamaldehyde

(c) C_6H_5—CH=CHC≡N

Cinnamonitrile

CHAPTER 23

Carbohydrates

POLYFUNCTIONAL COMPOUNDS IN NATURE

Carbohydrates, a very important class of naturally occurring chemicals, are materials that give structure to plants, flowers, vegetables, and trees. In addition, carbohydrates function as chemical energy-storage systems; they are metabolized to water, carbon dioxide, and heat or other energy. As such, they constitute an important source of food. Finally, they serve as the building units of fats (Sections 17-12 and 18-4) and nucleic acids (Section 27-7). Cellulose, starch, and ordinary table sugar are carbohydrates. Like glucose, $C_6(H_2O)_6$, many of the simple building blocks of complex carbohydrates have the general formula $C_n(H_2O)_n$.

In nature carbohydrates are produced primarily by a reaction sequence called **photosynthesis.** In this process, sunlight impinging on the chlorophyll of green plants is absorbed, and the photochemical energy thus obtained is used to convert carbon dioxide and water into oxygen and the polyfunctional structure of carbohydrates.

Photosynthesis of Glucose in Green Plants

$$6\ CO_2 + 6\ H_2O \underset{\text{Released metabolic energy}}{\overset{\text{Sunlight, chlorophyll}}{\rightleftharpoons}} \underset{\textbf{Glucose}}{C_6(H_2O)_6} + 6\ O_2$$

The detailed mechanism of this transformation is complicated and takes many steps, the first of which is the absorption of one quantum of light by the extended π system (Chapter 14) of chlorophyll.

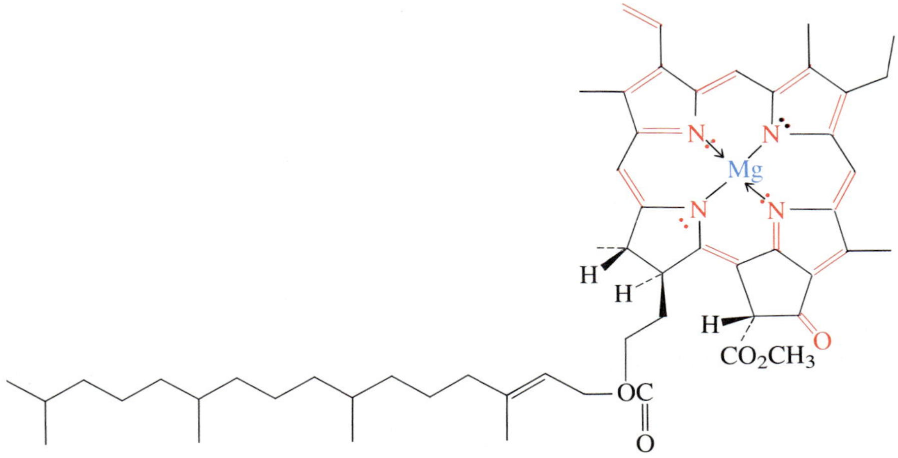

Chlorophyll a

The mechanism of the enzymatic degradation of carbohydrates has been and still is the subject of extensive research. The cycle of photosynthesis and carbohydrate metabolism is a beautiful example of how nature reuses its resources. First, CO_2 and water are consumed to convert solar energy into chemical energy and oxygen. When an organism needs some of the stored energy, it is generated by conversion of carbohydrate into CO_2 and water, using up roughly the same amount of oxygen originally liberated.

This chapter first describes the structure and naming of the most simple carbohydrates, the sugars. It then discusses their chemistry, which is governed by the presence of carbonyl and hydroxy functions along carbon chains of various lengths. Several methods useful in sugar synthesis and their structural analysis are then introduced, procedures that allow chains to be lengthened or shortened. Finally, it describes the various types of carbohydrates in nature.

23-1 The Names and Structures of Carbohydrates

The simplest carbohydrates are the sugars, or **saccharides.** With increasing chain length, sugars contain an increasing number of carbon-based stereocenters, giving rise to a multitude of diastereomers. Fortunately for us chemists, nature deals mainly with only one of the possible series of enantiomers. Because sugars are polyhydroxy carbonyl compounds, they form cyclic hemiacetals, affording additional structural and chemical variety.

Sugars Are Classified as Aldoses and Ketoses

Carbohydrate is the general name for the monomeric *(monosaccharides),* dimeric *(disaccharides),* trimeric *(trisaccharides),* oligomeric *(oligosaccharides),* and polymeric *(polysaccharides)* form of sugars (*saccharum,* Latin, sugar). A monosaccharide is an aldehyde or ketone containing at least two additional hydroxy groups. Thus, the two simplest members of this class of compounds are 2,3-dihydroxypropanal (glyceraldehyde) and 1,3-dihydroxypropanone (1,3-dihydroxyacetone).

$$\begin{array}{c} \mathrm{CHO} \\ | \\ \mathrm{H{-}C{-}OH} \\ | \\ \mathrm{CH_2OH} \end{array} \qquad \begin{array}{c} \mathrm{CH_2OH} \\ | \\ \mathrm{C{=}O} \\ | \\ \mathrm{CH_2OH} \end{array}$$

2,3-Dihydroxypropanal (Glyceraldehyde)
An aldotriose

1,3-Dihydroxypropanone (1,3-Dihydroxyacetone)
A ketotriose

Aldehydic sugars are classified as **aldoses;** those with a ketone function are called **ketoses.** Depending on their chain length, sugars are labeled *trioses* (3 carbons), *tetroses* (4 carbons), *pentoses* (5 carbons), *hexoses* (6 carbons), and so on. Therefore, 2,3-dihydroxypropanal (glyceraldehyde) is an aldotriose, whereas 1,3-dihydroxypropanone is a ketotriose.

Glucose, also known as dextrose, blood sugar, or grape sugar (*glykys,* Greek, sweet), is a pentahydroxyhexanal and hence belongs to the class of aldohexoses. It occurs naturally in many fruits and plants and in concentrations ranging from 0.08% to 0.1% in human blood. A corresponding isomeric ketohexose is *fructose,* the sweetest natural sugar (some synthetic sugars are sweeter), which also occurs in many fruits (*fructus,* Latin, fruit) and in honey. Another important natural sugar is the aldopentose *ribose,* which constitutes a building block of the ribonucleic acids (Section 27-7). The empirical formula for all these sugars is $C_n(H_2O)_n$, which is equivalent to hydrated carbon. This is one of the reasons why the compounds in this class are called carbohydrates.

A disaccharide is derived from two monosaccharides by formation of an ether (usually, acetal) bridge. Hydrolysis regenerates the monosaccharide. Ether formation between a mono- and a disaccharide results in a trisaccharide, and repetition of this process eventually produces a natural polymer (polysaccharide). Such polymeric carbohydrates constitute the basic framework of cellulose and starch (Section 23-4).

$$\begin{array}{c} \mathrm{CHO} \\ | \\ \mathrm{H{-}C{-}OH} \\ | \\ \mathrm{HO{-}C{-}H} \\ | \\ \mathrm{H{-}C{-}OH} \\ | \\ \mathrm{H{-}C{-}OH} \\ | \\ \mathrm{CH_2OH} \end{array}$$

Glucose
An aldohexose

$$\begin{array}{c} \mathrm{CH_2OH} \\ | \\ \mathrm{C{=}O} \\ | \\ \mathrm{HO{-}C{-}H} \\ | \\ \mathrm{H{-}C{-}OH} \\ | \\ \mathrm{H{-}C{-}OH} \\ | \\ \mathrm{CH_2OH} \end{array}$$

Fructose
A ketohexose

$$\begin{array}{c} \mathrm{CHO} \\ | \\ \mathrm{H{-}C{-}OH} \\ | \\ \mathrm{H{-}C{-}OH} \\ | \\ \mathrm{H{-}C{-}OH} \\ | \\ \mathrm{CH_2OH} \end{array}$$

Ribose
An aldopentose

EXERCISE 23-1

To which class of sugars do the following monosaccharides belong?

(a)
$$\begin{array}{c} \mathrm{CHO} \\ | \\ \mathrm{HCOH} \\ | \\ \mathrm{HCOH} \\ | \\ \mathrm{CH_2OH} \end{array}$$
Erythrose

(b)
$$\begin{array}{c} \mathrm{CHO} \\ | \\ \mathrm{HOCH} \\ | \\ \mathrm{HOCH} \\ | \\ \mathrm{HCOH} \\ | \\ \mathrm{CH_2OH} \end{array}$$
Lyxose

(c)
$$\begin{array}{c} \mathrm{CH_2OH} \\ | \\ \mathrm{C{=}O} \\ | \\ \mathrm{HOCH} \\ | \\ \mathrm{HCOH} \\ | \\ \mathrm{CH_2OH} \end{array}$$
Xylulose

Most Sugars Are Chiral and Optically Active

With the exception of 1,3-dihydroxypropanone, all the sugars mentioned so far contain carbon-based stereocenters. The simplest chiral sugar is 2,3-dihydroxypropanal (glyceraldehyde), with one asymmetric carbon. Its dextrorotatory form is found to be *R*, the levorotatory enantiomer *S*, as shown in the Fischer projections of the molecule. Such projections are used extensively to represent sugars. A review of Section 5-4 will remind you of this notation.

Fischer Projections of the Two Enantiomers of 2,3-Dihydroxypropanal (Glyceraldehyde)

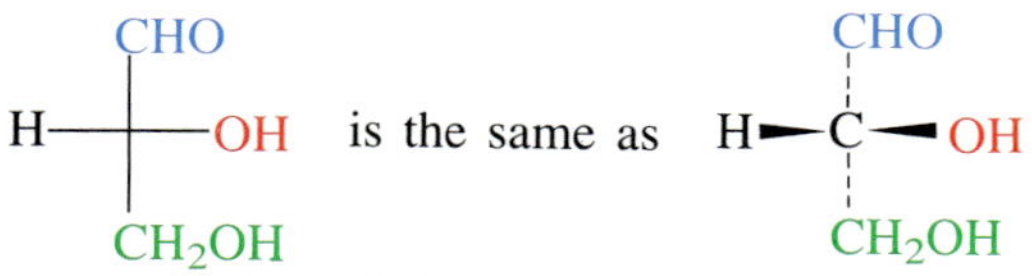

(*R*)-(+)-2,3-Dihydroxypropanal
[D-(+)-Glyceraldehyde]
$[\alpha]_D^{25°C} = +8.7°$

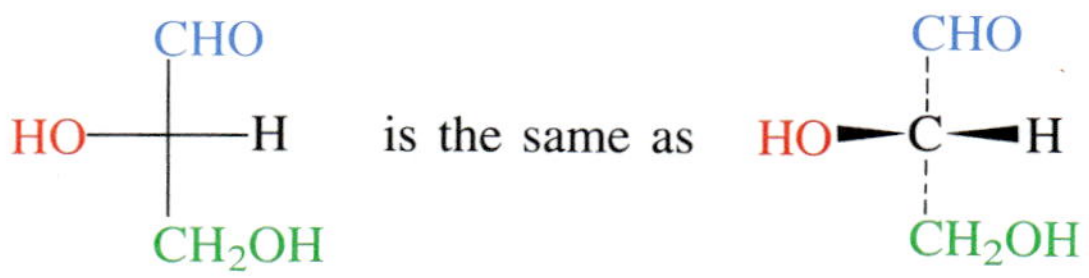

(*S*)-(−)-2,3-Dihydroxypropanal
[L-(−)-Glyceraldehyde]
$[\alpha]_D^{25°C} = -8.7°$

Even though *R* and *S* nomenclature is perfectly satisfactory for naming sugars, an older system is still in general use. It was developed before the absolute configuration of sugars had been established, and it relates all sugars to 2,3-dihydroxypropanal (glyceraldehyde). Instead of the letters *R* and *S*, it uses the prefixes D for the (+) enantiomer of glyceraldehyde and L for the (−) enantiomer (Section 5-3). Those monosaccharides whose highest-numbered (i.e., farthest from the aldehyde or keto group) stereocenter has the same absolute configuration as that of D-(+)-2,3-dihydroxypropanal [D-(+)-glyceraldehyde] are then labeled D, whereas those with opposite configuration are named L. Two diastereomers that differ only in this way at one stereocenter are also called **epimers.**

Designation of a D and an L Sugar

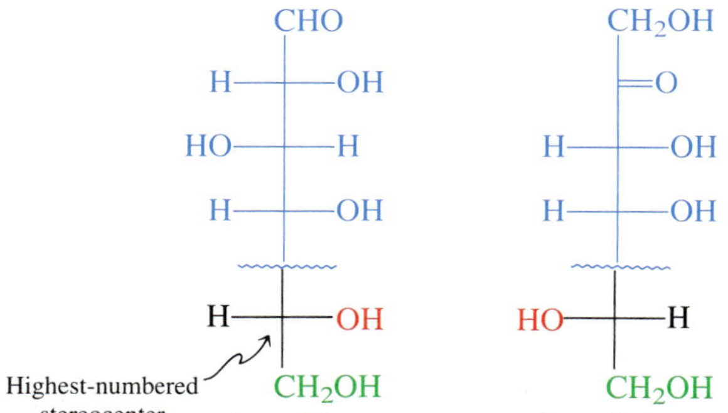

A D aldose

An L ketose

The D,L nomenclature divides the family of the sugars into two groups. As the number of stereocenters increases, there is an increasing number of stereoisomers. For example, the aldotetrose 2,3,4-trihydroxybutanal has two stereocenters and hence may exist as four stereoisomers: two diastereomers, each as a pair of enantiomers. Like many natural products, these diastereomers have common names that are often used in the literature, mainly because the complexity of these molecules leads to long systematic names. This chapter will therefore deviate from our usual procedure of labeling all molecules in a systematic manner. The isomer of 2,3,4-trihydroxybutanal with 2*R*,3*R* configuration is called *erythrose;* its diastereomer, *threose*. Note that each of these has two enantiomers, one belonging to the family of the D sugars, its mirror image to the L sugars. The sign of the optical rotation is not correlated with the D and L label (just as in the *R,S* notation; see Section 5-3). For example, D-glyceraldehyde is dextrorotatory, but D-erythrose is levorotatory.

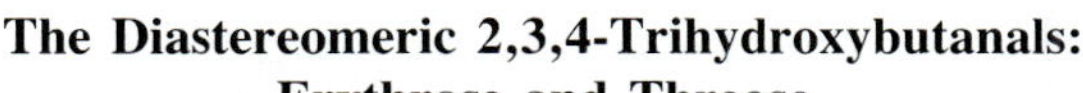

The Diastereomeric 2,3,4-Trihydroxybutanals: Erythrose and Threose

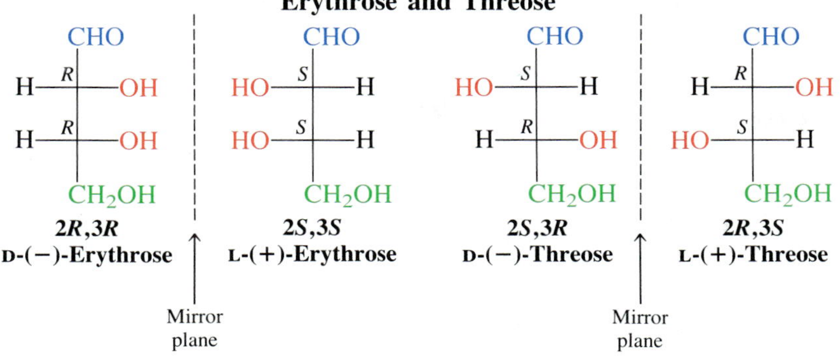

An aldopentose has three stereocenters and hence $2^3 = 8$ stereoisomers. There are $2^4 = 16$ such isomers in the group of aldohexoses. Why then use the D,L nomenclature even though it designates the absolute configuration of only one chiral center? Probably because *almost all naturally occurring sugars have the D configuration*. Evidently somewhere in the structural evolution of the sugar molecules, nature "chose" only one configuration for one end of the chain. The amino acids are another example of such selectivity (Chapter 27).

Figure 23-1 shows the series of D-aldoses up to the aldohexoses, with their signs of rotation and common names. Fischer projections are used throughout.

FIGURE 23-1
The D-aldoses (up to the aldohexoses).

D-(+)-Glyceraldehyde

D-(−)-Erythrose D-(−)-Threose

D-(−)-Ribose D-(−)-Arabinose D-(+)-Xylose D-(−)-Lyxose

D-(+)-Allose D-(+)-Altrose D-(+)-Glucose D-(+)-Mannose D-(−)-Gulose D-(−)-Idose D-(+)-Galactose D-(+)-Talose

FIGURE 23-2

The D-ketoses (up to the ketohexoses).

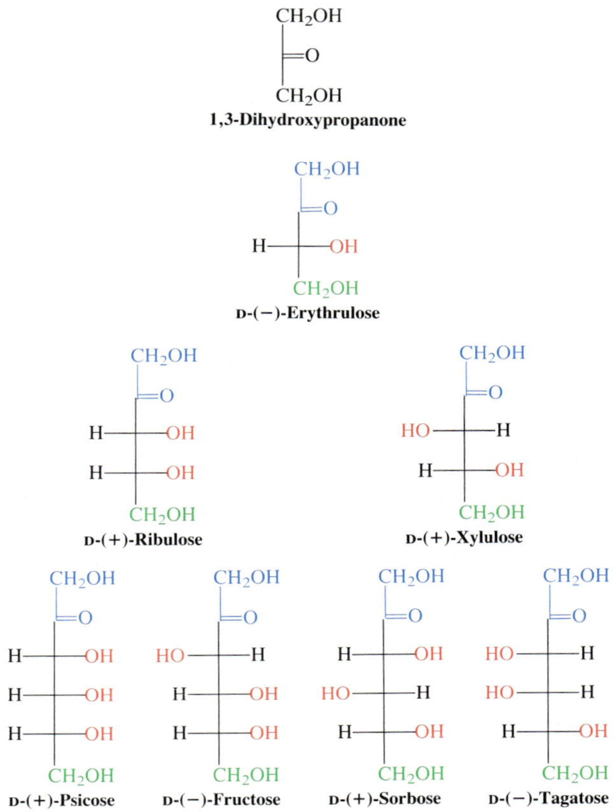

CH_2OH
=O
CH_2OH
1,3-Dihydroxypropanone

CH_2OH
=O
H—OH
CH_2OH
D-(−)-Erythrulose

CH_2OH
=O
H—OH
H—OH
CH_2OH
D-(+)-Ribulose

CH_2OH
=O
HO—H
H—OH
CH_2OH
D-(+)-Xylulose

CH_2OH
=O
H—OH
H—OH
H—OH
CH_2OH
D-(+)-Psicose

CH_2OH
=O
HO—H
H—OH
H—OH
CH_2OH
D-(−)-Fructose

CH_2OH
=O
H—OH
HO—H
H—OH
CH_2OH
D-(+)-Sorbose

CH_2OH
=O
HO—H
HO—H
H—OH
CH_2OH
D-(−)-Tagatose

To avoid confusion, these are drawn in a standard way: the carbon chain extends vertically, and the aldehyde terminus is placed at the top. In this convention, the hydroxy group at the highest-numbered stereocenter (at the bottom) points to the right. Figure 23-2 shows the analogous series of ketoses.

EXERCISE 23-2

(a) Give a systematic name for (a) D-(−)-ribose and (b) D-(+)-glucose. Remember to assign the *R* and *S* configuration at each stereocenter.

EXERCISE 23-3

Redraw the dashed-wedged line structure of sugar A as a Fischer projection and find its common name in Figure 23-1.

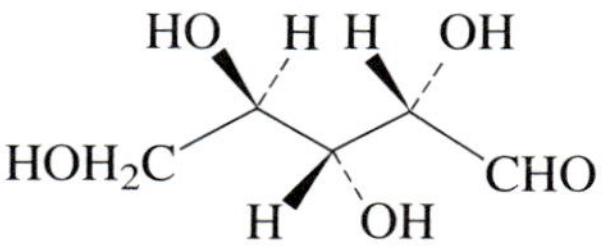

HO H H OH
HOH_2C CHO
H OH
A

Sugars Form Intramolecular Hemiacetals

So far, the structures of monosaccharides have been shown in the abstract form of Fischer projections. A more realistic way would be by the use of dashed and wedged line structures.

Fischer Projection and Dashed-Wedged Line Structures for D-(+)-Glucose

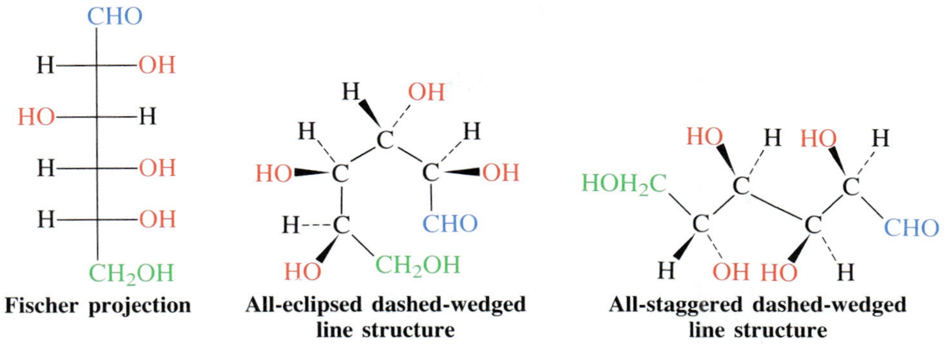

Fischer projection **All-eclipsed dashed-wedged line structure** **All-staggered dashed-wedged line structure**

It takes care to convert a Fischer projection into a dashed-wedged line structure (and vice versa). Recall (Section 5-4) that the Fischer projection represents the molecule in an *all-eclipsed* arrangement. It can be translated into an all-eclipsed dashed-wedged line picture as shown, by drawing the carbon chain in semicircular form. In the preceding illustration, the aldehyde group is on the right and we read in a counterclockwise direction along the stem. This convention positions the substituents on the right of the corresponding Fischer projection above the plane of the page pointing toward the viewer. Rearrange the structure to the all-staggered configuration, and note that the substituent groups originally all pointing to one side of the molecule are now directed alternately toward and away from the viewer. (Use molecular models to visualize the stereochemistry of D-glucose.)

Sugars are hydroxy carbonyl compounds that should be capable of intramolecular hemiacetal formation (see Section 15-5). Indeed, glucose and the other hexoses and pentoses exist as an equilibrium mixture with their cyclic hemiacetal isomers, in which the latter strongly predominate. Although in principle any one of the five hydroxy groups could add to the carbonyl group of the aldehyde, six-membered ring formation is usually preferred, but five-membered rings also are known.

Cyclic Hemiacetal Formation by Glucose

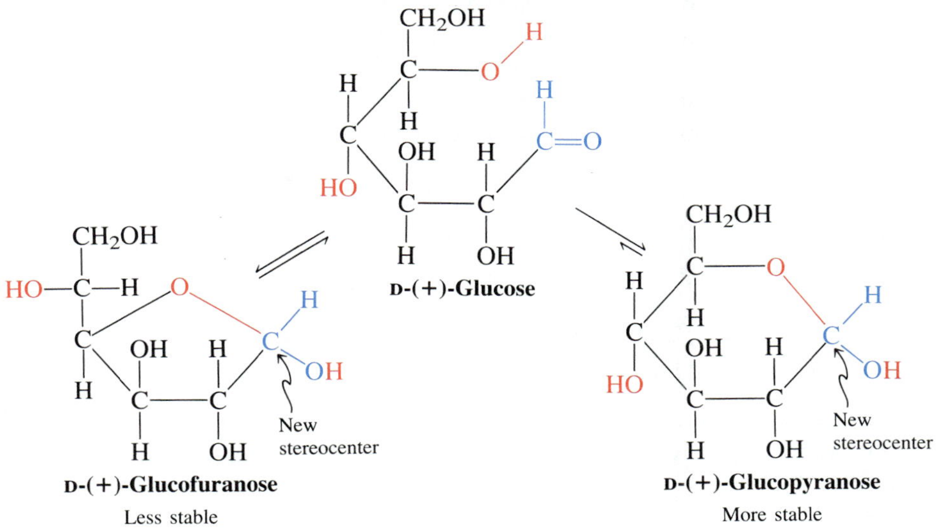

D-(+)-Glucose

D-(+)-Glucofuranose
Less stable

D-(+)-Glucopyranose
More stable

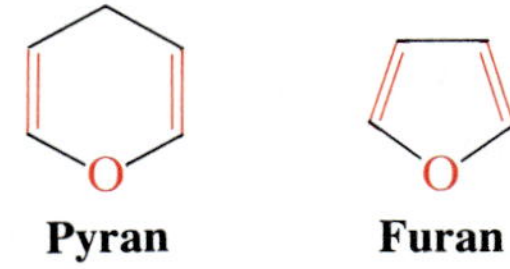

The hemiacetal thus formed is called a **pyranose,** a name derived from *pyran,* a six-membered cyclic ether (see Sections 9-5 and 26-1). Sugars in the five-membered ring form are called **furanoses,** from *furan*. In contrast with glucose, which exists primarily as the pyranose, fructose forms both fructopyranose and fructofuranose, in a 70:30 equilibrium mixture.

Cyclic Hemiacetal Formation by Fructose

D-(−)-Fructose

New stereocenter

30%

D-(−)-Fructofuranose

New stereocenter

70%

D-(−)-Fructopyranose

Inspection of the structures of the furanoses and pyranoses reveals that, on cyclization, the carbonyl carbon turns into a new stereocenter. As a consequence, hemiacetal formation leads to *two* new compounds, two diastereomers differing in the configuration of the hemiacetal group. If that configuration is S, the sugar is labeled α, when it is R, it is called β. Hence, for example, glucose may form α- or β-glucopyranose or -furanose. Because this type of diastereomer formation is unique to sugars, such isomers have been given a separate name: **anomers.** The new stereocenter is called the **anomeric carbon.**

EXERCISE 23-4

The anomers α- and β-glucopyranose should form in equal amounts because they are enantiomers. True or false? Explain your answer.

Fischer, Haworth, and Chair Cyclohexane Projections of Cyclic Sugar Structures

How can we best represent the stereochemistry of the cyclic forms of sugars? One approach is to adapt the Fischer projections, by simply drawing elongated lines to indicate the new bonds formed on cyclization, without changing the basic "grid" of the original formula.

Adapted Fischer Projections of Glucopyranoses

Cyclization

α-D-(+)-Glucopyranose
m.p. 146°C

β-D-(+)-Glucopyranose
m.p. 150°C

A projection that better reveals the real three-dimensional structure of the sugar molecule was designed by Haworth.* The cyclic ether is written in line notation as a pentagon or hexagon, the anomeric carbon placed on the right, and the ether oxygen on top. The substituents located above or below the ring are attached to vertical lines.

Haworth Projections

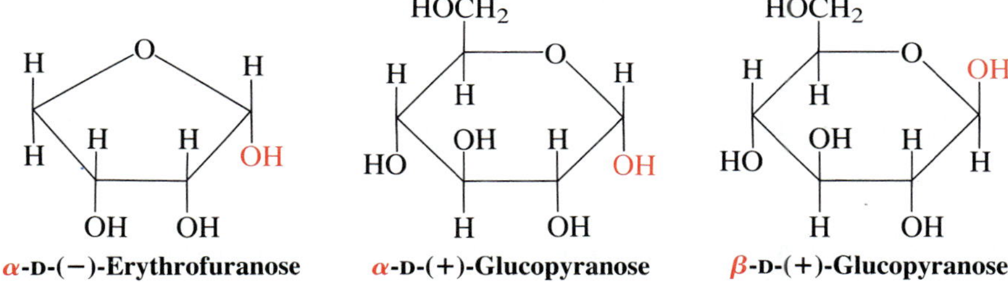

α-D-(−)-Erythrofuranose **α-D-(+)-Glucopyranose** **β-D-(+)-Glucopyranose**

In a Haworth projection, the α anomer has the OH group at the anomeric carbon pointing down, whereas the β anomer has it pointing up.

EXERCISE 23-5

Draw the structure of (a) α-D-fructofuranose, (b) β-D-glucofuranose, and (c) β-D-arabinopyranose.

Haworth projections have been and are used extensively in the literature of sugar chemistry, but here, to make use of our knowledge of conformation (Sections 4-3 and 4-4), the cyclic forms of sugars will be presented as envelope (for furanoses) or chair (for pyranoses) conformations. As in the Haworth notation, the ether oxygen usually will be placed top right, and the anomeric carbon at the right vertex of the envelope or chair.

*Sir Walter N. Haworth, 1883–1950, University of Birmingham, England, Nobel Prize 1937.

Conformational Pictures of Glucofuranose and -pyranose

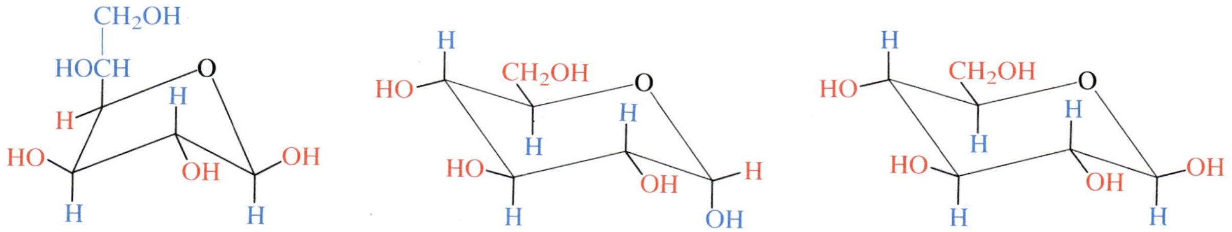

β-D-Glucofuranose **α-D-Glucopyranose** **β-D-Glucopyranose**

Although there are exceptions to this rule, most aldohexoses adopt the chair conformation that places the relatively bulky hydroxymethyl group at the C-5 terminus in the equatorial position. For glucose, this preference means that, in the α form, four of the five substituents are equatorial, and one is forced to lie axial; in the β form, *all* substituents can be equatorial. This situation is unique for glucose; the other seven aldohexoses (see Figure 23-1) contain one or more axial substituents.

EXERCISE 23-6

Using the values in Table 4-3, estimate the difference in free energy between the all-equatorial conformer of β-D-glucopyranose and that obtained by ring flip (assume that $\Delta G^\circ_{CH_2OH} = \Delta G^\circ_{CH_3} = 1.70$ kcal mole^{-1} and that the ring-oxygen mimics a methylene group).

Glucose Crystallizes as α-Glucopyranose

Glucose precipitates from concentrated solutions at room temperature to give crystals that melt at 146°C. Structural analysis by X-ray diffraction reveals that these crystals contain only the α-D-(+)-glucopyranose anomer (Figure 23-3).

FIGURE 23-3

Structure of α-D-(+)-glucopyranose, with some selected bond lengths and angles.

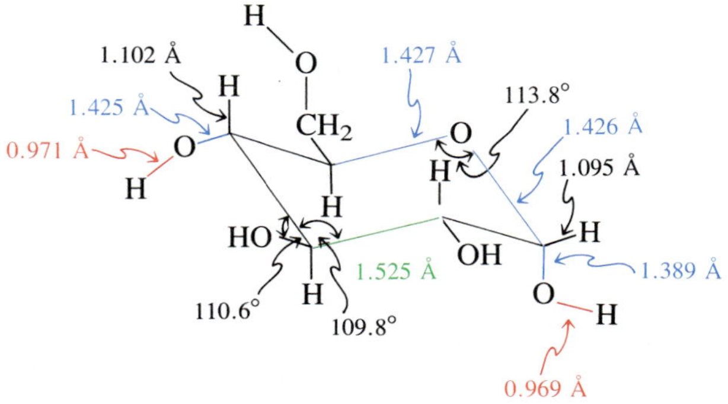

The Two Anomers of Glucopyranose Interconvert: Mutarotation

If crystalline α-D-(+)-glucopyranose is dissolved in water and its optical rotation measured immediately, a value $[\alpha]_D^{25^\circ C} = +112^\circ$ is obtained. Curiously, this value decreases with time until it reaches a constant $+52.7^\circ$. This change is accelerated by both acids and bases. Evidently, some chemical change alters the initial specific rotation of the sample. Indeed, in solution, the α-pyranose rapidly establishes an equilibrium (in a process that is catalyzed by acid and base; see Section 15-5) with a small amount of the open-chain aldehyde isomer, which in turn undergoes renewed and reversible ring closure to the β anomer.

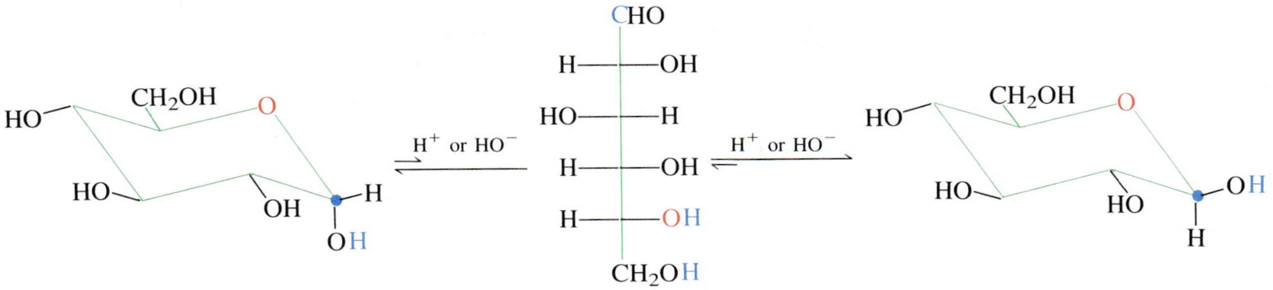

α-D-(+)-Glucopyranose
$[\alpha]_D^{25°C} = +112°$
36.4%

$\xrightleftharpoons{H^+ \text{ or } HO^-}$

Aldehyde form
0.003%

$\xrightleftharpoons{H^+ \text{ or } HO^-}$

β-D-(+)-Glucopyranose
$[\alpha]_D^{25°C} = +18.7°$
63.6%

The β form has a considerably lower specific rotation (+18.7°) than its anomer; therefore, the observed α value in solution decreases. Similarly, a solution of the pure β anomer [m.p. 150°C, obtainable by crystallization of glucose from ethanoic (acetic) acid] gradually increases its specific rotation from +18.7° to +52.7°. At this point, a final equilibrium has been reached, with 36.4% of the α anomer and 63.6% of the β anomer. The change in optical rotation observed when a sugar equilibrates with its anomer is called **mutarotation** (*mutare*, Latin, to change).

The presence of the two anomers in an aqueous solution of D-glucose can be detected in its ^{13}C NMR spectrum (Figure 23-4, on the next page).

EXERCISE 23-7

An alternative mechanism for mutarotation bypasses the aldehyde intermediate and proceeds through oxonium ions. Formulate it.

EXERCISE 23-8

Calculate the equilibrium ratio of α- and β-glucopyranose (that has been given in the text) from the specific rotations of the pure anomers and the observed specific rotation at mutarotational equilibrium.

EXERCISE 23-9

By using Table 4-3, estimate the difference in energy between α- and β-glucopyranose at room temperature (25°C). Then calculate it by using the equilibrium percentage.

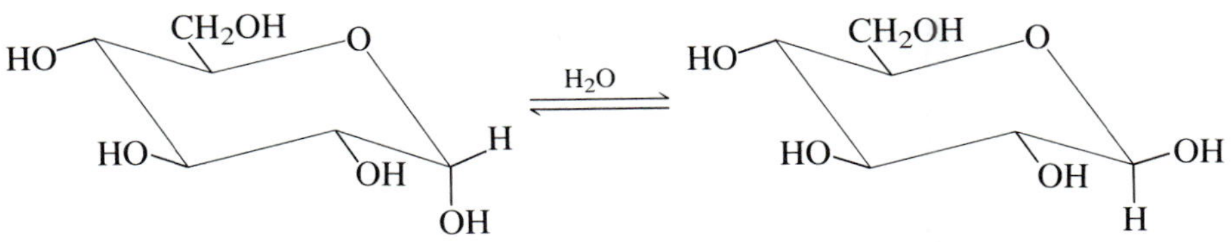

α-D-Glucopyranose $\xrightleftharpoons{H_2O}$ **β-D-Glucopyranose**

In summary, the simplest carbohydrates are sugars, which are polyhydroxy aldehydes (aldoses) and ketones (ketoses). They are classified as D when the highest-numbered stereocenter is *R*, L when it is *S*. Sugars related to each other by inversion at one stereocenter (as in D and L) are called epimers. Most of the naturally occurring sugars belong to the D family. The hexoses and pentoses can

FIGURE 23-4

63.07-MHz proton-decoupled ^{13}C NMR spectrum of glucose in water before reaching equilibrium. Two pairs of carbon atoms have the same chemical shift (one pair at 72.3 ppm, the other at 77.4 ppm); hence only ten signals are seen instead of twelve. The aldehyde form, because of its low concentration, cannot be detected. Assignments (from C-6 to C-1) for α-D-glucose are 62.3, 72.9, 71.3, 74.5, 73.2, and 93.7 ppm; for β-D-glucose, 62.4, 77.4, 71.3, 77.4, 75.8, and 97.5 ppm.

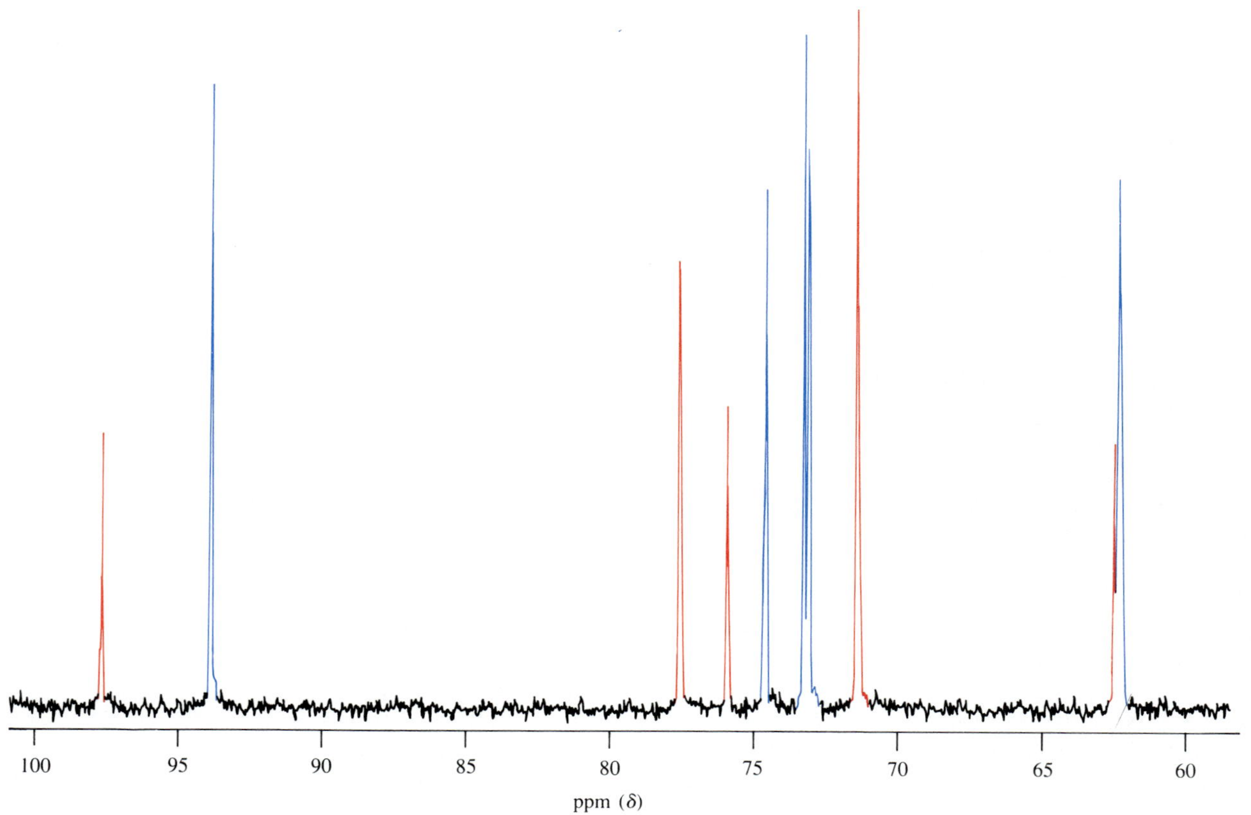

take the form of five- or six-membered cyclic hemiacetals. The acetal carbon (anomeric carbon) can have two configurations: α or β. In solution, the α and β forms of the sugars are in equilibrium with each other. The equilibration can be followed by starting with a pure anomer and observing the changes in specific rotation, a phenomenon also called mutarotation.

23-2 The Polyfunctional Chemistry of Sugars

Simple sugars exist in the form of various isomers: the open-chain carbonyl compound, and the α and β anomers of various cyclic forms. Because all of these are rapidly equilibrated, the relative rates of their individual reactions with various reagents will determine the product distribution of a particular transformation. We can therefore divide the reactions of sugars into two groups, those occurring on the linear form and those occurring on the cyclic forms, although sometimes both react competitively.

Oxidation of Sugars Leads to Carboxylic Acids or Degradation

The open-chain monosaccharides undergo the reactions typical of polyfunctional compounds. For example, aldoses contain the oxidizable aldehyde group and therefore respond to the classical oxidation tests such as exposure to Fehling's or Tollens's solutions (Section 15-8). The α-hydroxy substituent in ketoses is similarly oxidized.

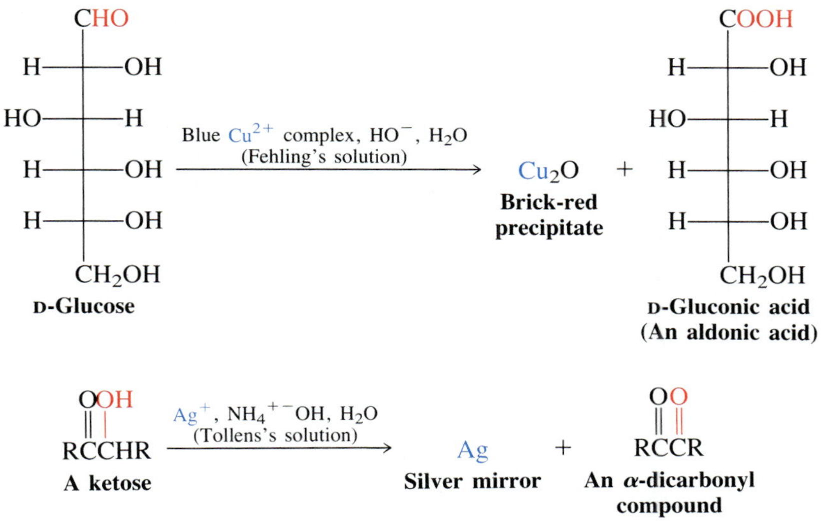

CHO
H—|—OH
HO—|—H
H—|—OH
H—|—OH
CH_2OH
D-Glucose

$\xrightarrow[\text{(Fehling's solution)}]{\text{Blue } Cu^{2+} \text{ complex, } HO^-,\ H_2O}$ Cu_2O **Brick-red precipitate** +

COOH
H—|—OH
HO—|—H
H—|—OH
H—|—OH
CH_2OH
D-Gluconic acid (An aldonic acid)

$$\underset{\textbf{A ketose}}{\overset{O\ \ OH}{\overset{\|\ \ |}{RCCHR}}} \xrightarrow[\text{(Tollens's solution)}]{Ag^+,\ NH_4{}^{+\,-}OH,\ H_2O} \underset{\textbf{Silver mirror}}{Ag} + \underset{\textbf{An }\alpha\textbf{-dicarbonyl compound}}{\overset{O\ O}{\overset{\|\ \|}{RCCR}}}$$

In these reactions, the aldoses are transformed into **aldonic acids,** whereas ketoses convert into α-dicarbonyl compounds. Sugars that respond positively to these tests are called **reducing sugars.**

Aldonic acids are made on a preparative scale by oxidation of aldoses with bromine in buffered aqueous solution (pH = 5–6). For example, D-mannose yields D-mannonic acid in this way.

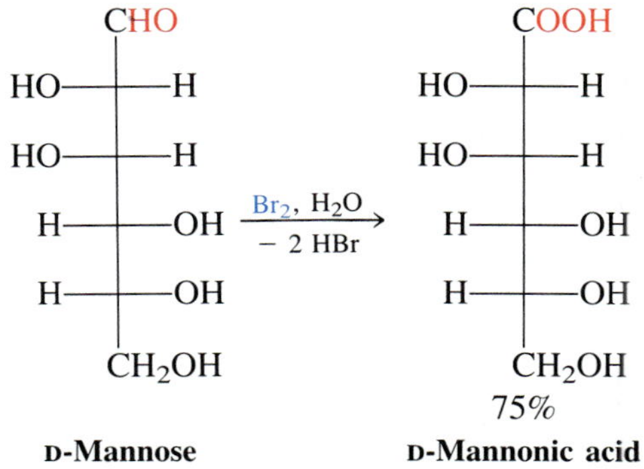

CHO
HO—|—H
HO—|—H
H—|—OH
H—|—OH
CH_2OH
D-Mannose

$\xrightarrow[-\ 2\ HBr]{Br_2,\ H_2O}$

COOH
HO—|—H
HO—|—H
H—|—OH
H—|—OH
CH_2OH
75%
D-Mannonic acid

On evaporation of solvent from the aqueous solution of an aldonic acid, the γ-lactone (Section 18-4) forms spontaneously.

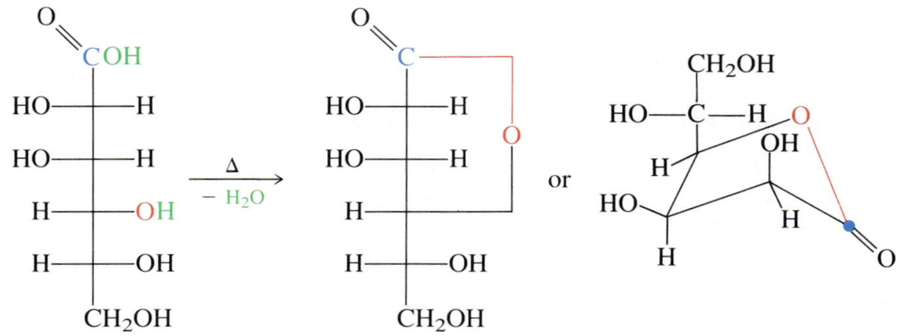

D-Mannonic acid $\xrightarrow[-\,H_2O]{\Delta}$ D-Mannono-γ-lactone (83%)

D-Mannonic acid

D-Mannono-γ-lactone

EXERCISE 23-10

A mechanism can be formulated for the conversion of mannose into mannono-γ-lactone by direct oxidation of the β-mannopyranose. Which group would have to be oxidized and what is the initial product? (Hint: Review the reactions of cyclic hemiacetals in Section 15-5 and transesterification in Section 18-4).

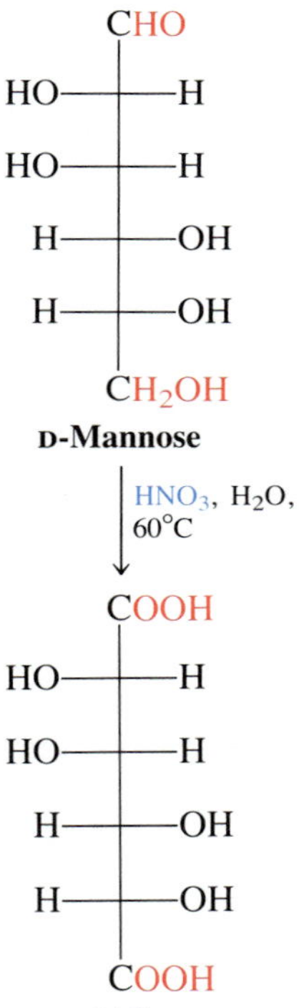

More-vigorous oxidation of an aldose leads to attack at the primary alcohol function as well as at the aldehyde group. The resulting dicarboxylic acid is called an **aldaric,** or **saccharic, acid.** This oxidation can be achieved with warm dilute aqueous nitric acid (see Section 17-5). For example, as shown in the margin, D-mannose is converted into D-mannaric acid under these conditions.

Because the aldaric acids have two carboxy groups, they can form double lactones. For example, D-mannaric acid is dehydrated twice on heating to give the double γ-lactone.

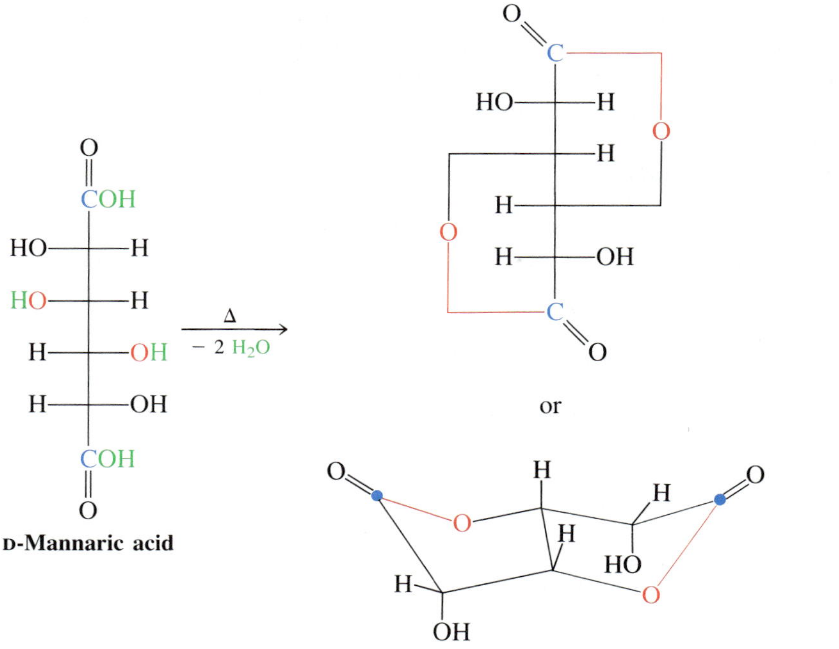

D-Mannose

HNO_3, H_2O, 60°C

44%

D-Mannaric acid

D-Mannaric acid $\xrightarrow[-\,2\,H_2O]{\Delta}$ D-Mannaric acid double γ-lactone (70%)

D-Mannaric acid

D-Mannaric acid double γ-lactone

EXERCISE 23-11

The two sugars D-allose and D-glucose (Figure 23-1) differ in configuration only at C-3. If you did not know which was which and you had samples of both, a polarimeter, and nitric acid at your disposal, how could you distinguish the two? Hint: Write the products of oxidation.

Periodic Acid Causes Oxidative Cleavage of Sugars

The methods for oxidation of sugars discussed so far leave the basic skeleton intact. A reagent that leads to C–C bond rupture is periodic acid, HIO_4. This compound oxidatively degrades vicinal diols to give carbonyl compounds.

Oxidative Cleavage of Vicinal Diols with Periodic Acid

$$\text{cis-1,2-Cyclohexanediol} \xrightarrow{HIO_4,\ H_2O} \text{Hexanedial } (OHC(CH_2)_4CHO),\ 77\%$$

The mechanism of this transformation proceeds through a cyclic periodic ester, which decomposes through an aromatic six-electron transition state.

Mechanism of Periodic Acid Cleavage of Vicinal Diols:

$$-\overset{|}{C}-OH / -\underset{|}{C}-OH + HIO_4 \xrightarrow{-H_2O} \text{Cyclic periodate ester} \longrightarrow 2\ \rangle C{=}O + HIO_3$$

Cyclic periodate ester

Because most sugars contain several pairs of vicinal diols, oxidation with HIO_4 can give complex mixtures. Sufficient oxidizing agent causes complete degradation of the chain to one-carbon compounds, a technique that has been applied in the structural elucidation of sugars. For example, treatment of glucose with five equivalents of HIO_4 results in the formation of five equivalents of methanoic (formic) acid and one of methanal (formaldehyde). Similar degradation of the isomeric fructose consumes an equal amount of oxidizing agent, but the products are three equivalents of the acid, two of the aldehyde, and one of carbon dioxide.

$$\text{D-Glucose} \xrightarrow{5\ HIO_4} 5\ HCOOH + HCHO$$

D-Glucose

$$\text{D-Fructose} \xrightarrow{5\ HIO_4} 3\ HCOOH + 2\ HCHO + CO_2$$

D-Fructose

How can these results be explained? First, various types of oxidizable carbon units may develop during the reaction. A terminal hydroxymethyl group next to another hydroxylated carbon (a typical arrangement at one end of the chain in a monosaccharide) can oxidize to methanal (formaldehyde) and the next lower aldose:

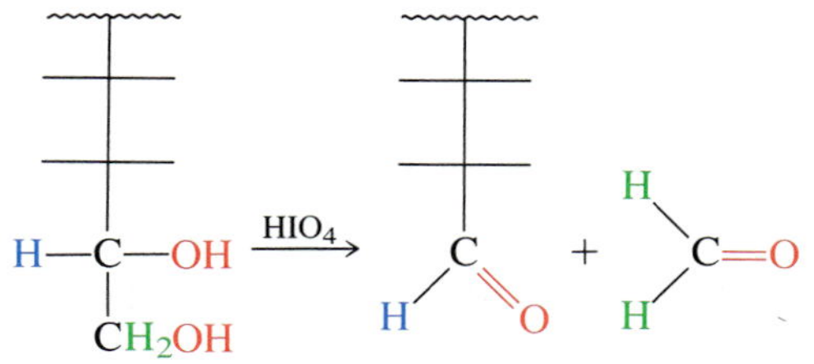

$$\text{H—C—OH (—CH}_2\text{OH)} \xrightarrow{\text{HIO}_4} \text{HC=O} + \text{H}_2\text{C=O}$$

The products of the oxidation of a sugar containing a terminal aldehyde unit (as found in an aldose or in an intermediate oxidation product) are derived from the corresponding hydrated form, the geminal diol, with which it is in equilibrium (Sections 15-3 and 15-5). Oxidative cleavage produces methanoic (formic) acid and a new, lower aldose.

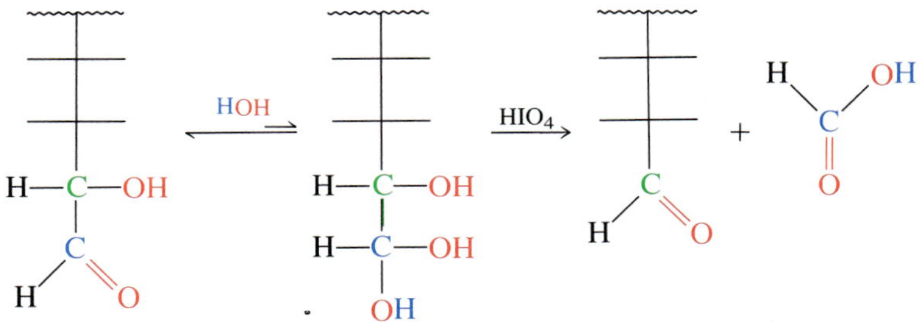

$$\text{H—C—OH (—CHO)} \underset{}{\overset{\text{HOH}}{\rightleftharpoons}} \text{H—C—OH (—H—C—OH, —OH)} \xrightarrow{\text{HIO}_4} \text{HC=O} + \text{HC(=O)OH}$$

Finally, a terminal hydroxymethyl group with an adjacent carbonyl function (as in fructose) can be recognized as the precursor of methanal (formaldehyde) and a carboxylic acid by the same mechanism.

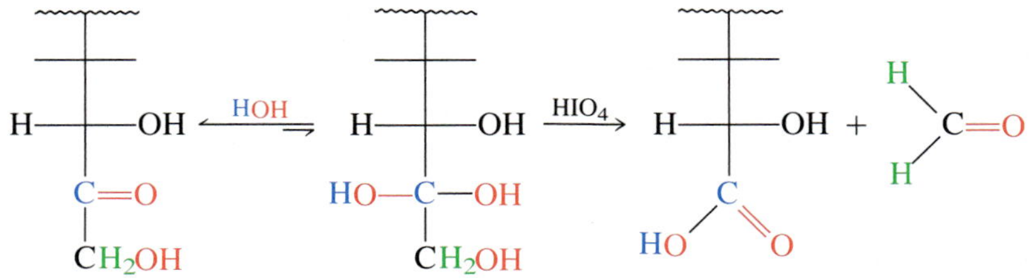

$$\text{H—C—OH (—C=O, —CH}_2\text{OH)} \overset{\text{HOH}}{\rightleftharpoons} \text{H—C—OH (—HO—C—OH, —CH}_2\text{OH)} \xrightarrow{\text{HIO}_4} \text{H—C—OH (—C(=O)OH)} + \text{H}_2\text{C=O}$$

If the resulting acid bears an α-hydroxy group, continued oxidation will furnish carbon dioxide and an aldehyde, ready for further degradation.

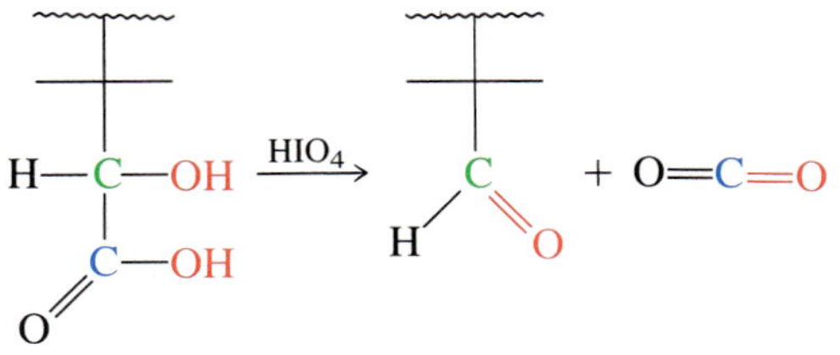

$$\text{H—C—OH (—C(=O)—OH)} \xrightarrow{\text{HIO}_4} \text{HC=O} + \text{O=C=O}$$

These considerations show that (1) the breaking of each C–C bond in the sugar consumes one molecule of HIO_4; (2) each aldehyde and geminal diol unit furnish an equivalent of methanoic (formic) acid; and (3) the primary hydroxy function gives methanal (formaldehyde), and the carbonyl group in ketoses gives CO_2. The number of equivalents of periodic acid consumed reveals the size of the sugar molecule, and the ratios of oxidation products are important clues to the number and arrangement of hydroxy and carbonyl functions.

EXERCISE 23-12

Write the expected products (and their ratios), if any, of the treatment of the following compounds with HIO_4: (a) 1,2-ethanediol (ethylene glycol); (b) 1,2-propanediol; (c) 1,2,3-propanetriol; (d) 1,3-propanediol; (e) 2,4-dihydroxy-3,3-dimethylcyclobutanone; (f) D-threose.

Reduction of Sugars Gives Alditols

Aldoses and ketoses are reduced by the same types of reducing agents that convert aldehydes and ketones into alcohols. The resulting polyhydroxy compounds are called **alditols.** For example, D-glucose gives D-glucitol (older name, D-sorbitol) on treatment with sodium borohydride. The hydride reducing agent traps the small amount of the open-chain form of the sugar, in this way shifting the equilibrium from the unreactive cyclic hemiacetal to the product.

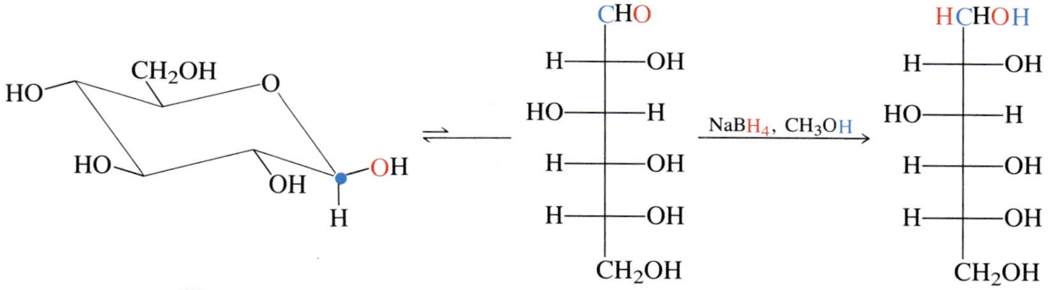

$$\text{D-Glucose (cyclic)} \rightleftharpoons \text{CHO–CH(OH)–CH(OH)–CH(OH)–CH(OH)–CH}_2\text{OH} \xrightarrow{NaBH_4,\ CH_3OH} \text{HCHOH–CH(OH)–CH(OH)–CH(OH)–CH(OH)–CH}_2\text{OH}$$

D-Glucose

D-Glucitol
(D-Sorbitol)

Many alditols occur in nature. D-Glucitol is found in red seaweed in concentrations as high as 14%, also in many berries (but not in grapes), in cherries, in plums, in pears, and in apples. It is prepared commercially by high-pressure hydrogenation of D-glucose or by electrochemical reduction.

EXERCISE 23-13

(a) Reduction of D-ribose with sodium borohydride gives a product without optical activity. Explain. (b) Similar reduction of D-fructose gives two optically active products. Explain.

The Carbonyl Group Undergoes Condensations: Phenylhydrazones and Osazones

As might be expected, the carbonyl function in aldoses and ketoses undergoes condensation reactions with amine derivatives (Section 15-6). For example, treatment of D-mannose with phenylhydrazine gives the corresponding D-mannose phenylhydrazone.

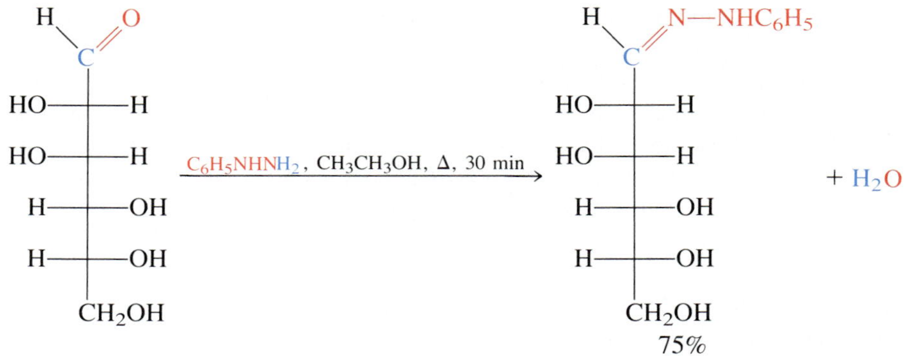

$$\text{D-Mannose (CHO; HO–H; HO–H; H–OH; H–OH; CH}_2\text{OH)} \xrightarrow{C_6H_5NHNH_2,\ CH_3CH_3OH,\ \Delta,\ 30\ \text{min}} \text{D-Mannose phenylhydrazone (HC=N–NHC}_6\text{H}_5\text{; HO–H; HO–H; H–OH; H–OH; CH}_2\text{OH)} + H_2O$$

75%

D-Mannose **D-Mannose phenylhydrazone**

Surprisingly, the reaction does not stop at this stage but can be induced to continue with additional phenylhydrazine (two extra equivalents). The final product is a double phenylhydrazone, also called a **phenylosazone.** In addition, one equivalent each of benzenamine (aniline), ammonia, and water is generated.

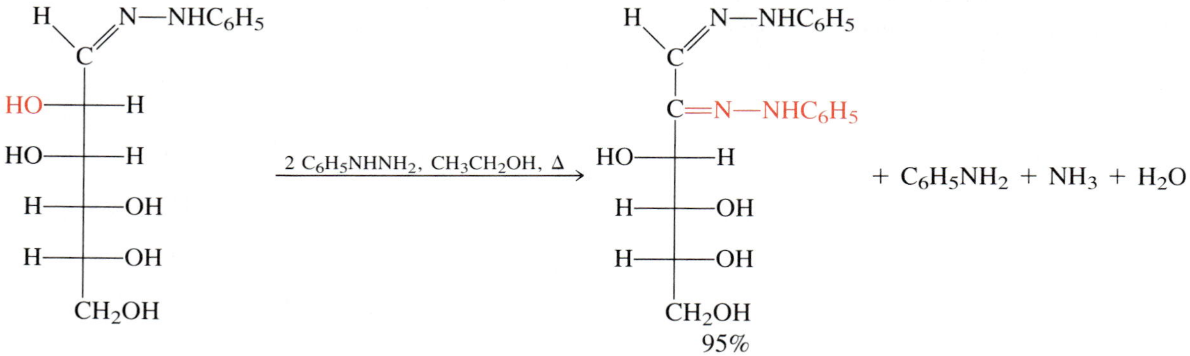

$$\text{HC=N–NHC}_6\text{H}_5\text{; HO–H; HO–H; H–OH; H–OH; CH}_2\text{OH} \xrightarrow{2\ C_6H_5NHNH_2,\ CH_3CH_2OH,\ \Delta} \text{HC=N–NHC}_6\text{H}_5\text{; C=N–NHC}_6\text{H}_5\text{; HO–H; H–OH; H–OH; CH}_2\text{OH} + C_6H_5NH_2 + NH_3 + H_2O$$

95%

A phenylosazone

Although the detailed mechanism of osazone synthesis is not completely understood, a possible pathway is shown in the following scheme. Once formed, the osazones do not continue to react with excess phenylhydrazine but are stable under the conditions of the reaction.

Mechanism of Phenylosazone Formation:

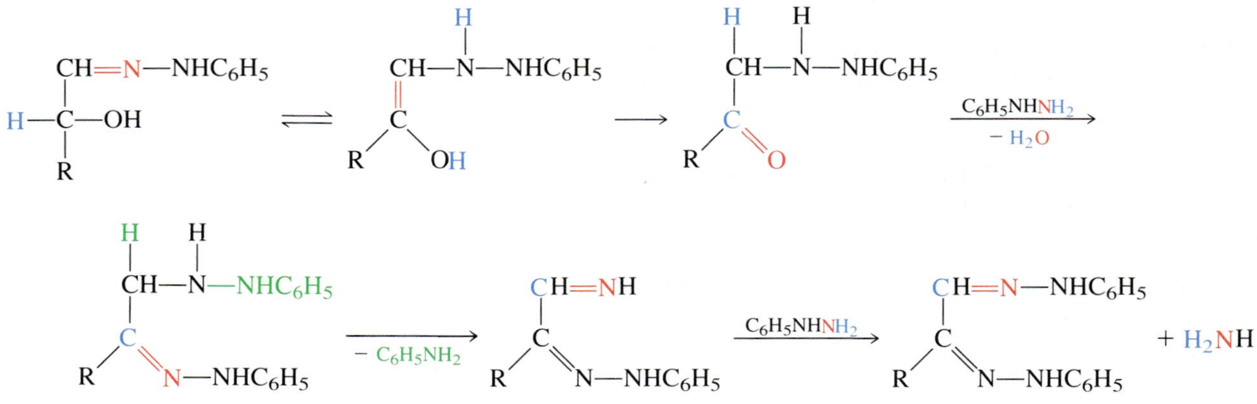

$$\text{CH=N–NHC}_6\text{H}_5\text{; H–C–OH; R} \rightleftharpoons \text{CH–N(H)–NHC}_6\text{H}_5\text{; C(=)(R)OH} \longrightarrow \text{CH(H)–N(H)–NHC}_6\text{H}_5\text{; C(R)=O} \xrightarrow[-\ H_2O]{C_6H_5NHNH_2}$$

$$\text{CH(H)–N(H)–NHC}_6\text{H}_5\text{; C(R)=N–NHC}_6\text{H}_5 \xrightarrow[-\ C_6H_5NH_2]{} \text{CH=NH; C(R)=N–NHC}_6\text{H}_5 \xrightarrow{C_6H_5NHNH_2} \text{CH=N–NHC}_6\text{H}_5\text{; C(R)=N–NHC}_6\text{H}_5 + H_2NH$$

Historically, the discovery of osazone formation marked a significant advance in the practical aspects of sugar chemistry. Sugars, like many other polyhydroxy compounds, are well known for their reluctance to crystallize from syrups. Their osazones, on the other hand, readily form yellow crystals with sharp melting points, simplifying the isolation and characterization of many sugars, particularly if they have been formed as mixtures or are impure.

EXERCISE 23-14

Compare the structures of the phenylosazones of D-glucose, D-mannose, and D-fructose. Do you notice anything unusual?

Hydroxy Groups in Sugars Form Esters and Ethers

As polyhydroxy compounds, sugars can be converted into several alcohol derivatives. Esters can be prepared by standard techniques (Section 17-8), usually employing excess reagent to completely convert all hydroxy functions, including the hemiacetal group. For example, treatment of β-D-glucopyranose with five equivalents of ethanoic (acetic) anhydride furnishes the pentaethanoate (pentaacetate).

$$\beta\text{-D-Glucopyranose} + 5\ CH_3\overset{O}{\overset{\|}{C}}O\overset{O}{\overset{\|}{C}}CH_3 \xrightarrow[-\ 5\ CH_3COOH]{\text{Pyridine, }0°C,\ 24\text{ h}} \beta\text{-D-Glucopyranose pentaethanoate (91\%)}$$

β-D-Glucopyranose

91%

β-D-Glucopyranose pentaethanoate

Similarly, complete methylation can be effected under the conditions of the Williamson ether synthesis (Section 9-5).

$$\beta\text{-D-Ribopyranose} \xrightarrow[-\ 4\ CH_3OSO_3Na,\ -\ 4\ HOH]{4\ CH_3OSO_2OCH_3,\ NaOH} \beta\text{-D-Ribopyranose tetramethyl ether (70\%)}$$

β-D-Ribopyranose

70%

β-D-Ribopyranose tetramethyl ether

The acetal function can be selectively hydrolyzed to the hemiacetal (see Section 15-5).

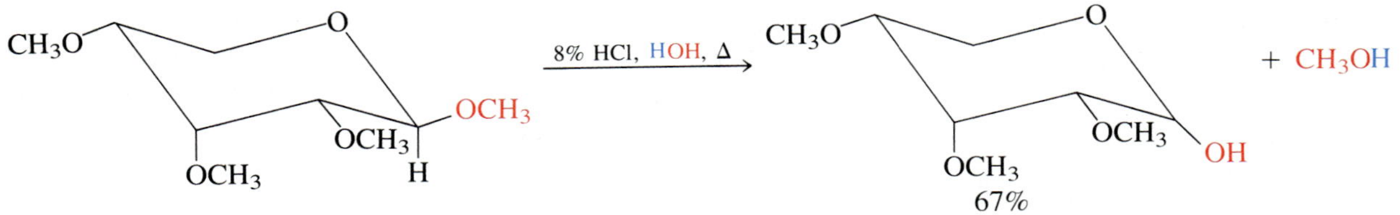

CH_3O … OCH_3 … OCH_3 … OCH_3 … H $\xrightarrow{8\%\ HCl,\ HOH,\ \Delta}$ CH_3O … OCH_3 … OCH_3 … OH + CH_3OH

67%

D-Ribopyranose trimethyl ether
Mixture of α and β forms

The reverse is also possible; the hemiacetal unit can be selectively converted into the acetal. For example, treatment of D-glucose with acidic methanol leads to the formation of the two methyl acetals. Sugar acetals are called **glycosides.** Thus, glucose forms *glucosides*.

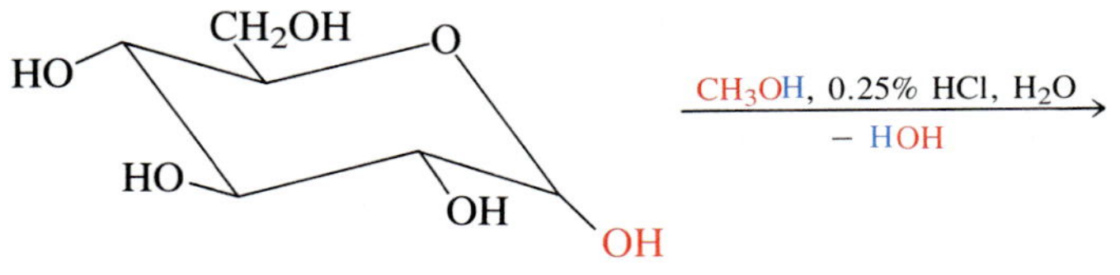

α- or β-D-Glucopyranose $\xrightarrow[-\ HOH]{CH_3OH,\ 0.25\%\ HCl,\ H_2O}$

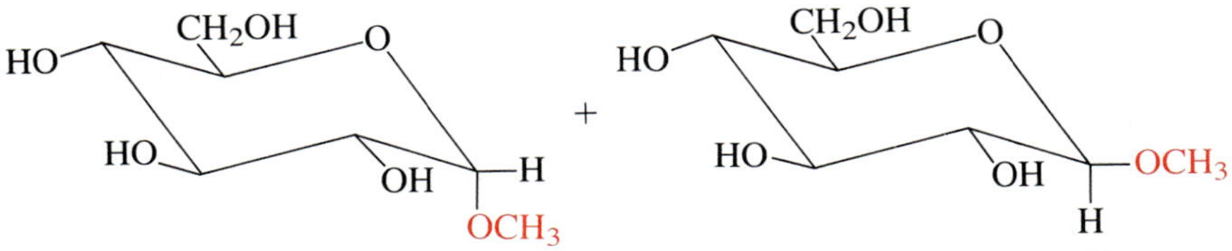

Methyl α-D-glucopyranoside
m.p. 166°C, $[\alpha]_D^{25°C} = +158°$

+

Methyl β-D-glucopyranoside
m.p. 105°C, $[\alpha]_D^{25°C} = -33°$

Because glycosides contain a blocked anomeric carbon atom, they do not show mutarotation in the absence of acid, they test negatively to Fehling's and Tollens' reagents (they are *non*reducing sugars), and they are indifferent to reagents that attack carbonyl groups. Such protection can be useful in synthesis and in structural analysis (see Exercise 23-17).

EXERCISE 23-15

The same mixture of glucosides is formed in the methylation of D-glucose with acidic methanol, regardless of whether you start with the α or β form. Why?

EXERCISE 23-16

Draw the structure of methyl α-D-arabinofuranoside.

EXERCISE 23-17

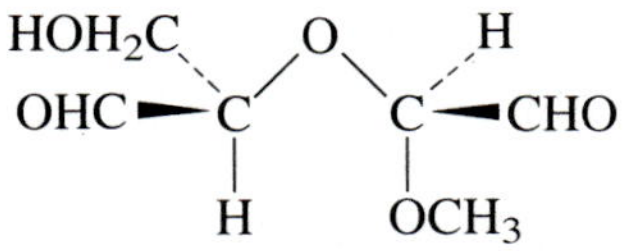

Methyl α-D-glucopyranoside consumes two equivalents of HIO_4 to give one equivalent each of methanoic (formic) acid and dialdehyde A (shown at the left). An unknown aldopentose methyl furanose reacted with one equivalent of HIO_4 to give A without the formation of methanoic (formic) acid. Suggest a structure for the unknown. Is there more than one solution to this problem?

HOH_2C, OHC, H, O, H, CHO, OCH_3

A

Neighboring Hydroxy Groups in Sugars Can Be Linked as Cyclic Ethers and Esters

The presence of neighboring pairs of hydroxy groups in the sugars allows for the formation of cyclic ether and ester derivatives. For example, it is possible to synthesize five- or six-membered cyclic sugar acetals from the vicinal (and also β-diol) units by treating them with carbonyl compounds (Section 15-5), and cyclic carbonates by exposing them to phosgene, $COCl_2$ (Section 18-4).

Cyclic Acetal and Carbonate Formation from Vicinal Diols

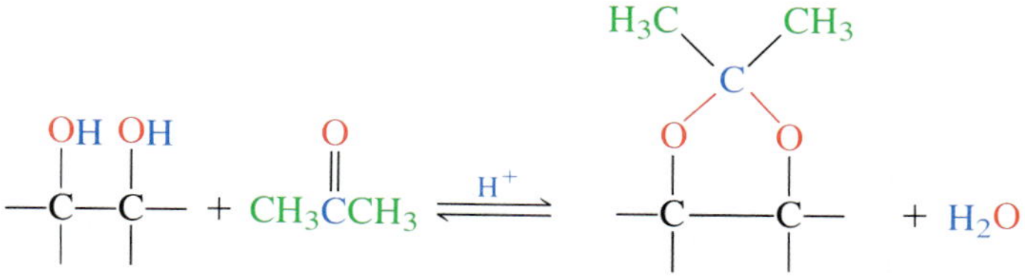

Propanone (acetone) acetal

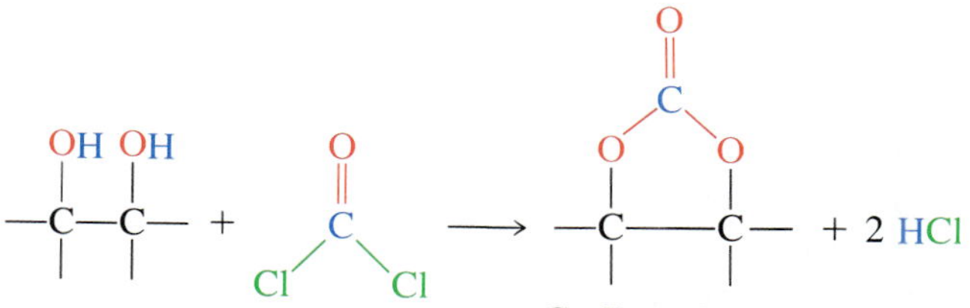

Cyclic carbonate

Such transformations work best if the two hydroxy groups are positioned cis to allow a relatively unstrained ring to form. For example, in the presence of excess acidic propanone (acetone), β-D-arabinopyranose is converted into the double acetal.

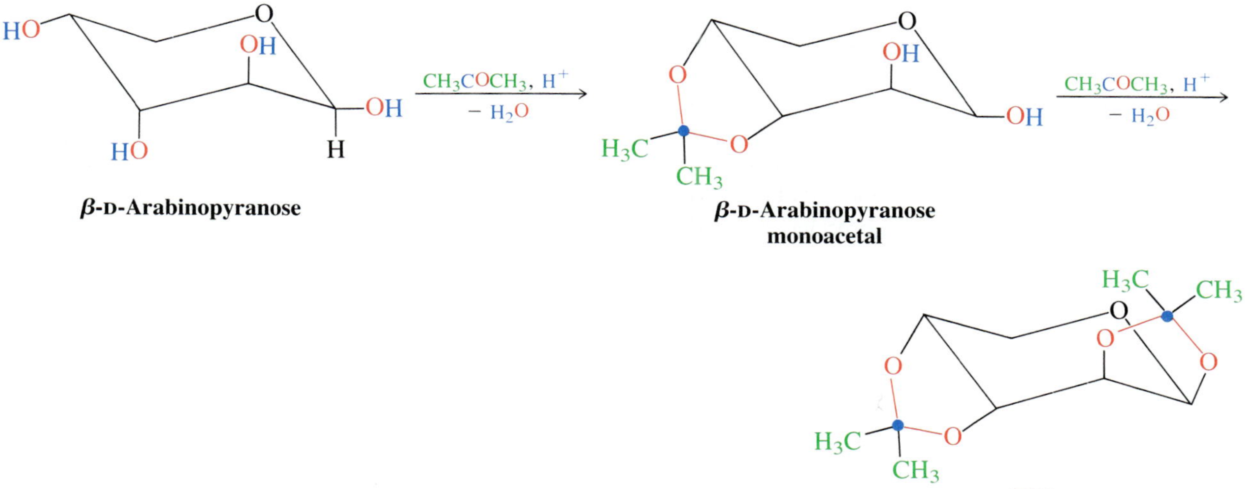

β-D-Arabinopyranose

β-D-Arabinopyranose monoacetal

55%

β-D-Arabinopyranose double acetal

Cyclic acetal and ester formation from sugars is often necessary to subject the remaining, unprotected alcohol functions to selective transformations, such as oxidation to carbonyl compounds, conversion into leaving groups, and eliminations.

EXERCISE 23-18

Suggest a feasible synthesis of compound A from D-galactose.

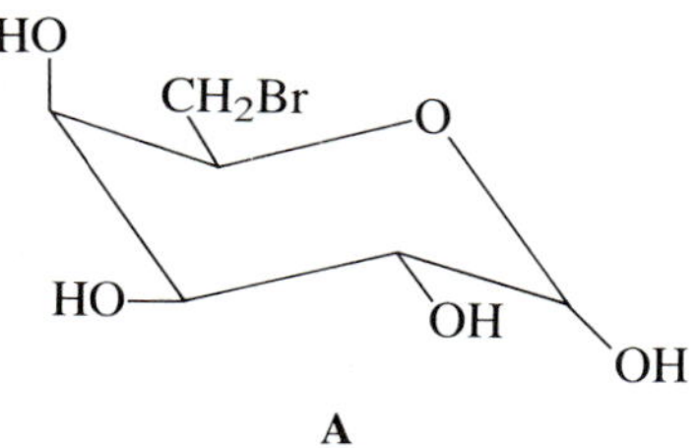

To summarize, the chemistry of the sugars is largely that expected for polyhydroxy carbonyl compounds. Oxidation (by Br_2) of the aldehyde group of aldoses gives aldonic acids; more-vigorous oxidation (by HNO_3) converts sugars into aldaric acids. Both types of product are readily dehydrated to γ-lactones. Oxidative cleavage with periodic acid degrades the sugar backbone to methanoic (formic) acid, methanal (formaldehyde), and CO_2, the ratio of these products depending on the structure of the sugar. Reduction of the carbonyl function (by $NaBH_4$) furnishes alditols. One equivalent of phenylhydrazine converts a sugar into the phenylhydrazone, but additional hydrazine reagent causes oxidation of the center adjacent to the hydrazone function to furnish the osazone. The various hydroxy groups can be esterified or converted into ethers. The hemiacetal unit can be selectively protected as the acetal, also called a glycoside. Finally, the various diol units in the sugar backbone may be linked as cyclic acetals or carbonates, depending on steric requirements.

23-3 The Step-by-Step Buildup and Degradation of Sugars: Proof of the Structure of the Aldoses

Larger sugars can be made from smaller ones and vice versa, by chain lengthening and chain shortening. These transformations can also be used to structurally correlate various sugars, a procedure applied by Fischer to prove the relative configuration of all the stereocenters in the aldoses shown in Figure 23-1.

Chain Lengthening: Kiliani-Fischer Synthesis

In the (modified) **Kiliani-Fischer* synthesis** of sugars, an aldose is first treated with hydrogen cyanide to give the corresponding cyanohydrin. Because this transformation forms a new stereocenter, two diastereomers appear. Separation

*Professor Heinrich Kiliani, 1855–1945, University of Freiburg, Germany; Professor Emil Fischer, see Section 5-4.

of the diastereomers and partial reduction of the nitrile group by catalytic hydrogenation in the presence of acidic water then gives the aldehyde groups of the chain-extended sugars. In this hydrogenation, a modified palladium catalyst (similar to Lindlar's catalyst, Section 13-6) allows selective reduction of the nitrile to the imine, which hydrolyzes under the reaction conditions. The special catalyst is necessary to prevent the reduction from proceeding all the way to the amine (Section 18-6).

General Kiliani-Fischer Synthesis of Sugars

STEP 1: Cyanohydrin formation

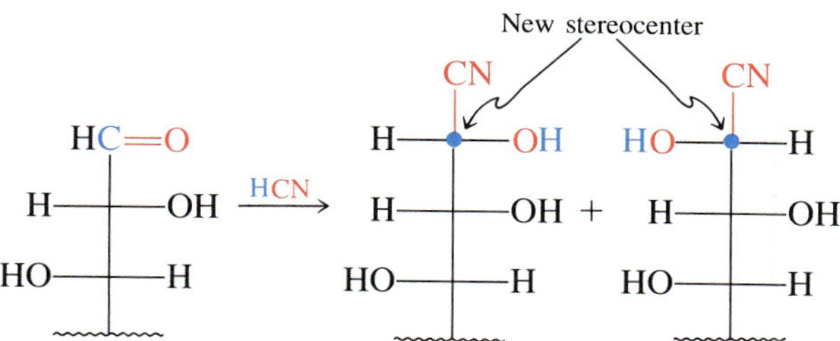

Two new diastereomeric nitriles

STEP 2: Reduction and hydrolysis (only one diastereomer is shown)

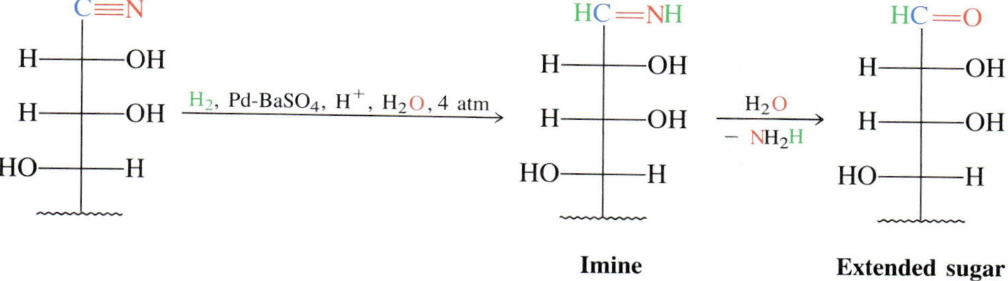

Imine **Extended sugar**

EXERCISE 23-19

What are the products of Kiliani-Fischer chain extension of (a) D-erythrose and (b) D-arabinose?

Chain Shortening: Wohl and Ruff Degradations

Whereas the Kiliani-Fischer approach synthesizes higher sugars, complementary strategies degrade higher to lower sugars one carbon at a time. These are the **Wohl*** and **Ruff† degradations.** Both procedures remove the carbonyl group of an aldose and simultaneously convert the neighboring carbon into the aldehyde function of the new sugar.

The Wohl degradation is essentially the reverse of the Kiliani-Fischer chain extension. The aldehyde group of the aldose is first converted into an oxime by condensation with hydroxylamine. Subsequent complete esterification with ethanoic (acetic) anhydride produces an intermediate polyethanoyl derivative.

---

*Professor Alfred Wohl, 1863–1933, University of Danzig, Germany.

†Professor Otto Ruff, 1871–1939, University of Danzig, Germany.

Wohl Degradation of D-Gulose

D-Gulose $\xrightarrow[-\,H_2O]{H_2NOH}$ D-Gulose oxime (69%) $\xrightarrow[-\,6\,CH_3COOH]{6\,CH_3COCCH_3 \text{ (O, O)},\ \text{pyridine},\ 100^\circ C,\ 1\ h}$ D-Gulose oxime hexaethanoate $\xrightarrow{-\,CH_3COOH}$ D-Xylose cyanohydrin pentaethanoate (60%) $\xrightarrow[-\,5\,CH_3COOCH_3]{CH_3OH,\ CH_3O^-Na^+}$ D-Xylose cyanohydrin $\xrightarrow{-\,HCN}$ D-Xylose (90%)

This compound is unstable; it eliminates ethanoic (acetic) acid to give the corresponding nitrile. The ester groups are then removed by transesterification with sodium methoxide; under these conditions, the resulting free cyanohydrin spontaneously decomposes into the lower sugar.

The Ruff degradation takes a different course—it is an oxidative decarboxylation procedure. The sugar is first oxidized to the aldonic acid under standard conditions. Exposure to hydrogen peroxide in the presence of ferric ion then leads to the loss of the carboxy group and oxidation of the new terminus to the aldehyde function of the lower aldose. The mechanism of this decarboxylation is related to the one discussed in Section 17-11.

Both degradation procedures give rather low yields because of the sensitivity of the products to the reaction conditions. Nevertheless, they may be useful in structural elucidations (Exercise 23-20). Such studies were carried out originally by Fischer to establish the relative configuration of the monosaccharides (the *Fischer proof*). The next section will describe some of the logic behind Fischer's approach to this problem.

General Ruff Degradation of Sugars

HC=O, H—C—OH → (Br_2, H_2O) → COOH, H—C—OH → (Fe^{3+}, H_2O_2, H_2O; $- CO_2$) → HC=O

EXERCISE 23-20

Wohl degradation of two D pentoses A and B gave two new sugars C and D. Oxidation of C with HNO_3 gave *meso*-2,3-dihydroxybutanedioic (tartaric) acid, that of D resulted in an optically active acid. Oxidation of either A or B with HNO_3 furnished an optically active aldaric acid. What are compounds A, B, C, and D?

How Do We Establish the Relative Configuration of the Aldoses? An Exercise in Logic

Imagine that you have been presented with fourteen jars, each filled with one of the tetroses, pentoses, and hexoses in Figure 23-1. How would you establish the structure of each compound?

This was the challenge faced by Fischer in the late nineteenth century, when chemists had no modern spectrometers at their disposal. Fischer showed that this problem can be solved, without spectral techniques, by using a combination of synthetic manipulations and the logical interpretation of their outcome. Only one assumption had to be made: that the dextrorotatory 2,3-dihydroxypropanal (glyceraldehyde) has the D configuration (and not that of its mirror image). This assumption can be verified only by X-ray structural analysis, and indeed it was in 1950, long after Fischer's days. [In fact, it was not verified for (*R*)-2,3-dihydroxypropanal (D-glyceraldehyde) itself, but for optically active 2,3-dihydroxybutanedioic (tartaric) acid, with which the aldehyde can be structurally correlated through D-threose.] Fischer guessed at this absolute configuration and was lucky to have been right; otherwise, all the compounds in Figure 23-1 would have had to be written as their mirror images. However, at the time it was more important to establish the *relative* configuration of all the stereocenters, to associate each unique sugar with a unique sequence of such centers in its backbone.

Given the correct structure of (*R*)-2,3-dihydroxypropanal (D-glyceraldehyde), we can now set out to unambiguously prove the structure of all the higher D aldoses. (Consult Figure 23-1 as required while following these arguments.) First, we perform a Kiliani-Fischer chain lengthening on D-glyceraldehyde, which gives two new isomeric sugars. Separation and oxidation with nitric acid leads to *meso*-2,3-dihydroxybutanedioic (tartaric) acid from one product and an optically active acid from the other. Therefore, the former must be D-erythrose; the latter, D-threose. Note that the absolute configuration at the next-to-last carbon is the same in both sugars and in our starting material, (*R*)-2,3-dihydroxypropanal (D-glyceraldehyde). Recall that this is a common property of all the D sugars. The difference is at C-2; D-erythrose is 2*R*; D-threose, 2*S*.

Let us now use D-erythrose as our new starting material, because we know its structure, and lengthen the chain further.

CHO, H—C(*R*)—OH, H—C—OH, CH_2OH

D-Erythrose

CHO, HO—C(*S*)—H, H—C—OH, CH_2OH

D-Threose

CHO
H—C—OH
H—C—OH
CH_2OH
D-Erythrose

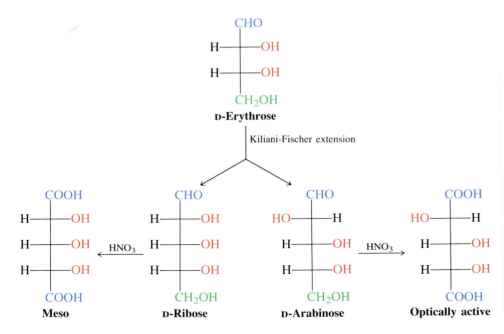

Kiliani-Fischer extension

COOH		CHO	CHO		COOH
H—C—OH		H—C—OH	HO—C—H		HO—C—H
H—C—OH	← HNO_3	H—C—OH	H—C—OH	HNO_3 →	H—C—OH
H—C—OH		H—C—OH	H—C—OH		H—C—OH
COOH		CH_2OH	CH_2OH		COOH
Meso		**D-Ribose**	**D-Arabinose**		**Optically active**

Again, two new sugars ensue (as they must, because we have again added a new stereocenter), two pentoses. We know their configuration at C-3 and C-4 (the same as that in C-2 and C-3 of their starting material) but not that at C-2. Their oxidation again produces one optically inactive dicarboxylic acid and one that is active. The former must therefore have the structure of D-ribose, the latter that of D-arabinose.

A very similar train of thought leads to the unambiguous assignment of the structures of D-xylose (oxidized to a meso dioic acid) and D-lyxose (oxidized to an optically active dioic acid), derived synthetically from D-threose, whose structure we ascertained at the very beginning.

COOH		CHO	CHO		COOH
H—C—OH		H—C—OH	HO—C—H		HO—C—H
HO—C—H	← HNO_3	HO—C—H	HO—C—H	HNO_3 →	HO—C—H
H—C—OH		H—C—OH	H—C—OH		H—C—OH
COOH		CH_2OH	CH_2OH		COOH
		D-Xylose	**D-Lyxose**		

We now know the structures of the four aldopentoses and may expose each of them to the chain-extension procedure. This gives us four pairs of aldohexoses, each pair distinguished from the other by the unique sequence of stereocenters at C-3, C-4, and C-5. The members of each pair differ only in their configuration at C-2.

The structural assignment for the four sugars obtained from D-ribose and D-lyxose, respectively, is again accomplished by oxidation to the corresponding aldaric acids. Both D-allose and D-galactose give optically inactive oxidation products, in contrast with their counterparts D-altrose and D-talose, which give optically active dicarboxylic acids.

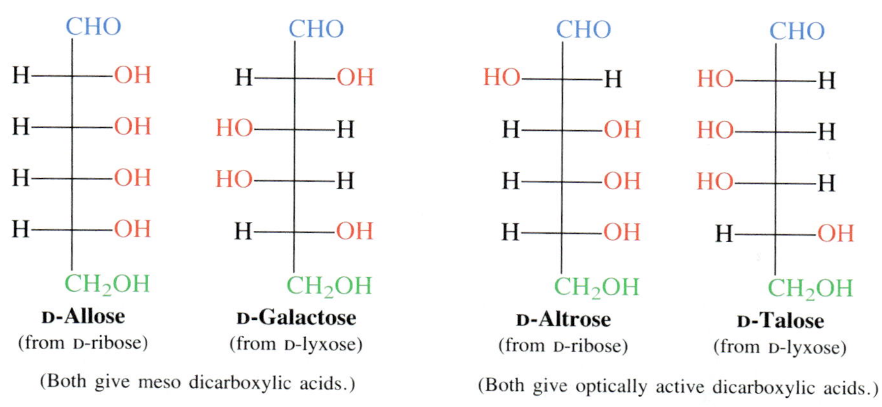

The structural assignment of the four remaining sugars cannot be based on the approach taken thus far, because all four give optically active dicarboxylic acids on oxidation.

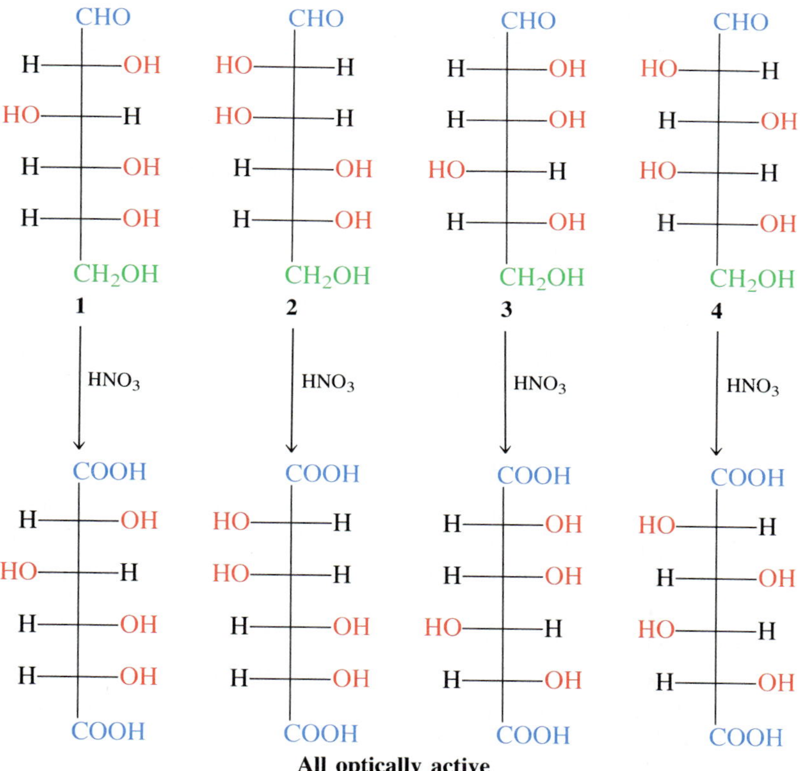

However, it is found that the two carboxylic acids derived from sugars 1 and 3 are enantiomers—that is, mirror images of each other. This result is possible only if 1 and 3 have the structures of D-glucose or D-gulose. You can verify this relation of the two aldaric acids by building molecular models.

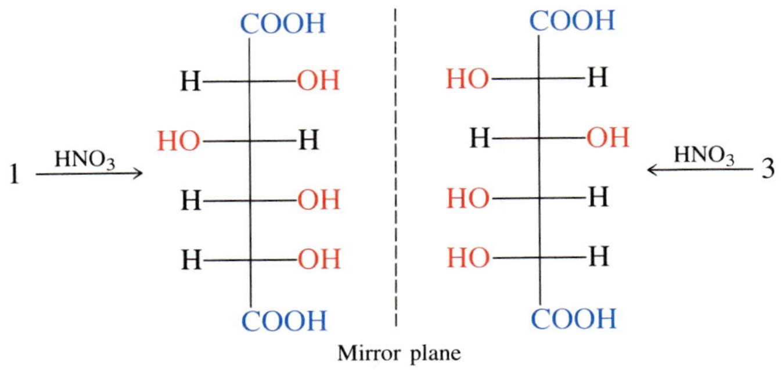

COOH | COOH
H—OH | HO—H
HO—H | H—OH
1 $\xrightarrow{HNO_3}$ H—OH | HO—H $\xleftarrow{HNO_3}$ 3
H—OH | HO—H
COOH | COOH

Mirror plane

We now proceed experimentally as follows. D-Arabinose is converted into a pair of new sugars, 1 and 2, by Kiliani-Fischer chain extension; D-xylose furnishes sugars 3 and 4. With these results in hand, the structural assignments fall into place. Sugar 1 must have the structure of D-glucose, and sugar 3 must have the structure of D-gulose. Therefore, 2 is assigned the structure of D-mannose and 4 that of D-idose.

EXERCISE 23-21

In the preceding discussion, we assigned the structures of D-ribose and D-arabinose by virtue of the fact that on oxidation the first gives a meso dioic acid, the second an optically active isomer. Could you arrive at the same result by ^{13}C NMR spectroscopy?

In summary, sugars can be made from other sugars by step-by-step one-carbon chain lengthening or shortening. These techniques, in conjunction with the symmetry properties of the various aldaric acids, allow the stereochemical assignments of the aldoses.

23-4 Disaccharides, Polysaccharides, and Other Sugars in Nature

A substantial fraction of the natural sugars occur in dimeric, trimeric, higher oligomeric (between two and ten sugar units), and polymeric form.

Sucrose Is a Disaccharide Derived from Glucose and Fructose

The sugar that is most familiar to us is ordinary table sugar, **sucrose.** The average yearly consumption of sucrose in the United States is about 100 pounds per person. Sucrose is one of the few natural chemicals consumed in unmodified form (water and sodium chloride are others). Sucrose is isolated from sugar cane and sugar beets, in which it is particularly abundant (about 14%–20% by weight), although it is present in many plants in smaller concentrations. World production is about 150 trillion pounds a year, and there are countries (e.g., Cuba) whose entire economy depends on the world price of sucrose.

Sucrose has not been discussed in this chapter so far, because it is not a simple monosaccharide but a disaccharide composed of two units: glucose and fructose. The structure of sucrose can be deduced from its chemical behavior: acidic hydrolysis splits it into glucose and fructose; it is a nonreducing sugar; it does not form an osazone; and it does not undergo mutarotation. These findings suggest that the component monosaccharide units are linked by an acetal bridge connecting the two anomeric carbons; in this way, the two cyclic hemiacetal functions protect each other. X-ray structural analysis confirms this hypothesis: sucrose is a disaccharide in which the α-D-glucopyranose form of glucose is attached to β-D-fructofuranose in this way.

Sucrose, an α-D-glucopyranosyl β-D-fructofuranoside

Two representations of the molecule are shown. At the left, both cyclic sugars are drawn in the usual way: the anomeric carbon on the right, the acetal oxygen on top. Another structure with more-favorable steric interactions is shown at the right, a rotamer in which the two sugar units point away from each other.

Sucrose has a specific rotation of $+66.5^\circ$. Treatment with aqueous acid decreases the rotation until it reaches a value of -20.0°. The same effect is observed with the enzyme *invertase*. The phenomenon is known as the **inversion of sucrose,** and its mechanism is related to mutarotation of monosaccharides. It includes three separate reactions: hydrolysis of the disaccharide to the component monosaccharides α-D-glucopyranose and β-D-fructofuranose; mutarotation of α-D-glucopyranose to the equilibrium mixture with the β form; and mutarotation of β-D-fructofuranose to the slightly more stable β-D-fructopyranose. Because the value for the specific rotation of fructose (-92°) is more negative than the value for glucose ($+52.7^\circ$) is positive, the resulting mixture has a net negative rotation, *inverted* from that of the original sucrose solution.

Inversion of Sucrose

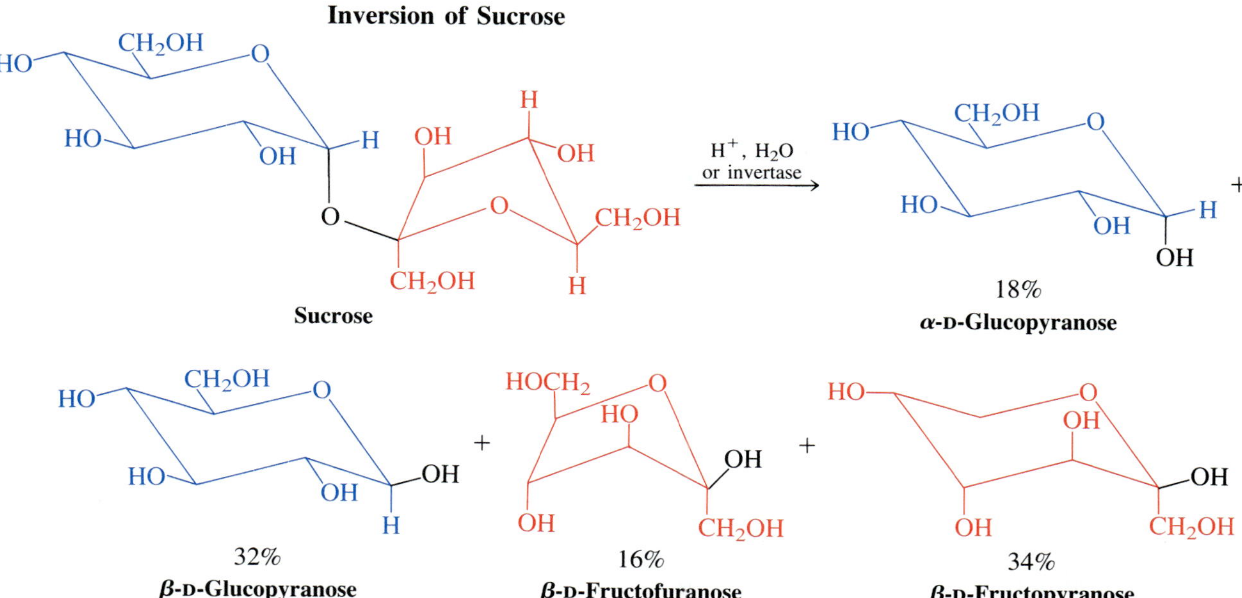

EXERCISE 23-22

Write the products (if any) of the reaction of sucrose with (a) excess $(CH_3)_2SO_4$, NaOH; (b) 1. H^+, H_2O, 2. $NaBH_4$; and (c) NH_2OH.

Sucrose contains an acetal linkage between the anomeric carbons of the component sugars. One could imagine other acetal linkages with other hydroxy groups. Indeed, **maltose** (malt sugar), which is obtained in 80% yield by enzymatic *(amylase)* degradation of starch (to be discussed later in this section), is a dimer of glucose in which the hemiacetal oxygen of one glucose molecule (in the α-anomeric form) is bound to C-4 of the second.

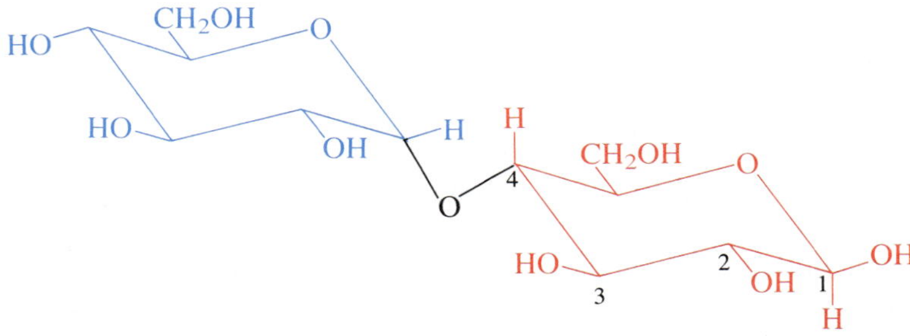

β-Maltose, an α-D-glucopyranosyl-β-D-glucopyranose

In this arrangement, the glucose unit maintains its unprotected hemiacetal structure, with its distinct chemistry. For example, maltose is a reducing sugar, it forms osazones, and it undergoes mutarotation. Maltose is hydrolyzed to two molecules of glucose by aqueous acid or by the enzyme *maltase*. It is about one-third as sweet as sucrose.

EXERCISE 23-23

Draw the structure of the initial product of β-maltose when it is subjected to (a) oxidation with Br_2; (b) phenylhydrazine (3 equivalents); (c) conditions that effect mutarotation.

Another common disaccharide is **cellobiose,** obtained by the hydrolysis of cellulose (to be discussed later in this section). Its chemical properties are almost identical with those of maltose, and so is its structure: the same except for the stereochemistry at the acetal linkage, β instead of α.

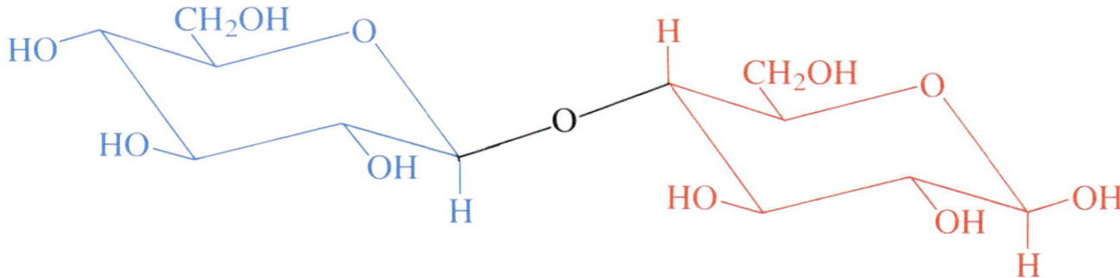

β-Cellobiose, a β-D-glucopyranosyl-β-D-glucopyranose

Aqueous acid cleaves cellobiose into its component glucose molecules just as efficiently as it hydrolyzes maltose. However, enzymatic hydrolysis requires a different enzyme, *β-glucosidase,* which specifically attacks only the β-acetal bridge. In contrast, maltase is specific for α-acetal units of the type found in maltose.

After sucrose, the most-abundant natural disaccharide is **lactose** (milk sugar). It is found in human and most animal milk (about 5% solution), constituting more than one-third of the solid residue remaining on evaporation of all volatiles. Its structure is made up of galactose and glucose units, connected in the form of a β-D-galactopyranosyl-D-glucopyranose. Crystallization from water furnishes only the α anomer.

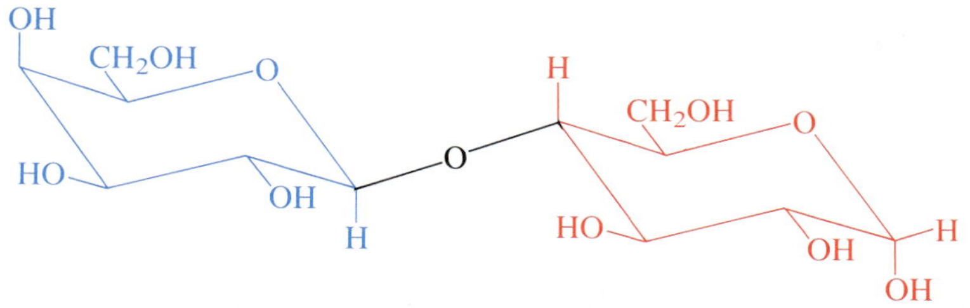

Crystalline α-lactose, a β-D-galactopyranosyl-α-D-glucopyranose

It is remarkable, considering the number of permutations in the number and kind of possible ether junctions, that the structural diversity of the natural disaccharides is relatively limited to the types discussed so far. Similarly, there are only a few natural tri- and tetrasaccharides that can be found in quantity. An example is **raffinose,** a nonreducing trisaccharide built from one molecule each of D-galactose, D-glucose, and D-fructose. The linkage between the last two components is identical with the one in sucrose, whereas that of the first two is new, being an acetal bridge between the α-anomeric carbon of galactose and the terminal hydroxymethyl group of glucose. Raffinose is found in certain plant materials, such as Australian manna and cotton seeds.

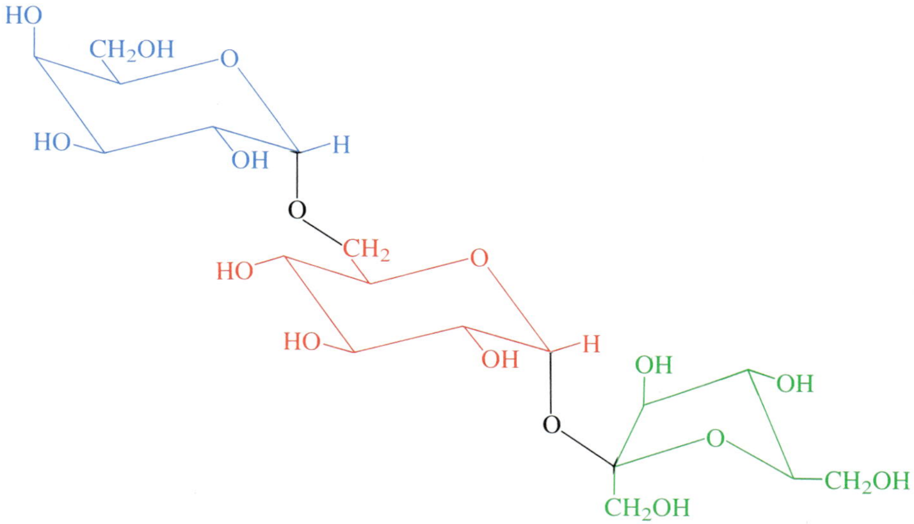

Raffinose

Polysaccharides are the polymers of monosaccharides. Their possible structural diversity is comparable to that of alkene polymers (Sections 12-7 and 12-8), particularly in variations of chain length and branching. Nature, however, has been remarkably conservative in its construction of such polymers. The three most-abundant natural polysaccharides, cellulose, starch, and glycogen, are derived from the same monomer: glucose.

Cellulose is a poly-β-glucopyranoside linked at C-4, containing about 3000 monomeric units and having a molecular weight of about 500,000. It is largely linear.

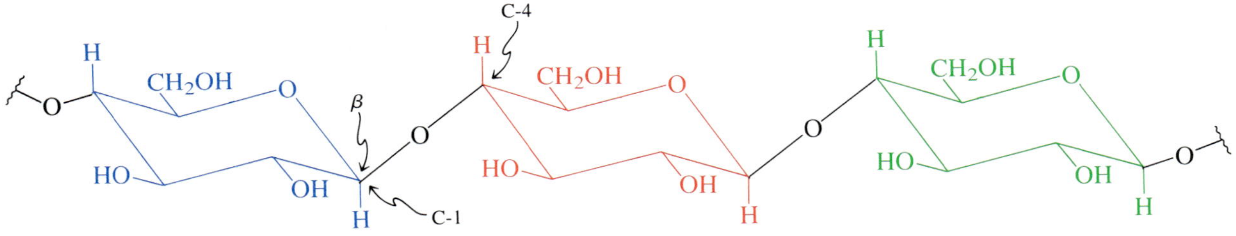

Cellulose

Individual strands of cellulose tend to align with each other and are connected by multiple hydrogen bonds. The development of so many hydrogen bonds is responsible for the highly rigid structure of cellulose and its effective use as the cell-wall material in organisms. Thus, cellulose is abundant in trees and other plants. Cotton fiber is almost pure cellulose, as in filter paper. Wood and straw contain about 50% of the polysaccharide.

X-ray data indicate that the cellulose chain is composed of repeating units of 10.3 Å in length, not 5.1 Å, the length of one glucose molecule. This result may be explained if alternate glucose units are rotated 180° with respect to each other. Thus, cellulose is best described as a polycellobiose. This arrangement allows for hydrogen bonding on both sides of the cellulose chain, making for even better alignment and rigidity (Figure 23-5).

FIGURE 23-5

Suggested hydrogen bonding between two cellulose strands. Note that such bonding may occur within one strand, as well as between two of them.

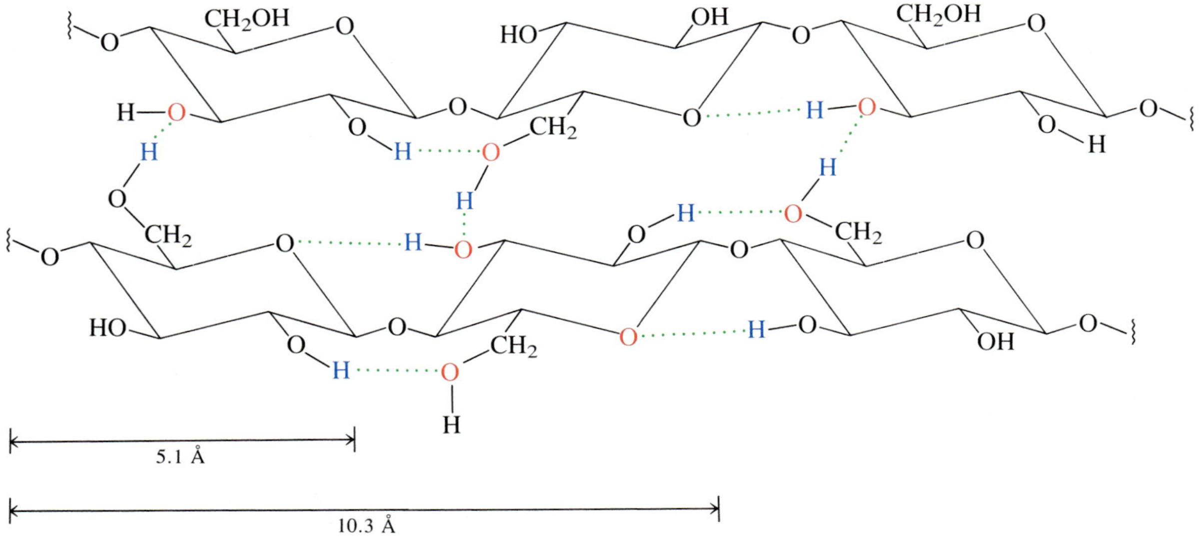

5.1 Å

10.3 Å

Several derivatives of cellulose have commercial uses. Conversion of the free hydroxy groups into nitrate esters with nitric acid results in *nitrocellulose*. If the nitrate content is high, this material is explosive and is used in smokeless gunpowder. A lower nitrate content gives a polymer that was important as one of the first commercial plastics: celluloid. For a long time, nitrocellulose was used extensively in the photographic and film industries. Unfortunately, it is highly flammable and gradually decomposes, it is now used only rarely.

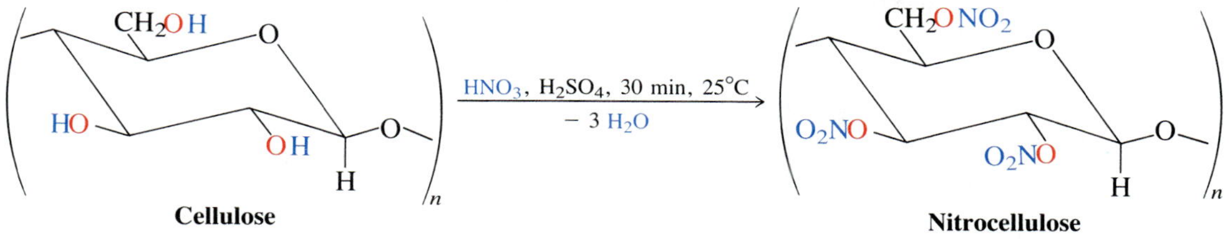

$$\text{Cellulose} \xrightarrow[-\,3\,H_2O]{HNO_3,\ H_2SO_4,\ 30\text{ min},\ 25°C} \text{Nitrocellulose}$$

Cellulose **Nitrocellulose**

Cellulose, which is insoluble in almost all solvents, can be solubilized by blocking the hydroxy groups as adducts to carbon disulfide, the sulfur analog of CO_2. The resulting functional group is called *xanthate*. Subsequent treatment with acid regenerates the insoluble polymer; this process may be controlled to give fibers *(rayon)* or sheets *(cellophane)*.

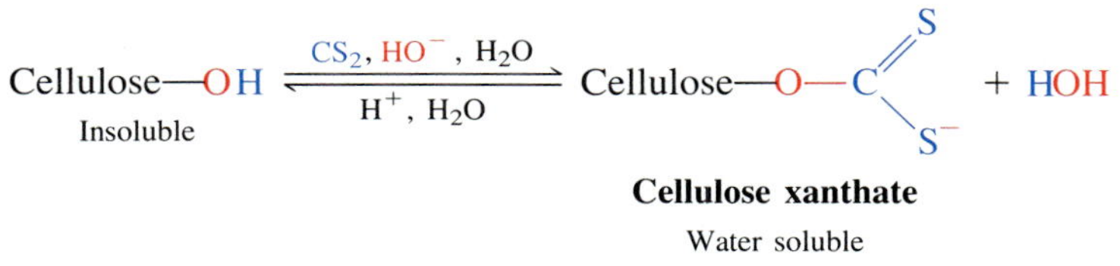

$$\underset{\text{Insoluble}}{\text{Cellulose—OH}} \underset{H^+,\ H_2O}{\overset{CS_2,\ HO^-,\ H_2O}{\rightleftharpoons}} \text{Cellulose—O—C(=S)S}^- + HOH$$

Cellulose xanthate
Water soluble

Unlike cellulose, **starch** is a polyglucose connected by α-acetal linkages in a variety of configurations. It functions as a food reserve in plants and (like cellulose) is readily cleaved by aqueous acid into glucose. Major sources of starch are corn, potatoes, wheat, and rice. Hot water swells granular starch and allows the separation of the two major components: **amylose** (~20%) and **amylopectin** (~80%). Both are soluble in hot water, but the former is less soluble in cold water. Amylose contains a few hundred glucose units per molecule (molecular weight, 150,000–600,000). It has a different structure from cellulose, even though both polymers are unbranched. The difference in the stereochemistry at the anomeric carbons leads to the preference of amylose to form a helical polymer arrangement (not the straight chain shown in the formula). Note that the disaccharide units in amylose are the same as those in maltose.

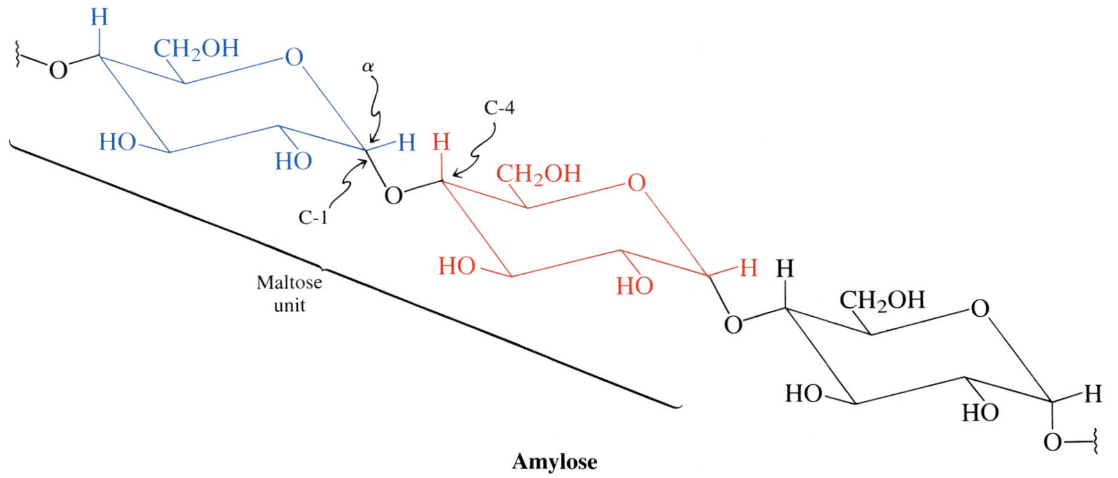

α
C-4
C-1
Maltose unit

Amylose

In contrast with amylose, the more-soluble amylopectin is branched, mainly at C-6, about once every twenty–to–twenty-five glucose units. Its molecular weight runs into the millions.

Amylopectin

Glycogen: A Source of Energy

Another polysaccharide with a structure very similar to amylopectin but with greater branching (one per ten glucose units) and of much larger size (as much as 100 million molecular weight) is **glycogen.** This compound is of considerable biological importance because it is one of the major energy-storage polysaccharides in humans and animals and because it provides an immediate source of glucose between meals and in periods of (strenuous) physical activity. It is accumulated especially in the liver and in rested skeletal muscle in relatively large amounts. The manner in which cells make use of this energy storage is a fascinating story in biochemistry.

A special enzyme, *phosphorylase,* first breaks glycogen down to give a derivative of glucose: α-D-glucopyranosyl 1-phosphate. This transformation takes place at one of the nonreducing terminal sugar groups of the glycogen molecule and proceeds in a step-by-step manner—one glucose molecule at a time. Because glycogen is so highly branched, there are many such end groups at which the enzyme can ''nibble'' away, making sure that, at a time of high energy requirements, a sufficient amount of glucose becomes quickly available.

Phosphorylase cannot break α-1,6-glycosidic bonds. As soon as it gets close to such a branching point (in fact, as soon as it reaches a terminal residue four units away from that point), it stops (Figure 23-6, on page 1074). At this stage, a different enzyme comes into play, *transferase,* which can shift blocks of three terminal glucosyl residues from one branch to another. This process leaves one residual glucose substituent at the branching point. Now a third enzyme is required to remove that last obstacle to obtaining a new straight chain. This

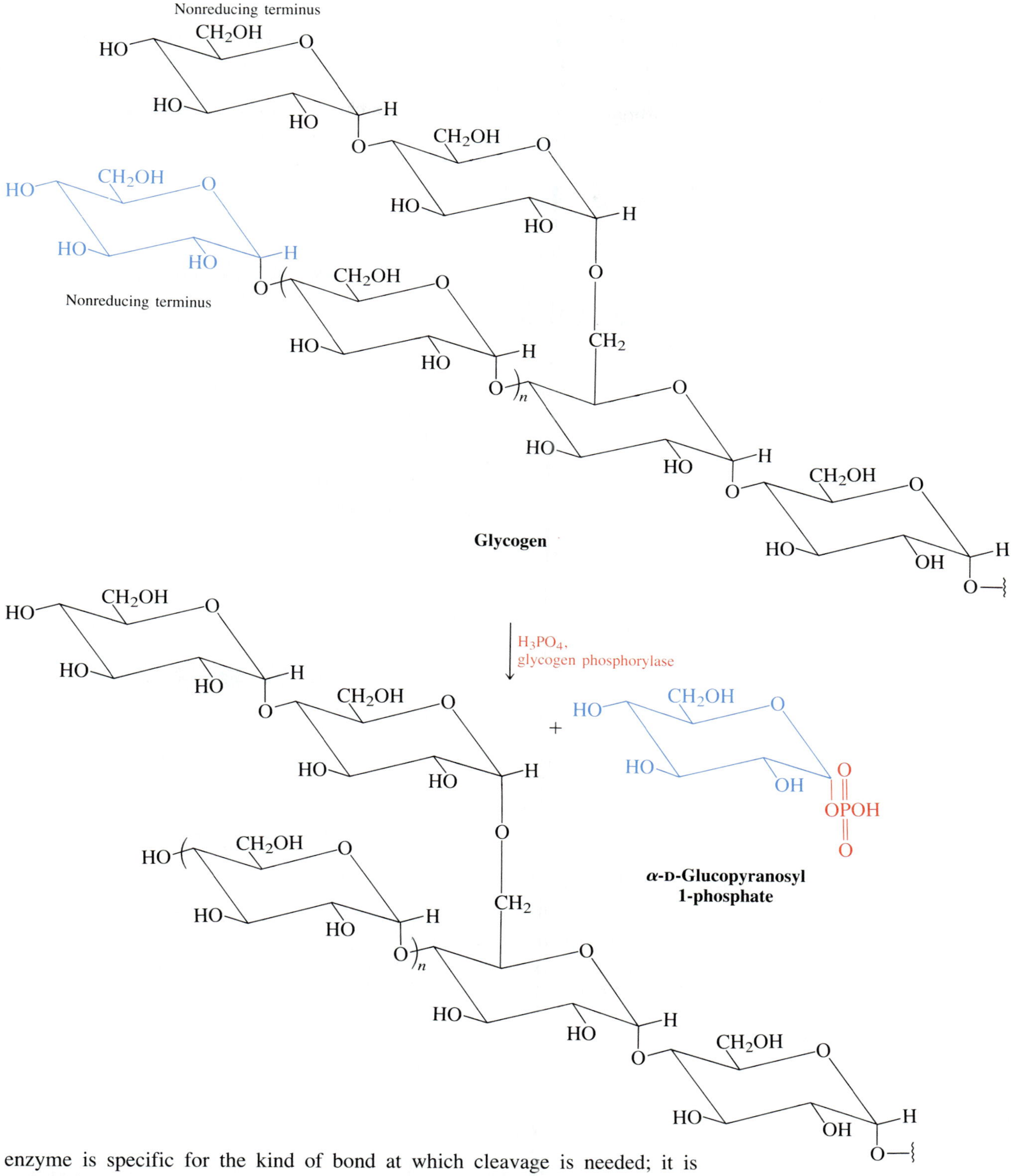

Nonreducing terminus

Nonreducing terminus

Glycogen

H_3PO_4,
glycogen phosphorylase

α-D-Glucopyranosyl 1-phosphate

enzyme is specific for the kind of bond at which cleavage is needed; it is *α-1,6-glucosidase,* also known as the *debranching enzyme*. Once this enzyme has completed its task, phosphorylase can continue degrading the glucose chain until it reaches another branch, and so on.

FIGURE 23-6

Steps in the degradation of a glycogen side chain. Initially, phosphorylase removes glucose units 1 through 5 and 15 through 17 step by step. The enzyme is now four sugar units away from a branching point (10). Transferase moves units 6 through 8 in one block and attaches them to unit 14. A third enzyme, α-1,6-glucosidase debranches the system at glucose unit 10, by removing glucose 9. A straight chain has been formed and phosphorylase can continue its degradation job.

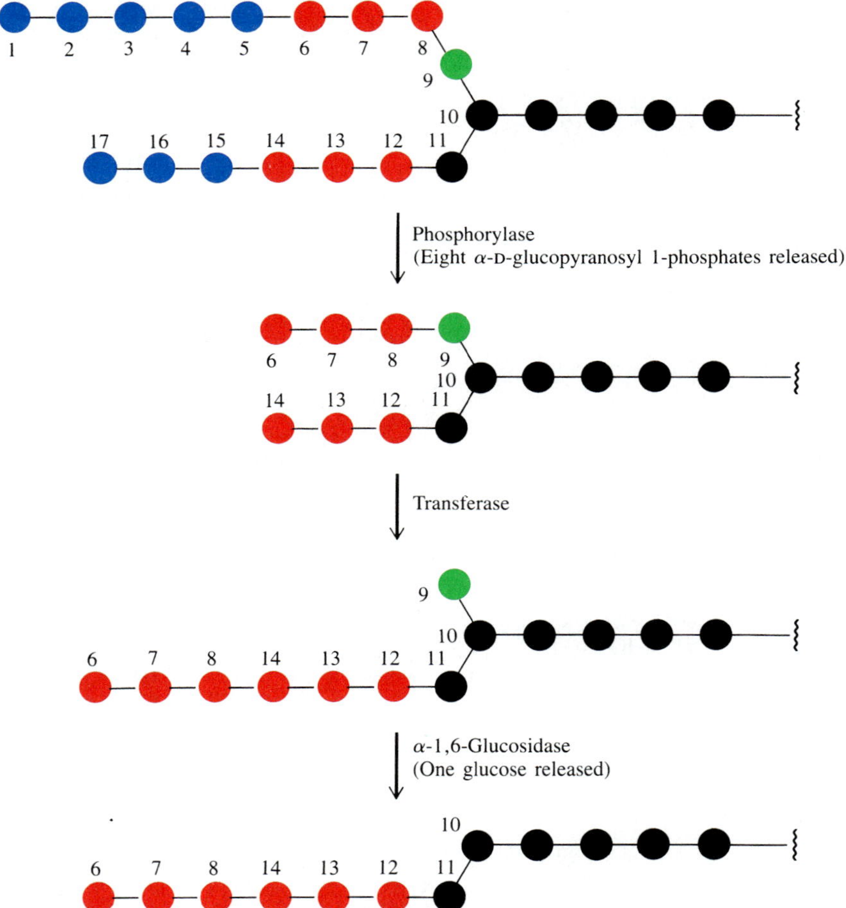

The glucose liberated from glycogen is converted into 2-oxopropanoic (pyruvic) acid by a complex pathway (glycolysis) that includes several enzymes. This acid then gives rise to various products, depending on the particular organism and conditions.

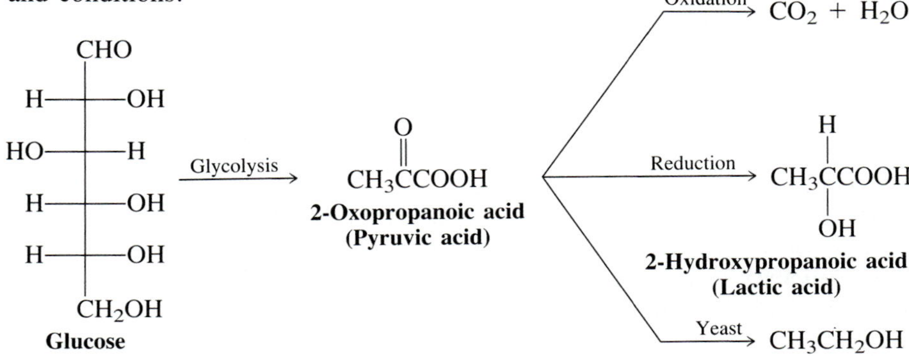

In an aerobic (oxygen-rich) environment, further oxidation results in CO_2 and H_2O, with a maximum gain in energy. If there is a poor supply of oxygen, as, for example, in an actively contracting muscle, incomplete reduction gives 2-hydroxypropanoic (lactic) acid (see Section 22-1). Some anaerobic organisms, such as yeast, convert 2-oxopropanoic (pyruvic) acid into ethanol.

The three polysaccharides cellulose, starch, and glycogen are also called *glucans* because they are composed almost exclusively of glucose molecules. There exist a number of other minor glucans in nature. In some organisms, the polysaccharide in their metabolism is based on other sugar units. *Inulin* is a fructosan, yielding mainly fructose on hydrolysis. Inulin, rather than starch, is the food reservoir in the roots of Compositae and can be isolated from the tubers of dahlias and Jerusalem artichokes. Other nonglucan polysaccharides found in nature are xylans, mannans, and galactans.

Modified Sugars in Nature

Many of the naturally occurring sugars have a modified structure or are attached to some other organic molecule. There is a large class of sugars in which at least one of the hydroxy groups has been replaced by an amine function. They are called **glycosylamines** if the nitrogen is attached to the anomeric carbon and **amino deoxy sugars** if it is located elsewhere.

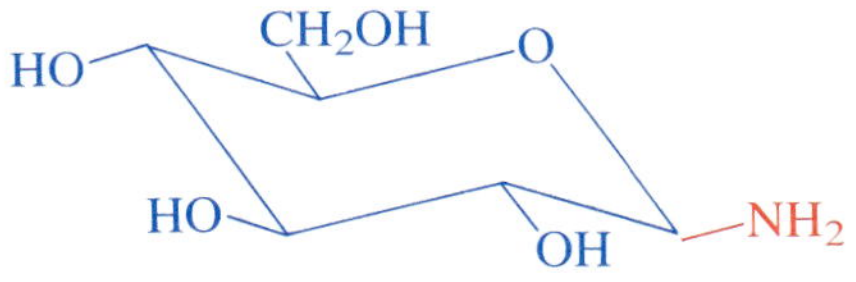

β-D-Glucopyranosylamine

(A glycosylamine)

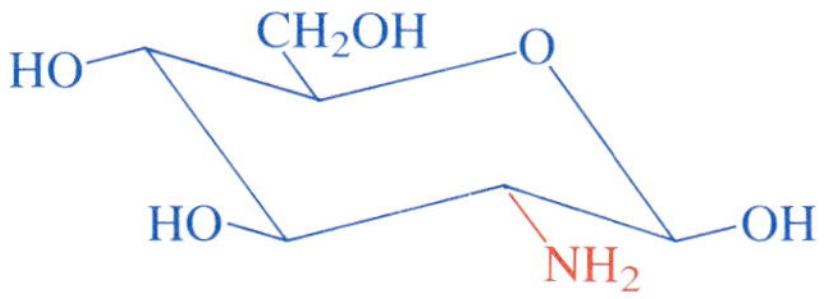

2-Amino 2-deoxy-D-glucopyranose

(An amino sugar)

Glycosylamines are present in other important biological polymers: the *nucleic acids* (Section 27-7), which contain the genetic code and are responsible for protein biosynthesis. Ribonucleic acid is a polymer made of repeating units called nucleotides, which are substituted glycosylamines. An example is uridylic acid.

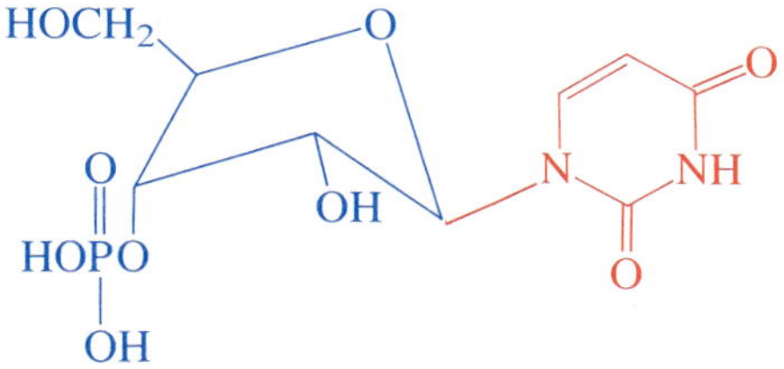

Uridylic acid

A repeating β-D-glucosamine unit is found in *chitin,* the building material (with calcium carbonate) of crab and lobster shells.

Chitin

If a sugar is attached by its anomeric carbon to the hydroxy group of another complex residue, it is called a **glycosyl group.** The remainder of the molecule (or the product of removing the sugar by hydrolysis) is called the **aglycon.** Examples are *amygdalin* and *adriamycin*. Amygdalin, a derivative of the β-1,6-glycoside gentiobiose, is isolated from apricot pits, along with its monosaccharide analog *Laetrile*.

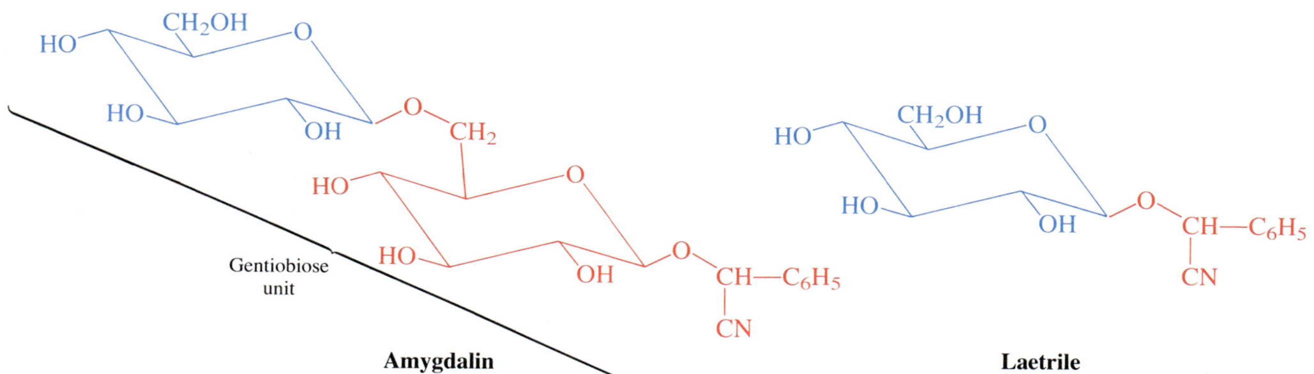

Amygdalin **Laetrile**

Laetrile is notorious for its controversial use in the treatment of cancer. Proponents claim that the molecule interacts with the enzyme β-glucosidase in such a way as to release cyanide ion, which destroys cancer cells. Noncancerous cells are said to be able to enzymatically deactivate cyanide to cyanate ion. As a consequence of these claims, many patients, particularly those for whom conventional therapy has failed, have sought refuge in Laetrile. Its curative effects have been disputed by the medical profession. The National Institutes of Health conducted a study that indicated the inefficacy of the drug. However, critics of this study assert that it was flawed by improper use and application of Laetrile.

Adriamycin is a member of the anthracycline family of antibiotics. Adriamycin and its deoxy analog daunomycin have been remarkably effective in the treatment of a wide variety of human cancers. They now constitute a cornerstone of combination cancer chemotherapy. The aglycon part of these systems is a linear tetracyclic framework incorporating an anthraquinone moiety (Section 25-4). The amino sugar is called daunosamine.

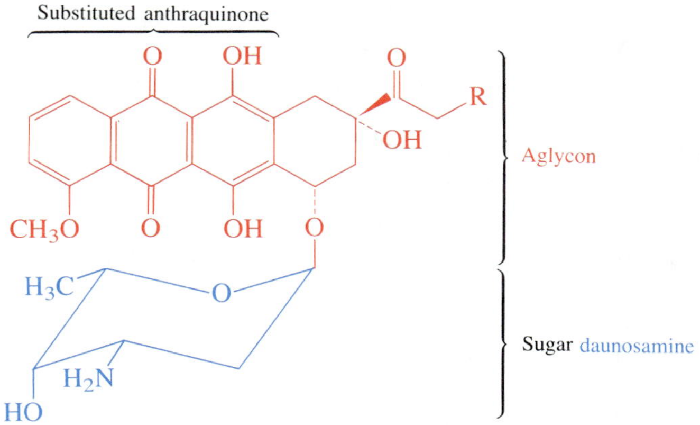

Substituted anthraquinone

Aglycon

Sugar daunosamine

Adriamycin (R = OH)
Daunomycin (R = H)

An unusual group of antibiotics, the *aminoglycoside antibiotics,* is based almost exclusively on oligosaccharide structures. Of particular therapeutic importance is streptomycin (an antituberculosis agent), isolated in 1944 from cultures of the mold *Streptomyces griseus*.

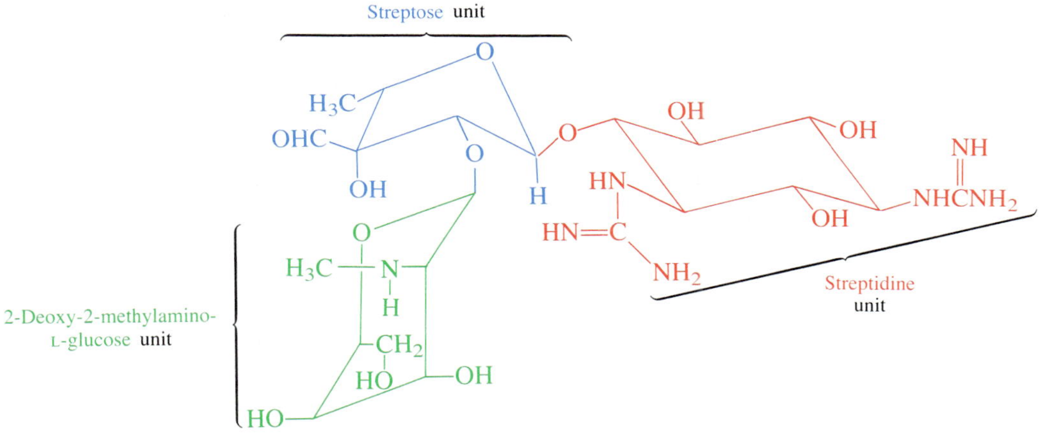

Streptose unit

2-Deoxy-2-methylamino-L-glucose unit

Streptidine unit

Streptomycin

The molecule consists of three subunits, two of which are sugars: the furanose streptose and the glucose derivative 2-deoxy-2-methylamino-L-glucose (an example of the rare L form). The third, streptidine, is actually a hexasubstituted cyclohexane.

In summary, sucrose is a dimer derived from linking α-D-glucopyranose with β-D-fructofuranose at the anomeric centers. It shows inversion of its optical rotation on hydrolysis to its mutarotating component sugars. The disaccharide maltose is a glucose dimer in which the components are linked by a carbon–oxygen bond between an α-anomeric carbon of a glucose molecule and C-4 of the second. Cellobiose is almost identical structurally with maltose, but has a β configuration at the acetal carbon. Lactose has a β-D-galactose linked to glucose in the same manner as in cellobiose. The polysaccharides cellulose, starch, and glycogen are all polyglucosides. Cellulose consists of repeating dimeric cellobiose units. Starch, on the other hand, may be regarded as a polymaltose derivative. Its occasional branching poses a challenge to enzymatic degradation, as in that of glycogen. Metabolism of these polymers first gives monomeric glucose, which is then oxidized to 2-oxopropanoic (pyruvic) acid. Depending on the organism and physiological conditions, this molecule may then be oxidized further to CO_2 and H_2O or reduced to 2-hydroxypropanoic (lactic) acid or ethanol. Finally, many sugars occur in nature in modified form or as simple appendages to other structures. Examples include amino sugars, glycosylamines, amygdalin, and adriamycin. The aminoglycoside antibiotics consist entirely of saccharide molecules, modified and unmodified.

Summary of New Reactions

1 CYCLIC HEMIACETAL FORMATION IN SUGARS

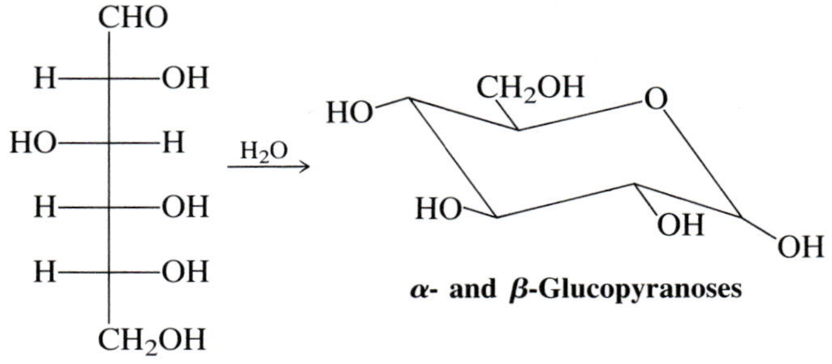

CHO, H—OH, HO—H, H—OH, H—OH, CH_2OH $\xrightarrow{H_2O}$ HO, CH_2OH, O, HO, OH, OH

α- and β-Glucopyranoses

2 MUTAROTATION

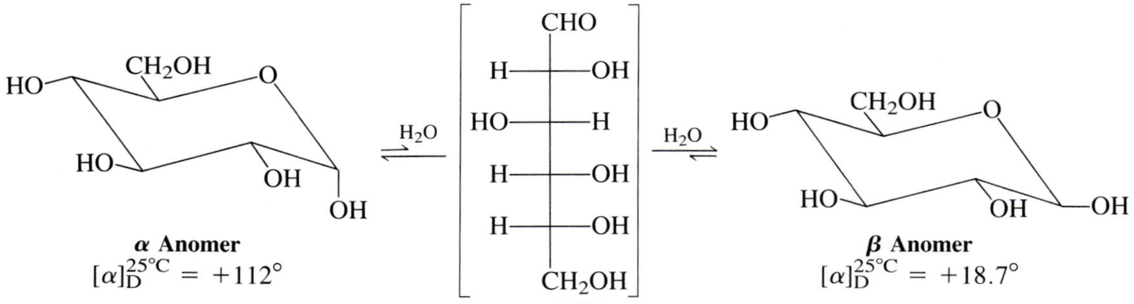

HO, CH_2OH, O, HO, OH, OH $\underset{}{\overset{H_2O}{\rightleftharpoons}}$ [CHO, H—OH, HO—H, H—OH, H—OH, CH_2OH] $\overset{H_2O}{\rightleftharpoons}$ HO, CH_2OH, O, HO, OH, OH

α Anomer
$[\alpha]_D^{25°C} = +112°$

Equilibrium $[\alpha]_D^{25°C} = +52.7°$

β Anomer
$[\alpha]_D^{25°C} = +18.7°$

3 OXIDATION

Tests for reducing sugars:

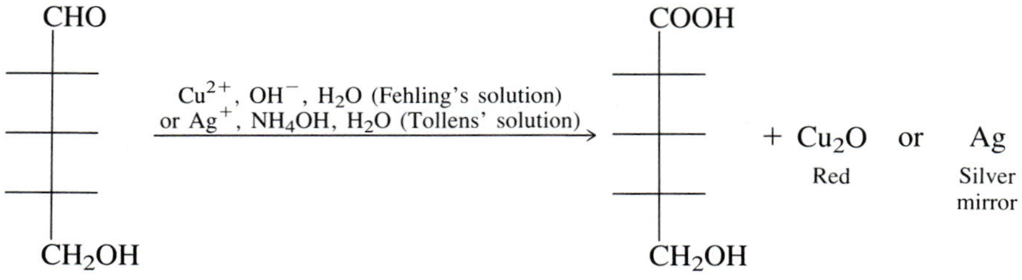

CHO, CH_2OH $\xrightarrow[\text{or } Ag^+,\ NH_4OH,\ H_2O \text{ (Tollens' solution)}]{Cu^{2+},\ OH^-,\ H_2O \text{ (Fehling's solution)}}$ COOH, CH_2OH + Cu_2O or Ag

Red — Silver mirror

Aldonic acid synthesis:

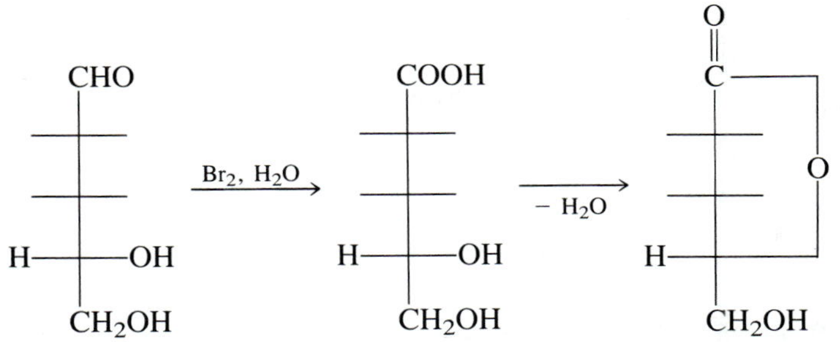

CHO / — / — / H—OH / CH_2OH $\xrightarrow{Br_2,\ H_2O}$ COOH / — / — / H—OH / CH_2OH $\xrightarrow[-\ H_2O]{}$ C(=O)—O— (ring) / — / — / H— / CH_2OH

Aldonic acid **γ-Lactone**

Aldaric acid synthesis:

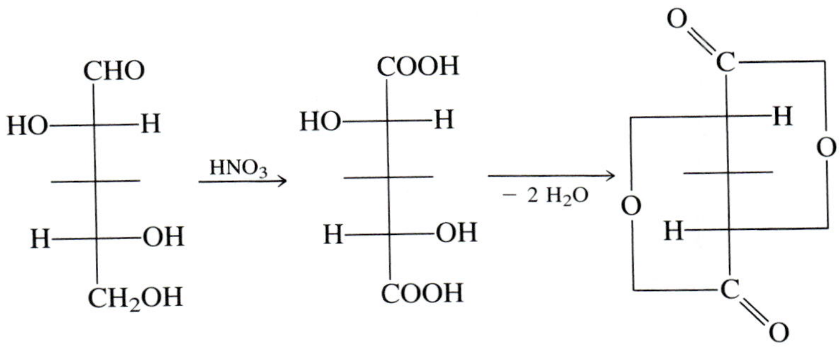

CHO / HO—H / — / H—OH / CH_2OH $\xrightarrow{HNO_3}$ COOH / HO—H / — / H—OH / COOH $\xrightarrow[-\ 2\ H_2O]{}$ O=C—O— (ring) / —H / — / H— / C=O (ring)

Aldaric acid **Double γ-lactone**

4 SUGAR DEGRADATION

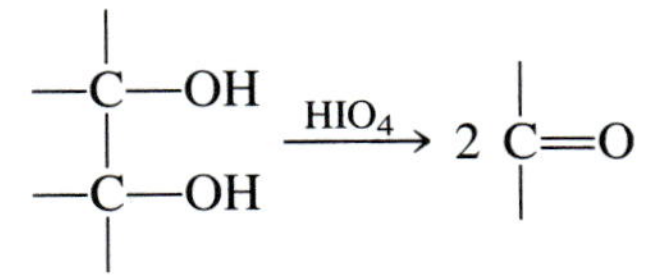

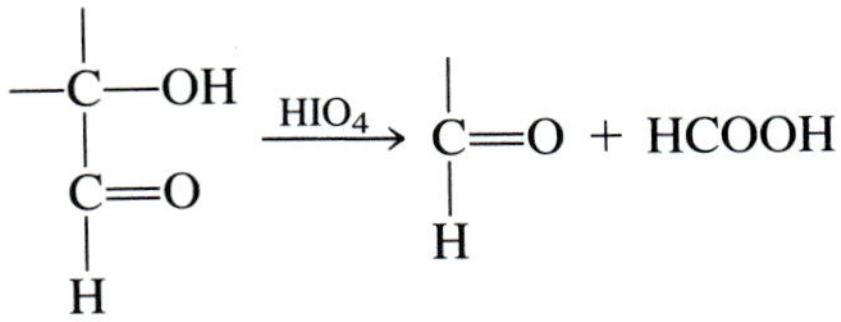

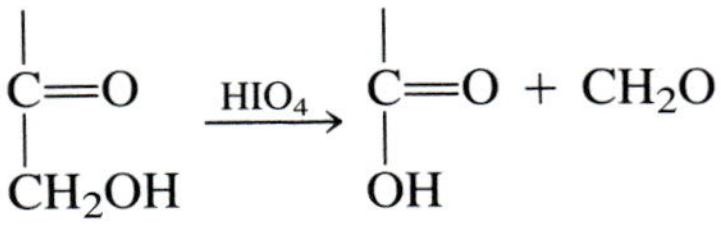

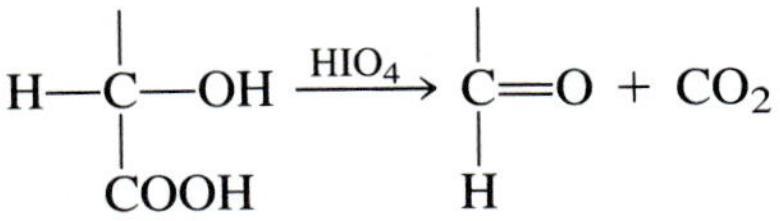

$$\begin{array}{c}|\\-\mathrm{C}-\mathrm{OH}\\|\\-\mathrm{C}-\mathrm{OH}\\|\end{array} \xrightarrow{HIO_4} 2\ \begin{array}{c}|\\\mathrm{C}{=}\mathrm{O}\\|\end{array}$$

$$\begin{array}{c}|\\-\mathrm{C}-\mathrm{OH}\\|\\\mathrm{C}{=}\mathrm{O}\\|\\\mathrm{H}\end{array} \xrightarrow{HIO_4} \begin{array}{c}|\\\mathrm{C}{=}\mathrm{O}\\|\\\mathrm{H}\end{array} + \mathrm{HCOOH}$$

$$\begin{array}{c}|\\\mathrm{C}{=}\mathrm{O}\\|\\\mathrm{CH_2OH}\end{array} \xrightarrow{HIO_4} \begin{array}{c}|\\\mathrm{C}{=}\mathrm{O}\\|\\\mathrm{OH}\end{array} + \mathrm{CH_2O}$$

$$\begin{array}{c}|\\\mathrm{H}-\mathrm{C}-\mathrm{OH}\\|\\\mathrm{COOH}\end{array} \xrightarrow{HIO_4} \begin{array}{c}|\\\mathrm{C}{=}\mathrm{O}\\|\\\mathrm{H}\end{array} + \mathrm{CO_2}$$

5 REDUCTION

CHO → ($NaBH_4$) → CH_2OH

H—C—OH

CH_2OH → CH_2OH

6 HYDRAZONES AND OSAZONES

CHO → ($C_6H_5NHNH_2$, 1 equivalent) → $CH{=}N{-}NHC_6H_5$ + H_2O

H—C—OH

CH_2OH

Phenylhydrazone

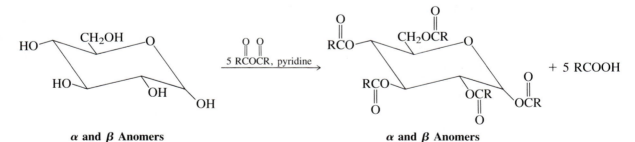

CHO, H—C—OH, H—C—OH, CH_2OH → (3 $C_6H_5NHNH_2$) → $HC{=}N{-}NHC_6H_5$, $C{=}N{-}NHC_6H_5$, H—C—OH, CH_2OH + $C_6H_5NH_2$ + NH_3 + 2 H_2O

Osazone

7 ESTERS

HO, CH_2OH, O, HO, OH, OH → (5 RCOCR (O O), pyridine) → RCO (O), CH_2OCR (O), O, RCO (O), OCR (O), OCR (O) + 5 RCOOH

***α* and *β* Anomers** → ***α* and *β* Anomers**

8 GLYCOSIDES

HO, CH_2OH, O, HO, OH, OH ⇌ (CH_3OH, H^+ / H_2O, H^+) HO, CH_2OH, O, HO, OH, OCH_3 + H_2O

***α* and *β* Anomers** ⇌ ***α* and *β* Anomers**

9 ETHERS

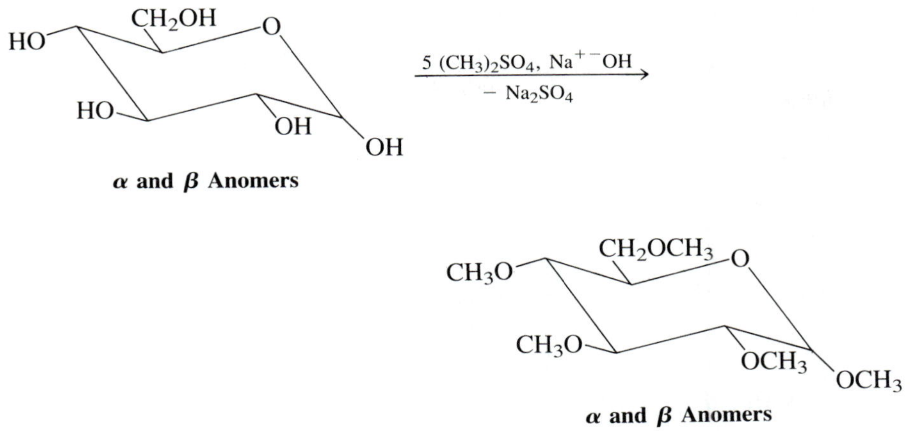

HO, CH_2OH, O, HO, OH, OH

α and β Anomers

$\xrightarrow[-\ Na_2SO_4]{5\ (CH_3)_2SO_4,\ Na^+\ ^-OH}$

CH_3O, CH_2OCH_3, O, CH_3O, OCH_3, OCH_3

α and β Anomers

10 CYCLIC ACETALS

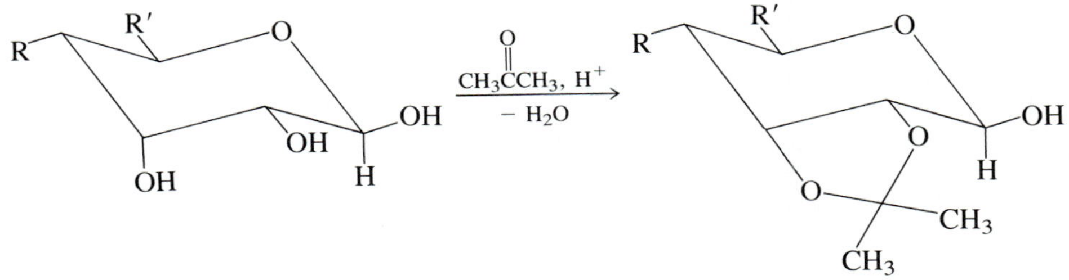

R, R′, O, OH, OH, OH, H

$\xrightarrow[-\ H_2O]{CH_3\overset{O}{\overset{\|}{C}}CH_3,\ H^+}$

R, R′, O, OH, O, H, O, CH_3, CH_3

11 CYCLIC CARBONATES

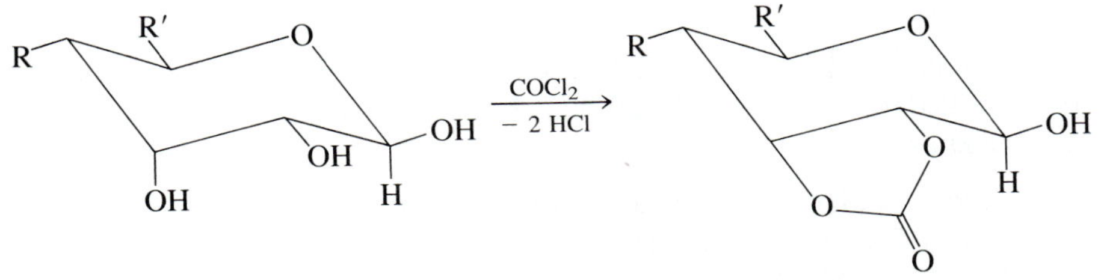

R, R′, O, OH, OH, OH, H

$\xrightarrow[-\ 2\ HCl]{COCl_2}$

R, R′, O, OH, O, H, O, O

12 KILIANI-FISCHER SYNTHESIS

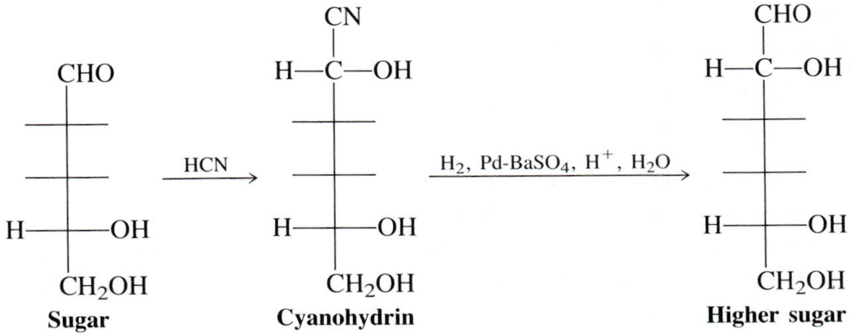

CHO; H—C—OH; CH_2OH
Sugar

$\xrightarrow{HCN}$

CN; H—C—OH; H—C—OH; CH_2OH
Cyanohydrin

$\xrightarrow{H_2,\ Pd\text{-}BaSO_4,\ H^+,\ H_2O}$

CHO; H—C—OH; H—C—OH; CH_2OH
Higher sugar

13 WOHL DEGRADATION

CHO / H—OH / — / H—OH / CH_2OH $\xrightarrow[-\,H_2O]{NH_2OH}$ CH═NOH / H—OH / — / H—OH / CH_2OH $\xrightarrow{CH_3\overset{O}{\overset{\|}{C}}O\overset{O}{\overset{\|}{C}}CH_3}$ N≡C / H—$O\overset{O}{\overset{\|}{C}}CH_3$ / — / H—$O\overset{O}{\overset{\|}{C}}CH_3$ / $CH_2O\underset{O}{\underset{\|}{C}}CH_3$ $\xrightarrow[-\,CH_3COO^-Na^+,\ -\,NaCN]{CH_3ONa,\ CH_3OH}$ CHO / — / H—OH / CH_2OH

Sugar — **Oxime** — **Protected cyanohydrin** — **Lower sugar**

14 RUFF DEGRADATION

CHO / H—OH / — / H—OH / CH_2OH $\xrightarrow{Br_2,\ H_2O}$ COOH / H—OH / — / H—OH / CH_2OH $\xrightarrow[-\,CO_2]{Fe^{3+},\ H_2O_2}$ CHO / — / H—OH / CH_2OH

Summary of Important Concepts

1 Carbohydrates are naturally occurring polyhydroxy carbonyl compounds that can exist as monomers, dimers, oligomers, and polymers.

2 Monosaccharides are called aldoses if they are aldehydes and ketoses if they are ketones. The chain length is indicated by the prefix tri-, tetr-, pent-, hex-, and so on.

3 Most natural carbohydrates belong to the D family; that is, the stereocenter farthest from the carbonyl group has the same configuration as that in (*R*)-(+)-2,3-dihydroxypropanal [D-(+)-glyceraldehyde].

4 The keto forms of carbohydrates exist in equilibrium with the corresponding five-membered (furanoses) or six-membered (pyranoses) cyclic hemiacetals. The new stereocenter formed by cyclization is called the anomeric carbon, and the two anomers are designated α and β.

5 Equilibration between anomers in solution gives rise to changes in the measured optical rotation called mutarotation.

6 The reactions of the saccharides are characteristic of carbonyl, alcohol, and hemiacetal groups. They include oxidation of the aldehyde to the carboxy function of aldonic acids, double oxidation to aldaric acids, oxidative cleavage of vicinal diol units, reduction to alditols, condensations, esterifications, and acetal formations.

7 The synthesis of higher sugars is based on Kiliani-Fischer chain lengthening, the new carbon being introduced by cyanide ion. The synthesis of lower sugars relies on Wohl or Ruff chain shortening, a terminal carbon being expelled either as cyanide ion or as CO_2.

8 The techniques of chain lengthening and shortening can be used with the symmetry properties of aldaric acids for structural correlations within the family of the aldoses.

9 Di- and higher saccharides are formed by ether formation between monomers; the ether bridge usually includes at least one hemiacetal hydroxy group.

10 The change in optical rotation observed in aqueous solutions of sucrose, called the inversion of sucrose, is due to the equilibration of the starting sugar with the various cyclic and anomeric forms of its component monomers.

11 Many natural sugars contain modified backbones. Amino groups may have replaced hydroxy groups, there may be substituents of various complexity (aglycons), the backbone carbon atoms of a sugar may lack oxygens, and (rarely) the sugar may adopt the L configuration.

Problems

1 To which classes of sugars do the following monosaccharides belong? Which are D and which are L?

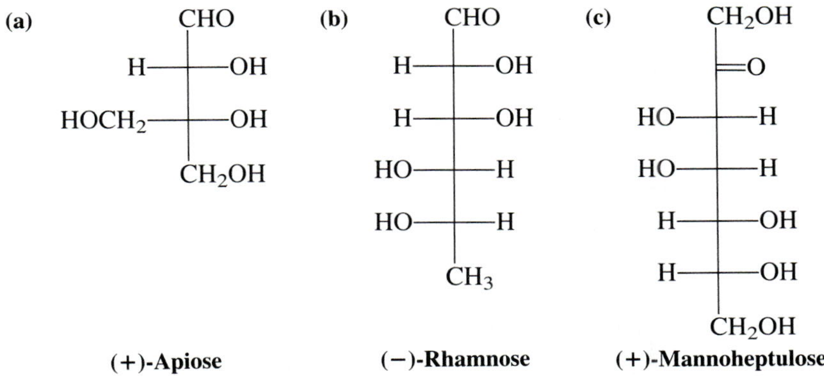

2 The designations D and L as applied to sugars refer to the configuration of the highest-numbered stereocenter. If the configuration of the highest-numbered stereocenter of D-ribose (Figure 23-1) is switched from D to L, is the product L-ribose? If not, what is the product? How is it related to D-ribose (i.e., what kind of isomers are they)?

3 Draw open-chain (Fischer-projection) structures for L-(+)-ribose and L-(−)-glucose (see Exercise 23-2). What are the systematic names of these sugars?

4 Identify the following sugars, which are represented by unconventionally drawn Fischer projections. Hint: It will be necessary to convert these into more-conventional representations *without* inverting any of the stereocenters.

(a) 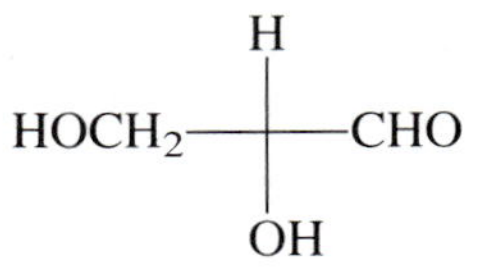 H; HOCH$_2$—CHO; OH

(b) 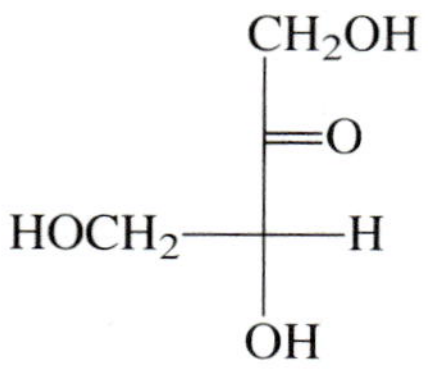 CH$_2$OH; =O; HOCH$_2$—H; OH

(c) 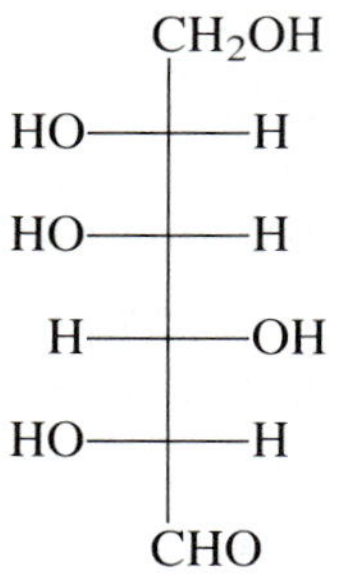 CH$_2$OH; HO—H; HO—H; H—OH; HO—H; CHO

(d) 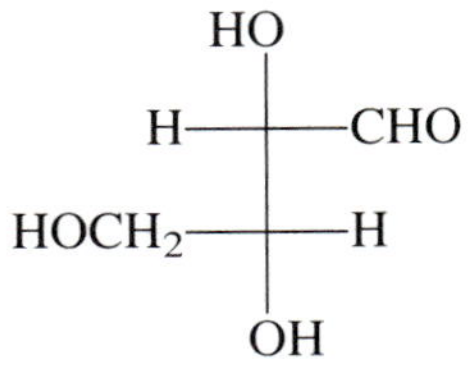 HO; H—CHO; HOCH$_2$—H; OH

(e) 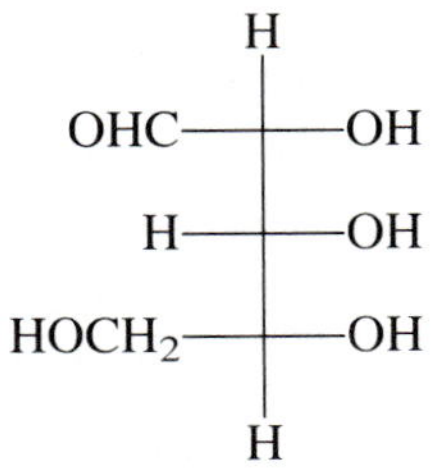 H; OHC—OH; H—OH; HOCH$_2$—OH; H

5 Redraw each of the following sugars as a Fischer projection, in open-chain form, and find its common name.

(a) 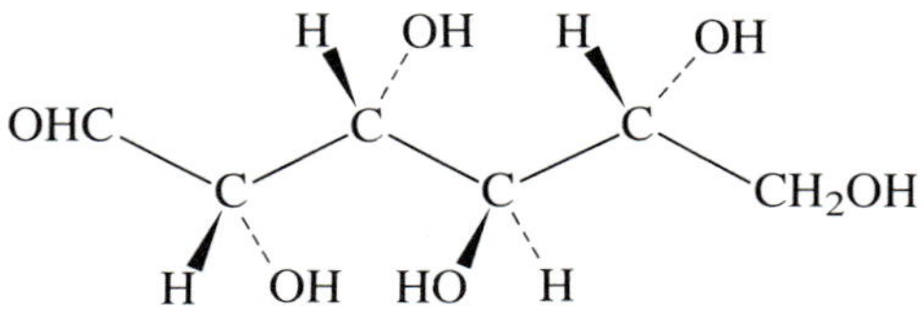 OHC, H, OH, H, OH, CH$_2$OH, H, OH, HO, H

(b) 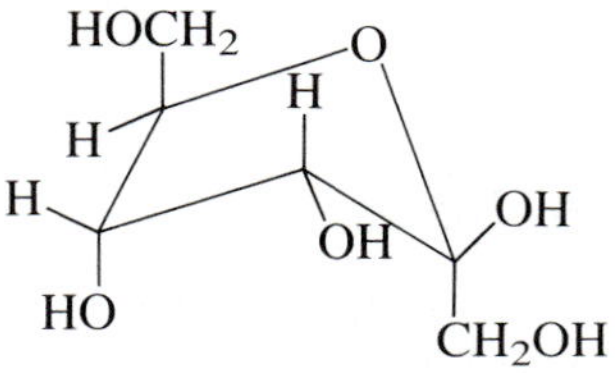 HOCH$_2$, O, H, H, H, OH, OH, HO, CH$_2$OH

(c) 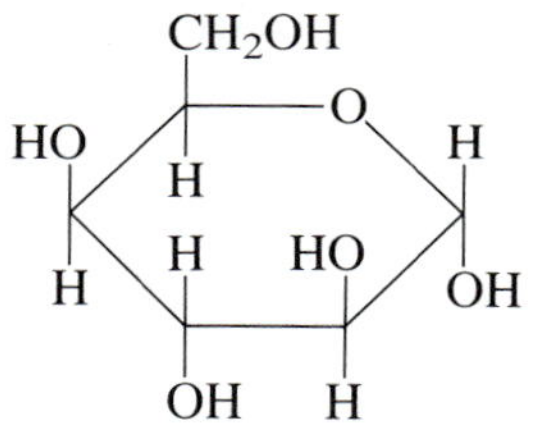 CH$_2$OH, O, HO, H, H, H, H, HO, OH, OH, H

(d) 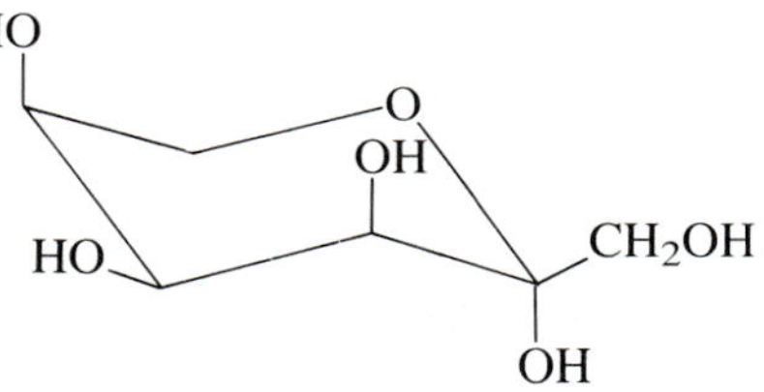 HO, O, OH, HO, CH$_2$OH, OH

6 For each of the following sugars, (i) draw all reasonable cyclic structures, using either Haworth or conformational formulas, (ii) indicate which structures are pyranoses and which are furanoses, and (iii) label α and β anomers.

(a) (−)-Threose.
(b) (−)-Allose.
(c) (−)-Ribulose.
(d) (+)-Sorbose.
(e) (+)-Mannoheptulose (Problem 1).

7 Are any of the sugars in Problem 5 incapable of mutarotation? Why?

8 Draw the most-stable conformation of each of the following sugars in its pyranose form.

(a) α-D-Arabinose.
(b) β-D-Galactose.
(c) β-D-Mannose.
(d) α-D-Idose.

9 Write the expected products of the reaction of each of the following sugars with (i) Br_2, H_2O; (ii) HNO_3, H_2O, 60°C; (iii) $NaBH_4$; and (iv) excess $C_6H_5NHNH_2$, CH_3CH_2OH, Δ. Find the common names of all the products.

(a) D-(−)-Threose.
(b) D-(+)-Xylose.
(c) D-(+)-Galactose.

10 Draw the Fischer projection of an aldohexose that will give the same osazone as **(a)** D-(−)-idose and **(b)** L-(−)-altrose.

11 **(a)** Which of the aldopentoses (Figure 23-1) would give optically active alditols upon reduction with $NaBH_4$?

(b) Using D-fructose, illustrate the results of $NaBH_4$ reduction of a ketose. Is the situation more complicated than reduction of an aldose? Explain.

12 Which of the following glucoses and glucose derivatives are capable of undergoing mutarotation?

(a) α-D-Glucopyranose.

(b) Methyl α-D-glucopyranoside.

(c) Methyl α-2,3,4,6-tetra-*O*-methyl-D-glucopyranoside (i.e., the tetramethyl ether at carbons 2, 3, 4, and 6).

(d) α-2,3,4,6-Tetra-*O*-methyl-D-glucopyranose.

(e) α-D-Glucopyranose 1,2-monopropanone acetal.

13 **(a)** Explain why the oxygen at C-1 of an aldopyranose can be methylated so much more easily than the other oxygens in the molecule.

(b) Explain why the methyl ether unit at C-1 of a fully methylated aldopyranose can be hydrolyzed so much more easily than the other methyl ether functions in the molecule.

(c) Write the expected product(s) of the following reaction:

$$\text{D-Fructose} \xrightarrow{CH_3OH,\ 0.25\%\ HCl,\ H_2O}$$

14 Of the four aldopentoses, two readily form double acetals when treated with excess acidic propanone (acetone), but the other two form only monoacetals. Explain.

15 **(a)** A mixture of (*R*)-2,3-dihydroxypropanal (D-glyceraldehyde) and 1,3-dihydroxypropanone (1,3-dihydroxyacetone) that is treated with aqueous NaOH rapidly yields a mixture of three sugars: D-fructose, D-sorbose, and racemic dendroketose (only one enantiomer is shown here). Explain this result by means of a detailed mechanism.

(b) The same mixture of products is also obtained if either the aldehyde or the ketone *alone* is treated with base. Explain. Hint: Closely examine the intermediates in your mechanism for part **a**.

CH_2OH
=O
H— —OH
$HOCH_2$— —OH
CH_2OH

Dendroketose

16 A fundamental transformation in carbohydrate metabolism is the enzyme-catalyzed conversion of D-fructose-1,6-diphosphate into the two smaller monophosphates shown:

$CH_2OP(O)(OH)_2$ / =O / HO— —H / H— —OH / H— —OH / $CH_2OP(O)(OH)_2$ $\xrightarrow{\text{Enzyme}}$ $CH_2OP(O)(OH)_2$ / =O / CH_2OH + CHO / H— —OH / $CH_2OP(O)(OH)_2$

There is a nonenzymic reaction that accomplishes the same type of bond cleavage. What is this version called? Illustrate the nonenzymic mechanism with the carbohydrate structures above. Hint: See Section 16-3.

17 Write or draw the missing reagents and structures **a** through **g.** What is the common name of **g?**

$$\text{D-(+)-Xylose} \xrightarrow{(a)} \underset{\text{D-Xylonic acid}}{\textbf{(b)}} \xrightarrow{(c)} \underset{\text{Methyl D-xylonate}}{\textbf{(d)}} \xrightarrow{NH_3,\ \Delta} \underset{\textbf{(e)}}{C_5H_{11}NO_5} \xrightarrow{Br_2,\ NaOH} CO_2 + \underset{\textbf{(f)}}{C_4H_{11}NO_4} \xrightarrow{\Delta} NH_3 + \underset{\textbf{(g)}}{C_4H_8O_4}$$

The preceding sequence (called the Weerman degradation) achieves the same end as what procedures described in this chapter?

18 D-Sedoheptulose is a sugar that plays a role in a metabolic cycle (the *pentose oxidation cycle*) that converts glucose into 2,3-dihydroxypropanal (glyceraldehyde) plus three equivalents of CO_2. Determine the structure of D-sedoheptulose from the following information.

(i) $\text{D-Sedoheptulose} \xrightarrow{6\ HIO_4} 4\ \overset{O}{\overset{\|}{H C}}OH + 2\ H\overset{O}{\overset{\|}{C}}H + CO_2$

(ii) $\text{D-Sedoheptulose} \xrightarrow{C_6H_5NHNH_2}$ an osazone identical with that formed by another sugar, aldoheptose A

(iii) $\text{Aldoheptose A} \xrightarrow{\text{Ruff degradation}}$ aldohexose B

(iv) $\text{Aldohexose B} \xrightarrow{HNO_3,\ H_2O,\ \Delta}$ an optically active product

(v) $\text{Aldohexose B} \xrightarrow{\text{Ruff degradation}}$ D-ribose

19 Illustrate the results of Kiliani-Fischer chain elongation of D-talose. How many products are formed? Draw their structure(s). On treatment with warm HNO_3 do the product(s) give optically active or inactive dicarboxylic acids?

20 Write a plausible mechanism for the decarboxylation step in the general Ruff degradation.

21 Fischer's solution to the problem of sugar structures was actually much more difficult to achieve experimentally than Section 23-3 implies. For one thing, the only sugars that he could readily obtain from natural sources were glucose, mannose, and arabinose. (Erythrose and threose were, in fact, not then available at all, either naturally or synthetically.) His ingenious solution required a source of gulose so that he could make the critical comparison of dicarboxylic acids described at the end of the section. Unfortunately, gulose does not occur in nature; so Fischer had to make it. His synthesis, from glucose, was difficult because at a key point he got a troublesome mixture of products. Nowadays the following synthesis might be used.

Write or draw the missing reagents and structures **a** through **g.** Use Fischer projections for all structures. Follow the instructions and hints in parentheses.

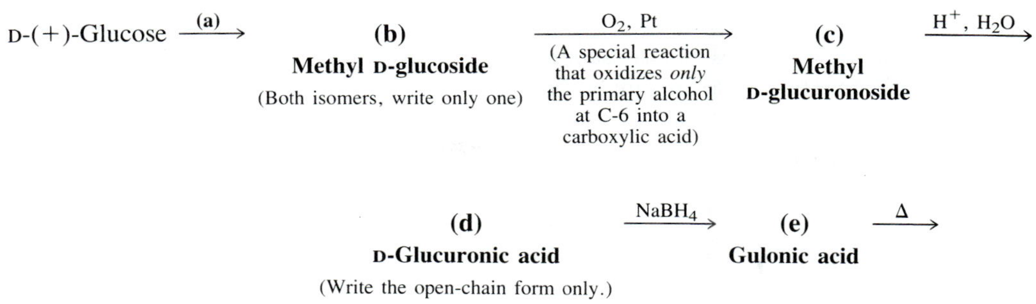

Is the gulose that Fischer synthesized from D-glucose an L sugar or a D sugar? (Be careful. Fischer himself got the wrong answer at first, and that confused *everybody* for years.)

22 **(a)** Write a detailed mechanism for the isomerization of β-D-fructofuranose from the hydrolysis of sucrose into an equilibrium mixture of the β-pyranose and β-furanose forms.

(b) Although fructose usually appears as a furanose when it is part of a polysaccharide, in the pure crystalline form, fructose adopts a β-pyranose structure. Draw β-D-fructopyranose in its most-stable conformation. In water at 20°C, the equilibrium mixture contains about 80% pyranose and 20% furanose.

(c) What is the free-energy difference between the pyranose and furanose forms at this temperature?

(d) Pure β-D-fructopyranose has $[\alpha]_D^{20°C} = -132°$. The equilibrium pyranose-furanose mixture has $[\alpha]_D^{20°C} = -92°$. Calculate $[\alpha]_D^{20°C}$ for pure β-D-fructofuranose.

23 Classify each of the following sugars and sugar derivatives as either reducing or nonreducing.

(a) D-Glyceraldehyde.
(b) D-Arabinose.
(c) β-D-Arabinopyranose 3,4-monopropanone acetal.
(d) β-D-Arabinopyranose double propanone acetal.
(e) D-Ribulose.
(f) D-Galactose.
(g) Methyl β-D-galactopyranoside.
(h) β-D-Galacturonic acid.
(i) β-Cellobiose.
(j) α-Lactose.

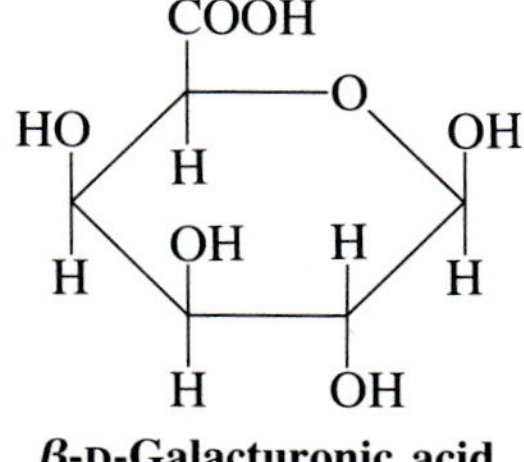

β-D-Galacturonic acid

24 Explain why raffinose (Section 23-4) is a nonreducing sugar.

25 Is α-lactose capable of mutarotation? Write an equation to illustrate the process.

26 Trehalose, sophorose, and turanose are disaccharides. Trehalose is found in the cocoons of some insects, sophorose turns up in a few bean varieties, and turanose is an ingredient in low-grade honey made by bees with indigestion from a diet of pine tree sap. Identify among the following structures those that correspond to trehalose, sophorose, and turanose on the basis of the information (i, ii, iii) given below.

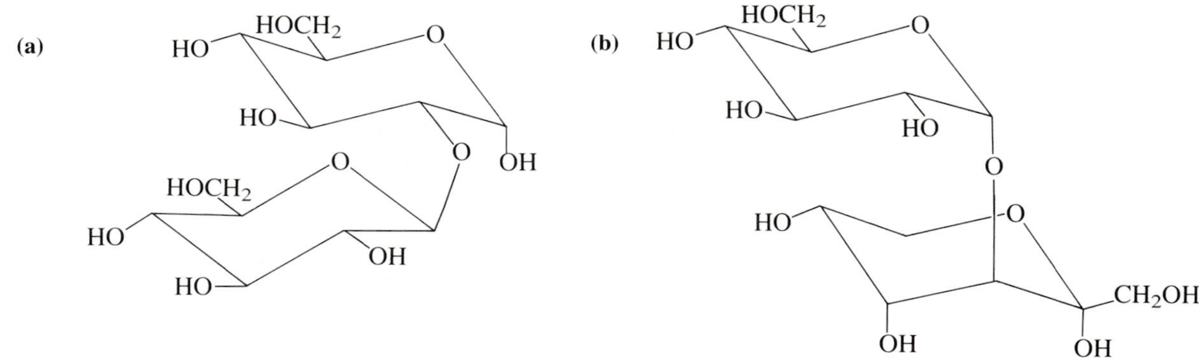

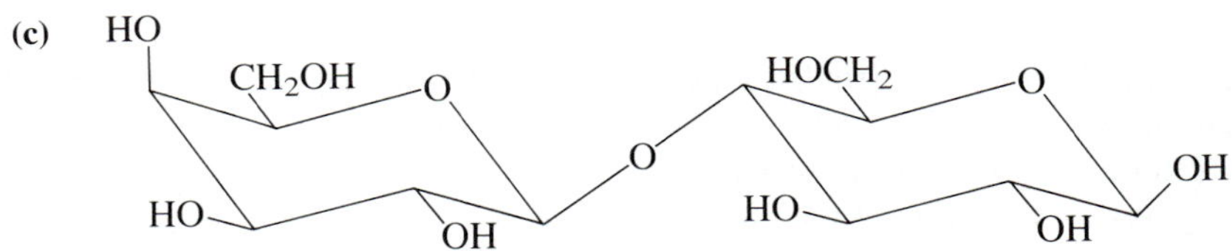

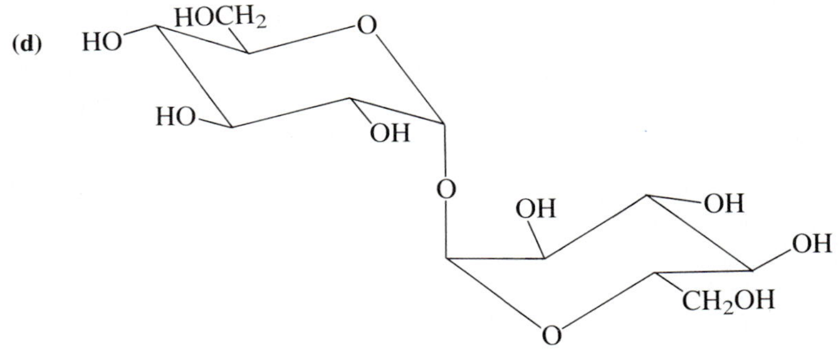

(i) Turanose and sophorose are reducing sugars. Trehalose is a nonreducing sugar.

(ii) On hydrolysis, sophorose and trehalose give two molecules each of aldoses. Turanose gives one molecule of an aldose and one molecule of a ketose.

(iii) The two aldoses that comprise sophorose are anomers of each other.

27 Rutinose is a reducing sugar that is part of several bioflavonoids, a group of compounds found in many plants. They have therapeutic value in maintaining the strength of blood-vessel walls. One rutinose-containing bioflavonoid is hesperidin, which is present in lemons and oranges.

On acid hydrolysis, rutinose gives one equivalent each of D-glucose and a sugar C with the formula $C_6H_{12}O_5$. Sugar C reacts with four equivalents of HIO_4 to give three equivalents of methanoic acid and one equivalent each of methanal and ethanal. Sugar C can be synthesized as shown at the left.

L-(−)-Mannose

1. $HSCH_2CH_2SH$, $ZnCl_2$
2. Raney Ni (Section 15-5)
3. Br_2, H_2O

↓

sugar C

(a) What is the structure of sugar C?

Complete methylation of rutinose (by excess dimethyl sulfate) followed by acid hydrolysis gives one equivalent of methyl 2,3,4-tri-*O*-methyl-D-glucoside and one equivalent of the 2,3,4-tri-*O*-methyl derivative of sugar C.

(b) What possible structure(s) of rutinose are consistent with these data?

28 Vitamin C (ascorbic acid) occurs almost universally in the plant and animal kingdoms. (According to Linus Pauling, mountain goats biosynthesize from 12 to 14 grams of it per day.) Animals produce it from D-glucose in the liver by the four-step sequence D-glucose → D-glucuronic acid (see Problem 21) → D-glucuronic acid γ-lactone → L-gulonic acid γ-lactone → vitamin C.

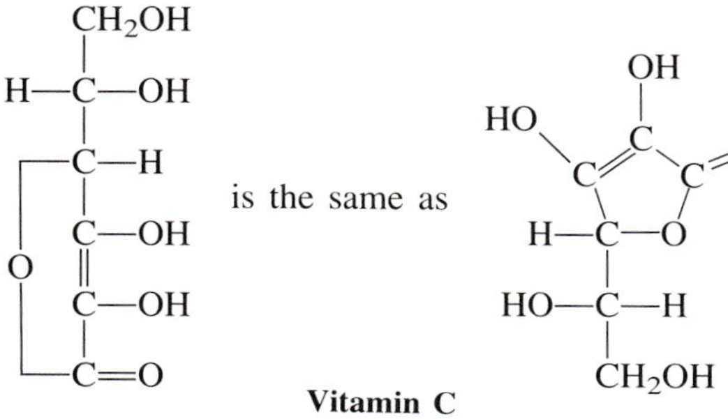

CH_2OH, H—C—OH, C—H, C—OH, C—OH, C=O (ring O) is the same as [ring structure: HO, OH, C=O, H—C—O, HO—C—H, CH_2OH]

Vitamin C

The enzyme that catalyzes the last reaction, L-gulonolactone oxidase, is absent from humans, some monkeys, guinea pigs, and birds, presumably because of a defective gene resulting from a mutation that may have occurred some 60 million years ago. As a result, we have to get our vitamin C from food or make it synthetically. In fact, the ascorbic acid in almost all vitamin supplements is synthetic. An outline of one of the major commercial syntheses follows. Write or draw the missing reagents and product structures **a** through **f.**

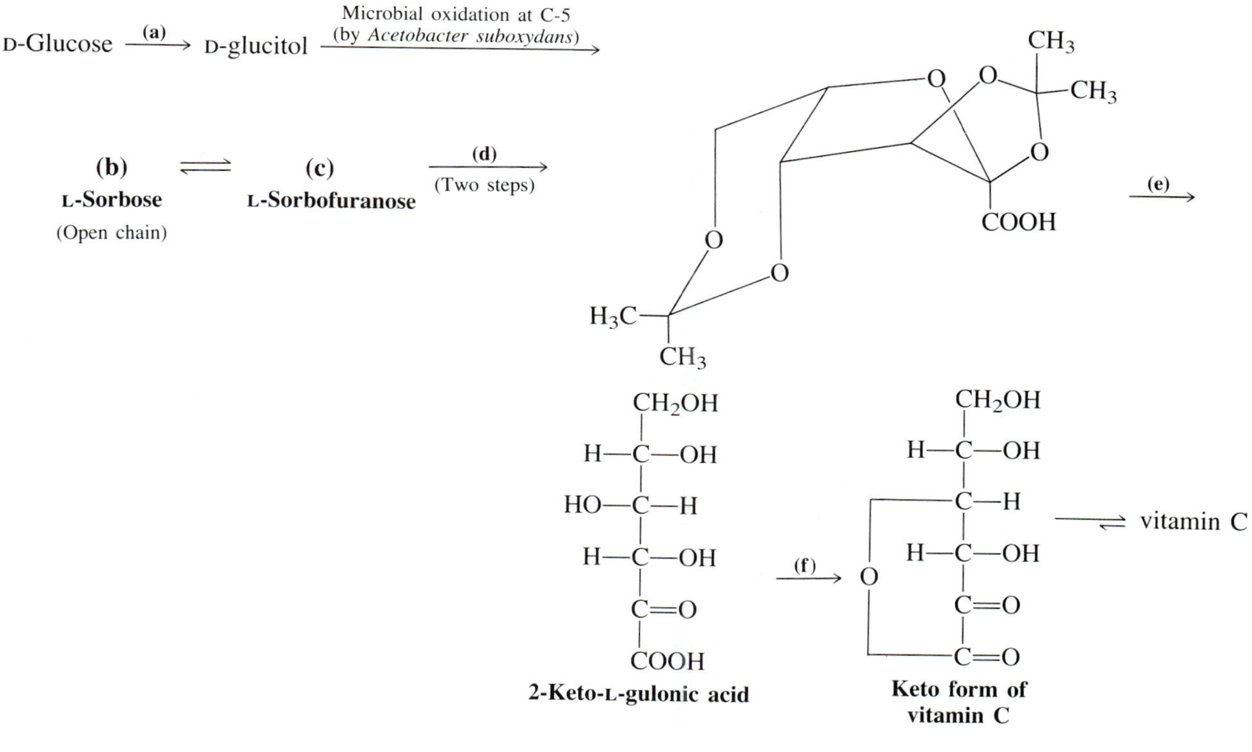

D-Glucose —(a)→ D-glucitol —Microbial oxidation at C-5 (by *Acetobacter suboxydans*)→

(b) **L-Sorbose** (Open chain) ⇌ **(c)** **L-Sorbofuranose** —(d) (Two steps)→ [diacetonide structure with CH_3, CH_3, COOH, H_3C, CH_3] —(e)→

CH_2OH, H—C—OH, HO—C—H, H—C—OH, C=O, COOH
2-Keto-L-gulonic acid —(f)→ CH_2OH, H—C—OH, C—H, H—C—OH, C=O, C=O (ring O)
Keto form of vitamin C ⇌ vitamin C

CHAPTER 24

Substituted Benzenes

Chapters 22 and 23 were concerned with the chemistry of mainly aliphatic molecules having two or more functional groups, particularly those containing oxygen. This chapter expands on this theme by taking a closer look at the reactivity of aromatic compounds having added functionality. The questions it will answer are these: How reactive is the benzene ring compared with other functional groups? How much does the presence of the aromatic ring modify the behavior of neighboring reactive centers?

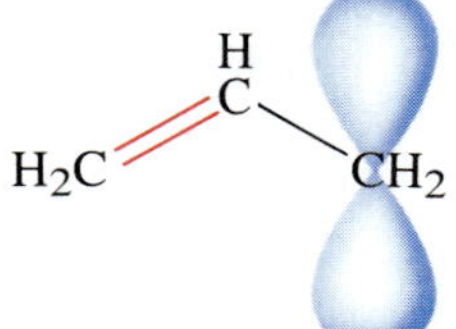

The 2-propenyl (allyl) system

24-1 Phenylmethyl (Benzyl) Is an Analog of 2-Propenyl (Allyl): Benzylic Resonance Stabilization

In orbital terms, the phenylmethyl (benzyl) group, $C_6H_5CH_2$, may be regarded as a benzene ring in which its π system interacts with an extra p orbital—in 2-propenyl (allyl, Section 14-1), a double bond is extended in a similar way. Because of this interaction, the benzene ring stabilizes adjacent radical, cationic, and anionic centers by resonance.

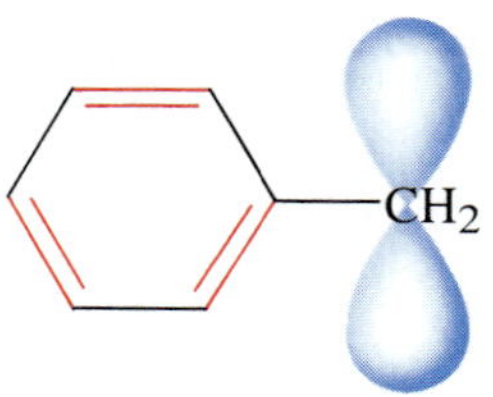

The phenylmethyl (benzyl) system

Side-Chain Halogenation of Alkylbenzenes Is Facilitated by the Resonance Stabilization of Benzylic Radicals

When benzene is exposed to chlorine or bromine, normally no reaction will occur unless a Lewis acid catalyst is added. The Lewis acid catalyzes halogenation of the ring (Section 19-5).

$$\text{no reaction} \xleftarrow{\text{Br—Br}} C_6H_6 \xrightarrow[-\,\text{HBr}]{\text{Br—Br, FeBr}_3} C_6H_5Br$$

In contrast, methylbenzene (toluene), when attacked by chlorine or bromine, gives halogenated products even in the absence of a catalyst (although heat or light facilitates reaction). Analysis of these products shows that reaction takes place at the methyl group, not at the aromatic ring. Excess halogen can lead to multiple substitutions.

$$C_6H_5CH_2H + Br_2 \xrightarrow{\Delta} C_6H_5CH_2Br + HBr$$

(Bromomethyl)-benzene

$$C_6H_5CH_3 \xrightarrow[-\,HCl]{Cl_2,\ h\nu} C_6H_5CH_2Cl \xrightarrow[-\,HCl]{Cl_2,\ h\nu} C_6H_5CHCl_2 \xrightarrow[-\,HCl]{Cl_2,\ h\nu} C_6H_5CCl_3$$

(Chloromethyl)-benzene **(Dichloromethyl)-benzene** **(Trichloromethyl)-benzene**

In these processes, one equivalent of hydrogen halide is generated for each substitution. As in the halogenation of alkanes (Sections 3-5 through 3-8) and the allylic halogenation of alkenes (Section 14-2), the mechanism of benzylic halogenation proceeds through free-radical intermediates.

Mechanism of the Halogenation of Methylbenzene (Toluene):

STEP 1: Halogen dissociation

$$X_2 \xrightarrow{\Delta \text{ or } h\nu} 2\ X\cdot$$

STEP 2: Phenylmethyl (benzyl) radical formation

$$X\cdot + C_6H_5CH_2H \longrightarrow C_6H_5\dot{C}H_2 + HX$$

Phenylmethyl (benzyl) radical

STEP 3: Phenylmethyl (benzyl) radical halogenation

$$C_6H_5\dot{C}H_2 + X_2 \longrightarrow C_6H_5CH_2X + X\cdot$$

The halogen molecule dissociates in a first step (which is induced by heat or light) to give two halogen atoms. One of them then abstracts a benzylic hydrogen, leaving a phenylmethyl (benzyl) radical. This intermediate reacts with another halogen molecule to give the product (halomethyl)benzene and another halogen atom, which reenters the cycle.

FIGURE 24-1
Three ways of depicting delocalization in the phenylmethyl (benzyl) radical.

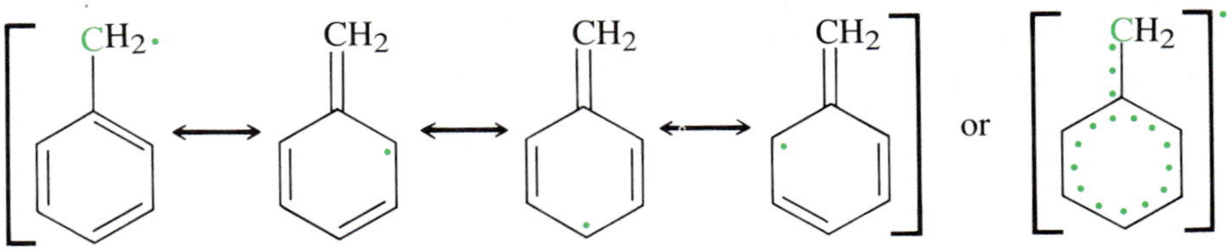

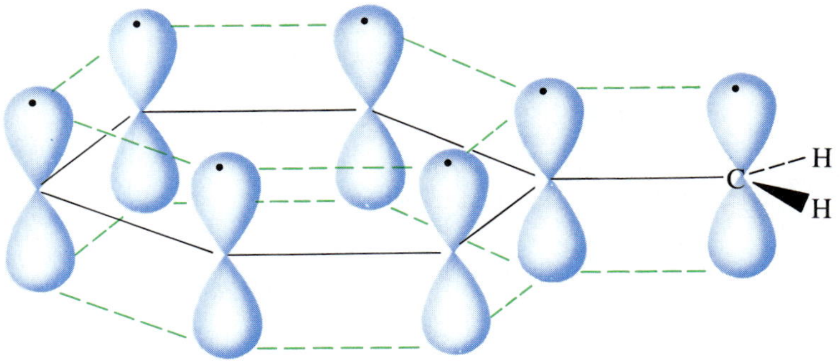

The reason for the ease of benzylic halogenations is the relative weakness of the benzylic C–H bond: $DH^\circ = 87$ kcal mole^{-1}, close to the value for the allylic C–H bond. This bond is weak because after its rupture the benzene π system can enter into resonance with the adjacent radical center, a phenomenon called *benzylic resonance*. As with 2-propenyl (allyl), the effect can be pictured in a variety of ways, such as resonance structures, dotted lines indicating the extent of delocalization, or atomic orbitals (Figure 24-1).

On inspection of the various resonance structures of the intermediate phenylmethyl (benzyl) radical, you may wonder why the halogen attacks only at the benzylic position, but not *ortho* or *para*. The answer is simple: reaction at any but the benzylic carbon would destroy the aromatic character of the benzene ring, making the transition state of such a reaction energetically unfavored.

EXERCISE 24-1

N-Bromobutanimide (*N*-bromosuccinimide, NBS) is frequently used in benzylic brominations (as in allylic brominations, Section 14-2). The following two brominations have been carried out with this reagent. Explain the difference in the rates. Hint: It has nothing to do with the structure of the reagent, but rather with the structures of the intermediate radicals.

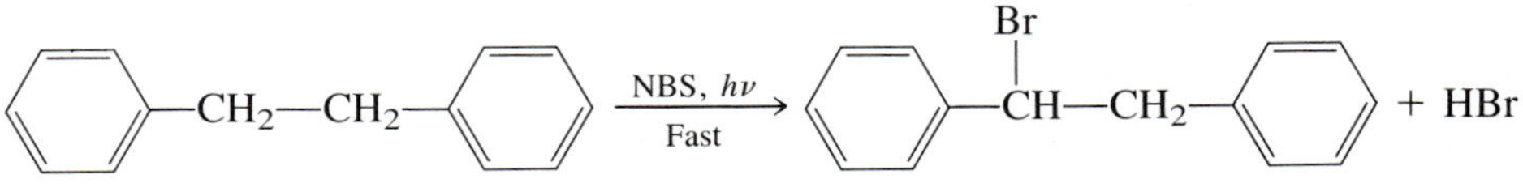

1,2-Diphenylethane

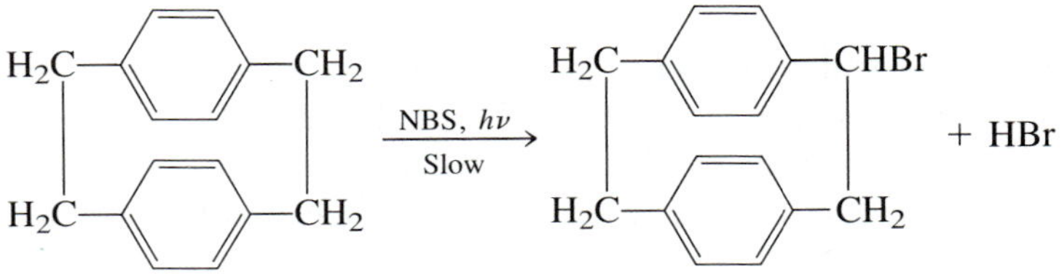

[2.2]Paracyclophane

BOX 24-1

Triphenylmethyl: A Stable Radical

Phenyl substitution at the benzylic center increases the stability of the corresponding radicals. The triphenylmethyl radical is stable enough to exist in solution even at room temperature, in equilibrium with a predominant dimer. This dimer is not (as you might have expected) hexaphenylethane but a molecule in which one triphenylmethyl group is linked to the *para* position of one of the phenyl substituents of another.

The Triphenylmethyl Radical and Its Dimer

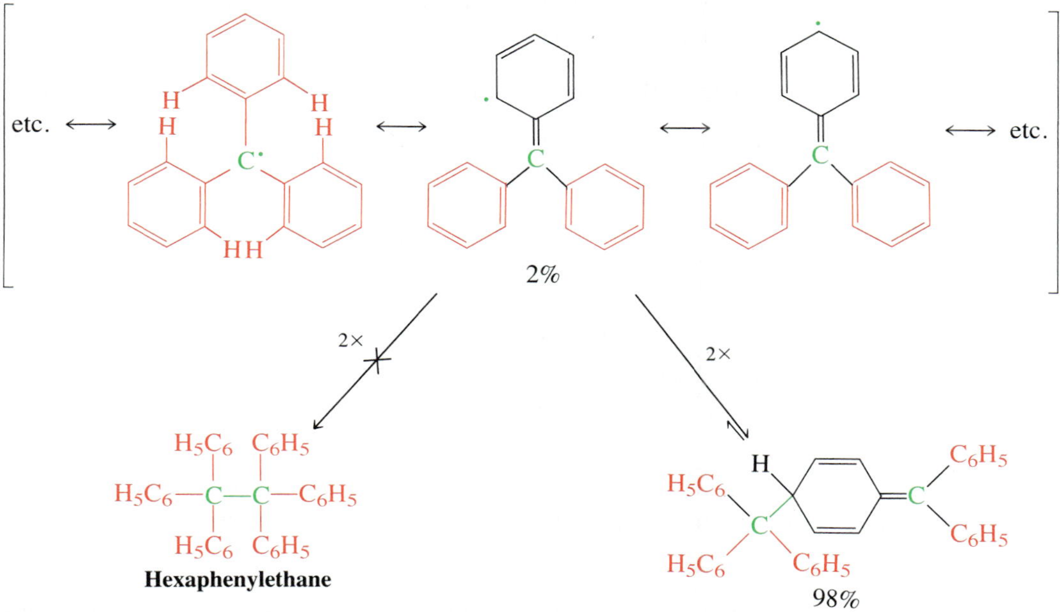

etc. ⟷ [triphenylmethyl radical resonance structures] ⟷ etc.

2%

2× (not formed): Hexaphenylethane, $(H_5C_6)_3C{-}C(C_6H_5)_3$

2×: dimer, $(H_5C_6)_2(C_6H_5)C$–(H)cyclohexadiene$=C(C_6H_5)_2$, 98%

Hexaphenylethane

Steric hindrance explains this unusual result: hexaphenylethane is much more crowded around the newly formed C–C bond than the alternative. This effect even outweighs the loss of the aromaticity of one benzene ring.

EXERCISE 24-2

Suggest a spectroscopic technique (and predict the results of its application) that would distinguish between the two suggested structures of the triphenylmethyl dimer.

The Positive Charge in Benzylic Cations Is Delocalized

Benzylic resonance also has a profound effect on the reactivity of benzylic halides and sulfonates. For example, the 4-methylbenzenesulfonate of phenylmethanol (benzyl alcohol) reacts with ethanol much more rapidly than the corresponding ethyl sulfonate does.

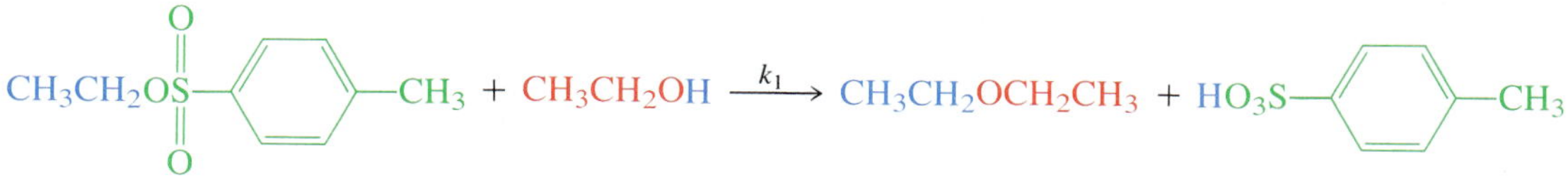

$$CH_3CH_2OSO_2{-}C_6H_4{-}CH_3 + CH_3CH_2OH \xrightarrow{k_1} CH_3CH_2OCH_2CH_3 + HO_3S{-}C_6H_4{-}CH_3$$

Ethyl 4-methylbenzenesulfonate

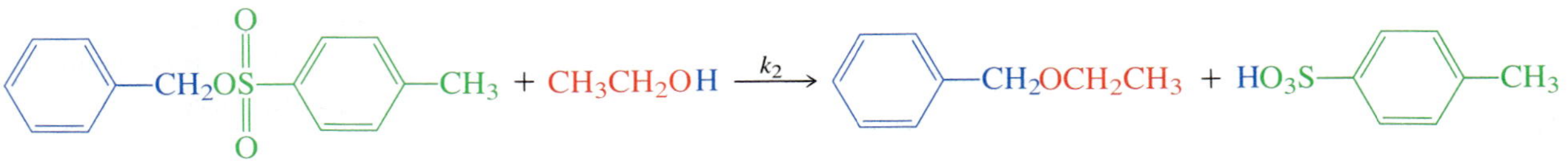

$$C_6H_5CH_2OSO_2C_6H_4CH_3 + CH_3CH_2OH \xrightarrow{k_2} C_6H_5CH_2OCH_2CH_3 + HO_3SC_6H_4CH_3$$

Phenylmethyl 4-methylbenzenesulfonate **(Ethoxymethyl)benzene**

$$k_2 : k_1 = 100:1$$

The reason for the rate difference is a difference in mechanism. Ethyl 4-methylbenzenesulfonate cannot undergo S_N1 reactions; it reacts with ethanol by an S_N2 process. This transformation is slow because alcohols are relatively poor nucleophiles. The phenylmethyl derivative, on the other hand, can dissociate to form a delocalized carbocation and therefore follow an S_N1 mechanism.

Mechanism of the Ethanolysis of Phenylmethyl 4-Methylbenzenesulfonate:

$$C_6H_5CH_2-OSO_2C_6H_4CH_3 \longrightarrow [C_6H_5\overset{+}{C}H_2 \longleftrightarrow \longleftrightarrow \longleftrightarrow] + {}^-O_3SC_6H_4CH_3 \xrightarrow{CH_3CH_2OH} C_6H_5CH_2OCH_2CH_3 + HO_3SC_6H_4CH_3$$

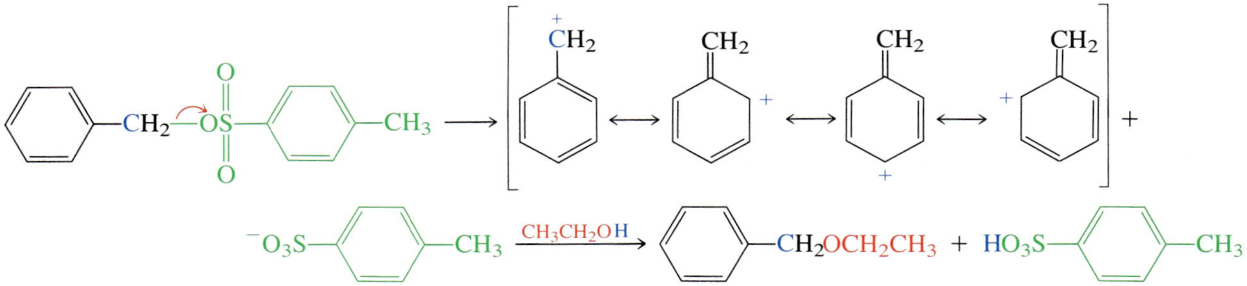

EXERCISE 24-3

Which one of the two chlorides will solvolyze more rapidly: (chloromethyl)benzene, $C_6H_5CH_2Cl$, or chlorodiphenylmethane, $(C_6H_5)_2CHCl$? Explain your answer.

EXERCISE 24-4

Phenylmethanol (benzyl alcohol) is converted into (chloromethyl)benzene in the presence of hydrogen chloride much more rapidly than ethanol is converted into chloroethane. Explain.

BOX 24-2

The Triphenylmethyl Cation Forms Stable Salts

The triphenylmethyl cation is sufficiently stable to be isolated in the form of salts.

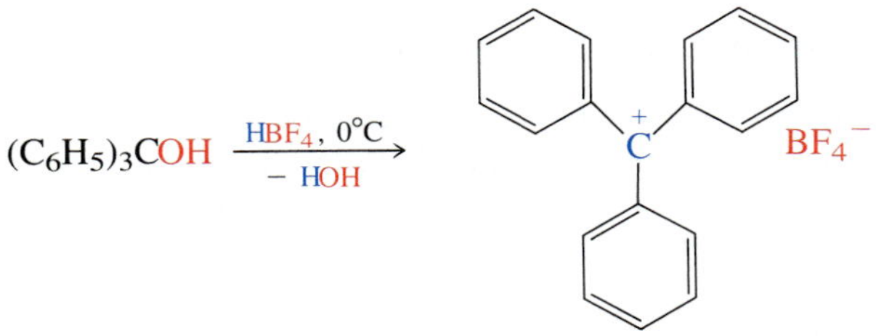

$$(C_6H_5)_3COH \xrightarrow[-\,HOH]{HBF_4,\ 0^\circ C} (C_6H_5)_3\overset{+}{C}\ BF_4^-$$

92%

Triphenylmethyl tetrafluoroborate (Stable orange salt)

FIGURE 24-2
Transition state for the S_N2 reaction of a (halomethyl)benzene.

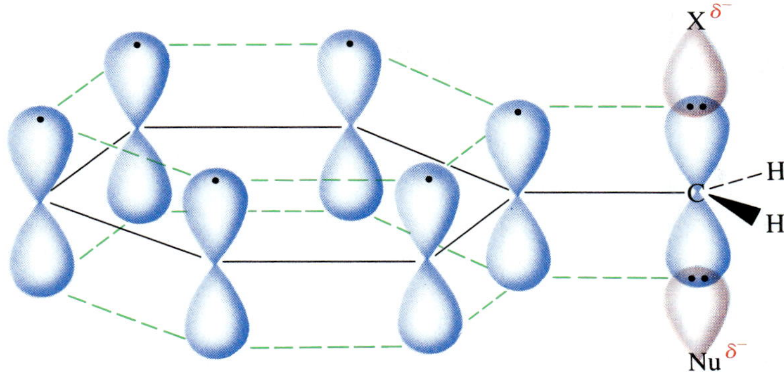

Benzylic resonance also enhances bimolecular substitution at benzylic carbons; such reactions are about two orders of magnitude as fast as the S_N2 reactions of primary haloalkanes. The transition state is stabilized by overlap of the benzenic with the adjacent p orbitals (Figure 24-2).

$C_6H_5CH_2Br + {}^{-}CN \xrightarrow{S_N2} C_6H_5CH_2CN + Br^{-}$

(Bromomethyl)benzene → 81% **Phenylethanenitrile**

The Methyl Group in Methylbenzene (Toluene) Is Relatively Acidic: Resonance in Phenylmethyl (Benzyl) Anions

A negative charge adjacent to a benzene ring, as in phenylmethyl (benzyl) anion, is stabilized by conjugation in much the same way that the corresponding radical and cation are stabilized.

$C_6H_5CH_3 \rightleftharpoons [C_6H_5\ddot{C}H_2^{-} \longleftrightarrow \longleftrightarrow \longleftrightarrow] + H^+$

$pK_a \sim 41$

The acidity of methylbenzene (toluene; $pK_a \sim 41$) is therefore considerably greater than that of ethane ($pK_a \sim 50$) and comparable to that of propene ($pK_a \sim 40$), which is deprotonated to produce the resonance-stabilized 2-propenyl (allyl) anion (Section 14-1). Consequently, methylbenzene (toluene) can be deprotonated by butyllithium. The corresponding Grignard reagent is made in the usual way from a (halomethyl)benzene and magnesium.

Deprotonation of Methylbenzene

$$C_6H_5CH_3 + CH_3CH_2CH_2CH_2Li \xrightarrow{(CH_3)_2NCH_2CH_2N(CH_3)_2,\ THF,\ \Delta} C_6H_5CH_2Li + CH_3CH_2CH_2CH_2H$$

Methylbenzene (Toluene) → **Phenylmethyllithium (Benzyllithium)**

BOX 24-3

Triphenylmethyl Anion: An Indicator

Triphenylmethane ($pK_a \sim 31$) is readily deprotonated by butyllithium to form the red triphenylmethyl anion. The color of this species can be used as an indicator in the deprotonation of compounds more acidic than triphenylmethane. For example, the end point of the deprotonation of methoxyethyne is detectable in this way.

$$\underset{\textbf{Methoxyethyne}}{CH_3OC{\equiv}CH} + (C_6H_5)_3CH\ (\text{trace}) \xrightarrow[-\ CH_3CH_2CH_2CH_2H]{CH_3CH_2CH_2CH_2Li\ (\text{slight excess})} CH_3OC{\equiv}C^-\ Li^+ + \underset{\text{Red}}{(C_6H_5)_3C^-\ Li^+}$$

EXERCISE 24-5

Which molecule in each of the following pairs is more reactive with the indicated reagents, and why?

(a) p-$CH_3OC_6H_4CH_2Br$ (ring bearing CH_2Br and OCH_3) or p-$CH_3OC_6H_4CH_2Cl$ (ring bearing CH_2Cl and OCH_3) with $NaOCH_3$ in CH_3OH

(b) $(C_6H_5)_2CH_2$ or $C_6H_5CH_3$ with $CH_3CH_2CH_2CH_2Li$

(c) $C_6H_5CH_2OH$ (ring bearing CH_2OH) or p-$O_2NC_6H_4CH_2OH$ (ring bearing CH_2OH and NO_2) with HCl

In summary, benzylic radicals, cations, and anions are stabilized by resonance with the benzene ring. This effect allows for relatively facile free-radical halogenations, S_N1 and S_N2 reactions, and benzylic anion formation.

24-2 Oxidation and Reduction of Benzene and Its Derivatives

Because of its aromatic character, the benzene ring is quite unreactive. Nevertheless, it does undergo certain transformations—in particular, electrophilic aromatic substitution (Chapters 19 and 20) and (under special conditions) even catalytic hydrogenation (Section 19-2). This section describes how certain other reagents oxidize and reduce the benzene nucleus and its substituents.

The Benzene Ring Undergoes Ozonolysis

Ozone cleaves alkenes at the double bonds (Section 12-5). Is ozone a strong enough reagent to attack benzene in a similar way? The answer is yes. Treatment of benzene with ozone followed by reductive work-up furnishes ethanedial (glyoxal). The same treatment of 1,2-dimethylbenzene (*o*-xylene) leads to a statistically predictable mixture of all possible oxidation products, indicating that all the double bonds in the aromatic ring are about equally reactive.

Ozonolysis of the Benzene Ring

Benzene $\xrightarrow{\text{1. } O_3\text{; 2. } (CH_3)_2S}$ 3 HC(=O)C(=O)H

Ethanedial (Glyoxal)

1,2-Dimethylbenzene $\xrightarrow{\text{1. } O_3\text{; 2. } (CH_3)_2S}$ $CH_3C(=O)C(=O)CH_3$ + $CH_3C(=O)C(=O)H$ + HC(=O)C(=O)H

2,3-Butanedione	**2-Oxopropanal**	**Ethanedial (Glyoxal)**
1 :	2 :	3

One-Electron Reduction of Benzene: A 1,4-Cyclohexadiene Synthesis

The catalytic hydrogenation of benzene cannot be stopped before the reduction of all three double bonds is complete. Selective reduction is possible, however, under the conditions of one-electron transfer, as in the conversion of alkynes into trans alkenes by sodium in liquid ammonia (Section 13-6), in the conversion of α,β-unsaturated carbonyl compounds into their saturated analogs (Section 16-5), in the pinacol coupling (Section 15-8), and in the acyloin condensation (Section 22-1). Thus, when benzene is dissolved in a mixture of liquid ammonia and ethanol and treated with an alkali metal, the resulting product is 1,4-cyclohexadiene. This reaction is named the **Birch* reduction.**

Birch Reduction of Benzene

Benzene $\xrightarrow{\text{Na, liquid } NH_3,\ CH_3CH_2OH}$ 1,4-Cyclohexadiene (H, H at C-1 and H, H at C-4)

1,4-Cyclohexadiene

*Professor Arthur J. Birch, b. 1915, Cambridge University, England.

Mechanism of the Birch Reduction:

STEP 1: One-electron transfer

+ Na ⟶ Na^+ +

Benzene radical anion

STEP 2: Protonation

+ CH_3CH_2OH ⟶ + $CH_3CH_2O^-$

Cyclohexadienyl radical

STEP 3: Second one-electron transfer

+ Na ⟶ Na^+ +

Cyclohexadienyl anion

STEP 4: Second protonation

+ $CH_3CH_2O^-$ + CH_3CH_2OH ⟶ + $CH_3CH_2O^-$

1,3-Cyclohexadiene

1,4-Cyclohexadiene

The mechanism of the Birch reduction starts with transfer of one electron to benzene, furnishing a radical anion. Ethanol then protonates this species to give a cyclohexadienyl radical, ready for further one-electron reduction. The resulting cyclohexadienyl anion is then protonated again. Interestingly, although this protonation could lead to the thermodynamically more stable conjugated 1,3-cyclohexadiene, that pathway is avoided, and the sole product is 1,4-cyclohexadiene. This kinetic result is due to the uneven electron distribution in the anion.

Substituents can control the direction of the Birch reduction. Electron donors usually end up bound to the alkene carbons, whereas acceptors are found at the saturated positions of the products. However, the range of such controlling substituents is limited to those that are stable under the reaction conditions. Thus, alkynes, carbonyls, esters, and other readily reduced functions are incompatible with the reagents used in the Birch reduction.

CH_3, CH_2CH_3

Na, liquid NH_3, CH_3CH_2OH →

CH_3, CH_2CH_3

85%

1-Ethyl-2-methylbenzene

1-Ethyl-2-methyl-1,4-cyclohexadiene

OCH_3

Li, liquid NH_3, CH_3CH_2OH →

OCH_3

84%

Methoxybenzene (Anisole)

1-Methoxy-1,4-cyclohexadiene

COOH

1. Na, liquid NH_3, CH_3CH_2OH
2. H^+, H_2O →

COOH

Benzenecarboxylic acid (Benzoic acid)

2,5-Cyclohexadienecarboxylic acid

EXERCISE 24-6

Predict the outcome of the Birch reduction of the following substrates: (a) methylbenzene (toluene); (b) 2-methylbenzenecarboxylic (*o*-toluic) acid; (c) [benzocyclobutene structure]

EXERCISE 24-7

An important transformation of steroid hormones is the change of the estrone derivative A to the enone B. Propose a series of reactions that would effect it.

H_3C, OH, H, H, H, CH_3O - - - - - → H_3C, OH, H, H, H, H, O

A **B**

Special Oxidations and Reductions of Substituents Attached to Benzene

Methylbenzenes are oxidized by such reagents as $KMnO_4$, $Na_2Cr_2O_7$, and HNO_3 to the corresponding benzenecarboxylic (benzoic) acids.

Benzylic Oxidations of Methylbenzenes

CH_3

1. $KMnO_4$, HO^-, Δ
2. H^+, H_2O →

COOH

100%

Methylbenzene (Toluene)

Benzenecarboxylic acid (Benzoic acid)

$$\text{4-Chloro-2-(2,2,2-trichloroethyl)-1-methylbenzene} \xrightarrow{Na_2Cr_2O_7,\ H_2SO_4,\ CH_3COOH} \text{4-Chloro-2-(2,2,2-trichloroethyl)benzenecarboxylic acid (93\%)}$$

4-Chloro-2-(2,2,2-trichloro-ethyl)-1-methylbenzene

93%

4-Chloro-2-(2,2,2-trichloro-ethyl)benzenecarboxylic acid

$$\text{4-Fluoro-1-methyl-2-nitrobenzene} \xrightarrow{15\%\ HNO_3,\ H_2O,\ 190^\circ C,\ 5\ h} \text{4-Fluoro-2-nitrobenzenecarboxylic acid (69\%)}$$

4-Fluoro-1-methyl-2-nitrobenzene

69%

4-Fluoro-2-nitrobenzenecarboxylic acid

The mechanisms of these oxidations apparently are complex; it is thought that phenylmethyl (benzyl) cations are intermediates. Longer-chain alkyl groups are oxidatively cleaved at the benzylic position.

$$\text{1-Butyl-4-methylbenzene} \xrightarrow[2.\ H^+,\ H_2O]{1.\ KMnO_4,\ HO^-,\ \Delta} \text{1,4-Benzenedicarboxylic acid (80\%)}$$

1-Butyl-4-methylbenzene

80%

1,4-Benzenedicarboxylic acid (Terephthalic acid)

Industry uses oxygen (because it is inexpensive) as an oxidizing agent in the large-scale preparation of benzenecarboxylic acids. Special catalysts have been developed for this purpose. For example, benzenecarboxylic (benzoic) acid can be made from methylbenzene (toluene), oxygen, and a mixed cobalt-manganese ethanoate (acetate) catalyst. 1,2-Benzenedicarboxylic (phthalic) anhydride is derived from 1,2-dimethylbenzene (*o*-xylene) by vapor-phase oxidation with air and vanadium pentoxide (for the use of this catalyst in the oxidation of cyclohexanol to hexanedioic acid, see Section 17-5).

$$C_6H_5CH_3 \xrightarrow{O_2,\ Co(OCCH_3)_2,\ Mn(OCCH_3)_2,\ \Delta} C_6H_5COOH + H_2O$$

$$\text{1,2-Dimethylbenzene} \xrightarrow{O_2,\ V_2O_5,\ 350^\circ\text{–}400^\circ C} \text{Phthalic anhydride} + 3\ H_2O$$

1,2-Benzenedicarboxylic anhydride (Phthalic anhydride)

The special reactivity of the benzylic position is manifest also in the mild conditions required for the oxidation of benzylic alcohols to the corresponding carbonyl compounds. For example, manganese dioxide, MnO_2, performs this oxidation selectively in the presence of other (nonbenzylic) hydroxy groups. (Recall that MnO_2 was used in the conversion of allylic alcohols into α,β-unsaturated aldehydes and ketones; see Section 16-4.)

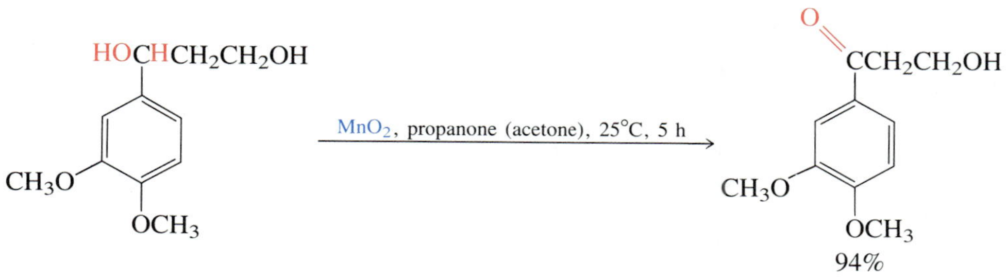

HOCHCH$_2$CH$_2$OH (3,4-dimethoxyphenyl) —[MnO_2, propanone (acetone), 25°C, 5 h]→ O=CCH$_2$CH$_2$OH (3,4-dimethoxyphenyl), 94%

Benzylic alcohols and ethers can also be catalytically hydrogenolized, a reaction in which the benzylic carbon–oxygen bond is broken.

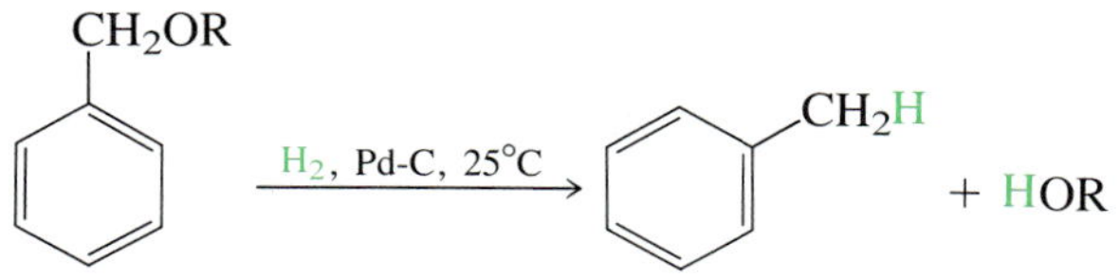

C$_6$H$_5$CH$_2$OR —[H_2, Pd-C, 25°C]→ C$_6$H$_5$CH$_2$H + HOR

This transformation is not possible for ordinary alcohols and ethers. As a result, the phenylmethyl (benzyl) substituent is a valuable protecting group for hydroxy functions, because it can be removed under neutral conditions.

EXERCISE 24-8

Write synthetic schemes that would connect the following starting materials with their products.

(a) 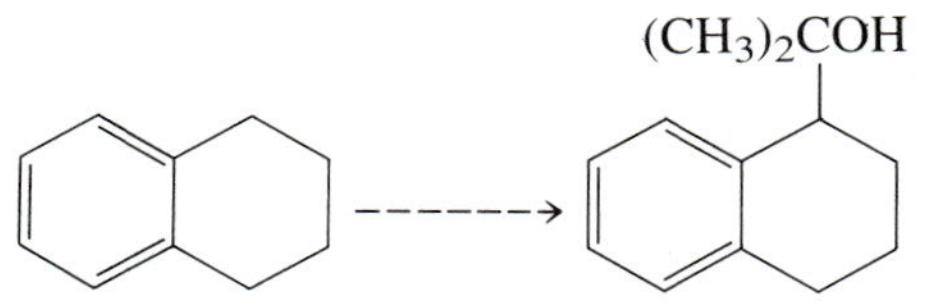 Tetrahydronaphthalene - - - - - -→ 1-[(CH$_3$)$_2$COH]-tetrahydronaphthalene

(b) C$_6$H$_5$CH$_2$CH$_3$ - - - - - -→ C$_6$H$_5$CH$_3$

(c) 1,2,4,5-tetramethylbenzene (H$_3$C, H$_3$C, CH$_3$, CH$_3$) - - - - - -→ benzenetetracarboxylic dianhydride (pyromellitic dianhydride)

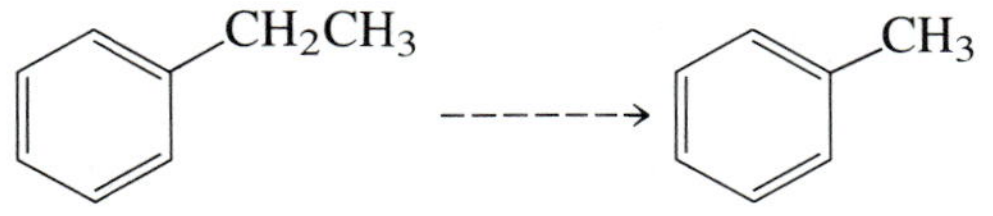

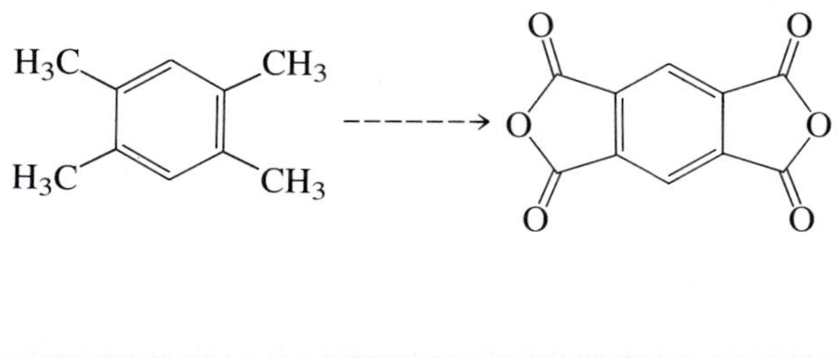

BOX 24-4

Phenylmethyl (Benzyl) as a Protecting Group for Alcohols

The use of a phenylmethyl (benzyl) protecting group in part of a synthesis of a sesquiterpene is shown in the following scheme, which also serves to review some key organic transformations.

HO, CH_3 → 1. NaH 2. $C_6H_5CH_2Br$, C_6H_6, DMSO; Protection of OH → $C_6H_5CH_2O$, CH_3 (80%) → CH_3COOH, H_2O; Deprotection of carbonyl group →

$C_6H_5CH_2O$, CH_3, O (93%) → $CH_3CH{=}P(C_6H_5)_3$, DMSO; Wittig reaction → $C_6H_5CH_2O$, CH_3, $CHCH_3$ (94%) → 1. BH_3 2. oxidation; Hydroboration-oxidation →

$C_6H_5CH_2O$, CH_3, $COCH_3$ (99%) → 1. CH_3Li 2. H^+, H_2O → $C_6H_5CH_2O$, CH_3, CH_3–C(CH_3)–OH (98%) → H_2, Pd-C; Deprotection of OH →

HO, CH_3, CH_3–C(CH_3)–OH (98%)

To summarize, the benzene nucleus, as well as its substituents, can be induced to undergo selective oxidations and reductions. Ozone oxidizes the ring; sodium in liquid ammonia reduces it. Benzylic oxidations of alkyl groups occur in the presence of permanganate, chromate, or nitric acid; benzylic alcohols are converted into the corresponding ketones by manganese dioxide. Industrial oxidations with oxygen use special catalysts (Co, Mn, V) for specific products, such as benzenecarboxylic (benzoic) acid from methylbenzene (toluene) and 1,2-benzenedicarboxylic (phthalic) anhydride from 1,2-dimethylbenzene (*o*-xylene). The benzylic ether function can be cleaved by hydrogenolysis in a transformation that allows the phenylmethyl (benzyl) substituent to be used as a protecting group for the hydroxy function in alcohols.

24-3 The Naming and Preparation of Benzenols (Phenols)

Arenes substituted by hydroxy groups are called **benzenols (phenols)** (Section 19-1). They are unusual because of their enolic structure. Remember that enols are usually unstable: they tautomerize easily to the corresponding ketones because of the relatively strong carbonyl bond. Benzenols (phenols), however, prefer the enol to the keto form because the aromatic character of the benzene ring is preserved.

Keto and Enol Forms of Benzenol (Phenol)

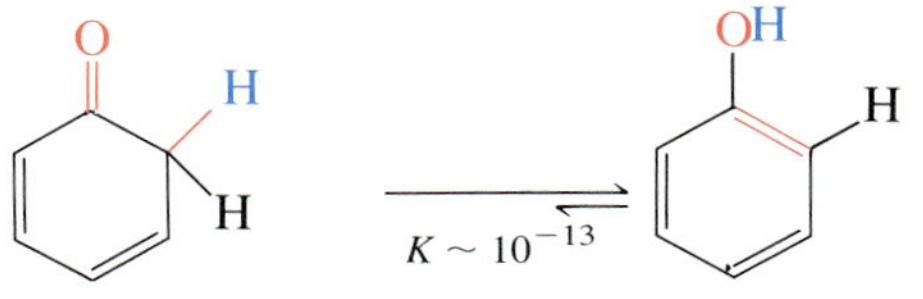

$$\text{2,4-Cyclohexadienone} \underset{K \sim 10^{-13}}{\rightleftharpoons} \text{Benzenol (Phenol)}$$

2,4-Cyclohexadienone **Benzenol (Phenol)**

Benzenols (phenols) and their ethers are ubiquitous in nature; some derivatives have medicinal and herbicidal applications, whereas others are important industrial materials. This section first explains the names of these compounds. It then describes an important difference between benzenols (phenols) and alkanols—the higher acidity of the former, a consequence of the presence of the neighboring aromatic ring. The aromatic ring is also the reason why benzenols (phenols) have to be made quite differently from alcohols, by aromatic substitution reactions.

How to Name Benzenols (Phenols)

Benzenol (phenol) itself was formerly known as *carbolic acid*. It forms colorless needles (m.p. 41°C), has a characteristic odor, and is somewhat soluble in water. Aqueous solutions of it (or its methyl-substituted derivatives) are used as disinfectants, but its main use is for the preparation of polymers (*phenolic resins*). Total U.S. production of benzenol (phenol) in 1984 was 2.9 billion pounds. Pure benzenol (phenol) causes severe skin burns and is poisonous; deaths have been reported from the ingestion of as little as 1 g. Fatal poisoning may also result from absorption through the skin.

Substituted benzenols (phenols) are named according to the system described in Section 19-1. The carboxy and carbonyl functional groups take precedence over the hydroxy group in the naming of such benzenes. Phenyl ethers are named as alkoxybenzenes. As a substituent, C_6H_5O is called **phenoxy.** Many benzenol (phenol) derivatives have common names.

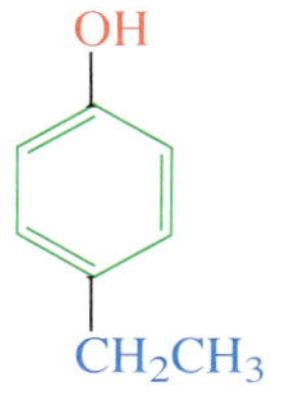

OH, CH_2CH_3	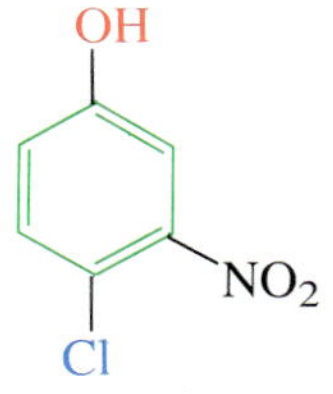 OH, NO_2, Cl	COOH, OH
4-Ethylbenzenol	**4-Chloro-3-nitrobenzenol**	**3-Hydroxybenzenecarboxylic acid**
(*p*-Ethylphenol)	**(4-Chloro-3-nitrophenol)**	**(*m*-Hydroxybenzoic acid)**

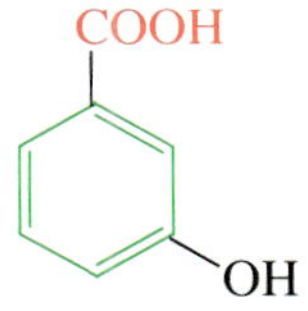

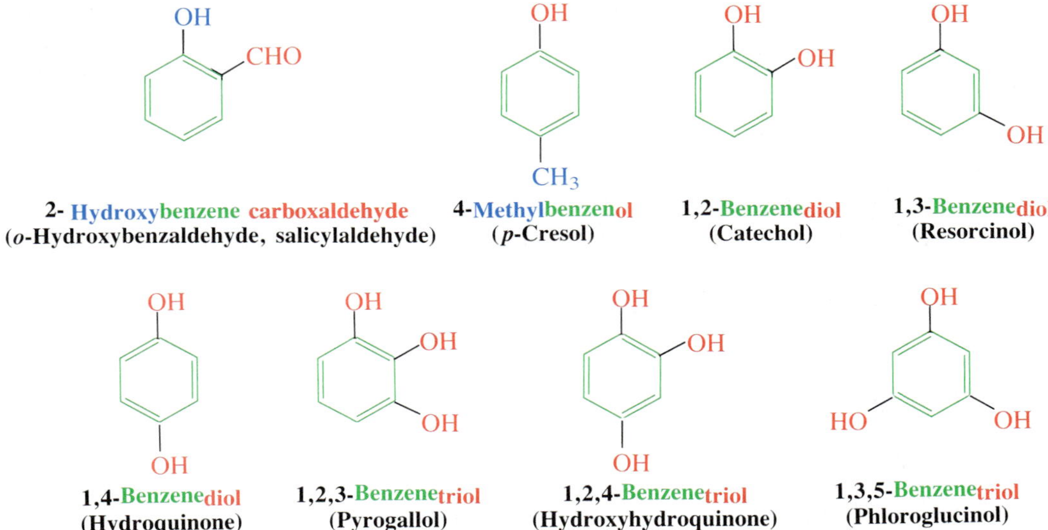

Benzenols (Phenols) Are Unusually Acidic

Benzenols (phenols) have $\mathrm{p}K_a$ values in the range from 8 to 10. Why are they so much more acidic than the alkanols? The reason is resonance: the negative charge in the corresponding anion is stabilized by delocalization into the aromatic ring.

Acidity of Benzenol (Phenol)

$\mathrm{p}K_a \sim 10$

The acidity of this system can be greatly affected by ring substituents that are capable of resonance. 4-Nitrobenzenol (*p*-nitrophenol), for example, has a $\mathrm{p}K_a$ of 7.15.

$\mathrm{p}K_a = 7.15$

The 2-isomer has similar acidity ($\mathrm{p}K_a = 7.22$), whereas nitrosubstitution at C-3 results in a $\mathrm{p}K_a$ of 8.39. Multiple nitration increases the acidity to that of carboxylic or even mineral acids. Electron-donating substituents have the opposite effect.

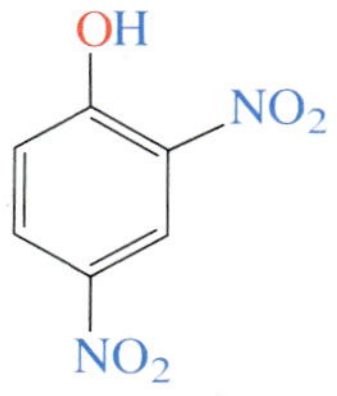

2,4-Dinitrobenzenol (2,4-Dinitrophenol)
$pK_a = 4.09$

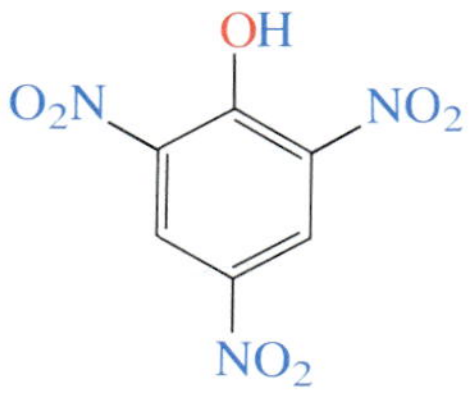

2,4,6-Trinitrobenzenol (2,4,6-Trinitrophenol, picric acid)
$pK_a = 0.25$

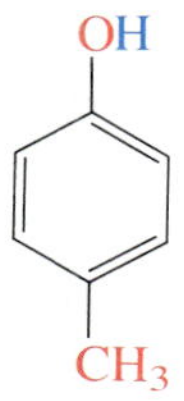

4-Methylbenzenol (*p*-Cresol)
$pK_a = 10.26$

As Section 24-4 will show, the oxygen in benzenol (phenol) and its ethers is also weakly basic, and its protonation gives rise to ether cleavage.

EXERCISE 24-9

Why is 3-nitrobenzenol (*m*-nitrophenol) less acidic than the other two isomers but more acidic than benzenol (phenol) itself?

EXERCISE 24-10

Rank in order of increasing acidity: benzenol (phenol), A; 3,4-dimethylbenzenol (3,4-dimethylphenol), B; 3-hydroxybenzenecarboxylic (*m*-hydroxybenzoic) acid, C; 4-(fluoromethyl)benzenol [*p*-(fluoromethyl)phenol], D.

Preparation of Benzenols (Phenols): Introduction of a Hydroxy Group into the Aromatic Ring by Ipso Substitution

Direct electrophilic addition of OH to arenes is difficult, owing to the scarcity of reagents that generate an electrophilic hydroxy group, such as HO^+. Because of these difficulties, benzenols (phenols) are usually prepared by various other methods, one of which is ipso substitution (Section 20-6) of an appropriate leaving group. For example, benzenol (phenol) can be made (and at one time was manufactured) from sodium benzenesulfonate by heating in molten sodium hydroxide.

Direct Nucleophilic Aromatic Hydroxylation

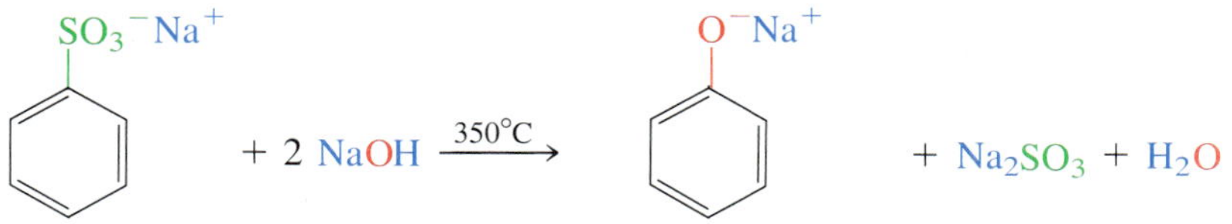

$$C_6H_5SO_3^-Na^+ + 2\ NaOH \xrightarrow{350^\circ C} C_6H_5O^-Na^+ + Na_2SO_3 + H_2O$$

Sodium salt of benzenol (phenol)

The reaction initially produces the benzenol (phenol) salt, which is subsequently treated with HCl. The nucleophilic substitution of chlorobenzene was also a commercial process; it very likely proceeds through a benzyne intermediate (Section 20-6).

$$C_6H_5Cl \xrightarrow[\text{2. HCl}]{\text{1. NaOH, 350}^\circ\text{C}} C_6H_5OH$$

EXERCISE 24-11

1-Chloro-4-methylbenzene (*p*-chlorotoluene) is not a good starting material for the preparation of 4-methylbenzenol (*p*-cresol) by direct reaction with hot NaOH because it forms a mixture of two products. Why does it do so, and what are the two products? Propose a synthesis from methylbenzene (toluene). Hint: Review the parts of Chapter 20 pertinent to this problem.

BOX 24-5

Chlorobenzenols (Chlorophenols) Are Powerfully Toxic Materials

The direct nucleophilic substitution of chloride in chloroarenes is a synthetic pathway to a number of herbicides, pesticides, and antibacterials. For example, hydroxylation of 1,2,4,5-tetrachlorobenzene gives 2,4,5-trichlorobenzenol (2,4,5-trichlorophenol, 2,4,5-TCP), an intermediate in the synthesis of 2,4,5-trichlorophenoxyethanoic acid (2,4,5-T).

1. NaOH, 150°C
2. H^+, H_2O
$-$ NaCl

$ClCH_2COOH$, NaOH, H_2O, Δ
$-$ NaCl

OCH_2COOH

85%

2,4,5-Trichlorobenzenol (2,4,5-Trichlorophenol, 2,4,5-TCP)

2,4,5-Trichlorophenoxy-ethanoic acid (2,4,5-Trichlorophenoxy-acetic acid, 2,4,5-T)

This acid is a powerful herbicide of particular value in brush control. A 1 : 1 mixture of the butyl esters of 2,4,5-T and its 2,4-dichloro analog (2,4-D) was used in large amounts (estimated at more than 10 million gallons from 1965 to 1970) as a defoliant (code name, Agent Orange) during the Vietnam war.

These chemicals are toxic irritants. 2,4,5-T is notorious because of a much more toxic impurity that forms in small quantities during its preparation: 2,3,7,8-tetrachlorodibenzo-*p*-dioxin (TCDD or, simply, *dioxin*). Heating 2,4,5-T to between 500° and 600°C has been shown to produce dioxin, and thus extreme care has to be taken to control the reaction temperatures in the preparation of 2,4,5-T. Dioxin can also be made directly from 2,4,5-trichlorobenzenol (2,4,5-trichlorophenol) by coupling through double dehydrochlorination.

2

0.5 *M* K_2CO_3, Cu powder, 240°–250°C
$-$ 2 KCl, $-$ H_2O

2,3,7,8-Tetrachlorodibenzo-*p*-dioxin (Dioxin)

The toxicity of dioxin (lethal dose, for test animals, in moles per kilogram of body weight) is about 500 times that of strychnine and more than 100,000 times that of sodium cyanide. It is embryotoxic, teratogenic (causing deformations of the fetus), and a suspected carcinogen in humans. In smaller than lethal concentrations, it causes severe skin rashes and lesions *(chloracne)*. In 1976, a runaway reaction in a chemical plant in Seveso, Italy, led to the accidental release of a cloud of overheated 2,4,5-trichlorobenzenol (2,4,5-trichlorophenol) contaminated with dioxin. It is estimated that more than 130 pounds of the poison was vaporized, causing numerous deaths among animals and severe skin irritations in many humans.

Other examples of chlorinated benzenols (phenols) with physiological activity are 2,3,4,5,6-pentachlorobenzenol (pentachlorophenol), a fungicide, and hexachlorophene (common name), a skin germicide once used in soaps and other toiletry products. It was banned when it was discovered to cause brain damage.

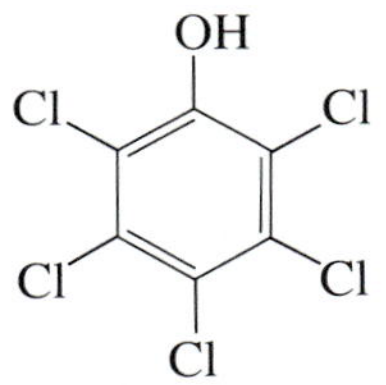

2,3,4,5,6-Pentachlorobenzenol
(Pentachlorophenol, a fungicide)

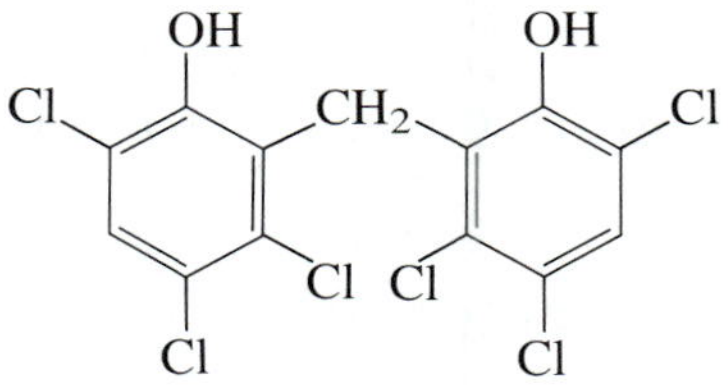

Hexachlorophene
(A skin germicide)

Benzenol (Phenol) by the Cumene Hydroperoxide Process: The Economics of Industrial Syntheses

Another industrial preparation of benzenol (phenol) highlights the economic constraints on any process that has commercial significance. In this approach, called the cumene hydroperoxide process, benzene and propene are oxidized in a series of steps by air to benzenol (phenol) and propanone (acetone). Although the goal of the sequence is to make the former product, it is the sales potential of the ketone byproduct that makes the process economically feasible.

General Cumene Hydroperoxide Process

$$CH_3CH{=}CH_2 + C_6H_6 + O_2 \longrightarrow CH_3\overset{O}{\overset{\|}{C}}CH_3 + C_6H_5OH$$

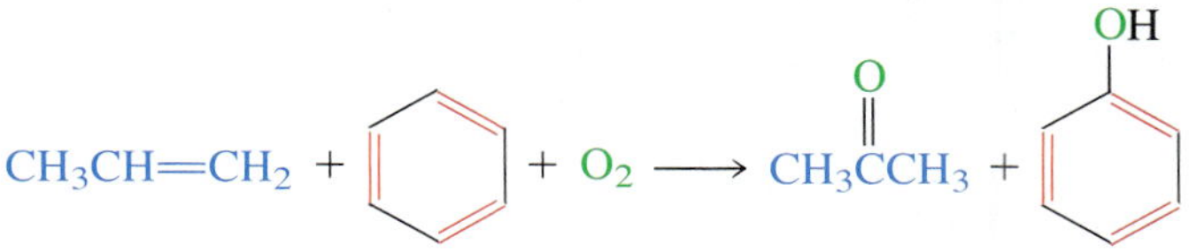

The synthesis proceeds through several separate reactions. In the first, benzene is converted into 1-methylethylbenzene (isopropylbenzene, or cumene) by Friedel-Crafts alkylation with propene under acidic conditions (Section 19-7).

REACTION 1: Alkylation of benzene

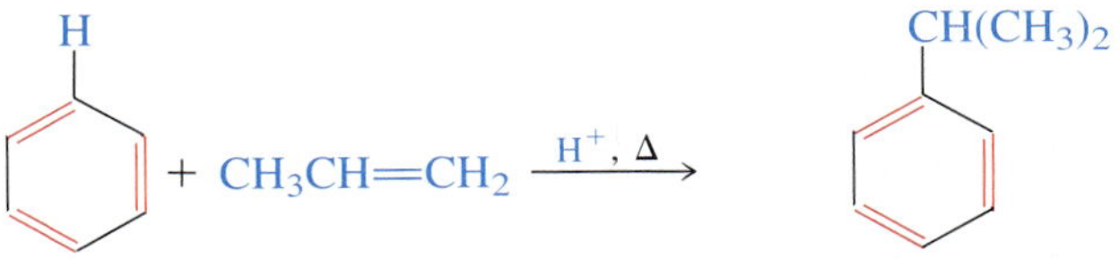

$$C_6H_5H + CH_3CH{=}CH_2 \xrightarrow{H^+,\ \Delta} C_6H_5CH(CH_3)_2$$

1-Methylethylbenzene
(Isopropylbenzene, or cumene)

REACTION 2: Oxidation to the hydroperoxide

$$C_6H_5CH(CH_3)_2 + O_2 \xrightarrow{\text{Initiator}} C_6H_5C(CH_3)_2\text{—O—O—H}$$

1-Methyl-1-phenylethyl hydroperoxide (Cumene hydroperoxide)

REACTION 3: Cleavage to products

$$C_6H_5C(CH_3)_2OOH \xrightarrow{10\%\ H_2SO_4,\ H_2O,\ \Delta} C_6H_5OH + CH_3\overset{O}{\overset{\|}{C}}CH_3$$

In the second reaction, the alkyl benzene is oxidized by air to the corresponding hydroperoxide. The ease with which this transformation occurs is due to the ready initiation of a free-radical chain process through the tertiary benzylic radical. Apparently, the initiator is oxygen.

Mechanism of Cumene Hydroperoxide Formation:

STEP 1: Initiation

$$C_6H_5CH(CH_3)_2 + O_2 \longrightarrow C_6H_5\dot{C}(CH_3)_2 + \cdot\ddot{O}\ddot{O}H$$

Tertiary benzylic radical

STEPS 2 AND 3: Propagation

$$C_6H_5\dot{C}(CH_3)_2 + O_2 \longrightarrow C_6H_5C(CH_3)_2\text{—}\ddot{O}\text{—}\ddot{O}\cdot$$

$$C_6H_5C(CH_3)_2\text{—}\ddot{O}\text{—}\ddot{O}\cdot + C_6H_5CH(CH_3)_2 \longrightarrow C_6H_5C(CH_3)_2\text{—}\ddot{O}\text{—}\ddot{O}H + C_6H_5\dot{C}(CH_3)_2$$

In the third and final reaction, the hydroperoxide is treated with dilute sulfuric acid to give the two products, benzenol (phenol) and propanone (acetone). In detail, this process resembles the acid-catalyzed rearrangement of 2,2-dimethyl-1-propanol (neopentyl alcohol, section 9-2). Protonation of the hydroxy group followed by phenyl migration with simultaneous extrusion of water gives a rearranged carbocation, stabilized by resonance with the oxygen of the phenoxy group. Addition of water followed by deprotonation and reprotonation of the two oxygens (labeled *proton shift*) gives an intermediate that rapidly decomposes to the products.

Mechanism of Cumene Hydroperoxide Decomposition:

Migration, $-H_2O$; H_2O; Proton shift; Decomposition

$$C_6H_5C(CH_3)_2\text{—}\ddot{O}\text{—}\ddot{O}H + H^+ \longrightarrow C_6H_5C(CH_3)_2\text{—}\ddot{O}\text{—}\overset{+}{O}H_2 \xrightarrow[-H_2O]{\text{Migration}} \left[(CH_3)_2\overset{+}{C}\text{—}\ddot{O}C_6H_5 \longleftrightarrow (CH_3)_2C{=}\overset{+}{\ddot{O}}C_6H_5\right] \xrightarrow{H_2O} (CH_3)_2C(\overset{+}{O}H_2)\text{—}\ddot{O}C_6H_5 \xrightarrow{\text{Proton shift}}$$

$$(CH_3)_2C(HO)\text{—}\overset{+}{O}(H)C_6H_5 \xrightarrow{\text{Decomposition}} C_6H_5\ddot{O}H + CH_3\overset{\overset{+}{O}H}{\overset{\|}{C}}CH_3 \longrightarrow C_6H_5\ddot{O}H + CH_3\overset{O}{\overset{\|}{C}}CH_3 + H^+$$

Benzenols (Phenols) from Arenediazonium Salts

The most-general laboratory procedure for making benzenols (phenols) is through arenediazonium salts, $ArN_2{}^+X^-$. Recall that primary alkanamines can be *N*-nitrosated but that the resulting species rearrange to diazonium ions, which are unstable—they lose nitrogen to give carbocations (Section 21-5). In contrast, primary benzenamines (anilines) are attacked by cold nitrous acid, in a reaction called *diazotization,* to give relatively stable arenediazonium salts.

Diazotization

$$R\text{—}C_6H_4\text{—}NH_2 \xrightarrow{NaNO_2,\ H^+,\ H_2O,\ 0°C} R\text{—}C_6H_4\text{—}\overset{+}{N}{\equiv}N\colon$$

An arenediazonium ion

The reactions of these species will be discussed in greater detail in Section 24-6. On heating, nitrogen is evolved with formation of reactive aryl cations. These are trapped by water to give the desired benzenols (phenols).

Decomposition of Arenediazonium Salts

$$\text{R-C}_6\text{H}_4\text{-N}^+\equiv\text{N:} \xrightarrow[-\,N_2]{\Delta} \text{R-C}_6\text{H}_4^{+} \xrightarrow[-\,H^+]{HOH} \text{R-C}_6\text{H}_4\text{-OH}$$

An aryl cation

EXERCISE 24-12

Propose a synthesis of (4-phenylmethyl)benzenol (*p*-benzylphenol) from benzene.

In summary, benzenols (phenols) exist in the enol form because of aromatic stabilization. They are named as benzenols (phenols) according to the rules for naming aromatic compounds explained in Section 19-1. Those derivatives bearing carboxy or methanoyl (formyl) groups on the aromatic ring are called hydroxybenzenecarboxylic (hydroxybenzoic) acids or hydroxybenzenecarboxaldehydes (hydroxybenzaldehydes). Benzenols (phenols) are acidic because the corresponding anions are resonance stabilized. They may be prepared by any of four methods: direct nucleophilic hydroxylation, hydroxylation of halobenzenes through benzynes, the cumene hydroperoxide process [for benzenol (phenol)], and decomposition of arenediazonium salts in water.

24-4 The Reactivity of Benzenols (Phenols): Alcohol and Aromatic Chemistry

The hydroxy group in benzenols (phenols) undergoes several of the reactions of alcohols (Chapter 9), such as protonation, Williamson ether synthesis, and esterification. The aromatic ring, on the other hand, is readily attacked by electrophiles to give substituted derivatives.

The Oxygen in Benzenols (Phenols) Is Only Weakly Basic

Benzenols (and their ethers) can be protonated by strong acid to give the corresponding oxonium ions. Thus, as with the alkanols, the hydroxy group imparts amphoteric character (Section 8-3). However, the basicity of benzenol (phenol) is even less than that of the alkanols, because the lone electron pairs on the oxygen are delocalized into the benzene ring (Section 20-1). The pK_a values for phenyloxonium ions are, therefore, lower than those of alkyloxonium ions.

pK_a Values of Methyl- and Phenyloxonium Ion

$$CH_3\ddot{O}^+H_2 \rightleftharpoons CH_3\ddot{O}H + H^+ \qquad pK_a = -2.2$$

$$C_6H_5\ddot{O}^+H_2 \rightleftharpoons C_6H_5\ddot{O}H + H^+ \qquad pK_a = -6.7$$

Unlike secondary and tertiary oxonium ions derived from alcohols, phenyloxonium derivatives do not dissociate to form phenyl cations because such ions have too high an energy content (see Section 24-6).

Protonated alkoxybenzenes are cleaved in the presence of nucleophiles such as Br^- or I^- (e.g., from HBr or HI) to give benzenol (phenol) and the corresponding haloalkane.

$$3\text{-}CH_3OC_6H_4COOH \xrightarrow{HBr,\ \Delta} 3\text{-}HOC_6H_4COOH + CH_3Br$$

3-Methoxybenzenecarboxylic acid (*m*-Methoxybenzoic acid) → **3-Hydroxybenzenecarboxylic acid (*m*-Hydroxybenzoic acid)** 90%

Some Lewis acids accomplish the same task.

$$(2\text{-}CH_3OC_6H_4CO)_2 \xrightarrow{1.\ AlCl_3,\ HCl,\ \text{nitrobenzene (solvent)};\ 2.\ H^+,\ H_2O} (2\text{-}HOC_6H_4CO)_2 + 2\ CH_3Cl$$

60%

EXERCISE 24-13

Why does cleavage of an alkoxybenzene by acid not produce a halobenzene and the alkanol?

Williamson Ether Synthesis with Benzenols (Phenols)

Phenoxide ion is a good nucleophile; thus, it displaces the leaving groups from haloalkanes and alkyl sulfonates to give alkoxybenzenes. These reactions can be carried out selectively even in the presence of alkanols because of the difference in acidity between the two respective hydroxy functions.

$$3\text{-}ClC_6H_4OH + CH_3CH_2CH_2Br \xrightarrow[-\ NaBr,\ -\ HOH]{NaOH,\ H_2O} 3\text{-}ClC_6H_4OCH_2CH_2CH_3$$

3-Chlorobenzenol (*m*-Chlorophenol) → **1-Chloro-3-propoxybenzene (*m*-Chlorophenyl propyl ether)** 63%

$$\text{HOC}_6\text{H}_4\text{CH}_2\text{OH} + (\text{CH}_3)_2\text{SO}_4 \xrightarrow[-\ \text{Na}_2\text{SO}_4,\ -\ \text{HOH}]{\text{NaOH}} \text{CH}_3\text{OC}_6\text{H}_4\text{CH}_2\text{OH}\ (40\%)$$

(4-Hydroxymethyl)benzenol
[(*p*-Hydroxymethyl)phenol]

4-Methoxyphenylmethanol
(*p*-Methoxybenzyl alcohol)

Benzenols (Phenols) Form Phenyl Alkanoates

The esterification of benzenols (phenols) usually requires the use of activated carboxylic acid derivatives, such as alkanoyl halides or carboxylic anhydrides, because reaction of the carboxylic acid itself (catalyzed by mineral acid, Section 17-8) is endothermic in this case.

$$\text{HOC}_6\text{H}_4\text{CH}_3 + \text{CH}_3\text{CH}_2\overset{\text{O}}{\overset{\|}{\text{C}}}\text{Cl} \xrightarrow[-\ \text{NaCl},\ -\ \text{HOH}]{\text{NaOH, H}_2\text{O}} \text{CH}_3\text{CH}_2\overset{\text{O}}{\overset{\|}{\text{C}}}\text{OC}_6\text{H}_4\text{CH}_3$$

4-Methylbenzenol
(*p*-Cresol)

Propanoyl chloride

4-Methylphenyl propanoate
(*p*-Methylphenyl propanoate)

BOX 24-6

Aspirin, a Physiologically Active Phenyl Alkanoate

Ethanoylation (acetylation) of 2-hydroxybenzenecarboxylic (*o*-hydroxybenzoic) acid, also known as salicylic acid, results in the formation of the drug *aspirin,* probably the most-consumed drug in the world. Production capacity in the United States alone is about 43 million pounds per year.

$$o\text{-HOC}_6\text{H}_4\text{COOH} \xrightarrow[-\ \text{CH}_3\text{COOH}]{\text{CH}_3\overset{\text{O}}{\overset{\|}{\text{C}}}\text{O}\overset{\text{O}}{\overset{\|}{\text{C}}}\text{CH}_3,\ \text{H}^+,\ \Delta} o\text{-CH}_3\overset{\text{O}}{\overset{\|}{\text{C}}}\text{OC}_6\text{H}_4\text{COOH}$$

2-Hydroxybenzenecarboxylic acid
(*o*-Hydroxybenzoic acid,
salicylic acid)

2-Ethanoyloxybenzenecarboxylic acid
(*o*-Acetoxybenzoic acid,
acetylsalicylic acid, aspirin)

Aspirin is used as an analgesic, antipyretic, and antirheumatic (antiinflammatory, particularly for arthritis). It is thought to interfere with the synthesis of prostaglandins (Section 16-5), thereby reducing fever, pain, and inflammation. Even though it is a popular medicine, it has some unwelcome side effects—particularly, gastric irritation and even bleeding. Recently, it has also been suspected to cause some childhood illnesses—notably, Reye's syndrome, a brain disease. Because of some of these drawbacks, there has been a gradual increase in the sales of other analgesics, such as Tylenol, another benzenol (phenol) derivative, prepared from 4-aminobenzenol (*p*-aminophenol) by ethanoylation.

$$\text{HO-C}_6\text{H}_4\text{-NH}_2 \xrightarrow[-\ CH_3COOH]{CH_3COCCH_3\ (\text{O O}),\ CH_3COOH} \text{HO-C}_6\text{H}_4\text{-NHCCH}_3\ (\text{C=O})$$

4-Aminobenzenol (*p*-Aminophenol) → ***N*-(4-Hydroxyphenyl)ethanamide [*N*-(*p*-Hydroxyphenyl)acetamide, Tylenol]**

Tylenol was much discussed in 1982 and 1986, when capsules of the drug laced with sodium cyanide caused the deaths of a total of eight people. These incidents have led to extensive revision of the procedures for packaging medications.

EXERCISE 24-14

Explain why, in the preparation of Tylenol, the amide is formed rather than the ester.

Benzenols (Phenols) Are Activated Substrates in Electrophilic Aromatic Substitutions

Benzenols (phenols) contain activated aromatic rings that undergo ready electrophilic substitution at the *ortho* and *para* positions (Sections 20-1 and 20-3). For example, even dilute aqueous nitric acid causes nitration.

$$\text{C}_6\text{H}_5\text{OH} \xrightarrow{HNO_3,\ CHCl_3,\ 15^\circ C} \text{2-NO}_2\text{-C}_6\text{H}_4\text{OH} + \text{4-NO}_2\text{-C}_6\text{H}_4\text{OH}$$

26%	61%
2-Nitrobenzenol (*o*-Nitrophenol)	**4-Nitrobenzenol (*p*-Nitrophenol)**

Friedel-Crafts alkanoylation (acylation) of benzenols (phenols) is complicated by ester formation and is better carried out on ether derivatives of benzenol.

$$\text{CH}_3\text{O-C}_6\text{H}_4\text{-H} + CH_3CCl\ (\text{C=O}) \xrightarrow[-\ HCl]{AlCl_3,\ CS_2} \text{CH}_3\text{O-C}_6\text{H}_4\text{-C(=O)CH}_3$$

70%

Methoxybenzene (Anisole) → **1-(4-Methoxyphenyl)ethanone (*p*-Methoxyacetophenone)**

Electrophilic attack at the *para* position is frequently predominant because of steric effects. However, it is normal to obtain mixtures resulting from both *ortho* and *para* substitution, and their composition is highly dependent on reagents and reaction conditions.

EXERCISE 24-15

Friedel-Crafts methylation of methoxybenzene (anisole) with chloromethane in the presence of $AlCl_3$ gives a 2:1 ratio of *ortho*:*para* products. Treatment of methoxybenzene with 2-chloro-2-methylpropane (*tert*-butyl chloride) under the same conditions furnishes only 1-methoxy-4-(1,1-dimethylethyl)benzene (*p*-*tert*-butylanisole). Explain.

BOX 24-7

Phenolphthalein: A pH Indicator

An unusual application of the Friedel-Crafts reaction is to the preparation of phenolphthalein, in which 1,2-benzenedicarboxylic (phthalic) anhydride attacks two molecules of benzenol (phenol).

2 (phenol: OH, H) + O O O $\xrightarrow[-\ H_2O]{H^+,\ \Delta}$ (HO, OH, O, O)

1,2-Benzenedicarboxylic anhydride (Phthalic anhydride)

Phenolphthalein

A probable mechanism (abbreviated) of this reaction is shown here.

Mechanism of Phenolphthalein Formation:

O O O $\xrightarrow{H^+}$ C+ O OH O + (phenol –OH) $\xrightarrow[\text{Crafts reaction, } -\ H^+]{\text{First Friedel-}}$ O COH O OH $\xrightarrow[\text{formation}]{\text{Cyclic hemiacetal}}$

HO OH O O $\xrightarrow[-\ HOH]{H^+}$ OH + O O + (phenol –OH) $\xrightarrow[\text{Crafts reaction, } -\ H^+]{\text{Second Friedel-}}$ phenolphthalein

Phenolphthalein is commonly used in the laboratory as a pH indicator. At pH below 8.5, the molecule exists in its lactone form, which is colorless. However, at pH ~ 9, the two hydroxy protons are removed, and the lactone ring opens up to give a deep red dianion. The color is due to the extensive delocalization of charge in the system.

Phenolphthalein as a pH Indicator

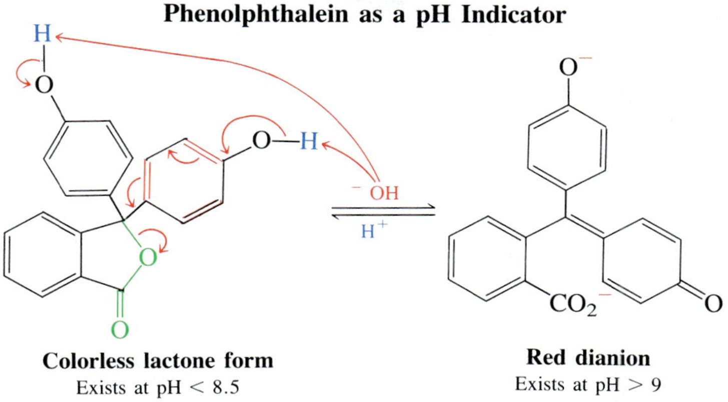

$^-$OH / H^+

Colorless lactone form
Exists at pH < 8.5

Red dianion
Exists at pH > 9

The properties of a good pH indicator of this type are (1) sharp color change between its two forms, (2) fast exchange between the acidic and the basic structures, and (3) stability in the pH range of interest.

The Special Reactivity of Phenoxide Ions

Under basic conditions, benzenols (phenols) can undergo electrophilic substitution, even with very mild electrophiles, through intermediate phenoxide ions. An industrially important application is the reaction with methanal (formaldehyde), which leads to *o*- and *p*-hydroxymethylation. Mechanistically, these processes may be considered enolate condensations, much like the aldol reaction (Section 16-3).

Hydroxymethylation of Benzenol (Phenol)

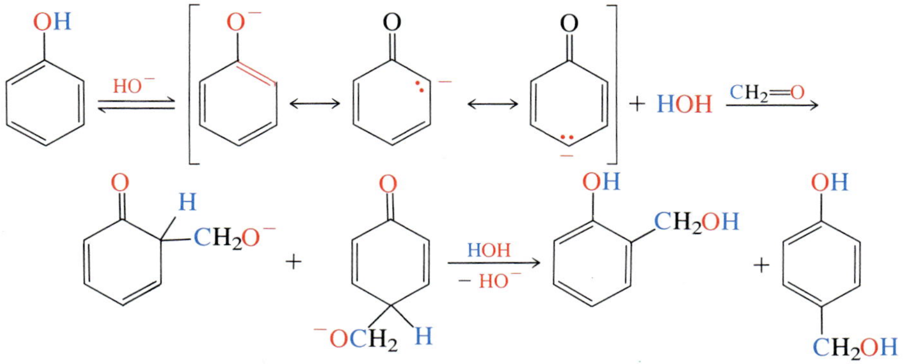

HO^-; + HOH; $CH_2{=}O$; CH_2O^-; $^-OCH_2$; HOH, $-HO^-$; CH_2OH

The initial aldol products are unstable: they dehydrate on heating, giving reactive intermediates called quinomethanes.

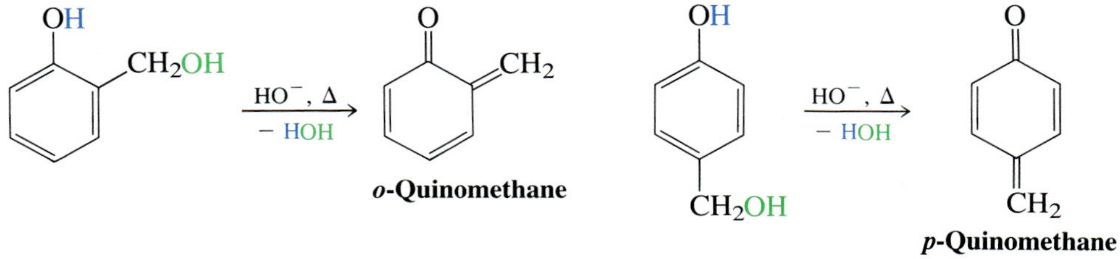

HO^-, Δ, $-$ HOH

***o*-Quinomethane**

***p*-Quinomethane**

Because these are α,β-unsaturated carbonyl compounds, they may undergo Michael additions (Section 16-5) with excess phenoxide ion. The resulting benzenols (phenols) can be hydroxymethylated again and the entire process repeated. Eventually, a complex benzenol-methanal (phenol-formaldehyde) copolymer, also called a phenolic resin, is formed. Total production of these resins in the United States in 1984 exceeded 1.6 billion pounds. Their major uses are in plywood (45%), insulation (14%), molding compounds (9%), fibrous and granulated wood (9%), laminates (8%), and foundry core binders (5%).

Phenolic Resin Synthesis

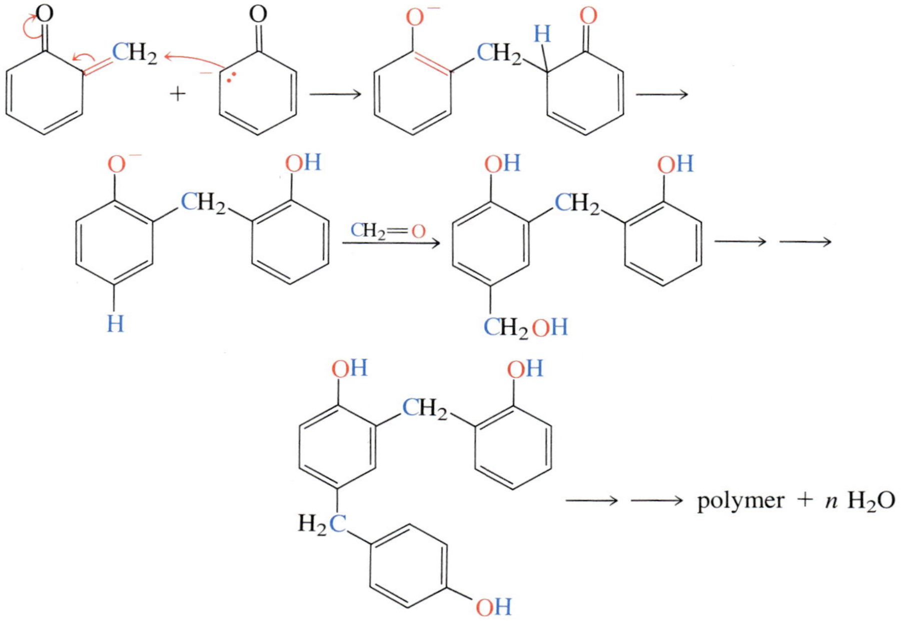

CH_2 + ⟶ O^- CH_2 H O ⟶

O^- CH_2 OH H $\xrightarrow{CH_2{=}O}$ OH CH_2 OH CH_2OH ⟶ ⟶

OH CH_2 OH H_2C OH ⟶ ⟶ polymer + n H_2O

In the **Kolbe* reaction,** phenoxide attacks carbon dioxide to furnish the salt of 2-hydroxybenzenecarboxylic acid (*o*-hydroxybenzoic acid, salicylic acid, precursor to aspirin; see Box 24-6):

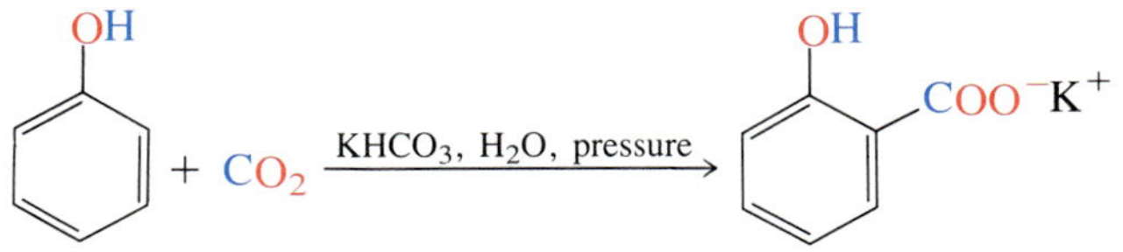

OH + CO_2 $\xrightarrow{KHCO_3,\ H_2O,\ \text{pressure}}$ OH COO^-K^+

EXERCISE 24-16

Formulate a mechanism for the Kolbe reaction.

EXERCISE 24-17

Hexachlorophene (see Section 24-3) is prepared in one step from 2,4,5-trichlorobenzenol (2,4,5-trichlorophenol) and methanal (formaldehyde) in the presence of sulfuric acid. How does this reaction proceed? (Hint: Formulate an acid-catalyzed hydroxymethylation for the first step).

---

*Professor Hermann Kolbe, 1818–1884, University of Leipzig, Germany.

2-Propenyloxybenzene (Allyl Phenyl Ether) Undergoes an Electrocyclic Reaction: The Claisen Rearrangement

At 200°C, 2-propenyloxybenzene (allyl phenyl ether) undergoes an unusual reaction that resembles electrophilic substitution but proceeds by a different mechanism: the starting material rearranges to 2-(2-propenyl)benzenol (*o*-allylphenol).

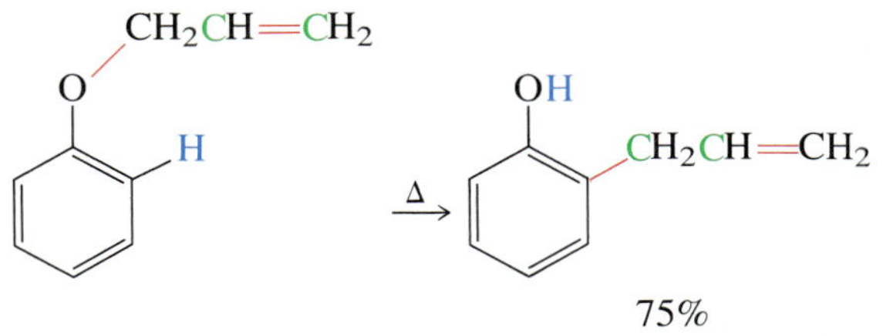

$$C_6H_5OCH_2CH{=}CH_2 \xrightarrow{\Delta} o\text{-}HOC_6H_4CH_2CH{=}CH_2 \quad 75\%$$

2-Propenyloxybenzene (Allyl phenyl ether) → **2-(2-Propenyl)benzenol (*o*-Allylphenol)**

This transformation, called the **Claisen* rearrangement,** is another concerted reaction with a transition state that accommodates the movement of six electrons. The initial intermediate is a high-energy isomer, 6-(2-propenyl)-2,4-cyclohexadienone, which enolizes to the final product.

Mechanism of the Claisen Rearrangement:

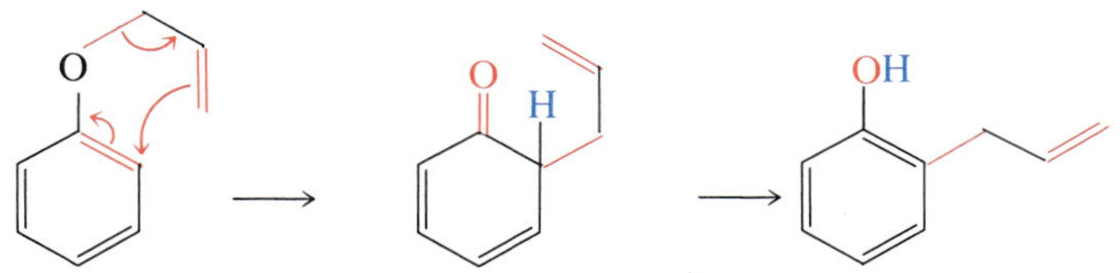

6-(2-Propenyl)-2,4-cyclohexadienone

With the nonaromatic 1-ethenoxy-2-propene (allyl vinyl ether), the Claisen rearrangement stops at the initial stage because the carbonyl group in this molecule is stable. This is called the *aliphatic Claisen rearrangement*.

Aliphatic Claisen Rearrangement

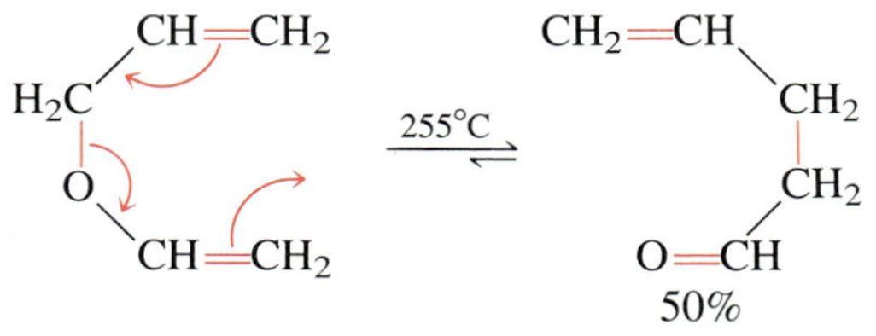

$$CH_2{=}CHOCH_2CH{=}CH_2 \underset{}{\overset{255°C}{\rightleftharpoons}} CH_2{=}CHCH_2CH_2CH{=}O \quad 50\%$$

1-Ethenoxy-2-propene (Allyl vinyl ether) → **4-Pentenal**

The carbon analog of the Claisen rearrangement is called the **Cope† rearrangement;** it takes place in compounds containing 1,5-diene units.

---

*This is the Professor Claisen of the Claisen condensation (Sections 18-4 and 22-3).

†This is the Professor Cope of the Cope elimination (Section 21-5).

Cope Rearrangement

H_5C_6 — 178°C ⇌ — H_5C_6

72%

3-Phenyl-1,5-hexadiene **trans-1-Phenyl-1,5-hexadiene**

EXERCISE 24-18

Design an experiment that would allow you to detect the degenerate Cope rearrangement in 1,5-hexadiene.

EXERCISE 24-19

Explain the following transformation by a mechanism.

HO — $Na^{+}\,^{-}OH$, H_2O → OHC

BOX 24-8

Exotic Degenerate Cope Rearrangements

An unusual degenerate Cope rearrangement is undergone by bicyclo[5.1.0]octa-2,5-diene. At 180°C, the compound interconverts with an identical structure, but having a different connectivity (as shown by the numbering of the carbons).

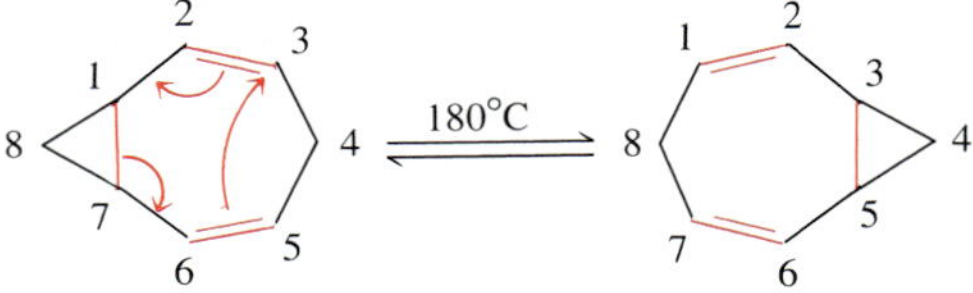

Bicyclo[5.1.0]octa-2,5-diene

An even more fascinating molecule is **bullvalene,** structurally derived from bicyclo[5.1.0]octa-2,5-diene by linking carbons 5 and 8 with an ethene bridge. This molecule has the option of undergoing more than 1.2 million degenerate Cope rearrangements, only two of which are shown here.

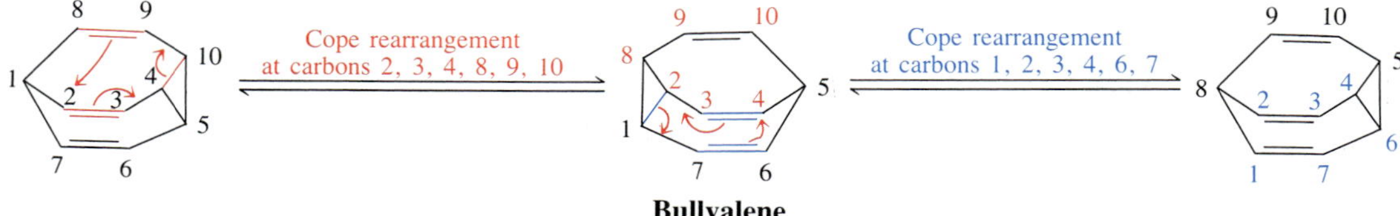

Bullvalene

Any three adjacent carbons can, at any time, be the component carbons of the cyclopropane ring. The degenerate rearrangement in bullvalene is fast on the NMR time scale. At 100°C, only *one* hydrogen resonance is detected, because all positions equilibrate rapidly. At −25°C, however, the spectrum shows two absorptions, one in the alkene region (6 H) and one in the saturated area (4 H), as expected for a frozen structure.

In summary, benzenols (phenols) and alkoxybenzenes react in a bifunctional manner. The oxygen may be protonated although it is less basic than the oxygen in the alkanols and alkoxyalkanes. Protonated benzenols and their derivatives do not ionize to phenyl cations, but the ethers can be cleaved to benzenols (phenols) and haloalkanes by HX. Alkoxybenzenes are made by Williamson ether synthesis, aryl alkanoates by alkanoylation. The benzene ring in benzenols (phenols) is subject to electrophilic aromatic substitution, particularly under basic conditions. Phenoxide ions can be hydroxymethylated and carboxylated. 2-Propenyloxybenzene rearranges to 2-(2-propenyl)benzenol (*o*-allylphenol) by an electrocyclic mechanism that moves six electrons (Claisen rearrangement). Similar concerted reactions are undergone by aliphatic unsaturated ethers (aliphatic Claisen rearrangement) and by hydrocarbons containing 1,5-diene units (Cope rearrangement).

24-5 The Oxidation of Benzenols (Phenols): Cyclohexadienediones (Quinones)

Benzenols (phenols) can be oxidized to carbonyl derivatives. Unlike the oxidation of alkanols, that of the benzene ring does not stop until it reaches the stage of a **cyclohexadienedione (benzoquinone).** This section describes the reactions of this new class of compounds.

Cyclohexadienediones (Benzoquinones) and Benzenediols (Hydroquinones) Are Redox Couples

1,2- and 1,4-Benzenediols (catechols and hydroquinones) are oxidized to the corresponding diketones, 3,5-cyclohexadiene-1,2-diones and 2,5-cyclohexadiene-1,4-diones (*o*-benzoquinones and *p*-benzoquinones), by a variety of oxidizing agents, such as sodium dichromate or silver oxide. Yields can be variable if the resulting diones are reactive, like the 1,2-dione, which partly decomposes under the conditions of its formation.

$$\text{1,2-Benzenediol} \xrightarrow{Ag_2O,\ (CH_3CH_2)_2O} \text{3,5-Cyclohexadiene-1,2-dione}$$

Low yield

1,2-Benzenediol (Catechol) → **3,5-Cyclohexadiene-1,2-dione (*o*-Benzoquinone)**

$$\text{1,4-Benzenediol} \xrightarrow{Na_2Cr_2O_7,\ H_2SO_4} \text{2,5-Cyclohexadiene-1,4-dione}$$

92%

1,4-Benzenediol (Hydroquinone) → **2,5-Cyclohexadiene-1,4-dione (*p*-Benzoquinone)**

Even simple benzenols (phenols) can be oxidized to such diones. The second oxygen is introduced at C-4 *(para)* by several oxidants. A special reagent that accomplishes this transformation is potassium nitrosodisulfonate, $\cdot ON(SO_3^-K^+)_2$, a radical species also known as **Frémy's* salt.** It is prepared by oxidation of sodium hydroxylaminedisulfonate with potassium permanganate, and it is stable and water soluble.

$$HON(SO_3^-Na^+)_2 \xrightarrow{KMnO_4} \cdot ON(SO_3^-K^+)_2 \quad 67\%$$

Sodium hydroxylamine-disulfonate → **Potassium nitrosodisulfonate (Frémy's salt)**

EXERCISE 24-20

Formulate a Lewis structure for Frémy's salt.

Oxidation of Benzenols (Phenols) by Fremy's Salt

$\xrightarrow{\cdot ON(SO_3K)_2}$ 82%

2-Methylbenzenol (*o*-Cresol) → **2-Methyl-2,5-cyclohexadiene-1,4-dione (*o*-Toluquinone)**

$\xrightarrow{\cdot ON(SO_3K)_2}$ 87%

2,3,5,6-Tetramethylbenzenol (2,3,5,6-Tetramethylphenol) → **2,3,5,6-Tetramethyl-2,5-cyclohexadiene-1,4-dione (Duroquinone)**

An important property of these diones is their capacity to act as mild oxidizing agents. During oxidation of an organic substrate, the dione unit is reduced to the diol. This reversible reduction-oxidation relation can be described by a general redox equation.

*Professor Edmond Frémy, 1814–1894, École Polytechnique, Paris.

Redox Relation Between 2,5-Cyclohexadiene-1,4-dione (*p*-Benzoquinone) and 1,4-Benzenediol (Hydroquinone)

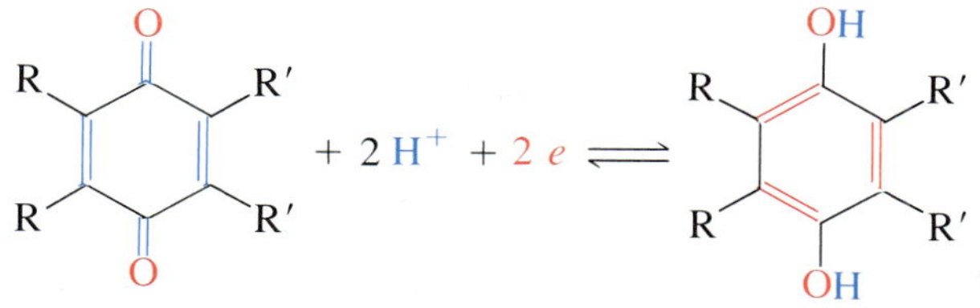

BOX 24-9

Natural Cyclohexadienediones (Quinones) as Reversible Oxidizing Agents

Nature makes use of the quinone-dihydroquinone redox couple in reversible oxidation reactions, which are ultimately part of the complicated cascade by which oxygen is used in biochemical degradations. An important series of compounds used for this purpose are the **ubiquinones** (a name coined to indicate their ubiquitous presence in nature), also collectively called **coenzyme Q (CoQ).** The ubiquinones are substituted 2,5-cyclohexadiene-1,4-dione (*p*-benzoquinone) derivatives bearing a side chain made up of 2-methylbutadiene units (isoprene, Sections 4-7 and 14-6).

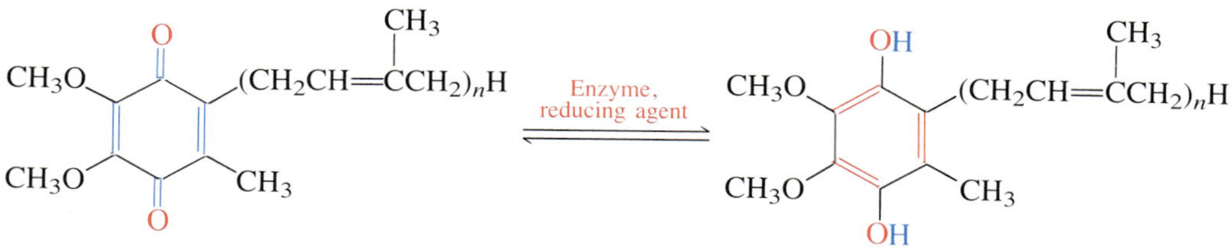

Ubiquinones (n = 6, 8, 10)
(Coenzyme Q)

A related compound is vitamin K_1, an antihemorrhagic agent in the blood (it stops the bleeding of wounds.)

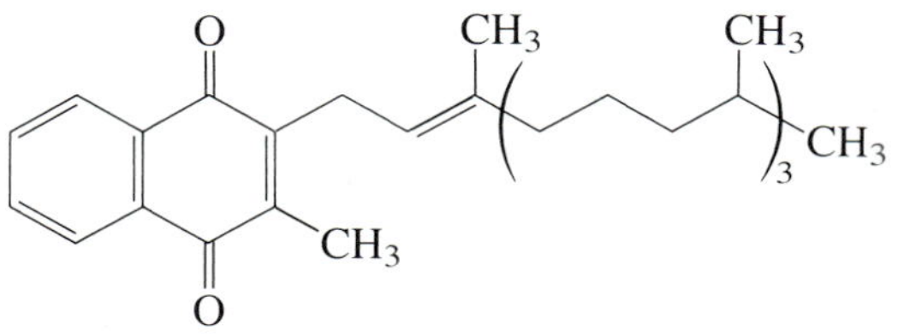

Vitamin K_1

2,5-Cyclohexadiene-1,4-diones (Benzoquinones) and Photography

1,4-Benzenediol (hydroquinone) can be used as a developer in the photographic process. Black-and-white photographic film contains tiny crystals of silver bromide. On exposure to light, the silver bromide becomes photoactivated. In this form, it is readily reduced by alkaline 1,4-benzenediol (hydroquinone) to metallic silver, which forms a black precipitate. Thus, areas on a photographic plate or sheet that have been exposed to light and such a developer show a black image.

Photographic Process

$$AgBr \xrightarrow[\text{Photoactivation}]{h\nu} AgBr^*$$

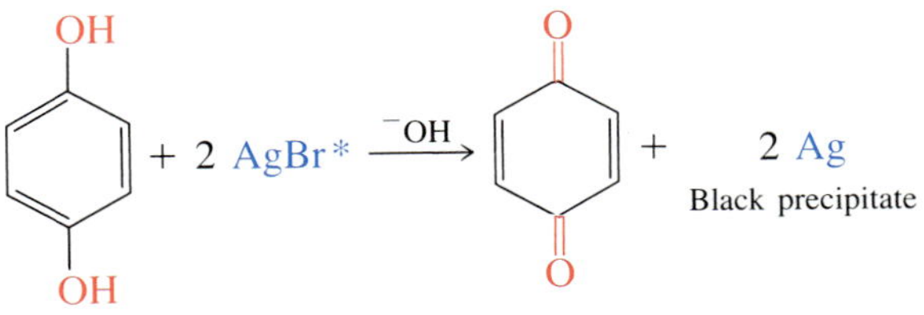

1,4-Benzenediol + 2 $AgBr^*$ $\xrightarrow{^-OH}$ 2,5-Cyclohexadiene-1,4-dione + 2 Ag (Black precipitate)

In practice, after exposure and development, the film is washed with a *fixer,* a solution of a compound such as sodium thiosulfate, capable of dissolving and removing unreduced silver bromide. Most commercial developers contain 4-aminobenzenols (*p*-aminophenols), not 1,4-benzenediol (hydroquinone).

Metol: A Commercial Developer for Photographic Films

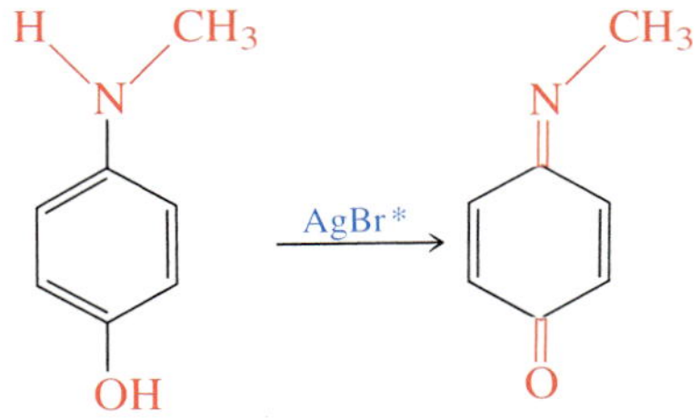

Metol $\xrightarrow{AgBr^*}$ N-methyl quinone imine

Metol

The Enone Units in 2,5-Cyclohexadiene-1,4-diones (*p*-Benzoquinones) Undergo Conjugate and Diels-Alder Additions.

2,5-Cyclohexadiene-1,4-diones (*p*-benzoquinones) function as reactive α,β-unsaturated ketones in conjugate additions (see Section 16-5). For example, hydrogen chloride adds to give an intermediate hydroxy dienone that enolizes to the aromatic 2-chloro-1,4-benzenediol.

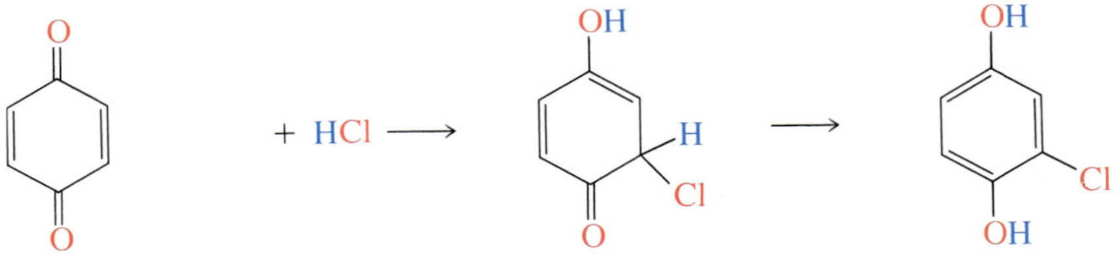

2,5-Cyclohexadiene-1,4-dione (*p*-Benzoquinone) + HCl ⟶ 6-Chloro-4-hydroxy-2,4-cyclohexadienone ⟶ 2-Chloro-1,4-benzenediol

2,5-Cyclohexadiene-1,4-dione (*p*-Benzoquinone)

6-Chloro-4-hydroxy-2,4-cyclohexadienone

2-Chloro-1,4-benzenediol

The double bonds also undergo cycloadditions to dienes (Section 14-5). The initial cycloadduct to 1,3-butadiene tautomerizes on heating with acid to the aromatic system.

Diels-Alder Reactions
of 2,5-Cyclohexadiene-1,4-dione (*p*-Benzoquinone)

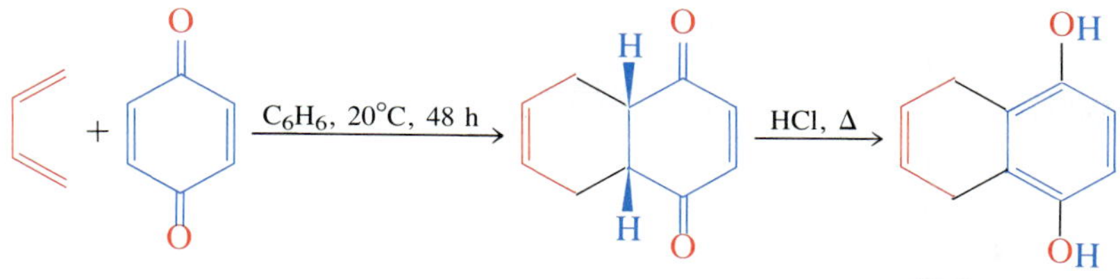

C_6H_6, 20°C, 48 h
HCl, Δ

88% overall

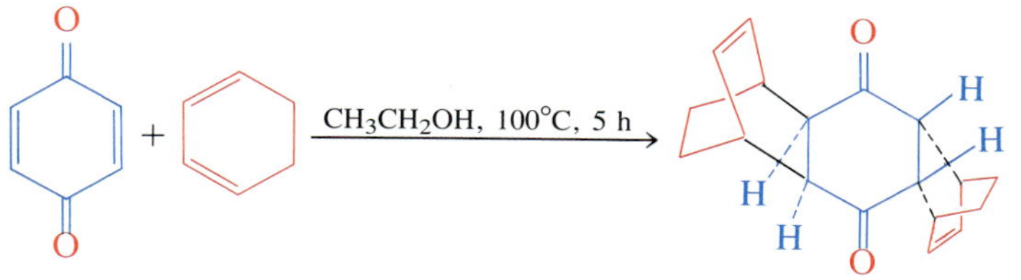

CH_3CH_2OH, 100°C, 5 h

Excess

80%
Double endo adduct

Diels-Alder additions to cyclic dienes follow the endo rule, as can be seen in the double addition to 1,3-cyclohexadiene.

EXERCISE 24-21

Explain the following result by a mechanism.

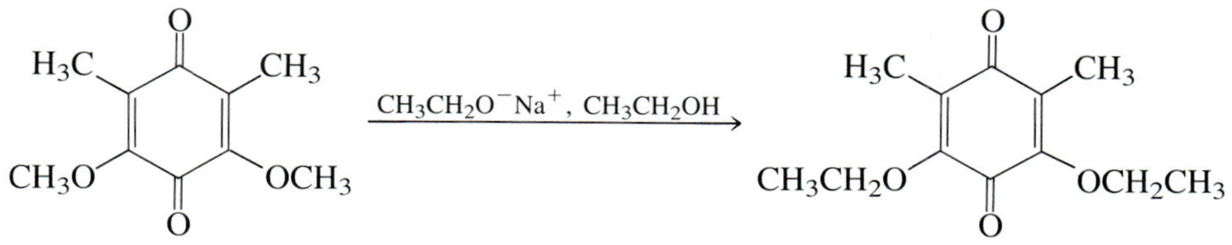

H_3C
CH_3
CH_3O
OCH_3
$CH_3CH_2O^-Na^+$, CH_3CH_2OH
CH_3CH_2O
OCH_2CH_3

In summary, benzenols (phenols) are oxidized to the corresponding diones (benzoquinones). The diones enter into reversible redox reactions that yield the corresponding diols and undergo conjugate additions and Diels-Alder additions to the double bonds.

24-6 Arenediazonium Salts, Useful Synthetic Intermediates

As mentioned earlier (Section 24-3), *N*-nitrosation of primary benzenamines (anilines) furnishes the relatively stable arenediazonium salts usable in the preparation of benzenols (phenols). This stability contrasts with the reactivity of alkanediazonium ions, which extrude nitrogen as fast as the diazonium salts are formed (Section 21-5). Why are the aromatic systems so much more stable? This section answers that question. Arenediazonium salts are precursors of haloarenes and arenenitriles by replacement of nitrogen and precursors of azo compounds by electrophilic substitution.

Arenediazonium Salts Are Stabilized by Resonance

The reason for the relative stability of arenediazonium salts is resonance and the high energy of the aryl cations formed on loss of nitrogen. One of the electron pairs making up the aromatic π system can be delocalized into the functional group, resulting in charge-separated resonance structures containing a double bond between the benzene ring and the attached nitrogen, as shown for the benzenediazonium cation.

Resonance in the Benzenediazonium Cation

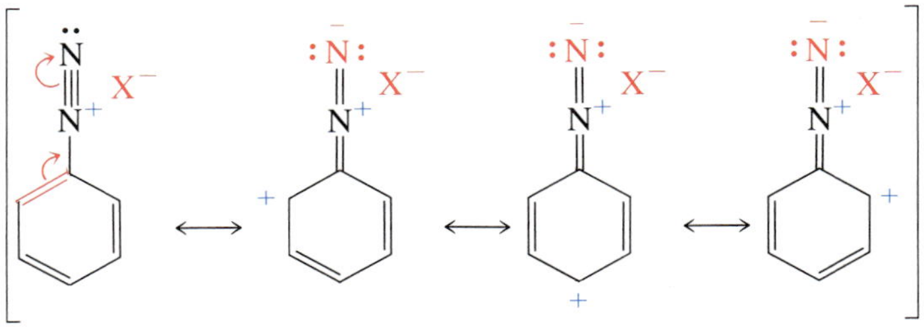

Only at high temperatures does nitrogen extrusion occur, to form the very reactive phenyl cation. When this is done in aqueous solution, benzenols (phenols) are produced (Section 24-3).

Why is the phenyl cation so reactive? After all, it is a carbocation that is part of a benzene ring. Should it not be resonance stabilized, like the phenylmethyl (benzyl) cation? The answer is no, as may be seen in the molecular-orbital picture of the phenyl cation (Figure 24-3). The empty orbital associated with the positive charge is one of the sp^2 hybrids aligned *perpendicular* to the π framework that normally produces aromatic resonance stabilization. Hence, this orbital cannot overlap with the π bonds and the positive charge cannot be delocalized.

FIGURE 24-3
Orbital picture of the phenyl cation.

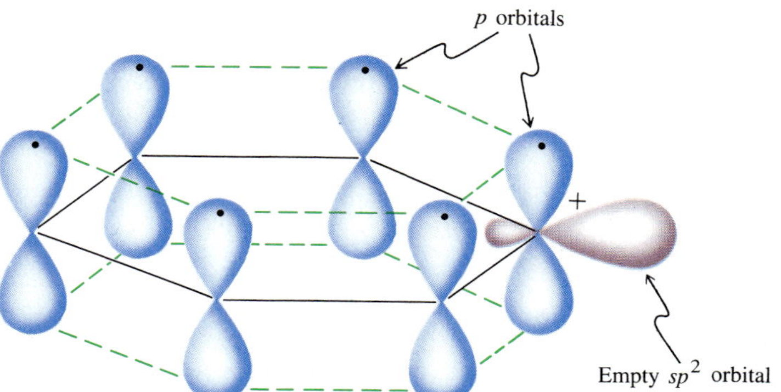

Arenediazonium Salts Can Be Converted into Other Substituted Benzenes

When arenediazonium salts are decomposed in the presence of nucleophiles other than water, the corresponding substituted benzenes are formed. For example, diazotization of arenamines (anilines) in the presence of hydrogen iodide results in the corresponding iodoarenes.

CHO, NH_2 → (CH_3COOH, HI, $NaNO_2$) → CHO, I + N_2
53%

Attempts to obtain other haloarenes in this way are frequently complicated by side reactions. One solution to this problem is the **Sandmeyer reaction*,** which makes use of the fact that the exchange of the nitrogen substituent for halogen is considerably facilitated by the presence of cuprous [Cu(I)] salts. The detailed mechanism of this decomposition reaction is not known, but is likely to proceed through radicals.

Sandmeyer Reactions

CH_3, NH_2 → (HCl, $NaNO_2$, 0°C) → CH_3, $N_2^+Cl^-$ → (CuCl, 60°C) → CH_3, Cl + N_2

2-Methylbenzenamine (*o*-Methylaniline)

79% overall
1-Chloro-2-methylbenzene (*o*-Chlorotoluene)

Cl, NH_2 → (1. HBr, $NaNO_2$, 0°C; 2. CuBr, 100°C) → Cl, Br

2-Chlorobenzenamine (*o*-Chloroaniline)

73%
1-Bromo-2-chlorobenzene (*o*-Bromochlorobenzene)

Cuprous ion catalysis is also used for the preparation of aromatic nitriles from arenamines (anilines). For this purpose, cuprous cyanide, CuCN, is added to the diazonium salt, in the presence of excess potassium cyanide.

CH_3, NH_2 → (1. HCl, $NaNO_2$, 0°C; 2. CuCN, KCN, 50°C) → CH_3, CN

70%
4-Methylbenzenenitrile (*p*-Tolunitrile)

Thermal decomposition of diazonium tetrafluoroborates yields the corresponding fluoroarenes. The transformation is known as the **Schiemann† reaction,** and no additional catalysts are required. The reaction is important, because direct electrophilic fluorination of benzenes with fluorine is difficult to control, owing to the exothermic character of the reaction.

---

*Dr. Traugott Sandmeyer, 1854–1922, Geigy Company, Basel.

†Professor Günther Schiemann, 1899–1969, Technical University of Hannover, West Germany.

COOH $N_2^+Cl^-$ $\xrightarrow{HBF_4}$ COOH $N_2^+BF_4^-$ $\xrightarrow{120°C}$ COOH F + N_2 + BF_3

A diazonium tetrafluoroborate

78%

2-Fluorobenzene-carboxylic acid (*o*-Fluorobenzoic acid)

The diazonium group can be removed reductively by reducing agents. The sequence diazotization–reduction is a way to replace the amino group in arenamines (anilines) with hydrogen. Among the reducing agents employed are aqueous hypophosphorous acid, H_3PO_2, and sodium borohydride in nonaqueous solution. This method is used to synthesize specifically substituted arenes in which the amino group is a directing substituent, to be removed in the last steps of a synthesis.

CH_3, Br, N_2^+ $\xrightarrow{H_3PO_2,\ H_2O,\ 25°C}$ CH_3, Br, H

85%

1-Bromo-3-methylbenzene (*m*-Bromotoluene)

Consider, for example, the synthesis of 1,3-dibromobenzene (*m*-dibromobenzene). Direct electrophilic bromination of benzene is not feasible for this purpose; after the first bromine has been introduced, the second will attack *ortho* or *para*. What is required is a *meta*-directing substituent, which can be transformed eventually into bromine. The nitro group is such a substituent. Double nitration of benzene furnishes 1,3-dinitrobenzene (*m*-dinitrobenzene). Reduction leads to the benzenediamine, which is then converted into the dihalo derivative.

$\xrightarrow{HNO_3,\ H_2SO_4,\ \Delta}$ NO_2, NO_2 $\xrightarrow{H_2,\ Pd}$

NH_2, NH_2 $\xrightarrow{1.\ NaNO_2,\ H^+,\ H_2O;\ 2.\ CuBr,\ 100°C}$ Br, Br

EXERCISE 24-22

Propose a synthesis of 1,3,5-tribromobenzene from benzene.

Electrophilic Substitution of Arenediazonium Salts: Azo Coupling

Being positively charged, arenediazonium ions have electrophilic character. Because the positive charge is delocalized, however, the salts are not very reactive. Nevertheless, they can accomplish electrophilic aromatic substitution if the substrate is an activated arene, such as benzenol (phenol) or benzenamine (aniline). This reaction is called **diazo coupling;** it leads to highly colored compounds, many of which are used as coloring agents *(azo dyes)*. For example, reaction of *N,N*-dimethylbenzenamine (*N,N*-dimethylaniline) with benzenediazonium chloride gives the brilliant orange dye *Butter Yellow*. This compound was once used as a food coloring agent, but has been declared a suspect carcinogen by the Food and Drug Administration.

Diazo Coupling

$$C_6H_5N_2^+\ Cl^- + C_6H_5\ddot{N}(CH_3)_2 \xrightarrow[pH\ =\ 3\text{–}5]{H^+,\ H_2O} C_6H_5\text{–N=N–}(H)C_6H_4\text{=}\overset{+}{N}(CH_3)_2\ Cl^- \xrightarrow{-\ HCl}$$

$$C_6H_5\text{–}\underbrace{N\text{=}N}_{\text{Azo function}}\text{–}C_6H_4\text{–}N(CH_3)_2$$

4-Dimethylaminoazobenzene
(*p*-Dimethylaminoazobenzene, Butter Yellow)

Azo dyes are sometimes employed as indicators because they change color on protonation. For example, Butter Yellow is red at and below pH 3, yellow at and above pH 4.0. At low pH, it is protonated at the azo function (not the dimethylamino group), because the corresponding cation is resonance stabilized.

$$\underset{\text{Yellow}}{(CH_3)_2\ddot{N}C_6H_4\ddot{N}\text{=}\ddot{N}C_6H_5} \xrightarrow{pH\ =\ 2.9} \underset{\text{Red}}{\left[(CH_3)_2\ddot{N}C_6H_4\ddot{N}\text{=}\overset{+}{N}(H)C_6H_5 \longleftrightarrow (CH_3)_2\ddot{N}C_6H_4\overset{+}{N}\text{–}\ddot{N}(H)C_6H_5 \longleftrightarrow (CH_3)_2\overset{+}{N}\text{=}C_6H_4\text{=}\ddot{N}\text{–}\ddot{N}(H)C_6H_5 \longleftrightarrow \text{etc.}\right]}$$

Dyes used in the clothing industry usually contain sulfonic acid groups that impart water solubility and allow the dye molecule to attach itself ionically to charged sites on the polymer framework of the textile.

Industrial Dyes

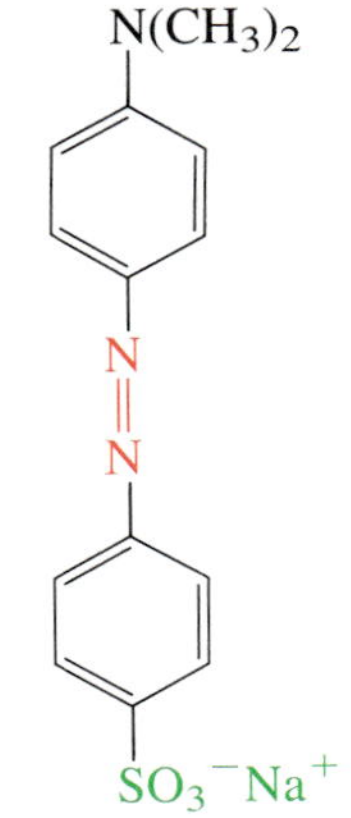

$N(CH_3)_2$ N=N $SO_3^-Na^+$

Methyl Orange
pH = 3.1, red
pH= 4.4, yellow

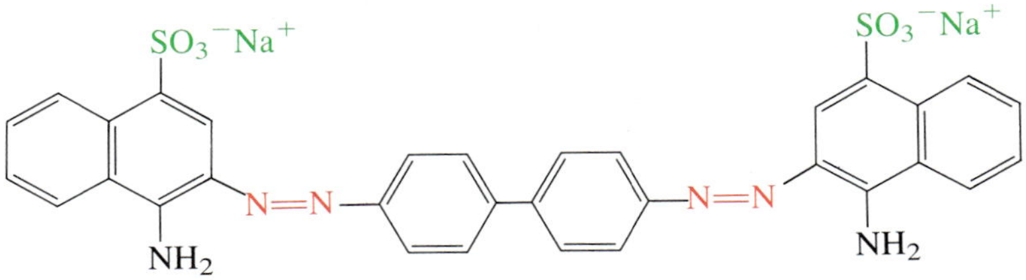

$SO_3^-Na^+$ N=N NH_2 N=N $SO_3^-Na^+$ NH_2

Congo Red
pH = 3.0, blue-violet
pH = 5.0, red

To summarize, arenediazonium salts, which are more stable than alkanediazonium salts because of resonance, are starting materials not only for benzenols (phenols), but also for haloarenes, arenenitriles, and reduced aromatics by displacement of nitrogen gas. The intermediates in some of these reactions may be aryl cations, highly reactive because of the absence of any electronically stabilizing features, but other, more-complicated mechanisms may be followed. The ability to transform arenediazonium salts in this way gives considerable scope to the regioselective construction of substituted benzenes. Finally, arenediazonium cations attack activated benzene rings by diazo coupling, a process that furnishes azobenzenes, which are often highly colored.

Summary of New Reactions

Benzylic Resonance

1 FREE-RADICAL HALOGENATION

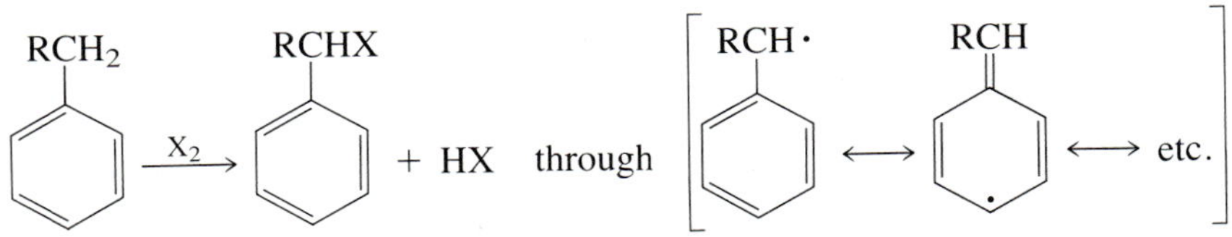

RCH_2 $\xrightarrow{X_2}$ RCHX + HX through [RCH· ⟷ RCH ⟷ etc.]

Benzylic radical

2 SOLVOLYSIS

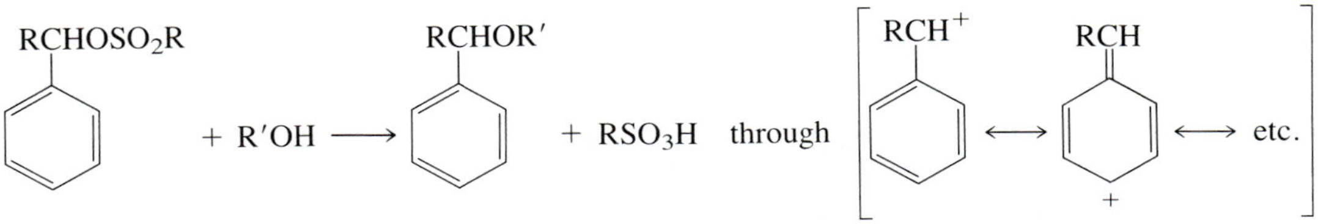

$RCHOSO_2R$ + R′OH ⟶ RCHOR′ + RSO_3H through [RCH^+ ⟷ RCH ⟷ etc.]

Benzylic cation

3 (HALOMETHYL)BENZENES FROM PHENYLMETHANOL

$C_6H_5CH_2OH + HX \longrightarrow C_6H_5CH_2X$

4 S_N2 REACTIONS OF (HALOMETHYL)BENZENES

$C_6H_5CH_2X + :Nu^- \longrightarrow C_6H_5CH_2Nu + X^-$

Through delocalized transition state

5 BENZYLIC DEPROTONATION

$C_6H_5CH_3 + RLi \longrightarrow C_6H_5CH_2Li + RH$

$pK_a \sim 41$

Phenylmethyllithium (Benzyllithium)

Oxidation and Reduction of the Aromatic Nucleus

6 OZONOLYSIS

$C_6H_6 \xrightarrow{\text{1. } O_3 \text{ 2. } (CH_3)_2S} 3\ HCCH$ (OO)

1,2-dialkylbenzene $\xrightarrow{\text{1. } O_3 \text{ 2. } (CH_3)_2S}$ RCCR + RCCH + HCCH (each OO)

7 BIRCH REDUCTION

$C_6H_6 \xrightarrow{\text{Na, liquid } NH_3,\ CH_3CH_2OH}$ 1,4-cyclohexadiene

Regioselectivity:

1,2-dialkylbenzene $\xrightarrow{\text{Na, liquid } NH_3,\ CH_3CH_2OH}$ 1,2-dialkyl-1,4-cyclohexadiene

R = alkyl

$$C_6H_5OR \xrightarrow{\text{Na, liquid } NH_3,\ CH_3CH_2OH} \text{1-alkoxy-1,4-cyclohexadiene}$$

$$C_6H_5COOH \xrightarrow[\text{2. } H^+,\ H_2O]{\text{1. Na, liquid } NH_3,\ CH_3CH_2OH} \text{2,5-cyclohexadiene-1-carboxylic acid}$$

Oxidation and Reduction Reactions on Aromatic Side Chains

8 OXIDATION

$$C_6H_5CH_2R \xrightarrow[\text{2. } H^+,\ H_2O]{\text{1. } KMnO_4,\ HO^-,\ \Delta} C_6H_5COOH$$

Other suitable oxidizing agents:

$Na_2Cr_2O_7$, H^+; HNO_3, H_2O; O_2, $Co(O\overset{O}{\overset{\|}{C}}CH_3)_2$, $Mn(O\overset{O}{\overset{\|}{C}}CH_3)_2$; O_2, V_2O_5.

Benzylic alcohols:

$$C_6H_5CHOHR \xrightarrow{MnO_2} C_6H_5C(=O)R$$

9 REDUCTION

$$C_6H_5CH_2OR \xrightarrow{H_2,\ Pd\text{-}C} C_6H_5CH_3 + ROH$$

$C_6H_5CH_2$ is a protecting group for ROH

Benzenols (Phenols)

10 ACIDITY

$$C_6H_5OH \rightleftharpoons H^+ + \left[C_6H_5\ddot{\underset{\cdot\cdot}{O}}\!:^- \longleftrightarrow \text{(cyclohexadienone, } \ddot{C}^- \text{ at para)} \longleftrightarrow \text{etc.} \right]$$

11 PREPARATION BY NUCLEOPHILIC AROMATIC SUBSTITUTION

$C_6H_5SO_3^-Na^+ \xrightarrow{\text{1. NaOH, }\Delta \text{; 2. }H^+, H_2O} C_6H_5OH$

$C_6H_5X \xrightarrow{\text{1. NaOH, }\Delta \text{; 2. }H^+, H_2O} C_6H_5OH$

12 CUMENE HYDROPEROXIDE PROCESS

$C_6H_6 \xrightarrow{CH_3CH{=}CH_2,\ H^+} C_6H_5CH(CH_3)_2 \xrightarrow{O_2} C_6H_5C(CH_3)_2OOH \xrightarrow{10\%\ H_2SO_4} C_6H_5OH + CH_3CCH_3$ (O above central C)

13 ARENEDIAZONIUM SALT HYDROLYSIS

$C_6H_5NH_2 \xrightarrow{NaNO_2,\ H^+,\ 0^\circ C} C_6H_5N_2^+ \xrightarrow{H_2O,\ \Delta} C_6H_5OH + N_2$

Benzenediazonium cation

Reactions of Benzenols (Phenols) and Alkoxybenzenes

14 ETHER CLEAVAGE

$C_6H_5OR \xrightarrow{HBr,\ \Delta} C_6H_5OH + RBr$

$C_6H_5OR \xrightarrow{\text{1. }AlCl_3,\ HCl\text{; 2. }H_2O} C_6H_5OH + RCl$

15 ETHER FORMATION

$C_6H_5OH + RX \xrightarrow{NaOH,\ H_2O} C_6H_5OR$

Alkoxybenzene

16 ESTERIFICATION

$C_6H_5OH + RCCl \xrightarrow{Base} C_6H_5OCR$ (O above the C of RCCl and OCR)

17 ELECTROPHILIC AROMATIC SUBSTITUTION

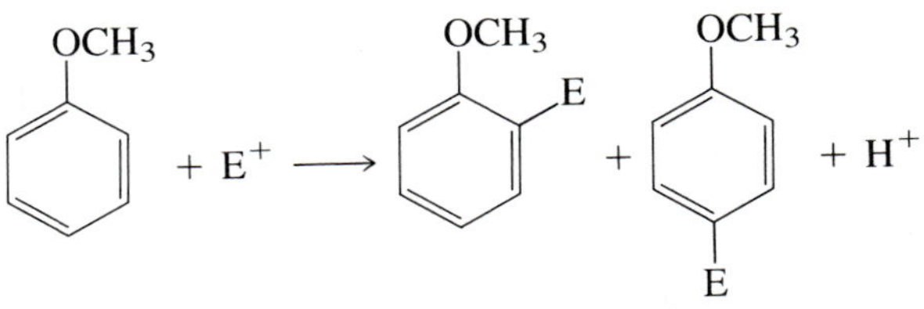

OCH$_3$ + E$^+$ ⟶ OCH$_3$, E + OCH$_3$, E + H$^+$

18 PHENOLIC RESINS

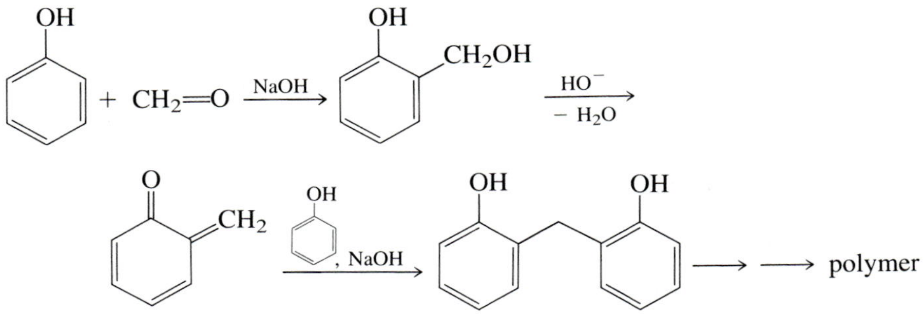

OH + $CH_2{=}O$ $\xrightarrow{\text{NaOH}}$ OH, CH_2OH $\xrightarrow[-\,H_2O]{HO^-}$

O, CH_2 $\xrightarrow{\text{OH, NaOH}}$ OH OH ⟶ ⟶ polymer

19 KOLBE REACTION

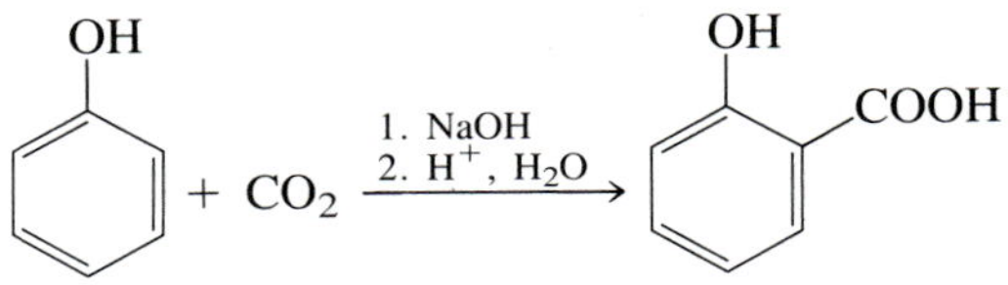

OH + CO_2 $\xrightarrow[\text{2. H}^+,\ H_2O]{\text{1. NaOH}}$ OH, COOH

20 CLAISEN REARRANGEMENT

Aromatic Claisen rearrangement:

O $\xrightarrow{\Delta}$ O, H ⟶ OH

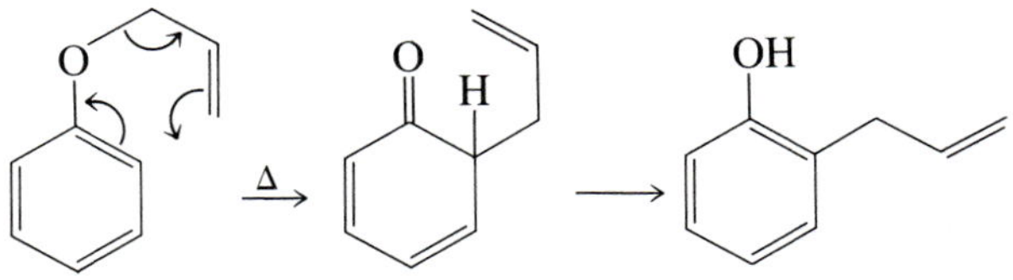

Aliphatic Claisen rearrangement:

O ⟶ O, H

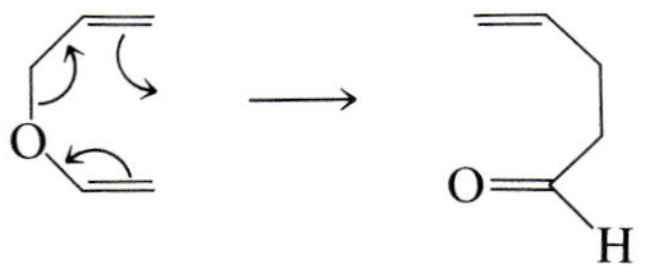

21 COPE REARRANGEMENT

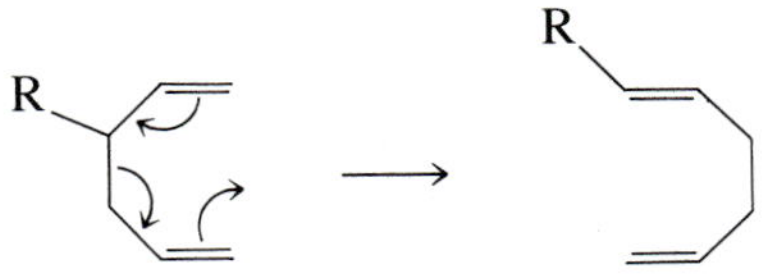

R ⟶ R

22 OXIDATION

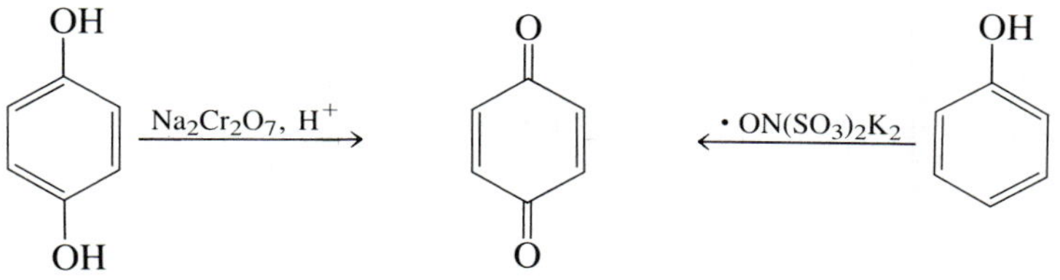

$$\text{HO-C}_6\text{H}_4\text{-OH} \xrightarrow{Na_2Cr_2O_7,\ H^+} \text{2,5-Cyclohexadiene-1,4-dione} \xleftarrow{\cdot ON(SO_3)_2K_2} \text{C}_6\text{H}_5\text{OH}$$

2,5-Cyclohexadiene-1,4-dione
(*p*-Benzoquinone)

23 CONJUGATE ADDITIONS
to 2,5-CYCLOHEXADIENE-1,4-DIONES (*p*-BENZOQUINONES)

$$\text{O=C}_6\text{H}_4\text{=O} + HCl \longrightarrow \text{2-chloro-1,4-dihydroxybenzene (OH, OH, Cl)}$$

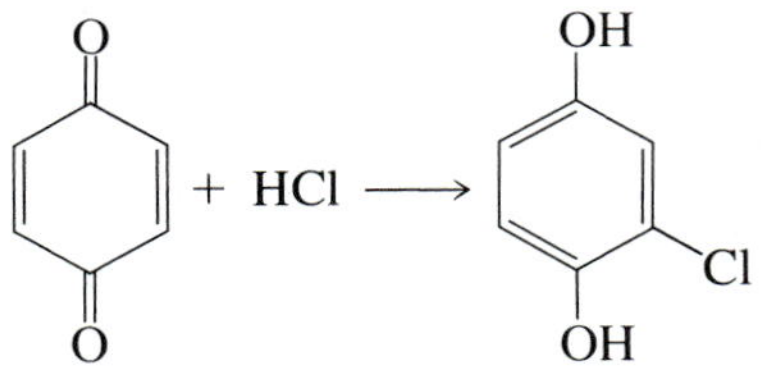

24 DIELS-ALDER CYCLOADDITIONS
to 2,5-CYCLOHEXADIENE-1,4-DIONES (*p*-BENZOQUINONES)

$$\text{O=C}_6\text{H}_4\text{=O} + \text{CH}_2\text{=CH-CH=CH}_2 \longrightarrow \text{cis-fused bicyclic dione (O, O, H, H)}$$

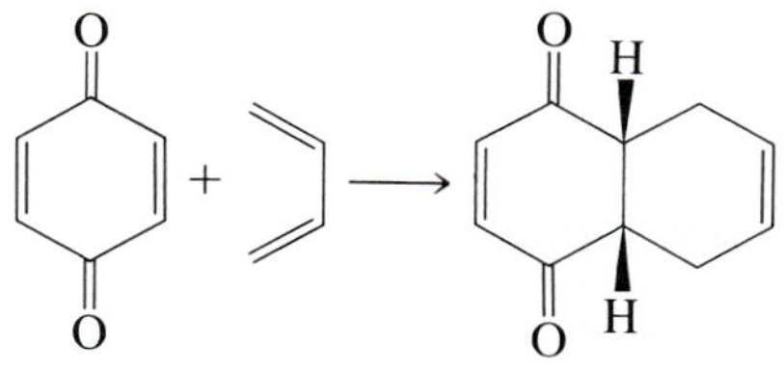

Arenediazonium Salts

25 RESONANCE

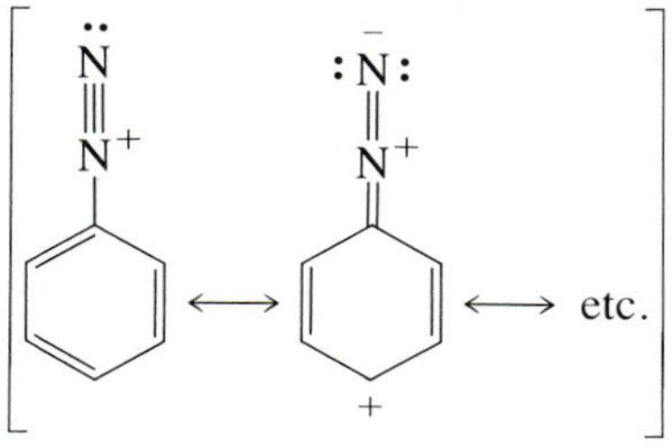

$$\left[\text{C}_6\text{H}_5\text{-}N^+\equiv\ddot{N} \longleftrightarrow \text{(+)C}_6\text{H}_5\text{=}N^+\text{=}\ddot{N}^-{:} \longleftrightarrow \text{etc.} \right]$$

26 SANDMEYER REACTIONS

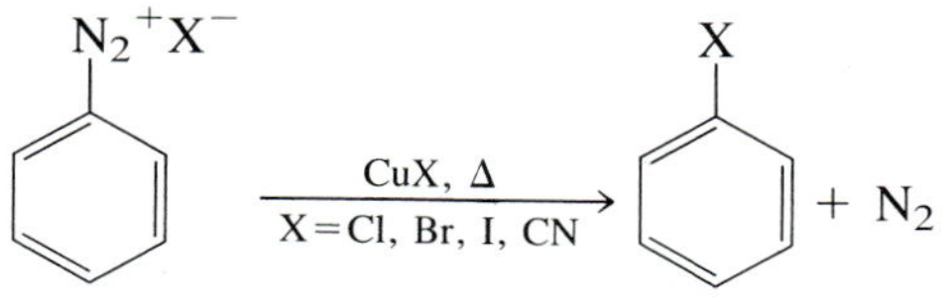

$$\text{C}_6\text{H}_5N_2{}^+X^- \xrightarrow[X=Cl,\ Br,\ I,\ CN]{CuX,\ \Delta} \text{C}_6\text{H}_5X + N_2$$

27 SCHIEMANN REACTION

$$C_6H_5N_2^+BF_4^- \xrightarrow{\Delta} C_6H_5F + N_2 + BF_3$$

28 REDUCTION

$$C_6H_5N_2^+ \xrightarrow{H_3PO_2 \text{ or } NaBH_4} C_6H_6 + N_2$$

29 DIAZO COUPLING

$$C_6H_5N^+{\equiv}N + C_6H_5OH \longrightarrow C_6H_5{-}N{=}N{-}C_6H_4{-}OH + H^+$$

Azo compound

Summary of Important Concepts

1 Benzylic radical, cation, and anion formations are facilitated by resonance of the resulting centers with a benzene π system. S_N2 reactions at benzylic carbons are accelerated by resonance in their transition states.

2 The aromatic π system in benzene derivatives can be disrupted by ozonolysis or Birch reduction.

3 Benzenols (phenols) are aromatic enols, undergoing reactions typical of the hydroxy function and the aromatic ring.

4 Cyclohexadienediones and benzenediols function as redox couples in laboratory reactions and in nature.

5 Arenediazonium ions are stabilized by resonance but furnish reactive aryl cations in which the positive charge cannot be delocalized into the aromatic ring.

6 The amino group can be used to direct electrophilic aromatic substitution, after which it is removable by diazotization and reduction.

Problems

1 Write the expected major product(s) of each of the following reactions.

(a) 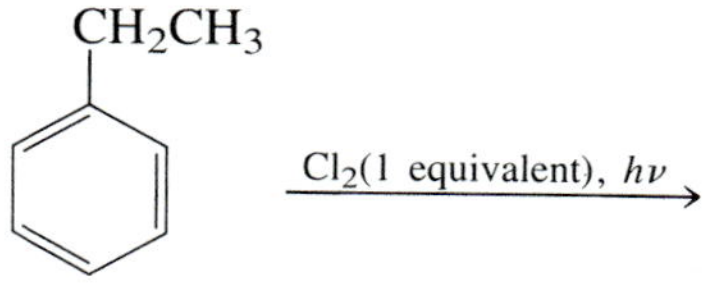 [ethylbenzene, CH_2CH_3] $\xrightarrow{Cl_2\text{(1 equivalent)}, h\nu}$

(b) 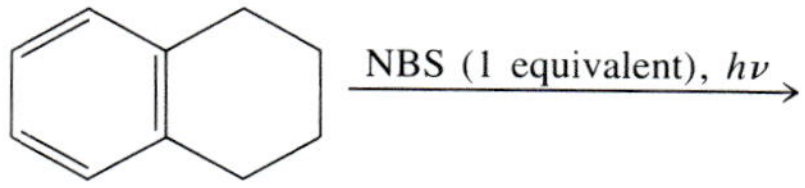 [tetrahydronaphthalene] $\xrightarrow{\text{NBS (1 equivalent)}, h\nu}$

(c) 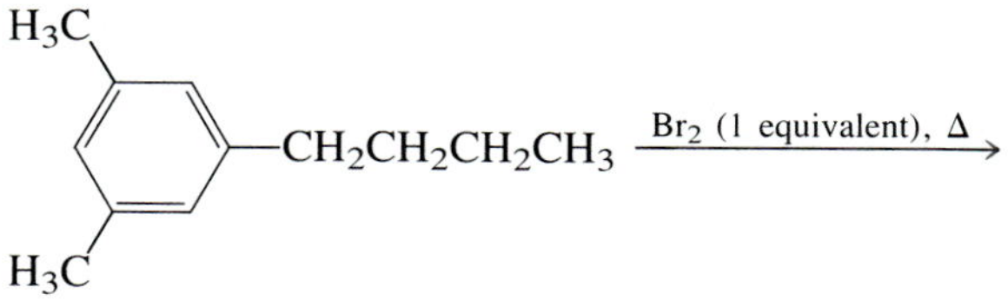 [3,5-dimethyl, H_3C, H_3C]$-CH_2CH_2CH_2CH_3$ $\xrightarrow{Br_2\text{ (1 equivalent)}, \Delta}$

2 By drawing appropriate resonance structures, illustrate why halogen atom attachment at the *para* position of phenylmethyl (benzyl) radical is unfavored compared with attachment at the benzylic position.

3 Is the site of reaction in the biosynthesis of norepinephrine from dopamine (see Chapter 5, Problem 27) consistent with the principles outlined in this chapter? Would it be easier or more difficult to duplicate this transformation nonenzymatically? Explain.

4 Write the expected products of the following reactions or reaction sequences.

(a) 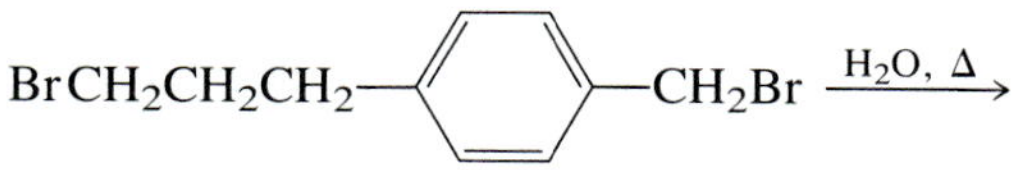 $BrCH_2CH_2CH_2$–[benzene]–CH_2Br $\xrightarrow{H_2O, \Delta}$

(b) 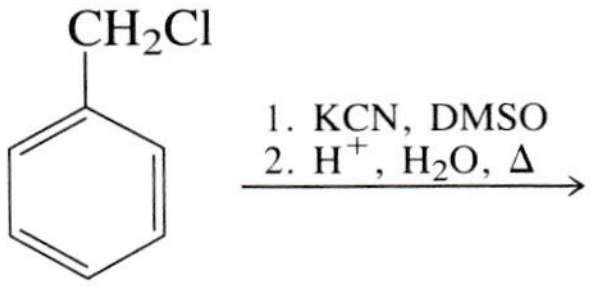 [benzene, CH_2Cl] $\xrightarrow[\text{2. } H^+, H_2O, \Delta]{\text{1. KCN, DMSO}}$

(c) 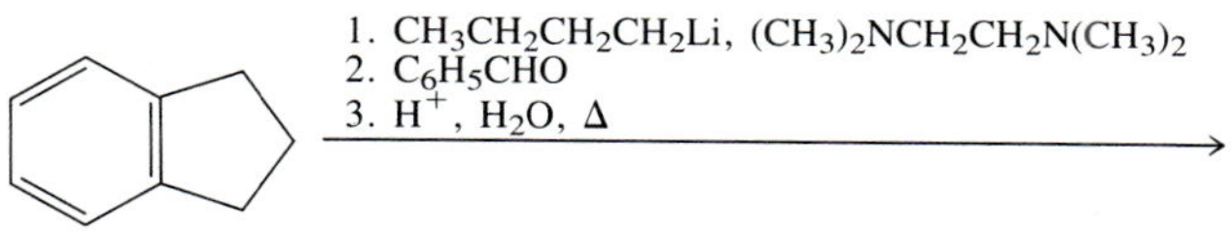 [indane]
1. $CH_3CH_2CH_2CH_2Li$, $(CH_3)_2NCH_2CH_2N(CH_3)_2$
2. C_6H_5CHO
3. H^+, H_2O, Δ
$\longrightarrow$

5 The hydrocarbon with the common name *fluorene* is, like triphenylmethane, acidic enough ($pK_a \sim 23$) to be a useful indicator in deprotonation reactions of compounds of greater acidity. Indicate the most-acidic hydrogen(s) in fluorene. Draw resonance structures to explain the relative stability of its conjugate base.

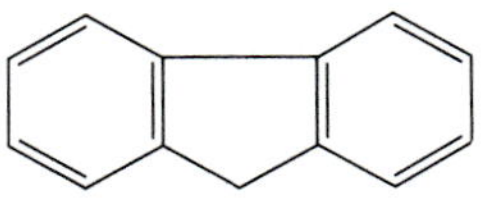

Fluorene

6 *N,N*,2-Trimethylbenzenamine is readily deprotonated by butyllithium at 0°C, a reaction that does not require addition of external amines (compare deprotonation of methylbenzene, Section 24-1). Explain, and draw the structure of the resulting lithium reagent.

7 **(a)** Is it possible to distinguish between the three isomers of dimethylbenzene (xylene) by means of ozonolysis alone? Illustrate your answer with reactions.

(b) Upon ozonolysis, an unidentified hydrocarbon A gives the two products shown in equimolar amounts. Write a structure for hydrocarbon A consistent with this result.

A $\xrightarrow{\text{1. } O_3 \text{; 2. } (CH_3)_2S}$ [cyclobutane-1,2-dione structure] + HCCCH$_2$CH$_2$CCH (with C=O groups)

1,2-Cyclobutanedione **2,5-Dioxohexanedial**

8 Write the product(s) that you expect to result from the following reaction sequence, which has been used to synthesize both steroids and "norhomosteroids" for biological testing.

[Structure: tetracyclic steroid-like compound bearing HO, H_3C, HO and an aromatic ring with OCH_3] $\xrightarrow{\text{1. Na, } NH_3\text{, } CH_3CH_2OH \text{; 2. } H^+\text{, } H_2O}$

9 Outline a straightforward, practical, and efficient synthesis of each of the following compounds. Start with benzene or methylbenzene. Assume that the *para* isomer (but *not* the *ortho* isomer) may be separated efficiently from any mixtures of *ortho* and *para* substitution products.

(a) $C_6H_5CH_2CH_2C{\equiv}CH$

(b) 4-Cl-C_6H_4-$CONH_2$

(c) C_6H_5-CO-C_6H_4-$COOCH_3$ (*para*)

(d) 2,6-dibromobenzoic acid (COOH flanked by Br, Br)

10 The synthesis of germanicol (begun in Problem 30 of Chapter 16) utilizes several reactions of arenes in its later stages. Fill in reagents **a** through **e** where they have been omitted from the following synthetic sequence. Each letter may correspond to between one and three steps.

[Structure: ketal-protected polycyclic compound with H_3C, CH_3, CH_3, CH_3, OCH_2–phenyl, and CH_2CH=O side chain] $\xrightarrow{\text{(a)}}$ [Structure: CH_3, CH_3, OCH_2–phenyl, $CH_2CH(OH)$–aryl bearing OCH_3] $\xrightarrow{\text{(b)}}$

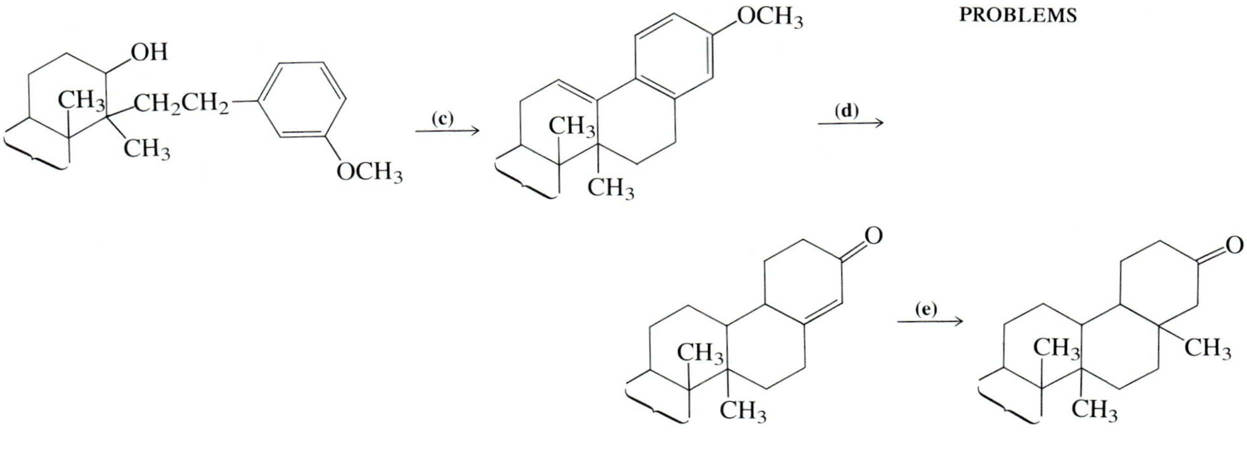

OH, CH_3, CH_2CH_2, CH_3, OCH_3 (c) → OCH_3, CH_3, CH_3 (d) →

O, CH_3, CH_3 (e) → O, CH_3, CH_3, CH_3

11 6,6-Dialkyl-2,4-cyclohexadienones do not spontaneously enolize to form benzenols. Why not?

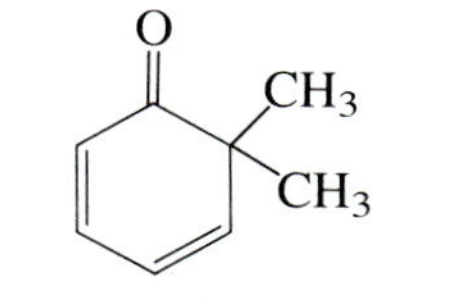

O, CH_3, CH_3

6,6-Dimethyl-2,4-cyclohexadienone

12 Write the expected product(s) of each of the following reactions and reaction sequences.

(a) 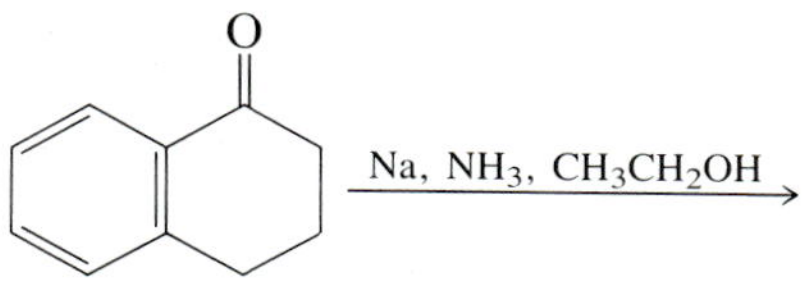 O; Na, NH_3, CH_3CH_2OH →

(b) 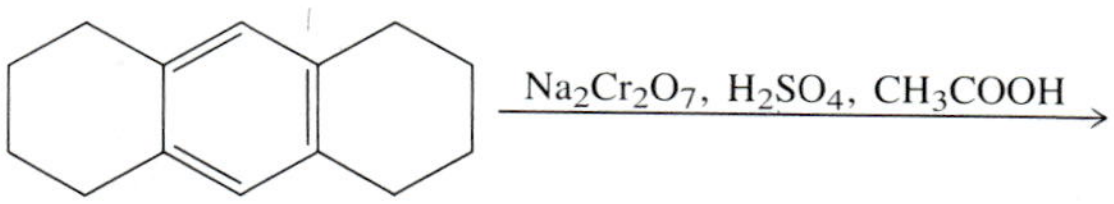 $Na_2Cr_2O_7$, H_2SO_4, CH_3COOH →

(c) 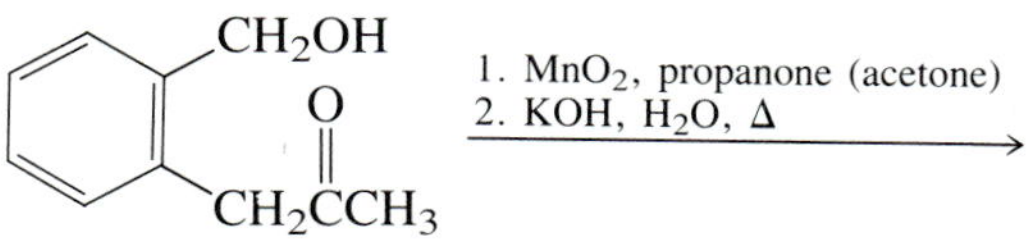 CH_2OH, O, $CH_2\overset{\|}{C}CH_3$; 1. MnO_2, propanone (acetone) 2. KOH, H_2O, Δ →

(d) 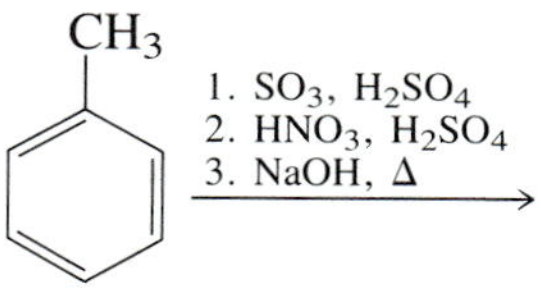 CH_3; 1. SO_3, H_2SO_4 2. HNO_3, H_2SO_4 3. NaOH, Δ →

13 Rank the following compounds in order of descending acidity.

(a) CH_3OH

(b) CH_3COOH

(c) 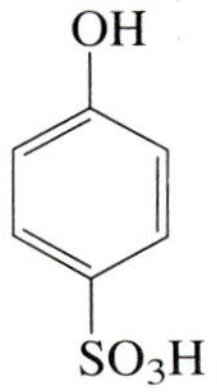 OH, SO_3H

(d) 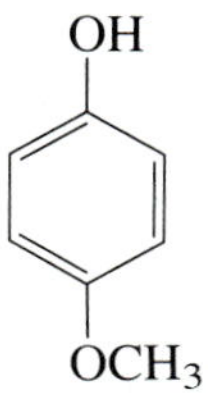 OH, OCH_3

(e) 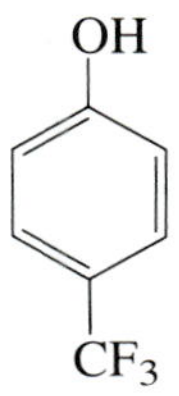 OH, CF_3

(f) 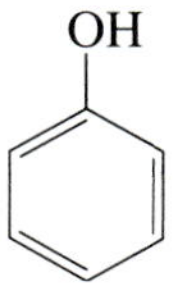 OH

14 Design a synthesis of each of the following benzenols (phenols). Begin each synthesis either with benzene or with any monosubstituted benzene derivative.

(a) OH, CH_3 (4-methylbenzenol)

(b) OH, Br, Br (2,6-dibromobenzenol)

(c) The three benzenediols

(d) OH, Cl, NO_2, NO_2

15 Starting with benzene, propose syntheses of each of the following benzenol (phenol) derivatives.

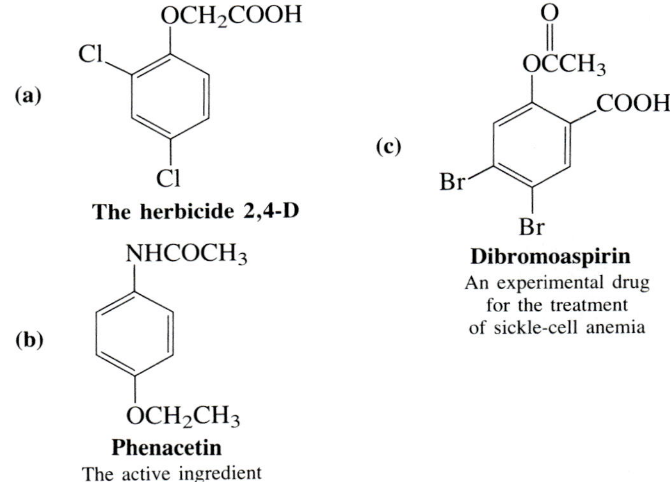

The herbicide 2,4-D

Phenacetin
The active ingredient in Midol

Dibromoaspirin
An experimental drug for the treatment of sickle-cell anemia

16 **(a)** Propose a mechanism for the cleavage of an alkoxyarene (e.g., methoxybenzene) by $AlCl_3$.

(b) A related process is the reaction of an aryl ester (e.g., a phenyl alkanoate) with $AlCl_3$, which results in alkanoylbenzenols:

$OCCH_3$ (phenyl ethanoate) $\xrightarrow{AlCl_3}$ 2-ethanoylbenzenol (CCH_3, OH) + 4-ethanoylbenzenol (OH, CCH_3)

Propose a mechanism for this reaction called the *Fries rearrangement*.

(c) Show how the Fries rearrangement can be applied to the synthesis of 2-ethanoyl-4-methylbenzenol.

OH, CCH_3, CH_3

2-Ethanoyl-4-methylbenzenol

17 Write the expected product(s) of each of the following reaction sequences.

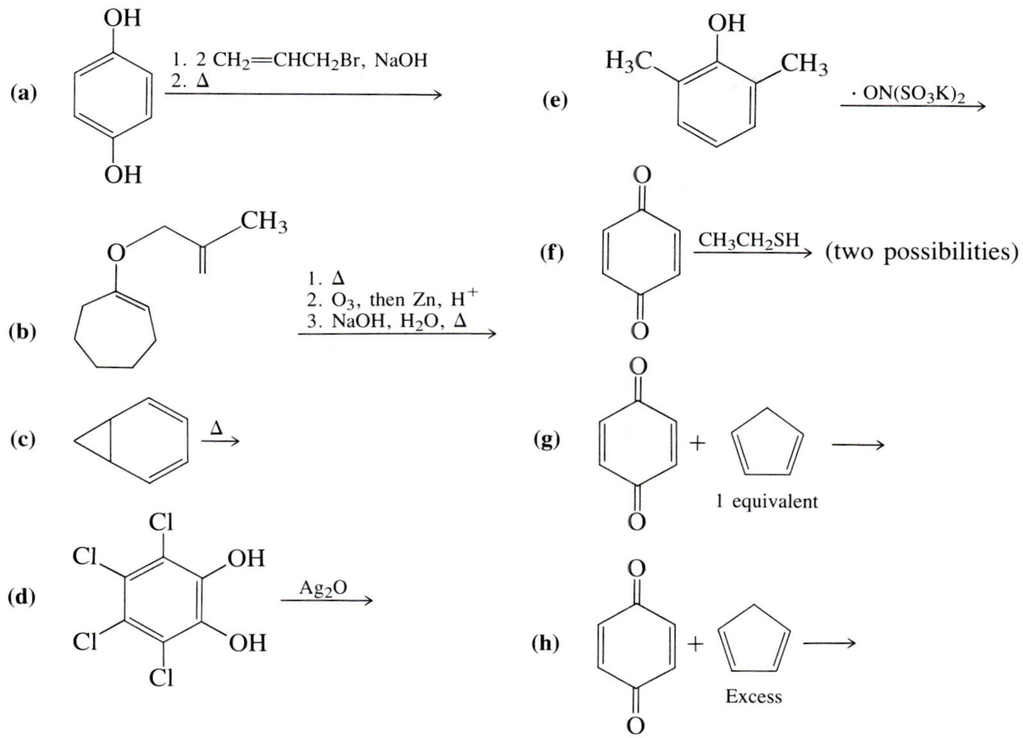

(a) [hydroquinone] 1. 2 $CH_2{=}CHCH_2Br$, NaOH 2. Δ →

(b) 1. Δ 2. O_3, then Zn, H^+ 3. NaOH, H_2O, Δ →

(c) Δ →

(d) Cl, Cl, Cl, Cl, OH, OH; Ag_2O →

(e) H_3C, OH, CH_3; $\cdot ON(SO_3K)_2$ →

(f) CH_3CH_2SH → (two possibilities)

(g) + [cyclopentadiene] → 1 equivalent

(h) + [cyclopentadiene] → Excess

18 As a children's medicine, Tylenol has a major marketing advantage over aspirin: *liquid Tylenol* preparations (essentially, Tylenol dissolved in flavored water) are stable, whereas comparable aspirin solutions are not. Explain.

19 Juvabione is a sesquiterpene found in the balsam fir. It has high juvenile-hormone activity (i.e., it prevents maturation of the larvae of certain insects). An outline of a synthesis is shown below. Supply the missing reagents.

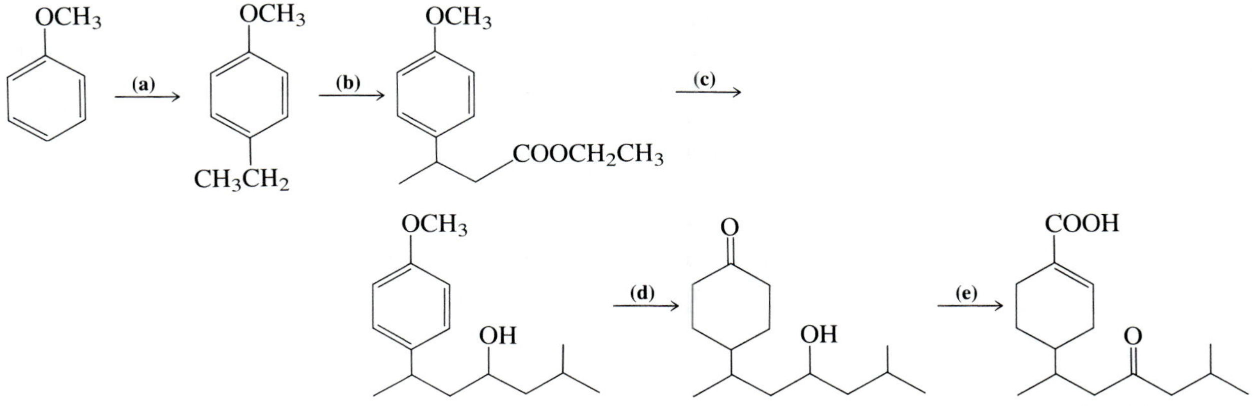

OCH_3 —(a)→ OCH_3, CH_3CH_2 —(b)→ OCH_3, $COOCH_2CH_3$ —(c)→

OCH_3, OH —(d)→ O, OH —(e)→ COOH, O

Juvabione

20 Biochemical oxidation of aromatic rings is catalyzed by a group of liver enzymes called aryl hydroxylases. Part of this chemical process is the conversion of toxic aromatic hydrocarbons such as benzene into water-soluble benzenols (phenols), which can be easily excreted. However, the primary purpose of the enzyme is to enable the synthesis of biologically useful compounds, such as the amino acid tyrosine:

$$\text{Phenylalanine (}C_6H_5CH_2CH(NH_2)COOH\text{)} \xrightarrow{O_2,\ \text{hydroxylase}} \text{Tyrosine (}p\text{-}HOC_6H_4CH_2CH(NH_2)COOH\text{)}$$

Phenylalanine **Tyrosine**

(a) Extrapolating from your knowledge of benzene chemistry, which of the three following possibilities seems most reasonable?
(i) The oxygen is introduced by electrophilic attack on the ring.
(ii) The oxygen is introduced by free-radical attack on the ring.
(iii) The oxygen is introduced by nucleophilic attack on the ring.

(b) It is widely suspected that oxacyclopropanes play a role in arene hydroxylation. Part of the evidence is the following observation: if the site to be hydroxylated is initially labeled with deuterium, a substantial proportion of the product still contains deuterium atoms, which have apparently migrated to the position *ortho* to the site of hydroxylation:

$$p\text{-}R\text{-}C_6H_4D \xrightarrow{O_2,\ \text{hydroxylase}} 4\text{-}R\text{-}2\text{-}D\text{-}C_6H_3OH$$ through (arene oxide bearing D and H on the epoxide carbons) as an intermediate

Suggest a plausible mechanism for the formation of the oxacyclopropane intermediate and its conversion into the observed product. Assume the availability of catalytic amounts of acids and bases, as necessary.

Note: In victims of the genetically transmitted disorder called phenylketonuria (PKU), the hydroxylase enzyme system described here does not function. Instead, phenylalanine in the brain is converted into 2-phenyl-2-oxopropanoic (phenylpyruvic) acid, the reverse of the process shown in Problem 18 of Chapter 21. The buildup of this compound in the brain can lead to severe retardation; thus people with PKU (which can be diagnosed at birth) must be restricted to diets low in phenylalanine.

21 A laboratory technique for the synthesis of the amino acid phenylalanine (Problem 20) utilizes the reagent *phthalimidomalonic ester,* prepared as shown.

$$\text{(phthalimide)}N^-K^+ + BrCH(COOCH_2CH_3)_2 \longrightarrow \text{(phthalimide)}N\text{—}CH(COOCH_2CH_3)_2$$

Potassium phthalimide **Diethyl 2-bromopropanedioate (Ethyl bromomalonate)** **Phthalimidomalonic ester**

Propose a synthesis of phenylalanine from this reagent and either benzene or methylbenzene (toluene). Hint: Review the material in Section 22-4.

22 Follow the instructions given in parts **a** through **d** about the scheme below, which is a clever synthesis of one of the *anthracyclines,* some of the most clinically useful anticancer drugs.

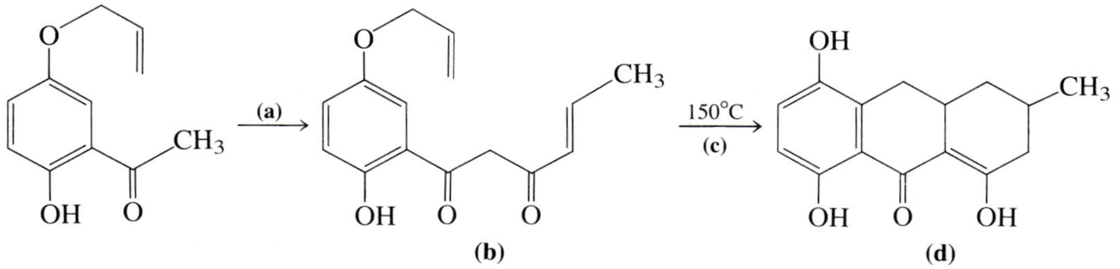

(a) → (b) → 150°C (c) → (d)

(a) Supply reagents for this step.

(b) Write another tautomer for this compound. (Hint: Section 16-2.)

(c) Two successive reactions take place during heating. What are they? Write their mechanisms.

(d) The product of reaction **c** has to undergo an isomerization to give this final product. Indicate how and why this isomerization occurs.

23 A common application of the Cope rearrangement is in ring-enlargement sequences. Fill in the reagents and products missing from the following scheme, which illustrates the construction of a ten-membered ring.

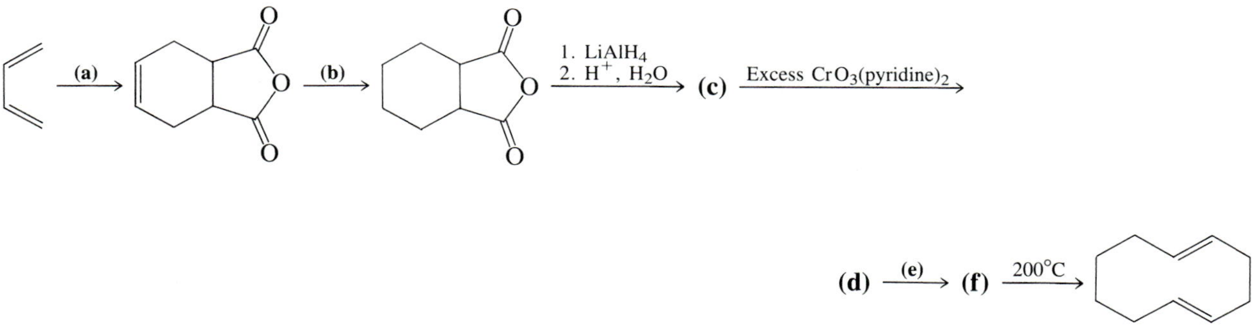

(a) → (b) → 1. $LiAlH_4$ 2. H^+, H_2O → **(c)** Excess $CrO_3(pyridine)_2$ →

(d) (e) → **(f)** 200°C →

24 Substituents can have a substantial effect on the redox relation between 2,5-cyclohexadiene-1,4-diones (*p*-benzoquinones) and 1,4-benzenediols (hydroquinones) by preferential resonance stabilization of one form or the other.

Compare the following three equilibria. Which of them lies farthest to the left? Which one lies farthest to the right? Draw resonance forms to indicate how the substituents affect the equilibrium by stabilizing one of the forms over the other.

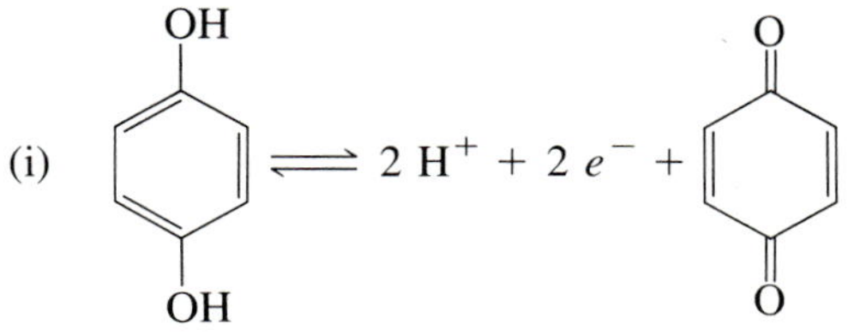

(i) OH, OH ⇌ $2\ H^+ + 2\ e^- +$ O, O

(ii) 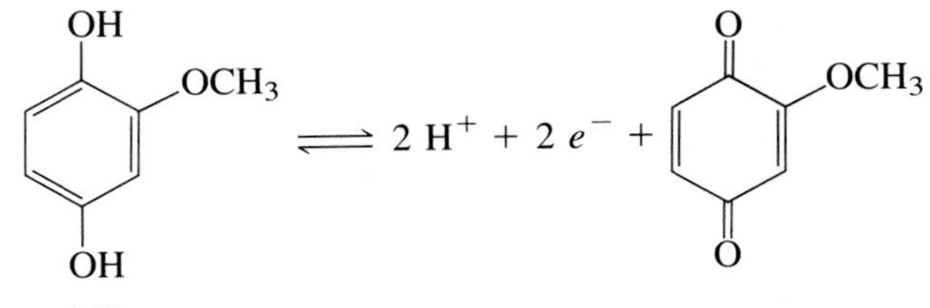 2-methoxy-1,4-benzenediol (OH, OCH_3, OH) $\rightleftharpoons$ $2\,H^+ + 2\,e^- +$ methoxy-1,4-benzoquinone (O, OCH_3, O)

(iii) 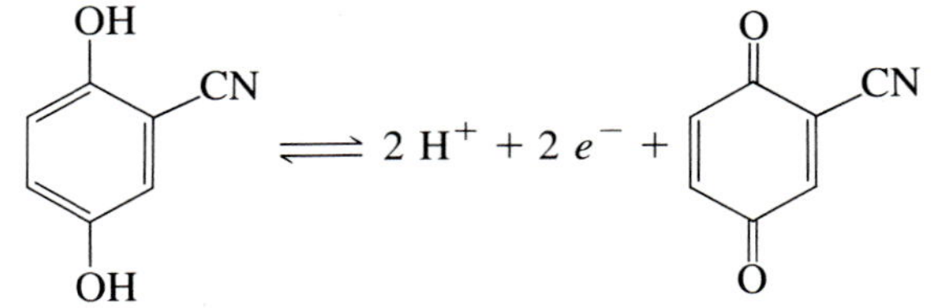 2-cyano-1,4-benzenediol (OH, CN, OH) $\rightleftharpoons$ $2\,H^+ + 2\,e^- +$ cyano-1,4-benzoquinone (O, CN, O)

25 Write a complete mechanism for the diazotization of benzenamine (aniline) in the presence of HCl and $NaNO_2$ and a plausible mechanism (based on what you have learned in Section 24-6) for its conversion into iodobenzene by treatment with aqueous iodide ion (e.g., from K^+I^-).

26 Devise a synthesis of each of the following substituted benzene derivatives. Start each synthesis with benzene.

(a) 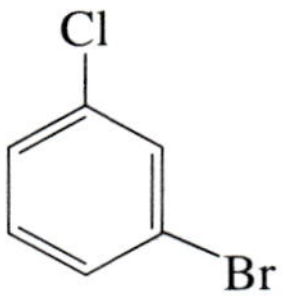 Cl, Br (1,3-)

(b) 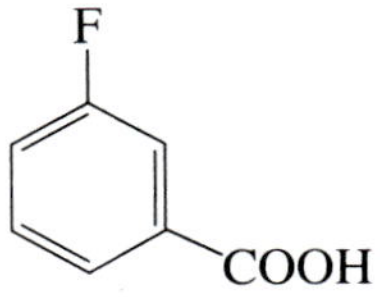 F, COOH (1,3-)

(c) 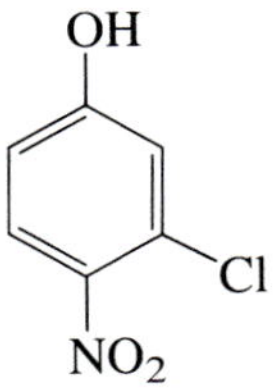 OH, Cl, NO_2

(d) 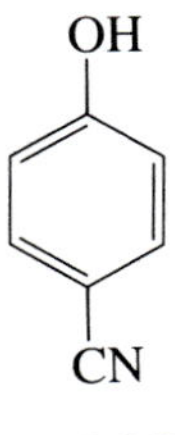 OH, CN (1,4-)

(e) 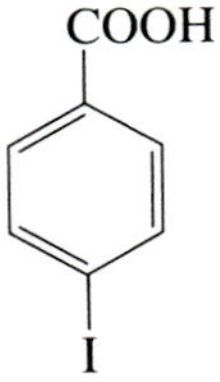 COOH, I (1,4-)

(f) Cl, Br, Cl, Br

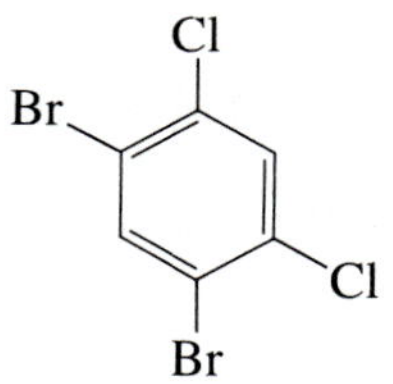

(g) 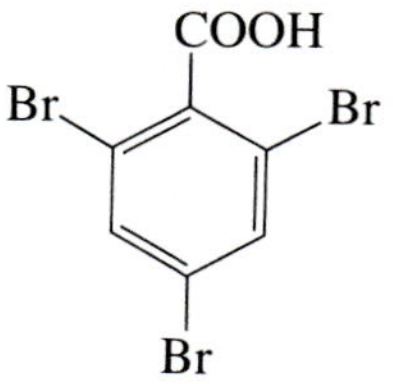 COOH, Br, Br, Br

(h) 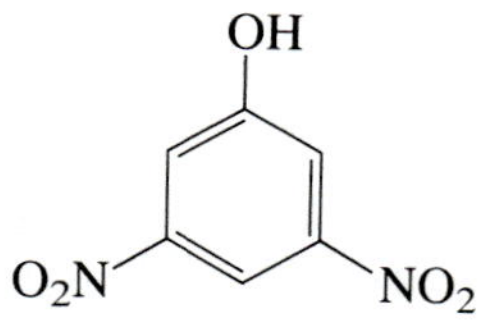 OH, O_2N, NO_2

27 Write the most-reasonable structure of the product of each of the following reaction sequences.

(a) p-aminobenzenesulfonic acid (NH_2, SO_3H) —— 1. $NaNO_2$, HCl, 5°C; 2. HO–(m-hydroxyphenyl)–OH ——→ Golden Yellow

(b) m-aminobenzenesulfonic acid (NH_2, SO_3H) —— 1. $NaNO_2$, HCl, 5°C; 2. C_6H_5–NH–C_6H_5 ——→ Metanil Yellow

For the following reaction, assume that electrophilic substitution is preferentially on the most-activated ring.

(c) p-aminobenzenesulfonic acid (NH_2, SO_3H) —— 1. $NaNO_2$, HCl, 5°C; 2. 1-naphthol (OH) ——→ Orange I

28 Show the reagents that would be necessary for the synthesis by diazo coupling of each of the three compounds listed below.

(a) Methyl Orange.
(b) Congo Red.
(c) Prontosil, H_2N–$C_6H_3(NH_2)$–N=N–C_6H_4–SO_2NH_2, which is converted microbially into sulfanilamide, H_2N–C_6H_4–SO_2NH_2. The accidental discovery of the antibacterial properties of prontosil in the 1930s led indirectly to the development of sulfa drugs as antibiotics in the 1940s.

29 A common natural flavoring substance A with the formula $C_{14}H_{18}O_8$ is hydrolyzed by dilute aqueous acid into one equivalent of D-glucose and one equivalent of a compound B with the formula $C_8H_8O_3$, which has ^1H NMR spectrum B (on the next page) and gives the following additional spectral data:
IR: $\tilde{\nu}$ = 3160, 1663, 853, 807 cm^{-1}.
^{13}C NMR: δ = 56.0 (q), 109.5 (d), 114.8 (d), 127.4 (d), 129.5 (s), 147.5 (s), 152.3 (s), and 191.3 (d) ppm.

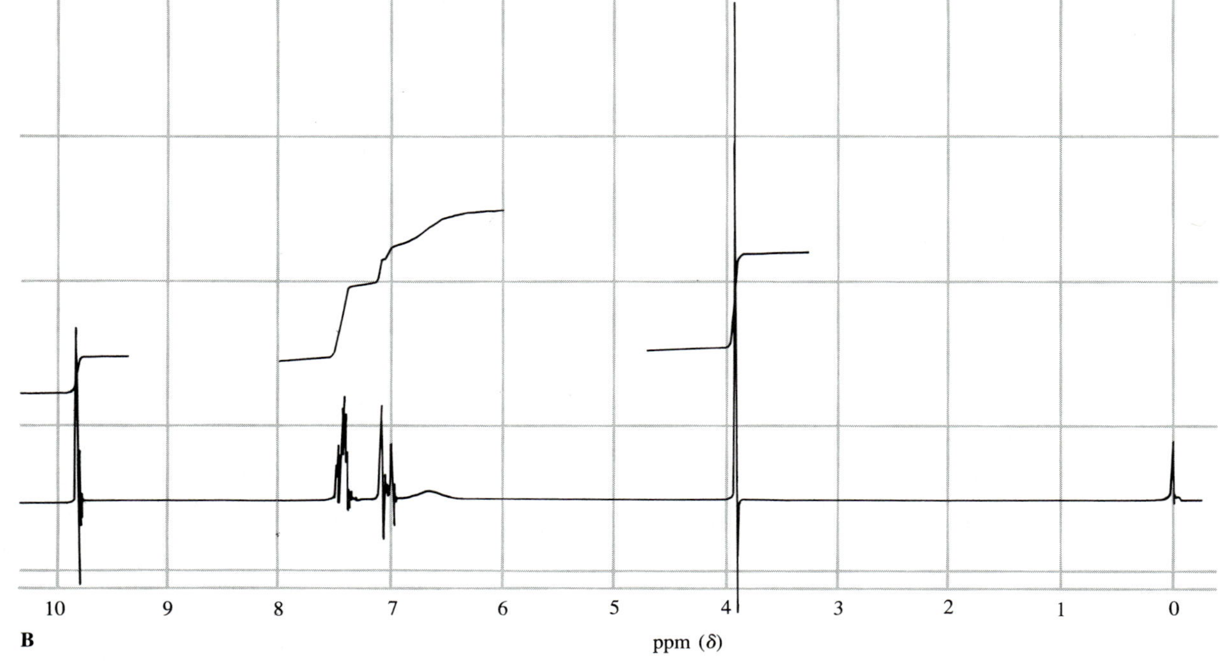

If B is treated with Ag_2O, it gives compound C, with the formula $C_8H_8O_4$, which can be independently synthesized from benzenecarboxylic acid as follows:

COOH (benzene ring) $\xrightarrow[\text{3. } CH_3I,\ NaOH]{\text{1. } SO_3,\ H_2SO_4;\ \text{2. } KOH,\ \Delta}$ $C_8H_8O_3$ **D** $\xrightarrow{\text{Conc. } H_2SO_4,\ \Delta^*}$ $C_8H_8SO_6$ **E** $\xrightarrow[\text{2. } H^+,\ H_2O,\ \Delta]{\text{1. } HNO_3,\ H_2SO_4,\ \Delta}$

$C_8H_7NO_5$ **F** $\xrightarrow{H_2,\ Pd}$ $C_8H_9NO_3$ **G** $\xrightarrow[\text{2. } H_2O,\ \Delta]{\text{1. } NaNO_2,\ H^+,\ H_2O,\ 0°C}$ C

(a) Identify the structures of compounds B through G.

(b) If A is treated with excess $(CH_3)_2SO_4$ and NaOH before acid hydrolysis, the products are B and the 2,3,4,6-tetramethyl ether of D-glucose. Write a reasonable structure for compound A.

30 The urushiols are the irritants in poison ivy and poison oak that give you rashes and make you itch after touching them. Use the following information to determine the structures of urushiols I and II, the two major members of this family of unpleasant compounds.

(i) Formulas: Urushiol I is $C_{21}H_{36}O_2$; urushiol II is $C_{21}H_{34}O_2$

* Substitutes *para* to most strongly activating group.

(ii) Urushiol II $\xrightarrow{H_2,\ Pd\text{-}C}$ urushiol I

(iii) Urushiol II $\xrightarrow{\text{Excess } CH_3I,\ NaOH}$ $C_{23}H_{38}O_2$ **Dimethylurushiol II** $\xrightarrow[\text{2. Zn, } H_2O]{\text{1. } O_3}$

$CH_3CH_2CH_2CH_2CH_2CH_2CHO$ + $C_{16}H_{24}O_3$ **Aldehyde H**

(iv) Synthesis of aldehyde H:

OCH_3 (methoxybenzene) $\xrightarrow[\text{2. } HNO_3,\ H_2SO_4]{\text{1. } SO_3,\ H_2SO_4}$ $C_7H_7NSO_6$ **I** $\xrightarrow{H^+,\ H_2O,\ \Delta}$ $C_7H_7NO_3$ **J** $\xrightarrow{\text{1. } H_2,\ Pd;\ \text{2. } NaNO_2,\ H^+,\ H_2O;\ \text{3. } H_2O,\ \Delta}$

$C_7H_8O_2$ **K** $\xrightarrow{\text{1. } CO_2,\ \text{pressure, } KHCO_3,\ H_2O;\ \text{2. } NaOH,\ CH_3I;\ \text{3. } H^+,\ H_2O}$ $C_9H_{10}O_4$ **L** $\xrightarrow{\text{1. } LiAlH_4;\ \text{2. } H^+,\ H_2O;\ \text{3. } MnO_2,\ \text{propanone (acetone)}}$

$C_9H_{10}O_3$ **M** $\xrightarrow{\text{1. } C_6H_5CH_2O(CH_2)_6CH{=}P(C_6H_5)_3;\ \text{2. Excess } H_2,\ Pd\text{-}C;\ \text{3. } CrO_3(\text{pyridine})_2}$ aldehyde **H**

CHAPTER 25

Polycyclic Benzenoid Hydrocarbons and Other Cyclic Polyenes

You now know how benzene and its derivatives react and to what extent their transformations are influenced by the aromatic array of π electrons. What happens if several benzene rings are fused to give a more-extended π system? Do these compounds also enjoy the special stability of benzene? Do they show similar chemical behavior, such as electrophilic aromatic substitution? Moreover, is the cyclic delocalization of six electrons and the resulting stabilization unique to benzene or are there other cyclic polyenes that have this property? This chapter answers these questions.

25-1 The Naming of Fused-Ring Benzenes: Polycyclic Benzenoid Hydrocarbons

The fusion (Section 4-6) of benzene rings to each other leads to a class of molecules called (somewhat loosely) **polycyclic benzenoid hydrocarbons.** In these structures, two or more benzene rings share two or more carbon atoms. There is no simple system for naming these structures; so we shall use their common names.

There is only one way to fuse one benzene ring to another. The resulting compound is called naphthalene. Further fusion can occur in a linear manner to give anthracene, tetracene, pentacene, and so on, a series called the **acenes.** Angular fusion, also called **peri fusion,** results in phenanthrene, which can be further annelated to a variety of other benzenoid polycycles.

Naphthalene **Anthracene** **Tetracene (Naphthacene)**

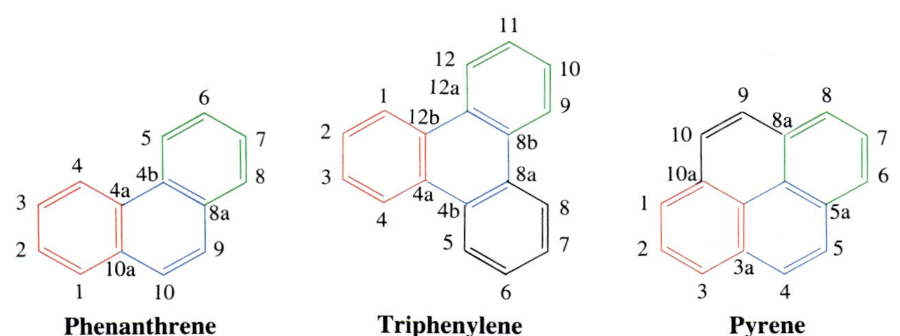

Phenanthrene **Triphenylene** **Pyrene**

Each has its own numbering system around the periphery. A quaternary carbon is given the number of the preceding carbon in the sequence followed by the letters "a," "b," and so on, depending on how close it is to that carbon.

EXERCISE 25-1

Name the following compounds or draw their structures: (a) 2,6-dimethylnaphthalene; (b) 1-bromo-6-nitrophenanthrene; (c) 9,10-diphenylanthracene; (d)

COOH

(extrapolate from the name of the corresponding benzene derivative); (e)

OCH_3

In summary, polycyclic benzenoid hydrocarbons are formed by fusion of benzene rings. These compounds have common names, and each has its own numbering system, including the labeling of the quaternary carbons by the letters of the alphabet.

25-2 Physical Properties of Naphthalene, the Smallest Fused Benzenoid Hydrocarbon

In contrast with benzene, which is a liquid, naphthalene is a colorless crystalline material with a melting point of 80°C. It is probably best known as a moth

FIGURE 25-1
Extended π conjugation in naphthalene is manifest in its UV spectrum (measured in 95% ethanol).

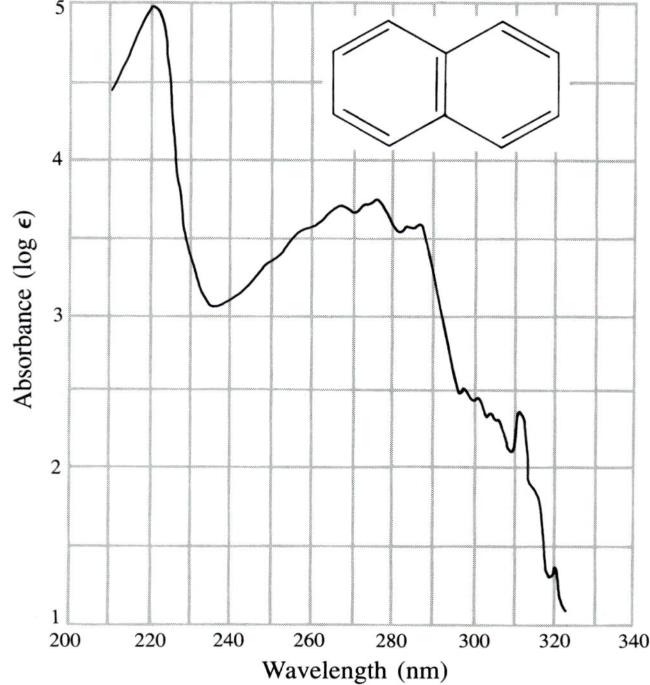

repellent and insecticide, although in these capacities it has been partly replaced by chlorinated compounds such as 1,4-dichlorobenzene (*p*-dichlorobenzene).

Naphthalene Is Aromatic: A Look at Spectra

Is naphthalene still aromatic? Let us consider some of its spectral properties. Particularly revealing are the ultraviolet and NMR spectra.

The ultraviolet spectrum of naphthalene (Figure 25-1) shows a pattern typical for an extended conjugated system, with peaks at wavelengths as long as 320 nm. On the basis of this observation, the electrons are delocalized and π conjugation is more extended than that in benzene (Section 19-2, Figure 19-7). Thus, it appears that the added four π electrons enter into efficient overlap with those of the attached benzene ring. In fact, it is possible to draw several resonance structures.

Resonance Structures in Naphthalene

Alternatively, the continuous overlap of the ten *p* orbitals can be shown as in Figure 25-2.

According to these representations, the structure of naphthalene should be symmetrical, with planar and almost hexagonal benzene rings and two perpendicular mirror planes bisecting the molecule. X-ray crystallographic measure-

FIGURE 25-2
Orbital picture of the overlap in naphthalene.

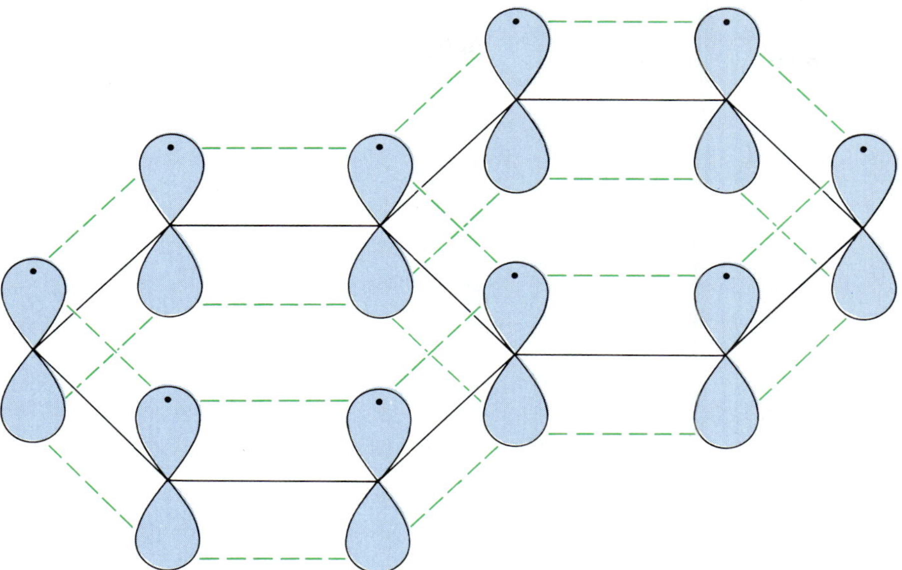

ments confirm this prediction (Figure 25-3). The C–C bonds deviate only slightly in length from those in benzene (1.39 Å), and they are clearly different from pure single (1.54 Å) and double bonds (1.33 Å).

What about the aromatic character of naphthalene? Here the ^{1}H NMR spectrum is helpful (Figure 25-4, on the next page). Two symmetrical multiplets can be observed at $\delta = 7.40$ and 7.77 ppm, characteristic of aromatic hydrogens deshielded by the ring-current effect of the π-electron loop (see Section 19-2, Figure 19-9). Coupling in the naphthalene nucleus is very similar to that in substituted benzenes: $J_{ortho} = 7.54$ Hz, $J_{meta} = 1.37$ Hz, and $J_{para} = 0.66$ Hz. The ^{13}C NMR spectrum shows three lines at $\delta = 126.5$, 128.5, and 134.4 ppm (quaternary carbons), chemical shifts that are in the range of other benzene derivatives. Thus, on the basis of structural and spectral criteria, naphthalene is aromatic.

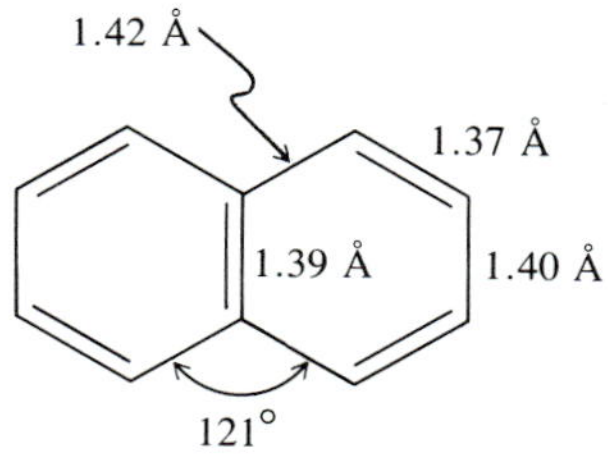

These properties of naphthalene hold for most of the other polycyclic benzenoid hydrocarbons. It appears that the cyclic delocalization in the individual benzene rings is not significantly perturbed by the fact that they have to share at least one π bond.

FIGURE 25-3
The molecular structure of naphthalene. The bond angles in the rings are 120°.

1.42 Å 1.37 Å 1.39 Å 1.40 Å 121°

EXERCISE 25-2

A substituted naphthalene $C_{10}H_8O_2$ gave the following spectral data: $m/z = 160$ (M^+); ^{1}H NMR $\delta = 6.92$ (dd, $J = 7.5$ Hz and 1.4 Hz, 2 H), 7.00 (s, 2 H), and 7.60 (d, $J = 7.5$ Hz, 2 H) ppm; ^{13}C NMR $\delta = 107.5$, 115.3, 123.0, 129.3, 136.8, and 155.8 ppm; IR $\tilde{\nu} = 3100$ cm^{-1}. What is its structure?

In summary, the physical properties of naphthalene are typical of an aromatic system. Its UV spectrum reveals that there is extensive delocalization of all π electrons, its molecular structure shows bond lengths and bond angles very similar to those in benzene, and its ^{1}H NMR spectrum reveals deshielded ring hydrogens indicative of an aromatic ring current. Other polycyclic benzenoid hydrocarbons have similar properties and are considered aromatic.

FIGURE 25-4
90-MHz ^{1}H NMR spectrum of naphthalene in CCl_4.

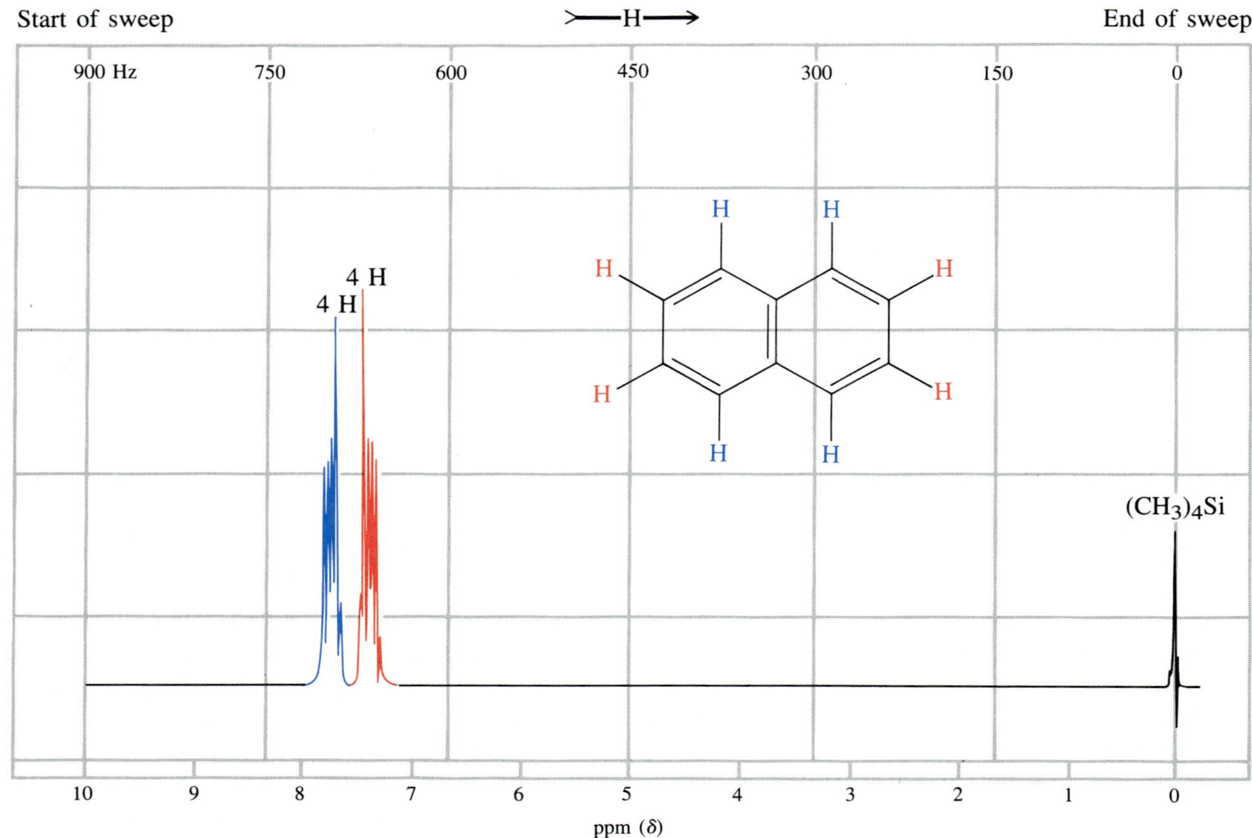

25-3 The Synthesis and Reactions of Naphthalenes

This section shows that naphthalenes can be prepared by reactions already described. Furthermore, the naphthalene nucleus undergoes electrophilic aromatic substitution reactions that follow the principles developed for the electrophilic chemistry of benzene and its derivatives.

Preparation of Naphthalenes: An Exercise in Annelation

Naphthalene itself and the methylnaphthalenes are obtained from coal tar, but specifically substituted derivatives must be synthesized. A general synthesis of the naphthalene nucleus starting with a substituted benzene uses a sequence of familiar reactions: Friedel-Crafts alkanoylation (Section 19-7) with butanedioic (succinic) anhydride, Clemmensen reduction (Section 15-8) of the resulting ketone, subsequent intramolecular Friedel-Crafts alkanoylation to form the second ring, and conversion of the resulting bicyclic system into the aromatic compound by Grignard addition (Section 8-6), acid-catalyzed dehydration (Section 7-4), and dehydrogenation (Section 20-5). This approach allows, for example, a simple preparation of 1,7-dialkylated naphthalenes.

A Simple Synthesis of 1,7-Dialkylnaphthalenes

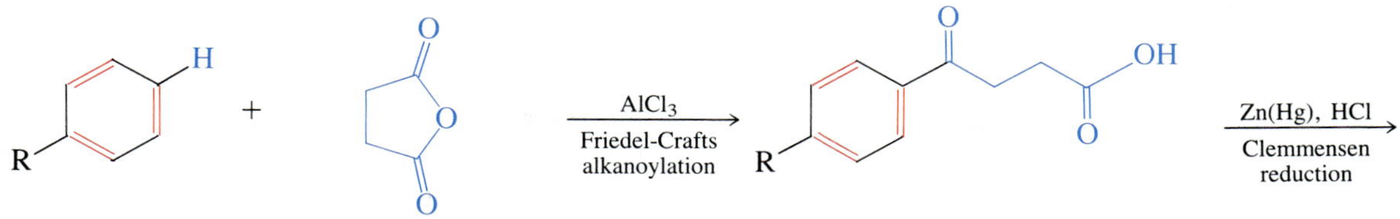

Benzene derivative (R activating) + **Butanedioic anhydride (Succinic anhydride)** $\xrightarrow[\text{Friedel-Crafts alkanoylation}]{AlCl_3}$ **4-Aryl-4-oxobutanoic acid** $\xrightarrow[\text{Clemmensen reduction}]{Zn(Hg),\ HCl}$

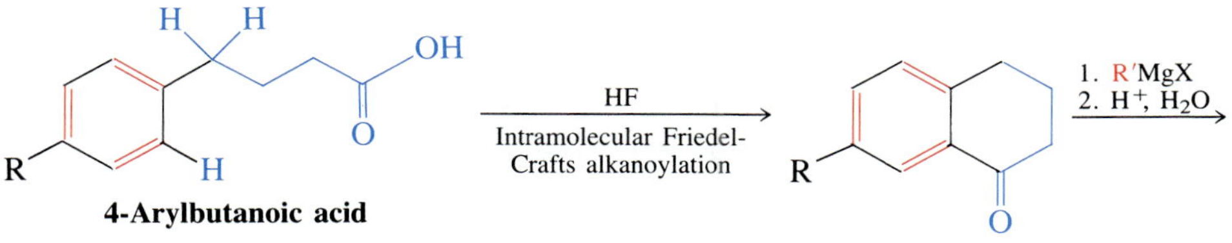

4-Arylbutanoic acid $\xrightarrow[\text{Intramolecular Friedel-Crafts alkanoylation}]{HF}$ $\xrightarrow{1.\ R'MgX\quad 2.\ H^+,\ H_2O}$

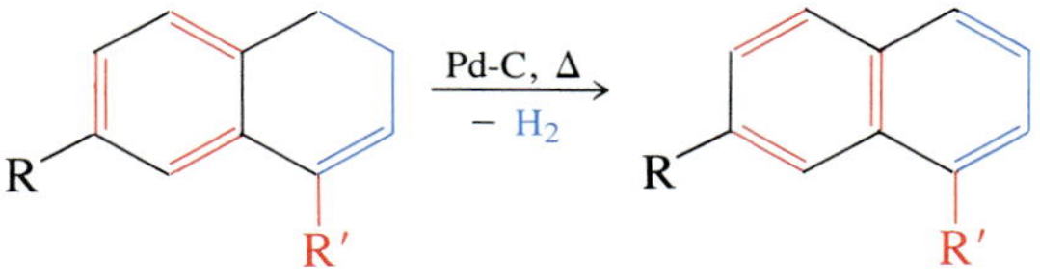

Substituted dihydronaphthalene $\xrightarrow[-\ H_2]{\text{Pd-C, }\Delta}$ **1,7-Disubstituted naphthalene**

EXERCISE 25-3

Propose a synthesis of 2-ethylnaphthalene starting from benzene and any other simple reagent.

Naphthalene Is Activated Toward Electrophilic Substitution

Treatment of naphthalene with bromine, even in the absence of a catalyst, results in smooth conversion into 1-bromonaphthalene. Thus, the aromatic character of the molecule is manifest also in its reactivity: electrophiles undergo substitution rather than addition.

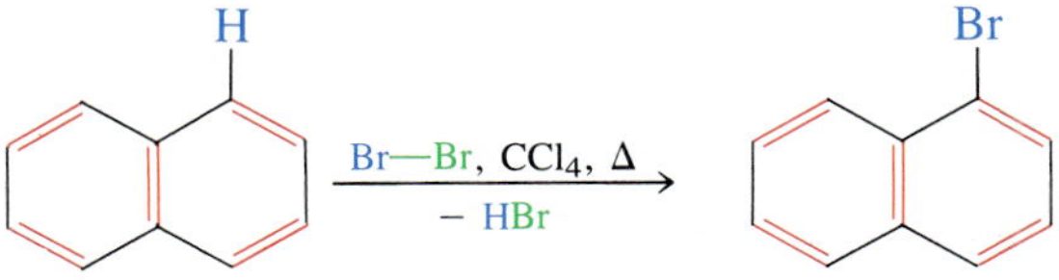

$\xrightarrow[-\ HBr]{Br{-}Br,\ CCl_4,\ \Delta}$ 75% **1-Bromonaphthalene**

The mild conditions required for this transformation also reveal that naphthalene is activated with respect to electrophilic aromatic substitution. The reason for the facility of electrophilic attack is the highly delocalized nature of the resulting cation. Its features can be nicely pictured by resonance structures. Note that two of these structures leave a benzene ring intact.

Electrophilic Reactivity of Naphthalene: Attack at C-1

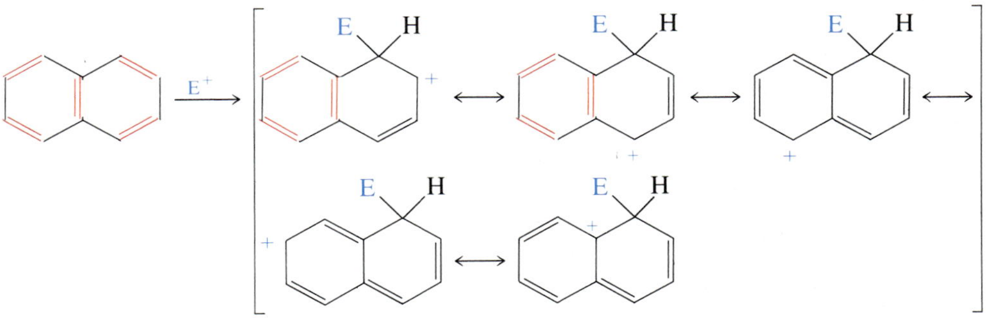

E^+

E H E H E H E H E H

Why does bromine substitute only at C-1 and not at C-2? An answer is found by inspection of the resonance structures corresponding to the second possibility. As with electrophilic attack at C-1, five such resonance descriptions can be formulated.

Consequences of Electrophilic Attack on Naphthalene at C-2

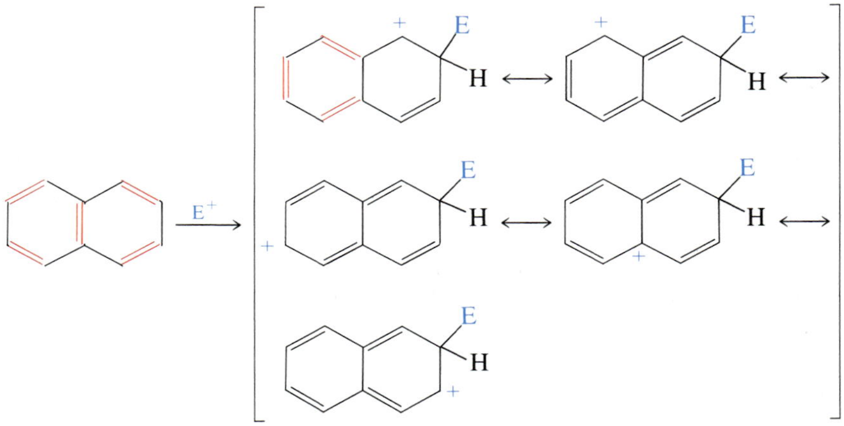

E^+

E H E H E H E H E H

Although this result might at first glance indicate that attack at either position is energetically similar, there is an important difference between the two modes: attack at C-1 allows *two* of the resonance structures to keep an intact benzene ring, with the full benefit of aromatic cyclic delocalization. On the other hand, attack at C-2 permits only *one* such structure. Because the first step in electrophilic aromatic substitution is rate determining (Section 19-4, Figure 19-16) and the energy level of its transition state corresponds to the relative stability of the resulting carbocation, attack at C-1 is faster than at C-2. The specificity of the bromination of naphthalene extends to other electrophilic substitutions, such as nitration and Friedel-Crafts alkanoylation.

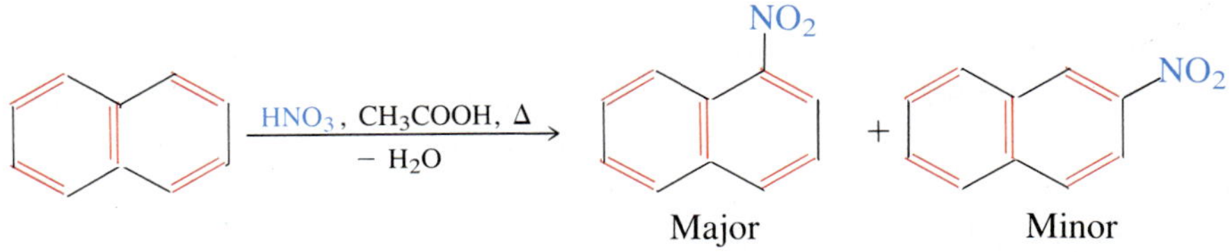

$$\xrightarrow[-\ H_2O]{HNO_3,\ CH_3COOH,\ \Delta}$$

NO_2 + NO_2

Major
1-Nitronaphthalene

Minor
2-Nitronaphthalene

H

O ‖ CCl

, $AlCl_3$, CS_2

− HCl

O

81%

1-Phenylmethanoylnaphthalene
(1-Benzoylnaphthalene;
1-naphthyl phenyl ketone)

An unusual case is the sulfonation of naphthalene. At 80°C, the product is mainly 1-naphthalenesulfonic acid. If this compound is treated with concentrated sulfuric acid at 165°C, however, it isomerizes to the more-stable 2-naphthalenesulfonic acid. The explanation is that initial sulfonation is reversible (Section 19-6) and is followed by attack at C-2 to give the thermodynamically favored product.

SO_3H

Conc. H_2SO_4, 80°C

165°C

Conc. H_2SO_4, 165°C

SO_3H

96%
1-Naphthalenesulfonic acid
(Kinetic product)

85%
2-Naphthalenesulfonic acid
(Thermodynamic product)

Why is 2-naphthalenesulfonic acid more stable than its isomer? The reason is mainly steric: substitution at C-1 places the incoming group in close proximity to the hydrogen at C-8, resulting in steric hindrance. This type of congestion is absent near C-2.

Steric Hindrance in 1-Substituted Naphthalenes

H R

Electrophilic Attack on Substituted Naphthalenes: Control of Regioselectivity

The rules of orientation in electrophilic attack on monosubstituted benzenes (Chapter 20) extend easily to the naphthalene nucleus. The substituted ring is the one most affected by the substituent already present: an activating group usually directs the incoming electrophile to the same ring, a deactivating group directs it away. For example, 1-naphthalenol (1-naphthol) undergoes electrophilic nitration at C-2 and C-4.

Nitration of 1-Naphthalenol (1-Naphthol)

OH

HNO_3, H_2SO_4
$- H_2O$

H
+O
H NO$_2$

and

H
+O
NO$_2$
H

Para attack

Ortho attack

$- H^+$

OH
NO$_2$

+

OH
NO$_2$

4-Nitro-1-naphthalenol
(4-Nitro-1-naphthol; major)

2-Nitro-1-naphthalenol
(2-Nitro-1-naphthol; minor)

An activating group at C-2 has two possibilities for *ortho* substitution, at C-1 and at C-3. The *para* position is quaternary and cannot be substituted. The resonance structures for attack at C-1 and C-3 show the former to be considerably more favorable in spite of steric hindrance, because it allows for one of the benzene rings to stay aromatic.

Activating Substituent at C-2

O
‖
$\ddot{N}HCCH_3$

HNO_3, CH_3COOH
$- H_2O$

O_2N H
O
‖
$\overset{+}{N}HCCH_3$

⟷ etc.

Attack at C-1 favored

HNO_3, CH_3COOH (✕)

$- H^+$

O
‖
$\overset{+}{N}HCCH_3$
NO$_2$
H

⟷ etc.

Attack at C-3 unfavored

NO$_2$
O
‖
$\ddot{N}HCCH_3$

71%

***N*-(1-Nitro-2-naphthyl)ethanamide**

Deactivating groups in one ring usually enforce electrophilic substitutions in the other ring and preferentially in the positions closest to the first ring (C-5 and C-8), as if the substituent were not there.

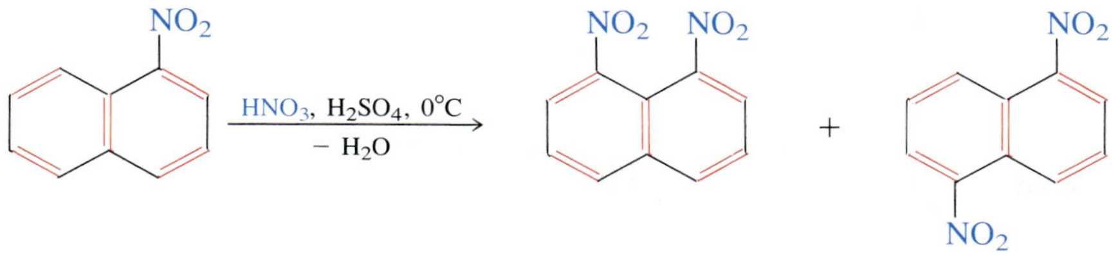

NO_2 $\xrightarrow[-\ H_2O]{HNO_3,\ H_2SO_4,\ 0°C}$ NO_2 NO_2 + NO_2 NO_2

30% **1,8-Dinitronaphthalene**

60% **1,5-Dinitronaphthalene**

EXERCISE 25-4

On the basis of the relative viability of the sets of resonance structures arising from initial electrophilic attack, predict the position of electrophilic aromatic nitration in (a) 1-ethylnaphthalene; (b) 2-nitronaphthalene; and (c) 5-methoxy-1-nitronaphthalene.

In summary, a general synthetic approach to substituted naphthalenes is to use butanedioic (succinic) anhydride as a reagent for double Friedel-Crafts alkanoylation, affording a means by which a substituted benzene ring can be annelated into a naphthalene framework. Naphthalene is activated with respect to electrophilic aromatic substitution; kinetically favored attack takes place at C-1. Sulfonation is reversible and eventually gives the derivative substituted at C-2, the thermodynamically favored product, which avoids steric hindrance with the hydrogen at C-8. Electrophilic attack on a substituted naphthalene takes place on an activated ring and away from a deactivated ring, with regioselectivity in accordance with the general rules developed for electrophilic aromatic substitution of benzene derivatives.

25-4 Tricyclic Benzenoid Hydrocarbons: Anthracene and Phenanthrene

Linear and angular fusion of a third benzene ring onto naphthalene results in the next two higher systems, anthracene and phenanthrene. Although isomeric and seemingly very similar, they have different thermodynamic stability: anthracene is about 6 kcal $mole^{-1}$ less stable than phenanthrene. Enumeration of the various resonance structures of the molecules explains why. The former has only four, and only two contain two fully aromatic benzene rings. The latter has five, three of which incorporate two aromatic benzenes, one even three.

Resonance in Anthracene

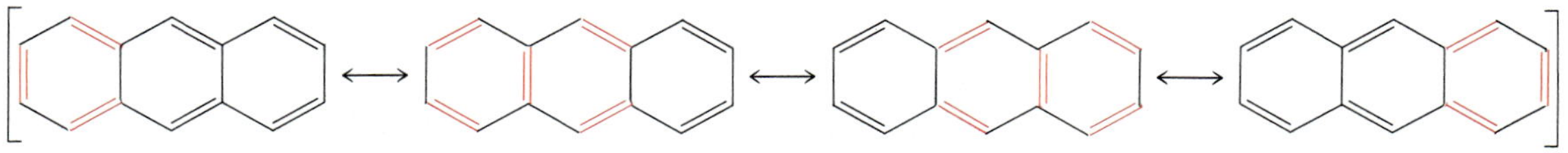

Resonance in Phenanthrene

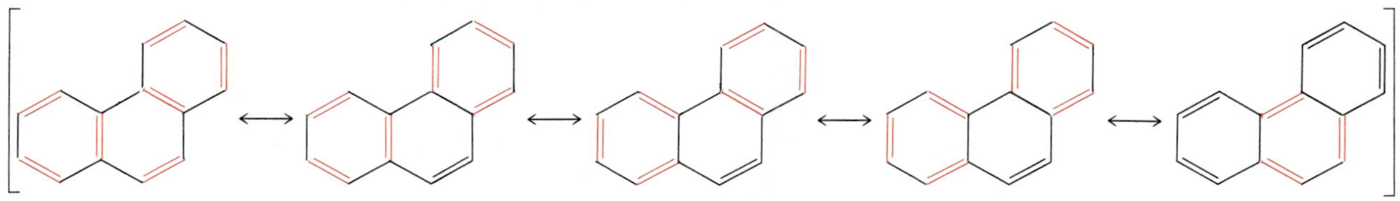

This section reviews the preparation and reactions of these compounds.

EXERCISE 25-5

Formulate all the possible resonance forms of tetracene (naphthacene, see Section 25-1). What is the maximum number of completely aromatic benzene rings in these structures?

Synthesis of Anthracenes and Phenanthrenes

As with naphthalene, the higher benzenoid hydrocarbons anthracene and phenanthrene can be constructed by ring-closure reactions. For example, Friedel-Crafts reaction of benzene with 1,2-benzenedicarboxylic (phthalic) anhydride, followed by Clemmensen reduction of the resulting acid, and ring closure by renewed (intramolecular) Friedel-Crafts reaction gives anthrone (common name), which is readily converted into anthracene and its 9-substituted derivatives as shown.

1,2-Benzenedicarboxylic anhydride (Phthalic anhydride) + benzene $\xrightarrow{AlCl_3,\ \Delta}$ **2-(Phenylmethanoyl)benzenecarboxylic acid (2-Benzoylbenzoic acid)** (CO_2H) 90% $\xrightarrow{Zn(Hg),\ HCl}$ (CO_2H) 95% $\xrightarrow[-H_2O]{HF}$

Anthrone 82% $\xrightarrow{1.\ CH_3CH_2MgBr\quad 2.\ H^+,\ H_2O}$ (CH_3CH_2, OH) $\xrightarrow[-H_2O]{H^+,\ \Delta}$ CH_3CH_2 80%–90% **9-Ethylanthracene**

Interestingly, anthrone is stable even though it is the keto form of the completely delocalized anthracenol. Evidently, the gain in resonance energy in going to the latter does not outweigh the loss of the strong C–O double bond.

Anthrone (H, H; O) $\rightleftharpoons$ **9-Anthracenol** (OH)

This suggests that the central ring in anthracene is not quite as aromatic as benzene itself, a fact also revealed by the resonance picture (only two of the four resonance structures depict this ring as aromatic) and, as we shall see, the reactions of anthracene.

An alternative way of producing the anthracene nucleus relies on Diels-Alder cycloadditions. Thus, 2,5-cyclohexadiene-1,4-dione (*p*-benzoquinone) undergoes the Diels-Alder reaction with 1,3-butadiene not only once (Section 24-5), but twice. This transformation was carried out by Diels and Alder themselves in

1928. The resulting adduct is readily oxidized to the corresponding dione (9,10-anthraquinone), which is then reduced by sodium borohydride and a Lewis acid to anthracene.

CH_3CH_2OH, 100°C

60%

$K^{+-}OH$, O_2

$NaBH_4$, BF_3, $(CH_3CH_2)_2O$

9,10-Anthraquinone

73% (for the last two steps)
Anthracene

Because the double Diels-Alder reaction can be carried out with substituted dienes, various substituted anthracenes are available by this route. Organometallic reactions at the carbonyl groups also allow modification of the 9 and 10 positions.

EXERCISE 25-6

Propose a reasonable mechanism for the reduction by sodium borohydride of 9,10-anthracenedione (9,10-anthraquinone) to anthracene. Hint: Consider the function of the Lewis acid.

EXERCISE 25-7

Suggest a synthesis of 2,3,6,7-tetramethyl-9,10-diphenylanthracene starting from 2,5-cyclohexadiene-1,4-dione (*p*-benzoquinone).

Ring closures also furnish phenanthrenes. For example, Friedel-Crafts reaction of naphthalene with butanedioic (succinic) anhydride in nitrobenzene causes substitution both at C-1 and at C-2. The two resulting 4-(1-naphthyl)- and 4-(2-naphthyl)-4-oxobutanoic acids are reduced under Clemmensen conditions and then cyclized to the phenanthrene skeleton.

$AlCl_3$, $C_6H_5NO_2$ — COOH, 36% + HOOC, 47% — Zn-Hg, HCl, Δ

COOH + HOOC, 90% — $AlCl_3$, $- H_2O$ — 88%

The product ketones can then be reduced, dehydrated, and dehydrogenated to phenanthrene, or they can be separated and individually transformed to substituted derivatives.

1. $CH_3CH_2CH_2CH_2Li$
2. H^+, H_2O
3. S, Δ

EXERCISE 25-8

Explain the absence of ketone B in the intramolecular Friedel-Crafts alkanoylation of A.

COOH $\xrightarrow{AlCl_3}$ (✗)

A **B**

An alternative approach starts with 1,2-diarylethenes (also called stilbenes) and makes use of the photochemical conrotatory electrocyclic ring closure of *cis*-1,3,5-hexatrienes to cyclohexadienes (Section 14-5). The resulting dihydrophenanthrene is reactive and cannot be isolated.

$h\nu$, Conrotatory ring closure; I_2, O_2

***cis*-1,2-Diphenylethene (*cis*-Stilbene)** ***trans*-4a,4b-Dihydrophenanthrene** 82%

However, it can be trapped by added oxidizing agents, such as oxygen in the presence of catalytic amounts of iodine. Under these conditions, good yields of the corresponding phenanthrenes are obtained.

Anthracene and Phenanthrene Have Reactive Central Rings

Additions to C-9 and C-10 in anthracene and phenanthrene furnish molecules with two unperturbed benzene units. Consequently, these two positions are much more reactive than would be expected for a benzene ring. For example, catalytic hydrogenation leads to the respective dihydroaromatic compounds.

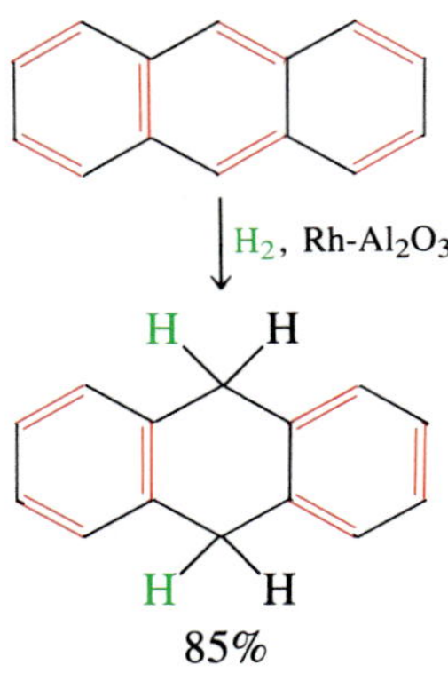

85%
9,10-Dihydroanthracene

H_2, Cu-Cr oxide

77%
9,10-Dihydrophenanthrene

BOX 25-1

The Helicenes

The oxidative photocyclization of 1,2-diarylethenes has been used in the synthesis of more-complex benzenoid hydrocarbons. A particularly interesting case is phenanthro[3,4-*c*]phenanthrene, in which six angularly fused benzene rings are arranged in an almost complete circle. This compound has been prepared by double photocyclization.

C_6H_6, $h\nu$, I_2, O_2 →

60%
Phenanthro[3,4-*c*]phenanthrene
(Hexahelicene)

The molecule has been given the common name *hexahelicene* because it is forced to adopt a helical arrangement—if the molecule were planar, parts of the two terminal benzene rings would be forced to occupy the same region in space (hence the distorted drawing shown above). Twisted from planarity, the molecule forms a helix (Figure 25-5), which spirals either clockwise or counterclockwise depending on the direction of the twist.

FIGURE 25-5

The spatial arrangement of hexahelicene. The helical structure can be described by the three planes shown. Each incorporates approximately one naphthalene unit. The angle between the two planes containing the terminal benzene rings is approximately 58.5°.

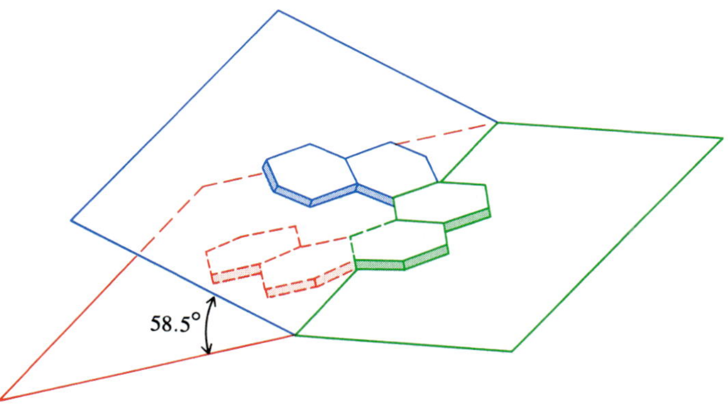

Because of its helical structure, hexahelicene is chiral, and its pure enantiomers have large specific rotations: $[\alpha]_D^{25°C} = 3640°$. Other higher helicenes are known that exhibit even more pronounced optical activity. Note that, even though helicenes are chiral, they lack stereocenters (Chapter 5).

Halogenation often gives addition products rather than substituted ones, although the resulting dihalides readily lose hydrogen halide to give the corresponding haloarenes.

Cl—Cl, CS_2, 0°C; 20°C, − HCl

9,10-Dichloro-9,10-dihydroanthracene

80%
9-Chloroanthracene

Br—Br, CCl_4; Δ, − HBr

9,10-Dibromo-9,10-dihydrophenanthrene

94%
9-Bromophenanthrene

The halogenation of anthracene may be regarded as a 1,4-addition to a 1,3-diene (Section 14-3). The diene character of this unit manifests itself also in Diels-Alder cycloadditions.

2-Butene-1,4-dioic anhydride (Maleic anhydride)

1,2-Dimethylbenzene (*o*-xylene), Δ

95%

This reaction is not possible for phenanthrene, because it has only one reactive double bond. However, this bond enters into photochemical [2+2]cycloadditions (Section 14-5).

CO_2CH_3 ; $h\nu$, $(CH_3CH_2)_2O$; CO_2CH_3 ; CO_2CH_3 ; CO_2CH_3

62%

EXERCISE 25-9

Explain the following transformation by a mechanism. Hint: Start by protonating the nitrogen.

CH_2CN ; Conc. H_2SO_4, 0°C ; NH_2

85%

9-Phenanthrenamine

Coal as a Source of Polycyclic Benzenoid Hydrocarbons

Many benzenoid hydrocarbons, including benzene, are derived from **coal.** Heating coal in the absence of air produces coal gas containing methane and other gaseous products. The remainder of the distillate is **coal tar,** which can be fractionally distilled to give benzene, methylbenzene (toluene), the dimethylbenzenes (xylenes), benzenol (phenol), naphthalene, and higher polycyclic aromatic hydrocarbons, as well as heterocyclic compounds (Chapter 26). The residue is **coke,** used in large quantities for the smelting of iron ore to steel.

Coal is not simply carbon, but an amorphous polymer consisting of layers of weakly linked polycyclic aromatic and hydroaromatic compounds (Figure 25-6, on the next page). On heating, the primary coal structure degrades to fragments of molecular weight between 300 and 1000, a large proportion of which are soluble in organic solvents. Coal solubilization and the conversion of coal into liquid fuels is of current interest in efforts to use coal as a source of new industrial chemical feedstocks.

Polycyclic Aromatic Hydrocarbons and Cancer

Many polycyclic benzenoid hydrocarbons are carcinogenic. The first observation of human cancer caused by such compounds was made in 1775 by Sir Percival Pott, a surgeon at London's St. Bartholomew's hospital, who recognized that chimney sweeps were prone to scrotal cancer. Since then, a great deal of research has gone into identifying which polycyclic benzenoid hydrocarbons

FIGURE 25-6
Suggested model for a partial structure of coal.

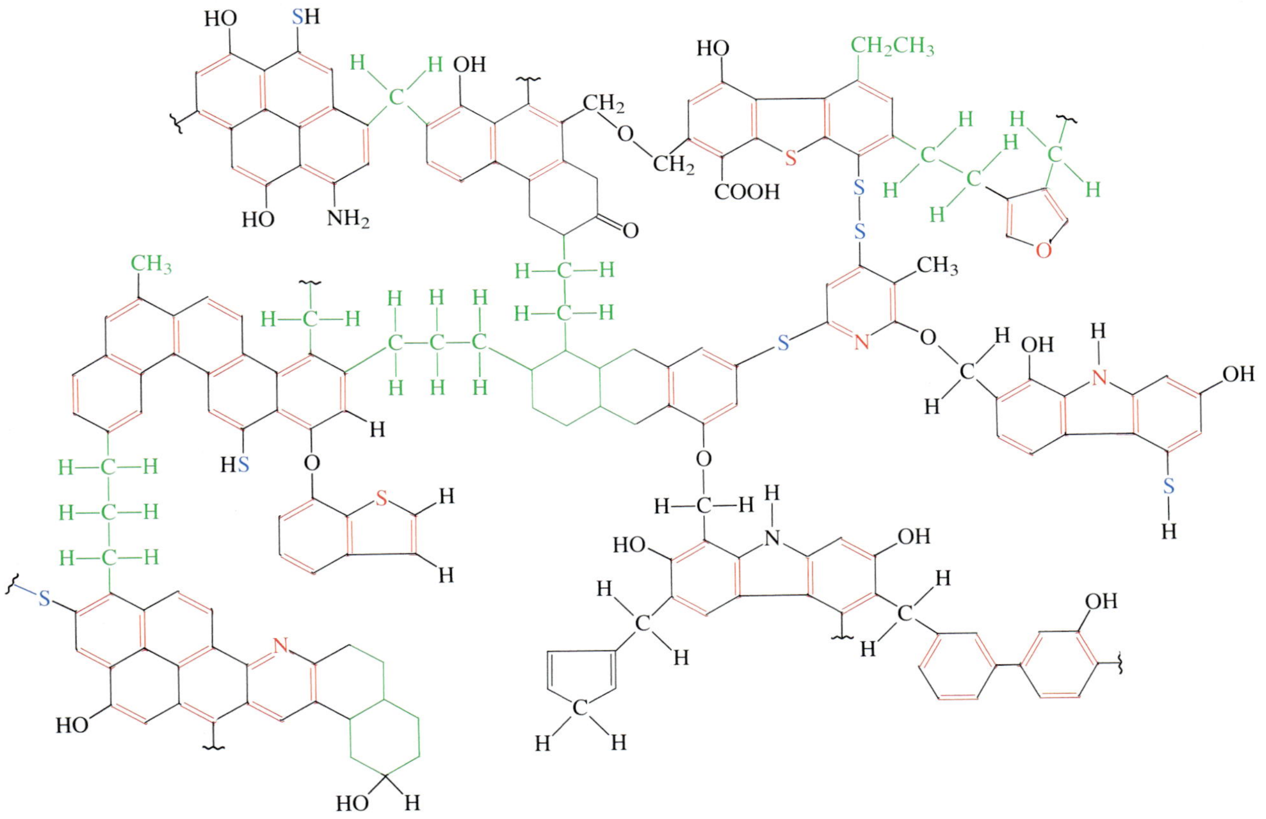

have this physiological property and how their structure correlates with activity. A particularly well studied molecule is benz[*a*]pyrene, a widely distributed environmental pollutant. It is frequently produced in the combustion of organic matter, such as automobile fuel and oil (for domestic heating and industrial power generation), in incineration of refuse, in forest fires, in burning cigarettes and cigars, and even in roasting meats. The annual release into the atmosphere in the United States alone has been estimated at 1,300 tons.

Carcinogenic Benzenoid Hydrocarbons

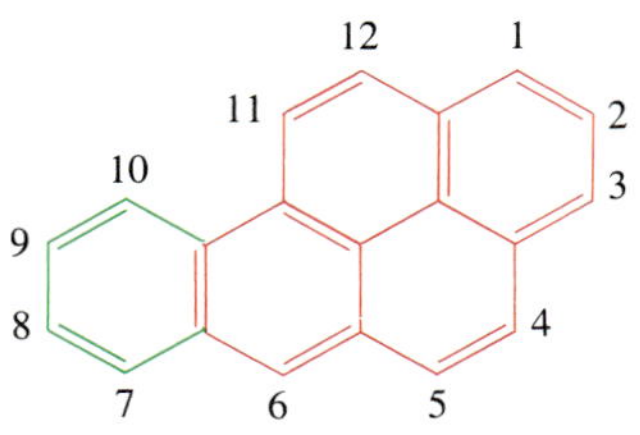

Benz[*a*]pyrene 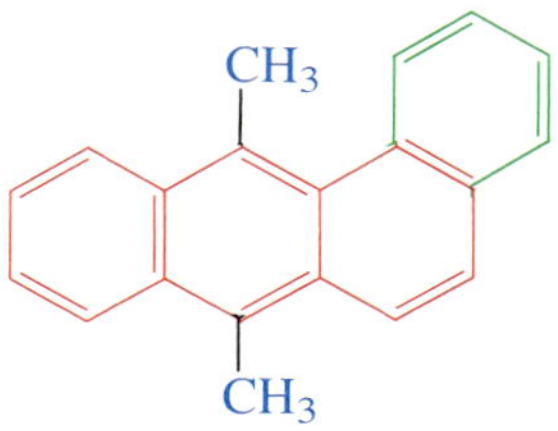 | **7,12-Dimethylbenz-[*a*]anthracene** 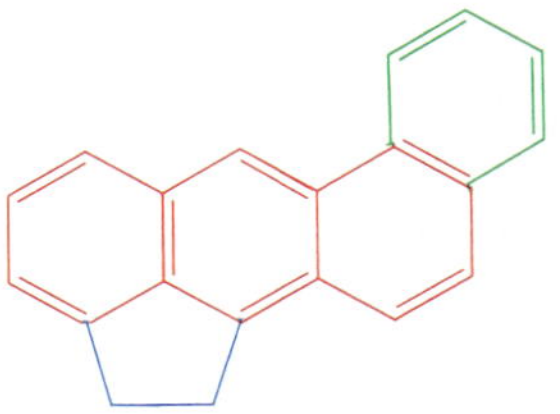 | **Cholanthrene**

What is the mechanism of carcinogenic action of benz[*a*]pyrene? The answer to this question is not complete. It is thought that an oxidizing enzyme (an *oxidase*) of the liver converts the hydrocarbon into the oxacyclopropane at C-7 and C-8. Another enzyme *(epoxide hydratase)* catalyses the hydration of the product to the trans diol. Further oxidation then results in the ultimate carcinogen, a new oxacyclopropane at C-9 and C-10.

Enzymatic Conversion of Benz[*a*]pyrene into the Ultimate Carcinogen

10
9
8
7
Oxidase
Hydratase

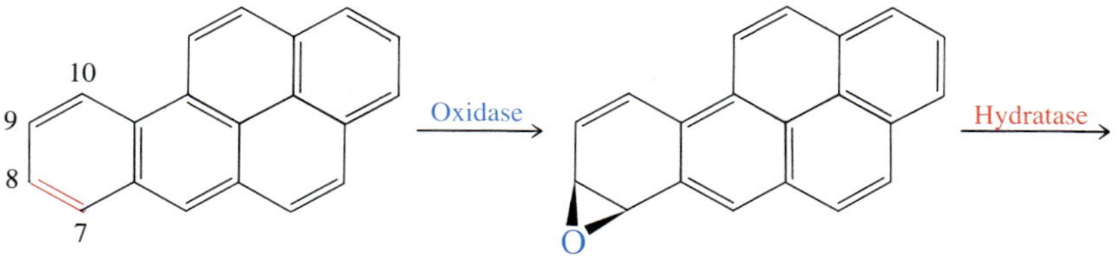

Benz[*a*]pyrene oxacyclopropane

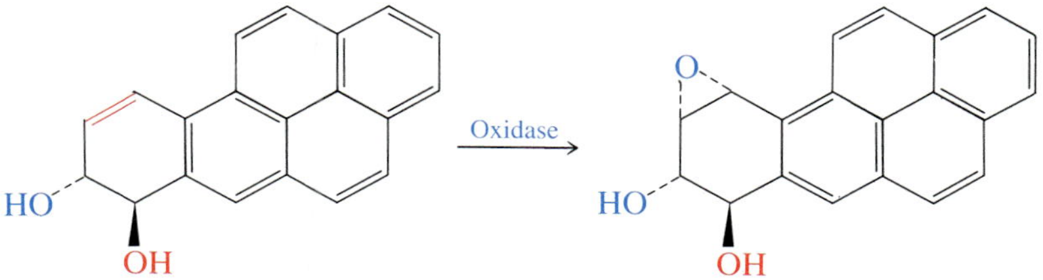

HO
OH
Oxidase
O
HO
OH

7,8-Dihydrobenz[*a*]pyrene-*trans*-7,8-diol

Carcinogen

The carcinogenic event is believed to occur when the amine nitrogen in guanine, one of the bases in the DNA strand (see Section 27-7), attacks the three-membered ring as a nucleophile. This reaction significantly alters the structure of one of the base pairs in DNA, leading to a mismatch during DNA replication.

The Carcinogenic Event

O
HO
OH
DNA
O
N
NH
HC
N
N
NH
DNA
HO
HO
OH

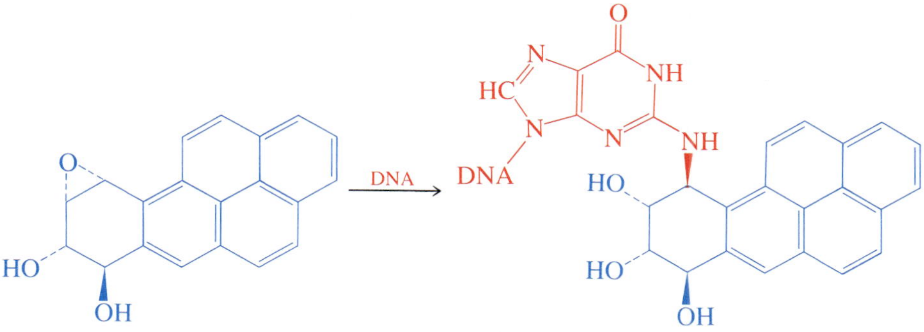

This change can lead to an alteration (mutation) of the genetic code, which may then generate a line of rapidly and indiscriminately proliferating cells typical of cancer. Not all mutations are carcinogenic; in fact, most of them lead to destruction of only the one affected cell. Exposure to the carcinogen only increases the likelihood of a carcinogenic event.

Notice that the carcinogen acts as an alkylating agent on DNA. This indicates that other alkylating agents could also be carcinogenic, and indeed that is found to be the case. The Occupational Safety and Health Administration (OSHA) has published a list of carcinogens that includes simple alkylating agents such as 1,2-dibromoethane, diazomethane, ethyl methanesulfonate, and β-propiolactone.

Carcinogenic Alkylating Agents and Sites of Reactivity

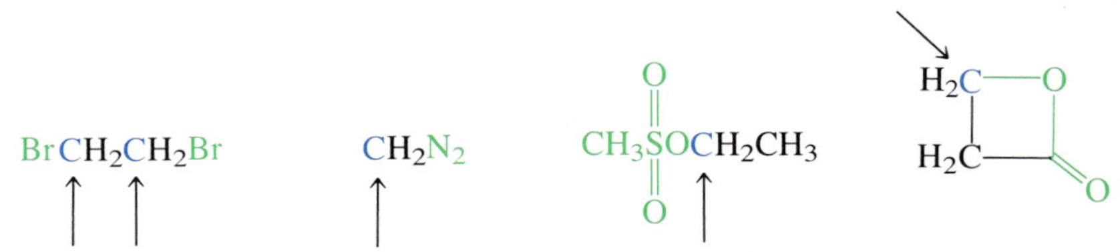

$BrCH_2CH_2Br$ — **1,2-Dibromoethane**

CH_2N_2 — **Diazomethane**

$CH_3SO_2OCH_2CH_3$ — **Ethyl methane-sulfonate**

H_2C—O / H_2C—C=O — **β-Propiolactone**

In summary, anthracenes and phenanthrenes can be made by Friedel-Crafts reactions of ring compounds with cyclic anhydrides. The anthracene framework is also readily constructed by double Diels-Alder cycloadditions to 2,5-cyclohexadiene-1,4-dione (*p*-benzoquinone), whereas an alternative route to phenanthrenes uses the oxidative photocyclization of 1,2-diarylethenes. Both systems have reactive central rings that undergo addition reactions with hydrogen (in the presence of a catalyst), halogens, and alkenes.

25-5 1,3-Cyclobutadiene, 1,3,5,7-Cyclooctatetraene, and Other Cyclic Polyenes: Hückel's Rule

Is the special stability and reactivity associated with cyclic delocalization unique to benzene and polycyclic benzenoids, or are there other cyclic π systems that have similar properties? Indeed, other cyclic conjugated polyenes are aromatic if they contain $4n + 2$ π electrons (n being zero or an integer). This pattern is known as Hückel's rule.

1,3-Cyclobutadiene, the Smallest Cyclic Polyene

1,3-Cyclobutadiene is an extremely reactive molecule, devoid of any aromatic properties. It can be observed only at very low temperatures and it must be prepared from special precursors that decompose on irradiation.

Preparation of 1,3-Cyclobutadiene

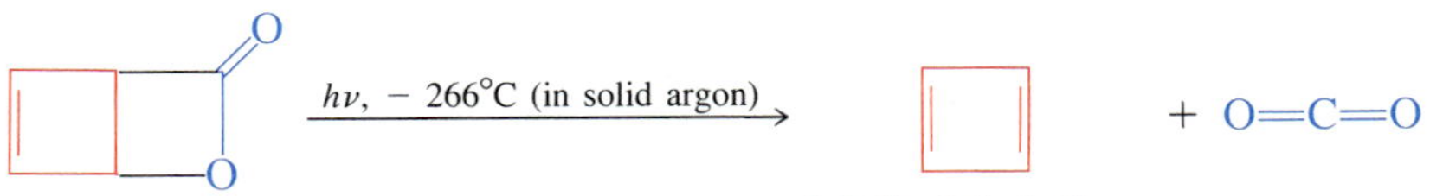

$$\xrightarrow{h\nu,\ -266^{\circ}\text{C (in solid argon)}} \quad + \ O{=}C{=}O$$

1,3-Cyclobutadiene

The reactivity of cyclobutadiene can be seen in its rapid Diels-Alder reactions, in which it can act as either diene or dienophile.

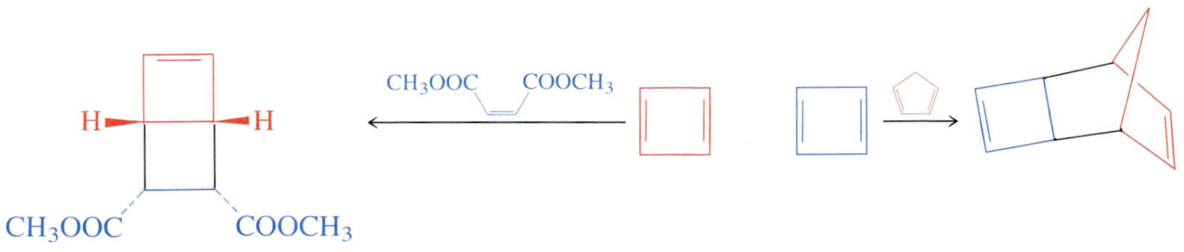

H H CH_3OOC $COOCH_3$ $\longleftarrow$ CH_3OOC $COOCH_3$

EXERCISE 25-10

1,3-Cyclobutadiene dimerizes at temperatures as low as −200°C to two isomeric products. Suggest structures for these dimers.

Substituted cyclobutadienes are less reactive, particularly if the substituents are bulky; they have been used to probe the spectroscopic features of the system of four π electrons. Particularly interesting is the ^1H NMR spectrum of 1,2,3-tris(1,1-dimethylethyl)cyclobutadiene (1,2,3-tri-*tert*-butylcyclobutadiene), in which the ring hydrogen resonates at $\delta = 5.38$ ppm, at much-higher field than expected for an aromatic system. This and other properties of cyclobutadiene show that it is quite unlike benzene. Let us now consider the next-higher cyclic polyene analog of benzene, 1,3,5,7-cyclooctatetraene.

1,3,5,7-Cyclooctatetraene Is Nonplanar and Nonaromatic

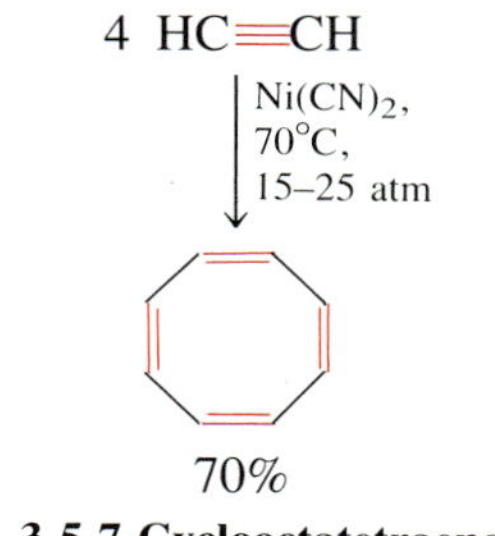

1,3,5,7-Cyclooctatetraene can be prepared in good yield by the nickel-catalyzed cyclotetramerization of ethyne (Section 13-7). It is a yellow liquid, b.p. 152°C, that is stable if kept cold but readily polymerizes when heated. It is oxidized by air, readily hydrogenated to cyclooctene and cyclooctane, and subject to electrophilic additions and to cycloadditions. This chemical reactivity is again not what would be expected if the molecule were aromatic. Spectral and structural data confirm the lack of aromaticity. Thus, the ^1H NMR spectrum shows a sharp singlet at $\delta = 5.68$ ppm, typical of an alkene. The molecular-structure determination reveals that cyclooctatetraene is actually *nonplanar* and tub-shaped (Figure 25-7). The double bonds are nearly orthogonal, and they alternate with single bonds.

FIGURE 25-7
The molecular structure of 1,3,5,7-cyclooctatetraene.

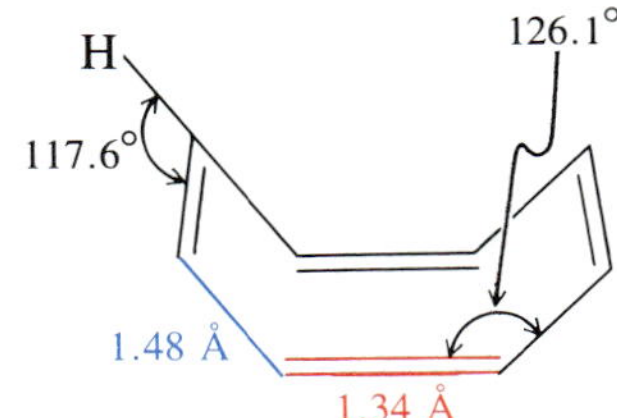

Only Cyclic Conjugated Polyenes Containing 4*n*+2 Pi Electrons Are Aromatic

Unlike cyclobutadiene and cyclooctatetraene, certain higher cyclic conjugated polyenes are aromatic. All of them have one property in common: they contain $4n + 2$ π electrons.

The first such system was prepared in 1956 by Sondheimer (see Section 13-6); it was 1,3,5,7,9,11,13,15,17-cyclooctadecanonaene, containing eighteen π electrons. To avoid such cumbersome names, Sondheimer introduced a simpler system of naming cyclic conjugated polyenes. He named completely conjugated monocyclic hydrocarbons $(CH)_N$ as **[*N*]annulenes,** in which *N* denotes the ring size. Thus, cyclobutadiene would be called [4]annulene, benzene [6]annulene, cyclooctatetraene [8]annulene. The first unstrained aromatic system in the series after benzene is [18]annulene.

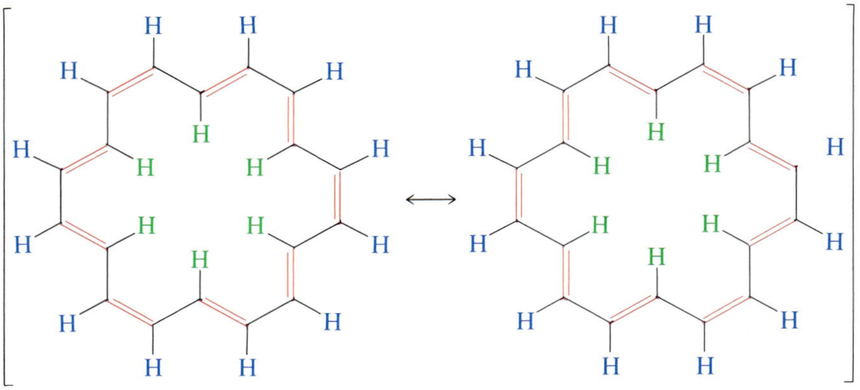

[18]Annulene
(1,3,5,7,9,11,13,15,17-Cyclooctadecanonaene)

[18]Annulene contains delocalized electrons, is fairly planar, and shows little alternation of the single and double bonds. Like benzene, therefore, it can be described by a set of two equal resonance structures. In accord with its aromatic character, the molecule is relatively stable and undergoes electrophilic aromatic substitution. It also exhibits a benzenelike ring-current effect in ^{1}H NMR.

EXERCISE 25-11

The ^{1}H NMR spectrum of [18]annulene shows two signals, at $\delta = 9.28$ (12 H) and -2.99 (6 H) ppm. Explain. Hint: Consult Figures 19-9 and 19-10.

Since the preparation of [18]annulene, many other annulenes have been made: those with $4n$ π electrons, such as cyclobutadiene and cyclooctatetraene, are not aromatic, but those with $4n + 2$ electrons, such as benzene and [18]annulene, are aromatic. This behavior had been predicted earlier by the theoretical chemist Hückel,* who formulated what is known as the $4n + 2$ rule, or Hückel's rule, in 1938. The rule expresses the regular molecular-orbital patterns in cyclic conjugated polyenes. The p orbitals mix to give an equal number of π molecular orbitals symmetrically spaced as shown in Figure 25-8. All levels are composed of degenerate pairs, except for the lowest bonding and highest antibonding orbitals. A closed-shell system is possible (i.e., all bonding molecular orbitals are occupied; see Section 1-6) only if there are $4n + 2$ π electrons.

EXERCISE 25-12

The unusual molecule 1,6-methano[10]annulene exhibits two sets of signals in the ^{1}H NMR spectrum at $\delta = 7.10$ (8 H) and -0.50 (2 H) ppm. Is this result a sign of aromatic character?

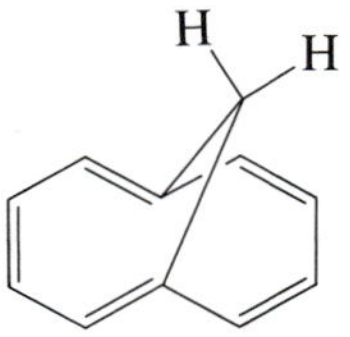

1,6-Methano[10]annulene

---

*Professor Erich Hückel, 1896–1984, University of Marburg, Germany.

FIGURE 25-8
A. The regular pattern of the π molecular orbitals in cyclic conjugated polyenes.
B. Molecular-orbital levels in 1,3-cyclobutadiene: four π electrons are not enough to result in a closed shell.
C. The six π electrons in benzene allow for a closed-shell configuration.

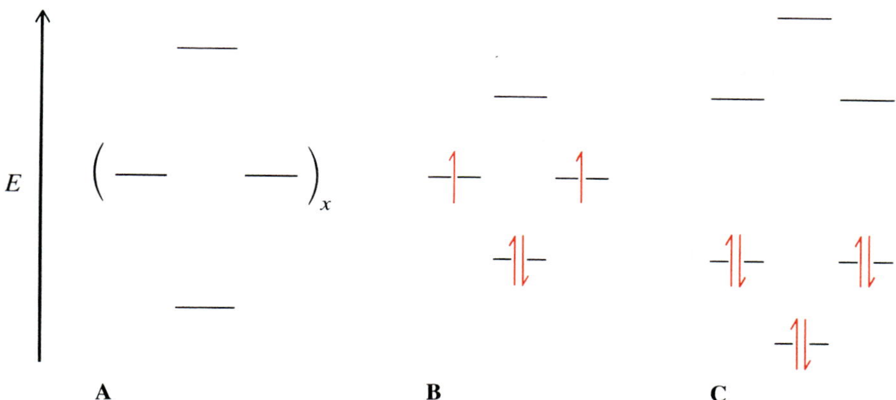

EXERCISE 25-13

On the basis of Hückel's rule, label the following molecules as aromatic or nonaromatic: (a) [30]annulene; (b) [16]annulene;

(c) ; (d) CH_3 H_3C

Hückel's Rule and Charged Molecules

Hückel's rule also applies to charged molecules, as long as there is cyclic delocalization. For example, 1,3-cyclopentadiene is unusually acidic (p$K_a \sim$ 16) because the resulting cyclopentadienyl anion forms a delocalized system of six π electrons in which the negative charge is equally distributed over all five carbon atoms. For comparison, the pK_a of propene (Section 14-1) is 40.

Aromatic Cyclopentadienyl Anion

H H $\rightleftharpoons$ H^+ + [$\longleftrightarrow$ $\longleftrightarrow$ $\longleftrightarrow$ etc.] or

pK_a = 16

The acidity of 1,3-cyclopentadiene can be increased even further by electron-withdrawing substituents. For example, the pK_a of cyclopentadiene bearing one cyano group is 9.8, that of the 1,3-dicyano derivative 2.52, lower than that of ethanoic (acetic) acid.

In contrast, the cyclopentadienyl cation, a system of four π electrons, can be produced only at low temperature, by treatment of 5-bromo-1,3-cyclopentadiene with SbF_5. The product is extremely reactive and is shown by spectroscopy to exist as a diradical.

Nonaromatic Cyclopentadienyl Cation

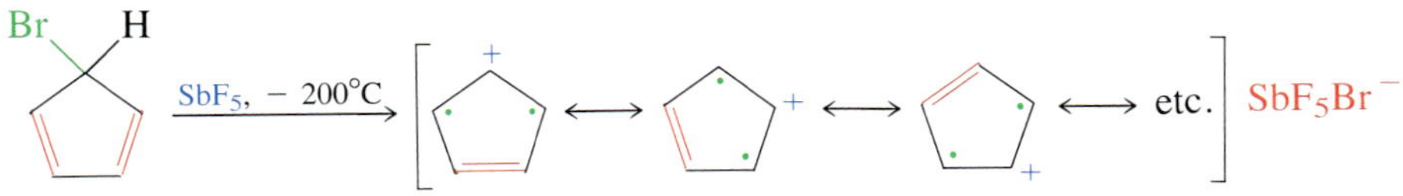

Br H; SbF_5, − 200°C → [+ ⟷ + ⟷ + ⟷ etc.] SbF_5Br^-

When 1,3,5-cycloheptatriene is treated with bromine, a stable salt is formed, cycloheptatrienyl bromide. In this molecule, the organic cation contains six delocalized π electrons, and the positive charge is equally distributed over seven carbons. Even though a carbocation, the system is remarkably unreactive, as expected for an aromatic system.

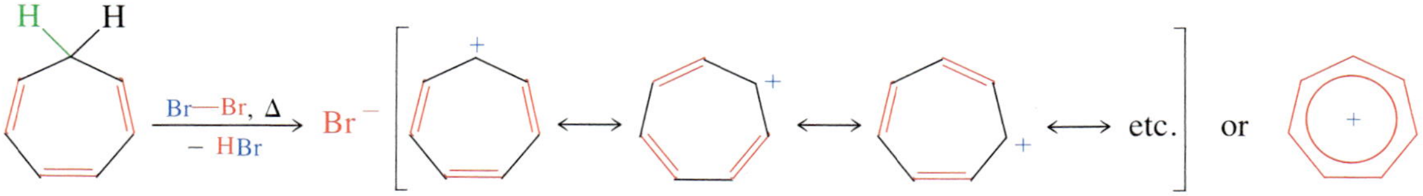

H H; Br—Br, Δ; − HBr → Br^- [+ ⟷ + ⟷ + ⟷ etc.] or +

EXERCISE 25-14

Draw an orbital picture of (a) the cyclopentadienyl anion and (b) cycloheptatrienyl cation (consult Figure 19-1).

EXERCISE 25-15

On the basis of Hückel's rule, label the following molecules aromatic or nonaromatic: (a) cyclopropenyl cation; (b) cyclononatetraenyl anion; (c)

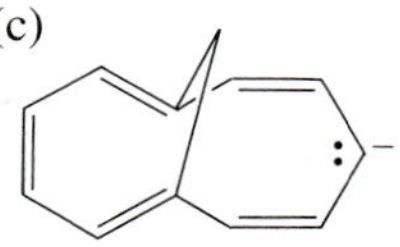

Two-Electron Reduction and Oxidation of Nonaromatic Cyclic Polyenes

Cyclic systems of $4n$ π electrons can be converted into their aromatic counterparts by two-electron oxidations and reductions. For example, cyclooctatetraene is reduced by alkali metals to the corresponding aromatic dianion. This species is planar, contains fully delocalized electrons, and is relatively stable. It also exhibits an aromatic ring current in ^{1}H NMR.

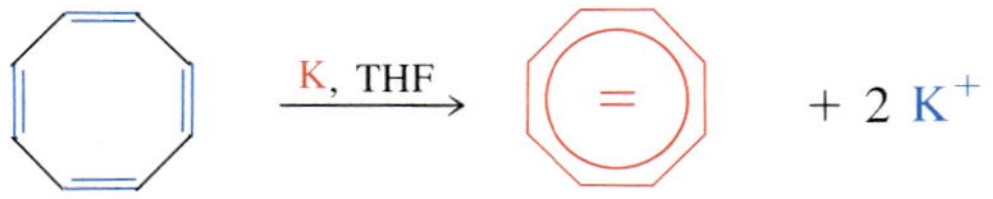

K, THF → = + 2 K^+

Eight π electrons, nonaromatic

Ten π electrons, aromatic

Similarly, [16]annulene can be either reduced to its dianion or oxidized to its dication, both products being aromatic. On formation of the dication, the configuration of the molecule changes.

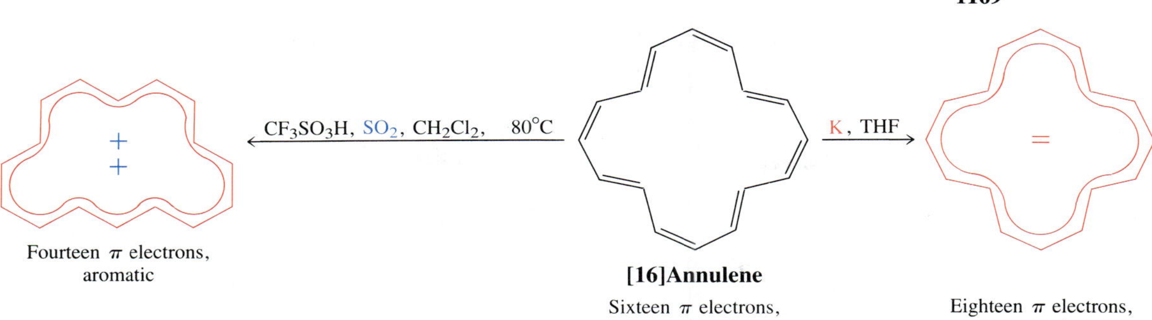

Fourteen π electrons, aromatic

[16]Annulene
Sixteen π electrons, nonaromatic

Eighteen π electrons, aromatic

EXERCISE 25-16

The triene A can be readily deprotonated twice to give the stable dianion B. However, the neutral analog of B, the tetraene C (pentalene), is extremely unstable. Explain.

Base

A **B** **C**

EXERCISE 25-17

Azulene is a deep blue (see Figure 14-18) aromatic hydrocarbon that is readily attacked by electrophiles at C-1, by nucleophiles at C-4. Explain.

Azulene

In summary, cyclic conjugated polyenes are aromatic if their π-electron count is $4n + 2$. This number corresponds to a completely filled set of bonding molecular orbitals. On the other hand, $4n$ π systems have open-shell structures that are unstable, reactive, and lack aromatic ring-current effects in ^{1}H NMR.

Summary of New Reactions

1 PREPARATION OF NAPHTHALENE

1. (succinic anhydride), $AlCl_3$
2. Zn (Hg), HCl
3. HF

1. $NaBH_4$
2. H^+
3. Pd-C, Δ

2 ELECTROPHILIC AROMATIC SUBSTITUTION OF NAPHTHALENE

E^+, $-H^+$

Thermodynamic product

Kinetic product

3 PREPARATION OF ANTHRACENE AND PHENANTHRENE

Anthracene:

1. $AlCl_3$, Δ
2. Zn (Hg), HCl, Δ
3. HF

1. $NaBH_4$
2. H^+

1. 2 $CH_2{=}CHCH{=}CH_2$
2. KOH, O_2

$NaBH_4$, BF_3, $(CH_3CH_2)_2O$

Phenanthrene:

1. O, $AlCl_3$
2. Zn (Hg), HCl
3. HF

\+

1. $NaBH_4$
2. H^+
3. Pd-C

$h\nu$, I_2, O_2

4 REACTIONS OF ANTHRACENE AND PHENANTHRENE

Reduction:

H_2, Rh-Al_2O_3

H_2, Cu-Cr oxide

Halogenation (addition-elimination):

1. X_2
2. Δ

X

1. X_2
2. Δ

X

Cycloadditions:

O O H H O O O O

CO_2CH_3 CO_2CH_3

CO_2CH_3 $h\nu$, $(CH_3CH_2)_2O$ CO_2CH_3

Hückel's Rule

5 1,3-CYCLOBUTADIENE

Preparation:

O O $h\nu$, $-266°C$, Ar $+ CO_2$

Diels-Alder cycloaddition:

H H CH_3OOC $COOCH_3$ CH_3OOC $COOCH_3$

6 1,3,5,7-CYCLOOCTATETRAENE

$$4\ HC{\equiv}CH \xrightarrow{Ni(CN)_2,\ 70°C,\ 15\text{–}25\ atm}$$

7 CYCLOPENTADIENYL ANION

$$\rightleftharpoons\ + H^+$$

$pK_a = 16$

8 CYCLOPENTADIENYL CATION

Br H

$$\xrightarrow{SbF_5,\ -200°C}\ + SbF_5Br^-$$

9 CYCLOHEPTATRIENYL CATION

$$\xrightarrow[-\ HBr]{Br_2}\ Br^-$$

10 CYCLOOCTATETRAENE DIANION

$$\xrightarrow{K,\ THF}\ + 2\ K^+$$

11 [16]ANNULENE DICATION AND DIANION

$$[16]^{2+} \xleftarrow{CF_3SO_3H,\ SO_2,\ CH_2Cl_2,\ -80°C} \quad \xrightarrow{K,\ THF} [16]^{2-}$$

[16]Annulene

Summary of Important Concepts

1 The polycyclic benzenoid hydrocarbons are composed of linearly or angularly fused benzene rings. The simplest members of this class of compounds are naphthalene, anthracene, and phenanthrene.

2 In these molecules, benzene rings share two (or more) carbon atoms, whose π electrons are delocalized over the entire ring system. Thus, naphthalene shows some of the properties characteristic of the aromatic ring in benzene: the electronic spectra reveal extended conjugation, ^{1}H NMR exhibits deshield-

ing ring-current effects, there is little bond alternation, and the π system undergoes electrophilic aromatic substitution.

3 Synthetic routes to the polycyclic benzenoid hydrocarbons, such as naphthalene, anthracene, and phenanthrene, rely on ring-closure reactions, such as intramolecular Friedel-Crafts alkanoylations, Diels-Alder cycloadditions, and oxidative photocyclizations.

4 Naphthalene undergoes kinetically favored electrophilic substitution at C-1 because of the relative stability of the intermediate carbocation. However, a substituent at this position suffers from steric hindrance exerted by whatever group is at C-8. Hence, if the initial substitution is reversible, the final, thermodynamically favored product will bear the electrophile at C-2.

5 Electron-donating substituents on one of the naphthalene rings direct electrophiles to the same ring, *ortho* and *para*. Electron-withdrawing substituents direct electrophiles away from that ring; substitution is mainly at C-5 and C-8.

6 Positions 9 and 10 in anthracene and phenanthrene have alkenelike reactivity, making addition reactions possible.

7 The ultimate carcinogen derived from benz[*a*]pyrene appears to be an oxacyclopropanediol in which C 7 and C-8 bear hydroxy groups and C-9 and C-10 are bridged by oxygen. This molecule alkylates one of the nitrogens of one of the DNA bases, thus causing mutations.

8 Benzene is the smallest member of the class of aromatic cyclic polyenes following Hückel's $4n + 2$ rule. Most of the $4n$ π systems are relatively reactive species devoid of aromatic properties. Hückel's rule also extends to aromatic charged systems, such as the cyclopentadienyl anion, cycloheptatrienyl cation, and cyclooctatetraene dianion.

Problems

1 Catalytic hydrogenation of naphthalene over Pd-C results in rapid addition of two moles of H_2. Propose a structure for this product.

2 The ^{1}H NMR spectrum of naphthalene shows two multiplets (Figure 25-4). The upfield absorption ($\delta = 7.40$ ppm) is due to the hydrogens at C-2, C-3, C-6, and C-7, and the downfield multiplet ($\delta = 7.77$ ppm) is due to the hydrogens at C-1, C-4, C-5, and C-8. Why do you suppose the latter hydrogens are more deshielded than the former?

3 An unknown compound whose mass-spectral parent ion appears at $m/z = 144$ has an IR spectrum with a broad, intense absorption at about 3290 cm^{-1}, a ^{13}C NMR spectrum with ten signals between $\delta = 105$ and 115 ppm, and ^{1}H NMR spectrum A (on the next page). Propose a structure for this compound. Hint: Note the presence of a one-proton singlet in the ^{1}H NMR spectrum at $\delta = 7.25$ ppm.

4 Draw the structure of *N,N,N′,N′*-tetramethyl-1,8-naphthalenediamine. This compound is such a strong base that it has been given the nickname "Proton Sponge." Suggest an explanation for the high base strength of Proton Sponge.

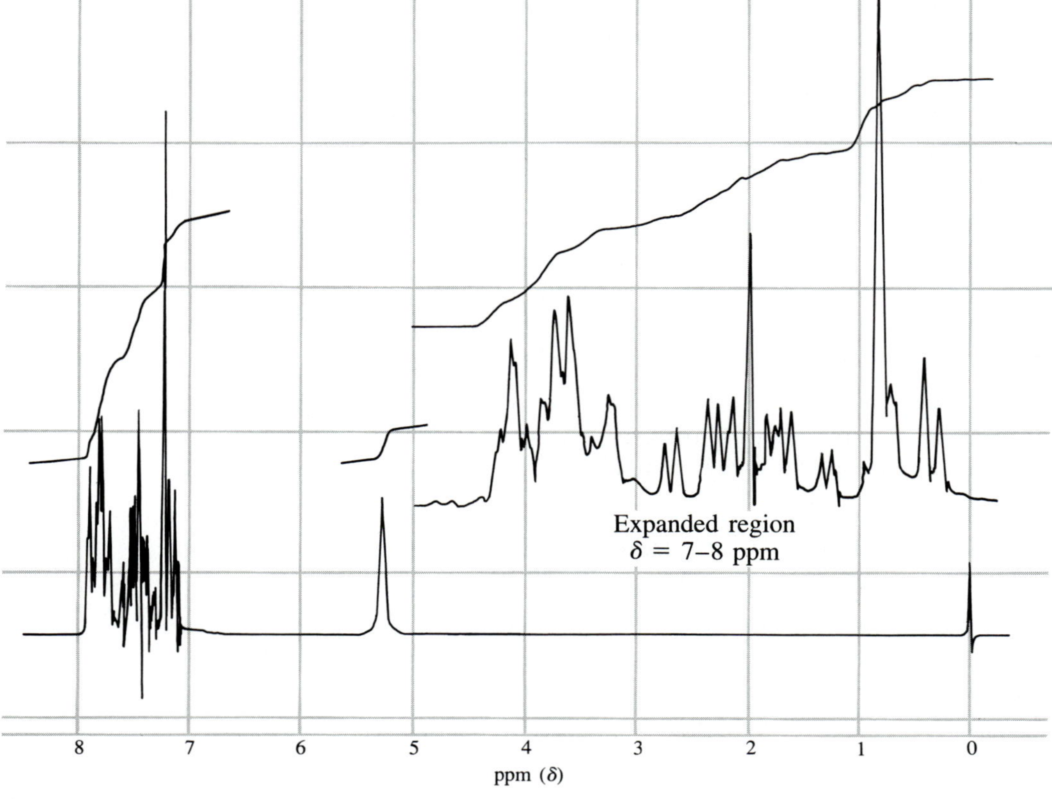

A

5 Predict the major mononitration product of each of the following disubstituted naphthalenes.

(a) 1,3-Dimethylnaphthalene.
(b) 1-Chloro-5-methoxynaphthalene.
(c) 1,7-Dinitronaphthalene.
(d) 1,6-Dichloronaphthalene.

6 Write the expected product(s) of each of the following reactions.

(a) 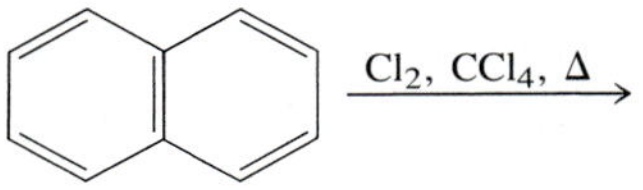 Naphthalene $\xrightarrow{Cl_2,\ CCl_4,\ \Delta}$

(b) 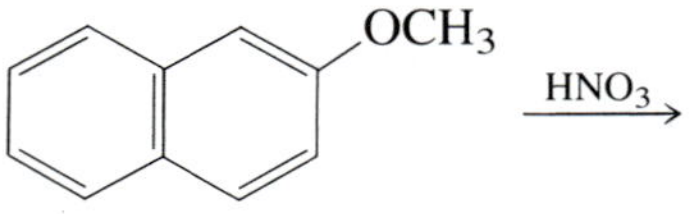 2-Methoxynaphthalene (OCH_3) $\xrightarrow{HNO_3}$

(c) 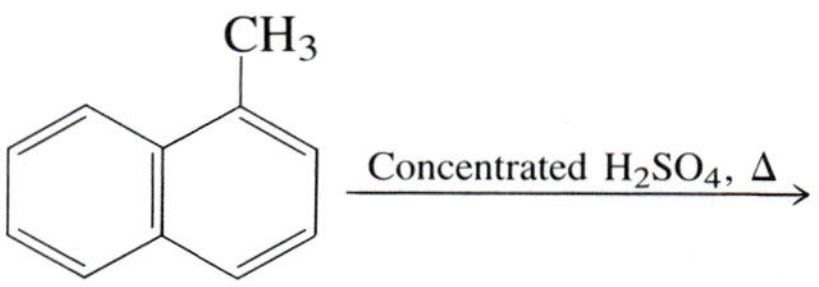 1-Methylnaphthalene (CH_3) $\xrightarrow{\text{Concentrated } H_2SO_4,\ \Delta}$

(d) 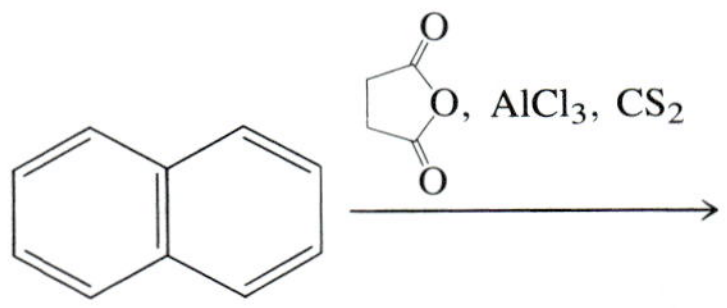 [naphthalene] $\xrightarrow{\text{succinic anhydride, } AlCl_3,\ CS_2}$

(e) 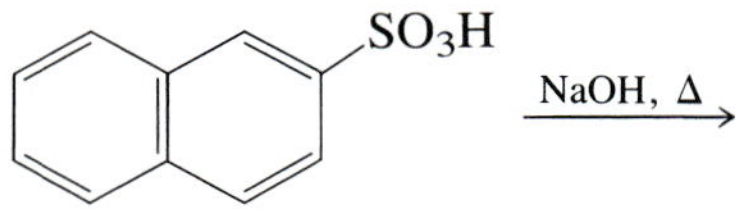 [2-naphthalenesulfonic acid, SO_3H] $\xrightarrow{NaOH,\ \Delta}$

(f) Product of part **e** $\xrightarrow{HO_3S-C_6H_4-N_2^+,\ NaOH,\ 5°C}$ Orange II (dye)

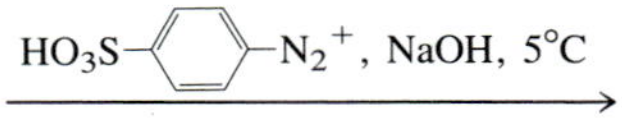

(g) 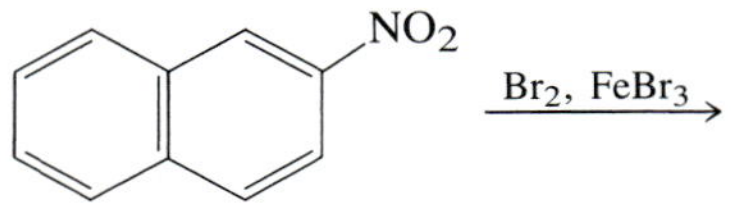 [2-nitronaphthalene, NO_2] $\xrightarrow{Br_2,\ FeBr_3}$

(h) 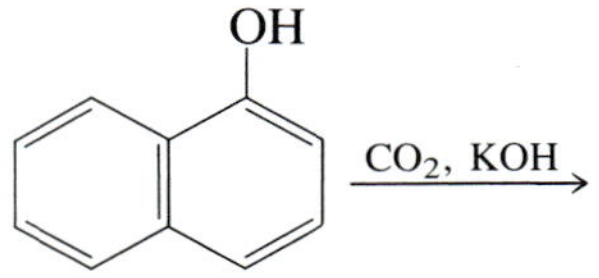 [1-naphthalenol, OH] $\xrightarrow{CO_2,\ KOH}$

7 The product of the following reaction sequence is a dye aptly named Allura Red. Draw its likely structure.

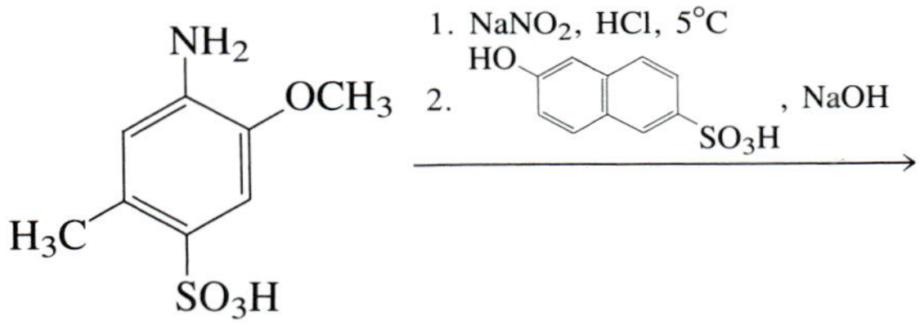

[Aromatic amine with NH_2, OCH_3, H_3C, SO_3H substituents] $\xrightarrow{\text{1. } NaNO_2,\ HCl,\ 5°C \quad \text{2. HO-naphthalene-}SO_3H,\ NaOH}$

8 Propose a simple synthesis of 1,4-naphthalenediol beginning with 2,5-cyclohexadiene-1,4-dione (*p*-benzoquinone).

9 Propose a synthesis of 2-(2-propenyl)-1-naphthalenol from naphthalene.

10 Write the expected product(s) of Friedel-Crafts alkanoylation (acylation) of naphthalene with 1,2-benzenedicarboxylic anhydride (phthalic anhydride). Starting with these products, illustrate the synthesis of one or more polycyclic benzenoid hydrocarbons, following the general methodology introduced in Sections 25-3 and 25-4.

11 The steroid equilenin is an estrogen that was first isolated from the urine of pregnant mares.

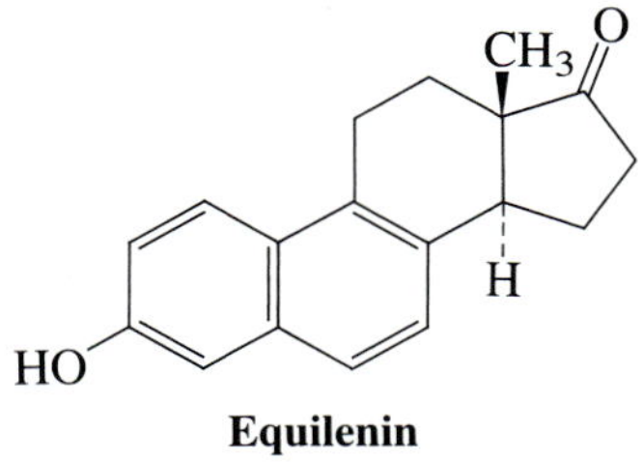

[Structure with labels: CH_3, O, H, HO]

Equilenin

Equilenin was, in fact, the very first naturally occurring steroid to be prepared by total synthesis (1939). An outline of the critical early stages of this synthesis is shown here. Fill in the missing reagents, as required, **a** through **e.**

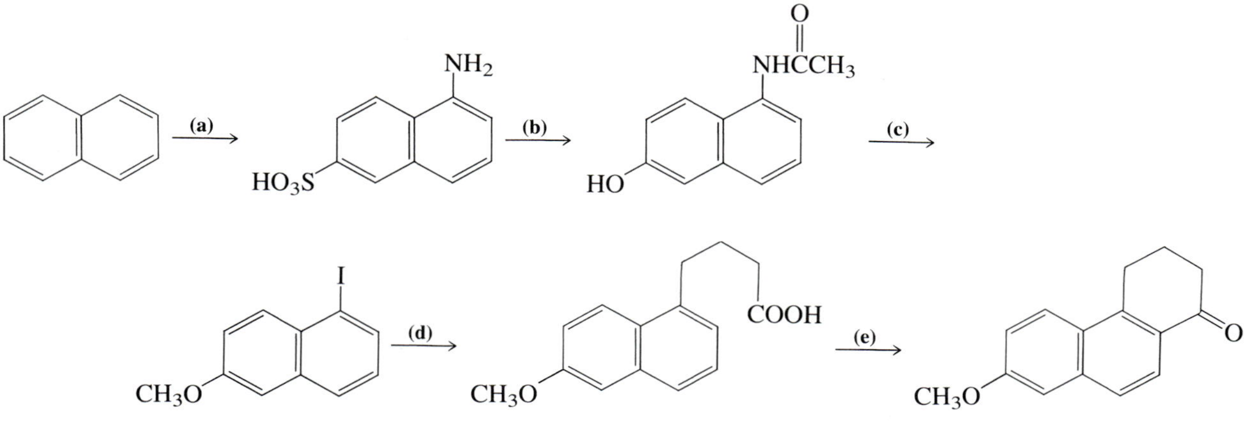

(a) → NH_2, HO_3S (b) → $NHCCH_3$ (O), HO (c) →

I, CH_3O (d) → COOH, CH_3O (e) → O, CH_3O

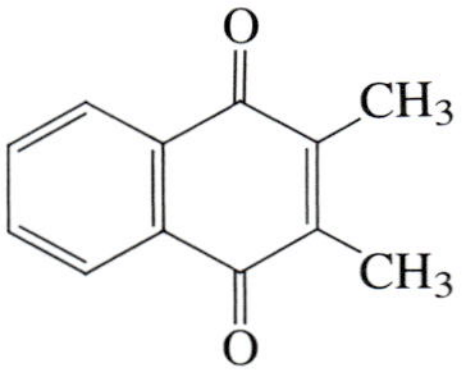

O, CH_3, CH_3, O

Vitamin K₃

12 The simplest of the vitamin K family (Section 24-5) is vitamin K_3. Propose a synthesis of this substance from benzene and any other organic compound(s) that you need.

13 Write the expected products of each of the following reactions.

(a) 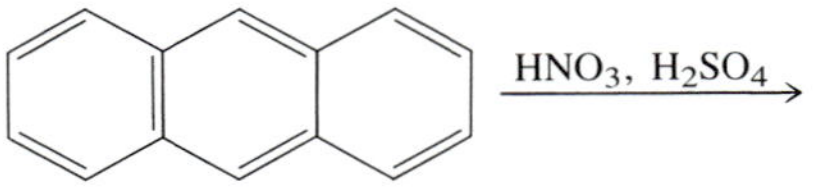 HNO_3, H_2SO_4 →

(b) 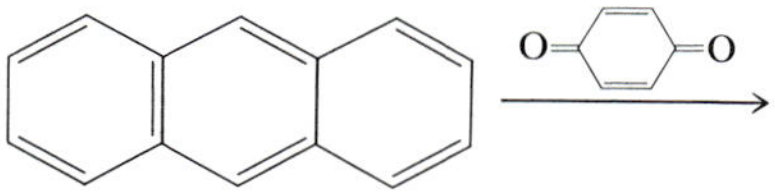 O= =O →

(c) 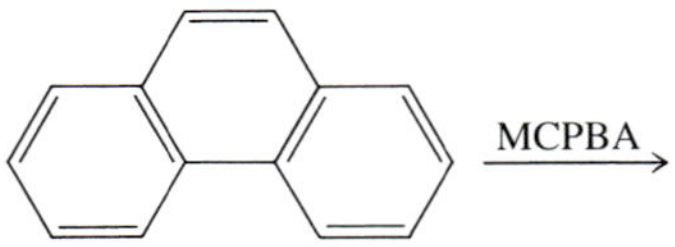 MCPBA →

(d) 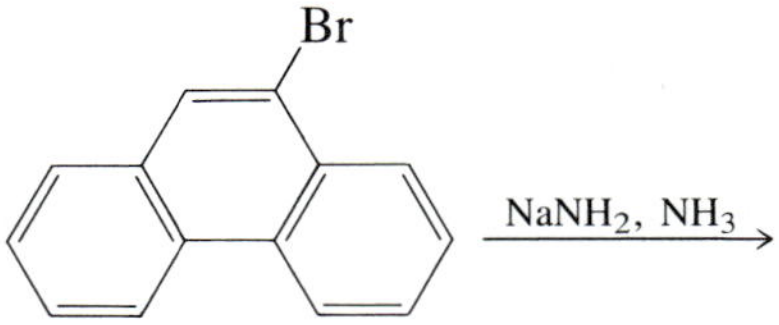 Br; $NaNH_2$, NH_3 →

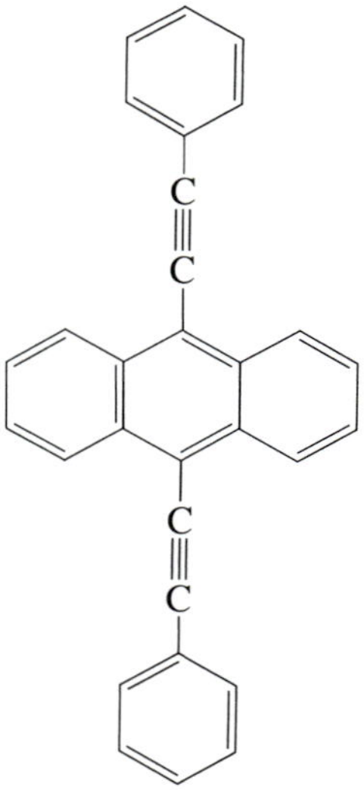

C≡C, C≡C

9,10-Di(phenylethynyl)-anthracene

14 The brilliant green fluorescence of 9,10-di(phenylethynyl)anthracene is the source of light in a variety of chemiluminescent products (e.g., Cyalume emergency light sticks), in which light is generated chemically, without heat. Devise a synthesis of this anthracene derivative from 2,5-cyclohexadiene-1,4-dione (*p*-benzoquinone).

15 Friedel-Crafts alkanoylation (acylation) of phenanthrene gives substitution in the 3-position, predominantly:

$CH_3\overset{O}{\overset{\|}{C}}Cl$, $AlCl_3$ → CH_3

Suggest an explanation for this observation. Hint: Both electronic and steric factors should be considered.

16 For each of the hydrocarbons depicted in Section 25-1, write a resonance form that illustrates the presence of $4n + 2$ π electrons around the periphery of the molecule. Is there anything about one of these six systems that sets it apart from the rest?

17 Which of the following structures qualify as being aromatic, according to Hückel's rule?

(a)

(b) $CH{=}CH_2$

(c)

(d)

(e)

(f) $[\ ]^{2-}$ $2\ K^+$

(g)

(h)

18 The ^1H NMR spectrum of the most-stable isomer of [14]annulene shows two signals, at $\delta = -0.61$ (4 H) and 7.88 (10 H) ppm. Two possible structures for [14]annulene are shown below. How do they differ? Which one corresponds to the NMR spectrum described?

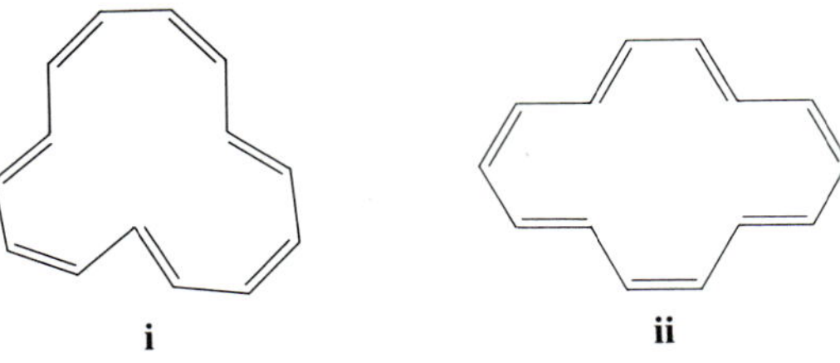

19 The unusual acidity of fluorene (pK_a ~ 23) was mentioned in Problem 5 of Chapter 24. Review your answer to that particular problem, and revise your explanation in light of Section 25-5.

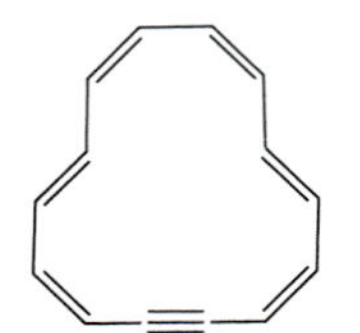

1,2-Dehydro[14]annulene

20 Treatment of 3,4-dibromo-1,2-diphenylcyclobutene with the powerful Lewis acid SbF_5 in an inert solvent gives a species showing the following ^1H NMR signals: δ = 8.78 (m, 4 H), 9.40 (m, 6 H), and 10.68 (s, 2 H) ppm. Suggest a structure for the species that is responsible for this spectrum.

21 How should the presence of a triple bond in a conjugated cyclic polyene affect the presence or absence of aromaticity? For example, would you expect 1,2-dehydro[14]annulene, shown at the left, to be aromatic?

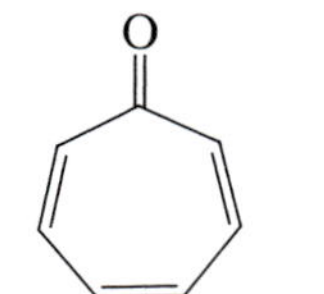

2,4,6-Cycloheptatrienone (Tropone)

22 **(a)** 2,4,6-Cycloheptatrienone (tropone) is a perfectly stable ketone, but its smaller relative 2,4-cyclopentadienone is an extremely reactive molecule capable of only fleeting existence. Suggest an explanation. Hint: Consider dipolar resonance forms involving the carbonyl group in each system.

(b) Explain the unexpected result of the following reaction.

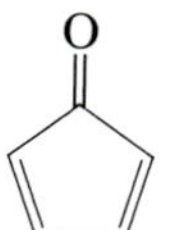

2,4-Cyclopentadienone

Tetraphenylcyclopentadienone (four C_6H_5 groups) $\xrightarrow{\text{1. } (CH_3)_3CLi \text{ 2. } CH_3I}$ product bearing $OC(CH_3)_3$, four C_6H_5 groups, and CH_3

23 Early work aimed at proving the structures of the powerfully active opium alkaloids was complicated by unusual results of standard structure-determining reactions. For example, two successive Hofmann eliminations on codeine yielded the phenanthrene derivative shown here. Propose mechanisms for the transformations that must have taken place. Suggest an explanation for the unusual carbon–carbon bond cleavage observed.

Codeine $\xrightarrow{\text{1. } CH_3I \text{ 2. } Ag_2O, \Delta \text{ 3. } CH_3I \text{ 4. } Ag_2O, \Delta}$ $(CH_3)_3N + CH_2{=}CH_2 +$ phenanthrene derivative (CH_3O, O bridge)

Codeine

24 Suggest a synthetic sequence for the conversion of compounds (i) and (ii) into tetracyclic diketone (iii), an intermediate in the synthesis of daunomycin, a potent anticancer agent.

i (CH_3O, CH_3O, $CH_2C{\equiv}N$) ii ($CH_3CH_2O_2C$, CH_3) iii (CH_3O, CH_3O, $C{\equiv}N$, CH_3)

CHAPTER 26

Heterocycles

HETEROATOMS IN CYCLIC ORGANIC COMPOUNDS

Carbocyclic compounds are cyclic molecules in which the rings are made up of only carbon atoms. In contrast, **heterocycles** are their analogs in which one or more ring carbons have been replaced by a heteroatom, such as nitrogen, oxygen, sulfur, phosphorus, silicon, a metal, and so on. The most-common heterocyclic systems contain nitrogen or oxygen or both. Several of their derivatives have appeared in the discussion of cyclic ethers—for example, oxacyclopentane (tetrahydrofuran) and crown ethers (Section 9-5), cyclic acetals (Sections 15-5 and 23-2), cyclic dicarboxylic acid derivatives (Sections 17-7, 17-8, 17-9, 18-3, 18-4, and 18-5), halonium ions (Section 12-3), and 1,3-dithiacyclohexanes (dithianes; Section 22-2).

It has been estimated that more than 65% of all published chemical studies deal in one way or another with heterocyclic systems. More than half of natural compounds are heterocyclic, and a high percentage of drugs contain heterocycles. Several earlier chapters have described representatives of these natural products. The following are some additional examples.

Cocaine

(Stimulant, topical anesthetic; found in coca leaves)

Pyridoxine, vitamin B_6

(Enzyme cofactor vitamin)

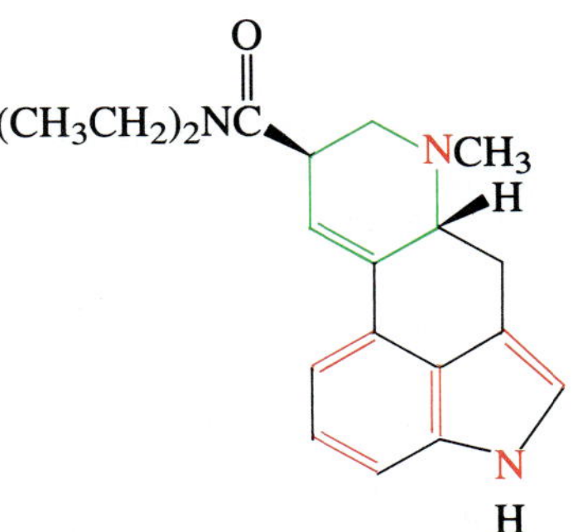

Lysergic acid diethylamide

(LSD; psychotomimetic)

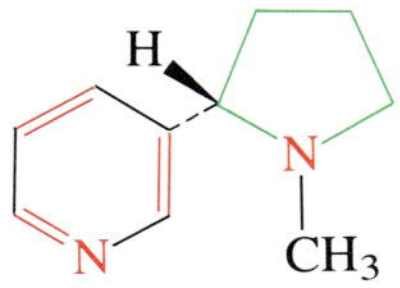

Nicotine

(Found in dried tobacco leaves in 2%–8% concentration)

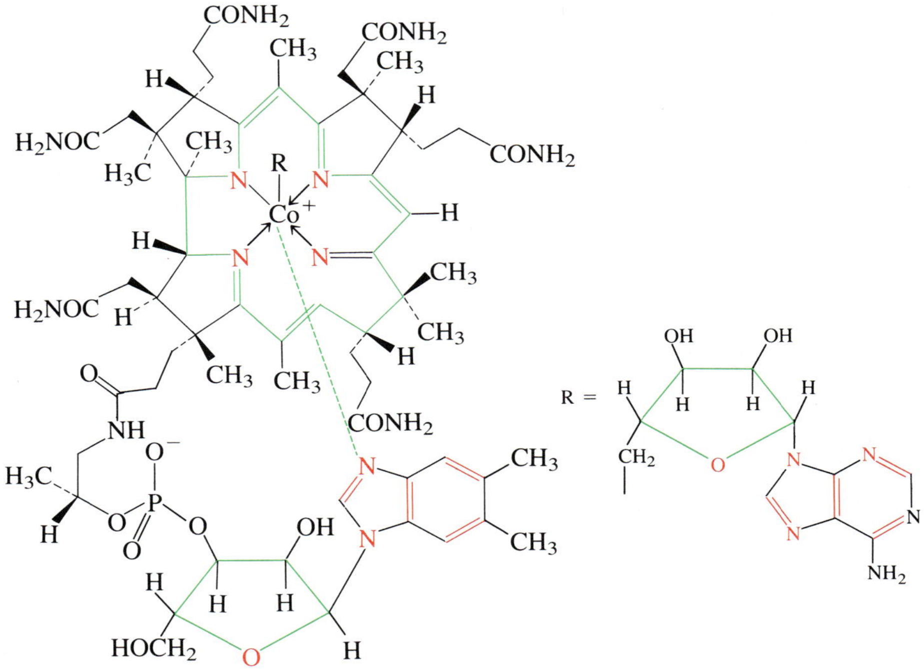

Vitamin B_{12}
(Cobalamin)

(Catalyzes biological rearrangements and methylations)

This chapter describes the naming, synthesis, and reactions of some saturated and aromatic heterocyclic compounds in order of increasing ring size, starting with the heterocyclopropanes. Some of this chemistry is not new, but a simple extension of transformations discussed earlier for carbocycles. However, the presence of the heteroatom does impart special reactivity, which causes the chemical behavior of some of the heterocyclic compounds to differ from that of their carbocyclic analogs.

Oxacyclopropane
(Oxirane, ethylene oxide)

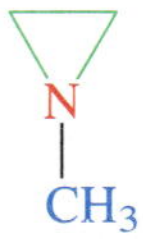

***N*-Methylazacyclopropane**
(*N*-Methylaziridine)

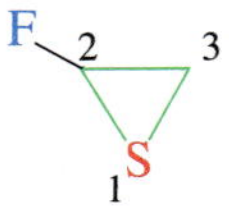

2-Fluorothiacyclopropane
(2-Fluorothiirane)

26-1 The Naming of Heterocycles

Like all the other classes of compounds described in this book, this last one contains many members bearing common names. Moreover, there are several competing systems for naming heterocycles that require memorization, are not always applicable, and are sometimes confusing. This chapter will adhere to the simplest system, one that regards saturated heterocycles as being derived from the related carbocycles but uses a prefix to denote the presence and identity of the heteroatom: **aza-** for nitrogen, **oxa-** for oxygen, **thia-** for sulfur, **phospha-** for phosphorus, and so on. However, other names will be given in parentheses, particularly those that are widely used. The location of substituents is indicated by numbering the ring atoms starting with the heteroatom.

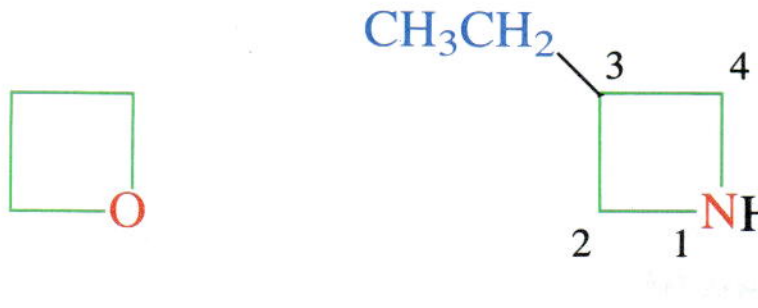

Oxacyclobutane
(Oxetane)

3-Ethylazacyclobutane
(3-Ethylazetidine)

2,2-Dimethylthiacyclobutane
(2,2-Dimethylthietane)

***trans*-3,4-Dibromooxacyclopentane**
(*trans*-3,4-Dibromotetrahydrofuran)

Azacyclopentane
(Pyrrolidine)

Thiacyclopentane
(Tetrahydrothiophene)

3-Methyloxacyclohexane
(3-Methyltetrahydropyran)

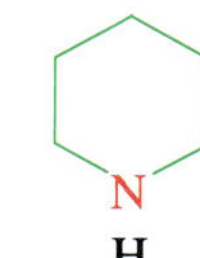

Azacyclohexane
(Piperidine)

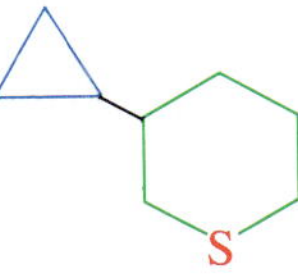

3-Cyclopropylthiacyclohexane
(3-Cyclopropyltetrahydrothiopyran)

Although the unsaturated heterocycles could in principle also be named according to this system (e.g., furan should be called 1-oxa-2,4-cyclopentadiene, and pyridine should be called azabenzene or, better, aza-1,3,5-cyclohexatriene), common names are so firmly entrenched in the literature that we shall use them here.

Pyrrole

Furan

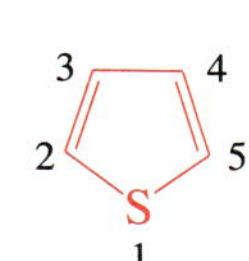

Thiophene

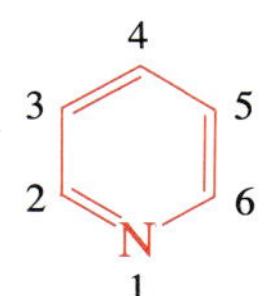

Pyridine

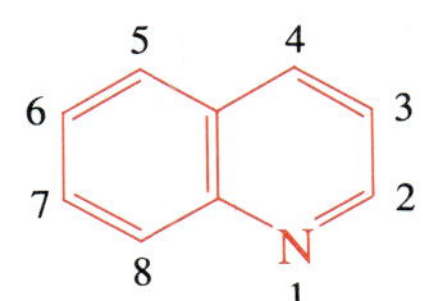

Quinoline

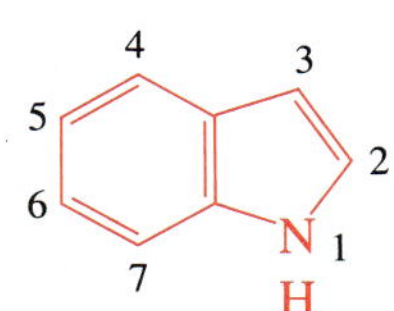

Indole

Adenine
(see Section 27-5)

EXERCISE 26-1

Name or draw the following compounds: (a) *trans*-2,4-dimethyloxacyclopentane (*trans*-2,4-dimethyltetrahydrofuran); (b) *N*-ethylazacyclopropane;

(c) O_2N–(pyridine)–NO_2 ; (d) (4-bromoindole structure, Br, N, H)

26-2 Three-membered Heterocycles: Strain Imparts Reactivity

This section describes some of the ways in which heterocyclopropanes are made, primarily those methods employing ring-closure reactions and electrophilic additions to alkenes. As in the extensive chemistry of the oxacyclopropanes (Section 9-6), ring strain allows these molecules to undergo nucleophilic ring-opening readily.

Preparative Routes to Heterocyclopropanes

Azacyclopropanes (aziridines) can be prepared by direct addition of nitrenes (Section 18-5), the nitrogen analogs of carbenes (Section 21-5), to alkenes. For example, irradiation or heating of ethyl azidocarboxylate leads to a reactive nitrene that is trapped by an alkene to furnish the corresponding azacyclopropane in moderate yields.

$$\left[:N{\equiv}\overset{+}{N}{-}\ddot{\overset{-}{N}}{-}\overset{O}{\overset{\|}{C}}OCH_2CH_3 \longleftrightarrow {}^{-}:\ddot{N}{=}\overset{+}{N}{=}\ddot{N}{-}\overset{O}{\overset{\|}{C}}OCH_2CH_3 \longleftrightarrow :N{\equiv}\overset{+}{N}{-}\ddot{N}{=}\overset{{}^{-}:\ddot{O}:}{\overset{|}{C}}OCH_2CH_3\right] \xrightarrow{h\nu \text{ or } \Delta}$$

$$:N{\equiv}N: + :\ddot{N}{-}\overset{O}{\overset{\|}{C}}OCH_2CH_3 \xrightarrow[-\,N_2]{\text{cyclohexene}} \text{(bicyclic aziridine)}N{-}CO_2CH_2CH_3$$

A nitrene

50%

Better results are obtained by allowing the azide to first undergo cycloaddition to a double bond. This reaction is analogous to the addition of ozone to alkenes to give molozonides (Section 12-5). However, the products, called triazolines, are more stable.

$$\text{Bicyclo[2.2.1]heptene} + \ddot{N}{\equiv}\overset{+}{N}{-}\ddot{N}:^{-}{-}CO_2CH_2CH_3 \xrightarrow{\text{Pentane, } 25^{\circ}C} \text{triazoline} \xrightarrow{h\nu,\ \text{propanone (acetone)}}$$

94%

A triazoline

$$\text{(tricyclic aziridine)}\ddot{N}CO_2CH_2CH_3 + :N{\equiv}N:$$

95%

Most azacyclopropane (aziridine) syntheses employ ring-closure reactions. For example, iodine isocyanate, an electrophilic reagent with a positively polarized iodine (e.g., $I^{+}NCO^{-}$; Section 12-3), adds to double bonds to give the iodo isocyanate derivative.

Iodine isocyanate

trans-2-Iodocyclohexyl isocyanate

A carbamate ester

80%

An *N*-methoxycarbonyl-azacyclopropane

90%

Isocyanates are found also in the Hofmann rearrangement, in which they are readily converted into the corresponding carbamate esters by reaction with alcohols. In the present case, treatment with methanol gives the desired intermediate. The relatively acidic amide proton ($pK_a \sim 15$) of this species is removed by base, and the amidate ion then undergoes an intramolecular S_N2 reaction yielding an *N*-methoxycarbonylazacyclopropane. Base hydrolysis at slightly elevated temperature removes the carboxy group (see Section 18-5) to furnish the free cyclic amine.

EXERCISE 26-2

Azacyclopropanes (aziridines) can also be made from oxacyclopropane by the sequence (1) RNH_2, (2) HCl, (3) base. How does this reaction sequence work?

Intramolecular S_N2 reactions can also be employed in the preparation of oxacyclopropanes from vicinal chloro alcohols (Section 9-5). An alternative approach to the preparation of oxacyclopropanes is oxidation of alkenes by peroxycarboxylic acid (Section 12-5). Recall that this reaction proceeds by an electrophilic mechanism.

General Oxacyclopropane Formation from Alkenes

As a result, double bonds bearing electron-withdrawing groups, such as α,β-unsaturated carbonyl compounds, undergo oxacyclopropane formation under these conditions only very sluggishly. However, they can be oxidized to the corresponding oxacyclopropanes by a nucleophilic source of oxygen. Thus, when α,β-unsaturated aldehydes or ketones are treated with basic hydrogen peroxide, a Michael-type addition (Section 16-5) takes place, with hydroperoxide ion as the nucleophile.

Oxacyclopropane Formation from Propenal with Basic Hydrogen Peroxide

$$H_2C{=}CHCHO \xrightarrow{30\%\ H_2O_2,\ NaOH,\ CH_3OH} \text{2-methanoyloxacyclopropane (85\%)}$$

Propenal → **2-Methanoyloxacyclopropane**

Mechanism:

Michael addition

Ring closure

EXERCISE 26-3

Treatment of either *cis*- or *trans*-3-penten-2-one with basic H_2O_2 results in the same product, *trans*-2-ethanoyl-3-methyloxacyclopropane. Explain by a mechanism.

Thiacyclopropanes (thiiranes) are best synthesized from the readily prepared corresponding oxacyclopropanes. A reagent that accomplishes the direct conversion of one into the other is potassium thiocyanate, $K^{+-}SCN$. The transformation is stereospecific and proceeds with inversion at both reacting carbons.

Direct Conversion of an Oxacyclopropane into a Thiacyclopropane

C_6H_6, $NaHCO_3$; $K^{+-}SCN$, $-\ K^{+-}OCN$

56% (overall)

***trans*-2,3-Dideuteriothiacyclopropane**

The mechanism of this reaction begins with nucleophilic opening of the oxacyclopropane ring. The alkoxide oxygen thus produced then adds to the nitrile group to give an intermediate heterocyclic anion.

Mechanism:

$^{-}$:S̈—C≡N: **Thiocyanate ion** → First inversion → Rotation → → Rotation → Second inversion → + $^{-}$:Ö—C≡N: **Cyanate ion**

Ring-opening in the opposite direction, followed by intramolecular displacement of the cyanate function, gives the final thiacyclopropane product.

EXERCISE 26-4

Propose a synthesis of compound A (shown at the right) from cyclohexane.

A

Ring Strain Governs Heterocyclopropane Reactivity

Heterocyclopropanes are relatively reactive, because ring strain is released by nucleophilic ring-opening. Under basic conditions, reaction occurs at the less-substituted center and causes inversion (Section 9-6).

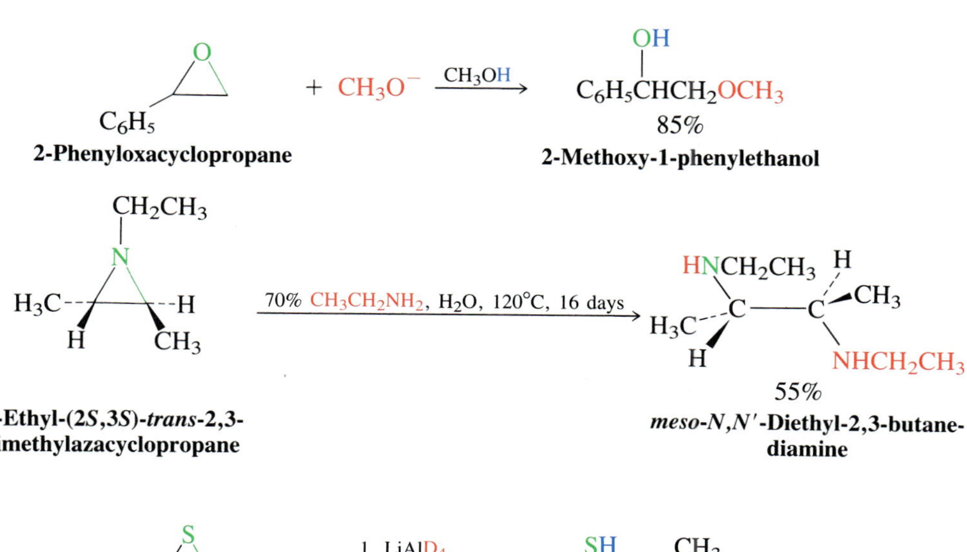

2-Phenyloxacyclopropane → **2-Methoxy-1-phenylethanol**

***N*-Ethyl-(2*S*,3*S*)-*trans*-2,3-dimethylazacyclopropane** → ***meso*-*N*,*N*′-Diethyl-2,3-butanediamine**

1. $LiAlD_4$
2. H^+, H_2O

SH, CH_3, H, D — 35%

(2*R*,3*R*)-*trans*-2,3-Dimethylthiacyclopropane → **(2*R*,3*S*)-3-Deuterio-2-butanethiol**

EXERCISE 26-5

Treatment of target A of Exercise 26-4 with hydrogen chloride gives a thiol product. Draw its structure, including stereochemistry.

EXERCISE 26-6

Explain the following result by a mechanism. Hint: Try a ring-opening catalyzed by a Lewis acid.

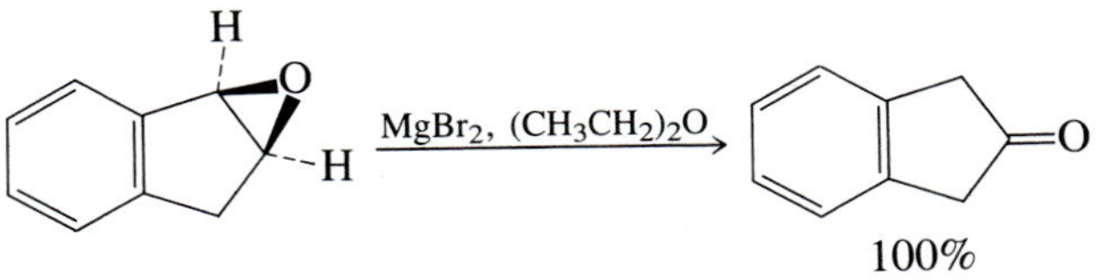

$\xrightarrow{MgBr_2,\ (CH_3CH_2)_2O}$ 100%

In summary, the heterocyclopropanes are made by direct addition of electrophilic heteroatom-containing reagents such as nitrenes or peroxycarboxylic acids, by ring-closure reactions, by nucleophilic addition of hydroperoxide ion to α,β-unsaturated carbonyl compounds, and by exchange of one heteroatom for another (as in the conversion of oxa- into thiacyclopropanes). Their reactivity results from the release of strain by ring-opening.

26-3 Preparation and Reactions of Four- and Five-membered Heterocycloalkanes

This section briefly surveys the methods used to prepare some of the four- and five-membered heterocycloalkanes, and their reactions as well. The heterocyclobutanes are available by ring closure, although the rates of cyclization are low (Section 9-5), and by [2+2]cycloadditions (Section 14-5). Because of ring strain, they are more reactive than the corresponding heterocyclopentanes. The latter also can be made by intramolecular S_N2 reactions or from the corresponding unsaturated systems by hydrogenation.

Preparation of Four- and Five-membered Heterocycloalkanes

Although ring closures leading to four-membered heterocycloalkanes by intramolecular S_N2 reactions are relatively slow, useful yields can be obtained in a variety of cases.

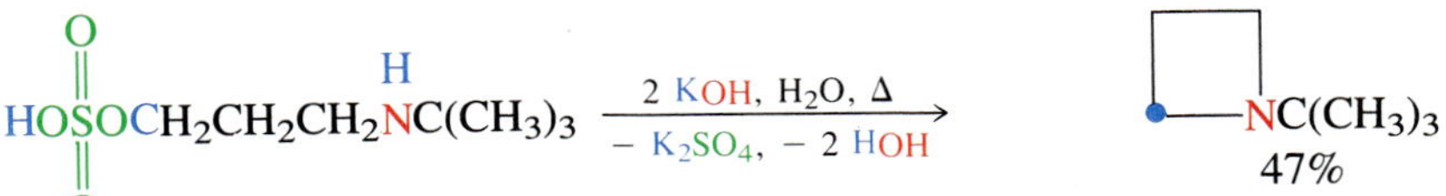

$$HOS(=O)_2OCH_2CH_2CH_2N(H)C(CH_3)_3 \xrightarrow[-\ K_2SO_4,\ -\ 2\ HOH]{2\ KOH,\ H_2O,\ \Delta} \text{(azacyclobutane)}-NC(CH_3)_3$$

47%

***N*-(1,1-Dimethylethyl)azacyclobutane**

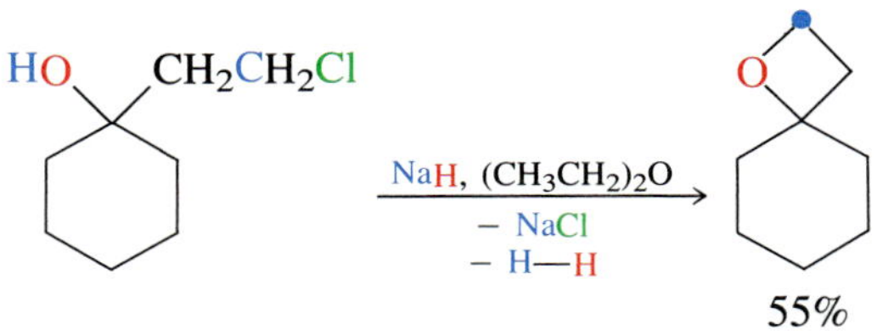

HO, CH_2CH_2Cl (cyclohexane)

$$\xrightarrow[-\ NaCl \quad -\ H{-}H]{NaH,\ (CH_3CH_2)_2O}$$

55%

EXERCISE 26-7

(2-Chloromethyl)oxacyclopropane reacts with hydrogen sulfide ion to give thiacyclobutan-3-ol. Explain by a mechanism.

Four-membered ring heterocycles are also available by direct [2+2]cycloaddition of appropriate double bonds, by the use of photochemical conditions, or special catalysts, or reactive substrates such as chlorosulfonyl isocyanate.

[2+2]Cycloadditions in Heterocyclobutane Synthesis

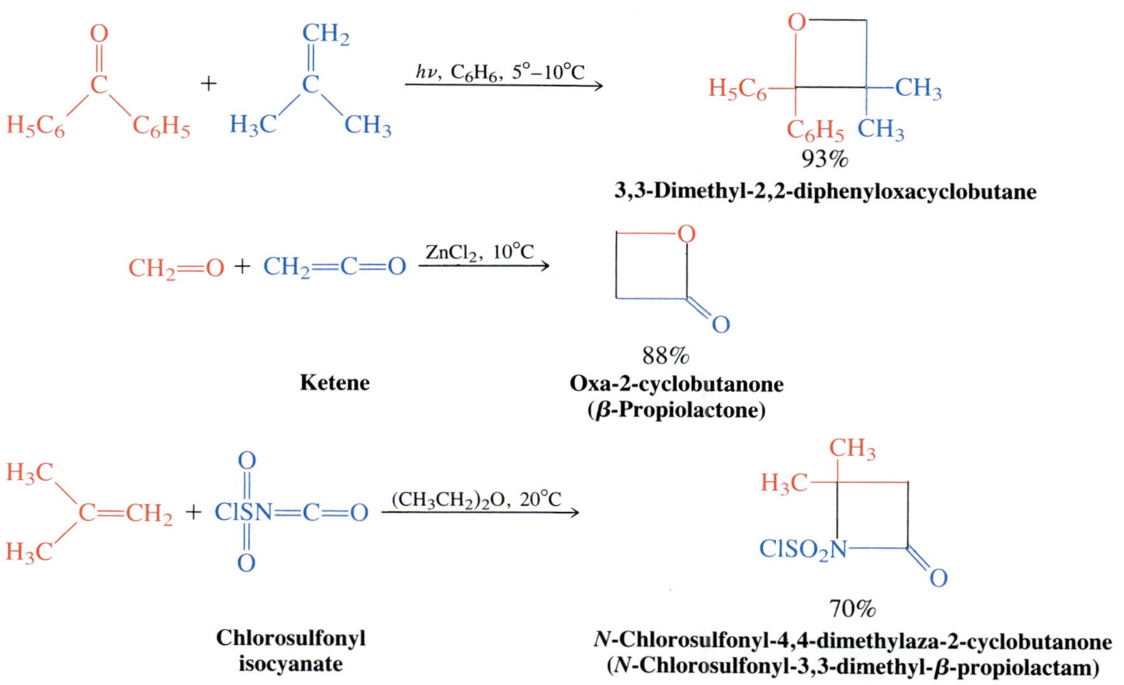

EXERCISE 26-8

The mechanisms of the preceding cycloadditions of ketene are thought to include dipolar intermediates. Formulate such mechanisms.

Heterocycloalkanes containing five-membered (and larger) rings are most frequently made by intramolecular S_N2 reactions. An alternative is the catalytic hydrogenation of the corresponding unsaturated derivatives, if they are readily available. Thus, for example, pyrroles have been converted in this way into azacyclopentanes, and furans into oxacyclopentanes. Thiophene is difficult to reduce in this manner because sulfur is a catalyst poison. However, with an excess of catalyst, good yields of thiacyclopentane are obtained.

N–CO_2CH_3 (pyrrole) $\xrightarrow{H_2,\ Pt,\ CH_3CO_2H}$ N–CO_2CH_3 (pyrrolidine), 95%

***N*-Methoxycarbonylpyrrole** → ***N*-Methoxycarbonylazacyclopentane (*N*-Methoxycarbonylpyrrolidine)**

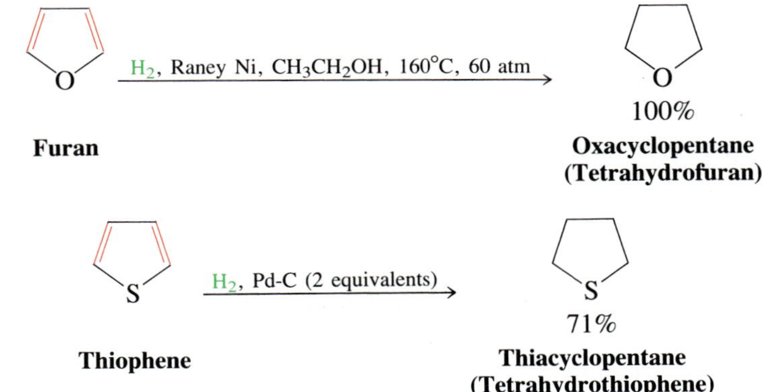

Heterocyclobutanes React by Release of Ring Strain. Heterocyclopentanes Are Unreactive

The reactivity of the four- and five-membered heterocycloalkanes bears out expectations based on ring strain: only the strained-ring systems are reactive, and their reactions usually open the rings, as in the reaction of oxacyclobutane with methanamine (methylamine).

$$\text{(oxacyclobutane)} + CH_3NH_2 \xrightarrow{150^\circ C} CH_3NH(CH_2)_3OH$$

45%

N-Methyl-3-amino-1-propanol

EXERCISE 26-9

Treatment of thiacyclobutane with chlorine in $CHCl_3$ at $-70^\circ C$ gives $ClCH_2CH_2CH_2SCl$ in 30% yield. Suggest a mechanism for this transformation. Hint: The sulfur in sulfides is nucleophilic (Section 9-7).

EXERCISE 26-10

2-Methyloxacyclobutane reacts with hydrogen chloride to give two products. Write their structures.

The unstrained heterocyclopentanes are relatively inert; recall that oxacyclopentane (tetrahydrofuran, THF) is used as a solvent. The heteroatoms in aza- and thiacyclopentane allow these species to undergo their own characteristic transformations (see Sections 9-7, 15-6, 16-1, and 21-5).

EXERCISE 26-11

Treatment of azacyclopentane (pyrrolidine) with sodium nitrite in ethanoic (acetic) acid gives a liquid, b.p. 99°–100°C (15 mm Hg), which has the composition $C_4H_8N_2O$. Propose a structure for this compound.

In summary, the four- and five-membered heterocycloalkanes are available by intramolecular S_N2 reactions and (in the former case) by [2+2]cycloadditions. Predictably, the strained homologs are more reactive than their five-membered counterparts in nucleophilic ring-opening reactions.

26-4
The Aromatic Heterocyclopentadienes: Pyrrole, Furan, and Thiophene

The 1-hetero-2,4-cyclopentadienes contain a butadiene unit bridged by a heteroatom bearing lone electron pairs. Do these systems delocalize their electrons to form an aromatic six-π-electron framework?

This section will answer the question in the affirmative, review the methods used to prepare these compounds and their derivatives, and describe some of their reactions, particularly electrophilic aromatic substitutions.

Pyrrole, Furan, and Thiophene Contain Delocalized Lone Electron Pairs

The electronic structure of the three heterocycles pyrrole, furan, and thiophene is similar to that of the cyclopentadienyl anion (Section 25-5). The cyclopentadienyl anion may be viewed as a butadiene bridged by a negatively charged carbon whose electron pair is delocalized over the other four carbons. The heterocyclic analogs contain a neutral atom in that place, bearing lone electron pairs.

Analogy Between Cyclopentadienyl Anion and the Aromatic Heterocyclopentadienes

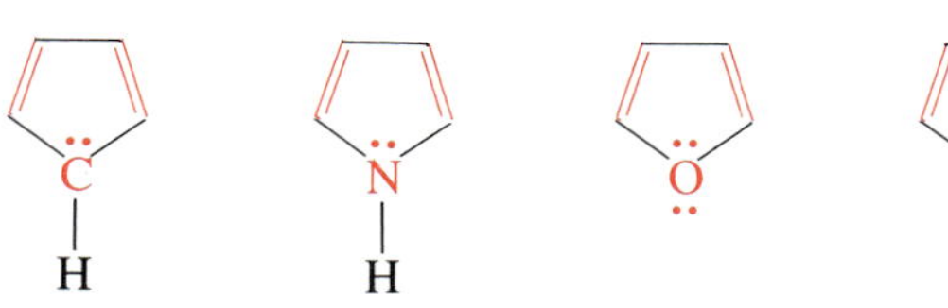

One of these pairs is similarly delocalized, furnishing the two electrons necessary to satisfy the $4n + 2$ rule. To maximize overlap, the heteroatoms are hybridized to sp^2 (Figure 26-1), the delocalized electron pair being assigned to the remaining p orbital. In pyrrole, the sp^2-hybridized nitrogen bears a hydrogen substituent in the plane of the molecule. For furan and thiophene, the second lone electron pair is placed into one of the sp^2 hybrid orbitals, again in the plane and therefore with no opportunity to achieve overlap. The situation is much like the arrangement in the phenyl anion (Section 20-6) or the phenyl cation (in which this orbital is empty; Section 24-6, Figure 24-3).

FIGURE 26-1
Molecular-orbital pictures of (A) pyrrole and (B) furan and thiophene (X = O or S).

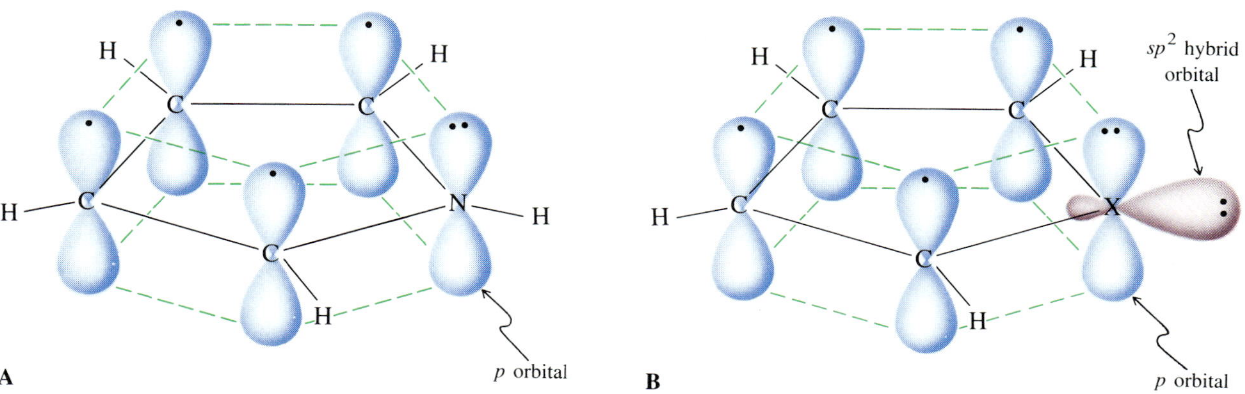

The delocalization of the lone electron pair in the 1-hetero-2,4-cyclopentadienes can also be described by charge-separated resonance structures, as shown for pyrrole.

Resonance Structures of Pyrrole

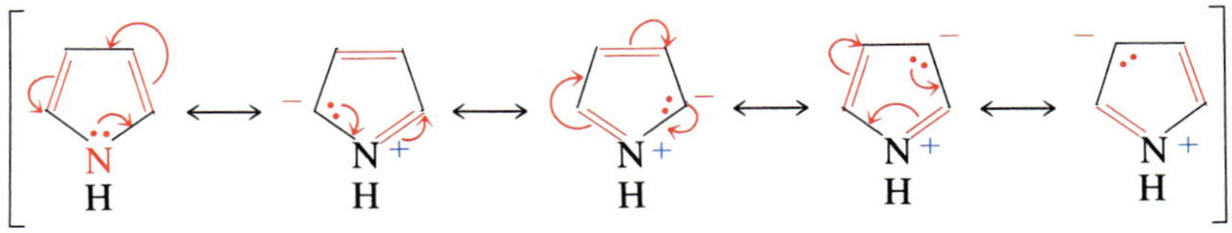

Notice that there are four dipolar structures in which a positive charge is placed on the heteroatom and a negative charge successively on each of the carbons. This picture suggests that the heteroatom should be relatively electron poor and that the carbons should be relatively electron rich. Indeed, as we shall see, the reactivity of these compounds bears out that expectation.

EXERCISE 26-12

Azacyclopentane has a dipole moment of 1.57 D, pyrrole has one of 1.80 D. However, the dipole vectors in the two molecules point in opposite directions. What is the sense of direction of this vector in each structure? Explain your answer.

Preparation of Pyrroles, Furans, and Thiophenes

Syntheses of the heterocyclopentadienes use a variety of cyclization strategies. A general approach is the **Paal-Knorr* synthesis** (for pyrroles) and its variations (for the other heterocycles). The target molecule is approached from an enolizable γ-dicarbonyl compound that is treated with an amine derivative (for pyrroles), or P_2O_5 (for furans), or P_2S_5 (for thiophenes). Formally, this process may be regarded as a dehydration of an intermediate double enol (or its nitrogen or sulfur equivalent) to the heterocycle.

General Cyclization of a γ-Dicarbonyl Compound to a 1-Hetero-2,4-Cyclopentadiene

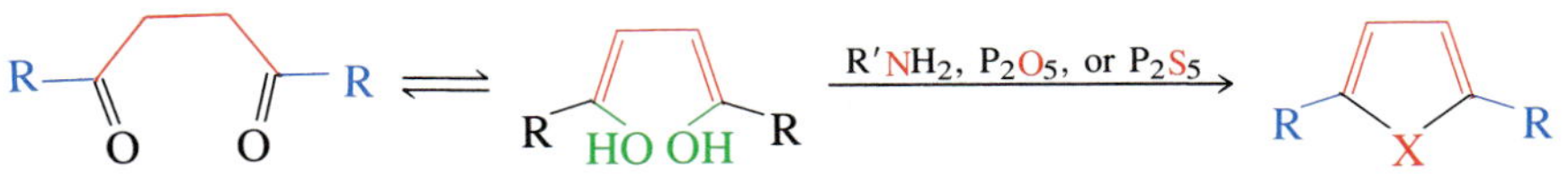

R'NH_2, P_2O_5, or P_2S_5

HO OH

X = NR', O, or S

Examples:

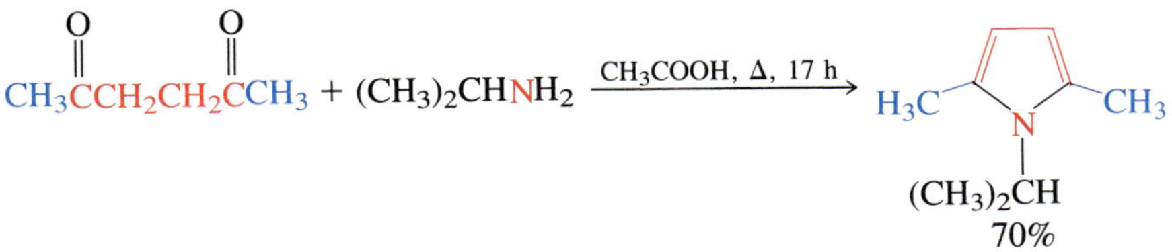

$CH_3CCH_2CH_2CCH_3$ + $(CH_3)_2CHNH_2$ $\xrightarrow{CH_3COOH,\ \Delta,\ 17\ h}$

H_3C N CH_3

$(CH_3)_2CH$

70%

***N*-1-Methylethyl-2,5-dimethylpyrrole**

*Professor Karl Paal, 1860–1935, University of Erlangen, Germany; Professor Ludwig Knorr, 1859–1921, University of Jena, Germany.

C_6H_5 C_6H_5 $\xrightarrow{P_2O_5,\ 150°C}$ C_6H_5 O C_6H_5

62%

$$CH_3\overset{O}{\overset{\|}{C}}CH_2CH_2\overset{O}{\overset{\|}{C}}CH_3 \xrightarrow{P_2S_5,\ 140°\text{–}150°C} H_3C \text{ S } CH_3$$

60%

2,5-Dimethylthiophene

EXERCISE 26-13

The following equation is an example of another synthesis of pyrroles. Write a mechanism for this transformation.

$$CH_3\overset{O}{\overset{\|}{C}}\underset{NH_2}{\underset{|}{C}}HCO_2CH_2CH_3 + CH_3\overset{O}{\overset{\|}{C}}CH_2CO_2CH_2CH_3 \xrightarrow{\Delta}$$

Ethyl 2-amino-3-oxobutanoate **Ethyl 3-oxobutanoate**

H_3C, $CO_2CH_2CH_3$, $CH_3CH_2O_2C$, N, H, CH_3

Diethyl 3,5-dimethylpyrrole-2,4-dicarboxylate

Aromatic Heterocyclopentadienes Undergo Electrophilic Aromatic Substitution

As expected for aromatic systems, the 1-hetero-2,4-cyclopentadienes undergo electrophilic substitution. There are two sites of possible attack: at C-2 and at C-3. Which one should be more reactive? An answer may be found by the same procedure used to predict the regioselectivity of electrophilic aromatic substitution of substituted benzenes (Chapter 20): enumeration of all the possible resonance structures for the two modes of reaction.

Consequences of Electrophilic Attack at C-2 and C-3 in the Aromatic Heterocyclopentadienes

Attack at C-2:

X $\xrightarrow{E^+}$ [X(H)(E) $^+$ $\longleftrightarrow$ $^+$ X(H)(E) $\longleftrightarrow$ X^+(H)(E)]

Attack at C-3:

X $\xrightarrow{E^+}$ [X(H)(E) $^+$ $\longleftrightarrow$ X^+(H)(E)]

Both modes benefit from the presence of the resonance-contributing heteroatom, but attack at C-2 leads to an intermediate with an additional resonance structure, thus indicating this position to be the preferred center of substitution. Indeed, such selectivity is generally observed. However, because C-3 also is activated to electrophilic attack, mixtures of products can form, depending on conditions, substrates, and electrophiles.

Electrophilic Aromatic Substitution of Pyrrole, Furan, and Thiophene

Pyrrole $\xrightarrow[-\ CH_3COOH]{CH_3\overset{O}{\overset{\|}{C}}ONO_2,\ -10°C}$ 2-Nitropyrrole (50%) + 3-Nitropyrrole (13%)

Furan $\xrightarrow[-\ HCl]{Cl{-}Cl,\ CH_2Cl_2}$ 2-Chlorofuran (64%)

2-Methylthiophene (H_3C–) $\xrightarrow[-\ HCl]{CH_3\overset{O}{\overset{\|}{C}}Cl,\ SnCl_4}$ 2-Ethanoyl-5-methylthiophene (64%) (H_3C–, –$\overset{O}{\overset{\|}{C}}CH_3$)

The relative nucleophilic reactivity of the three heterocycles decreases in the order pyrrole > furan > thiophene (≫ benzene).

EXERCISE 26-14

Protonation of pyrrole furnishes a species that exhibits five signals in the ^1H NMR spectrum and four peaks in the ^{13}C NMR spectrum. Assign a structure to this species.

EXERCISE 26-15

The monobromination of thiophene-3-carboxylic acid gives only one product. What is its structure and why is it the only product formed?

1-Hetero-2,4-Cyclopentadienes Can Undergo Ring-Opening and Cycloaddition Reactions

Furans are masked γ-dicarbonyl compounds that can be hydrolyzed under mild conditions. The reaction may be viewed as the reverse of the Paal-Knorr-type synthesis of furans. Pyrrole polymerizes under these reaction conditions, whereas thiophene is stable.

Unmasking a Furan by Hydrolysis to a γ-Dicarbonyl Compound

$$\text{2,5-Dimethylfuran} \xrightarrow{CH_3COOH,\ H_2SO_4,\ H_2O,\ \Delta} CH_3\overset{O}{\overset{\|}{C}}CH_2CH_2\overset{O}{\overset{\|}{C}}CH_3$$

90%

2,5-Hexanedione

Mechanism:

$$H_3C\text{-furan-}CH_3 + H^+ \longrightarrow H_3C\text{-(protonated furan)-}CH_3 \xrightarrow[-\ H^+]{H\ddot{O}H}$$

$$H_3C\text{-(2-hydroxy-dihydrofuran)-}CH_3 \xrightarrow{H^+} H_3C\text{-(protonated)-}CH_3 \longrightarrow$$

$$H_3C\text{-}\overset{+}{O}H\text{=C}\cdots\text{C=O-}CH_3 \xrightarrow{-\ H^+} H_3C\overset{:O:}{\overset{\|}{C}}CH_2CH_2\overset{:O:}{\overset{\|}{C}}CH_3$$

Pyrrole undergoes similar ring-opening in the presence of hydroxylamine hydrochloride to give the dioxime (Section 15-6) of butanedial.

$$\text{Pyrrole} + H_3\overset{+}{N}OHCl^- \xrightarrow[-\ NH_3]{CH_3CH_2OH,\ Na_2CO_3,\ \Delta} HON{=}CHCH_2CH_2CH{=}NOH$$

80%

Butanedial dioxime

Raney-nickel desulfurization (Section 15-5) of thiophene derivatives results in sulfur-free acyclic saturated compounds.

$$\text{2-thienyl-}CH(OCH_2CH_3)_2 \xrightarrow[-\ NiS]{\text{Raney Ni},\ (CH_3CH_2)_2O,\ \Delta} CH_3(CH_2)_3CH(OCH_2CH_3)_2$$

50%

The sulfur in thiophene is subject to oxidation by peroxycarboxylic acids to give the corresponding highly reactive intermediates thiophene sulfoxide and sulfone, which undergo Diels-Alder addition to each other, although in low yield.

30% H_2O_2, CH_3COOH, 23°C, 7 days

Thiophene sulfoxide **Thiophene sulfone** (Not isolable)

15%

Diels-Alder cycloadditions are undergone also by the other aromatic heterocycles, an indication of the diene character of the π system.

$(CH_3CH_2)_2O$, 25°C

90°C

95%
Endo adduct
Kinetic product

90%
Exo adduct
Thermodynamic product

EXERCISE 26-16

Explain the following result.

CO_2CH_3 C≡C CO_2CH_3, NCO_2CH_3 +

200°C, 1 h

CH_3O_2C CO_2CH_3 N CO_2CH_3

40%
Trimethyl *N*,3,4-pyrroletricarboxylate

There is another product formed in this reaction. What is it?

Indole, a Benzpyrrole

Among the benzannelated derivatives of the 1-hetero-2,4-cyclopentadienes, indole is probably the most important, forming part of many natural products. Indole is related to pyrrole in the same manner that naphthalene is related to benzene. Its electronic makeup is indicated by the various possible resonance structures that can be formulated for the molecule.

Resonance in Indole

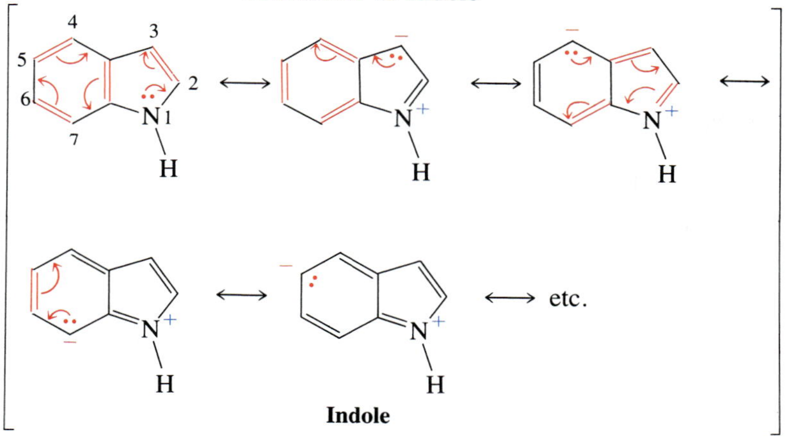

4 3 5 2 6 N1 7 H ⟷ N^+ H ⟷ N^+ H ⟷ N^+ H ⟷ N^+ H ⟷ etc.

Indole

Although those resonance structures that disturb the cyclic six-π-electron system of the fused benzene ring are less important, they indicate the electron-donating effect of the heteroatom.

Indoles are most generally available by the **Fischer* indole synthesis.** In this procedure, an arylhydrazone of an aldehyde or ketone (Section 15-6) is treated with polyphosphoric acid (PPA), or a mineral acid, or a Lewis acid; this causes the extrusion of ammonia with simultaneous ring closure to give the desired heterocycle.

General Fischer Indole Synthesis

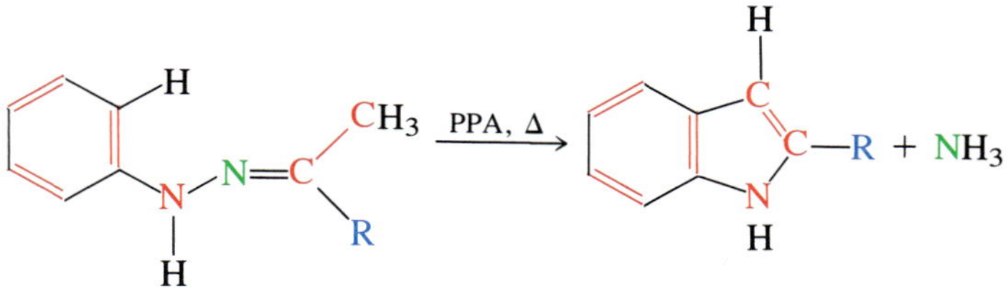

H N N=C CH_3 R H $\xrightarrow{\text{PPA, }\Delta}$ H C C—R N H + NH_3

Examples:

H_3C C N N CH_3 $\xrightarrow[-\,NH_3]{\text{PPA, 100°C}}$ N CH_3

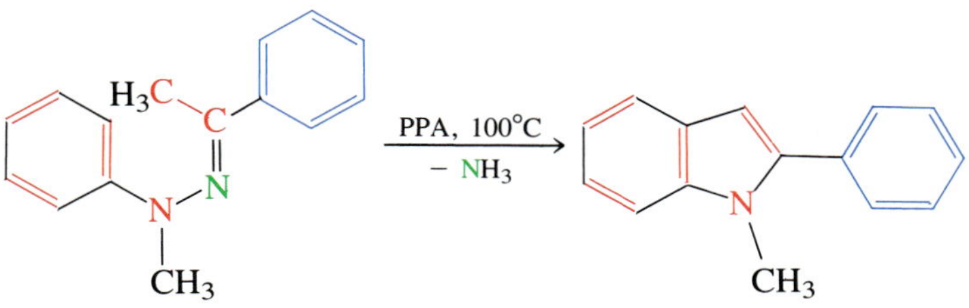

73%

***N*-Methyl-2-phenylindole**

*Professor Emil Fischer, see Section 5-4.

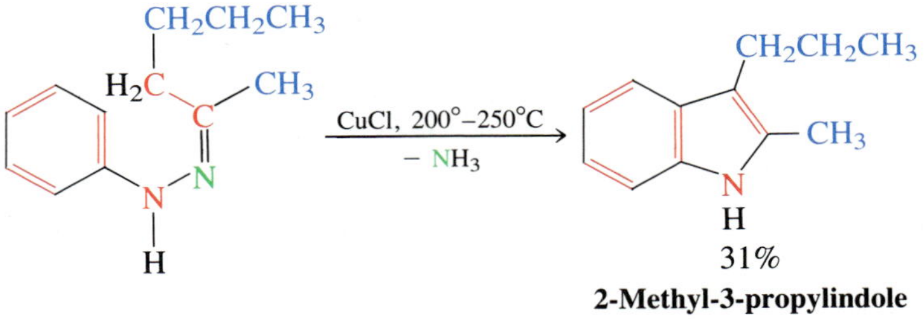

CuCl, 200°–250°C
$-NH_3$

31%

2-Methyl-3-propylindole

The mechanism of the Fischer indole synthesis is thought to begin with acid-catalyzed rearrangement of the arylhydrazone from an imine to an enamine form (Section 15-6).

Mechanism of the Fischer Indole Synthesis:

Imine ⇌ (Catalytic H^+) **Enamine** ⇌ ($+2\,H^+$ / $-2\,H^+$)

"Diaza-Cope" rearrangement

$-H^+$ / $+H^+$

H^+ shift

$-\ddot{N}H_3,\ -H^+$

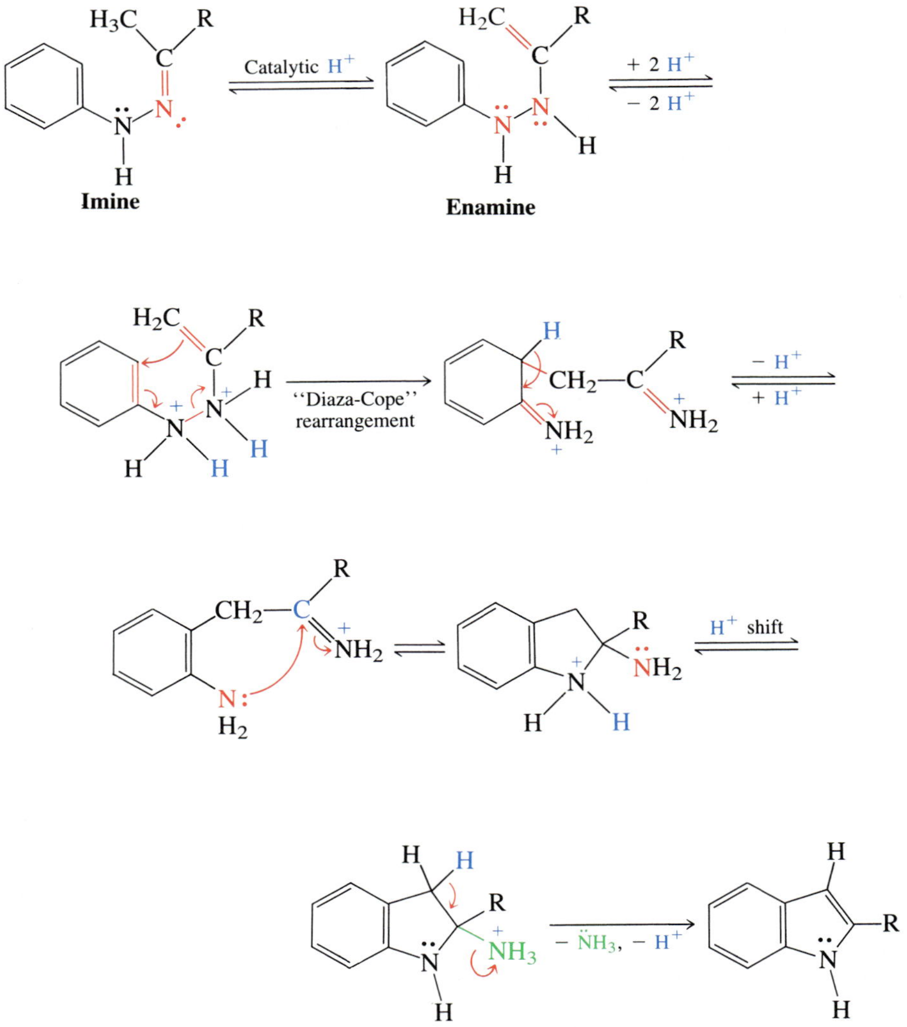

The two basic nitrogen atoms are probably protonated in the strongly acidic medium to form an activated system that undergoes a diaza analog of the Cope rearrangement (Section 24-4) incorporating one of the aromatic π bonds and the double bond of the enamine. Subsequent deprotonation furnishes a benzenamine (aniline) capable of intramolecular nucleophilic attack on the appended iminium group. This ring closure is followed by loss of ammonia and a proton to give the aromatic indole nucleus.

EXERCISE 26-17

Give the products of the treatment of the following arylhydrazones with acid: (a) 2-methylcyclohexanone phenylhydrazone; (b) 1-phenyl-2-propanone phenylhydrazone (two products); (c) 2-oxopropanoic (pyruvic) acid phenylhydrazone.

EXERCISE 26-18

Explain the following result:

H_3C H $CH(CH_3)_2$ N N H $\xrightarrow[2.\ H^+,\ H_2O]{1.\ ZnCl_2}$ H_3C CH_3 CH_3 N H

EXERCISE 26-19

Predict the preferred site of electrophilic aromatic substitution in indole. Explain your choice.

In summary, pyrrole, furan, and thiophene contain delocalized aromatic π systems analogous to that of the cyclopentadienyl anion. A general method for the preparation of 1-hetero-2,4-cyclopentadienes is based on the cyclization of enolizable 1,4-dicarbonyl compounds. The donation of the lone electron pair on the heteroatom to the diene unit makes the carbon atoms in these systems electron rich and therefore more susceptible to electrophilic aromatic substitution than those in benzene. Electrophilic attack is frequently favored at C-2, but substitution at C-3 is also observed, depending on conditions, substrates, and electrophiles. Some rings can be opened by hydrolysis or by desulfurization (for thiophenes). The diene unit can be reactive enough to undergo Diels-Alder cycloadditions. Indole is a benzpyrrole containing a delocalized π system. Indoles are made by treatment of arylhydrazones with acid, which ultimately leads to ring closure and extrusion of one molecule of ammonia (Fischer indole synthesis).

26-5
Pyridine, an Azabenzene

Conceptual replacement of a CH unit in benzene by an sp^2-hybridized nitrogen atom gives pyridine. Pyridine is aromatic, as judged by its physical and chemical properties. This section describes the electronic structure of pyridine, the preparation of some of its derivatives, and its electrophilic and nucleophilic substitution reactions. Finally, as an exercise in pyridine chemistry, the synthesis of nicotine is described.

FIGURE 26-2
Molecular-orbital picture of pyridine.

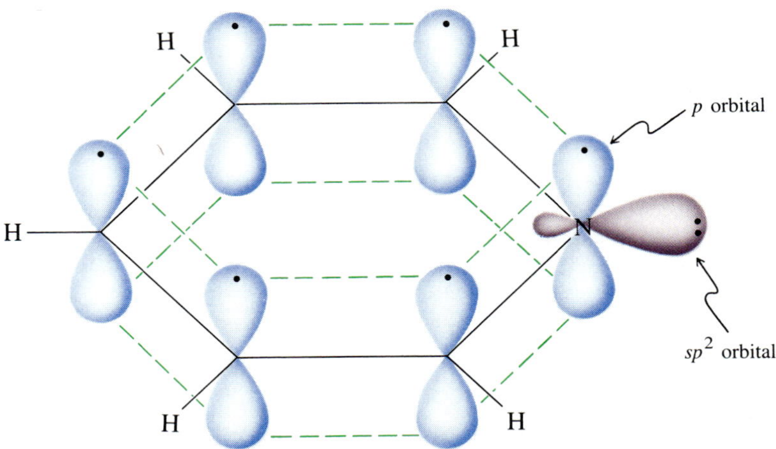

Pyridine Is a Cyclic Aromatic Imine

Pyridine contains an sp^2-hybridized nitrogen atom as in an imine (Section 15-6). In contrast with pyrrole, there is only one electron in the *p* orbital that completes the aromatic π-electron arrangement of the aromatic ring; as in the phenyl anion, the lone electron pair is located in one of the sp^2 hybrid atomic orbitals in the molecular plane (Figure 26-2). Therefore, in pyridine, the heteroatom does not donate excess electron density to the remainder of the molecule. Quite the contrary: because nitrogen is more electronegative than carbon (Table 1-4), it withdraws electron density from the ring, both inductively and by resonance.

Resonance in Pyridine

EXERCISE 26-20

The dipole moment in azacyclohexane (piperidine) is 1.57 D. In which direction does the dipole vector point? Answer the same question for pyridine. Do you expect the dipole moment in pyridine to be larger or smaller than that in azacyclohexane (piperidine)? Explain your answer.

Aromatic delocalization in pyridine is signaled by the ^1H NMR spectrum, which reveals the presence of a ring current. The electron-withdrawing capability of the nitrogen is manifest in larger chemical shifts (more deshielding) at C-2 and C-4, as expected from the resonance picture.

^{1}H NMR Chemical Shifts (ppm) in Pyridine and Benzene

H 7.27

7.46 H, H 7.06, H 8.50, N

Because the lone electron pair on nitrogen is not tied up by conjugation (as in pyrrole, Exercise 26-14), pyridine acts as a weak base. (It is used as such in numerous organic transformations.) Compared with alkanamines (pK_a of ammonium salts ~ 10), the pyridinium ion has a low pK_a, because the nitrogen is sp^2 and not sp^3 hybridized (review Section 11-2 for the effect of hybridization on acidity).

Pyridine is the smallest member of the azabenzene family. Some of its higher aza-analogs are shown here. They behave similarly to pyridine, but show the increasing effect of aza-substitution—in particular, increasing electron deficiency.

Pyridine Is a Weak Base

N + H^+ ⇌ N$^+$–H

Pyridinium ion
pK_a = 5.29

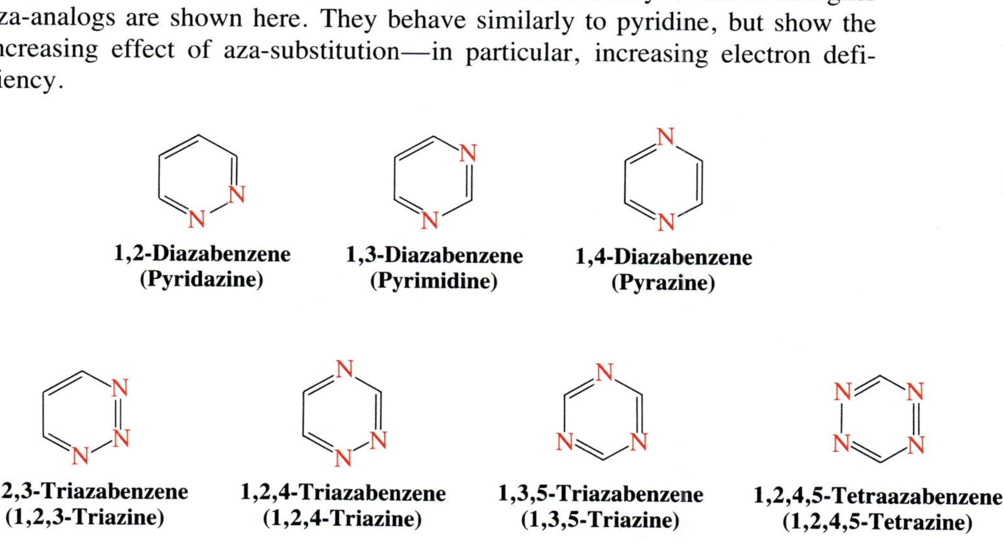

1,2-Diazabenzene (Pyridazine) **1,3-Diazabenzene (Pyrimidine)** **1,4-Diazabenzene (Pyrazine)**

1,2,3-Triazabenzene (1,2,3-Triazine) **1,2,4-Triazabenzene (1,2,4-Triazine)** **1,3,5-Triazabenzene (1,3,5-Triazine)** **1,2,4,5-Tetraazabenzene (1,2,4,5-Tetrazine)**

Pyridines Are Made by Condensation Reactions

Pyridine and simple alkylpyridines are obtained from coal tar. Many of the more-highly-substituted pyridines are in turn made by both electrophilic and nucleophilic substitution of the simpler derivatives.

Pyridines can be made by condensation reactions of acyclic starting materials such as carbonyl compounds with ammonia. The most general of these methods is the **Hantzsch* pyridine synthesis.** In this reaction, two molecules of a β-dicarbonyl compound, an aldehyde, and ammonia combine in several steps to give a substituted dihydropyridine, which is readily oxidized by nitric acid to the aromatic system.

*Professor Arthur Hantzsch, 1857–1935, University of Leipzig, Germany.

Hantzsch Synthesis of 2,6-Dimethylpyridine

$$CH_2{=}O + 2\ CH_3COCH_2CO_2CH_2CH_3 + NH_3 \xrightarrow{-3\ H_2O}$$

Diethyl 1,4-dihydro-2,6-dimethyl-3,5-pyridinedicarboxylate (89%) $\xrightarrow{HNO_3,\ H_2SO_4}$

Diethyl 2,6-dimethyl-3,5-pyridinedicarboxylate (65%) $\xrightarrow[-\ 2\ CH_3CH_2OH,\ -\ 2\ CaCO_3]{1.\ KOH,\ H_2O\quad 2.\ CaO,\ \Delta}$ 2,6-Dimethylpyridine (65%)

89%
Diethyl 1,4-dihydro-2,6-dimethyl-3,5-pyridinedicarboxylate

65%
Diethyl 2,6-dimethyl-3,5-pyridinedicarboxylate

65%
2,6-Dimethylpyridine

If the β-dicarbonyl compound is a 3-keto ester, the resulting product is a 3,5-pyridinedicarboxylic ester. Hydrolysis followed by pyrolysis of the calcium salt of the acid causes decarboxylation.

In a likely mechanism for the Hantzsch synthesis, the aldehyde [in this example, methanal (formaldehyde)] undergoes a Knoevenagel condensation (Section 22-5) with the 3-keto ester. At the same time, the ester forms its enamine in equilibrium concentrations (Section 15-6) by condensation with ammonia. The enamine then functions as a Michael donor (Section 16-5) to the activated Knoevenagel condensation product.

Mechanism of the Hantzsch Pyridine Synthesis:

STEP 1: Knoevenagel condensation of the aldehyde with the 3-keto ester

$$H_2C{=}O + CH_3\overset{O}{\overset{\|}{C}}CH_2CO_2CH_2CH_3 \longrightarrow H_2C{=}C(CO_2CH_2CH_3)(\overset{O}{\overset{\|}{C}}CH_3) + H_2O$$

STEP 2: Enamine formation of the 3-keto ester with ammonia

$$CH_3\overset{O}{\overset{\|}{C}}CH_2CO_2CH_2CH_3 + NH_3 \rightleftharpoons CH_3\overset{NH_2}{\overset{|}{C}}{=}CHCO_2CH_2CH_3 + H_2O$$

Enamine

STEP 3: Michael addition of the enamine to the Knoevenagel condensation product and proton shift

$CH_3CH_2O_2C$ H H_2C $CO_2CH_2CH_3$ ⟶ H_3C NH_2 :O CH_3

$CH_3CH_2O_2C$ H $CO_2CH_2CH_3$ H_3C $\overset{+}{N}H_2$ $:O^-$ CH_3 $\xrightarrow{H^+ \text{ shift}}$ $CH_3CH_2O_2C$ H $CO_2CH_2CH_3$ H_3C NH_2 O CH_3

Keto enamine

STEP 4: Intramolecular condensation of the keto enamine and rearrangement

$CH_3CH_2O_2C$ $CO_2CH_2CH_3$ H_3C NH_2 O CH_3

Keto enamine

$\xrightarrow{-H_2O}$ $CH_3CH_2O_2C$ H $CO_2CH_2CH_3$ H_3C N CH_3 ⟶

Diethyl 3,4-dihydro-2,6-dimethyl-3,5-pyridinedicarboxylate

$CH_3CH_2O_2C$ $CO_2CH_2CH_3$ H_3C N H CH_3

Diethyl 1,4-dihydro-2,6-dimethyl-3,5-pyridinedicarboxylate

The resulting dipolar species undergoes a proton shift to form an intermediate keto enamine that is perfectly set up for an intramolecular condensation. The 3,4-dihydropyridine derivative produced in this way is less stable than the corresponding 1,4-dihydro isomer and therefore rearranges to the latter by a simple proton shift.

EXERCISE 26-21

What starting materials would you use in the Hantzsch synthesis of the following pyridines?

(a) $CH_3CH_2O_2C$, NO_2, CN, H_3C, N, CH_3 (b) $C(CH_3)_3$, NC, CN, H_3C, N, CH_3 (c) CH_3, CH_3CH_2, N, CH_2CH_3

Pyridine Undergoes Electrophilic Aromatic Substitution Only Under Extreme Conditions

Because the pyridine ring is electron poor, the system undergoes electrophilic aromatic substitution only with great difficulty, several orders of magnitude more slowly than benzene.

Electrophilic Aromatic Substitution of Pyridine

Pyridine $\xrightarrow[-\ H_2O]{NaNO_3,\ \text{fuming}\ H_2SO_4,\ 300°C}$ 3-Nitropyridine (NO_2), 4.5%

3-Nitropyridine

Pyridine $\xrightarrow[-\ HBr]{Br—Br,\ H_2SO_4,\ SO_3}$ 3-Bromopyridine (Br), 86%

3-Bromopyridine

EXERCISE 26-22

Explain why electrophilic aromatic substitution of pyridine, even though sluggish, is at C-3.

Activating substituents allow for milder reaction conditions or improved yields.

H_3C, N, CH_3 $\xrightarrow[-\ H_2O]{KNO_3,\ \text{fuming}\ H_2SO_4,\ 100°C}$ H_3C, N, CH_3, NO_2; 81%

2,6-Dimethylpyridine → **2,6-Dimethyl-3-nitropyridine**

N, NH_2 $\xrightarrow[-\ HBr]{Br—Br,\ CH_3COOH,\ 20°C}$ Br, N, NH_2; 90%

2-Aminopyridine → **2-Amino-5-bromopyridine**

Pyridine Undergoes Nucleophilic Substitution

Because the pyridine ring is relatively electron deficient, it undergoes nucleophilic substitution much more readily than benzene (Section 20-6). Attack at C-2 and C-4 is preferred because it leads to intermediates in which the negative charge may be placed on nitrogen. To understand nucleophilic substitutions of the pyridine nucleus, it is helpful to regard the system as a cyclic imine. Attack at C-2 is like 1,2-addition to the imine function, and attack at C-4 can be regarded as conjugate addition to an α,β-unsaturated system.

Consequences of Nucleophilic Attack on Pyridine at C-2, C-3, and C-4

Attack at C-2:

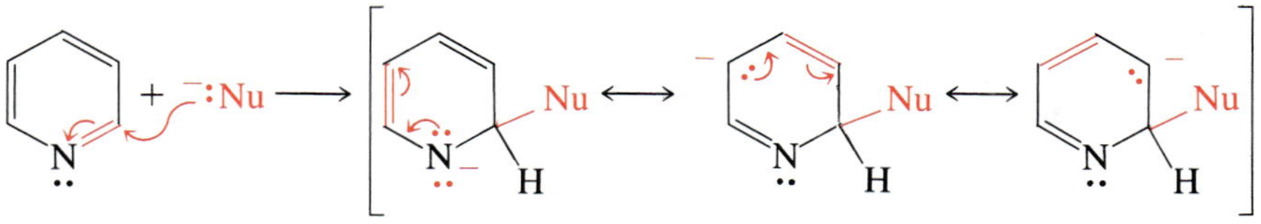

Attack at C-3:

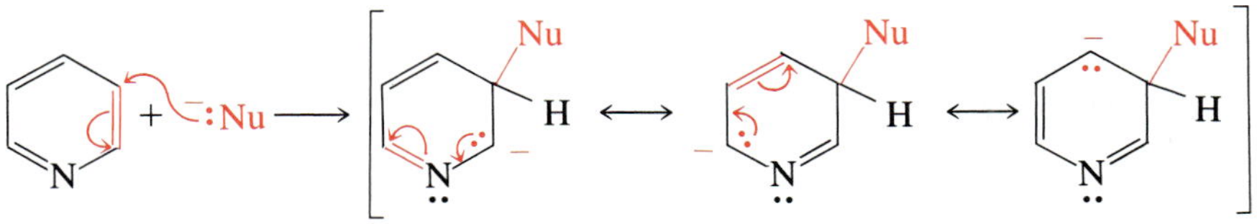

Attack at C-4:

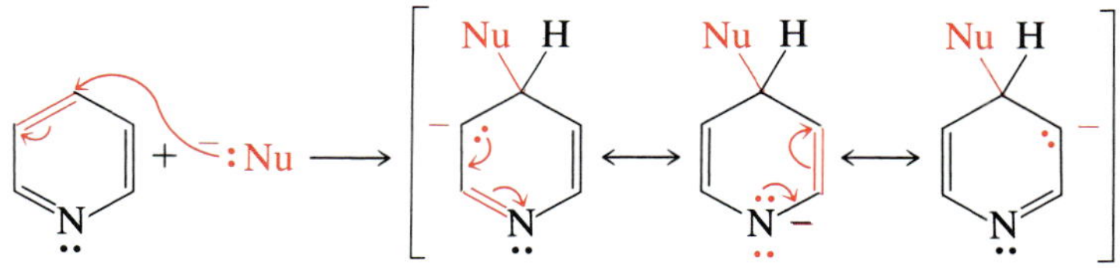

An example of nucleophilic substitution of pyridine is the **Chichibabin* reaction,** in which the heterocycle is converted into 2-aminopyridine by treatment with sodium amide in liquid ammonia. The product of this reaction before aqueous work-up is the resonance-stabilized 2-pyridineamide ion. Note the contrast with electrophilic substitutions, which include deprotonations, not loss of hydride.

The Chichibabin Reaction

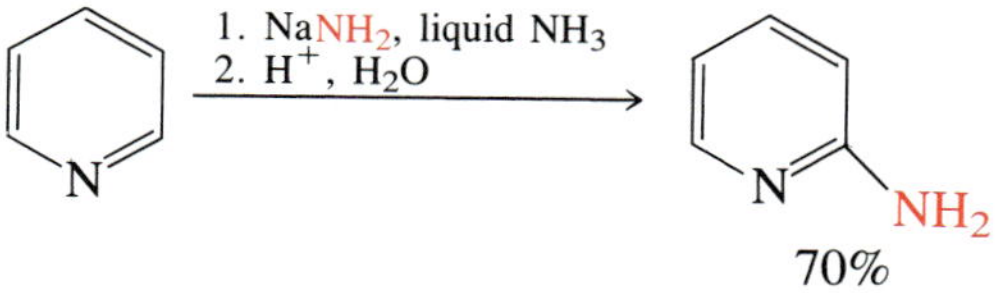

1. NaNH_2, liquid NH_3
2. H^+, H_2O

NH_2
70%
2-Aminopyridine

*Professor Alexei E. Chichibabin, 1871–1945, University of Moscow.

Mechanism:

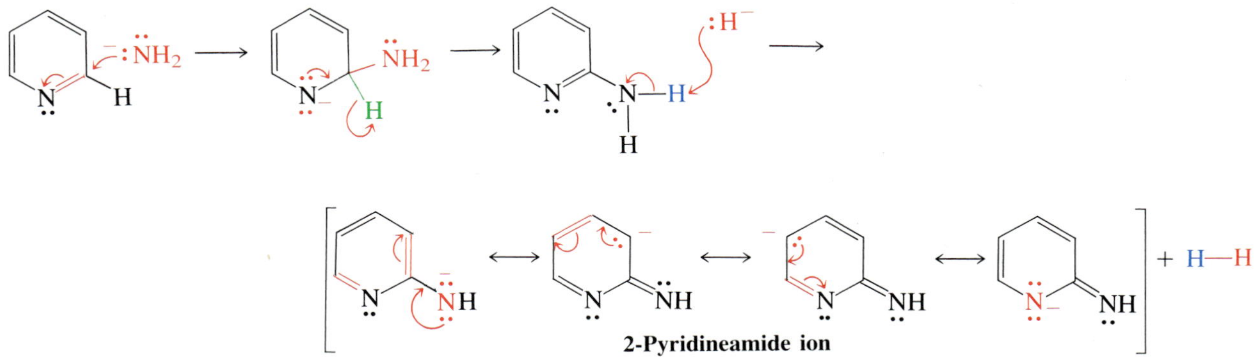

2-Pyridineamide ion

$+ H—H$

Transformations related to the Chichibabin reaction occur on treatment of pyridines with Grignard or organolithium reagents.

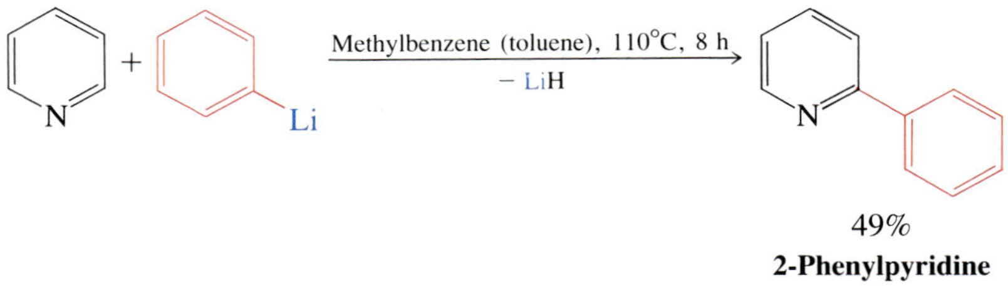

Methylbenzene (toluene), 110°C, 8 h
− LiH

49%
2-Phenylpyridine

In most nucleophilic substitutions of pyridines, halides are leaving groups, the 2- and 4-halopyridines being particularly reactive.

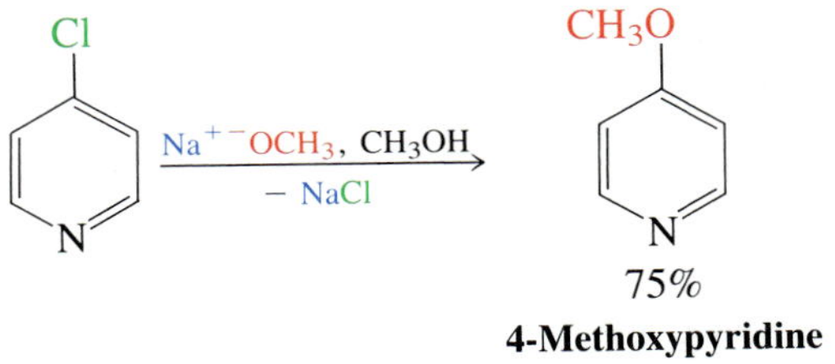

$Na^{+}\ ^{-}OCH_3$, CH_3OH
− NaCl

75%
4-Methoxypyridine

EXERCISE 26-23

The relative rates of the reactions of 2-, 3-, and 4-chloropyridine with sodium methoxide in methanol are 3,000 : 1 : 81,000. Explain.

BOX 26-1

Pyridinium Salts in Nature: Nicotinamide Adenine Dinucleotide

A complex pyridinium derivative, **nicotinamide adenine dinucleotide** (NAD^+), is an important biological oxidizing agent. The structure consists of a pyridine ring [derived from 3-pyridinecarboxylic (nicotinic) acid], two ribose molecules (Section 23-1) linked by a pyrophosphate bridge, and the base adenine (Section 27-1).

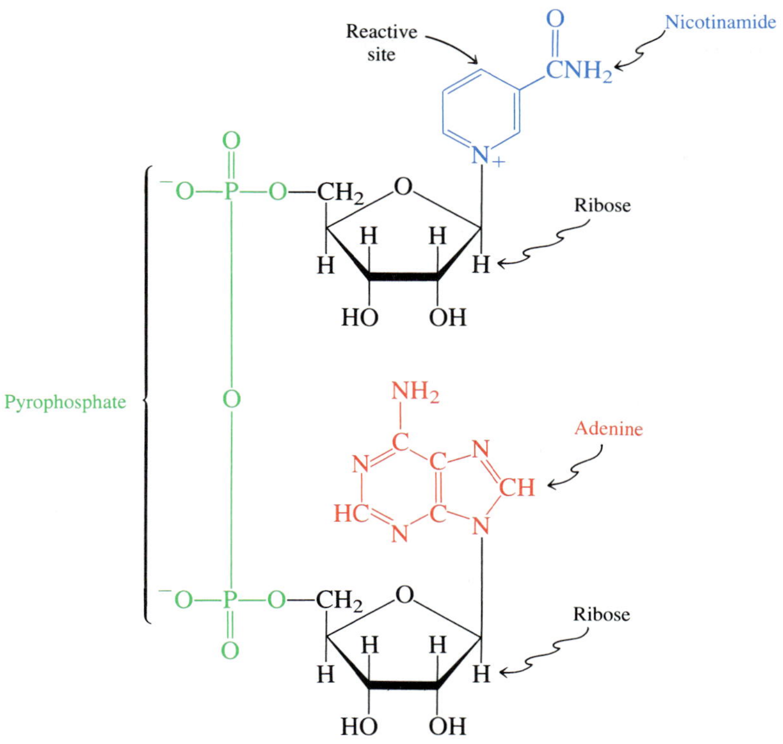

Reactive site
Nicotinamide
Ribose
Pyrophosphate
Adenine
Ribose

Nicotinamide adenine dinucleotide

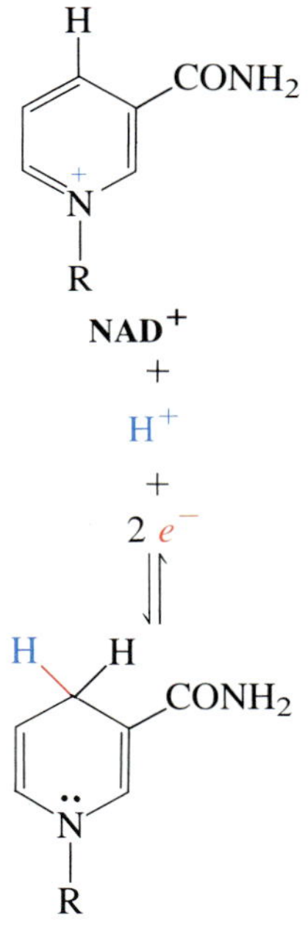

Most organisms derive their energy from the oxidation (removal of electrons) of fuel molecules, such as glucose or fatty acids; the ultimate oxidant (electron acceptor) is oxygen, which gives water. Such biological oxidations proceed through a cascade of electron-transfer reactions requiring the intermediacy of special redox reagents. NAD^+ is one such molecule. In the oxidation of a substrate, the pyridinium ring in NAD^+ undergoes a two-electron reduction with simultaneous protonation.

Reduction of NAD^+

NAD^+ + H^+ + $2e^-$ ⇌ NADH

NAD^+ is the electron acceptor in many biological oxidations of alcohols to aldehydes (including the conversion of vitamin A into retinal, Section 16-4). This reaction can be seen as a transfer of hydride from C-1 of the alcohol to the pyridinium nucleus with simultaneous deprotonation.

Oxidation of Alcohols by NAD^+

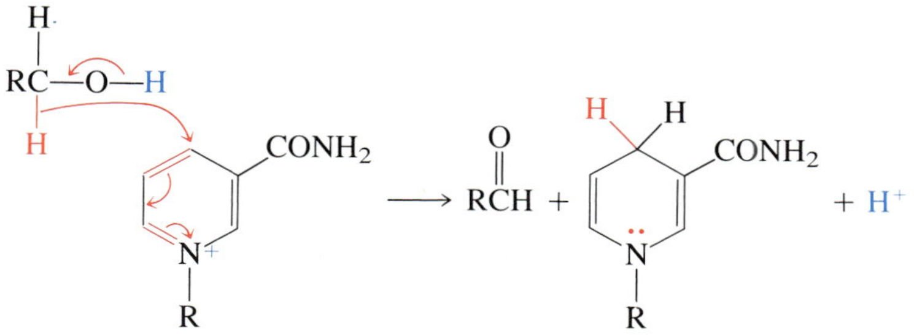

RCH_2OH + NAD^+ ⟶ $RCHO$ + NADH + H^+

The Total Synthesis of Nicotine: An Exercise in Heterocyclic Chemistry

Nicotine is a simple pyridine derivative, whose synthesis demonstrates some of the principles of heterocyclic chemistry. The first total synthesis was reported in 1928; it started from ethyl 3-pyridinecarboxylate (ethyl β-nicotinate). Claisen condensation (Section 22-3) with *N*-methyl-γ-butyrolactam (*N*-methylpyrrolidinone) was followed by acid hydrolysis of the lactam ring, producing a 3-keto carboxylic acid, which underwent decarboxylation as expected (Section 22-4). The resulting ketone was reduced by catalytic hydrogenation, and the alcohol was treated with hydrogen iodide to obtain the corresponding iodide. Subsequent exposure to dilute base caused ring closure to racemic nicotine.

Total Synthesis of Nicotine

Ethyl 3-pyridinecarboxylate (Ethyl β-nicotinate) + ***N*-Methyl-γ-butyrolactam (*N*-Methylpyrrolidinone)** $\xrightarrow[-\ CH_3CH_2OH]{CH_3CH_2O^-Na^+,\ C_6H_6,\ \Delta}$ 70% $\xrightarrow{\text{Conc. HCl, 130°C}}$

$\xrightarrow[-\ CO_2,\ -\ H^+]{}$ 38% **3-[4-(*N*-Methylamino)butanoyl]pyridine** $\xrightarrow{H_2,\ \text{Pd-C},\ H_2O,\ 24\ h}$

$\xrightarrow[-\ HOH]{HI,\ 100°C,\ 4\ h}$ $\xrightarrow[-\ HOH,\ -\ KI]{K_2CO_3,\ KOH,\ H_2O,\ \Delta,\ 24\ h}$

31% from 3-[4-(*N*-methylamino)butanoyl]pyridine
(*R*,*S*)-Nicotine

26-6
Quinoline and Isoquinoline: The Benzpyridines

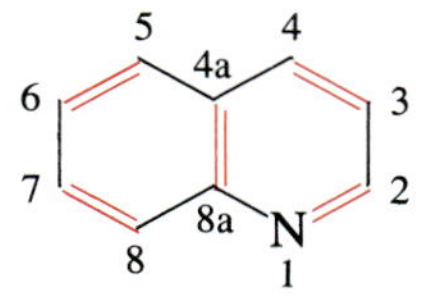

Benzene can be fused to pyridine in two ways to give the two azanaphthalenes, quinoline and isoquinoline.* Both are liquids with high boiling points. Many of their derivatives are found in nature or have been synthesized in search of medicinal activity. Like pyridine, quinoline and isoquinoline are readily available from coal tar. Substituted derivatives may be synthesized by condensation reactions.

5 4 4a 6 3 7 2 8a N 8 1

Quinoline

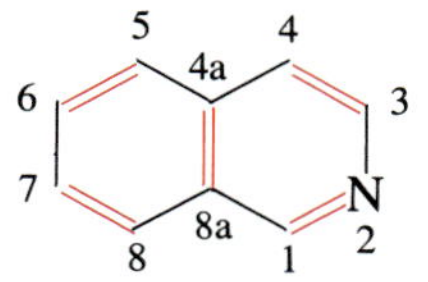

5 4 4a 6 3 7 N 8a 8 1 2

Isoquinoline

Preparation of Quinolines and Isoquinolines

Quinoline derivatives can be made by the **Friedländer† synthesis,** which uses a 2-aminobenzenecarboxaldehyde and an enolizable carbonyl derivative.

Friedländer Synthesis of Quinolines

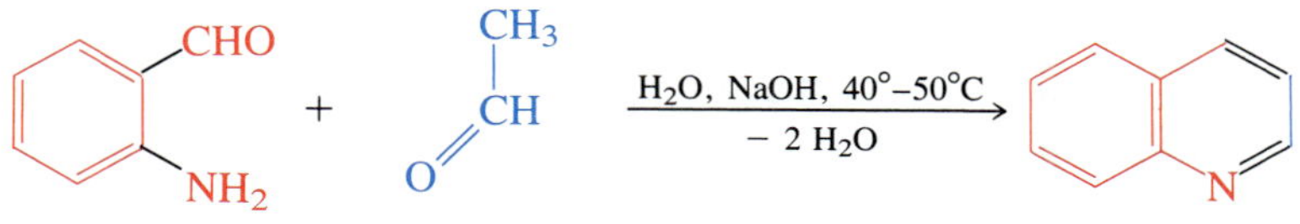

CHO, NH_2 + CH_3–CH=O $\xrightarrow[-\,2\,H_2O]{H_2O,\ NaOH,\ 40°\text{–}50°C}$ N

2-Aminobenzene-carboxaldehyde **Ethanal (Acetaldehyde)** 85% **Quinoline**

H, C=O, CH_3O, NH_2, OCH_3 + $CO_2CH_2CH_3$, CH_2, C=O, $CO_2CH_2CH_3$ $\xrightarrow[-\,2\,H_2O]{NaOH,\ H_2O,\ \Delta}$ $CO_2CH_2CH_3$, CH_3O, N, $CO_2CH_2CH_3$, OCH_3

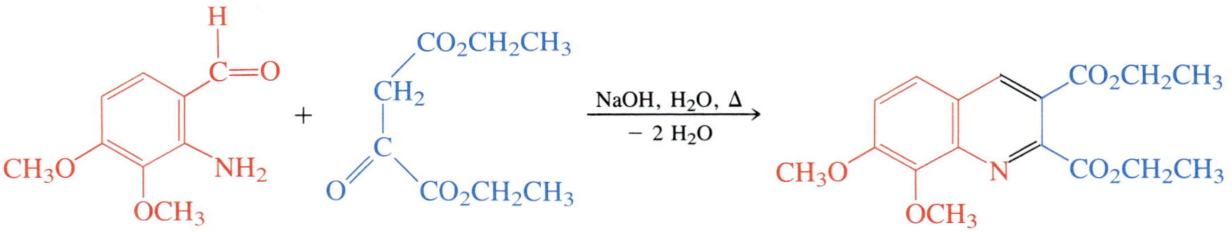

2-Amino-3,4-dimethoxy-benzenecarboxaldehyde **Diethyl 2-oxo-butanedioate** 65% **Diethyl 7,8-dimethoxyquinoline-2,3-dicarboxylate**

EXERCISE 26-24

Formulate a mechanism for the Friedländer synthesis of quinoline.

A source of the isoquinoline nucleus is the **Bischler-Napieralski‡ synthesis,** which incorporates an intramolecular Friedel-Crafts-type cyclization step. In this approach, the amide of a 2-phenylethanamine (β-phenethylamine) is treated with phosphorus pentoxide or a similar dehydrating agent to furnish a

*According to our systematic way of naming heterocycles, these compounds should be called 1-aza- and 2-azanaphthalene.

†Professor Paul Friedländer, 1857–1923, Technical University of Darmstadt, Germany.

‡A. Bischler and B. Napieralski published their synthesis in 1893, University of Zürich.

3,4-dihydroisoquinoline. This product can then be dehydrogenated to the fully aromatic system.

Bischler-Napieralski Synthesis of 1-Phenylisoquinoline

$POCl_3$, P_2O_5, 1,2-dimethylbenzene (*o*-xylene), Δ, 1 h → 100% **1-Phenyl-3,4-dihydroisoquinoline**; Pd, 200°C, $- H_2$ → 90% **1-Phenylisoquinoline**

EXERCISE 26-25

Formulate a mechanism for the Bischler-Napieralski synthesis of 1-phenyl-3,4-dihydroisoquinoline.

Reactions of Quinolines and Isoquinolines: Electrophiles Attack the Benzene Ring, Nucleophiles the Pyridine Ring

As might be expected, because pyridine is electron poor compared with benzene, electrophilic substitutions on quinoline and isoquinoline take place at the benzene ring. As with naphthalene, substitution at the carbons next to the other ring predominates.

H_2SO_4, SO_3, fuming HNO_3, 15°–20°C, 5 h, $- H_2O$ → 35% **5-Nitroquinoline** + 43% **8-Nitroquinoline**

H_2SO_4, HNO_3, 0°C, 30 min, $- H_2O$ → 72% **5-Nitroisoquinoline** + 8% **8-Nitroisoquinoline**

In contrast with electrophiles, nucleophiles prefer reaction at the electron-poor pyridine nucleus. These reactions are quite analogous to those with pyridine.

Chichibabin Reaction of Quinoline and Isoquinoline

1. $BaNH_2$, liquid NH_3, 20°C, 20 days
2. H^+, H_2O

NH_2

80%

2-Aminoquinoline

COOH

1. KNH_2, liquid NH_3
2. CH_3COOH

COOH

NH_2

71%

1-Aminoisoquinoline-4-carboxylic acid

EXERCISE 26-26

Quinoline and isoquinoline react with organometallic reagents exactly as pyridine does (Section 26-5). Give the products of their reaction with 2-propenylmagnesium bromide (allylmagnesium bromide).

The structures in the margin are representative of higher aza analogs of naphthalene.

1,2-Diazanaphthalene (Cinnoline)

2,3-Diazanaphthalene (Phthalazine)

1,3-Diazanaphthalene (Quinazoline)

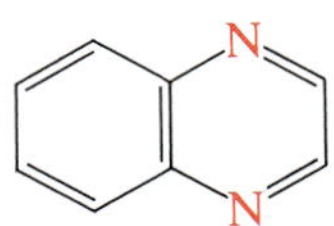

1,4-Diazanaphthalene (Quinoxaline)

1,3,8-Triazanaphthalene (Pyrido[2,3-*d*]pyrimidine)

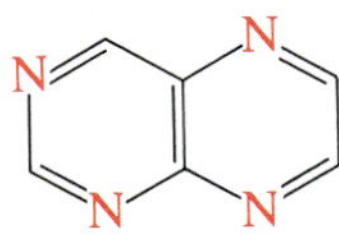

1,3,5,8-Tetraazanaphthalene (Pteridine)

BOX 26-2

Natural Distribution of 1,3,5,8-Tetraazanaphthalene (Pteridine)

The 1,3,5,8-tetraazanaphthalene (pteridine) ring system is present in a number of interesting natural products. Xanthopterin and leucopterin are insect pigments.

OH, OH, H_2N

Xanthopterin
(Yellow butterfly and other insect pigment)

OH, OH, OH, H_2N

Leucopterin
(Colorless substance found in white butterfly wings)

Folic acid (Section 19-6) is a biologically important molecule incorporating a 1,3,5,8-tetraazanaphthalene (pteridine) ring, 4-aminobenzenecarboxylic acid, and (*S*)-2-aminopentanedioic (glutamic) acid (Section 27-1). Mammals have to obtain this substance from their diet.

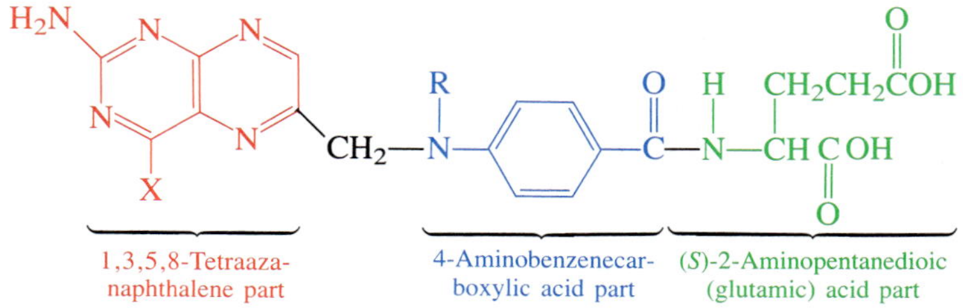

1,3,5,8-Tetraaza-naphthalene part

4-Aminobenzenecarboxylic acid part

(*S*)-2-Aminopentanedioic (glutamic) acid part

Folic acid (X = OH, R = H)
Methotrexate (X = NH_2, R = CH_3)

Tetrahydrofolic acid functions as a biological carrier of one-carbon units. The reactive part of the molecule is at nitrogens N-5 and N-10.

Tetrahydrofolic Acid as a Carrier of One-Carbon Units

Tetrahydrofolic acid

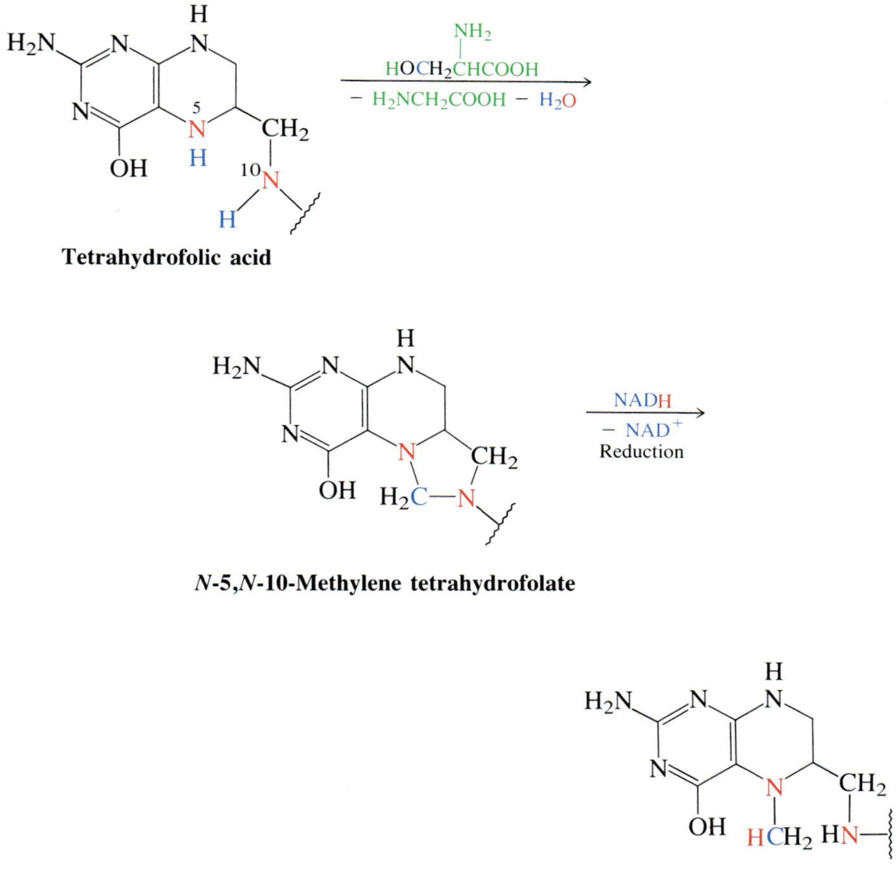

$HOCH_2CH(NH_2)COOH$, $- H_2NCH_2COOH - H_2O$

***N*-5,*N*-10-Methylene tetrahydrofolate**

NADH, $- NAD^+$, Reduction

***N*-5-Methyl tetrahydrofolate**

A derivative of folic acid, methotrexate, is sufficiently similar structurally that it can enter into some of the reactions of folic acid. It also acts as an inhibitor in some of the processes of cell division that are mediated by folic acid. As a result, it is a useful drug in cancer chemotherapy. Because cancer cells divide much more rapidly than normal cells, they are strongly affected by the presence of this compound.

Riboflavin (vitamin B_2) is a benzfused analog of 1,3,5,8-tetraazanaphthalene (pteridine) bearing a ribose unit; it is found in animal and plant tissues.

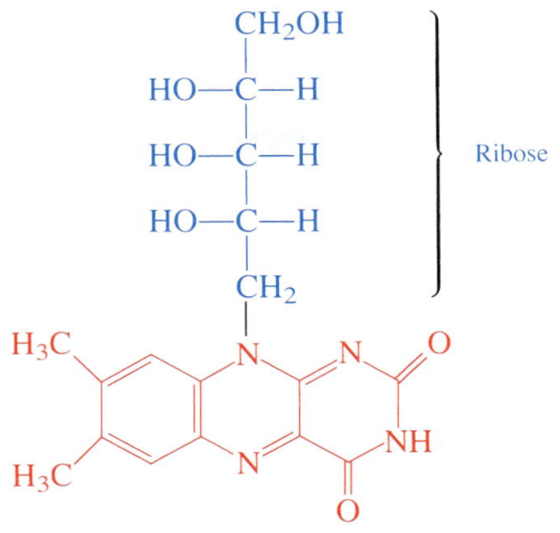

Riboflavin
(Vitamin B_2)

In summary, the azanaphthalenes quinoline and isoquinoline may be regarded as benzpyridines. They are synthesized by condensation reactions (Friedländer and Bischler-Napieralski syntheses) in which the heterocycle is fused onto an existing substituted benzene ring. Electrophiles attack the benzene ring of azanaphthalenes, nucleophiles the pyridine ring.

26-7 Nitrogen Heterocycles in Nature: Alkaloids

The alkaloids are natural nitrogen-containing compounds, found particularly in plants. The name is derived from their basic characteristic properties (alkali-like), which are induced by the lone electron pair of nitrogen. A variety of alkaloids have been described in the earlier sections of this book.

Many alkaloids have potent pharmacological properties. One of the most-potent and most-abused alkaloids is *heroin,* the ethanoyl (acetyl) derivative of morphine (Section 9-8). Morphine and related alkaloids are responsible for the physiological effect of opium poppies. The danger of these drugs is their physiological (as well as psychological) addictiveness.

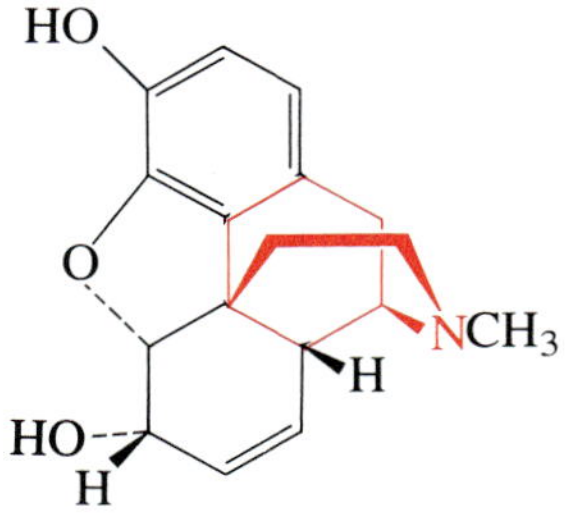

Morphine 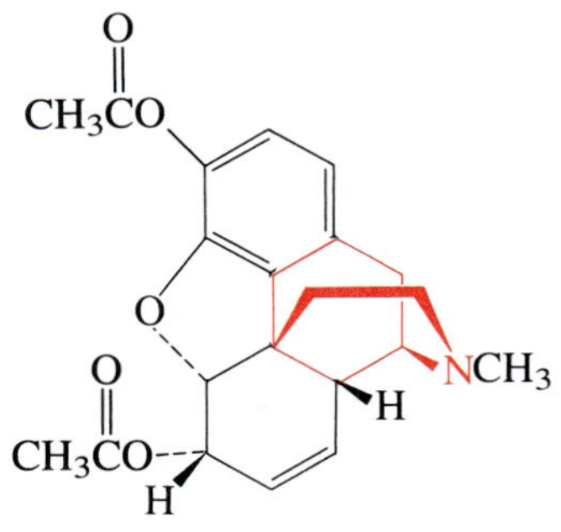 **Heroin** 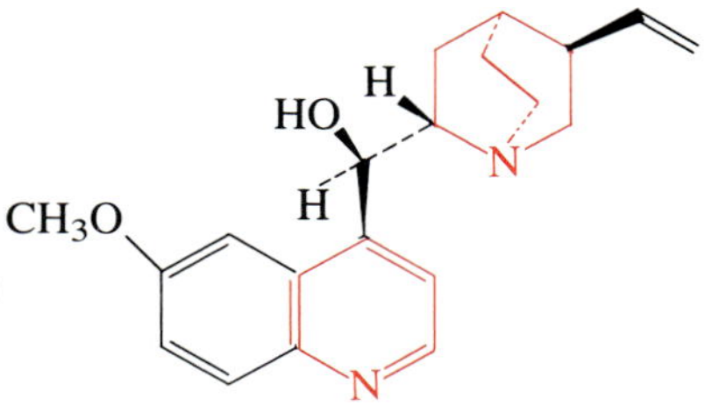 **Quinine**

Quinine, isolated from cinchona bark (as much as 8% concentration) is the oldest-known effective antimalarial agent. The name malaria is derived from the Italian *malo,* bad, and *aria,* air, referring to the old theory that the disease is caused by noxious effluent gases from marsh land. A malaria attack consists of a chill accompanied or followed by a fever, which terminates in a sweating stage. Such attacks may recur regularly. Malaria is caused by a protozoan parasite (*Plasmodium* species) transmitted by the bite of an infected female mosquito of the genus *Anopheles*. It is estimated that as many as 10 million people are affected by the disease, and its incidence is increasing.

Strychnine and *brucine* are powerful poisons (the lethal dose in animals is approximately 5–8 mg kg^{-1}), the lethal ingredients of many a detective novel.

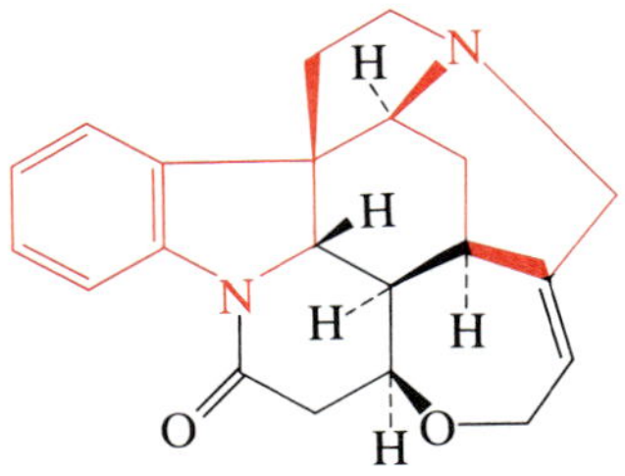

Strychnine 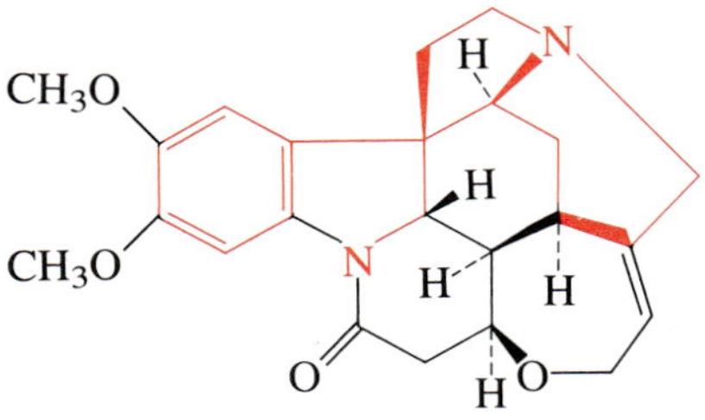 **Brucine**

The isoquinoline and 1,2,3,4-tetrahydroisoquinoline nuclei are abundant among the alkaloids, and their derivatives are physiologically active, for example, as hallucinogens, central-nervous-system agents (depressants and stimulants), and hypotensives.

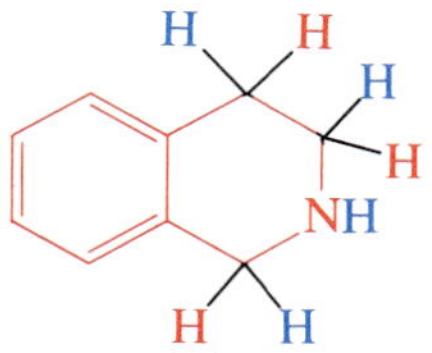

1,2,3,4-Tetrahydro-isoquinoline

BOX 26-3

Tetrahydroisoquinolines and Alcoholism

It has been hypothesized that tetrahydroisoquinolines formed after ingestion of alcohol are responsible for *delirium tremens,* a state of hyperexcitability, tremulousness, hallucinosis, and seizures characteristic of alcoholics. This hypothesis is based on the finding that ethanol in the blood increases the amount of ethanal (acetaldehyde) in the brain, by enzymatic oxidation. Ethanal (acetaldehyde) has been shown to inhibit an enzyme *(aldehyde dehydrogenase)* that is responsible for the oxidation of 3,4-dihydroxyphenylethanal (3,4-dihydroxyphenylacetaldehyde), another brain metabolite, to the corresponding acid. This aldehyde is derived from 4-(2-aminoethyl)-1,2-benzenediol (dopamine), also by enzymatic oxidation. When present in larger than normal concentrations, the aldehyde and the amine may condense to form the biological molecule *N*-norlaudanosoline. It is known that this tetrahydroisoquinoline is a biosynthetic precursor of morphine, which, when generated in the brain in this way, might be responsible for delirium tremens.

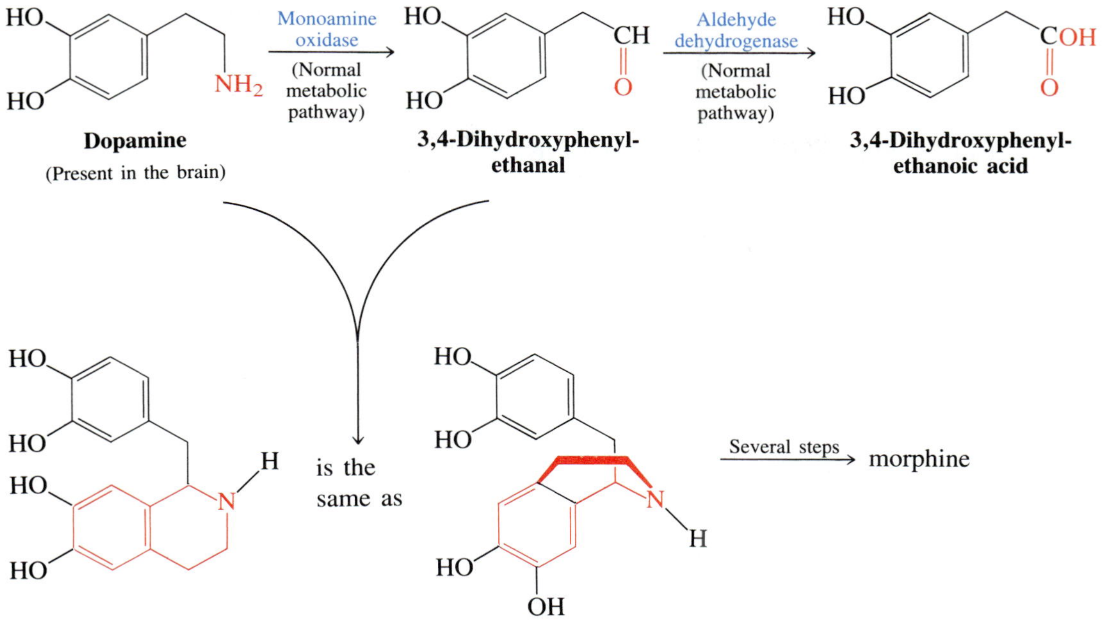

HO HO NH2 Dopamine (Present in the brain) —Monoamine oxidase (Normal metabolic pathway)→ 3,4-Dihydroxyphenylethanal —Aldehyde dehydrogenase (Normal metabolic pathway)→ 3,4-Dihydroxyphenylethanoic acid

Dopamine
(Present in the brain)

3,4-Dihydroxyphenyl-ethanal

3,4-Dihydroxyphenyl-ethanoic acid

is the same as

Several steps → morphine

***N*-Norlaudanosoline**

The following sequence shows how *N*-norlaudanosoline is biologically transformed into morphine.

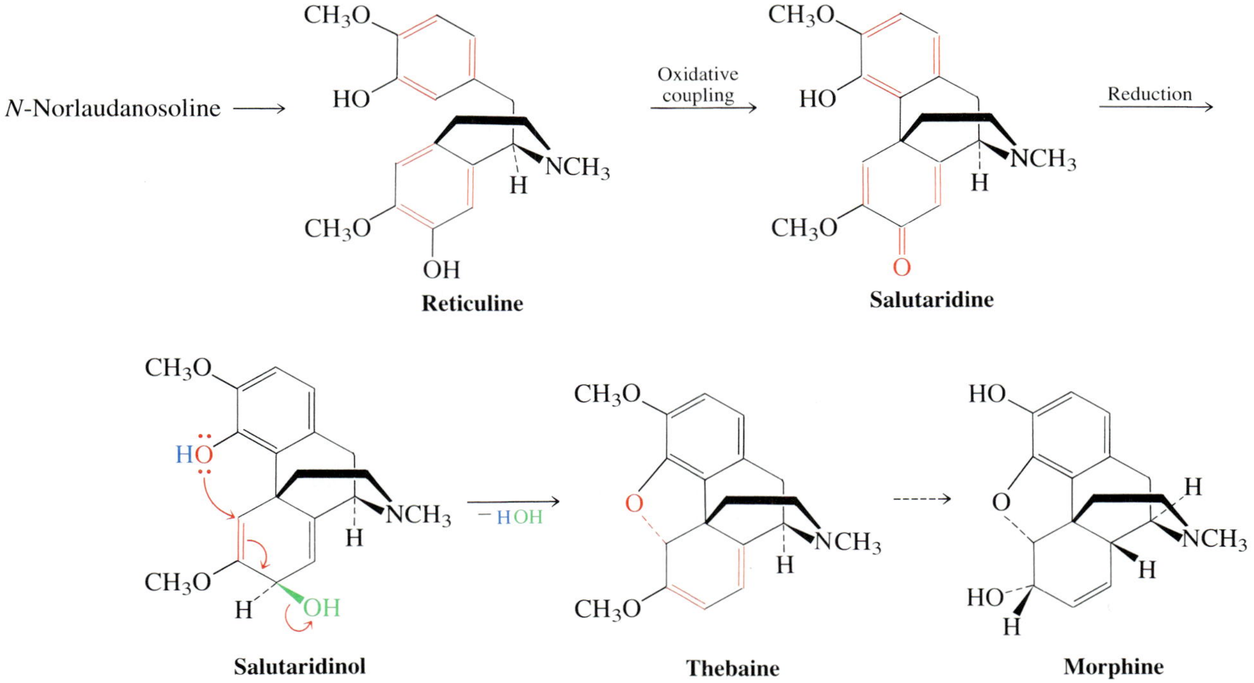

N-Norlaudanosoline ⟶ Reticuline —Oxidative coupling→ Salutaridine —Reduction→

Reticuline

Salutaridine

Salutaridinol —(−HOH)→ Thebaine ⇢ Morphine

Salutaridinol

Thebaine

Morphine

Initial methylation gives reticuline, which undergoes an oxidative coupling reaction *ortho* with respect to one phenolic hydroxy group, *para* with respect to the other, yielding salutaridine. Enzymatic reduction to salutaridinol is followed by formation of an ether bridge to give thebaine, a close precursor to morphine.

In summary, many natural nitrogen-containing compounds are alkaloids, which are physiologically active in various ways.

Summary of New Reactions

1 SYNTHESIS OF AZACYCLOPROPANES (AZIRIDINES)

$$RCH{=}CHR' + N_3COCH_2CH_3 \xrightarrow{h\nu \text{ or } \Delta} \text{(aziridine)}NCOCH_2CH_3 + N_2$$

$$RCH{=}CHR' + :\ddot{N}{=}\overset{+}{N}{-}\overset{-}{\ddot{N}}CO_2CH_2CH_3 \longrightarrow \text{(triazoline, } N{-}CO_2CH_2CH_3) \xrightarrow{h\nu} \text{(aziridine)}NCOCH_2CH_3 + N_2$$

$$RCH{=}CHR' + IN{=}C{=}O \longrightarrow RCH(I){-}CH(R')NCO \xrightarrow{CH_3OH} RCH(I){-}CH(R')HNCOCH_3 \xrightarrow{50\%\ NaOH}$$

$$\text{(aziridine)}NCOCH_3 \xrightarrow{KOH,\ \Delta} \text{(aziridine)}NCOH \longrightarrow \text{(aziridine)}NH + CO_2$$

2 SYNTHESIS OF OXACYCLOPROPANES

$$RR'C{=}CR''R''' + RCOOH \longrightarrow \text{(oxacyclopropane: R, R', R'', R''')} + RCOH$$

From α,β-unsaturated carbonyl compounds:

H_2O_2, HO^-

3 SYNTHESIS OF THIACYCLOPROPANES (THIIRANES)

KSCN

4 REACTIONS OF HETEROCYCLOPROPANES

1. $^-$:Nu or HNu
2. H_2O

5 SYNTHESIS OF HETEROCYCLOBUTANES

$h\nu$

$CH_2{=}O + CH_2{=}C{=}O \xrightarrow{ZnCl_2}$

$ClSO_2N{=}C{=}O$

6 RING-OPENING OF HETEROCYCLOBUTANES

$:\ddot{X}(CH_2)_3Nu$

7 SYNTHESIS OF HETEROCYCLOPENTANES

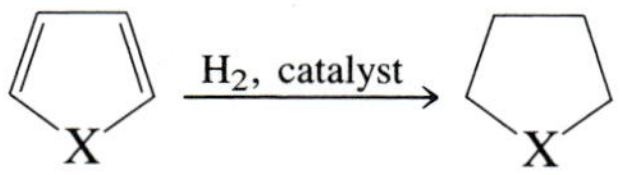

H_2, catalyst

X X

8 PAAL-KNORR SYNTHESIS OF 1-HETERO-2,4-CYCLOPENTADIENES

R R O O P_2O_5, Δ R O R

P_2S_5 $(NH_4)_2CO_3$, Δ

R S R R N H R

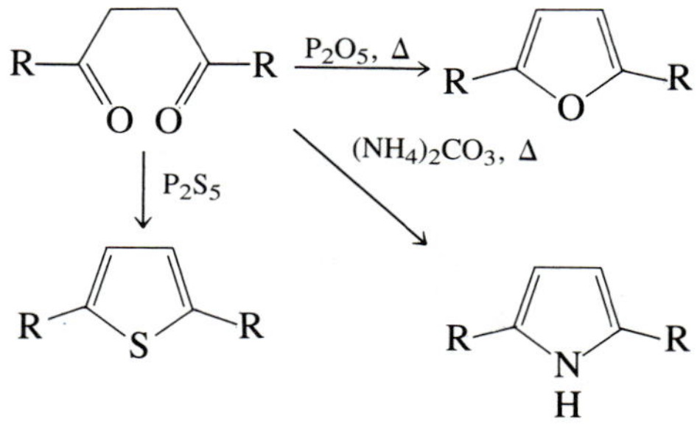

9 REACTIONS OF 1-HETERO-2,4-CYCLOPENTADIENES

Electrophilic substitution:

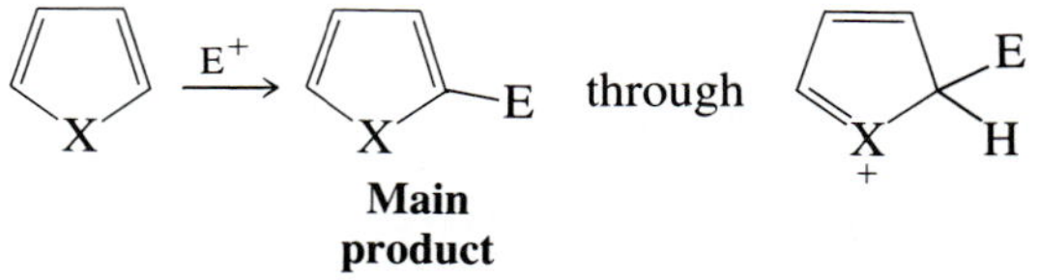

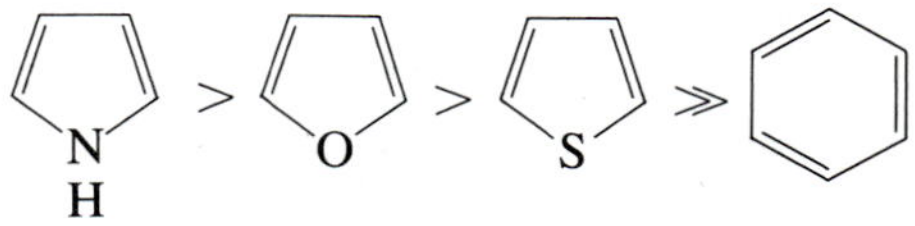

X E^+ X E through X + E H

Main product

N H > O > S ≫

Relative rates

Ring-opening:

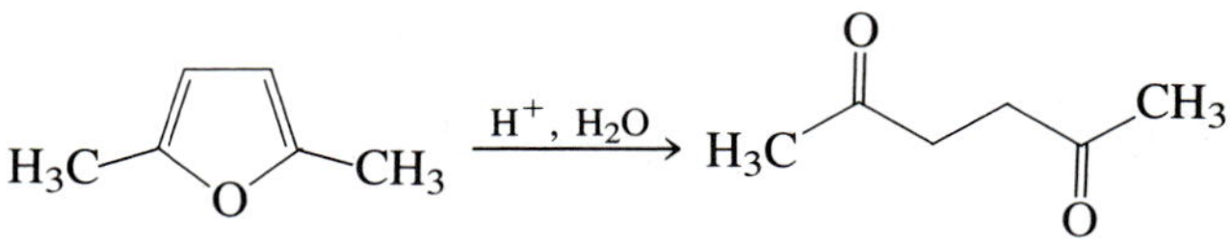

H_3C O CH_3 H^+, H_2O H_3C O O CH_3

Cycloaddition:

S RCO_3H S O + S O O ⟶ O S S O O

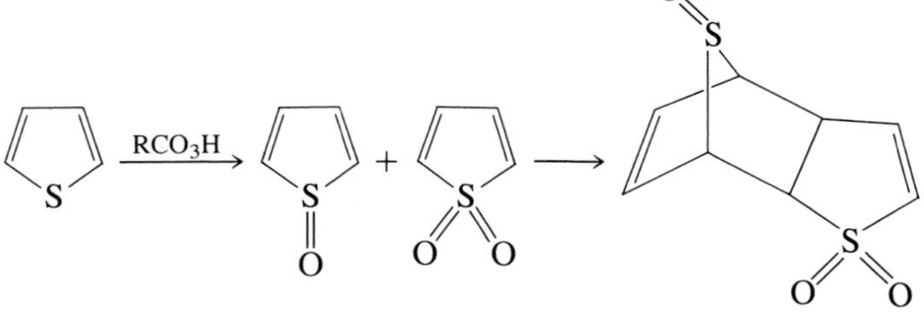

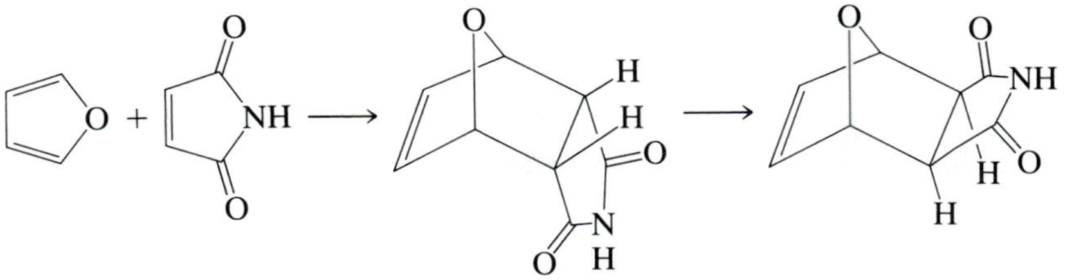

$$\text{Furan} + \text{Maleimide} \longrightarrow \text{Endo adduct} \longrightarrow \text{Exo adduct}$$

Endo adduct
Kinetic product

Exo adduct
Thermodynamic product

10 FISCHER INDOLE SYNTHESIS

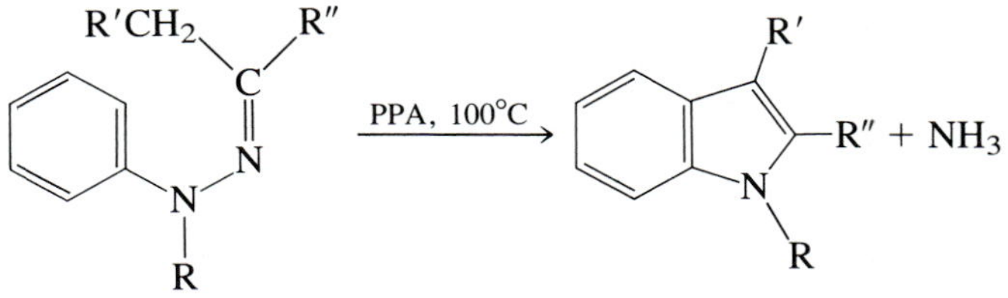

$$\text{PhN(R)N=C(CH_2R')R''} \xrightarrow{\text{PPA, 100°C}} \text{1-R-3-R'-2-R''-indole} + NH_3$$

11 HANTZSCH SYNTHESIS OF PYRIDINES

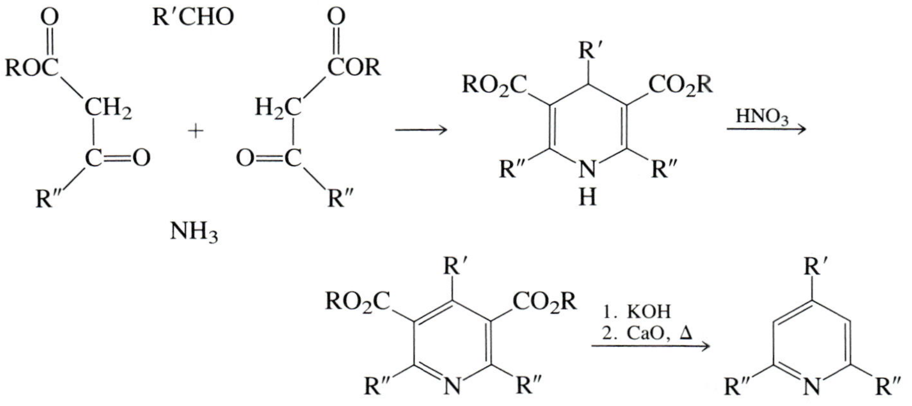

$$2\ ROC(=O)CH_2C(=O)R'' + R'CHO + NH_3 \longrightarrow \text{dialkyl 1,4-dihydro-4-R'-2,6-di-R''-pyridine-3,5-dicarboxylate} \xrightarrow{HNO_3}$$

$$\text{dialkyl 4-R'-2,6-di-R''-pyridine-3,5-dicarboxylate} \xrightarrow[\text{2. CaO, }\Delta]{\text{1. KOH}} \text{4-R'-2,6-di-R''-pyridine}$$

12 REACTIONS OF PYRIDINE

Protonation:

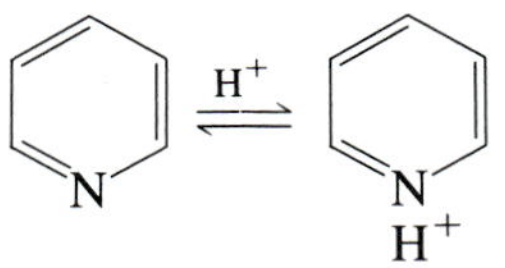

$$\text{Pyridine} \underset{}{\overset{H^+}{\rightleftharpoons}} \text{Pyridinium (N–}H^+\text{)}$$

$pK_a = 5.29$
Pyridinium ion

Electrophilic substitution:

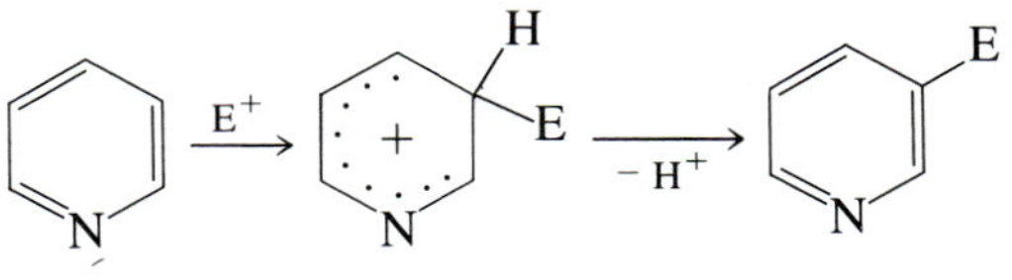

$$\text{Pyridine} \xrightarrow{E^+} \text{cation intermediate (H, E at C-3)} \xrightarrow{-H^+} \text{3-E-pyridine}$$

Nucleophilic substitution:

$\xrightarrow[\text{Chichibabin reaction}]{NaNH_2,\ \text{liquid } NH_3}$ $\xrightarrow{-H^-}$ $\xrightarrow{H^+,\ H_2O}$ N NH$_2$ H NH$_2$

$\xrightarrow[\text{Organometallic reagent}]{RLi}$ N R

X Nu $\xrightarrow{:Nu^-}$ + X^-

Halopyridine

13 FRIEDLÄNDER SYNTHESIS OF QUINOLINES

CHO NH$_2$ + H_2C R C O R′ ⟶ R N R′

14 BISCHLER-NAPIERALSKI SYNTHESIS OF ISOQUINOLINES

R NH O $\xrightarrow[-\ H_2O]{POCl_3,\ P_2O_5,\ \Delta}$ N R $\xrightarrow[-\ H_2]{Pd,\ 200°C}$ N R

15 REACTIONS OF QUINOLINE AND ISOQUINOLINE

Electrophilic substitution:

↓ N ↑ ↓ N ↑

Nucleophilic substitution:

N ↖ N ↑

Summary of Important Concepts

1 The heterocycloalkanes can be named by using cycloalkane nomenclature and indicating the presence of the heteroatom by the prefix aza- for nitrogen, oxa- for oxygen, thia- for sulfur, and so on. Other systematic and common names abound and are often used in the literature, particularly for the aromatic heterocycles.

2 The strained three- and four-membered heterocycloalkanes react easily with nucleophiles to undergo ring-opening.

3 The 1-hetero-2,4-cyclopentadienes are aromatic and have an arrangement of six π electrons similar to that in the cyclopentadienyl anion. The heteroatom is sp^2 hybridized, the p orbital contributing two electrons to the π system. As a consequence, the diene unit is electron rich and reactive in electrophilic aromatic substitutions. The heteroatoms can also undergo their typical reactions, such as nitrosation (pyrroles) and oxidation (thiophenes).

4 Replacement of one (or more) of the CH units in benzene by an sp^2-hybridized nitrogen gives rise to pyridine (and other azabenzenes). The p orbital on the heteroatom contributes one electron to the π system; the lone electron pair is located in an sp^2 hybrid atomic orbital in the molecular plane. Azabenzenes are electron poor, because the electronegative nitrogen withdraws electron density from the ring by induction and by resonance. Electrophilic aromatic substitution of azabenzenes is sluggish. On the other hand, nucleophilic aromatic substitution is facilitated, as shown by the Chichibabin reaction, substitutions by organometallic reagents next to the nitrogen, and the displacement of halide ion from halopyridines by nucleophiles.

5 The azanaphthalenes (benzpyridines) quinoline and isoquinoline contain an electron-poor pyridine ring, susceptible to nucleophilic attack, and a relatively electron rich benzene ring that enters into electrophilic aromatic substitution reactions, usually at the positions closest to the heterocyclic unit.

Problems

1 Name or draw the following compounds.

(a) *cis*-2,3-Diphenyloxacyclopropane

(b) 3-Azacyclobutanone

(c) 1,3-Oxathiacyclopentane

(d) 2-Butanoyl-1,3-dithiacyclohexane

(e) 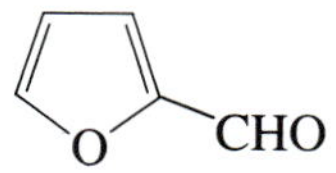 O CHO

(f) 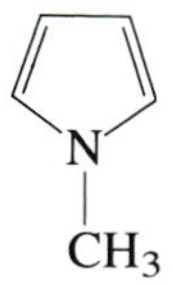 N CH_3

(g) 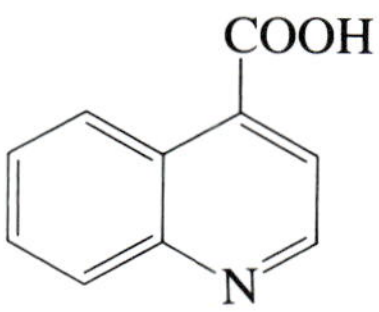 COOH N

(h) 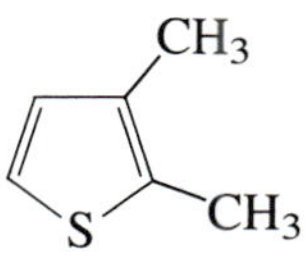 CH_3 S CH_3

2 Write the likely product(s) of each of the following reaction sequences.

(a) 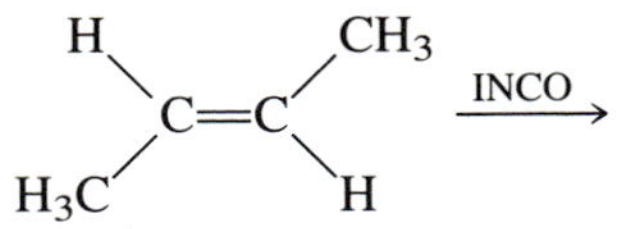 $H(H_3C)C{=}C(CH_3)H$ $\xrightarrow{\text{INCO}}$

(b) Product of part **a** $\xrightarrow[\text{3. KOH, }\Delta]{\text{1. }CH_3OH\text{; 2. NaOH, }H_2O}$

(c) 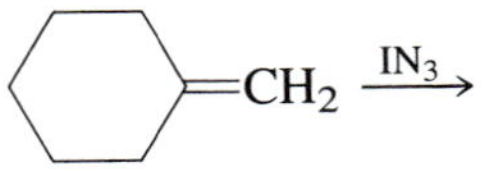 $=CH_2$ $\xrightarrow{IN_3}$ (Hint: Write a Lewis structure for IN_3.)

(d) Product of part **c** $\xrightarrow[\text{2. }H_2O]{\text{1. }LiAlH_4}$ (Hint: Section 21-4.)

3 The following two reactions give steroidal products with the same formula but completely different structures. Draw the structures and explain the different results.

(a) 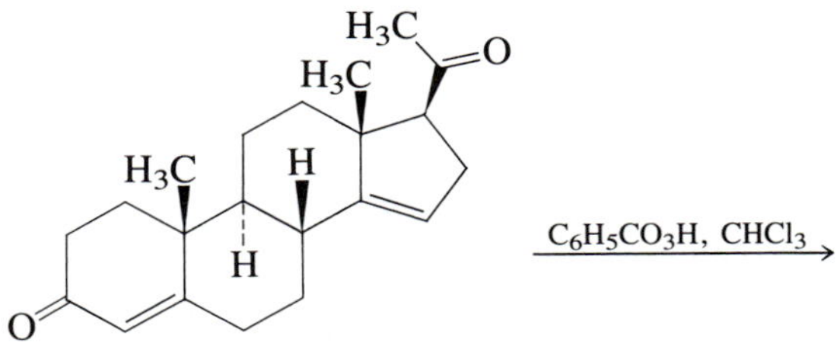 [steroid structure: H₃C, H₃C, H₃C, H, H, =O, O] $\xrightarrow{C_6H_5CO_3H,\ CHCl_3}$

(b) 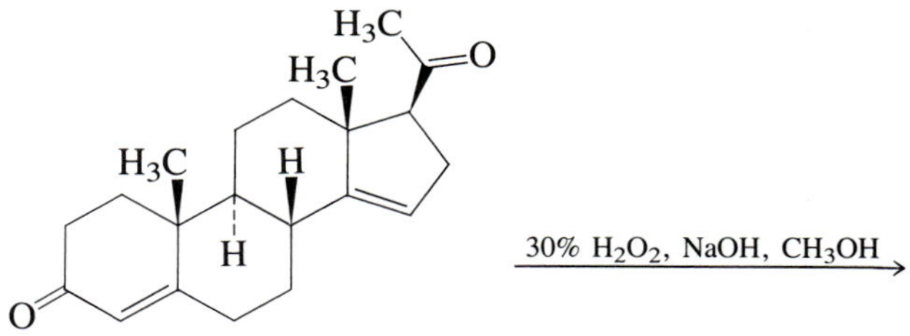 [steroid structure: H₃C, H₃C, H₃C, H, H, =O, O] $\xrightarrow{30\%\ H_2O_2,\ NaOH,\ CH_3OH}$

4 The following two reactions give stereoisomeric products. Write their structures and suggest a reason for the contrasting results. Hint: Make a model.

(a) 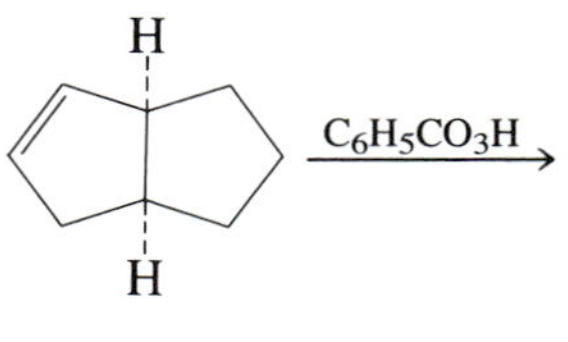 [bicyclic alkene, H, H] $\xrightarrow{C_6H_5CO_3H}$

(b) 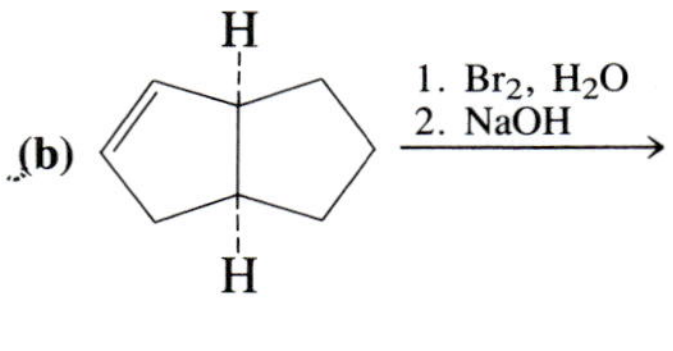 [bicyclic alkene, H, H] $\xrightarrow[\text{2. NaOH}]{\text{1. }Br_2,\ H_2O}$

5 The Darzens condensation is one of the older methods (1904) for the synthesis of three-membered heterocycles. It is most commonly the reaction of a 2-halo ester with a carbonyl derivative in the presence of base. The following examples of the Darzens condensation show how it is applied to the synthesis of oxacyclopropane and azacyclopropane. Suggest a reasonable mechanism for each of these reactions.

(a) $C_6H_5CHO + C_6H_5CHClCOOCH_2CH_3 \xrightarrow{KOC(CH_3)_3,\ (CH_3)_3COH}$ H$_5$C$_6$—C(H)(–O–)C(C$_6$H$_5$)(COOCH$_2$CH$_3$)

(b) $C_6H_5CH{=}NC_6H_5 + ClCH_2COOCH_2CH_3 \xrightarrow{KOC(CH_3)_3,\ CH_3OCH_2CH_2OCH_3}$ C_6H_5CH—(N–C_6H_5)—$CHCOOCH_2CH_3$

6 Propose a synthesis of each of the following heterocyclic compounds. In each case, construct the heterocyclic ring from suitable nonheterocyclic precursors.

(a) CH_3CH—CH_2 (N–CH_3 aziridine)

(c)

(b) NCOOCH$_3$ (H, H, H, H)

(d) O

7 Write the expected product of each of the following reaction sequences.

(a) $\xrightarrow{KSCN,\ CH_3CH_2OH,\ \Delta}$

(R = alkyl chain)

(b) N–CH_3 $\xrightarrow{NaOCH_2CH_3,\ CH_3CH_2OH,\ \Delta}$

(c) H, O, H $\xrightarrow{1.\ LiAlH_4;\ 2.\ H^+,\ H_2O}$

8 Propose a method for the transformation of diol (i), shown in the margin, into oxacyclobutane (ii).

HO, CH$_3$, OH

i

?

CH$_3$, O

ii

9 Intramolecular versions of [2+2]cycloadditions are commonly used to prepare polycyclic structures containing heterocyclobutanes. Under photochem-

ical conditions, 5-hexen-2-one cyclizes to an isomeric product that exhibits the following ^{1}H NMR data. Propose a structure for this product.

$$\underset{\textbf{5-Hexen-2-one}}{CH_3\overset{\overset{\displaystyle O}{\|}}{C}CH_2CH_2CH{=}CH_2} \xrightarrow{h\nu} {}^1\text{H NMR: } \delta = 1.35 \text{ (s, 3 H), } 2.29 \text{ (m, 4 H), } 2.85 \text{ (m, 1 H), } 4.39 \text{ (d of d, 1 H), and } 4.81 \text{ (d of d, 1 H) ppm.}$$

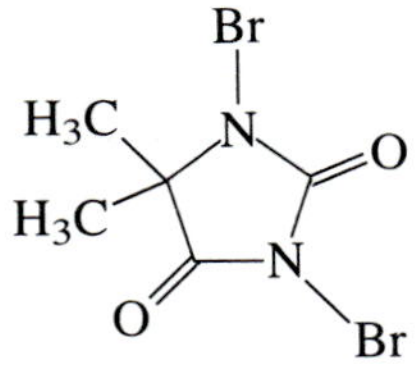

10 **(a)** The compound shown in the margin, with the common name 1,3-dibromo-5,5-dimethylhydantoin, is useful as a source of electrophilic bromine (Br^+) for addition reactions. Give a more-systematic name for this heterocyclic compound.

(b) An even more remarkable heterocyclic compound (B) is prepared by the following reaction sequence. Using the given information, deduce structures for A and B, and name the latter.

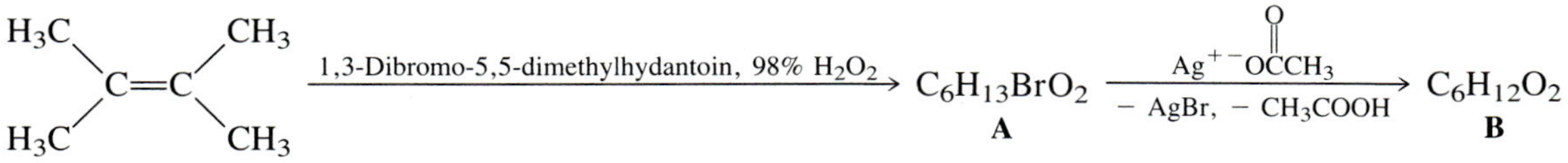

$$(CH_3)_2C{=}C(CH_3)_2 \xrightarrow{\text{1,3-Dibromo-5,5-dimethylhydantoin, 98\% } H_2O_2} \underset{\textbf{A}}{C_6H_{13}BrO_2} \xrightarrow[-\text{ AgBr, } -\text{ } CH_3COOH]{Ag^{+}\,{}^{-}O\overset{\overset{\displaystyle O}{\|}}{C}CH_3} \underset{\textbf{B}}{C_6H_{12}O_2}$$

Heterocycle B is a yellow, crystalline, sweet-smelling compound that decomposes on gentle heating to two molecules of propanone (acetone), one of which is formed directly in its $n \rightarrow \pi^*$ excited state (Sections 14-7 and 15-2). This electronically excited product is chemiluminescent.

$$B \longrightarrow CH_3\overset{\overset{\displaystyle O}{\|}}{C}CH_3 + \left[CH_3\overset{\overset{\displaystyle O}{\|}}{C}CH_3\right]^{n \rightarrow \pi^*} \longrightarrow h\nu + 2\ CH_3\overset{\overset{\displaystyle O}{\|}}{C}CH_3$$

Heterocycles similar to B are responsible for the chemiluminescence produced by a number of species (e.g., fireflies and several deep-sea fish); they also serve as the energy sources in commercial chemiluminescent products (Problem 14 of Chapter 25).

11 The penicillins are a class of antibiotics containing two heterocyclic rings that interfere with the construction of cell walls by bacteria. The interference results from reaction of the penicillin with an amino group of a protein that closes gaps that develop during construction of the cell wall. The insides of the cell leak out, and the organism dies.

(a) Suggest a reasonable product for the reaction of penicillin G with the amino group of a protein (Protein-NH_2). Hint: First identify the most-reactive electrophilic site in the penicillin molecule.

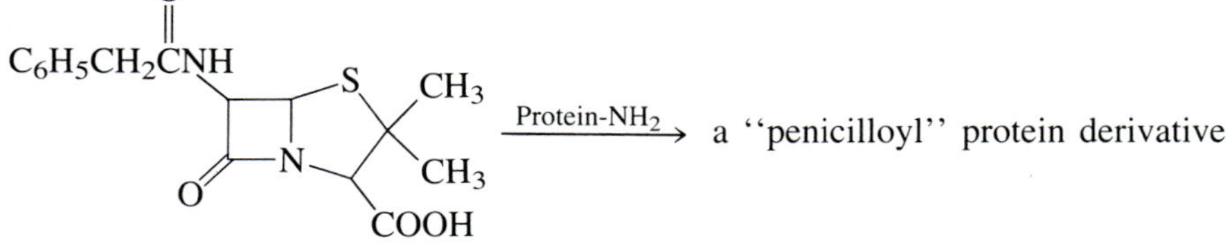

Penicillin G $\xrightarrow{\text{Protein-}NH_2}$ a "penicilloyl" protein derivative

(b) Penicillin-resistant bacteria secrete an enzyme (penicillinase) that catalyzes hydrolysis of the antibiotic faster than the antibiotic can attack the cell-wall proteins. Propose a structure for the product of this hydrolysis reaction, and suggest a reason why hydrolysis destroys the antibiotic properties of penicillin.

$$\text{Penicillin G} \xrightarrow{H_2O,\ \text{penicillinase}} \text{penicilloic acid}$$

(Hydrolysis product; no antibiotic acitivity)

12 Azacyclohexanes (piperidines) can be synthesized by reaction of ammonia with *cross-conjugated dienones:* ketones conjugated on both sides with double bonds. Propose a mechanism for the following synthesis of 2,2,6,6-tetramethylaza-4-cyclohexanone.

$$(CH_3)_2C{=}CH\overset{\overset{\displaystyle O}{\|}}{C}CH{=}C(CH_3)_2 \xrightarrow{NH_3}$$

(2,2,6,6-tetramethyl-4-piperidinone ring: O, H_3C, CH_3, H_3C, CH_3, N–H)

13 Compound C, C_8H_8O, exhibits ^1H NMR spectrum C. On treatment with concentrated aqueous HCl, it is converted almost instantaneously into a compound that exhibits spectrum D (on the next page). Approximate integrated intensities of the signals in spectrum D are as follows: $\delta = 7.1$–7.4 (5 H), 4.8 (1 H), 4.2 (2 H), and 3.8 (2 H) ppm. What is compound C, and what is the product of its treatment with aqueous acid?

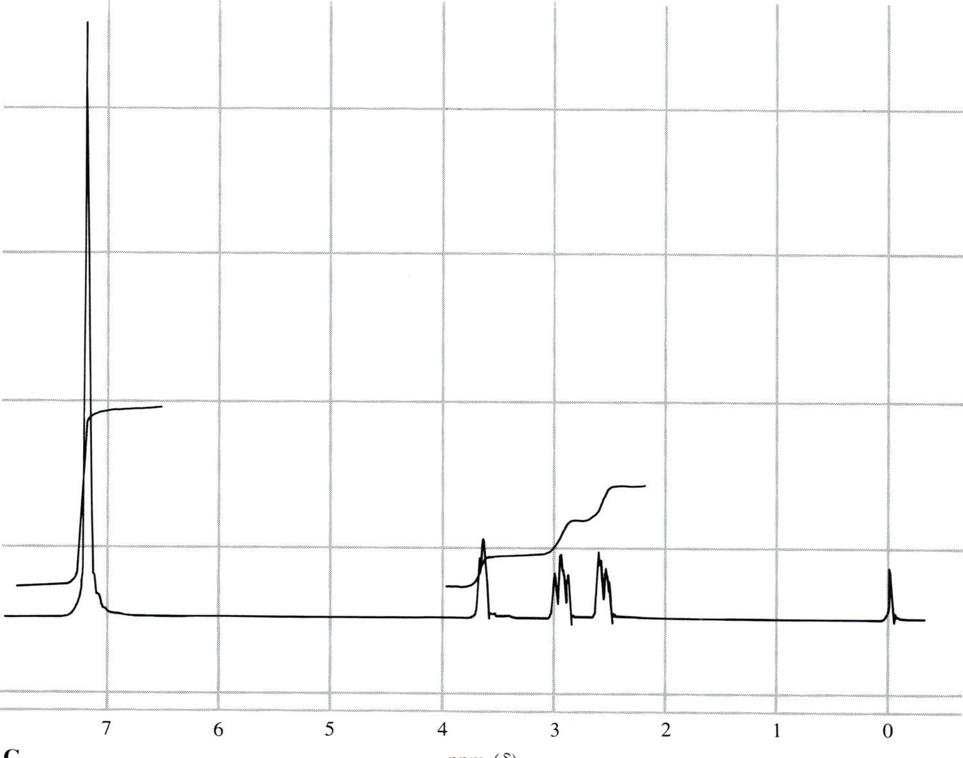

C

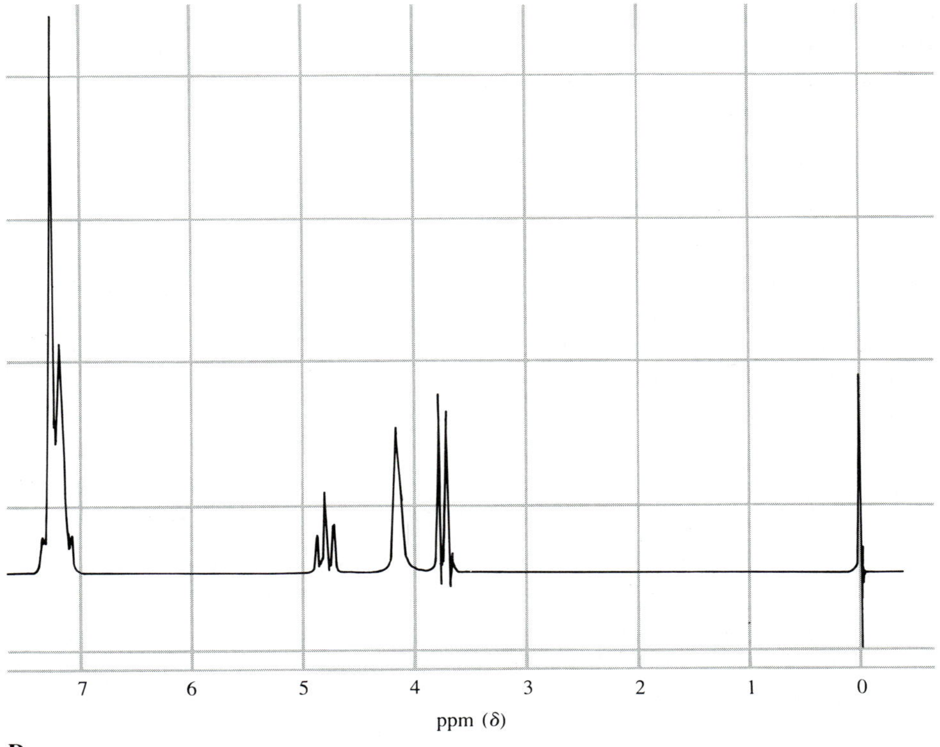

D

14 The following heterocyclopentadienes contain more than one heteroatom. For each one, identify the orbitals occupied by all lone electron pairs on the heteroatoms, and determine whether the molecule qualifies as aromatic.

Pyrazole **Imidazole** **Thiazole** **Isoxazole**

Would you expect any of these heterocycles to be a stronger base than pyrrole?

15 Write the product of each of the following reactions.

(a) $\xrightarrow{CH_3NH_2}$

(b) $\xrightarrow{P_2O_5,\ \Delta}$

16 1-Hetero-2,4-cyclopentadienes can be prepared by condensation of an α-dicarbonyl compound and certain heteroatom-containing diesters. Propose a mechanism for the following pyrrole synthesis.

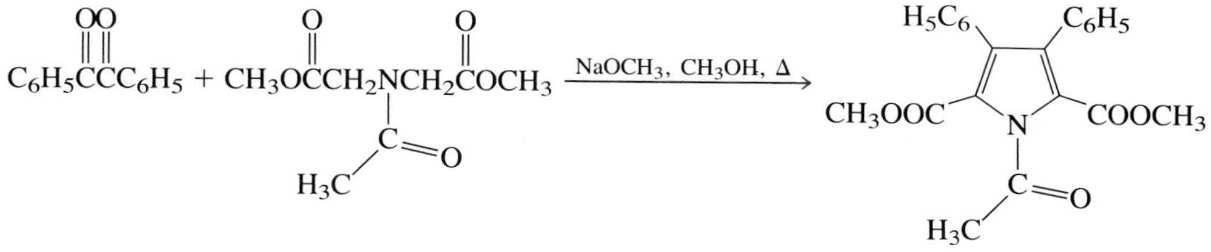

$$C_6H_5\overset{O}{\overset{\|}{C}}\overset{O}{\overset{\|}{C}}C_6H_5 + CH_3O\overset{O}{\overset{\|}{C}}CH_2\underset{\underset{H_3C}{C=O}}{N}CH_2\overset{O}{\overset{\|}{C}}OCH_3 \xrightarrow{NaOCH_3,\ CH_3OH,\ \Delta}$$

H_5C_6 C_6H_5 CH_3OOC N $COOCH_3$ C=O H_3C

How would you use a similar approach to synthesize 2,5-thiophenedicarboxylic acid?

17 Write the expected major product(s) of each of the following reactions. Explain how you chose the position of substitution in each case.

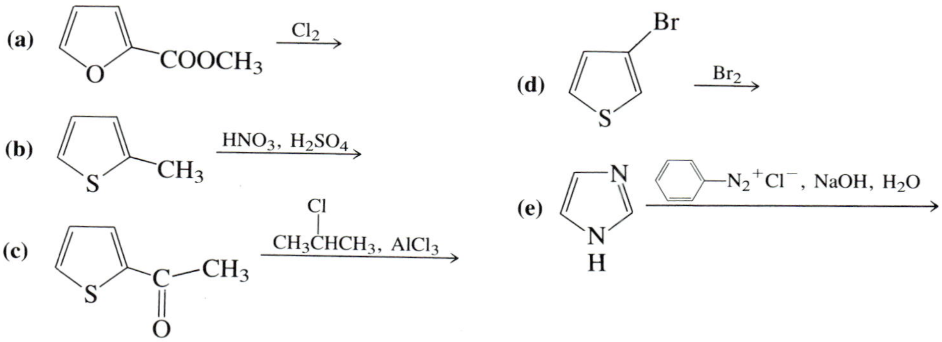

(a) O, $COOCH_3$ $\xrightarrow{Cl_2}$

(b) S, CH_3 $\xrightarrow{HNO_3,\ H_2SO_4}$

(c) S, $\overset{O}{\overset{\|}{C}}—CH_3$ $\xrightarrow{CH_3\overset{Cl}{\overset{|}{C}}HCH_3,\ AlCl_3}$

(d) S, Br $\xrightarrow{Br_2}$

(e) N, N, H $\xrightarrow{C_6H_5—N_2^+Cl^-,\ NaOH,\ H_2O}$

18 Thiophene and its selenium analog, selenophene, are not equally reactive toward electrophiles. In fact, one is brominated fifty times as fast as the other. Which is faster, and why?

19 Write the products expected of each of the following reactions.

(a) H_3C CH_3 N, N $\xrightarrow{\text{Fuming } H_2SO_4,\ 270°C}$

(b) O + S, O, O, O $\xrightarrow{\Delta,\ \text{pressure}}$

(c) N, Br $\xrightarrow{KSH,\ CH_3OH,\ \Delta}$

(d) S $\xrightarrow[2.\ \text{Raney Ni},\ \Delta]{1.\ C_6H_5COCl,\ SnCl_4}$

(e) N $\xrightarrow{(CH_3)_3CLi,\ \Delta}$

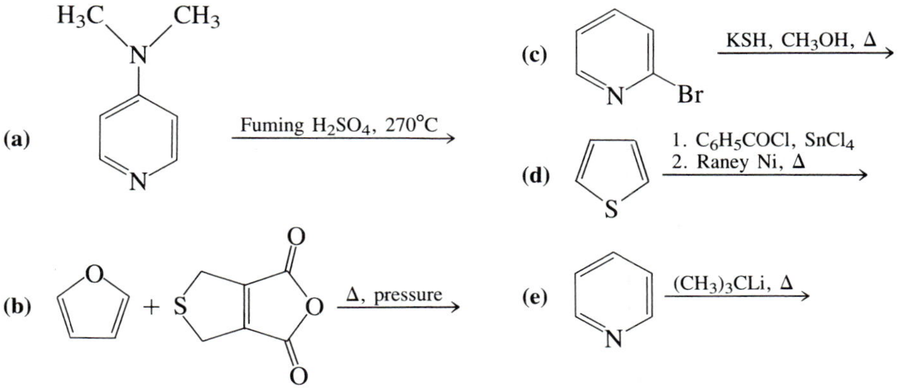

20 Propose a synthesis of each of the following substituted heterocycles, using synthetic sequences presented in this chapter.

(a) 3-acetyl-2-methylquinoline: O, $\overset{\|}{C}CH_3$, N, CH_3

(b) H_5C_6, N, C_6H_5

(c) H_3C, CH_3, N, H

(d) CH_3O, CH_3, CH_3, N, H

21 Heterocycle E, C_5H_6O, exhibits ^1H NMR spectrum E and is converted by H_2 and Raney nickel into F, $C_5H_{10}O$, with spectrum F. Identify compounds E and F. Note: The coupling constants in the spectra of the compounds in this problem and the next one are rather small; they are therefore not nearly as useful in structure elucidation as the coupling constants around a benzene ring.

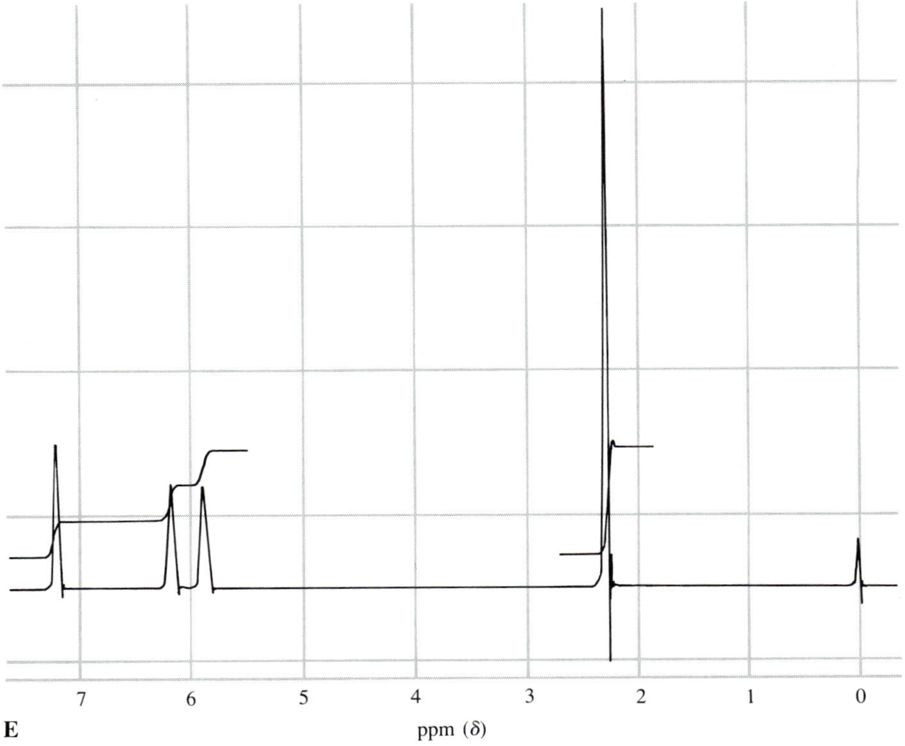

22 The commercial synthesis of a certain useful heterocyclic derivative requires treatment of a mixture of aldopentoses (derived from corncobs, straw, etc.) with hot acid under dehydrating conditions. The product, G, has ^1H NMR spectrum G, shows a strong IR band at 1670 cm^{-1}, and is formed in nearly quantitative yield. Identify G and formulate a mechanism for its formation.

$$\text{Aldopentoses} \xrightarrow{H^+,\ \Delta} \underset{\textbf{G}}{C_5H_4O_2}$$

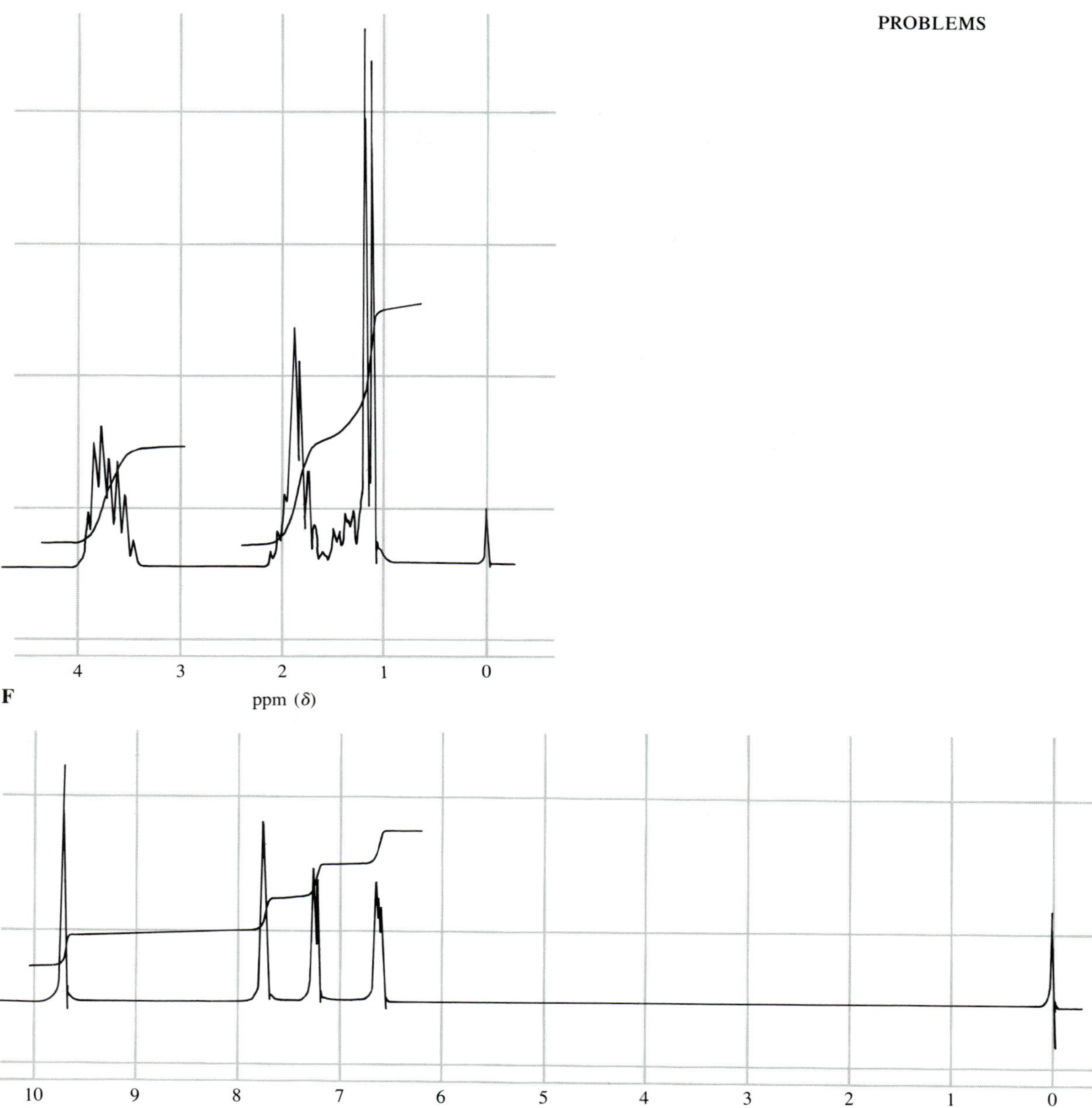

Compound G is a valuable synthetic starting material. The following sequence converts it into *furethonium,* which is useful in the treatment of glaucoma. What is the structure of furethonium?

$$\text{G} \xrightarrow[\text{2. Excess CH}_3\text{I}]{\text{1. NH}_3\text{, NaBH}_3\text{CN}} \text{furethonium}$$

23 Treatment of a 3-alkanoylindole with $LiAlH_4$ (shown on the next page) reduces the carbonyl all the way to a CH_2 group. Explain by a plausible mechanism. Note: Direct nucleophilic displacement of alkoxide by hydride is *not* plausible.

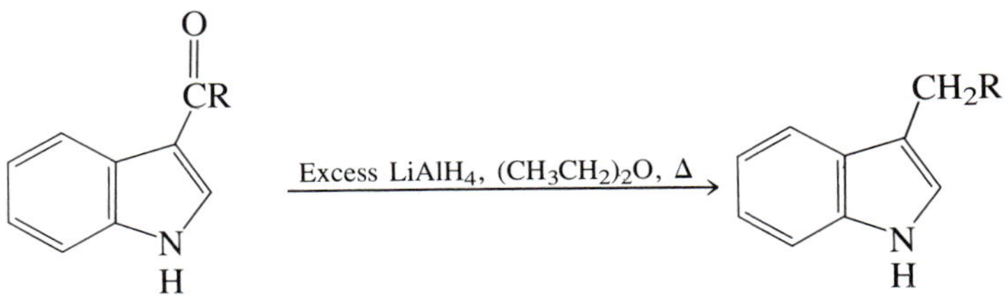

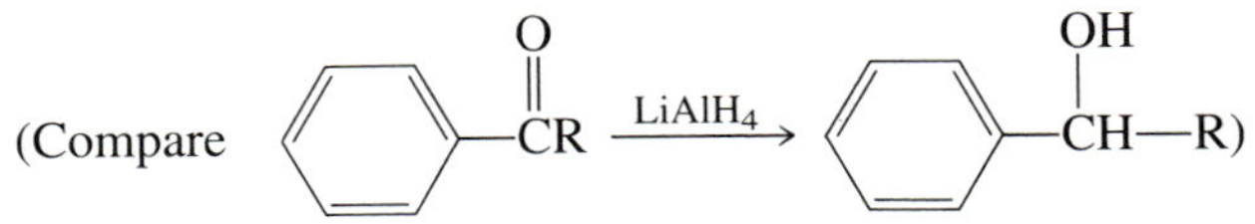

$$\text{3-acylindole (C(=O)R at C-3)} \xrightarrow{\text{Excess LiAlH}_4,\ (CH_3CH_2)_2O,\ \Delta} \text{3-(CH}_2\text{R)indole}$$

$$\text{(Compare } C_6H_5\overset{O}{\overset{\|}{C}}R \xrightarrow{\text{LiAlH}_4} C_6H_5\overset{OH}{\overset{|}{C}}H\text{—R)}$$

24 Chelidonic acid, an oxa-4-cyclohexanone (common name, γ-pyrone), is found in a number of plants and is synthesized from propanone (acetone) and diethyl ethanedioate. Formulate a mechanism for this transformation.

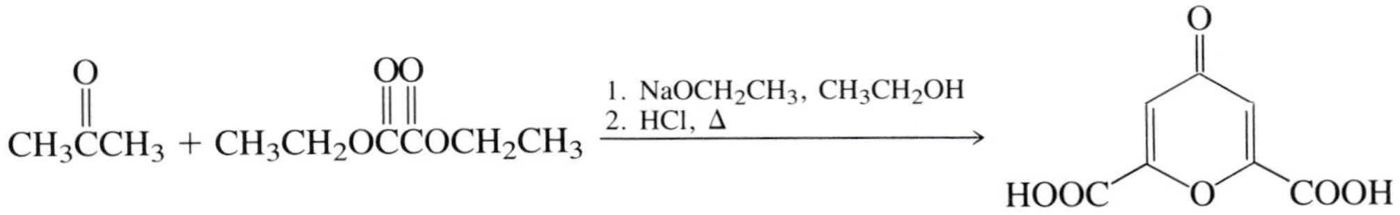

$$CH_3\overset{O}{\overset{\|}{C}}CH_3 + CH_3CH_2O\overset{O}{\overset{\|}{C}}\overset{O}{\overset{\|}{C}}OCH_2CH_3 \xrightarrow[\text{2. HCl, }\Delta]{\text{1. NaOCH}_2CH_3,\ CH_3CH_2OH} \text{Chelidonic acid (4-oxo-4H-pyran-2,6-dicarboxylic acid; HOOC, COOH)}$$

Chelidonic acid

25 Reserpine is a naturally occurring indole alkaloid with powerful tranquilizing and antihypertensive activity. Many such compounds possess a characteristic structural feature: one nitrogen atom at a ring fusion separated by two carbons from another nitrogen atom.

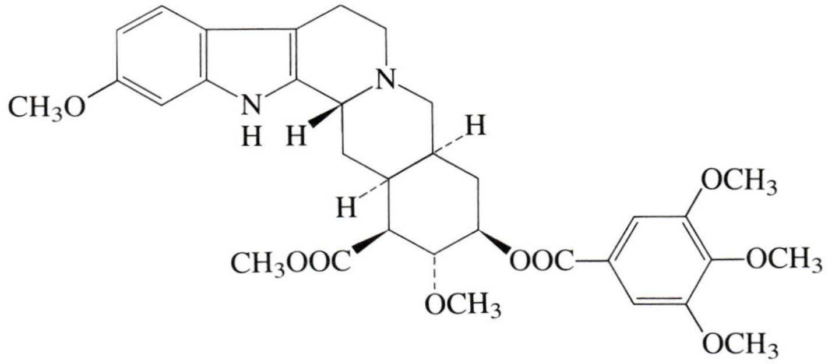

Reserpine structure (CH_3O on indole ring, N–H, H, H, H; CH_3OOC, OCH_3, OOC–3,4,5-trimethoxyphenyl (OCH_3, OCH_3, OCH_3))

Reserpine

A series of compounds with modified versions of this structural feature have been synthesized and shown also to have antihypertensive activity, as well as antifibrillatory activity. One such synthesis is shown here. Name or draw the missing reagents and products **a** through **c.**

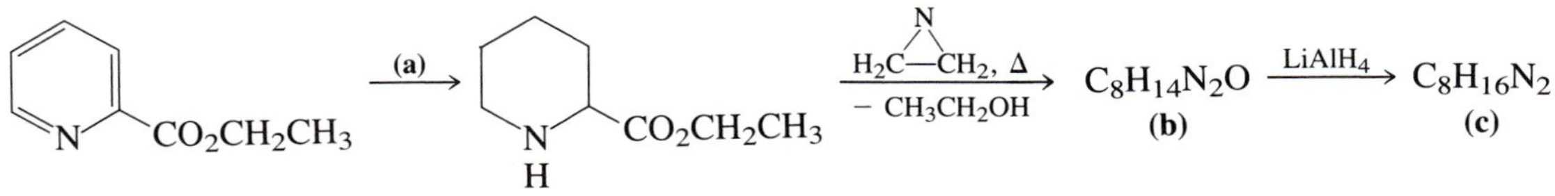

$$\text{2-pyridyl-}CO_2CH_2CH_3 \xrightarrow{\textbf{(a)}} \text{2-piperidyl(N–H)-}CO_2CH_2CH_3 \xrightarrow[-\ CH_3CH_2OH]{\text{aziridine (H}_2C\text{—CH}_2\text{, NH)},\ \Delta} \underset{\textbf{(b)}}{C_8H_{14}N_2O} \xrightarrow{\text{LiAlH}_4} \underset{\textbf{(c)}}{C_8H_{16}N_2}$$

26 Starting with benzenamine (aniline) and pyridine, propose a synthesis for the antimicrobial sulfa drug sulfapyridine:

Sulfapyridine

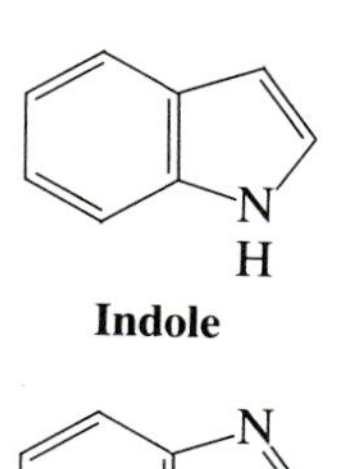

Indole

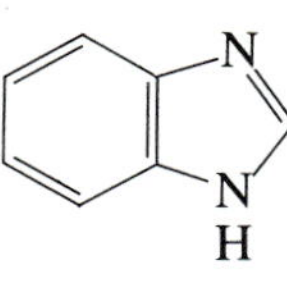

Benzimidazole

27 Write detailed mechanisms for each of the first two reactions in the total synthesis of nicotine (Section 26-5).

28 Derivatives of benzimidazole possess biological activity somewhat like that of indoles and purines (of which adenine, Section 26-1, is an example). Benzimidazoles are commonly prepared from benzene-1,2-diamine. Devise a short synthesis of 2-methylbenzimidazole from benzene-1,2-diamine.

Benzene-1,2-diamine → 2-Methylbenzimidazole

Purine

2,3-Dimethylquinoxaline

29 Benzene-1,2-diamine (Problem 28) is also the precursor of choice for the synthesis of quinoxaline derivatives. Propose a simple method for the conversion of benzene-1,2-diamine into 2,3-dimethylquinoxaline, a precursor of *mediquox,* an agent used to treat respiratory infections in poultry.

Mediquox

30 The following is a rapid synthesis of one of the heterocycles in this chapter. Draw the structure of the product, which has ^{1}H NMR spectrum H.

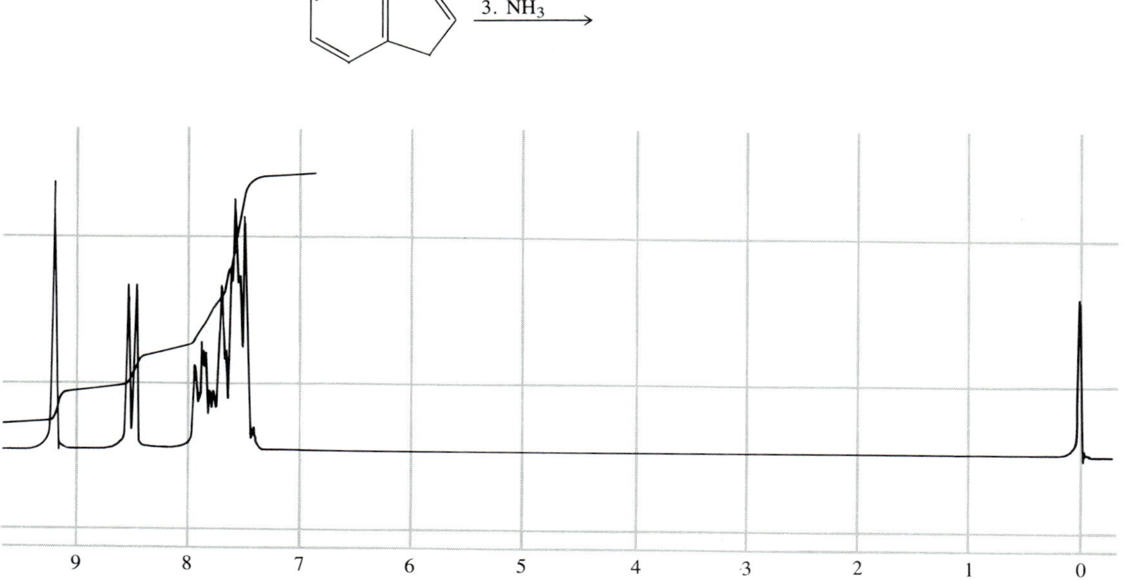

H

CHAPTER 27

Amino Acids, Peptides, and Proteins

NITROGEN-CONTAINING MONOMERS AND POLYMERS IN NATURE

Chapter 23 showed that monosaccharides can form polymers by making ether bonds to other monosaccharides. Nature makes use of the resulting polysaccharides for energy storage and for building cell walls. This chapter will consider a second type of natural polymer: the **polypeptide.**

The monomer unit in a polypeptide is a **2-amino** (or α-amino) **acid.** The polymer forms by repeated reaction of the carboxylic acid function of one amino acid with the amino group in another to make a chain of amides (Section 18-5). The amide bond joining amino acids is also called a **peptide bond.**

$$2n\ \mathrm{HN(H)}\!-\!\mathrm{C(R)(H)}\!-\!\overset{\displaystyle O}{\overset{\|}{\mathrm{C}}}\mathrm{OH} \longrightarrow -\!\left(\mathrm{NH}\!-\!\mathrm{C(R)(H)}\!-\!\overset{\displaystyle O}{\overset{\|}{\mathrm{C}}}\!-\!\mathrm{NH}\!-\!\mathrm{C(R)(H)}\!-\!\overset{\displaystyle O}{\overset{\|}{\mathrm{C}}}\right)_n\!- + 2n\ \mathrm{H_2O}$$

2-Amino acid (α-Amino acid) — **Polyamino acid (Polypeptide)**

The oligomers formed by linking amino acids in this way are called **peptides.** For example, two amino acids give rise to a dipeptide, three to a tripeptide, and so on. Certain large natural polypeptides are called **proteins.** Some protein molecules contain more than 8,000 amino acid units and have a molecular weight greater than 1,000,000.

Proteins have diverse functions in biological systems. For example, they act as catalysts **(enzymes)** for many natural chemical reactions. The catalyzed transformations range in complexity from a simple hydration of carbon dioxide to the replication of an entire chromosome. Enzymes can accelerate certain reactions many millionfold.

Proteins also act as transport and storage facilities. Thus, oxygen is transported by the protein *hemoglobin;* iron is carried in the blood by *transferrin* and stored in the liver by *ferritin*. Proteins play a crucial role in coordinated motion, such as muscle contraction. They give mechanical support to skin and bone; they are antibodies responsible for our immune protection; they generate and transmit nerve impulses (e.g., *rhodopsin* is the photoreceptor protein in retinal rod cells, see Section 16-4); and they control growth and differentiation—that is, how much and what part of the genetic information stored in DNA is to be used at any given time.

27-1 The Structure and Acid-Base Properties of Amino Acids

This chapter will begin by describing the structure of the twenty most-common natural amino acids, the building blocks of proteins. Although all of them can be readily named in a systematic manner, they rarely are; so we shall use the common names found in the literature. Because of the presence of both amino and carboxy functions, amino acids are amphoteric—both acidic and basic.

The Naming and Structure of the Common 2-Amino Acids: The Stereocenter Has the *S* Configuration

Although there are more than five hundred naturally occurring amino acids, the proteins in all species, from bacteria to humans, consist mainly of only twenty different amino acids. Adult humans can synthesize all but eight. These eight, called *essential amino acids,* must be included in our diet (Table 27-1, footnote a). In the twenty common amino acids used in nature, an amino group is located at C-2, the α-carbon. Amino acids may be drawn as dashed-wedged line structures or as Fischer projections.

How to Draw (2*S*)-Amino Acids and Their Relation to the L-Sugars

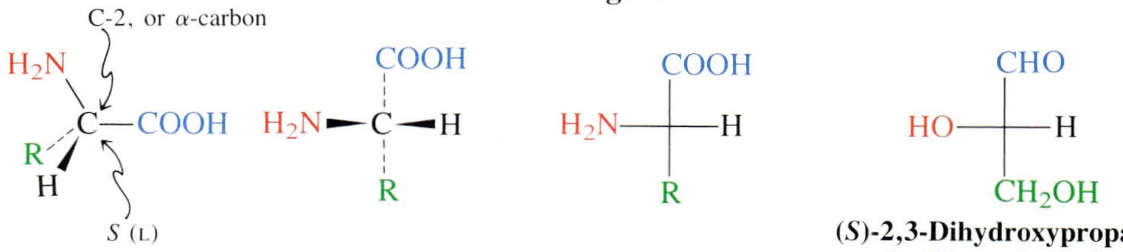

C-2, or α-carbon
H_2N, C—COOH, R, H
S (L)
COOH, H_2N—C—H, R
COOH, H_2N—H, R
CHO, HO—H, CH_2OH
(*S*)-2,3-Dihydroxypropanal (L-Glyceraldehyde)

Dashed-wedged line structures **Fischer projections**

In all but glycine, C-2 is a stereocenter and usually adopts the *S* configuration. As with the sugars (Section 23-1), the older amino acid nomenclature employs D and L prefixes, relating all (2*S*)-amino acids to (*S*)-2,3-dihydroxypropanal (L-glyceraldehyde).

The basic amino acids have common names, which are used in the literature almost to the exclusion of systematic nomenclature. Table 27-1 gives their structures and names, as well as pK_a data. A mnemonic three-letter code is based on abbreviated forms of these names. We shall see later (Section 27-3) that these codes are convenient for the description of peptides. The R group in $RCH(NH_2)COOH$ can be alkyl or aryl, and it can contain hydroxy, amino, mercapto or sulfide, and carboxy groups.

TABLE 27-1
The basic natural (2*S*)-amino acids

(Fischer projection: COOH at top, H_2N at left, H at right, R at bottom)

R	Name	Three-letter code	pK_a of COOH	pK_a of $^+NH_3$	pK_a of acidic function in R
—H	Glycine	Gly	2.4	9.8	—
Alkyl Group:					
$—CH_3$	Alanine	Ala	2.4	9.9	—
$—CH(CH_3)_2$	Valine[a]	Val	2.3	9.7	—
$—CH_2CH(CH_3)_2$	Leucine[a]	Leu	2.3	9.7	—
$—CHCH_2CH_3$ (*S*) with CH_3 branch	Isoleucine[a]	Ile	2.3	9.7	—
$—CH_2C_6H_5$	Phenylalanine[a]	Phe	2.6	9.2	—
HN—C(COOH[b])(H)—CH_2 (ring)	Proline	Pro	2.0	10.6	—
Hydroxy-containing:					
$—CH_2OH$	Serine	Ser	2.2	9.4	—
—CHOH (*R*) with CH_3 branch	Threonine[a]	Thr	2.1	9.1	—
$—CH_2—C_6H_4—OH$	Tyrosine	Tyr	2.2	9.1	10.1
Amino-containing:					
$—CH_2CNH_2$ (C=O)	Asparagine	Asn	2.0	8.8	—
$—CH_2CH_2CNH_2$ (C=O)	Glutamine	Gln	2.2	9.1	—
$—(CH_2)_4NH_2$	Lysine[a]	Lys	2.2	9.2	10.8[c]
$—(CH_2)_3NHCNH_2$ (C=NH)	Arginine	Arg	1.8	9.0	13.2[c]
$—CH_2$—(3-indolyl, N—H)	Tryptophan[a]	Trp	2.4	9.4	—
$—H_2C$—(4-imidazolyl, NH, N)	Histidine	His	1.8	9.2	6.1[c]

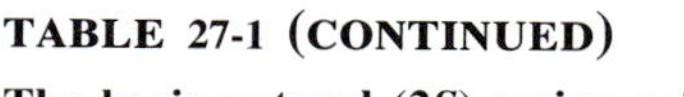

TABLE 27-1 (CONTINUED)
The basic natural (2S)-amino acids

R	Name	Three-letter code	pK_a of COOH	pK_a of $^+NH_3$	pK_a of acidic function in R
Mercapto- or Sulfide-containing:					
$—CH_2SH$	Cysteine[d]	Cys	1.9	10.3	8.4
$—CH_2CH_2SCH_3$	Methionine[a]	Met	2.2	9.3	—
Carboxy-containing:					
$—CH_2COOH$	Aspartic acid	Asp	2.0	10.0	3.9
$—CH_2CH_2COOH$	Glutamic acid	Glu	2.1	10.0	4.3

[a]Essential amino acids. [b]Entire structure. [c]pK_a of conjugate acid. [d]The stereocenter is R.

As emphasized in the discussion of the natural D-sugars (Section 23-1), the fact that a molecule belongs to the L-family does not mean that it must be levorotatory. Thus, although leucine is levorotatory ($[\alpha]_D^{25°C} = -10.8°$), alanine ($[\alpha]_D^{25°C} = +8.5°$), valine ($[\alpha]_D^{25°C} = +13.9°$), and isoleucine ($[\alpha]_D^{25°C} = +11.3°$) are dextrorotatory.

EXERCISE 27-1

Give the systematic names of alanine, valine, leucine, isoleucine, phenylalanine, serine, tyrosine, lysine, cysteine, methionine, aspartic acid, and glutamic acid.

Amino Acids Are Acidic and Basic

Because they contain both carboxylic acid and amine functions, it is not surprising that amino acids are amphoteric (Section 8-3)—that is, both acidic and basic. An ammonium ion (pK_a ~ 10–11) is significantly less acidic than a carboxylic acid (pK_a ~ 2–5); consequently, amino acids actually exist as **zwitterionic ammonium carboxylates.** The highly polar nature of this structure allows amino acids to form particularly strong crystal lattices. Therefore, most of them are fairly insoluble and do not melt, but decompose on heating.

Zwitterion

In aqueous solution, various acid-base equilibria involving the functional groups are established. Consider, for example, the simplest member in the series, glycine. Depending on the pH of the medium, it may exist predominantly as the diprotonated cation (pH < 1), the monoprotonated zwitterion (pH ~ 6), or the deprotonated 2-amino carboxylate ion (pH > 13).

$$H_2\overset{+}{N}(H)CH_2COOH \underset{H^+}{\overset{HO^-}{\rightleftharpoons}} H_2\overset{+}{N}(H)CH_2COO^- \underset{H^+}{\overset{HO^-}{\rightleftharpoons}} H_2NCH_2COO^-$$

Predominates at pH < 1 | Predominates at pH ~ 6 | Predominates at pH > 13

Table 27-1 records pK_a values corresponding to the various equilibria. The first pK_a (2.4) refers to the equilibrium

$$\overset{+}{H_3N}CH_2COOH + H_2O \rightleftharpoons \overset{+}{H_3N}CH_2COO^- + H_2\overset{+}{O}H$$
$$pK_a = 2.4$$

$$K_1 = \frac{[\overset{+}{H_3N}CH_2COO^-][H_2\overset{+}{O}H]}{[\overset{+}{H_3N}CH_2COOH]} = 10^{-2.4}$$

Note that this pK_a is more than two units lower than that of an ordinary carboxylic acid (pK_a $CH_3COOH = 4.74$). This difference is a consequence of the electron-withdrawing effect of the protonated amino group. The second pK_a value (9.8) describes the second deprotonation step:

$$\overset{+}{H_3N}CH_2COO^- + H_2O \rightleftharpoons H_2NCH_2COO^- + H_2\overset{+}{O}H$$
$$pK_a = 9.8$$
$$K_2 = \frac{[H_2NCH_2COO^-][H_2\overset{+}{O}H]}{[\overset{+}{H_3N}CH_2COO^-]} = 10^{-9.8}$$

It is important to know at what pH the concentration of charge-neutralized zwitterionic form is the largest. This would be at a point at which the extent of protonation equals that of deprotonation and hence $[\overset{+}{H_3N}CH_2COOH] = [H_2NCH_2COO^-]$. To solve for the pH at this point, we can rewrite the equations describing K_1 and K_2 in the following way.

$$\text{from } K_1\text{:}\quad [\overset{+}{H_3N}CH_2COO^-] = \frac{[\overset{+}{H_3N}CH_2COOH] \times 10^{-2.4}}{[H_2\overset{+}{O}H]}$$

$$\text{from } K_2\text{:}\quad [\overset{+}{H_3N}CH_2COO^-] = \frac{[H_2NCH_2COO^-][H_2\overset{+}{O}H]}{10^{-9.8}}$$

Therefore,

$$\frac{[H_2NCH_2COO^-][H_2\overset{+}{O}H]}{10^{-9.8}} = \frac{[\overset{+}{H_3N}CH_2COOH] \times 10^{-2.4}}{[H_2\overset{+}{O}H]}$$

and

$$[H_2\overset{+}{O}H]^2 = \frac{10^{-12.2} \times [\overset{+}{H_3N}CH_2COOH]}{[H_2NCH_2COO^-]}$$

Because $[\overset{+}{H_3N}CH_2COOH] = [H_2NCH_2COO^-]$, it follows that $[H_2\overset{+}{O}H] = 10^{-6.1}$; thus, pH = 6.1.

This point is also called the **isoelectric point** (p*I*), because the number of positively charged amino acid molecules equals the number of their negatively charged counterparts. As can be verified readily, the isoelectric point is the average of the two pK_a values of the amino acid.

$$pI = \frac{pK_{COOH} + pK_{\overset{+}{N}H_2H}}{2}$$

The situation is a little more complicated if the side chain of the acid bears an additional acidic or basic function. Table 27-1 shows seven entries in which this is the case. The first is tyrosine, bearing an acidic 4-hydroxyphenylmethyl substituent with pK_a = 10.1, typical of benzenols (phenols; Section 24-4).

Assignment of pK_a Values in Selected Amino Acids

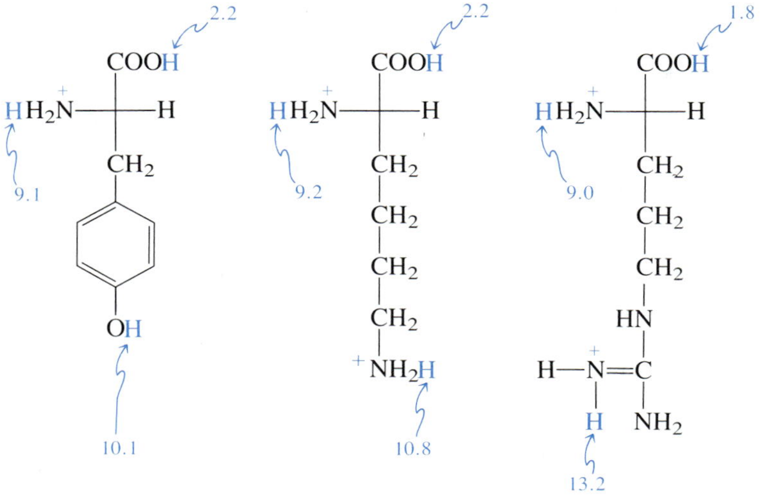

Lysine has an additional amino group that can be protonated in a strongly acidic medium to furnish a dication. On raising the pH of the solution, deprotonation of the carboxy group occurs first, to be followed by proton loss from the nitrogen at C-2, and finally from the remote ammonium function. The isoelectric point is located halfway between the last two pK_as, at $pI = 10$.

Arginine bears a substituent new to us: the relatively basic **guanidino group** (pK_a of conjugate acid ~ 13).

Guanidine

EXERCISE 27-2

Guanidine is found in turnip juice, mushrooms, corn germ, rice hulls, mussels, and earthworms. Its basicity is due to the formation of a highly resonance stabilized protonated form. Draw its resonance structures. Hint: Review Section 18-1.

Histidine contains another new substituent: an **imidazole.** Imidazoles are members of the class of diheterocyclopentadienes. Others include oxazole, thiazole, and pyrazole. In these rings, all of which are aromatic, one of the nitrogen atoms is hybridized as in pyridine, and the other nitrogen (or heteroatom) is hybridized as in pyrrole.

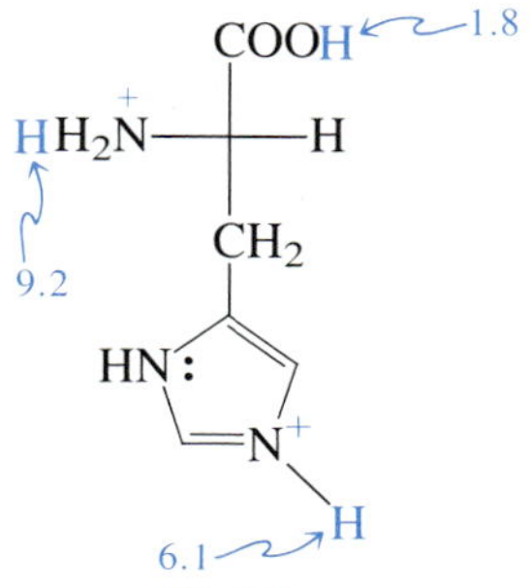

Histidine

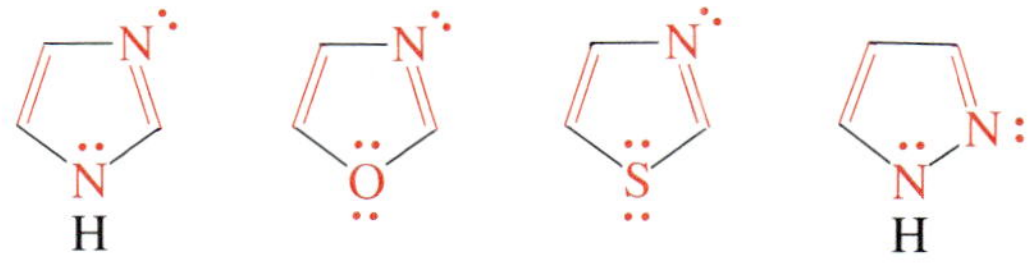

Imidazole **Oxazole** **Thiazole** **Pyrazole**

EXERCISE 27-3

Draw an orbital picture of imidazole (use Figure 26-1 as a model).

The imidazole nucleus is relatively basic because the protonated species is stabilized by resonance and can be described by two equivalent resonance structures.

Resonance in Protonated Imidazole

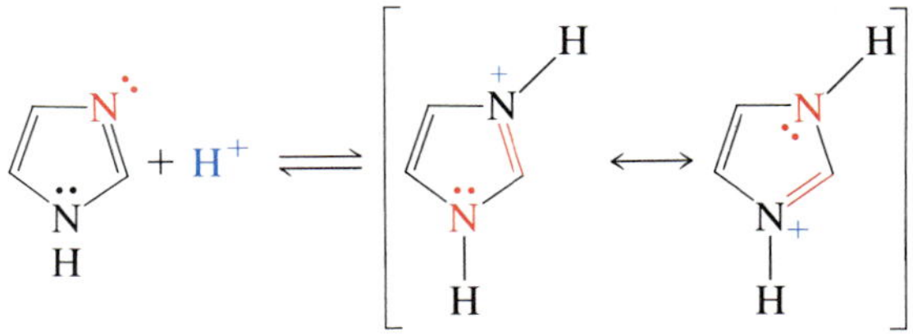

$$\text{imidazole} + H^+ \rightleftharpoons [\text{resonance structures}]$$

$pK_a = 7.0$

This resonance stabilization is related to that in amides (Section 18-1). Imidazole is significantly protonated at physiological pH. It can therefore function as a proton acceptor and donor at the active site of a variety of enzymes.

The amino acid cysteine bears a mercapto substituent. Recall that, apart from their acidic character, thiols can be oxidized to disulfides under mild conditions (Section 9-7). In nature, various enzymes are capable of oxidatively coupling the mercapto groups in the cysteines of proteins and peptides, leading to reversible linking of peptide strands (Section 9-7).

Aspartic acid and glutamic acid are amino dicarboxylic acids. At physiological pH, both of the carboxy functions are deprotonated, and the molecules exist as the zwitterionic anions aspartate and glutamate. Monosodium glutamate (MSG) is used as a flavor enhancer in various foods.

In summary, there are twenty elementary (2*S*)-amino acids, all of which have common names. Unless there are additional acid-base functions in the chain, their acid-base behavior is governed by two pK_a values, the lower one describing the deprotonation of the protonated amino function. At the isoelectric point, which is the average of the two pK_a values, the number of amino acid molecules with net zero charge is maximized. Some amino acids contain additional acidic or basic functions, such as hydroxy, amino, guanidino, imidazolyl, mercapto, and carboxy.

27-2 The Preparation of Amino Acids: A Combination of Amine and Carboxylic Acid Chemistry

The chemistry of amines was dealt with in Chapter 21, that of carboxylic acids and their derivatives in Chapters 17 and 18. This section will apply the combined information in those chapters to the preparation of 2-amino acids. Optically active amino acids can be made by resolution or by biological methods.

From Carboxylic to 2-Amino Acids: Hell-Volhard-Zelinsky Bromination Followed by Amination

What would be the quickest way of introducing a 2-amino substituent into a carboxylic acid? Section 17-11 pointed out that simple 2-functionalization of an acid is possible by the Hell-Volhard-Zelinsky bromination. Furthermore, the bromine in the product can be displaced by nucleophiles, such as ammonia. In this way, propanoic acid is converted into racemic alanine in 45% overall yield in two steps.

$$CH_3CH_2COOH \xrightarrow[-\ HBr]{Br—Br,\ catalytic\ PBr_3} CH_3CH(Br)COOH \xrightarrow[-\ HBr]{NH_3,\ H_2O,\ 25°C,\ 4\ days} CH_3CH(\overset{+}{N}H_3)COO^-$$

Propanoic acid — **2-Bromopropanoic acid** (80%) — **(*R*,*S*)-Alanine** (56%)

Unfortunately, this approach frequently suffers from relatively low yields. A better synthesis utilizes Gabriel's procedure for the preparation of primary amines (Section 21-4).

Applications of the Gabriel Synthesis to the Preparation of Amino Acids

In the Gabriel synthesis of amines, 1,2-benzenedicarboxylic imide (phthalimide) anion is *N*-alkylated and the resulting product hydrolyzed with acid to furnish the amine and 1,2-benzenedicarboxylic (phthalic) acid. If, in the first step of this sequence, diethyl 2-bromopropanedioate (diethyl 2-bromomalonate; readily available by bromination of diethyl propanedioate) is employed as the alkylating reagent, the product can be hydrolyzed and then decarboxylated by acid (Section 22-4). Further hydrolysis then liberates an amino acid.

Gabriel Synthesis of Glycine

$$HCBr(COOCH_2CH_3)_2 + \text{phthalimide } :\ddot{N}:^-\ K^+ \xrightarrow{-\ KBr} \text{phthalimido-}N—CH(COOCH_2CH_3)_2 \xrightarrow[-\ 2\ CH_3CH_2OH]{H^+,\ H_2O,\ \Delta}$$

Diethyl 2-bromopropanedioate (Diethyl 2-bromomalonate) + **Potassium 1,2-benzenedicarboxylic imide (Potassium phthalimide)** → 85%

$$\text{phthalimido-}N—CH(COOH)_2 \xrightarrow{-\ CO_2} \xrightarrow[-\ C_6H_4(COOH)_2]{H^+,\ H_2O} H_3\overset{+}{N}CH_2COO^-$$

85% **Glycine**

One of the advantages of this approach is the versatility of the initially formed 2-substituted propanedioate, which can itself be alkylated, thus allowing for the preparation of a variety of substituted amino acids.

$$\text{phthalimido-}N—CH(CO_2CH_2CH_3)_2 \xrightarrow{1.\ CH_3CH_2O^-Na^+\quad 2.\ RX\quad 3.\ H^+,\ H_2O,\ \Delta} RCH(\overset{+}{N}H_3)COO^-$$

EXERCISE 27-4

Propose Gabriel syntheses of methionine, aspartic acid, and glutamic acid.

A variation of the Gabriel synthesis utilizes diethyl *N*-ethanoyl-2-aminopropanedioate (acetamidomalonic ester) instead of the imide derivative.

$$CH_3\overset{\overset{\displaystyle O}{\|}}{C}NHCH(CO_2CH_2CH_3)_2 \xrightarrow[-\,HOH]{NaOH,\ CH_2{=}O} \xrightarrow{H^+,\ H_2O,\ \Delta} HOCH_2\overset{\overset{\displaystyle ^+NH_3}{|}}{C}HCOO^-$$

Diethyl *N*-ethanoyl-2-aminopropanedioate (Acetamidomalonic ester) → 65% **Serine**

Amino Acids from Aldehydes: Strecker Synthesis

The **Strecker* synthesis** employs in its crucial step a variation of the cyanohydrin formation from aldehydes and hydrogen cyanide (Section 15-7).

$$R\overset{\overset{\displaystyle O}{\|}}{C}H + HCN \rightleftharpoons R{-}\underset{\underset{\displaystyle H}{|}}{\overset{\overset{\displaystyle OH}{|}}{C}}{-}CN$$

Cyanohydrin

When the same reaction is carried out in the presence of ammonia, it is the intermediate imine that undergoes addition of hydrogen cyanide, to furnish the corresponding 2-amino nitriles. Subsequent acidic or basic hydrolysis results in the desired amino acids.

Strecker Synthesis of Alanine

$$CH_3\overset{\overset{\displaystyle O}{\|}}{C}H \xrightarrow[-\,H_2O]{NH_3} CH_3\overset{\overset{\displaystyle NH}{\|}}{C}H \xrightarrow{HCN} H_3C{-}\underset{\underset{\displaystyle H}{|}}{\overset{\overset{\displaystyle NH_2}{|}}{C}}{-}CN \xrightarrow{H^+,\ H_2O} CH_3\overset{\overset{\displaystyle ^+NH_3}{|}}{C}HCOO^-$$

Ethanal (Acetaldehyde) — **Imine** — **2-Aminopropanenitrile** — 55% **Alanine**

EXERCISE 27-5

Propose Strecker syntheses of glycine and methionine.

How to Synthesize Enantiomerically Pure Amino Acids

All the preceding methods produce amino acids in racemic form. However, for many synthetic purposes—in particular, peptide and protein syntheses—enantiomerically pure compounds are required; in most syntheses of natural peptides, they must have the *S* configuration. To solve this problem, the racemic amino acids have to be either resolved (Section 5-7) or prepared as one enantiomer by enantioselective reactions.

---

*Professor Adolf Strecker, 1822–1871, University of Würzburg, Germany.

A conceptually straightforward approach to the preparation of pure enantiomers of amino acids would be resolution of their diastereomeric salts. Typically, the amine group is first protected as an amide, and the resulting product then treated with an optically active amine (Section 21-6), such as the alkaloid brucine (Section 26-7). The two diastereomers formed can be separated by fractional crystallization. Unfortunately, in practice, this method can be tedious and can suffer from poor yields.

Resolution of Racemic Valine

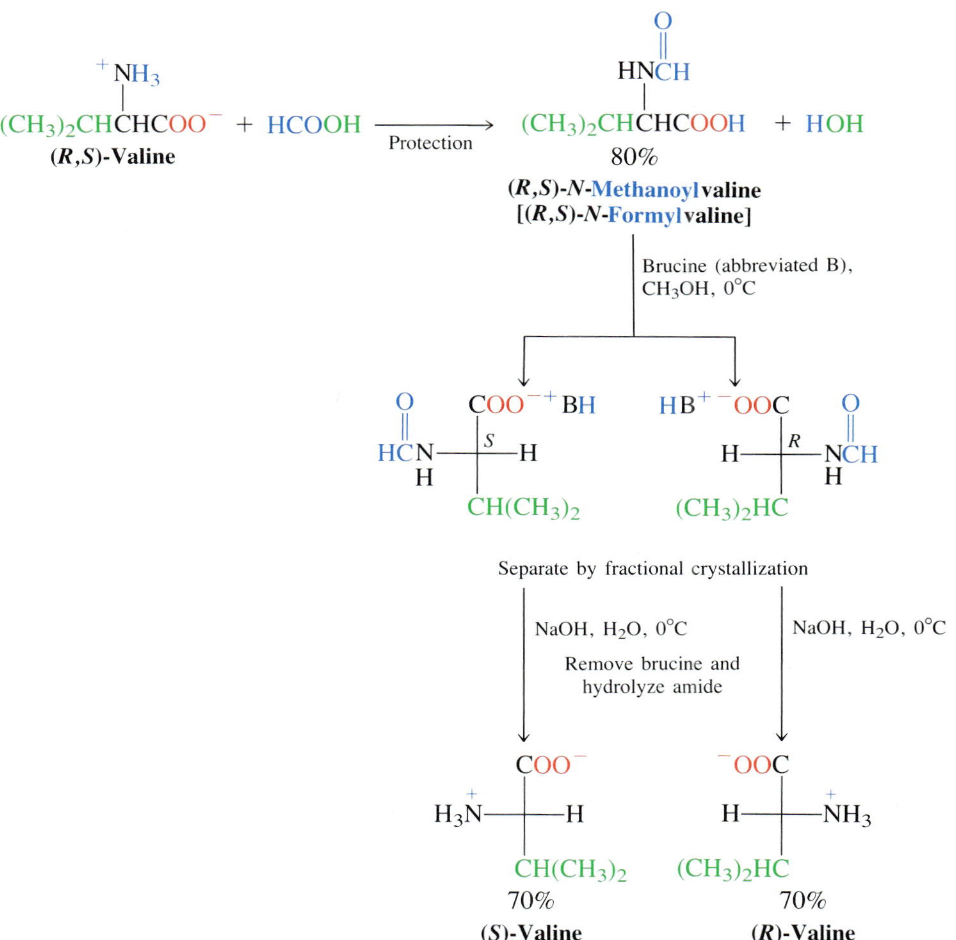

In an alternative approach, the stereocenter at C-2 is formed enantioselectively. Nature makes use of this strategy in the biosynthesis of amino acids. Thus, the enzyme *glutamate dehydrogenase* converts the prochiral carbonyl group in 2-oxopentanedioic acid into the amine substituent in (*S*)-glutamic acid by a biological reductive amination (for chemical reductive aminations, see Section 21-4). The reducing agent is NADH (Section 26-5).

$$HOOCCH_2CH_2\overset{O}{\overset{\|}{C}}COOH + NH_3 + H^+ \xrightarrow[-\,NAD^+]{NADH,\ \text{glutamate dehydrogenase}} HOOCCH_2CH_2\overset{^+NH_3}{\overset{|}{C}}HCOO^- + H_2O$$

2-Oxopentanedioic acid → **(*S*)-Glutamic acid**

(*S*)-Glutamic acid is the direct biosynthetic precursor of glutamine, proline, and arginine. Moreover, it functions as an aminating agent for other 2-oxo acids, with the help of another enzyme, *transaminase,* to make further amino acids available.

$$\mathrm{H{-}\overset{\overset{+}{N}H_3}{\underset{R}{C}}{-}COO^- + R'\overset{O}{\overset{\|}{C}}COO^- \xrightleftharpoons{\text{Transaminase}} R\overset{O}{\overset{\|}{C}}COO^- + H{-}\overset{\overset{+}{N}H_3}{\underset{R'}{C}}{-}COO^-}$$

Chemists have used the catalytic activity of enzymes to prepare optically pure amino acids in an approach called **kinetic resolution.** In this process, the handedness of the enzyme allows it to select only one of the enantiomers to be converted into some other product. The remaining enantiomer can then be separated from the mixture.

An example of this principle is the enzymatic kinetic resolution of racemic alanine by *hog renal acylase I,* isolated from hog kidneys. This enzyme catalytically hydrolyzes the amide linkages of only *N*-alkanoylated (2*S*)-amino acids. In practice, the racemic acid is first ethanoylated (acetylated) with ethanoic (acetic) anhydride and the resulting *N*-ethanoylalanine (*N*-acetylalanine) is subsequently exposed to the enzyme in aqueous solution.

Biological Kinetic Resolution of Racemic Alanine

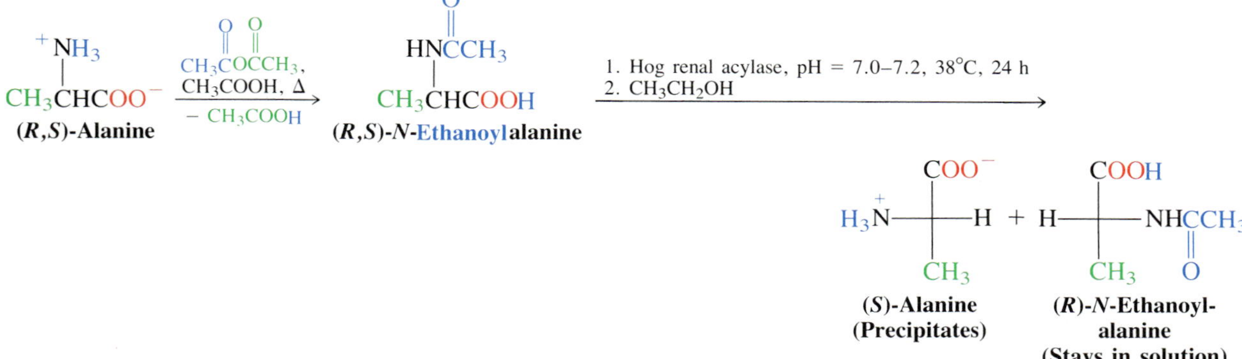

Addition of ethanol to this mixture causes crystallization of the natural (*S*)-amino acid; the more-soluble (*R*)-*N*-ethanoylalanine does not undergo reaction and remains in the mother liquors. The procedure is very effective: only about 150 mg of enzyme is required to resolve 600 g of amino acid.

For the preparation of small quantities of enantiomerically pure product, the selective enzymatic oxidation of a racemic amino acid has been used. This procedure destroys one enantiomer, leaving the other behind. In this way, rattlesnake venom, containing *L-amino acid oxidase,* yields the pure *R* form, whereas kidney *D-amino acid oxidase* furnishes the pure *S* form of the acid. This method is not useful for large preparations, however, because one of the two enantiomers cannot be recovered.

In summary, racemic amino acids are made by the amination of 2-bromo carboxylic acids, applications of the Gabriel synthesis of amines, and the

Strecker synthesis, which proceeds through an imine variation of the preparation of cyanohydrins, followed by hydrolysis. Enantiomerically pure amino acids can be obtained by resolution of a racemic mixture or by biological methods.

27-3 Amino Acid Oligomers and Polymers: The Structure of Peptides and Proteins

As mentioned in the introduction to this chapter, amino acids can be polymerized by multiple formation of peptide bonds to give polypeptides. This section will show how such polypeptide chains can be depicted, the structure of polypeptides and proteins, and how the sequence of amino acids is established in such polymers.

A Short Notation for Depicting Polypeptides

In drawing a polypeptide chain, the amino end is placed at the left, the carboxy end at the right. The configuration at all stereocenters is presumed to be *S*.

How to Draw the Structure of a Tripeptide

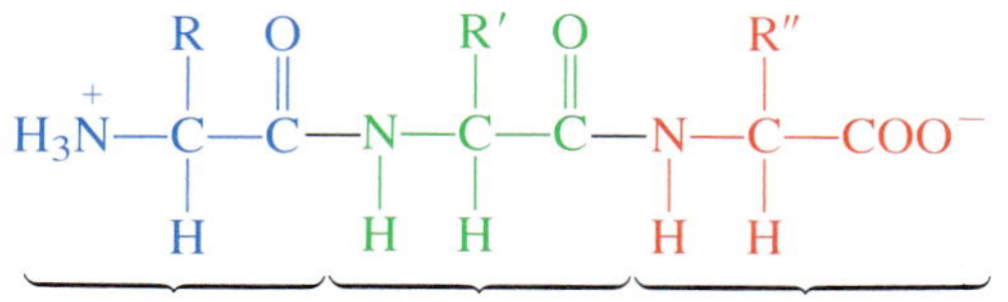

$$\overset{+}{H_3N}-\underset{\underset{H}{|}}{\overset{\overset{R}{|}}{C}}-\overset{\overset{O}{\|}}{C}-\underset{\underset{H}{|}}{N}-\underset{\underset{H}{|}}{\overset{\overset{R'}{|}}{C}}-\overset{\overset{O}{\|}}{C}-\underset{\underset{H}{|}}{N}-\underset{\underset{H}{|}}{\overset{\overset{R''}{|}}{C}}-COO^-$$

Amino acid 1 — Amino acid 2 — Amino acid 3

The chain incorporating the amide (peptide) bonds is called the **main chain,** the substituents R, R′, and so on are the **side chains.** The individual amino acid units forming the peptide are referred to as **residues.** In some proteins, two or more polypeptide chains are linked by disulfide bridges (Sections 9-7 and 27-1) or hydrogen bonds.

The naming of peptides is straightforward. Starting from the amino-terminal end, the names of the individual residues are simply connected in sequence, each regarded as a substituent to the next amino acid, ending with the carboxy-terminal residue. Because this procedure rapidly becomes cumbersome, the abbreviations listed in Table 27-1 are used for larger peptides.

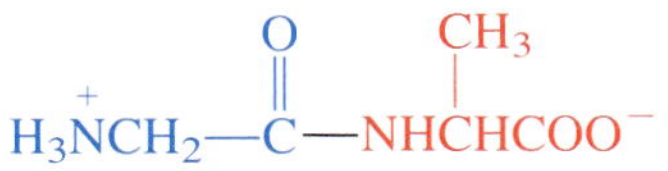

$$\overset{+}{H_3N}CH_2-\overset{\overset{O}{\|}}{C}-NH\overset{\overset{CH_3}{|}}{C}HCOO^-$$

Glycylalanine
Gly-Ala

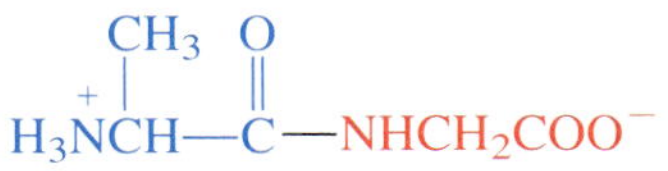

$$\overset{+}{H_3N}\overset{\overset{CH_3}{|}}{C}H-\overset{\overset{O}{\|}}{C}-NHCH_2COO^-$$

Alanylglycine
Ala-Gly

$$\overset{+}{H_3N}\overset{\overset{C_6H_5CH_2}{|}}{C}H-\overset{\overset{O}{\|}}{C}-NH\overset{\overset{(CH_3)_2CH}{|}}{\overset{CH_2}{|}}{C}H-\overset{\overset{O}{\|}}{C}-NH\overset{\overset{CH_3}{|}}{\overset{HCOH}{|}}{C}HCOO^-$$

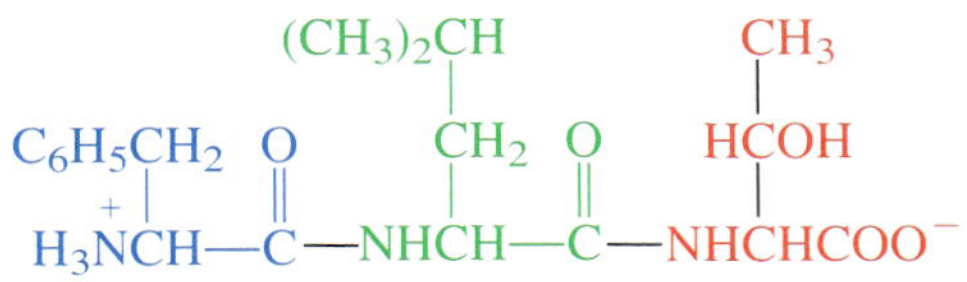

Phenylalanylleucylthreonine
Phe-Leu-Thr

Let us look at some examples of peptides and their structural variety. A dipeptide ester, aspartame, is a low-calorie artificial sweetener (Nutrasweet). In the three-letter notation, the ester end is denoted by —OCH_3. Glutathione, a tripeptide, is found in all living cells, and in particularly high concentrations in the lens of the eye. It is unusual in that its glutamic acid residue is linked at the γ-carboxy group (denoted γ-Glu) to the rest of the peptide.

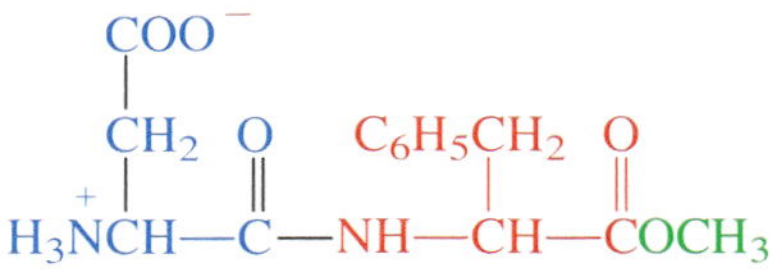

$$H_3\overset{+}{N}CH(CH_2COO^-)—\overset{O}{\overset{\|}{C}}—NH—CH(CH_2C_6H_5)—\overset{O}{\overset{\|}{C}}OCH_3$$

Aspartylphenylalanine methyl ester
Asp-Phe-OCH_3
(Aspartame)

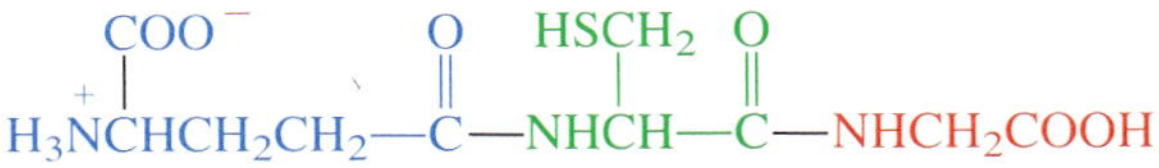

$$H_3\overset{+}{N}CH(COO^-)CH_2CH_2—\overset{O}{\overset{\|}{C}}—NHCH(CH_2SH)—\overset{O}{\overset{\|}{C}}—NHCH_2COOH$$

γ-Glutamylcysteinylglycine
γ-Glu-Cys-Gly
(Glutathione)

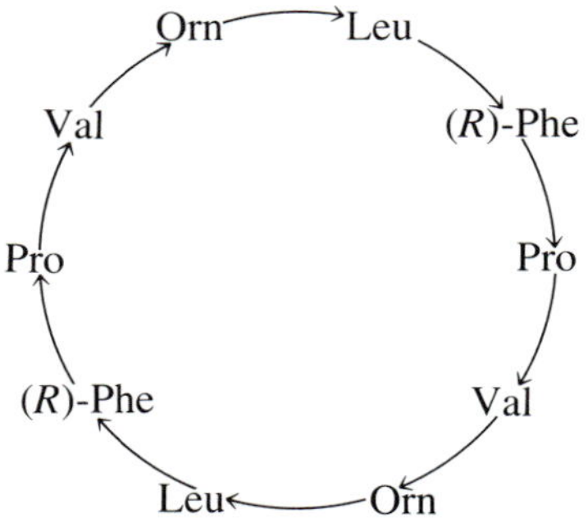

Orn → Leu → (*R*)-Phe → Pro → Val → Orn → Leu → (*R*)-Phe → Pro → Val → Orn (cyclic)

Gramicidin S

It functions as a biological reducing agent by being readily oxidized enzymatically at the cysteine mercapto unit to the disulfide-bridged dimer. Gramicidin S is a cyclic peptide antibiotic constructed out of two identical pentapeptides that have been joined head to tail. It contains phenylalanine in the *R* configuration and a rare amino acid, ornithine [Orn, a lower homolog (one CH_2 group less) of lysine]. In the short notation in which gramicidin S is shown, the sense in which the amino acids are linked (amino to carboxy direction) is indicated by arrows.

Although short notations are practical, they do not reveal some of the noncovalent bonding interactions in peptides. A dashed-wedged line picture of gramicidin S shows several important hydrogen bonds contributing substantially to the stereochemical rigidity of the ring.

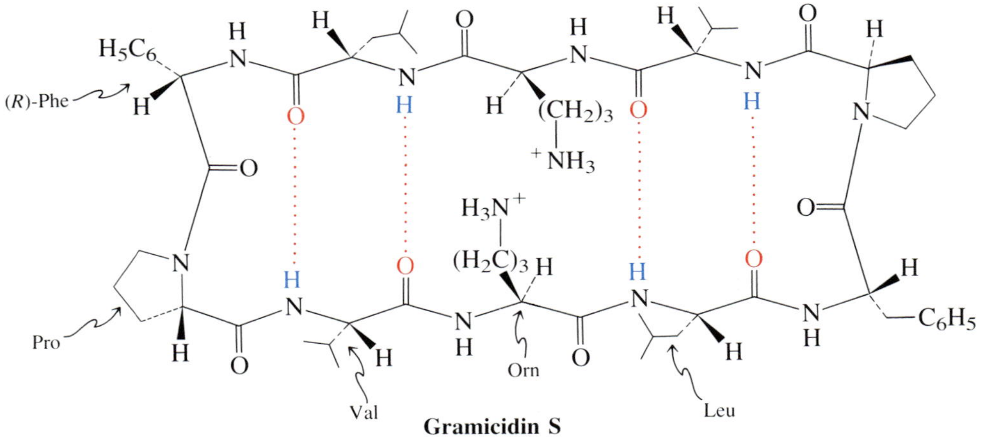

(*R*)-Phe, Pro, Val, Orn, Leu

Gramicidin S

(Dashed-wedged line notation)

A second feature inducing rigidity is the strongly dipolar nature of the amide functional group (Section 18-1), which causes an appreciable barrier to rotation around the relatively short carbonyl–nitrogen bond.

FIGURE 27-1

Amino acid sequence of bovine (cattle) insulin. The amino terminus is at the left in both chains.

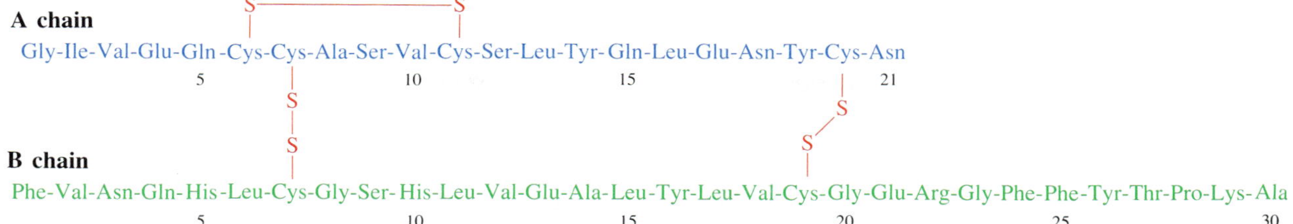

A more-complex sequence of amino acids is found in the protein hormone insulin (Figure 27-1), a drug important in the treatment of diabetes because of its ability to regulate glucose metabolism. Because most synthetic methods give only low yields, a major source of insulin, until recently, has been the pancreas of slaughtered animals. An exciting new development is the efficient preparation of the protein by genetic-engineering methods.

Insulin contains fifty-one amino acid residues incorporated into two chains, labeled A and B. The chains are connected by two disulfide bridges, and there is an additional such linkage connecting the cysteine residues at positions 6 and 11 of the A chain, causing it to loop. Both chains fold up in a way that minimizes steric interference and maximizes electrostatic, London, and hydrogen-bonding attractions. These forces give rise to a fairly condensed three-dimensional structure (Figure 27-2).

FIGURE 27-2

Three-dimensional structure of insulin. Residues on chain A are blue, those on B green. The disulfide bridges are indicated in red. (After *Biochemistry*, 2d ed., by Lubert Stryer. W. H. Freeman and Company. Copyright © 1975, 1981.)

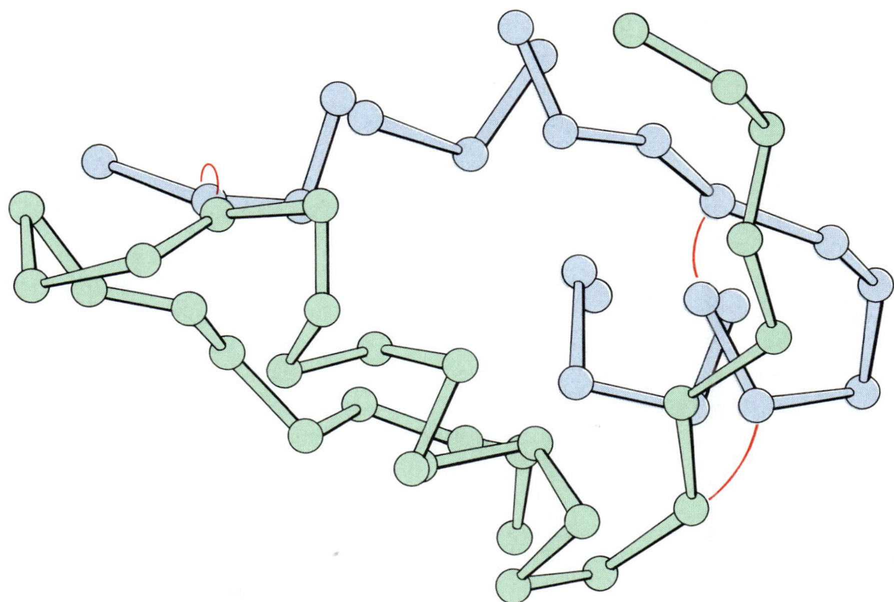

EXERCISE 27-6

Vasopressin, also known as antidiuretic hormone, controls the excretion of water from the body. Write its full structure. Note that there is an intramolecular disulfide bridge between the two cysteine molecules.

```
S—————————————S
|             |
Cys-Tyr-Phe-Gln-Asn-Cys-Pro-Arg-Gly-NH2
```

Vasopressin

Proteins Fold in Space: Secondary and Tertiary Structures

Like insulin, other polypeptide chains adopt well-defined three-dimensional structures rather than the loose, random arrangement exemplified by Figure 27-3A. Whereas the sequence of amino acids in the chain defines the *primary structure,* its folding pattern gives rise to the *secondary structure* of the polypeptide. The secondary structure is due in part to the presence of disulfide bridges but is mainly a result of the rigidity of the amide bond and the maximization of hydrogen and other noncovalent bonding along the chain(s). Two important arrangements are the *pleated sheet,* or *β configuration,* and the *α helix* (Figure 27-3B and C).

In the pleated sheet (Figure 27-3B), two chains line up so as to place the amino groups of one peptide opposite the carbonyl groups of a second, allowing for hydrogen bonds to form. Such bonds can also develop within a single chain if it loops back on itself. Multiple hydrogen bonding of this type can impart considerable rigidity to a system. In the β configuration, adjacent planes defined by the three atoms (C, O, N) of the amide linkages form a specific angle, giving rise to a pleated-sheet structure (Figure 27-4, on page 1246).

The α helix (Figure 27-3C) allows for intramolecular hydrogen bonding between nearby amino acids in the chain. There are 3.6 amino acids per turn of the helix, two equivalent points in neighboring turns being about 5.4 Å apart.

Not all polypeptides adopt idealized structures such as those shown in Figure 27-3B and C. Some (or parts of some) exhibit random arrangements (Figure 27-3A), sometimes in equilibrium with one or both of the more-ordered alternatives. Such equilibria may depend strongly on the pH of the solution surrounding the polymer. If too much charge of the same kind builds up along the chain, charge repulsion will enforce a more-random orientation. In addition, the bulky proline, because its amino nitrogen is also part of the substituent ring, can cause a kink or bend in an α helix.

The further folding, coiling, and other aggregation of polypeptides give rise to their *tertiary structure*. A variety of forces, all arising from the R group, come into play to stabilize such molecules, including disulfide bridges, hydrogen bonds, London forces, and electrostatic attraction and repulsion. There are also *micellar effects* (Section 18-4, Figure 18-1): the polymer adopts a structure that maximizes exposure of polar groups to the aqueous environment, while minimizing exposure of hydrophobic groups (e.g., alkyl and phenyl). Pronounced folding is observed in the **globular proteins,** many of which perform chemical transport and catalysis (e.g., myoglobin and hemoglobin, Section 27-6). In the **fibrous proteins,** such as *myosin* (in muscle) and *α-keratin* (in hair, nails, and wool), several α helices are coiled in such a way as to produce a **superhelix,** a structure made out of a helical polypeptide chain (Figure 27-5).

FIGURE 27-3

Various structural arrangements of a polypeptide: (A) random structure; (B) the pleated sheet, or β configuration, which is held in place by hydrogen bonds (dotted lines) between two polypeptide strands; (C) the α helix, in which the polymer chain is arranged as a right-handed spiral held rigidly in shape by intramolecular hydrogen bonds. (After "Proteins," by Paul Doty, *Scientific American*, September 1957. Copyright © 1957. Scientific American, Inc.)

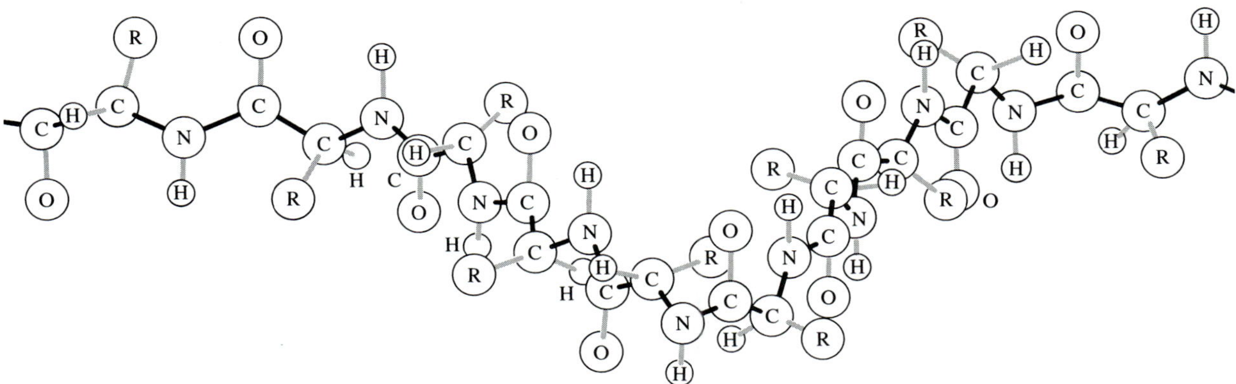

A

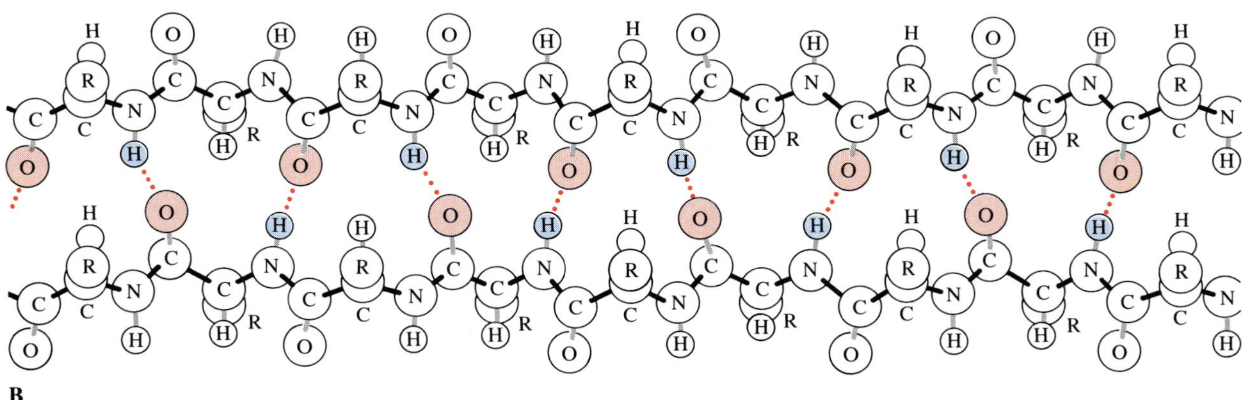

B

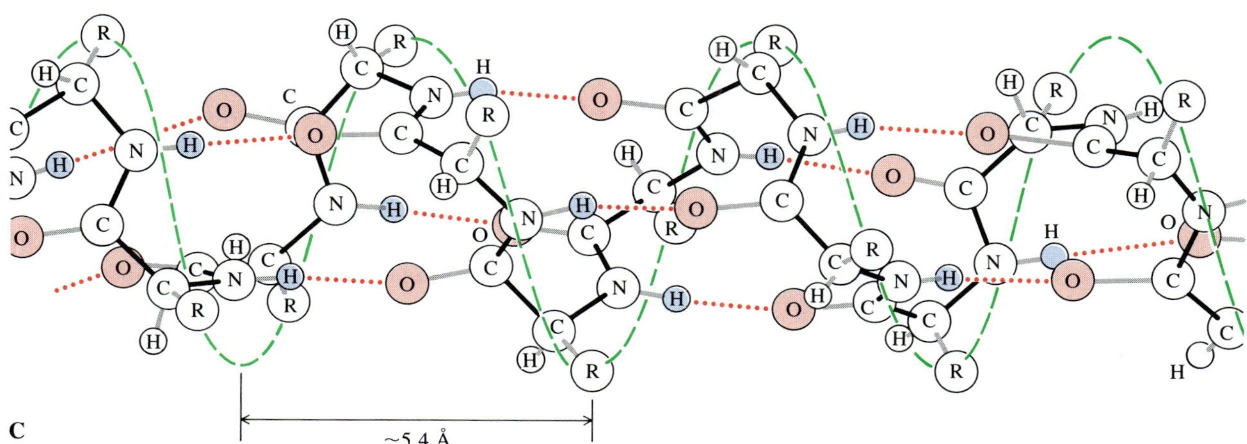

C

FIGURE 27-4

Pleated-sheet structure of a polypeptide. The colored peptide bonds define the individual pleats; the positions of the side chains, R, are alternately above and below the planes of the sheets. The dotted lines indicate hydrogen bonds to a neighboring chain or to water.

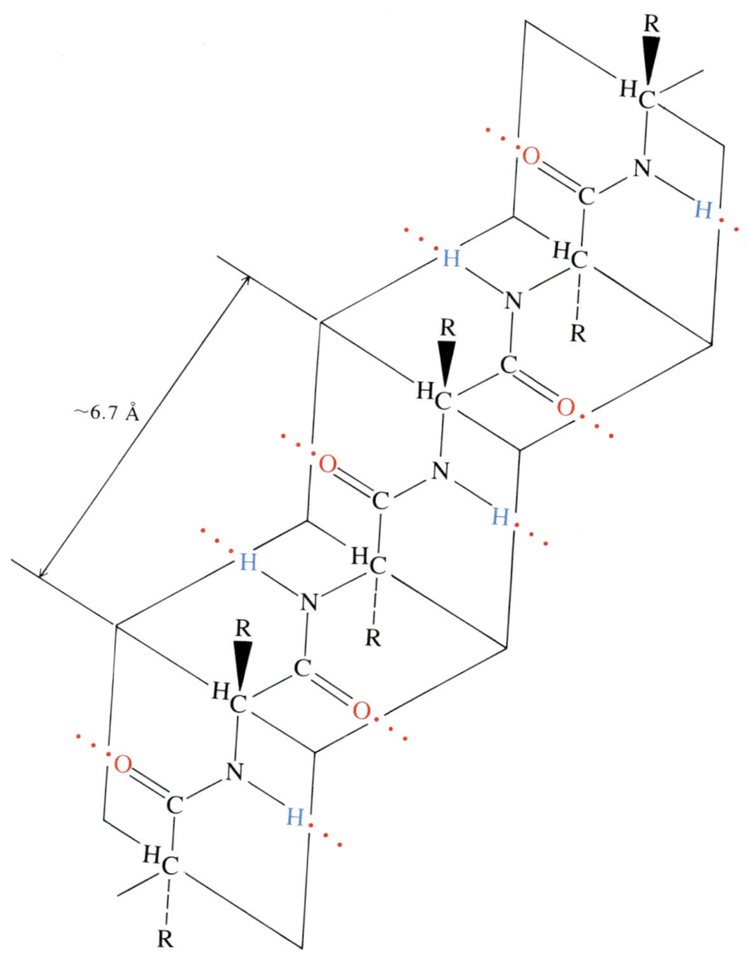

FIGURE 27-5

Idealized picture of a superhelix, a helix within a helix.

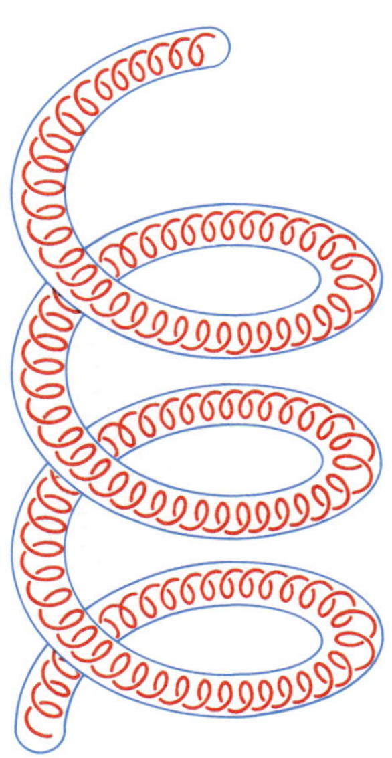

The tertiary structure of a protein is important because it is likely to define a pocket, or active site, specifically designed to fit a substrate of the protein. In enzymes, this specificity facilitates substrate complexation, activation, and turnover. In other proteins, it simply may allow for attachment of a molecule that is to be transported to some other site. Denaturation, or breakdown of the tertiary structure of a protein, usually causes precipitation of the protein and destroys any catalytic activity. Denaturation is caused by exposure to excessive heat or extreme pH values. Think, for example, of what happens to clear egg white when it is poured into a hot frying pan or to milk when it is added to lemon tea.

Some molecules, such as hemoglobin (Section 27-6), also adopt a *quaternary structure,* in which two or more amino acid chains, each with its own tertiary structure, combine to form a larger assembly.

In summary, polypeptides are polymers of amino acids linked by amide bonds. Their amino acid sequence can be described in a shorthand notation using the three-letter abbreviations compiled in Table 27-1. The amino end group is placed at the left, the carboxy end at the right. Polypeptides can be cyclic and can also be linked by disulfide and hydrogen bonds. The sequence of amino acids is the primary structure of a polypeptide, folding gives rise to its secondary structure, and further folding, coiling, and aggregation to its tertiary structure.

27-4 Determination of Polypeptide Primary Structure: Amino Acid Sequencing

Biological function requires definite amino acid sequences in polypeptides and proteins. In the late 1950s and early 1960s, it was discovered that such sequences are genetically predetermined by DNA, the molecule containing our hereditary material (see Section 27-7). The identification of the primary structure of polypeptides, called **polypeptide sequencing,** therefore strikes at the heart of the question of how DNA expresses its genetic material. Moreover, knowledge of the amino acid sequence of a protein helps to understand its mechanism of action. The function of a protein may be completely changed by the incorporation of a "monkey wrench" residue into a normal structure. For example, sickle-cell anemia, a potentially lethal disease, is the result of changing a single amino acid in the protein hemoglobin (Section 27-6). Polypeptide sequencing is also important in efforts to understand the evolution of life. Thus, functionally similar proteins in related species might be expected to have similar amino acid sequences. The closeness of relationship might be revealed by establishing primary structure. This section shows how chemical means, in conjunction with analytical techniques, allow us to do just that.

The First Step: Break the Sulfide Bridges and Purify

Many polypeptides are made up of two or more chains linked by disulfide bridges. The first step in the analysis of such structures is to break these bonds and separate the resulting subunits. One way of doing this is to oxidize the disulfide bridges to sulfonic acids with peroxymethanoic (performic) acid.

$$\text{Chain A}-S-S-\text{Chain B} \xrightarrow{HCOOH\ (O=)} \text{Chain A}-SO_3H \quad HO_3S-\text{Chain B}$$

The problem of subsequent purification is a great one, and attempts at its solution consume many hours in the laboratory. Several techniques are available that separate polypeptides on the basis of size, solubility in a particular solvent, charge, and ability to bind to a support.

In *dialysis,* the polypeptide is separated from smaller fragments literally by filtration through a semipermeable membrane. A second method, *gel-filtration chromatography,* uses a carbohydrate polymer in the form of a column of beads as a support. Smaller molecules diffuse on average more easily into the beads, spending a longer time on the column than large ones do, thus eluting more slowly. In *ion-exchange chromatography,* a charged support separates molecules according to the amount of charge they carry. For example, a column containing carboxylate groups binds positively charged (e.g., amine-protonated) polypeptides but not neutral and negative species. Separation can be effected by changing the pH of the eluting solvent, which ultimately controls the charge level on the support and on the polypeptides to be separated.

A powerful application of this principle is *electrophoresis*. Here the mobility of a polypeptide in an electric field is measured. A spot of the mixture to be separated is placed on a piece of absorbing paper that is moistened with a buffered aqueous solution and attached to two electrodes. When the voltage is turned on, positively charged species move toward the cathode, negatively charged molecules toward the anode. The velocity, V, of each migrating species is proportional to the strength of the field, E, and the net charge, Z, on the molecule (in turn, dependent on pH) and inversely proportional to a shape-related constant, f, the *frictional resistance*.

$$V = \frac{EZ}{f}$$

The separating power of this technique is extraordinary. More than a thousand different proteins from one species of bacterium have been resolved in a single experiment.

Finally, *affinity chromatography* exploits the tendency of polypeptides to bind weakly to certain supports by hydrogen bonds and other attractive forces. Peptides of differing sizes and shapes have differing retention times in a column containing such a support.

Once the individual polypeptide strands obtained by breaking the disulfide bridges are purified, the next step in structural analysis is to establish their amino acid composition.

The Second Step: Which Amino Acids Are Present?

To determine which amino acids and how much of each is present in the polypeptide, the entire chain is degraded by amide hydrolysis (6 N HCl, 110°C, 24 h) to give a mixture of the free amino acids. The mixture is separated on an automated **amino acid analyzer.** This instrument consists of an ion-exchange column bearing a negatively charged support, usually containing carboxylate or sulfonate ions. The amino acids are passed through the column in slightly acidic solution. Depending on their structure, they are more or less protonated and therefore are more or less retained on the column; thus they separate. Further control is achieved by slowly increasing the pH of the eluting solvent; this causes the eventual deprotonation, and therefore elution, of the amino acids, beginning with the most acidic and ending with the most basic.

At the end of the column is an analyzer, a reservoir containing a special indicator. Eluted amino acids produce a violet color whose intensity is propor-

FIGURE 27-6

The result, recorded as a chromatogram, of separating various amino acids on an amino acid analyzer using a polysulfonated ion-exchange resin. The more-acidic products (e.g., aspartic acid) are generally eluted first. Ammonia is included for comparison.

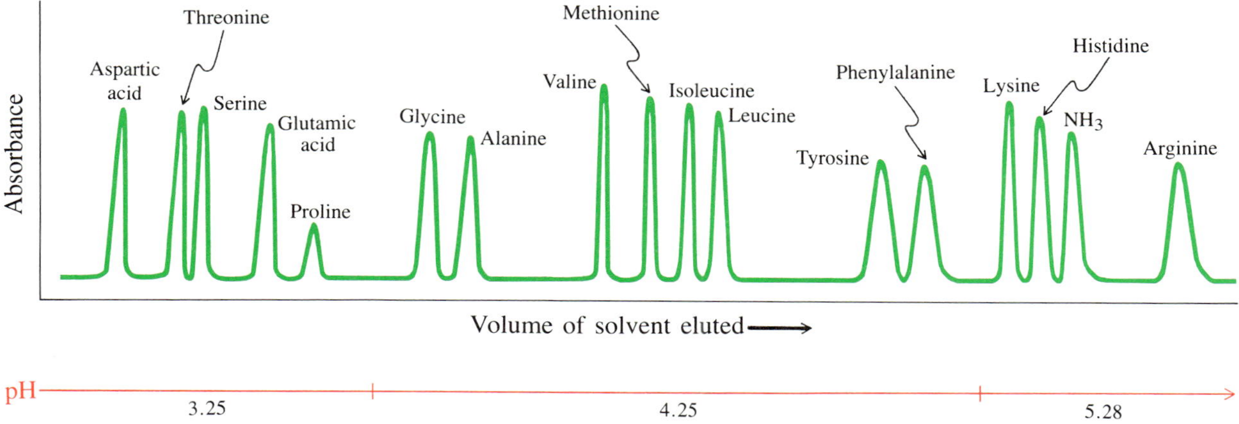

tional to the amount of acid present. These color changes can be recorded as a function of time or of the volume of solvent eluted, resulting in a characteristic chromatogram (Figure 27-6). The area under each peak is a measure of the relative amount of a specific amino acid in the mixture.

The amino acid analyzer can readily establish the composition of a polypeptide. For example, the chromatogram of hydrolyzed glutathione (Section 27-3) gives three equally sized peaks, corresponding to glutamic acid, glycine, and cysteine.

EXERCISE 27-7

Give the expected results of the amino acid analysis of the A chain in insulin (Figure 27-1).

Given the gross make-up of a polypeptide, other methods are used to reveal the order in which the individual amino acids are bound to each other: the amino acid sequence.

The Third Step: Amino Acid Sequencing from the Amino Terminus

Several different chemical methods can reveal the identity of the residue at the amino terminus in a polypeptide. They all exploit the uniqueness of this group: it is the only one bearing a free amino substituent. This property allows for specific chemical reactions that serve to tag the terminal amino acid.

One such procedure is the **Sanger* degradation.** In this process, the peptide is first exposed to 1-fluoro-2,4-dinitrobenzene. The amino group of the amino-terminal residue attacks this reagent to effect nucleophilic aromatic substitution (Section 20-6), picking up the dinitrophenyl tag. After complete hydrolysis of the polypeptide, the tagged amino acid is readily identified in the mixture by its chromatographic behavior.

---

*Professor Frederick Sanger, b. 1918, Cambridge University, England, Nobel Prize 1958 and 1980.

TABLE 27-2
The specificity of hydrolytic enzymes in polypeptide cleavage

Enzyme	Site of cleavage
Trypsin	Lys, Arg, carboxy end
Clostripain	Arg, carboxy end
Chymotrypsin	Phe, Trp, Tyr, carboxy end
Pepsin	Asp, Glu, Leu, Phe, Trp, Tyr, carboxy end
Thermolysin	Leu, Ile, Val, amino end

EXERCISE 27-9

A polypeptide containing twenty-one amino acids was hydrolyzed by thermolysin. The products of this treatment were Gly, Ile, Val-Cys-Ser, Leu-Tyr-Gln, Val-Glu-Gln-Cys-Cys-Ala-Ser, and Leu-Glu-Asn-Tyr-Cys-Asn. When the same polypeptide was hydrolyzed by chymotrypsin, the products were Cys-Asn, Gln-Leu-Glu-Asn-Tyr, and Gly-Ile-Val-Glu-Gln-Cys-Cys-Ala-Ser-Val-Cys-Ser-Leu-Tyr. Give the amino acid sequence of this molecule.

In summary, the structure of polypeptides is established by various degradation schemes. First, disulfide bridges are broken, and then the kind and relative abundance of the component amino acids in each polypeptide are determined by complete hydrolysis and amino acid analysis. The amino-terminal residues can be identified by Sanger or Edman degradation. Repeated Edman degradation gives the sequence of shorter polypeptides. Such shorter pieces are made from longer polypeptides by specific enzymatic hydrolysis.

27-5 Synthesis of Polypeptides: A Challenge in the Application of Protecting Groups

In a sense, the topic of peptide synthesis is a trivial one: only one type of bond, the amide linkage, has to be made. The formation of this linkage was described in Section 17-9; why discuss it further? This section shows that, in fact, achieving selectivity poses great problems, for which specific solutions have to be found. Consider even as simple a target as the dipeptide glycylalanine. Just heating glycine and alanine to make the peptide bond by dehydration would result in a complex mixture of di-, tri-, and higher peptides with random sequences. This is so because both reactive ends of the two starting materials would randomly form bonds to their own kind and to each other and because there is no way to prevent uncontrolled and random oligomerization.

An Attempt at the Synthesis of Glycylalanine by Thermal Dehydration

$$\text{Gly} + \text{Ala} \xrightarrow[-\,H_2O]{\Delta} \text{Gly-Gly} + \text{Ala-Gly} + \underset{\substack{\text{Desired}\\ \text{product}}}{\text{Gly-Ala}} + \text{Ala-Ala} + \text{Gly-Gly-Ala} + \text{Ala-Gly-Ala}\ \text{etc.}$$

Selective Peptide Synthesis Requires Protecting Groups

To form peptide bonds selectively, the functional groups of the amino acids have to be protected. There are amino- and carboxy-protecting groups. The amino group is frequently blocked by a **phenylmethoxycarbonyl group** (abbreviated **carbobenzoxy** or **Cbz**), introduced by reaction of an amino acid with phenylmethyl chloromethanoate (benzyl chloroformate).

Protection of the Amino Group in Glycine

$$\overset{+}{H_3N}CH_2COO^- + C_6H_5CH_2O\overset{O}{\overset{\|}{C}}Cl \xrightarrow[-\ NaCl,\ -\ HOH]{NaOH} C_6H_5CH_2O\overset{O}{\overset{\|}{C}}NHCH_2COOH\ \ 80\%$$

Glycine; Phenylmethyl chloromethanoate (Benzyl chloroformate); Phenylmethoxycarbonylglycine (Carbobenzoxyglycine, Cbz-Gly)

The amino group is deprotected by hydrogenolysis (Section 24-2), which initially furnishes the carbamic acid as a reactive intermediate (Section 18-5). Decarboxylation occurs instantly to restore the amino function.

Deprotection of the Amino Group in Glycine

$$C_6H_5CH_2O\overset{O}{\overset{\|}{C}}NHCH_2COOH \xrightarrow{H_2,\ Pd\text{-}C} C_6H_5CH_2H + \underbrace{HO\overset{O}{\overset{\|}{C}}NH}_{\text{Carbamic acid function}}CH_2COOH \longrightarrow CO_2 + \overset{+}{H_3N}CH_2COO^-\ \ 95\%$$

Another amino-protecting group is **1,1-dimethylethoxycarbonyl (*tert*-butoxycarbonyl, Boc),** introduced by reaction with bis(1,1-dimethylethyl) dicarbonate (di-*tert*-butyl dicarbonate).

Protection of the Amino Group in Amino Acids as the Boc Derivative

$$\overset{+}{H_3N}\overset{R}{\overset{|}{C}}HCOO^- + (CH_3)_3CO\overset{O}{\overset{\|}{C}}O\overset{O}{\overset{\|}{C}}OC(CH_3)_3 \xrightarrow[-\ CO_2,\ -\ (CH_3)_3COH]{(CH_3CH_2)_3N} (CH_3)_3CO\overset{O}{\overset{\|}{C}}NH\overset{R}{\overset{|}{C}}HCOOH\ \ 70\%\text{–}100\%$$

Bis(1,1-dimethylethyl) dicarbonate (Di-*tert*-butyl dicarbonate); 1,1-Dimethylethoxycarbonylamino acid (*tert*-Butoxycarbonylamino acid, Boc-amino acid)

Deprotection in this case is achieved by treatment with acid under conditions mild enough to leave other peptide bonds untouched.

General Deprotection of Boc-Amino Acids

$$(CH_3)_3CO\overset{O}{\overset{\|}{C}}NH\overset{R}{\overset{|}{C}}HCOOH \xrightarrow{HCl\ \text{or}\ CF_3COOH,\ 25°C} \overset{+}{H_3N}\overset{R}{\overset{|}{C}}HCOO^- + CO_2 + CH_2{=}C(CH_3)_2$$

EXERCISE 27-10

The mechanism of the deprotection of Boc-amino acids is different from that of the normal ester hydrolysis (Section 18-4): it proceeds through the intermediate 1,1-dimethylethyl (*tert*-butyl) cation. Formulate this mechanism.

The carboxy terminus of an amino acid is protected by the formation of a simple ester such as methyl or ethyl. Deprotection results from treatment with base; this reaction is selective because esters are more reactive toward nucleophiles than amides are. Phenylmethyl alkanoates can be cleaved by hydrogenolysis under neutral conditions.

With the ability to protect either end of the amino acid, it is clear how a selective peptide synthesis may be achieved: by coupling an amino-protected with a carboxy-protected unit.

Formation of the Peptide Bond Proceeds by Carboxy Activation

Because the protecting groups are sensitive to acid and base, the peptide bond must be formed under the mildest possible conditions. Special carboxy-activating reagents are used.

Perhaps the most general of these reagents is **dicyclohexylcarbodiimide (DCC).** The electrophilic reactivity of this molecule is similar to that of ketene (Section 18-3) or an isocyanate; it is ultimately hydrated to *N,N′*-dicyclohexylurea.

Peptide Bond Formation with Dicyclohexylcarbodiimide

RCOOH + R′NH₂ + C₆H₁₁N=C=NC₆H₁₁ ⟶ RC(=O)NHR′ + O=C(NHC₆H₁₁)₂

Dicyclohexylcarbodiimide **N,N′-Dicyclohexylurea**

The mechanism of this transformation is thought to begin with addition of the carboxylic acid to DCC, as it would add to ketene to give an anhydride (Section 18-3). Like an anhydride, the product contains an activated carbonyl group that is susceptible to attack by the amine by an addition-elimination mechanism.

Mechanism of Peptide-Bond Formation with DCC:

RC(=O)OH + C₆H₁₁N=C=NC₆H₁₁ ⟶ ⟶ RC(=O)OC(=NC₆H₁₁)NHC₆H₁₁ —(H₂N—R′)⟶ ⟶ R—C(OH)(HNR′)—O—C(=NC₆H₁₁)NHC₆H₁₁ ⟶ RC(=O)NHR′ + O=C(NHC₆H₁₁)₂

Armed with this knowledge, let us return to the problem of the synthesis of glycylalanine. It is solved by adding amino-protected glycine to an alanyl ester in the presence of DCC. The resulting product is then deprotected to give the desired dipeptide.

Preparation of Gly-Ala

$$\underset{\textbf{Boc-Gly}}{(CH_3)_3C\overset{O}{\overset{\|}{C}}NHCH_2COOH} + \underset{\textbf{Ala-OCH}_2\textbf{C}_6\textbf{H}_5}{H_2N\overset{H_3C}{\overset{|}{C}}H\overset{O}{\overset{\|}{C}}OCH_2C_6H_5} \xrightarrow{DCC}$$

$$\underset{\textbf{Boc-Gly-Ala-OCH}_2\textbf{C}_6\textbf{H}_5}{(CH_3)_3CO\overset{O}{\overset{\|}{C}}NHCH_2\overset{O}{\overset{\|}{C}}NH\overset{H_3C}{\overset{|}{C}}H\overset{O}{\overset{\|}{C}}OCH_2C_6H_5} \xrightarrow[2.\ H_2,\ Pd\text{-}C]{1.\ H^+,\ H_2O}$$

$$\underset{\textbf{Gly-Ala}}{H_3\overset{+}{N}CH_2\overset{O}{\overset{\|}{C}}NH\overset{CH_3}{\overset{|}{C}}HCOO^-} + C_6H_5CH_3 + CO_2 + CH_2{=}C(CH_3)_2$$

For the preparation of a higher peptide, deprotection of only one end is required, followed by renewed coupling, and so on.

EXERCISE 27-11

Propose a synthesis of Leu-Ala-Val from the component amino acids.

Polypeptide Preparation Can Be Automated: The Merrifield Solid-Phase Synthesis

An ingenious approach to the synthesis of polypeptides has been developed by Merrifield.* It relies on the use of a solid support of polystyrene. The technique is known as **solid-phase synthesis.**

Polystyrene beads, although insoluble, swell considerably in certain organic solvents, such as dichloromethane. The swollen material allows reagents to move in and out of the polymer matrix easily. Thus, the phenyl rings can be functionalized by electrophilic aromatic substitution.

Electrophilic Chloromethylation of Polystyrene

$$\underset{\textbf{Polystyrene}}{\text{–}CH_2\text{–}\underset{C_6H_5}{CH}\text{–}CH_2\text{–}\underset{C_6H_5}{CH}\text{–}} \xrightarrow[-\ CH_3CH_2OH]{ClCH_2OCH_2CH_3,\ SnCl_4} \underset{\textbf{Functionalized polystyrene}}{\text{–}CH_2\text{–}\underset{C_6H_4CH_2Cl}{CH}\text{–}CH_2\text{–}\underset{C_6H_5}{CH}\text{–}}$$

For the purposes of solid-phase polypeptide synthesis, polystyrene is 1%–10% chloromethylated. This low level of incorporation ensures that the reacting centers on the bead surface do not interfere with each other.

*Professor Robert B. Merrifield, b. 1921, Rockefeller University, New York, Nobel Prize 1984.

EXERCISE 27-12

Formulate a plausible mechanism for the chloromethylation of the benzene rings in polystyrene.

A dipeptide synthesis on chloromethylated polystyrene would proceed as follows.

Solid-Phase Synthesis of a Dipeptide

$$(CH_3)_3CO\overset{O}{\overset{\|}{C}}NH\overset{R}{\overset{|}{C}}HCOO^- + ClCH_2\text{—C}_6\text{H}_4\text{—}\sim\text{polystyrene chain}$$

1. Attachment of protected amino aid — $- Cl^-$

$$(CH_3)_3CO\overset{O}{\overset{\|}{C}}NH\overset{R}{\overset{|}{C}}H\overset{O}{\overset{\|}{C}}OCH_2\text{—C}_6\text{H}_4\text{—}\sim\text{polystyrene chain}$$

2. Deprotection of amino terminus — CF_3CO_2H, CH_2Cl_2

$$H_2N\overset{R}{\overset{|}{C}}H\overset{O}{\overset{\|}{C}}OCH_2\text{—C}_6\text{H}_4\text{—}\sim\text{polystyrene chain}$$

3. Coupling to protected second amino acid — $(CH_3)_3CO\overset{O}{\overset{\|}{C}}NH\overset{R'}{\overset{|}{C}}HCOOH$, DCC

$$(CH_3)_3CO\overset{O}{\overset{\|}{C}}NH\overset{R'}{\overset{|}{C}}H\overset{O}{\overset{\|}{C}}NH\overset{R}{\overset{|}{C}}H\overset{O}{\overset{\|}{C}}OCH_2\text{—C}_6\text{H}_4\text{—}\sim\text{polystyrene chain}$$

4. Deprotection of amino terminus — CF_3CO_2H, CH_2Cl_2

$$H_2N\overset{R'}{\overset{|}{C}}H\overset{O}{\overset{\|}{C}}NH\overset{R}{\overset{|}{C}}H\overset{O}{\overset{\|}{C}}OCH_2\text{—C}_6\text{H}_4\text{—}\sim\text{polystyrene chain}$$

5. Disconnection of dipeptide from polymer — HF

$$\overset{+}{H_3}N\overset{R'}{\overset{|}{C}}H\overset{O}{\overset{\|}{C}}NH\overset{R}{\overset{|}{C}}HCOO^- + FCH_2\text{—C}_6\text{H}_4\text{—}\sim\text{polystyrene chain}$$

Dipeptide

First an amino-protected amino acid is anchored on the polystyrene by nucleophilic substitution of the benzylic chloride by carboxylate. Deprotection is then followed by coupling with a second amino-protected amino acid. Renewed deprotection and final removal of the dipeptide by treatment with hydrogen fluoride completes the sequence.

The great advantage of the solid-phase approach is the ease with which products may be isolated. Because all the intermediates are immobilized on the polymer, they can be purified by simple filtration and washing. Obviously, it is not necessary to stop at the dipeptide stage. Repetition of the deprotection-coupling sequence leads to larger and larger peptides. Merrifield designed a machine that would carry out the required series of manipulations automatically, each cycle requiring only a few hours. In this way, the first total synthesis of the protein insulin was accomplished. More than five thousand separate operations were required to assemble the fifty-one amino acids in the two separate chains, but, thanks to the automated procedure, this took only several days.

Automated protein synthesis opens up exciting possibilities. First, it can be used to confirm the structure of polypeptides that have been analyzed by chain degradation and sequencing. Second, it can be used to construct unnatural proteins that might be more active and more specific than natural ones. Such proteins could be invaluable in the treatment of disease or in the understanding of biological function and activity.

In summary, polypeptides are made by coupling an amino-protected amino acid with another in which the carboxy end is protected. Typical protecting groups are readily cleaved esters and related functions. Coupling proceeds under mild conditions with dicyclohexylcarbodiimide as a dehydrating agent. Solid-phase synthesis is an automated procedure in which a carboxy-anchored peptide chain is built up from amino-protected monomers by cycles of coupling and deprotection.

27-6 Polypeptides in Nature: Oxygen Transport by the Proteins Myoglobin and Hemoglobin

This section deals with two natural polypeptides that function as oxygen carriers in vertebrates: the proteins myoglobin and hemoglobin. Myoglobin is active in the muscle, where it stores oxygen and releases it when needed. Hemoglobin is contained in red blood cells and facilitates oxygen transport. Without its presence, blood would be able to absorb only a fraction (about 2%) of the oxygen needed by the body. How is the oxygen bound in these proteins?

Active Sites in Myoglobin and Hemoglobin: The Heme Group

The secret of the oxygen-carrying ability of myoglobin and hemoglobin is a special nonpolypeptide unit, called a **heme group,** attached to the protein. Heme is a cyclic organic ligand (called a **porphyrin**) made out of four linked, substituted pyrrole units surrounding an iron atom (Figure 27-7). The complex is red, giving blood its characteristic color.

The iron in the heme is attached to four nitrogens but can accommodate two additional groups above and below the plane of the porphyrin ring. In myoglobin, one of these groups is the imidazole ring of a histidine unit attached to one of the α-helical segments of the protein (Figure 27-8A). The other is most important for the protein's function: oxygen. Close to the oxygen-binding site, there is a second imidazole of a histidine unit, which appears to protect this side of the heme by steric hindrance. For example, carbon monoxide, which also binds to the iron in the heme group, and thus blocks oxygen transport, is prevented from binding as strongly as it normally would because of the presence of the second imidazole group. Consequently, carbon monoxide poisoning can be reversed by administering oxygen to a patient who has been exposed to the gas. The two imidazole substituents in the neighborhood of the iron atom in the heme group are brought into close proximity by the unique folding pattern of the

FIGURE 27-7

Porphine is the simplest porphyrin. Note that the system forms an aromatic ring of eighteen delocalized π electrons. The heme group is responsible for binding oxygen. Two of the bonds to iron are dative (coordinate covalent), indicated by arrows.

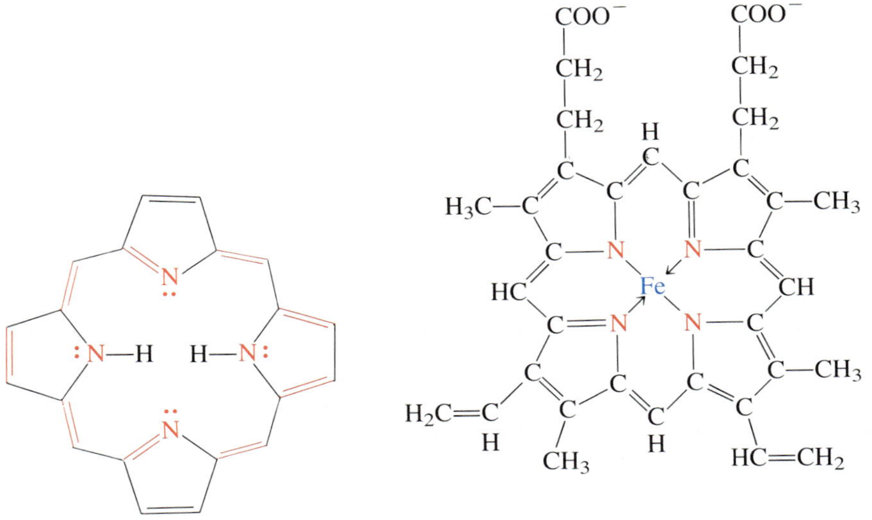

Porphine **Heme**

protein. The rest of the polypeptide chain serves as a mantle, shielding and protecting the active site from unwanted intruders and controlling the kinetics of its action (Figures 27-8B and C).

Myoglobin and hemoglobin offer excellent examples of the four structural levels in proteins. The primary structure of myoglobin consists of 153 amino acid residues of known sequence. Myoglobin has eight α-helical segments that

FIGURE 27-8

A. Schematic representation of the active site in myoglobin.

B. Schematic representation of the tertiary structure of myoglobin and its heme.

C. Secondary and tertiary structure of myoglobin (after "The Hemoglobin Molecule," by M. F. Perutz, *Scientific American,* November, 1964. Copyright © 1964. Scientific American, Inc.).

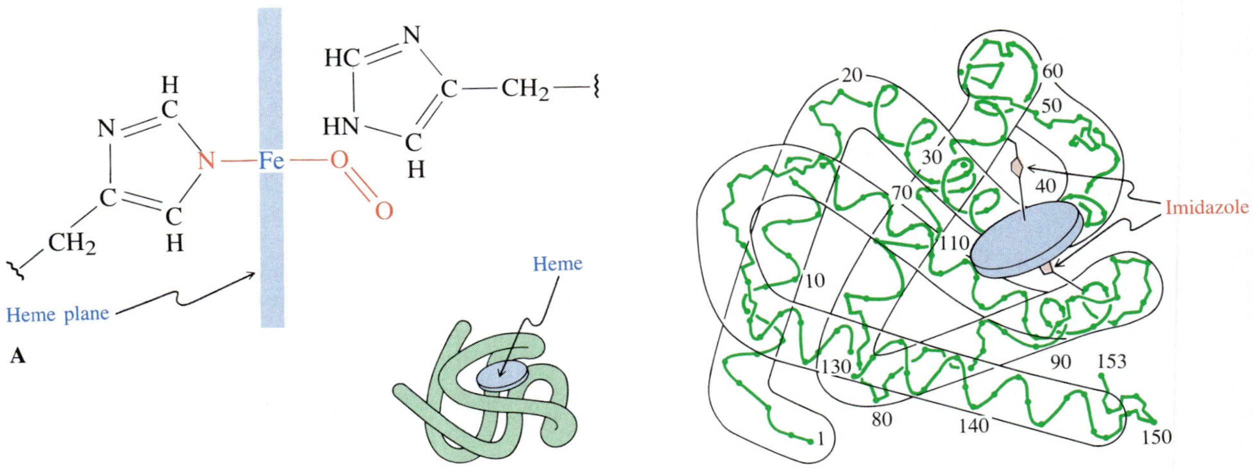

FIGURE 27-9

The quaternary structure of hemoglobin. Each α and β chain has its own heme group. (After R. E. Dickerson and I. Geis, 1969, *The Structure and Action of Proteins,* Benjamin-Cummings, p. 56. Copyright 1969 by Irving Geis.)

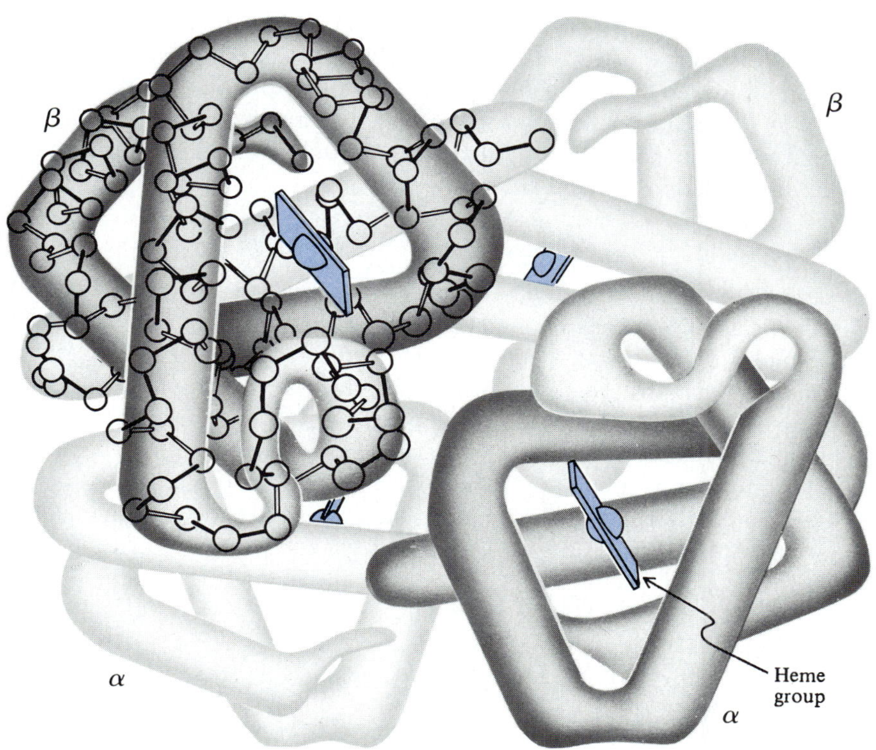

constitute its secondary structure, the longest having 23 residues. The tertiary structure has the bends that give myoglobin its three-dimensional shape.

Hemoglobin contains four protein chains: two ***α* chains** of 141 residues each, and two ***β* chains** of 146 residues each. Each chain has its own heme group and a tertiary structure similar to that of myoglobin. Although there is little interaction between the two α chains or the two β chains, there are many contacts between them. Furthermore, α_1 is closely attached to β_1, as is α_2 to β_2. These interactions give hemoglobin its quaternary structure (Figure 27-9), in addition to its secondary and tertiary structures.

The folding of the hemoglobin and myoglobin of several living species is strikingly similar even though the amino acid sequences differ. This finding appears to indicate that this particular tertiary structure is an optimal configuration around the heme group. It allows the heme to absorb oxygen as it is introduced through the lung, hang on to it as long as necessary for safe transport, and release it when required.

27-7 The Biosynthesis of Proteins: Nucleic Acids

How does nature assemble the structure of proteins? The answer to this question is based on one of the most-exciting discoveries in science, the nature and

FIGURE 27-10
Part of a DNA chain.

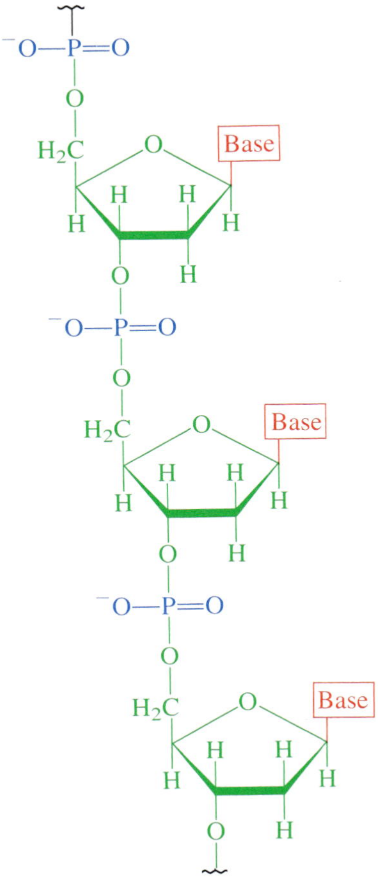

workings of the genetic code. All hereditary information is embedded in the **deoxyribonucleic acids (DNA).** The expression of this information in the synthesis of the many enzymes necessary for cell function is carried out by the **ribonucleic acids (RNA).** After the carbohydrates and polypeptides, the nucleic acids are the third major type of biological polymer. Their repeating units are called **nucleotides.** Just like the amino acids in a polypeptide chain, nucleotides have characteristic structural features. Thus, instead of the amino and the carboxy group linked by a variably substituted carbon atom, nucleotides contain a *phosphate group* linked through a *sugar moiety* to a *nitrogen heterocycle,* the **DNA** or **RNA base.** The final section of this book will give you a glimpse of the structure and function of these biological polymers.

Four Heterocycles Define the Structure of Nucleic Acids

Considering the structural diversity of natural products, the structures of DNA and RNA are simple. All their component units are polyfunctional, and it is one of the wonders of nature that evolution has selected for a very few specific structural combinations. Nucleic acids are polymers in which phosphate units link sugar units bearing various bases (Figure 27-10).

In DNA, the sugar moieties are 2-deoxyriboses, and only four bases are present: cytosine (C), thymine (T), adenine (A), and guanine (G). The sugar characteristic of RNA is ribose, and again there are four bases, but the nucleic acid incorporates uracil (U) instead of thymine.

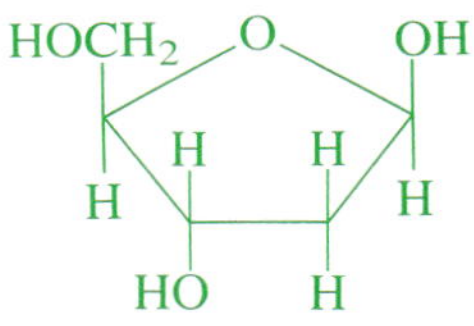

2-Deoxyribose 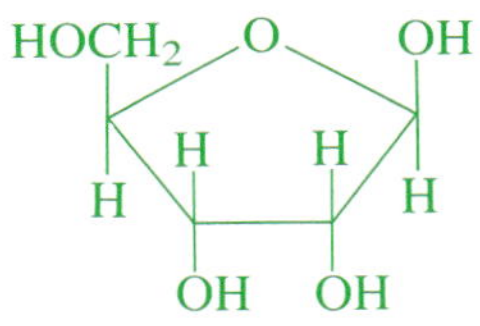 **Ribose**

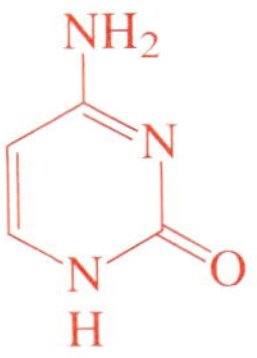

Cytosine (C) 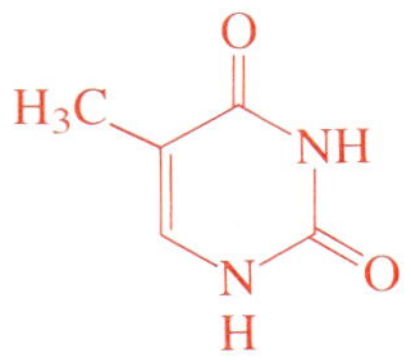 **Thymine (T)**  **Adenine (A)**

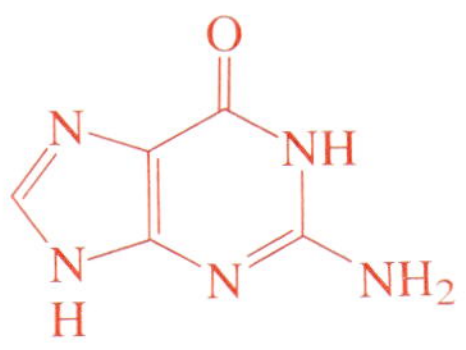

Guanine (G) 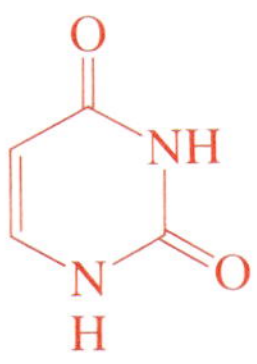 **Uracil (U)**

The nucleotide may be formally assembled by first replacing the hydroxy group at C-1 in the sugar with one of the base nitrogens. The resulting molecular unit is called a **nucleoside.** Second, a phosphate substituent is introduced at C-5. The four bases present in DNA and RNA thus give rise to four nucleotides.

The Four Nucleotides of DNA

2-Deoxyadenylic acid **2-Deoxyguanidylic acid**

2-Deoxycytidylic acid **2-Deoxythymidylic acid**

The Four Nucleotides of RNA

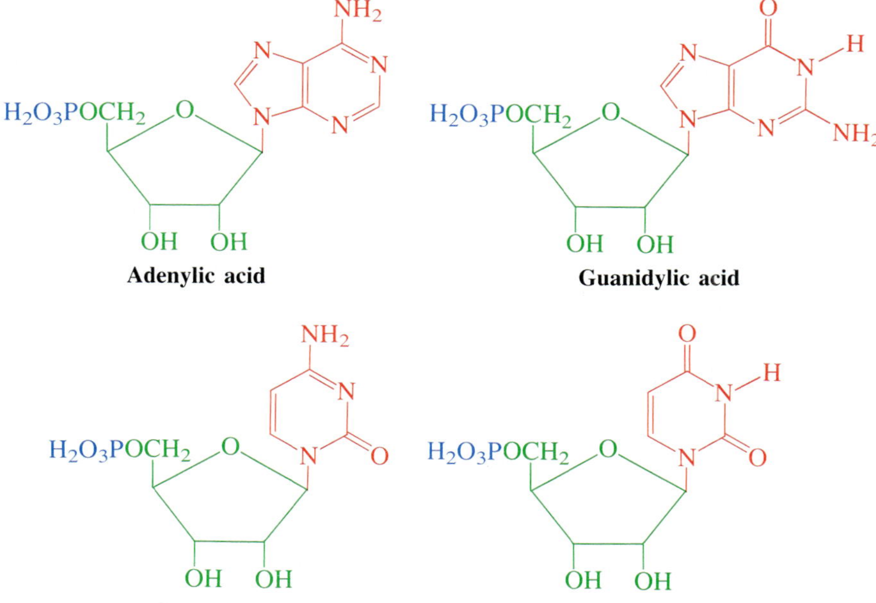

Adenylic acid **Guanidylic acid**

Cytidylic acid **Uridylic acid**

The polymeric chain shown in Figure 27-10 is then readily derived by repeatedly forming a phosphate ester bridge from C-5 of the sugar unit of one nucleotide to C-3 of another. In this polymer, the bases adopt the same role as that of the 2-substituent in the amino acids of a polypeptide: their sequence may vary from one nucleic acid to the other. This base sequence has fundamental biological importance.

FIGURE 27-11

Hydrogen bonding between the complementary base pairs adenine–thymine and guanine–cytosine.

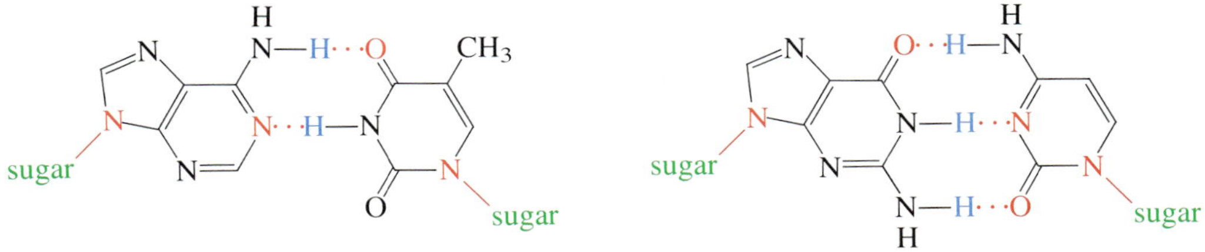

Adenine–thymine **Guanine–cytosine**

The Structure of Nucleic Acids Reveals a Double Helix

Nucleic acids can form extraordinarily long chains with molecular weights in the billions. Like proteins, they adopt secondary and tertiary structures. In 1953, Watson and Crick* made their well-known proposal that DNA had a double-helical structure composed of two strands with complementary base sequences. A crucial piece of information leading to this proposal was that, in the DNA of various species, the ratio of adenine to thymine, like that of guanine to cytosine, was always one to one. This led to the suggestion that two DNA chains are held together by hydrogen bonding in such a way that adenine and guanine in one chain always face thymine and cytosine in the other (Figure 27-11). Thus, if a piece of DNA in one strand has the sequence -A-G-C-T-A-C-G-A-T-C-, this entire segment is hydrogen bonded to a complementary strand -T-C-G-A-T-G-C-T-A-G-, as shown.

```
~~~~—A—G—C—T—A—C—G—A—T—C—~~~~
     :  :  :  :  :  :  :  :  :  :
~~~~—T—C—G—A—T—G—C—T—A—G—~~~~
```

Because of other structural constraints, the arrangement that maximizes hydrogen bonding and minimizes steric repulsion is the double helix (Figure 27-12).

DNA Replicates by Unwinding and the Assembly of New Complementary Strands

There is no restriction on the variety of sequences of the bases in the nucleic acids. Watson and Crick proposed that the specific base sequence of a particular DNA contained all genetic information necessary for the duplication of a cell and, indeed, the growth and development of the organism as a whole. Moreover, the exact complementarity of the double-helical structure led them to suggest a way in which DNA might replicate itself and pass on the genetic code.

*Professor James D. Watson, b. 1928, Harvard University, Nobel Prize 1960 (medicine); Professor Francis H. F. C. Crick, b. 1916, Cambridge University, England, Nobel Prize 1960 (medicine).

FIGURE 27-12

A. The two nucleic acid strands of a DNA double helix are held together by hydrogen bonding between the complementary sets of bases. Note that the two chains run in opposite directions and that all the bases are on the inside of the double helix. The diameter of the helix is 20 Å, base–base separation across the strands is ~3.4 Å, and the helical turn repeats every 34 Å.

B. One of the strands of a DNA double helix, in a view down the axis of the molecule (after *Biochemistry,* 2d ed., by Lubert Stryer. W. H. Freeman and Company. Copyright © 1975, 1981).

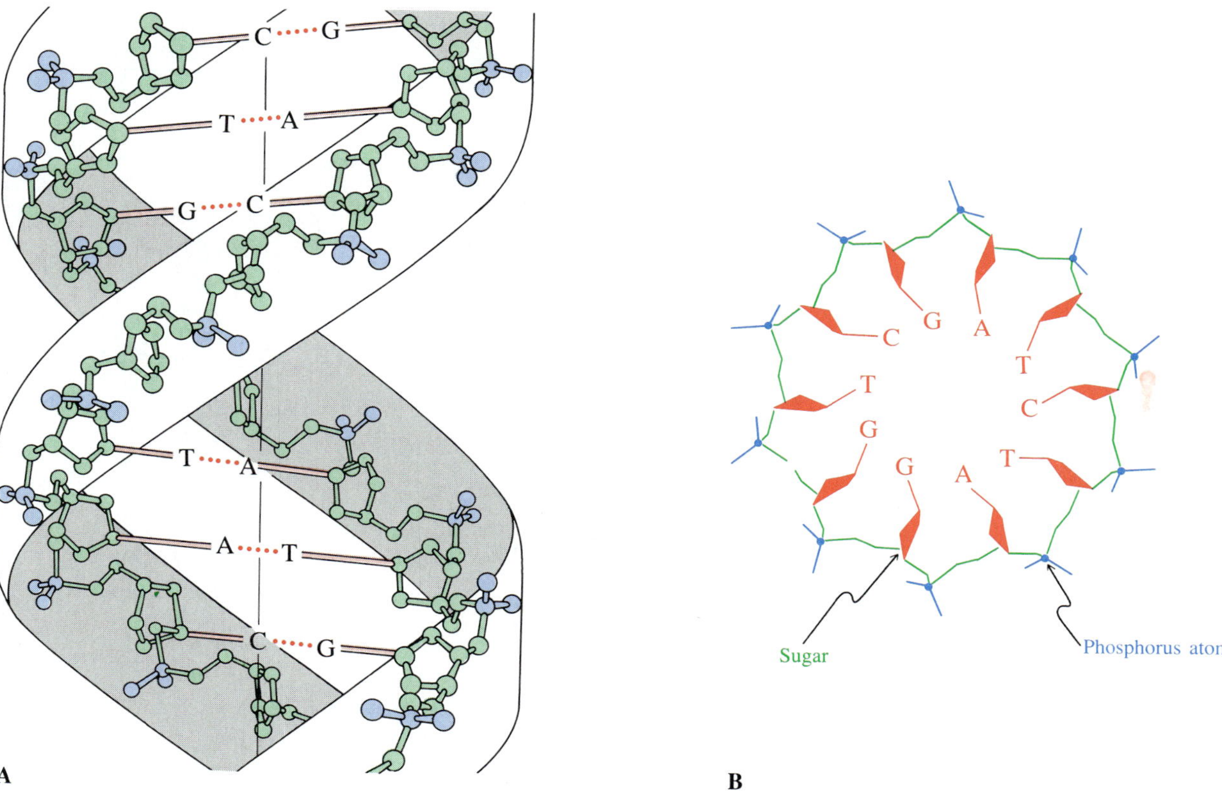

In this mechanism, each of the two strands of DNA functions as a template. The double helix partly unwinds, and enzymes then begin to assemble the new DNA by coupling nucleotides to each other in a sequence complementary to that in the template, always juxtaposing C to G and A to T (Figure 27-13). Eventually, two complete double helices are produced from the original.

Protein Synthesis Through RNA

DNA gives rise to RNA by a replication process that is very similar to that described for DNA. In RNA, however, ribose takes the place of deoxyribose as the repeating sugar unit, and uracil is incorporated instead of thymine. The resulting nucleic acid is called **messenger RNA** (**mRNA;** Figure 27-14). Its chain is much shorter than that of DNA, and it does not stay bound to the template but breaks away as its synthesis is finished.

The mRNA is the template responsible for the correct sequencing of the amino acid units in proteins. Each sequence of three bases, called a **codon,**

FIGURE 27-13

Model for DNA replication. The double helix initially unwinds to two single strands, each of which is used as a template for reconstruction of the complementary nucleic acid sequence.

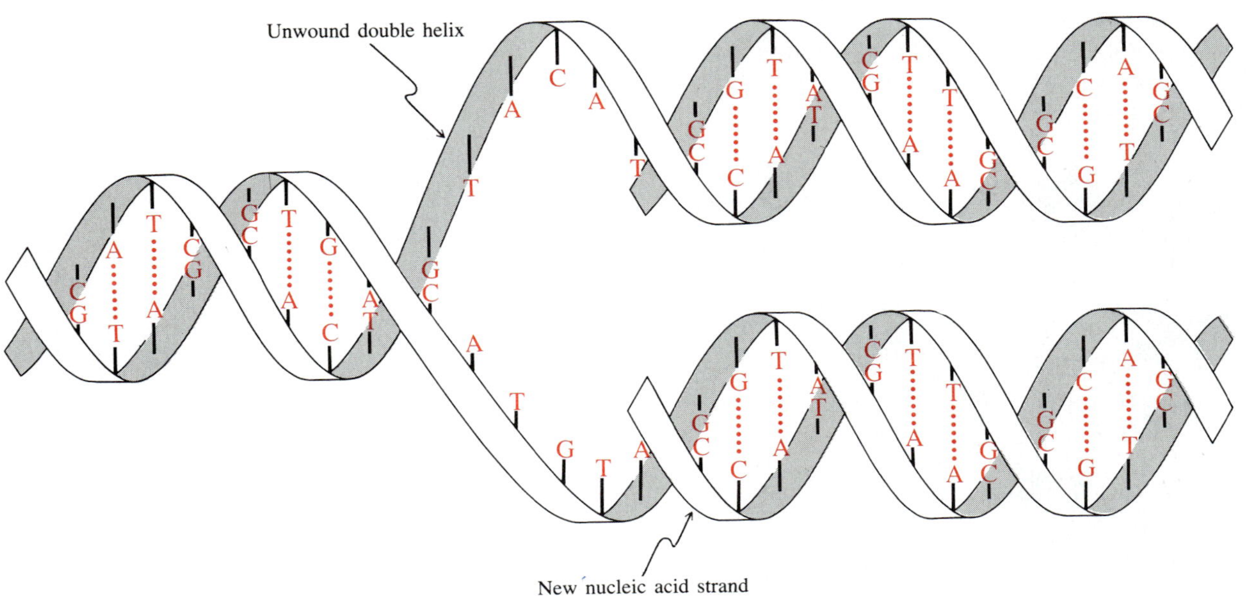

FIGURE 27-14

Simplified picture of messenger RNA synthesis.

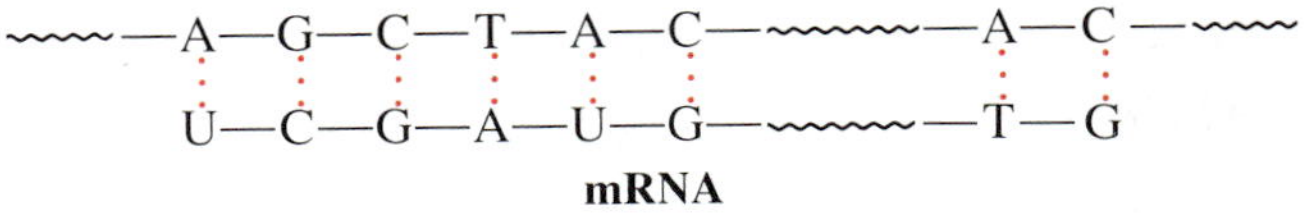

~~~—A—G—C—T—A—C—~~~~—A—C—~~~

U—C—G—A—U—G—~~~~—T—G

mRNA

specifies a particular amino acid (Table 27-3). Simple permutation of this three-base code with a total of four bases gives $4^3 = 64$ possible distinct sequences. That number is more than enough, because only twenty different amino acids are needed for protein synthesis. This might seem like overkill, but consider that the next lower alternative—namely, a two-base code—would give only $4^2 = 16$ combinations, too few for the number of different amino acids found in natural proteins.

Codons do not overlap; in other words, the three bases specifying one amino acid are not part of another preceding or succeeding codon. Moreover, the "reading" of the base sequence is consecutive; each codon immediately follows the next, uninterrupted by genetic "commas" or "hyphens." Nature also makes full use of all 64 codons, by allowing for several of them to describe the same amino acid (Table 27-3). Only tryptophan and methionine are characterized by single three-base codes. Some codons act as signals to initiate or terminate production of a polypeptide chain. Note that the initiator codon (AUG) is also the codon for methionine. Thus, if the codon AUG appears *after* a chain has been initiated, methionine will be produced. The complete base sequence of the DNA in a cell defines its **genetic code.**
~~~

TABLE 27-3

The three-base code for the common amino acids used in protein synthesis

Amino acid	Base sequence	Amino acid	Base sequence	Amino acid	Base sequence
Ala	GCA	His	CAC	Ser	AGC
	GCC		CAU		AGU
	GCG	Ile	AUA		UCA
	GCU		AUC		UCG
Arg	AGA		AUU		UCC
	AGG	Leu	CUA		UCU
	CGA		CUC	Thr	ACA
	CGC		CUG		ACC
	CGG		CUU		ACG
	CGU		UUA		ACU
Asn	AAC		UUG	Trp	UGG
	AAU	Lys	AAA	Tyr	UAC
Asp	GAC		AAG		UAU
	GAU	Met	AUG	Val	GUA
Cys	UGC	Phe	UUU		GUG
	UGU		UUC		GUC
Gln	CAA	Pro	CCA		GUU
	CAG		CCC	Chain initiation	AUG
Glu	GAA		CCG	Chain termination	UGA
	GAG		CCU		UAA
Gly	GGA				UAG
	GGC				
	GGG				
	GGU				

Mutations in the base sequence of DNA can be caused by physical (radiation) or chemical (carcinogens; see, e.g., Section 25-4) interference. Mutations can either replace one base with another or add or delete one base or more. Here is some of the potential value of redundant codons. If, for example, the sequence CCG (proline) were mutated to the sequence CCC, proline would still be synthesized as normal.

EXERCISE 27-13

Given the following RNA sequence, what amino acid sequence would be produced? Remember that the chain must have both an initiating and a terminating codon. What would happen if the first U in the sequence were eliminated by radiation?

G-G-A-U-G-A-A-G-U-A-U-G-C-A-U-C-A-U-G-C-U-U-A-A-G-C-U-A-G-C-A-A-U

Proteins are synthesized along the mRNA template with the help of a set of other important nucleic acids, called **transfer ribonucleic acids (tRNAs).** These are molecules of relatively low molecular weight, containing from 70 to about 90 nucleotides. Each tRNA is specifically designed to carry one of the 20 amino acids to the mRNA in the course of protein buildup. As the polypeptide chain grows longer, it begins to develop its characteristic secondary and tertiary structure (α helix, pleated sheets, etc.), helped by enzymes that form the necessary disulfide bridges. All of this happens with remarkable speed. It is estimated that a protein made up of approximately 150 amino acid residues can be biosynthesized in less than one minute. Clearly, nature still has the edge over the synthetic organic chemist, at least in this domain.

In summary, the nucleic acids DNA and RNA are polymers containing monomeric units called nucleotides. There are four nucleotides for each, varying only in the structure of the base: cytosine (C), thymine (T), adenine (A), and guanine (G) for DNA; cytosine, uracil (U), adenine, and guanine for RNA. The two nucleic acids differ also in the identity of the sugar unit: deoxyribose for DNA, ribose for RNA. DNA replication and RNA synthesis from DNA is facilitated by the complementary character of the base pairs A–T, G–C, and A–U. The double helix partly unwinds and functions as a template for replication. RNA is responsible for protein biosynthesis; each three-base sequence, or codon, specifies a particular amino acid. Codons do not overlap, and more than one can specify the same amino acid.

Summary of New Reactions

1 ACIDITY OF AMINO ACIDS

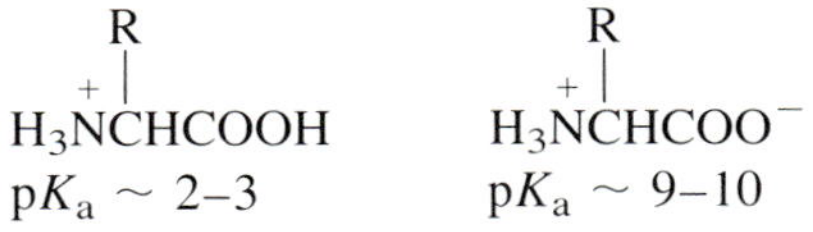

$$\overset{+}{H_3N}CH(R)COOH \quad pK_a \sim 2\text{–}3 \qquad\qquad \overset{+}{H_3N}CH(R)COO^- \quad pK_a \sim 9\text{–}10$$

$$\text{Isoelectric point } pI = \frac{pK_{COOH} + pK_{\overset{+}{N}H_3}}{2}$$

2 THE RELATIVELY BASIC GUANIDINO SUBSTITUENT IN ARGININE

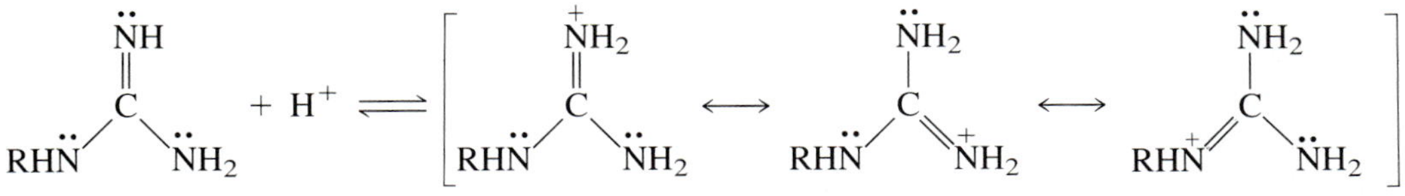

$$RHN\text{–}C(=NH)\text{–}NH_2 + H^+ \rightleftharpoons \left[RHN\text{–}C(=\overset{+}{N}H_2)\text{–}NH_2 \longleftrightarrow RHN\text{–}C(NH_2)=\overset{+}{N}H_2 \longleftrightarrow R\overset{+}{H}N=C(NH_2)\text{–}NH_2 \right]$$

$$pK_a \sim 13$$

3 THE BASICITY OF IMIDAZOLE IN HISTIDINE

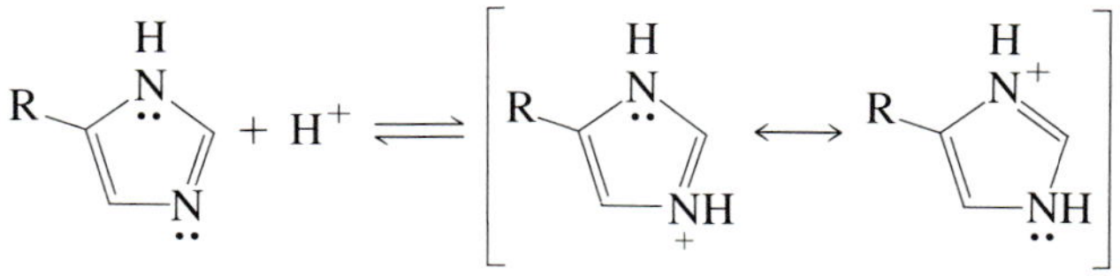

$$\text{R-imidazole} + H^+ \rightleftharpoons \left[\text{imidazolium (N}^+\text{H at N-3)} \longleftrightarrow \text{imidazolium (N}^+\text{H at N-1)} \right]$$

$$pK_a \sim 7.0$$

Preparation of Amino Acids

4 HELL-VOLHARD-ZELINSKY BROMINATION FOLLOWED BY AMINATION

$$RCH_2COOH \xrightarrow[\text{2. } NH_3]{\text{1. } Br_2\text{, catalytic } PBr_3} R\underset{}{\overset{\overset{+}{N}H_3}{|}}{C}HCOO^-$$

5 GABRIEL SYNTHESIS

$$\text{Phthalimide } N^-K^+ + BrCH(CO_2CH_3)_2 \xrightarrow[-\,KBr]{} \text{Phthalimide-}NCH(CO_2CH_3)_2 \xrightarrow[\text{3. } H^+,\, H_2O,\, \Delta]{\text{1. } RO^-Na^+ \text{ 2. } R'X} R'CH(\overset{+}{N}H_3)COO^-$$

Variation:

$$CH_3\overset{O}{\overset{\|}{C}}NHCH(CO_2R)_2 \xrightarrow[\text{2. } H^+,\, H_2O,\, \Delta]{\text{1. NaOH, } R'X} R'CH(\overset{+}{N}H_3)COO^-$$

6 STRECKER SYNTHESIS

$$R\overset{O}{\overset{\|}{C}}H \xrightarrow{HCN,\ NH_3} RCH(NH_2)CN \xrightarrow{H^+,\ H_2O,\ \Delta} RCH(\overset{+}{N}H_3)COO^-$$

7 ENZYMATIC KINETIC RESOLUTION

$$RCH(\overset{+}{N}H_3)COO^- \xrightarrow[-\,CH_3COOH]{CH_3\overset{O}{\overset{\|}{C}}O\overset{O}{\overset{\|}{C}}CH_3} RCH(HN\overset{O}{\overset{\|}{C}}CH_3)COOH \xrightarrow{\text{Hog renal acylase}} H_3\overset{+}{N}\text{—}C(COO^-)(R)\text{—}H + H\text{—}C(COOH)(R)\text{—}NH\overset{O}{\overset{\|}{C}}CH_3$$

(*S*)-Amino acid — **Unreacted (*R*)-enantiomer**

Polypeptide Sequencing

8 DISULFIDE CLEAVAGE

$$\text{—S—S—} \xrightarrow{HCO_3H} \text{—}SO_3H \quad \text{—}SO_3H$$

9 HYDROLYSIS

$$\text{Peptide} \xrightarrow{6\ \text{N HCl, } 110^\circ\text{C, 24 h}} \text{amino acids}$$

10 SANGER DEGRADATION

$$O_2N\text{-}C_6H_3(NO_2)\text{-}F + H_2N\overset{R}{\underset{}{C}}H\overset{O}{\overset{\|}{C}}\text{—}\underset{H}{N}\text{—}\sim \xrightarrow[-\ HF]{} \xrightarrow{6\ \text{N HCl},\ \Delta} O_2N\text{-}C_6H_3(NO_2)\text{-}NH\overset{R}{C}HCO_2H + \text{amino acids}$$

Amino terminus of peptide

11 EDMAN DEGRADATION

$$C_6H_5\text{—}N{=}C{=}S + H_2N\overset{R}{C}H\overset{O}{\overset{\|}{C}}\text{—}\underset{H}{N}\text{—}\sim \longrightarrow \xrightarrow{H^+,\ H_2O} \text{Phenylthiohydantoin} + H_2N\text{—}\sim$$

Phenylthiohydantoin **Lower polypeptide**

Preparation of Polypeptides

12 PROTECTING GROUPS

$$R\overset{\overset{+}{N}H_3}{C}HCOO^- + C_6H_5CH_2O\overset{O}{\overset{\|}{C}}Cl \xrightarrow[-\ NaCl]{NaOH} C_6H_5CH_2O\overset{O}{\overset{\|}{C}}NH\overset{R}{C}HCOOH \xrightarrow[\text{Deprotection}]{H_2,\ \text{Pd-C}} R\overset{\overset{+}{N}H_3}{C}HCOO^- + C_6H_5CH_3 + CO_2$$

Phenylmethyl chloromethanoate (Benzyl chloroformate)

Cbz-protected amino acid

$$R\overset{\overset{+}{N}H_3}{C}HCOO^- + (CH_3)_3CO\overset{O}{\overset{\|}{C}}O\overset{O}{\overset{\|}{C}}OC(CH_3)_3 \xrightarrow{(CH_3CH_2)_3N} (CH_3)_3CO\overset{O}{\overset{\|}{C}}NH\overset{R}{C}HCOOH \xrightarrow[\text{Deprotection}]{H^+,\ H_2O} R\overset{\overset{+}{N}H_3}{C}HCOO^- + CO_2 + CH_2{=}C(CH_3)_2$$

Bis(1,1-Dimethylethyl) dicarbonate (Di-*tert*-butyl dicarbonate)

Boc-protected amino acid

13 PEPTIDE-BOND FORMATION WITH DICYCLOHEXYLCARBODIIMIDE

$$\text{Cbz-Gly} + \text{Ala-}CH_2C_6H_5 + \underset{\textbf{DCC}}{C_6H_{11}N{=}C{=}NC_6H_{11}} \longrightarrow \text{Cbz-Gly-Ala-}CH_2C_6H_5 + C_6H_{11}NH\overset{O}{\overset{\|}{C}}NHC_6H_{11}$$

$$\text{(P)} \xrightarrow[-\ CH_3CH_2OH]{ClCH_2OCH_2CH_3,\ SnCl_4} \text{(P)}\!-\!CH_2Cl \xrightarrow[2.\ H^+,\ H_2O]{1.\ (CH_3)_3CO\overset{O}{\overset{\|}{C}}NH\overset{R}{\overset{|}{C}}HCOO^-}$$

(P) = polystyrene

$$\text{(P)}\!-\!CH_2O\overset{O}{\overset{\|}{C}}-\overset{R}{\overset{|}{C}}HNH_2 \xrightarrow[2.\ H^+,\ H_2O]{1.\ (CH_3)_3CO\overset{O}{\overset{\|}{C}}NH\overset{R'}{\overset{|}{C}}HCOOH,\ DCC} \text{(P)}\!-\!CH_2O\overset{O}{\overset{\|}{C}}-\overset{R}{\overset{|}{C}}HNH\overset{O}{\overset{\|}{C}}-\overset{R'}{\overset{|}{C}}HNH_2 \xrightarrow{HF}$$

$$\text{(P)}\!-\!CH_2F + H_3\overset{+}{N}\overset{R'}{\overset{|}{C}}H\overset{O}{\overset{\|}{C}}NH\overset{R}{\overset{|}{C}}HCOO^-$$

Summary of Important Concepts

1 Polypeptides are poly(amino acids) linked by amide bonds. Most natural polypeptides are made from only nineteen different (2*S*)-amino acids and glycine, all of which have common names and three-letter abbreviations.

2 Amino acids are amphoteric; they can be protonated and deprotonated.

3 Besides fractional crystallization of diastereomers, one way of obtaining enantiomerically pure amino acids is by kinetic resolution of a racemic mixture by enzymes.

4 The structures of polypeptides are varied; they can be linear, cyclic, disulfide bridged, pleated sheet, α helical or superhelical, or disordered, depending on size, composition, hydrogen bonding, and electrostatic and London forces.

5 Amino acids are separated mainly by virtue of their pH-dependent differences in ability to bind to solid supports.

6 Polypeptide sequencing entails a combination of selective chain cleavage and amino acid analysis of the resulting shorter polypeptide fragments.

7 Polypeptide synthesis requires end-protected amino acids that are coupled by dicyclohexylcarbodiimide. The product can be selectively deprotected at either end to allow for further extension of the chain. The use of solid supports, as in the Merrifield synthesis, can be automated.

8 The proteins myoglobin and hemoglobin are polypeptides in which the amino acid chain envelops the active site, heme. The heme contains an iron atom that reversibly binds oxygen, allowing for oxygen uptake, transport, and delivery.

9 The nucleic acids are biopolymers made of phosphate-linked base-bearing sugars. Only four different bases and one sugar are used for DNA and RNA. Because the base pairs adenine–thymine, guanine–cytosine, and adenine–uracil pair up by particularly favorable hydrogen bonding, a nucleic acid can

adopt a dimeric helical structure containing complementary base sequences. In DNA, this arrangement unwinds and functions as a template during DNA replication and RNA synthesis. In protein synthesis, each amino acid is specified by a set of three consecutive RNA bases, called a codon. Thus, the base sequence (genetic code) in a strand of RNA translates into a specific amino acid sequence in a protein.

Problems

1 Draw stereochemically correct structural formulas for isoleucine and threonine (Table 27-1). What is a systematic name for threonine?

2 The abbreviation *allo* means *diastereomer* in amino acid terms. Draw allo-L-isoleucine and give it a systematic name.

3 Draw the structure that each of the following amino acids would have in aqueous solution at the indicated pH values. Calculate the isoelectric point for each amino acid.

- **(a)** Alanine at pH = 1, 7, and 12.
- **(b)** Serine at pH = 1, 7, and 12.
- **(c)** Tyrosine at pH = 1, 7, 9.5, and 12.
- **(d)** Histidine at pH = 1, 5, 7, and 12.
- **(e)** Cysteine at pH = 1, 7, 9, and 12.
- **(f)** Aspartic acid at pH = 1, 3, 7, and 12.
- **(g)** Arginine at pH = 1, 7, 12, and 14.

4 Group the amino acids in Problem 3 according to whether they are **(a)** positively charged, **(b)** neutral, or **(c)** negatively charged at pH = 7.

5 Using either one of the methods in Section 27-2 or a route of your own devising, propose a reasonable synthesis of each of the following amino acids in racemic form.

- **(a)** Valine.
- **(b)** Leucine.
- **(c)** Proline.
- **(d)** Threonine.
- **(e)** Lysine.

6 **(a)** Illustrate the Strecker synthesis of phenylalanine. Is the product chiral? Does it exhibit optical activity?

(b) It has been found that replacement of NH_3 by an optically active amine in the Strecker synthesis of phenylalanine leads to an excess of one enantiomer of the product.

Assign the *R* or *S* configuration to each stereocenter in the following structures, and explain why the use of a chiral amine causes preferential formation of one stereoisomer of the final product.

$C_6H_5CH_2CHO$ —[$C_6H_5C(H)(CH_3)NH_2$, HCN; $- H_2O$]→ $C_6H_5CH_2$–C(H)(CN)–NH–C(H)(CH_3)C_6H_5 (Mainly) —[1. H^+, H_2O; 2. H_2, Pd-C]→ $C_6H_5CH_2$–C(H)(COO^-)$\overset{+}{N}H_3$ (Mainly)

7 The antibacterial agent in garlic, allicin (recall Problem 36 of Chapter 9), is synthesized from the unusual amino acid alliin by the action of the enzyme *allinase*. Because allinase is an extracellular enzyme, this process takes place only when garlic cells are crushed. Propose a reasonable synthesis for the amino acid alliin. Hint: Begin by designing a synthesis of an amino acid from Table 27-1 that is structurally related to alliin.

$$H_2C{=}CHCH_2\overset{O}{\overset{\|}{S}}CH_2\overset{\overset{+}{N}H_3}{\overset{|}{C}}HCOO^-$$

Alliin

8 Devise a procedure for separating a mixture of the four stereoisomers of isoleucine into its four components: (+)-isoleucine, (−)-isoleucine, (+)-alloisoleucine, and (−)-alloisoleucine (Problem 2). Make use of the fact that alloisoleucine is much more soluble in 80% ethanol at all temperatures than is isoleucine.

9 Identify each of the following structures as a dipeptide, tripeptide, and so on, and point out all the peptide bonds:

(a) $H_3\overset{+}{N}-CH[(CH_3)_2CH]-C(=O)-NH-CH(CH_3)-C(=O)-NH-CH(HSCH_2)-COO^-$

(b) $H_3\overset{+}{N}-CH(HOCH_2)-C(=O)-NH-CH(CH_2COO^-)-COO^-$

(c) $H_3\overset{+}{N}-CH(CH_2\text{-imidazolyl})-C(=O)-NH-CH(CH_3CHOH)-C(=O)-N\!\!<\!(CH_2CH_2CH_2 \text{ ring})\!>\!CH-C(=O)-NH-CH[CH_2(CH_2)_3\overset{+}{N}H_3]-COO^-$

(d) $H_3\overset{+}{N}-CH(CH_2C_6H_4OH\text{-}p)-C(=O)-NH-CH_2-C(=O)-NH-CH_2-C(=O)-NH-CH(CH_2C_6H_5)-C(=O)-NH-CH[CH_2CH(CH_3)_2]-COO^-$

10 Using the standard three-letter abbreviations for amino acids, write the peptide structures in Problem 9 in short notation.

11 Indicate which of the amino acids in Problem 3 and the peptides in Problem 9 would migrate in an electrophoresis apparatus at pH = 7 **(a)** toward the anode or **(b)** toward the cathode.

12 Silk consists of β sheets whose polypeptide chains are made up of the repeating amino acid sequence Gly-Ser-Gly-Ala-Gly-Ala. What characteristics of amino acid side chains appear to favor the β-sheet configuration? Do the illustrations of β-sheet structures (Figures 27-3B and 27-4) suggest an explanation for this preference?

13 Identify as many stretches of α helix as you can in the structure of myoglobin (Figure 27-8C). Prolines are located in myoglobin at positions 37, 88, 100, and 120. How does each of these prolines affect the tertiary structure of the molecule?

14 Of the 153 amino acids in myoglobin, 78 contain polar side chains (i.e., Arg, Asn, Asp, Gln, Glu, His, Lys, Ser, Thr, Trp, and Tyr). When myoglobin adopts its natural folded conformation, 76 of these 78 polar side chains (all but those of two histidines) project outward from its surface. Meanwhile, in addition to the two histidines, the interior of myoglobin contains only Gly, Val, Leu, Ala, Ile, Phe, Pro, and Met. Explain.

15 Explain the following three observations.

(a) Silk, like most polypeptides with sheet structures, is water insoluble.

(b) Globular proteins like myoglobin generally dissolve readily in water.

(c) Disruption of the tertiary structure of a globular protein (denaturation) leads to precipitation from aqueous solution.

16 In your own words, outline the procedure that might have been followed by the researchers who determined which amino acids were present in vasopressin (Exercise 27-6).

17 Write the products of Sanger degradation of the peptides in Problem 9.

18 What would be the outcome of reaction of gramicidin S with 1-fluoro-2,4-dinitrobenzene (Sanger's reagent)? With phenyl isothiocyanate (Edman degradation)?

19 The polypeptide bradykinin is a tissue hormone that can function as a potent pain-producing agent. Treatment of bradykinin with 1-fluoro-2,4-dinitrobenzene followed by complete acid hydrolysis produces *N*-(2,4-dinitrophenyl)arginine together with free Arg, Gly, Phe, Pro, and Ser. Incomplete acid hydrolysis causes random cleavage of many bradykinin molecules into an assortment of peptide fragments that includes Arg-Pro-Pro-Gly, Phe-Arg, Ser-Pro-Phe, and Gly-Phe-Ser. Complete hydrolysis followed by amino acid analysis indicates a ratio of 3 Pro, 2 Phe, 2 Arg, and one each of Gly and Ser. Deduce the amino acid sequence of bradykinin.

20 Somatostatin is a polypeptide hormone with several functions, including regulation of the secretion of insulin by the pancreas. It is useful in the treatment of certain kinds of diabetes. Somatostatin contains one disulfide linkage. After its cleavage by HCO_3H, just a single polypeptide chain is present.

(a) What does this tell you?

Treatment of this polypeptide chain with trypsin yields the following three peptides: Ala-Gly-Cys(SO_3H)-Lys, Thr-Phe-Thr-Ser-Cys(SO_3H), and Asn-Phe-Phe-Trp-Lys.

(b) Now what do you know about the structure of somatostatin?

Hydrolysis of the polypeptide by chymotrypsin leads to Lys-Thr-Phe, Thr-Ser-Cys(SO_3H), Ala-Gly-Cys(SO_3H)-Lys-Asn-Phe, free Phe, and free Trp.

(c) Write the entire amino acid sequence of somatostatin.

21 The amino acid sequence of met-enkephalin, a brain peptide with powerful opiatelike biological activity, is Tyr-Gly-Gly-Phe-Met. What would be the products of step-by-step Edman degradation of met-enkephalin?

The peptide shown in part **d** of Problem 9 is leu-enkephalin, a relative of met-enkephalin with similar properties. How would the results of Edman degradation of leu-enkephalin differ from those of met-enkephalin?

22 Secreted by the pituitary gland, corticotropin is a hormone that stimulates the adrenal cortex. Determine its primary structure from the following information.

(i) Hydrolysis by chymotrypsin produces six peptides: Arg-Trp, Ser-Tyr, Pro-Leu-Glu-Phe, Ser-Met-Glu-His-Phe, Pro-Asp-Ala-Gly-Glu-Asp-Gln-Ser-Ala-Glu-Ala-Phe, and Gly-Lys-Pro-Val-Gly-Lys-Lys-Arg-Arg-Pro-Val-Lys-Val-Tyr.

(ii) Hydrolysis by trypsin produces free lysine, free arginine, and the following five peptides: Trp-Gly-Lys, Pro-Val-Lys, Pro-Val-Gly-Lys, Ser-Tyr-Ser-Met-Glu-His-Phe-Arg, and Val-Tyr-Pro-Asp-Ala-Gly-Glu-Asp-Gln-Ser-Ala-Glu-Ala-Phe-Pro-Leu-Glu-Phe.

23 Glucagon is a pancreatic hormone whose function opposes that of insulin: it causes an increase in glucose levels in the blood. It consists of a polypeptide chain with 29 amino acid units. Treatment of glucagon with thermolysin produces four fragments, including the tripeptide Val-Gln-Tyr, the tetrapeptide Leu-Met-Asn-Thr, a nine-amino-acid peptide A and a 13-amino-acid peptide B. Sanger degradation of A yields *N*-(2,4-dinitrophenyl)leucine, and Sanger degradation of B yields *N*-(2,4-dinitrophenyl)histidine.

Peptide A is not cleaved by chymotrypsin, but clostripain breaks it down into Leu-Asp-Ser-Arg, Ala-Gln-Asp-Phe, and a free Arg.

Peptide B is cleaved by chymotrypsin into Ser-Lys-Tyr, Thr-Ser-Asp-Tyr, and His-Ser-Gln-Gly-Thr-Phe.

(a) At this stage, how much do you know for certain about the structure of glucagon? What uncertainties still remain?

(b) One of the products of trypsin hydrolysis of the intact glucagon molecule is the peptide Tyr-Leu-Asp-Ser-Arg. Does this help?

(c) One of the products of chymotrypsin hydrolysis of the intact hormone is Leu-Met-Asn-Thr, the same tetrapeptide released by thermolysin. Now can you piece together the entire molecule?

24 Propose a synthesis of leu-enkephalin (part **d** of Problem 9) from the component amino acids.

25 The following molecule is thyrotropin-releasing hormone (TRH). It is secreted by the hypothalamus, causing the release of thyrotropin from the pituitary gland, which, in turn, stimulates the thyroid gland. The thyroid produces hormones, such as thyroxine, that control metabolism in general.

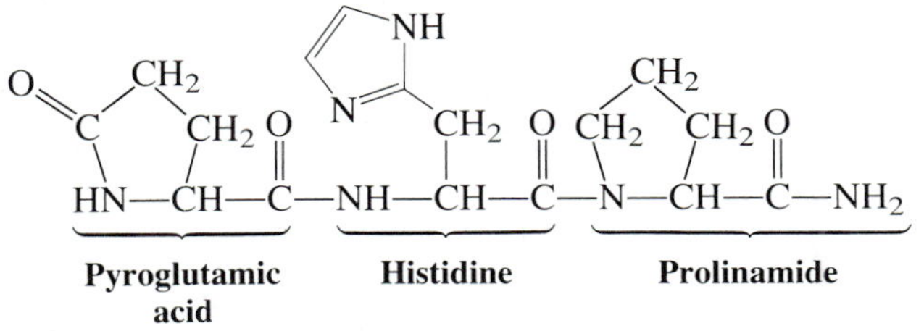

```
                                 NH
      CH2                 N           CH2
O=C       CH2 O              CH2  O CH2   CH2 O
   \     /    ||              |   || \    /   ||
    HN—CH—C—NH—CH—C—N—CH—C—NH2
```

Pyroglutamic acid — **Histidine** — **Prolinamide**

The initial isolation of TRH required the processing of four tons of hypothalamic tissue, from which 1 mg of the hormone was obtained. Needless to say, it is a bit more convenient to synthesize TRH in the laboratory than to extract it from natural sources. Devise a synthesis of TRH from Glu, His, and Pro. Note that pyroglutamic acid is just the lactam of Glu and may be readily obtained by heating Glu to between 135° and 140°C.

26 Consider the synthesis of aspartame (Section 27-3). Is there a structural feature in one of its component amino acids that might make the synthesis difficult? What other amino acids contain groups that might cause problems in synthesizing peptides containing them?

27 **(a)** The structures illustrated for the four DNA bases (Section 27-7) represent only the most-stable tautomers. Draw one or more alternative tautomers for each of these heterocycles (review tautomerism, Sections 13-6, 16-1, and 16-2).

(b) In certain cases, the presence of a small amount of one of these less-stable tautomers can lead to an error in DNA replication or mRNA synthesis due to faulty base pairing. One example is the imine tautomer of adenine, which pairs with cytosine instead of thymine. Draw a possible structure for this hydrogen-bonded base pair (see Figure 27-11).

(c) Using Table 27-3, derive a possible nucleic acid sequence for an mRNA that would code for the five amino acids in met-enkephalin (Problem 21). If the mispairing described in part **b** were at the first possible position in the synthesis of this mRNA sequence, what would be the consequence in the amino acid sequence of the peptide? (Ignore the initiation codon.)

28 Factor VIII is one of the proteins participating in the formation of blood clots. A defect in the gene whose DNA sequence codes for Factor VIII is responsible for classic hemophilia *(bleeders' disease)*. Factor VIII contains 2,332 amino acids. How many nucleotides are needed to code for its synthesis?

29 Hydroxyproline (Hyp), like many other amino acids that are not "officially" classified as essential, is nonetheless a very necessary biological substance. It constitutes about 14% of the amino acid content of the protein *collagen*. Collagen is the main constituent of skin and connective tissue. It is also present, together with inorganic substances, in nails, bones, and teeth.

(a) The systematic name for hydroxyproline is (2*S*,4*R*)-4-hydroxy-azacyclopentane-2-carboxylic acid. Draw a stereochemically correct structural formula for this amino acid.

(b) Hydroxyproline is synthesized in the body in peptide-bound form from peptide-bound proline and O_2, in an enzyme-catalyzed process that requires Vitamin C. In the absence of Vitamin C, only a defective, Hyp-deficient collagen can be produced. Vitamin C deficiency causes *scurvy,* a condition characterized by bleeding of the skin and swollen, bleeding gums.

In the following reaction sequence, an efficient laboratory synthesis of hydroxyproline, fill in the necessary reagents (i) and (ii), and formulate detailed mechanisms for the steps marked with an asterisk.

$ClCH_2CH$—CH_2 (epoxide, O) $\xrightarrow{\text{i}}$ (phthalimido)N—CH_2CH—CH_2 (epoxide, O) $\xrightarrow{Na^{+}\,{}^{-}CH(COOCH_2CH_3)_2{}^{*}}$

(phthalimido)NCH_2— (lactone, O, O) —$COOCH_2CH_3$ $\xrightarrow{SO_2Cl_2}$

(phthalimido)NCH_2— (lactone, O, O) —C(Cl)—$COOCH_2CH_3$ $\xrightarrow{\text{ii*}}$ hydroxyproline

(c) Gelatin, which is partly hydrolyzed collagen, is rich in hydroxyproline and, as a result, is often touted as a remedy for split or brittle nails. Like most proteins, however, gelatin is almost completely broken down into individual amino acids in the stomach and small intestine before absorption. Is the free hydroxyproline thus introduced into the bloodstream of any use to the body in the synthesis of collagen? Hint: Does Table 27-3 list a three-base code for hydroxyproline?

30 Sickle-cell anemia is an often fatal genetic condition caused by a single error in the DNA gene that codes for the β chain of hemoglobin. The correct nucleic acid sequence (read from the mRNA template) begins with AUG-GUGCACCUGACUCCUG**A**GGAGAAG. . . , and so forth.

(a) Translate this into the corresponding amino acid sequence of the protein.

(b) The mutation that gives rise to the sickle-cell condition is replacement of the boldface A in the preceding sequence by U. What is the consequence of this error in the corresponding amino acid sequence?

(c) This amino acid substitution alters the properties of the hemoglobin molecule—in particular, its polarity and its shape. Suggest reasons for both these effects. (Refer to Table 27-1 for amino acid structures and to Figure 27-8C for the structure of myoglobin, which is similar to that of hemoglobin. Note the location of the amino acid substitution in the tertiary structure of the protein.)

Answers to Exercises

CHAPTER 1

1-1

(a)

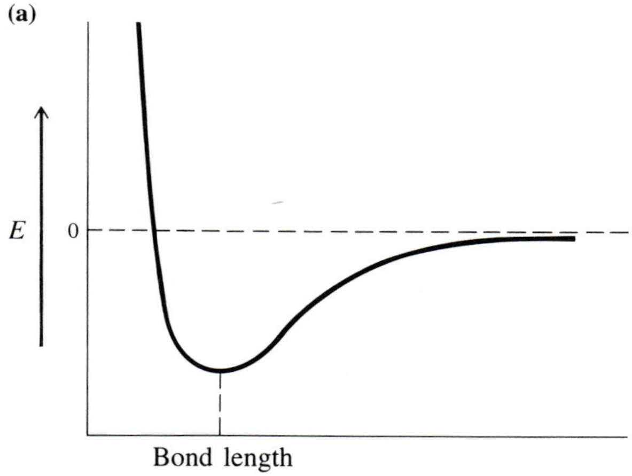

(b) Self-explanatory.

1-2

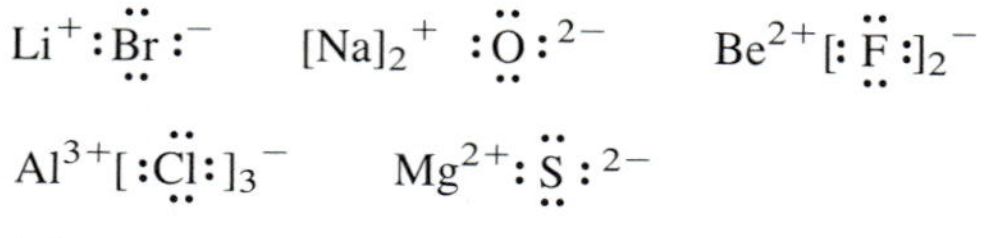

1-3

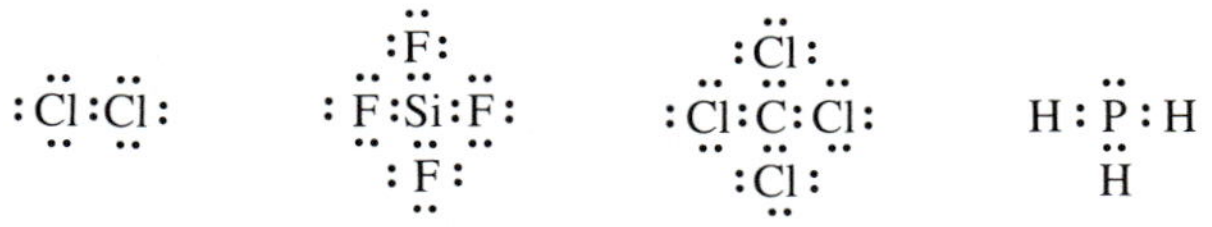

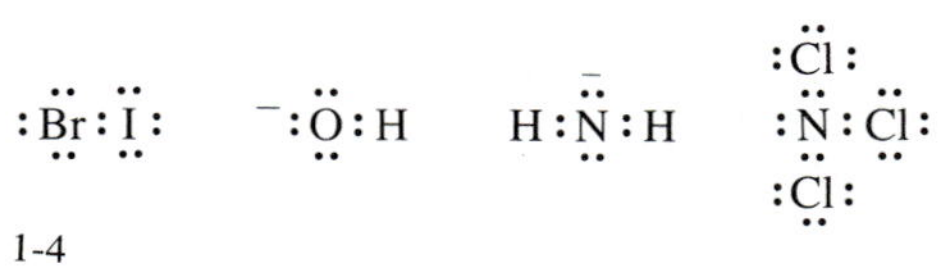

1-4

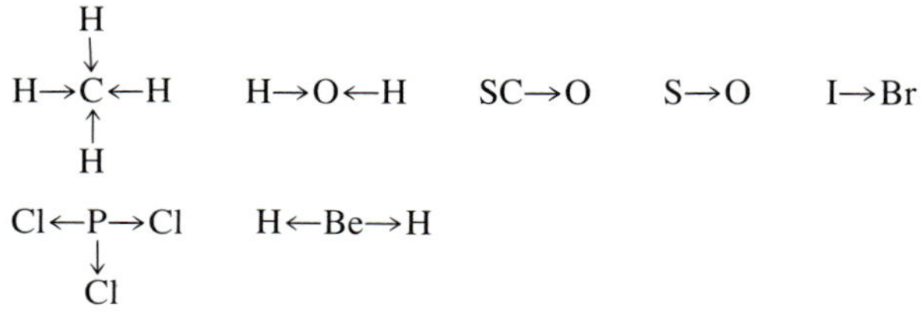

1-5

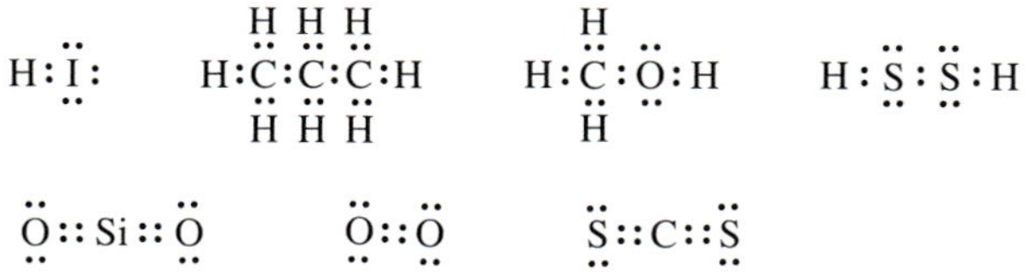

1-6

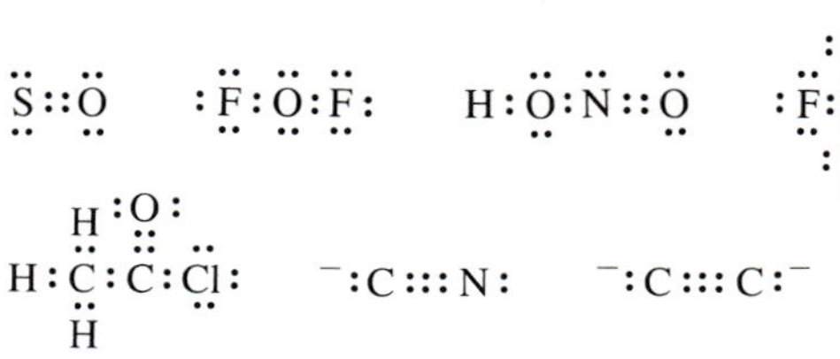

1-7

It should be close to trigonal (counting the lone electron

pair), with equal N–O bond lengths and one-half of a negative charge on each oxygen atom.

$$\left[:\ddot{O}\overset{\large N}{\diagdown}\ddot{O}:^- \longleftrightarrow {}^-:\ddot{O}\diagdown N = \ddot{O}: \right]$$

1-8

(a) $\left[{}^-:C\equiv\overset{+}{N}—\ddot{O}:^- \longleftrightarrow \cdot{}^{2-}\ddot{C}=\overset{+}{N}=\ddot{O} \right]$

The left-hand structure is preferred, because the charges are more evenly distributed and the negative charge resides on the relatively electronegative oxygen.

(b) $\left[{}^-\ddot{N}=\ddot{O} \longleftrightarrow \ddot{N}:\ddot{O}:^- \right]$

The left-hand structure is preferred, because the right-hand one has no octet on nitrogen.

1-9

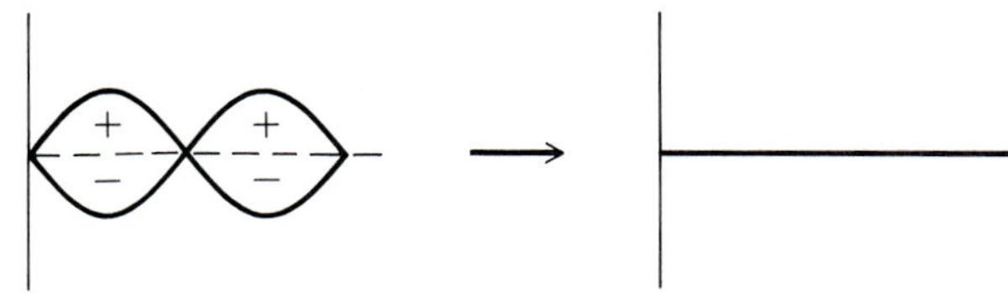

1-10

S $(1s)^2(2s)^2(2p)^6(3s)^2(3p)^4$
P $(1s)^2(2s)^2(2p)^6(3s)^2(3p)^3$

1-11

The molecular-orbital picture is similar to that shown for the bonding in H_2 (Figures 1-11 and 1-12). However, the presence of only one antibonding but two bonding electrons results in net bonding.

He ⥮ ⟶ ↿ (antibonding), ⥮ (bonding) ⟵ ↿ He^+

1-12

CH_3^+ or $H:\overset{+}{\underset{..}{C}}:H$ with H below — No octet

CH_3^- or $H:\overset{-}{\underset{..}{\ddot{C}}}:H$ with H below

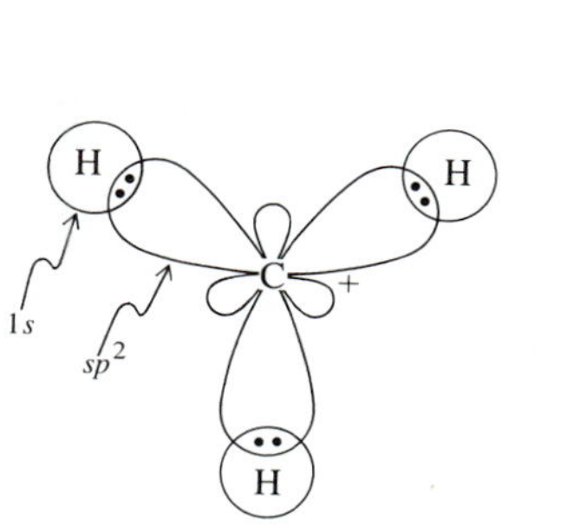

Trigonal, sp^2 hybridized, electron deficient like BH_3.

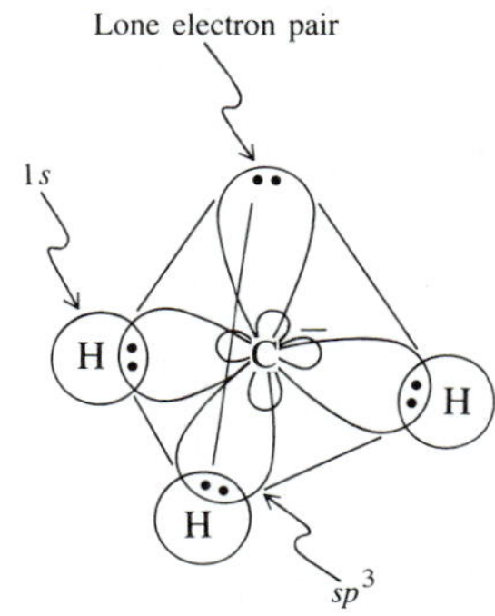

Tetrahedral, sp^3 hybridized, closed shell.

1-13

$C_6H_{12}O_6$. Molecular weight

6 C:	72.067
12 H:	12.096
6 O:	95.996
Total:	180.159

$$\%\ C: \frac{72.067}{180.159} \times 100 = 40.0$$

$$\%\ H: \frac{12.096}{180.159} \times 100 = 6.71$$

$$\%\ O: \frac{95.996}{180.159} \times 100 = 53.28$$

1-14

```
   H  H  H  H              H
   |  |  |  |              |
H—C—C—C—C—H         H  H—C—H
   |  |  |  |          |    |
   H  H  H  H       H—C————C—H
                       |    |
                       H  H—C—H
                            |
                            H
```

Butane **Isobutane**

CHAPTER 2

2-1

(a)

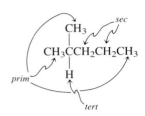

(b) Higher homologs:

CH_3 CH_3
CH_3CH $CHCH_3$

CH_3
$CH_3CH_2CH_2CH$
CH_3

CH_3
CH_2
CH_3CH_2CH
CH_3

CH_3
$CH_3CH_2CCH_3$
CH_3

Lower homologs:

CH_3
CH_3CH
CH_3

$CH_3CH_2CH_2CH_3$

2-2

CH_3
$CH_3CHCH_2CH_2CH_3$
Isohexane

CH_3
CH_3CCH_3
CH_3
Neopentane

2-3

CH_3
$CH_3CCH_2CH_2CH_3$
H

prim *sec* *tert*

2-4

Self-explanatory.

2-5

H_3C H

2-Methylbutane

H CH_3
H_3C H

2,3-Dimethylbutane

2-6

In this example (graphed below), the energy difference between the two staggered conformers turns out to be quite small.

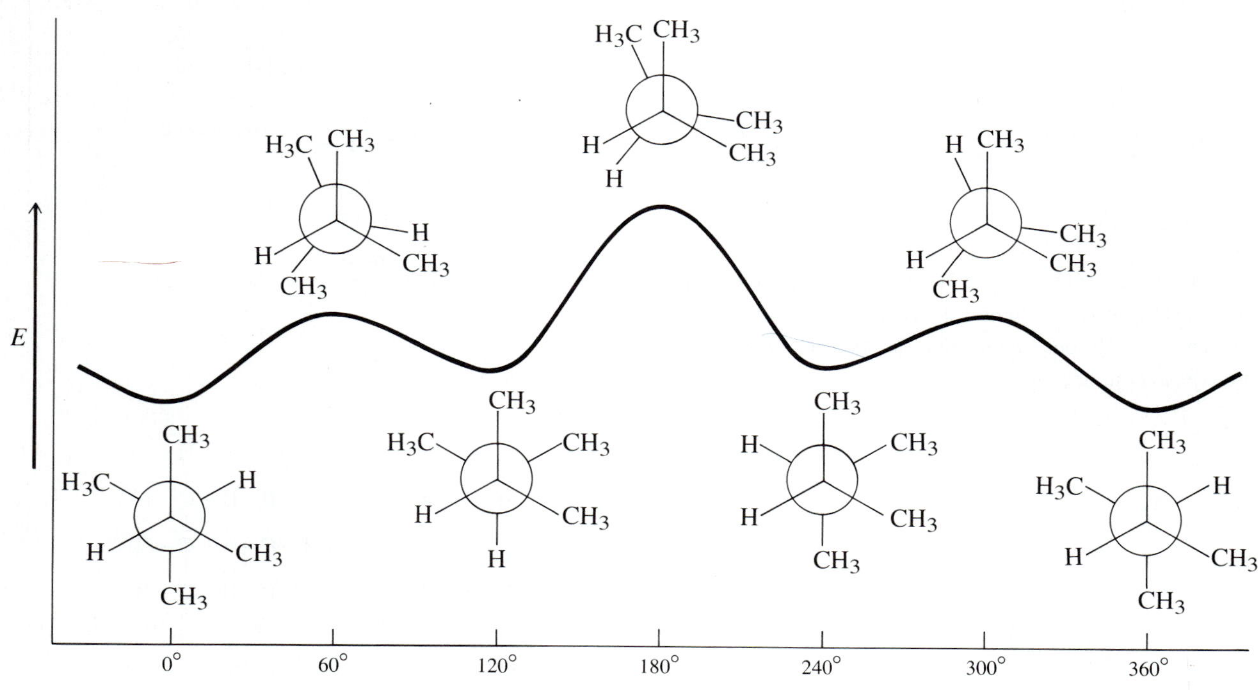

2-7
$0.9 = -1.36 \log K$ kcal mole^{-1} at 25°C
$K = 0.219$; *anti*:*gauche* = 82:18
$0.9 = -RT \ln K = -1.71 \log K$ kcal mole^{-1} at 100°C
$K = 0.297$; *anti*:*gauche* = 77:23

2-8
$$\Delta G^\circ = \Delta H^\circ - T\Delta S^\circ$$
$$= -15.5 \text{ kcal mole}^{-1} - (298 \text{ deg} \times -31.3 \text{ cal deg}^{-1} \text{ mole}^{-1})$$
$$= -6.17 \text{ kcal mole}^{-1}$$
The entropy is negative because two molecules are converted into one in this reaction.

2-9
After 50% conversion, only 1/2 molar concentration of starting materials is present. Hence, for first order: rate = k[A]. At 50% conversion, rate will be 1/2 initial rate. For second order: rate = k[A][B]. At 50% conversion, rate = k(1/2)[A](1/2)[B] = 1/4 initial rate.

2-10
$k = 10^{14} e^{-58.4/1.53} = 3.03 \times 10^{-3} \text{ sec}^{-1}$
For the reverse reaction:
$\Delta G^\circ = \Delta H^\circ - T\Delta S^\circ = +15.5 - 24.2 = -8.7$ kcal mole^{-1}
Hence, the dissociation equilibrium lies on the side of ethene and HCl at this high temperature, where the entropy factor overrides the ΔH° term.

CHAPTER 3

3-1
A simple answer would be that the strength of a bond depends not only on the size and energies of the orbitals, but also on Coulombic contributions. Thus, in going from N to O to F, nuclear charge increases in the core, allowing for more nuclear-core–electronic attraction in binding to CH_3. The bond in question becomes more and more polar along the series.

3-2
First: CH_3–$C(CH_3)_3$ $DH^\circ = 84$ kcal mole^{-1}
Second: CH_3–CH_3 $DH^\circ = 90$ kcal mole^{-1}

3-3
$3\,C + 3\,H_2 \longrightarrow$ cyclopropane
$\Delta H_f^\circ = +12.7$ kcal mole^{-1}

Cyclopropane + 4.5 $O_2 \longrightarrow 3\,CO_2 + 3\,H_2O$
ΔH° (3 CO_2) = −282.3 kcal mole^{-1}
ΔH° (3 H_2O) = −204.9 kcal mole^{-1}
−487.2 kcal mole^{-1}

$\Delta H^\circ_{comb} = -487.2 - (+12.7) = -499.9$ kcal mole^{-1}

3-4
$$CH_3CH_3 + Cl_2 \xrightarrow{h\nu} CH_3CH_2Cl + HCl$$
$\Delta H^\circ = 98 + 58 - 80 - 103 = -27$ kcal mole^{-1}

Mechanism

Initiation:

$Cl_2 \xrightarrow{h\nu} 2\,:\ddot{Cl}\cdot$ $\Delta H^\circ = +58$ kcal mole^{-1}

Propagation:

$CH_3CH_3 + :\ddot{Cl}\cdot \longrightarrow CH_3CH_2\cdot + H\ddot{Cl}:$
$\Delta H^\circ = -5$ kcal mole^{-1}

$CH_3CH_2\cdot + Cl_2 \longrightarrow CH_3CH_2\ddot{Cl}: + :\ddot{Cl}\cdot$
$\Delta H^\circ = -22$ kcal mole^{-1}

Termination:

$:\ddot{Cl}\cdot + :\ddot{Cl}\cdot \longrightarrow Cl_2$ $\Delta H^\circ = -58$ kcal mole^{-1}

$CH_3CH_2\cdot + :\ddot{Cl}\cdot \longrightarrow CH_3CH_2\ddot{Cl}:$
$\Delta H^\circ = -80$ kcal mole^{-1}

$CH_3CH_2\cdot + \cdot CH_2CH_3 \longrightarrow CH_3CH_2CH_2CH_3$
$\Delta H^\circ = -82$ kcal mole^{-1}

3-5
$CH_4 + Cl_2 + Br_2 \longrightarrow$
$CH_3Cl + CH_4 + Cl_2 + Br_2 + HCl$

Cl_2 is more reactive than Br_2.

3-6

$$CH_3CH_2CH_2CH_3 + Cl_2 \xrightarrow{h\nu}$$

$$CH_3CH_2CH_2CH_2Cl + CH_3CH_2\underset{}{\overset{Cl}{C}}HCH_3 + HCl$$

Ratio of primary to secondary product:

$$(6 \times 1):(4 \times 4) = 6:16 = 3:8$$

In other words, 2-chlorobutane:1-chlorobutane = 8:3.

3-7

A **B** **C** **D** **E**

3 primary, 3 types (2 each) secondary, 1 tertiary hydrogen. Relative amounts of A, B, C, D, E:

$$A:B:C:D:E = (3 \times 1):(1 \times 5):(4 \times 4):(4 \times 4):(2 \times 4) = 3:5:16:16:8$$

This problem is actually more complicated because of the possibility of forming cis and trans isomers (see Section 4-1).

3-8

(a) $RH + SO_2Cl_2 \xrightarrow{\text{Initiator}} RCl + SO_2 + HCl$

Possible mechanism:

Initiation:

$$Init^{\cdot} + SO_2Cl_2 \longrightarrow Init\text{–}Cl + {}^{\cdot}SO_2Cl$$

Propagation:

$${}^{\cdot}SO_2Cl + RH \longrightarrow R^{\cdot} + HSO_2Cl \quad (\downarrow HCl + SO_2)$$

$$R^{\cdot} + SO_2Cl_2 \longrightarrow RCl + {}^{\cdot}SO_2Cl$$

Termination (only one is shown):

$$R^{\cdot} + {}^{\cdot}SO_2Cl \longrightarrow RSO_2Cl$$

An alternative to this scheme would be a second initiation step generating $Cl^{\cdot}$:

$${}^{\cdot}SO_2Cl \longrightarrow SO_2 + {}^{\cdot}Cl$$

${}^{\cdot}Cl$ could then enter the familiar propagation cycle when Cl_2 is used as the chlorinating agent. This possibility is unlikely, because the selectivity in chlorinations with SO_2Cl_2 as reagent is different from that found with Cl_2, suggesting a different chain carrier.

(b)

NCl + RH —Initiator→ NH + RCl

Initiation:

Init· + N—Cl ⟶ Init–Cl + N·

Propagation:

N· + RH ⟶ NH + R·

R· + N—Cl ⟶ RCl + N·

Termination (only one is shown):

R· + N· ⟶ R—N

3-9 $CH_3CH_2CH_3$
↑ ↑

Will give mixture, because of competitive primary and secondary chlorination (indicated by arrows).

$$H_3C{-}C(CH_3)_2{-}CH_3 \longrightarrow (CH_3)_3CCH_2Cl$$

Only one type of C–H; so 2,2-dimethylpropane should give good selectivity.

Same situation as in 2,2-dimethylpropane.

Will give a bad mixture.

CHAPTER 4

4-1
Aspects of ring strain and conformational analysis are discussed in Sections 4-2 through 4-5.

Note that the cycloalkanes are much less flexible than the straight-chain alkanes and thus have less conformational freedom. Cyclopropane must be flat and all hydrogens eclipsed. The higher cycloalkanes have increasingly more flexibility, hydrogens being in staggered positions and the carbon atoms of the ring eventually being able to adopt *anti* configurations.

4-2

trans-1-Bromo-2-methylcyclohexane **cis-1-Bromo-3-methylcyclohexane**

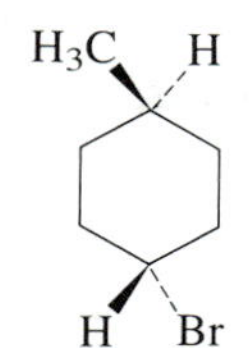

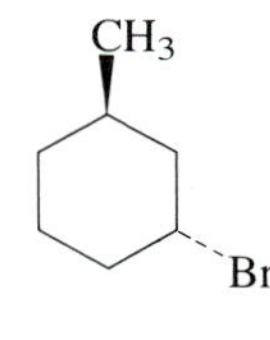

***trans*-1-Bromo-4-Methylcyclohexane** ***trans*-1-Bromo-3-methylcyclohexane**

***cis*-1-Bromo-4-methylcyclohexane**

4-3
ΔG° (*gauche*-eclipsed) butane = 3.6 kcal mole^{-1}. Flattening the chair cyclohexane results in six eclipsing "butane segments." Thus, ΔG° (chair-flat) cyclohexane = $6 \times 3.6 = 21.6$ kcal mole^{-1}.

4-4
(a) ΔG° = energy difference between an axial methyl and axial ethyl group: $1.75 - 1.7$ = about 0.05 kcal mole^{-1}; that is, very small.
(b) Same as part a.
(c) $1.75 + 1.7 = 3.45$ kcal mole^{-1}.

4-5
Decreasing order of stability:

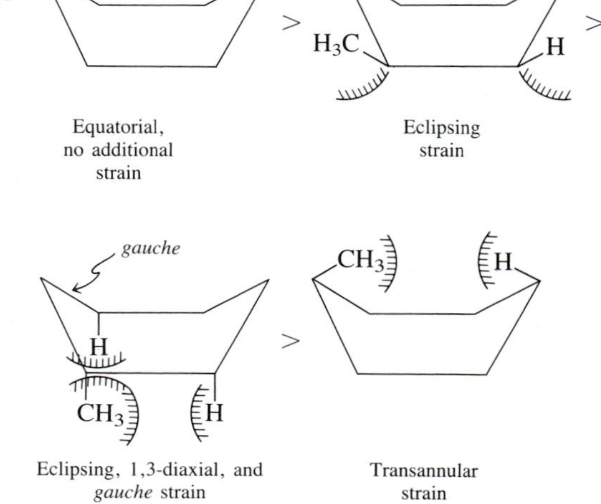

4-6
There are three, shown in decreasing order of stability:

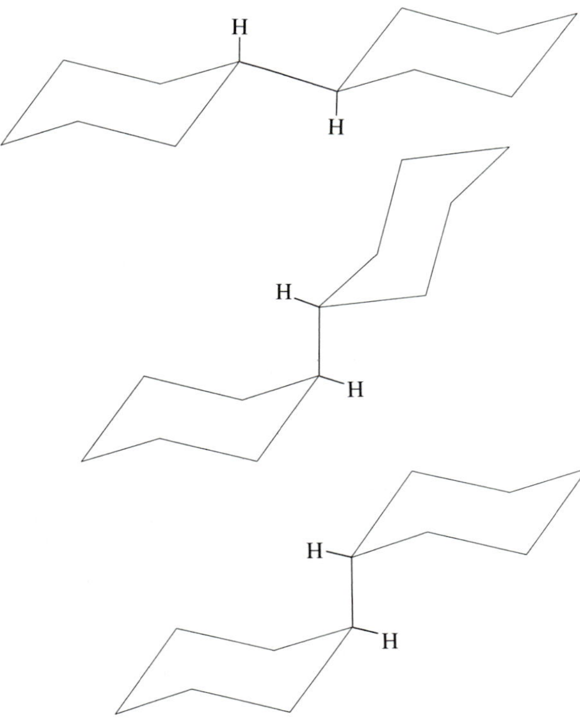

4-7
trans-Bicyclo[4.4.0]decane is fairly rigid. Full chair–chair conformational ''flipping'' is not possible. On the other hand, the axial and equatorial positions in the cis isomer can be interchanged by conformational isomerization of both rings. The barrier to this exchange is small (E_a = 14 kcal mole^{-1}). Because one of the appended bonds is always axial, the cis isomer is less stable than the trans isomer by 2 kcal mole^{-1} (as measured by combustion experiments).

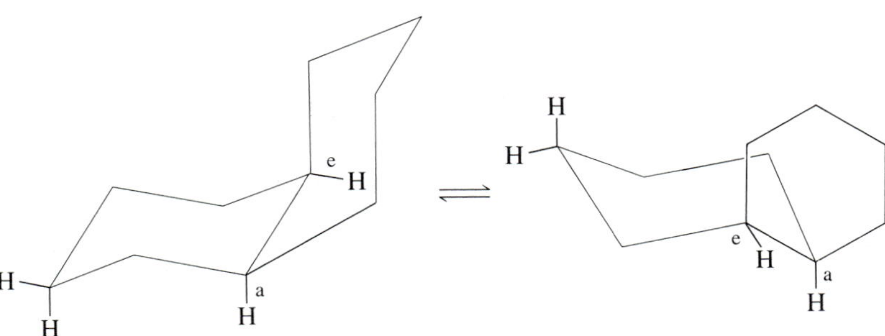

Ring flip in *cis*-bicyclo[4.4.0]decane

4-8

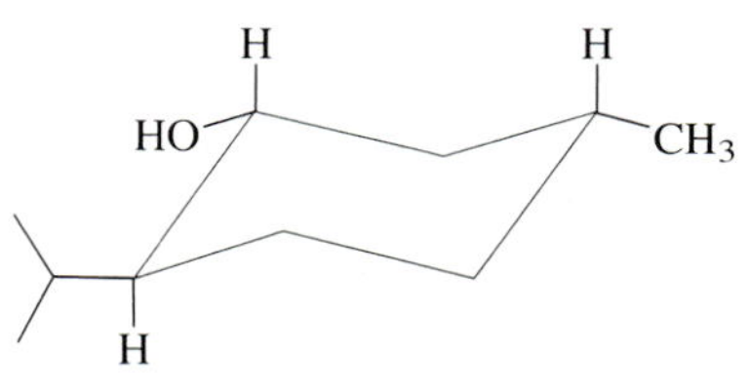

All equatorial

4-9
Chrysanthemic acid: alkene, carboxylic acid, ester.
Grandisol: alkene, alcohol.
Menthol: alcohol.
Camphor: ketone.
β-Cadinene: alkene.

CHAPTER 5

5-1

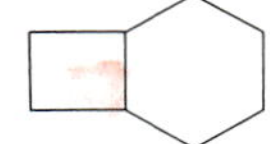
Bicyclo[4.2.0]octane

Bicyclo[3.3.0]octane

Both hydrocarbons have the same molecular formula: C_8H_{14}. Therefore, they are (structural) isomers.

5-2
There are several boat and twist-boat forms of methylcyclohexane, some of which are shown:

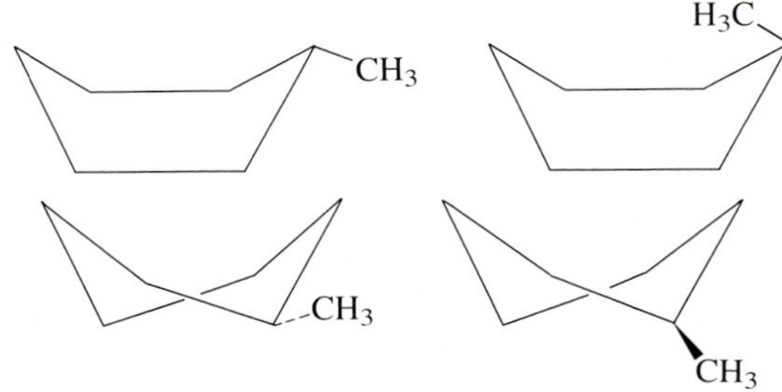

5-3

All are chiral. Note, however, that 2-methylbutadiene (isoprene) itself is achiral. Number of stereocenters: chrysanthemic acid, 2; grandisol, 2; menthol, 3; camphor, 2; β-cadinene, 3; cholesterol, 8; cholic acid, 11; cortisone, 6; testosterone, 6; estradiol, 5; progesterone, 6; norethynodrel, 5; mestranol, 5.

5-4

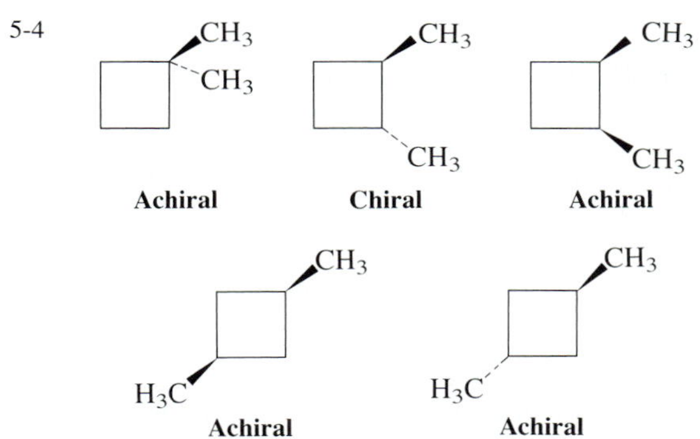

trans-1,2-Dimethylcyclobutane is the only molecule in the series lacking both a mirror plane and a center of symmetry.

5-5
(+)-Bromochlorofluoromethane: *R*
(−)-Bromochlorofluoromethane: *S*
(−)-2-Bromobutane: *R*
(+)-2-Bromobutane: *S*
(+)-2-Aminopropanoic acid: *S*
(−)-2-Hydroxypropanoic acid: *R*

5-6

H, Cl, CH_2CH_3, CH_3 on C: ***S***

CH_3, F, Cl, CH_2CH_3 on C: ***R***

Br, HC≡C, CH_3, CH=CH_2 on C: ***S***

5-7

H (top), Br (left), Cl (right), F (bottom)

H (top), Cl (left), Br (right), F (bottom)

CH_2CH_3 (top), H (left), Br (right), CH_3 (bottom)

CH_2CH_3 (top), Br (left), H (right), CH_3 (bottom)

H (top), H_2N (left), CH_3 (right), COOH (bottom)

H (top), HO (left), COOH (right), CH_3 (bottom)

H (top), H_3C (left), CH_2CH_3 (right), Cl (bottom)

CH_3 (top), CH_3CH_2 (left), Cl (right), F (bottom)

Br (top), CH_2=CH (left), CH_3 (right), C≡CH (bottom)

5-8

Br *a* (top), H *d* (left), D *c* (right), CH_3 *b* (bottom) → *d* (top), *a* (left), *b* (right), *c* (bottom): ***R***

Cl *c* (top), F *d* (left), Br *b* (right), I *a* (bottom) → *d* (top), *c* (left), *a* (right), *b* (bottom): ***R***

CH_3 *c* (top), H_2N *a* (left), COOH *b* (right), H *d* (bottom) → *d* (top), *b* (left), *a* (right), *c* (bottom): ***S***

5-9
1: (2*S*,3*S*)-2-Fluoro-3-methylpentane.
2: (2*R*,3*S*)-2-Fluoro-3-methylpentane.
3: (2*R*,3*R*)-2-Fluoro-3-methylpentane.
4: (2*S*,3*S*)-2-Fluoro-3-methylpentane.
1 and 2 are diastereomers; 1 and 3 are enantiomers; 1 and 4 are identical; 2, 3, and 4 are diastereomers; 3 and 4 are enantiomers.

5-10

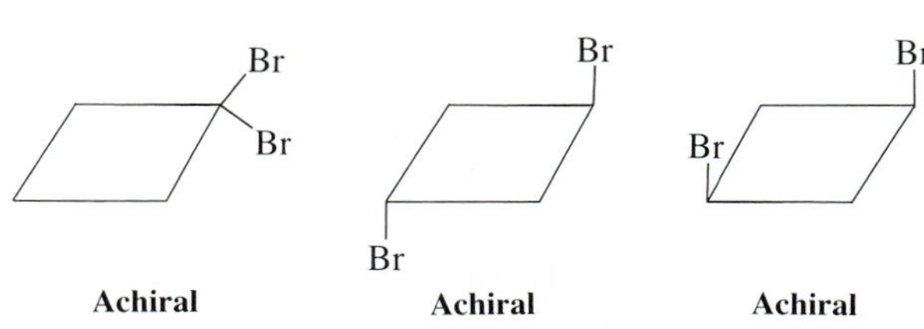

All other isomeric dibromocyclobutanes have, in fact, no stereocenters.

Br Br — **Achiral**; Br Br — **Chiral**; Br Br — **Meso**

Mirror plane; Br Br — **Meso**; Br Br — **Chiral**

cis-1,3-Dibromocyclopentane is meso because a mirror plane bisects the molecule; it passes through the methylene group connecting the two stereocenters.

5-11

CH_3 / H—Br / H—Cl / H—F / CH_3

CH_3 / Br—H / H—Cl / H—F / CH_3

CH_3 / Br—H / Cl—H / H—F / CH_3

CH_3 / Br—H / H—Cl / F—H / CH_3

With the four mirror images, there are four enantiomeric pairs of diastereomers.

5-12

Almost any halogenation at C-2 gives a racemate; the exception is bromination, which results in achiral 2,2-dibromobutane. In addition, bromination at C-3 gives the two 2,3-dibromobutane diastereomers, one of which, 2*R*,3*S*, is meso.

5-13

Attack at C-1:

CH_2Br / H—Br / H—H / H—H / CH_3

(*R*)-1,2-Dibromopentane

Chiral, optically active

Attack at C-2:

CH_3 / Br—Br / H—H / H—H / CH_3

2,2-Dibromopentane

Achiral

Attack at C-3:

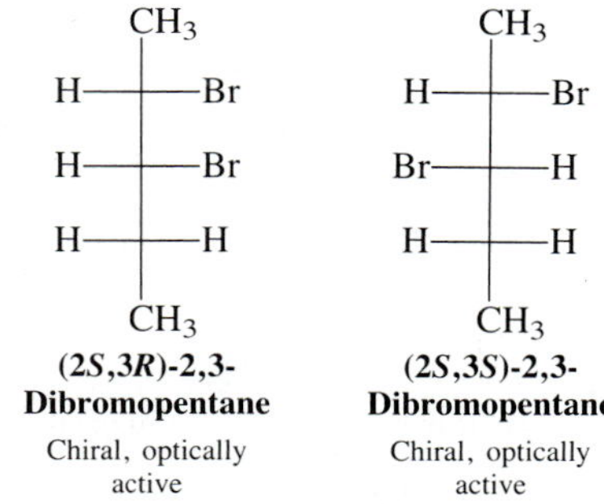

(2*S*,3*R*)-2,3-Dibromopentane
Chiral, optically active

(2*S*,3*S*)-2,3-Dibromopentane
Chiral, optically active

Diastereomers, formed in unequal amounts

Attack at C-4:

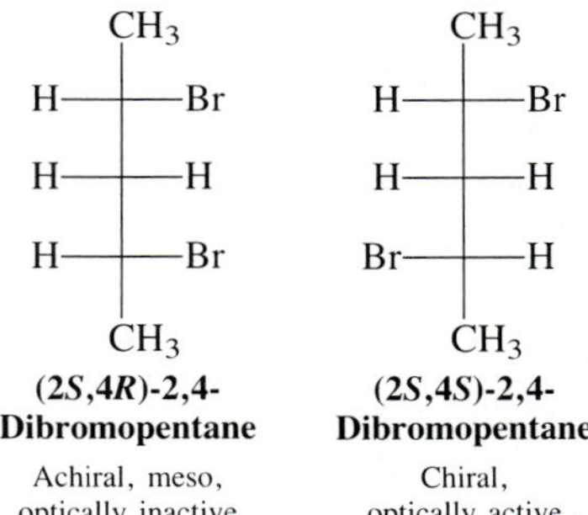

(2*S*,4*R*)-2,4-Dibromopentane
Achiral, meso, optically inactive

(2*S*,4*S*)-2,4-Dibromopentane
Chiral, optically active

Diastereomers, formed in unequal amounts

Attack at C-5:

CH_3 / H—Br / H—H / H—H / CH_2Br

(*S*)-1,4-Dibromopentane

Chiral, optically active

5-14

Attack at C-1:

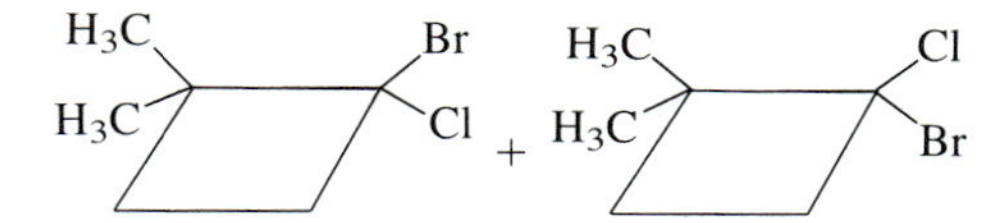

[Structures: two cyclobutane derivatives with H_3C, H_3C on one carbon and Br, Cl on the adjacent carbon (Br/Cl interchanged), joined by +]

Both are chiral but formed in equal amounts; hence, a racemate, optically inactive.

Attack at the methyl groups:

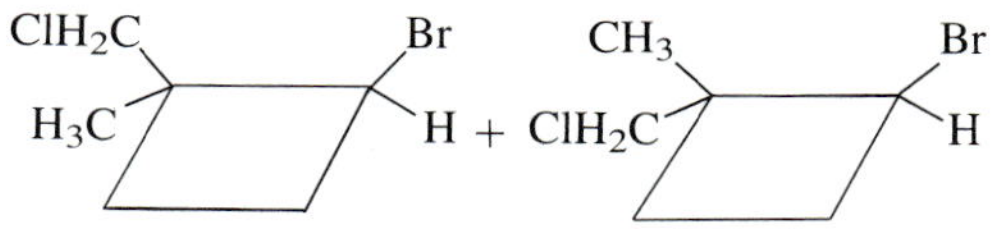

[Structures: ClH_2C, H_3C / Br, H + CH_3, ClH_2C / Br, H]

Both are diastereomers, formed in unequal amounts and in optically active form.

Attack at C-3:

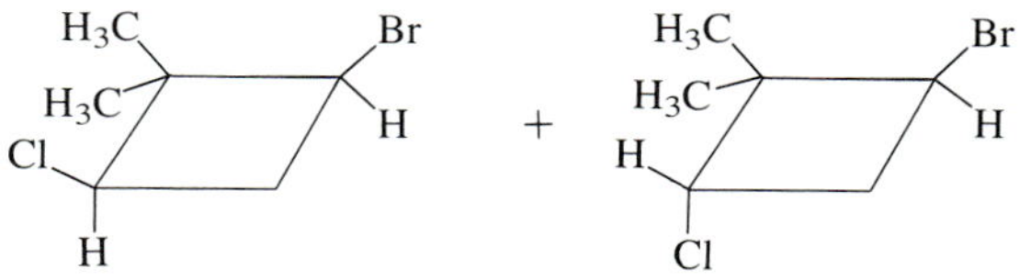

[Structures: H_3C, H_3C / Br, H with Cl, H at C-3 + H_3C, H_3C / Br, H with H, Cl at C-3]

Chiral cis diastereomer

Formed in amounts different from the trans dihalide, optically active

Chiral trans diastereomer

Formed in amounts different from the cis isomer, optically active

Attack at C-4:

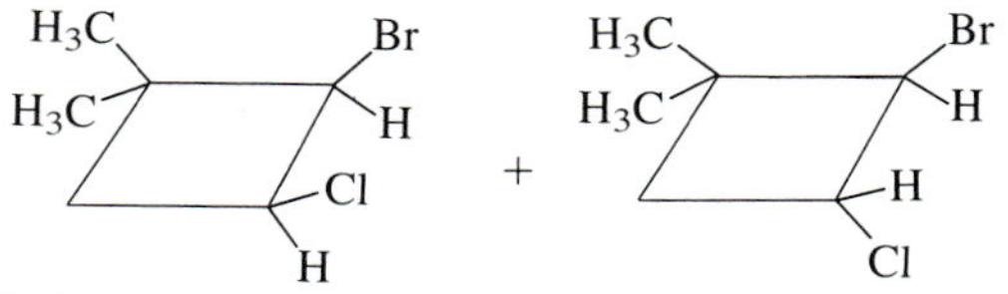

[Structures: H_3C, H_3C / Br, H with Cl, H at C-4 + H_3C, H_3C / Br, H with H, Cl at C-4]

Chiral cis diastereomer

Formed in amounts different from the trans dihalide, optically active

Chiral trans diastereomer

Formed in amounts different from the cis isomer, optically active

CHAPTER 6

6-1

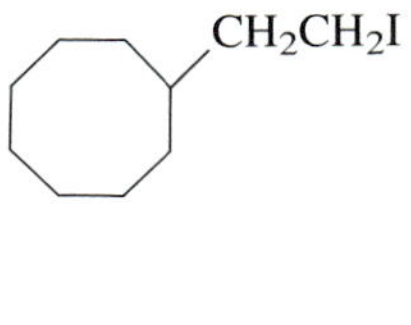

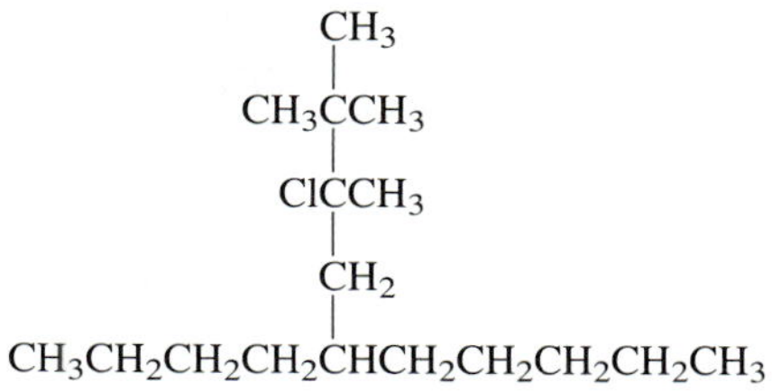

CH_2CH_2I

$$\begin{array}{c} CH_3 \\ | \\ CH_3CCH_3 \\ | \\ ClCCH_3 \\ | \\ CH_2 \\ | \\ CH_3CH_2CH_2CH_2CHCH_2CH_2CH_2CH_2CH_3 \end{array}$$

Note the similarity of this structure to that of 6-(2-chloro-2,3,3-trimethylbutyl)undecane. Why is it named so differently?

6-2

(a) $CH_3CH_2CH_2CH_2\ddot{\underset{..}{I}}:$

(b) $CH_3CH_2CH_2CH_2\ddot{\underset{..}{O}}CH_2CH_3$

(c) $CH_3CH_2CH_2CH_2\ddot{N}{=}\overset{+}{N}{=}\ddot{N}:^-$

(d) $\left[CH_3CH_2CH_2CH_2\overset{CH_3}{\underset{CH_3}{As}}CH_3\right]^+ \; :\ddot{\underset{..}{Br}}:^-$

(e) $\left[CH_3CH_2CH_2CH_2\ddot{\underset{CH_3}{Se}}CH_3\right]^+ \; :\ddot{\underset{..}{Br}}:^-$

6-3

(a) $CH_3I + :N(CH_3)_3$

(b) There are two approaches:

$CH_3\ddot{\underset{..}{S}}:^- + CH_3CH_2\ddot{\underset{..}{I}}:$ or $CH_3\ddot{\underset{..}{I}}: + CH_3CH_2\ddot{\underset{..}{S}}:^-$

6-4

$$H_3C{-}\ddot{\underset{..}{Cl}}: \xrightarrow{\text{Slow}} \cdot CH_3 + :\ddot{\underset{..}{Cl}}\cdot$$

$$:\ddot{\underset{..}{Cl}}\cdot + {}^-:\ddot{\underset{..}{O}}H \xrightarrow{\text{Fast}} :\ddot{\underset{..}{Cl}}:^- + \cdot\ddot{\underset{..}{O}}H$$

$$\cdot CH_3 + \cdot\ddot{\underset{..}{O}}H \xrightarrow{\text{Fast}} H_3C{-}\ddot{\underset{..}{O}}H$$

The first step requires the dissociation energy of the carbon–chlorine bond and is slow. The second (electron transfer) and third steps (radical recombination) are fast.

6-5

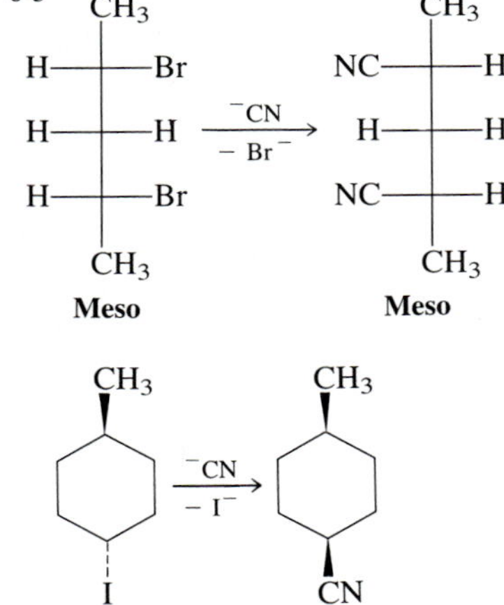

6-6
I^- is a better leaving group than Cl^-. Hence the product is $Cl(CH_2)_6SeCH_3$.

6-7

$$pK_b = 14 - pK_a$$

I^-	Br^-	Cl^-	HSO_4^-	$CH_3SO_3^-$	F^-	CH_3COO^-
19.2	18.7	16.2	19	15.2	10.8	9.3

CN^-	CH_3S^-	CH_3O^-	HO^-	H_2N^-	CH_3^-
4.8	4	−1.5	−1.7	−21	~ −36

6-8

6-9

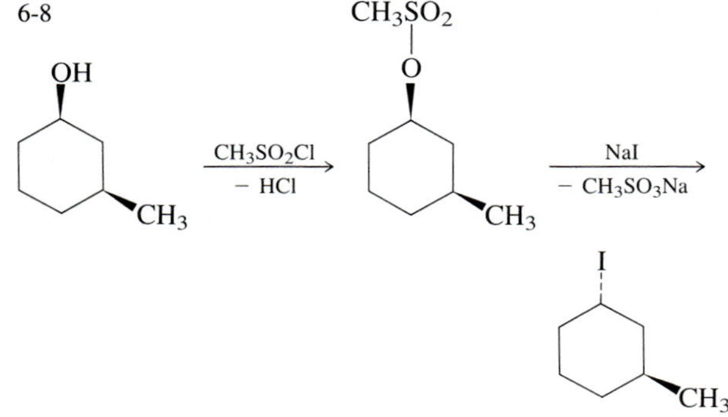

6-10
(a) CH_3S^- (b) CH_3NH^- (c) HSe^-

6-11
(a) $P(CH_3)_3$ (b) $CH_3CH_2Se^-$ (c) H_2O

6-12
(a) CH_3SeH (b) $(CH_3)_2PH$

6-13
(a) CH_3S^- (b) $(CH_3)_2NH$

6-14
The more-reactive substrates are **(a)** [bromocyclohexane structure: Br]

and **(b)** $CH_3CH_2CH_2Br$

6-15 [structures: Br > H3C, Br]

CHAPTER 7

7-1
Compound A is a 2,2-dialkyl-1-halopropane (neopentyl halide) derivative. The carbon bearing the potential leaving group is primary and very hindered and therefore very unreactive with respect to any substitution reactions. Compound B is a 1,1-dialkyl-1-haloethane (*tert*-alkyl halide) derivative and undergoes solvolysis.

7-2
Bonds broken: 67 + 119 = 186 kcal $mole^{-1}$
Bonds made: 93 + 87 = 180 kcal $mole^{-1}$
ΔH° = +6 kcal $mole^{-1}$

By this calculation, the reaction should actually be endothermic. It still proceeds because of the excess water employed and the favorable solvation energies of the products.

7-3

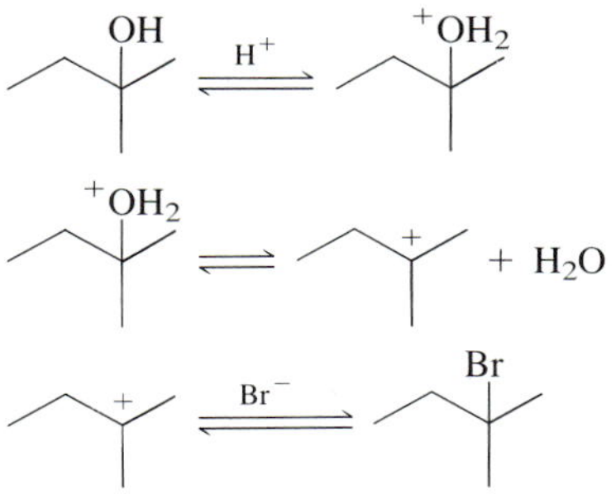

OH $\xrightarrow{H^+}$ $^+OH_2$

$^+OH_2$ ⇌ + H_2O

$\xrightarrow{Br^-}$ Br

This mechanism is exactly the reverse of that formulated for the hydrolysis of 2-bromo-2-methylpropane.

7-4

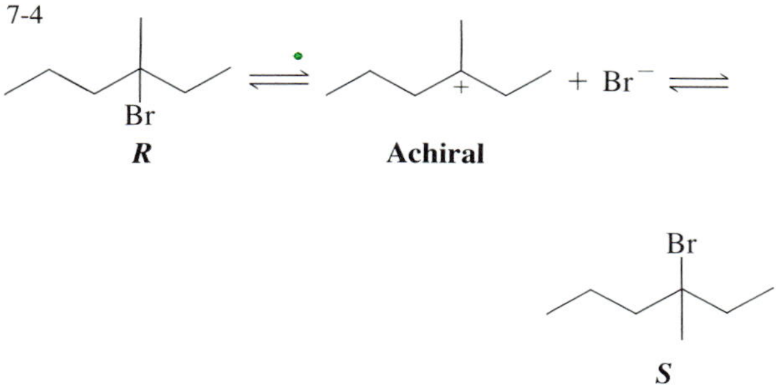

Br **R** ⇌ **Achiral** + Br^- ⇌ Br **S**

The molecule dissociates to the achiral tertiary carbocation. Recombination gives a 1:1 mixture of *R* and *S* product.

7-5

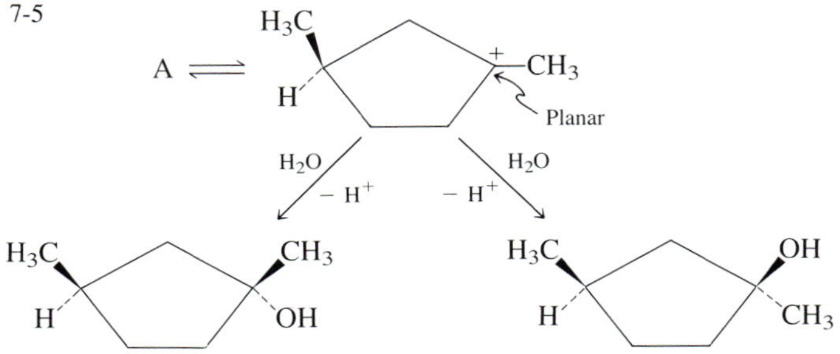

A ⇌ H_3C, H, CH_3 (+), Planar

H_2O, $-H^+$ → H_3C, H, CH_3, OH

H_2O, $-H^+$ → H_3C, H, OH, CH_3

7-6

$$(CH_3)_3COH + H^+ \rightleftharpoons (CH_3)_3C^+ + H_2O$$

Br^- Faster → $(CH_3)_3CBr$

Cl^- Slower → $(CH_3)_3CCl$

7-7

The bicyclic system cannot easily form a planar carbocation intermediate required for an S_N1 process. Backside attack as in the S_N2 reaction is completely blocked.

7-8

(a) This is an S_N2 reaction that occurs with inversion.

(b) In a weakly nucleophilic protic solvent, mainly solvolysis occurs through the intermediacy of an achiral carbocation.

7-9

$$(CH_3)_3CBr \rightleftharpoons (CH_3)_3C^+ + Br^-$$

CH_3CH_2OH, S_N1 → $(CH_3)_3COCH_2CH_3$ + H^+

H_2O, S_N1 → $(CH_3)_3COH$ + H^+

E1 → $H_2C{=}C(CH_3)_2$ + H^+

7-10

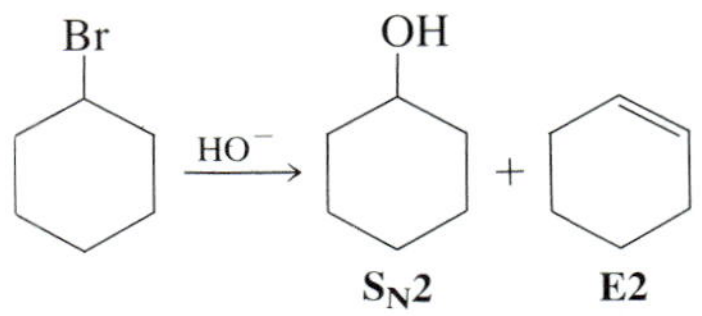

Br $\xrightarrow{HO^-}$ OH (S_N2) + (**E2**)

S_N2 **E2**

7-11

$CH_2{=}CH_2$; no E2 possible; $CH_2{=}C(CH_3)_2$; no E2 possible.

7-12

I^- is a better leaving group, allowing for selective elimination of HI.

7-13

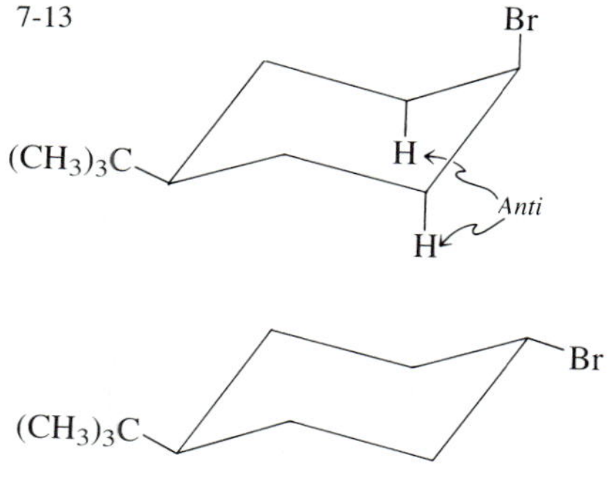

$(CH_3)_3C$, Br, H, H, *Anti*

$(CH_3)_3C$, Br

The 1,1-dimethylethyl group in both cases prefers the equatorial position. In the cis isomer, this preference places the leaving group *anti* to two hydrogens. In the trans isomer, the bromine is forced to be in the equatorial position without any *anti* hydrogens. In order for elimination to occur, the molecule has to first undergo conformational flip at an energy cost of about 5.5 kcal mole^{-1} (Table 4-3).

7-14

(a) $N(CH_3)_3$, stronger base, worse nucleophile

(b) $(CH_3\overset{\displaystyle CH_3}{\overset{|}{C}}H)_2N^-$, more-hindered base

(c) Cl^-, stronger base, worse nucleophile

7-15

(a) The second example will give more E2 product, because a stronger base is present.

(b) The first example will give E2 product, mainly because of the presence of a strong, hindered base.

CHAPTER 8

8-1

(a) HO CH3 **(b)** OH Br

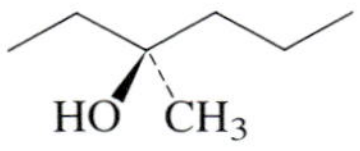

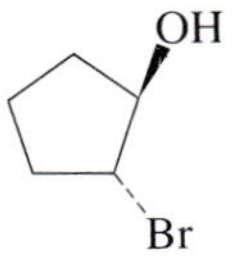

(c) $(CH_3)_3CCH_2OH$

8-2

(a) 4-Methyl-2-pentanol; **(b)** *cis*-4-ethylcyclohexanol; **(c)** 3-bromo-2-chloro-1-butanol.

8-3 OH < OH Cl < OH Cl < OH Cl

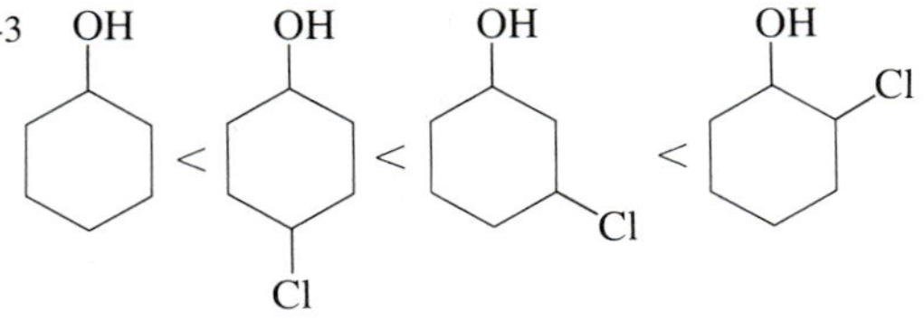

8-4

In solution, $(CH_3)_3COH$ is a weaker acid than CH_3OH. The equilibrium lies to the right.

8-5

The first step is an S_N2 reaction that results in inversion. The second step leaves the stereocenter unchanged. Hence the product is (*S*)-2-butanol.

8-6

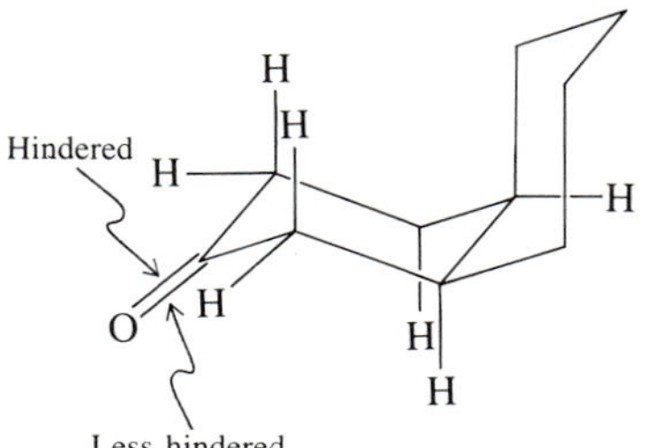

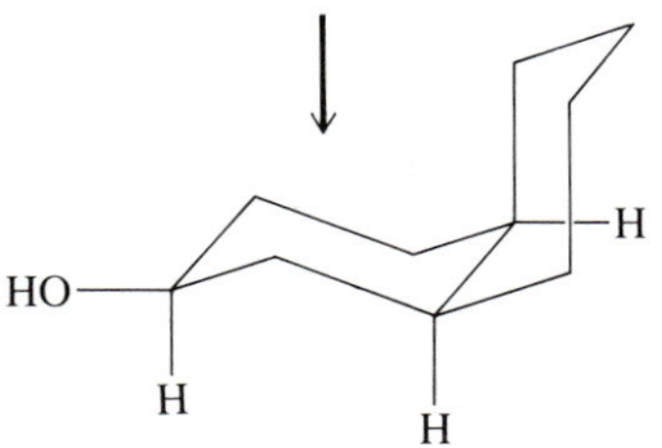

Predominant alcohol diastereomer

8-7

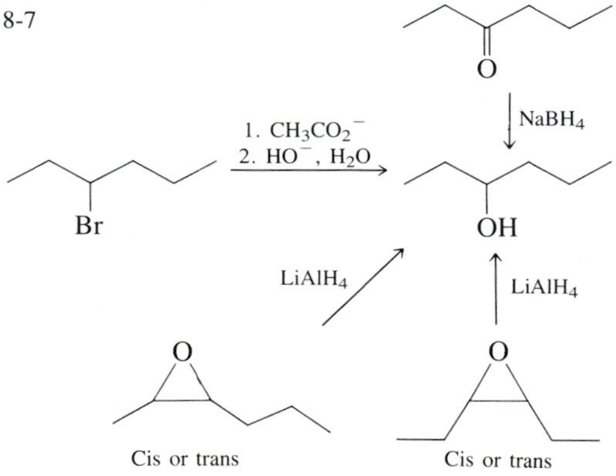

O

Br → (1. $CH_3CO_2^-$ 2. HO^-, H_2O) → OH; ketone (O) → ($NaBH_4$) → OH

$LiAlH_4$ (from epoxide O, Cis or trans); $LiAlH_4$ (from epoxide O, Cis or trans)

Cis or trans — Cis or trans

Note: This starting material will also give 2-hexanol on reduction

8-8

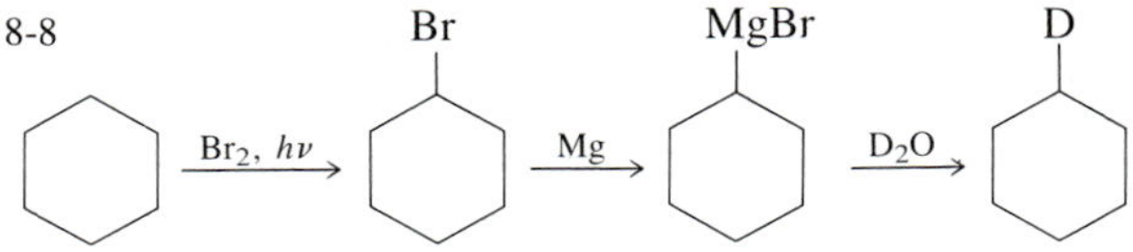

cyclohexane $\xrightarrow{Br_2,\ h\nu}$ Br $\xrightarrow{Mg}$ MgBr $\xrightarrow{D_2O}$ D

8-9

$(CH_3)_4C \xrightarrow{Br_2,\ h\nu} (CH_3)_3CCH_2Br \xrightarrow{Li}$

$(CH_3)_3CCH_2Li \xrightarrow{CuI}$

$[(CH_3)_3CCH_2]_2CuLi \xrightarrow{(CH_3)_3CCH_2Br}$

$(CH_3)_3CCH_2CH_2C(CH_3)_3$

8-10

$$(CH_3)_2CHOH \xrightarrow{HBr} (CH_3)_2CHBr \xrightarrow{Mg} (CH_3)_2CHMgBr \xrightarrow{CH_2=O} (CH_3)_2CHCH_2OH$$

8-11

(a) $CH_3CH_2CH_2MgBr + CH_3\overset{O}{\overset{\|}{C}}H$

(b) Cyclobutanone $+ (CH_3)_3CLi$

(c) $CH_3\overset{O}{\overset{\|}{C}}OCH_3 + 2\ CH_3MgBr$

(d) $CH_3CH_2CH_2CH_2Li + H_2C\overset{O}{—}CH_2$ (ethylene oxide)

8-12

$$CH_3CH_2CH_2CH_2C(CH_3)_2—C(OH)(CH_3)CH_3 \Longrightarrow$$

2,3,3-Trimethyl-2-heptanol

$$CH_3\overset{O}{\overset{\|}{C}}CH_3 + CH_3CH_2CH_2CH_2C(CH_3)_2Li \Longrightarrow$$

$$CH_3CH_2CH_2CH_2C(CH_3)_2Br \Longrightarrow$$

2-Bromo-2-methylhexane

$$CH_3CH_2CH_2CH_2C(CH_3)_2OH \Longrightarrow$$

2-Methyl-2-hexanol

$$CH_3\overset{O}{\overset{\|}{C}}CH_3 + CH_3CH_2CH_2CH_2MgBr$$

8-13

$$CH_3C(CH_3)_2OH \xrightarrow[2.\ Mg]{1.\ HBr} CH_3C(CH_3)_2MgBr \xrightarrow{H_2C\overset{O}{—}CH_2}$$

$$CH_3C(CH_3)_2CH_2CH_2OH \xrightarrow{(CH_3\overset{CH_3}{C}H)_2N^-Li^+}$$

$$CH_3C(CH_3)_2CH_2CH_2O^-Li^+ \xrightarrow{CH_3I} CH_3C(CH_3)_2CH_2CH_2OCH_3$$

CHAPTER 9

9-1

$$\underset{pK_a\ =\ 15.5}{CH_3OH} + {}^-CN \xrightleftharpoons{K\ =\ 10^{-6.3}} CH_3O^- + \underset{pK\ =\ 9.2}{HCN}$$

Answer: No.

9-2

2-Methylcyclohexanol ($\ddot{O}H$) $\xrightleftharpoons{H^+}$ protonated alcohol ($\overset{+}{\ddot{O}}H_2$) $\xrightleftharpoons{-\ H_2\ddot{O}:}$

Secondary carbocation (H₃C, H; hydride shift) $\longrightarrow$ Tertiary carbocation $\xrightarrow{:\ddot{Br}:^-}$ 1-bromo-1-methylcyclohexane (H_3C, $:\ddot{Br}:$)

Secondary carbocation **Tertiary carbocation**

9-3

(a) $CH_3C(OCH_3)(CH_3)CH_2CH_2CH_3$

(b) 1-chloro-1-ethylcyclohexane (CH_3CH_2, Cl on ring carbon)

9-4

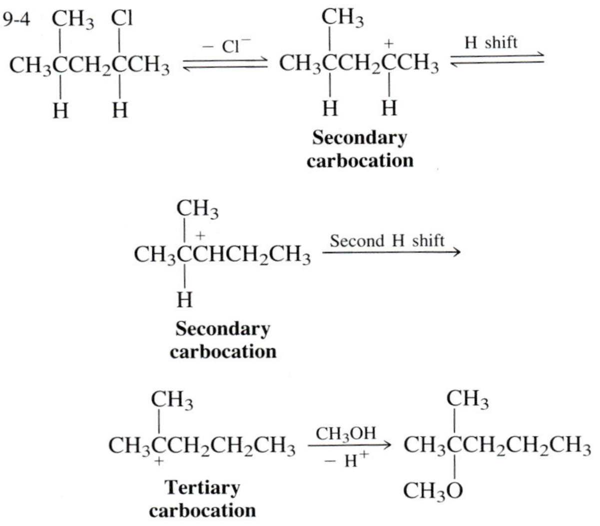

$$CH_3CH(CH_3)CH_2CHClCH_3 \xrightleftharpoons{-Cl^-} CH_3CH(CH_3)CH_2\overset{+}{C}HCH_3 \xrightleftharpoons{\text{H shift}}$$

Secondary carbocation

$$CH_3\overset{+}{C}HCH(CH_3)CH_2CH_3 \xrightarrow{\text{Second H shift}}$$

Secondary carbocation

$$CH_3\overset{+}{C}(CH_3)CH_2CH_2CH_3 \xrightarrow[-H^+]{CH_3OH} CH_3C(CH_3)(OCH_3)CH_2CH_2CH_3$$

Tertiary carbocation

9-5

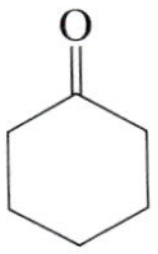

90%
Cyclohexanone

9-6

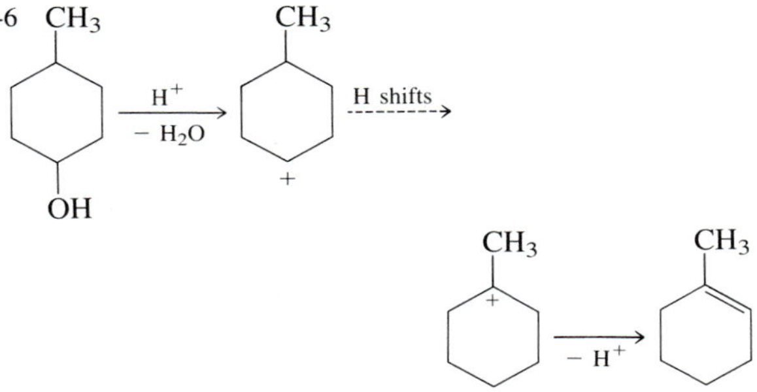

H shifts; $-H^+$

9-7

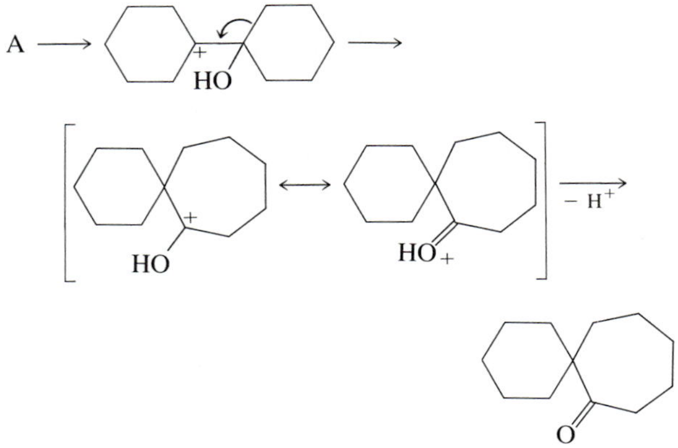

A →

$-H^+$

9-8

(a) P, I_2; **(b)** HCl; **(c)** PBr_3.

9-9

CH_3OH → 1. PBr_3 2. Mg → CH_3MgBr

CH_3OH → 1. $K_2Cr_2O_7$, H^+ 2. CH_3OH, H^+ → $HCOCH_3$ (O double bonded)

→ $CH_3CH(OH)CH_3$ → 1. $K_2Cr_2O_7$, H^+ 2. CH_3MgBr 3. HBr → $(CH_3)_3CBr$

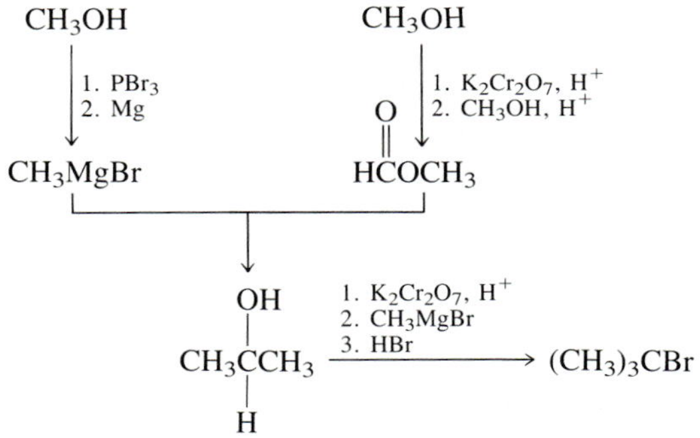

9-10

This is an example of the intramolecular Williamson synthesis, to be discussed in the next subsection.

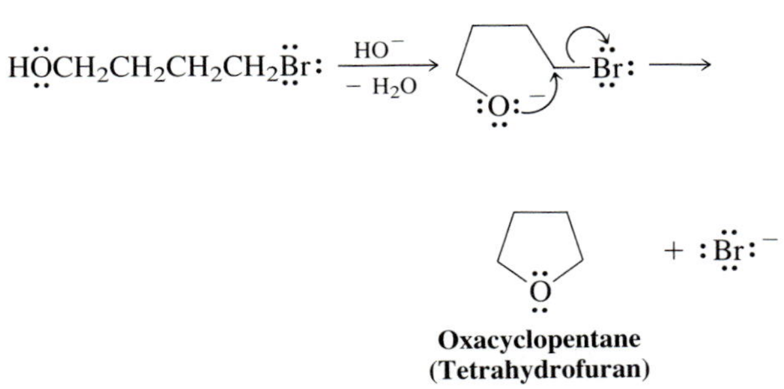

$$HOCH_2CH_2CH_2CH_2Br \xrightarrow[-H_2O]{HO^-} \longrightarrow$$

$+ :Br:^-$

Oxacyclopentane (Tetrahydrofuran)

9-11

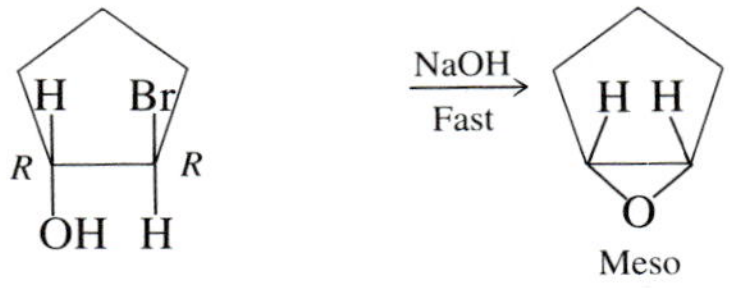

NaOH, Fast

(1*R*,2*R*)-2-Bromocyclopentanol

The nucleophilic oxygen and the leaving group are trans (*anti*).

Meso

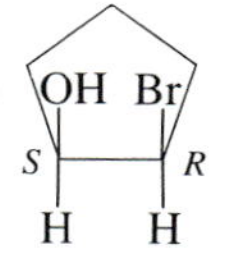

(1*S*,2*R*)-2-Bromocyclopentanol

Here nucleophile and leaving group are cis (*syn*).

$\xrightarrow{NaOH}$ no epoxide formation, relatively slow E2 and S_N2

9-12

$$BrCH_2CH_2CH_2OH \xrightarrow[4.\ H^+,\ H_2O]{1.\ (CH_3)_3COH,\ H^+ \quad 2.\ Mg \quad 3.\ D_2O} DCH_2CH_2CH_2OH$$

9-13

Mechanism 1:

$$H\ddot{O}CH_2CH_2CH_2CH_2\ddot{O}H + H^+ \rightleftharpoons$$

$$H_2C(CH_2)_3\overset{+}{O}H_2 \text{ (cyclizing: } H\ddot{O}: \text{ attacks } CH_2\text{)} \xrightarrow{-H_2\ddot{O}} \text{(oxolane ring, } \ddot{O}H^+) \rightleftharpoons$$

$$\text{(oxolane, } \ddot{O}) + H^+$$

Mechanism 2:

$$CH_3\underset{:\ddot{O}H}{\overset{CH_3}{|}}\!\!CCH_2CH_2CH_2CH_2\ddot{O}H + H^+ \underset{+H_2\ddot{O}}{\overset{-H_2\ddot{O}}{\rightleftharpoons}}$$

$$(CH_3)_2\overset{+}{C}CH_2CH_2CH_2CH_2\ddot{O}H \longrightarrow \text{(oxane ring, } \overset{+}{\ddot{O}}\text{–H, } C(CH_3)_2) \rightleftharpoons$$

$$\text{(2,2-dimethyloxane, } \ddot{O}) + H^+$$

9-14

(a) This ether is best synthesized by solvolysis:

$$CH_3CH_2C(CH_3)_2Br + CH_3CH(CH_3)OH \longrightarrow CH_3CH_2C(CH_3)_2\text{—O—}CH(CH_3)CH_3$$

Solvent (the $CH_3CH(CH_3)OH$); product: **2-Methyl-2-(1-methylethoxy)butane**

The alternative, an S_N2 reaction, would give elimination:

$$CH_3CH_2C(CH_3)_2O^- + CH_3CH(CH_3)Br \longrightarrow$$

$$CH_3CH{=}CH_2 + CH_3CH_2C(CH_3)_2OH$$

(b) This target is best prepared by an S_N2 reaction with a halomethane, because such an alkylating agent cannot undergo elimination. The alternative would be nucleophilic substitution of a 1-halo-2,2-dimethylpropane, a reaction that is normally too slow.

$$CH_3C(CH_3)_2CH_2O^- + CH_3Cl \longrightarrow$$

$$CH_3C(CH_3)_2CH_2OCH_3 \quad + Cl^-$$

1-Methoxy-2,2-dimethylpropane

$$CH_3C(CH_3)_2CH_2Br + CH_3O^- \longrightarrow \text{slow reaction, impractical}$$

9-15

$$CH_3OCH_3 + 2\ HI \xrightarrow{\Delta} 2\ CH_3I + H_2O$$

Mechanism:

$$CH_3\ddot{O}CH_3 + H\ddot{I}: \rightleftharpoons CH_3\overset{+}{\underset{H}{O}}CH_3 + :\ddot{I}:^-$$

$$:\ddot{I}:^- + CH_3\text{—}\overset{H}{\overset{|}{O^+}}\text{—}CH_3 \longrightarrow CH_3\ddot{I}: + H\ddot{O}CH_3$$

(Continued on next page.)

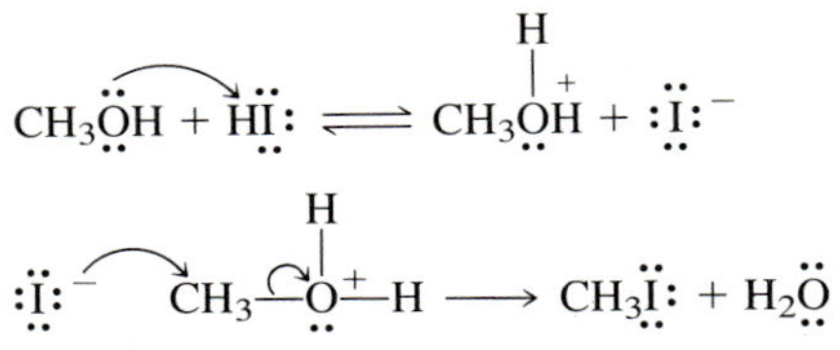

$CH_3\ddot{O}H + H\ddot{I}: \rightleftharpoons CH_3\overset{+}{\ddot{O}}H_2 + :\ddot{I}:^-$

$:\ddot{I}:^- \quad CH_3-\overset{+}{\ddot{O}}-H \longrightarrow CH_3\ddot{I}: + H_2\ddot{O}$

9-16

Methyllithium reacts with compound A to form initially the less-stable trans-diaxial 2-methylcyclohexanol conformer, which undergoes ring flip to the more-stable trans-diequatorial form.

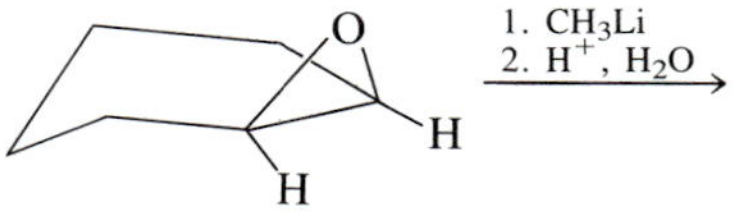

A $\xrightarrow{\text{1. } CH_3Li \text{ 2. } H^+, H_2O}$

Ring flip

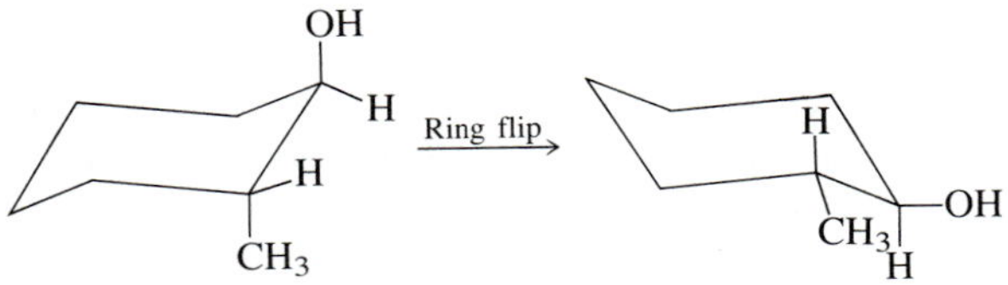

Trans-diaxial **Trans-diequatorial**

***trans*-2-Methylcyclohexanol**

9-17

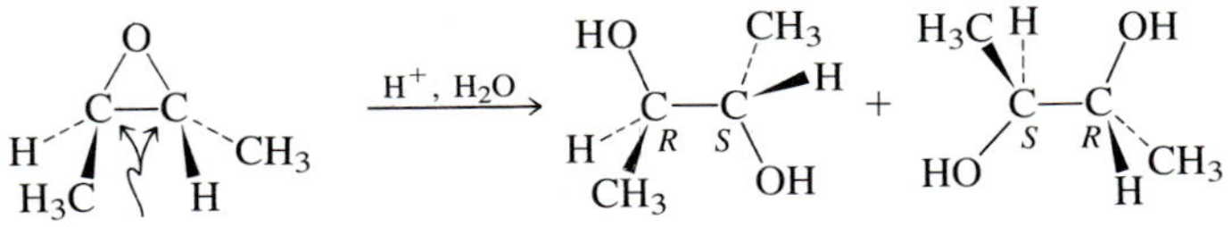

$\xrightarrow{H^+, H_2O}$ +

Attack by water

(2*R*,3*R*)-*trans*-2,3-Dimethyloxacyclopropane

Both products are the same as

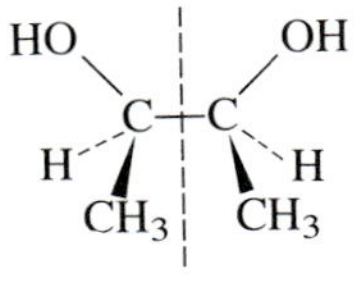

Mirror plane

***meso*-2,3-Butanediol (Not optically active)**

Because the starting material is symmetric, attack at either oxacyclopropane carbon is equally likely, giving the *same* molecule in either case.

9-18

(a) oxacyclopropane $+ HS^- \longrightarrow HOCH_2CH_2S^- \xrightarrow{\text{oxacyclopropane}} \xrightarrow{H^+, H_2O} HOCH_2CH_2SCH_2CH_2OH \xrightarrow{SOCl_2} ClCH_2CH_2SCH_2CH_2Cl$

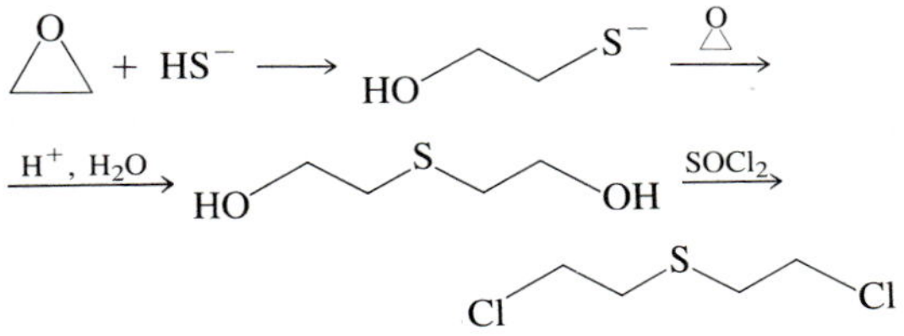

(b) Intramolecular sulfonium salt formation:

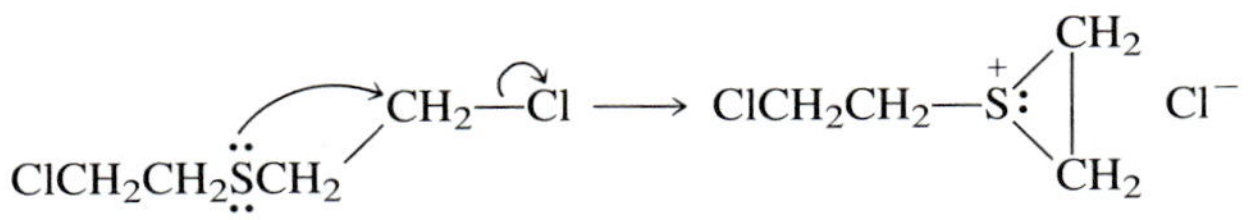

$ClCH_2CH_2\ddot{S}CH_2CH_2-Cl \longrightarrow ClCH_2CH_2-\overset{+}{S}(CH_2CH_2) \; Cl^-$

Nucleophiles attack by ring-opening:

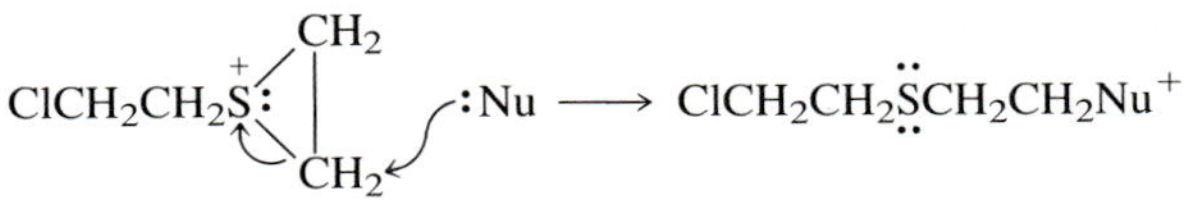

$ClCH_2CH_2\overset{+}{S}(CH_2CH_2) + :Nu \longrightarrow ClCH_2CH_2\ddot{S}CH_2CH_2Nu^+$

CHAPTER 10

10-1

$CH_3(CH_2)_4CH(CH_3)OH \qquad (CH_3)_3C-C(CH_3)_2OH$

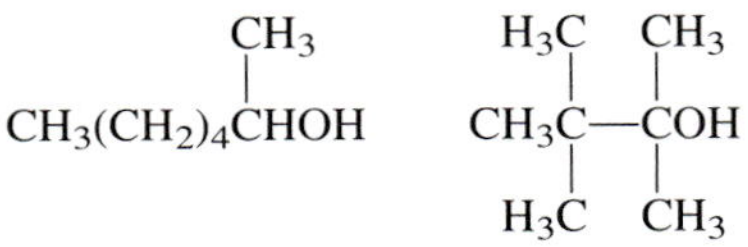

10-2

$DH^\circ_{Cl_2} = 58 \text{ kcal mole}^{-1}.$

$\Delta E = 28{,}600/\lambda$

$\lambda = 28{,}600/\Delta E = 490$ nm, in the ultraviolet-visible range.

10-3

$\delta = 288/90 = 3.20$ ppm and $\delta = 297/90 = 3.30$ ppm. In a 100-MHz spectrometer, the signals would appear 320 and 330 Hz downfield from $(CH_3)_4Si$.

10-4
In both cases, the methyl group resonates at higher field. In $ClCH_2OCH_3$, the methylene hydrogens are relatively deshielded because of the cumulative electron-withdrawing effect of the two heteroatoms. In $CH_3OCH_2CH_2OCH_3$, the difference is that between the primary and secondary hydrogens.

10-5
(a) $(H_3C)_2(CH_3)C—C(CH_3)(CH_3)_2$ (2,2,3,3-tetramethylbutane: $CH_3C—CCH_3$ with H_3C, CH_3 above and H_3C, CH_3 below)
One peak

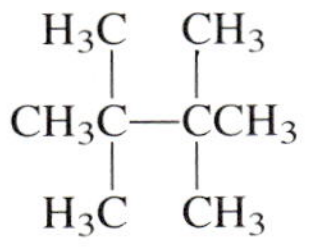

(b) $CH_3OCH_2CH_2OCH_2CH_2OCH_3$
Three peaks

(c) Oxirane ($H_2C—CH_2$ bridged by O)
One peak

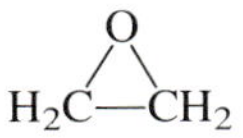

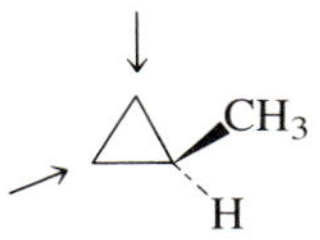

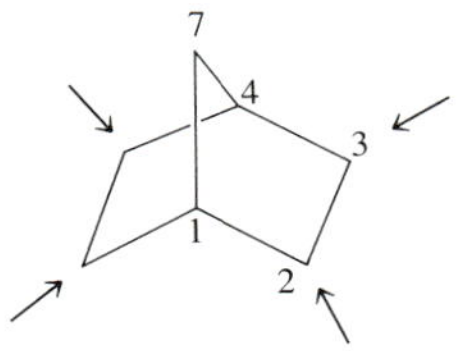

10-6
(a) None. **(b)** (methylcyclopropane with CH3 and H; arrows) **(c)** (bicyclic structure numbered 1, 2, 3, 4, 7 with arrows) but not at C-7

10-7

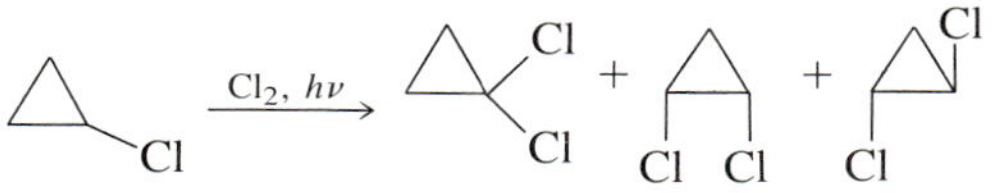

(chlorocyclopropane) $\xrightarrow{Cl_2,\ h\nu}$ (1,1-dichlorocyclopropane) + (cis-1,2-dichlorocyclopropane) + (trans-1,2-dichlorocyclopropane)
Cis Trans

1,1-Dichlorocyclopropane shows only one signal, whereas *cis*-1,2-dichlorocyclopropane exhibits three (the two hydrogens next to the chlorine atoms at C-1 and C-2 are equivalent, whereas those at C-3 are diastereotopic). In contrast, in the trans isomer, the hydrogens at C-3 are not diastereotopic, as shown by a 180° rotational symmetry operation:

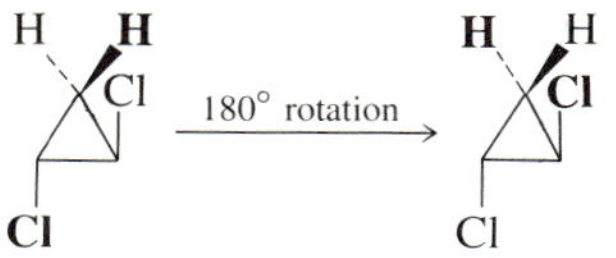

$\xrightarrow{180^\circ \text{ rotation}}$

Therefore, this compound reveals only two signals.

10-8

The following δ values were recorded in CCl_4 solution:
(a) $\delta = 3.38$ (q, $J = 7.1$ Hz, 4 H) and 1.12 (t, $J = 7.1$ Hz, 6 H) ppm;
(b) $\delta = 3.53$ (t, $J = 6.2$ Hz, 4 H) and 2.34 (quin, $J = 6.2$ Hz, 2 H) ppm;
(c) $\delta = 3.19$ (s, 1 H), 1.48 (q, $J = 6.7$ Hz, 2 H), 1.14 (s, 6 H), and 0.90 (t, $J = 6.7$ Hz, 3 H) ppm.

10-9
The highest-field doublet ($J = 6.5$ Hz) of relative intensity 6 is due to the two equivalent methyl groups split by the adjacent tertiary hydrogen. The lowest-field doublet ($J = 6$ Hz) of intensity 2 is generated by the two hydrogens of the chloromethyl group, again split by the neighboring tertiary nucleus. Thus, the tertiary position is simultaneously coupled to the six methyl and the two methylene hydrogens by two distinct coupling constants. The resulting multiplet should consist of a septet of triplets or a triplet of septets: a maximum of twenty-one peaks. However, because the two coupling constants to the respective sets of nonequivalent hyrogens are similar, only nine peaks are seen, in conformity with the $N + 1$ rule.

10-10 $H_3C—CH_2—CH_2—Br$
↑ qt ↑ ttq ↑ tt

10-11
(a) 3; **(b)** 3; **(c)** 7; **(d)** 2.

11-16

The formation of A is through an acid-catalyzed rearrangement, *not* an oxidation.

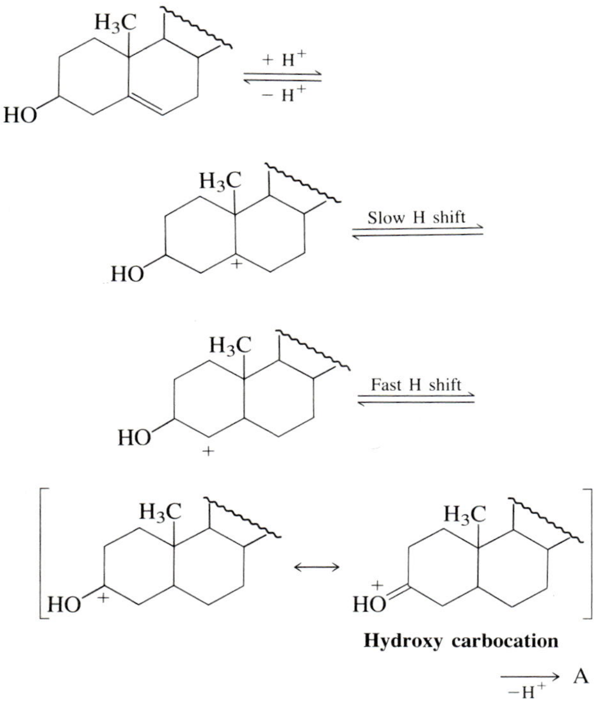
H_3C
HO
$+ H^+$
$- H^+$
Slow H shift
Fast H shift
Hydroxy carbocation
$-H^+$
A

Note that the first hydrogen shift is uphill (from a tertiary to a secondary carbocation), but this unfavorable equilibrium is driven by the favorable nature of the subsequent steps.

CHAPTER 12

12-1

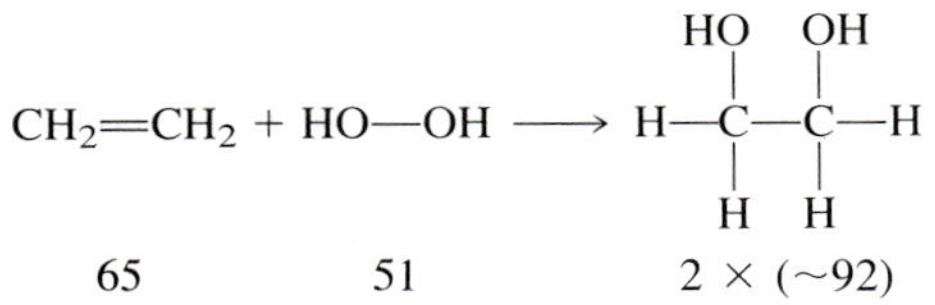
$CH_2{=}CH_2 + HO{-}OH \longrightarrow$
HO OH
H—C—C—H
H H
65
51
2 × (~92)

Then $\Delta H° = -68$ kcal mole^{-1}. Even though very exothermic, this reaction requires a catalyst. (See, e.g., Section 12-5.)

12-2

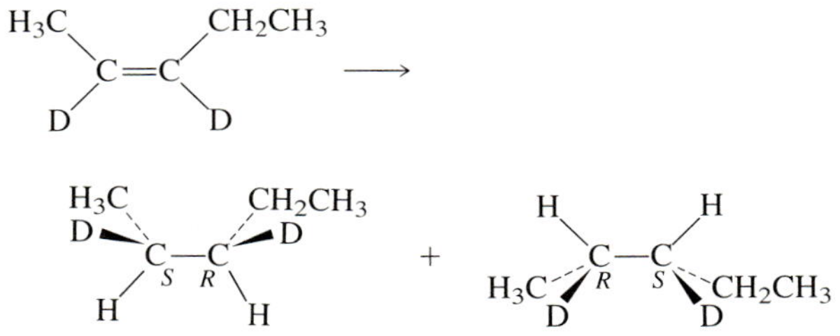
H_3C
CH_2CH_3
D
H
S
R
+

Racemic

12-3

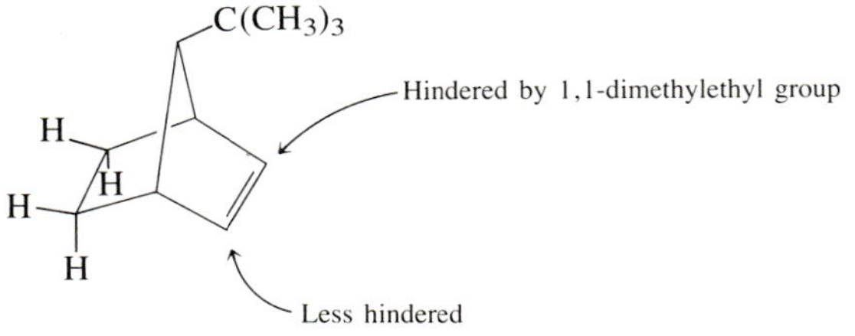
$C(CH_3)_3$
Hindered by 1,1-dimethylethyl group
Less hindered
H

12-4

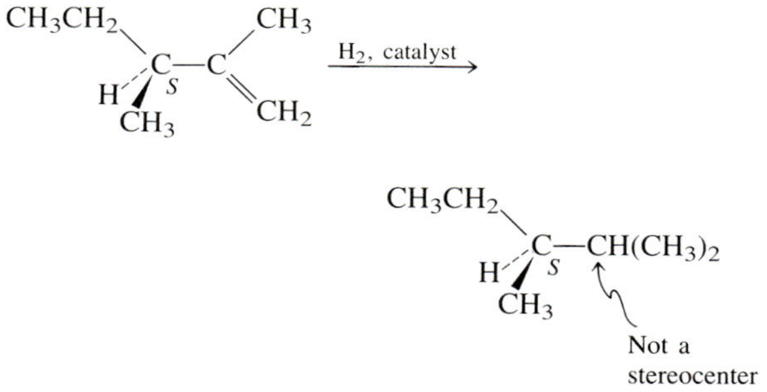
CH_3CH_2
CH_3
CH_2
H
S
H_2, catalyst
$CH(CH_3)_2$
Not a stereocenter

12-5

(a)

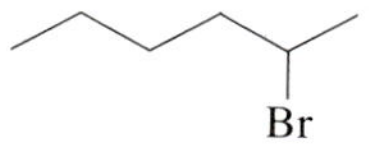
Br

(b)

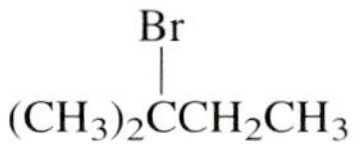
Br
$(CH_3)_2CCH_2CH_3$

(c)

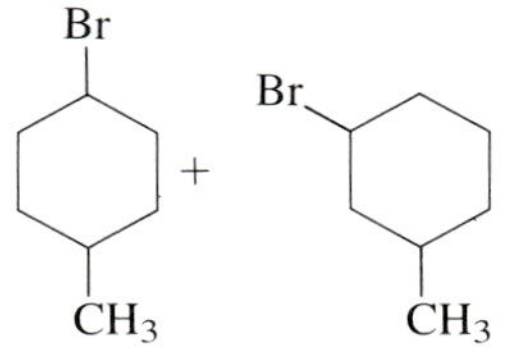
Br
CH_3
+

Both cis and trans

12-6

Protonation to the 1,1-dimethylethyl (*tert*-butyl) cation is reversible. With D^+, fast exchange of all hydrogens for deuterium will occur.

$$CH_2{=}C(CH_3)_2 \underset{-D^+}{\overset{+D^+}{\rightleftharpoons}} DCH_2\overset{+}{C}(CH_3)_2 \underset{+H^+}{\overset{-H^+}{\rightleftharpoons}}$$

$$DCH{=}C(CH_3)_2 \underset{-D^+}{\overset{+D^+}{\rightleftharpoons}} D_2CH\overset{+}{C}(CH_3)_2 \underset{+H^+}{\overset{-H^+}{\rightleftharpoons}}$$

$$D_2C{=}C(CH_3)_2 \underset{-D^+}{\overset{+D^+}{\rightleftharpoons}} D_3C\overset{+}{C}(CH_3)_2 \underset{+H^+}{\overset{-H^+}{\rightleftharpoons}}$$

$$(D_3C)(H_3C)C{=}CH_2 \underset{-D^+}{\overset{+D^+}{\rightleftharpoons}} \text{and so on} \dashrightarrow$$

$$(CD_3)_3C^+ \underset{-D_2O}{\overset{D_2O}{\rightleftharpoons}} (CD_3)_3COD + D^+$$

12-7

$$\underset{65}{CH_2{=}CH_2} + \underset{37}{F{-}F} \longrightarrow FCH_2{-}CH_2F$$

$2 \times (\sim 107)$ kcal mole^{-1}

$\Delta H^\circ = -112$ kcal mole^{-1}

$$\underset{65}{CH_2{=}CH_2} + \underset{36}{I{-}I} \longrightarrow ICH_2{-}CH_2I$$

$2 \times (\sim 53)$ kcal mole^{-1}

$\Delta H^\circ = -5$ kcal mole^{-1}

12-8

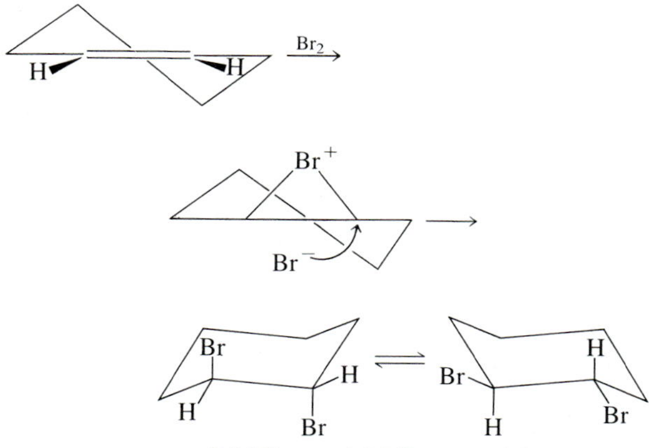

$\xrightarrow{Br_2}$ Br$^+$ / Br$^-$ $\longrightarrow$

(1*S*,2*S*)-*trans*-1,2-Dibromocyclohexane

$\xrightarrow{Br_2}$ Br$^+$ / Br$^-$ $\longrightarrow$

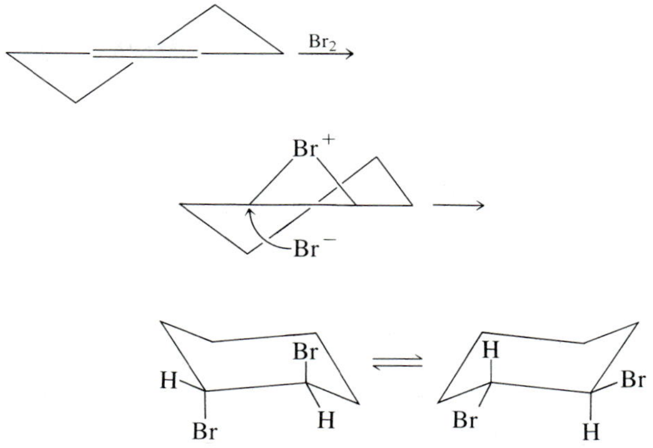

(1*R*,2*R*)-*trans*-1,2-Dibromocyclohexane

Anti addition to either conformation initially gives the trans-diaxial conformer.

12-9

(a) Only one diastereomer is formed (as a racemate):

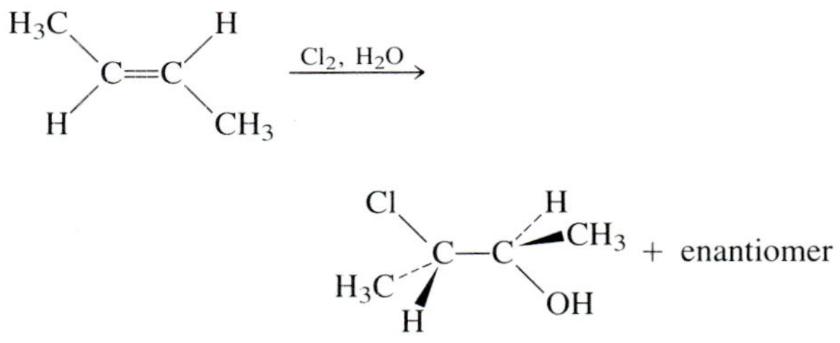

$$(H_3C)(H)C{=}C(H)(CH_3) \xrightarrow{Cl_2,\ H_2O} Cl(H_3C)(H)C{-}C(H)(CH_3)(OH) + \text{enantiomer}$$

(b) Two isomers are formed, but only one diastereomer of each (as racemates):

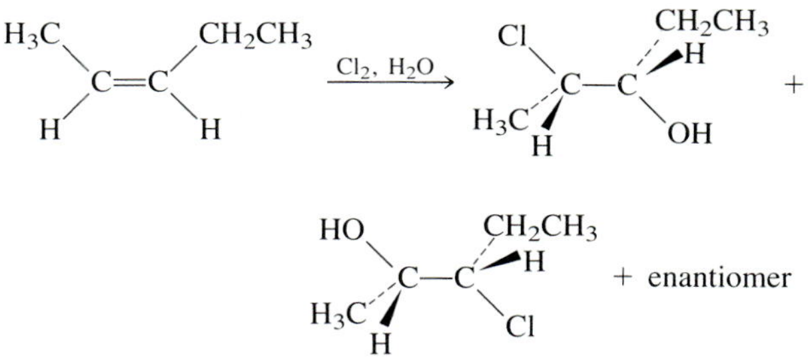

$$(H_3C)(H)C{=}C(CH_2CH_3)(H) \xrightarrow{Cl_2,\ H_2O} Cl(H_3C)(H)C{-}C(CH_2CH_3)(H)(OH) +$$

$$HO(H_3C)(H)C{-}C(CH_2CH_3)(H)(Cl) + \text{enantiomer}$$

12-10

(a) $CH_3CH(OCH_3)CH_2Cl$ (both enantiomers)

(b) [Structures: four bromohydrin stereoisomers of methylcyclohexane with Br and OH] + all enantiomers

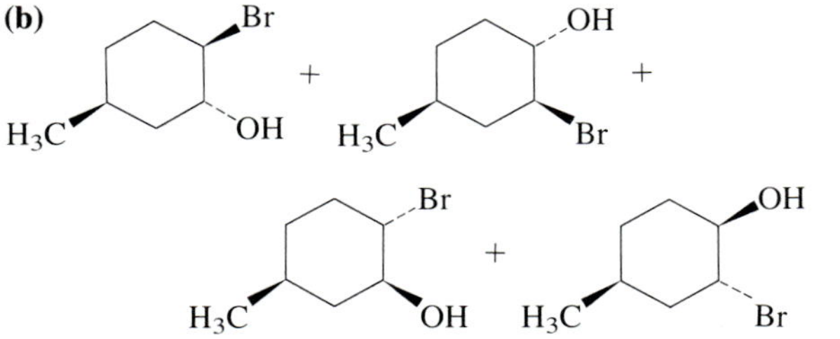

12-11

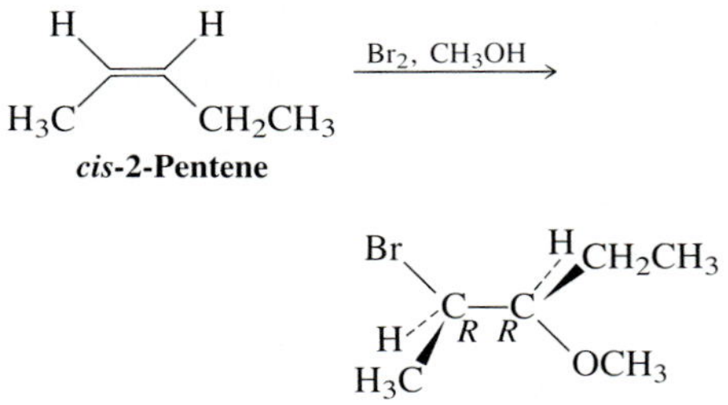

H_3C–CH=CH–CH_2CH_3 (*cis*-2-Pentene) $\xrightarrow{Br_2,\ CH_3OH}$ (2*R*,3*R*) product (CH_3CHBr–CH(OCH_3)CH_2CH_3) + enantiomer

Opening of the bromonium ion can also give (3*R*,2*R*)- and (3*S*,2*S*)-3-bromo-2-methoxypentane.

12-12

Mercuration is followed by *intramolecular* trapping of the mercurinium ion by one of the hydroxy groups.

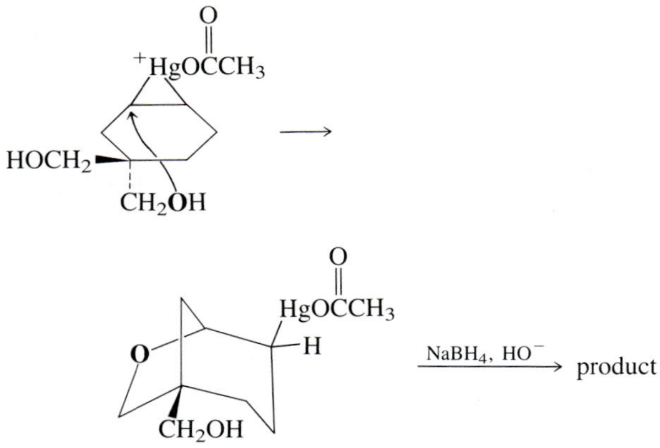

$^+HgOCCH_3$ (with C=O), $HOCH_2$, CH_2OH ⟶ $HgOCCH_3$, H, O, CH_2OH $\xrightarrow{NaBH_4,\ HO^-}$ product

12-13

(a) $CH_3CH_2CH_2OH$

(b) H, D, H_3C (*S*) C—C (*R*) OH, CH_3, D + enantiomer

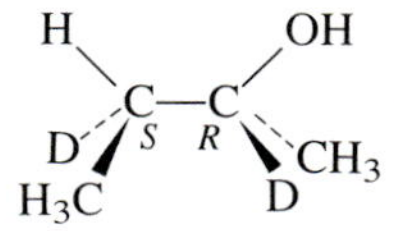

12-14

(a) Cyclopentane with H, D, Br, D + enantiomer

(b) H, H_3C, D (*R*) C—C (*R*) Br, CH_3, D + enantiomer

12-15

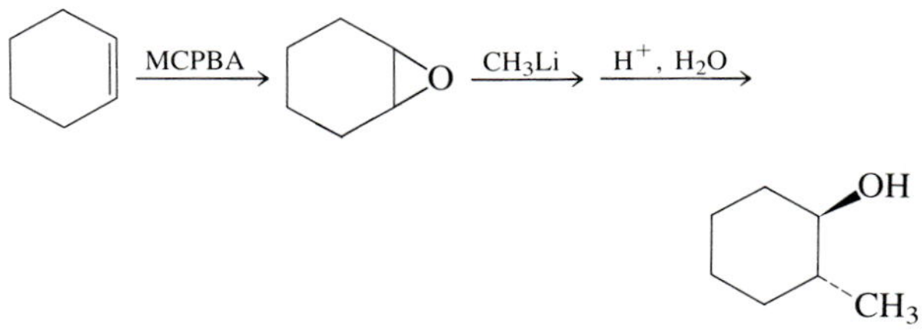

Cyclohexene $\xrightarrow{MCPBA}$ cyclohexene oxide $\xrightarrow{CH_3Li}$ $\xrightarrow{H^+,\ H_2O}$ *trans*-2-methylcyclohexanol (OH, CH_3)

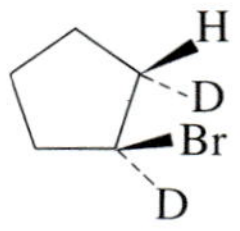

12-16

(a) [chain] OH + enantiomer (HO, H)

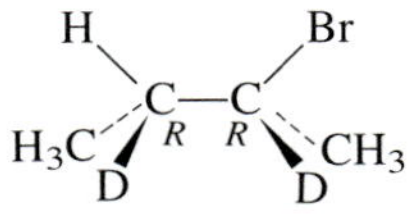

(b) Cyclohexane-1,2-diol (OH, OH) + enantiomer

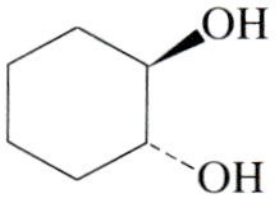

70%

(c) HO, H, H_3C C—C H, CH_2CH_3, OH + enantiomer

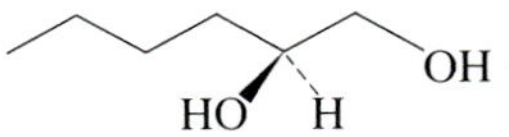

12-17

H_3C, H C=C CH_3, H $\xrightarrow{H_2O_2,\ \text{catalytic}\ OsO_4}$

HO, H_3C, H C—C OH, CH_3, H same as HO, H_3C, H C—C H, CH_3, OH

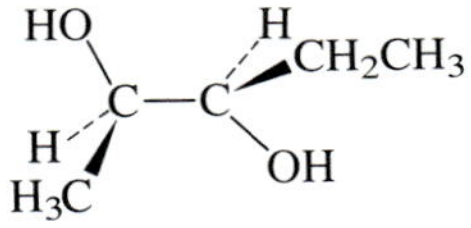

Eclipsed **Staggered**

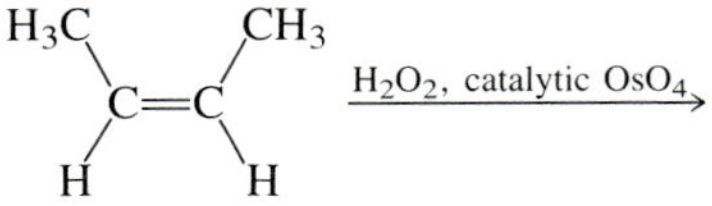

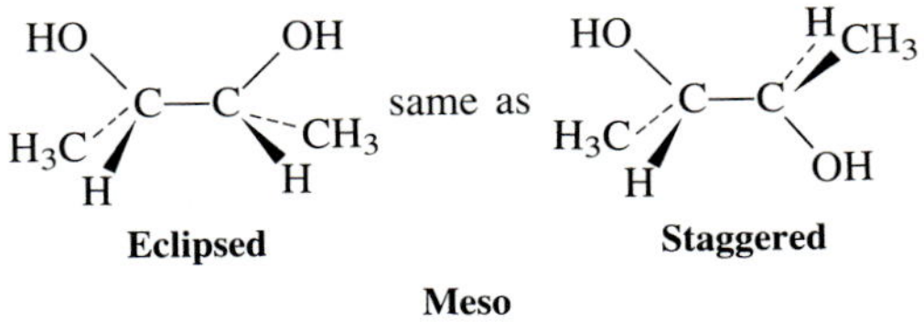

Meso

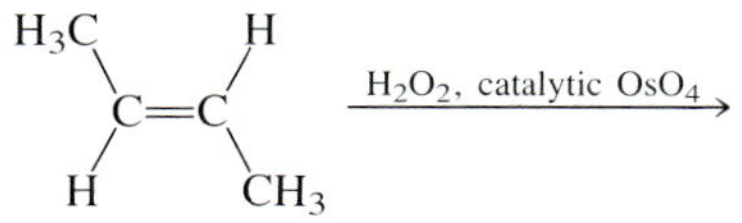

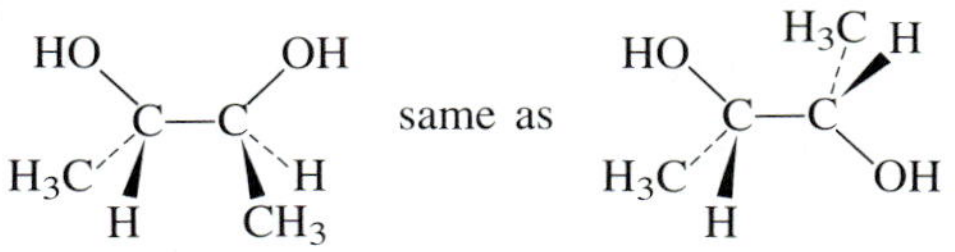

H_2O_2, catalytic OsO_4

Eclipsed same as Staggered

(R,R), (S,S)

12-18

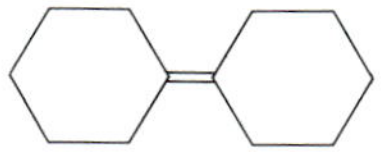

$\mathbf{C_{12}H_{20}}$

12-19

(a) **(b)** 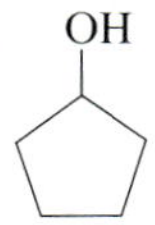 + CH_3OH

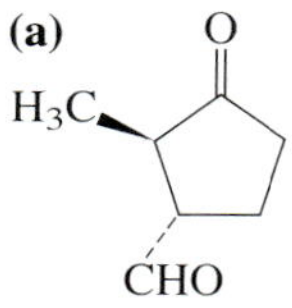

12-20

Do not be fooled by the way structures are drawn.

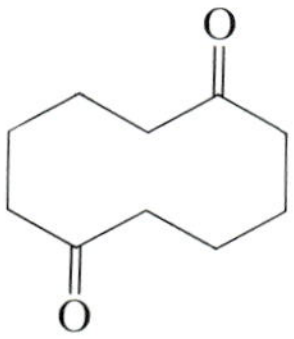

is the same as 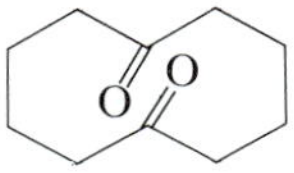 Therefore

the starting material is

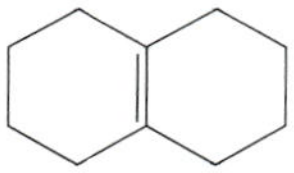

12-21

HI Reaction: $65 - 53 = +12$ kcal mole^{-1}

HCl Reaction: $103 - 81 = +22$ kcal mole^{-1}

12-22

Initiation:

$Br—CBr_3 + RO^{\cdot} \longrightarrow ROBr + {\cdot}CBr_3$

↑ Weak bond

Propagation:

$CH_3CH{=}CH_2 + {\cdot}CBr_3 \longrightarrow CH_3\dot{C}H—CH_2CBr_3$

$CH_3\dot{C}HCH_2CBr_3 + BrCBr_3 \longrightarrow CH_3CH(Br)CH_2CBr_3 + {\cdot}CBr_3$

12-23

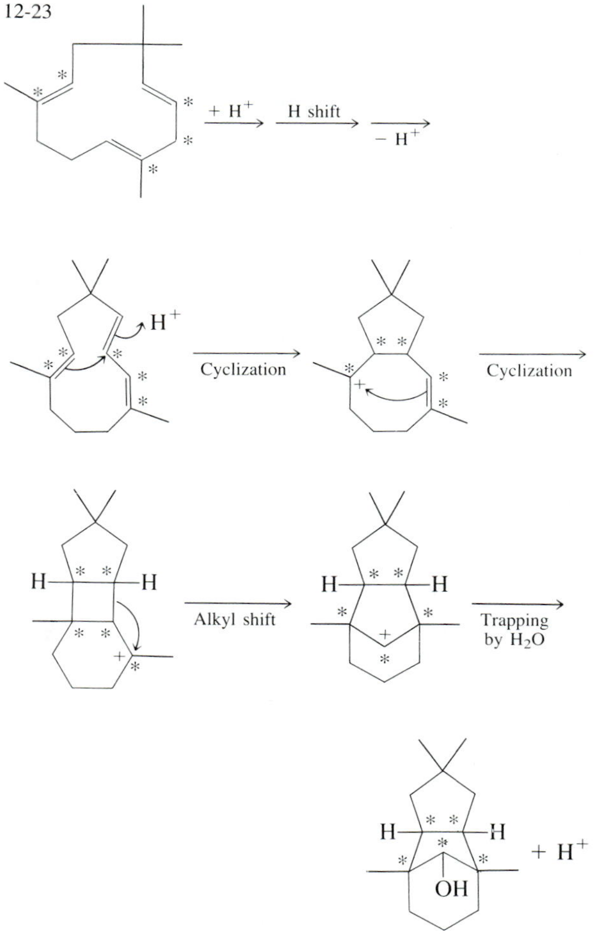

+ H^+ → H shift → $-H^+$ →

Cyclization → Cyclization →

Alkyl shift → Trapping by H_2O →

+ H^+

12-24

This is an irregular copolymer with both monomers incorporated in random numbers but regioselectively along the chain. Write a mechanism for its formation.

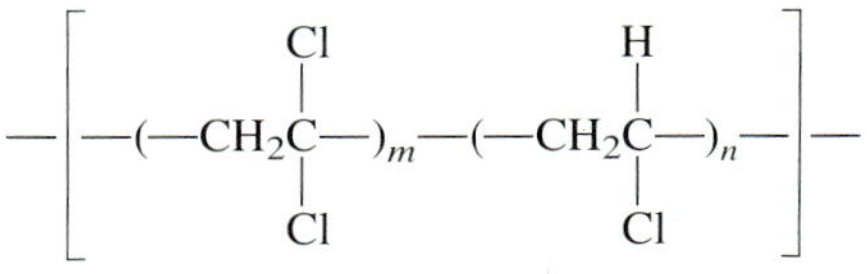

$$—[—(—CH_2CCl_2—)_m—(—CH_2CHCl—)_n—]—$$

CHAPTER 13

13-1

(a)

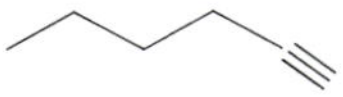

1-Hexyne  **2-Hexyne**

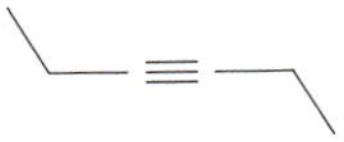

3-Hexyne 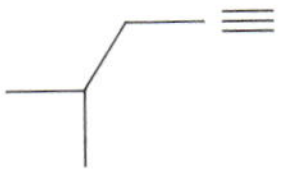 **4-Methyl-1-pentyne**

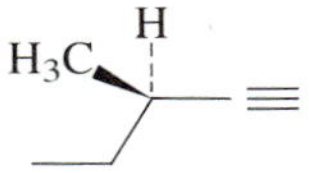

(*R*)-3-Methyl-1-pentyne 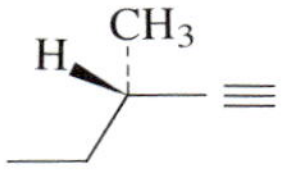 **(*S*)-3-Methyl-1-pentyne**

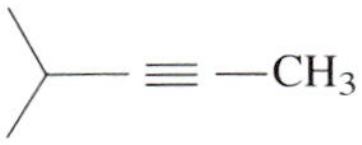

4-Methyl-2-pentyne **3,3-Dimethyl-1-butyne**

$(CH_3)_3C—≡$

(b) (*R*)-3-Methyl-1-penten-4-yne.

(c)

3-Butyn-1-ol 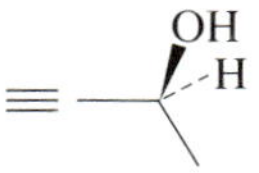 **(*S*)-3-Butyn-2-ol** 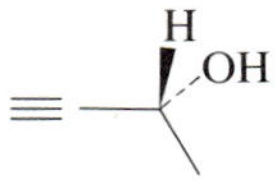 **(*R*)-3-Butyn-2-ol**

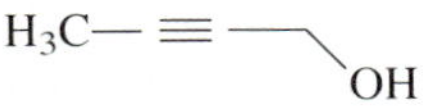

2-Butyn-1-ol 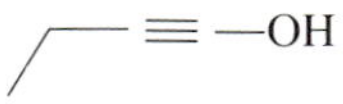 **1-Butyn-1-ol**

(This compound is highly unstable and does not exist in solution.)

13-2

Only those bases will deprotonate ethyne ($pK_a = 25$) for which the pK_a of the conjugate acid is higher: $(CH_3)_3COH$, $pK_a \sim 18$, and so $(CH_3)_3CO^-$ is too weak; but $[(CH_3)_2CH]_2NH$, $pK_a \sim 40$, and therefore LDA is a suitable base.

13-3

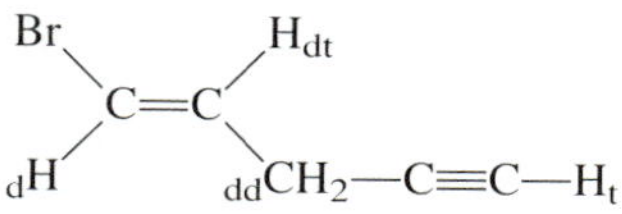

$$\text{Br(H}_{dt}\text{)C=C}\ ;\ {}_{d}\text{H—C=C—}{}_{dd}\text{CH}_2\text{—C}\equiv\text{C—H}_t$$

This answer ignores allylic coupling (Table 11-1), which is very small.

13-4

$\Delta G^\circ = -RT \ln K = 3.78$ kcal mole^{-1}.

13-5

Order of acidity:

$$CH_3CH{=}C{=}\mathbf{CH_2} < CH_3\mathbf{CH_2}C{\equiv}CH$$

Both deprotonations furnish the same anion. Because the allene is more stable than 1-butyne, it is less readily deprotonated. Draw a potential-energy diagram.

13-6

Yes. There are no elements of symmetry, such as a mirror plane or a point of inversion, that would render them achiral.

13-7

The hydrogen alpha to fluorine is more acidic and Br^- is a better leaving group than F^-.

$$\underset{F}{CH_2}—\underset{Br}{CH_2} \xrightarrow{CH_3O^-} FHC{=}CH_2$$

13-8

(a) $HC{\equiv}CLi \xrightarrow{CH_3CH_2CH_2Br}$

$$HC{\equiv}CCH_2CH_2CH_3 \xrightarrow[\text{2. } CH_3CH_2\overset{O}{\overset{\|}{C}}H]{\text{1. } CH_3CH_2CH_2CH_2Li}$$

$$CH_3CH_2\overset{OH}{\overset{|}{C}}HC{\equiv}CCH_2CH_2CH_3$$

(b)

$$CH_3(CH_2)_3C{\equiv}CH \xrightarrow[\text{2. } H\overset{O}{\overset{\|}{C}}OCH_3]{\text{1. } CH_3CH_2MgBr}$$

$$CH_3(CH_2)_3C{\equiv}C\underset{H}{\overset{OH}{C}}C{\equiv}C(CH_2)_3CH_3$$

13-9

In the presence of sodium amide, the terminal alkyne unit is deprotonated. Electron transfer to a negatively charged alkynyl group is not favored.

$$CH_3(CH_2)_2C{\equiv}C(CH_2)_4C{\equiv}CH \xrightarrow{NaNH_2,\ \text{liquid } NH_3}$$

$$CH_3(CH_2)_2C{\equiv}C(CH_2)_4C{\equiv}C{:}^- \xrightarrow{Na,\ \text{liquid } NH_3} \xrightarrow{H^+,\ H_2O}$$

$$CH_3(CH_2)_2CH{=}CH(CH_2)_4C{\equiv}CH$$

75%

13-10

$$CH_3(CH_2)_3C{\equiv}CH \xrightarrow[\text{2. } CH_3COOD]{\text{1. Diisoamylborane}} \begin{matrix} CH_3(CH_2)_3 & & H \\ & C{=}C & \\ H & & D \end{matrix}$$

$$CH_3(CH_2)_3C{\equiv}CH \xrightarrow{\text{1. } CH_3CH_2MgBr;\ \text{2. } D_2O;\ \text{3. Diisoamylborane};\ \text{4. } CH_3COOH} \begin{matrix} CH_3(CH_2)_3 & & D \\ & C{=}C & \\ H & & H \end{matrix}$$

13-11

$$(CH_3)_3CC{\equiv}CH \xrightarrow[\text{2. } H_2O_2,\ HO^-]{\text{1. Dicyclohexylborane}} (CH_3)_3CCH_2\overset{O}{\overset{\|}{C}}H$$

13-12

Protonation generates a bromine-stabilized carbocation:

$$(H_3C)_2C{=}C(\ddot{B}r{:})CH_3 \xrightarrow{H^+}$$

$$\left[H_3C-\underset{H}{\overset{H_3C}{C}}-\overset{:\ddot{B}r:}{\underset{+}{C}}-CH_3 \longleftrightarrow H_3C-\underset{H}{\overset{H_3C}{C}}-\overset{+\ddot{B}r:}{\overset{\|}{C}}-CH_3 \right]$$

13-13

$$CH_3\overset{O}{\overset{\|}{C}}CH_3 \xrightarrow[\text{2. } H^+,\ H_2O]{\text{1. } LiC{\equiv}CH} HC{\equiv}C-\underset{CH_3}{\overset{CH_3}{C}}OH \xrightarrow{H^+,\ H_2O,\ Hg^{2+}} CH_3\overset{O}{\overset{\|}{C}}-\underset{CH_3}{\overset{CH_3}{C}}OH$$

13-14

[Scheme: tertiary allylic alcohol bearing a $CH_3C{\equiv}C$-containing side chain $\xrightarrow[-\ H_2O]{H^+}$ allylic cation; cyclization to a six-membered ring cation; attack of the alkyne to form a bicyclic vinyl cation (with H_3C, CH_3, H labels) $\xrightarrow{Cl^-}$ vinyl chloride product with CH_3, Cl, H_3C, H.]

13-15

$$\equiv \xrightarrow{\text{1. } CH_3CH_2CH_2CH_2Li;\ \text{2. ethylene oxide}} {\equiv}\!-\!CH_2CH_2-OH \xrightarrow{Cu^+,\ O_2}$$

$$HO-CH_2CH_2-C{\equiv}C-C{\equiv}C-CH_2CH_2-OH \xrightarrow{H_2,\ Pd\text{-}C} HO(CH_2)_8OH$$

13-16

$$H-C{\equiv}C-H \xrightarrow[-\ Cl^-]{HgCl_2} \text{(mercurinium ion, } \overset{+}{Hg}Cl\text{)} \xrightarrow[-\ H^+]{H_2O} \begin{matrix} HO & & H \\ & C{=}C & \\ H & & HgCl \end{matrix} \xrightarrow{H^+}$$

$$\begin{matrix} HO & & H \\ & \overset{+}{C}-C & \\ H & H & HgCl \end{matrix} \xrightarrow[-\ HgCl_2]{+\ Cl^-} \begin{matrix} HO & & H \\ & C{=}C & \\ H & & H \end{matrix} \longrightarrow CH_3\overset{O}{\overset{\|}{C}}H$$

13-17

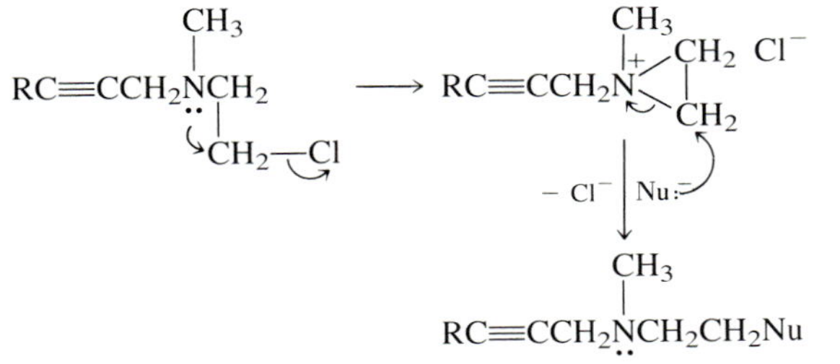

$$RC{\equiv}CCH_2\ddot{N}(CH_3)CH_2CH_2{-}Cl \longrightarrow RC{\equiv}CCH_2\overset{+}{N}(CH_3)(CH_2CH_2)\ Cl^- \xrightarrow[-\,Cl^-]{Nu:^-} RC{\equiv}CCH_2\ddot{N}(CH_3)CH_2CH_2Nu$$

(c) [structure] + [structure]

Br Br

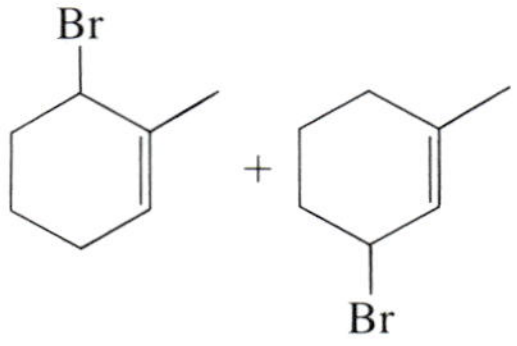

Bromination at the primary allylic position is too slow.

CHAPTER 14

14-1

The intermediate allylic cation is achiral.

14-2

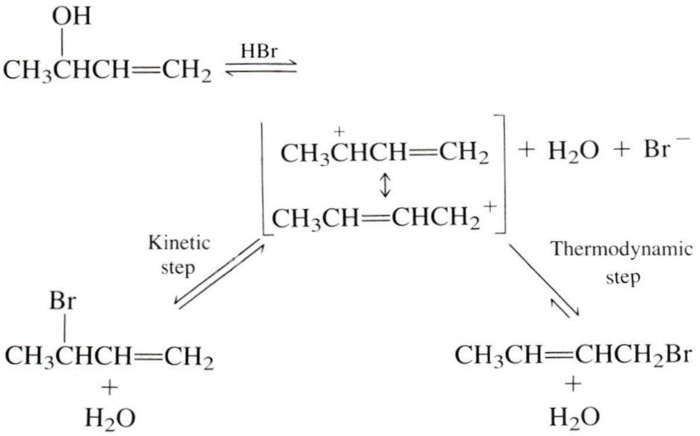

$$CH_3CH(OH)CH{=}CH_2 \overset{HBr}{\rightleftharpoons} \left[CH_3\overset{+}{C}HCH{=}CH_2 \leftrightarrow CH_3CH{=}CHCH_2^+\right] + H_2O + Br^-$$

Kinetic step: $CH_3CH(Br)CH{=}CH_2 + H_2O$

Thermodynamic step: $CH_3CH{=}CHCH_2Br + H_2O$

14-3

[structure] $\longrightarrow$ [structure] $+ RCOO^-$

CH_3 O

OCR

CH_3

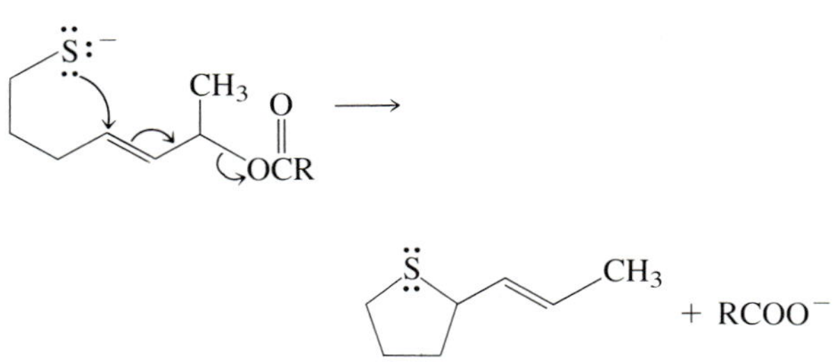

14-4

(a) 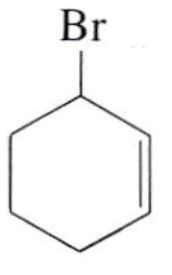 [structure] Br

(b) 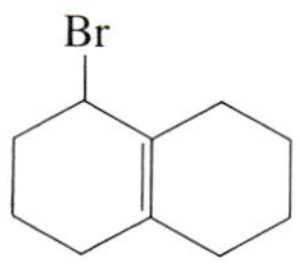 [structure] Br

14-5

[structure] + [structure] MgBr $\longrightarrow$ [structure] $\xrightarrow{1.\ O_3\quad 2.\ NaBH_4}$ [structure]

HO HO OH

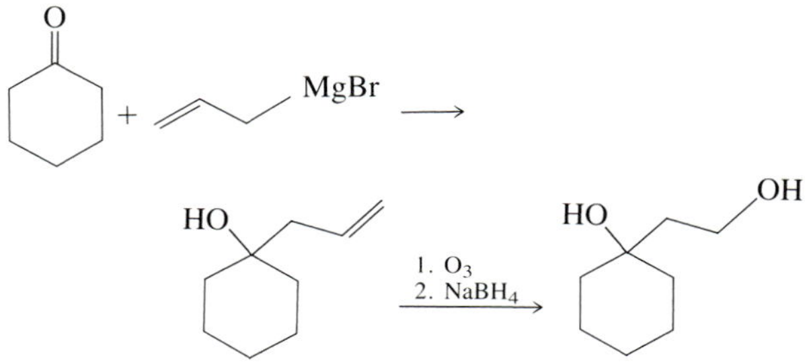

14-6

(a) 5-Bromo-1,3-cycloheptadiene

(b) (*E*)-2,3-Dimethyl-1,3-pentadiene

(c) 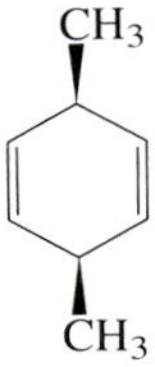 [structure] CH_3 CH_3

(d)  [structure] Br Br

14-7

An internal trans double bond is more stable than a terminal double bond by about 2.7 kcal mole^{-1} (see Figure 11-12). This difference plus the expected resonance energy of 3.5 kcal mole^{-1} add up to 6.2 kcal mole^{-1}, pretty close to the observed value.

14-8

The effect of the two allylic double bonds is roughly additive. The DH° of the central methylene bond in 1,4-pentadiene may be estimated by taking the DH° of a secondary C–H bond (95 kcal mole^{-1}) and subtracting twice the amount of allylic stabilization (in this case a little less than expected, about 2 × 12 kcal mole^{-1}).

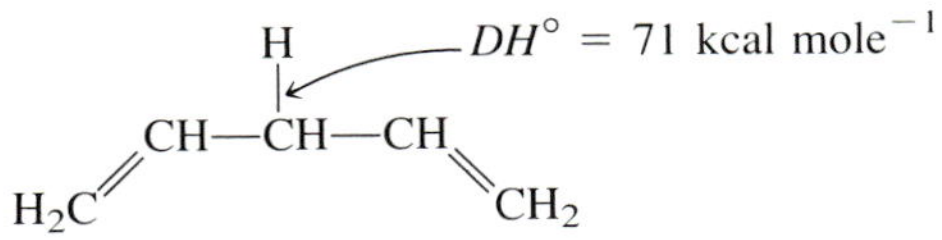

$$H_2C{=}CH{-}CH(H){-}CH{=}CH_2 \qquad DH^\circ = 71\ \text{kcal mole}^{-1}$$

14-9

(a)

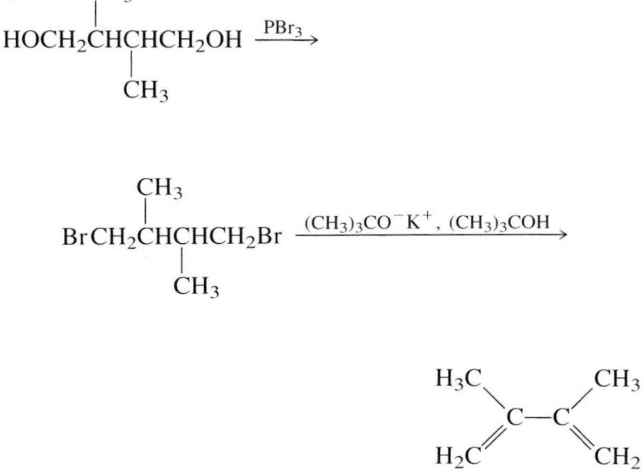

$$HOCH_2CH(CH_3)CH(CH_3)CH_2OH \xrightarrow{PBr_3}$$

$$BrCH_2CH(CH_3)CH(CH_3)CH_2Br \xrightarrow{(CH_3)_3CO^-K^+,\ (CH_3)_3COH}$$

$$H_2C{=}C(CH_3){-}C(CH_3){=}CH_2$$

(b)

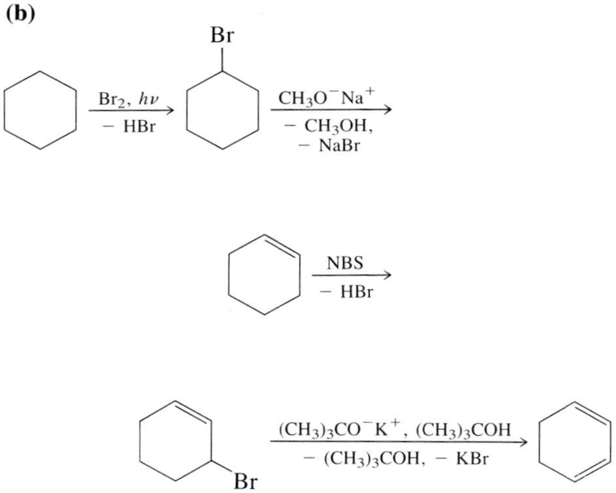

Cyclohexane $\xrightarrow[-\ HBr]{Br_2,\ h\nu}$ bromocyclohexane $\xrightarrow[-\ CH_3OH,\ -\ NaBr]{CH_3O^-Na^+}$

Cyclohexene $\xrightarrow[-\ HBr]{NBS}$

3-Bromocyclohexene $\xrightarrow[-\ (CH_3)_3COH,\ -\ KBr]{(CH_3)_3CO^-K^+,\ (CH_3)_3COH}$ 1,3-cyclohexadiene

14-10

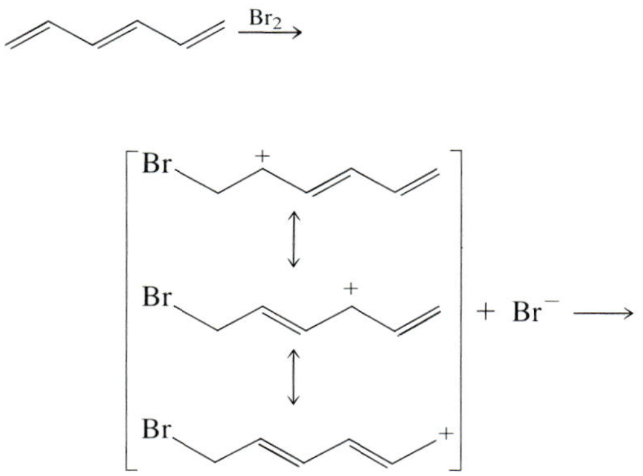

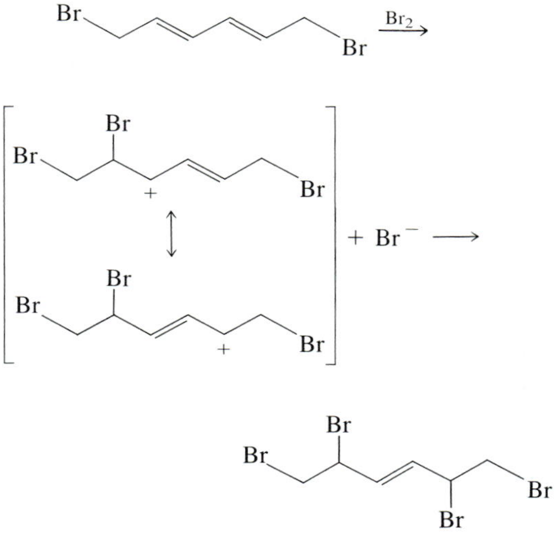

Hexatriene $\xrightarrow{Br_2}$

[Br–CH₂–CH⁺–CH=CH–CH=CH₂ ↔ Br–CH₂–CH=CH–CH⁺–CH=CH₂ ↔ Br–CH₂–CH=CH–CH=CH–CH₂⁺] + Br^- ⟶

Br–CH₂–CH=CH–CH=CH–CH₂–Br $\xrightarrow{Br_2}$

[Br–CH₂–CH(Br)–CH⁺–CH=CH–CH₂–Br ↔ Br–CH₂–CH(Br)–CH=CH–CH⁺–CH₂–Br] + Br^- ⟶

Br–CH₂–CH(Br)–CH=CH–CH(Br)–CH₂–Br

14-11

(a) Butadiene + $(NC)_2C{=}C(CN)_2$ ⟶ 4,4,5,5-tetracyanocyclohexene

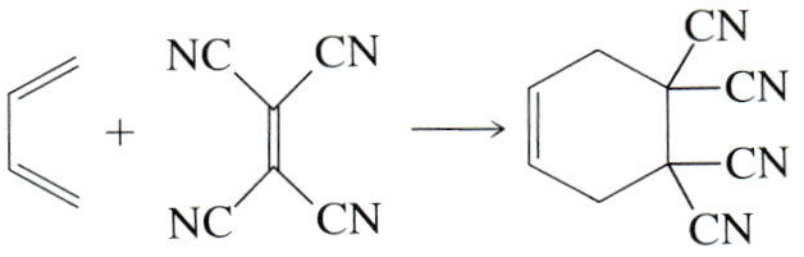

(b) Cyclopentadiene + $(NC)_2C{=}C(CN)_2$ ⟶ bicyclic tetracyano adduct

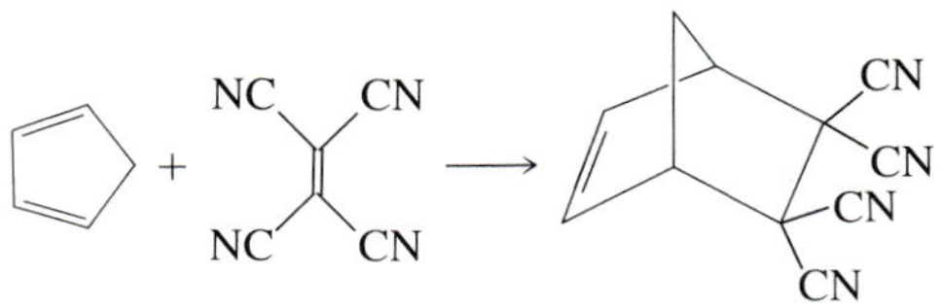

Make a model of this product.

(c) 1,2-Dimethylenecyclohexane + $(NC)_2C{=}C(CN)_2$ ⟶ tetracyano octahydronaphthalene

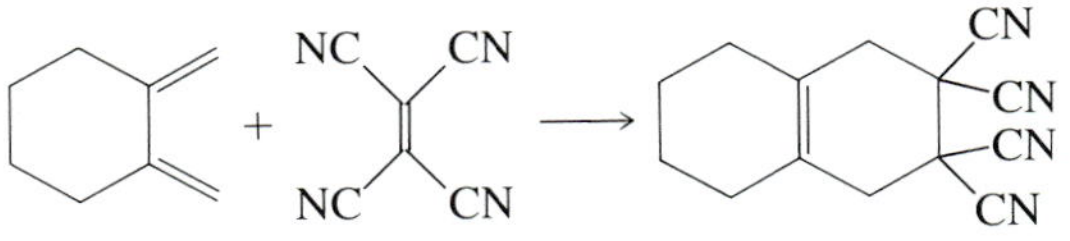

14-12

(a) 4,5-Dimethyl-cyclohexene with trans CN groups (H_3C, H_3C, CN, CN)

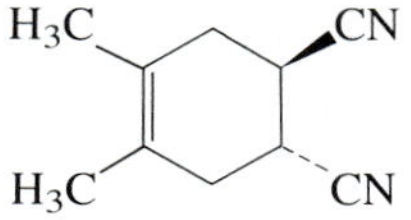

(b) H_3C–C(=CH–CH_3)–CH=CH–CH_3 or H_3C–C(=CH–CH_3)–CH=CH–CH_3 (cis) + $F_2C{=}CF_2$

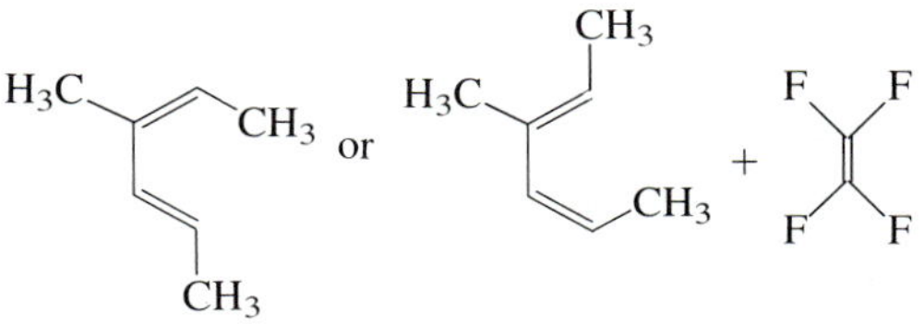

14-13

The cis-trans isomer cannot readily reach the *s*-cis conformation because of steric hindrance.

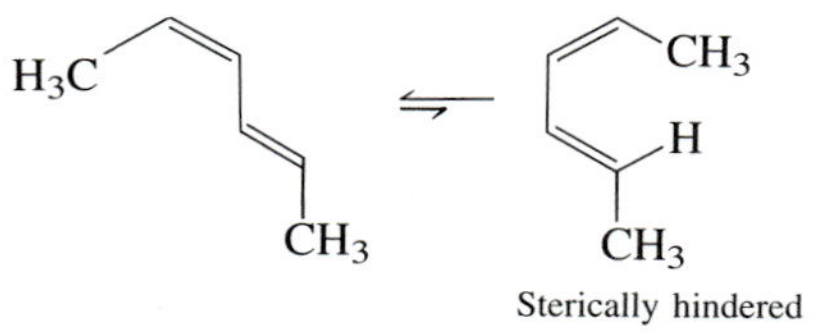

H_3C CH_3 ⇌ CH_3 H CH_3

Sterically hindered

14-14

(a) CH_3 CH_3 + CO_2CH_3 ⟶ CH_3 CO_2CH_3 CH_3

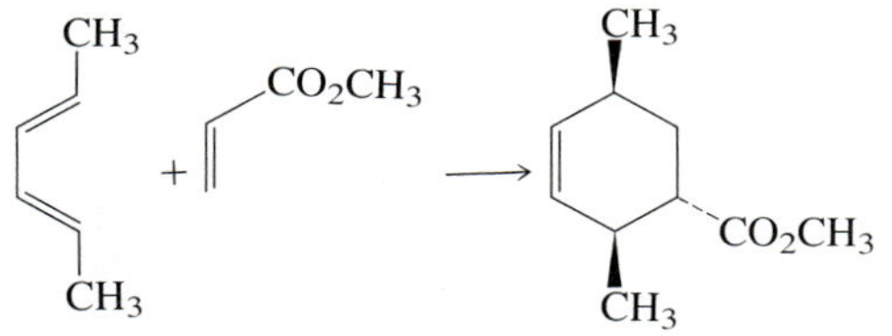

(b) CH_3 + O O O ⟶ CH_3 O O O

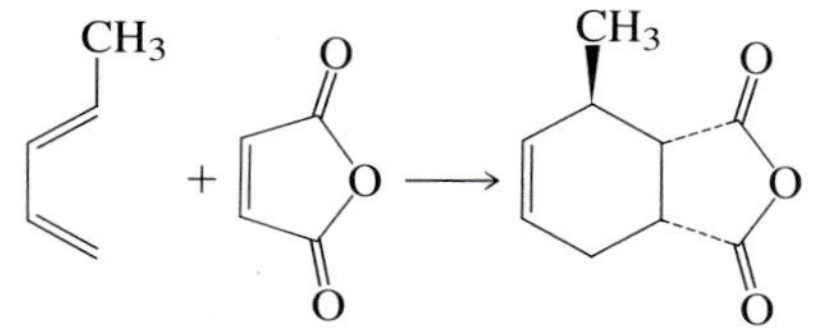

(c) + CO_2CH_3 CH_3O_2C ⟶

CO_2CH_3 H H CO_2CH_3

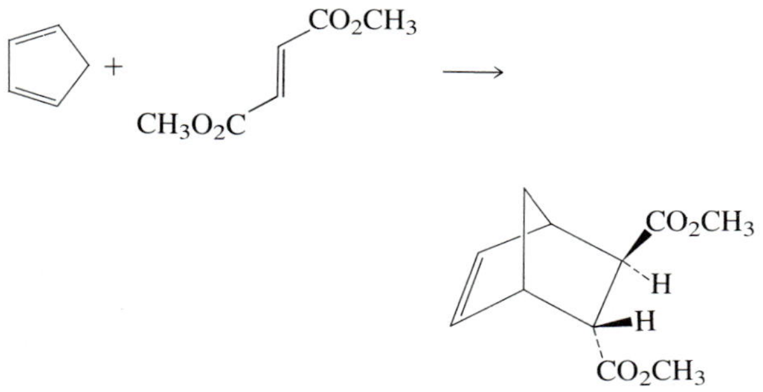

14-15

The first product is the result of exo addition, the second product the outcome of endo addition.

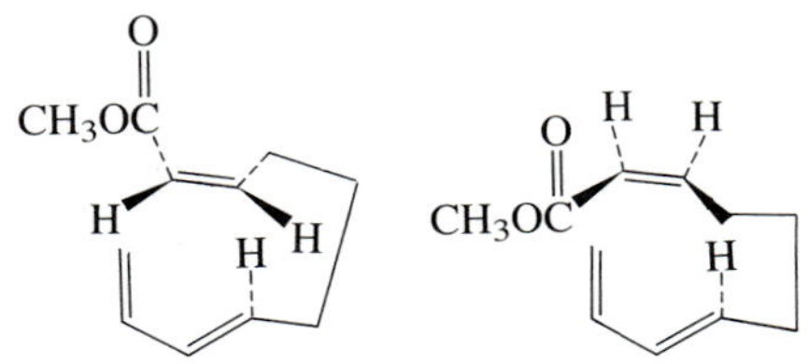

O CH_3OC H H H

O H H CH_3OC H

14-16

This reaction has been called a domino Diels-Alder cycloaddition.

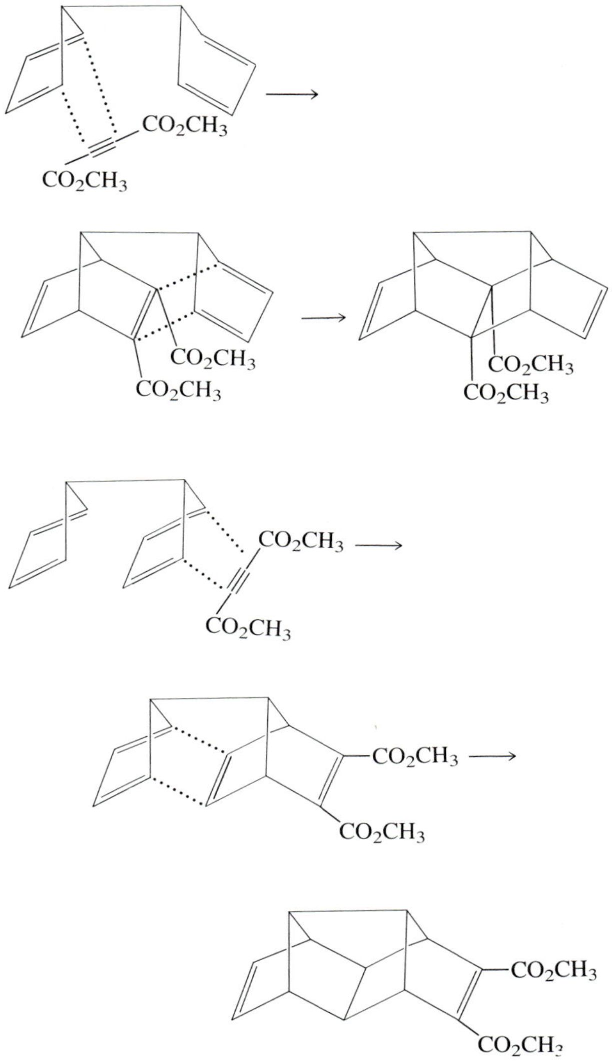

CO_2CH_3 CO_2CH_3 ⟶

CO_2CH_3 CO_2CH_3 ⟶ CO_2CH_3 CO_2CH_3

CO_2CH_3 ⟶ CO_2CH_3

CO_2CH_3 ⟶ CO_2CH_3

CO_2CH_3 CO_2CH_3

14-17

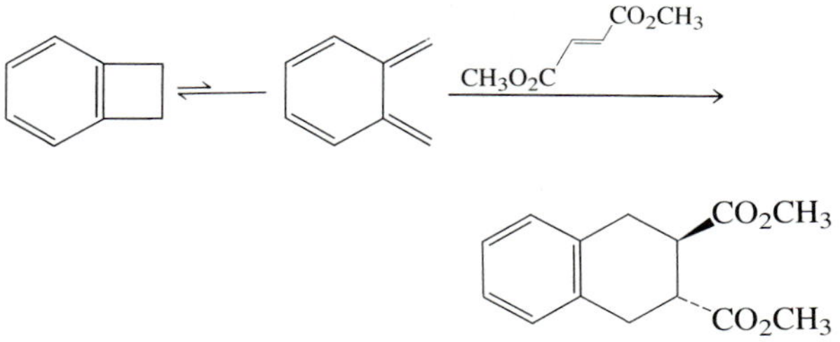

⇌ CO_2CH_3 CH_3O_2C ⟶

CO_2CH_3 CO_2CH_3

14-18
Conrotatory.

14-19

λ_{max} = 217 nm ⇢ Calculated 222 nm (measured 222.5 nm) ⇢

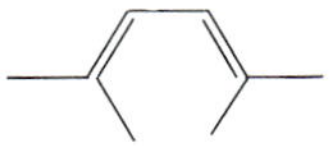

Calculated 237 nm (measured 241.5 nm)

CHAPTER 15

15-1
(a) 2-Cyclohexenone
(b) (*E*)-4-Methyl-4-hexenal

(c)

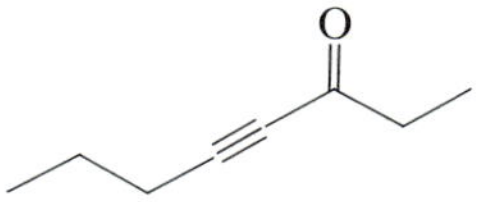

(d)

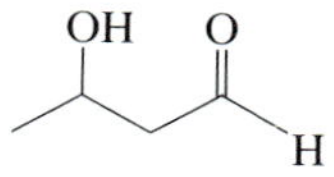

15-2 $CH_3COCH_2CH_2COCH_3$, an oxidation product.

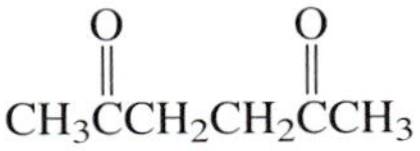

15-3 J = 6.7 Hz; J = 7.7 Hz; J_{trans} = 16.1 Hz
$J(CH_3—H\text{-}2)$ = 1.6 Hz

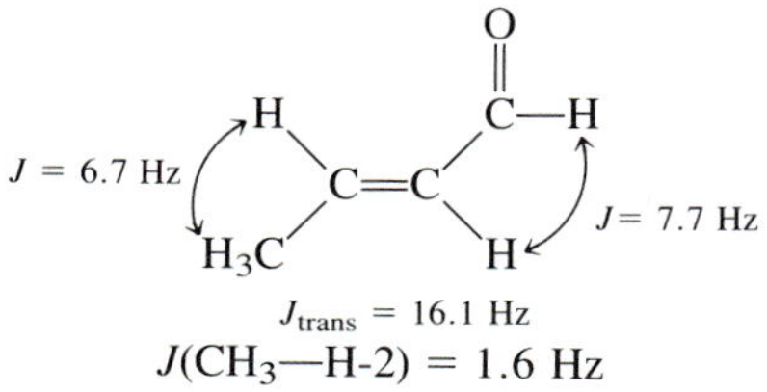

15-4

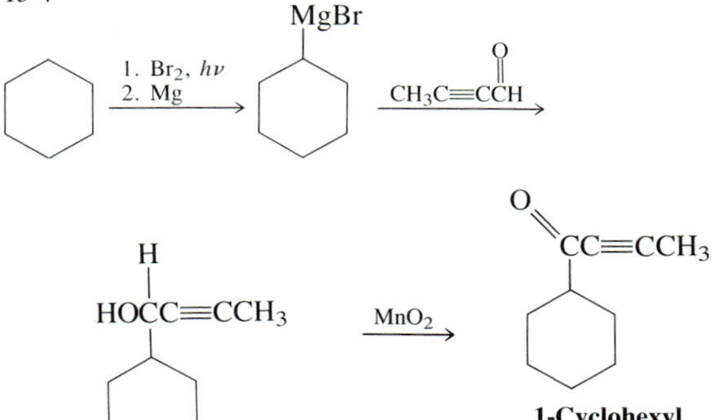

Cyclohexane $\xrightarrow{\text{1. } Br_2,\ h\nu;\ \text{2. Mg}}$ Cyclohexyl-MgBr $\xrightarrow{CH_3C\equiv CCHO}$ cyclohexyl-$CH(OH)C\equiv CCH_3$ $\xrightarrow{MnO_2}$

1-Cyclohexyl 1-propynyl ketone

15-5
(a) 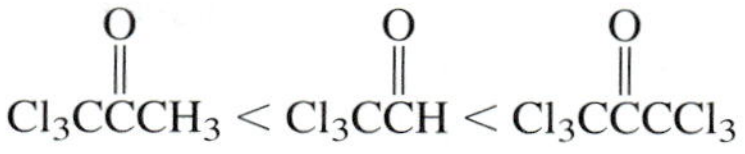 $Cl_3CCOCH_3 < Cl_3CCHO < Cl_3CCOCCl_3$

(b) 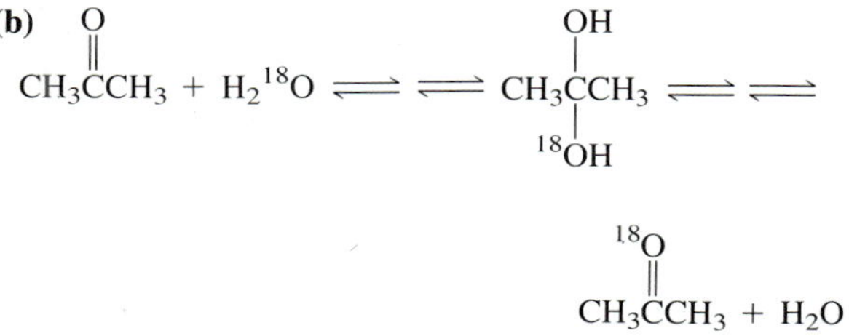 $CH_3COCH_3 + H_2{}^{18}O \rightleftharpoons \rightleftharpoons CH_3C(OH)(^{18}OH)CH_3 \rightleftharpoons \rightleftharpoons CH_3C(^{18}O)CH_3 + H_2O$

15-6 $CH_3CO(CH_2)_4Br$ (**A**) $\xrightarrow[\text{Protection}]{HOCH_2CH_2OH,\ H^+}$ dioxolane of $CH_3C(CH_2)_4Br$ $\xrightarrow{\text{1. Mg; 2. } CH_2{=}O}$ dioxolane of $CH_3C(CH_2)_4CH_2OH$ $\xrightarrow[\text{Deprotection}]{H^+,\ H_2O}$ $CH_3CO(CH_2)_4CH_2OH$ (**B**)

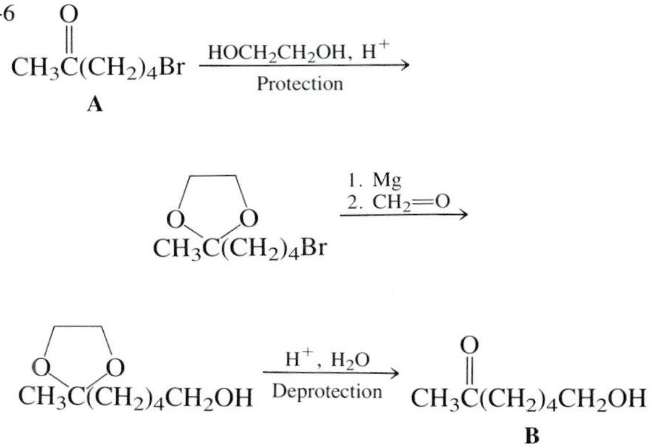

15-7

C $\xrightarrow{H^+}$ ⟶ ⟶ $\xrightarrow{-H^+}$ **D**

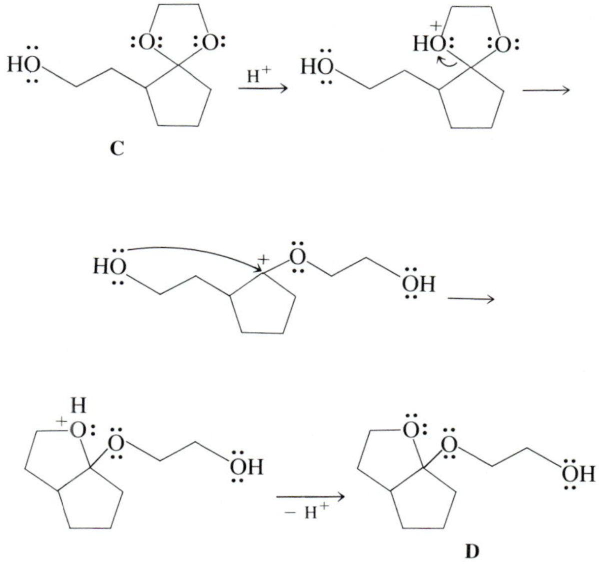

15-8

Sorbose is in equilibrium with its cyclic hemiacetal form, which incorporates a five-membered ring. Double propanone acetal formation completes the transformation.

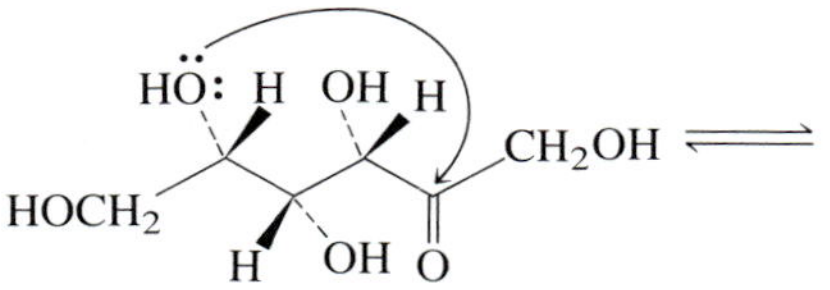

HÖ: H OH H
HOCH$_2$ CH$_2$OH ⇌
H OH O
A

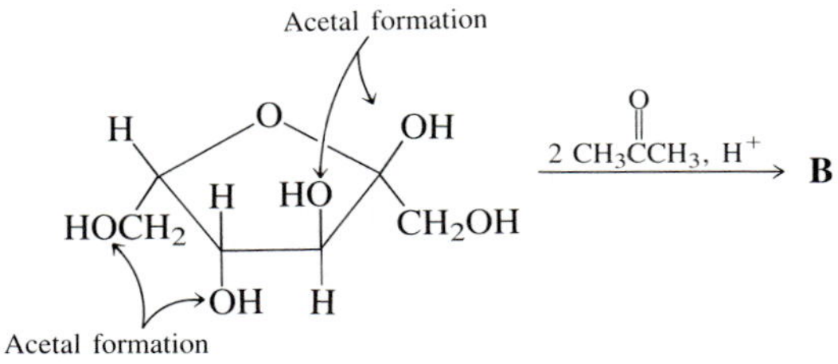

Acetal formation
H O OH
H HO
HOCH$_2$ CH$_2$OH
OH H
Acetal formation

$\xrightarrow{2\ CH_3\overset{O}{\overset{\|}{C}}CH_3,\ H^+}$ B

15-9

$\xrightarrow[2.\ CH_3SCH_3]{1.\ O_3}$

$\xrightarrow[2.\ \text{Raney Ni}]{1.\ HSCH_2CH_2SH}$

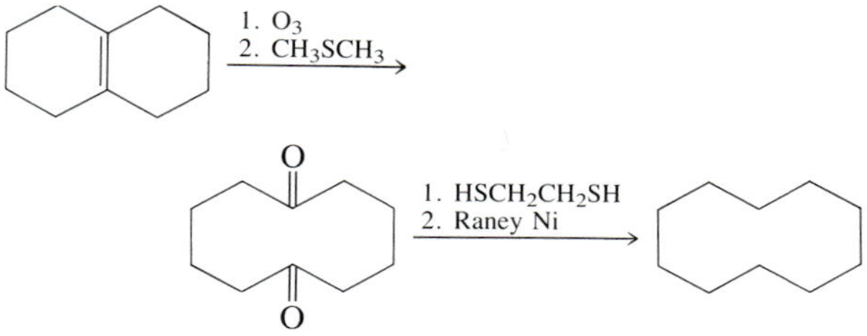

15-10

The mechanism of imidazolidine formation is similar to that formulated for cyclic acetal synthesis.

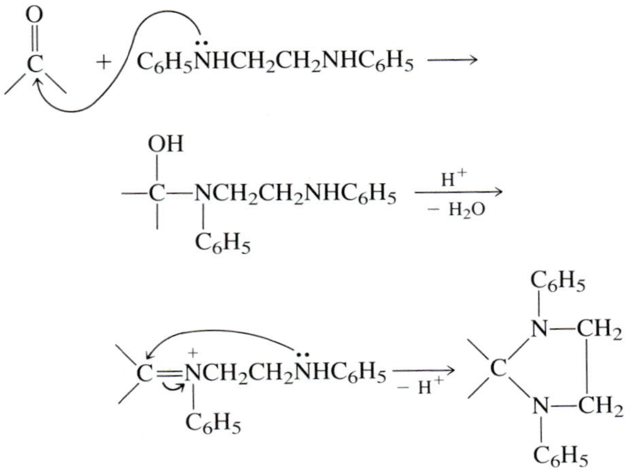

$>C{=}O + C_6H_5\ddot{N}HCH_2CH_2NHC_6H_5 \longrightarrow$

$-\overset{OH}{\underset{|}{\overset{|}{C}}}-\underset{C_6H_5}{\underset{|}{N}}CH_2CH_2NHC_6H_5 \xrightarrow[-\ H_2O]{H^+}$

$>C{=}\overset{+}{\underset{C_6H_5}{\underset{|}{N}}}CH_2CH_2\ddot{N}HC_6H_5 \xrightarrow{-\ H^+}$ (imidazolidine: C bonded to N(C_6H_5)—CH_2 and N(C_6H_5)—CH_2)

15-11

(a)

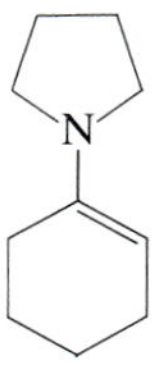

N

(b)

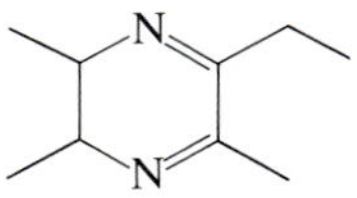

N

N

(c) HO H

N

NHC$_6$H$_5$

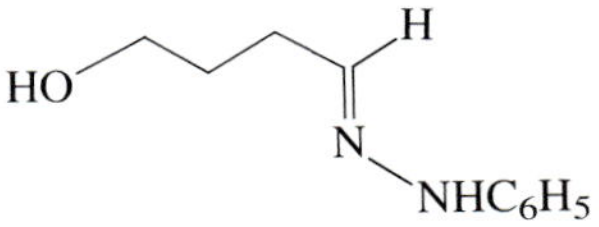

15-12

(a)

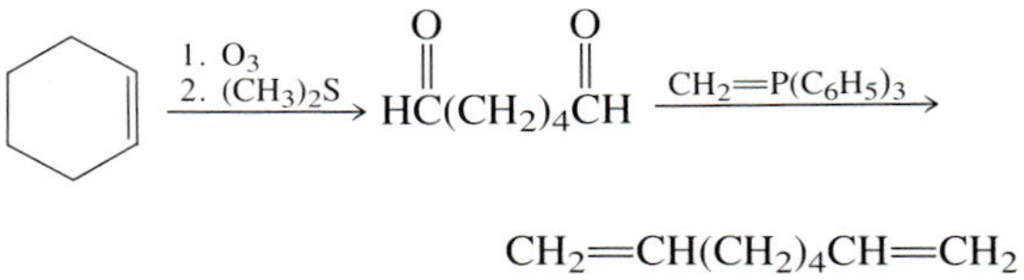

$\xrightarrow[2.\ (CH_3)_2S]{1.\ O_3}$ $H\overset{O}{\overset{\|}{C}}(CH_2)_4\overset{O}{\overset{\|}{C}}H$ $\xrightarrow{CH_2{=}P(C_6H_5)_3}$

$CH_2{=}CH(CH_2)_4CH{=}CH_2$

(b)

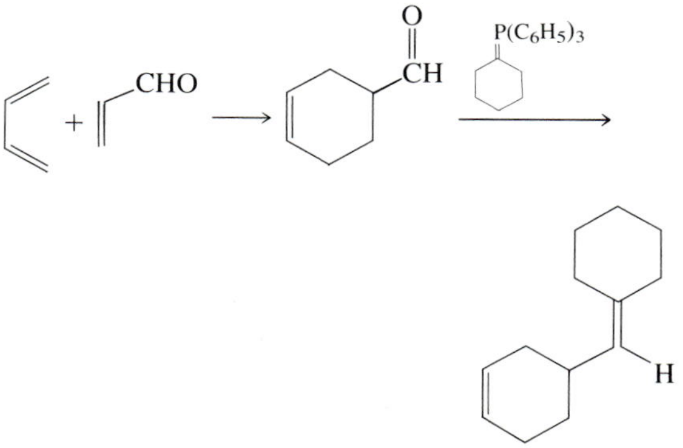

+ CHO ⟶ $\overset{O}{\overset{\|}{C}}H$ $\xrightarrow{P(C_6H_5)_3}$

H

15-13

(a)

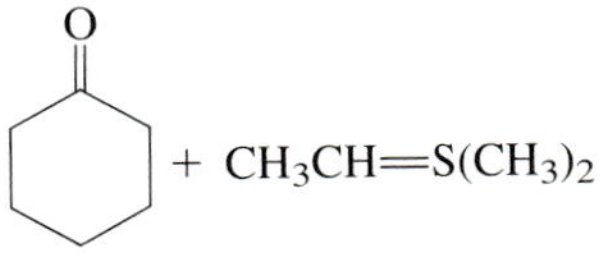

O

+ $CH_3CH{=}S(CH_3)_2$

(b)

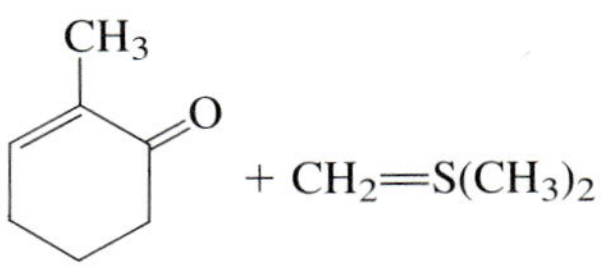

CH$_3$

O

+ $CH_2{=}S(CH_3)_2$

(c) $CH_3CH_2\overset{O}{\overset{\|}{C}}H + CH_2{=}S(CH_3)_2$

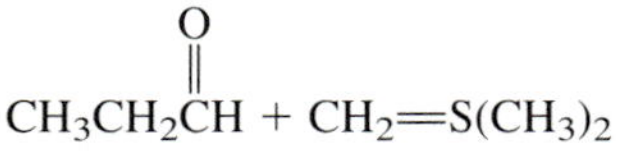

15-14

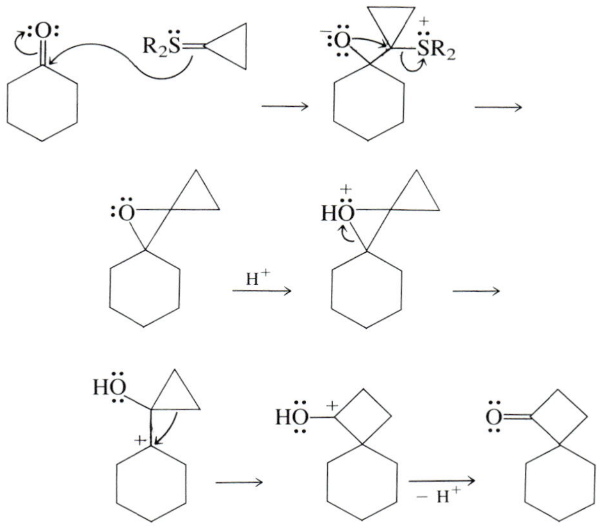

$\longrightarrow$ $\longrightarrow$

$\xrightarrow{H^+}$ $\longrightarrow$

$\longrightarrow$ $\xrightarrow{-\,H^+}$

15-15

$CH_3\overset{O}{\overset{\|}{C}}CH_3 + :\overset{-}{C}H_2—\overset{+}{N}{\equiv}N:$

$CH_3C(O^-)(CH_3)—CH_2—\overset{+}{N}{\equiv}N: \xrightarrow{-\,N_2} CH_3\overset{O}{\overset{\|}{C}}CH_2CH_3 + N_2$

$CH_3C(O^-)(CH_3)—CH_2—\overset{+}{N}{\equiv}N: \xrightarrow{-\,N_2}$ $(H_3C)_2C—CH_2$ (epoxide)

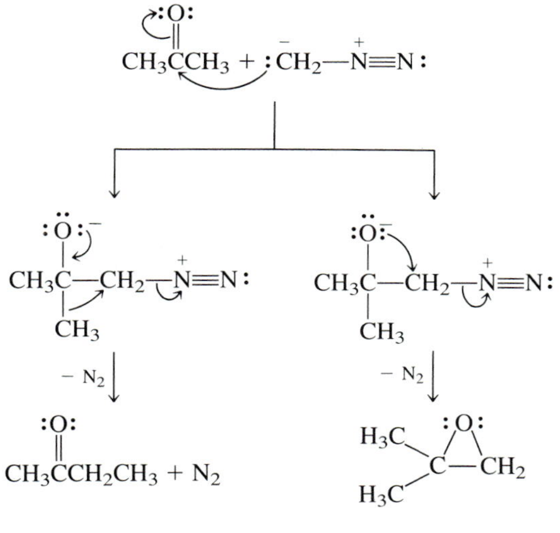

15-16

(a) $CH_2{=}CHCH_2CH_2O\overset{O}{\overset{\|}{C}}CH_3$

(b)

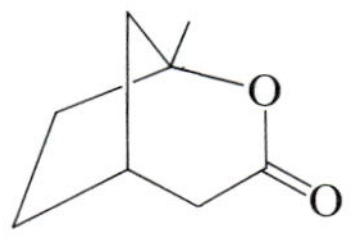

(c) $(CH_3)_3CO\overset{O}{\overset{\|}{C}}CH_3$

15-17

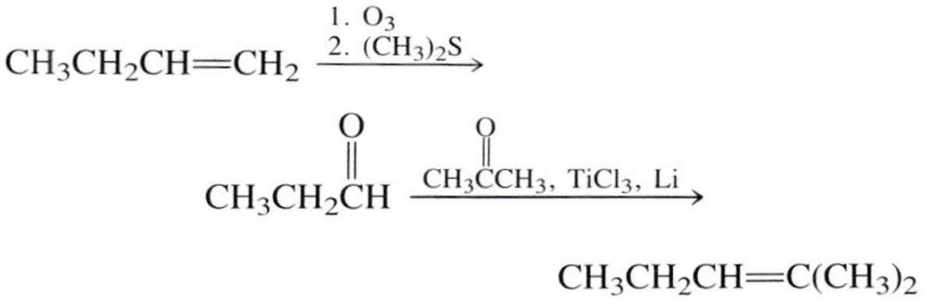

$CH_3CH_2CH{=}CH_2 \xrightarrow[2.\ (CH_3)_2S]{1.\ O_3} CH_3CH_2\overset{O}{\overset{\|}{C}}H \xrightarrow{CH_3\overset{O}{\overset{\|}{C}}CH_3,\ TiCl_3,\ Li} CH_3CH_2CH{=}C(CH_3)_2$

CHAPTER 16

16-1

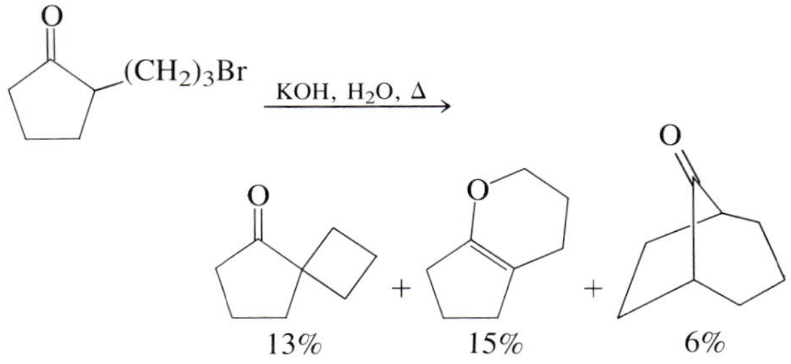

$(CH_2)_3Br$ $\xrightarrow{KOH,\ H_2O,\ \Delta}$

13% + 15% + 6%

16-2

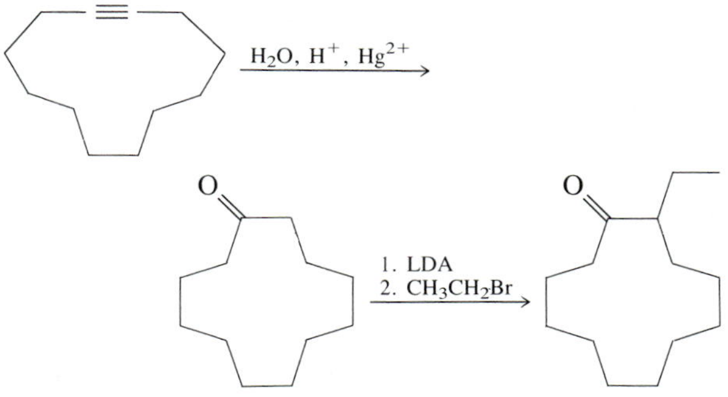

$\xrightarrow{H_2O,\ H^+,\ Hg^{2+}}$

$\xrightarrow[2.\ CH_3CH_2Br]{1.\ LDA}$

16-3

First product:

$\xrightarrow{CH_3—I}$ CH_3 I^-

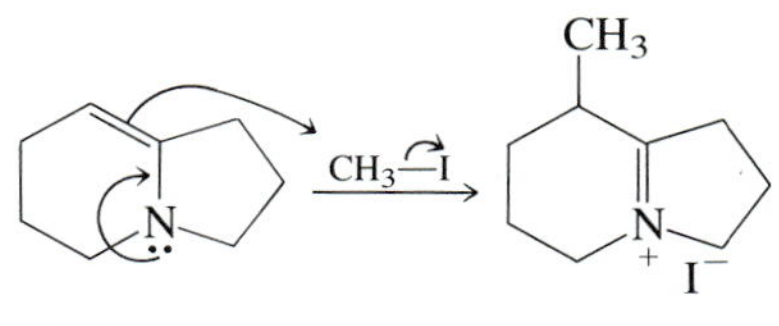

Second product:

CH_3 $\xrightarrow{CH_3—I}$ H_3C CH_3 I^-

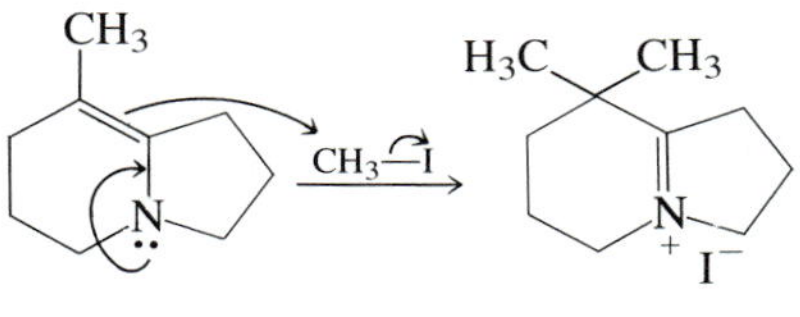

Third product:

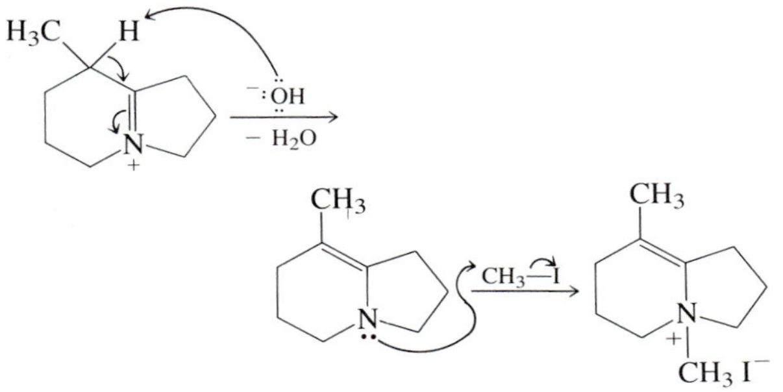

$\xrightarrow[-\ H_2O]{^-:\ddot{O}H}$ $\xrightarrow{CH_3\text{—}I}$ $CH_3\ I^-$

In this case, *C*-alkylation is relatively hindered; therefore, *N*-alkylation becomes competitive.

16-4

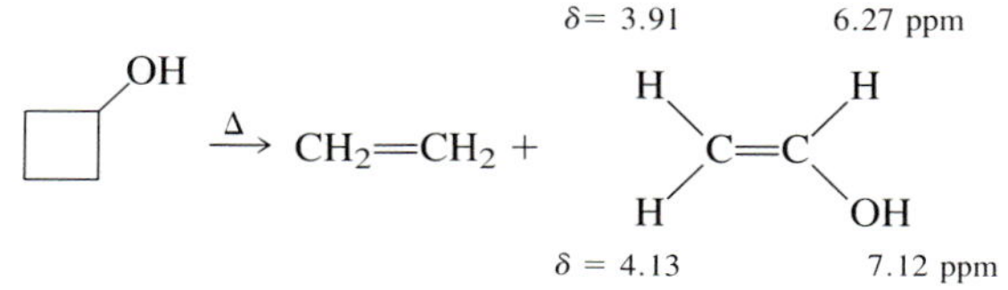

$\xrightarrow{\Delta}$ $CH_2{=}CH_2$ + (H)(H)C=C(H)(OH)

$\delta = 3.91$ 6.27 ppm

$\delta = 4.13$ 7.12 ppm

$J_{trans} = 14.0$ Hz
$J_{cis} = 6.5$ Hz
$J_{gem} = 1.8$ Hz

Two of the alkenyl hydrogens resonate at unusually high field. This phenomenon is generally observed in the ^{1}H NMR spectra of enols and their derivatives. Can you explain it? (Hint: Draw the dipolar resonance structure of ethenol.)

16-5

(a)

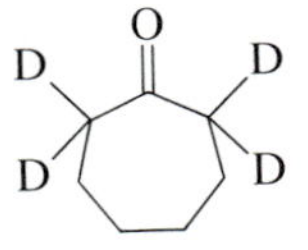

(b) $(CH_3)_3CCH$ (C=O)
No enolizable hydrogen

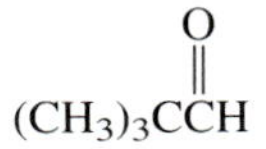

(c) $(CH_3)_3CCCD_3$ (C=O)

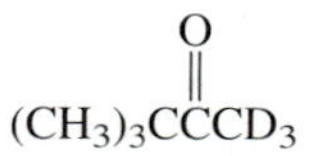

(d) CHO

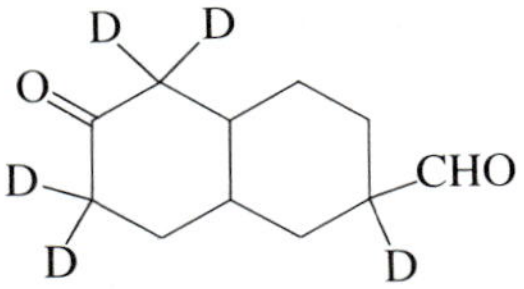

16-6

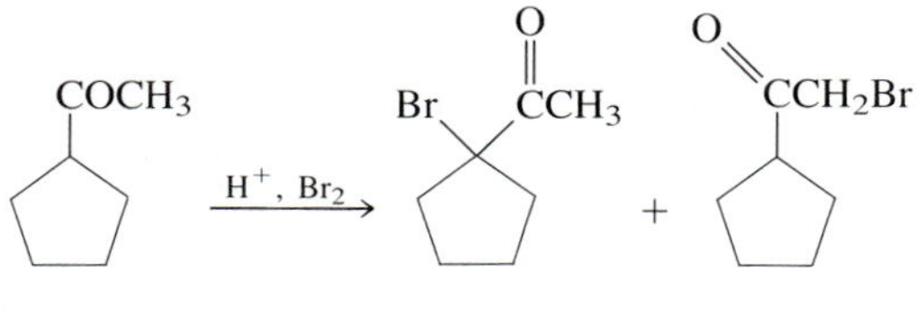

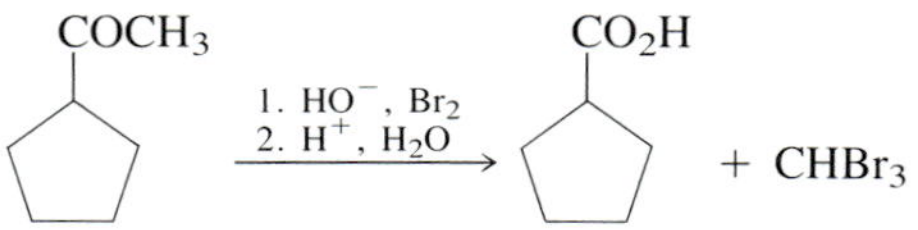

$COCH_3$ $\xrightarrow{H^+,\ Br_2}$ Br CCH_3 + CCH_2Br

$COCH_3$ $\xrightarrow[2.\ H^+,\ H_2O]{1.\ HO^-,\ Br_2}$ CO_2H + $CHBr_3$

16-7

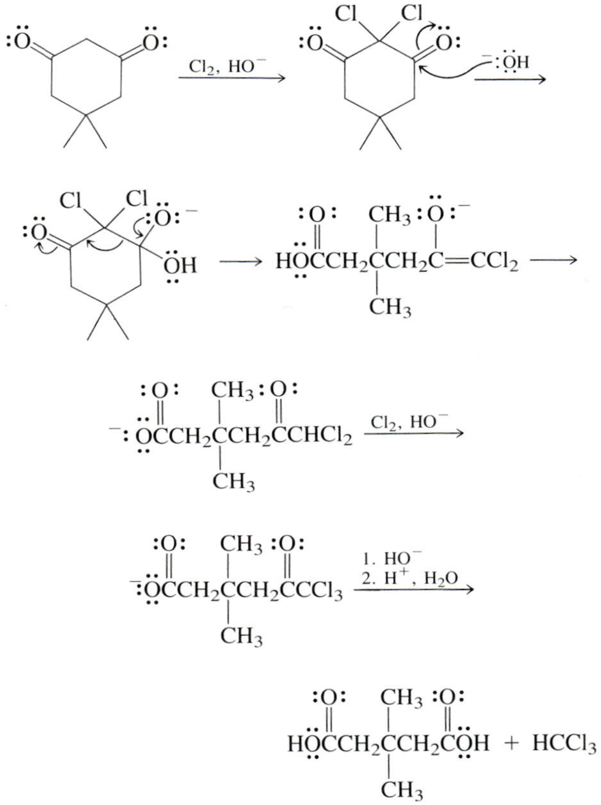

$\xrightarrow{Cl_2,\ HO^-}$ Cl Cl $\xrightarrow{^-:\ddot{O}H}$

Cl Cl $\ddot{O}H$ $\longrightarrow$ $H\ddot{O}CCH_2\overset{CH_3}{C}CH_2\overset{:O:^-}{C}{=}CCl_2$ $\longrightarrow$

$^-:\ddot{O}CCH_2CCH_2CCHCl_2$ $\xrightarrow{Cl_2,\ HO^-}$

$^-:\ddot{O}CCH_2CCH_2CCCl_3$ $\xrightarrow[2.\ H^+,\ H_2O]{1.\ HO^-}$

$H\ddot{O}CCH_2CCH_2C\ddot{O}H$ + $HCCl_3$

16-8

CH_3CHCH_2CH (:ÖH, :O:) **Aldol** $\overset{HO^-}{\rightleftharpoons}$ CH_3CHCH_2CH (:Ö:⁻) + $HCCH_3$ $\rightleftharpoons$

CH_3CHCH_2CH (:Ö, CHCH₃, :Ö:⁻, :O:) $\rightleftharpoons$ $\overset{H^+}{\rightleftharpoons}$ H₃C, H, :Ö, Ö:, H, H₃C, C, CH₂, C, H, ÖH

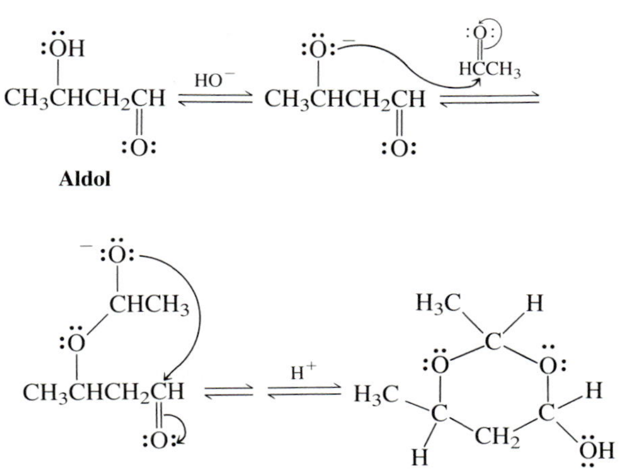

16-9

(a) 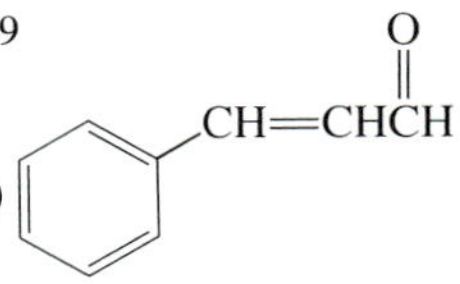 $C_6H_5CH{=}CHCHO$

(b)

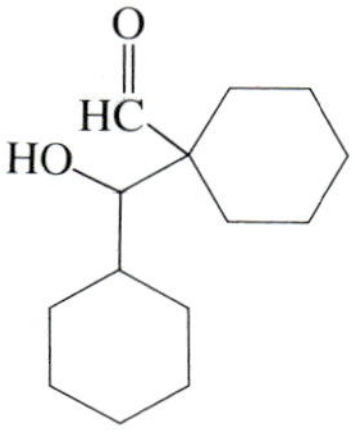

(c) 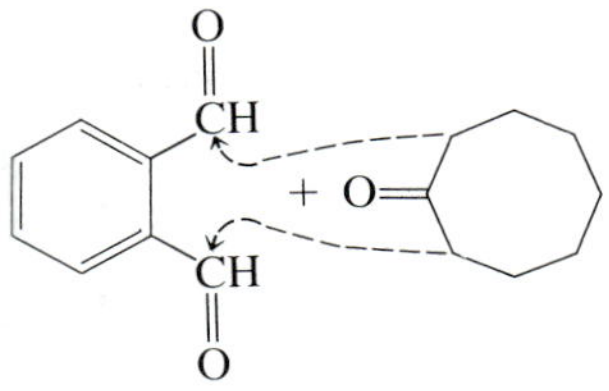 $CH_2{=}CHCH{=}C(CH_3)CHO$

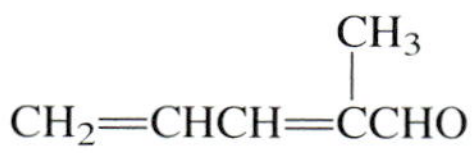

16-10

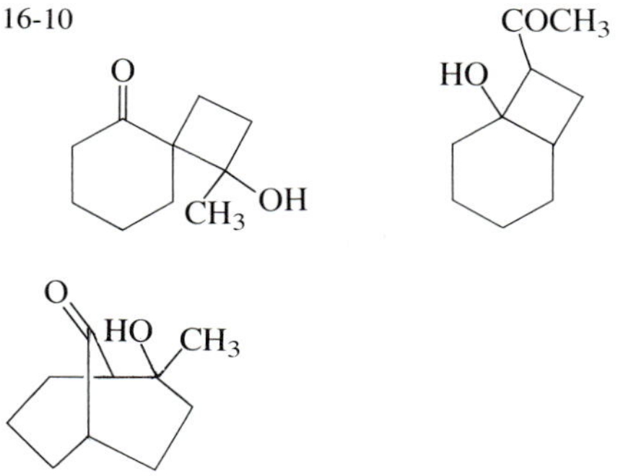

16-11

(a) Na$_2$CO$_3$, 100°C

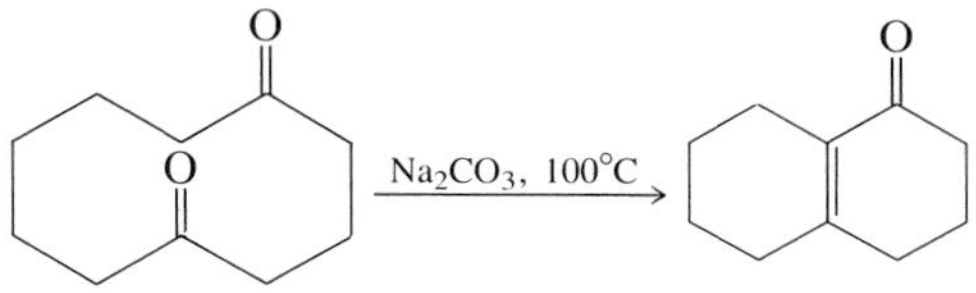

(b) 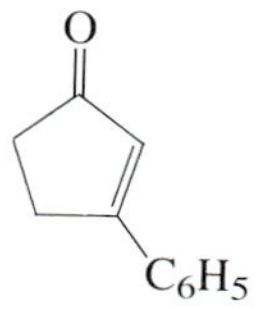 C_6H_5

(c)

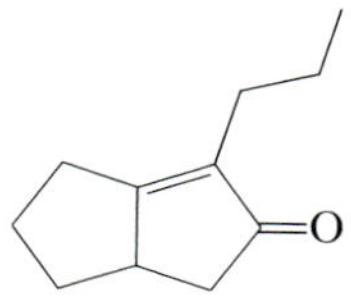

16-12

(a) HO^- ; H_2, Pd-C

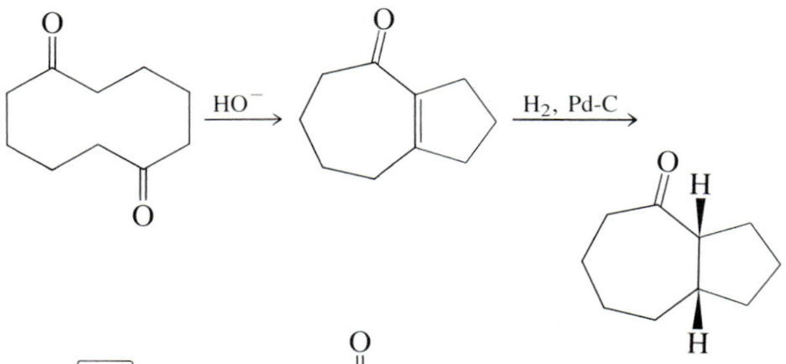

(b) 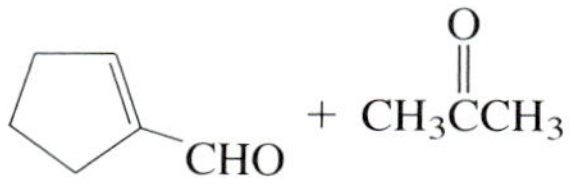 CHO + CH_3CCH_3 (O)

(c) CH, CH, + O=

16-13

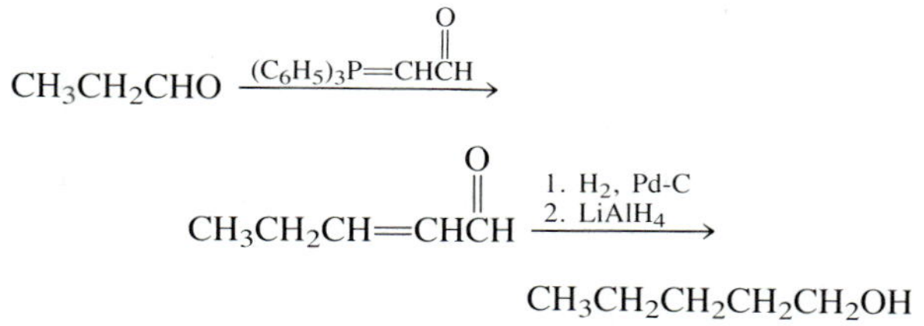

$CH_3CH_2CHO \xrightarrow{(C_6H_5)_3P{=}CHCH(O)} CH_3CH_2CH{=}CHCHO \xrightarrow{1.\ H_2,\ Pd\text{-}C;\ 2.\ LiAlH_4} CH_3CH_2CH_2CH_2CH_2OH$

16-14

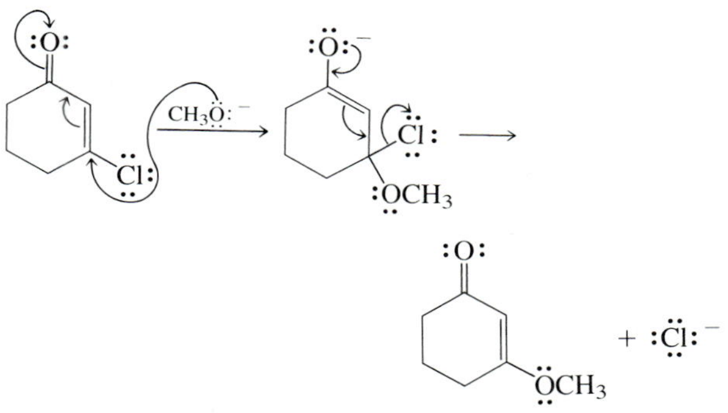

$CH_3\ddot{O}:^-$; :Cl: ; :OCH$_3$; + :Cl:$^-$

16-15

$H_2C{=}CHCCH_3 \rightleftharpoons H_2C{=}CH{-}C(\overset{+}{O}H)CH_3 \xrightarrow{H_2\ddot{N}\ddot{N}H_2}$

$H_2C{-}CH{=}C(OH)CH_3$ with $H_2N^+{-}\ddot{N}H_2$ $\xrightarrow{-H^+}$

$CH_2CH_2CCH_3$ (O), HN–NH$_2$ $\longrightarrow$ CH$_3$, :O:$^-$, HN–NH$_2^+$ $\longrightarrow$

CH$_3$, OH, HN–NH $\xrightarrow[-H_2O]{H^+}$ CH$_3$, HN–N

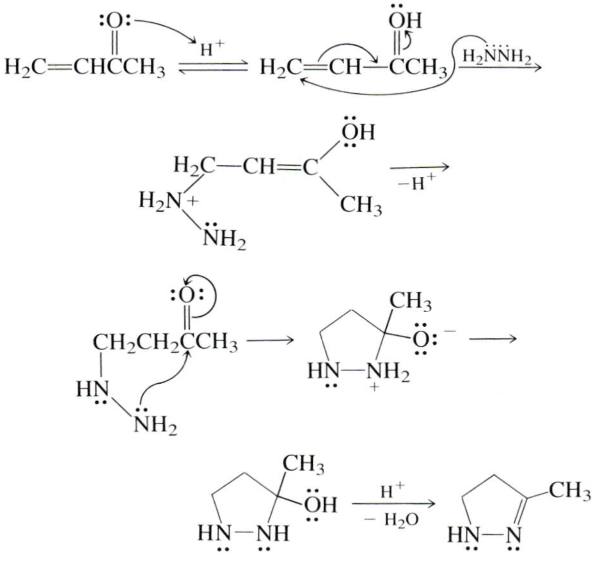

An alternative mechanism would be hydrazone formation followed by intramolecular 1,4-addition of the second amino group. Formulate it.

16-16

(a) 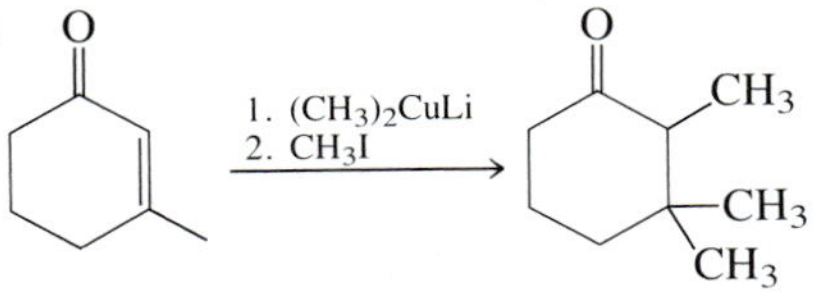 3-methylcyclohex-2-enone $\xrightarrow[\text{2. } CH_3I]{\text{1. } (CH_3)_2CuLi}$ 2,3,3-trimethylcyclohexanone (CH_3, CH_3, CH_3)

(b) 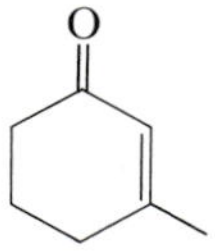 3-methylcyclohex-2-enone + $(CH_2{=}CHCH_2CH_2)_2CuLi \longrightarrow$

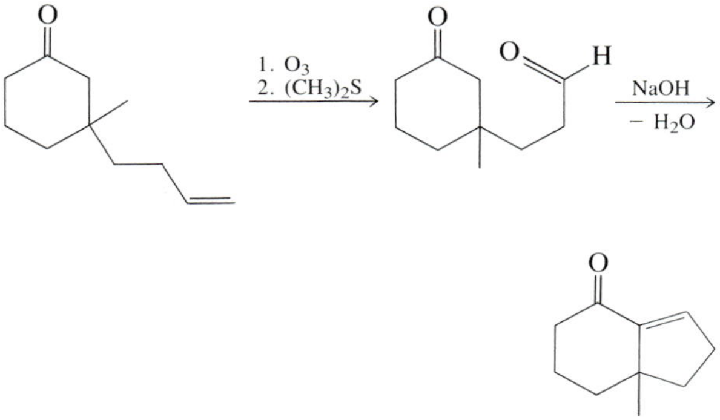

3-(3-butenyl)-3-methylcyclohexanone $\xrightarrow[\text{2. } (CH_3)_2S]{\text{1. } O_3}$ aldehyde (O, H) $\xrightarrow[-\,H_2O]{NaOH}$ bicyclic enone (O)

16-17

1-pyrrolidinyl-3,4-dihydronaphthalene (enamine, $\ddot{N}$) + $CH_2{=}CH{-}\overset{\overset{\textstyle :\ddot{O}:}{\|}}{C}CH_3 \longrightarrow$

iminium ($\overset{+}{N}$) $CH_2CH{=}C(\ddot{O}:^-)CH_3$ $\xrightarrow{\text{Hydrolysis}}$ 2-(3-oxobutyl)-1-tetralone (:O:, :O:)

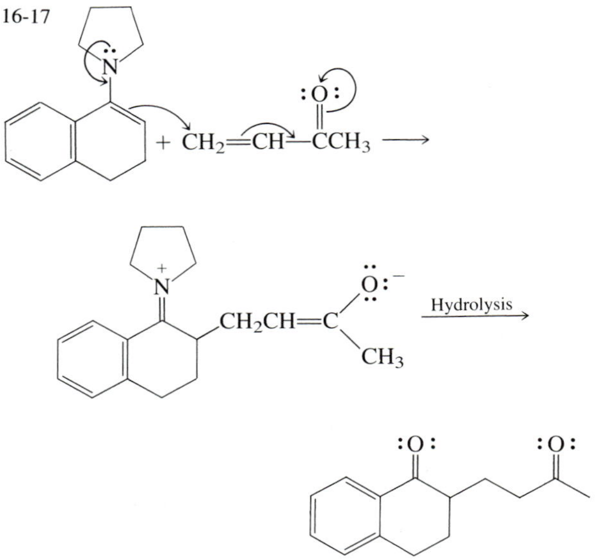

16-18

(a) $CH_3\overset{\overset{\textstyle :\ddot{O}:^-}{|}}{C}{=}CH_2 + CH_2{=}CH\overset{\overset{\textstyle O}{\|}}{C}CH_3$

(b) 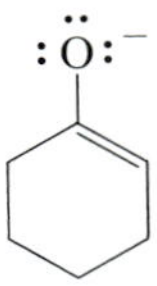 cyclohexenolate ($:\ddot{O}:^-$) $+ CH_2{=}CH\overset{\overset{\textstyle O}{\|}}{C}CH_3$

(c) $C_6H_5CH{=}\overset{\overset{\textstyle :\ddot{O}:^-}{|}}{C}CH_3 + CH_2{=}CH\overset{\overset{\textstyle O}{\|}}{C}CH_3$

CHAPTER 17

17-1

(a) 5-Bromo-3-chloroheptanoic acid

(b) 4-Oxocyclohexanecarboxylic acid

(c) $HOOCCH_2CH_2CH_2CBr_2COOH$ (Br above and below the C)

(d) $CH_3\overset{\overset{\textstyle OH}{|}}{C}HCH_2CH_2\overset{\overset{\textstyle O}{\|}}{C}OH$

17-2

$CH_3CH_2CH_2COOH$, butanoic (butyric) acid

17-3

$ClCH_2\overset{\overset{\textstyle O}{\|}}{C}OH$

17-4

(a) $CH_3CBr_2COOH > CH_3CHBrCOOH > CH_3CH_2COOH$

(b) $CH_3\overset{\overset{\textstyle F}{|}}{C}HCH_2COOH > CH_3\overset{\overset{\textstyle Br}{|}}{C}HCH_2COOH$

(c) 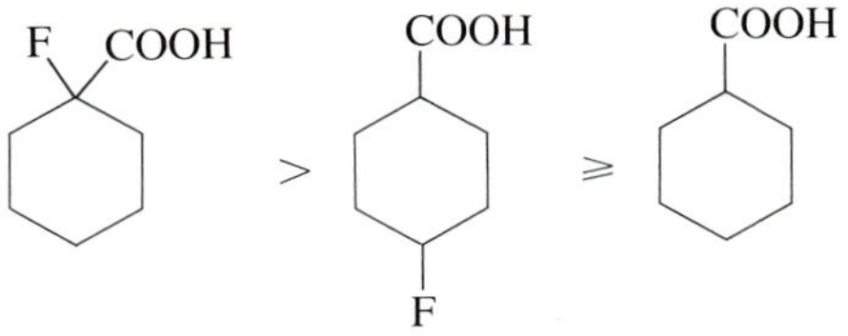 1-fluorocyclohexanecarboxylic acid (F, COOH) > 4-fluorocyclohexanecarboxylic acid (COOH, F) ≥ cyclohexanecarboxylic acid (COOH)

17-5

Protonated propanone (acetone) has fewer resonance forms:

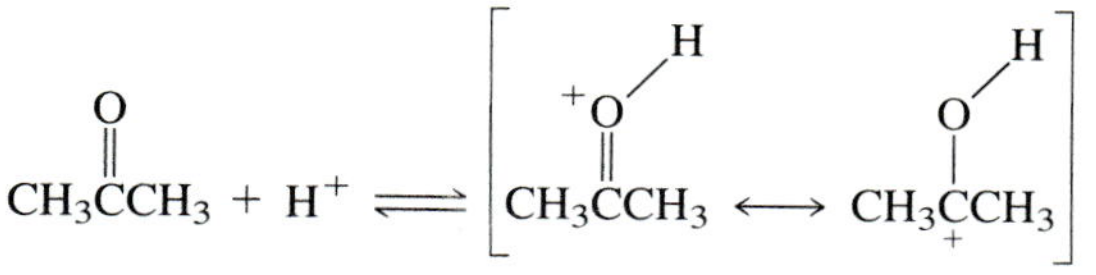

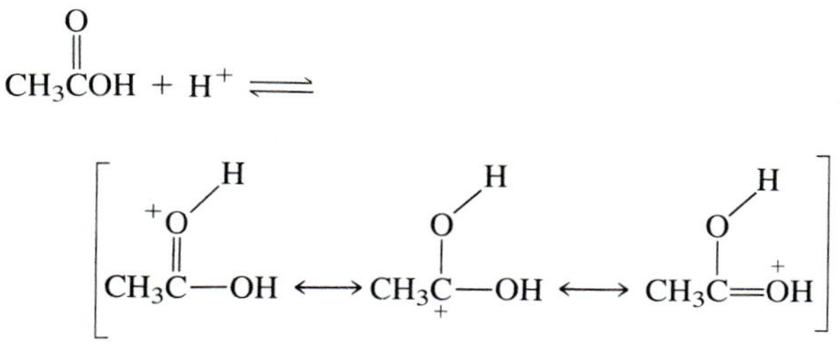

$$CH_3\overset{O}{\overset{\|}{C}}CH_3 + H^+ \rightleftharpoons \left[CH_3\overset{{}^{+}O-H}{\overset{\|}{C}}CH_3 \longleftrightarrow CH_3\overset{O-H}{\overset{|}{\underset{+}{C}}}CH_3 \right]$$

$$CH_3\overset{O}{\overset{\|}{C}}OH + H^+ \rightleftharpoons$$

$$\left[CH_3\overset{{}^{+}O-H}{\overset{\|}{C}}-OH \longleftrightarrow CH_3\overset{O-H}{\overset{|}{\underset{+}{C}}}-OH \longleftrightarrow CH_3\overset{O-H}{\overset{|}{C}}\overset{+}{=}OH \right]$$

17-6

(a) 1. HCN, 2. H^+, H_2O.

(b) Methylenecyclohexane $\xrightarrow{HBr}$ 1-bromo-1-methylcyclohexane (H_3C, Br) $\xrightarrow{1.\ Mg\quad 2.\ CO_2\quad 3.\ H^+,\ H_2O}$ 1-methylcyclohexanecarboxylic acid (H_3C, COOH)

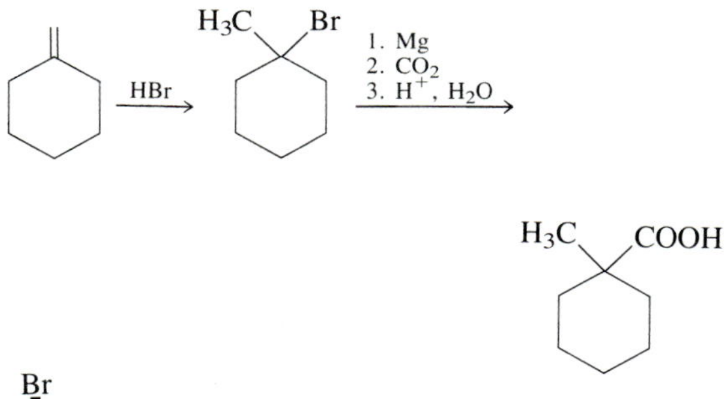

(c) Br / OCH_3 cyclohexane $\xrightarrow[\ S_N2]{^-CN,\ -\,Br^-}$ N≡C / OCH_3 cyclohexane $\xrightarrow{1.\ HO^-,\ H_2O\quad 2.\ H^+,\ H_2O}$ COOH / OCH_3 cyclohexane

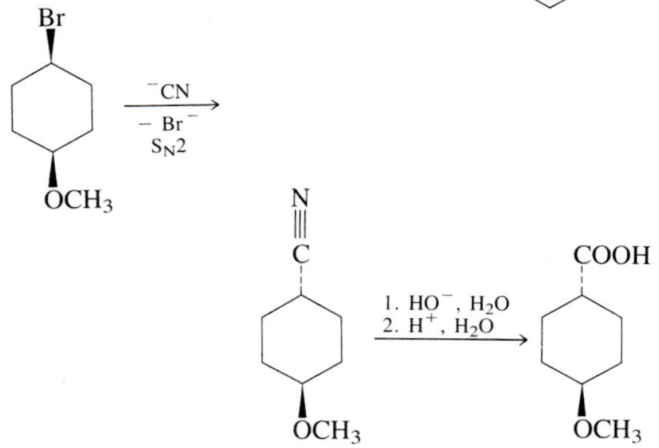

17-7

Both anhydrides are produced along with HCl, which catalyzes their decomposition.

(a)

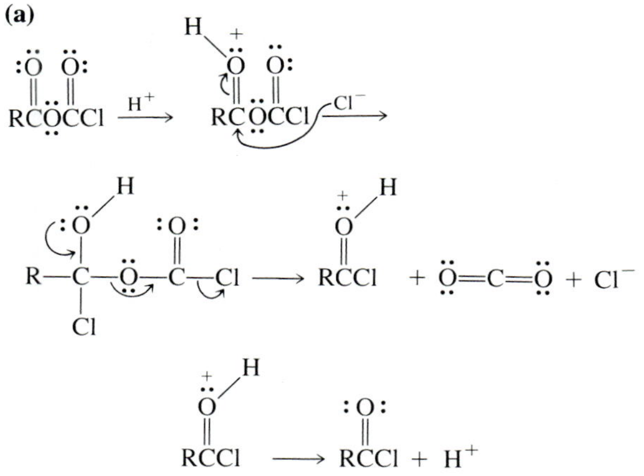

$$R\overset{:\ddot{O}}{\overset{\|}{C}}\ddot{O}\overset{\ddot{O}:}{\overset{\|}{C}}Cl \xrightarrow{H^+} R\overset{H-\overset{+}{O}}{\overset{\|}{C}}\ddot{O}\overset{\ddot{O}:}{\overset{\|}{C}}Cl \xrightarrow{Cl^-}$$

$$R-\underset{Cl}{\overset{:\ddot{O}-H}{\overset{|}{\underset{|}{C}}}}-\ddot{O}-\overset{:O:}{\overset{\|}{C}}-Cl \longrightarrow R\overset{\overset{+}{O}-H}{\overset{\|}{C}}Cl + \ddot{O}=C=\ddot{O} + Cl^-$$

$$R\overset{\overset{+}{O}-H}{\overset{\|}{C}}Cl \longrightarrow R\overset{:O:}{\overset{\|}{C}}Cl + H^+$$

(b)

$$R\overset{:\ddot{O}}{\overset{\|}{C}}\ddot{O}\overset{:\ddot{O}}{\overset{\|}{C}}-\overset{\ddot{O}:}{\overset{\|}{C}}Cl \xrightarrow{HCl} R-\underset{Cl}{\overset{:\ddot{O}H}{\overset{|}{\underset{|}{C}}}}-\ddot{O}-\overset{:\ddot{O}}{\overset{\|}{C}}-\overset{\ddot{O}:}{\overset{\|}{C}}-Cl \longrightarrow$$

$$R\overset{:O:}{\overset{\|}{C}}Cl + \ddot{O}=C=\ddot{O} + \overset{:O:}{\overset{\|}{\ddot{C}}} + H^+ + Cl^-$$

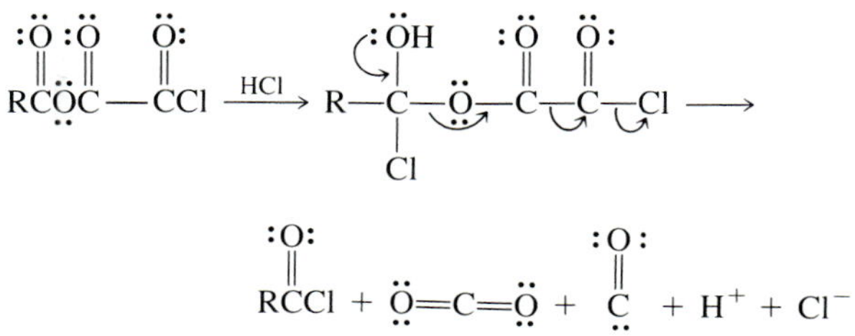

17-8

(a)

$$1.\ CH_3\overset{O}{\overset{\|}{C}}Cl + Na^{+\,-}O\overset{O}{\overset{\|}{C}}CH_2CH_3$$

$$2.\ CH_3\overset{O}{\overset{\|}{C}}O^-Na^+ + Cl\overset{O}{\overset{\|}{C}}CH_2CH_3$$

(b)

$$CH_3\overset{H_3C}{\overset{|}{C}}H\overset{O}{\overset{\|}{C}}OH + SOCl_2 \text{ or } COCl_2$$

17-9

The reaction is self-catalyzed by acid.

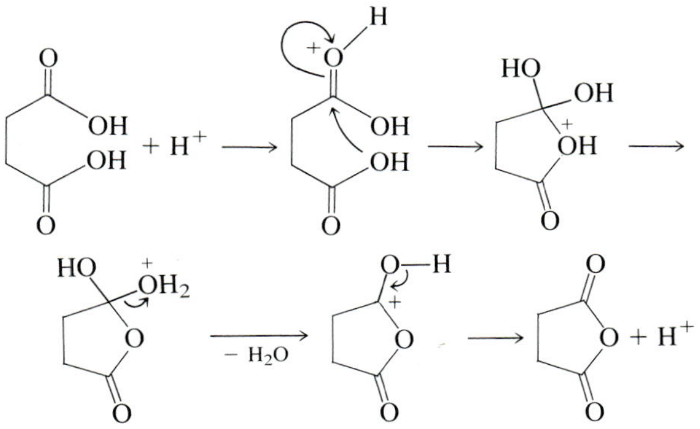

Butanedioic acid + H^+ ⟶ protonated acid (carbonyl $^{+}$O–H attacked by the other OH) ⟶ cyclic intermediate (HO, OH, $\overset{+}{O}H$) ⟶

cyclic intermediate (HO, $\overset{+}{O}H_2$) $\xrightarrow{-\,H_2O}$ ring cation ($\overset{+}{}$O–H) ⟶ succinic anhydride + H^+

17-10

The tetrahedral intermediate formed after hydration of the ester bears two equivalent hydroxy groups. Because its formation is reversible, the reverse process of eliminating water and regenerating the ester has an equal chance of proceeding through labeled or unlabeled species.

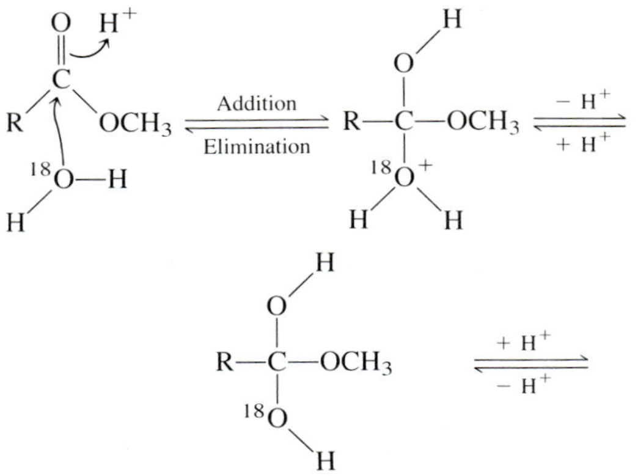

Addition / Elimination; $-H^+$ / $+H^+$; $+H^+$ / $-H^+$

Tetrahedral intermediate, both OH groups equivalent

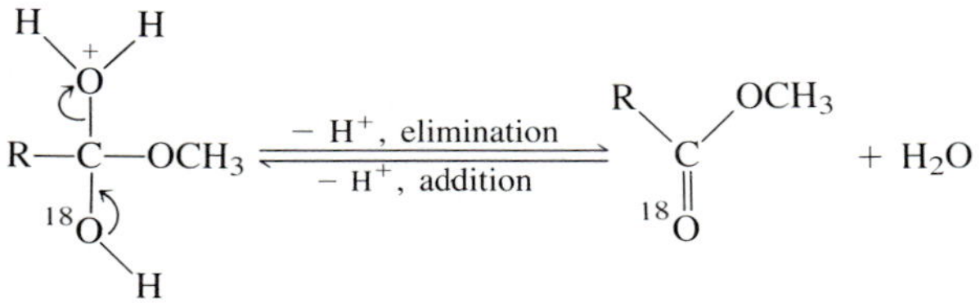

$-H^+$, elimination / $-H^+$, addition; $+ H_2O$

Regenerated labeled ester

17-11

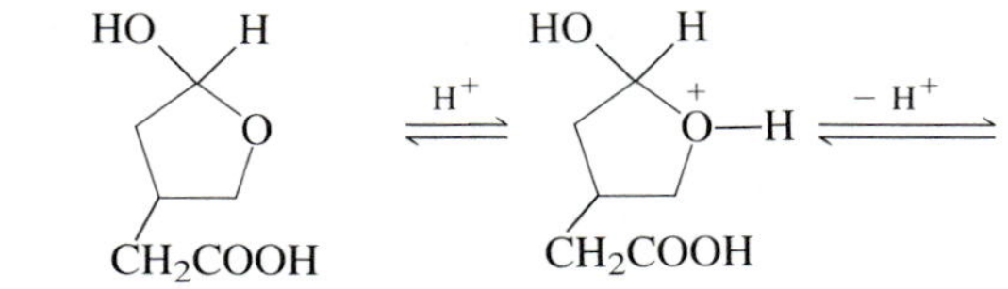

H^+; $-H^+$

This is an intramolecular hemiacetal (see Section 15-5)

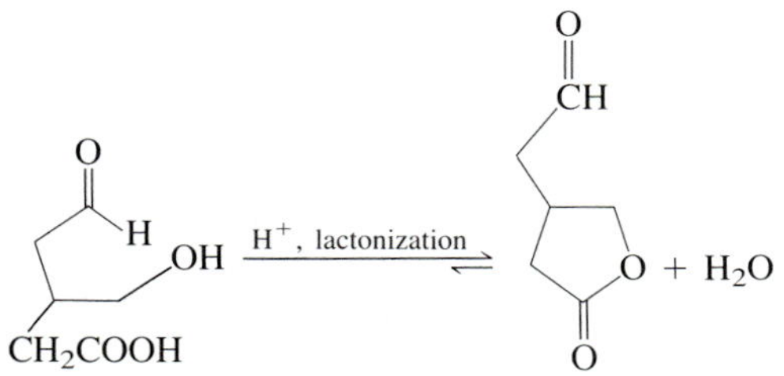

H^+, lactonization; $+ H_2O$

17-12

$\xrightarrow[-Br^-]{CN}$

$\xrightarrow{H^+, H_2O, \text{ hydrolysis}}$

$\xrightarrow[-H_2O]{NaOH, \Delta}$

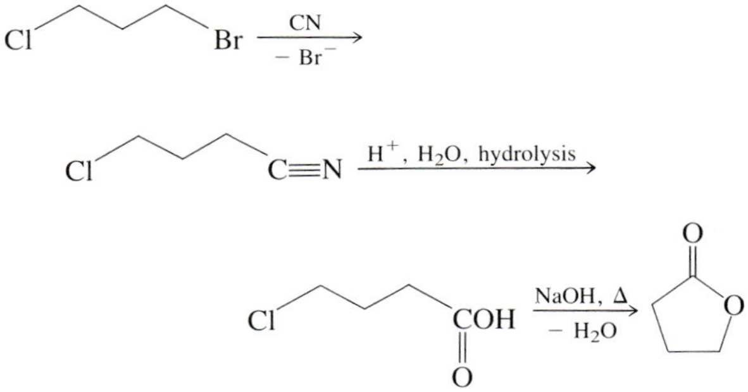

17-13

(a) Ester hydrolysis does not affect the stereocenter; **(b)** the S_N2 reaction with carboxylate ions inverts the stereocenter.

17-14

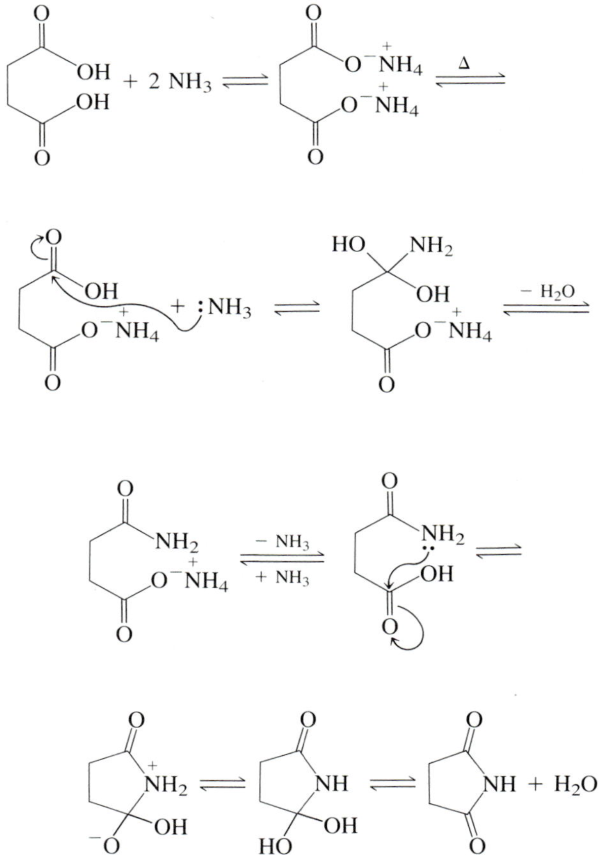

$+ 2 NH_3$; Δ

$+ :NH_3$; $-H_2O$

$-NH_3$ / $+NH_3$

$+ H_2O$

17-15

$$CH_3CH_2CH_2CH_2Li \xrightarrow{CO_2} CH_3CH_2CH_2CH_2\overset{O}{\overset{\|}{C}}O^-Li^+ \xrightarrow{CH_3CH_2CH_2CH_2Li}$$

$$CH_3CH_2CH_2CH_2C(O^-Li^+)(CH_2CH_2CH_2CH_3)O^-Li^+ \xrightarrow{H^+,\ H_2O} (CH_3CH_2CH_2CH_2)_2C{=}O$$

17-16

(a) 1. H^+, H_2O, 2. $LiAlH_4$, 3. H^+, H_2O;
(b) 1. $LiAlD_4$, 2. H^+, H_2O.

17-17

$$CH_3COOH \xrightarrow{2\ LDA} {}^-{:}CH_2\ddot{C}(=O){\ddot{O}}{:}^- \quad \mathbf{A}$$

(a) A + H_2C—CH_2 (ethylene oxide) **(b)** A + CH_3COCH_3
(c) A + CH_3I

17-18

(a) $CH_3CH_2CH_2COOH \xrightarrow{1.\ 2\ LDA;\ 2.\ CH_3I;\ 3.\ H^+,\ H_2O} CH_3CH_2CH(CH_3)COOH \xrightarrow{1.\ AgNO_3,\ KOH;\ 2.\ Br_2} CH_3CH_2CH(CH_3)Br$

(b) $CH_3CH_2CH_2COOH \xrightarrow{1.\ AgNO_3,\ KOH;\ 2.\ Br_2} CH_3CH_2CH_2Br \xrightarrow{1.\ (CH_3)_3CO^-K^+;\ 2.\ KMnO_4,\ HO^-} CH_3COOH$

CHAPTER 18

18-1

At room temperature, rotation around the amide bond is slow on the NMR time scale and two distinct rotamers can be observed:

$$H_3C\text{–}C(=O)\text{–}N(H)\text{–}NHC_6H_5 \rightleftharpoons H_3C\text{–}C(=O)\text{–}N(H)(NHC_6H_5)$$

Heating makes the equilibration so fast that the NMR technique can no longer distinguish between the two species.

18-2

$$H_3C\text{–}C(={:}\ddot{O}{:})\text{–}\ddot{C}l{:} + H^+ \longrightarrow$$

$$\left[H_3C\text{–}C(={}^+{:}O\text{–}H)\text{–}\ddot{C}l{:} \longleftrightarrow H_3C\text{–}C^+(\text{–}\ddot{O}\text{–}H)\text{–}\ddot{C}l{:} \longleftrightarrow H_3C\text{–}C(\text{–}\ddot{O}\text{–}H)={Cl}{:}^+ \right]$$

Not a strong contributor

$$H_3C\text{–}C(={:}O{:})\text{–}\ddot{N}H_2 + H^+ \longrightarrow$$

$$\left[H_3C\text{–}C(={}^+{:}O\text{–}H)\text{–}\ddot{N}H_2 \longleftrightarrow H_3C\text{–}C^+(\text{–}\ddot{O}\text{–}H)\text{–}\ddot{N}H_2 \longleftrightarrow H_3C\text{–}C(\text{–}\ddot{O}\text{–}H)={}^+NH_2 \right]$$

Strong contributor

strength and the electronic structure of the resulting radical cations. (Draw resonance structures.) Do these cations fragment by loss of water?

$m/z = 85$, $CH_3CH_2CH_2CH_2C{\equiv}O^+$

(α cleavage)

19-17

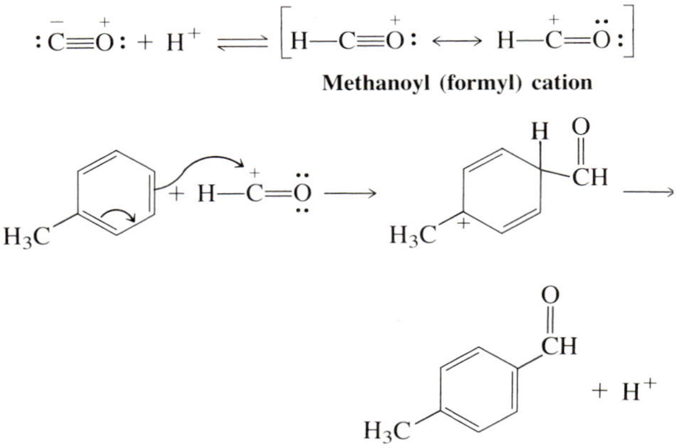

$:\overset{-}{C}\equiv\overset{+}{O}: + H^+ \rightleftharpoons [H—C\equiv\overset{+}{O}: \longleftrightarrow H—\overset{+}{C}=\ddot{O}:]$

Methanoyl (formyl) cation

19-18

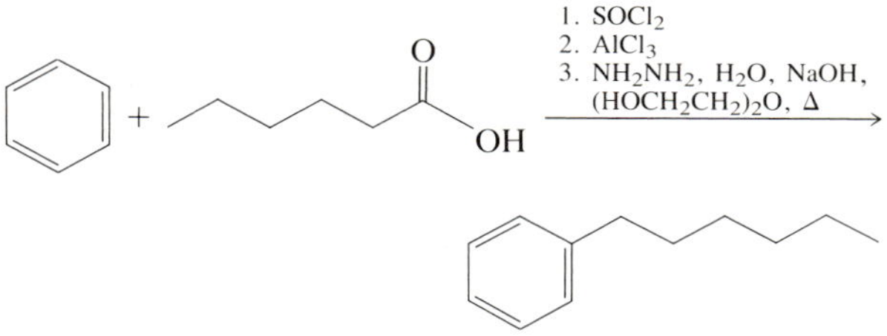

1. $SOCl_2$
2. $AlCl_3$
3. NH_2NH_2, H_2O, NaOH, $(HOCH_2CH_2)_2O$, Δ

19-19

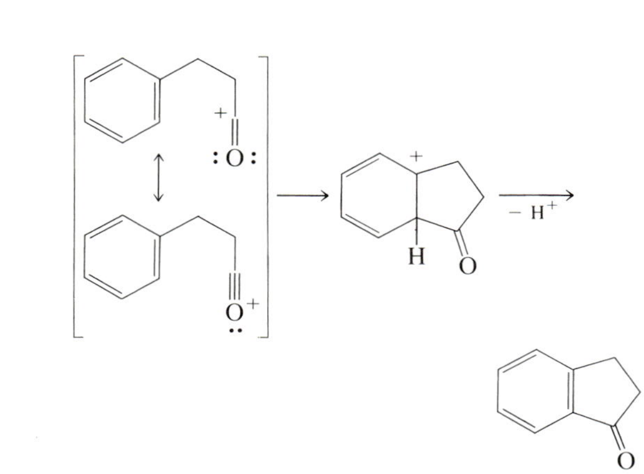

$-CH_3OH$

$-H^+$

CHAPTER 20

20-1

D, B, A, C. The fused ring in tetralin, D, can be considered to have the effect of two alkyl substituents.

20-2

Three of the resonance structures written for benzenol (phenol) place a positive charge on oxygen as in a protonated carbonyl compound. Moreover, the corresponding anion is unusually stable because the negative charge can be delocalized into the ring.

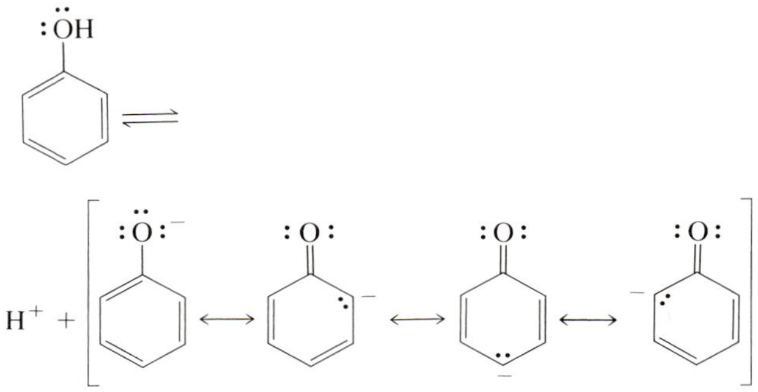

The lone electron pair on the nitrogen in benzenamine (aniline) is tied up by resonance with the benzene ring and is therefore less available for protonation.

20-3

Activated: A, D Deactivated: B, C

20-4

(a)

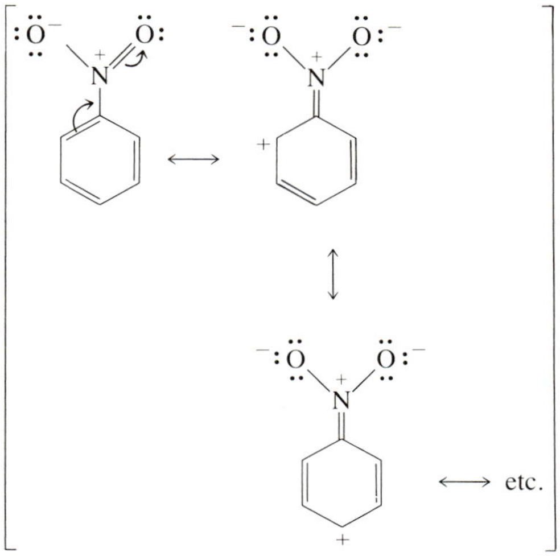

⟷ etc.

The nitro group is deactivating, both by induction (positive charge on nitrogen) as well as by resonance with the ring. The latter contribution is minor because it disrupts the resonance in the nitro group itself, which is similar to that in allyl:

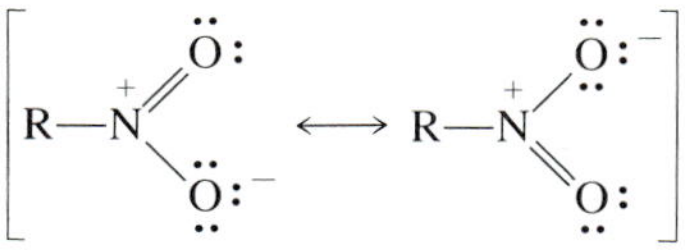

$$\left[R{-}\overset{+}{N}(=\ddot{O}:)(-\ddot{O}:^-) \longleftrightarrow R{-}\overset{+}{N}(-\ddot{O}:^-)(=\ddot{O}:) \right]$$

(b) In $C_6H_5\overset{+}{N}R_3$, the positively charged ammonium group acts as a strong inductive electron withdrawer. The lone electron pair of the original amino group is used for bond formation to one of the alkyl substituents.

(c) [Resonance structures of benzenesulfonic acid: $C_6H_5SO_2OH \longleftrightarrow \longleftrightarrow$ … $\longleftrightarrow$ etc.]

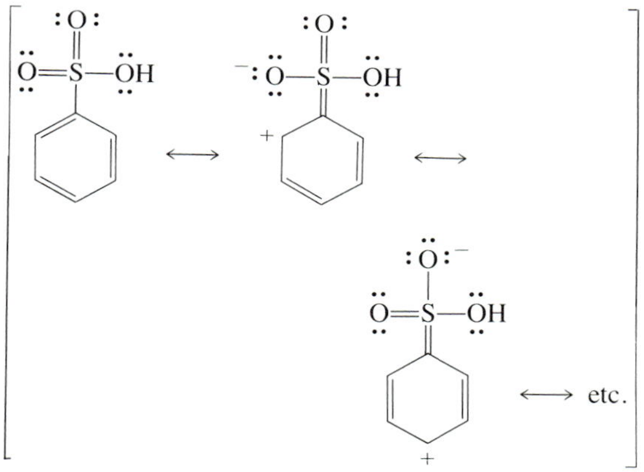

Benzenesulfonic acid is deactivated by resonance, like benzenecarboxylic (benzoic) acid.

(d) [Resonance structures of biphenyl $\longleftrightarrow \longleftrightarrow \longleftrightarrow$ etc.; labels: Deactivated ring, Activated ring]

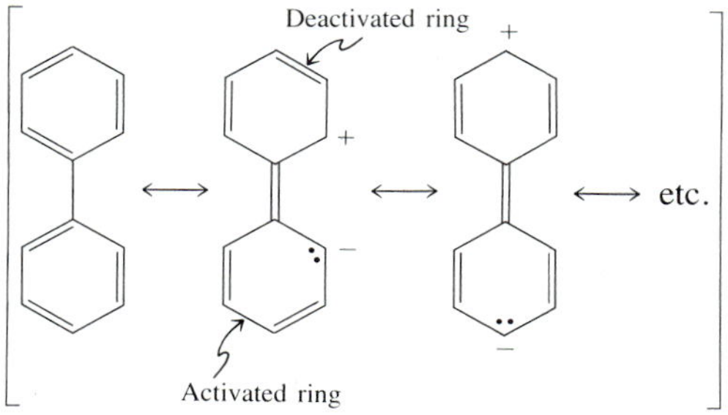

Note that, in phenylbenzene (the IUPAC name is biphenyl), resonance structures appear to indicate that one benzene ring will be activated to the extent that the other is deactivated, the net result being no effect. It turns out that this is not true; the abundance of π electrons in the system render it more electron-rich than benzene. Therefore, phenyl is a weak activator.

20-5

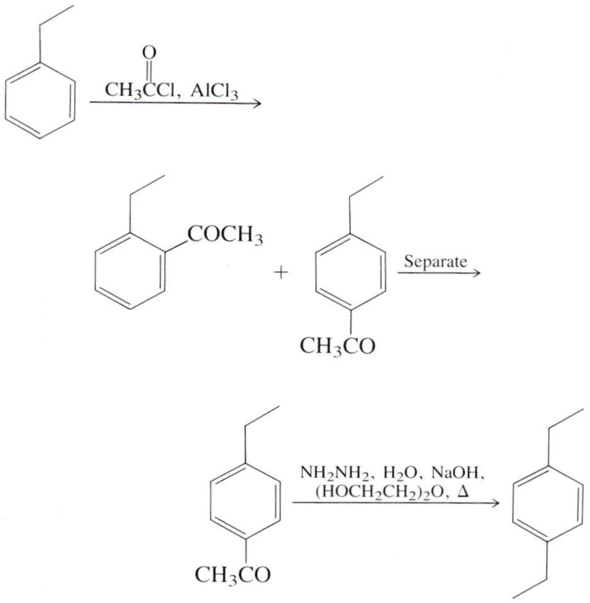

Ethylbenzene $\xrightarrow{CH_3\overset{O}{\overset{\|}{C}}Cl,\ AlCl_3}$ 1-ethyl-2-acetylbenzene ($COCH_3$) + 1-ethyl-4-acetylbenzene (CH_3CO) $\xrightarrow{\text{Separate}}$

4-Ethylacetophenone (CH_3CO) $\xrightarrow{NH_2NH_2,\ H_2O,\ NaOH,\ (HOCH_2CH_2)_2O,\ \Delta}$ 1,4-diethylbenzene

20-6

Methylbenzene (toluene) is activated and will consume all of the electrophile before the latter has a chance to attack the deactivated ring of (trifluoromethyl)benzene.

20-7

Ortho attack:

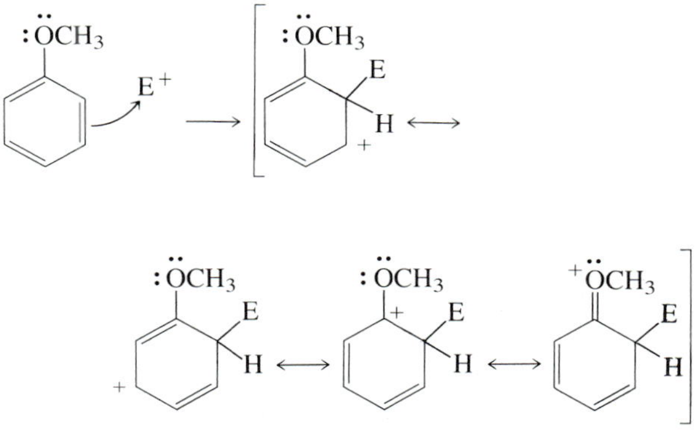

$:\ddot{O}CH_3$-benzene + E^+ $\longrightarrow$ [cation with $:\ddot{O}CH_3$, E, H, + $\longleftrightarrow$ cation with $:\ddot{O}CH_3$, E, H, + $\longleftrightarrow$ cation with $:\ddot{O}CH_3$, + , E, H $\longleftrightarrow$ cation with $^+\ddot{O}CH_3$ (C=O double bond), E, H]

Meta attack:

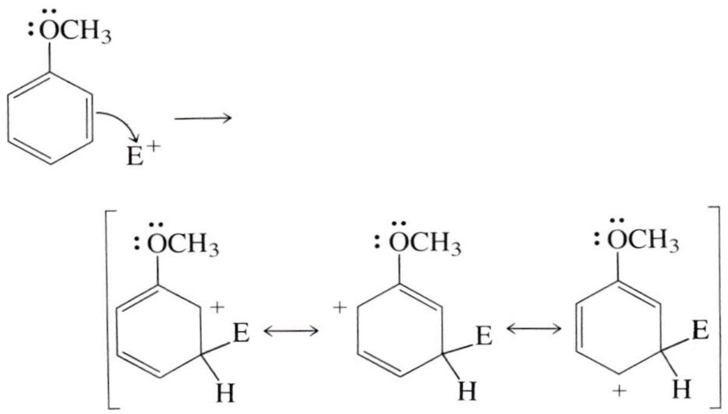

$:\ddot{O}CH_3$ … E^+ ⟶

$:\ddot{O}CH_3$ … E ⟷ $:\ddot{O}CH_3$ … E ⟷ $:\ddot{O}CH_3$ … E H

Para attack:

$:\ddot{O}CH_3$ … E^+ ⟶ [$:\ddot{O}CH_3$ … E H ⟷ $:\ddot{O}CH_3$ … E H ⟷ $^+\ddot{O}CH_3$ … E H ⟷ $:\ddot{O}CH_3$ … E H]

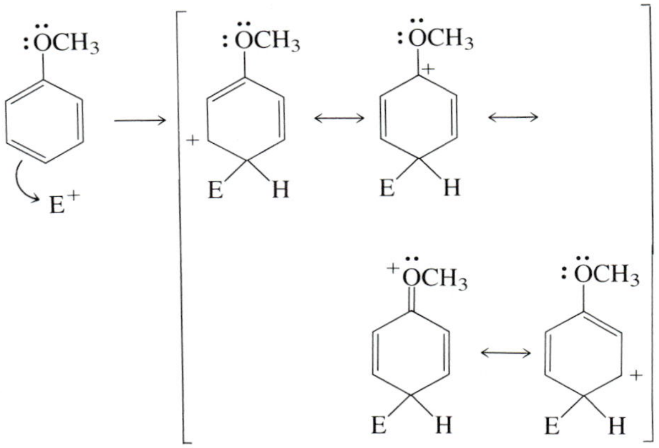

20-8
The lone electron pair is already tied up by resonance in the amide bond and is therefore less available to the ring.

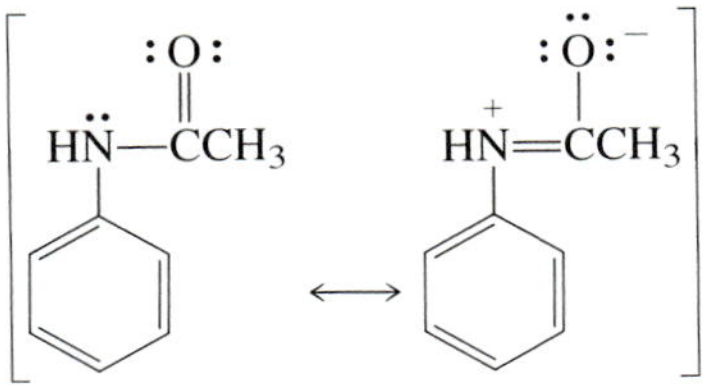

[$:\ddot{O}:$ $H\ddot{N}{-}CCH_3$ ⟷ $:\ddot{O}:^-$ $H\overset{+}{N}{=}CCH_3$]

20-9
Benzenamine (aniline) is completely protonated in strong acid. The lone electron pair is no longer available for resonance with the ring. Hence, the ammonium substituent is an inductive deactivator and *meta* director.

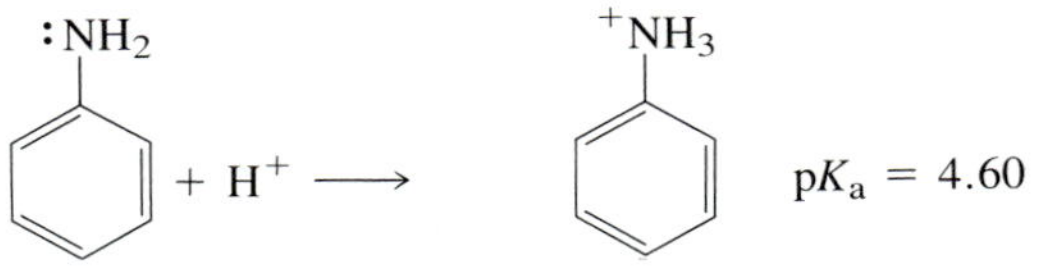

$:NH_2$ + H^+ ⟶ $^+NH_3$ $pK_a = 4.60$

Benzenammonium ion
(Anilinium ion)

20-10
(a) C-1 (= C-4) and C-2 (= C-3); **(b)** C-4 (= C-6) and C-5; **(c)** mainly C-4, with only some C-2, because C-2, although doubly activated, is also sterically hindered; **(d)** C-2 (= C-3 = C-5 = C-6).

20-11

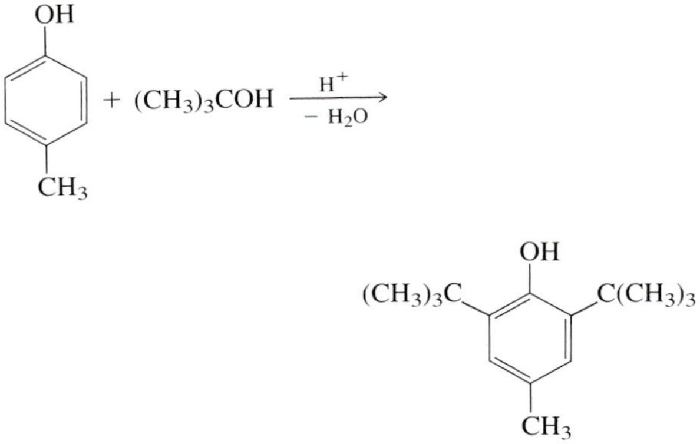

OH, CH_3 + $(CH_3)_3COH$ $\xrightarrow[-\,H_2O]{H^+}$

OH, $(CH_3)_3C$, $C(CH_3)_3$, CH_3

20-12
(a) C-3 and C-4; **(b)** C-5; **(c)** mainly C-3, with only some C-2, because the nitro group is a more-powerful deactivator than the ester group (Table 20-3); **(d)** mainly C-4 because NO_2 is a *meta* director and Br is an *ortho* and *para* director; for steric reasons, there is only a small amount of C-6 substitution.

20-13
(a) C-5 and C-7; **(b)** C-4 and C-6; **(c)** C-4 and C-6.

20-14

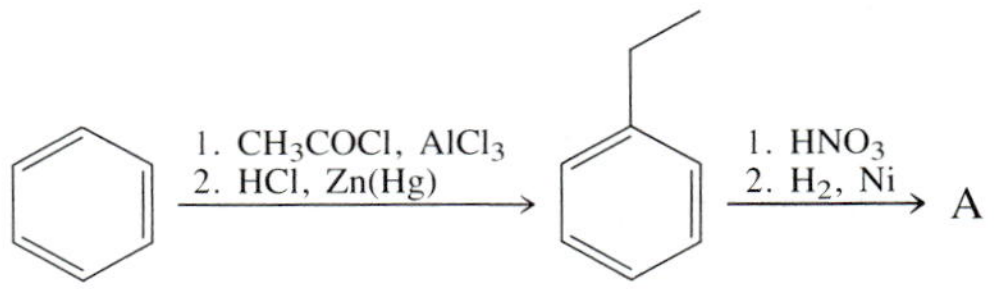

$\xrightarrow[\text{2. HCl, Zn(Hg)}]{\text{1. }CH_3COCl,\ AlCl_3}$ $\xrightarrow[\text{2. }H_2,\ Ni]{\text{1. }HNO_3}$ A

20-15

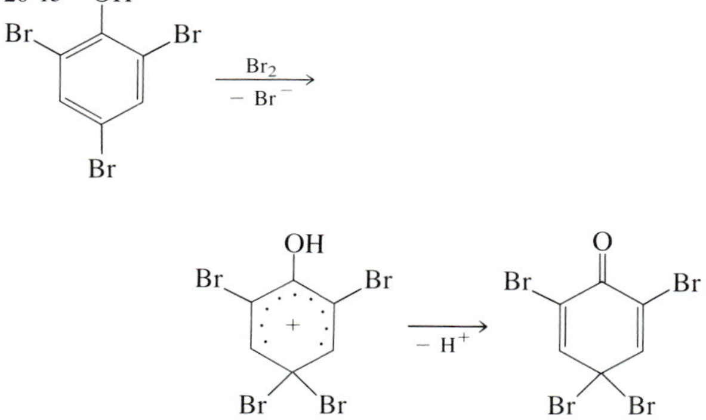

$\xrightarrow[-\,Br^-]{Br_2}$ $\xrightarrow{-\,H^+}$

20-16

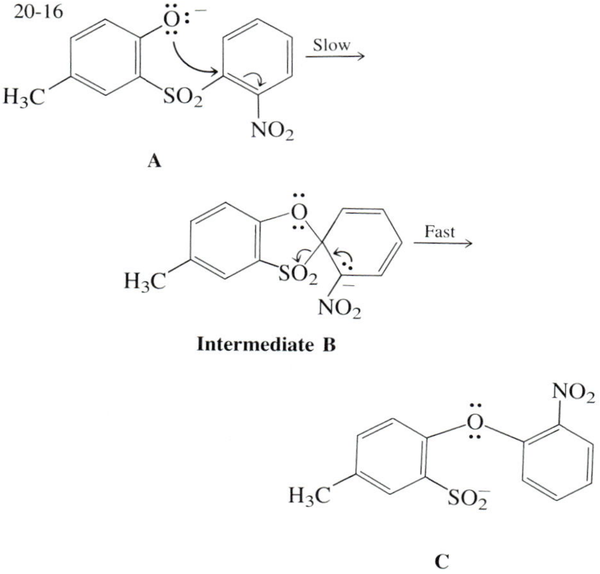

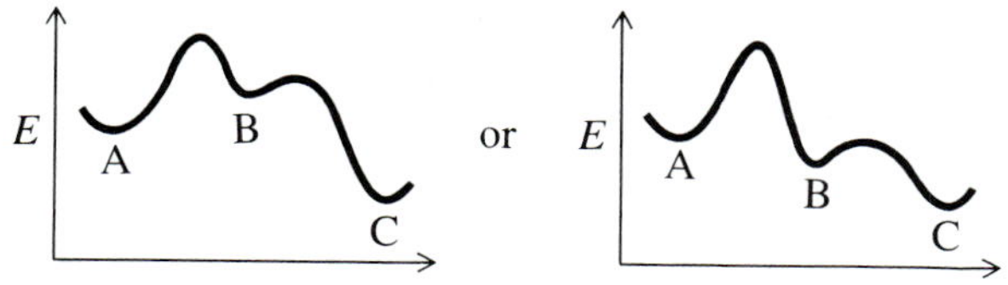

A $\xrightarrow{\text{Slow}}$

Intermediate B $\xrightarrow{\text{Fast}}$

C

E A B C or E A B C

20-17

$\xrightarrow[-\,NH_3,\ -\,Br^-]{^-:\ddot{N}H_2}$ $\xrightarrow{^-:\ddot{N}H_2}$

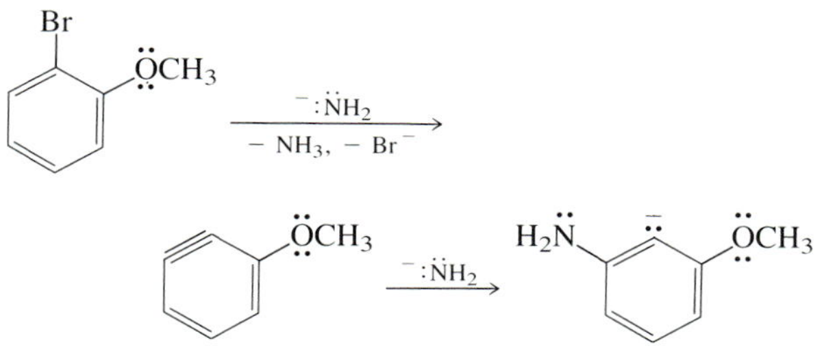

The product of amide addition to the first intermediate benzyne is stabilized by the inductively electron withdrawing effect of the methoxy oxygen; therefore, it is formed regioselectively. Protonation gives the major product. Note that there is no possibility for resonance in this system because the reactive electron pair is located in an sp^2 orbital that lies perpendicular to the π system.

CHAPTER 21

21-1

(a) 2-Butanamine, *sec*-butylamine

(b) *N,N*-Dimethylbenzenamine, *N,N*-dimethylaniline

(c) 6-Bromo-2-hexanamine, 5-bromo-2-methylpentylamine

21-2

(a) $HC{\equiv}CCH_2NH_2$; **(b)** 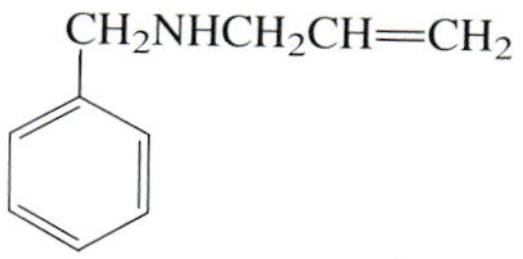 $C_6H_5CH_2NHCH_2CH{=}CH_2$

(c) $(CH_3)_3CNHCH_3$;

(d) same as **c.**

21-3

The lesser electronegativity of nitrogen, compared with oxygen, allows for slightly more diffuse orbitals and hence longer bonds to other nuclei.

21-4

No. Inversion leads to the rapid interchange of the two different environments.

21-5

The enantiomer on the left is S, the other R.

21-6

Less, because, again, nitrogen is less electronegative than oxygen. See Tables 10-2 and 10-3 for the effect of the electronegativity of substituent atoms on chemical shifts.

21-7

IR: secondary amine, hence a weak band at 3400 cm^{-1}.

^{1}H NMR: s for the 1,1-dimethylethyl (*tert*-butyl) group at high field,
s for the attached methylene group at $\delta \sim 2.7$,
q for the second methylene unit close to the first,

(Continued on the next page.)

t for the unique methyl group at high field, closest to the 1,1-dimethylethyl (*tert*-butyl) signal.

^{13}C NMR: five signals, two at low field about 45–50 ppm.

Mass: m/z = 115 (M^+), 100 $[(CH_3)_3CCH_2NH{=}CH_2]^+$, and 58 $(CH_2{=}NHCH_2CH_3)^+$. In this case, two different iminium ions can be formed by fragmentation.

21-8

As discussed in Section 20-1, the nitrogen lone electron pair is tied up by resonance with the benzene ring. Therefore, the nitrogen is less nucleophilic than the one in an alkanamine.

21-9

The nitromethane anion is resonance stabilized.

$CH_3NO_2 + {}^-{:}B \rightleftharpoons$
pK_a = 10.21

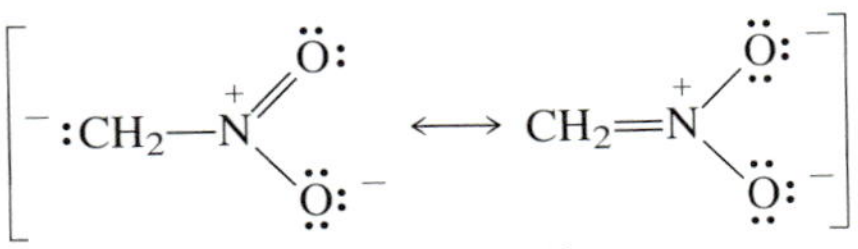

$[{}^-{:}CH_2{-}N^+(=O)O^- \longleftrightarrow CH_2{=}N^+(O^-)O^-]$

Nitromethane anion

It can be alkylated, and it enters into aldol reactions.

21-10

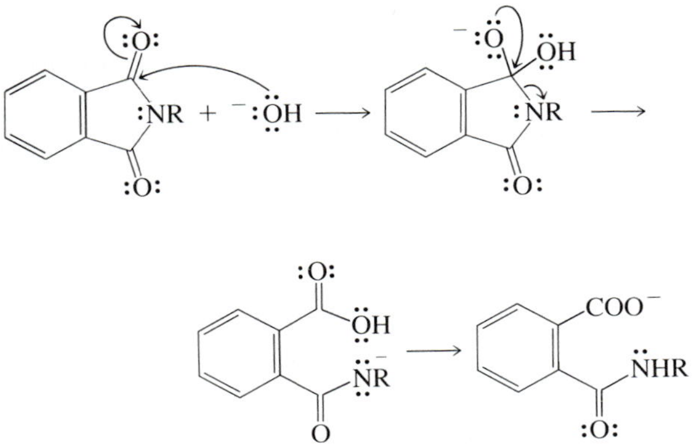

:NR + $^-$:ÖH ⟶ ⟶

Continue as in the normal amide hydrolysis by base (Section 18-5).

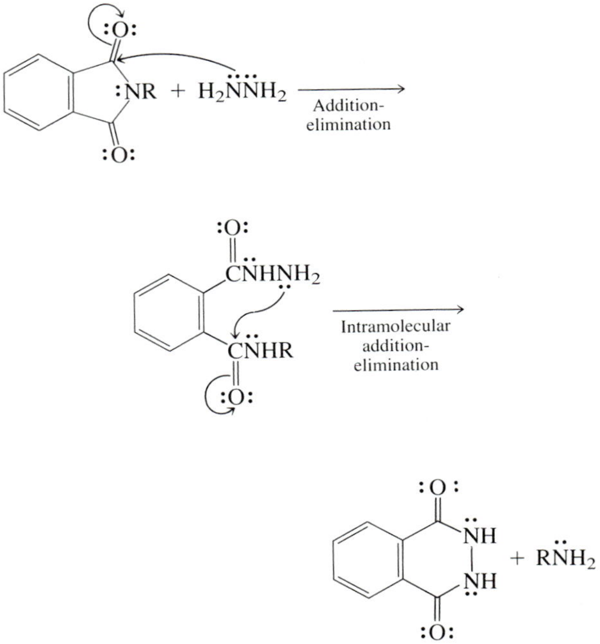

:NR + $H_2\ddot{N}\ddot{N}H_2$ ⟶ (Addition-elimination)

ĊNHNH$_2$ / ĊNHR ⟶ (Intramolecular addition-elimination)

NH / NH + $R\ddot{N}H_2$

21-11

$\ddot{O}^+$–H + $H\ddot{N}(CH_3)_2$ ⟶

$R{-}C(\ddot{O}H)(R'){-}N^+H(CH_3)_2$ ⟶ $R{-}C(\overset{+}{O}H_2)(R'){-}\ddot{N}(CH_3)_2$ $\xrightarrow{-H_2\ddot{O}}$

$RR'C{=}\overset{+}{N}(CH_3)_2$ $\xrightarrow{NaBH_3CN}$

$R{-}CH(R'){-}\ddot{N}(CH_3)_2$

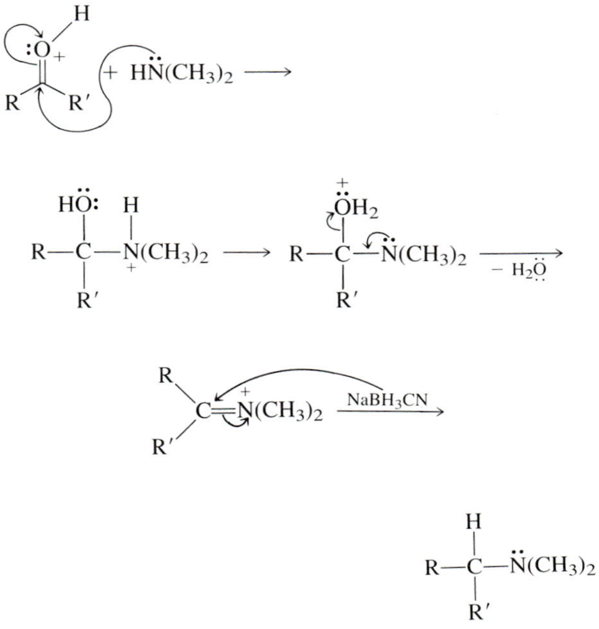

21-12

Not all intermediates are shown.

NH_2 (C=O, H; CH=O) $\xrightarrow[-H_2O]{H^+}$ (N, C, H; C=O, H) $\xrightarrow{NaBH_3CN}$

(NH; CH=O) $\xrightarrow[-H_2O]{H^+}$

(N^+) $\xrightarrow{NaBH_3CN}$ (N)

35%

21-13

(a) NH_2 + CH_3I

Excess

(b) NH_2 + $CH_2{=}O$ + $NaBH_3CN$

(c) $NHCH$ (O) + $LiAlH_4$

21-14

(a) $CH_3CH{=}CH_2$ and $CH_2{=}CH_2$;

(b) $CH_3CH_2CH{=}CH_2$ and $CH_3CH{=}CHCH_3$ (cis and trans). It is interesting that in both cases the terminal alkene predominates. This reaction is kinetically controlled according to the Hofmann rule (Section 11-5). Thus, the base prefers attack at the less-bulky end of the sterically encumbered quaternary ammonium ion.

21-15

(N) $\xrightarrow{\text{1. } CH_3I \text{ 2. } Ag_2O,\ H_2O \text{ 3. } \Delta}$ ($N{-}CH_3$) $\xrightarrow{\text{1. } CH_3I \text{ 2. } Ag_2O,\ H_2O \text{ 3. } \Delta}$

$(CH_3)_2N$ + double-bond isomer (as a side product) $\xrightarrow{\text{1. } CH_3I \text{ 2. } Ag_2O,\ H_2O \text{ 3. } \Delta}$

+ (CH_3)

21-16

H_3C, C_6H_5, H, CH_3, H, $\overset{+}{N}(CH_3)_2$, $^-$:Ö:

21-17

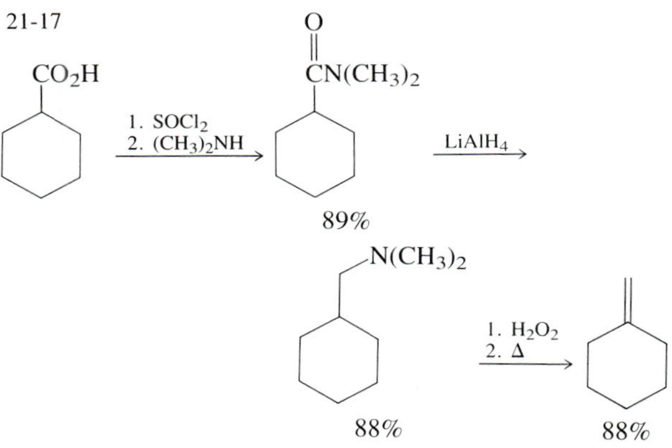

21-18

H, H, H, H, H, H

Bicyclo[1.1.0]butane

21-19

(a) 1. $SOCl_2$, 2. CH_2N_2 (1 equivalent);

(b) 1. $SOCl_2$, 2. CH_2N_2 (2 equivalents);

(c) 1. $SOCl_2$, 2. CH_2N_2 (2 equivalents), 3. (cyclohexene), $CuSO_4$, Δ.

CHAPTER 22

22-1

(a) Diels-Alder reaction; **(b)** Michael addition; **(c)** cyclic acetal formation.

22-2

(a) $CH_3CH_2CH_2CO_2CH_3$, 1. Na, 2. Cu^{2+}

(b) CO_2CH_3 / CO_2CH_3 + Na

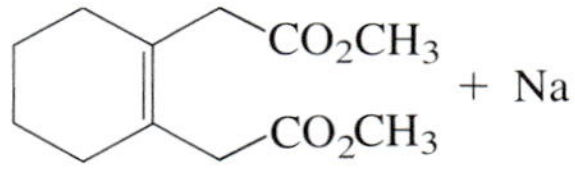

(c) CO_2CH_3 / $CH_2CO_2CH_3$ / CH_3O + Na

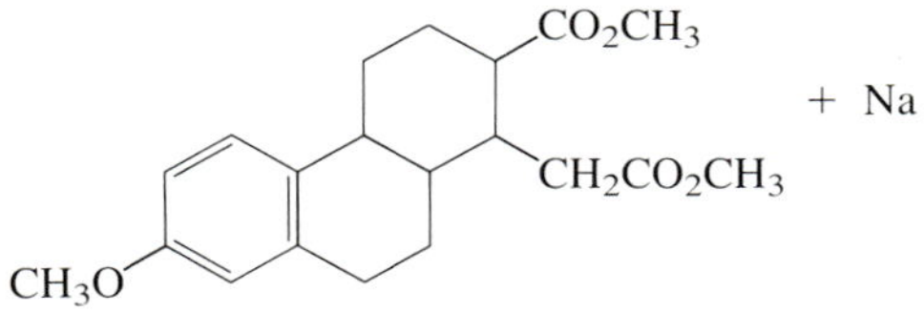

22-3

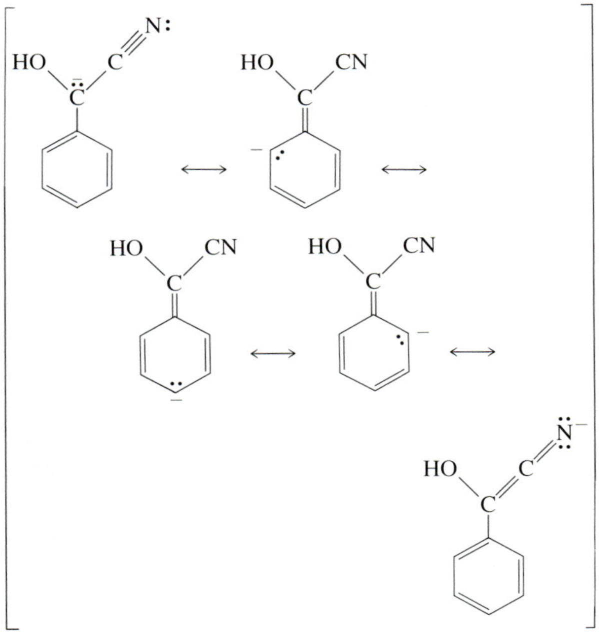

HO, C, N: ; HO, C, CN ; HO, C, CN ; HO, C, CN ; HO, C, C, N

22-4

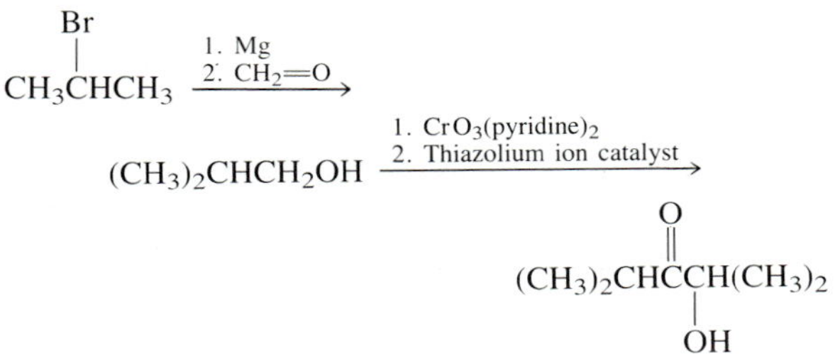

CH_3CHCH_3 (Br) $\xrightarrow{\text{1. Mg; 2. } CH_2{=}O}$ $(CH_3)_2CHCH_2OH$ $\xrightarrow{\text{1. } CrO_3(\text{pyridine})_2\text{; 2. Thiazolium ion catalyst}}$ $(CH_3)_2CHCCH(CH_3)_2$ (O, OH)

22-5

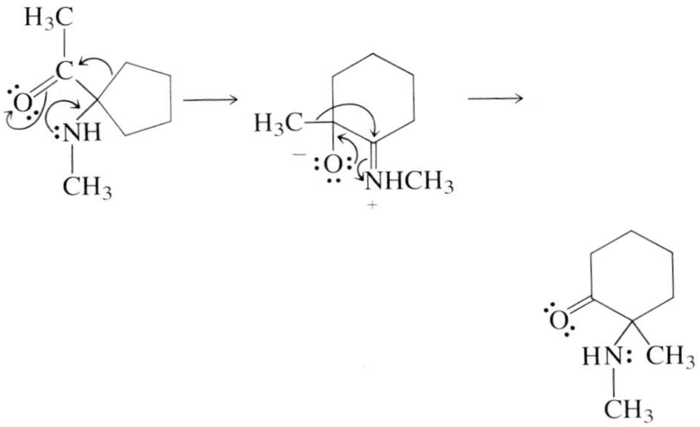

H_3C, C, O, NH, CH_3 → H_3C, O, $NHCH_3$ → O, HN, CH_3, CH_3

22-6

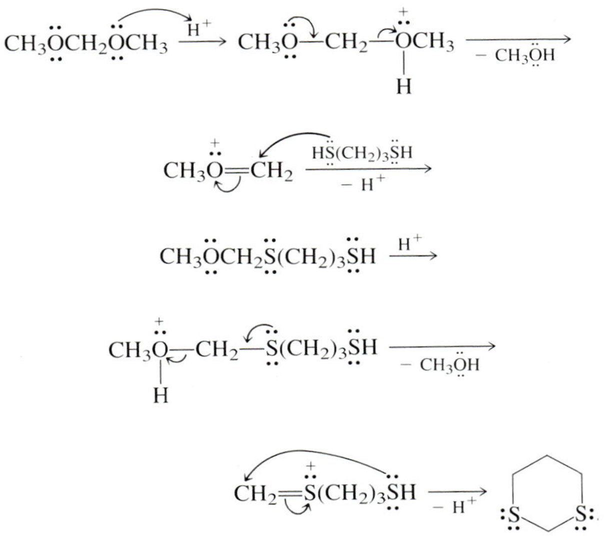

$CH_3OCH_2OCH_3 \xrightarrow{H^+} CH_3O{-}CH_2{-}\overset{+}{O}CH_3$ (H) $\xrightarrow{-CH_3OH}$

$CH_3\overset{+}{O}{=}CH_2 \xrightarrow[-H^+]{HS(CH_2)_3SH}$

$CH_3OCH_2S(CH_2)_3SH \xrightarrow{H^+}$

$CH_3\overset{+}{O}{-}CH_2{-}S(CH_2)_3SH$ (H) $\xrightarrow{-CH_3OH}$

$CH_2{=}\overset{+}{S}(CH_2)_3SH \xrightarrow{-H^+}$ S, S

22-7

O

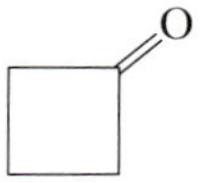

22-8

The starting material undergoes a retro-Claisen condensation. The methyl ethanoate (acetate) so formed then reacts by a forward Claisen condensation.

22-9

This mechanism (top of following page) is abbreviated, showing only the most important steps.

$$CH_3CH_2O_2C(CH_2)_3\overset{O}{\overset{\|}{C}}OCH_2CH_3 \xrightarrow[-CH_3CH_2OH]{CH_3CH_2O^-} CH_3CH_2O_2CCH_2CH_2\ddot{\bar{C}}HCO_2CH_2CH_3 \xrightarrow[-\ CH_3CH_2O^-]{CH_3CH_2O\overset{O}{\overset{\|}{C}}\overset{O}{\overset{\|}{C}}OCH_2CH_3}$$

$$CH_3CH_2O_2CCH_2CH_2CH(\overset{O}{\overset{\|}{C}}\overset{O}{\overset{\|}{C}}OCH_2CH_3)CO_2CH_2CH_3 \xrightarrow[-CH_3CH_2OH]{CH_3CH_2O^-} CH_3CH_2O_2C\ddot{\bar{C}}HCH_2CH_2CH(\overset{O}{\overset{\|}{C}}\overset{O}{\overset{\|}{C}}OCH_2CH_3)CO_2CH_2CH_3 \xrightarrow[-\ CH_3CH_2O^-]{}$$

(cyclopentane-1,2-dione bearing $CH_3CH_2O_2C$ and $CO_2CH_2CH_3$ substituents)

22-10

This mechanism also is abbreviated.

$$\text{diethyl phthalate (}C_6H_4(CO\,OCH_2CH_3)_2\text{)} + {}^-{:}CH_2CO_2CH_2CH_3 \xrightarrow[-\ CH_3CH_2O^-,\ -\ H^+]{}$$

$$C_6H_4(CO\ddot{\bar{C}}HCO_2CH_2CH_3)(CO\,OCH_2CH_3) \xrightarrow[-\ CH_3CH_2O^-]{} \text{1,3-indandione-2-}CO_2CH_2CH_3$$

22-11

(a) cyclohexanone + $CH_3CH_2O_2CCO_2CH_2CH_3$

1. $CH_3CH_2O^-$, 2. H^+, H_2O

(b) $CH_3\overset{O}{\overset{\|}{C}}CH_3 + HCO_2CH_2CH_3$

1. $CH_3CH_2O^-$, 2. H^+, H_2O

(c) cyclooctanone + $CH_3CH_2O\overset{O}{\overset{\|}{C}}OCH_2CH_3$

1. NaH, 2. H^+, H_2O

(d) $H_3C\overset{O}{\overset{\|}{C}}CH_2CH_2CH_2CH_2CO_2CH_2CH_3$

1. $CH_3CH_2O^-$, 2. H^+, H_2O

22-12

$$CH_3\overset{O}{\overset{\|}{C}}CH_2CH_2CH_2CO_2CH_3 \xrightarrow[\text{See Section 22-4}]{1.\ (C_6H_5)_3CO^-K^+\quad 2.\ H^+,\ H_2O}$$

1,3-cyclohexanedione, 100%

$$\xrightarrow{2\ NaOCH_3,\ CH_3I,\ CH_3OH}$$

80%

2,2-Dimethyl-1,3-cyclohexanedione

22-13

$$\text{2-methyl-1,3-cyclohexanedione} + {}^{-}\!:\ddot{O}H \longrightarrow$$

$$\text{tetrahedral intermediate} \longrightarrow HO\!-\!C(=O)(CH_2)_3C(=C\!-\!CH_3)\!-\!\ddot{O}:^{-} \longrightarrow$$

The enolate ion is a relatively good leaving group

$$CH_3CH_2\overset{\overset{\displaystyle :O:}{\|}}{C}(CH_2)_3CO_2^{-} \xrightarrow{H^+,\ H_2O} CH_3CH_2\overset{\overset{\displaystyle :O:}{\|}}{C}(CH_2)_3COOH$$

22-14

(a) 2-Butylcyclohexanone

(b) $CH_3CH_2CO_2H$
Propanoic acid

(c) $CH_3\overset{O}{\overset{\|}{C}}\underset{\underset{CH_3}{|}}{C}H\overset{O}{\overset{\|}{C}}\!-\!C_6H_5$
2-Methyl-1-phenyl-1,3-butanedione

(d) This sequence is general for 2-halo esters:

$$CH_3\overset{O}{\overset{\|}{C}}CH_2\overset{O}{\overset{\|}{C}}OCH_2CH_3 \xrightarrow[2.\ BrCH_2CO_2CH_2CH_3]{1.\ CH_3CH_2O^-Na^+} CH_3\overset{O}{\overset{\|}{C}}\underset{\underset{CH_2\overset{\|}{\underset{O}{C}}OCH_2CH_3}{|}}{C}H\overset{O}{\overset{\|}{C}}OCH_2CH_3 \xrightarrow[2.\ H^+,\ H_2O,\ \Delta]{1.\ NaOH}$$

$$CH_3\overset{O}{\overset{\|}{C}}CH_2CH_2\overset{O}{\overset{\|}{C}}OH$$
4-Oxopentanoic acid

Note that only the carboxy group located β to the ketone carbonyl can undergo decarboxylation.

(e) $HO\overset{O}{\overset{\|}{C}}\underset{\underset{CH_3CH_2}{|}}{C}HCH_2\overset{O}{\overset{\|}{C}}OH$
2-Ethylbutanedioic acid

Excessive heating may dehydrate this product to the anhydride (Section 17-7).

(f) $CH_3\overset{O}{\overset{\|}{C}}CH_2CH_2\overset{O}{\overset{\|}{C}}CH_3$
2,5-Hexanedione

22-15

$$CH_2(CO_2CH_2CH_3)_2 + Br(CH_2)_5Cl \xrightarrow[-\ CH_3CH_2OH,\ NaBr]{CH_3CH_2O^-Na^+} Cl(CH_2)_5\overset{\overset{H}{|}}{\underset{\underset{CO_2CH_2CH_3}{|}}{C}}CO_2CH_2CH_3 \xrightarrow[-\ CH_3CH_2OH]{CH_3CH_2O^-Na^+}$$

$$\text{Cl}\!-\!(CH_2)_5\!-\!{}^{-}\ddot{C}(CO_2CH_2CH_3)_2\ Na^+ \xrightarrow[-\ NaCl]{} \text{1,1-cyclohexane}(CO_2CH_2CH_3)_2 \xrightarrow[2.\ H^+,\ H_2O,\ \Delta]{1.\ KOH,\ CH_3CH_2OH} \text{C}_6\text{H}_{10}(\text{H})\text{COOH}$$

Cyclohexanecarboxylic acid

22-16

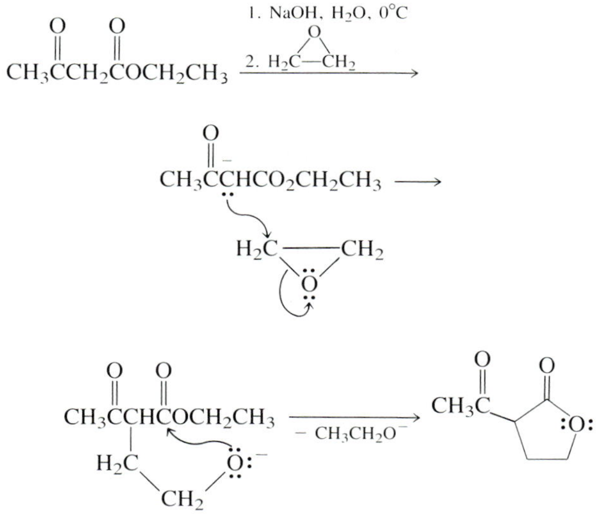

$$CH_3\overset{O}{\overset{\|}{C}}CH_2\overset{O}{\overset{\|}{C}}OCH_2CH_3 \xrightarrow[\text{2. } H_2C\text{—}CH_2 \text{ (epoxide)}]{\text{1. NaOH, } H_2O,\ 0^\circ C}$$

$$CH_3\overset{O}{\overset{\|}{C}}\ddot{C}^-HCO_2CH_2CH_3 \longrightarrow \quad H_2C\text{—}CH_2 \text{ (epoxide, ring-opening)}$$

$$CH_3\overset{O}{\overset{\|}{C}}\underset{H_2C\text{—}CH_2\text{—}\ddot{O}:^-}{CHCOCH_2CH_3} \xrightarrow{-\ CH_3CH_2O^-} CH_3\overset{O}{\overset{\|}{C}}\text{-(2-oxotetrahydrofuran-3-yl)}$$

22-17

All of these reactions are reported in the literature with the yields indicated.

(a) 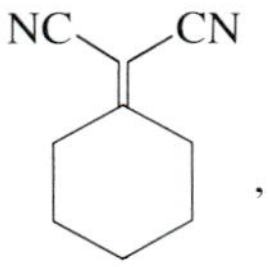 $(NC)_2C{=}C_6H_{10}$ (dicyanomethylenecyclohexane), 80%

(b) 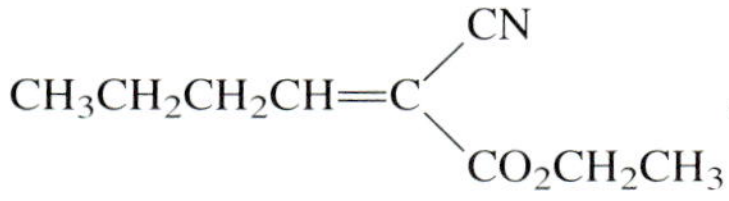 $CH_3CH_2CH_2CH{=}C(CN)(CO_2CH_2CH_3)$, 74%

(c) 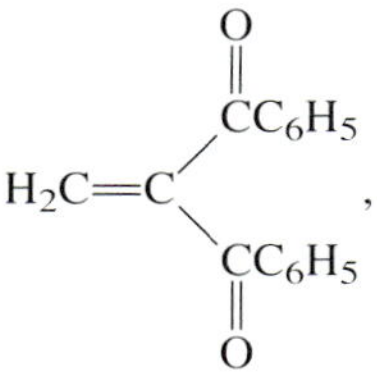 $H_2C{=}C(\overset{O}{\overset{\|}{C}}C_6H_5)_2$, 98%

(d) 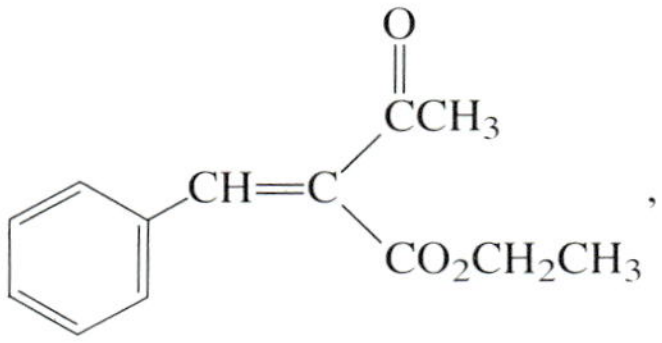 $C_6H_5CH{=}C(\overset{O}{\overset{\|}{C}}CH_3)(CO_2CH_2CH_3)$, 95%

22-18

A = activating group

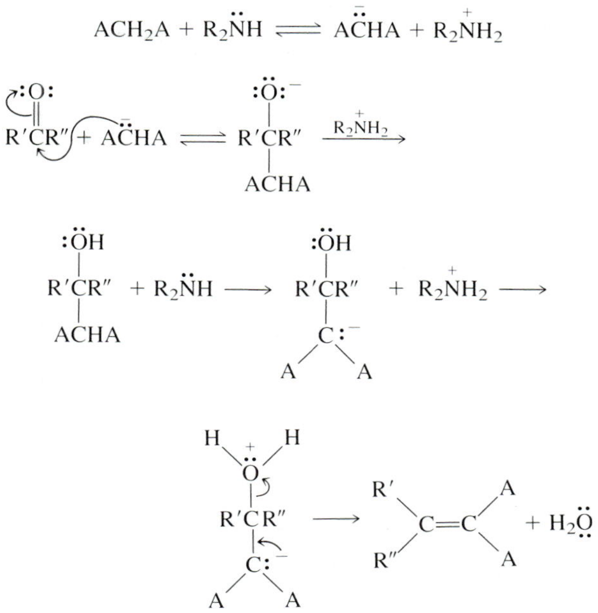

$$ACH_2A + R_2\ddot{N}H \rightleftharpoons A\ddot{C}^-HA + R_2\overset{+}{N}H_2$$

$$R'\overset{:O:}{\overset{\|}{C}}R'' + A\ddot{C}^-HA \rightleftharpoons R'\underset{ACHA}{\overset{:\ddot{O}:^-}{C}}R'' \xrightarrow{R_2\overset{+}{N}H_2}$$

$$R'\underset{ACHA}{\overset{:\ddot{O}H}{C}}R'' + R_2\ddot{N}H \longrightarrow R'\underset{\ddot{C}^-A_2}{\overset{:\ddot{O}H}{C}}R'' + R_2\overset{+}{N}H_2 \longrightarrow$$

$$R'\underset{\ddot{C}^-A_2}{\overset{H_2\overset{+}{O}}{C}}R'' \longrightarrow R'R''C{=}CA_2 + H_2\ddot{O}$$

22-19

(a) 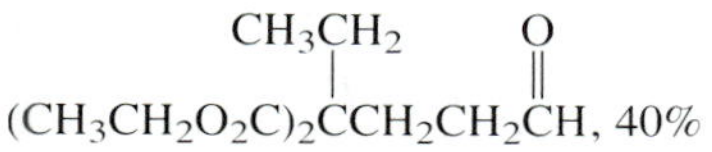 $(CH_3CH_2O_2C)_2\underset{}{\overset{CH_3CH_2}{C}}CH_2CH_2\overset{O}{\overset{\|}{C}}H$, 40%

(b) 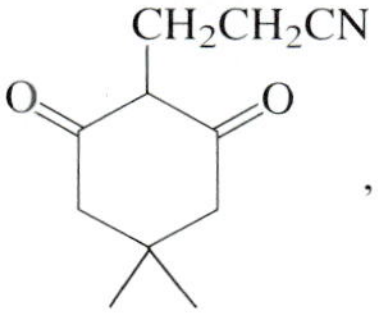 2-(CH_2CH_2CN)-5,5-dimethylcyclohexane-1,3-dione, 56%

(c) 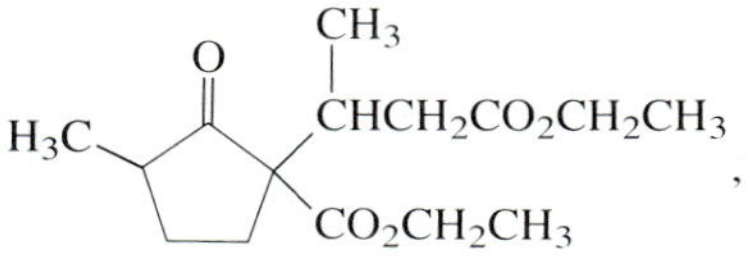 2-[$\overset{CH_3}{CH}CH_2CO_2CH_2CH_3$]-2-($CO_2CH_2CH_3$)-5-($H_3C$)-cyclopentanone, 66%

22-20

O ·· O + $CH_2{=}CHCN$ ⟶

NC :⁻ H Acidic O O ⟶

CN O ·· O $\xrightarrow[\text{Second Michael addition}]{CH_2{=}CHCN}$ ⟶ NC CN O O

22-21

First step: Michael addition

O $CO_2CH_2CH_3$ + $CH_2{=}CH\overset{O}{\overset{\|}{C}}CH_3$ ⟶

H_3C O O $CO_2CH_2CH_3$

Second step: aldol condensation

$H_2\ddot{C}^-$ O O $CO_2CH_2CH_3$ ---→ O $CO_2CH_2CH_3$

CHAPTER 23

23-1

(a) Aldotetrose; **(b)** aldopentose; **(c)** ketopentose.

23-2

(a) (2*R*,3*R*,4*R*)-2,3,4,5-Tetrahydroxypentanal

(b) (2*R*,3*S*,4*R*,5*R*)-2,3,4,5,6-Pentahydroxyhexanal

23-3

CHO
HO—H
H—OH
H—OH
CH_2OH

D-(−)-Arabinose

23-4

False. They should form in unequal amounts because two diastereomers are formed. In fact the ratio of α to β is 36:64. Similarly, the relative amounts of α-D-, β-D-fructopyranose, α-D-, and β-D-fructofuranose at equilibrium in aqueous solution are 3%, 57%, 9%, and 31%.

23-5

(a) $HOCH_2$ O CH_2OH *S* OH OH OH

(b) CH_2OH HO—C—H O OH *R* OH H OH

(c) O OH *R* OH H HO OH

23-6
Four axial OH groups, $4 \times 0.94 = 3.76$ kcal mole^{-1}; 1 axial CH_2OH, 1.70 kcal mole^{-1}; $\Delta G = 5.46$ kcal mole^{-1}. The concentration of this conformer in solution is therefore negligible.

23-7
Only the anomeric carbon and its vicinity are shown:

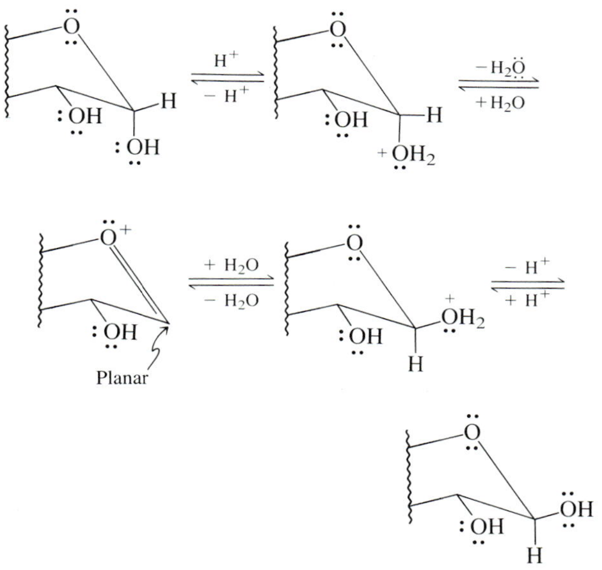

Planar

23-8
Pure α form, $+112^\circ$; pure β form, $+18.7^\circ$ ($\Delta\alpha = 93.3^\circ$). After equilibration, $+52.7^\circ$. Mole fraction of α: $(52.7 - 18.7)/93.3 = 0.364$. Hence, the mole fraction of $\beta = 0.636$; thus, the equilibrium mole ratio $\beta:\alpha = 0.636/0.364 = 1.75:1$.

23-9

$\Delta G^\circ_{\text{estimated}} = -0.94$ kcal mole^{-1} (one axial OH);

$\Delta G^\circ = -RT \ln K = -1.36 \log 63.6/36.4 = -0.33$ kcal mole^{-1}.

The difference between the two values is due to the fact that the six-membered ring is a cyclic ether (not a cyclohexane).

23-10
Oxidation at the anomeric center gives the δ-lactone, which then isomerizes by intramolecular transesterification to the γ-lactone, which happens to be more stable.

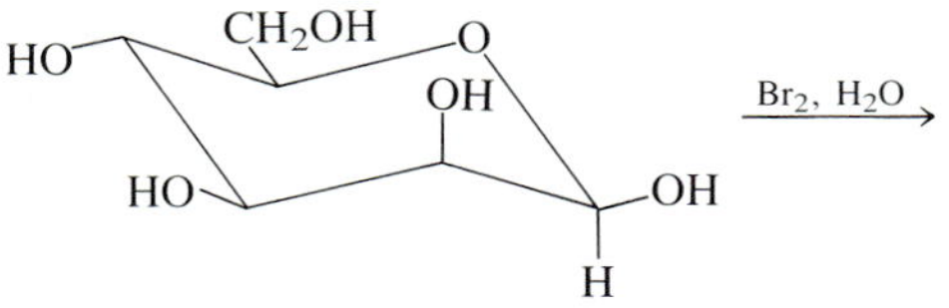

β-D-Mannopyranose

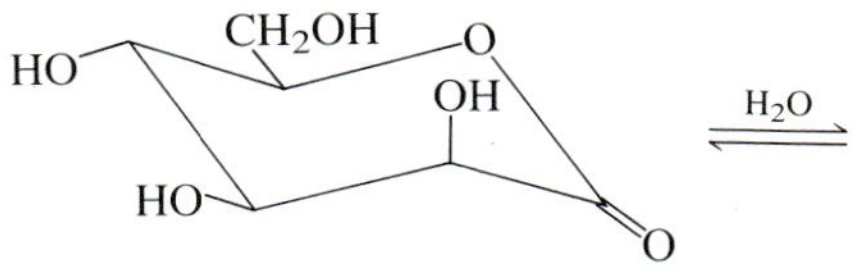

$\xrightarrow{Br_2,\ H_2O}$

D-Mannono-δ-lactone

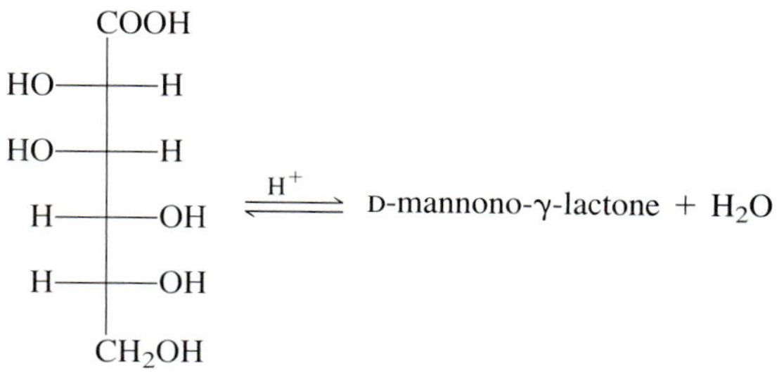

$\xrightarrow{H_2O}$

D-Mannonic acid

$\xrightarrow{H^+}$ D-mannono-γ-lactone + H_2O

23-11
Oxidation of D-glucose should give an optically active aldaric acid, whereas that of D-allose leads to loss of optical activity. This result is a consequence of turning the two end groups along the sugar chain into the same substituent.

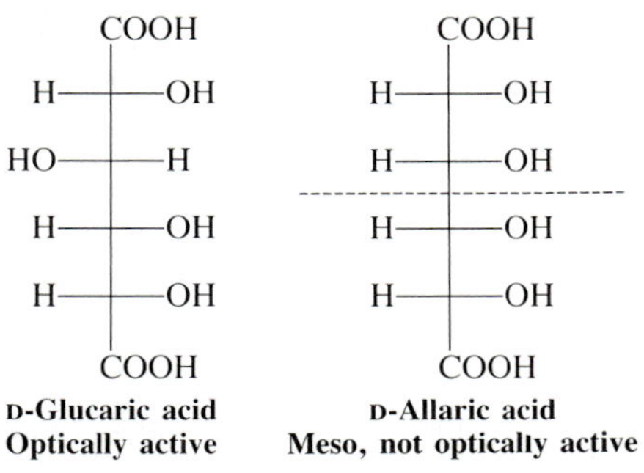

D-Glucaric acid Optically active **D-Allaric acid Meso, not optically active**

This operation may cause important changes in the symmetry of the molecule. Thus, D-allaric acid has a mirror plane. It is therefore a meso compound and not optically active. (This also means that D-allaric acid is identical with L-allaric acid.) On the other hand, D-glucaric acid is still optically active.

Other simple aldoses that turn into meso-aldaric acids are D-erythrose, D-ribose, D-xylose, and D-galactose (see Figure 23-1).

23-12

(a) 2 $CH_2{=}O$; **(b)** $CH_3CH{=}O + CH_2{=}O$; **(c)** 2 $CH_2{=}O$ + HCOOH; **(d)** no reaction; **(e)** $OHCC(CH_3)_2CHO + CO_2$; **(f)** 3 HCOOH + $CH_2{=}O$.

23-13

(a) Ribitol is a meso compound

(b) D-Mannitol (major) and D-glucitol

23-14

All of them are the same.

23-15

The mechanism of acetal formation proceeds through the same intermediate cation in both cases:

23-16

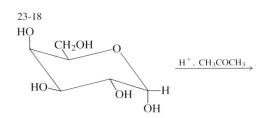

23-17

Same structure as that in Exercise 23-16 or its diastereomers with respect to C-2 and C-3.

23-18

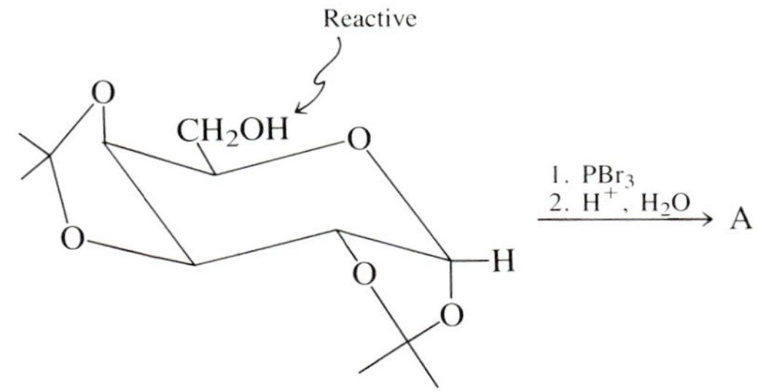

23-19

(a) D-Ribose and D-arabinose

(b) D-Glucose and D-mannose

23-20

A, D-arabinose; B, D-lyxose; C, D-erythrose; D, D-threose.

23-21

^{13}C NMR would show only three lines for ribaric acid, but five for arabinaric acid.

23-22

(a)

(b) CH_2OH / H—OH / HO—H / H—OH / H—OH / CH_2OH + CH_2OH / HO—H / HO—H / H—OH / H—OH / CH_2OH

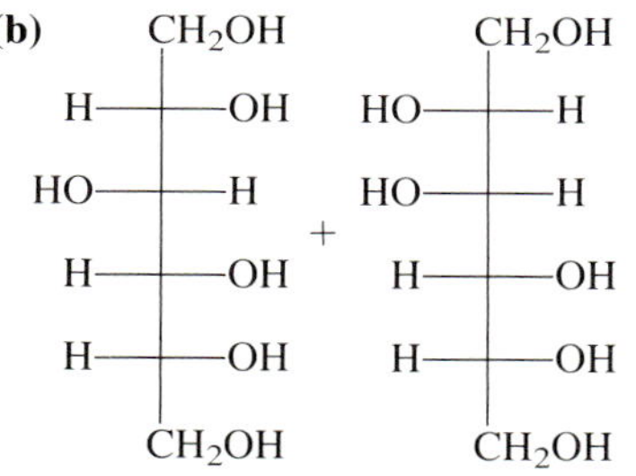

(c) No reaction.

23-23

(a)

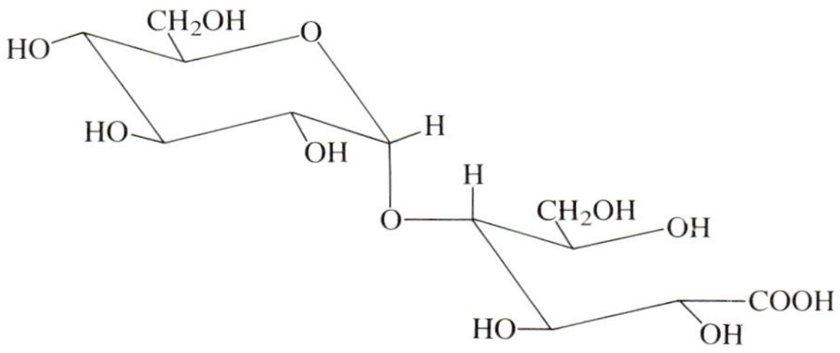

HO, CH_2OH, O, HO, OH, H, O, H, CH_2OH, OH, HO, COOH, OH

(b)

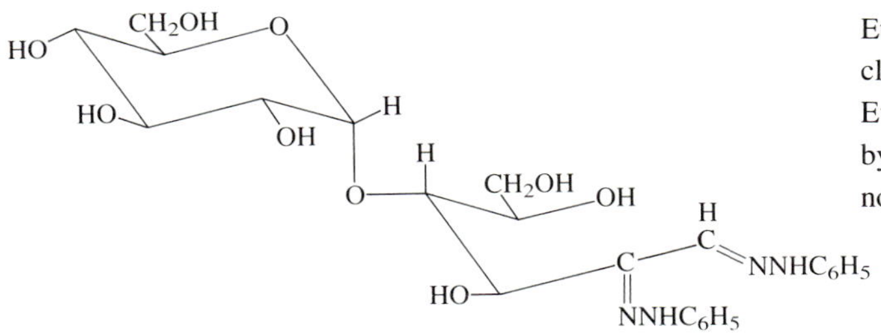

HO, CH_2OH, O, HO, OH, H, O, H, CH_2OH, OH, H, C, $NNHC_6H_5$, C, HO, $NNHC_6H_5$

(c)

HO, CH_2OH, O, HO, OH, H, O, H, 4, CH_2OH, O, HO, 3, 2, OH, 1, H, OH

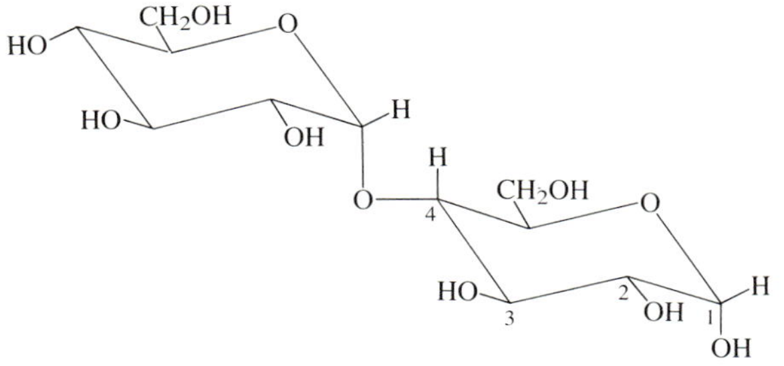

α-Maltose

CHAPTER 24

24-1

As your models will show, the orbitals of the benzylic radical derived from [2.2]paracyclophane are sterically prohibited from efficient overlap with those of the adjacent benzene ring.

24-2

The ^{13}C NMR spectrum of hexaphenylethane would be expected to show 5 lines; observed dimer, 14 lines.

24-3

$(C_6H_5)_2CHCl$ solvolyzes faster because the additional phenyl group causes extra resonance stabilization of the intermediate carbocation. This molecule is even more reactive under S_N1 conditions than 2-chloro-2-methylpropane (*tert*-butyl chloride).

24-4

$$C_6H_5CH_2OH \xrightarrow[-\,H_2O]{H^+} C_6H_5CH_2{}^+ \xrightarrow{Cl^-} C_6H_5CH_2Cl \qquad (S_N1 \text{ mechanism})$$

Ethanol has to react through an S_N2 mechanism that includes chloride attack on the protonated hydroxy group. Even if the conversion of phenylmethanol were to proceed by this pathway, it would be accelerated relative to ethanol because of a delocalized transition state.

24-5

(a) $4\text{-}CH_3OC_6H_4CH_2Br$, because it contains a better leaving group; **(b)** $(C_6H_5)_2CH_2$, because the corresponding anion is better resonance stabilized; **(c)** $C_6H_5CH_2OH$, because the corresponding phenylmethyl (benzyl) cation is not destabilized by the extra nitro group (draw resonance structures).

24-6

(a) CH_3 **(b)** COOH, CH_3 **(c)**

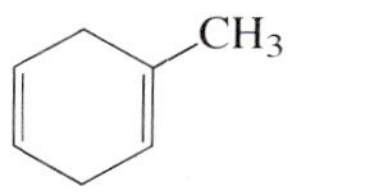

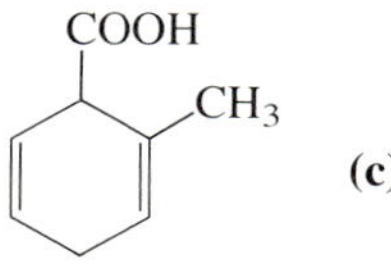

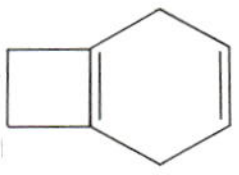

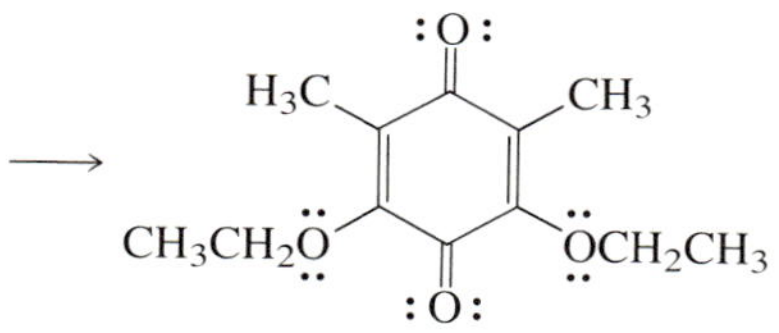

$\longrightarrow$ H_3C, CH_3, $CH_3CH_2\ddot{O}$, $\ddot{O}CH_2CH_3$, :O:, :O:

24-22

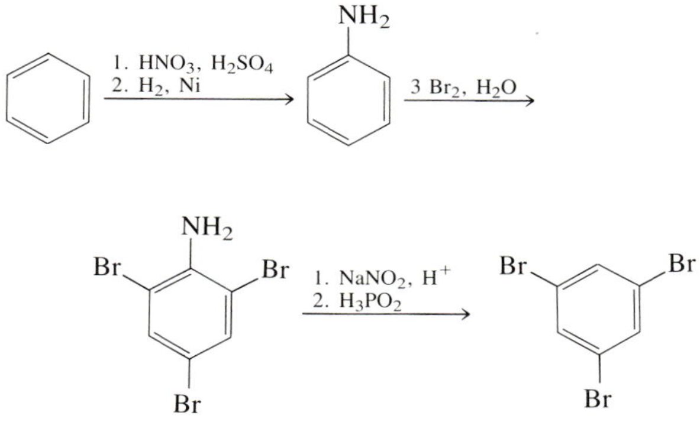

1. HNO_3, H_2SO_4
2. H_2, Ni

NH_2

3 Br_2, H_2O

NH_2, Br, Br, Br

1. $NaNO_2$, H^+
2. H_3PO_2

Br, Br, Br

CHAPTER 25

25-1

(a) H_3C, CH_3

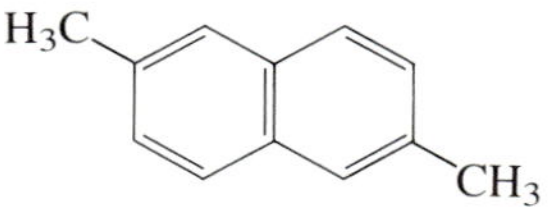

(b) NO_2, Br

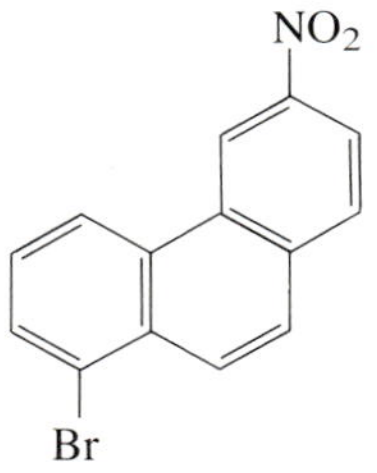

(c) C_6H_5, C_6H_5

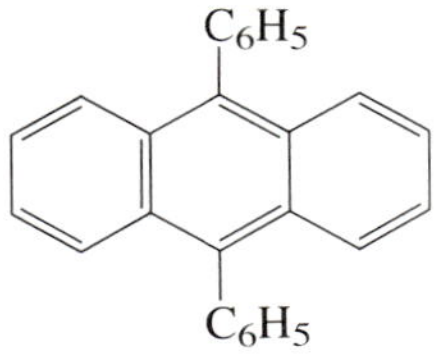

(d) 1-Naphthalenecarboxylic acid (1-naphthoic acid)
(e) 2-Methoxytriphenylene

25-2

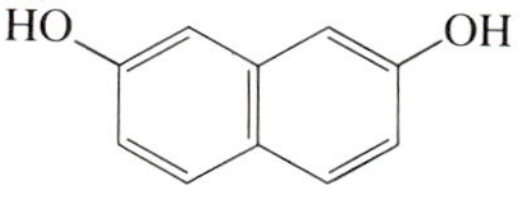

HO, OH

25-3

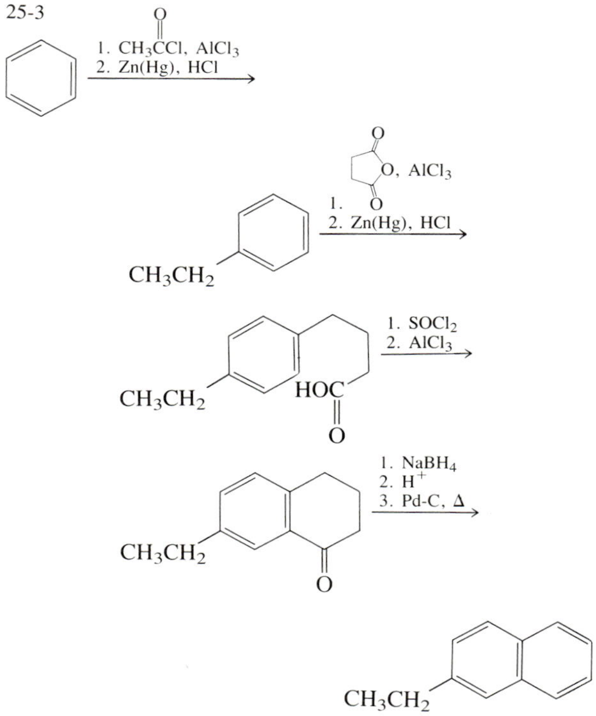

1. $CH_3\overset{O}{\overset{\|}{C}}Cl$, $AlCl_3$
2. Zn(Hg), HCl

CH_3CH_2

1. (succinic anhydride) O, $AlCl_3$
2. Zn(Hg), HCl

CH_3CH_2, $HO\overset{}{C}$=O

1. $SOCl_2$
2. $AlCl_3$

CH_3CH_2, O

1. $NaBH_4$
2. H^+
3. Pd-C, Δ

CH_3CH_2

25-4

(a) at C-4; **(b)** at C-5 and C-8; **(c)** at C-8.

25-5

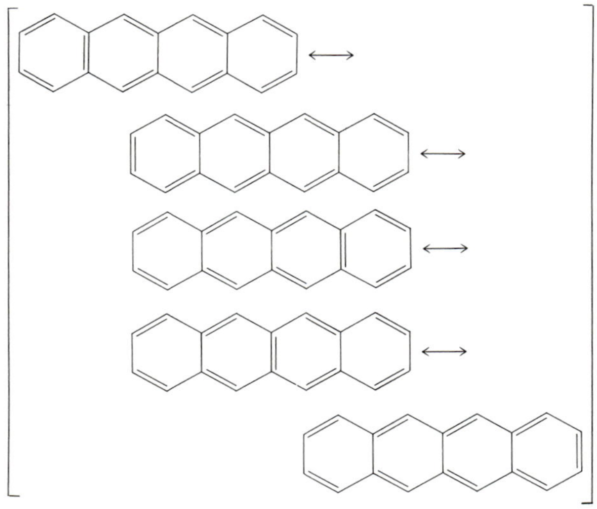

The maximum number of aromatic benzene Kekulé rings is two, in three of the resonance structures (the first, third, and fourth).

25-6

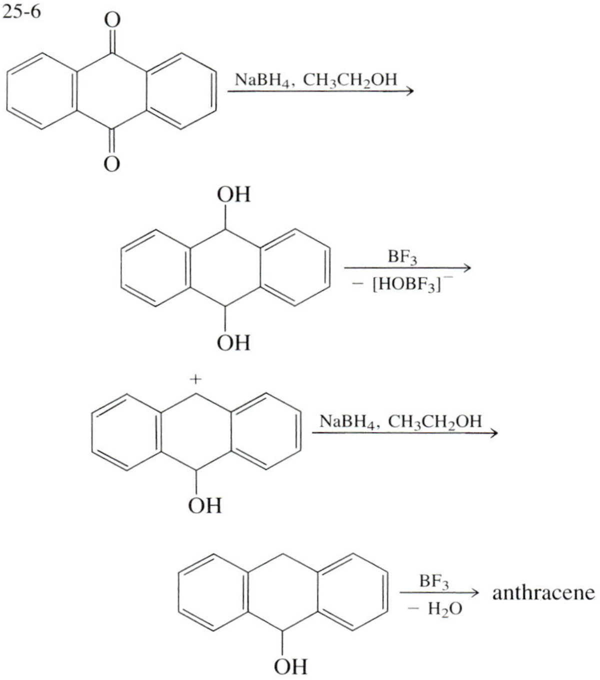

O, O

$\xrightarrow{NaBH_4,\ CH_3CH_2OH}$

OH, OH

$\xrightarrow[-\ [HOBF_3]^-]{BF_3}$

+, OH

$\xrightarrow{NaBH_4,\ CH_3CH_2OH}$

OH

$\xrightarrow[-\ H_2O]{BF_3}$ anthracene

25-7

2 H_3C, H_3C + O, O $\longrightarrow$

H_3C, H_3C, O, O, CH_3, CH_3 $\xrightarrow{S,\ \Delta}$

H_3C, H_3C, O, O, CH_3, CH_3 $\xrightarrow[2.\ NaBH_4,\ BF_3]{1.\ 2\ C_6H_5MgBr}$

H_3C, H_3C, C_6H_5, C_6H_5, CH_3, CH_3

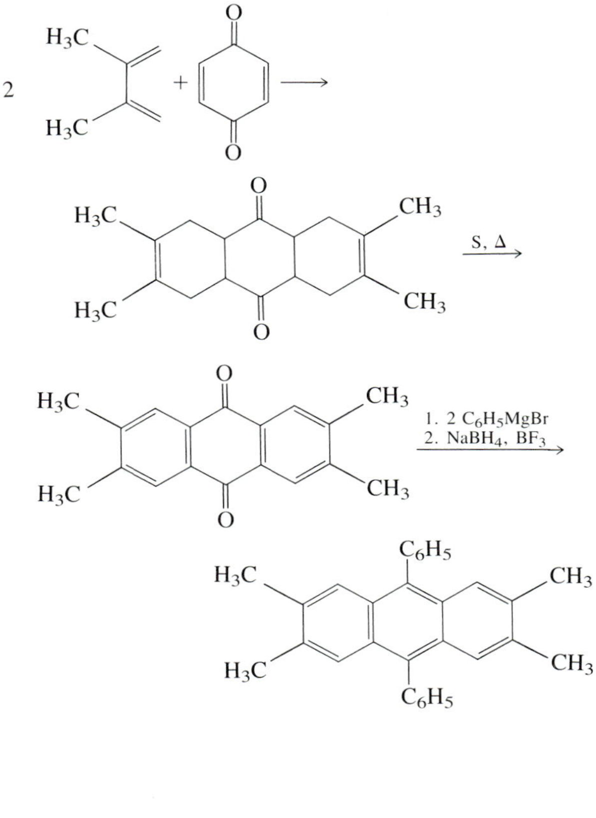

25-8

Attack at C-3 is not favored, for reasons of resonance (Section 25-3).

25-9

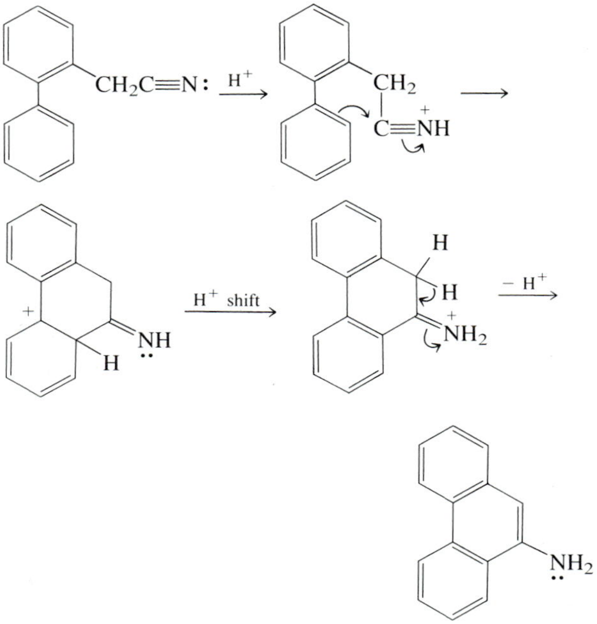

$CH_2C{\equiv}N:$ $\xrightarrow{H^+}$ CH_2, $C{\equiv}\overset{+}{N}H$ $\longrightarrow$

+, NH, H $\xrightarrow{H^+\ shift}$ H, H, $\overset{+}{N}H_2$ $\xrightarrow{-\ H^+}$

NH_2

25-10

This is an unusual Diels-Alder reaction in which one molecule acts as a diene, the other as a dienophile.

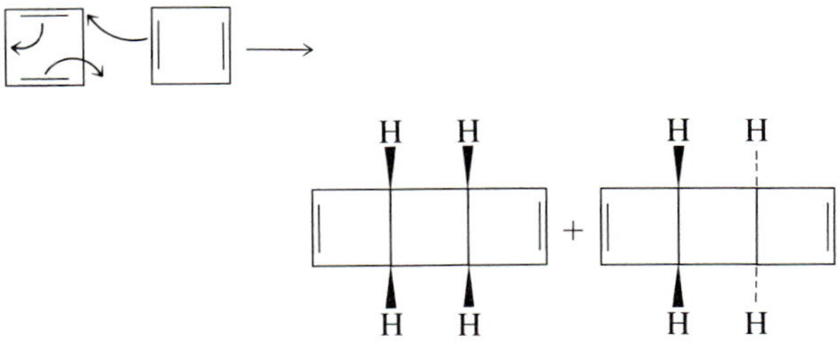

$\longrightarrow$

H H H H, H H H H, +

25-11

The ring-current effect deshields the outside twelve hydrogens, but shields the inner six.

25-12

Yes, the eight hydrogens in the π periphery are deshielded, and the bridge hydrogens over the ring are shielded.

25-13
(a), (c), (d) Aromatic; **(b)** nonaromatic.

25-14

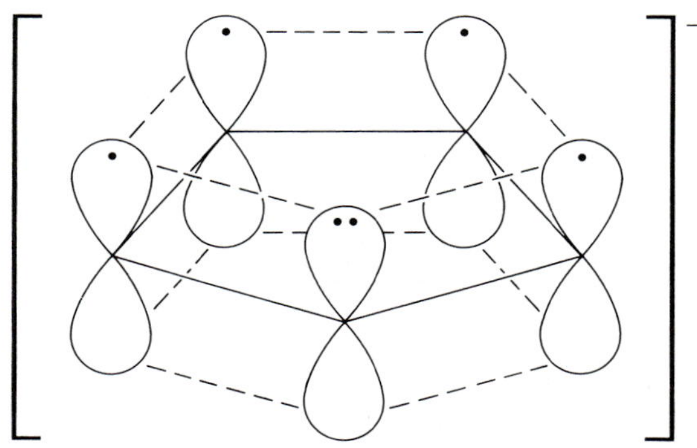

(a)

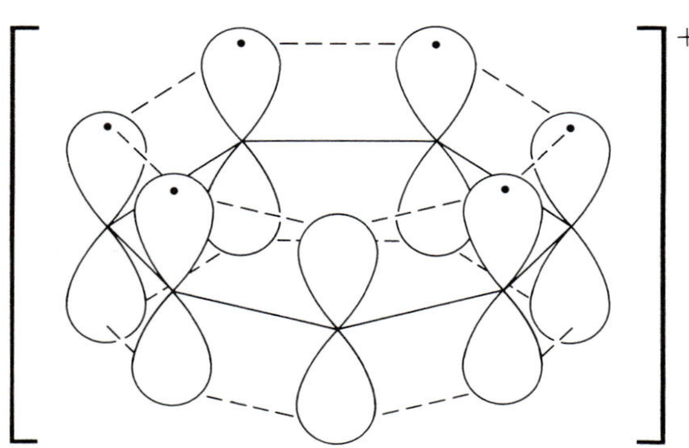

(b)

25-15
(a), (b) Aromatic; **(c)** nonaromatic.

25-16
The dianion is an aromatic system of ten π electrons, but pentalene has $4n$ π electrons.

25-17
Attack by an electrophile at C-1 allows for a cycloheptatrienyl cation resonance structure. Similarly, nucleophilic attack at C-4 gives an intermediate with a cyclopentadienyl anion ring.

E H + H Nu

CHAPTER 26

26-1
(a) H_3C, O, CH_3 **(b)** N, CH_2CH_3

(c) 2,6-Dinitropyridine **(d)** 4-Bromoindole

26-2

$\ddot{O}$ + $R\ddot{N}H_2$ ⟶ ⟶ $H\ddot{O}$: $\ddot{N}HR$ $\xrightarrow[-\,H_2O]{HCl}$

Cl :NHR $\xrightarrow[-\,HCl]{Base}$ N R

26-3

$CH_3CH{=}CH{-}\overset{:O:}{\overset{\|}{C}}CH_3$ $H\ddot{O}\ddot{O}:^-$ $\xrightarrow{\text{Michael addition}}$

[$CH_3CH{-}CH{=}C(O^-)CH_3$ with $HO{-}O$ ↔ $CH_3CH{-}\bar{C}H{-}\overset{:O:}{\overset{\|}{C}}CH_3$ with $HO{-}O$] $\xrightarrow{\text{Ring closure}}$

Free rotation in this intermediate destroys stereochemistry

$CH_3CH{-}CH\overset{O}{\overset{\|}{C}}CH_3$ (epoxide O) + $^-:\ddot{O}H$

Cis and trans

26-4

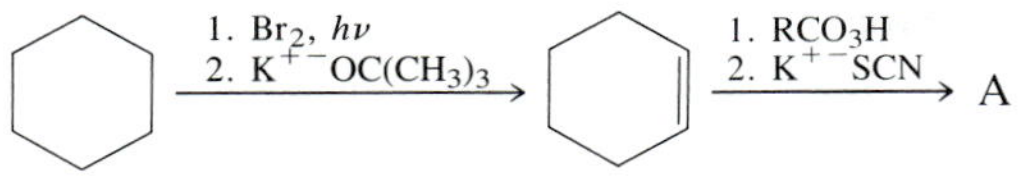

1. Br_2, $h\nu$ 2. $K^{+}\,^{-}OC(CH_3)_3$ → 1. RCO_3H 2. $K^{+}\,^{-}SCN$ → A

26-5

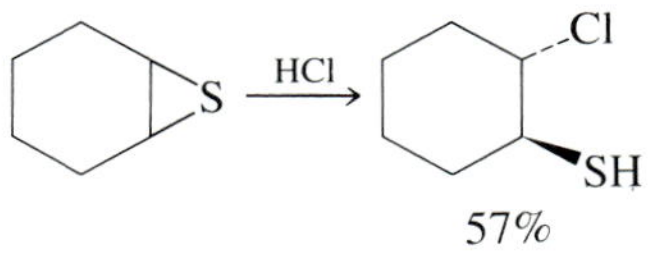

HCl → Cl, SH

57%

26-6

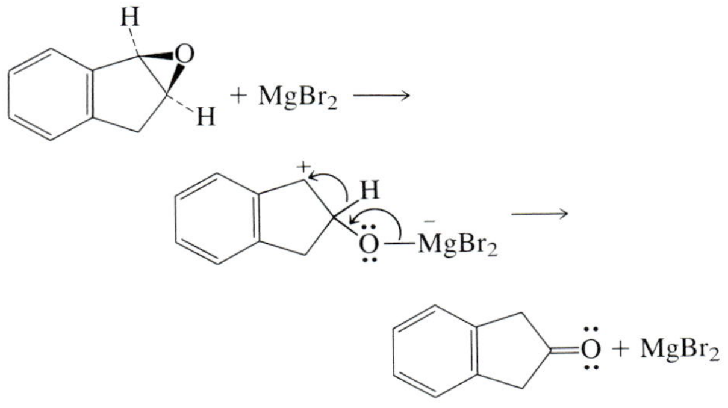

H, O, H + $MgBr_2$ →

H, $\ddot{O}$—$MgBr_2$ →

$\ddot{O}$ + $MgBr_2$

26-7

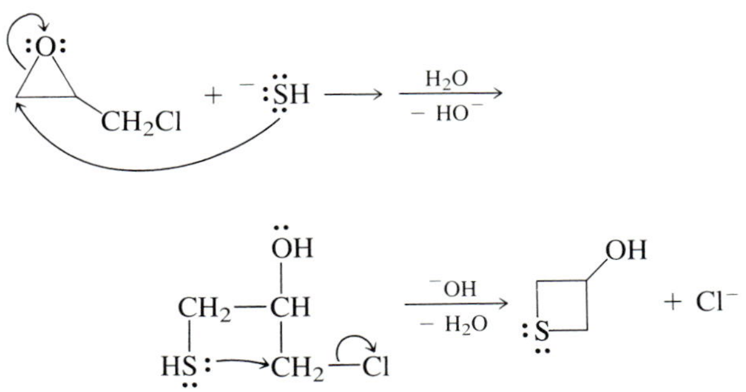

:O: CH_2Cl + $^{-}$:SH → $\frac{H_2O}{-\,HO^-}$

OH
CH_2—CH
HS: → CH_2—Cl

$\frac{^{-}OH}{-\,H_2O}$ OH :S: + Cl^-

26-8

$CH_2=\ddot{O}$

(a) $CH_2=C=\ddot{O}^{+}$—$\bar{Zn}Cl_2$ →

$CH_2=\ddot{O}^{+}$
$CH_2=C$—$\ddot{O}$—$Zn^{-}Cl_2$ → $\ddot{O}$: $\ddot{O}$: + $ZnCl_2$

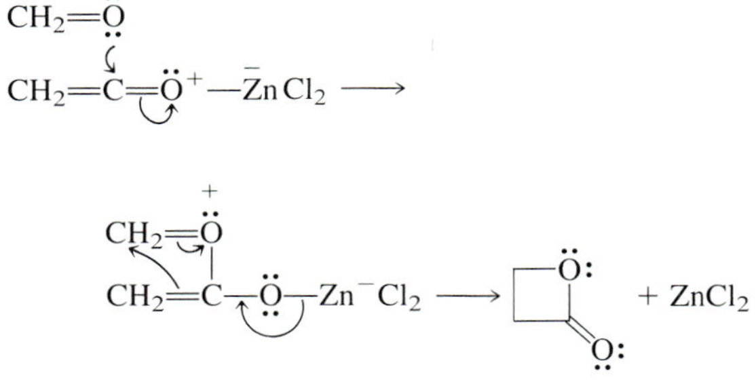

(b) H_3C, H_3C $C=CH_2$ →
$ClSO_2$—$\ddot{N}=C=O$

[$(CH_3)_2\overset{+}{C}$ CH_2 C O $ClSO_2$—$\ddot{N}^{-}$] → H_3C H_3C N O $ClSO_2$

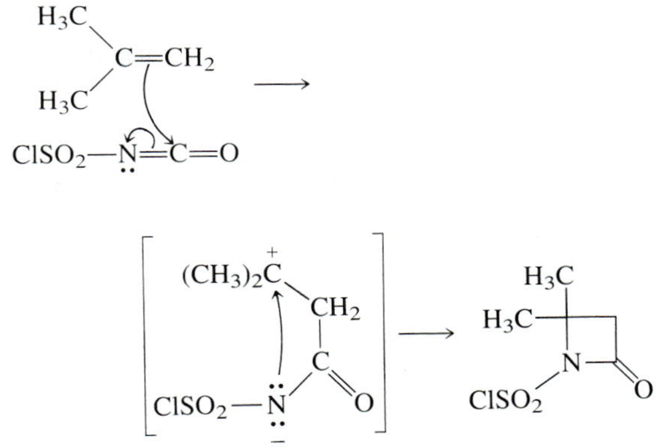

26-9

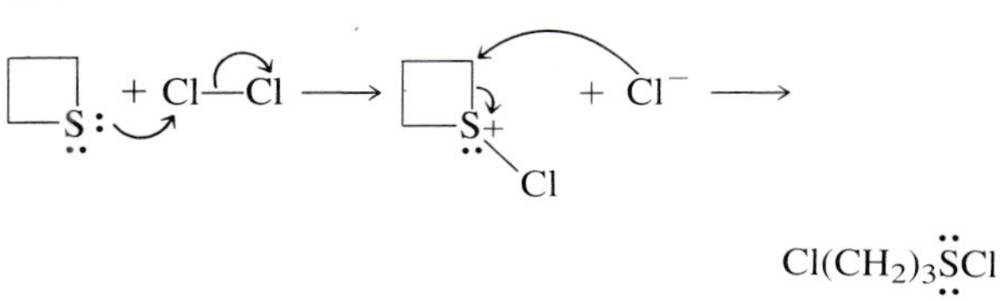

S: + Cl—Cl → S^{+} Cl + Cl^- →

$Cl(CH_2)_3\ddot{S}Cl$

26-10

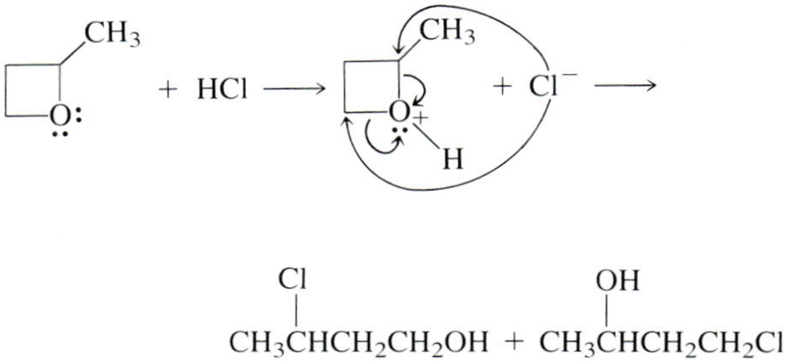

CH_3 $\ddot{O}$: + HCl → CH_3 O^{+} H + Cl^- →

$CH_3\overset{Cl}{C}HCH_2CH_2OH$ + $CH_3\overset{OH}{C}HCH_2CH_2Cl$

26-11

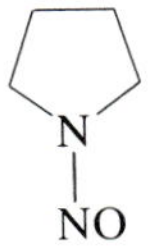

N
NO

26-12

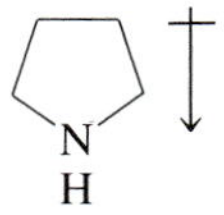

N H 1.57 D

Nitrogen is more electronegative than carbon

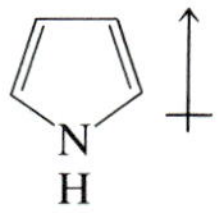

N H 1.80 D

Because of resonance, the molecule is now polarized in the opposite direction

26-13

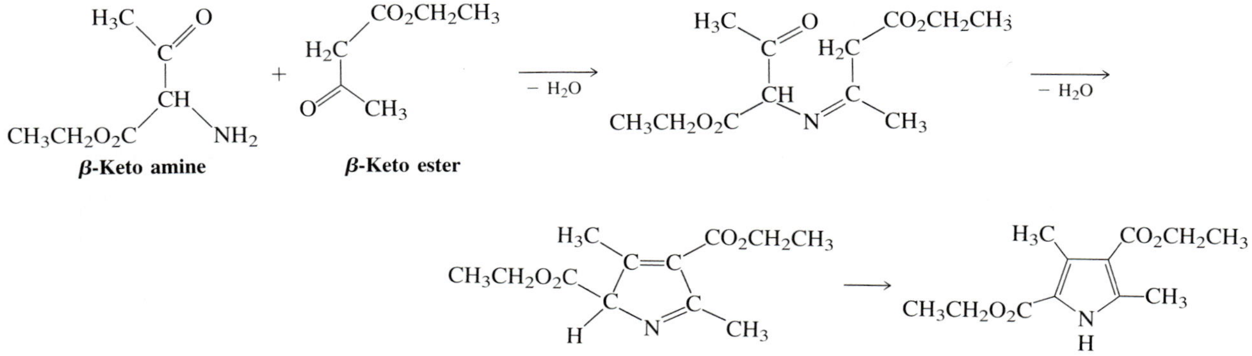

H_3C, O, C, CH, $CH_3CH_2O_2C$, NH_2 + $CO_2CH_2CH_3$, H_2C, C, O, CH_3 $\xrightarrow{-H_2O}$ H_3C, O, C, CH, $CH_3CH_2O_2C$, N, C, H_2C, $CO_2CH_2CH_3$, CH_3 $\xrightarrow{-H_2O}$

β-Keto amine **β-Keto ester**

H_3C, $CO_2CH_2CH_3$, C=C, $CH_3CH_2O_2C$, C, H, N=C, CH_3 ⟶ H_3C, $CO_2CH_2CH_3$, $CH_3CH_2O_2C$, N, H, CH_3

26-14

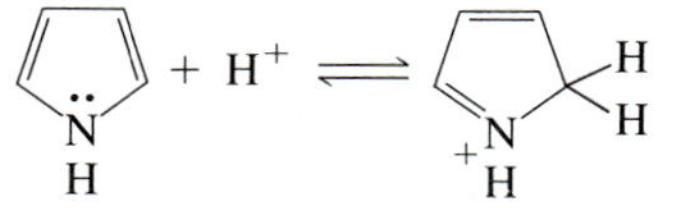

N, H + H^+ ⇌ N, H, +, H, H

$pK_a = -4.4$

26-15

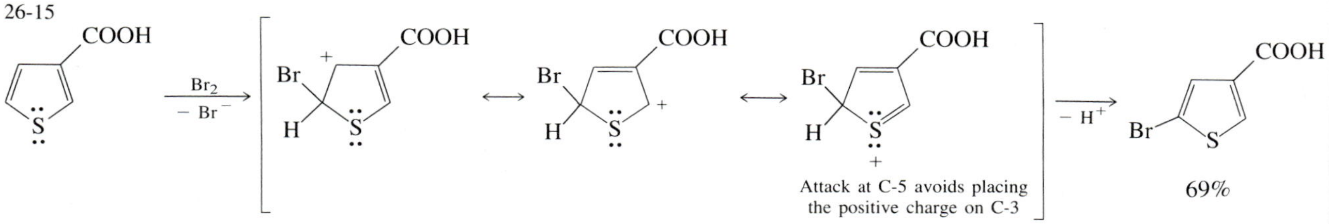

COOH, S $\xrightarrow[-Br^-]{Br_2}$ [Br, +, H, S, COOH ⟷ Br, H, S, +, COOH ⟷ Br, H, S, +, COOH] $\xrightarrow{-H^+}$ Br, S, COOH

Attack at C-5 avoids placing the positive charge on C-3

69%

26-16

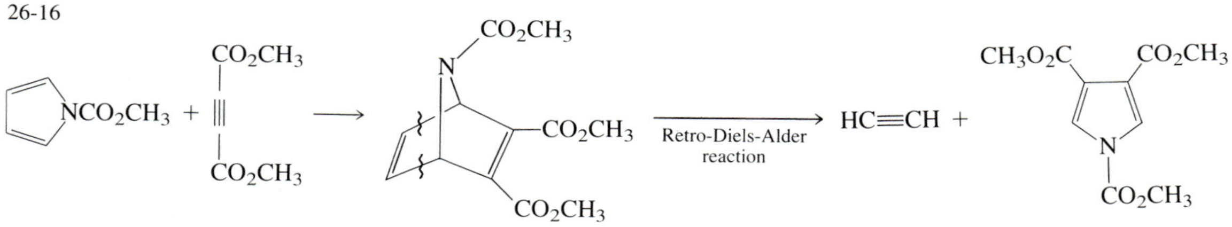

NCO_2CH_3 + CO_2CH_3, CO_2CH_3 ⟶ CO_2CH_3, N, CO_2CH_3, CO_2CH_3 $\xrightarrow{\text{Retro-Diels-Alder reaction}}$ $HC{\equiv}CH$ + CH_3O_2C, CO_2CH_3, N, CO_2CH_3

26-17

(a) N, H, N, CH_3 $\xrightarrow{H_2SO_4}$ N, H, CH_3

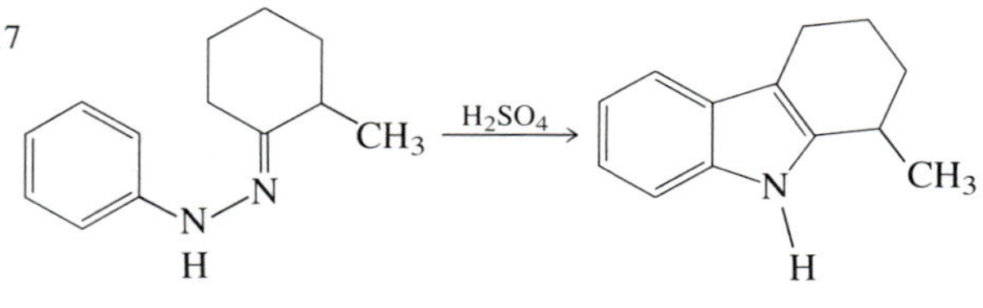

(b) N, H, N, H_3C, C_6H_5 $\xrightarrow{PPA}$ C_6H_5, CH_3, N, H + N, H, C_6H_5

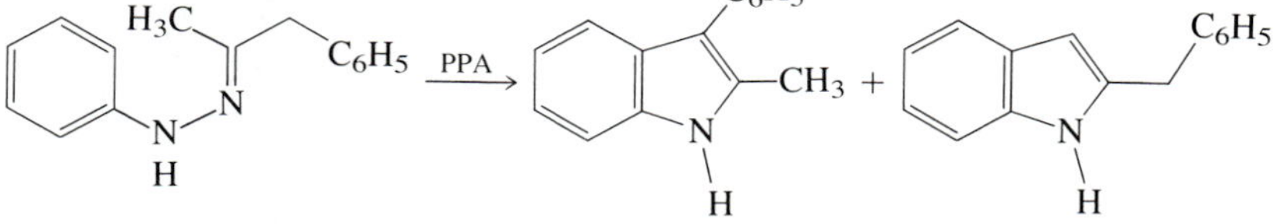

2-Methyl-3-phenylindole- **2-(Phenylmethyl)-indole**

(c) H_3C, COOH; N–N(H) hydrazone $\xrightarrow{PCl_5}$

—COOH

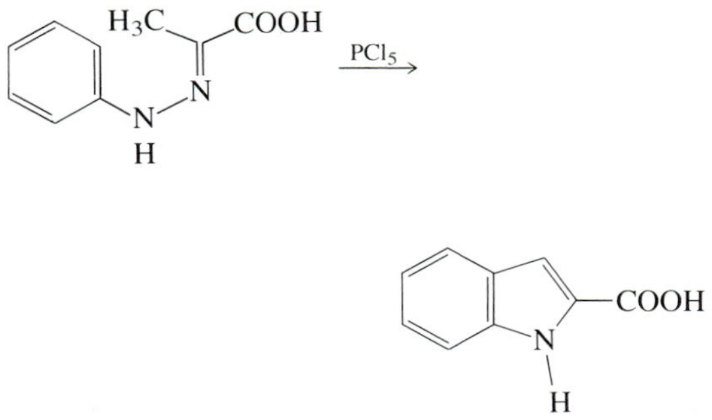

2-Indolecarboxylic acid

26-18

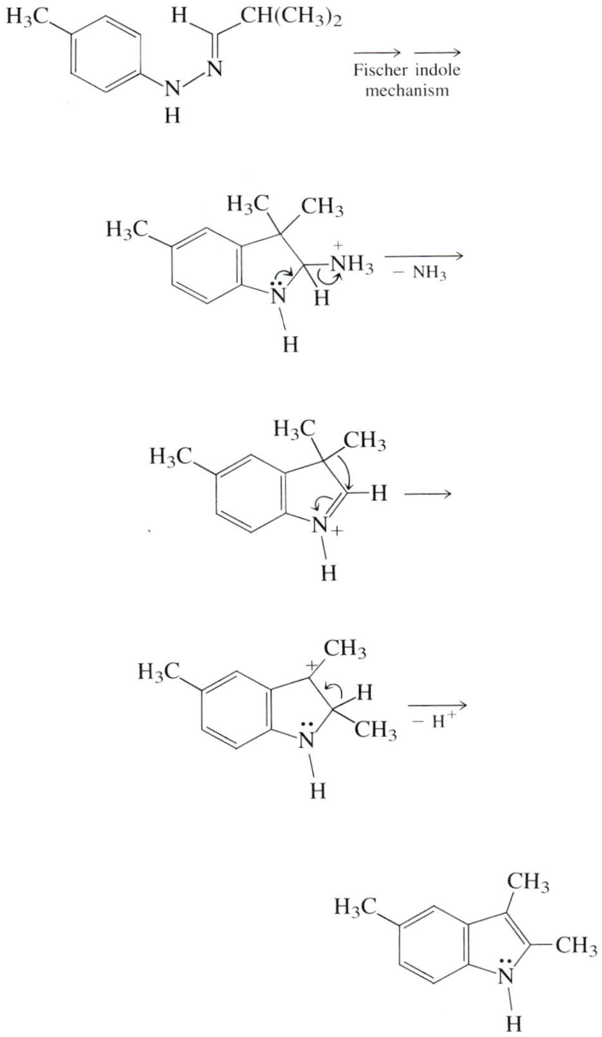

H_3C … H, $CH(CH_3)_2$ … N–N(H) $\longrightarrow \longrightarrow$ Fischer indole mechanism

H_3C, H_3C, CH_3, $\overset{+}{N}H_3$, H, H $\xrightarrow{-\,NH_3}$

H_3C, H_3C, CH_3, —H, N^+, H $\longrightarrow$

H_3C, CH_3, H, CH_3, N, H $\xrightarrow{-\,H^+}$

H_3C, CH_3, —CH_3, N, H

26-19

+ E^+ $\longrightarrow$

E, H; + $\longleftrightarrow$ E, H; $+N$, H

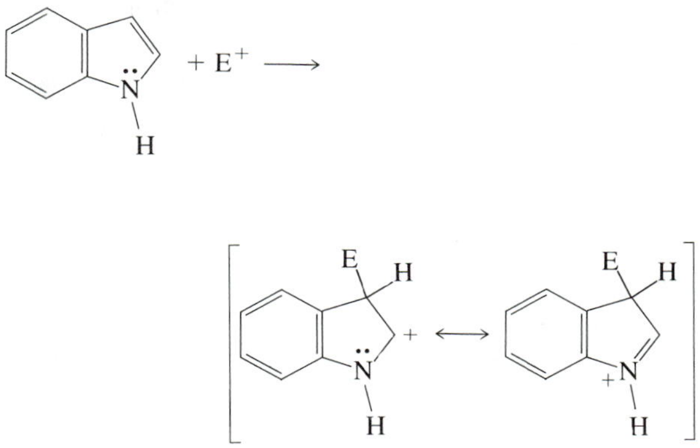

Only attack at C-3 produces the iminium resonance structure without disrupting the benzene ring

26-20
Because of the electronegativity of nitrogen, the dipole vector in both compounds points toward the heteroatom. The dipole moment of pyridine is 2.26 D, larger than that in azacyclohexane (piperidine) because the nitrogen is sp^2 hybridized. (See Section 11-2 for the effects of hybridization on electron-withdrawing power.)

26-21

(a) $CH_3\overset{O}{\overset{\|}{C}}CH_2CO_2CH_2CH_3$, NH_3, (2-nitrobenzaldehyde: NO_2, CHO),

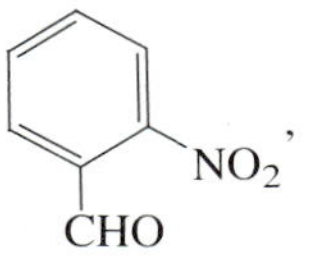

$CH_3\overset{O}{\overset{\|}{C}}CH_2CN$;

(b) $CH_3\overset{O}{\overset{\|}{C}}CH_2CN$, NH_3, $(CH_3)_3CCHO$;

(c) $CH_3CH_2\overset{O}{\overset{\|}{C}}CH_2CO_2CH_2CH_3$, NH_3, CH_3CHO.

26-22

C-3 is the least-deactivated position in the ring. Attack at C-2 or C-4 generates intermediate cations with resonance structures that place the positive charge on the electronegative nitrogen.

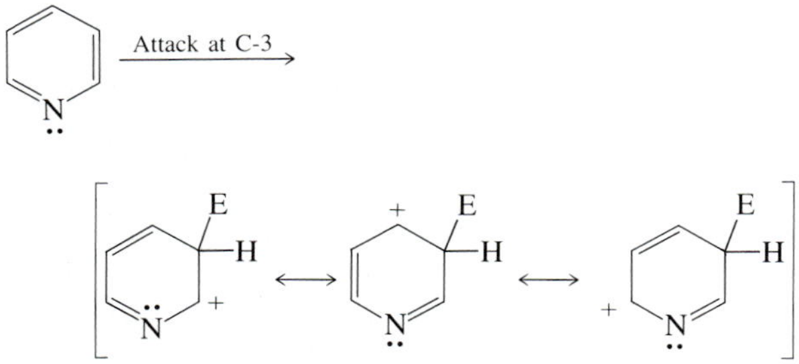

Attack at C-3

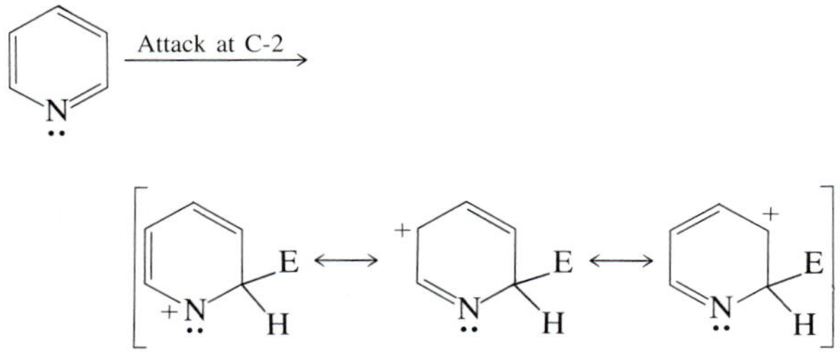

Attack at C-2

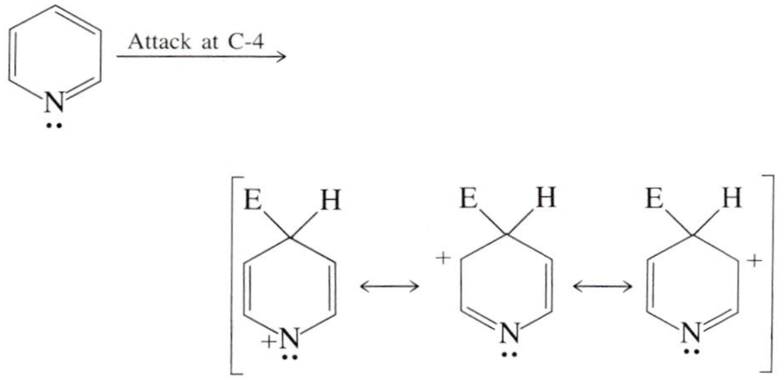

Attack at C-4

26-23

Attack at C-2 and C-4 produces the more-highly-resonance-stabilized anions (only the most-important resonance structures are shown):

2-chloropyridine,

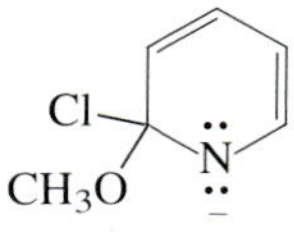

3-chloropyridine,

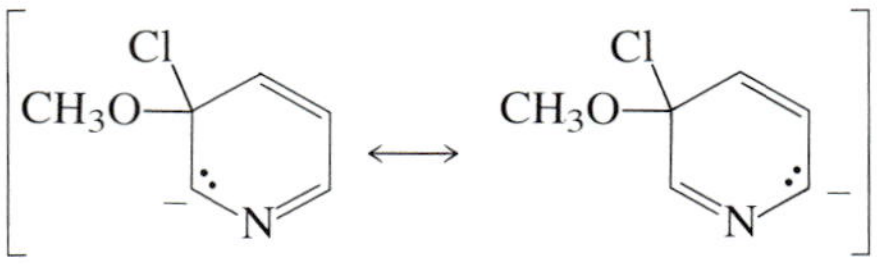

4-chloropyridine,

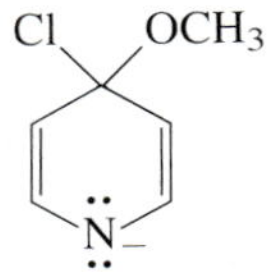

26-24

Abbreviated mechanism:

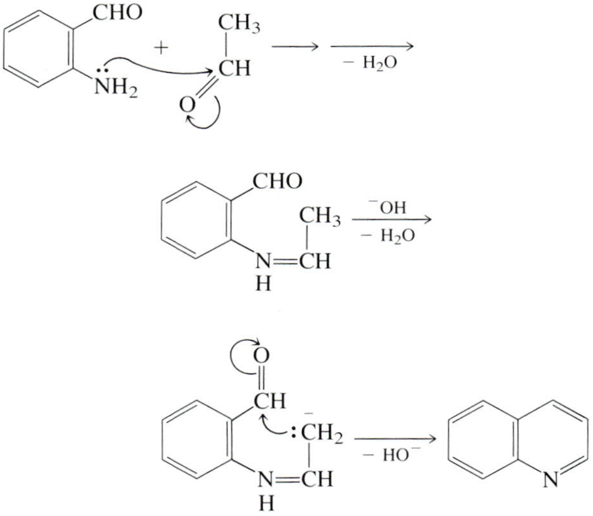

26-25

Abbreviated mechanism:

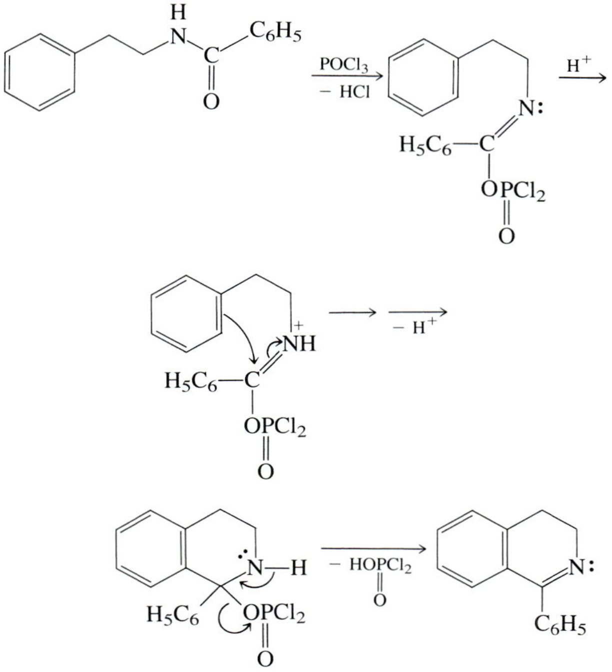

$\xrightarrow[-\ HCl]{POCl_3}$ $\xrightarrow{H^+}$ $\longrightarrow$ $\xrightarrow[-\ H^+]{}$ $\xrightarrow[-\ HOPCl_2(=O)]{}$

26-26

1. $(CH_3CH_2)_2O$, 18 h, Δ
2. NH_4Cl

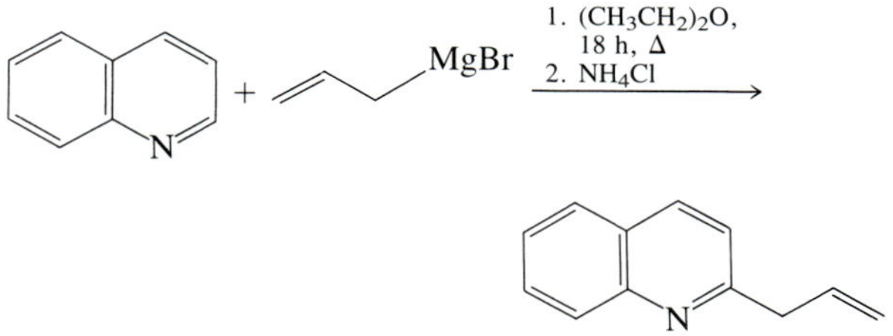

56%

2-(2-Propenyl)quinoline

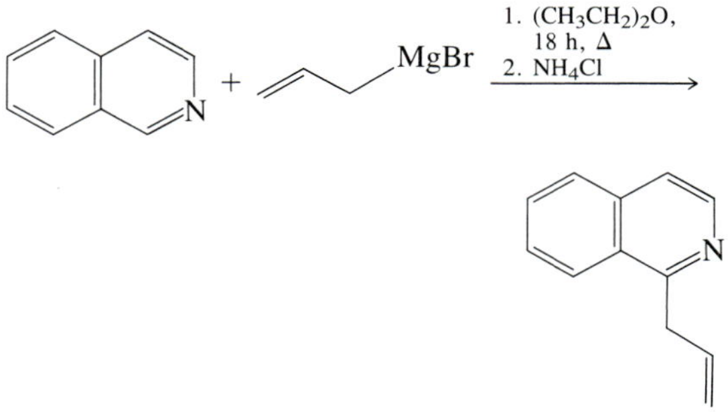

1. $(CH_3CH_2)_2O$, 18 h, Δ
2. NH_4Cl

57%

1-(2-Propenyl)isoquinoline

CHAPTER 27

27-1

(2*S*)-Aminopropanoic acid; (2*S*)-amino-3-methylbutanoic acid; (2*S*)-amino-4-methylpentanoic acid; (2*S*)-amino-3-methylpentanoic acid; (2*S*)-amino-3-phenylpropanoic acid; (2*S*)-amino-3-hydroxypropanoic acid; (2*S*)-amino-3-(4-hydroxyphenyl)propanoic acid; (2*S*,6)-diaminohexanoic acid; (2*R*)-amino-3-mercaptopropanoic acid; (2*S*)-amino-4-(methylthio)butanoic acid; (2*S*)-aminobutanedioic acid; (2*S*)-aminopentanedioic acid.

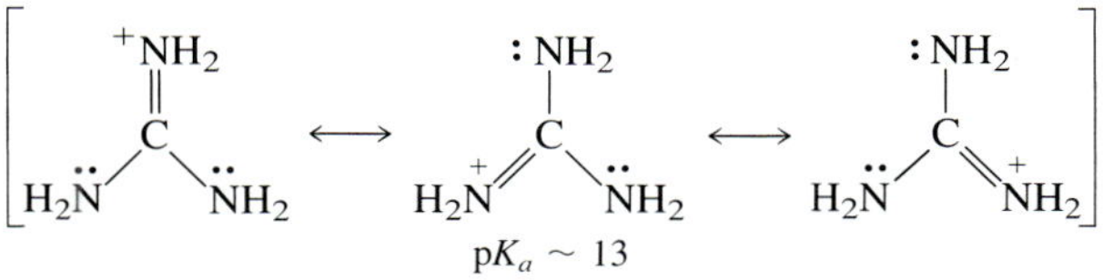

$pK_a \sim 13$

27-3

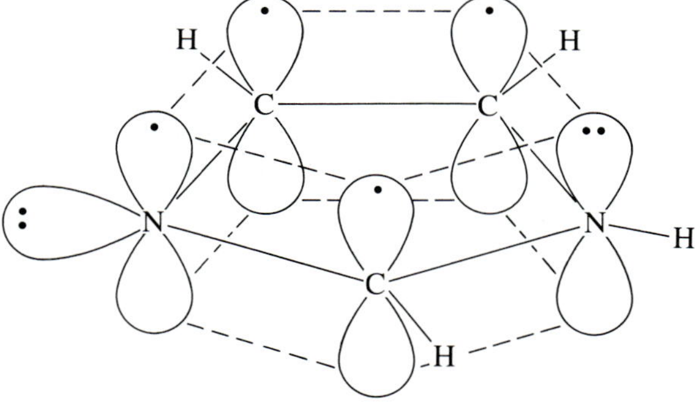

27-4

The yields given are those found in the literature.

Phthalimidomalonic ester anion, $N\text{—}\ddot{C}^{-}(CO_2CH_2CH_3)_2$:

$\xrightarrow{ClCH_2CH_2SCH_3} \xrightarrow{H^+,\ H_2O,\ \Delta} CH_3SCH_2CH_2CH(\overset{+}{N}H_3)CO_2^-$ 85% **Methionine**

$\xrightarrow{ClCH_2COOCH_2CH_3} \xrightarrow{H^+,\ H_2O,\ \Delta} HOOCCH_2CH(\overset{+}{N}H_3)CO_2^-$ 33% **Aspartic acid**

$\xrightarrow[\text{Michael addition}]{CH_2{=}CHCOOCH_2CH_3} \xrightarrow{H^+,\ H_2O,\ \Delta} HOOCH_2CH_2CH(\overset{+}{N}H_3)CO_2^-$ 75% **Glutamic acid**

27-5

These syntheses are found in the literature:

$CH_2{=}O \xrightarrow{NH_4^{+\,-}CN,\ H_2SO_4} H_2NCH_2CN$ (**2-Aminoethanenitrile**) $\xrightarrow{BaO,\ H_2O,\ \Delta} H_3\overset{+}{N}CH_2COO^-$ 42% **Glycine**

$CH_3SH + CH_2{=}CHCH{=}O \xrightarrow[\text{Michael Addition}]{} CH_3SCH_2CH_2CH{=}O$ 84% (**3-(Methylthio)propanal**) $\xrightarrow{1.\ Na^{+\,-}CN,\ (NH_4)_2CO_3;\ 2.\ NaOH} CH_3SCH_2CH_2CH(\overset{+}{N}H_3)COO^-$ 58% **Methionine**

27-6

$H_3\overset{+}{N}CH(CH_2S\text{—})C(O)\text{—}NHCH(CH_2C_6H_4OH)\text{—}C(O)\text{—}NHCH(CH_2C_6H_5)\text{—}C(O)\text{—}NHCH(CH_2CH_2C(O)NH_2)\text{—}C(O)\text{—}NHCH(CH_2C(O)NH_2)\text{—}C(O)\text{—}NHCH(CH_2S\text{—})\text{—}C(O)\text{—}N(\text{pyrrolidine})CH\text{—}C(O)\text{—}NHCH((CH_2)_3NHC(=NH)NH_2)\text{—}C(O)\text{—}NHCH_2C(O)NH_2$

(The two CH_2 groups of the first and sixth residues are joined: $H_2C\text{—}S\text{—}S\text{—}CH_2$.)

27-7

Hydrolysis of the A chain in insulin produces one equivalent each of Gly, Ile, and Ala, two each of Val, Glu, Gln, Ser, Leu, Tyr, and Asn, and four of Cys.

27-8

$$C_6H_5\ddot{N}{=}C{=}S + H_2\ddot{N}CH_2CONH_2 \longrightarrow C_6H_5\ddot{N}^{-}{-}C({=}S){-}\overset{+}{N}H_2CH_2CONH_2 \xrightarrow{H^+\ \text{shift}} \ldots \xrightarrow{H^+} \ldots \xrightarrow{H^+\ \text{shift}} \ldots \xrightarrow{-H^+} \ldots + NH_3$$

27-9

It's the A chain of insulin.

27-10

$$(CH_3)_3CO\overset{O}{\overset{\|}{C}}NH\overset{R}{\overset{|}{C}}HCOOH \underset{}{\overset{H^+}{\rightleftharpoons}} (CH_3)_3C{-}O{-}\overset{+OH}{\overset{\|}{C}}NH\overset{R}{\overset{|}{C}}HCOOH \rightleftharpoons (CH_3)_3C^+ + O{=}\overset{OH}{\overset{|}{C}}NH\overset{R}{\overset{|}{C}}HCOOH \xrightarrow[-H^+]{}$$

$$H_3\overset{+}{N}\overset{R}{\overset{|}{C}}HCOO^- + CO_2 + CH_2{=}C(CH_3)_2$$

27-11

1. Ala + $(CH_3)_3CO\overset{O}{\overset{\|}{C}}O\overset{O}{\overset{\|}{C}}OC(CH_3)_3 \longrightarrow$ Boc-Ala + CO_2 + $(CH_3)_3COH$
2. Val + $CH_3OH \xrightarrow{H^+}$ Val-OCH_3 + H_2O
3. Boc-Ala + Val-$OCH_3 \xrightarrow{DCC}$ Boc-Ala-Val-OCH_3
4. Boc-Ala-Val-$OCH_3 \xrightarrow{H^+}$ Ala-Val-OCH_3 + CO_2 + $CH_2{=}C(CH_3)_2$
5. Leu + $(CH_3)_3CO\overset{O}{\overset{\|}{C}}O\overset{O}{\overset{\|}{C}}OC(CH_3)_3 \longrightarrow$ Boc-Leu + CO_2 + $(CH_3)_3COH$
6. Boc-Leu + Ala-Val-$OCH_3 \xrightarrow{DCC}$ Boc-Leu-Ala-Val-OCH_3
7. Boc-Leu-Ala-Val-$OCH_3 \xrightarrow{1.\ H^+,\ H_2O;\ 2.\ HO^-,\ H_2O}$ Leu-Ala-Val

Periodic Table of the Elen

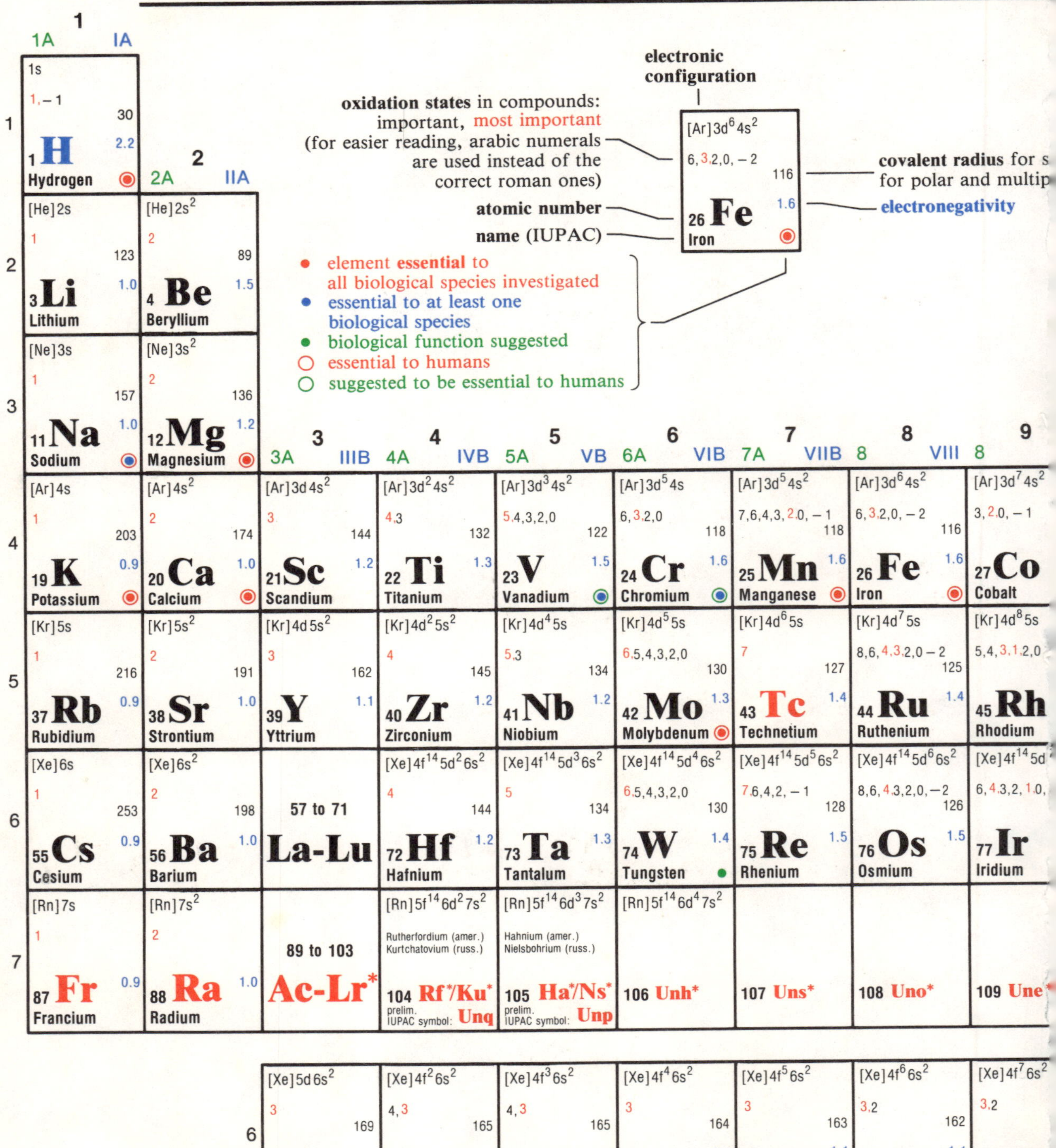

6	$[Xe]5d6s^{2}$ 3 169 1.1 57 La Lanthanum	$[Xe]4f^{2}6s^{2}$ 4, 3 165 1.1 58 Ce Cerium	$[Xe]4f^{3}6s^{2}$ 4, 3 165 1.1 59 Pr Praseodymium	$[Xe]4f^{4}6s^{2}$ 3 164 1.1 60 Nd Neodymium	$[Xe]4f^{5}6s^{2}$ 3 163 1.1 61 Pm Promethium	$[Xe]4f^{6}6s^{2}$ 3, 2 162 1.1 62 Sm Samarium	$[Xe]4f^{7}6s^{2}$ 3, 2 63 Eu Europium
7	$[Rn]6d7s^{2}$ 3 1.0 89 Ac Actinium	$[Rn]6d^{2}7s^{2}$ 4 165 1.1 90 Th Thorium	$[Rn]5f^{2}6d7s^{2}$ 5, 4 1.1 91 Pa Protactinium	$[Rn]5f^{3}6d7s^{2}$ 6, 5, 4, 3 142 1.2 92 U Uranium	$[Rn]5f^{4}6d7s^{2}$ 6, 5, 4, 3 1.2 93 Np Neptunium	$[Rn]5f^{6}7s^{2}$ 6, 5, 4, 3 1.2 94 Pu Plutonium	$[Rn]5f^{7}7s^{2}$ 6, 5, 4, 3 95 Am* Americium